Spotlight Figures

Get Ready for a Whole New Mastering Experience

NEW! Ready-to-Go Teaching Modules help instructors find the best assets to use before, during, and after class to teach the toughest topics in A&P. Created by teachers for teachers, these curated sets of teaching tools save you time by highlighting the most effective and engaging animations, videos, quizzing, coaching and active learning activities from MasteringA&P.

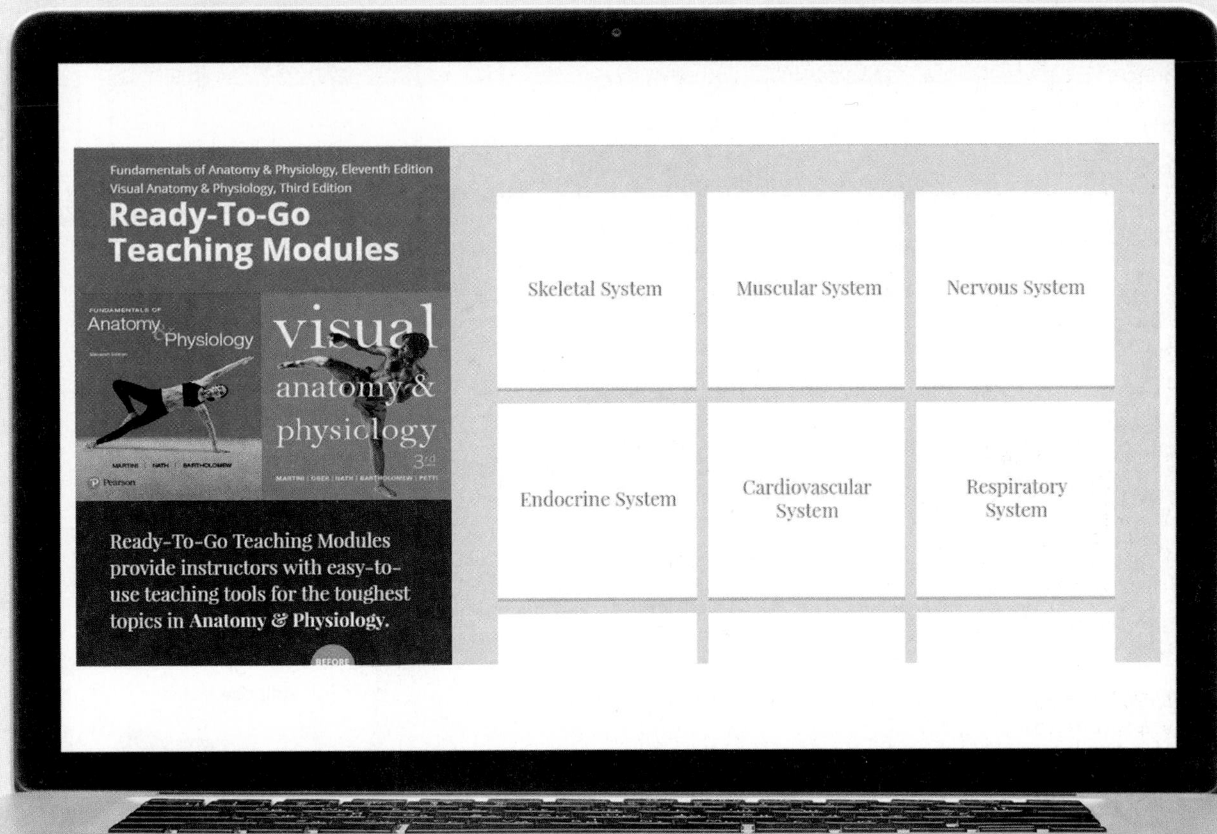
Fundamentals of Anatomy & Physiology, Eleventh Edition
Visual Anatomy & Physiology, Third Edition
Ready-To-Go Teaching Modules
Anatomy & Physiology
visual anatomy & physiology
Ready-To-Go Teaching Modules provide instructors with easy-to-use teaching tools for the toughest topics in Anatomy & Physiology.
Skeletal System
Muscular System
Nervous System
Endocrine System
Cardiovascular System
Respiratory System

Help Students Use Art More Effectively

NEW! SmartArt Videos help students navigate select, complex pieces of art for some of the toughest topics in A&P. Author Kevin Petti walks students through several figures and provides additional background and detail. The videos can be accessed via QR codes in the book and offer accompanying assignments through Mastering A&P.

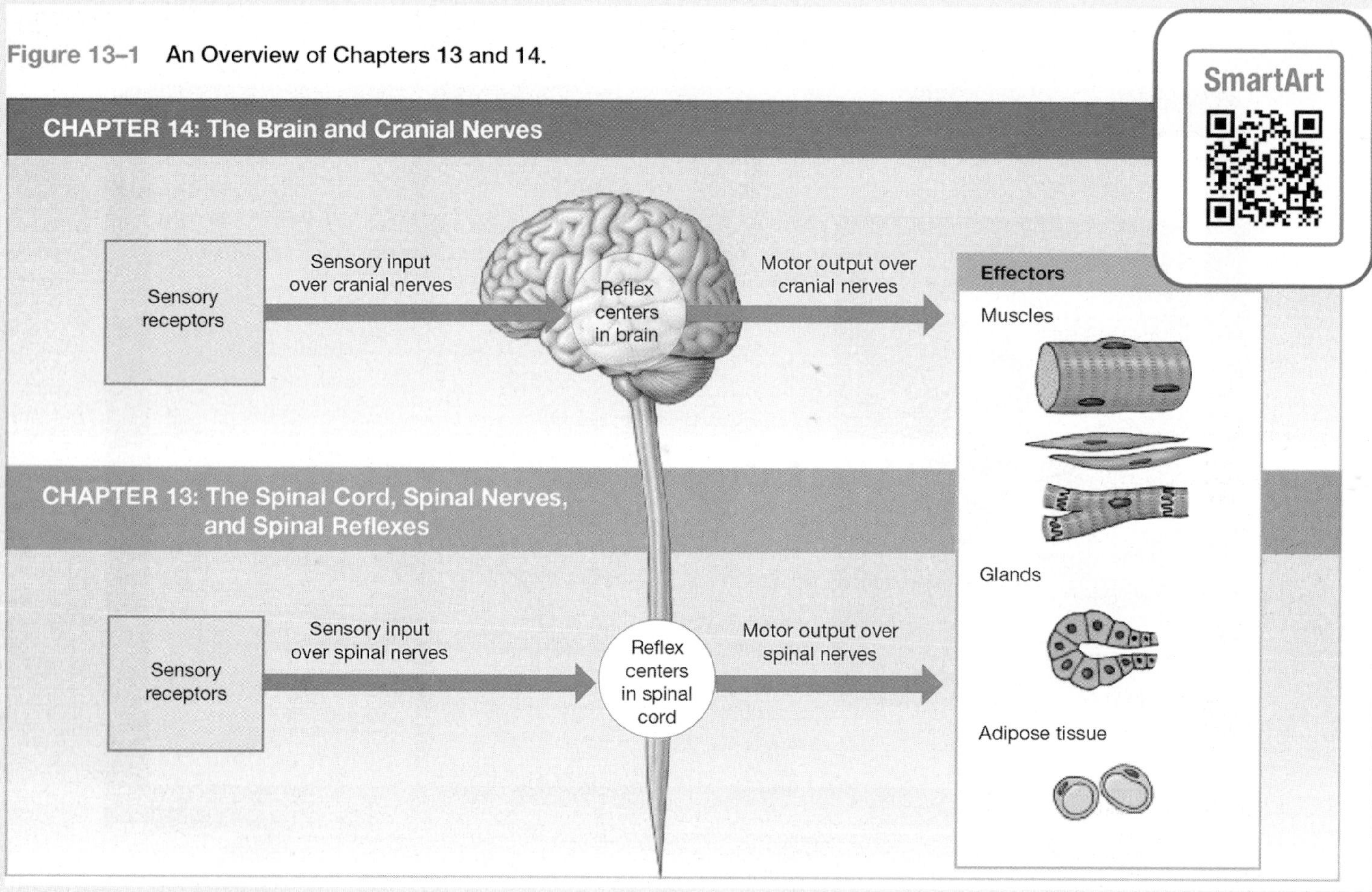

SmartArt
CHAPTER 14: The Brain and Cranial Nerves
Sensory receptors
Sensory input over cranial nerves
Reflex centers in brain
Motor output over cranial nerves
Effectors
Muscles
CHAPTER 13: The Spinal Cord, Spinal Nerves, and Spinal Reflexes
Sensory receptors
Sensory input over spinal nerves
Reflex centers in spinal cord
Motor output over spinal nerves
Glands
Adipose tissue

Figure 13–1 An Overview of Chapters 13 and 14.

Spotlight Figures provide highly visual one- and two-page presentations of tough topics in the book, with a particular focus on physiology.

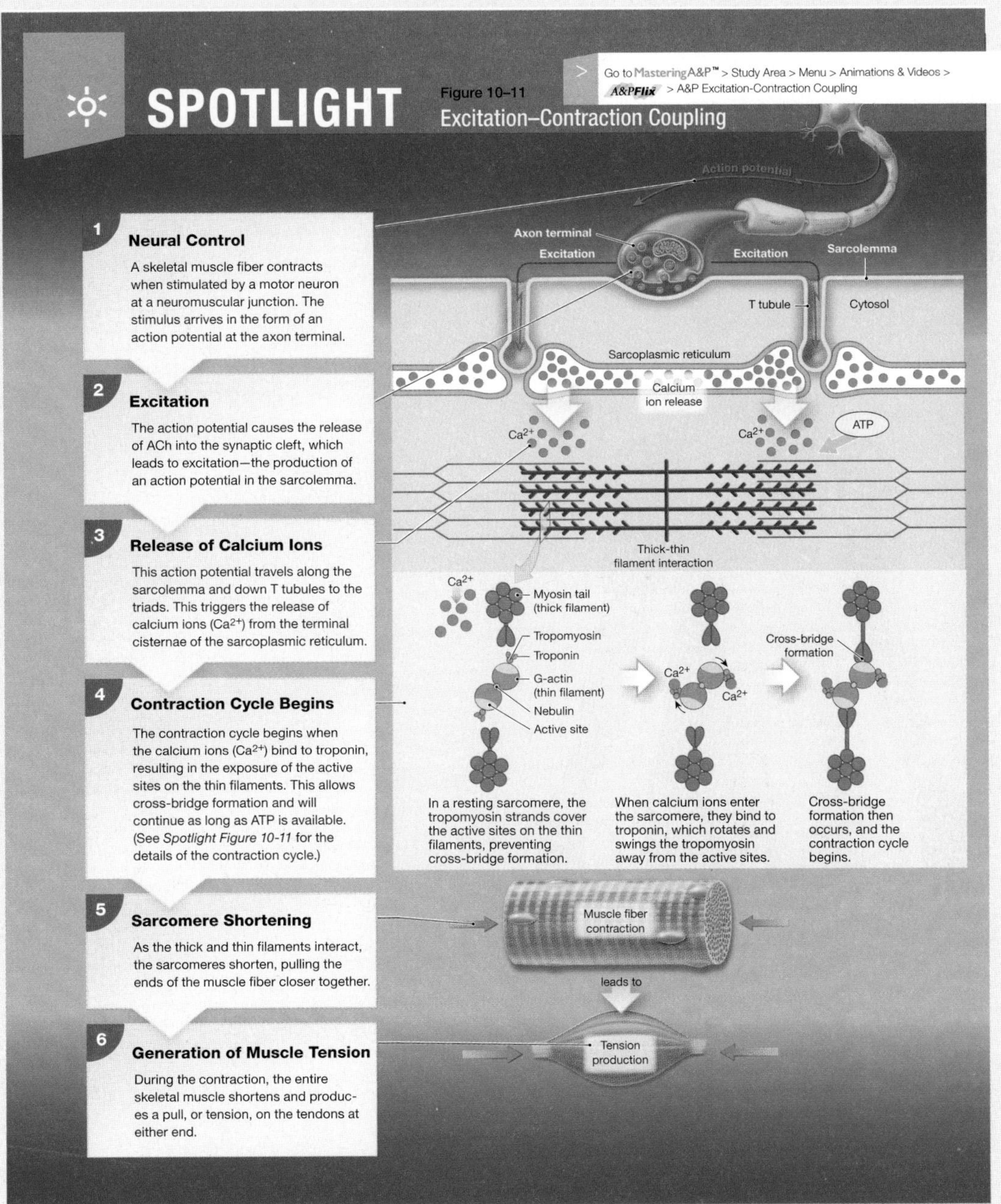

SPOTLIGHT
Figure 10–11
Excitation–Contraction Coupling
Go to MasteringA&P™ > Study Area > Menu > Animations & Videos > A&PFlix > A&P Excitation-Contraction Coupling
Action potential
1 Neural Control
A skeletal muscle fiber contracts when stimulated by a motor neuron at a neuromuscular junction. The stimulus arrives in the form of an action potential at the axon terminal.
2 Excitation
The action potential causes the release of ACh into the synaptic cleft, which leads to excitation—the production of an action potential in the sarcolemma.
3 Release of Calcium Ions
This action potential travels along the sarcolemma and down T tubules to the triads. This triggers the release of calcium ions (Ca^{2+}) from the terminal cisternae of the sarcoplasmic reticulum.
4 Contraction Cycle Begins
The contraction cycle begins when the calcium ions (Ca^{2+}) bind to troponin, resulting in the exposure of the active sites on the thin filaments. This allows cross-bridge formation and will continue as long as ATP is available. (See *Spotlight Figure 10-11* for the details of the contraction cycle.)
5 Sarcomere Shortening
As the thick and thin filaments interact, the sarcomeres shorten, pulling the ends of the muscle fiber closer together.
6 Generation of Muscle Tension
During the contraction, the entire skeletal muscle shortens and produces a pull, or tension, on the tendons at either end.
Axon terminal
Excitation
Excitation
Sarcolemma
T tubule
Cytosol
Sarcoplasmic reticulum
Calcium ion release
Ca^{2+}
ATP
Thick-thin filament interaction
Myosin tail (thick filament)
Tropomyosin
Troponin
G-actin (thin filament)
Nebulin
Active site
Cross-bridge formation
In a resting sarcomere, the tropomyosin strands cover the active sites on the thin filaments, preventing cross-bridge formation.
When calcium ions enter the sarcomere, they bind to troponin, which rotates and swings the tropomyosin away from the active sites.
Cross-bridge formation then occurs, and the contraction cycle begins.
Muscle fiber contraction
leads to
Tension production

NEW! MasteringA&P references within the chapter direct students to specific digital resources, such as tutorials, animations, and videos, that will help further their understanding of key concepts in the course.

Systems Integration in the Classroom

NEW! Build Your Knowledge features show how each body system influences the others. As students progress through the book, they will build their knowledge about how the body systems work together to maintain homeostasis.

Build Your Knowledge

Figure 11–24 Integration of the MUSCULAR system with the other body systems presented so far.

Integumentary System

- The Integumentary System removes excess body heat, synthesizes vitamin D_3 for calcium and phosphate absorption, and protects underlying muscles.
- The muscular system includes facial muscles that pull on the skin of the face to produce facial expressions

Skeletal System

- The Skeletal System provides mineral reserves for maintaining normal calcium and phosphate levels in body fluids, supports skeletal muscles, and provides sites of muscle attachment.
- The muscular system provides skeletal movement and support, and stabilizes bones and joints. Stresses exerted by tendons maintain normal bone structure and bone mass.

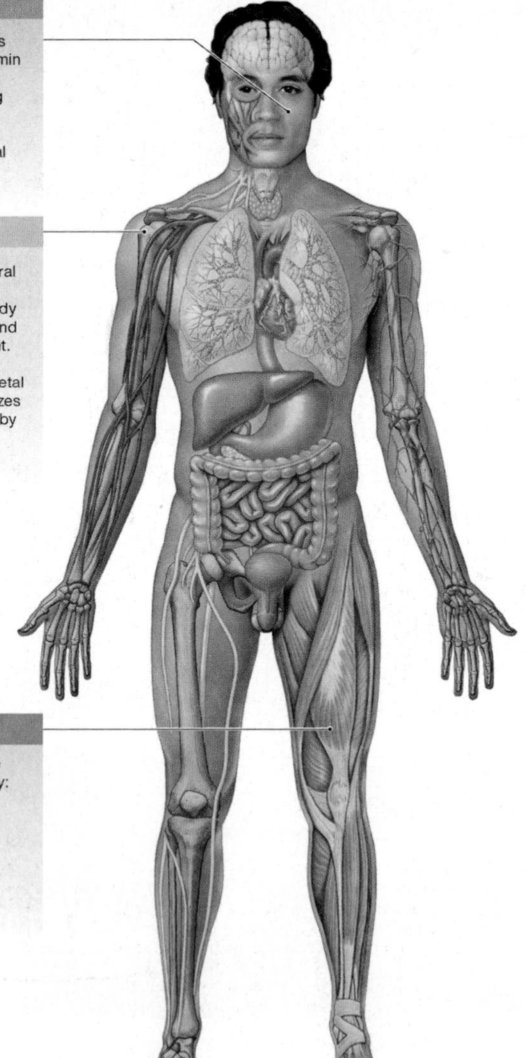

Muscular System

The muscular system performs these primary functions for the human body:

- It produces skeletal movement
- It helps maintain posture and body position
- It supports soft tissues
- It guards entrances and exits to the body
- It helps maintain body temperature

and Beyond

Clinical Cases get students motivated for their future careers. Each chapter opens with a story-based Clinical Case related to the chapter content and ends with a Clinical Case Wrap-Up.

+ CLINICAL CASE He Has Fish Skin!

I shook his hand and immediately I knew something was different about him. When Will clasped my hand between both of his, I felt like my hand was sandwiched between two sheets of thick, shaggy sandpaper. There was none of the moistness or warmth of a usual handshake. These hands belonged to Grandpa Will.

Grandpa Will's grandsons adored him, and the feeling was mutual! They lured him into chasing them around the backyard. Because it was a hot summer day, play

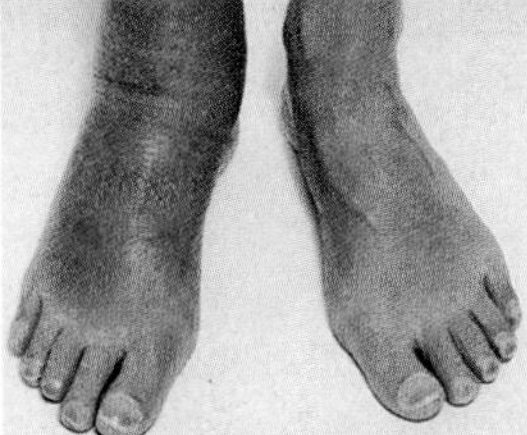

lasted all of 15 minutes and then Grandpa brought the gang back to the air-conditioned comfort of the house. He sank back into the recliner. He was flushed and breathing hard, but his shirt stayed dry and crisp—there wasn't a bead of sweat visible on him. The boys climbed onto his lap, laughing, as he encircled them with those coarse hands. "Oh, Grandpa, you feel like a fish!" **What is happening with Grandpa Will's integumentary system? To find out, turn to the Clinical Case Wrap-Up on p. 179.**

+ CLINICAL CASE Wrap-Up He Has Fish Skin!

Grandpa Will has *ichthyosis vulgaris* (IK-thē-oh-sis vōl-GĀR-is). Ichthyosis literally means "fish-like condition," a feature his grandsons noticed right away. Will inherited this skin condition from his parents, who carried a gene mutation for a structural protein. The lack of this protein impairs keratinization. With this condition, skin cells also have fewer desmosomes and tight junctions, so the epidermis becomes flaky and resembles fish scales.

There is no cure. Every day Grandpa must tend to his skin by applying moisturizers to draw and retain moisture and emollients to soften the scales. He avoids soap, which further dries his skin, and uses rubber gloves to wash the dishes. In winter, he runs a humidifier in the bedroom. If his skin care gets lax, his skin can break into deep, painful cracks. Because the natural skin barrier is disrupted, secondary infection can move in.

1. Why is ichthyosis called a "disorder of cornification"? Which cells are involved?
2. Grandpa has trouble when it's hot outside. What's wrong?

See the blue Answers tab at the back of the book.

+ Clinical Note Abnormal Bone Development

A variety of endocrine or metabolic problems can result in characteristic skeletal changes. In **pituitary growth failure**, inadequate production of growth hormone leads to reduced epiphyseal cartilage activity and abnormally short bones. This condition is becoming increasingly rare in the United States, because children can be treated with synthetic human growth hormone.

Gigantism results from an overproduction of growth hormone before puberty (**Photo a**). (The world record for height is 272 cm, or 8 ft, 11 in. It was reached by Robert Wadlow, of Alton, Illinois, who died at age 22 in 1940. Wadlow weighed 216 kg, or 475 lb.) If the growth hormone level rises abnormally after epiphyseal cartilages close, the skeleton does not grow longer. Instead, bones get thicker, especially in the face, jaw, and hands. Cartilage growth and alterations in soft-tissue structure lead to changes in physical features, such as the contours of the face. These physical changes take place in the disorder called **acromegaly** (ak-roh-MEG-ah-lē).

Several inherited metabolic conditions that affect many systems influence the growth and development of the skeletal system. These conditions produce characteristic variations in body proportions. For example, many individuals with **Marfan's syndrome** are very tall and have long, slender limbs (**Photo b**). The cause is excessive cartilage formation at the epiphyseal cartilages. The underlying mutation affects the structure of connective tissue throughout the body, and commonly causes cardiovascular events such as the sudden death of athletes during strenuous athletic contests.

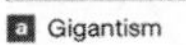

a Gigantism

b Marfan's syndrome

Clinical Notes appear within every chapter and expand upon topics just discussed. They present diseases and pathologies along with their relationship to normal function.

Related Clinical Terms

carbuncle: A skin infection that often involves a group of hair follicles. The infected material forms a lump, which occurs deep in the skin; the medical term for multiple boils.

cold sore: A lesion that typically occurs in or around the mouth and is caused by a dormant herpes simplex virus that may be reactivated by factors such as stress, fever, or sunburn. Also called *fever blister.*

comedo: The primary sign of acne consisting of an enlarged pore filled with skin debris, bacteria, and sebum (oil); the medical term for a blackhead.

dermatology: The branch of medicine concerned with the diagnosis, treatment, and prevention of diseases of the skin, hair, and nails.

eczema: Rash characterized by inflamed, itchy, dry, scaly, or irritated skin.

frostbite: Injury to body tissues caused by exposure to below-freezing temperatures, typically affecting the nose, fingers, or toes and sometimes resulting in gangrene.

furuncle: A skin infection involving an entire hair follicle and nearby skin tissue; the medical term for a boil.

gangrene: A term that describes dead or dying body tissue that occurs because the local blood supply to the tissue is either lost or is inadequate to keep the tissue alive.

Clinical Terms end every chapter with a list of relevant clinical terms and definitions.

Continuous Learning Before, During, and After Class

Dynamic Study Modules enable students to study more effectively on their own. With the Dynamic Study Modules mobile app, students can quickly access and learn the concepts they need to be more successful on quizzes and exams.

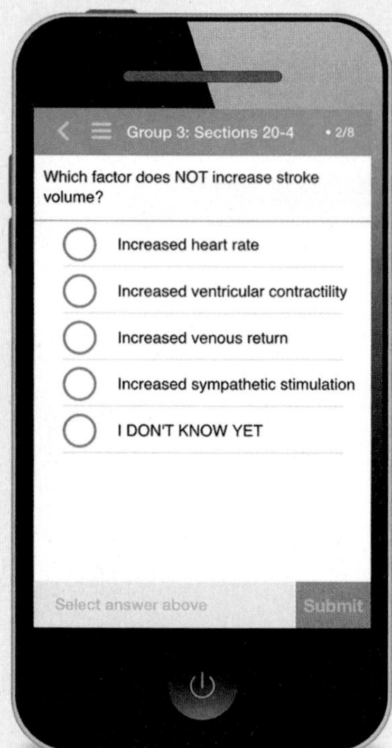

Group 3: Sections 20-4
• 2/8
Which factor does NOT increase stroke volume?
Increased heart rate
Increased ventricular contractility
Increased venous return
Increased sympathetic stimulation
I DON'T KNOW YET
Select answer above
Submit

NEW! Instructors can now select which questions to assign to students.

NEW! SmartArt Videos help students navigate some of the complex figures in the text. They are accessible via QR code in the book and are assignable in MasteringA&P.

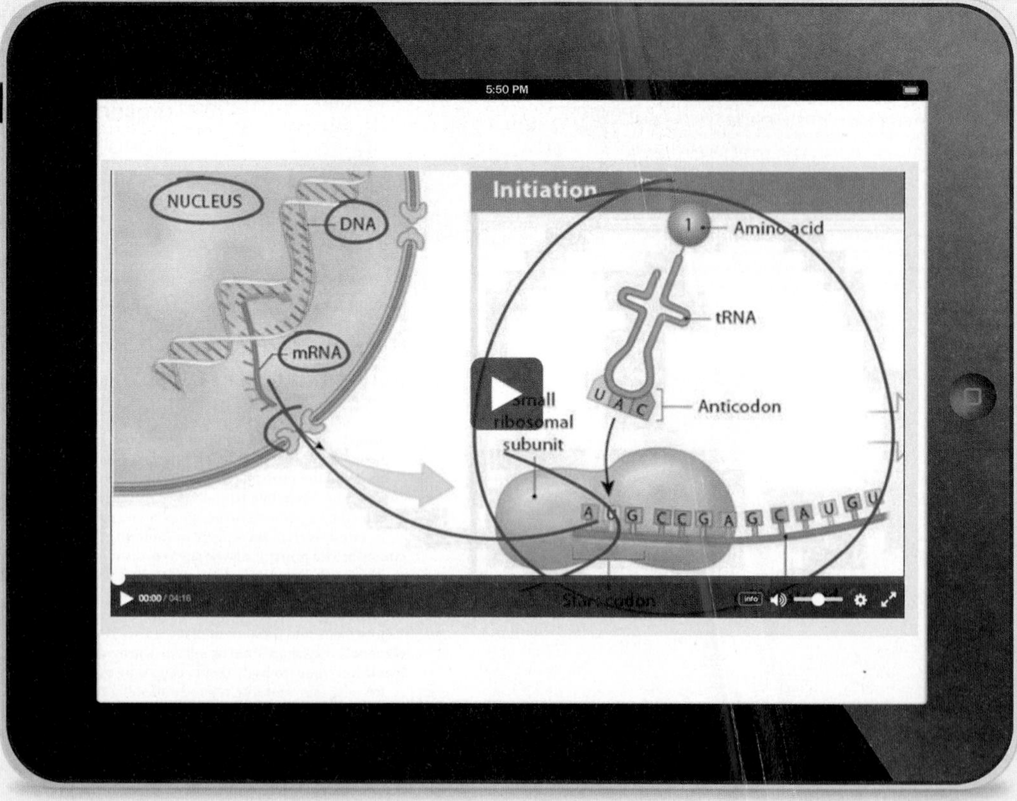

5:50 PM
NUCLEUS
DNA
mRNA
Initiation
Amino acid
tRNA
Anticodon
Small ribosomal subunit
Start codon

with MasteringA&P™

Learning Catalytics is a "bring your own device" (laptop, smartphone, or tablet) engagement, assessment, and classroom intelligence system. Students use their device to respond to open-ended questions and then discuss answers in class based on their responses.

> ***"My students are so busy and engaged answering Learning Catalytics questions during lecture that they don't have time for Facebook."***
> —Declan De Paor, Old Dominion University

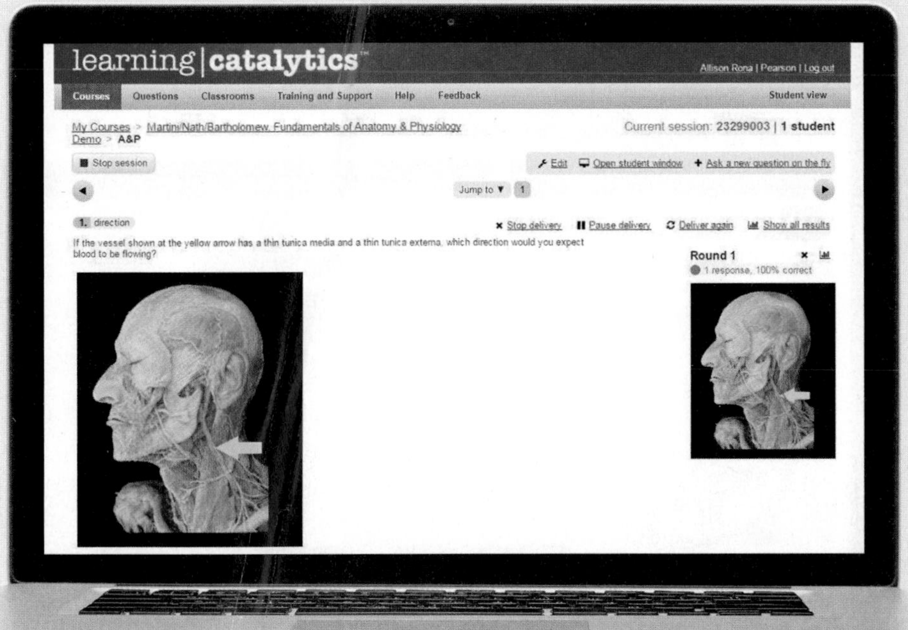
learning | catalytics
Allison Rona | Pearson | Log out
Courses
Questions
Classrooms
Training and Support
Help
Feedback
Student view
My Courses > Martini/Nath/Bartholomew, Fundamentals of Anatomy & Physiology Demo > A&P
Current session: 23299003 | 1 student
Stop session
Edit
Open student window
Ask a new question on the fly
Jump to
1. direction
Stop delivery
Pause delivery
Deliver again
Show all results
If the vessel shown at the yellow arrow has a thin tunica media and a thin tunica externa, which direction would you expect blood to be flowing?
Round 1
1 response, 100% correct

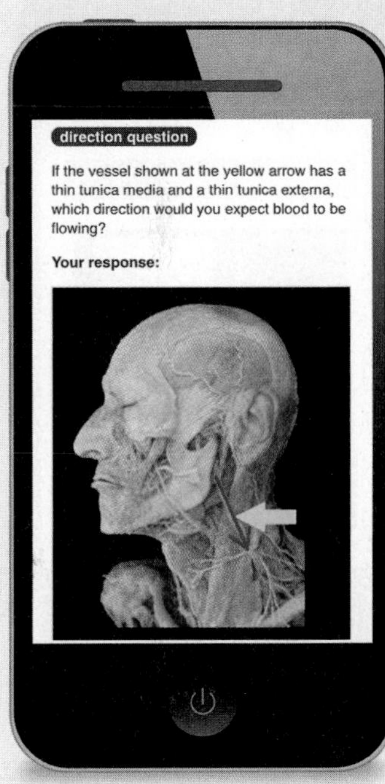
direction question
If the vessel shown at the yellow arrow has a thin tunica media and a thin tunica externa, which direction would you expect blood to be flowing?
Your response:

MasteringA&P™

NEW! Interactive Physiology 2.0 helps students advance beyond memorization to a genuine understanding of complex physiological processes. Fun, interactive tutorials, games, and quizzes give students additional explanations to help them grasp difficult concepts. IP 2.0 features brand-new graphics, quicker navigation, and more robust interactivity.

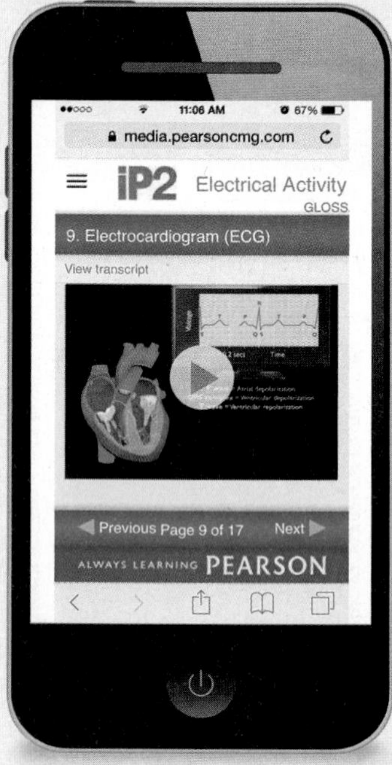
11:06 AM
67%
media.pearsoncmg.com
iP2
Electrical Activity
GLOSS
9. Electrocardiogram (ECG)
View transcript
Previous Page 9 of 17
Next
ALWAYS LEARNING PEARSON

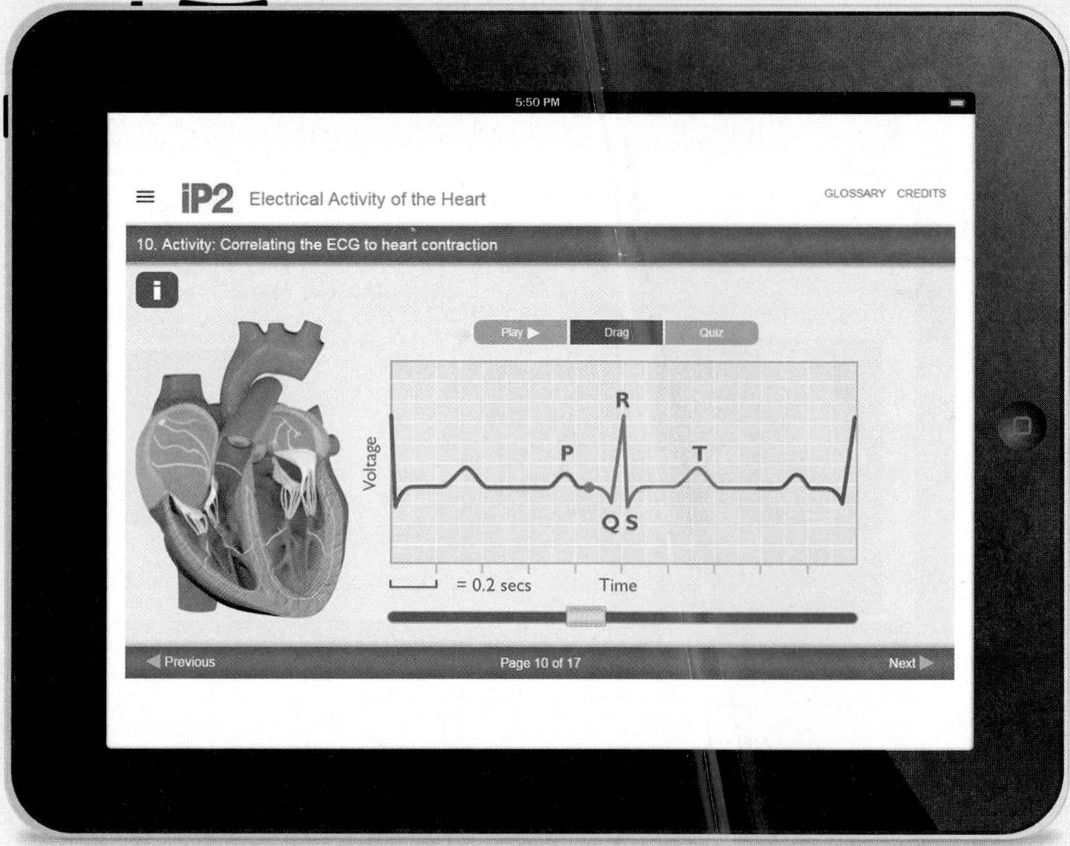
5:50 PM
iP2
Electrical Activity of the Heart
GLOSSARY
CREDITS
10. Activity: Correlating the ECG to heart contraction
Play
Drag
Quiz
R
P
T
Q S
Voltage
= 0.2 secs
Time
Previous
Page 10 of 17
Next

NEW IP 2.0 modules include:

- Resting Membrane Potential
- Electrical Activity of the Heart
- Cardiac Output
- Factors Affecting Blood Pressure
- Generation of an Action Potential

Coming Soon:

- Cardiac Cycle
- Glomerular Filtration
- Neuromuscular Junction
- Tubular Reabsorption and Secretion
- Excitation Contraction Coupling

More Practice, More Learning

A&P Flix Coaching Activities bring interactivity to these popular 3D movie-quality animations by asking students to answer questions related to the video.

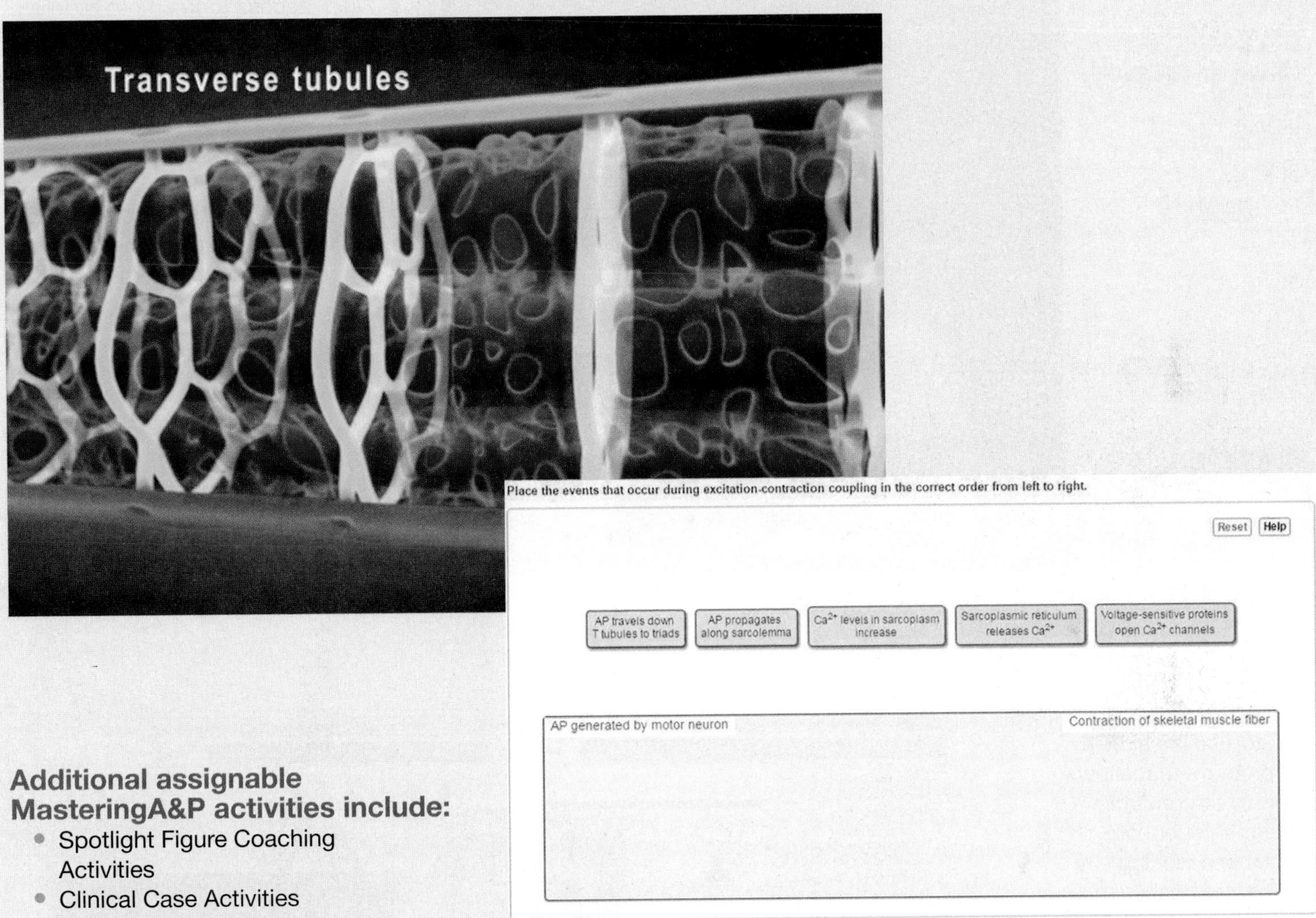
Transverse tubules
Place the events that occur during excitation-contraction coupling in the correct order from left to right.
Reset
Help
AP travels down T tubules to triads
AP propagates along sarcolemma
Ca2+ levels in sarcoplasm increase
Sarcoplasmic reticulum releases Ca2+
Voltage-sensitive proteins open Ca2+ channels
AP generated by motor neuron
Contraction of skeletal muscle fiber
Submit
Hints
My Answers
Give Up
Review Part
Incorrect; Try Again
The AP causes the voltage-sensitive proteins located in the T tubules to change shape.

Additional assignable MasteringA&P activities include:

- Spotlight Figure Coaching Activities
- Clinical Case Activities
- Clinical Note Activities
- Bone & Dissection Video Coaching Activities
- And More!

NEW! Beginning Fall 2017, all of the assignments from Wood's *Laboratory Manual for A&P featuring Martini Art*, 6e can be accessed in your *Fundamentals of A&P* Mastering course! Only one MasteringA&P code is needed to access these assignments.

MasteringA&P™ Lab Practice

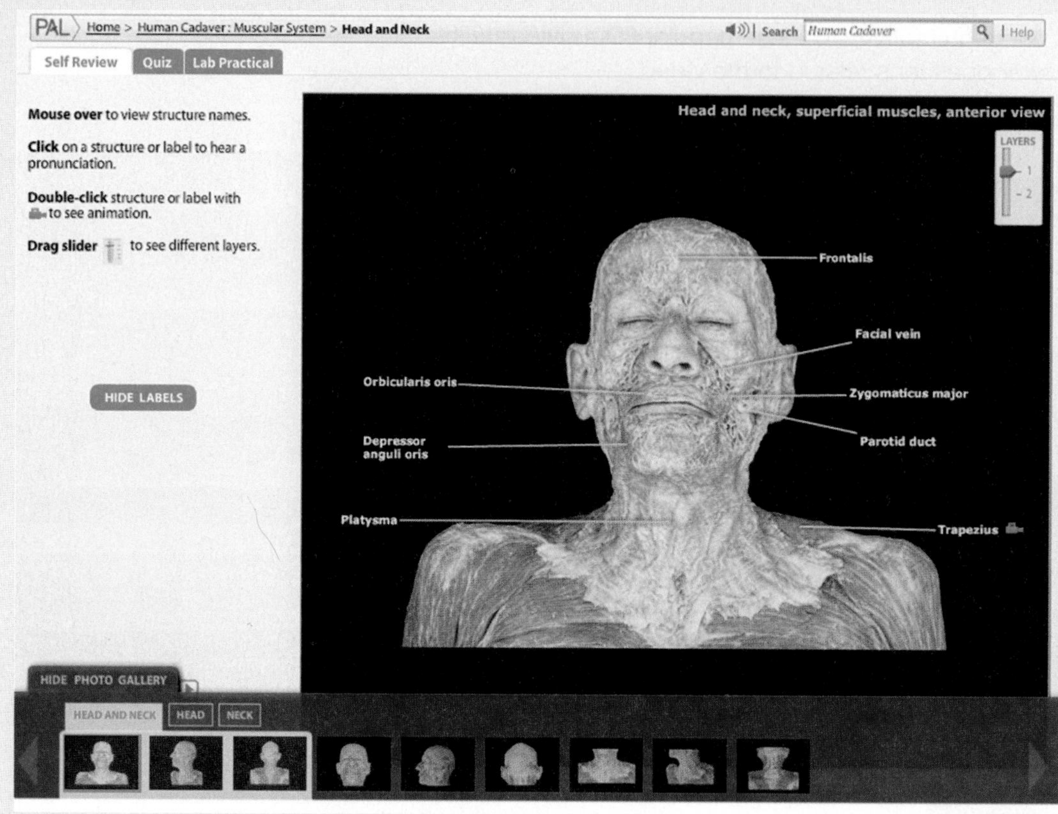
PAL
Home > Human Cadaver: Muscular System > Head and Neck
Search
Human Cadaver
Help
Self Review
Quiz
Lab Practical
Mouse over to view structure names.
Click on a structure or label to hear a pronunciation.
Double-click structure or label with to see animation.
Drag slider to see different layers.
HIDE LABELS
Head and neck, superficial muscles, anterior view
LAYERS
Frontalis
Facial vein
Orbicularis oris
Zygomaticus major
Depressor anguli oris
Parotid duct
Platysma
Trapezius
HIDE PHOTO GALLERY
HEAD AND NECK
HEAD
NECK

Practice Anatomy Lab (PAL™ 3.0) is a virtual anatomy study and practice tool that gives students 24/7 access to the most widely used lab specimens, including the human cadaver, anatomical models, histology, cat, and fetal pig. PAL 3.0 is easy to use and includes built-in audio pronunciations, rotatable bones, and simulated fill-in-the-blank lab practical exams.

PhysioEx 9.1 is an easy-to-use lab simulation program that allows students to conduct experiments that are difficult in a wet lab environment because of time, cost, or safety concerns. Students are able to repeat labs as often as they like, perform experiments without animals, and are asked to frequently stop and predict within the labs.

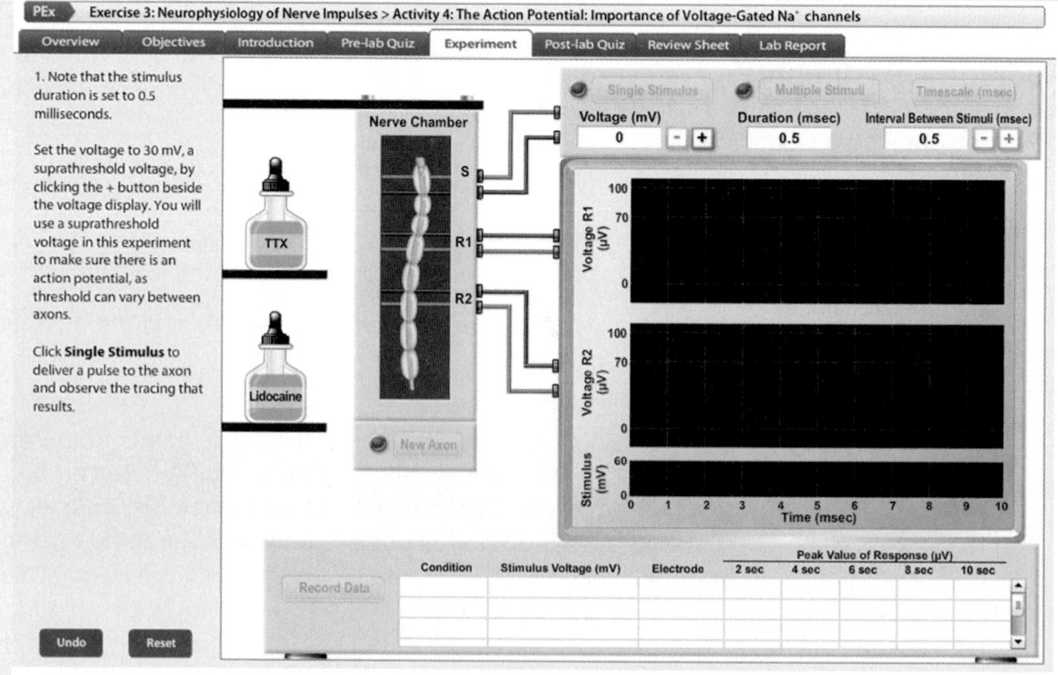
PEx
Exercise 3: Neurophysiology of Nerve Impulses > Activity 4: The Action Potential: Importance of Voltage-Gated Na+ channels
Overview
Objectives
Introduction
Pre-lab Quiz
Experiment
Post-lab Quiz
Review Sheet
Lab Report
1. Note that the stimulus duration is set to 0.5 milliseconds.
Set the voltage to 30 mV, a suprathreshold voltage, by clicking the + button beside the voltage display. You will use a suprathreshold voltage in this experiment to make sure there is an action potential, as threshold can vary between axons.
Click Single Stimulus to deliver a pulse to the axon and observe the tracing that results.
Undo
Reset
Nerve Chamber
TTX
Lidocaine
S
R1
R2
New Axon
Single Stimulus
Multiple Stimuli
Timescale (msec)
Voltage (mV)
Duration (msec)
Interval Between Stimuli (msec)
Voltage R1 (μV)
Voltage R2 (μV)
Stimulus (mV)
Time (msec)
Record Data
Condition
Stimulus Voltage (mV)
Electrode
Peak Value of Response (μV)
2 sec
4 sec
6 sec
8 sec
10 sec

Access the Complete Textbook On or Offline with eText 2.0

NEW! The **Eleventh Edition** is available in Pearson's fully-accessible eText 2.0 platform.*

NEW! The eText 2.0 mobile app offers offline access and can be downloaded for most iOS and Android phones and tablets from the iTunes or Google Play stores.

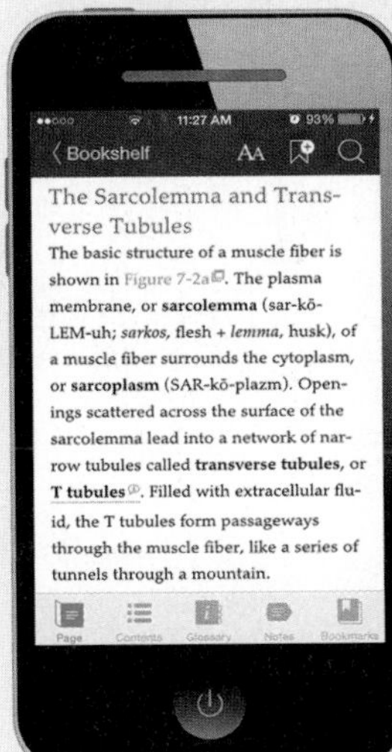
11:27 AM
Bookshelf
The Sarcolemma and Transverse Tubules
The basic structure of a muscle fiber is shown in Figure 7-2a. The plasma membrane, or sarcolemma (sar-kō-LEM-uh; sarkos, flesh + lemma, husk), of a muscle fiber surrounds the cytoplasm, or sarcoplasm (SAR-kō-plazm). Openings scattered across the surface of the sarcolemma lead into a network of narrow tubules called transverse tubules, or T tubules. Filled with extracellular fluid, the T tubules form passageways through the muscle fiber, like a series of tunnels through a mountain.

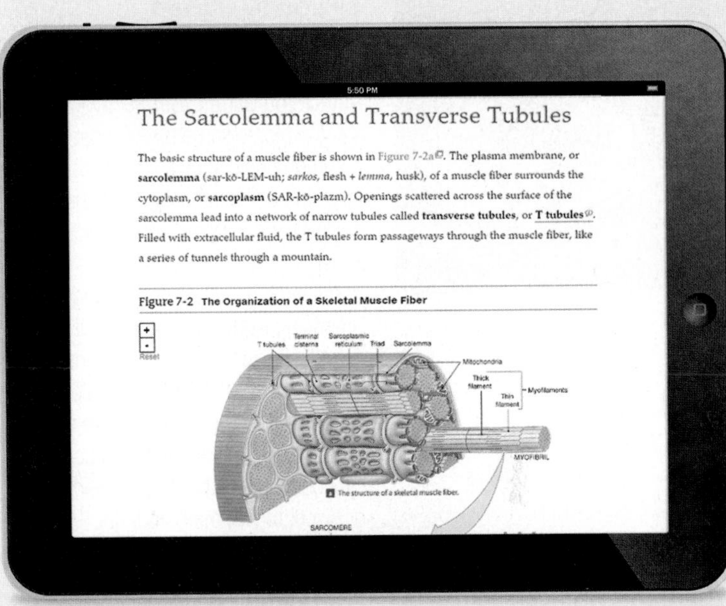
5:50 PM
The Sarcolemma and Transverse Tubules
The basic structure of a muscle fiber is shown in Figure 7-2a. The plasma membrane, or sarcolemma (sar-kō-LEM-uh; sarkos, flesh + lemma, husk), of a muscle fiber surrounds the cytoplasm, or sarcoplasm (SAR-kō-plazm). Openings scattered across the surface of the sarcolemma lead into a network of narrow tubules called transverse tubules, or T tubules. Filled with extracellular fluid, the T tubules form passageways through the muscle fiber, like a series of tunnels through a mountain.
Figure 7-2 The Organization of a Skeletal Muscle Fiber

Powerful interactive and customization functions include instructor and student note-taking, highlighting, bookmarking, search, and links to glossary terms.

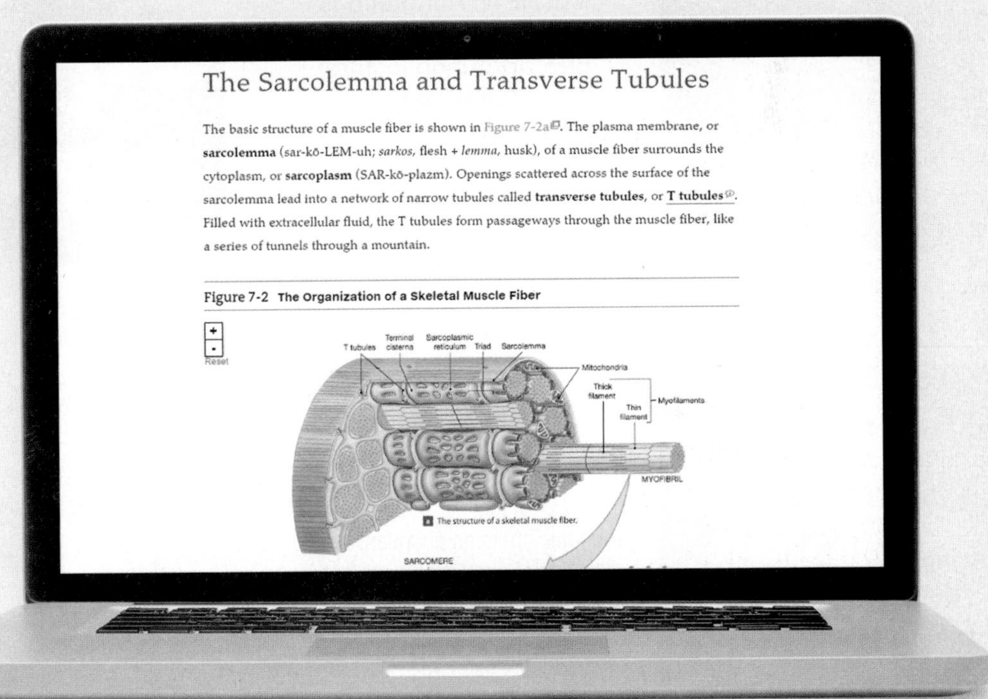
The Sarcolemma and Transverse Tubules
The basic structure of a muscle fiber is shown in Figure 7-2a. The plasma membrane, or sarcolemma (sar-kō-LEM-uh; sarkos, flesh + lemma, husk), of a muscle fiber surrounds the cytoplasm, or sarcoplasm (SAR-kō-plazm). Openings scattered across the surface of the sarcolemma lead into a network of narrow tubules called transverse tubules, or T tubules. Filled with extracellular fluid, the T tubules form passageways through the muscle fiber, like a series of tunnels through a mountain.
Figure 7-2 The Organization of a Skeletal Muscle Fiber
Reset
T tubules
Terminal cisterna
Sarcoplasmic reticulum
Triad
Sarcolemma
Mitochondria
Thick filament
Thin filament
Myofilaments
MYOFIBRIL
The structure of a skeletal muscle fiber.
SARCOMERE

*The eText 2.0 edition will be live for Fall 2017 classes.

Instructor and Student Support

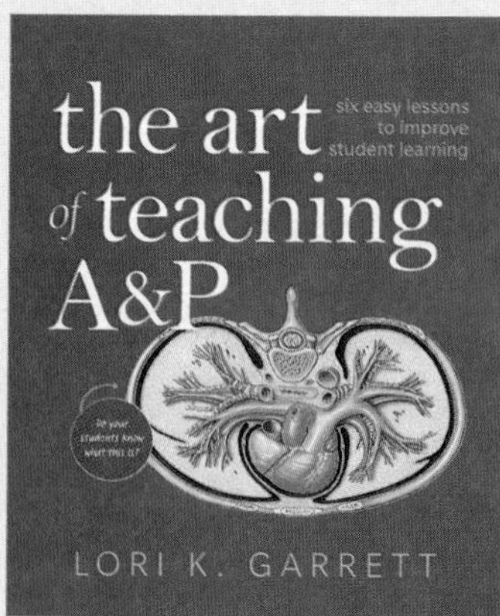
the art of teaching A&P
six easy lessons to improve student learning
LORI K. GARRETT

NEW! The Art of Teaching A&P: Six Easy Lessons to Improve Student Learning by Lori K. Garrett
978-0-13-446951-5
0-13-446951-8
Author Lori Garrett *(Get Ready for A&P)* explores some of the most common challenges she's encountered in her classroom when using art to teach anatomy and physiology. From describing the challenge to researching why it occurs and proposing solutions to address it, Lori provides insight into how students look at images. She presents ideas for how educators can best use figures and illustrations to teach complex concepts without overwhelming or discouraging their students.

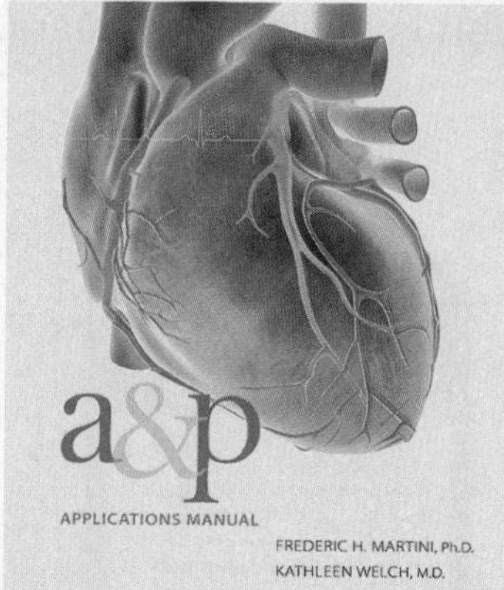
a&p
APPLICATIONS MANUAL
FREDERIC H. MARTINI, Ph.D.
KATHLEEN WELCH, M.D.

A&P Applications Manual by Frederic H. Martini and Kathleen Welch
978-0-32-194973-8
0-32-194973-0
This manual contains extensive discussions on clinical topics and disorders to help students apply the concepts of anatomy and physiology to daily life and their future health professions. Free when packaged with the textbook.

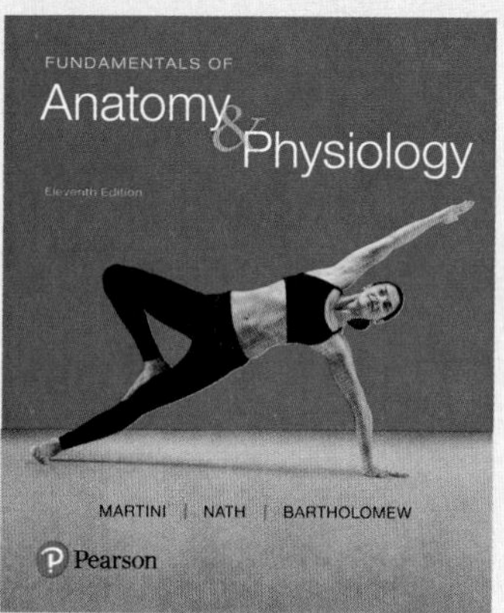
FUNDAMENTALS OF
Anatomy & Physiology
MARTINI | NATH | BARTHOLOMEW
Pearson

A complete package of instructor resources includes:

- Customizable PowerPoint slides (with **NEW! Annotations on how to present complex art during lecture)**
- All figures from the book in JPEG format
- A&P Flix 3D movie-quality animations on tough topics
- Test Bank
- And more!

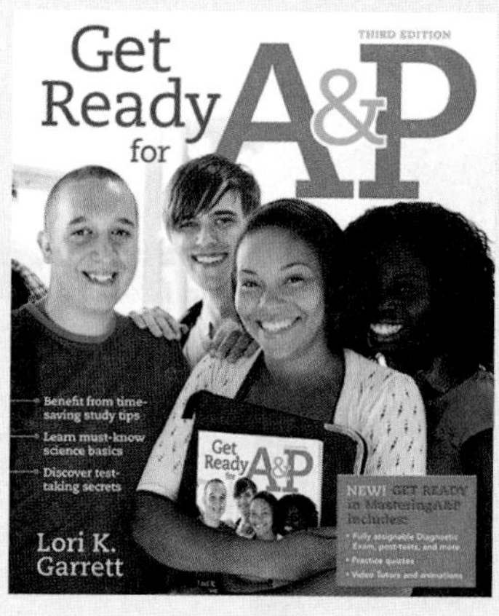
THIRD EDITION
Get Ready for A&P
Lori K. Garrett

Get Ready for A&P by Lori K. Garrett
978-0-32-181336-7
0-32-181336-7
This book and online component were created to help students be better prepared for their A&P course. Features include pre-tests, guided explanations followed by interactive quizzes and exercises, and end-of-chapter cumulative tests. Also available in the Study Area of MasteringA&P. Free when packaged with the textbook.

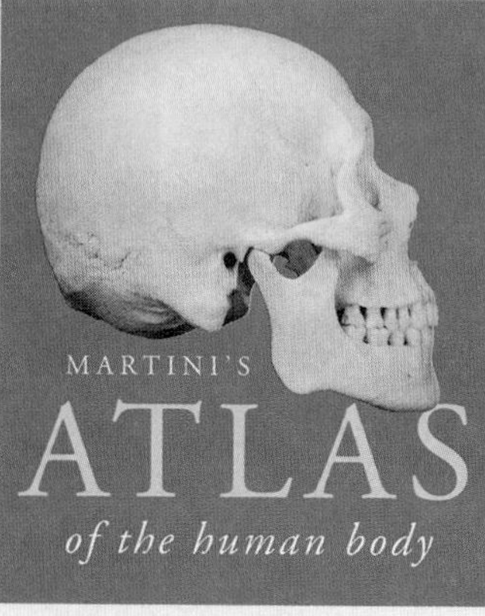
MARTINI'S
ATLAS
of the human body

Martini's Atlas of the Human Body by Frederic H. Martini
978-0-32-194072-8
0-32-194072-5
The Atlas offers an abundant collection of anatomy photographs, radiology scans, and embryology summaries, helping students visualize structures and become familiar with the types of images seen in a clinical setting. Free when packaged with the textbook.

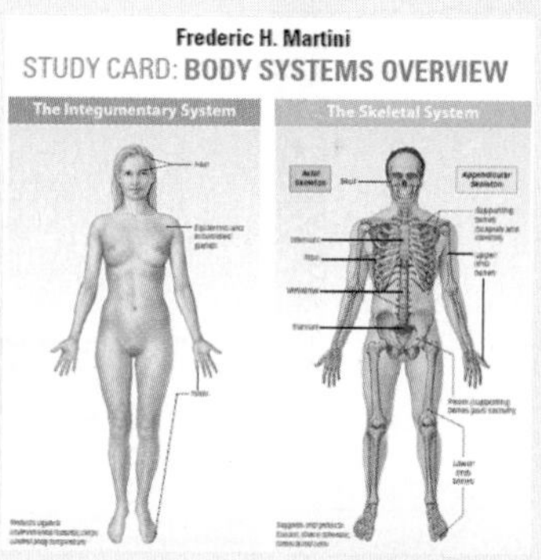
Frederic H. Martini
STUDY CARD: BODY SYSTEMS OVERVIEW
The Integumentary System
The Skeletal System

Study Card for Martini: Body Systems Overview
978-0-13-460995-9
0-13-460995-6
A six-panel laminated card showing all body systems and their organs and functions. Free when packaged with the textbook.

FUNDAMENTALS OF

Anatomy & Physiology

Eleventh Edition

Frederic H. Martini, Ph.D.
University of Hawaii at Manoa

Judi L. Nath, Ph.D.
Lourdes University, Sylvania, Ohio

Edwin F. Bartholomew, M.S.

William C. Ober, M.D.
Art Coordinator and Illustrator

Claire E. Ober, R.N.
Illustrator

Kathleen Welch, M.D.
Clinical Consultant

Ralph T. Hutchings
Biomedical Photographer

Kevin Petti
San Diego Miramar College
SmartArt Videos

Christine Boudrie, M.D.
Lourdes University, Sylvania, Ohio
Clinical Cases

Ruth Anne O'Keefe, M.D.

Pearson

Editor-in-Chief: *Serina Beauparlant*
Courseware Portfolio Manager: *Cheryl Cechvala*
Content Producer: *Jessica Picone*
Managing Producer: *Nancy Tabor*
Courseware Director, Content Development: *Barbara Yien*
Courseware Sr. Analyst: *Suzanne Olivier*
Courseware Editorial Assistant: *Kimberly Twardochleb*
Rich Media Content Producer: *Lauren Chen*
Associate Mastering Producer: *Kristen Sanchez*
Full-Service Vendor: *Cenveo® Publisher Services*
Copyeditor: *Lorretta Palagi*
Art Coordinator: *Lisa Torri*
Design Manager: *Mark Ong*
Interior Designer: *tani hasegawa*
Cover Designer: *tani hasegawa*
Rights & Permissions Project Manager: *Kathleen Zander, Jason Perkins*
Rights & Permissions Management: *Cenveo® Publisher Services*
Photo Researcher: *Kristin Piljay*
Manufacturing Buyer: *Stacey Weinberger*
Executive Marketing Manager: *Allison Rona*
Cover Photo Credit: *RGB Ventures/SuperStock/Alamy Stock Photo*

Notice: Our knowledge in clinical sciences is constantly changing. The authors and the publisher of this volume have taken care that the information contained herein is accurate and compatible with the standards generally accepted at the time of the publication. Nevertheless, it is difficult to ensure that all information given is entirely accurate for all circumstances. The authors and the publisher disclaim any liability, loss, or damage incurred as a consequence, directly or indirectly, of the use and application of any of the contents of this volume.

Acknowledgements of third party content appear on page C-1, which constitutes an extension of this copyright page.

Library of Congress Cataloging-in-Publication Data

Names: Martini, Frederic, author. | Nath, Judi Lindsley, author. | Bartholomew, Edwin F., author. | Ober, William C., illustrator | Ober, Claire E., illustrator. | Welch, Kathleen (Kathleen Martini), consultant. | Hutchings, R. T., illustrator.
Title: Fundamentals of anatomy & physiology / Frederic H. Martini, Judi L. Nath, Edwin F. Bartholomew; William C. Ober, art coordinator and illustrator; Claire E. Ober, illustrator; Kathleen Welch, clinical consultant; Ralph T. Hutchings, biomedical photographer; clinical cases by Christine Boudrie, Ruth Anne O'Keefe.
Other titles: Fundamentals of anatomy and physiology
Description: Eleventh edition. | New York : Pearson Education, Inc., [2018] | Includes bibliographical references and index.
Identifiers: LCCN 2016041391| ISBN 9780134396026 (student edition : alk. paper) | ISBN 0134396022 (student edition : alk. paper)
Subjects: | MESH: Anatomy | Physiology
Classification: LCC QP34.5 | NLM QS 4 | DDC 612—dc23 LC record available at https://lccn.loc.gov/2016041391

4 18

ISBN 10: 0-13-439602-2; ISBN 13: 978-0-13-439602-6 (Student edition)
ISBN 10: 0-13-447727-8; ISBN 13: 978-0-13-447727-5 (Instructor's Review Copy)
ISBN 10: 0-13-457600-4; ISBN 13: 978-0-13-457600-8 (NASTA)

Text and Illustration Team

Frederic (Ric) H. Martini, Ph.D.

Author

Dr. Martini received his Ph.D. from Cornell University in comparative and functional anatomy for work on the pathophysiology of stress. In addition to professional publications that include journal articles and contributed chapters, technical reports, and magazine articles, he is the lead author of 10 undergraduate texts on anatomy and physiology. Dr. Martini is currently affiliated with the University of Hawaii at Manoa and has a long-standing bond with the Shoals Marine Laboratory, a joint venture between Cornell University and the University of New Hampshire. He has been active in the Human Anatomy and Physiology Society (HAPS) for over 24 years and was a member of the committee that established the course curriculum guidelines for A&P. He is now a President Emeritus of HAPS after serving as President-Elect, President, and Past-President over 2005–2007. Dr. Martini is also a member of the American Physiological Society, the American Association of Anatomists, the Society for Integrative and Comparative Biology, the Australia/New Zealand Association of Clinical Anatomists, the Hawaii Academy of Science, the American Association for the Advancement of Science, and the International Society of Vertebrate Morphologists.

Edwin F. Bartholomew, M.S.

Author

Edwin F. Bartholomew received his undergraduate degree from Bowling Green State University and his M.S. from the University of Hawaii. Mr. Bartholomew has taught human anatomy and physiology at both the secondary and undergraduate levels. In addition, he has taught courses ranging from botany to zoology at Maui Community College (now the University of Hawaii Maui College). For many years, he taught at historic Lahainaluna High School, the oldest high school west of the Rockies, where he assisted in establishing a Health Occupations Students of America (HOSA) chapter. He is a coauthor of *Fundamentals of Anatomy & Physiology, Visual Anatomy & Physiology, Essentials of Anatomy & Physiology, Visual Essentials of Anatomy & Physiology, Structure and Function of the Human Body,* and *The Human Body in Health and Disease* (all published by Pearson). Mr. Bartholomew is a member of the Human Anatomy and Physiology Society (HAPS), the National Association of Biology Teachers, the National Science Teachers Association, and the American Association for the Advancement of Science.

Judi L. Nath, Ph.D.

Author

Dr. Judi Nath is a biology professor and the writer-in-residence at Lourdes University, where she teaches at both the undergraduate and graduate levels. Primary courses include anatomy, physiology, pathophysiology, medical terminology, and science writing. She received her bachelor's and master's degrees from Bowling Green State University, which included study abroad at the University of Salzburg in Austria. Her doctoral work focused on autoimmunity, and she completed her Ph.D. from the University of Toledo. Dr. Nath is devoted to her students and strives to convey the intricacies of science in captivating ways that are meaningful, interactive, and exciting. She has won the Faculty Excellence Award—an accolade recognizing effective teaching, scholarship, and community service—multiple times and in 2013 was named as an Ohio Memorable Educator. She is active in many professional organizations, notably the Human Anatomy and Physiology Society (HAPS), where she has served several terms on the board of directors. Dr. Nath is a coauthor of *Visual Anatomy & Physiology, Visual Essentials of Anatomy & Physiology, Anatomy & Physiology,* and *Human Anatomy* (published by Pearson), and she is the sole author of *Using Medical Terminology* and *Stedman's Medical Terminology* (published by Wolters Kluwer). Her favorite charities are those that have significantly affected her life, including the local Humane Society, the Cystic Fibrosis Foundation, and the ALS Association. In 2015, she and her husband established the Nath Science Scholarship at Lourdes University to assist students pursuing science-based careers. When not working, days are filled with family life, bicycling, and hanging with the dogs.

William C. Ober, M.D.

Art Coordinator and Illustrator

Dr. Ober received his undergraduate degree from Washington and Lee University and his M.D. from the University of Virginia. He also studied in the Department of Art as Applied to Medicine at Johns Hopkins University. After graduation, Dr. Ober completed a residency in Family Practice and later was on the faculty at the University of Virginia in the Department of Family Medicine and in the Department of Sports Medicine. He also served as Chief of Medicine of Martha Jefferson Hospital in Charlottesville, Virginia. He is currently a Visiting Professor of Biology at Washington and Lee University, where he has taught several courses and led student trips to the Galapagos Islands. He was on the Core Faculty at Shoals Marine Laboratory for 24 years, where he taught Biological Illustration every summer. Dr. Ober has collaborated with Dr. Martini on all of his textbooks in every edition.

Claire E. Ober, R.N.

Illustrator

Claire E. Ober, R.N., B.A., practiced family, pediatric, and obstetric nursing before turning to medical illustration as a full-time career. She returned to school at Mary Baldwin College, where she received her degree with distinction in studio art. Following a 5-year apprenticeship, she has worked as Dr. Ober's partner in Medical & Scientific Illustration since 1986. She was on the Core Faculty at Shoals Marine Laboratory and co-taught the Biological Illustration course with Dr. Ober for 24 years. The textbooks illustrated by Medical & Scientific Illustration have won numerous design and illustration awards.

Kathleen Welch, M.D.

Clinical Consultant

Dr. Welch received her B.A. from the University of Wisconsin–Madison, her M.D. from the University of Washington in Seattle, and did her residency in Family Practice at the University of North Carolina in Chapel Hill. Participating in the Seattle WWAMI rural medical education program, she studied in Fairbanks, Anchorage, and Juneau, Alaska, with time in Boise, Idaho, and Anacortes, Washington, as well. For 2 years, she served as Director of Maternal and Child Health at the LBJ Tropical Medical Center in American Samoa and subsequently was a member of the Department of Family Practice at the Kaiser Permanente Clinic in Lahaina, Hawaii, and on the staff at Maui Memorial Hospital. She was in private practice from 1987 until her retirement in 2012. Dr. Welch is a Fellow of the American Academy of Family Practice and a member of the Hawaii Medical Association, the Maui County Medical Association, and the Human Anatomy and Physiology Society (HAPS). With Dr. Martini, she has coauthored both a textbook on anatomy and physiology and the *A&P Applications Manual*. She and Dr. Martini were married in 1979, and they have one son.

Ralph T. Hutchings

Biomedical Photographer

Mr. Hutchings was associated with the Royal College of Surgeons for 20 years. An engineer by training, he has focused for years on photographing the structure of the human body. The result has been a series of color atlases, including the *Color Atlas of Human Anatomy*, the *Color Atlas of Surface Anatomy*, and *The Human Skeleton* (all published by Mosby-Yearbook Publishing). For his anatomical portrayal of the human body, the International Photographers Association has chosen Mr. Hutchings as the best photographer of humans in the 20th century. He lives in North London, where he tries to balance the demands of his photographic assignments with his hobbies of early motor cars and airplanes.

Christine Boudrie, M.D.

Clinical Contributor

Dr. Boudrie studied at Brown University in Providence, Rhode Island, for her B.S. in biology, and also obtained her M.D. there. After graduation she served in the National Health Service Corps, a program of the U.S. Public Health Service, which sponsored her last 2 years of medical school. She was assigned to provide health education to the rural communities of southeast Michigan with a special focus on seniors. She has had the great pleasure of working with a variety of undergraduate and graduate students in the Northeast and Midwest, earning teaching excellence awards and a nomination for Carnegie Foundation's U.S. Professor of the Year in 2014. Currently, she chairs the Department of Biology and Health Sciences at Lourdes University, a small Franciscan liberal arts school in northwest Ohio.

Ruth Anne O'Keefe, M.D.

Clinical Contributor

Dr. O'Keefe did her undergraduate studies at Marquette University, attended graduate school at the University of Wisconsin, and received her M.D. from George Washington University. She was the first woman to study orthopedics at The Ohio State University during her residency. She did fellowship training in trauma surgery at Loma Linda University in California. She serves on the board of Global Health Partnerships, a group that partners with a clinic serving 35,000 people in remote Kenya. She lives in Albuquerque with her Sweet Ed. She is mother of four, grandmother of nine, and foster mother to many.

Kevin Petti, PhD

Smart Art Video Contributor

Dr. Petti is a professor at San Diego Miramar College, and teaches courses in human anatomy and physiology, human dissection, and health education. He is President Emeritus of the Human Anatomy and Physiology Society (HAPS) and holds a doctorate from the University of San Diego. As a dual U.S./Italian citizen, he also teaches courses in Italy that focus on the genesis of anatomy as a science and its influence on the Renaissance masters.

Preface

The Eleventh Edition of *Fundamentals of Anatomy & Physiology* is a comprehensive textbook that fulfills the needs of today's students while addressing the concerns of their teachers. We focused our attention on the question "How can we make this information meaningful, manageable, and comprehensible?" During the revision process, we drew upon our content knowledge, research skills, artistic talents, and years of classroom experience to make this edition the best yet.

The broad changes to this edition are presented in the **New to the Eleventh Edition** section below, and the specific changes are presented in the **Chapter-by-Chapter Changes in the Eleventh Edition** section that follows.

New to the Eleventh Edition

In addition to the many technical changes in this edition, such as updated statistics and anatomy and physiology descriptions, we have made the following key changes:

- **NEW SmartArt Videos** help students better navigate key, complex pieces of art. Author Kevin Petti walks students through select pieces of art from the book, providing additional background and detail.
- **NEW design for homeostasis figures** replaces former Tenth Edition figures in various chapters.
- **NEW Questions have been added to selected figures in all chapters to reinforce text–art integration.**
- **Easier narrative leads to improved clarity of text.** Clearly organized text uses simpler, shorter, more active sentences, with a reading level that makes reading and studying easier for students.
- **Anatomical terms** have been updated based on *Terminologia Anatomica, Terminologia Histologica*, and *Terminologia Embryologica*. Eponyms continue to be included within the narrative.

Hallmark Features of This Text

- **50 Spotlight Figures** provide highly visual one- and two-page presentations of tough topics in the book, with a particular focus on physiology.
- **29 Clinical Cases** get students motivated for their future careers. Each chapter opens with a story-based Clinical Case related to the chapter content and ends with a Clinical Case Wrap-Up.
- **The repetition of the chapter-opening Learning Outcomes below the coordinated section headings within the chapters** underscores the connection between the HAPS-based Learning Outcomes and the associated teaching points. Author Judi Nath sat on the Human Anatomy and Physiology Society (HAPS) committee that developed the HAPS Learning Outcomes recommended to A&P teachers, and the Learning Outcomes in this book are based on them.

Chapter-by-Chapter Changes in the Eleventh Edition

This annotated Table of Contents provides examples of revision highlights in each chapter of the Eleventh Edition. For a more complete list of changes, please contact the publisher.

Chapter 1: An Introduction to Anatomy and Physiology

- Added a new Section 1–1 on using the text and art in tandem.
- New separate section (1-4) on medical terminology.
- Reorganized the chapter to start with simpler anatomical topics and build to more complex physiological ones. Homeostasis and the roles of negative feedback now conclude the chapter as Sections 1–7 and 1–8, respectively.
- NEW Figure 1–1 A Conceptual Framework for Learning
- NEW Clinical Note: *Habeas Corpus* ("You Shall Have the Body")
- NEW Clinical Note: The Sounds of the Body
- Figure 1–8 The Control of Room Temperature (new homeostasis design)
- Figure 1–9 Negative Feedback: Control of Body Temperature (new homeostasis design)
- Former Spotlight Figure 1–10 Diagnostic Imaging Techniques is now a Clinical Note.
- Questions added to Figures 1–3, 1–4, 1–5, 1–6, and 1–9.

Chapter 2: The Chemical Level of Organization

- Clinical Case: What Is Wrong with My Baby? revised
- Clinical Note: Radiation Sickness revised
- NEW Figure 2–1 Hydrogen Atom with Electron Cloud
- NEW Section 2–9 gathers together coverage of monomers, polymers, and functional groups to provide an overview to the organic compounds.
- Table 2–8.Turnover Times moved to the Appendix as Turnover Times of Organic Components of Four Cell Types.
- NEW Clinical Note: Too Sweet on Sugar?
- Questions added to Figures 2–3, 2–8, 2–9, 2–12, 2–15, 2–17, 2–24, and 2–26.

Chapter 3: The Cellular Level of Organization

- Clinical Case: The Beat Must Go On! revised (new title)
- Figure 3–2 The Plasma Membrane revised (new added part b)
- Figure 3–8 Lysosome Functions revised
- NEW Clinical Note: Lysosomal Storage Disease
- NEW Clinical Note: Free Radicals
- Figure 3–13 The Process of Translation revised

- NEW Clinical Note: Drugs and the Plasma Membrane
- Figure 3–21 Receptor–Mediated Endocytosis revised
- Spotlight Figure 3–23 Stages of a Cell's Life Cycle revised
- Questions added to Figures 3–3, 3–9, 3–11, 3–15, 3–17, 3–18, and 3–19.

Chapter 4: The Tissue Level of Organization

- NEW Figure 4–1 An Orientation to the Body's Tissues
- Figure 4–2 Cell Junctions revised (*basal lamina* replaces *clear layer* and *reticular lamina* replaces *dense layer*)
- Table 4–1.Classifying Epithelia revised
- Connective tissue proper has been separated out into its own section, Section 4–5. This section now also includes the discussion of fasciae.
- Figure 4–9 The Cells and Fibers of Connective Tissue Proper revised (added fibrocyte)
- Figure 4–10 Embryonic Connective Tissues revised (now share labels)
- The fluid connective tissues blood and lymph now have their own section, Section 4–6.
- Questions added to Figures 4–3, 4–14, 4–16, 4–18, and 4–19.

Chapter 5: The Integumentary System

- NEW Clinical Case: He Has Fish Skin!
- Figure 5–1 The Components of the Integumentary System revised
- The dermis and hypodermis sections have been moved up to become Sections 5–2 and 5–3, respectively, to give students more anatomical background to understand the later physiological sections.
- Spotlight Figure 5–3 The Epidermis revised (matched SEM and art)
- NEW Clinical Note: Nips, Tucks, and Shots
- Figure 5–12 Hair Follicles and Hairs revised (new part b)
- Figure 5–14 Sweat Glands revised (uses *eccrine sweat glands* as primary term)
- NEW Clinical Note: Your Skin, A Mirror of Your Health
- NEW Clinical Note: Burns and Grafts
- NEW Build Your Knowledge Figure 5–15 Integration of the INTEGUMENTARY system with the other body systems presented so far (replaces System Integrator)
- Questions added to Figures 5–1, 5–6, 5–8, 5–10, and 5–13.

Chapter 6: Bones and Bone Structure (formerly called Osseous Tissue and Bone Structure)

- NEW Figure 6–4 Bone Lacking a Calcified Matrix
- Figure 6–5 Types of Bone Cells revised (art and layout to parallel text)
- NEW Figure 6–6 Osteons of Compact Bone (former part a removed)
- We now clarify in the section titles that Section 6–5 covers both interstitial and appositional growth, while remodeling is covered in Section 6–6.
- Spotlight Figure 6–17 Types of Fractures and Steps in Repair revised (tibia replaces humerus to better match photograph)
- Questions added to Figures 6–3, 6–5, 6–7, and 6–10.

Chapter 7: The Axial Skeleton

- Figure 7–2 Cranial and Facial Subdivisions of the Skull revised
- Figure 7–3 The Adult Skull revised (hyphenates the terms *supra-orbital* and *infra-orbital*)
- Figure 7–9 The Ethmoid revised (*ethmoidal labyrinth* replaces *lateral mass*)
- Spotlight Figure 7–4 Sectional Anatomy of the Skull revised (updated trigeminal nerve [V] terminology)
- Figure 7–14 The Orbital Complex revised (art and photograph now share labels)
- Figure 7–15 The Nasal Complex revised (part b new art)
- Figure 7–17 The Vertebral Column revised (new color-coded vertebral regions)
- Figure 7–22 Sacrum and Coccyx revised (new coccyx label configuration)
- Questions added to Figures 7–16, 7–17, and 7–23.

Chapter 8: The Appendicular Skeleton

- NEW Clinical Case: Timber!!
- Figure 8–6 Bones of the Right Wrist and Hand revised (carpal bones separated out into proximal and distal carpals)
- NEW Clinical Note: Shin Splints
- Clinical Note: Carpal Tunnel Syndrome includes new illustration
- Questions added to Figures 8–1, 8–6, 8–8, and 8–12.

Chapter 9: Joints

- NEW Clinical Note: Bursitis and Bunions
- NEW Clinical Note: Dislocation
- Spotlight Figure 9–2 Joint Movement revised (headings labeled as parts a, b, and c; *plane joint* replaces *gliding joint*)
- Figure 9–5 Special Movements (part labels added; arrows moved onto photographs in new parts d and e)
- Section 9–5 now covers the hinge joints of the elbow and knee, while Section 9–6 covers the ball-and-socket shoulder and hip joints.
- NEW Build Your Knowledge Figure 9–11 Integration of the SKELETAL system with the other body systems presented so far (replaces System Integrator)
- Questions added to Figures 9–1, 9–3, 9–6, and 9–9.

Chapter 10: Muscle Tissue

- NEW Clinical Case: Keep on Keepin' On
- Figure 10–1 The Organization of Skeletal Muscles revised (added tendon attachment to bone)
- Figure 10–5 Sarcomere Structure, Superficial and Cross-Sectional Views revised (new figure icon)
- Figure 10–6 Levels of Functional Organization in a Skeletal Muscle revised (new grouping of art)
- Figure 10–7 Thin and Thick Filaments revised (new art for parts b, c, and d)
- Spotlight Figure 10–9 Events at the Neuromuscular Junction revised (art now shows Na^+ flow through membrane channels)
- Spotlight Figure 10–11 The Contraction Cycle and Cross-Bridge Formation revised (improved step boxes visibility)
- Figure 10–16 Effects of Repeated Stimulations revised (new art organization and explanatory text)
- Information about tension production at the level of skeletal muscles has been separated out into a new section, Section 10–6.
- Figure 10–20 Muscle Metabolism revised (text and art in bottom box)

- Figure 10–21 Fast versus Slow Fibers revised (micrograph is a TEM not LM)
- Coverage of muscle fatigue has been moved from the muscle metabolism section to the muscle performance section, Section 10–8.
- NEW Clinical Note: Electromyography
- Discussion on the effects of skeletal muscle aging has been moved from Chapter 11 and included with muscle hypertrophy and atrophy in Section 10–8.
- Questions added to Figures 10–3, 10–6, 10–14, and 10–21.

Chapter 11: The Muscular System

- NEW Clinical Case: Downward-Facing Dog
- Figure 11–1 Muscle Types Based on Pattern of Fascicle Organization revised
- Figure 11–2 The Three Classes of Levers revised (new icons for each lever)
- Spotlight Figure 11–3 Muscle Action revised (new art in part c)
- The introduction to axial and appendicular muscles has been made into a separate section, Section 11–5, to provide an overview before we cover the muscles in detail.
- NEW Clinical Note: Signs of Stroke
- Figure 11–12 Oblique and Rectus Muscles and the Diaphragm revised (added *transversus thoracis* label to part c)
- Figure 11–17 Muscles That Move the Forearm and Hand revised (corrected leader for *triceps brachii, medial head*)
- Figure 11–18 Muscles That Move the Hand and Fingers revised
- Figure 11–21 Muscles That Move the Leg revised (*quadriceps femoris* replaces *quadriceps muscles*)
- NEW Build Your Knowledge Figure 11–24 Integration of the MUSCULAR system with the other body systems presented so far (replaces System Integrator)
- Questions added to Figures 11–5, 11–6, 11–10, 11–17, 11–19, and 11–21.

Chapter 12: Nervous Tissue

- Chapter title changed from Neural Tissue to Nervous Tissue
- Section 12–1 includes discussion of the Enteric Nervous System (ENS) as a third division of the nervous system
- Figure 12–1 A Functional Overview of the Nervous System revised (added a body figure to support text-art integration)
- Moved coverage of synapse structures from Section 12–2 into Section 12–7 so it is now right before students need it to understand synaptic function.
- Figure 12–3 Structural Classification of Neurons revised (moved part labels and text above art)
- Figure 12–5 Neuroglia in the CNS revised (deleted micrograph; label grouping for neuroglia)
- Schwann cell text updated (*neurolemmocytes* replaces *neurilemma cells* and *neurolemma* replaces *neurilemma*).
- Figure 12–7 Peripheral Nerve Regeneration after Injury revised
- Spotlight Figure 12–8 Resting Membrane Potential revised (text revised in first two columns)
- Figure 12–9 Electrochemical Gradients for Potassium and Sodium Ions revised (text revised in part c)
- Figure 12–11 Graded Potentials revised (text in step 2)
- NEW Spotlight Figure 12–13 Generation of an Action Potential revised (text in step boxes)
- Figure 12–14 Propagation of an Action Potential revised (added part labels)
- NEW Figure 12–16 Events in the Functioning of a Cholinergic Synapse revised (now runs across two pages; text in steps revised)
- Table 12–4 Representative Neurotransmitters and Neuromodulators revised (endorphins separated from opioids)
- Figure 12–17 Mechanisms of Neurotransmitter and Receptor Function revised (chemically gated ion channel art now matches that in previous figures)
- Questions added to Figures 12–2, 12–4, and 12–16.

Chapter 13: The Spinal Cord, Spinal Nerves, and Spinal Reflexes

- Figure 13–1 An Overview of Chapters 13 and 14 revised
- Figure 13–2 Gross Anatomy of the Adult Spinal Cord revised (added new part b)
- Uses the term *posterior* and *anterior* in reference to spinal roots, ganglion, and rami instead of *dorsal* and *ventral* (e.g., Figure 13–3, 13–4, 13–5, and Spotlight Figure 13–8)
- Figure 13–6 A Peripheral Nerve revised (corrected magnified section in part a)
- NEW Figure 13–9 Nerve Plexuses and Peripheral Nerves revised (labels grouped and boxed)
- Figure 13–10 The Cervical Plexus revised (corrected cranial nerve designation, e.g., *accessory nerve [XI]* replaces *accessory nerve [N XI]*)
- Figure 13–12 The Lumbar and Sacral Plexuses revised (removed Clinical Note)
- Spotlight Figure 13–14 Spinal Reflexes revised (added part labels to better coordinate with text)
- Figure 13–15 The Classification of Reflexes revised (reorganized categories within inclusive boxes)
- Figure 13–17 The Plantar Reflex and Babinski Reflex revised (*Babinski reflex* replaces *Babinski sign/positive Babinski reflex* and *plantar reflex* replaces *negative Babinski reflex*)
- Questions added to Figures 13–3, 13–5, 13–9, and 13–15.

Chapter 14: The Brain and Cranial Nerves

- Figure 14–1 An Introduction to Brain Structures and Functions revised (added part labels a–f to better coordinate with text)
- Figure 14–2 Ventricular System revised (*ventricular system of the brain* replaces *ventricles of the brain*)
- Figure 14–3 The Relationships among the Brain, Cranium, and Cranial Meninges revised *periosteal cranial dura* replaces *dura mater [periosteal layer]* and *meningeal cranial dura* replaces *dura mater [meningeal layer]*)
- Figure 14–5 The Diencephalon and Brainstem revised (corrected cranial nerve designation, e.g., in Cranial Nerves box, CN replaces N for nerve designations.)
- The sections on the midbrain (now Section 14–5) and cerebellum (now Section 14–6) have been switched, so that we now cover all of the brainstem together.
- Figure 14–10 The Thalamus revised (thalamic nuclei labels now color coded to clarify brain regions that receive thalamic input; *medial geniculate body* and *lateral geniculate body* replace *medial geniculate nucleus* and *lateral geniculate nucleus*)
- Figure 14–18 Origins of the Cranial Nerves revised (new brain cadaver photograph; cranial nerve labels boxed together)
- Questions added to Figures 14–1, 14–3, 14–9, 14–13, 14–15, 14–22, and 14–26.

Chapter 15: Sensory Pathways and the Somatic Nervous System

- Figure 15–1 An Overview of Events Occurring Along the Sensory and Motor Pathways revised
- Figure 15–2 Receptors and Receptive Fields revised (different colors for each receptive field and added Epidermis and Free nerve endings labels)
- Figure 15–3 Tonic and Phasic Sensory Receptors revised (new background colors for graphs)
- Figure 15–4 Tactile Receptors in the Skin revised (added *myelin sheath* to afferent nerve fiber in part c; part d, *bulbous corpuscle* replaces *Ruffini corpuscle*; part e, *lamellar [pacinian] corpuscle* replaces *lamellated [pacinian] corpuscle*)
- NEW Figure 15–6 Locations and Functions of Chemoreceptors
- Figure 15–7 Sensory Pathways and Ascending Tracts in the Spinal Cord revised (*gracile fasciculus* replaces *fasciculus gracilis*, *cuneate fasciculus* replaces *fasciculus* cuneate)
- Spotlight Figure 15–8 Somatic Sensory Pathways revised (introduced "somatotopy" in Sensory Homunculus boxed text)
- Questions added to Figures 15–1, 15–2, 15–4, 15–7, and 15–10.

Chapter 16: The Autonomic Nervous System and Higher-Order Functions

- NEW Clinical Case: Remember Me?
- NEW Spotlight Figure 16–2 The Autonomic Nervous System (incorporates old Figures 16–4 and 16–6. added Pons and Medulla oblongata labels on the art)
- A new summary Section 16–6 called "The differences in the organization of sympathetic and parasympathetic structures lead to widespread sympathetic effects and specific parasympathetic effects" has been created.
- The sections on memory, states of consciousness, and behavior have been combined into Section 16–9.
- Figure 16–11 The Reticular Activating System (RAS) revised (*CN II* and *CN VIII* replace *N II* and *N VIII*, respectively)
- NEW Build Your Knowledge Figure 16–12 Integration of the NERVOUS system with the other body systems presented so far (replaces System Integrator)
- Questions added to Figures 16–1, 16–3, 16–4, 16–7, and 16–11.

Chapter 17: The Special Senses

- Figure 17–1 The Olfactory Organs revised (I replaces N I)
- Spotlight Figure 17–2 Olfaction and Gustation revised (added part a and b labels)
- Figure 17–3 Papillae, Taste Buds, and Gustatory Receptor Cells revised (new figure title; added *Midline groove* label to part a)
- Figure 17–4 External Features and Accessory Structures of the Eye revised (*lateral angle* replaces *lateral canthus*, *medial angle* replaces *medial canthus*, *bulbar conjunctiva* replaces *ocular conjunctiva*, *eyelid* replaces *palpebrae*)
- Figure 17–5 The Sectional Anatomy of the Eye revised (*corneoscleral junction* replaces *corneal limbus*)
- Figure 17–6 The Pupillary Muscles revised (*dilator pupillae* replaces *pupillary dilator muscles*; *sphincter pupillae* replaces *pupillary constrictor*)
- Figure 17–7 The Organization of the Retina revised (*pigmented layer of retina* replaces *pigmented part of retina*; switched parts b and c to parallel new sequence in the text)
- A new overview section, Section 17–4, called "The focusing of light on the retina leads to the formation of a visual image" has been created in the text.
- Figure 17–10 Factors Affecting Focal Distance revised (clarified text within figure; added Focal point label to all the art)
- Figure 17–11 Accommodation revised (*fovea centralis* replaces *fovea*)
- Figure 17–14 Structure of Rods, Cones, and the Rhodopsin Molecule revised (*pigmented epithelium* replaces *pigment epithelium*)
- Figure 17–23 The Internal Ear revised (*ampullary crest* replaces *crista ampullaris*; clarified position of membranous labyrinth in part a art)
- Figure 17–24 The Semicircular Ducts revised (*ampullary cupula* replaces *cupula*; *vestibular nerve* replaces *vestibular branch* in part a)
- Figure 17–26 Pathways for Equilibrium Sensations revised (*cochlear nerve* replaces *cochlear branch*)
- Figure 17–30 Sound and Hearing revised (added new art to illustrate step 4)
- Figure 17–32 Pathways for Auditory Sensations revised (auditory replaces sound and acoustic in steps 2 and 5)
- Questions added to Figures 17–4, 17–7, 17–21, and 17–28.

Chapter 18: The Endocrine System

- Figure 18–1 Organs and Tissues of the Endocrine System revised (clarified hormones in Gonads box)
- Table 18–1 Mechanisms of Intercellular Communication revised (added autocrine communication)
- Spotlight Figure 18–3 G Proteins and Second Messengers revised (added positive feedback involving protein kinase C; clarified calcium ion sources for binding with calmodulin)
- Figure 18–6 Three Mechanisms of Hypothalamic Control over Endocrine Function revised (removed numbers and added color coding to enhance links between hypothalamic structures and functions)
- Figure 18–7 The Hypophyseal Portal System and the Blood Supply to the Pituitary Gland revised (*regulatory hormones* replaces *regulatory factors*)
- Figure 18–8 Feedback Control of Endocrine Secretion revised (added two banners to separate part a from parts b and c; incorporated old part d with a new color-coded table within part a)
- Figure 18–9 Pituitary Hormones and Their Targets revised (added color codes to correlate with Figure 18–6)
- Figure 18–11 Synthesis and Regulation of Thyroid Hormones (added step art to part a that describes synthesis, storage, and secretion of thyroid hormones; added new homeostasis design to part b that illustrates the regulation of thyroid secretion)
- Figure 18–12 Anatomy of the Parathyroid Glands revised (*principal cells* replaces *chief cells*)
- Figure 18–13 Homeostatic Regulation of the Blood Calcium Ion Concentration revised (new homeostasis design)
- Figure 18–14 The Adrenal Gland and Adrenal Hormones revised (added new micrograph and new design for part c)
- Figure 18–17 Homeostatic Regulation of the Blood Glucose Concentration revised (new homeostasis design)
- Figure 18–19 Endocrine Functions of the Kidneys revised (new homeostasis design in part b)
- NEW Build Your Knowledge Figure 18–21 Integration of the ENDOCRINE system with the other body systems presented so far (replaces System Integrator)
- Questions added to Figures 18–6, 18–8, 18–9, 18–14, and 18–17.

Chapter 19: Blood

- NEW Clinical Case: Crisis in the Blood
- Section 19–1 now covers the main functions and characteristics of blood, as well as an introduction to both plasma and formed elements (combined with the old Section 19–2).
- Figure 19–4 Stages of RBC Maturation: Erythropoiesis and Figure 19–5 Recycling of Red Blood Cell Components sequence changed because of chapter reorganization.
- Figure 19–6 Blood Types and Cross-Reactions revised (corrected shapes of anti-A and anti-B antibodies)
- Figure 19–7 Blood Type Testing revised (anti-Rh replaces anti-D; added "clumping" or "no clumping" under test results for clarification)
- Figure 19–11 The Phases of Hemostasis (Vascular, Platelet, and Coagulation) and Clot Retraction revised (*clotting factors* replaces *platelet factors* in step 2; new blood clot SEM)
- Table 19–2.Differences in Blood Group Distribution revised
- Questions added to Figures 19–3, 19–5, 19–6, and 19–10.

Chapter 20: The Heart

- Figure 20–1 An Overview of the Cardiovascular System revised (new art and boxed labels)
- Figure 20–2 The Location of the Heart in the Thoracic Cavity revised (*parietal layer of serous pericardium* replaces *parietal pericardium*)
- Figure 20–4 The Heart Wall revised (*visceral layer of serous pericardium* replaces *epicardium [visceral pericardium]*)
- Figure 20–5 The Sectional Anatomy of the Heart revised (*tricuspid valve* replaces *right AV [tricuspid] valve*; *mitral valve* replaces *left AV [mitral] valve*)
- Figure 20–7 Valves of the Heart and Blood Flow revised (red arrows replace black arrows in part a; black arrows deleted in part b)
- Figure 20–10 The Conducting System of the Heart and the Pacemaker Potential revised (*pacemaker potential* replaces *prepotential*)
- Figure 20–11 Impulse Conduction through the Heart and Accompanying ECG Tracings revised (added ECG tracings next to the step art)
- Figure 20–12 An Electrocardiogram (ECG) revised (*QRS complex* replaces *QRS interval* in part b)
- Figure 20–14 Cardiac Contractile Cells revised (cardiac contractile cells replaces *cardiac muscle cells*; former Figure 20–5 moved because of chapter reorganization to provide structural information right before functional information)
- Figure 20–15 Action Potentials in Cardiac Contractile Cells and Skeletal Muscle Fibers revised (*ventricular contractile cell* replaces *ventricular muscle cell*)
- Figure 20–16 Phases of the Cardiac Cycle revised (moved labels for Atrial systole, Atrial diastole, Ventricular systole, and Ventricular diastole to perimeter of art for increased correlation)
- Figure 20–17 Pressure and Volume Relationships in the Cardiac Cycle revised (modified colors of banners to match the perimeter art of Figure 20–16 Phases of the Cardiac Cycle for increased correlation)
- Figure 20–19 Factors Affecting Cardiac Output revised (added EDV and ESV)
- Figure 20–23 Factors Affecting Stroke Volume revised (added key)
- Figure 20–24 A Summary of the Factors Affecting Cardiac Output revised (deleted arrow from Preload to End-systolic volume box)
- Table 20–1 Structural and Functional Differences between Cardiac Contractile Cells and Skeletal Muscle Fibers revised (*cardiac contractile cells* replaces *cardiac muscle cells*)
- Questions added to Figures 20–1, 20–5, 20–11, 20–15, 20–21, and 20–24.

Chapter 21: Blood Vessels and Circulation

- Figure 21–2 Histological Structures of Blood Vessels revised (added luminal diameters for all vessels)
- Figure 21–4 The Organization of a Capillary Bed revised (deleted metarterioles)
- Figure 21–8 Relationships among Vessel Luminal Diameter, Cross-Sectional Area, Blood Pressure, and Blood Velocity within the Systemic Circuit revised (*vessel luminal diameter* replaces *vessel diameter* in part a; *vessel lumens* replaces *vessels* in part b)
- Figure 21–11 Forces Acting across Capillary Walls revised (added tissue cells background)
- The discussion of vasomotion has been moved from Section 21–1 to Section 21–3, to cover this process with other vessel physiology.
- Figure 21–12 Short-Term and Long-Term Cardiovascular Responses revised (new homeostasis design)
- Figure 21–13 Baroreceptor Reflexes of the Carotid and Aortic Sinuses revised (new homeostasis design)
- Figure 21–14 The Chemoreceptor Reflexes revised (new homeostasis design)
- Figure 21–15 The Hormonal Regulation of Blood Pressure and Blood Volume revised (new homeostasis design)
- Figure 21–16 Cardiovascular Responses to Blood Loss revised (new homeostasis design)
- Figure 21–24 Arteries Supplying the Abdominopelvic Organs revised
- Figure 21–27 Major Veins of the Head, Neck, and Brain revised (added confluence of sinuses to parts a, b and c)
- Figure 21–28 The Venous Drainage of the Abdomen and Chest revised (*median sacral* replaces *medial sacral*; *hemi-azygos* replaces *hemiazygos*)
- Figure 21–29 Flowchart of Circulation to the Superior and Inferior Venae Cavae revised
- Figure 21–31 The Hepatic Portal System revised
- **NEW** Build Your Knowledge Figure 21–34 Integration of the CARDIOVASCULAR system with the other body systems presented so far (replaces System Integrator)
- Questions added to Figures 21–2, 21–7, 21–12, 21–15, 21–21, and 21–29.

Chapter 22: The Lymphatic System and Immunity

- The coverage of the lymphatic system is now Section 22–1.
- Figure 22–1 The Components of the Lymphatic System revised (Other Lymphoid Tissues and Organs heading replaces Lymphoid Tissues and Organs heading because lymph nodes are organs)
- Figure 22–5 Lymphoid Nodules moved (formerly Figure 22–7, moved due to chapter reorganization)
- Figure 22–6 The Structure of a Lymph Node revised and moved (*cortex* replaces *outer cortex*; *paracortex* replaces *deep cortex*; formerly Figure 22–8, moved due to chapter reorganization)
- Figure 22–7 The Thymus moved (formerly Figure 22–9, moved due to chapter reorganization)
- Figure 22–8 The Spleen moved (formerly Figure 22–10, moved due to chapter reorganization)
- The original Section 22–1 has been moved to become Section 22–2 and adapted so that it is now titled "Lymphocytes are important to the innate (nonspecific) and adaptive (specific) defenses that protect the body."
- We have broadened the definition of the term "immune response" from a "defense against specific antigens" to "the body's reaction to infectious agents and abnormal substances."

- Figure 22–9 The Origin and Distribution of Lymphocytes revised and moved (hemocytoblasts replaces multipotent hemopoietic stem cell; formerly Figure 22–10, moved due to chapter reorganization)
- Figure 22–10 Innate Defenses revised
- Figure 22–11 How Natural Killer Cells Kill Cellular Targets moved (formerly Figure 22–12, moved due to chapter reorganization)
- Figure 22–12 Interferons revised
- NEW Figure 22–13 Pathways of Complement Activation revised (added the Lectin Pathway)
- Figure 22–14 Inflammation and the Steps in Tissue Repair moved (formerly Figure 22–15, moved due to chapter reorganization)
- Figure 22–15 Classes of Lymphocytes revised and moved (*regulatory T cells* replaces *suppressor T cells*; formerly Figure 22–5, moved due to chapter reorganization)
- Figure 22–16 An Overview of Adaptive Immunity revised and moved (former title: An Overview of the Immune Response; formerly Figure 22–17, moved due to chapter reorganization)
- Figure 22–17 Forms of Immunity revised and moved (*acquired* replaces *induced*; formerly Figure 22–16, moved due to chapter reorganization)
- Figure 22–18 Antigens and MHC Proteins revised
- Spotlight Figure 22–21 Cytokines of the Immune System revised and moved (formerly Figure 22–28, moved due to chapter reorganization)
- Figure 22–22 A Summary of the Pathways of T Cell Activation revised and moved (*regulatory T cells* replaces *suppressor T cells*; formerly Figure 22–21, moved due to text reorganization)
- Figure 22–23 The Sensitization and Activation of B Cells moved (formerly Figure 22–22, moved due to chapter reorganization)
- Figure 22–24 Antibody Structure and Function moved (formerly Figure 22–23, moved due to chapter reorganization)
- Figure 22–27 An Integrated Summary of the Immune Response revised and moved (*regulatory T cells* replaces *suppressor T cells*; formerly Figure 22–26, moved due to chapter reorganization
- NEW Build Your Knowledge Figure 22–30 Integration of the LYMPHATIC system with the other body systems presented so far (replaces System Integrator)
- Questions added to Figures 22–3, 22–8, 22–12, 22–17, 22–25, and 22–26.

Chapter 23: The Respiratory System

- NEW Clinical Case: No Rest for the Weary
- Figure 23–3 The Structures of the Upper Respiratory System revised (*epithelial surface* replaces *superficial view* in micrograph of part a)
- Figure 23–3 The Structures of the Upper Respiratory System revised (*pharyngeal opening of auditory tube* replaces *nasopharyngeal meatus*)
- Original Sections 23–3 and 23–4 have been combined into a new Section 23–3 on the conducting portion of the lower respiratory system. This section now includes coverage of the bronchial tree.
- Figure 23–6 The Anatomy of the Trachea revised (cross-sectional diagram of trachea and esophagus replaces micrograph to better highlight trachealis)
- NEW Section 23–4 has been added titled "The respiratory portion of the lower respiratory system is where gas exchange occurs." This covers the respiratory bronchioles, alveolar ducts and alveoli, and the blood air barrier.
- Figure 23–7 The Bronchi, Lobules, and Alveoli of the Lung revised and moved (new art in part c; formerly Figure 23–9, moved due to chapter reorganization)
- Figure 23–8 Alveolar Organization revised and moved (*pneumocyte type I* and *type II* replaces *type I* and *type II pneumocyte*; *blood air barrier* replaces *respiratory membrane*; formerly Figure 23–10, moved due to chapter reorganization
- Figure 23–9 The Gross Anatomy of the Lungs revised and moved (formerly Figure 23–7, moved due to chapter reorganization)
- Figure 23–10 The Relationship between the Lungs and Heart revised (labeled Anterior border in part b; formerly Figure 23–8, moved due to chapter reorganization)
- Figure 23–11 An Overview of the Key Steps in Respiration revised
- NEW Figure 23–13 Primary and Accessory Respiratory Muscles
- NEW Spotlight Figure 23–14 Pulmonary Ventilation
- Figure 23–15 Pressure and Volume Changes during Inhalation and Exhalation revised and moved (outlined boxes with same color as respective line graphs for better correlation; formerly Figure 23–14, moved due to chapter reorganization)
- Figure 23–16 Pulmonary Volumes and Capacities revised
- Figure 23–18 An Overview of Respiratory Processes and Partial Pressures in Respiration revised (added new icon art)
- Figure 23–23 A Summary of the Primary Gas Transport Mechanisms revised (added oxygen and carbon dioxide partial pressure values)
- Spotlight Figure 23–25 Control of Respiration revised
- Figure 23–26 The Chemoreceptor Response to Changes in P_{CO_2} revised (new homeostasis design)
- NEW Build Your Knowledge Figure 23–28 Integration of the RESPIRATORY system with the other body systems presented so far (replaces System Integrator)
- Questions added to Figures 23–2, 23–7, 23–8, 23–13, 23–16, 23–20, and 23–26.

Chapter 24: The Digestive System

- Figure 24–1 Components of the Digestive System revised (*mechanical digestion* replaces *mechanical processing*)
- Figure 24–2 The Mesenteries revised (added Visceral peritoneum label to part d)
- Figure 24–3 Histological Organization of the Digestive Tract revised (*muscular layer* replaces *muscularis externa*; *intestinal glands* replaces *mucosal glands*; *submucosal neural plexus* replaces *submucosal plexus*)
- Figure 24–4 Peristalsis revised (Initial State now step 1)
- Figure 24–6 Anatomy of the Oral Cavity revised (*oral vestibule* replaces *vestibule*; *frenulum of tongue* replaces *lingual frenulum*)
- Figure 24–7 The Teeth moved (formerly Figure 24–8, moved due to chapter reorganization)
- Figure 24–8 Deciduous and Permanent Dentitions revised (new title; *deciduous* replaces *primary*; *permanent* replaces *secondary*; *canine* replaces *cuspid*; formerly Figure 24–9, moved due to chapter reorganization)
- Figure 24–9 Anatomy of the Salivary Glands moved (formerly Figure 24–7, moved due to chapter reorganization)
- Section 24–3, titled "The pharynx and esophagus are passageways that transport the food bolus from the oral cavity to the stomach," now combines coverage of the pharynx, esophagus, and deglutition.
- Figure 24–12 Gross Anatomy of the Stomach revised (new title; *pyloric part* replaces *pylorus*)
- Figure 24–14 The Secretion of Hydrochloric Acid Ions revised (new title; *anion countertransport mechanism* replaces *countertransport mechanism*; added Dissociation label for clarification)

- Spotlight Figure 24–15 The Regulation of Gastric Activity revised (clarified Key in steps 1 and 2)
- The new Section 24–5 called "Accessory digestive organs, such as the pancreas and liver, produce secretions that aid in chemical digestion" now covers these accessory organs all in one place.
- Figure 24–16 Anatomy of the Pancreas moved (formerly Figure 24–18, moved due to chapter reorganization)
- Figure 24–17 Gross Anatomy of the Liver revised and moved (new title; added Peritoneal cavity label to part a; formerly Figure 24–19, moved due to chapter reorganization)
- Figure 24–18 Histology of the Liver revised and moved (*portal triad* replaces *portal area*; reoriented micrograph to better correlate with art in part b; renamed portal triad structures to *interlobular bile duct, interlobular vein*, and *interlobular artery; stellate macrophage* replaces *Kupffer cells*; formerly Figure 24–20, moved due to chapter reorganization)
- Figure 24–19 The Anatomy and Physiology of the Gallbladder and Bile Ducts revised (*bile duct* replaces *common bile duct*; formerly Figure 24–21, moved due to chapter reorganization)
- Figure 24–20 Gross Anatomy and Segments of the Intestine moved (new title; formerly Figure 24–16, moved due to chapter reorganization)
- Figure 24–21 Histology of the Intestinal Wall revised (new title; added new part c showing Paneth cells; *intestinal gland* replaces *intestinal crypt*; formerly Figure 24–17, moved due to chapter reorganization)
- Figure 24–22 The Secretion and Effects of Major Duodenal Hormones revised (new title; clarified secretin's primary effect)
- Figure 24–23 The Secretion and Effects of Major Digestive Tract Hormones revised (new title; added new pancreas art)
- Figure 24–25 Histology of the Colon revised (new title; added two more teniae coli to the icon art to show general positions of all three teniae coli)
- Added coverage of the microbiome under Section 24–7 on the large intestine.
- NEW Figure 24–26 The Defecation Reflex
- Spotlight Figure 24–27 The Chemical Events of Digestion revised
- Figure 24–27 Digestive Secretion and Water Reabsorption in the Digestive Tract revised (added new art next to Dietary Input box)
- **NEW** Build Your Knowledge Figure 23–28 Integration of the DIGESTIVE system with the other body systems presented so far (replaces System Integrator)
- Questions added to Figures 24–4, 24–9, 24–12, 24–23, and 24–26.

Chapter 25: Metabolism, Nutrition, and Energetics (title changed to include nutrition)

- NEW Figure 25–1 Metabolism of Organic Nutrients and Nutrient Pools
- We now cover oxidation–reduction reactions in Section 25–1.
- Figure 25–2 Glycolysis moved (formerly Figure 25–3)
- Figure 25–3 The Citric Acid Cycle revised and moved (*electron transport chain* replaces *electron transport system*; formerly Figure 25–4)
- NEW Spotlight Figure 25–4 The Electron Transport Chain and ATP Formation
- Figure 25–5 A Summary of the Energy Yield of Glycolysis and Aerobic Metabolism revised (total ATP yield from a glucose molecule based on new values of ATP yield per NADH [2.5 ATP vs. previous 3 ATP] and $FADH_2$ [1.5 ATP vs. previous 2 ATP]).
- Figure 25–6 Glycolysis and Gluconeogenesis revised (added NADH $\rightarrow$ NAD to show pyruvate is reduced to form lactate when oxygen is lacking)
- Figure 25–7 Lipolysis and Beta-Oxidation revised (new title; lowered total ATP yield)
- Figure 25–8 Lipid Transport and Use revised (formerly Figure 25–9)
- Spotlight Figure 25–10 Absorptive and Postabsorptive States revised (*membrane receptor* replaces *carrier protein*; formerly Spotlight Figure 25–11)
- Figure 25–11 MyPlate, MyWins revised (new title)
- Questions added to Figures 25–2, 25–5, 25–7, 25–8, and 25–14.

Chapter 26: The Urinary System

- Figure 26–6 The Anatomy of a Representative Nephron and the Collecting System revised (new figure title; removed functional anatomy descriptions; *descending thin limb* replaces *thin descending limb* in all relevant figures)
- Figure 26–7 The Functional Anatomy of a Representative Nephron and the Collecting System revised (added *Extraglomerular mesangial cells* label in part a to clarify their distinction from juxtaglomerular cells; *intraglomerular mesangial cell* replaces *mesangial cell*)
- Figure 26–8 The Locations and Structures of Cortical and Juxtamedullary Nephrons moved (formerly Figure 26–7, renumbered because of chapter reorganization)
- Figure 26–9 An Overview of Urine Formation revised (added functional anatomy descriptions from former Figure 26–6)
- Figure 26–11 The Response to a Reduction in the GFR revised (new homeostasis design)
- There is a new section called Principles of Reabsorption and Secretion at the beginning of Section 26–5 to provide an overview of this process before we get into its details.
- Figure 26–12 Transport Activities at the PCT revised (corrected color of cotransport mechanism symbol in the art)
- A new Section 26–6 called "Countercurrent multiplication allows the kidneys to regulate the volume and concentration of urine" has been added to emphasize this content, especially the role of the medullary osmotic gradient. This also includes a more complete kidney function testing section.
- Spotlight Figure 26–16 Summary of Renal Function revised (added new step 8 discussing papillary duct permeability to urea and art showing urea transporter)
- Figure 26–18 Organs for Conducting and Storing Urine revised (deleted "[in urogenital diaphragm]" in part b)
- NEW Figure 26–20 The Control of Urination
- NEW Build Your Knowledge Figure 26–21 Integration of the URINARY system with the other body systems presented so far (replaces System Integrator)
- Questions added to Figures 26–5, 26–6, 26–11, 26–14, and 26–18.

Chapter 27: Fluid, Electrolyte, and Acid–Base Balance

- Figure 27–5 Homeostatic Regulation of Sodium Ion Concentration in Body Fluids revised (new homeostasis design)
- Figure 27–6 Integration of Fluid Volume Regulation and Sodium Ion Concentration in Body Fluids revised (new homeostasis design)
- Figure 27–7 Major Factors Involved in Disturbances of Potassium Ion Balance revised (new homeostasis design)
- Figure 27–8 Three Classes of Acids Found in the Body revised (*metabolic acids* replaces *organic acids*)

- Figure 27–13 pH Regulation of Tubular Fluid by Kidney Tubule Cells revised (incorporated buffer system type next to relevant chemical reactions for better art–text integration)
- Figure 27–15 Homeostatic Regulation of Acid–Base Balance revised (new homeostasis design)
- Figure 27–16 Responses to Metabolic Acidosis revised (new homeostasis design)
- Figure 27–17 Responses to Metabolic Alkalosis revised (new homeostasis design)
- Questions added to Figures 27–2, 27–7, 27–10, 27–14, and 27–16.

Chapter 28: The Reproductive System

- NEW Clinical Case: And Baby Makes Three?
- Section 28–2, retitled "The structures of the male reproductive system consist of the testes and scrotum, duct system, accessory glands, and penis," is now focused on male reproductive anatomy.
- FAP10 Figure 28–2 The Descent of the Testes deleted
- Figure 28–4 Anatomy of the Seminiferous Tubules revised (includes only parts a and b of former Figure 28–5)
- Figure 28–5 Anatomy of the Epididymis revised (former Figure 28–9 moved due to chapter reorganization)
- Figure 28–6 Anatomy of the Ductus Deferens and Accessory Glands revised and reorganized (former Figure 28–10 moved due to chapter reorganization)
- Figure 28–7 Anatomy of the Penis revised and reorganized (former Figure 28–11 moved due to chapter reorganization; new erectile tissue box)
- There is now a Section 28–3 called "Spermatogenesis occurs in the testes, and hormones from the hypothalamus, pituitary gland, and testes control male reproductive functions" that covers male reproductive physiology.
- Section 28–3 now starts with an Overview of Mitosis and Meiosis.
- NEW Figure 28–8 A Comparison of Chromosomes in Mitosis and Meiosis
- Figure 28–9 The Process of Spermatogenesis revised (former Figure 28–7 moved due to chapter reorganization; *sperm* replaces *spermatozoa*)
- Figure 28–10 Spermatogenesis in a Seminiferous Tubule revised (includes only parts c and d of former Figure 28–5; moved due to chapter reorganization)
- Figure 28–11 The Process of Spermiogenesis and Anatomy of a Sperm revised (former Figure 28–8 moved due to chapter reorganization; *sperm* replaces *spermatozoa*)
- The reworked Section 28–4 is now titled "The structures of the female reproductive system consist of the ovaries, uterine tubes, uterus, vagina, and external genitalia" and focuses on presenting the female reproductive anatomy.
- Figure 28–15 Anatomy of the Uterine Tubes revised (former Figure 28–17 moved due to chapter reorganization; new epithelial surface SEM)
- Figure 28–19 Anatomy of the Female External Genitalia revised (former Figure 28–22 moved due to chapter reorganization)
- The reworked Section 28–5 titled "Oogenesis occurs in the ovaries, and hormones from the hypothalamus, pituitary gland, and ovaries control female reproductive functions" presents female reproductive physiology. This section now gathers information on oogenesis, the ovarian cycle, and the uterine cycle, as well as their coordination.
- Figure 28–21 The Process of Oogenesis revised (new title; former Figure 28–15 moved due to chapter reorganization)
- Figure 28–22 Follicle Development and the Ovarian Cycle revised (former Figure 28–16 moved due to chapter reorganization; new ovary art)
- Figure 28–23 A Comparison of the Structure of the Endometrium during the Phases of the Uterine Cycle revised (new title; former Figure 28–20 moved due to chapter reorganization)
- Spotlight Figure 28–24 Hormonal Regulation of Female Reproduction revised (text in Follicle Phase of the Ovarian Cycle box changed to reflect that one tertiary follicle from a group becomes dominant; *Tertiary ovarian follicle development* label replaces *Follicle development* label; temperature ranges changed for both Celsius and Fahrenheit scales; and Menses label changed to Menstrual Phase)
- Under Section 28–6, there are new discussions of contraception and infertility, and sexually transmitted diseases.
- Under Section 28–7, there is a new discussion of development of internal reproductive organs, with a new Figure 28–26 The Development of Male and Female Internal Reproductive Organs.
- NEW Build Your Knowledge Figure 28–27 Integration of the REPRODUCTIVE system with the other body systems presented so far (replaces System Integrator)
- Questions added to Figures 28–7, 28–9, 28–11, 28–22, 28–23, and 28–25.

Chapter 29: Development and Inheritance

- Figure 29–1 Fertilization revised (changed some titles and text in step art; clarified when DNA synthesis occurs)
- Figure 29–3 Stages in Implantation revised (*cytotrophoblast* replaces *cellular trophoblast*; *syncytiotrophoblast* replaces *syncytial trophoblast*)
- Figure 29–4 The Inner Cell Mass and Gastrulation revised (changed Gastrulation from Day 12 to Day 15)
- Spotlight Figure 29–5 Extra-Embryonic Membranes and Placenta Formation revised (added cervical plug to Week 10/step 5 art)
- Figure 29–6 Anatomy of the Placenta after the First Trimester revised (replaced first sentence of part a text)
- Figure 29–7 The First 12 Weeks of Development revised (new art at 3 weeks of development replaces Week 2 SEM)
- Section 29–5, now called "During the second and third trimesters, fetal development primarily involves growth and organ function," focuses on the fetal development during this period.
- Section 29–6, called "During gestation, maternal organ systems support the developing fetus; the reproductive system in particular undergoes structural and functional changes" now presents the maternal changes, including hormonal effects.
- Figure 29–12 The Milk Ejection Reflex revised (new title)
- Figure 29–17 Inheritance of an X-Linked Trait revised (former Figure 29–18 moved due to chapter reorganization)
- Figure 29–18 Crossing Over and Recombination revised (clarified text in part b; former Figure 29–17 moved due to chapter reorganization)
- Questions added to Figures 29–2, 29–4, 29–10, 29–14, and 29–15.

Appendix

- NEW Table 3 Four Common Methods of Reporting Gas Pressure
- NEW Table 4 Turnover Times of Organic Components of Four Cell Types

Acknowledgments

This textbook represents a group effort, and we would like to acknowledge the people who worked together with us to create this Eleventh Edition.

Foremost on our thank-you list are the instructors who offered invaluable suggestions throughout the revision process. We thank them for their participation and list their names and affiliations below.

Lois Borek, *Georgia State University*
Angela Bruni, *Mississippi Gulf Coast Community College*
Marien Cendon, *Miami Dade College, Kendall Campus*
Jose Chestnut, *Essex County College*
James E. Clark, *Manchester Community College*
Ferdinand Esser, *Mercy College*
Robert S. Kellar, *Northern Arizona University*
Beth A. Kersten, *State College of Florida*
Mary Katherine Lockwood, *University of New Hampshire*
Naomi Machell, *Delaware County Community College*
Russell Nolan
Amanda R. Pendleton, *Amarillo College*
Louise Petroka, *Gateway Community College*
Courtney B. Ross, *Gwinnett Technical College*
Natalia Schmidt, *Leeward Community College*
Scott L. Simerlein, *Purdue University North Central*
Patricia Steinke, *San Jacinto College Central*
Diane G. Tice, *Morrisville State College*
Pauline Ward, *Houston Community College*
Sarah Ward, *Colorado Northwestern Community College*
Mary Weis, *Collin College*
Colleen Winters, *Towson University*

The accuracy and currency of the clinical material in this edition reflects the work of our Clinical Case contributors, Christine Boudrie, M.D. and Ruth Anne O'Keefe, M.D., who provided constant, useful feedback on each chapter.

Virtually without exception, reviewers stressed the importance of accurate, integrated, and visually attractive illustrations in helping readers understand essential material. The revision of the art program was directed by Bill Ober, M.D., and Claire E. Ober, R.N. Their suggestions about presentation sequence, topics of clinical importance, and revisions to the proposed art were of incalculable value to us and to the project. The illustration program for this edition was further enhanced by the efforts of several other talented individuals. Jim Gibson designed many of the Spotlight Figures in the art program and consulted on the design and layout of the individual figures. His talents have helped produce an illustration program that is dynamic, cohesive, and easy to understand. Anita Impagliazzo helped create the photo/art combinations that have resulted in clearer presentations and a greater sense of realism in important anatomical figures. We are also grateful to the talented team at Imagineering (imagineeringart.com) for their dedicated and detailed illustrative work on key figures. The color micrographs in this edition were provided by Dr. Robert Tallitsch, and his assistance is much appreciated. Many of the striking anatomy photos in the text and in *Martini's Atlas of the Human Body* are the work of biomedical photographer Ralph Hutchings; his images played a key role in the illustration program.

We also express our appreciation to the editors and support staff at Pearson Science.

We owe special thanks to Senior Acquisitions Editor Cheryl Cechvala for shepherding this project from start to finish. Her ability to manage every detail with such fervor and interest is commendable. She also has an incredible command of the English language coupled with exceptional oratory ability—it's always fun having candid conversations. Although this was her first year as our editor, she possesses the skills of a seasoned veteran. She is our biggest advocate and is always willing to champion our cause—despite the challenges of working with authors. We are appreciative of all her efforts on our behalf.

Content Producer Jessica Picone was extremely skilled at keeping this project moving forward. Throughout every iteration, she kept track of the files, ensured we were on task, and maintained her high standards. Working with authors can be challenging, and Jessica was up for the task! Editorial Coordinator Kimberly Twardochleb was always available and answered every question we had with speed and accuracy. Having two highly skilled professionals working with us eased our burden. Thanks for not only preparing our material for publication, but making sure it was the best it could possibly be. This past year could not have happened without them.

Suzanne Olivier, our Development Editor, is the absolute best in the business. Suzanne's ability to look at science material in new ways was astonishing. Moreover, she skillfully encouraged us to think about presenting science information a bit differently, too. She played an essential part in revising this Eleventh Edition. Her unfailing attention to readability, consistency, and quality was indispensable to the authors in meeting our goal of delivering complex A&P content in a more student-friendly, learner-centered way.

We are grateful to Lorretta Palagi for her very careful attention to detail and consistency in her copyedit of the text and art.

This book would not exist without the extraordinary dedication of the Production team who solved many problems under pressure with unfailing good cheer. Norine Strang skillfully led her excellent team at Cenveo to move the book smoothly through composition.

The striking cover and clear, navigable interior design were created by tani hasegawa. Thanks also to Mark Ong, Design Manager, who devised innovative solutions for several complex design challenges.

Thanks to our Photo Researcher, Kristin Piljay, and the permissions team at Cenveo for finding, obtaining, and coordinating all the photos in the photo program.

Thanks are also due to Kate Abderholden, Editorial Assistant, who served as project editor for the print supplements for instructors. Thanks also to Stacey Weinberger for handling the physical manufacturing of the book.

We are also grateful to Lauren Chen, Rich Media Content Producer, and Kristin Sanchez, Associate Mastering Producer, for their creative efforts on the media package, most especially MasteringA&P.

We would also like to express our gratitude to the following people at Pearson Science: Paul Corey, President, who continues to support all our texts; Barbara Yien, Director of Development, who kindly kept all phases moving forward under all circumstances; Allison Rona, Executive Marketing Manager; and the dedicated Pearson Science sales representatives for their continuing support of this project. Special thanks to Editor-in-Chief Serina Beauparlant, who took over the reins and worked closely with our new editor, Cheryl Cechvala, to ensure we had the resources necessary to publish what students need to succeed and what professors want in a textbook. And, a round of applause and a backflip go out to Derek Perrigo, Senior A&P Specialist, our biggest cheerleader.

To help improve future editions, we encourage you to send any pertinent information, suggestions, or comments about the organization or content of this textbook to us directly, using the e-mail addresses below. We warmly welcome comments and suggestions and will carefully consider them in the preparation of the Twelfth Edition.

Frederic (Ric) H. Martini
Haiku, Hawaii
martini@pearson.com

Judi L. Nath
Sandusky, Ohio
nath@pearson.com

Edwin F. Bartholomew
Lahaina, Hawaii
bartholomew@pearson.com

Contents

UNIT 2 SUPPORT AND MOVEMENT

5 The Integumentary System 152

8 The Appendicular Skeleton 243

9 Joints 265

10 Muscle Tissue 291

UNIT 3 CONTROL AND REGULATION

14 The Brain and Cranial Nerves 465

15 Sensory Pathways and the Somatic Nervous System 512

16 The Autonomic Nervous System and Higher-Order Functions 535

17 The Special Senses 565

UNIT 4 FLUIDS AND TRANSPORT

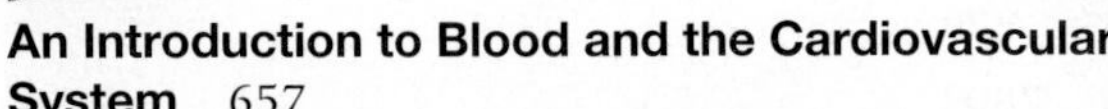

19 Blood 656

20 The Heart 688

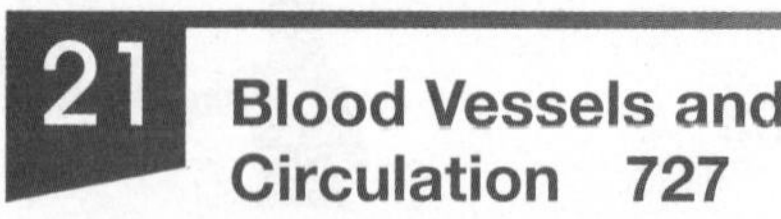

UNIT 5 ENVIRONMENTAL EXCHANGE

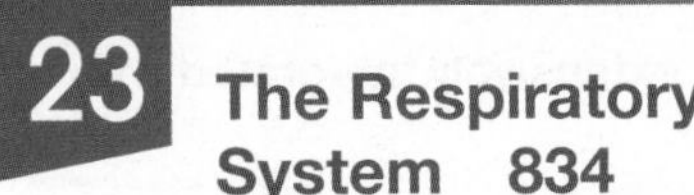

23

23 The Respiratory System 834

UNIT 6 CONTINUITY OF LIFE

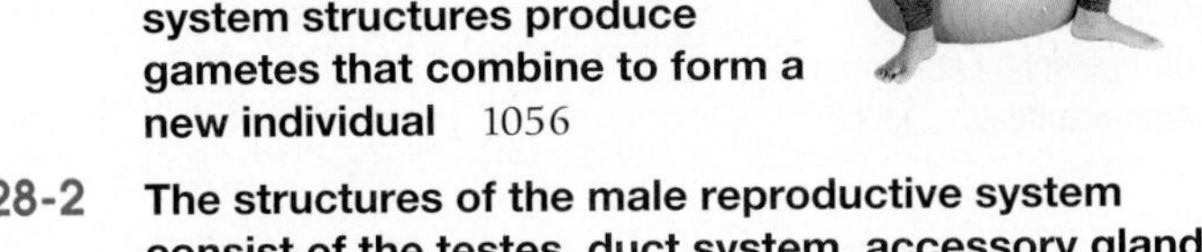

28 The Reproductive System 1055

29 Development and Inheritance 1103

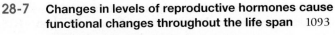

1 An Introduction to Anatomy and Physiology

Learning Outcomes

These Learning Outcomes correspond by number to this chapter's sections and indicate what you should be able to do after completing the chapter.

- 1-1 Describe how to use the text and art to master learning. p. 2
- 1-2 Define anatomy and physiology, explain the relationship between these sciences, and describe various specialties of each discipline. p. 3
- 1-3 Identify the major levels of organization in organisms, from the simplest to the most complex, and identify the major components of each organ system. p. 6
- 1-4 Describe the origins of anatomical and physiological terms, and explain the significance of *Terminologia Anatomica*. p. 7
- 1-5 Use anatomical terms to describe body regions, body sections, and relative positions. p. 7
- 1-6 Identify the major body cavities of the trunk and their subdivisions, and describe the functions of each. p. 14
- 1-7 Explain the concept of homeostasis. p. 18
- 1-8 Describe how negative feedback and positive feedback are involved in homeostatic regulation. p. 19

CLINICAL CASE Using A&P to Save a Life

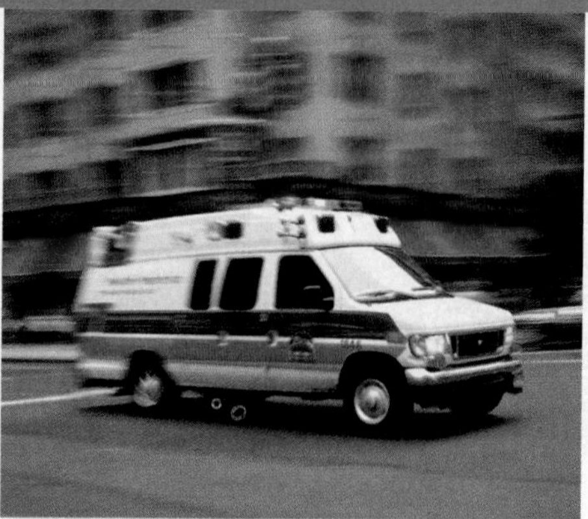

An emergency medical technician (EMT) is on the way to the emergency department with a young victim of street violence. A knife with a 6-inch blade had been found next to the bleeding, unconscious man.

"We have a young male with multiple stab wounds. He has lost a lot of blood and we can barely get a blood pressure," the EMT radios to the triage nurse in the emergency department as the ambulance squeals through traffic. "We started an IV and we are pouring in fluid as fast as we can."

"Where are the wounds?" asks the receiving nurse.

"He has a deep wound in his right upper quadrant, just inferior to the diaphragm. I can see bruising from the hub of the knife around the wound, and there is another wound in his anterior right thigh. His pulse is 120 and thready (weak). His blood pressure is 60 over 30."

"How long has he been down?" questions the nurse.

"Less than a half hour. We intubated him (inserted a breathing tube) and started a large-bore IV as soon as we got there. We are 10 minutes out now."

"Keep the fluids going wide open, keep pressure on the thigh, and take him directly to Trauma Room 1," come the instructions. Meanwhile, the nurse orders the trauma team to Trauma Room 1, orders X-Ray to be on standby in the room, and requests 4 units of type O negative whole blood—the universal donor blood—from the blood bank. **Will the team be ready to save this young man?** **To find out, turn to the Clinical Case Wrap-Up on p. 26.**

An Introduction to Studying the Human Body

Welcome to the field of human anatomy and physiology—known simply as A&P! In this textbook we will introduce you to the inner workings of the human body, giving information about both its structure (anatomy) and its function (physiology). Many students who use this book are preparing for jobs in health-related fields; but regardless of your career choice, you will find the information within these pages relevant to your future.

We will focus on the human body, but the principles you will learn apply to other living things as well. Our world contains an enormous diversity of living organisms, which vary widely in appearance and lifestyle. One aim of *biology*—the study of life—is to discover the unity and the patterns that underlie this diversity. As we study human anatomy and physiology, three main concepts will emerge: (1) the principle of complementarity of structure and function, (2) the hierarchy of structural relationships, and (3) homeostasis, the tendency toward internal balance. These principles are the foundation for learning about the human organism.

Before we begin with the science of human anatomy and physiology, let's turn our attention to the science of learning and learning strategies. To make the most of your learning experience, apply these strategies, which were collected from academic research.

1-1 To make the most of your learning, read the text and view the art together

Learning Outcome Describe how to use the text and art to master learning.

Getting to Know Your Textbook

This first section of the book sets the stage for your success in this course and introduces you to the basic principles of learning. Just as there are three underlying concepts in A&P, there are two basic principles to using your textbook effectively to learn A&P. Practicing these principles will help you throughout your college career.

Let's start. Think back to your first childhood book. You most likely began with a "picture book." Then, as you learned the alphabet and developed speech, you progressed to "word books." The next step was "chapter books." Somewhere along the way, you quit looking at pictures and focused solely on the words (text). Maybe the shift in focus to text-based reading without looking at the pictures happened in high school. You began reading words—paragraph upon paragraph, page upon page of words. Now, you are in college, and we need to realign your focus.

In college, you are faced with lots of new terms, abstract concepts, and unfamiliar images. That's great, because college is intended to increase your knowledge and expand your horizons. However, research has shown that undergraduate students have a tendency to simply read the text (also called the *narrative*), without paying attention to the pictures (referred to as visuals, art, diagrams, illustrations, figures, or images). While you can certainly learn from this approach, further research demonstrates that when students *read the text and then look at the corresponding picture, they actually learn the material better*!

Although this may sound quite intuitive, most students do not do that. So, we wrote a book that truly integrates text with art to help you learn A&P. Please continue reading as we walk you through the process of using a textbook to enhance your learning. Much of what we're about to tell you applies to most of your college textbooks, but we'll focus on this book, since it was designed for students such as you.

Anatomy of a Chapter

Your book is broken down into *sections* with *text–art integration*, and specific *learning outcomes* for each section based on a *learning classification scheme*. A **section** is a unit about a topic that continues to build on previously learned topics. The sectional layout promotes logical, efficient navigation through the material, while callouts to figures integrate the text with the art. **Text–art integration** implies that the figures are close to the lines of text and that the figure legends are adjacent to the art. Look at that figure when you see a callout for it. The figure callouts look like this: (Figure 1–1). They are color-coded on purpose so you can stop reading, look at the figure, and then find your place again when you go back to reading the text. So, *strategy #1 is to read the text and then study the image that goes along with the narrative.*

Learning outcomes are educational objectives that use key verbs and target specific skills, goals, aims, and achievements. The learning outcomes appear at the beginning of each chapter and within the chapter under the sentence-based headings. *Strategy #2 is to pay attention to these learning outcomes because they are tied directly to testing and tell you what you should be able to do after reading that specific section and studying the images.* These learning outcomes are based on a **learning classification scheme**, which identifies the fundamental levels of learning from lower order skills to those of higher-order skills.

Figure 1–1 A Conceptual Framework for Learning.

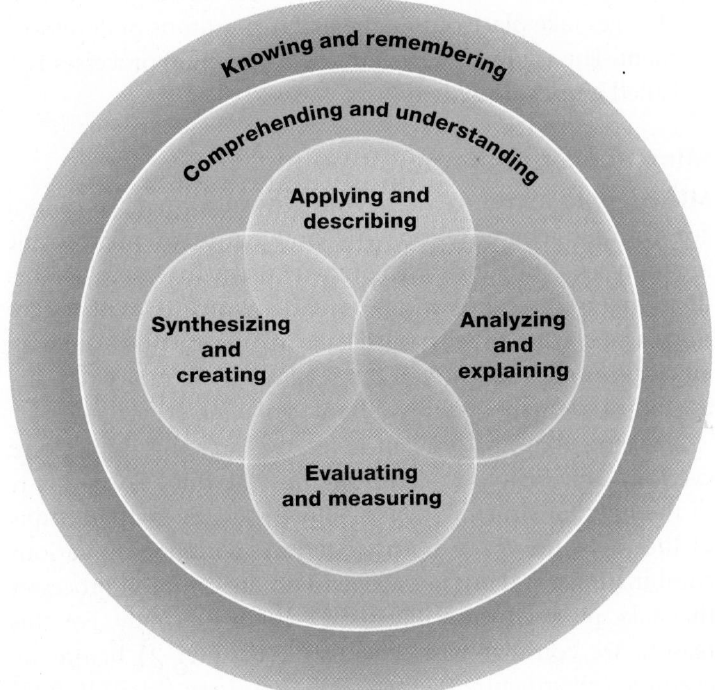
Knowing and remembering
Comprehending and understanding
Applying and describing
Synthesizing and creating
Analyzing and explaining
Evaluating and measuring

You'll see key verbs in your learning outcomes and you'll notice some overlap among them within the levels of learning. From lower to higher, these levels are (1) knowing and remembering, (2) comprehending and understanding, (3) applying and describing, (4) analyzing and explaining, (5) evaluating and measuring, and (6) synthesizing and creating (Figure 1–1). (Here is where you can practice using what you just learned: Look at that figure, think about it, and then return to this text.) If you practice these basic strategies—(1) read the narrative and study the image and (2) pay attention to the learning outcomes—you are well on your way to success!

✓ Checkpoint

1. **Describe a learning outcome.**
2. **Explain how to use your textbook most effectively to enhance your learning.**

See the blue Answers tab at the back of the book.

1-2 Anatomy (structure) and physiology (function) are closely integrated

Learning Outcome Define anatomy and physiology, explain the relationship between these sciences, and describe various specialties of each discipline.

Anatomy is the study of internal and external body structures and their physical relationships among other body parts. In contrast, **physiology** is the study of how living organisms perform their vital functions. Someone studying anatomy might, for example, examine where a particular muscle attaches to the skeleton. Someone studying physiology might consider how a muscle contracts or what forces a contracting muscle exerts on the skeleton. You will be studying both anatomy and physiology in this text, so let's look at the relationships between these sciences.

Anatomy and physiology are closely integrated, both in theory and practice. Anatomical information provides clues about functions, and physiological processes can be explained only in terms of the underlying anatomy. This is a very important concept in living systems:

> All specific functions are performed by specific structures, and the form of a structure relates to its function. This is known as the *principle of complementarity of structure and function*.

The link between structure and function is always present, but not always understood. For example, the anatomy of the heart was clearly described in the 15th century, but almost 200 years passed before the heart's pumping action was demonstrated.

Anatomists and physiologists approach the relationship between structure and function from different perspectives. To understand the difference, suppose you asked an anatomist and a physiologist to examine a car and report their findings. The anatomist might begin by measuring and photographing the various parts of the car and, if possible, taking it apart and putting it back together. The anatomist could then explain its key *structural relationships*—for example, how the pistons are seated in the engine cylinders, how the crankshaft is connected to the pistons, and how the transmission links the drive shaft to the axles and, thus, to the wheels. The physiologist also would note the relationships among the car's parts, but he or she would focus mainly on its *functional* characteristics, such as how the combustion of gasoline in the cylinders moves the pistons up and down and makes the drive shaft rotate, and how the transmission conveys this motion to the axles and wheels so that the car moves. Additionally, he or she might also study the amount of power that the engine could generate, the amount of force transmitted to the wheels in different gears, and so forth.

Our basic approach in this textbook will be to start with the descriptive anatomy of body structures (appearance, size, shape, location, weight, and color) before considering the related functions. Sometimes the groups of organs that make up an *organ system* perform very diverse functions, and in those cases we consider the functions of each individual organ separately. A good example is our discussion of the digestive system and its organs. You will learn about the functions of the salivary glands in one section, and the functions of the tongue in another. In other systems, the organs work together so extensively that we present an overall discussion of their physiology, after we describe the system's anatomy. The lymphatic system (which contains a network of vessels) and the cardiovascular system, for example, are treated using this approach.

Anatomy

When you look at something, how far away you are from it often determines what you see. You get a very different view of your neighborhood from a satellite photo than from your front yard. Similarly, your method of observation has a dramatic effect on your understanding of the structure of the human body. Based on the degree of structural detail being considered, we divide **human anatomy**, the study of the structure of the human body, into *gross (macroscopic) anatomy* and *microscopic anatomy.*

Gross Anatomy

Gross anatomy, or *macroscopic anatomy*, involves examining fairly large structures. Gross anatomy (from the Latin term *grossus*, meaning "thick" or "massive") can be conducted without using a microscope and can involve the study of anatomy by dissecting a cadaver. There are many different forms of gross anatomy:

- *Surface anatomy*, or superficial anatomy, is the study of the general form of the body's surface, especially in relation to its deeper parts.
- *Regional anatomy* focuses on the anatomical organization of specific areas of the body, such as the head, neck, or trunk. Many advanced courses in anatomy stress a regional approach, because it emphasizes the spatial relationships among structures already familiar to students.
- *Sectional anatomy* is the study of the relationship of the body's structures by examining cross sections of the tissue or organ.
- *Systemic anatomy* is the study of the structure of *organ systems*, which are groups of organs that function together in a coordinated manner. Examples include the *skeletal system*, composed primarily of bones; the *muscular system*, made up of skeletal muscles; and the *cardiovascular system*, consisting of the heart, blood, and vessels. We take a systemic anatomy approach in this book because this format works better to clarify the functional relationships among the component organs. We introduce the 11 organ systems in the human body later in the chapter.
- *Clinical anatomy* includes a number of subspecialties important in clinical practice. Examples include *pathological anatomy* (anatomical features that change during illness), *radiographic anatomy* (anatomical structures seen using specialized imaging techniques), and *surgical anatomy* (anatomical landmarks important in surgery).
- *Developmental anatomy* describes the changes in form that take place between conception and adulthood. The techniques of developmental anatomists are similar to those used in gross anatomy and in microscopic anatomy (discussed next) because developmental anatomy considers anatomical structures over a broad range of sizes—from a single cell to an adult human. The most extensive structural changes take place during the first two months of development. The study of these early developmental processes is called **embryology** (em-brē-OL-ō-jē).

Microscopic Anatomy

Microscopic anatomy deals with structures that we cannot see without magnification. The boundaries of microscopic anatomy are set by the limits of the equipment we use. With a dissecting microscope you can see tissue structure. With a light microscope, you can see basic details of cell structure. And with an electron microscope, you can see individual molecules that are only a few nanometers (billionths of a meter) across.

Microscopic anatomy includes two major subdivisions: cytology and histology. **Cytology** (sī-TOL-ō-jē) is the study of the internal structure of individual *cells*, the simplest units of life. Cells are made up of chemical substances in various combinations, and our lives depend on the chemical processes that take place in the trillions of cells in the body. For this reason, we consider basic chemistry (Chapter 2) before we examine cell structure (Chapter 3). **Histology** (his-TOL-ō-jē)

+ Clinical Note *Habeas Corpus* ("You Shall Have the Body")

It is the first day of Anatomy. Students await the arrival of their white-coated and gloved professor. Anxiety mounts as a stretcher covered in surgical drapes is wheeled in. This is the cadaver. Who will faint? *Will it be me?* For many students in the health professions, cadaver dissection is a cornerstone of their training. These students are following in a revered tradition that began 2300 years ago with the first examinations of the body after death by Greek royal physicians. The expression "a skeleton in your closet" dates from a later era when medical students had to procure bodies on their own for study (and keep them hidden in the closet). There is much to be learned from death. Cadaver dissections also reveal much about life. After working closely on a cadaver for months, students develop an attachment to "their" body, often naming it. The intimate revelations of the scalpel, the highly personal variations of human anatomy, the Rubik's cube of disease, and the stark reality of death combine to leave a deep intellectual and emotional mark on the student. Students and faculty may end the course with a ceremony to pay their respects to this human body and to this privileged experience.

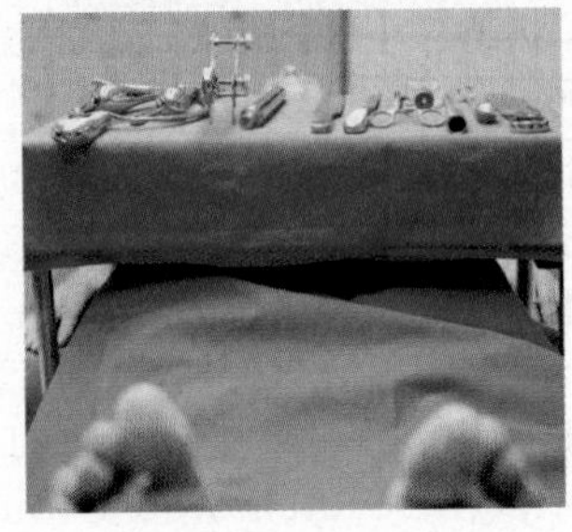

is the examination of *tissues*—groups of specialized cells that work together to perform specific functions (Chapter 4). Tissues combine to form *organs*, such as the heart, kidney, liver, or brain, each with specific functions. Many organs are easy to examine without a microscope, so at the organ level we cross the boundary from microscopic anatomy to gross anatomy. As we proceed through the text, we will consider details at all levels, from microscopic to macroscopic.

Physiology

Human physiology is the study of the functions, or workings, of the human body. These functions are complex processes and much more difficult to examine than most anatomical structures. As a result, there are even more specialties in physiology than in anatomy. Examples include the following:

- *Cell physiology*, the study of the functions of cells, is the cornerstone of human physiology. Cell physiology looks at the chemistry of the cell. It includes both chemical processes within cells and chemical interactions among cells.
- *Organ physiology* is the study of the function of specific organs. An example is *cardiac physiology*, the study of heart function—how the heart works.
- *Systemic physiology* includes all aspects of the functioning of specific organ systems. Cardiovascular physiology, respiratory physiology, and reproductive physiology are examples.
- *Pathological physiology* is the study of the effects of diseases on organ functions or system functions. Modern medicine depends on an understanding of both normal physiology and pathological physiology.

Physicians normally use a combination of anatomical, physiological, chemical, and psychological information when they evaluate patients. When a patient presents with **signs** (an objective disease indication like a fever) and **symptoms** (a subjective disease indication, such as tiredness), the physician will look at the structures affected (gross anatomy), perhaps collect a fluid or tissue sample (microscopic anatomy) for analysis, and ask questions to find out what changes from normal functioning the patient is experiencing. Think back to your last trip to a doctor's office. Not only did the physician examine your body, noting any anatomical abnormalities, but he or she also evaluated your physiological processes by asking questions, observing your movements, listening to your body sounds, taking your temperature, and perhaps requesting chemical analyses of fluids such as blood or urine.

In evaluating all these observations to reach a diagnosis, physicians rely on a logical framework based on the scientific method. The **scientific method** is a system of advancing knowledge that begins by proposing a hypothesis to answer a question, and then testing that hypothesis with data collected through observation and experimentation. This method is at the core of all scientific thought, including medical diagnosis.

√ Checkpoint

3. **Define *anatomy*.**
4. **Define *physiology*.**
5. **Describe how anatomy and physiology are closely related.**
6. **What is the difference between gross anatomy and microscopic anatomy?**
7. **Identify several specialties of physiology.**
8. **Why is it difficult to separate anatomy from physiology?**

See the blue Answers tab at the back of the book.

1-3 Levels of organization progress from chemicals to a complete organism

Learning Outcome **Identify the major levels of organization in organisms, from the simplest to the most complex, and identify major components of each organ system.**

Our understanding of how the human body works is based on investigations of its different levels of organization. Higher levels of organization are more complex and more variable than lower levels. Chapters 2, 3, and 4 consider the chemical, cellular, and tissue levels of organization of the human body. These levels are the foundations of more complex structures and vital processes, as we describe in Chapters 5–29. The six levels of organization of the human body are shown in **Spotlight Figure 1–2** and include:

- *The Chemical Level.* **Atoms** are the smallest stable units of matter. They can combine to form **molecules** with complex shapes. The atomic components and unique three-dimensional shape of a particular molecule determine its function. For example, complex protein molecules form filaments that produce the contractions of muscle cells in the heart. We explore this level of organization in Chapter 2.
- *The Cellular Level.* **Cells** are the smallest living units in the body. Complex molecules can form various types of larger structures called *organelles*. Each organelle has a specific function in a cell. Energy-producing organelles provide the energy needed for heart muscle cell contractions. We examine the cellular level of organization in Chapter 3.
- *The Tissue Level.* A **tissue** is a group of cells working together to perform one or more specific functions. Heart muscle cells, also called cardiac muscle cells (*cardium*, heart), interact with other types of cells and with materials outside the cell to form cardiac muscle tissue. We consider the tissue level of organization in Chapter 4.
- *The Organ Level.* **Organs** are made of two or more tissues working together to perform specific functions. Layers of cardiac muscle tissue, in combination with another type of tissue called connective tissue, form the bulk of the wall of the heart, which is a hollow, three-dimensional organ.
- *The Organ System Level.* A group of organs interacting to perform a particular function forms an **organ system**. Each time the heart contracts, for example, it pushes blood into a network of blood vessels. Together, the heart, blood, and blood vessels make up the cardiovascular system, one of 11 organ systems in the body. This system functions to distribute oxygen and nutrients throughout the body.
- *The Organism Level.* An individual life form is an **organism**. In our case, an individual human is the highest level of organization that we consider. All of the body's organ systems must work together to maintain the life and health of the organism.

The organization at each level determines not only the structural characteristics but also the functions of higher levels. For example, the arrangement of atoms and molecules at the chemical level creates the protein filaments and organelles at the cellular level that give individual cardiac muscle cells the ability to contract. At the tissue level, these cells are linked, forming cardiac muscle tissue. The structure of the tissue ensures that the contractions are coordinated, producing a powerful heartbeat. When that beat occurs, the internal anatomy of the heart, an organ, enables it to function as a pump. The heart is filled with blood and connected to the blood vessels, and its pumping action circulates blood through the vessels of the cardiovascular system. Through interactions with the respiratory, digestive, urinary, and other systems, the cardiovascular system performs a variety of functions essential to the survival of the organism.

Something that affects a system will ultimately affect each of the system's parts. For example, after massive blood loss, the heart cannot pump blood effectively. When the heart cannot pump and blood cannot flow, oxygen and nutrients cannot be distributed to the heart or around the body. Very soon, the cardiac muscle tissue begins to break down as individual muscle cells die from oxygen and nutrient starvation. These changes will not be restricted to the cardiovascular system. All cells, tissues, and organs in the body will be damaged. **Spotlight Figure 1–2** illustrates the levels of organization and introduces the 11 interdependent, interconnected organ systems in the human body.

The cells, tissues, organs, and organ systems of the body coexist in a relatively small, shared environment, much like the residents of a large city. Just as city dwellers breathe the same air and drink the water supplied by the local water company, cells in the human body absorb oxygen and nutrients from the fluids that surround them. If a city is blanketed in smog or its water supply is contaminated, its inhabitants will become ill. Similarly, if the body fluid composition becomes abnormal, cells will be injured or destroyed. For example, suppose the temperature or salt content of the blood changes. The effect on the heart could range from the need for a minor adjustment (heart muscle tissue contracts more often, raising the heart rate) to a total disaster (the heart stops beating, so the individual dies).

✓ Checkpoint

9. **Identify the major levels of organization of the human body from the simplest to the most complex.**
10. **Identify the organ systems of the body and cite some major structures of each.**
11. **At which level of organization does a histologist investigate structures?**

See the blue Answers tab at the back of the book.

1-4 Medical terminology is important to understanding anatomy and physiology

Learning Outcome **Describe the origins of anatomical and physiological terms, and explain the significance of *Terminologia Anatomica*.**

Early anatomists faced serious problems when trying to communicate. Saying that a bump is "on the back," for example, does not give very precise information about its location. So anatomists created illustrated maps of the human body and gave each structure a specific name. They used prominent anatomical structures as landmarks, measured distances in centimeters or inches, and discussed these subjects in specialized directional terms. Modern anatomists continue and build on these practices. In effect, anatomy uses a special language that you must learn almost at the start of your study.

That special language, called **medical terminology**, involves using word roots, prefixes, suffixes, and combining forms to build terms related to the body in health and disease. Many of the anatomical and physiological terms you will encounter in this textbook are derived from Greek or Latin roots that originated more than 1500 years ago. In fact, the term *anatomy* is derived from Greek roots that mean "a cutting open"; the term *physiology* also comes from Greek. Learning the word parts used in medical terminology will greatly assist in your study of anatomy and physiology and in your preparation for any health-related career.

There are four basic building blocks—or word parts—of medical terms. *Word roots* are the basic, meaningful parts of a term that cannot be broken down into another term with another definition. *Prefixes* are word elements that are attached to the beginning of words to modify their meaning but cannot stand alone. *Suffixes* are similar to prefixes, except they are word elements or letters added to the end of a word or word part to form another term. *Combining forms* are independent words or word roots that are used in combination with words, prefixes, suffixes, or other combining forms to build a new term. As we introduce new terms, we will provide notes on pronunciation and relevant word parts. In addition, the table inside the back cover of your textbook lists many commonly used word roots, prefixes, suffixes, and combining forms.

To illustrate the building of medical terms, consider the word *pathology* (puh-THOL-ō-jē). Breaking this word into its basic parts reveals its meaning. The prefix *path-* refers to disease (the Greek term for "disease" is *pathos*). The suffix *-ology* means "study of." So pathology is the study of disease.

Latin and Greek terms are not the only ones that have been imported into the anatomical vocabulary over the centuries, and this vocabulary continues to expand. Many anatomical structures and clinical conditions were first named after either the discoverer or, in the case of diseases, the most famous victim. During the past 100 years, most of these commemorative names, or **eponyms** (EH-pō-nimz), have been replaced by more precise terms. Where appropriate, we will give both the eponym and the more precise term, because in clinical medicine, both terms may be used.

To avoid the miscommunication that plagued the early anatomists, it is important for scientists throughout the world to use the same name for each body structure. In 1998, two scientific organizations—the Federative Committee on Anatomical Terminology (FCAT) and the International Federation of Associations of Anatomists (IFAA)—published *Terminologia Anatomica* (*TA*). *Terminologia Anatomica* established the worldwide standard for human anatomical terminology. The successor of FCAT is the Federative International Programme on Anatomical Terminologies (FIPAT). In April 2011, FIPAT published *TA* online. Latin continues to be the language of anatomy, but this reference provides an English equivalent term for each anatomical structure. For example, the *tendo calcaneus* (Latin) is also called the calcaneal tendon (English). You may know the structure better by its eponym, the Achilles tendon. Eponyms are not found in *TA*. We have used *TA* as our standard in preparing this textbook.

✓ Checkpoint

12. **Describe medical terminology.**
13. **Define *eponym*.**
14. **Name the book that serves as the international standard for anatomical terms.**

See the blue Answers tab at the back of the book.

1-5 Anatomical terms describe body regions, anatomical positions and directions, and body sections

Learning Outcome **Use anatomical terms to describe body regions, body sections, and relative positions.**

Anatomists use anatomical terms to describe body regions, relative positions and directions, and body sections, as well as major body cavities and their subdivisions. In the following sections we introduce the terms used in superficial anatomy and sectional anatomy.

Surface Anatomy

Surface anatomy involves locating structures on or near the body surface. A familiarity with anatomical landmarks (structures that can be felt or palpated), anatomical regions (specific areas used for reference purposes), and terms for anatomical directions will make the material in subsequent chapters easier to understand.

SPOTLIGHT

Figure 1–2
Levels of Organization

Interacting atoms form molecules that combine to form the protein filaments of a heart muscle cell. Such cells interlock, forming heart muscle tissue, which makes up most of the walls of the heart, a three-dimensional organ. The heart is only one component of the cardiovascular system, which also includes the blood and blood vessels. The various organ systems must work together to maintain life at the organism level.

Chemical Level

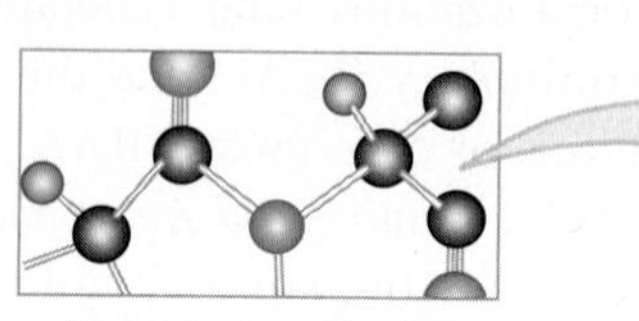

Atoms in combination

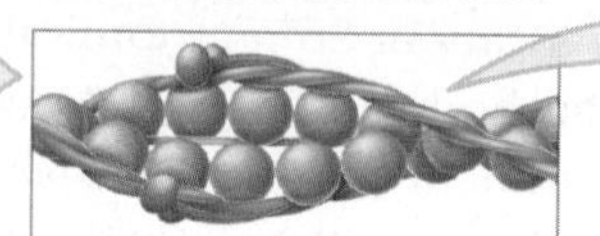

Complex protein molecule

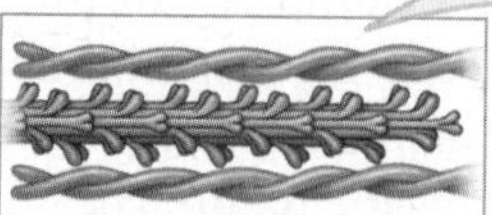

Protein filaments

Cellular Level

Heart muscle cell

THE ORGAN SYSTEMS

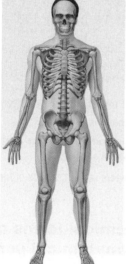

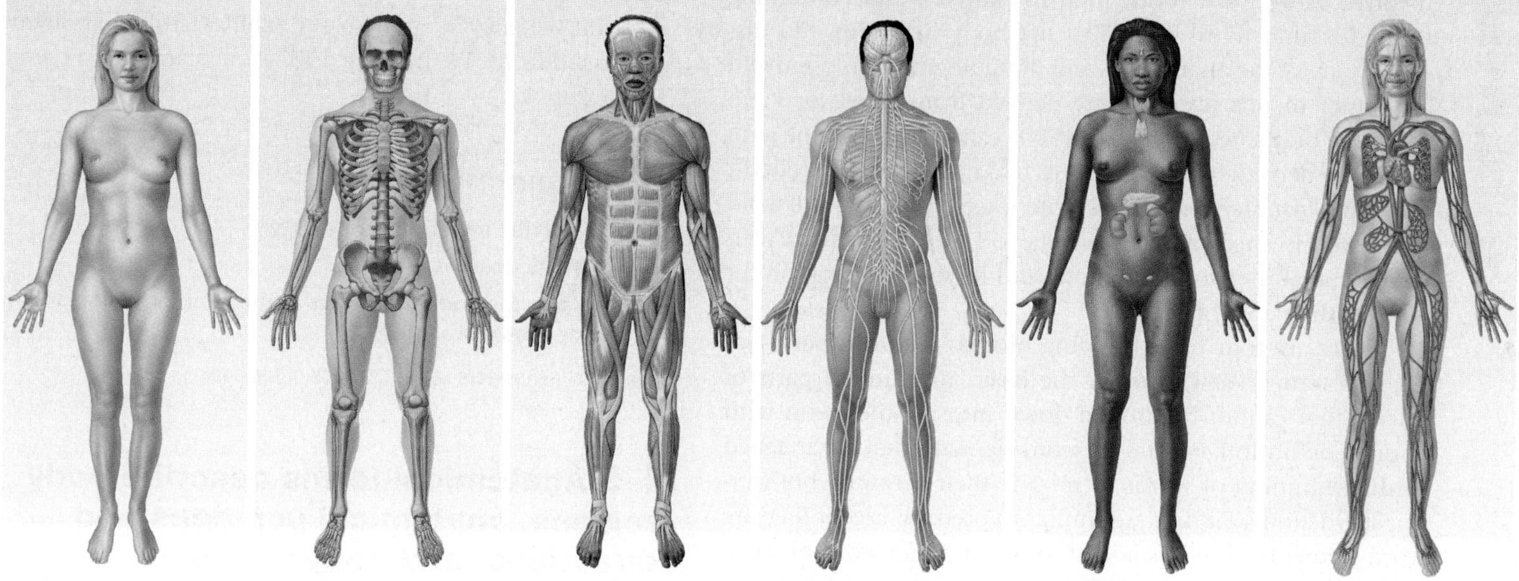

Integumentary

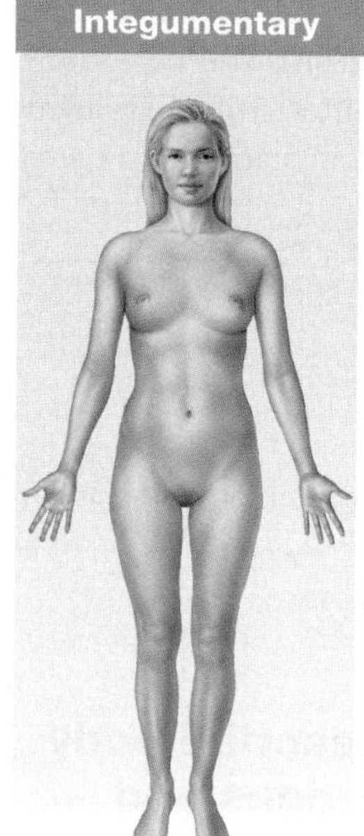

Major Organs
- Skin
- Hair
- Sweat glands
- Nails

Functions
- Protects against environmental hazards
- Helps regulate body temperature
- Provides sensory information

Skeletal

Major Organs
- Bones
- Cartilages
- Associated ligaments
- Bone marrow

Functions
- Provides support and protection for other tissues
- Stores calcium and other minerals
- Forms blood cells

Muscular

Major Organs
- Skeletal muscles and associated tendons

Functions
- Provides movement
- Provides protection and support for other tissues
- Generates heat that maintains body temperature

Nervous

Major Organs
- Brain
- Spinal cord
- Peripheral nerves
- Sense organs

Functions
- Directs immediate responses to stimuli
- Coordinates or moderates activities of other organ systems
- Provides and interprets sensory information about external conditions

Endocrine

Major Organs
- Pituitary gland
- Thyroid gland
- Pancreas
- Adrenal glands
- Gonads
- Endocrine tissues in other systems

Functions
- Directs long-term changes in the activities of other organ systems
- Adjusts metabolic activity and energy use by the body
- Controls many structural and functional changes during development

Cardiovascular

Major Organs
- Heart
- Blood
- Blood vessels

Functions
- Distributes blood cells, water, and dissolved materials including nutrients, waste products, oxygen, and carbon dioxide
- Distributes heat and assists in control of body temperature

Tissue Level

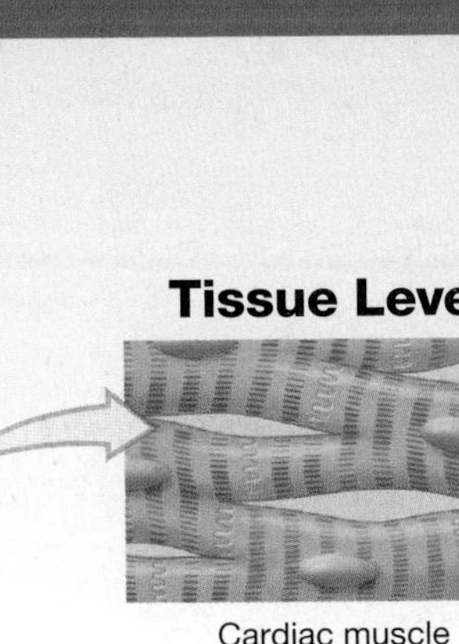

Cardiac muscle tissue

Organ Level

The heart

Organ system level

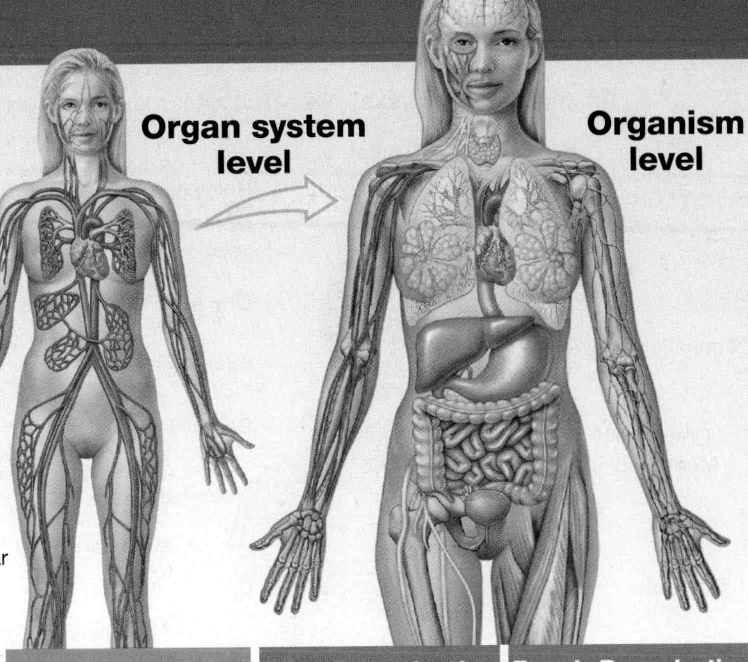

The cardiovascular system

Organism level

Lymphatic

Major Organs
- Spleen
- Thymus
- Lymphatic vessels
- Lymph nodes
- Tonsils

Functions
- Defends against infection and disease
- Returns tissue fluids to the bloodstream

Respiratory

Major Organs
- Nasal cavities
- Sinuses
- Larynx
- Trachea
- Bronchi
- Lungs
- Alveoli

Functions
- Delivers air to alveoli (sites in lungs where gas exchange occurs)
- Provides oxygen to bloodstream
- Removes carbon dioxide from bloodstream
- Produces sounds for communication

Digestive

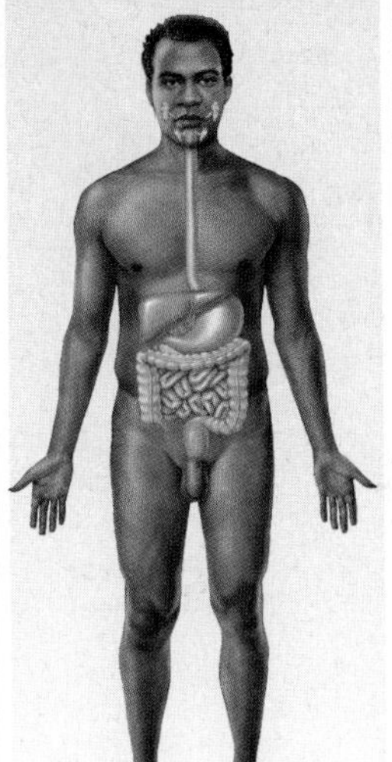

Major Organs
- Teeth
- Tongue
- Pharynx
- Esophagus
- Stomach
- Small intestine
- Large intestine
- Liver
- Gallbladder
- Pancreas

Functions
- Processes and digests food
- Absorbs and conserves water
- Absorbs nutrients
- Stores energy reserves

Urinary

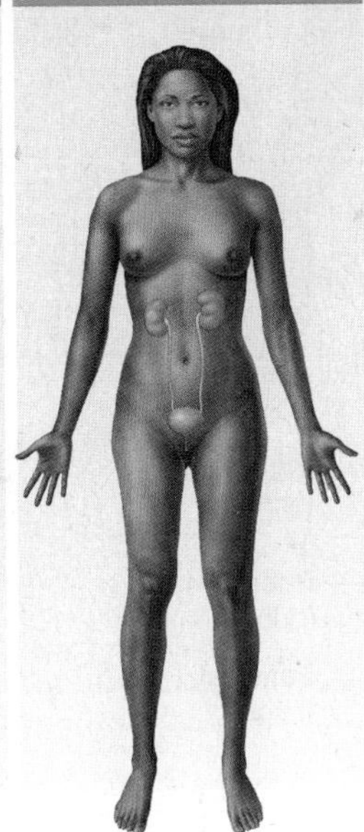

Major Organs
- Kidneys
- Ureters
- Urinary bladder
- Urethra

Functions
- Excretes waste products from the blood
- Controls water balance by regulating volume of urine produced
- Stores urine prior to voluntary elimination
- Regulates blood ion concentrations and pH

Male Reproductive

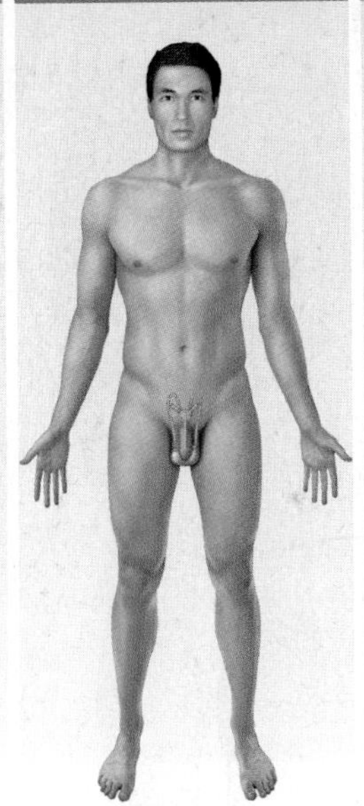

Major Organs
- Testes
- Epididymides
- Ductus deferentia
- Seminal vesicles
- Prostate gland
- Penis
- Scrotum

Functions
- Produces male sex cells (sperm), seminal fluids, and hormones
- Sexual intercourse

Female Reproductive

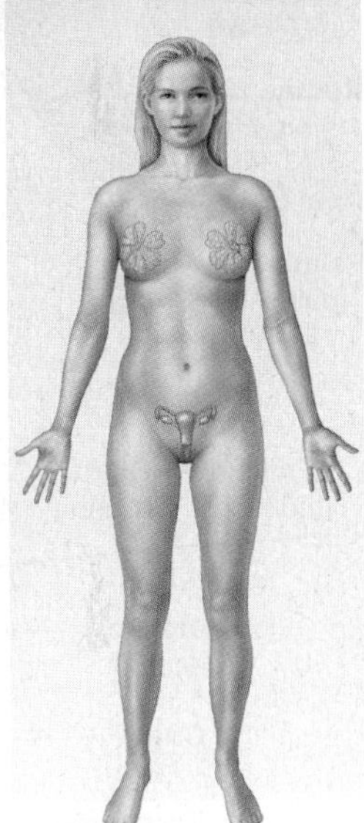

Major Organs
- Ovaries
- Uterine tubes
- Uterus
- Vagina
- Labia
- Clitoris
- Mammary glands

Functions
- Produces female sex cells (oocytes) and hormones
- Supports developing embryo from conception to delivery
- Provides milk to nourish newborn infant
- Sexual intercourse

Figure 1–3 Anatomical Landmarks. Anatomical terms are shown in boldface type and common names are in plain type.

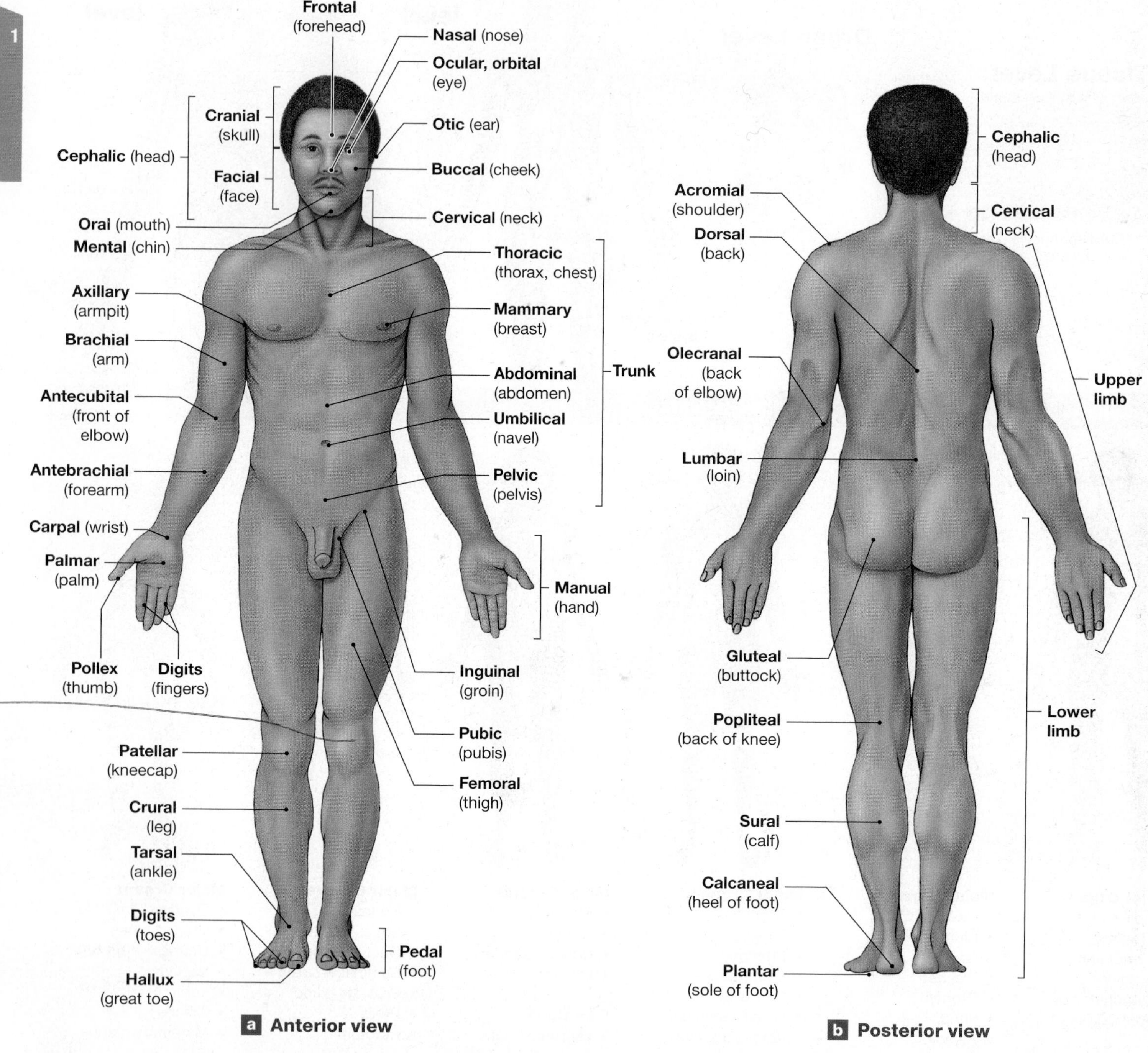
Frontal
(forehead)
Nasal (nose)
Ocular, orbital
(eye)
Cranial
(skull)
Cephalic (head)
Facial
(face)
Otic (ear)
Buccal (cheek)
Oral (mouth)
Mental (chin)
Cervical (neck)
Thoracic
(thorax, chest)
Axillary
(armpit)
Mammary
(breast)
Brachial
(arm)
Abdominal
(abdomen)
Trunk
Antecubital
(front of
elbow)
Umbilical
(navel)
Antebrachial
(forearm)
Pelvic
(pelvis)
Carpal (wrist)
Palmar
(palm)
Manual
(hand)
Pollex
(thumb)
Digits
(fingers)
Inguinal
(groin)
Pubic
(pubis)
Patellar
(kneecap)
Femoral
(thigh)
Crural
(leg)
Tarsal
(ankle)
Digits
(toes)
Pedal
(foot)
Hallux
(great toe)
a Anterior view
Cephalic
(head)
Acromial
(shoulder)
Cervical
(neck)
Dorsal
(back)
Olecranal
(back
of elbow)
Upper
limb
Lumbar
(loin)
Gluteal
(buttock)
Popliteal
(back of knee)
Lower
limb
Sural
(calf)
Calcaneal
(heel of foot)
Plantar
(sole of foot)
b Posterior view

? Are the following anatomical landmarks visible from the anterior or posterior view: dorsal, gluteal, calcaneal?

Anatomical Landmarks

Figure 1–3 presents important anatomical landmarks. Understanding the terms and their origins will help you remember both the location of a particular structure and its name. For example, *brachial* refers to the arm, and later we will consider the *brachialis muscle* and the *brachial artery*, which are in the arm, as their names suggest.

The standard anatomical reference for the human form is the **anatomical position**. This is also called the *anatomic position*. When the body is in this position, the hands are at the sides with the palms facing forward, and the feet are together. Figure 1–3a shows an individual in the anatomical position as seen from the front, called an *anterior view*. Figure 1–3b shows the body as seen from the back, called a *posterior view*. Unless otherwise noted, all descriptions in this text refer to the body in the anatomical position. A person lying down is said to be **supine** (sū-PĪN) when face up, and **prone** when face down.

Tips & Tools

Supine means "up." In order to carry a bowl of *soup*, your hand must be in the *supine* position.

Anatomical Regions

To describe a general area of interest or injury, clinicians and anatomists often need broader terms in addition to specific landmarks. They use two methods—dividing into quadrants and dividing into regions—to map the surface of the abdomen and pelvis.

Clinicians refer to four **abdominopelvic quadrants** (Figure 1–4a) formed by a pair of imaginary perpendicular lines that intersect at the umbilicus (navel). This simple method of dividing into quadrants provides useful references for describing the location of aches, pains, and injuries. Knowing the location can help the clinician determine the possible cause. For example, tenderness in the right lower quadrant (RLQ) is a symptom of appendicitis. Tenderness in the right upper quadrant (RUQ), however, may indicate gallbladder or liver problems.

Anatomists prefer more precise terms to describe the location and orientation of internal organs. They recognize nine **abdominopelvic regions** (Figure 1–4b). Figure 1–4c shows the relationships among quadrants, regions, and internal organs.

Tips & Tools

The imaginary lines dividing the abdominopelvic regions resemble a tic-tac-toe game.

Anatomical Directions

Figure 1–5 introduces the main directional terms and some examples of their use. There are many different terms, and some can be used interchangeably. For example, *anterior* refers to the front of the body when viewed in the anatomical position. In humans, this term is equivalent to *ventral*, which refers to the belly. *Posterior* refers to the back of the body; this term is equivalent to *dorsal*. When reading anatomical descriptions, remember that the terms *left* and *right* always refer to the left and right sides of the *subject*, not of the observer.

Before you read further, analyze the image in detail, and practice using the terms. We start using these terms in the rest of this chapter. If you are familiar with the basic vocabulary, the anatomical descriptions throughout this textbook will be easier to follow.

Figure 1–4 Abdominopelvic Quadrants and Regions.

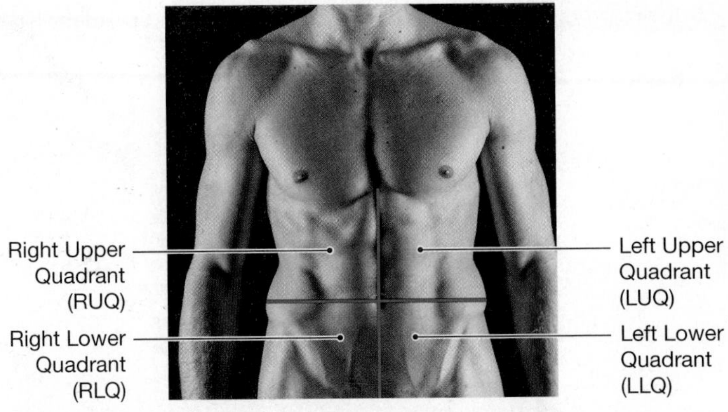
Right Upper Quadrant (RUQ)
Right Lower Quadrant (RLQ)
Left Upper Quadrant (LUQ)
Left Lower Quadrant (LLQ)

a **Abdominopelvic quadrants**. The four abdominopelvic quadrants are formed by two perpendicular lines that intersect at the navel. The terms for these quadrants, or their abbreviations, are most often used in clinical discussions.

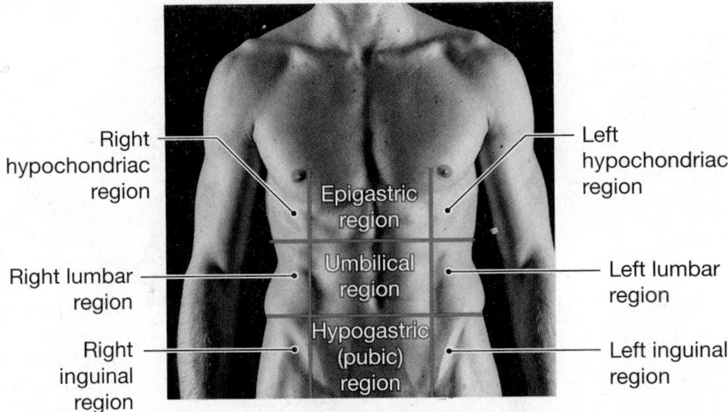
Right hypochondriac region
Epigastric region
Left hypochondriac region
Right lumbar region
Umbilical region
Left lumbar region
Right inguinal region
Hypogastric (pubic) region
Left inguinal region

b **Abdominopelvic regions**. The nine abdominopelvic regions provide more precise regional descriptions.

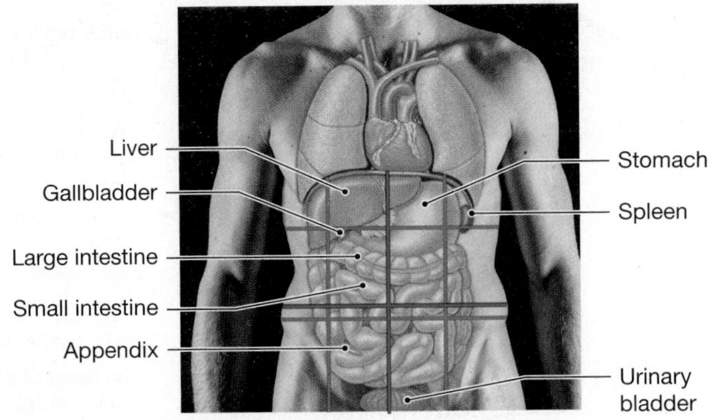
Liver
Gallbladder
Large intestine
Small intestine
Appendix
Stomach
Spleen
Urinary bladder

c **Anatomical relationships**. The relationship between the abdominopelvic quadrants and regions and the locations of the internal organs are shown here.

? In which abdominopelvic quadrant and region is the stomach predominantly found?

Figure 1–5 **Directional References.**

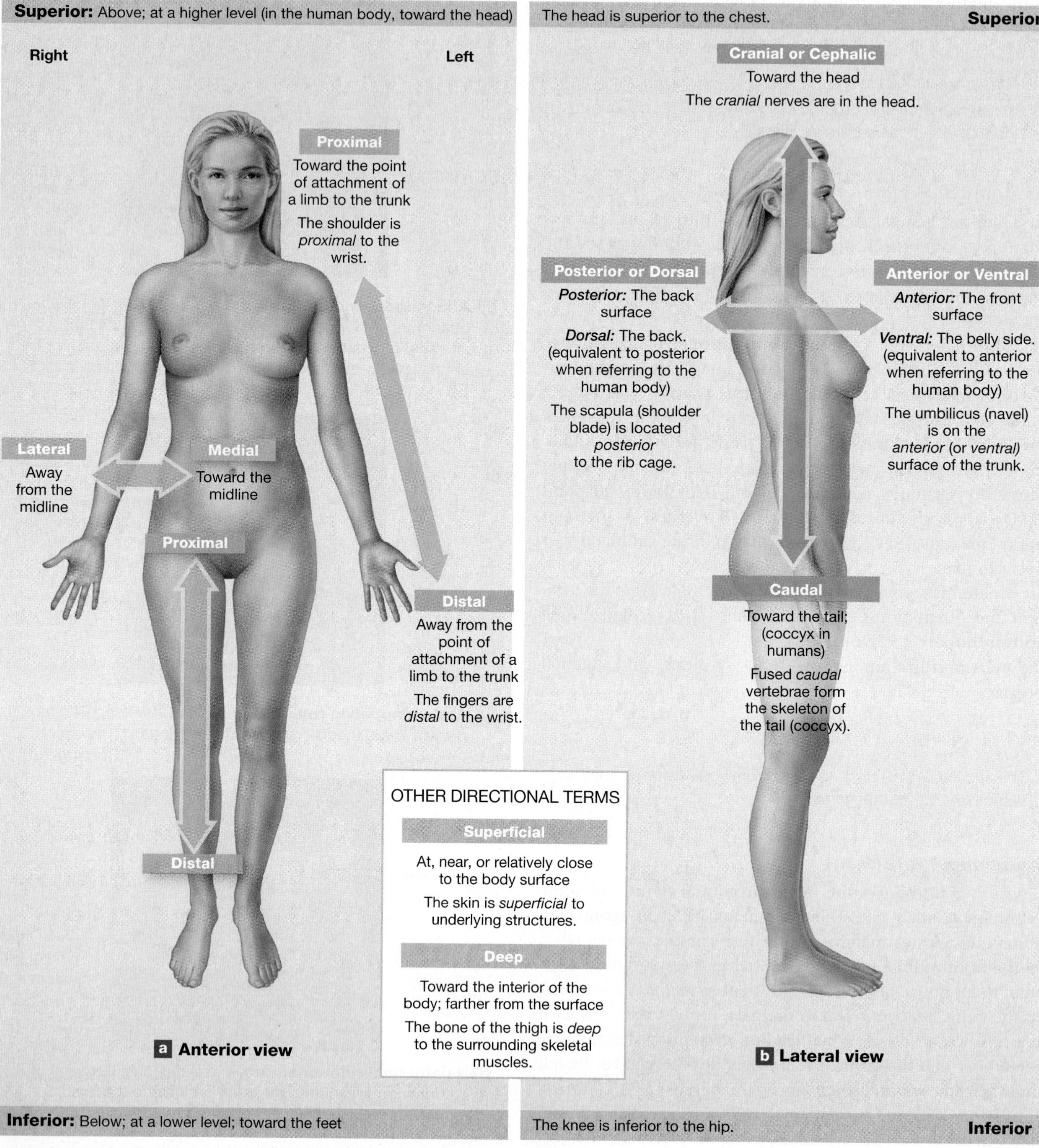
Superior: Above; at a higher level (in the human body, toward the head)
The head is superior to the chest.
Superior
Right
Left
Proximal
Toward the point of attachment of a limb to the trunk
The shoulder is proximal to the wrist.
Lateral
Away from the midline
Medial
Toward the midline
Proximal
Distal
Away from the point of attachment of a limb to the trunk
The fingers are distal to the wrist.
Distal
a Anterior view
Cranial or Cephalic
Toward the head
The cranial nerves are in the head.
Posterior or Dorsal
Posterior: The back surface
Dorsal: The back. (equivalent to posterior when referring to the human body)
The scapula (shoulder blade) is located posterior to the rib cage.
Anterior or Ventral
Anterior: The front surface
Ventral: The belly side. (equivalent to anterior when referring to the human body)
The umbilicus (navel) is on the anterior (or ventral) surface of the trunk.
Caudal
Toward the tail; (coccyx in humans)
Fused caudal vertebrae form the skeleton of the tail (coccyx).
OTHER DIRECTIONAL TERMS
Superficial
At, near, or relatively close to the body surface
The skin is superficial to underlying structures.
Deep
Toward the interior of the body; farther from the surface
The bone of the thigh is deep to the surrounding skeletal muscles.
b Lateral view
Inferior: Below; at a lower level; toward the feet
The knee is inferior to the hip.
Inferior

? Using directional references for a person in the anatomical position, how would you describe the relationship of the hand compared to the elbow? To the groin?

+ Clinical Note The Sounds of the Body

We pay attention to the sounds of the body because they provide evidence of normal function. Midwives and doctors used to place their ears directly on the patient's body to listen. Then a piece of new-fangled technology, called the *stethoscope*, was invented in 1816 that gave the patient more privacy and the practitioner a better listen. *Auscultation* (aws-kul-TĀ-shun) is the practice of listening to the various sounds made by body organs with a stethoscope. After a surgical operation, the recovery room nurse auscultates the patient's abdomen to listen for bowel sounds (called by the imitative term *borborygmi* [bor-bō-RIG-mī]). These sounds confirm that the intestine is resuming its characteristic motility after anesthesia. When students get their college physicals, the practitioner auscultates the lungs over the dorsal surface of the body and superior to the clavicles (where the tips of the lungs lie). It is a thrilling moment when a pregnant woman hears her baby's heartbeat for the first time with the help of an ultrasound technician. A Doppler ultrasound device bounces sound waves off of a fetus's heart that are detected at the mother's skin surface in her pubic region (see Clinical Note: Diagnostic Imaging Techniques, pp. 16–17). The heartbeat is usually first heard when the fetus is about 12 weeks old. The sounds of an adult heart are heard at the general locations labeled on the anterior thoracic region of the body shown here. The heart sounds of "lubb-dupp" (which you will study in Chapter 20) tell us that heart valves are closing correctly during the heart's cycle. However, an unexpected "whoosh" can alert us to the possibility of a medical problem called a *heart murmur*.

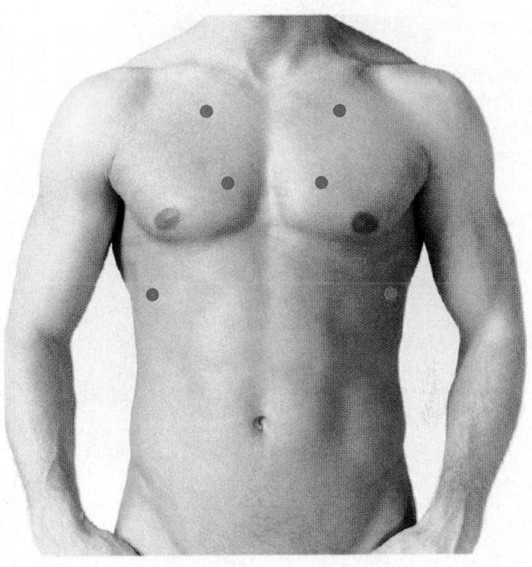

Sectional Anatomy

Sometimes the only way to understand the relationships among the parts of a three-dimensional object is to slice through it and look at the internal organization. A slice through a three-dimensional object is called a *section*.

An understanding of sectional views is particularly important now that imaging techniques enable us to see inside the living body. These views are sometimes difficult to interpret, but it is worth spending the time required to understand what they show. Once you are able to interpret sectional views, you will have a good mental model for studying the anatomy and physiology of a particular region or system. Radiologists and other medical professionals responsible for interpreting medical scans spend much of their time analyzing sectional views of the body.

Any section through a three-dimensional object can be described in reference to a **sectional plane**, as indicated in Figure 1–6. A *plane* is a two-dimensional flat surface, and a section is a single view or slice along a plane. Common planes are frontal (coronal), sagittal, and transverse (horizontal).

- The **frontal (coronal) plane** is a vertical plane that divides the body or organ into anterior and posterior portions. A cut in this plane is called a **frontal section**, or *coronal section*.
- The **sagittal plane** is a vertical plane that divides the body into left and right portions. A cut in this plane is called a **sagittal section**. If the plane lies in the middle, it is called a **midsagittal plane**, and if it is offset from the middle, it is called a **parasagittal plane**.
- The **transverse plane** divides the body into *superior* and *inferior* portions. A cut in this plane is called a **transverse section**, or *cross section*. Unless otherwise noted, in this textbook all anatomical diagrams that present cross-sectional views of the body are oriented as though the subject were supine with you, the observer, standing at the subject's feet and looking toward the head.

The atlas that accompanies this text contains images of sections taken through the body in various planes.

Checkpoint

15. **What is the purpose of anatomical terms?**
16. **For a body in the anatomical position, describe an anterior view and a posterior view.**

See the blue Answers tab at the back of the book.

Figure 1–6 Sectional Planes.

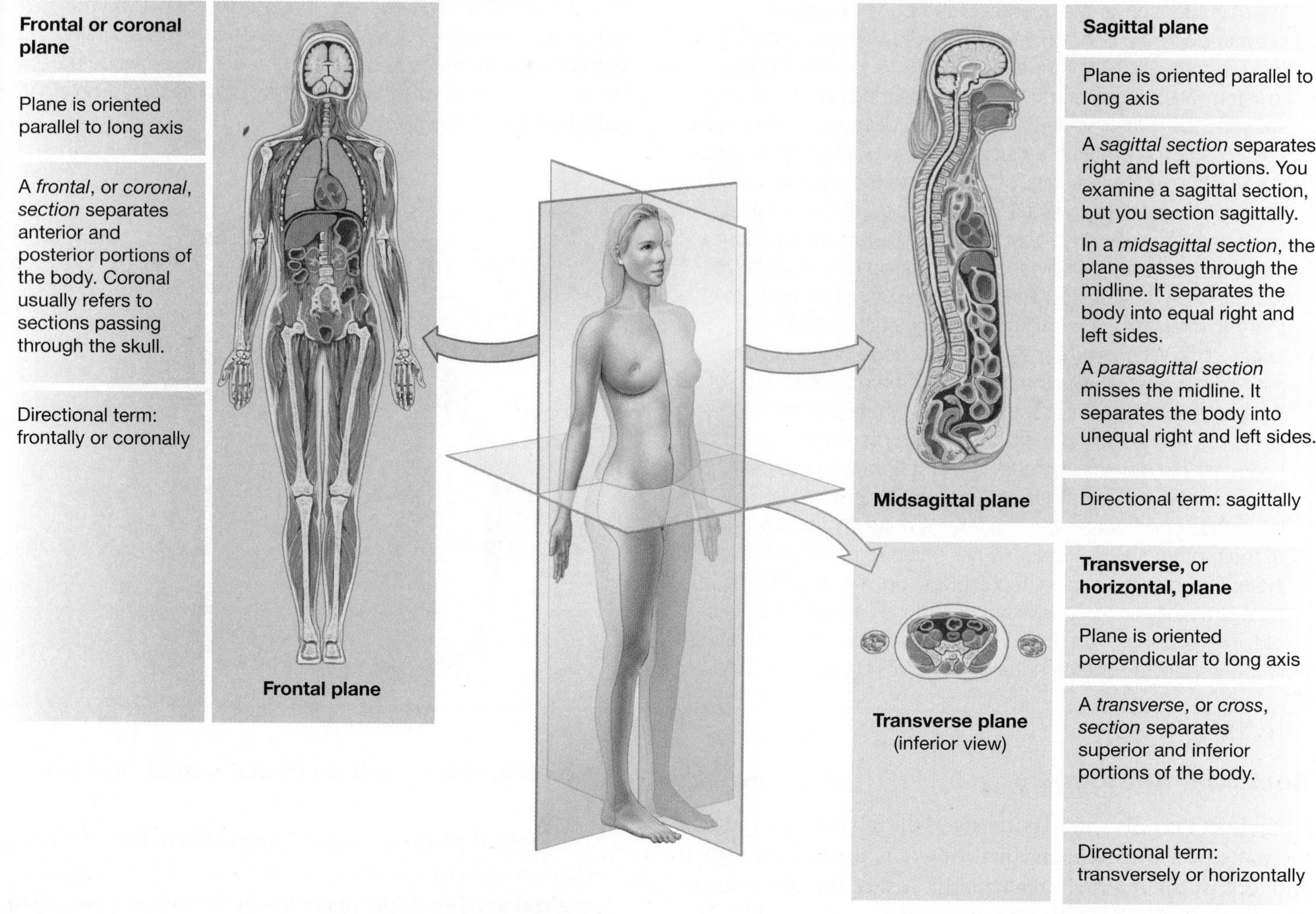
Frontal or coronal plane
Plane is oriented parallel to long axis
A *frontal*, or *coronal*, *section* separates anterior and posterior portions of the body. Coronal usually refers to sections passing through the skull.
Directional term: frontally or coronally
Frontal plane
Sagittal plane
Plane is oriented parallel to long axis
A *sagittal section* separates right and left portions. You examine a sagittal section, but you section sagittally.
In a *midsagittal section*, the plane passes through the midline. It separates the body into equal right and left sides.
A *parasagittal section* misses the midline. It separates the body into unequal right and left sides.
Midsagittal plane
Directional term: sagittally
Transverse, or horizontal, plane
Plane is oriented perpendicular to long axis
A *transverse*, or *cross*, *section* separates superior and inferior portions of the body.
Transverse plane (inferior view)
Directional term: transversely or horizontally

? Which plane separates the body into superior and inferior portions? Which plane separates the body into anterior and posterior portions?

1-6 Body cavities of the trunk protect internal organs and allow them to change shape

Learning Outcome **Identify the major body cavities of the trunk and their subdivisions, and describe the functions of each.**

The body's trunk is subdivided into three major regions established by the body wall: the thoracic, abdominal, and pelvic regions. Most of our vital organs are located within these regions of the trunk. The true **body cavities** are closed, fluid filled, and lined by a thin tissue layer called a *serous membrane,* or *serosa*. The vital organs of the trunk are suspended within these body cavities; they do not simply lie there. Early anatomists used the term *cavity* when referring to internal regions. For example, everything deep to the chest wall of the thoracic region is considered to be within the **thoracic cavity**, and all of the structures deep to the abdominal and pelvic walls are said to lie within the **abdominopelvic cavity**. Internally, the **diaphragm** (DĪ-uh-fram), a flat muscular sheet, separates these anatomical regions.

The boundaries of the true body cavities and the regional "cavities" are not identical. For example, the thoracic cavity contains two *pleural cavities* (each surrounding a lung), the *pericardial cavity* (surrounding the heart), and the *mediastinum* (a large tissue mass). The *peritoneal cavity* (surrounding abdominal organs) extends only partway into the pelvic cavity (surrounding pelvic organs). Figure 1–7 shows the boundaries between the subdivisions of the thoracic cavity and the abdominopelvic cavity.

The body cavities of the trunk have two essential functions: (1) They protect delicate organs from shocks and impacts, and (2) they permit significant changes in the size and shape of

Figure 1–7 Relationships among the Subdivisions of the Body Cavities of the Trunk.

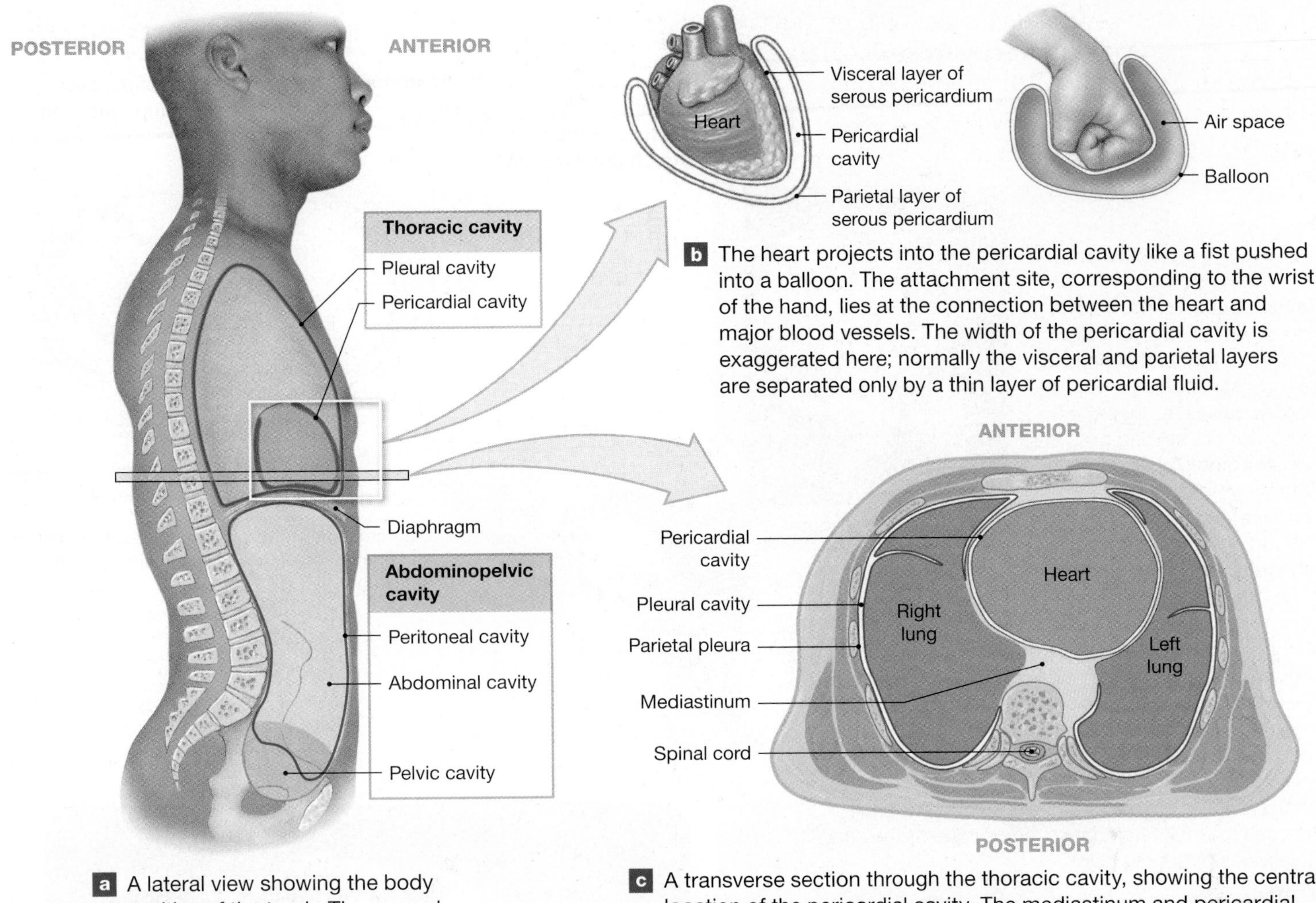
POSTERIOR
ANTERIOR
Thoracic cavity
Pleural cavity
Pericardial cavity
Diaphragm
Abdominopelvic cavity
Peritoneal cavity
Abdominal cavity
Pelvic cavity
Heart
Visceral layer of serous pericardium
Pericardial cavity
Parietal layer of serous pericardium
Air space
Balloon
ANTERIOR
Pericardial cavity
Pleural cavity
Parietal pleura
Mediastinum
Spinal cord
Heart
Right lung
Left lung
POSTERIOR

a A lateral view showing the body cavities of the trunk. The muscular diaphragm subdivides them into a superior thoracic cavity and an inferior abdominopelvic cavity. Three of the four adult true body cavities are shown and outlined in red; only one of the two pleural cavities can be shown in a sagittal section.

b The heart projects into the pericardial cavity like a fist pushed into a balloon. The attachment site, corresponding to the wrist of the hand, lies at the connection between the heart and major blood vessels. The width of the pericardial cavity is exaggerated here; normally the visceral and parietal layers are separated only by a thin layer of pericardial fluid.

c A transverse section through the thoracic cavity, showing the central location of the pericardial cavity. The mediastinum and pericardial cavity lie between the two pleural cavities. Note that this transverse or cross-sectional view is oriented as though you were standing at the feet of a supine person and looking toward that person's head. This inferior view of a transverse section is the standard presentation for clinical images. Unless otherwise noted, transverse or cross-sectional views in this text use this same orientation (*see Clinical Note: Diagnostic Imaging Techniques*).

internal organs. For example, the lungs, heart, stomach, intestines, urinary bladder, and many other organs can expand and contract without distorting surrounding tissues or disrupting the activities of nearby organs because they project into body cavities.

The internal organs that are enclosed by these cavities are known as **viscera** (VIS-e-ruh). A delicate serous membrane lines the walls of these internal cavities and covers the surfaces of the enclosed viscera. A watery fluid, called *serous fluid*, moistens serous membranes, coats opposing surfaces, and reduces friction. The portion of a serous membrane that directly covers a visceral organ is called the *visceral serosa*. The opposing layer that lines the inner surface of the body wall or chamber is called the *parietal serosa*. The parietal and visceral membranes are one membrane: The parietal serosa folds back onto itself, forming the visceral serosa. Because the moist parietal and visceral serosae are usually in close contact, the body cavities are called *potential spaces*. In some clinical conditions, however, excess fluid can accumulate within these potential spaces, increasing their volume and exerting pressure on the enclosed viscera.

The Thoracic Cavity

The thoracic cavity contains the lungs and heart; associated organs of the respiratory, cardiovascular, and lymphatic systems; the inferior portions of the esophagus; and the thymus (Figure 1–7a, c). The thoracic cavity is subdivided into the left and right **pleural cavities** (holding the lungs), separated by

Clinical Note Diagnostic Imaging Techniques

During the past several decades, rapid progress has been made in discovering more accurate and more detailed ways to image the human body, both in health and disease.

X-rays

X-rays are the oldest and still the most common method of imaging. X-rays are a form of high-energy radiation that can penetrate living tissues. An x-ray beam travels through the body before striking a photographic plate. Not all of the projected x-rays arrive at the film. The body absorbs or deflects some of those x-rays. The ability to stop the passage of x-rays is referred to as **radiopacity**. When taking an x-ray, these areas that are impenetrable by x-rays appear light or white on the exposed film and are said to be **radiopaque**. In the body, air has the lowest radiopacity. Fat, liver, blood, muscle, and bone are increasingly radiopaque. As a result, radiopaque tissues look white, and less radiopaque tissues are in shades of gray to black.

An x-ray of the skull, taken from the left side

To use x-rays to visualize soft tissues, a very radiopaque substance must be introduced. To study the upper digestive tract, a radiopaque barium solution is ingested by the patient. The resulting x-ray shows the contours of the stomach and intestines.

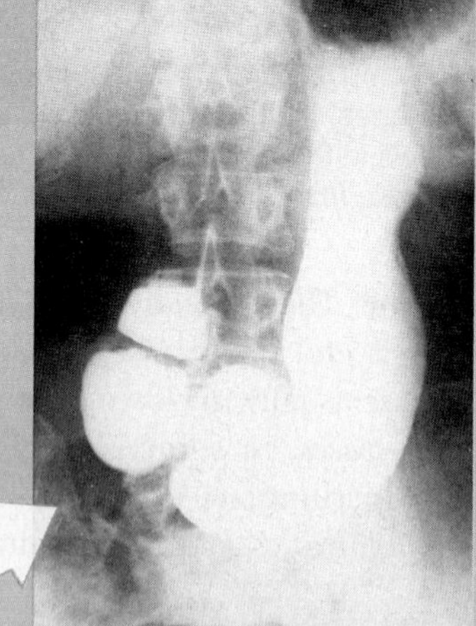

A **barium-contrast x-ray** of the upper digestive tract

Standard Scanning Techniques

More recently, a variety of **scanning techniques** dependent on computers have been developed to show the less radiopaque, soft tissues of the body in much greater detail.

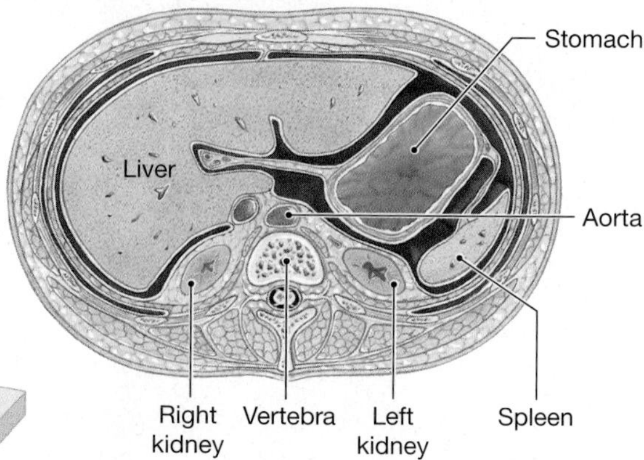
Stomach
Liver
Aorta
Right kidney
Vertebra
Left kidney
Spleen

Diagrammatic views showing the relative position and orientation of the CT scan below and the MRI to the right.

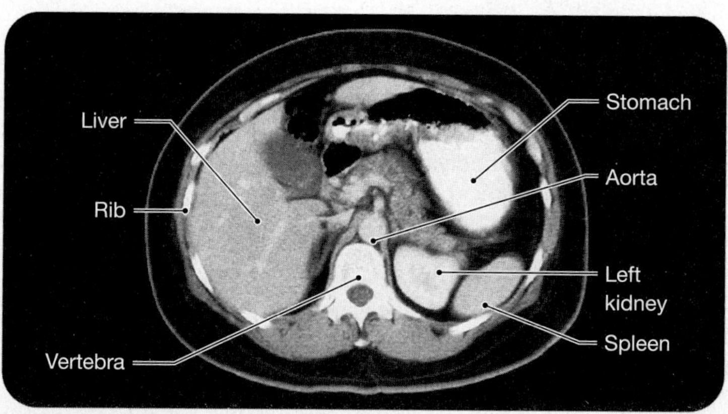
Liver
Rib
Vertebra
Stomach
Aorta
Left kidney
Spleen

CT scan of the abdomen

CT (computed tomography) **scans** use computers to reconstruct sectional views. A single x-ray source rotates around the body, and the x-ray beam strikes a sensor monitored by the computer. The x-ray source completes one revolution around the body every few seconds. It then moves a short distance and repeats the process. The result is usually displayed as a sectional view in black and white, but it can be colorized for visual effect. CT scans show three-dimensional relationships and soft tissue structures more clearly than do standard x-rays.

- Note that when anatomical diagrams or scans present cross-sectional views, the sections are presented from an inferior perspective, as though the observer were standing at the feet of a person in the supine position and looking toward the head of the subject.

Special Scanning Methods

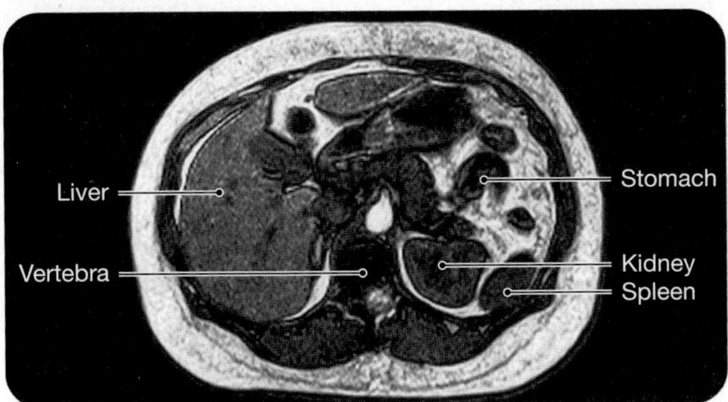
Liver
Vertebra
Stomach
Kidney
Spleen

MRI scan of the abdomen

An **MRI** of the same region (in this case, the abdomen) can show soft tissue structure in even greater detail than a CT scan. Magnetic resonance imaging surrounds part or all of the body with a magnetic field 3000 times as strong as that of Earth. This field causes particles within atoms throughout the body to line up in a uniform direction. Energy from pulses of radio waves are absorbed and released by the different atoms. The released energy is used to create a detailed image of the soft tissue structure.

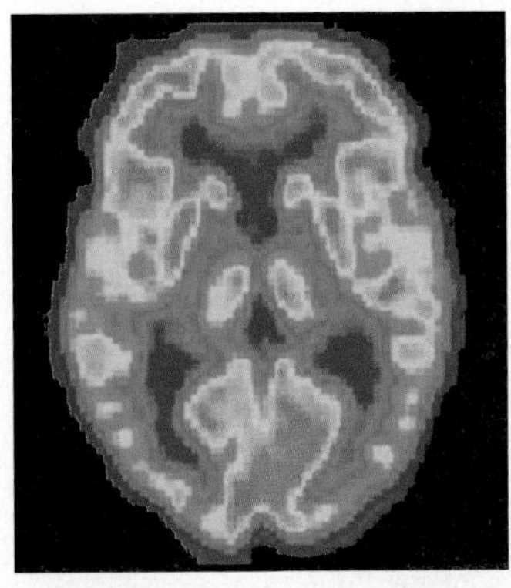

PET scan of the brain

Positron emission tomography (**PET**) is an imaging technique that assesses metabolic and physiological activity of a structure. A PET scan is an important tool in evaluating healthy and diseased brain function.

Ultrasound of the uterus

In **ultrasound** procedures, a small transmitter contacting the skin broadcasts a brief, narrow burst of high-frequency sound and then detects the echoes. The sound waves are reflected by internal structures, and a picture, or **echogram**, is assembled from the pattern of echoes. These images lack the clarity of other procedures, but no adverse effects have been reported, and fetal development can be monitored without a significant risk of birth defects. Special methods of transmission and processing permit analysis of the beating heart without the complications that can accompany dye injections.

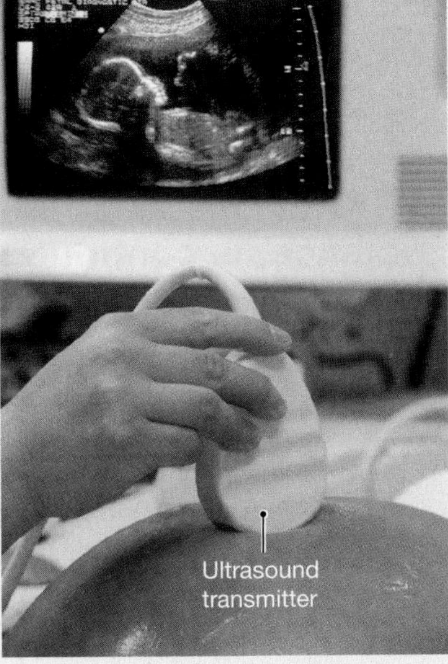
Ultrasound transmitter

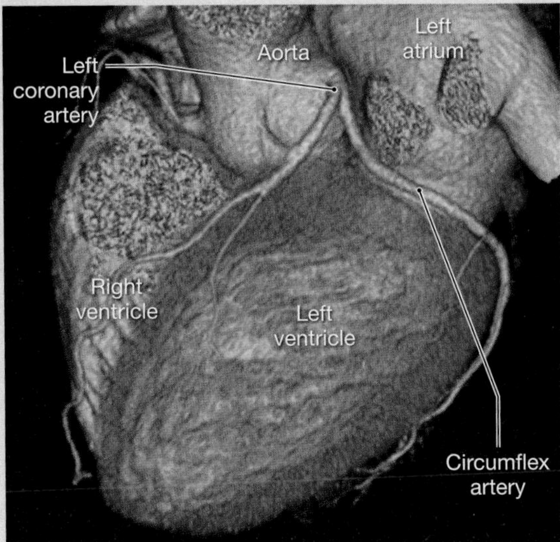
Left coronary artery
Aorta
Left atrium
Right ventricle
Left ventricle
Circumflex artery

Spiral scan of the heart

A **spiral CT scan** is a form of three-dimensional imaging technology that is becoming increasingly important in clinical settings. During a spiral CT scan, the patient is on a platform that advances at a steady pace through the scanner while the imaging source, usually x-rays, rotates continuously around the patient. Because the x-ray detector gathers data quickly and continuously, a higher quality image is generated, and the patient is exposed to less radiation as compared to a standard CT scanner, which collects data more slowly and only one slice of the body at a time.

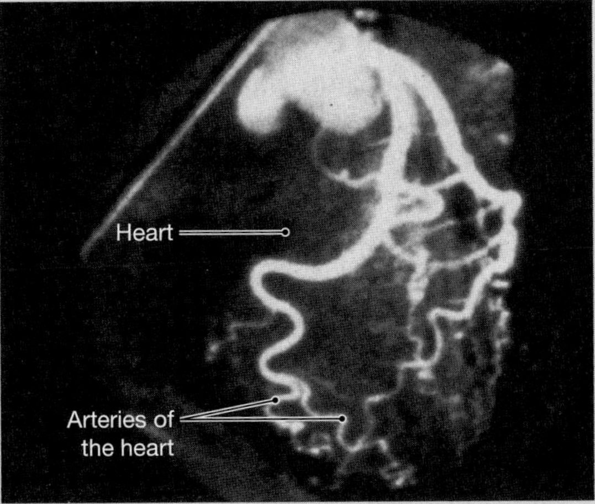
Heart
Arteries of the heart

Digital subtraction angiography of coronary arteries

Digital subtraction angiography (**DSA**) is used to monitor blood flow through specific organs, such as the brain, heart, lungs, and kidneys. X-rays are taken before and after radiopaque dye is administered, and a computer "subtracts" details common to both images. The result is a high-contrast image showing the distribution of the dye.

a mass of tissue called the **mediastinum** (mē-dē-a-STĪ-num). Each pleural cavity surrounds a lung and is lined by a slippery serous membrane that reduces friction as the lung expands and recoils during breathing. The serous membrane lining a pleural cavity is called a *pleura* (PLOOR-ah). The *visceral pleura* covers the outer surfaces of a lung, and the *parietal pleura* covers the mediastinal surface and the inner body wall.

The mediastinum consists of a mass of connective tissue that surrounds, stabilizes, and supports the esophagus, trachea, and thymus, as well as the major blood vessels that originate or end at the heart. The mediastinum also contains the **pericardial cavity**, a small chamber that surrounds the heart. The relationship between the heart and the pericardial cavity resembles that of a fist pushing into a balloon (Figure 1–7b). The wrist corresponds to the *base* (attached portion) of the heart, and the balloon corresponds to the serous membrane that lines the pericardial cavity. The serous membrane associated with the heart is called the *pericardium* (*peri-*, around + *cardium*, heart). The layer covering the heart is the *visceral layer of serous pericardium*, and the opposing surface is the *parietal layer of serous pericardium*. During each beat, the heart changes in size and shape. The pericardial cavity permits these changes, and the slippery pericardial serous membrane lining prevents friction between the heart and nearby structures in the thoracic cavity.

The Abdominopelvic Cavity

The abdominopelvic cavity extends from the diaphragm to the pelvis. It is subdivided into a superior *abdominal cavity* and an inferior *pelvic cavity* (Figure 1–7a). The abdominopelvic cavity contains the **peritoneal** (per-i-tō-NĒ-al) **cavity**, a potential space lined by a serous membrane known as the *peritoneum* (per-i-tō-NĒ-um). The *parietal peritoneum* lines the inner surface of the body wall. A narrow space containing a small amount of fluid separates the parietal peritoneum from the *visceral peritoneum*, which covers the enclosed organs. You are probably already aware of the movements of the organs in this cavity. Most of us have had at least one embarrassing moment when a digestive organ contracted, producing a movement of liquid or gas and a gurgling or rumbling sound. The peritoneum allows the organs of the digestive system to slide across one another without damage to themselves or the walls of the cavity.

The **abdominal cavity** extends from the inferior (toward the feet) surface of the diaphragm to the level of the superior (toward the head) margins of the pelvis. This cavity contains the liver, stomach, spleen, small intestine, and most of the large intestine. (Look back at Figure 1–4c that shows the positions of most of these organs.) The organs are partially or completely enclosed by the peritoneal cavity, much as the heart and lungs are enclosed by the pericardial and pleural cavities, respectively. A few organs, such as the kidneys and pancreas, lie between the peritoneal lining and the muscular wall of the abdominal cavity. Those organs are said to be *retroperitoneal* (*retro*, behind).

The **pelvic cavity** is inferior to the abdominal cavity. The bones of the pelvis form the walls of the pelvic cavity, and a layer of muscle forms its floor. The pelvic cavity contains the urinary bladder, various reproductive organs, and the distal (farthest) portion of the large intestine. In females, the pelvic cavity contains the ovaries, uterine tubes, and uterus. In males, it contains the prostate gland and seminal glands (seminal vesicles). The pelvic cavity also contains the inferior portion of the peritoneal cavity. The peritoneum covers the ovaries and the uterus in females, as well as the superior portion of the urinary bladder in both sexes. Organs such as the urinary bladder and the distal portions of the ureters and large intestine, which extend inferior to the peritoneal cavity, are said to be *infraperitoneal*.

The true body cavities of the trunk in the adult share a common embryological origin. The term "dorsal body cavity" is sometimes used to refer to the internal chamber of the skull (cranial cavity) and the space enclosed by the vertebrae (vertebral cavity). These chambers, which are defined by bony structures, are anatomically and embryonically distinct from true body cavities, and the term "dorsal body cavity" is not encountered in either clinical anatomy or comparative anatomy. For these reasons, we have avoided using that term in our discussion of body cavities.

A partial list of chambers, or spaces, within the body that are not true body cavities would include the cranial cavity, vertebral cavity, oral cavity, digestive cavity, orbits (eye sockets), tympanic cavity of each middle ear, nasal cavities, and paranasal sinuses (air-filled chambers within some cranial bones that are connected to the nasal cavities). These structures will be discussed in later chapters.

The Clinical Note: Diagnostic Imaging Techniques on pp. 16–17 highlights some clinical tests commonly used for viewing the interior of the body.

Checkpoint

17. **Name two essential functions of the body cavities of the trunk.**
18. **Describe the various body cavities of the trunk.**

See the blue Answers tab at the back of the book.

1-7 Homeostasis, the state of internal balance, is continuously regulated

Learning Outcome **Explain the concept of homeostasis.**

Homeostasis (hō-mē-o-STĀ-sis; from the Greek *homeo*, similar + *stasis*, state of standing) refers to the existence of a stable internal environment. Various physiological processes act to prevent harmful changes in the composition of body fluids and the environment inside our cells. Maintaining homeostasis is absolutely vital to an organism's survival. Failure to maintain homeostasis soon leads to illness or even death. The principle of homeostasis is the central theme of this text and the foundation of all modern physiology.

Mechanisms of Homeostatic Regulation

Homeostatic regulation is the adjustment of physiological systems to preserve homeostasis. Physiological systems have evolved to maintain homeostasis in an environment that is often inconsistent, unpredictable, and potentially dangerous. An understanding of homeostatic regulation is crucial to making accurate predictions about the body's responses to both normal and abnormal conditions.

Homeostatic regulation involves two general mechanisms: autoregulation and extrinsic regulation.

1. **Autoregulation** is a process that occurs when a cell, tissue, organ, or organ system adjusts in response to some environmental change. For example, when the oxygen level decreases in a tissue, the cells release chemicals that widen, or dilate, blood vessels. This dilation increases the blood flow and provides more oxygen to the region.
2. **Extrinsic regulation** is a process that results from the activities of the nervous system or endocrine system. These organ systems detect an environmental change and send an electrical signal (nervous system) or chemical messenger (endocrine system) to control or adjust the activities of another or many other systems simultaneously. For example, when you exercise, your nervous system issues commands that increase your heart rate so that blood will circulate faster. Your nervous system also causes blood flow to be reduced to less active organs, such as the digestive tract. The oxygen in circulating blood is then available to the active muscles, which need it most.

In general, the nervous system directs rapid, short-term, and very specific responses. For example, if you accidentally set your hand on a hot stove, the heat would produce a painful, localized disturbance of homeostasis. Your nervous system would respond by ordering specific muscles to contract and pull your hand away from the stove. These contractions last only as long as the neural activity continues, usually a matter of seconds.

In contrast, the endocrine system releases chemical messengers called *hormones* into the bloodstream. These molecular messengers can affect tissues and organs throughout the body. The responses may not be immediately apparent, but they may persist for days or weeks. Examples of homeostatic regulation dependent on endocrine function include the long-term regulation of blood volume and composition, and the adjustment of organ system function during starvation.

An Overview of the Process of Homeostatic Regulation

Regardless of the system involved, homeostatic regulation always works to keep the internal environment within certain limits, or a range. A homeostatic regulatory mechanism consists of three parts: (1) a **receptor**, a sensor that is sensitive to a particular stimulus or environmental change; (2) a **control center**, which receives and processes the information supplied by the receptor and sends out commands; and (3) an **effector**, a cell or organ that responds to the commands of the control center and whose activity either opposes or enhances the stimulus. You are probably already familiar with similar mechanical regulatory mechanisms, such as the one involving the thermostat in your house or apartment (Figure 1–8a).

The thermostat is the control center. It receives information about room temperature from an internal or remote thermometer (a receptor). The setting on the thermostat establishes the **set point**, or desired value, which in this case is the temperature you select. (In our example, the set point is 22°C, or about 72°F.) The function of the thermostat is to keep room temperature within acceptable limits, usually within a degree or so of the set point. In summer, the thermostat performs this function by controlling an air conditioner (an effector). When the temperature at the thermometer rises above the set point, the thermostat turns on the air conditioner, which then cools the room. Then, when the temperature at the thermometer returns to the set point, the thermostat turns off the air conditioner. The control is not precise, especially if the room is large, and the thermostat is located on just one wall. Over time, the temperature in the center of the room fluctuates in a range above and below the set point (Figure 1–8b).

We can summarize the essential feature of temperature control by a thermostat very simply: A variation outside the set point triggers an automatic response that corrects the situation. In this way, variation in temperature is kept within an acceptable range. Now let's explore how the body uses a similar method of regulation called negative feedback.

✓ Checkpoint

19. **Define *homeostasis*.**
20. **Which general mechanism of homeostatic regulation always involves the nervous or endocrine system?**
21. **Why is homeostatic regulation important to an organism?**

See the blue Answers tab at the back of the book.

1-8 Negative feedback opposes variations from normal, whereas positive feedback enhances them

Learning Outcome **Describe how negative feedback and positive feedback are involved in homeostatic regulation.**

To keep variation in key body systems within ranges that are compatible with our long-term survival, the body uses a method of homeostatic regulation called *negative feedback.* In this process, an effector activated by the control center opposes,

Figure 1–8 The Control of Room Temperature.

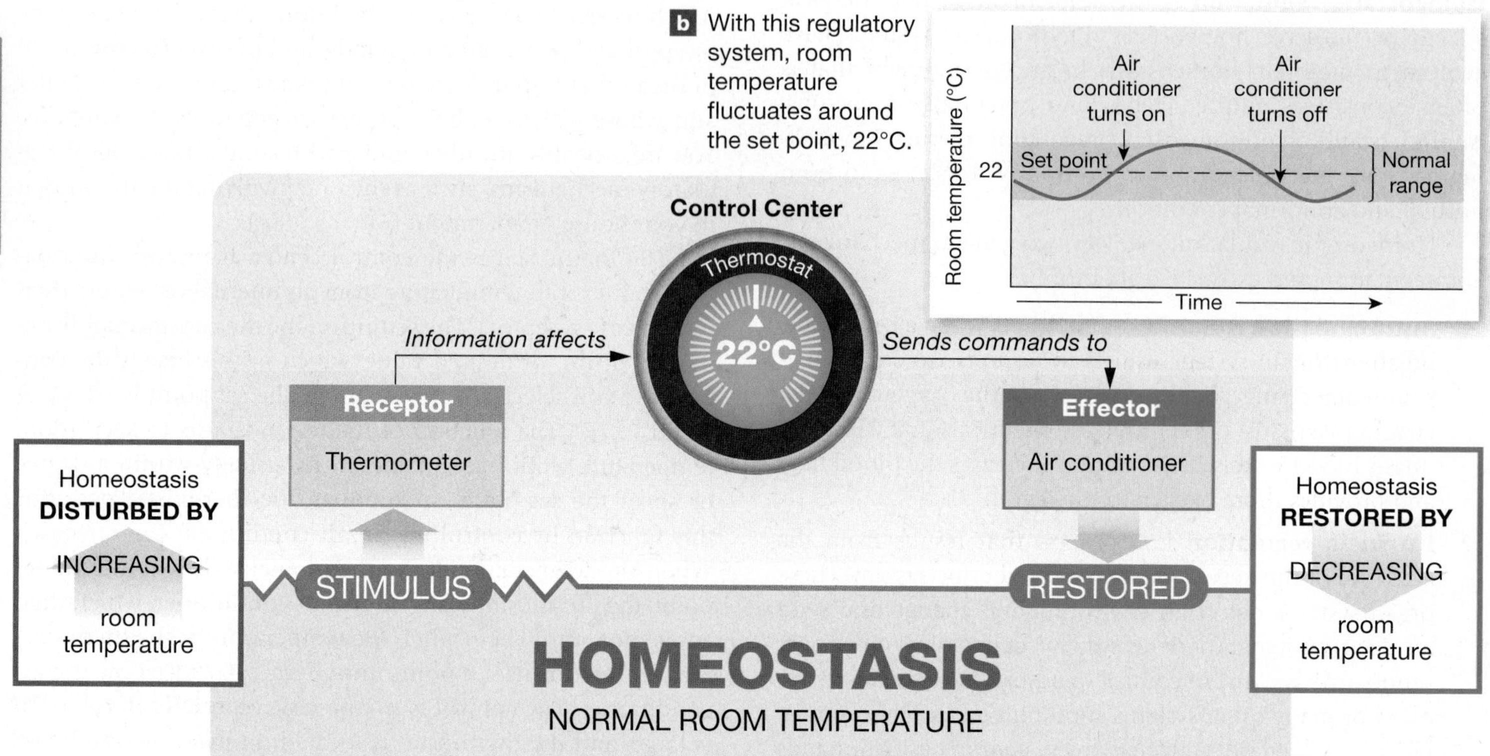
b With this regulatory system, room temperature fluctuates around the set point, 22°C.
Room temperature (°C)
22
Set point
Air conditioner turns on
Air conditioner turns off
Normal range
Time
Control Center
Thermostat
22°C
Information affects
Sends commands to
Receptor
Thermometer
Effector
Air conditioner
Homeostasis DISTURBED BY INCREASING room temperature
STIMULUS
RESTORED
Homeostasis RESTORED BY DECREASING room temperature
HOMEOSTASIS
NORMAL ROOM TEMPERATURE

a In response to input from a receptor (a thermometer), a control center (a thermostat) triggers an effector response (either an air conditioner or a heater) that restores normal temperature. In this case, when room temperature rises above the set point, the thermostat turns on the air conditioner, and the temperature returns to normal.

or *negates*, the original stimulus. In this way, negative feedback tends to minimize change. The body also has another method of homeostatic regulation called *positive feedback*, which instead tends to enhance or *increase* the change that triggered it. However, most homeostatic regulatory mechanisms involve negative feedback. Let's examine the roles of negative and positive feedback in homeostasis before considering the roles of organ systems in regulating homeostasis.

The Role of Negative Feedback in Homeostasis

An important example of **negative feedback**, a way of counteracting a change, is the control of body temperature, a process called *thermoregulation*. In thermoregulation, the relationship between heat loss, which takes place mainly at the body surface, and heat production, which takes place in all active tissues, is altered.

In the homeostatic control of body temperature (Figure 1–9a), the thermoregulatory control center is in the *hypothalamus*, a region of the brain. This control center receives information from two sets of temperature receptors, one in the skin and the other within the hypothalamus. At the normal set point, body temperature (as measured with an oral thermometer) is approximately 37°C (98.6°F).

If body temperature rises above 37.2°C, activity in the control center targets two effectors: (1) muscle tissue lining the walls of blood vessels supplying blood to the skin and (2) sweat glands. The muscle tissue relaxes so the blood vessels dilate (widen), increasing blood flow through vessels near the body surface, and the sweat glands speed up their secretion of sweat. The skin then acts like a radiator by losing heat to the environment, and the evaporation of sweat speeds the process.

As body temperature returns to normal, temperature in the hypothalamus decreases, and the thermoregulatory center becomes less active. Blood flow to the skin and sweat gland activity then decrease to previous levels. Body temperature drops below the set point as the secreted sweat evaporates.

Negative feedback is the primary mechanism of homeostatic regulation, and it provides long-term control over the body's internal conditions and systems. Homeostatic mechanisms using negative feedback normally ignore minor variations. They maintain a normal *range* rather than a fixed value. In our example, body temperature fluctuated around the set-point temperature (Figure 1–9b). The regulatory process itself

Figure 1–9 Negative Feedback: Control of Body Temperature. In negative feedback, a stimulus produces a response that opposes or negates the original stimulus.

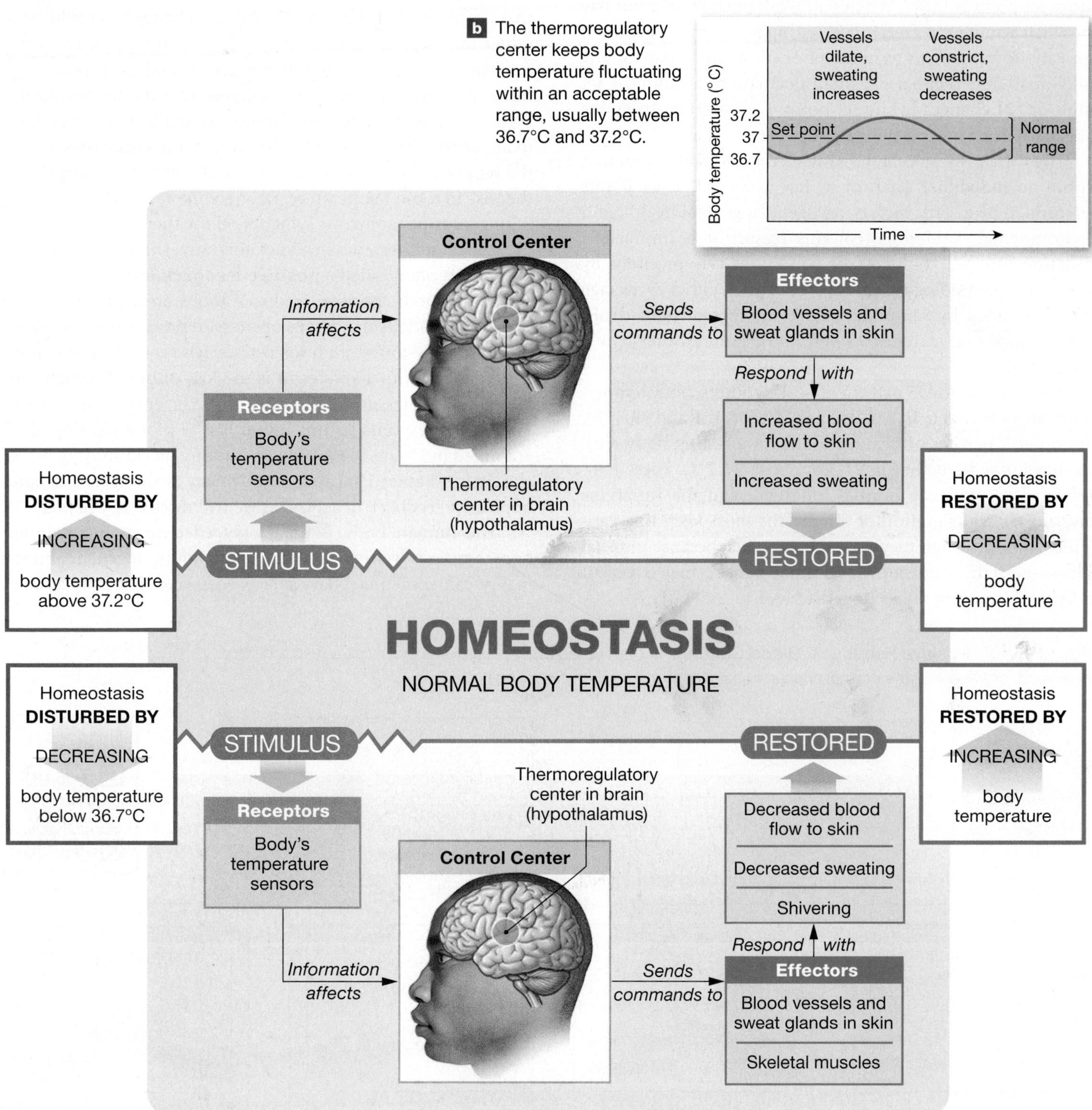
b The thermoregulatory center keeps body temperature fluctuating within an acceptable range, usually between 36.7°C and 37.2°C.
Vessels dilate, sweating increases
Vessels constrict, sweating decreases
Body temperature (°C)
37.2
37
36.7
Set point
Normal range
Time
Control Center
Information affects
Sends commands to
Effectors
Blood vessels and sweat glands in skin
Respond with
Increased blood flow to skin
Increased sweating
Receptors
Body's temperature sensors
Thermoregulatory center in brain (hypothalamus)
Homeostasis DISTURBED BY INCREASING body temperature above 37.2°C
STIMULUS
RESTORED
Homeostasis RESTORED BY DECREASING body temperature
HOMEOSTASIS
NORMAL BODY TEMPERATURE
Homeostasis DISTURBED BY DECREASING body temperature below 36.7°C
STIMULUS
RESTORED
Homeostasis RESTORED BY INCREASING body temperature
Thermoregulatory center in brain (hypothalamus)
Receptors
Body's temperature sensors
Control Center
Decreased blood flow to skin
Decreased sweating
Shivering
Respond with
Effectors
Information affects
Sends commands to
Blood vessels and sweat glands in skin
Skeletal muscles

a Events in the regulation of body temperature, which are comparable to those shown in *Figure 1–8*. A control center in the brain (the hypothalamus) functions as a thermostat with a set point of 37°C. If body temperature exceeds 37.2°C, heat loss is increased through increased blood flow to the skin and increased sweating.

? If a person's body temperature gets too high, the body will respond by decreasing its temperature to restore homeostasis. What are the homeostatic responses to an increase in body temperature?

is dynamic. That is, it is constantly changing because the set point may vary with changing environments or differing activity levels. For example, when you are asleep, your thermoregulatory set point is lower. When you work outside on a hot day (or when you have a fever), it is set higher. Body temperature can vary from moment to moment or from day to day for any individual, due to either (1) small fluctuations around the set point or (2) changes in the set point. Comparable variations take place in all other aspects of physiology.

The variability among individuals is even greater than that within an individual. Each of us has homeostatic set points determined by genetic factors, age, gender, general health, and environmental conditions. For this reason, it is impractical to define "normal" homeostatic conditions very precisely. By convention, physiological values are reported either as average values obtained by sampling a large number of individuals, or as a range that includes 95 percent or more of the sample population.

For example, for 95 percent of healthy adults, body temperature ranges between 36.7°C and 37.2°C (98.1°F and 98.9°F). The other 5 percent of healthy adults have resting body temperatures that are below 36.7°C or above 37.2°C. These temperatures are perfectly normal for them, and the variations have no clinical significance. Physicians must keep this variability in mind when they review lab reports, because unusual values—even those outside the "normal" range—may represent individual variation rather than disease.

The Role of Positive Feedback in Homeostasis

In **positive feedback**, an initial stimulus produces a response that amplifies or enhances the original change in conditions, rather than opposing it. You seldom encounter positive feedback in your daily life, simply because it tends to produce extreme responses. For example, suppose that the thermostat in Figure 1–8a was accidentally connected to a heater rather than to an air conditioner. Now when room temperature rises above the set point, the thermostat turns on the heater, causing a further rise in room temperature. Room temperature will continue to increase until someone switches off the thermostat, turns off the heater, or intervenes in some other way. This kind of escalating cycle is often called a **positive feedback loop**.

In the body, positive feedback loops are typically found when a potentially dangerous or stressful process must be completed quickly to restore homeostasis. For example, the immediate danger from a severe cut is the loss of blood, which can lower blood pressure and reduce the efficiency of the heart. The body's response to this blood loss is blood clotting, diagrammed in Figure 1–10. We will examine blood clotting more closely in Chapter 19. Labor and delivery are another example of positive feedback in action, as we will discuss in Chapter 29.

The human body is amazingly effective at maintaining homeostasis. Nevertheless, an infection, injury, or genetic abnormality can sometimes have effects so severe that homeostatic

Figure 1–10 Positive Feedback: Blood Clotting. In positive feedback, a stimulus produces a response that accelerates or enhances the original change in conditions, rather than opposing it.

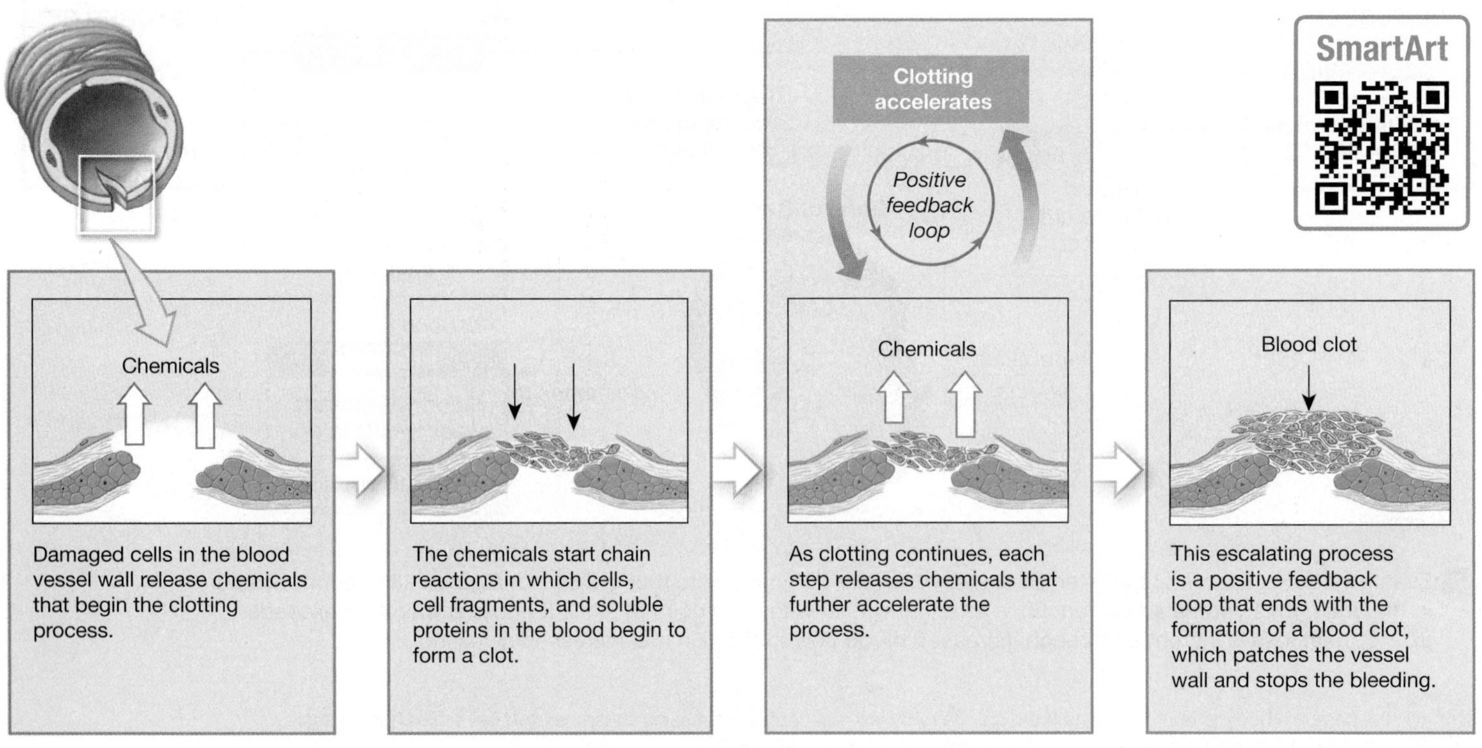

Chemicals
Damaged cells in the blood vessel wall release chemicals that begin the clotting process.
The chemicals start chain reactions in which cells, cell fragments, and soluble proteins in the blood begin to form a clot.
Clotting accelerates
Positive feedback loop
Chemicals
As clotting continues, each step releases chemicals that further accelerate the process.
Blood clot
This escalating process is a positive feedback loop that ends with the formation of a blood clot, which patches the vessel wall and stops the bleeding.
SmartArt

mechanisms cannot fully compensate for them. One or more variables within the internal environment may then be pushed outside their normal range of values. When this happens, organ systems begin to malfunction, producing a state known as illness, or **disease**. In Chapters 5–29, we devote much attention to the mechanisms that bring about a variety of human diseases.

Systems Integration, Equilibrium, and Homeostasis

Homeostatic regulation controls aspects of the internal environment that affect every cell in the body. No single organ system has total control over any of these aspects. Instead, such control requires the coordinated efforts of multiple organ systems. In later chapters we will explore the functions of each organ system and see how the systems interact to preserve homeostasis. **Table 1–1** lists the roles of various organ systems in regulating several important functions that are subject to homeostatic control. Note that in each case such regulation involves several organ systems.

A **state of equilibrium** exists when opposing processes or forces are in balance. In the case of body temperature, a state of equilibrium exists when the rate of heat loss equals the rate of heat production. Each physiological system functions to maintain a state of equilibrium that keeps vital conditions within a normal range of values. This is often called a state of **dynamic equilibrium** because physiological systems are continually adapting and adjusting to changing conditions. For example, when muscles become more active, more heat is produced. More heat must then be lost at the skin surface to reestablish a state of equilibrium before body temperature rises outside normal ranges. Yet the adjustments made to control body temperature have other consequences. The sweating that increases heat loss at the skin surface increases losses of both water and salts. Other systems must then compensate for these losses and reestablish an equilibrium state for water and salts.

Note this general pattern: Any adjustments made by one physiological system have direct and indirect effects on a variety of other systems. Maintaining homeostasis is like a juggling act that keeps lots of different objects in the air.

Each organ system interacts with and depends on other organ systems, but introductory students may often find it easier to learn the basics of anatomy and physiology one system at a time. Chapters 5–29 are organized around individual systems, but remember that these systems all work together. The 11 *Build Your Knowledge Figures* in later chapters will help reinforce this message. Each provides an overview of one system's functions and summarizes its functional relationships with systems covered in previous chapters.

Table 1–1 The Roles of Organ Systems in Homeostatic Regulation

Internal Stimulus	Primary Organ Systems Involved	Functions of the Organ Systems
Body temperature	Integumentary system	Heat loss
	Muscular system	Heat production
	Cardiovascular system	Heat distribution
	Nervous system	Coordination of blood flow, heat production, and heat loss
Body fluid composition		
Nutrient concentration	Digestive system	Nutrient absorption, storage, and release
	Cardiovascular system	Nutrient distribution
	Urinary system	Control of nutrient loss in the urine
	Skeletal system	Mineral storage and release
Oxygen, carbon dioxide levels	Respiratory system	Absorption of oxygen, elimination of carbon dioxide
	Cardiovascular system	Internal transport of oxygen and carbon dioxide
Levels of toxins and pathogens	Lymphatic system	Removal, destruction, or inactivation of toxins and pathogens
Body fluid volume	Urinary system	Elimination or conservation of water from the blood
	Digestive system	Absorption of water; loss of water in feces
	Integumentary system	Loss of water through perspiration
	Cardiovascular system and lymphatic system	Distribution of water throughout body tissues
Waste concentration	Urinary system	Excretion of wastes from the blood
	Digestive system	Elimination of wastes from the liver in feces
	Cardiovascular system	Transport of wastes products to sites of excretion
Blood pressure	Cardiovascular system	Pressure generated by the heart moves blood through blood vessels
	Nervous system and endocrine system	Adjustments in heart rate and blood vessel diameter can raise or lower blood pressure

1

Checkpoint

22. Explain the function of negative feedback systems.
23. What happens to the body when homeostasis breaks down?
24. Explain how a positive feedback system works.
25. Why is positive feedback helpful in blood clotting but unsuitable for the regulation of body temperature?
26. Define *equilibrium.*
27. When the body continuously adapts by using homeostatic mechanisms, it is said to be in a state of_____ equilibrium.

See the blue Answers tab at the back of the book.

1 Chapter Review

Study Outline

MasteringA&P™ Access more chapter study tools online in the MasteringA&P Study Area:

- Chapter Quizzes, Chapter Practice Test, MP3 Tutor Sessions, and Clinical Case Studies
- Practice Anatomy Lab PAL 3.0
- A&P Flix A&PFlix
- Interactive Physiology iP2
- PhysioEx PhysioEx 9.1

An Introduction to Studying the Human Body p. 2

1. Biology is the study of life. One of its goals is to discover the unity and the patterns that underlie the diversity of organisms.

1-1 To make the most of your learning, read the text and view the art together p. 2

2. This text is divided into **sections**, which are units about a topic that continues to build on previously learned topics.
3. To enhance learning, the text and art should be read and studied together. For this reason, there is **text–art integration** whereby the figures are placed in proximity to the narrative.
4. **Learning outcomes** are tied to testing and tell the reader what educational goals should be achieved after reading each section. Learning outcomes are based on a **learning classification scheme**, which identifies the fundamental levels of learning from lower order skills to those of higher order skills. *(Figure 1–1)*

1-2 Anatomy (structure) and physiology (function) are closely integrated p. 3

5. **Anatomy** is the study of internal and external structures of the body and the physical relationships among body parts. **Physiology** is the study of how living organisms perform their vital functions. All physiological functions are performed by specific structures.
6. All specific functions are performed by specific structures.
7. In **gross** *(macroscopic)* **anatomy**, we consider features that are visible without a microscope. This field includes *surface anatomy* (general form and superficial markings), *regional anatomy* (anatomical organization of specific areas of the body), *sectional anatomy* (relationship of the body's structures by examining cross sections of tissues or organs), and *systemic anatomy* (structure of organ systems). *Clinical anatomy* includes anatomical subspecialties important to the practice of medicine. In *developmental anatomy*, we examine the changes in form that occur between conception and physical maturity. In *embryology*, we study developmental processes that occur during the first two months of development.
8. The equipment used determines the limits of *microscopic anatomy*. In **cytology**, we analyze the internal structure of individual cells. In **histology**, we examine **tissues**, groups of cells that perform specific functions. Tissues combine to form **organs**, anatomical structures with multiple functions. *(Spotlight Figure 1–2)*
9. Human physiology is the study of the functions of the human body. It is based on *cell physiology*, the study of the functions of cells. In *organ physiology*, we study the physiology of specific organs. In *systemic physiology*, we consider all aspects of the functioning of specific organ systems. In *pathological physiology*, we study the effects of diseases on organ or system functions.

1-3 Levels of organization progress from chemicals to a complete organism p. 6

10. Anatomical structures and physiological mechanisms occur in a series of interacting levels of organization. *(Spotlight Figure 1–2)*
11. The 11 **organ systems** of the body are the integumentary, skeletal, muscular, nervous, endocrine, cardiovascular, lymphatic, respiratory, digestive, urinary, and reproductive systems. *(Spotlight Figure 1–2)*

1-4 Medical terminology is important to understanding anatomy and physiology p. 7

12. **Medical terminology** is the use of word roots, prefixes, suffixes, and combining forms to construct anatomical, physiological, or medical terms.
13. *Terminologia Anatomica* was used as the standard in preparing your text.

1-5 Anatomical terms describe body regions, anatomical positions and directions, and body sections p. 7

14. The standard arrangement for anatomical reference is called the **anatomical position**. A person who is lying down is either **supine** (face up) or **prone** (face down). *(Figure 1–3)*
15. **Abdominopelvic quadrants** and **abdominopelvic regions** represent two approaches to describing anatomical regions of that portion of the body. *(Figure 1–4)*
16. The use of special directional terms provides clarity for the description of anatomical structures. *(Figure 1–5)*
17. The three **sectional planes** (**transverse**, or *horizontal*, **plane**; **frontal**, or *coronal*, **plane**; and **sagittal plane**) describe relationships among the parts of the three-dimensional human body. *(Figure 1–6)*

1-6 Body cavities of the trunk protect internal organs and allow them to change shape p. 14

18. **Body cavities** protect delicate organs and permit significant changes in the size and shape of internal organs. The body cavities of the trunk surround organs of the respiratory, cardiovascular, digestive, urinary, and reproductive systems. *(Figure 1–7)*
19. The **diaphragm** divides the (superior) **thoracic** and (inferior) **abdominopelvic cavities**. The thoracic cavity consists of two **pleural cavities** (each surrounding a lung) with a central tissue mass known as the **mediastinum**. Within the mediastinum is the **pericardial cavity**, which surrounds the heart. The abdominopelvic cavity consists of the **abdominal cavity** and the **pelvic cavity** and contains the *peritoneal cavity*, a chamber lined by the *peritoneum*, a *serous membrane*. *(Figure 1–7)*
20. Diagnostic imaging techniques are used in clinical medicine to view the body's interior.

1-7 Homeostasis, the state of internal balance, is continuously regulated p. 18

21. **Homeostasis** is the existence of a stable environment within the body.
22. Physiological systems preserve homeostasis through **homeostatic regulation**.
23. **Autoregulation** occurs when a cell, tissue, organ, or organ system adjusts its activities automatically in response to some environmental change. **Extrinsic regulation** results from the activities of the nervous system or endocrine system.
24. Homeostatic regulation mechanisms usually involve a **receptor** that is sensitive to a particular stimulus; a **control center**, which receives and processes the information supplied by the receptor and then sends out commands; and an **effector** that responds to the commands of the control center and whose activity either opposes or enhances the stimulus. *(Figure 1–8)*

1-8 Negative feedback opposes variations from normal, whereas positive feedback enhances them p. 19

25. **Negative feedback** is a corrective mechanism involving an action that directly opposes a variation from normal limits. *(Figure 1–9)*
26. In **positive feedback**, an initial stimulus produces a response that exaggerates or enhances the change in the original conditions, creating a **positive feedback** loop. *(Figure 1–10)*
27. No single organ system has total control over the body's internal environment; all organ systems work together. *(Table 1–1)*

Review Questions

See the blue Answers tab at the back of the book.

LEVEL 1 Reviewing Facts and Terms

1. Label the directional terms in the figures below.

(a) ________	**(b)** ________
(c) ________	**(d)** ________
(e) ________	**(f)** ________
(g) ________	**(h)** ________
(i) ________	**(j)** ________

Match each numbered item with the most closely related lettered item. Use letters for answers in the spaces provided.

____ **2.** cytology	**(a)** study of tissues
____ **3.** physiology	**(b)** constant internal environment
____ **4.** histology	**(c)** face-up position
____ **5.** anatomy	**(d)** study of functions
____ **6.** homeostasis	**(e)** positive feedback
____ **7.** muscle	**(f)** organ system
____ **8.** heart	**(g)** study of cells
____ **9.** endocrine	**(h)** negative feedback
____ **10.** temperature regulation	**(i)** serous membrane
____ **11.** labor and delivery	**(j)** study of internal and external body structures
____ **12.** supine	**(k)** diaphragm
____ **13.** prone	**(l)** tissue
____ **14.** divides thoracic and abdominopelvic body cavities	**(m)** peritoneal cavity
____ **15.** abdominopelvic cavity	**(n)** organ
____ **16.** pericardium	**(o)** face-down position

17. The following is a list of six levels of organization that make up the human body:

(1) tissue	**(2)** cell
(3) organ	**(4)** chemical
(5) organism	**(6)** organ system

The correct order, from the simplest to the most complex level, is **(a)** 2, 4, 1, 3, 6, 5, **(b)** 4, 2, 1, 3, 6, 5, **(c)** 4, 2, 1, 6, 3, 5, **(d)** 4, 2, 3, 1, 6, 5, **(e)** 2, 1, 4, 3, 5, 6.

18. The study of the structure of tissues is called **(a)** gross anatomy, **(b)** cytology, **(c)** histology, **(d)** organology.

19. The increasingly forceful labor contractions during childbirth are an example of **(a)** receptor activation, **(b)** effector shutdown, **(c)** negative feedback, **(d)** positive feedback.
20. Failure of homeostatic regulation in the body results in **(a)** autoregulation, **(b)** extrinsic regulation, **(c)** disease, **(d)** positive feedback.
21. A plane through the body that passes perpendicular to the long axis of the body and divides the body into a superior and an inferior section is a **(a)** sagittal section, **(b)** transverse section, **(c)** coronal section, **(d)** frontal section.
22. Which body cavity would enclose each of the following organs? **(a)** heart, **(b)** small intestine, large intestine, **(c)** lung, **(d)** kidneys.
23. The mediastinum is the region between the **(a)** lungs and heart, **(b)** two pleural cavities, **(c)** chest and abdomen, **(d)** heart and pericardium.
24. A learning outcome is best described as **(a)** a goal of learning after reading a section based on a learning classification scheme, **(b)** an abstract concept linking anatomy to physiology, **(c)** a type of homeostatic mechanism, **(d)** the same thing as text–art integration.

LEVEL 2 Reviewing Concepts

25. **(a)** Define *anatomy*.
    (b) Define *physiology*.
26. The two major body cavities of the trunk are the **(a)** pleural cavity and pericardial cavity, **(b)** pericardial cavity and peritoneal cavity, **(c)** pleural cavity and peritoneal cavity, **(d)** thoracic cavity and abdominopelvic cavity.
27. What distinguishes autoregulation from extrinsic regulation?
28. Describe the anatomical position.
29. Which sectional plane could divide the body so that the face remains intact? **(a)** sagittal plane, **(b)** frontal (coronal) plane, **(c)** equatorial plane, **(d)** midsagittal plane, **(e)** parasagittal plane.
30. Which the following is *not* an example of negative feedback? **(a)** Increased pressure in the aorta triggers mechanisms to lower blood pressure. **(b)** A rise in blood calcium levels triggers the release of a hormone that lowers blood calcium levels. **(c)** A rise in estrogen during the menstrual cycle increases the number of progesterone receptors in the uterus. **(d)** Increased blood sugar stimulates the release of a hormone from the pancreas that stimulates the liver to store blood sugar.

LEVEL 3 Critical Thinking and Clinical Applications

31. The hormone *insulin* is released from the pancreas in response to an increased level of glucose (sugar) in the blood. If this hormone is controlled by negative feedback, what effect would insulin have on the blood glucose level?
32. A stroke occurs when blood flow to the brain is disrupted, causing brain cells to die. Why might a stroke result in a rise or fall of normal body temperature?

+ CLINICAL CASE Wrap-Up Using A&P to Save a Life

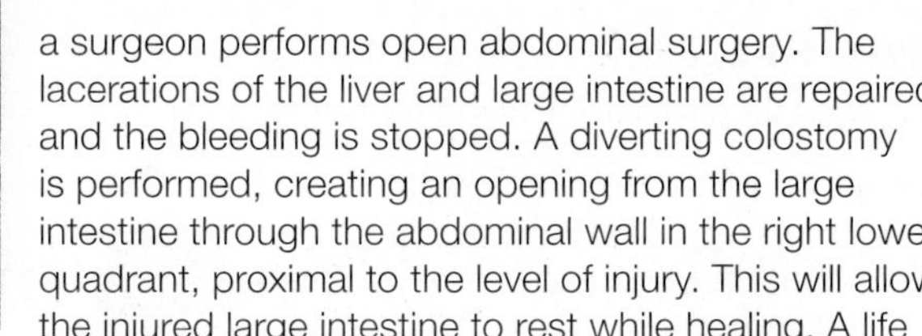

The patient is wheeled through the door and into Trauma Room 1. He is barely alive. Because of the location of the abdominal wound, in the right upper quadrant, and the likely depth of the wound, more than 6 inches, it is probable that the liver has been lacerated. Depending on the angle of the knife, sections of large intestine, containing bacteria that are deadly when released into the abdominal cavity, have probably also been lacerated. Any blood spilled onto the street or into the patient's abdominal cavity is lost from the cardiovascular system and cannot carry oxygen to body tissues. For this reason, the trauma nurse is ready with blood for transfusion.

An x-ray shows free air in the abdominal cavity, a sure sign the large intestine is cut. It also shows a fluid level indicating free blood in the abdominal cavity. Both conditions need immediate surgical attention. The patient is taken directly to the operating room where a surgeon performs open abdominal surgery. The lacerations of the liver and large intestine are repaired and the bleeding is stopped. A diverting colostomy is performed, creating an opening from the large intestine through the abdominal wall in the right lower quadrant, proximal to the level of injury. This will allow the injured large intestine to rest while healing. A life is saved because an EMT, trauma nurse, and surgeon know their anatomy and physiology.

1. Besides the liver and most of the large intestine, what other major organs are in the abdominal cavity?
2. If the deep knife wound had been superior to the diaphragm, what body cavity would the knife have entered, and what organs does it contain?

See the blue Answers tab at the back of the book.

Related Clinical Terms

acute: A disease of short duration but typically severe.

chemotherapy: The treatment of disease or mental disorder by the use of chemical substances, especially the treatment of cancer by cytotoxic and other drugs.

chronic: Illness persisting for a long time or constantly recurring. Often contrasted with *acute*.

epidemiology: The branch of science that deals with the incidence, distribution, and possible control of diseases and other factors relating to health.

etiology: The science and study of the cause of diseases.

idiopathic: Denoting any disease or condition of unknown cause.

morbidity: The state of being diseased or unhealthy, or the incidence of disease in a population.

pathophysiology: The functional changes that accompany a particular syndrome or disease.

syndrome: A condition characterized by a group of associated symptoms.

2 The Chemical Level of Organization

Learning Outcomes

These Learning Outcomes correspond by number to this chapter's sections and indicate what you should be able to do after completing the chapter.

✚ CLINICAL CASE What Is Wrong with My Baby?

Sean is Maureen's first baby. Maureen and her husband, Conner, had enjoyed an uncomplicated pregnancy and delivery. Maureen had felt healthy throughout the pregnancy, but something was wrong with her baby.

Sean is 1 month old, and not thriving. He seems to have a good appetite and breast-feeds as if he were starving; yet he has dropped 20 percent of his normal birth weight of 7 pounds, 8 ounces. He is down to only 6 pounds at his 1-month checkup, and his skin looks "wrinkly." His stools appear greasy and foamy. Maureen also notices that Sean's skin tastes salty.

Most alarmingly, he seems to be having some difficulty breathing. His breathing is wheezy. Maureen and Conner are both from big families, and none of the babies has ever been sickly like this. **What is wrong with baby Sean?** **To find out, turn to the Clinical Case Wrap-Up on p. 64.**

An Introduction to the Chemical Level of Organization

In this chapter we consider the structure of *atoms*, the basic chemical building blocks. You will also learn how atoms can combine to form increasingly complex structures, and how those types of complex structures function in the human body.

2-1 Atoms are the basic particles of matter

Learning Outcome Describe an atom and how atomic structure affects interactions between atoms.

Our study of the human body begins at the chemical level of organization. **Chemistry** is the science that deals with the structure of **matter**, defined as anything that takes up space and has mass. **Mass**, the amount of material in matter, is a physical property that determines the weight of an object in Earth's gravitational field. For our purposes, the mass of an object is the same as its weight. However, the two are not always equivalent: In orbit you would be weightless, but your mass would remain unchanged.

The smallest stable units of matter are called **atoms**. Air, elephants, oranges, oceans, rocks, and people are all composed of atoms in varying combinations. The unique characteristics of each object, living or nonliving, result from the types of atoms involved and the ways those atoms combine and interact.

Atoms are composed of **subatomic particles**. Many different subatomic particles exist, but only three—*protons, neutrons*, and *electrons*—are important for understanding the chemical properties of matter. Protons and neutrons are similar in size and mass, but **protons** (p^+) have a positive electrical charge. **Neutrons** (n or n^0) are electrically *neutral*, or uncharged. **Electrons** (e^-) are much lighter than protons—only 1/1836 as massive—and have a negative electrical charge. For this reason, the mass of an atom is determined primarily by the number of protons and neutrons in the **nucleus**, the central region of an atom. The mass of a large object, such as your body, is the sum of the masses of all its component atoms.

Atomic Structure

Atoms normally contain equal numbers of protons and electrons. The number of protons in an atom is known as its **atomic number**. *Hydrogen* (represented as H) is the simplest atom, with an atomic number of 1. An atom of hydrogen contains one proton and one electron. Hydrogen's proton is located in the center of the atom and forms the nucleus. Hydrogen atoms seldom contain neutrons, but when neutrons are present, they are also located in the nucleus. All atoms other than hydrogen have both neutrons and protons in their nuclei.

Electrons travel around the nucleus at high speed, within a spherical area called the **electron cloud** (Figure 2–1). We often illustrate atomic structure in the simplified form shown for hydrogen in Figure 2–2a (see p. 30). In this two-dimensional representation, the electrons occupy a circular **electron shell**. One reason an electron tends to remain in its electron shell is that the negatively charged electron is attracted to the positively charged proton. The attraction between opposite electrical charges is an example of an *electrical force*. As you will see in later chapters, electrical forces are involved in many physiological processes.

The dimensions of the electron cloud determine the overall size of the atom. To get an idea of the scale involved, consider that

> Go to MasteringA&P™ > Study Area > Get Ready for A&P > 2. Get Ready for A&P Website> Chapter 5: Chemistry > Atomic Structure: Video Tutor: Atomic Structure

Figure 2–1 **Hydrogen Atom with Electron Cloud.** This space-filling model of a hydrogen atom depicts the three-dimensional electron cloud formed by the single electron orbiting the nucleus.

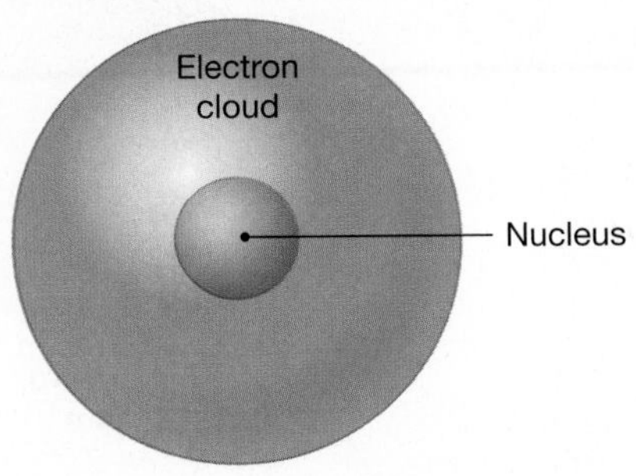
Electron cloud
Nucleus

if the nucleus were the size of a tennis ball, the electron cloud of a hydrogen atom would have a radius of 10 km (about 6 miles!). In reality, atoms are so small that atomic measurements are reported in nanometers (NAN-ō-mē-terz) (nm). One nanometer is 10^{-9} meter (0.000000001 m), or one billionth of a meter. The very largest atoms approach 0.5 nm in diameter.

Elements and Isotopes

An **element** is a pure substance composed of atoms of only one kind. Atoms are the smallest particles of an element that still retain the characteristics of that element. As a result, each element has uniform composition and properties. Each element includes all the atoms with the same number of protons, and thus the same atomic number. Only 92 elements exist in nature. Researchers have created about two dozen additional elements through nuclear reactions in laboratories.

Every element has a chemical symbol, an abbreviation recognized by scientists everywhere. Most of the symbols are easy to connect with the English names of the elements (O for oxygen, N for nitrogen, C for carbon, and so on), but a few are abbreviations of their Latin names. For example, the symbol for sodium, Na, comes from the Latin word *natrium*.

2

Elements cannot be changed or broken down into simpler substances, whether by chemical processes, heating, or other ordinary physical means. For example, an atom of carbon always remains an atom of carbon, regardless of the chemical events in which it may take part.

Our bodies consist of many elements, and the 13 most abundant elements are listed in **Table 2–1**. Our bodies also contain atoms of another 14 elements—called *trace elements*—that are present in very small amounts.

The atoms of a single element all have the same number of protons, but they can differ in the number of neutrons in the nucleus. For example, most hydrogen nuclei consist of just a single proton, but 0.015 percent also contain 1 neutron, and a very small percentage contain 2 neutrons (**Figure 2–2**). Atoms of the same element whose nuclei contain different numbers of neutrons are called **isotopes**.

Different isotopes of an element have essentially identical chemical properties, and are alike except in mass. The **mass number**—the total number of protons plus neutrons in the nucleus of an atom—is used to designate isotopes. For example, hydrogen has 3 isotopes, distinguished by their mass numbers (1, 2, or 3). Hydrogen-1, or ^{1}H, has 1 proton and 1 electron (**Figure 2–2a**). Hydrogen-2, or ^{2}H, also known as *deuterium*, has 1 proton, 1 electron, and 1 neutron (**Figure 2–2b**). Hydrogen-3,

Table 2–1 Principal Elements in the Human Body

Element (% of total body weight)	Significance
Oxygen, O (65)	A component of water and other compounds; gaseous form is essential for respiration
Carbon, C (18.6)	Found in all organic molecules
Hydrogen, H (9.7)	A component of water and most other compounds in the body
Nitrogen, N (3.2)	Found in proteins, nucleic acids, and other organic compounds
Calcium, Ca (1.8)	Found in bones and teeth; important for membrane function, nerve impulses, muscle contraction, and blood clotting
Phosphorus, P (1.0)	Found in bones and teeth, nucleic acids, and high-energy compounds
Potassium, K (0.4)	Important for proper membrane function, nerve impulses, and muscle contraction
Sodium, Na (0.2)	Important for blood volume, membrane function, nerve impulses, and muscle contraction
Chlorine, Cl (0.2)	Important for blood volume, membrane function, and water absorption
Magnesium, Mg (0.06)	A cofactor for many enzymes
Sulfur, S (0.04)	Found in many proteins
Iron, Fe (0.007)	Essential for oxygen transport and energy capture
Iodine, I (0.0002)	A component of hormones of the thyroid gland
Trace elements: silicon (Si), fluorine (F), copper (Cu), manganese (Mn), zinc (Zn), selenium (Se), cobalt (Co), molybdenum (Mo), cadmium (Cd), chromium (Cr), tin (Sn), aluminum (Al), boron (B), and vanadium (V)	Some function as cofactors; the functions of many trace elements are poorly understood

Figure 2–2 The Structure of Hydrogen Atoms. Three forms of hydrogen atoms are shown using the two-dimensional electron-shell model, which indicates the spherical electron cloud surrounding the nucleus.

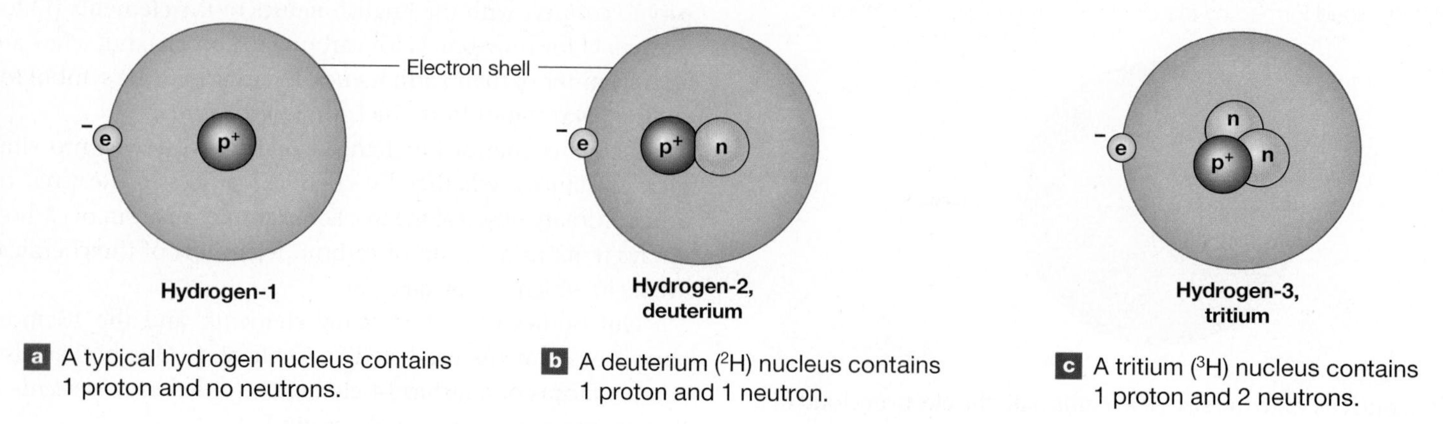
Electron shell
e
p+
Hydrogen-1
a A typical hydrogen nucleus contains 1 proton and no neutrons.
e
p+
n
Hydrogen-2, deuterium
b A deuterium (^2H) nucleus contains 1 proton and 1 neutron.
e
n
p+
n
Hydrogen-3, tritium
c A tritium (^3H) nucleus contains 1 proton and 2 neutrons.

or ^3H, also known as *tritium*, has 1 proton, 1 electron, and 2 neutrons (Figure 2–2c).

The nuclei of some isotopes are unstable, or radioactive. That is, they spontaneously break down and give off *radiation* (energy in the form of moving subatomic particles or waves) in measurable amounts. Such isotopes are called **radioisotopes**. The breakdown process is called *radioactive decay*. The decay rate of a radioisotope is commonly expressed as its **half-life**: the time required for half of a given amount of the isotope to decay. Radioisotopes differ radically in how rapidly they decay; their half-lives range from fractions of a second to billions of years.

Weakly radioactive isotopes are sometimes used in diagnostic procedures to monitor the structural or functional characteristics of internal organs. However, strongly radioactive isotopes are dangerous, because the radiation they give off can alter the number of electrons in an atom, break apart molecules, and destroy cells and tissues.

Atomic Weights

A typical *oxygen* atom has an atomic number of 8 and contains 8 protons and 8 neutrons. The mass number of this isotope is therefore 16. The mass numbers of other isotopes of oxygen depend on the number of neutrons present. Mass numbers are useful because they tell us the number of subatomic particles in the nuclei of different atoms. However, they do not tell us the *actual* mass of the atoms. For example, they do not take into account the masses of the electrons or the slight difference between the mass of a proton and that of a neutron. The actual mass of an atom of a specific isotope is known as its *atomic mass*.

The unit used to express atomic mass is the *atomic mass unit* (amu), or *dalton*. By international agreement, 1 amu is equal to one-twelfth the mass of a carbon-12 atom. One atomic mass unit is very close to the mass of a single proton or neutron. Thus, the atomic mass of an atom of the most common isotope of hydrogen is very close to 1, and that of the most common isotope of oxygen is very close to 16.

The **atomic weight** of an element is an average of the different atomic masses and proportions of its different isotopes. This results in the atomic weight of an element being very close to the mass number of the most common isotope of that element. For example, the mass number of the most common isotope of hydrogen is 1, but the atomic weight of hydrogen is closer to 1.01, primarily because some hydrogen atoms (0.02 percent) have a mass number of 2, and even fewer have a mass number of 3. (The periodic table of the elements in the Appendix at the back of this book shows the atomic weight of each element.)

Atoms take part in chemical reactions in fixed numerical ratios. To form water, for example, exactly 2 atoms of hydrogen combine with 1 atom of oxygen. But individual atoms are far too small and too numerous to be counted, so to determine the number of atoms chemists use a unit called the *mole*. For any element, a **mole** (abbreviated *mol*) is a specific quantity with a weight in grams equal to that element's atomic weight.

The mole is useful because 1 mol of a given element always contains the same number of atoms as 1 mol of any other element (just as we use *dozen* to stand for 12 items). That number (called *Avogadro's number*) is 6.023×10^{23}, or about 600 billion trillion. Expressing relationships in moles rather than in grams makes it much easier to keep track of the relative numbers of atoms in chemical samples and processes. For example, if a report stated that a sample contains 0.5 mol of hydrogen atoms and 0.5 mol of oxygen atoms, you would know immediately that the 2 elements were present in equal numbers. That would not be so evident if the report stated that there were 0.505 g of hydrogen atoms and 8.00 g of oxygen atoms. Most chemical analyses and clinical laboratory tests report data in moles (mol), millimoles (mmol—1/1000 mol, or 10^{-3} mol), or micromoles (μmol—1/1,000,000 mol, or 10^{-6} mol).

Electrons and Energy Levels

Atoms are electrically neutral. In other words, every positively charged proton is balanced by a negatively charged electron. Thus, each increase in the atomic number has a comparable increase in the number of electrons traveling around the nucleus.

Within the electron cloud, electrons occupy an orderly series of *energy levels*. The electrons in an energy level may travel in complex patterns around the nucleus, but for our purposes the patterns can be diagrammed as a series of concentric electron shells. The first electron shell is the one closest to the nucleus, and it corresponds to the lowest energy level of electrons. Note that the terms *electron shell* and *energy level* can generally be used interchangeably.

There are up to eight energy levels in atoms, depending on their atomic number, but let's just look at the first three here. Each energy level is limited in the number of electrons it can hold. The first energy level can hold at most 2 electrons. The next two levels can each hold up to 8 electrons, an observation called the *octet rule*. Note that the maximum number of electrons that may occupy shells 1 through 3 corresponds to the number of elements in rows 1 through 3 of the periodic table of the elements (see Appendix). The electrons in an atom occupy successive shells in an orderly manner: The first energy level fills before any electrons enter the second, and the second energy level fills before any electrons enter the third.

The outermost energy level, or electron shell, forms the "surface" of the atom and is called the **valence shell**. (The *valence* of an element refers to its combining power with other atoms.) The number of electrons in this level determines the chemical properties of the element. Atoms with unfilled valence shells are unstable—that is, they will react with other atoms, usually in ways that result in full valence shells. In contrast, atoms with a filled valence shell are stable and therefore do not readily react with other atoms.

As indicated in **Figure 2–3a**, a hydrogen atom has 1 electron in the first energy level, the valence shell, so that level is unfilled. Therefore, a hydrogen atom readily reacts with other atoms. A helium atom has 2 electrons in its first energy level, which means its valence shell is filled (**Figure 2–3b**). This makes the helium atom very stable; it will not ordinarily react with other atoms. A lithium atom has 3 electrons (**Figure 2–3c**). Its first energy level can hold only 2 of them, so lithium has a single electron in a second, unfilled energy level. Thus, like hydrogen, lithium is unstable and reactive. The valence shell is filled in a neon atom, which has an atomic number of 10 (**Figure 2–3d**). Neon atoms, like helium atoms and other elements in the far right column of the periodic table, are very stable. The atoms of elements that are most important to biological systems are unstable (see **Table 2–1**, p. 29). Their instability promotes atomic interactions to form larger structures.

Figure 2–3 The Arrangement of Electrons into Energy Levels. The electron-shell model is also used to indicate the relative energy of electrons in an atom.

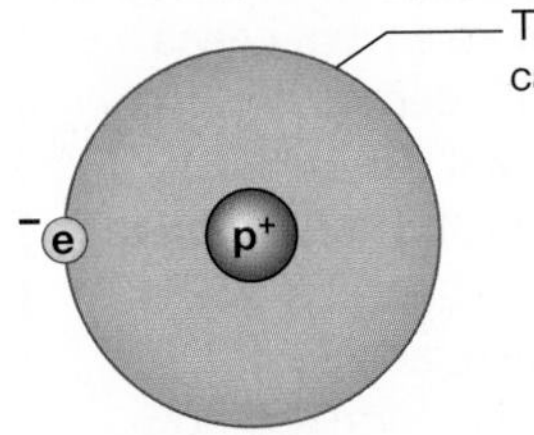
The first energy level can hold a maximum of 2 electrons.
e
p+
n

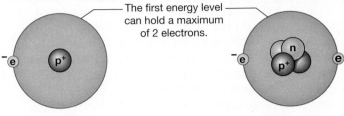

Hydrogen, H
Atomic number: 1
Mass number: 1
1 electron

Helium, He
Atomic number: 2
Mass number: 4
(2 protons + 2 neutrons)
2 electrons

a Hydrogen (H). A typical hydrogen atom has 1 proton and 1 electron. The electron orbiting the nucleus occupies the first, or lowest, energy level, diagrammed as an electron shell.

b Helium (He). An atom of helium has 2 protons, 2 neutrons, and 2 electrons. The 2 electrons orbit in the same energy level.

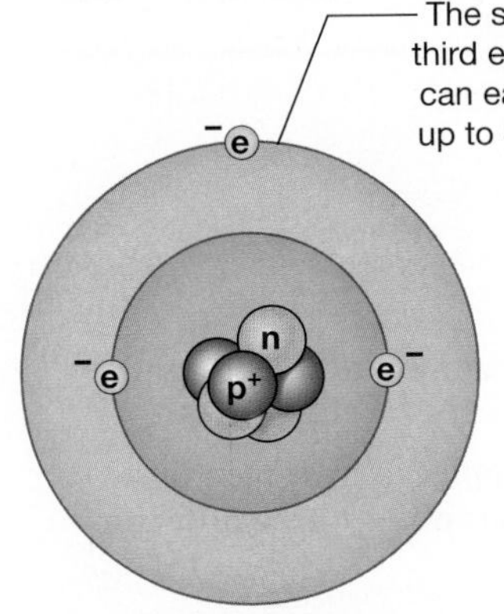
The second and third energy levels can each contain up to 8 electrons.
e
p+
n

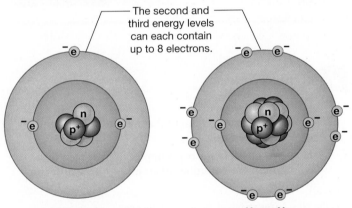

Lithium, Li
Atomic number: 3
Mass number: 6
(3 protons + 3 neutrons)
3 electrons

Neon, Ne
Atomic number: 10
Mass number: 20
(10 protons + 10 neutrons)
10 electrons

c Lithium (Li). A lithium atom has 3 protons, 3 neutrons, and 3 electrons. The first energy level can hold only 2 electrons, so the third electron occupies the second energy level.

d Neon (Ne). A neon atom has 10 protons, 10 neutrons, and 10 electrons. The second level can hold up to 8 electrons; thus, both the first and second energy levels are filled.

? How many electrons can the second energy level hold when it is completely filled?

+ Clinical Note Radiation Sickness

Radiation sickness results from excessive exposure to ionizing radiation. It is characterized by fatigue, nausea, vomiting, and loss of teeth and hair. More severe cases cause anemia (low red blood cell count), central nervous damage, and death. In a clinical setting, it can occur when large doses of cancer-treating radiation are given to a person over a short time period. (X-rays and CT scans use low-dose radiation and do not cause radiation sickness.) The amount of radiation received determines how sick a person can become. In 2011, a tsunami damaged a nuclear reactor in Fukushima, Japan, releasing a lethal radioactive form of iodine, I-131. Since then, public health officials have made potassium iodide (KI) available to people who live near such reactors. This benign salt saturates the thyroid gland with stable (nonradioactive) iodine so that it cannot absorb I-131.

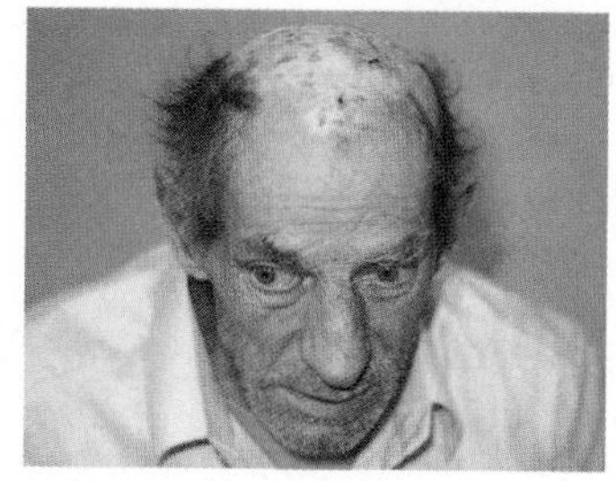

✓ Checkpoint

1. **Define *atom*.**
2. **Atoms of the same element that have different numbers of neutrons are called _____.**
3. **How is it possible for two samples of hydrogen to contain the same number of atoms, yet have different weights?**

See the blue Answers tab at the back of the book.

2-2 Chemical bonds are forces formed by interactions between atoms

Learning Outcome **Compare the ways in which atoms combine to form molecules and compounds.**

Elements without active chemical properties are said to be *inert*. The noble gases helium, neon, and argon have filled valence shells and are called *inert gases*, because their atoms do not undergo chemical reactions.

Elements with unfilled valence shells, such as hydrogen and lithium, are called *reactive*, because they readily interact or combine with other atoms. Reactive atoms become stable by gaining, losing, or sharing electrons to fill their valence shells. The interactions often involve the formation of **chemical bonds**, which hold the participating atoms together once the chemical reaction has ended.

When chemical bonding takes place, new chemical entities called *molecules* and *compounds* are created. Chemical bonding may or may not involve the sharing of electrons. The term **molecule** refers to any chemical structure consisting of atoms held together by shared electrons. A **compound** is a pure chemical substance made up of atoms of two or more different elements in a fixed proportion, regardless of whether electrons are shared or not. The two categories overlap, but they are not the same. Not all molecules are compounds, because some molecules consist of atoms of only one element. (For example, two oxygen atoms can be joined by sharing electrons to form a molecule of oxygen.) And not all compounds consist of molecules, because some compounds, such as ordinary table salt (sodium chloride) are held together by bonds that do not involve shared electrons.

Many substances, however, fit both categories. Take water, for example. Water is a compound because it contains two different elements—hydrogen and oxygen—in a fixed proportion of two hydrogen atoms to one oxygen atom. It also consists of molecules, because the two hydrogen atoms and one oxygen atom are held together by shared electrons. As we will see in later sections, most biologically important compounds, from carbohydrates to DNA, are molecules.

Regardless of the type of bonding involved, a chemical compound has properties that can be quite different from those of its components. For example, a mixture of hydrogen gas and oxygen gas can explode, but the explosion is a chemical reaction that produces liquid water, a compound used to put out fires.

The **molecular weight** of a molecule or compound is the sum of the atomic weights of its component atoms. It follows from the definition of the mole given previously (p. 30) that the molecular weight of a molecule or compound in grams is equal to the mass of 1 mol of molecules or a compound. Molecular weights are important because we cannot handle individual molecules or parts of a compound nor can we easily count the billions involved in chemical reactions in the body.

As an example of how to calculate a molecular weight, let's use water, or H_2O. The atomic weight of hydrogen is 1.0079. To simplify our calculations, we round that value to 1. Then, 1 hydrogen molecule (H_2) with its 2 atoms has a molecular weight of $2(1 \times 2 = 2)$. One oxygen atom has an atomic weight of 16. Summing up, the molecular weight of 1 molecule of H_2O is 18. One mole of water molecules would then have a mass of 18 grams.

The human body consists of countless molecules and compounds, so it is a challenge to describe these substances and their varied interactions. Chemists simplify such descriptions through a standardized system of *chemical notation*. The basic rules of this system for atoms and molecules are described in **Spotlight Figure 2–4a,b.**

SPOTLIGHT

Figure 2–4

Chemical Notation

In order to discuss the specific compounds that occur in the human body, we must be able to describe chemical compounds and reactions clearly. To do this, we use a simple form of "chemical shorthand" known as **chemical notation**. Chemical notation enables us to describe complex events briefly and precisely; its rules are summarized below.

VISUAL REPRESENTATION | CHEMICAL NOTATION

a Atoms

The symbol of an element indicates 1 atom of that element. A number preceding the symbol of an element indicates more than 1 atom of that element.

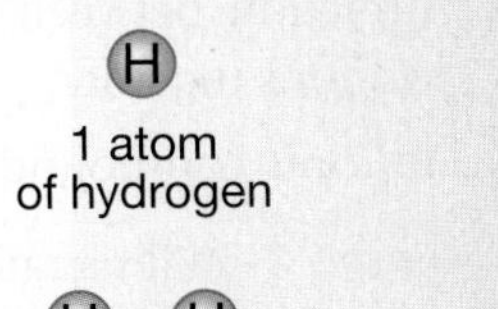
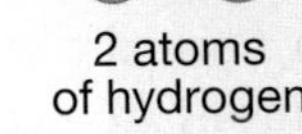
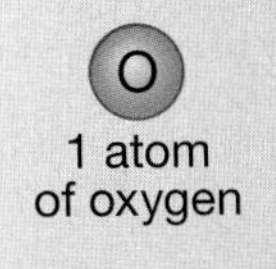
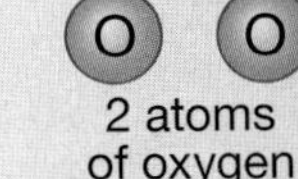

1 atom of hydrogen; 1 atom of oxygen; 2 atoms of hydrogen; 2 atoms of oxygen

=

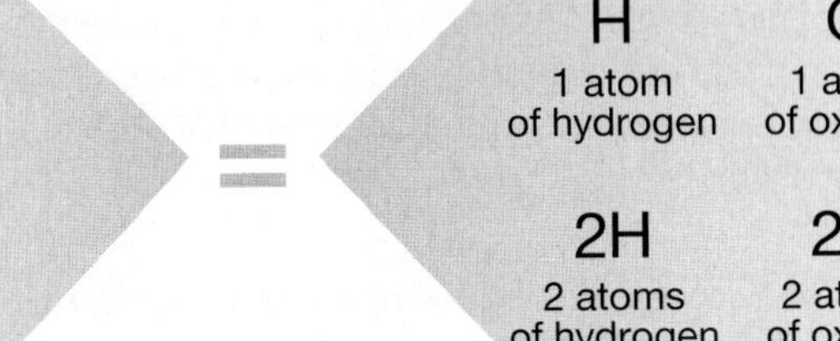

H — 1 atom of hydrogen; O — 1 atom of oxygen

2H — 2 atoms of hydrogen; 2O — 2 atoms of oxygen

b Molecules

A numerical subscript following the symbol of an element indicates the number of atoms of that element in a molecule.

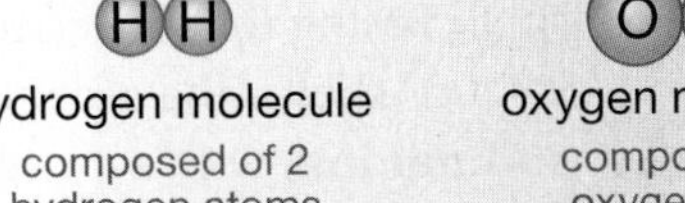
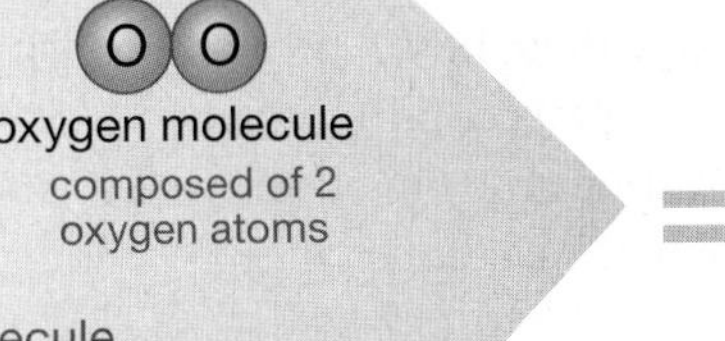

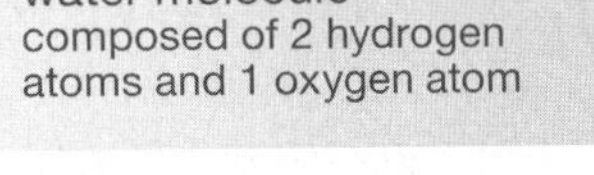

hydrogen molecule composed of 2 hydrogen atoms; oxygen molecule composed of 2 oxygen atoms

water molecule composed of 2 hydrogen atoms and 1 oxygen atom

=

H_2 hydrogen molecule; O_2 oxygen molecule

H_2O water molecule

c Reactions

In a description of a chemical reaction, the participants at the start of the reaction are called reactants, and the reaction generates one or more products. Chemical reactions are represented by chemical equations. An arrow indicates the direction of the reaction, from reactants (usually on the left) to products (usually on the right). In the following reaction, 2 atoms of hydrogen combine with 1 atom of oxygen to produce a single molecule of water.

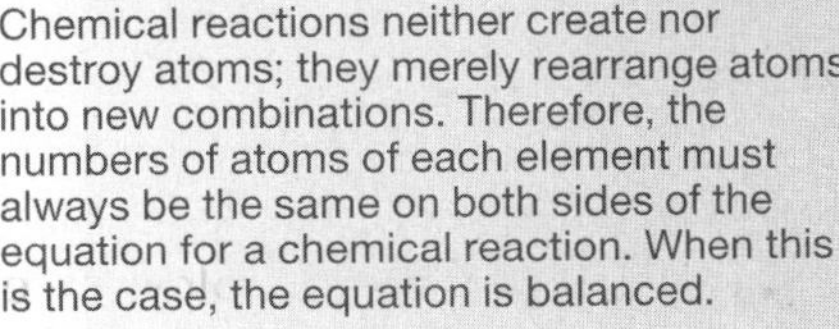

Reactants **Product**

Chemical reactions neither create nor destroy atoms; they merely rearrange atoms into new combinations. Therefore, the numbers of atoms of each element must always be the same on both sides of the equation for a chemical reaction. When this is the case, the equation is balanced.

=

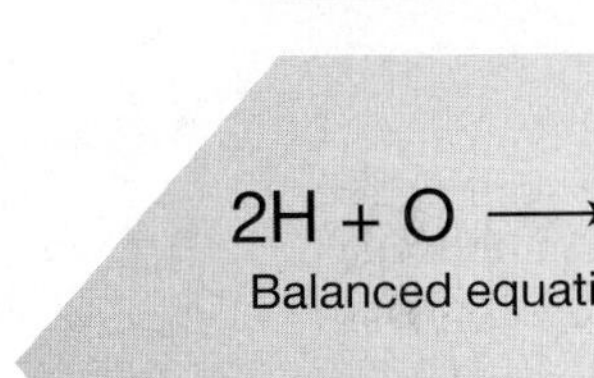

$2H + O \longrightarrow H_2O$

Balanced equation

$2H + 2O \longrightarrow H_2O$

Unbalanced equation

d Ions

A superscript plus or minus sign following the symbol of an element indicates an ion. A single plus sign indicates a cation with a charge of +1. (The original atom has given up 1 electron.) A single minus sign indicates an anion with a charge of –1. (The original atom has gained 1 electron.) If more than 1 electron is involved, the charge on the ion is indicated by a number preceding the plus or minus sign.

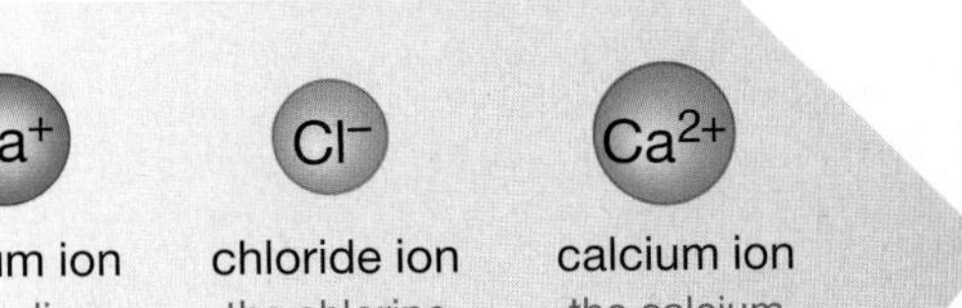

sodium ion — the sodium atom has given up 1 electron; chloride ion — the chlorine atom has gained 1 electron; calcium ion — the calcium atom has given up 2 electrons

=

Na^+ sodium ion; Cl^- chloride ion; Ca^{2+} calcium ion

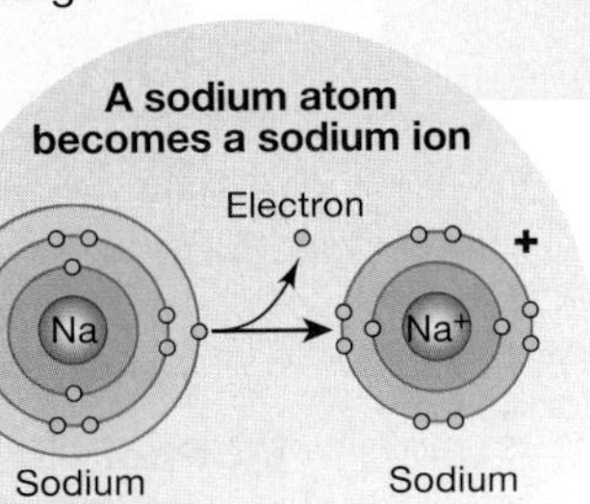
A sodium atom becomes a sodium ion
Electron
Na
Na+
+
Sodium atom (Na)
Sodium ion (Na+)

In the sections that follow, we consider three basic types of chemical bonds: *ionic bonds*, *covalent bonds*, and *hydrogen bonds*.

Ionic Bonds

An **ion** is an atom or group of atoms that has an electric charge, either positive or negative. Ions with a positive charge (+) are called **cations** (KAT-ī-onz). Ions with a negative charge (−) are called **anions** (AN-ī-onz).

Tips & Tools

Think of the ***t*** in ca***t***ion as a plus sign (+) to remember that a ca***t***ion has a positive charge, and think of the ***n*** in a***n***ion as standing for ***n***egative (−) to remember that a***n***ions have a ***n***egative charge.

Atoms become ions by losing or gaining electrons, so ions have an unequal number of protons and electrons. We assign a value of +1 to the charge on a proton, while the charge on an electron is −1. An atom that loses an electron becomes a cation with a charge of +1, because it then has 1 proton that lacks a corresponding electron. Losing a second electron would give the cation a charge of +2. Adding an extra electron to a neutral atom produces an anion with a charge of −1. Adding a second electron gives the anion a charge of −2 (**Spotlight Figure 2–4d**).

Ionic bonds are chemical bonds created by the electrical attraction between anions and cations. In the formation of an ionic bond, the following steps occur:

1. One atom—the *electron donor*—loses one or more electrons and becomes a cation, with a positive (+) charge. Another atom—the *electron acceptor*—gains those same electrons and becomes an anion, with a negative (−) charge.
2. Attraction between the opposite charges then draws the two ions together.
3. An ionic compound is formed.

Figure 2–5a illustrates the formation of an ionic bond. The sodium atom diagrammed in ❶ has an atomic number of 11, so this atom normally contains 11 protons and 11 electrons. (Because neutrons are electrically neutral, they do not affect the formation of ions or ionic bonds.) Electrons fill the first and second energy levels, and a single electron occupies the valence shell. Losing that 1 electron would give the sodium atom a full valence shell—the second energy level—and would produce a **sodium ion**, with a charge of +1. (The chemical shorthand for a sodium ion is Na^+.) But a sodium atom cannot simply throw

Figure 2–5 The Formation of Ionic Bonds.

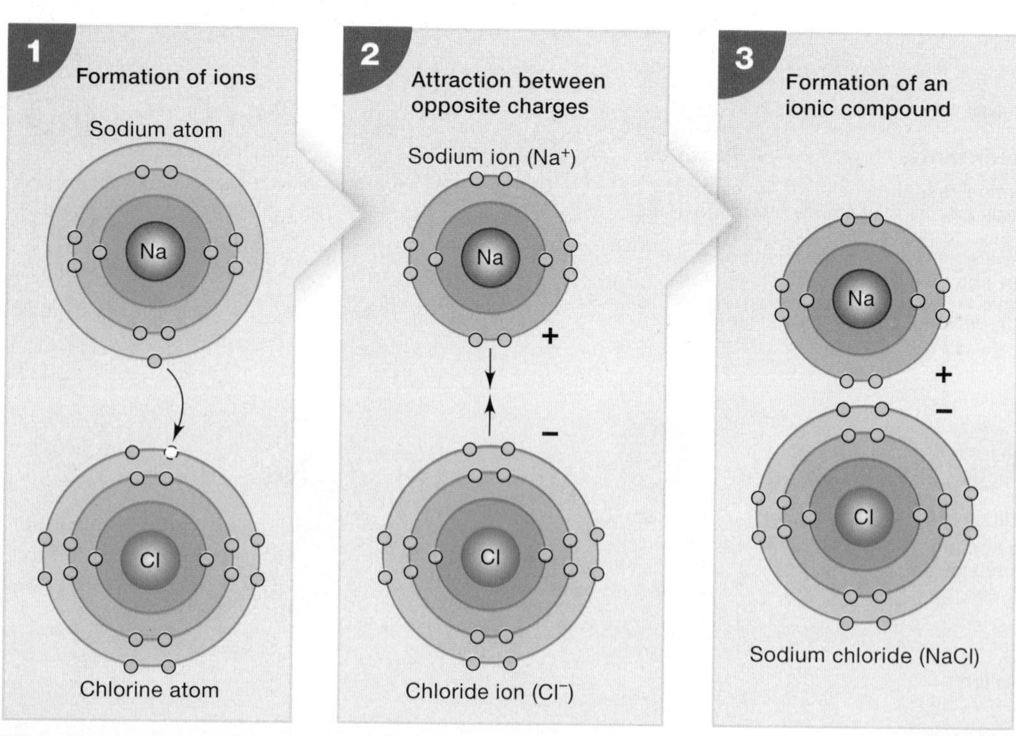
1 Formation of ions
Sodium atom
Na
Cl
Chlorine atom
2 Attraction between opposite charges
Sodium ion (Na+)
Na
+
−
Cl
Chloride ion (Cl−)
3 Formation of an ionic compound
Na
+
−
Cl
Sodium chloride (NaCl)

a Formation of an ionic bond. ❶ A sodium (Na) atom gives up an electron, which is gained by a chlorine (Cl) atom. ❷ Because the sodium ion (Na^+) and chloride ion (Cl^-) have opposite charges, they are attracted to one another. ❸ The association of sodium and chloride ions forms the ionic compound sodium chloride.

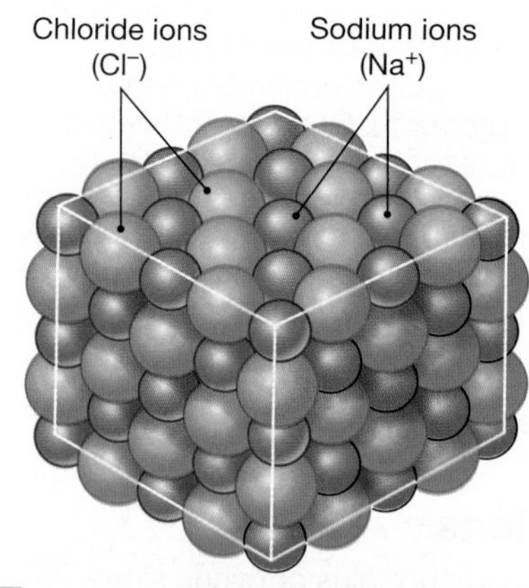
Chloride ions (Cl−)
Sodium ions (Na+)

b Sodium chloride crystal. Large numbers of sodium and chloride ions form a crystal of sodium chloride (table salt).

c Photo of sodium chloride crystals

Go to MasteringA&P™ > Study Area > Get Ready for A&P > 2. Get Ready for A&P Website> Chapter 5: Chemistry > Chemical Bonding: Video Tutor: Introduction to Chemical Bonding away the electron: The electron must be donated to an electron acceptor. A chlorine atom has seven electrons in its valence shell, so it needs only one electron to achieve stability. A sodium atom can provide the extra electron. In the process (1), the chlorine atom becomes a **chloride ion** (Cl^-) with a charge of -1.

Both atoms have now become stable ions with filled outermost energy levels. But the two ions do not move apart after the electron transfer, because the positively charged sodium ion is attracted to the negatively charged chloride ion (2). The combination of oppositely charged ions forms an *ionic compound*—in this case, **sodium chloride**, or NaCl (3). Large numbers of sodium and chloride ions interact to form highly structured crystals, held together by the strong electrical attraction of oppositely charged ions (Figure 2–5b,c).

Note that ionic compounds are not molecules. That is because they consist of a group of ions rather than atoms bonded by shared electrons. Sodium chloride and other ionic compounds are common in body fluids, but they are not present as intact crystals. When placed in water, many ionic compounds dissolve, and some or all of the component anions and cations separate.

Covalent Bonds

Some atoms can complete their valence shells not by gaining or losing electrons, but by sharing electrons with other atoms. Such sharing creates **covalent** (kō-VĀ-lent) **bonds** between the atoms involved.

Individual hydrogen atoms, as diagrammed in Figure 2–2a, do not exist in nature. Instead, we find hydrogen molecules. Molecular hydrogen consists of a pair of hydrogen atoms (Figure 2–6). In chemical shorthand, molecular hydrogen is H_2, where H is the chemical symbol for hydrogen, and the subscript 2 indicates the number of atoms. Molecular hydrogen is a gas that is present in the atmosphere in very small quantities. When the two hydrogen atoms share their electrons, each electron whirls around both nuclei. The sharing of one pair of electrons creates a **single covalent bond**, which can be represented by a single line (—) in the *structural formula* of a molecule.

Oxygen, with an atomic number of 8, has 2 electrons in its first energy level and 6 in its second. The oxygen atoms diagrammed in Figure 2–6 become stable by sharing two pairs of electrons, forming a **double covalent bond**, represented by two lines (=) in its structural formula. Molecular oxygen (O_2) is an atmospheric gas that most organisms need to survive. Our cells would die without a relatively constant supply of oxygen.

In our bodies, chemical processes that consume oxygen generally also produce **carbon dioxide** (CO_2) as a waste product. Each of the oxygen atoms in a carbon dioxide molecule forms double covalent bonds with the carbon atom, as Figure 2–6 shows.

A triple covalent bond is the sharing of three pairs of electrons, and is indicated by three lines (≡) in a structural formula. A triple covalent bond joins 2 nitrogen atoms to form molecular nitrogen (N_2) (see Figure 2–6). Molecular nitrogen accounts for about 79 percent of our planet's atmosphere, but our cells ignore it completely. In fact, deep-sea divers live for long periods while breathing artificial air that does not contain nitrogen. (We discuss the reasons for eliminating nitrogen under these conditions in the Decompression Sickness Clinical Note in Chapter 23.)

Figure 2–6 Covalent Bonds in Five Common Molecules.

Molecule	Electron-Shell Model and Structural Formula
Hydrogen (H_2)	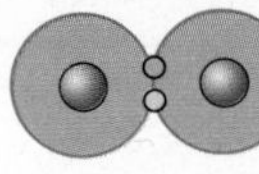 H–H
Oxygen (O_2)	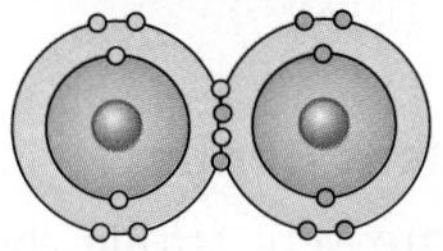 O=O
Carbon dioxide (CO_2)	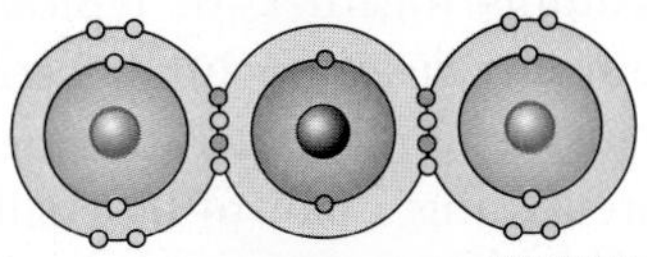 O=C=O
Nitrogen (N_2)	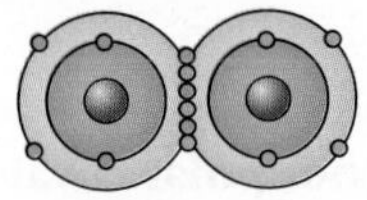 N≡N
Nitric oxide (NO)	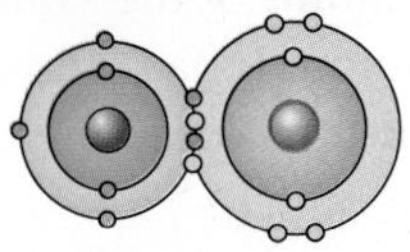 N=O

Covalent bonds usually form molecules in which the valence shells of the atoms involved are filled. An atom, ion, or molecule that contains unpaired electrons in its valence shell is called a *free radical*. Free radicals are highly reactive. Almost as fast as it forms, a free radical enters additional reactions that are typically destructive. For example, free radicals can damage or destroy vital compounds, such as proteins. Evidence suggests that the cumulative damage from free radicals inside and outside our cells is a major factor in the aging process. Free radicals sometimes form in the course of normal metabolism, but cells have several methods of removing or inactivating them.

Nitric oxide (NO), however, is a free radical with several important functions in the body (see Figure 2–6). It reacts readily with other atoms or molecules, but it is involved in chemical communication in the nervous system, in the control of blood vessel diameter, in blood clotting, and in the defense against bacteria and other pathogens (disease-causing organisms).

> Go to MasteringA&P™ > Study Area > Get Ready for A&P > 2. Get Ready for A&P Website> Chapter 5: Chemistry > Chemical Bonding: Video Tutor: Ionic Bonding

Tips & Tools

Remember this mnemonic for the covalent bonding of hydrogen, oxygen, nitrogen, and carbon atoms: HONC 1234. ***H****ydrogen shares* *1* pair of electrons (H—), ***o****xygen shares* *2* pairs (—O—), ***n****itrogen shares 3 pairs* (—N—), and ***c****arbon shares* *4* pairs (—C—).

Nonpolar Covalent Bonds

Covalent bonds are very strong, because the shared electrons hold the atoms together. In typical covalent bonds the atoms remain electrically neutral, because each shared electron spends just as much time "at home" as away. (If you and a friend were tossing a pair of baseballs back and forth as fast as you could, on average, each of you would have just one baseball.)

Many covalent bonds involve an equal sharing of electrons. Such bonds are called **nonpolar covalent bonds**. They occur, for instance, between 2 atoms of the same type. Nonpolar covalent bonds are very common. In fact, those involving carbon atoms form most of the structural components of the human body.

Polar Covalent Bonds

Covalent bonds involving different types of atoms may involve an unequal sharing of electrons, because the elements differ in how strongly they attract electrons. An unequal sharing of electrons creates a **polar covalent bond**. In chemistry, *polarity* refers to a separation of positive and negative electric charge. It can apply to a bond or entire molecule. For example, in a molecule of water, an oxygen atom forms covalent bonds with 2 hydrogen atoms (Figure 2–7a). The oxygen nucleus (with its 8 protons) has a much stronger attraction for the shared electrons than the hydrogen atoms (each with a single proton). As a result, the electrons spend more time orbiting the oxygen nucleus than orbiting the hydrogen nuclei.

Because the oxygen atom has two extra electrons most of the time, it develops a slight (partial) negative charge, indicated by δ^-, as shown in Figure 2–7b. At the same time, each hydrogen atom develops a slight (partial) positive charge, δ^+, because its electron is away much of the time. (Suppose you and a friend were tossing a pair of baseballs back and forth, but one of you returned them as fast as possible while the other held onto them for a while before throwing them back. One of you would now, on average, have more than one baseball, and the other would have less than one.)

The unequal sharing of electrons makes polar covalent bonds somewhat weaker than nonpolar covalent bonds. Polar covalent bonds often create *polar molecules*—molecules that have positive and negative ends. Water is the most important polar molecule in the body.

Figure 2–7 Water Molecules Contain Polar Covalent Bonds.

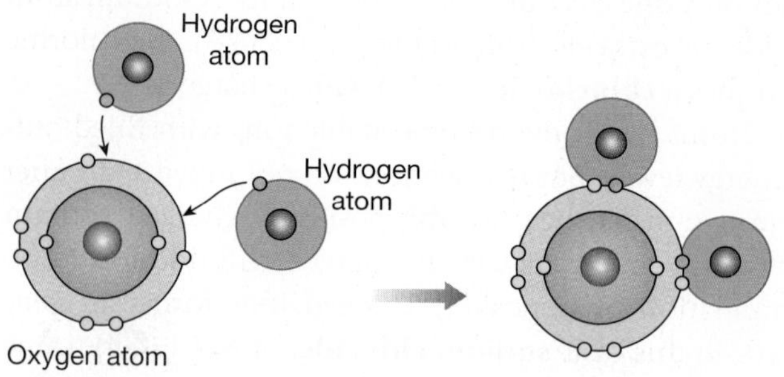
Hydrogen atom
Hydrogen atom
Oxygen atom

a Formation of a water molecule. In forming a water molecule, an oxygen atom completes its valence shell by sharing electrons with a pair of hydrogen atoms. The sharing is unequal, because the oxygen atom holds the electrons more tightly than do the hydrogen atoms.

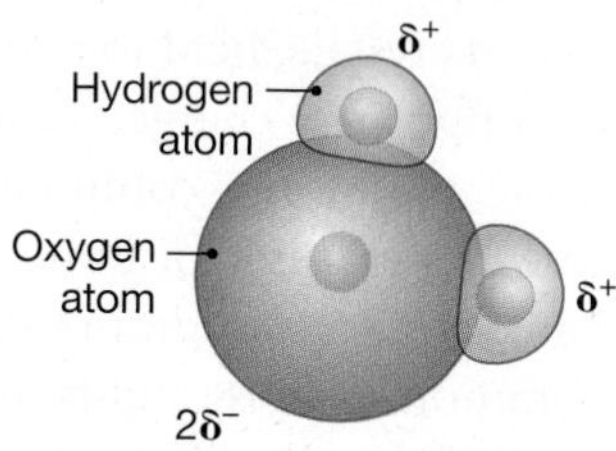
δ+
Hydrogen atom
Oxygen atom
δ+
2δ−

b Electric charges on a water molecule. Because the oxygen atom has 2 extra electrons much of the time, it develops a slight negative charge, and the hydrogen atoms become weakly positive. The bonds in a water molecule are polar covalent bonds.

Hydrogen Bonds

Covalent and ionic bonds tie atoms together to form molecules and/or compounds. Other, comparatively weak forces also act between adjacent molecules, and even between atoms within a large molecule. The most important of these weak attractive forces is the hydrogen bond. A **hydrogen bond** is the attraction between a slight positive charge (δ^+) on the hydrogen atom of a polar covalent bond and a slight negative charge (δ^-) on an oxygen, nitrogen, or fluorine atom of another polar covalent bond.

Hydrogen bonds are too weak to create molecules, but they can change the shapes of molecules or pull molecules closer together. For example, hydrogen bonding occurs between water molecules, forming clumps, or groups, of interconnected water molecules (Figure 2–8). At a water surface, this attraction between water molecules slows the rate of evaporation and creates what is known as *surface tension*. Surface tension acts as a barrier that keeps lightweight objects that cannot break through the top layer of water molecules from entering the water. For example, it allows insects to walk across the surface of a pond or puddle. Similarly, the surface tension in a layer of tears on the eye prevents dust particles from touching the surface of the

Go to MasteringA&P™ > Study Area > Get Ready for A&P > 2. Get Ready for A&P Website> Chapter 5: Chemistry > Chemical Bonding: Video Tutor: Covalent Bonding

Figure 2–8 Hydrogen Bonds Form between Water Molecules. The hydrogen atoms of a water molecule have a slight positive charge, and the oxygen atom has a slight negative charge (see *Figure 2–7b*). The distances between these molecules have been exaggerated for clarity.

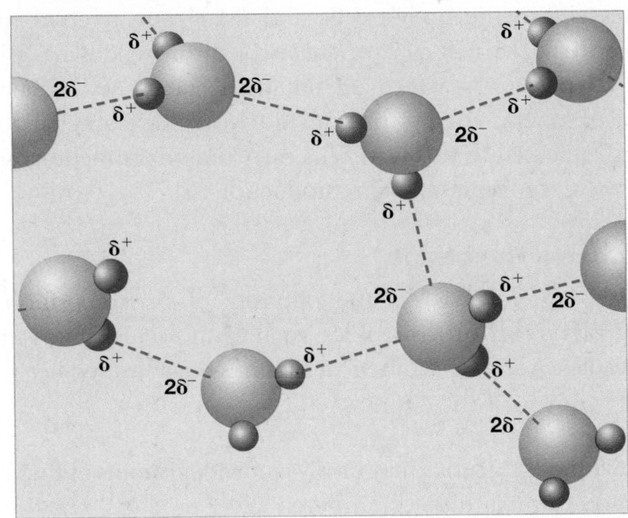
δ+
2δ−

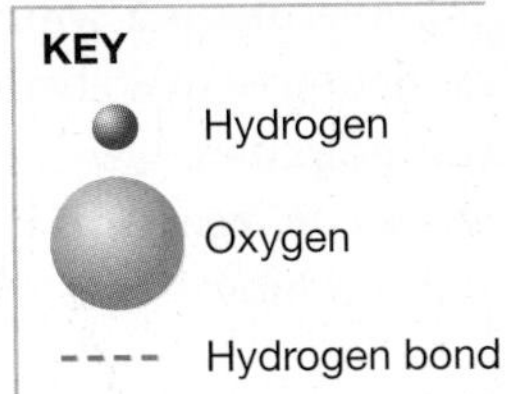
KEY
Hydrogen
Oxygen
Hydrogen bond

? Hydrogen bonds do not form between two hydrogen atoms. Where do hydrogen bonds form?

eye. At the cellular level, hydrogen bonds affect the shapes and properties of large, complex molecules, such as proteins and nucleic acids (including DNA).

States of Matter

Most matter in our environment exists in one of three states: solid, liquid, or gas. *Solids* keep their volume and their shape at ordinary temperatures and pressures. A lump of granite, a brick, and a textbook are solid objects. *Liquids* have a constant volume, but no fixed shape. The shape of a liquid is determined by the shape of its container. Water, coffee, and soda are liquids. A *gas* has no constant volume and no fixed shape. Gases can be compressed or expanded, and unlike liquids they will fill a container of any size. The most familiar example is the air of our atmosphere.

What determines whether a substance is a solid, liquid, or gas? A matter's state depends on the degree of interaction among its atoms or molecules. The particles of a solid are held tightly together, while those of a gas are very far apart. Water is the only substance that occurs as a solid (ice), a liquid (water), and a gas (water vapor) at temperatures compatible with life. Water exists as a liquid over a broad range of temperatures primarily because of hydrogen bonding among the water molecules. We talk more about water's unusual properties in Section 2-6.

Checkpoint

4. **Define *chemical bond* and identify several types of chemical bonds.**
5. **Which kind of bond holds atoms in a water molecule together? What attracts water molecules to one another?**
6. **Both oxygen and neon are gases at room temperature. Oxygen combines readily with other elements, but neon does not. Why?**

See the blue Answers tab at the back of the book.

2-3 Decomposition, synthesis, and exchange reactions are important types of chemical reactions in physiology

Learning Outcome Distinguish among the major types of chemical reactions that are important for studying physiology.

Cells stay alive and functional by controlling chemical reactions. In a **chemical reaction**, new chemical bonds form between atoms, or existing bonds between atoms are broken. These changes take place as atoms in the reacting substances, called **reactants**, are rearranged to form different substances, or **products** (see **Spotlight Figure 2–4c**).

In effect, each cell is a chemical factory. Cells use chemical reactions to provide the energy they need to maintain homeostasis and to perform essential functions such as growth, maintenance and repair, secretion (discharging from a cell), and contraction. All of the reactions under way in the cells and tissues of the body at any given moment make up its **metabolism** (me-TAB-ō-lizm).

Basic Energy Concepts

An understanding of some basic relationships between matter and energy is helpful for any discussion of chemical reactions. **Work** is the movement of an object or a change in the physical structure of matter. In your body, work includes movements such as walking or running, and also the synthesis of *organic* (carbon-containing) molecules and the conversion of liquid water to water vapor (evaporation). **Energy** is the capacity to do work, and movement or physical change cannot take place without energy. The two major types of energy are kinetic energy and potential energy:

- **Kinetic energy** is the energy of motion—energy that can be transferred to another object and do work. When you fall off a ladder, it is kinetic energy that does the damage.

- **Potential energy** is stored energy—energy that has the potential to do work. It may derive from an object's position (you standing on a ladder) or from its physical or chemical structure (a stretched spring or a charged battery).

Kinetic energy must be used in climbing the ladder, in stretching the spring, or in charging the battery. The resulting potential energy is converted back into kinetic energy when you descend, the spring recoils, or the battery discharges. The kinetic energy can then be used to perform work. For example, in an MP3 player, the chemical potential energy stored in small batteries is converted to kinetic energy that vibrates the sound-producing membranes in headphones or external speakers.

Energy cannot be destroyed: It can only be converted from one form to another. A conversion between potential energy and kinetic energy is never 100 percent efficient. Each time an energy exchange occurs, some of the energy is released in the form of heat. *Heat* is an increase in random molecular motion, and the temperature of an object is proportional to the average kinetic energy of its molecules. Heat can never be completely converted to work or any other form of energy, and cells cannot capture it or use it to do work.

Cells do work as they use energy to synthesize complex molecules and move materials into, out of, and within the cell. The cells of a skeletal muscle at rest, for example, contain potential energy in the form of the positions of protein filaments and the covalent bonds between molecules inside the cells. When a muscle contracts, it performs work. Potential energy is converted into kinetic energy, and heat is released. The amount of heat is proportional to the amount of work done. As a result, when you exercise, your body temperature rises.

Types of Chemical Reactions

Three types of chemical reactions are important to the study of physiology: decomposition reactions, synthesis reactions, and exchange reactions. Many of these chemical reactions are also *reversible reactions*.

Decomposition Reactions

A **decomposition reaction** breaks a molecule into smaller fragments. Here is a diagram of a simple *decomposition reaction*:

$$AB \longrightarrow A + B$$

Decomposition reactions take place outside cells as well as inside them. For example, a typical meal contains molecules of fats, sugars, and proteins that are too large and too complex to be absorbed and used by your body. Decomposition reactions in the digestive tract break these molecules down into smaller fragments that can be absorbed.

Decomposition reactions involving water are important in the breakdown of complex molecules in the body. In this process, which is called a **hydrolysis reaction** (hī-DROL-i-sis; *hydro-*, water + *lysis*, a loosening), one of the bonds in a complex molecule is broken, and the components of a water molecule (H and OH) are added to the resulting fragments:

$$AB + H_2O \longrightarrow AH + BOH$$

Collectively, the decomposition reactions of complex molecules within the body's cells and tissues are referred to as **catabolism** (ka-TAB-ō-lizm; *katabole*, a throwing down). When a covalent bond—a form of potential energy—is broken, it releases kinetic energy that can do work. By harnessing the energy released in this way, cells carry out vital functions such as growth, movement, and reproduction.

Synthesis Reactions

Synthesis (SIN-the-sis) is the opposite of decomposition. A **synthesis reaction** assembles smaller molecules into larger molecules. A simple synthetic reaction is diagrammed here:

$$A + B \longrightarrow AB$$

Synthesis reactions may involve individual atoms or the combination of molecules to form even larger products. The formation of water from hydrogen and oxygen molecules is a synthesis reaction. Synthesis always involves the formation of new chemical bonds, whether the reactants are atoms or molecules.

A **dehydration synthesis**, or *condensation*, **reaction** is the formation of a complex molecule by the removal of a water molecule:

$$AH + BOH \longrightarrow AB + H_2O$$

Dehydration synthesis is the opposite of hydrolysis. We look at examples of both reactions in later sections (Sections 2-9, 2-10, and 2-11).

Collectively, the synthesis of new molecules within the body's cells and tissues is known as **anabolism** (a-NAB-ō-lizm; *anabole*, a throwing upward). Anabolism is usually considered an "uphill" process because it takes energy to create a chemical bond (just as it takes energy to push something uphill). Cells must balance their energy budgets, with catabolism providing the energy to support anabolism and other vital functions.

Tips & Tools

To remember the difference between anabolism (synthesis) and catabolism (breakdown), relate the terms to words you already know: *Anabolic* steroids are used to build up muscle tissue, while both *catastrophe* and *catabolism* involve destruction (breakdown).

Exchange Reactions

In an **exchange reaction**, parts of the reacting molecules are shuffled around to produce new products:

$$AB + CD \longrightarrow AD + CB$$

The reactants and products contain the same components (A, B, C, and D), but those components are present in different combinations. In an exchange reaction, the reactant molecules AB and CD must break apart (a decomposition) before they can interact with each other to form AD and CB (a synthesis).

Reversible Reactions

At least in theory, chemical reactions are reversible, so if $A + B \longrightarrow AB$, then $AB \longrightarrow A + B$. Many important biological reactions are freely reversible. Such reactions can be represented as an equation:

$$A + B \rightleftharpoons AB$$

This equation indicates that, in a sense, two reactions are taking place at the same time. One is a synthesis reaction ($A + B \rightarrow AB$) and the other is a decomposition reaction ($AB \rightarrow A + B$).

Recall from Chapter 1 that a state of *equilibrium* exists when opposing processes or forces are in balance. At equilibrium, the rates of the two reactions are in balance. As fast as one molecule of AB forms, another degrades into A + B.

What happens when equilibrium is disturbed—say, if you add more AB? In our example, the rate of the synthesis reaction is directly proportional to the frequency of encounters between A and B. In turn, the frequency of encounters depends on the degree of crowding. (You are much more likely to bump into another person in a crowded room than in a room that is almost empty.) So adding more AB molecules will increase the rate of conversion of AB to A and B. The amounts of A and B will then increase, leading to an increase in the rate of the reverse reaction—the formation of AB from A and B. Eventually, a balance, or equilibrium, is again established.

Tips & Tools

Jell-O provides an example of a physical reversible reaction. Once Jell-O has been refrigerated, the gelatin sets up and forms a solid, but if it sits without refrigeration for too long, it turns back into a liquid again.

✓ Checkpoint

7. **Using the rules for chemical notation, how is an ion's electrical charge represented?**
8. **Using the rules for chemical notation, write the molecular formula for glucose, a compound composed of 6 carbon atoms, 12 hydrogen atoms, and 6 oxygen atoms.**
9. **Identify and describe three types of chemical reactions important to human physiology.**
10. **In cells, glucose, a six-carbon molecule, is converted into two three-carbon molecules by a reaction that releases energy. What type of reaction is this?**

See the blue Answers tab at the back of the book.

2-4 Enzymes speed up reactions by lowering the energy needed to start them

Learning Outcome Describe the crucial role of enzymes in metabolism.

Most **biochemical reactions** (those that happen in living organisms) do not take place spontaneously, or if they do, they occur so slowly that they would be of little value to living cells. Before a reaction can proceed, enough energy must be provided to activate the reactants. The amount of energy required to start a reaction is called the **activation energy**. Many reactions can be activated by changes in temperature or acidity, but such changes are deadly to cells. For example, every day your cells break down complex sugars as part of your normal metabolism. Yet to break down a complex sugar in a laboratory, you must boil it in an acidic solution. Your cells don't have that option! Temperatures that high and solutions that corrosive would immediately destroy living tissues. Instead, your cells use special proteins called *enzymes* to catalyze (speed up) most of the complex synthesis and decomposition reactions in your body.

Enzymes promote chemical reactions by lowering their required activation energy (Figure 2–9). In doing so, they make it possible for chemical reactions, such as the breakdown of sugars, to proceed under conditions compatible with life. Cells make enzyme molecules, each of which promotes a specific reaction.

Enzymes belong to a class of substances called **catalysts** (KAT-uh-lists; *katalysis*, dissolution), compounds that speed up

Figure 2–9 Enzymes Lower Activation Energy. Enzymes lower the activation energy required for a chemical reaction to proceed readily (in order, from 1–4) under conditions in the body.

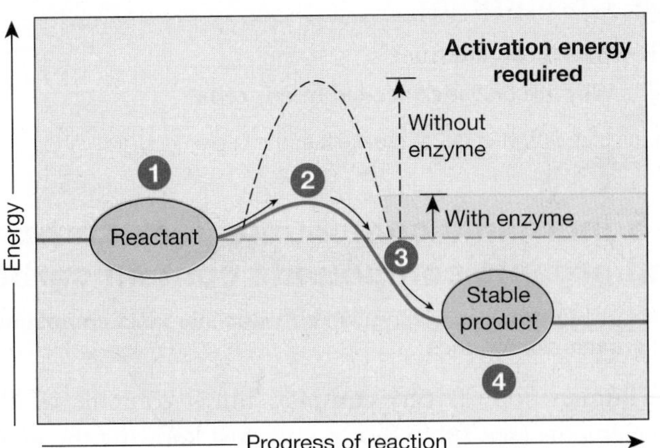
Activation energy required
Without enzyme
With enzyme
1
2
3
4
Reactant
Stable product
Energy
Progress of reaction

? Which number represents the greatest amount of energy that must be overcome during the reaction? Which number represents the lowest amount of reaction energy?

chemical reactions without themselves being permanently changed or consumed. Enzymatic reactions, which are reversible, can be written as

$$A + B \overset{\text{enzyme}}{\rightleftharpoons} AB$$

An appropriate enzyme can accelerate, or speed up, a reaction, but an enzyme affects only the *rate* of the reaction, not its direction or the products that are formed. An enzyme cannot bring about a reaction that would otherwise be impossible. Enzymatic reactions are generally reversible, and they proceed until equilibrium is reached.

The complex reactions that support life take place in a series of interlocking steps, each controlled by a specific enzyme. Such a reaction sequence is called a *metabolic pathway*. A synthetic pathway can be diagrammed as

$$A \xrightarrow[\text{Step 1}]{\text{enzyme 1}} B \xrightarrow[\text{Step 2}]{\text{enzyme 2}} C \xrightarrow[\text{Step 3}]{\text{enzyme 3}} \text{and so on.}$$

In many cases, the steps in the synthetic pathway differ from those in the decomposition pathway, and separate enzymes are often involved.

It takes activation energy to start a chemical reaction, but once it has begun, the reaction as a whole may absorb or release energy as it proceeds to completion. If the amount of energy released is greater than the activation energy needed to start the reaction, there will be a net release of energy. Reactions that release energy are said to be **exergonic** (*exo-*, outside + *ergon*, work). Exergonic reactions are relatively common in the body. They generate the heat that maintains your body temperature.

If more energy is required to begin the reaction than is released as it proceeds, the reaction as a whole will absorb energy. Such reactions are called **endergonic** (*endo-*, inside). The synthesis of molecules such as fats and proteins results from endergonic reactions.

Checkpoint

11. **What is an enzyme?**
12. **Why are enzymes needed in our cells?**

See the blue Answers tab at the back of the book.

2-5 Inorganic compounds lack carbon, and organic compounds contain carbon

Learning Outcome **Distinguish between inorganic compounds and organic compounds.**

The human body is very complex, but it contains relatively few elements (see **Table 2–1**, p. 29). Just knowing the identity and quantity of each element in the body will not help you understand the body any more than memorizing the alphabet will help you understand this text. Just as 26 letters can be combined to form thousands of different words in this text, only about 26 elements combine to form thousands of different chemical compounds in our bodies. As we saw in Chapter 1, these compounds make up the living cells that form the body's framework and carry on all its life processes. Learning about the major classes of chemical compounds will help you to understand the structure and function of the human body.

Two of the major classes of compounds are nutrients and metabolites. **Nutrients** are the substances from food that are necessary for normal physiological functions. Nutrients include carbohydrates, proteins, lipids (fats), vitamins, minerals, and water. **Metabolites** (me-TAB-ō-lītz; *metabole*, change) are substances that are involved in, or are a by-product of, metabolism. We can broadly categorize nutrients and metabolites as either inorganic or organic. **Inorganic compounds** generally do not contain carbon and hydrogen atoms as their primary structural components. (If present, they do not form C—H bonds.) In contrast, carbon and hydrogen always form the basis for **organic compounds**. Their molecules can be much larger and more complex than inorganic compounds. *Carbohydrates*, *proteins*, and *lipids* are organic nutrients used by the body—we cover these in later sections.

The most important inorganic compounds in the body are as follows:

- *carbon dioxide*, a by-product of cell metabolism;
- *oxygen*, an atmospheric gas required in important metabolic reactions;
- *water*, which accounts for more than half of our body weight; and
- *acids, bases,* and *salts*—compounds held together partially or completely by ionic bonds.

In the next section, we focus on water, its properties, and how those properties establish the conditions necessary for life. Most of the other inorganic molecules and compounds in the body exist in association with water, the primary component of our body fluids. Both carbon dioxide and oxygen, for example, are gas molecules that are transported in body fluids. Also, all the inorganic acids, bases, and salts we will discuss are dissolved in body fluids.

Checkpoint

13. **Compare inorganic compounds to organic compounds.**

See the blue Answers tab at the back of the book.

2-6 Physiological systems depend on water

Learning Outcome **Explain how the chemical properties of water make life possible.**

Water (H_2O) is the most important substance in the body. It makes up to two-thirds of total body weight. A change in the body's water content can be fatal, because virtually all physiological systems will be affected.

Although water is familiar to everyone, it has some highly unusual properties: universal solvent, reactivity, high heat capacity, and lubrication. All are important to the human body.

- *Universal Solvent.* A remarkable number of inorganic and organic molecules and compounds are water *soluble,* meaning they will dissolve or break up in water. The individual particles become distributed within the water, and the result is a **solution**—a uniform mixture of two or more substances. The liquid in which other atoms, ions, or molecules are distributed is called the **solvent**. The dissolved substances are the **solutes**. In *aqueous* (AK-wē-us) *solutions,* water is the solvent. Water is often called a "universal solvent" because more substances dissolve in it than any other liquid.
- *Reactivity.* In our bodies, chemical reactions take place in water, but water molecules are also reactants in some reactions. Hydrolysis and dehydration synthesis are two examples noted earlier in the chapter.
- *High Heat Capacity.* **Heat capacity** is the quantity of heat required to raise the temperature of a unit mass of a substance 1°C. Water has an unusually high heat capacity, because water molecules in the liquid state are attracted to one another through hydrogen bonding. Important consequences of this attraction include the following:
    - The temperature of water must be quite high (it requires a lot of energy) to break all of the hydrogen bonds between individual water molecules, and allow them to escape and become water vapor, a gas. Therefore, water remains a liquid over a broad range of environmental temperatures, and the freezing and boiling points of water are far apart.
    - Water carries a great deal of heat away with it when it changes from a liquid to a gas. This feature explains the cooling effect of perspiration on the skin.
    - An unusually large amount of heat energy is required to change the temperature of 1 g of water by 1°C. As a result, a large mass of water changes temperature slowly. This property is called *thermal inertia*. Thermal inertia helps stabilize body temperature because water accounts for up to two-thirds of the weight of the human body.
- *Lubrication.* Water is an effective lubricant because there is little friction between water molecules. So even a thin layer of water between two opposing surfaces will greatly reduce friction between them. (That is why driving on wet roads can be tricky. Your tires may start sliding on a layer of water rather than maintaining contact with the road.) Within joints such as the knee, an aqueous solution prevents friction between the opposing surfaces. Similarly, a small amount of fluid in the body cavities prevents friction between internal organs, such as the heart or lungs, and the body wall. ⮌ p. 15

The Properties of Aqueous Solutions

Water's chemical structure makes it an unusually effective solvent. The covalent bonds in a water molecule are oriented so that the hydrogen atoms are fairly close together. As a result, the water molecule has positive- and negative-charged ends, or poles (Figure 2–10a). For this reason, a water molecule is called a **polar molecule**.

2

Many inorganic compounds are held together partly or completely by ionic bonds. In water, these compounds undergo **dissociation** (di-sō-sē-Ā-shun)—the splitting of a compound into smaller molecules. **Ionization** (ī-on-i-ZĀ-shun) is the dissociation into ions. In this process, ionic bonds are broken as the individual ions interact with the positive or negative ends of polar water molecules (Figure 2–10b). The result is a mixture of cations and anions surrounded by water molecules. The water molecules around each ion form a *hydration sphere* that isolates the ions from each other, thus preventing the formation of ionic bonds.

An aqueous solution containing anions and cations will conduct an electrical current. When this happens, cations (+) move toward the negative side, and anions (−) move toward the positive side. Electrical forces across plasma membranes affect the functioning of all cells, and small electrical currents carried by ions are essential to muscle contraction and nerve function. (We will discuss these processes in more detail in Chapters 10 and 12.)

Electrolytes and Body Fluids

Soluble inorganic substances whose ions will conduct an electrical current in solution are called **electrolytes** (e-LEK-trō-lītz). Sodium chloride in solution is an electrolyte. The dissociation of electrolytes in blood and other body fluids releases a variety of ions. Table 2–2 lists important electrolytes and the ions released when they dissociate.

Changes in the concentrations of electrolytes in body fluids will disturb almost every vital function. For example, a declining potassium ion (K^+) level will lead to a general muscular paralysis, and a rising concentration will cause weak and irregular heartbeats. The concentrations of ions in body fluids are

Table 2–2 Important Electrolytes That Dissociate in Body Fluids

Electrolyte	Ions Released
$NaCl$ (sodium chloride)	→ $Na^+ + Cl^-$
KCl (potassium chloride)	→ $K^+ + Cl^-$
$CaPO_4$ (calcium phosphate)	→ $Ca^{2+} + PO_4^{2-}$
$NaHCO_3$ (sodium bicarbonate)	→ $Na^+ + HCO_3^-$
$MgCl_2$ (magnesium chloride)	→ $Mg^{2+} + 2Cl^-$
Na_2HPO_4 (sodium hydrogen phosphate)	→ $2Na^+ + HPO_4^{2-}$
Na_2SO_4 (sodium sulfate)	→ $2Na^+ + SO_4^{2-}$

Figure 2–10 Water Molecules Surround Solutes in Aqueous Solutions.

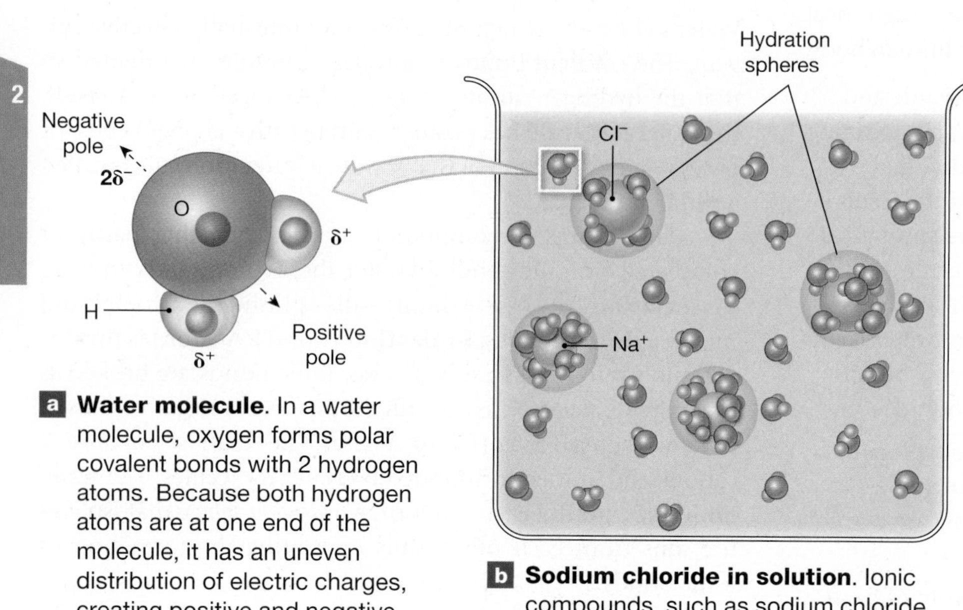
Negative pole
2δ⁻
O
δ⁺
H
δ⁺
Positive pole
Hydration spheres
Cl⁻
Na⁺

a **Water molecule.** In a water molecule, oxygen forms polar covalent bonds with 2 hydrogen atoms. Because both hydrogen atoms are at one end of the molecule, it has an uneven distribution of electric charges, creating positive and negative poles.

b **Sodium chloride in solution.** Ionic compounds, such as sodium chloride, dissociate in water as the polar water molecules break the ionic bonds in the large crystal structure. Each ion in solution is surrounded by water molecules, creating hydration spheres.

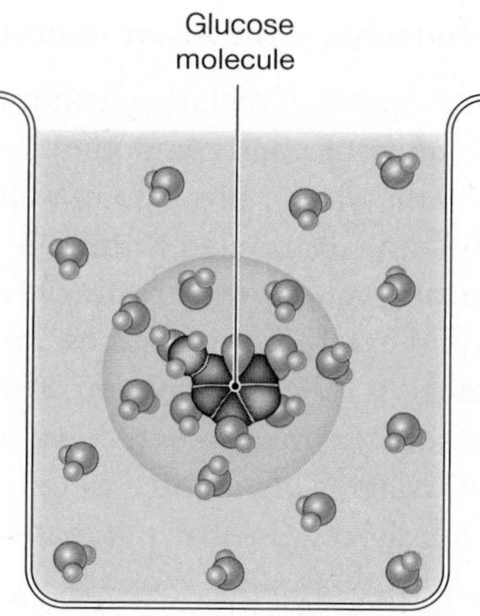
Glucose molecule

c **Glucose in solution.** Hydration spheres also form around an organic molecule containing polar covalent bonds. If the molecule binds water strongly, as does glucose, it will be carried into solution—in other words, it will dissolve. Note that the molecule does not dissociate, as occurs for ionic compounds.

carefully regulated, mostly by the coordination of activities at the kidneys (ion excretion), the digestive tract (ion absorption), and the skeletal system (ion storage or release).

Hydrophilic and Hydrophobic Compounds

Some organic molecules contain polar covalent bonds, which also attract water molecules. The hydration spheres that form may then carry these molecules into solution. Molecules that interact readily with water molecules in this way are called **hydrophilic** (hī-drō-FIL-ik; *hydro-*, water + *philos*, loving). Glucose, an important soluble sugar, is one example (Figure 2–10c).

When nonpolar molecules are exposed to water, hydration spheres do not form and the molecules do not dissolve. Molecules that do not readily interact with water are called **hydrophobic** (hī-drō-FŌB-ik; *hydro-*, water + *phobos*, fear). Fats and oils of all kinds are some of the most familiar hydrophobic molecules. For example, body fat deposits consist of large, hydrophobic droplets trapped in the watery interior of cells. Gasoline and heating oil are hydrophobic molecules not found in the body. When accidentally spilled into lakes or oceans, they form long-lasting oil slicks instead of dissolving.

Tips & Tools

To distinguish between hydrophobic and hydrophilic, remember that a phobia is a fear of something, and that *-philic* ends with "lic," which resembles "like."

Colloids and Suspensions

Body fluids may contain large and complex organic molecules, such as proteins, that are held in solution by their association with water molecules (see Figure 2–10c). A solution containing dispersed proteins or other large molecules is called a **colloid** (KOL-oyd). Liquid Jell-O is a familiar colloid. The particles or molecules in a colloid will remain in solution indefinitely.

In contrast, a **suspension** contains large particles in solution, but if undisturbed, its particles will settle out of solution due to the force of gravity. For example, stirring beach sand into a bucket of water creates a temporary suspension that will last only until the sand settles to the bottom. Whole blood is another temporary suspension, because the blood cells are suspended in the blood plasma. If clotting is prevented, the cells in a blood sample will gradually settle to the bottom of the container. Measuring that settling rate, or "sedimentation rate," is a common laboratory test.

✓ Checkpoint

14. **Explain how the chemical properties of water make life possible.**

See the blue Answers tab at the back of the book.

2-7 Body fluid pH is vital for homeostasis

Learning Outcome Explain what pH is and discuss its importance.

A hydrogen atom involved in a chemical bond or participating in a chemical reaction can easily lose its electron to become a **hydrogen ion (H^+)**. Hydrogen ions are extremely reactive in solution. In excessive numbers, they will disrupt cell and tissue functions. As a result, the concentration of hydrogen ions in body fluids must be regulated precisely.

A few hydrogen ions are normally present even in a sample of pure water, because some of the water molecules dissociate spontaneously, releasing cations and anions. The dissociation of water is a reversible reaction. We can represent it as

$$H_2O \rightleftharpoons H^+ + OH^-$$

Notice that the dissociation of one water molecule yields a hydrogen ion (H^+) and a **hydroxide** (hī-DROK-sīd) **ion (OH^-)**.

However, very few water molecules ionize in pure water, so the number of hydrogen and hydroxide ions is small. The quantities are usually reported in moles, making it easy to keep track of the numbers of hydrogen and hydroxide ions. One liter of pure water contains about 0.0000001 mol of hydrogen ions and an equal number of hydroxide ions. In other words, the concentration of hydrogen ions in a solution of pure water is 0.0000001 mol per liter. This can be written as

$$[H^+] = 1 \times 10^{-7} \text{mol/L}$$

The brackets around the H^+ signify "the concentration of," another example of chemical notation.

The hydrogen ion concentration in body fluids is so important to physiological processes that we use a special shorthand to express it. The **pH** of a solution is defined as the negative logarithm of the hydrogen ion concentration in moles per liter. So instead of using the equation $[H^+] = 1 \times 10^{-7}$ mol/L, we say that the pH of pure water is $-(-7)$, or 7.

Using pH values saves space, but always remember that the pH number is an *exponent* and that the pH scale is logarithmic. For instance, a pH of 6 ($[H^+] = 1 \times 10^{-6}$, or 0.000001 mol/L) means that the concentration of hydrogen ions is *10 times greater than* it is at a pH of 7 ($[H^+] = 1 \times 10^{-7}$, or 0.0000001 mol/L). The pH scale ranges from 0 to 14 (**Figure 2–11**).

Pure water has a pH of 7, but as **Figure 2–11** indicates, solutions display a wide range of pH values, depending on the nature of the solutes involved:

- A solution with a pH of 7 is said to be **neutral**, because it contains equal numbers of hydrogen and hydroxide ions.
- A solution with a pH below 7 is **acidic** (a-SI-dik), meaning that it contains more hydrogen ions than hydroxide ions.
- A solution with a pH above 7 is **basic**, or *alkaline* (AL-kuh-lin), meaning that it has more hydroxide ions than hydrogen ions.

The normal pH of blood ranges from 7.35 to 7.45. Abnormal fluctuations in pH can damage cells and tissues by breaking chemical bonds, changing the shapes of proteins, and altering cellular functions. *Acidosis* is an abnormal physiological state caused by low blood pH (below 7.35). A pH below 7 can produce coma. Likewise, *alkalosis* results from an abnormally high pH (above 7.45). A blood pH above 7.8 generally causes uncontrollable and sustained skeletal muscle contractions.

Figure 2–11 The pH Scale Indicates Hydrogen Ion Concentration. The pH scale is logarithmic; an increase or decrease of one unit corresponds to a tenfold change in H^+ concentration.

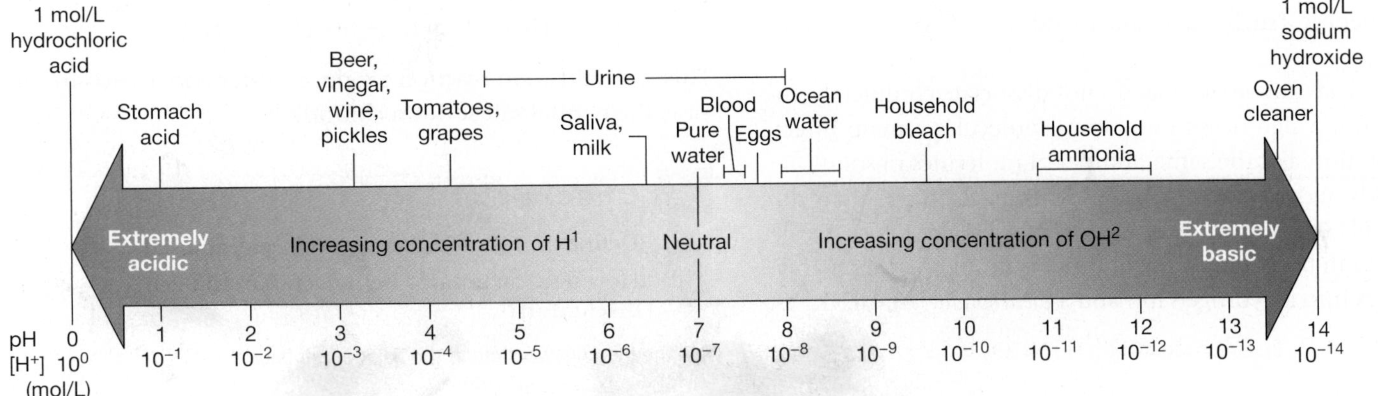
1 mol/L hydrochloric acid
Stomach acid
Beer, vinegar, wine, pickles
Tomatoes, grapes
Urine
Saliva, milk
Blood
Pure water
Eggs
Ocean water
Household bleach
Household ammonia
Oven cleaner
1 mol/L sodium hydroxide
Extremely acidic
Increasing concentration of H1
Neutral
Increasing concentration of OH2
Extremely basic
pH 0 1 2 3 4 5 6 7 8 9 10 11 12 13 14
[H+] 10^0 10^-1 10^-2 10^-3 10^-4 10^-5 10^-6 10^-7 10^-8 10^-9 10^-10 10^-11 10^-12 10^-13 10^-14
(mol/L)

2

Checkpoint

15. **Define pH, and explain how the pH scale relates to acidity and alkalinity.**
16. **What is the significance of pH in physiological systems?**

See the blue Answers tab at the back of the book.

2-8 Acids, bases, and salts have important physiological roles

Learning Outcome Describe the physiological roles of acids, bases, and salts and the role of buffers in body fluids.

Acids and Bases

The body contains *acids* and *bases* that may cause acidosis or alkalosis, respectively. An **acid** (*acere*, sour) is any solute that dissociates in solution and releases hydrogen ions, lowering the pH. A hydrogen atom that loses its electron consists solely of a proton, so we often refer to hydrogen ions simply as protons, and to acids as *proton donors*. In contrast, a **base** is a solute that removes hydrogen ions from a solution, raising the pH. It acts as a *proton acceptor*. Acids and bases are often identified in terms of *strength*. Strength is a measure of the degree of dissociation of either an acid or a base when it is in solution. There are both strong and weak acids and bases.

A *strong acid* dissociates completely in solution, and the reaction occurs essentially in one direction only. *Hydrochloric acid* (HCl) is a representative strong acid. In water, it ionizes as follows:

$$HCl \longrightarrow H^+ + Cl^-$$

The stomach produces this powerful acid to help break down food. Hardware stores sell HCl under the name muriatic acid, for cleaning concrete and swimming pools.

In solution, many bases release a hydroxide ion (OH^-). Hydroxide ions have an attraction for hydrogen ions and react quickly with them to form water molecules. A *strong base* dissociates completely in solution. *Sodium hydroxide*, NaOH, is a strong base. In solution, it releases sodium ions and hydroxide ions:

$$NaOH \longrightarrow Na^+ + OH^-$$

Strong bases have a variety of industrial and household uses. Drain openers (such as Drāno) and lye are two familiar examples.

Weak acids and *weak bases* do not dissociate completely. At equilibrium, a significant number of molecules remain intact in the solution. For the same number of molecules in solution, weak acids and weak bases have less impact on pH than do strong acids and strong bases. *Carbonic acid* (H_2CO_3) is a weak acid found in body fluids. In solution, carbonic acid reversibly dissociates into a hydrogen ion and a *bicarbonate ion*, HCO_3^-:

$$H_2CO_3 \rightleftharpoons H^+ + HCO_3^-$$

Salts

A **salt** is an ionic compound containing any cation except a hydrogen ion, and any anion except a hydroxide ion. Because they are held together by ionic bonds, many salts dissociate completely in water, releasing cations and anions. For example, sodium chloride (NaCl) dissociates immediately in water, releasing Na^+ and Cl^-. Sodium and chloride ions are the most abundant ions in body fluids. However, many other ions are present in lesser amounts as a result of the dissociation of other ionic compounds. Ionic concentrations in the body are regulated in ways we describe in Chapters 26 and 27.

The dissociation of sodium chloride does not affect the local concentrations of hydrogen ions or hydroxide ions. For this reason, NaCl, like many salts, is a "neutral" solute. It does not make a solution more acidic or more basic. In contrast, some salts may interact with water molecules and indirectly affect the concentrations of H^+ and OH^- ions. Thus, in some cases, the dissociation of salts makes a solution slightly acidic or slightly basic.

Buffers and pH Control

Buffers are compounds that stabilize the pH of a solution by removing or replacing hydrogen ions. *Buffer systems* usually involve a weak acid and its related salt, which functions as a weak base. For example, the carbonic acid–bicarbonate buffer system (detailed in Chapter 27) consists of carbonic acid and sodium bicarbonate, $NaHCO_3$, otherwise known as baking soda. Buffers and buffer systems in body fluids help maintain the pH within normal limits. The pH of several body fluids is included in **Figure 2–11**.

The use of antacids is one example of the type of reaction that takes place in buffer systems. Antacids use sodium bicarbonate to neutralize excess hydrochloric acid in the stomach.

Note that the effects of neutralization are most evident when you add a strong acid to a strong base. For example, by adding hydrochloric acid to sodium hydroxide, you neutralize both the strong acid and the strong base:

$$HCl + NaOH \longrightarrow H_2O + NaCl$$

This neutralization reaction produces water and a salt—in this case, the neutral salt sodium chloride.

Checkpoint

17. **Define the following terms: *acid*, *base*, and salt.**
18. **How does an antacid help decrease stomach discomfort?**

See the blue Answers tab at the back of the book.

2-9 Living things contain organic compounds made up of monomers, polymers, and functional groups

Learning Outcome: **Describe monomers and polymers, and the importance of functional groups in organic compounds.**

The macromolecules of living things are all organic compounds. Many of these large organic molecules are made up of long chains of carbon atoms linked by covalent bonds. The carbon atoms typically form additional covalent bonds with hydrogen or oxygen atoms and, less commonly, with nitrogen, phosphorus, sulfur, iron, or other elements.

The *macromolecules* of life are complex structures with varied functions and properties. They include carbohydrates, lipids, proteins, and nucleic acids. Each macromolecule is made up of monomer subunits. A **monomer** (MON-ō-mer; *mono-* single + *-mer,* member of a group) is a molecule that can be bonded to other identical molecules to form a **polymer**. Repeating monomers join together through *dehydration synthesis* reactions to form polymers, sometimes called *mers.*

The monomers of carbohydrates, lipids, proteins, and nucleic acids are separated, or released, through *hydrolysis reactions.* As a result of such reactions, carbohydrates release monosaccharides, lipids (fats) release fatty acids and glycerol, proteins release amino acids, and nucleic acids release nucleotides. We discuss the structures and functions of these molecules in the next four sections.

Organic compounds are diverse, but certain groupings of atoms, known as *functional groups,* are responsible for the characteristic reactions of a particular compound. Furthermore, these functional groups greatly influence the properties of any molecule of which they are a part. **Table 2–3** details the functional groups you will study in this chapter.

✓ Checkpoint

19. **What macromolecules are important to living things?**
20. **Which functional group acts as an acid?**

See the blue Answers tab at the back of the book.

2-10 Carbohydrates contain carbon, hydrogen, and oxygen in a 1:2:1 ratio

Learning Outcome **Discuss the structures and functions of carbohydrates.**

A **carbohydrate** is an organic molecule that contains carbon, hydrogen, and oxygen in a ratio near 1:2:1. Familiar carbohydrates include the sugars and starches that make up about half of the typical U.S. diet. Carbohydrates typically account for less than 1 percent of total body weight. Carbohydrates are most important as energy sources that are catabolized. In the following sections, we focus on *monosaccharides, disaccharides,* and *polysaccharides.*

Monosaccharides

A **monosaccharide** (mon-ō-SAK-uh-rīd; *mono-,* single + *sakcharon,* sugar), or *simple sugar,* is a carbohydrate with three to seven carbon atoms. Depending on how many carbons it contains, a monosaccharide can be called a *triose* (three carbon atoms), *tetrose* (four carbon atoms), *pentose* (five carbon atoms), *hexose* (six carbon atoms), or *heptose* (seven carbon atoms). The *-ose* ending indicates a sugar. The hexose **glucose** (GLŪ-kōs), $C_6H_{12}O_6$, is the most important metabolic "fuel" in the body. Monosaccharides such as glucose dissolve readily in water and are rapidly distributed throughout the body by blood and other body fluids.

Table 2–3 Important Functional Groups of Organic Compounds

Functional Group	Structural Formula*	Importance	Examples
Amino group $—NH_2$	R—N(—H)—H	Acts as a base, accepting H^+, depending on pH; can form bonds with other molecules	Amino acids
Carboxyl group —COOH	R—C(—OH)=O	Acts as an acid, releasing H^+ to become $R—COO^-$	Fatty acids, amino acids
Hydroxyl group —OH	R—O—H	May link molecules through dehydration synthesis (condensation); hydrogen bonding between hydroxyl groups and water molecules; affects solubility	Carbohydrates, fatty acids, amino acids
Phosphate group $—PO_4$	R—O—P(=O)(—O^-)—O^-	May link other molecules to form larger structures; may store energy in high-energy bonds	Phospholipids, nucleic acids, high-energy compounds

*A structural formula shows the covalent bonds within a molecule or functional group (see Figure 2–6). The letter R represents the term *R group* and is used to denote the rest of the molecule to which a functional group is attached.

Figure 2–12 The Structures of Glucose. Note that the ring form, the most common form of glucose, is represented with a kind of shorthand in later figures: We leave out the carbon atom at five corners of the hexagon, although we do show the oxygen atom at the remaining corner.

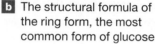

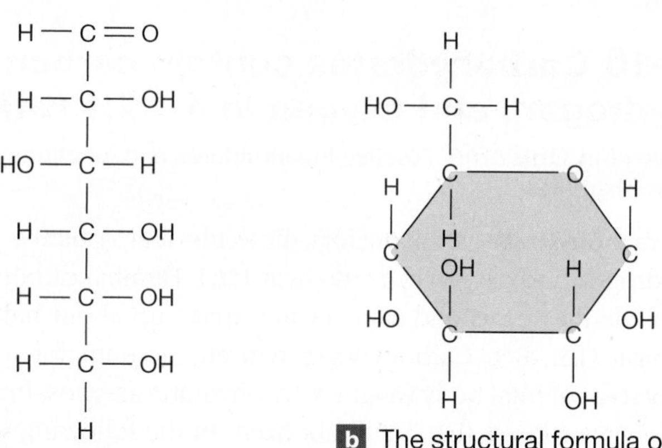
H—C=O
H—C—OH
HO—C—H
H—C—OH
H—C—OH
H—C—OH
H
HO—C—H
C
O
H
OH
HO
OH

a The structural formula of the straight-chain form

b The structural formula of the ring form, the most common form of glucose

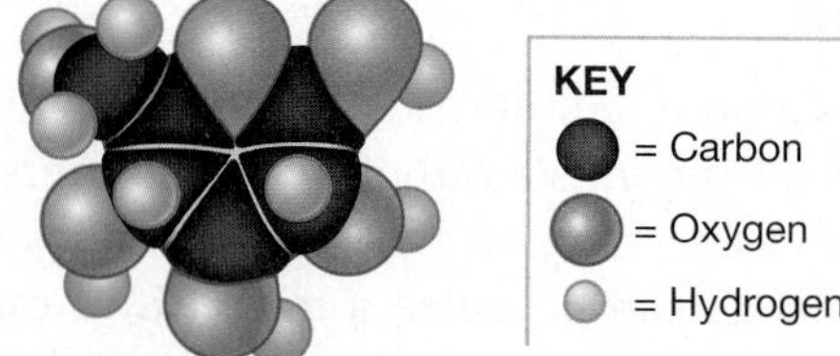
KEY
= Carbon
= Oxygen
= Hydrogen

c A three-dimensional model showing the organization of atoms in the ring form

? How many oxygen atoms are shown in each glucose structure?

As shown by two-dimensional diagrams known as *structural formulas*, the atoms in a glucose molecule may form either a straight chain (**Figure 2–12a**) or a ring (**Figure 2–12b**). In the body, the ring form is more common. A three-dimensional model shows the arrangement of atoms in the ring form most accurately (**Figure 2–12c**).

The three-dimensional structure of an organic molecule is an important characteristic, because it usually determines the molecule's fate or function. Some molecules have the same molecular formula—in other words, the same types and numbers of atoms—but different structures. Such molecules are called **isomers**. The body usually treats different isomers as distinct molecules. For example, the monosaccharides glucose and fructose are isomers. *Fructose* is a hexose found in many fruits and in secretions of the male reproductive tract. It has the same chemical formula as glucose, $C_6H_{12}O_6$, but the arrangement of its atoms differs from that of glucose. As a result, separate enzymes and reaction sequences control its breakdown and synthesis.

Disaccharides and Polysaccharides

Carbohydrates other than simple sugars are complex molecules composed of monosaccharide building blocks, or monomers. Two monosaccharide monomers joined together form a **disaccharide** (dī-SAK-uh-rīd; *di-*, two). Disaccharides such as *sucrose* (table sugar) have a sweet taste and, like monosaccharides, are quite soluble in water.

The formation of sucrose involves a dehydration synthesis reaction (**Figure 2–13a**). Recall that dehydration synthesis reactions link molecules together by the removal of a water

Figure 2–13 The Formation and Breakdown of Complex Sugars.

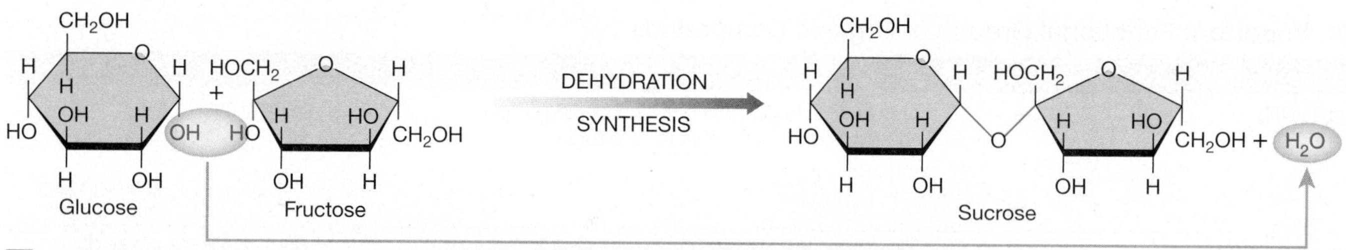
CH_2OH
$HOCH_2$
Glucose
Fructose
DEHYDRATION
SYNTHESIS
Sucrose
CH_2OH + H_2O

a Formation of the disaccharide sucrose through dehydration synthesis. During dehydration synthesis, 2 molecules are joined by the removal of a water molecule.

b Breakdown of sucrose into simple sugars by hydrolysis. Hydrolysis reverses the steps of dehydration synthesis; a complex molecule is broken down by the addition of a water molecule.

molecule. The breakdown of sucrose into simple sugars is an example of hydrolysis, or breakdown by the addition of a water molecule (Figure 2–13b). Hydrolysis is the functional opposite of dehydration synthesis.

Many foods contain disaccharides, but all carbohydrates except monosaccharides must be broken apart through hydrolysis before they can provide useful energy. Most popular junk foods (high in calories but otherwise lacking in nutritional content), such as candies and sodas, are full of monosaccharides (commonly fructose) and disaccharides (generally sucrose). Some people cannot tolerate sugar for medical reasons. Others avoid it in an effort to control their weight (because excess sugars are converted to fat for long-term storage). Many of these people use *artificial sweeteners* in their foods and beverages. These compounds have a very sweet taste, but they either cannot be broken down in the body or are used in insignificant amounts.

More complex carbohydrates result when repeated dehydration synthesis reactions add additional monosaccharides or disaccharides. These large molecules (polymers) are called **polysaccharides** (pol-ē-SAK-uh-rīdz; *poly-*, many). Polysaccharide chains can be straight or highly branched. *Cellulose*, a structural component of many plants, is a polysaccharide that our bodies cannot digest because the particular linkages between the glucose molecules cannot be cleaved by enzymes in the body. Foods such as celery, which contains cellulose, water, and little else, contribute fiber to digestive wastes but do not provide a source of energy.

Starches are large polysaccharides formed from glucose molecules. Most starches are manufactured by plants. Your digestive tract can break these molecules into monosaccharides. Starches such as those in potatoes and grains are a major dietary energy source.

The polysaccharide **glycogen** (GLĪ-kō-jen), or *animal starch*, has many side branches consisting of chains of glucose molecules (Figure 2–14). Like most other starches, glycogen does not dissolve in water or other body fluids. Muscle cells make and store glycogen. When muscle cells have a high demand for glucose, glycogen molecules are broken down. When the need is low, these cells absorb glucose from the bloodstream and rebuild glycogen reserves. Table 2–4 summarizes information about carbohydrates.

Figure 2–14 The Structure of the Polysaccharide Glycogen. Liver and muscle cells store glucose as the polysaccharide glycogen, a long, branching chain of glucose molecules.

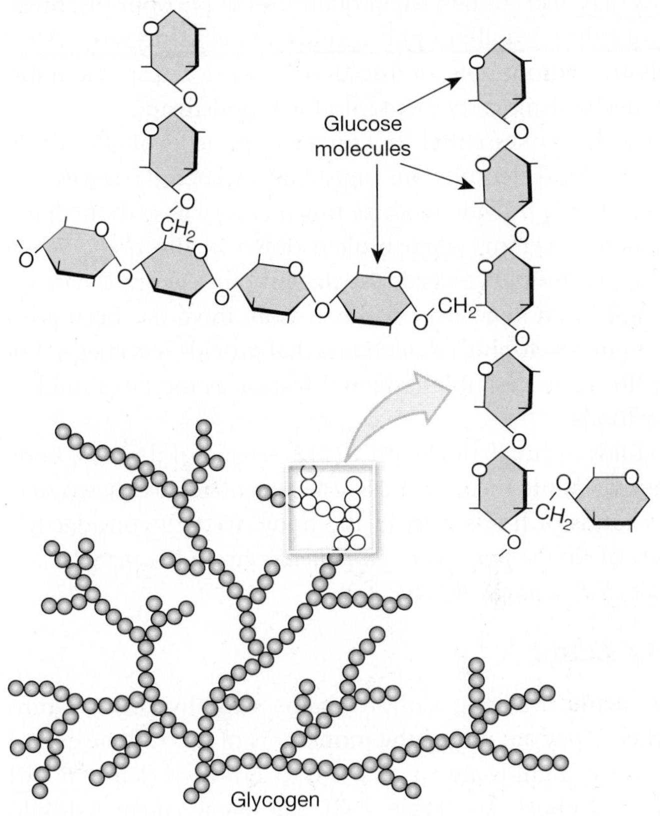
Glucose molecules
O
CH_2
Glycogen

2

✓ Checkpoint

21. **Plant starch and glycogen are both polysaccharides. What monomer do they have in common?**

See the blue Answers tab at the back of the book.

2-11 Lipids often contain a carbon-to-hydrogen ratio of 1:2

Learning Outcome Discuss the structures and functions of lipids.

Like carbohydrates, **lipids** (*lipos*, fat) contain carbon, hydrogen, and oxygen, and the carbon-to-hydrogen ratio is near 1:2. However, lipids contain much less oxygen than do carbohydrates with the same number of carbon atoms.

Table 2–4 Carbohydrates in the Body

Structural Class	Examples	Primary Function	Remarks
Monosaccharides (simple sugars)	Glucose, fructose	Energy source	Manufactured in the body and obtained from food; distributed in body fluids
Disaccharides	Sucrose, lactose, maltose	Energy source	Sucrose is table sugar, lactose is in milk, and maltose is malt sugar found in germinating grain; all must be broken down to monosaccharides before absorption
Polysaccharides	Glycogen	Storage of glucose	Glycogen is in animal cells; other starches and cellulose are within or around plant cells

The hydrogen-to-oxygen ratio is therefore very large. For example, a representative lipid, such as lauric acid (found in coconut, laurel, and palm kernel oils), has a formula of $C_{12}H_{24}O_2$. Lipids may also contain small quantities of phosphorus, nitrogen, or sulfur. Familiar lipids include *fats, oils,* and *waxes*. Most lipids are hydrophobic, or insoluble in water, but special transport mechanisms carry them into the bloodstream.

Lipids form essential structural components of all cells. In addition, lipid deposits are important as energy reserves. On average, lipids provide twice as much energy as carbohydrates do, gram for gram, when broken down in the body. When the supply of lipids exceeds the demand for energy, the excess is stored in fat deposits. For this reason, there has been great interest in developing *fat substitutes* that provide less energy, but have the same desirable taste and texture as the fats found in many foods.

Lipids normally make up 12–18 percent of the total body weight of adult men, and 18–24 percent for adult women. Many kinds of lipids exist in the body. We will consider five classes of lipids: *fatty acids, eicosanoids, glycerides, steroids,* and *phospholipids and glycolipids*.

Fatty Acids

Fatty acids are long carbon chains with hydrogen atoms attached. They are one of the monomers of lipids. One end of the carbon chain is always attached to a *carboxyl* (kar-BOK-sil) *group*, —COOH (see **Table 2–3**). The name *carboxyl* should help you remember that a carbon and a hydroxyl (—OH) group are the important structural features of fatty acids. The carbon chain attached to the carboxyl group is known as the *hydrocarbon tail* of the fatty acid. **Figure 2–15a** shows a representative fatty acid, *lauric acid*.

Fatty acids have a very limited solubility in water. When a fatty acid is in solution, only the carboxyl end associates with water molecules, because that is the only hydrophilic portion of the molecule. The hydrocarbon tail is hydrophobic. In general, the longer the hydrocarbon tail, the lower the solubility of the molecule.

Fatty acids may be either saturated or unsaturated (**Figure 2–15b**). These terms refer to the number of hydrogen atoms bound to the carbon atoms in the hydrocarbon tail. In a *saturated* fatty acid, each carbon atom in the tail has four single covalent bonds (see **Figure 2–15a**). Within the tail, two of those bonds bind adjacent carbon atoms, and the other two bind hydrogen atoms. The carbon atom at the end of the tail binds three hydrogen atoms. In an *unsaturated* fatty acid, one or more of the single covalent bonds between the carbon atoms have been replaced by a double covalent bond. As a result, the carbon atoms involved will each bind only one hydrogen atom rather than two. This changes the shape of the hydrocarbon tail, giving it a sharp bend, as you can see in **Figure 2–15b**. The change also affects the way the fatty acid is metabolized.

Figure 2–15 Fatty Acids.

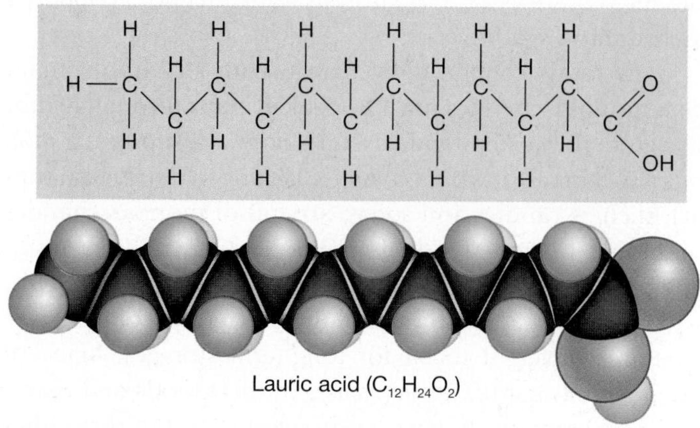
Lauric acid ($C_{12}H_{24}O_2$)

a Lauric acid shows two structural characteristics common to all fatty acids: a long chain of carbon atoms and a carboxyl group (—COOH) at one end.

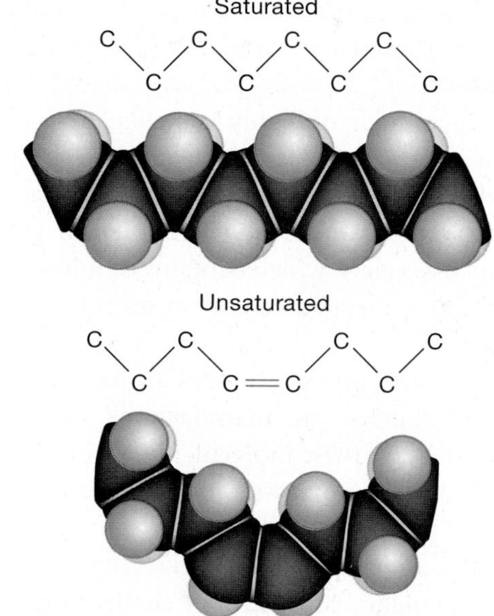
Saturated
Unsaturated

b A fatty acid is either saturated (has single covalent bonds only) or unsaturated (has one or more double covalent bonds). The presence of a double bond causes a sharp bend in the molecule.

What type of bond does an unsaturated fatty acid contain that a saturated fatty acid does not?

A *monounsaturated* fatty acid has a single double bond in the hydrocarbon tail. A *polyunsaturated* fatty acid contains two or more double bonds.

Eicosanoids

Eicosanoids (ī-KŌ-sa-noydz) are lipids derived from *arachidonic* (ah-rak-i-DON-ik) *acid*, a fatty acid that must be absorbed in the diet because the body cannot synthesize it. The two

Figure 2–16 Prostaglandins. Prostaglandins contain 20 carbon atoms and a 5-carbon ring.

major classes of eicosanoids are *leukotrienes* and *prostaglandins*. **Leukotrienes** (lū-kō-TRĪ-ēnz) are produced mostly by cells involved with coordinating the responses to injury or disease. We consider leukotrienes in Chapters 18 and 22.

Prostaglandins (pros-tuh-GLAN-dinz) are short-chain fatty acids in which five of the carbon atoms are joined in a ring Figure 2–16). These compounds are released by cells to coordinate or direct local cellular activities, and they are extremely powerful even in small quantities. Virtually all tissues synthesize and respond to them. The effects of prostaglandins vary with their structure and their release site. Prostaglandins released by damaged tissues, for example, stimulate nerve endings and produce the sensation of pain (Chapter 15). Those released in the uterus help trigger the start of labor contractions (Chapter 29).

The body uses several types of chemical messengers. Those that are produced in one part of the body and have effects on distant parts are called *hormones*. Hormones are distributed throughout the body in the bloodstream, but most prostaglandins affect only the area in which they are produced. As a result, prostaglandins are often called *local hormones*. The distinction is not a rigid one, however, as some prostaglandins also enter the bloodstream and affect other areas. We discuss hormones and prostaglandins in Chapter 18.

Glycerides

Unlike monosaccharides, individual fatty acids cannot be strung together in a chain by dehydration synthesis to form a polymer. But they can be attached to a modified simple sugar, **glycerol** (GLIS-er-ol), through a similar reaction. The result is a lipid known as a **glyceride** (GLIS-er-īd). Dehydration synthesis reactions can produce a **monoglyceride** (mon-ō-GLIS-er-īd), consisting of glycerol plus one fatty acid. Subsequent reactions can yield a **diglyceride** (glycerol + two fatty acids) and then a **triglyceride** (glycerol + three fatty acids), as in Figure 2–17. Hydrolysis breaks the glycerides into fatty acids and glycerol (monomers). Comparing Figure 2–17 with Figure 2–13 shows that dehydration synthesis and hydrolysis operate the same way, whether the molecules involved are carbohydrates or lipids.

Triglycerides, also known as *triacylglycerols* or *neutral fats*, are important lipid polymers. They have the following important functions:

- *Energy Source.* Fat deposits in the body represent a significant energy reserve. In times of need, the triglycerides are taken apart by hydrolysis, yielding fatty acids that can be broken down to provide energy.

+ Clinical Note Too Sweet on Sugar?

A baby's first and favorite taste is sweet: mother's milk is rich in *lactose* (milk sugar), a disaccharide of glucose and galactose. This preference for sweet persists throughout life. It is easier to tempt the poor appetite of a frail, elderly person with a bowl of pudding than with a bowl of steamed kale. Manufacturers of processed foods know this.

When heart disease became endemic in the United States, holding first place as the killer of Americans, the medical community advocated a low-fat diet for heart health. Artery-clogging fats were removed from manufactured foods—such as cookies, soups, and other boxed, bagged, and frozen products—and replaced with sugar for flavor and mouth appeal.

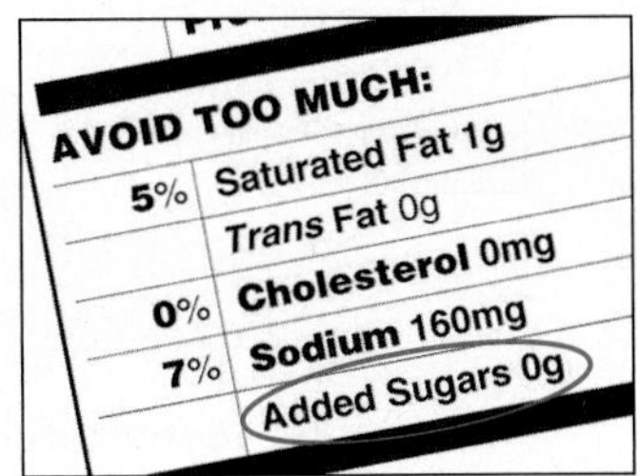
AVOID TOO MUCH:
5% Saturated Fat 1g
Trans Fat 0g
0% Cholesterol 0mg
7% Sodium 160mg
Added Sugars 0g

The sweetening of the American diet has wreaked a new kind of havoc on American health. Dental hygienists see more *dental caries* (cavities). Obesity is climbing at an alarming rate. Grade-school children are developing more *type 2 diabetes*, formerly called "adult-onset diabetes." These serious and potentially fatal diseases generally did not appear until a person had lived several decades with a poor lifestyle.

Glucose is a necessary nutrient. Our body's cells depend on it for fuel; our neurons (brain cells) require it. However, we should meet our glucose needs through complex carbohydrates, or polysaccharides (such as glycogen). In contrast to simple sugars, complex carbohydrates are digested slowly by decomposition reactions in the digestive tract. The component monosaccharides of glucose are released and absorbed gradually, maintaining a steady blood glucose level. In turn, the pancreas is signaled only as needed to make the protein hormone *insulin*, which stimulates the transport of glucose into the body's cells. Complex carbohydrates promote satiety (fullness) and support healthy sugar metabolism.

Figure 2–17 Triglyceride Formation. The formation of a triglyceride involves the attachment of fatty acids to a glycerol molecule through dehydration synthesis. In this example, a triglyceride is formed by the attachment of one unsaturated and two saturated fatty acids to a glycerol molecule.

Glycerol
Fatty acids
Fatty acid 1 Saturated
Fatty acid 2 Saturated
Fatty acid 3 Unsaturated
DEHYDRATION SYNTHESIS
HYDROLYSIS
+ H_2O
+ H_2O
+ H_2O
Triglyceride

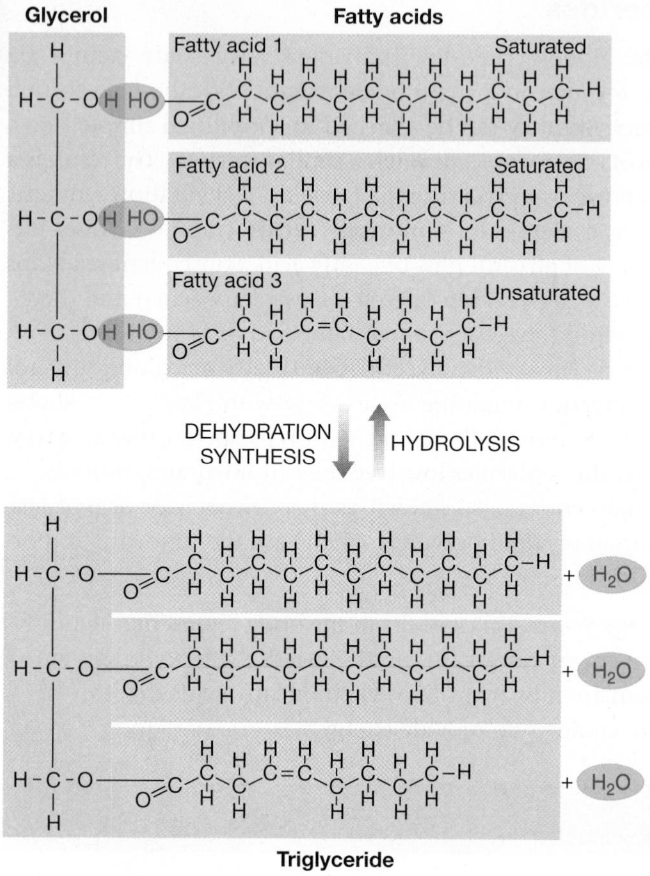

? What makes fatty acid 3 an unsaturated fatty acid?

- *Insulation*. Fat deposits under the skin serve as insulation, slowing heat loss to the environment. Heat loss across a layer of lipids is only about one-third of the heat loss through other tissues.
- *Protection*. A fat deposit around a delicate organ such as a kidney provides a cushion that protects against bumps or jolts.

Triglycerides are stored in the body as lipid droplets within cells. The droplets absorb and accumulate lipid-soluble vitamins, drugs, or toxins that appear in body fluids. This accumulation has both positive and negative effects. For example, the body's lipid reserves retain both valuable lipid-soluble vitamins (A, D, E, K) and potentially dangerous lipid-soluble pesticides, such as the now-banned DDT.

Steroids

Steroids are large lipid molecules that share a distinctive four-ring carbon structure (Figure 2–18). They differ in the functional groups that are attached to this basic framework. The steroid **cholesterol** (kō-LES-ter-ol; *chole-*, bile + *stereos*, solid) and related steroids are important for several reasons:

- The outer boundary of all animal cells, called a plasma membrane, contains cholesterol (Figure 2–18a). Cells need cholesterol to maintain their plasma membranes, as well as for cell growth and division.
- Steroid hormones are involved in the regulation of sexual function. Examples include the sex hormones *estrogen* and *testosterone* (Figure 2–18b,c).
- Steroid hormones are important in the regulation of tissue metabolism and mineral balance. Examples include *corticosteroids* from the adrenal cortex, which play a role in carbohydrate and protein metabolism, and *calcitriol* from the kidneys, a hormone important in the regulation of the body's calcium ion concentrations.
- Steroid derivatives called *bile salts* are required for the normal processing of dietary fats. The liver produces bile salts and secretes them in bile. They interact with lipids in the intestinal tract and assist with the digestion and absorption of lipids.

The body obtains cholesterol in two ways: (1) by absorbing it from animal products in the diet and (2) by synthesizing it. Liver,

Figure 2–18 Steroids Have a Complex Four-Ring Structure. Individual steroids differ in the side chains attached to the carbon rings.

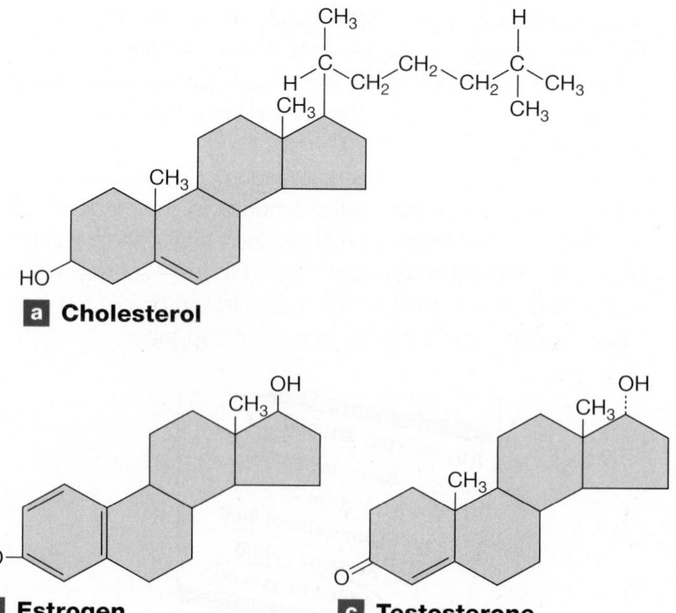

a Cholesterol
b Estrogen
c Testosterone

meat, shellfish, and egg yolks are especially rich dietary sources of cholesterol. People with *hypercholesterolemia*, a condition characterized by very high levels of blood cholesterol, have an increased risk of developing a form of heart disease called coronary artery disease (CAD). In CAD, excess cholesterol deposits on arterial walls, forming plaques that obstruct blood flow to the heart. Currently, it is suggested that a healthy person consume no more than 300 mg of cholesterol per day, and others with diabetes, high cholesterol, or heart disease should consume no more than 200 mg per day. Unfortunately, the blood cholesterol level can be difficult to control by dietary restriction alone because the body can synthesize cholesterol as well. In fact, the body makes more than enough, so strict vegetarians do not need to eat animal products to ensure adequate amounts of cholesterol.

Phospholipids and Glycolipids

Phospholipids (FOS-fō-lip-idz) and **glycolipids** (GLĪ-kō-lip-idz) are structurally related, and our cells can synthesize both types of lipids, primarily from fatty acids. In a *phospho*lipid, a *phosphate group* (PO_4^{3-}) links a diglyceride to a nonlipid group (**Figure 2–19a**). There are different types of phospholipids; the one shown in this figure is lecithin. In a *glyco*lipid, a carbohydrate is attached to a diglyceride (**Figure 2–19b**). Note that placing *-lipid* last in these names indicates that the molecule consists primarily of lipid.

The long hydrocarbon tails of phospholipids and glycolipids are hydrophobic, but the opposite ends, the nonlipid *heads*, are hydrophilic. In water, large numbers of these molecules tend to form droplets, or *micelles* (mī-SELZ), with the hydrophilic portions on the outside (**Figure 2–19c**). Most meals contain a mixture of lipids and other organic molecules, and micelles form as the food breaks down in your digestive tract. In addition to phospholipids and glycolipids, micelles may contain other insoluble lipids, such as steroids, glycerides, and long-chain fatty acids.

Phospholipids and glycolipids (as well as cholesterol) are called *structural lipids*, because they help form and maintain intracellular structures called membranes. At the cellular level, *membranes* are sheets or layers composed mainly of lipids. For example, the plasma membrane surrounding each cell is composed primarily of phospholipids. It separates the aqueous solution inside the cell from the aqueous solution outside the cell. Also, various internal membranes subdivide the interior of the cell into specialized compartments, each with a distinctive chemical nature and, as a result, a different function.

The five types of lipids and their characteristics are summarized in **Table 2–5**.

Checkpoint

22. **Describe lipids.**
23. **Which lipids would you find in human plasma membranes?**

See the blue Answers tab at the back of the book.

2-12 Proteins contain carbon, hydrogen, oxygen, and nitrogen and are formed from amino acids

Learning Outcome **Discuss the structures and functions of proteins.**

Proteins are the most abundant organic molecules in the human body and in many ways the most important. The human body contains many different proteins, and they account for about 20 percent of total body weight. All proteins contain carbon, hydrogen, oxygen, and nitrogen. Smaller quantities of sulfur and phosphorus may also be present. *Amino acids* are simple organic compounds (monomers) that combine to form proteins (polymers). p. 45

Proteins carry out a variety of essential functions, which we can group into the following major categories:

- *Support. Structural proteins* create a three-dimensional framework for the body. They provide strength, organization, and support for cells, tissues, and organs.

Table 2–5 Representative Lipids and Their Functions in the Body

Lipid Type	Example(s)	Primary Functions	Remarks
Fatty acids	Lauric acid	Energy source	Absorbed from food or synthesized in cells; transported in the blood
Eicosanoids	Prostaglandins, leukotrienes	Chemical messengers coordinating local cellular activities	Prostaglandins are produced in most body tissues
Glycerides	Monoglycerides, diglycerides, triglycerides	Energy source, energy storage, insulation, and physical protection	Stored in fat deposits; must be broken down to fatty acids and glycerol before they can be used as an energy source
Steroids	Cholesterol	Structural component of plasma membranes, hormones, digestive secretions in bile	All have the same four-carbon ring framework
Phospholipids, glycolipids	Lecithin (a phospholipid)	Structural components of plasma membranes	Derived from fatty acids and nonlipid components

Figure 2–19 Phospholipids and Glycolipids.

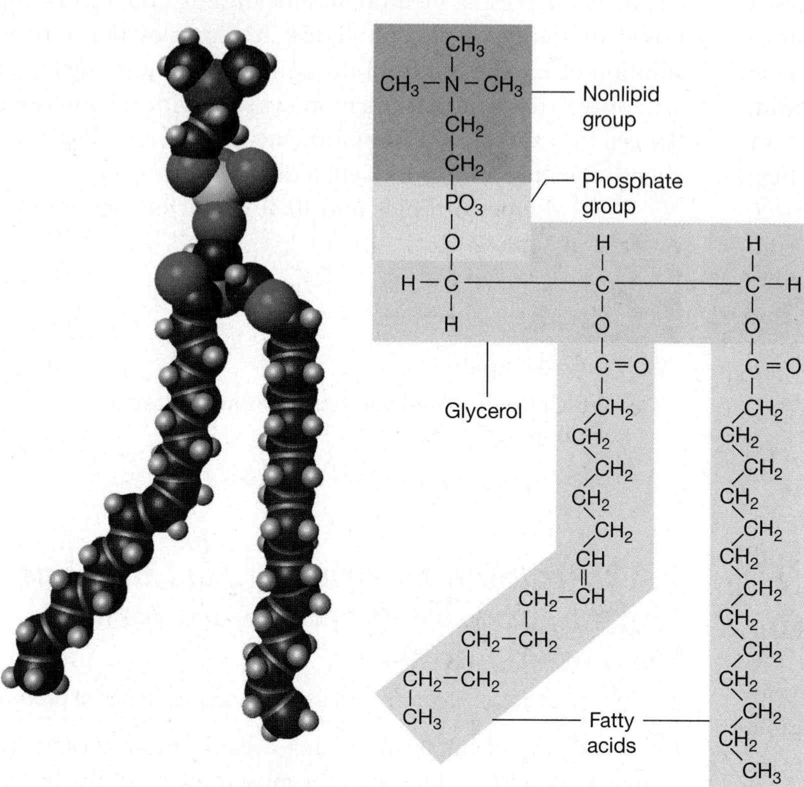
CH_3
CH_3-N-CH_3
CH_2
CH_2
PO_3
Nonlipid group
Phosphate group
Glycerol
Fatty acids

a The phospholipid *lecithin*. In a phospholipid, a phosphate group links a nonlipid molecule to a diglyceride.

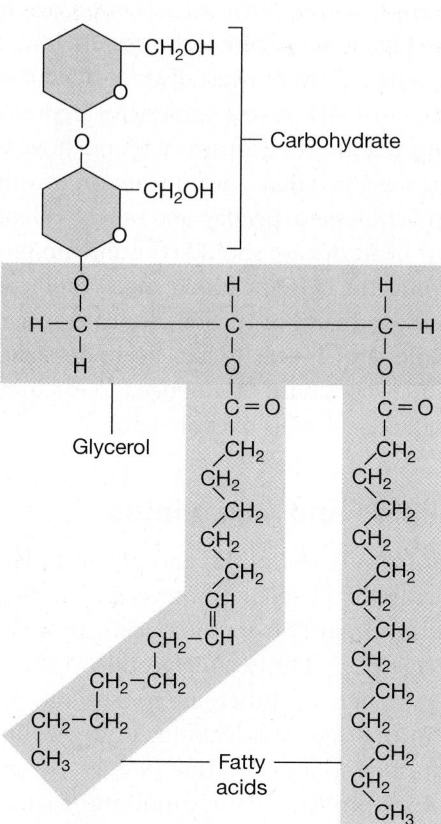
CH_2OH
CH_2OH
Carbohydrate
Glycerol
Fatty acids

b In a glycolipid, a carbohydrate is attached to a diglyceride.

c When large numbers of phospholipids and glycolipids are in water, they form micelles, with the hydrophilic heads facing the water molecules, and the hydrophobic tails on the inside of each droplet.

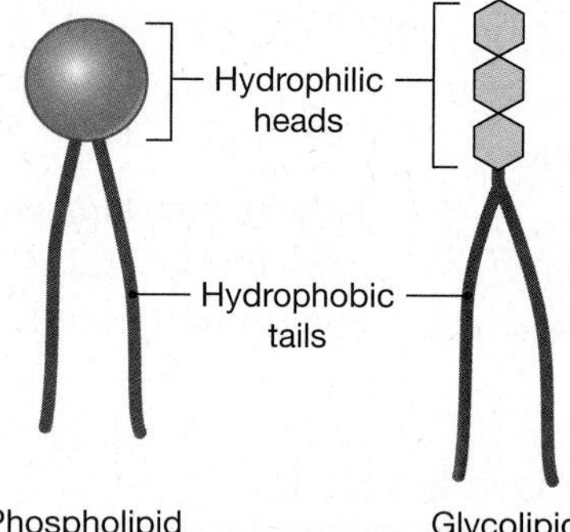
Hydrophilic heads
Hydrophobic tails
Phospholipid
Glycolipid

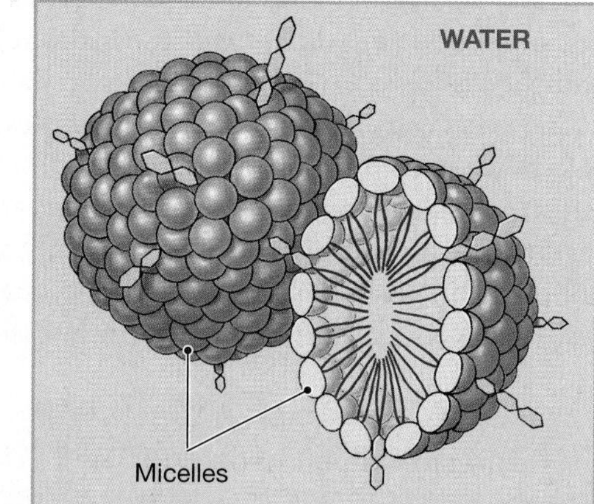
WATER
Micelles

- *Movement. Contractile proteins* bring about muscular contraction. Related proteins are responsible for the movement of individual cells.
- *Transport.* Special *transport proteins* bind many substances for transport in the blood, including insoluble lipids, respiratory gases, special minerals such as iron, and several hormones. These substances would not otherwise be transported in the blood. Other specialized proteins move materials from one part of a cell to another.
- *Buffering.* Proteins provide a buffering action and in this way help prevent dangerous changes in the pH of body fluids.

- *Metabolic Regulation*. Many proteins are enzymes, which as you may recall speed up chemical reactions in cells. The sensitivity of enzymes to environmental factors such as temperature and pH is extremely important in controlling the pace and direction of metabolic reactions.
- *Coordination and Control*. Protein hormones can influence the metabolic activities of every cell in the body or affect the function of specific organs or organ systems.
- *Defense*. Proteins defend the body in many ways. The tough, waterproof proteins of the skin, hair, and nails protect the body from environmental hazards. Proteins called *antibodies* help protect us from disease by taking part in the *immune response*. Special *clotting proteins* restrict bleeding after an injury.

Protein Structure

Proteins are organic polymers that consist of long chains of similar organic molecules called **amino acids**. Twenty different amino acid monomers occur in significant quantities in the body. All 20 amino acids are small, water-soluble molecules. A typical protein contains 1000 amino acids, while the largest protein complexes have 100,000 or more.

Each amino acid consists of five parts (Figure 2–20):

- a central carbon atom,
- a hydrogen atom,
- an *amino group* ($—NH_2$),
- a carboxyl group ($—COOH$),
- an *R group* (a variable *side chain* of one or more atoms that identifies a specific amino acid).

The name *amino acid* refers to the presence of the amino group and the carboxyl group, which all amino acids have in common. At physiological pH levels, the carboxyl group can act as an acid by releasing a hydrogen ion to become a *carboxyl ion* (COO^-). The amino group can act as a base by accepting a hydrogen ion, to become an amino ion ($—NH_3^+$). The result is a molecule that has both positive and negative charges, but a net charge of zero. Such molecules are called *zwitterions,* derived from the German word that means "hybrid."

Figure 2–20 Amino Acids. Each amino acid consists of a central carbon atom to which four different groups are attached: a hydrogen atom, an amino group ($—NH_2$), a carboxyl group ($—COOH$), and a variable side group designated as R.

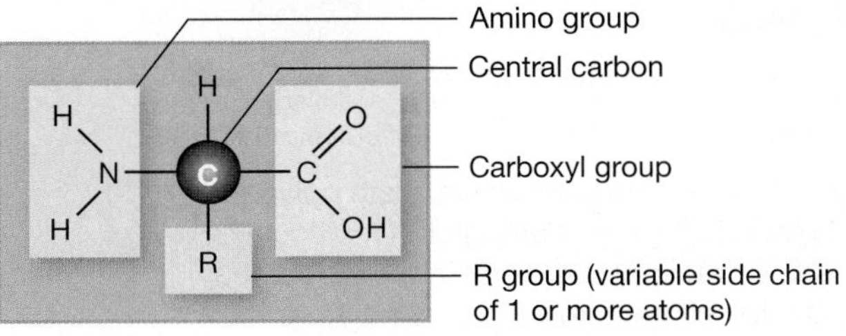
Structure of an Amino Acid
Amino group
Central carbon
Carboxyl group
R group (variable side chain of 1 or more atoms)
H
N
C
O
OH
R

Figure 2–21 The Formation of Peptide Bonds. Peptides form as dehydration synthesis creates a peptide bond between the carboxyl group of one amino acid and the amino group of another. In this example, a peptide bond links the amino acids glycine (for which R = H) and alanine ($R = CH_3$) to form a dipeptide.

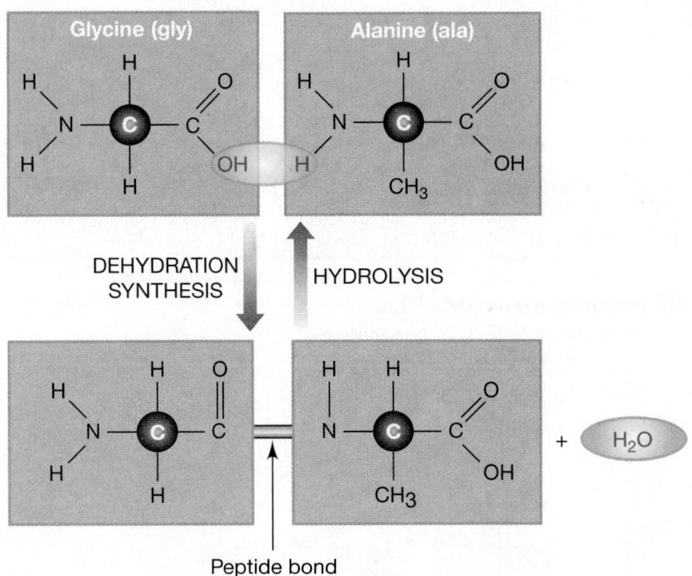
Peptide Bond Formation
Glycine (gly)
Alanine (ala)
DEHYDRATION SYNTHESIS
HYDROLYSIS
Peptide bond
CH3
OH
H2O

A protein begins to form as amino acids are strung together into long chains. Figure 2–21 shows how dehydration synthesis can link two representative amino acids: *glycine* and *alanine.* This reaction creates a covalent bond between the carboxyl group of one amino acid and the amino group of another. Such a bond is known as a **peptide bond**. Molecules consisting of amino acids held together by peptide bonds are called **peptides**. The molecule created in this example is called a *dipeptide,* because it contains two amino acids.

The chain can be lengthened by the addition of more amino acids. Attaching a third amino acid produces a *tripeptide.* Tripeptides and larger peptide chains are called **polypeptides**. Polypeptides with more than 100 amino acids are usually called proteins. Familiar proteins include *hemoglobin* in red blood cells, *collagen* in skin, bones, and muscles, and *keratin* in fingernails and hair.

The different atoms of the R groups distinguish one amino acid from another, giving each its own chemical properties. For example, different R groups are polar, nonpolar, or electrically charged. Amino acids with nonpolar R groups are hydrophobic, whereas amino acids with polar R groups that form hydrogen bonds with water are hydrophilic. Amino acids with electrically charged R groups are strongly hydrophilic.

The properties of the R groups contribute to the overall shape and function of proteins.

Protein Shape

The characteristics of a particular protein are determined in part by the R groups on its amino acids. But the properties of a protein are more than just the sum of the properties of its parts, for polypeptides can have highly complex shapes that are important to their function. Proteins can have four levels of structural complexity (Figure 2–22):

1. **Primary structure** is the sequence of amino acids along the length of a single polypeptide (Figure 2–22a).

Figure 2–22 Protein Structure.

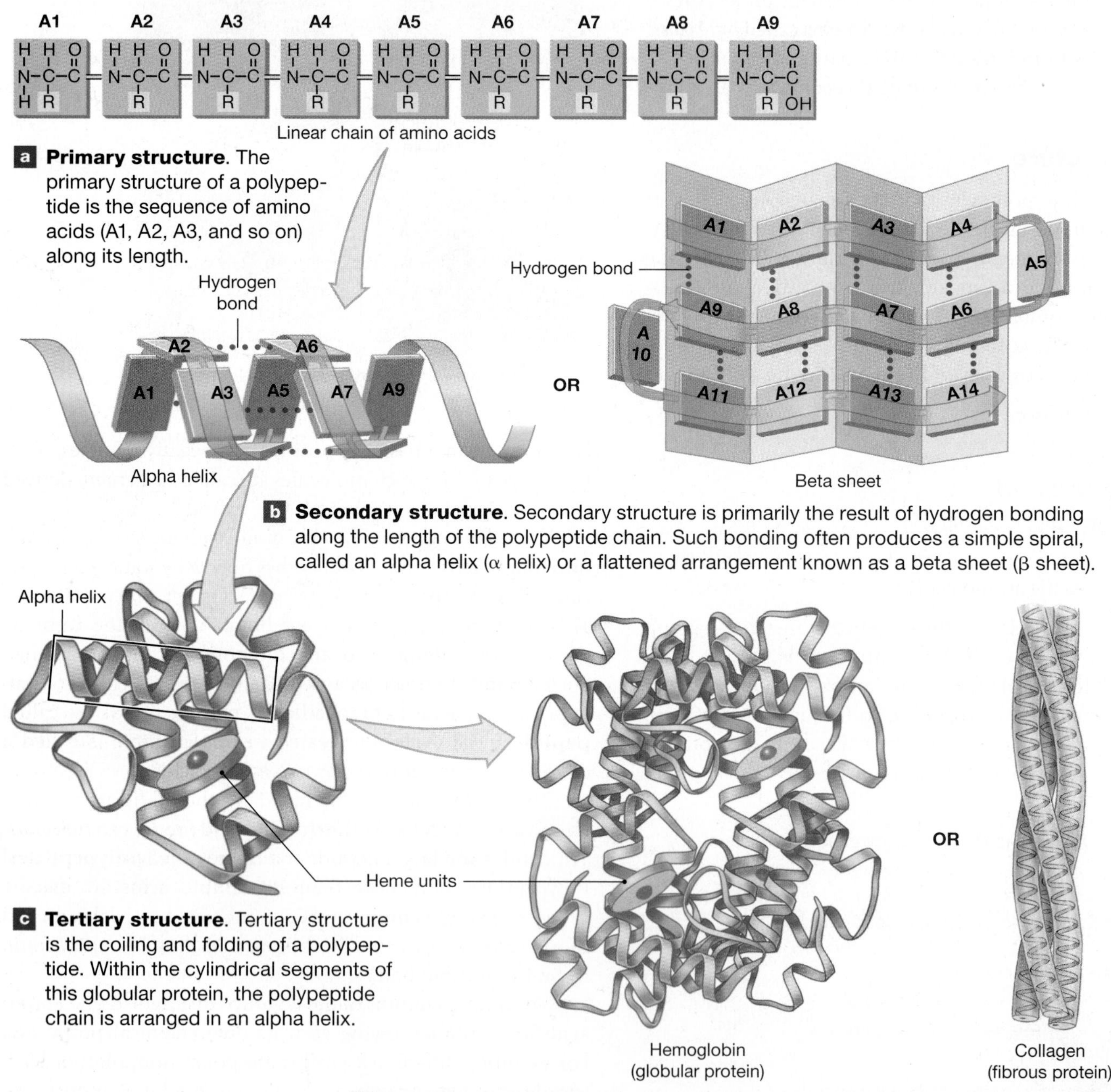
A1
A2
A3
A4
A5
A6
A7
A8
A9
Linear chain of amino acids
Hydrogen bond
Alpha helix
Hydrogen bond
OR
A10
A11
A12
A13
A14
Beta sheet
Alpha helix
Heme units
OR
Hemoglobin (globular protein)
Collagen (fibrous protein)

a **Primary structure**. The primary structure of a polypeptide is the sequence of amino acids (A1, A2, A3, and so on) along its length.

b **Secondary structure**. Secondary structure is primarily the result of hydrogen bonding along the length of the polypeptide chain. Such bonding often produces a simple spiral, called an alpha helix (α helix) or a flattened arrangement known as a beta sheet (β sheet).

c **Tertiary structure**. Tertiary structure is the coiling and folding of a polypeptide. Within the cylindrical segments of this globular protein, the polypeptide chain is arranged in an alpha helix.

d **Quaternary structure**. Quaternary structure develops when separate polypeptide subunits interact to form a larger molecule. A single hemoglobin molecule contains four globular subunits. Hemoglobin transports oxygen in the blood; the oxygen binds reversibly to the heme units. In collagen, three helical polypeptide subunits intertwine. Collagen is the principal extracellular protein in most organs.

Peptide bonds are responsible for the primary structure of proteins.

2. **Secondary structure** is the shape that results from the presence of hydrogen bonds between atoms at different parts of the polypeptide chain. Hydrogen bonding may create either an *alpha helix* (simple spiral) or *beta sheet* (a flat pleated sheet) (Figure 2–22b). Which one forms depends on where hydrogen bonding takes place between the sequence of amino acids in the polypeptide chain. The alpha helix is the more common form, but a given polypeptide chain may have both helical and pleated sections. Ribbon diagrams are used to represent the three-dimensional structure of proteins. An alpha helix appears as a coiled ribbon, beta sheets as arrows, and less-structured polypeptide chains as narrow ribbons or tubes.
3. **Tertiary structure** is the complex coiling and folding that gives a protein its final three-dimensional shape (Figure 2–22c). Tertiary structure results primarily from hydrophobic and hydrophilic interactions between the R groups of the polypeptide chain and the surrounding water molecules, and to a lesser extent from interactions between the R groups of amino acids in different parts of the molecule. Most such interactions are relatively weak. One, however, is very strong: the *disulfide bond,* a covalent bond that may form between the sulfur atoms of two molecules of the amino acid *cysteine* located at different sites along the chain (not shown). Disulfide bonds create permanent loops or coils in a polypeptide chain.
4. **Quaternary structure** is the interaction between individual polypeptide chains to form a protein complex (Figure 2–22d). Each of the polypeptide subunits has its own secondary and tertiary structures. For example, the protein *hemoglobin* contains four subunits. In Figure 2–22d, the polypeptide subunits with the same color (blue or purple) have the same structure. The hemoglobin in red blood cells binds and transports oxygen. *Collagen,* composed of three windings of alpha helical polypeptides, is the most abundant structural protein and is found in skin, bones, muscles, cartilages, and tendons. Collagen fibers form the framework that supports cells in most tissues. In *keratin,* two alpha helical polypeptides are wound together like the strands of a rope. Keratin is the tough, water-resistant protein at the surface of the skin and in nails and hair.

Fibrous and Globular Proteins

Proteins fall into two general structural classes on the basis of their overall shape and properties:

- **Globular proteins** are compact, generally rounded, and soluble in water. Many enzymes, hormones, and other molecules that circulate in the bloodstream are globular proteins. These proteins can function only if they remain in solution. The unique shape of each globular protein comes from its tertiary structure. Hemoglobin and *myoglobin,* a protein in muscle cells, are both globular proteins. The enzymes that control chemical reactions inside cells are also globular proteins.
- **Fibrous proteins** form extended sheets or strands. Fibrous proteins are tough, durable, and generally insoluble in water. They usually play structural roles in the body. Their shapes are usually due to secondary structure (for proteins with the pleated-sheet form) or quaternary structure (as we just described for collagen and keratin).

Protein Shape and Function

The shape of a protein determines its functional characteristics, and the sequence of amino acids ultimately determines its shape. The 20 amino acids can be linked in an astonishing number of combinations, creating proteins of enormously varied shape and function. Changing only 1 of the 10,000 or more amino acids in a protein can significantly alter the way the protein functions. For example, several cancers and *sickle cell anemia,* a blood disorder, result from changing just a single amino acid in the amino acid sequences of complex proteins.

The tertiary and quaternary shapes of complex proteins depend not only on their amino acid sequence, but also on the local environmental conditions. Small changes in the ionic composition, temperature, or pH of their surroundings can affect the function of proteins. Protein shape can also be affected by hydrogen bonding to other molecules in solution. The significance of these factors is most striking when we consider enzymes, for these proteins are essential to the metabolic operations in every one of our cells.

Enzyme Function

Enzymes are among the most important of all the body's proteins. As noted earlier in this chapter, enzymes catalyze the chemical reactions that sustain life. Almost everything that happens inside the human body does so because a specific enzyme makes it possible.

The reactants in enzymatic reactions are called **substrates**. As in other types of chemical reactions, the interactions among substrates yield specific products. Before an enzyme can function as a catalyst—accelerating a chemical reaction without itself being permanently changed or consumed—the substrates must bind to a special region of the enzyme. This region is called the **active site**. It is typically a groove or pocket into which one or more substrates nestle, like a key fitting into a lock. Weak electrical attractive forces, such as hydrogen bonding, reinforce the physical fit. The tertiary or quaternary structure of the enzyme molecule determines the shape of the active site. Although enzymes are proteins, any organic or inorganic compound that will bind to the active site can be a substrate.

Figure 2–23 **A Simplified View of Enzyme Structure and Function.** Each enzyme contains a specific active site somewhere on its exposed surface.

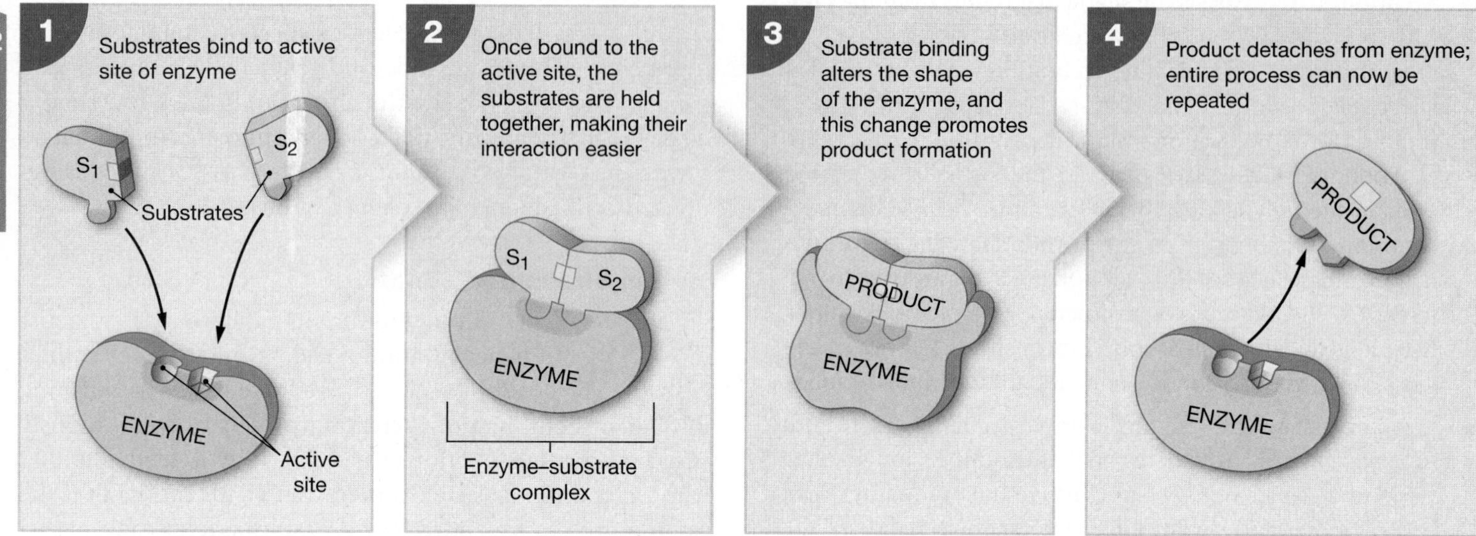
1
Substrates bind to active site of enzyme
S_1
S_2
Substrates
ENZYME
Active site
2
Once bound to the active site, the substrates are held together, making their interaction easier
S_1
S_2
ENZYME
Enzyme–substrate complex
3
Substrate binding alters the shape of the enzyme, and this change promotes product formation
PRODUCT
ENZYME
4
Product detaches from enzyme; entire process can now be repeated
PRODUCT
ENZYME

Figure 2–23 presents one example of enzyme structure and function. Substrates bind to the enzyme at its active site (1). Substrate binding produces an *enzyme–substrate complex* (2). Substrate binding typically produces a temporary, reversible change in the shape of the enzyme that may place physical stresses on the substrate molecules, leading to product formation (3). The product is then released, freeing the enzyme to repeat the process (4).

Enzymes work quickly, cycling rapidly between substrates and products. For example, an enzyme providing energy during a muscular contraction performs its reaction sequence 100 times per second. Hydrolytic enzymes can work even faster, breaking down almost 20,000 molecules a second!

Figure 2–23 shows an enzyme that catalyzes a synthesis reaction. Other enzymes may catalyze decomposition reactions, reversible reactions, or exchange reactions. Regardless of the reaction they catalyze, all enzymes share the basic characteristics of specificity, saturation limits, and regulation:

- *Specificity*. Each enzyme catalyzes only one type of reaction, a characteristic called **specificity**. An enzyme's specificity is due to the ability of its active sites to bind only to substrates with particular shapes and charges. For this reason, differences in enzyme structure that do not affect the active site and do not change the response of the enzyme to substrate binding do not affect enzyme function. Such enzyme variants are called *isozymes*.
- *Saturation Limits*. The rate of an enzymatic reaction is directly related to the concentrations of substrate molecules and enzymes. An enzyme molecule must encounter appropriate substrates before it can catalyze a reaction. The higher the substrate concentration, the more frequent these encounters. When substrate concentrations are high enough that every enzyme molecule is cycling through its reaction sequence at top speed, further increases in substrate concentration will not affect the rate of reaction unless additional enzyme molecules are provided. The substrate concentration required to reach the maximum rate of reaction is called the *saturation limit*. An enzyme that has reached its saturation limit is said to be **saturated**. To increase the reaction rate further, the cell must increase the number of enzyme molecules available. This is one important way that cells promote specific reactions.
- *Regulation*. Each cell contains an assortment of enzymes, and any particular enzyme may be active under one set of conditions and inactive under another. Virtually anything that changes the tertiary or quaternary shape of an enzyme can turn it "on" or "off" and in this way control reaction rates inside the cell. Because the shape change is immediate, enzyme activation or inactivation is an important method of short-term control over reaction rates and metabolic pathways. Here we will consider only one example of enzyme regulation: the presence or absence of *cofactors*.

Cofactors and Enzyme Function

A **cofactor** is an ion or a molecule that must bind to an enzyme before substrates can also bind. Without a cofactor, the enzyme is intact but nonfunctional. With the cofactor, the enzyme can catalyze a specific reaction. Examples of ionic cofactors include calcium ion (Ca^{2+}) and magnesium ion (Mg^{2+}), which bind at the enzyme's active site. Cofactors may also bind at other sites, as long as they produce a change in the shape of the active site that makes substrate binding possible.

Coenzymes are nonprotein organic molecules that function as cofactors. Our bodies convert many vitamins into essential coenzymes. *Vitamins*, detailed in Chapter 25, are organic nutrients structurally related to lipids or carbohydrates, but have unique functional roles. Because the human body cannot synthesize most of the vitamins it needs, you must obtain them from your diet.

Temperature and pH Affect Enzyme Function

Each enzyme works best at specific temperatures and pH values. As temperatures rise, protein shape changes and enzyme function deteriorates. Eventually the protein undergoes **denaturation**, a change in tertiary or quaternary structure that makes it nonfunctional. You see permanent denaturation when you fry an egg. As the temperature rises, the proteins in the egg white denature. Eventually, the proteins become completely and irreversibly denatured, forming an insoluble white mass. In the body, death occurs at very high body temperatures (above 43°C, or 110°F) because structural proteins and enzymes soon denature, causing irreparable damage to organs and organ systems.

Enzymes are equally sensitive to changes in pH. *Pepsin*, an enzyme that breaks down food proteins in the stomach, works best at a pH of 2.0 (strongly acidic). Your small intestine contains *trypsin*, another protein-degrading enzyme. Trypsin works only in an alkaline environment, with an optimum pH of 7.7 (weakly basic).

Glycoproteins and Proteoglycans

Glycoproteins (GLĪ-kō-prō-tēnz) and **proteoglycans** (prō-tē-ō-GLĪ-kanz) are combinations of protein and carbohydrate molecules. Glyco*proteins* are large proteins with small carbohydrate groups attached. These molecules may function as enzymes, antibodies, hormones, or protein components of plasma membranes. Glycoproteins in plasma membranes play a major role in identifying normal versus abnormal cells. They are also important in the immune response (Chapter 22). Glycoprotein secretions called *mucins* absorb water to form **mucus**. Mucus coats and lubricates the surfaces of the reproductive and digestive tracts.

Proteo*glycans* are large polysaccharide molecules linked by polypeptide chains. Proteoglycans bind adjacent cells together, and give tissue fluids a viscous (syrupy) consistency.

Checkpoint

24. **Describe a protein.**
25. **How does boiling a protein affect its structural and functional properties?**

See the blue Answers tab at the back of the book.

2-13 DNA and RNA are nucleic acids

Learning Outcome Discuss the structures and functions of nucleic acids.

Nucleic (nū -KLĀ-ik) **acids** are large organic molecules composed of carbon, hydrogen, oxygen, nitrogen, and phosphorus. Nucleic acids store and process information at the molecular level inside cells. The two classes of nucleic acid molecules are **deoxyribonucleic** (dē-oks-ē-rī-bō-nū -KLĀ-ik) **acid (DNA)** and **ribonucleic** (rī-bō-nū-KLĀ-ik) **acid (RNA)**. As we will see, these two classes of nucleic acids differ in composition, structure, and function.

The primary role of nucleic acids is to store and transfer information—specifically, information essential to the synthesis of proteins within our cells. The DNA in our cells encodes the information needed to build proteins, while several forms of RNA cooperate to build specific proteins using the information provided by DNA.

DNA contains the instructions for making proteins with correct shapes and therefore correct functions. Those proteins then control our inherited characteristics. For example, by directing the synthesis of structural proteins, DNA determines all of our physical characteristics, including eye color, hair color, and blood type. By directing the synthesis of many functional proteins, including enzymes, DNA regulates all aspects of cellular metabolism, including the creation and destruction of lipids, carbohydrates, and other vital molecules.

In this section, we look at how nucleic acids are structured, and the similarities and differences between DNA and RNA. In Chapter 3 we detail the ways that DNA and RNA work together.

Structure of Nucleic Acids

A nucleic acid consists of one or two long chains of repeating subunits. The individual subunits (monomers) of the chain are called **nucleotides**. Each nucleotide has three parts (Figure 2–24a): (1) a pentose (five-carbon sugar) attached to both, (2) a phosphate group and (3) a **nitrogenous** (nitrogen-containing) **base**. The pentose is either **ribose** (in RNA) or **deoxyribose** (in DNA).

Five nitrogenous bases occur in nucleic acids: **adenine (A)**, **guanine (G)**, **cytosine (C)**, **thymine (T)**, and **uracil (U)**. Adenine and guanine are double-ringed molecules called *purines* (Figure 2–24b). The other three bases are single-ringed molecules called *pyrimidines* (Figure 2–24c). Both RNA and DNA contain adenine, guanine, and cytosine. Uracil occurs only in RNA and thymine occurs only in DNA.

A nucleotide forms when a phosphate group binds to a pentose already attached to a nitrogenous base. A nucleic acid forms when nucleotides are joined by dehydration synthesis, which attaches the phosphate group of one nucleotide to the sugar of another. The "backbone" of a nucleic acid molecule is a linear sugar-to-phosphate-to-sugar sequence, with the nitrogenous bases projecting to one side.

Comparison of RNA and DNA

In both DNA and RNA, it is the sequence of nitrogenous bases that carries the information about how to make proteins.

Figure 2–24 **Nucleotides and Nitrogenous Bases.**

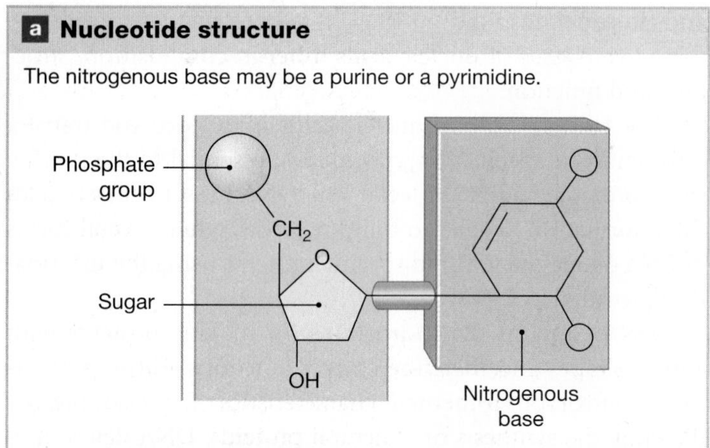
a Nucleotide structure
The nitrogenous base may be a purine or a pyrimidine.
Phosphate group
CH_2
O
Sugar
OH
Nitrogenous base

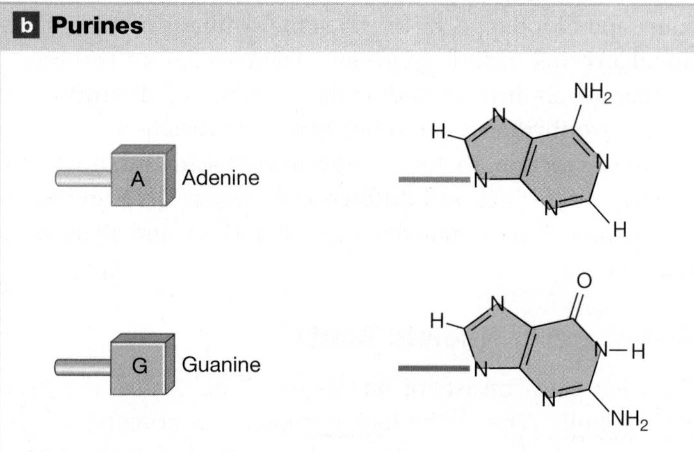
b Purines
A
Adenine
NH_2
H
N
G
Guanine
O
N—H
NH_2

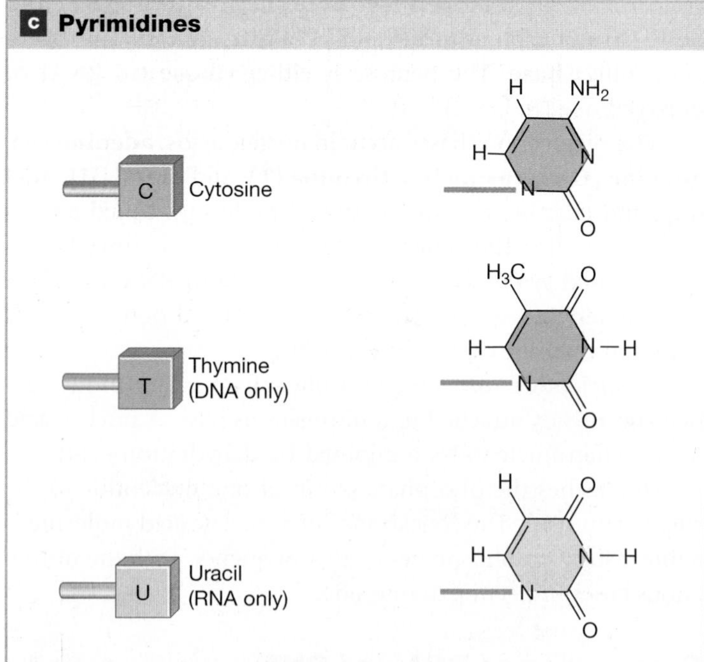
c Pyrimidines
C
Cytosine
H
NH_2
N
O
T
Thymine (DNA only)
H_3C
N—H
U
Uracil (RNA only)

? What structural differences make adenine and guanine different from cytosine, thymine, and uracil?

Figure 2–25 **The Structure of Nucleic Acids.** The nucleic acids RNA and DNA are long chains of nucleotides.

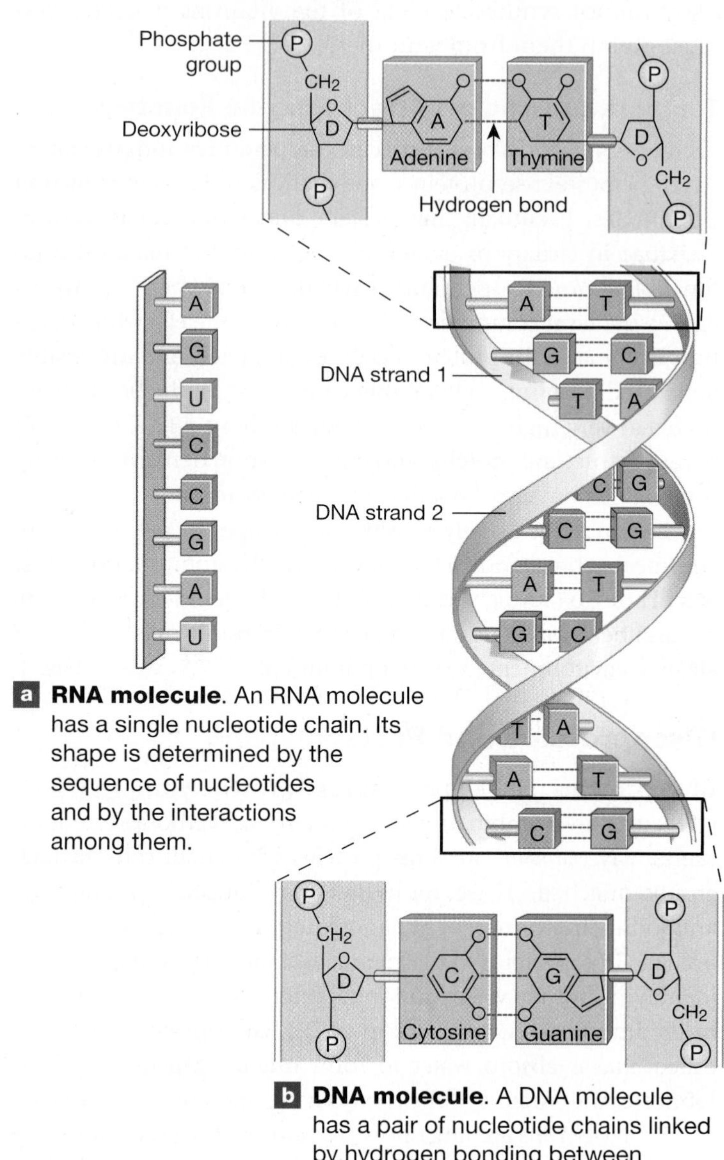
Phosphate group
Deoxyribose
CH_2
P
D
A
T
Adenine
Thymine
Hydrogen bond
DNA strand 1
DNA strand 2
G
C
U
C
Cytosine
Guanine

a RNA molecule. An RNA molecule has a single nucleotide chain. Its shape is determined by the sequence of nucleotides and by the interactions among them.

b DNA molecule. A DNA molecule has a pair of nucleotide chains linked by hydrogen bonding between complementary base pairs.

However, there are other important structural differences between the two types of nucleic acids. A molecule of RNA consists of a single chain of nucleotides (**Figure 2–25a**). Its shape depends on the order of the nucleotides and the interactions among them. Our cells have various forms of RNA with different shapes and functions. As we will see in Chapter 3, protein synthesis requires three types of RNA: (1) *messenger RNA (mRNA)*, (2) *transfer RNA (tRNA)*, and (3) *ribosomal RNA (rRNA)*. There are also other types of RNA whose roles are under active research.

A DNA molecule consists of a *pair* of nucleotide chains, with two sugar-phosphate backbones on the outside and the nitrogenous bases projecting inward (**Figure 2–25b**). Hydrogen bonding between opposing nitrogenous bases

Table 2–6 Comparison of RNA and DNA

Characteristic	RNA	DNA
Sugar	Ribose	Deoxyribose
Nitrogenous bases	Adenine (A) Guanine (G) Cytosine (C) Uracil (U)	Adenine Guanine Cytosine Thymine (T)
Number of nucleotides in typical molecule	Varies from fewer than 100 nucleotides to about 50,000	Always more than 45 million
Shape of molecule	Varies with hydrogen bonding along the length of the strand; three main types (mRNA, rRNA, tRNA)	Paired strands coiled in a double helix
Function	Performs protein synthesis as directed by DNA	Stores genetic information that controls protein synthesis

holds the two strands together. The shapes of the nitrogenous bases allow adenine to bond only to thymine, and cytosine to bond only to guanine. As a result, the combinations adenine–thymine (A–T) and cytosine–guanine (C–G) are known as **complementary base pairs**, and the two nucleotide chains of the DNA molecule are known as **complementary strands**. The two strands of DNA twist around one another in a double helix that resembles a spiral staircase. Each step of the staircase corresponds to one complementary base pair.

Through a sequence of events described in Chapter 3 , the cell uses one of the two complementary DNA strands to provide the information needed to synthesize a specific protein. Table 2–6 summarizes our comparison of RNA and DNA.

Checkpoint

26. **Describe a nucleic acid.**
27. **A large organic molecule made of the sugar ribose, nitrogenous bases, and phosphate groups is which kind of nucleic acid?**

See the blue Answers tab at the back of the book.

2-14 ATP is a high-energy compound used by cells

Learning Outcome **Discuss the structures and functions of high-energy compounds.**

To perform their vital functions, cells must use energy, which they obtain by breaking down organic substrates (catabolism). To be useful, that energy must be transferred from molecule to molecule or from one part of the cell to another.

The usual method of energy transfer involves the creation and breakdown of high-energy bonds by enzymes within cells. A *high-energy bond* is a covalent bond whose breakdown releases energy the cell can use directly. In our cells, a high-energy bond generally binds a phosphate group (PO_4^{3-}) to an organic molecule. The product with such a bond is called a **high-energy compound**. Most high-energy compounds are derived from nucleotides, the building blocks of nucleic acids (Figure 2–26).

The process of attaching a phosphate group to another molecule is called **phosphorylation** (fos-for-i-LĀ-shun). This

Figure 2–26 **The Structure of ATP.** A molecule of ATP is formed by attaching two phosphate groups to the nucleotide adenosine monophosphate. These two phosphate groups are connected by high-energy bonds incorporating energy released by catabolism. Cells most often obtain quick energy to power cellular operations by removing one phosphate group from ATP, forming ADP (adenosine diphosphate). ADP can later be reconverted to ATP, and the cycle repeated.

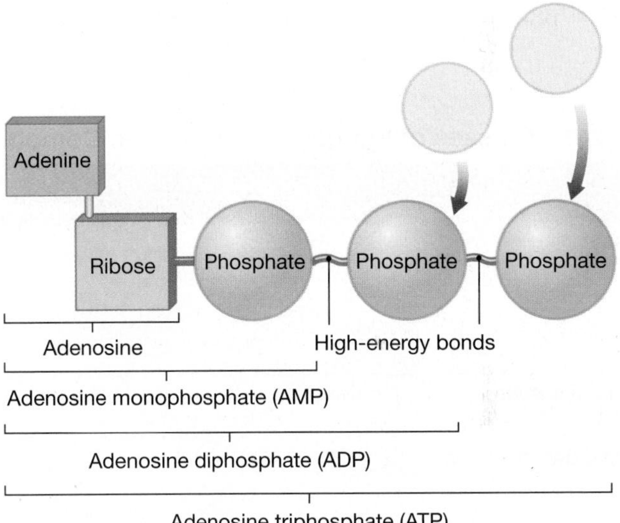
Adenine
Ribose
Phosphate
Phosphate
Phosphate
Adenosine
High-energy bonds
Adenosine monophosphate (AMP)
Adenosine diphosphate (ADP)
Adenosine triphosphate (ATP)

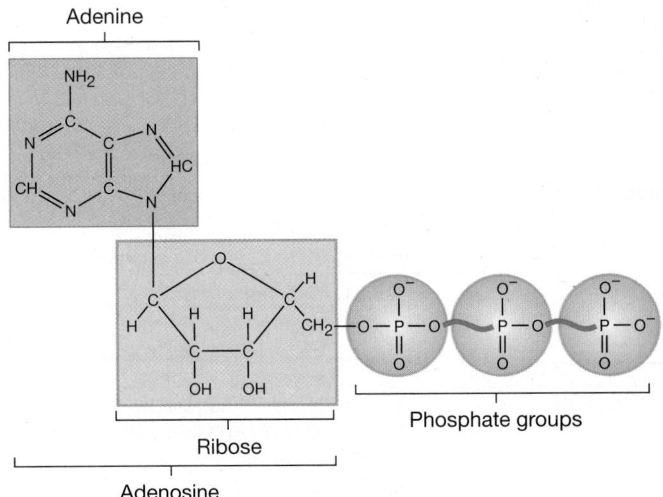
Adenine
NH2
Ribose
Phosphate groups
Adenosine

? How many phosphorylations does AMP undergo to become ATP?

process does not necessarily produce high-energy bonds. For example, in the synthesis of sucrose, a phosphate group is first attached to glucose. The creation of a high-energy compound requires (1) a phosphate group, (2) enzymes capable of catalyzing the reactions involved, and (3) suitable organic substrates to which the phosphate can be added.

The most important such substrate is the nucleotide **adenosine monophosphate (AMP)**, which already contains one phosphate group. Attaching a second phosphate group produces **adenosine diphosphate (ADP)**. A significant energy input is required to convert AMP to ADP, and the second phosphate is attached by a high-energy bond. Even more energy is required to add a third phosphate and create the high-energy compound **adenosine triphosphate (ATP)** (Figure 2–26).

The conversion of ADP to ATP is the most important method of storing energy in our cells. The breakdown of ATP to ADP is the most important method of releasing energy. The relationships involved in this energy transfer can be diagrammed as

$$\text{ADP} + \text{phosphate group} + \text{energy} \rightleftharpoons \text{ATP} + H_2O$$

The hydrolytic breakdown of ATP to ADP requires an enzyme known as **adenosine triphosphatase (ATPase)**, as well as a molecule of water. Throughout life, cells continuously generate ATP from ADP and then use the energy provided by the breakdown of ATP to perform vital functions, such as the synthesis of proteins or the contraction of muscles.

ATP is our most abundant high-energy compound, but there are others. They are typically other nucleotides that have undergone phosphorylation. For example, *guanosine triphosphate (GTP)* is a nucleotide-based high-energy compound that transfers energy in specific enzymatic reactions.

Table 2–7 summarizes the inorganic and organic compounds covered in this chapter.

Checkpoint

28. **Describe ATP.**
29. **What molecule is produced by the phosphorylation of ADP?**

See the blue Answers tab at the back of the book.

Table 2–7 Classes of Inorganic and Organic Compounds

Class	Building Blocks (Elements and/or Monomers)	Sources	Functions
INORGANIC			
Water	Hydrogen and oxygen atoms	Absorbed from the diet or generated by metabolism	Solvent; transport medium for dissolved materials and heat; cooling through evaporation; medium for chemical reactions; reactant in hydrolysis
Acids, bases, salts	H^+, OH^-, various anions and cations	Obtained from the diet or generated by metabolism	Structural components; buffers; sources of ions
Dissolved gases	O, C, N, and other atoms	Atmosphere, metabolism	O_2: required for cellular metabolism CO_2: generated by cells as a waste product NO: chemical messenger in cardiovascular, nervous, and lymphatic systems
ORGANIC			
Carbohydrates	C, H, O, in some cases N; CHO in a 1:2:1 ratio Monosaccharide monomers	Obtained from the diet or manufactured in the body	Energy source; some structural role when attached to lipids or proteins; energy storage
Lipids	C, H, O, in some cases N or P; CHO not in 1:2:1 ratio Fatty acids and glycerol monomers	Obtained from the diet or manufactured in the body	Energy source; energy storage; insulation; structural components; chemical messengers; protection
Proteins	C, H, O, N, commonly S Amino acid monomers	20 common amino acids; roughly half can be manufactured in the body, others must be obtained from the diet	Catalysts for metabolic reactions; structural components; movement; transport; buffers; defense; control and coordination of activities
Nucleic acids	C, H, O, N, and P; nucleotides composed of phosphates, sugars, and nitrogenous bases Nucleotide monomers	Obtained from the diet or manufactured in the body	Storage and processing of genetic information
High-energy compounds	Nucleotides joined to phosphates by high-energy bonds	Synthesized by all cells	Storage or transfer of energy

2 Chapter Review

Study Outline

An Introduction to the Chemical Level of Organization p. 28

1. Chemicals combine to form complex structures.

2-1 Atoms are the basic particles of matter p. 28

2. Atoms are the smallest units of matter. They consist of **protons**, **neutrons**, and **electrons**. Protons and neutrons reside in the **nucleus** of an atom.
3. The number of protons in an atom is its **atomic number**. Each **element** includes all the atoms that have the same number of protons and thus the same atomic number.
4. Within an atom, an **electron cloud** surrounds the nucleus. *(Figure 2–1)*
5. The **mass number** of an atom is the total number of protons and neutrons in its nucleus. **Isotopes** are atoms of the same element whose nuclei contain different numbers of neutrons.
6. Electrons occupy an orderly series of **energy levels**, commonly illustrated as **electron shells**. The electrons in the outermost energy level, or **valence shell**, determine an element's chemical properties. *(Figures 2–2, 2–3)*

2-2 Chemical bonds are forces formed by interactions between atoms p. 32

7. Atoms can combine through chemical reactions that create **chemical bonds**. A **molecule** is any chemical structure consisting of atoms held together by covalent bonds. A **compound** is a chemical substance made up of atoms of two or more elements in a fixed proportion.
8. The rules of **chemical notation** are used to describe chemical compounds and reactions. *(Spotlight Figure 2–4)*
9. An **ionic bond** results from the attraction between **ions**, atoms that have gained or lost electrons. **Cations** are positively charged; **anions** are negatively charged. *(Figure 2–5)*
10. Atoms that share electrons to form a molecule are held together by **covalent bonds**. A sharing of one pair of electrons is a **single covalent bond**; a sharing of two pairs is a **double covalent bond**; and a sharing of three pairs is a **triple covalent bond**. A bond with equal sharing of electrons is a **nonpolar covalent bond**; a bond with unequal sharing of electrons is a **polar covalent bond**. *(Figures 2–6, 2–7)*
11. A **hydrogen bond** is a weak, but important, electrical attraction that can affect the shapes and properties of molecules. *(Figure 2–8)*
12. Matter can exist as a *solid*, a *liquid*, or a *gas*, depending on the nature of the interactions among the component atoms or molecules.
13. The **molecular weight** of a molecule or a compound is the sum of the atomic weights of its component atoms.

2-3 Decomposition, synthesis, and exchange reactions are important types of chemical reactions in physiology p. 37

14. A chemical reaction occurs when **reactants** are rearranged to form one or more **products**. Collectively, all the **chemical reactions** in the body constitute its **metabolism**. Through metabolism, cells capture, store, and use energy to maintain homeostasis and to perform essential functions.
15. **Work** is the movement of an object or a change in the physical structure of matter. **Energy** is the capacity to perform work.
16. **Kinetic energy** is the energy of motion. **Potential energy** is stored energy that results from the position or structure of an object. Conversions from potential to kinetic energy (or vice versa) are not 100 percent efficient. Every such energy conversion releases *heat*.
17. A chemical reaction is classified as a **decomposition**, a **synthesis**, or an **exchange reaction**.
18. Cells gain energy to power their functions by **catabolism**, the breakdown of complex molecules. Much of this energy supports **anabolism**, the synthesis of new molecules.
19. All chemical reactions are theoretically reversible. At **equilibrium**, the rates of two opposite reactions are in balance.

2-4 Enzymes speed up reactions by lowering the energy needed to start them p. 39

20. **Activation energy** is the amount of energy required to start a reaction. **Enzymes** are **catalysts**—compounds that accelerate chemical reactions without themselves being permanently changed or consumed. Enzymes promote chemical reactions by lowering the activation energy needed. *(Figure 2–9)*
21. **Exergonic** reactions release energy. **Endergonic** reactions absorb energy.

2-5 Inorganic compounds lack carbon, and organic compounds contain carbon p. 40

22. **Nutrients** are the essential elements and molecules normally obtained from the diet. **Metabolites**, on the other hand, are molecules that can be synthesized or broken down by chemical reactions inside our bodies. Nutrients and metabolites can be broadly categorized as either **inorganic** or **organic compounds**.

2-6 Physiological systems depend on water p. 40

23. Water is the most important constituent of the body.
24. A **solution** is a uniform mixture of two or more substances. It consists of a medium, or **solvent**, in which atoms, ions, or molecules of another substance, or **solute**, are individually dispersed. In *aqueous solutions*, water is the solvent. *(Figure 2–10)*

> MasteringA&P™ Access more chapter study tools online in the MasteringA&P Study Area:
> - Chapter Quizzes, Chapter Practice Test, MP3 Tutor Sessions, and Clinical Case Studies
> - Practice Anatomy Lab PAL 3.0
> - Interactive Physiology iP2
> - A&P Flix A&PFlix
> - PhysioEx PhysioEx 9.1

25. Many inorganic substances, called **electrolytes**, undergo **dissociation**, or **ionization**, in water to form ions. Molecules that interact readily with water molecules are called **hydrophilic**. Those that do not are called **hydrophobic**. *(Figure 2–10; Table 2–2)*

2-7 Body fluid pH is vital for homeostasis p. 43

26. The **pH** of a solution indicates the concentration of hydrogen ions it contains. Solutions are classified as **neutral**, **acidic**, or **basic** *(alkaline)* on the basis of pH. *(Figure 2–11)*

2-8 Acids, bases, and salts have important physiological roles p. 44

27. An **acid** releases hydrogen ions. A **base** removes hydrogen ions from a solution. *Strong acids* and *strong bases* ionize completely. In the case of *weak acids* and *weak bases*, only some of the molecules ionize.
28. A **salt** is an electrolyte whose cation is not a hydrogen ion (OH^+) and whose anion is not a hydroxide ion (OH^-).
29. **Buffers** remove or replace hydrogen ions in solution. Buffers and *buffer systems* in body fluids maintain the pH within normal limits.

2-9 Living things contain organic compounds made up of monomers, polymers, and functional groups p. 45

30. Carbon and hydrogen are the main constituents of **organic compounds**, which generally contain oxygen as well. Identical **monomers** join together through *dehydration synthesis* reactions and form long complex chains called **polymers**. Organic polymers include carbohydrates, lipids, proteins, and nucleic acids.
31. The properties of the different classes of organic monomers and polymers are a result of the presence of *functional groups* of atoms. *(Table 2–3)*

2-10 Carbohydrates contain carbon, hydrogen, and oxygen in a 1:2:1 ratio p. 45

32. **Carbohydrates** are most important as an energy source for metabolic processes. The three major types of carbohydrates are **monosaccharides** *(simple sugars)*, **disaccharides**, and **polysaccharides**. Disaccharides and polysaccharides form from monosaccharide monomers by **dehydration synthesis**. *(Figures 2–12 to 2–14; Table 2–4)*

2-11 Lipids often contain a carbon-to-hydrogen ratio of 1:2 p. 47

33. **Lipids** include *fats, oils,* and *waxes*. Most are insoluble in water. The five important classes of lipids are **fatty acids**, **eicosanoids**, **glycerides**, **steroids**, and **phospholipids** and **glycolipids**. *(Figures 2–15 to 2–19; Table 2–5)*
34. **Triglycerides** *(neutral fats)* consist of three fatty acid molecules attached by dehydration synthesis to a molecule of **glycerol**. **Diglycerides** consist of two fatty acids and glycerol. **Monoglycerides** consist of one fatty acid plus glycerol. Fatty acids and glycerol are lipid monomers. *(Figure 2–17)*
35. Steroids (1) are components of plasma membranes, (2) include sex hormones and hormones regulating metabolic activities, and (3) are important in lipid digestion. *(Figure 2–18)*
36. **Phospholipids** and **glycolipids** are structural lipids that are components of *micelles* and plasma membranes *(Figure 2–19)*.

2-12 Proteins contain carbon, hydrogen, oxygen, and nitrogen and are formed from amino acids p. 51

37. **Proteins** perform a variety of essential functions in the body. Seven important types of proteins are *structural proteins, contractile proteins, transport proteins, buffering proteins, enzymes, hormones,* and *antibodies*.
38. Proteins are organic polymers made up of chains of **amino acids**. Each amino acid monomer consists of an *amino group*, a *carboxyl group*, a *hydrogen atom*, and an *R group (side chain)* attached to a central carbon atom. A **polypeptide** is a linear sequence of amino acids held together by **peptide bonds**; **proteins** are polypeptides containing over 100 amino acids. *(Figures 2–20, 2–21)*
39. The four levels of protein structure are **primary structure** (amino acid sequence), **secondary structure** (amino acid interactions, such as hydrogen bonds), **tertiary structure** (complex folding, disulfide bonds, and interaction with water molecules), and **quaternary structure** (formation of protein complexes from individual subunits). **Globular proteins**, such as *myoglobin* and *hemoglobin*, are generally rounded and water soluble. **Fibrous proteins**, such as *collagen* and *keratin*, are elongated, tough, durable, and generally insoluble. *(Figure 2–22)*
40. The reactants in an enzymatic reaction, called **substrates**, interact to yield a product by binding to the enzyme's **active site**. **Cofactors** are ions or molecules that must bind to the enzyme before the substrates can bind. **Coenzymes** are organic cofactors commonly derived from *vitamins*. *(Figure 2–23)*
41. The shape of a protein determines its functional characteristics. Each protein works best at an optimal combination of temperature and pH and will undergo temporary or permanent **denaturation**, or change in shape, at temperatures or pH values outside the normal range.

2-13 DNA and RNA are nucleic acids p. 57

42. **Nucleic acids** store and process information at the molecular level. The two kinds of nucleic acids are **deoxyribonucleic acid (DNA)** and **ribonucleic acid (RNA)**. *(Figures 2–24, 2–25; Table 2–6)*
43. Nucleic acids are organic polymers made up of chains of **nucleotides**. Each nucleotide contains a sugar, a phosphate group, and a **nitrogenous base**. The sugar is *ribose* in RNA and *deoxyribose* in DNA. DNA is a two-stranded double helix containing the nitrogenous bases **adenine**, **guanine**, **cytosine**, and **thymine**. RNA consists of a single strand and contains **uracil** instead of thymine.

2-14 ATP is a high-energy compound used by cells p. 59

44. Cells store energy in the *high-energy bonds* of **high-energy compounds**. The most important high-energy compound is **ATP (adenosine triphosphate)**. Cells make ATP by adding a phosphate group to **ADP (adenosine diphosphate)** through **phosphorylation**. When ATP is broken down to ADP and phosphate, energy is released. Cells can use this energy to power essential activities. *(Figure 2–26; Table 2–7)*

Review Questions

See the blue Answers tab at the back of the book.

LEVEL 1 Reviewing Facts and Terms

1. An oxygen atom has eight protons. **(a)** Sketch in the arrangement of electrons around the nucleus of the oxygen atom in the following diagram. **(b)** How many more electrons will it take to fill the outermost energy level?

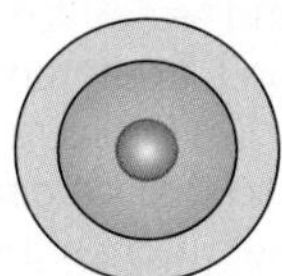

Oxygen atom

2. What is the following type of decomposition reaction called?

$$ABCD + H_2O \longrightarrow ABCH + DOH$$

3. The subatomic particle with the least mass **(a)** carries a negative charge, **(b)** carries a positive charge, **(c)** plays no part in the atom's chemical reactions, **(d)** is found only in the nucleus.
4. Isotopes of an element differ from each other in the number of **(a)** protons in the nucleus, **(b)** neutrons in the nucleus, **(c)** electrons in the outer shells, **(d)** a, b, and c are all correct.
5. The number and arrangement of electrons in an atom's outer energy level (valence shell) determine the atom's **(a)** atomic weight, **(b)** atomic number, **(c)** molecular weight, **(d)** chemical properties.
6. All organic compounds in the human body contain all of the following elements *except* **(a)** hydrogen, **(b)** oxygen, **(c)** carbon, **(d)** calcium, **(e)** both a and d.
7. A substance containing atoms of different elements that are bonded together is called a(n) **(a)** molecule, **(b)** compound, **(c)** mixture, **(d)** isotope, **(e)** solution.
8. All the chemical reactions that occur in the human body are collectively referred to as **(a)** anabolism, **(b)** catabolism, **(c)** metabolism, **(d)** homeostasis.
9. Which of the following chemical equations illustrates a typical decomposition reaction?
   (a) $A + B \longrightarrow AB$
   (b) $AB + CD \longrightarrow AD + CB$
   (c) $2A_2 + B_2 \longrightarrow 2A_2B$
   (d) $AB \longrightarrow A + B$
10. The speed, or rate, of a chemical reaction is influenced by **(a)** the presence of catalysts, **(b)** the temperature, **(c)** the concentration of the reactants, **(d)** a, b, and c are all correct.
11. A pH of 7.8 in the human body typifies a condition referred to as **(a)** acidosis, **(b)** alkalosis, **(c)** dehydration, **(d)** homeostasis.
12. A(n) _____ is a solute that dissociates to release hydrogen ions, and a(n) _____ is a solute that removes hydrogen ions from solution. **(a)** base, acid, **(b)** salt, base, **(c)** acid, salt, **(d)** acid, base.
13. Special catalytic molecules called _____ speed up chemical reactions in the human body. **(a)** enzymes, **(b)** cytozymes, **(c)** cofactors, **(d)** activators, **(e)** cytochromes.
14. Which of the following is *not* a function of a protein? **(a)** support, **(b)** transport, **(c)** metabolic regulation, **(d)** storage of genetic information, **(e)** movement.
15. Complementary base pairing in DNA includes the pairs **(a)** adenine–uracil and cytosine–guanine, **(b)** adenine–thymine and cytosine–guanine, **(c)** adenine–guanine and cytosine–thymine, **(d)** guanine–uracil and cytosine–thymine.
16. What are the three subatomic particles in atoms?
17. What four major classes of organic compounds (polymers) are found in the body?
18. List three important functions of triglycerides (neutral fats) in the body.
19. List seven major functions performed by proteins.
20. **(a)** What three basic components make up a nucleotide of DNA? **(b)** What three basic components make up a nucleotide of RNA?
21. What three components are required to create the high-energy compound ATP?

LEVEL 2 Reviewing Concepts

22. If a polypeptide contains 10 peptide bonds, how many amino acids does it contain? **(a)** 9, **(b)** 10, **(c)** 11, **(d)** 12.
23. A dehydration synthesis reaction between glycerol and a single fatty acid would yield a(n) **(a)** micelle, **(b)** omega-3 fatty acid, **(c)** triglyceride, **(d)** monoglyceride, **(e)** diglyceride.
24. Explain how enzymes function in chemical reactions.
25. What is a salt? How does a salt differ from an acid or a base?
26. Explain the differences among nonpolar covalent bonds, polar covalent bonds, and ionic bonds.
27. In an exergonic reaction, **(a)** large molecules are broken down into smaller ones, **(b)** small molecules are assembled into larger ones, **(c)** molecules are rearranged to form new molecules, **(d)** molecules move from reactants to products and back, **(e)** energy is released during the reaction.
28. The hydrogen bonding that occurs in water is responsible for all of the following *except* **(a)** the high boiling point of water, **(b)** the low freezing point of water, **(c)** the ability of water to dissolve nonpolar substances, **(d)** the ability of water to dissolve inorganic salts, **(e)** the surface tension of water.
29. A sample that contains an organic molecule has the following constituents: carbon, hydrogen, oxygen, nitrogen, and phosphorus. Is the molecule more likely to be a carbohydrate, a lipid, a protein, or a nucleic acid?

LEVEL 3 Critical Thinking and Clinical Applications

30. An atom of the element calcium has 20 protons and 20 neutrons. Determine the following information about calcium: **(a)** number of electrons, **(b)** atomic number, **(c)** atomic weight, **(d)** number of electrons in each energy level.
31. A certain reaction pathway consists of four steps. How would decreasing the amount of enzyme that catalyzes the second step affect the amount of product produced at the end of the pathway?
32. An important buffer system in the human body involves carbon dioxide (CO_2) and bicarbonate ion (HCO_3^-) in the reversible reaction

$$CO_2 + H_2O \rightleftharpoons H_2CO_3 \rightleftharpoons H^+ + HCO_3^-$$

If a person becomes excited and exhales large amounts of CO_2, how will the pH of the person's body be affected?

2

+ CLINICAL CASE Wrap-Up What Is Wrong with My Baby?

Baby Sean has *cystic fibrosis (CF)*, a life-threatening genetic disease. He inherited a defective gene from each parent. This faulty gene adversely affects the transport of salt—sodium chloride (NaCl)—into and out of cells. As a result, thick, sticky secretions are produced in both the digestive and respiratory systems.

In someone with CF, the digestive juices produced are so thick they clog in the pancreas and cannot get into the small intestine. Digestive juices contain enzymes that break down carbohydrates, lipids, and proteins in food so that they can be absorbed and used by cells. Without these enzymes, Sean's food passes right through him without being digested, leading to weight loss and fatty stools. In addition, the abnormal secretions clog the lungs, making Sean short of breath, wheezy, and susceptible to repeated lung infections. Finally, CF makes his skin taste very salty because of salt being lost in the sweat.

The diagnosis of cystic fibrosis is made with a sweat test. A sweat-producing chemical is applied to an area of Sean's skin, and the sweat is collected and tested for chloride concentration.

Sean's treatment starts immediately. His parents give him digestive enzymes before each feeding so that he can absorb nutrients. Sean's parents learn chest physical therapy: For 20–60 minutes, twice per day, they percuss (clap forcefully with a cupped hand) and vibrate (shake with an open hand) his chest to loosen the thick mucus so he can cough it up and out. Soon Sean gains weight and begins to thrive, but this therapy will be lifelong.

1. What undigested substances would have made Sean's stools greasy and foamy?
2. Why are digestive enzymes necessary for life?

See the blue Answers tab at the back of the book.

Related Clinical Terms

artificial sweetener: Organic molecules that can stimulate taste buds and provide a sweet taste to foods without adding substantial amounts of calories to the diet.

heavy metal: The term used for a group of elements on the "heavier" end of the periodic table of elements. Some heavy metals—cobalt, copper, iron, manganese, molybdenum, vanadium, strontium, and zinc—are essential to health in trace amounts. Others are nonessential and can be harmful to health in excessive amounts. These include cadmium, antimony, chromium, mercury, lead, and arsenic.

3 The Cellular Level of Organization

Learning Outcomes

These Learning Outcomes correspond by number to this chapter's sections and indicate what you should be able to do after completing the chapter.

3-1 ▪ **List the functions of the plasma membrane and the structural features that enable it to perform those functions.** p. 66

3-2 ▪ **Describe the organelles of a typical cell, and indicate the specific functions of each.** p. 71

3-3 ▪ **Explain the functions of the cell nucleus, and discuss the nature and importance of the genetic code.** p. 83

3-4 ▪ **Summarize the role of DNA in protein synthesis, cell structure, and cell function.** p. 85

3-5 ▪ **Describe the processes of cellular diffusion and osmosis, and explain their role in physiological systems.** p. 90

3-6 ▪ **Describe carrier-mediated transport and vesicular transport mechanisms used by cells to facilitate the absorption or removal of specific substances.** p. 94

3-7 ▪ **Explain the origin and significance of the cell membrane potential.** p. 99

3-8 ▪ **Describe the stages of the cell life cycle, including interphase, mitosis, and cytokinesis, and explain their significance.** p. 102

3-9 ▪ **Discuss the regulation of the cell life cycle.** p. 103

3-10 ▪ **Discuss the relationship between cell division and cancer.** p. 107

3-11 ▪ **Define cellular differentiation, and explain its importance.** p. 109

CLINICAL CASE The Beat Must Go On!

3

Five-year-old Jackson comes home from kindergarten with yet another ear infection. He is a happy child with a great appetite, but he can't seem to stay healthy. His nose is constantly plugged, and he always has a cough. Sometimes he coughs so hard he vomits.

Jackson tested negative for cystic fibrosis, a genetic disease that produces thick, sticky mucus. There are no pets or cigarette smokers in the home.

Because this cough has gone on so long, Jackson is now getting a chest x-ray. The x-ray technician asks if she can take another film, carefully checking to be sure she has the "right" marker on Jackson's right side.

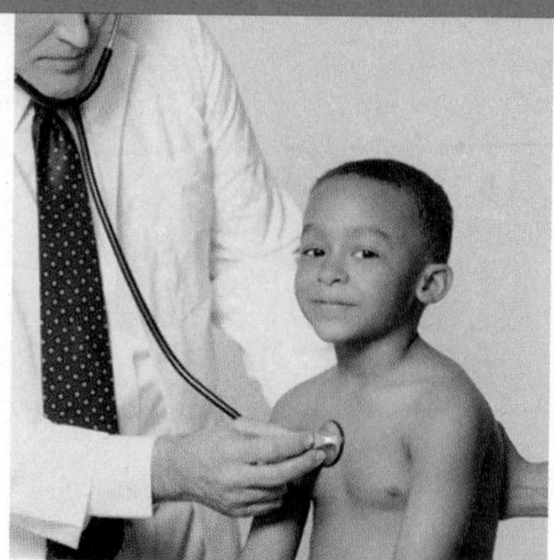

"Now that's funny," the tech says. "It looks like this x-ray is backward, but I know I took it correctly. Jackson's heart must be on the wrong side." **What clue has the technician discovered that might explain Jackson's chronic infections and cough? To find out, turn to the Clinical Case Wrap-Up on p. 113.**

An Introduction to Cells

In this chapter we will see how combinations of chemicals form *cells*, the smallest living units in the human body. We will also look at the chemical events that sustain life, which occur mostly inside cells.

Much of anatomy and physiology cannot be seen with the naked (unaided) eye, because cells are very small. The unaided human eye can see objects only about 0.1 mm in diameter. A typical cell is much smaller than that. As a result, no one could actually examine the structure of a cell until effective microscopes were invented about 400 years ago and the field of *microscopy* (investigating objects using a microscope) was born. Microscopes have evolved from the compound light microscope, which magnifies objects up to 1000 times, to the scanning electron microscope, which can magnify an object from 10 to 500,000 times!

Over time, the work of scientists in this area has led to the development of the *cell theory* in its current form. We can summarize its basic concepts as follows:

- Cells are the building blocks of all organisms.
- All cells come from the division of preexisting cells.
- Cells are the smallest units that carry out life's essential physiological functions.
- Each cell maintains homeostasis at the cellular level. Homeostasis at the level of the tissue, organ, organ system, and organism reflects the combined and coordinated actions of many cells.

The human body contains trillions of cells. All our activities—from running to thinking—result from these combined and coordinated actions of millions or even billions of cells. Many insights into human physiology arose from studies of the functioning of individual cells. What we have learned over the years has given us a new understanding of cellular physiology and the mechanisms of homeostatic control. Today, the study of cellular structure and function, or **cytology**, is part of the broader discipline of **cell biology**, which integrates aspects of biology, chemistry, and physics.

The human body contains two general classes of cells: sex cells and somatic cells. **Sex cells** (also called *germ cells* or *reproductive cells*) are either the *sperm* of males or the *oocytes* (Ō-ō-sītz), or immature ova ("eggs"), of females. The fusion of a sperm and an oocyte at fertilization is the first step in the development of a new individual. **Somatic cells** (*soma*, body), or body cells, include all the other cells. In this chapter, we focus on somatic cells. We discuss sex cells in Chapters 28 and 29.

In the rest of this chapter, we describe the structure of a typical somatic cell, consider some of the ways in which cells interact with their environment, and discuss how somatic cells reproduce. Keep in mind that the "typical" somatic cell is like the "average" person: Any description masks many individual variations. **Spotlight Figure 3–1** summarizes the anatomy of a typical (representative) cell.

3-1 The plasma membrane separates the cell from its surrounding environment and performs various functions

Learning Outcome List the functions of the plasma membrane and the structural features that enable it to perform those functions.

When you view a cell through a microscope, the first structure you encounter is its outer boundary, called the **plasma membrane**, or **cell membrane**. Membranes are neither rigid nor uniform in structure. At each location, the inner and outer surfaces of the plasma membrane may differ in important ways. For example, some enzymes are found only on the inner surface of

Go to MasteringA&P™ > Study Area > Menu > Additional Study Tools > Get Ready for A&P > Chapter 6: Cell Biology > Cell Theory: Video Tutor: General Cell Structure the membrane, and some receptor molecules are found only on the outer surface. In general, plasma membranes have these functions:

- *Physical Isolation.* The plasma membrane is a physical barrier that separates the inside of the cell, or *cytoplasm,* from the surrounding extracellular fluid. For example, the plasma membrane keeps enzymes and structural proteins inside the cell. Conditions inside and outside the cell are very different, and those differences must be maintained to preserve homeostasis.
- *Regulation of Exchange with the Environment.* The plasma membrane acts as a kind of gatekeeper. It controls the entry of ions and nutrients, the elimination of wastes, and the release of secretions.
- *Sensitivity to the Environment.* The plasma membrane is the first part of the cell affected by changes in the composition, concentration, or pH of the extracellular fluid. It also contains a variety of special structural molecules known as *receptors* that allow the cell to recognize and respond to specific molecules in its environment. For instance, the plasma membrane may receive chemical signals from other cells. The binding of just one molecule to a receptor may trigger the activation or deactivation of enzymes that affect many cellular activities.
- *Structural Support.* Specialized connections between plasma membranes, or between membranes and extracellular materials, give tissues stability. For example, the cells at the surface of the skin are tightly bound together, while those in the deepest layers are attached to extracellular protein fibers in underlying tissues.

The plasma membrane is extremely thin, ranging from 6 to 10 nm in thickness. The membrane contains lipids, proteins, and carbohydrates.

Membrane Lipids

Lipids form most of the surface area of the plasma membrane, but they make up only about 42 percent of its weight. The plasma membrane is called a **phospholipid bilayer**, because the phospholipid molecules in it form two layers (Figure 3–2).

Recall from Chapter 2 that a phospholipid has both a hydrophilic end (the phosphate portion) and a hydrophobic end (the lipid portion). ⮌ p. 51 In each half of the bilayer, the phospholipids lie with their hydrophilic heads at the membrane surface and their hydrophobic tails on the inside. In this arrangement, the hydrophilic ("water-loving") heads of the two layers are in contact with the watery environments on both sides of the membrane—the extracellular fluid outside the cell and the intracellular fluid, or *cytosol,* inside the cell. (The cytosol is the fluid component of the cytoplasm.) The hydrophobic ("water-fearing") tails form the interior of the membrane.

The lipid bilayer also contains cholesterol and other steroids, small quantities of other lipids, proteins, and glycolipids. Cholesterol is an important component of plasma membranes, with almost one cholesterol molecule for each phospholipid molecule. ⮌ p. 50 Cholesterol "stiffens" the plasma membrane, making it less fluid and less permeable.

Note the similarities in lipid organization between the plasma membrane and a micelle (look back at Figure 2–19c, p. 52). Ions and water-soluble compounds cannot enter a micelle, because the lipid tails of the phospholipid molecules are hydrophobic and will not associate with water molecules. For the same reason, such substances cannot cross the lipid portion of the plasma membrane. In this way, the hydrophobic center of the membrane isolates the cytoplasm from the surrounding fluid environment.

Membrane Proteins

Proteins, which are much denser than lipids, account for about 55 percent of the weight of a plasma membrane. There are two general structural classes of membrane proteins based on their location, integral and peripheral (see Figure 3–2). **Integral proteins** are part of the plasma membrane structure and cannot be easily separated from it without damaging or destroying the membrane. Most integral proteins span the width of the membrane one or more times, and are known as *transmembrane proteins.* These proteins contain hydrophobic portions embedded within the hydrophobic lipid bilayer, with their hydrophilic portions extended into the extracellular environment and cytosol. **Peripheral proteins** are bound to the inner or outer surface of the membrane and (like sticky notes) are easily separated from it. Integral proteins greatly outnumber peripheral proteins.

Various types of membrane proteins carry out particular specialized functions. Here are some examples of important types of membrane proteins:

- *Anchoring Proteins.* **Anchoring proteins** attach the plasma membrane to other structures and stabilize its position. Inside the cell, membrane proteins are bound to the *cytoskeleton,* a network of supporting filaments in the cytoplasm. Outer membrane proteins may attach the cell to extracellular protein fibers or to another cell.
- *Recognition Proteins (Identifiers).* The cells of the immune system recognize other cells as normal or abnormal based on the presence or absence of characteristic **recognition proteins**. Many important recognition proteins are glycoproteins. ⮌ p. 57 (We discuss one group, the MHC proteins involved in the immune response, in Chapter 22.)

SPOTLIGHT Figure 3–1 Anatomy of a Model Cell

In our model cell, a *plasma membrane* separates the cell contents, called the *cytoplasm*, from its surroundings. The cytoplasm can be subdivided into the *cytosol*, a fluid, and intracellular structures collectively known as *organelles* (or-ga-NELZ). Organelles are structures suspended within the cytosol that perform specific functions within the cell. They can be subdivided into membranous and nonmembranous organelles. Cells are surrounded by a watery medium known as the **extracellular fluid**. The extracellular fluid in most tissues is called **interstitial** (in-ter-STISH-ul) **fluid**.

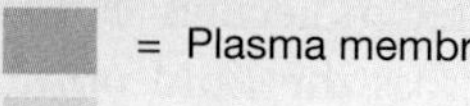

= Plasma membrane
= Nonmembranous organelles
= Membranous organelles

Microvilli

Microvilli are extensions of the plasma membrane containing microfilaments.

Function
Increase surface area to facilitate absorption of extracellular materials

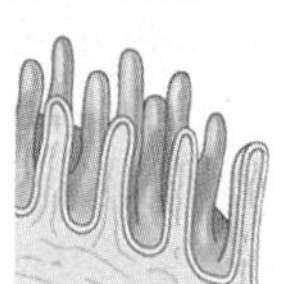

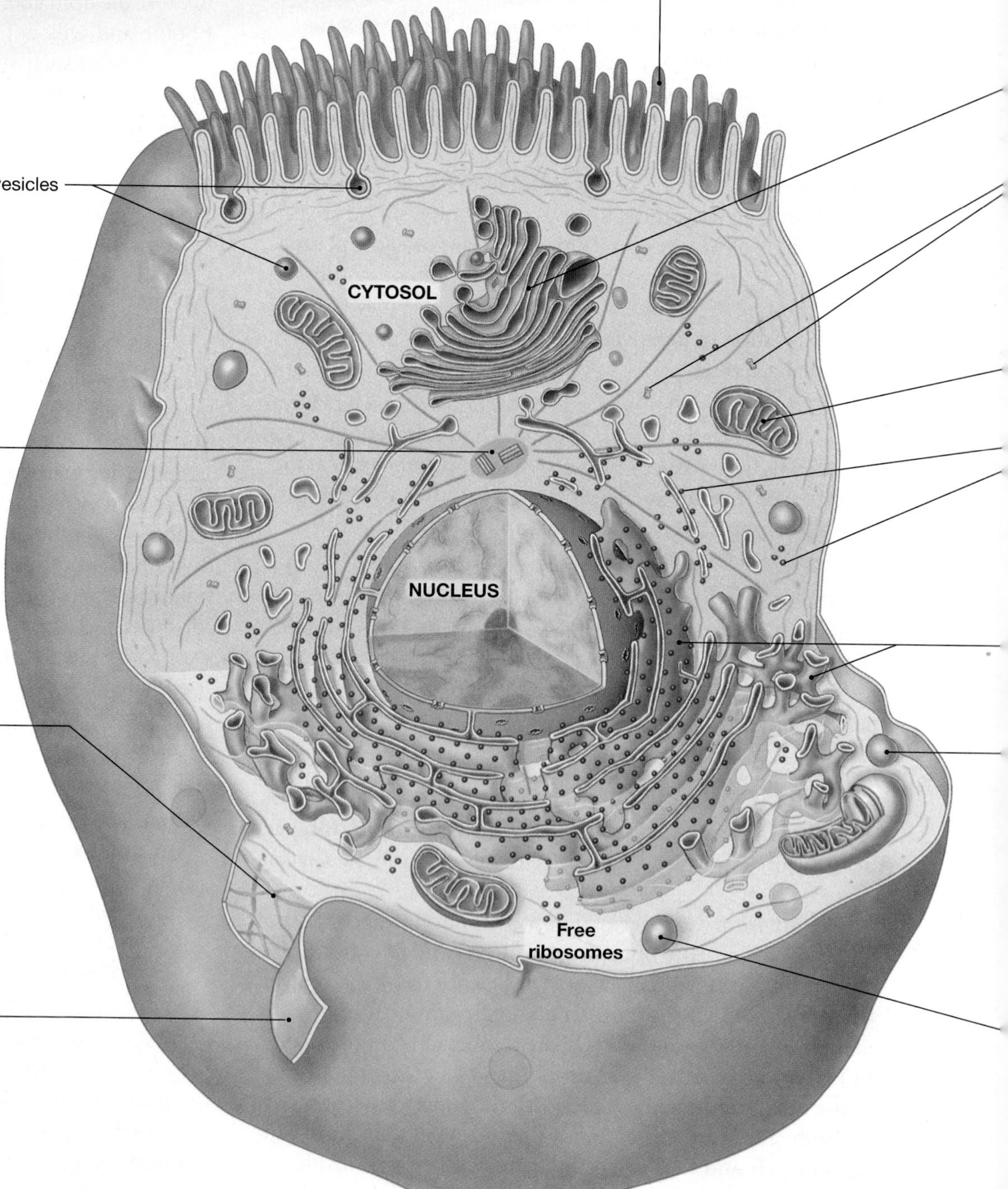

Secretory vesicles

CYTOSOL

NUCLEUS

Free ribosomes

Centrosome and Centrioles

Cytoplasm containing two centrioles at right angles; each centriole is composed of 9 microtubule triplets in a 9 + 0 array

Functions
Essential for movement of chromosomes during cell division; organization of microtubules in cytoskeleton

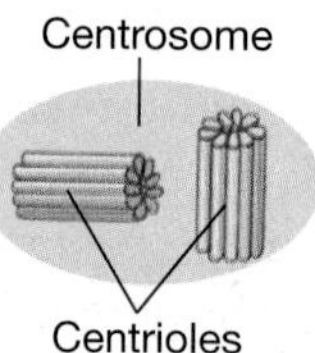

Centrosome
Centrioles

Cytoskeleton

Proteins organized in fine filaments or slender tubes

Functions
Strength and support; movement of cellular structures and materials

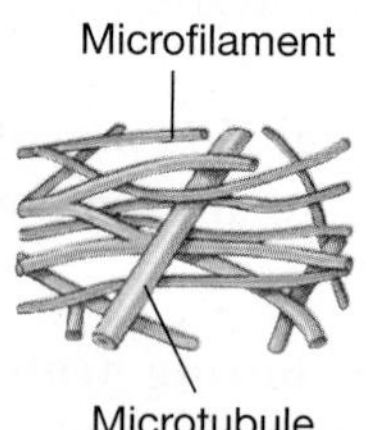

Microfilament
Microtubule

Plasma Membrane

Lipid bilayer containing phospholipids, steroids, proteins, and carbohydrates

Functions
Isolation; protection; sensitivity; support; controls entry and exit of materials

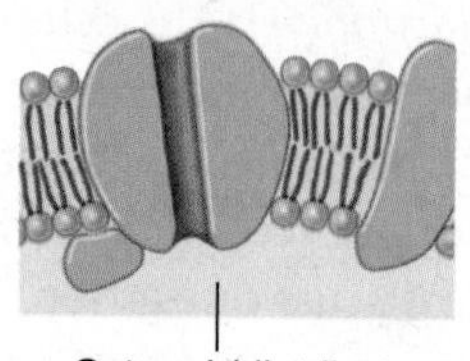

Cytosol (distributes materials by diffusion)

Cilia

Cilia are long extensions of the plasma membrane containing microtubules. There are two types: primary and motile.

Functions
A primary cilium acts as a sensor. Motile cilia move materials over cell surfaces

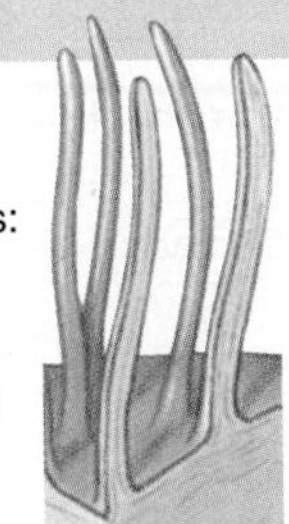

Proteasomes

Hollow cylinders of proteolytic enzymes with regulatory proteins at their ends

Functions
Breakdown and recycling of damaged or abnormal intracellular proteins

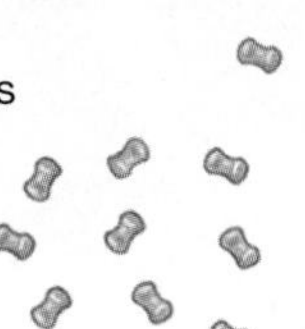

Ribosomes

RNA + proteins; fixed ribosomes bound to rough endoplasmic reticulum; free ribosomes scattered in cytoplasm

Function
Protein synthesis

Peroxisomes

Vesicles containing degradative enzymes

Functions
Catabolism of fats and other organic compounds; neutralization of toxic compounds generated in the process

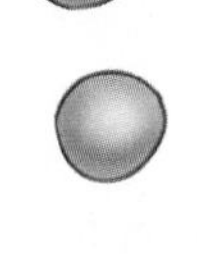

Lysosomes

Vesicles containing digestive enzymes

Function
Intracellular removal of damaged organelles or pathogens

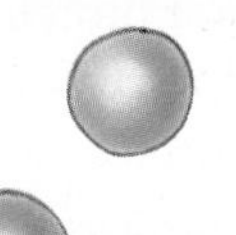

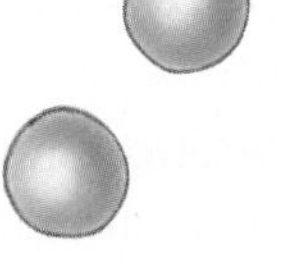

Golgi apparatus

Stacks of flattened membranes (cisternae) containing chambers

Functions
Storage, alteration, and packaging of secretory products and lysosomal enzymes

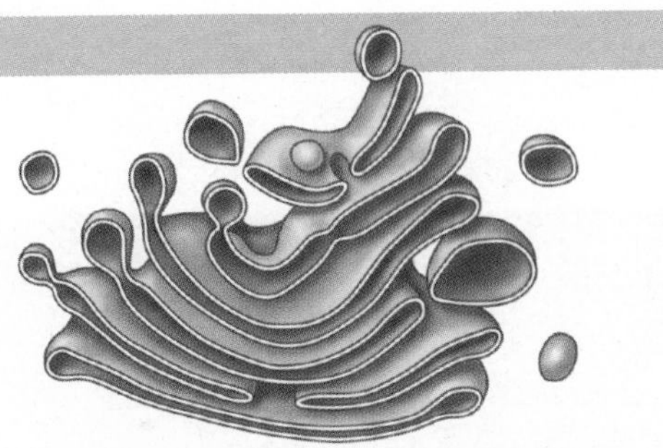

Mitochondria

Double membrane, with inner membrane folds (cristae) enclosing important metabolic enzymes

Function
Produce 95% of the ATP required by the cell

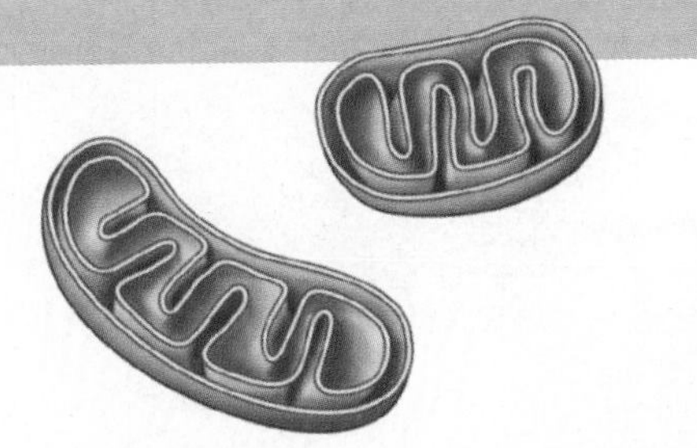

Endoplasmic reticulum (ER)

Network of membranous channels extending throughout the cytoplasm

Functions
Synthesis of secretory products; intracellular storage and transport; detoxification of drugs or toxins

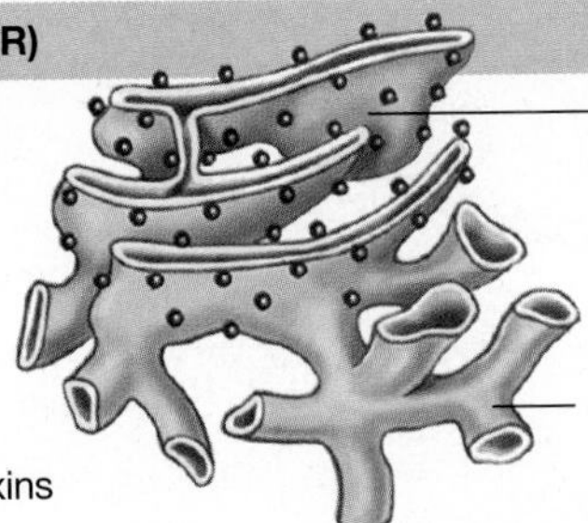
Rough ER has ribosomes, and it modifies and packages newly synthesized proteins
Smooth ER does not have ribosomes and synthesizes lipids and carbohydrates

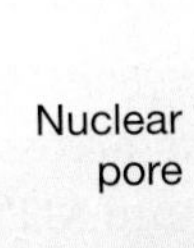

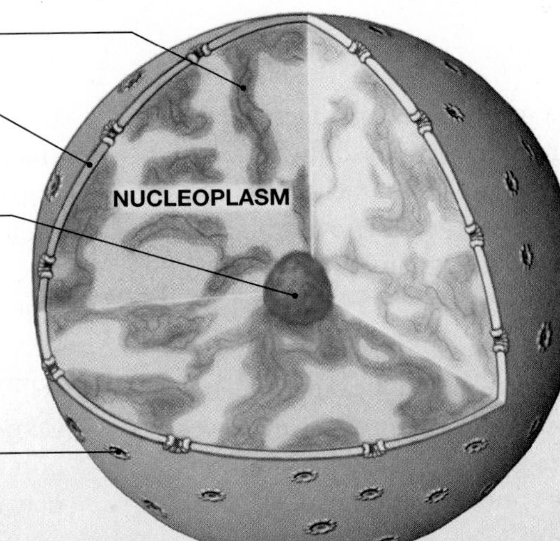
Chromatin
Nuclear envelope
NUCLEOPLASM
Nucleolus (site of rRNA synthesis and assembly of ribosomal subunits)
Nuclear pore

NUCLEUS

Nucleoplasm containing nucleotides, enzymes, nucleoproteins, and chromatin; surrounded by a double membrane, the nuclear envelope

Functions
Control of metabolism; storage and processing of genetic information; control of protein synthesis

Figure 3–2 The Plasma Membrane.

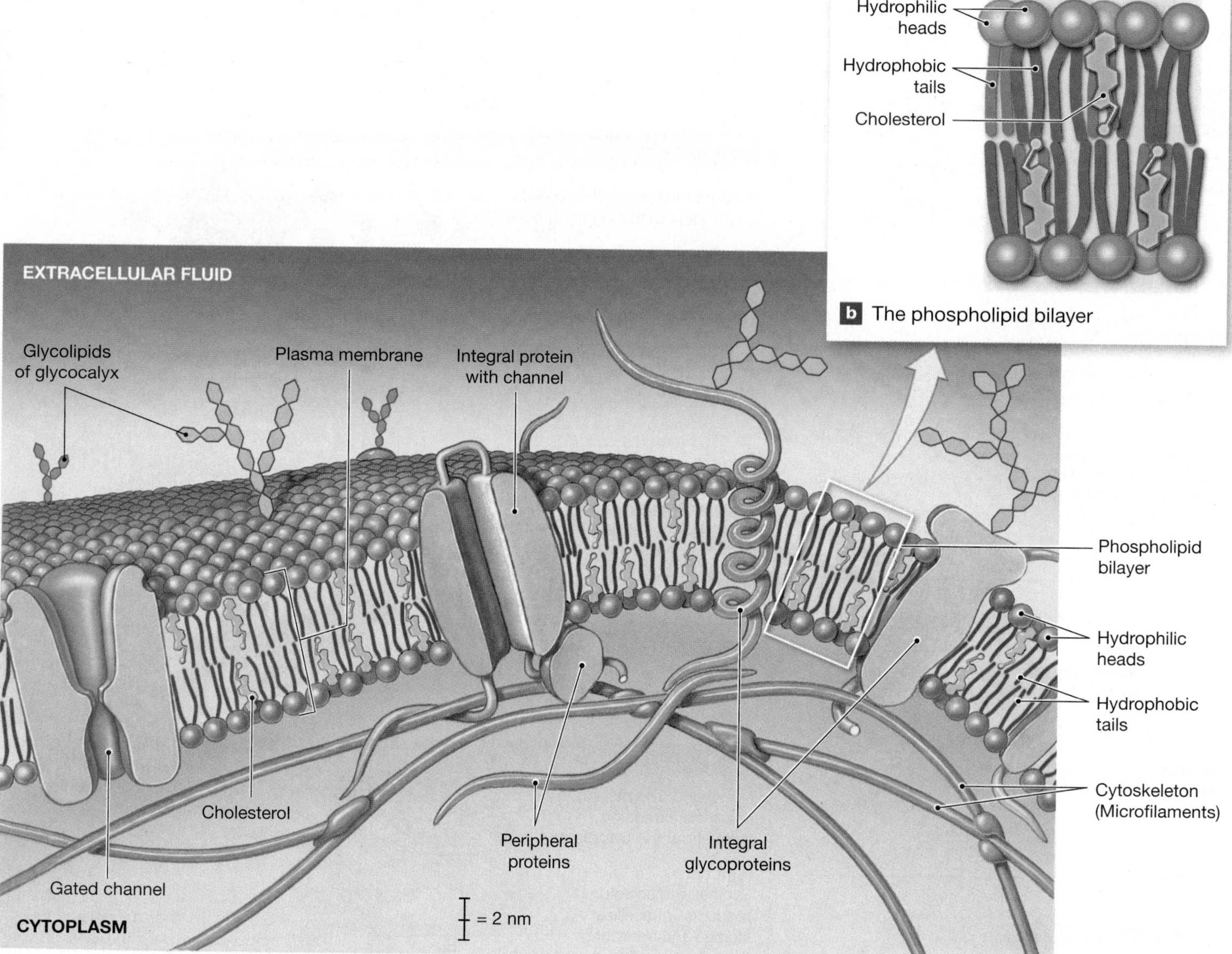
Hydrophilic heads
Hydrophobic tails
Cholesterol
b The phospholipid bilayer
EXTRACELLULAR FLUID
Glycolipids of glycocalyx
Plasma membrane
Integral protein with channel
Phospholipid bilayer
Hydrophilic heads
Hydrophobic tails
Cytoskeleton (Microfilaments)
Cholesterol
Peripheral proteins
Integral glycoproteins
Gated channel
CYTOPLASM
= 2 nm

a The plasma membrane

- *Enzymes.* Enzymes in plasma membranes may be integral or peripheral proteins. They catalyze reactions in the extracellular fluid or in the cytosol, depending on their location. For example, dipeptides are broken down into amino acids by enzymes on the extracellular membrane surfaces of cells that line the intestinal tract.
- *Receptor Proteins.* **Receptor proteins** in the plasma membrane are sensitive to the presence of specific extracellular ions or molecules called **ligands** (LĪ-gandz). A ligand can be anything from a small ion, such as a calcium ion, to a relatively large and complex hormone. When a ligand binds to the appropriate receptor, that binding may trigger changes in the activity of the cell. For example, the binding of the hormone *insulin* to a specific membrane receptor protein is the key step that leads to an increase in the cell's rate of glucose absorption. Plasma membranes differ in the type and number of receptor proteins they contain, and these differences account for a cell's sensitivity to specific hormones and other ligands.
- *Carrier Proteins.* **Carrier proteins** bind solutes and transport them across the plasma membrane. Carrier proteins may require ATP as an energy source. p. 59 For example,

virtually all cells have carrier proteins that bring glucose into the cytoplasm without expending ATP. Yet these cells must use ATP to transport ions such as sodium (Na^+) and calcium (Ca^{2+}) out of the cytoplasm and across the plasma membrane.

- *Channels.* Some integral proteins contain a **channel**, or central pore, that forms a passageway completely through the plasma membrane. Such channels permit water and small solutes to move across the plasma membrane. Ions do not dissolve in lipids, so they cannot cross the phospholipid bilayer. For this reason, ions and other small water-soluble substances can cross the membrane only by passing through channels. Most of the communication between the interior and exterior of the cell occurs through these channels. Some of the channels are called **gated channels** because they can open or close to regulate the passage of substances. Channels account for about 0.2 percent of the total surface area of the plasma membrane, but they are extremely important in physiological processes such as muscle contraction and nerve impulse transmission, as we discuss in Chapters 10 and 12.

Some integral and peripheral proteins are always confined to specific areas of the plasma membrane. These areas, called *rafts*, mark the location of anchoring proteins and some kinds of receptor proteins. Yet because membrane phospholipids are fluid at body temperature, many other integral proteins drift across the surface of the membrane like ice cubes in a bowl of punch. In addition, the composition of the entire plasma membrane can change over time as large areas of the membrane surface are removed and recycled in the ongoing process of metabolic turnover. A table listing the turnover times of the organic components of representative cells is found in the Appendix.

Membrane Carbohydrates

Carbohydrates account for about 3 percent of the weight of a plasma membrane. The carbohydrates in the plasma membrane are parts of complex molecules such as *proteoglycans*, *glycoproteins*, and *glycolipids*. pp. 51, 57 The carbohydrate portions of these large molecules extend beyond the outer surface of the membrane, forming a layer known as the **glycocalyx** (glī-kō-KĀ-liks; *calyx*, cup).

The glycocalyx has a variety of important functions, including the following:

- *Lubrication and Protection.* The glycoproteins and glycolipids form a viscous (thick and sticky) layer that lubricates and protects the plasma membrane.
- *Anchoring and Locomotion.* Because its components are sticky, the glycocalyx can help anchor the cell in place. It also takes part in the movement of specialized cells.
- *Specificity in Binding.* Glycoproteins and glycolipids can function as receptors, binding specific extracellular compounds. Such binding can change the properties of the cell surface and indirectly affect the cell's behavior.
- *Recognition.* Cells involved with the immune response recognize glycoproteins and glycolipids as normal or abnormal. The characteristics of the glycocalyx are genetically determined. The body's immune system recognizes its own membrane glycoproteins and glycolipids as "self" rather than as "foreign." This recognition system keeps your immune system from attacking your cells, and allows it to recognize and destroy invading pathogens (disease-causing agents).

The plasma membrane forms a barrier between the cytosol and the extracellular fluid. If the cell is to survive, dissolved substances and larger compounds must be permitted to cross this barrier. Nutrients must be able to enter the cell, and metabolic wastes must be able to leave. The structure of the plasma membrane is ideally suited to this need for selective transport. We will discuss selective transport and other membrane functions further, after we have completed our overview of cellular anatomy.

Checkpoint

1. **List the general functions of the plasma membrane.**
2. **Identify the components of the plasma membrane that allow it to carry out its functions.**
3. **Which component of the plasma membrane is primarily responsible for the membrane's ability to form a physical barrier between the cell's internal and external environments?**
4. **Which type of integral protein allows water, ions, and small water-soluble solutes to pass through the plasma membrane?**

See the blue Answers tab at the back of the book.

3-2 Organelles within the cytoplasm perform particular functions

Learning Outcome **Describe the organelles of a typical cell, and indicate the specific functions of each.**

Cytoplasm is a general term for the material between the plasma membrane and the membrane that surrounds the nucleus. The cytoplasm has three major subdivisions: cytosol, organelles, and inclusions. **Cytosol** is also known as *intracellular fluid.* It is a mixture of water and various dissolved and insoluble materials, in which the organelles and inclusions are suspended. Cytosol is a colloid with a consistency that varies between that of thin maple syrup and almost-set gelatin. Cytosol contains many more proteins than does extracellular fluid, the watery medium that surrounds cells. p. 42 These proteins

Go to MasteringA&P™ > Study Area > Menu > Additional Study Tools > Get Ready for A&P > Chapter 6: Cell Biology > The Cell Membrane: Video Tutor: Cell Membrane Structure are so important to the cell that they make up about 30 percent of a typical cell's weight.

Organelles ("little organs") are the internal structures of cells that perform most of the tasks that keep a cell alive and functioning normally. Each organelle has specific functions related to cell structure, growth, maintenance, and metabolism. We can divide cellular organelles into two broad categories, nonmembranous and membranous. **Nonmembranous organelles** are not completely enclosed by membranes, and all of their components are in direct contact with the cytosol. **Membranous organelles** are isolated from the cytosol by phospholipid membranes, just as the plasma membrane isolates the cytosol from the extracellular fluid. The *nucleus* is also surrounded by a membranous envelope—and is, strictly speaking, a membranous organelle. It has so many vital functions that we will consider it in a separate section.

Inclusions are masses of insoluble materials. Some inclusions are stored nutrients, such as glycogen granules in liver or in skeletal muscle cells and lipid droplets in fat cells. Others are pigment granules such as the brown skin pigment *melanin*, found in hair and skin cells.

The Cytosol

Cytosol is different from extracellular fluid. Some important differences between cytosol and extracellular fluid are as follows:

- *Sodium and potassium concentrations differ.* The concentration of potassium ions (K^+) is higher in the cytosol than in the extracellular fluid, whereas the concentration of sodium ions (Na^+) is lower in the cytosol than in the extracellular fluid.
- *Suspended protein concentrations differ.* The cytosol contains a much higher concentration of suspended proteins than does extracellular fluid. Many of the proteins are enzymes that regulate metabolic operations. Others are associated with the various organelles. The consistency of the cytosol is determined in large part by the enzymes and cytoskeletal proteins.
- *Nutrient concentrations differ.* The cytosol usually contains smaller quantities of carbohydrates and lipids, and smaller reserves of amino acids than does the extracellular fluid. The extracellular fluid is a transport medium only, and no materials are stored there. The carbohydrates in the cytosol are broken down to provide energy. When carbohydrates are unavailable, lipids, in particular triglycerides, are used instead as a source of energy. The amino acids are primarily used to manufacture proteins.

Nonmembranous Organelles

Nonmembranous organelles do not have a definite boundary. The cell's nonmembranous organelles include the *cytoskeleton*, *centrosome* with *centrioles*, *ribosomes*, and *proteasomes*. Some cells also have what are called cellular extensions, which include *microvilli*, *cilia*, and *flagella*.

The Cytoskeleton

The **cytoskeleton** serves as the cell's skeleton. It is an internal protein framework that gives the cytosol strength and flexibility. The cytoskeleton of all cells is made of *microfilaments*, *intermediate filaments*, and *microtubules* (Figure 3–3). Muscle cells contain these cytoskeletal elements plus *thick filaments*. The filaments and microtubules of the cytoskeleton form a dynamic network whose continual reorganization affects cell shape and function. This network provides support for organelles, and keeps them in their proper positions. Interactions between cytoskeletal components are also important in moving organelles and in changing the shape of the cell.

We will consider only a few of the many functions of the cytoskeleton in this section. In addition to the functions we describe here, the cytoskeleton plays a role in the metabolic organization of the cell. It determines where in the cytoplasm key enzymatic reactions take place and where specific proteins are synthesized. For example, many enzymes (especially those involved with metabolism and energy production), ribosomes, and RNA molecules responsible for the synthesis of proteins are attached to the microfilaments and microtubules of the cytoskeleton. The varied metabolic functions of the cytoskeleton are now a subject of intensive research.

Microfilaments. The smallest cytoskeletal structures are the rod-shaped **microfilaments**. These protein strands are generally about 5 nm in diameter. Typical microfilaments are made of the protein **actin**. In skeletal muscle cells, the thin actin filaments interact with other protein strands of thick *myosin* filaments to cause contraction. In cells that form a layer or lining, such as the lining of the intestinal tract, actin filaments also form a layer, called the *terminal web*, just inside the plasma membrane at the exposed surface of the cell (Figure 3–3a).

Microfilaments have the following major functions:

- *Microfilaments anchor the cytoskeleton* to integral proteins of the plasma membrane. They give the cell additional mechanical strength and attach the plasma membrane to the enclosed cytoplasm.
- *Microfilaments, interacting with other proteins, determine the consistency of the cytosol.* Where microfilaments form a dense, flexible network, the cytosol has a gelatinous consistency. Where they are widely dispersed, the cytosol is more fluid.

Intermediate Filaments. The protein composition of **intermediate filaments** varies among cell types. These filaments range from 9 to 11 nm in diameter. They are so named because they

Figure 3–3 The Cytoskeleton.

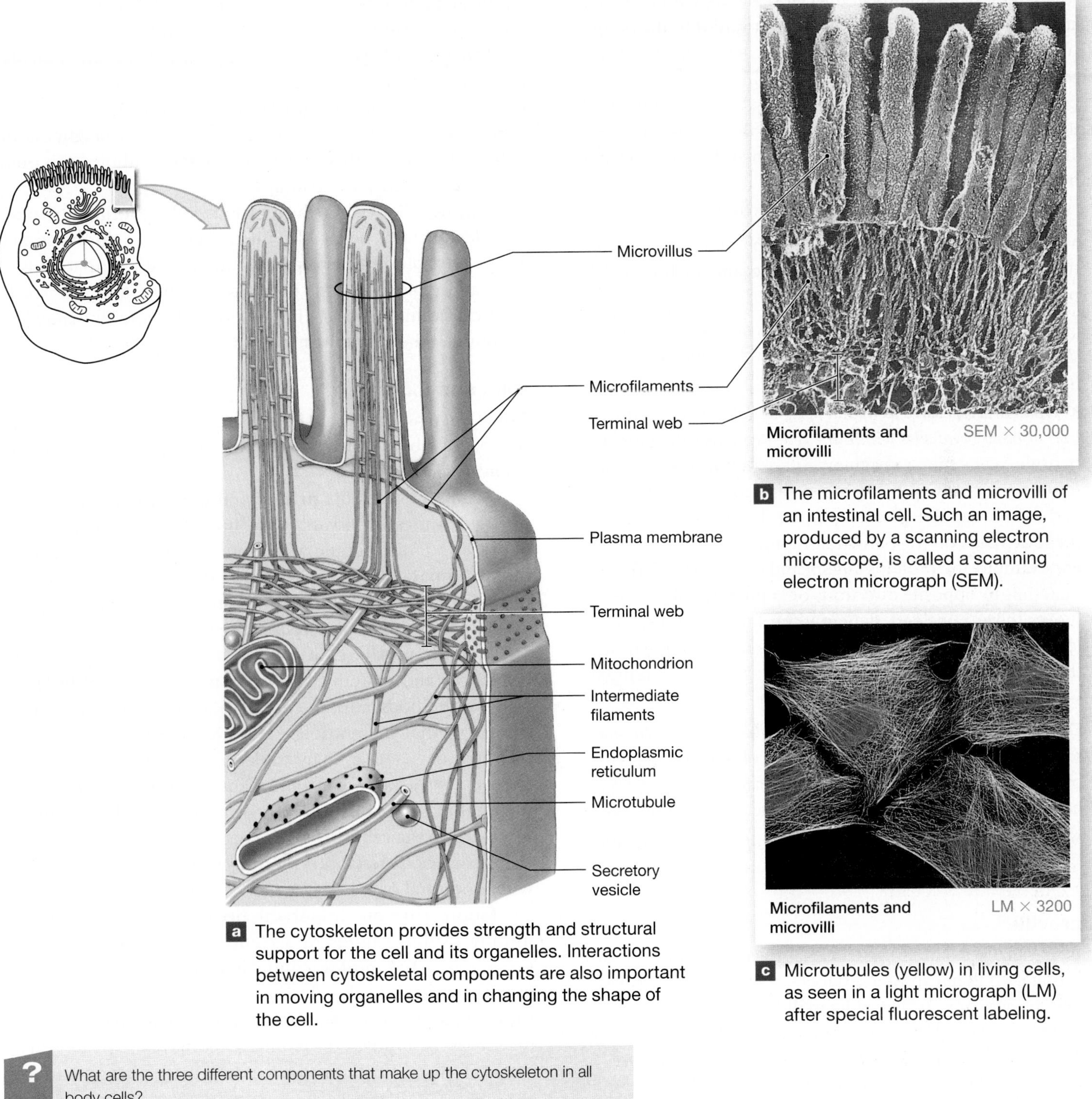
Microvillus
Microfilaments
Terminal web
Plasma membrane
Terminal web
Mitochondrion
Intermediate filaments
Endoplasmic reticulum
Microtubule
Secretory vesicle
Microfilaments and microvilli
SEM × 30,000
Microfilaments and microvilli
LM × 3200

a The cytoskeleton provides strength and structural support for the cell and its organelles. Interactions between cytoskeletal components are also important in moving organelles and in changing the shape of the cell.

b The microfilaments and microvilli of an intestinal cell. Such an image, produced by a scanning electron microscope, is called a scanning electron micrograph (SEM).

c Microtubules (yellow) in living cells, as seen in a light micrograph (LM) after special fluorescent labeling.

? What are the three different components that make up the cytoskeleton in all body cells?

are intermediate in size between microfilaments and microtubules. Intermediate filaments, which are insoluble in the watery medium, are the most durable of the cytoskeletal elements.

Intermediate filaments (1) *strengthen the cell* and help maintain its shape, (2) *stabilize the positions of organelles,* and (3) *stabilize the position of the cell with respect to surrounding cells* through specialized attachments to the plasma membrane. Many cells contain specialized intermediate filaments with unique functions. For example, the keratin fibers in superficial layers of the skin are intermediate filaments that make these layers strong and able to resist stretching.

Microtubules. Most cells contain **microtubules**, hollow tubes built from the globular protein **tubulin**. Microtubules are the largest components of the cytoskeleton, with diameters of about 25 nm. Microtubules extend outward into the periphery of the cell from a region near the nucleus called the *centrosome* (see **Spotlight Figure 3–1**).

Each microtubule forms by the aggregation of tubulin molecules, growing out from its origin at the centrosome. The entire structure persists for a time and then disassembles into individual tubulin molecules again.

Microtubules have the following functions:

- *Microtubules form the main portions of the cytoskeleton,* giving the cell strength, maintaining its shape, and anchoring the position of major organelles.
- *Microtubules change the shape of the cell, and may assist in cell movement,* through the assembly and disassembly of microtubules.
- *Microtubules can serve as a kind of monorail system to move vesicles or other organelles within the cell.* Proteins called **motor proteins** create the movement. These motor proteins bind to both the structure being moved and to a microtubule and then move along its length. The direction of movement depends on the particular motor protein involved. For example, the proteins *kinesin* and *dynein* (DĪ-nēn) carry materials in opposite directions on a microtubule: Kinesin moves toward one end, dynein toward the other. Regardless of the direction of transport or the nature of the motor, the process requires ATP and is essential to normal cellular function.
- *During cell division, microtubules distribute duplicated chromosomes containing DNA* to opposite ends of the dividing cell by forming a network of microtubules called the *spindle apparatus.* We look at this process in more detail in Section 3-8.
- *Microtubules form structural components of organelles,* such as *centrioles* and *cilia.*

Microvilli

Many cells have small, finger-shaped projections of the plasma membrane on their exposed surfaces (**Figure 3–3b**). These nonmotile projections, called **microvilli** (singular, *microvillus*), greatly increase the surface area of the cell exposed to the extracellular environment. Accordingly, they cover the surfaces of cells that are actively absorbing materials, such as the cells lining the digestive tract. Microvilli have extensive connections with the cytoskeleton. A core of microfilaments stiffens each microvillus and anchors it to the cytoskeleton at the terminal web.

Centrosome and Centrioles

The **centrosome** is a region of cytoplasm located next to the nucleus in a cell. It is the *microtubule-organizing center* of animal cells, and the heart of the cytoskeletal system. Microtubules of the cytoskeleton generally begin at the centrosome and radiate out through the cytoplasm. In all animal cells capable of undergoing cell division, the centrosome surrounds a pair of cylindrical structures called **centrioles**. The centrioles lie perpendicular to each other and are composed of short microtubules. The microtubules form nine groups, three in each group. Each of these nine "triplets" is connected to its nearest neighbors on either side. Because there are no central microtubules, this organization is called a *9 + 0 array* (**Figure 3–4a**).

During cell division, the centrioles aid the formation of the spindle apparatus needed for the movement of chromosomes. Mature red blood cells, skeletal muscle cells, cardiac muscle cells, and typical neurons have no centrioles, and as a result, these cells cannot divide. The centrioles also form the *basal bodies* found at the base of some cellular extensions.

Cilia and Flagella

Cilia (singular, *cilium*) are fairly long, slender extensions of the plasma membrane. Two types of cilia are found in human cells, *nonmotile* and *motile.*

A single, nonmotile **primary cilium** is found on the cells of a wide variety of tissues in the body. Its structure is similar to the 9 + 0 microtubule organization of a centriole. The nonmotile primary cilium acts as a signal sensor, detecting environmental stimuli and coordinating activities such as embryonic development and homeostasis at the tissue level.

Multiple **motile cilia** are found on cells lining both the respiratory and reproductive tracts, and at various other locations in the body. Motile cilia have an internal arrangement similar to that of centrioles. However, in cilia, nine *pairs* of microtubules (rather than triplets) surround a central pair (**Figure 3–4b**)—an organization known as a *9 + 2 array.* The microtubules are anchored to a compact **basal body** situated just beneath the cell surface. The organization of microtubules in the basal body resembles the array of a centriole: nine triplets with no central pair.

Motile cilia are important because they "beat" rhythmically to move fluids or secretions across the cell surface (**Figure 3–4c**). A motile cilium is relatively stiff during the *power stroke* and flexible during the *return stroke.* For example, the ciliated cells along your trachea beat their cilia in synchronized waves to move sticky mucus and trapped dust particles toward the throat and away from delicate respiratory surfaces. If the cilia are damaged or immobilized by heavy smoking or a metabolic problem, this cleansing action is lost. When irritants are no longer removed, a chronic cough and respiratory infections develop. Ciliated cells also move oocytes along the uterine tubes, and sperm from the testes into the male reproductive tract. Defective primary cilia are known to be responsible for a wide range of human disorders, collectively known as *ciliopathies.*

Figure 3–4 Centrioles and Cilia.

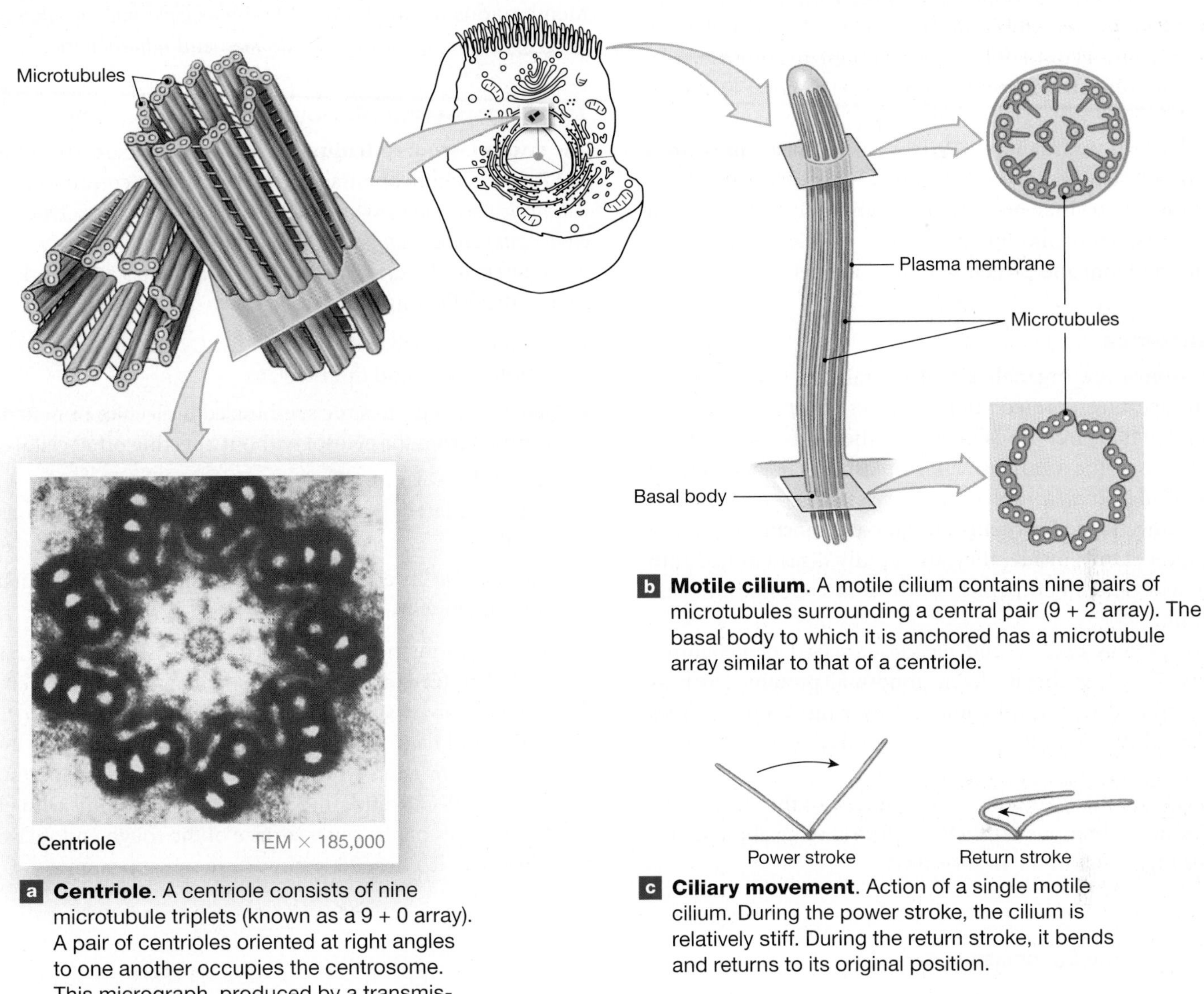
Microtubules
Plasma membrane
Microtubules
Basal body
Centriole
TEM × 185,000
Power stroke
Return stroke

a **Centriole.** A centriole consists of nine microtubule triplets (known as a 9 + 0 array). A pair of centrioles oriented at right angles to one another occupies the centrosome. This micrograph, produced by a transmission electron microscope, is called a TEM.

b **Motile cilium.** A motile cilium contains nine pairs of microtubules surrounding a central pair (9 + 2 array). The basal body to which it is anchored has a microtubule array similar to that of a centriole.

c **Ciliary movement.** Action of a single motile cilium. During the power stroke, the cilium is relatively stiff. During the return stroke, it bends and returns to its original position.

A **flagellum** (plural, *flagella*) is a whip-like extension of the plasma membrane. Flagella have the same 9 + 2 microtubule organization as motile cilia, but are much longer and beat in a wavelike fashion. The only human cell with a flagellum is a sperm, and there is only one flagellum per cell. Sperm with more than one flagellum are abnormal and cannot fertilize an oocyte.

Ribosomes

Now let's turn to how proteins are produced within cells, using information provided by the DNA of the nucleus. **Ribosomes** are the organelles responsible for protein synthesis. The number of ribosomes in a particular cell varies with the type of cell and its demand for new proteins. For example, liver cells, which manufacture blood proteins, contain far more ribosomes than do fat cells, which primarily synthesize lipids.

Individual ribosomes are not visible with a light microscope. In an electron micrograph, they appear as dense granules approximately 25 nm in diameter. Each ribosome is about 60 percent RNA and 40 percent protein.

A functional ribosome consists of two subunits that are normally separate and distinct. One is called a **small ribosomal subunit** and the other a **large ribosomal subunit**. These subunits contain special proteins and **ribosomal RNA (rRNA)**, one of the RNA types introduced in Chapter 2. Before protein synthesis can begin, a small and a large ribosomal subunit in the cytoplasm must join together with a strand of *messenger RNA* (*mRNA*), another type of RNA. Protein synthesis then begins in the cytoplasm.

3

The two major types of functional ribosomes in cells are free ribosomes and fixed ribosomes. **Free ribosomes** are scattered throughout the cytoplasm. The proteins they manufacture directly enter the cytosol. Ribosomes synthesizing proteins with destinations other than the cytosol become temporarily bound, or fixed, to the *endoplasmic reticulum (ER)*, a membranous organelle. Proteins manufactured by such **fixed ribosomes** enter the ER, where they are modified and packaged for use within the cell or they are secreted from the cell. We look at ribosomal structure and functions when we discuss the endoplasmic reticulum and protein synthesis in later sections.

Proteasomes

Proteasomes are organelles that contain an assortment of protein-digesting (proteolytic) enzymes, or *proteases*. They are smaller than free ribosomes and their job is to remove proteins from the cytoplasm. Cytoplasmic enzymes attach chains of *ubiquitin*, a molecular "tag," to proteins destined for recycling. Tagged proteins are quickly transported into a proteasome. Once inside, they are rapidly disassembled into amino acids and small peptides, which are released into the cytoplasm.

Proteasomes remove and recycle damaged or denatured proteins. They also break down abnormal proteins, such as those produced within cells infected by viruses. Proteasomes also play a key role in the immune response, as we will see in Chapter 22.

Check **Spotlight Figure 3–1** for a review of the characteristics of nonmembranous organelles before reading about membranous organelles in the next section.

Membranous Organelles

Membranous organelles include the *endoplasmic reticulum*, the *Golgi apparatus*, *lysosomes*, *peroxisomes*, and *mitochondria*.

The Endoplasmic Reticulum

The **endoplasmic reticulum** (en-dō-PLAZ-mik re-TIK-ū-lum), or **ER**, is a network of intracellular membranes continuous with the *nuclear envelope*, which surrounds the nucleus. The name *endoplasmic reticulum* is very descriptive. *Endo-* means "within," *plasm* refers to the cytoplasm, and a *reticulum* is a network. The ER has the following major functions:

- *Synthesis.* Specialized regions of the ER synthesize proteins, carbohydrates, and lipids.
- *Storage.* The ER can store synthesized molecules or materials absorbed from the cytosol without affecting other cellular operations.
- *Transport.* Materials can travel from place to place within the ER.
- *Detoxification.* The ER can absorb drugs or toxins and neutralize them with enzymes.

The ER forms hollow tubes, flattened sheets, and chambers called **cisternae** (sis-TUR-nē; singular, *cisterna*, a reservoir for water). Two types of ER exist: *smooth endoplasmic reticulum (SER)* and *rough endoplasmic reticulum (RER)* (**Figure 3–5**). The term *smooth* refers to the fact that no fixed ribosomes are associated with the smooth endoplasmic reticulum, while the fixed ribosomes on the outer surface of the rough endoplasmic reticulum give it a beaded, grainy, or rough appearance.

Figure 3–5 The Endoplasmic Reticulum.

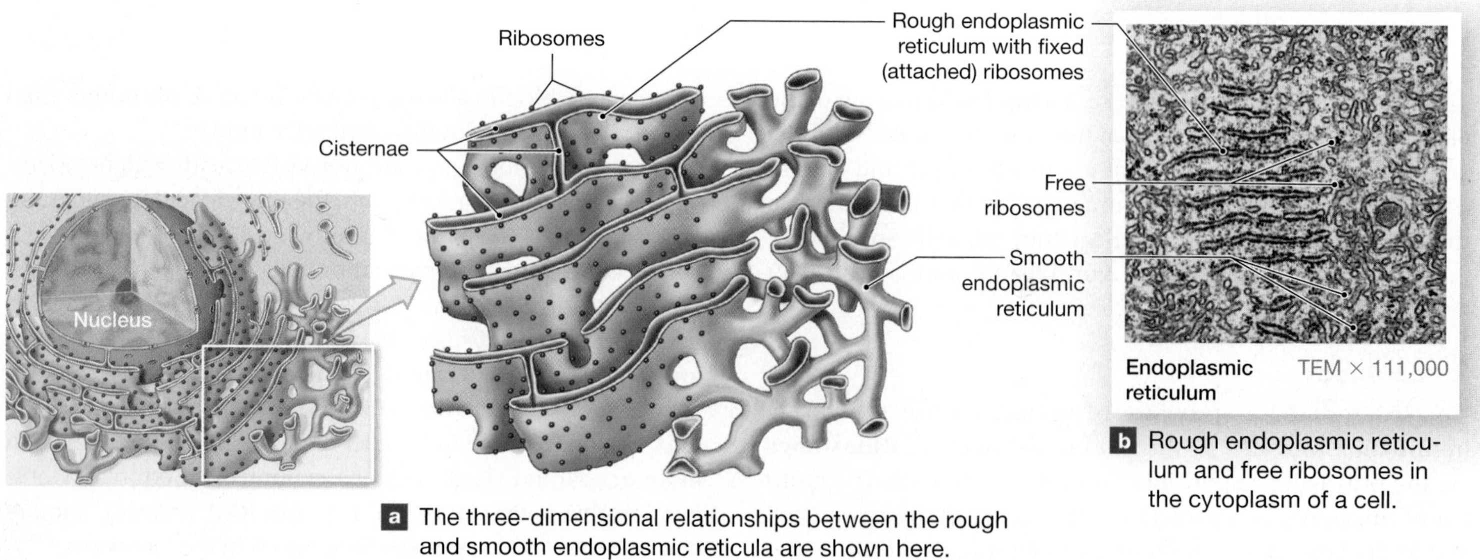
Ribosomes
Rough endoplasmic reticulum with fixed (attached) ribosomes
Cisternae
Free ribosomes
Smooth endoplasmic reticulum
Nucleus
Endoplasmic reticulum
TEM × 111,000

a The three-dimensional relationships between the rough and smooth endoplasmic reticula are shown here.

b Rough endoplasmic reticulum and free ribosomes in the cytoplasm of a cell.

The amount of endoplasmic reticulum and the proportion of RER to SER vary with the type of cell and its ongoing activities. For example, pancreatic cells that manufacture digestive enzymes contain an extensive RER, but the SER is relatively small. The situation is just the reverse in the cells of reproductive organs that synthesize steroid hormones.

Smooth Endoplasmic Reticulum. The **smooth endoplasmic reticulum (SER)** is involved with the synthesis of lipids, fatty acids, and carbohydrates; the sequestering of calcium ions; and the detoxification of drugs. Important functions of the SER include the following:

- *synthesis of the phospholipids and cholesterol* needed for maintenance and growth of the plasma membrane, ER, nuclear membrane, and Golgi apparatus in all cells;
- *synthesis of steroid hormones,* such as *androgens* and *estrogens* (the dominant sex hormones in males and in females, respectively) in the reproductive organs;
- *synthesis and storage of glycerides, especially triacylglycerides (triglycerides),* in liver cells and fat cells; and
- *synthesis and storage of glycogen* in skeletal muscle and liver cells.

In muscle cells, neurons, and many other types of cells, the SER also adjusts the composition of the cytosol by absorbing and storing ions, such as Ca^{2+}, or large molecules. In addition, the SER in liver and kidney cells detoxifies or inactivates drugs.

Rough Endoplasmic Reticulum. The **rough endoplasmic reticulum (RER)** functions as a combination workshop and shipping warehouse, because the fixed ribosomes on the RER synthesize proteins. Many of these newly synthesized proteins are then chemically modified in the RER and packaged for export to their next destination, the Golgi apparatus.

The new polypeptide chains produced at fixed ribosomes are released into the cisternae of the RER. Inside the RER, each protein assumes its secondary and tertiary structures. ⮌ p. 55 Some of the proteins are enzymes that will function inside the endoplasmic reticulum. Other proteins are chemically modified by the attachment of carbohydrates, creating glycoproteins.

Most of the proteins and glycoproteins produced by the RER are packaged into small membranous sacs that pinch off from the tips of the cisternae. These **transport vesicles** then deliver their contents to the Golgi apparatus.

The Golgi Apparatus

When a transport vesicle carries a newly synthesized protein or glycoprotein that is destined for export from the cell, it travels from the ER to the **Golgi** (GŌL-jē) **apparatus**, or **Golgi complex**, an organelle that looks a bit like a stack of dinner plates (Figure 3–6). This organelle typically consists of five or six flattened membranous discs called *cisternae.* A single cell may contain several of these organelles, most often near the nucleus.

The Golgi apparatus has the following major functions (**Spotlight Figure 3–7**). It:

- *modifies and packages secretions,* such as hormones or enzymes, for release from the cell;
- *adds or removes carbohydrates to or from proteins* to change protein structure and thus function,
- *renews or modifies the plasma membrane*; and
- *packages special enzymes within vesicles (lysosomes)* for use in the cytoplasm.

Figure 3–6 The Golgi Apparatus.

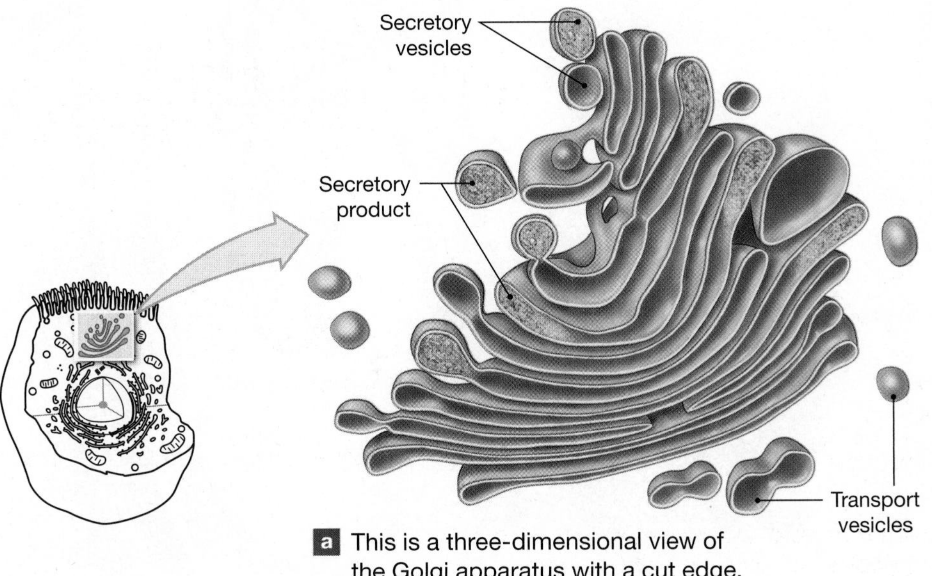
Secretory vesicles
Secretory product
Transport vesicles

a This is a three-dimensional view of the Golgi apparatus with a cut edge.

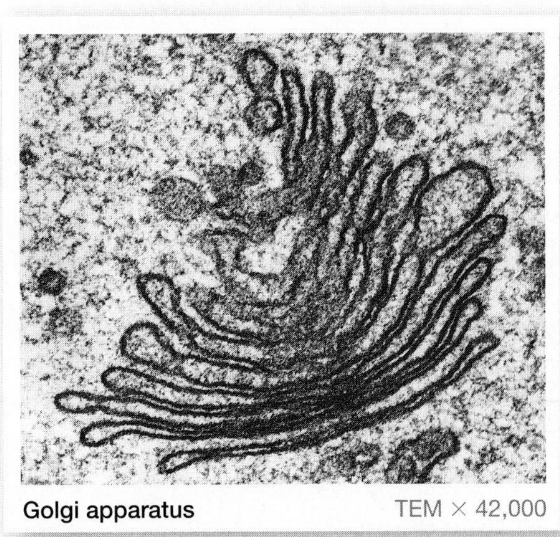
Golgi apparatus
TEM × 42,000

b This is a sectional view of the Golgi apparatus of an active secretory cell.

SPOTLIGHT

Figure 3–7

Protein Synthesis, Processing, and Packaging

The Golgi apparatus plays a major role in modifying and packaging newly synthesized proteins. Some proteins and glycoproteins synthesized in the rough endoplasmic reticulum (RER) are delivered to the Golgi apparatus by transport vesicles. Here's a summary of the process, beginning with DNA.

1 Protein synthesis begins when a gene on DNA produces messenger RNA (mRNA), the template for protein synthesis.

2 The mRNA leaves the nucleus and attaches to a free ribosome in the cytoplasm, or a fixed ribosome on the RER.

3a Proteins constructed on free ribosomes are released into the cytosol for use within the cell.

3b Protein synthesis on fixed ribosomes occurs at the RER. The newly synthesized protein folds into its three-dimensional shape.

4 The proteins are then modified within the ER. Regions of the ER then bud off, forming transport vesicles containing modified proteins and glycoproteins.

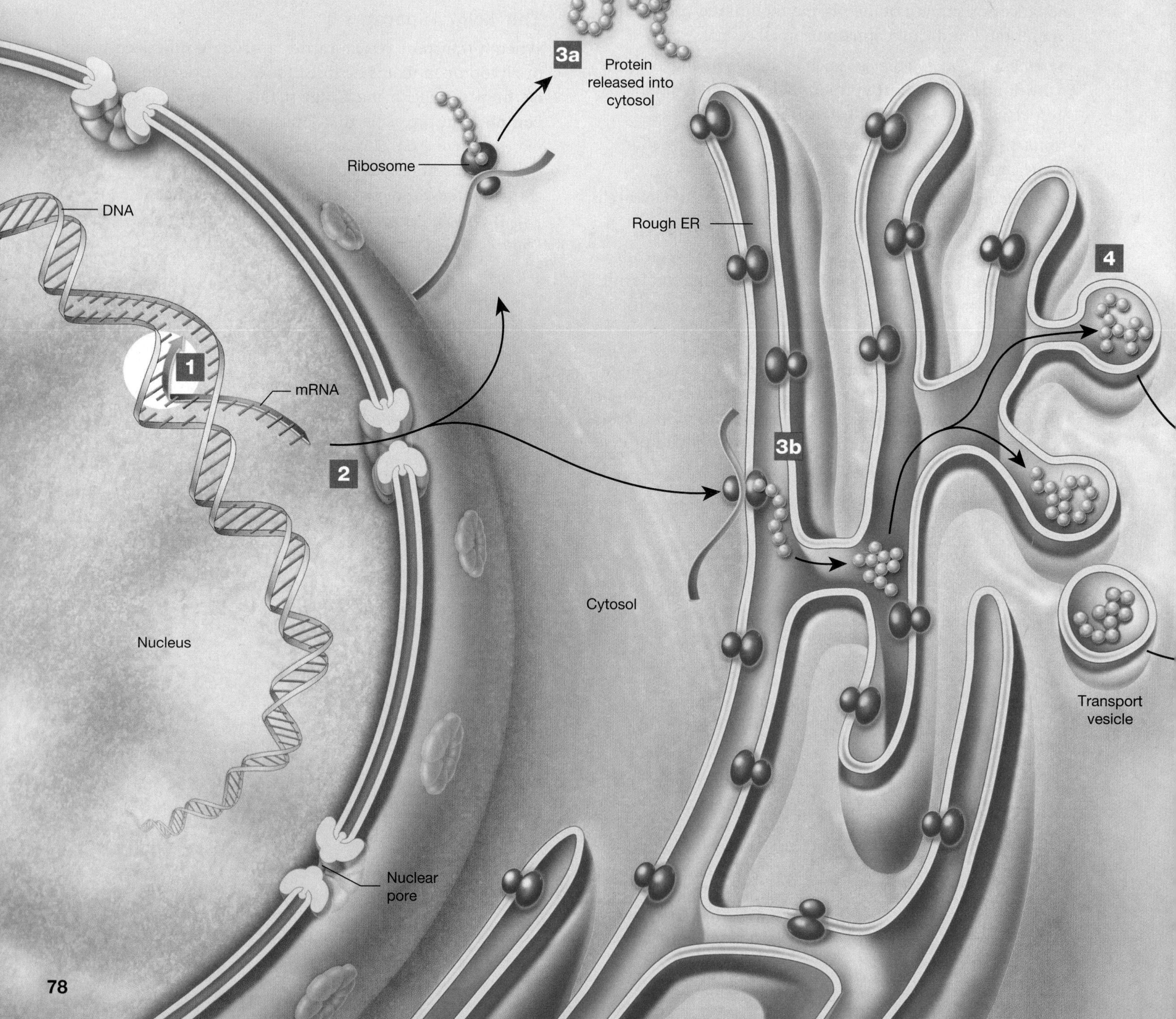
3a
Protein released into cytosol
Ribosome
DNA
Rough ER
4
1
mRNA
2
3b
Cytosol
Nucleus
Transport vesicle
Nuclear pore

5 The transport vesicles carry the proteins and glycoproteins generated in the ER toward the Golgi apparatus. The transport vesicles then fuse to create the forming *cis* face ("receiving side") of the Golgi apparatus.

6 Multiple transport vesicles combine to form cisternae on the *cis* face. Further protein and glycoprotein modification and packaging occur as the cisternae move toward the maturing (*trans*) face. Small transport vesicles return resident Golgi proteins to the forming cis face for reuse.

7 The maturing *trans* face ("shipping side") generates vesicles that carry modified proteins away from the Golgi apparatus. One type of vesicle becomes a lysosome, which contains digestive enzymes.

8 Two other types of vesicles proceed to the plasma membrane: secretory and membrane renewal. **Secretory vesicles** fuse with the plasma membrane and empty their products outside the cell by exocytosis. **Membrane renewal vesicles** add new lipids and proteins to the plasma membrane.

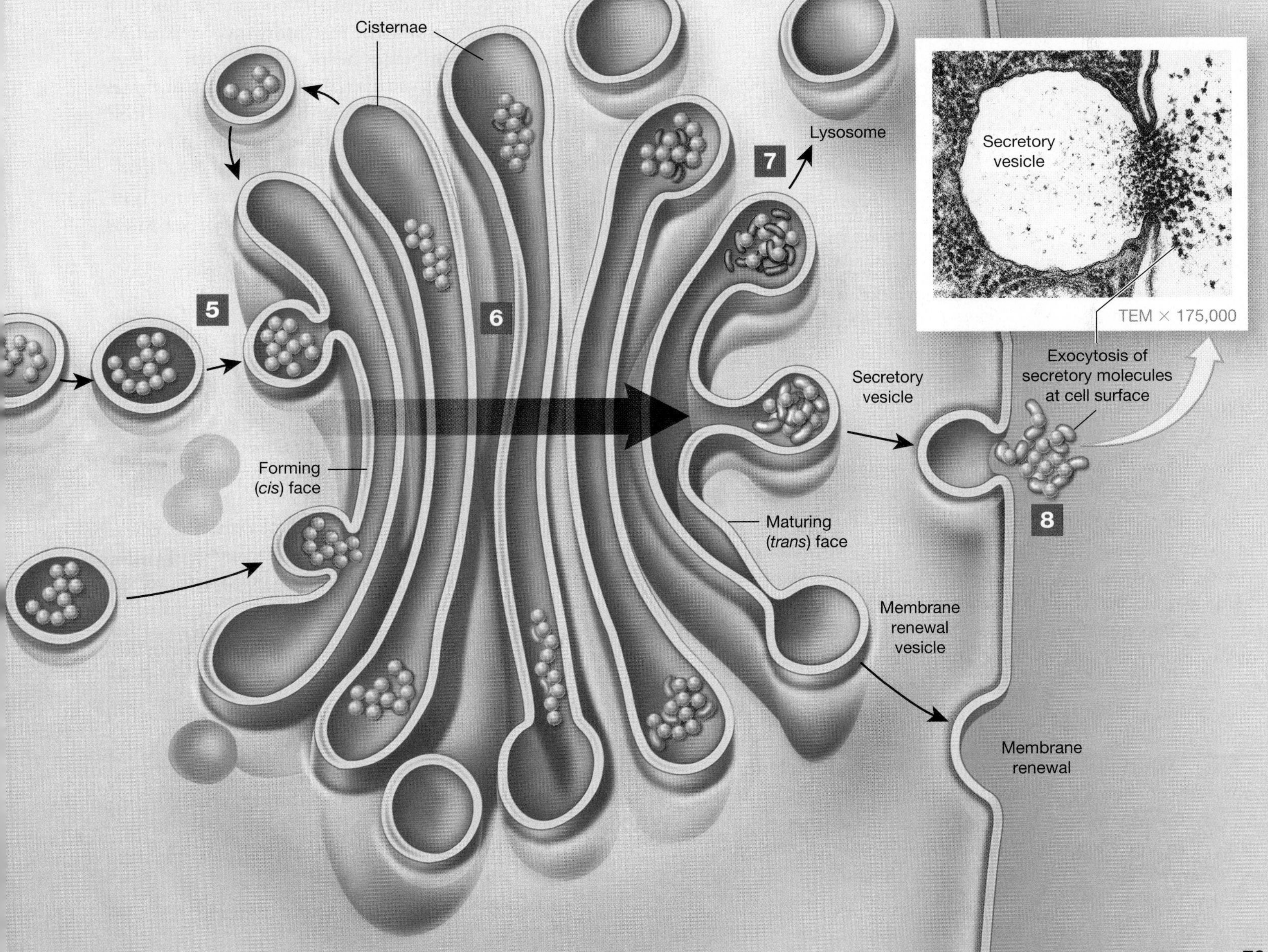
Cisternae
Lysosome
7
Secretory vesicle
TEM × 175,000
5
6
Forming (cis) face
Secretory vesicle
Exocytosis of secretory molecules at cell surface
8
Maturing (trans) face
Membrane renewal vesicle
Membrane renewal

Figure 3–8 Lysosome Functions. Primary lysosomes, formed by the Golgi apparatus, contain inactive enzymes. They may be activated under any of the three basic conditions indicated here.

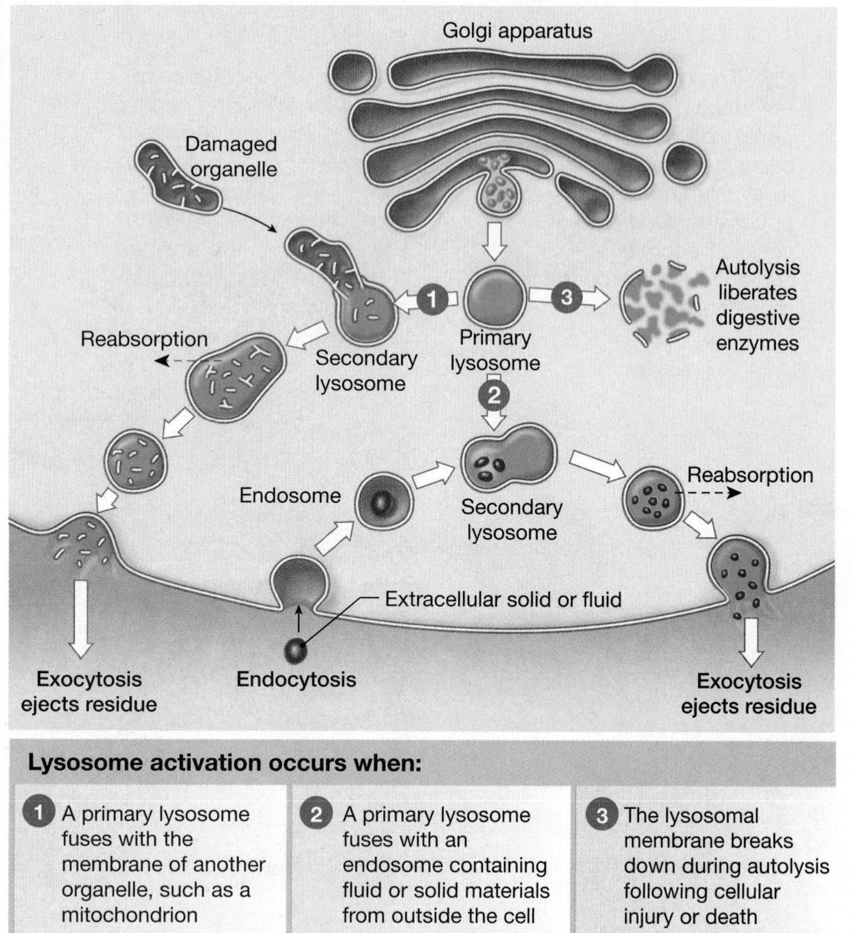
Golgi apparatus
Damaged organelle
Autolysis liberates digestive enzymes
1
3
Primary lysosome
Reabsorption
Secondary lysosome
2
Endosome
Secondary lysosome
Reabsorption
Extracellular solid or fluid
Exocytosis ejects residue
Endocytosis
Exocytosis ejects residue

Lysosome activation occurs when:

1 A primary lysosome fuses with the membrane of another organelle, such as a mitochondrion	2 A primary lysosome fuses with an endosome containing fluid or solid materials from outside the cell	3 The lysosomal membrane breaks down during autolysis following cellular injury or death

Lysosomes

Cells often need to break down and recycle large organic molecules and even complex structures like organelles. The breakdown process requires powerful enzymes, and it often generates toxic chemicals that could damage or kill the cell. **Lysosomes** (LĪ-sō-sōmz; *lyso-*, a loosening + *soma*, body) are vesicles that provide an isolated environment for potentially dangerous chemical reactions. These vesicles, produced by the Golgi apparatus, contain digestive enzymes that break organic polymers into monomers. Lysosomes are small, often spherical bodies with contents that look dense and dark in electron micrographs.

Lysosomes have several functions (Figure 3–8). One is to remove damaged organelles. *Primary lysosomes* contain inactive enzymes. When these lysosomes fuse with the membranes of damaged organelles (such as mitochondria or fragments of the ER), the enzymes are activated and *secondary lysosomes* are formed. The enzymes then break down the contents. The cytosol reabsorbs released nutrients, and the remaining material is expelled from the cell.

Lysosomes also destroy bacteria (as well as liquids and organic debris) that enter the cell from the extracellular fluid. The cell encloses these substances in a small portion of the plasma membrane, which is then pinched off to form a transport vesicle, or *endosome*, in the cytoplasm. (We discuss this method of transporting substances into the cell, called *endocytosis*, in Section 3-6.) Then a primary lysosome fuses with the vesicle, forming a secondary lysosome. Activated enzymes inside break down the contents and release usable substances, such as sugars or amino acids. In this way, the cell both protects itself against harmful substances and obtains valuable nutrients.

Lysosomes also do essential cleanup and recycling inside the cell. For example, when muscle cells are inactive, lysosomes gradually break down their contractile proteins. (This mechanism accounts for the reduction in muscle mass that accompanies aging.) The process is usually precisely controlled, but in a damaged or dead cell, the regulatory mechanism fails as lysosome membranes become increasingly permeable. Lysosomes then disintegrate, releasing enzymes that become activated within the cytosol. These enzymes rapidly destroy the cell's proteins and organelles in a process called **autolysis** (aw-TOL-i-sis; *auto-*, self). Although many factors appear to increase lysosome membrane permeability, we do not yet know how to control lysosomal activities.

Peroxisomes

Peroxisomes are smaller than lysosomes and carry a different group of enzymes. In contrast to lysosomes, which are produced at the Golgi apparatus, new peroxisomes are produced by the growth and subdivision of existing peroxisomes. Their enzymes are produced at free ribosomes and transported from the cytosol into the peroxisomes by carrier proteins.

Peroxisomes absorb and break down fatty acids and other organic compounds. As they do so, peroxisomes generate hydrogen peroxide (H_2O_2), a potentially dangerous free radical. p. 35 Catalase, the most abundant enzyme within

+ Clinical Note Lysosomal Storage Diseases

Problems in producing lysosomal enzymes cause more than 30 serious diseases affecting children. In these conditions, called *lysosomal storage diseases*, the lack of a specific lysosomal enzyme results in the buildup of waste products and debris that lysosomes normally remove and recycle. Affected individuals may die when vital cells, such as those of the heart, can no longer function.

the peroxisome, then breaks down the hydrogen peroxide to oxygen and water. In this way, peroxisomes protect the cell from the potentially damaging effects of the free radicals produced during catabolism. Peroxisomes are present in all cells, but their numbers are highest in metabolically active cells, such as liver cells.

Mitochondria

The cells of all living things require energy to carry out the functions of life. The organelles that produce energy in the form of ATP molecules are the **mitochondria** (mī-tō-KON-drē-ŭh; singular, *mitochondrion*; *mitos*, thread + *chondrion*, granule). The number of mitochondria in a particular cell varies with the cell's energy demands. These organelles may account for 30 percent of the volume of a heart muscle cell, yet are absent in red blood cells.

Mitochondria have an unusual double membrane (**Figure 3–9a**). The outer membrane surrounds the organelle. The inner membrane contains numerous folds called **cristae** (KRIS-tē), which surround the fluid contents, or **matrix**, of the mitochondrion. Cristae increase the membrane surface area

Figure 3–9 Mitochondria.

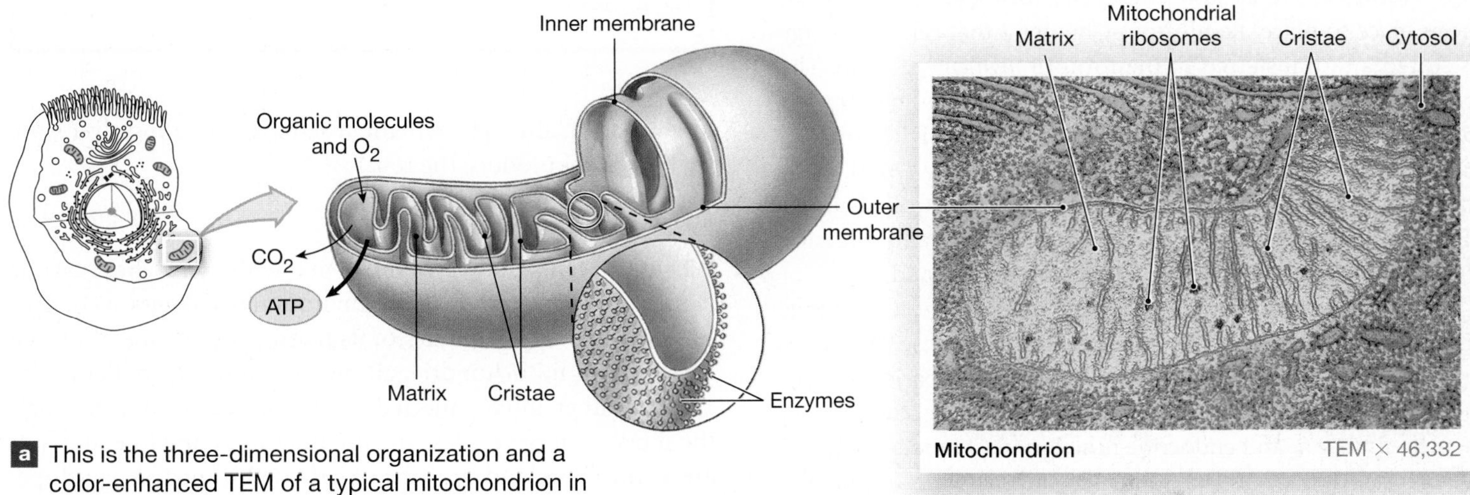
Inner membrane
Organic molecules and O_2
CO_2
ATP
Matrix
Cristae
Enzymes
Outer membrane
Mitochondrial ribosomes
Cytosol
Mitochondrion
TEM × 46,332

a This is the three-dimensional organization and a color-enhanced TEM of a typical mitochondrion in longitudinal section.

b This is an overview of the role of mitochondria in energy production. Mitochondria absorb oxygen and short carbon chains, such as pyruvate, and they generate carbon dioxide, ATP, and water.

? What are the two reactants shown here that are necessary for energy production? What are the three products shown here as a result of this reaction?

in contact with the matrix and so allow more attached protein complexes and enzymes involved in making ATP from ADP and P_i (inorganic phosphate). Metabolic enzymes in the matrix catalyze reactions that release carbon dioxide and provide some additional energy for cellular functions.

Mitochondria contain their own DNA (mtDNA) and ribosomes. The mtDNA codes for small numbers of RNA and polypeptide molecules. The polypeptides are used in enzymes required for energy production. Although mitochondria contain their own genetic system, their functions depend on imported proteins coded by nuclear DNA.

Most of the chemical reactions that release energy take place in the mitochondria, yet most of the cellular activities that require energy occur in the surrounding cytoplasm. For this reason, cells must store energy in a form that can be moved from place to place. Recall from Chapter 2 that cellular energy is stored and transferred in the form of *high-energy bonds.* The best example is the high-energy bond that attaches an inorganic phosphate ion (P_i) to adenosine diphosphate (ADP), forming the high-energy compound *adenosine triphosphate (ATP)*. Cells can then break this high-energy bond under controlled conditions, reconverting ATP to ADP and phosphate and releasing energy for the cell's use when and where it is needed.

Many cells generate ATP and other high-energy compounds as they break down carbohydrates, especially glucose. We examine the entire process in Chapter 25, but a few basic concepts now will help you follow discussions of muscle contraction, neuron function, and endocrine function in Chapters 10–18.

Most ATP is produced inside mitochondria, but the first steps take place in the cytosol (Figure 3–9b). A reaction sequence called **glycolysis** (*glycos*, sugar + *-lysis*, a loosening) breaks down a glucose molecule into two molecules of *pyruvate.* Mitochondria then absorb the pyruvate molecules.

In the mitochondrial matrix, a carbon dioxide (CO_2) molecule is removed from each absorbed pyruvate molecule. The remainder then enters the **citric acid cycle** (also known as the *Krebs cycle* and the *tricarboxylic acid cycle* or *TCA cycle*). The citric acid cycle is an enzymatic pathway that breaks down the absorbed pyruvate.

The remnants of pyruvate molecules contain carbon, oxygen, and hydrogen atoms. The carbon and oxygen atoms are released as carbon dioxide, which diffuses out of the cell. The hydrogen atoms are delivered to carrier protein complexes in the cristae. There the electrons are removed from the hydrogen atoms and passed along a chain of coenzymes and ultimately transferred to oxygen atoms. The energy released during these steps indirectly supports the enzymatic conversion of ADP to ATP. p. 59

Because mitochondrial activity requires oxygen, this method of ATP production is known as **aerobic metabolism** (*aer*, air + *bios*, life), or *cellular respiration.* Aerobic metabolism in mitochondria produces about 95 percent of the ATP needed to keep a cell alive. (Enzymatic reactions, including glycolysis, in the cytosol produce the rest.)

+ Clinical Note Free Radicals

Throughout a typical day or after exposure to pollution, cells generate free radicals. *Free radicals* are highly reactive atoms or molecules that contain an unpaired electron "seeking" the electrochemical stability of another electron. Common free radicals containing oxygen are known as reactive oxygen species (ROS). Free radicals such as ROS can damage proteins, DNA, and lipids. They prevent proteins from assuming their functional quaternary structure, DNA becomes cross-linked and unable to replicate, and the phospholipid bilayer of membranous organelles and the plasma membrane itself is pierced. Oxidative damage by free radicals underlies aging and numerous diseases such as Alzheimer's.

Membrane Flow

When the temperature changes markedly, you change your clothes. Similarly, when a cell's environment changes, it changes the structure and properties of its plasma membrane. With the exception of mitochondria, all membranous organelles in the cell are either interconnected or in communication through the movement of vesicles. The membranes of the RER and SER are continuous and are connected to the nuclear envelope. Transport vesicles connect the ER with the Golgi apparatus, and secretory vesicles link the Golgi apparatus with the plasma membrane. Finally, vesicles forming at the exposed surface of the cell remove and recycle segments of the plasma membrane. This continuous movement and exchange of membrane segments is called **membrane flow**, or *membrane trafficking*. In an actively secreting cell, an area equal to the entire membrane surface may be replaced each hour. This process was shown in **Spotlight Figure 3–7**.

Membrane flow is an example of the dynamic nature of cells. It provides a way for cells to change the characteristics of their plasma membranes by altering their lipids and the proteins serving as receptors, channels, anchors, and enzymes—as they grow, mature, or respond to a specific environmental stimulus.

✓ Checkpoint

5. **Explain the difference between the cytoplasm and the cytosol.**
6. **What are the major differences between cytosol and extracellular fluid?**
7. **Identify the nonmembranous organelles, and cite a function of each.**

8. **Identify the membranous organelles, and cite their functions.**
9. **Explain why certain cells in the ovaries and testes contain large amounts of smooth endoplasmic reticulum.**
10. **What does the presence of many mitochondria imply about a cell's energy requirements?**

See the blue Answers tab at the back of the book.

3-3 The nucleus contains DNA and enzymes essential for controlling cellular activities

Learning Outcome **Explain the functions of the cell nucleus, and discuss the nature and importance of the genetic code.**

The **nucleus** is usually the largest and most conspicuous structure in a cell. If you look at a human cell under a light microscope, it may be the only organelle you can see. The nucleus is the control center for cellular operations. A single nucleus stores all the information needed to direct the synthesis of more than 100,000 different proteins in the human body. The nucleus determines the structure of the cell and what functions it can perform by controlling which proteins are synthesized, under what circumstances, and in what amounts. A cell without a nucleus cannot repair itself, so it will disintegrate within three or four months.

Structure of the Nucleus

Most cells contain a single nucleus, but exceptions exist. For example, skeletal muscle cells have many nuclei, but mature red blood cells have none. Figure 3–10 details the structure of a typical nucleus. The nucleus is surrounded by a membranous nuclear envelope, which encloses its contents, including DNA.

Figure 3–10 The Nucleus.

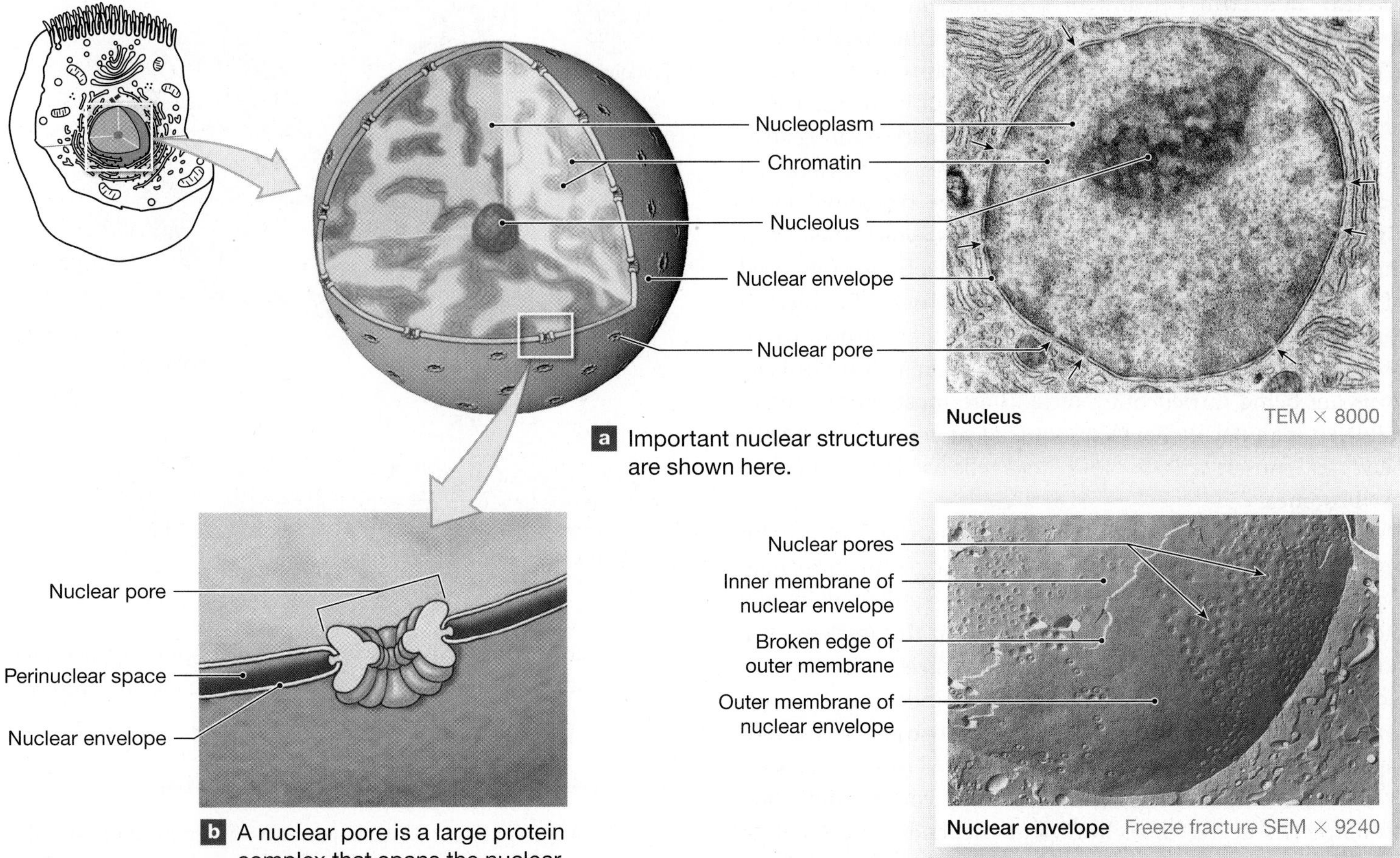
Nucleoplasm
Chromatin
Nucleolus
Nuclear envelope
Nuclear pore
Nucleus
TEM × 8000
Nuclear pore
Perinuclear space
Nuclear envelope
Nuclear pores
Inner membrane of nuclear envelope
Broken edge of outer membrane
Outer membrane of nuclear envelope
Nuclear envelope
Freeze fracture SEM × 9240

a Important nuclear structures are shown here.

b A nuclear pore is a large protein complex that spans the nuclear envelope.

c This cell was frozen and then broken apart to make the double membrane of its nuclear envelope visible. The technique, called *freeze fracture* or *freeze-etching*, provides a unique perspective on the internal organization of cells. The nuclear envelope and nuclear pores are visible. The fracturing process broke away part of the outer membrane of the nuclear envelope, and the broken edge of this membrane can be seen.

Nuclear Envelope

3

Surrounding the nucleus and separating it from the cytosol is a **nuclear envelope**, a double membrane with its two layers separated by a narrow *perinuclear space* (*peri-*, around). At several locations, the nuclear envelope is connected to the rough endoplasmic reticulum (see **Spotlight Figure 3–1**).

To direct processes that take place in the cytoplasm, the nucleus must receive information about conditions and activities in other parts of the cell. Chemical communication between the nucleus and the cytoplasm takes place through openings in the nuclear envelope called **nuclear pores**. Each pore has about 50 associated proteins, forming a *nuclear pore complex*. Each nuclear pore complex regulates the transport of materials, such as RNA and other proteins, between the nucleus and the cytoplasm. They are large enough to allow ions and small molecules to enter or leave, but are too small for DNA to pass freely.

Contents of the Nucleus

The fluid portion of the nucleus is called the *nucleoplasm* or *karyolymph* (*karyo-*, nucleus). The nucleoplasm contains the **nuclear matrix**, a network of fine filaments that provides structural support and may be involved in the regulation of genetic activity. The nucleoplasm also contains ions, enzymes, RNA and DNA nucleotides, small amounts of RNA, and DNA.

In addition, most nuclei contain several dark-staining areas called **nucleoli** (nū -KLĒ-ō-lī; singular, *nucleolus*). Nucleoli are transient nuclear organelles that synthesize ribosomal RNA. They also assemble the ribosomal subunits, which then enter the cytoplasm through nuclear pores. Nucleoli are composed of RNA, enzymes, and proteins called **histones**. The nucleoli form around portions of DNA that contain the instructions for producing ribosomal proteins and RNA when those instructions are being carried out. Nucleoli are most prominent in cells that manufacture large amounts of proteins, such as liver, nerve, and muscle cells, because those cells need large numbers of ribosomes.

The DNA in the nucleus stores the instructions for protein synthesis. Interactions between the DNA and the histones help determine which information is available to the cell at any moment. The organization of DNA within the nucleus of a nondividing cell and one preparing for cell division is shown in **Figure 3–11**. At intervals, the DNA strands wind around the histones, forming a complex known as a **nucleosome**. Such winding allows a great deal of DNA to be packaged in a small space.

The entire chain of nucleosomes may coil around other proteins. The degree of coiling varies, depending on whether cell division is under way. In cells that are not dividing, the nucleosomes are loosely coiled within the nucleus, forming a tangle of fine filaments known as **chromatin**. Chromatin gives the nucleus a clumped, grainy appearance. Just before cell division begins, the coiling becomes tighter, forming distinct structures called **chromosomes** (*chroma*, color). In humans, the nuclei of somatic cells contain 23 pairs of chromosomes. One member of each pair is derived from the mother, and one from the father.

Figure 3–11 The Organization of DNA within the Nucleus. DNA strands are coiled around histones to form nucleosomes. Nucleosomes form coils that may be very tight or rather loose. In cells that are not dividing, the DNA is loosely coiled, forming a tangled network known as chromatin. When the coiling becomes tighter, as it does in preparation for cell division, the DNA becomes visible as distinct structures called chromosomes. (Terms associated with cell division are highlighted with an asterisk [*] and discussed in **Spotlight Figure 3–23**.)

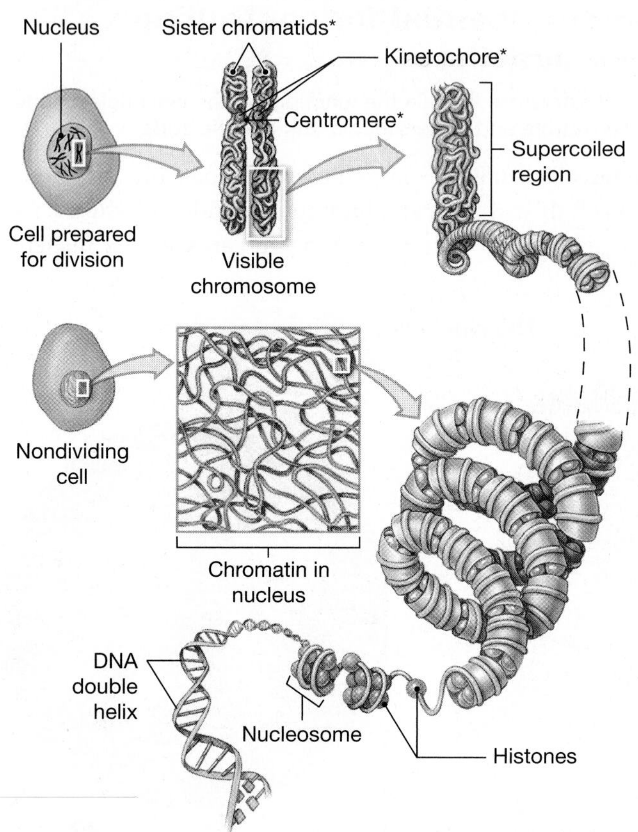
Nucleus
Sister chromatids*
Kinetochore*
Centromere*
Supercoiled region
Cell prepared for division
Visible chromosome
Nondividing cell
Chromatin in nucleus
DNA double helix
Nucleosome
Histones

? How is DNA organized in the nucleus when the cell is prepared for division? How is DNA organized in the nucleus when the cell is not dividing?

Information Storage in the Nucleus

As we saw in Chapter 2, each protein molecule consists of a unique sequence of amino acids. p. 53 Any "recipe" for a protein must specify the order of amino acids in the polypeptide chain. This information is stored in the chemical structure of the DNA strands in the nucleus. The chemical "language" the cell uses is known as the **genetic code**. An understanding of the genetic code has enabled researchers to learn how cells build proteins and how various structural and functional characteristics are inherited from generation to generation.

How does the genetic code work? Recall the basic structure of nucleic acids described in Chapter 2. p. 57 A single DNA molecule consists of a pair of DNA strands held together by hydrogen bonding between complementary nitrogenous bases. Information is stored in the sequence of nitrogenous bases along the length of the DNA strands. Those nitrogenous bases are adenine (A), thymine (T), cytosine (C), and guanine (G).

The genetic code is called a *triplet code,* because a sequence of three nitrogenous bases represents a single amino acid. Thus, the information encoded in the sequence of nitrogenous bases must be read in groups of three. For example, the triplet thymine–guanine–thymine (TGT) on one DNA strand, called the *coding strand,* is the code for the amino acid cysteine. More than one triplet may represent the same amino acid, however. For example, the DNA triplet thymine–guanine–cytosine (TGC) also codes for cysteine.

A **gene** is the functional unit of heredity. It is the sequence of nucleotides of a DNA strand that specifies the amino acids needed to produce a specific protein. The needed number of triplet nucleotides, each coding for an amino acid, depends on the size of the polypeptide represented. A relatively short polypeptide chain might need fewer than 100 triplets, but the instructions for building a large protein might involve 1000 or more triplets.

Not all nucleotide sequences of a DNA strand carry instructions for assembling proteins. Some sequences contain instructions for synthesizing ribosomal RNA or another type that we will discuss shortly, called *transfer RNA.* Others have a regulatory function, and still others have as yet no known function.

✓ Checkpoint

11. **Describe the contents and structure of the nucleus.**
12. **What is a gene?**

See the blue Answers tab at the back of the book.

3-4 DNA controls protein synthesis, cell structure, and cell function

Learning Outcome **Summarize the role of DNA in protein synthesis, cell structure, and cell function.**

The DNA of the nucleus controls cell structure and function through the synthesis of specific proteins. **Protein synthesis** is the assembling of functional polypeptides in the cytoplasm. The major events of protein synthesis are transcription and translation.

Regulation of Transcription by Gene Activation

Each DNA molecule contains thousands of genes—the information needed to synthesize thousands of proteins. Normally, the genes are tightly coiled, and histones bound to the DNA keep the genes inactive. Before a gene can affect a cell, the portion of the DNA molecule containing that gene must be uncoiled and the histones temporarily removed in a process called **gene activation**.

The factors controlling this process are only partially understood. We know, however, that every gene contains segments responsible for regulating its own activity. In effect, these are triplets of nucleotides that say "read this message" or "do not read this message," "message starts here," or "message ends here." These "read me," "don't read me," and "start" signals form a special region of DNA called the *promoter,* or control segment, at the start of each gene. Each gene ends with a "stop" signal. Gene activation begins with the temporary disruption of the weak hydrogen bonds between the nitrogenous bases of the two DNA strands and the removal of the histone that guards the promoter.

Transcription of DNA into mRNA

Transcription is the synthesis of RNA from a DNA template. The use of the term *transcription* makes sense, as it means "to copy" or "rewrite."

All three types of RNA are formed through the transcription of DNA, but we will focus here on the transcription of **messenger RNA (mRNA)**, which carries the information

+ Clinical Note DNA Fingerprinting

Every nucleated somatic cell in the body carries a set of 46 chromosomes (23 pairs) that are copies of the set formed at fertilization. Not all the DNA of these chromosomes codes for proteins. Within the non–protein coding DNA are regions called short tandem repeats (STRs), which contain the same nucleotide sequence repeated over and over. The number of STRs and the number of repetitions vary among individuals.

In the United States, 13 core STRs are used for identifying individuals by *DNA fingerprinting*. The individual differences are reflected in the different lengths of these STRs. The chance that any two individuals, other than identical twins, will have the same length and pattern of repeating DNA segments in 13 STRs is approximately 1 in 575 trillion. For this reason, individuals can be identified on the basis of their DNA pattern, just as they can on the basis of a fingerprint. Skin cells, semen, hair root cells, or cheek cells can serve as the DNA source. Information from DNA fingerprinting has been used to convict or acquit people accused of violent crimes, such as rape or murder. The U.S. Crime Act of 1994 allowed the FBI to establish and maintain the National DNA Index System.

needed to synthesize proteins. The synthesis of mRNA is essential, because DNA is too large to leave the nucleus. Instead, its information is copied to messenger RNA, which is smaller and *can* leave the nucleus and carry the information to the cytoplasm, where protein synthesis takes place.

Recall that the two DNA strands in a gene are complementary. The strand containing the triplets that specify the sequence of amino acids in the polypeptide is the **coding strand**. The other strand, called the **template strand**, contains complementary triplets that will be used as a template for mRNA production. The resulting mRNA will have a nucleotide sequence identical to that of the coding strand, but with uracil substituted for thymine.

The Process of Transcription

The steps in transcription are illustrated in Figure 3-12:

1. **RNA Polymerase Binding.** Once the DNA strands have separated and the histones have been removed to expose the promoter, transcription can begin. The key event is the binding of the enzyme **RNA polymerase** to the promoter of a gene on the template strand.

2. **RNA Polymerase Nucleotide Linking.** RNA polymerase promotes hydrogen bonding between the nitrogenous bases of the template strand nucleotides and complementary nucleotides in the nucleoplasm. This enzyme begins at a "start" signal in the promoter region. It then strings nucleotides together by covalent bonding. The RNA polymerase travels along the DNA strand, interacting with only a small portion of the template strand at any one time. The complementary DNA strands separate in front of the enzyme as it moves one nucleotide at a time, and the strands then reassociate behind it.

The enzyme collects additional nucleotides and attaches them one by one to the growing chain. The nucleotides involved are those characteristic of RNA, not of DNA. In other words, RNA polymerase can attach adenine, guanine, cytosine, or uracil, but never thymine. For this reason, wherever an A occurs in the DNA strand, the enzyme attaches a U rather than a T to the

Figure 3–12 mRNA Transcription. In this figure, a small portion of a single DNA molecule, containing a single gene, is undergoing transcription. (1) The two DNA strands separate, and RNA polymerase binds to the promoter of the gene. (2) The RNA polymerase moves from one nucleotide to another along the length of the template strand. At each site, complementary RNA nucleotides form hydrogen bonds with the DNA nucleotides of the template strand. The RNA polymerase then strings the arriving nucleotides together into a strand of mRNA. (3) On reaching the stop signal at the end of the gene, the RNA polymerase and the mRNA strand detach, and the two DNA strands reassociate.

SmartArt

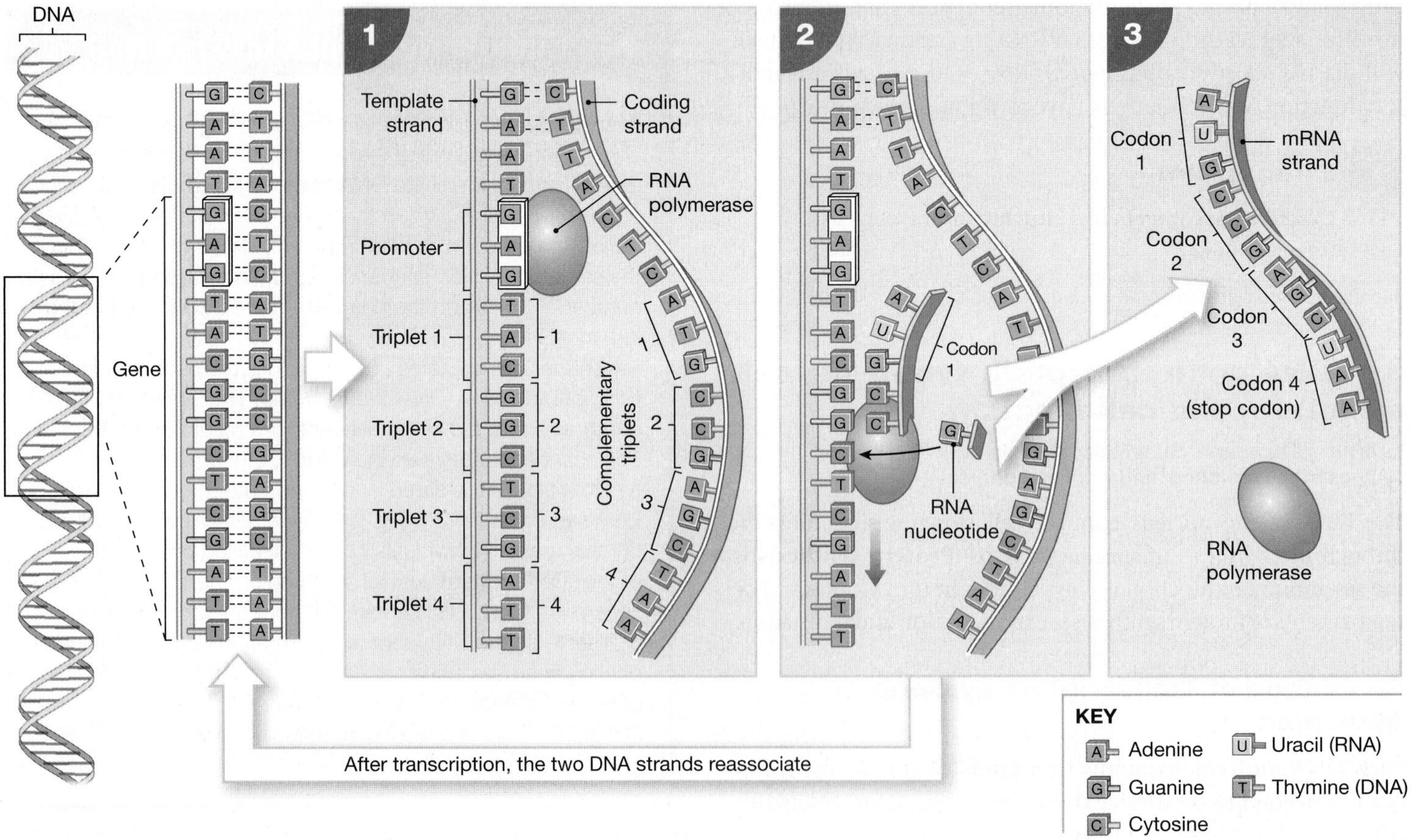

DNA; Gene; Template strand; Coding strand; RNA polymerase; Promoter; Triplet 1; Triplet 2; Triplet 3; Triplet 4; Complementary triplets; Codon 1; RNA nucleotide; Codon 2; Codon 3; Codon 4 (stop codon); mRNA strand

After transcription, the two DNA strands reassociate

KEY: A Adenine; G Guanine; C Cytosine; U Uracil (RNA); T Thymine (DNA)

growing mRNA strand. In this way, RNA polymerase assembles a complete strand of mRNA.

Notice that the nucleotide sequence of the DNA template strand determines the nucleotide sequence of the mRNA strand. Each DNA triplet corresponds to a sequence of three nucleotides in the mRNA strand. Such a three-base mRNA sequence is called a **codon** (KŌ-don). Codons contain nitrogenous bases that are complementary to those of the triplets in the DNA template strand. For example, if the DNA triplet is TCG, the corresponding mRNA codon will be AGC. This method of copying ensures that the mRNA exactly matches the coding strand of the gene (except for the substitution of uracil for thymine).

3 **Detachment of mRNA.** At the "stop" signal, the enzyme and the mRNA strand detach from the DNA strand, and transcription ends. The complementary DNA strands now complete their reassociation as hydrogen bonding takes place again between complementary base pairs.

RNA Processing

The nucleotide sequence of each gene includes a number of triplets that are not needed to build a functional protein. As a result, the mRNA strand assembled during transcription, sometimes called immature mRNA or *pre-mRNA*, must be "edited" before it leaves the nucleus to direct protein synthesis. In this **RNA processing**, noncoding intervening sequences, called **introns**, are snipped out, and the remaining coding segments, or **exons**, are spliced together. The process creates a much shorter, functional strand of mRNA that then enters the cytoplasm through a nuclear pore.

Intron removal is extremely important and tightly regulated. An error in the editing will produce an abnormal protein with potentially disastrous results. Moreover, we now know that by changing the editing instructions and removing different introns, a single gene can produce mRNAs that code for several different proteins. Some introns, however, act as enzymes to catalyze their own removal. How this variable editing is regulated is unknown.

Translation from mRNA into a Polypeptide

Translation is the formation of a linear chain of amino acids, a polypeptide, using the information from an mRNA strand. Again, the name makes sense: To *translate* is to present the same information in a different language. In this case, a message written in the "language" of nucleic acids (the sequence of nitrogenous bases) is translated into the "language" of proteins (the sequence of amino acids in a polypeptide chain). Each mRNA codon designates a particular amino acid to be inserted into the polypeptide chain.

The amino acids are provided by **transfer RNA (tRNA)**, a relatively small and mobile type of RNA. Each tRNA molecule binds and delivers a specific type of amino acid. More than 20 kinds of transfer RNA exist—at least one for each of the 20 amino acids used in protein synthesis.

Table 3–1 Examples of the Genetic Code

DNA Triplet				
Template Strand	Coding Strand	mRNA Codon	tRNA Anticodon	Amino Acid
AAA	TTT	UUU	AAA	Phenylalanine
AAT	TTA	UUA	AAU	Leucine
ACA	TGT	UGU	ACA	Cysteine
CAA	GTT	GUU	CAA	Valine
TAC	ATG	AUG	UAC	Methionine
TCG	AGC	AGC	UCG	Serine
GGC	CCG	CCG	GGC	Proline
CGG	GCC	GCC	CGG	Alanine

A tRNA molecule has a "tail" that binds an amino acid. Roughly midway along its length, the nucleotide chain of the tRNA forms a tight loop that can interact with an mRNA strand. The loop contains three nitrogenous bases that form an **anticodon**. During translation, the anticodon bonds with a complementary mRNA codon. The anticodon base sequence indicates the type of amino acid carried by the tRNA. For example, a tRNA with the anticodon GGC always carries the amino acid *proline*, whereas a tRNA with the anticodon CGG carries *alanine*. **Table 3–1** lists examples of several codons and

+ Clinical Note Mutations

Mutations are permanent changes in a cell's DNA that affect the nucleotide sequence of one or more genes. The simplest is a *point mutation*, a change in a single nucleotide that affects one codon. The triplet code has some flexibility, because several different codons can specify the same amino acid. But a point mutation that produces a codon that specifies a different amino acid will usually change the structure of the completed protein. A single change in the amino acid sequence of a structural protein can prove fatal. Certain cancers and two potentially lethal blood disorders, *thalassemia* and *sickle cell anemia*, result from changes in a single nucleotide.

Several hundred inherited disorders have been traced to abnormalities in enzyme or protein structure that reflect single changes in the nucleotide sequence. More extensive mutations, such as additions or deletions of nucleotides, can affect multiple codons in one gene or in several adjacent genes. They can also affect the structure of one or more chromosomes.

Most mutations occur during DNA replication, when cells are duplicating their DNA in preparation for cell division. A single cell, a group of cells, or an entire individual may be affected. Entire organisms are affected when the changes are made early in development.

Figure 3–13 The Process of Translation.

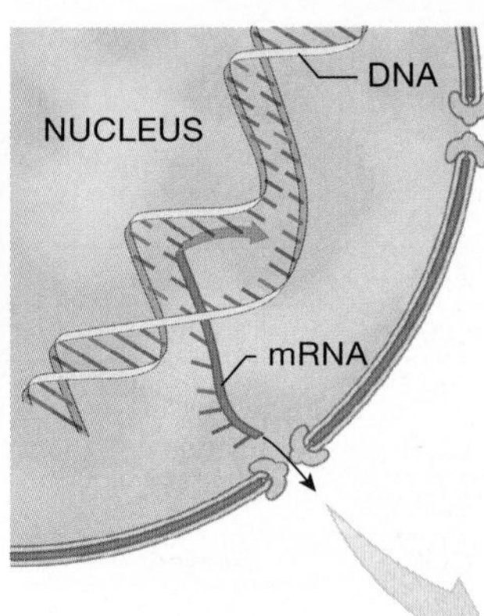
DNA
NUCLEUS
mRNA

SmartArt

Initiation

Translation begins as the mRNA strand binds to a small ribosomal subunit at the start codon. The attachment of tRNA and the subsequent binding of the large ribosomal subunit creates a functional ribosome, with the mRNA nestled in the gap between the small and the large ribosomal subunits.

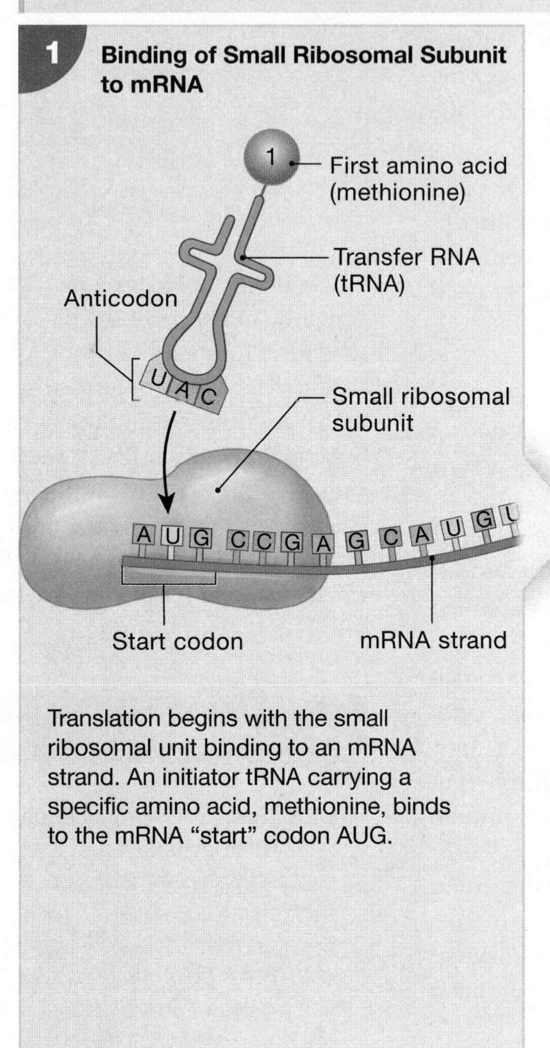
1
Binding of Small Ribosomal Subunit to mRNA
1
First amino acid (methionine)
Transfer RNA (tRNA)
Anticodon
U A C
Small ribosomal subunit
A U G C C G A G C A U G
Start codon
mRNA strand

Translation begins with the small ribosomal unit binding to an mRNA strand. An initiator tRNA carrying a specific amino acid, methionine, binds to the mRNA "start" codon AUG.

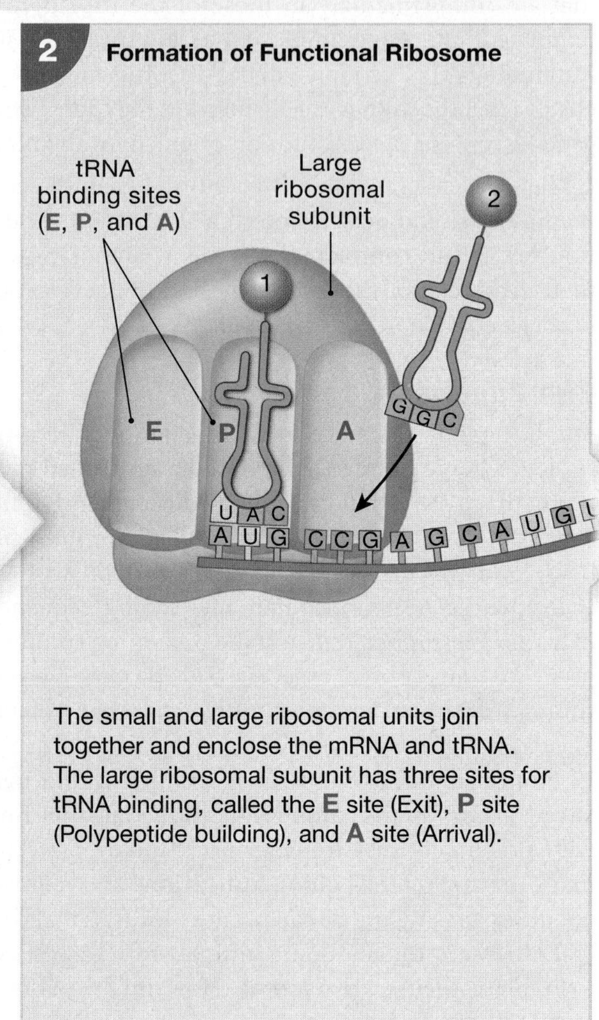
2
Formation of Functional Ribosome
tRNA binding sites (E, P, and A)
Large ribosomal subunit
2
1
G G C
E
P
A
U A C
A U G C C G A G C A U G

The small and large ribosomal units join together and enclose the mRNA and tRNA. The large ribosomal subunit has three sites for tRNA binding, called the **E** site (Exit), **P** site (Polypeptide building), and **A** site (Arrival).

KEY

- A Adenine
- G Guanine
- C Cytosine
- U Uracil

anticodons that specify individual amino acids and summarizes the relationships among DNA template and coding strand triplets, codons, anticodons, and amino acids.

During translation, each codon along the mRNA strand binds a complementary anticodon on a tRNA molecule. For example, if the mRNA has the three-codon sequence AUG–CCG–AGC, it will bind to three tRNAs with anticodons UAC, GGC, and UCG, respectively. The amino acid sequence of the polypeptide chain depends on the resulting sequence of codons along the mRNA strand. In this case, the amino acid sequence in the resulting polypeptide would be methionine–proline–serine. (You can follow the correspondence among codons, anticodons, and amino acids in **Table 3–1**. For a complete table of the mRNA codons and which amino acids and start or stop signals they represent, see the Appendix at the back of the book.)

Recall that ribosomes are organelles that play a key role in protein synthesis. You may recall that a functional ribosome consists of two subunits, a small ribosomal subunit and a large ribosomal subunit. At the beginning of the translation process, a small and a large ribosomal subunit must join together with an mRNA strand. **Figure 3–13** illustrates the three phases of the translation process: *initiation*, *elongation*, and *termination*. Please study this figure carefully before going on.

Elongation

In elongation, amino acids are added one by one to the growing polypeptide chain, as the ribosome moves along the mRNA strand. Translation proceeds swiftly, producing a typical protein in about 20 seconds. Although only two mRNA codons are "read" by a ribosome at any one time, many ribosomes can bind to a single mRNA strand, and a multitude of identical polypeptides may be quickly and efficiently produced.

Termination

Elongation ends when the ribosome reaches the stop codon at the end of the mRNA strand. The mRNA strand remains intact, and it can interact with other ribosomes to create additional copies of the same polypeptide chain.

3 Formation of Peptide Bond

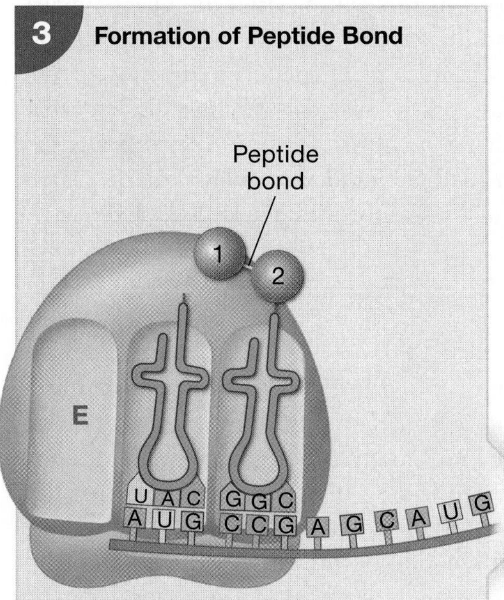
Peptide bond
1
2
E
U A C
G G C
A U G
C C G
A G C A U G

A second tRNA then arrives at the A site of the ribosome, carrying amino acid 2. Its anticodon binds to the second codon of the mRNA strand. Ribosomal enzymes now remove amino acid 1 from the first tRNA in the P site and attach it to amino acid 2 with a peptide bond.

4 Extension of Polypeptide

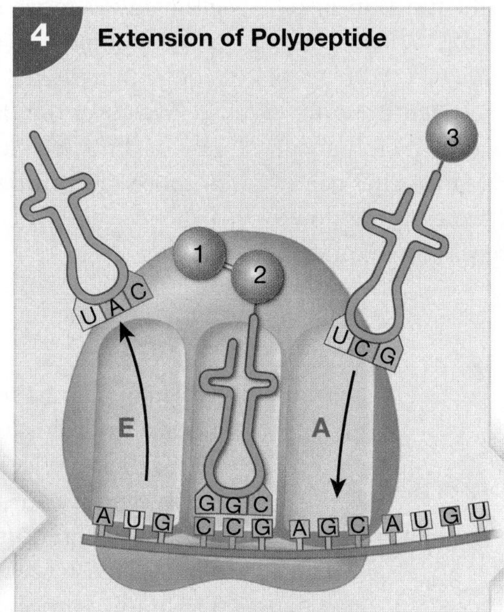
3
1
2
U A C
U C G
E
A
A U G
G G C
C C G
A G C
A U G U

The ribosome now moves one codon farther along the length of the mRNA strand. A third tRNA arrives at the A site, bearing amino acid 3. The first tRNA then detaches from the E site of the ribosome and reenters the cytosol. It can pick up another amino acid molecule in the cytosol and repeat the process.

5 Completion and Release of Polypeptide

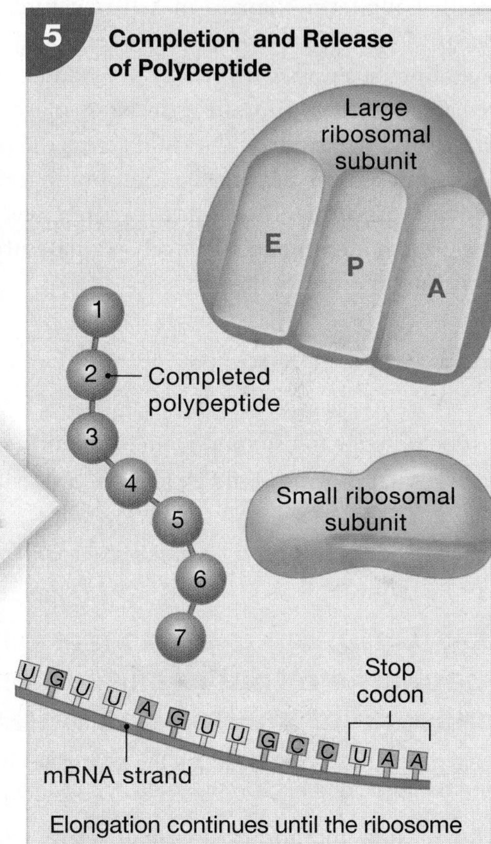
Large ribosomal subunit
E
P
A
1
2
Completed polypeptide
3
4
5
6
7
Small ribosomal subunit
Stop codon
U G U U A G U U G C C U A A
mRNA strand

Elongation continues until the ribosome reaches a stop codon. Termination occurs as a protein-release factor bonds with the stop codon and the completed polypeptide is released. The ribosomal subunits then separate, freeing the mRNA strand.

How DNA Controls Cell Structure and Function

As we noted previously, the DNA of the nucleus controls the cell by directing the synthesis of specific proteins. By controlling protein synthesis, virtually every aspect of cell structure and function can be regulated.

How is cell structure and function regulated at the DNA level? How does the nucleus "know" which genes to activate or deactivate (silence) to maintain cellular homeostasis? Changes in the extracellular environment may result in ions or molecules entering the cell directly through membrane channels or by binding to membrane receptors that then alter the intracellular (cytoplasmic) environment. Such direct entry into the cell or binding of ligands can initiate chemical *signaling pathways* that form intracellular signaling molecules. Either situation may trigger the activation or inactivation of cytoplasmic enzymes and provide short-term adjustments that maintain homeostasis.

Alternatively, substances that cross the plasma membrane, such as chemical messengers, hormones, or products that form from a signaling pathway, may enter the nucleus through nuclear pores and bind to specific receptors or promoters along the DNA strands. The result is a change in genetic activity that provides long-term homeostatic adjustments affecting the cell's biochemical and physical characteristics. For example, synthesizing additional enzymes, fewer enzymes, or different enzymes can change the biochemical processes under way in the cell. Changes in the physical structure of the cell result from alterations in the rates or types of structural proteins that are synthesized. Long-term changes in cell structure and function also occur as part of growth, development, and aging. (We discuss signaling pathways within cells in Chapter 18.)

Continual communication takes place between the cytoplasm and the extracellular fluid across the plasma membrane; and what crosses the plasma membrane today may alter gene activity tomorrow. In the next two sections, we examine how the plasma membrane selectively regulates the passage of materials into and out of the cell.

Checkpoint

13. **Define *gene activation*.**
14. **Describe transcription and translation.**
15. **What process would be affected by the lack of the enzyme RNA polymerase?**

See the blue Answers tab at the back of the book.

3-5 Diffusion is a passive transport mechanism that assists membrane passage of solutes and water

Learning Outcome **Describe the processes of cellular diffusion and osmosis, and explain their role in physiological systems.**

The plasma membrane is an effective barrier between the cytoplasm and the extracellular fluid. A consequence is that conditions inside the cell can be much different from conditions outside the cell. However, the barrier cannot be perfect, because cells are not self-sufficient. Each day they require nutrients to provide the energy they need to stay alive and function normally. They also generate waste products that must be removed. Your body has passageways and openings for nutrients, gases, and wastes, but a continuous, relatively uniform membrane surrounds the cell. So how do materials—whether nutrients or waste products—get across the plasma membrane without damaging it or reducing its effectiveness as a barrier?

The key to answering this question is **permeability**, the property of the plasma membrane that determines precisely which substances can enter or leave the cytoplasm. A membrane through which nothing can pass is **impermeable**. A membrane through which any substance can pass without difficulty is **freely permeable**. Plasma membranes are called **selectively permeable**, however, because their permeability lies somewhere between those extremes.

A selectively permeable membrane permits the free passage of some materials and restricts the passage of others. The distinction may be based on size, electrical charge, molecular shape, lipid solubility, or other factors. Cells differ in their permeabilities, depending on the lipids and proteins in the plasma membrane and how these components are arranged.

Passage across the membrane is either passive or active. *Passive processes* move ions or molecules across the plasma membrane with no expenditure of energy by the cell. *Active processes* require that the cell expend energy, generally in the form of ATP.

We cover the main passive process, *diffusion*, and a special case of diffusion called *osmosis*, in this section. In the next section, we cover carrier-mediated transport, which can be passive or active, and vesicular transport, which is always an active process.

Diffusion

Driven by thermal energy, ions and molecules are constantly in motion, colliding and bouncing off one another and off obstacles in their paths. The movement is both passive and random: A molecule can bounce in any direction. One result of this continuous random motion is that, over time, the molecules in any given space will tend to become evenly distributed. This distribution process, the net movement of a substance from an area of higher concentration to an area of lower concentration, is called **diffusion**. The difference between the high and low concentrations of a substance is a **concentration gradient** (and thus a potential energy gradient). Because of its net movement, diffusion is often described as proceeding "down a concentration gradient" or "downhill."

Diffusion tends to eliminate concentration gradients. After a gradient has been eliminated, the molecular motion continues, but net movement no longer occurs in any particular direction. (For simplicity, we restrict use of the term *diffusion* to the directional movement that eliminates concentration gradients—a process sometimes called *net diffusion.*)

Diffusion in air and water is a slow process, and it is most important over very short distances. A simple, everyday example can give you a mental image of how diffusion works. Consider what happens when you drop a colored sugar cube into water (Figure 3–14). Placing the cube in a large volume of clear water sets up a steep concentration gradient for both ingredients as they dissolve: The sugar and dye concentrations are higher near the cube and negligible elsewhere. As time passes, the dye and sugar molecules spread through the solution until they are distributed evenly. However, compared to a cell, a beaker of water is enormous, and additional factors (which we will ignore) account for dye and sugar distribution over distances of centimeters as opposed to micrometers.

Go to MasteringA&P™ > Study Area > Menu > Clinical > Clinical Case Studies > Chapter 3: Dangerously Thin: A Case Study on the Genetic Code

Figure 3–14 **Diffusion.**

1
Placing a colored sugar cube into a water-filled beaker establishes a steep concentration gradient.
2
As the cube begins to dissolve, many sugar and dye molecules are in one location, and none are elsewhere.
3
With time, the sugar and dye molecules spread through the water.
4
Eventually, the concentration gradient is eliminated and the molecules are evenly distributed throughout the solution.

Diffusion is important in body fluids because it tends to eliminate local concentration gradients. For example, every cell in the body generates carbon dioxide (CO_2), so the intracellular CO_2 concentration is fairly high. The CO_2 concentration is lower in the surrounding interstitial fluid, and lower still in the circulating blood. Plasma membranes are freely permeable to CO_2, so it can diffuse down its concentration gradient. It travels from the cell's interior into the interstitial fluid and then into the bloodstream, for eventual delivery to the lungs.

+ Clinical Note Drugs and the Plasma Membrane

Many clinically important drugs affect the plasma membrane. For some anesthetics, such as chloroform, ether, halothane, and nitrous oxide, potency is directly correlated with their lipid solubility. High lipid solubility speeds the drug's entry into cells and enhances its ability to block ion channels or change other properties of plasma membranes. These changes reduce the sensitivity of neurons and muscle cells. However, some common anesthetics have relatively low lipid solubility. For example, the local anesthetics *procaine* and *lidocaine* affect nerve cells by blocking sodium ion channels in their plasma membranes. This blockage reduces or eliminates the responsiveness of these cells to painful (or any other) stimuli.

To be effective, the diffusion of nutrients, waste products, and dissolved gases must keep pace with the demands of active cells. Several important factors influence diffusion rates:

- *Distance.* The shorter the distance, the more quickly concentration gradients are eliminated. In the human body, few cells are farther than 25 μm from a blood vessel.
- *Ion and Molecule Size.* The smaller the ion and molecule size, the faster its rate of diffusion. Ions and small organic molecules, such as glucose, diffuse more rapidly than do large proteins.
- *Temperature.* The higher the temperature, the faster the diffusion rate. Diffusion proceeds somewhat more quickly at human body temperature (about 37°C, or 98.6°F) than at cooler environmental temperatures.
- *Concentration Gradient.* The steeper the concentration gradient, the faster diffusion proceeds. For example, when cells become more active, they use more oxygen, so the intracellular concentration of oxygen declines. This change increases the concentration gradient of oxygen between the inside of the cell (somewhat low) and the interstitial fluid outside (somewhat high). The rate of oxygen diffusion into the cell then increases.
- *Electrical Forces.* Opposite electrical charges (+ and −) attract each other, and like charges (+ and + or − and −) repel each other. The cytoplasmic (inner) surface of the plasma membrane has a net negative charge relative to the extracellular (outer) surface. This negative charge tends to attract positive ions from the extracellular fluid into the cell, while repelling the entry of negative ions.

3

An ion's or molecule's concentration gradient, or *chemical gradient*, can work to enhance or detract from the electrical gradient. For example, interstitial fluid contains higher concentrations of sodium ions (Na^+) and chloride ions (Cl^-) than does cytosol. Both the concentration gradient and the electrical gradient favor diffusion of the positively charged sodium ions into the cell. In contrast, diffusion of the negatively charged chloride ions into the cell is favored by the chemical gradient, but opposed by the electrical gradient. For any ion, the net result of the chemical and electrical forces acting on it is called the *electrochemical gradient.*

Diffusion across Plasma Membranes

In extracellular fluids, water and dissolved solutes diffuse freely. A plasma membrane, however, acts as a barrier that selectively restricts diffusion: Some substances pass through easily, but others cannot penetrate the membrane. An ion or a molecule can diffuse across a plasma membrane only by (1) crossing the lipid portion of the membrane by simple diffusion or (2) passing through a membrane channel (Figure 3–15).

Simple Diffusion. Alcohol, fatty acids, and steroids can enter cells easily, because they can diffuse through the lipid portions of the membrane. Lipid-soluble drugs, dissolved gases such as oxygen and carbon dioxide, and water molecules also enter and leave our cells by diffusing through the phospholipid bilayer.

Channel-Mediated Diffusion. The situation is more complicated for ions and water-soluble compounds, which are not lipid soluble. To enter or leave the cytoplasm, these substances must pass through a membrane channel.

Membrane channels are very small passageways created by transmembrane proteins. An average channel is about 0.8 nm in diameter. Water molecules can enter or exit freely, but even a small organic molecule, such as glucose, is too big to fit through the channels. Whether an ion can cross a particular membrane channel depends on many factors, including the size and charge of the ion, the size of the hydration sphere, and interactions between the ion and the channel walls. p. 41 Membrane channels that are always open and allow ions to pass across the plasma membrane are called *leak channels*, or passive channels.

The mechanics of diffusion through membrane channels is more complex than simple diffusion. For example, the rate at which a particular ion diffuses across the membrane can be limited by the availability of suitable channels. However, many ions, including

Figure 3–15 **Diffusion across the Plasma Membrane.** The path a substance takes in crossing a plasma membrane depends on the substance's size and lipid solubility.

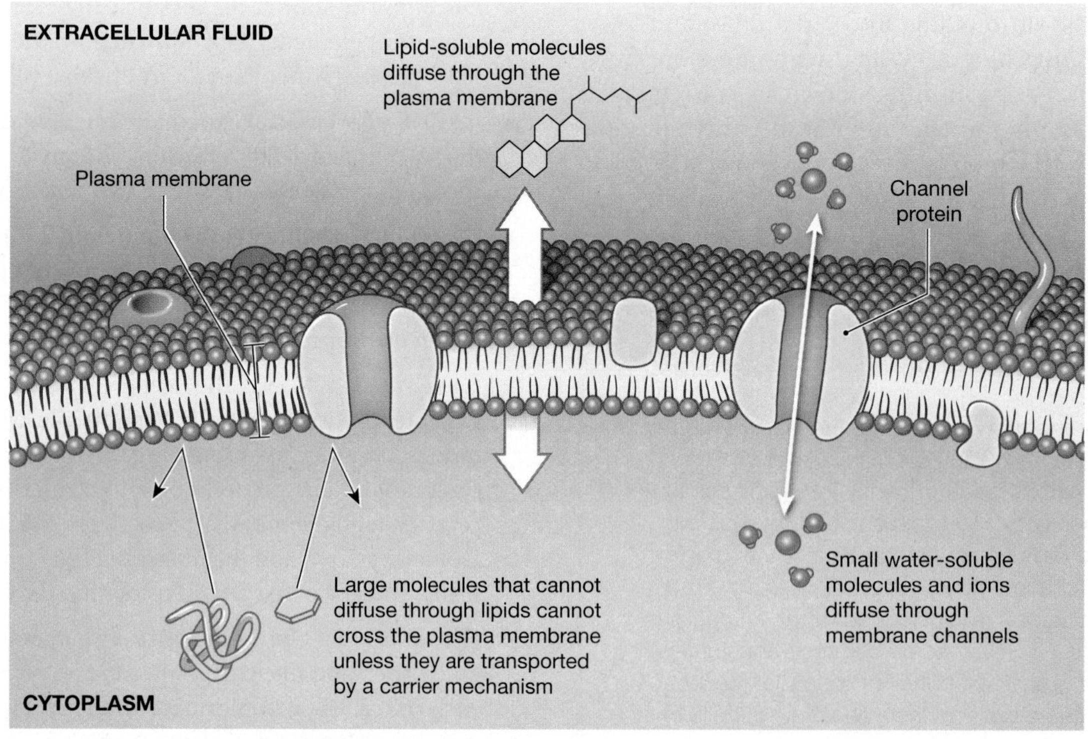
EXTRACELLULAR FLUID
Lipid-soluble molecules diffuse through the plasma membrane
Plasma membrane
Channel protein
Small water-soluble molecules and ions diffuse through membrane channels
Large molecules that cannot diffuse through lipids cannot cross the plasma membrane unless they are transported by a carrier mechanism
CYTOPLASM

? How do small water-soluble molecules and ions diffuse across the plasma membrane? How do lipid-soluble molecules diffuse across the plasma membrane?

> Go to MasteringA&P™ > Study Area> Menu > Animations & Videos > A&PFlix > A&P > Membrane Transport

sodium, potassium, and chloride, move across the plasma membrane at rates comparable to those for simple diffusion.

Osmosis: Diffusion of Water across Selectively Permeable Membranes

The net diffusion of water across a membrane is so important that it has a special name: **osmosis** (oz-MŌ-sis; *osmos*, a push). We will always use the term *osmosis* for the movement of water, and the term *diffusion* for the movement of solutes.

Intracellular and extracellular fluids are solutions with a variety of dissolved materials. Each solute diffuses as though it were the only material in solution. The diffusion of sodium ions, for example, takes place only in response to a concentration gradient for sodium. A concentration gradient for another ion has no effect on the rate or direction of sodium ion diffusion.

Some solutes diffuse into the cytoplasm, others diffuse out, and a few (such as proteins) are unable to diffuse across the plasma membrane at all. Yet if we ignore the individual identities and simply count ions and molecules, we find that the *total* concentration of dissolved ions and molecules on either side of the plasma membrane stays the same. This state of equilibrium persists because a typical plasma membrane is freely permeable to water.

To understand the basis for such equilibrium, consider that whenever a solute concentration gradient exists, a concentration gradient for *water* also exists. Dissolved solute molecules occupy space that would otherwise be taken up by water molecules, so the higher the solute concentration, the lower the water concentration. As a result, *water molecules tend to flow across a selectively permeable membrane toward the solution with the higher solute concentration,* because this movement is *down* the concentration gradient for water. Water will move until the water concentrations—and thus solute concentrations—are the same on either side of the membrane; in other words, water moves until a state of equilibrium is reached. Remember these basic characteristics of osmosis:

- Osmosis is the diffusion of water molecules across a selectively permeable membrane.
- Osmosis takes place across a selectively permeable membrane that is freely permeable to water, but not freely permeable to all solutes.
- In osmosis, water flows across a selectively permeable membrane toward the solution that has the higher concentration of solutes, because that is where the concentration of water is lower.

Osmosis and Osmotic Pressure

The process of osmosis and its pressure effect are diagrammed in Figure 3–16. **1** shows two solutions (A and B), with different solute concentrations, separated by a selectively permeable membrane. As osmosis takes place, water molecules cross the

Figure 3–16 Osmosis. The osmotic pressure of solution B is equal to the amount of hydrostatic pressure required to stop the osmotic flow from solution A to solution B.

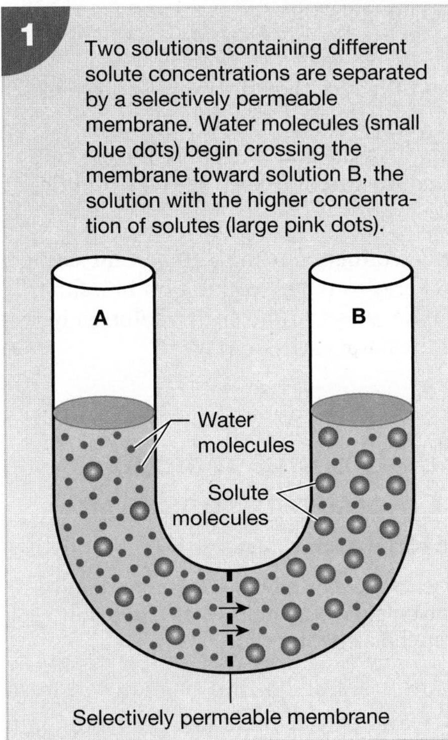
1
Two solutions containing different solute concentrations are separated by a selectively permeable membrane. Water molecules (small blue dots) begin crossing the membrane toward solution B, the solution with the higher concentration of solutes (large pink dots).
A
B
Water molecules
Solute molecules
Selectively permeable membrane

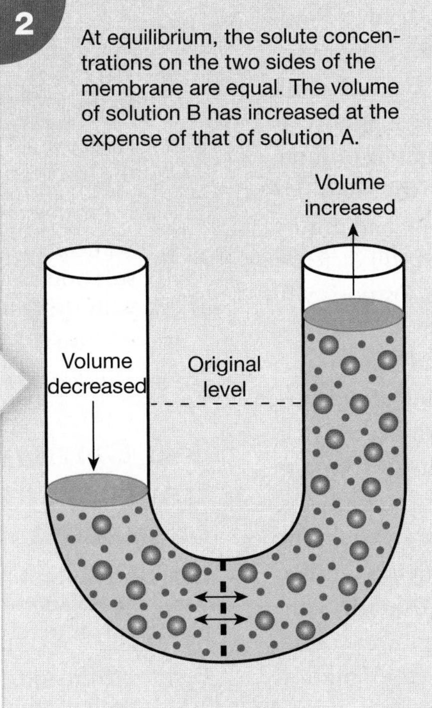
2
At equilibrium, the solute concentrations on the two sides of the membrane are equal. The volume of solution B has increased at the expense of that of solution A.
Volume increased
Volume decreased
Original level

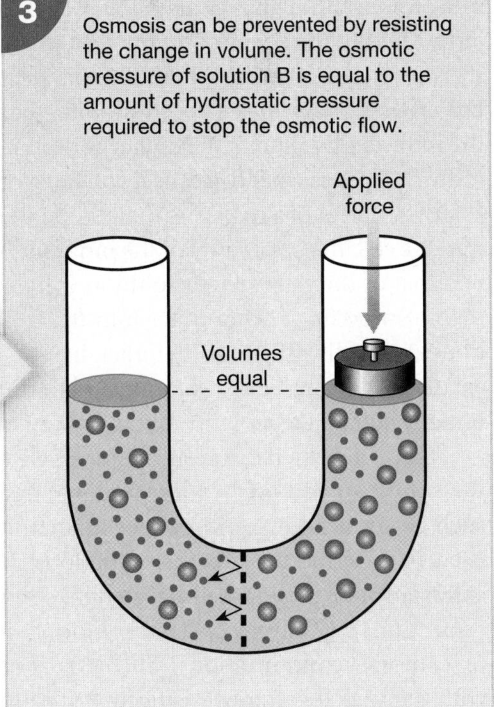
3
Osmosis can be prevented by resisting the change in volume. The osmotic pressure of solution B is equal to the amount of hydrostatic pressure required to stop the osmotic flow.
Applied force
Volumes equal

Go to MasteringA&P™ > Study Area > Menu > Additional Study Tools > Get Ready for A&P > Chapter 6: Cell Biology > Movement Processes: Video Tutor: Diffusion

Go to MasteringA&P™ > Study Area > Menu > Additional Study Tools > Get Ready for A&P > Chapter 6: Cell Biology > Movement Processes: Video Tutor: Osmosis membrane until the solute concentrations in the two solutions are identical (2). For this reason, the volume of solution B increases while that of solution A decreases. The greater the initial difference in solute concentrations, the stronger is the osmotic flow.

The **osmotic pressure** of a solution is an indication of the force with which pure water moves into that solution as a result of its solute concentration. We can measure a solution's osmotic pressure in several ways. For example, an opposing pressure can prevent the osmotic flow of water into the solution. Pushing against a fluid generates **hydrostatic pressure**. In 3 hydrostatic pressure opposes the osmotic pressure of solution B, so no net osmotic flow occurs.

Osmosis eliminates solute concentration differences more rapidly than solute diffusion. Why? In large part this happens because water molecules can also cross a membrane through abundant water channels called **aquaporins**, which exceed the number of solute channels, through which water can also pass. This difference results in a higher membrane permeability for water than for solutes.

Osmolarity and Tonicity

The total solute concentration in an aqueous solution is the solution's **osmolarity**, or **osmotic concentration**. The nature of the solutes, however, is often as important as the total osmolarity. Therefore, when we describe the effects of various osmotic solutions on cells, we usually use the term **tonicity** instead of osmolarity. A solution that does not cause an osmotic flow of water into or out of a cell is called isotonic (*iso-*, same + *tonos*, tension).

The terms *osmolarity* and *tonicity* do not always mean the same thing, although they are often used interchangeably. Osmolarity refers to the solute concentration of the solution, but tonicity is a description of how the solution affects the shape of a cell. For example, consider a solution with the same osmolarity as the intracellular fluid, but a higher concentration of one or more individual ions. What happens if you put a cell into this solution? If any of those ions can cross the plasma membrane and diffuse into the cell, the osmolarity of the intracellular fluid will increase, and that of the extracellular solution will decrease. Osmosis will then also occur, moving water into the cell. If the process continues, the cell will gradually inflate like a water balloon. In this example, the extracellular solution and the intracellular fluid were initially equal in osmolarity, but they were not isotonic.

However, when discussing osmosis we commonly compare the solute, or osmotic, concentrations between two solutions, such as intracellular and extracellular fluids. Three terms are used to describe the possible situations: (1) *isotonic* (the two solutions being compared have equal solute, or osmotic, concentrations), (2) *hypotonic* (the solution with the lower solute, or osmotic, concentration), and (3) *hypertonic* (the solution with the higher solute, or osmotic, concentration)

What visible effects do various osmotic solutions have on cells? Figure 3–17a shows a red blood cell (RBC) in an **isotonic** solution, where no osmotic flow takes place and the size and shape of the cell look normal. But if you put a red blood cell into a **hypotonic** (*hypo-*, lower) solution, water will flow into the cell, causing it to swell up like a balloon (Figure 3–17b). The cell may eventually burst, releasing its contents. This event is **hemolysis** (*hemo-*, blood + *lysis*, a loosening). In contrast, a cell in a **hypertonic** (*hyper-*, higher) solution will lose water by osmosis. As it does, the cell shrivels and dehydrates. The shrinking of red blood cells is called **crenation** (Figure 3–17c).

In the emergency department, it is often necessary to give patients large volumes of fluid to combat severe blood loss or dehydration. The procedure used in many cases is called an intravenous, or IV, drip and the fluids enter a patient's vein. One fluid frequently administered is a 0.9 percent (0.9 g/dL) solution of sodium chloride (NaCl). This solution, called *normal saline,* approximates the normal osmotic concentration of extracellular fluids. It is used because sodium and chloride are the most abundant ions in the extracellular fluid. Little net movement of either ion across plasma membranes takes place, so normal saline is essentially isotonic to body cells. An alternative IV treatment involves the use of an isotonic saline solution containing *dextran,* a carbohydrate that cannot cross plasma membranes. The dextran molecules elevate the osmolarity and osmotic pressure of the blood, and as water enters the blood vessels from the surrounding tissue fluid by osmosis, blood volume increases.

✓ Checkpoint

16. **What is meant by the term *selectively permeable* when referring to a plasma membrane?**
17. **Define *diffusion*.**
18. **List five factors that influence the diffusion of substances in the body.**
19. **How would a decrease in the concentration of oxygen in the lungs affect the diffusion of oxygen into the blood?**
20. **Define *osmosis*.**
21. **Some pediatricians recommend using a 10 percent salt solution as a nasal spray to relieve congestion in infants with stuffy noses. What effect would such a solution have on the cells lining the nasal cavity, and why?**

See the blue Answers tab at the back of the book.

3-6 Carrier-mediated and vesicular transport assist membrane passage of specific substances

Learning Outcome Describe carrier-mediated transport and vesicular transport mechanisms used by cells to facilitate the absorption or removal of specific substances.

Besides diffusion, substances are taken into or removed from cells in two additional ways: carrier-mediated transport and vesicular transport. *Carrier-mediated transport* requires specialized integral membrane proteins. This type of transport can

> Go to MasteringA&P™ > Study Area> Menu > Lab Tools > PhysioEx 9.1 > Exercise 1: Cell Transport Mechanisms and Permeability > Activity 3: Simulating Osmotic Pressure

Figure 3–17 **Osmotic Flow across a Plasma Membrane.** The smaller paired arrows indicate an equilibrium with no net water movement. The larger arrows indicate the direction of osmotic water movement.

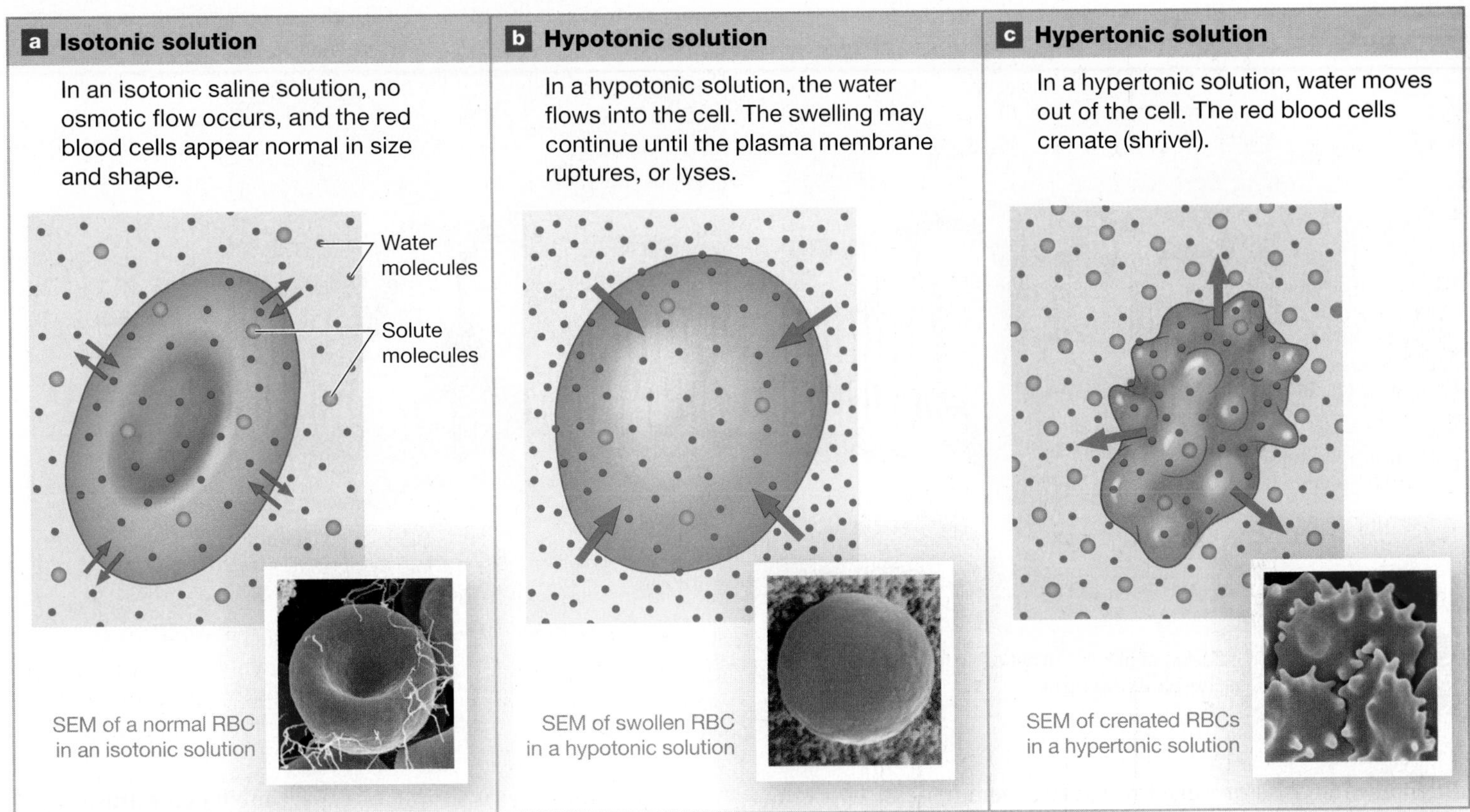

a Isotonic solution
In an isotonic saline solution, no osmotic flow occurs, and the red blood cells appear normal in size and shape.
Water molecules
Solute molecules
SEM of a normal RBC in an isotonic solution
b Hypotonic solution
In a hypotonic solution, the water flows into the cell. The swelling may continue until the plasma membrane ruptures, or lyses.
SEM of swollen RBC in a hypotonic solution
c Hypertonic solution
In a hypertonic solution, water moves out of the cell. The red blood cells crenate (shrivel).
SEM of crenated RBCs in a hypertonic solution

? Describe the concentration of the intracellular fluid relative to the extracellular fluid in parts b and c.

be passive or active, depending on the substance transported and the nature of the transport mechanism. *Vesicular transport* involves moving materials within small membranous sacs, or *vesicles*. Vesicular transport is always an active process.

Carrier-Mediated Transport

In **carrier-mediated transport**, integral proteins bind specific ions or organic substrates and carry them across the plasma membrane. All forms of carrier-mediated transport have the following characteristics, which they share with enzymes:

- *Specificity.* Each carrier protein in the plasma membrane binds and transports only certain substances. For example, the carrier protein that transports glucose will not transport other simple sugars.
- *Saturation Limits.* The availability of substrate molecules and carrier proteins limits the rate of transport into or out of the cell, just as the availability of substrates and enzymes limits enzymatic reaction rates. When all the available carrier proteins are operating at maximum speed, the carriers are said to be *saturated*. The rate of transport cannot increase further, regardless of the size of the concentration gradient.
- *Regulation.* Just as enzyme activity often depends on the presence of cofactors, the binding of other molecules, such as hormones, can affect the activity of carrier proteins. For this reason, hormones provide an important means of coordinating carrier protein activity throughout the body. We examine the interplay between hormones and plasma membranes when we study the endocrine system (Chapter 18) and metabolism (Chapter 25).

Many examples of carrier-mediated transport involve the movement of a single substrate molecule across the plasma membrane. However, a few carrier proteins may move more than one substrate at a time. In **symport**, or **cotransport**, mechanisms, the common carrier protein (symporter) transports two different molecules or ions through a membrane in the same direction, either into or out of the cell. In **antiport**, or **countertransport**, mechanisms, the carrier protein (antiporter) transports two

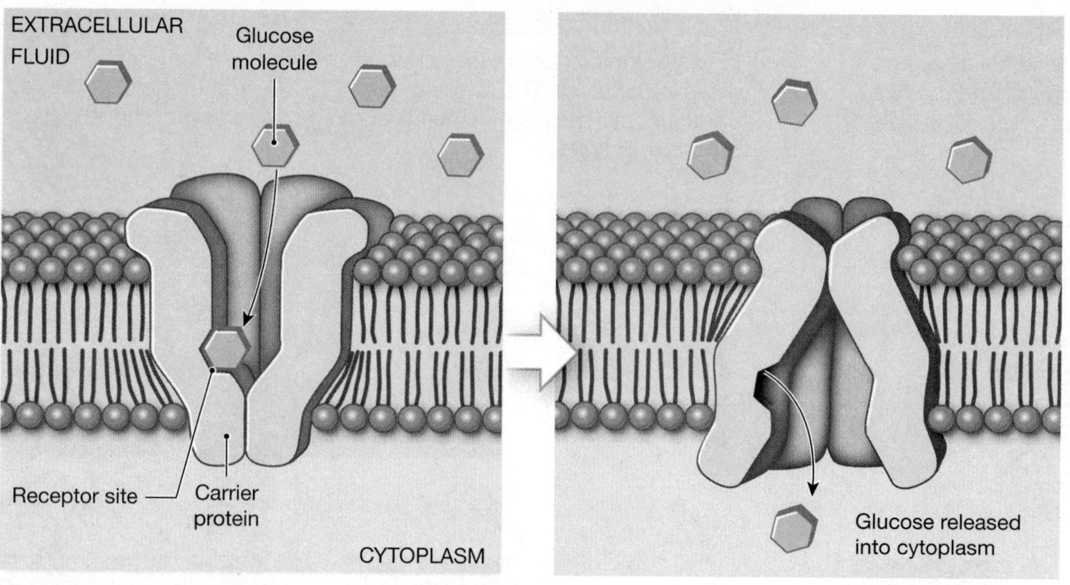
EXTRACELLULAR FLUID
Glucose molecule
Receptor site
Carrier protein
CYTOPLASM
Glucose released into cytoplasm

Figure 3–18 Facilitated Diffusion. In facilitated diffusion, a molecule, such as glucose, binds to a specific receptor site on a carrier protein. This binding permits the molecule to diffuse across the plasma membrane. If the direction of the glucose concentration gradient is reversed, then facilitated diffusion would occur in the opposite direction.

? In the facilitated diffusion of glucose, what determines the direction in which glucose molecules will be transported?

different molecules or ions through the membrane in opposite directions.

Let's look at two examples of carrier-mediated transport: *facilitated diffusion* and *active transport.*

Facilitated Diffusion

Many essential nutrients, such as glucose and amino acids, are insoluble in lipids and too large to fit through membrane channels. These substances can be passively transported across the membrane by carrier proteins in a process called **facilitated diffusion** (Figure 3–18). The molecule to be transported must first bind to a **receptor site** on the carrier protein. The shape of the protein then changes, moving the molecule across the plasma membrane and releasing it into the cytoplasm. This takes place without ever creating a continuous open channel between the cell's exterior and interior.

Tips & Tools

To understand the mechanism of cellular channels, think about entering a store that has a double set of automated sliding doors. When you near the first set of automated doors, it opens. As you step onto the mat between the sets of doors, the doors behind you close, trapping you in the vestibule. When you take another step forward, the second set of doors opens, enabling you to enter the store.

As in the case of simple or carrier-mediated diffusion, no ATP is expended in facilitated diffusion: The molecules simply move from an area of higher concentration to one of lower concentration. However, once the carrier proteins are saturated, the rate of transport cannot increase, regardless of further increases in the concentration gradient. Facilitated diffusion may occur in either direction across a plasma membrane.

All cells move glucose across their membranes through facilitated diffusion. However, several different carrier proteins are involved. In muscle cells, fat cells, and many other types of cells, the glucose transporter functions only when stimulated by the hormone *insulin.* Inadequate production of this hormone is one cause of *diabetes mellitus,* a metabolic disorder characterized by a high blood glucose level, which we discuss in Chapter 18.

Active Transport

In **active transport**, a high-energy bond (in ATP or another high-energy compound) provides the energy needed to move ions or molecules across the membrane. Despite the energy cost, active transport offers one great advantage: It does not depend on a concentration gradient. As a result, the cell can import or export specific substrates, *regardless of their intracellular or extracellular concentrations.*

All cells have carrier proteins called **ion pumps**, which actively transport the cations sodium (Na^+), potassium (K^+), calcium (Ca^{2+}), and magnesium (Mg^{2+}) across their plasma membranes. Specialized cells can transport additional ions, such as iodide (I^-), chloride (Cl^-), and iron (Fe^{2+}).

Many of these carrier proteins move a specific cation or anion in one direction only, either into or out of the cell. Sometimes one carrier protein will move more than one kind of ion at the same time. If a countertransport, or antiporter, mechanism occurs, the carrier protein is called an **exchange pump**.

There are two main types of active transport, primary and secondary.

Primary Active Transport: The Sodium–Potassium Exchange Pump. **Primary active transport** is the process of pumping solutes against a concentration gradient using the energy from ATP. One type of primary active transport involves sodium and potassium ions, which are the main cations in body fluids. The sodium ion concentration is high in the extracellular fluid, but low in the cytoplasm. The distribution of potassium ions in the body is just the opposite: low in the extracellular fluid and high in the cytoplasm. Because of leak channels in plasma membranes, sodium ions slowly diffuse into the cell, and potassium ions diffuse out. Homeostasis within the cell depends on removing sodium ions and recapturing lost potassium ions. This exchange takes place by a **sodium–potassium exchange pump**. The carrier protein involved in the process is called *sodium–potassium ATPase.*

The sodium–potassium exchange pump exchanges intracellular sodium ions for extracellular potassium ions (Figure 3–19). On average, for each ATP molecule hydrolyzed to ADP and P_i (inorganic phosphate), three sodium ions are ejected and the cell reclaims two potassium ions. If ATP is readily available, the rate of transport depends on the concentration of sodium ions in the cytoplasm. When the concentration rises, the pump becomes more active. The energy demands are impressive: Sodium–potassium ATPase may use up to 40 percent of the ATP produced by a resting cell!

Secondary Active Transport. In **secondary active transport**, the transport mechanism itself does not require energy from ATP, but the cell often needs to expend ATP at a later time to preserve homeostasis. As with facilitated diffusion, a secondary active transport mechanism moves a specific substrate down its concentration gradient. Unlike the carrier proteins in facilitated diffusion, however, these carrier proteins can also move another substrate at the same time, without regard to its concentration gradient. In effect, the concentration gradient for one substance provides the driving force needed by the carrier protein, and the second substance gets a "free ride."

The concentration gradient for sodium ions most often provides the driving force for symport mechanisms that move materials into the cell. For example, sodium-linked cotransport is important in the absorption of glucose and amino acids along the intestinal tract. The initial transport activity proceeds without direct energy expenditure, but the cell must then expend ATP to pump the arriving sodium ions out of the cell using the sodium–potassium exchange pump (Figure 3–20).

Sodium ions are also involved with many antiport mechanisms. For example, sodium–calcium countertransport keeps the intracellular calcium ion concentration very low.

Figure 3–19 The Sodium–Potassium Exchange Pump. The operation of the sodium–potassium exchange pump is an example of active transport. For each ATP converted to ADP, this carrier protein pump, also called sodium–potassium ATPase, carries three Na^+ out of the cell and two K^+ into the cell.

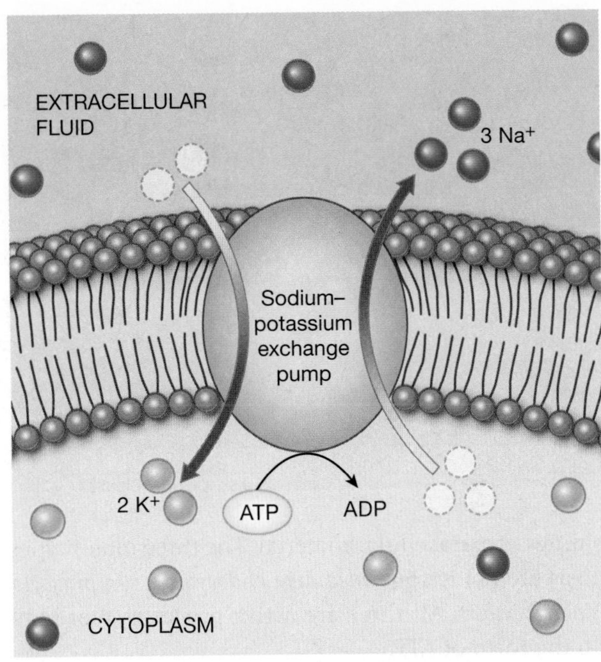
EXTRACELLULAR
FLUID
3 Na+
Sodium–
potassium
exchange
pump
2 K+
ATP
ADP
CYTOPLASM

? What provides the energy for the sodium–potassium exchange pump?

Vesicular Transport

In **vesicular transport**, the traffic of materials into or out of the cell takes place in **vesicles**, small membranous sacs that form at, or fuse with, the plasma membrane. Because these vesicles move tiny droplets of fluid and solutes rather than single molecules, this process is also known as *bulk transport.* The two major types of vesicular transport are *endocytosis* and *exocytosis.*

Endocytosis

We saw earlier in this chapter that extracellular materials can be packaged in vesicles at the cell surface and imported into the cell. This process, called **endocytosis**, involves relatively

Figure 3–20 **Secondary Active Transport.** In secondary active transport, glucose transport by a carrier protein will take place only after the carrier has bound two sodium ions. In three cycles, three glucose molecules and six sodium ions are transported into the cytoplasm. The cell then pumps the sodium ions across the plasma membrane by the sodium–potassium exchange pump, at a cost of two ATP molecules.

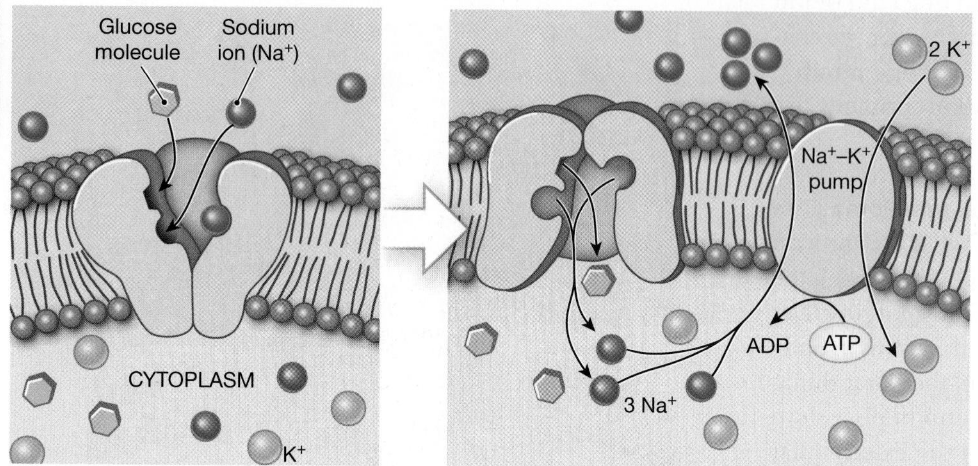
Glucose molecule
Sodium ion (Na+)
CYTOPLASM
K+
2 K+
Na+–K+ pump
ADP
ATP
3 Na+

large volumes of extracellular material. The three major types of endocytosis are (1) *receptor-mediated endocytosis,* (2) *pinocytosis,* and (3) *phagocytosis.* All three are active processes that require energy in the form of ATP.

The vesicles of endocytosis are generally known as *endosomes.* Their contents remain isolated from the cytoplasm, trapped within the vesicle. The enclosed materials may move into the surrounding cytoplasm by active transport, simple or facilitated diffusion, or the destruction of the vesicle membrane.

Receptor-Mediated Endocytosis. A highly selective process, **receptor-mediated endocytosis** produces vesicles that contain a specific target molecule in high concentrations. Most receptor molecules are glycoproteins, and each binds a specific ligand, or target molecule, such as a transport protein or a hormone. Cholesterol and iron ions (Fe^{2+}) are too large to pass through membrane pores, but they are able to enter cells by receptor-mediated endocytosis.

Receptor-mediated transport begins as receptors and their ligands migrate to specialized areas of the plasma membrane called **clathrin-coated pits**. These areas of plasma membrane are inwardly depressed and are "coated" with high concentrations of the protein *clathrin* on their cytoplasmic surface. See Figure 3–21 for an illustration and description of the complete process.

Another form of endocytosis that targets specific molecules involves receptors associated with membrane lipids and small flask-shaped indentations called **caveolae** (kav-ē-Ō-lē), a term that means "little caves." Specialized membrane proteins other than clathrin are also involved.

Pinocytosis. "Cell drinking," or **pinocytosis** (pi-nō-sī-TŌ-sis), is the formation of endosomes filled with extracellular fluid. Most cells carry out pinocytosis. This process is not as selective as receptor-mediated endocytosis, because no receptor proteins are involved. The target appears to be the fluid contents in general, rather than specific bound ligands. In pinocytosis, a deep groove or pocket forms in the plasma membrane and then pinches off. The steps involved are similar to the steps in receptor-mediated endocytosis, except that ligand binding is not involved.

Phagocytosis. "Cell eating," or **phagocytosis** (fag-ō-sī-TŌ-sis), produces endosomes called *phagosomes* containing solid objects that may be as large as the cell itself. In this process, cytoplasmic extensions called **pseudopodia** (sū-dō-PŌ-dē-ah; singular *pseudopodium; pseudo-*, false + *podon*, foot) surround the object, and their membranes fuse to form a phagosome. This vesicle then fuses with many lysosomes, and lysosomal enzymes digest its contents. Only specialized cells, such as the macrophages, that protect tissues by engulfing bacteria, cell debris, and other abnormal materials, perform phagocytosis.

Exocytosis

Exocytosis (ek-sō-sī-TŌ-sis), introduced in **Spotlight Figure 3–7** (p. 78) in our discussion of the Golgi apparatus, is the functional reverse of endocytosis. In exocytosis, a vesicle formed inside the cell fuses with, and becomes part of, the plasma membrane. When this takes place, the vesicle contents are released into the extracellular environment. The ejected materials may be secretory products, such as mucins or hormones, or waste products, such as those accumulating in endocytic vesicles.

Please now take a long look at **Spotlight Figure 3–22.** This figure provides a summary of all the passive and active transport processes in cells, in a way that will allow you to compare and recall them.

Transcytosis

In a few specialized cells, endocytosis produces vesicles on one side of the cell that are discharged through exocytosis on the opposite side. This method of bulk transport is common in cells lining capillaries, which use a combination of pinocytosis and exocytosis to transfer fluid and solutes from the bloodstream into the surrounding tissues. This overall transport process is called *transcytosis* (tranz-sī-TŌ-sis).

Figure 3–21 Receptor-Mediated Endocytosis.

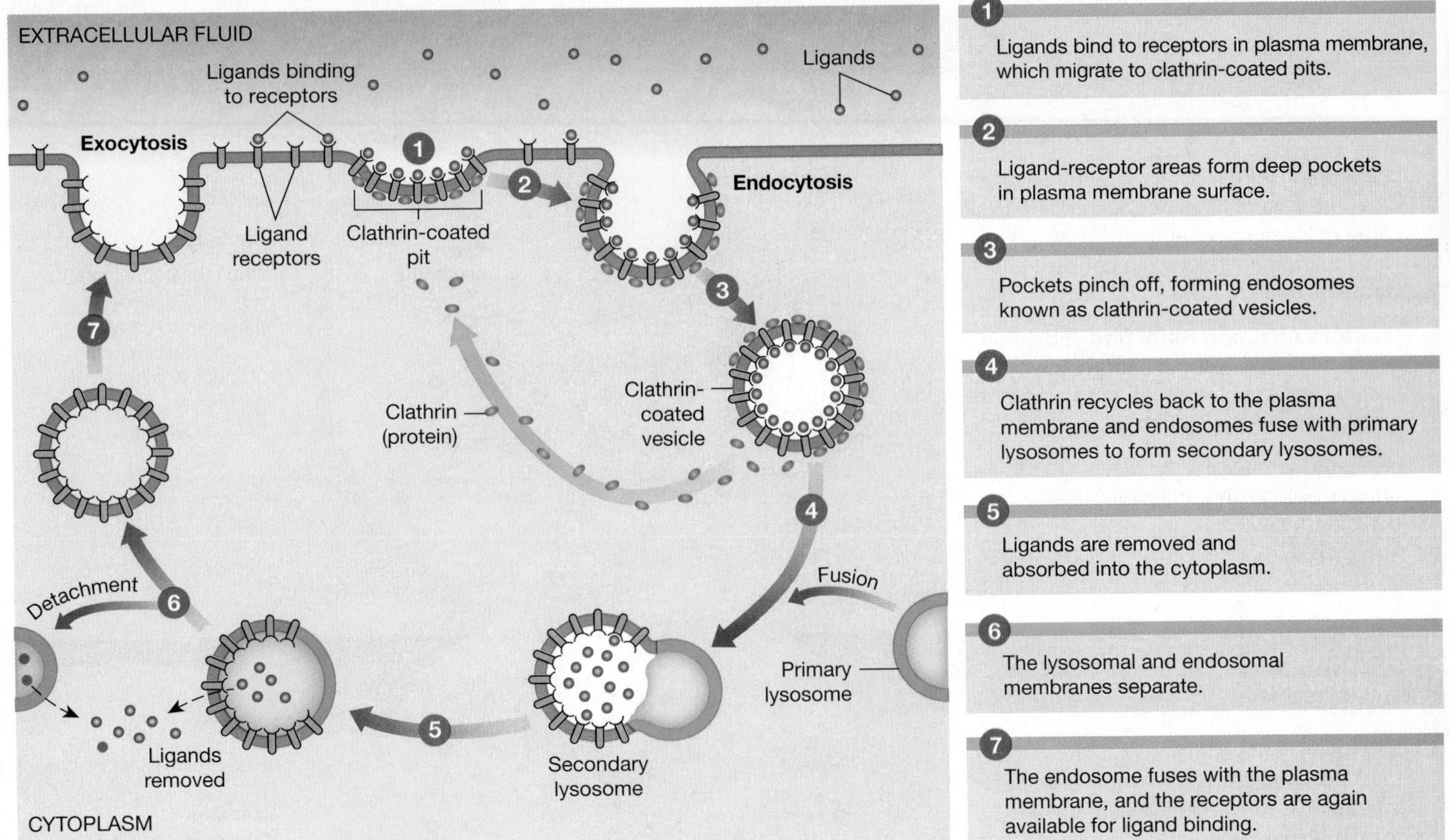
EXTRACELLULAR FLUID
Ligands binding to receptors
Ligands
Exocytosis
Endocytosis
Ligand receptors
Clathrin-coated pit
Clathrin (protein)
Clathrin-coated vesicle
Fusion
Primary lysosome
Detachment
Ligands removed
Secondary lysosome
CYTOPLASM
1 Ligands bind to receptors in plasma membrane, which migrate to clathrin-coated pits.
2 Ligand-receptor areas form deep pockets in plasma membrane surface.
3 Pockets pinch off, forming endosomes known as clathrin-coated vesicles.
4 Clathrin recycles back to the plasma membrane and endosomes fuse with primary lysosomes to form secondary lysosomes.
5 Ligands are removed and absorbed into the cytoplasm.
6 The lysosomal and endosomal membranes separate.
7 The endosome fuses with the plasma membrane, and the receptors are again available for ligand binding.

Checkpoint

22. **Describe the process of carrier-mediated transport.**
23. **The concentration of hydrogen ions (H^+) in the stomach contents during digestion rises to a higher level than in the cells lining the stomach. Which transport process must be operating?**
24. **Describe endocytosis.**
25. **Describe exocytosis.**
26. **What is the process called whereby certain types of white blood cells engulf bacteria?**

See the blue Answers tab at the back of the book.

3-7 The membrane potential of a cell results from the unequal distribution of positive and negative charges across the plasma membrane

Learning Outcome **Explain the origin and significance of the cell membrane potential.**

Recall that the cytoplasm-facing surface of a cell's plasma membrane has a slight negative charge with respect to the membrane's extracellular surface. (Keep in mind that this difference in charge occurs only at the plasma membrane. The electrical charge within the cytoplasm itself is neutral when compared to the extracellular fluid.) The slight excess of positive charges on the exterior of the plasma membrane is due to cations, namely sodium. The negative charge along the cytoplasmic side of the plasma membrane is due to negatively charged protein anions that are too large to pass through any membrane ion channels. A membrane's unequal charge distribution is created by differences in the permeability of the membrane to various ions, as well as by active transport mechanisms.

The positive and negative charges are attracted to each other and would normally rush together, but the plasma membrane keeps them apart. When positive and negative charges are held apart, a **potential difference** exists between them. *Potential* refers to the energy contained by something relative to its position to something else. For example, the water behind a dam has *potential energy*—stored energy that can be released to do work. We refer to the potential difference across a plasma membrane as the **membrane potential**.

The unit of measurement of potential difference is the *volt* (V). Most cars, for example, have 12-V batteries. The membrane potentials of cells are much smaller, typically in the vicinity of 0.07 V. Such a value is usually expressed as 70 *millivolts* (mV), or thousandths of a volt.

SPOTLIGHT

Figure 3–22
Overview of Membrane Transport

Passive Processes

No cellular energy expended

Diffusion

Diffusion is the movement of molecules from an area of higher concentration to an area of lower concentration. That is, the movement occurs down a concentration gradient (high to low).

Factors Affecting Rate: Size of gradient, molecular size, electric charge, lipid solubility, temperature, and the presence of membrane channel proteins

Substances Involved: Gases, small inorganic ions and molecules, lipid-soluble materials

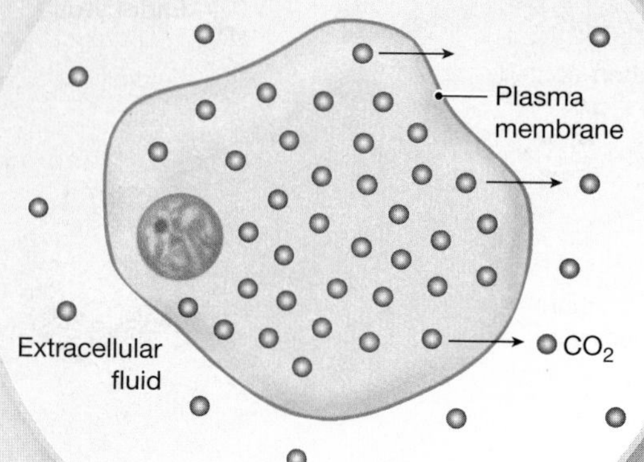

Plasma membrane
Extracellular fluid
CO_2

Example:
When the concentration of CO_2 inside a cell is greater than outside the cell, the CO_2 diffuses out of the cell and into the extracellular fluid.

Osmosis

Osmosis is the diffusion of water molecules across a selectively permeable membrane. Movement occurs toward higher solute concentration because that is where the concentration of water is lower. Osmosis continues until the concentration gradient is eliminated.

Factors Affecting Rate: Concentration gradient; opposing pressure

Substances Involved: Water only

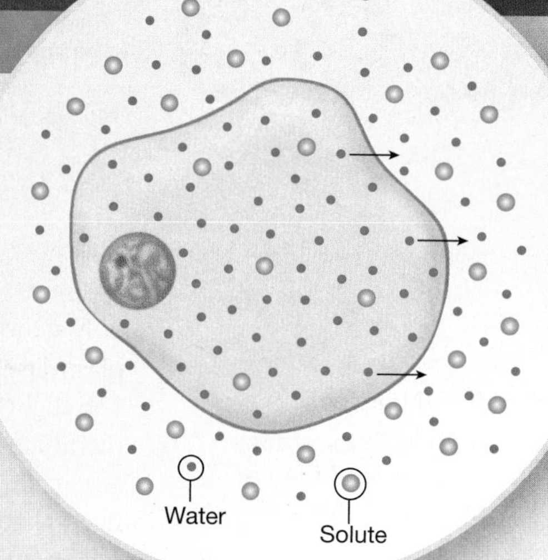

Water
Solute

Example:
If the solute concentration outside a cell is greater than inside the cell, water molecules will move across the plasma membrane into the extracellular fluid.

Facilitated diffusion

Facilitated diffusion is the movement of materials across a membrane by a carrier protein. Movement follows the concentration gradient.

Factors Affecting Rate: Concentration gradient, availability of carrier proteins

Substances Involved: Simple sugars and amino acids

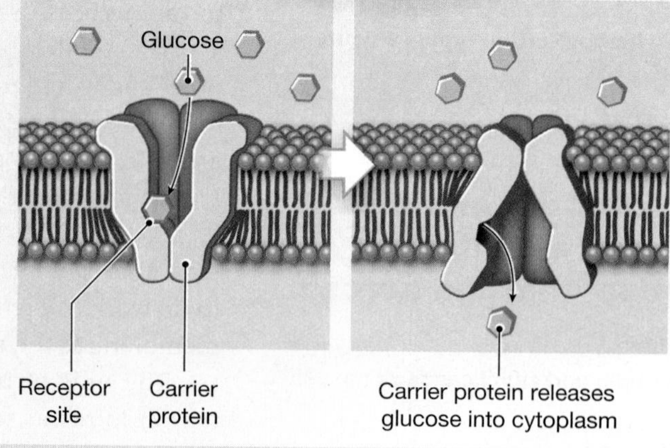

Glucose
Receptor site
Carrier protein
Carrier protein releases glucose into cytoplasm

Example:
Nutrients, such as glucose, that are insoluble in lipids and too large to fit through membrane channels may be transported across the plasma membrane by carrier proteins. Many carrier proteins move a specific substance in one direction only, either into or out of the cell.

Go to MasteringA&P™ > Study Area > Menu > Animations & Videos > A&PFlix > A&P > Membrane Transport

Active Processes

Require ATP

Active transport

Active transport requires carrier proteins that move specific substances across a membrane against their concentration gradient. If the carrier moves one solute in one direction and another solute in the opposite direction, it is called an **exchange pump**.

Factors Affecting Rate: Availability of carrier protein, substrate, and ATP

Substances Involved: Na^+, K^+, Ca^{2+}, Mg^{2+} ; other solutes in special cases

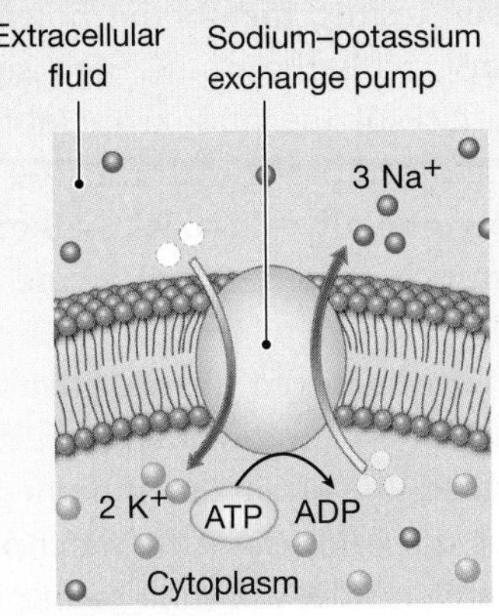

Extracellular fluid
Sodium–potassium exchange pump
3 Na+
2 K+
ATP
ADP
Cytoplasm

Example:
The **sodium–potassium exchange pump**. One ATP molecule is hydrolyzed to ADP for each 3 sodium ions that are ejected from the cell, while 2 potassium ions are reclaimed.

Endocytosis

Endocytosis is the packaging of extracellular materials into a vesicle for transport into the cell.

Receptor-Mediated Endocytosis

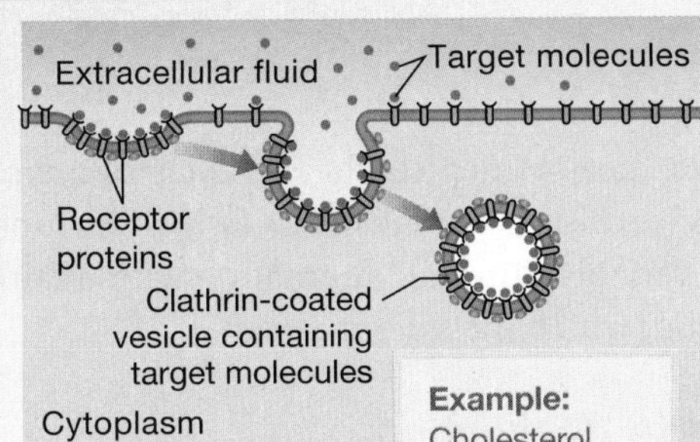

Extracellular fluid
Target molecules
Receptor proteins
Clathrin-coated vesicle containing target molecules
Cytoplasm

Example:
Cholesterol and iron ions are transported this way.

In **receptor-mediated endocytosis**, target molecules called ligands bind to receptor proteins on the membrane surface, triggering vesicle formation.

Factors Affecting Rate: Number of receptors on the plasma membrane and the concentration of ligands

Substances Involved: Ligands

Pinocytosis

In **pinocytosis**, vesicles form at the plasma membrane and bring fluids and small molecules into the cell. This process is often called "cell drinking."

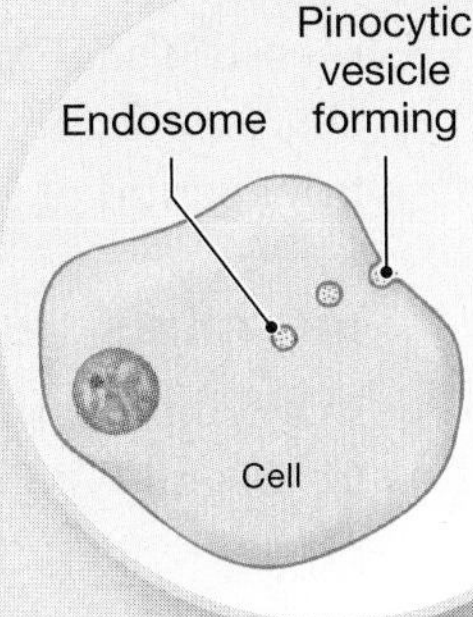

Pinocytic vesicle forming
Endosome
Cell

Example:
Once the vesicle is inside the cytoplasm, water and small molecules enter the cell across the vesicle membrane.

Factors Affecting Rate: Stimulus and mechanism not understood

Substances Involved: Extracellular fluid, with dissolved molecules such as nutrients

Phagocytosis

In **phagocytosis**, vesicles called phagosomes form at the plasma membrane to bring particles into the cell. This process is often called "cell eating."

Factors Affecting Rate: Presence of pathogens and cellular debris

Substances Involved: Bacteria, viruses, cellular debris, and other foreign material

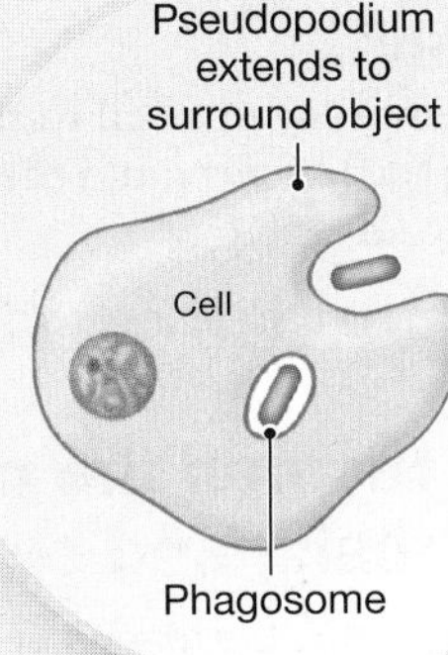

Pseudopodium extends to surround object
Cell
Phagosome

Example:
Large particles are brought into the cell by cytoplasmic extensions (called **pseudopodia**) that engulf the particle and pull it into the cell.

Exocytosis

In **exocytosis**, intracellular vesicles fuse with the plasma membrane to release fluids and/or solids from the cells.

Factors Affecting Rate: Stimulus and mechanism incompletely understood

Substances Involved: Fluid and cellular wastes; secretory products from some cells

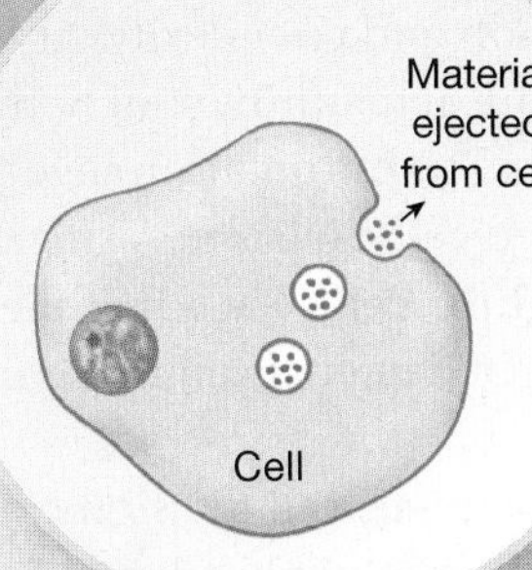

Material ejected from cell
Cell

Example:
Cellular wastes in vesicles are ejected from the cell.

The membrane potential in an unstimulated, or undisturbed, cell is called the **resting membrane potential**. Each type of cell has a characteristic resting membrane potential between −10 mV and −100 mV, with the minus sign signifying that the cytoplasmic surface of the plasma membrane has an excess of negative charges compared with the outside. Examples include fat cells (−40 mV), thyroid cells (−50 mV), neurons (−70 mV), skeletal muscle cells (−85 mV), and cardiac muscle cells (−90 mV).

If the cell's lipid barrier were removed, the positive and negative charges would rush together and the potential difference would be eliminated. Instead, the plasma membrane acts like a dam across a stream. Just as a dam resists the water pressure that builds up on the upstream side, a plasma membrane resists electrochemical forces that would otherwise drive ions into or out of the cell. So, like water behind a dam, the ions held on either side of the plasma membrane have potential energy. People have designed many ways to use the potential energy stored behind a dam—for example, turning a mill wheel or a turbine. Similarly, cells have ways of using the potential energy stored in the membrane potential. For example, changes in the membrane potential make possible the transmission of information in the nervous system, and thus our perceptions and thoughts. As we will see in later chapters, changes in the membrane potential also trigger the contractions of muscles and the secretions of glands.

✓ Checkpoint

27. **What is the membrane potential of a cell, and in what units is it expressed?**
28. **If the plasma membrane of a cell were freely permeable to sodium ions (Na^+), how would the membrane potential be affected?**

See the blue Answers tab at the back of the book.

3-8 Stages of the cell life cycle include interphase, mitosis, and cytokinesis

Learning Outcome Describe the stages of the cell life cycle, including interphase, mitosis, and cytokinesis, and explain their significance.

The period between fertilization and physical maturity involves tremendous changes in organization and complexity. At fertilization, a single cell is all there is, but at maturity, your body has about 75 trillion cells. This amazing transformation involves a form of cellular reproduction called **cell division**. A single cell divides to produce a pair of **daughter cells**, each half the size of the original. Before dividing, each daughter cell will grow to the size of the original cell.

Even when development is complete, cell division continues to be essential to survival. Cells are highly adaptable, but physical wear and tear, toxic chemicals, temperature changes, and other environmental stresses can damage them. And, like individuals, cells age. The life span of a cell varies from hours to decades, depending on the type of cell and the stresses involved. Many cells apparently self-destruct after a certain period of time because specific "suicide genes" in the nucleus are activated. This genetically controlled death of cells is called **apoptosis** (ap-op-TŌ-sis; *apo-*, separated from + *ptosis,* a falling). Researchers have identified several genes involved in the regulation of this process. For example, a gene called *bcl-2* encodes a regulator protein that appears to prevent apoptosis and to keep a cell alive and functional. If something interferes with the function of this gene, the cell self-destructs.

Because a typical cell does not live nearly as long as a typical person, cell populations must be maintained over time by cell division. For cell division to be successful, the genetic material in the nucleus must be duplicated accurately, and one copy must be distributed to each daughter cell. The duplication of the cell's genetic material is called *DNA replication*, and nuclear division is called *mitosis* (mī-TŌ-sis). Mitosis occurs during the division of somatic cells. The production of sex cells involves a different process, *meiosis* (mī-Ō-sis), described in Chapter 28.

The Cell Life Cycle

Spotlight Figure 3–23 (p. 104) depicts the life cycle of a typical cell. That life cycle consists of an *interphase* (which includes DNA replication) of variable duration alternating with a fairly brief period of *M phase* (which includes mitosis).

Interphase and DNA Replication

In a cell preparing to divide, interphase can be divided into the G_1, S, and G_2 phases. DNA replication takes place during the S phase.

Interphase. Interphase is the stage of the cell life cycle between mitotic divisions. Somatic cells spend the majority of their lives in interphase. During this time, they perform all their normal functions. An interphase cell in a nondividing state is said to be in the G_0 phase. Cells of different tissues vary in the amount of time spent in the G_0 phase. Interphase cells preparing to divide exit the G_0 phase and move on to the G_1, S, and G_2 phases.

The Process of DNA Replication. Recall that each DNA molecule consists of a pair of DNA strands joined by hydrogen bonds between complementary nitrogenous bases. ↺ p. 57 Duplication of DNA, or **DNA replication**, begins when enzymes called *helicases* unwind the strands and disrupt the weak hydrogen bonds between the bases. As the strands unwind, molecules of the enzyme **DNA polymerase** bind to the exposed nitrogenous bases. This enzyme (1) promotes bonding between the nitrogenous bases of the DNA strand and complementary DNA nucleotides in the nucleoplasm

> Go to MasteringA&P™ > Study Area > Menu > Additional Study Tools > Get Ready for A&P > Chapter 6: Cell Biology > The Cell Cycle: Video Tutor: The Cell Cycle

and (2) links the nucleotides by covalent bonds. **Spotlight Figure 3–24** (p. 106) diagrams DNA replication.

M Phase

The M phase includes mitosis and cytokinesis. Mitosis refers to nuclear division and *cytokinesis* refers to cytoplasmic division.

Mitosis. **Mitosis** is the duplication of the chromosomes in the nucleus and their separation into two identical sets in the process of somatic cell division. Although we describe mitosis in stages called *prophase, metaphase, anaphase,* and *telophase,* it is really one continuous process. See **Spotlight Figure 3–23** for a description of this process.

Tips & Tools

To remember the correct sequence of events during mitosis, imagine the U-shaped contour rug surrounding your toilet as the P-MAT, for ***p***rophase, ***m***etaphase, ***a***naphase, and ***t***elophase.

Cytokinesis. **Cytokinesis** is the division of the cytoplasm during mitosis. This process usually begins in late anaphase and continues throughout telophase, producing two daughter cells. The completion of cytokinesis marks the end of cell division, creating two separate and complete cells, each surrounded by its own plasma membrane.

The Mitotic Rate and Energy Use

The preparations for cell division that take place between the G_1 and M phases are difficult to recognize in a light micrograph. However, the start of mitosis is easy to recognize, because the chromatin condenses to form highly visible chromosomes. The frequency of cell division in a tissue can be estimated by the number of cells in mitosis at any time. As a result, we often use the term **mitotic rate** when we discuss rates of cell division. In general, the longer the life expectancy of a cell type, the slower the mitotic rate. Long-lived cells, such as muscle cells and neurons, either never divide or do so only under special circumstances. Other cells, such as those covering the surface of the skin or the lining of the digestive tract, are subject to attack by chemicals, pathogens, and abrasion. They survive for only days or even hours. Special cells called **stem cells** maintain these cell populations through repeated cycles of cell division.

Stem cells are relatively unspecialized. Their only function is to produce daughter cells. Each time a stem cell divides, one of its daughter cells develops functional specializations while the other prepares for further stem cell divisions. The rate of stem cell division can vary with the type of tissue and the demand for new cells. In heavily abraded skin, stem cells may divide more than once a day, but stem cells in adult connective tissues may remain inactive for years.

Dividing cells use an unusually large amount of energy. For example, they must synthesize new organic materials and move organelles and chromosomes within the cell. All these processes require ATP in substantial amounts. Cells that do not have adequate energy sources cannot divide. In a person who is starving, normal cell growth and maintenance grind to a halt. For this reason, prolonged starvation stunts childhood growth, slows wound healing, lowers resistance to disease, thins the skin, and changes the lining of the digestive tract.

✓ Checkpoint

29. **Give the biological terms for (a) cellular reproduction and (b) cellular death.**
30. **What enzymes must be present for DNA replication to proceed normally?**
31. **Describe interphase, and identify its stages.**
32. **A cell is actively manufacturing enough organelles to serve two functional cells. This cell is probably in what phase of its life cycle?**
33. **Define *mitosis*, and list its four stages.**
34. **What would happen if spindle fibers failed to form in a cell during mitosis?**

See the blue Answers tab at the back of the book.

Tips & Tools

When considering the relative length of time that a cell spends in interphase compared to mitosis, think about an upcoming test. You prepare a long time (interphase) for something that happens quickly (mitosis).

3-9 Several factors regulate the cell life cycle

Learning Outcome **Discuss the regulation of the cell life cycle.**

In normal adult tissues, the rate of cell division balances the rate of cell loss or destruction. Mitotic rates are genetically controlled, and many different stimuli may be responsible for activating genes that promote cell division. Some of the stimuli are internal, and many cells set their own pace of mitosis and cell division.

An important internal trigger is the level of **M-phase promoting factor (MPF)**, also known as *maturation-promoting factor.* MPF is assembled from two parts: a cell division cycle protein called *Cdc2* and a second protein called *cyclin.* The cyclin level climbs as the cell life cycle proceeds. When the level is high enough, MPF appears in the cytoplasm and mitosis gets under way.

Go to MasteringA&P™ > Study Area > Menu > Animations & Videos > A&PFlix > A&P > Mitosis

INTERPHASE

Most cells spend only a small part of their time actively engaged in cell division. Somatic cells spend the majority of their functional lives in a state known as **interphase**. During interphase, a cell performs all its normal functions and, if necessary, prepares for cell division.

A cell that is ready to divide first enters the **G_1** phase. In this phase, the cell makes enough mitochondria, cytoskeletal elements, endoplasmic reticula, ribosomes, Golgi apparatus membranes, and cytosol for two functional cells. Centriole replication begins in G_1 and commonly continues until G_2. In cells dividing at top speed, G_1 may last just 8–12 hours. Such cells use all their energy on mitosis, and all other activities cease. When G_1 lasts for days, weeks, or months, preparation for mitosis occurs as the cells perform their normal functions.

When the activities of **G_1** have been completed, the cell enters the **S phase**. Over the next 6–8 hours, the cell duplicates its chromosomes. This involves DNA replication and the synthesis of histones and other proteins in the nucleus.

Once DNA replication has ended, there is a brief (2–5-hour) **G_2 phase** devoted to last-minute protein synthesis and to the completion of centriole replication.

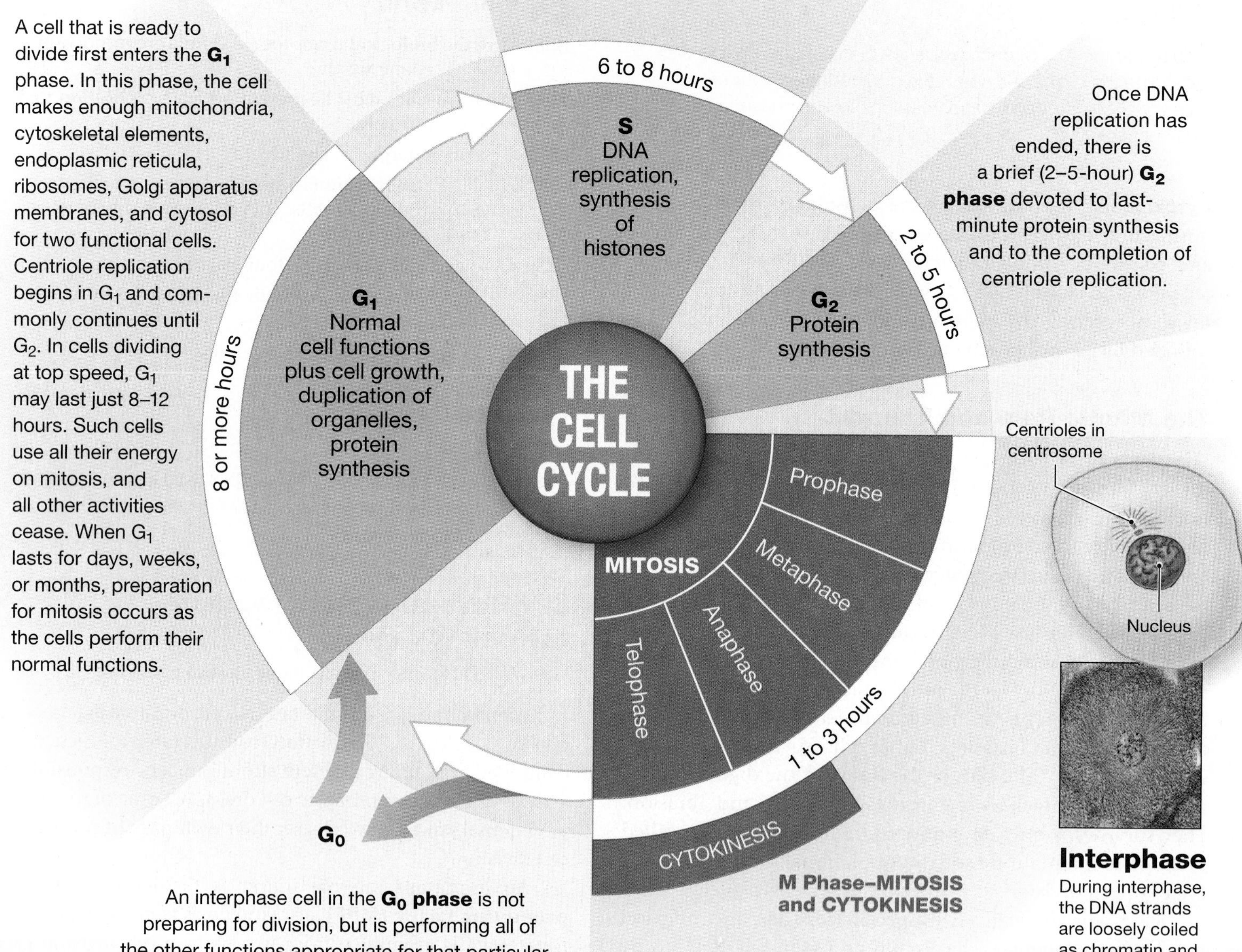

6 to 8 hours
S
DNA replication, synthesis of histones
2 to 5 hours
G_2
Protein synthesis
8 or more hours
G_1
Normal cell functions plus cell growth, duplication of organelles, protein synthesis
THE CELL CYCLE
MITOSIS
Prophase
Metaphase
Anaphase
Telophase
1 to 3 hours
CYTOKINESIS
G_0
M Phase–MITOSIS and CYTOKINESIS
Centrioles in centrosome
Nucleus

Interphase
During interphase, the DNA strands are loosely coiled as chromatin and chromosomes cannot be seen.

An interphase cell in the **G_0 phase** is not preparing for division, but is performing all of the other functions appropriate for that particular cell type. Some mature cells, such as skeletal muscle cells and most neurons, remain in G_0 indefinitely and never divide. In contrast, stem cells, which divide repeatedly with very brief interphase periods, never enter G_0.

Go to MasteringA&P™ > Study Area > Menu > Animations & Videos > A&PFlix > A&P > Mitosis

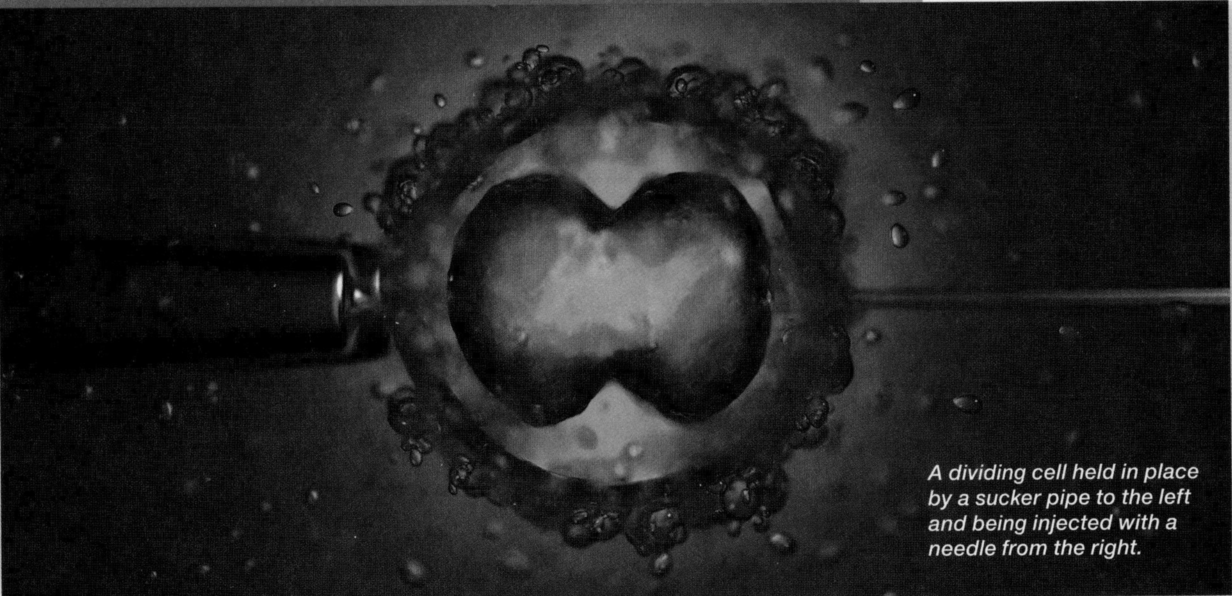
A dividing cell held in place by a sucker pipe to the left and being injected with a needle from the right.

MITOSIS AND CYTOKINESIS

The M phase of the cell cycle includes mitosis and cytokinesis. Mitosis separates the duplicated chromosomes of a cell into two identical nuclei. The term **mitosis** specifically refers to the division and duplication of the cell's nucleus; division of the cytoplasm to form two distinct new cells involves a separate but related process known as **cytokinesis**.

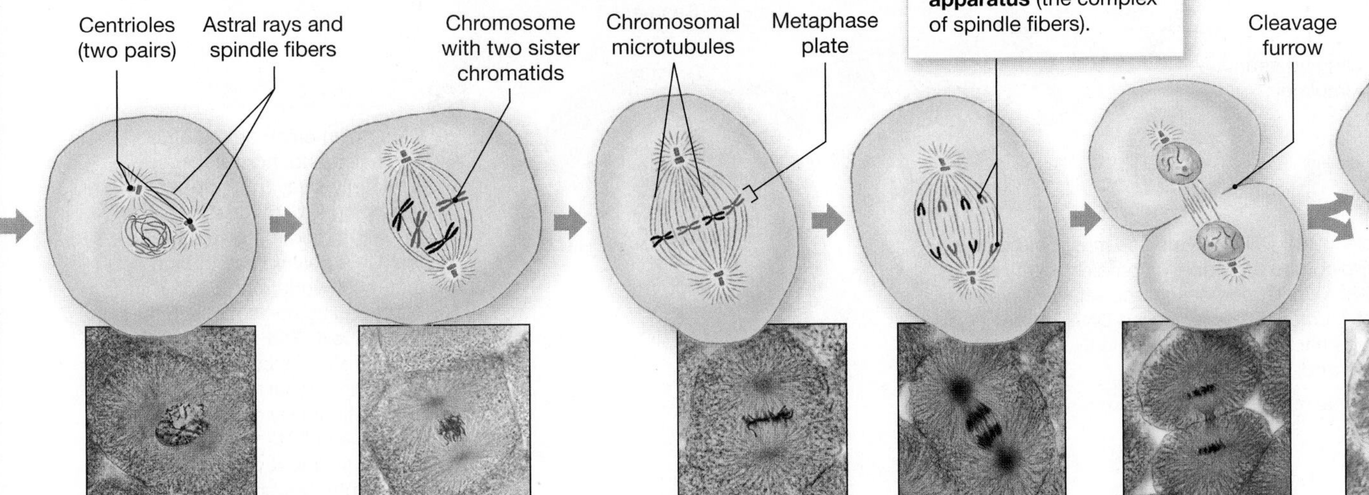
Centrioles (two pairs)
Astral rays and spindle fibers
Chromosome with two sister chromatids
Chromosomal microtubules
Metaphase plate
The two daughter chromosomes are pulled apart and toward opposite ends of the cell along the spindle apparatus (the complex of spindle fibers).
Cleavage furrow
Daughter cells

Early prophase

Prophase (PRŌ-fāz; *pro*-, before) begins when the chromatin condenses and chromosomes become visible as single structures under a light microscope. An array of microtubules called **spindle fibers** extends between the centriole pairs. Smaller microtubules called **astral rays** radiate into the cytoplasm.

Late prophase

As a result of DNA replication during the S phase, two copies of each chromosome now exist. Each copy, called a **chromatid** (KRŌ-ma-tid), is connected to its duplicate copy at a single point, the **centromere** (SEN-trō-mēr). **Kinetochores** (ki-NĒ-tō-korz) are the protein-bound areas of the centromere; they attach to spindle fibers forming **chromosomal microtubules**.

Metaphase

Metaphase (MET-a-fāz; *meta*-, after) begins as the chromatids move to a narrow central zone called the metaphase plate. Metaphase ends when all the chromatids are aligned in the plane of the metaphase plate.

Anaphase

Anaphase (AN-a-fāz; *ana*-, apart) begins when the centromere of each chromatid pair splits and the chromatids separate. The two **daughter chromosomes** are now pulled toward opposite ends of the cell along the chromosomal microtubules.

Telophase

During telophase (TĒL- ō-fāz; *telo*-, end), each new cell prepares to return to the interphase state. The nuclear membranes re-form, the nuclei enlarge, and the chromosomes gradually uncoil. This stage marks the end of mitosis.

Cytokinesis

Cytokinesis is the division of the cytoplasm into two daughter cells. Cytokinesis usually begins with the formation of a **cleavage furrow** and continues throughout telophase. The completion of cytokinesis marks the end of cell division.

SPOTLIGHT

Figure 3–24 DNA Replication

Go to MasteringA&P™ > Study Area > Menu > Animations & Videos > A&PFlix > A&P > DNA Replication

For cell division to be successful, the genetic material in the nucleus must be duplicated accurately, and one copy must be distributed to each daughter cell. This duplication of the cell's genetic material is called **DNA replication**.

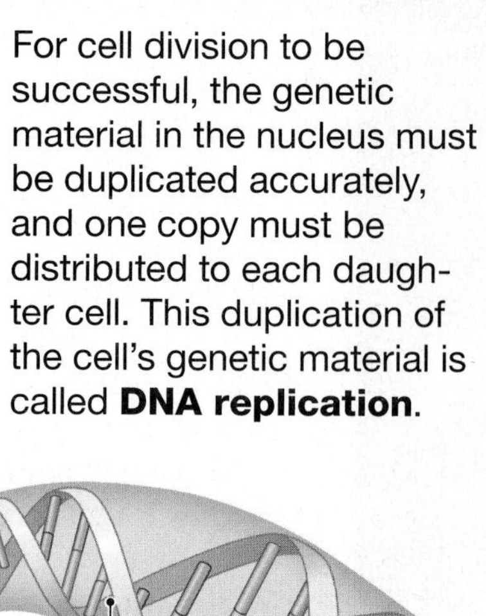

Each DNA molecule consists of two nucleotide strands joined by hydrogen bonds between their complementary nitrogenous bases.

1 DNA replication begins when **helicase** enzymes unwind the strands and disrupt the hydrogen bonds between the bases. As the strands unwind, molecules of **DNA polymerase** bind to the exposed nitrogenous bases. This enzyme (1) promotes bonding between the nitrogenous bases of each separated strand with complementary DNA nucleotides in the nucleoplasm and (2) links the nucleotides by covalent bonds.

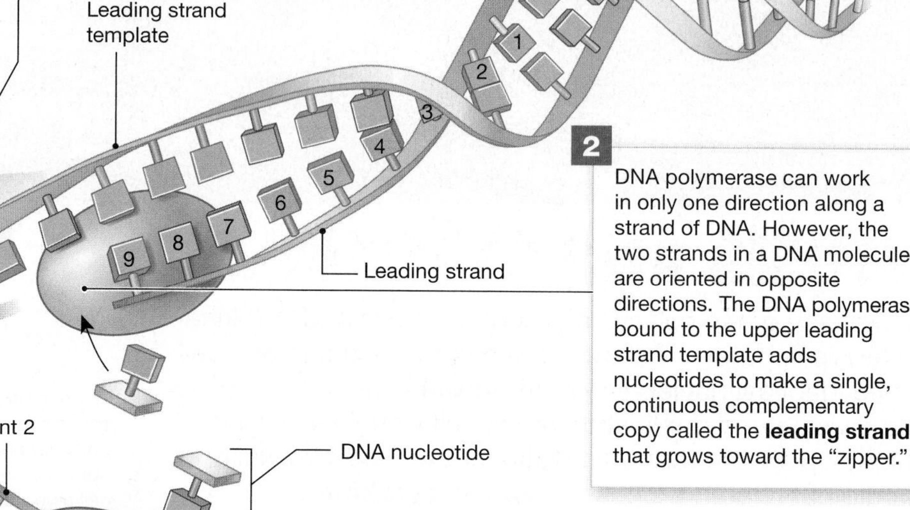

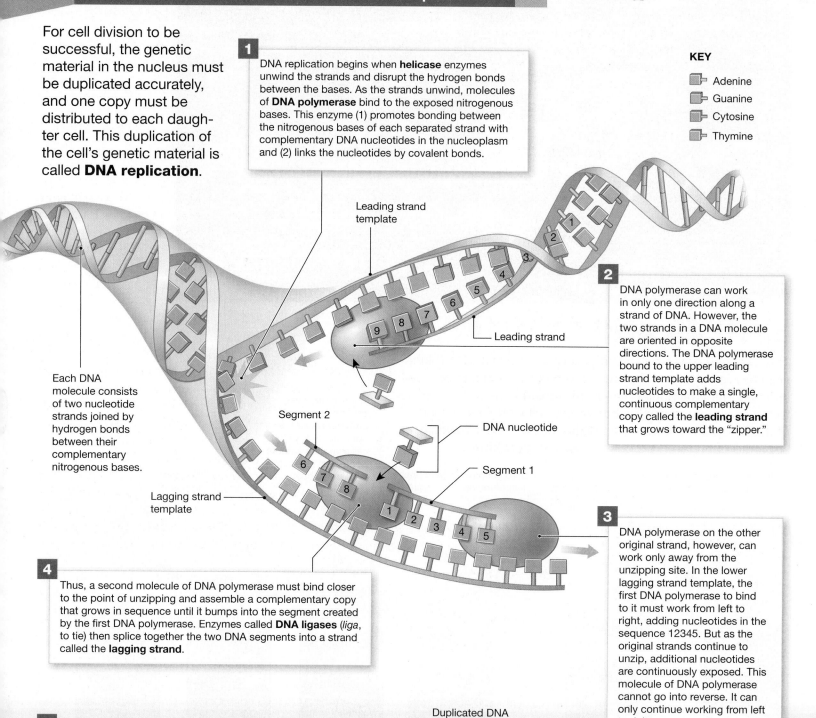

KEY
Adenine
Guanine
Cytosine
Thymine
Leading strand template
Leading strand
Segment 2
DNA nucleotide
Segment 1
Lagging strand template

2 DNA polymerase can work in only one direction along a strand of DNA. However, the two strands in a DNA molecule are oriented in opposite directions. The DNA polymerase bound to the upper leading strand template adds nucleotides to make a single, continuous complementary copy called the **leading strand** that grows toward the "zipper."

3 DNA polymerase on the other original strand, however, can work only away from the unzipping site. In the lower lagging strand template, the first DNA polymerase to bind to it must work from left to right, adding nucleotides in the sequence 12345. But as the original strands continue to unzip, additional nucleotides are continuously exposed. This molecule of DNA polymerase cannot go into reverse. It can only continue working from left to right.

4 Thus, a second molecule of DNA polymerase must bind closer to the point of unzipping and assemble a complementary copy that grows in sequence until it bumps into the segment created by the first DNA polymerase. Enzymes called **DNA ligases** (*liga*, to tie) then splice together the two DNA segments into a strand called the **lagging strand**.

5 Eventually, the unzipping completely separates the original strands. The copying and last splicing are completed, and 2 identical DNA molecules are formed. Once the DNA has been replicated, the centrioles duplicated, and the necessary enzymes and proteins have been synthesized, the cell leaves interphase and is ready to proceed to mitosis.

Duplicated DNA double helices

Various natural body substances, such as hormones, peptides, or nutrients, derived from food can stimulate the division of specific types of cells. These substances are known as **growth factors**. Table 3–2 lists some of these chemical factors and their target tissues. We discuss them in later chapters.

Researchers have identified genes that inhibit cell division. Such genes are known as *repressor genes*. One gene, named *TP53*, encodes a protein called *p53*, that binds to DNA in the nucleus. The binding, in turn, activates another gene that directs the production of growth-inhibiting factors inside the cell. Roughly half of all cancers are associated with abnormal forms of the *p53* gene.

There are indications that in humans, the *number* of cell divisions that a cell and its descendants undergo is regulated at the chromosome level by structures called **telomeres**. Telomeres are terminal segments of DNA with associated proteins. These DNA–protein complexes bend and fold repeatedly to form caps at the ends of chromosomes, much like the plastic sheaths on the tips of shoestrings. Telomeres have several functions, notably to attach chromosomes to the nuclear matrix and to protect the ends of the chromosomes from damage during mitosis. The telomeres themselves, however, are subject to wear and tear over the years. Each time a cell divides during adult life, some of the repeating segments break off, and the telomeres get shorter. When they get too short, repressor gene activity signals the cell to stop dividing.

Checkpoint

35. **Define *growth factor*, and identify several growth factors that affect cell division.**

See the blue Answers tab at the back of the book.

3-10 Abnormal cell growth and division characterize tumors and cancers

Learning Outcome **Discuss the relationship between cell division and cancer.**

When the rates of cell division and growth exceed the rate of cell death, a tissue begins to enlarge. A **tumor**, or *neoplasm*, is a mass or swelling produced by abnormal cell growth and division. In a *benign tumor*, the cells usually remain within the tissue where it originated. Such a tumor seldom threatens an individual's life and can usually be surgically removed if its size or position disturbs tissue function. Cells in a **malignant tumor** no longer respond to normal controls. These cells do not remain confined within the epithelium or a connective tissue capsule, but spread into surrounding tissues. The tumor of origin is called the *primary tumor*, and the spreading process is called **invasion**. Malignant cells may also travel to distant tissues and organs and establish *secondary tumors*.

Cancer is an illness that results from the abnormal proliferation of any of the cells in the body. It is characterized by mutations that disrupt normal cell regulatory controls and produce potentially malignant cells. Normal cells begin the path to becoming malignant when a mutation occurs in a gene involved with cell growth, differentiation, or division. The modified genes are called **oncogenes** (ON-kō-gēnz; *oncos*, tumor).

Agents that cause a mutation (a change in DNA) are called **mutagens**. Cancer-causing agents, which include many mutagens, are known as **carcinogens**. Examples of carcinogens include radiation, chemicals, bacteria, and viruses. Some chemical carcinogens, such as alcohol and estrogen, are not directly mutagenic. Instead, they promote cancer by promoting rapid cell divisions, which increase the chances that a mutation will occur.

Table 3–2 Chemical Factors Affecting Cell Division

Factor	Sources	Effects	Targets
M-phase promoting factor (maturation-promoting factor) (MPF)	Forms within cytoplasm from Cdc2 and cyclin	Initiates mitosis	Regulatory mechanism active in all dividing cells
Growth hormone	Anterior lobe of the pituitary gland	Stimulation of growth, cell division, differentiation	All cells, especially in epithelial and connective tissues
Prolactin	Anterior lobe of the pituitary gland	Stimulation of cell growth, division, development	Gland and duct cells of mammary glands
Nerve growth factor (NGF)	Salivary glands; other sources suspected	Stimulation of nerve cell repair and development	Neurons and neuroglia
Epidermal growth factor (EGF)	Duodenal glands; other sources suspected	Stimulation of stem cell divisions and epithelial repairs	Epidermis
Fibroblast growth factor (FGF)	Wide range of cells	Division and differentiation of fibroblasts and related cells	Connective tissues
Erythropoietin (EPO)	Kidneys (primary source)	Stimulation of stem cell divisions and maturation of red blood cells	Red bone marrow
Thymosins and related compounds	Thymus	Stimulation of division and differentiation of lymphocytes (especially T cells)	Thymus and other lymphoid tissues and organs
Chalones	Many tissues	Inhibition of cell division	Cells in the immediate area

Figure 3–25 The Development and Metastasis of Cancer.

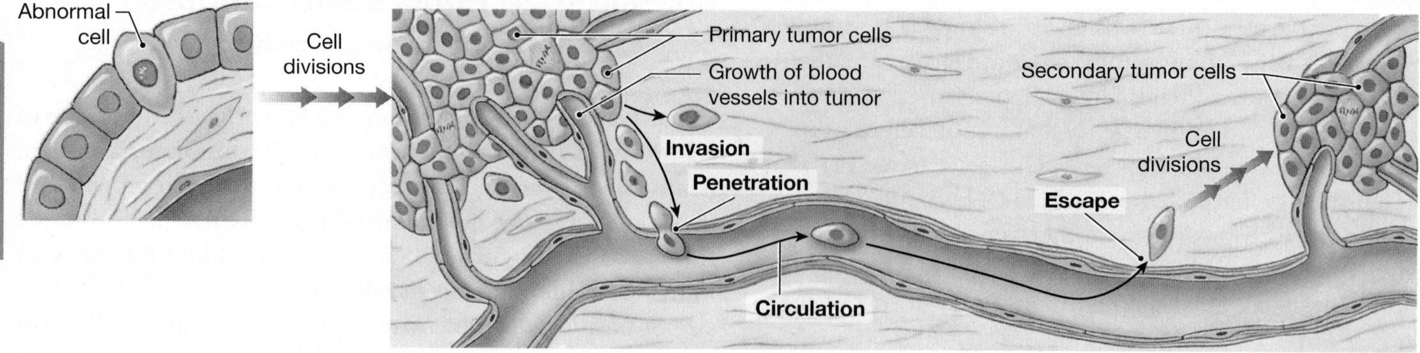
Abnormal cell
Cell divisions
Primary tumor cells
Growth of blood vessels into tumor
Invasion
Penetration
Circulation
Secondary tumor cells
Cell divisions
Escape

Normal cells become malignant in a multistep process. For example, a mutation may occur in one cell, which is passed on to its daughter cells. One of those cells may undergo another mutation that speeds its rate of cell division. Further growth-promoting mutations in one of those cells and its proliferating descendants may produce a primary tumor. The tumor now contains descendants of the original abnormal cell that have been selected for rapid growth. While these tumor cells are ultimately all descended from one abnormal cell, they are not identical to it due to their accumulated mutations.

Cancer cells do not resemble normal cells. They have a different shape and typically become abnormally larger or smaller. Malignant tumor cells are also less subject to controls limiting their adhesion and movement between adjacent cells and extracellular tissues. At first, the growth of the primary tumor distorts the tissue, but the basic tissue organization remains intact. **Metastasis** (me-TAS-ta-sis; *meta-*, after + *stasis*, standing still), or the spread of the cancer to other areas, begins with invasion as malignant cells "break out" of the primary tumor and invade the surrounding tissue. They may then enter the lymphatic system and accumulate in nearby lymph nodes. When metastasis involves the penetration of blood vessels, the cancer cells circulate throughout the body. Metastatic cancer progresses in the steps diagrammed in Figure 3–25.

Responding to cues that are as yet unknown, cancer cells in the bloodstream ultimately escape out of blood vessels to

+ Clinical Note Telomerase, Aging, and Cancer

Each telomere contains a sequence of about 8000 nitrogenous bases, but they are multiple copies of the same six-base sequence, TTAGGG, repeated over and over again. Telomeres are not formed by DNA polymerase. Instead, they are created by the enzyme *telomerase.* Telomerase is functional early in life, but by adulthood it becomes inactive. As a result, the telomere segments lost during each mitotic division are not replaced. Eventually, the telomeres become so short that the cell no longer divides.

This mechanism is a factor in the aging process, since many of the signs of age result from the gradual loss of functional stem cell populations. Experiments are in progress to determine whether activating telomerase (or a suspected alternative repair enzyme) can forestall or reverse the effects of aging. This would seem to be a very promising area of research. In fact, the 2009 Nobel Prize in Physiology or Medicine was awarded for telomerase research. Activate telomerase, and halt aging—sounds good, doesn't it? Unfortunately, there's a catch: In adults, telomerase activation is a key step in the development of cancer.

If for some reason a cell with short telomeres does *not* respond normally to repressor genes, it will continue to divide. The result is mechanical damage to the DNA strands, chromosomal abnormalities, and mutations. Interestingly, one of the first consequences of such damage is the abnormal activation of telomerase. Once this occurs, the abnormal cells can continue dividing indefinitely. Telomerase is active in at least 90 percent of all cancer cells. Research is under way to find out how to turn off telomerase that has been improperly activated.

+ Clinical Note Breakthroughs with Stem Cells

In most cases, cell differentiation is irreversible. Once genes are turned off, they won't be turned back on. However, some cells, such as stem cells, are relatively undifferentiated. These cells can differentiate into any of several different cell types, depending on local conditions. For example, if nutrients are abundant, stem cells can differentiate into fat cells. Researchers are gradually discovering what chemical cues and genes control the differentiation of specific cell types. Such research has resulted in the ability to "turn back the clock" in some types of adult somatic cells and reprogram them into a form of stem cell called *induced pluripotent stem (iPS) cells.* The ability to take a person's stem cells or somatic cells and create new cells or neurons to treat disease may one day revolutionize the practice of medicine.

The actor Michael J. Fox, of *Back to the Future* fame, has made **Parkinson's Disease (PD)** research his personal mission since he was diagnosed with PD at age 30. PD is a neurodegenerative disease characterized by progressive degeneration and loss of dopamine-producing neurons. (Dopamine is one kind of *neurotransmitter*, a substance that one neuron releases to communicate with other neurons.) PD may be the first disorder suited to treatment by implanting stem cells. Several laboratories have induced either iPS cells or *embryonic stem cells* to differentiate into cells that function as dopamine-producing neurons. In rodent studies, both iPS and embryonic stem cell–derived dopamine neurons re-innervated the brains of rats with Parkinson's disease. These cells also released dopamine and improved motor function.

establish secondary tumors at other sites. The cancer cells in these tumors are extremely active metabolically, and their presence stimulates the growth of new blood vessels (*angiogenesis*) into the area. The increased blood supply provides additional nutrients to the cancer cells and further accelerates tumor growth and metastasis.

As malignant tumors grow, organ function begins to deteriorate. The malignant cells may no longer perform their original functions, or they may perform normal functions in an abnormal way. For example, endocrine cancer cells may produce normal hormones, but in excessively large amounts.

Cancer cells do not use energy very efficiently. They grow and multiply at the expense of healthy tissues, competing for space and nutrients with normal cells. This competition contributes to the starved appearance of many patients in the late stages of cancer. Death may result from the compression of vital organs when nonfunctional cancer cells have killed or replaced the healthy cells in those organs, or when the cancer cells have starved normal tissues of essential nutrients. We discuss cancer further in later chapters that deal with specific systems.

√ Checkpoint

36. **An illness characterized by mutations that disrupt normal control mechanisms and produce potentially malignant cells is termed _____.**
37. **Define *metastasis*.**

See the blue Answers tab at the back of the book.

3-11 Cellular differentiation is cellular specialization as a result of gene activation or repression

Learning Outcome **Define cellular differentiation, and explain its importance.**

An individual's liver cells, fat cells, and neurons all contain the same chromosomes and genes, but in each case a different set of genes has been turned *off*. In other words, your liver cells and fat cells differ because liver cells have one set of genes accessible for transcription, or activation, and fat cells have another.

When a gene is functionally eliminated, the cell loses the ability to produce a particular protein—and to perform any functions involving that protein. Each time another gene switches off, the cell's functional abilities become more restricted. This development of specific cellular characteristics and functions that are different from the original cell is called **cellular differentiation**.

Fertilization produces a single cell with all its genetic potential intact. Repeated cell divisions follow, and cellular differentiation begins as the number of cells increases. Cellular differentiation produces specialized cells with limited capabilities. These cells form organized collections known as *tissues*, each with discrete functional roles.In Chapter 4, we examine the structure and function of tissues and consider the role of tissue interactions in maintaining homeostasis.

Checkpoint

38. **Define *cellular differentiation*.**

See the blue Answers tab at the back of the book.

3 Chapter Review

Study Outline

An Introduction to Cells p. 66

1. Contemporary *cell theory* incorporates several basic concepts: (1) Cells are the building blocks of all plants and animals; (2) cells are produced by the division of preexisting cells; (3) cells are the smallest units that perform all vital physiological functions; and (4) each cell maintains homeostasis at the cellular level. *(Spotlight Figure 3–1)*
2. **Cytology**, the study of cellular structure and function, is part of **cell biology**.
3. The human body contains two types of cells: **sex cells** (*sperm* and *oocytes*) and **somatic cells** (all other cells). *(Spotlight Figure 3–1)*

3-1 The plasma membrane separates the cell from its surrounding environment and performs various functions p. 66

4. A typical cell is surrounded by **extracellular fluid**—specifically, the **interstitial fluid** of the tissue. The cell's outer boundary is the **plasma membrane** (**cell membrane**).
5. The plasma membrane's functions include physical isolation, regulation of exchange with the environment, sensitivity to the environment, and structural support. *(Figure 3–2)*
6. The plasma membrane is a **phospholipid bilayer**, which contains other lipids, proteins, and carbohydrates.
7. **Integral proteins** are part of the membrane itself; **peripheral proteins** are attached to, but can separate from, the membrane.
8. Membrane proteins can act as anchors (**anchoring proteins**), identifiers (**recognition proteins**), enzymes, receptors (**receptor proteins**), carriers (**carrier proteins**), or **channels**.
9. The carbohydrate portions of *proteoglycans, glycoproteins,* and *glycolipids* form the **glycocalyx** on the outer cell surface. Functions include lubrication and protection, anchoring and locomotion, specificity in binding, and recognition.

3-2 Organelles within the cytoplasm perform particular functions p. 71

10. The **cytoplasm** contains the fluid **cytosol** and the **organelles** suspended in the cytosol.
11. Cytosol (intracellular fluid) differs from extracellular fluid in composition and in the presence of **inclusions**.
12. **Nonmembranous organelles** are not completely enclosed by membranes, and all of their components are in direct contact with the cytosol. They include the *cytoskeleton, microvilli, centrioles, cilia, ribosomes,* and *proteasomes. (Spotlight Figure 3–1)*
13. **Membranous organelles** are surrounded by phospholipid membranes that isolate them from the cytosol. They include the *endoplasmic reticulum,* the *Golgi apparatus, lysosomes, peroxisomes, mitochondria,* and *nucleus. (Spotlight Figure 3–1)*
14. The **cytoskeleton** gives the cytoplasm strength and flexibility. All cells have three cytoskeletal components: **microfilaments** (typically made of **actin**), **intermediate filaments**, and **microtubules** (made of **tubulin**). Muscle cells also have *thick filaments* (made of *myosin*). *(Figure 3–3)*
15. **Microvilli** are small, nonmotile projections of the plasma membrane that increase the surface area exposed to the extracellular environment. *(Figure 3–3)*

> **MasteringA&P™** Access more chapter study tools online in the MasteringA&P Study Area:
> - Chapter Quizzes, Chapter Practice Test, MP3 Tutor Sessions, and Clinical Case Studies
> - Practice Anatomy Lab PAL 3.0
> - A&P Flix A&PFlix
> - Interactive Physiology iP2
> - PhysioEx PhysioEx 9.1

16. The **centrosome** is a region of the cytoplasm near the nucleus. It is the microtubule organizing center surrounding the paired centrioles. **Centrioles** direct the movement of chromosomes during cell division. *(Figure 3–4)*
17. A single, nonmotile **primary cilium** is sensitive to environmental stimuli. Multiple **motile cilia**, each anchored by a **basal body**, beat rhythmically to move fluids or secretions across the cell surface. *(Figure 3–4)*
18. **Ribosomes**, responsible for manufacturing proteins, are composed of a **small** and a **large ribosomal subunit**, both of which contain **ribosomal RNA (rRNA)**. **Free ribosomes** are in the cytoplasm, and **fixed ribosomes** are attached to the endoplasmic reticulum. *(Spotlight Figure 3–1)*
19. **Proteasomes** remove and break down damaged or abnormal proteins that have been tagged with *ubiquitin.*
20. The **endoplasmic reticulum (ER)** is a network of intracellular membranes that function in synthesis, storage, transport, and detoxification. The ER forms hollow tubes, flattened sheets, and chambers called **cisternae**. **Smooth endoplasmic reticulum (SER)** is involved in lipid synthesis; **rough endoplasmic reticulum (RER)** contains ribosomes on its outer surface and forms **transport vesicles**. *(Figure 3–5; Spotlight Figure 3–7)*
21. The **Golgi apparatus**, or **Golgi complex**, forms **secretory vesicles** and new plasma membrane components, and packages *lysosomes.* Secretions exit the cell by exocytosis. *(Figure 3–6; Spotlight Figure 3–7)*
22. **Lysosomes**, vesicles filled with digestive enzymes, are responsible for the **autolysis** of injured cells. *(Figures 3–6, 3–8; Spotlight Figure 3–7)*
23. **Peroxisomes** carry enzymes that neutralize potentially dangerous free radicals.
24. **Membrane flow**, or *membrane trafficking,* refers to the continuous movement and recycling of the membrane among the ER, vesicles, the Golgi apparatus, and the plasma membrane.
25. **Mitochondria** are responsible for ATP production through aerobic metabolism. The **matrix**, or fluid contents of a mitochondrion, lies inside the **cristae**, or folds of an inner membrane. *(Figure 3–9)*

3-3 The nucleus contains DNA and enzymes essential for controlling cellular activities p. 83

26. The **nucleus** is the control center of cellular operations. It is surrounded by a **nuclear envelope** (a double membrane with a **perinuclear space**), through which it communicates with the cytosol by way of **nuclear pores**. *(Spotlight Figure 3–1; Figure 3–10)*
27. The nucleus contains a supportive **nuclear matrix**; one or more **nucleoli** typically are present.
28. The nucleus controls the cell by directing the synthesis of specific proteins, using information stored in **chromosomes**,

which consist of DNA bound to **histones**. In nondividing cells, DNA and associated proteins form a tangle of filaments called **chromatin**. *(Figure 3–11)*

29. The cell's information storage system, the **genetic code**, is called a *triplet code* because a sequence of three nitrogenous bases specifies the identity of a single amino acid. Each **gene** contains all the DNA triplets needed to produce a specific polypeptide chain.

3-4 DNA controls protein synthesis, cell structure, and cell function p. 85

30. As **gene activation** begins, **RNA polymerase** must bind to the gene.
31. **Transcription** is the production of RNA from a DNA template. After transcription, a strand of **messenger RNA (mRNA)** carries instructions from the nucleus to the cytoplasm. *(Figure 3–12)*
32. During **translation**, a functional polypeptide is constructed using the information contained in the sequence of **codons** along an mRNA strand. The mRNA codons correspond to DNA triplets. The sequence of codons determines the sequence of amino acids in the polypeptide.
33. During translation, complementary base pairing of **anticodons** to mRNA codons occurs, and **transfer RNA (tRNA)** molecules bring amino acids to the ribosomal complex. Translation includes three phases: **initiation**, **elongation**, and **termination**. *(Spotlight Figure 3–13; Table 3–1)*
34. Through protein synthesis, the DNA of the nucleus controls the synthesis of enzymes and structural proteins that provide short- and long-term homeostatic adjustments affecting the cell's biochemical and physical characteristics.

3-5 Diffusion is a passive transport mechanism that assists membrane passage of solutes and water p. 90

35. The **permeability** of a barrier such as the plasma membrane is an indication of the barrier's effectiveness. Nothing can pass through an **impermeable** barrier; anything can pass through a **freely permeable** barrier. Plasma membranes are **selectively permeable**.
36. **Diffusion** is the net movement of a substance from an area of higher concentration to an area of lower concentration. Diffusion occurs until the **concentration gradient** is eliminated. *(Figures 3–14, 3–15)*
37. Most lipid-soluble materials, water, and gases freely diffuse across the phospholipid bilayer of the plasma membrane. Small water-soluble molecules and ions rely on channel-mediated diffusion through a passageway within a transmembrane protein. **Leak channels** are passive channels that allow ions across the plasma membrane.
38. **Osmosis** is the net flow of water across a selectively permeable membrane in response to differences in solute concentration. **Osmotic pressure** of a solution is the force of water movement into that solution resulting from its solute concentration. **Hydrostatic pressure** can oppose osmotic pressure. *(Figure 3–16)*
39. **Tonicity** describes the effects of osmotic solutions on cells. A solution that does not cause an osmotic flow is **isotonic**. A solution that causes water to flow into a cell is **hypotonic** and can lead to **hemolysis** of red blood cells. A solution that causes water to flow out of a cell is **hypertonic** and can lead to **crenation**. *(Figure 3–17)*

3-6 Carrier-mediated and vesicular transport assist membrane passage of specific substances p. 94

40. Carrier-mediated transport involves the binding and transporting of specific ions by integral proteins. **Symport**, or **cotransport**, moves two substances in the same direction; **antiport**, or **countertransport**, moves them in opposite directions.
41. In **facilitated diffusion**, compounds are transported across a membrane after binding to a **receptor site** within the channel of a carrier protein. *(Figure 3–18)*
42. **Active transport** mechanisms consume ATP and are not dependent on concentration gradients. Some **ion pumps** are **exchange pumps**. **Secondary active transport** may involve symport (cotransport) or antiport (countertransport). *(Figures 3–19, 3–20)*
43. In **vesicular transport**, materials move into or out of the cell in membranous **vesicles**. Movement into the cell is accomplished through **endocytosis**, an active process that can take three forms: **receptor-mediated endocytosis** (by means of **clathrin-coated vesicles**), **pinocytosis**, or **phagocytosis** (using **pseudopodia**). The ejection of materials from the cytoplasm is accomplished by **exocytosis**. *(Figure 3–21; Spotlight Figure 3–22)*

3-7 The membrane potential of a cell results from the unequal distribution of positive and negative charges across the plasma membrane p. 99

44. The **potential difference**, measured in volts, between the two sides of a plasma membrane is a **membrane potential.** The membrane potential in an unstimulated, or undisturbed, cell is the cell's **resting membrane potential**.

3-8 Stages of the cell life cycle include interphase, mitosis, and cytokinesis p. 102

45. **Cell division** is the reproduction of cells. **Apoptosis** is the genetically controlled death of cells. **Mitosis** is the nuclear division of somatic cells. Sex cells are produced by **meiosis**. *(Spotlight Figure 3–23)*
46. Most somatic cells spend the majority of their time in **interphase**, which includes the **G_1, S (DNA replication)**, and **G_2 phases**. *(Spotlight Figures 3–23, 3–24)*
47. Mitosis proceeds in four stages: **prophase**, **metaphase**, **anaphase**, and **telophase**. *(Spotlight Figure 3–23)*
48. During **cytokinesis**, the cytoplasm is divided to form two daughter cells and cell division ends. *(Spotlight Figure 3–23)*
49. In general, the longer the life expectancy of a cell type, the slower the **mitotic rate**. **Stem cells** undergo frequent mitosis to replace other, more specialized cells.

3-9 Several factors regulate the cell life cycle p. 103

50. A variety of hormones and **growth factors** can stimulate cell division and growth. *(Table 3–2)*

3-10 Abnormal cell growth and division characterize tumors and cancers p. 107

51. Produced by abnormal cell growth and division, a **tumor**, or *neoplasm*, can be *benign* or *malignant*. Malignant cells may spread locally (by **invasion**) or to distant tissues and organs (through **metastasis**). The resultant illness is called **cancer**. Modified genes called **oncogenes** often cause malignancy. *(Figure 3–25)*

3-11 Cellular differentiation is cellular specialization as a result of gene activation or repression p. 109

52. **Cellular differentiation**, a process of cellular specialization, results from the inactivation of particular genes in different cells, producing populations of cells with limited capabilities. Specialized cells form organized collections called *tissues*, each of which has certain functional roles.

Review Questions

See the blue Answers tab at the back of the book.

LEVEL 1 Reviewing Facts and Terms

1. In the following diagram, identify the type of solution (hypertonic, hypotonic, or isotonic) in which the red blood cells are immersed.

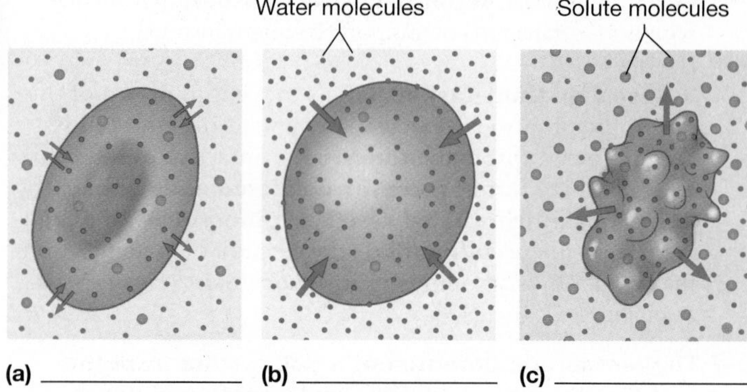
Water molecules
Solute molecules

(a) __________ (b) __________ (c) __________

2. The process that transports solid objects such as bacteria into the cell is called **(a)** pinocytosis, **(b)** phagocytosis, **(c)** exocytosis, **(d)** receptor-mediated endocytosis, **(e)** channel-mediated transport.
3. Plasma membranes are said to be **(a)** impermeable, **(b)** freely permeable, **(c)** selectively permeable, **(d)** actively permeable, **(e)** slightly permeable.
4. _____ ion concentration is high in extracellular fluids, and _____ ion concentration is high in the cytoplasm. **(a)** Calcium; magnesium, **(b)** Chloride; sodium, **(c)** Potassium; sodium, **(d)** Sodium; potassium.
5. At resting membrane potential, the cytoplasmic side of the membrane surface of the cell is _____, and the cell extracellular membrane surface is _____. **(a)** slightly negative; slightly positive, **(b)** slightly positive; slightly negative, **(c)** slightly positive; neutral; **(d)** slightly negative; neutral.
6. The organelle responsible for a variety of functions centering on the synthesis of lipids and carbohydrates is **(a)** the Golgi apparatus, **(b)** the rough endoplasmic reticulum, **(c)** the smooth endoplasmic reticulum, **(d)** mitochondria.
7. The synthesis of a functional polypeptide using the information in an mRNA strand is **(a)** translation, **(b)** transcription, **(c)** replication, **(d)** gene activation.
8. Our somatic cell nuclei contain _____ pairs of chromosomes. **(a)** 8, **(b)** 16, **(c)** 23, **(d)** 46.
9. The movement of water across a membrane from an area of low solute concentration to an area of higher solute concentration is known as **(a)** osmosis, **(b)** active transport, **(c)** diffusion, **(d)** facilitated transport, **(e)** filtration.
10. The interphase of the cell life cycle is divided into **(a)** prophase, metaphase, anaphase, and telophase, **(b)** G_0, G_1, S, and G_2, **(c)** mitosis and cytokinesis, **(d)** all of these.
11. List the four basic concepts that make up the cell theory.
12. What are four general functions of the plasma membrane?
13. What are the primary functions of membrane proteins?
14. By what three major transport mechanisms do substances get into and out of cells?
15. List five important factors that influence diffusion rates.
16. What are the four major functions of the endoplasmic reticulum?

LEVEL 2 Reviewing Concepts

17. Diffusion is important in body fluids, because it tends to **(a)** increase local concentration gradients, **(b)** eliminate local concentration gradients, **(c)** move substances against concentration gradients, **(d)** create concentration gradients.
18. Microvilli are found **(a)** mostly in muscle cells, **(b)** on the inside of plasma membranes, **(c)** in large numbers on cells that secrete hormones, **(d)** in cells that are actively engaged in absorption, **(e)** only on cells lining the reproductive tract.
19. When a cell is placed in a(n) _____ solution, the cell will lose water through osmosis. This process results in the _____ of red blood cells **(a)** hypotonic; crenation, **(b)** hypertonic; crenation, **(c)** isotonic; hemolysis, **(d)** hypotonic; hemolysis.
20. Suppose that a DNA segment has the following nucleotide sequence: CTC–ATA–CGA–TTC–AAG–TTA. Which nucleotide sequences would a complementary mRNA strand have? **(a)** GAG–UAU–GAU–AAC–UUG–AAU, **(b)** GAG–TAT–GCT–AAG–TTC–AAT, **(c)** GAG–UAU–GCU–AAG–UUC–AAU, **(d)** GUG–UAU–GGA–UUG–AAC–GGU.
21. How many amino acids are coded in the DNA segment in Review Question 20? **(a)** 18, **(b)** 9, **(c)** 6, **(d)** 3.
22. The sodium–potassium exchange pump **(a)** is an example of facilitated diffusion, **(b)** does not require the input of cellular energy in the form of ATP, **(c)** moves the sodium and potassium ions along their concentration gradients, **(d)** is composed of a carrier protein located in the plasma membrane, **(e)** is not necessary for the maintenance of homeostasis.
23. If a cell lacked ribosomes, it would not be able to **(a)** move, **(b)** synthesize proteins, **(c)** produce DNA, **(d)** metabolize sugar, **(e)** divide.
24. List, in sequence, the phases of the interphase stage of the cell life cycle, and briefly describe what happens in each.
25. List the stages of mitosis, and briefly describe the events that occur in each.
26. **(a)** What is cytokinesis? **(b)** What is the role of cytokinesis in the cell cycle?

LEVEL 3 Critical Thinking and Clinical Applications

27. The transport of a certain molecule exhibits the following characteristics: (1) The molecule moves down its concentration gradient; (2) at concentrations above a given level, the rate of transport does not increase; and (3) cellular energy is not required for transport to occur. Which transport process is at work?
28. Solutions A and B are separated by a selectively permeable barrier. Over time, the level of fluid on side A increases. Which solution initially had the higher concentration of solute?
29. A molecule that blocks the ion channels lining integral proteins in the plasma membrane would interfere with **(a)** cell recognition, **(b)** the movement of lipid-soluble molecules, **(c)** producing changes in the electrical charges across a plasma membrane, **(d)** the ability of protein hormones to stimulate the cell, **(e)** the cell's ability to divide.
30. What is the benefit of having some of the cellular organelles enclosed by a membrane similar to the plasma membrane?

+ CLINICAL CASE Wrap-Up The Beat Must Go On!

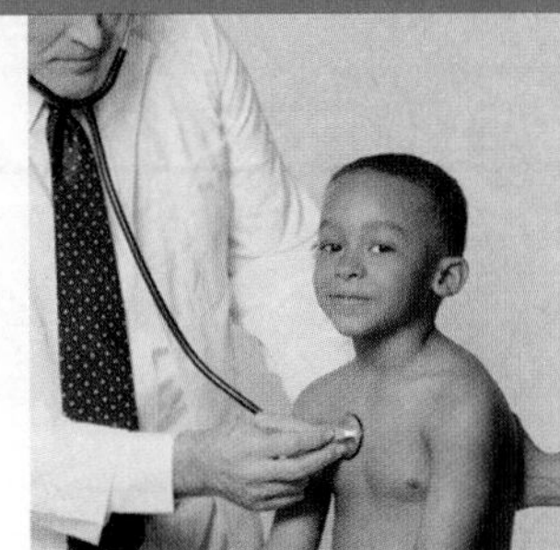

Jackson is diagnosed with *primary ciliary dyskinesia*, or *PCD (dys*, something wrong with; *kinesia*, movement), an inherited disorder that affects motile cilia. For this reason, his ears, sinuses, airways, and reproductive tract are all affected. Without the normal sweeping, wavelike motion of cilia to keep the lining of these areas clear of debris and infecting organisms, frequent infections of the ears, nose, throat, sinuses, and lungs occur.

The motion of cilia is apparently important in determining the left–right arrangement of organs during embryonic and fetal development. The tech initially thought she had reversed the right/left markers on Jackson's chest x-ray, but it turns out that his heart really is on the wrong side. In fact, all of Jackson's chest organs are arranged in a mirror-image of normal, a condition known as *situs inversus*, which occurs in 50% of patients with PCD.

Because there are only about 25,000 people with primary ciliary dyskinesia in the United States, it is classified as an "orphan" disease (any disease that affects fewer than 200,000 people). Many orphan diseases are genetic and therefore chronic (long lasting).

1. What other cell, with a "tail" that has the same microtubule structure as motile cilia, might also be affected as a result of Jackson's primary ciliary dyskinesia?
2. What other consequences could result from Jackson's disorder once he reaches adulthood?

See the blue Answers tab at the back of the book.

Related Clinical Terms

anaplasia: An irreversible change in the size and shape of tissue cells.

dysplasia: A reversible change in the normal shape, size, and organization of tissue cells.

genetic engineering: A general term that encompasses attempts to change the genetic makeup of cells or organisms, including humans.

hyperplasia: An increase in the number of normal cells (not tumor formation) in a tissue or organ, thus enlarging that tissue or organ.

hypertrophy: The enlargement of an organ or tissue due to an increase in the size of its cells.

liposome: A minute spherical sac of lipid molecules enclosing a water droplet. Often formed artificially to carry drugs into the tissues.

necrosis: Death of one or more cells in an organ or tissue due to disease, injury, or inadequate blood supply.

oncologist: Physician who specializes in the identification and treatment of cancers.

prion: A protein particle that is not visible microscopically, contains no nucleic acid, is resistant to destruction, and is thought to be the cause of some brain diseases such as bovine spongiform encephalopathy (BSE), scrapie, and Creutzfeldt-Jakob disease.

scanning electron micrograph (SEM): An image produced by an electron microscope in which a beam of focused electrons moves across an object with that object producing secondary electrons that are scattered and formatted into a three-dimensional image on a cathode-ray tube—also called *scanning microscope.*

transmission electron micrograph (TEM): A cross-sectional image produced by an electron microscope that passes a beam of electrons through an extremely small specimen. After passing through the sample the electrons are focused to form a magnified sectional view.

4 The Tissue Level of Organization

Learning Outcomes

These Learning Outcomes correspond by number to this chapter's sections and indicate what you should be able to do after completing the chapter.

CLINICAL CASE The Rubber Girl

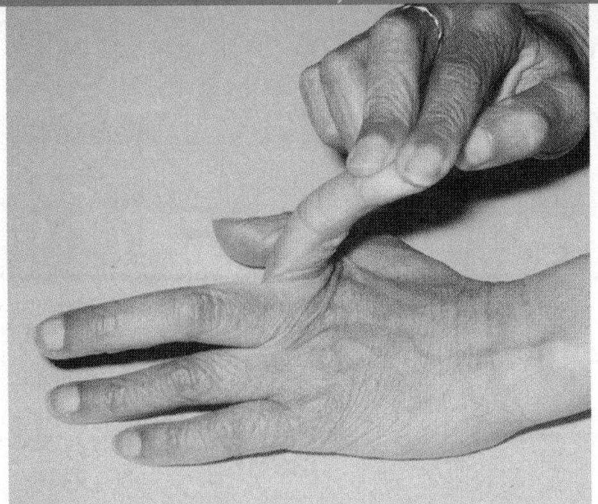

All her life Anne Marie has felt "delicate." She is a horrible athlete. She once dislocated both shoulders attempting to play high school volleyball. And she has an unstable feeling, as if her kneecaps will dislocate, when she tries dancing or any other twisting activity. She sprains one ankle or the other at least once a year. She has always been "double jointed," with fingers that she can bend back until they nearly touch her posterior forearm. Her elbows and knees all hyperextend, moving beyond their normal anatomical position.

Anne Marie's skin has always been delicate, pale, and very stretchy. She seems to get abrasions and cuts easily. These heal slowly, leaving huge scars that are also fragile. Once, she had stitches in her face and they pulled right through the skin. Now she has a wide scar over her eyebrow. She bruises easily, too.

Anne Marie thinks she is a lot like her dad. He has the same easy-bruising skin that seems to tear with minimal trauma and heal poorly. He was never very athletic and is also loose jointed. He has twisted both ankles many times. Luckily he has a desk job. **What is the cause of Anne Marie's unusual physical traits? To find out, turn to the Clinical Case Wrap-Up on p. 151.**

An Introduction to the Tissue Level of Organization

In this chapter we discuss how a variety of cell types arranged in various combinations form *tissues*, which are structures with discrete structural and functional properties. Tissues in combination form organs, such as the heart or liver, and in turn organs can be grouped into 11 organ systems. ⮌ p. 6 Therefore, learning about tissues will help you understand the descriptions of organs and organ systems in later chapters. It will also help you make the connections between anatomical structures and their physiological functions. ATLAS: Embryology Summary 1: The Formation of Tissues

4-1 The four tissue types are epithelial, connective, muscle, and nervous

Learning Outcome **Identify the four major types of tissues in the body and describe their roles.**

The human body contains trillions of cells, but only about 200 *types* of cells. To work efficiently, several different types of cells must coordinate their efforts. Cells working together form **tissues**—collections of specialized cells and cell products that carry out a limited number of functions (Figure 4–1). The study of tissues is called **histology**. Histologists recognize the following four basic types of tissue:

- **Epithelial tissue** covers exposed surfaces, lines internal passageways and chambers, and forms glands.
- **Connective tissue** fills internal spaces, provides structural support for other tissues, transports materials within the body, and stores energy.
- **Muscle tissue** is specialized for contraction and includes the skeletal muscles of the body, the muscle of the heart, and the muscular walls of hollow organs.
- **Nervous tissue** carries information from one part of the body to another in the form of electrical impulses.

In this chapter we introduce the basic characteristics of these tissues.

Checkpoint

1. **Define *histology*.**
2. **Identify the four major types of tissues in the body.**

See the blue Answers tab at the back of the book.

4-2 Epithelial tissue covers body surfaces, lines internal surfaces, and serves other essential functions

Learning Outcome **Discuss the types and functions of epithelial tissue.**

Let's begin our discussion with **epithelial tissue**, because it includes a very familiar feature—the surface of your skin. Epithelial tissue includes epithelia and glands. **Epithelia** (ep-i-THĒ-lē-a; singular, *epithelium*) are layers of cells that cover external or line internal surfaces. *Glands* are structures that produce fluid secretions. They are either attached to or derived from epithelia.

Epithelia cover every exposed surface of your body. Epithelia form the surface of the skin and line the digestive, respiratory, reproductive, and urinary tracts. In fact, they line all passageways that communicate with the outside world. The more delicate epithelia line internal cavities and passageways, such as the chest cavity, fluid-filled spaces in the brain, the inner surfaces of blood vessels, and the chambers of the heart.

We now discuss epithelial tissue functions, characteristics, specializations, and how it maintains its effectiveness as a barrier between the external environment and the body's interior.

Figure 4–1 An Orientation to the Body's Tissues.

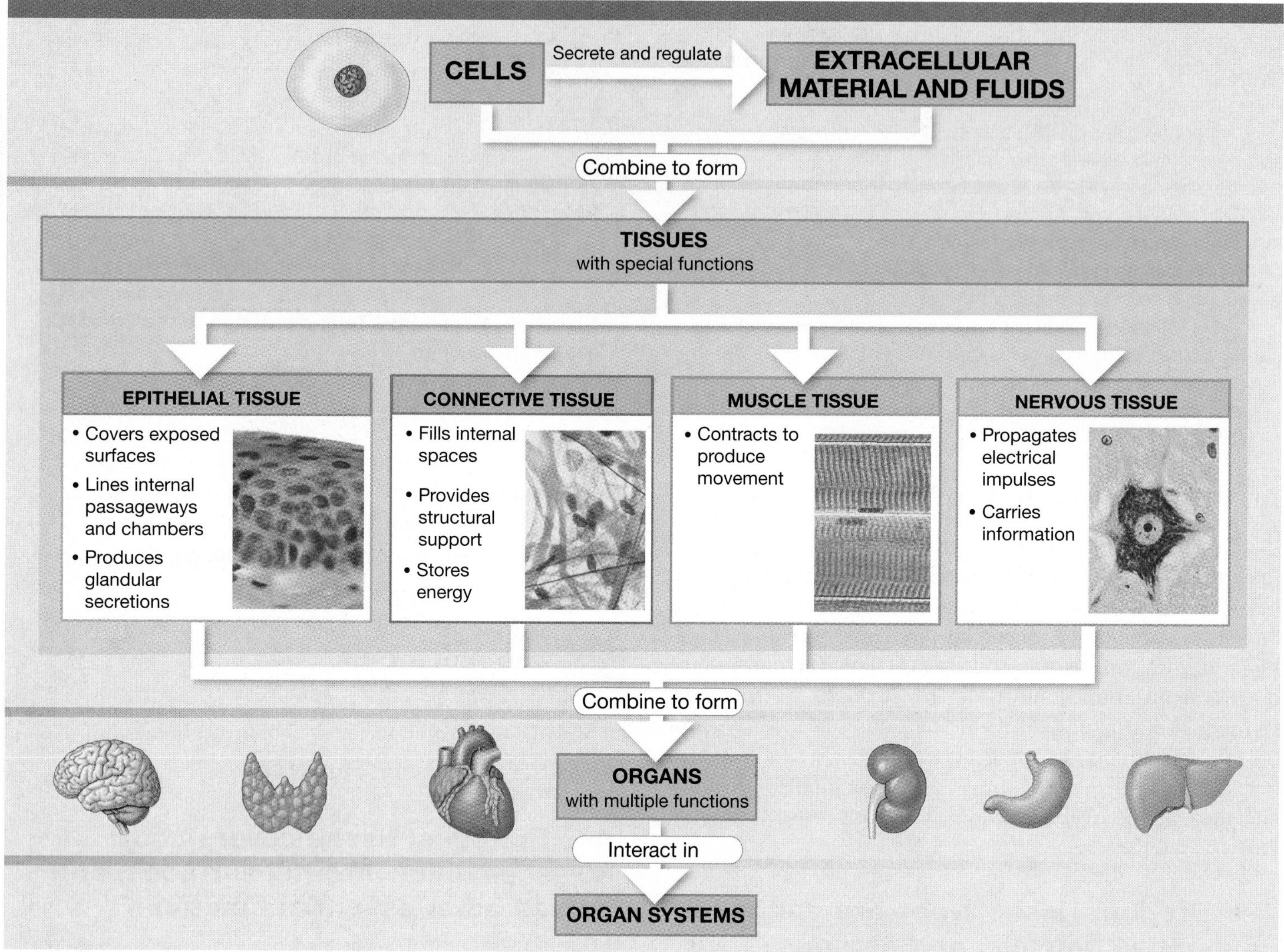
CELLS
Secrete and regulate
EXTRACELLULAR MATERIAL AND FLUIDS
Combine to form
TISSUES
with special functions
EPITHELIAL TISSUE
• Covers exposed surfaces
• Lines internal passageways and chambers
• Produces glandular secretions
CONNECTIVE TISSUE
• Fills internal spaces
• Provides structural support
• Stores energy
MUSCLE TISSUE
• Contracts to produce movement
NERVOUS TISSUE
• Propagates electrical impulses
• Carries information
Combine to form
ORGANS
with multiple functions
Interact in
ORGAN SYSTEMS

Functions of Epithelial Tissue

Epithelia perform the following four essential functions:

- *Provide Physical Protection.* Epithelia protect exposed and internal surfaces from abrasion, dehydration, and destruction by chemical or biological agents.
- *Control Permeability.* Any substance that enters or leaves your body must cross an epithelium. Some epithelia are relatively impermeable. Others are easily crossed by compounds as large as proteins. Many epithelia contain the molecular "machinery" needed for absorbing or secreting specific substances. The epithelial barrier can also be regulated and modified in response to stimuli. For example, hormones can affect the transport of ions and nutrients through epithelial cells. Even physical stress can alter the structure and properties of epithelia. For example, calluses form on your hands when you do manual labor for some time.
- *Provide Sensation.* Most epithelia are extremely sensitive to stimulation, because they have a large sensory nerve supply. These sensory nerves continually provide information about the external and internal environments. For example, the lightest touch of a mosquito will stimulate sensory neurons that tell you where to swat. A *neuroepithelium* is an epithelium that is specialized to perform a particular sensory function. Our neuroepithelia contain sensory cells that provide the sensations of smell, taste, sight, equilibrium, and hearing.
- *Produce Specialized Secretions.* Epithelial cells that produce secretions are called *gland cells.* Individual gland cells are usually scattered among other cell types in an epithelium.

In a *glandular epithelium*, most or all of the epithelial cells produce secretions. These cells either discharge their secretions onto the surface of the epithelium (to provide physical protection or temperature regulation) or release them into the surrounding interstitial fluid and blood (to act as chemical messengers).

Characteristics of Epithelial Tissue

Epithelia have several important features:

- *Polarity*. An epithelium has an exposed surface, either facing the external environment or an internal space, and a base, which is attached to underlying tissues. The term **polarity** refers to the presence of structural and functional differences between the exposed and attached surfaces. An epithelium consisting of a single layer of cells has an exposed *apical surface* and an attached *basal surface*. The two surfaces differ in plasma membrane structure and function. Often, the apical surface has microvilli; sometimes it has cilia. The functional polarity is also evident in the uneven distribution of organelles between the exposed surface and the basement membrane. In some epithelia, such as the lining of the kidney tubules, mitochondria are concentrated near the base of the cell, where energy is in high demand for the cell's transport activities (Figure 4–2).
- *Cellularity*. Epithelia are made almost entirely of cells bound closely together by interconnections known as *cell junctions*. In other tissue types, the cells are often widely separated by extracellular materials.

Figure 4–2 **The Polarity of Epithelial Cells.**

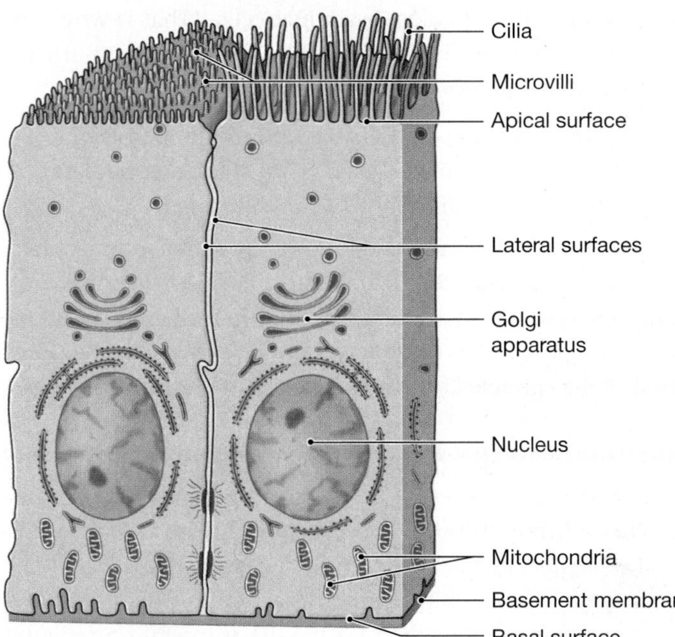
Cilia
Microvilli
Apical surface
Lateral surfaces
Golgi apparatus
Nucleus
Mitochondria
Basement membrane
Basal surface

- *Attachment*. The base of an epithelium is bound to a thin, noncellular **basement membrane**. This basement membrane is formed from the fusion of several successive layers (the basal lamina and reticular lamina), a collagen matrix, and proteoglycans (intercellular cement). The basement membrane adheres to the basal surface and to the underlying tissues to establish the cell's border and resist stretching.
- *Avascularity*. Epithelia are **avascular** (ā-VAS-kū-lar; *a-*, without + *vas*, vessel), which means that they lack blood vessels. Epithelial cells get nutrients by diffusion or absorption across either the exposed or the attached epithelial surface.
- *Regeneration*. Epithelial cells that are damaged or lost at the exposed surface are continuously replaced through stem cell divisions in the epithelium. Regeneration is a characteristic of other tissues as well, but the rates of cell division and replacement are typically much higher in epithelia than in other tissues.

Specializations of Epithelial Cells

How are epithelial cells different from other body cells? They have several structural specializations. For the epithelium as a whole to perform the functions just listed, individual epithelial cells may be specialized for (1) the movement of fluids over the epithelial surface, providing protection and lubrication; (2) the movement of fluids through the epithelium, to control permeability; or (3) the production of secretions that provide physical protection or act as chemical messengers.

The specialized epithelial cell is often divided into two functional regions, which means the cell has a strong polarity. One is the **apical surface**, where the cell is exposed to an internal or external environment. The other consists of the **basolateral surfaces**, which include both the base (basal surface), where the cell attaches to underlying epithelial cells or deeper tissues, and the sides (lateral surfaces), where the cell contacts its neighbors (see Figure 4–2).

Many epithelial cells that line internal passageways have *microvilli* on their exposed surfaces. ⮌ p. 74 Just a few may be present, or microvilli may carpet the entire surface. Microvilli are especially abundant on epithelial surfaces where absorption and secretion take place, such as along portions of the digestive system and kidneys. The epithelial cells in these locations are transport specialists. Each cell has at least 20 times more surface area to transport substances than it would have without microvilli.

Motile cilia are characteristic of surfaces covered by a **ciliated epithelium**. ⮌ p. 74 A typical ciliated cell contains about 250 cilia that beat in a coordinated manner. The synchronized beating of the cilia moves substances over the epithelial surface. For example, the ciliated epithelium that lines the respiratory tract moves mucus up from the lungs and toward the

4

throat, as though on an escalator. The sticky mucus traps inhaled particles, including dust, pollen, and pathogens. The ciliated epithelium then carries the mucus and the trapped debris to the throat. There they are swallowed or expelled by coughing. If abrasion or toxic compounds such as nicotine and carbon monoxide in cigarette smoke injure the cilia or the epithelial cells, ciliary movement can stop. Without the protective upward flow of mucus, infection or disease becomes more likely.

Maintaining the Integrity of Epithelia

To be effective as a barrier, an epithelium must form a complete cover or lining. Three factors help maintain the physical integrity of an epithelium: (1) intercellular connections, (2) attachment to the basement membrane, and (3) epithelial maintenance and repair.

Intercellular Connections

Cells in an epithelium are firmly attached to one another, and the epithelium as a unit is attached to extracellular fibers of the superficial basal lamina of the basement membrane. Many cells in your body form permanent or temporary bonds with other cells or extracellular material. Epithelial cells, however, are specialists in intercellular connection.

Intercellular connections involve either extensive areas of opposing plasma membranes or specialized attachment sites called *cell junctions*, discussed shortly. Large areas of opposing plasma membranes are interconnected by transmembrane proteins called **cell adhesion molecules (CAMs)**, which bind to each other and to extracellular materials. For example, CAMs on the basolateral surface of an epithelium help bind the cell to the underlying basement membrane. The membranes of adjacent cells may also be bonded by a thin layer of proteoglycans that contain polysaccharide derivatives known as *glycosaminoglycans (GAGs)*, most notably, *hyaluronan* (hyaluronic acid).

Cell junctions are specialized areas of the plasma membrane that attach a cell to another cell or to extracellular materials. The three most common types of cell junctions are (1) gap junctions, (2) tight junctions, and (3) desmosomes (Figure 4–3a).

Gap Junctions. Some epithelial functions require rapid intercellular communication. At a **gap junction** (Figure 4–3b), two cells are held together by two embedded interlocking transmembrane proteins called *connexons*. Each connexon is composed of six *connexin* proteins that form a cylinder with a central pore. Two aligned connexons form a narrow passageway that lets small molecules and ions pass from cell to cell. Gap junctions are common among epithelial cells, where the movement of ions helps coordinate functions such as the beating of cilia. Gap junctions are also common in other tissues. For example, gap junctions in cardiac muscle tissue and smooth muscle tissue are essential in coordinating muscle cell contractions. A number of human diseases are caused by mutations in the genes that code for different connexin proteins.

Tight Junctions. Tight junctions encircle the apical regions of epithelial cells. At a **tight junction** (also known as an *occluding junction*), the lipid portions of the two plasma membranes are tightly bound together by interlocking membrane proteins (Figure 4–3c). Inferior to the tight junctions, a continuous *adhesion belt* forms a band that encircles cells and binds them to their neighbors. The bands are attached to the microfilaments of the terminal web. ↺ p. 72

Tight junctions largely prevent water and solutes from passing between the cells. When the epithelium lines a tube, such as the intestinal tract, the apical surfaces of the epithelial cells are exposed to the space inside the tube, a passageway called the **lumen** (LŪ-men). Tight junctions effectively isolate the contents of the lumen from the basolateral surfaces of the cell. For example, tight junctions near the apical surfaces of cells that line the digestive tract help keep enzymes, acids, and wastes in the lumen from reaching the basolateral surfaces and digesting or otherwise damaging the underlying tissues and organs.

Desmosomes. Most epithelial cells are subject to mechanical stresses—stretching, bending, twisting, or compression—so they must have durable interconnections. At a **desmosome** (DEZ-mō-sōm; *desmos*, ligament + *soma*, body), CAMs and proteoglycans link the opposing plasma membranes. Desmosomes are very strong and can resist stretching and twisting.

A typical desmosome is formed by components from two cells. Within each cell, a complex known as a *dense area* is connected to the cytoskeleton. This connection to the cytoskeleton gives the desmosome—and the epithelium—its strength. For example, desmosomes are abundant between cells in the superficial layers of the skin. As a result, damaged skin cells are usually lost in sheets rather than as individual cells. (That is why your skin peels rather than sheds off as a powder after a sunburn.)

There are two types of desmosomes:

- *Spot desmosomes* are small discs connected to bands of intermediate filaments (Figure 4–3d). The intermediate filaments stabilize the shape of the cell.
- *Hemidesmosomes* resemble half of a spot desmosome. Rather than attaching one cell to another, a hemidesmosome attaches a cell to extracellular filaments in the basement membrane (Figure 4–3e). This attachment helps stabilize the position of the epithelial cell and anchors it to underlying tissues.

Attachment to the Basement Membrane

Not only do epithelial cells hold onto one another, but they also remain firmly connected to the rest of the body. In fact, epithelial cells must be attached to other cells, or they will die. The inner surface of each epithelium is attached to a two-part basement membrane composed of a basal lamina and a reticular

Figure 4–3 **Cell Junctions.**

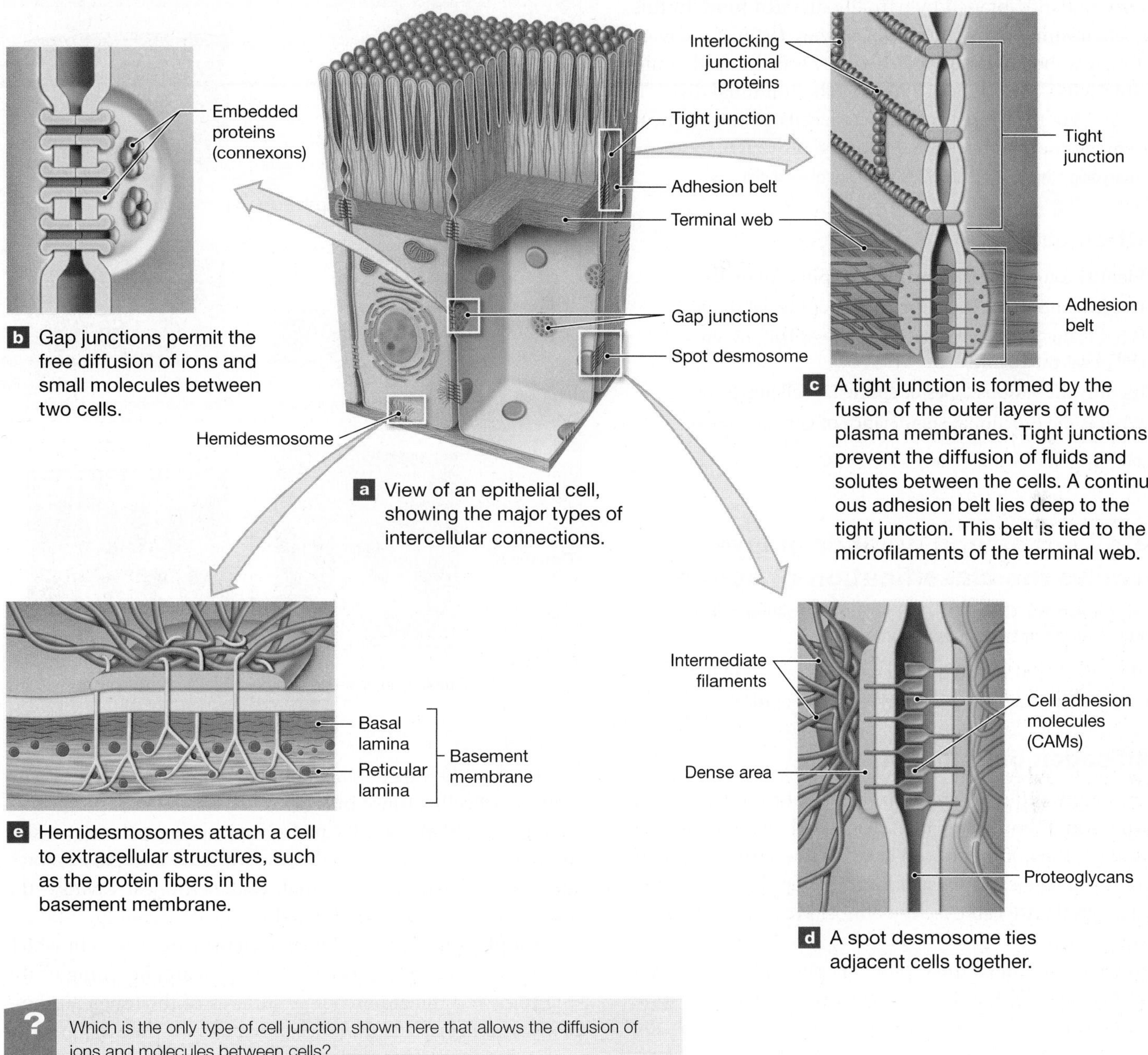

Embedded proteins (connexons)
Interlocking junctional proteins
Tight junction
Adhesion belt
Terminal web
Gap junctions
Spot desmosome
Hemidesmosome
Tight junction
Adhesion belt
Intermediate filaments
Cell adhesion molecules (CAMs)
Dense area
Proteoglycans
Basal lamina
Reticular lamina
Basement membrane

a View of an epithelial cell, showing the major types of intercellular connections.

b Gap junctions permit the free diffusion of ions and small molecules between two cells.

c A tight junction is formed by the fusion of the outer layers of two plasma membranes. Tight junctions prevent the diffusion of fluids and solutes between the cells. A continuous adhesion belt lies deep to the tight junction. This belt is tied to the microfilaments of the terminal web.

d A spot desmosome ties adjacent cells together.

e Hemidesmosomes attach a cell to extracellular structures, such as the protein fibers in the basement membrane.

? Which is the only type of cell junction shown here that allows the diffusion of ions and molecules between cells?

lamina. The layer closer to the epithelium, the *basal lamina,* is an amorphous, ill-organized layer thought to function as a selective filter (Figure 4–3e). It is secreted by the adjacent layer of epithelial cells. The basal lamina restricts the movement of proteins and other large molecules from the underlying connective tissue into the epithelium.

The deeper portion of the basement membrane, the *reticular lamina,* consists mostly of reticular fibers and ground substance. These connective tissue components are discussed later on page 130. This layer gives the basement membrane its strength. Attachments between the fibers of the basal lamina and those of the reticular lamina hold the two layers together, and hemidesmosomes attach the epithelial cells to the composite basement membrane. The reticular lamina also acts as a filter that determines what substances can diffuse between the adjacent tissues and the epithelium.

Epithelial Maintenance and Repair

Epithelial cells lead hard lives. They are exposed to disruptive enzymes, toxic chemicals, pathogenic bacteria, physical

4

distortion, and mechanical abrasion. Consider the lining of your small intestine, where epithelial cells are exposed to a variety of enzymes and abraded by partially digested food. In this extreme environment, an epithelial cell may last just a day or two before it is shed or destroyed. The only way the epithelium can maintain its structure over time is by the continual division of *stem cells*. ↺ p. 103 Most epithelial stem cells are located near the basement membrane, in a relatively protected location. ATLAS: Embryology Summary 2: The Development of Epithelia

✓ Checkpoint

3. **Identify four essential functions of epithelial tissue.**
4. **List five important characteristics of epithelial tissue.**
5. **What is the probable function of an epithelium whose cells bear microvilli?**
6. **Identify the various types of epithelial cell junctions.**
7. **What is the functional significance of gap junctions?**

See the blue Answers tab at the back of the book.

4-3 Cell shape and number of layers determine the classification of epithelia

Learning Outcome Describe the relationship between structure and function for each type of epithelium.

Recall that epithelial tissue includes epithelia and glands. The glands are either attached to or derived from epithelia.

Classification of Epithelia

There are many different specialized types of epithelia. You can easily sort these into categories based on (1) the cell shape and (2) the number of cell layers between the basement membrane and the exposed surface of the epithelium. Using these two criteria, we can describe almost every epithelium in the body.

Epithelial cells have three basic shapes: *squamous*, *cuboidal*, and *columnar* (**Table 4–1**). To classify by shape, we look at the superficial cells in a section perpendicular to both the apical surface and the basement membrane. In sectional view, squamous cells appear thin and flat like scales, cuboidal cells look like little boxes, and columnar cells are tall and relatively slender rectangles.

Once you have determined whether the superficial cells are squamous, cuboidal, or columnar, then look at the number of cell layers. There are only two options: *simple* or *stratified*.

If only one layer of cells covers the basement membrane, that layer is a **simple epithelium**. Simple epithelia are necessarily thin. All the cells have the same polarity, so the distance from the nucleus to the basement membrane does not change from one cell to the next.

Table 4–1 Classifying Epithelia

	Simple	Stratified
Squamous	Simple squamous epithelium	Stratified squamous epithelium
Cuboidal	Simple cuboidal epithelium	Stratified cuboidal epithelium
Columnar	Simple columnar epithelium	Stratified columnar epithelium

Because they are so thin, simple epithelia are fragile. A single layer of cells cannot provide much mechanical protection, so simple epithelia are located only in protected areas inside the body. They line internal compartments and passageways, such as the pleural, pericardial, and peritoneal cavities; the heart chambers; and blood vessels.

Simple epithelia are also characteristic of regions in which secretion or absorption occurs. Examples are the lining of the intestines and the gas-exchange surfaces of the lungs. In these places, thinness is an advantage. It reduces the time required for materials to cross the epithelial barrier.

In a **stratified epithelium**, several layers of cells cover the basement membrane. Stratified epithelia are generally located in areas that are exposed to mechanical or chemical stresses. The surface of the skin and the lining of the mouth are examples.

Tips & Tools

To help you remember the meanings of the terms *squamous* and *stratified*, associate the word "**squa**mous" with "**sca**ly," and the word "**strat**ified" with "**strat**osphere," an upper **layer** of Earth's atmosphere.

Go to MasteringA&P™ > Study Area > Menu > Lab Tools > PAL™ 3.0 > Histology: Epithelial Tissue

Figure 4–4 Squamous Epithelia.

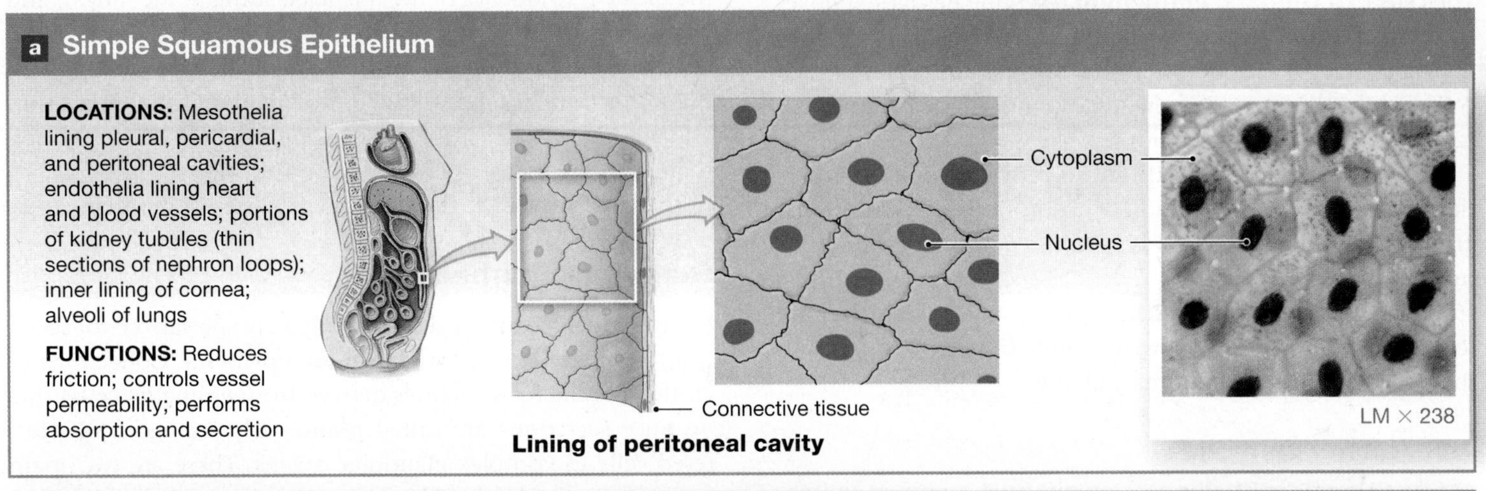
a Simple Squamous Epithelium
LOCATIONS: Mesothelia lining pleural, pericardial, and peritoneal cavities; endothelia lining heart and blood vessels; portions of kidney tubules (thin sections of nephron loops); inner lining of cornea; alveoli of lungs
FUNCTIONS: Reduces friction; controls vessel permeability; performs absorption and secretion
Cytoplasm
Nucleus
Connective tissue
Lining of peritoneal cavity
LM × 238

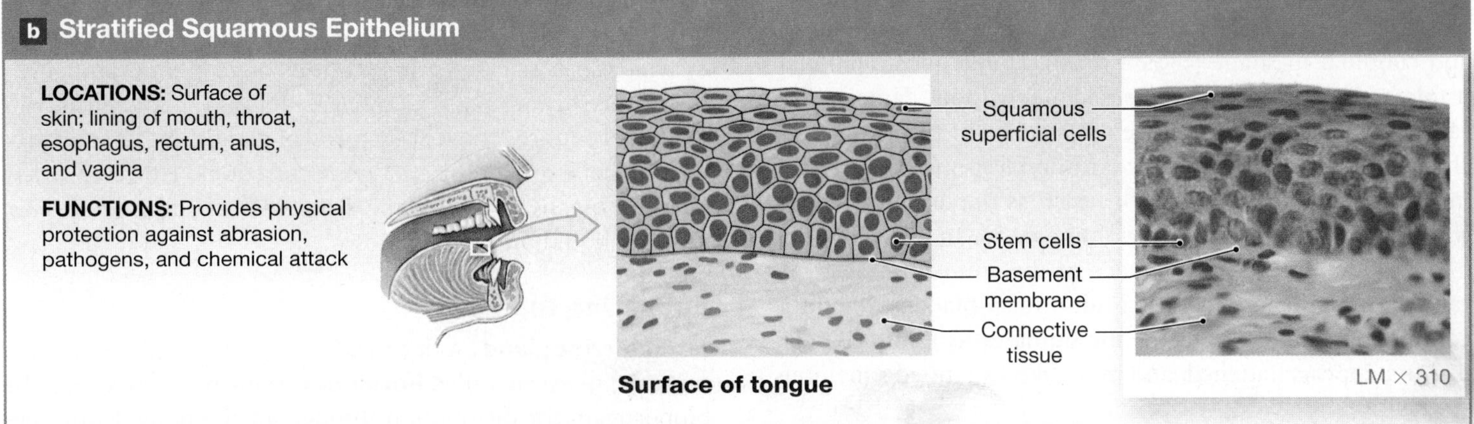
b Stratified Squamous Epithelium
LOCATIONS: Surface of skin; lining of mouth, throat, esophagus, rectum, anus, and vagina
FUNCTIONS: Provides physical protection against abrasion, pathogens, and chemical attack
Squamous superficial cells
Stem cells
Basement membrane
Connective tissue
Surface of tongue
LM × 310

Squamous Epithelia

The cells in a **squamous epithelium** (SKWĀ-mus; *squama*, plate or scale) are thin, flat, and somewhat irregular in shape, like pieces of a jigsaw puzzle (Figure 4–4). From the surface, the cells resemble fried eggs laid side by side. In sectional view, the disc-shaped nucleus occupies the thickest portion of each cell (see Table 4–1).

A **simple squamous epithelium** is the body's most delicate type of epithelium. This type of epithelium is located in protected regions where absorption or diffusion takes place, or where a slick, slippery surface reduces friction. Examples are the respiratory exchange surfaces (*alveoli*) of the lungs, the lining of the thoracic and abdominopelvic body cavities of the trunk (Figure 4–4a), and the lining of the heart and blood vessels. Smooth linings are extremely important. For example, any irregularity in the lining of a blood vessel may result in the formation of a potentially dangerous blood clot.

Special names have been given to the simple squamous epithelia that line chambers and passageways that do not communicate with the outside world. The simple squamous epithelium that lines the body cavities enclosing the lungs, heart, and abdominal organs is called a **mesothelium** (mez-ō-THĒ-lē-um; *mesos*, middle). The pleura, pericardium, and peritoneum each contain a superficial layer of mesothelium. The simple squamous epithelium lining the inner surface of the heart and all blood vessels is called an **endothelium** (en-dō-THĒ-lē-um; *endo-*, inside).

A **stratified squamous epithelium** (Figure 4–4b) is generally located where mechanical stresses are severe. The cells form a series of layers, like the layers in a sheet of plywood. The surface of the skin and the lining of the mouth, esophagus, and anus are areas where this type of epithelium protects against physical and chemical attacks.

On exposed body surfaces, where mechanical stress and dehydration are potential problems, superficial layers of epithelial cells are packed with filaments of the protein *keratin*. As a result, these layers are both tough and water resistant. Such an epithelium is said to be *keratinized*. A *nonkeratinized* stratified squamous epithelium resists abrasion, but will dry out and deteriorate unless kept moist. Nonkeratinized stratified squamous epithelia are found in the oral cavity, pharynx, esophagus, anus, and vagina.

Cuboidal Epithelia

The cells of a **cuboidal epithelium** resemble hexagonal boxes from their apical surfaces. (In typical sectional views they appear square.) The spherical nuclei are near the center of each cell, and the distance between adjacent nuclei is roughly equal to the height of the epithelium.

A **simple cuboidal epithelium** provides limited protection and occurs where secretion or absorption takes place. Such an epithelium makes up glands and lines portions of the kidney tubules (Figure 4–5a).

Stratified cuboidal epithelia are relatively rare. They are located along the ducts of sweat glands (Figure 4–5b) and in the larger ducts of the mammary glands.

Transitional Epithelia

A **transitional epithelium** is an unusual stratified epithelium because its cells can change between being squamous and cuboidal in shape (Figure 4–5c). Unlike most epithelia, it tolerates repeated cycles of stretching without damage. It is called transitional because the appearance of the epithelium changes as stretching occurs. A transitional epithelium is found in regions of the urinary system, such as the urinary bladder, where large changes in urine volume occur. In an empty urinary bladder, the superficial cells of the epithelium are typically plump and cuboidal. In a full urinary bladder, when the volume of urine has stretched the lining to its limits, the epithelium appears flattened, and more like a stratified squamous epithelium.

Columnar Epithelia

In a typical sectional view, **columnar epithelial cells** appear rectangular. In reality, the densely packed cells are hexagonal, but they are taller and more slender than cells in a cuboidal epithelium (Figure 4–6). The elongated nuclei are crowded into a narrow band close to the basement membrane. The height of the epithelium is several times the distance between adjacent nuclei.

A **simple columnar epithelium** is typically found where absorption or secretion takes place, such as in the small intestine (Figure 4–6a). In the stomach and large intestine, the secretions of simple columnar epithelia protect against chemical stresses.

Portions of the respiratory tract contain a **pseudostratified columnar epithelium**, a columnar epithelium that includes several types of cells with varying shapes and functions. The distances between the cell nuclei and the exposed surface vary, so the epithelium appears to be layered, or stratified (Figure 4–6b). It is not truly stratified, though, because every epithelial cell contacts the basement membrane. Pseudostratified columnar epithelial cells typically have cilia. Epithelia of this type line most of the nasal cavity, the trachea (windpipe), the bronchi (branches of the trachea leading to the lungs), and portions of the male reproductive tract.

Stratified columnar epithelia are relatively rare, providing protection along portions of the pharynx, epiglottis, anus, and urethra, as well as along a few large excretory ducts. The epithelium has either two layers or multiple layers (Figure 4–6c). In the latter case, only the superficial cells are columnar.

Glandular Epithelia

Many epithelia contain gland cells that are specialized for secretion; these are known as **glandular epithelia**. Collections of epithelial cells (or structures derived from epithelial cells) that produce secretions are called **glands**. They range from scattered cells to complex glandular organs. There are two main types of glands, classified by where they deliver their secretions. *Endocrine glands* release their secretions into the blood, whereas *exocrine glands* release their secretions directly onto an epithelial surface or into passageways called *ducts* that open onto an epithelial surface. The vast majority of glands in the body produce either exocrine or endocrine secretions. However, a few complex organs, including the digestive tract and the pancreas, produce both kinds of secretions.

Endocrine Glands

An **endocrine gland** produces *endocrine* (*endo-*, inside + *krino*, to separate) *secretions*, called **hormones**. These hormones enter the bloodstream for distribution throughout the body. Hormones regulate or coordinate the activities of various tissues, organs, and organ systems. Because their secretions are not released into ducts, endocrine glands are often called *ductless glands.*

Examples of endocrine glands include the thyroid gland, thymus, and pituitary gland. Endocrine glands, which are separate organs, are made up of endocrine cells. Endocrine cells may also be present outside of a gland as part of an epithelial surface, such as the lining of the digestive tract. We consider endocrine glands, cells, and hormones further in Chapter 18.

Exocrine Glands

Exocrine glands produce *exocrine* (*exo-*, outside) *secretions*, which are discharged onto an epithelial surface. Most exocrine secretions reach the surface through tubular **ducts**, which empty onto the skin surface or onto an epithelium lining an internal passageway that communicates with the exterior. Examples of exocrine secretions are enzymes entering the digestive tract, perspiration on the skin, tears in the eyes, and milk produced by mammary glands.

We classify exocrine glands from three perspectives: (1) by the structure of the glands, (2) by how those glands secrete their products, and (3) by what those products are.

Figure 4–5 Cuboidal and Transitional Epithelia.

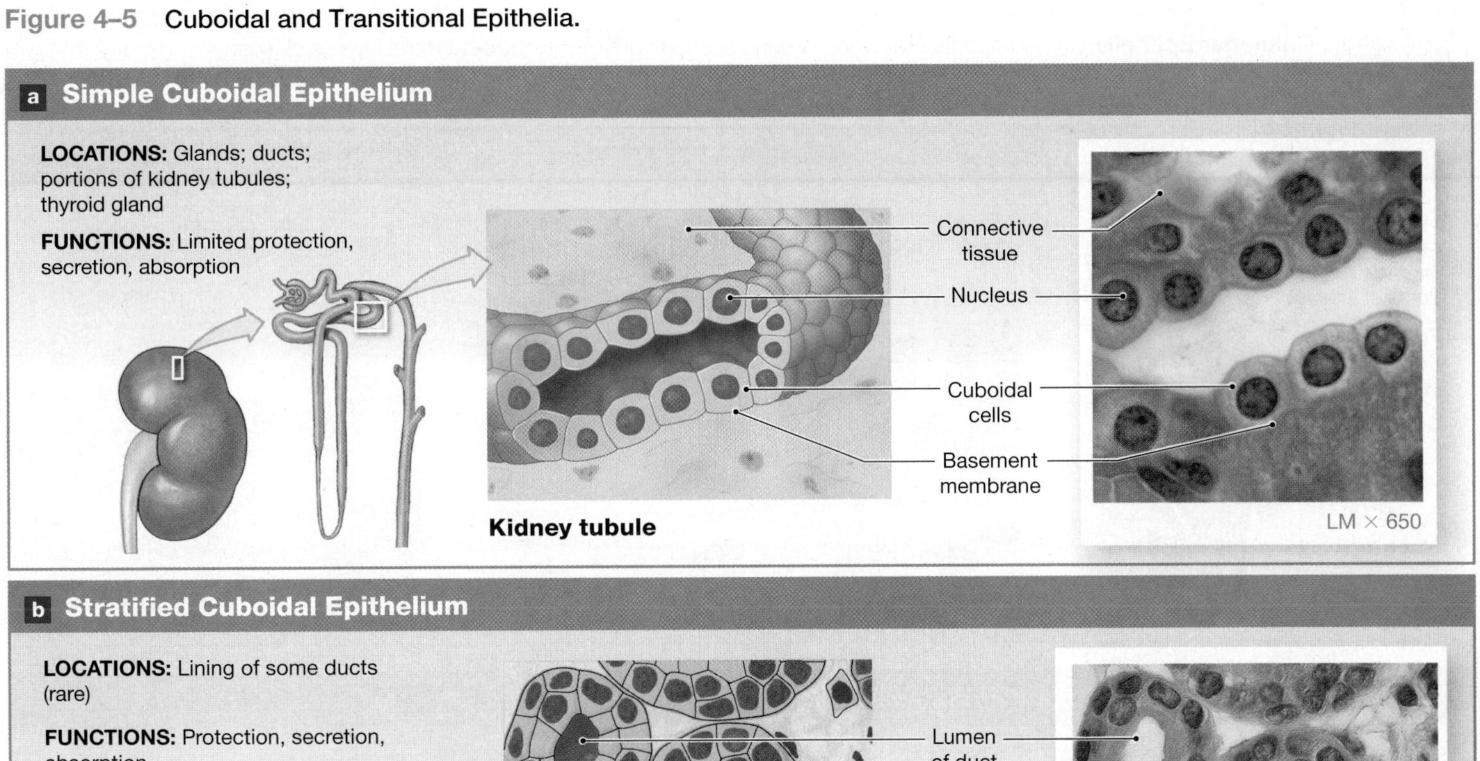
a Simple Cuboidal Epithelium
LOCATIONS: Glands; ducts; portions of kidney tubules; thyroid gland
FUNCTIONS: Limited protection, secretion, absorption
Connective tissue
Nucleus
Cuboidal cells
Basement membrane
Kidney tubule
LM × 650

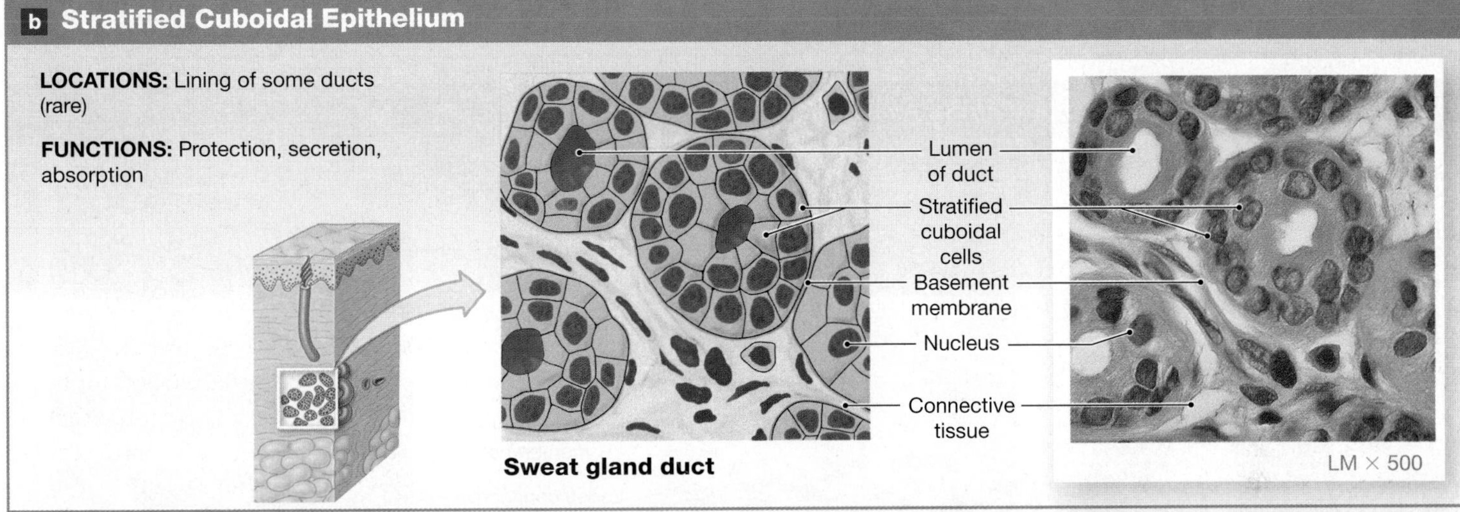
b Stratified Cuboidal Epithelium
LOCATIONS: Lining of some ducts (rare)
FUNCTIONS: Protection, secretion, absorption
Lumen of duct
Stratified cuboidal cells
Basement membrane
Nucleus
Connective tissue
Sweat gland duct
LM × 500

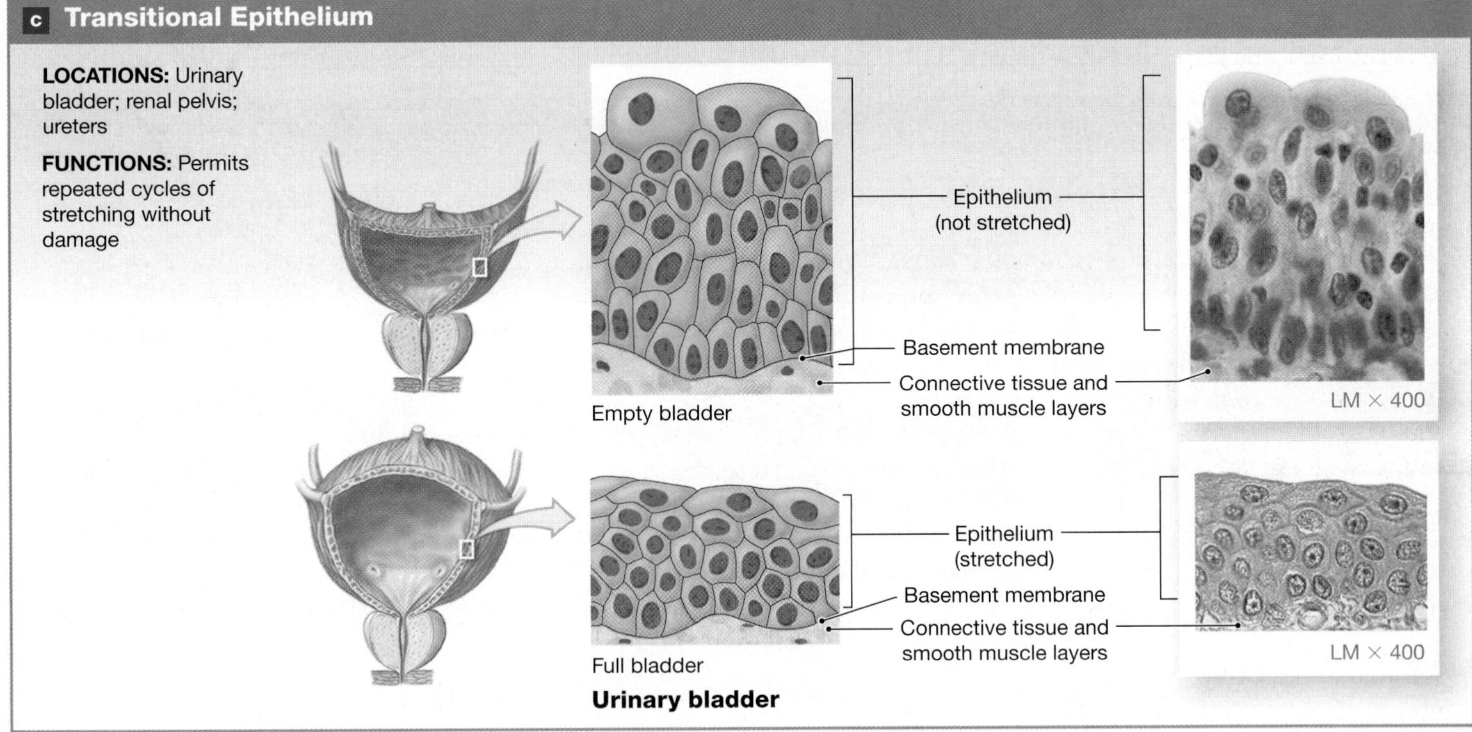
c Transitional Epithelium
LOCATIONS: Urinary bladder; renal pelvis; ureters
FUNCTIONS: Permits repeated cycles of stretching without damage
Epithelium (not stretched)
Basement membrane
Connective tissue and smooth muscle layers
Empty bladder
LM × 400
Epithelium (stretched)
Basement membrane
Connective tissue and smooth muscle layers
Full bladder
Urinary bladder
LM × 400

Figure 4–6 **Columnar Epithelia.** In part c, note that when viewing sections of tissues, the superficial layer is closest to the lumen.

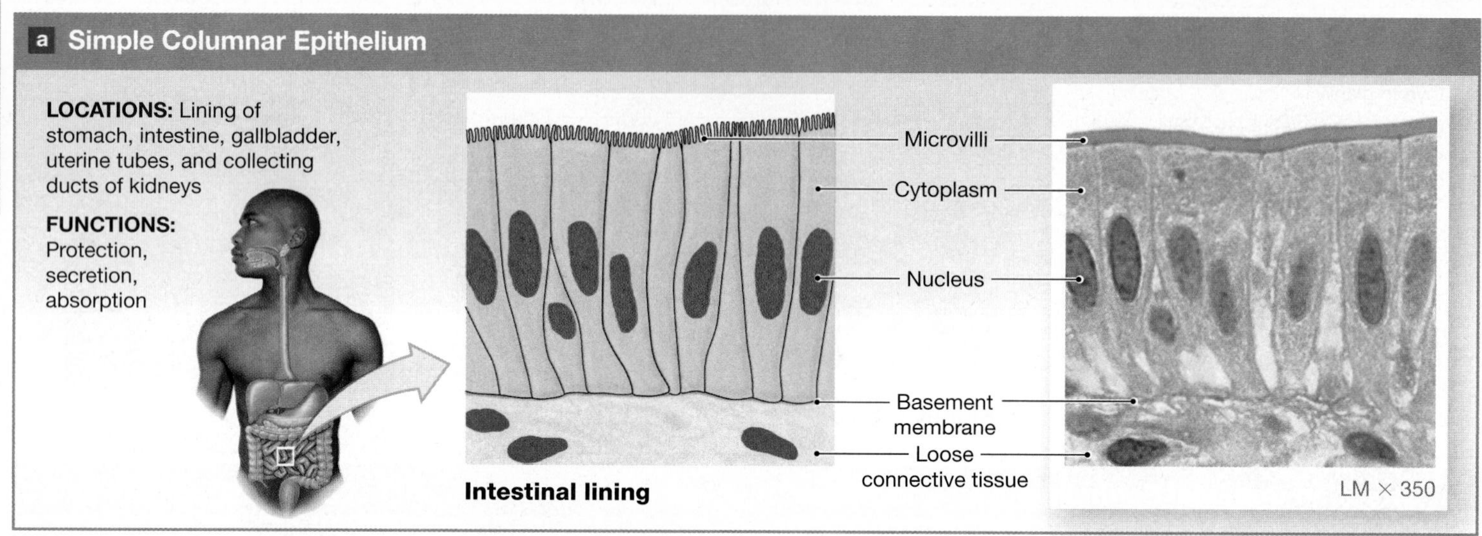
a Simple Columnar Epithelium
LOCATIONS: Lining of stomach, intestine, gallbladder, uterine tubes, and collecting ducts of kidneys
FUNCTIONS: Protection, secretion, absorption
Microvilli
Cytoplasm
Nucleus
Basement membrane
Loose connective tissue
Intestinal lining
LM × 350

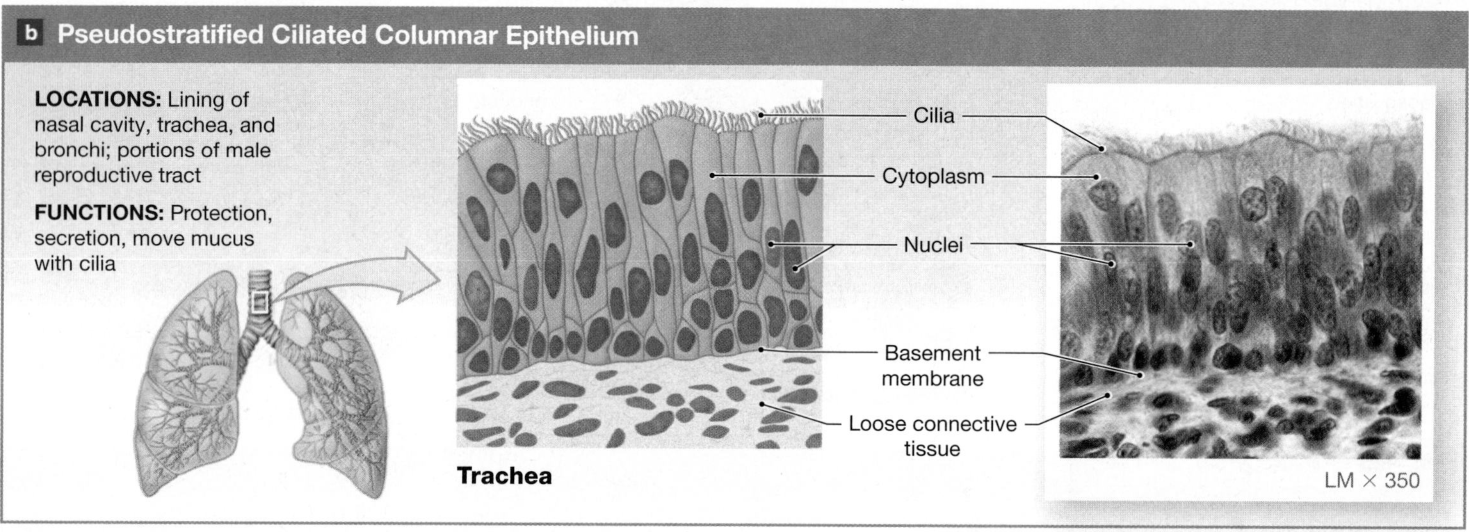
b Pseudostratified Ciliated Columnar Epithelium
LOCATIONS: Lining of nasal cavity, trachea, and bronchi; portions of male reproductive tract
FUNCTIONS: Protection, secretion, move mucus with cilia
Cilia
Cytoplasm
Nuclei
Basement membrane
Loose connective tissue
Trachea
LM × 350

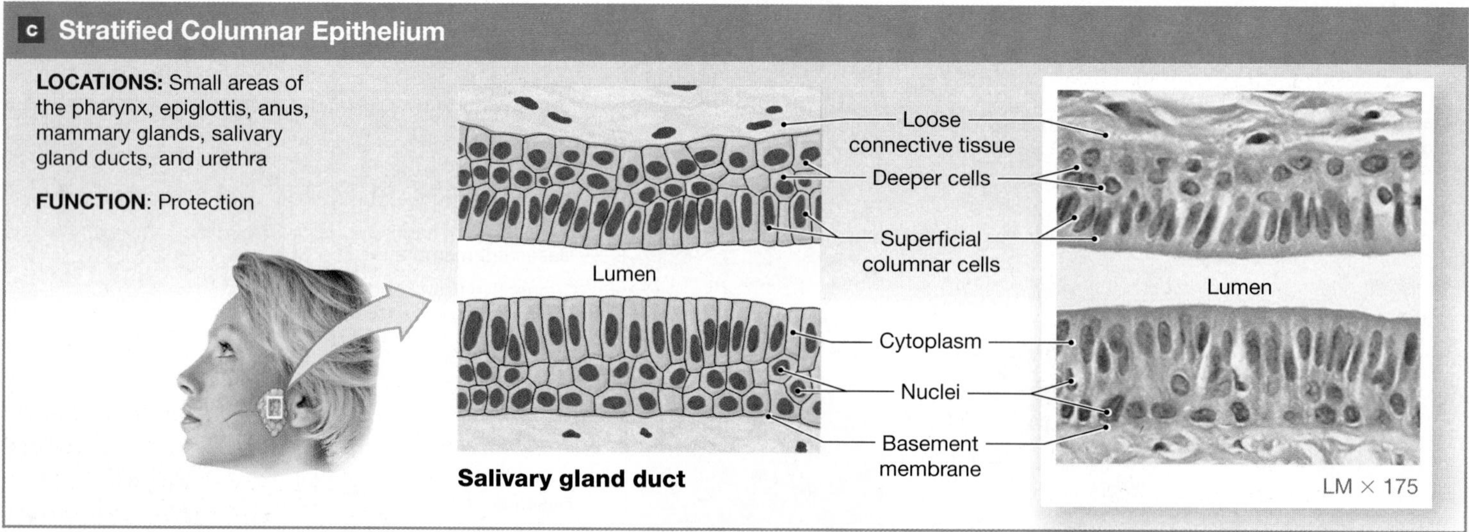
c Stratified Columnar Epithelium
LOCATIONS: Small areas of the pharynx, epiglottis, anus, mammary glands, salivary gland ducts, and urethra
FUNCTION: Protection
Loose connective tissue
Deeper cells
Superficial columnar cells
Lumen
Lumen
Cytoplasm
Nuclei
Basement membrane
Salivary gland duct
LM × 175

? In "simple columnar epithelium," which word describes cell shape and which word describes the number of cell layers?

Figure 4–7 A Structural Classification of Exocrine Glands.

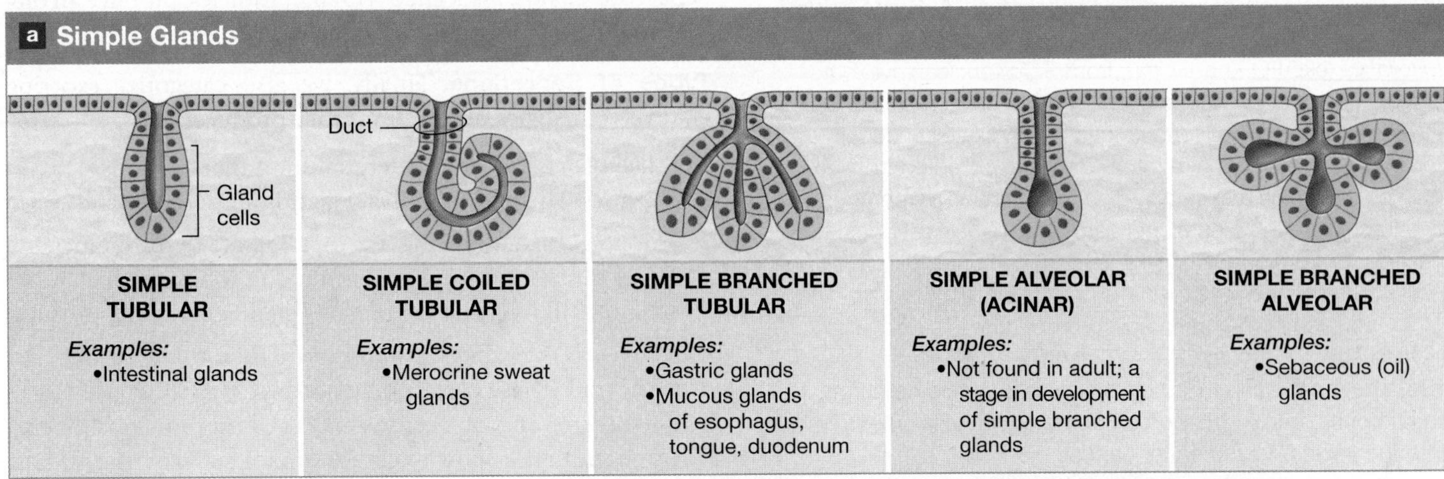
a Simple Glands
Duct
Gland cells
SIMPLE TUBULAR
Examples:
•Intestinal glands
SIMPLE COILED TUBULAR
Examples:
•Merocrine sweat glands
SIMPLE BRANCHED TUBULAR
Examples:
•Gastric glands
•Mucous glands of esophagus, tongue, duodenum
SIMPLE ALVEOLAR (ACINAR)
Examples:
•Not found in adult; a stage in development of simple branched glands
SIMPLE BRANCHED ALVEOLAR
Examples:
•Sebaceous (oil) glands

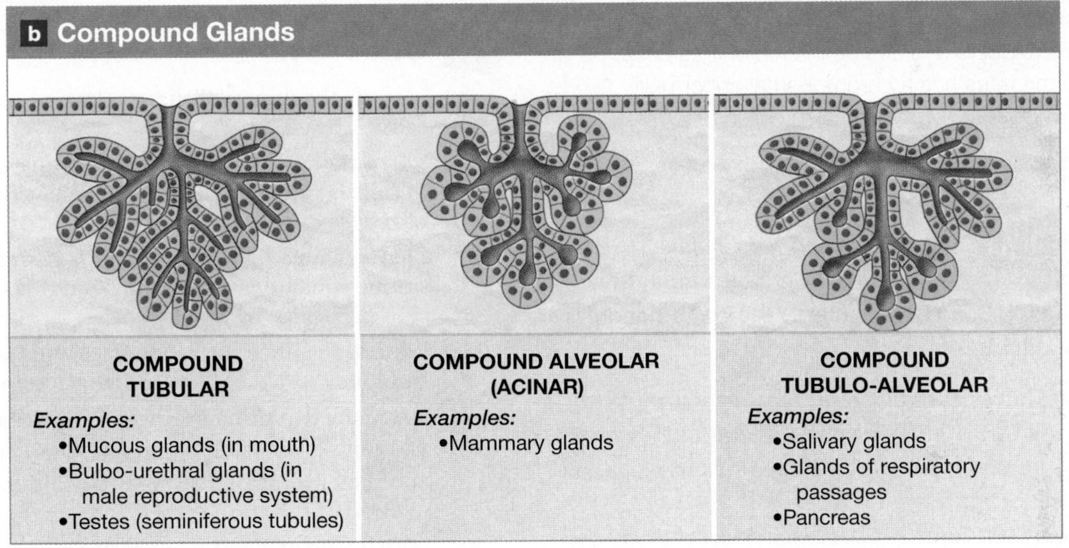
b Compound Glands
COMPOUND TUBULAR
Examples:
•Mucous glands (in mouth)
•Bulbo-urethral glands (in male reproductive system)
•Testes (seminiferous tubules)
COMPOUND ALVEOLAR (ACINAR)
Examples:
•Mammary glands
COMPOUND TUBULO-ALVEOLAR
Examples:
•Salivary glands
•Glands of respiratory passages
•Pancreas

Gland Structure. The first method of classifying exocrine glands is by structure. In epithelia that have independent, scattered gland cells, the individual secretory cells are called **unicellular glands**. **Multicellular glands** include glandular epithelia and aggregations of gland cells that produce exocrine or endocrine secretions.

Unicellular exocrine glands consist of single cells called **goblet cells** that are specialized for secretion. Like mucous cells, goblet cells secrete *mucin*, which mixes with water to form a sticky lubricant called **mucus**. Their location determines whether they are termed *goblet cells* or *mucous cells*. Goblet cells are scattered among the absorptive cells in the columnar epithelium of the small and large intestines. Mucous cells are scattered among other epithelial tissues.

The simplest **multicellular exocrine gland** is a *secretory sheet*, in which gland cells form an epithelium that releases secretions into an inner compartment. Mucin-secreting cells that line the stomach, for instance, protect that organ from its own acids and enzymes by continuously secreting mucin.

Most other multicellular exocrine glands are in pockets set back from the epithelial surface. Their secretions travel through one or more ducts to the surface. Examples include the salivary glands, which produce mucins and digestive enzymes.

Three characteristics are useful in describing the structure of multicellular exocrine glands (Figure 4–7):

1. *The Structure of the Duct.* A gland is *simple* if it has a single duct that does not divide on its way to the gland cells (Figure 4–7a). The gland is *compound* if the duct divides one or more times on its way to the gland cells (Figure 4–7b).
2. *The Shape of the Secretory Portion of the Gland.* Glands whose glandular cells form tubes are *tubular*; the tubes may be straight or coiled. Those that form blind pockets are *alveolar* (al-VĒ-ō-lar; *alveolus*, sac) or *acinar* (AS-i-nar; *acinus*, chamber). Glands whose secretory cells form both tubes and pockets are called *tubulo-alveolar* and *tubulo-acinar*.

4

+ Clinical Note Exfoliative Cytology

Exfoliative (eks-FŌ-lē-a-tiv; *ex*-, from + *folium*, leaf) *cytology* is the study of cells shed or removed from epithelial surfaces. Cells can be examined for changes that indicate cancer or for genetic screening of a fetus. Cells are collected from epithelia lining the respiratory, digestive, urinary, or reproductive tract; from the pleural, pericardial, or peritoneal body cavities; or from an epithelial surface. The harvesting of epithelial cells interrupts their intercellular junctions.

One such common sampling procedure is a *Pap test*, named after Dr. George Papanicolaou, who pioneered its use. The most familiar Pap test involves scraping cells from the tip of the *cervix*, the portion of the uterus that projects into the vagina, to screen for cancer.

Amniocentesis (am-nē-oh-sen-TĒ-sis) collects shed epithelial cells from a sample of *amniotic fluid*, which surrounds and protects a developing fetus. Examination of these cells can determine if the fetus has a chromosomal abnormality, such as occurs in *Down's syndrome*, also called *trisomy 21*.

3. *The Relationship between the Ducts and the Glandular Areas.* A gland is *branched* if several secretory areas (tubular or acinar) share a duct. ("Branched" refers to the glandular areas and not to the duct.)

Methods of Secretion. Exocrine glands are made up of exocrine cells, which use one of three different methods to secrete their products: (1) *merocrine secretion*, (2) *apocrine secretion*, or (3) *holocrine secretion*.

In **merocrine** (MER-ō-krin; *meros*, part) **secretion**, the product is released from an exocrine cell by secretory vesicles through exocytosis (**Figure 4–8a**). ↺ p. 98 This is the most common method of exocrine secretion. For example, mucin is a merocrine secretion. The mucous secretions of the salivary glands coat food and reduce friction during swallowing. In the skin, merocrine sweat glands produce the watery perspiration that helps cool you on a hot day.

Apocrine (AP-ō-krin; *apo*-, off) **secretion** from an exocrine cell involves the loss of cytoplasm as well as the secretory product (**Figure 4–8b**). The apical portion of the cytoplasm becomes packed with secretory vesicles and is then shed. Milk production in the mammary glands involves a combination of merocrine and apocrine secretions.

Merocrine and apocrine secretions leave a cell relatively intact and able to continue secreting. **Holocrine** (HOL-ō-krin; *holos*, entire) **secretion**, by contrast, destroys the gland cell. During holocrine secretion, a superficial cell in a stratified glandular epithelium becomes packed with secretory vesicles and then bursts, releasing the secretion, but killing the cell (**Figure 4–8c**). Further secretion depends on replacing destroyed gland cells by the division of underlying stem cells. Sebaceous glands, associated with hair follicles, produce an oily hair coating by means of holocrine secretion.

Types of Secretions. Finally, we also categorize exocrine glands by the types of secretions they produce:

- *Serous glands* secrete a watery solution that contains enzymes. The parotid salivary glands are serous glands.
- *Mucous glands* secrete mucins that hydrate to form mucus. The sublingual salivary glands and the submucosal glands of the small intestine are mucous glands.
- *Mixed exocrine glands* contain more than one type of gland cell. These glands may produce two different exocrine secretions, one serous and the other mucous. The submandibular salivary glands are mixed exocrine glands.

✓ Checkpoint

8. **Identify the three cell shapes characteristic of epithelial cells.**
9. **When classifying epithelial tissue, one with a single layer of cells is called ____, whereas one with multiple layers of cells is called ____.**
10. **Viewed with a light microscope, a tissue appears as a simple squamous epithelium. Can this be a sample of the skin surface? Why or why not?**
11. **Why do the pharynx, esophagus, anus, and vagina have a similar stratified epithelial organization?**
12. **Name the two primary types of glandular epithelia.**
13. **The secretory cells of sebaceous glands fill with vesicles and then rupture, releasing their contents. Which method of secretion is this?**
14. **Which type of gland releases its secretions directly into the interstitial fluid?**

See the blue Answers tab at the back of the book.

4-4 Connective tissue has varied roles in the body that reflect the physical properties of its three main types

Learning Outcome **List the specific functions of connective tissue, and describe the three main categories of connective tissue.**

It is impossible to discuss epithelial tissue without mentioning an associated type of tissue: **connective tissue**, a diverse group of supporting tissues. Recall that the reticular lamina layer of the basement membrane of all epithelial tissues is made up of connective tissue components. In essence, connective tissue connects the epithelium to the rest of the body. Other types of connective tissue include bone, fat, and blood.

Connective tissues vary widely in appearance and function, but they all share three basic components: (1) specialized cells, (2) extracellular protein fibers, and (3) a fluid known

Figure 4–8 Methods of Glandular Secretion.

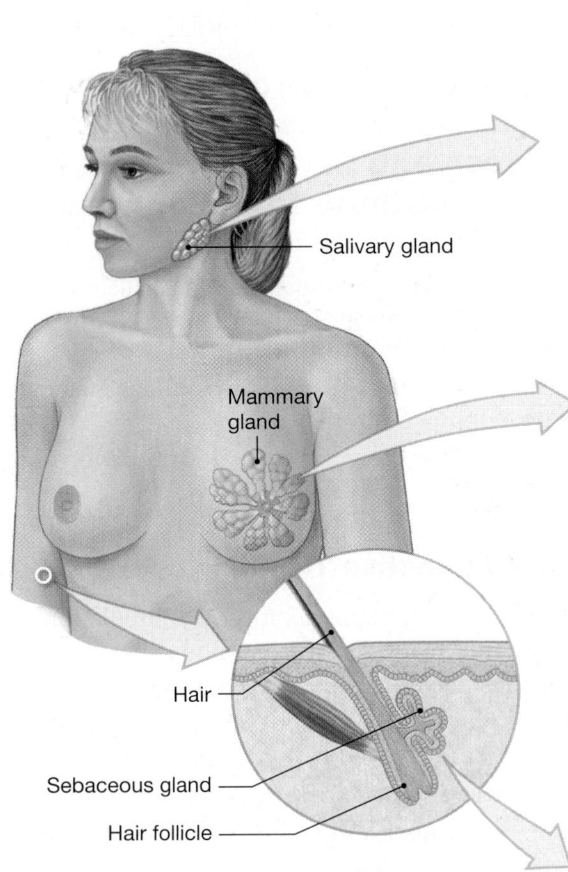
Salivary gland
Mammary gland
Hair
Sebaceous gland
Hair follicle

a Merocrine secretion

In merocrine secretion, the product is released from secretory vesicles at the apical surface of the gland cell by exocytosis.

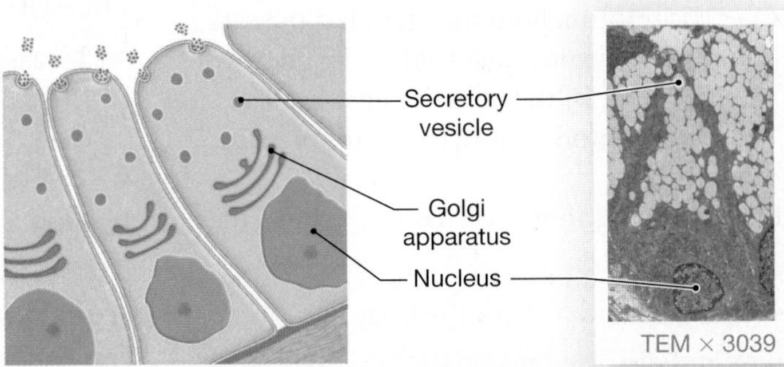
Secretory vesicle
Golgi apparatus
Nucleus
TEM × 3039

b Apocrine secretion

Apocrine secretion involves the loss of apical cytoplasm. Inclusions, secretory vesicles, and other cytoplasmic components are shed in the process. The gland cell then grows and repairs itself before it releases additional secretions.

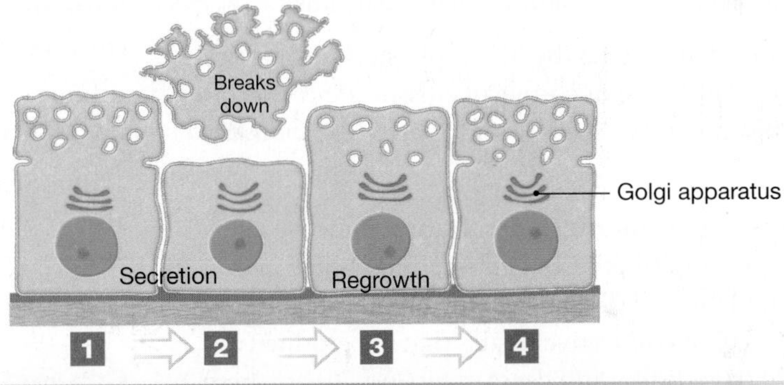
Breaks down
Golgi apparatus
Secretion
Regrowth
1
2
3
4

c Holocrine secretion

Holocrine secretion occurs as superficial gland cells burst. Continued secretion involves the replacement of these cells through the mitotic divisions of underlying stem cells.

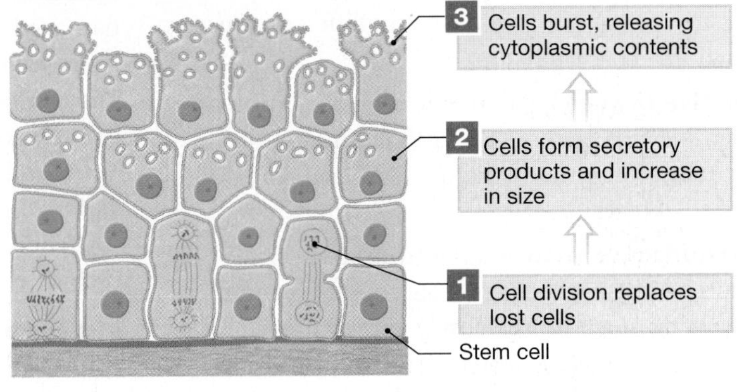
3 Cells burst, releasing cytoplasmic contents
2 Cells form secretory products and increase in size
1 Cell division replaces lost cells
Stem cell

? Which method of secretion destroys the cell?

as *ground substance*. Together, the extracellular fibers and ground substance make up the **matrix**, which surrounds the cells. While cells make up the bulk of epithelial tissue, the matrix typically accounts for most of the volume of connective tissues. ATLAS: Embryology Summary 3: The Origins of Connective Tissues

Connective tissues occur throughout the body, but they are never exposed to the outside environment. Many connective tissues are highly vascular (that is, they have many blood vessels) and contain sensory receptors that detect pain, pressure, temperature, and other stimuli.

The specific functions of connective tissues include the following:

- Establishing a structural framework for the body.
- Transporting fluids and dissolved materials.
- Protecting delicate organs.
- Supporting, surrounding, and interconnecting other types of tissue.
- Storing energy, especially in the form of triglycerides.
- Defending the body from invading microorganisms.

We classify connective tissues by their physical properties. The three general categories of connective tissue are *connective tissue proper*, *fluid connective tissues*, and *supporting connective tissues*.

- **Connective tissue proper** includes those connective tissues with many types of cells and extracellular fibers in a viscous (syrupy) ground substance. This broad category contains a variety of connective tissues that are grouped into *loose connective tissues* and *dense connective tissues*. These groupings are based on the number of cell types present, and on the relative properties and proportions of fibers and ground substance. Both *adipose tissue*, or fat (a loose connective tissue), and *tendons* (a dense connective tissue) are connective tissue proper, but they have very different structural and functional characteristics.
- **Fluid connective tissues** have distinctive populations of cells suspended in a watery matrix that contains dissolved proteins. The only two types are *blood* and *lymph*.
- **Supporting connective tissues** differ from connective tissue proper in having a less diverse cell population and a matrix containing much more densely packed fibers. Supporting connective tissues protect soft tissues and support the weight of part or all of the body. The two types of supporting connective tissues are *cartilage* and *bone*. (The matrix of cartilage is a gel with characteristics that vary with the predominant type of fiber. The matrix of bone contains mineral deposits, primarily calcium salts, that provide rigidity.)

Now that you have an overview of connective tissue, the next three sections explore these three types in more detail.

Checkpoint

15. **Identify several functions of connective tissues.**
16. **List the three categories of connective tissues.**

See the blue Answers tab at the back of the book.

4-5 Connective tissue proper includes loose connective tissues that fill internal spaces and dense connective tissues that contribute to the internal framework of the body

Learning Outcome Compare the structures and functions of the various types of connective tissue proper, and the layers of connective tissue called fasciae.

In this section we describe the cells and extracellular materials of connective tissue proper and the structure and function of loose and dense connective tissues. We also describe embryonic connective tissues.

Structure of Connective Tissue Proper

Connective tissue proper contains a varied cell population. It also contains a matrix made up of extracellular fibers and a viscous ground substance (Figure 4–9). Let's look at each of these components.

Connective Tissue Proper Cell Populations

Some cells in connective tissue proper function in local maintenance, repair, and energy storage. These cells include *fibroblasts*, *fibrocytes*, *adipocytes*, and *mesenchymal cells*. All are permanent residents of the connective tissue. Other cells defend and repair damaged tissues. These cells are more mobile and include *macrophages*, *mast cells*, *lymphocytes*, *plasma cells*, and *microphages*. They are not permanent residents. Instead they migrate through healthy connective tissues and collect at sites of tissue injury. The number of cells and cell types in a tissue at any moment varies with local conditions.

The fixed and migratory cells of connective tissue proper include the following:

- **Fibroblasts** (FĪ-brō-blasts) are the only cells that are *always* present in connective tissue proper. They are also the most abundant fixed residents of connective tissue proper. Fibroblasts secrete hyaluronan (a polysaccharide derivative) and proteins. (Recall that hyaluronan is one of the ingredients that helps lock epithelial cells together.) In connective tissue proper, extracellular fluid, hyaluronan, and proteins interact to form the proteoglycans that make ground

Figure 4–9 The Cells and Fibers of Connective Tissue Proper. A diagrammatic view of the cell types and fibers of connective tissue proper. (Microphages, not shown, are common only in damaged or abnormal tissues.)

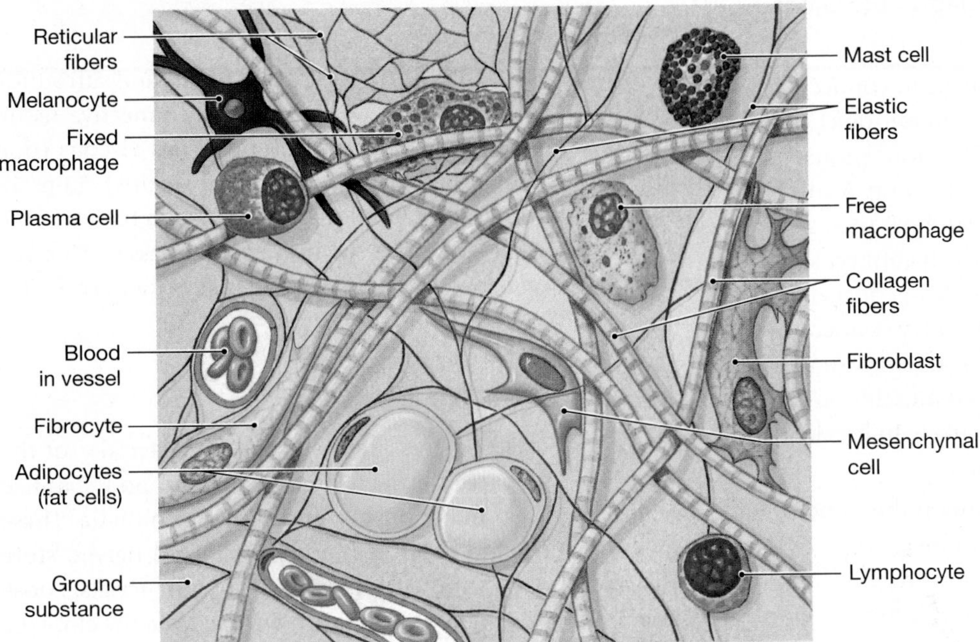
Reticular fibers
Melanocyte
Fixed macrophage
Plasma cell
Blood in vessel
Fibrocyte
Adipocytes (fat cells)
Ground substance
Mast cell
Elastic fibers
Free macrophage
Collagen fibers
Fibroblast
Mesenchymal cell
Lymphocyte

substance viscous. Each fibroblast also secretes protein subunits that assemble to form large extracellular fibers. p. 55

- **Fibrocytes** (FĪ-brō-sītz) are the second most abundant fixed cell in connective tissue proper. They differentiate from fibroblasts. Fibrocytes are spindle-shaped cells that maintain the connective tissue fibers of connective tissue proper.
- **Adipocytes** (AD-i-pō-sītz) are also known as fat cells. A typical adipocyte contains a single, enormous lipid droplet. The nucleus, other organelles, and cytoplasm are squeezed to one side, making a sectional view of the cell resemble a class ring. The number of adipocytes varies from one type of connective tissue to another, from one region of the body to another, and among individuals.
- **Mesenchymal cells** are stem cells that are present in many connective tissues. These cells respond to local injury or infection by dividing to produce daughter cells that differentiate into fibroblasts, macrophages, or other connective tissue cells.
- **Melanocytes** (me-LAN-ō-sītz) synthesize and store the brown pigment **melanin** (MEL-a-nin), which gives tissues a dark color. Melanocytes are common in the epithelium of the skin, where they play a major role in determining skin color. Melanocytes are also abundant in connective tissues of the eye and the dermis of the skin, although the number present differs by body region and among individuals.
- **Macrophages** (MAK-rō-fā-jez) are large phagocytic cells scattered throughout the matrix. These scavengers engulf damaged cells or pathogens that enter the tissue. (Their name literally means "big eater.") Although not abundant, macrophages are important in mobilizing the body's defenses. When stimulated, they release chemicals that activate the immune system and attract large numbers of additional macrophages and other cells involved in tissue defense.

  Macrophages are either *fixed macrophages*, which spend long periods in a tissue, or *free macrophages*, which migrate rapidly through tissues. In effect, fixed macrophages provide a "frontline" defense that can be reinforced by the arrival of free macrophages and other specialized cells.
- **Mast cells** circulate in the blood in an immature form before they migrate to other vascularized tissues (tissues with blood vessels) and undergo final maturation. The cytoplasm of a mast cell is filled with granules containing **histamine** (HIS-tuh-mēn) and **heparin** (HEP-uh-rin). Histamine, released after injury or infection, stimulates local inflammation. (You are probably familiar with the inflammatory effects of histamine if you've had a cold. People often take antihistamines to reduce cold symptoms.) *Basophils*, blood cells that enter damaged tissues and enhance the inflammation process, also contain granules of histamine and heparin. The level of heparin in the blood is normally low. It is an anticoagulant that enhances local blood flow during inflammation and reduces the development of blood clots in areas of slow-moving blood.
- **Lymphocytes** (LIM-fō-sītz) migrate throughout the body, traveling through connective tissues and other tissues. Their numbers increase markedly wherever tissue damage occurs. Some lymphocytes may develop into **plasma cells**, which produce *antibodies*—proteins involved in defending the body against disease.
- **Microphages** [*neutrophils* and *eosinophils* (ē-ō-SIN-ō-filz)] are phagocytic blood cells that normally move through connective tissues in small numbers. They are attracted to the site of an infection or injury by chemicals released by macrophages and mast cells.

Connective Tissue Fibers

Three types of fibers occur in connective tissue proper: *collagen, reticular*, and *elastic fibers* (see Figure 4–9). Fibroblasts form all three by secreting protein subunits that interact in the matrix. Fibrocytes then maintain these fibers.

- **Collagen fibers** are the most common fibers in connective tissue proper. They are long, straight, and unbranched. Each collagen fiber consists of a bundle of fibrous protein subunits wound together like the strands of a rope. A collagen fiber is flexible like a rope, but it is stronger than steel when pulled from either end. Tendons, which connect skeletal muscles to bones, consist almost entirely of collagen fibers. *Ligaments* are similar to tendons, but they connect one bone to another bone. Tendons and ligaments can withstand tremendous forces. Uncontrolled muscle contractions or skeletal movements are more likely to break a bone than to snap a tendon or a ligament.
- **Reticular fibers** (*reticulum*, network) contain the same protein subunits as do collagen fibers, but they are arranged differently. ⮌ p. 55 Thinner than collagen fibers, reticular fibers form a branching, interwoven framework that is tough, yet flexible. Reticular fibers resist forces applied from many directions because they form a network rather than share a common alignment. This interwoven network, called a *stroma*, stabilizes the relative positions of the functional cells, or *parenchyma* (pa-RENG-ki-ma), of organs such as the liver. Reticular fibers also stabilize the positions of an organ's blood vessels, nerves, and other structures, despite changing positions and the pull of gravity.
- **Elastic fibers** contain the protein *elastin*. Elastic fibers are branched and wavy. After stretching, they return to their original length. *Elastic ligaments*, which are dominated by elastic fibers, are rare but have important functions, such as interconnecting vertebrae.

Ground Substance

Ground substance fills the spaces between cells and surrounds connective tissue fibers (see Figure 4–9). In connective tissue proper, ground substance is clear, colorless, and viscous (due to the presence of proteoglycans and glycoproteins). ⮌ p. 57 Ground substance is viscous enough that bacteria have trouble moving through it—imagine swimming in molasses. This viscosity slows the spread of pathogens and makes them easier for phagocytes to catch.

Loose Connective Tissues

Loose connective tissues are the "packing materials" of the body. They fill spaces between organs, cushion and stabilize specialized cells in many organs, and support epithelia. These tissues surround and support blood vessels and nerves, store lipids, and provide a route for the diffusion of materials. Loose connective tissues include *mucous connective tissue* in embryos and *areolar tissue, adipose tissue*, and *reticular tissue* in adults.

Embryonic Connective Tissues

Mesenchyme (MEZ-en-kīm), or *embryonic connective tissue*, is the first connective tissue to appear in a developing embryo. Mesenchyme contains an abundance of star-shaped stem cells (mesenchymal cells) separated by a matrix with very fine protein filaments (Figure 4–10a). Mesenchyme gives rise to all other connective tissues. **Mucous connective tissue** (Figure 4–10b), or *Wharton's jelly*, is a loose connective tissue found in many parts of the embryo, including the umbilical cord.

Figure 4–10 Embryonic Connective Tissues.

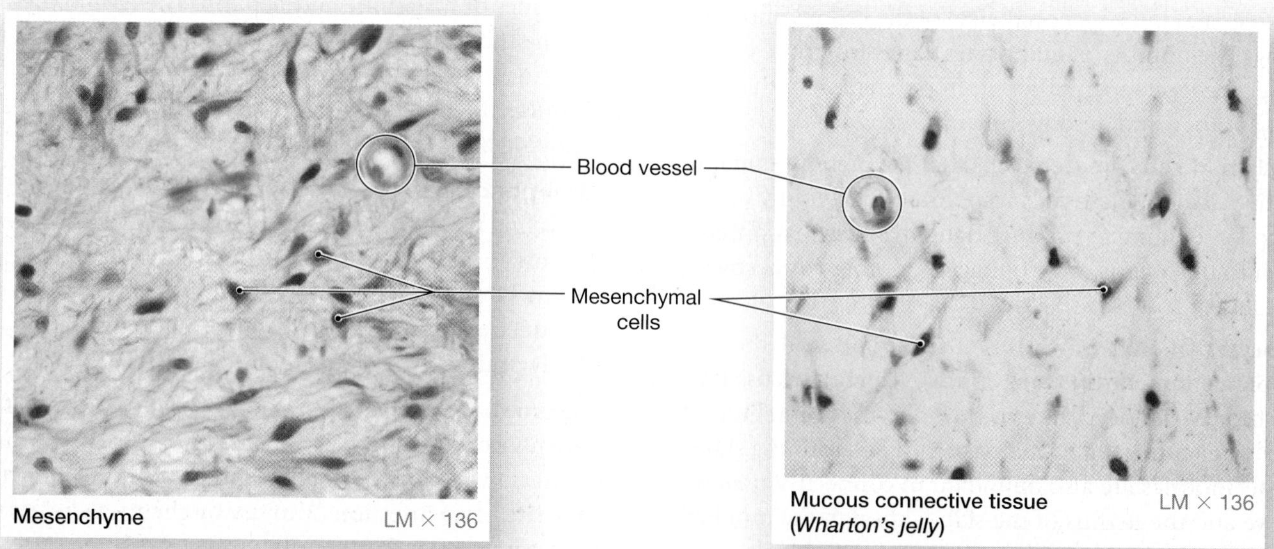
Blood vessel
Mesenchymal cells
Mesenchyme
LM × 136
Mucous connective tissue
(Wharton's jelly)
LM × 136

a This is the first connective tissue to appear in an embryo.

b This sample was taken from the umbilical cord of a fetus.

Adults have neither form of embryonic connective tissue. Instead, many adult connective tissues contain scattered mesenchymal stem cells that can assist in tissue repair after an injury.

Areolar Tissue

Areolar tissue (*areola*, little space) is the least specialized connective tissue in adults. It may contain all the cells and fibers of any connective tissue proper in a very loosely organized array (Figure 4–11a). Areolar tissue has an open framework. A viscous ground substance provides most of its volume and absorbs shocks. Areolar tissue can distort without damage because its fibers are loosely organized. Elastic fibers make it resilient, so areolar tissue returns to its original shape after external pressure is relieved.

Areolar tissue forms a layer that separates the skin from deeper structures. In addition to providing padding, the elastic properties of this layer allow a considerable amount of independent movement. For this reason, if you pinch the skin of your arm, you will not affect the underlying muscle. Conversely, contractions of the underlying muscle do not pull against your skin: As the muscle bulges, the areolar tissue stretches. The areolar tissue layer under the skin is a common injection site for drugs because it has an extensive blood supply.

The capillaries (thin-walled blood vessels) in areolar tissue deliver oxygen and nutrients and remove carbon dioxide and waste products. They also carry wandering cells to and from the tissue. Epithelia commonly cover areolar tissue, and fibrocytes maintain the reticular lamina of the basement membrane that separates the two kinds of tissue. The epithelial cells rely on the oxygen and nutrients that diffuse across the basement membrane from capillaries in the underlying connective tissue.

Adipose Tissue

The distinction between areolar tissue and fat, or **adipose tissue**, is a matter of degree. Adipocytes (fat cells) account for most of the volume of adipose tissue (Figure 4–11b), but only a fraction of the volume of areolar tissue. Adipose tissue provides padding, absorbs shocks, acts as an insulator to slow heat loss through the skin, and serves as packing or filler around structures. Adipose tissue is common under the skin of the sides (between the last rib and the hips), buttocks, and breasts. It fills the bony sockets behind the eyes and surrounds the kidneys. It is also common beneath the mesothelial lining of the pericardial and peritoneal cavities.

Most of the adipose tissue in the body is called **white fat**, because it has a pale, yellow-white color. In infants and young children, however, the adipose tissue between the shoulder blades, around the neck, and possibly elsewhere in the upper body is highly vascularized, and the individual adipocytes contain numerous mitochondria. Together, these characteristics give the tissue a deep, rich color, leading to the name **brown fat**. When these cells are stimulated by the nervous system, lipid breakdown speeds up. The cells do not capture the energy that is released. Instead, the surrounding tissues absorb it as heat. The heat warms the circulating blood, which distributes the heat throughout the body. In this way, an infant can double its metabolic heat generation very quickly. Until recently, it was thought that adults had little, if any, active brown fat tissue. However, recent evidence has indicated that it is most likely present in adults, and that its absence may be connected to obesity.

Adipocytes are not simply fat storage bins. Instead, they are metabolically active cells. Their lipids are constantly being broken down and replaced. When nutrients are scarce, adipocytes deflate like collapsing balloons as their lipids are broken down and the fatty acids are released to support metabolism.

What happens if nutrients become plentiful again? The lost weight can easily be regained in the same areas of the body because the cells are not killed but merely reduced in size. In adults, adipocytes cannot divide. The number of fat cells in peripheral tissues is established in the first few weeks of a newborn's life, perhaps in response to the amount of fats in the diet. However, that is not the end of the story, because loose connective tissues also contain mesenchymal cells. If circulating lipid levels are chronically elevated, the mesenchymal cells will divide, giving rise to cells that differentiate into fat cells. As a result, areas of areolar tissue can become adipose tissue in times of nutritional plenty, even in adults.

Liposuction is a cosmetic surgical procedure that removes excess or unwanted adipose tissue. However, liposuction provides only a temporary and potentially risky solution to the problem of excess weight. Once again, the reason is that adipose tissue can regenerate through the differentiation of mesenchymal cells.

+ Clinical Note Marfan Syndrome

Marfan syndrome is an inherited condition caused by the production of an abnormally weak form of *fibrillin*, a glycoprotein that gives strength and elasticity to connective tissues. The effects of this defect are widespread because most organs contain connective tissues. The most visible sign of Marfan syndrome involves the skeleton. Most individuals with the condition are tall and have abnormally long limbs and fingers. The most serious consequences involve the cardiovascular system. About 90 percent of people with Marfan syndrome have structural abnormalities in their cardiovascular system. The most dangerous possibility is that the weakened elastic connective tissues in the walls of major arteries, such as the aorta, may burst, causing a sudden, fatal loss of blood.

4

Figure 4–11 Loose Connective Tissues.

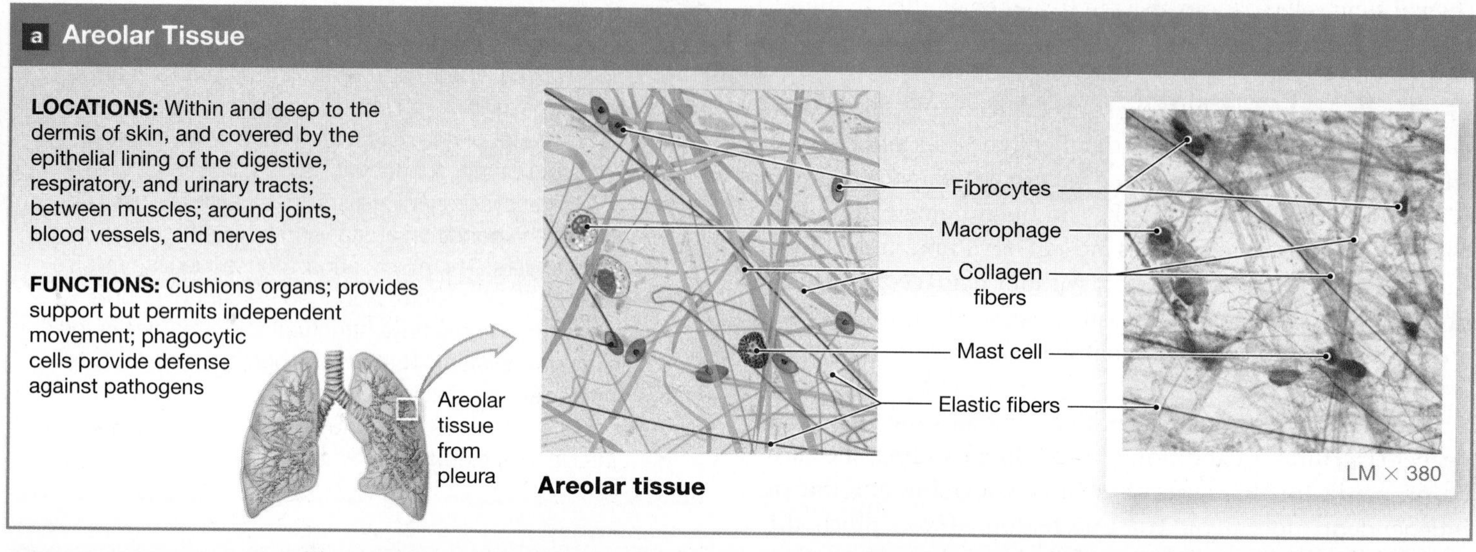
a Areolar Tissue
LOCATIONS: Within and deep to the dermis of skin, and covered by the epithelial lining of the digestive, respiratory, and urinary tracts; between muscles; around joints, blood vessels, and nerves
FUNCTIONS: Cushions organs; provides support but permits independent movement; phagocytic cells provide defense against pathogens
Areolar tissue from pleura
Areolar tissue
Fibrocytes
Macrophage
Collagen fibers
Mast cell
Elastic fibers
LM × 380

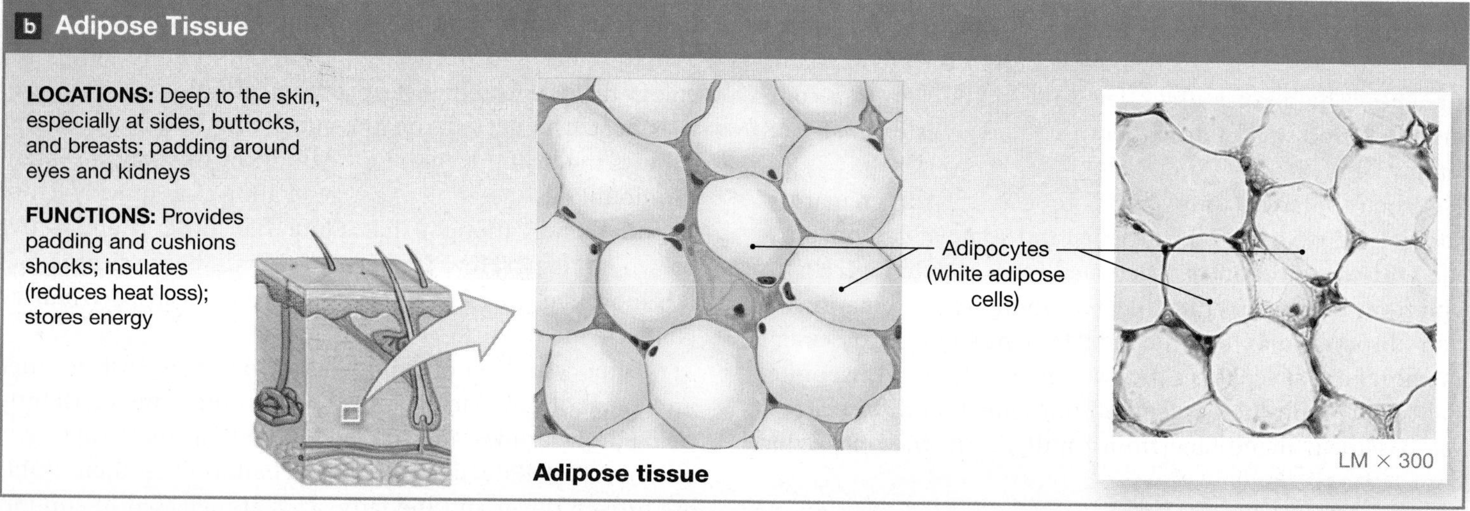
b Adipose Tissue
LOCATIONS: Deep to the skin, especially at sides, buttocks, and breasts; padding around eyes and kidneys
FUNCTIONS: Provides padding and cushions shocks; insulates (reduces heat loss); stores energy
Adipose tissue
Adipocytes (white adipose cells)
LM × 300

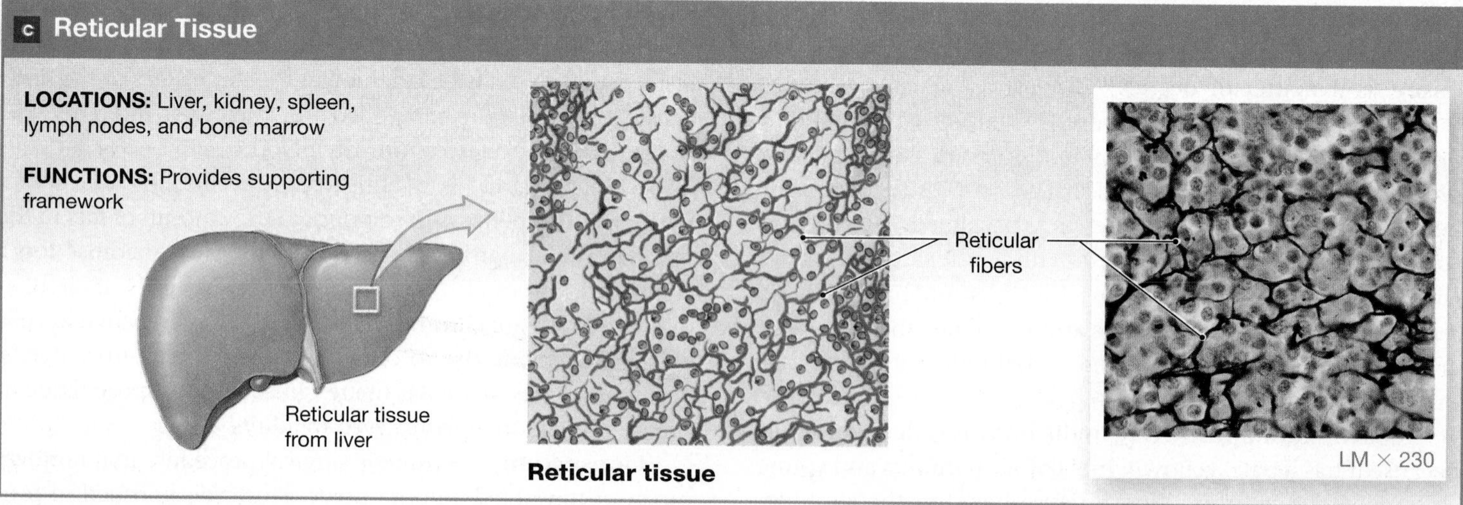
c Reticular Tissue
LOCATIONS: Liver, kidney, spleen, lymph nodes, and bone marrow
FUNCTIONS: Provides supporting framework
Reticular tissue from liver
Reticular tissue
Reticular fibers
LM × 230

? Which type of loose connective tissue provides padding, cushions shock, reduces heat loss, and stores energy? Where are some places in the body this type of tissue is commonly found?

Reticular Tissue

As mentioned earlier, organs such as the liver, kidney, and spleen contain **reticular tissue**, in which reticular fibers form a complex three-dimensional stroma (Figure 4–11c). The stroma supports the functional cells of these organs. This fibrous framework is also found in the lymph nodes and bone marrow. Fixed macrophages, fibroblasts, and fibrocytes are associated with the reticular fibers, but these cells are seldom visible in microscopic sections, because specialized cells with other functions dominate the organs.

Dense Connective Tissues

Fibers create most of the volume of **dense connective tissues**. These tissues are often called **collagenous** (ko-LAJ-e-nus) **tissues**, because collagen fibers are the dominant type of fiber in them. The body has two types of dense connective tissues: *dense regular connective tissue* and *dense irregular connective tissue*. *Elastic tissue* is a specialized type of dense regular connective tissue.

Dense Regular Connective Tissue

In **dense regular connective tissue**, the collagen fibers are parallel to each other, packed tightly, and aligned with the forces applied to the tissue. **Tendons** are cords of dense regular connective tissue that attach skeletal muscles to bones (Figure 4–12a). The collagen fibers run longitudinally along the tendon and transfer the pull of the contracting muscle to the bone. **Ligaments** resemble tendons, but connect one bone to another or stabilize the positions of internal organs.

An **aponeurosis** (AP-ō-noo-RŌ-sis; plural, *aponeuroses*) is a tendinous sheet that attaches a broad, flat muscle to another muscle or to several bones of the skeleton. It can also stabilize the positions of tendons and ligaments. Aponeuroses are associated with large muscles of the skull, lower back, and abdomen, and with the tendons and ligaments of the palms of the hands and the soles of the feet. Large numbers of fibroblasts are scattered among the collagen fibers of tendons, ligaments, and aponeuroses.

Dense Irregular Connective Tissue

In contrast to dense regular connective tissue, the fibers in **dense irregular connective tissue** form an interwoven meshwork in no consistent pattern (Figure 4–12b). These tissues strengthen and support areas subjected to stresses from many directions. A layer of dense irregular connective tissue gives the deep layer of the skin, called the *dermis*, its strength. Cured leather (animal skin) is an excellent example of the interwoven nature of this tissue. Dense irregular connective tissue forms a thick fibrous layer called a *capsule*, which surrounds internal organs such as the liver, kidneys, and spleen and encloses the cavities of joints. Except at joints, dense irregular connective tissue also forms a sheath around cartilages (the *perichondrium*) and bones (the *periosteum*).

Elastic Tissue

Dense regular and dense irregular connective tissues contain variable amounts of elastic fibers. When elastic fibers outnumber collagen fibers, the tissue has a springy, resilient nature that allows it to tolerate cycles of extension and recoil. Abundant elastic fibers are present in the connective tissue that supports transitional epithelia, in the walls of large blood vessels such as the aorta, between erectile tissue and the skin of the penis, and around the respiratory passageways.

Elastic tissue is a dense regular connective tissue made up mainly of elastic fibers. Elastic ligaments are almost completely dominated by elastic fibers. They help stabilize the positions of the vertebrae of the spinal column (Figure 4–12c).

Fasciae: Layers of Connective Tissue Proper

As discussed earlier, connective tissue provides the internal structure of the body. Layers of connective tissue surround and support the organs within the trunk cavities and connect them to the rest of the body. These layers (1) provide strength and stability, (2) maintain the relative positions of internal organs, and (3) supply a route for the distribution of blood vessels, lymphatic vessels, and nerves.

Fasciae (FASH-ē-ē; singular, *fascia*; FASH-ē-uh) are connective tissue layers and wrappings that support and surround organs. We can divide the fasciae into three layers: the superficial fascia, the deep fascia, and the subserous fascia (Figure 4–13).

1. The **superficial fascia** refers to the *subcutaneous layer*. This layer of areolar and adipose tissue separates the skin from underlying tissues and organs. It also provides insulation and padding, and lets the skin and underlying structures move independently.
2. The **deep fascia** consists of sheets of dense regular connective tissue. The organization of the fibers is like that of a sheet of plywood: In each layer making up the sheet, all fibers run in the same direction, but the orientation of the fibers changes from layer to layer. This arrangement helps the tissue resist forces coming from many directions. The tough capsules that surround most organs, including the kidneys and the organs enclosed by the thoracic and peritoneal cavities, are bound to this fascia layer. The perichondrium around cartilages, the periosteum around bones and the ligaments that interconnect them, and the connective tissues of muscle (including tendons) are also connected to the deeper fascia layer. Because the function of deep fascia differs by location, you may see it referred to more specifically as *muscular fascia*, *parietal fascia*, and *visceral fascia*.

Figure 4–12 Dense Connective Tissues.

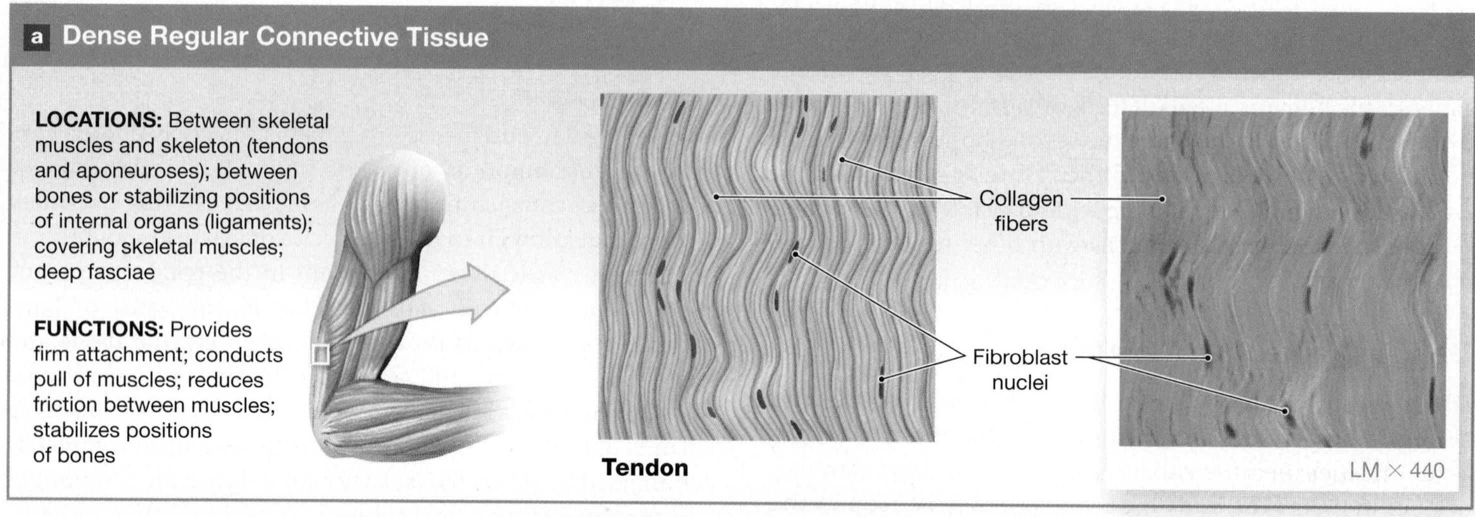
a Dense Regular Connective Tissue
LOCATIONS: Between skeletal muscles and skeleton (tendons and aponeuroses); between bones or stabilizing positions of internal organs (ligaments); covering skeletal muscles; deep fasciae
FUNCTIONS: Provides firm attachment; conducts pull of muscles; reduces friction between muscles; stabilizes positions of bones
Collagen fibers
Fibroblast nuclei
Tendon
LM × 440

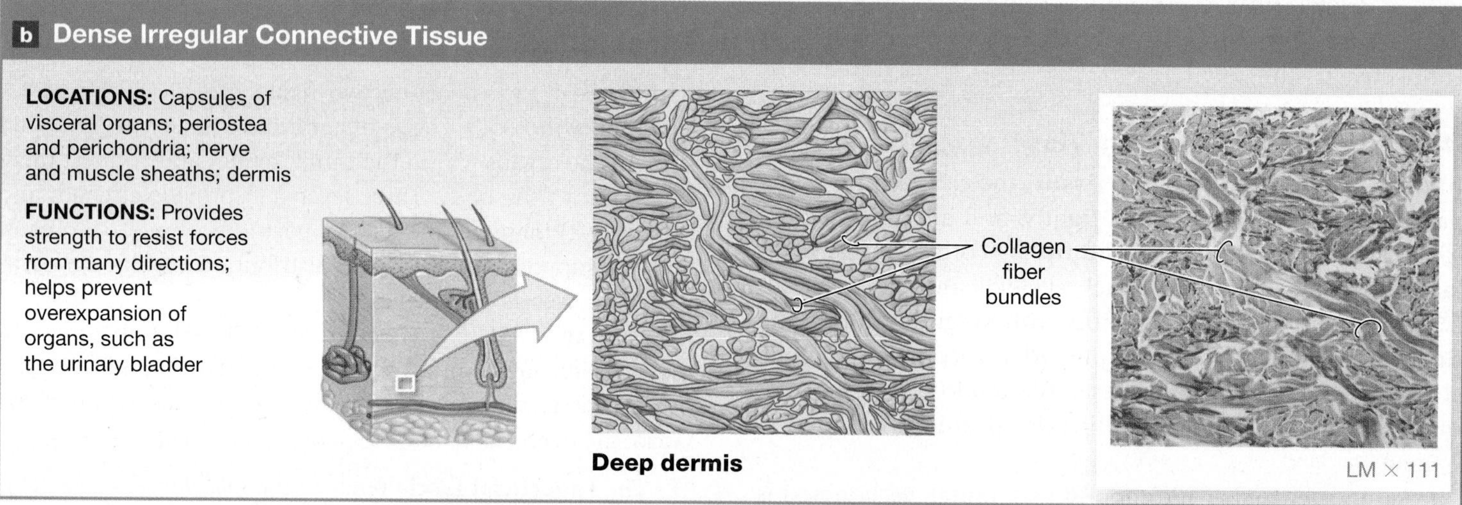
b Dense Irregular Connective Tissue
LOCATIONS: Capsules of visceral organs; periostea and perichondria; nerve and muscle sheaths; dermis
FUNCTIONS: Provides strength to resist forces from many directions; helps prevent overexpansion of organs, such as the urinary bladder
Collagen fiber bundles
Deep dermis
LM × 111

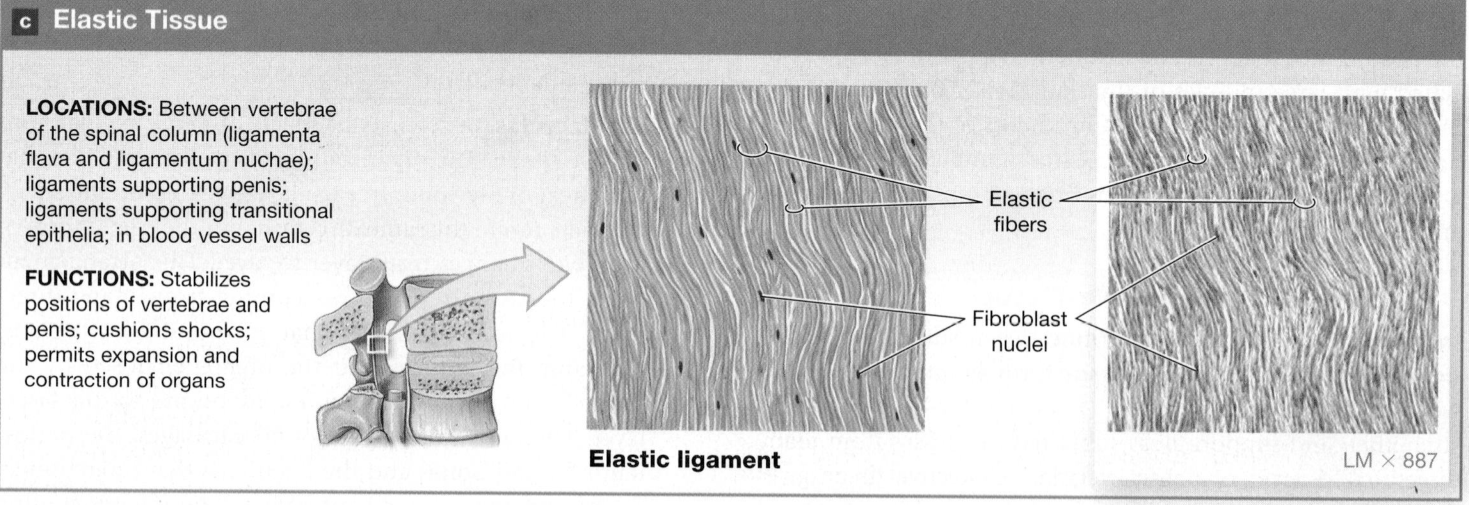
c Elastic Tissue
LOCATIONS: Between vertebrae of the spinal column (ligamenta flava and ligamentum nuchae); ligaments supporting penis; ligaments supporting transitional epithelia; in blood vessel walls
FUNCTIONS: Stabilizes positions of vertebrae and penis; cushions shocks; permits expansion and contraction of organs
Elastic fibers
Fibroblast nuclei
Elastic ligament
LM × 887

Figure 4–13 **The Fasciae.** The relationships among the connective tissue layers in the body.

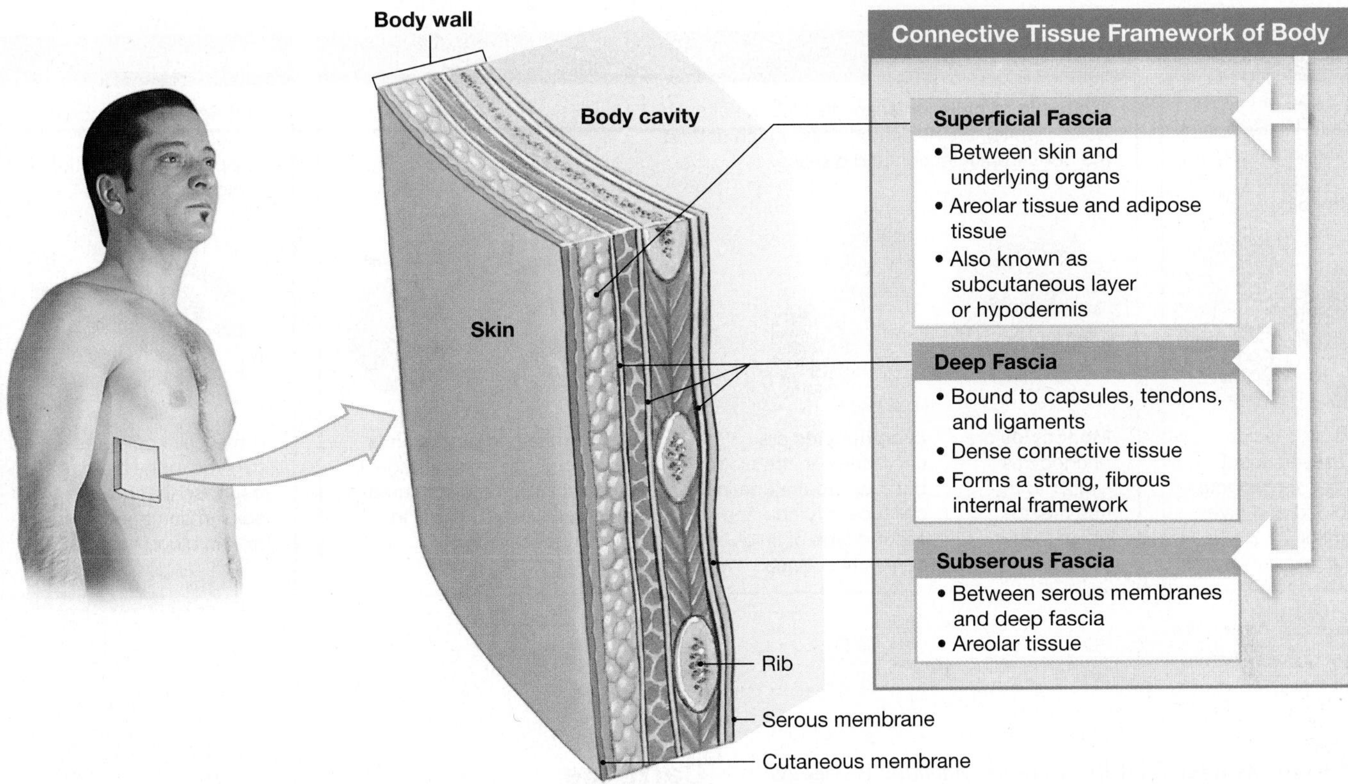

Body wall
Body cavity
Skin
Rib
Serous membrane
Cutaneous membrane
Connective Tissue Framework of Body
Superficial Fascia
• Between skin and underlying organs
• Areolar tissue and adipose tissue
• Also known as subcutaneous layer or hypodermis
Deep Fascia
• Bound to capsules, tendons, and ligaments
• Dense connective tissue
• Forms a strong, fibrous internal framework
Subserous Fascia
• Between serous membranes and deep fascia
• Areolar tissue

The dense connective tissue components are interwoven. For example, fascia around a muscle blends into the tendon, whose fibers intermingle with those of the periosteum. This arrangement creates a strong, fibrous network and binds structural elements together.

3. The **subserous fascia** is a layer of areolar tissue that lies between the deep fascia and the serous membranes that line true body cavities. p. 14 Because this layer separates the serous membranes from the deep fascia, movements of muscles or muscular organs do not severely distort the delicate body cavity linings.

Checkpoint

17. **Identify the cells found in connective tissue proper.**
18. **Lack of vitamin C in the diet interferes with the ability of fibroblasts to produce collagen. What effect might this interference have on connective tissue?**
19. **Many allergy sufferers take antihistamines to relieve their allergy symptoms. Which cells produce the substance that this medication blocks?**
20. **Which type of connective tissue contains primarily triglycerides?**
21. **Name the three layers of fascia and their types of connective tissue.**

See the blue Answers tab at the back of the book.

4-6 Blood and lymph are fluid connective tissues that transport cells and dissolved materials

Learning Outcome Describe the fluid connective tissues blood and lymph, and explain their relationship with interstitial fluid in maintaining homeostasis.

Blood and lymph are fluid connective tissues with distinctive collections of cells. The fluid matrix that surrounds the cells also includes many types of suspended proteins that do not form insoluble fibers under normal conditions.

In **blood**, the watery matrix is called **plasma**. Plasma contains blood cells and fragments of cells, collectively known as *formed elements*. There are three types of formed elements: red blood cells, white blood cells, and platelets (Figure 4–14).

Recall from Chapter 3 that an extracellular fluid called interstitial fluid surrounds the cells of the human body. p. 68 In terms of decreasing volume, the three major subdivisions of extracellular fluid are *interstitial fluid*, *plasma*, and *lymph*.

Plasma is normally confined to the blood vessels (arteries, capillaries, and veins) of the cardiovascular system. Contractions of the heart keep it in motion. *Arteries* carry blood away from the heart and into the tissues of the body. In those tissues, blood pressure forces water and small solutes out of the bloodstream across the walls of *capillaries*, the smallest blood vessels. This process is the origin of the interstitial fluid that bathes the body's cells. The remaining blood flows from the capillaries into *veins* that return it to the heart.

Figure 4–14 **Formed Elements in the Blood.**

Red blood cells

Red blood cells, or erythrocytes (e-RITH-rō-sīts), transport oxygen (and, to a lesser degree, carbon dioxide) in the blood.

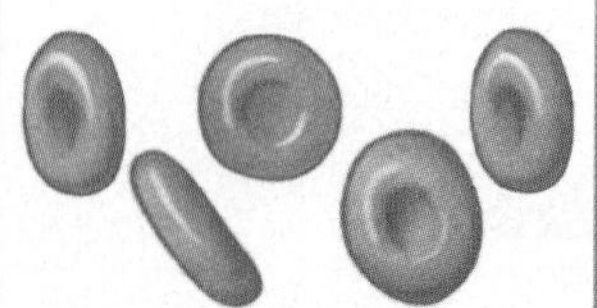

Red blood cells lack a nucleus. They account for about half the volume of whole blood and give blood its color.

White blood cells

White blood cells, or leukocytes (LŪ-kō-sīts; *leuko-*, white), are nucleated cells, which defend the body from infection and disease.

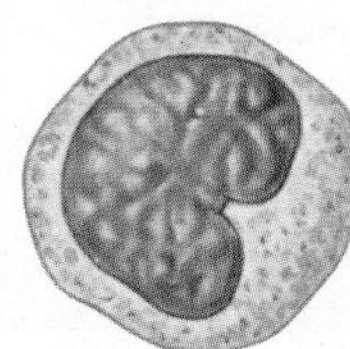

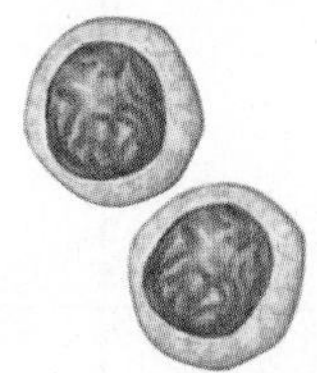

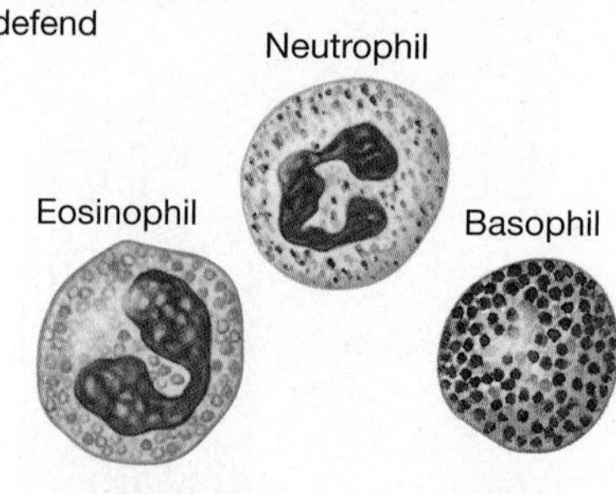
Neutrophil
Eosinophil
Basophil

Monocytes are phagocytes similar to the free macrophages in other tissues.

Lymphocytes are uncommon in the blood but they are the dominant cell type in lymph, the second type of fluid connective tissue.

Eosinophils and **neutrophils** are phagocytes (microphages). **Basophils** promote inflammation much like mast cells in other connective tissues.

Platelets

Platelets are membrane-enclosed packets of cytoplasm that function in blood clotting.

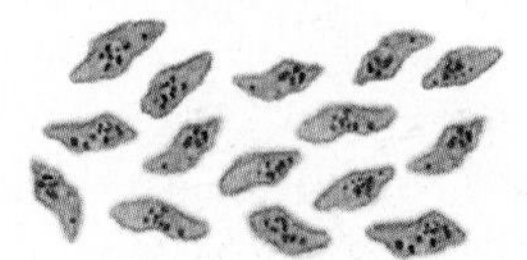

These cell fragments are involved in the clotting response that seals leaks in damaged or broken blood vessels.

? Which formed element of the blood contains a nucleus?

Lymph forms as interstitial fluid enters *lymphatic vessels.* As fluid passes along the lymphatic vessels, cells of the immune system (lymphocytes) monitor its composition and respond to signs of injury or infection. The lymphatic vessels ultimately return the lymph to large veins near the heart. This recirculation of fluid—from the cardiovascular system, through the interstitial fluid, to the lymph, and then back to the cardiovascular system—is a continuous process that is essential to homeostasis. It helps eliminate local differences in the levels of nutrients, wastes, or toxins; maintains blood volume; and alerts the immune system to infections that may be under way in peripheral tissues.

Checkpoint

22. **Which two types of connective tissue have a fluid matrix?**
23. **Describe the recirculation of fluid in the body.**

See the blue Answers tab at the back of the book.

4-7 The supporting connective tissues cartilage and bone provide a strong framework

Learning Outcome Describe how cartilage and bone function as supporting connective tissues.

Cartilage and *bone* are called supporting connective tissues because they provide a strong framework that supports the rest of the body. The matrix of each tissue contains numerous fibers. In bone, it also contains deposits of insoluble calcium salts.

Cartilage

The matrix of **cartilage** is a firm gel that contains polysaccharide derivatives called **chondroitin sulfates** (kon-DROY-tin; *chondros*, cartilage). Chondroitin sulfates form complexes with proteins in the ground substance, producing proteoglycans. Cartilage cells, or **chondrocytes** (KON-drō-sītz), are the only cells in the cartilage matrix. They occupy small chambers known as **lacunae** (la-KOO-nē; *lacus*, lake). The physical properties of cartilage depend on the proteoglycans of the matrix, and on the type and abundance of extracellular fibers.

Unlike other connective tissues, cartilage is avascular, so all exchange of nutrients and waste products must take place by diffusion through the matrix. Because of this situation, cartilage heals poorly. Blood vessels do not grow into cartilage because chondrocytes produce a chemical that discourages their formation. This chemical inhibitor of new blood vessel formation is called *antiangiogenesis factor* (*anti-*, against + *angeion*, vessel + *genesis*, production). Other angiogenesis inhibitors have also been identified in the body or developed as drugs. Various inhibitors are now used in cancer treatments to discourage the formation of new blood vessels to tumors, thereby slowing their growth.

Cartilage is generally set apart from surrounding tissues by a fibrous **perichondrium** (per-i-KON-drē-um; *peri-*, around). The perichondrium contains two distinct layers: an outer, fibrous region of dense irregular connective tissue, and an inner, cellular layer. The fibrous layer gives mechanical support and protection and attaches the cartilage to other structures.

The cellular layer is important to the growth and maintenance of the cartilage. Blood vessels in the perichondrium provide oxygen and nutrients to the underlying chondrocytes.

Types of Cartilage

The body contains three major types of cartilage: hyaline cartilage, elastic cartilage, and fibrocartilage.

- **Hyaline cartilage** (HĪ-uh-lin; *hyalos*, glass) is the most common type of cartilage. Except inside joint cavities, a dense perichondrium surrounds hyaline cartilages. Hyaline cartilage is tough but somewhat flexible because its matrix contains closely packed collagen fibers. Because these fibers are not in large bundles and do not stain darkly, they are not always apparent in light microscopy (Figure 4–15a). Examples in adults include the connections between the ribs and the sternum; the nasal cartilages and the supporting cartilages along the conducting passageways of the respiratory tract; and the *articular cartilages*, which cover opposing bone surfaces within many joints, such as the elbow and knee.
- **Elastic cartilage** (Figure 4–15b) is extremely resilient and flexible because it contains numerous elastic fibers. These cartilages usually have a yellowish color on gross dissection. Elastic cartilage forms the external flap (the *auricle*) of the outer ear; the *epiglottis*, which prevents food and liquids from entering the windpipe when swallowing; a passageway to the middle ear cavity (the *auditory canal*); and small cartilages (the *cuneiform cartilages*) in the larynx, or voice box.
- **Fibrocartilage** is extremely durable and tough because it has little ground substance and its matrix is dominated by densely interwoven collagen fibers (Figure 4–15c). Pads of fibrocartilage (*intervertebral discs*) lie between the spinal vertebrae, around tendons and within or around joints, and between the pubic bones of the pelvis. In these positions, fibrocartilage resists compression, absorbs shocks, prevents damaging bone-to-bone contact, and limits movement.

Cartilage Growth

Cartilage grows by two mechanisms: *interstitial growth* and *appositional growth*. **Interstitial growth** enlarges the cartilage from within (Figure 4–16a). Chondrocytes in the cartilage matrix divide, and the daughter cells produce additional matrix. Interstitial growth is most important during development. It begins early in embryonic development and continues through adolescence.

Appositional growth gradually increases the size of the cartilage by adding to its outer surface (Figure 4–16b). In this process, cells of the inner layer of the perichondrium divide repeatedly and become *chondroblasts* (immature chondrocytes). These cells begin producing cartilage matrix. As they become surrounded by and embedded in new matrix, they differentiate into mature chondrocytes, becoming part of the cartilage and so enlarging it.

Both interstitial and appositional growth take place during development, but interstitial growth contributes more to the mass of adult cartilage. Neither interstitial nor appositional growth takes place in the cartilages of normal adults. However, appositional growth may occur in unusual circumstances, such as after cartilage has been damaged or excessively stimulated by *growth hormone* from the pituitary gland. Minor damage to cartilage can be repaired by appositional growth at the damaged surface. After more severe damage, a dense fibrous patch will replace the injured portion of the cartilage.

Several complex joints, including the knee, contain both hyaline cartilage and fibrocartilage. The hyaline cartilage covers bony surfaces, and fibrocartilage pads in the joint prevent contact between bones during movement. Because injuries to these pads do not heal, such injuries can interfere with normal movements. After repeated or severe damage, joint mobility is severely reduced. Surgery generally produces only a temporary or incomplete repair.

Bone

Here we focus only on significant differences between cartilage and bone. We examine the detailed histology of **bone**, or **osseous** (OS-ē-us; *os*, bone) **tissue**, in Chapter 6. The volume of ground substance in bone is very small. About two-thirds of the matrix of bone is **calcified**, consisting of a mixture of calcium salts—primarily calcium phosphate, with lesser amounts of calcium carbonate. The rest of the matrix is dominated by collagen fibers. This combination gives bone truly remarkable properties. By themselves, calcium salts are hard but rather brittle, whereas collagen fibers are strong but relatively flexible. In bone, the presence of the minerals surrounding the collagen fibers produces a strong, somewhat flexible combination that is highly resistant to shattering. In its overall properties, bone can compete with the best steel-reinforced concrete. In essence, the collagen fibers in bone act like the steel reinforcing rods, and the mineralized matrix acts like the concrete.

Figure 4–17 (p. 139) shows the general organization of bone. Lacunae in the matrix contain **osteocytes** (OS-tē-ō-sītz), or bone cells. The lacunae are typically organized around blood vessels in tubes called *central canals* that branch through the bony matrix. Diffusion cannot take place through the hard matrix, but osteocytes communicate with the blood vessels and with one another by means of slender cytoplasmic extensions. These extensions run through long, slender passageways in the matrix called **canaliculi** (kan-a-LIK-ū-lē; little canals). These passageways form a branching network for the exchange of materials between blood vessels, interstitial fluid, and osteocytes.

Except in joint cavities, where a layer of hyaline cartilage covers bone, the surfaces are sheathed by a **periosteum** (per-ē-OS-tē-um), a covering composed of fibrous (outer) and cellular (inner) layers. The periosteum helps to attach a bone to surrounding tissues and to associated tendons and ligaments. The cellular layer takes part in bone growth and helps with repairs after an injury.

Figure 4–15 Types of Cartilage.

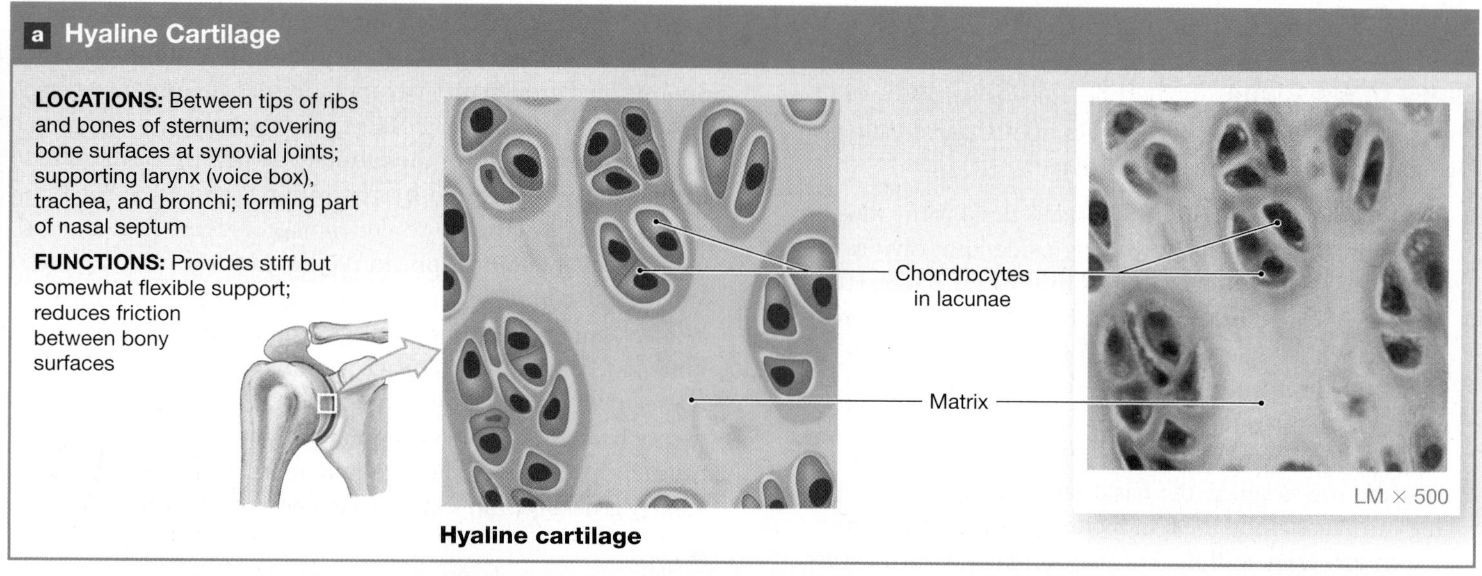
a Hyaline Cartilage
LOCATIONS: Between tips of ribs and bones of sternum; covering bone surfaces at synovial joints; supporting larynx (voice box), trachea, and bronchi; forming part of nasal septum
FUNCTIONS: Provides stiff but somewhat flexible support; reduces friction between bony surfaces
Chondrocytes in lacunae
Matrix
Hyaline cartilage
LM × 500

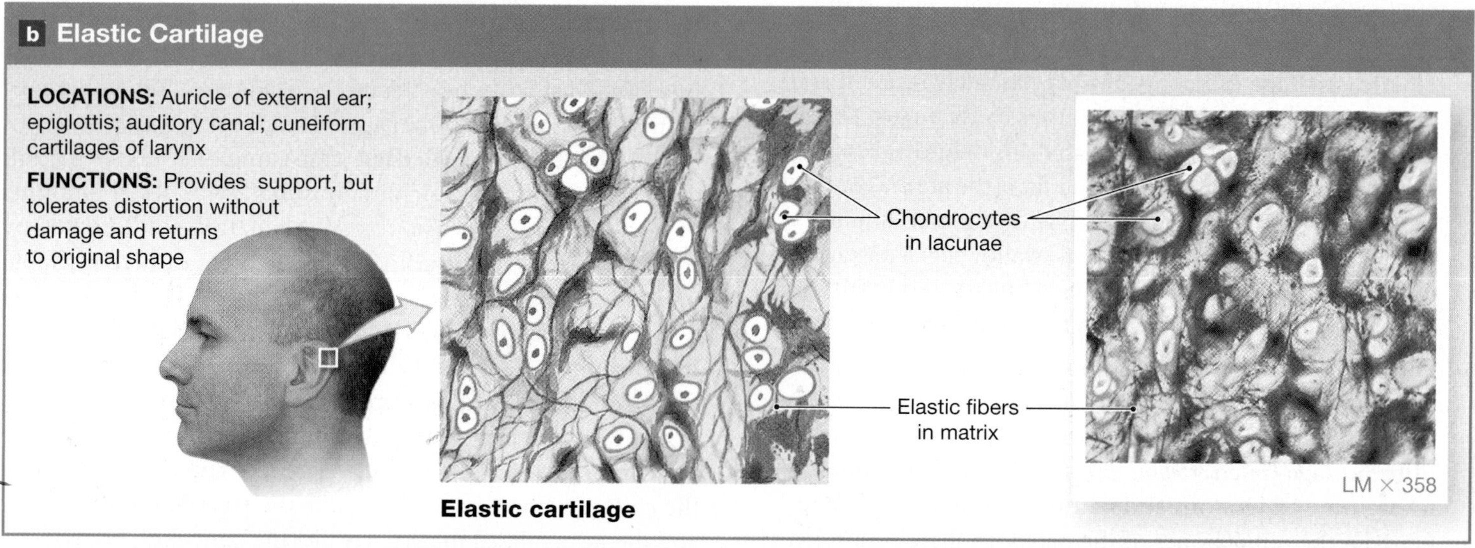
b Elastic Cartilage
LOCATIONS: Auricle of external ear; epiglottis; auditory canal; cuneiform cartilages of larynx
FUNCTIONS: Provides support, but tolerates distortion without damage and returns to original shape
Chondrocytes in lacunae
Elastic fibers in matrix
Elastic cartilage
LM × 358

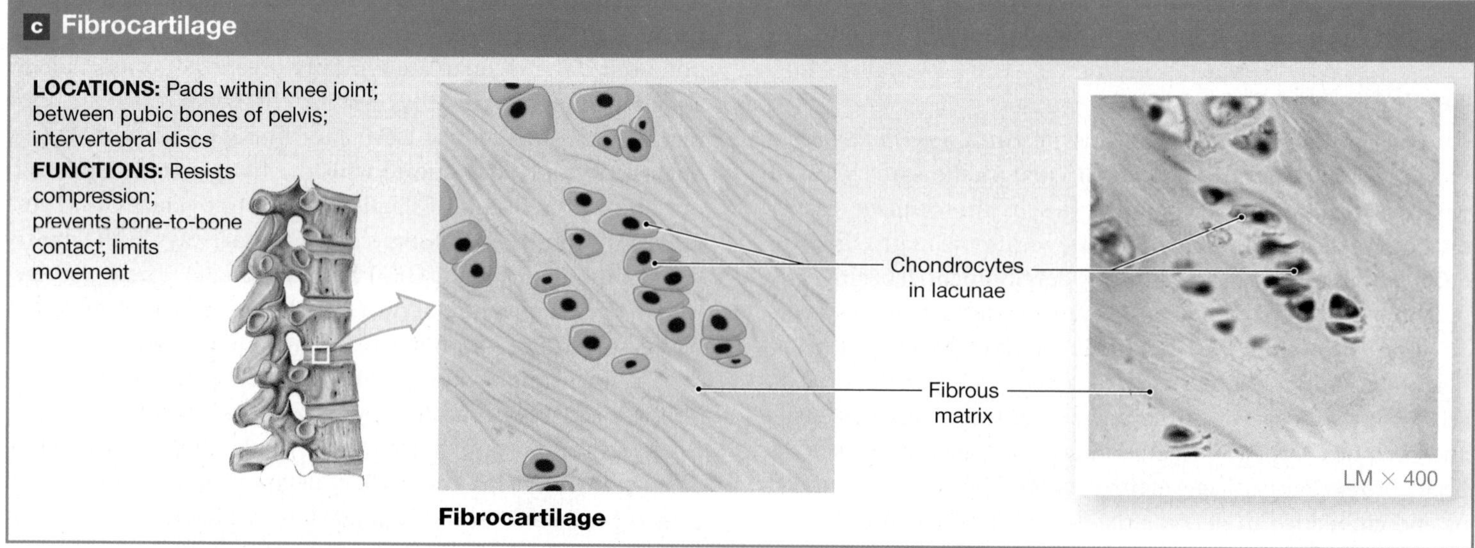
c Fibrocartilage
LOCATIONS: Pads within knee joint; between pubic bones of pelvis; intervertebral discs
FUNCTIONS: Resists compression; prevents bone-to-bone contact; limits movement
Chondrocytes in lacunae
Fibrous matrix
Fibrocartilage
LM × 400

Figure 4–16 The Growth of Cartilage.

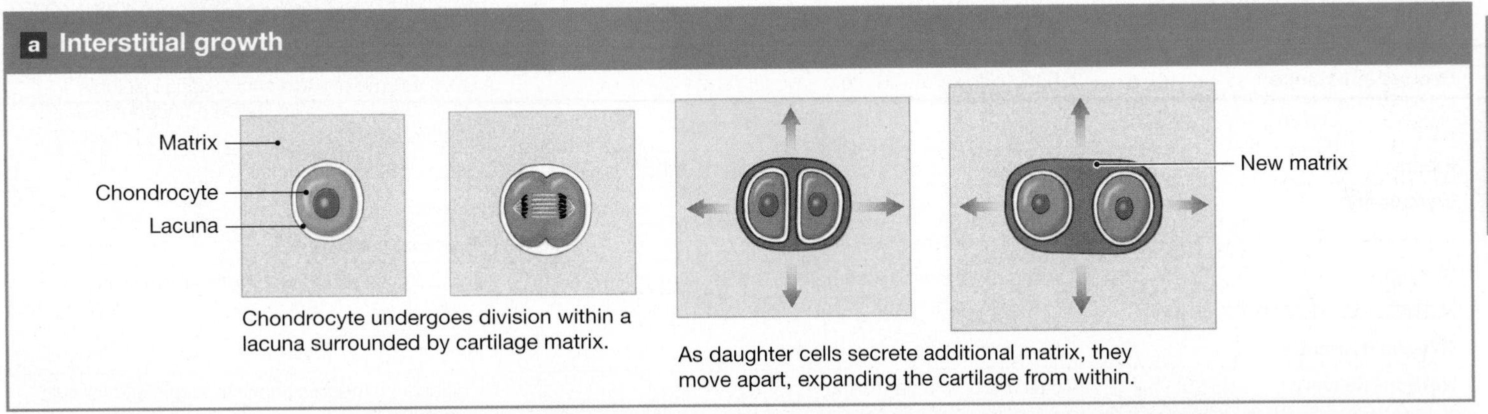
a Interstitial growth
Matrix
Chondrocyte
Lacuna
New matrix
Chondrocyte undergoes division within a lacuna surrounded by cartilage matrix.
As daughter cells secrete additional matrix, they move apart, expanding the cartilage from within.

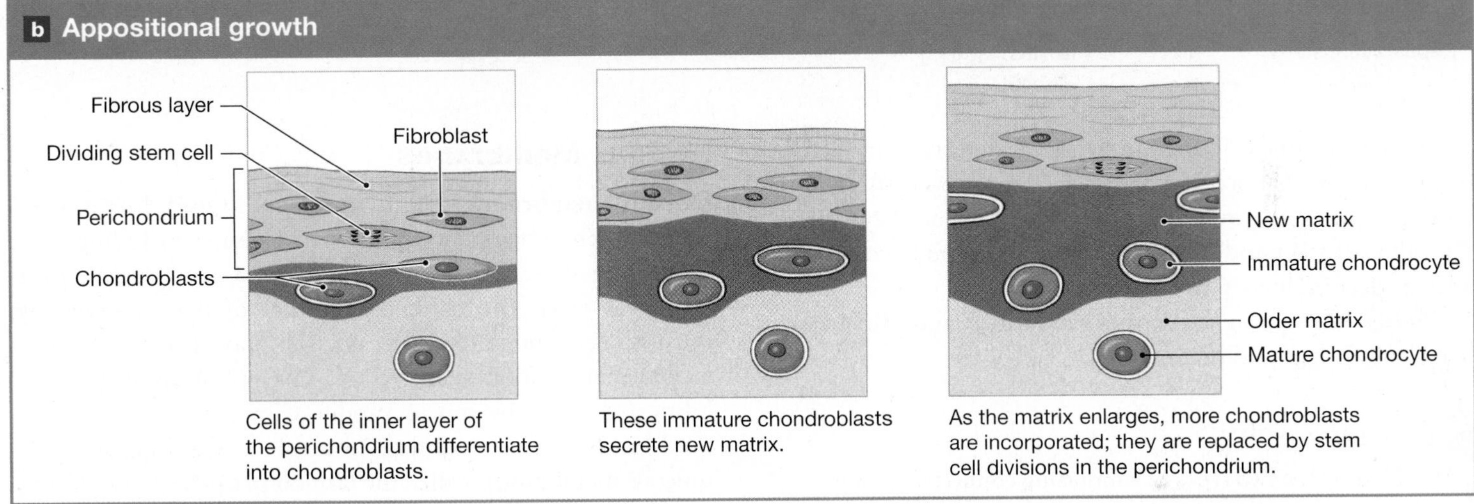
b Appositional growth
Fibrous layer
Dividing stem cell
Perichondrium
Chondroblasts
Fibroblast
New matrix
Immature chondrocyte
Older matrix
Mature chondrocyte
Cells of the inner layer of the perichondrium differentiate into chondroblasts.
These immature chondroblasts secrete new matrix.
As the matrix enlarges, more chondroblasts are incorporated; they are replaced by stem cell divisions in the perichondrium.

? What is the only type of cell found in cartilage matrix?

Figure 4–17 **Bone.** The osteocytes in bone are generally organized in groups around a central canal that contains blood vessels.

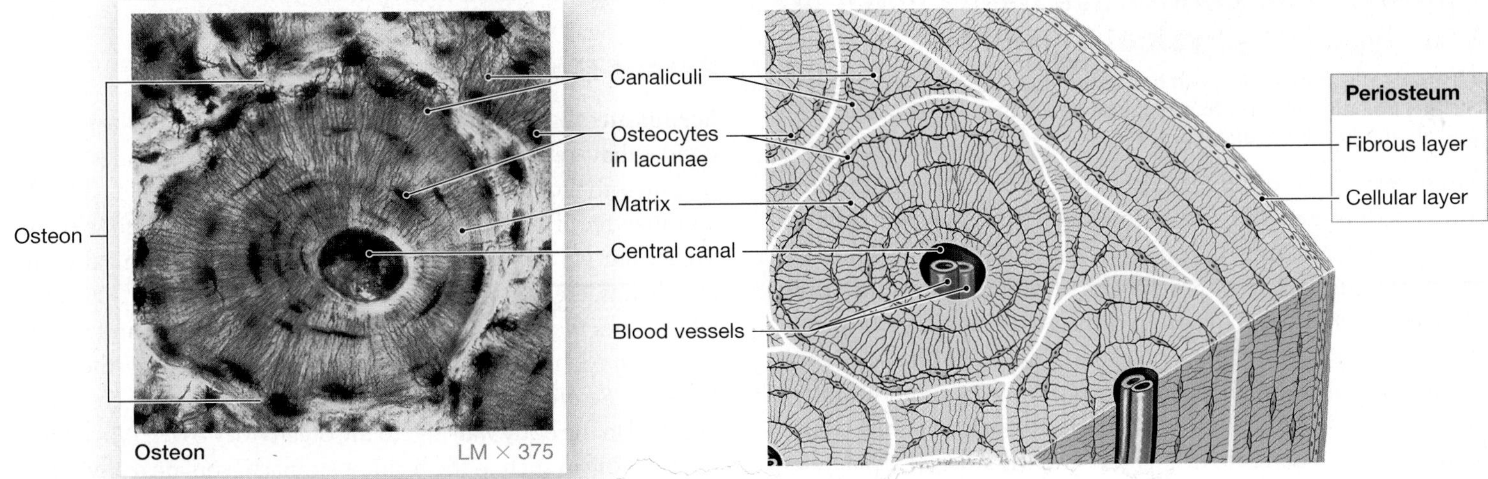
Osteon
Canaliculi
Osteocytes in lacunae
Matrix
Central canal
Blood vessels
Periosteum
Fibrous layer
Cellular layer
Osteon
LM × 375

Table 4–2 A Comparison of Cartilage and Bone

Characteristic	Cartilage	Bone
STRUCTURAL FEATURES		
Cells	Chondrocytes in lacunae	Osteocytes in lacunae
Ground substance	Chondroitin sulfate (in proteoglycans) and water	A small volume of liquid surrounding insoluble crystals of calcium salts (calcium phosphate and calcium carbonate)
Fibers	Collagen, elastic, and reticular fibers (proportions vary)	Collagen fibers predominate
Vascularity	None	Extensive
Covering	Perichondrium (two layers)	Periosteum (two layers)
Strength	Limited: bends easily, but hard to break	Strong: resists distortion until breaking point
METABOLIC FEATURES		
Oxygen demands	Low	High
Nutrient delivery	By diffusion through matrix	By diffusion through cytoplasm and interstitial fluid in canaliculi
Growth	Interstitial and appositional	Appositional only
Repair capabilities	Limited	Extensive

Unlike cartilage, bone undergoes extensive remodeling throughout life. Complete repairs can be made even after severe damage. Bones also respond to the stresses placed on them. They grow thicker and stronger with exercise and become thin and brittle with inactivity.

Table 4–2 summarizes the similarities and differences between cartilage and bone.

✓ Checkpoint

24. **Identify the two types of supporting connective tissue.**
25. **Why does bone heal faster than cartilage?**
26. **If a person has a herniated intervertebral disc, characterized by displacement of the pad of cartilage between the vertebrae, which type of cartilage has been damaged?**

See the blue Answers tab at the back of the book.

4-8 Tissue membranes made from epithelia and connective tissue make up four types of physical barriers

Learning Outcome **Explain how epithelial and connective tissues combine to form four types of tissue membranes, and specify the functions of each.**

There are many different types of anatomical membranes. You encountered plasma membranes that enclose cells in Chapter 3, and you will find many other kinds of membranes in later chapters. Here we are concerned with **tissue membranes** that form physical barriers by lining or covering body surfaces. Each such membrane consists of an epithelium supported by connective tissue. Four types of tissue membranes are found in the body: (1) *mucous membranes*, (2) *serous membranes*, (3) the *cutaneous membrane*, and (4) *synovial membranes* (Figure 4–18).

Mucous Membranes

Mucous membranes, or **mucosae** (mū-KŌ-sē), line passageways and chambers that open to the exterior, including those in the digestive, respiratory, urinary, and reproductive tracts (Figure 4–18a). The epithelial surfaces of these passageways must be kept moist to reduce friction and, in many cases, to facilitate absorption or secretion. The epithelial surfaces are lubricated either by mucus produced by mucous cells, or by fluids, such as urine or semen. The areolar tissue component of a mucous membrane is called the **lamina propria** (PRŌ-prē-uh). We also look at the organization of specific mucous membranes in later chapters.

Many mucous membranes contain simple epithelia that perform absorptive or secretory functions, such as the simple columnar epithelium of the digestive tract. However, other types of epithelia may be involved. For example, a stratified squamous epithelium is part of the mucous membrane of the mouth, and the mucous membrane along most of the urinary tract contains a transitional epithelium.

Serous Membranes

Serous membranes line the sealed, internal cavities of the trunk—cavities that are not open to the exterior. These membranes consist of a mesothelium supported by areolar tissue (Figure 4–18b). As you may recall from Chapter 1, the three types of serous membranes are (1) the *peritoneum*, which lines the peritoneal cavity and covers the surfaces of the enclosed organs; (2) the *pleura*, which lines the pleural cavities and covers the lungs; and (3) the *pericardium*, which lines the pericardial cavity and covers the heart. ⮌ p. 18 Serous membranes are very thin, but they are firmly attached to the body wall and to the organs they cover. When looking at an organ such as the heart or stomach, you are really seeing the tissues of the organ through a transparent serous membrane.

Figure 4–18 Types of Membranes.

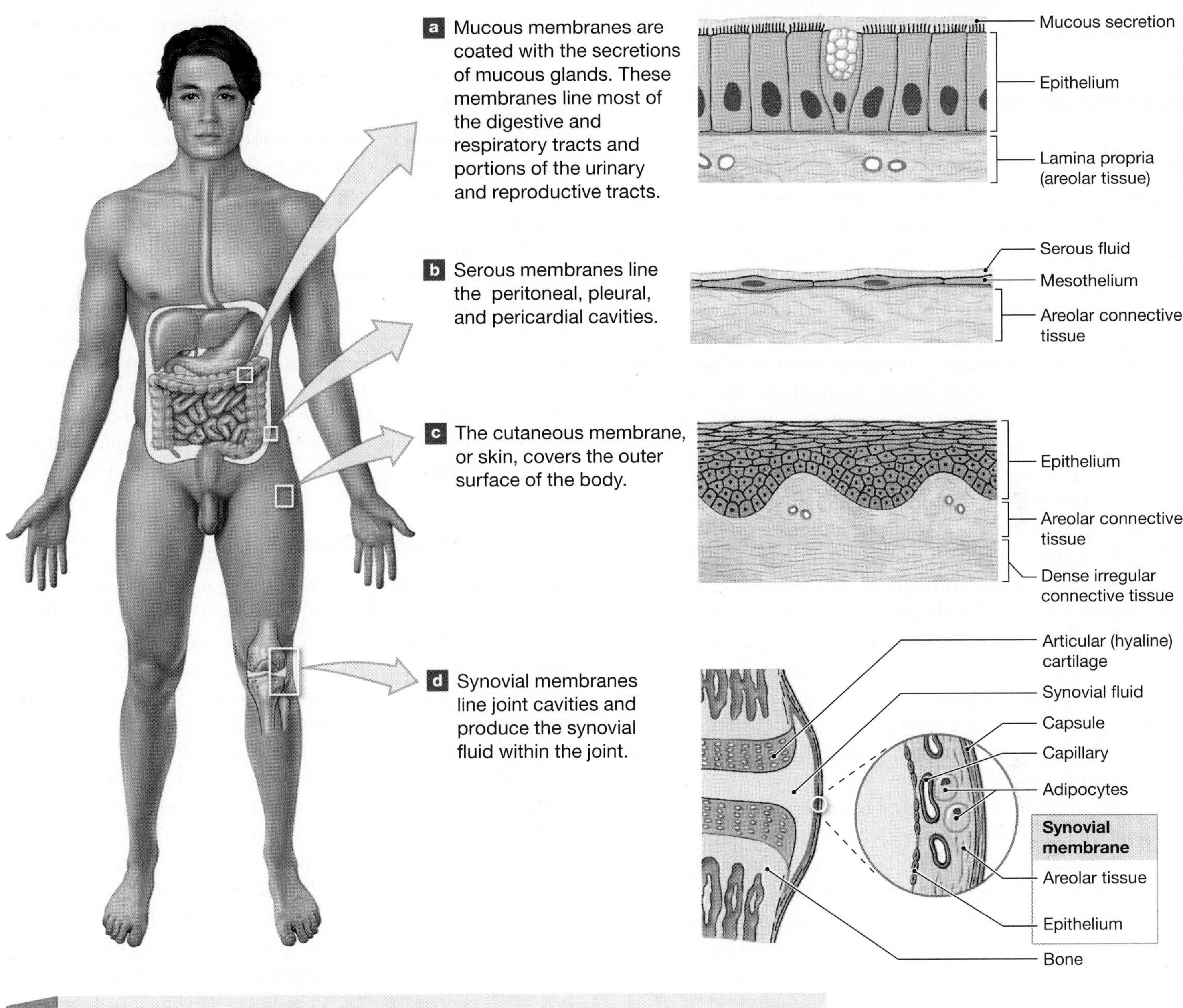
a Mucous membranes are coated with the secretions of mucous glands. These membranes line most of the digestive and respiratory tracts and portions of the urinary and reproductive tracts.
Mucous secretion
Epithelium
Lamina propria (areolar tissue)
b Serous membranes line the peritoneal, pleural, and pericardial cavities.
Serous fluid
Mesothelium
Areolar connective tissue
c The cutaneous membrane, or skin, covers the outer surface of the body.
Epithelium
Areolar connective tissue
Dense irregular connective tissue
d Synovial membranes line joint cavities and produce the synovial fluid within the joint.
Articular (hyaline) cartilage
Synovial fluid
Capsule
Capillary
Adipocytes
Synovial membrane
Areolar tissue
Epithelium
Bone

? What is the name of the epithelium that lines the peritoneal cavity?

The primary function of any serous membrane is to minimize friction between the surfaces that it covers. Each serous membrane can be divided into a *parietal portion*, which lines the inner surface of the cavity, and an opposing *visceral portion*, or **serosa**, which covers the outer surfaces of visceral organs. These organs often move or change shape as they carry out their various functions, and the parietal and visceral surfaces of a serous membrane are in close contact at all times. Friction between these surfaces is kept to a minimum because mesothelia are very thin and permeable, and tissue fluids continuously diffuse onto the exposed surfaces, keeping them moist and slippery.

The slippery fluid secreted by the mesothelium of a serous membrane is called *serous fluid*. However, after an injury or in certain disease states where there is an imbalance of hydrostatic and osmotic forces, the volume of serous fluid, now called *transudate*, may increase dramatically. This change may complicate existing medical problems or produce new ones.

The Cutaneous Membrane

The **cutaneous membrane** is the skin that covers the surface of your body. It consists of a keratinized stratified squamous epithelium and a layer of areolar tissue reinforced by underlying dense irregular connective tissue (Figure 4–18c). In contrast to serous and mucous membranes, the cutaneous membrane is thick, relatively waterproof, and usually dry. We take a closer look at the cutaneous membrane in Chapter 5.

Synovial Membranes

Adjacent bones often interact at joints, or *articulations*. There the two articulating bones are very close together or in contact. Joints that permit significant amounts of movement are complex structures. A fibrous capsule surrounds such a joint, and the ends of the articulating bones lie within a *joint cavity* filled with **synovial** (si-NŌ-vē-ul) **fluid**. The synovial fluid is produced by a **synovial membrane**, which lines the joint cavity (Figure 4–18d).

A synovial membrane consists of two layers: areolar tissue and an atypical or unusual epithelium. The extensive area of areolar tissue contains a matrix of interwoven collagen fibers, proteoglycans, and glycoproteins. An incomplete layer, or sheet, of macrophages and specialized fibroblasts overlies and separates the areolar tissue from the joint cavity. These cells regulate the composition of the synovial fluid. This incomplete layer of cells is often called an epithelium, but it differs from true epithelia in four respects: (1) It develops within a connective tissue; (2) no basement membrane is present; (3) gaps of up to 1 mm may separate adjacent cells; and (4) fluid and solutes are continuously exchanged between the synovial fluid and capillaries in the underlying connective tissue.

Even though a smooth layer of articular (hyaline) cartilage covers the ends of the bones, the surfaces must be lubricated to keep friction from damaging the opposing surfaces. Synovial fluid provides the much-needed lubrication. This fluid is similar in composition to the ground substance in loose connective tissues. Synovial fluid circulates from the areolar tissue into the joint cavity and percolates through the articular cartilages, supplying oxygen and nutrients to the chondrocytes. Joint movement is important in stimulating the formation and circulation of synovial fluid: If a **synovial joint** is immobilized for long periods, the articular cartilages and the synovial membrane undergo degenerative changes.

✓ Checkpoint

27. **Identify the four types of tissue membranes found in the body.**
28. **Which cavities in the body are lined by serous membranes?**
29. **Which type of tissue membrane typically lines body passageways (such as the urinary tract) that open to the exterior?**

See the blue Answers tab at the back of the book.

4-9 The three types of muscle tissue are skeletal, cardiac, and smooth

Learning Outcome Describe the three types of muscle tissue and the special structural features of each type.

Several vital functions involve movement of one kind or another—materials move along the digestive tract, blood flows within the vessels of the cardiovascular system, and the body itself moves from place to place. **Muscle tissue**, which is specialized for contraction, produces movement. Muscle cells have organelles and properties distinct from those of other cells.

There are three types of muscle tissue: (1) *skeletal muscle* forms the large muscles that produce gross body movements; (2) *cardiac muscle*, found in the heart, is responsible for circulating the blood; and (3) *smooth muscle* is found in the walls of visceral organs and a variety of other locations, where it contracts and provides support. The contraction mechanism is similar in all three types of muscle tissue, but the muscle cells differ in internal organization. We examine only general characteristics in this chapter and then discuss each type of muscle more fully in Chapter 10.

Skeletal Muscle Tissue

Skeletal muscle tissue contains very long muscle cells—up to 0.3 m (1 ft) or more in length. The individual muscle cells are usually called **muscle fibers** because they are relatively long and cylindrical (slender). Each muscle fiber has several hundred nuclei just inside the plasma membrane (Figure 4–19a). For this reason, the fibers are described as *multinucleate*.

Skeletal muscle fibers are incapable of dividing. New muscle fibers are produced instead through the divisions of **myosatellite cells** (*muscle satellite cells*), stem cells that persist in adult skeletal muscle tissue. As a result, skeletal muscle tissue can at least partially repair itself after an injury.

As noted in Chapter 3, the cytoskeleton contains actin and myosin filaments. ⮌ p. 72 Actin and myosin interactions produce contractions. In skeletal muscle fibers, however, these filaments are organized into repeating patterns that give the cells a *striated*, or banded, appearance. The *striations*, or bands, are easy to see in light micrographs. Skeletal muscle fibers do not usually contract unless stimulated by nerves, and the nervous system provides voluntary control over their activities. Thus, skeletal muscle is called **striated voluntary muscle**.

Tips & Tools

Associate the sound of the word **stri**ated with the sound of the word **stri**ped.

A *skeletal muscle* is an organ of the muscular system. Muscle tissue predominates, but a muscle contains all four types of body tissue. Within a skeletal muscle, adjacent skeletal muscle fibers are tied together by collagen and elastic fibers that blend

Figure 4–19 **Types of Muscle Tissue.**

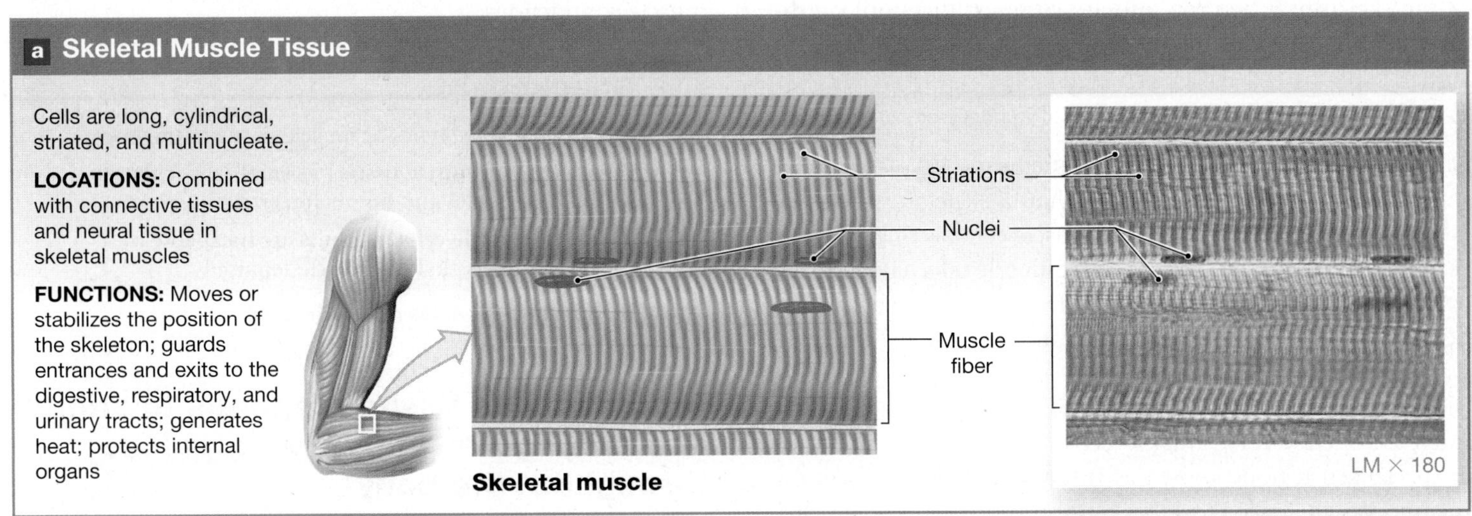
a Skeletal Muscle Tissue
Cells are long, cylindrical, striated, and multinucleate.
LOCATIONS: Combined with connective tissues and neural tissue in skeletal muscles
FUNCTIONS: Moves or stabilizes the position of the skeleton; guards entrances and exits to the digestive, respiratory, and urinary tracts; generates heat; protects internal organs
Striations
Nuclei
Muscle fiber
Skeletal muscle
LM × 180

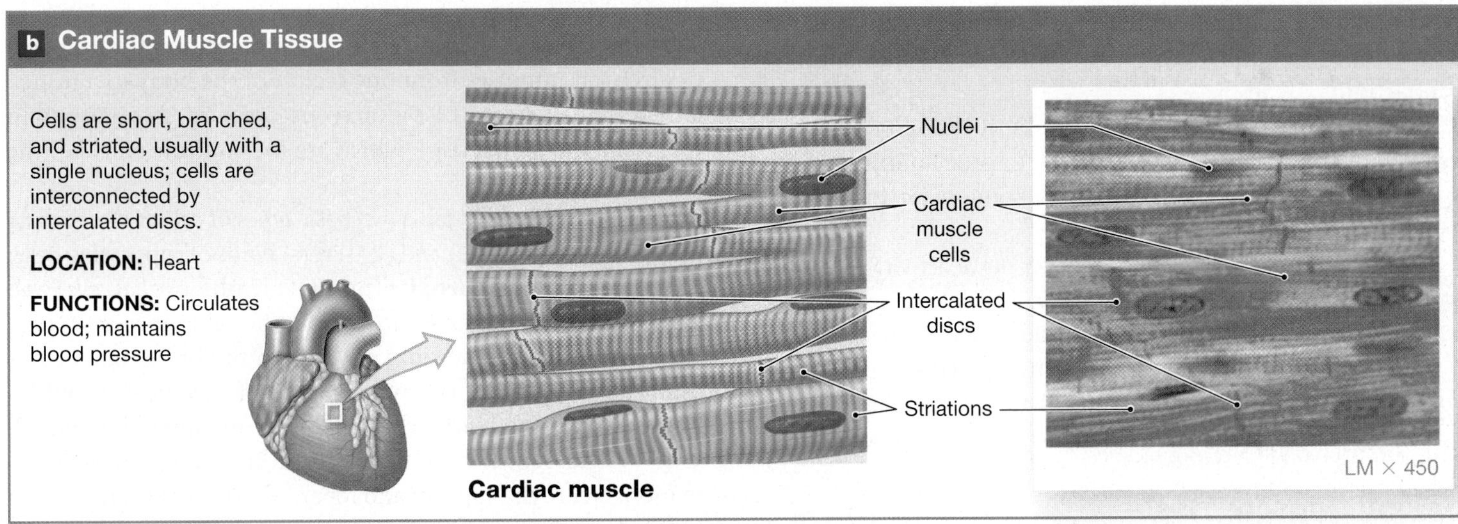
b Cardiac Muscle Tissue
Cells are short, branched, and striated, usually with a single nucleus; cells are interconnected by intercalated discs.
LOCATION: Heart
FUNCTIONS: Circulates blood; maintains blood pressure
Nuclei
Cardiac muscle cells
Intercalated discs
Striations
Cardiac muscle
LM × 450

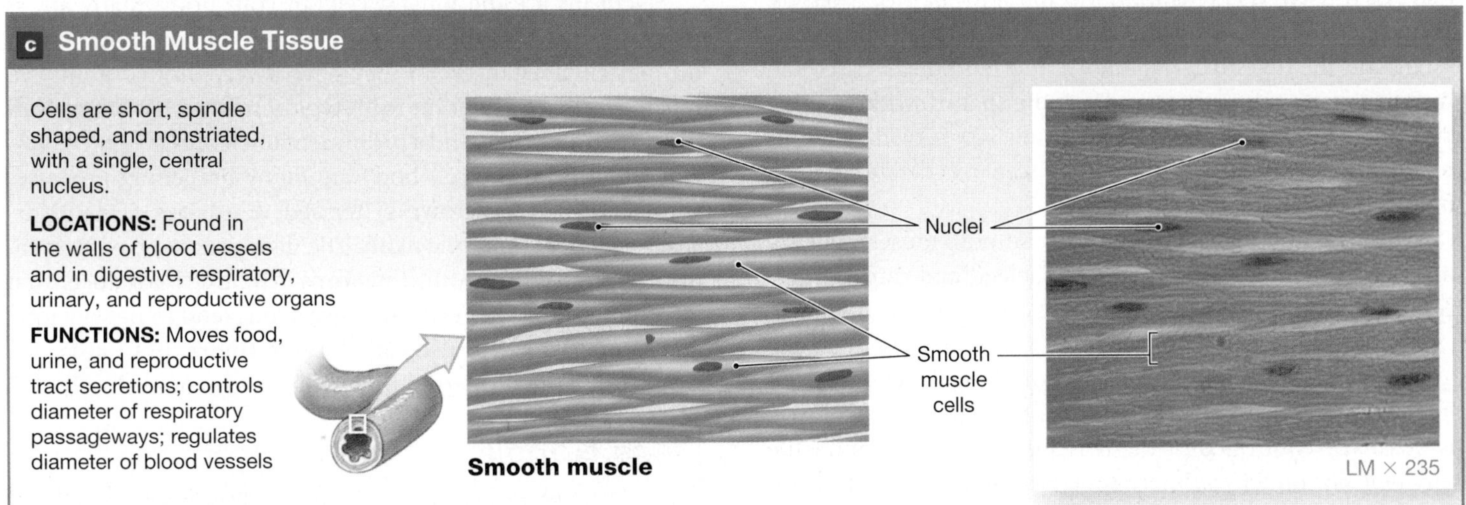
c Smooth Muscle Tissue
Cells are short, spindle shaped, and nonstriated, with a single, central nucleus.
LOCATIONS: Found in the walls of blood vessels and in digestive, respiratory, urinary, and reproductive organs
FUNCTIONS: Moves food, urine, and reproductive tract secretions; controls diameter of respiratory passageways; regulates diameter of blood vessels
Nuclei
Smooth muscle cells
Smooth muscle
LM × 235

? Which muscle tissue consists of multinucleate cells?

into the attached tendon or aponeurosis. The tendon or aponeurosis conducts the force of contraction, often to a bone of the skeleton. When the muscles contract, they pull on the attached bone, producing movement.

Cardiac Muscle Tissue

Cardiac muscle tissue is found in the heart. A typical cardiac muscle cell is thinner and shorter than a skeletal muscle cell (Figure 4–19b). A typical cardiac muscle cell has one centrally positioned nucleus, but some cardiac muscle cells have as many as five. Prominent striations resemble those of skeletal muscle because the actin and myosin filaments are arranged the same way in the cells of both tissue types.

Cardiac muscle tissue consists of a branching network of interconnected cardiac muscle cells. The cells are connected at specialized regions known as **intercalated discs**. There the membranes are locked together by desmosomes, proteoglycans, and gap junctions. Ion movement through gap junctions helps synchronize the contractions of the cardiac muscle cells, and the desmosomes and proteoglycans lock the cells together during a contraction.

Cardiac muscle tissue has a very limited ability to repair itself. Some cardiac muscle cells do divide after an injury to the heart, but the repairs are incomplete. Some heart function is usually lost.

Cardiac muscle cells do not rely on nerve activity to start a contraction. Instead, specialized cardiac muscle cells called *pacemaker cells* set a regular rate of contraction. The nervous system can alter the rate of pacemaker cell activity, but it does not provide voluntary control over individual cardiac muscle cells. For this reason, cardiac muscle is called **striated involuntary muscle**.

Smooth Muscle Tissue

Smooth muscle tissue is found in the walls of blood vessels, around hollow organs such as the urinary bladder, and in layers around the digestive, respiratory, and reproductive tracts. A smooth muscle cell is short and spindle shaped, with tapering ends and a single, central nucleus (Figure 4–19c). These cells can divide, so smooth muscle tissue can regenerate after an injury.

The actin and myosin filaments in smooth muscle cells are organized differently from those of skeletal and cardiac muscles. One result is that smooth muscle tissue has no striations.

Smooth muscle cells may contract on their own, with gap junctions between adjacent cells coordinating the contractions of individual cells. The contraction of some smooth muscle tissue can be controlled by the nervous system, but contractile activity is not under voluntary control. [Imagine the effort you would need to exert conscious control over the smooth muscles along the 8 m (26 ft) of digestive tract, not to mention the miles of blood vessels!] Smooth muscle is known as **nonstriated involuntary muscle** because of its appearance and the fact that the nervous system usually does not voluntarily control smooth muscle contractions.

✓ Checkpoint

30. **Identify the three types of muscle tissue in the body.**
31. **Which type of muscle tissue has small, tapering cells with single nuclei and no obvious striations?**
32. **If skeletal muscle cells in adults are incapable of dividing, how is skeletal muscle repaired?**

See the blue Answers tab at the back of the book.

4-10 Nervous tissue responds to stimuli and propagates electrical impulses throughout the body

Learning Outcome Discuss the basic structure and role of nervous tissue.

Nervous tissue is specialized for the propagation (movement) of electrical impulses from one region of the body to another. Ninety-eight percent of the nervous tissue in the body is in the brain and spinal cord, which are the control centers of the nervous system.

Nervous tissue contains two basic types of cells: (1) **neurons** (NŪ-ronz; *neuro*, nerve) and (2) several kinds of supporting cells, collectively called **neuroglia** (nū-ROG-lē-uh), or *glial cells* (*glia*, glue). Our conscious and unconscious thought processes reflect the communication among neurons in the brain. Such communication involves the propagation of electrical impulses, in the form of reversible changes in the membrane potential. ⮌ p. 99 The frequency of the electrical impulses convey information. Neuroglia support and repair nervous tissue and supply nutrients to neurons.

Neurons are the longest cells in your body. Many are as long as a meter (39 in.)! Most neurons cannot divide under normal circumstances, so they have a very limited ability to repair themselves after injury. A typical neuron has a large **cell body** with a nucleus and a prominent nucleolus (Figure 4–20). Extending from the cell body are many branching processes (projections or outgrowths) termed **dendrites** (DEN-drītz; *dendron*, a tree), and one **axon**. The dendrites receive information, typically from other neurons. The axon conducts that information to other cells. Because axons tend to be very long and slender, they are also called **nerve fibers**. In Chapter 12, we examine the properties of nervous tissue more closely.

Tips & Tools

To remember the direction of information flow in a neuron, associate the "t" in dendri**t**e with "**t**o" and the "a" in **a**xon with "**a**way."

Go to MasteringA&P™ > Study Area > Menu > Lab Tools > PAL 3.0 > Histology: Muscle Tissue

Figure 4–20 **Nervous Tissue.**

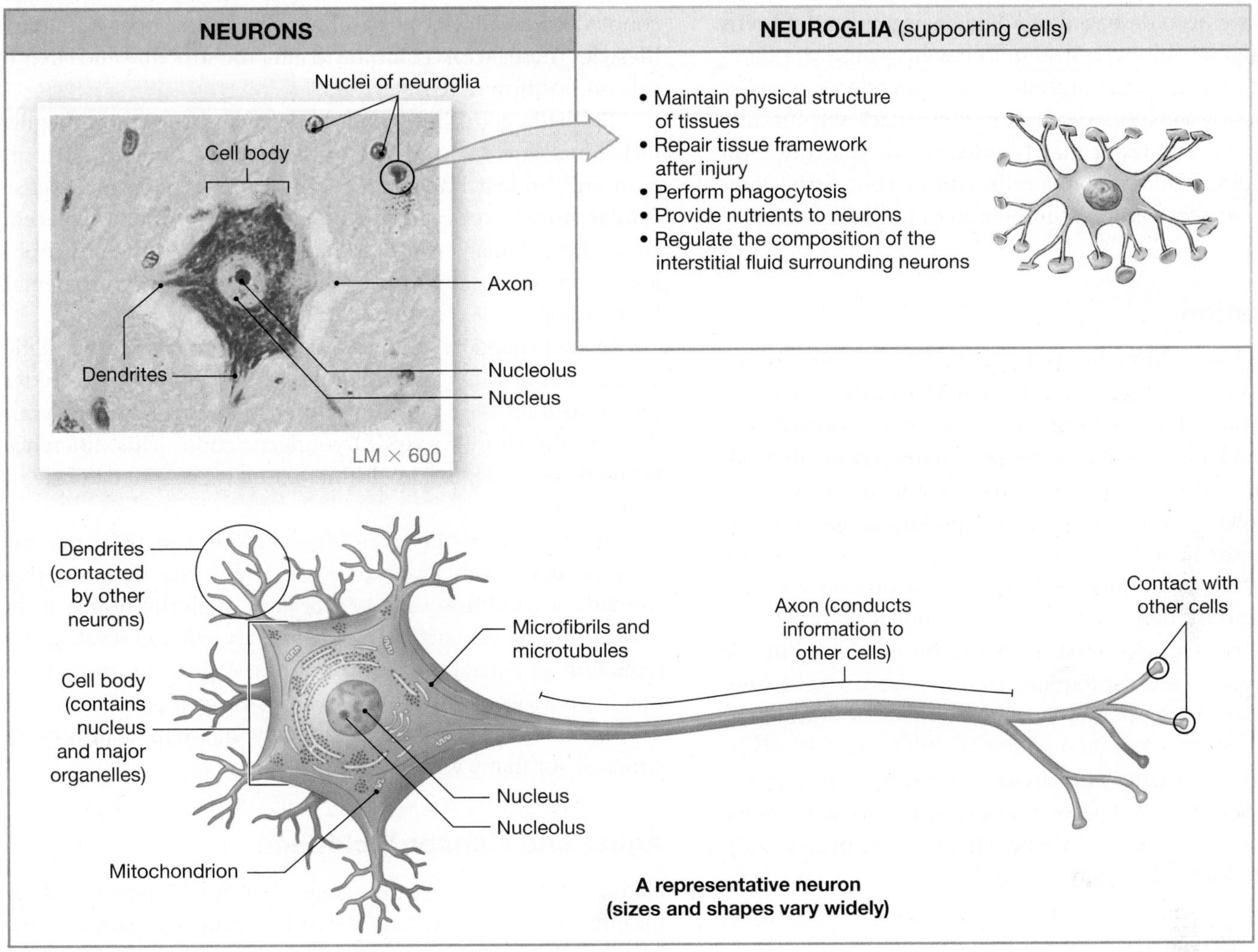

NEURONS
Nuclei of neuroglia
Cell body
Axon
Dendrites
Nucleolus
Nucleus
LM × 600
NEUROGLIA (supporting cells)
• Maintain physical structure of tissues
• Repair tissue framework after injury
• Perform phagocytosis
• Provide nutrients to neurons
• Regulate the composition of the interstitial fluid surrounding neurons
Dendrites (contacted by other neurons)
Cell body (contains nucleus and major organelles)
Mitochondrion
Microfibrils and microtubules
Nucleus
Nucleolus
Axon (conducts information to other cells)
Contact with other cells
A representative neuron (sizes and shapes vary widely)

Checkpoint

33. **A tissue contains irregularly shaped cells with many fibrous projections, some several centimeters long. These are probably which type of cell?**

See the blue Answers tab at the back of the book.

4-11 The response to tissue injury involves inflammation and regeneration

Learning Outcome **Describe how injuries affect the tissues of the body.**

Now let's consider what happens after an injury, focusing on the interaction among different tissues. Our example includes connective tissue (blood), epithelium (the endothelia of blood vessels), muscle tissue (smooth in the vessel walls), and nervous tissue (sensory nerve endings). **Inflammation**, or the **inflammatory response**, is a process that isolates the injured area while damaged cells, tissue components, and any dangerous microorganisms, which could cause *infection*, are cleaned up. **Regeneration** is the repair process that restores normal function after inflammation has subsided. In later chapters, especially Chapters 5 and 22, we consider inflammation and regeneration in more detail.

Inflammation

Stimuli that can produce inflammation include impact, abrasion, distortion, chemical irritation, infection by pathogenic organisms (such as bacteria or viruses), and extreme temperatures (hot or cold). Each of these stimuli kills cells, damages connective tissue fibers, or injures the tissue in some way. Such changes alter the chemical composition of the interstitial fluid: Damaged cells release prostaglandins, proteins, and potassium ions, and the injury itself may have introduced foreign proteins or pathogens into the body. Damaged connective tissue activates mast cells, which release chemicals that stimulate the inflammatory response.

Go to MasteringA&P™ > Study Area > Menu > Lab Tools > PAL 3.0 > Histology: Nervous Tissue

Tissue conditions soon become even more abnormal. **Necrosis** (ne-KRŌ-sis) is the tissue destruction that takes place after cells have been damaged or killed. It begins several hours after the original injury. Lysosomal enzymes cause the damage. Through widespread autolysis, lysosomes release enzymes that first destroy the injured cells and then attack surrounding tissues. ⮌ p. 80 The result may be **pus**, which is a collection of debris, fluid, dead and dying cells, and necrotic tissue components. An **abscess** is an accumulation of pus in an enclosed tissue space.

Regeneration

Each organ has a different ability to regenerate after injury. This ability is directly linked to the pattern of tissue organization in the injured organ. Epithelia, connective tissues (except cartilage), and smooth muscle tissue usually regenerate well, but skeletal and cardiac muscle tissues and nervous tissue regenerate relatively poorly, if at all. The skin, which is dominated by epithelia and connective tissues, regenerates rapidly and completely after injury. (We consider the process in Chapter 5.) In contrast, damage to the heart is much more serious. The connective tissues of the heart can be repaired, but the majority of damaged cardiac muscle cells are replaced only by fibrous tissue. The permanent replacement of normal tissue by fibrous tissue is called *fibrosis* (fī-BRŌ-sis). Fibrosis in muscle and other tissues may occur in response to injury, disease, or aging. **Spotlight Figure 4–21** shows the tissue response to injury and the process of tissue regeneration; please study this figure before going on.

Checkpoint

34. **Identify the two phases in the response to tissue injury.**
35. **List the four common signs of inflammation.**

See the blue Answers tab at the back of the book.

4-12 With advancing age, tissue regeneration decreases and cancer rates increase

Learning Outcome Describe how aging affects the tissues of the body.

Aging has two important effects on tissues: The body's ability to repair damage to tissues decreases, and cancer is more likely to occur.

Aging and Tissue Structure

Tissues change with age, and the speed and effectiveness of tissue regeneration decrease. Repair and maintenance activities throughout the body slow down, and the rate of energy consumption in general decreases. All these changes reflect various hormonal alterations that take place with age, often coupled with reduced physical activity and a more sedentary lifestyle. These factors combine to alter the structure and chemical composition of many tissues.

Epithelia get thinner and connective tissues more fragile. Individuals bruise more easily and bones become brittle. Joint pain and broken bones are common in the elderly. Because cardiac muscle cells and neurons are not normally replaced, cumulative damage can eventually cause major health problems, such as cardiovascular disease or deterioration in mental functioning.

In later chapters, we consider the effects of aging on specific organs and systems. For example, the chondrocytes of older individuals produce a slightly different form of proteoglycan than do the chondrocytes of younger people. This difference probably accounts for the thinner and less resilient cartilage of older people.

In some cases, the tissue degeneration can be temporarily slowed or even reversed. Age-related reduction in bone strength, a condition called *osteoporosis*, typically results from a combination of inactivity, a low dietary calcium level, and a reduction in circulating sex hormones. A program of exercise that includes weight-bearing activity, calcium supplements, and in some cases medication can generally maintain healthy bone structure for many years.

Aging and Cancer Incidence

Cancer rates increase with age, and about 25 percent of all people in the United States develop cancer at some point in their lives. An estimated 70–80 percent of cancer cases result from chemical exposure, environmental factors, or some combination of the two, and 40 percent of those cancers are caused by cigarette smoke. Each year in the United States, more than 500,000 individuals die of cancer, making it second only to heart disease as a cause of death.We discussed the development and growth of cancer in Chapter 3. ⮌ p. 107

With this chapter, you have concluded the introductory portion of this text. In combination, the four basic tissue types described here form all of the organs and body systems we discuss in subsequent chapters.

Checkpoint

36. **Identify some age-related factors that affect tissue repair and structure.**
37. **What would account for the observed increase in cancer rates with age?**

See the blue Answers tab at the back of the book.

SPOTLIGHT

Figure 4–21

Inflammation and Regeneration

Tissues are not isolated, and they combine to form organs with diverse functions. Therefore, any injury affects several types of tissue simultaneously. To preserve homeostasis, the tissues must respond in a coordinated way. The restoration of homeostasis involves two related processes: inflammation and regeneration.

Mast Cell Activation

When an injury damages connective tissue, mast cells release a variety of chemicals. This process, called **mast cell activation**, stimulates inflammation.

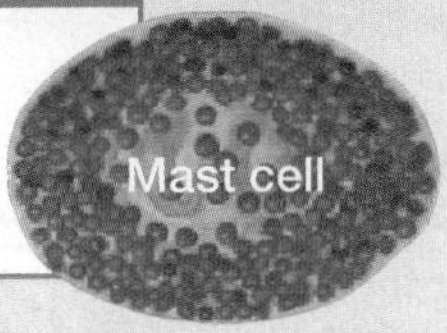

Mast cell

Histamine
Heparin
Prostaglandins

stimulates

Exposure to Pathogens and Toxins

Injured tissue contains an abnormal concentration of pathogens, toxins, wastes, and the chemicals from injured cells.

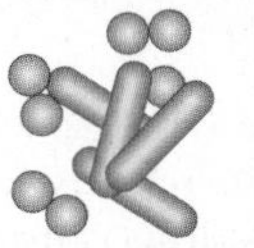

When a tissue is injured, a general defense mechanism is activated.

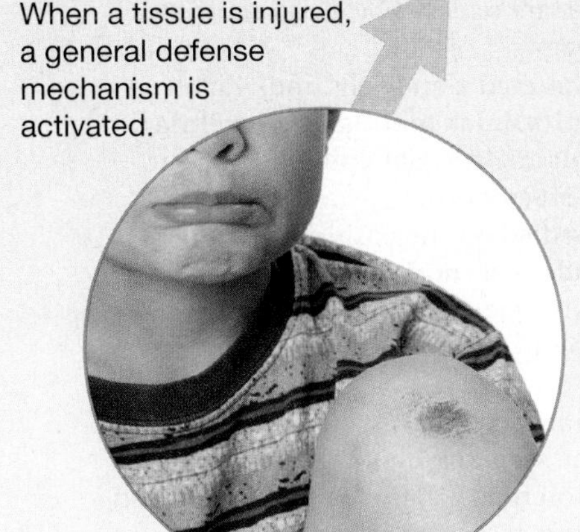

INFLAMMATION

Inflammation produces several familiar indications of injury. These indications are the so-called *cardinal signs of inflammation*: redness, heat (warmth), swelling, pain, and sometimes loss of function. Inflammation may also result from the presence of pathogens, such as harmful bacteria, within the tissues. The presence of these pathogens constitutes an **infection**.

Increased Blood Flow	Increased Vessel Permeability	Pain
In response to the released chemicals, blood vessels dilate, increasing blood flow through the damaged tissue. 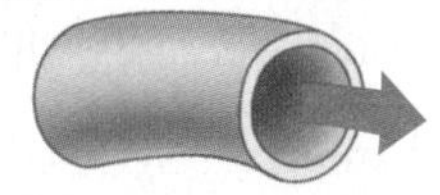	Vessel dilation is accompanied by an increase in the permeability of the capillary walls. Plasma now diffuses into the injured tissue, so the area becomes swollen. 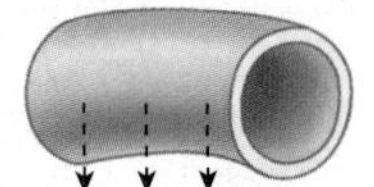	The abnormal conditions within the tissue and the chemicals released by mast cells stimulate nerve endings that produce the sensation of pain. PAIN

Increased Local Temperature	Increased Oxygen and Nutrients	Increased Phagocytosis	Removal of Toxins and Wastes
The increased blood flow and permeability cause the tissue to become warm and red. 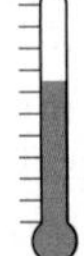	Vessel dilation, increased blood flow, and increased vessel permeability result in enhanced delivery of oxygen and nutrients. 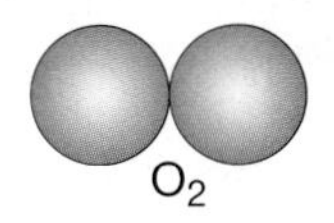 O_2	Phagocytes in the tissue are activated, and they begin engulfing tissue debris and pathogens. 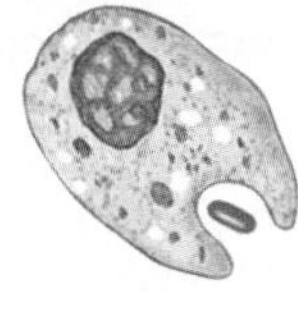 	Enhanced circulation carries away toxins and wastes, distributing them to the kidneys for excretion, or to the liver for inactivation.  Toxins and wastes

Inflammation Subsides

Over a period of hours to days, the cleanup process generally succeeds in eliminating the inflammatory stimuli.

Regeneration

Regeneration is the repair that occurs after the damaged tissue has been stabilized and the inflammation has subsided. Fibroblasts move into the area, laying down a collagenous framework known as **scar tissue**. Over time, scar tissue is usually "remodeled" and gradually assumes a more normal appearance.

Inhibits mast cell activation

Normal tissue conditions restored

4 Chapter Review

Study Outline

MasteringA&P™ Access more chapter study tools online in the MasteringA&P Study Area:
- Chapter Quizzes, Chapter Practice Test, MP3 Tutor Sessions, and Clinical Case Studies
- Practice Anatomy Lab PAL 3.0
- A&P Flix A&PFlix
- Interactive Physiology iP2
- PhysioEx PhysioEx 9.1

An Introduction to the Tissue Level of Organization p. 115

1. Tissues are structures with discrete structural and functional properties that combine to form organs.

4-1 The four tissue types are epithelial, connective, muscle, and nervous p. 115

2. **Tissues** are collections of specialized cells and cell products that perform a relatively limited number of functions. The four *tissue types* are *epithelial tissue, connective tissue, muscle tissue,* and *nervous tissue. (Figure 4–1)*
3. **Histology** is the study of tissues.

4-2 Epithelial tissue covers body surfaces, lines internal surfaces, and serves essential functions p. 115

4. **Epithelial tissue** includes epithelia and glands. An **epithelium** is an **avascular** layer of cells that forms a barrier that provides protection and regulates permeability. **Glands** are secretory structures derived from epithelia. Epithelial cells may show **polarity**, the presence of structural and functional differences between their exposed (apical) and attached (basal) surfaces.
5. A **basement membrane** attaches epithelia to underlying connective tissues.
6. Epithelia provide physical protection, control permeability, provide sensation, and produce specialized secretions. Gland cells are epithelial cells that produce secretions. In **glandular epithelia**, most cells produce secretions.
7. Epithelial cells are specialized to perform secretory or transport functions and to maintain the physical integrity of the epithelium. *(Figure 4–2)*
8. Many epithelial cells have microvilli.
9. The coordinated beating of motile cilia on a **ciliated epithelium** moves materials across the epithelial surface.
10. Cells can attach to other cells or to extracellular protein fibers by means of **cell adhesion molecules (CAMs)** or at specialized attachment sites called **cell junctions**. The three major types of cell junctions are **tight junctions**, **gap junctions**, and **desmosomes**. *(Figure 4–3)*
11. The base of each epithelium is connected to a two-part basement membrane consisting of a *basal lamina* and a *reticular lamina*. Divisions by stem cells continually replace the short-lived epithelial cells.

4-3 Cell shape and number of layers determine the classification of epithelia p. 120

12. Epithelia are classified on the basis of the number of cell layers and the shape of the cells at the apical surface.
13. A **simple epithelium** has a single layer of cells covering the basement membrane; a **stratified epithelium** has several layers. The cells in a **squamous epithelium** are thin and flat. Cells in a **cuboidal epithelium** resemble hexagonal boxes; those in a **columnar epithelium** are taller and more slender. *(Table 4–1; Figures 4–4 to 4–6)*
14. Epithelial cells (or structures derived from epithelial cells) that produce secretions are called *glands*. **Exocrine glands** discharge secretions onto the body surface or into **ducts**, which communicate with the exterior. *Hormones*, the secretions of **endocrine glands**, are released by gland cells into the surrounding bloodstream.
15. In epithelia that contain scattered gland cells, individual secretory cells are called **unicellular glands**. **Multicellular glands** are organs that contain glandular epithelia that produce exocrine or endocrine secretions.
16. Exocrine glands can be classified on the basis of structure as **unicellular exocrine glands** or as **multicellular exocrine glands**. Multicellular exocrine glands can be further classified according to structure. **Goblet cells** and **mucous cells** secrete mucin. *(Figure 4–7)*
17. A glandular epithelial cell may release its secretions by merocrine, apocrine, or holocrine secretory methods. In **merocrine secretion**, the most common method, the product is released through exocytosis. **Apocrine secretion** involves the loss of both the secretory product and cytoplasm. Unlike the other two methods, **holocrine secretion** destroys the gland cell, which becomes packed with secretions and then bursts. *(Figure 4–8)*

4-4 Connective tissue has varied roles in the body that reflect the physical properties of its three main types p. 126

18. **Connective tissues** are internal tissues with many important functions: establishing a structural framework; transporting fluids and dissolved materials; protecting delicate organs; supporting, surrounding, and interconnecting tissues; storing energy reserves; and defending the body from microorganisms.
19. All connective tissues contain specialized cells and a **matrix**, composed of extracellular protein fibers and a **ground substance**.
20. **Connective tissue proper** is connective tissue that contains varied cell populations and fiber types surrounded by a syrupy ground substance.
21. **Fluid connective tissues** have distinctive populations of cells suspended in a watery matrix that contains dissolved proteins. The two types of fluid connective tissues are *blood* and *lymph*.
22. **Supporting connective tissues** have a less diverse cell population than connective tissue proper and a dense matrix with closely packed fibers. The two types of supporting connective tissues are *cartilage* and *bone*.

4-5 Connective tissue proper includes loose connective tissues that fill internal spaces and dense connective tissues that contribute to the internal framework of the body p. 128

23. Connective tissue proper contains fibers, a viscous ground substance, and a varied population of cells, including **fibroblasts**, **fibrocytes**, **macrophages**, **adipocytes**, **mesenchymal cells**, **melanocytes**, **mast cells**, **lymphocytes**, and **microphages**. *(Figure 4–9)*
24. The three types of fibers in connective tissue are **collagen fibers**, **reticular fibers**, and **elastic fibers**.
25. The first connective tissue to appear in an embryo is **mesenchyme**, or *embryonic connective tissue. (Figure 4–10)*
26. Connective tissue proper is classified as either **loose connective tissue** or **dense connective tissue**. Loose connective tissues are mesenchyme and **mucous connective tissues** in the embryo; **areolar tissue**; **adipose tissue**, including **white fat** and **brown fat**; and **reticular tissue**. Most of the volume in dense connective tissue consists of fibers. The two types of dense connective tissue are **dense regular connective tissue** and **dense irregular connective tissue**. **Elastic tissue** is a dense regular connective tissue. *(Figures 4–11, 4–12)*
27. A network of connective tissue proper ties internal organs and systems together. This network consists of the **superficial fascia** (*subcutaneous layer*, separating the skin from underlying tissues and organs), the **deep fascia** (dense regular connective tissue), and the **subserous fascia** (the layer between the deep fascia and the serous membranes that line true body cavities). *(Figure 4–13)*

4-6 Blood and lymph are fluid connective tissues that transport cells and dissolved materials p. 135

28. **Blood** and **lymph** are connective tissues that contain distinctive collections of cells in a fluid matrix.
29. Blood contains *formed elements:* **red blood cells** (*erythrocytes*), **white blood cells** (*leukocytes*), and **platelets**. The watery matrix of blood is called **plasma**. *(Figure 4–14)*
30. *Arteries* carry blood away from the heart and toward *capillaries*, where water and small solutes move into the interstitial fluid of surrounding tissues. *Veins* return blood to the heart.
31. **Lymph** forms as interstitial fluid enters the lymphatic vessels, which return lymph to the cardiovascular system.

4-7 The supporting connective tissues cartilage and bone provide a strong framework p. 136

32. Cartilage and bone are called supporting connective tissues because they support the rest of the body.
33. **Chondrocytes**, or cartilage cells, rely on diffusion through the avascular matrix to obtain nutrients.
34. The matrix of cartilage is a firm gel that contains **chondroitin sulfates** (used to form proteoglycans) and chondrocytes. Chondrocytes occupy chambers called **lacunae**. A fibrous **perichondrium** separates cartilage from surrounding tissues. The three types of cartilage are **hyaline cartilage**, **elastic cartilage**, and **fibrocartilage**. *(Figure 4–15)*
35. **Cartilage** grows by two mechanisms: **interstitial growth** and **appositional growth**. *(Figure 4–16)*
36. **Bone**, or **osseous tissue**, consists of **osteocytes**, little ground substance, and a dense, mineralized matrix. Osteocytes are situated in lacunae. The matrix consists of calcium salts and collagen fibers, giving it unique properties. *(Figure 4–17; Table 4–2)*
37. Osteocytes depend on diffusion within **canaliculi** for nutrient intake.
38. Each bone is surrounded by a **periosteum** with fibrous and cellular layers.

4

4-8 Tissue membranes made from epithelia and connective tissue make up four types of physical barriers p. 140

39. Membranes form a barrier or interface. Epithelia and connective tissues combine to form membranes that cover and protect other structures and tissues. *(Figure 4–18)*
40. **Mucous membranes** line cavities that communicate with the exterior. They contain areolar tissue called the **lamina propria**.
41. **Serous membranes** line the body's sealed internal cavities of the trunk. They secrete a fluid called serous fluid.
42. The **cutaneous membrane**, or skin, covers the body surface.
43. **Synovial membranes** form an incomplete lining within the cavities of synovial joints and secrete **synovial fluid**.

4-9 The three types of muscle tissue are skeletal, cardiac, and smooth p. 142

44. **Muscle tissue** is specialized for contraction. *(Figure 4–19)*
45. The cells of **skeletal muscle tissue** are *multinucleate*. Skeletal muscle, or **striated voluntary muscle**, produces new fibers by the division of **myosatellite cells**, a kind of stem cell.
46. The **cardiac muscle cells** of **cardiac muscle tissue** occur only in the heart. Cardiac muscle, or **striated involuntary muscle**, relies on *pacemaker cells* to stimulate regular contraction.
47. **Smooth muscle tissue**, or **nonstriated involuntary muscle**, is not striated. Smooth muscle cells can divide and therefore regenerate after injury.

4-10 Nervous tissue responds to stimuli and propagates electrical impulses throughout the body p. 144

48. **Nervous tissue** propagates electrical impulses, which convey information from one area of the body to another.
49. Cells in nervous tissue are either neurons or neuroglia. **Neurons** transmit information as electrical impulses. Several kinds of **neuroglia** (*glial cells*) exist, and their basic functions include supporting nervous tissue and helping supply nutrients to neurons. *(Figure 4–20)*
50. A typical neuron has a **cell body**, **dendrites**, and an **axon**, or **nerve fiber**. The axon carries information to other cells.

4-11 The response to tissue injury involves inflammation and regeneration p. 145

51. Any injury affects several types of tissue simultaneously, and the tissues respond in a coordinated manner. Two processes restore homeostasis: inflammation and regeneration.
52. **Inflammation**, or the **inflammatory response**, isolates the injured area while damaged cells, tissue components, and any dangerous microorganisms (which could cause **infection**) are cleaned up. **Regeneration** is the repair process that restores normal function. *(Spotlight Figure 4–21)*

4-12 With advancing age, tissue regeneration decreases and cancer rates increase p. 146

53. Tissues change with age. Repair and maintenance become less efficient, and the structure and chemical composition of many tissues are altered.
54. The incidence of cancer increases with age, with roughly three-quarters of all cases caused by exposure to chemicals or by other environmental factors, such as cigarette smoke.

Review Questions

See the blue Answers tab at the back of the book.

LEVEL 1 Reviewing Facts and Terms

1. Identify the six types of epithelial tissue shown in the drawing below.

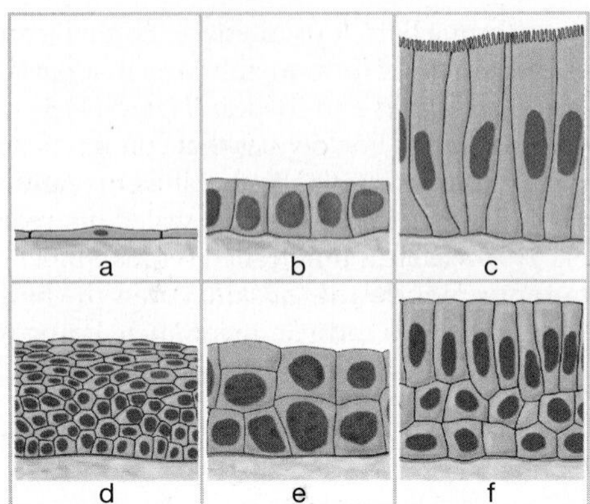
a
b
c
d
e
f

(a)________ (b)________
(c)________ (d)________
(e)________ (f)________

2. Collections of specialized cells and cell products that perform a relatively limited number of functions are called **(a)** cellular aggregates, **(b)** tissues, **(c)** organs, **(d)** organ systems, **(e)** organisms.
3. Tissue that is specialized for contraction is **(a)** epithelial tissue, **(b)** muscle tissue, **(c)** connective tissue, **(d)** nervous tissue.
4. A type of cell junction common in cardiac and smooth muscle tissues is the **(a)** hemidesmosome, **(b)** basal junction, **(c)** tight junction, **(d)** gap junction.
5. The most abundant connections between cells in the superficial layers of the skin are **(a)** connexons, **(b)** gap junctions, **(c)** desmosomes, **(d)** tight junctions.
6. A ________ membrane has an epithelium that is stratified and supported by dense connective tissue. **(a)** synovial, **(b)** serous, **(c)** cutaneous, **(d)** mucous.
7. Mucous secretions that coat the passageways of the digestive and respiratory tracts result from ________ secretion. **(a)** apocrine, **(b)** merocrine, **(c)** holocrine, **(d)** endocrine.
8. Matrix is a characteristic of which type of tissue? **(a)** epithelial, **(b)** nervous, **(c)** muscle, **(d)** connective.
9. Which of the following epithelia most easily permits diffusion? **(a)** stratified squamous, **(b)** simple squamous, **(c)** transitional, **(d)** simple columnar.
10. Functions of connective tissue include **(a)** establishing a structural framework for the body, **(b)** storing energy reserves, **(c)** providing protection for delicate organs, **(d)** all of these, **(e)** a and c only.
11. The three major types of cartilage in the body are **(a)** collagen, reticular, and elastic, **(b)** areolar, adipose, and reticular, **(c)** hyaline, elastic, and fibrous, **(d)** tendons, reticular, and elastic.
12. The primary function of serous membranes in the body is to **(a)** minimize friction between opposing surfaces, **(b)** line cavities that communicate with the exterior, **(c)** perform absorptive and secretory functions, **(d)** cover the surface of the body.
13. The type of cartilage growth characterized by adding new layers of cartilage to the surface is **(a)** interstitial growth, **(b)** appositional growth, **(c)** intramembranous growth, **(d)** longitudinal growth.
14. Tissue changes with age can result from **(a)** hormonal changes, **(b)** increased need for sleep, **(c)** improper nutrition, **(d)** all of these, **(e)** a and c only.
15. Axons, dendrites, and a cell body are characteristic of cells located in **(a)** nervous tissue, **(b)** muscle tissue, **(c)** connective tissue, **(d)** epithelial tissue.
16. The repair process necessary to restore normal function after inflammation subsides in damaged tissues is **(a)** isolation, **(b)** regeneration, **(c)** reconstruction, **(d)** all of these.
17. What are the four essential functions of epithelial tissue?
18. Differentiate between endocrine and exocrine glands.
19. By what three methods do exocrine glandular epithelial cells release their secretions?
20. List three basic components of connective tissues.
21. What are the four kinds of membranes composed of epithelial and connective tissue that cover and protect other structures and tissues in the body?
22. What two cell populations make up nervous tissue? What is the function of each?

LEVEL 2 Reviewing Concepts

23. What is the difference between an exocrine secretion and an endocrine secretion?
24. A significant structural feature in the digestive system is the presence of tight junctions near the exposed surfaces of cells lining the digestive tract. Why are these junctions so important?
25. Describe the fluid connective tissues in the human body. What are the main differences between fluid connective tissues and supporting connective tissues?
26. Why are infections always a serious threat after a severe burn or abrasion?
27. A layer of glycoproteins and a network of fine protein filaments that prevents the movement of proteins and other large molecules from the connective tissue to the epithelium describe **(a)** interfacial canals, **(b)** the basement membrane, **(c)** the reticular lamina, **(d)** areolar tissue, **(e)** squamous epithelium.
28. Why does damaged cartilage heal slowly? **(a)** Chondrocytes cannot be replaced if killed, and other cell types must take their place. **(b)** Cartilage is avascular, so nutrients and other molecules must diffuse to the site of injury. **(c)** Damaged cartilage becomes calcified, thus blocking the movement of materials required for healing. **(d)** Chondrocytes divide more slowly than other cell types, delaying the healing process. **(e)** Damaged collagen cannot be quickly replaced, thereby slowing the healing process.
29. List the similarities and differences among the three types of muscle tissue.

LEVEL 3 Critical Thinking and Clinical Applications

30. Assuming that you had the necessary materials to perform a detailed chemical analysis of body secretions, how could you determine whether a secretion was merocrine or apocrine?
31. During a lab practical, a student examines a tissue that is composed of densely packed protein fibers that run parallel to each other and form a cord. There are no striations, but small nuclei are visible. The student identifies the tissue as skeletal muscle. Why is the student's choice wrong, and what tissue is he probably observing?
32. While in a chemistry lab, Jim accidentally spills a small amount of a caustic (burning) chemical on his arm. What changes in the characteristics of the skin would you expect to observe, and what would cause these changes?

4

+ CLINICAL CASE Wrap-Up The Rubber Girl

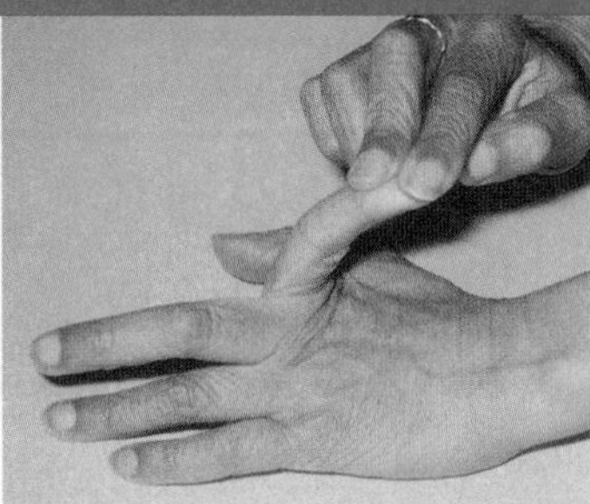

Anne Marie suffers from a variant of *Ehlers-Danlos* (Ā-lerz DAHN-los) *syndrome*—a rare inherited disorder of connective tissue. Specifically, the collagen fibers in the matrix of connective tissue are defective. Because connective tissue and collagen occur throughout the body, many tissues are involved, particularly the joints and skin.

Anne Marie's loose joints and dislocations are due to deficient dense connective tissue. Because of abnormal collagen fibers, both her ligaments (which connect bone to bone) and her tendons (which connect muscle to bone) are abnormally loose and heal poorly when overstretched. When she twists her knee and feels as if her kneecap is dislocating, it is, indeed, sliding out of its groove. The ligaments that protect her ankles have been repeatedly sprained.

Thin, translucent skin that is soft, velvety, and stretchy is also typical of Ehlers-Danlos syndrome. Scars appear wide, fragile, and abnormal. Bruises and skin tears occur easily.

1. Connective tissues are made up of specialized cells and a matrix of extracellular protein fibers and a viscous fluid known as ground substance. What part of connective tissue proper is defective in Ehlers-Danlos syndrome?
2. Do you think this genetic syndrome affects just Anne Marie's joints and skin, or would you suspect other tissues are affected as well?

See the blue Answers tab at the back of the book.

Related Clinical Terms

autopsy: An examination of a dead body to discover the cause of death or the extent of disease.

biopsy: An examination of tissue removed from a living body to discover the presence, cause, or extent of disease.

cachexia: Weakness and wasting of the body due to severe chronic illness.

carcinoma: A cancer arising in the epithelial tissue of the skin or of the lining of the internal organs.

immunotherapy: The prevention or treatment of disease with substances that stimulate the immune response.

lesion: A region in an organ or tissue that has suffered damage from injury or disease; a wound, ulcer, abscess, or tumor, for example.

metaplasia: A reversible structural change that alters the character of a tissue.

pathologist: A physician who specializes in the study of disease processes in tissues and body fluids.

remission: Abatement, ending, or lessening in severity of the signs and symptoms of a disease.

sarcoma: A malignant tumor of connective or other nonepithelial tissue.

tissue engineering: The process of growing tissue either inside or outside of a body to be transplanted into a patient or used for testing.

tissue rejection: Occurs when a transplant recipient's immune system attacks a transplanted organ or tissue.

tissue transplantation: Moving tissues (or organs) from one body and placing them into another body via medical procedures for the purpose of replacing the recipient's damaged or failing tissue (or organ).

tumor grading: A system used to classify cancer cells in terms of how abnormal they look under a microscope and how quickly the tumor is likely to grow and spread.

tumor staging: Defining at what point a patient is in the development of a malignant disease when the diagnosis is made.

xenotransplant: The process of grafting or transplanting organs or tissues between members of different species.

5 The Integumentary System

Learning Outcomes

These Learning Outcomes correspond by number to this chapter's sections and indicate what you should be able to do after completing the chapter.

5-1 ■ Describe the main structural features of the epidermis, and explain the functional significance of each. p. 153

5-2 ■ Describe the structures and functions of the dermis. p. 158

5-3 ■ Describe the structures and functions of the subcutaneous layer. p. 160

5-4 ■ Explain what accounts for individual differences in skin color, and discuss the response of melanocytes to sunlight exposure. p. 161

5-5 ■ Describe the interaction between sunlight and vitamin D_3 production. p. 163

5-6 ■ Describe the mechanisms that produce hair, and explain the structural basis for hair texture and color. p. 165

5-7 ■ Discuss the various kinds of glands in the skin, and list the secretions of those glands. p. 168

5-8 ■ Describe the anatomical structures of nails, and explain how they are formed. p. 171

5-9 ■ Explain how the skin responds to injury and repairs itself. p. 172

5-10 ■ Summarize the effects of aging on the skin. p. 175

CLINICAL CASE He Has Fish Skin!

I shook his hand and immediately I knew something was different about him. When Will clasped my hand between both of his, I felt like my hand was sandwiched between two sheets of thick, shaggy sandpaper. There was none of the moistness or warmth of a usual handshake. These hands belonged to Grandpa Will.

Grandpa Will's grandsons adored him, and the feeling was mutual! They lured him into chasing them around the backyard. Because it was a hot summer day, play lasted all of 15 minutes and then Grandpa brought the gang back to the air-conditioned comfort of the house. He sank back into the recliner. He was flushed and breathing hard, but his shirt stayed dry and crisp—there wasn't a bead of sweat visible on him. The boys climbed onto his lap, laughing, as he encircled them with those coarse hands. "Oh, Grandpa, you feel like a fish!" **What is happening with Grandpa Will's integumentary system? To find out, turn to the Clinical Case Wrap-Up on p. 179.**

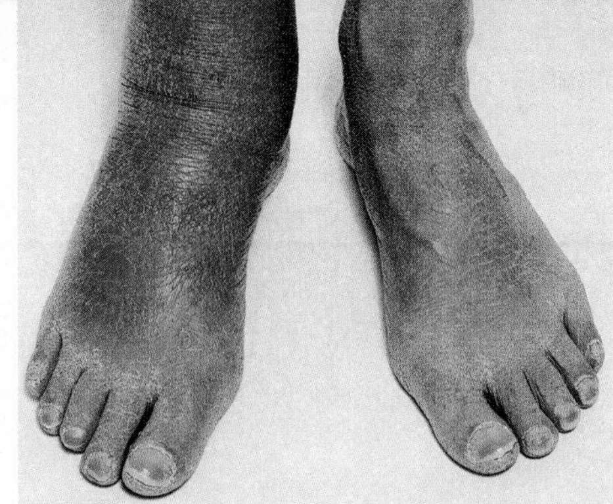

An Introduction to the Integumentary System

You are probably more familiar with the skin than with any other organ system. No other organ system is as accessible, large, and underappreciated as the **integumentary system**. Often referred to simply as the **integument** (in-TEG-ū-ment), this system makes up about 16 percent of your total body weight. Its surface, 1.5–2 m^2 (16.1–21.5 sq. ft.) in area, is continually abraded, attacked by microorganisms, irradiated by sunlight, and exposed to environmental chemicals. The integumentary system is your body's first line of defense against an often hostile environment. It's the place where you and the outside world meet.

The integumentary system has two major parts: the **cutaneous membrane**, or skin, and the **accessory structures** (Figure 5–1).

1. The cutaneous membrane has two components: the *epidermis* (*epi-*, above), or superficial epithelium, and the *dermis*, an underlying area of connective tissues.
2. The accessory structures include *hair* and *hair follicles*, *exocrine glands*, and *nails*. They are embedded in the dermis and project up to or above the surface of the epidermis.

The integument does not function in isolation. An extensive network of blood vessels branches through the dermis. Nerve fiber endings and *sensory receptors* monitor touch, pressure, temperature, and pain, providing valuable information to the central nervous system about the state of the body.

Deep to the dermis is a layer of loose connective tissue called the *subcutaneous layer* (*hypodermis*). The subcutaneous layer separates the integument from the deep fascia around other organs, such as muscles and bones. ⮌ p. 133 Although the subcutaneous layer is often considered separate from the integument, we will consider it in this chapter because its connective tissue fibers are interwoven with those of the dermis.

The general functions of the integumentary system, which are summarized in Figure 5–2, include the following:

- *Protection* of underlying tissues and organs against impact, abrasion, fluid loss, and chemical attack.
- *Excretion* of salts, water, and organic wastes by glands.
- *Maintenance of normal body temperature* through either insulation or evaporative cooling, as needed.
- *Production of melanin*, which protects underlying tissue from ultraviolet (UV) radiation.
- *Production of keratin*, which protects against abrasion and repels water.
- *Synthesis of vitamin D_3*, a steroid that is converted to *calcitriol*, a hormone important to normal calcium ion metabolism.
- *Storage of lipids* in adipocytes in the dermis and in adipose tissue in the subcutaneous layer.
- *Detection* of touch, pressure, pain, vibration, and temperature stimuli, and the relaying of that information to the nervous system. (We consider these *general senses*, which provide information about the external environment, in Chapter 15.)
- *Coordination of the immune response* to pathogens and cancers in the skin.

5-1 The epidermis is a protective covering composed of layers with various functions

Learning Outcome **Describe the main structural features of the epidermis, and explain the functional significance of each.**

The epidermis is a stratified squamous epithelium. Recall from Chapter 4 that such an epithelium provides physical protection for the dermis, prevents water loss, and helps keep microorganisms outside the body. ⮌ p. 121

Figure 5–1 The Components of the Integumentary System.

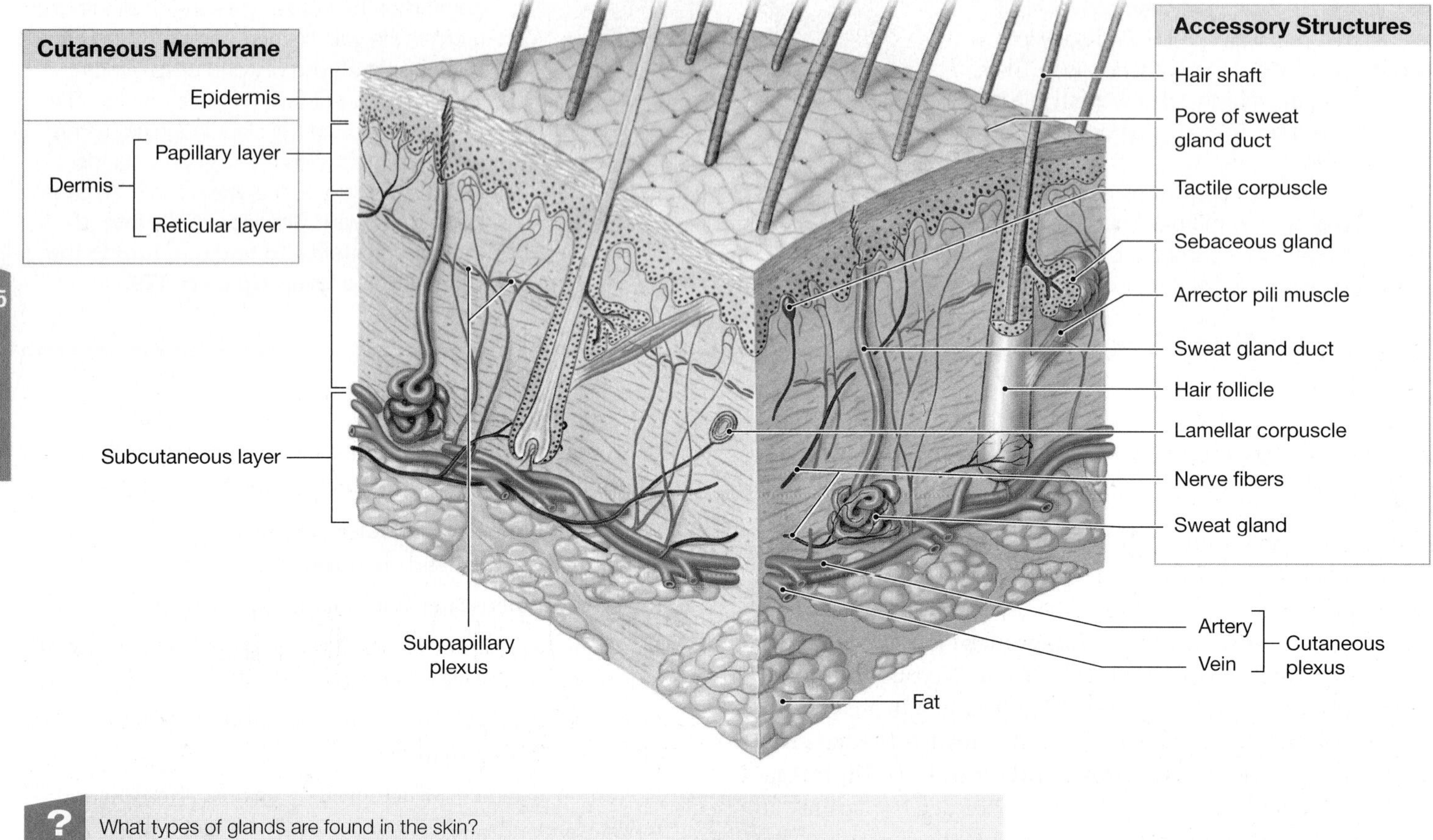
Cutaneous Membrane
Epidermis
Dermis
Papillary layer
Reticular layer
Subcutaneous layer
Subpapillary plexus
Fat
Accessory Structures
Hair shaft
Pore of sweat gland duct
Tactile corpuscle
Sebaceous gland
Arrector pili muscle
Sweat gland duct
Hair follicle
Lamellar corpuscle
Nerve fibers
Sweat gland
Artery
Vein
Cutaneous plexus

? What types of glands are found in the skin?

Like all epithelia, the epidermis is avascular. Because there are no local blood vessels, epidermal cells rely on the diffusion of nutrients and oxygen from capillaries within the dermis. As a result, the epidermal cells with the highest metabolic demands are found close to the basement membrane, where the diffusion distance is short. The superficial cells, far removed from the source of nutrients, are dead. **Spotlight Figure 5–3a** shows and describes the basic organization of the epidermis.

Cells of the Epidermis: Keratinocytes

Keratinocytes (ke-RAT-i-nō-sītz), the body's most abundant epithelial cells, dominate the epidermis. These cells form several layers and contain large amounts of the protein **keratin** (KER-a-tin; *keros*, horn). Keratin is a tough, fibrous protein that is also the basic structural component of hair and nails in humans. ⮌ p. 55

Layers of the Epidermis

The two types of skin, thin and thick, differ in their number of layers of keratinocytes. **Thin skin**, which contains four layers of keratinocytes, covers most of the body surface. **Thick skin**, which has five layers, is found on the palms of the hands and the soles of the feet. Note that the terms *thin* and *thick* refer to the relative thickness of the epidermis, not to the cutaneous membrane as a whole. **Spotlight Figure 5–3b** shows the five layers of keratinocytes in a section of the epidermis in an area of thick skin.

The boundaries between the layers are often difficult to see in a standard light micrograph. The various layers have Latin names beginning with the word *stratum* (plural, *strata*), which means "layer." The second part of the name refers to the function or appearance of the layer. The strata, in order from the basement membrane toward the free surface, are the *stratum basale*, the *stratum spinosum*, the *stratum granulosum*, the *stratum lucidum*, and the *stratum corneum*.

Stratum Basale

The deepest layer of the epidermis is the **stratum basale** (STRA-tum buh-SAL-āy, "basal layer") or *stratum germinativum* (jer-mi-na-TĒ-vum). Hemidesmosomes attach the cells of this layer to the basement membrane that separates the epidermis from the areolar tissue of the dermis. ⮌ p. 118 The stratum basale and the underlying dermis interlock, increasing the

Figure 5–2 **The Cutaneous Membrane and Accessory Structures of the Integumentary System.**

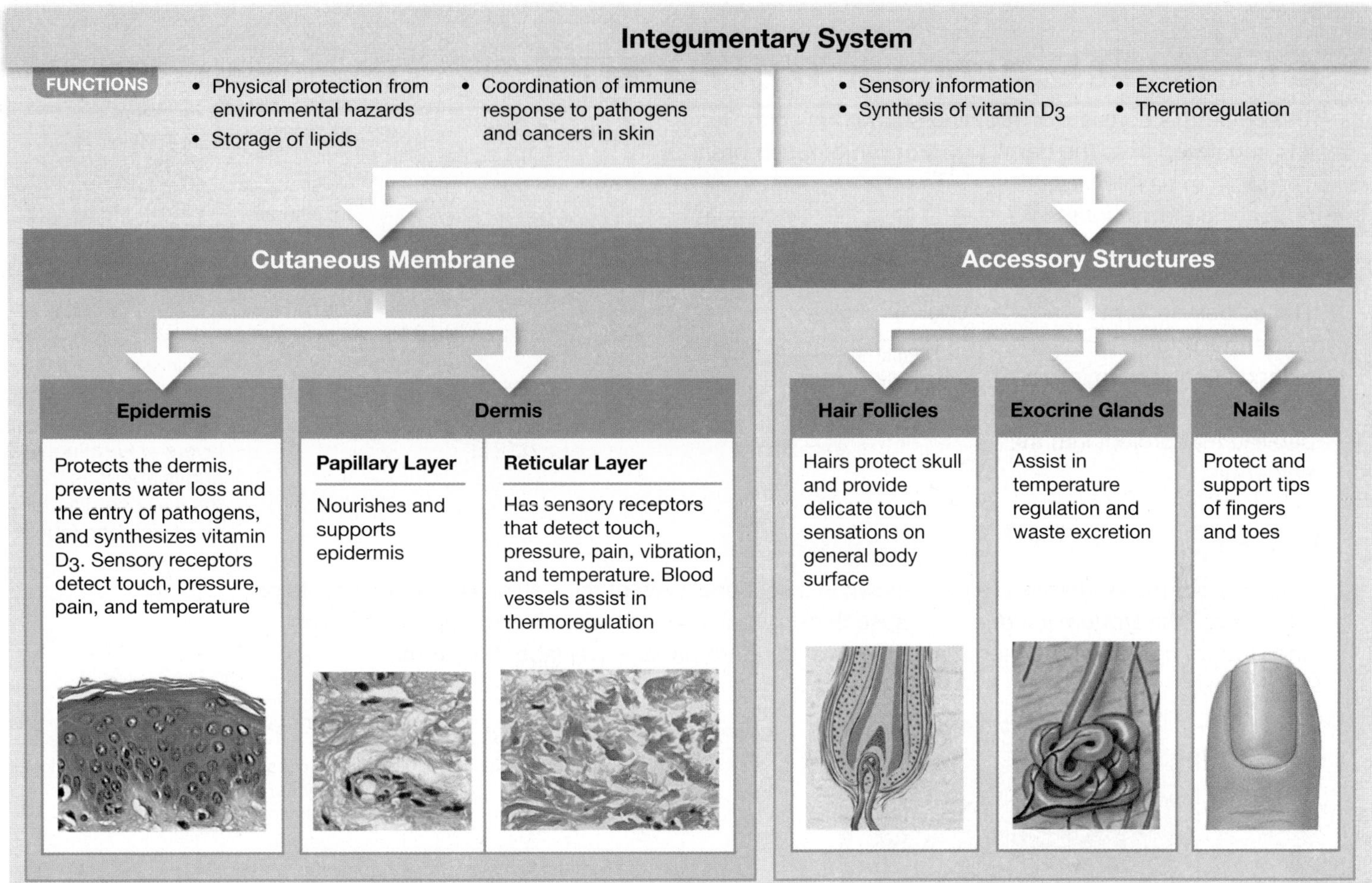
Integumentary System
FUNCTIONS
• Physical protection from environmental hazards
• Storage of lipids
• Coordination of immune response to pathogens and cancers in skin
• Sensory information
• Synthesis of vitamin D3
• Excretion
• Thermoregulation
Cutaneous Membrane
Accessory Structures
Epidermis
Protects the dermis, prevents water loss and the entry of pathogens, and synthesizes vitamin D3. Sensory receptors detect touch, pressure, pain, and temperature
Dermis
Papillary Layer
Nourishes and supports epidermis
Reticular Layer
Has sensory receptors that detect touch, pressure, pain, vibration, and temperature. Blood vessels assist in thermoregulation
Hair Follicles
Hairs protect skull and provide delicate touch sensations on general body surface
Exocrine Glands
Assist in temperature regulation and waste excretion
Nails
Protect and support tips of fingers and toes

strength of the bond between the epidermis and dermis. The stratum basale forms **epidermal ridges**, which extend into the dermis and are adjacent to dermal projections called **dermal papillae** (pa-PIL-ē; singular, *papilla*; a nipple-shaped mound) that extend into the epidermis (see **Spotlight Figure 5–3a**). These ridges and papillae are important because the strength of the attachment is proportional to the surface area of the basement membrane. In other words, the more numerous and deeper the folds, the larger the area of attachment becomes.

Basal cells, or *germinative cells*, dominate the stratum basale. Basal cells are stem cells that divide to replace the more superficial keratinocytes that are shed at the epithelial surface.

Skin surfaces that lack hair also contain specialized epithelial sensory cells known as *tactile (Merkel) cells* scattered among the cells of the stratum basale. Each tactile cell together with a sensory nerve ending is called a **tactile disc**. The tactile cells are sensitive to touch and, when compressed, release chemicals that stimulate their associated sensory nerve endings. (The skin contains many other kinds of sensory receptors, as we will see in later sections.)

The brown tones of skin result from the synthesis of pigment by cells called *melanocytes.* ⮌ p. 129 These pigment cells are distributed throughout the stratum basale, with cell processes extending into more superficial layers.

Stratum Spinosum

Each time a stem cell divides, one of the daughter cells is pushed superficial to the stratum basale into the **stratum spinosum**. This stratum consists of 8 to 10 layers of keratinocytes bound together by desmosomes. ⮌ p. 118 The name *stratum spinosum*, which means "spiny layer," refers to the fact that the cells look like miniature pincushions in standard histological sections. They look that way because the keratinocytes were processed with chemicals that shrank the cytoplasm but left the cytoskeletal elements and desmosomes intact. Some of the cells entering this layer from the stratum basale continue to divide, further increasing the thickness of the epithelium.

The stratum spinosum also contains cells that participate in the immune response. These cells are called *dendritic cells* (*dendron*, tree) because of their branching projections. They are also known as *Langerhans cells.* They stimulate defense against

SPOTLIGHT

Figure 5–3
The Epidermis

a BASIC ORGANIZATION OF THE EPIDERMIS

The epidermis consists of stratified squamous epithelium. It is separated from the dermis by a basement membrane. The stratum basale and the underlying dermis interlock, strengthening the bond between the two. The epidermis forms epidermal ridges, which extend into the dermis and are adjacent to dermal papillae that project into the epidermis.

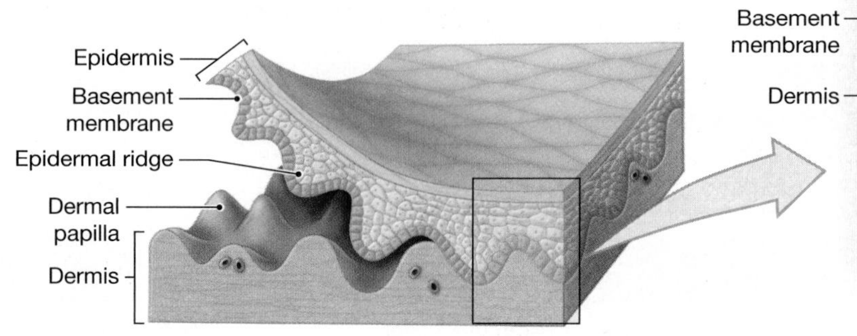

Epidermis
Basement membrane
Epidermal ridge
Dermal papilla
Dermis

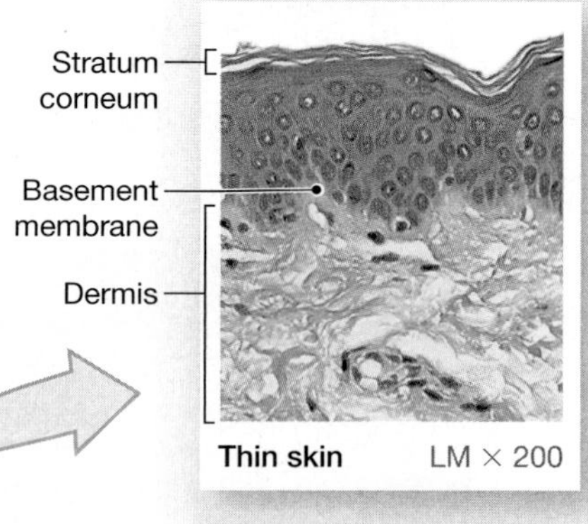

Stratum corneum

Basement membrane

Dermis

Thin skin LM × 200

Thin skin contains four layers of keratinocytes, and is about as thick as the wall of a plastic sandwich bag (about 0.08 mm).

b LAYERS OF THE EPIDERMIS

The layers of the epidermis are best shown in a sectional view of thick skin. **Thick skin** contains a fifth layer, the *stratum lucidum*. Because thick skin also has a much thicker superficial layer (the *stratum corneum*), it is about as thick as a standard paper towel (about 0.5 mm).

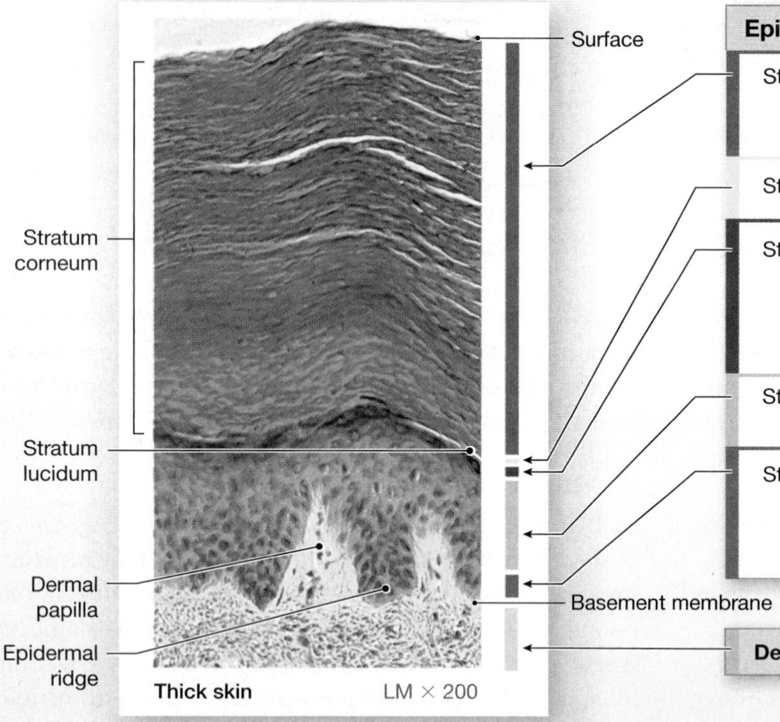

Surface

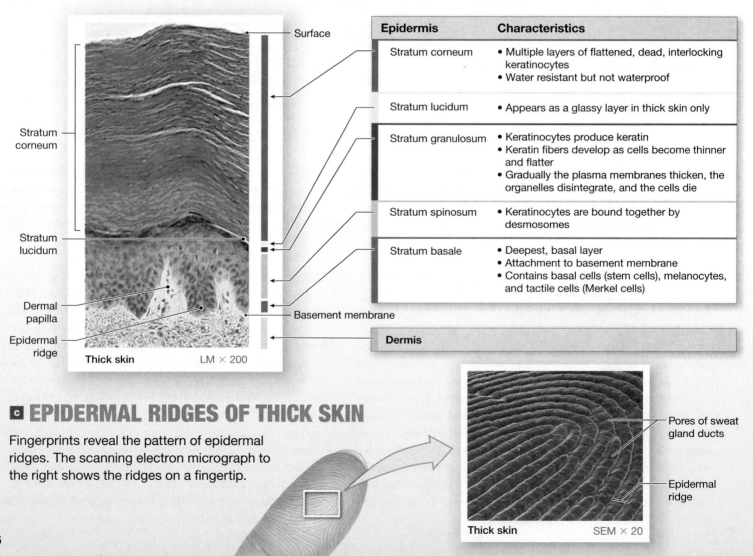

Stratum corneum

Stratum lucidum

Dermal papilla

Epidermal ridge

Basement membrane

Thick skin LM × 200

Epidermis	Characteristics
Stratum corneum	• Multiple layers of flattened, dead, interlocking keratinocytes • Water resistant but not waterproof
Stratum lucidum	• Appears as a glassy layer in thick skin only
Stratum granulosum	• Keratinocytes produce keratin • Keratin fibers develop as cells become thinner and flatter • Gradually the plasma membranes thicken, the organelles disintegrate, and the cells die
Stratum spinosum	• Keratinocytes are bound together by desmosomes
Stratum basale	• Deepest, basal layer • Attachment to basement membrane • Contains basal cells (stem cells), melanocytes, and tactile cells (Merkel cells)
Dermis	

c EPIDERMAL RIDGES OF THICK SKIN

Fingerprints reveal the pattern of epidermal ridges. The scanning electron micrograph to the right shows the ridges on a fingertip.

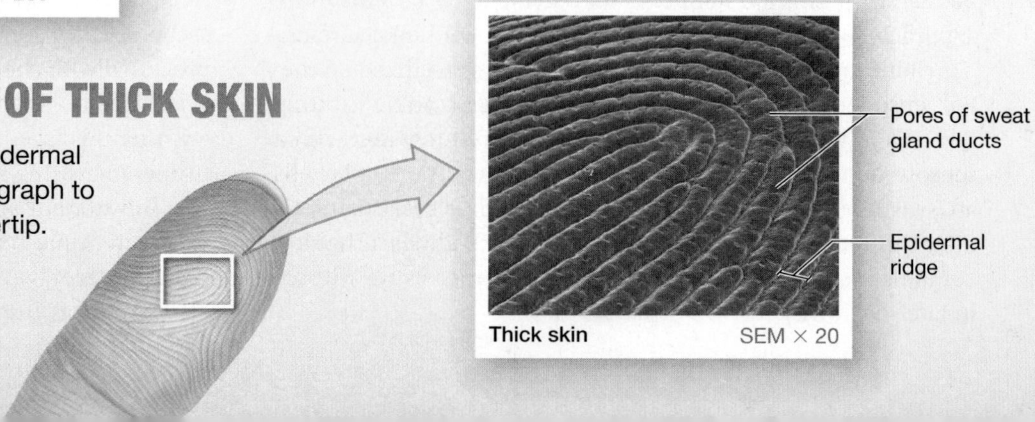

Pores of sweat gland ducts

Epidermal ridge

Thick skin SEM × 20

(1) microorganisms that manage to penetrate the superficial layers of the epidermis and (2) superficial skin cancers. We consider dendritic cells and other cells of the immune response in Chapter 22.

Stratum Granulosum

Superficial to the stratum spinosum is the **stratum granulosum**, or "granular layer." It consists of three to five layers of keratinocytes derived from the stratum spinosum. By the time cells are pushed into this layer, most have stopped dividing and have started making large amounts of keratin. As keratin fibers accumulate, the cells grow thinner and flatter, and their plasma membranes thicken and become less permeable. The keratinocytes also make a protein called **keratohyalin** (ker-a-tō-HĪ-a-lin), which forms dense cytoplasmic granules that promote dehydration of the cell as well as aggregation and cross-linking of the keratin fibers. The nuclei and other organelles then disintegrate, and the cells die. Further dehydration creates a tightly interlocked layer of cells that consists of keratin fibers surrounded by keratohyalin.

Stratum Lucidum

In the thick skin of the palms and soles, a glassy **stratum lucidum** ("clear layer") covers the stratum granulosum. The cells in the stratum lucidum are flattened, densely packed, largely without organelles, and filled with keratin.

Stratum Corneum

At the exposed surface of both thick skin and thin skin is the **stratum corneum** (KOR-nē-um; *corneus*, horny). It normally contains 15 to 30 layers of keratinized cells. **Keratinization**, or *cornification*, is the formation of protective, superficial layers of cells filled with keratin. This process takes place on all exposed skin surfaces except the anterior surfaces of the eyes. The dead cells in each layer of the stratum corneum remain tightly interconnected by desmosomes. The connections are so secure that keratinized cells are generally shed in large groups or sheets rather than individually (think about how you peel after a sunburn). It's also interesting to note, that every minute, we shed about 30,000 to 40,000 skin cells. That adds up to about 9 pounds of skin cells every year!

It takes 7 to 10 days for a cell to move from the stratum basale to the stratum corneum. The dead cells generally remain in the exposed stratum corneum for an additional two weeks before they are shed or washed away. This arrangement places the deeper portions of the epithelium and underlying tissues beneath a protective barrier of dead, durable, and expendable cells. Normally, the surface of the stratum corneum is relatively dry, so it is unsuitable for the growth of many microorganisms. Lipid secretions from sebaceous glands coat the surface, helping to prevent water loss or gain.

The contours of the skin surface follow the ridge patterns of the dermal papillae in the stratum basale, which vary from small conical pegs in thin skin to the complex whorls seen on the thick skin of the palms and soles. Ridges on the palms and soles increase the surface area of the skin and increase friction, ensuring a secure grip. The ridge patterns on the tips of the fingers are the basis of fingerprints (**Spotlight Figure 5–3c**). These ridge shapes are determined partially by the interaction of genes and partially by the intrauterine environment. That is, during fetal development our genes determine general characteristics, but the contact with amniotic fluid, another fetus (in the case of twins), or the uterine wall affects the ultimate pattern of a fetus's fingerprints. So identical twins do not have identical fingerprints. However, the pattern of your epidermal ridges is unique and does not change during your lifetime.

The stratum corneum is water resistant, but not waterproof. Water from interstitial fluids slowly diffuses to the surface and evaporates into the surrounding air. You lose about 500 mL (about 1 pt) of water in this way each day. The process is called **insensible perspiration**, because you are unable to see or feel (sense) the water loss. In contrast, you are usually very aware of the **sensible perspiration** produced by active sweat glands.

Damage to the epidermis can increase the rate of insensible perspiration. If the damage breaks connections between superficial and deeper layers of the epidermis, fluid accumulates in pockets, or *blisters*, within the epidermis. (Blisters also form between the epidermis and dermis if the basement membrane is damaged.) If damage to the stratum corneum makes it less effective as a water barrier, the rate of insensible perspiration skyrockets, and a potentially dangerous fluid loss occurs. This loss is a serious consequence of severe burns and a complication in the condition known as *xerosis* (excessively dry skin).

Why do wrinkles form in the skin on the fingers and toes after a long soak in the bath? Previously, this effect was commonly explained as being due to the osmotic movement of water across the epithelium. That is, the fingertips absorb water and swell, making the skin wrinkle. However, such a process failed to explain why wrinkling did not occur if there was nerve damage in the fingers. It turns out that fingertips do not swell in water but actually shrink when they wrinkle, an effect caused by the constriction of underlying blood vessels. The effect is due to the autonomic nervous system, which functions mostly outside our awareness and controls blood vessel diameter. Is there any benefit to having wrinkled fingertips? Recent studies have shown that the wrinkles improve the ability to grip wet or submerged objects, much like the treads of a tire improve traction on a wet road.

Epidermal Growth Factor

Epidermal growth factor (EGF) is one of the peptide growth factors introduced in Chapter 3. ⊃ p. 107 EGF is produced by the salivary glands and glands of the duodenum (initial segment of the small intestine). It was named for its effects on the epidermis, but we now know that EGF has widespread effects on epithelia throughout the body. Here are some roles of EGF:

- Promoting the divisions of basal cells in the stratum basale and stratum spinosum.
- Accelerating the production of keratin (keratinization) in differentiating keratinocytes.
- Stimulating epidermal development and epidermal repair after injury.
- Stimulating secretory product synthesis and secretion by epithelial glands.

In the procedure known as *tissue culture,* cells are grown under laboratory conditions for experimental or therapeutic use. Epidermal growth factor has such a pronounced effect that it can be used in tissue culture to stimulate the growth and division of epidermal cells (or other epithelial cells). Using human epidermal growth factor (hEGF) produced through recombinant DNA technology, it is now possible to grow sheets of epidermal cells for use in treating severe or extensive burns. The burned areas can be covered by epidermal sheets "grown" from a small sample of intact skin from another part of the burn victim's body. (We consider this treatment in Clinical Note: Burns and Grafts on p. 174).

Checkpoint

1. **Identify the layers of the epidermis.**
2. **Dandruff is caused by excessive shedding of cells from the outer layer of skin on the scalp. So dandruff is composed of cells from which epidermal layer?**
3. **A splinter that penetrates to the third layer of the epidermis of the palm is lodged in which layer?**
4. **Why does taking a bath cause wrinkly fingertips and toes?**
5. **Some criminals sand the tips of their fingers so as not to leave recognizable fingerprints. Would this practice permanently remove fingerprints? Why or why not?**
6. **Name the sources of epidermal growth factor in the body.**
7. **Identify some roles of epidermal growth factor related to the epidermis.**

See the blue Answers tab at the back of the book.

5-2 The dermis is the tissue layer that supports the epidermis

Learning Outcome **Describe the structures and functions of the dermis.**

The dermis lies between the epidermis and the subcutaneous layer (see Figure 5–1). Accessory structures of epidermal origin, such as hair follicles and sweat glands, extend into the dermis. In addition, the dermis contains networks of blood vessels and nerve fibers.

Layers of the Dermis

The dermis is made up of two major layers: (1) a superficial *papillary layer* and (2) a deeper *reticular layer.* The **papillary layer** consists of areolar tissue. It contains the capillaries, lymphatic vessels, and sensory nerve fibers that supply the surface of the skin. p. 131 The papillary layer gets its name from the dermal papillae that project between the epidermal ridges (see Spotlight Figure 5–3a).

Because of the abundance of sensory nerve endings and receptors in the skin, regional infection or inflammation of the skin can be very painful. **Dermatitis** (der-muh-TĪ-tis) is an inflammation of the skin that primarily involves the papillary layer. The inflammation typically begins in a part of the skin exposed to infection or irritated by chemicals, radiation, or mechanical stimuli. Dermatitis may cause no discomfort, or it may produce an annoying itch, as in poison ivy. Other forms of the condition can be quite painful, and the inflammation can spread rapidly across the entire integument.

The **reticular layer** lies deep to the papillary layer. It consists of an interwoven meshwork of dense irregular connective tissue containing both collagen and elastic fibers (Figure 5–4). Bundles of collagen fibers extend superficially beyond the reticular layer to blend into those of the papillary layer, so the boundary between the two layers is indistinct. Collagen fibers of the reticular layer also extend into the deeper subcutaneous layer. In addition to extracellular protein fibers, the dermis contains all the cells of connective tissue proper. p. 128

Dermal Strength and Elasticity

Collagen and elastic fibers give the dermis strength and elasticity. *Collagen fibers* are very strong and resist stretching, but they are easily bent or twisted. *Elastic fibers* permit stretching and then recoil to their original length. The elastic fibers provide flexibility, and the collagen fibers limit that flexibility to prevent damage to the tissue.

Figure 5–4 **Reticular Layer of Dermis.**

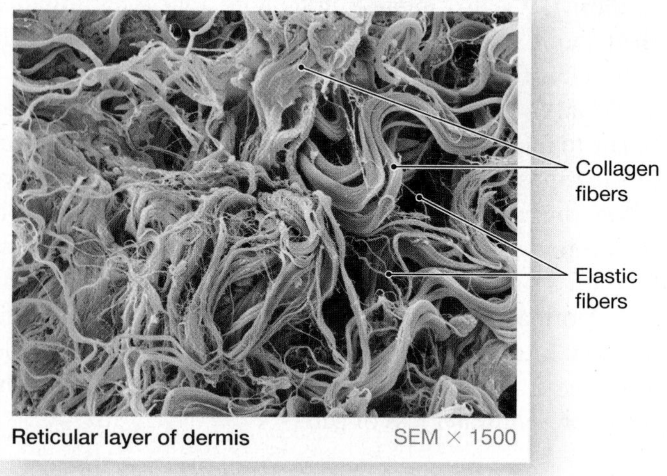
Collagen fibers
Elastic fibers
Reticular layer of dermis
SEM × 1500

The water content of the skin also helps maintain its flexibility and resilience. These properties are collectively known as *skin turgor.* One of the signs of dehydration is the loss of skin turgor, revealed by pinching the skin on the back of the hand. A dehydrated dermis will remain peaked when pinched, but hydrated skin will flatten out.

Aging, hormones, and the destructive effects of UV radiation permanently reduce the amount of elastin in the dermis. Wrinkles and sagging skin result. The extensive distortion of the dermis that occurs over the abdomen during pregnancy or after substantial weight gain can exceed the elastic limits of the skin. The resulting damage to the dermis prevents it from recoiling to its original size after delivery or weight loss. The skin then wrinkles and creases, creating a network of **stretch marks**.

To treat reduced skin elasticity, *tretinoin (Retin-A)* is a derivative of vitamin A that can be applied to the skin as a cream or gel. This drug was originally developed to treat acne, but it also increases blood flow to the dermis and stimulates dermal repair. As a result, the rate of wrinkle formation decreases, and existing wrinkles become smaller. The degree of improvement varies among individuals.

Tension Lines

Most of the collagen and elastic fibers at any location are arranged in parallel bundles oriented to resist the forces applied to the skin during normal movement. The resulting pattern of fiber bundles in the skin establishes **tension lines**, also called *cleavage lines.* Tension lines are clinically significant, because a cut made parallel to a tension line will usually remain closed and heal with little scarring, whereas a cut made at right angles to a tension line will be pulled open as severed elastic fibers recoil, resulting in greater scarring. When possible, surgeons make incisions parallel to tension lines (Figure 5–5).

The Dermal Blood Supply

Arteries supplying the skin lie deep within the subcutaneous layer. Branches of these arteries form two networks, or *plexuses,* in the dermis. The deeper network lies along the border of the subcutaneous layer with the reticular layer of the dermis. This network is called the **cutaneous plexus** (Figure 5–6). Tributaries of these arteries supply both the adipose tissues of the subcutaneous layer and the tissues of the integument. As small arteries travel toward the epidermis, branches supply the hair follicles, sweat glands, and other structures in the dermis.

Figure 5–5 Tension Lines of the Skin. Tension lines follow the pattern of collagen fiber bundles in the dermis of the skin.

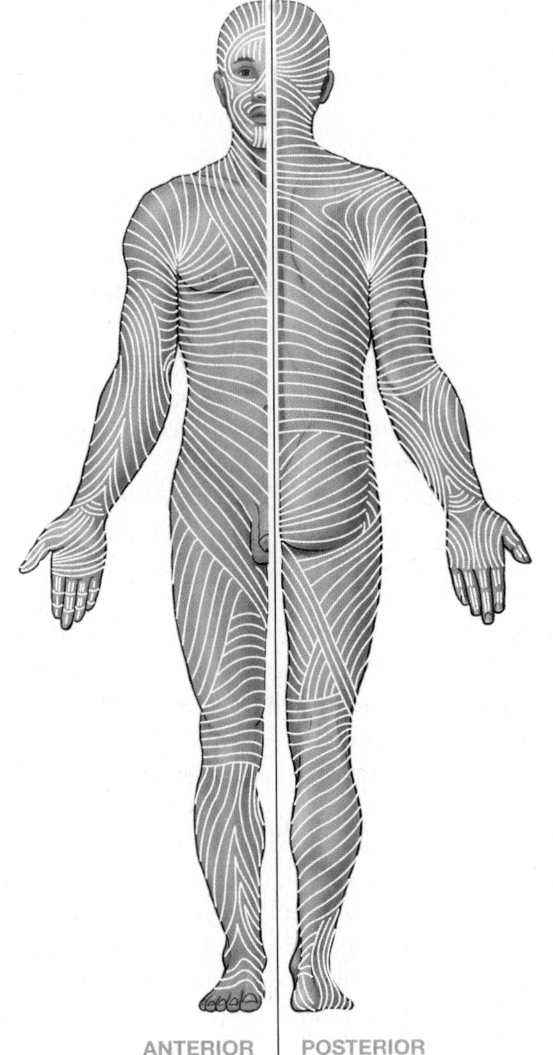
ANTERIOR
POSTERIOR

Figure 5–6 Dermal Circulation. The cutaneous and subpapillary plexuses are shown.

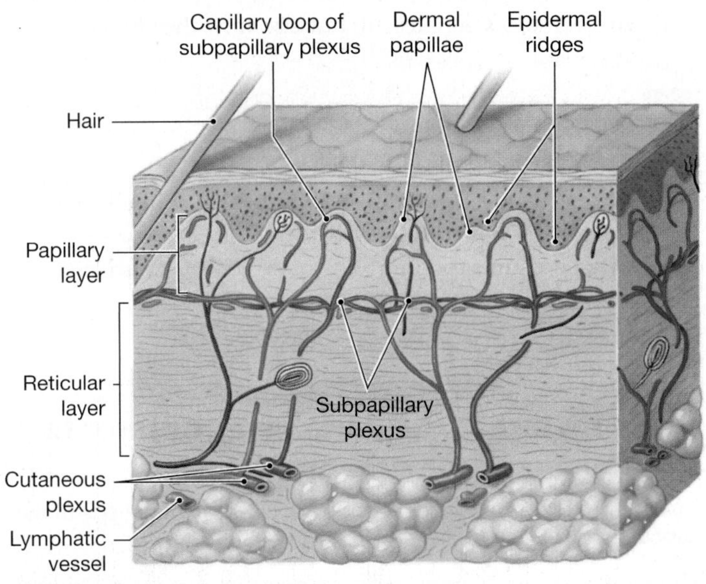
Capillary loop of subpapillary plexus
Dermal papillae
Epidermal ridges
Hair
Papillary layer
Reticular layer
Subpapillary plexus
Cutaneous plexus
Lymphatic vessel

? Which plexus supplies blood to the capillary loops that follow the epidermis–dermis boundary?

5

On reaching the papillary layer, the small arteries form another branching network, the **subpapillary plexus** (see Figure 5–6). It provides arterial blood to capillary loops that follow the contours of the epidermis–dermis boundary. These capillaries empty into small veins of the subpapillary plexus, which drain into veins accompanying the arteries of the cutaneous plexus. This network in turn connects to larger veins in the subcutaneous layer.

Trauma to the skin often results in a *contusion*, or bruise. When dermal blood vessels rupture, blood leaks into the dermis, and the area develops the familiar "black and blue" color.

5

Innervation of the Skin

The integument is filled with sensory nerve fiber endings and sensory receptors. Anything that comes in contact with the skin—from the lightest touch of a mosquito to the weight of a loaded backpack—initiates a nerve impulse that can reach our conscious awareness. Nerve fibers in the skin control blood flow, adjust gland secretion rates, and monitor sensory receptors in the dermis and the deeper layers of the epidermis.

We have already noted that the deeper layers of the epidermis contain tactile discs. They are fine touch and pressure receptors. The epidermis also contains sensory nerve endings that provide sensations of pain and temperature. The dermis contains similar nerve endings, as well as other, more specialized receptors. Examples shown in Figure 5–1 include receptors sensitive to light touch—*tactile corpuscles* (*Meissner corpuscles*), located in dermal papillae—and receptors sensitive to deep pressure and vibration—*lamellar corpuscles* (*pacinian corpuscles*), in the reticular layer.

Even this partial list of the receptors in the skin is enough to highlight the importance of the integument as a sensory structure. We will return to this topic in Chapter 15, where we consider not only what receptors are present, but also how they function.

Checkpoint

8. **Describe the location of the dermis.**
9. **Where are the capillaries and sensory nerve fibers that supply the epidermis located?**
10. **What accounts for the ability of the dermis to undergo repeated stretching?**

See the blue Answers tab at the back of the book.

5-3 The subcutaneous layer connects the dermis to underlying tissues

Learning Outcome Describe the structures and functions of the subcutaneous layer.

The **subcutaneous layer**, or *hypodermis*, lies deep to the dermis (see Figure 5–1). The boundary between the two is generally indistinct because the connective tissue fibers of the reticular layer are extensively interwoven with those of the subcutaneous layer. The subcutaneous layer is not a part of the integument, but it is important in stabilizing the position of the skin in relation to underlying tissues, such as skeletal muscles or other organs, while permitting independent movement.

Figure 5–7 The Subcutaneous Layer.

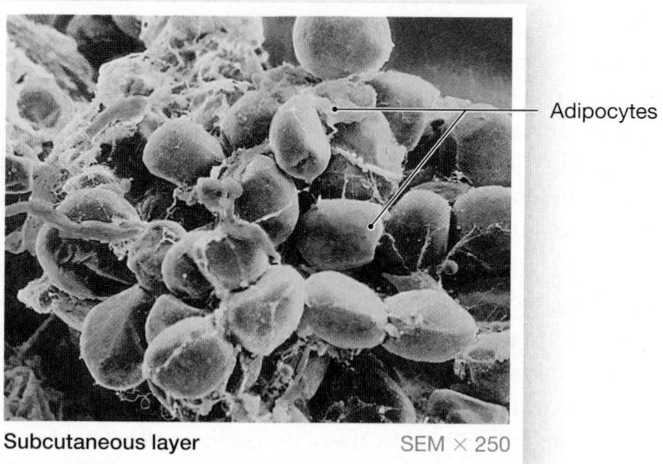
Adipocytes
Subcutaneous layer
SEM × 250

The subcutaneous layer is quite elastic. It consists primarily of adipose tissue (areolar tissue dominated by adipocytes) (Figure 5–7). Also called subcutaneous fat, it makes up about 80 percent of all body fat. (The remaining amount is visceral fat that is associated with visceral organs.) Only the superficial region of the subcutaneous layer contains large arteries and veins. The venous circulation of this region contains a substantial amount of blood, and much of this volume will shift to the general circulation if these veins constrict. For that reason, we often describe the skin as a *blood reservoir*. The rest of the subcutaneous layer contains a limited number of capillaries and no vital organs. This last characteristic makes **subcutaneous injection**—by means of a **hypodermic needle**—a useful method of administering drugs.

Most infants and small children have extensive "baby fat," which provides insulation and helps reduce heat loss. This subcutaneous fat also serves as a substantial energy reserve and as a shock absorber for the rough-and-tumble activities of our early years.

As we grow, the distribution of subcutaneous fat changes. The greatest changes take place in response to circulating sex hormones. Beginning at puberty, men build up subcutaneous fat at the neck, on the arms, along the lower back, and over the buttocks. In contrast, women accumulate subcutaneous fat at the breasts, buttocks, hips, and thighs. In adults of either gender, the subcutaneous layer of the backs of the hands and the upper surfaces of the feet contains few fat cells, but distressing amounts of adipose tissue can accumulate in the abdominal region, producing a "potbelly." An excessive amount of such abdominal fat, also called *central adiposity*, is strongly correlated with cardiovascular disease.

Clinical Note Nips, Tucks, and Shots

According to the American Society of Plastic Surgeons, millions of cosmetic procedures are performed every year in the United States. The most common cosmetic surgery sought by men is rhinoplasty (nose reshaping); for women, it is breast augmentation.

Rhinoplasty can improve the appearance of the nose or relieve breathing problems. Remodeling of bone and cartilage can be done through an open (external) incision or endonasally (inside the nose).

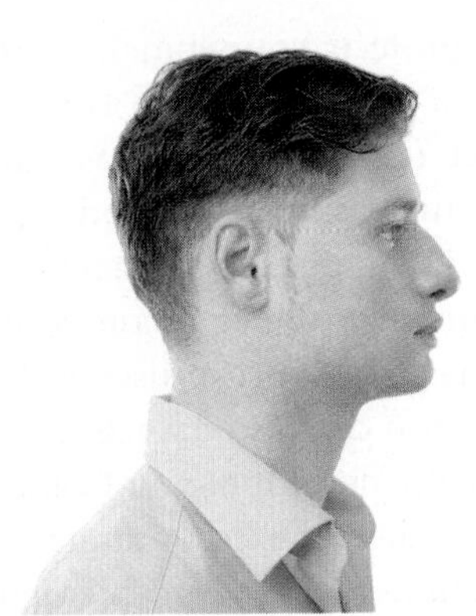

Before

After

Breast augmentation and reconstruction following breast cancer surgery now use shaped implants made of silicone or saline that more closely resemble the look and feel of the natural breast. However, implants can make it harder to obtain a clear image of the breast during mammography screening.

"**Nips and tucks**" remove excess skin and underlying fat in the face, chin, back of the arm, belly, buttocks, and thighs. *Abdominoplasty* (tummy tuck) reduces loose abdominal skin after multiple pregnancies or significant weight loss. A visible scar remains across the lower abdomen. *Liposuction* removes subcutaneous adipose tissue through a tube inserted deep to the skin while the person is anesthetized. Liposuction might sound like an easy way to remove unwanted fat, but in practice, it can be dangerous. There are risks from anesthesia, and bleeding (adipose tissue is quite vascular), sensory deficits, infection, and fluid loss can also occur.

Other popular cosmetic procedures are minimally invasive and can be done in the outpatient setting with local anesthesia. For example, millions of **botulinum toxin A** (Botox) shots are given annually to reduce wrinkles. Botox weakens muscles by blocking the signals from nerves to those muscles. **Dermal fillers**, such as hyaluronan (hyalurronic acid) or collagen, are injected to plump up sagging skin. However, the effects of these fillers are short lived because the fillers are cleared by dermal macrophages.

Checkpoint

11. **List the two terms for the tissue that connects the dermis to underlying tissues.**
12. **Describe the subcutaneous layer.**
13. **Identify several functions of subcutaneous fat.**

See the blue Answers tab at the back of the book.

5-4 Epidermal pigmentation and dermal circulation influence skin color

Learning Outcome Explain what accounts for individual differences in skin color, and discuss the response of melanocytes to sunlight exposure.

One of the most obvious distinguishing characteristics among humans is skin color. What gives skin its various colors? In this section we examine how pigments in the epidermis and blood flow in the dermis influence skin color. We also discuss clinical implications of changes in skin color.

The Role of Epidermal Pigmentation

The epidermis contains variable quantities of two pigments: melanin and carotene. Both of these pigments contribute to skin color, but melanin, which is made in the skin, is the primary determinant.

Melanin

Melanin is a pigment produced by **melanocytes**, pigment-producing cells, introduced in Chapter 4. There are two types of melanin, a red-yellow form (pheomelanin) and a brown-black form (eumelanin). The melanocytes involved are located in the stratum basale, squeezed between or deep to the epithelial cells (Figure 5–8). Melanocytes manufacture both types of melanin from the amino acid *tyrosine* and package it into intracellular vesicles called *melanosomes.* These vesicles, which can contain either type of melanin, travel within the processes of melanocytes and are transferred intact to nearby basal keratinocytes. The transfer of pigmentation colors the keratinocytes temporarily, until the melanosomes are destroyed by fusion with lysosomes. In individuals with pale skin, this transfer takes place in the stratum basale and stratum spinosum, and the cells of more superficial layers lose their pigmentation. In dark-skinned people, the melanosomes are larger and more numerous, and the transfer may occur in the stratum granulosum as well, making skin pigmentation darker and more persistent.

Go to MasteringA&P™ > Study Area > Menu > Lab Tools > PAL 3.0 > Histology: Integumentary System

5

Figure 5–8 Melanocytes.

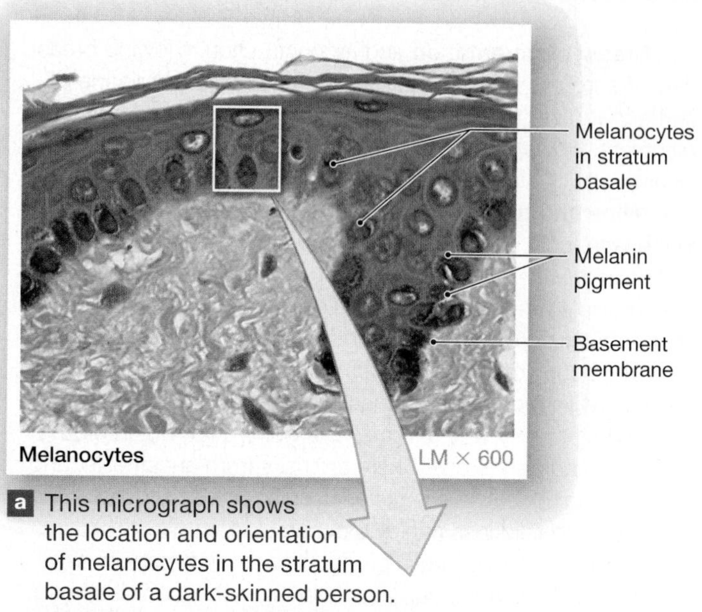
Melanocytes in stratum basale
Melanin pigment
Basement membrane
Melanocytes
LM × 600

a This micrograph shows the location and orientation of melanocytes in the stratum basale of a dark-skinned person.

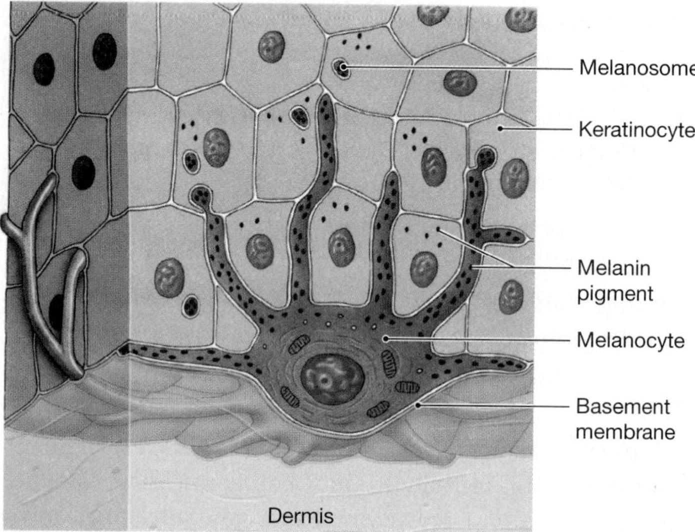
Melanosome
Keratinocyte
Melanin pigment
Melanocyte
Basement membrane
Dermis

b Melanocytes produce and store melanin.

? In what layer of the skin are melanocytes found?

The ratio of melanocytes to basal cells ranges between 1:4 and 1:20, depending on the region of the body. The skin covering most areas of the body has about 1000 melanocytes per square millimeter. The cheeks and forehead, the nipples, and the genital region (the scrotum of males and the labia majora of females) have higher concentrations (about 2000 per square millimeter). The differences in skin pigmentation among individuals do not reflect different numbers of melanocytes, but rather, different levels of melanin synthesis. A deficiency or absence of melanin production leads to a disorder known as *albinism*. Individuals with this condition have a normal distribution of melanocytes, but the cells are incapable of producing melanin.

There can also be localized differences in the rates of melanin production by melanocytes. *Freckles* are small, pigmented areas on relatively pale skin. These spots typically have an irregular border. They represent the areas serviced by melanocytes that are producing larger-than-average amounts of melanin. Freckles tend to be most abundant on surfaces such as the face, probably due to its greater exposure to the sun. *Lentigos* are similar to freckles but have regular borders and contain abnormal melanocytes. *Senile lentigos*, or *liver spots*, are variably pigmented areas that develop on sun-exposed skin in older individuals with pale skin.

The melanin in keratinocytes protects your epidermis and dermis from the harmful effects of sunlight, which contains significant amounts of **ultraviolet (UV) radiation**. Melanocytes respond to UV exposure by increasing their production of melanin. A small amount of UV radiation is beneficial, because it stimulates the epidermis to produce a compound required for calcium ion homeostasis (a process discussed in a later section). However, UV radiation can also damage DNA, causing mutations and promoting the development of cancer. Within keratinocytes, melanosomes become concentrated in the region around the nucleus, where the melanin pigments act like a sunshade to provide some UV protection for the DNA in those cells.

UV radiation can also produce some immediate effects, such as burns. If severe, sunburns can damage both the epidermis and the dermis. Thus, the pigment layers in the epidermis help protect both epidermal and dermal tissues. However, the production of melanin is not rapid enough to prevent sunburn the first day you spend at the beach. Melanin synthesis accelerates slowly, peaking about 10 days after the initial exposure. Individuals of any skin color can suffer sun damage to the skin, but dark-skinned individuals have greater initial protection.

Over time, cumulative UV damage to the skin can harm fibroblasts, impairing maintenance of the dermis. ⮌ p. 128 The result is premature wrinkling. In addition, skin cancers can develop from chromosomal damage in basal cells or melanocytes.

Carotene

Carotene (KAR-uh-tēn) is an orange-yellow pigment that normally accumulates in epidermal cells. It is most apparent in cells of the stratum corneum of light-skinned individuals, but it also accumulates in fatty tissues in the deep dermis and subcutaneous layer. Carotene is also found in a variety of orange vegetables, such as carrots and squashes. The skin of someone who eats lots of carrots can actually turn orange from an overabundance of carotene. The color change is very striking in pale-skinned individuals, but less obvious in people with

> Go to MasteringA&P™ > Study Area > Menu > Lab Tools > PAL 3.0 > Anatomical Models: Integumentary System

darker skin pigmentation. Carotene can be converted to vitamin A, which is required for both the normal maintenance of epithelia and the synthesis of photoreceptor (light receptor) pigments in the eye.

The Role of Dermal Circulation: Hemoglobin

Blood contains red blood cells filled with the pigment *hemoglobin*, which binds and transports oxygen in the bloodstream. When bound to oxygen, hemoglobin is bright red, giving capillaries in the dermis a reddish tint that is most apparent in lightly pigmented individuals. If those vessels are dilated, the red tones become much more pronounced. For example, your skin becomes flushed and red when your body temperature rises, because the superficial blood vessels dilate so that the skin can act like a radiator and lose heat. ⮌ p. 20

The skin becomes relatively pale when its blood supply is temporarily reduced. A light-skinned person who is frightened may "turn white" due to a sudden drop in blood supply to the skin. During a sustained reduction in circulatory supply, the oxygen level in the tissues declines. Under these conditions, hemoglobin releases oxygen and turns a much darker red. Seen from the surface, the skin then takes on a bluish coloration called **cyanosis** (sī-uh-NŌ-sis; *kyanos*, blue). In individuals of any skin color, cyanosis is easiest to see in areas of very thin skin, such as the lips or beneath the nails. It can also occur in response to extreme cold or as a result of cardiovascular or respiratory disorders, such as heart failure or severe asthma.

Disease-Related Changes in Skin Color

Because the skin is easy to observe, changes in the skin's appearance can be useful in diagnosing diseases that primarily affect other body systems. Several diseases can produce secondary effects on skin color and pigmentation:

- In *jaundice* (JAWN-dis), the liver is unable to excrete bile, so a yellowish pigment accumulates in body fluids. In advanced stages, the skin and whites of the eyes turn yellow.
- Some tumors affecting the pituitary gland result in the secretion of large amounts of *melanocyte-stimulating hormone (MSH)*. This hormone causes melanocytes to overproduce melanin, leading to a darkening of the skin, as if the person has an extremely deep bronze tan.
- In *Addison's disease*, the pituitary gland secretes large quantities of *adrenocorticotropic hormone (ACTH)*, which is structurally similar to MSH. The effect of ACTH on skin color is similar to that of MSH.
- In *vitiligo* (vit-i-LĪ-gō), individuals lose their melanocytes, causing white patches on otherwise normal skin (Figure 5–9). The condition develops in about 1 percent of the population. Its incidence increases among individuals with thyroid gland disorders, Addison's disease, or several other disorders. It is suspected that vitiligo develops when the immune defenses malfunction and antibodies attack normal melanocytes. The primary problem with vitiligo is cosmetic, especially for individuals with darkly pigmented skin.

Figure 5–9 Vitiligo.

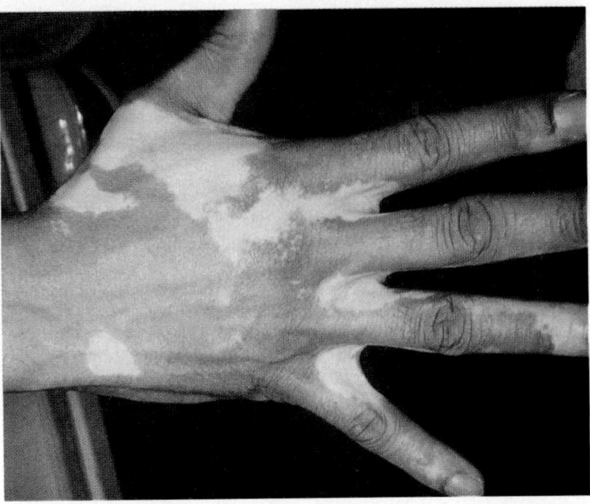

Checkpoint

14. **Name the two major pigments in the epidermis.**
15. **Why does exposure to sunlight darken skin?**
16. **Why does the skin of a fair-skinned person appear red during exercise in hot weather?**

See the blue Answers tab at the back of the book.

5-5 Sunlight causes epidermal cells to convert a steroid into vitamin D_3

Learning Outcome Describe the interaction between sunlight and vitamin D_3 production.

Too much sunlight can damage epithelial cells and deeper tissues, but limited exposure to sunlight is beneficial. When exposed to UV radiation, epidermal cells in the stratum spinosum and stratum basale convert a cholesterol-related steroid compound into *cholecalciferol* (kō-le-kal-SIF-er-ol), or **vitamin D_3**. The liver then converts cholecalciferol into an intermediary product used by the kidneys to synthesize the hormone **calcitriol** (kal-si-TRĪ-ol) (Figure 5–10). Calcitriol is essential for the normal absorption of calcium and phosphate ions in the small intestine. An inadequate supply leads to impaired bone maintenance and growth.

The term *vitamin* is usually reserved for essential organic nutrients that we must get from the diet because the body either cannot make them or makes them in insufficient amounts. Cholecalciferol can be absorbed by the digestive tract, and if the skin cannot make enough cholecalciferol, a dietary supply

5

Clinical Note Skin Cancer

Almost everyone has several *benign* (be-NĪN; noncancerous) tumors of the skin. Moles and warts are common examples. However, **skin cancers**, which can be dangerous, are the most common form of cancer.

An *actinic keratosis*, a precancerous scaly area on the skin, is an indication that sun damage has occurred. In untreated individuals, it can lead to squamous cell carcinoma. In contrast, *basal cell carcinoma*, a cancer that originates in the stratum basale is the most common skin cancer (Photo a). About two-thirds of these cancers appear in body areas subjected to chronic UV exposure. Researchers have identified genetic factors that predispose people to this condition. *Squamous cell carcinomas* are less common, but are almost totally restricted to areas of sun-exposed skin. *Metastasis* (spread to distant body sites) seldom occurs in treated squamous cell and basal cell carcinomas, and most people survive these cancers. The usual treatment involves the surgical removal of the tumor, and 95 percent of patients survive for five years or longer after treatment. (This statistic, the *five-year survival rate*, is a common method of reporting long-term outcomes.)

Unlike these common and seldom life-threatening cancers, *malignant melanomas* (mel-ah-NŌ-muz) are extremely dangerous (Photo b). In this condition, cancerous melanocytes grow rapidly and metastasize through the lymphatic system and/or blood vessels. The outlook for long-term survival is in many cases determined by how early the condition is diagnosed. If the cancer is detected early, while it is still localized, the five-year survival rate is 99 percent. If it is not detected until extensive metastasis has occurred, the survival rate drops to 14 percent.

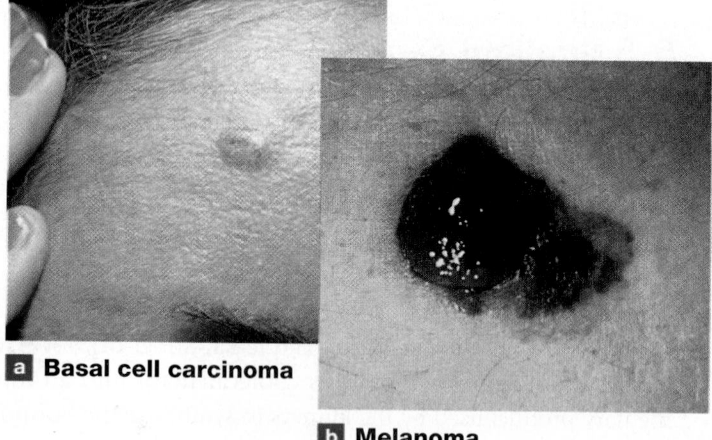

a Basal cell carcinoma

b Melanoma

To detect melanoma at an early stage, you must examine your skin, and you must know what to look for. The mnemonic ABCDE makes it easy to remember this cancer's key characteristics:

- **A** is for *asymmetry:* One half of a melanoma is unlike the other half. Typically, they are raised; they may also ooze or bleed.
- **B** is for *border:* The border of a melanoma is generally irregular, and in some cases notched.
- **C** is for *color:* A melanoma is generally mottled, with any combination of tan, brown, black, red, pink, white, and blue tones.
- **D** is for *diameter:* Melanomas are usually more than 6 mm in diameter, or approximately the area covered by a pencil eraser.
- **E** is for *evolving:* Benign lesions look the same over time. Any change in size, shape, or color signals possible melanoma.

Prevent sun damage by avoiding exposure to the sun during the middle of the day and by using a sunscreen (not a tanning oil) before any sun exposure. This practice also delays the cosmetic problems of aging and wrinkling. *Everyone* who spends any time out in the sun should choose a broad-spectrum sunscreen with a sun protection factor (SPF) of at least 15. Blondes, redheads, and people with very pale skin are better off with an SPF of 20 to 30. Fair-skinned individuals who live in the tropics are most susceptible to all forms of skin cancer, because their melanocytes are unable to shield them from UV radiation. (The risks are the same for those who spend time in a tanning salon or tanning bed.) The protection offered by these sunscreens comes both from organic molecules that absorb UV radiation and from inorganic pigments that absorb, scatter, and reflect UV rays. The higher the SPF, the more of these chemicals the product contains; hence, fewer UV rays are able to penetrate to the skin's surface. Wearing a hat with a brim and panels to shield the neck and face, long pants, and long-sleeved shirts provides added protection.

The use of sunscreens will be even more important as the ozone gas in the upper atmosphere is further destroyed by our industrial emissions. Ozone absorbs UV radiation before it reaches Earth's surface. In doing so, ozone assists the melanocytes in preventing skin cancer.

(from foods such as egg yolks, fish oil, and fortified milk) will maintain normal bone development. Under these circumstances, dietary cholecalciferol acts like a vitamin. For this reason, cholecalciferol is also called *vitamin* D_3.

If cholecalciferol cannot be produced by the skin and is not included in the diet, bone development is abnormal and bone maintenance is inadequate. For example, children who live in areas with overcast skies and whose diet lacks cholecalciferol can have abnormal bone development. This condition, called **rickets** (Figure 5–11), has largely been eliminated in the United States because dairy companies are required to add cholecalciferol, usually identified as "vitamin D," to the milk

Figure 5–10 Sources of Vitamin D_3.

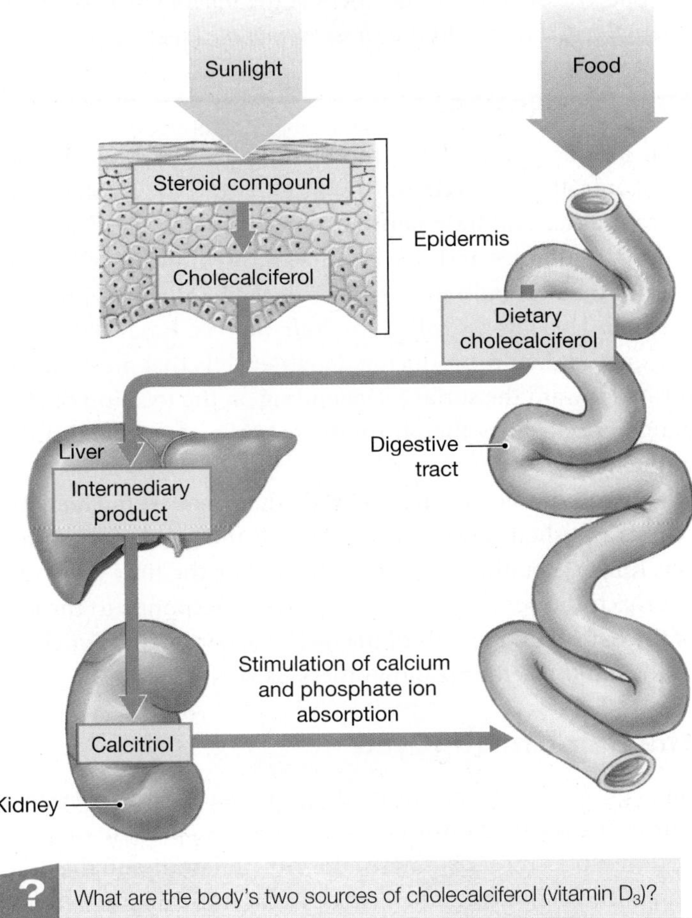
Sunlight
Food
Steroid compound
Cholecalciferol
Epidermis
Dietary cholecalciferol
Liver
Intermediary product
Digestive tract
Stimulation of calcium and phosphate ion absorption
Calcitriol
Kidney

? What are the body's two sources of cholecalciferol (vitamin D_3)?

Figure 5–11 Rickets. Rickets, a disease caused by vitamin D_3 deficiency, results in the bending of abnormally weak and flexible bones under the weight of the body, plus other structural changes.

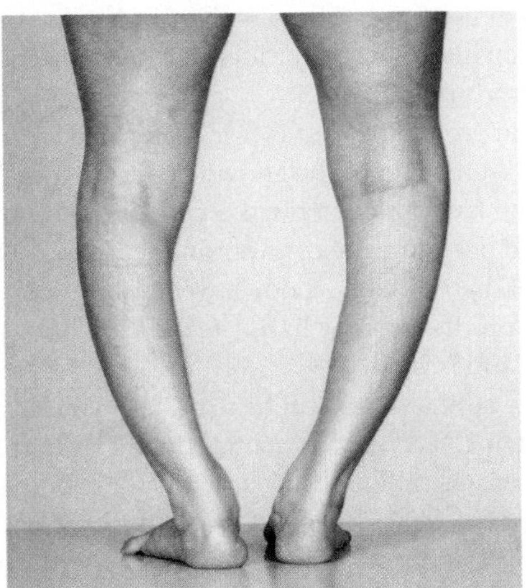

sold in grocery stores. In Chapter 6, we consider the hormonal control of bone growth in greater detail.

 **Checkpoint**

17. **Explain the relationship between sunlight exposure and vitamin D_3 synthesis.**
18. **In some cultures, women must be covered completely, except for their eyes, when they go out in public. Explain why these women may develop bone problems later in life.**

See the blue Answers tab at the back of the book.

5-6 Hair is made of keratinized dead cells pushed to the skin surface where it has protecting and insulating roles

Learning Outcome **Describe the mechanisms that produce hair, and explain the structural basis for hair texture and color.**

Hair and several other structures—hair follicles, sebaceous and sweat glands, and nails—are considered accessory structures of the integument. During embryonic development, these structures originate from the epidermis, so they are also known as *epidermal derivatives.* Although located in the dermis, they project to the surface of the skin. ATLAS: Embryology Summary 5: The Development of the Integumentary System

Hairs project above the surface of the skin almost everywhere, except over the sides and soles of the feet, the palms of the hands, the sides of the fingers and toes, the lips, and portions of the external genitalia. The human body has about 2.5 million hairs, and 75 percent of them are on the general body surface, not on the head. Hairs are nonliving structures produced in organs called **hair follicles**.

The hairs and hair follicles on your body have important functions. The 500,000 or so hairs on your head protect your scalp from ultraviolet radiation, help cushion light impacts to the head, and insulate the skull. The hairs guarding the entrances to your nostrils and external ear canals help keep out foreign particles and insects. Your eyelashes do the same for the surfaces of the eyes. Eyebrows are important because they help keep sweat out of your eyes. Hairs also play a critical role as sensory receptors.

Hair and Hair Follicle Structure

Figure 5–12 illustrates important details about the structure of hairs and hair follicles. Each hair follicle opens onto the surface of the epidermis but extends deep into the dermis and usually into the hypodermis. Deep to the epidermis, each follicle is wrapped in a dense connective tissue sheath. A **root hair plexus** of sensory nerves surrounds the base of each hair follicle (Figure 5–12a). As a result, you can feel the movement of the

5

+ Clinical Note Decubitus Ulcers

Problems with dermal circulation affect both the epidermis and the dermis. An *ulcer* is a localized area where the epithelium has been eroded. *Decubitus* (dē-KYŪ-bih-tus) *ulcers*, or *bedsores*, affect patients whose circulation is restricted, especially when a splint, a cast, or lying in bed continuously compresses superficial blood vessels. Such sores most commonly affect the skin covering joints or bony prominences, where dermal blood vessels are pressed against deeper structures. The chronic lack of circulation kills epidermal cells, removing a barrier to bacterial infection. Eventually, dermal tissues deteriorate as well. The prevention and care of bedsores is an expertise of nursing. Prevention entails frequently changing the patient's position. Treatment involves the use of hyperbaric (high-pressure) oxygen to deliver maximum oxygen to blood traveling through compressed tissues.

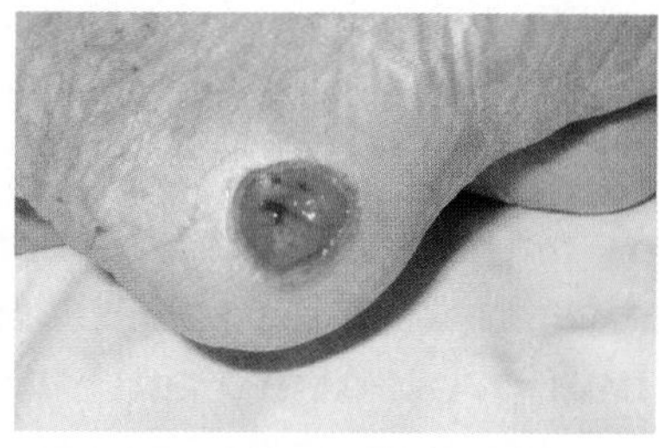

shaft of even a single hair. This sensitivity provides an early-warning system that may help prevent injury. For example, you may be able to swat a mosquito before it reaches your skin.

A bundle of smooth muscle cells forms the **arrector pili** (a-REK-tor PĪ-lī) **muscle**, also called *arrector muscle of hair*. It extends from the papillary layer of the dermis to the connective tissue sheath surrounding the hair follicle. When stimulated, the arrector pili muscle contracts, pulling on the follicle and forcing the hair to stand erect. Contraction may be the result of emotional states, such as fear or rage, or a response to cold, producing "goose bumps." In a furry mammal, this action increases the thickness of its insulating coat. Humans do not receive any comparable insulating benefits, but the response persists.

Each hair is a long, cylindrical structure that extends outward, past the epidermal surface (Figure 5–12a). The **hair root** is the portion that anchors the hair into the skin. The root begins at the base of the hair, at the *hair bulb*, and extends distally to the point at which the internal organization of the hair is complete, about halfway to the skin surface. The **hair shaft** extends from this halfway point to the exposed tip of the hair. The hair we see is the protruding part of the hair shaft.

Seen in cross section, a hair has three layers (Figure 5–12b). The cells in the core of the hair make up what is called the **medulla**. Cells farther from the center of the hair matrix form the **cortex**, an intermediate layer. Those at the edges of the hair matrix form the **cuticle**, which will be the surface of the hair.

Around the hair, the epithelial cells of the follicle walls are organized into several concentric layers. Moving outward from the hair cuticle, these layers include the *internal root sheath*, the *external root sheath*, and the *glassy membrane* (see Figure 5–12b).

Hair Production

Hair production begins at the bulging base of a hair follicle, a region called the **hair bulb** (Figure 5–12c,d). The hair bulb surrounds a small **hair papilla**, a peg of connective tissue containing capillaries and nerves. A layer of epithelial cells at the base of the hair bulb in contact with the hair papilla produces the hair. This layer is called the **hair matrix**. Basal cells of the hair matrix divide, producing daughter cells that are gradually pushed toward the surface. Depending on the location of these daughter cells, they then form the medulla, cortex, and cuticle of the hair.

As cell divisions continue at the hair matrix, the daughter cells are pushed toward the surface of the skin, and the hair gets longer. Keratinization is completed by the time these cells approach the surface. At the level that corresponds to the start of the hair shaft, the cells of the medulla, cortex, and cuticle are dead, and the keratinization process is at an end.

The Hair Growth Cycle

Hairs grow and are shed according to a **hair growth cycle**. A hair in the scalp grows for two to five years, at a rate of about 0.33 mm per day. Variations in the growth rate and in the duration of the hair growth cycle account for individual differences in the length of uncut hair.

While hair is growing, the cells of the hair root absorb nutrients and incorporate them into the hair structure. As a result, clipping or collecting hair for analysis can be helpful in diagnosing several disorders. For example, hairs of individuals with lead poisoning or other heavy metal poisoning contain high levels of those metal ions. Hair samples containing nucleated cells can also be used for identification purposes through DNA fingerprinting. ⮌ p. 85

As a hair grows, the root is firmly attached to the matrix of the follicle. At the end of the growth cycle, the follicle becomes inactive. The hair is now termed a **club hair**. The follicle gets smaller, and over time the connections between the hair matrix and the club hair root break down. When another cycle begins, the follicle produces a new hair. The old club hair is pushed to the surface and is shed.

Healthy adults with a full head of hair typically lose about 100 head hairs each day. Sustained losses of more than 100 hairs per day generally indicate that a net loss of hairs is under way, and hair loss will eventually be noticeable. Temporary increases in hair loss can result from drugs, dietary factors, radiation, an excess of vitamin A, high fever, stress, or hormonal factors related to pregnancy.

Figure 5–12 Hair Follicles and Hairs.

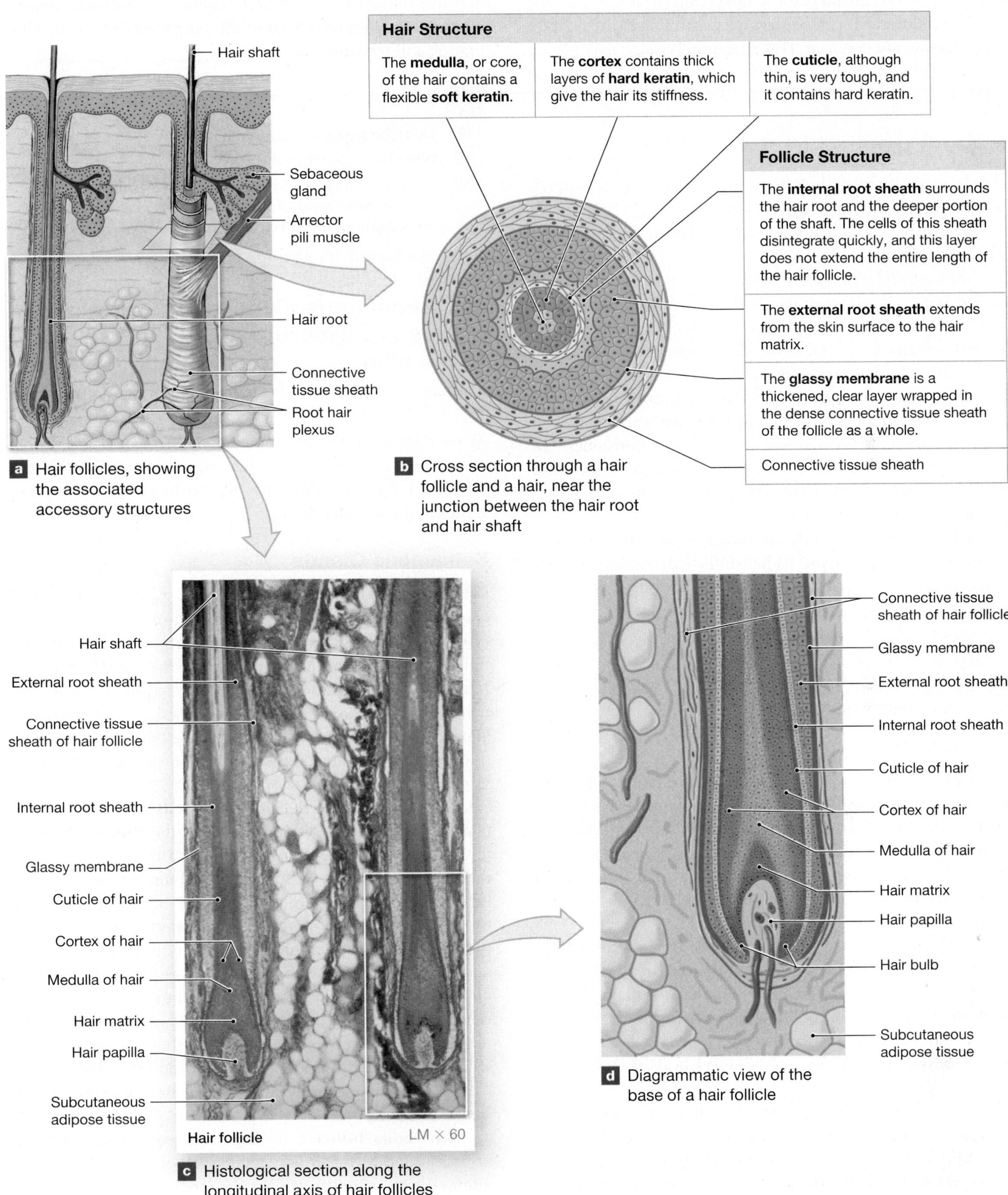
Hair Structure
The medulla, or core, of the hair contains a flexible soft keratin.
The cortex contains thick layers of hard keratin, which give the hair its stiffness.
The cuticle, although thin, is very tough, and it contains hard keratin.
Follicle Structure
The internal root sheath surrounds the hair root and the deeper portion of the shaft. The cells of this sheath disintegrate quickly, and this layer does not extend the entire length of the hair follicle.
The external root sheath extends from the skin surface to the hair matrix.
The glassy membrane is a thickened, clear layer wrapped in the dense connective tissue sheath of the follicle as a whole.
Connective tissue sheath
Hair shaft
Sebaceous gland
Arrector pili muscle
Hair root
Connective tissue sheath
Root hair plexus
Hair shaft
External root sheath
Connective tissue sheath of hair follicle
Internal root sheath
Glassy membrane
Cuticle of hair
Cortex of hair
Medulla of hair
Hair matrix
Hair papilla
Subcutaneous adipose tissue
Hair follicle
LM × 60
Connective tissue sheath of hair follicle
Glassy membrane
External root sheath
Internal root sheath
Cuticle of hair
Cortex of hair
Medulla of hair
Hair matrix
Hair papilla
Hair bulb
Subcutaneous adipose tissue

a Hair follicles, showing the associated accessory structures

b Cross section through a hair follicle and a hair, near the junction between the hair root and hair shaft

c Histological section along the longitudinal axis of hair follicles

d Diagrammatic view of the base of a hair follicle

5

In males, changes in the level of the sex hormones circulating in the blood can affect the scalp, causing a shift in the type of hair produced (discussed shortly), beginning at the temples and the crown of the head. This alteration is called *male pattern baldness.* Some cases of male pattern baldness respond to drug therapies, such as the topical application of *minoxidil (Rogaine).*

Types of Hairs

Hairs first appear after about three months of embryonic development. These hairs, collectively known as *lanugo* (la-NŪ-gō), are extremely fine and unpigmented. Most lanugo hairs are shed before birth.

They are replaced by one of two types of hairs in the adult integument: vellus hairs or terminal hairs. **Vellus hairs** are the fine "peach fuzz" hairs located over much of the body surface. **Terminal hairs** are heavy, more deeply pigmented, and sometimes curly. The hairs on your head, including your eyebrows and eyelashes, are terminal hairs that are present throughout life. However, in other areas hair follicles may alter the type of the hairs produced in response to circulating hormones. For example, vellus hairs are present at the armpits, pubic area, and limbs until puberty. After puberty, the follicles produce terminal hairs, in response to circulating sex hormones.

What makes hair curly or straight? The cross-sectional shape of the hair shaft and its hair follicle determine whether hairs are curly or straight. A curly hair and its follicle are oval; a straight hair and its follicle are round.

Tips & Tools

Associate the word *vellus* ("peach fuzz") with "velvet."

Hair Color

What gives your hair its color? Variations in hair color reflect differences in structure and variations in the pigment produced by melanocytes at the hair matrix. Different forms and amounts of melanin give a dark brown, yellow-brown, or red color to the hair. Genes determine these structural and biochemical characteristics, but hormonal and environmental factors also influence the condition of your hair.

As pigment production decreases with age, hair color lightens. White hair results from the combination of a lack of pigment and the presence of air bubbles in the medulla of the hair shaft. As the proportion of white hairs increases, the person's overall hair color is described as gray or even white. Because hair itself is dead and inert, any changes in its coloration are gradual.

What happens when we dye our hair? We are able to change our hair color by using chemicals that disrupt the cuticle and permit dyes to enter and stain the cortex and medulla. These color treatments damage the hair's protective cuticle layer and dehydrate and weaken the hair shaft. As a result, the hair becomes more thin and brittle. Conditioners and oil treatments attempt to reduce the effects of this structural damage by rehydrating and recoating the shaft.

Checkpoint

19. **Describe a typical strand of hair.**
20. **What happens when the arrector pili muscle contracts?**
21. **Once a burn on the forearm that destroys the epidermis and extensive areas of the deep dermis heals, will hair grow again in the affected area?**

See the blue Answers tab at the back of the book.

5-7 Sebaceous glands and sweat glands are exocrine glands found in the skin

Learning Outcome **Discuss the various kinds of glands in the skin, and list the secretions of those glands.**

As we have seen, the integumentary system contains various accessory structures that produce exocrine secretions. In this section we focus on the structure and function of sebaceous glands and sweat glands.

Sebaceous Glands

Sebaceous (se-BĀ-shus) **glands**, or *oil glands,* are holocrine glands that discharge an oily lipid secretion into hair follicles (Figure 5–13). The gland cells produce large quantities of these lipids, called **sebum** (SĒ-bum), as they mature. These glands release the sebum through holocrine secretion, a process that involves the rupture of the secretory cells. ⮌ p. 126 Sebaceous glands that communicate with a single hair follicle share a duct and thus are classified as simple branched alveolar glands. ⮌ p. 125

The sebum released from gland cells enters the lumen (open passageway around the hair) of the gland. The arrector pili muscles that erect the hair then contract, squeezing the sebaceous gland and forcing the sebum into the hair follicle and onto the surface of the skin. Sebum is a mixture of triglycerides, cholesterol, proteins, and electrolytes. It inhibits the growth of bacteria, lubricates and protects the hair shaft, and conditions the surrounding skin. The hair shaft is made of keratin, which is a tough protein, but dead, keratinized cells become dry and brittle once exposed to the environment. Interestingly, when we shampoo and wash, we remove the oily secretions of sebaceous glands, only to add other lipids to our hair in the form of conditioners, and to our skin in the form of creams and lotions.

Sebaceous follicles are large sebaceous glands that are not associated with hair follicles. Instead, their ducts discharge sebum directly onto the epidermis (see Figure 5–13). We have

Figure 5–13 The Structure of Sebaceous Glands and Sebaceous Follicles.

Sebaceous follicle
Sebaceous gland
Epidermis
Dermis
Subcutaneous layer
Lumen (hair removed)
Wall of hair follicle
Basement membrane
Discharge of sebum
Lumen of duct
Breakdown of cell membranes
Mitosis and growth
Basal cells

Sebaceous gland LM × 150

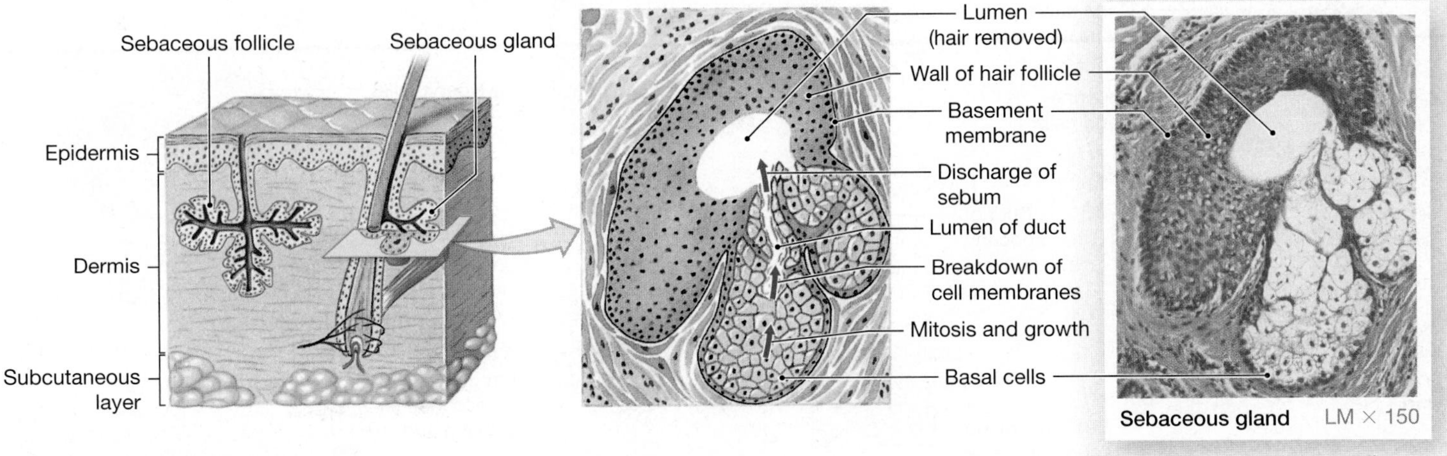

? Where do sebaceous glands and sebaceous follicles discharge their secretions?

sebaceous follicles on the face, back, chest, nipples, and external genitalia.

Surprisingly, sebaceous glands are very active during the last few months of fetal development. Their secretions, mixed with shed epidermal cells, form a protective superficial layer—the *vernix caseosa*—that coats the skin surface. Sebaceous gland activity all but stops after birth, but it increases again at puberty in response to rising levels of sex hormones.

Seborrheic (seb-ō-RĒ-ik) *dermatitis* is an inflammation around abnormally active sebaceous glands, most often in the scalp. The affected area becomes red and oily, with increased epidermal scaling. In infants, mild cases are called *cradle cap*. Seborrheic dermatitis is a common cause of dandruff in adults. Anxiety, stress, and fungal or bacterial infections can aggravate the problem.

Sweat Glands

The skin contains two types of **sweat glands**, or *sudoriferous glands* (*sudor*, sweat): *apocrine sweat glands* and *eccrine sweat glands* (Figure 5–14).

Apocrine Sweat Glands

In the armpits, around the nipples, and in the pubic region, **apocrine sweat glands** secrete their products into hair follicles (Figure 5–14a). These coiled, tubular glands produce a sticky, cloudy, and potentially odorous secretion. The name *apocrine* was originally chosen because it was thought the gland cells used an apocrine method of secretion. ↩ p. 126 We now know that they rely on merocrine secretion, but the name has not changed.

Apocrine sweat glands begin secreting at puberty. This sweat is a nutrient source for bacteria, whose growth and breakdown intensify its odor. Surrounding the secretory cells in these glands are special myoepithelial cells that contract and squeeze the gland, causing the accumulated sweat to discharge into the hair follicles. The secretory activities of the gland cells and the contractions of myoepithelial cells are controlled by the nervous system and by circulating hormones.

Eccrine Sweat Glands

Eccrine (EK-rin) **sweat glands** are also known as *merocrine sweat glands*. These coiled, tubular glands discharge their secretions directly onto the surface of the skin (Figure 5–14b). Eccrine sweat glands are far more numerous and widely distributed than apocrine sweat glands. The adult integument contains 2–5 million eccrine sweat glands, which are smaller than apocrine sweat glands and do not extend as deeply into the dermis. The palms and soles have the highest numbers; the palms have an estimated 500 eccrine sweat glands per square centimeter (3000 per square inch).

As noted earlier, the sweat from eccrine sweat glands is called sensible perspiration. Sweat is 99 percent water, but it also contains some other components, including sodium chloride, which gives sweat a salty taste. It has a pH range of 4.0–6.8. (See the Appendix for a complete analysis of the composition of normal sweat.)

Figure 5–14 Sweat Glands.

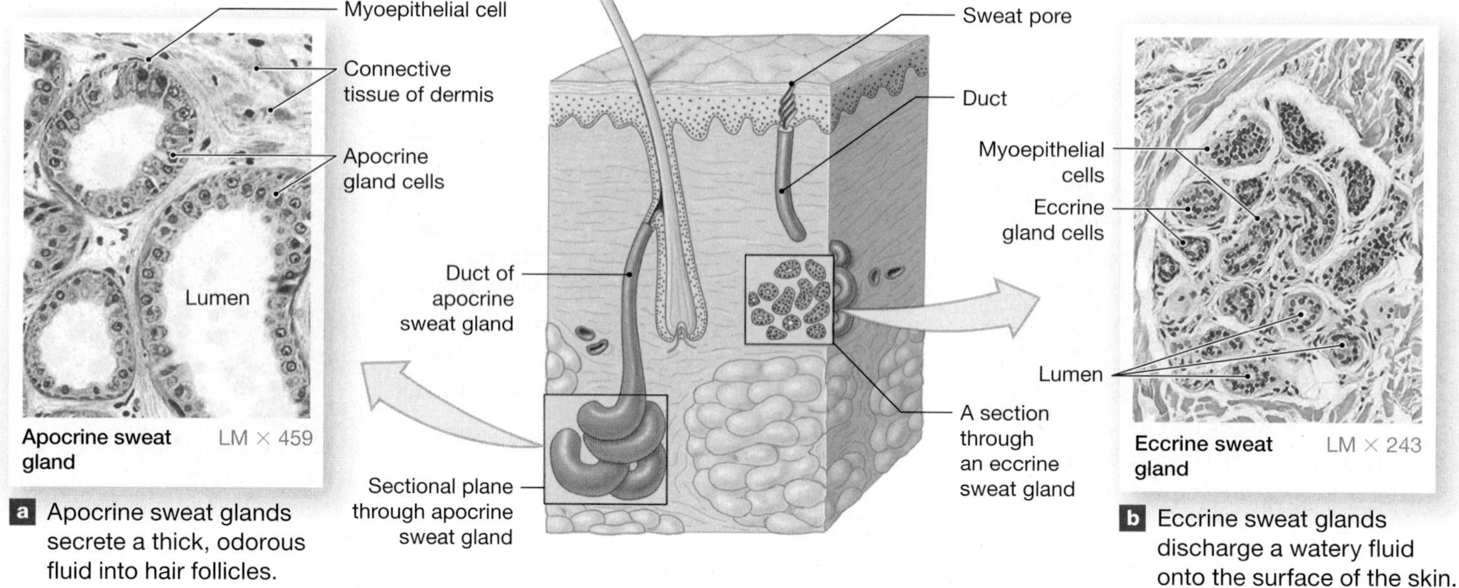
Myoepithelial cell
Connective tissue of dermis
Apocrine gland cells
Lumen
Apocrine sweat gland
LM × 459
Duct of apocrine sweat gland
Sectional plane through apocrine sweat gland
Sweat pore
Duct
Myoepithelial cells
Eccrine gland cells
Lumen
A section through an eccrine sweat gland
Eccrine sweat gland
LM × 243

a Apocrine sweat glands secrete a thick, odorous fluid into hair follicles.

b Eccrine sweat glands discharge a watery fluid onto the surface of the skin.

The functions of eccrine sweat glands include the following:

- *Cooling the Surface of the Skin to Reduce Body Temperature.* This is the primary function of sensible perspiration (p. 157). The degree of secretory activity is regulated by neural and hormonal mechanisms. When all the eccrine sweat glands are working at their maximum, we can perspire more than a gallon per hour, and dangerous fluid and electrolyte losses can occur. For this reason, athletes in endurance sports must drink fluids at regular intervals.
- *Excreting Water and Electrolytes.* Water, salts (mostly sodium chloride), and a number of metabolized drugs are excreted.
- *Providing Protection from Environmental Hazards.* Sweat dilutes harmful chemicals in contact with the skin. It also discourages the growth of microorganisms in two ways: (1) by either flushing them from the surface or making it difficult for them to adhere to the epidermal surface and (2) through the action of *dermicidin,* a small peptide with powerful antibiotic properties.

The armpits provide consistent heat and moisture, ideal conditions for bacterial growth on the body surface. Deodorants and antiperspirants, singly or in combination, are commonly used to combat underarm bacteria and odor. Deodorants work by temporarily reducing bacteria numbers and by masking the body odor they produce. Antiperspirants target the sweat glands. Antiperspirants contain aluminum compounds that reduce the amount of moisture available to bacteria in two ways: (1) They interact with sweat to plug up sweat gland ducts, and (2) they cause the sweat gland pores to constrict. However, the plugs are eventually lost through the normal shedding of cells at the surface of the skin.

Other Integumentary Glands

As we have seen, sebaceous glands are located wherever there are hair follicles, apocrine sweat glands are located in relatively restricted areas, and eccrine sweat glands are widely distributed across the body surface. The skin also contains a variety of specialized glands that are restricted to specific locations. Here are two important examples:

1. The **mammary glands** of the breasts are anatomically related to apocrine sweat glands. A complex interaction between sex hormones and pituitary hormones controls their development and secretion.We discuss mammary gland structure and function in Chapter 28.
2. **Ceruminous** (se-RŪ-mi-nus) **glands** are modified sweat glands in the passageway of the external ear. Their secretions combine with those of nearby sebaceous glands, forming a mixture called **cerumen**, or earwax. Together with tiny hairs along the ear canal, earwax helps trap foreign particles, preventing them from reaching the eardrum.

Control of Glandular Secretions and Thermoregulation

The autonomic nervous system (ANS) is the portion of the nervous system that adjusts our basic life support systems without our conscious control (Chapter 16). It controls the activation and deactivation of sebaceous glands and apocrine sweat glands. Regional control is not possible. The commands from the ANS affect all the glands of that type, everywhere on the body surface.

Eccrine sweat glands are much more precisely controlled, and the amount of secretion and the area of the body involved can vary independently. For example, when you are nervously awaiting an anatomy and physiology exam, only your palms may begin to sweat.

As we noted earlier, the primary function of sensible perspiration is to cool the surface of the skin and to reduce body temperature. When the environmental temperature is high, this is a key component of *thermoregulation*, the homeostatic process of maintaining temperature. You may remember that Chapter 1 introduced the negative feedback mechanisms of thermoregulation. p. 20 When you sweat in the hot sun, all your eccrine glands are working together. The blood vessels beneath your epidermis are dilated and filled with blood, your skin reddens, and the surface of your skin is warm and wet. As the moisture evaporates, your skin cools. If your body temperature then falls below normal, sensible perspiration stops, blood flow to the skin is reduced, and the skin surface cools and dries, releasing little heat into the environment.We provide additional details of this process in Chapter 25.

Checkpoint

22. **Identify two types of exocrine glands found in the skin.**
23. **What are the functions of sebaceous secretions?**
24. **Deodorants are used to mask the effects of secretions from which type of skin gland?**
25. **Which type of skin gland is most affected by the hormonal changes that occur during puberty?**

See the blue Answers tab at the back of the book.

5-8 Nails are keratinized epidermal cells that protect the tips of fingers and toes

Learning Outcome Describe the anatomical structures of nails, and explain how they are formed.

Nails protect the exposed dorsal surfaces of the tips of the fingers and toes (Figure 5–15a). They also help limit distortion of the digits from mechanical stress—for example, when you run or grasp objects. The **nail body**, the visible portion of the nail, covers an area of epidermis called the **nail bed** (Figure 5–15b). The nail body is recessed deep to the level of the surrounding epithelium and is bordered on either side by **lateral nail grooves** (depressions) and **lateral nail folds**. The **free edge** of the nail—the distal portion that continues past the nail bed—extends over the **hyponychium** (hī-pō-NIK-ē-um), an area of thickened stratum corneum (Figure 5–15c).

Nail production takes place at the **nail root**, an epidermal fold not visible from the surface. The deepest portion of the nail root lies very close to the bone of the fingertip. A portion of the stratum corneum of the nail root extends over the exposed nail, forming the **eponychium** (ep-ō-NIK-ē-um; *epi-*, over + *onyx*, nail), or **cuticle**. Underlying blood vessels give the nail its characteristic pink color. Near the root, these vessels

Figure 5–15 The Structure of a Nail.

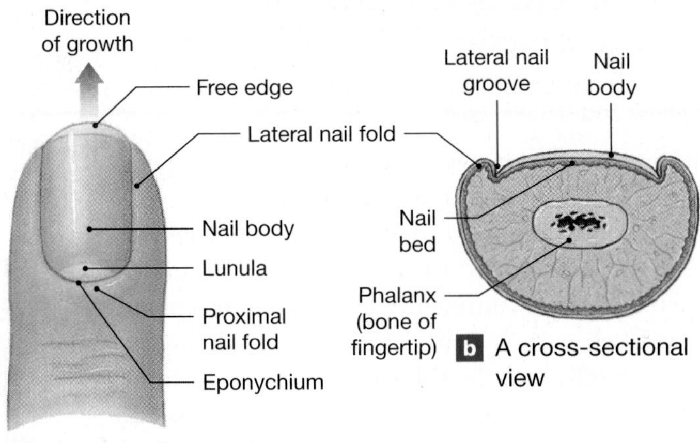
Direction of growth
Free edge
Lateral nail fold
Nail body
Lunula
Proximal nail fold
Eponychium
Lateral nail groove
Nail body
Nail bed
Phalanx (bone of fingertip)

a A superficial view

b A cross-sectional view

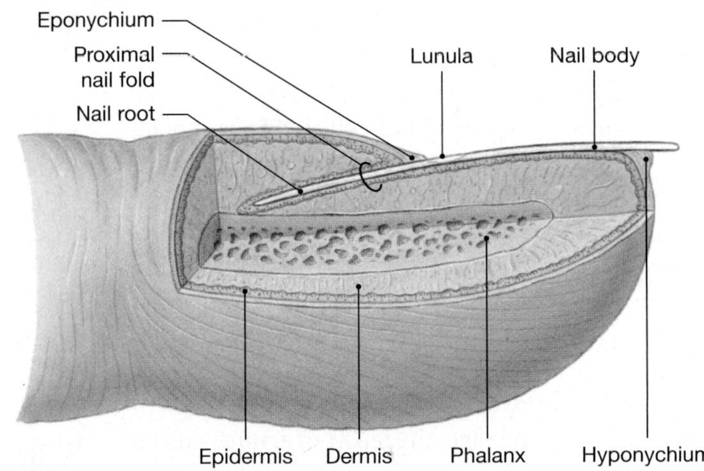
Eponychium
Proximal nail fold
Nail root
Lunula
Nail body
Epidermis
Dermis
Phalanx
Hyponychium

c A longitudinal section

may be obscured, leaving a pale crescent known as the **lunula** (LŪ-nū-la; *luna*, moon) (see Figure 5–15a).

The body of the nail consists of dead, tightly compressed cells packed with keratin. The cells producing the nails can be affected by conditions that alter body metabolism, so changes in the shape, structure, or appearance of the nails can provide useful diagnostic information. For example, the nails may turn yellow in individuals who have chronic respiratory disorders, thyroid gland disorders, or AIDS. Nails may become pitted and distorted as a result of *psoriasis* (so-RĪ-a-sis), a condition marked by rapid stem cell division in the stratum basale, and concave as a result of some blood disorders.

Checkpoint

26. **What substance makes fingernails hard?**
27. **What term is used to describe the thickened stratum corneum underlying the free edge of a nail?**
28. **Where does nail growth occur?**

See the blue Answers tab at the back of the book.

5-9 After an injury, the integument is repaired in several phases

Learning Outcome Explain how the skin responds to injury and repairs itself.

The integumentary system can function independently. It often responds directly and automatically to local influences without the involvement of the nervous or endocrine systems. For example, when the skin is continually subjected to mechanical stresses, stem cells in the stratum basale divide more rapidly, and the thickness of the epithelium increases. That is why calluses form on your palms when you perform work with your hands.

A more dramatic display of local regulation can be seen after an injury to the skin. The skin can regenerate effectively, even after considerable damage, because stem cells persist in both the epithelial and connective tissue. Basal cell divisions replace lost epidermal cells, and mesenchymal cell divisions replace lost dermal cells.

The process can be slow. When large surface areas are involved, infection and fluid loss complicate the situation. The speed and effectiveness of skin repair vary with the type of wound. For example, a slender, straight cut, or *incision*, may heal fairly quickly compared with a deep scrape, or *abrasion*, which involves the repair of a much greater surface area.

Figure 5–16 illustrates the four phases in the regeneration of the skin after an injury: inflammation, migration, proliferation, and scarring.

When damage extends through the epidermis and into the dermis, bleeding, along with swelling, redness, and pain occur. Mast cells in the region trigger an inflammatory response. This is the *inflammation phase* (1).

A dried blood clot, or **scab**, forms at the surface as part of the *migration phase* (2). The scab temporarily restores the integrity of the epidermis and restricts the entry of additional microorganisms into the area.

The bulk of the clot consists of an insoluble network of *fibrin*, a fibrous protein that forms from blood proteins during the clotting response. The clot's color is due to trapped red blood cells.

If the wound occupies an extensive area or involves a region covered by thin skin, dermal repairs must be under way before epithelial cells can cover the surface. Divisions by fibroblasts and mesenchymal cells produce mobile epithelial cells

Figure 5–16 Repair of Injury to the Integument.

1 Inflammation Phase

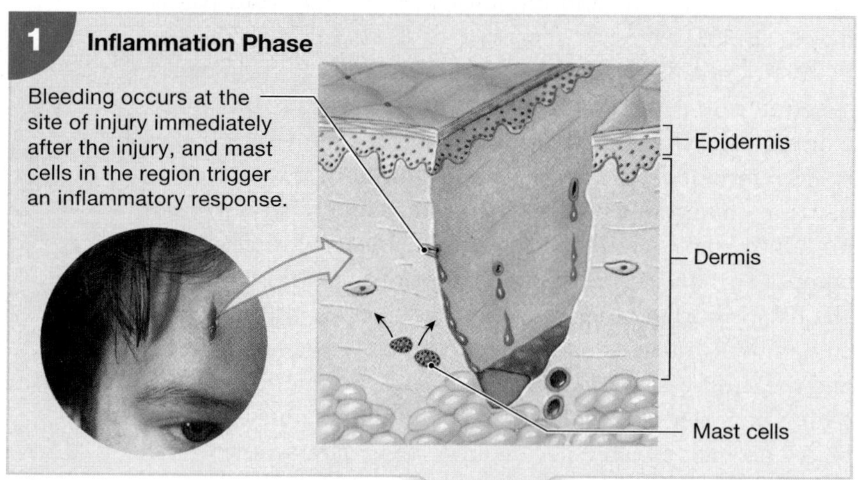
Bleeding occurs at the site of injury immediately after the injury, and mast cells in the region trigger an inflammatory response.
Epidermis
Dermis
Mast cells

2 Migration Phase

After several hours, a scab has formed and cells of the stratum basale are migrating along the edges of the wound. Phagocytic cells are removing debris, and more of these cells are arriving with the enhanced circulation in the area. Clotting around the edges of the affected area partially isolates the region.

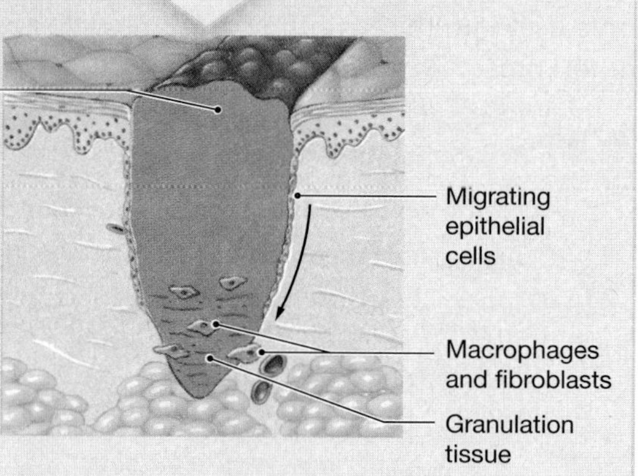
Migrating epithelial cells
Macrophages and fibroblasts
Granulation tissue

3 Proliferation Phase

About a week after the injury, the scab has been undermined by epidermal cells migrating over the collagen fiber meshwork produced by fibroblast proliferation and activity. Phagocytic activity around the site has almost ended, and the fibrin clot is dissolving.

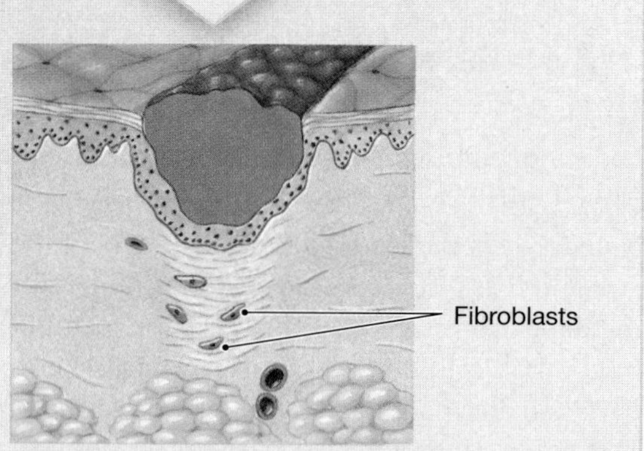
Fibroblasts

4 Scarring Phase

After several weeks, the scab has been shed, and the epidermis is complete. A shallow depression marks the injury site, but fibroblasts in the dermis continue to create scar tissue that will gradually elevate the overlying epidermis.

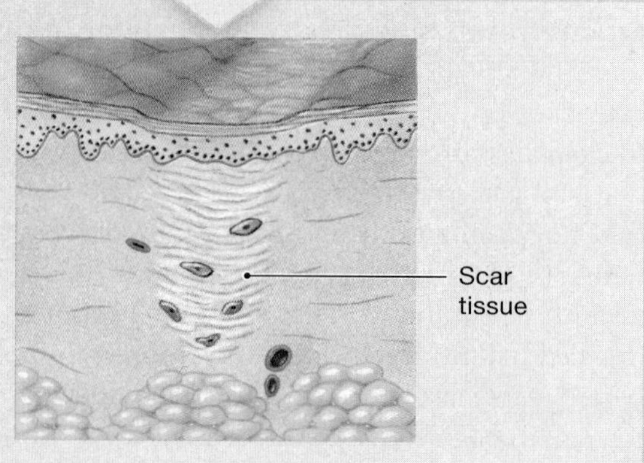
Scar tissue

that invade the deeper areas of injury. Endothelial cells of damaged blood vessels also begin to divide, and capillaries follow the fibroblasts, enhancing circulation. Phagocytic cells such as macrophages remove debris in the area. The combination of blood clot, fibroblasts, and an extensive capillary network at the base of a wound is called **granulation tissue**.

Over time, deeper portions of the clot dissolve, and the number of capillaries declines in the *proliferation phase* ③. Fibroblasts produce a collagen fiber meshwork, the proliferation mentioned in the title of the phase. The repairs do not restore the integument to its original condition, however, because the dermis will contain an abnormally large number of collagen fibers and a few blood vessels. Severely damaged hair follicles, sebaceous or sweat glands, muscle cells, and nerves are seldom repaired. They, too, are replaced by fibrous tissue.

In the *scarring phase*, the formation of rather inflexible, fibrous, noncellular **scar tissue** completes the repair process, but without restoring the tissue to its original condition ④.

We do not know what regulates the extent of scar tissue formation, and the process is highly variable. For example, surgical procedures performed on a fetus do not leave scars, perhaps because damaged fetal tissues do not produce the same types of growth factors that adult tissues do. In some adults, most often those with dark skin, scar tissue formation may continue beyond the requirements of tissue repair. The result is a thickened mass of scar tissue that begins at the site of injury and grows into the surrounding dermis. This thick, raised area of scar tissue, called a **keloid** (KĒ-loyd), is covered by a shiny, smooth epidermal surface (Figure 5–17). Keloids can develop anywhere on the body where the tissue has been injured, including as a result of surgeries. They are harmless and considered by some to be a cosmetic problem. However, some keloids are intentional and some cultures produce keloids as a form of body decoration.

Figure 5–17 A Keloid. Keloids are areas of raised fibrous scar tissue.

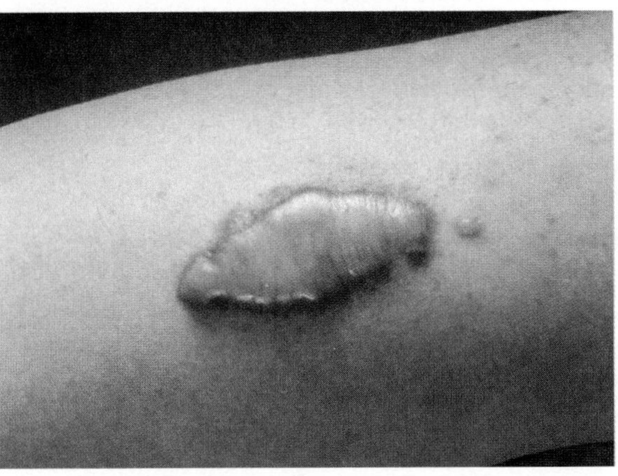

People in societies around the world adorn the skin with culturally significant markings of one kind or another. Tattoos, piercings, keloids and other scar patterns, and makeup are all used to "enhance" the appearance of the integument. Several African cultures perform scarification, creating a series of complex, raised scars on the skin. Polynesian cultures have long used tattoos as a sign of status and beauty. A dark pigment is inserted deep within the dermis of the skin by tapping on a needle, shark tooth, or bit of bone. Because the pigment is inert, the markings remain for life, clearly visible through the overlying epidermis. American popular culture has rediscovered tattoos as a fashionable way of adorning the body. Some of the risks associated with tattooing include allergic reactions, bloodborne infections from the use of contaminated equipment, and skin infections.

Tattoos can now be partially or completely removed by laser surgery. The removal process takes time (10 or more sessions may be required to remove a large tattoo), and scars often remain. To remove the tattoo, an intense, narrow beam of light from a laser breaks down the ink molecules in the dermis. Each blast of the laser that destroys the ink also burns the surrounding dermal tissue. Although the burns are minor, they accumulate and result in the formation of localized scar tissue.

Checkpoint

29. **What term describes the combination of fibrin clots, fibroblasts, and the extensive network of capillaries in healing tissue?**
30. **Why can skin regenerate effectively even after considerable damage?**

See the blue Answers tab at the back of the book.

Clinical Note
Your Skin, A Mirror of Your Health

Changes in skin color, tone, and overall condition commonly reveal disease. Such changes can assist in the diagnosis of conditions involving other systems. A *contusion*, or bruise, for example, is a swollen, discolored area where blood has leaked through vessel walls. Extensive bruising without any obvious cause may indicate a blood-clotting disorder. Yellowish skin and mucous membranes may signify jaundice, which generally indicates some type of liver disorder. The general condition of the skin can also be significant. In addition to color changes, changes in the flexibility, elasticity, dryness, or sensitivity of the skin commonly follow malfunctions in other organ systems.

Clinical Note Burns and Grafts

Burns are common injuries that result from skin exposure to heat, friction, radiation, electrical shock, or strong chemical agents. In evaluating burns in a clinical setting, two key factors must be determined: the depth of the burn and the percentage of skin surface area that has been burned.

Classification by Depth of Burn

Partial-Thickness Burns

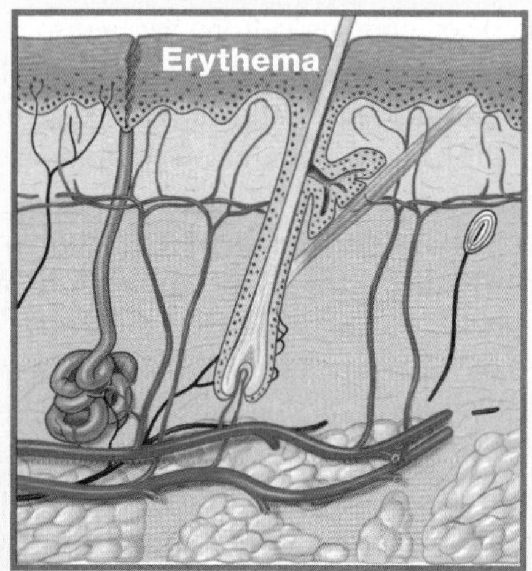
Erythema

In a **first-degree burn** or *partial-thickness burn*, only the surface of the epidermis is damaged. In this type of burn, which includes most sunburns, the skin reddens and can be painful. The redness, a sign called **erythema** (er-i-THĒ-muh), results from inflammation of the sun-damaged tissues.

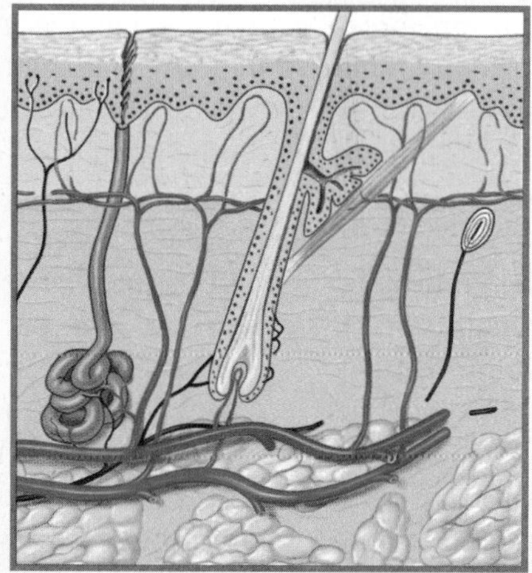

In a **second-degree burn** or *partial-thickness burn*, the entire epidermis and perhaps some of the dermis are damaged. Hair follicles and glands are usually not affected, but blistering, pain, and swelling occur. If the blisters rupture, infection can easily develop. Healing typically takes one to two weeks, and some scar tissue may form.

Full-Thickness Burns

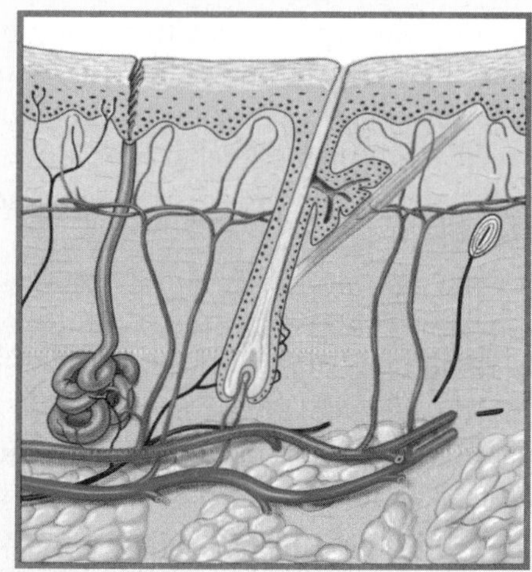

Third-degree burns, or *full-thickness burns*, destroy the epidermis and dermis, extending into the subcutaneous layer. Despite swelling, these burns are less painful than second-degree burns, because sensory nerves are destroyed. These burns cannot repair themselves, because granulation tissue cannot form and epithelial cells cannot cover the injury site. *Skin grafting* is usually necessary.

Estimation of Surface Burn Area

A simple method for estimating burn area is the **rule of nines**. The surface area in adults is divided into multiples of 9 and then the damaged regions are totaled. This rule is modified for children because their body proportions are different.

The depth of the burn can be assessed quickly with a pin. Because loss of sensation indicates a full-thickness burn, no reaction to a pin prick indicates a third-degree burn.

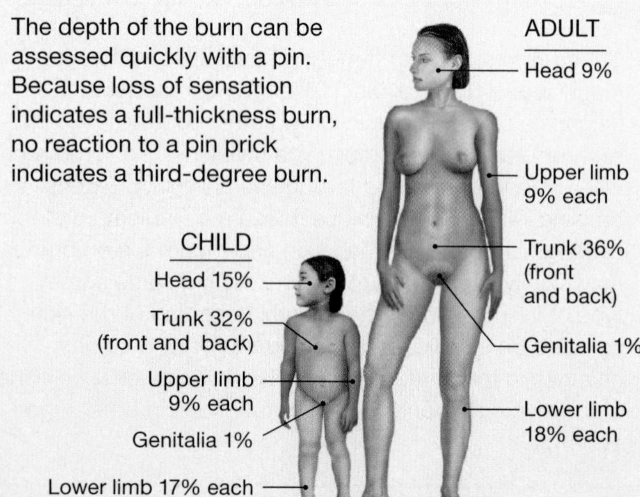
ADULT
Head 9%
Upper limb 9% each
Trunk 36% (front and back)
Genitalia 1%
Lower limb 18% each
CHILD
Head 15%
Trunk 32% (front and back)
Upper limb 9% each
Genitalia 1%
Lower limb 17% each

Burns and Skin Function

Burns that cover more than 20 percent of the skin surface threaten life, because they affect the following functions.

Skin Functions Affected by Burns

- **Fluid and Electrolyte Balance**. Fluid and electrolytes are lost because the skin is no longer an effective barrier. In second-degree burns, fluid loss may reach five times the normal level.
- **Thermoregulation**. Increased fluid loss causes increased cooling by evaporation. As a result, more energy must be used to maintain normal body temperature.
- **Protection from Infection**. Widespread infection (*sepsis*) can result. Infection is the leading cause of death in burn victims.

Skin Grafts

In a **skin graft**, areas of intact skin are transplanted to cover the site of the burn. A *split-thickness graft* involves a transfer of the epidermis and superficial portions of the dermis. A *full-thickness graft* involves the epidermis and both layers of the dermis.

5-10 Effects of aging on the skin include thinning, wrinkling, and reduced melanocyte activity

Learning Outcome **Summarize the effects of aging on the skin.**

Aging affects all the components of the integumentary system:

- The epidermis thins as basal cell activity declines. The connections between the epidermis and dermis weaken, making older people more prone to skin injury, tears, and infections.
- The number of dendritic cells (Langerhans cells) decreases to about 50 percent of levels seen at about age 21. This decrease may reduce the sensitivity of the immune system and further encourage skin damage and infection.
- Vitamin D_3 production declines by about 75 percent. The result can be reduced calcium ion and phosphate ion absorption, eventually leading to muscle weakness and a reduction in bone strength and density.
- Melanocyte activity declines. In light-skinned individuals the skin becomes very pale. With less melanin in the skin, people become more sensitive to sunlight and more likely to experience sunburn.
- Glandular activity declines. The skin becomes dry and often scaly, because sebum production is reduced. Eccrine sweat glands are also less active. With impaired perspiration, older people cannot lose heat as fast as younger people can. So the elderly are at greater risk of overheating in warm environments.
- The blood supply to the dermis is reduced. Reduction in blood flow makes the skin become cool, which in turn can stimulate thermoreceptors, making a person feel cold even in a warm room. However, reduced circulation and sweat gland function in the elderly lessens their ability to lose body heat. For this reason, overexertion or exposure to high temperatures (such as those in a sauna or hot tub) can cause body temperatures to soar dangerously high.
- Hair follicles stop functioning or produce thinner, finer hairs. With decreased melanocyte activity, these hairs are gray or white.
- The dermis thins, and the elastic fiber network decreases in size. As a result, the integument becomes weaker and less resilient, and sagging and wrinkling occur. These effects are most noticeable in areas of the body that have been exposed to the sun.
- With changes in levels of sex hormones, secondary sexual characteristics in hair and body fat distribution begin to fade.
- Skin repairs proceed more slowly. Repairs to an uninfected blister might take three to four weeks in a young adult, but the same repairs could take six to eight weeks at age 65–75. Because healing takes place more slowly, recurring infections may result.

Build Your Knowledge Figure 5–18 reviews the integumentary system. Build Your Knowledge figures will appear after each body system has been covered to help build your understanding of the interconnections among all other body systems. Since we have covered only one body system thus far, Figure 5–18 does not include references to body systems covered in upcoming chapters.

Checkpoint

31. **Older people do not tolerate the summer heat as well as they did when they were young, and they are more prone to heat-related illnesses. What accounts for these changes?**
32. **Why does hair turn white or gray with age?**

See the blue Answers tab at the back of the book.

Build Your Knowledge

Figure 5–18 Integration of the INTEGUMENTARY system with the other body systems presented so far.

Integumentary System

The integumentary system performs a variety of functions for the human body. It:

- covers and protects underlying tissues and organs from impacts, abrasions, chemicals, and infections, and it prevents the loss of body fluids
- maintains normal body temperature by regulating heat exchange with the environment
- synthesizes vitamin D_3 in the epidermis, which aids calcium uptake and stores large reserves of lipids in the dermis
- detects touch, pressure, pain, and temperature stimuli
- excretes salts, water, and organic wastes
- coordination of the immune response to pathogens and skin cancers

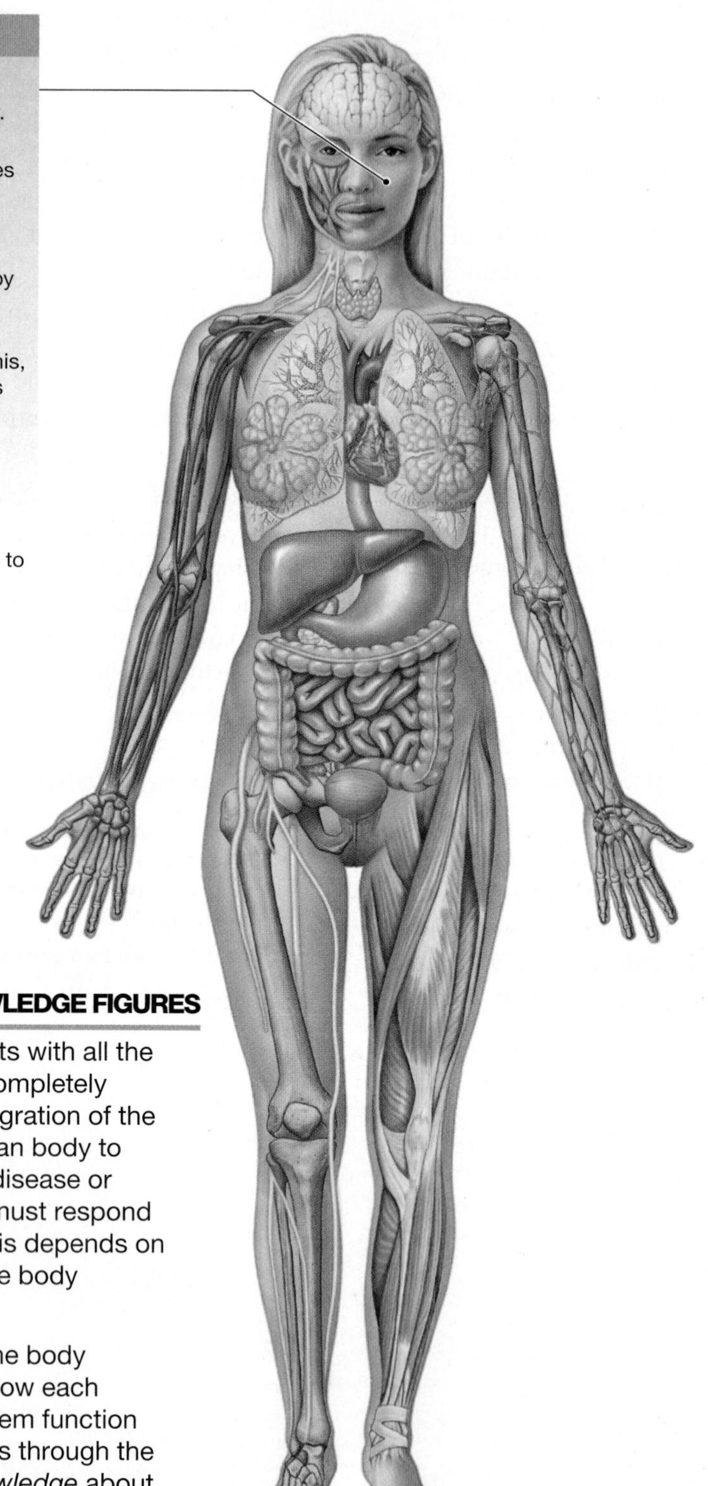

ABOUT THE BUILD YOUR KNOWLEDGE FIGURES

Since each body system interacts with all the others, no one system can be completely understood in isolation. The integration of the various systems allows the human body to function seamlessly, and when disease or injury strikes, multiple systems must respond for healing to occur. Homeostasis depends on the thorough integration of all the body systems working as one.

These figures will introduce the body systems one by one and show how each influences the others to make them function more effectively. As we progress through the systems, you will *build your knowledge* about the complementary nature of the interactions between the various body systems.

5 Chapter Review

Study Outline

> MasteringA&P™ Access more chapter study tools online in the MasteringA&P Study Area:
> - Chapter Quizzes, Chapter Practice Test, MP3 Tutor Sessions, and Clinical Case Studies
> - Practice Anatomy Lab PAL 3.0
> - A&P Flix A&PFlix
> - Interactive Physiology iP2
> - PhysioEx PhysioEx 9.1

An Introduction to the Integumentary System p. 153

1. The **integument**, or **integumentary system**, consists of the **cutaneous membrane**, or *skin* (which includes the **epidermis** and the **dermis**) and the **accessory structures**. Beneath the dermis lies the **subcutaneous layer** (*hypodermis*). *(Figures 5–1, 5–2)*
2. Functions of the integument include *protection, excretion, temperature maintenance, vitamin D_3 synthesis, nutrient storage, sensory detection*, and *coordination of the immune response. (Figure 5–2)*

5-1 The epidermis is a protective covering composed of layers with various functions p. 153

3. **Thin skin**, consisting of four layers of **keratinocytes**, covers most of the body. Heavily abraded body surfaces may be covered by **thick skin** containing five layers of keratinocytes. *(Spotlight Figure 5–3)*
4. The epidermis provides physical protection, prevents fluid loss, and helps keep microorganisms out of the body.
5. Cell divisions in the **stratum basale**, the deepest epidermal layer, replace more superficial cells. *(Spotlight Figure 5–3)*
6. As epidermal cells age, they pass through the stratum basale, the **stratum spinosum**, the **stratum granulosum**, the **stratum lucidum** (in thick skin), and the **stratum corneum.** In the process, they accumulate large amounts of **keratin.** Ultimately, the cells are shed. *(Spotlight Figure 5–3)*
7. The epidermis and dermis are bonded together by **epidermal ridges** interlocked with **dermal papillae**. *(Spotlight Figure 5–3)*
8. *Dendritic (Langerhans) cells* in the stratum spinosum are part of the immune system. *Tactile (Merkel) cells* in the stratum basale provide sensory information about objects that touch the skin.
9. **Epidermal growth factor (EGF)** promotes growth, division, and repair of the epidermis and epithelial gland synthesis and secretion of secretory product.

5-2 The dermis is the tissue layer that supports the epidermis p. 158

10. The dermis consists of the superficial **papillary layer** and the deeper **reticular layer**. *(Figures 5–1, 5–2, 5–4)*
11. The papillary layer of the dermis contains blood vessels, lymphatic vessels, and sensory nerves that supply the epidermis. The reticular layer consists of a meshwork of collagen and elastic fibers oriented to resist tension in the skin.
12. Extensive distention of the dermis can cause **stretch marks**.
13. The pattern of collagen and elastic fiber bundles forms **tension lines** (cleavage lines). *(Figure 5–5)*
14. Arteries to the skin form the **cutaneous plexus** in the subcutaneous layer. The **subpapillary plexus** is at the base of the papillary layer of the dermis. *(Figure 5–6)*
15. Integumentary sensory receptors detect both light touch and pressure.

5-3 The subcutaneous layer connects the dermis to underlying tissues p. 160

16. The **subcutaneous layer** (*hypodermis*), stabilizes the skin's position against underlying organs and tissues. *(Figures 5–1, 5–7)*

5-4 Epidermal pigmentation and dermal circulation influence skin color p. 161

17. The color of the epidermis depends on two factors: epidermal pigmentation and dermal blood supply.
18. The epidermis contains the pigments **melanin** and **carotene**. **Melanocytes**, which produce melanin, protect us from **ultraviolet (UV) radiation**. *(Figure 5–8)*
19. Interruptions of the dermal blood supply or poor oxygenation in the lungs can lead to **cyanosis**. **Vitiligo** is the appearance of nonpigmented white patches of various sizes on otherwise normally pigmented skin. *(Figure 5–9)*

5-5 Sunlight causes epidermal cells to convert a steroid into vitamin D_3 p. 163

20. Epidermal cells synthesize **vitamin D_3**, or **cholecalciferol**, when exposed to the UV radiation in sunlight. **Rickets** is a disease caused by vitamin D_3 deficiency. *(Figures 5–10, 5–11)*

5-6 Hair is made of keratinized dead cells pushed to the skin surface where it has protecting and insulating roles p. 165

21. **Hairs** originate in complex organs called **hair follicles**. Each hair has a **root** and a **shaft**. At the base of the root are a **hair papilla**, surrounded by a **hair bulb**, and a **root hair plexus** of sensory nerves. Hairs have a **medulla**, or core of soft keratin, surrounded by a **cortex** of hard keratin. The **cuticle** is a superficial layer of dead cells that protects the hair. *(Figure 5–12)*
22. Our bodies have both **vellus hairs** ("peach fuzz") and heavy **terminal hairs**. A hair that has stopped growing is called a **club hair**. A curly hair and its follicle are oval; a straight hair and its follicle are round.
23. Each **arrector pili muscle** (*arrector muscle of hair*) can erect a single hair. *(Figure 5–12)*
24. Our hairs grow and are shed according to the **hair growth cycle**. A typical hair on the head grows for two to five years and is then shed.

5-7 Sebaceous glands and sweat glands are exocrine glands found in the skin p. 168

25. A typical **sebaceous gland** discharges waxy **sebum** into a lumen and, ultimately, into a hair follicle. **Sebaceous follicles** are large sebaceous glands that discharge sebum directly onto the epidermis. *(Figure 5–13)*
26. The two types of **sweat glands**, or *sudoriferous glands*, are apocrine and eccrine (*merocrine*) sweat glands. **Apocrine sweat glands** produce an odorous secretion. The more numerous

eccrine sweat glands produce a watery secretion known as sensible perspiration. *(Figure 5–14)*

27. **Mammary glands** of the breasts are structurally similar to apocrine sweat glands. **Ceruminous glands** in the ear produce a waxy substance called **cerumen**.

5-8 Nails are keratinized epidermal cells that protect the tips of fingers and toes p. 171

28. The **nail body** of a **nail** covers the **nail bed**. Nail production occurs at the **nail root**, which is covered by the **cuticle**, or **eponychium**. The **free edge** of the nail extends over the **hyponychium**. *(Figure 5–15)*

5-9 After an injury, the integument is repaired in several phases p. 172

29. Based on the division of stem cells, the skin can regenerate effectively even after considerable damage. The process begins with bleeding and inflammation (*inflammation phase*), the formation of a **scab** and **granulation tissue** (*migration phase*), loss of granulation tissue and undermining of the scab (*proliferation phase*), and ends with the creation of **scar tissue** (*scarring phase*). *(Figure 5–16)*
30. **Keloids** are raised areas of fibrous scar tissue. *(Figure 5–17)*

5-10 Effects of aging on the skin include thinning, wrinkling, and reduced melanocyte activity p. 175

31. With aging, the integument thins, blood flow decreases, cellular activity decreases, and repairs take place more slowly.
32. The integumentary system provides physical protection for all other body systems. *(Figure 5–18)*

Review Questions

See the blue Answers tab at the back of the book.

LEVEL 1 Reviewing Facts and Terms

1. Identify the different portions **(a–d)** of the cutaneous membrane and the underlying layer of loose connective tissue **(e)** in the diagram to the right.
   (a) ________
   (b) ________
   (c) ________
   (d) ________
   (e) ________
2. The two major components of the integumentary system are **(a)** the cutaneous membrane and the accessory structures, **(b)** the epidermis and the subcutaneous layer, **(c)** the hair and the nails, **(d)** the dermis and the subcutaneous layer.
3. Beginning at the basement membrane and traveling toward the free surface, the epidermis includes the following strata: **(a)** corneum, lucidum, granulosum, spinosum, basale, **(b)** granulosum, lucidum, spinosum, basale, corneum, **(c)** basale, spinosum, granulosum, lucidum, corneum, **(d)** lucidum, granulosum, spinosum, basale, corneum.
4. Each of the following is a function of the integumentary system, *except* **(a)** protection of underlying tissue, **(b)** excretion of salts and wastes, **(c)** maintenance of body temperature, **(d)** synthesis of vitamin C, **(e)** storage of nutrients.
5. Exposure of the skin to ultraviolet (UV) radiation **(a)** can result in increased numbers of melanocytes forming in the skin, **(b)** can result in decreased melanin production in melanocytes, **(c)** can cause destruction of vitamin D_3, **(d)** can result in damage to the DNA of cells in the stratum basale, **(e)** has no effect on the skin cells.
6. The two major components of the dermis are the **(a)** superficial fascia and cutaneous membrane, **(b)** epidermis and subcutaneous layer, **(c)** papillary layer and reticular layer, **(d)** stratum basale and stratum corneum.
7. The cutaneous plexus and subpapillary plexus consist of **(a)** blood vessels providing the dermal blood supply, **(b)** a network of nerves providing dermal sensations, **(c)** specialized cells for cutaneous sensations, **(d)** gland cells that release cutaneous secretions.
8. The accessory structures of the integument include the **(a)** blood vessels, glands, muscles, and nerves, **(b)** tactile discs, lamellar corpuscles, and tactile corpuscles, **(c)** hair, skin, and nails, **(d)** hair follicles, nails, sebaceous glands, and sweat glands.
9. The portion of the hair follicle where cell divisions occur is the **(a)** shaft, **(b)** matrix, **(c)** root hair plexus, **(d)** cuticle.
10. The two types of exocrine glands in the skin are **(a)** eccrine and sweat glands, **(b)** sebaceous and sweat glands, **(c)** apocrine and sweat glands, **(d)** sebaceous and eccrine glands.
11. Apocrine sweat glands can be controlled by **(a)** the autonomic nervous system, **(b)** regional control mechanisms, **(c)** the endocrine system, **(d)** both a and c.
12. The primary function of sensible perspiration is to **(a)** get rid of wastes, **(b)** protect the skin from dryness, **(c)** maintain electrolyte balance, **(d)** reduce body temperature.
13. The stratum corneum of the nail root, which extends over the exposed nail, is called the **(a)** hyponychium, **(b)** eponychium, **(c)** lunula, **(d)** cerumen.
14. Muscle weakness and a reduction in bone strength in the elderly result from decreased **(a)** vitamin D_3 production, **(b)** melanin production, **(c)** sebum production, **(d)** dermal blood supply.
15. In which layer(s) of the epidermis does cell division occur?
16. What is the function of the arrector pili muscles?
17. What widespread effects does epidermal growth factor (EGF) have on the integument?
18. What two major layers constitute the dermis, and what components are in each layer?
19. List the four phases in the regeneration of the skin after an injury.

LEVEL 2 Reviewing Concepts

20. Contrast insensible perspiration and sensible perspiration.
21. In clinical practice, drugs can be delivered by diffusion across the skin. This delivery method is called *transdermal administration*. Why are fat-soluble drugs more suitable for transdermal administration than drugs that are water soluble?
22. In our society, a tanned body is associated with good health. However, medical research constantly warns about the dangers of excessive exposure to the sun. Describe how a tan develops and cite a benefit of minimal sun exposure.
23. Why is it important for a surgeon to choose—when possible—an incision pattern according to the skin's tension lines?
24. The fibrous protein that is responsible for the strength and water resistance of the skin surface is **(a)** collagen, **(b)** eleidin, **(c)** keratin, **(d)** elastin, **(e)** keratohyalin.

25. The darker a person's skin color, **(a)** the more melanocytes she has in her skin, **(b)** the more layers she has in her epidermis, **(c)** the more melanin her melanocytes produce, **(d)** the more superficial her blood vessels.
26. In order for bacteria on the skin to cause an infection in the skin, they must accomplish all of the following, *except* **(a)** survive the bactericidal components of sebum, **(b)** avoid being flushed from the surface of the skin by sweat, **(c)** penetrate the stratum corneum, **(d)** penetrate to the level of the capillaries, **(e)** escape the dendritic cells.

LEVEL 3 Critical Thinking and Clinical Applications

27. In an elderly person, blood supply to the dermis is reduced and sweat glands are less active. This combination of factors would most affect **(a)** the ability to thermoregulate, **(b)** the ability to heal injured skin, **(c)** the ease with which the skin is injured, **(d)** the physical characteristics of the skin, **(e)** the ability to grow hair.
28. Two patients arrive at the emergency department. One has cut his finger with a knife; the other has stepped on a nail. Which wound has a greater chance of becoming infected? Why?
29. Exposure to optimum amounts of sunlight is necessary for proper bone maintenance and growth in children. **(a)** What does sunlight do to promote bone maintenance and growth? **(b)** If a child lives in an area where exposure to sunlight is rare because of pollution or overcast skies, what can be done to minimize impaired maintenance and growth of bone?
30. One of the factors to which lie detectors respond is an increase in electrical skin conductance due to the presence of moisture. Explain the physiological basis for the use of this indicator.
31. Many people change the natural appearance of their hair, either by coloring it or by altering the degree of curl in it. Which layers of the hair do you suppose are affected by the chemicals added during these procedures? Why are the effects of the procedures not permanent?

5

+ CLINICAL CASE Wrap-Up He Has Fish Skin!

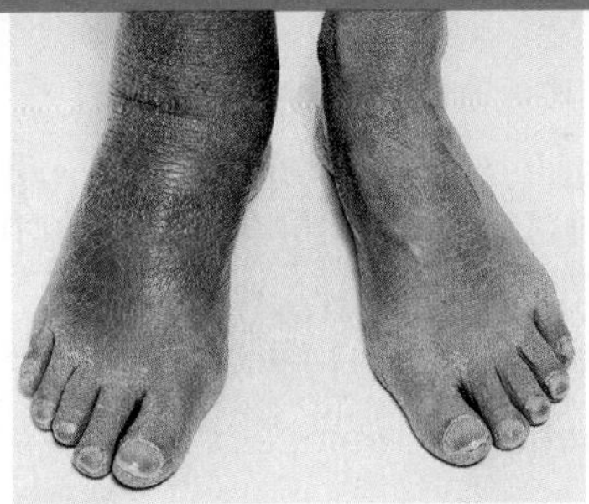

Grandpa Will has *ichthyosis vulgaris* (IK-thē-oh-sis vōl-GĀR-is). Ichthyosis literally means "fish-like condition," a feature his grandsons noticed right away. Will inherited this skin condition from his parents, who carried a gene mutation for a structural protein. The lack of this protein impairs keratinization. With this condition, skin cells also have fewer desmosomes and tight junctions, so the epidermis becomes flaky and resembles fish scales.

There is no cure. Every day Grandpa must tend to his skin by applying moisturizers to draw and retain moisture and emollients to soften the scales. He avoids soap, which further dries his skin, and uses rubber gloves to wash the dishes. In winter, he runs a humidifier in the bedroom. If his skin care gets lax, his skin can break into deep, painful cracks. Because the natural skin barrier is disrupted, secondary infection can move in.

1. Why is ichthyosis called a "disorder of cornification"? Which cells are involved?
2. Grandpa has trouble when it's hot outside. What's wrong?

See the blue Answers tab at the back of the book.

Related Clinical Terms

carbuncle: A skin infection that often involves a group of hair follicles. The infected material forms a lump, which occurs deep in the skin; the medical term for multiple boils.

cold sore: A lesion that typically occurs in or around the mouth and is caused by a dormant herpes simplex virus that may be reactivated by factors such as stress, fever, or sunburn. Also called *fever blister.*

comedo: The primary sign of acne consisting of an enlarged pore filled with skin debris, bacteria, and sebum (oil); the medical term for a blackhead.

dermatology: The branch of medicine concerned with the diagnosis, treatment, and prevention of diseases of the skin, hair, and nails.

eczema: Rash characterized by inflamed, itchy, dry, scaly, or irritated skin.

frostbite: Injury to body tissues caused by exposure to below-freezing temperatures, typically affecting the nose, fingers, or toes and sometimes resulting in gangrene.

furuncle: A skin infection involving an entire hair follicle and nearby skin tissue; the medical term for a boil.

gangrene: A term that describes dead or dying body tissue that occurs because the local blood supply to the tissue is either lost or is inadequate to keep the tissue alive.

impetigo: An infection of the surface of the skin, caused by staphylococcus ("staph") and streptococcus ("strep") bacteria.

nevus: A benign pigmented spot on the skin such as a mole.

onycholysis: A nail disorder characterized by a spontaneous separation of the nail bed starting at the distal free margin and progressing proximally.

pallor: An unhealthy pale appearance.

porphyria: A rare hereditary disease in which the blood pigment hemoglobin is abnormally metabolized. Porphyrins are excreted in the urine, which becomes dark; other symptoms include mental disturbances and extreme sensitivity of the skin to light.

rosacea: A condition in which certain facial blood vessels enlarge, giving the cheeks and nose a flushed appearance.

scleroderma: An idiopathic chronic autoimmune disease characterized by hardening and contraction of the skin and connective tissue, either locally or throughout the body.

tinea: A skin infection caused by a fungus; also called *ringworm.*

urticaria: Skin condition characterized by red, itchy, raised areas that appear in varying shapes and sizes; commonly called *hives.*

6 Bones and Bone Structure

Learning Outcomes

These Learning Outcomes correspond by number to this chapter's sections and indicate what you should be able to do after completing the chapter.

- 6-1 ▪ Describe the major functions of the skeletal system. p. 181
- 6-2 ▪ Classify bones according to shape and structure, giving examples of each type, and explain the functional significance of each of the major types of bone markings. p. 182
- 6-3 ▪ Identify the cell types in bone, and list their major functions. p. 183
- 6-4 ▪ Compare the structures and functions of compact bone and spongy bone. p. 187
- 6-5 ▪ Compare the mechanisms of endochondral ossification and intramembranous ossification. p. 190
- 6-6 ▪ Describe the remodeling and homeostatic mechanisms of the skeletal system. p. 194
- 6-7 ▪ Discuss the effects of exercise, nutrition, and hormones on bone development and on the skeletal system. p. 196
- 6-8 ▪ Explain the role of calcium as it relates to the skeletal system. p. 198
- 6-9 ▪ Describe the types of fractures, and explain how fractures heal. p. 201
- 6-10 ▪ Summarize the effects of the aging process on the skeletal system. p. 204

CLINICAL CASE A Case of Child Abuse?

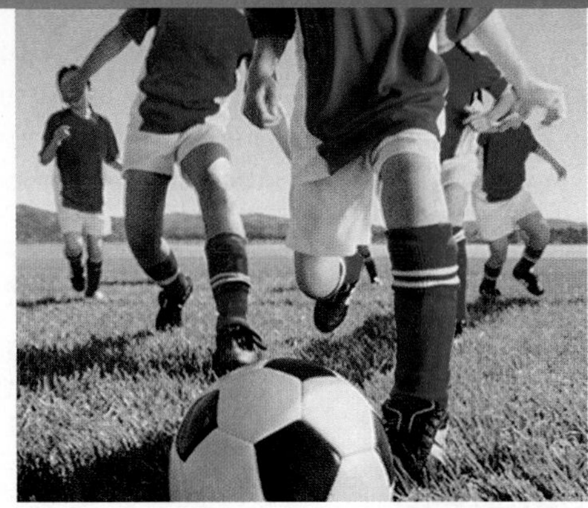

Alex's mom has been waiting in the emergency room for over 2 hours since her son was wheeled off to get an x-ray. Poor Alex had been carried home from third-grade soccer practice. The coach said she had not witnessed the injury, but it seemed like Alex had twisted and fallen without even having contact with another player. He was unable to bear weight on his left leg.

A handsome young man, *not* wearing a white coat, approaches her. "Mrs. Otero, I am Mr. Wang, the hospital social worker. I am here to speak with you about your son, Alex. Alex has an acute fracture of his tibia and fibula—the bones in his leg—but he also has two healed rib fractures and a healing fracture of his upper arm as well. We fear he has been the victim of child abuse, so he is being admitted until we can investigate further."

Mrs. Otero is dumbfounded by the suggestion of child abuse. She knows that no one at home has hurt him. Alex had complained about shoulder pain after playing on the monkey bars a few weeks ago, but she had not suspected his arm was broken. She considered Alex to be a lot like her. Both were somewhat fragile and prone to breaking bones, although she had not had any more fractures since she had married. They both had soft, discolored teeth. And, oddly enough, the whites of their eyes had a blue tint. **If Alex is not a victim of child abuse, what else could explain his multiple fractures? To find out, turn to the Clinical Case Wrap-Up on p. 206.**

An Introduction to Bones and Bone Tissue

Many people think of the skeleton as being rather boring, but this is far from the truth. The bones of the skeleton are more than just racks from which muscles hang; they have a variety of vital functions. Our bones are complex, dynamic organs that constantly change to adapt to the demands we place on them. In addition to other functions, bones support the weight of the body, and work with muscles to maintain body position and to produce controlled, precise movements. Without the skeleton to pull against, contracting muscle fibers could not make us sit, stand, walk, or run.

Chapters 6–9 describe the structure and function of the skeletal system. This chapter starts with a brief introduction to the skeletal system, and then expands on the introduction to bone presented in Chapter 4 by describing bone, or osseous tissue, a supporting connective tissue. ⮌ p. 137 We also examine the mechanisms involved in the growth, remodeling, and repair of the skeleton. Throughout the chapter we will explore how the structure of bones allows them to perform their functions.

6-1 The skeletal system has several major functions

Learning Outcome **Describe the major functions of the skeletal system.**

Your **skeletal system** includes the bones of the skeleton and the cartilages, ligaments, and other connective tissues that stabilize or interconnect the bones. This system has the following major functions:

- *Support.* The skeletal system provides structural support for the entire body. Individual bones or groups of bones provide a framework for the attachment of soft tissues and organs.
- *Storage of Minerals and Lipids.* Minerals are inorganic ions that contribute to the osmotic concentration of body fluids, as you will learn in Chapter 25. Minerals also take part in various physiological processes, and several are important as enzyme cofactors. Calcium is the most abundant mineral in the human body. The calcium salts of bone are a valuable mineral reserve that maintains normal concentrations of calcium and phosphate ions in body fluids. In addition, the bones of the skeleton store energy in the form of lipids in *yellow bone marrow*, an adipose tissue that is found in certain internal bone cavities.
- *Blood Cell Production.* Red blood cells, white blood cells, and other blood elements are produced in *red bone marrow*, which fills the internal cavities of many bones.We describe blood cell formation when we examine the cardiovascular and lymphatic systems (Chapters 19 and 22).
- *Protection.* Skeletal structures surround many soft tissues and organs. The ribs protect the heart and lungs, the skull encloses the brain, the vertebrae shield the spinal cord, and the pelvis cradles digestive and reproductive organs.
- *Leverage.* Many bones function as levers that can change the magnitude and direction of the forces generated by skeletal muscles. The movements produced range from the precise motion of a fingertip to changes in the position of the entire body.

All of the features and properties of the skeletal system ultimately depend on the unique and dynamic properties of bone. The bone specimens that you study in lab or that you may have seen in the skeletons of dead animals are only the dry remains of this living tissue. They have the same relationship to the bone in a living organism as a kiln-dried 2-by-4 board has to a living oak tree.

Checkpoint

1. **Name the major functions of the skeletal system.**

See the blue Answers tab at the back of the book.

6-2 Bones are classified according to shape and structure, and they have a variety of bone markings

Learning Outcome Classify bones according to shape and structure, giving examples of each type, and explain the functional significance of each of the major types of bone markings.

Bones are categorized by their general shape and structure. We begin our discussion with bone shapes, turn our attention to the bumps, grooves, and openings associated with bones, and then close with bone structure.

Bone Shapes

The adult skeleton typically contains 206 major bones. We can divide these into the following six broad categories according to their individual shapes (Figure 6–1):

- **Sutural bones**, or *Wormian bones*, are small, flat, irregularly shaped bones between the flat bones of the skull (Figure 6–1a). There are individual variations in the number, shape, and position of the sutural bones.
- **Irregular bones** have complex shapes (Figure 6–1b).

Figure 6–1 A Classification of Bones by Shape.

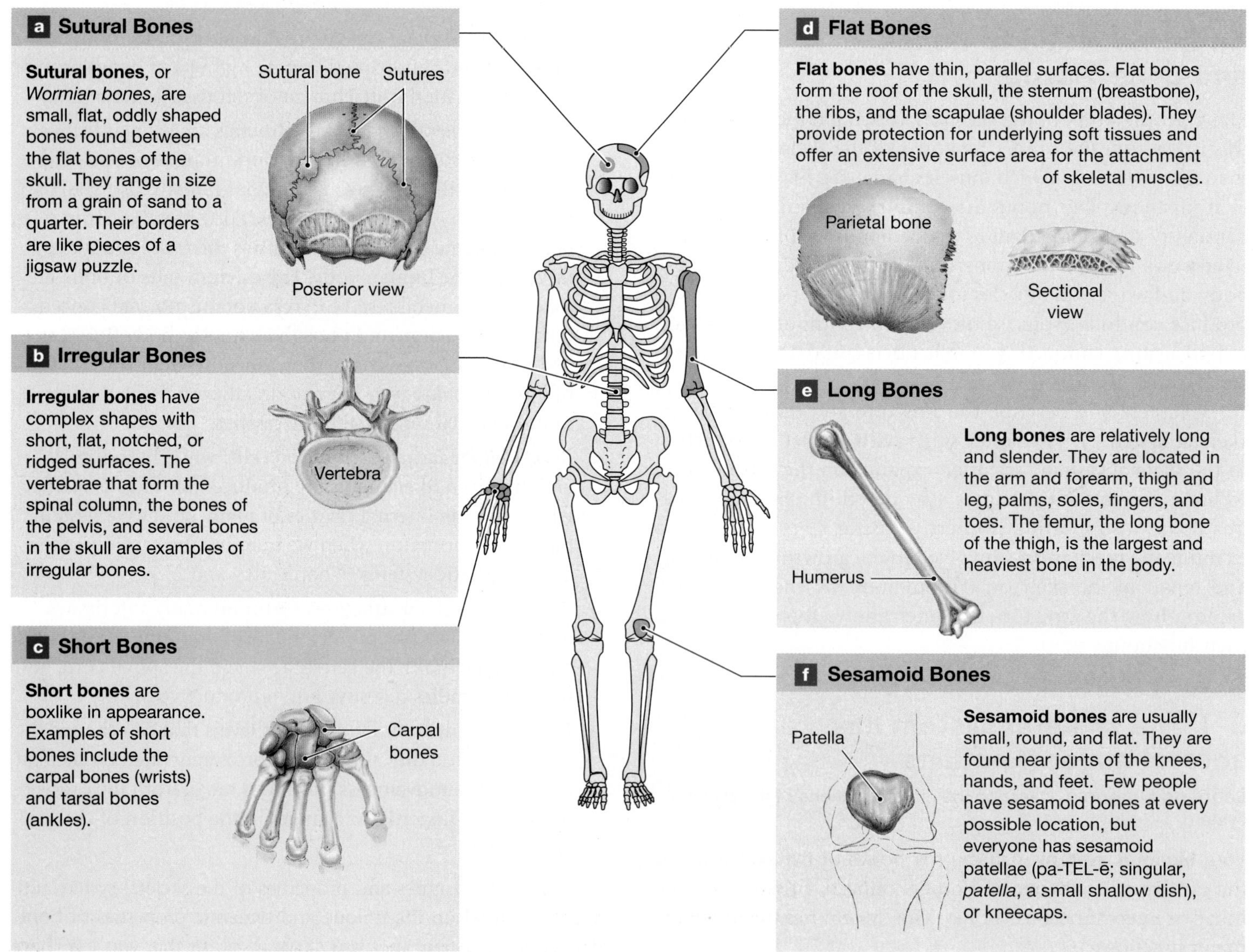
a Sutural Bones
Sutural bones, or Wormian bones, are small, flat, oddly shaped bones found between the flat bones of the skull. They range in size from a grain of sand to a quarter. Their borders are like pieces of a jigsaw puzzle.
Sutural bone
Sutures
Posterior view
b Irregular Bones
Irregular bones have complex shapes with short, flat, notched, or ridged surfaces. The vertebrae that form the spinal column, the bones of the pelvis, and several bones in the skull are examples of irregular bones.
Vertebra
c Short Bones
Short bones are boxlike in appearance. Examples of short bones include the carpal bones (wrists) and tarsal bones (ankles).
Carpal bones
d Flat Bones
Flat bones have thin, parallel surfaces. Flat bones form the roof of the skull, the sternum (breastbone), the ribs, and the scapulae (shoulder blades). They provide protection for underlying soft tissues and offer an extensive surface area for the attachment of skeletal muscles.
Parietal bone
Sectional view
e Long Bones
Long bones are relatively long and slender. They are located in the arm and forearm, thigh and leg, palms, soles, fingers, and toes. The femur, the long bone of the thigh, is the largest and heaviest bone in the body.
Humerus
f Sesamoid Bones
Sesamoid bones are usually small, round, and flat. They are found near joints of the knees, hands, and feet. Few people have sesamoid bones at every possible location, but everyone has sesamoid patellae (pa-TEL-ē; singular, patella, a small shallow dish), or kneecaps.
Patella

- **Short bones** are boxy, with approximately equal dimensions (Figure 6–1c).
- **Flat bones** have thin, parallel surfaces, which produce a flattened shape (Figure 6–1d).
- **Long bones** are relatively long and slender, consisting of a shaft with two ends that are wider than the shaft (Figure 6–1e).
- **Sesamoid bones** are usually small, round, and flat, and shaped somewhat like a sesame seed (Figure 6–1f). They tend to develop within tendons. Except for the patellae (pa-TEL-ē; singular, *patella*, a small shallow dish), or kneecaps, there are individual variations in the location and number of sesamoid bones. These variations, along with varying numbers of sutural bones, account for individual differences in the total number of bones in the skeleton. (Sesamoid bones may form in at least 26 locations.)

Bone Markings

The surfaces of each bone in your body have characteristic **bone markings** or *surface features*. Examples include projections, openings, and depressions. Projections form where muscles, tendons, and ligaments attach and where adjacent bones form joints. Openings and depressions in bone are sites where blood vessels or nerves lie alongside or penetrate the bone. Detailed examination of bone markings can yield an abundance of anatomical information. For example, anthropologists, criminologists, and pathologists can often determine the size, age, sex, and general appearance of an individual on the basis of incomplete skeletal remains.

Figure 6–2 introduces the prominent bone markings. It illustrates specific anatomical terms for the various projections, depressions, and openings. These bone markings provide fixed landmarks that can help us determine the position of the soft-tissue components of other organ systems.

Bone Structure

Figure 6–3a introduces the anatomy of the femur, the long bone of the thigh. This representative long bone has an extended tubular shaft, or **diaphysis** (dī-AF-i-sis). At each end is an expanded area known as the **epiphysis** (ē-PIF-i-sis). The diaphysis is connected to each epiphysis at a narrow zone known as the **metaphysis** (me-TAF-i-sis; *meta*, between).

The wall of the diaphysis consists of a layer of compact bone. **Compact bone** is relatively dense and solid. Compact bone forms a sturdy protective layer that surrounds a central space called the **medullary cavity** (*medulla*, innermost part), or *marrow cavity*.

The epiphyses consist largely of spongy bone. **Spongy bone**, also called *trabecular* (tra-bek-YŪ-lar) *bone*, consists of an open network of struts and plates that resembles a three-dimensional garden lattice.

Figure 6–3b shows the structure of a flat bone from the skull, one of the *parietal* (pa-RĪ-e-tul) *bones*. A flat bone resembles a spongy bone sandwich, with layers of compact bone covering a core of spongy bone. Within the cranium, the layer of spongy bone between the layers of compact bone is called the *diploë* (DIP-lō-ē; *diplous*, twofold). Red bone marrow is present within spongy bone, but there is no large medullary cavity as in the diaphysis of a long bone.

We now consider the histology of a typical bone.

✓ Checkpoint

2. **Identify the categories used for classifying a bone according to shape.**
3. **Define bone markings.**

See the blue Answers tab at the back of the book.

6

6-3 Bone is composed of matrix and several types of cells: osteogenic cells, osteoblasts, osteocytes, and osteoclasts

Learning Outcome **Identify the cell types in bone, and list their major functions.**

Bone tissue is a supporting connective tissue. Like other connective tissues, it contains specialized cells and a matrix consisting of extracellular protein fibers and a ground substance. (You may want to review the sections on dense connective tissues, cartilage, and bone in Chapter 4 now.) pp. 133, 136

Recall these characteristics of bone:

- The matrix of bone is very dense due to deposits of calcium salts around the protein fibers.
- The matrix contains bone cells, or *osteocytes* (OS-tē-ō-sītz), within pockets called *lacunae* (la-KŪ-nē). (The same term is used for the spaces that chondrocytes occupy in cartilage. p. 136) The lacunae of bone are typically organized around blood vessels that branch through the bony matrix.
- *Canaliculi* (kan-a-LIK-yū-lī), narrow passageways through the matrix, extend between the lacunae and nearby blood vessels, forming a branching network through which osteocytes exchange nutrients, wastes, and gases.
- Except at joints, a *periosteum* (per-ē-OS-tē-um) covers the outer surfaces of bones. It consists of an outer fibrous layer and an inner cellular layer.

We now take a closer look at the organization of the matrix and cells of bone.

Bone Matrix

Calcium phosphate, $Ca_3(PO_4)_2$, makes up almost two-thirds of the weight of bone. Calcium phosphate interacts with calcium

Figure 6–2 An Introduction to Bone Markings.

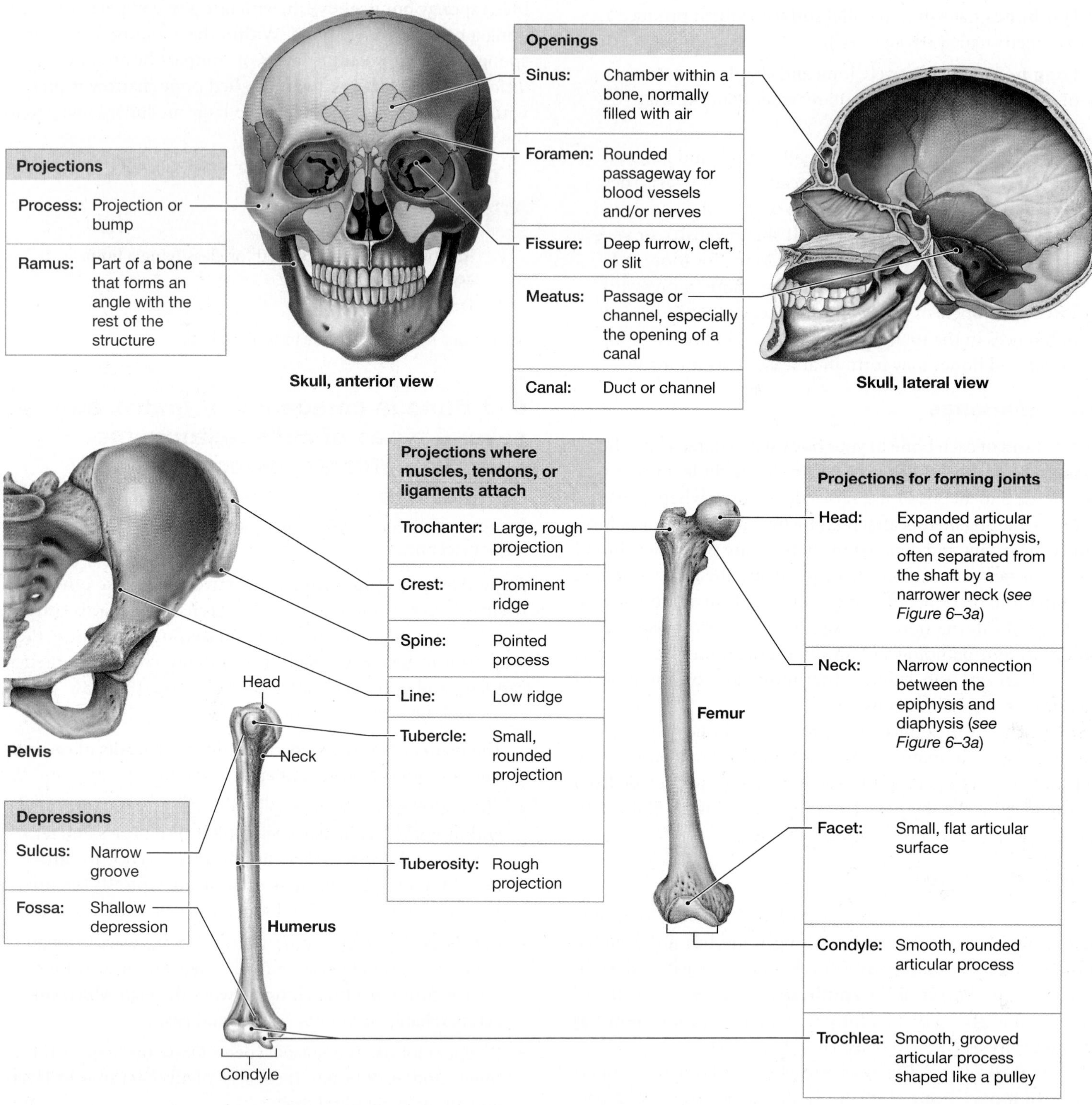
Projections
Process: Projection or bump
Ramus: Part of a bone that forms an angle with the rest of the structure
Skull, anterior view
Openings
Sinus: Chamber within a bone, normally filled with air
Foramen: Rounded passageway for blood vessels and/or nerves
Fissure: Deep furrow, cleft, or slit
Meatus: Passage or channel, especially the opening of a canal
Canal: Duct or channel
Skull, lateral view
Pelvis
Projections where muscles, tendons, or ligaments attach
Trochanter: Large, rough projection
Crest: Prominent ridge
Spine: Pointed process
Line: Low ridge
Tubercle: Small, rounded projection
Tuberosity: Rough projection
Head
Neck
Humerus
Condyle
Depressions
Sulcus: Narrow groove
Fossa: Shallow depression
Femur
Projections for forming joints
Head: Expanded articular end of an epiphysis, often separated from the shaft by a narrower neck (see Figure 6–3a)
Neck: Narrow connection between the epiphysis and diaphysis (see Figure 6–3a)
Facet: Small, flat articular surface
Condyle: Smooth, rounded articular process
Trochlea: Smooth, grooved articular process shaped like a pulley

hydroxide, $Ca(OH)_2$, to form crystals of **hydroxyapatite** (hī-drok-sē-AP-a-tīt), $Ca_{10}(PO_4)_6(OH)_2$. As these crystals form, they incorporate other calcium salts, such as calcium carbonate ($CaCO_3$), and ions such as sodium, magnesium, and fluoride. A bone without a calcified matrix looks normal, but is very flexible (Figure 6–4). About one-third of the weight of bone is collagen fibers. Cells make up only 2 percent of the mass of a typical bone.

Calcium phosphate crystals are very hard, but relatively inflexible and quite brittle. They can withstand compression but are likely to shatter with twisting, bending, or sudden impacts. Collagen fibers, by contrast, are remarkably strong.

Figure 6–3 Bone Structure.

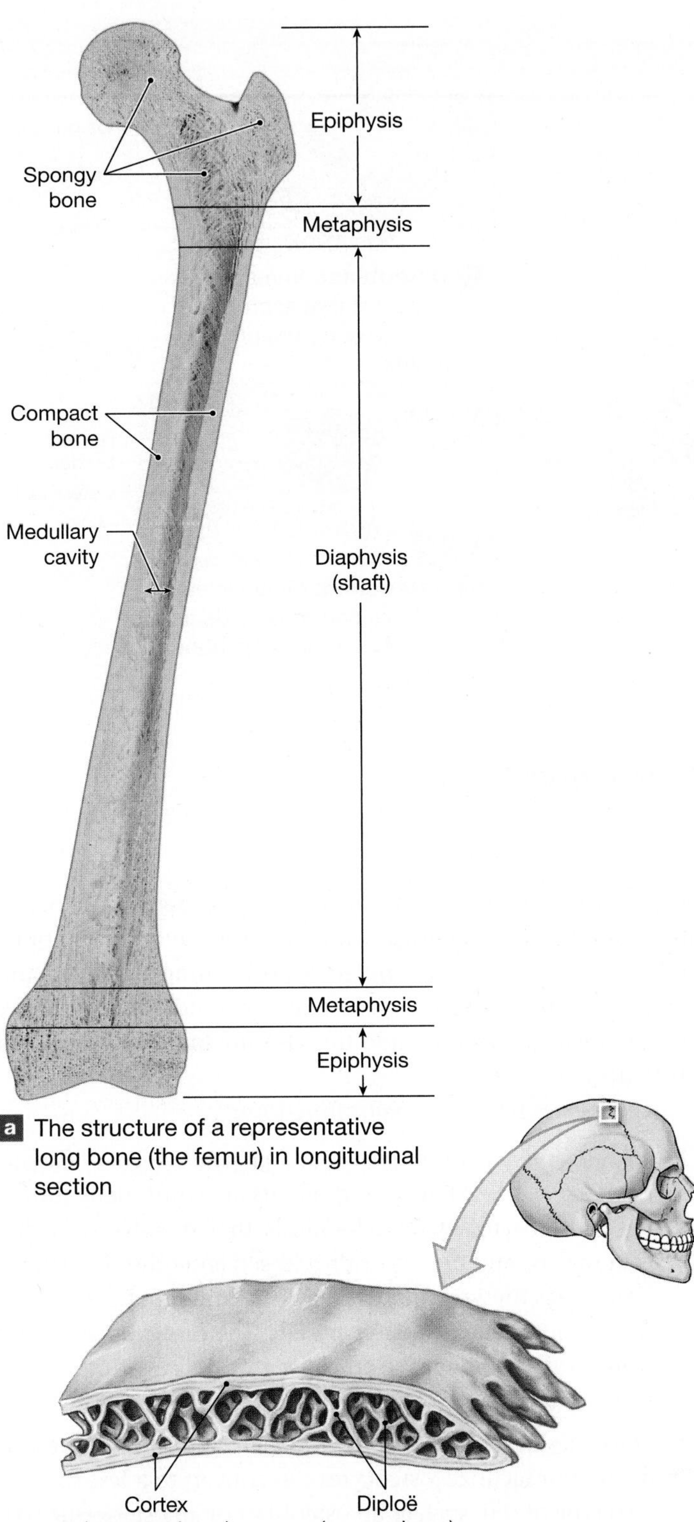
Epiphysis
Spongy bone
Metaphysis
Compact bone
Medullary cavity
Diaphysis (shaft)
Metaphysis
Epiphysis
Cortex (compact bone)
Diploë (spongy bone)

a The structure of a representative long bone (the femur) in longitudinal section

b The structure of a flat bone (the parietal bone)

? Where is spongy bone found in long bones? Where is spongy bone found in flat bones?

Figure 6–4 Bone Lacking a Calcified Matrix.

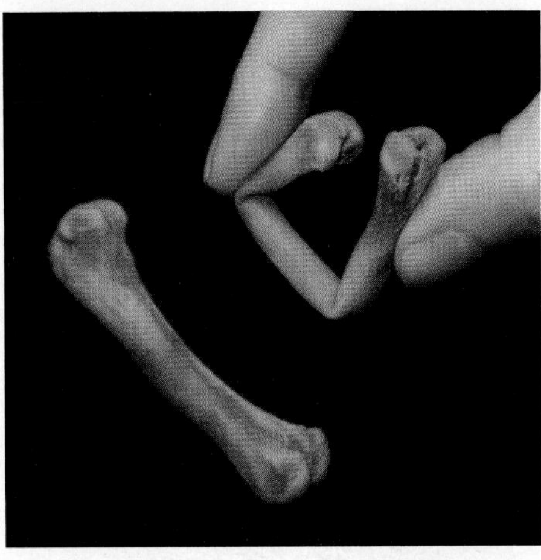

When subjected to tension (pull), they are stronger than steel. They are flexible as well as tough and can easily tolerate twisting and bending, but offer little resistance to compression. When compressed, they simply bend out of the way.

The composition of the matrix in compact bone is the same as that in spongy bone. The collagen fibers provide an organic framework for the formation of hydroxyapatite crystals. These crystals form small plates and rods that are locked into the collagen fibers at regular angles. The result is a protein–crystal combination that has the flexibility of collagen and the compressive strength of hydroxyapatite crystals. The protein–crystal interactions allow bone to be strong, somewhat flexible, and highly resistant to breaking or shattering under compression.

In its overall properties, bone is on par with the best steel-reinforced concrete. In fact, bone is far superior to concrete, because it can undergo remodeling (cycles of bone formation and resorption) as needed and can repair itself after injury.

Bone Cells

Bone contains four types of cells: osteogenic cells, osteoblasts, osteocytes, and osteoclasts (Figure 6–5). Refer back to the bone discussion in Chapter 4, p. 137.

Osteogenic Cells

Bone contains small numbers of mesenchymal cells called **osteogenic** (os-tē-ō-JEN-ik) **cells**, or *osteoprogenitor* (os-tē-ō-prō-JEN-i-tor) *cells.* These stem cells divide to produce daughter cells that differentiate into osteoblasts (Figure 6–5a). Osteogenic cells maintain populations of osteoblasts and so are important in the repair of a *fracture* (a break or a crack in a bone). Osteogenic cells are found in the inner, cellular layer of the periosteum. They are also found in a cellular layer, or *endosteum*, that lines medullary cavities and passageways for blood vessels that penetrate the matrix of compact bone.

Figure 6–5 Types of Bone Cells.

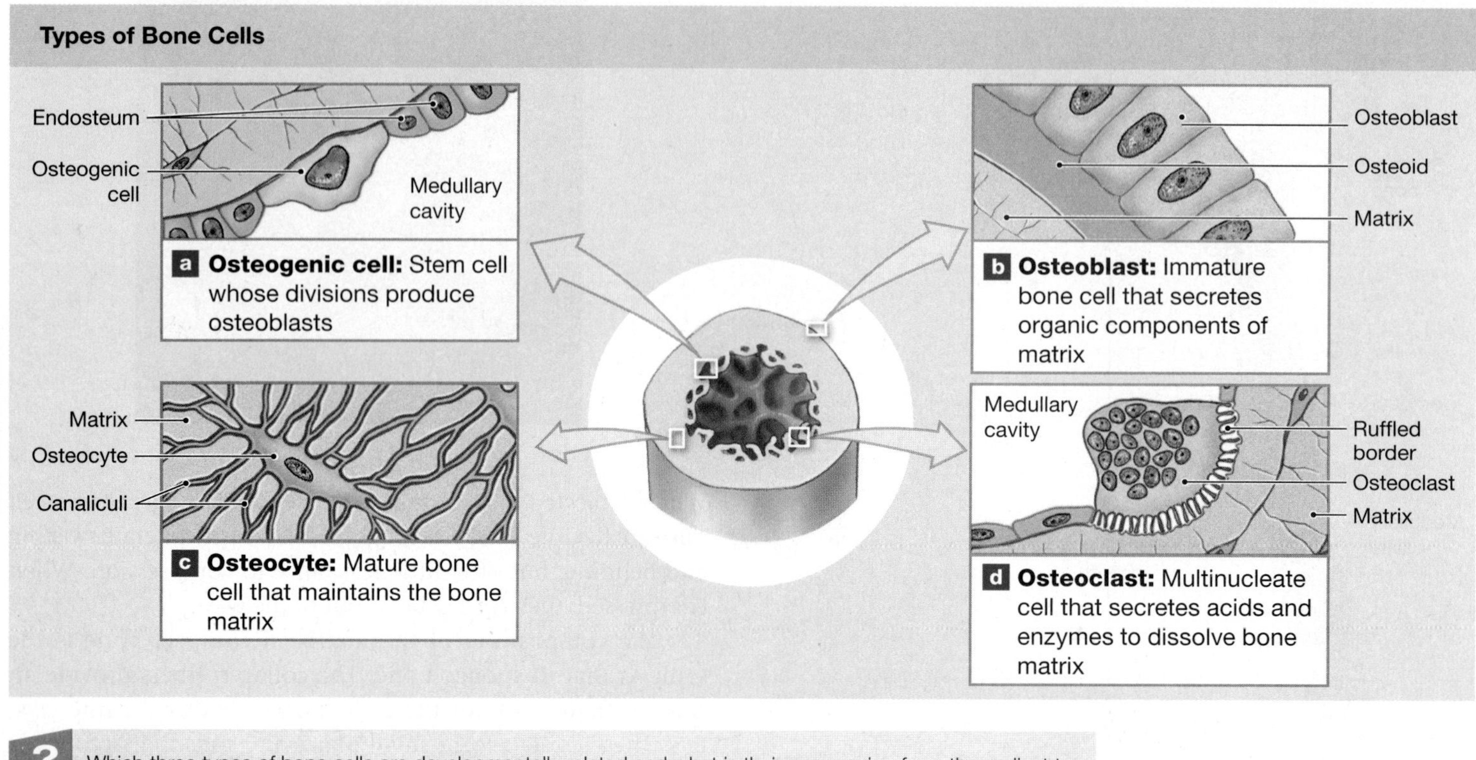

Types of Bone Cells
Endosteum
Osteogenic cell
Medullary cavity
a Osteogenic cell: Stem cell whose divisions produce osteoblasts
Matrix
Osteocyte
Canaliculi
c Osteocyte: Mature bone cell that maintains the bone matrix
Osteoblast
Osteoid
Matrix
b Osteoblast: Immature bone cell that secretes organic components of matrix
Medullary cavity
Ruffled border
Osteoclast
Matrix
d Osteoclast: Multinucleate cell that secretes acids and enzymes to dissolve bone matrix

? Which three types of bone cells are developmentally related and what is their progression from the earliest to the latest stage?

Osteoblasts

Osteoblasts (OS-tē-ō-blasts; *-blast,* immature precursor cell) produce new bone matrix in a process called **osteogenesis** (os-tē-ō-JEN-e-sis; *genesis,* production) or **ossification** (os-i-fi-KĀ-shun). Osteoblasts make and release the proteins and other organic components of the matrix. Before calcium salts are deposited, this organic matrix is called **osteoid** (OS-tē-oyd) (Figure 6–5b). Osteoblasts also help increase local concentrations of calcium phosphate above its solubility limit, triggering the deposition of calcium salts in the organic matrix. This process converts osteoid to bone. Osteocytes develop from osteoblasts that have become completely surrounded by bone matrix.

Osteocytes

Osteocytes (*osteo-,* bone + *-cyte,* cell) are mature bone cells that make up most of the cell population. Each osteocyte occupies a lacuna, a pocket sandwiched between layers of matrix (Figure 6–5c). Osteocytes cannot divide, and a lacuna never contains more than one osteocyte.

Narrow passageways called **canaliculi** radiate through the matrix. Canaliculi contain cytoplasmic extensions of osteocytes, thus supporting cell-to-cell communication between osteocytes in different lacunae and access to nutrients supplied by blood vessels in the central canal. Neighboring osteocytes are linked by gap junctions, which permit the cells to rapidly exchange ions and small molecules, including nutrients and hormones. The interstitial fluid that surrounds the osteocytes and their extensions is an additional route for the diffusion of nutrients and wastes.

Osteocytes have two major functions:

1. *Osteocytes maintain the protein and mineral content of the surrounding matrix.* Matrix components are continually turned over. Osteocytes secrete chemicals that dissolve the adjacent matrix, and the minerals released enter the circulation. Osteocytes then rebuild the matrix, stimulating the deposition of new hydroxyapatite crystals. The turnover rate varies from bone to bone. We consider this process further in a later section.
2. *Osteocytes take part in the repair of damaged bone.* If released from their lacunae, osteocytes can convert to a less specialized type of cell, such as an osteoblast or an osteogenic cell. These additional cells assist bone repair.

Osteoclasts

Osteoclasts (OS-tē-ō-clasts; *klastos,* broken) are cells that absorb and remove bone matrix (Figure 6–5d). They are large cells with 50 or more nuclei. The osteoclasts generally occur in shallow depressions called *osteoclastic crypts* (*Howship's lacunae*) that they have eroded into the matrix. Osteoclasts secrete acids and protein-digesting enzymes that dissolve the matrix and release the stored minerals. During this process, fingerlike processes, called the *ruffled border*, increase the secretory surface area of the cell in contact with the surrounding matrix. This erosion process is called **osteolysis** (os-tē-OL-i-sis; *osteo-*, bone + *lysis*, a loosening), or *resorption*. The released products are resorbed at this border. Osteolysis is an important part of the regulation of calcium and phosphate ion concentrations in body fluids.

Osteoclasts are not related to osteogenic cells or their descendants. Instead, they are derived from the same stem cells that produce monocytes and macrophages, which are cells involved in the body's defense mechanisms.

We look at how these bone cells function together in Section 6-6.

Checkpoint

4. **Mature bone cells are known as ____, bone-building cells are called ____, and ____ are bone-resorbing cells.**
5. **How would the compressive strength of a bone be affected if its ratio of collagen to hydroxyapatite increased?**
6. **If the activity of osteoclasts exceeds the activity of osteoblasts in a bone, how will the mass of the bone be affected?**

See the blue Answers tab at the back of the book.

6-4 Compact bone contains parallel osteons, and spongy bone contains trabeculae

Learning Outcome **Compare the structures and functions of compact bone and spongy bone.**

Compact bone functions to protect, support, and resist stress. Spongy bone provides some support and stores marrow. In this section, we examine in detail the structures of compact and spongy bone, which allow these types of bone to perform their functions. We also look at the outer covering and inner lining of bone, called the periosteum and endosteum, respectively.

Compact Bone Structure

At the microscopic level, the basic functional unit of mature compact bone is the **osteon** (OS-tē-on), or *Haversian* (ha-VER-jhun) *system*. In an osteon, the osteocytes are arranged in concentric layers around a vascular **central canal**, or *Haversian canal* (Figure 6–6). This canal contains one or more blood vessels (normally a capillary and a *venule*, a very small vein) that carry blood to and from the osteon (Figure 6–7a). Central canals generally run parallel to the surface of the bone. Other passageways, known as

Figure 6–6 **Osteons of Compact Bone.**

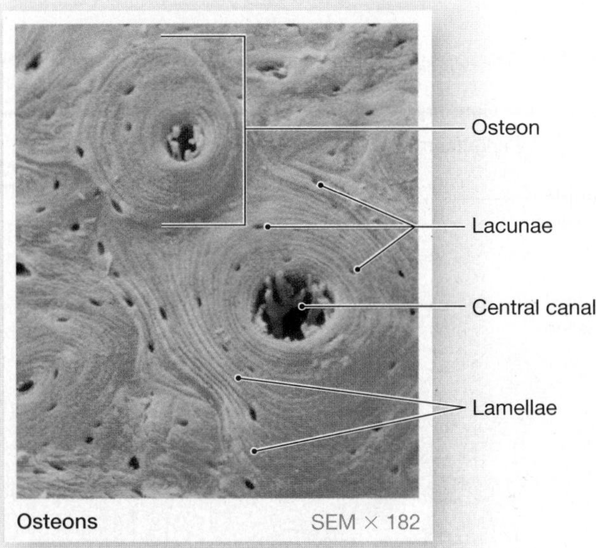
Osteon
Lacunae
Central canal
Lamellae
Osteons
SEM × 182

perforating canals, or *Volkmann's* (FŌLK-manz) *canals*, extend perpendicular to the surface. Blood vessels in these canals supply blood both to osteons deeper in the bone and to tissues of the medullary cavity.

Bone matrix forms layers called **lamellae** (lah-MEL-lē; singular, *lamella*, a thin plate). The lamellae of each osteon form a series of nested cylinders around the central canal. In transverse section, these *concentric lamellae* create a targetlike pattern, with the central canal as the bull's-eye. Collagen fibers within each lamella form a spiral pattern that adds strength and resiliency (Figure 6–7b).

Interstitial lamellae fill in the spaces between the osteons in compact bone. These lamellae are remnants of osteons whose matrix components have been almost completely recycled by osteoclasts. *Circumferential lamellae* (*circum-*, around + *ferre*, to bear) are found at the outer and inner surfaces of the bone, where they are covered by the periosteum and endosteum, respectively (see Figure 6–7). These lamellae are produced during the growth of the bone, as we will describe in Section 6-6.

Tips & Tools

To understand the relationship of collagen fibers to bone matrix, think about the reinforcing steel rods (rebar) in concrete. Like the rebar in concrete, collagen adds tensional strength and flexibility to bone.

Compact bone is thickest where stresses are applied from a limited number of directions. All osteons in compact bone are aligned the same, making such bones very strong when stressed along the axis of alignment. You might think of a single osteon as a drinking straw with very thick walls. When you try to push the ends of the straw together or to pull them apart, the straw

6

Figure 6–7 The Structure of Compact Bone.

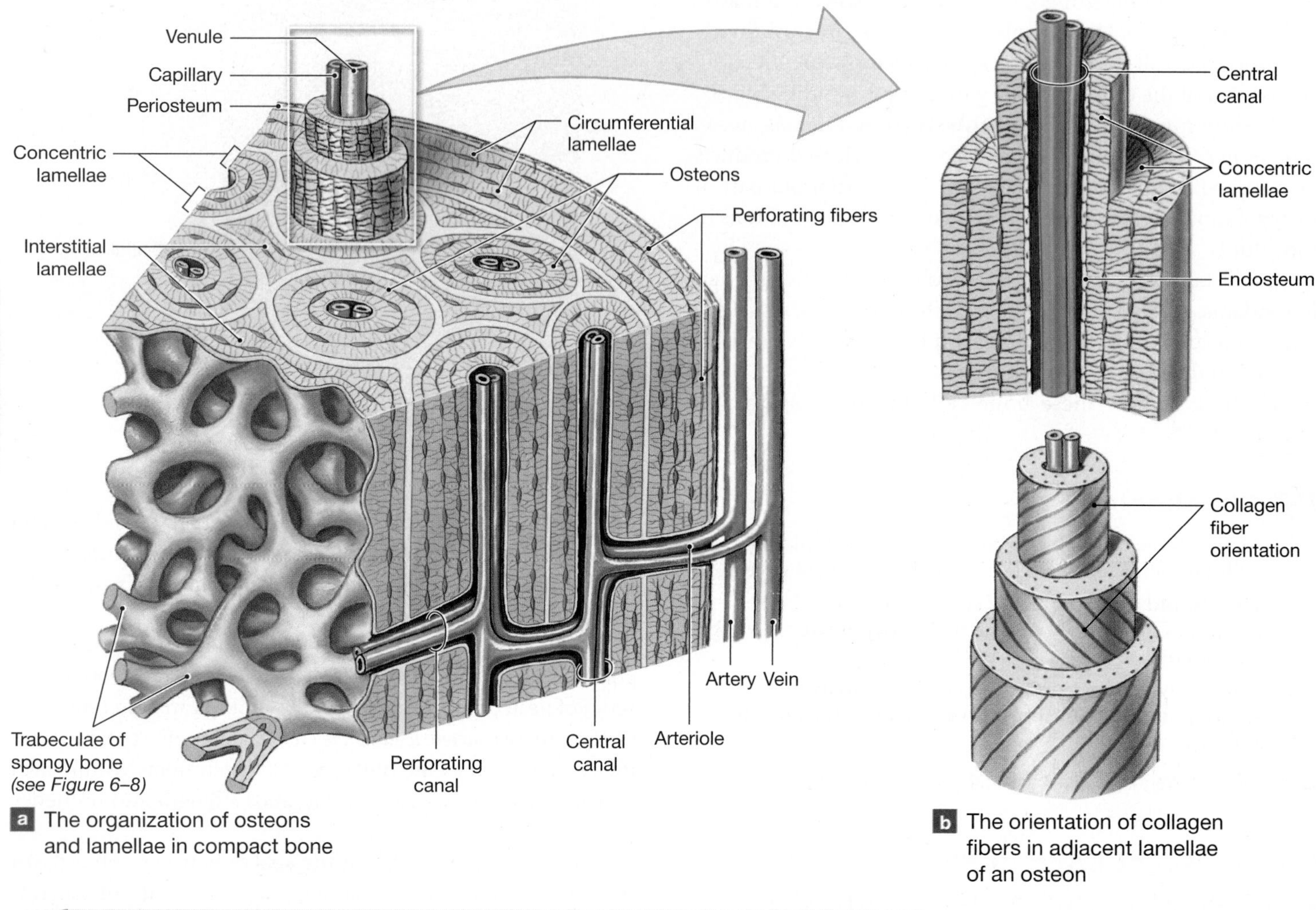
Venule
Capillary
Periosteum
Concentric lamellae
Interstitial lamellae
Circumferential lamellae
Osteons
Perforating fibers
Trabeculae of spongy bone (see Figure 6–8)
Perforating canal
Central canal
Arteriole
Artery
Vein
Central canal
Concentric lamellae
Endosteum
Collagen fiber orientation

a The organization of osteons and lamellae in compact bone

b The orientation of collagen fibers in adjacent lamellae of an osteon

? What are the three ways in which lamellae are organized in compact bone?

quite strong. But if you hold the ends and push from the side, the straw will easily bend at a sharp angle.

The osteons in the diaphysis of a long bone are parallel to the long axis of the shaft. For this reason, the shaft does not bend, even when extreme forces are applied to either end. (The femur can withstand 10–15 times the body's weight without breaking.) Yet a much smaller force applied to the side of the shaft can break the femur. A sudden sideways force, as in a fall or auto accident, causes the majority of breaks in this bone.

Spongy Bone Structure

In spongy bone, lamellae are not arranged in osteons. No osteons are present. The matrix in spongy bone forms a meshwork of supporting bundles of fibers called **trabeculae** (tra-BEK-ū-lē) (Figure 6–8). These thin trabeculae branch, creating an open network. The trabeculae are oriented along stress lines and extensively cross-braced.

Spongy bone is found where bones are not heavily stressed or where stresses originate from many directions. In addition, spongy bone is less dense than compact bone. Spongy bone reduces the weight of the skeleton, making it easier for muscles to move the bones.

There are no capillaries or venules in the matrix of spongy bone. Instead, spongy bone within the epiphyses of long bones, such as the femur, and the interior of other large bones such as the sternum and ilium (a large bone of the pelvis) contains **red bone marrow**, which forms blood cells. The framework of trabeculae supports and protects the cells of the bone marrow. Blood vessels within this tissue deliver nutrients to the osteocytes by diffusion along canaliculi that

Figure 6–8 **The Structure of Spongy Bone.**

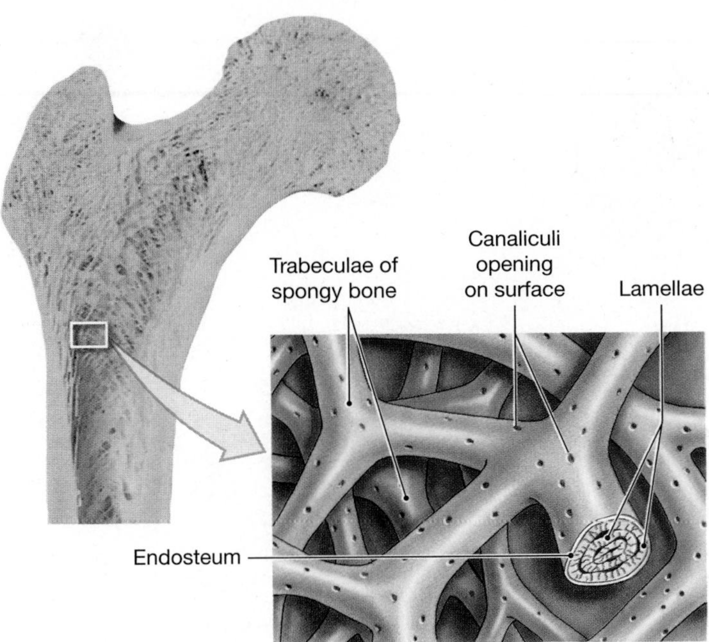
Trabeculae of
spongy bone
Canaliculi
opening
on surface
Lamellae
Endosteum

open onto the surfaces of trabeculae; they also remove wastes generated by the osteocytes. At other sites, spongy bone may contain **yellow bone marrow**—adipose tissue important as an energy reserve.

Coordinated Functions of Compact and Spongy Bone

How do compact bone and spongy bone work together to support the body's weight? Figure 6–9 shows the distribution of forces applied to the femur at the hip joint, and illustrates the functional relationship between compact bone and spongy bone. The head of the femur articulates with a socket on the lateral surface of the pelvis. At the proximal epiphysis of the femur, trabeculae transfer forces from the pelvis to the compact bone of the femoral shaft, across the hip joint. At the distal epiphysis, trabeculae transfer weight from the shaft to the leg, across the knee joint. The femoral head projects medially, and the body weight compresses the medial side of the shaft. However, because the force is applied off center, the bone must also resist the tendency to bend into a lateral bow. So while the medial portion of the shaft is under compression, the lateral portion of the shaft, which resists this bending, is placed under a stretching load, or *tension*. The medullary cavity does not reduce the bone's strength because the center of the bone is not subjected to compression or tension.

Surface Coverings of Bone

Bones have two kinds of coverings: periosteum on the outside and endosteum lining the inside. Let's look in more detail at each of these coverings.

Figure 6–9 **The Distribution of Forces on a Long Bone.** The femur, or thigh bone, has a diaphysis (shaft) with walls of compact bone and epiphyses filled with spongy bone. The body weight is transferred to the femur at the hip joint. Because the hip joint is off center relative to the axis of the shaft, the body weight is distributed along the bone in a way that compresses the medial (inner) portion of the shaft and stretches the lateral (outer) portion.

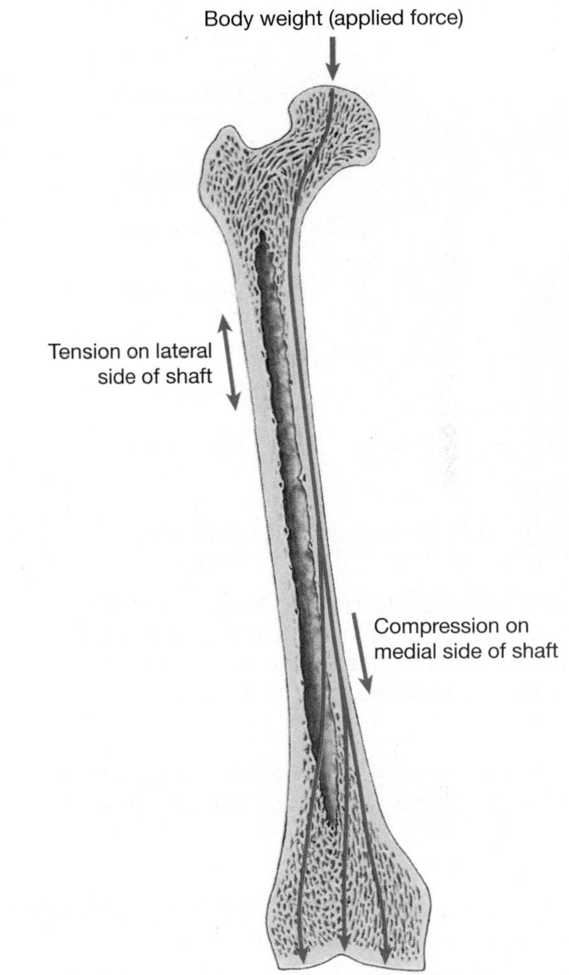
Body weight (applied force)
Tension on lateral
side of shaft
Compression on
medial side of shaft

The Periosteum

Except within joint cavities, the superficial layer of compact bone that covers all bones is wrapped by a **periosteum** (*peri*, around), a membrane with a fibrous outer layer and a cellular inner layer (Figure 6–10a). The periosteum (1) isolates the bone from surrounding tissues, (2) provides a route for the blood vessels and nerves, and (3) takes part in bone growth and repair.

Near joints, the periosteum becomes continuous with the connective tissues that hold the bones in place. At a joint that permits significant amounts of movement, called a *synovial* (si-NŌ-vē-ul) *joint*, the periosteum is continuous with the joint capsule. The fibers of the periosteum are also interwoven with those of the tendons attached to the bone. As the bone grows, these tendon fibers are cemented into the circumferential lamellae by osteoblasts from the cellular layer of the periosteum. Collagen fibers incorporated into bone tissue from

Figure 6–10 The Periosteum and Endosteum.

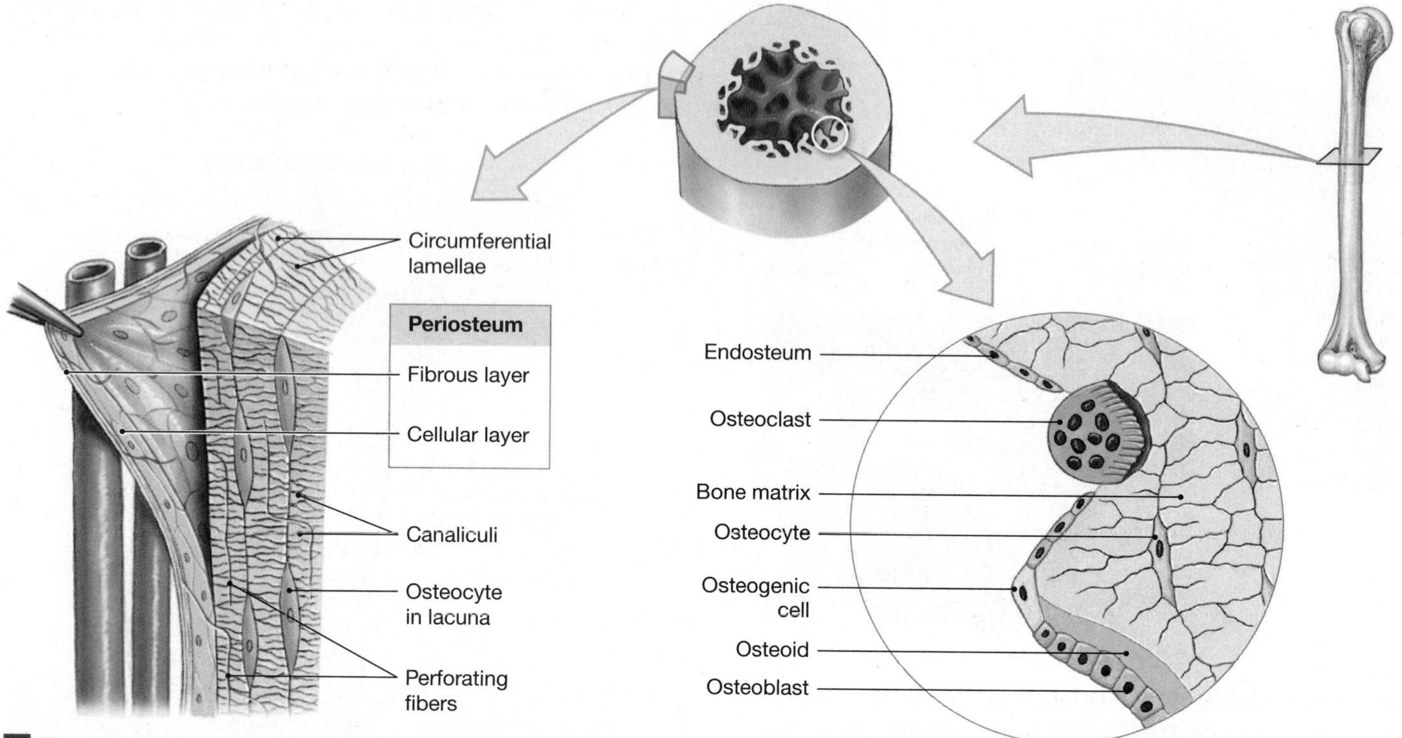

Circumferential lamellae
Periosteum
Fibrous layer
Cellular layer
Canaliculi
Osteocyte in lacuna
Perforating fibers
Endosteum
Osteoclast
Bone matrix
Osteocyte
Osteogenic cell
Osteoid
Osteoblast

a The periosteum contains outer (fibrous) and inner (cellular) layers. Collagen fibers of the periosteum are continuous with those of the bone, adjacent joint capsules, and attached tendons and ligaments.

b The endosteum is an incomplete cellular layer containing osteoblasts, osteogenic cells, and osteoclasts.

? Select the correct words for the following sentence: The (endosteum/periosteum) covers the superficial layer of compact bone and is made up of (the fibrous and cellular layers/osteoblasts, osteoclasts, and osteocytes).

tendons and ligaments, as well as from the superficial periosteum, are called *perforating (Sharpey's) fibers.*

This method of attachment bonds the tendons and ligaments into the general structure of the bone, providing a much stronger attachment than would otherwise be possible. An extremely powerful pull on a tendon or ligament will usually break a bone rather than snap the collagen fibers at the bone surface.

The Endosteum

The **endosteum** (*endo-*, inside), an incomplete cellular layer, lines the medullary cavity (Figure 6–10b). This layer is active during bone growth, repair, and remodeling. It covers the trabeculae of spongy bone and lines the inner surfaces of the central canals of compact bone. The endosteum consists of a simple flattened layer of osteogenic cells that covers the bone matrix, generally without any intervening connective tissue fibers. Where the cellular layer is not complete, the matrix is exposed. At these exposed sites, osteoclasts and osteoblasts can remove or deposit matrix components.

Checkpoint

7. **Compare the structures and functions of compact bone and spongy bone.**
8. **A sample of long bone has lamellae, which are not arranged in osteons. Is the sample most likely taken from the epiphysis or diaphysis?**

See the blue Answers tab at the back of the book.

6-5 Bones form through ossification and enlarge through interstitial and appositional growth

Learning Outcome Compare the mechanisms of endochondral ossification and intramembranous ossification.

The growth of the skeleton determines the size and proportions of your body. In this section, we consider the physical processes of bone formation, or **ossification**, and bone growth. Ossification, or osteogenesis, refers specifically to the formation of

> **+ Clinical Note** Heterotopic Bone Formation
>
> **Heterotopic** (*hetero-*, different + *topos*, place) **ossification** (**HO**) is the abnormal development of bone in non-skeletal tissues. This new bone grows at a rate that is three times faster than normal bone formation and is classified by three main types: myositis ossificans, traumatic myositis ossificans, and neurogenic heterotopic ossification. *Myositis ossificans* is a rare genetic form that causes ossification of the muscles. *Traumatic myositis ossificans* occurs in an area that has experienced repeated trauma or a single hard hit to soft tissue, or as a result of hip surgery. *Neurogenic heterotopic ossification* occurs frequently as a complication of spinal cord injury, but the pathophysiology is not as yet understood. All forms are characterized by pain and decreased range of motion; treatment is problematic because any surgical intervention may trigger more ossification.

bone. The process of **calcification**—the deposition of calcium salts—takes place during ossification, but it can also occur in other tissues. When calcification occurs in tissues other than bone, the result is a calcified tissue (such as calcified cartilage) that does not resemble bone.

Ossification occurs in two ways: endochondral and intramembranous. In *endochondral ossification*, bone replaces existing cartilage. Then bone growth occurs through interstitial growth (in length) and appositional growth (in width). In *intramembranous ossification*, bone develops directly from mesenchyme (loosely organized embryonic connective tissue) or fibrous connective tissue.

The bony skeleton begins to form about six weeks after fertilization, when the embryo is approximately 12 mm (0.5 in.) long. (At this stage, the existing skeletal elements are made of cartilage.) During fetal (beyond the eighth week) development, these cartilages are then replaced by bone, by either endochondral or intramembranous ossification. Endochondral ossification occurs mostly in long bones, while intramembranous ossification occurs mostly in flat bones.

During development after birth, the bones undergo a tremendous increase in size. Bone growth continues through adolescence, and portions of the skeleton generally do not stop growing until about age 25.

Endochondral Ossification

During development, most bones originate as hyaline cartilages that are miniature models of the corresponding bones of the adult skeleton. These cartilage models are gradually replaced by bone through the process of **endochondral** (en-dō-KON-drul) (*chondros*, cartilage) **ossification**. The steps in limb bone development provide a good example of this process (**Spotlight Figure 6–11**). Please study steps 1–4 of this figure thoroughly now. Note that this process includes the development of a **primary ossification center** *inside* the cartilage model.

Growth in Length: Interstitial Growth

The process of interstitial growth is shown in steps 5–7 of **Spotlight Figure 6–11**. Please read through these steps carefully. Note that this process involves the development of **secondary ossification centers**.

The completion of epiphyseal growth is called **epiphyseal closure**. The timing of epiphyseal closure differs from bone to bone and from individual to individual. The toes may complete ossification by age 11, but parts of the pelvis or the wrist may continue to enlarge until about age 25. The timing of endochondral ossification can be monitored by comparing the width of the **epiphyseal cartilages** in successive x-rays. In adults, the former location of this cartilage can often be seen in x-rays as a narrow *epiphyseal line*, which remains after epiphyseal closure (**Figure 6–12**).

Growth in Width: Appositional Growth

A superficial layer of bone, or bone collar, forms early in endochondral ossification (see **Spotlight Figure 6–11**, step 2). After that, the developing bone increases in diameter through **appositional growth** at the outer surface. In this process, cells of the inner layer of the periosteum differentiate into osteoblasts and deposit superficial layers of bone matrix. Eventually, these osteoblasts become surrounded by matrix and differentiate into osteocytes.

Over much of the surface, appositional growth adds a series of layers that form circumferential lamellae. In time, the deepest circumferential lamellae are recycled and replaced by osteons typical of compact bone. However, blood vessels and collagen fibers of the periosteum can sometimes become enclosed within the matrix produced by osteoblasts. Osteons may then form around the smaller vessels. While bone matrix is being added to the outer surface of the growing bone, osteoclasts are removing bone matrix at the inner surface, but at a slower rate. As a result, the medullary cavity gradually enlarges as the bone gets larger in diameter.

Intramembranous Ossification

Intramembranous (in-tra-MEM-bra-nus) **ossification** begins when osteoblasts differentiate within a mesenchymal or fibrous connective tissue. This type of ossification is also called *dermal ossification* because it normally takes place in the deeper layers of the dermis. The bones that result are called *dermal bones*. Examples of dermal bones are the flat bones of the skull, the mandible (lower jaw), and the clavicles (collarbones).

SPOTLIGHT

Figure 6–11 Endochondral Ossification

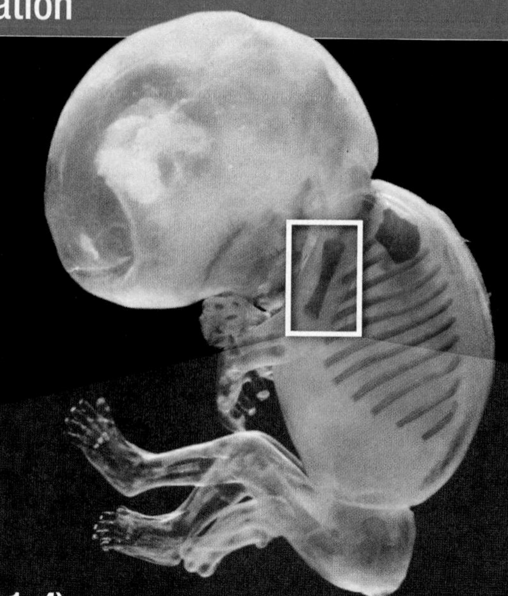

Endochondral ossification begins with the formation of a hyaline cartilage model. This model serves as the pattern for almost all bone development. Limb bone development is a good example of this process. By the time an embryo is six weeks old, the proximal bones of the limbs, the humerus (upper limb) and femur (lower limb), have formed, but they are composed entirely of cartilage. These cartilage models continue to grow by expansion of the cartilage matrix (interstitial growth) and the production of more cartilage at the outer surface (appositional growth).

Initiation of Ossification in the Developing Bone (Steps 1–4)

1 As the cartilage enlarges, chondrocytes near the center of the shaft increase greatly in size. The matrix is reduced to a series of small struts that soon begin to calcify. The enlarged chondrocytes then die and disintegrate, leaving cavities within the cartilage.

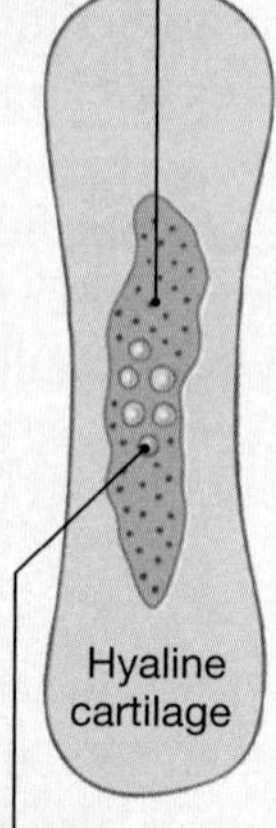
Enlarging chondrocytes within calcifying matrix
Hyaline cartilage
Disintegrating chondrocytes of the cartilage model

2 Blood vessels grow around the edges of the cartilage, and the cells of the perichondrium convert to osteoblasts. The shaft of the cartilage then becomes ensheathed in a superficial layer of bone.

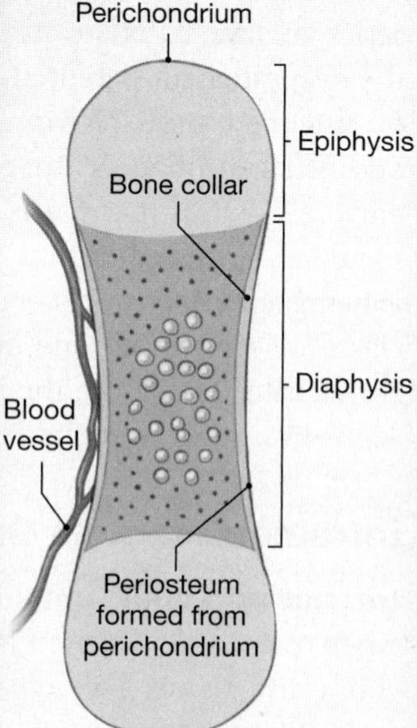
Perichondrium
Epiphysis
Bone collar
Diaphysis
Blood vessel
Periosteum formed from perichondrium

3 Blood vessels penetrate the cartilage and invade the central region. Fibroblasts migrating with the blood vessels differentiate into osteoblasts and begin producing spongy bone at a primary ossification center. Bone formation then spreads along the shaft toward both ends of the former cartilage model.

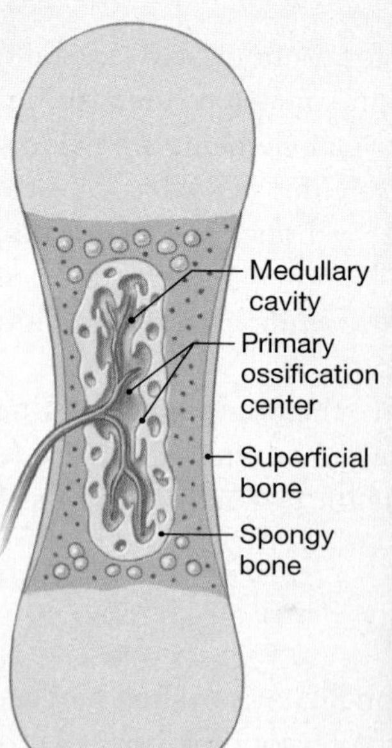
Medullary cavity
Primary ossification center
Superficial bone
Spongy bone

4 Remodeling occurs as growth continues, creating a medullary cavity. The osseous tissue of the shaft becomes thicker, and the cartilage near each epiphysis is replaced by shafts of bone. Further growth involves increases in length and diameter.

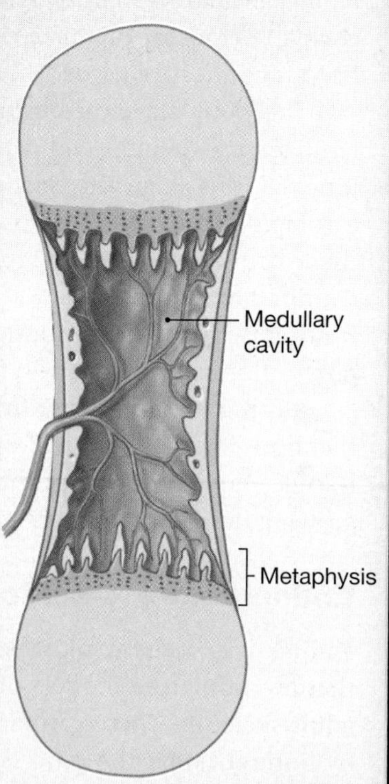
Medullary cavity
Metaphysis

SmartArt

Increasing the Length of a Developing Long Bone (Steps 5–7)

During the initial stages of ossification, osteoblasts move away from the primary ossification center toward the epiphyses. But they do not complete the ossification of the model immediately, because the cartilages of the epiphyses continue to grow. The region where the cartilage is being replaced by bone lies at the metaphysis, the junction between the diaphysis and an epiphysis. On the diaphyseal (shaft) side of the metaphysis, osteoblasts continually invade the cartilage and replace it with bone, while on the epiphyseal side, new cartilage is produced at the same rate. The situation is like a pair of joggers, one in front of the other. As long as they are running at the same speed, they can run for miles without colliding. In this case, the osteoblasts and the epiphysis are both "running away" from the primary ossification center. As a result, the osteoblasts never catch up with the epiphysis, and the bone continues to grow longer and longer.

5 Capillaries and osteoblasts migrate into the epiphyses, creating secondary ossification centers.

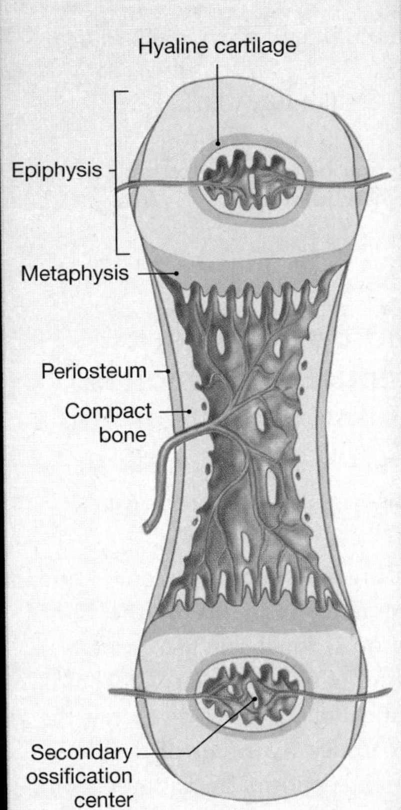

Hyaline cartilage
Epiphysis
Metaphysis
Periosteum
Compact bone
Secondary ossification center

6 The epiphyses eventually become filled with spongy bone. The metaphysis, a relatively narrow cartilaginous region called the **epiphyseal cartilage**, or *epiphyseal plate*, now separates the epiphysis from the diaphysis. On the shaft side of the metaphysis, osteoblasts continuously invade the cartilage and replace it with bone. New cartilage is produced at the same rate on the epiphyseal side.

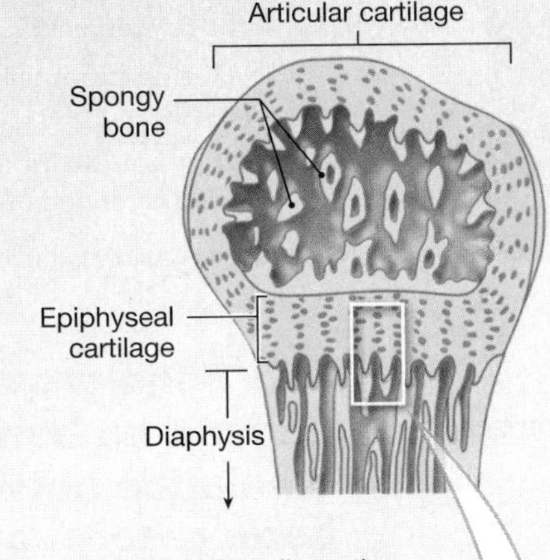

Articular cartilage
Spongy bone
Epiphyseal cartilage
Diaphysis

Within the epiphyseal cartilage, the chondrocytes are organized into zones.

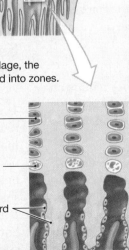

Chondrocytes at the epiphyseal side of the cartilage continue to divide and enlarge.

Chondrocytes degenerate at the diaphyseal side.

Osteoblasts migrate upward from the diaphysis and cartilage is gradually replaced by bone.

7 At puberty, the rate of epiphyseal cartilage production slows and the rate of osteoblast activity accelerates. As a result, the epiphyseal cartilage gets narrower and narrower, until it ultimately disappears. This event is called **epiphyseal closure**. The former location of the epiphyseal cartilage becomes a distinct **epiphyseal line** that remains after epiphyseal growth has ended.

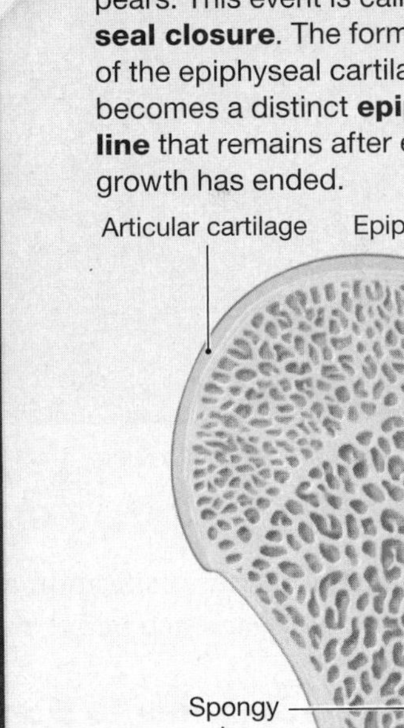

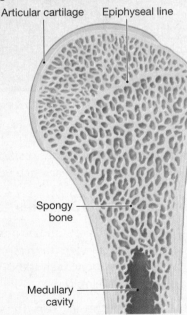

Articular cartilage
Epiphyseal line
Spongy bone
Medullary cavity

A thin cap of the original cartilage model remains exposed to the joint cavity as the **articular cartilage**. This cartilage prevents damaging the joint from bone-to-bone contact.

Figure 6–12 Bone Growth at Epiphyseal Cartilages.

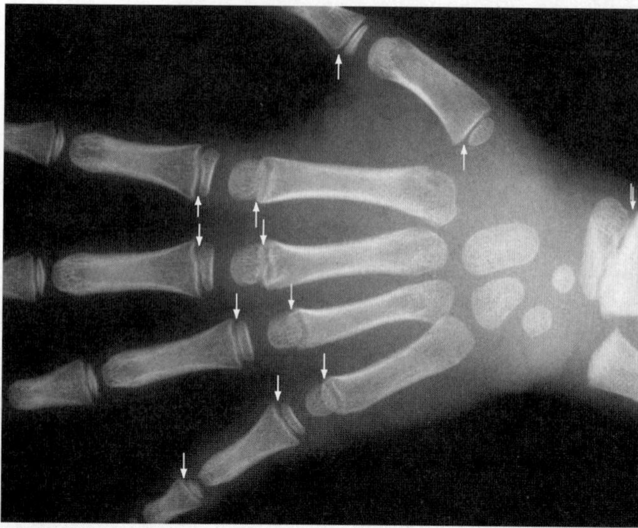

a An x-ray of growing epiphyseal cartilages (arrows)

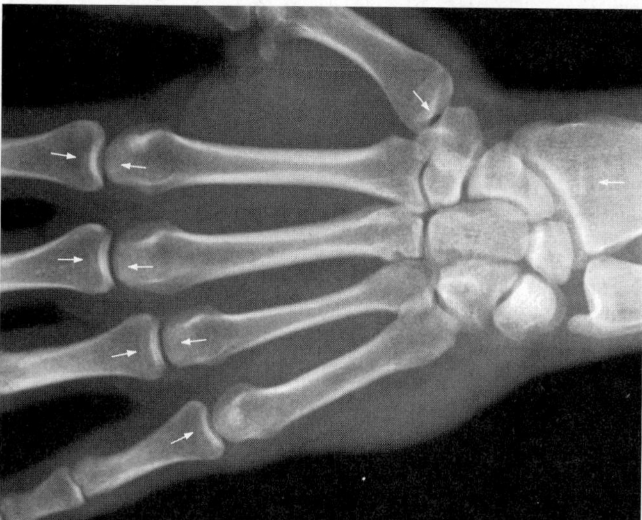

b Epiphyseal lines in an adult (arrows)

The steps of intramembranous ossification are summarized in Figure 6–13. Please study each step before moving on.

Blood and Nerve Supplies to Bone

In order for bones to grow and be maintained, they require an extensive blood supply. For this reason, osseous tissue is highly vascular. In a typical bone such as the humerus, three major sets of blood vessels develop (Figure 6–14):

1. *The Nutrient Artery and Vein.* The blood vessels that supply the diaphysis form by invading the cartilage model as endochondral ossification begins. Most bones have only one *nutrient artery* and one *nutrient vein*, but a few bones, including the femur, have more than one of each. The vessels enter the bone through one or more round passageways called *nutrient foramina* in the diaphysis. Branches of these large vessels form smaller perforating canals and extend along the length of the shaft into the osteons of the surrounding compact bone.
2. *Metaphyseal Vessels.* The *metaphyseal* (met-a-FIZ-ē-ul) *vessels* supply blood to the inner (diaphyseal) surface of each epiphyseal cartilage, where that cartilage is being replaced by bone.
3. *Periosteal Vessels.* Blood vessels from the periosteum provide blood to the superficial osteons of the shaft. During endochondral bone formation, branches of periosteal vessels also enter the epiphyses, providing blood to the secondary ossification centers.

Following the closure of the epiphyses, all three sets of vessels become extensively interconnected.

The periosteum also contains a network of lymphatic vessels (lymphatics) and sensory nerves. The lymphatics collect lymph from branches that enter the bone and reach individual osteons by the perforating canals. The sensory nerves penetrate the compact bone with the nutrient artery to innervate the endosteum, medullary cavity, and epiphyses. Because of the rich sensory innervation, injuries to bones are usually very painful.

In the next section, we examine the maintenance and replacement of mineral reserves in the adult skeleton.

✓ Checkpoint

9. **In endochondral ossification, what is the original source of osteoblasts?**
10. **During intramembranous ossification, which type of tissue is replaced by bone?**
11. **How could x-rays of the femur be used to determine whether a person has reached full height?**

See the blue Answers tab at the back of the book.

6-6 Bone growth and development depend on bone remodeling, which is a balance between bone formation and bone resorption

Learning Outcome Describe the remodeling and homeostatic mechanisms of the skeletal system.

The process of **bone remodeling** continuously recycles and renews the organic and mineral components of the bone matrix. Bone remodeling goes on throughout life, as part of normal bone maintenance. Remodeling can replace the matrix but leave the bone as a whole unchanged, or it may change the shape, internal architecture, or mineral content of the bone. Through remodeling, older mineral deposits are removed from bone and released into the circulation at the same time that circulating minerals are being absorbed and deposited.

Figure 6–13 Intramembranous Ossification.

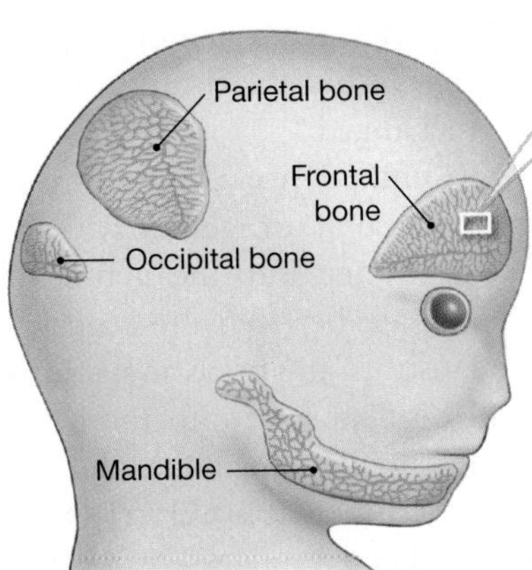
Parietal bone
Frontal bone
Occipital bone
Mandible

Intramembranous ossification starts about the eighth week of embryonic development. This type of ossification occurs in the deeper layers of the dermis, forming **dermal bones**.

1 Mesenchymal cells cluster together, differentiate into osteoblasts, and start to secrete the organic components of the matrix. The resulting osteoid then becomes mineralized with calcium salts forming bone matrix.

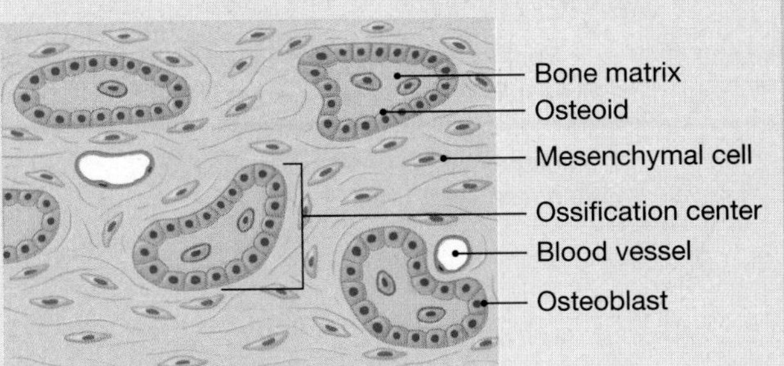
Bone matrix
Osteoid
Mesenchymal cell
Ossification center
Blood vessel
Osteoblast

2 As ossification proceeds, some osteoblasts are trapped inside bony pockets where they differentiate into osteocytes. The developing bone grows outward from the ossification center in small struts called **spicules**.

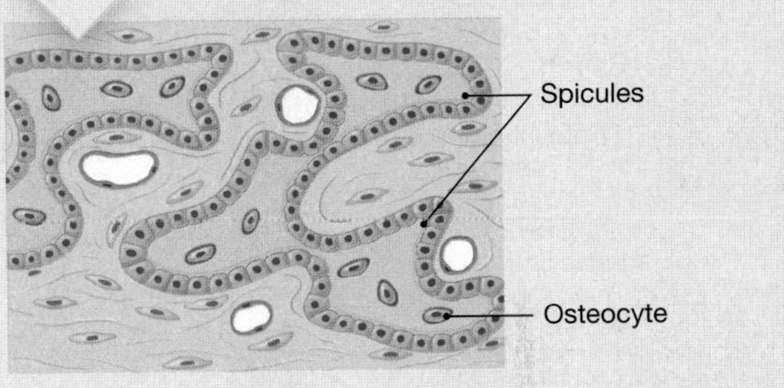
Spicules
Osteocyte

3 Blood vessels begin to branch within the region and grow between the spicules. The rate of bone growth accelerates with oxygen and a reliable supply of nutrients. As spicules interconnect, they trap blood vessels within the bone.

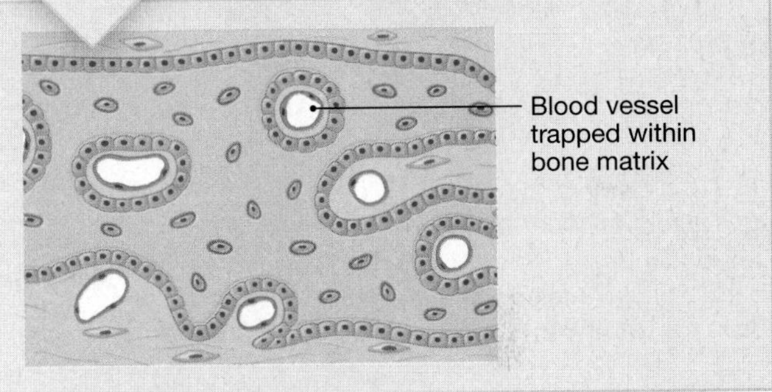
Blood vessel trapped within bone matrix

4 Continued deposition of bone by osteoblasts located close to blood vessels results in a plate of spongy bone with blood vessels weaving throughout.

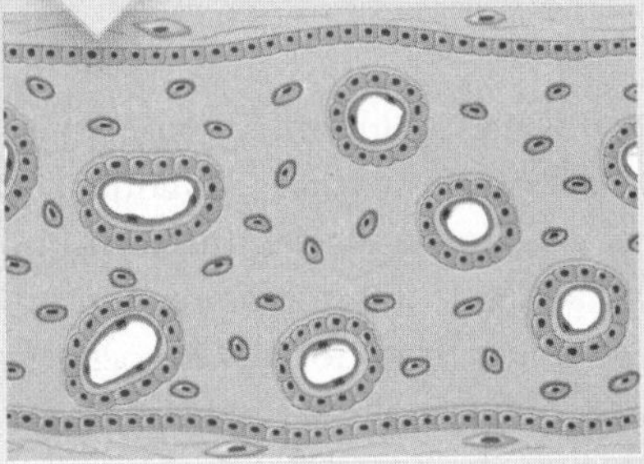

5 Subsequent remodeling around blood vessels produces osteons typical of compact bone. Osteoblasts on the bone surface along with connective tissue around the bone become the periosteum.

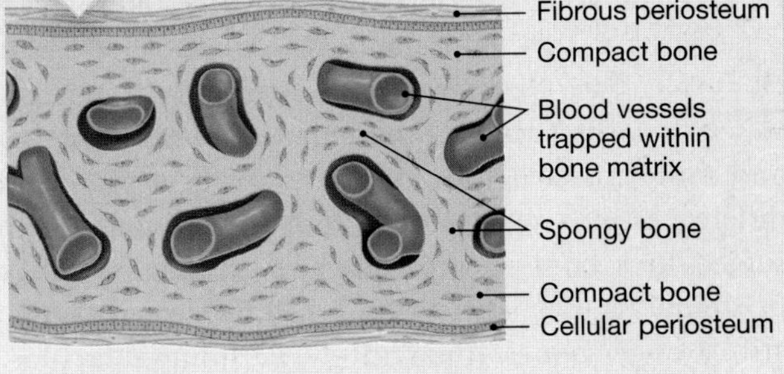
Fibrous periosteum
Compact bone
Blood vessels trapped within bone matrix
Spongy bone
Compact bone
Cellular periosteum

Areas of spongy bone are remodeled forming the diploë and a thin covering of compact bone.

Figure 6–14 The Blood Supply to a Mature Bone.

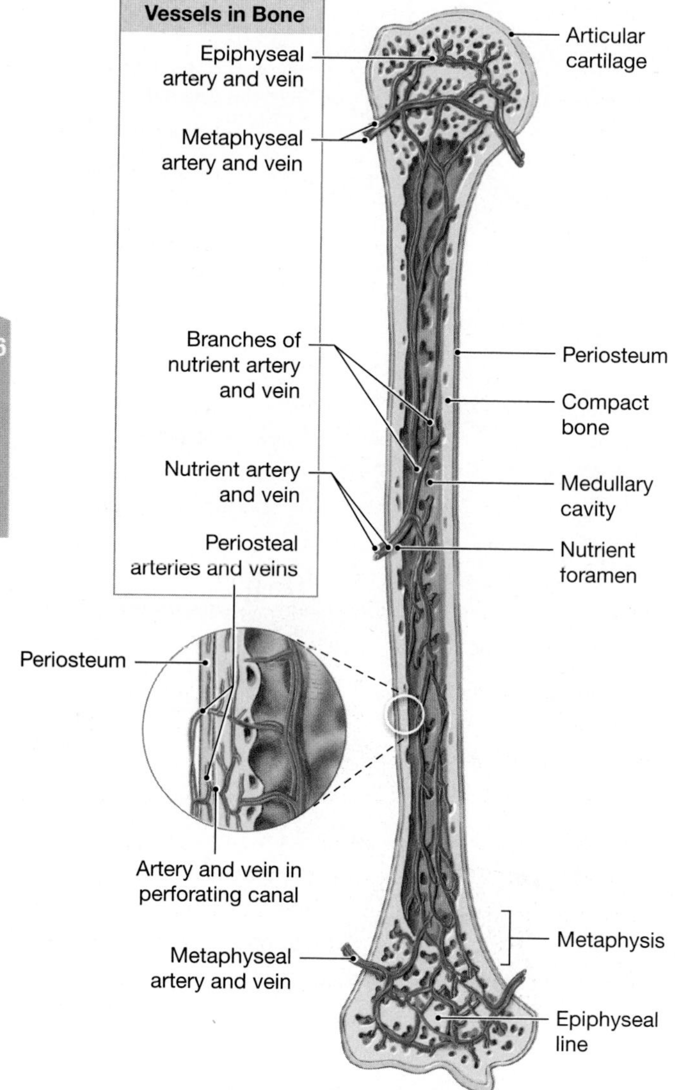
Vessels in Bone
Epiphyseal artery and vein
Metaphyseal artery and vein
Branches of nutrient artery and vein
Nutrient artery and vein
Periosteal arteries and veins
Periosteum
Artery and vein in perforating canal
Metaphyseal artery and vein
Articular cartilage
Periosteum
Compact bone
Medullary cavity
Nutrient foramen
Metaphysis
Epiphyseal line

Bone remodeling involves an interplay among the activities of osteocytes, osteoblasts, and osteoclasts. In adults, osteocytes are continuously removing and replacing the surrounding calcium salts. Osteoclasts and osteoblasts also remain active even after the epiphyseal cartilages have closed. Osteoclasts are constantly removing matrix, and osteoblasts are always adding to it. Normally, their activities are balanced: As quickly as osteoblasts form one osteon, osteoclasts remove another. The homeostatic balance between the opposing activities of osteoclasts and osteoblasts is very important. When osteoclasts remove calcium salts faster than osteoblasts deposit them, bones weaken. When osteoblast activity predominates, bones become stronger and more massive. This opposition causes some interesting differences in bone structure among individuals. People who subject their bones to muscular stress through weight training or strenuous exercise develop not only stronger muscles, but also stronger bones. Alternatively, declining muscular activity due to immobility leads to reduced bone mass at sites of muscle attachment.

The turnover rate of bone is quite high. In young adults, almost one-fifth of the skeleton is recycled and replaced each year. However, not every part of every bone is affected equally. The rate of turnover differs regionally and even locally. For example, the spongy bone in the head of the femur may be replaced two or three times each year, but the compact bone along the shaft remains largely unchanged.

Because of their biochemical similarity to calcium, heavy-metal ions such as lead, strontium, cobalt, or radioactive uranium or plutonium can be incorporated into the matrix of bone. Osteoblasts do not differentiate between these heavy-metal ions and calcium. This means that any heavy metal ions present in the bloodstream will be deposited into the bone matrix. Some of these ions are potentially dangerous, and the turnover of bone matrix can have detrimental health effects as ions that are absorbed and accumulated are released into the circulation over a period of years. This was one of the major complications in the aftermath of the Ukrainian Chernobyl nuclear reactor incident in 1986. Radioactive compounds released in the meltdown of the reactor were deposited into the bones of exposed individuals. Over time, the radiation released by their own bones has caused thyroid cancers, leukemia (cancer of the blood cells, which starts in the red bone marrow), and other potentially fatal cancers. Many of these same events are expected to appear in the aftermath of the Japanese Fukushima Daiichi nuclear power plant disaster that occurred in March 2011.

✓ Checkpoint

12. **Describe bone remodeling.**
13. **Explain how heavy-metal ions could be incorporated into bone matrix.**

See the blue Answers tab at the back of the book.

6-7 Exercise, nutrition, and hormones affect bone development and the skeletal system

Learning Outcome Discuss the effects of exercise, nutrition, and hormones on bone development and on the skeletal system.

What factors have the most important effects on the processes of bone remodeling? Exercise, nutrition, and hormones are key.

The Effects of Exercise on Bone

The turnover and recycling of minerals give each bone the ability to adapt to new stresses. The sensitivity of osteoblasts to electrical events has been theorized as the mechanism that controls the internal organization and structure of bone. Whenever a bone is stressed, the mineral crystals generate minute electrical

Figure 6–15 A Chemical Analysis of Bone.

Composition of Bone	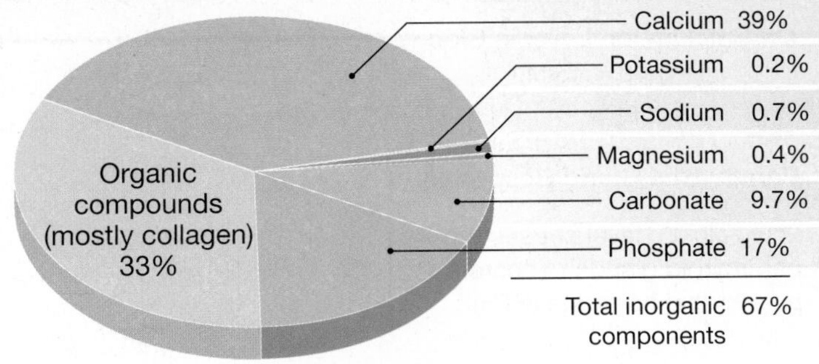	Bone Contains
Calcium	39%	99% of the body's calcium
Potassium	0.2%	4% of the body's potassium
Sodium	0.7%	35% of the body's sodium
Magnesium	0.4%	50% of the body's magnesium
Carbonate	9.7%	80% of the body's carbonate
Phosphate	17%	99% of the body's phosphate
Total inorganic components	67%	
Organic compounds (mostly collagen)	33%	

Hormones and Calcium Ion Balance

Calcium ions play a role in a variety of physiological processes, so the body must tightly control the calcium ion concentration in order to prevent damage to essential physiological systems. Even small variations from the normal concentration affect cellular operations, and larger changes can cause a clinical crisis. Calcium ions are particularly important to both the plasma membranes and the intracellular activities of neurons and muscle cells, especially cardiac muscle cells. If the calcium ion concentration of body fluids increases by 30 percent, neurons and muscle cells become unresponsive. If the calcium ion level decreases by 35 percent, neurons become so excitable that convulsions can occur. A 50 percent reduction in calcium ion concentration generally causes death. Calcium ion concentration is so closely regulated, however, that daily fluctuations of more than 10 percent are highly unusual.

Two hormones with opposing effects maintain calcium ion homeostasis. These hormones, parathyroid hormone and calcitonin, coordinate the storage, absorption, and excretion of calcium ions. Three target sites and functions are involved: (1) bones (storage), (2) digestive tract (absorption), and (3) kidneys (excretion).

Figure 6–16a indicates factors that increase the calcium ion level in the blood. When the calcium ion concentration in the blood falls below normal, cells of the **parathyroid glands**, embedded in the thyroid gland in the neck, release **parathyroid hormone (PTH)** into the bloodstream. Parathyroid hormone has the following three major effects, all of which *increase* the blood calcium ion level:

- *Stimulating osteoclast activity (indirectly).* Osteoclasts do not have PTH receptors, but PTH binds to receptors on adjacent osteoblasts. This binding causes the osteoblasts to release an osteoclast differentiation factor called RANKL (receptor activator of nuclear factor-kB). RANKL activates receptors on pre-osteoclast cells, which in turn causes them to differentiate into mature osteoclasts. These mature osteoclasts release enzymes that promote bone resorption, thereby releasing calcium ions into the bloodstream.
- *Increasing the amount of calcium ions absorbed by the intestines by enhancing calcitriol secretion by the kidneys.* Under normal circumstances, calcitriol is always present, and parathyroid hormone controls its effect on the intestinal epithelium.
- *Decreasing the amount of calcium ions excreted by the kidneys.* This causes more calcium to remain in the bloodstream.

Under these conditions, *more* calcium ions enter body fluids, and *fewer* are lost. The calcium ion concentration increases to a normal level, and homeostasis is restored.

Calcitonin was discovered more than 50 years ago, but its role in human physiology remains unclear. Although calcitonin secretion is stimulated by increased blood calcium, it does not appear to cause bone deposition of calcium by osteoblasts. Its effects are restricted to osteoclasts, whose activity is reduced. The kidneys respond to calcitonin by excreting more calcium and secreting less calcitriol, which lessens calcium absorption by the intestines (Figure 6–16b). In pregnant and nursing women, both calcitonin and vitamin D_3 are increased, suggesting that these increases may be important in transferring maternal calcium to the fetus and for preserving and recovering bone mineral density in the mother. Yet, its exact mechanism of action remains elusive. Calcitonin is used as a treatment for osteoporosis in postmenopausal women. Pharmacological calcitonin is administered as a nasal spray, takes about 15 minutes to be absorbed into the blood, and can inhibit osteoclast activity in these women.

Changes in mineral content do not necessarily affect the shape of the bone because the bone matrix contains protein fibers as well as mineral deposits. In *osteomalacia* (os-tē-ō-ma-LĀ-shē-uh; *malakia*, softness), the bones appear normal, although they are weak and flexible due to poor

Figure 6–16 Factors That Increase the Blood Calcium Ion Level.

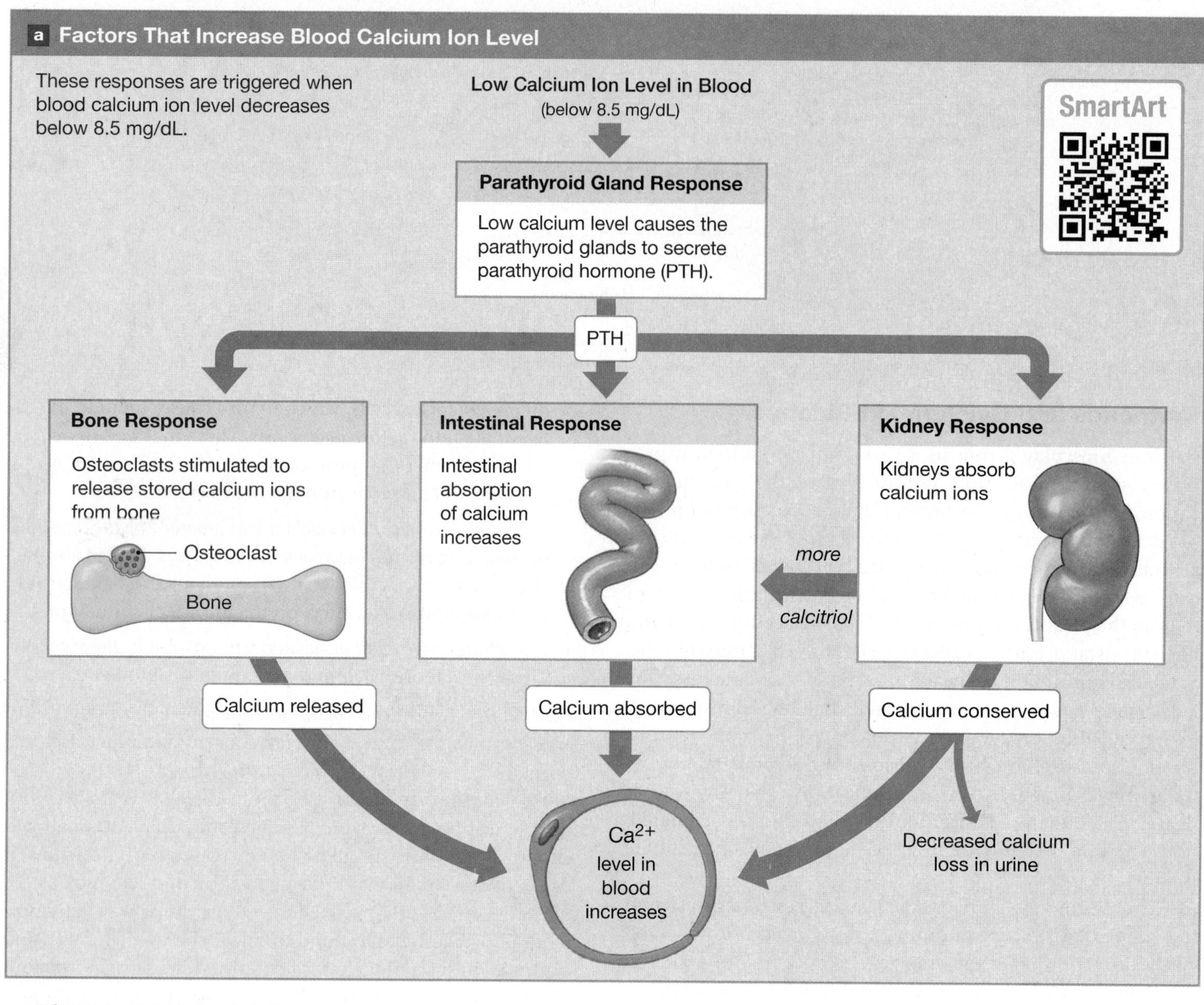
a Factors That Increase Blood Calcium Ion Level
These responses are triggered when blood calcium ion level decreases below 8.5 mg/dL.
Low Calcium Ion Level in Blood
(below 8.5 mg/dL)
SmartArt
Parathyroid Gland Response
Low calcium level causes the parathyroid glands to secrete parathyroid hormone (PTH).
PTH
Bone Response
Osteoclasts stimulated to release stored calcium ions from bone
Osteoclast
Bone
Intestinal Response
Intestinal absorption of calcium increases
Kidney Response
Kidneys absorb calcium ions
more
calcitriol
Calcium released
Calcium absorbed
Calcium conserved
Decreased calcium loss in urine
Ca2+ level in blood increases

? What gland responds when the blood calcium level is too low?

mineralization. *Rickets*, a form of osteomalacia affecting children, generally results from a vitamin D_3 deficiency caused by inadequate skin exposure to sunlight and an inadequate dietary supply of the vitamin. p. 164 The bones of children with rickets are so poorly mineralized that they become very flexible. The bones bend laterally, and affected individuals develop a bowlegged appearance because the walls of each femur can no longer resist the tension and compression forces applied by the body weight (see **Figure 6–9**). In the United States, milk is fortified with vitamin D; this practice began in the 1930s to prevent rickets.

Checkpoint

17. **Explain the role of PTH and the interaction between PTH and calcitriol on blood calcium ion level.**
18. **Why does a child who has rickets have difficulty walking?**

See the blue Answers tab at the back of the book.

Figure 6–16 Factors That Increase the Blood Calcium Ion Level. (*continued*)

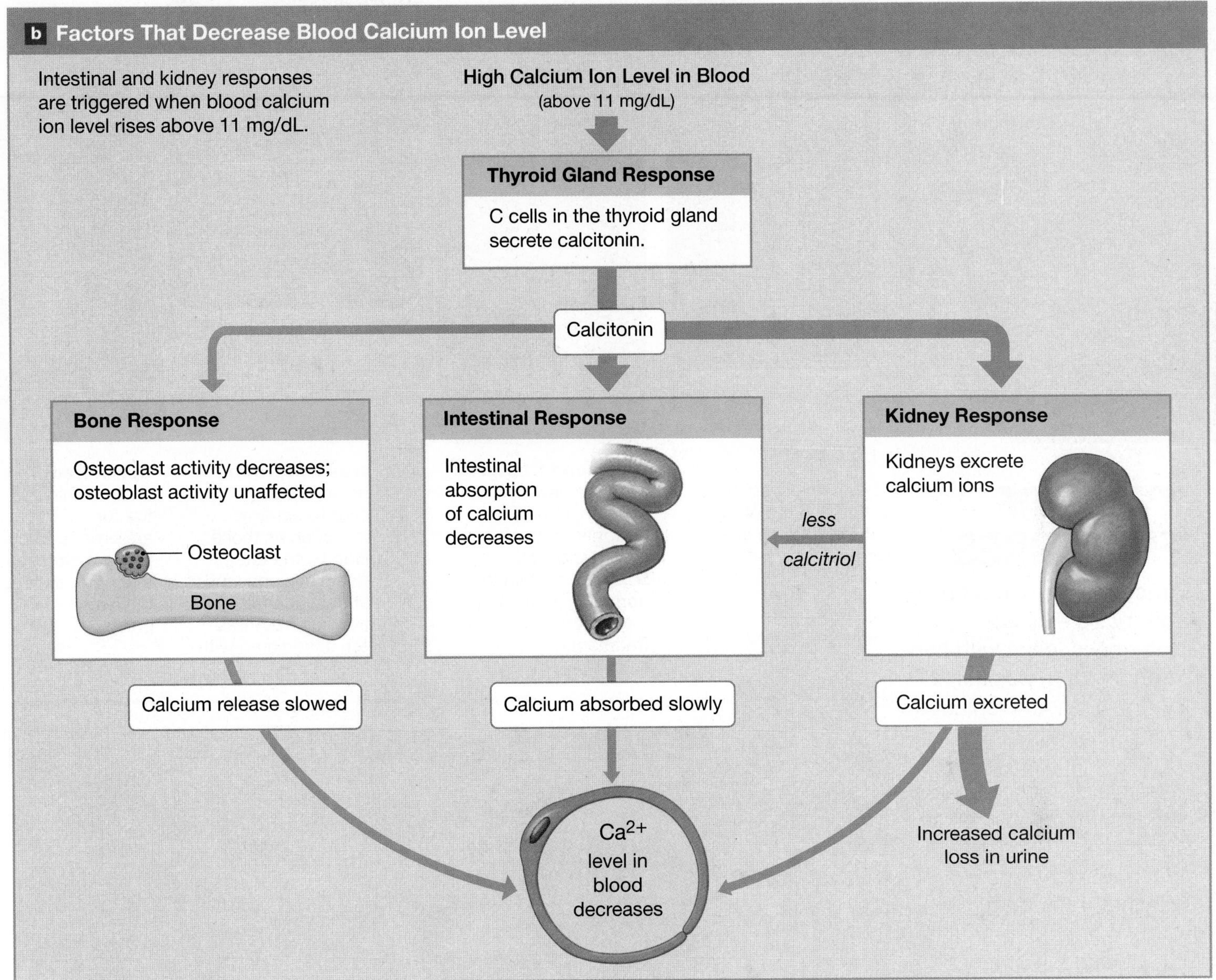
b Factors That Decrease Blood Calcium Ion Level
Intestinal and kidney responses are triggered when blood calcium ion level rises above 11 mg/dL.
High Calcium Ion Level in Blood
(above 11 mg/dL)
Thyroid Gland Response
C cells in the thyroid gland secrete calcitonin.
Calcitonin
Bone Response
Osteoclast activity decreases; osteoblast activity unaffected
Osteoclast
Bone
Intestinal Response
Intestinal absorption of calcium decreases
less
calcitriol
Kidney Response
Kidneys excrete calcium ions
Calcium release slowed
Calcium absorbed slowly
Calcium excreted
Ca2+ level in blood decreases
Increased calcium loss in urine

6-9 A fracture is a crack or break in a bone

Learning Outcome Describe the types of fractures, and explain how fractures heal.

Despite its mineral strength, bone can crack or even break if it is subjected to extreme loads, sudden impacts, or stresses from unusual directions. The damage that results is called a **fracture**. Most fractures heal even after severe damage, if the blood supply and the cellular components of the endosteum and periosteum survive. **Spotlight Figure 6–17** illustrates the different types of fractures and the repair process. Please study this figure carefully before reading the next section. Note that the two main types of fractures are **open** (or compound) and **closed** (or simple).

Checkpoint

19. **List the four steps involved in fracture repair, beginning at the onset of the bone break.**
20. **At which point in fracture repair would you find an external callus?**

See the blue Answers tab at the back of the book.

Go to MasteringA&P™ > Study Area > Menu > Clinical > Clinical Case Studies > Chapter 6: Look Out Below: A Case Study on Bone Tissue Structure and Repair

SPOTLIGHT

Figure 6–17
Types of Fractures and Steps in Repair

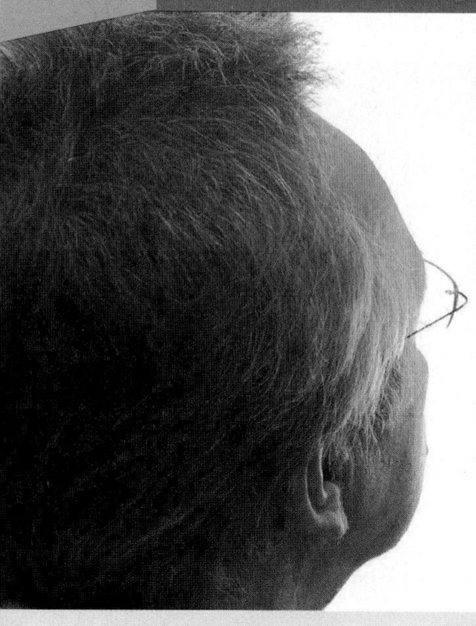

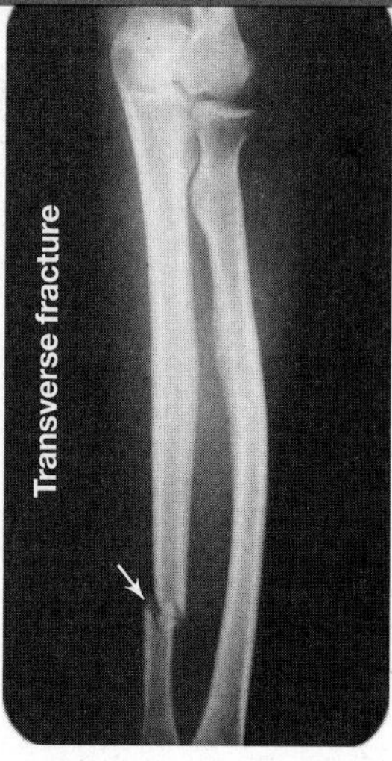
Transverse fracture

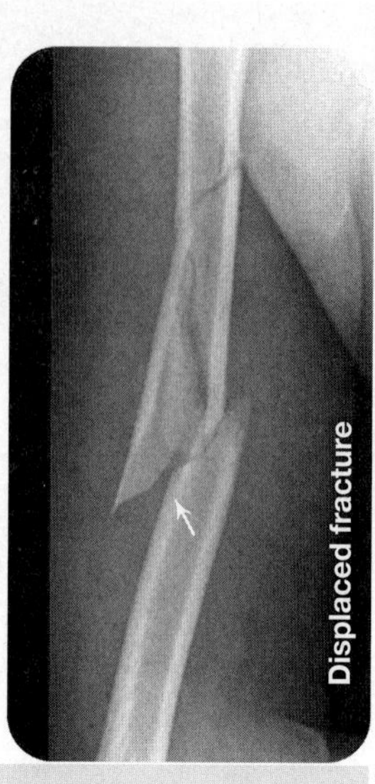
Displaced fracture

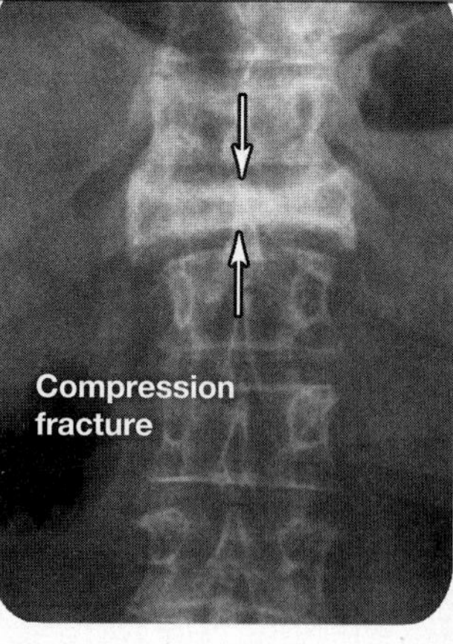
Compression fracture

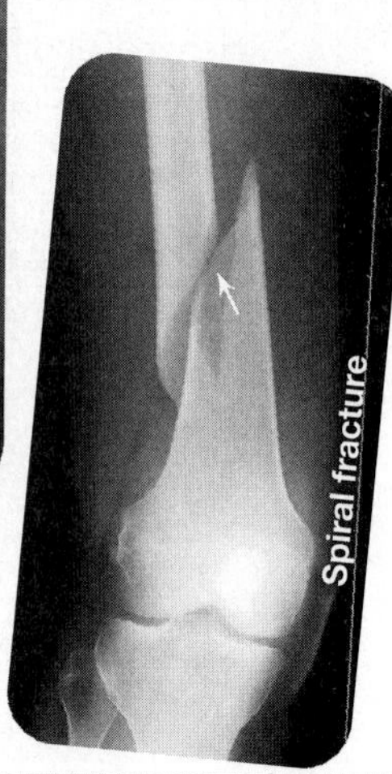
Spiral fracture

TYPES OF FRACTURES

Fractures are named by four basic either/or categories, each with two groupings: (1) position of bone ends (nondisplaced or displaced); (2) completeness of break (complete or incomplete); (3) orientation of break to the bone's long axis (linear or transverse); and (4) whether or not the broken bone penetrates the skin (open or closed). **Open** (*compound*) fractures project through the skin. **Closed** (*simple*) fractures are completely internal. Open fractures tend to be more dangerous than closed due to an increased risk of infection or uncontrolled bleeding. In closed fractures, the bone ends can only be seen on x-ray because the skin remains intact.

Transverse fractures, such as this fracture of the ulna, break a bone shaft across its long axis.

Displaced fractures produce new and abnormal bone arrangements. **Nondisplaced fractures** retain the normal alignment of the bones or fragments.

Compression fractures occur in vertebrae subjected to extreme stresses, such as those produced by the forces that arise when you land on your seat in a fall. Compression fractures are often associated with osteoporosis.

Spiral fractures, such as this fracture of the tibia, are produced by twisting stresses that spread along the length of the bone.

REPAIR OF A FRACTURE

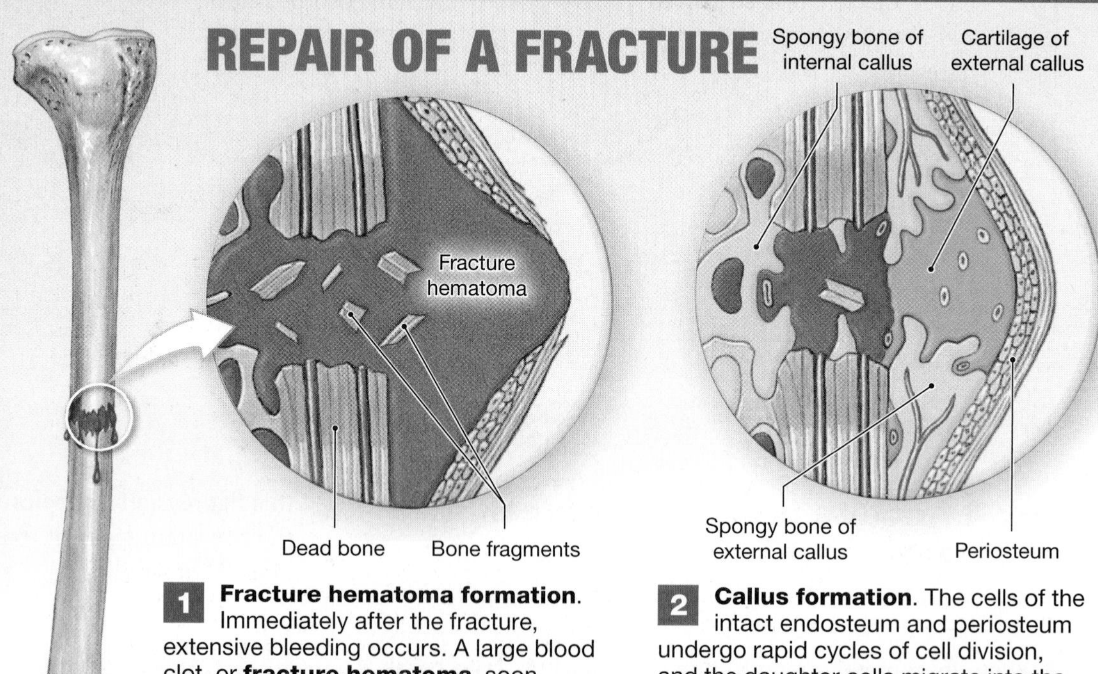
Fracture hematoma
Dead bone
Bone fragments
Spongy bone of internal callus
Cartilage of external callus
Spongy bone of external callus
Periosteum

1 **Fracture hematoma formation.** Immediately after the fracture, extensive bleeding occurs. A large blood clot, or **fracture hematoma**, soon closes off the injured vessels and leaves a fibrous meshwork in the damaged area. The disruption of circulation kills local osteocytes, broadening the area affected. Dead bone soon extends along the shaft in either direction.

2 **Callus formation.** The cells of the intact endosteum and periosteum undergo rapid cycles of cell division, and the daughter cells migrate into the fracture zone. An **internal callus** forms as a network of spongy bone unites the inner edges of the fracture. An **external callus** of cartilage and bovne encircles and stabilizes the outer edges of the fracture.

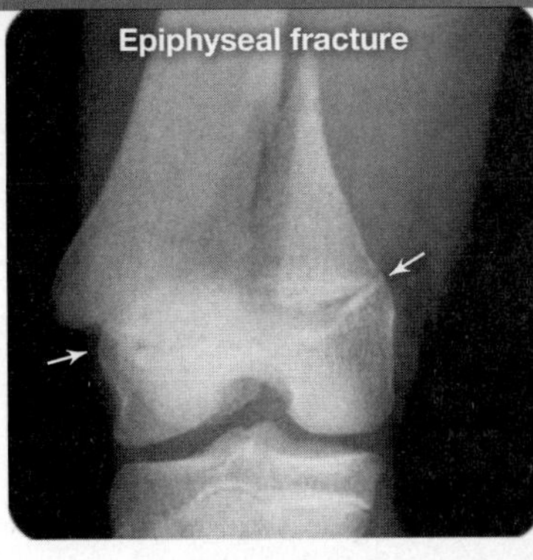
Epiphyseal fracture

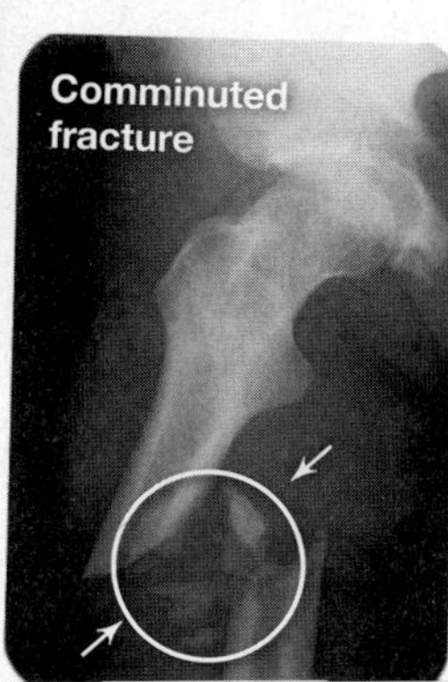
Comminuted fracture

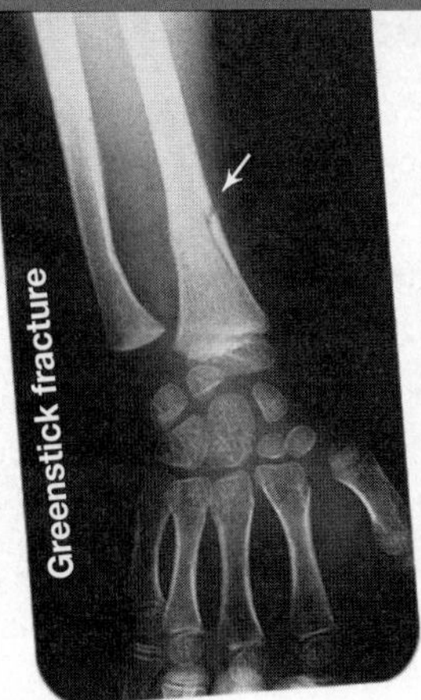
Greenstick fracture

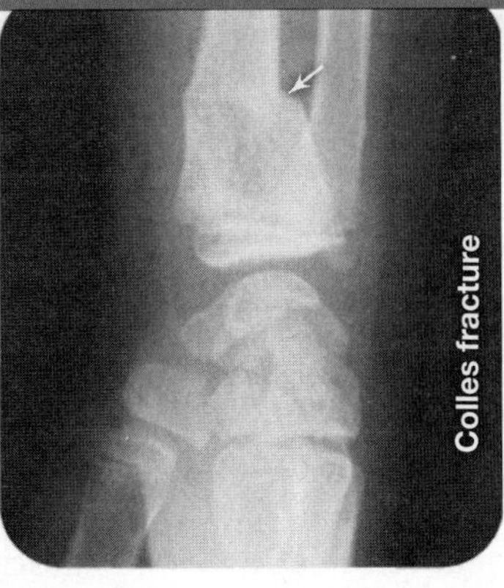
Colles fracture

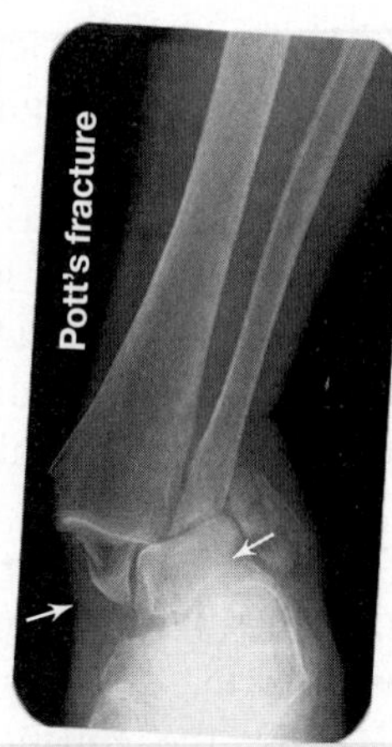
Pott's fracture

Epiphyseal fractures, such as this fracture of the femur, tend to occur where the bone matrix is undergoing calcification and chondrocytes are dying. A clean transverse fracture along this line generally heals well. Unless carefully treated, fractures between the epiphysis and the epiphyseal cartilage can permanently stop growth at this site.

Comminuted fractures, such as this fracture of the femur, shatter the affected area into a multitude of bony fragments.

In a **greenstick fracture**, such as this fracture of the radius, only one side of the shaft is broken, and the other is bent. This type of fracture generally occurs in children, whose long bones have yet to ossify fully.

A **Colles fracture**, a break in the distal portion of the radius, is typically the result of reaching out to cushion a fall.

A **Pott's fracture**, also called a bimalleolar fracture, occurs at the ankle and affects both the medial malleolus of the distal tibia and the lateral malleolus of the distal fibula.

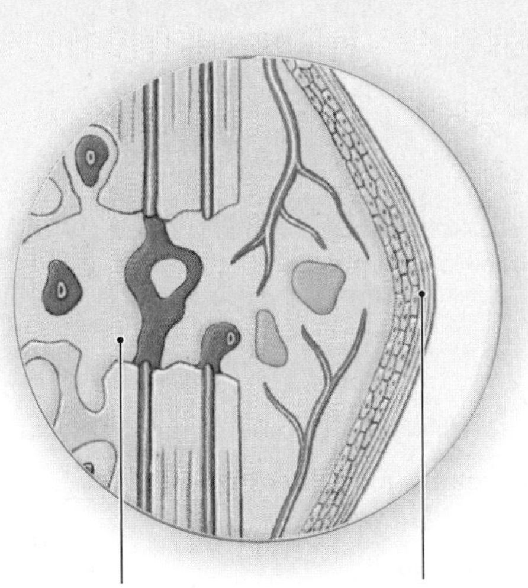
Internal callus
External callus
External callus

3 **Spongy bone formation**. As the repair continues, osteoblasts replace the central cartilage of the external callus with spongy bone, which then unites the broken ends. Fragments of dead bone and the areas of bone closest to the break are resorbed and replaced. The ends of the fracture are now held firmly in place and can withstand normal stresses from muscle contractions.

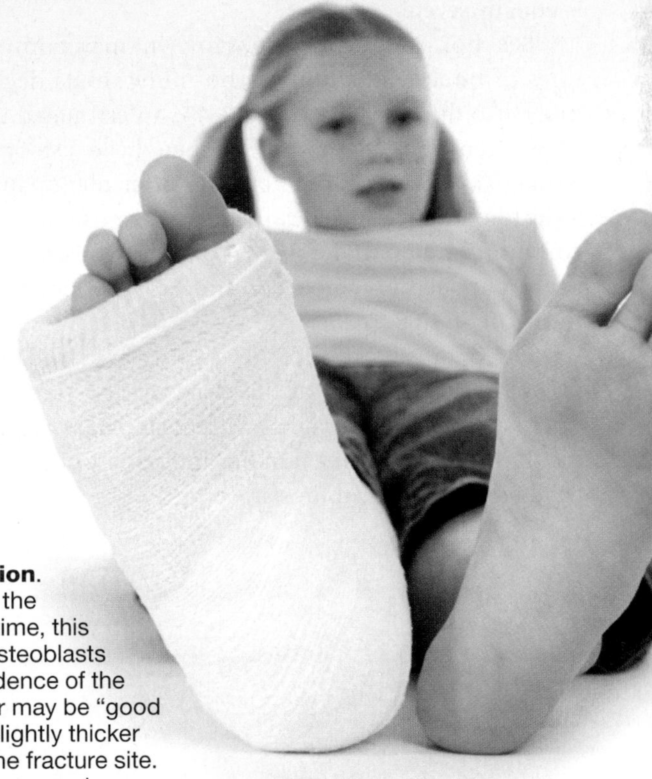

4 **Compact bone formation**. A swelling initially marks the location of the fracture. Over time, this region will be remodeled by osteoblasts and osteoclasts, and little evidence of the fracture will remain. The repair may be "good as new" or the bone may be slightly thicker and stronger than normal at the fracture site. Under comparable stresses, a second fracture will generally occur at a different site.

6-10 Osteopenia has widespread effects on aging bones

Learning Outcome Summarize the effects of the aging process on the skeletal system.

6

The bones of the skeleton become thinner and weaker as a normal part of the aging process. Inadequate ossification is called **osteopenia** (os-tē-ō-PĒ-nē-uh; *penia*, lacking). All of us become slightly osteopenic as we age. This reduction in bone mass begins between ages 30 and 40. At this age, osteoblast activity begins to decrease, while osteoclast activity continues at the previous level. Once the reduction begins, women lose about 8 percent of their skeletal mass every decade. Men lose less—about 3 percent per decade. Not all parts of the skeleton are equally affected. Epiphyses, vertebrae, and the jaws lose more mass than other sites, resulting in fragile limbs, reduction in height, and loss of teeth.

When the reduction in bone mass is sufficient to compromise normal function, the condition is known as **osteoporosis** (os-tē-ō-po-RŌ-sis; *porosus*, porous). The brittle, fragile bones that result are likely to break when exposed to stresses that younger individuals could easily tolerate. For example, a hip fracture can occur when a 90-year-old person simply tries to stand. Any fractures in older individuals lead to a loss of independence and immobility that further weakens the skeleton. The extent of the loss of spongy bone mass due to osteoporosis is shown in Figure 6–18. The reduction in compact bone mass is equally severe.

Sex hormones are important in maintaining normal rates of bone deposition. As hormone levels decline, bone mass also diminishes. Over age 45, an estimated 29 percent of women and 18 percent of men have osteoporosis. In women, the condition accelerates after menopause, due to a decline in circulating estrogen. Severe osteoporosis is less common in men under age 60 than in women of that same age group because men continue to produce testosterone until late in life.

Osteoporosis can also develop as a secondary effect of many cancers. Cancers of the bone marrow, breast, or other tissues release a chemical known as **osteoclast-activating factor**. It increases both the number and activity of osteoclasts and produces severe osteoporosis.

Figure 6–18 **The Effects of Osteoporosis on Spongy Bone.**

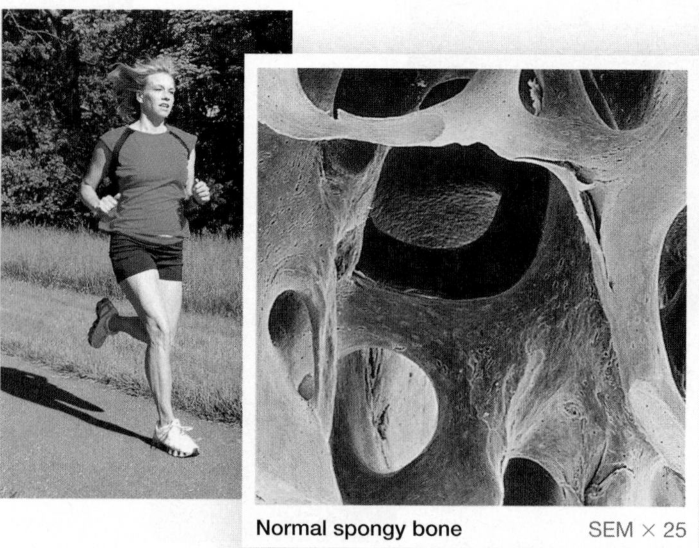

Normal spongy bone SEM × 25

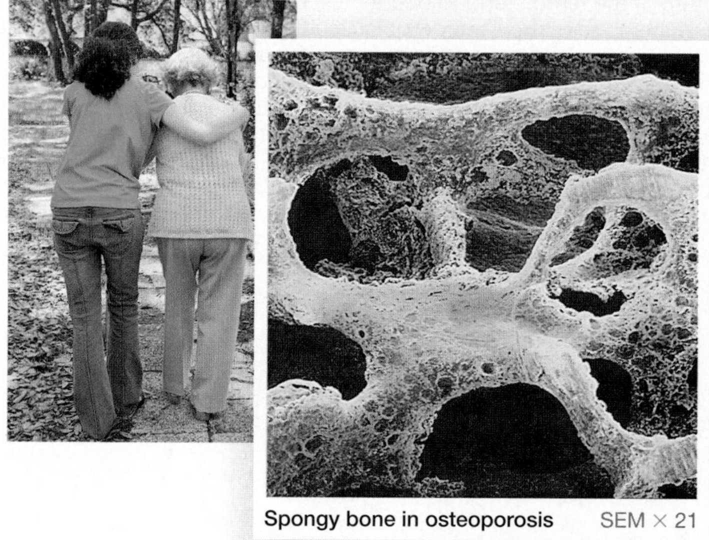

Spongy bone in osteoporosis SEM × 21

Checkpoint

21. Define *osteopenia*.
22. Why is osteoporosis more common in older women than in older men?

See the blue Answers tab at the back of the book.

6 Chapter Review

Study Outline

> MasteringA&P™ Access more chapter study tools online in the MasteringA&P Study Area:
> - Chapter Quizzes, Chapter Practice Test, MP3 Tutor Sessions, and Clinical Case Studies
> - Practice Anatomy Lab PAL 3.0
> - A&P Flix A&PFlix
> - Interactive Physiology iP2
> - PhysioEx PhysioEx 9.1

An Introduction to Bones and Bone Tissue p. 181

1. The skeletal system has a variety of purposes, such as providing a framework for body posture and allowing for precise movements.

6-1 The skeletal system has several major functions p. 181

2. The skeletal system includes the bones of the skeleton and the cartilages, ligaments, and other connective tissues that stabilize or connect the bones. Major functions of the skeletal system include support, storage of minerals and lipids, blood cell production, protection, and leverage.

6-2 Bones are classified according to shape and structure, and they have a variety of bone markings p. 182

3. Bones may be categorized as **sutural bones** (*Wormian bones*), **irregular bones**, **short bones**, **flat bones**, **long bones**, and **sesamoid bones**. *(Figure 6–1)*
4. Each bone has characteristic **bone markings**, or *surface features*, including projections, openings, and depressions. *(Figure 6–2)*
5. A representative long bone has a **diaphysis**, **epiphyses**, **metaphyses**, and a **medullary cavity**. *(Figure 6–3)*
6. The two types of bone tissue are **compact bone** and **spongy bone** (*trabecular bone*).
7. The medullary cavity and spaces within spongy bone contain either **red bone marrow** (for blood cell formation) or **yellow bone marrow** (for lipid storage).

6-3 Bone is composed of matrix and several types of cells: osteogenic cells, osteoblasts, osteocytes, and osteoclasts p. 183

8. **Bone tissue** is a supporting connective tissue with a solid matrix and is ensheathed by a *periosteum*.
9. Bone matrix consists largely of calcium salts that form crystals of **hydroxyapatite**. A bone without a calcified matrix is very flexible. *(Figure 6–4)*
10. **Osteogenic cells** differentiate into osteoblasts. **Osteoblasts** synthesize new bone matrix by **osteogenesis** or **ossification**. **Osteocytes** are mature bone cells located in lacunae. Adjacent osteocytes are interconnected by **canaliculi**. They maintain the surrounding matrix and assist bone repair. **Osteoclasts** dissolve the bony matrix through **osteolysis.** *(Figure 6–5)*

6-4 Compact bone contains parallel osteons, and spongy bone contains trabeculae p. 187

11. The basic functional unit of compact bone is the **osteon**, containing osteocytes arranged around a **central canal**. **Perforating canals** extend perpendicularly to the bone surface. *(Figures 6–6, 6–7)*
12. Bone matrix forms layers called **lamellae**. *Concentric lamellae* surround an osteon. *(Figure 6–7)*
13. Spongy bone contains **trabeculae**, typically in an open network. *(Figure 6–8)*
14. Compact bone is located where stresses come from a limited range of directions, such as along the diaphysis of long bones. Spongy bone is located where stresses are few or come from many directions, such as at the epiphyses of long bones. *(Figure 6–9)*
15. A bone is covered by a **periosteum** and lined with an **endosteum**. *(Figure 6–10)*

6-5 Bones form through ossification and enlarge through interstitial and appositional growth p. 190

16. **Ossification** (or *osteogenesis*) is the process of bone formation. **Calcification** is the process of depositing calcium salts within a tissue.
17. **Endochondral ossification** begins with a cartilage model that is gradually replaced by bone at the metaphysis. In this way, bone length increases. *(Spotlight Figure 6–11)*
18. The ends of the cartilage model remain as **articular cartilages**. *(Spotlight Figure 6–11)*
19. The timing of *epiphyseal closure* of the **epiphyseal cartilage** and the formation of an **epiphyseal line** differs among bones and among individuals. *(Spotlight Figure 6–11, Figure 6–12)*
20. Bone diameter increases through **appositional growth.**
21. **Intramembranous ossification** begins when osteoblasts differentiate within connective tissue. The process produces dermal bones. Such ossification begins at an **ossification center**. *(Figure 6–13)*
22. Three major sets of blood vessels provide an extensive supply of blood to bone. *(Figure 6–14)*

6-6 Bone growth and development depend on bone remodeling, which is a balance between bone formation and bone resorption p. 194

23. The organic and mineral components of bone are continuously recycled and renewed through **bone remodeling**, which involves interplay among the activities of osteocytes, osteoblasts, and osteoclasts.

6-7 Exercise, nutrition, and hormones affect bone development and the skeletal system p. 196

24. The shapes and thicknesses of bones reflect the stresses applied to them. For this reason, regular exercise is important in maintaining normal bone structure.
25. Normal ossification requires a reliable source of minerals, vitamins, and hormones.
26. *Growth hormone* and *thyroxine* stimulate bone growth. Parathyroid hormone and, to a lesser extent, calcitonin control blood calcium levels. *(Table 6–1)*

6-8 Calcium plays a critical role in bone physiology p. 198

27. Calcium is the most abundant mineral in the human body; about 99 percent of it is located in the skeleton. *(Figure 6–15)*

28. Interactions among the bones, intestines of the digestive tract, and kidneys affect the calcium ion concentration. *(Figure 6–16)*
29. **Parathyroid hormone (PTH)** is the main hormone that regulates calcium ion homeostasis. PTH increases calcium ion level in the blood. In pregnant and nursing women, calcitonin aids calcium ion transfer to the fetus and preserves bone mineral density in the mother. *(Figure 6–16)*

6-9 A fracture is a crack or break in a bone p. 201

30. A break or crack in a bone is a **fracture**. The repair of a fracture involves the formation of a **fracture hematoma**, an **external callus**, and an **internal callus**. *(Spotlight Figure 6–17)*

6-10 Osteopenia has widespread effects on aging bones p. 204

31. The effects of aging on the skeleton include **osteopenia** and **osteoporosis.** *(Figure 6–18)*

Review Questions

See the blue Answers tab at the back of the book.

LEVEL 1 Reviewing Facts and Terms

1. Blood cell formation occurs in **(a)** yellow bone marrow, **(b)** red bone marrow, **(c)** the matrix of bone tissue, **(d)** the ground substance of bones.
2. Two-thirds of the weight of bone is accounted for by **(a)** crystals of calcium phosphate, **(b)** collagen fibers, **(c)** osteocytes, **(d)** calcium carbonate.
3. The membrane found wrapping the bones, except within the joint cavity, is the **(a)** periosteum, **(b)** endosteum, **(c)** perforating fibers, **(d)** a, b, and c are correct.
4. The basic functional unit of compact bone is the Haversian system or **(a)** osteocyte, **(b)** osteoclast, **(c)** osteon, **(d)** osseous matrix, **(e)** osseous lamellae.
5. The vitamins essential for normal adult bone maintenance and repair are **(a)** A and E, **(b)** C and D_3, **(c)** B and E, **(d)** B complex and K.
6. The hormones that coordinate the storage, absorption, and excretion of calcium ions are **(a)** growth hormone and thyroxine, **(b)** calcitonin and parathyroid hormone, **(c)** calcitriol and cholecalciferol, **(d)** estrogens and androgens.
7. Classify the bones in the following diagram according to their shape.

(a) ________ **(b)** ________
(c) ________ **(d)** ________
(e) ________ **(f)** ________

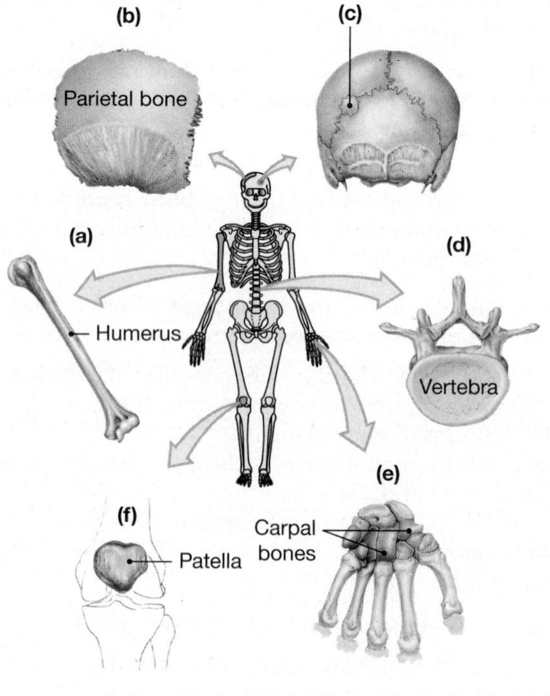

8. The presence of an epiphyseal line indicates **(a)** epiphyseal growth has ended, **(b)** epiphyseal growth is just beginning, **(c)** growth of bone diameter is just beginning, **(d)** the bone is fractured at the location, **(e)** no particular event.
9. The *primary* reason that osteoporosis accelerates after menopause in women is **(a)** reduced levels of circulating estrogens, **(b)** reduced levels of vitamin C, **(c)** diminished osteoclast activity, **(d)** increased osteoblast activity.
10. The nonpathologic loss of bone that occurs with aging is called **(a)** osteomyelitis, **(b)** osteoporosis, **(c)** osteopenia, **(d)** osteitis, **(e)** osteomalacia.
11. Name the major functions of the skeletal system.
12. List the four distinctive cell populations of bone tissue.
13. What are the primary parts of a typical long bone?
14. What is the primary difference between endochondral ossification and intramembranous ossification?
15. List the organic and inorganic components of bone matrix.
16. **(a)** What nutritional factors are essential for normal bone growth and maintenance?
    (b) What hormonal factors are necessary for normal bone growth and maintenance?
17. Which three organs or tissues interact to assist in the regulation of calcium ion concentration in body fluids?
18. What are the major effects of parathyroid hormone?

LEVEL 2 Reviewing Concepts

19. If spongy bone has no osteons, how do nutrients reach the osteocytes?
20. Why are stresses or impacts to the side of the shaft in a long bone more dangerous than stress applied to the long axis of the shaft?
21. Why do extended periods of inactivity cause degenerative changes in the skeleton?
22. What are the functional relationships between the skeleton, on the one hand, and the digestive and urinary systems, on the other?
23. Why would a physician be concerned about the growth patterns of a young child request an x-ray of the hand?
24. Why does a second fracture in the same bone tend to occur at a site different from that of the first fracture?
25. The process of bone growth at the epiphyseal cartilage is similar to **(a)** intramembranous ossification, **(b)** endochondral ossification, **(c)** the process of osteopenia, **(d)** the process of healing a fracture, **(e)** the process of calcification.
26. How might bone markings be useful in identifying the remains of an individual who was shot and killed years ago?

LEVEL 3 Critical Thinking and Clinical Applications

27. While playing on her swing set, 10-year-old Sally falls and breaks her right leg. At the emergency room, the doctor tells her parents that the proximal end of the tibia where the epiphysis meets the diaphysis is fractured. The fracture is properly set and eventually heals. During a routine physical when she is 18, Sally learns that her right leg is 2 cm shorter than her left, probably because of her accident. What might account for this difference?
28. Which of the following conditions would you possibly observe in a child who is suffering from rickets? **(a)** abnormally short limbs, **(b)** abnormally long limbs, **(c)** oversized facial bones, **(d)** bowed legs, **(e)** weak, brittle bones
29. Frank does not enter puberty until he is 16. What effect would you predict this will have on his stature? **(a)** Frank will probably be taller than if he had started puberty earlier. **(b)** Frank will probably be shorter than if he had started puberty earlier. **(c)** Frank will probably be a dwarf. **(d)** Frank will have bones that are heavier than normal. **(e)** The late onset of puberty will have no effect on Frank's stature.
30. In physical anthropology, cultural conclusions can be inferred from a thorough examination of the skeletons of ancient peoples. What sorts of clues might bones provide as to the lifestyles of those individuals?

+ CLINICAL CASE Wrap-Up A Case of Child Abuse?

6

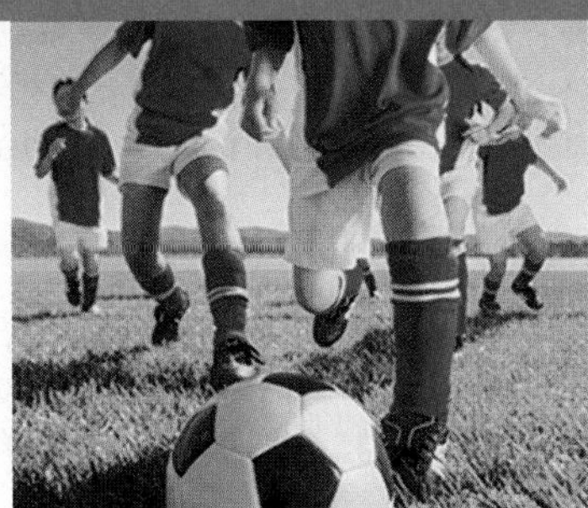

Alex and his mother both suffer from *osteogenesis imperfecta (OI)*, which means the imperfect formation of bone. Osteogenesis imperfecta is the most commonly inherited connective tissue disorder in the United States. The inherited defective gene codes for collagen that is made by osteoblasts and released into the bone matrix. This collagen is critical for bone strength and also important in forming teeth, ligaments, and the whites of the eyes. There are several types of OI with varying severities, but the Otero family suffers from the most common, relatively mild type.

Without enough normal collagen in the bone matrix, the bones are brittle and fracture easily, often with minimal trauma. In more severe cases, the skeleton is not strong enough to support the body's weight. Patients with OI have characteristic heart-shaped faces, and the whites of their eyes have a blue tint. They often suffer hearing loss.

Osteogenesis imperfecta seems to have affected people throughout history. An Egyptian mummy dating from 1000 BCE has been found with evidence of osteogenesis imperfecta.

Because of the way that this bone disorder is inherited, each of Alex's children will have a 50 percent chance of inheriting osteogenesis imperfecta. At least Alex will be aware of the problem and hopefully will never be accused of child abuse.

1. Imagine a piece of Alex's tibia under the microscope. What does it look like before the fracture?
2. Should Alex continue playing soccer? Explain.

See the blue Answers tab at the back of the book.

Related Clinical Terms

achondroplasia: A disorder of bone growth that causes the most common type of dwarfism.

bone marrow transplant: Transferring healthy bone marrow stem cells from one person into another, replacing bone marrow that is either dysfunctional or has been destroyed by chemotherapy or radiation.

bone mineral density test (BMD): A test to predict the risk of bone fractures by measuring how much calcium and other types of minerals are present in the patient's bones.

bone scan: A nuclear scanning test that identifies new areas of bone growth or breakdown. Used to evaluate damage, find cancer in the bones, and/or monitor the bone's conditions (including infection and trauma).

closed reduction: The correction of a bone fracture by manipulation without incision into the skin.

dual-energy x-ray absorptiometry (DEXA): Procedure that uses very small amounts of radiation to measure changes in bone density as small as 1 percent; the test monitors bone density in osteoporosis and osteopenia.

open reduction: The correction of a bone fracture by making an incision into the skin and rejoining the fractured bone parts, often by mechanical means such as a rod, plate, or screw.

orthopedics: The branch of medicine dealing with the correction of deformities of bones or muscles.

osteogenesis imperfecta (OI): An inherited (genetic) disorder characterized by extreme fragility of the bones; also called *brittle bone disease*.

osteomyelitis: An acute or chronic bone infection.

osteopetrosis: A rare hereditary bone disorder in which the bones become overly dense; it presents in one of three forms: osteopetrosis tarda, osteopetrosis congenita, and "marble bone" disease.

osteosarcoma: A type of cancer that starts in the bones; also called osteogenic sarcoma.

Paget's disease: A chronic disorder that can result in enlarged and misshapen bones due to abnormal bone destruction and regrowth.

traction: The application of a sustained pull on a limb or muscle in order to maintain the position of a fractured bone until healing occurs or to correct a deformity.

7 The Axial Skeleton

Learning Outcomes

These Learning Outcomes correspond by number to this chapter's sections and indicate what you should be able to do after completing the chapter.

- 7-1 ■ Identify the bones of the axial skeleton, and specify their functions. p. 209
- 7-2 ■ Identify the bones, foramina, and fissures of the cranium and face, and explain the significance of the markings on the individual bones. p. 209
- 7-3 ■ Describe the structure and functions of the orbital complex, nasal complex, and paranasal sinuses. p. 225
- 7-4 ■ Describe the key structural differences among the skulls of infants, children, and adults. p. 226
- 7-5 ■ Identify and describe the curves of the spinal column, and indicate the function of each. p. 228
- 7-6 ■ Identify the five vertebral regions, and describe the distinctive structural and functional characteristics of the vertebrae in each region. p. 229
- 7-7 ■ Explain the significance of the joints between the thoracic vertebrae and ribs, and between the ribs and sternum. p. 237

CLINICAL CASE Knocked Out

Dr. Deb Grady, a member of the American Board of Ringside Medicine, is covering a Golden Gloves boxing event as the ringside physician. She is closely watching a match between two older teenage boys. Up to the third round the fighters seem to be evenly matched. Suddenly, Dr. Grady witnesses a crushing punch delivered to the face of one fighter, knocking him out cold.

Dr. Grady jumps into the ring and carefully rolls the fighter over. She sees immediate and alarming swelling and bruising of the soft tissues around his left eye and profuse bleeding from a cut over the eye. As the young boxer regains consciousness, he sputters and coughs. Dr. Grady feels crackling subcutaneous air (like popping bubble wrap) in the expanding soft tissues around his eye. "How many fingers am I holding up?" she asks the fighter, holding up her index finger. "Two," the confused fighter answers. He is seeing double. "Follow my finger," she instructs, moving her finger up to the level of his forehead. The fighter can track her finger with his right eye, but his injured left eye remains looking straight ahead.

Dr. Grady calls the fight and accompanies the young boxer to the closest emergency room. **What has happened to the boxer's left eye? To find out, turn to the** **Clinical Case Wrap-Up on p. 242.**

An Introduction to the Divisions of the Skeleton

The human skeleton has two main divisions, axial and appendicular. The **axial skeleton**, the bones of the head and trunk, forms the longitudinal axis of the body (Figure 7–1). The **appendicular skeleton** includes the bones that support and form the limbs. In this chapter, we turn our attention to the functional anatomy of the axial skeleton; in Chapter 8 we examine the appendicular skeleton.

7-1 The 80 bones of the head and trunk make up the axial skeleton

Learning Outcome Identify the bones of the axial skeleton, and specify their functions.

The axial skeleton has 80 bones, about 40 percent of the bones in the human body:

- the *skull* (8 *cranial bones* and 14 *facial bones*),
- the bones associated with the skull (6 *auditory ossicles* and the *hyoid bone*),
- the *thoracic cage* (the *sternum* and 24 *ribs*), and
- the *vertebral column* (24 *vertebrae*, the *sacrum*, and the *coccyx*).

The axial skeleton provides a framework that supports and protects the brain, the spinal cord, and the thoracic and abdominal organs. It also provides an extensive surface area for the attachment of muscles that (1) adjust the positions of the head, neck, and trunk; (2) perform breathing movements; and (3) stabilize or position parts of the appendicular skeleton. Note that the axial skeleton does not include the clavicles (collar bones) or scapulae (shoulder blades), or the bones of the pelvic girdle (hip bones). The joints of the axial skeleton allow limited movement, but they are very strong and heavily reinforced with ligaments.

Checkpoint

1. **Identify the bones of the axial skeleton.**
2. **List the primary functions of the axial skeleton.**

See the blue Answers tab at the back of the book.

7-2 The skull's 8 cranial bones protect the brain, and its 14 facial bones form the mouth, nose, and orbits

Learning Outcome Identify the bones, foramina, and fissures of the cranium and face, and explain the significance of the markings on the individual bones.

The **skull** contains 22 bones: 8 form the **cranium**, and 14 are associated with the face (Figures 7–2, 7–3).

Cranial, Facial, and Associated Bones

The eight **cranial bones** enclose the **cranial cavity**, a chamber that supports the brain (Figure 7–2a, **Spotlight Figure 7–4**). The *occipital* (ok-SIP-e-tul), right and left *parietal* (pa-RĪ-e-tul), and *frontal bones* form the **calvaria** (kal-VA-rē-uh), or *skullcap*. Blood vessels, nerves, and membranes that stabilize the brain's position are attached to the inner surface of the cranium. Its outer surface provides an extensive area for the attachment of muscles that move the eyes, jaws, and head.

If you think of the cranium as the house where the brain resides, then the **facial bones** are the front porch (Figure 7–2b). Muscles that control facial expressions and manipulate food are attached to the nine superficial facial bones.

Figure 7–1 **The Axial Skeleton.** ATLAS: Plates 1a,b

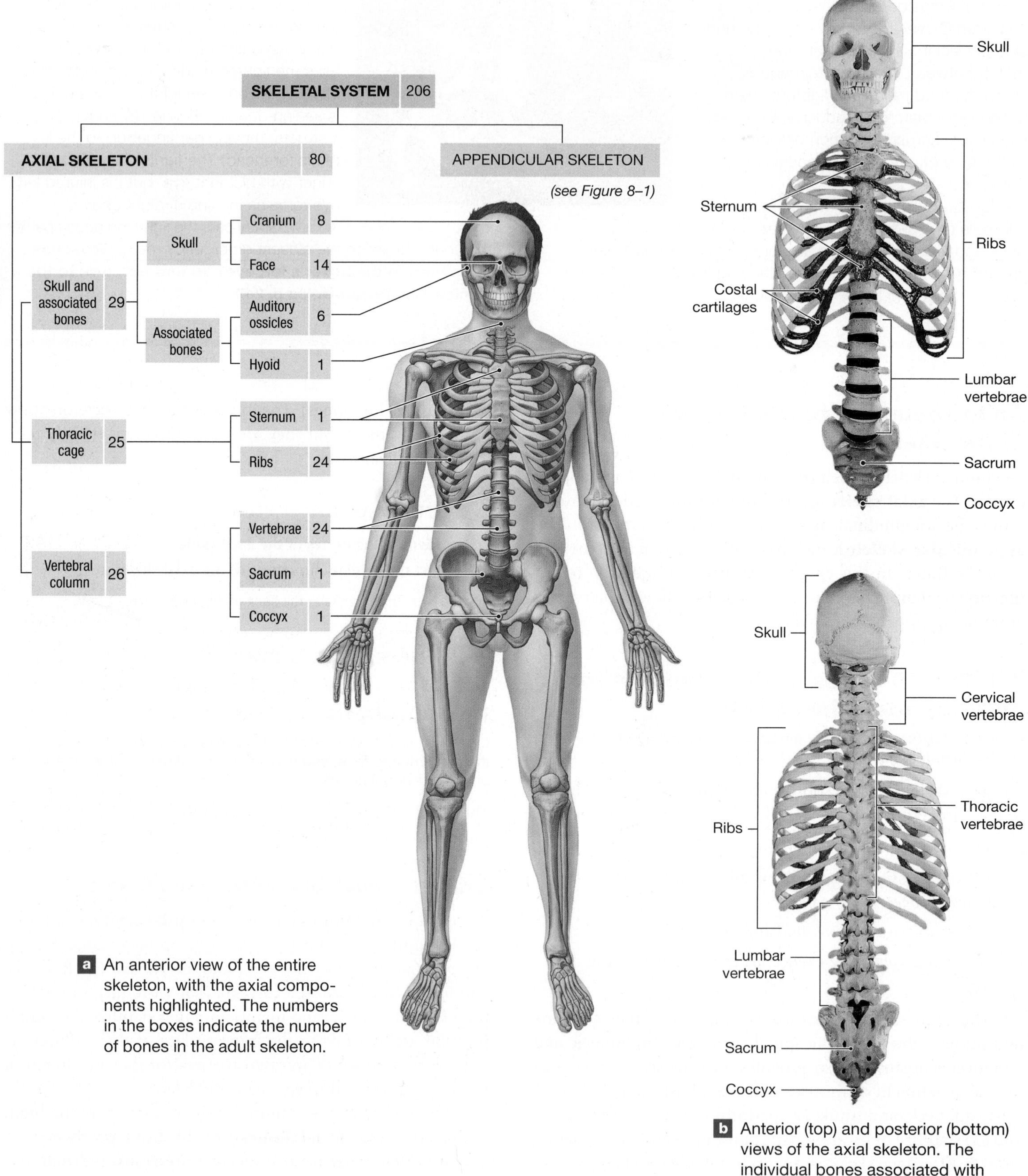
SKELETAL SYSTEM 206
AXIAL SKELETON 80
APPENDICULAR SKELETON
(see Figure 8–1)
Skull and associated bones 29
Skull
Cranium 8
Face 14
Associated bones
Auditory ossicles 6
Hyoid 1
Thoracic cage 25
Sternum 1
Ribs 24
Vertebral column 26
Vertebrae 24
Sacrum 1
Coccyx 1
Skull
Sternum
Ribs
Costal cartilages
Lumbar vertebrae
Sacrum
Coccyx
Skull
Cervical vertebrae
Ribs
Thoracic vertebrae
Lumbar vertebrae
Sacrum
Coccyx

a An anterior view of the entire skeleton, with the axial components highlighted. The numbers in the boxes indicate the number of bones in the adult skeleton.

b Anterior (top) and posterior (bottom) views of the axial skeleton. The individual bones associated with the skull are not visible.

Figure 7–2 Cranial and Facial Subdivisions of the Skull.

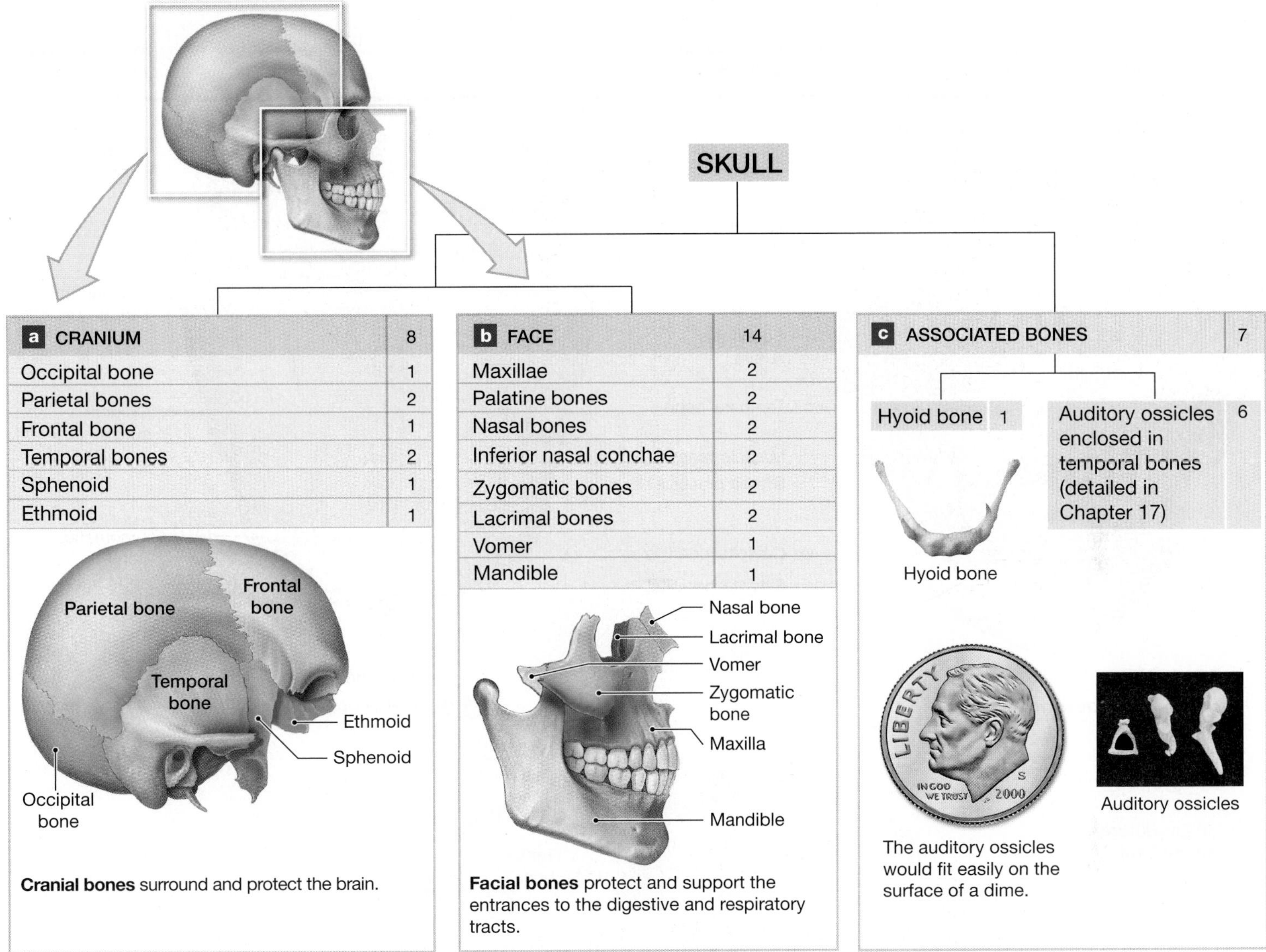
SKULL
a CRANIUM 8
Occipital bone 1
Parietal bones 2
Frontal bone 1
Temporal bones 2
Sphenoid 1
Ethmoid 1
Parietal bone
Frontal bone
Temporal bone
Ethmoid
Sphenoid
Occipital bone
Cranial bones surround and protect the brain.
b FACE 14
Maxillae 2
Palatine bones 2
Nasal bones 2
Inferior nasal conchae 2
Zygomatic bones 2
Lacrimal bones 2
Vomer 1
Mandible 1
Nasal bone
Lacrimal bone
Vomer
Zygomatic bone
Maxilla
Mandible
Facial bones protect and support the entrances to the digestive and respiratory tracts.
c ASSOCIATED BONES 7
Hyoid bone 1
Auditory ossicles enclosed in temporal bones (detailed in Chapter 17) 6
Hyoid bone
LIBERTY
IN GOD WE TRUST
2000
Auditory ossicles
The auditory ossicles would fit easily on the surface of a dime.

The five deeper facial bones help separate the oral and nasal cavities, increase the surface area of the nasal cavities, or help form the **nasal septum** (*septum*, wall) that divides the **nasal cavity** into left and right halves.

In addition, seven other bones are associated with the skull: six *auditory ossicles*, involved with hearing, and the *hyoid bone*, which supports the voice box (Figure 7–2c).

Joints, or *articulations*, form where two or more bones connect. A joint between the occipital bone of the skull and the first vertebra of the neck stabilizes the positions of the brain and spinal cord. The joints between the vertebrae of the neck permit a wide range of head movements.

Sutures

The connections between most of the skull bones of adults are immovable joints called **sutures** (SŪ-cherz). (An exception is the freely movable joint where the mandible contacts the cranium.) At a suture, bones are tied firmly together with dense fibrous connective tissue. The names of the four major sutures are as follows:

- *Lambdoid Suture.* The **lambdoid** (LAM-doyd) **suture** arches across the posterior surface of the skull (Figure 7–3a). This suture connects the occipital bone with the two parietal bones. One or more **sutural bones** (*Wormian bones*) may be present along the lambdoid suture. ⮌ p. 182
- *Coronal Suture.* The **coronal suture** attaches the frontal bone to the parietal bones of either side (Figure 7–3b). A cut through the body that parallels the coronal suture produces a *frontal*, or *coronal*, *section* (see Figure 1–6, ⮌ p. 14).
- *Sagittal Suture.* The **sagittal suture** extends from the lambdoid suture to the coronal suture, between the parietal bones (see Figure 7–3b). A cut along the midline of this

Figure 7–3 The Adult Skull. ATLAS: Plates 4a,b; 5a–e

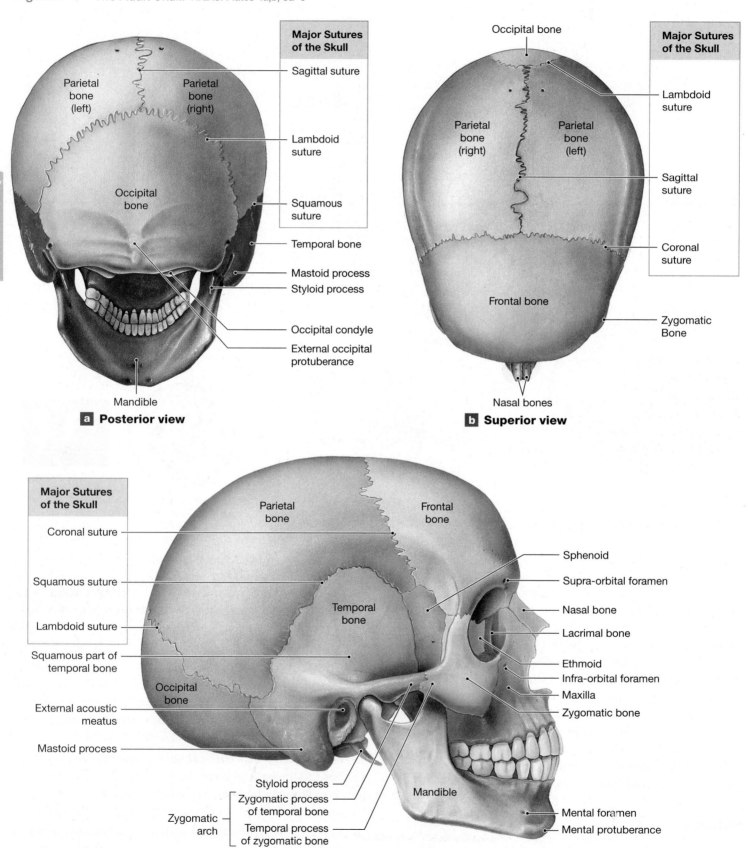

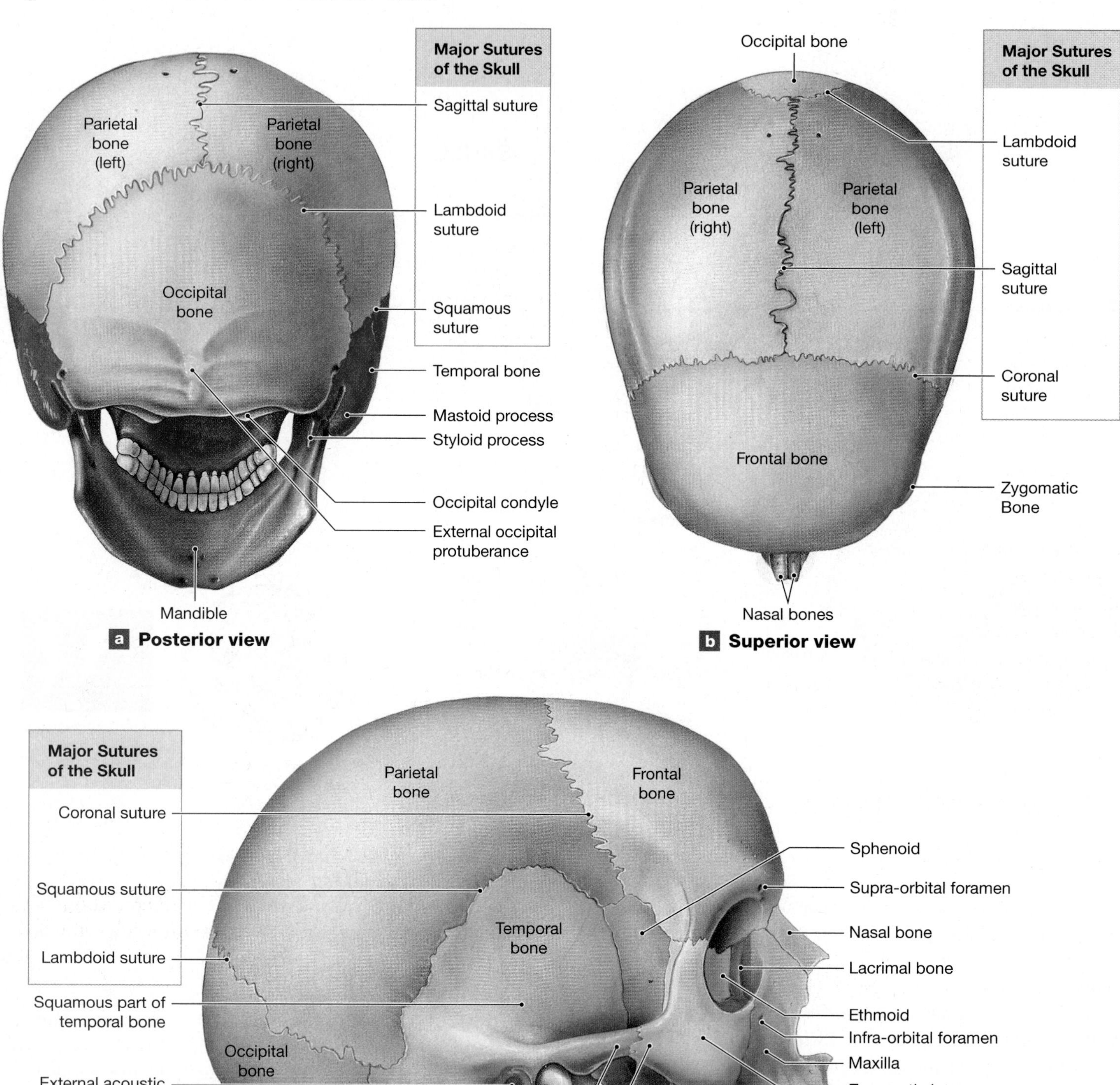
Major Sutures of the Skull
Sagittal suture
Lambdoid suture
Squamous suture
Parietal bone (left)
Parietal bone (right)
Occipital bone
Temporal bone
Mastoid process
Styloid process
Occipital condyle
External occipital protuberance
Mandible
a Posterior view
Occipital bone
Major Sutures of the Skull
Lambdoid suture
Parietal bone (right)
Parietal bone (left)
Sagittal suture
Coronal suture
Frontal bone
Zygomatic Bone
Nasal bones
b Superior view
Major Sutures of the Skull
Coronal suture
Squamous suture
Lambdoid suture
Squamous part of temporal bone
External acoustic meatus
Mastoid process
Parietal bone
Frontal bone
Temporal bone
Occipital bone
Sphenoid
Supra-orbital foramen
Nasal bone
Lacrimal bone
Ethmoid
Infra-orbital foramen
Maxilla
Zygomatic bone
Styloid process
Zygomatic process of temporal bone
Zygomatic arch
Temporal process of zygomatic bone
Mandible
Mental foramen
Mental protuberance
c Lateral view

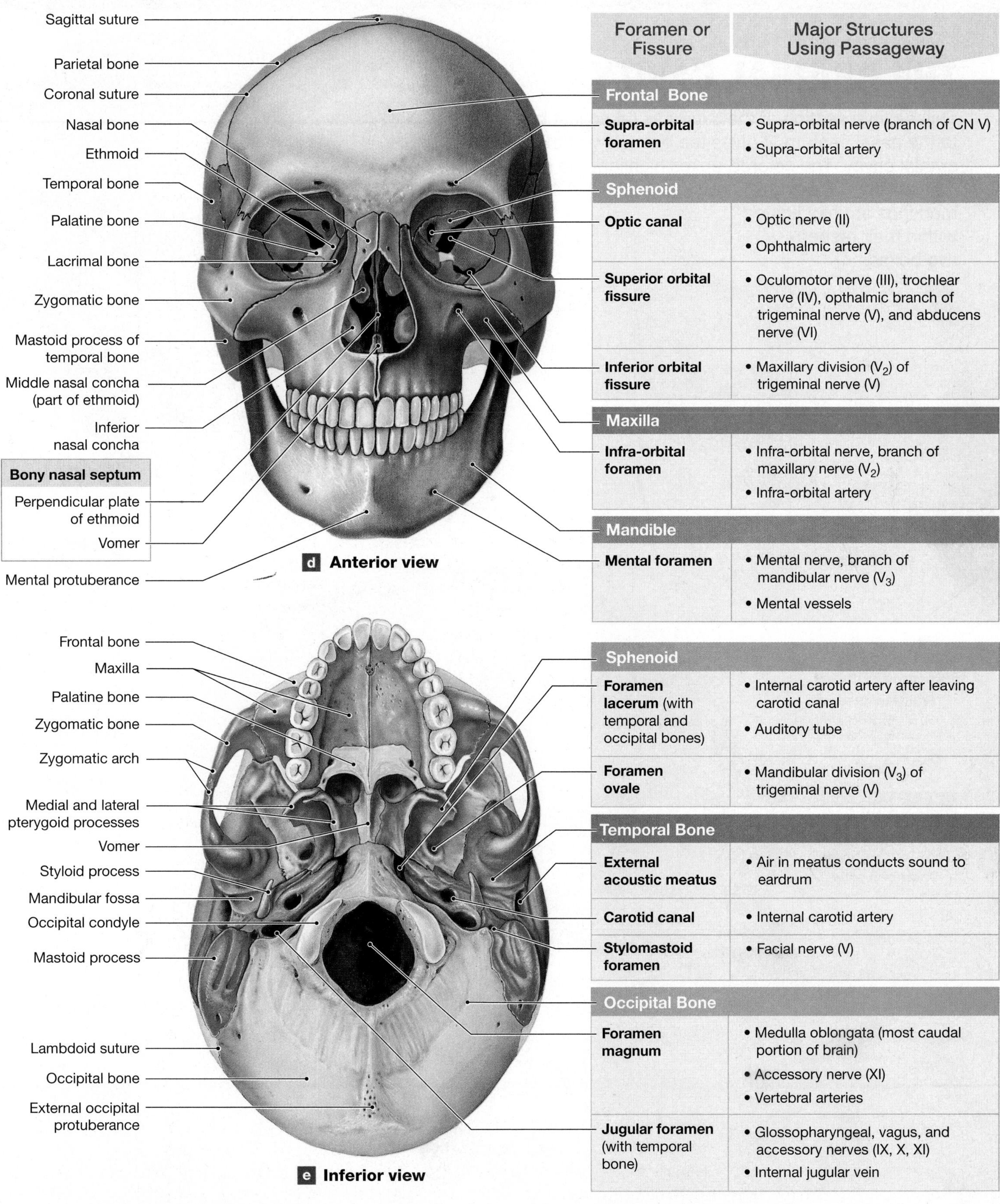
Sagittal suture
Parietal bone
Coronal suture
Nasal bone
Ethmoid
Temporal bone
Palatine bone
Lacrimal bone
Zygomatic bone
Mastoid process of temporal bone
Middle nasal concha (part of ethmoid)
Inferior nasal concha
Bony nasal septum
Perpendicular plate of ethmoid
Vomer
Mental protuberance
Frontal bone
Maxilla
Palatine bone
Zygomatic bone
Zygomatic arch
Medial and lateral pterygoid processes
Vomer
Styloid process
Mandibular fossa
Occipital condyle
Mastoid process
Lambdoid suture
Occipital bone
External occipital protuberance

d **Anterior view**

e **Inferior view**

Foramen or Fissure	Major Structures Using Passageway
Frontal Bone	
Supra-orbital foramen	• Supra-orbital nerve (branch of CN V) • Supra-orbital artery
Sphenoid	
Optic canal	• Optic nerve (II) • Ophthalmic artery
Superior orbital fissure	• Oculomotor nerve (III), trochlear nerve (IV), opthalmic branch of trigeminal nerve (V), and abducens nerve (VI)
Inferior orbital fissure	• Maxillary division (V_2) of trigeminal nerve (V)
Maxilla	
Infra-orbital foramen	• Infra-orbital nerve, branch of maxillary nerve (V_2) • Infra-orbital artery
Mandible	
Mental foramen	• Mental nerve, branch of mandibular nerve (V_3) • Mental vessels

Foramen or Fissure	Major Structures Using Passageway
Sphenoid	
Foramen lacerum (with temporal and occipital bones)	• Internal carotid artery after leaving carotid canal • Auditory tube
Foramen ovale	• Mandibular division (V_3) of trigeminal nerve (V)
Temporal Bone	
External acoustic meatus	• Air in meatus conducts sound to eardrum
Carotid canal	• Internal carotid artery
Stylomastoid foramen	• Facial nerve (V)
Occipital Bone	
Foramen magnum	• Medulla oblongata (most caudal portion of brain) • Accessory nerve (XI) • Vertebral arteries
Jugular foramen (with temporal bone)	• Glossopharyngeal, vagus, and accessory nerves (IX, X, XI) • Internal jugular vein

7

? Looking at all the views of the skull, name the major sutures.

SPOTLIGHT

Figure 7–4
Sectional Anatomy of the Skull

THE CRANIAL CAVITY

This sagittal section of the skull was taken just lateral to the nasal septum. It gives the best view of the cranial cavity. The colored boxes highlight the bones that surround the cranial cavity, and some of their key markings are highlighted within their respective boxes.

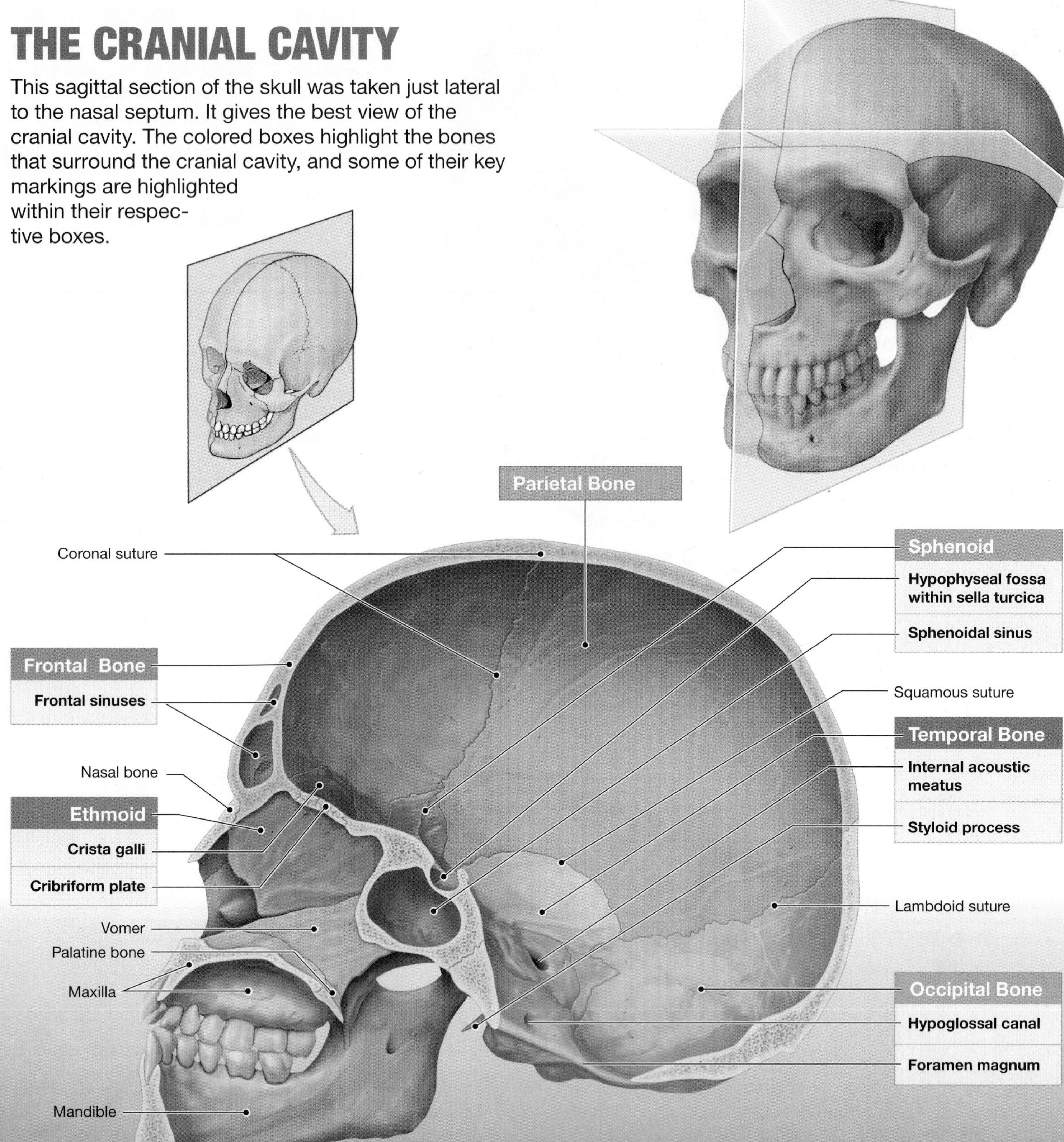

Parietal Bone
Coronal suture
Sphenoid
Hypophyseal fossa within sella turcica
Sphenoidal sinus
Frontal Bone
Frontal sinuses
Squamous suture
Temporal Bone
Internal acoustic meatus
Styloid process
Nasal bone
Ethmoid
Crista galli
Cribriform plate
Lambdoid suture
Vomer
Palatine bone
Maxilla
Occipital Bone
Hypoglossal canal
Foramen magnum
Mandible

a **Sagittal section**

THE FLOOR OF THE CRANIUM

This superior view of a horizontal section through the skull shows the floor of the cranial cavity. This is where most of the nerves and blood vessels enter or leave the cranial cavity. Compare this figure and chart with *Figure 7–3e* showing the inferior view of the skull and *Figure 14–18* (p. 495) showing the origin of the cranial nerves.

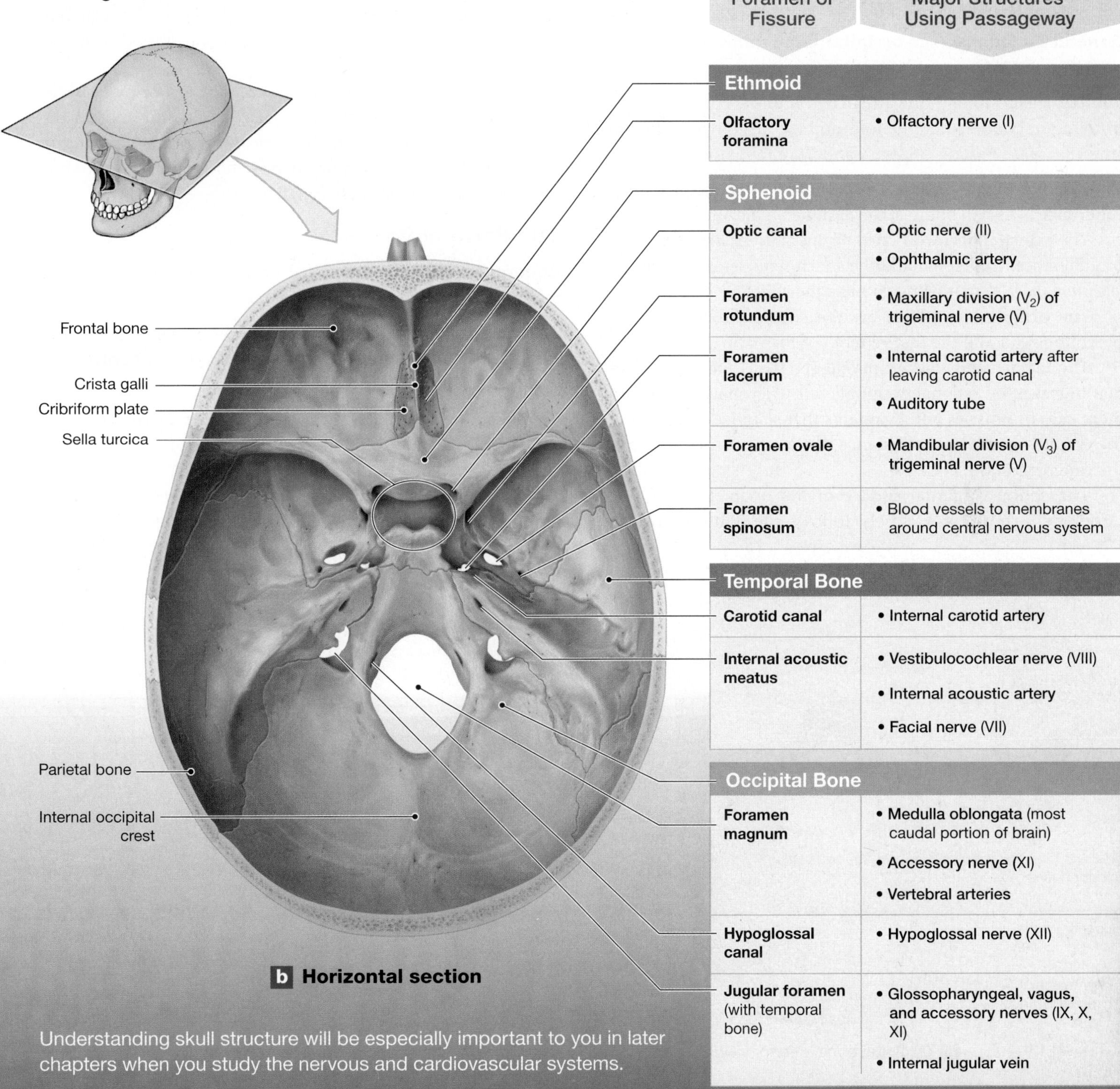

Frontal bone
Crista galli
Cribriform plate
Sella turcica
Parietal bone
Internal occipital crest

b Horizontal section

Foramen or Fissure	Major Structures Using Passageway
Ethmoid	
Olfactory foramina	• **Olfactory nerve** (I)
Sphenoid	
Optic canal	• **Optic nerve** (II) • **Ophthalmic artery**
Foramen rotundum	• **Maxillary division** (V_2) **of trigeminal nerve** (V)
Foramen lacerum	• **Internal carotid artery** after leaving carotid canal • **Auditory tube**
Foramen ovale	• **Mandibular division** (V_3) **of trigeminal nerve** (V)
Foramen spinosum	• Blood vessels to membranes around central nervous system
Temporal Bone	
Carotid canal	• **Internal carotid artery**
Internal acoustic meatus	• **Vestibulocochlear nerve** (VIII) • **Internal acoustic artery** • **Facial nerve** (VII)
Occipital Bone	
Foramen magnum	• **Medulla oblongata** (most caudal portion of brain) • **Accessory nerve** (XI) • **Vertebral arteries**
Hypoglossal canal	• **Hypoglossal nerve** (XII)
Jugular foramen (with temporal bone)	• **Glossopharyngeal, vagus, and accessory nerves** (IX, X, XI) • **Internal jugular vein**

Understanding skull structure will be especially important to you in later chapters when you study the nervous and cardiovascular systems.

FOCUS The Individual Bones of the Skull

We can explore the surface features of these bones further, using the related images in the *Atlas*. Foramina and fissures create passages for vessels and nerves. The vessels are detailed in Chapter 21. The nerves are described in the Focus on cranial nerves in Chapter 14.

Cranial Bones

The Occipital Bone (Figure 7–5a)

General Functions: The **occipital bone** forms much of the posterior and inferior surfaces of the cranium.

Joints: The occipital bone articulates with the parietal bones, the temporal bones, the sphenoid, and the first cervical vertebra (the atlas) (see Figure 7–3a–c, e and Spotlight Figure 7–4).

Regions/Markings (Figure 7–5a): The **external occipital protuberance** is a small bump at the midline on the inferior surface.

The **external occipital crest** begins at the external occipital protuberance. This crest marks the attachment of a ligament (*ligamentum nuchae*) that helps stabilize the vertebrae of the neck.

The **occipital condyles** are the sites of articulation between the skull and the first vertebra of the neck.

The *inferior* and *superior nuchal* (NŪ-kul) *lines* are ridges that intersect the external occipital crest. They mark the attachment sites of muscles and ligaments that stabilize the joints at the occipital condyles and balance the weight of the head over the vertebrae of the neck.

The concave internal surface of the occipital bone (see Spotlight Figure 7–4a) closely follows the contours of the brain. The grooves follow the paths of major blood vessels, and the ridges mark the attachment sites of membranes that stabilize the position of the brain.

Foramina: The **foramen magnum** (see Spotlight Figure 7–4b) connects the cranial cavity with the vertebral canal, which is enclosed by the vertebral column. This foramen surrounds the connection between the brain and spinal cord.

The **jugular foramen** lies between the occipital bone and the temporal bone (see Figure 7–3e). The *internal jugular vein* passes through this foramen, carrying venous blood from the brain.

The **hypoglossal canals** (see Spotlight Figure 7–4b) begin at the lateral base of each occipital condyle and end on the inner surface of the occipital bone near the foramen magnum. The *hypoglossal nerves*, cranial nerves that control the tongue muscles, pass through these canals.

The Parietal Bones (Figure 7–5b)

General Functions: The paired **parietal bones** form part of the superior and lateral surfaces of the cranium.

Joints: The parietal bones articulate with one another and with the occipital, temporal, frontal, and sphenoid bones (see Figure 7–3a–d and Spotlight Figure 7–4).

Regions/Markings: The *superior* and *inferior temporal lines* are low ridges that mark the attachment sites of the *temporalis*, a large muscle that closes the mouth (Figure 7–5b).

Grooves on the inner surface of the parietal bones mark the paths of cranial blood vessels (see Spotlight Figure 7–4a).

Figure 7–5 The Occipital and Parietal Bones.

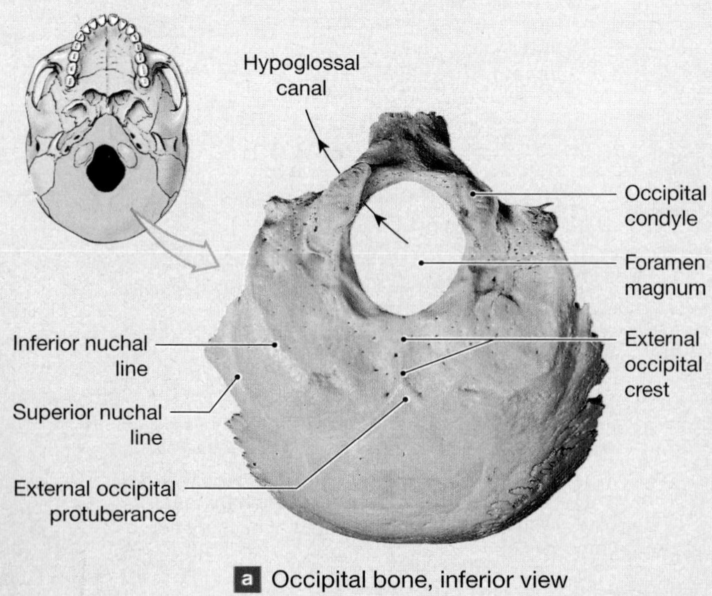
Hypoglossal canal
Occipital condyle
Foramen magnum
External occipital crest
Inferior nuchal line
Superior nuchal line
External occipital protuberance

a Occipital bone, inferior view

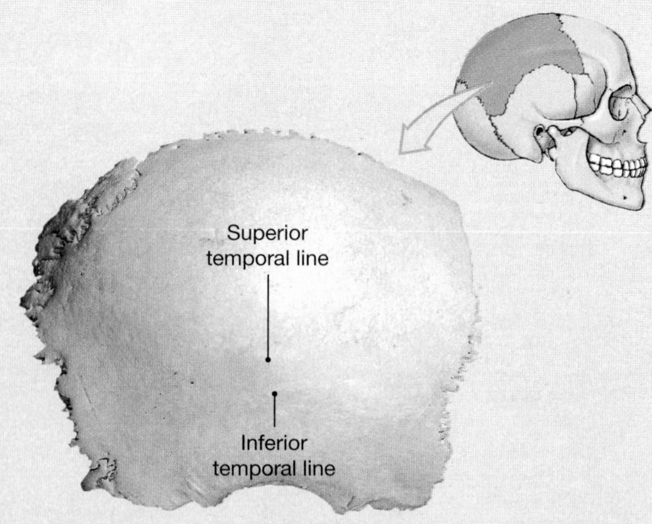
Superior temporal line
Inferior temporal line

b Right parietal bone, lateral view

The Frontal Bone (Figure 7–6)

General Functions: The **frontal bone** forms the anterior portion of the cranium and the roof of the *orbits* (eye sockets). Mucus from the *frontal sinuses* within this bone helps to flush the surfaces of the nasal cavities.

Joints: The frontal bone articulates with the parietal, sphenoid, ethmoid, nasal, lacrimal, maxillary, and zygomatic bones (see Figure 7–3b–e and **Spotlight Figure 7–4**).

Regions/Markings: The **forehead** forms the anterior, superior portion of the cranium. It provides surface area for the attachment of facial muscles. The *superior temporal line* is continuous with the superior temporal line of the parietal bone.

The **supra-orbital margin** is a thickening of the frontal bone that helps protect the eye. The **glabella** (*glabellus*, smooth) is the area that is between the supra-orbital margins and superior to the nasal bones. It marks the most forward projection of the frontal bone.

The **lacrimal fossa** on the superior and lateral surface of the orbit is a shallow depression that marks the location of the *lacrimal* (tear) *gland*. Tears from this gland lubricate the surface of the eye.

The **frontal sinuses** are extremely variable in size and time of appearance. They generally appear after age 6, but some people never develop them. We describe the frontal sinuses and other sinuses of the cranium and face in Section 7-3.

Foramina: The **supra-orbital foramen** provides passage for blood vessels that supply the eyebrow, eyelids, and frontal sinuses. In some cases, this foramen is incomplete. The blood vessels then cross the orbital rim within a **supra-orbital notch.**

Remarks: During development, the bones of the cranium form by the fusion of separate centers of ossification. At birth, these fusion processes are not yet complete: Two frontal bones articulate along the *frontal suture.* The suture generally disappears by age 8 as the bones fuse. In rare cases, it persists as the *metopic suture* and runs through the midline of the forehead. It is sometimes mistaken for a frontal bone fracture.

The Temporal Bones (Figure 7–7)

General Functions: The paired **temporal bones** (1) form part of both the lateral walls of the cranium and the *zygomatic arches*, (2) form the only joints with the mandible, (3) surround and protect the sense organs of the internal ear, and (4) are attachment sites for muscles that close the jaws and move the head.

Joints: The temporal bones articulate with the zygomatic, sphenoid, parietal, and occipital bones of the cranium and with the mandible (see Figure 7–3 and **Spotlight Figure 7–4**).

Regions/Markings: The **squamous part** of the temporal bone is the lateral portion that borders the squamous suture.

Figure 7–6 The Frontal Bone.

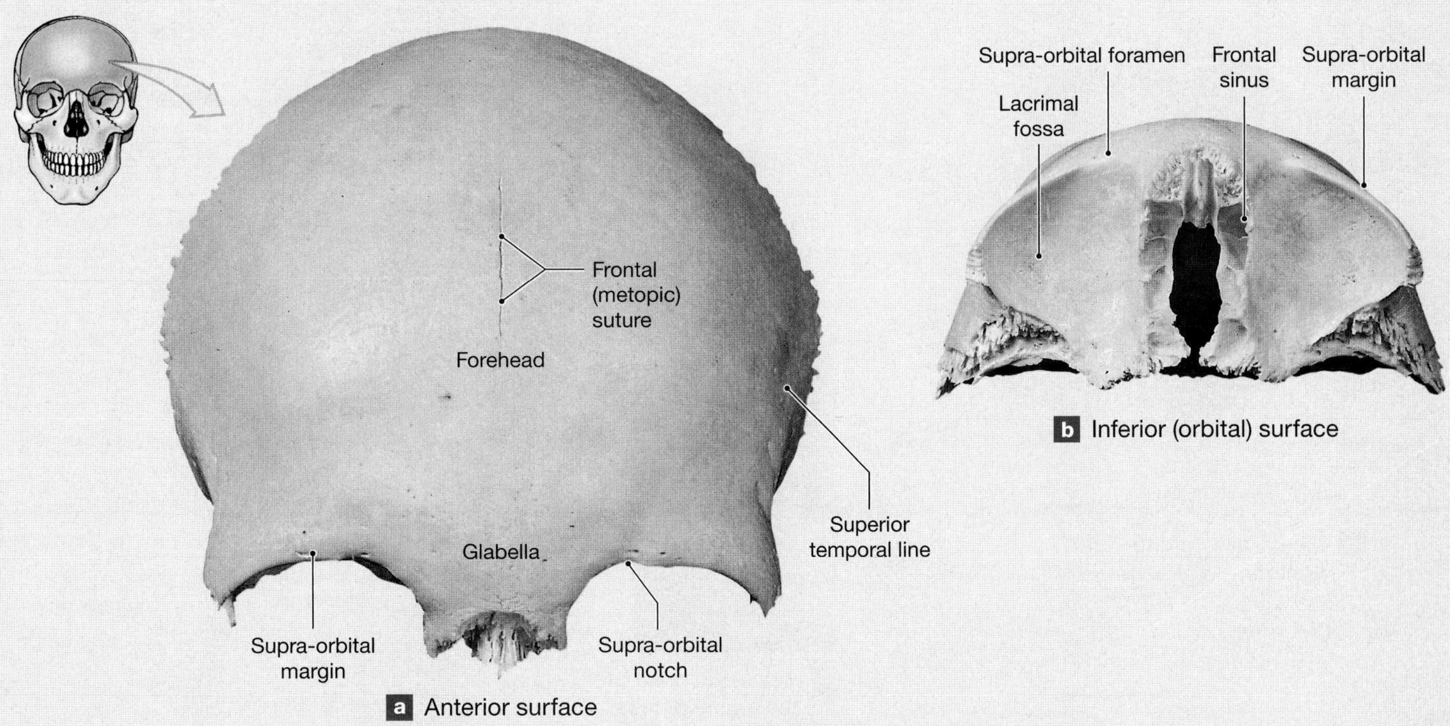

Frontal (metopic) suture
Forehead
Glabella
Superior temporal line
Supra-orbital margin
Supra-orbital notch
a Anterior surface
Supra-orbital foramen
Frontal sinus
Supra-orbital margin
Lacrimal fossa
b Inferior (orbital) surface

The concave, internal surface parallels the surface of the brain and is called the *cerebral surface.*

The **zygomatic process**, inferior to the temporal squamous part (Figure 7–7a), articulates with the *temporal process* of the zygomatic bone.

Together, these processes form the **zygomatic arch** (see Figure 7–3c,e).

The **mandibular fossa** on the inferior surface marks the site of articulation with the mandible.

The **mastoid process** (Figure 7–7b,c) is an attachment site for muscles that rotate or extend the head. It contains *mastoid cells* (also called mastoid air cells), small, interconnected cavities that connect to the middle ear cavity. *Mastoiditis*, inflammation and infection in the air cell system, develops if pathogens reach the mastoid process. Signs and symptoms include severe earaches, fever, and swelling behind the ear.

The **styloid** (STĪ-loyd; *stylos*, pillar) **process**, near the base of the mastoid process, is attached to ligaments that support the hyoid bone and to the tendons of several muscles associated with the hyoid bone, the tongue, and the pharynx (throat).

The **petrous part** of the temporal bone, located on its internal surface, encloses the structures of the *internal ear*—sense organs that provide information about hearing and balance (see Figure 7–7a).

The auditory ossicles are located in the *tympanic cavity*, or *middle ear*, a cavity within the petrous part. These tiny bones—three on each side—transfer sound vibrations from the delicate *tympanic membrane*, or eardrum, to the internal ear. (We discuss these bones and their functions in Chapter 17.)

Foramina (see Figure 7–3e): The jugular foramen, between the temporal and occipital bones, provides passage for the internal jugular vein.

The **carotid canal** provides passage for the internal carotid artery, a major artery to the brain. As it leaves the carotid canal, the internal carotid artery passes through the anterior portion of the foramen lacerum.

The **foramen lacerum** (LAS-er-um; *lacerare*, to tear) is a jagged slit extending between the sphenoid and the petrous part of the temporal bone. This slit contains hyaline cartilage and small arteries that supply the inner surface of the cranium.

The *auditory tube* is an air-filled passageway that connects the pharynx (throat) to the middle ear. This tube passes through the posterior portion of the foramen lacerum.

The **external acoustic meatus** on the lateral surface, ends at the tympanic membrane, or eardrum (which disintegrates during the preparation of a dried skull).

Figure 7–7 The Temporal Bones.

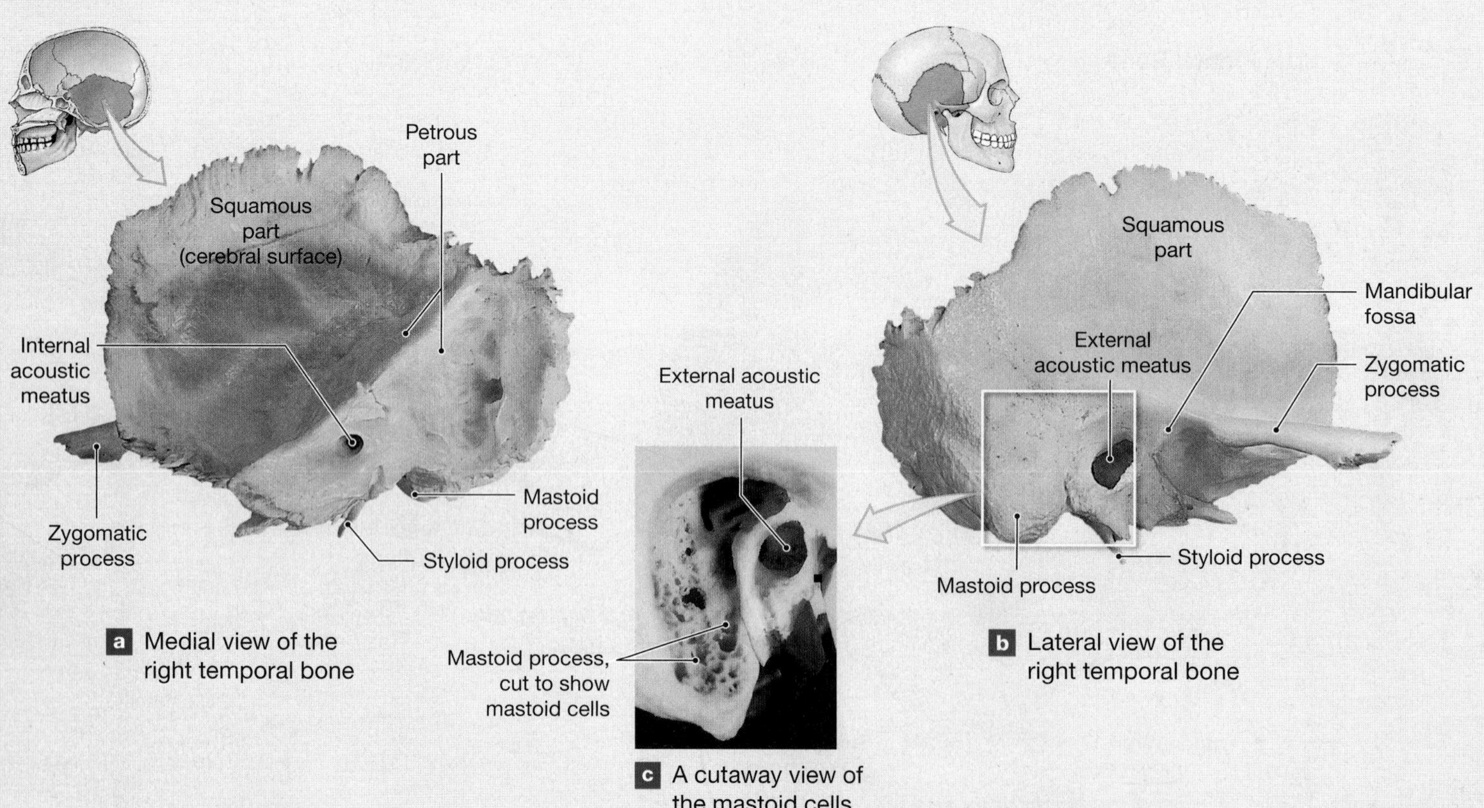

Petrous part
Squamous part (cerebral surface)
Internal acoustic meatus
Zygomatic process
Mastoid process
Styloid process
External acoustic meatus
Mastoid process, cut to show mastoid cells
Squamous part
Mandibular fossa
External acoustic meatus
Zygomatic process
Styloid process
Mastoid process

a Medial view of the right temporal bone

c A cutaway view of the mastoid cells

b Lateral view of the right temporal bone

The **stylomastoid foramen** lies posterior to the base of the styloid process. The *facial nerve* passes through this foramen to control the facial muscles.

The **internal acoustic meatus**, begins on the medial surface of the petrous part of the temporal bone. It carries blood vessels and nerves to the internal ear and conveys the facial nerve to the stylomastoid foramen.

The Sphenoid (Figure 7–8)

General Functions: The **sphenoid** forms part of the floor of the cranium, unites the cranial and facial bones, and acts as a cross-brace that strengthens the sides of the skull. Mucus from the *sphenoidal sinuses* within this bone helps clean the surfaces of the nasal cavities.

Joints: The sphenoid articulates with the ethmoid and the frontal, occipital, parietal, and temporal bones of the cranium and the palatine bones, zygomatic bones, maxillae, and vomer of the face (see Figure 7–3c–e and Spotlight Figure 7–4).

Regions/Markings (see Figure 7–8): The shape of the sphenoid can be compared to a bat with its wings extended. This bone is relatively large, but more superficial bones hide much of it.

The **body** forms the central axis of the sphenoid.

The **sella turcica** (TUR-si-kuh; "Turkish saddle") is a bony, saddle-shaped enclosure on the superior surface of the body (Figure 7–8a). Within the sella turcica is a depression known as the **hypophyseal** (hī-pō-FIZ-ē-al) **fossa** (see Spotlight Figure 7–4a). This fossa contains the *pituitary gland.*

Figure 7–8 The Sphenoid.

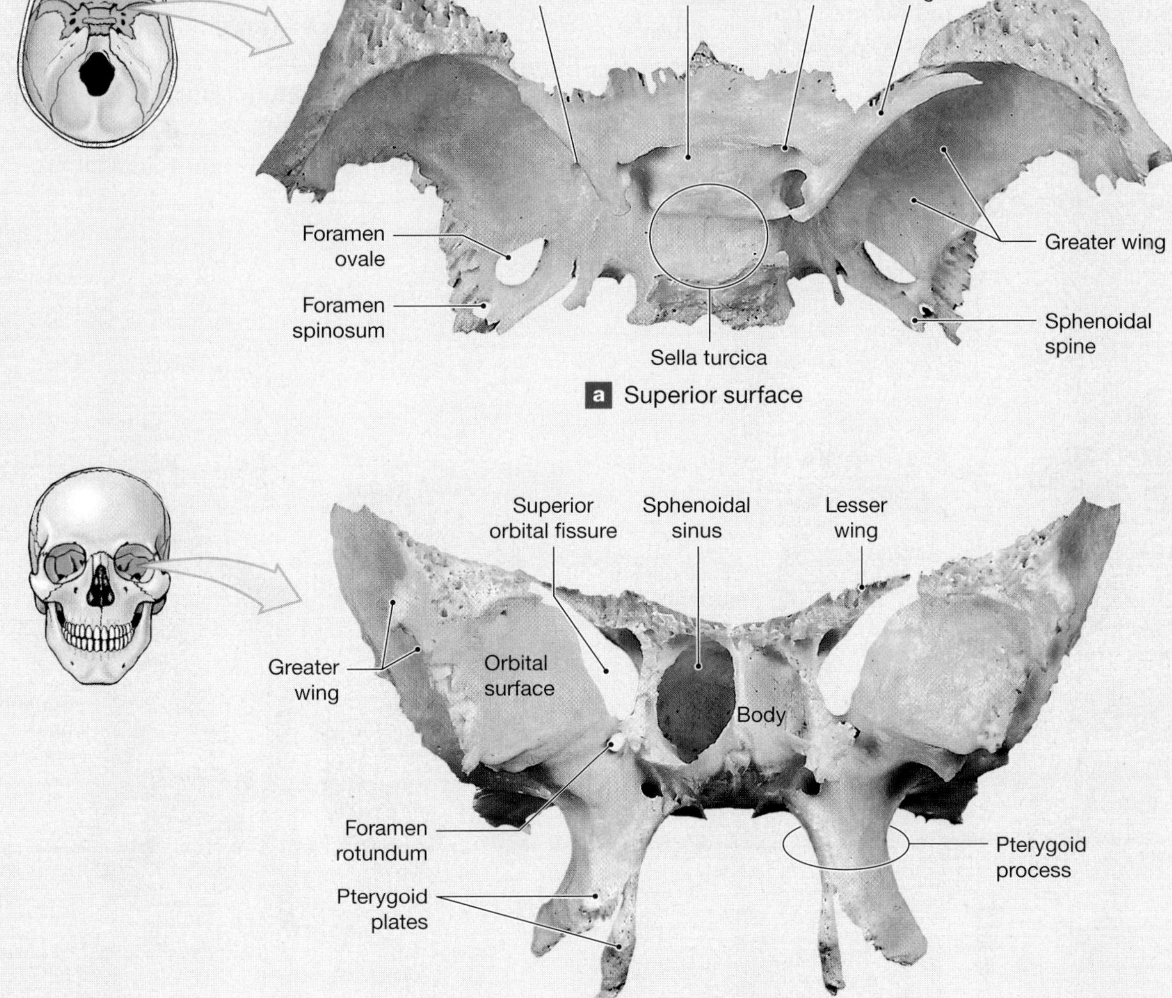

Foramen rotundum
Prechiasmatic sulcus
Optic canal
Lesser wing
Foramen ovale
Foramen spinosum
Greater wing
Sphenoidal spine
Sella turcica
a Superior surface
Superior orbital fissure
Sphenoidal sinus
Lesser wing
Greater wing
Orbital surface
Body
Foramen rotundum
Pterygoid plates
Pterygoid process
b Anterior surface

The **sphenoidal sinuses** (Figure 7–8b) are on either side of the body, inferior to the sella turcica.

The **lesser wings** extend horizontally anterior to the sella turcica.

The **greater wings** extend laterally from the body and form part of the cranial floor. A sharp *sphenoidal spine* lies at the posterior, lateral corner of each greater wing. Anteriorly, each greater wing forms part of the posterior wall orbit.

The **pterygoid** (TER-i-goyd; *pterygion*, wing) **processes** are vertical projections that originate on either side of the body. Each pterygoid process forms a pair of *pterygoid plates*, which are attachment sites for muscles that move the mandible and soft palate.

Foramina (see Spotlight Figure 7–4b and Figure 7–8): The **optic canals** permit passage of the optic nerves from the eyes to the brain. The **prechiasmatic sulcus**, also known as the *optic groove*, is a narrow grove anterior to the optic chiasm, the point where two optic nerves cross.

Penetrating each greater wing are a **superior orbital fissure**, **foramen rotundum**, **foramen ovale** (ō-VAH-lē), and **foramen spinosum**. These passages carry blood vessels and nerves to the orbit, face, jaws, and membranes of the cranial cavity, respectively.

The Ethmoid (Figure 7–9)

General Functions: The **ethmoid** forms the anteromedial floor of the cranium, the roof of the nasal cavity, and part of the nasal septum (which subdivides the nasal cavity) and medial orbital wall. Mucus from a network of sinuses, or *ethmoidal cells*, within this bone flushes the surfaces of the nasal cavities.

Joints: The ethmoid articulates with the frontal bone and sphenoid of the cranium and with the maxillae, nasal, lacrimal, and palatine bones, and the inferior nasal conchae and vomer of the face (see Figure 7–3c,d and Spotlight Figure 7–4).

Regions/Markings: The ethmoid has three parts: (1) the cribriform plate, (2) the ethmoidal labyrinth, and (3) the perpendicular plate (Figure 7–9a).

The **cribriform plate** (*cribrum*, sieve) forms the anteromedial floor of the cranium and the roof of the nasal cavity. The **crista galli** (*crista*, crest + *gallus*, rooster, cock; "cock's comb") is a bony ridge that projects superior to the cribriform plate. A membrane that stabilizes the position of the brain, the *falx cerebri*, attaches to this ridge.

An **ethmoidal labyrinth** opens into the nasal cavity on each side. Each labyrinth consists of **ethmoidal cells**, interconnected air-filled cavities. The cells are arranged in three groups, anterior, middle, and posterior, and are closed laterally by the orbital plate, which forms part of the wall of the orbit. The **superior nasal conchae** (KONG-kē; singular, *concha* a snail shell) and the **middle nasal conchae** are delicate projections that form part of the ethmoidal labyrinth (Figure 7–9b).

Figure 7–9 The Ethmoid.

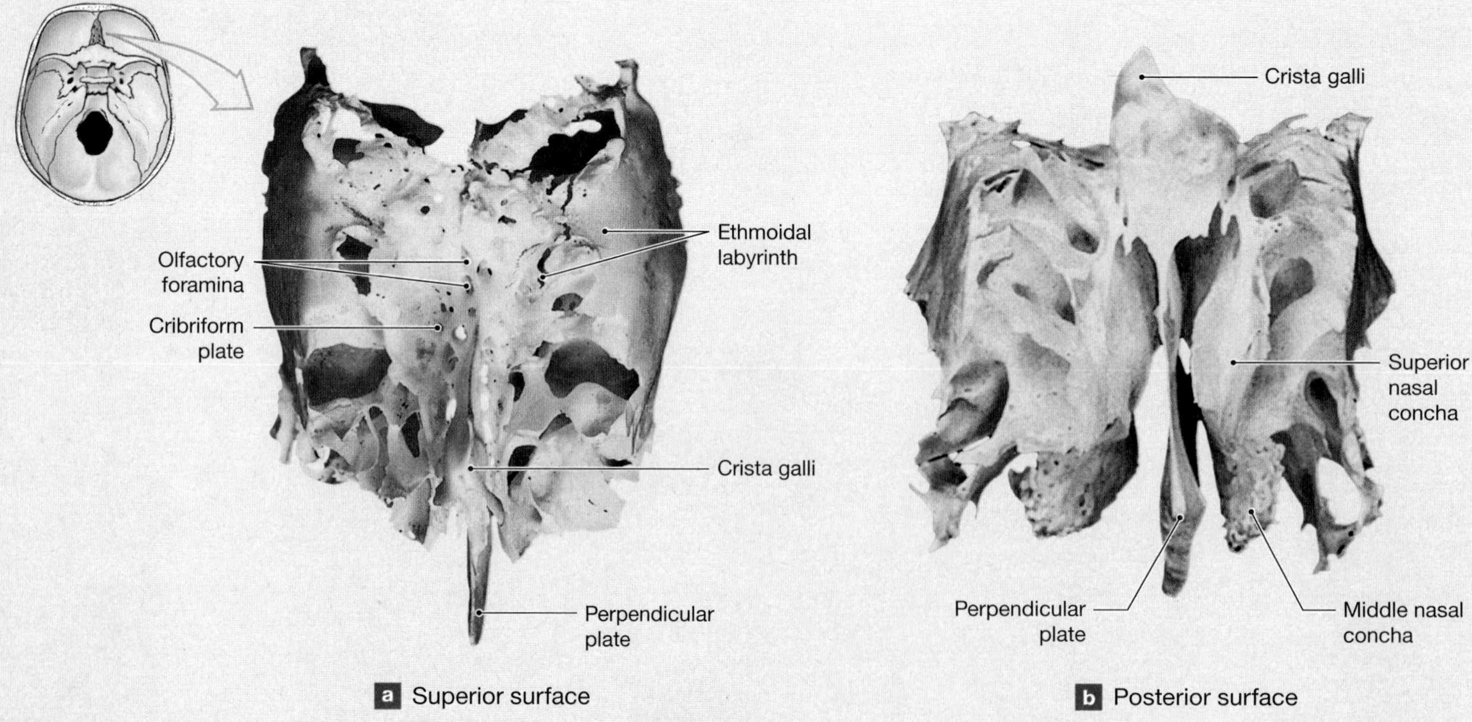

Olfactory foramina
Cribriform plate
Ethmoidal labyrinth
Crista galli
Perpendicular plate
a Superior surface
Crista galli
Superior nasal concha
Perpendicular plate
Middle nasal concha
b Posterior surface

The perpendicular plate forms part of the nasal septum, along with the vomer and a piece of hyaline cartilage.

Foramina: The **olfactory foramina** in the cribriform plate permit passage of the olfactory nerves, which provide the sense of smell.

Remarks: *Olfactory* (smell) *receptors* are located in the epithelium that covers the inferior surfaces of the cribriform plate, the medial surfaces of the superior nasal conchae, and the superior portion of the perpendicular plate.

The nasal conchae break up the airflow in the nasal cavity. They create swirls, turbulence, and eddies that have three major functions: (1) Particles in the air are thrown against the sticky mucus that covers the walls of the nasal cavity; (2) air movement is slowed, providing time for warming, humidifying, and removing dust before the air reaches more delicate portions of the respiratory tract; and (3) air is directed toward the superior portion of the nasal cavity, adjacent to the cribriform plate, where the olfactory receptors are located.

Facial Bones

The Maxillae (Figure 7–10a,b)

General Functions: The paired **maxillae**, or *upper jawbones*, support the upper teeth and form the inferior orbital rims, the lateral margins of the external nares, the upper jaw, and most of the hard palate. The *maxillary sinuses* in these bones produce mucus that flushes the inferior surfaces of the nasal cavities. The maxillae are the largest facial bones, and the maxillary sinuses are the largest sinuses.

Joints: The maxillae articulate with the frontal bones and ethmoid, with one another, and with all the other facial bones except the mandible (see Figure 7–3c–e and **Spotlight Figure 7–4a**).

Regions/Markings: The **orbital rim** protects the eye and other structures in the orbit (Figure 7–10a). The *anterior nasal spine* is found at the anterior portion of the maxilla, at its articulation with the maxilla of the other side. It is an attachment point for the cartilaginous anterior portion of the nasal septum.

Figure 7–10 The Maxillae and Palatine Bones. ATLAS: Plates 8a–d; 12d

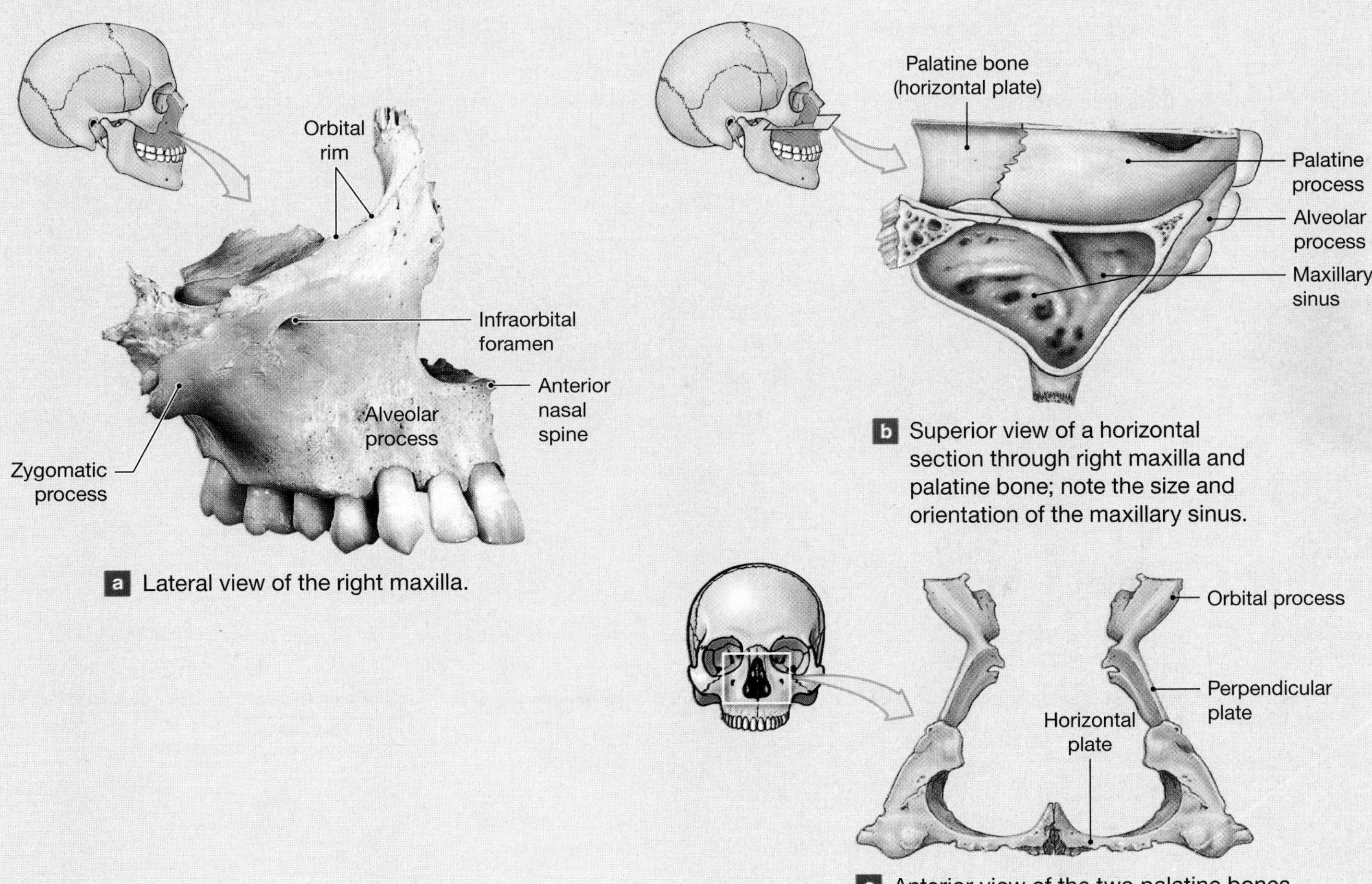

Orbital rim
Infraorbital foramen
Anterior nasal spine
Alveolar process
Zygomatic process
Palatine bone (horizontal plate)
Palatine process
Alveolar process
Maxillary sinus
Orbital process
Perpendicular plate
Horizontal plate

a Lateral view of the right maxilla.

b Superior view of a horizontal section through right maxilla and palatine bone; note the size and orientation of the maxillary sinus.

c Anterior view of the two palatine bones.

The **alveolar process** of the maxilla is a projecting ridge that contains the tooth sockets for the upper teeth.

The **palatine processes** form most of the **hard palate**, or bony roof of the mouth. A developmental disorder known as a *cleft palate* occurs when the maxillae fail to meet along the midline of the hard palate, resulting in a split in the roof of the mouth. ATLAS: Embryology Summary 6: The Development of the Skull

The **maxillary sinuses** lighten the portion of the maxillae superior to the teeth (Figure 7–10b).

The **nasolacrimal canal** is formed by a maxilla and a lacrimal bone. It protects the *lacrimal sac* and the **nasolacrimal duct**, which carries tears from the orbit to the nasal cavity.

Foramina: The **infra-orbital foramen** marks the path of a major sensory nerve that reaches the brain by the foramen rotundum of the sphenoid.

The **inferior orbital fissure** (see Figure 7–3d) lies between the maxilla and the sphenoid. It permits passage of cranial nerves and blood vessels.

The Palatine Bones (Figure 7–10b,c)

General Functions: The paired **palatine bones** form the posterior portions of the hard palate. They also contribute to the floor of each orbit.

Joints: The palatine bones articulate with one another, with the maxillae, with the sphenoid and ethmoid, with the inferior nasal conchae, and with the vomer (see Figure 7–3e and Spotlight Figure 7–4a).

Regions/Markings: The palatine bones are shaped like an L (Figure 7–10c). The **horizontal plate** forms the posterior part of the hard palate. The **perpendicular plate** extends from the horizontal plate to the **orbital process**, which forms part of the floor of the orbit. This process contains a small sinus that usually opens into the sphenoidal sinus.

Foramina: Small blood vessels and nerves supplying the roof of the mouth penetrate the lateral portion of the horizontal plate.

The Nasal Bones (Figure 7–11)

General Functions: The paired **nasal bones** support the superior portions of the bridge of the nose. They are connected to cartilages that support the distal portions of the nose. These flexible cartilages, and associated soft tissues, extend to the superior border of the **external nares** (NA-rēz; singular, *naris*), the entrances to the nasal cavity.

Joints: The paired nasal bones articulate with one another, with the ethmoid, and with the frontal bone and maxillae (see Figure 7–3b–d and Spotlight Figure 7–4a).

The Vomer (Figure 7–11)

General Functions: The **vomer** forms the inferior portion of the bony nasal septum.

Figure 7–11 The Smaller Bones of the Face.

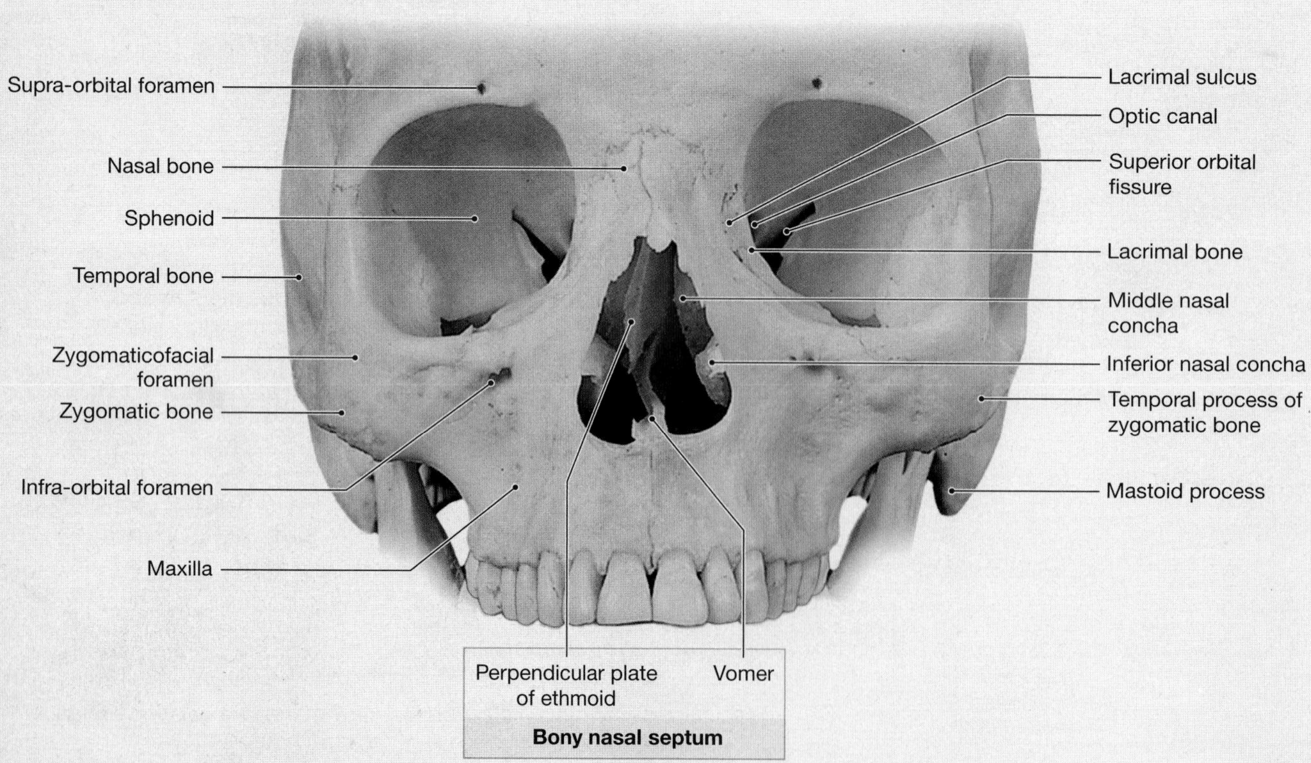

Supra-orbital foramen
Nasal bone
Sphenoid
Temporal bone
Zygomaticofacial foramen
Zygomatic bone
Infra-orbital foramen
Maxilla
Lacrimal sulcus
Optic canal
Superior orbital fissure
Lacrimal bone
Middle nasal concha
Inferior nasal concha
Temporal process of zygomatic bone
Mastoid process
Perpendicular plate of ethmoid
Vomer
Bony nasal septum

Joints: The vomer articulates with the maxillae, sphenoid, ethmoid, and palatine bones, and with the cartilaginous part of the nasal septum, which extends into the fleshy part of the nose (see Figure 7–3d,e and Spotlight Figure 7–4a).

The Inferior Nasal Conchae (Figure 7–11)

General Functions: The paired **inferior nasal conchae** slow inhaled air and create turbulence as the air passes through the nasal cavity. They also increase the epithelial surface area to warm and humidify inhaled air.

Joints: The inferior nasal conchae articulate with the maxillae, ethmoid, palatine, and lacrimal bones (see Figure 7–3d).

The Zygomatic Bones (Figure 7–11)

General Functions: The paired **zygomatic bones** contribute to the rims and lateral walls of the orbits and form the prominent cheekbones and parts of the zygomatic arches.

Joints: The zygomatic bones articulate with the maxillae, and the sphenoid, frontal, and temporal bones (see Figure 7–3b–e).

Regions/Markings: The **temporal process** curves posteriorly to meet the zygomatic process of the temporal bone.

Foramina: The **zygomaticofacial foramen** on the anterior surface of each zygomatic bone carries a sensory nerve that innervates the cheek.

The Lacrimal Bones (Figure 7–11)

General Functions: The paired **lacrimal bones** form parts of the medial walls of the orbits.

Joints: The lacrimal bones are the smallest facial bones. They articulate with the frontal bone and maxillae, and with the ethmoid (see Figure 7–3c,d).

Regions/Markings: The **lacrimal sulcus**, a groove along the anterior, lateral surface of the lacrimal bone, marks the location of the lacrimal sac. The lacrimal sulcus leads to the nasolacrimal canal, which begins at the orbit and opens into the nasal cavity. As noted earlier, the lacrimal bone and the maxilla form this canal.

The Mandible (Figure 7–12)

General Functions: The **mandible** forms the lower jaw.

Joints: The mandible articulates with the mandibular fossae of the temporal bones (see Figures 7–3c,e and 7–7b).

Figure 7–12 The Mandible.

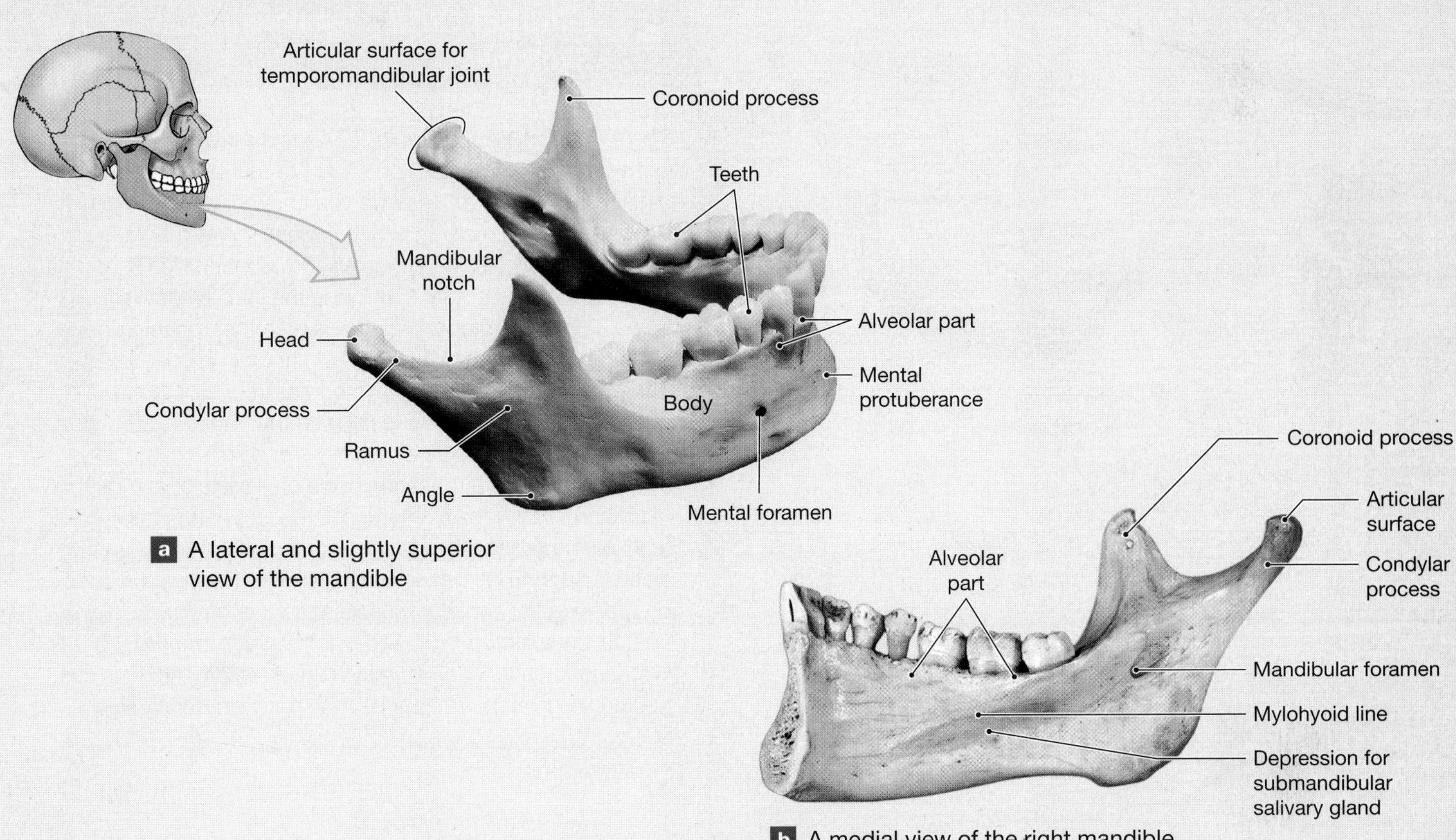
Articular surface for temporomandibular joint
Coronoid process
Teeth
Mandibular notch
Head
Alveolar part
Mental protuberance
Condylar process
Body
Ramus
Angle
Mental foramen
a A lateral and slightly superior view of the mandible
Coronoid process
Articular surface
Alveolar part
Condylar process
Mandibular foramen
Mylohyoid line
Depression for submandibular salivary gland
b A medial view of the right mandible

Regions/Markings: The **body** of the mandible is the horizontal portion of that bone (Figure 7–12a).

The **alveolar part** of the mandible is the portion of the mandible that surrounds and supports the lower teeth. The *dental alveoli* are the sockets for the roots of the teeth.

The *mental protuberance* (*mentalis*, chin) is the attachment site for several facial muscles.

The **ramus** of the mandible is the ascending part that begins at the *angle* of the mandible on either side.

On each ramus:

- The **condylar process** articulates with the temporal bone at the *temporomandibular joint*.
- The **coronoid** (KOR-ō-noyd) **process** is the insertion point for the *temporalis*, a powerful muscle that closes the jaws.
- The **mandibular notch** is the depression that separates the condylar and coronoid processes.

A prominent depression on the mandible's medial surface marks the position of the *submandibular salivary gland* (Figure 7–12b).

The *mylohyoid line* marks the insertion of the *mylohyoid*, which supports the floor of the mouth.

Foramina: The **mental foramina** are openings for nerves that carry sensory information from the lips and chin to the brain.

The **mandibular foramen** is the entrance to the *mandibular canal*, a passageway for blood vessels and nerves that service the lower teeth. Dentists typically anesthetize the sensory nerve that enters this canal before they work on the lower teeth.

The Hyoid Bone (Figure 7–13)

General Functions: The **hyoid bone** supports the larynx (voicebox) and is the attachment site for muscles of the larynx, pharynx, and tongue.

Joints: *Stylohyoid ligaments* connect the *lesser horns* to the styloid processes of the temporal bones.

Regions/Processes: The **body** of the hyoid is an attachment site for muscles of the larynx, pharynx, and tongue.

The larger and more lateral **greater horns** help support the larynx and are attached to muscles that move the tongue.

The shorter and more medial **lesser horns** are attached to the stylohyoid ligaments. From these ligaments, the hyoid and larynx hang beneath the skull like a swing from the limb of a tree.

Figure 7–13 A Superior View of the Hyoid Bone.

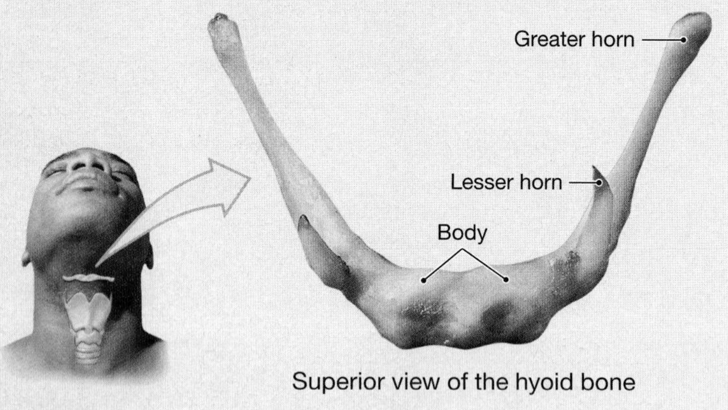
Greater horn
Lesser horn
Body
Superior view of the hyoid bone

Clinical Note Temporomandibular Joint Syndrome

The *temporomandibular joint (TMJ)* is a movable joint between each temporal bone and the mandible. This joint allows your jaw to move while you chew or talk. The disadvantage of such mobility is that forceful forward or lateral movement can dislocate the jaw. The connective tissue sheath, or *capsule*, that surrounds the joint is relatively loose. A pad of fibrocartilage separates the opposing bone surfaces. In **TMJ syndrome**, the mandible is pulled slightly out of alignment, generally by spasms in one of the jaw muscles. The person has facial pain that radiates around the ear on the affected side and cannot open the mouth fully. TMJ syndrome is a repeating cycle of muscle spasm → misalignment → pain → muscle spasm. It has been linked to involuntary behaviors, such as grinding the teeth during sleep (*bruxism*), and to emotional stress. Treatment focuses on breaking the cycle of muscle spasm and pain. Applying heat to the affected joint, and using anti-inflammatory drugs, local anesthetics, or both, may help. If teeth grinding is suspected, special mouth guards may be worn during sleep.

> Go to MasteringA&P™ > Study Area > Menu > Lab Tools > Bone & Dissection Videos > Bone Videos > Cranium

the sagittal suture produces a *midsagittal section.* A slice that parallels the sagittal suture produces a *parasagittal section.* p. 13

- *Squamous Sutures.* A **squamous** (SKWĀ-mus) **suture** on each side of the skull joins the temporal bone and the parietal bone of that side. Figure 7–3a shows the intersection between the squamous sutures and the lambdoid suture. Figure 7–3c shows the path of the squamous suture on the right side of the skull.

Sinuses, Foramina, and Fissures

Recall that sinuses, foramina, and fissures are bone markings that describe openings in bones. p. 184 Several bones of the skull contain air-filled chambers called **sinuses**. You can see these in the sagittal section of the cranial cavity in **Spotlight Figure 7–4a**. Sinuses have three major functions: (1) They lessen the weight of the bone; (2) they are lined with mucous membranes, which produce mucus that moistens and cleans the air in and near the sinus; and (3) they serve as resonating chambers in speech production. We consider the sinuses as we discuss specific bones (Figures 7–5 to 7–12). Figure 7–3d,e and **Spotlight Figure 7–4b** show important foramina and fissures in cranial and facial bones, along with the major structures using these passageways.

Checkpoint

3. **In which bone is the foramen magnum located?**
4. **Tomás suffers a hit to the skull that fractures the right superior lateral surface of his cranium. Which bone is fractured?**
5. **Which bone contains the depression called the sella turcica? What is located in this depression?**
6. **Identify the facial bones.**
7. **Identify the bone containing the mental foramen, and list the structures using this passageway.**
8. **Identify the bone containing the optic canal, and cite the structures using this passageway.**
9. **Name the foramina found in the ethmoid bone.**

See the blue Answers tab at the back of the book.

7-3 Each orbital complex contains and protects an eye, and the nasal complex encloses the nasal cavities

Learning Outcome Describe the structure and functions of the orbital complex, nasal complex, and paranasal sinuses.

The facial bones not only protect and support the openings of the digestive and respiratory systems, but also protect the sense organs responsible for vision and smell. Together, certain cranial bones and facial bones form an *orbital complex,* which surrounds each eye, and the *nasal complex,* which surrounds the nasal cavities.

The Orbital Complexes

The **orbits** are the bony sockets that contain and protect the eyes. Seven bones of the **orbital complex** form each orbit (Figure 7–14). The frontal bone forms the roof, and the maxilla provides most of the orbital floor. The orbital rim and the first portion of the medial wall are formed by the maxilla, the lacrimal bone, and the ethmoidal labyrinth. The ethmoidal labyrinth articulates with the sphenoid and a small process of the palatine bone. Several prominent foramina and fissures penetrate the sphenoid or lie between it and the maxilla. The *superior orbital fissure* is a cleft through which nerves controlling eye movement pass. Laterally, the sphenoid and maxilla articulate with the zygomatic bone, which forms the lateral wall and rim of the orbit.

Figure 7–14 The Orbital Complex. The right orbital region. ATLAS: Plate 5f

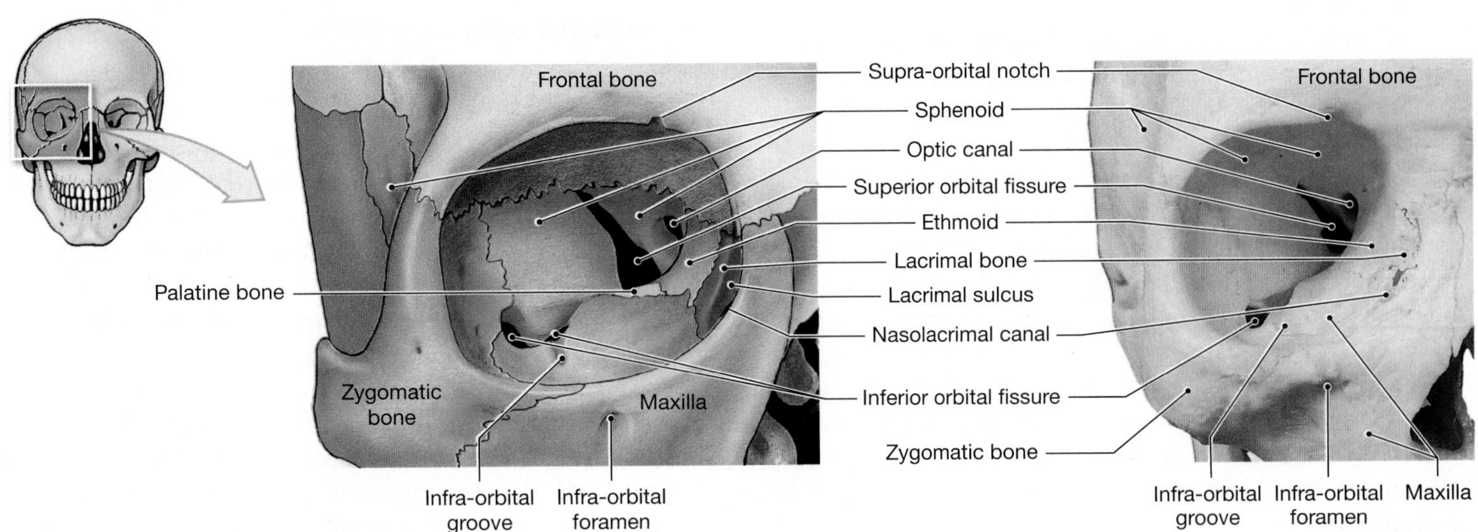

Frontal bone
Palatine bone
Zygomatic bone
Maxilla
Infra-orbital groove
Infra-orbital foramen
Supra-orbital notch
Sphenoid
Optic canal
Superior orbital fissure
Ethmoid
Lacrimal bone
Lacrimal sulcus
Nasolacrimal canal
Inferior orbital fissure
Zygomatic bone
Frontal bone
Infra-orbital groove
Infra-orbital foramen
Maxilla

Go to Mastering A&P™ > Study Area > Menu > Lab Tools > Bone & Dissection Videos > Bone Videos > Facial Bone

Figure 7–15 **The Nasal Complex.** ATLAS: Plates 11b; 12d; 13b,g

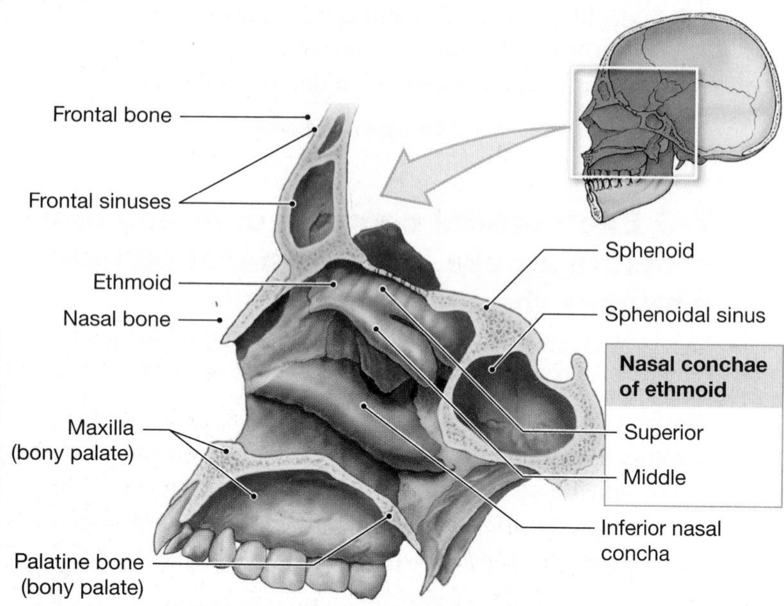

Frontal bone
Frontal sinuses
Ethmoid
Nasal bone
Maxilla (bony palate)
Palatine bone (bony palate)
Sphenoid
Sphenoidal sinus
Nasal conchae of ethmoid
Superior
Middle
Inferior nasal concha

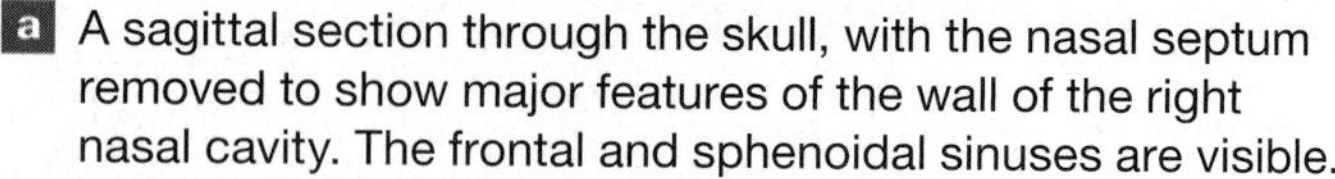

a A sagittal section through the skull, with the nasal septum removed to show major features of the wall of the right nasal cavity. The frontal and sphenoidal sinuses are visible.

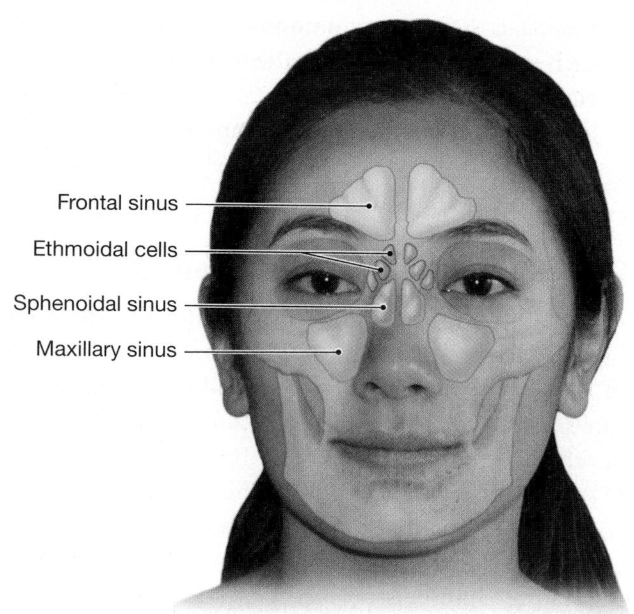

Frontal sinus
Ethmoidal cells
Sphenoidal sinus
Maxillary sinus

b Locations of the paranasal sinuses.

The Nasal Complex

The **nasal complex** (Figure 7–15) includes the bones that enclose the nasal cavities and the *paranasal sinuses,* air-filled chambers connected to the nasal cavities. The frontal bone, sphenoid, and ethmoid form the superior wall of the nasal cavities (Figure 7–15a). The lateral walls are formed by the maxillae and the lacrimal bones (not shown), the ethmoid (the superior and middle nasal conchae), and the inferior nasal conchae. The soft tissues of the nose form much of the anterior margin of the nasal cavity, but the maxillae and nasal bones support the bridge of the nose.

The sphenoid, ethmoid, frontal bone, and paired palatine bones and maxillae contain the **paranasal sinuses.** Figure 7–15b shows the locations of the *frontal* and *sphenoidal sinuses, ethmoidal cells,* and *maxillary sinuses.* (The tiny palatine sinuses, not shown, generally open into the sphenoidal sinuses.) The paranasal sinuses lighten the skull bones and house an extensive area of mucous epithelium, which releases mucus into the nasal cavities. This ciliated epithelium moves the mucus back toward the throat, where it is eventually swallowed or expelled by coughing. Incoming air is humidified and warmed as it flows across this thick carpet of mucus. Foreign particulates, such as dust or microorganisms, become trapped in the sticky mucus and are then swallowed or expelled. This mechanism helps protect the more delicate portions of the respiratory tract.

Checkpoint

10. **Identify the bones of the orbital complex.**
11. **Identify the bones of the nasal complex.**
12. **Identify the bones containing the paranasal sinuses.**

See the blue Answers tab at the back of the book.

7-4 Fontanelles are non-ossified fibrous areas between cranial bones that ease birth and allow for rapid brain growth in infants and children

Learning Outcome Describe the key structural differences among the skulls of infants, children, and adults.

Many different centers of ossification are involved in forming the skull. As development proceeds, the centers fuse, producing a smaller number of composite bones. For example, the sphenoid begins as 14 separate ossification centers.

The skull organizes around the developing brain. As the time of birth approaches, the brain enlarges rapidly. The bones of the skull are also growing, but they fail to keep pace. At birth, fusion is not yet complete: There are two frontal bones, four occipital bones, and several sphenoid and temporal elements. At this time, the cranial bones are connected by areas of fibrous connective tissue (Figure 7–16). The connections are quite flexible, so the skull can withstand the distortion that normally takes place during delivery.

+ Clinical Note Sinusitis

When an irritant or allergen, such as pepper, pollen, or dust, is introduced into the nasal passages, our bodies work to remove it, often by sneezing. The sneeze expels the irritant out with the air. Small particles or milder irritants may trigger mucus production by the epithelium of the paranasal sinuses. The mucus stream flushes the nasal surfaces, often removing the irritants. In *allergic rhinitis* (rī-NĪ-tus), an overactive immune response to an allergen causes excessive mucosal swelling, mucus production, and sneezing.

A sinus infection is another matter entirely. A viral, bacterial, or fungal infection produces an inflammation of the mucous membrane of the nasal cavity. As the membrane swells, the small passageways called *ostia* within the paranasal sinuses narrow. Mucus drains more slowly, the sinuses fill, and the person experiences headaches and pressure within the facial bones. This condition is called **sinusitis**. The maxillary sinuses are commonly involved.

Sinusitis pain relief is the basis of a large over-the-counter drug market in the United States. The active ingredients in these drugs dry the epithelial linings, reduce pain, and restrict further swelling. A nondrug remedy is the neti (NEH-tē) pot. To use this small teapot-shaped vessel, it is filled with warm salt water and gently poured into your nostrils while your head is tilted over the sink. The salt water flows through your nasal cavity, irrigating the nasal passages, and out the other nostril. The process is repeated on the other side.

Chronic (long-term) sinusitis may accompany chronic allergies. It may also occur as the result of a **deviated nasal septum**, in which the cartilaginous structure dividing the nasal is bent**.** Septal deviation often blocks the drainage of one or more sinuses, producing chronic cycles of infection and inflammation. A deviated septum can result from injuries to the nose and can usually be corrected or improved by surgery.

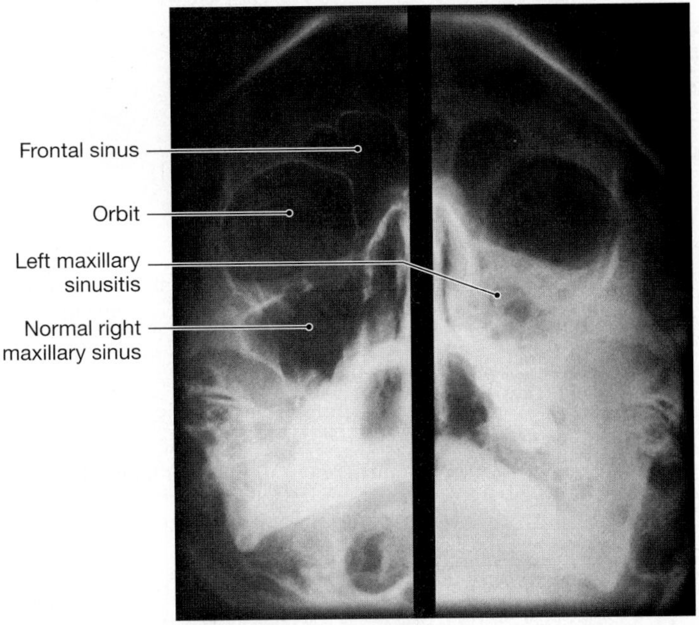
Frontal sinus
Orbit
Left maxillary sinusitis
Normal right maxillary sinus

The largest fibrous areas between the cranial bones are known as **fontanelles** (fon-tuh-NELZ; sometimes spelled *fontanels*). There are six such fontanelles:

- The paired *sphenoidal fontanelles* are at the junctions between the squamous sutures and the coronal suture (Figure 7–16a).
- The paired *mastoid fontanelles* are at the junctions between the squamous sutures and the lambdoid suture.
- The *anterior fontanelle* is the largest fontanelle. It lies at the intersection of the frontal, sagittal, and coronal sutures in the anterior portion of the skull (Figure 7–16b).
- The *posterior fontanelle* is at the junction between the lambdoid and sagittal sutures.

The anterior fontanelle is often called the "soft spot" on newborns. It is usually the only fontanelle easily seen and felt by new parents. The anterior fontanelle pulses as the heart beats because it covers a major blood vessel. This fontanelle is sometimes used to determine whether an infant is dehydrated, because the surface becomes indented when blood volume is low.

The sphenoidal, mastoid, and posterior fontanelles disappear within a month or two after birth. The anterior fontanelle generally persists until the child is nearly 2 years old. Even after the fontanelles disappear, the bones of the skull remain separated by fibrous connections.

The skulls of infants and adults differ in other ways as well, such as in the shape and structure of cranial elements. This difference accounts for variations in proportions as well as in size. The cranium of a young child, compared with the skull as a whole, is relatively larger than that of an adult. The most significant growth in the skull takes place before age 5, because at that age the brain stops enlarging and the cranial sutures develop. The growth of the cranium is generally coordinated with the expansion of the brain. If one or more sutures form before the brain stops growing, the skull will be abnormal in shape, size, or both.

✓ Checkpoint

13. **Define *fontanelle*, and identify the major fontanelles.**
14. **What purpose does a fontanelle serve?**

See the blue Answers tab at the back of the book.

Figure 7–16 Infant Skull and Fontanelles.

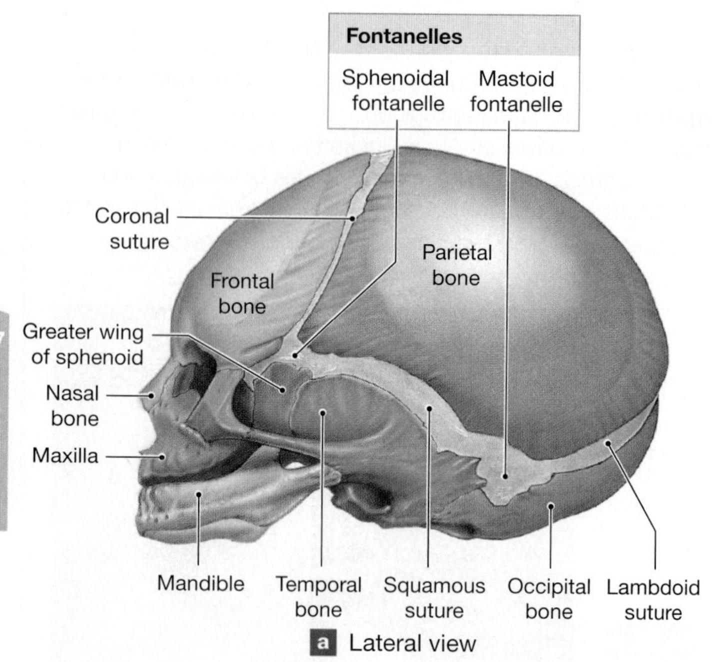
Fontanelles
Sphenoidal fontanelle
Mastoid fontanelle
Coronal suture
Parietal bone
Frontal bone
Greater wing of sphenoid
Nasal bone
Maxilla
Mandible
Temporal bone
Squamous suture
Occipital bone
Lambdoid suture

a Lateral view

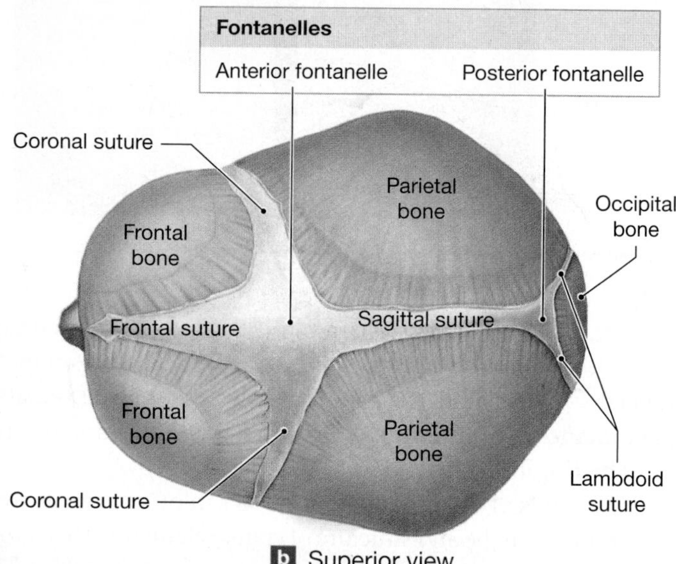
Fontanelles
Anterior fontanelle
Posterior fontanelle
Coronal suture
Parietal bone
Occipital bone
Frontal bone
Frontal suture
Sagittal suture
Frontal bone
Parietal bone
Lambdoid suture
Coronal suture

b Superior view

? Name the fontanelles found on the infant skull. Why aren't they labeled on the adult skull?

7-5 The vertebral column has four flexible and supportive spinal curves

Learning Outcome Identify and describe the curves of the spinal column, and indicate the function of each.

The rest of the axial skeleton consists of the vertebral column, ribs, and sternum. The adult **vertebral column**, or *spine*, consists of 26 bones: the **vertebrae** (24), the **sacrum** (1), and the **coccyx** (KOK-siks), or tailbone (1). The vertebrae provide a column of support, bearing the weight of the head, neck, and trunk, and ultimately transferring the weight to the appendicular skeleton of the lower limbs. The vertebrae also protect the spinal cord and help maintain an upright body position, as when we sit or stand. The average length of the vertebral column of an adult male and female is about 71 cm (28 in.) and 61 cm (24 in.), respectively.

The vertebral column is not straight and rigid. A lateral view shows four **spinal curves**: (1) the **cervical curve**, (2) the **thoracic curve**, (3) the **lumbar curve**, and (4) the **sacral curve** (Figure 7–17).

You may have noticed that the central line of an infant's body, called the axis, forms a C-shape, with the back curving posteriorly. The C-shape results from the thoracic and sacral curves. These are called **primary curves**, because they appear in fetal development. They are also called *accommodation curves*, because they accommodate the thoracic and abdominopelvic viscera (internal organs). The primary curves are present at birth.

The cervical and lumbar curves, known as **secondary curves**, appear several months after birth. They are also called *compensation curves*, because they help shift the weight to permit an upright posture. The cervical and lumbar secondary curves become accentuated as the toddler learns to walk and run. All four curves are fully developed by age 10. ATLAS: Embryology Summary 7: The Development of the Vertebral Column

When you stand, the weight of your body must be transmitted through the vertebral column to the hips and ultimately to the lower limbs. Yet most of your body weight lies anterior to the vertebral column. The various curves bring that weight in line with the body axis. Think about what you do automatically when standing with a heavy object hugged to your chest. You avoid toppling forward by exaggerating your lumbar curve and by keeping your weight back toward the body axis. This posture can lead to discomfort at the base of the spinal column.

+ Clinical Note Craniostenosis

Malformations of the skull can result from *craniostenosis* (krā-nē-ō-sten-Ō-sis; *stenosis*, narrowing), the premature closure of cranial fontanelles. As the brain continues to enlarge, the rest of the skull changes shape to accommodate it. A long and narrow head is produced by early closure of the sagittal suture. A very broad skull results if the coronal suture forms prematurely. Early closure of all cranial sutures restricts the development of the brain, and surgery must be performed to prevent brain damage. If brain enlargement stops due to genetic or developmental abnormalities, however, skull growth ceases as well. This condition, which results in an undersized head, is called *microcephaly* (mī-krō-SEF-uh-lē; *micro-*, small + *cephalon*, head).

For example, many women in the last three months of pregnancy develop chronic back pain from the changes in lumbar curvature that adjust for the increasing weight of the fetus. In many parts of the world, people often carry heavy objects balanced on their head. This practice increases the load on the vertebral column but does not affect the spinal curves because the weight is aligned with the axis of the spine.

Checkpoint

15. **What is the importance of the secondary curves of the spine?**
16. **List the components of the vertebral column.**

See the blue Answers tab at the back of the book.

Figure 7–17 **The Vertebral Column.** ATLAS: Plate 2b

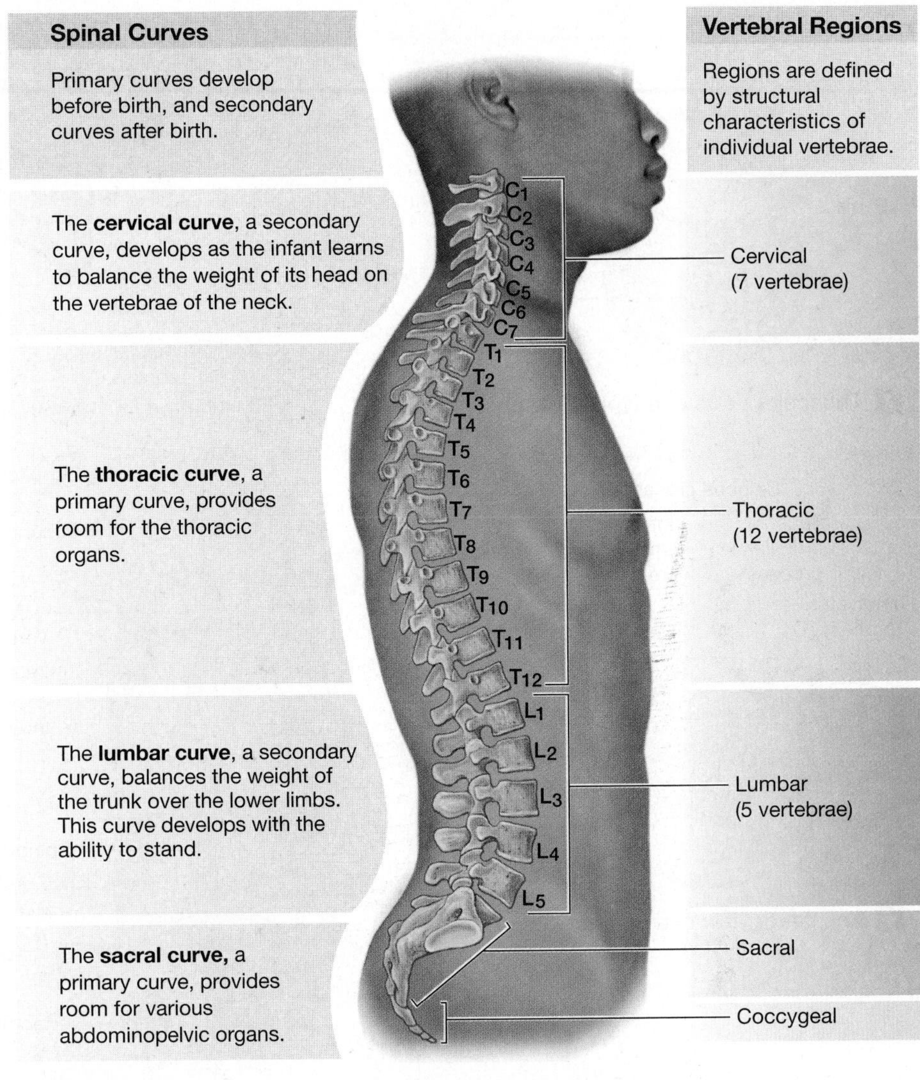
Spinal Curves
Primary curves develop before birth, and secondary curves after birth.
The **cervical curve**, a secondary curve, develops as the infant learns to balance the weight of its head on the vertebrae of the neck.
The **thoracic curve**, a primary curve, provides room for the thoracic organs.
The **lumbar curve**, a secondary curve, balances the weight of the trunk over the lower limbs. This curve develops with the ability to stand.
The **sacral curve,** a primary curve, provides room for various abdominopelvic organs.
Vertebral Regions
Regions are defined by structural characteristics of individual vertebrae.
C1
C2
C3
C4
C5
C6
C7
Cervical (7 vertebrae)
T1
T2
T3
T4
T5
T6
T7
T8
T9
T10
T11
T12
Thoracic (12 vertebrae)
L1
L2
L3
L4
L5
Lumbar (5 vertebrae)
Sacral
Coccygeal

? Which two regions of the spine form primary curves?

7-6 The five vertebral regions—cervical, thoracic, lumbar, sacral, and coccygeal—each have characteristic vertebrae

Learning Outcome Identify the five vertebral regions, and describe the distinctive structural and functional characteristics of vertebrae in each region.

As Figure 7–17 shows, the vertebral column is divided into *cervical, thoracic, lumbar, sacral,* and *coccygeal regions*. The cervical, thoracic, and lumbar regions consist of individual vertebrae (singular, *vertebra*). Seven **cervical vertebrae** (C_1–C_7 or CI–CVII) make up the neck and extend inferiorly to the trunk. Twelve **thoracic vertebrae** (T_1–T_{12} or TI–TXII) form the superior portion of the back. Each articulates with one or more pairs of ribs. Five **lumbar vertebrae** (L_1–L_5 or LI–LV) form the inferior portion of the back. The fifth articulates with the sacrum, which in turn articulates with the coccyx.

Let's look first at the general characteristics of vertebra, and then at the differences among vertebrae in each of these regions.

Vertebral Anatomy

Each vertebra consists of three basic parts: (1) a *vertebral body,* (2) a *vertebral arch,* and (3) *articular processes* (Figure 7–18a). Most vertebrae also have several types of processes that do not articulate with other bones.

The **vertebral body** is the part of a vertebra that transfers weight along the axis of the vertebral column. The body forms the anterior margin of each **vertebral foramen**, while the **vertebral arch** forms the posterior margin (Figure 7–18a,b). The vertebral arch has walls, called **pedicles** (PED-i-kulz), and a roof, formed by flat layers called **laminae** (**LAM-i-nē**; singular, *lamina*, a thin plate) (Figure 7–18a). The pedicles arise along the posterior and lateral margins of the body. The laminae on either side extend dorsally and medially to complete the roof.

A **spinous process** projects posteriorly from the point where the vertebral laminae fuse to complete the vertebral arch (Figure 7–18b,c). You can see—and feel—the spinous processes through the skin of the back when the spine is flexed. **Transverse processes** project laterally or dorsolaterally on both sides from the point where the laminae join the pedicles. These processes are sites of muscle attachment, and they may also articulate with the ribs.

Figure 7–18 Vertebral Anatomy.

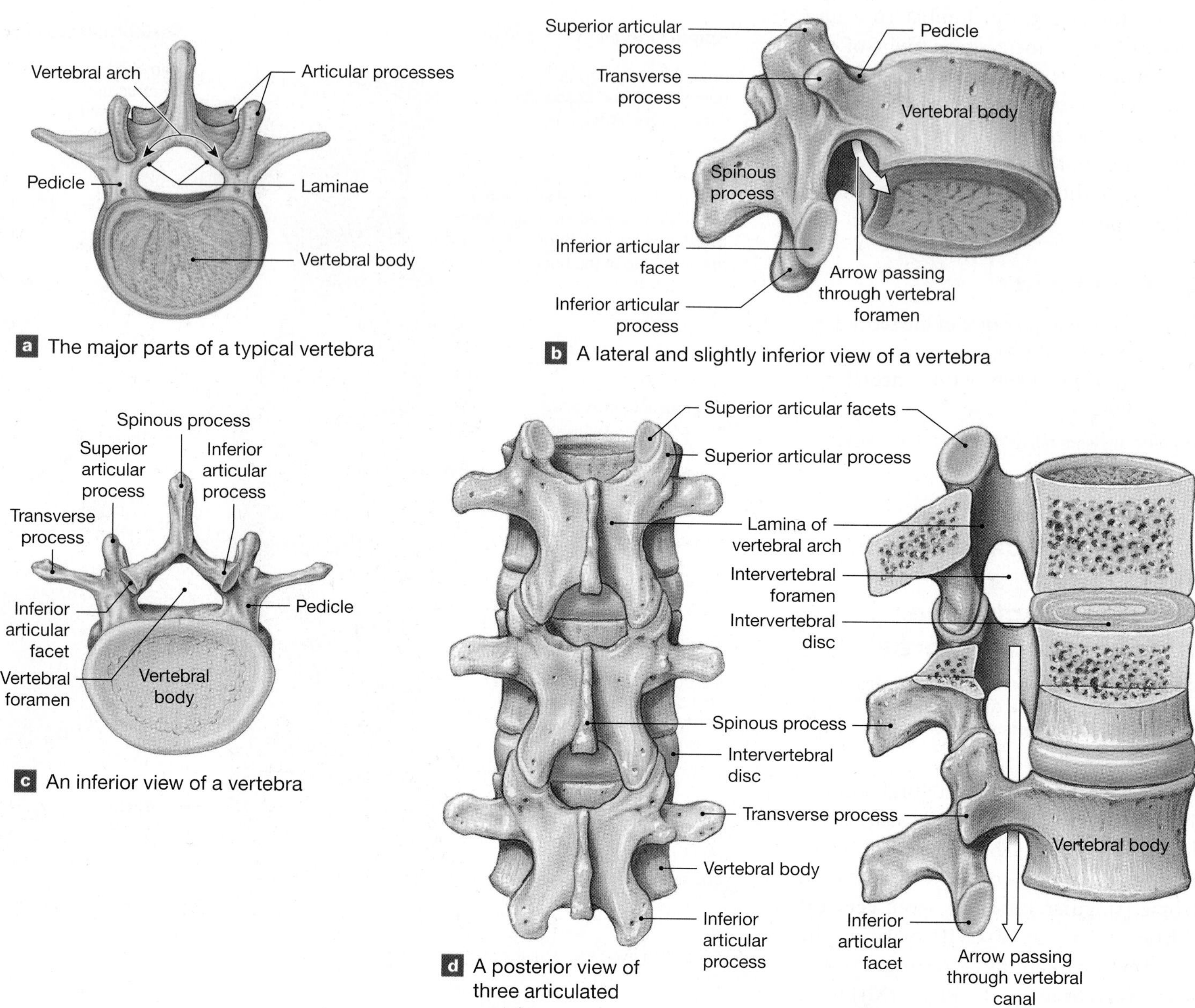
Vertebral arch
Articular processes
Pedicle
Laminae
Vertebral body
Superior articular process
Transverse process
Pedicle
Vertebral body
Spinous process
Inferior articular facet
Inferior articular process
Arrow passing through vertebral foramen
Spinous process
Superior articular process
Inferior articular process
Transverse process
Inferior articular facet
Pedicle
Vertebral foramen
Vertebral body
Superior articular facets
Superior articular process
Lamina of vertebral arch
Intervertebral foramen
Intervertebral disc
Spinous process
Intervertebral disc
Transverse process
Vertebral body
Vertebral body
Inferior articular process
Inferior articular facet
Arrow passing through vertebral canal

a The major parts of a typical vertebra

b A lateral and slightly inferior view of a vertebra

c An inferior view of a vertebra

d A posterior view of three articulated vertebrae

e A lateral and sectional view of three articulated vertebrae

Like the transverse processes, the **articular processes** arise at the junction between the pedicles and the laminae (Figure 7–18c,d). A **superior** and an **inferior articular process** lie on each side of the vertebra. The superior articular processes articulate with the inferior articular processes of a more superior vertebra (or the occipital condyles, in the case of the first cervical vertebra). The inferior articular processes articulate with the superior articular processes of a more inferior vertebra (or the sacrum, in the case of the last lumbar vertebra). Each articular process has a smooth concave surface called an **articular facet**. The superior processes have articular facets on their dorsal surfaces, whereas the inferior processes articulate along their ventral surfaces.

The bodies of adjacent vertebrae are interconnected by ligaments, but are separated by pads of fibrocartilage, the **intervertebral discs**. Gaps called **intervertebral foramina** separate the pedicles of successive vertebrae. These gaps permit the passage of nerves running to or from the enclosed spinal cord. Together, the vertebral foramina of successive vertebrae form the **vertebral canal**, which encloses the spinal cord (Figure 7–18e).

Characteristics of Regional Vertebrae

Each vertebra has characteristic markings and articulations, but here we focus on the general characteristics of each region. Note that when we refer to a specific vertebra, we use a capital

Figure 7–19 The Cervical Vertebrae. ATLAS: Plates 20b; 21a–e

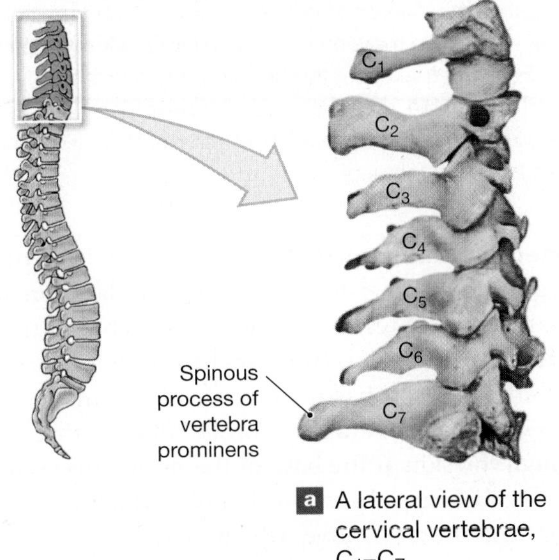

C1
C2
C3
C4
C5
C6
C7
Spinous process of vertebra prominens

a A lateral view of the cervical vertebrae, C_1–C_7.

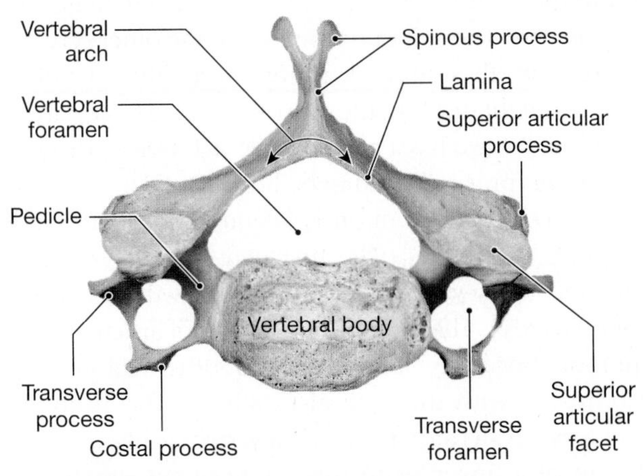

Vertebral arch
Vertebral foramen
Pedicle
Spinous process
Lamina
Superior articular process
Vertebral body
Transverse process
Costal process
Transverse foramen
Superior articular facet

b A superior view of a representative cervical vertebra (C_3–C_6). Note the typical features listed in *Table 7–1*.

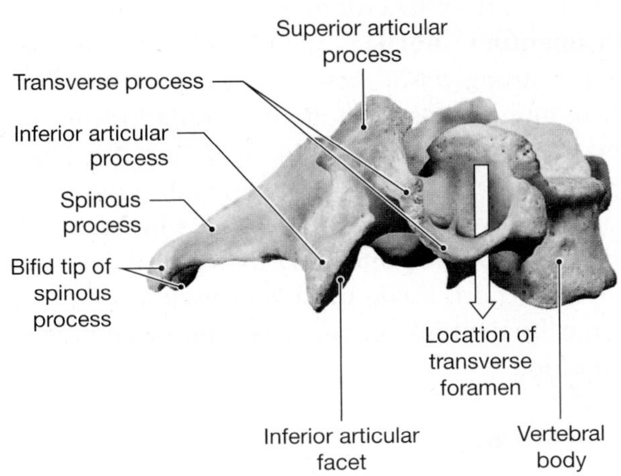

Superior articular process
Transverse process
Inferior articular process
Spinous process
Bifid tip of spinous process
Location of transverse foramen
Inferior articular facet
Vertebral body

c A lateral view of the same vertebra as in part b.

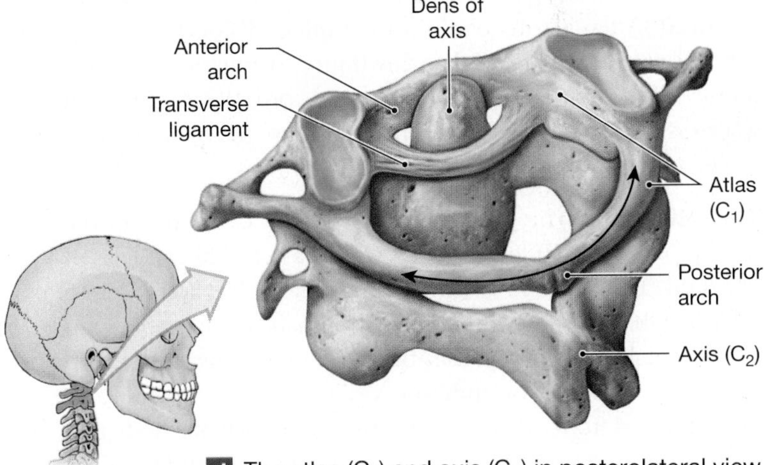

Dens of axis
Anterior arch
Transverse ligament
Atlas (C1)
Posterior arch
Axis (C2)

d The atlas (C_1) and axis (C_2) in posterolateral view.

letter for the vertebral region: C, T, L, S, and Co represent the cervical, thoracic, lumbar, sacral, and coccygeal regions, respectively. In addition, we use a numerical subscript to indicate the relative position of the vertebra within that region, with 1 indicating the vertebra closest to the skull. For example, C_1 is in contact with the skull and C_3 is the third cervical vertebra. Similarly, T_{12} is in contact with L_1, and L_4 is the fourth lumbar vertebra (see Figure 7–17). We will use this shorthand throughout the text.

Keep in mind that the diameter of the spinal cord decreases as you proceed caudally along the vertebral canal, and so does the diameter of the vertebral arch. As you continue toward the sacrum, the loading increases and the vertebral bodies gradually enlarge. Let's see how regional variations determine each vertebral group's function.

Cervical Vertebrae

Most mammals—whether giraffes, whales, mice, or humans—have seven cervical vertebrae, C_1 to C_7 (Figure 7–19a). The cervical vertebrae extend from the occipital bone of the skull to the thorax. These vertebrae are the smallest in the vertebral column. In addition, the body of a cervical vertebra is small compared with the size of its vertebral foramen (Figure 7–19b). Because cervical vertebrae support only the weight of the head, the vertebral body can be relatively small and light.

Laterally, the transverse processes are fused to the **costal processes**, which originate near the ventrolateral portion of the vertebral body (Figure 7–19b). The costal and transverse processes encircle prominent, round **transverse foramina**. These passageways protect the *vertebral arteries* and *vertebral veins*, important blood vessels that service the brain.

In a typical cervical vertebra, the superior surface of the body is concave from side to side, and it slopes, with the anterior edge inferior to the posterior edge (Figure 7–19c). Vertebra C_1 has no spinous process. The spinous processes of the other cervical vertebrae are relatively stumpy, generally shorter than the diameter of the vertebral foramen. In the case of vertebrae C_2–C_6, the tip of each spinous process has a prominent notch (see Figure 7–19b). A split spinous process is said to be **bifid** (BĪ-fid).

The preceding description is adequate for identifying the cervical vertebrae C_3–C_6. The first two cervical vertebrae are unique, and the seventh is modified (we will describe these vertebrae shortly). The interlocking bodies of articulated C_3–C_7 permit more flexibility than do those of other regions.

Compared with the cervical vertebrae, your head is relatively massive. It sits atop the cervical vertebrae like a soup bowl on the tip of a finger. With this arrangement, small muscles can produce significant effects by tipping the balance one way or another. But if you change position suddenly, as in a fall or during rapid acceleration (a jet takeoff) or deceleration (a car crash), the balancing muscles are not strong enough to stabilize the head. A dangerous partial or complete dislocation of the cervical vertebrae can result, with injury to muscles and ligaments and potential injury to the spinal cord. We use the term **whiplash** to describe such an injury, because the movement of the head resembles the cracking of a whip.

The Atlas (C_1). The **atlas**, cervical vertebra C_1, holds up the head (Figure 7–19d). It articulates with the occipital condyles of the skull. This vertebra is named after Atlas, who, according to Greek myth, holds the world on his shoulders. The joint between the occipital condyles and the atlas is a joint that permits you to nod (such as when you indicate "yes"). The atlas can easily be distinguished from other vertebrae by (1) the lack of a body and spinous process, and (2) the presence of a large, round vertebral foramen bounded by **anterior** and **posterior arches**.

The atlas articulates with the second cervical vertebra, the *axis*. This joint permits rotation (as when you shake your head to indicate "no").

The Axis (C_2). During development, the body of the atlas fuses to the body of the second cervical vertebra, called the **axis** (C_2) (see Figure 7–19d). This fusion creates the prominent **dens** (DENZ; *dens*, tooth), or *odontoid* (ō-DON-toyd; *odoous*, tooth) *process*, of the axis. A transverse ligament binds the dens to the inner surface of the atlas, forming a pivot for rotation of the atlas and skull. Important muscles controlling the position of the head and neck attach to the especially robust spinous process of the axis.

In children, the fusion between the dens and axis is incomplete. Impacts or even severe shaking can cause dislocation of the dens and severe damage to the spinal cord. In adults, a hit to the base of the skull can be equally dangerous, because a dislocation of the atlas–axis joint can force the dens into the base of the brain, with fatal results.

Tips & Tools

To remember the difference between the atlas and the axis and their respective movements, consider Greek mythology and the Earth. In this case, the head (Earth) is held up by Atlas and is capable of nodding "yes" movements. The Earth rotates on its axis, and the axis allows for us to shake our heads in a "no" movement.

The Vertebra Prominens (C_7). The transition from one vertebral region to another is not abrupt. The last vertebra of one region generally resembles the first vertebra of the next. The **vertebra prominens** (prominent vertebra), or seventh cervical vertebra (C_7), has a long, slender spinous process (see Figure 7–19a) that ends in a broad tubercle that you can feel through the skin at the base of the neck. This vertebra is the interface between the cervical curve, which arches anteriorly, and the thoracic curve, which arches posteriorly (see Figure 7–17a). The transverse processes of C_7 are large, providing additional surface area for muscle attachment.

The **ligamentum nuchae** (lig-uh-MEN-tum NŪ-kē; *nucha*, nape), a strong elastic ligament, begins at the vertebra prominens and extends to an insertion (attachment site) along the external occipital crest of the skull. ⮌ p. 216 Along the way, it attaches to the spinous processes of the other cervical vertebrae. When your head is upright, this ligament acts like the string on a bow, maintaining the cervical curvature without muscular effort. If you bend your neck forward, the elasticity in the ligamentum nuchae helps return your head to an upright position.

Thoracic Vertebrae

There are 12 thoracic vertebrae, T_1 to T_{12} (Figure 7–20a). A typical thoracic vertebra has a distinctive heart-shaped body that is more massive than that of a cervical vertebra (Figure 7–20b). The vertebral foramen is smaller in relation to the vertebral body than in the cervical vertebrae. The long, slender spinous process projects posteriorly and inferiorly.

Each thoracic vertebra articulates with ribs along the dorsolateral surfaces of the body. The **costal facets** on the vertebral bodies articulate with the heads of the ribs. The location and structure of the joints vary somewhat among thoracic vertebrae (see Figure 7–20a). Vertebrae T_1–T_8 each articulate with two pairs of ribs, so their vertebral bodies have two costal facets (*superior* and *inferior*) on each side. Vertebrae T_9–T_{11} have a single costal facet on each side, and each vertebra articulates with a single pair of ribs. The transverse processes of vertebrae T_1–T_{10} are relatively thick and contain **transverse costal facets** for rib articulation (Figure 7–20b,c). Thus, rib pairs 1 through 10 contact their vertebrae at two points: a costal facet and a transverse costal facet.

The spinous processes of T_{10}, T_{11}, and T_{12} increasingly resemble those of the lumbar region as the transition between the

Figure 7–20 The Thoracic Vertebrae. ATLAS: Plates 22a–c

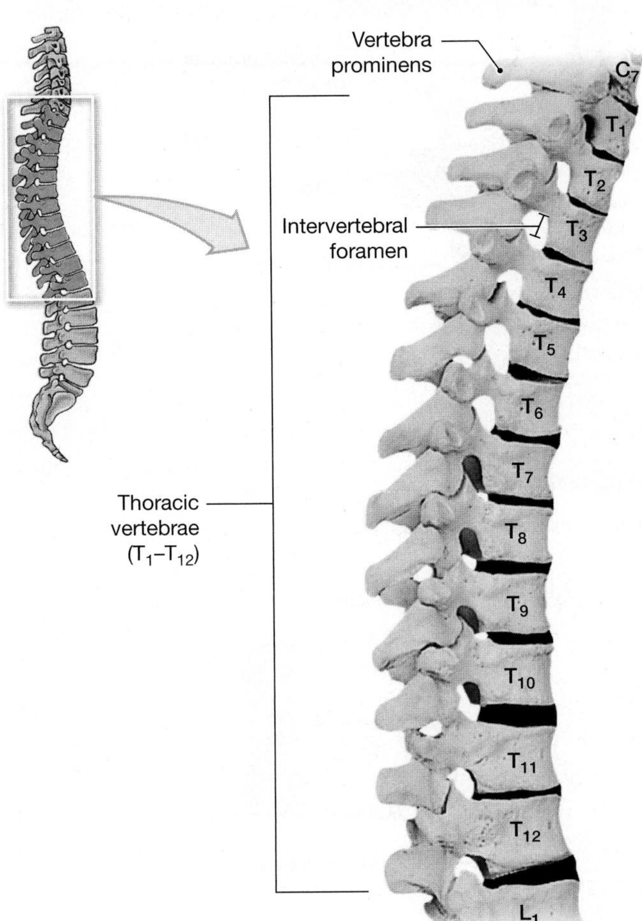
Vertebra prominens
C_7
T_1
T_2
Intervertebral foramen
T_3
T_4
T_5
T_6
T_7
Thoracic vertebrae (T_1–T_{12})
T_8
T_9
T_{10}
T_{11}
T_{12}
L_1

a A lateral view of the thoracic region of the vertebral column. The vertebra prominens (C_7) resembles T_1, but lacks facets for rib articulation. Vertebra T_{12} resembles the first lumbar vertebra (L_1) but has a facet for rib articulation.

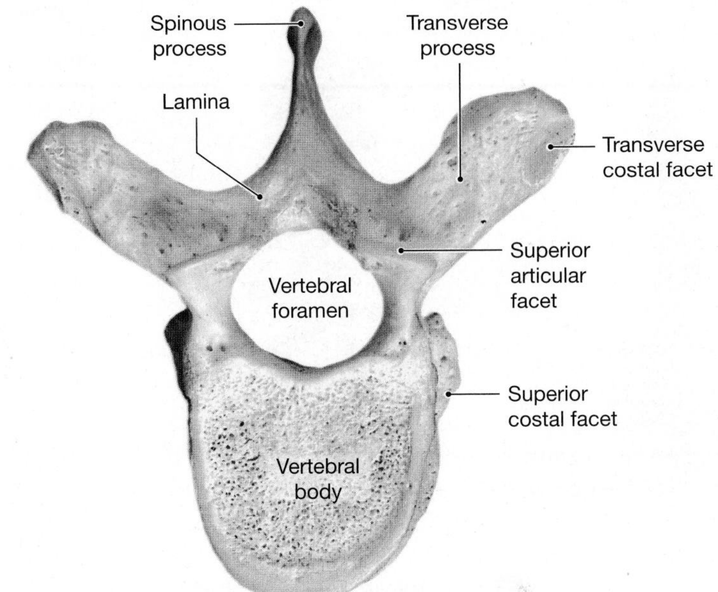
Spinous process
Transverse process
Lamina
Transverse costal facet
Superior articular facet
Vertebral foramen
Superior costal facet
Vertebral body

b Thoracic vertebra, superior view.

Transverse costal facet for tubercle of superior rib
Superior costal facet for head of superior rib
Transverse process
Vertebral body
Spinous process
Inferior costal facet for head of inferior rib

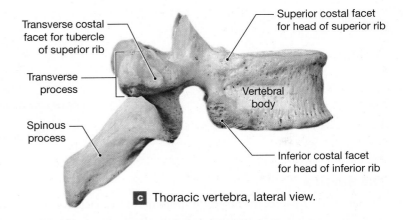

c Thoracic vertebra, lateral view.

thoracic and lumbar curves approaches. Because the inferior thoracic and lumbar vertebrae carry so much weight, the transition between the thoracic and lumbar curves is difficult to stabilize. As a result, compression fracture dislocations from a hard fall tend to involve the last thoracic and first two lumbar vertebrae.

Lumbar Vertebrae

The five lumbar vertebrae (L_1–L_5) are the largest vertebrae (**Figure 7–21a**). The body of a typical lumbar vertebra (**Figure 7–21b,c**) is thicker than that of a thoracic vertebra, and the superior and inferior surfaces are oval rather than heart shaped. Also note that lumbar vertebrae do not have costal facets or transverse costal facets. The slender transverse processes project dorsolaterally, while the stumpy spinous processes project dorsally. The vertebral foramen is triangular. Finally, the superior articular processes face medially ("up and in"), and the inferior articular processes face laterally ("down and out").

The lumbar vertebrae withstand the most weight. Their massive spinous processes provide surface area for the attachment of lower back muscles that reinforce or adjust the lumbar curve.

Table 7–1 summarizes the features of cervical, thoracic, and lumbar vertebrae.

Tips & Tools

To remember the number of bones in the first three spinal curves, think about mealtimes. You eat breakfast at 7 a.m. (7 cervical vertebrae), lunch at 12 p.m. (12 thoracic vertebrae), and dinner at 5 p.m. (5 lumbar vertebrae).

Go to MasteringA&P™ > Study Area > Menu > Lab Tools > Bone & Dissection Videos > Bone Videos > Typical Vertebra

Figure 7–21 **The Lumbar Vertebrae.** ATLAS: Plates 23a–c

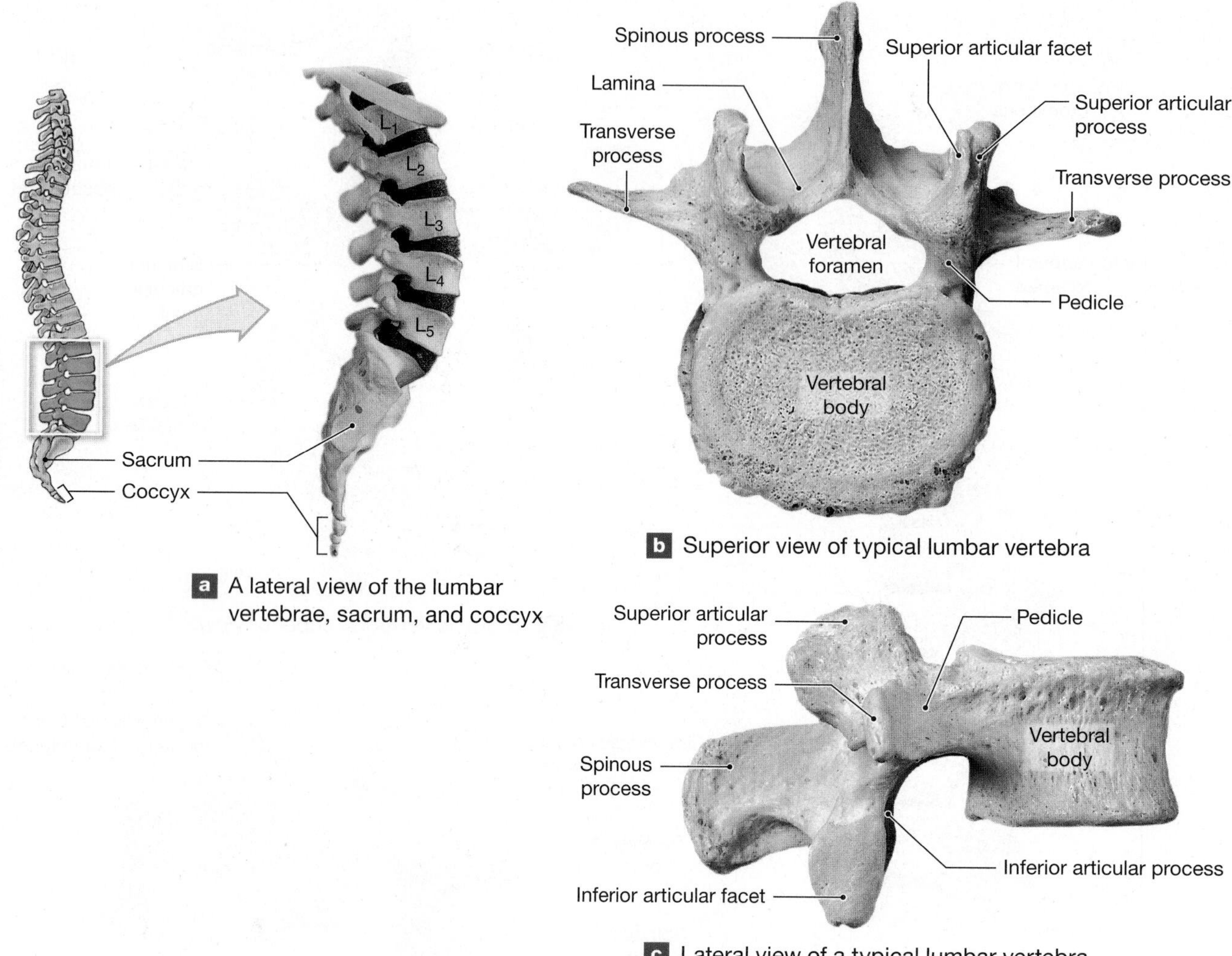

L1
L2
L3
L4
L5
Sacrum
Coccyx
Spinous process
Lamina
Transverse process
Superior articular facet
Superior articular process
Transverse process
Vertebral foramen
Pedicle
Vertebral body
Superior articular process
Transverse process
Spinous process
Inferior articular facet
Pedicle
Vertebral body
Inferior articular process

a A lateral view of the lumbar vertebrae, sacrum, and coccyx

b Superior view of typical lumbar vertebra

c Lateral view of a typical lumbar vertebra

The Sacrum

The sacrum consists of the fused components of five sacral vertebrae. These vertebrae begin fusing shortly after puberty, and are generally completely fused at ages 25–30. The sacrum protects the reproductive, digestive, and urinary organs. Through paired joints, it attaches the axial skeleton to the paired hip bones, or pelvic girdle, of the appendicular skeleton (see Figure 7–1a).

The superior articular processes of the first sacral vertebra articulate with the last lumbar vertebra, L_5. The **sacral canal** is a passageway that begins between these articular processes and extends the length of the sacrum (Figure 7–22a). Nerves and membranes that line the vertebral canal in the spinal cord continue into the sacral canal.

The broad, posterior sacral surface has an extensive area for muscle attachment, especially for muscles that move the thigh. Three sacral crests, the *median* and *lateral sacral crests*, and the *sacral cornu*, provide these attachment sites (Figure 7–22b). The **median sacral crest** is a ridge formed by the fused spinous processes of the sacral vertebrae. The laminae of the fifth sacral vertebra fail to contact one another at the midline. They form the **sacral cornua** (KOR-nū-uh; singular, *cornu; cornua*, horns). These ridges form the margins of the **sacral hiatus** (hī-Ā-tus), the opening at the inferior end of the sacral canal. This opening is covered by connective tissues. Four pairs of **sacral foramina** open on either side of the median sacral crest. The intervertebral foramina of the fused sacral vertebrae open into these passageways. Each **lateral sacral crest** is a ridge that represents the fused transverse processes of the sacral vertebrae.

The sacrum is curved, with a convex posterior surface (see Figure 7–22b). The degree of curvature is more pronounced in males than in females. The **auricular surface** is a thickened, flattened area lateral and anterior to the superior part of the lateral sacral crest. The auricular surface is the site of articulation with the pelvic girdle (the *sacro-iliac joint*). The **sacral tuberosity** (see Figure 7–22a) is a roughened area between the lateral sacral crest and the auricular surface. It marks the attachment site of ligaments that stabilize the sacro-iliac joint.

Table 7–1 Regional Differences in Vertebral Structure and Function

	Type (Number)		
Feature	**Cervical Vertebrae (7)**	**Thoracic Vertebrae (12)**	**Lumbar Vertebrae (5)**
Location	Neck	Chest	Inferior portion of back
Body	Small, oval, curved faces	Medium, heart-shaped, flat faces; facets for rib articulations	Massive, oval, flat faces
Vertebral foramen	Large	Medium	Small
Spinous process	Long; split tip; points inferiorly	Long, slender; not split; points inferiorly	Blunt, broad; points posteriorly
Transverse processes	Have transverse foramina	All but two (T_{11}, T_{12}) have facets for rib joints	Short; no articular facets or transverse foramina
Functions	Support skull, stabilize relative positions of brain and spinal cord, and allow controlled head movement	Support weight of head, neck, upper limbs, and chest; articulate with ribs to allow changes in volume of thoracic cage	Support weight of head, neck, upper limbs, and trunk
Typical appearance (superior view)			

Figure 7–22 The Sacrum and Coccyx.

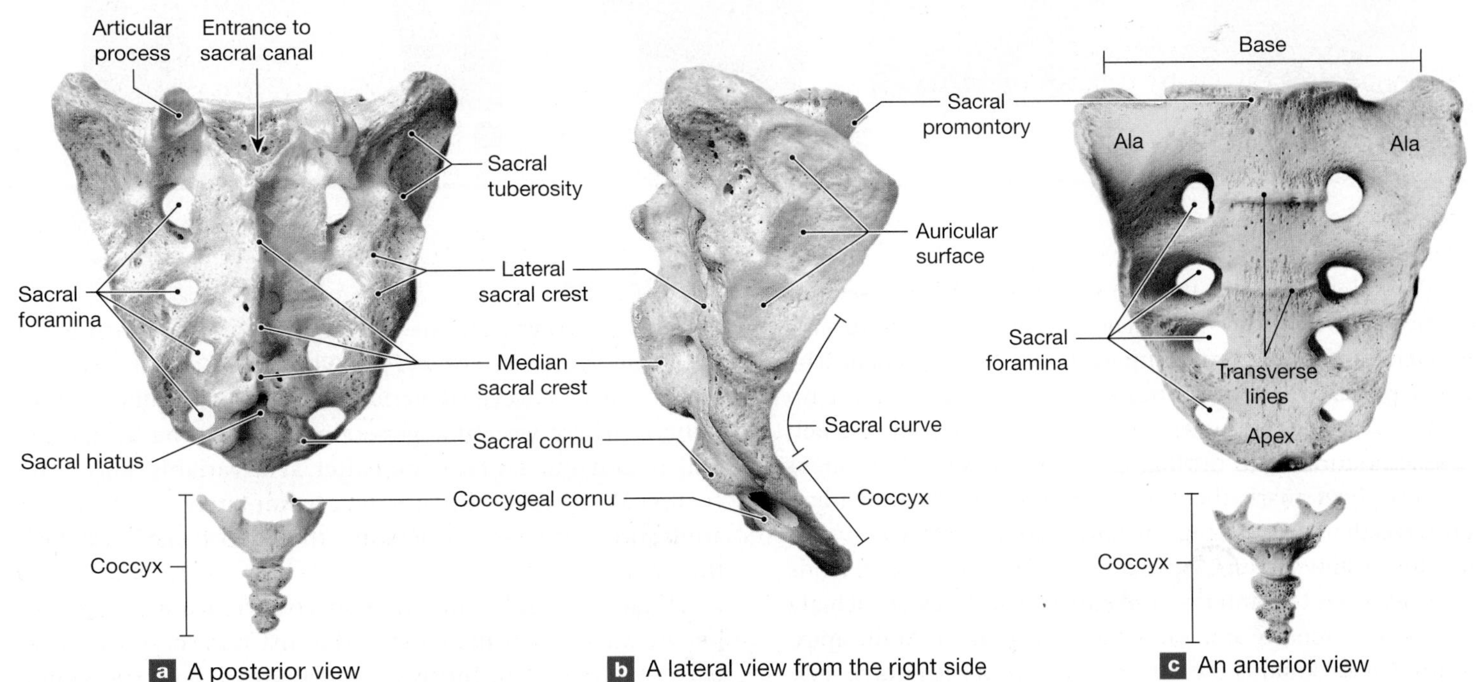

Articular process
Entrance to sacral canal
Sacral tuberosity
Sacral foramina
Lateral sacral crest
Median sacral crest
Sacral hiatus
Sacral cornu
Coccygeal cornu
Coccyx
Sacral promontory
Auricular surface
Sacral curve
Coccyx
Base
Ala
Ala
Sacral foramina
Transverse lines
Apex
Coccyx

a A posterior view

b A lateral view from the right side

c An anterior view

+ Clinical Note Kyphosis, Lordosis, and Scoliosis

The vertebral column must move, balance, and support the trunk and head. Conditions that damage the bones, muscles, and/or nerves can result in distorted curves and impaired function. **Kyphosis** (kī-FŌ-sis; *kyphos*, hump-backed, bent) is an abnormal posterior exaggeration of the thoracic curvature, producing a "round-back" appearance (Photo a). This condition can be caused by (1) osteoporosis with compression fractures affecting the anterior portions of vertebral bodies, (2) chronic contractions in muscles that insert on the vertebrae, or (3) abnormal vertebral growth. In **lordosis** (lor-DŌ-sis; *lordosis*, a bending backward), or "swayback," both the abdomen and buttocks protrude abnormally (Photo b). This causes an anterior exaggeration of the lumbar curvature. This may occur during pregnancy or result from abdominal obesity or abdominal wall muscle weakness. **Scoliosis** (skō-lē-Ō-sis; *scoliosis*, crookedness) is an abnormal lateral curvature of the spine in one or more of the movable vertebrae (Photo c). Scoliosis is the most common spinal curvature abnormality. This condition may result from damage to vertebral bodies during development, or from muscular paralysis affecting one side of the back (as in some cases of polio). In many cases, the cause of abnormal spinal curvatures is not known. Idiopathic (of no known cause) scoliosis generally appears in girls during adolescence, when periods of growth are most rapid. Small curves may later stabilize once growth is complete. For larger curves, bracing may prevent progression. Severe cases can be treated through surgical straightening with implanted metal rods or cables. School nurses provide scoliosis clinics to screen grade-school children.

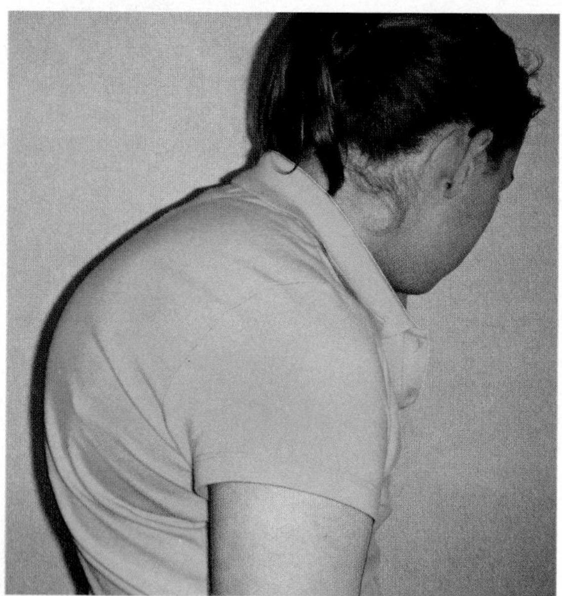
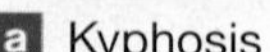
a Kyphosis

b Lordosis

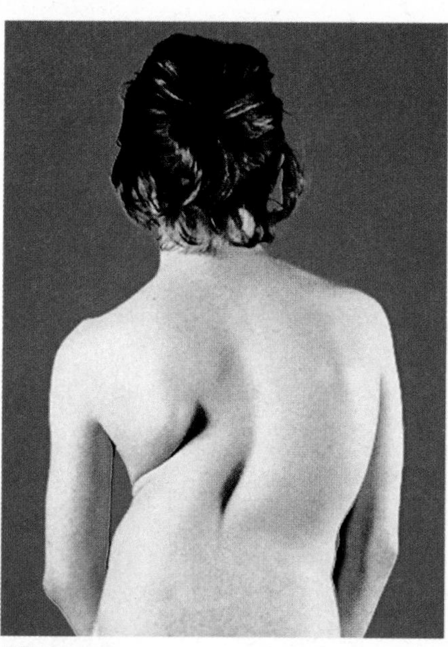
c Scoliosis

The subdivisions of the sacrum are most clearly seen in anterior view (Figure 7–22c). The narrow, inferior portion is the sacral **apex**. The broad superior surface is the **base**. The **sacral promontory**, a prominent bulge at the anterior tip of the base, is an important landmark in females during pelvic examinations and during labor and delivery. Prominent *transverse lines* mark the former boundaries of individual vertebrae that fuse during the formation of the sacrum. At the base of the sacrum, a broad sacral **ala**, or *wing*, extends on either side. The anterior and superior surfaces of each ala provide an extensive area for muscle attachment. At the apex, a flattened area marks the site of articulation with the coccyx.

The Coccyx

The small coccyx consists of four (in some cases three or five) coccygeal vertebrae (see Figure 7–22). Ossification of the distal coccygeal vertebrae is not complete before puberty. These vertebrae generally begin fusing by age 26, and fusion takes place thereafter at a variable pace. The coccygeal vertebrae do not fuse completely until late in adulthood. In very old persons, the coccyx may fuse with the sacrum.

Ligaments and a muscle that constricts the anal opening are attached to the coccyx. The first two coccygeal vertebrae have transverse processes and unfused vertebral arches.

The prominent lamina of the first coccygeal vertebrae is known as the **coccygeal cornu**. Each lamina curves to meet the sacral cornu.

✓ Checkpoint

17. **Why does the vertebral column of an adult have fewer vertebrae than that of a newborn?**
18. **Joe suffered a hairline fracture (fracture without separation of the fragments) at the base of the dens. Which bone is fractured, and where is it located?**
19. **Examining a human vertebra, you notice that, in addition to the large foramen for the spinal cord, two smaller foramina are on either side of the bone in the region of the transverse processes. From which region of the vertebral column is this vertebra?**
20. **Why are the bodies of the lumbar vertebrae larger than those of the cervical and thoracic vertebrae?**

See the blue Answers tab at the back of the book.

7-7 The thoracic cage protects organs in the chest and provides sites for muscle attachment

Learning Outcome **Explain the significance of the joints between the thoracic vertebrae and the ribs, and between the ribs and sternum.**

The skeleton of the chest, or **thoracic cage** (Figure 7–23), provides bony support to the walls of the thoracic cavity. It consists of the thoracic vertebrae, the ribs, costal cartilages, and the sternum (breastbone). The thoracic cage as a whole serves the following main functions:

- It protects the heart, lungs, thymus, and other structures in the thoracic cavity.
- It serves as an attachment point for muscles involved in (1) breathing, (2) maintaining the position of the vertebral column, and (3) moving the pectoral girdles (both clavicles and scapulae) and upper limbs.

Figure 7–23 The Thoracic Cage. ATLAS: Plate 22b

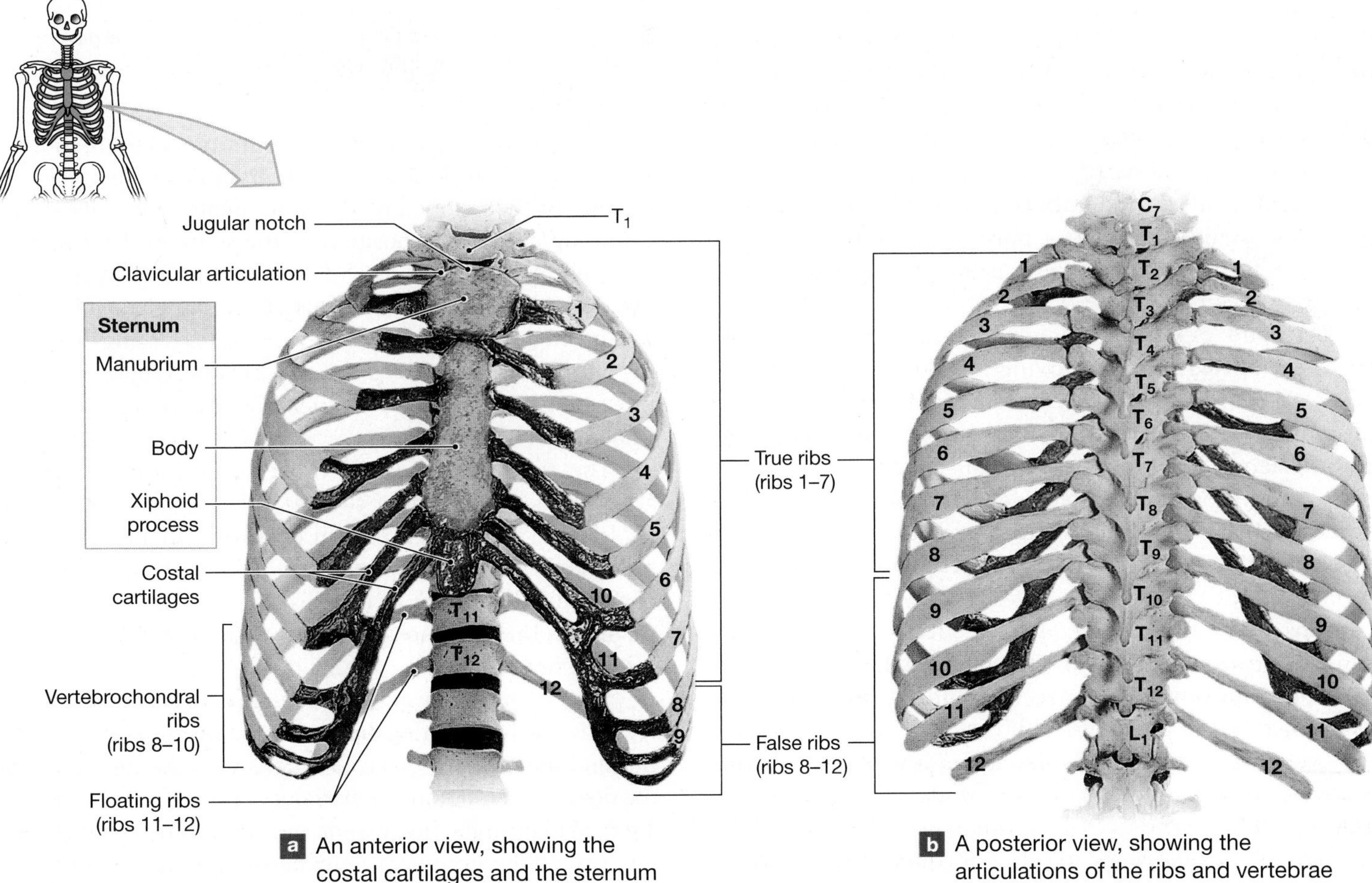

Jugular notch
Clavicular articulation
Sternum
Manubrium
Body
Xiphoid process
Costal cartilages
Vertebrochondral ribs (ribs 8–10)
Floating ribs (ribs 11–12)
T1
T11
T12
True ribs (ribs 1–7)
False ribs (ribs 8–12)
C7
T1
T2
T3
T4
T5
T6
T7
T8
T9
T10
T11
T12
L1
1
2
3
4
5
6
7
8
9
10
11
12

a An anterior view, showing the costal cartilages and the sternum

b A posterior view, showing the articulations of the ribs and vertebrae

? Which two pairs of ribs are the floating ribs? What distinguishes them from the other false ribs?

The ribs and the sternum form the *rib cage*, whose movements are important in breathing. Let's look at the components of the rib cage more closely.

The Ribs

Ribs are long, curved, flattened bones that originate on or between the thoracic vertebrae and end in the wall of the thoracic cavity. Each of us, regardless of sex, has 12 pairs of ribs (Figure 7–23a). The first seven pairs (ribs 1–7; I–VII) are called **true ribs**, or *vertebrosternal ribs*. They reach the anterior body wall and are connected to the sternum by separate cartilaginous extensions, the **costal cartilages**. Beginning with the upper first rib, the vertebrosternal ribs gradually increase in length and in radius of curvature.

Ribs 8–12 (VIII–XII) are called **false ribs**, because they do not attach directly to the sternum. The costal cartilages of ribs 8–10, the *vertebrochondral ribs*, fuse together and merge with the cartilages of rib pair 7 before they reach the sternum. The last two pairs of ribs (11–12; XI–XII) are called *floating ribs* because they have no connection with the sternum. They are also called *vertebral ribs* because they are attached only to the vertebrae (Figure 7–23b) and muscles of the body wall.

Figure 7–24a shows the superior surface of a typical rib. The *vertebral end* of the rib articulates with the vertebral column at the **head**, or *capitulum* (ka-PIT-ū-lum). A ridge divides the articular surface of the head into superior and inferior articular facets (Figure 7–24b). From the head, a short **neck** leads to the **tubercle**, a small elevation that projects dorsally. The inferior portion of the tubercle contains an articular facet that contacts the transverse process of the thoracic vertebra.

The bend, or *angle*, of the rib is the site where the tubular **body**, or *shaft*, begins curving toward the sternum (see Figure 7–24b). The internal rib surface is concave, and a prominent *costal groove* along its inferior border marks the path of nerves and blood vessels. The superficial surface is convex and provides an attachment site for muscles of the pectoral girdle and trunk. The *intercostal muscles*, which move the ribs, are attached to the superior and inferior surfaces.

Ribs 1 and 10 originate at costal facets on vertebrae T_1 and T_{10}, respectively, and their tubercular facets articulate with the transverse costal facets on those vertebrae. The heads of ribs 2–9 articulate with costal facets on two adjacent vertebrae. Their tubercular facets articulate with the transverse costal facets of the inferior member of the vertebral pair. Ribs 11 and 12, which originate at T_{11} and T_{12}, do not have tubercular facets and do not contact the transverse processes of T_{11} or T_{12}.

The ribs are quite mobile, due to their complex musculature, dual articulations at the vertebrae, and flexible connection to the sternum. Note how the ribs angle inferiorly as they curve away from the vertebral column (see Figure 7–23b). Because of their curvature, the movements of the ribs during breathing change the position of the sternum. Lowering the ribs pulls the sternum inward, and raising the ribs moves it outward. As a result, movements of the ribs change both the width and the depth of the thoracic cage, decreasing or increasing its volume accordingly as we breathe out or in. These movements are important in breathing, as we see in detail in Chapter 23.

The ribs can bend and move to cushion shocks and absorb hits, but severe or sudden impacts can cause painful rib fractures. A cracked rib can heal without a cast or splint because the ribs are tightly bound in connective tissues. But comminuted fractures of the ribs can send bone splinters or fragments into the thoracic cavity, with potential damage to internal organs.

Surgery on the heart, lungs, or other organs in the thorax typically involves entering the thoracic cavity. The mobility of the ribs and the cartilaginous connections with the sternum allow the ribs to be temporarily moved out of the way. "Rib spreaders" are used to push the ribs apart in much the same way that a jack lifts a car off the ground for a tire change. If more extensive access is required, the cartilages of the sternum can be cut and the entire sternum folded out of the way. Once the sternum is replaced, scar tissue reunites the cartilages, and the ribs heal fairly rapidly.

Figure 7–24 The Ribs. ATLAS: Plates 22a,b

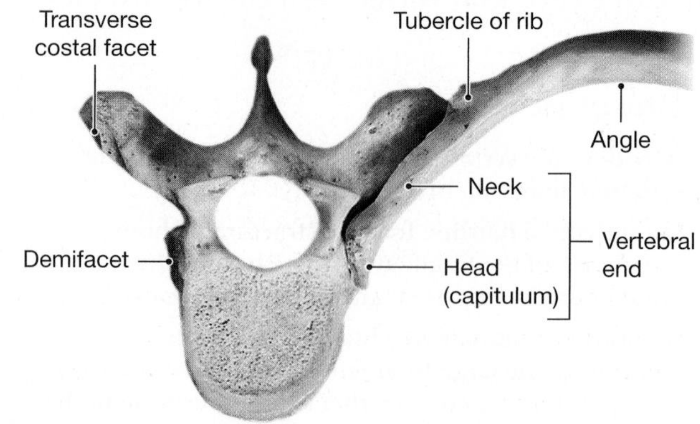
Transverse costal facet
Tubercle of rib
Angle
Neck
Vertebral end
Demifacet
Head (capitulum)

a A superior view of a rib showing the joints between it and a thoracic vertebra

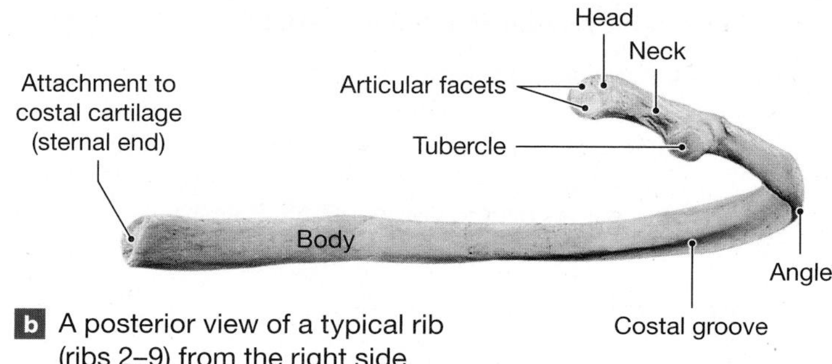
Head
Neck
Attachment to costal cartilage (sternal end)
Articular facets
Tubercle
Body
Angle
Costal groove

b A posterior view of a typical rib (ribs 2–9) from the right side

The Sternum

The adult **sternum**, or breastbone, is a flat bone that forms in the anterior midline of the thoracic wall (see Figure 7–23a). The sternum has the following three parts:

- The broad, triangular **manubrium** (ma-NŪ-brē-um) articulates with the *clavicles* (collarbones) and the cartilages of the first pair of ribs. The manubrium is the widest and most superior portion of the sternum. Only the first pair of ribs is attached by cartilage to this portion of the sternum. The **jugular notch**, located between the clavicular articulations, is a shallow indentation on the superior surface of the manubrium.
- The tongue-shaped **body** attaches to the inferior surface of the manubrium and extends inferiorly along the midline. Individual costal cartilages from rib pairs 2–7 are attached to this portion of the sternum.
- The **xiphoid** (ZI-foyd) **process** is the smallest part of the sternum. It is attached to the inferior surface of the body. The *diaphragm* (breathing muscle) and *rectus abdominis* muscle attach to the xiphoid process.

Ossification of the sternum begins at 6 to 10 ossification centers, and fusion is not complete until at least age 25. Before that age, the sternal body consists of four separate bones. In adults, their boundaries appear as a series of transverse lines crossing the sternum. The xiphoid process is generally the last part to ossify and fuse. Its connection to the sternal body can be broken by impact or strong pressure, creating a spear of bone that can severely damage the liver. The xiphoid process is used as a palpable landmark during the administration of cardiopulmonary resuscitation (CPR), and CPR training strongly emphasizes proper hand positioning to reduce the chances of breaking ribs or the xiphoid process.

Checkpoint

21. **Describe false ribs.**
22. **Improper administration of cardiopulmonary resuscitation (CPR) can result in a fracture of which bones?**
23. **What are the main differences between true ribs and floating ribs?**

See the blue Answers tab at the back of the book.

7 Chapter Review

Study Outline

MasteringA&P™ Access more chapter study tools online in the MasteringA&P Study Area:

Chapter Quizzes, Chapter Practice Test, MP3 Tutor Sessions, and Clinical Case Studies

- Practice Anatomy Lab PAL 3.0
- A&P Flix A&PFlix
- Interactive Physiology iP2
- PhysioEx PhysioEx 9.1

An Introduction to the Divisions of the Skeleton p. 209

1. The human skeleton has two main divisions: the **axial skeleton** and the **appendicular skeleton**. The axial skeleton forms the longitudinal axis of the body and the appendicular skeleton includes the pectoral and pelvic girdles, which support the limbs. *(Figure 7–1)*

7-1 The 80 bones of the head and trunk make up the axial skeleton p. 209

2. The **axial skeleton** is divided into the **skull** and **associated bones** (**auditory ossicles** and **hyoid**), the **thoracic cage**, and the **vertebral column**. *(Figures 7–1, 7–3)*

7-2 The skull's 8 cranial bones protect the brain, and its 14 facial bones form the mouth, nose, and orbits p. 209

3. The **skull** consists of the **cranium** and the bones of the face. The cranium, composed of **cranial bones**, encloses the **cranial cavity**. The **facial bones** protect and support the entrances to the digestive and respiratory tracts. *(Figures 7–2, 7–3)*
4. Prominent superficial landmarks on the skull include the **lambdoid**, **coronal**, **sagittal**, and **squamous sutures**. *(Figure 7–3, Spotlight Figure 7–4)*
5. The bones of the cranium are the **occipital bone**, the two **parietal bones**, the **frontal bone**, the two **temporal bones**, the **sphenoid**, and the **ethmoid**. *(Figures 7–2 to 7–9)*
6. The occipital bone surrounds the **foramen magnum**. *(Figures 7–3 to 7–5)*
7. The frontal bone contains the **frontal sinuses**. *(Spotlight Figure 7–4, Figure 7–6)*
8. The **auditory ossicles** are located in a cavity within the temporal bone. *(Figure 7–7)*
9. The bones of the face are the **maxillae**, the **palatine bones**, the **nasal bones**, the **vomer**, the **inferior nasal conchae**, the **zygomatic bones**, the **lacrimal bones**, and the **mandible**. *(Figures 7–2 to Spotlight Figure 7–4, 7–10 to 7–12)*

10. The left and right maxillae, or *maxillary bones,* are the largest facial bones; they form the upper jaw and most of the **hard palate**. *(Figures 7–3, Spotlight Figure 7–4, 7–10)*
11. The palatine bones are small L-shaped bones that form the posterior portions of the hard palate and contribute to the floor of the orbital cavities. *(Figures 7–3, Spotlight Figure 7–4, 7–10)*
12. The paired nasal bones extend to the superior border of the **external nares**. *(Figures 7–3, Spotlight Figure 7–4, 7–11)*
13. The vomer forms the inferior portion of the **nasal septum**. *(Figures 7–3, Spotlight Figure 7–4, 7–11)*
14. The **temporal process** of the zygomatic bone articulates with the **zygomatic process** of the temporal bone to form the **zygomatic arch**. *(Figures 7–3, 7–7, 7–11)*
15. The paired lacrimal bones, the smallest bones of the face, are situated medially in each **orbit**. *(Figures 7–3, 7–11)*
16. The mandible is the bone of the lower jaw. *(Figures 7–3, Spotlight Figure 7–4, 7–12)*
17. The **hyoid bone**, suspended by *stylohyoid ligaments,* supports the larynx. *(Figure 7–13)*
18. The foramina and fissures of the adult skull are summarized in *Spotlight Figure 7–4*.

7-3 Each orbital complex contains and protects an eye, and the nasal complex encloses the nasal cavities p. 225

19. Seven bones form each **orbital complex.** *(Figure 7–14)*
20. The **nasal complex** includes the bones that enclose the nasal cavities and the **paranasal sinuses**, hollow airways that connect with the nasal passages. *(Figure 7–15)*

7-4 Fontanelles are non-ossified fibrous areas between cranial bones that ease birth and allow for rapid brain growth in infants and children p. 226

21. Fibrous connective tissue **fontanelles** permit the skulls of infants and children to continue growing after birth. *(Figure 7–16)*

7-5 The vertebral column has four flexible and supportive spinal curves p. 228

22. The **vertebral column** consists of the vertebrae, sacrum, and coccyx. We have 7 **cervical vertebrae** (the first articulates with the skull), 12 **thoracic vertebrae** (which articulate with the ribs), and 5 **lumbar vertebrae** (the last articulates with the sacrum). The **sacrum** and **coccyx** consist of fused vertebrae. *(Figure 7–17)*
23. The spinal column has four **spinal curves**. The **thoracic** and **sacral curves** are called **primary** or **accommodation curves**; the **lumbar** and **cervical curves** are known as **secondary** or **compensation curves**. *(Figure 7–17)*
24. A typical vertebra has a **vertebral body** and a **vertebral arch**, and articulates with adjacent vertebrae at the **superior** and **inferior articular processes**. *(Figure 7–18)*
25. **Intervertebral discs** separate adjacent vertebrae. Spaces between successive **pedicles** form the **intervertebral foramina**. *(Figure 7–18)*

7-6 The five vertebral regions—cervical, thoracic, lumbar, sacral, and coccygeal—each have characteristic vertebrae p. 229

26. Cervical vertebrae are distinguished by the shape of the body, the relative size of the vertebral foramen, the presence of **costal processes** with **transverse foramina**, and notched **spinous processes**. These vertebrae include the **atlas**, **axis**, and **vertebra prominens**. *(Figure 7–19, Table 7–1)*
27. Thoracic vertebrae have a distinctive heart-shaped body; long, slender spinous processes; and articulations for the ribs. *(Figures 7–20, 7–23, Table 7–1)*
28. The lumbar vertebrae are the most massive and least mobile of the vertebrae; they are subjected to the greatest strains. *(Figure 7–21, Table 7–1)*
29. The sacrum protects reproductive, digestive, and urinary organs and articulates with the pelvic girdle and with the fused elements of the coccyx. *(Figure 7–22)*

7-7 The thoracic cage protects organs in the chest and provides sites for muscle attachment p. 237

30. The skeleton of the **thoracic cage** consists of the thoracic vertebrae, the ribs, and the sternum. The **ribs** and **sternum** form the *rib cage. (Figure 7–23)*
31. Ribs 1–7 are **true ribs**, or *vertebrosternal ribs.* Ribs 8–12 are called **false ribs**; they include the *vertebrochondral ribs* (ribs 8–10) and two pairs of *floating (vertebral) ribs* (ribs 11–12). A typical rib has a **head**, or *capitulum;* a **neck**; a **tubercle**; an *angle;* and a **body**, or *shaft.* A *costal groove* marks the path of nerves and blood vessels. *(Figures 7–23, 7–24)*
32. The sternum consists of the **manubrium**, **body**, and **xiphoid process**. *(Figure 7–23)*

Review Questions

See the blue Answers tab at the back of the book.

LEVEL 1 Reviewing Facts and Terms

1. Identify the cranial and facial bones in the diagram below.

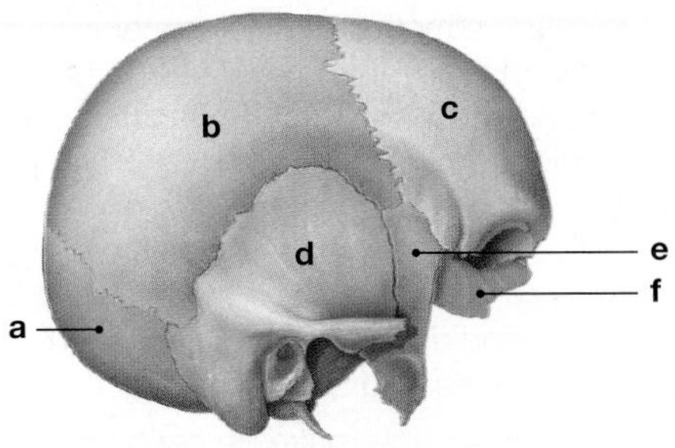

a
b
c
d
e
f

Cranial bones

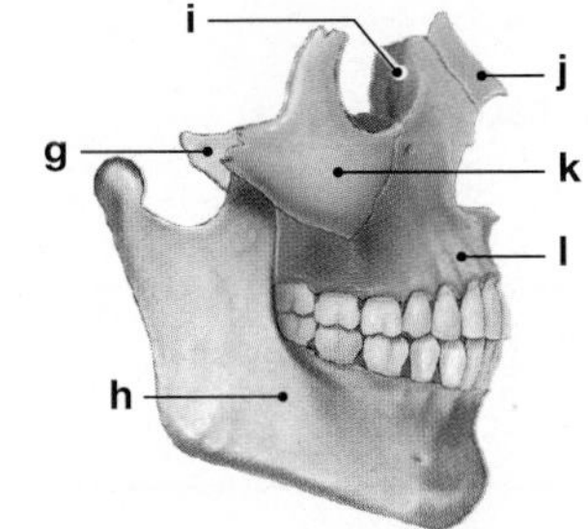

g
h
i
j
k
l

Facial bones

(a) ________	**(b)** ________
(c) ________	**(d)** ________
(e) ________	**(f)** ________
(g) ________	**(h)** ________
(i) ________	**(j)** ________
(k) ________	**(l)** ________

2. Which of the following lists contains *only* facial bones? **(a)** mandible, maxilla, nasal, zygomatic, **(b)** frontal, occipital, zygomatic, parietal, **(c)** occipital, sphenoid, temporal, lacrimal, **(d)** frontal, parietal, occipital, sphenoid.
3. The unpaired facial bones include the **(a)** lacrimal and nasal, **(b)** vomer and mandible, **(c)** maxilla and mandible, **(d)** zygomatic and palatine.
4. The boundaries between skull bones are immovable joints called **(a)** foramina, **(b)** fontanelles, **(c)** lacunae, **(d)** sutures.
5. The joint between the frontal and parietal bones is correctly called the ______ suture. **(a)** parietal, **(b)** lambdoid, **(c)** squamous, **(d)** coronal.
6. Blood vessels that drain blood from the head pass through the **(a)** jugular foramina, **(b)** hypoglossal canals, **(c)** stylomastoid foramina, **(d)** mental foramina, **(e)** lateral canals.
7. For each of the following vertebrae, indicate its vertebral region.

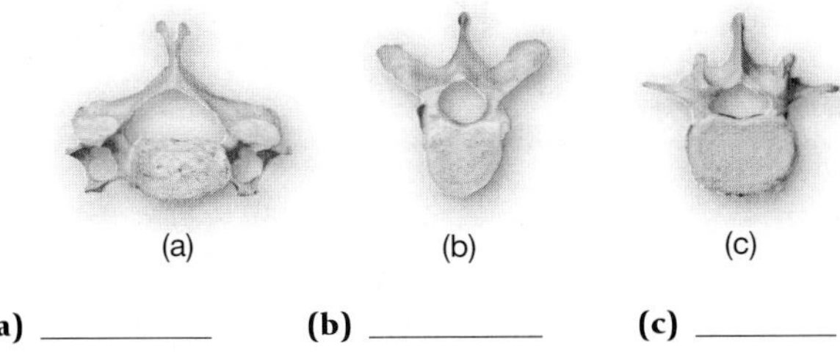

(a)
(b)
(c)

(a) ________ **(b)** ________ **(c)** ________

8. Cervical vertebrae can usually be distinguished from other vertebrae by the presence of **(a)** transverse processes, **(b)** transverse foramina, **(c)** demifacets on the body, **(d)** the vertebra prominens, **(e)** large spinous processes.
9. The lateral walls of the vertebral foramen are formed by the **(a)** body of the vertebra, **(b)** spinous process, **(c)** pedicles, **(d)** laminae, **(e)** transverse processes.
10. The part of the vertebra that transfers weight along the axis of the vertebral column is the **(a)** vertebral arch, **(b)** lamina, **(c)** pedicles, **(d)** body.
11. Which eight bones make up the cranium?
12. What seven bones constitute the orbital complex?
13. What is the primary function of the vomer?
14. In addition to the spinal curves, what skeletal element contributes to the flexibility of the vertebral column?

LEVEL 2 Reviewing Concepts

15. What is the relationship between the temporal bone and the ear?
16. What is the relationship between the ethmoid and the nasal cavity?
17. Describe how movement of the ribs functions in breathing.
18. Why is it important to keep your back straight when you lift a heavy object?
19. The atlas (C_1) can be distinguished from the other vertebrae by **(a)** the presence of anterior and posterior vertebral arches, **(b)** the lack of a body, **(c)** the presence of superior facets and inferior articular facets, **(d)** all of these.
20. What purpose do the fontanelles serve during birth?
21. The secondary spinal curves **(a)** help position the body weight over the legs, **(b)** accommodate the thoracic and abdominopelvic viscera, **(c)** include the thoracic curvature, **(d)** do all of these, **(e)** do only a and c.
22. When you rotate your head to look to one side, **(a)** the atlas rotates on the occipital condyles, **(b)** C_1 and C_2 rotate on the other cervical vertebrae, **(c)** the atlas rotates on the dens of the axis, **(d)** the skull rotates the atlas, **(e)** all cervical vertebrae rotate.
23. Improper administration of cardiopulmonary resuscitation (CPR) can force the ______ into the liver. **(a)** floating ribs, **(b)** lumbar vertebrae, **(c)** manubrium of the sternum, **(d)** costal cartilage, **(e)** xiphoid process.

LEVEL 3 Critical Thinking and Clinical Applications

24. Jane has an upper respiratory infection and begins to feel pain in her teeth. This is a good indication that the infection is located in the **(a)** frontal sinuses, **(b)** sphenoid bone, **(c)** temporal bone, **(d)** maxillary sinuses, **(e)** zygomatic bones.
25. While working at an excavation, an archaeologist finds several small skull bones. She examines the frontal, parietal, and occipital bones and concludes that the skulls are those of children not yet 1 year old. How can she tell their ages from an examination of these bones?
26. Mary is in her last month of pregnancy and is suffering from lower back pains. Since she is carrying excess weight in front of her, she wonders why her back hurts. What would you tell her?

CLINICAL CASE Wrap-Up Knocked Out

The young boxer has sustained a "blowout" fracture of his left orbital complex. The orbit that contains the eye is formed by seven facial bones and is surrounded by four air-filled sinuses. The orbital rim is strong, composed of thick, protective bone. But the very thin medial wall and floor of the orbit were literally blown out by the intact eyeball itself when he was struck with such a forceful punch.

7

A computerized tomography (CT) scan reveals fractures of the ethmoid, lacrimal, palatine, and maxillary bones that make up the delicate medial wall and floor of the left orbit. Many tiny fracture fragments can be seen within the maxillary sinus.

The opening to the air-filled sinus allowed the air to escape into the soft tissues surrounding the eye when the fighter coughed, causing the subcutaneous air expansion around the eye. The muscles that move the eye had become entrapped within the fracture fragments, preventing the fighter from gazing upward and giving rise to his double vision.

The young boxer is taken to the operating room. A surgical repair of the orbital floor is performed, using an implant to support the eyeball until fracture healing is complete.

1. What are the seven bones that make up each orbital complex?
2. What are the four paranasal sinuses surrounding each orbital complex?

See the blue Answers tab at the back of the book.

Related Clinical Terms

craniotomy: The surgical removal of a section of bone (bone flap) from the skull for the purpose of operating on the underlying tissues.

deviated nasal septum: A bent nasal septum (cartilaginous structure dividing the left and right nasal cavities) that slows or prevents sinus drainage.

herniated disc: A disc (fibrocartilage pad) between the vertebrae that slips out of place or ruptures; if it presses on a nerve, it can cause back pain or sciatica.

laminectomy: A surgical operation to remove the posterior vertebral arch on a vertebra, usually to give access to the spinal cord or to relieve pressure on nerves.

spina bifida: A condition resulting from the failure of the vertebral laminae to unite during development; commonly associated with developmental abnormalities of the brain and spinal cord.

spinal fusion: A surgical procedure that stabilizes the spine by joining together (fusing) two or more vertebrae using bone grafts, metal rods, or screws.

8 The Appendicular Skeleton

Learning Outcomes

These Learning Outcomes correspond by number to this chapter's sections and indicate what you should be able to do after completing the chapter.

- 8-1 ■ Identify the bones that form the pectoral girdles, their functions, and their surface features. p. 244
- 8-2 ■ Identify the bones of the upper limbs, their functions, and their surface features. p. 247
- 8-3 ■ Identify the bones that form the pelvic girdle, their functions, and their surface features. p. 252
- 8-4 ■ Identify the bones of the lower limbs, their functions, and their surface features. p. 254
- 8-5 ■ Summarize sex differences and age-related changes in the human skeleton. p. 260

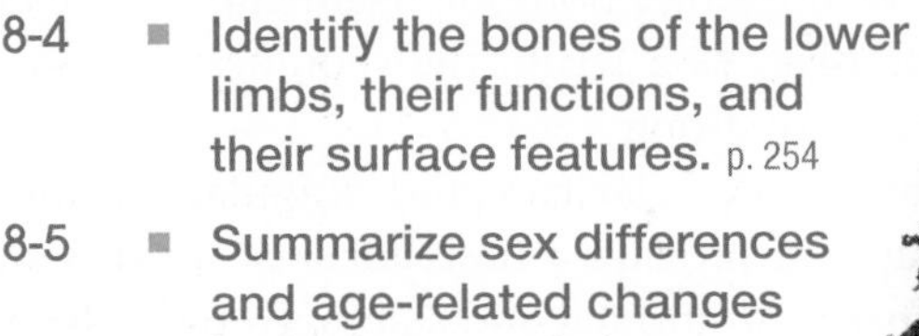

CLINICAL CASE Timber!!

"Unh!" The air is knocked out of Ken, a 25-year-old apprentice tree trimmer. He has slipped and fallen from a large branch 12 feet up. He hit the ground sideways, landing on his right arm, which he instinctively thrust out to the side to try and break his fall. He lies there, dazed, a vicious pain lancing across his upper chest. He has trouble catching his breath.

Doug, his coworker, reaches him a moment later. Sizing up a bad situation, Doug punches 9-1-1 into his cell phone. Shortly after, the EMTs arrive. They take Ken's vital signs and transfer him to a stretcher for transport to the local emergency room (ER).

At the ER, the doctor cuts Ken's shirt off and reveals a sagging right shoulder. Ken is curling his arm into a fetal position to guard it from any movement, which is excruciating. On his upper chest, close to the sternum, there is a knob of swollen tissue. The doctor orders an x-ray to confirm his suspicion. **Where was Ken injured?** **To find out, turn to the Clinical Case Wrap-Up on p. 264.**

An Introduction to the Appendicular Skeleton

In the previous chapter, we discussed the 80 bones of the axial skeleton. In this chapter we focus on the remaining 60 percent of the bones of the skeletal system, which are appended to the axial skeleton. The **appendicular skeleton** includes the bones of the limbs and the supporting bone (pectoral and pelvic) girdles that connect them to the trunk (Figure 8–1).

To appreciate the role of the appendicular skeleton in your life, make a mental list of all the things you have done with your arms or legs today. Standing, walking, writing, turning pages, eating, dressing, shaking hands, texting—the list quickly becomes unwieldy. Your axial skeleton protects and supports internal organs and takes part in vital functions, such as breathing. But your appendicular skeleton, with attached skeletal muscles, lets you manipulate objects and move from place to place.

The appendicular skeleton is dominated by the long bones that form the limbs. The upper limbs allow us to move freely, and the lower limbs are adapted for movement and support. ⮌ p. 209 Long bones share common features. For example, one epiphysis is usually called the *head*, the diaphysis is called the *shaft*, and the narrower metaphysis between the head and shaft is called the *neck*.

8-1 The pectoral (shoulder) girdles attach the upper limbs to the axial skeleton

Learning Outcome **Identify the bones that form the pectoral girdles, their functions, and their surface features.**

Each arm *articulates* (that is, forms a joint) with the trunk at a **pectoral girdle**, or *shoulder girdle* (Figures 8–1, 8–2a). The pectoral girdles consist of four bones—two S-shaped *clavicles* (KLAV-i-kulz; collarbones) and two broad, flat *scapulae* (SKAP-ū-lē; singular, *scapula*, SKAP-ū-luh; shoulder blades).

The medial, anterior end of each clavicle articulates with the manubrium of the sternum. ⮌ p. 239 These joints are the *only* direct connections between the pectoral girdles and the axial skeleton. Skeletal muscles support and position the scapulae, which have no direct bony or ligamentous connections to the thoracic cage. As a result, the shoulders are extremely mobile, but not very strong.

Movements of the clavicles and scapulae position the shoulder joints and provide a basis for arm movement. The shoulder joints are positioned and stabilized by skeletal muscles that extend between the axial skeleton and the pectoral girdles. Once the joints are in position, other skeletal muscles, including several that originate on the pectoral girdle, move the upper limbs.

The surfaces of the scapulae and clavicles are extremely important as sites for muscle attachment. Bony ridges and projections mark the attachment sites of major muscles. Other bone markings, such as sulci or foramina, indicate the positions of nerves that control the muscles or the passage of blood vessels that nourish the muscles and bones. Note that in the descriptions of bones in this chapter, we emphasize surface markings that either have functional importance (such as the attachment sites for skeletal muscles and the paths of major nerves and blood vessels) or that serve as anatomical landmarks (such as the xiphoid process for administering CPR).

The Clavicles

The paired **clavicles** are S-shaped bones that originate at the superior, lateral border of the manubrium of the sternum, lateral to the jugular notch (Figure 8–2a). From this pyramid-shaped **sternal end**, each clavicle curves laterally and posteriorly for about half its length. It then forms a smooth posterior

Figure 8–1 An Anterior View of the Appendicular Skeleton. The numbers in the boxes indicate the total number of bones of each type or within each category. ATLAS: Plates 1a,b

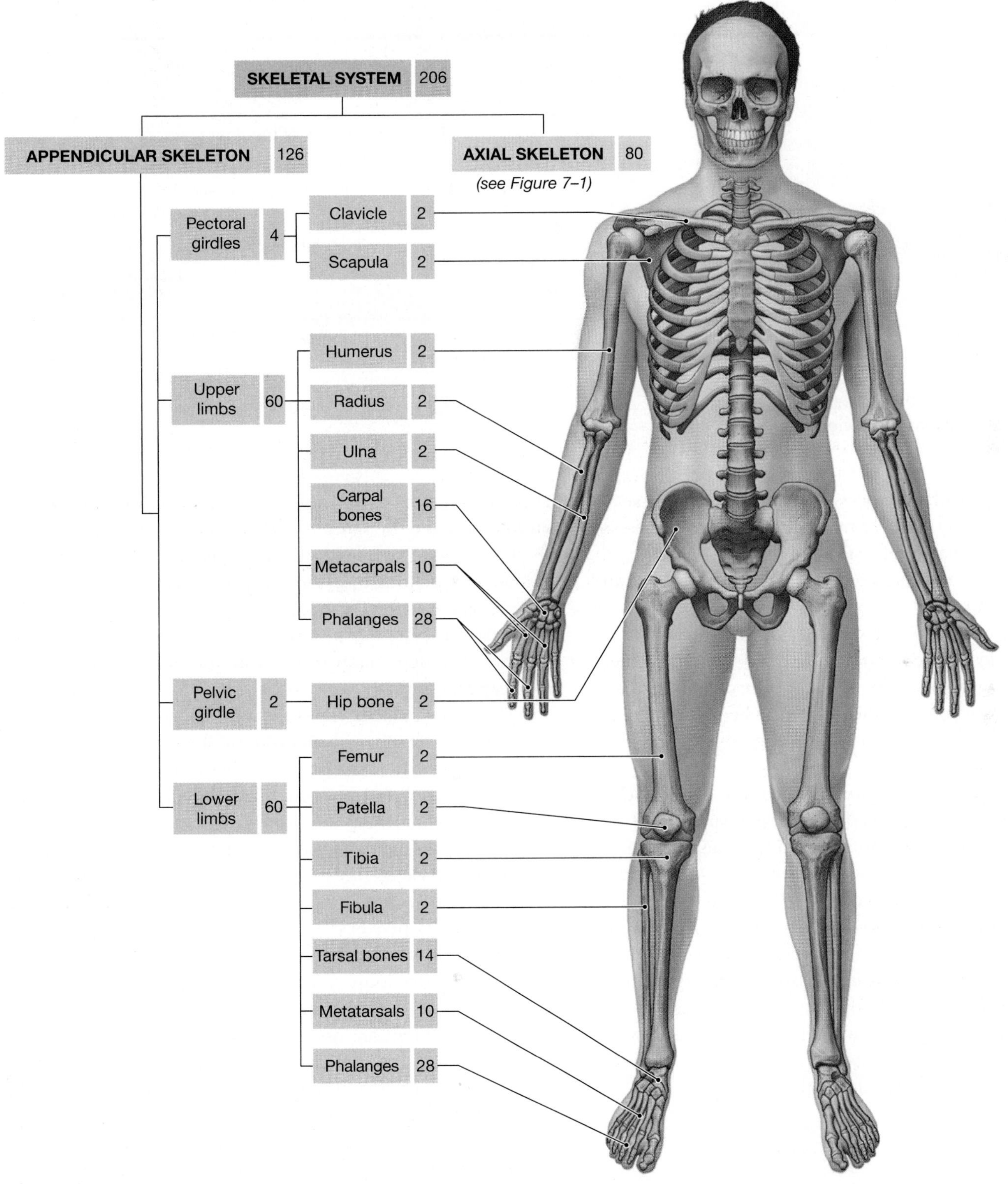
SKELETAL SYSTEM 206
APPENDICULAR SKELETON 126
AXIAL SKELETON 80
(see Figure 7–1)
Pectoral girdles 4
Clavicle 2
Scapula 2
Upper limbs 60
Humerus 2
Radius 2
Ulna 2
Carpal bones 16
Metacarpals 10
Phalanges 28
Pelvic girdle 2
Hip bone 2
Lower limbs 60
Femur 2
Patella 2
Tibia 2
Fibula 2
Tarsal bones 14
Metatarsals 10
Phalanges 28

? What structures comprise the appendicular skeleton?

Figure 8–2 The Right Clavicle. ATLAS: Plates 26a,b

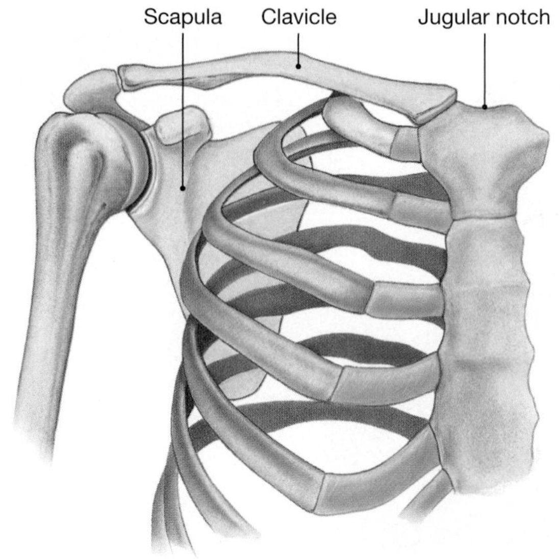
Scapula
Clavicle
Jugular notch

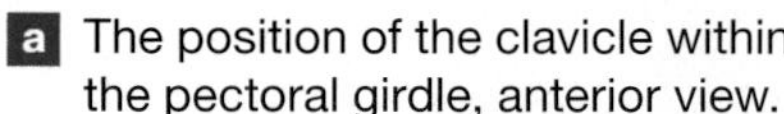

a The position of the clavicle within the pectoral girdle, anterior view.

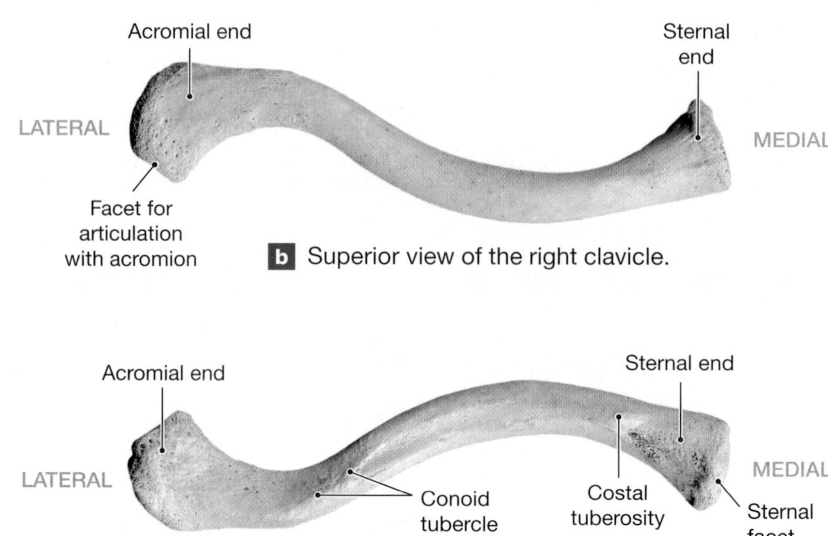
Acromial end
Sternal end
LATERAL
MEDIAL
Facet for articulation with acromion
Acromial end
Sternal end
LATERAL
MEDIAL
Conoid tubercle
Costal tuberosity
Sternal facet

b Superior view of the right clavicle.

c Inferior view of the right clavicle.

curve to articulate with a process of the scapula, the *acromion* (a-KRŌ-mē-on). The flat, **acromial end** of the clavicle is broader than the sternal end (Figure 8–2b,c).

The smooth, superior surface of the clavicle lies just beneath the skin. The acromial end has a rough inferior surface that has prominent lines and tubercles (Figure 8–2c). The *conoid tubercle* and *costal tuberosity* on the inferior surface are attachment sites for muscles and ligaments of the shoulder. The combination of the direction of curvature and the differences between superior and inferior surfaces makes it easy to distinguish a left clavicle from a right clavicle.

You can explore the connection between your own clavicles and sternum. With your fingers in your jugular notch (the depression just superior to the manubrium), locate the clavicle on either side and find the *sternoclavicular joints* where the sternum articulates with the clavicles. When you move your shoulders, you can feel the sternal ends of the clavicles change their positions.

The clavicles are small and fragile, so it's no surprise that fractures of the clavicle are fairly common. For example, a simple fall can fracture a clavicle if you land on your hand with your arm outstretched. Fortunately, most clavicular fractures heal rapidly with a simple clavicle strap.

The Scapulae

The paired **scapulae** are flat bones found on the posterior side of the rib cage. Each triangular-shaped scapula is made up of two large processes for muscular attachment and the *glenoid cavity* for articulation with an arm bone (Figure 8–3). The three sides of each scapula are the **superior border**; the **medial border**, or *vertebral border*; and the **lateral border** (Figure 8–3a,c) or *axillary border*. Muscles that position the scapula attach along these sides. The corners of the triangle are called the *superior angle*, the *inferior angle*, and the *lateral angle*. The anterior, or *costal*, surface of the scapula is smooth and concave. The depression in the anterior surface is called the **subscapular fossa**.

The lateral angle of the scapula forms a broad process that supports the cup-shaped **glenoid cavity** (Figure 8–3a,b). At the glenoid cavity, the scapula articulates with the *humerus*, the proximal bone of the upper limb. This articulation is the shoulder joint, also known as the *glenohumeral joint*.

Two large scapular processes extend beyond the margin of the glenoid cavity (see Figure 8–3b) superior to the head of the humerus. The smaller, anterior projection is the **coracoid** (KOR-uh-koyd) **process**. The **acromion** is the larger, posterior process. If you run your fingers along the superior surface of your shoulder joint, you will feel this process. The acromion articulates with the clavicle at the *acromioclavicular joint*. Ligaments and tendons associated with the shoulder joint are attached to both the coracoid process and the acromion.

The acromion is continuous with the **spine** of the scapula, a ridge that crosses the posterior surface before ending at the medial border (Figure 8–3c). The scapular spine divides the convex posterior surface into two regions. The area superior to this spine is the **supraspinous fossa** (*supra*, above). The region inferior to the spine is the **infraspinous fossa** (*infra*, beneath). Small ridges and lines mark the entire posterior surface where smaller muscles attach to the scapula.

> Go to MasteringA&P™ > Study Area > Menu > Lab Tools > Bone & Dissection Videos > Bone Videos > Scapula Bone

Figure 8–3 **The Right Scapula.** ATLAS: Plates 26a,b

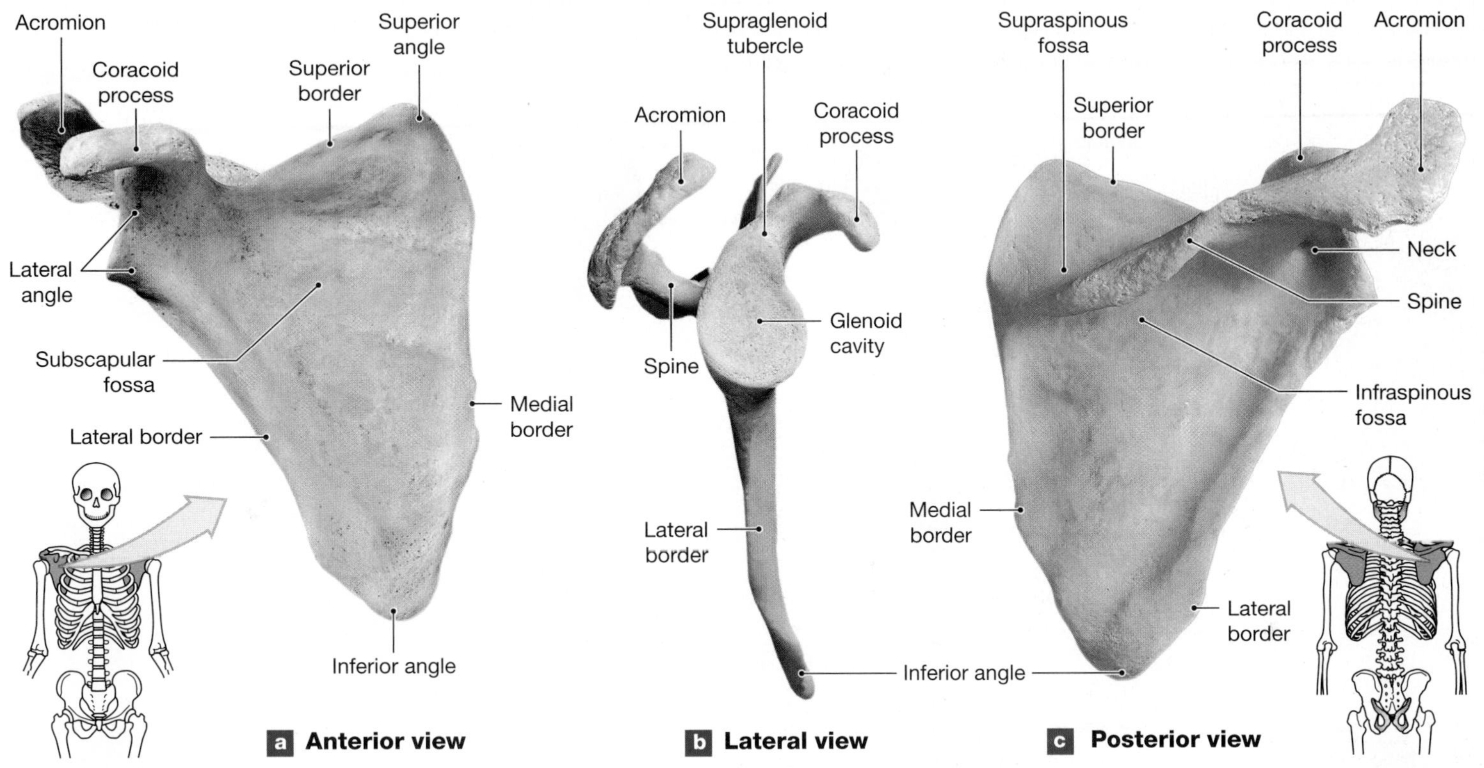

Acromion
Coracoid process
Superior border
Superior angle
Lateral angle
Subscapular fossa
Lateral border
Medial border
Inferior angle
a Anterior view
Supraglenoid tubercle
Acromion
Coracoid process
Glenoid cavity
Spine
Lateral border
b Lateral view
Supraspinous fossa
Superior border
Coracoid process
Acromion
Neck
Spine
Infraspinous fossa
Medial border
Lateral border
Inferior angle
c Posterior view

Checkpoint

1. **Name the bones of the pectoral girdles.**
2. **How would a broken clavicle affect the mobility of the scapula?**
3. **Which bone articulates with the scapula at the glenoid cavity?**

See the blue Answers tab at the back of the book.

8-2 The bones of the upper limbs are adapted for free movement

Learning Outcome **Identify the bones of the upper limbs, their functions, and their surface features.**

The skeleton of the upper limbs consists of the bones of the arms, forearms, wrists, and hands. Note that in anatomical descriptions, the term *arm* refers only to the proximal portion of the upper limb (from shoulder to elbow), not to the entire limb. The distal portion of the upper limb (from elbow to wrist) is the *forearm*. Let's examine the bones of the right upper limb.

Arm Bone: The Humerus

The arm, or *brachium*, contains only one long bone, the **humerus**, which extends from the scapula to the elbow. From proximal to distal, the main parts of the humerus are the *head*, *shaft*, and *condyle*, where it articulates with the forearm bones.

At the proximal end of the humerus, the round **head** articulates with the scapula (Figure 8–4a,b). The prominent **greater tubercle** is a rounded projection on the lateral surface of the epiphysis, near the margin of the humeral head. The greater tubercle establishes the lateral contour of the shoulder. You can verify its position by feeling for a bump located a few centimeters from the tip of the acromion. The **lesser tubercle** is a smaller projection that lies on the anterior, medial surface of the epiphysis, separated from the greater tubercle by the **intertubercular sulcus**, or *bicipital groove*. A large tendon runs along the groove. Both tubercles are important sites for muscle attachment.

The **anatomical neck** lies between the tubercles and the articular surface of the head. This neck marks the extent of the joint capsule. The narrower distal **surgical neck** corresponds to the metaphysis of the growing bone. The name reflects the fact that fractures typically occur at this site.

The proximal shaft of the humerus is round in section. The **deltoid tuberosity** is a large, rough elevation on the lateral surface of the shaft, approximately halfway along its length. It is named after the *deltoid*, a muscle that attaches to it.

On the posterior surface, the deltoid tuberosity ends at the **radial groove** (see Figure 8–4b). This depression marks the path of the *radial nerve*, a large nerve for both sensory information from the posterior surface of the limb and motor control over the large muscles that straighten (extend) the elbow.

> Go to MasteringA&P™ > Study Area > Menu > Lab Tools > Bone & Dissection Videos > Bone Videos > Humerus

Figure 8–4 The Right Humerus and Elbow Joint. ATLAS: Plates 31; 34a–d Note that for simplicity, an illustration of a single bone throughout this chapter will have labels that do not include the name of the bone. For example, these photographs of a humerus have the label *Head* rather than *Head of the humerus*, or *Humeral head.* In figures where more than one bone is shown, labels will use complete names to avoid confusion.

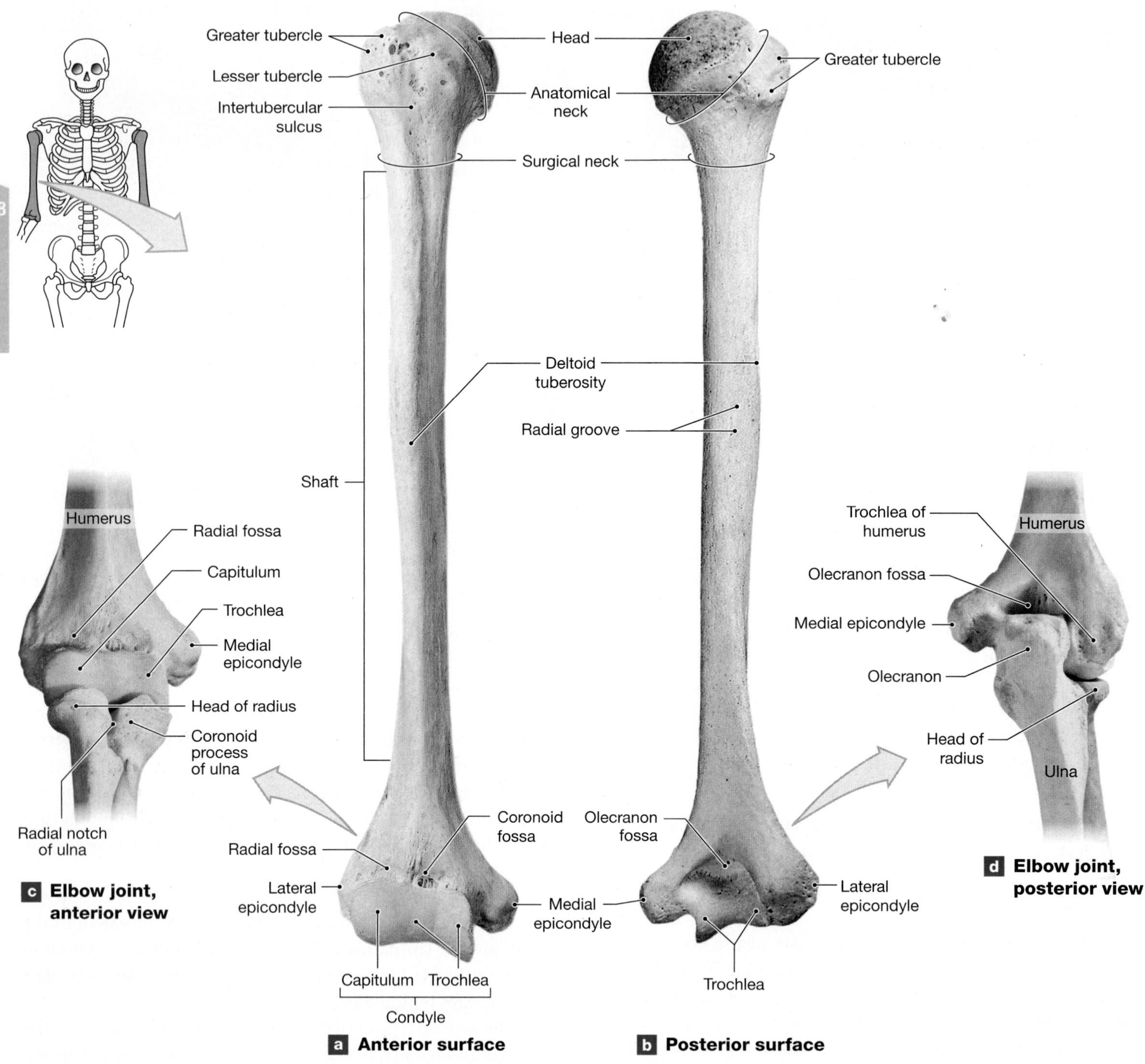
Greater tubercle
Lesser tubercle
Intertubercular sulcus
Head
Anatomical neck
Surgical neck
Greater tubercle
Deltoid tuberosity
Radial groove
Shaft
Humerus
Radial fossa
Capitulum
Trochlea
Medial epicondyle
Head of radius
Coronoid process of ulna
Radial notch of ulna
Radial fossa
Lateral epicondyle
Coronoid fossa
Olecranon fossa
Medial epicondyle
Capitulum
Trochlea
Condyle
Trochlea
Lateral epicondyle
Trochlea of humerus
Humerus
Olecranon fossa
Medial epicondyle
Olecranon
Head of radius
Ulna

c **Elbow joint, anterior view**

a **Anterior surface**

b **Posterior surface**

d **Elbow joint, posterior view**

Distal to the radial groove, the posterior surface of the humerus is fairly flat.

Near the distal articulation with the bones of the forearm, the humeral shaft expands to either side at the **medial** and **lateral epicondyles**. *Epicondyles* are processes that develop proximal to an articulation. They provide additional surface area for muscle attachment. The prominent medial head and the differences between the medial and lateral epicondyles make it easy to tell a left humerus from a right humerus.

The *ulnar nerve* crosses the posterior surface of the medial epicondyle. A hit at the posteromedial surface of the elbow joint can strike this nerve and produce a temporary numbness and paralysis of muscles on the anterior surface of the forearm. Because of the odd sensation, this area is sometimes called the *funny bone.*

At the **condyle**, the humerus articulates with the *ulna* and the *radius*, the bones of the forearm. The condyle is divided into two articular regions: the trochlea and the capitulum (see Figure 8–4a,b). The **trochlea** is the spool-shaped medial portion of the condyle. It extends from the base of the **coronoid fossa** on the anterior surface to the **olecranon** (ō-LEK-ruh-non) **fossa** on the posterior surface. The rounded **capitulum** forms the lateral surface of the condyle (Figure 8–4c).

A shallow **radial fossa** superior to the capitulum accommodates a portion of the radial head when the elbow bends (see Figure 8–4a). Projections from the ulnar surface fit into the coronoid fossa (when bending the elbow) and the olecranon fossa (when straightening the elbow) as the elbow nears its total range of motion (Figure 8–4d).

Bones of the Forearm

The *ulna* and *radius* are parallel long bones that support the forearm, or *antebrachium.* In the anatomical position, the ulna lies medial to the radius. The **interosseous membrane**, a fibrous sheet, connects the lateral margin of the ulna to the radius (Figure 8–5a,b).

The Ulna

From proximal to distal, the **ulna** is made up of the *olecranon*, the shaft, and the *ulnar head*. The **olecranon**, the proximal end of the ulna, is the point of the elbow (see Figure 8–5a). On the anterior surface of the proximal epiphysis, the **trochlear notch** of the ulna (see Figure 8–5b) articulates with the trochlea of the humerus at the elbow joint. Figure 8–5c shows the trochlear notch of the ulna in a lateral view.

Figure 8–5 The Right Radius and Ulna. ATLAS: Plates 31; 35f; 36a,b

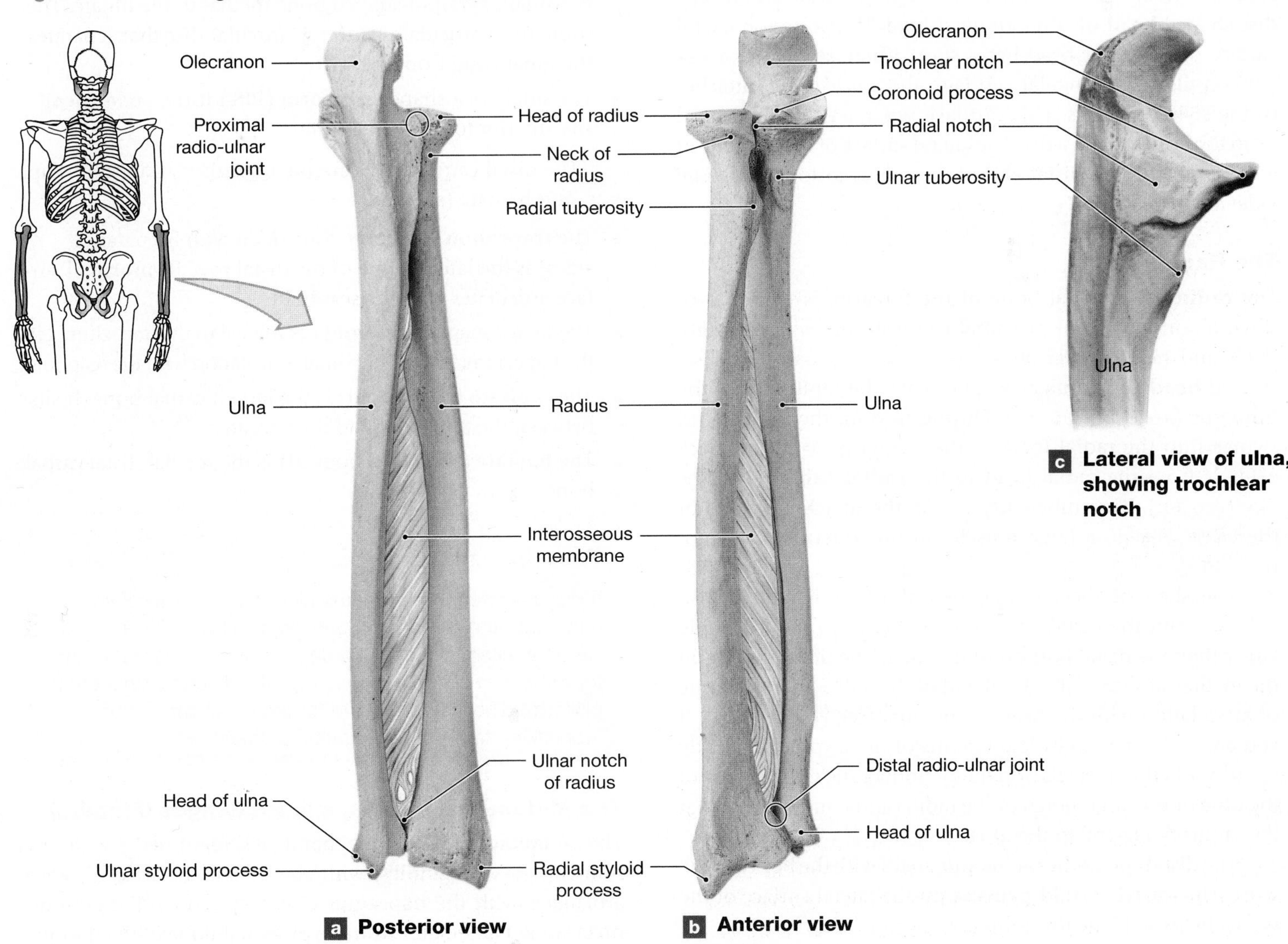
Olecranon
Proximal radio-ulnar joint
Head of radius
Neck of radius
Radial tuberosity
Ulna
Radius
Interosseous membrane
Ulnar notch of radius
Head of ulna
Ulnar styloid process
Radial styloid process
a Posterior view
Olecranon
Trochlear notch
Coronoid process
Radial notch
Ulnar tuberosity
Ulna
Distal radio-ulnar joint
Head of ulna
b Anterior view
Ulna
c Lateral view of ulna, showing trochlear notch

Tips & Tools

The memory tool for olecranon involves the letter **l**: The knob you feel at the e**l**bow is the o**l**ecranon of the u**l**na. You can remember that the trochlear notch is a feature of the **u**lna because this notch forms a U in lateral view.

The olecranon forms the superior lip of the trochlear notch. The **coronoid process** forms its inferior lip. At the limit of *extension*, movement that straightens the forearm and arm, the olecranon swings into the olecranon fossa on the posterior surface of the humerus (see Figure 8–4d). At the limit of *flexion*, a movement that decreases the angle between the articulating bones, the arm and forearm form a tight V, and the coronoid process projects into the coronoid fossa on the anterior humeral surface. Lateral to the coronoid process, a smooth **radial notch** accommodates the head of the radius at the *proximal radio-ulnar joint* (see Figure 8–5a).

Viewed in cross section, the shaft of the ulna is triangular. Near the wrist, the shaft of the ulna narrows before ending at a disc-shaped **head of ulna**, or *ulnar head*. The posterior, lateral surface of the ulnar head has a short **ulnar styloid process** (*stylos*, pillar). A triangular cartilage, the *articular disc*, attaches to the styloid process. This cartilage separates the ulnar head from the bones of the wrist. The lateral surface of the ulnar head articulates with the distal end of the radius to form the *distal radio-ulnar joint* (see Figure 8–5b).

The Radius

The **radius**, the lateral bone of the forearm, is made up of three main parts from proximal to distal: the *head of radius*, *shaft*, and *radial styloid process* (see Figure 8–5a). The disc-shaped **head of radius** articulates with the capitulum of the humerus (see Figure 8–4c). During flexion, the radial head swings into the radial fossa of the humerus. A narrow neck extends from the radial head to the **radial tuberosity** (see Figure 8–5b). This tuberosity marks the attachment site of the *biceps brachii*, a large muscle on the anterior surface of the arm.

The shaft of the radius curves along its length. It also enlarges, and the distal portion of the radius is considerably larger than the distal portion of the ulna. The **ulnar notch** on the medial surface of the distal end of the radius marks the site of articulation with the head of the ulna (see Figure 8–5a). If you are looking at an isolated radius or ulna, you can quickly identify whether it is left or right by finding the radial notch of the ulna or the ulnar notch of the radius and remembering that the radius lies lateral to the ulna.

The distal end of the radius articulates with the bones of the wrist. The **radial styloid process** on the lateral surface of the radius helps stabilize this joint (see Figure 8–5a,b).

Bones of the Wrist and Hand

The *wrist*, or carpus, contains eight *carpal bones*. Some of the carpal bones articulate with the five *metacarpal* (met-uh-KAR-pul; *metacarpus*, hand) *bones* to support the hand (Figure 8–6). Each hand has 14 finger bones, or *phalanges* (fa-LAN-jēz; singular, *phalanx*, FĀ-lanks).

The Carpal (Wrist) Bones

The **carpal bones** in the wrist form two rows, one with four **proximal carpal bones** and the other with four **distal carpal bones**.

The proximal carpal bones are the scaphoid, lunate, triquetrum, and pisiform (see Figure 8–6):

- The **scaphoid** (*skaphe*, boat) is the proximal carpal bone on the lateral border of the wrist. It is the carpal bone closest to the styloid process of the radius.
- The comma-shaped **lunate** (*luna*, moon) lies medial to the scaphoid. Like the scaphoid, it articulates with the radius.
- The **triquetrum** (trī-KWĒ-trum; *triquetrus*, three-cornered) is a small, pyramid-shaped bone medial to the lunate. The triquetrum articulates with the articular disc that separates the ulnar head from the wrist.
- The small, pea-shaped **pisiform** (PIS-i-form; *pisum*, pea) sits anterior to the triquetrum.

The distal carpal bones are the trapezium, trapezoid, capitate, and hamate (see Figure 8–6):

- The **trapezium** (*trapezion*, four-sided with no parallel sides) is the lateral bone of the distal row. Its proximal surface articulates with the scaphoid.
- The wedge-shaped **trapezoid** lies medial to the trapezium. Like the trapezium, it has a proximal articulation with the scaphoid.
- The **capitate** (*caput*, head) is the largest carpal bone. It sits between the trapezoid and the hamate.
- The **hamate** (*hamatum*, hooked) is the medial distal carpal bone.

Tips & Tools

It may help you to identify the eight carpal bones if you remember the sentence "Sam Likes To Push The Toy Car Hard." In lateral-to-medial order, the first four words stand for the proximal carpal bones (scaphoid, lunate, triquetrum, pisiform). The last four stand for the distal carpal bones (trapezium, trapezoid, capitate, hamate).

The Metacarpals (Palm) and Phalanges (Fingers)

The 5 **metacarpals** of the hand are identified by Roman numerals I–V, beginning with the lateral metacarpal, which articulates with the trapezium (see Figure 8–6). For example, metacarpal I articulates with the proximal bone of the thumb.

Figure 8–6 Bones of the Right Wrist and Hand. ATLAS: Plates 38a,b

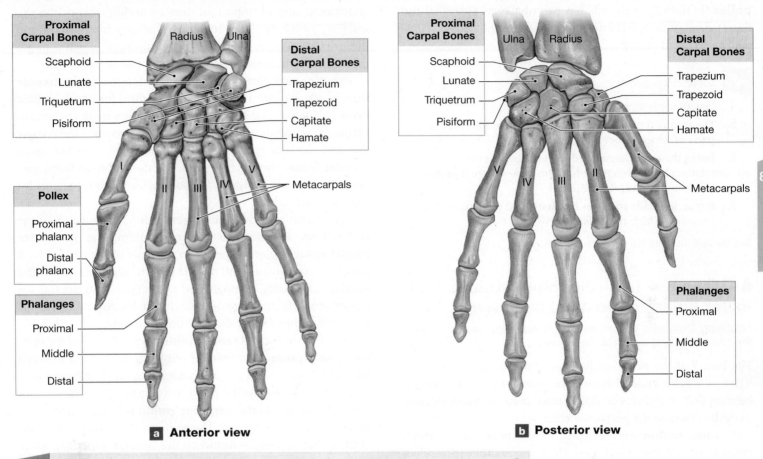

Proximal Carpal Bones
Scaphoid
Lunate
Triquetrum
Pisiform

Radius
Ulna

Distal Carpal Bones
Trapezium
Trapezoid
Capitate
Hamate

I II III IV V Metacarpals

Pollex
Proximal phalanx
Distal phalanx

Phalanges
Proximal
Middle
Distal

a **Anterior view**

Proximal Carpal Bones
Scaphoid
Lunate
Triquetrum
Pisiform

Ulna
Radius

Distal Carpal Bones
Trapezium
Trapezoid
Capitate
Hamate

V IV III II I Metacarpals

Phalanges
Proximal
Middle
Distal

b **Posterior view**

? With which carpal bones does metacarpal IV articulate?

+ Clinical Note Carpal Tunnel Syndrome

The carpal bones articulate with one another at joints that permit limited sliding and twisting. Ligaments interconnect the carpal bones and help stabilize the wrist joint. The tendons of muscles that flex the fingers pass across the anterior surface of the wrist. These tendons are sandwiched between the intercarpal ligaments and a broad, superficial transverse ligament called the *flexor retinaculum.* Inflammation of the connective tissues between the flexor retinaculum and the carpal bones can compress the tendons and the adjacent *median nerve.* The result is pain, weakness, and reduced wrist mobility. This condition, called *carpal tunnel syndrome,* is common in occupations that require repetitive wrist motion.

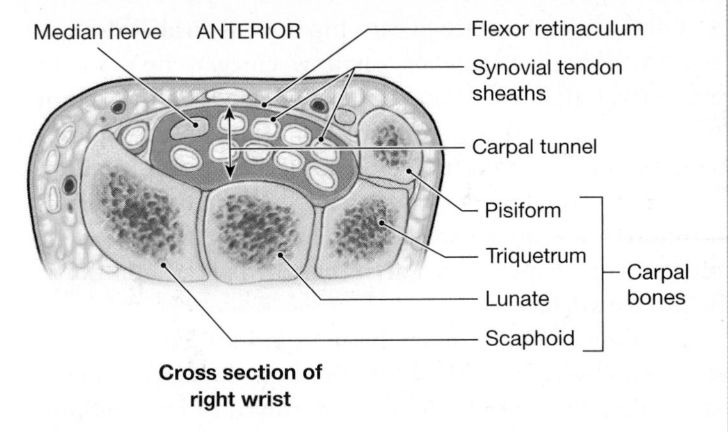
Median nerve
ANTERIOR
Flexor retinaculum
Synovial tendon sheaths
Carpal tunnel
Pisiform
Triquetrum
Lunate
Scaphoid
Carpal bones

Cross section of right wrist

Distally, the metacarpals articulate with the proximal **phalanges**, the finger bones. The first finger, known as the **pollex** (POL-eks), or thumb, has two phalanges (*proximal* and *distal*). Each of the other four fingers has three phalanges (*proximal, middle,* and *distal*).

Checkpoint

4. **List all the bones of an upper limb.**
5. **The rounded projections on either side of the elbow are parts of which bone?**
6. **Using the terms *lateral* and *medial*, describe the relationship between the forearm bones when a person is in the anatomical position.**
7. **Bill accidentally fractures his first distal phalanx with a hammer. Which finger is broken?**

See the blue Answers tab at the back of the book.

8-3 The pelvic girdle (hips) attaches the lower limbs to the axial skeleton

Learning Outcome Identify the bones that form the pelvic girdle, their functions, and their surface features.

The bones of the pelvic girdle attach to the lower limbs (see Figure 8–1) and must withstand the stresses involved in weight bearing and mobility. For this reason, they are more massive than the bones of the pectoral girdles.

In this section we first consider the *pelvic girdle,* which consists of the two *hip bones.* Then we examine the *pelvis,* a composite structure that includes the pelvic girdle of the appendicular skeleton plus the sacrum and coccyx of the axial skeleton. pp. 234–237

The Pelvic Girdle (Hip Bones)

The **pelvic girdle** is made up of the paired **hip bones**, also called the **coxal bones**, or **pelvic bones**. Each hip bone forms by the fusion of three bones: an **ilium** (IL-ē-um; plural, *ilia*), an **ischium** (IS-kē-um; plural, *ischia*), and a **pubis** (PŪ-bis) (Figure 8–7).

On the lateral surface of each hip bone is an **acetabulum** (as-e-TAB-ū-lum; *acetabulum,* shallow vinegar cup), a concave socket that articulates with the head of the femur (Figure 8–7a). A ridge of bone forms the lateral and superior margins of the acetabulum, which has a diameter of about 5 cm (2 in.). The anterior and inferior portion of the ridge is incomplete, leaving a gap called the **acetabular notch**. The smooth, cup-shaped articular surface of the acetabulum is the **lunate surface**.

The ilium, ischium, and pubis meet inside the acetabulum, as though it were a pie sliced into three pieces. Superior to the acetabulum, the ilium forms a broad, curved surface that provides an extensive area for the attachment of muscles, tendons, and ligaments (see Figure 8–7a). Bone markings along the margin of the ilium include the *iliac spines,* which mark the attachment sites of important muscles and ligaments; the *gluteal lines,* which mark the attachment of large hip muscles; and the **greater sciatic** (sī-AT-ik) **notch**, through which a major nerve (the *sciatic nerve*) reaches the lower limb.

The broadest part of the ilium extends between the **arcuate line**, which is continuous with the pectineal line, and the **iliac crest** (Figure 8–7b). These prominent ridges mark the attachments of ligaments and muscles. The area between the arcuate line and the iliac crest forms a shallow depression known as the **iliac fossa**. The concave surface of the iliac fossa helps support the abdominal organs. It also provides additional area for muscle attachment.

The ischium forms the posterior, inferior portion of the acetabulum. Posterior to the acetabulum, the prominent **ischial spine** projects superior to the *lesser sciatic notch.* Blood vessels, nerves, and a small muscle pass through this notch. The **ischial tuberosity**, a roughened projection, is located at the posterior and lateral edge of the ischium. When you are sitting, the ischial tuberosities bear your body's weight.

The narrow **ischial ramus** continues until it meets the **inferior pubic ramus**. The inferior pubic ramus extends between the ischial ramus and the *pubic tubercle,* a small, elevated area anterior and lateral to the pubic symphysis. There the inferior pubic ramus meets the **superior pubic ramus**, which originates near the acetabulum. The *pectineal line,* a ridge that ends at the pubic tubercle, is found on the anterior, superior surface of the superior ramus (see Figure 8–7b).

The pubic ramus and ischial ramus encircle the **obturator** (OB-tū-rā-tor) **foramen** (see Figure 8–7a). This space is closed by a sheet of collagen fibers whose inner and outer surfaces provide a firm base for the attachment of muscles of the hip.

In a medial view of a hip bone, the anterior and medial surface of the pubis has a roughened area that marks the site of articulation with the pubis of the opposite side (see Figure 8–7b). At this articulation—the **pubic symphysis**—the two pubic bones are attached to a middle pad of fibrocartilage.

The Pelvis (Pelvic Girdle, Sacrum, and Coccyx)

Figure 8–8 shows anterior and posterior views of the **pelvis**, which consists of the two hip bones, the sacrum, and the coccyx. Posteriorly, the **auricular surfaces** of the ilia have a sturdy articulation with the auricular surfaces of the sacrum, attaching the pelvic girdle to the axial skeleton at the *sacroiliac joints.* p. 234 Ligaments arising at the **iliac tuberosity** (see Figure 8–7), a roughened area superior to the auricular surface, stabilize this joint. An extensive network of ligaments

Figure 8–7 The Right Hip Bone. The left and right hip bones make up the pelvic girdle.

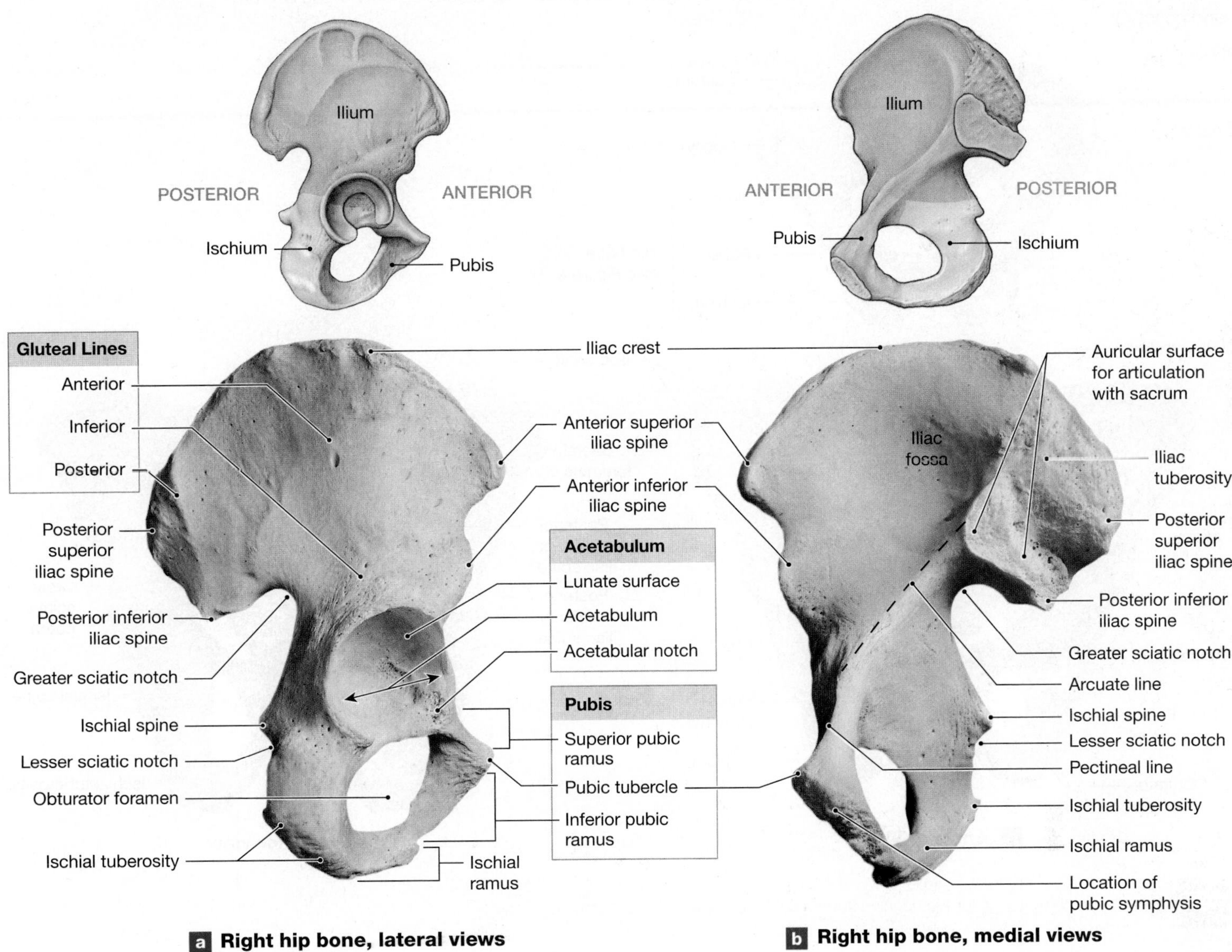
Ilium
POSTERIOR
ANTERIOR
Ischium
Pubis
Ilium
ANTERIOR
POSTERIOR
Pubis
Ischium
Gluteal Lines
Anterior
Inferior
Posterior
Posterior superior iliac spine
Posterior inferior iliac spine
Greater sciatic notch
Ischial spine
Lesser sciatic notch
Obturator foramen
Ischial tuberosity
Iliac crest
Anterior superior iliac spine
Anterior inferior iliac spine
Acetabulum
Lunate surface
Acetabulum
Acetabular notch
Pubis
Superior pubic ramus
Pubic tubercle
Inferior pubic ramus
Ischial ramus
Iliac fossa
Auricular surface for articulation with sacrum
Iliac tuberosity
Posterior superior iliac spine
Posterior inferior iliac spine
Greater sciatic notch
Arcuate line
Ischial spine
Lesser sciatic notch
Pectineal line
Ischial tuberosity
Ischial ramus
Location of pubic symphysis

a Right hip bone, lateral views

b Right hip bone, medial views

connects the lateral borders of the sacrum with the iliac crest, the ischial tuberosity, the ischial spine, and the arcuate line. Other ligaments tie the ilia to the posterior lumbar vertebrae. These interconnections increase the stability of the pelvis.

The pelvic cavity is the space enclosed by the bones of the pelvis. ⮌ p. 18 The pelvis may be divided into the *true (lesser) pelvis* and the *false (greater) pelvis* that is separated by the pelvic brim (Figure 8–9a,b). The **true pelvis** is inferior to the pelvic brim. The superior limit of the true pelvis is a line that extends from either side of the base of the sacrum, along the arcuate line and pectineal line to the pubic symphysis. The bony edge of the true pelvis is called the **pelvic brim**, or *linea terminalis*. The enclosed space is the **pelvic inlet**. The **false pelvis** is the upper portion of the pelvic cavity surrounded by the broad, bladelike portions of each ilium superior to the pelvic brim (see Figure 8–9a).

The **pelvic outlet** is the opening bounded by the coccyx, the ischial tuberosities, and the inferior border of the pubic symphysis (Figure 8–9b,c). The surface region bordered by the inferior edges of the pelvis is called the *perineum* (per-i-NĒ-um). Perineal muscles form the floor of the pelvic cavity and support the organs in the true pelvis.

The shape of the pelvis of a female is somewhat different from that of a male. Some of the differences are the result of variations in body size and muscle mass. For example, in females, the pelvis is generally smoother and lighter and has less-prominent markings. Look ahead to **Spotlight Figure 8–14** for a comparison of some major overall differences between adult male and female pelves.

> Go to MasteringA&P™ > Study Area > Menu > Lab Tools > Bone & Dissection Videos > Bone Videos > Hip Bone

Figure 8–8 The Pelvis of an Adult Male. (Look back at Figure 7–22, p. 235, for a detailed view of the sacrum and coccyx.)

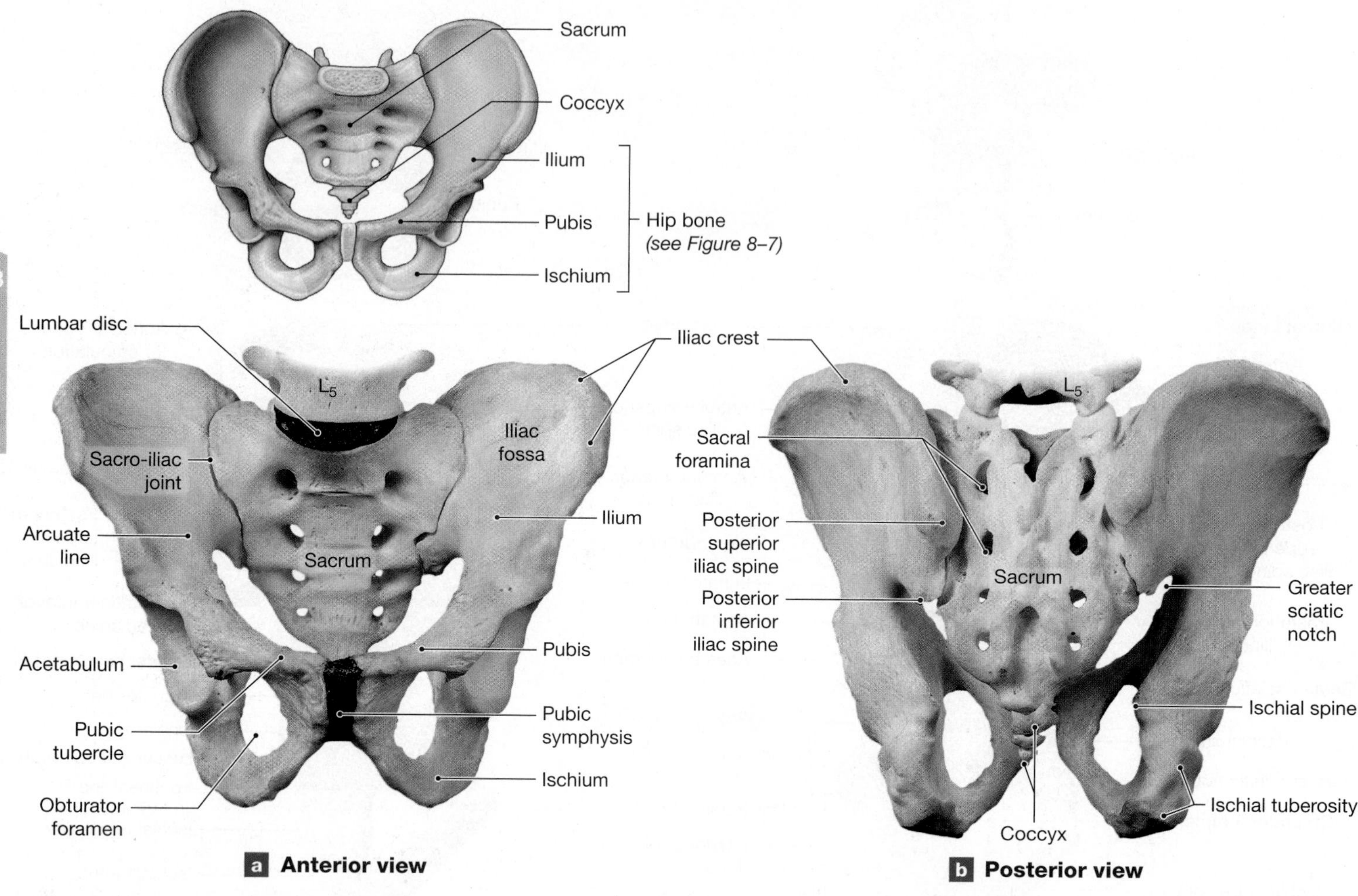
Sacrum
Coccyx
Ilium
Pubis
Ischium
Hip bone
(see Figure 8–7)
Lumbar disc
L5
Iliac crest
Iliac fossa
Sacro-iliac joint
Arcuate line
Sacrum
Ilium
Acetabulum
Pubis
Pubic tubercle
Pubic symphysis
Ischium
Obturator foramen
a Anterior view
Sacral foramina
Posterior superior iliac spine
Posterior inferior iliac spine
Greater sciatic notch
Ischial spine
Ischial tuberosity
Coccyx
b Posterior view

? What is the correct term for the opening formed by the pubis and the ischium?

Checkpoint

8. **Name the bones of the pelvic girdle.**
9. **Which three bones make up a hip bone?**
10. **When you are seated, which part of the pelvis supports your body's weight?**

See the blue Answers tab at the back of the book.

8-4 The bones of the lower limbs are adapted for movement and support

Learning Outcome **Identify the bones of the lower limbs, their functions, and their surface features.**

The skeleton of each lower limb consists of a *femur* (thigh), a *patella* (kneecap), a *tibia* and a *fibula* (leg), and the *tarsal bones, metatarsals,* and *phalanges* of the foot. Once again, anatomical terminology differs from common usage. Notice that in anatomical terms, *leg* refers only to the distal portion of the limb, not to the entire lower limb. We will use *thigh* and *leg,* rather than *upper leg* and *lower leg.*

The functional anatomy of the lower limbs differs from that of the upper limbs, primarily because the lower limbs transfer the body weight to the ground. For this reason, the bones of the lower limbs are more massive than those of the upper limbs. We now examine the bones of the right lower limb.

The Femur (Thighbone)

The **femur** is the longest and heaviest bone in the body. It is made up of three major parts (from proximal to distal): the *femoral head,* the *shaft,* and the *condyles* (Figure 8–10a,b). It articulates with the hip bone at the hip joint and with the tibia of the leg at the knee joint. The rounded epiphysis, or **head**

Figure 8–9 Divisions of the Pelvis.

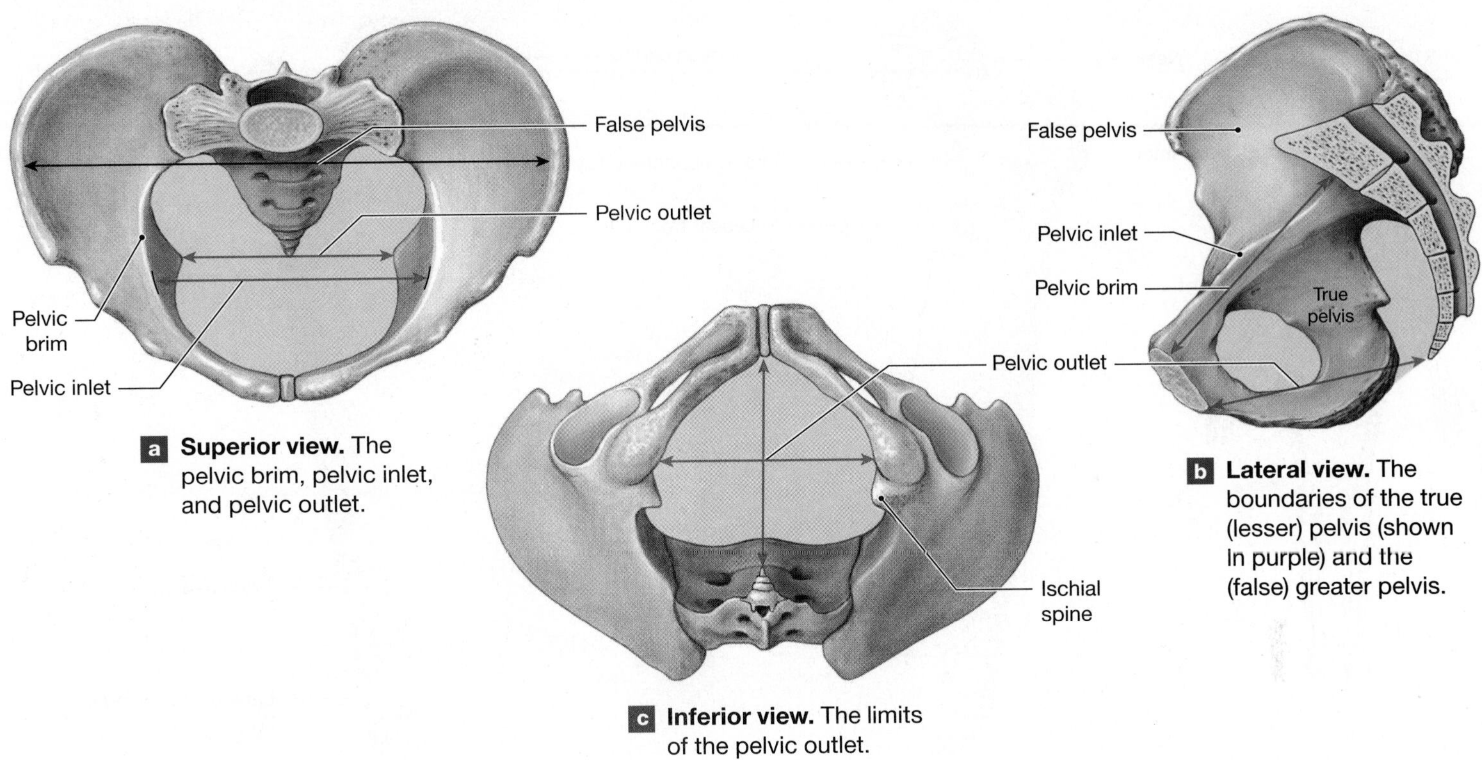
False pelvis
Pelvic outlet
Pelvic brim
Pelvic inlet
False pelvis
Pelvic inlet
Pelvic brim
True pelvis
Pelvic outlet
Ischial spine

a Superior view. The pelvic brim, pelvic inlet, and pelvic outlet.

b Lateral view. The boundaries of the true (lesser) pelvis (shown in purple) and the (false) greater pelvis.

c Inferior view. The limits of the pelvic outlet.

of the femur, articulates with the hip bone at the acetabulum. A ligament attaches the acetabulum to the femur at the **fovea capitis**, a small pit in the center of the femoral head. The **neck** of the femur joins the **shaft** at an angle of about 125°.

The **greater** and **lesser trochanters** are large, rough projections that originate at the junction of the neck and shaft. The greater trochanter projects laterally; the lesser trochanter projects posteriorly and medially. These trochanters develop where large tendons attach to the femur. On the anterior surface of the femur, the raised **intertrochanteric** (in-ter-trō-kan-TER-ik) **line** marks the edge of the articular capsule (see Figure 8–10a). This line continues around to the posterior surface as the **intertrochanteric crest**.

The **linea aspera** is a rough ridge that runs along the center of the posterior surface of the femoral shaft, marking the attachment site of powerful hip muscles (see Figure 8–10b). As it nears the knee joint, the linea aspera divides into a pair of ridges (the *medial* and *lateral supracondylar ridges*) that continue to the projections of the **medial** and **lateral epicondyles** (see Figure 8–10a,b).

Inferior to these epicondyles are the smoothly rounded **medial** and **lateral condyles**, which are part of the knee joint. The two condyles are separated by a deep **intercondylar fossa**. The medial and lateral condyles extend across the inferior surface of the femur, but the intercondylar fossa does not reach the anterior surface. The anterior and inferior surfaces of the two condyles are separated by the **patellar surface**, a smooth articular surface over which the patella glides. The adductor tubercle is a surface marking for muscle attachment.

+ Clinical Note Hip Fracture

A *hip fracture*, or a "broken hip," actually involves the femur, not a hip bone. The two types of hip fractures are femoral neck fractures and intertrochanteric fractures. In a femoral neck fracture, the head of the femur separates from the femur (and its blood supply). There is no disconnection to the blood supply in an intertrochanteric fracture. As a result, such fractures may be repaired with a metal plate and screws.

The Patella (Kneecap)

The **patella**, or kneecap, is a large sesamoid bone that forms within the tendon of the *quadriceps femoris*, a group of muscles that extends (straightens) the knee. The patella has a rough, convex anterior surface and a broad **base** (Figure 8–11a). The roughened surface reflects the attachment of the *quadriceps tendon* (anterior and superior surfaces) and the *patellar ligament* (anterior and inferior surfaces). The patellar ligament connects the **apex** of the patella to the tibia.

> Go to MasteringA&P™ > Study Area > Menu > Lab Tools > Bone & Dissection Videos > Bone Videos > Femur

Figure 8–10 Bone Markings on the Right Femur. ATLAS: Plates 32; 75a–d; 77

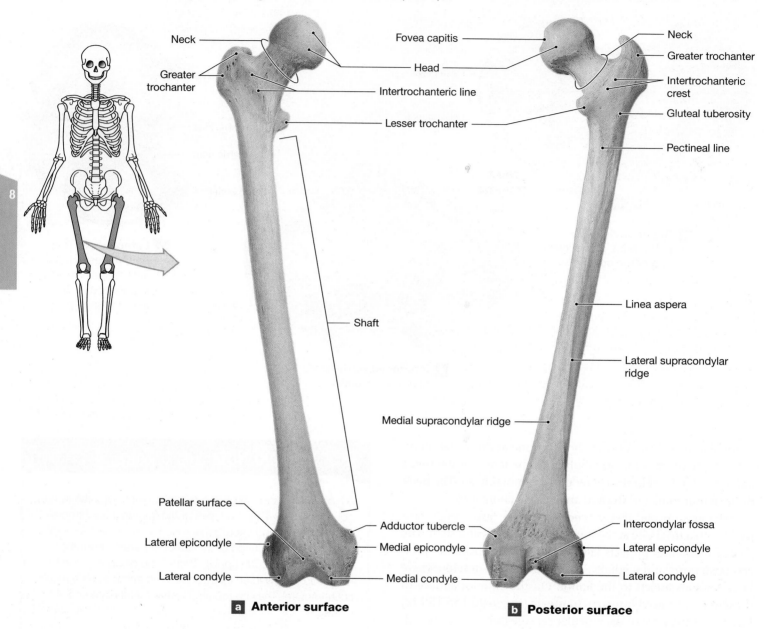

Neck
Greater trochanter
Fovea capitis
Head
Intertrochanteric line
Lesser trochanter
Shaft
Patellar surface
Lateral epicondyle
Lateral condyle
Adductor tubercle
Medial epicondyle
Medial condyle

a Anterior surface

Neck
Greater trochanter
Intertrochanteric crest
Gluteal tuberosity
Pectineal line
Linea aspera
Lateral supracondylar ridge
Medial supracondylar ridge
Intercondylar fossa
Lateral epicondyle
Lateral condyle

b Posterior surface

Figure 8–11 The Right Patella and Patella with Femur.

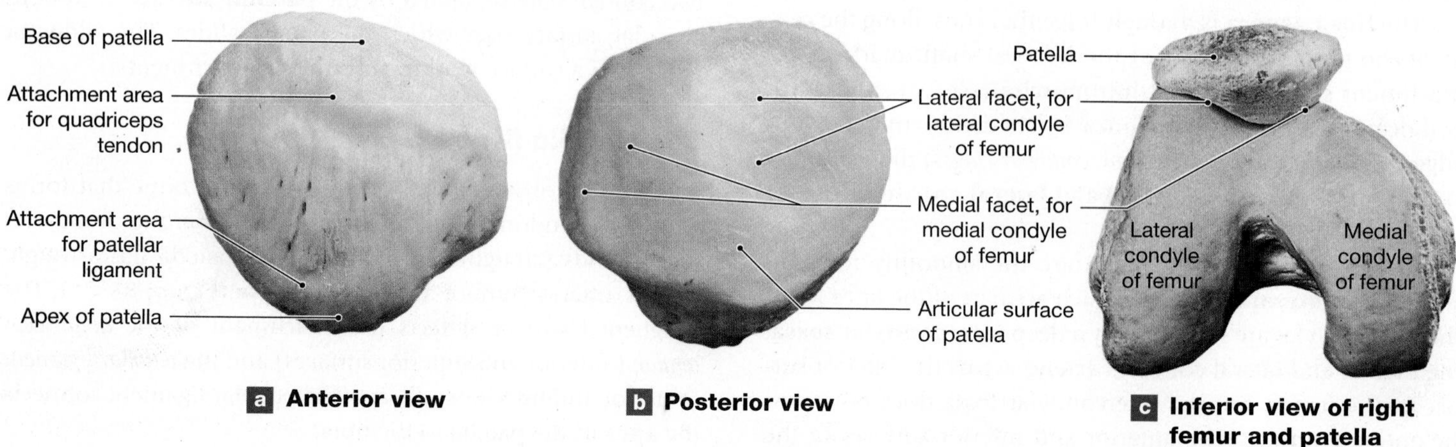
Base of patella
Attachment area for quadriceps tendon
Attachment area for patellar ligament
Apex of patella
Lateral facet, for lateral condyle of femur
Medial facet, for medial condyle of femur
Articular surface of patella
Patella
Lateral condyle of femur
Medial condyle of femur

a Anterior view **b Posterior view** **c Inferior view of right femur and patella**

+ Clinical Note Shin Splints

What do a runner, dancer, and military recruit have in common? The possibility of a painful stress injury of the tibia called *shin splints*. Muscles and tendons of the lower leg pull on bone when workout intensity increases, and inflammation is often the result. The tender, mildly swollen shin responds well to rest, ice, and over-the-counter pain relievers. Proper stretching and new shoes also help, as does cross training to vary the impact on the leg. Too much too soon is too hard on the tibia!

The posterior patellar surface (Figure 8–11b) has two concave facets (called the medial and lateral facets). These facets articulate with the medial and lateral condyles of the femur (Figure 8–11c). Normally, the patella glides across the patellar surface of the femur. Its direction of movement is superior–inferior (up and down), not medial–lateral (side to side).

The patellae are cartilaginous at birth. They start to ossify after the person begins walking, as thigh and leg movements become more powerful. Ossification usually begins at age 2 or 3 and ends around the time of puberty.

Bones of the Leg

The *tibia* (TIB-ē-uh) and *fibula* (FIB-ū-luh) together support the leg. The slender fibula parallels the lateral border of the tibia, with the tibia medial to the fibula in anatomical position. The medial border of the thin fibula is bound to the tibia by the **interosseous membrane**. This membrane helps stabilize the positions of the two bones. It also provides additional surface area for muscle attachment. A cross section of the tibia and fibula is shown in Figure 8–12.

The Tibia

The **tibia**, or shinbone, is the large medial bone of the leg; its main parts (from proximal to distal) are the *condyles*, the *shaft*, and a process called the *medial malleolus* (Figure 8–12a). The medial and lateral condyles of the femur articulate with the **medial** and **lateral tibial condyles** at the proximal end of the tibia. The **intercondylar eminence** is a ridge that separates the condyles (Figure 8–12b). The anterior surface of the tibia near the condyles has a prominent, rough **tibial tuberosity** (see Figure 8–12a), which you can feel through the skin. This tuberosity marks the attachment of the patellar ligament.

The **anterior margin** is a ridge that begins at the tibial tuberosity and extends distally along the anterior tibial surface. You can also easily feel the anterior margin of the tibia through the skin. As it nears the ankle joint, the tibia broadens, and the medial border ends in the **medial malleolus** (ma-LĒ-ō-lus; *malleolus*, hammer), a large process familiar to you as the medial bump at the ankle. The inferior surface of the tibia articulates with the proximal bone of the ankle. The medial malleolus supports this joint medially.

The Fibula

From proximal to distal, the **fibula** is made up of a *head, shaft*, and *lateral malleolus* process (see Figure 8–12a,b). The **head** of the fibula articulates with the tibia. The articular facet is located on the anterior, inferior surface of the lateral tibial condyle.

As its small diameter suggests, the fibula does not help transfer weight to the ankle and foot. In fact, it does not even articulate with the femur. However, the fibula is an important site for the attachment of muscles that move the foot and toes. In addition, the distal tip of the fibula extends laterally to the ankle joint. This fibular process, the **lateral malleolus**, gives lateral stability to the ankle. However, forceful movement of the foot outward and backward can dislocate the ankle, breaking both the lateral malleolus of the fibula and the medial malleolus of the tibia. This injury is called a *Pott's fracture*. ⮌ p. 203

Tips & Tools

To remember how to distinguish the fibula from the tibia, consider the **fib** and **l** parts of the word *fibula*. To tell a small **l**ie is to **fib**. The **fib**ula is sma**ll**er than the tibia, and is **l**ateral to it.

Bones of the Ankle and Foot

The ankle, or *tarsus*, consists of seven *tarsal bones* (Figure 8–13). The *metatarsals* are five long bones that form the distal portion of the foot, or *metatarsus*. The toes of the foot contain 14 *phalanges*, or toe bones.

The Tarsal Bones

The **tarsal bones** make up the ankle. The large **talus** transfers the weight of the body from the tibia toward the toes. The articulation between the talus and the tibia occurs across the superior and medial surfaces of the **trochlea**, a spool- or pulley-shaped articular process. The lateral surface of the trochlea articulates with the lateral malleolus of the fibula.

Tips & Tools

To remember where the talus is located, think that the **t**alus is on **t**op of the foot and articulates with the **t**ibia.

The **calcaneus** (kal-KĀ-nē-us), or heel bone, is the largest of the tarsal bones (Figure 8–13a,b). When you stand, most of your weight is transferred from the tibia through the talus to the calcaneus, and then to the ground. The posterior portion of the calcaneus is a rough, knob-shaped projection. This is the attachment site for the *calcaneal tendon (Achilles tendon)*, which

Figure 8–12 **The Right Tibia and Fibula.** ATLAS: Plates 32; 80a,b; 83a,b

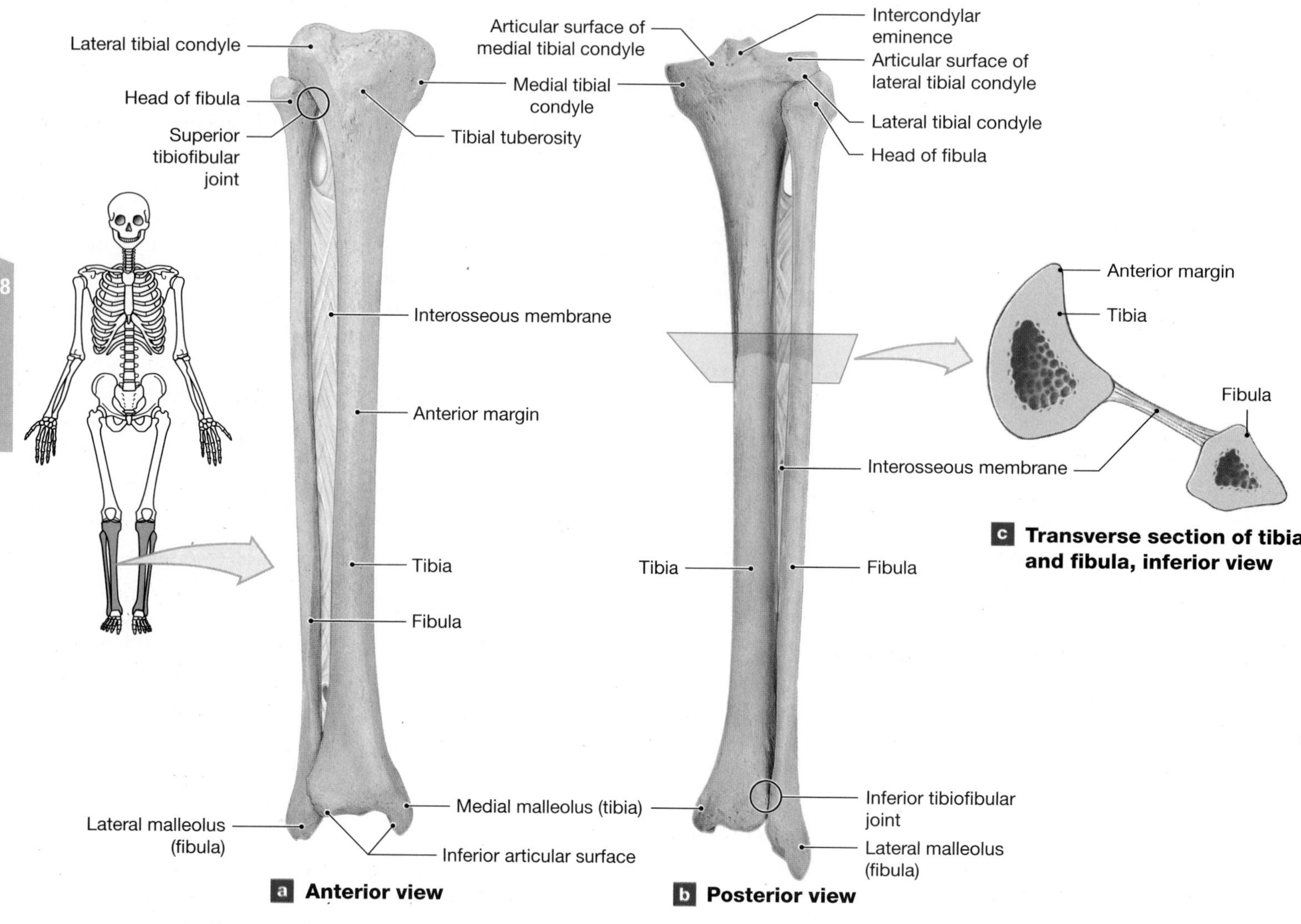

Lateral tibial condyle
Head of fibula
Superior tibiofibular joint
Articular surface of medial tibial condyle
Medial tibial condyle
Tibial tuberosity
Interosseous membrane
Anterior margin
Tibia
Fibula
Lateral malleolus (fibula)
Medial malleolus (tibia)
Inferior articular surface
a Anterior view
Intercondylar eminence
Articular surface of lateral tibial condyle
Lateral tibial condyle
Head of fibula
Interosseous membrane
Tibia
Fibula
Inferior tibiofibular joint
Lateral malleolus (fibula)
b Posterior view
Anterior margin
Tibia
Fibula
Interosseous membrane
c Transverse section of tibia and fibula, inferior view

? Which of the leg bones does not articulate with the femur?

extends from the calf muscles. When you are standing, these strong muscles can lift your heels off the ground so that you stand on tiptoes. The superior and anterior surfaces of the calcaneus have smooth facets for articulation with other tarsal bones.

The **cuboid** articulates with the anterior surface of the calcaneus. The **navicular** is anterior to the talus, on the medial side of the ankle. It articulates with the talus and with the three *cuneiform* (kū-NĒ-i-form) *bones.* These wedge-shaped bones are arranged in a row, with articulations between them. They are named according to their position: **medial cuneiform, intermediate cuneiform**, and **lateral cuneiform**. Proximally, the cuneiform bones articulate with the anterior surface of the navicular. The lateral cuneiform also articulates with the medial surface of the cuboid. The distal surfaces of the cuboid and the cuneiform bones articulate with the metatarsals of the foot.

Tips & Tools

To remember the names of the tarsal bones in the order presented in this text, try the memory aid "Tom Can Control Not Much In Life."

The Metatarsals and Phalanges

The five **metatarsals** in the foot are identified by Roman numerals I–V, proceeding from medial to lateral across the sole (see Figure 8–13a,b). Proximally, metatarsals I–III articulate with the three cuneiform bones, and metatarsals IV and V articulate with the cuboid. Distally, each metatarsal articulates with a different proximal phalanx.

Figure 8–13 Bones of the Ankle and Foot. ATLAS: Plates 32; 85a; 86a,c; 87a–c; 88

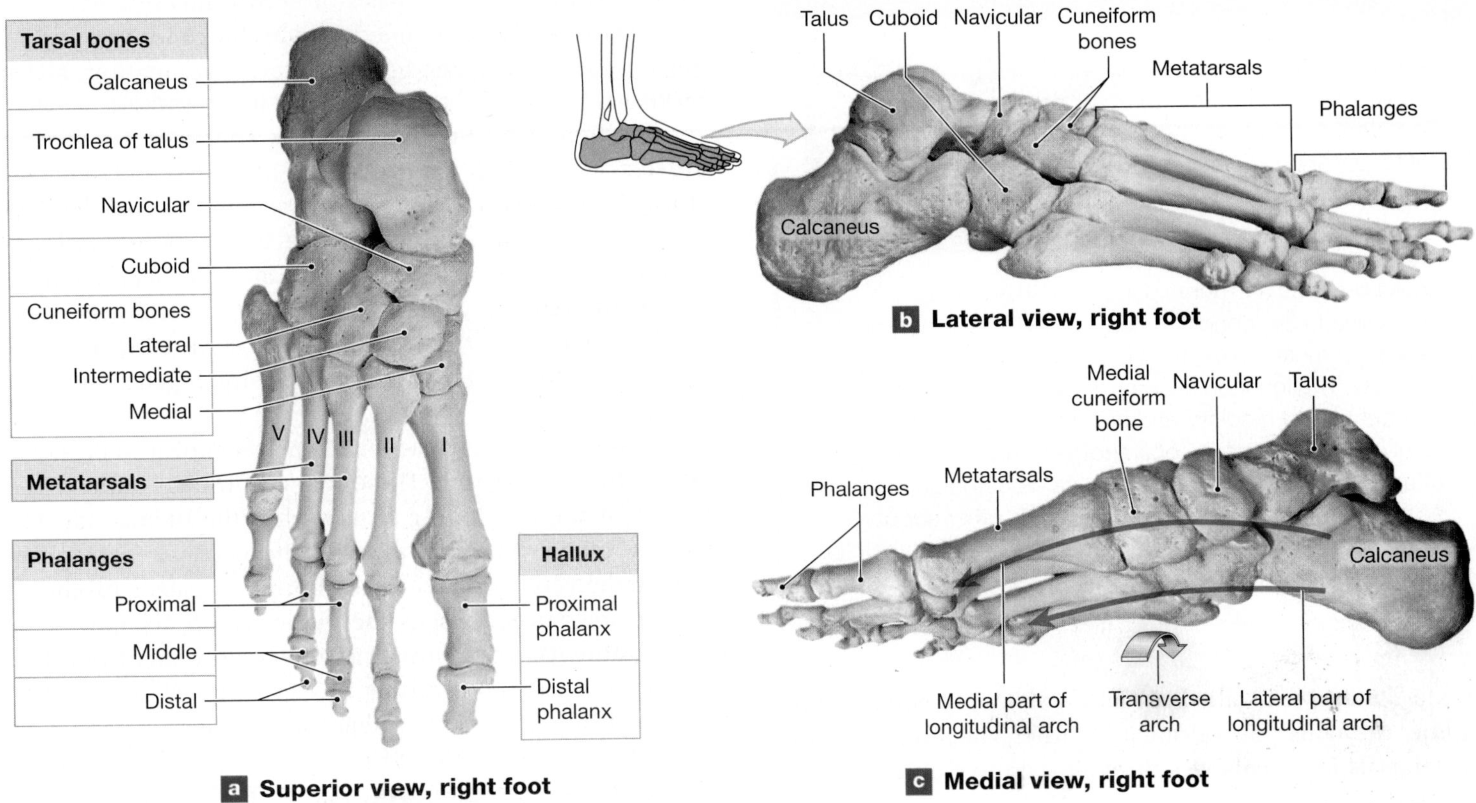

Tarsal bones
Calcaneus
Trochlea of talus
Navicular
Cuboid
Cuneiform bones
Lateral
Intermediate
Medial
V IV III II I
Metatarsals
Phalanges
Proximal
Middle
Distal
Hallux
Proximal phalanx
Distal phalanx
Talus
Cuboid
Navicular
Cuneiform bones
Metatarsals
Phalanges
Calcaneus
Medial cuneiform bone
Navicular
Talus
Metatarsals
Phalanges
Calcaneus
Medial part of longitudinal arch
Transverse arch
Lateral part of longitudinal arch

a Superior view, right foot

b Lateral view, right foot

c Medial view, right foot

The **phalanges** of the toes have the same anatomical organization as do those of the fingers (see Figure 8–13a,b). The **hallux**, or great toe, has two phalanges (*proximal* and *distal*). The other four toes have three phalanges apiece (*proximal, middle,* and *distal*).

Arches of the Foot

When you are standing or walking, the bones of your feet do not rest flat on the ground. Instead, weight transfer occurs along arches in the foot (Figure 8–13c) that are maintained by ligaments and tendons. There are two main types of arches, *longitudinal* and *transverse*.

The main **longitudinal arch** ties the calcaneus to the distal portions of the metatarsals. As a result, the medial plantar surface of the foot remains elevated, so that the muscles, nerves, and blood vessels that supply the inferior surface are not squeezed between the metatarsals and the ground. Note that the *lateral part* of the longitudinal arch has much less curvature than the *medial part*. Part of the reason is that the medial part has more elasticity. This elasticity absorbs the shocks from sudden changes in weight loading, for example, from the stresses that running or ballet dancing place on the toes. The **transverse arch** describes the degree of curvature change from one side of the foot to the other. In the condition known as *flatfeet*, normal arches are lost ("fall") or never form.

When you stand normally, your body weight is distributed evenly between the calcaneus and the distal ends of the metatarsals. The amount of weight transferred forward depends on the position of the foot and the placement of your body weight. During flexion at the ankle, a movement called *dorsiflexion*, all your body weight rests on the calcaneus—as when you "dig in your heels." During extension at the

+ Clinical Note Stress Fractures

Running is beneficial to overall health but places the foot bones under more stress than does walking. *Stress fractures* are hairline fractures that develop in bones subjected to repeated shocks or impacts. Stress fractures of the foot usually involve one of the metatarsals. These fractures are caused either by improper placement of the foot while running or by poor arch support. In a fitness regime that includes street running, proper support for the bones of the foot is essential. An entire running-shoe market has arisen around the amateur and professional runner's need for good arch support.

8

Clinical Note Club Foot

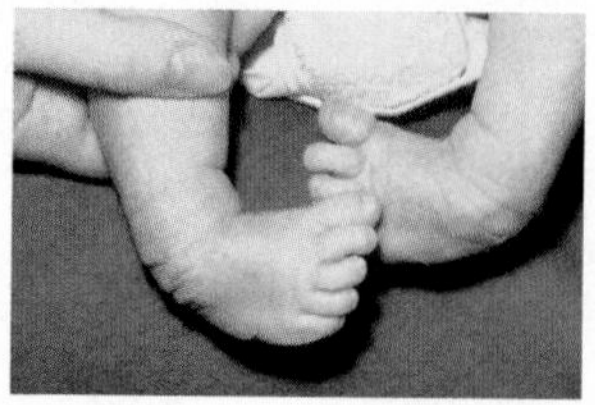

The arches of the foot are usually present at birth. Sometimes, however, they fail to develop properly. In **club foot** (*congenital talipes equinovarus*), abnormal muscle development distorts growing bones and joints. One or both feet may be involved. The condition can be mild, moderate, or severe. In most cases, the tibia, ankle, and foot are affected. The longitudinal arch is exaggerated, and the feet are turned medially and inverted. If both feet are involved, the soles face one another. This condition affects 1 in 1000 births and is twice as common in boys as girls. Prompt treatment with casts or other supports in infancy helps alleviate the problem. Fewer than half the cases require surgery.

ankle, a movement called *plantar flexion*, when you stand on tiptoe, the talus and calcaneus transfer your weight to the metatarsals and phalanges through the more anterior tarsal bones.

Checkpoint

11. **Identify the bones of the lower limb.**
12. **The fibula is not part of the knee joint and does not bear weight. However, when it is fractured, walking is difficult. Why?**
13. **While jumping off the back steps at his house, 10-year-old Joey lands on his right heel and breaks his foot. Which foot bone is most likely broken?**
14. **Which foot bone transfers the weight of the body from the tibia toward the toes?**

See the blue Answers tab at the back of the book.

8-5 Differences in sex and age account for individual skeletal variation

Learning Outcome **Summarize sex differences and age-related changes in the human skeleton.**

A comprehensive study of a human skeleton can reveal important information about a person. We can estimate a person's muscular development and muscle mass from the appearance of various ridges and from the general bone mass. Details such as the condition of the teeth or the presence of healed fractures give an indication of the person's medical history. Two important facts, sex and age, can be determined or closely estimated on the basis of known measurements. In some cases, the skeleton may provide clues about a person's nutritional state, handedness, and even occupation. ATLAS: Embryology Summary 8: The Development of the Appendicular Skeleton

Spotlight Figure 8–14 identifies characteristic relative differences between the skeletons of males and females. However, not every skeleton shows every feature detailed in this figure. Many differences, including bone markings on the skull, cranial capacity, and general skeletal features, reflect differences in average body size, muscle mass, and muscular strength. Note that differences in cranial capacity do not indicate any differences in intelligence between the sexes.

Some variations in the female pelvis appear to be adaptations for childbearing. These adaptations help support the weight of the developing fetus within the uterus, and the passage of the newborn through the pelvic outlet during vaginal delivery. In addition, the hormone *relaxin*, produced during pregnancy, loosens the pubic symphysis and sacroiliac ligaments. This loosening allows movement between the hip bones that can further increase the size of the pelvic inlet and outlet. In cases where the fetus is too large to fit through the pelvic outlet, delivery is accomplished by cesarean section or C-section. A C-section is a surgical incision made through the abdominal wall and uterus to deliver the newborn.

The general changes in the skeletal system that take place with age are summarized in **Table 8–1**, p. 262. Note that these changes begin at age 3 months and continue throughout life. The epiphyseal cartilages, for example, begin to fuse at about age 3. The timing of epiphyseal closure is a key factor determining adult body size. Young people whose long bones are still growing should avoid heavy weight training, because they risk crushing the epiphyseal cartilages and thus shortening their stature (look back at **Spotlight Figure 6–11**, p. 193). Degenerative changes in the normal skeletal system, such as a reduction in mineral content in the bony matrix, typically do not begin until age 30–45.

Checkpoint

15. **Compare and contrast the bones of males and females with respect to weight and bone markings.**
16. **An anthropologist discovered several bones in a deep grave. After close visual inspection and careful measurements, what sort of information could the bones reveal?**
17. **How is the female pelvis adapted for childbearing?**

See the blue Answers tab at the back of the book.

Go to MasteringA&P™ > Study Area > Menu > Lab Tools > PAL 3.0 > Human Cadaver > Appendicular Skeleton

SPOTLIGHT

Figure 8–14 Sex Differences in the Human Skeleton

The sex of an adult human skeleton can be determined by many of the details seen in the bones. The skull and pelvis are particularly helpful. Not every skeleton shows every feature clearly, but this chart contrasts the basic differences between the sexes.

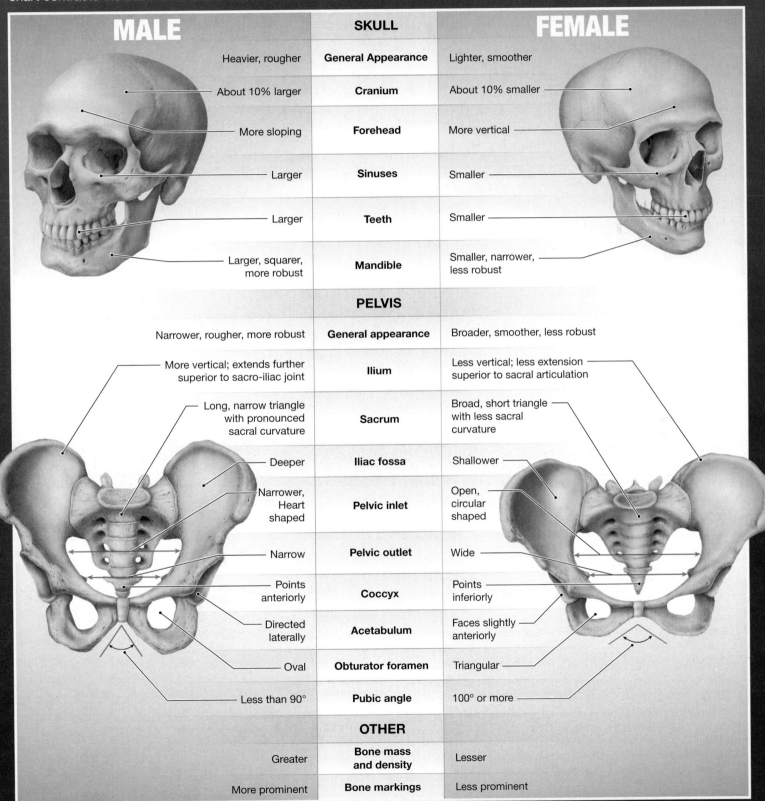

MALE		FEMALE
	SKULL	
Heavier, rougher	**General Appearance**	Lighter, smoother
About 10% larger	**Cranium**	About 10% smaller
More sloping	**Forehead**	More vertical
Larger	**Sinuses**	Smaller
Larger	**Teeth**	Smaller
Larger, squarer, more robust	**Mandible**	Smaller, narrower, less robust
	PELVIS	
Narrower, rougher, more robust	**General appearance**	Broader, smoother, less robust
More vertical; extends further superior to sacro-iliac joint	**Ilium**	Less vertical; less extension superior to sacral articulation
Long, narrow triangle with pronounced sacral curvature	**Sacrum**	Broad, short triangle with less sacral curvature
Deeper	**Iliac fossa**	Shallower
Narrower, Heart shaped	**Pelvic inlet**	Open, circular shaped
Narrow	**Pelvic outlet**	Wide
Points anteriorly	**Coccyx**	Points inferiorly
Directed laterally	**Acetabulum**	Faces slightly anteriorly
Oval	**Obturator foramen**	Triangular
Less than 90°	**Pubic angle**	100° or more
	OTHER	
Greater	**Bone mass and density**	Lesser
More prominent	**Bone markings**	Less prominent

Table 8–1 Age-Related Changes in the Skeleton

Region and Feature	Events	Age in Years
GENERAL SKELETON		
Bony matrix	Reduction in mineral content; increased risk of osteoporosis	Begins at ages 30–45; values differ for males versus females between ages 45 and 65; similar reductions occur in both sexes after age 65
Bone markings	Reduction in size, roughness	Gradual reduction with increasing age and decreasing muscular strength and mass
SKULL		
Fontanelles	Closure	Completed by age 2
Frontal suture	Fusion	2–8
Occipital bone	Fusion of ossification centers	1–4
Styloid process	Fusion with temporal bone	12–16
Hyoid bone	Complete ossification and fusion	25–30
Teeth	Loss of "baby teeth"; appearance of secondary dentition; eruption of permanent molars	Detailed in Chapter 24 (digestive system)
Mandible	Loss of teeth; reduction in bone mass; change in angle at mandibular notch	Accelerates in later years (60+)
VERTEBRAE		
Curvature	Development of major curves	3 months–10 years
Intervertebral discs	Reduction in size, percentage contribution to height	Accelerates in later years (60+)
LONG BONES		
Epiphyseal cartilages	Fusion	Begins about age 3; ranges vary, but general analysis permits determination of approximate age
PECTORAL AND PELVIC GIRDLES		
Epiphyses	Fusion	Relatively narrow ranges of ages (e.g., 14–16, 16–18, 22–25) increase accuracy of age estimates

8 Chapter Review

Study Outline

An Introduction to the Appendicular Skeleton p. 244

1. The **appendicular skeleton** includes the bones of the upper and lower limbs and the pectoral and pelvic girdles, which connect the limbs to the trunk. *(Figure 8–1)*

8-1 The pectoral (shoulder) girdles attach the upper limbs to the axial skeleton p. 244

2. Each upper limb articulates with the trunk through a **pectoral girdle**, or *shoulder girdle*, which consists of a **scapula** and a **clavicle**.
3. On each side, a clavicle and scapula position the shoulder joint, help move the upper limb, and provide a base for muscle attachment. *(Figures 8–2, 8–3)*
4. Both the **coracoid process** and the **acromion** of the scapula have attached ligaments and tendons associated with the shoulder joint. *(Figure 8–3)*

> MasteringA&P™ Access more chapter study tools online in the MasteringA&P Study Area:
> - Chapter Quizzes, Chapter Practice Test, MP3 Tutor Sessions, and Clinical Case Studies
> - Practice Anatomy Lab PAL 3.0
> - A&P Flix A&PFlix
> - Interactive Physiology iP2
> - PhysioEx PhysioEx 9.1

8-2 The bones of the upper limbs are adapted for free movement p. 247

5. The scapula articulates with the **humerus** at the shoulder (*glenohumeral*) joint. The **greater** and **lesser tubercles** of the humerus are important sites of muscle attachment. *(Figure 8–4)*
6. The humerus articulates with the **radius** and **ulna**, the bones of the forearm, at the elbow joint. *(Figure 8–5)*

7. The **carpal bones** of the wrist form two rows. The distal row articulates with the five **metacarpals**. Four of the fingers contain three **phalanges**; the **pollex** (thumb) has only two phalanges. *(Figure 8–6)*

8-3 The pelvic girdle (hips) attaches the lower limbs to the axial skeleton p. 252

8. The bones of the **pelvic girdle** are more massive than those of the pectoral girdle because of their role in weight bearing and locomotion.
9. The pelvic girdle consists of two **hip bones** (**coxal bones** or **pelvic bones**). Each hip bone forms through the fusion of three bones: the **ilium**, **ischium**, and **pubis**. *(Figure 8–7)*
10. The ilium is the largest component of the hip bone. Inside the **acetabulum**, the ilium is fused to the ischium (posteriorly) and the pubis (anteriorly). The *pubic symphysis* limits movement between the pubic bones of the left and right hip bones. *(Figures 8–7, 8–8)*
11. The **pelvis** consists of the hip bones, the sacrum, and the coccyx. It is subdivided into the **false** *(greater)* **pelvis** and the **true** *(lesser)* **pelvis.** *(Figures 8–8, 8–9)*

8-4 The bones of the lower limbs are adapted for movement and support p. 254

12. The **femur** is the longest and heaviest bone in the body. It articulates with the **tibia** at the knee joint. *(Figures 8–10, 8–12)*
13. The **patella** is a large sesamoid bone. *(Figure 8–11)*
14. The **fibula** parallels the tibia laterally. *(Figure 8–12)*
15. The *tarsus*, or ankle, has seven **tarsal bones**. *(Figure 8–13)*
16. The basic organizational pattern of the **metatarsals** and **phalanges** of the foot resembles that of the hand. All the toes have three phalanges, except for the **hallux** (great toe), which has two. *(Figure 8–13)*
17. When a person stands, most of the body weight is transferred to the **calcaneus**, and the rest is passed on to the five metatarsals through the **talus**. Weight transfer occurs along the **longitudinal arch**. There is also a **transverse arch** that runs from one side of the foot to the other. *(Figure 8–13)*

8-5 Differences in sex and age account for individual skeletal variation p. 260

18. Studying a human skeleton can reveal important information, such as the person's weight, sex, body size, muscle mass, and age. *(Spotlight Figure 8–14; Table 8–1)*
19. Age-related changes take place in the skeletal system. These changes begin at about age 3 months and continue throughout life. *(Table 8–1)*

Review Questions

See the blue Answers tab at the back of the book.

LEVEL 1 Reviewing Facts and Terms

1. In the following photographs of the scapula, identify the three views (a–c) and the indicated bone markings (d–g).

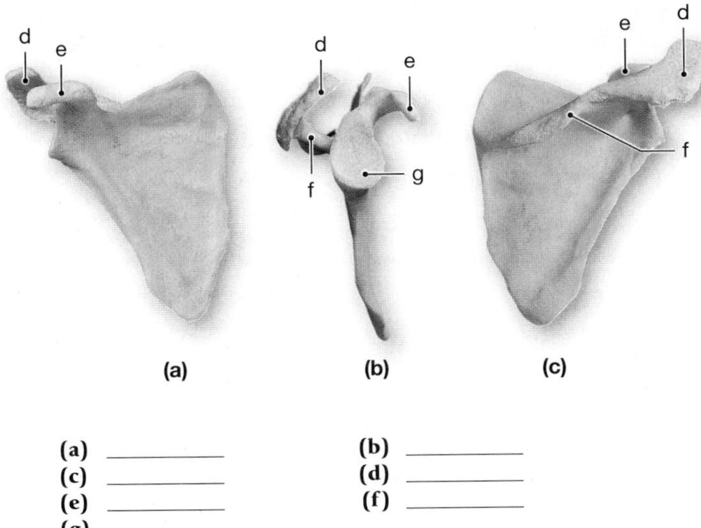

d
e
d
e
e
d
f
g
f
(a)
(b)
(c)

(a) ________ (b) ________
(c) ________ (d) ________
(e) ________ (f) ________
(g) ________

2. Which of the following is primarily responsible for stabilizing, positioning, and bracing the pectoral girdles? **(a)** tendons, **(b)** ligaments, **(c)** the joint shape, **(d)** muscles, **(e)** the shape of the bones within the joint.
3. In the following drawing of the pelvis, label the structures indicated.

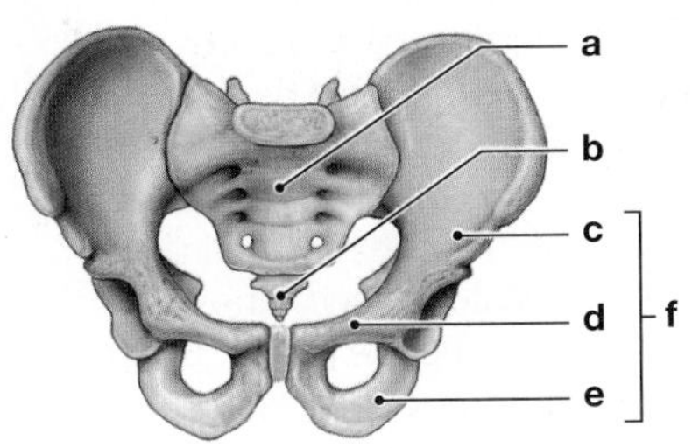

a
b
c
d
e
f

(a) ________ (b) ________
(c) ________ (d) ________
(e) ________ (f) ________

4. In anatomical position, the ulna lies **(a)** medial to the radius, **(b)** lateral to the radius, **(c)** inferior to the radius, **(d)** superior to the radius.
5. The point of the elbow is actually the ________ of the ulna. **(a)** styloid process, **(b)** olecranon, **(c)** coronoid process, **(d)** trochlear notch.
6. The bones of the hand articulate distally with the **(a)** carpal bones, **(b)** ulna and radius, **(c)** metacarpals, **(d)** phalanges.
7. The epiphysis of the femur articulates with the pelvis at the **(a)** pubic symphysis, **(b)** acetabulum, **(c)** sciatic notch, **(d)** obturator foramen.
8. What is the name of the flexible sheet that interconnects the radius and ulna (and the tibia and fibula)?
9. Name the bones that make up each hip bone.
10. Which seven bones make up the ankle (tarsus)?

LEVEL 2 Reviewing Concepts

11. The presence of tubercles on bones indicates the positions of **(a)** tendons and ligaments, **(b)** muscle attachments, **(c)** ridges and flanges, **(d)** a and b.
12. At the glenoid cavity, the scapula articulates with the proximal end of the **(a)** humerus, **(b)** radius, **(c)** ulna, **(d)** femur.
13. All of the following are structural characteristics of the female pelvic girdle compared to the male pelvic girdle with either one or two exceptions. Identify the exception(s). **(a)** The female pelvis is adapted for childbearing. **(b)** The female pelvic girdle is lighter than the male pelvic girdle. **(c)** Relaxin produced by the male pelvic girdle loosens the pubic symphysis and sacro-iliac joints. **(d)** The shape of the female pelvic girdle is the same as that of the male. **(e)** Answers c and d are exceptions.
14. The large foramen between the pubic ramus and ischial ramus is the **(a)** foramen magnum, **(b)** suborbital foramen, **(c)** acetabulum, **(d)** obturator foramen.
15. Which of the following skeletal characteristics is an adaption for childbearing? **(a)** inferior angle of 100° or more between the pubic bones, **(b)** a relatively broad, low pelvis, **(c)** less curvature of the sacrum and coccyx, **(d)** All of these are correct.
16. The fibula **(a)** forms an important part of the knee joint, **(b)** articulates with the femur, **(c)** helps to bear the weight of the body, **(d)** provides lateral stability to the ankle, **(e)** does a and b.
17. The tarsal bone that transfers and distributes weight to the heel or toes is the **(a)** cuneiform, **(b)** calcaneus, **(c)** talus, **(d)** navicular.
18. What is the difference in skeletal structure between the pelvic girdle and the pelvis?
19. Why would a self-defense instructor advise a student to strike an attacker's clavicle?
20. Jack injures himself playing hockey, and the physician who examines him informs him that he has dislocated his pollex. What part of Jack's body did he injure? **(a)** his arm, **(b)** his leg, **(c)** his hip, **(d)** his thumb, **(e)** his shoulder.
21. The pelvis **(a)** protects the upper abdominal organs, **(b)** contains bones from both the axial and appendicular skeletons, **(c)** is composed of the coxal bones, sacrum, and coccyx, **(d)** does all of these, **(e)** does b and c.
22. Why is the tibia, but not the fibula, involved in the transfer of weight to the ankle and foot?
23. In determining the age of a skeleton, all of the following pieces of information would be helpful *except* **(a)** the number of cranial sutures, **(b)** the size and roughness of the markings of the bones, **(c)** the presence or absence of fontanelles, **(d)** the presence or absence of epiphyseal cartilages, **(e)** the types of minerals deposited in the bones.

LEVEL 3 Critical Thinking and Clinical Applications

24. Why would a person suffering from osteoporosis be more likely to suffer a hip fracture than a broken shoulder?
25. While fireman Fred is fighting a fire in a building, part of the ceiling collapses, and a beam strikes him on his left shoulder. He is rescued, but has a great deal of pain in his shoulder. He cannot move his arm properly, especially in the anterior direction. His clavicle is not broken, and his humerus is intact. What is the probable nature of Fred's injury?
26. Investigators find the pelvis of a human and are able to identify the sex, the relative age, and some physical characteristics of the person. How is this possible from only the pelvis?

+ CLINICAL CASE Wrap-Up Timber!

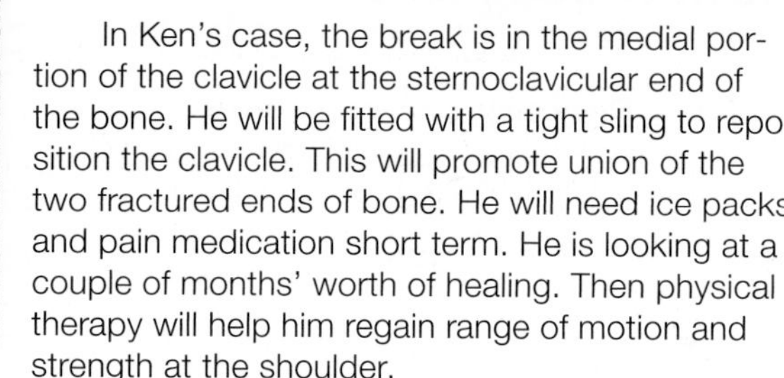

Ken's diagnosis is a fractured right clavicle. This is a very common injury involving people of all ages. Most breaks occur in the middle of the clavicle, but they can involve the rib cage or the scapula.

Several vital vascular and neural structures pass near the clavicle. Just deep to this bone run the subclavian artery and vein (*subclavian*, below the clavicle), which can be torn when the clavicle is damaged. Nerves supplying the arm may be injured with a clavicle fracture on the lateral end. The sharp end of a broken clavicle can puncture a lung. ER personnel will monitor for these potential, grave complications that would require surgical repair.

In Ken's case, the break is in the medial portion of the clavicle at the sternoclavicular end of the bone. He will be fitted with a tight sling to reposition the clavicle. This will promote union of the two fractured ends of bone. He will need ice packs and pain medication short term. He is looking at a couple of months' worth of healing. Then physical therapy will help him regain range of motion and strength at the shoulder.

1. What is the role of the clavicle?
2. What functional deficits will Ken have with such a fracture?

See the blue Answers tab at the back of the book.

Related Clinical Terms

bone bruise: Bleeding within the periosteum of a bone.

bone graft: A surgical procedure that transplants bone tissue to repair and rebuild diseased or damaged bone.

genu valgum: Deformity in which the knees angle medially and touch one another while standing; commonly called knock-knee.

pelvimetry: Measurement of the dimensions of the female pelvis.

9 Joints

Learning Outcomes

These Learning Outcomes correspond by number to this chapter's sections and indicate what you should be able to do after completing the chapter.

CLINICAL CASE What's the Matter with the Birthday Girl?

Jessica just woke up from her nap at preschool. Earlier this morning the class celebrated her fourth birthday with cupcakes and singing. Now she cries when she tries to roll over. Her best friend coaxes her to get up. "Come on, Jessie, let's play tag," she says. Jessica shakes her head "no," and struggles to sit up on the side of the cot. She lets her feet dangle but won't put her weight on her legs.

The other children are now buzzing around the room. In contrast, Jessica's withdrawal attracts the attention of her teacher, who comes over to check on her. She reaches an arm out to help pull Jessica up and Jessica winces with pain.

A short while later, when Jessica's mother comes to pick her up, the teacher draws her aside for a quick conference. "Jessica has not been participating in any play activities this whole week," reports the teacher. "Even today, on her birthday, she refused to play tag. She takes a really long time getting up from the floor after story circle. And her legs seemed to be hurting her after naptime. I wonder what's going on?" **What's the matter with Jessica? To find out, turn to the Clinical Case Wrap-Up on p. 290.**

An Introduction to Joints

In this chapter we consider the ways bones interact wherever they meet at **joints**, or **articulations**. In the last two chapters, you learned the individual bones of the skeleton and that they provide strength, support, and protection for softer tissues of the body. However, your daily life demands more of the skeleton—it must also facilitate and adapt to body movements. Think of your activities in a typical day: You breathe, talk, walk, sit, stand, and change positions countless times. In each case, your skeleton is directly involved. Movements can occur only at joints because the bones of the skeleton are fairly inflexible. The characteristic structure of a joint determines the type and amount of movement that may take place. Each joint reflects a compromise between the need for strength and stability and the need for mobility.

In this chapter we compare the relationships between articular form and function. We consider several examples that range from relatively immobile but very strong joints (the intervertebral joints) to a highly mobile but relatively weak joint (the shoulder).

9-1 Joints are categorized according to their structure or range of motion

Learning Outcome **Contrast the major categories of joints, and explain the relationship between structure and function for each category.**

We use two classification schemes to categorize joints: structural or functional. The structural classification scheme relies solely on the anatomy of the joint, whereas the functional scheme is based on the amount of movement possible at a joint, a property known as the *range of motion*.

Using the structural scheme, we classify joints as *fibrous*, *cartilaginous*, *bony*, or *synovial*. Bony joints form when fibrous or cartilaginous joints ossify. The ossification may be normal or abnormal, and may occur at various times in life. (We discuss synovial joints in more detail shortly.)

In the functional scheme, we first take into account the range of motion for a joint. Such descriptions produce groups of immovable, slightly movable, and freely movable joints. Then each of these groups is further subdivided primarily on the basis of the anatomy of the joint, giving us this final classification (shown in more detail in **Table 9–1**):

- An *immovable joint* is a **synarthrosis** (sin-ar-THRŌ-sis; *syn*, together + *arthros*, joint). A synarthrosis can be *fibrous* or *cartilaginous*, depending on the type of connection formed by the two bones in the joint. Over time, these two bones may fuse (ossify), forming a bony joint.
- A *slightly movable joint* is an **amphiarthrosis** (am-fē-ar-THRŌ-sis; *amphi*, on both sides). An amphiarthrosis is either *fibrous* or *cartilaginous*, depending on the type of the connection between the opposing bones.
- A *freely movable joint* is a **diarthrosis** (dī-ar-THRŌ-sis; *dia*, through), or *synovial joint*. Diarthroses are subdivided functionally, according to their planes of movement (which we cover in Section 9-2).

The two classification schemes are loosely correlated. We see many anatomical patterns among immovable or slightly movable joints, but there is only one type of freely movable joint—synovial joints. All synovial joints are diarthroses, and the two terms are interchangeable. As you will see, synarthroses and amphiarthroses are found largely in the axial skeleton, whereas diarthroses predominate in the appendicular skeleton. We will use the functional classification rather than the anatomical one in this chapter, because our primary interest is in how joints work.

Table 9–1 Functional and Structural Classifications of Joints

Functional Category	Structural Category and Type	Description
SYNARTHROSIS (no movement)		
At a synarthrosis, the bony edges are quite close together and may even interlock. These extremely strong joints are located where movement between the bones must be prevented.	FIBROUS: **Suture**	A **suture** (*sutura*, a sewing together) is a synarthrotic joint located only between the bones of the skull. The edges of the bones are interlocked and bound together at the suture by dense fibrous connective tissue.
	FIBROUS: **Gomphosis**	A **gomphosis** (gom-FŌ-sis; *gomphos*, bolt) is a synarthrosis that binds the teeth to bony sockets in the maxillae and mandible. The fibrous connection between a tooth and its socket is a *periodontal* (per-ē-ō-DON-tal; *peri*, around + *odontos*, tooth) *ligament*.
	CARTILAGINOUS: **Synchondrosis**	A **synchondrosis** (sin-kon-DRŌ-sis; *syn*, together + *chondros*, cartilage) is a rigid, cartilaginous bridge between two articulating bones. The cartilaginous connection between the ends of the first pair of vertebrosternal ribs and manubrium of the sternum is a synchondrosis. Another example is the epiphyseal cartilage, which connects the diaphysis to the epiphysis in a growing long bone.
	BONY: **Synostosis**	A **synostosis** (sin-os-TŌ-sis) is a totally rigid, immovable joint created when two bones fuse and the boundary between them disappears. The rare frontal (metopic) suture of the frontal bone, the fusion of an infant's left and right mandibular bones, and the epiphyseal lines of mature long bones are synostoses.
AMPHIARTHROSIS (little movement)		
An amphiarthrosis permits more movement than a synarthrosis, but is much stronger than a freely movable joint. The articulating bones are connected by collagen fibers or cartilage.	FIBROUS: **Syndesmosis**	At a **syndesmosis** (sin-dez-MŌ-sis; *syndesmos*, ligament), bones are connected by a ligament. One example is the distal joint between the tibia and fibula.
	CARTILAGINOUS: **Symphysis**	At a **symphysis**, the articulating bones are connected by a wedge or pad of fibrocartilage. The joint between the two pubic bones (the *pubic symphysis*) is an example of a symphysis.
DIARTHROSIS (free movement)		
Planes of Movement • **Monaxial:** movement in one plane; elbow, ankle • **Biaxial:** movement in two planes; ribs and wrist • **Triaxial:** movement in three planes; shoulder, hip	SYNOVIAL	**Synovial** (si-NŌ-ve-ul) **joints** permit a wider range of motion than do other types of joints. They are typically located at the ends of long bones, such as those of the upper and lower limbs.

Checkpoint

1. Name and describe the three types of joints as classified by range of motion.
2. What characteristics do typical synarthrotic and amphiarthrotic joints share?
3. In a newborn, the large bones of the skull are joined by fibrous connective tissue. The bones later grow, interlock, and form immovable joints. Structurally, which type of joints are these?

See the blue Answers tab at the back of the book.

9

9-2 Diarthroses: Synovial joints contain synovial fluid and are surrounded by a joint capsule and stabilizing accessory structures

Learning Outcome Describe the basic structure of a synovial joint, and describe common synovial joint accessory structures and their functions.

Synovial joints are freely movable diarthroses. A two-layered **joint capsule**, also called an **articular capsule**, surrounds the synovial joint (Figure 9–1). This capsule is continuous with the periostea of the articulating bones. The joint capsule contains an inner *synovial membrane* and an outer *fibrous capsule.* This membrane does not cover the articulating surfaces within the joint, which are protected by *articular cartilage*. Recall that a synovial membrane consists of areolar tissue covered by an incomplete epithelial layer. The *synovial fluid* that fills the *joint cavity* originates in the areolar tissue of the synovial membrane. p. 142

Remember that a joint cannot be both highly mobile and very strong. The greater the range of motion at a joint, the weaker it becomes. A synarthrosis is the strongest type of joint, but it permits no movement. In contrast, a diarthrosis, such as the shoulder joint, is far weaker but has a broad range of motion. Because of this relative weakness, synovial joints are surrounded by a variety of stabilizing *accessory structures*.

Now let's consider the major features of synovial joints.

Articular Cartilage

Under normal conditions, the bony surfaces at a synovial joint cannot contact one another, because special **articular cartilage** covers the articulating surfaces. Articular cartilage resembles hyaline cartilage elsewhere in the body. p. 137

Figure 9–1 The Structure of a Synovial Joint.

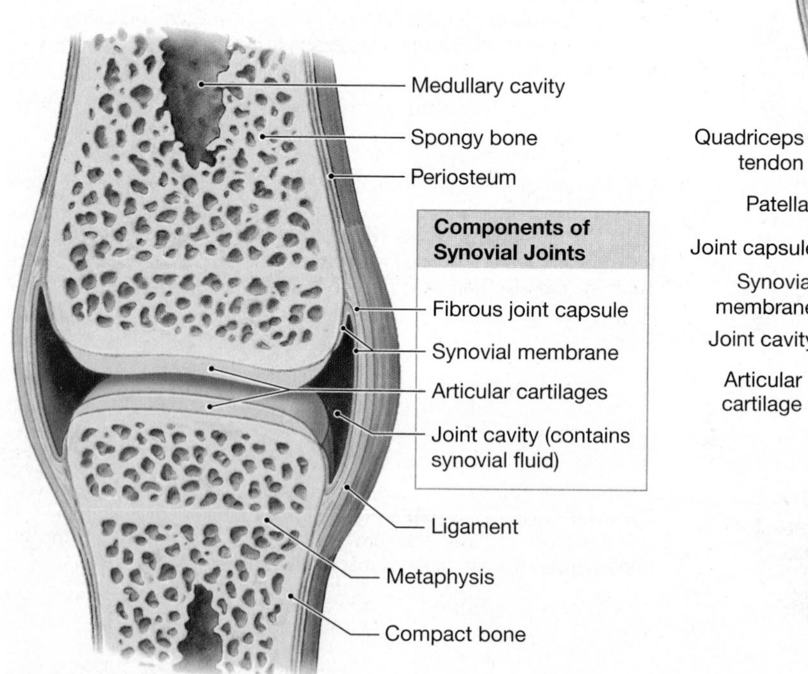
Medullary cavity
Spongy bone
Periosteum
Components of Synovial Joints
Fibrous joint capsule
Synovial membrane
Articular cartilages
Joint cavity (contains synovial fluid)
Ligament
Metaphysis
Compact bone

a Synovial joint, sagittal section

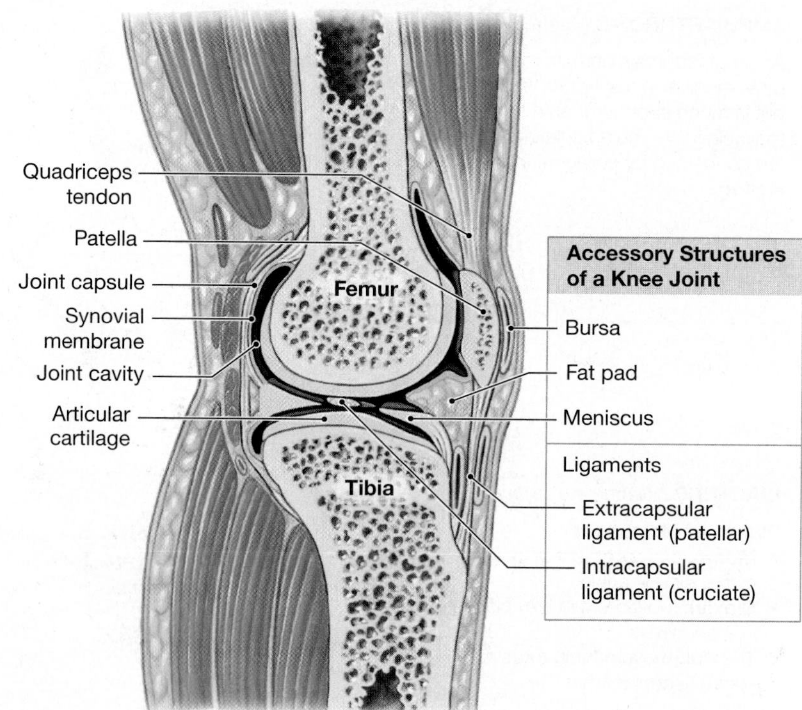
Quadriceps tendon
Patella
Joint capsule
Synovial membrane
Joint cavity
Articular cartilage
Femur
Tibia
Accessory Structures of a Knee Joint
Bursa
Fat pad
Meniscus
Ligaments
Extracapsular ligament (patellar)
Intracapsular ligament (cruciate)

b Knee joint, sagittal section

? What structure covers the bony surfaces of a synovial joint to prevent the surfaces from touching?

However, articular cartilage has no perichondrium (a fibrous sheath described in Chapter 4), and the matrix contains more water than that of other cartilage.

The surface of articular cartilage is slick and smooth. This feature alone can reduce friction during movement at the joint. However, even when pressure is applied across a joint, the smooth articular cartilages do not touch one another because a thin film of synovial fluid within the joint cavity separates them (Figure 9–1a). This fluid acts as a lubricant, minimizing friction.

Synovial joints cannot function normally if the articular cartilages are damaged. When such damage occurs, the cartilage matrix may begin to break down. The exposed surface will then change from a slick, smooth-gliding surface to a rough abrasive surface of bristly collagen fibers. This abrasive surface drastically increases friction at the joint.

Synovial Fluid

What is synovial fluid? It is a clear, viscous solution with the consistency of egg yolk or heavy molasses. **Synovial fluid** resembles interstitial fluid, but contains proteoglycans with a high concentration of hyaluronan secreted by fibroblasts of the synovial membrane. p. 128 Even in a large joint such as the knee, the total quantity of synovial fluid in a joint is small—normally less than 3 mL.

The synovial fluid within a joint has the following primary functions:

- *Lubrication.* Articular cartilages act like sponges filled with synovial fluid. When part of an articular cartilage is compressed, some of the synovial fluid is squeezed out of the cartilage and into the space between the opposing surfaces. This thin layer of fluid greatly reduces friction between moving surfaces, just as a thin film of water makes a highway slick by reducing friction between a car's tires and the road. When the compression stops, synovial fluid is pulled back into the articular cartilage.
- *Nutrient Distribution.* The synovial fluid in a joint must circulate continuously to provide nutrients and a waste disposal route for the chondrocytes of the avascular articular cartilage. Whenever the joint moves, the compression and reexpansion of the articular cartilage pumps synovial fluid into and out of the cartilage matrix. As the synovial fluid flows through the areolar tissue of the synovial membrane, additional nutrients are delivered and wastes are absorbed by diffusion across capillary walls.
- *Shock Absorption.* Synovial fluid cushions joints that are subjected to compression from shocks and sudden impacts. This occurs because the viscosity of synovial fluid increases with increased pressure. For example, your hip, knee, and ankle joints are compressed as you walk. They are even more compressed when you jog or run. When the pressure across a joint suddenly increases, the resulting shock is lessened as the viscosity of synovial fluid increases. As the pressure lessens, viscosity decreases and it regains its lubricating function across the articular surfaces.

Accessory Structures

Synovial joints may have a variety of protective and stabilizing accessory structures, including pads of cartilage or fat, ligaments, tendons, and *bursae* (protective pockets). Figure 9–1b shows these structures in the knee joint.

Cartilages and Fat Pads

In several joints, including the knee, pads made of fibrocartilage or fat may lie between the opposing articular surfaces. A fibrocartilage pad called a **meniscus** (me-NIS-kus; a crescent; plural, *menisci;* me-NIS-kē) is located between opposing bones within a synovial joint. Menisci may subdivide a synovial cavity, channel the flow of synovial fluid, or allow for variations in the shapes of the articular surfaces.

Fat pads are localized masses of adipose tissue covered by a layer of synovial membrane. They are commonly superficial to the joint capsule. Fat pads protect the articular cartilages and act as packing material for the joint. When the bones move, the fat pads fill in the spaces created as the joint cavity changes shape.

Ligaments

Accessory **ligaments** are localized thickenings that support, strengthen, and reinforce synovial joints. *Extracapsular ligaments* are outside the joint capsule, while *intracapsular ligaments* are inside the joint capsule.

In a **sprain**, a ligament is stretched so much that some of the collagen fibers are torn, but the ligament as a whole survives and the joint is not damaged. With excessive force, one of the attached bones often breaks before the ligament tears. In general, a broken bone heals much more quickly and effectively than does a torn ligament. The reason is that ligaments have no direct blood supply and thus must obtain essential substances by the slow process of diffusion.

Tendons

Tendons are not part of the joint itself, but they do play a role in stabilizing the joint, because tendons connect the fleshy part of muscles to bones that make up the joint. Furthermore, tendons passing across or around a joint may limit the joint's range of motion, yet provide mechanical support for it. For example, tendons associated with the muscles of the arm help brace the shoulder joint.

Bursae

Bursae (BUR-sē; singular, *bursa;* a pouch) are small, thin, fluid-filled pockets in connective tissue that reduce friction and act as shock absorbers. They contain synovial fluid and are lined by a synovial membrane. Bursae may be connected to the joint cavity or may be separate from it. They form where a tendon or ligament rubs against other tissues. Bursae are located around most synovial joints, including the shoulder joint. *Synovial tendon sheaths* are tubular bursae that surround tendons where they cross bony surfaces. Bursae may also appear deep to the skin, covering a bone or lying within other connective tissues exposed to friction or pressure. Bursae that develop in abnormal locations, or because of abnormal stresses, are called *adventitious bursae.*

Factors That Stabilize Synovial Joints

9

For any synovial joint, movement beyond its normal range of motion will cause damage. Several factors are responsible for limiting a joint's range of motion, stabilizing it, and reducing the chance of it being injured:

- The collagen fibers of the joint capsule and any accessory, extracapsular, or intracapsular ligaments
- The shapes of the articulating surfaces and menisci, which may prevent movement in specific directions
- The presence of other bones, skeletal muscles, or fat pads around the joint
- Tension in tendons attached to the articulating bones.

When a skeletal muscle contracts and pulls on a tendon on a bone in a joint, movement in a specific direction is either encouraged or opposed. The pattern of stabilizing structures versus degree of mobility varies among joints. In general, the more stable the joint, the more restricted its range of motion. For example, the hip joint is stabilized by the shapes of the bones (the head of the femur projects into the acetabulum of the hip bone), a heavy capsule, intracapsular and extracapsular ligaments, tendons, and massive muscles. It is therefore very strong and stable. In contrast, the elbow is another stable joint, but it gains its stability from the interlocking shapes of the articulating bones. The capsule and associated ligaments provide additional support. Finally, the shoulder joint, as the most mobile synovial joint, relies primarily on the surrounding ligaments, muscles, and tendons for stability. For this reason, it is also fairly weak.

+ Clinical Note Bursitis and Bunions

Bursae are normally flat and contain a small amount of synovial fluid. When injured, however, they become inflamed and swell with synovial fluid. This causes pain in the affected area whenever the tendon or ligament moves, resulting in a condition called **bursitis**. Inflammation can result from the friction due to repetitive motion, pressure over the joint, infection, or trauma. Bursitis associated with repetitive motion typically occurs at the shoulder. Musicians, golfers, baseball pitchers, and tennis players may develop bursitis there. The most common pressure-related bursitis is a **bunion**. Bunions may form over the base of the great toe as a result of friction and distortion of the first metatarsophalangeal joint. High heels and tight, pointy-toed shoes are often the culprit!

+ Clinical Note Dislocation

When reinforcing structures cannot protect a joint from extreme stresses, a **dislocation** results. In a dislocation, there is complete loss of contact between the articulating surfaces. Articular cartilages are damaged, ligaments tear, and the joint capsule is distorted. The *inside* of a joint has no pain receptors, but nerves that monitor the capsule, ligaments, and tendons are quite sensitive, so dislocations are very painful.

✓ Checkpoint

4. **Describe the components of a synovial joint, and identify the functions of each.**
5. **Why would improper circulation of synovial fluid lead to the degeneration of articular cartilages in the affected joint?**
6. **Define *dislocation*.**

See the blue Answers tab at the back of the book.

9-3 Diarthroses: The different types of synovial joints allow a wide range of skeletal movements

Learning Outcome **Describe how the anatomical and functional properties of synovial joints permit movements of the skeleton.**

Range of motion (ROM) refers to the full movement at a particular joint. This range is usually the distance between flexion (decreasing the angle) and extension (increasing the angle). If a person has limited ROM, that means a joint cannot move through its normal range of movement. In this section, the terms *movement* and *motion* are used interchangeably.

Types of Movements at Synovial Joints

To describe human movement, we need a frame of reference that enables accurate and precise communication. In descriptions of movement at synovial joints, phrases such as "bend the leg" or "raise the arm" are imprecise. Anatomists use descriptive terms that have specific meanings. **Spotlight Figure 9–2a** (p. 272) provides a simple way to model these different types of movements with a pencil and representative articular surface.

Gliding Movement

In a gliding movement, two opposing surfaces slide past one another in one plane. Gliding occurs between the flat or nearly flat surfaces of articulating carpal bones, between tarsal bones, and between the clavicles and the sternum. The movement can occur in almost any direction, but the amount of movement is slight, and rotation is generally prevented by the capsule and associated ligaments.

Angular Movement

Examples of angular movement include *flexion, extension, abduction*, and *adduction* (Figure 9–3). Descriptions of these movements are based on reference to an individual in the anatomical position. ⮌ p. 10

Figure 9–3 Angular Movements and Circumduction. The red dots indicate the locations of the joints involved in the movements illustrated.

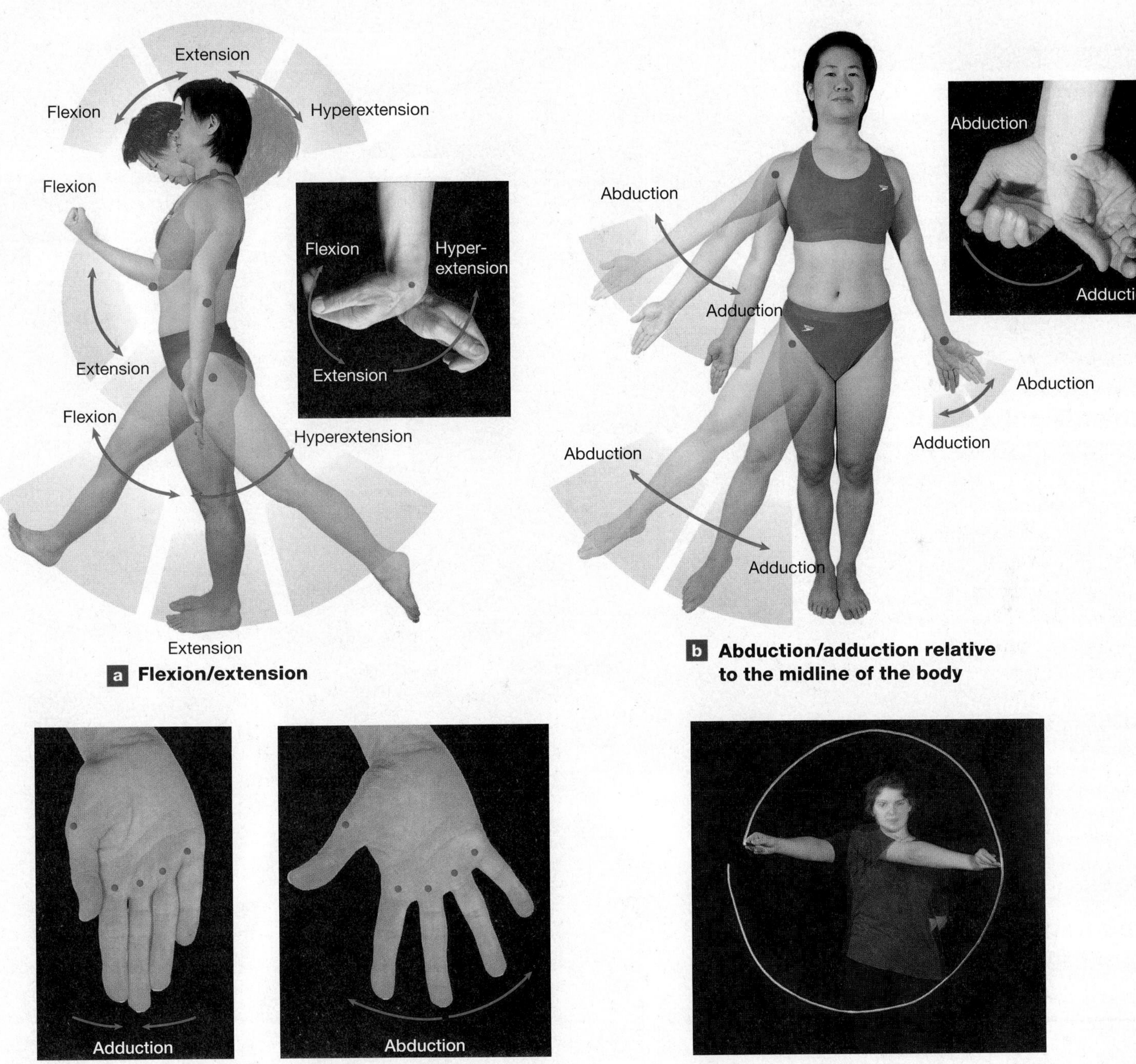
Extension
Flexion
Hyperextension
Flexion
Flexion
Hyper-
extension
Extension
Extension
Flexion
Hyperextension
Extension
a Flexion/extension
Abduction
Abduction
Adduction
Adduction
Abduction
Adduction
Abduction
Adduction
Abduction
b Abduction/adduction relative to the midline of the body
Adduction
Abduction
c Adduction/abduction relative to central digit
d Circumduction

? List the various types of angular movements.

SPOTLIGHT

Figure 9–2

Joint Movement

a Simple Model of Articular Motion

Take a pencil as your model and stand it upright on the surface of a desk. The pencil represents a bone, and the desk is an articular surface. A lot of twisting, pushing, and pulling will demonstrate that there are only three ways to move the pencil.

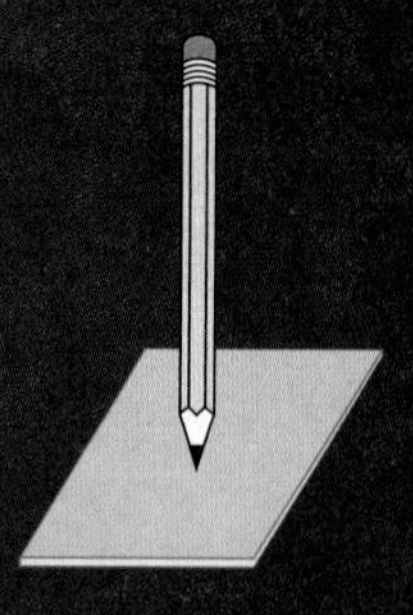

Moving the Point

Linear motion

Possible movement 1 shows the pencil can move. If you hold the pencil upright, without securing the point, you can push the pencil across the surface. This kind of motion is called **gliding**. You could slide the point forward or backward, from side to side, or diagonally.

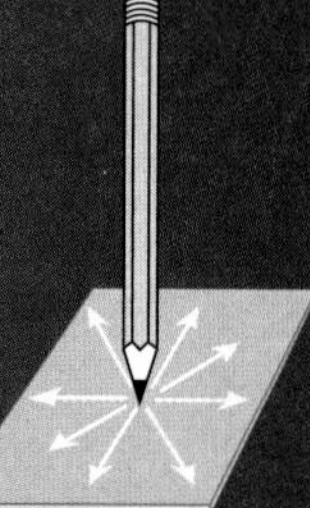

Changing the Shaft Angle

Angular motion

Possible movement 2 shows the pencil shaft can change its angle with the surface. With the tip held in position, you can move the eraser end of the pencil forward and backward, from side to side, or at some intermediate angle.

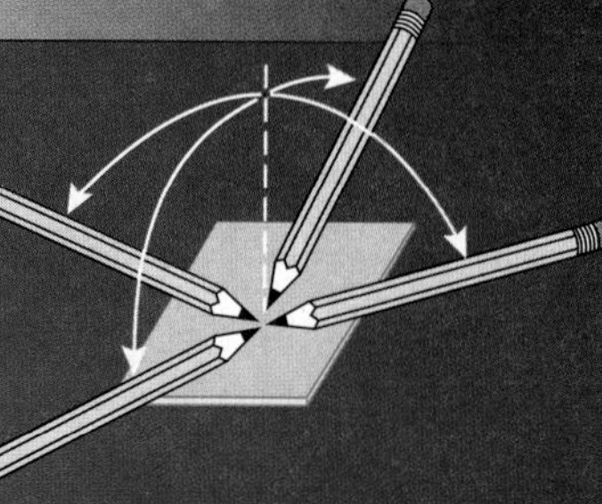

Circumduction

A more complex angular motion is possible. Grasp the pencil eraser and move the pencil in any direction until it is no longer vertical. Now, swing the eraser through a complete circle in a movement called **circumduction**.

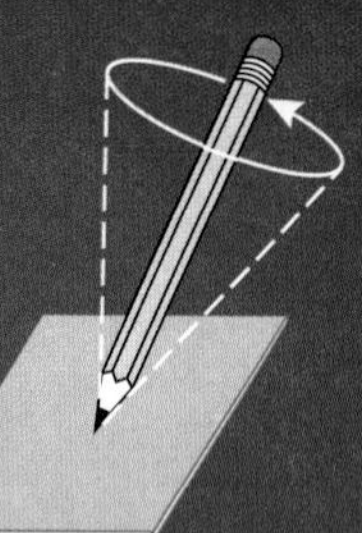

Rotating the Shaft

Rotation

Possible movement 3 shows that the pencil shaft can rotate. If you keep the shaft vertical and the point at one location, you can still spin the pencil around its longitudinal axis in a movement called **rotation**. No joint can freely rotate because this would tangle blood vessels, nerves, and muscles, as they crossed the joint.

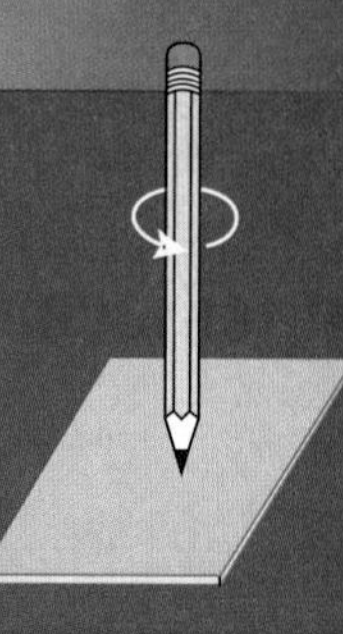

b Axes of Motion

Movement of joints can also be described by the number of axes that they can rotate around. A joint that permits movement around one axis is called **monaxial**, a joint that permits movement around two axes is called **biaxial**, and one that permits movement around three axes, is called **triaxial**.

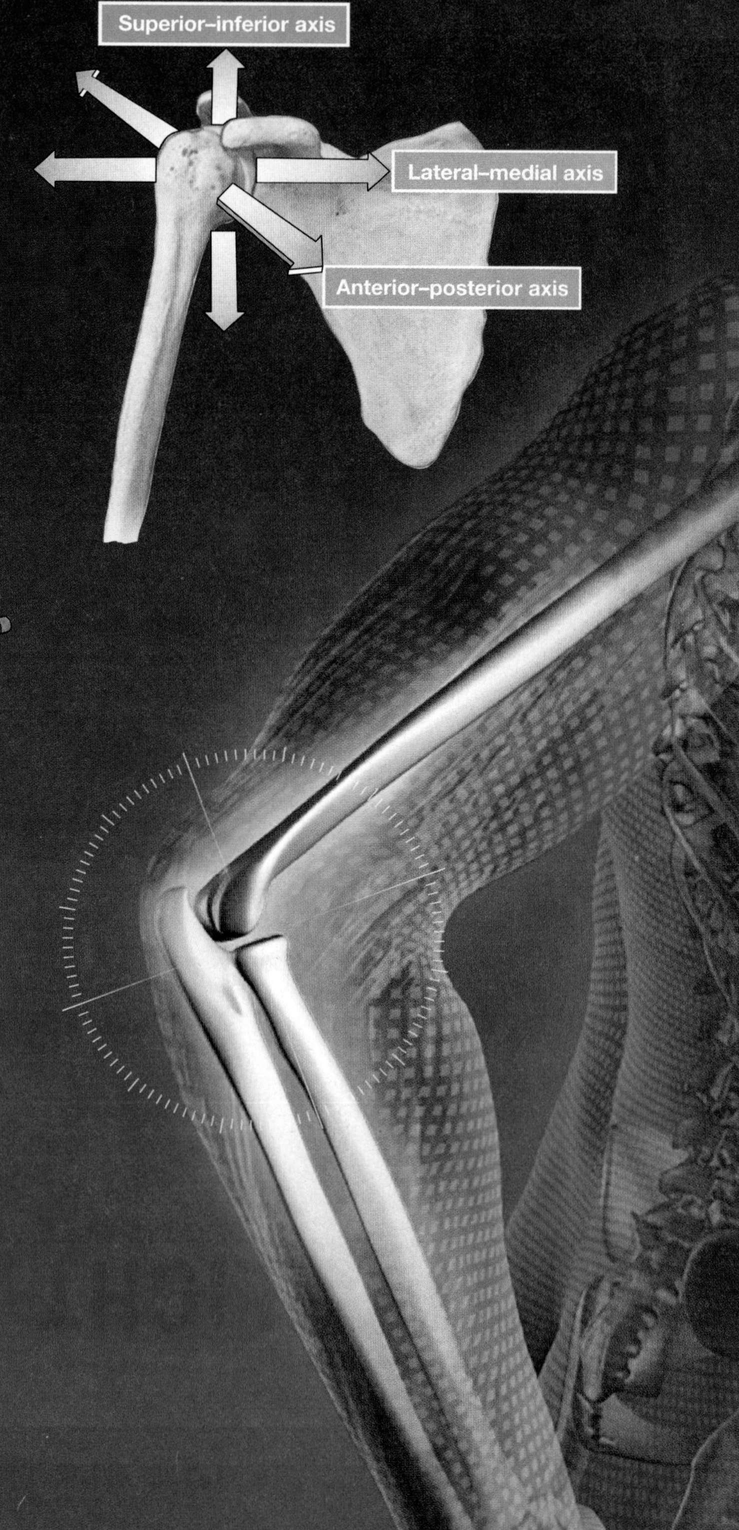
Superior–inferior axis
Lateral–medial axis
Anterior–posterior axis

C Classification of Synovial Joints

Synovial joints are freely movable diarthrotic joints, and they are classified by the type and range of motion permitted. Synovial joints are described as plane, hinge, condylar, saddle, pivot, or ball-and-socket on the basis of the shapes of the articulating surfaces, which in turn determines the joint movement.

Plane joint

Plane joints, or gliding joints, have flattened or slightly curved surfaces that slide across one another, but the amount of movement is very slight.

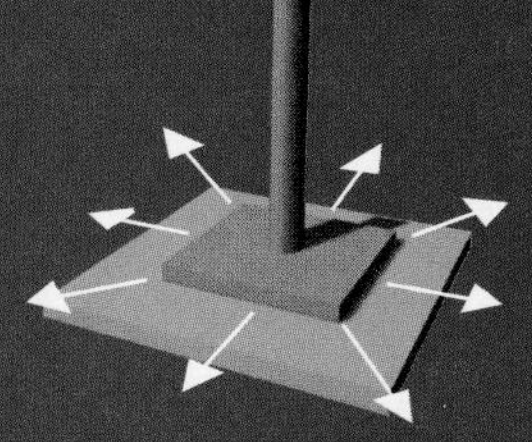

Movement:
Gliding.
Slight nonaxial

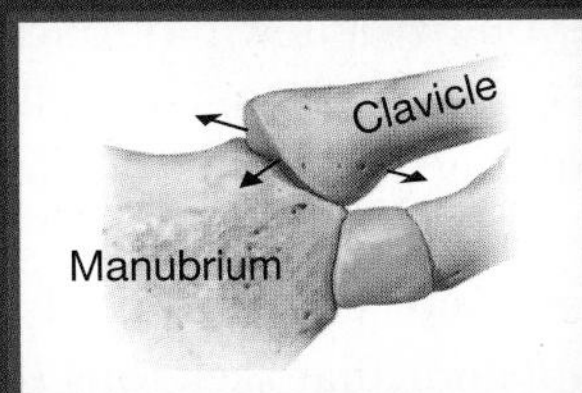
Clavicle
Manubrium

Examples:
- Acromioclavicular and claviculosternal joints
- Intercarpal joints
- Vertebrocostal joints
- Sacro-iliac joints

Hinge joint

Hinge joints permit angular motion in a single plane, like the opening and closing of a door.

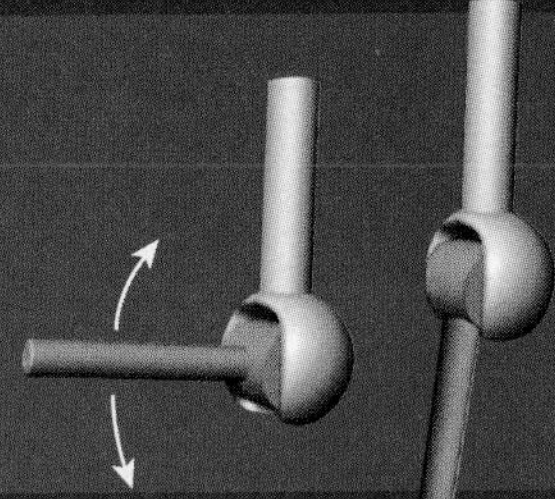

Movement:
Angular.
Monaxial

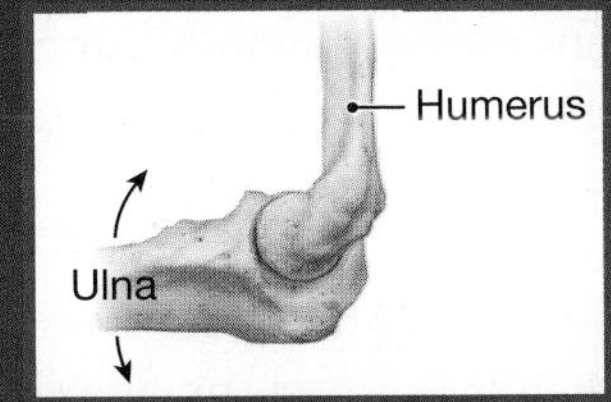
Humerus
Ulna

Examples:
- Elbow joint
- Knee joint
- Ankle joint
- Interphalangeal joint

Condylar joint

Condylar joints, or ellipsoid joints, have an oval articular face nestled within a depression on the opposing surface.

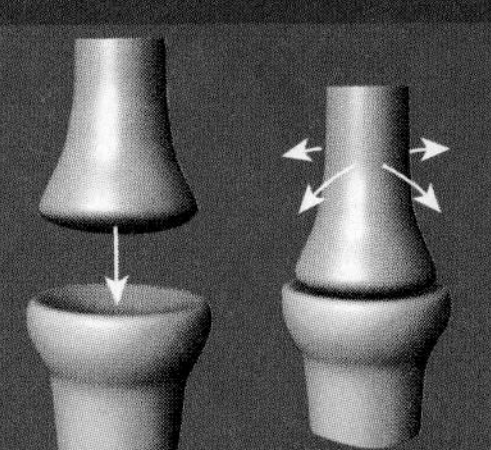

Movement:
Angular.
Biaxial

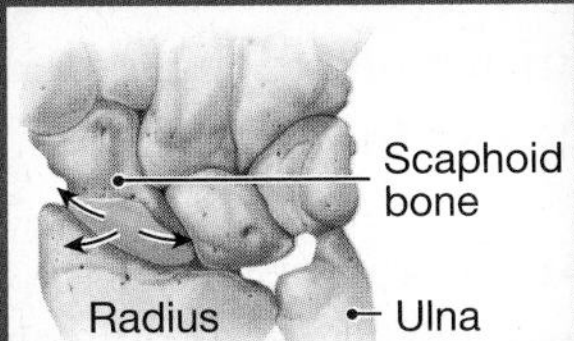
Scaphoid bone
Radius
Ulna

Examples:
- Radiocarpal joint
- Metacarpophalangeal joints 2–5
- Metatarsophalangeal joints

Saddle joint

Saddle joints have complex articular faces and fit together like a rider in a saddle. Each face is concave along one axis and convex along the other.

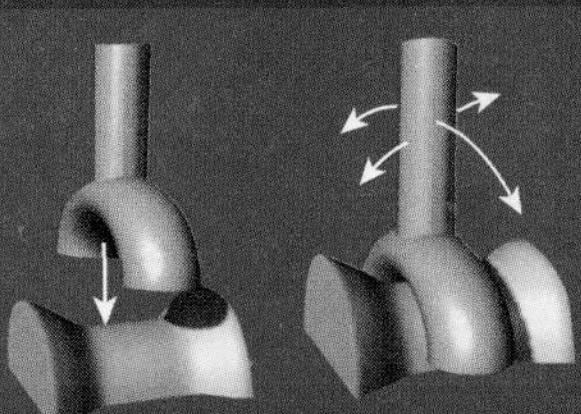

Movement:
Angular.
Biaxial

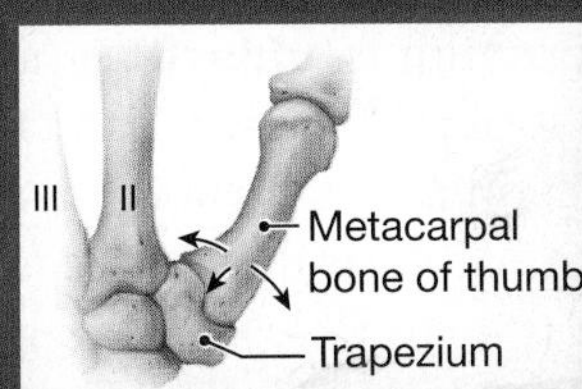
III
II
Metacarpal bone of thumb
Trapezium

Examples:
- First carpometacarpal joint

Pivot joint

Pivot joints only permit rotation.

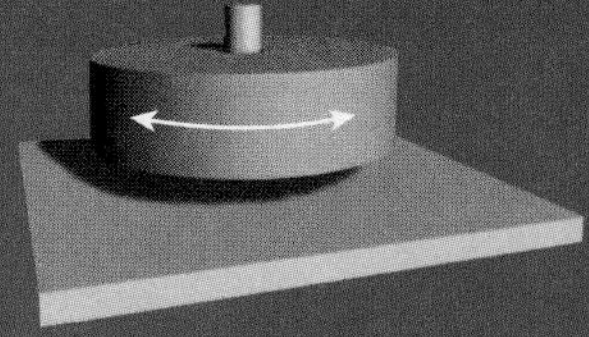

Movement:
Rotation.
Monaxial

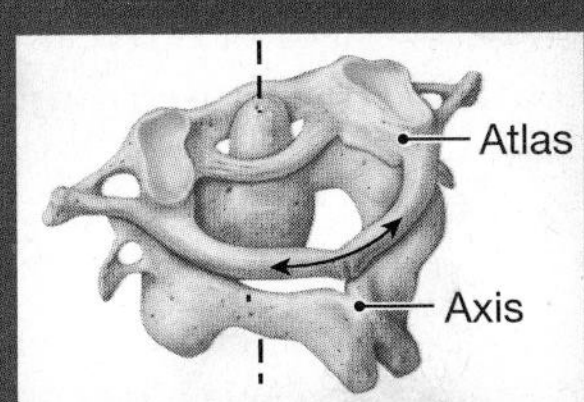
Atlas
Axis

Examples:
- Atlanto-axial joint
- Proximal radio-ulnar joint

Ball-and-socket joint

In a ball-and-socket joint, the round head of one bone rests within a cup-shaped depression in another.

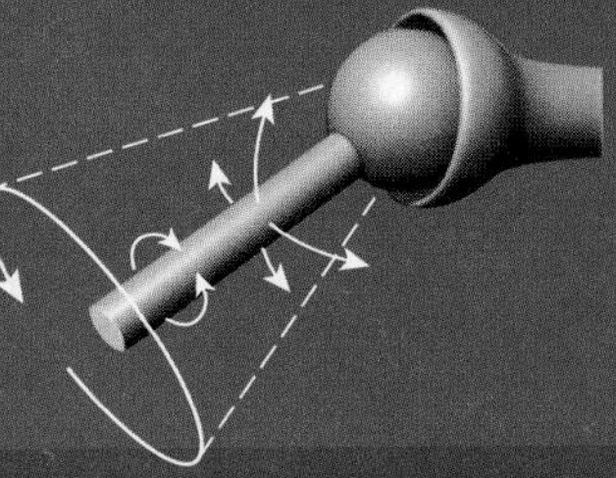

Movement:
Angular, circumduction, and rotation.
Triaxial

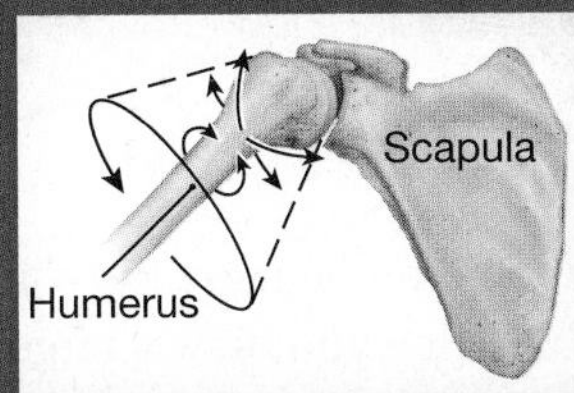
Scapula
Humerus

Examples:
- Shoulder joint
- Hip joint

Flexion and Extension. **Flexion** (FLEK-shun) is movement in the anterior–posterior (sagittal) plane that decreases the angle between articulating bones. **Extension** occurs in the same plane, but it increases the angle between articulating bones (Figure 9–3a). We usually apply these terms to the movements of the long bones of the limbs, but we can also use them to describe movements of the axial skeleton. For example, when you bring your head toward your chest, you flex the intervertebral joints of the neck. When you bend down to touch your toes, you flex the intervertebral joints of the spine. Extension reverses these movements, returning you to the anatomical position.

When a person is in the anatomical position, all of the major joints of the axial and appendicular skeletons (except the ankle) are at full extension. (Shortly we will introduce special terms to describe movements of the ankle joint.) Flexion of the shoulder joint or hip joint moves the limbs anteriorly, whereas extension moves them posteriorly. Flexion of the wrist joint moves the hand anteriorly, and extension moves it posteriorly.

In each of these examples, extension can be continued past the anatomical position. Extension past the anatomical position is called **hyperextension** (see Figure 9–3a). When you hyperextend your neck, you can gaze at the ceiling. Ligaments, bony processes, or soft tissues prevent hyperextension of many joints, such as the elbow or the knee.

Abduction and Adduction. **Abduction** (*ab*, from) is movement *away* from the longitudinal axis (midline) of the body in the frontal plane (Figure 9–3b). For example, swinging the upper limb to the side is abduction of the limb. Moving it back to the anatomical position is **adduction** (*ad*, to). Adduction of the wrist moves the heel of the hand and fingers *toward* the body, whereas abduction moves them farther away. Spreading the fingers or toes apart abducts them, because they move *away from* a central digit (Figure 9–3c). Bringing them together is called adduction. (Fingers move toward or away from the middle finger; toes move toward or away from the second toe.) Abduction and adduction always refer to movements of the appendicular skeleton, not to those of the axial skeleton.

Tips & Tools

When someone is **abduct**ed, they are taken away, just as **abduct**ion takes the limb away from the body. During **add**uction, the limb is **add**ed to the body.

Circumduction

Circumduction is a complete circular movement. Moving your arm in a loop is circumduction, as when you draw a large circle on a whiteboard (Figure 9–3d). Your hand moves in a circle, but your arm does not rotate.

Figure 9–4 Rotational Movements.

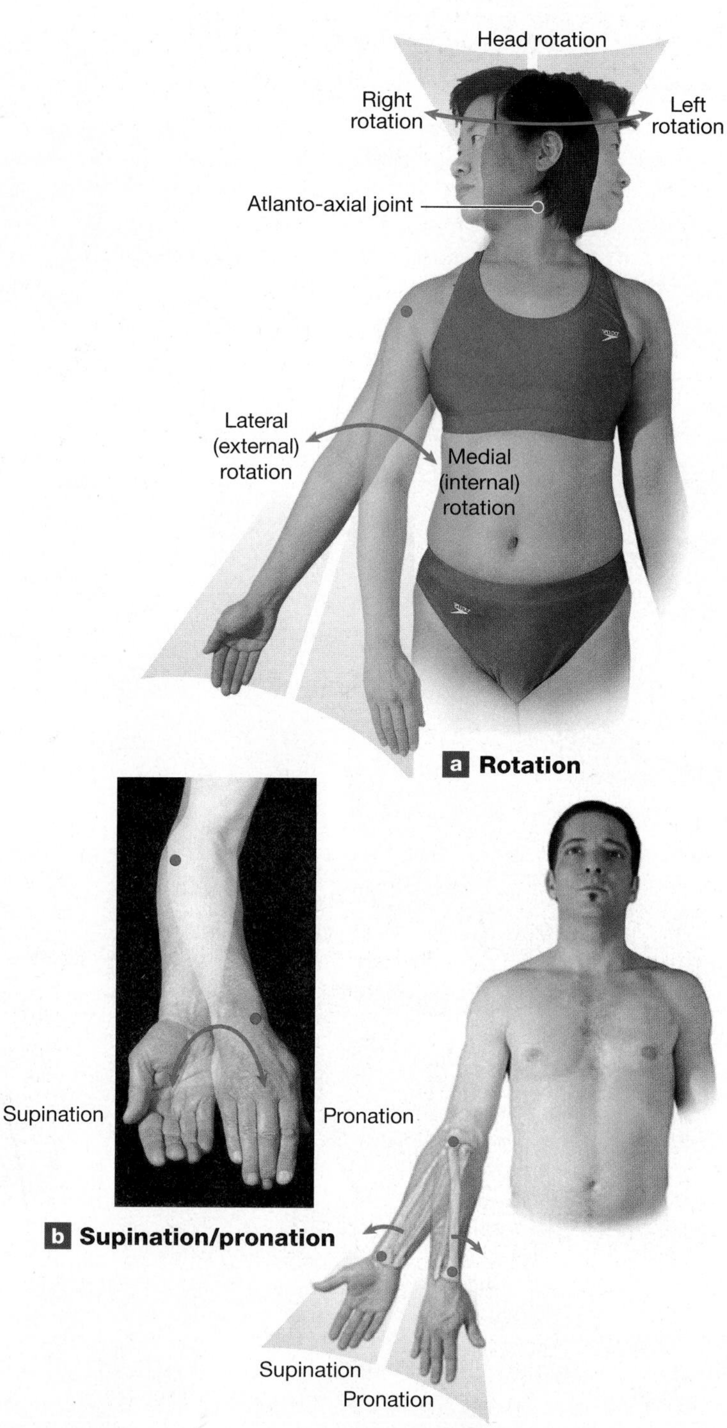
Head rotation
Right rotation
Left rotation
Atlanto-axial joint
Lateral (external) rotation
Medial (internal) rotation
a Rotation
Supination
Pronation
b Supination/pronation
Supination
Pronation

Rotational Movement

We also describe rotational movements with reference to a person in the anatomical position. Rotation of the head may involve *left rotation* or *right rotation* (Figure 9–4a). We describe limb rotation by reference to the longitudinal axis of the trunk. During **medial rotation**, also known as *internal rotation*, the anterior surface of a limb turns toward the long axis of the trunk (see Figure 9–4a). The reverse movement is called **lateral rotation**, or *external rotation*.

The proximal joint between the radius and the ulna permits rotation of the radial head. As the shaft of the radius rotates, the distal epiphysis of the radius rolls across the anterior surface of the ulna. This movement, called **pronation** (prō-NĀ-shun), turns the wrist and hand from palm facing front to palm facing back (Figure 9–4b). The opposing movement, in which the palm is turned anteriorly, is **supination** (sū-pi-NĀ-shun). The forearm is supinated in the anatomical position. This view makes it easier to follow the path of the blood vessels, nerves, and tendons, which rotate with the radius during pronation.

Tips & Tools

In order to carry a bowl of *soup*, the hand must be *sup*inated.

Special Movements

Several special terms apply to specific joints or unusual types of movement (Figure 9–5):

- **Inversion** (*in*, into + *vertere*, to turn) is a twisting movement of the foot that turns the sole inward, elevating the medial edge of the sole. The opposite movement is called **eversion** (ē-VER-zhun) (Figure 9–5a).
- **Dorsiflexion** is flexion at the ankle joint and elevation of the sole, as when you dig in your heel. **Plantar flexion** (*plantaris*, sole), the opposite movement, extends the ankle joint and elevates the heel, as when you stand on tiptoe (Figure 9–5b). However, it is also acceptable (and simpler) to use "flexion and extension at the ankle," rather than "dorsiflexion and plantar flexion."
- **Opposition** is movement of the thumb toward the surface of the palm or the pads of other fingers (Figure 9–5c). Opposition enables you to grasp and hold objects between your thumb and palm. It involves movement at the first

9

Figure 9–5 Special Movements.

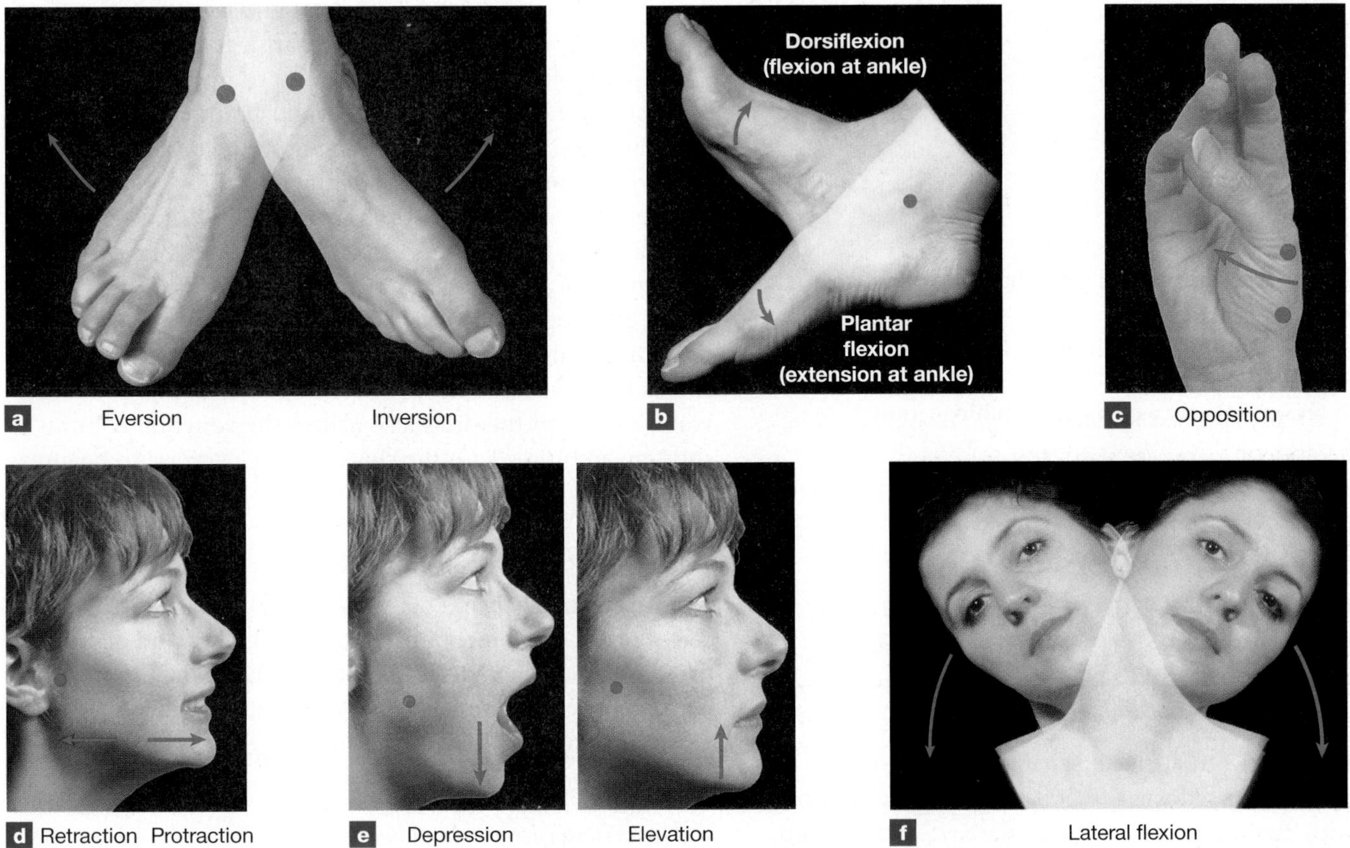

Dorsiflexion
(flexion at ankle)
Plantar
flexion
(extension at ankle)
a Eversion
Inversion
b
c Opposition
d Retraction Protraction
e Depression
Elevation
f Lateral flexion

carpometacarpal and metacarpophalangeal joints. Flexion at the fifth metacarpophalangeal joint can assist this movement. **Reposition** is the movement that returns the thumb and fingers from opposition.

- **Protraction** is movement of a body part anteriorly in the horizontal plane. **Retraction** is the reverse movement. You protract your jaw when you jut your chin forward, and you retract your jaw when you return it to its normal position (Figure 9–5d).
- **Depression** and **elevation** take place when we move a structure inferiorly and superiorly, respectively. You depress your mandible when you open your mouth, and you elevate your mandible as you close your mouth (Figure 9–5e). Another familiar elevation takes place when you shrug your shoulders.
- **Lateral flexion** occurs when your vertebral column bends to the side (Figure 9–5f). This movement is most pronounced in the cervical and thoracic regions.

Classification of Synovial Joints

Spotlight Figure 9–2b (p. 272) shows how the movement of synovial joints can be described by one to three axes of motion.

Synovial joints are classified by the type and ranges of motion they permit. Spotlight Figure 9–2c (p. 273) shows the six basic categories of synovial joints, which are based on the structure of the articular surfaces. These categories are *plane (gliding), hinge, condylar (ellipsoid), saddle, pivot,* and *ball-and-socket joints*. Please study this figure before going on in the chapter.

✓ Checkpoint

7. **Identify the types of synovial joints based on the shapes of the articulating surfaces.**
8. **When you do jumping jacks, which lower limb movements are necessary?**
9. **Which movements are associated with hinge joints?**

See the blue Answers tab at the back of the book.

9-4 Intervertebral joints contain intervertebral discs and ligaments that allow for vertebral movements

Learning Outcome Describe the joints between the vertebrae of the vertebral column.

The articulations between adjacent vertebrae form two types of joints—symphyses and synovial—that permit small movements associated with flexion and rotation of the vertebral column (Figure 9–6a). Little movement occurs between adjacent vertebral bodies. From axis to sacrum, the vertebrae are separated and cushioned by pads of fibrocartilage called *intervertebral discs*. Thus, the bodies of vertebrae form symphyses.

Intervertebral discs and symphyses are not found between the first and second cervical vertebrae or in the sacrum or coccyx, where vertebrae have fused. Recall that the first cervical vertebra has no vertebral body and no intervertebral disc. The only joint between the first two cervical vertebrae is a synovial joint that allows much more rotation than the symphyses between other cervical vertebrae.

Structure of Intervertebral Joints

Intervertebral joints are joints between adjacent vertebral bodies (these are symphyses) and joints between adjacent articular processes (these are synovial) along the vertebral column. Let's look at the structure of the intervertebral discs and ligaments that support these joints.

Intervertebral Discs

Each **intervertebral disc** has a tough outer layer of fibrocartilage, the **anulus fibrosus** (AN-ū-lus fī-BRŌ-sus). The collagen fibers of this layer attach the disc to the bodies of adjacent vertebrae. The anulus fibrosus surrounds the **nucleus pulposus** (pul-PŌ-sus), a soft, elastic, gelatinous core. The nucleus pulposus gives the disc resiliency and enables it to absorb shocks. The superior and inferior surfaces of the disc are almost completely covered by thin **vertebral end plates** that are composed of hyaline cartilage and fibrocartilage (Figure 9–6b).

Movement of the vertebral column compresses the nucleus pulposus and displaces it in the opposite direction. This displacement permits smooth gliding movements between vertebrae while maintaining their alignment.

The intervertebral discs make a significant contribution to a person's height: They account for about one-quarter of the length of the vertebral column superior to the sacrum. As we grow older, the water content of the nucleus pulposus in each disc decreases. The discs gradually become less effective as cushions, and the chances of vertebral injury increase. Water loss from the discs also causes the vertebral column to shorten, accounting for the characteristic decrease in height as we get older.

Intervertebral Ligaments

Numerous strong ligaments are attached to the bodies and processes of all vertebrae, binding them together and stabilizing the vertebral column (see Figure 9–6b). Ligaments interconnecting adjacent vertebrae include the following:

- The *ligamenta flava* are paired ligaments that connect the laminae of adjacent vertebrae.
- The *posterior longitudinal ligament* is a fibrous band that parallels the anterior longitudinal ligament and connects the posterior surfaces of adjacent vertebral bodies.

Figure 9–6 **Intervertebral Joints.** ATLAS: Plates 20b; 23c

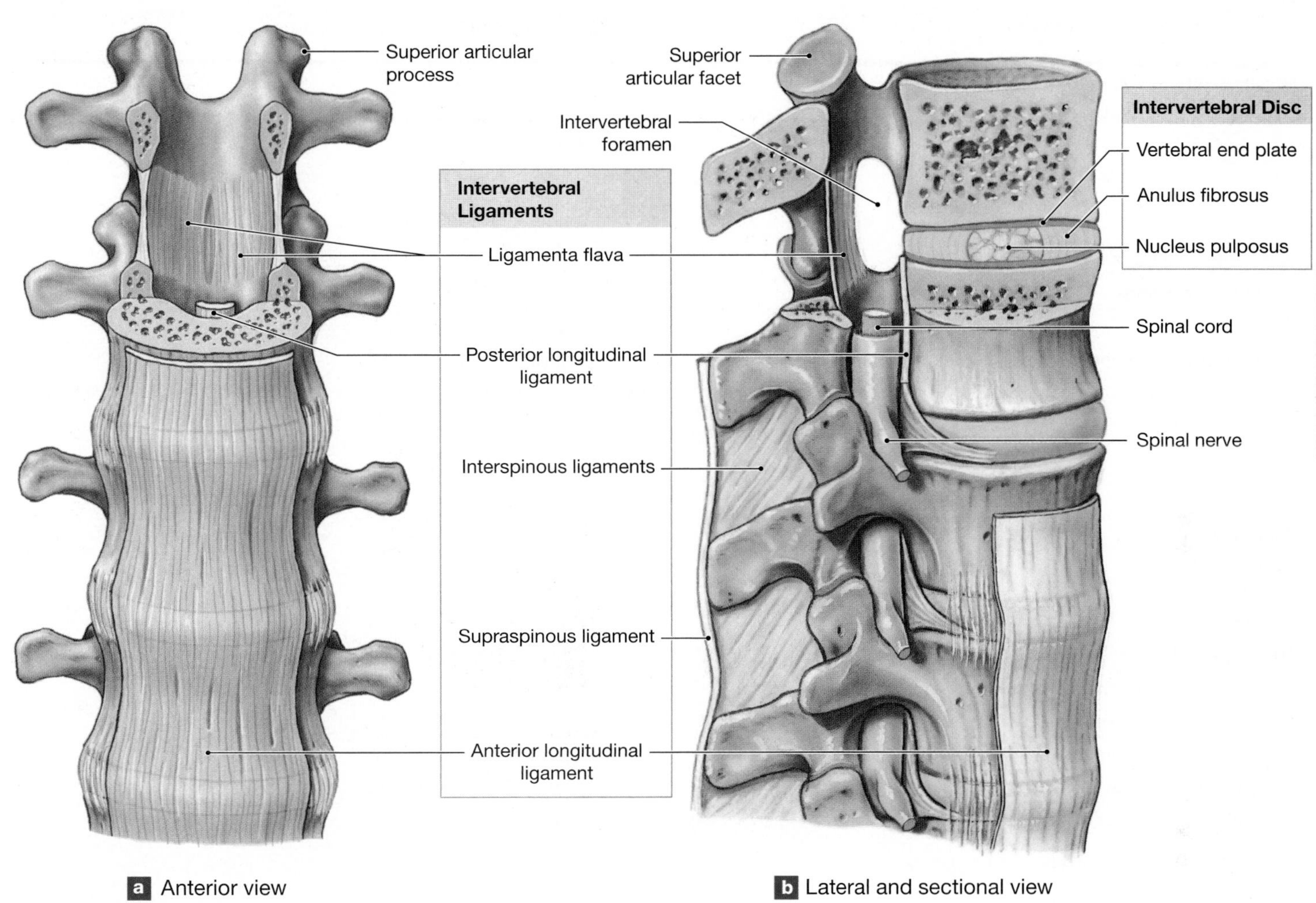
Superior articular process
Superior articular facet
Intervertebral foramen
Intervertebral Ligaments
Ligamenta flava
Posterior longitudinal ligament
Interspinous ligaments
Supraspinous ligament
Anterior longitudinal ligament
Intervertebral Disc
Vertebral end plate
Anulus fibrosus
Nucleus pulposus
Spinal cord
Spinal nerve
a Anterior view
b Lateral and sectional view

? Within an intervertebral disc, which structure provides the tough outer layer of fibrocartilage and which structure provides the soft inner core for resiliency and shock absorption?

- The *interspinous ligaments* are bands of fibrous tissue that connect the spinous processes of adjacent vertebrae.
- The *supraspinous ligament* is a longitudinal fibrous band that is attached to the tips of the spinous processes of the vertebrae from C_7 to the sacrum. The *ligamentum nuchae,* which extends from vertebra C_7 to the base of the skull, is continuous with the supraspinous ligament. p. 232
- The *anterior longitudinal ligament* is a wide fibrous band that connects the anterior surfaces of adjacent vertebral bodies.

Tips & Tools

The inter**spinous** ligaments and supra**spinous** ligament get their names because they are attached to the **spinous** processes of the vertebrae.

Vertebral Movements

The following movements can take place across the intervertebral joints of the vertebral column: (1) flexion, or bending anteriorly; (2) extension, or bending posteriorly; (3) lateral flexion, or bending laterally; and (4) rotation. Table 9–2 summarizes information about intervertebral and other joints of the axial skeleton.

Checkpoint

10. **What type of joint only permits slight movements?**
11. **Which regions of the vertebral column lack intervertebral discs? Explain why the absence of discs is significant.**
12. **Which vertebral movements are involved in (a) bending forward, (b) bending to the side, and (c) moving the head to signify "no"?**

See the blue Answers tab at the back of the book.

Table 9–2 Joints of the Axial Skeleton

Element	Joint	Type of Joint	Movement
SKULL			
Cranial and facial bones of skull	Various	Synarthroses (suture or synostosis)	None
Maxilla/teeth and mandible/teeth	Alveolar	Synarthrosis (gomphosis)	None
Temporal bone/mandible	Temporomandibular	Combined plane joint and hinge diarthrosis	Elevation, depression, and lateral gliding
VERTEBRAL COLUMN			
Occipital bone/atlas	Atlanto-occipital	Condylar diarthrosis	Flexion/extension
Atlas/axis	Atlanto-axial	Pivot diarthrosis	Rotation
Other vertebral elements	Intervertebral (between vertebral bodies)	Amphiarthrosis (symphysis)	Slight movement
	Intervertebral (between articular processes)	Plane diarthrosis	Slight rotation and flexion/extension
L_5/sacrum	Between L_5 body and sacral body	Amphiarthrosis (symphysis)	Slight movement
	Between inferior articular processes of L_5 and articular processes of sacrum	Plane diarthrosis	Slight flexion/extension
Sacrum/hip bone	Sacro-iliac	Plane diarthrosis	Slight movement
Sacrum/coccyx	Sacrococcygeal	Plane diarthrosis *(may become fused)*	Slight movement
Coccygeal bones		Synarthrosis (synostosis)	No movement
THORACIC CAGE			
Bodies of T_1–T_{12} and heads of ribs	Costovertebral	Plane diarthrosis	Slight movement
Transverse processes of T_1–T_{10}	Costovertebral	Plane diarthrosis	Slight movement
Ribs and costal cartilages		Synarthrosis (synchondrosis)	No movement
Sternum and first costal cartilage	Sternocostal (1st)	Synarthrosis (synchondrosis)	No movement
Sternum and costal cartilages 2–7	Sternocostal (2nd–7th)	Plane diarthrosis*	Slight movement

*Commonly converts to synchondrosis in elderly individuals.

9-5 The elbow and knee are both hinge joints

Learning Outcome **Describe the structure and function of the elbow joint and the knee joint.**

In this section we consider the structure and function of two major hinge joints: the elbow joint and the knee joint.

The Elbow Joint

The elbow joint is a complex hinge joint that involves the humerus, radius, and ulna (Figure 9–7). The largest and strongest articulation at the elbow is the *humero-ulnar joint,* where the trochlea of the humerus articulates with the trochlear notch of the ulna. This joint works like a door hinge, with physical limitations imposed on the range of motion. In the case of the elbow, the shape of the trochlear notch of the ulna determines the plane of movement, and the combination of the notch and the olecranon limits the degree of extension permitted. At the smaller *humeroradial joint,* the capitulum of the humerus articulates with the head of the radius.

Muscles that attach to the rough surface of the olecranon produce extension of the elbow. These muscles are mainly under the control of the *radial nerve,* which passes along the radial groove of the humerus. ⮌ p. 247 The large *biceps brachii* covers the anterior surface of the arm. Its distal tendon is attached to the radius at the radial tuberosity. Contraction of this muscle produces supination of the forearm and flexion at the elbow.

The elbow joint is extremely stable because (1) the bony surfaces of the humerus and ulna interlock, (2) a single, thick joint capsule surrounds both the humero-ulnar and proximal radio-ulnar joints, and (3) strong ligaments reinforce the joint capsule.

The *radial collateral ligament* stabilizes the lateral surface of the elbow joint (Figure 9–7a). It extends between the lateral epicondyle of the humerus and the *anular ligament,* which binds the head of the radius to the ulna. The *ulnar collateral ligament* stabilizes the medial surface of the elbow joint. This ligament extends from the medial epicondyle of the humerus anteriorly to the coronoid process of the ulna and posteriorly to the olecranon (Figure 9–7b).

Figure 9–7 **The Right Elbow Joint Showing Stabilizing Ligaments.** ATLAS: Plates 35a–g

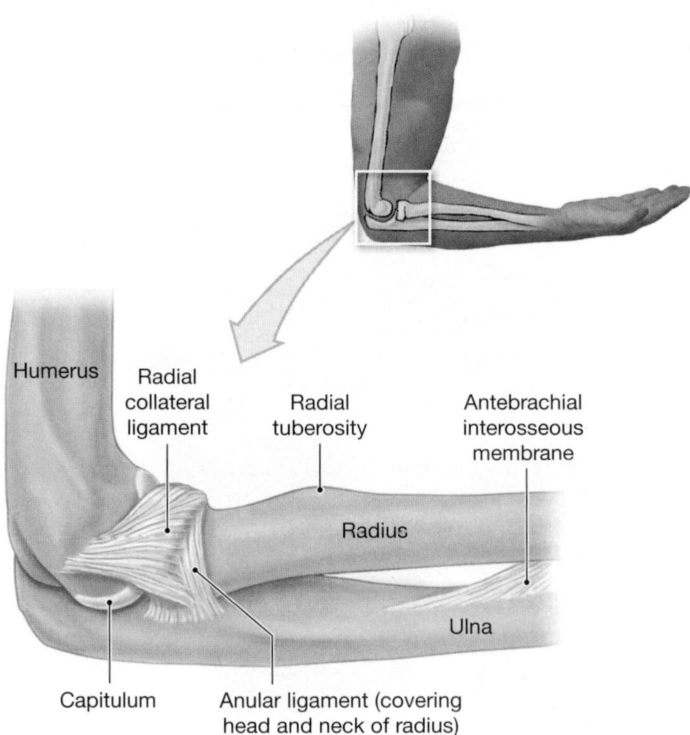

Humerus
Radial collateral ligament
Radial tuberosity
Antebrachial interosseous membrane
Radius
Ulna
Capitulum
Anular ligament (covering head and neck of radius)
a Lateral view

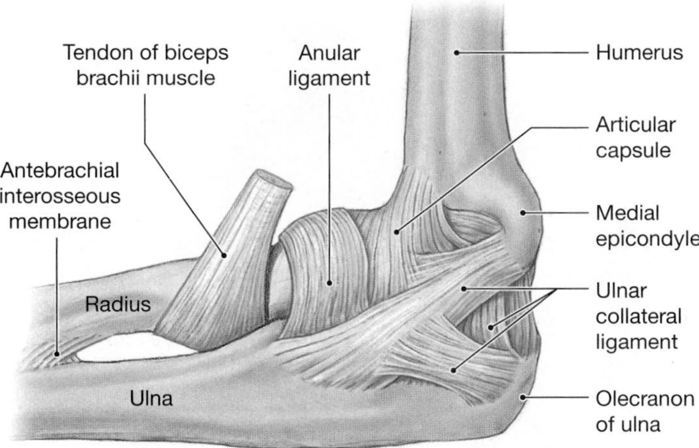

Tendon of biceps brachii muscle
Anular ligament
Humerus
Articular capsule
Antebrachial interosseous membrane
Medial epicondyle
Radius
Ulnar collateral ligament
Ulna
Olecranon of ulna
b Medial view

Despite the strength of the joint capsule and ligaments, the elbow can be damaged by severe impacts or unusual stresses. For example, if you fall on your hand with a partially flexed elbow, contractions of muscles that extend the elbow may break the ulna at the center of the trochlear notch. Less violent stresses can produce dislocations or other injuries to the elbow, especially if epiphyseal growth has not been completed. For example, parents in a rush may drag a toddler along behind them, exerting an upward, twisting pull on the elbow joint that can result in a partial dislocation of the radial head from the annular ligament known as *nursemaid's elbow.*

The Knee Joint

The knee joint transfers the weight of the body from the femur to the tibia. The knee joint functions as a hinge, but the joint is far more complex than the elbow or even the ankle. The rounded condyles of the femur roll across the superior surface of the tibia, so the points of contact are constantly changing. The joint permits flexion, extension, and very limited rotation.

The knee joint contains three separate articulations. Two are between the femur and tibia (medial condyle to medial condyle, and lateral condyle to lateral condyle). One is between the patella and the patellar surface of the femur (Figure 9–8a). Let's look at the joint itself and then its supporting accessory structures.

9

Joint Capsule and Joint Cavity

The joint capsule at the knee joint is thin and in some areas incomplete, but various ligaments and tendons of associated muscles strengthen it. A pair of fibrocartilage pads, the *medial* and *lateral menisci,* lies between the femoral and tibial surfaces (see Figures 9–1b, 9–8b). The menisci (1) act as cushions, (2) conform to the shape of the articulating surfaces as the femur changes position, and (3) provide lateral stability to the joint. Prominent fat pads cushion the margins of the joint. These fat pads also assist the many bursae in reducing friction between the patella and other tissues.

Accessory Structures

A complete dislocation of the knee is very rare, mainly because seven major ligaments stabilize the knee joint:

1. The tendon from the *quadriceps femoris* passes over the anterior surface of the joint and is responsible for extending the knee when the muscle contracts. The patella is embedded in this tendon, and the *patellar ligament* continues to its attachment on the anterior surface of the tibia (see Figure 9–8a). The patellar ligament and two ligamentous bands known as the *patellar retinaculae* support the anterior surface of the knee joint.

2,3. Two *popliteal ligaments* extend between the femur and the heads of the tibia and fibula (Figure 9–8c). These ligaments reinforce the knee joint's posterior surface.

4,5. Inside the joint capsule, the *anterior cruciate* (KRŪ-she-at) *ligament (ACL)* and *posterior cruciate ligament (PCL)* attach the intercondylar area of the tibia to the condyles of the femur (Figure 9–8d). The "anterior" and "posterior" in these terms refer to the sites of origin of these ligaments on the tibia. They cross one another as they proceed to their destinations on the femur. (The term *cruciate* is derived from the Latin word *cruciatus,* meaning "a cross.") The ACL and the PCL limit the anterior and posterior movement of the tibia and maintain the alignment of the femoral and tibial condyles.

9

Figure 9–8 The Right Knee Joint. ATLAS: Plates 78a–i; 79a,b; 80a,b

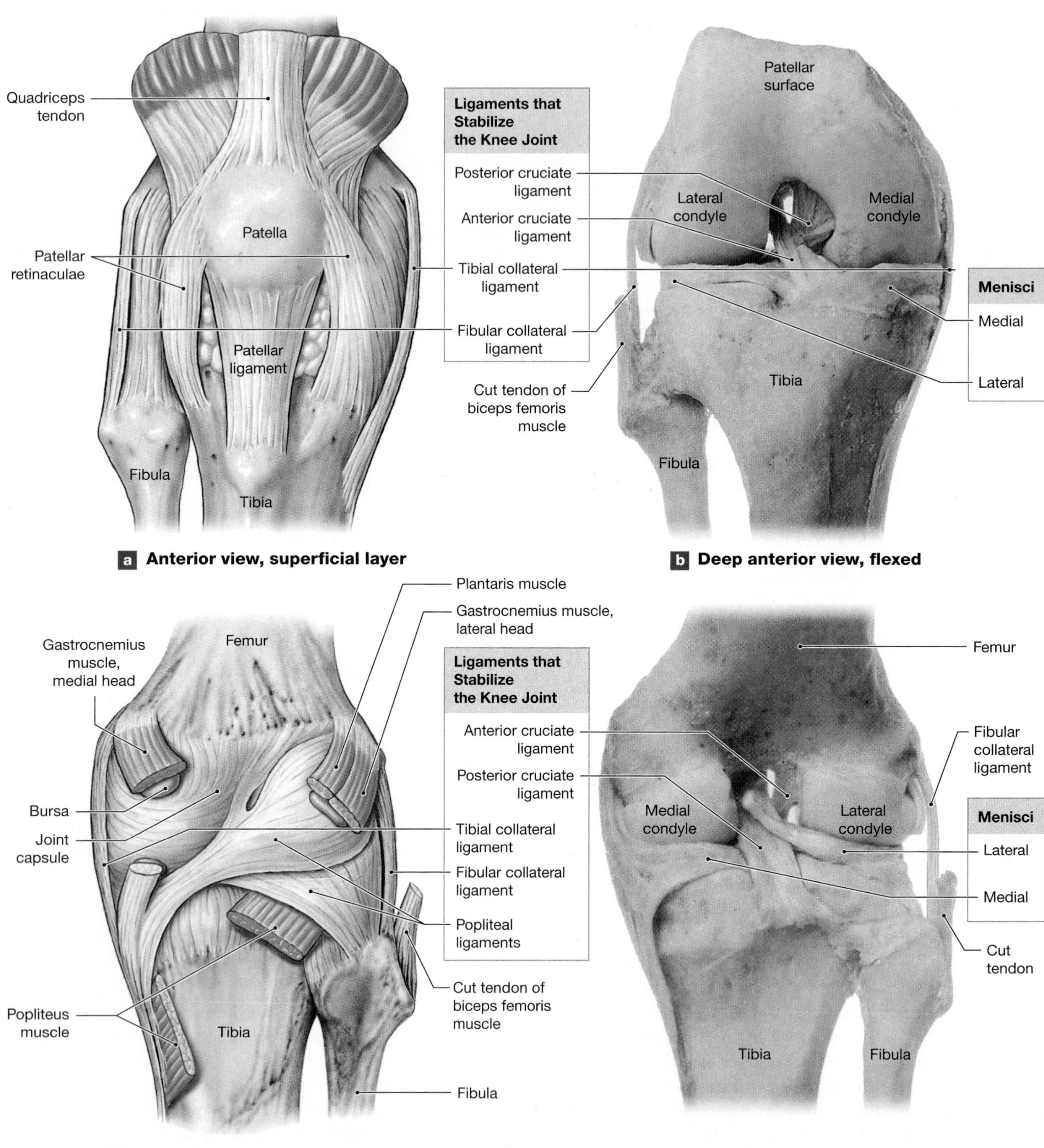

Quadriceps tendon
Patella
Patellar retinaculae
Patellar ligament
Fibula
Tibia
Ligaments that Stabilize the Knee Joint
Posterior cruciate ligament
Anterior cruciate ligament
Tibial collateral ligament
Fibular collateral ligament
Cut tendon of biceps femoris muscle
Patellar surface
Lateral condyle
Medial condyle
Menisci
Medial
Lateral
Tibia
Fibula
Plantaris muscle
Gastrocnemius muscle, lateral head
Gastrocnemius muscle, medial head
Femur
Bursa
Joint capsule
Ligaments that Stabilize the Knee Joint
Anterior cruciate ligament
Posterior cruciate ligament
Tibial collateral ligament
Fibular collateral ligament
Popliteal ligaments
Cut tendon of biceps femoris muscle
Popliteus muscle
Tibia
Fibula
Femur
Medial condyle
Lateral condyle
Fibular collateral ligament
Menisci
Lateral
Medial
Cut tendon
Tibia
Fibula

a Anterior view, superficial layer

b Deep anterior view, flexed

c Posterior view, superficial layer

d Deep posterior view, extended

Clinical Note Damage to Intervertebral Discs

If the posterior longitudinal ligament is weakened, which often occurs as we age, intervertebral discs become more vulnerable to damage. The compressed nucleus pulposus of a disc may distort the anulus fibrosus, forcing it partway into the vertebral canal. This condition, seen here in lateral view, is called a **bulging disc** (Figure a). If the nucleus pulposus breaks through the anulus fibrosus, it too may protrude into the vertebral canal. This condition, shown here in superior view, is called a **herniated disc** (Figure b). When a disc herniates, it compresses spinal nerves and may cause pain. Lumbar pain and disc disease are potential hazards for nurses, who lift and transfer patients as part of their job.

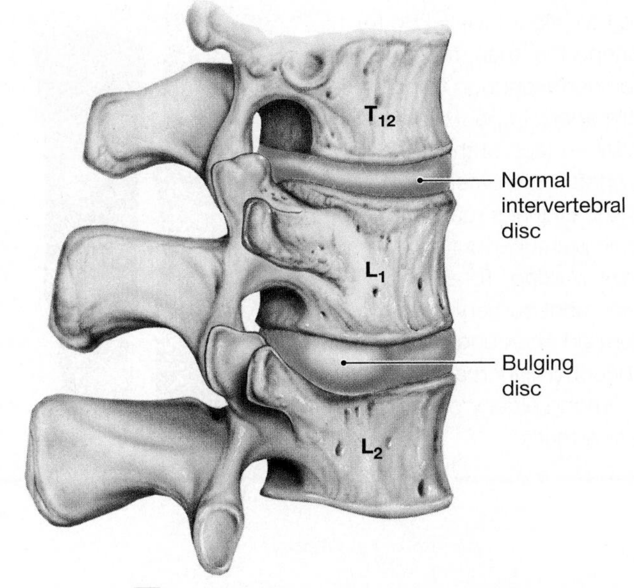
T_{12}
Normal intervertebral disc
L_1
Bulging disc
L_2

a A lateral view of the lumbar region of the spinal column, showing a bulging intervertebral disc

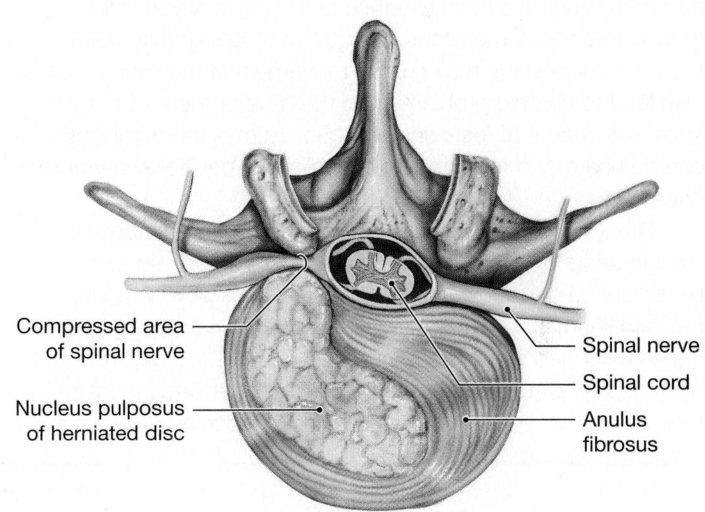
Compressed area of spinal nerve
Nucleus pulposus of herniated disc
Spinal nerve
Spinal cord
Anulus fibrosus

b A sectional view through a herniated disc, showing the release of the nucleus pulposus and its effect on the spinal cord and adjacent spinal nerves

6,7. The *tibial collateral ligament* (clinically called the *medial collateral ligament [MCL]*), reinforces the medial surface of the knee joint, and the *fibular collateral ligament* (clinically called the *lateral collateral ligament [LCL]*) reinforces the lateral surface (see Figure 9–8). These ligaments tighten only at full extension, the position in which they stabilize the joint.

At full extension, a slight lateral rotation of the tibia tightens the anterior cruciate ligament and wedges the lateral meniscus between the tibia and femur. The knee joint is essentially locked in the extended position. With the joint locked, a person can stand for prolonged periods without using (and tiring) the muscles that extend the knee. Unlocking the knee joint requires muscular contractions that medially rotate the tibia or laterally rotate the femur.

The knee joint is subjected to severe stresses in the course of normal activities. Painful knee injuries are all too familiar to both amateur and professional athletes. Treatment is often costly and prolonged, and repairs seldom make the joint "good as new."

Checkpoint

13. **Terry suffers an injury to his forearm and elbow. After the injury, he notices an unusually large degree of motion between the radius and the ulna at the elbow. Which ligament did Terry most likely damage?**
14. **What signs and symptoms would you expect in a person who has damaged the menisci of the knee joint?**
15. **Why is "clergyman's knee" (a type of bursitis) common among carpet layers and roofers?**
16. **Name the bones making up the knee joint.**

See the blue Answers tab at the back of the book.

9-6 The shoulder and hip are both ball-and-socket joints

Learning Outcome **Describe the structure and function of the shoulder joint and the hip joint.**

Next we examine the ball-and-socket joints of the body: the shoulder joint and the hip joint.

9

+ Clinical Note Knee Injuries

Athletes place tremendous stresses on their knees. Ordinarily, the medial and lateral menisci move as the position of the femur changes. The knee is most stable when it is straight. However, if a locked knee is struck from the side, the lateral meniscus can tear and the supporting ligaments can be seriously damaged. Placing a lot of weight on the knee while it is partially flexed can trap a meniscus between the tibia and femur, resulting in a break or tear in the cartilage. In the most common injury, the lateral surface of the leg is driven medially, tearing the medial meniscus. In addition to being quite painful, the torn cartilage may restrict movement at the joint. It can also lead to chronic problems and the development of a "trick knee"—a knee that feels unstable. Sometimes the meniscus can be heard and felt popping in and out of position when the knee is extended.

Other knee injuries involve tearing one or more stabilizing ligaments or damaging the patella. Torn ligaments can be difficult to correct surgically, and healing is slow. Anterior cruciate ligament (ACL) injuries are frequently caused by twisting an extended weight-bearing knee. Soccer, football, and basketball players are more likely than other athletes to injure their ACLs. Nonsurgical treatment with exercise and braces is possible, but requires a change in activity. Reconstructive surgery using part of the patellar tendon or a graft from a cadaver tendon may allow a return to active sports. A strong hit or force to the lateral surface of the knee, as in a football tackle or flying hockey puck, may damage both the ACL and the medial collateral ligament (MCL).

The patella can be injured in a number of ways. If the leg is immobilized (as it might be in a football pileup) while you try to extend the knee, the muscles are powerful enough to pull the patella apart. Impacts to the anterior surface of the knee can also shatter the patella.

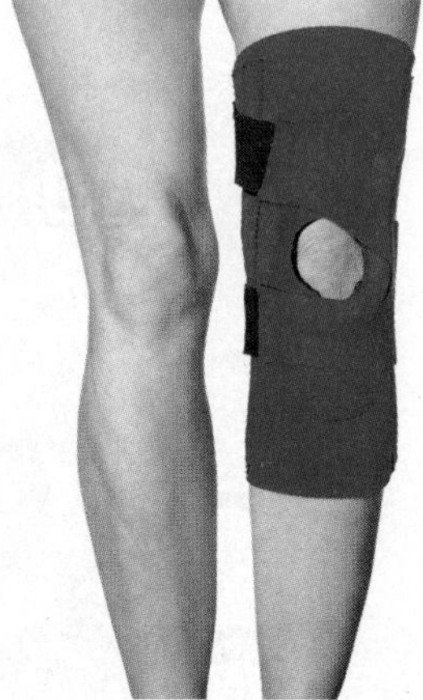

Knee injuries may lead to chronic painful arthritis that impairs walking. Total knee replacement surgery is rarely performed on young people, but it is becoming increasingly common among elderly patients with severe arthritis.

The Shoulder Joint

The **shoulder joint**, or *glenohumeral joint*, permits the greatest range of motion of any joint. It is also the most frequently dislocated joint, demonstrating the principle that stability must be sacrificed to obtain mobility. The shoulder is a ball-and-socket diarthrosis formed by the articulation of the head of the humerus with the glenoid cavity of the scapula (Figure 9–9a).

Joint Capsule and Joint Cavity

The area of the glenoid cavity is increased by a fibrocartilaginous **glenoid labrum** (*labrum*, lip or edge), which continues beyond the bony rim and deepens the socket (see Figure 9–9a). The relatively loose joint capsule extends from the scapula, proximal to the glenoid labrum, to the anatomical neck of the humerus. Somewhat oversized, the joint capsule permits an extensive range of motion. The bones of the pectoral girdle provide some stability to the superior surface, because the acromion and coracoid process of the scapula project laterally superior to the head of the humerus. However, the surrounding skeletal muscles provide most of the stability at this joint, with help from their associated tendons and various ligaments.

Accessory Structures

The major ligaments that help stabilize the shoulder joint are the *acromioclavicular, coracoclavicular, coraco-acromial, coracohumeral,* and *glenohumeral ligaments* (Figure 9–9b). The acromioclavicular ligament reinforces the capsule of the acromioclavicular joint and supports the superior surface of the shoulder. A **shoulder separation** is a relatively common injury involving partial or complete dislocation of the acromioclavicular joint. This injury can result from a hit to the superior surface of the shoulder. The acromion is forcibly depressed while the clavicle is held back by strong muscles. A dislocation of the shoulder joint is different, and involves the glenohumeral joint.

The muscles that stabilize the shoulder joint originate on the trunk, pectoral girdle, and humerus, and cover the anterior, superior, and posterior surfaces of the capsule. The tendons of the *supraspinatus, infraspinatus, teres minor,* and *subscapularis muscles* reinforce the joint capsule and limit range of movement. These muscles, known as the muscles of the **rotator cuff**, are the primary structures supporting the shoulder joint and limiting its range of motion (see Figure 9–9b). Damage to the rotator cuff typically occurs in sports that place severe strains on the shoulder. Whitewater kayakers, baseball pitchers, and quarterbacks are all at high risk for rotator cuff injuries.

Figure 9–9 The Shoulder Joint. ATLAS: Plate 27d

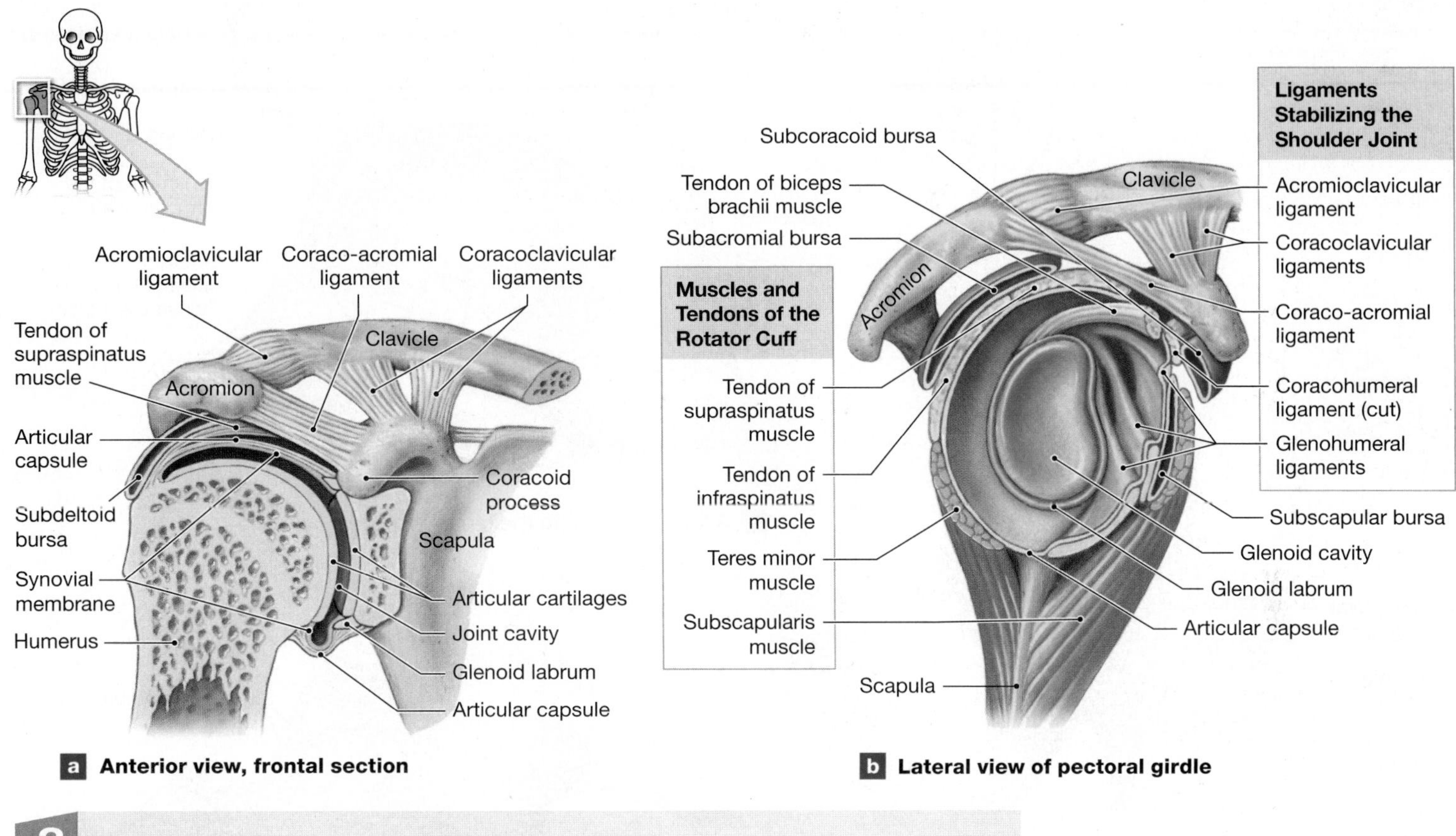
Acromioclavicular ligament
Coraco-acromial ligament
Coracoclavicular ligaments
Tendon of supraspinatus muscle
Clavicle
Acromion
Articular capsule
Coracoid process
Subdeltoid bursa
Scapula
Synovial membrane
Articular cartilages
Humerus
Joint cavity
Glenoid labrum
Articular capsule
Subcoracoid bursa
Tendon of biceps brachii muscle
Subacromial bursa
Muscles and Tendons of the Rotator Cuff
Tendon of supraspinatus muscle
Tendon of infraspinatus muscle
Teres minor muscle
Subscapularis muscle
Acromion
Clavicle
Scapula
Ligaments Stabilizing the Shoulder Joint
Acromioclavicular ligament
Coracoclavicular ligaments
Coraco-acromial ligament
Coracohumeral ligament (cut)
Glenohumeral ligaments
Subscapular bursa
Glenoid cavity
Glenoid labrum
Articular capsule

a Anterior view, frontal section

b Lateral view of pectoral girdle

? What two bones does the acromioclavicular ligament connect?

Tips & Tools

The rotator cuff muscles can be remembered by using the acronym **SITS**, for **s**upraspinatus, **i**nfraspinatus, **t**eres minor, and **s**ubscapularis.

The anterior, superior, and posterior surfaces of the shoulder joint are reinforced by ligaments, muscles, and tendons, but the inferior capsule is poorly reinforced. As a result, a dislocation caused by an impact or a violent muscle contraction is most likely to occur at this site. Such a dislocation can tear the inferior capsular wall and the glenoid labrum. The healing process typically leaves a weakness that increases the chances for future dislocations.

As at other joints, bursae at the shoulder reduce friction where large muscles and tendons pass across the joint capsule. The shoulder has several important bursae, such as the *subdeltoid bursa*, the *subcoracoid bursa*, the *subacromial bursa*, and the *subscapular bursa* (see Figure 9–9a,b). A tendon of the biceps brachii muscle runs through the shoulder joint. As it passes through the joint capsule, it is surrounded by a synovial tendon sheath that is continuous with the joint cavity. Inflammation of any of these extracapsular bursae can restrict motion and produce the painful symptoms of bursitis. p. 269

The Hip Joint

The hip joint is a sturdy ball-and-socket diarthrosis that permits flexion, extension, adduction, abduction, circumduction, and rotation. Figure 9–10 introduces the structure of the hip joint. The acetabulum, a deep fossa in the hip bone, accommodates the head of the femur. p. 252 Within the acetabulum, articular cartilage extends like a horseshoe to either side of the acetabular notch (Figure 9–10a). The *acetabular labrum*, a projecting rim of rubbery fibrocartilage, increases the depth of the joint cavity and helps to seal in synovial fluid.

The joint capsule of the hip joint is extremely dense and strong. It extends from the lateral and inferior surfaces of the hip bone to the intertrochanteric line and intertrochanteric crest of the femur, enclosing both the head and neck of the femur. p. 255 This arrangement helps to keep the femoral head from moving too far from the acetabulum.

Figure 9–10 The Right Hip Joint. ATLAS: Plates 71a,b; 72a

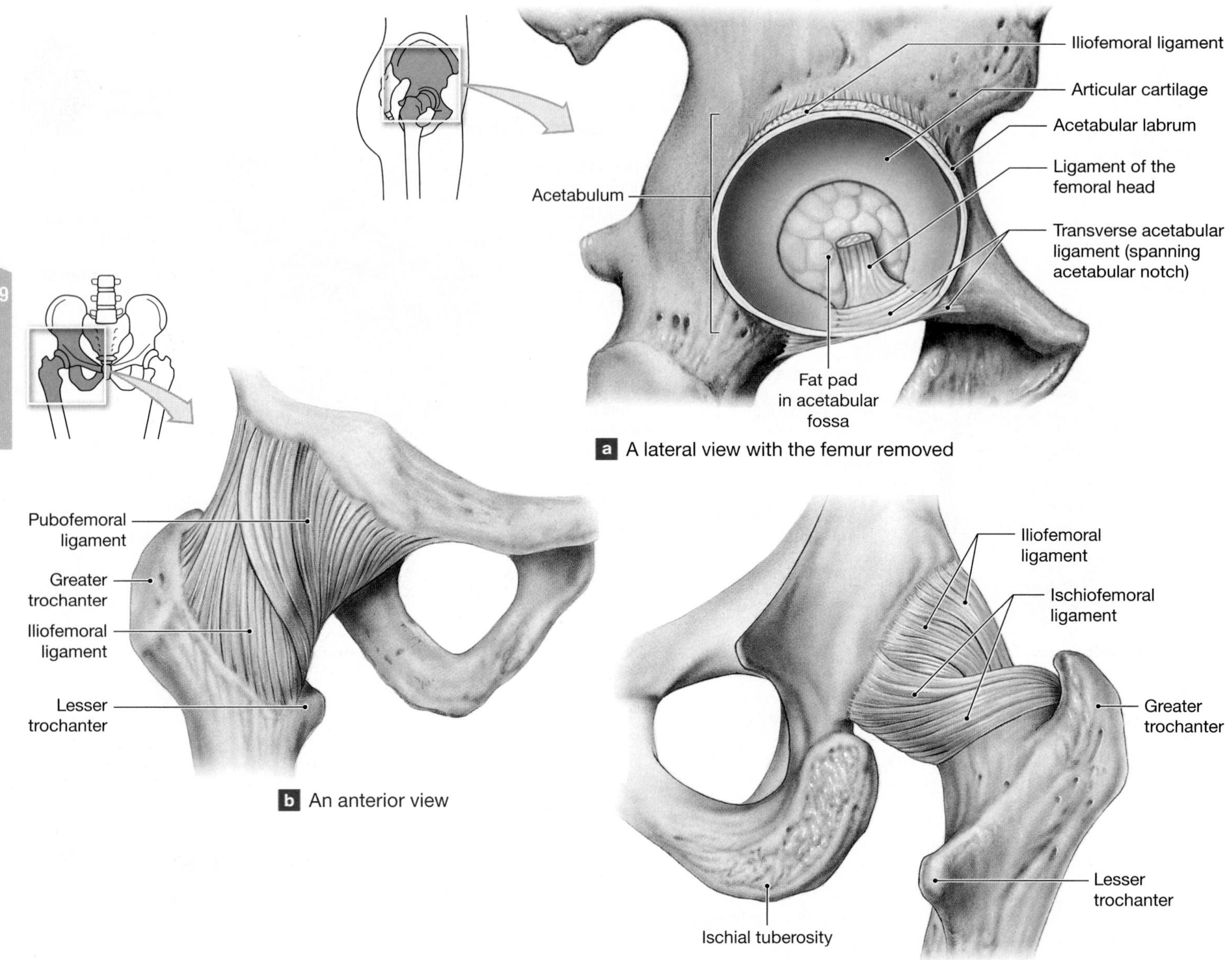
Iliofemoral ligament
Articular cartilage
Acetabular labrum
Ligament of the femoral head
Transverse acetabular ligament (spanning acetabular notch)
Acetabulum
Fat pad in acetabular fossa
Pubofemoral ligament
Greater trochanter
Iliofemoral ligament
Lesser trochanter
Iliofemoral ligament
Ischiofemoral ligament
Greater trochanter
Lesser trochanter
Ischial tuberosity

a A lateral view with the femur removed

b An anterior view

c A posterior view, showing additional ligaments that add strength to the capsule

Four broad ligaments reinforce the joint capsule (**Figure 9–10b,c**). Three of them—the *iliofemoral, pubofemoral,* and *ischiofemoral ligaments*—are regional thickenings of the capsule. The *transverse acetabular ligament* crosses the acetabular notch, filling in the gap in the inferior border of the acetabulum. A fifth ligament, the *ligament of the femoral head,* or *ligamentum teres* (*teres,* long and round), originates along the transverse acetabular ligament (see **Figure 9–10a**) and attaches to the fovea capitis, a small pit at the center of the femoral head. ⮌ p. 255 This ligament tenses only when the hip is flexed and the thigh is undergoing lateral rotation.

Much more important stabilization is provided by the bulk of the surrounding muscles, aided by ligaments and capsular fibers. The combination of an almost complete bony socket, a strong joint capsule, supporting ligaments, and muscular padding makes the hip joint extremely stable. The head of the femur is well supported, but the ball-and-socket joint is not directly aligned with the weight distribution along the shaft. Stress must be transferred at an angle from the joint, along the thin femoral neck to the length of the femur. ⮌ p. 189 *Hip fractures* (fractures of the femoral neck or between the greater and lesser trochanters of the femur) are more common than hip dislocations. As we noted in Chapter 6, femoral fractures at the hip are common in elderly individuals with severe osteoporosis. ⮌ p. 204

Table 9–3 summarizes information about the joints of the appendicular skeleton, the great majority of which are diarthroses.

Table 9–3 Joints of the Appendicular Skeleton

Articulating Bones	Joint	Type of Joint	Movement
JOINTS OF EACH PECTORAL GIRDLE AND UPPER LIMB			
Sternum/clavicle	Sternoclavicular	Plane diarthrosis*	Protraction/retraction, elevation/depression, slight rotation
Scapula/clavicle	Acromioclavicular	Plane diarthrosis	Slight movement
Scapula/humerus	Shoulder, or glenohumeral	Ball-and-socket diarthrosis	Flexion/extension, adduction/abduction, circumduction, rotation
Humerus/ulna and humerus/radius	Elbow (humero-ulnar and humeroradial)	Hinge diarthrosis	Flexion/extension
Radius/ulna	Proximal radio-ulnar	Pivot diarthrosis	Rotation
	Distal radio-ulnar	Pivot diarthrosis	Pronation/supination
Radius/carpal bones	Radiocarpal	Condylar diarthrosis	Flexion/extension, adduction/abduction, circumduction
Carpal bone to carpal bone	Intercarpal	Plane diarthrosis	Slight movement
Carpal bone to metacarpal bone (I)	Carpometacarpal of thumb	Saddle diarthrosis	Flexion/extension, adduction/abduction, circumduction, opposition
Carpal bone to metacarpal bone (II–V)	Carpometacarpal	Plane diarthrosis	Slight flexion/extension, adduction/abduction
Metacarpal bone to phalanx	Metacarpophalangeal	Condylar diarthrosis	Flexion/extension, adduction/abduction, circumduction
Phalanx/phalanx	Interphalangeal	Hinge diarthrosis	Flexion/extension
JOINTS OF THE PELVIC GIRDLE AND LOWER LIMBS			
Sacrum/ilium of hip bone	Sacro-iliac	Plane diarthrosis	Slight movement
Hip bone/hip bone	Pubic symphysis	Amphiarthrosis	None†
Hip bone/femur	Hip	Ball-and-socket diarthrosis	Flexion/extension, adduction/abduction, circumduction, rotation
Femur/tibia	Knee	Complex, functions as hinge	Flexion/extension, limited rotation
Tibia/fibula	Tibiofibular (proximal)	Plane diarthrosis	Slight movement
	Tibiofibular (distal)	Plane diarthrosis and amphiarthrotic syndesmosis	Slight movement
Tibia and fibula with talus	Ankle, or talocrural	Hinge diarthrosis	Flexion/extension (dorsiflexion/plantar flexion)
Tarsal bone to tarsal bone	Intertarsal	Plane diarthrosis	Slight movement
Tarsal bone to metatarsal bone	Tarsometatarsal	Plane diarthrosis	Slight movement
Metatarsal bone to phalanx	Metatarsophalangeal	Condylar diarthrosis	Flexion/extension, adduction/abduction
Phalanx/phalanx	Interphalangeal	Hinge diarthrosis	Flexion/extension

*A "double-plane joint," with two joint cavities separated by an articular cartilage.

†During pregnancy, hormones weaken the pubic symphysis and permit movement important to childbirth; see Chapter 29.

Checkpoint

17. **Which tissues or structures provide most of the stability for the shoulder joint?**
18. **Would a tennis player or a jogger be more likely to develop inflammation of the subscapular bursa? Why?**
19. **A football player received a hit to the upper surface of his shoulder, causing a shoulder separation. What does this mean?**
20. **Name the bones making up the shoulder joint.**
21. **At what site are the iliofemoral ligament, pubofemoral ligament, and ischiofemoral ligament located?**

See the blue Answers tab at the back of the book.

9-7 With advancing age, arthritis and other degenerative changes often impair joint mobility

Learning Outcome Describe the effects of aging on joints, and discuss the most common age-related clinical problems for joints.

Joints are subjected to h eavy wear and tear throughout our lifetimes. Problems with joint function are common, especially in older individuals. **Rheumatism** (RŪ-muh-tiz-um) is a general term for pain and stiffness affecting the *musculoskeletal system* (bones, muscles, and joints). **Arthritis** (ar-THRĪ-tis) means *joint inflammation,* and encompasses all the rheumatic diseases that affect synovial joints. Many types of arthritis are recognized.

Go to MasteringA&P™ > Study Area > Menu > Lab Tools > PAL 3.0 > Human Cadaver: Joints

Arthritis always involves damage to the articular cartilages, but the specific cause can vary. For example, arthritis can result from bacterial or viral infection, injury to the joint, decreased viscosity or lack of synovial fluid, metabolic problems, or severe physical stresses. The most common type of arthritis is *osteoarthritis* (os-tē-ō-ar-THRĪ-tis), also known as *degenerative joint disease (DJD)*. It generally affects people age 60 or older. Osteoarthritis can result from cumulative wear and tear at the joint surfaces or from genetic factors affecting collagen formation. In the U.S. population, 25 percent of women and 15 percent of men over age 60 show signs of this disease.

Rheumatoid arthritis is an inflammatory condition that affects about 0.5–1.0 percent of the adult population, generally those between ages 40 and 60, and women more than men. Rheumatoid arthritis occurs when the immune response mistakenly attacks the joint tissues. Such a condition, in which the body attacks its own tissues, is called an *autoimmune disease.* Allergies, bacteria, viruses, and genetic factors have all been proposed as contributing to or triggering the destructive inflammation.

In *gouty arthritis*, crystals of uric acid form within the synovial fluid of joints. The accumulation of crystals of uric acid over time eventually interferes with normal movement. This form of arthritis is named after the metabolic disorder known as *gout*, discussed further in Chapter 25. In gout, the crystals are derived from uric acid (a metabolic waste product), and the joint most often affected is the metatarsal–phalangeal joint of the great toe. Gout is relatively rare, but other forms of gouty arthritis are much more common. Some degree of calcium salt deposition occurs in the joints in 30–60 percent of those over age 85. The cause is unknown, but the condition appears to be linked to age-related changes in the articular cartilages.

Regular exercise, physical therapy, and drugs that reduce inflammation (such as aspirin) can often slow the progress of osteoarthritis. Surgical procedures can realign or redesign the affected joint. In extreme cases involving the hip, knee, elbow, or shoulder, the defective joint can be replaced by an artificial one.

Degenerative changes comparable to those seen in arthritis may result from joint immobilization. When motion ceases, so does the circulation of synovial fluid, and the cartilages begin to degenerate. **Continuous passive motion (CPM)** of any injured joint appears to encourage the repair process by improving the circulation of synovial fluid. A physical therapist or a machine often performs the movement during the recovery process.

With age, bone mass decreases and bones become weaker, so the risk of fractures increases. ⮌ p. 204 If osteoporosis develops, the bones may weaken to the point at which fractures occur in response to stresses that could easily be tolerated by normal bones. Hip fractures are among the most dangerous fractures seen in elderly people, with or without osteoporosis. These fractures, most often involving persons over age 60, may be accompanied by hip dislocation or pelvic fractures. Although severe hip fractures are most common among those over age 60, in recent years the frequency of hip fractures has increased dramatically among young, healthy professional athletes.

Checkpoint

22. **Define *rheumatism*.**
23. **Define *arthritis*.**

See the blue Answers tab at the back of the book.

9-8 The skeletal system supports and stores energy and minerals for other body systems

Learning Outcome **Explain the functional relationships between the skeletal system and other body systems.**

We have discussed the skeletal system in this and the last three chapters. Let's look now at how the bones of the skeletal system function with other body systems. As we discussed in Chapter 6, the living skeleton is dynamic and undergoes continuous remodeling. The balance between osteoblast and osteoclast activity is subject to change. When osteoblast activity predominates, bones thicken and strengthen. When osteoclast activity predominates, bones get thinner and weaker. The balance between bone formation and bone recycling varies with (1) the age of the person, (2) the physical stresses applied to the bone, (3) circulating hormone levels, (4) amounts of calcium and phosphorus absorption and excretion, and (5) genetic or environmental factors. Most of these variables involve some interaction between the skeletal system and other body systems.

In fact, the skeletal system is intimately associated with other systems. For instance, the bones of the skeleton are attachment sites for the muscular system, extensively connected with the cardiovascular and lymphatic systems, and largely under the physiological control of the endocrine system. The digestive and urinary systems also play important roles in providing the calcium and phosphate minerals needed for bone growth. In return, the skeleton represents a reserve of calcium, phosphate, and other minerals that can compensate for reductions in their supply from the diet. Build Your Knowledge Figure 9–11 reviews the components and functions of the skeletal system, and diagrams the major functional relationships between that system and other systems studied so far.

Checkpoint

24. **Explain why there must be a balance between osteoclast activity and osteoblast activity.**
25. **Describe the functional relationship between the skeletal system and the integumentary system.**

See the blue Answers tab at the back of the book.

Build Your Knowledge

Figure 9–11 Integration of the SKELETAL system with the other body systems presented so far.

Integumentary System

- The Integumentary System removes excess body heat, synthesizes vitamin D_3 for calcium and phosphate absorption, and protects underlying bones and joints.
- The skeletal System provides structural support for the skin.

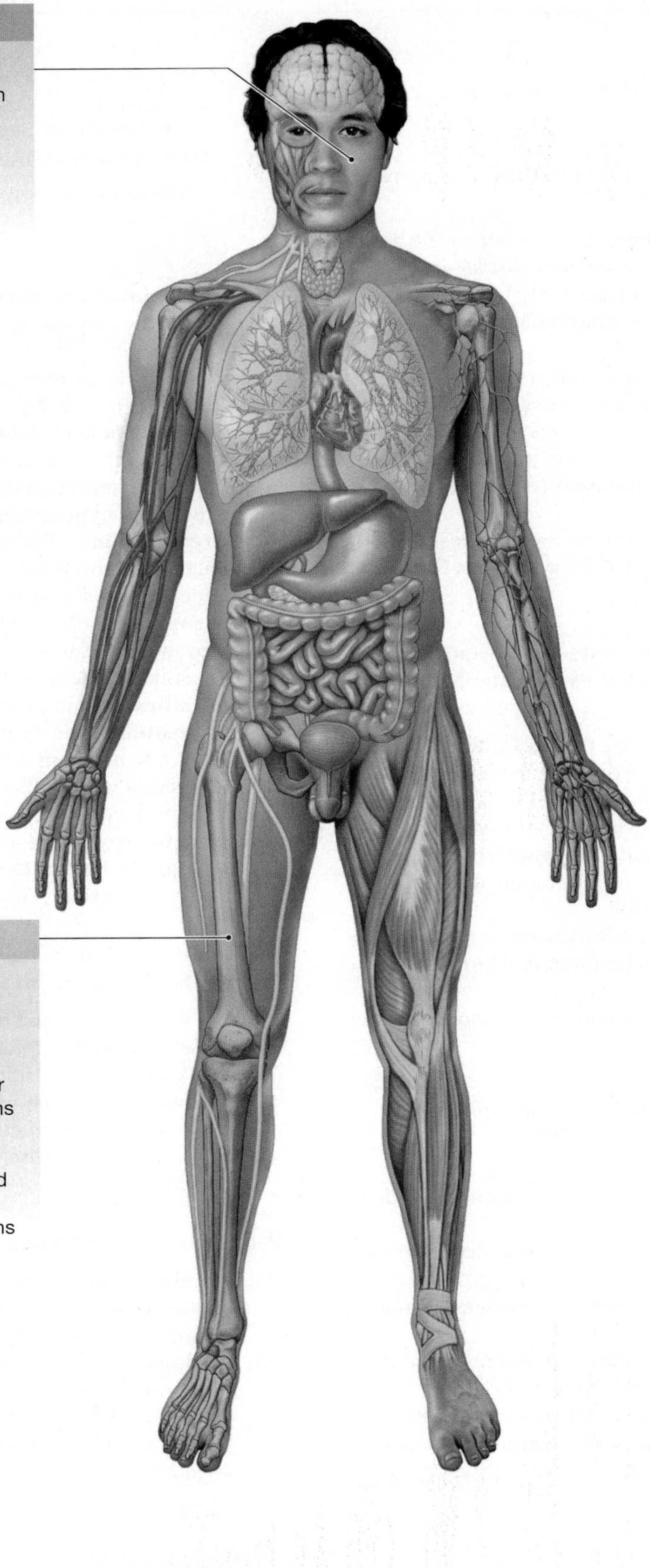

Skeletal System

The skeletal system performs several major functions for the human body. It:

- provides structural support for the body
- stores calcium, phosphate, and other minerals necessary for many functions in other organ systems, and lipids as energy reserves
- produces blood cells and other blood elements in red bone marrow
- protects many soft tissues and organs
- provides leverage for movements generated by skeletal muscles

9 Chapter Review

Study Outline

> MasteringA&P™ Access more chapter study tools online in the MasteringA&P Study Area:
> - Chapter Quizzes, Chapter Practice Test, MP3 Tutor Sessions, and Clinical Case Studies
> - Practice Anatomy Lab PAL 3.0
> - A&P Flix A&PFlix
> - Interactive Physiology iP2
> - PhysioEx PhysioEx 9.1

An Introduction to Joints p. 266

1. **Joints,** or **articulations**, exist wherever two bones interconnect.

9-1 Joints are categorized according to their structure or range of motion p. 266

2. *Immovable joints* are **synarthroses**; *slightly movable joints* are **amphiarthroses**; and joints that are *freely movable* are called **diarthroses** or synovial joints. *(Table 9–1)*
3. Alternatively, joints are classified structurally as *fibrous, cartilaginous, bony,* or *synovial. (Table 9–1)*
4. The four major types of synarthroses are a **suture** (skull bones bound together by dense connective tissue), a **gomphosis** (teeth bound to bony sockets by *periodontal ligaments*), a **synchondrosis** (two bones joined by a rigid cartilaginous bridge), and a **synostosis** (two bones completely fused).
5. The two major types of amphiarthroses are a **syndesmosis** (bones connected by a ligament) and a **symphysis** (bones connected by fibrocartilage).

9-2 Diarthroses: Synovial joints contain synovial fluid and are surrounded by a joint capsule and stabilizing accessory structures p. 268

6. The bony surfaces at diarthroses are enclosed within a **joint capsule**, also called an **articular capsule**, that is lined by a synovial membrane.
7. The bony surfaces within a synovial joint are covered by **articular cartilages**, and lubricated by **synovial fluid**. Functions of synovial fluid include lubrication, nutrient distribution, and shock absorption.
8. Accessory synovial structures include **menisci**, or *articular discs*; **fat pads**; **accessory ligaments**; **tendons**; and **bursae**. *(Figure 9–1)*
9. A **dislocation** occurs when articulating surfaces are forced out of position.

9-3 Diarthroses: The different types of synovial joints allow a wide range of skeletal movements p. 270

10. The possible types of articular motion are **linear motion** (**plane** or **gliding**), **angular motion**, and **rotation**. *(Spotlight Figure 9–2a)*
11. Joints are called **monaxial**, **biaxial**, or **triaxial**, depending on the axes of movement they allow. *(Spotlight Figure 9–2b)*
12. **Plane (gliding) joints** permit limited movement, generally in a single plane. *(Spotlight Figure 9–2c)*
13. **Hinge joints** are monaxial joints that permit only angular movement in one plane. *(Spotlight Figure 9–2c)*
14. **Condylar joints** (ellipsoid joints) are biaxial joints with an oval articular face that nestles within a depression in the opposing articular surface. *(Spotlight Figure 9–2c)*
15. **Saddle joints** are biaxial joints with articular faces that are concave on one axis and convex on the other. *(Spotlight Figure 9–2c)*
16. **Pivot joints** are monaxial joints that permit only rotation. *(Spotlight Figure 9–2c)*
17. **Ball-and-socket joints** are triaxial joints that permit rotation as well as other movements. *(Spotlight Figure 9–2c)*
18. Important terms that describe angular movement are **flexion**, **extension**, **hyperextension**, **abduction**, **adduction**, and **circumduction**. *(Figure 9–3)*
19. Rotational movement can be **left** or **right**, **medial** (*internal*) or **lateral** (*external*), or, in the bones of the forearm, **pronation** or **supination**. *(Figure 9–4)*
20. Movements of the foot include **eversion** and **inversion**. The ankle undergoes flexion and extension, also known as **dorsiflexion** and **plantar flexion**, respectively. *(Figure 9–5)*
21. **Opposition** is the thumb movement that enables us to grasp objects. **Reposition** is the opposite of opposition. *(Figure 9–5)*
22. **Retraction** involves moving a structure posteriorly; **protraction** is moving a structure anteriorly. **Depression** and **elevation** occur when we move a structure inferiorly and superiorly, respectively. **Lateral flexion** occurs when the vertebral column bends to one side. *(Figure 9–5)*

9-4 Intervertebral joints contain intervertebral discs and ligaments that allow for vertebral movements p. 276

23. The articular processes of vertebrae form gliding joints with those of adjacent vertebrae. The bodies form symphyses that are separated and cushioned by **intervertebral discs**, which contain an outer **anulus fibrosus** and an inner **nucleus pulposus**. **Vertebral end plates** cover the superior and inferior surfaces of the disc. Several ligaments stabilize the vertebral column. *(Figure 9–6; Table 9–2)*

9-5 The elbow and knee are both hinge joints p. 278

24. The **elbow joint** permits only flexion–extension. It is a hinge diarthrosis whose capsule is reinforced by strong ligaments. *(Figure 9–7; Table 9–3)*
25. The **knee joint** is a hinge joint made up of three articulations: two formed between the femur and tibia and one between the patella and femur. The joint permits flexion–extension and limited rotation, and it has various supporting ligaments. *(Figure 9–8; Table 9–3)*

9-6 The shoulder and hip are both ball-and-socket joints p. 281

26. The **shoulder joint**, or *glenohumeral joint*, is formed by the glenoid cavity and the head of the humerus. This articulation permits the greatest range of motion of any joint. It is a ball-and-socket diarthrosis with various stabilizing ligaments. Strength and stability are sacrificed in favor of mobility. *(Figure 9–9; Table 9–3)*
27. The **hip joint** is a ball-and-socket diarthrosis formed between the acetabulum of the hip bone and the head of the femur. The joint permits flexion–extension, adduction–abduction, circumduction, and rotation. Numerous ligaments stabilize the joint. *(Figure 9–10; Table 9–3)*

9-7 With advancing age, arthritis and other degenerative changes often impair joint mobility p. 285

28. Problems with joint function are relatively common, especially in older individuals. **Rheumatism** is a general term for pain and stiffness affecting the musculoskeletal system; several major forms exist. **Arthritis** encompasses all the rheumatic diseases that affect synovial joints. Arthritis becomes increasingly common with age.

9-8 The skeletal system supports and stores energy and minerals for other body systems p. 286

29. Growth and maintenance of the skeletal system are supported by the integumentary system. The skeletal system also interacts with the muscular, cardiovascular, lymphatic, digestive, urinary, and endocrine systems. *(Figure 9–11)*

9

Review Questions

See the blue Answers tab at the back of the book.

LEVEL 1 Reviewing Facts and Terms

1. Label the structures in the following illustration of a synovial joint.

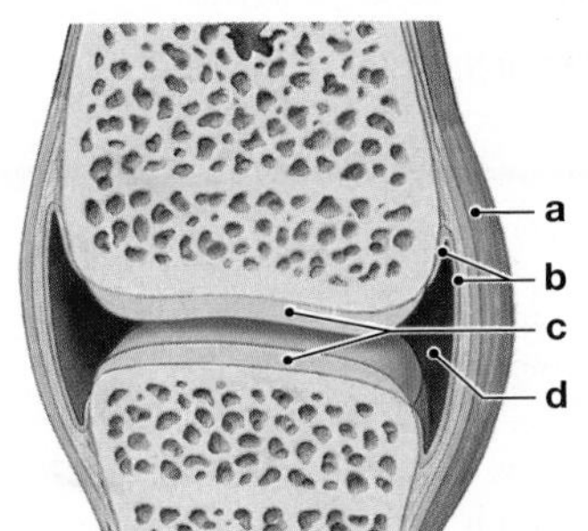
a
b
c
d

(a) ________ (b) ________
(c) ________ (d) ________

2. A synarthrosis located between the bones of the skull is a **(a)** symphysis, **(b)** syndesmosis, **(c)** synchondrosis, **(d)** suture.
3. The joint between adjacent vertebral bodies is a **(a)** syndesmosis, **(b)** symphysis, **(c)** synchondrosis, **(d)** synostosis.
4. The anterior joint between the two pubic bones is a **(a)** synchondrosis, **(b)** synostosis, **(c)** symphysis, **(d)** synarthrosis.
5. Joints typically located between the ends of adjacent long bones are **(a)** synarthroses, **(b)** amphiarthroses, **(c)** diarthroses, **(d)** symphyses.
6. The function of the articular cartilage is **(a)** to reduce friction, **(b)** to prevent bony surfaces from contacting one another, **(c)** to provide lubrication, **(d)** both a and b.
7. Which of the following is *not* a function of synovial fluid? **(a)** shock absorption, **(b)** nutrient distribution, **(c)** maintenance of ionic balance, **(d)** lubrication of the articular surfaces, **(e)** waste disposal.
8. The structures that limit the range of motion of a joint and provide mechanical support across or around the joint are **(a)** bursae, **(b)** tendons, **(c)** menisci, **(d)** all of these.
9. Complete loss of contact between two articulating surfaces is a **(a)** circumduction, **(b)** hyperextension, **(c)** dislocation, **(d)** supination.
10. Abduction and adduction always refer to movements of the **(a)** axial skeleton, **(b)** appendicular skeleton, **(c)** skull, **(d)** vertebral column.
11. Rotation of the forearm that makes the palm face posteriorly is **(a)** supination, **(b)** pronation, **(c)** proliferation, **(d)** projection.
12. A saddle joint permits ________ movement but prevents ________ movement. **(a)** rotational; gliding, **(b)** angular; gliding, **(c)** gliding; rotational, **(d)** angular; rotational.
13. Standing on tiptoe is an example of ________ at the ankle. **(a)** elevation, **(b)** flexion, **(c)** extension, **(d)** retraction.
14. Examples of monaxial joints, which permit angular movement in a single plane, are **(a)** the intercarpal and intertarsal joints, **(b)** the shoulder and hip joints, **(c)** the elbow and knee joints, **(d)** all of these.
15. Decreasing the angle between bones is termed **(a)** flexion, **(b)** extension, **(c)** abduction, **(d)** adduction, **(e)** hyperextension.
16. Movements that occur at the shoulder and the hip represent the actions that occur at a ________ joint. **(a)** hinge, **(b)** ball-and-socket, **(c)** pivot, **(d)** plane.
17. The anulus fibrosus and nucleus pulposus are structures associated with the **(a)** intervertebral discs, **(b)** knee and elbow, **(c)** shoulder and hip, **(d)** carpal and tarsal bones.
18. Subacromial, subcoracoid, and subscapular bursae reduce friction in the ________ joint. **(a)** hip, **(b)** knee, **(c)** elbow, **(d)** shoulder.
19. Although the knee joint is only one joint, it resembles ________ separate joints. **(a)** two, **(b)** three, **(c)** four, **(d)** five, **(e)** six.

LEVEL 2 Reviewing Concepts

20. Dislocations involving synovial joints are usually prevented by all of the following *except* **(a)** structures such as ligaments that stabilize and support the joint, **(b)** the position of bursae that limits the degree of movement, **(c)** the presence of other bones that prevent certain movements, **(d)** the position of muscles and fat pads that limits the degree of movement, **(e)** the shape of the articular surface.
21. The hip is an extremely stable joint because it has **(a)** a complete bony socket, **(b)** a strong joint capsule, **(c)** supporting ligaments, **(d)** all of these.
22. How does a meniscus (articular disc) function in a joint?
23. Partial or complete dislocation of the acromioclavicular joint is called a(n) ________.
24. How do articular cartilages differ from other cartilages in the body?
25. Differentiate between a bulging disc and a herniated disc.
26. How would you explain to your grandmother the characteristic decrease in height with advancing age?
27. List the six different types of diarthroses, and give an example of each.

LEVEL 3 Critical Thinking and Clinical Applications

28. While playing tennis, Dave "turns his ankle." He experiences swelling and pain. After being examined, he is told that he has no ruptured ligaments and that the structure of the ankle is not affected. On the basis of the signs and symptoms and the examination results, what happened to Dave's ankle?
29. Joe injures his knee during a football practice such that the synovial fluid in the knee joint no longer circulates normally. The physician who examines him tells him that they have to reestablish circulation of the synovial fluid before the articular cartilages become damaged. Why?
30. When playing a contact sport, which injury would you expect to occur more frequently, a dislocated shoulder or a dislocated hip? Why?

+ CLINICAL CASE Wrap-Up What's the Matter with the Birthday Girl?

9

Jessica's mother takes her to see her pediatrician. After hearing about Jessica's symptoms, the pediatrician refers her to a rheumatologist, a doctor who specializes in rheumatic diseases.

Jessica now has warm, swollen, tender knees with limited motion. The rheumatologist withdraws some synovial fluid from one of her knees using a small needle (a procedure known as *arthrocentesis*). Instead of a small amount of clear, viscous fluid with the consistency of heavy molasses, there is a large amount of thin, cloudy synovial fluid. This fluid cannot do its important jobs of joint lubrication, nutrient distribution, and shock absorption.

The doctor says that Jessica is suffering from juvenile rheumatoid arthritis, the most common type of childhood arthritis. Unlike the more common osteoarthritis seen in the elderly, rheumatoid arthritis is an autoimmune, inflammatory disease of the synovial membrane seen in people of all ages. For unknown reasons, the immune system mistakenly attacks the synovial membrane, causing it to grow abnormally and produce large amounts of unhealthy synovial fluid. Jessica may outgrow her joint disease, or the synovitis may destroy her joint cartilage.

1. What category of joints is likely to be affected by juvenile rheumatoid arthritis?
2. If you could look inside Jessica's knee joint, what do you think the synovial membrane would look like?

See the blue Answers tab at the back of the book.

Related Clinical Terms

ankylosing spondylitis: A chronic, progressive inflammatory disease of the intervertebral spaces that causes abnormal fusion of the vertebrae.

arthroplasty: The surgical reconstruction or creation of an artificial joint.

arthroscopy: Insertion of a narrow tube containing optical fibers and a tiny camera (arthroscope) directly into the joint for visual examination.

Bouchard nodes: Bony enlargements on the proximal interphalangeal joints due to osteoarthritis.

chondromalacia: Softening of cartilage as a result of strenuous activity or an overuse injury.

Heberden nodes: Bony overgrowths on the distal interphalangeal joints due to osteoarthritis that cause the patient to have knobby fingers.

joint mice: Small fibrous, cartilaginous, or bony loose bodies in the synovial cavity of a joint.

Lyme disease: An infectious disease transmitted to humans from the bite of a tick infected with *Borrelia burgdorferi* or *Borrelia mayonii*, causing flu-like symptoms and joint pain.

pannus: Granulation tissue (combination of fibrous connective tissue and capillaries) forming within a synovial membrane, which releases cartilage-destroying enzymes.

prepatellar bursitis: Inflammation of the bursa over the front of the knee just above the kneecap; also known as housemaid's knee.

prosthesis: An artificial substitute for a body part.

synovitis: Inflammation of the synovial membrane.

tophi: Deposits of uric acid crystals often found around joints and usually associated with gout.

10 Muscle Tissue

Learning Outcomes

These Learning Outcomes correspond by number to this chapter's sections and indicate what you should be able to do after completing the chapter.

CLINICAL CASE Keep on Keepin' on

www.mona-lila

Rhetta has been drumming in an all-girl band for five years. At their last gig two weeks ago, the crowd went wild. The band played an encore and she gave it her all. Afterwards, she could barely hang onto her drumsticks. This was not the first time that her arms "wimped out," but it was the worst time. It seems that lately, she can play great for the first few songs, but then she completely loses her strength. This week, she had trouble swallowing. She also complains of having double vision. Her eyelids droop, especially when she wakes up. She feels like a mess. Her band members have encouraged her to see a doctor and now she's ready to make it happen.

Rhetta sees a neurologist who takes a complete medical history and does a thorough physical exam. She realizes how weak her legs have gotten when he asks her to stand up from her chair without using her arms. Walking on tiptoes is impossible for her. When the doctor lifts her eyelids, she's not able to track his finger with her gaze, and she still sees double.

"I would like to perform a test," says the neurologist, starting an IV. He injects a small amount of a drug. Suddenly, Rhetta's strength returns. She opens both eyes wide and no longer sees double. She can toe walk, heel walk, and get up unassisted many times in a row. She feels like she could play her drums all night. But after 30 minutes, her weakness returns. **What is going on with Rhetta? To find out, turn to the Clinical Case Wrap-Up on p. 335.**

10

An Introduction to Muscle Tissue

In this chapter we discuss muscle tissue, one of the four primary tissue types. As we introduced in Chapter 4, our bodies contain three types of muscle tissue: (1) *skeletal muscle*, (2) *cardiac muscle*, and (3) *smooth muscle*. ⮌ p. 142 In this chapter we pay particular attention to skeletal muscle tissue. We examine the structural and physiological characteristics of skeletal muscle cells, and relate those features to the functions of the entire tissue. We conclude the chapter with an overview of the differences among skeletal, cardiac, and smooth muscle tissues. Reading this chapter will prepare you for our discussion of the muscular system in Chapter 11.

10-1 The primary function of muscle tissue is to produce movement

Learning Outcome **Identify the common properties of muscle tissue and the primary functions of skeletal muscles.**

Muscle tissue consists chiefly of muscle cells that are highly specialized for contraction. This contraction at the cellular level produces many types of movement in the body. Without these muscle tissues, nothing in the body would move, and the body itself could not move. Skeletal muscle tissue moves the body by pulling on our bones, making it possible for us to walk, dance, bite an apple, or play the guitar. Cardiac muscle tissue pumps blood through the cardiovascular system. Smooth muscle tissue pushes fluids and solids along the digestive tract and other internal organs, and regulates the diameters of small arteries and respiratory bronchioles.

Common Properties of Muscle Tissue

Muscle tissue shares these common properties: excitability, contractility, extensibility, and elasticity. **Excitability**, also known as responsiveness or irritability, is the ability to receive and respond to a stimulus. Muscle tissue responds to a chemical stimulus from a nerve cell with a change in membrane potential. **Contractility** refers to the ability of a muscle cell to shorten when it is stimulated. **Extensibility** is stretching movement of a muscle, while **elasticity** is the ability of a muscle to recoil (spring back) to its resting length.

Functions of Skeletal Muscle

Our skeletal muscles have the following functions:

- *Producing Movement.* In general, skeletal muscle contractions pull on tendons to move our bones. Skeletal muscle actions range from simple motions, such as extending the arm or breathing, to the highly coordinated movements of swimming or piano playing.
- *Maintaining Posture and Body Position.* Tension in our skeletal muscles maintains body posture—for example, holding your head still when you read a book or balancing your body weight above your feet when you walk. Without constant muscular activity, we could neither sit upright nor stand.
- *Supporting Soft Tissues.* Layers of skeletal muscle make up the abdominal wall and the pelvic floor cavity. These muscles support the weight of our visceral organs and shield our internal tissues from injury.
- *Guarding Body Entrances and Exits.* Skeletal muscles called sphincters (SFINK-terz) encircle the openings of the digestive and urinary tracts. These muscles give us voluntary control over swallowing, defecating, and urinating. (Involuntary sphincters have smooth muscle.)
- *Maintaining Body Temperature.* Muscle contractions use energy, and whenever energy is used in the body, some of it

is converted to heat. The heat released by working muscles keeps body temperature in the range needed for normal functioning.

- *Storing Nutrients.* When our diet contains too few proteins or calories, the contractile proteins in skeletal muscles are broken down, and their amino acids released into the circulation. The liver can use some of these amino acids to synthesize glucose, and others can be broken down to provide energy.

Let's now turn to the functional anatomy of a typical skeletal muscle.

Checkpoint

1. **Identify the three types of muscle tissue and cite their major functions.**
2. **Identify the common properties shared by muscle tissues.**
3. **Identify the primary functions of skeletal muscle.**

See the blue Answers tab at the back of the book.

10-2 Skeletal muscle contains muscle tissue, connective tissues, blood vessels, and nerves

Learning Outcome **Describe the organization of muscle at the tissue level.**

Figure 10–1 illustrates the organization of a representative skeletal muscle. **Skeletal muscles** are organs composed mainly of skeletal muscle tissue, but they also contain connective tissues, blood vessels, and nerves. Each cell in skeletal muscle tissue is a single muscle *fiber*. Skeletal muscles attach directly or indirectly to bones.

Here we examine how muscle tissue is organized in a skeletal muscle by connective tissues, and how skeletal muscles are supplied with blood vessels and nerves.

Organization of Connective Tissues and Muscle Tissue

As you can see in Figure 10–1a, the muscle tissue in a skeletal muscle is surrounded by three layers of connective tissue: (1) an epimysium, (2) a perimysium, and (3) an endomysium.

The **epimysium** (ep-i-MIZ-ē-um; *epi-*, on + *mys*, muscle) is a dense layer of collagen fibers that surrounds the entire muscle. It separates the muscle from nearby tissues and organs. The epimysium is connected to the deep *fascia* (FASH-ē-uh), a dense connective tissue layer. p. 133

The **perimysium** (per-i-MIZ-ē-um; *peri-*, around) divides the skeletal muscle into a series of compartments. Each compartment contains a bundle of muscle fibers called a **fascicle** (FAS-i-kl; *fasciculus*, a bundle; Figure 10–1b). The perimysium contains collagen and elastic fibers as well as the blood vessels and nerves that supply the muscle fibers within the fascicles.

Within a fascicle, the delicate connective tissue of the **endomysium** (en-dō-MIZ-ē-um; *endo-*, inside) surrounds the individual skeletal muscle cells, called *muscle fibers* (Figure 10–1c), and loosely interconnects adjacent muscle fibers. This flexible, elastic connective tissue layer contains (1) *capillary networks* (the smallest blood vessels) that supply blood to the muscle fibers; (2) **myosatellite cells**, stem cells that help repair damaged muscle tissue; and (3) nerve fibers that control the muscle. All these structures are in direct contact with the individual muscle fibers. p. 142

The collagen fibers of the perimysium and endomysium are interwoven and blend into one another. At each end of the muscle, the collagen fibers of the epimysium, perimysium, and endomysium come together to form either a bundle known as a **tendon**, or a broad sheet called an **aponeurosis** (ap-ō-nū-RŌ-sis). Tendons and aponeuroses usually attach skeletal muscles to bones. Where they contact the bone, the collagen fibers extend into the bone matrix, providing a firm attachment. As a result, any contraction of the muscle pulls on the attached bone.

Function of Skeletal Muscle Components

As we have seen, the connective tissues of the endomysium and perimysium contain the blood vessels and nerves that supply the muscle fibers. Muscle contraction requires tremendous quantities of energy. An extensive vascular network delivers the necessary oxygen and nutrients and carries away the metabolic wastes generated by active skeletal muscles. The blood vessels and the nerves generally enter the muscle together and follow the same branching course through the perimysium. Each fascicle receives branches of these blood vessels and nerves. Within the endomysium, arterioles (small arteries) supply blood to a capillary network that services the individual muscle fiber.

Skeletal muscles contract only when the central nervous system stimulates them. Axons, *nerve fibers* extending from neurons, penetrate the epimysium, branch through the perimysium, and enter the endomysium to innervate individual muscle fibers. Skeletal muscles are often called voluntary muscles, because we have voluntary control over their contractions. We may also control many skeletal muscles at a subconscious level. For example, skeletal muscles involved with breathing, such as the *diaphragm* (DĪ-a-fram), usually work outside our conscious awareness.

Next, let's examine the microscopic structure of a typical skeletal muscle fiber and relate that microstructure to the physiology of the contraction process.

Checkpoint

4. **Describe the three connective tissue layers associated with skeletal muscle tissue.**
5. **How would severing the tendon attached to a muscle affect the muscle's ability to move a body part?**

See the blue Answers tab at the back of the book.

Figure 10–1 The Organization of Skeletal Muscles. A skeletal muscle consists of fascicles (bundles of muscle fibers) enclosed by the epimysium. The bundles are separated by connective tissue fibers of the perimysium, and within each bundle, each of the muscle fibers is surrounded by an endomysium. Each muscle fiber has many superficial nuclei, as well as mitochondria and other organelles (see Figure 10–3).

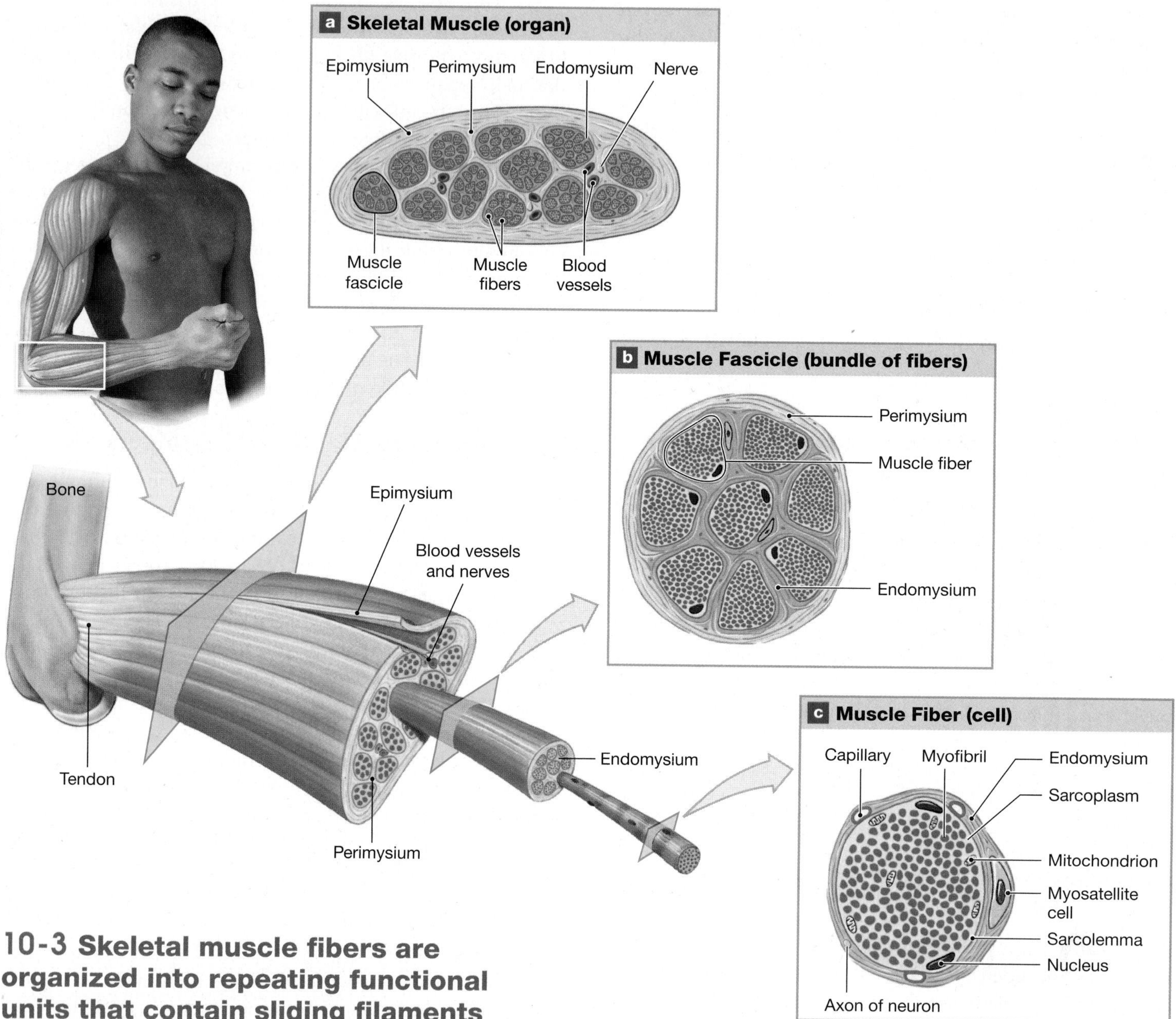

a Skeletal Muscle (organ)
Epimysium
Perimysium
Endomysium
Nerve
Muscle fascicle
Muscle fibers
Blood vessels
b Muscle Fascicle (bundle of fibers)
Perimysium
Muscle fiber
Endomysium
Bone
Epimysium
Blood vessels and nerves
Tendon
Endomysium
Perimysium
c Muscle Fiber (cell)
Capillary
Myofibril
Endomysium
Sarcoplasm
Mitochondrion
Myosatellite cell
Sarcolemma
Nucleus
Axon of neuron

10-3 Skeletal muscle fibers are organized into repeating functional units that contain sliding filaments

Learning Outcome Describe the characteristics of skeletal muscle fibers, and identify the components of a sarcomere.

Skeletal muscle fibers are quite different from the "typical" cells we described in Chapter 3. One obvious difference is size: Compared to other cells, skeletal muscle fibers are enormous. A muscle fiber from a thigh muscle could have a diameter of 100 μm and a length up to 30 cm (12 in.). Another major difference is that skeletal muscle fibers are *multinucleate:* Each contains hundreds of nuclei just internal to the plasma membrane. The genes in these nuclei control the production of enzymes and structural proteins required for normal muscle contraction. The more copies of these genes, the faster these proteins can be produced.

The distinctive features of size and multiple nuclei are related. During development, groups of embryonic cells called **myoblasts** (*myo-*, muscle + *blastos*, formative cell) fuse, forming individual multinucleate skeletal muscle fibers (Figure 10–2a). Each nucleus in a skeletal muscle fiber reflects the contribution of a single myoblast. Some myoblasts do not fuse with developing muscle fibers. These unfused cells remain in adult skeletal

Figure 10–2 The Formation of a Multinucleate Skeletal Muscle Fiber.

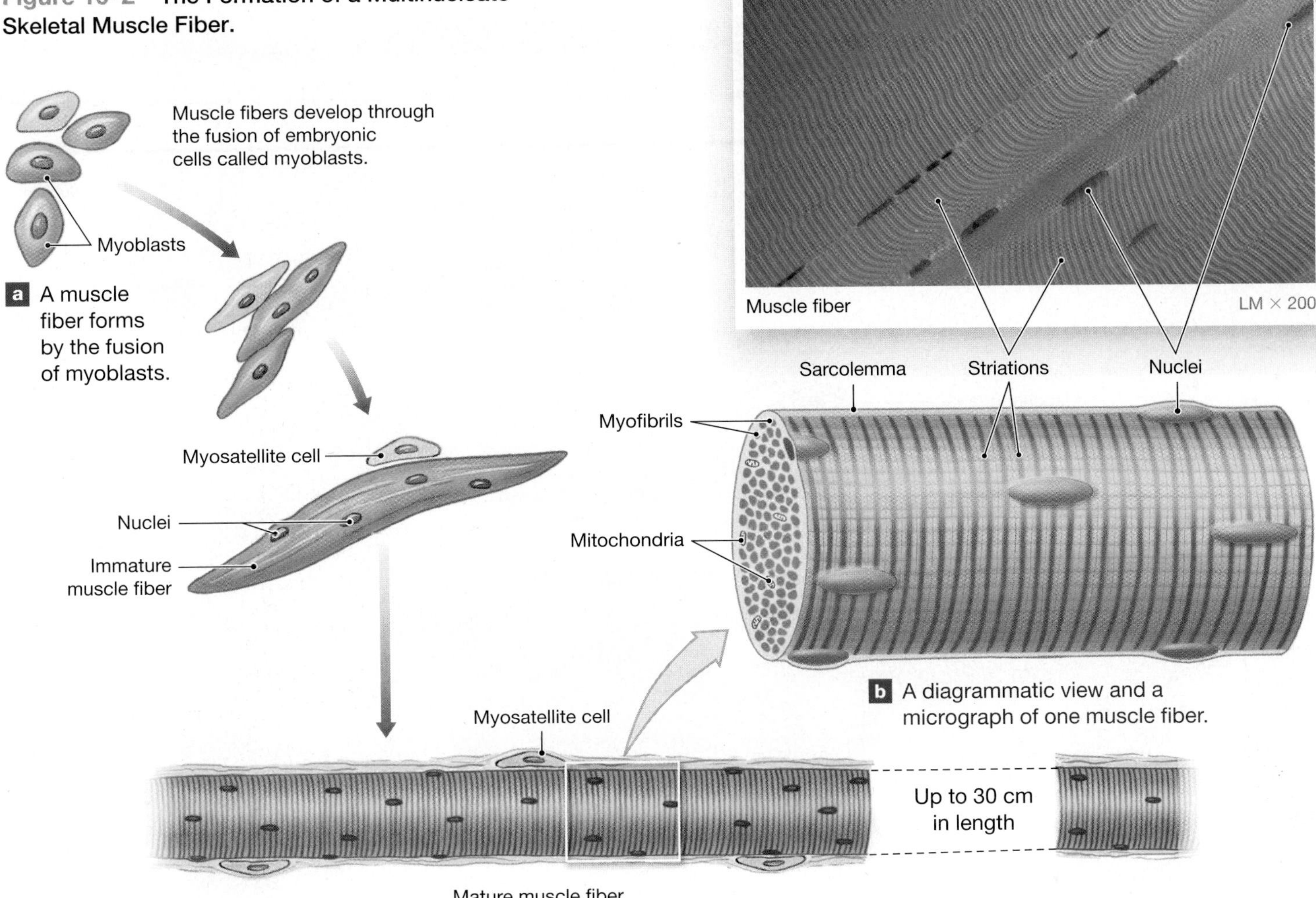
Muscle fibers develop through the fusion of embryonic cells called myoblasts.
Myoblasts
Myosatellite cell
Nuclei
Immature muscle fiber
Myofibrils
Mitochondria
Muscle fiber
LM × 200
Sarcolemma
Striations
Nuclei
Myosatellite cell
Up to 30 cm in length
Mature muscle fiber

a A muscle fiber forms by the fusion of myoblasts.

b A diagrammatic view and a micrograph of one muscle fiber.

10

muscle tissue as the myosatellite cells seen in Figures 10–1c and 10–2a. After an injury, myosatellite cells may enlarge, divide, and fuse with damaged muscle fibers, thereby assisting in the repair of the tissue.

A final way that skeletal muscle fibers are different from other cells is that they are banded, or *striated*. These striations are visible with a light microscope, so skeletal muscle tissue is also known as **striated muscle**. p. 142 Each muscle fiber contains hundreds to thousands of cylindrical structures called *myofibrils* (Figure 10–2b). As we will see, the striations are due to the precise arrangements of *thin* (*actin*) and *thick* (*myosin*) *filaments* in these myofibrils. The arrangement of these filaments forms the repeating functional unit called a *sarcomere*; the sliding of these filaments allows a muscle fiber to contract. In addition, the fiber contains other unique components that support the function of the sarcomeres. Let's take a look at these components.

The Sarcolemma and Transverse Tubules

The **sarcolemma** (sar-kō-LEM-uh; *sarkos*, flesh + *lemma*, husk), or plasma membrane of a muscle fiber, surrounds the **sarcoplasm** (SAR-kō-plazm), or cytoplasm of the muscle fiber (Figure 10–3). Like the plasma membranes of other body cells, the sarcolemma has a characteristic membrane potential. p. 99 In a skeletal muscle fiber, a sudden change in the membrane potential is the first step that leads to a contraction.

Even though a skeletal muscle fiber is very large, all regions of the cell must contract at the same time. For this reason, the signal to contract must be distributed quickly throughout the interior of the cell. This signal is propagated through the transverse tubules. **Transverse tubules**, or **T tubules**, are narrow tubes whose surfaces are continuous with the sarcolemma and extend deep into the sarcoplasm (see Figure 10–3). They are filled with extracellular fluid and form passageways through the muscle fiber, like a network of tunnels through a mountain. As extensions of the sarcolemma, T tubules share the same special property of being able to propagate an electrical impulse. As a result, electrical impulses generated by the sarcolemma also travel along the T tubules into the cell interior. These impulses, called *action potentials*, trigger muscle fiber contraction.

Figure 10–3 **The Structure and Internal Organization of a Skeletal Muscle Fiber.**

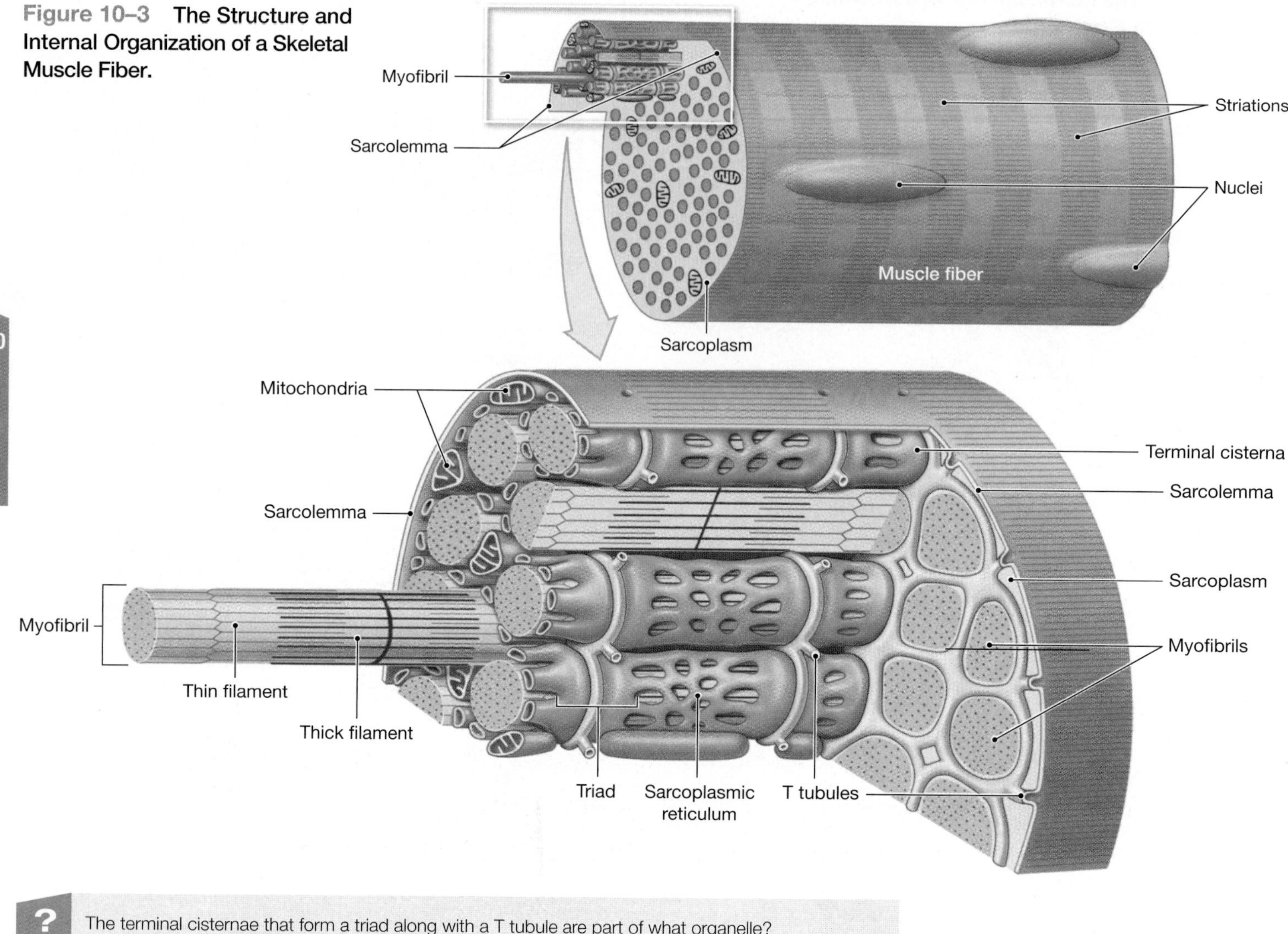
Myofibril
Sarcolemma
Striations
Nuclei
Muscle fiber
Sarcoplasm
Mitochondria
Terminal cisterna
Sarcolemma
Sarcolemma
Sarcoplasm
Myofibril
Myofibrils
Thin filament
Thick filament
Triad
Sarcoplasmic reticulum
T tubules

? The terminal cisternae that form a triad along with a T tubule are part of what organelle?

The Sarcoplasmic Reticulum

In skeletal muscle fibers, a membrane complex called the **sarcoplasmic reticulum (SR)** forms a tubular network around each myofibril, fitting over it like lacy shirtsleeves (see Figure 10–3). The SR is similar to the smooth endoplasmic reticulum of other cells.

Wherever a T tubule encircles a myofibril, the tubule is tightly bound to the membranes of the SR. On either side of a T tubule, the tubules of the SR enlarge, fuse, and form expanded chambers called **terminal cisternae** (sis-TUR-nē). This combination of a pair of terminal cisternae plus a T tubule is known as a **triad**. Although the membranes of the triad are tightly bound together, their fluid contents are separate and distinct.

The SR is specialized for the storage and release of calcium ions. In Chapter 3, we noted that special ion pumps keep the intracellular concentration of calcium ions (Ca^{2+}) very low. p. 71 Most cells pump calcium ions out across their plasma membranes and into the extracellular fluid. Although skeletal muscle fibers do export Ca^{2+} in this way, they also remove calcium ions from the cytosol by actively transporting them into the terminal cisternae of the SR. The cytosol of a resting skeletal muscle fiber contains a very low concentration of Ca^{2+}, around 10^{-7} mmol/L. The free Ca^{2+} concentration level inside the terminal cisternae may be as much as 1000 times higher. In addition, terminal cisternae contain the protein *calsequestrin*, which reversibly binds Ca^{2+}. Including both the free calcium and the bound calcium, the total concentration of Ca^{2+} inside terminal cisternae can be 40,000 times that of the surrounding cytosol.

Myofibrils

A **myofibril** is 1–2 μm in diameter and as long as the entire muscle fiber. Branches of the T tubules and the sarcoplasmic reticulum encircle each myofibril. The active shortening of myofibrils is responsible for skeletal muscle fiber contraction.

Myofibrils consist of bundles of protein filaments called **myofilaments**. Myofibrils contain two types of myofilaments

Figure 10–4 Sarcomere Structure, Longitudinal Views.

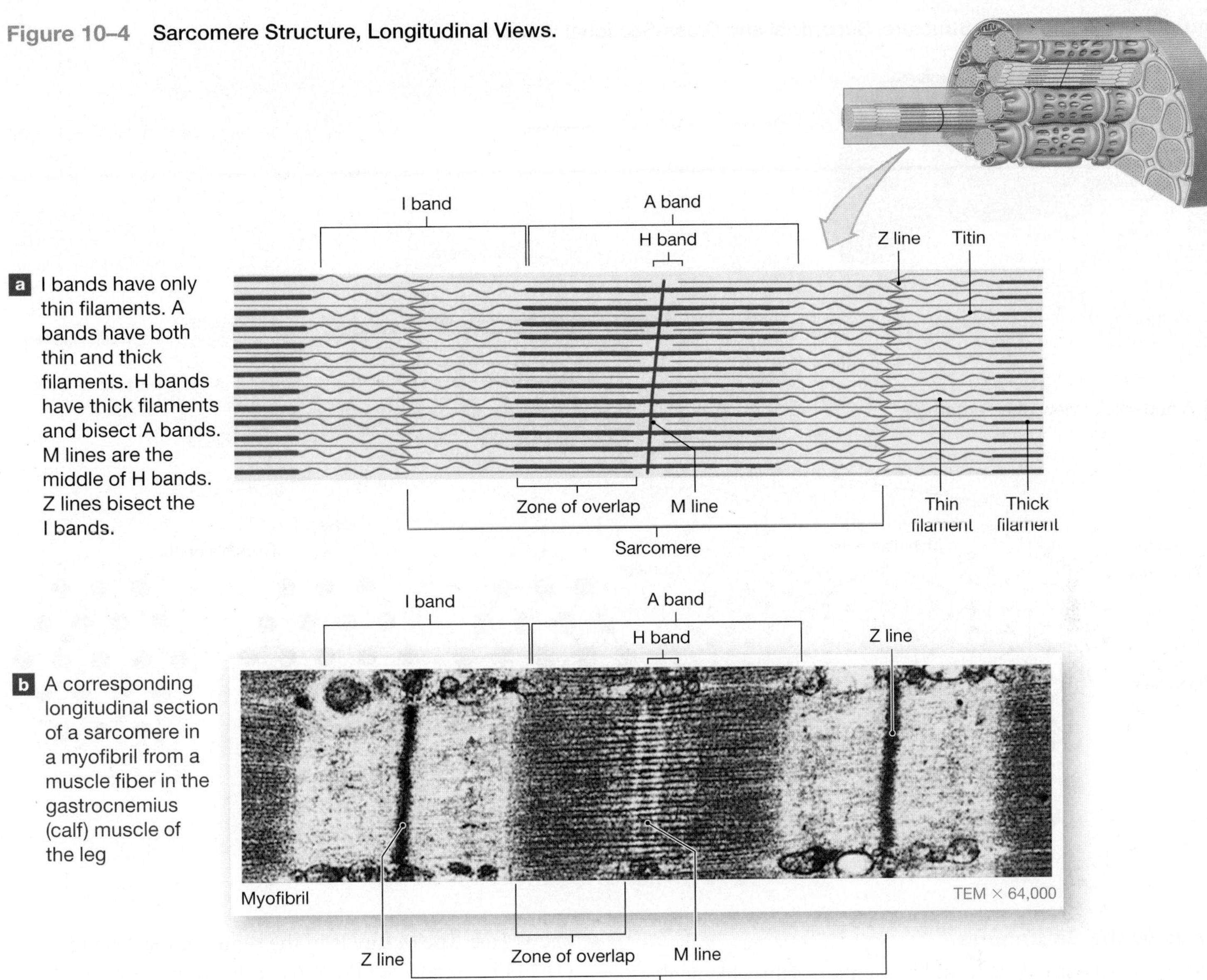
I band
A band
H band
Z line
Titin
Zone of overlap
M line
Thin filament
Thick filament
Sarcomere
I band
A band
H band
Z line
Myofibril
TEM × 64,000
Z line
Zone of overlap
M line
Sarcomere

a I bands have only thin filaments. A bands have both thin and thick filaments. H bands have thick filaments and bisect A bands. M lines are the middle of H bands. Z lines bisect the I bands.

b A corresponding longitudinal section of a sarcomere in a myofibril from a muscle fiber in the gastrocnemius (calf) muscle of the leg

10

that we introduced in Chapter 3: *thin filaments* composed primarily of *actin*, and *thick filaments* composed primarily of *myosin*. ⮌ p. 72 In addition, myofibrils contain *titin*, elastic myofilaments associated with the thick filaments. (We consider the role of titin later in the chapter.)

Myofibrils are anchored to the inner surface of the sarcolemma at each end of a skeletal muscle fiber. In turn, the outer surface of the sarcolemma is attached to collagen fibers of the tendon or aponeurosis of the skeletal muscle. As a result, when the myofibrils shorten or contract, the entire cell shortens and pulls on the tendon.

Scattered among the myofibrils are mitochondria and granules of glycogen, the storage form of glucose. Mitochondrial activity and glucose breakdown by glycolysis provide energy in the form of ATP for short-duration, maximum-intensity muscular contractions. ⮌ p. 81

Sarcomeres

As we have discussed, thin and thick myofilaments are organized into repeating functional units called **sarcomeres** (SAR-kō-mērz; *sarkos*, flesh + *meros*, part) (Figure 10–4). Sarcomeres are the smallest functional units of the muscle fiber. Interactions between the thick and thin filaments within sarcomeres are responsible for muscle contraction.

A myofibril consists of approximately 10,000 sarcomeres, end to end. Each sarcomere has a resting length of about 2 μm. A sarcomere contains (1) thick filaments, (2) thin filaments, (3) proteins that stabilize the positions of the thick and thin filaments, and (4) proteins that regulate the interactions between thick and thin filaments. Let's look at these components, including how their arrangement produces the characteristic banding of myofibrils and the sarcomeres they contain.

Figure 10–5 Sarcomere Structure, Superficial and Cross-Sectional Views.

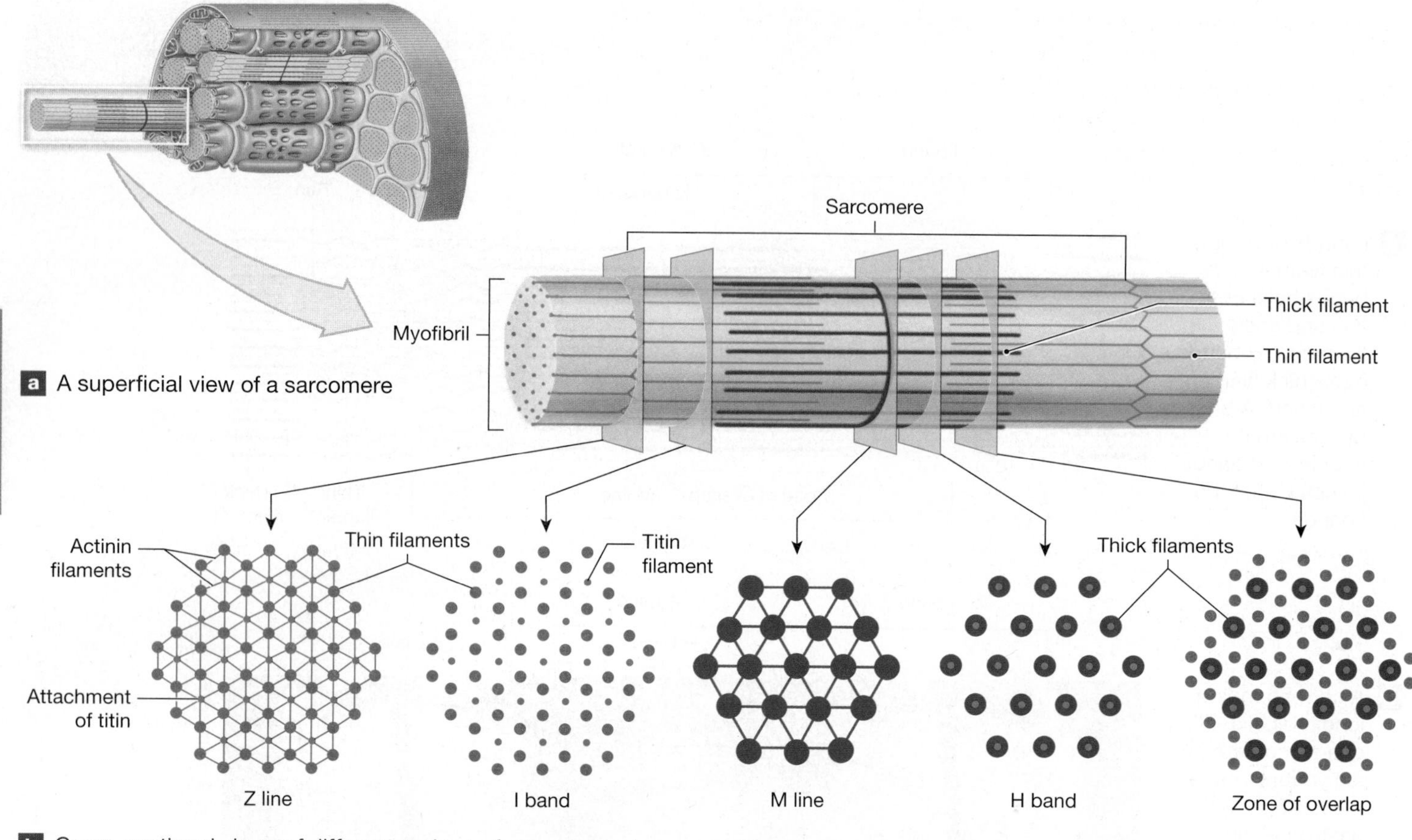
Sarcomere
Myofibril
Thick filament
Thin filament
a A superficial view of a sarcomere
Actinin filaments
Attachment of titin
Thin filaments
Titin filament
Thick filaments
Z line
I band
M line
H band
Zone of overlap
b Cross-sectional views of different regions of a sarcomere

Bands of the Sarcomere

Differences in the size, density, and distribution of thick filaments and thin filaments account for the banded appearance of each myofibril (Figure 10–4). Each sarcomere has dark bands called **A bands** and light bands called **I bands**. The names of these bands are derived from *anisotropic* and *isotropic*, which refer to their appearance when viewed using polarized light microscopy.

Tips & Tools

You can remember that actin occurs in thin filaments by associating the "tin" in ac**tin** with the word *thin*. Then remember that th**I**n filaments look l**I**ght and form the **I** band. Similarly, the **A** bands are d**A**rk.

The A Band. The A bands contain regions of overlapping thick (myosin) and thin (actin) filaments. The thick filaments are at the center of each sarcomere, in the A band. The A band is about as long as a typical thick filament. The A band also includes portions of thin filaments and contains these three subdivisions (see Figure 10–4):

- *The M Line.* The **M line** is in the center of the A band. (The M is for *middle.*) Proteins of the M line connect the central portion of each thick filament to neighboring thick filaments. These dark-staining proteins help stabilize the positions of the thick filaments.
- *The H Band.* In a resting sarcomere, the **H band** is a lighter region on either side of the M line. The H band contains thick filaments, but no thin filaments.
- *The Zone of Overlap.* The **zone of overlap** is a dark region where thin filaments are located between the thick filaments. Here three thick filaments surround each thin filament, and six thin filaments surround each thick filament.

The longitudinal views in Figure 10–4 and the cross-sectional views in Figure 10–5 should help you to visualize a three-dimensional image of these regions of a sarcomere.

Two T tubules encircle each sarcomere, and the triads containing them are located in the zones of overlap, at the edges of the A band (see Figure 10–3). As a result, calcium ions released by the SR enter the regions where thick and thin filaments can interact.

The I Band. The I band is a region of the sarcomere that contains thin filaments but no thick filaments. It extends from the A band of one sarcomere to the A band of the next sarcomere (see Figure 10–4). Z lines bisect the I bands and mark the boundary between adjacent sarcomeres. The Z lines consist of proteins called *actinins*, which interconnect thin filaments of adjacent sarcomeres. At both ends of the sarcomere, thin filaments extend from the Z lines toward the M line and into the zone of overlap.

Strands of the elastic protein **titin** extend from the tips of the thick filaments to attachment sites at the Z line (see Figures 10–4a, 10–5). Titin helps keep the thick and thin filaments in proper alignment and aids in restoring resting sarcomere length after contraction. It also helps the muscle fiber resist extreme stretching that would otherwise disrupt the contraction mechanism.

Each Z line is surrounded by a meshwork of intermediate filaments that interconnect adjacent myofibrils. The myofibrils closest to the sarcolemma, in turn, are bound to attachment sites on the interior surface of the membrane. Because the Z lines of all the myofibrils are aligned in this way, the alternating A bands and I bands form the banded appearance of the muscle fiber as a whole (see Figure 10–2b).

Figure 10–6 reviews the levels of skeletal muscle organization we have discussed so far.

Figure 10–6 **Levels of Functional Organization in a Skeletal Muscle.**

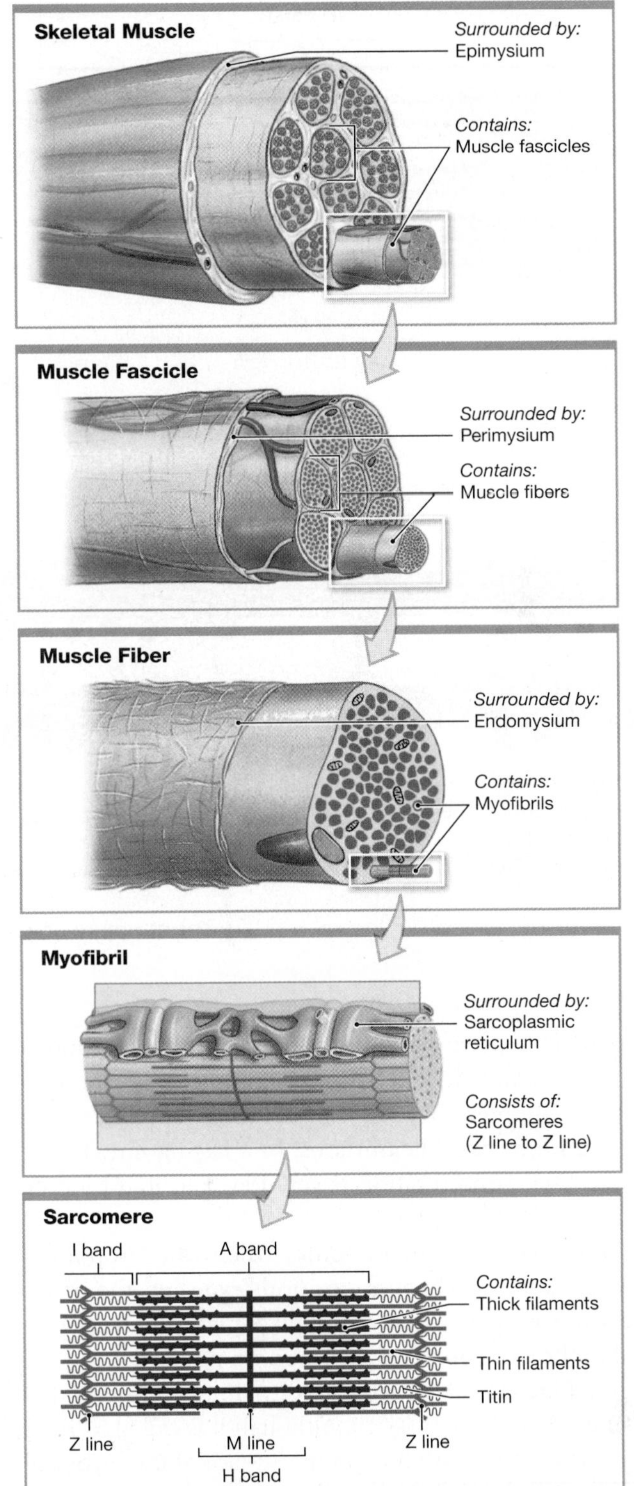
Skeletal Muscle
Surrounded by: Epimysium
Contains: Muscle fascicles
Muscle Fascicle
Surrounded by: Perimysium
Contains: Muscle fibers
Muscle Fiber
Surrounded by: Endomysium
Contains: Myofibrils
Myofibril
Surrounded by: Sarcoplasmic reticulum
Consists of: Sarcomeres (Z line to Z line)
Sarcomere
I band
A band
Contains: Thick filaments
Thin filaments
Titin
Z line
M line
H band
Z line

? Match the level of organization with the structure that surrounds it: *Level of organization:* skeletal muscle, muscle fascicle, muscle fiber, myofibril *Surrounding structure:* endomysium, epimysium, perimysium, sarcoplasmic reticulum

Myofilaments

Now let's consider the molecular structure of the thin and thick myofilaments responsible for muscle contraction.

Thin Filaments. A typical **thin filament** is 5–6 nm in diameter and is 1 μm long (Figure 10–7a). A single thin filament contains four main proteins: F-actin, nebulin, tropomyosin, and troponin (Figure 10–7b).

Filamentous actin, or **F-actin**, is a twisted strand composed of two rows of 300–400 individual globular molecules (monomers) of **G-actin** (Figure 10–7b). A long strand of **nebulin** extends along the F-actin strand in the cleft between the rows of G-actin molecules. Nebulin holds the F-actin strand together. Each G-actin molecule contains an **active site**, where myosin (in the thick filaments) can bind. This is similar to enzymatic reactions in which a substrate molecule binds to the active site of an enzyme. (Recall that a substrate is a substance that is acted on by an enzyme.) p. 55

Under resting conditions, however, the *troponin–tropomyosin complex* prevents myosin binding. Strands of **tropomyosin** (trō-pō-MĪ-ō-sin; *trope*, turning) cover the active sites on G-actin and prevent actin–myosin interaction. A tropomyosin molecule is a double-stranded protein that covers seven active sites. It is bound to one molecule of troponin midway along its length. A **troponin** (TRŌ-pō-nin) molecule consists of three globular subunits. One subunit binds to tropomyosin, locking

Figure 10–7 Thin and Thick Filaments.

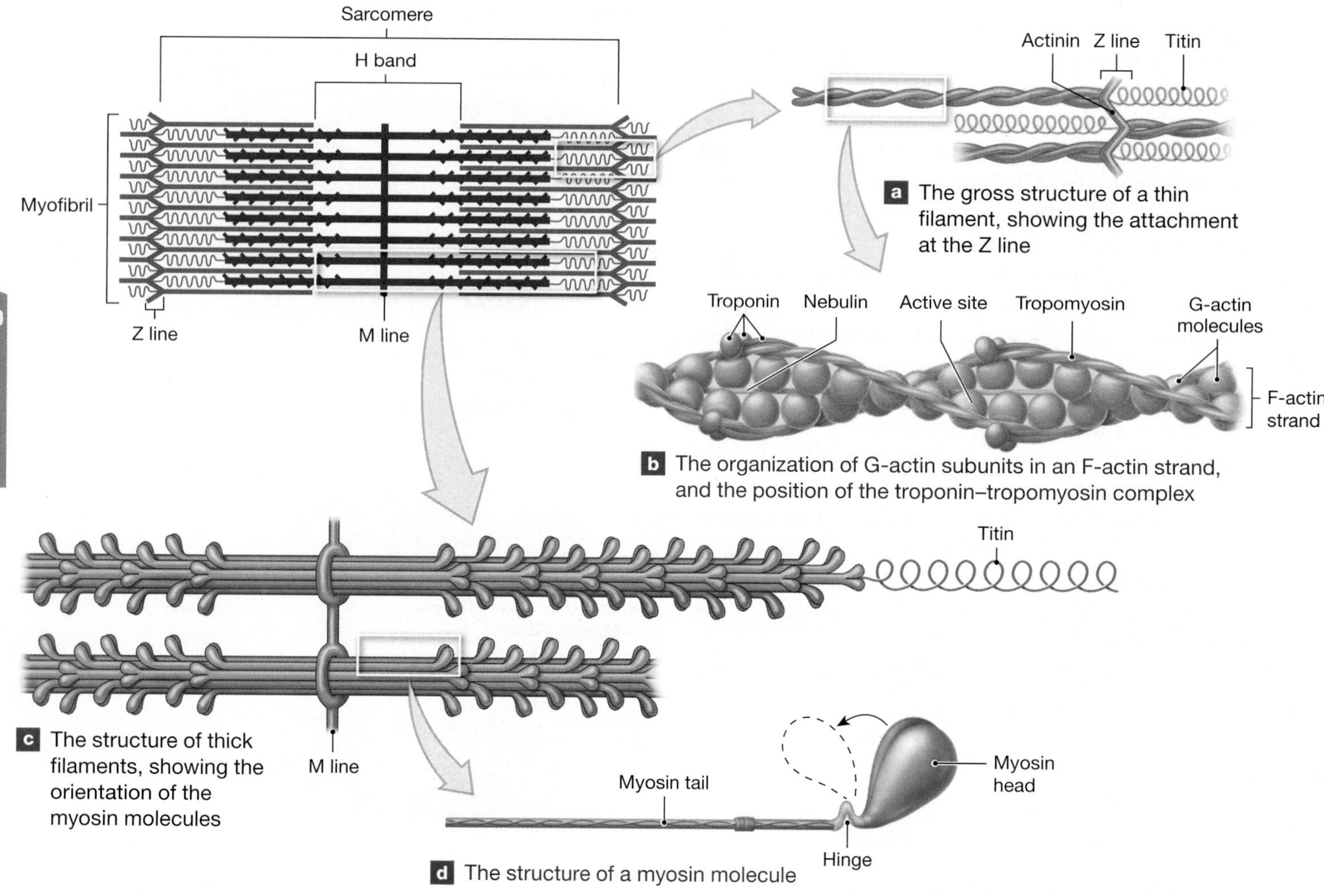
Sarcomere
H band
Myofibril
Z line
M line
Actinin
Z line
Titin
a The gross structure of a thin filament, showing the attachment at the Z line
Troponin
Nebulin
Active site
Tropomyosin
G-actin molecules
F-actin strand
b The organization of G-actin subunits in an F-actin strand, and the position of the troponin–tropomyosin complex
Titin
c The structure of thick filaments, showing the orientation of the myosin molecules
M line
Myosin tail
Myosin head
Hinge
d The structure of a myosin molecule

them together as a troponin–tropomyosin complex, and a second subunit binds to one G-actin, holding the troponin–tropomyosin complex in position. The third subunit has a receptor that binds two calcium ions. In a resting muscle, the intracellular Ca^{2+} concentration is very low, and that binding site is empty.

A contraction can occur only when the troponin–tropomyosin complex changes position, exposing the active sites on actin. The change in position takes place when calcium ions bind to their receptors on the troponin molecules.

The thin filaments are attached to the Z lines at either end of the sarcomere (see Figure 10–7a). Although it is called a "line" because it looks like a dark line on the surface of the myofibril under a microscope, the Z line appears as a zigzag line in sectional view and is more like a disc with a waffle surface (see Figure 10–5b). For this reason, the Z line is often called the *Z disc.*

Thick Filaments. Thick filaments are 10–12 nm in diameter and 1.6 μm long (Figure 10–7c). A thick filament contains about 300 **myosin** molecules, each made up of a pair of myosin subunits twisted around one another (Figure 10–7d). The long tail is bound to other myosin molecules in the thick filament. The free head, which projects outward toward the nearest thin filament, has two globular protein subunits.

All the myosin molecules are arranged with their tails pointing toward the M line (see Figure 10–7c). The H band includes a central region where there are no myosin heads. Elsewhere on the thick filaments, the myosin heads are arranged in a spiral, each facing one of the surrounding thin filaments.

Each thick filament has a core of titin. On either side of the M line, a strand of titin extends the length of the thick filament and then continues across the I band to the Z line on that side. Titin has been described as a "molecular spring." Titin has *elastic* properties, which means that it will recoil after stretching. In the normal resting sarcomere, the titin strands are completely relaxed. They become tense only when some external force stretches the sarcomere.

Figure 10–8 **Changes in the Appearance of a Sarcomere during the Contraction of a Skeletal Muscle Fiber.**

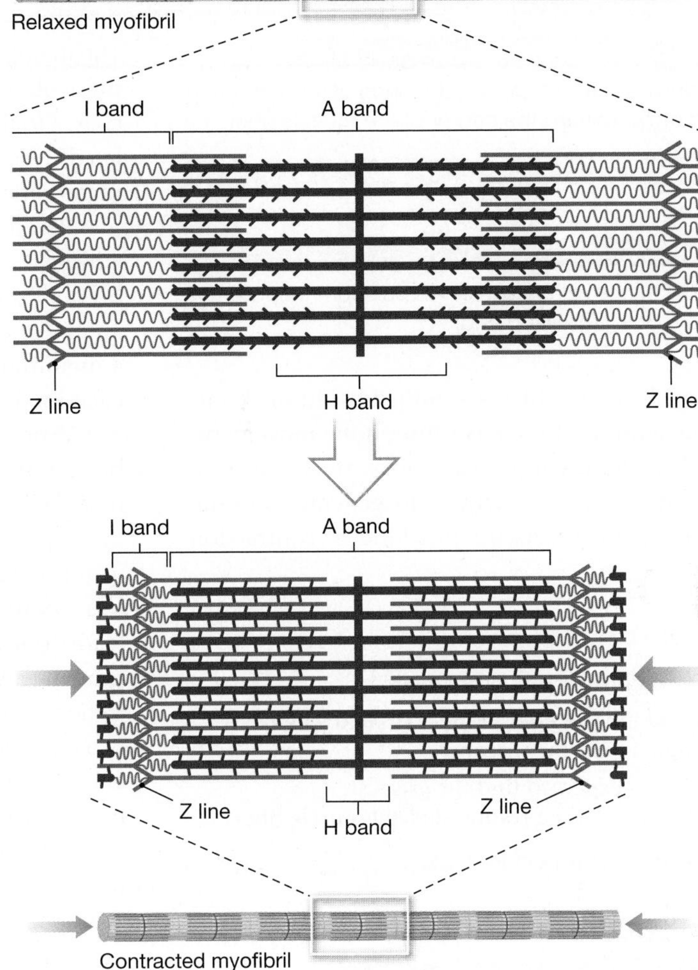
Relaxed myofibril
I band
A band
Z line
H band
Z line
I band
A band
Z line
H band
Z line
Contracted myofibril

a A relaxed sarcomere showing location of the A band, Z lines, and I band.

b During a contraction, the A band stays the same width, but the Z lines move closer together and the I band gets smaller. When the ends of a myofibril are free to move, the sarcomeres shorten simultaneously and the ends of the myofibril are pulled toward its center.

The Sliding-Filament Theory of Muscle Contraction

What happens to a sarcomere when a skeletal muscle fiber contracts? As you can see in Figure 10–8, (1) the H bands and I bands of the sarcomeres narrow, (2) the zones of overlap widen, (3) the Z lines move closer together, and (4) the width of the A band remains constant. These observations make sense only if the thin filaments are sliding toward the center of each sarcomere, alongside the thick filaments. This explanation is known as the **sliding-filament theory**. The contraction weakens with the disappearance of the I bands, at which point the Z lines are in contact with the ends of the thick filaments.

During a contraction, sliding occurs in every sarcomere along the myofibril. As a result, the myofibril gets shorter. Because myofibrils are attached to the sarcolemma at each Z line and at either end of the muscle fiber, when myofibrils get shorter, so does the muscle fiber.

Tips & Tools

To better understand the actions of sliding filaments during a muscle contraction, hold your hands in front of you, palms toward your body and thumbs sticking straight up. Now move your hands together, so that the fingers of one hand move in between the fingers of the other hand. Your fingers represent thin and thick filaments, and your thumbs the Z lines. Notice that finger length stays the same, but your thumbs move closer together.

If neither end of the myofibril is held in position, both ends move toward the middle, as you can see in Figure 10–9a. Such a contraction seldom occurs in an intact skeletal muscle, because one end of the muscle (the *origin*) is usually fixed in position during a contraction, while the other end (the *insertion*) moves. In that case, the free end moves toward the fixed end (Figure 10–9b). As the muscle fiber contracts, it pulls on

Figure 10–9 Shortening during a Contraction.

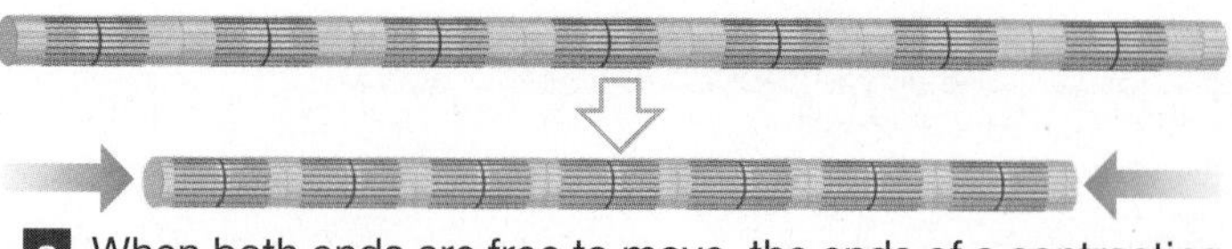

a When both ends are free to move, the ends of a contracting muscle fiber move toward the center of the muscle fiber.

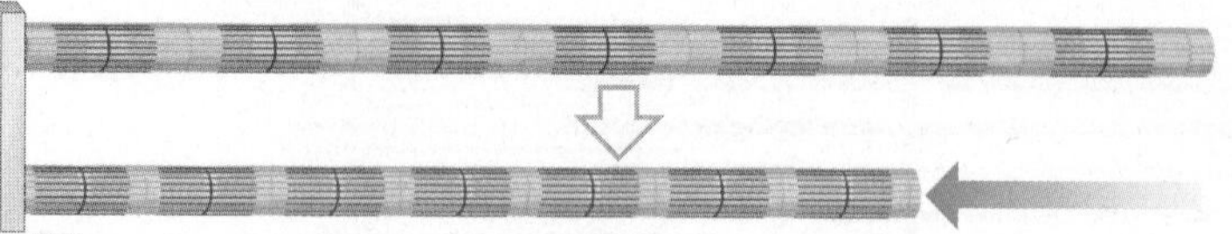

b When one end of a myofibril is fixed in position, the free end is pulled toward the fixed end.

the tendon's collagen fibers, which are embedded in the bone. This pulling on the tendon ultimately causes bone movement.

If neither end of the myofibril can move, thick and thin filament interactions consume energy and generate tension, but sliding cannot occur. We discuss this kind of contraction in a later section.

Checkpoint

6. **Describe the components of a sarcomere.**
7. **Why do skeletal muscle fibers appear striated when viewed through a light microscope?**
8. **Where would you expect to find the greatest concentration of Ca^{2+} in a resting skeletal muscle fiber?**

See the blue Answers tab at the back of the book.

10-4 Motor neurons stimulate skeletal muscle fibers to contract at the neuromuscular junction

Learning Outcome **Identify the components of the neuromuscular junction, and summarize the events involved in the neural control of skeletal muscle contraction and relaxation.**

You have seen that the myofilaments in a sarcomere change position during a contraction, but not *how* these changes occur. In this section we first discuss how the plasma membranes of neurons and skeletal muscle fibers propagate electrical impulses. (This topic is explored more fully in Chapter 12.) We then look at how skeletal muscle activity is under neural control (as you will see in **Spotlight Figure 10–10**), how neural stimulation of a muscle fiber is coupled to the contraction of the fiber (presented in **Spotlight Figure 10–11**), and how actin and myosin interact to cause sarcomere contraction (shown in **Spotlight Figure 10–12**). Read the following text about excitable membranes before studying these three Spotlight figures.

Electrical Impulses and Excitable Membranes

As we learned in Chapter 3, all the cells of the body maintain a membrane potential due to an unequal distribution of positive and negative charges across their plasma membrane. p. 99 The unequal charge distribution means cells are *polarized*, much like miniature batteries. In cells, the inner surface of the plasma membrane is slightly negative compared to the outer surface. The membrane potential is a measure of cellular polarization (in millivolt units) that compares the cytoplasmic membrane surface charge to the extracellular membrane surface charge. In unstimulated (resting) neurons and skeletal muscle fibers, typical resting membrane potentials are −70 mV and −85 mV, respectively.

Recall that the main contributors of negative charges within a cell are proteins that cannot cross the plasma membrane. There is also an excess of sodium ions outside a cell and an excess of potassium ions inside a cell. These ions can cross a membrane through membrane channels that are specific for each ion or by the use of ion pumps.

Various stimuli can lead to a temporary change in the distribution of electrical charges across the plasma membranes of all body cells. Differences in sodium ion and potassium ion membrane permeability underlie such changes. An influx of sodium ions leads to *depolarization* as the membrane potential becomes less negative. The movement of potassium ions out of a cell leads to *hyperpolarization* as the membrane potential becomes more negative. A return to the resting potential is called *repolarization*. In most cells, the depolarization or hyperpolarization of a plasma membrane is a localized change limited by the presence or absence of stimulation. Called a *graded potential*, it does not continue to spread over the plasma membrane.

Neurons and skeletal muscle fibers, however, have electrically **excitable membranes**. Excitable membranes permit rapid communication between different parts of a cell. In neurons and skeletal muscle fibers, the depolarization and repolarization events produce an electrical impulse, or **action potential**, that is propagated along their plasma membranes. Unlike the plasma membranes of most cells, excitable membranes contain *voltage-gated ion channels* that are activated and inactivated by changes in the membrane potential. These electrical channels become activated when the membranes of neurons and skeletal muscle fibers first depolarize from the resting potential to a *threshold potential* (from −70 to −60 mV for neurons and from −85 to −55 mV for skeletal muscle fibers). Upon reaching the threshold potential, *voltage-gated sodium ion channels* open and there is a rapid influx of positively charged sodium ions into the cell.

The inside of the membrane reverses from a negative to a positive charge, and the depolarization peaks at a membrane potential of +30 mV with the closure of the voltage-gated sodium ion channels. *Repolarization* of the membrane then begins as voltage-gated potassium ion channels open and positively charged potassium ions leave the cell. The loss of more positive charges than entered the cell causes the membrane potential to become negative again. A "resting" *refractory period* follows and the former concentrations of sodium and potassium ions across the plasma membrane are restored through the action of sodium–potassium ion pumps. p. 97 A further depolarization cannot occur until the refractory period is over. As a result, the action potential travels in one direction because the refractory period prevents it from propagating back in the direction from which it was initiated.

The Control of Skeletal Muscle Activity

Skeletal muscle fibers cannot begin contraction until they receive instructions from *motor neurons* of the central nervous system (brain and spinal cord). The motor neurons carry the instructions in the form of action potentials. A contraction begins when the sarcoplasmic reticulum releases stored calcium ions (Ca^{2+}) into the cytosol of the muscle fiber. That release is under the control of the nervous system. As the intracellular calcium ions are reabsorbed, the contraction ends and muscle relaxation occurs. This next section details the events that lead up to contraction and then cause relaxation.

Events at the Neuromuscular Junction

Communication between a neuron and another cell occurs at a *synapse.* When the other cell is a skeletal muscle fiber, the synapse is known as a **neuromuscular junction (NMJ)**. The NMJ is made up of an *axon terminal* (synaptic terminal) of a neuron, a specialized region of the sarcolemma called the *motor end plate,* and, in between, a narrow space called the *synaptic cleft.*

A motor neuron stimulates a muscle fiber through the series of steps shown in **Spotlight Figure 10–10** (pp. 304–305). In this process, the axon terminal of a neuron releases a chemical signal called a **neurotransmitter** into the synaptic cleft. This signal is *acetylcholine,* or *ACh.* ACh binds to an ACh membrane channel receptor on the *motor end plate* of the muscle fiber. The ACh receptor is a *chemically gated* Na^+ *channel.* ACh binding opens the channel, which allows Na^+ to enter the cell. The inflow of positive charges depolarizes the motor end plate and generates an action potential that sweeps across the sarcolemma. Make sure to study **Spotlight Figure 10–10** before reading on.

Excitation–Contraction Coupling

The link between the generation of an action potential in the sarcolemma and the start of a muscle contraction is called **excitation–contraction coupling**. This coupling occurs at the triads. On reaching a triad, an action potential triggers the release of Ca^{2+} from the terminal cisternae of the sarcoplasmic reticulum. The change in the permeability of the SR to Ca^{2+} is temporary, lasting only about 0.03 seconds. Yet within a millisecond, the Ca^{2+} concentration in and around the sarcomere reaches 100 times its resting level. Because the terminal cisternae are located at the zones of overlap, where the thick and thin filaments interact, the effect of calcium ion release on the sarcomere is almost instantaneous.

Troponin is the lock that keeps the active sites inaccessible; calcium ions are the key to that lock. Recall from **Figure 10–7b** that troponin binds to both actin and tropomyosin, and that the tropomyosin molecules cover the active sites and prevent interactions between thick filaments and thin filaments. Each troponin molecule also has a binding site for calcium ions, and this site is empty when the muscle fiber is at rest. Calcium ion binding changes the shape of the troponin molecule and weakens the bond between troponin and actin. The troponin molecule then changes position, rolling the attached tropomyosin strand away from the active sites (**Spotlight Figure 10–11**, p. 306). With this change, the *contraction cycle* begins.

The Contraction Cycle and Generation of Muscle Tension

The **contraction cycle** is a series of molecular events that enable muscle contraction. After the active sites are exposed, the myosin heads bind to them, forming **cross-bridges**. The connection between the head and the tail functions as a hinge that lets the head pivot. When it pivots, using the energy released from the hydrolysis of ATP, the head swings toward the M line, a movement known as the **power stroke**. This pivoting is the key step in muscle contraction. **Spotlight Figure 10–12**, pp. 308–309, shows the interlocking steps of the contraction cycle.

When muscle cells contract, they pull on the attached tendon fibers the way a line of people might pull on a rope. Let's consider what is happening in light of some basic physical principles that apply to muscle cells. The pull, called *tension,* is an active force: Energy must be expended to produce it. Tension is applied *to* some object, whether a rope, a rubber band, or a book on a tabletop. Tension applied to an object tends to pull the object toward the source of the tension. However, before movement can occur, the applied tension must overcome the object's *load* (or *resistance*), a passive force that opposes movement. The amount of load can depend on the weight of the object, its shape, friction, and other factors. When the applied tension exceeds the load, the object moves.

In contrast, *compression,* or a push applied to an object, tends to force the object away from the source of the compression. Again, no movement can occur until the applied compression exceeds the load of the object. Muscle cells can use energy to shorten and generate tension through interactions between thick and thin filaments, but not to lengthen and generate compression. In other words, muscle cells can pull, but they cannot push.

Each power stroke shortens the sarcomere by about 0.5 percent. The entire muscle shortens at the same rate because all the sarcomeres contract together. The speed of the shortening

SPOTLIGHT

Figure 10–10

Events at the Neuromuscular Junction

A single axon may branch to control more than one skeletal muscle fiber, but each muscle fiber has only one neuromuscular junction (NMJ). At the NMJ, the axon terminal of the neuron lies near the motor end plate of the muscle fiber.

Motor neuron
Path of electrical impulse (action potential)
Axon
Neuromuscular junction
Axon terminal
SEE BELOW
Sarcoplasmic reticulum
Motor end plate
Myofibril
Motor end plate

1 The cytoplasm of the axon terminal contains vesicles filled with molecules of acetylcholine, or ACh. Acetylcholine is a **neurotransmitter**, a chemical released by a neuron to change the permeability or other properties of another cell's plasma membrane. The synaptic cleft and the motor end plate contain molecules of the enzyme acetylcholinesterase (AChE), which breaks down ACh.

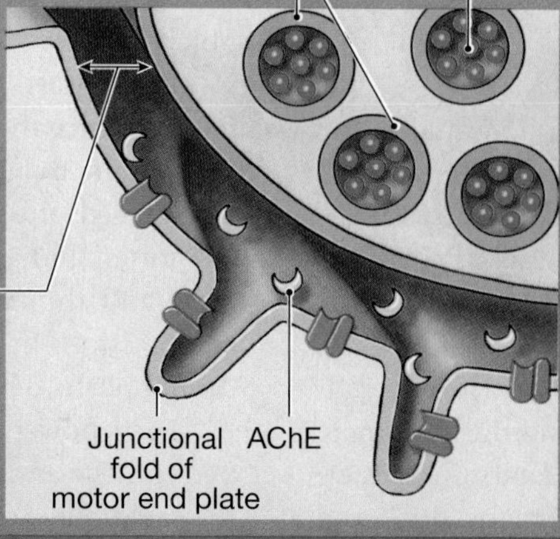

Vesicles
ACh
Junctional fold of motor end plate
AChE

The **synaptic cleft** is a narrow space that separates the axon terminal of the neuron from the opposing motor end plate.

2 The stimulus for ACh release is the arrival of an electrical impulse, or action potential, at the axon terminal. An action potential is a sudden change in the membrane potential that travels along the length of the axon.

Arriving action potential

Go to MasteringA&P™ > Study Area > Menu > Animations & Videos > A&PFlix > A&P > Events at the Neuromuscular Junction

Muscle Fiber

The action potential generated at the motor end plate now sweeps across the entire membrane surface. The effects are almost immediate because an action potential is an electrical event that flashes like a spark across the sarcolemmal surface. The effects are brief because the ACh has been removed, and no further stimulus acts upon the motor end plate until another action potential arrives at the axon terminal.

3 When the action potential reaches the neuron's axon terminal, permeability changes in its membrane trigger the exocytosis of ACh into the synaptic cleft. Exocytosis occurs as vesicles fuse with the neuron's plasma membrane.

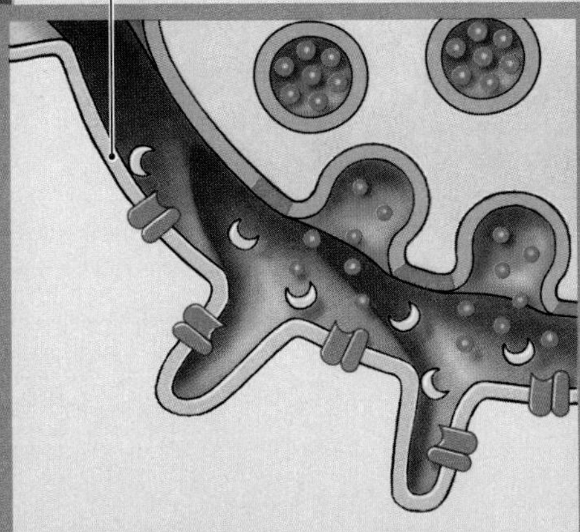
Motor
end plate

4 ACh molecules diffuse across the synaptic cleft and bind to ACh receptor membrane channels. ACh binding opens the membrane channel on the surface of the motor end plate. Because the extracellular fluid contains a high concentration of sodium ions, and sodium ion concentration inside the cell is very low, sodium ions rush into the cytosol.

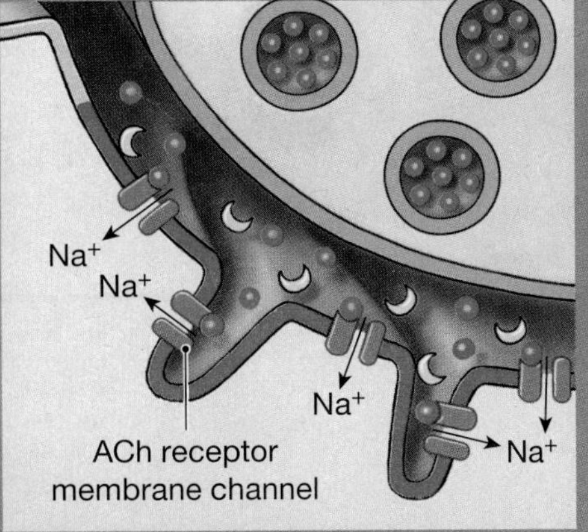
Na+
Na+
Na+
Na+
ACh receptor
membrane channel

5 The sudden inrush of sodium ions results in the generation of an action potential in the sarcolemma. ACh is removed from the synaptic cleft in two ways. ACh either diffuses away from the synapse, or it is broken down by AChE into acetic acid and choline. This removal closes the ACh receptor membrane channels. The muscle fiber pictured above indicates the propagation of the action potential along the sarcolemma.

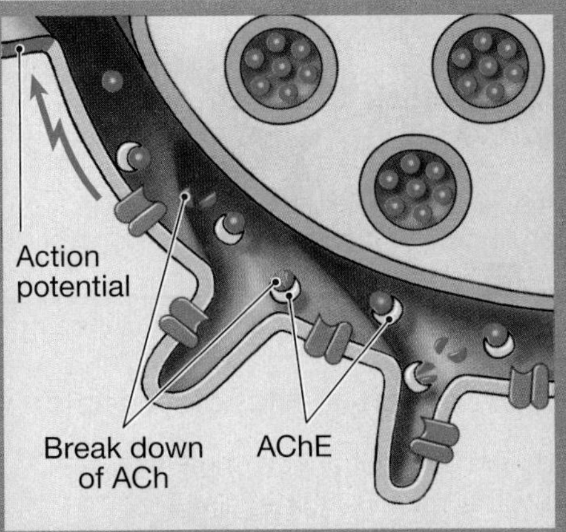
Action
potential
Break down
of ACh
AChE

SPOTLIGHT

Figure 10–11

Excitation–Contraction Coupling

Go to MasteringA&P™ > Study Area > Menu > Animations & Videos > A&PFlix > A&P > Excitation-Contraction Coupling

Action potential

Axon terminal

Excitation

Excitation

Sarcolemma

T tubule

Cytosol

Sarcoplasmic reticulum

Calcium ion release

Ca^{2+}

Ca^{2+}

ATP

Thick-thin filament interaction

Ca^{2+}

Myosin tail (thick filament)

Tropomyosin

Troponin

G-actin (thin filament)

Nebulin

Active site

Ca^{2+}

Ca^{2+}

Cross-bridge formation

In a resting sarcomere, the tropomyosin strands cover the active sites on the thin filaments, preventing cross-bridge formation.

When calcium ions enter the sarcomere, they bind to troponin, which rotates and swings the tropomyosin away from the active sites.

Cross-bridge formation then occurs, and the contraction cycle begins.

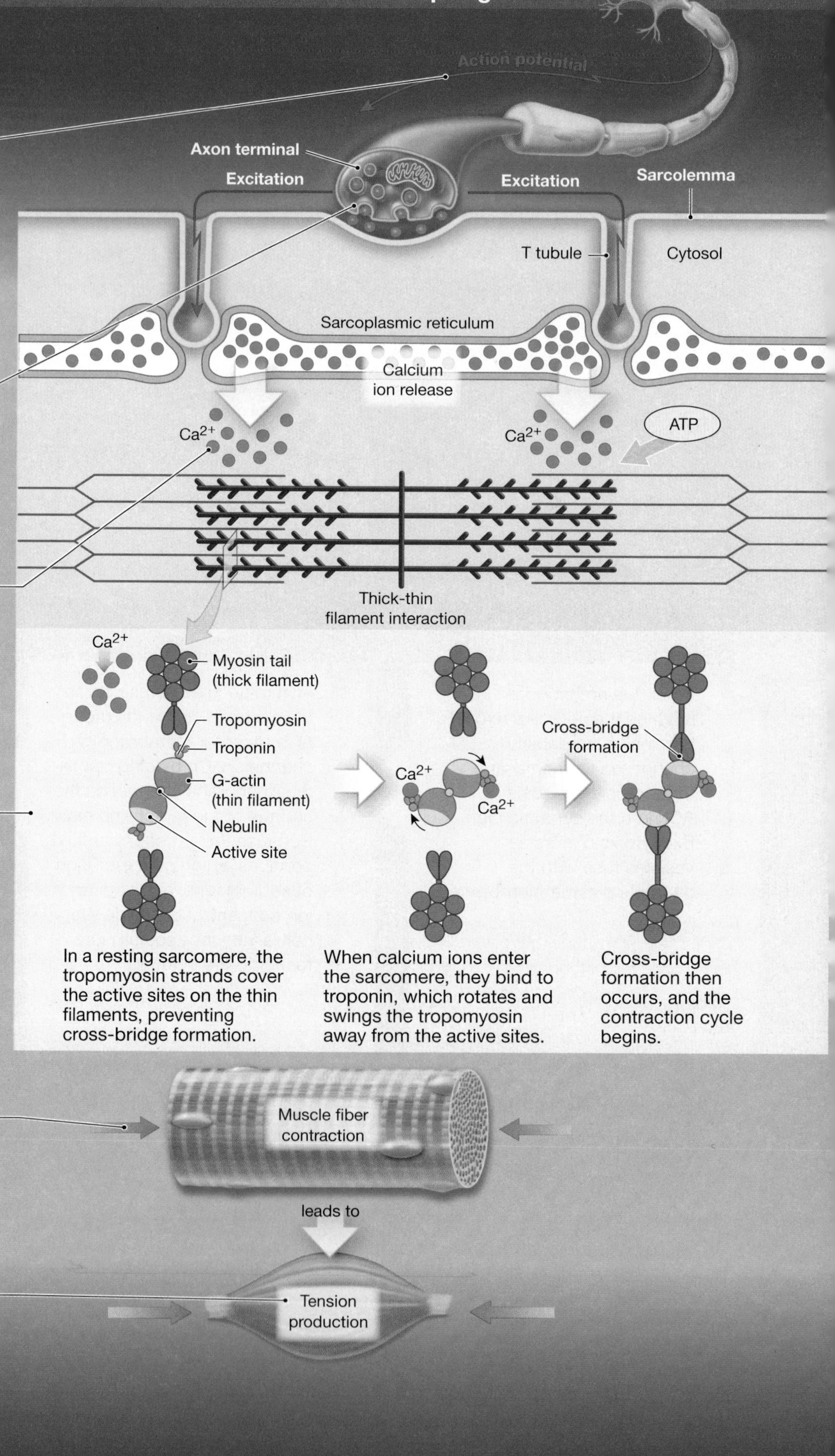

1

Neural Control

A skeletal muscle fiber contracts when stimulated by a motor neuron at a neuromuscular junction. The stimulus arrives in the form of an action potential at the axon terminal.

2

Excitation

The action potential causes the release of ACh into the synaptic cleft, which leads to excitation—the production of an action potential in the sarcolemma.

3

Release of Calcium Ions

This action potential travels along the sarcolemma and down T tubules to the triads. This triggers the release of calcium ions (Ca^{2+}) from the terminal cisternae of the sarcoplasmic reticulum.

4

Contraction Cycle Begins

The contraction cycle begins when the calcium ions (Ca^{2+}) bind to troponin, resulting in the exposure of the active sites on the thin filaments. This allows cross-bridge formation and will continue as long as ATP is available. (See *Spotlight Figure 10-11* for the details of the contraction cycle.)

5

Sarcomere Shortening

As the thick and thin filaments interact, the sarcomeres shorten, pulling the ends of the muscle fiber closer together.

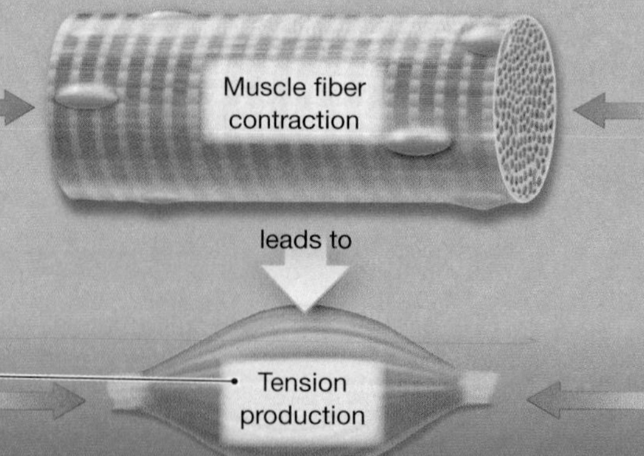
Muscle fiber contraction
leads to
Tension production

6

Generation of Muscle Tension

During the contraction, the entire skeletal muscle shortens and produces a pull, or tension, on the tendons at either end.

depends on the cycling rate (the number of power strokes per second). The greater the load, the slower the cycling rate.

To understand how the contraction cycle produces tension in a muscle fiber, imagine that you are on a tug-of-war team. You reach forward, grab the rope with both hands, and pull it toward you. This grab-and-pull corresponds to cross-bridge attachment and pivoting. You then let go of the rope, reach forward and grab it, and pull once again. Your actions are not synchronized with the rest of your team. At any given time, some people are reaching and grabbing, some are pulling, and others are letting go. (If everyone let go at the same time, your opponents would pull the rope away.) The amount of tension produced depends on how many people are pulling at the same time.

The situation is similar in a muscle fiber. The myosin heads along a thick filament work together to pull a thin filament toward the center of the sarcomere. Each myofibril consists of a string of sarcomeres, and in a contraction all of the thin filaments are pulled toward the centers of the sarcomeres.

+ Clinical Note Tetanus

Beware rusty nails! However, it's not the rust or the nail, but instead an infection by the very common bacterium *Clostridium tetani* that is dangerous. The resulting disease is called **tetanus**. *Clostridium* bacteria are found in soil and virtually everywhere else in the environment, but they can thrive only in tissues with a low oxygen level. For this reason, a deep puncture wound, such as that from a nail, is much more likely to result in tetanus than a shallow, open cut that bleeds freely. When active in body tissues, these bacteria release a powerful neurotoxin that affects the central nervous system. Motor neurons, which control skeletal muscles throughout the body, are particularly sensitive to it. The toxin suppresses the mechanism that inhibits motor neuron activity. The result is sustained, powerful contractions of skeletal muscles throughout the body. The disease is also called *lockjaw* because patients find it difficult to open the mouth.

Severe tetanus has a 40–60 percent mortality rate. In other words, for every 100 people who develop severe tetanus, 40 to 60 die. Approximately 500,000 cases of tetanus occur worldwide each year, but only about 100 of them occur in the United States, thanks to an effective immunization program beginning in childhood. (Recommended immunization involves a "tetanus shot" followed by a booster shot every 10 years.) In unimmunized patients, early administration of an antitoxin can prevent severe symptoms. However, this treatment does not reduce signs and symptoms that have already appeared.

Figure 10–13 reviews the contraction process. Steps 1–5 of this figure show the release of ACh at the neuromuscular junction, continuing through excitation–contraction coupling to the beginning of the contraction cycle.

Relaxation

How long does a contraction last? Its duration depends on three factors: (1) the period of stimulation at the neuromuscular junction (NMJ), (2) the presence of free calcium ions in the cytosol, and (3) the availability of ATP. A single action potential has only a brief effect on a muscle fiber, so a contraction will continue only if additional action potentials arrive at the NMJ in rapid succession. When they do, the continual release of ACh produces a series of action potentials in the sarcolemma that keeps the Ca^{2+} level elevated in the cytosol. Under these conditions, the contraction cycle will be repeated over and over.

10

Steps 6–10 of Figure 10–13 show what happens if just one action potential arrives at the neuromuscular junction. This starts with the breakdown of released ACh in the synaptic cleft by *acetylcholinesterase*, or *AChE*. Inside the muscle fiber, the permeability changes in the sarcoplasmic reticulum (SR) are also very brief. For these reasons, the Ca^{2+} concentration in the cytosol will quickly return to its normal resting level. Two mechanisms are involved in this process: (1) active Ca^{2+} transport into the SR, and (2) active Ca^{2+} transport across the

+ Clinical Note Rigor Mortis

When death occurs, circulation ceases and the skeletal muscles are deprived of nutrients and oxygen. Within a few hours, the skeletal muscle fibers run out of ATP and the sarcoplasmic reticulum becomes unable to pump calcium ions out of the cytosol. Calcium ions diffusing into the cytosol from the extracellular fluid or leaking out of the SR then trigger a sustained contraction. Without ATP, the cross-bridges cannot detach from the active sites, so skeletal muscles throughout the body become locked in the contracted position. Because all the skeletal muscles are involved, the body becomes "stiff as a board." This physical state—**rigor mortis**—lasts until the lysosomal enzymes released by autolysis (the process of self-digestion) break down the Z lines and titin filaments. Rigor mortis begins two to seven hours after death and ends after one to six days or when decomposition begins. The timing depends on environmental factors, such as temperature. Forensic pathologists (scientists who study the causes of disease and death) can estimate the time of death on the basis of the degree of rigor mortis and environmental conditions.

SPOTLIGHT

Figure 10–12
The Contraction Cycle and Cross-Bridge Formation

1
Contraction Cycle Begins

The contraction cycle involves a series of interrelated steps. It begins with the arrival of calcium ions (Ca^{2+}) within the zone of overlap in a sarcomere.

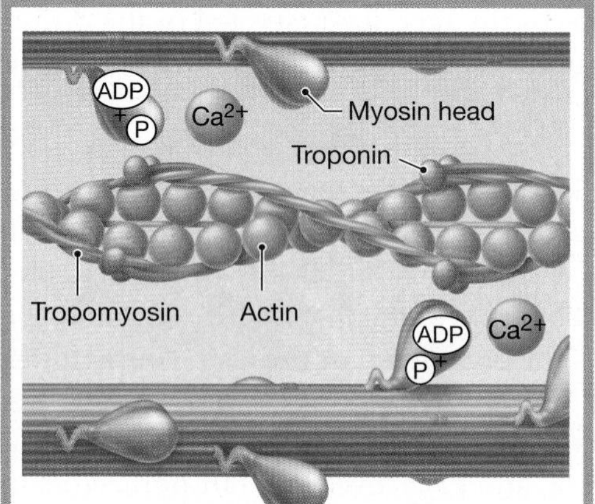

ADP
P
Ca^{2+}
Myosin head
Troponin
Tropomyosin
Actin
ADP
P
Ca^{2+}

2
Active-Site Exposure

Calcium ions bind to troponin, weakening the bond between actin and the troponin–tropomyosin complex. The troponin molecule then changes position, rolling the tropomyosin molecule away from the active sites on actin and allowing interaction with the energized myosin heads.

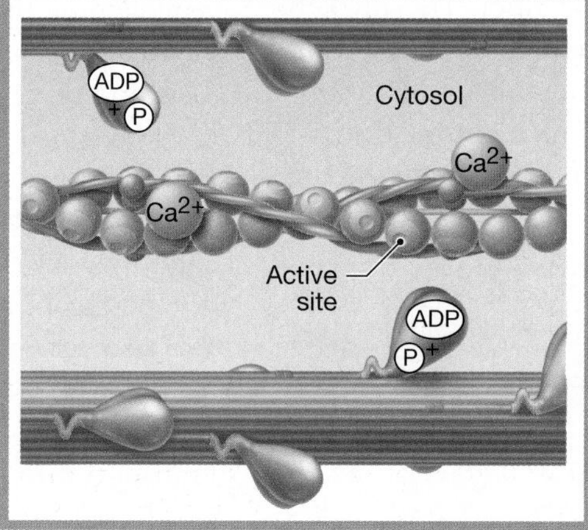

ADP
P
Cytosol
Ca^{2+}
Ca^{2+}
Active site
ADP
P

3
Cross-Bridge Formation

Once the active sites are exposed, the energized myosin heads bind to them, forming cross-bridges.

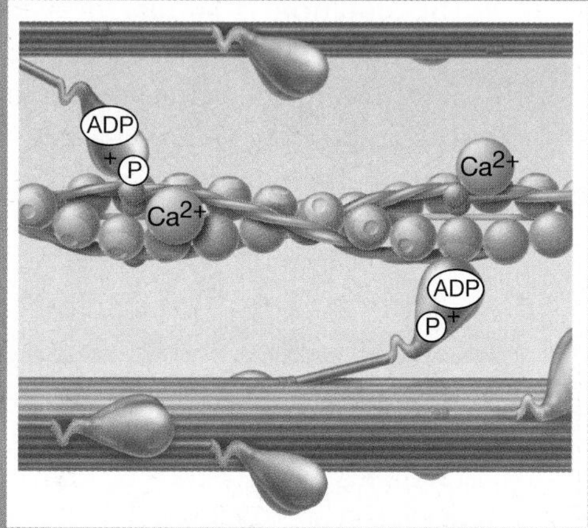

ADP
P
Ca^{2+}
Ca^{2+}
ADP
P

RESTING SARCOMERE

In the resting sarcomere, each myosin head is already "energized"—charged with the energy that will be used to power a contraction. Each myosin head points away from the M line. In this position, the myosin head is "cocked" like the spring in a mousetrap. Cocking the myosin head requires energy, which is obtained by breaking down ATP; in doing so, the myosin head functions as ATPase, an enzyme that breaks down ATP. At the start of the contraction cycle, the breakdown products, ADP and phosphate (represented as P), remain bound to the myosin head.

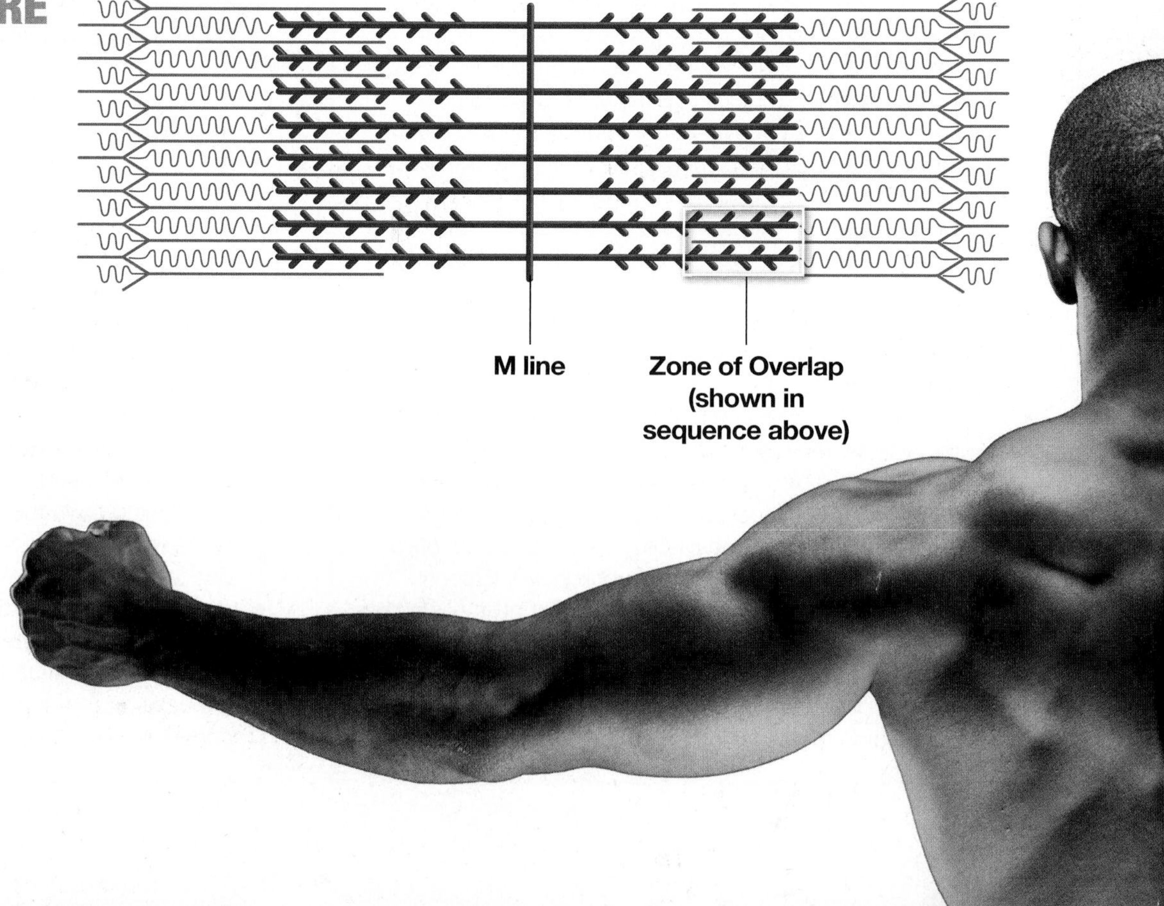

M line
Zone of Overlap (shown in sequence above)

Go to MasteringA&P™ > Study Area > Menu > Animations & Videos > A&PFlix > A&P > The Cross Bridge Cycle

4

Myosin Head Pivoting

After cross-bridge formation, the energy that was stored in the resting state is released as the myosin head pivots toward the M line. This action is called the power stroke; when it occurs, the bound ADP and phosphate group are released.

5

Cross-Bridge Detachment

When another ATP binds to the myosin head, the link between the myosin head and the active site on the actin molecule is broken. The active site is now exposed and able to form another cross-bridge.

6

Myosin Reactivation

Myosin reactivation occurs when the free myosin head splits ATP into ADP and P. The energy released is used to recock the myosin head.

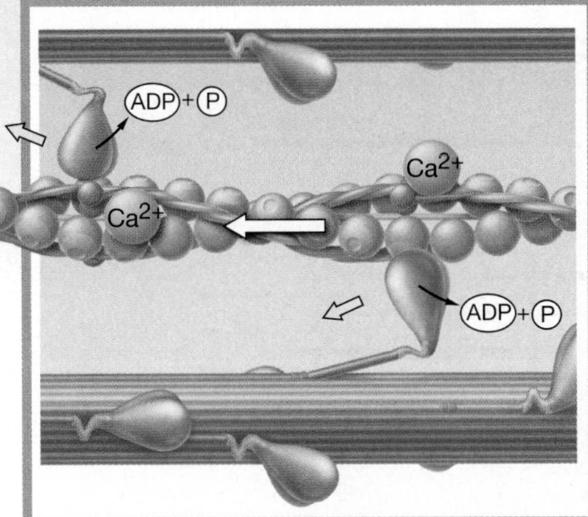
ADP+P
Ca^{2+}
Ca^{2+}
ADP+P

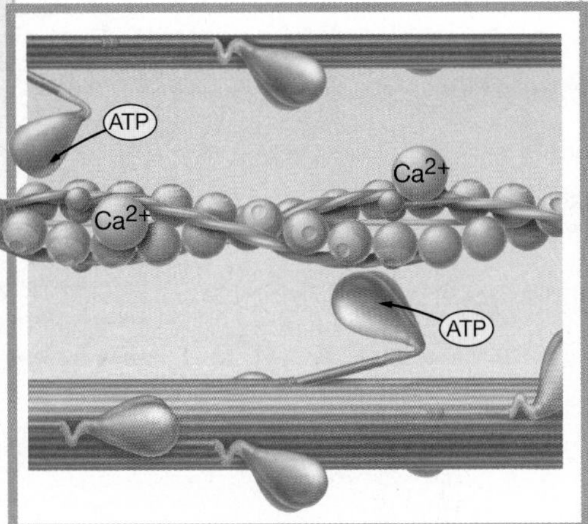
ATP
Ca^{2+}
Ca^{2+}
ATP

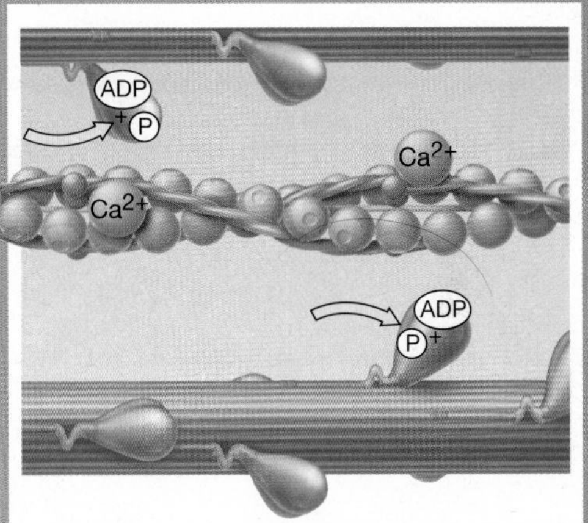
ADP + P
Ca^{2+}
Ca^{2+}
ADP P +

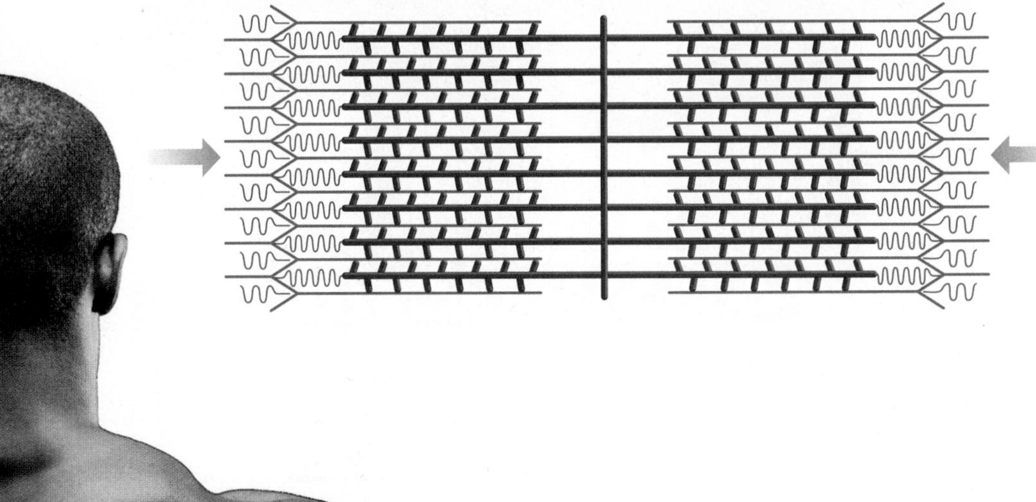

CONTRACTED SARCOMERE

The entire cycle is repeated several times each second, as long as Ca^{2+} concentrations remain elevated and ATP reserves are sufficient. Calcium ion levels will remain elevated only as long as action potentials continue to pass along the T tubules and stimulate the terminal cisternae. Once that stimulus is removed, the calcium channels in the SR close and calcium ion pumps pull Ca^{2+} from the cytosol and store it within the terminal cisternae. Troponin molecules then shift position, swinging the tropomyosin strands over the active sites and preventing further cross-bridge formation.

Figure 10–13 Steps Involved in Skeletal Muscle Contraction and Relaxation.

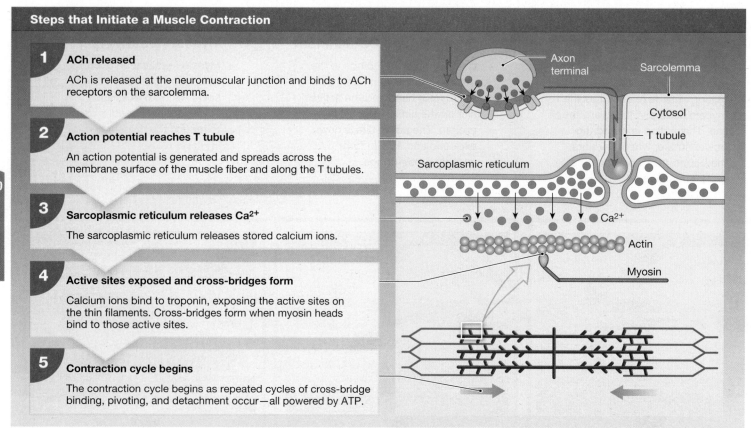

Steps that Initiate a Muscle Contraction

1 ACh released

ACh is released at the neuromuscular junction and binds to ACh receptors on the sarcolemma.

2 Action potential reaches T tubule

An action potential is generated and spreads across the membrane surface of the muscle fiber and along the T tubules.

3 Sarcoplasmic reticulum releases Ca^{2+}

The sarcoplasmic reticulum releases stored calcium ions.

4 Active sites exposed and cross-bridges form

Calcium ions bind to troponin, exposing the active sites on the thin filaments. Cross-bridges form when myosin heads bind to those active sites.

5 Contraction cycle begins

The contraction cycle begins as repeated cycles of cross-bridge binding, pivoting, and detachment occur—all powered by ATP.

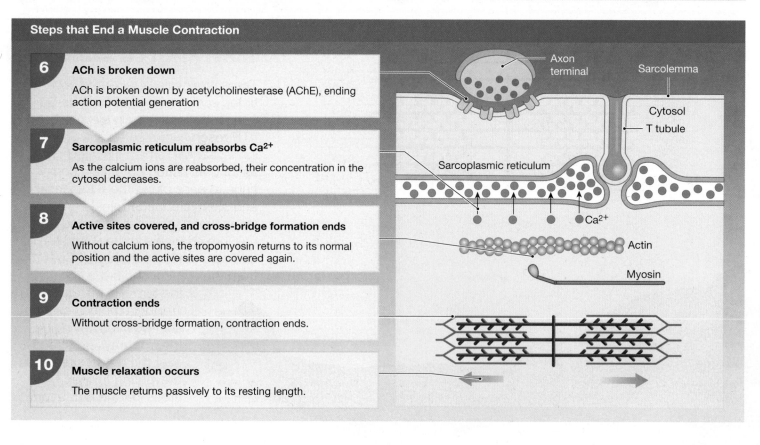

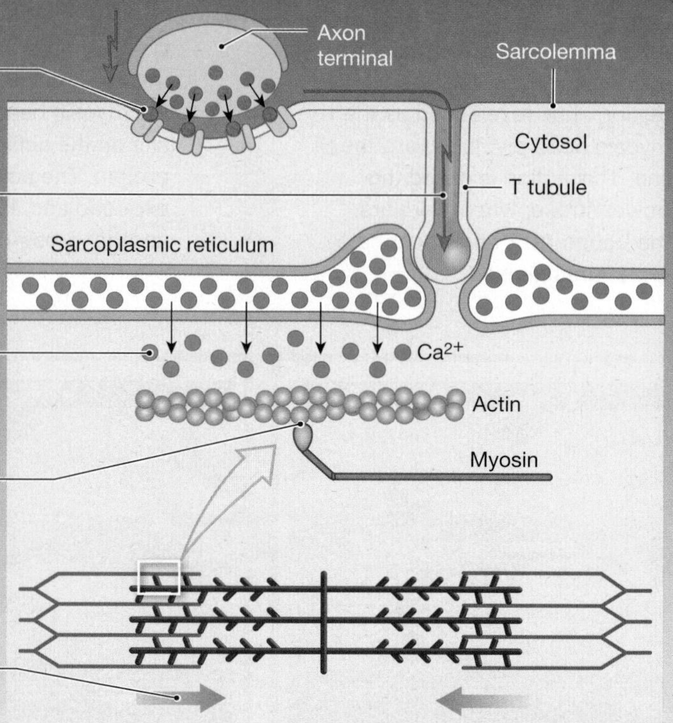

Axon terminal
Sarcolemma
Cytosol
T tubule
Sarcoplasmic reticulum
Ca2+
Actin
Myosin

Steps that End a Muscle Contraction

6 ACh is broken down

ACh is broken down by acetylcholinesterase (AChE), ending action potential generation

7 Sarcoplasmic reticulum reabsorbs Ca^{2+}

As the calcium ions are reabsorbed, their concentration in the cytosol decreases.

8 Active sites covered, and cross-bridge formation ends

Without calcium ions, the tropomyosin returns to its normal position and the active sites are covered again.

9 Contraction ends

Without cross-bridge formation, contraction ends.

10 Muscle relaxation occurs

The muscle returns passively to its resting length.

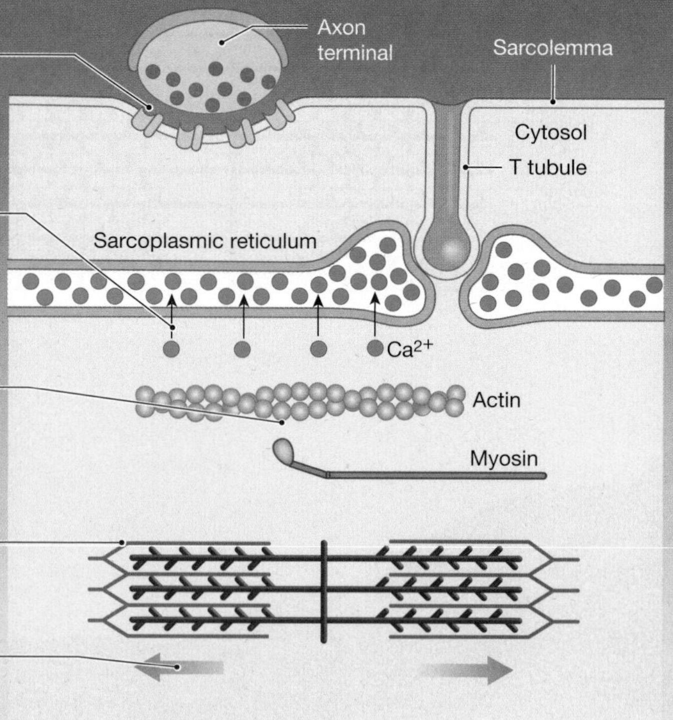

Axon terminal
Sarcolemma
Cytosol
T tubule
Sarcoplasmic reticulum
Ca2+
Actin
Myosin

sarcolemma into the extracellular fluid. Of the two, transport into the SR is far more important. Almost as soon as the calcium ions have been released, the SR returns to its normal permeability and begins to actively take up these ions from the surrounding cytosol. As the Ca^{2+} concentration in the cytosol falls, earlier events reverse themselves: (1) Calcium ions detach from troponin, (2) troponin returns to its original position, and (3) tropomyosin once again covers the active sites on actin. This ends the contraction.

The sarcomeres do not automatically return to their original length. The sarcomeres in a muscle fiber can shorten and develop tension, but there is no active mechanism for reversing the process—the power stroke cannot be reversed to push the Z lines farther apart. External forces must act on the contracted muscle fiber to stretch the myofibrils and sarcomeres to their original dimensions. We describe those forces in Section 10–5.

✓ Checkpoint

9. **Describe the neuromuscular junction.**
10. **How would a drug that blocks acetylcholine release affect muscle contraction?**
11. **What would happen to a resting skeletal muscle if the sarcolemma suddenly became very permeable to calcium ions?**
12. **Predict what would happen to a muscle if the motor end plate lacked acetylcholinesterase (AChE).**

See the blue Answers tab at the back of the book.

10-5 Muscle fibers produce different amounts of tension depending on sarcomere length and frequency of stimulation

Learning Outcome **Describe the mechanism responsible for the different amounts of tension produced in a muscle fiber.**

When sarcomeres shorten in a contraction, they shorten the muscle fiber. The tension produced by an individual muscle fiber can vary, and in this section we consider the specific factors involved in that variation.

The amount of tension produced by an individual muscle fiber in a contraction ultimately depends on the number of power strokes performed by cross-bridges. The number of contracting sarcomeres within the muscle fiber is fixed, because calcium ions are released from all triads in the muscle fiber at the same time. For this reason, a muscle fiber is either all "on" (producing tension) or all "off" (relaxed).

The tension produced by an individual muscle fiber *does* vary, however, for other reasons. Tension production is greatest when a muscle is stimulated at its optimal length. A couple of factors are important:

- Tension production depends on the fiber's resting length at the time of stimulation, which determines the amount of overlap between thick and thin filaments.
- It also depends on the frequency of stimulation, which affects the internal concentration of calcium ions and thus the amount bound to troponin.

Length–Tension Relationships

Figure 10–14 shows the effect of sarcomere length on active tension. When many people pull on a rope, the amount of tension produced is proportional to the number of people pulling. Similarly, in a skeletal muscle fiber, the amount of tension generated during a contraction depends on the number of power strokes performed by cross-bridges in each of the myofibrils. The number of cross-bridges that can form, in turn, depends on the amount of overlap between thick filaments and thin filaments within these sarcomeres. When the muscle fiber is stimulated to contract, only myosin heads in the zones of overlap can bind to active sites and produce tension. For these reasons, we can relate the tension produced by the entire muscle fiber to the length of individual sarcomeres, a concept known as the **length–tension relationship**.

A sarcomere works most efficiently within an optimal range of lengths. When the resting sarcomere length is within this range, the maximum number of cross-bridges can form, and the maximum amount of tension is produced. If the resting sarcomere length falls outside the range—if the sarcomere is stretched and lengthened, or compressed and shortened—it cannot produce as much tension when stimulated. An increase in sarcomere length reduces the tension produced by reducing the zone of overlap and, hence, the number of potential cross-bridges.

A decrease in the resting sarcomere length reduces efficiency because the stimulated sarcomere cannot shorten very much before the thin filaments extend across the center of the sarcomere (M line) and collide with or overlap the thin filaments of the opposite side. This interferes with the binding of myosin heads to active sites and the propagation of the action potential along the T tubules. Because the number of cross-bridges is reduced, tension declines in the stimulated muscle fiber. Tension production falls to zero when the resting sarcomere is at its shortest length. At this point, the thick filaments are jammed against the Z lines and no further shortening is possible. Although cross-bridge binding can still occur, the myosin heads cannot produce a power stroke and generate tension because the thin filaments cannot move.

Figure 10–14 The Effect of Sarcomere Length on Active Tension.

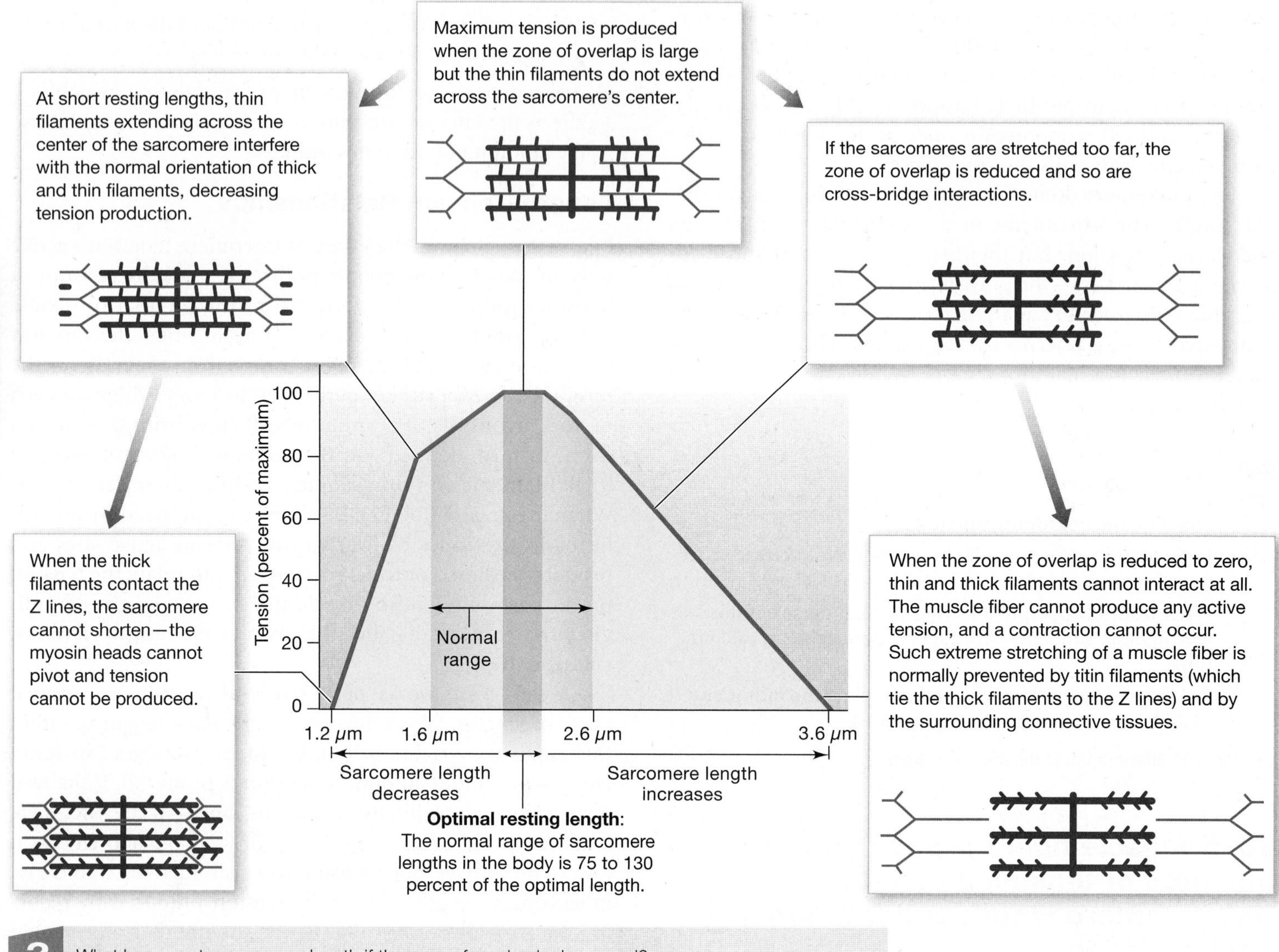

Maximum tension is produced when the zone of overlap is large but the thin filaments do not extend across the sarcomere's center.
At short resting lengths, thin filaments extending across the center of the sarcomere interfere with the normal orientation of thick and thin filaments, decreasing tension production.
If the sarcomeres are stretched too far, the zone of overlap is reduced and so are cross-bridge interactions.
When the thick filaments contact the Z lines, the sarcomere cannot shorten—the myosin heads cannot pivot and tension cannot be produced.
When the zone of overlap is reduced to zero, thin and thick filaments cannot interact at all. The muscle fiber cannot produce any active tension, and a contraction cannot occur. Such extreme stretching of a muscle fiber is normally prevented by titin filaments (which tie the thick filaments to the Z lines) and by the surrounding connective tissues.
Tension (percent of maximum)
100
80
60
40
20
0
Normal range
1.2 μm
1.6 μm
2.6 μm
3.6 μm
Sarcomere length decreases
Sarcomere length increases
Optimal resting length:
The normal range of sarcomere lengths in the body is 75 to 130 percent of the optimal length.

? What happens to sarcomere length if the zone of overlap is decreased?

In summary, skeletal muscle fibers contract most forcefully when stimulated within the range of their optimal resting lengths (see Figure 10–14). The arrangement of skeletal muscles, connective tissues, and bones normally prevents extreme compression or excessive stretching of muscle fibers. For example, straightening your elbow stretches your *biceps brachii*, but the bones and ligaments of the elbow stop this movement before the muscle fibers stretch too far. During an activity such as walking, in which muscles contract and relax cyclically, muscle fibers are stretched to a length very close to "ideal" before they are stimulated to contract. When muscles must contract over a larger range of resting lengths, they often "team up" to improve efficiency. (We discuss the mechanical principles involved in Chapter 11.)

Frequency of Stimulation

A single stimulation produces a single contraction, or *twitch*, that may last 7–100 milliseconds, depending on the muscle stimulated. A single twitch is so brief that there isn't enough time to activate a significant percentage of the available cross-bridges. For this reason, although muscle twitches can be produced by electrical stimulation in a laboratory, they are ineffective in performing useful work. However, if a second twitch occurs before the tension returns to zero, tension will peak at a higher level, because additional cross-bridges will form. Think of pushing a child on a swing: You push gently to start the swing moving, but if you push harder the second time, the child swings higher because the energy of the second push is added to the energy remaining from the first. Likewise, each successive contraction begins before the tension has fallen to the resting level, so the tension continues to increase until it peaks. To understand this process, we need to follow the changes that occur as the rate of stimulation increases. Although measuring twitches is useful experimentally, all consciously and subconsciously directed muscular activities—standing, walking, running, reaching, and so forth—involve sustained muscular contractions rather than twitches.

Twitches

A **twitch** is a single stimulus–contraction–relaxation sequence in a muscle fiber. Twitches vary in duration depending on the type of muscle, its location, internal and external environmental conditions, and other factors. For example, twitches in an eye muscle fiber can be as brief as 7.5 msec, but a twitch in a muscle fiber from the *soleus* (a small calf muscle) lasts about 100 msec. Figure 10–15a is a **myogram**, or graphic representation, showing tension development in muscle fibers from various skeletal muscles. As you can see here, the twitch in the *gastrocnemius muscle*, a prominent calf muscle, is intermediate in duration.

Figure 10–15b details the phases of a 40-msec twitch in a muscle fiber from the gastrocnemius. We can divide a single twitch into a *latent period*, a *contraction phase*, and a *relaxation phase:*

1. The **latent period** begins at stimulation and lasts about 2 msec. During this period, the action potential sweeps across the sarcolemma, and the SR releases calcium ions. The muscle fiber does not produce tension during the latent period, because the contraction cycle has yet to begin.
2. In the **contraction phase**, tension increases to a peak. As the tension increases, calcium ions are binding to troponin, active sites on thin filaments are being exposed, and cross-bridge interactions are occurring. For this muscle fiber, the contraction phase ends roughly 15 msec after stimulation.
3. The **relaxation phase** lasts about 25 msec. During this period, the Ca^{2+} level is decreasing as calcium ions are pumped back into the SR, active sites are being covered by tropomyosin, and the number of active cross-bridges is declining as they detach. As a result, tension decreases to the resting level.

Treppe

If a skeletal muscle fiber is stimulated a second time immediately after the relaxation phase of a twitch has ended, the resulting contraction will develop a slightly higher maximum tension than did the first contraction. The increase in maximum tension shown in Figure 10–16a will continue over the first 30–50 stimulations. After that, the amount of tension produced will remain constant. This pattern is called **treppe** (TREP-eh), which is German for "staircase," because the tension rises in stages, like the steps in a staircase. The rise is thought to result from a gradual increase in the concentration of Ca^{2+} in the cytosol, in part because the calcium ion pumps in the SR have too little time to recapture the ions between stimulations. Most skeletal muscles don't undergo treppe. However, treppe is a phenomenon in cardiac muscle. It occurs if stimuli of the same intensity are sent to the muscle fiber after a latent period.

Figure 10–15 The Development of Tension in a Twitch.

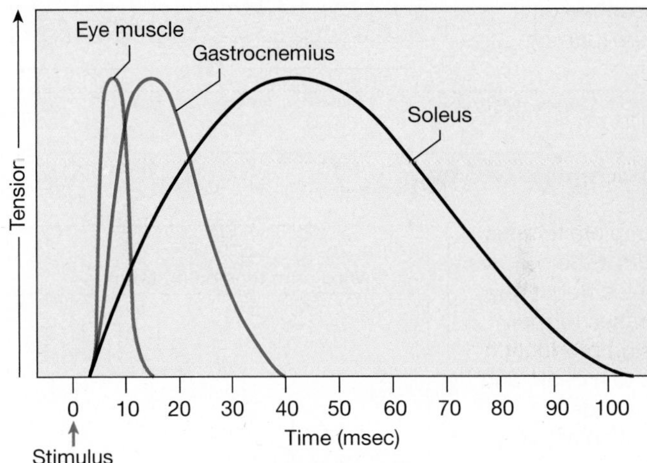
Eye muscle
Gastrocnemius
Soleus
Tension
0
10
20
30
40
50
60
70
80
90
100
Time (msec)
Stimulus

a A myogram showing differences in tension over time for a twitch in the fibers from different skeletal muscles.

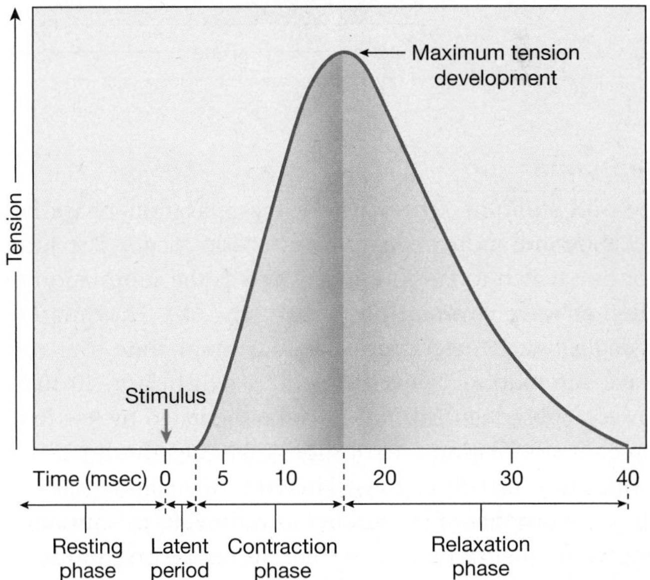
Maximum tension development
Tension
Stimulus
Time (msec) 0
5
10
20
30
40
Resting phase
Latent period
Contraction phase
Relaxation phase

b The details of tension over time for a single twitch in a fiber of the gastrocnemius muscle. Notice the presence of a latent period. It corresponds to the time needed for the propagation of an action potential and the subsequent release of calcium ions by the sarcoplasmic reticulum.

10

Figure 10–16 Effects of Repeated Stimulations.

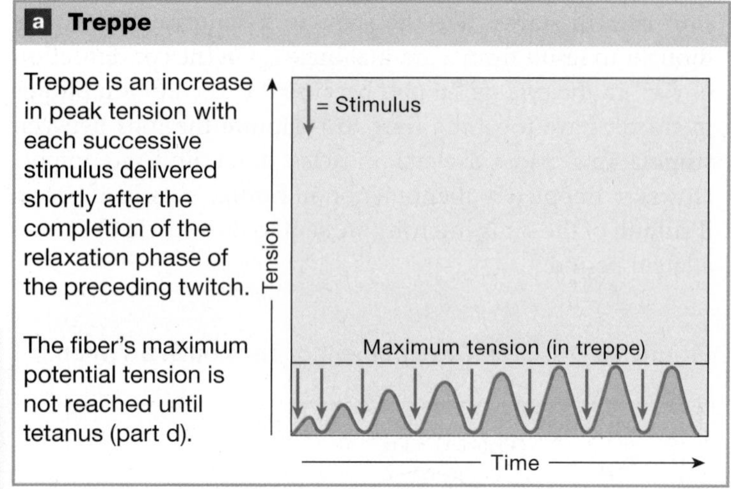
a Treppe
Treppe is an increase in peak tension with each successive stimulus delivered shortly after the completion of the relaxation phase of the preceding twitch.
The fiber's maximum potential tension is not reached until tetanus (part d).
= Stimulus
Tension
Maximum tension (in treppe)
Time

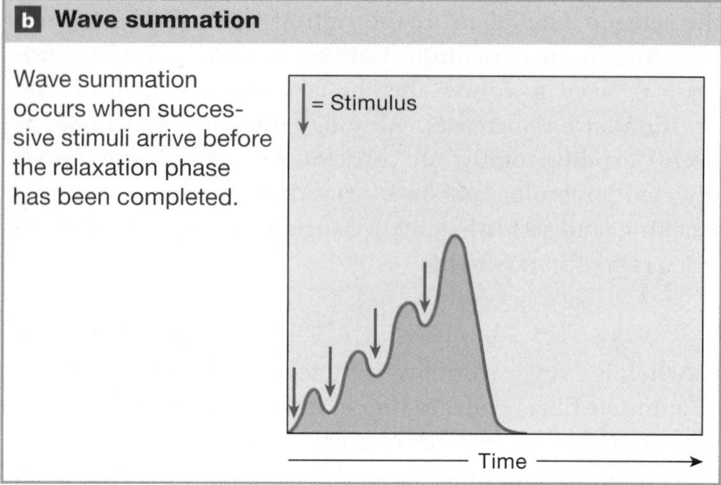
b Wave summation
Wave summation occurs when successive stimuli arrive before the relaxation phase has been completed.
= Stimulus
Time

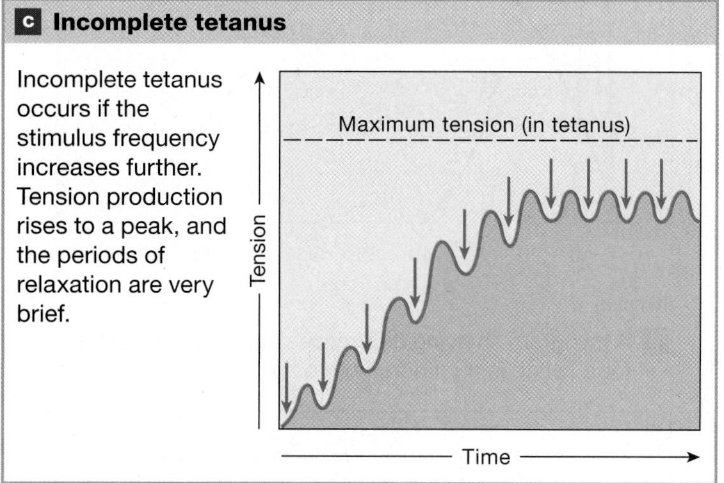
c Incomplete tetanus
Incomplete tetanus occurs if the stimulus frequency increases further. Tension production rises to a peak, and the periods of relaxation are very brief.
Tension
Maximum tension (in tetanus)
Time

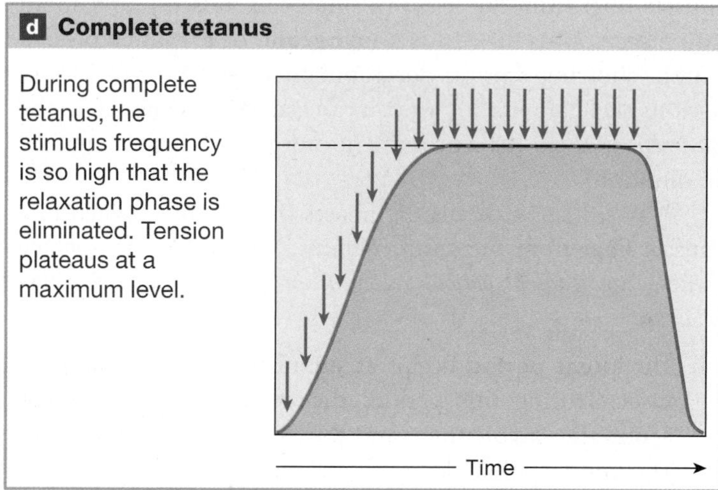
d Complete tetanus
During complete tetanus, the stimulus frequency is so high that the relaxation phase is eliminated. Tension plateaus at a maximum level.
Time

Wave Summation

If a second stimulus arrives before the relaxation phase has ended, a second, more powerful contraction occurs. The addition of one twitch to another in this way is the summation of twitches, or **wave summation** (Figure 10–16b). The duration of a single twitch determines the maximum time available for wave summation. For example, if a twitch lasts 20 msec (1/50 sec), subsequent stimuli must be separated by less than 20 msec—a stimulation rate of more than 50 stimuli per second. This rate is usually expressed in terms of *stimulus frequency*, which is the number of stimuli per unit time. In this instance, a stimulus frequency of greater than 50 per second produces wave summation, whereas a stimulus frequency of less than 50 per second produces individual twitches and treppe.

Incomplete and Complete Tetanus

If the stimulation continues and the muscle fiber is never allowed to relax completely, tension will increase until it reaches its maximum potential tension, called *tetanus* (*tetanos*, convulsive tension), which is roughly four times the maximum produced by treppe. A muscle producing almost maximum tension during rapid cycles of contraction and relaxation is in **incomplete tetanus** (Figure 10–16c).

When a higher stimulation frequency eliminates the relaxation phase, **complete tetanus** occurs (Figure 10–16d). In this situation, action potentials arrive so rapidly that the SR does not have time to reclaim the calcium ions. The high Ca^{2+} concentration in the cytosol prolongs the contraction, making it continuous. During a tetanic contraction, there is enough time for essentially all of the potential cross-bridges to form, and tension peaks.

Checkpoint

13. **Why does a muscle that has been overstretched produce minimal tension?**

See the blue Answers tab at the back of the book.

10-6 Skeletal muscles produce increased tension by recruiting additional motor units

Learning Outcome Compare the different types of skeletal muscle contraction.

Now that you are familiar with the basic mechanisms of muscle contraction at the level of the individual muscle fiber, we can begin to examine the performance of skeletal muscles—the organs of the muscular system. In this section, we consider the coordinated contractions of an entire population of skeletal muscle fibers.

The amount of tension produced by a muscle as a whole is the sum of the tensions generated by its individual muscle fibers, since they are all pulling together. For this reason, regulating the number of stimulated muscle fibers will control the tension of a skeletal muscle. This regulation involves both the number of muscle fibers innervated by a single motor neuron and how many of these motor neurons are stimulated.

The ability to control the amount of tension exerted by our skeletal muscles is critical during normal movements, so that our muscles contract smoothly, not jerkily. As we will see, muscle contractions differ. In *isotonic contractions*, the muscle length changes with a constant force. In *isometric contractions*, the length of the muscle does not change with a constant force. Furthermore, isotonic contractions are further divided into either concentric (muscle shortens) or eccentric (muscle lengthens) contractions. We discuss these different types below.

As muscle fibers actively shorten, they pull on the attached connective tissue fibers and tendons, which become stretched. The tension is transferred in turn to bones, which are moved against an external load. (We will look at the interactions between the muscular and skeletal systems in Chapter 11.)

Motor Units

A typical skeletal muscle contains thousands of muscle fibers. Some motor neurons control a few muscle fibers, but most control hundreds or thousands of them through multiple axon terminals. A **motor unit** is a motor neuron and all the muscle fibers that it controls. The size of a motor unit indicates how fine, or precise, a movement can be. Small motor units, with 4–6 muscle fibers, are found in the muscles of the eyes and fingers, where precise control is extremely important. Large motor units, with 1000–2000 muscle fibers, are found in weight-bearing muscles of the thighs and hips, where precise control is less urgent. The muscle fibers of each motor unit are intermingled with those of other motor units (**Figure 10–17a**). Because of this intermingling, the direction of pull exerted on the tendon does not change when the number of activated motor units changes.

Figure 10–17 The Arrangement and Activity of Motor Units in a Skeletal Muscle.

SmartArt

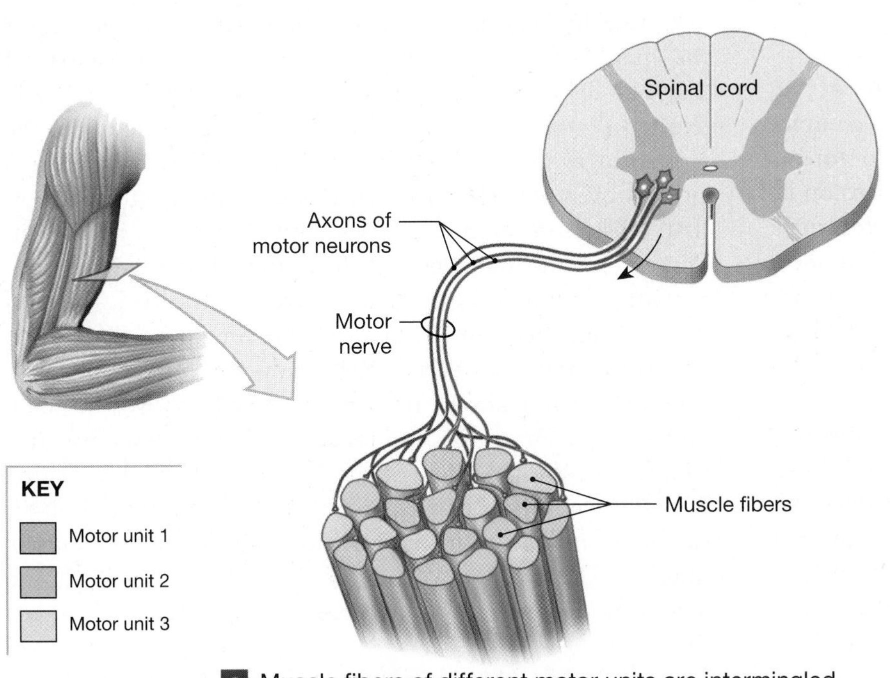
Spinal cord
Axons of motor neurons
Motor nerve
Muscle fibers
KEY
Motor unit 1
Motor unit 2
Motor unit 3

a Muscle fibers of different motor units are intermingled, so the forces applied to the tendon remain balanced regardless of which motor units are stimulated.

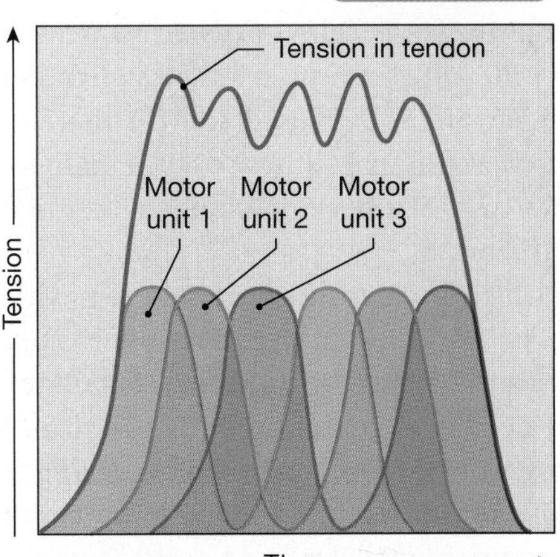
Tension in tendon
Motor unit 1
Motor unit 2
Motor unit 3
Tension
Time

b The tension applied to the tendon remains fairly constant, even though individual motor units cycle between contraction and relaxation.

10

We have all probably seen or felt an involuntary "muscle twitch" under the skin. Such a muscle twitch is called a **fasciculation** (fa-sik-ū-LĀ-shun). Unlike the twitches discussed earlier, a fasciculation involves more than one muscle fiber. It is the synchronous contraction of a motor unit.

Recruitment

When you decide to move your arm, specific groups of motor neurons in the spinal cord are stimulated. The contraction begins with the activation of the smallest motor units in the stimulated muscle. These motor units generally contain muscle fibers that contract fairly slowly. As the movement continues, larger motor units containing faster and more powerful muscle fibers are activated, and tension rises steeply. The smooth but steady increase in muscular tension produced by increasing the number of active motor units is called **recruitment**.

Peak tension occurs when all motor units in the muscle contract in a state of complete tetanus. Such powerful contractions do not last long, however, because the individual muscle fibers soon use up their available energy reserves. During a sustained contraction, motor units are activated on a rotating basis, so some of them are resting and recovering while others are actively contracting. In this "relay team" approach, called *asynchronous motor unit summation*, each motor unit can recover somewhat before it is stimulated again (Figure 10–17b). As a result, when your muscles contract for sustained periods, they produce slightly less than maximum tension.

Muscle Tone

In any skeletal muscle, some motor units are always active, even when the entire muscle is not contracting. Their contractions do not produce enough tension to cause movement, but they do tense and firm the muscle. This resting tension in a skeletal muscle is called **muscle tone**. A muscle with little muscle tone appears limp and soft, whereas one with moderate muscle tone is firm and solid. Different motor units are stimulated at different times so some individual muscle fibers can relax while others maintain a constant tension in the attached tendon.

Resting muscle tone stabilizes the positions of bones and joints. For example, in muscles involved with balance and posture, enough motor units are stimulated to produce the tension needed to maintain body position. Muscle tone also helps prevent sudden, uncontrolled changes in the positions of bones and joints. In addition to bracing the skeleton, the elastic nature of muscles and tendons lets skeletal muscles act as shock absorbers that cushion the impact of a sudden bump or shock. Heightened muscle tone accelerates the recruitment process during a voluntary contraction, because some of the motor units are already stimulated. Strong muscle tone also makes skeletal muscles appear firm and well defined, even at rest.

Activated muscle fibers use energy, so the greater the muscle tone, the higher the "resting" rate of *metabolism* (all the chemical reactions occurring in the body). Increasing this rate is one of the significant effects of exercise in a weight-loss program. You lose weight when your daily energy use exceeds your daily energy intake in food. Although exercise consumes energy very quickly, the period of activity is usually quite brief. In contrast, elevated muscle tone increases resting energy consumption by a small amount, but the effects are cumulative, and they continue 24 hours per day.

Types of Muscle Contractions

We can classify muscle contractions as *isotonic* or *isometric* on the basis of their pattern of tension production.

Isotonic Contractions

In an **isotonic contraction** (*iso-*, equal + *tonos*, tension), tension increases and the skeletal muscle's length changes. Lifting an object off a desk, walking, and running involve isotonic contractions. There are two types of isotonic contractions: concentric and eccentric.

Isotonic Concentric Contractions. In an **isotonic concentric contraction**, the muscle tension *exceeds* the load and the muscle *shortens*. Consider an experiment with a skeletal muscle that is 1 cm^2 in cross-sectional area and can produce roughly 4 kg (8.8 lb) of tension in complete tetanus. That is, 4 kg is the muscle's maximum potential tension. If we hang a load of 2 kg (4.4 lb) from that muscle and stimulate it, the muscle will shorten (Figure 10–18a). Before the muscle can shorten, the cross-bridges in the muscle fibers must produce enough tension to overcome the load—in this case, the 2-kg weight. At first, tension in the muscle rises until the tension in the tendon exceeds the load. Then, as the muscle shortens, its tension remains constant at a value that just exceeds the load. The term *isotonic* originated from this type of experiment.

In the body, however, the situation is more complicated. Muscles are not always positioned directly above the load, and they are attached to bones rather than to static weights. Changes in the positions of the muscle and the articulating bones, the effects of gravity, and other mechanical and physical factors interact to increase or decrease the load the muscle must overcome as a movement proceeds. Nevertheless, during a concentric contraction, the peak tension produced exceeds that load.

The speed of shortening varies with the difference between the amount of tension produced and the size of the load. If

Figure 10–18 **Concentric, Eccentric, and Isometric Contractions.**

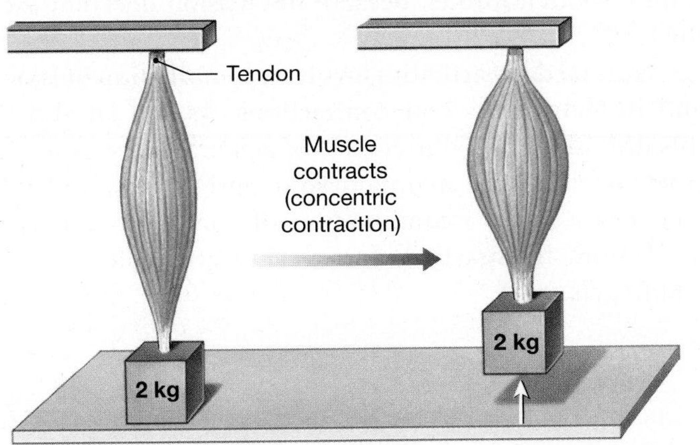
Tendon
Muscle contracts (concentric contraction)
2 kg
2 kg

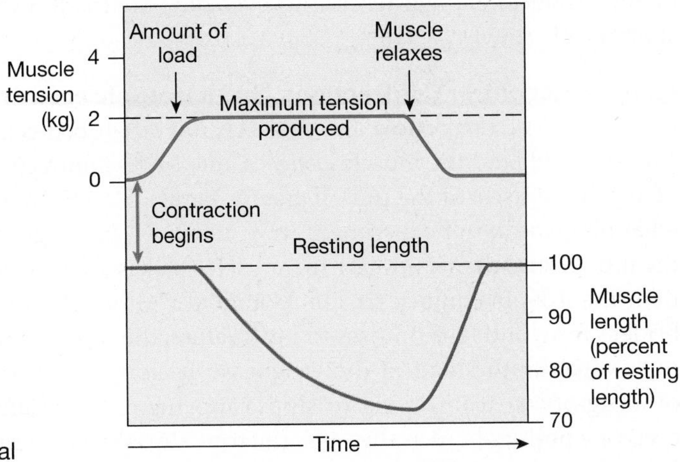
Muscle tension (kg)
4
2
0
Amount of load
Muscle relaxes
Maximum tension produced
Contraction begins
Resting length
100
90
80
70
Muscle length (percent of resting length)
Time

a **Isotonic concentric contraction.**
A muscle is attached to a weight (2 kg) that is ½ its maximum potential tension (4 kg). When stimulated, it develops enough tension to lift the weight. The tension remains constant, but the muscle shortens.

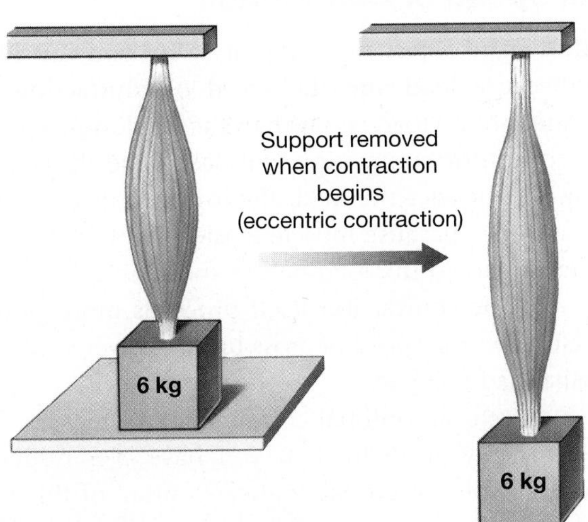
Support removed when contraction begins (eccentric contraction)
6 kg
6 kg

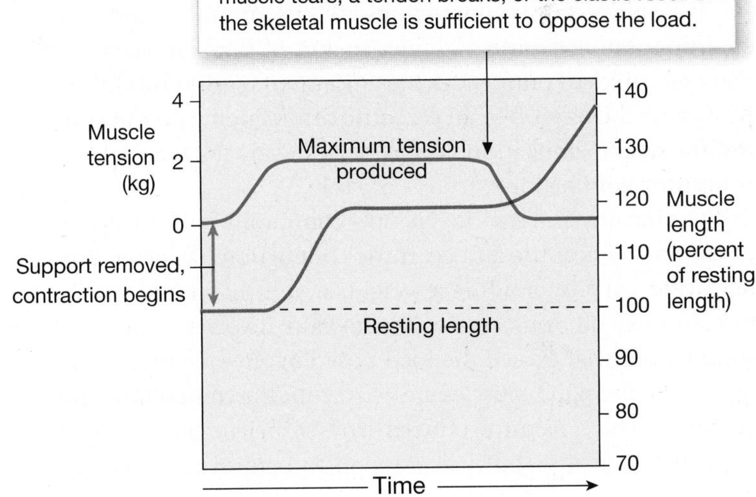
When the eccentric contraction ends, the unopposed load stretches the muscle until either the muscle tears, a tendon breaks, or the elastic recoil of the skeletal muscle is sufficient to oppose the load.
Muscle tension (kg)
4
2
0
Support removed, contraction begins
Maximum tension produced
Resting length
140
130
120
110
100
90
80
70
Muscle length (percent of resting length)
Time

b **Isotonic eccentric contraction.**
The tension remains constant, but the muscle lengthens.

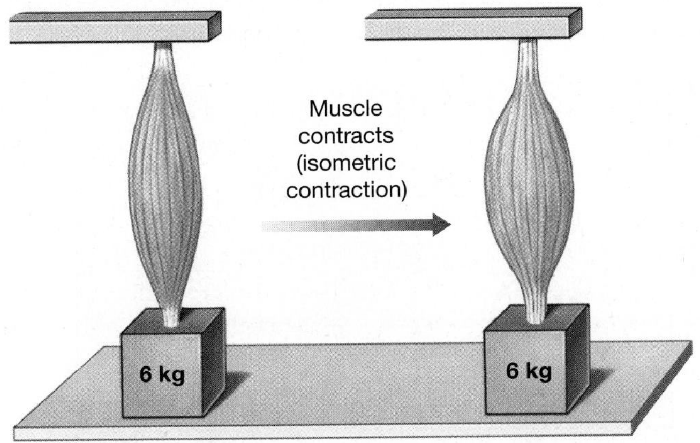
Muscle contracts (isometric contraction)
6 kg
6 kg

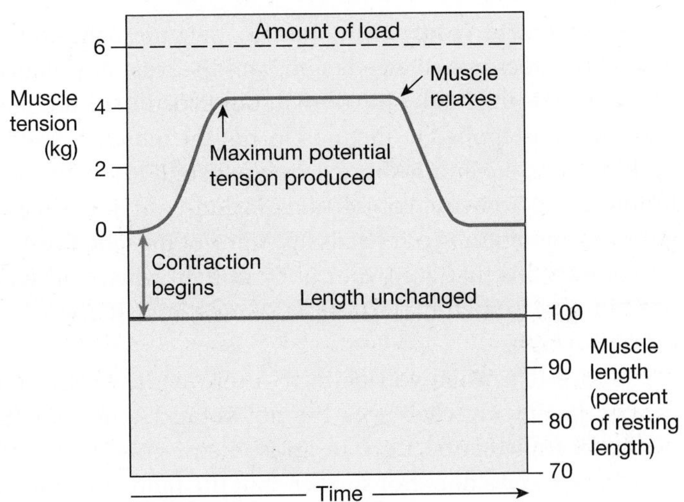
Amount of load
Muscle tension (kg)
6
4
2
0
Muscle relaxes
Maximum potential tension produced
Contraction begins
Length unchanged
100
90
80
70
Muscle length (percent of resting length)
Time

c **Isometric contraction.**
When stimulated, the tension rises, but the muscle length stays the same.

all the motor units are stimulated and the load is fairly small, the muscle will shorten very quickly. In contrast, if the muscle barely produces enough tension to overcome the load, it will shorten very slowly.

Isotonic Eccentric Contractions. In an **isotonic eccentric** (ek-SEN-trik) **contraction**, the peak tension developed is *less than* the load, and the muscle *elongates* due to the contraction of another muscle or the pull of gravity. Figure 10–18b shows what happens in our experiment if we attach a 6-kg weight to the muscle and then stimulate the muscle (6 kg is 2 kg greater than the 4-kg maximum tension potential). Although cross-bridges form and tension rises to peak values, the muscle cannot overcome the load of the weight, so it elongates. Think of a tug-of-war team trying to stop a moving car. Although everyone pulls as hard as they can, the rope slips through their fingers. The speed of elongation depends on the difference between the amount of tension developed by the active muscle fibers and the size of the load. In our analogy, the team might slow down a small car, but would have little effect on a large truck.

10

Isotonic eccentric contractions are very common, and they are important in a variety of movements. In these movements, you exert precise control over the amount of tension produced. By varying the tension in an eccentric contraction, you can control the rate of elongation, just as you can vary the tension in a concentric contraction.

During physical training, people commonly perform cycles of isotonic concentric and eccentric contractions, as when you do biceps curls by holding a weight in your hand and slowly flex and extend your elbow. The flexion involves concentric contractions that exceed the load posed by the weight. During extension, the same muscles are active, but the contractions are eccentric. The tension produced isn't sufficient to overcome the force of gravity, but it is enough to control the speed of movement.

Isometric Contractions

In an **isometric contraction** (*metric*, measure), the muscle as a whole does not change length, and the tension produced never exceeds the load. This time in our experiment the same 6-kg weight is applied to the muscle, but the muscle is fixed at both ends, so it cannot shorten or elongate (Figure 10–18c). Examples of isometric contractions include carrying a bag of groceries and holding our heads up. Many of the reflexive muscle contractions that keep your body upright when you stand or sit involve isometric contractions of muscles that oppose the force of gravity.

Notice that when you perform an isometric contraction, the contracting muscle bulges, but not as much as it does during an isotonic contraction. In an isometric contraction, the muscle *as a whole* does not shorten, but the individual muscle fibers shorten as connective tissues stretch. The muscle fibers cannot shorten further, because the tension does not exceed the load.

Normal daily activities involve a combination of isotonic and isometric muscular contractions. As you sit and read this text, isometric contractions of postural muscles stabilize your vertebrae and maintain your upright position. When you turn a page, a combination of concentric and eccentric isotonic contractions moves your arm, forearm, hand, and fingers.

Tips & Tools

During an *isotonic* contraction, such as lifting a baby, muscle tension remains constant while muscle length changes. During an *isometric* contraction, such as holding a baby at arm's length, muscle length remains constant but muscle tension changes.

Load and Speed of Contraction

You can lift a light object more rapidly than you can lift a heavy one because load and the speed of contraction are inversely related. If the load is less than the tension produced, an isotonic concentric contraction will occur, and the muscle will shorten. The heavier the load, the longer it takes for the movement to begin, because muscle tension (which increases gradually) must exceed the load before shortening can occur (Figure 10–19). The contraction itself proceeds more slowly. In the muscle fiber, the speed of cross-bridge power strokes is reduced as the load increases.

For each muscle, an optimal combination of tension and speed exists for any given load. If you have ever ridden a 21-speed bicycle, you are probably already aware of this fact. When you are cruising along comfortably, your thigh and leg

Figure 10–19 Load and Speed of Contraction. The heavier the load on a muscle, the longer it will take for the muscle to begin to shorten and the less the muscle will shorten.

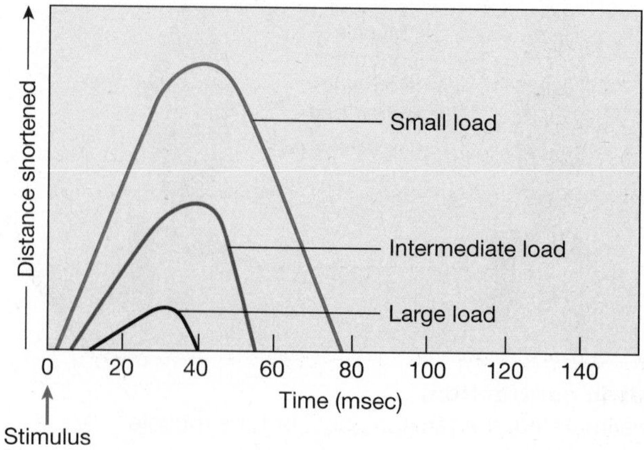
Distance shortened
Small load
Intermediate load
Large load
0
20
40
60
80
100
120
140
Time (msec)
Stimulus

muscles are working at an optimal combination of speed and tension. When you start up a hill, the load increases. Your muscles must now develop more tension, and they move more slowly; they are no longer working at optimal efficiency. If you then shift to a lower gear, the load on your muscles decreases and their speed increases, and the muscles are once again working efficiently.

Muscle Relaxation and the Return to Resting Length

As we noted earlier, there is no active mechanism for returning shortened muscle fibers to their resting length. After a contraction, a muscle returns to its original length through a combination of elastic forces, opposing muscle contractions, and gravity.

Elastic Forces

When a contraction ends, some of the energy initially "spent" in stretching the tendons and distorting intracellular organelles is recovered as they recoil or rebound to their original dimensions. This elasticity gradually helps return the muscle fibers in the muscle to their optimal resting length.

Opposing Muscle Contractions

The contraction of *opposing muscles*, those that perform opposite actions, can return a muscle to its resting length more quickly than elastic forces can. Consider the muscles of the arm that flex or extend the elbow. Contraction of the *biceps brachii* (BRĀ-kē-ī) on the anterior part of the arm flexes the elbow, and contraction of the *triceps brachii* on the posterior part of the arm extends the elbow. When the biceps brachii contracts, the triceps brachii is stretched. When the biceps brachii relaxes, contraction of the triceps brachii extends the elbow and stretches the muscle fibers of the biceps brachii to their original length.

Gravity

Gravity may assist opposing muscle groups in quickly returning a muscle to its resting length after a contraction. For example, imagine the biceps brachii fully contracted with the elbow pointed at the ground. When the muscle relaxes, gravity will pull the forearm down and stretch the muscle. However, some active muscle tension is needed to control the rate of movement and to prevent damage to the joint. In this example, isotonic eccentric contraction of the biceps brachii can be used to control the movement.

Checkpoint

14. **Can a skeletal muscle contract without shortening? Explain.**

See the blue Answers tab at the back of the book.

10-7 To maintain regular muscle fiber activity, energy and recovery are required

Learning Outcome **Describe the mechanisms by which muscle fibers obtain the energy to power contractions.**

Normal muscle function requires (1) substantial intracellular energy reserves, (2) a normal circulatory supply, (3) a normal blood oxygen level, and (4) a blood pH within normal limits. Anything that interferes with any of these factors will reduce the efficiency of muscle activity and cause premature muscle fatigue. For example, reduced blood flow from tight clothing, heart problems, or blood loss slows the delivery of oxygen and nutrients, accelerates the buildup of lactate and hydrogen ions, and leads to muscle fatigue.

Muscle metabolism is the sum of the chemical and physical changes that occur within muscle tissue. It involves generating the energy for contraction, using that energy, and then recovering from energy expenditure. The only energy source used directly for muscle contraction is ATP, and available stores are depleted within 4–6 seconds in active muscles. ATP is generated by direct phosphorylation of ADP by creatine phosphate (CP), anaerobic metabolism (glycolysis), and aerobic metabolism (citric acid cycle and electron transport chain). This section discusses these processes.

ATP Generation and Muscle Fiber Contraction

A single muscle fiber may contain 15 billion thick filaments. When that muscle fiber is actively contracting, each thick filament breaks down around 2500 ATP molecules per second. Even a small skeletal muscle contains thousands of muscle fibers, so the ATP demands of a contracting skeletal muscle are enormous.

In practical terms, the demand for ATP in a contracting muscle fiber is so high that it would be impossible to have all the necessary energy available as ATP before the contraction begins. Instead, a resting muscle fiber contains only enough ATP and other high-energy compounds to sustain a contraction until additional ATP can be generated. Throughout the rest of the contraction, the muscle fiber will generate ATP at roughly the same rate as it is used. Let's see how this works.

ATP and Creatine Phosphate Reserves

The primary function of ATP is to transfer energy from one location to another rather than to store it long term. At rest, a skeletal muscle fiber produces more ATP than it needs. Under these conditions, ATP transfers energy to creatine. **Creatine** (KRĒ-uh-tēn) is a small molecule that muscle cells assemble

Table 10–1 Sources of Energy in a Typical Muscle Fiber

Energy Source	Utilization Process	Initial Available Quantity	Number of Twitches Supported by Each Energy Source Alone	Duration of Isometric Tetanic Contraction Supported by Each Energy Source Alone
ATP	ATP → ADP + P	3 mmol	10	2 sec
CP	ADP + CP → ATP + C	20 mmol	70	15 sec
Glycogen	Glycolysis (anaerobic metabolism)	100 mmol	670	130 sec
Glycogen	Aerobic metabolism	100 mmol	12,000	2400 sec (40 min)

from fragments of amino acids. The energy transfer creates another high-energy compound, **creatine phosphate (CP)**, also known as *phosphocreatine*:

$$\text{ATP} + \text{creatine} \rightarrow \text{ADP} + \text{creatine phosphate}$$

During a contraction, each myosin head breaks down ATP, producing ADP and phosphate. The energy stored in creatine phosphate is then used to "recharge" ADP, converting it back to ATP through the reverse reaction:

$$\text{ADP} + \text{creatine phosphate} \rightarrow \text{ATP} + \text{creatine}$$

The enzyme that catalyzes this reaction is **creatine kinase** (KĪ-nāz; **CK**). When muscle cells are damaged, CK leaks across the plasma membranes and into the bloodstream. For this reason, a high blood concentration of CK usually indicates serious muscle damage.

Table 10–1 compares the energy sources of a representative muscle fiber. A resting skeletal muscle fiber contains about six times as much creatine phosphate as readily available ATP, but when a muscle fiber is undergoing a sustained contraction, these energy reserves are exhausted in only about 15 seconds. The muscle fiber must then rely on other mechanisms to generate ATP from ADP.

ATP Generation

As we saw in Chapter 3, most cells in the body generate ATP through (1) glycolysis in the cytoplasm and (2) aerobic metabolism in mitochondria. p. 82

Glycolysis. Glycolysis is the anaerobic breakdown of glucose to pyruvate in the cytosol of a cell. It is an **anaerobic process** because it does not require oxygen. Glycolysis provides a net gain of 2 ATP molecules and generates 2 *pyruvate molecules* from each glucose molecule.

The glucose broken down under these conditions comes primarily from the reserves of glycogen in the sarcoplasm. *Glycogen* is a chain of glucose molecules. p. 47 It is stored in granules close to the I bands of sarcomeres. Typical skeletal muscle fibers contain large glycogen reserves, which may account for 1.5 percent of the total muscle weight. When the muscle fiber begins to run short of ATP and CP, enzymes split the glycogen molecules, releasing glucose, which can be used to generate more ATP.

The ATP produced by glycolysis in the cytoplasm is only a small fraction of that produced by aerobic metabolism. Thus, when energy demands are relatively low and oxygen is readily available, glycolysis is important only because it provides the substrates for aerobic metabolism in the mitochondria. Yet, because it can proceed without oxygen, glycolysis becomes an important source of energy when energy demands are at a maximum and the available supply of oxygen limits how quickly the mitochondria can produce ATP.

Aerobic Metabolism. Aerobic metabolism normally provides 95 percent of the ATP demands of a resting cell. In this process, mitochondria absorb oxygen, ADP, phosphate ions, and organic substrates (such as pyruvate) from the surrounding cytoplasm. The molecules then enter the *citric acid cycle* (also known as the *tricarboxylic acid cycle* or the *Krebs cycle*), an enzymatic pathway that breaks down organic molecules. The carbon atoms of the organic substrate are released as carbon dioxide. The hydrogen atoms are shuttled to respiratory complexes in the inner mitochondrial membrane, where their electrons are removed. After a series of intermediate steps involving the *electron transport chain*, the protons and electrons are combined with oxygen to form water. Along the way, large amounts of energy are released and used to make ATP. The entire process is very efficient: For each molecule of pyruvate "fed" into the citric acid cycle, the cell gains 15 ATP molecules.

Resting skeletal muscle fibers rely almost exclusively on the aerobic metabolism of fatty acids to generate ATP. These fatty acids are absorbed from the circulation. When the muscle starts contracting, the mitochondria begin breaking down molecules of pyruvate from glycolysis instead of fatty acids. Because a typical skeletal muscle fiber contains large amounts of glycogen, the shift from fatty acid metabolism to glucose metabolism makes it possible for the cell to continue contracting for an extended period, even without an external source of nutrients.

Muscle Metabolism and Varying Activity Levels

As physical exertion and muscular activity increase, the pattern of muscle fiber energy production and use changes. Figure 10–20 illustrates **muscle metabolism** in a muscle fiber at rest and at

Figure 10–20 **Muscle Metabolism.**

SmartArt

Muscle Metabolism in a Resting Muscle Fiber

- The demand for ATP is low and sufficient oxygen is available for mitochondria to meet that demand.
- Fatty acids are absorbed and broken down in the mitochondria creating a surplus of ATP.
- Some mitochondrial ATP is used to convert absorbed glucose to glycogen.
- Mitochondrial ATP is also used to convert creatine to creatine phosphate (CP).
- This results in the buildup of energy reserves (glycogen and CP) in the muscle.

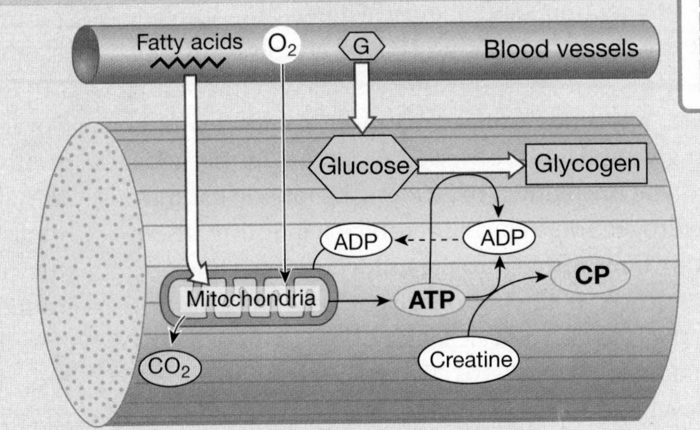

Fatty acids
O_2
G
Blood vessels
Glucose
Glycogen
ADP
ADP
Mitochondria
ATP
CP
CO_2
Creatine

Muscle Metabolism during Moderate Activity

- The demand for ATP increases.
- There is still enough oxygen for the mitochondria to meet the increased demand, but no excess ATP is produced.
- ATP is generated primarily by aerobic metabolism of glucose from stored glycogen.
- If the glycogen reserves are low, the muscle fiber can also break down other substrates, such as fatty acids.
- All of the ATP being generated is used to power muscle contraction.

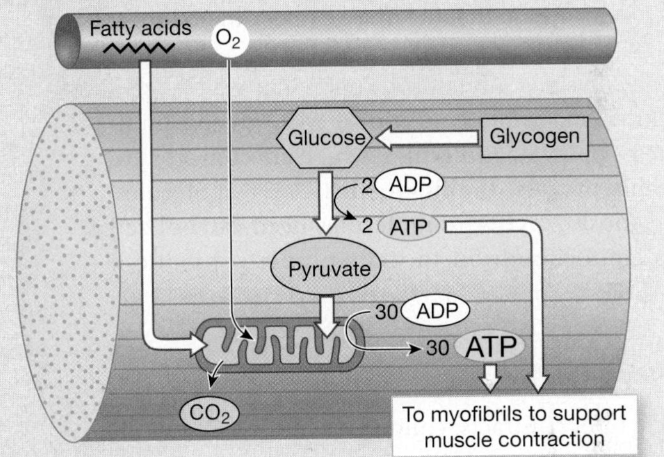

Fatty acids
O_2
Glucose
Glycogen
2 ADP
2 ATP
Pyruvate
30 ADP
30 ATP
CO_2
To myofibrils to support muscle contraction

Muscle Metabolism during Peak Activity

- The demand for ATP is enormous. Oxygen cannot diffuse into the fiber fast enough for the mitochondria to meet that demand. Only a third of the cell's ATP needs can be met by the mitochondria (not shown).
- The rest of the ATP comes from glycolysis, and when this produces pyruvate faster than the mitochondria can utilize it, the pyruvate builds up in the cytosol.
- The pyruvate is converted to lactate. Hydrogen ions from ATP hydrolysis are not absorbed by the mitochondria.
- The buildup of hydrogen ions increases cytosol acidity, which inhibits muscle contraction, leading to rapid fatigue.

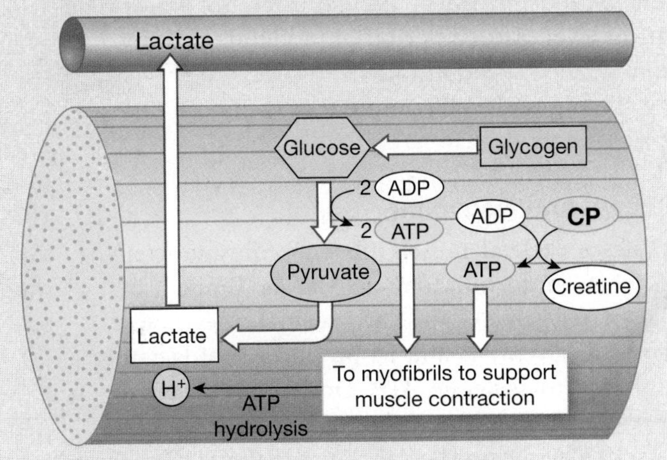

Lactate
Glucose
Glycogen
2 ADP
2 ATP
ADP
CP
Pyruvate
ATP
Creatine
Lactate
H^+
ATP hydrolysis
To myofibrils to support muscle contraction

10

moderate and peak levels of activity. Notice that this figure shows the results of the fact that when muscles become active, their energy consumption increases dramatically.

Although, as mentioned, the muscle fiber can generate additional ATP by the anaerobic process of glycolysis during times of peak energy demands, this process has two important drawbacks. First, any underutilized pyruvate molecules produced through glycolysis are converted to *lactate*. At the same time, the hydrolytic breakdown of ATP during muscle contraction releases hydrogen ions. During aerobic metabolism, these hydrogen ions are transported into the mitochondria. However, when ATP production is relying on glycolysis, the hydrogen ions remain in the cytosol. At peak activity, both hydrogen ions and lactate build up (**Figure 10–20**). This results in an increased lactate blood level and metabolic acidosis, often referred to as *lactic acidosis*. The hydrogen ions can then lower the intracellular pH. Buffers in the sarcoplasm resist pH shifts, but these mechanisms are limited. Eventually, changes in pH affect the workings of key enzymes so that the muscle fiber can no longer contract.

Second, glycolysis is a relatively inefficient way to generate ATP. Under anaerobic conditions, each glucose molecule generates 2 pyruvate molecules, which are converted to 2 lactate. In this process, the fiber gains 2 ATP molecules. However, had those 2 pyruvate molecules been catabolized aerobically in a mitochondrion, the cell would have produced 30 additional ATP.

The Recovery Period

When a muscle fiber contracts, conditions in the sarcoplasm change. Energy reserves are consumed, heat is released, and, if the contraction was at a maximum level, oxygen is less available and lactate is generated. In the **recovery period**, the conditions in muscle fibers are returned to a normal, pre-exertion level. After a period of moderate activity, muscle fibers may need several hours to recover. After sustained activity at a higher level, complete recovery can take a week.

Lactate Removal and Recycling

Glycolysis enables a skeletal muscle fiber to continue contracting even when mitochondrial activity is limited by the availability of oxygen. As we have seen, however, the anaerobic process of glycolysis and its production of lactate is not an ideal way to generate ATP. During the recovery period, when oxygen is available in abundance, lactate can be converted back to pyruvate. The pyruvate can then be used either by mitochondria to generate ATP or as a substrate for enzyme pathways that synthesize glucose and rebuild glycogen reserves. During exertion, lactate diffuses out of muscle fibers and into the bloodstream. The process continues after the exertion has ended, because the intracellular lactate concentration is still relatively high. The liver absorbs the lactate and converts it to pyruvate. About 20–30 percent of these new pyruvate molecules enter liver mitochondria where they are catabolized, providing the ATP needed to convert the other 70–80 percent of the pyruvate molecules to glucose. (We will cover these processes more fully in Chapter 25.) The glucose molecules are then released into the circulation, where they are absorbed by skeletal muscle fibers and used to rebuild their glycogen reserves. This shuffling of lactate produced in the muscles to the liver and glucose back to muscle cells is called the *Cori cycle*.

The Oxygen Debt

During the recovery period, the oxygen demand of the body's cells remains elevated above the normal resting level. The more ATP required for recovery, the more oxygen will be needed. The amount of oxygen required to restore normal, pre-exertion conditions is called the **oxygen debt**, or **excess postexercise oxygen consumption (EPOC)**.

Most of the additional oxygen consumption occurs in skeletal muscle fibers, which must restore ATP, CP, and glycogen concentrations to their former levels, and in liver cells, which generate the ATP needed to convert excess lactate to glucose. However, several other tissues also increase their rate of oxygen consumption and ATP generation during the recovery period. For example, sweat glands increase their secretory activity until normal body temperature is restored. While the oxygen debt is being repaid, breathing rate and depth are increased. As a result, you continue to breathe heavily after you stop exercising.

Heat Production and Loss

Muscular activity generates considerable heat. During a catabolic process, such as the breakdown of glycogen or the reactions of glycolysis, a muscle fiber captures only a portion of the released energy—the rest escapes as heat. A resting muscle fiber relying on aerobic metabolism captures about 42 percent of the energy released in catabolism. The other 58 percent warms the sarcoplasm, interstitial fluid, and circulating blood. Active skeletal muscles release roughly 85 percent of the heat needed to maintain normal body temperature.

When anaerobic metabolism is the primary method of ATP generation, muscle fibers become less efficient at capturing energy. At peak levels of exertion, only about 30 percent of the released energy is captured as ATP, and the remaining 70 percent warms the muscle and surrounding tissues. Body temperature soon climbs, and heat loss at the skin accelerates through mechanisms introduced in Chapters 1 and 5. pp. 20, 170 Heat loss continues during the recovery period.

Hormones and Muscle Metabolism

Hormones of the endocrine system adjust metabolic activities in skeletal muscle fibers. *Growth hormone* from the pituitary gland and *testosterone* (the primary sex hormone in males) stimulate the synthesis of contractile proteins, such as actin and myosin, and the enlargement of skeletal muscles. *Thyroid hormones* elevate the rate of energy consumption in resting and active skeletal muscles. During a sudden crisis, hormones of the adrenal gland, notably *epinephrine* (adrenaline), stimulate muscle metabolism and increase both the duration of stimulation and the force of contraction. We further examine the effects of hormones on muscle and other tissues in Chapter 18.

Checkpoint

15. **How do muscle fibers continuously synthesize ATP?**
16. **Define *oxygen debt*.**

See the blue Answers tab at the back of the book.

10-8 Muscle performance depends on muscle fiber type and physical conditioning

Learning Outcome **Relate the types of muscle fibers to muscle performance, discuss muscle hypertrophy, atrophy, and aging, and describe how physical conditioning affects muscle tissue.**

We can consider muscle performance in terms of **force**, the maximum amount of tension produced by a particular muscle or muscle group, and **endurance**, the amount of time during which a person can perform a particular activity. Here we consider the factors that determine the performance capabilities of any skeletal muscle: the type, distribution, and size of muscle fibers in the muscle, and physical conditioning or training.

Types of Skeletal Muscle Fibers

The human body has three major types of skeletal muscle fibers: *fast fibers*, *slow fibers*, and *intermediate fibers*.

Fast Fibers

Most of the skeletal muscle fibers in the body are called **fast fibers**, because they can reach peak twitch tension in 0.01 sec or less after stimulation. Fast fibers are large in diameter and contain densely packed myofibrils, large glycogen reserves, and relatively few mitochondria. Muscles dominated by fast fibers produce powerful contractions because the tension produced by a muscle fiber is directly proportional to the number of myofibrils. However, fast fibers fatigue rapidly because their contractions use ATP in massive amounts, and they have relatively few mitochondria to generate ATP. As a result, prolonged activity is supported primarily by anaerobic metabolism.

Slow Fibers

Slow fibers have only about half the diameter of fast fibers and take three times as long to reach peak tension after stimulation. These fibers are specialized in ways that enable them to continue contracting long after a fast fiber would have become fatigued. The most important specializations support aerobic metabolism in the numerous mitochondria.

One of the main characteristics of slow muscle fibers is that they are surrounded by a more extensive network of capillaries than fast muscle tissue. For this reason, they have a dramatically higher oxygen supply to support mitochondrial activity. Slow fibers also contain the red pigment **myoglobin** (MĪ-ō-glō-bin). This globular protein is similar to hemoglobin, the red oxygen-carrying pigment in blood. Both myoglobin and hemoglobin reversibly bind oxygen molecules. Other muscle fiber types contain small amounts of myoglobin, but it is most abundant in slow fibers. As a result, resting slow fibers hold substantial oxygen reserves that can be used during a contraction. Skeletal muscles dominated by slow fibers are dark red because slow fibers have both an extensive capillary supply and a high concentration of myoglobin.

With oxygen reserves and a more efficient blood supply, the mitochondria of slow fibers can contribute more ATP during contractions. In addition, demand for ATP is reduced because the cross-bridges in slow fibers cycle more slowly than those of fast fibers. Thus, slow fibers are less dependent on anaerobic metabolism than are fast fibers. Glycogen reserves of slow fibers are also smaller than those of fast fibers, because some of the mitochondrial energy production involves the breakdown of stored lipids rather than glycogen. Figure 10–21 compares the appearance of fast and slow fibers.

Intermediate Fibers

Most properties of **intermediate fibers** are intermediate between those of fast fibers and slow fibers. In appearance, intermediate fibers most closely resemble fast fibers because they contain little myoglobin and are relatively pale. They have an intermediate capillary network and mitochondrial supply around them and are more resistant to fatigue than are fast fibers.

The characteristics of these three types of muscle fibers are summarized in Table 10–2.

Muscle Performance and the Distribution of Muscle Fibers

The percentages of fast, intermediate, and slow fibers in a skeletal muscle can be quite variable. In muscles that contain a mixture of fast and intermediate fibers, the proportion can change with physical conditioning. For example, if a muscle is used repeatedly for endurance events, some of the fast fibers will develop the appearance and functional capabilities of

Figure 10–21 **Fast versus Slow Fibers.** The TEM on the right is a longitudinal section of skeletal muscle, showing more mitochondria (M) and a more extensive capillary supply in a slow fiber (R, for red) than in a fast fiber (W, for white).

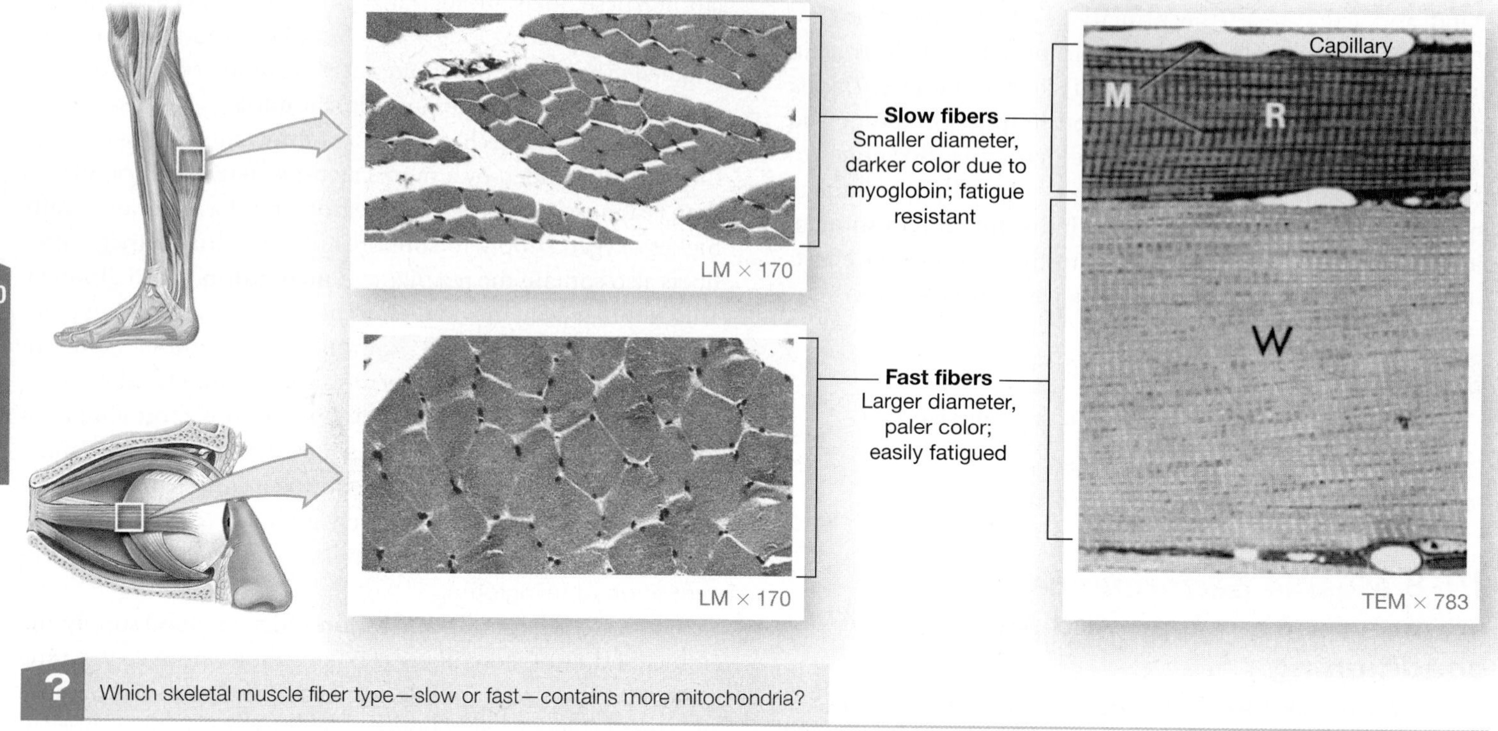
Slow fibers
Smaller diameter, darker color due to myoglobin; fatigue resistant
Fast fibers
Larger diameter, paler color; easily fatigued
LM × 170
LM × 170
Capillary
M
R
W
TEM × 783

? Which skeletal muscle fiber type—slow or fast—contains more mitochondria?

Table 10–2 Properties of Skeletal Muscle Fiber Types

Property	Fast Fibers	Slow Fibers	Intermediate Fibers
Cross-sectional diameter	Large	Small	Intermediate
Time to peak tension	Rapid	Prolonged	Medium
Contraction speed	Fast	Slow	Fast
Fatigue resistance	Low	High	Intermediate
Color	White	Red	Pink
Myoglobin content	Low	High	Low
Capillary supply	Scarce	Dense	Intermediate
Mitochondria	Few	Many	Intermediate
Glycolytic enzyme concentration in sarcoplasm	High	Low	High
Sources of substrates for ATP generation during contraction (metabolism)	Carbohydrates (anaerobic)	Lipids (fatty acids), carbohydrates, proteins (amino acids) (aerobic)	Primarily carbohydrates (anaerobic)
Alternative names	Type II-B, FF (fast fatigue), white, fast-twitch glycolytic	Type I, S (slow), red, SO (slow oxidative), slow-twitch oxidative	Type II-A, FR (fast resistant), fast-twitch oxidative

intermediate fibers. The muscle as a whole will become more resistant to fatigue.

Muscles dominated by fast fibers appear pale and are often called **white muscles**. Chicken breasts contain "white meat" because chickens use their wings only for brief intervals, as when fleeing a predator, and the power for flight comes from the anaerobic process of glycolysis in the fast fibers of their breast muscles. As noted earlier, extensive blood vessels and myoglobin give slow fibers a reddish color, so muscles dominated by slow fibers are known as **red muscles**. Chickens walk around all day, and these movements are powered by aerobic metabolism in the slow fibers of the "dark meat" of their legs.

What about human muscles? Most of our muscles contain a mixture of fiber types and so appear pink. However, there are no slow fibers in muscles of the eye or hand, where swift, but brief, contractions are required. Many of our back and calf muscles are dominated by slow fibers, and these muscles contract almost continuously to maintain an upright posture. Our genes determine the percentage of fast versus slow fibers in each muscle.

Muscle Hypertrophy, Atrophy, and Effects of Aging

As a result of repeated, exhaustive stimulation, muscle fibers develop more mitochondria, a higher concentration of glycolytic enzymes, and larger glycogen reserves. Such muscle fibers have more myofibrils than do less-stimulated fibers, and each myofibril contains more thick and thin filaments. The net effect is **hypertrophy** (hī-PER-trō-fē), an enlargement of the stimulated muscle. The number of muscle fibers does not change significantly, but the muscle as a whole enlarges because each muscle fiber increases in diameter.

Hypertrophy occurs in muscles that have been repeatedly stimulated to produce near-maximal tension. The intracellular changes that occur increase the amount of tension produced when these muscles contract. The muscles of a bodybuilder are excellent examples of muscular hypertrophy.

In contrast, a skeletal muscle that is not regularly stimulated by a motor neuron loses muscle tone and mass. The muscle becomes soft, and the muscle fibers become smaller and weaker. This reduction in muscle size, tone, and power is called **atrophy**. Individuals paralyzed by spinal cord injuries or other damage to the nervous system gradually lose muscle tone and size in the areas affected. In *polio*, a virus attacks motor neurons in the spinal cord and brain, causing muscular paralysis and atrophy. Even a temporary reduction in muscle use can lead to muscular atrophy, as you can easily observe by comparing "before and after" limb muscles in someone who has worn a cast. Muscle atrophy is reversible at first, but dying muscle fibers are not replaced. In extreme atrophy, the functional losses are permanent. That is why physical therapy is crucial for people who are temporarily unable to move normally. Direct electrical stimulation by an external device can substitute for nerve stimulation and prevent or reduce muscle atrophy.

As we age, muscle tissue also changes. We can summarize the effects of aging on the muscular system as follows:

- *Skeletal Muscle Fibers Become Smaller in Diameter.* This reduction in size reflects a decrease in the number of myofibrils. In addition, the muscle fibers contain smaller ATP, CP, and glycogen reserves and less myoglobin. The overall effect is a reduction in skeletal muscle size, strength, and endurance, combined with a tendency to fatigue more quickly. Because cardiovascular performance also decreases with age, blood flow to active muscles does not increase with exercise as rapidly as it does in younger people. These factors interact to produce decreases of 30–50 percent in anaerobic and aerobic performance by age 65.
- *Skeletal Muscles Become Less Elastic.* Aging skeletal muscles develop increasing amounts of fibrous connective tissue, a process called **fibrosis.** Fibrosis makes the muscle less flexible, and the collagen fibers can restrict movement and circulation.
- *Tolerance for Exercise Decreases.* A lower tolerance for exercise comes in part from tiring quickly and in part from reduced thermoregulation, as described in Chapter 5. p. 170 Individuals over age 65 cannot eliminate the heat their muscles generate during contraction as effectively as younger people can. For this reason, they are subject to overheating.
- *The Ability to Recover from Muscular Injuries Decreases.* The number of satellite cells steadily decreases with age, and the amount of fibrous tissue increases. As a result, when an injury occurs, repair capabilities are limited. Scar tissue formation is the usual result.

Regular exercise helps control body weight, strengthens bones, and generally improves the quality of life at all ages. Extremely demanding exercise is not as important as regular exercise. In fact, extreme exercise in the elderly can damage tendons, bones, and joints.

Muscle Fatigue

We say an active skeletal muscle is **fatigued** when it can no longer perform at the required level of activity. Many factors are involved in muscle fatigue. For example, muscle fatigue has been correlated with (1) depletion of metabolic reserves within the muscle fibers; (2) damage to the sarcolemma and sarcoplasmic reticulum; (3) a decline in pH within the muscle fibers and the muscle as a whole, which decreases calcium ion binding to troponin and alters enzyme activities; and (4) a sense of weariness and a reduction in the desire to continue the activity, due to the effects of low blood pH and sensations of pain. Muscle fatigue is cumulative—the effects become more pronounced as more neurons and muscle fibers are affected. The result is a gradual reduction in the capabilities and performance of the entire skeletal muscle.

If a muscle fiber is contracting at a moderate level and ATP demands can be met through aerobic metabolism, fatigue will not occur until glycogen, fatty acids, and protein reserves are depleted. This type of fatigue affects the muscles of endurance athletes, such as marathon runners, after hours of exertion. In contrast, sprinters get a different type of muscle fatigue. When a muscle produces a sudden, intense burst of activity at peak levels, glycolysis provides most of the ATP. After just seconds to minutes, tissue pH decreases due to lactic acidosis and the buildup of hydrogen ions, and the muscle can no longer function normally.

Physical Conditioning

Physical conditioning or training enables athletes and ordinary people of all ages to improve both power and endurance. In practice, the training schedule varies, depending on whether the activity is supported primarily by anaerobic or aerobic energy production.

Energy Use in Training

10

Anaerobic endurance is the length of time muscular contraction can continue to be supported by the existing energy reserves of ATP and CP and by glycolysis. Conditioning for anaerobic endurance improves an individual's power. Anaerobic endurance is limited by (1) the amount of ATP and CP available, (2) the amount of glycogen available for breakdown, and (3) the ability of the muscle to tolerate lactate and the buildup of hydrogen ions generated during the period of anaerobic metabolism. Typically, the onset of muscle fatigue occurs within 2 minutes of the start of maximal activity.

Activities that require above-average levels of anaerobic endurance include a 50-meter race or swim, pole vaulting, and competitive weight lifting. These activities involve the contractions of fast fibers. The energy for the first 10–20 seconds of activity comes from the ATP and CP reserves of the cytoplasm. As these energy sources decrease, an increase in glycogen breakdown and glycolysis provides additional energy. Athletes training to improve anaerobic endurance perform frequent, brief, intensive workouts that stimulate muscle hypertrophy.

Aerobic endurance is the length of time a muscle can continue to contract while supported by mitochondrial activities. Aerobic activities do not promote muscle hypertrophy. Conditioning for aerobic endurance improves an individual's ability to continue an activity for longer periods of time. Aerobic endurance is determined primarily by the availability of substrates for aerobic respiration, which muscle fibers can obtain by breaking down carbohydrates, lipids, or proteins. Initially, many of the nutrients broken down by muscle fibers come from reserves in the sarcoplasm. Prolonged aerobic activity, however, must be supported by nutrients provided by the circulating blood. Training to improve aerobic endurance generally involves sustained low levels of muscular activity. Examples include jogging, distance swimming, and other exercises that do not require peak tension production.

Physical training, using a combination of anaerobic and aerobic exercises, should be a lifelong pursuit. The recommended schedule is to alternate an anaerobic activity such as weight lifting or resistance training, with an aerobic activity, such as swimming or brisk walking. This combination, known as *cross-training,* enhances health by both increasing muscle mass and improving aerobic endurance.

Effects of Training

During exercise, blood vessels in the skeletal muscles dilate, which increases blood flow and the supply of oxygen and nutrients to the active muscle tissue. Warm-up periods are important because they both stimulate circulation in the muscles before the serious workout begins and activate the enzyme (glycogen phosphorylase) that breaks down glycogen to release glucose.

+ Clinical Note Delayed-Onset Muscle Soreness

Have you ever experienced muscle soreness the next day or several days after a period of physical exertion? Considerable controversy exists over the source and significance of this pain, which is known as *delayed-onset muscle soreness* (*DOMS*). You may have noticed a few of its characteristics:

- DOMS is distinct from the soreness you experience immediately after you stop exercising. The initial short-term soreness is probably related to the biochemical events associated with muscle fatigue.
- DOMS generally begins several hours after exercise and may last three or four days.
- The amount of DOMS is highest when the activity involves isotonic eccentric contractions (in which a muscle elongates despite producing tension). Activities dominated by concentric or isometric contractions produce less soreness.
- Levels of creatine kinase and myoglobin are elevated in the blood, indicating damage to muscle plasma membranes. The nature of the activity (eccentric, concentric, or isometric) has no effect on these levels, and the levels cannot be used to predict the degree of soreness.

Three mechanisms have been proposed to explain DOMS:

- Small tears may occur in the muscle tissue, leaving muscle fibers with damaged membranes. The sarcolemma of each damaged muscle fiber permits the loss of enzymes, myoglobin, and other chemicals that may stimulate nearby pain receptors.
- The pain may result from muscle spasms in the affected skeletal muscles. In some studies, stretching the muscle after exercise reduces the degree of soreness.
- The pain may result from tears in the connective tissue and tendons of the skeletal muscle.

Warm-ups also increase muscle temperature and thus accelerate the contraction process. Because glucose is a preferred energy source, endurance athletes such as marathon runners typically "load" or "bulk up" on carbohydrates for the three days before an event. They may also consume glucose-rich "sports drinks" during a competition. (We consider the risks and benefits of these practices in Chapter 25.)

Improvements in aerobic endurance result from two factors:

- *Alterations in the Characteristics of Muscle Fibers.* The proportion of fast and slow fibers in each muscle is genetically determined, and individual differences are significant. These variations affect aerobic endurance, because a person with more slow fibers in a particular muscle will be better able to perform under aerobic conditions than will a person with fewer. However, skeletal muscle cells respond to changes in the pattern of neural stimulation. Fast fibers trained for aerobic competition develop the characteristics of intermediate fibers, and this change improves aerobic endurance.
- *Improvements in Cardiovascular Performance.* Cardiovascular activity affects muscular performance by delivering oxygen and nutrients to active muscles. Physical training alters cardiovascular function by accelerating blood flow, thus improving oxygen and nutrient availability. Another important benefit of endurance training is an increase in capillaries that serve exercising muscles, providing better blood flow at the cellular level. (We examine factors involved in improving cardiovascular performance in Chapter 21.)

Checkpoint

17. **Identify the three types of skeletal muscle fibers.**
18. **Describe muscle fatigue.**
19. **Describe general age-related effects on skeletal muscle tissue.**
20. **Why would a sprinter experience muscle fatigue before a marathon runner would?**
21. **Which activity would be more likely to create an oxygen debt: swimming laps or lifting weights?**
22. **Which type of muscle fibers would you expect to predominate in the leg muscles of someone who excels at endurance activities, such as cycling or long-distance running?**

See the blue Answers tab at the back of the book.

10-9 Cardiac muscle tissue, found in the heart, produces coordinated and automatic contractions

Learning Outcome Identify the structural and functional differences between skeletal muscle fibers and cardiac muscle cells.

We introduced **cardiac muscle tissue** in Chapter 4, and briefly compared its properties with those of other types of muscle. **Cardiac muscle cells** are found only in the heart. We begin a more detailed examination of cardiac muscle tissue by considering its structural characteristics, and then go on to a brief discussion of its function.

Structural Characteristics of Cardiac Muscle Tissue

Like neurons and skeletal muscle fibers, cardiac muscle cells have excitable membranes. p. 302 Cardiac muscle cells also contain organized myofibrils, and the presence of many aligned sarcomeres gives the cells a striated appearance.

However, there are important structural differences between skeletal muscle fibers and cardiac muscle cells, including the following:

- Cardiac muscle cells are relatively small, averaging 10–20 μm in diameter and 50–100 μm in length.
- A typical cardiac muscle cell has a single, centrally placed nucleus (Figure 10–22a,b), although a few may have two or more nuclei. Unlike skeletal muscle, cardiac muscle cells are typically branched.
- The T tubules in a cardiac muscle cell are short and broad, and there are no triads (Figure 10–22c). The T tubules encircle the sarcomeres at the Z lines rather than at the zones of overlap.
- The SR of a cardiac muscle cell lacks terminal cisternae and stores fewer calcium ions. Its tubules contact the plasma membrane as well as the T tubules (see Figure 10–22c).
- Cardiac muscle cells are almost totally dependent on aerobic metabolism for the energy they need to continue contracting. They have energy reserves in the form of glycogen and lipid inclusions. The sarcoplasm of a cardiac muscle cell contains large numbers of mitochondria as well as abundant reserves of myoglobin, which store the oxygen needed to break down energy reserves during times of peak activity.
- Each cardiac muscle cell contacts several others at specialized sites known as *intercalated* (in-TER-ka-lā-ted) *discs.* p. 144

At an **intercalated disc**, the sarcolemmas of two adjacent cardiac muscle cells are extensively intertwined and bound together by gap junctions and desmosomes (see Figure 10–22a,b). p. 118 These connections help stabilize the positions of adjacent cells and maintain the three-dimensional structure of the tissue. They also allow ions to move from one cell to another so that cardiac muscle cells beat in rhythm. Therefore, two cardiac muscle cells can "pull together" with maximum efficiency because their myofibrils are essentially

10

Figure 10–22 Cardiac Muscle Tissue.

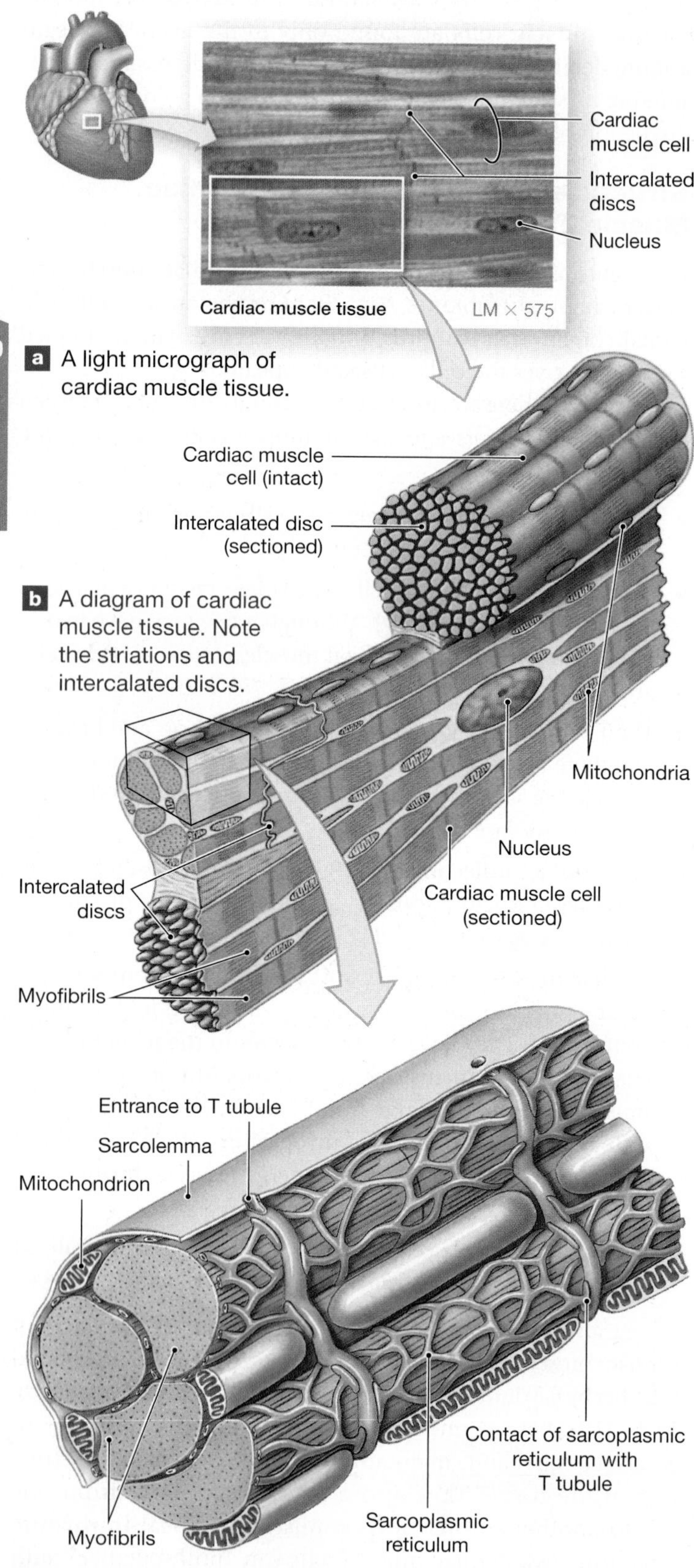
Cardiac muscle cell
Intercalated discs
Nucleus
Cardiac muscle tissue
LM × 575
Cardiac muscle cell (intact)
Intercalated disc (sectioned)
Mitochondria
Nucleus
Cardiac muscle cell (sectioned)
Intercalated discs
Myofibrils
Entrance to T tubule
Sarcolemma
Mitochondrion
Contact of sarcoplasmic reticulum with T tubule
Sarcoplasmic reticulum
Myofibrils

a A light micrograph of cardiac muscle tissue.

b A diagram of cardiac muscle tissue. Note the striations and intercalated discs.

c Cardiac muscle tissue has short, broad T tubules and an SR that lacks terminal cisternae.

locked together at the intercalated disc. There the myofibrils of two interlocking muscle cells are firmly anchored to their plasma membranes.

Functional Characteristics of Cardiac Muscle Tissue

The gap junctions allow ions and small molecules to move from one cell to another. These junctions create a direct electrical connection between two cardiac muscle cells. An action potential can travel across an intercalated disc, moving quickly from one cardiac muscle cell to another.

Because cardiac muscle cells are physically, chemically, and electrically connected to one another, the entire tissue resembles a single, enormous muscle cell. For this reason, cardiac muscle has been called a *functional syncytium* (sin-SISH-ē-um; a fused mass of cells).

In Chapter 20, we examine cardiac muscle physiology in detail, but here we briefly summarize the four major functional specialties of cardiac muscle:

- Cardiac muscle tissue contracts without neural stimulation. This property is called **automaticity**. Specialized cardiac muscle cells called **pacemaker cells** normally determine the timing of contractions.
- The nervous system can alter the pace or rate set by the pacemaker cells and adjust the amount of tension produced during a contraction.
- Cardiac muscle cell contractions last about 10 times as long as do those of skeletal muscle fibers. They also have longer refractory periods and do not readily fatigue.
- The properties of cardiac muscle sarcolemmas differ from those of skeletal muscle fibers. As in skeletal muscle fibers, an action potential triggers the release of calcium ions from the SR and the contraction of sarcomeres. However, an action potential in a cardiac muscle cell also makes the sarcolemma more permeable to extracellular calcium ions. Because their contractions require both intracellular and extracellular calcium ions, cardiac muscle cells are more sensitive to changes in the extracellular calcium ion concentration than are skeletal muscle fibers. As a result, individual twitches cannot undergo wave summation, and cardiac muscle tissue cannot produce tetanic contractions. This difference is important, because a heart in a sustained tetanic contraction could not pump blood.

Checkpoint

23. **Compare and contrast skeletal muscle tissue and cardiac muscle tissue.**
24. **What feature of cardiac muscle tissue allows the heart to act as a functional syncytium?**

See the blue Answers tab at the back of the book.

Go to MasteringA&P™ > Study Area > Menu > Lab Tools > PAL 3.0 > Histology > Muscular System

10-10 Smooth muscle tissue contracts to move substances within internal passageways

Learning Outcome Identify the structural and functional differences between skeletal muscle fibers and smooth muscle cells, and discuss the roles of smooth muscle tissue in systems throughout the body.

Smooth muscle tissue forms sheets, bundles, or sheaths around other tissues in almost every organ. In general, smooth muscle functions to coordinate the movement of substances through internal passageways. Smooth muscles play a variety of roles in various body systems:

- *Integumentary System:* Smooth muscles around blood vessels regulate the flow of blood to the superficial dermis; smooth muscles of the arrector pili elevate hairs. ⮌ p. 166
- *Cardiovascular System:* Smooth muscles around blood vessels control blood flow through vital organs and help regulate blood pressure.
- *Respiratory System:* Smooth muscles contract or relax to alter the diameters of the respiratory passageways and change their resistance to airflow.
- *Digestive System:* Extensive layers of smooth muscle in the walls of the digestive tract play an essential role in moving materials along the tract. Smooth muscle in the walls of the gallbladder contracts to eject bile into the digestive tract. In both the digestive and urinary systems, rings of smooth muscle called *sphincters* regulate the movement of materials along internal passageways.
- *Urinary System:* Smooth muscle tissue in the walls of small blood vessels alters the rate of filtration in the kidneys. Layers of smooth muscle in the walls of the ureters transport urine to the urinary bladder, and the contraction of the smooth muscle in the wall of the urinary bladder forces urine out of the body.
- *Reproductive System:* In males, layers of smooth muscle help move sperm along the reproductive tract and cause the ejection of glandular secretions from the accessory glands into the reproductive tract. In females, layers of smooth muscle help move oocytes (and perhaps sperm) along the reproductive tract, and contractions of the smooth muscle in the walls of the uterus expel the fetus at delivery.

Structural Characteristics of Smooth Muscle Tissue

Figure 10–23a shows typical smooth muscle tissue as seen under a light microscope.

Smooth muscle tissue, like skeletal and cardiac muscle tissues, contains actin and myosin. In skeletal and cardiac muscle cells, these proteins are organized in sarcomeres, with thin and thick filaments. However, the internal organization of a smooth muscle cell is very different:

- Smooth muscle cells are relatively long and slender, ranging from 5 to 10 μm in diameter and from 30 to 200 μm in length.
- Each cell is spindle shaped (tapered at both ends) and has a single, centrally located nucleus.
- A smooth muscle fiber has no T tubules, and the sarcoplasmic reticulum forms a loose network throughout the sarcoplasm.
- Smooth muscle cells lack myofibrils and sarcomeres. As a result, this tissue also has no striations and is called **nonstriated muscle**.
- Thick filaments are scattered throughout the sarcoplasm of a smooth muscle cell. The myosin proteins are organized differently than in skeletal or cardiac muscle cells, and smooth muscle cells have more myosin heads per thick filament.
- The thin filaments in a smooth muscle cell are attached to **dense bodies**, structures distributed throughout the sarcoplasm in a network of intermediate filaments composed of the protein *desmin* (Figure 10–23b). Some of the dense bodies are firmly attached to the sarcolemma. The dense bodies and intermediate filaments anchor the thin filaments such that when sliding occurs between thin and thick filaments, the cell shortens. Dense bodies are not arranged in straight lines, so when a contraction occurs, the muscle cell twists like a corkscrew.
- Adjacent smooth muscle cells are bound together at dense bodies, transmitting the contractile forces from cell to cell throughout the tissue.
- Although smooth muscle cells are surrounded by connective tissue, the collagen fibers never unite to form tendons or aponeuroses, as they do in skeletal muscles.

Functional Characteristics of Smooth Muscle Tissue

Smooth muscle tissue differs from other muscle tissue in (1) excitation–contraction coupling, (2) length–tension relationships, (3) control of contractions, and (4) smooth muscle tone.

Excitation–Contraction Coupling

In skeletal and cardiac muscles, the trigger for contraction is the binding of calcium ions to troponin. In contrast, the trigger for smooth muscle contraction is the appearance of free calcium ions in the cytoplasm. On stimulation, a surge of calcium ions enters the cell from the extracellular fluid, and

Figure 10–23 **Smooth Muscle Tissue.**

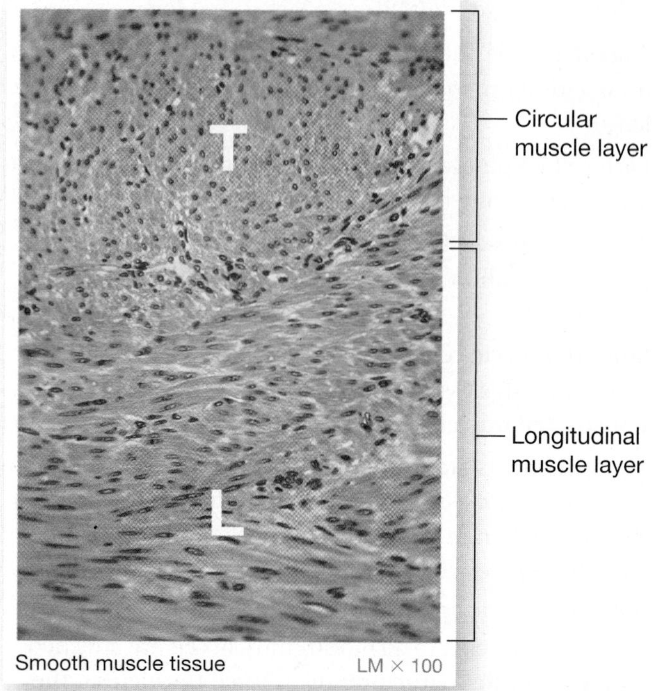
T
L
Circular muscle layer
Longitudinal muscle layer
Smooth muscle tissue
LM × 100

a Many visceral organs contain several layers of smooth muscle tissue oriented in different directions. Here, a single sectional view shows smooth muscle cells in both longitudinal (L) and transverse (T) sections.

Relaxed (sectional view)
Dense body
Actin
Myosin

Relaxed (superficial view)
Intermediate filaments (desmin)
Adjacent smooth muscle cells are bound together at dense bodies, transmitting the contractile forces from cell to cell throughout the tissue.

Contracted (superficial view)

b A single relaxed smooth muscle cell is spindle shaped and has no striations. Note the changes in cell shape as contraction occurs.

the sarcoplasmic reticulum releases additional calcium ions. The net result is a rise in the Ca^{2+} concentration throughout the cell.

Once in the sarcoplasm, the calcium ions interact with **calmodulin**, a calcium-binding protein. Calmodulin then activates the enzyme **myosin light chain kinase**, which in turn enables myosin heads to attach to actin.

Length–Tension Relationships

Because the thick and thin filaments are scattered and are not organized into sarcomeres in smooth muscle, tension development and resting length are not directly related. A stretched smooth muscle soon adapts to its new length and retains the ability to contract on demand. This ability to function over a wide range of lengths is called **plasticity**. Smooth muscle can contract over a range of lengths four times greater than that of skeletal muscle. Plasticity is especially important in digestive organs, such as the stomach, that change greatly in volume.

Despite the lack of sarcomeres, smooth muscle contractions can be just as powerful as those of skeletal muscles. Like skeletal muscle fibers, smooth muscle cells can undergo sustained contractions (as anyone who has experienced the uterine contractions that happen during labor can confirm).

Control of Contractions

Many smooth muscle cells are not innervated by motor neurons, and the neurons that do innervate smooth muscles are not under voluntary control. We categorize smooth muscle cells as either multiunit or visceral. **Multiunit smooth muscle cells** are innervated in motor units comparable to those of

> **+ Clinical Note** Electromyography
>
> Electromyography (EMG) is a low-risk test used to diagnose diseases of skeletal muscle, or of the motor neurons that serve it. The neurologist places a surface electrode over the weak or cramped muscle to find out if there is abnormal activity in the muscle when it should be resting. A small needle may also be inserted directly into the affected muscle to find out how the muscle reacts when a small electrical current is applied.

Table 10–3 A Comparison of Skeletal, Cardiac, and Smooth Muscle Tissues

Property	Skeletal Muscle	Cardiac Muscle	Smooth Muscle
Fiber dimensions (diameter × length)	100 μm × up to 30 cm	10–20 μm × 50–100 μm	5–10 μm × 30–200 μm
Nuclei	Multiple, near sarcolemma	Generally single, centrally located	Single, centrally located
Filament organization	In sarcomeres along myofibrils	In sarcomeres along myofibrils	Scattered throughout sarcoplasm
Sarcoplasmic reticulum (SR)	Terminal cisternae in triads at zones of overlap	SR tubules contact T tubules at Z lines	Dispersed throughout sarcoplasm, no T tubules
Control mechanism	Neural, at single neuromuscular junction on each muscle fiber	Automaticity (pacemaker cells), neural control limited to moderating pace and tension	Automaticity (pacesetter cells), neural or hormonal control (neural control limited to moderating pace and tension)
Ca^{2+} source	Release from SR	Extracellular fluid and release from SR	Extracellular fluid and release from SR
Ca^{2+} regulation	Troponin on thin filaments	Troponin on thin filaments	Calmodulin on myosin heads
Contraction	Rapid onset; may be tetanized; rapid fatigue	Slower onset; cannot be tetanized; resistant to fatigue	Slow onset; may be tetanized; resistant to fatigue
Energy source	Aerobic metabolism at moderate levels of activity; anaerobic during maximum activity (glycolysis)	Aerobic metabolism, usually lipid or carbohydrate substrates	Primarily aerobic metabolism

skeletal muscles, but each smooth muscle cell may be connected to more than one motor neuron. In contrast, many **visceral smooth muscle cells** lack a direct contact with any motor neuron.

Multiunit smooth muscle cells have excitable plasma membranes. They resemble skeletal muscle fibers and cardiac muscle cells in that neural activity produces an action potential that travels over the sarcolemma. However, these smooth muscle cells contract more slowly than do skeletal or cardiac muscle cells. We find multiunit smooth muscle cells in the iris of the eye, where they regulate the diameter of the pupil; within the walls of large arteries; in the arrector pili muscles of the skin; and along portions of the male reproductive tract. Multiunit smooth muscle cells do not typically occur in the digestive tract.

Visceral smooth muscle cells are arranged in sheets or layers. Within each layer, adjacent muscle cells are connected by gap junctions. As a result, whenever one muscle cell contracts, the electrical impulse that triggered the contraction can travel to adjacent smooth muscle cells. For this reason, the contraction spreads in a wave that soon involves every smooth muscle cell in the layer. The initial stimulus may be the activation of a motor neuron that contacts one of the muscle cells in the region. But smooth muscle cells also contract or relax in response to chemicals, hormones, the local concentration of oxygen or carbon dioxide, or a physical factor such as extreme stretching or irritation.

Many visceral smooth muscle layers show rhythmic cycles of activity in the absence of neural stimulation. For example, these cycles are characteristic of the smooth muscle cells in the wall of the digestive tract, where **pacesetter cells** spontaneously trigger the contraction of entire muscular sheets. Visceral smooth muscle cells are located in the walls of the digestive tract, the gallbladder, the urinary bladder, and many other internal organs.

Smooth Muscle Tone

Both multiunit and visceral smooth muscle tissues have a normal background level of activity, or smooth muscle tone. Neural, hormonal, or chemical factors can also stimulate smooth muscle relaxation, producing a decrease in muscle tone. For example, smooth muscle cells at the entrances to capillaries regulate the amount of blood flow into each vessel. If the tissue becomes starved for oxygen, the smooth muscle cells relax, allowing blood flow to increase and, hence, delivering additional oxygen. As conditions return to normal, the smooth muscle regains its normal muscle tone.

Table 10–3 summarizes how smooth muscle tissue differs from both skeletal and cardiac muscle tissues in structure and function.

✓ Checkpoint

25. **Identify the structural characteristics of smooth muscle tissue.**
26. **Which type of muscle tissue is least affected by changes in extracellular Ca^{2+} concentration during contraction?**
27. **Why can smooth muscle contract over a wider range of resting lengths than skeletal muscle can?**

See the blue Answers tab at the back of the book.

10 Chapter Review

Study Outline

An Introduction to Muscle Tissue p. 292

1. The three types of muscle tissue are *skeletal muscle, cardiac muscle,* and *smooth muscle.*

10-1 The primary function of muscle tissue is to produce movement p. 292

2. Common properties of muscle tissue: **excitability, contractility, extensibility**, and **elasticity**.
3. **Skeletal muscles** attach to bones directly or indirectly. Their functions are to (1) produce skeletal movement, (2) maintain posture and body position, (3) support soft tissues, (4) guard body entrances and exits, (5) maintain body temperature, and (6) store nutrient reserves.

10-2 Skeletal muscle contains muscle tissue, connective tissues, blood vessels, and nerves p. 293

4. The entire muscle is covered by an **epimysium**. Bundles of muscle fibers (cells) are sheathed by a **perimysium**, and each muscle fiber is surrounded by an **endomysium**. At the ends of the muscle are tendons or aponeuroses that attach the muscle to bones. (Figure 10–1)
5. The perimysium and endomysium contain the blood vessels and nerves that supply the muscle fibers.

10-3 Skeletal muscle fibers are organized into repeating functional units that contain sliding filaments p. 294

6. A skeletal muscle fiber has a **sarcolemma** (plasma membrane); **sarcoplasm** (cytoplasm); and **sarcoplasmic reticulum (SR)**, similar to the smooth endoplasmic reticulum of other cells. **Transverse (T) tubules** and **myofibrils** aid in contraction. Filaments in a myofibril are organized into repeating functional units called **sarcomeres.** *(Figures 10–2 to 10–6)*
7. **Myofibrils** contain **myofilaments** called **thin filaments** and **thick filaments.** *(Figures 10–2 to 10–6)*
8. Thin filaments consist of **F-actin, nebulin, tropomyosin,** and **troponin.** Tropomyosin molecules cover **active sites** on the **G-actin** subunits that form the F-actin strand. Troponin binds to G-actin and tropomyosin and holds the tropomyosin in position. *(Figure 10–7)*
9. Thick filaments consist of a bundle of myosin molecules around a **titin** core. Each **myosin** molecule has a long **tail** and a globular **head**, which forms **cross-bridges** with a thin filament during contraction. In a resting muscle cell, tropomyosin prevents the myosin heads from attaching to active sites on G-actin. *(Figure 10–7)*
10. The relationship between thick and thin filaments changes as a muscle fiber contracts. *(Figure 10–8)*
11. During a contraction, the free end of a myofibril moves toward the end that is attached or fixed. This occurs in an intact skeletal muscle, when one end of the muscle (the *origin*) is fixed in position during a contraction and the other end (the *insertion*) moves. *(Figure 10–9)*

> MasteringA&P™ Access more chapter study tools online in the MasteringA&P Study Area:
> - Chapter Quizzes, Chapter Practice Test, MP3 Tutor Sessions, and Clinical Case Studies
> - Practice Anatomy Lab PAL™ 3.0
> - A&P Flix A&PFlix™
> - Interactive Physiology iP2™
> - PhysioEx PhysioEx™ 9.1

10-4 Motor neurons stimulate skeletal muscle fibers to contract at the neuromuscular junction p. 302

12. Neurons and skeletal muscle fibers have **excitable membranes**, plasma membranes that can propagate electrical impulses, or action potentials.
13. A neuron controls the activity of a muscle fiber at a **neuromuscular junction (NMJ)**. *(Spotlight Figure 10–10)*
14. At the neuromuscular junction, when an **action potential** arrives at the neuron's **axon** (synaptic**) terminal**, **acetylcholine (ACh)** is released into the **synaptic cleft**. The binding of ACh to ACh membrane channel receptors on the motor end plate (and its junctional folds) of the muscle fiber leads to the generation of an action potential in the sarcolemma. *(Spotlight Figure 10–10)*
15. **Excitation–contraction coupling** occurs as the passage of an action potential along a T tubule triggers the release of Ca^{2+} from the cisternae of the SR at triads. *(Spotlight Figure 10–11)*
16. Release of Ca^{2+} initiates a **contraction cycle** of myosin head active-site exposure, cross-bridge formation, pivoting of the myosin head, cross-bridge detachment, and myosin reactivation. The calcium ions bind to troponin, which changes position and moves tropomyosin away from the active sites of actin. Cross-bridges of myosin heads then bind to actin. Next, each myosin head pivots, pulling the actin filament toward the center of the sarcomere, and then detaches. *(Spotlight Figure 10–12)*
17. When muscle cells contract, they create *tension* and pull on the attached tendons.
18. Acetylcholinesterase (AChE) breaks down ACh and limits the duration of muscle stimulation. *(Figure 10–13)*

10-5 Muscle fibers produce different amounts of tension depending on sarcomere length and frequency of stimulation p. 311

19. The amount of tension produced by a muscle fiber depends on the number of cross-bridges formed.
20. Skeletal muscle fibers can contract most forcefully when stimulated over a narrow range of resting lengths. *(Figure 10–14)*
21. A **twitch** is a cycle of contraction and relaxation produced by a single stimulus. *(Figure 10–15)*

22. Repeated stimulation at a slow rate produces **treppe**, a progressive increase in twitch tension. *(Figure 10–16)*
23. Repeated stimulation before the relaxation phase ends may produce **wave summation**, in which one twitch is added to another; **incomplete tetanus**, in which tension peaks and falls at intermediate stimulus rates; or **complete tetanus**, in which the relaxation phase is eliminated by very rapid stimuli. *(Figure 10–16)*

10-6 Skeletal muscles produce increased tension by recruiting additional motor units p. 315

24. The number and size of a muscle's motor units determine how precisely controlled its movements are. *(Figure 10–17)*
25. Resting **muscle tone** stabilizes bones and joints.
26. Normal activities generally include both **isotonic contractions** (in which the tension in a muscle rises and the length of the muscle changes) and **isometric contractions** (in which tension rises, but the length of the muscle stays the same). There are two types of isotonic contractions: **concentric** and **eccentric**. *(Figure 10–18)*
27. Load and speed of contraction are inversely related. *(Figure 10–19)*
28. The return to resting length after a contraction may involve elastic forces, opposing muscle contractions, and gravity.

10-7 To maintain regular muscle fiber activity, energy and recovery are required p. 319

29. Muscle contractions require large amounts of energy. *(Table 10–1)*
30. **Creatine phosphate (CP)** can release stored energy to convert ADP to ATP. *(Table 10–1)*
31. At rest or at moderate levels of activity, **aerobic metabolism** (the citric acid cycle and the electron transport chain in the mitochondria) can provide most of the ATP needed to support muscle contractions.
32. At peak levels of activity, the muscle cell relies heavily on **anaerobic metabolism (glycolysis)** to generate ATP, because the mitochondria cannot obtain enough oxygen to meet the existing ATP demands.
33. As muscular activity changes, the pattern of energy production and use changes. *(Figure 10–20)*
34. A fatigued muscle can no longer contract, because of a drop in pH due to the buildup of hydrogen ions, the exhaustion of energy resources, or other factors.
35. The **recovery period** begins immediately after a period of muscle activity and continues until conditions inside the muscle have returned to pre-exertion levels. The **oxygen debt**, or **excess postexercise oxygen consumption (EPOC)**, created during exercise is the amount of oxygen required during the recovery period to restore the muscle to its normal condition.
36. Circulating hormones may alter metabolic activities in skeletal muscle fibers.

10-8 Muscle performance depends on muscle fiber type and physical conditioning p. 323

37. The three types of skeletal muscle fibers are **fast fibers**, **slow fibers**, and **intermediate fibers**. *(Figure 10–21; Table 10–2)*
38. Fast fibers, which are large in diameter, contain densely packed myofibrils, large glycogen reserves, and relatively few mitochondria. They produce rapid and powerful contractions of relatively brief duration. *(Figure 10–21)*
39. Slow fibers are about half the diameter of fast fibers and take three times as long to contract after stimulation. Specializations such as abundant mitochondria, an extensive capillary supply, and high concentrations of **myoglobin** enable slow fibers to continue contracting for extended periods. *(Figure 10–21)*
40. Intermediate fibers are very similar to fast fibers, but have a greater resistance to fatigue.
41. Muscles dominated by fast fibers appear pale and are known as **white muscles**.
42. Muscles dominated by slow fibers are rich in myoglobin, appear red, and are known as **red muscles**.
43. Training to develop anaerobic endurance can lead to **hypertrophy** (enlargement) of the stimulated muscles.
44. With advanced age, skeletal muscles undergo **fibrosis**, the tolerance for exercise decreases, and repair of injuries slows.
45. **Anaerobic endurance** is the time over which muscular contractions can be sustained by glycolysis and reserves of ATP and CP.
46. **Aerobic endurance** is the time over which a muscle can continue to contract while supported by mitochondrial activities.

10-9 Cardiac muscle tissue, found in the heart, produces coordinated and automatic contractions p. 327

47. **Cardiac muscle tissue** is located only in the heart. **Cardiac muscle cells** are small; have one centrally located nucleus; have short, broad T tubules; and are dependent on aerobic metabolism. **Intercalated discs** bind the sarcolemmas of neighboring cardiac muscle cells. *(Figure 10–22; Table 10–3)*
48. Cardiac muscle cells contract without neural stimulation (**automaticity**), and their contractions last longer than those of skeletal muscle.
49. Because cardiac muscle twitches do not exhibit wave summation, cardiac muscle tissue cannot produce tetanic contractions.

10-10 Smooth muscle tissue contracts to move substances within internal passageways p. 329

50. **Smooth muscle tissue** is nonstriated, involuntary muscle tissue.
51. Smooth muscle cells lack sarcomeres and the resulting striations. The thin filaments are anchored to **dense bodies**. *(Figure 10–23; Table 10–3)*
52. Smooth muscle contracts when calcium ions interact with **calmodulin**, which activates **myosin light chain kinase**.
53. Smooth muscle functions over a wide range of lengths (**plasticity**).
54. In **multiunit smooth muscle cells,** each smooth muscle cell acts relatively independently of other smooth muscle cells in the organ. **Visceral smooth muscle cells** are not always innervated by motor neurons. Neurons that innervate smooth muscle cells are not under voluntary control.

10

Review Questions

See the blue Answers tab at the back of the book.

10

LEVEL 1 Reviewing Facts and Terms

1. Identify the structures in the following figure.

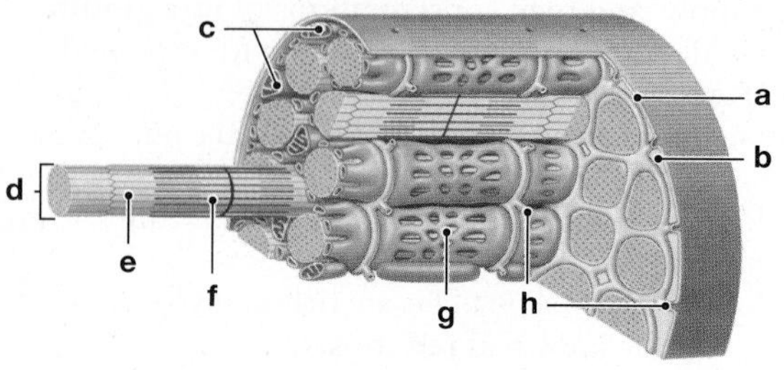

c
a
b
d
e
f
g
h

(a) ________ (b) ________
(c) ________ (d) ________
(e) ________ (f) ________
(g) ________ (h) ________

2. The connective tissue coverings of a skeletal muscle, listed from superficial to deep, are **(a)** endomysium, perimysium, and epimysium, **(b)** endomysium, epimysium, and perimysium, **(c)** epimysium, endomysium, and perimysium, **(d)** epimysium, perimysium, and endomysium.
3. The signal to contract is distributed deep into a muscle fiber by the **(a)** sarcolemma, **(b)** sarcomere, **(c)** transverse tubules, **(d)** myotubules, **(e)** myofibrils.
4. The detachment of the myosin cross-bridges is directly triggered by **(a)** the repolarization of T tubules, **(b)** the attachment of ATP to myosin heads, **(c)** the hydrolysis of ATP, **(d)** calcium ions.
5. A muscle producing near-peak tension during rapid cycles of contraction and relaxation is said to be in **(a)** incomplete tetanus, **(b)** treppe, **(c)** complete tetanus, **(d)** a twitch.
6. The type of contraction in which the tension rises, but the load does not move, is **(a)** a wave summation, **(b)** a twitch, **(c)** an isotonic contraction, **(d)** an isometric contraction.
7. Which of the following statements about myofibrils is *not* correct? **(a)** Each skeletal muscle fiber contains hundreds to thousands of myofibrils. **(b)** Myofibrils contain repeating units called sarcomeres. **(c)** Myofibrils extend the length of a skeletal muscle fiber. **(d)** Filaments consist of bundles of myofibrils. **(e)** Myofibrils are attached to the plasma membrane at both ends of a muscle fiber.
8. An action potential can travel quickly from one cardiac muscle cell to another because of the presence of **(a)** gap junctions, **(b)** tight junctions, **(c)** intercalated discs, **(d)** both a and c.
9. List the three types of muscle tissue in the body.
10. What three layers of connective tissue are part of each muscle? What functional role does each layer play?
11. The __________ contains vesicles filled with acetylcholine. **(a)** axon terminal, **(b)** motor end plate, **(c)** neuromuscular junction, **(d)** synaptic cleft, **(e)** transverse tubule.
12. What structural feature of a skeletal muscle fiber propagates action potentials into the interior of the cell?
13. What five interlocking steps are involved in the contraction process?
14. What two factors affect the amount of tension produced when a skeletal muscle contracts?
15. What forms of energy reserves do resting skeletal muscle fibers contain?
16. What two mechanisms are used to generate ATP from glucose in muscle cells?
17. What is the calcium-binding protein in smooth muscle tissue?

LEVEL 2 Reviewing Concepts

18. An activity that would require anaerobic endurance is **(a)** a 50-meter dash, **(b)** a pole vault, **(c)** a weight-lifting competition, **(d)** all of these.
19. Areas of the body where you would *not* expect to find slow fibers include the **(a)** back and calf muscles, **(b)** eye and hand, **(c)** chest and abdomen, **(d)** all of these.
20. During relaxation, muscles return to their original length because of all of the following *except* **(a)** actin and myosin actively pushing away from one another, **(b)** the contraction of opposing muscles, **(c)** the pull of gravity, **(d)** the elastic nature of the sarcolemma, **(e)** elastic forces.
21. According to the length–tension relationship, **(a)** longer muscles can generate more tension than shorter muscles, **(b)** the greater the zone of overlap in the sarcomere, the greater the tension the muscle can develop, **(c)** the greatest tension is achieved in sarcomeres where actin and myosin initially do not overlap, **(d)** there is an optimum range of actin and myosin overlap that will produce the greatest amount of tension, **(e)** both b and d are correct.
22. For each portion of a myogram tracing a twitch in a stimulated calf muscle fiber, describe the events that occur within the muscle fiber.
23. What three processes are involved in repaying the oxygen debt during a muscle's recovery period?
24. How does cardiac muscle tissue contract without neural stimulation?
25. Atracurium is a drug that blocks the binding of ACh to ACh receptors. Give an example of a site where such binding normally occurs, and predict the physiological effect of this drug.
26. Explain why a murder victim's time of death can be estimated according to the flexibility or rigidity of the body.
27. Which of the following activities would employ isometric contractions? **(a)** flexing the elbow, **(b)** chewing food, **(c)** maintaining an upright posture, **(d)** running, **(e)** writing.

LEVEL 3 Critical Thinking and Clinical Applications

28. Many potent insecticides contain toxins, called organophosphates, that interfere with the action of the enzyme acetylcholinesterase. Ivan is using an insecticide containing organophosphates and is very careless. Because he does not use gloves or a dust mask, he absorbs some of the chemical through his skin and inhales a large amount as well. What signs would you expect to observe in Ivan as a result of organophosphate poisoning?
29. Linda's father suffers an apparent heart attack and is rushed to the emergency room of the local hospital. The doctor on call tells her that he has ordered some blood work and that he will be able to tell if her father actually had a heart attack by looking at the blood levels of CK and cardiac troponin. Why would knowing the level of CK and cardiac troponin help to indicate if a person suffered a heart attack?
30. Bill broke his leg in a football game, and after six weeks the cast is finally removed. As he steps down from the examination table, he loses his balance and falls. Why?

CLINICAL CASE Wrap-Up Keep on Keepin' on

www.mona-lila

Rhetta is suffering from myasthenia gravis, a disease of the neuromuscular junction (NMJ). In people with this autoimmune disease, the immune system mistakenly makes antibodies that damage the acetylcholine (ACh) receptor sites on the surface of the motor end plates of skeletal muscle fibers. The ACh that is released by the axon terminal of the motor neuron cannot find enough intact ACh receptors on which to bind. For this reason, the muscle fiber membrane permeability to sodium ions does not change enough, and the cascade of events leading to muscular contraction is feeble. The result is weakness and fatigability. This weakness improves with rest as the ACh store is built up again. Often the muscles of the eyes, including the eyelids, and other facial muscles are affected first. All voluntary, skeletal muscles can be affected.

The neurologist injected Rhetta with Tensilon (edrophonium chloride), a short-acting acetylcholinesterase (AChE) inhibitor. The drug blocks AChE, the enzyme that breaks down ACh, and so prolongs muscle stimulation. Sadly, the dramatic effects, such as suddenly being able to open and move both eyes, are short lived. There are, however, many treatments available for myasthenia gravis, including long-acting AChE inhibitors. Rhetta responded well to treatment and was able to resume playing with her band.

10

1. How would a drug that inhibits AChE make a myasthenia gravis patient stronger?
2. Would myasthenia gravis affect Rhetta's heart muscle? Explain.

See the blue Answers tab at the back of the book.

Related Clinical Terms

botulism: A severe, potentially fatal paralysis of skeletal muscles, resulting from the consumption of a bacterial toxin.

Duchenne muscular dystrophy (DMD): One of the most common and best understood of the muscular dystrophies.

fibromyalgia: A chronic disorder characterized by widespread musculoskeletal pain, fatigue, and localized tenderness.

muscular dystrophies: A varied collection of inherited diseases that produce progressive muscle weakness and deterioration.

myasthenia gravis: A general muscular weakness resulting from a reduction in the number of ACh receptors on the motor end plate.

myopathy: Disease of muscle tissue.

RICE (rest, ice, compression, and elevation): Acronym for the standard treatment for muscle injuries, bruises, strains, and sprains.

11 The Muscular System

Learning Outcomes

These Learning Outcomes correspond by number to this chapter's sections and indicate what you should be able to do after completing the chapter.

11-1 ■ Describe the arrangement of fascicles in the various types of muscles, and explain the resulting functional differences. p. 337

11-2 ■ Describe the classes of levers, and explain how they make muscles more efficient. p. 339

11-3 ■ Predict the actions of a muscle on the basis of its origin and insertion, and explain how muscles interact to produce or oppose movements. p. 339

11-4 ■ Explain how the name of a muscle can help identify its location, appearance, or function. p. 343

11-5 ■ Compare and contrast the axial and appendicular muscles. p. 344

11-6 ■ Identify the principal axial muscles of the body, plus their origins, insertions, actions, and innervation. p. 347

11-7 ■ Identify the principal appendicular muscles of the body, plus their origins, insertions, actions, and innervation, and compare the major functional differences between the upper and lower limbs. p. 362

11-8 ■ Explain the functional relationship between the muscular system and other body systems, and explain the role of exercise in producing various responses in other body systems. p. 382

CLINICAL CASE Downward-Facing Dog

"Breathe and do what *you* can do," the instructor called out to the class in soothing tones. Rick concentrated on his yoga pose. He'd been coming to class with the encouragement of his nurse practitioner. "I think yoga could help you move and feel better. Your arthritis medications can only do so much for you. Give yoga the old college try. Let's see you again in three months," she urged.

Rick was impressed with the gentle practice of yoga. The first week in class his body felt all locked up, like the Tin Man in *The Wizard of Oz*. But he stayed with the program and practiced a little between classes. By now, three months later, he could stretch his arms overhead and balance on one foot for a few seconds in Tree Pose. His leg muscles felt stronger as he stepped sideways into Warrior Pose. His next challenge was to tackle the most famous yoga pose of all, Downward-Facing Dog. He thought that he might be up to trying it. Slow and easy, he told himself. **What will the nurse practitioner test on Rick's return visit?** **To find out, turn to the Clinical Case Wrap-Up on p. 387.**

An Introduction to the Muscular System

In this chapter we describe the gross anatomy of the muscular system and consider functional relationships between muscles and bones of the body. Most skeletal muscle fibers contract at similar rates and shorten to the same degree, but variations in their microscopic and macroscopic organization can dramatically affect the power, range, and speed of movement produced when a whole muscle contracts.

11-1 Fascicle arrangement is correlated with muscle power and range of motion

Learning Outcome Describe the arrangement of fascicles in the various types of muscles, and explain the resulting functional differences.

Muscle fibers in a skeletal muscle form bundles called *fascicles* (FAS-ih-kulz). ⮌ p. 293 The muscle fibers in a single fascicle are parallel, but the arrangement of fascicles in skeletal muscles can vary, as can the relationship between the fascicles and the associated tendon. Based on the patterns of fascicle arrangement, we can classify skeletal muscles as *parallel muscles*, *convergent muscles*, *pennate muscles*, and *circular muscles* (Figure 11–1).

Parallel Muscles

In a **parallel muscle**, the fascicles are parallel to the long axis of the muscle. Most of the skeletal muscles in the body are parallel muscles. Some are flat bands with broad attachments (*aponeuroses*; ap-ō-nū-RŌ-sēz) at each end. Others are plump and cylindrical, with tendons at one or both ends. Such a muscle is spindle shaped, with a central **body**, also known as the *belly*. For example, the *biceps brachii* of the arm is a parallel muscle with a central body (Figure 11–1a). When a parallel muscle contracts, it shortens and gets larger in diameter. When you flex your elbow, you can see the bulge of the contracting biceps brachii on the anterior surface of your arm. Other parallel muscles include the *rectus abdominis* (Figure 11–1b) and *supinator* (Figure 11–1c).

A skeletal muscle fiber can contract until it has shortened about 30 percent. The entire parallel muscle shortens by this amount when its fibers contract together, because these fibers are parallel to the long axis of the muscle. Thus, if the muscle is 10 cm (3.9 in.) long and one end is held in place, the other end will move 3 cm when the muscle contracts. The tension developed during this contraction depends on the total number of myofibrils the muscle contains. ⮌ p. 315

Convergent Muscles

In a **convergent muscle**, muscle fascicles extending over a broad area come together, or converge, on a common attachment site. The muscle may pull on a tendon, an aponeurosis, or a slender band of collagen fibers known as a **raphe** (RĀ-fē; seam). The muscle fibers typically spread out, like a fan or a broad triangle, with a tendon at the apex. Examples include the prominent *pectoralis muscles* of the chest (Figure 11–1d).

A convergent muscle can adapt to different activities because the stimulation of different portions of the muscle can change the direction it pulls. However, when the entire muscle contracts, the muscle fibers do not pull as hard on the attachment site as would a parallel muscle of the same size. Why? The reason is that convergent muscle fibers pull in different directions, rather than all pulling in the same direction as in parallel muscles.

Pennate Muscles

In a **pennate muscle** (PEN-āt; *penna*, feather), the fascicles form a common angle with the tendon. Because the muscle fibers pull at an angle, contracting pennate muscles do not move their tendons as far as parallel muscles do. But

Figure 11–1 Muscle Types Based on Pattern of Fascicle Arrangement.

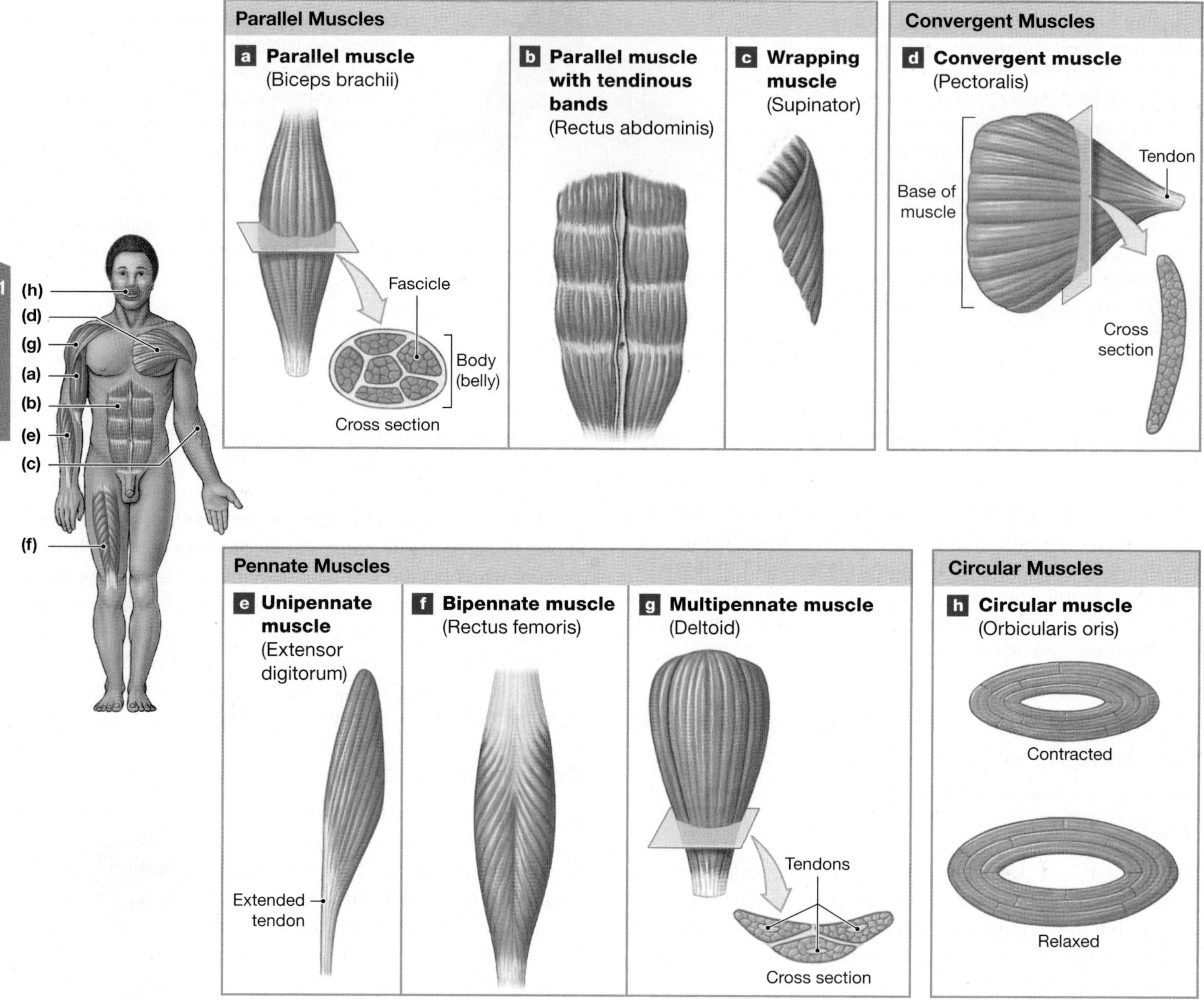

Parallel Muscles
a Parallel muscle (Biceps brachii)
Fascicle
Body (belly)
Cross section
b Parallel muscle with tendinous bands (Rectus abdominis)
c Wrapping muscle (Supinator)
Convergent Muscles
d Convergent muscle (Pectoralis)
Base of muscle
Tendon
Cross section
(h)
(d)
(g)
(a)
(b)
(e)
(c)
(f)
Pennate Muscles
e Unipennate muscle (Extensor digitorum)
Extended tendon
f Bipennate muscle (Rectus femoris)
g Multipennate muscle (Deltoid)
Tendons
Cross section
Circular Muscles
h Circular muscle (Orbicularis oris)
Contracted
Relaxed

a pennate muscle contains more muscle fibers—and thus more myofibrils—than does a parallel muscle of the same size. For this reason, the pennate muscle produces more tension.

There are three types of pennate muscles, depending on the arrangement of the fascicles with the tendon. If all the muscle fascicles are on the same side of the tendon, the pennate muscle is *unipennate.* The *extensor digitorum,* a forearm muscle that extends the finger joints, is unipennate (Figure 11–1e). More commonly, a pennate muscle has fascicles on both sides of a central tendon. Such a muscle is called *bipennate.* The *rectus femoris,* a prominent muscle that extends the knee, is bipennate (Figure 11–1f). If the tendon branches within a pennate muscle, the muscle is said to be *multipennate.* The triangular *deltoid* of the shoulder is multipennate (Figure 11–1g).

Circular Muscles

In a **circular muscle**, or **sphincter** (SFINK-ter), the fascicles are concentrically arranged around an opening. When the muscle contracts, the diameter of the opening becomes smaller. Circular muscles surround body openings or hollow organs and act as valves in the digestive and urinary tracts. An example is the *orbicularis oris* of the mouth (Figure 11–1h).

Checkpoint

1. **Based on patterns of fascicle arrangement, name the four types of skeletal muscle.**
2. **Why does a pennate muscle generate more tension than does a parallel muscle of the same size?**
3. **Which type of fascicle arrangement would you expect in a muscle guarding the anal opening between the large intestine and the exterior?**

See the blue Answers tab at the back of the book.

11-2 The use of bones as levers increases muscle efficiency

Learning Outcome **Describe the classes of levers, and explain how they make muscles more efficient.**

Skeletal muscles do not work in isolation. For muscles attached to the skeleton, the nature and site of their connections determine the force, speed, and range of the movement they produce. These characteristics are interdependent, and the relationships can explain a great deal about the general organization of the muscular and skeletal systems.

Attaching the muscle to a lever can modify the force, speed, or direction of movement produced by muscle contraction. A *lever* is a rigid structure—such as a board, a crowbar, or a bone—that moves on a fixed point called a *fulcrum*. A lever moves when pressure, called an *applied force*, is sufficient to overcome any load that would otherwise oppose or prevent such movement.

In the body, each bone is a lever and each joint is a fulcrum. Muscles provide the applied force. The load can vary from the weight of an object held in the hand to the weight of a limb or the weight of the entire body, depending on the situation. The important thing about levers is that they can change (1) the direction of an applied force, (2) the distance and speed of movement produced by an applied force, and (3) the effective strength of an applied force.

There are three classes of levers. A lever is classified according to the relative position of three elements: applied force, fulcrum, and load. Regardless of the class of lever, all follow the same mechanical principles: A mechanical advantage occurs when the applied force is farther from the fulcrum than the load. A mechanical disadvantage occurs when the applied force is closer to the load than the fulcrum.

We find examples of each class of lever in the human body (Figure 11–2). A pry bar or crowbar is an example of a **first-class lever**. In such a lever, the fulcrum (F) lies between the applied force (AF) and the load (L). We can describe the positions as L–F–AF. The body has few first-class levers. One, involved with extension of the neck and lifting the head, is shown in Figure 11–2a.

A familiar example of a **second-class lever** is a loaded wheelbarrow. The load lies between the applied force and the fulcrum (Figure 11–2b). The weight is the load, and the upward lift on the handle is the applied force. We can describe the positions as F–L–AF. In this arrangement, a small force can move a larger weight because the force is always farther from the fulcrum than the load is. That is, the effective force is increased. Notice, however, that when a force moves the handle, the load moves more slowly and covers a shorter distance. In other words, the effective force is increased at the expense of speed and distance. The body has few second-class levers. Ankle extension (plantar flexion) by the calf muscles involves a second-class lever (see Figure 11–2b).

In a **third-class lever**, such as a pair of tongs, the applied force is between the load and the fulcrum (Figure 11–2c). We can describe the positions as F–AF–L. Third-class levers are the most common levers in the body. The effect is the reverse of that for a second-class lever: Speed and distance traveled are increased at the expense of effective force.

Not every muscle is part of a lever system, but the presence of levers provides speed and versatility far in excess of what we would predict for the body on the basis of muscle physiology alone.

Checkpoint

4. **Define a *lever*, and describe the three classes of levers.**
5. **The joint between the occipital bone of the skull and the first cervical vertebra (atlas) is a part of which class of lever?**

See the blue Answers tab at the back of the book.

11-3 The origins and insertions of muscles determine their actions

Learning Outcome **Predict the actions of a muscle on the basis of its origin and insertion, and explain how muscles interact to produce or oppose movements.**

To understand the actions of skeletal muscles, we need to understand where they are connected to the bones that act as levers and which joints they cross.

Origins and Insertions

In Chapter 10 we noted that when both ends of a myofibril are free to move, the ends move toward the center during a contraction. In the body, the ends of a skeletal muscle are always attached to other structures that limit their movement. In most cases one end is fixed in position, and during a contraction the other end moves toward the fixed end. The less movable end is called the **origin** of the muscle. The more movable end is called the **insertion** of the muscle. The origin is typically proximal to the insertion. Almost all skeletal muscles either originate or insert on the skeleton.

Figure 11–2 The Three Classes of Levers.

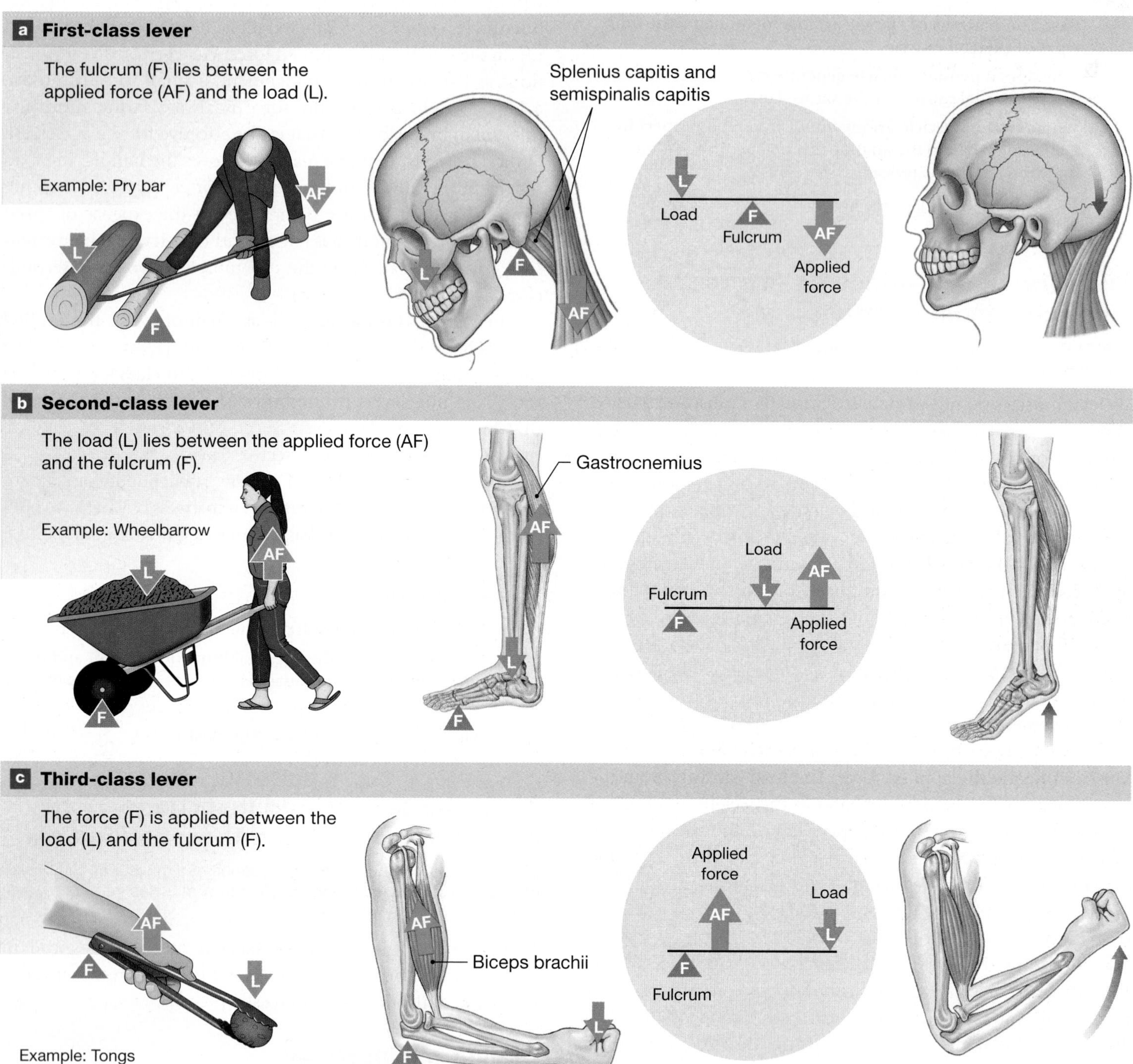
a First-class lever
The fulcrum (F) lies between the applied force (AF) and the load (L).
Example: Pry bar
AF
L
F
Splenius capitis and semispinalis capitis
Load
Fulcrum
Applied force
b Second-class lever
The load (L) lies between the applied force (AF) and the fulcrum (F).
Example: Wheelbarrow
Gastrocnemius
Load
Fulcrum
Applied force
c Third-class lever
The force (F) is applied between the load (L) and the fulcrum (F).
Biceps brachii
Applied force
Load
Fulcrum
Example: Tongs

The decision as to which end is the origin and which is the insertion is usually based on movement from the anatomical position. As an example, consider the *gastrocnemius*, a calf muscle that extends from the distal portion of the femur to the calcaneus. As Figure 11–2b shows, when the gastrocnemius contracts, it pulls the calcaneus toward the knee. As a result, we say that the gastrocnemius has its origin at the femur and its insertion at the calcaneus. Part of the fun of studying the muscular system is that you can actually do the movements and think about the muscles involved. As a result, laboratory activities focusing on muscle actions are often like disorganized aerobics classes.

When we cannot easily determine the origins and insertions on the basis of movement from the anatomical position, we have other rules to use. If a muscle extends between a broad aponeurosis and a narrow tendon, the aponeurosis is the origin and the tendon is the insertion. If several tendons are at one end and just one is at the other, the muscle has multiple origins and a single insertion. However, these simple rules cannot cover every situation. Knowing which end is the origin and which is the insertion is ultimately less important than knowing where the two ends attach and what the muscle accomplishes when it contracts.

Most muscles originate at a bone, but some originate at a connective tissue sheath or band. Examples of these sheaths or bands include *intermuscular septa* (components of the deep fascia that may separate adjacent skeletal muscles), *tendinous inscriptions* that join muscle fibers to form long muscles such as the *rectus abdominis*, the interosseous membranes of the forearm or leg, and the fibrous sheet that spans the obturator foramen of the pelvis.

Actions

When a muscle contracts, it produces a specific **action**, or movement. As introduced in Chapter 9, when a muscle moves a portion of the skeleton, that movement may involve flexion, extension, adduction, abduction, protraction, retraction, elevation, depression, rotation, circumduction, pronation, supination, inversion, eversion, lateral flexion, opposition, or reposition. (Before proceeding, you may want to review the discussions of planes of motion and **Spotlight Figure 9–2** and **Figures 9–3** to **9–5**.) pp. 272–276

We can describe actions in two ways, one focused on the bone and one on the joint. The first way describes actions in terms of the bone or region affected. For example, we say a muscle such as the biceps brachii performs "flexion of the forearm." However, specialists such as kinesiologists and physical therapists increasingly use the second way, which identifies the joint involved. In this approach, we say the action of the biceps brachii is "flexion at (or of) the elbow." Examples of muscle action are presented in **Spotlight Figure 11–3**.

In complex movements, muscles commonly work in groups rather than individually. Their cooperation improves the efficiency of a particular movement. For example, large muscles of the limbs produce flexion or extension over an extended range of motion. These muscles cannot produce powerful movements at full extension due to the positions of the articulating bones, but they are usually paired with one or more smaller muscles that provide assistance until the larger muscle can perform at maximum efficiency. At the start of the movement, the smaller muscle produces maximum tension, while the larger muscle produces minimum tension. The importance of the smaller "assistant" decreases as the movement proceeds and the effectiveness of the primary muscle increases.

To describe how muscles work together, we can use the following four functional types: agonist, antagonist, synergist, and fixator.

- An **agonist**, or **prime mover**, is a muscle whose contraction is mostly responsible for producing a particular movement. The biceps brachii is an agonist that produces flexion at the elbow.
- An **antagonist** is a muscle whose action opposes that of a particular agonist. The *triceps brachii* is an agonist that extends the elbow. For this reason, it is an antagonist of the biceps brachii. Likewise, the biceps brachii is an antagonist of the triceps brachii.

Agonists and antagonists are functional opposites. If one produces flexion, the other produces extension. When an agonist contracts to produce a particular movement, the corresponding antagonist is stretched, but it usually does not relax completely. Instead, it contracts eccentrically, with just enough tension to control the speed of the movement and ensure its smoothness. p. 318

You may find it easiest to learn about muscles in agonist–antagonist pairs (flexors–extensors, abductors–adductors) that act at a specific joint. This method highlights the functions of the muscles involved, and it can help organize the information into a logical framework. The tables in this chapter are arranged to support such an approach.

- When a **synergist** (*syn-*, together + *ergon*, work) contracts, it helps a larger agonist work efficiently. Synergists may provide additional pull near the insertion or may stabilize the point of origin. Their importance in assisting a particular movement may change as the movement progresses. In many cases, they are most useful at the start, when the agonist is stretched and unable to develop maximum tension. For example, the *latissimus dorsi* is a large trunk muscle that extends, adducts, and medially rotates the arm at the shoulder joint. A much smaller muscle, the *teres* (TER-ēz) *major*, assists in starting such movements when the shoulder joint is at full flexion.
- A **fixator** is a synergist that assists an agonist by preventing movement at another joint, thereby stabilizing the origin of the agonist. Recall that the biceps brachii is an agonist that produces flexion at the elbow. It has two tendons that originate on the scapula and one that inserts on the radius. During flexion, the trapezius and rhomboid act as fixators by stabilizing and preventing the movement of the scapula.

Checkpoint

6. **The *gracilis* attaches to the anterior surface of the tibia at one end, and to the pubis and ischium of the pelvis at the other. When the muscle contracts, flexion occurs at the hip. Which attachment point is considered the muscle's origin?**
7. **Muscle A abducts the humerus, and muscle B adducts the humerus. What is the relationship between these two muscles?**
8. **Define the term *synergist* as it relates to muscle action.**

See the blue Answers tab at the back of the book.

Go to MasteringA&P™ > Study Area > Menu > Animations & Videos > A&PFlix > Origins, Insertions, Actions, Innervations

11

SPOTLIGHT

Figure 11–3
Muscle Action

Go to MasteringA&P® > Study Area > A&PFlix > Group Muscle Actions and Joints

The action produced by a muscle at any one joint is largely dependent upon the structure of the joint and the location of the insertion of the muscle relative to the axis of movement at the joint. The direction, or geometric paths, of the action produced by a muscle—*called lines of action*—is often represented by an arrow (or more than one arrow in fan-shaped muscles).

Flexion and Extension

At joints that permit flexion and extension, muscles whose lines of action cross the anterior side of a joint are flexors of that joint, and muscles whose lines of action cross the posterior side of a joint are extensors of that joint.

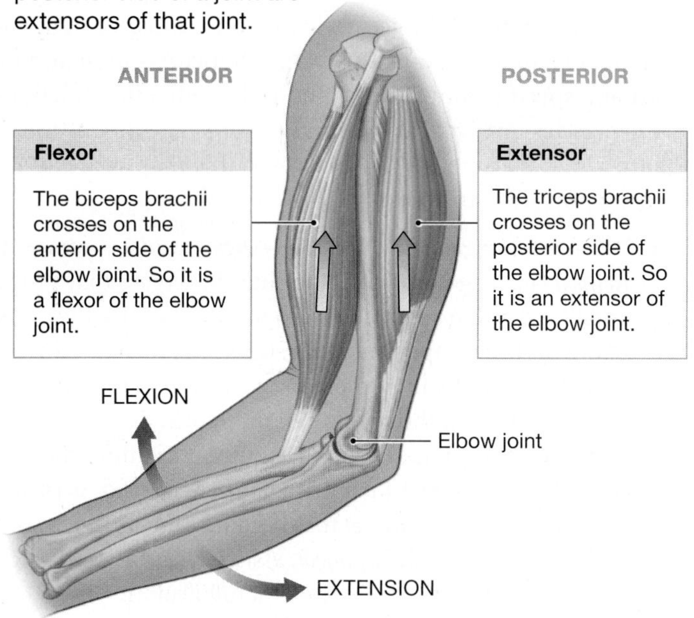
ANTERIOR
POSTERIOR
FLEXION
Elbow joint
EXTENSION

Flexor

The biceps brachii crosses on the anterior side of the elbow joint. So it is a flexor of the elbow joint.

Extensor

The triceps brachii crosses on the posterior side of the elbow joint. So it is an extensor of the elbow joint.

Abduction and Adduction

At joints that permit adduction and abduction, muscles whose lines of action cross the medial side of a joint are adductors of that joint, and muscles whose lines of action cross the lateral side of a joint are abductors of that joint.

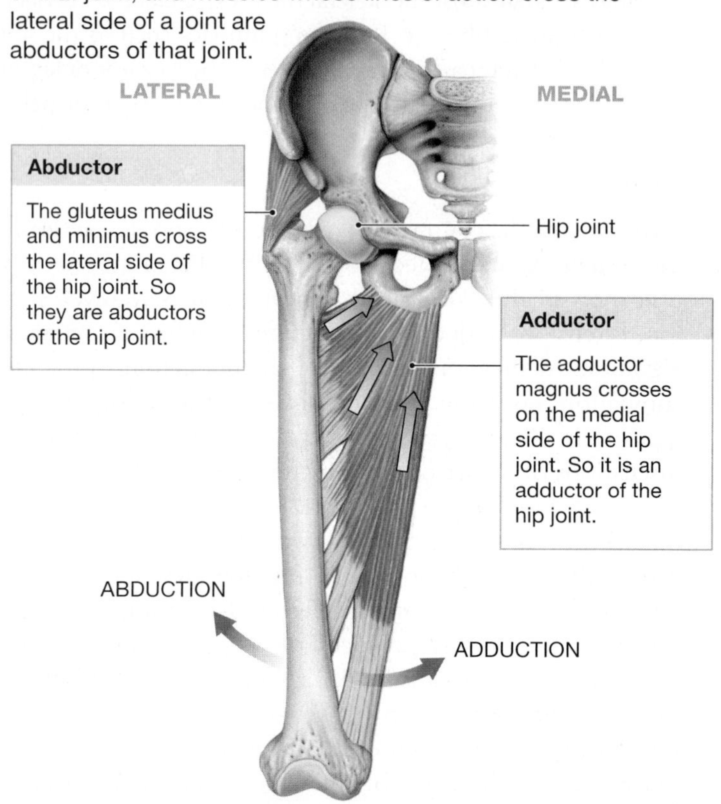
LATERAL
MEDIAL
Hip joint
ABDUCTION
ADDUCTION

Abductor

The gluteus medius and minimus cross the lateral side of the hip joint. So they are abductors of the hip joint.

Adductor

The adductor magnus crosses on the medial side of the hip joint. So it is an adductor of the hip joint.

Medial and Lateral Rotation

At joints that permit rotation, movement or turning of the body part occurs around its axis. The shoulder joint is a ball-and-socket joint that permits rotation. The subscapularis has lines of action that cross the anterior aspect of the shoulder joint. When the subscapularis contracts it produces medial rotation at the joint. The teres minor has lines of action that cross the posterior aspect of the shoulder joint. When the teres minor contracts, it produces lateral rotation at the shoulder.

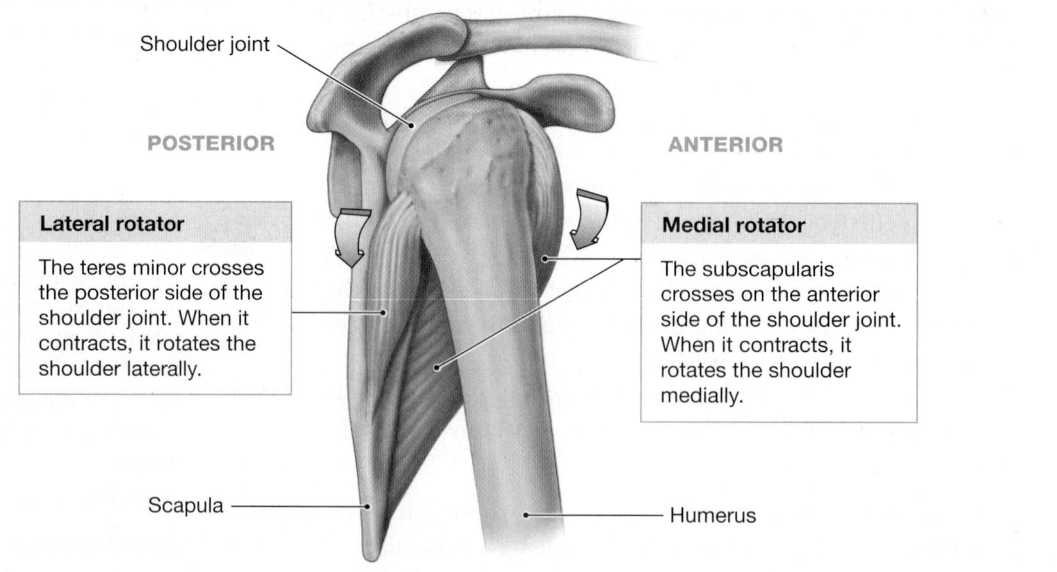
Shoulder joint
POSTERIOR
ANTERIOR
Scapula
Humerus

Lateral rotator

The teres minor crosses the posterior side of the shoulder joint. When it contracts, it rotates the shoulder laterally.

Medial rotator

The subscapularis crosses on the anterior side of the shoulder joint. When it contracts, it rotates the shoulder medially.

11-4 Descriptive terms are used to name skeletal muscles

Learning Outcome Explain how the name of a muscle can help identify its location, appearance, or function.

The human body has approximately 700 muscles. You do not need to learn every one of their names, but you will have to become familiar with many of them. Fortunately, the names anatomists assigned to the muscles include descriptive terms that can help you remember the names and identify the muscles. When you are faced with a new muscle name, it is helpful to first identify the descriptive portions of the name. The name of a muscle may include descriptive information about its region of the body; position, direction, and fascicle arrangement; structural characteristics; and action. **Table 11–1** includes a useful summary of muscle terminology.

Region of the Body

Regional terms are most common as modifiers that help identify individual muscles. In a few cases, a muscle is such a prominent feature of a body region that a name referring to the region alone will identify it. Examples include the *temporalis* of the head and the *brachialis* (brā-kē-A-lis) of the arm (**Figure 11–4a**).

Table 11–1 Muscle Terminology

Terms Indicating Specific Regions of the Body	Terms Indicating Position, Direction, or Fascicle Arrangement	Terms Indicating Structural Characteristics of the Muscle	Terms Indicating Actions
Abdominal (abdomen)	Anterior (front)	**NATURE OF ORIGIN**	**GENERAL**
Ancon (elbow)	External (on the outside)	Biceps (two heads)	Abductor (movement away)
Auricular (ear)	Extrinsic (outside the structure)	Triceps (three heads)	Adductor (movement toward)
Brachial (arm)	Inferior (below)	Quadriceps (four heads)	Depressor (lowering movement)
Capitis (head)	Internal (away from the surface)		Extensor (straightening movement)
Carpi (wrist)	Intrinsic (within the structure)	**SHAPE**	Flexor (bending movement)
Cervicis (neck)	Lateral (on the side)	Deltoid (triangle)	Levator (raising movement)
Coccygeal (coccyx)	Medial (middle)	Orbicularis (circle)	Pronator (turning into prone position)
Costal (rib)	Oblique (slanting)	Pectinate (comblike)	Supinator (turning into supine position)
Cutaneous (skin)	Posterior (back)	Piriformis (pear shaped)	Tensor (tensing movement)
Femoris (thigh)	Profundus (deep)	Platysma (flat plate)	
Glossal (tongue)	Rectus (straight)	Pyramidal (pyramid)	**SPECIFIC**
Hallux (great toe)	Superficial (toward the surface)	Rhomboid (parallelogram)	Buccinator (trumpeter)
Ilium (groin)	Superior (toward the head)	Serratus (serrated)	Risorius (laugher)
Inguinal (groin)	Transverse (crosswise)	Splenius (bandage)	Sartorius (like a tailor)
Lumbar (lumbar region)		Teres (round and long)	
Nasalis (nose)		Trapezius (trapezoid)	
Nuchal (back of neck)			
Ocular (eye)		**OTHER STRIKING FEATURES**	
Oris (mouth)		Alba (white)	
Palpebra (eyelid)		Brevis (short)	
Pollex (thumb)		Gracilis (slender)	
Popliteal (posterior to knee)		Latae (wide)	
Psoas (loin)		Latissimus (widest)	
Radial (forearm)		Longissimus (longest)	
Scapular (scapula)		Longus (long)	
Temporal (temple)		Magnus (large)	
Thoracic (thorax)		Major (larger)	
Tibial (tibia; shin)		Maximus (largest)	
Ulnar (ulna)		Minimus (smallest)	
		Minor (smaller)	
		Vastus (great)	

Figure 11–4 An Overview of the Major Skeletal Muscles.

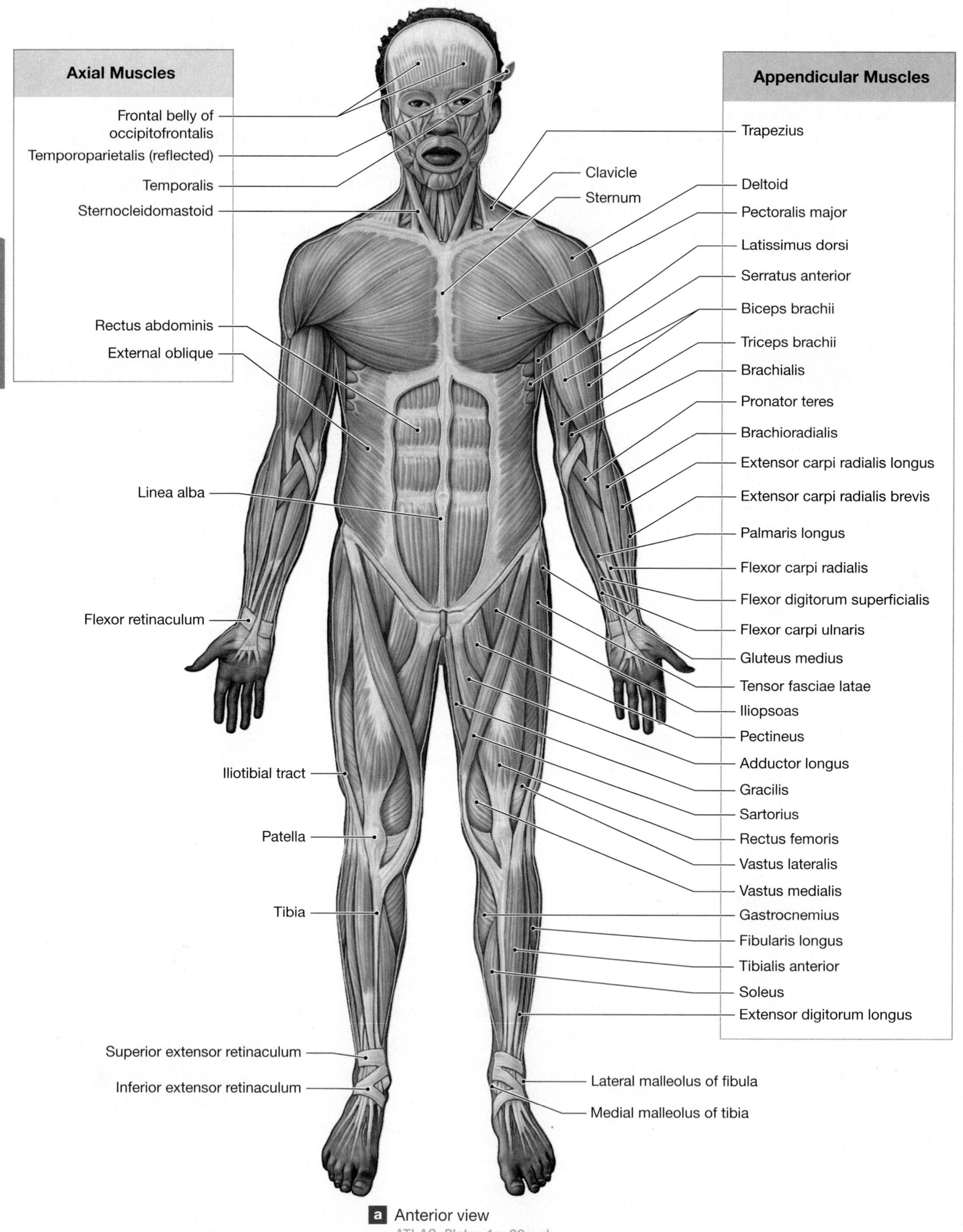
Axial Muscles
Frontal belly of occipitofrontalis
Temporoparietalis (reflected)
Temporalis
Sternocleidomastoid
Rectus abdominis
External oblique
Linea alba
Flexor retinaculum
Iliotibial tract
Patella
Tibia
Superior extensor retinaculum
Inferior extensor retinaculum
Clavicle
Sternum
Appendicular Muscles
Trapezius
Deltoid
Pectoralis major
Latissimus dorsi
Serratus anterior
Biceps brachii
Triceps brachii
Brachialis
Pronator teres
Brachioradialis
Extensor carpi radialis longus
Extensor carpi radialis brevis
Palmaris longus
Flexor carpi radialis
Flexor digitorum superficialis
Flexor carpi ulnaris
Gluteus medius
Tensor fasciae latae
Iliopsoas
Pectineus
Adductor longus
Gracilis
Sartorius
Rectus femoris
Vastus lateralis
Vastus medialis
Gastrocnemius
Fibularis longus
Tibialis anterior
Soleus
Extensor digitorum longus
Lateral malleolus of fibula
Medial malleolus of tibia

a Anterior view
ATLAS: Plates 1a; 39a–d

Figure 11–4 An Overview of the Major Skeletal Muscles. (*continued*)

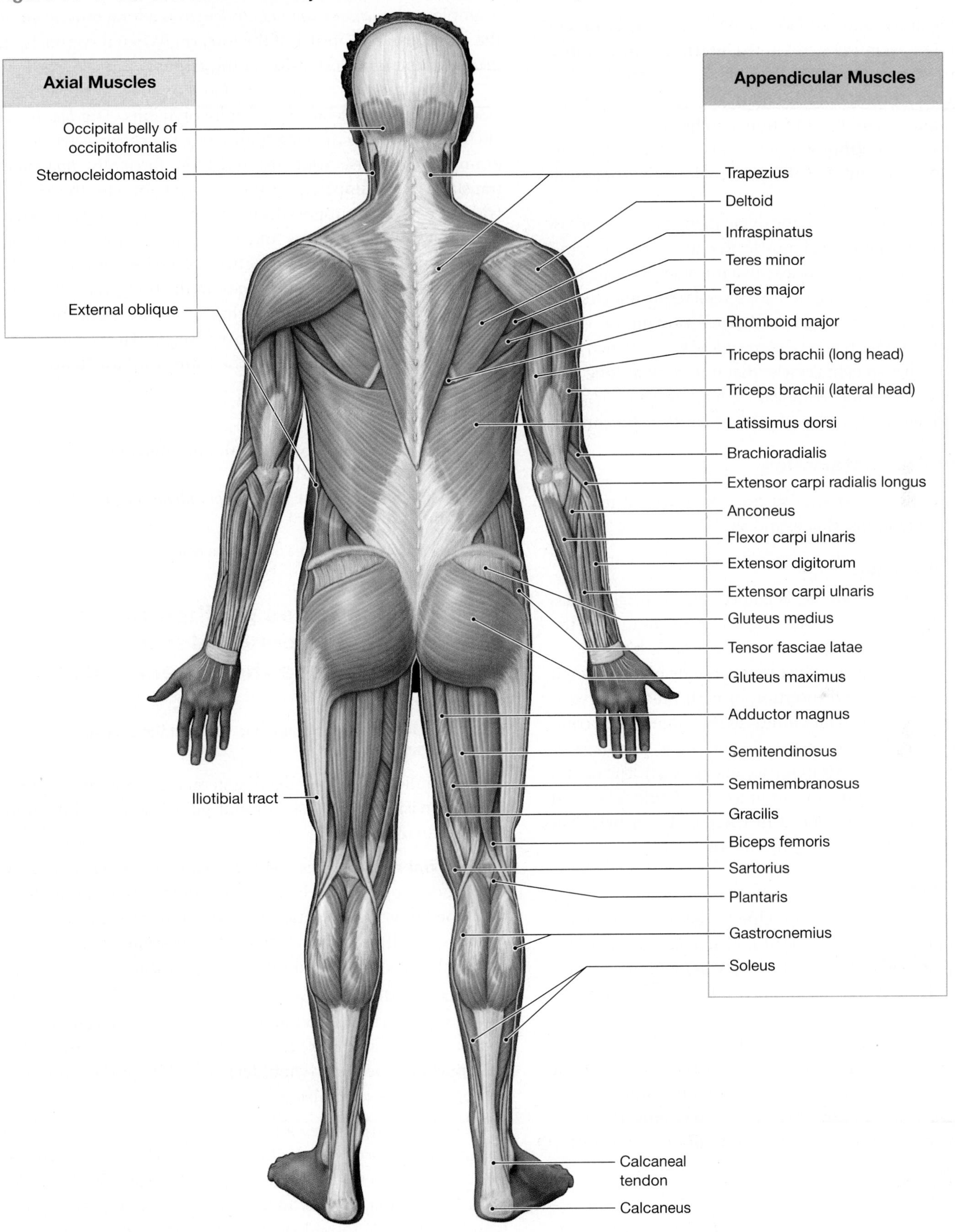
Axial Muscles
Occipital belly of occipitofrontalis
Sternocleidomastoid
External oblique
Iliotibial tract
Appendicular Muscles
Trapezius
Deltoid
Infraspinatus
Teres minor
Teres major
Rhomboid major
Triceps brachii (long head)
Triceps brachii (lateral head)
Latissimus dorsi
Brachioradialis
Extensor carpi radialis longus
Anconeus
Flexor carpi ulnaris
Extensor digitorum
Extensor carpi ulnaris
Gluteus medius
Tensor fasciae latae
Gluteus maximus
Adductor magnus
Semitendinosus
Semimembranosus
Gracilis
Biceps femoris
Sartorius
Plantaris
Gastrocnemius
Soleus
Calcaneal tendon
Calcaneus

b Posterior view
ATLAS: Plates 1b; 40a,b

Position, Direction, or Fascicle Arrangement

Muscles visible at the body surface are often called **externus** or **superficialis**. Deeper muscles are termed **internus** or **profundus**. Superficial muscles that position or stabilize an organ are called **extrinsic**. Muscles located entirely within an organ are **intrinsic**.

Muscle names may be directional indicators. For example, **transversus** and **oblique** indicate muscles that run across (transversus) or at a slanting (oblique) angle to the longitudinal axis of the body.

A muscle name may refer to the orientation of the muscle fascicles within a particular skeletal muscle. **Rectus** means "straight," and most rectus muscles have fascicles that run along the longitudinal axis of the muscle. Because we have several rectus muscles, the name typically includes a second term that refers to a precise region of the body. For example, the *rectus abdominis* of the abdomen is an axial muscle that has straight fascicles that run along its long axis. However, in the case of the *rectus femoris*, *rectus* refers to "straight muscle of the thigh" and not to its fascicles (which are bipennate).

Structural Characteristics

Some muscles are named after distinctive structural features, such as multiple tendons, shape, and size.

Origin and Insertion

The biceps brachii, for example, is named after its origin. It has two tendons of origin (*bi-*, two + *caput*, head). Similarly, the triceps brachii has three, and the *quadriceps femoris* has four.

Many muscle names include terms for body places that tell you the specific origin and insertion of each muscle. In such cases, the first part of the name indicates the origin, the second part the insertion. The *genioglossus*, for example, originates at the chin (*geneion*) and inserts in the tongue (*glossus*). The names may be long and difficult to pronounce, but **Table 11–1** and the anatomical terms introduced in Chapter 1 can help you identify and remember them. pp. 10–14

Shape and Size

Shape is sometimes an important clue to the name of a muscle. For example, the *trapezius* (tra-PĒ-zē-us), *deltoid*, *rhomboid* (ROM-boyd), and *orbicularis* (or-bik-ū-LĀ-ris) look like a trapezoid, a triangle (like the Greek letter delta, Δ), a rhomboid, and a circle, respectively.

Many terms refer to muscle size. Long muscles are called **longus** (long) or **longissimus** (longest). **Teres** muscles are both long and round. Short muscles are called **brevis**. Large ones are called **magnus** (big), **major** (bigger), or **maximus** (biggest). Small ones are called **minor** (smaller) or **minimus** (smallest).

Action

Many muscles are named *flexor*, *extensor*, *pronator*, *abductor*, *adductor*, and *rotator* (see **Spotlight Figure 11–3**). These are such common actions that the names almost always include other clues as to the appearance or location of the muscle. For example, the *extensor carpi radialis longus* is a long muscle along the radial (lateral) border of the forearm. When it contracts, its primary function is extension at the carpus (wrist).

A few muscles are named after the specific movements associated with special occupations or habits. The *buccinator* (BUK-si-nā-tor) on the face compresses the cheeks—when, for example, you purse your lips and blow forcefully. *Buccinator* translates as "trumpeter." Another facial muscle, the *risorius* (ri-SOR-ē-us), was supposedly named after the mood expressed: The Latin word *risor* means "one who laughs." However, a more appropriate description for the effect would be "a grimace." The *sartorius* (sar-TOR-ē-us), the longest in the body, is active when you cross your legs. Before sewing machines were invented, a tailor would sit on the floor cross-legged. The name of this muscle was derived from *sartor*, the Latin word for "tailor."

✓ Checkpoint

9. **Identify the kinds of descriptive information used to name skeletal muscles.**
10. **What does the name *flexor carpi radialis longus* tell you about this muscle?**

See the blue Answers tab at the back of the book.

11-5 Axial muscles position the axial skeleton, and appendicular muscles support and move the appendicular skeleton

Learning Outcome Compare and contrast the axial and appendicular muscles.

The separation of the skeletal system into axial and appendicular divisions serves as a useful guideline for subdividing the muscular system:

- The **axial muscles** arise on the axial skeleton. This category includes approximately 60 percent of the skeletal muscles in the body. They position the head and vertebral column; move the rib cage, which assists the movements that make breathing possible; and form the pelvic floor.
- The **appendicular muscles** stabilize or move structures of the appendicular skeleton. Forty percent of skeletal muscles are appendicular muscles, including those that move and support the pectoral (shoulder) and pelvic girdles and the upper and lower limbs.

Figure 11–4 provides an overview of the major axial and appendicular muscles of the human body. These are superficial muscles, which tend to be rather large. The superficial muscles cover deeper, smaller muscles that we cannot see unless the overlying muscles are removed, or *reflected*—that is, cut and pulled out of the way. Later figures that show deep muscles in specific regions will indicate whether superficial muscles have been reflected.

Next we study examples of both muscular divisions. Pay attention to patterns of origin, insertion, and action. In the figures in this chapter, you will find that some bony and cartilaginous landmarks are labeled to provide orientation.

The tables that follow also contain information about the innervation of the individual muscles. **Innervation** is the distribution of nerves to a region or organ. The tables indicate the nerves that control each muscle. Many of the muscles of the head and neck are innervated by *cranial nerves*, which originate at the brain and pass through the foramina of the skull. In addition, *spinal nerves* are connected to the spinal cord and pass through the intervertebral foramina. For example, spinal nerve L_1 passes between vertebrae L_1 and L_2. Spinal nerves may form a complex network called a *plexus* after exiting the spinal cord. One branch of this network may contain axons from several spinal nerves. Many tables identify the spinal nerves involved as well as the names of their specific branches.

Checkpoint

11. **Describe the location and general functions of axial muscles.**
12. **Describe the location and general functions of appendicular muscles.**

See the blue Answers tab at the back of the book.

11-6 Axial muscles are muscles of the head and neck, vertebral column, trunk, and pelvic floor

Learning Outcome **Identify the principal axial muscles of the body, plus their origins, insertions, actions, and innervation.**

The axial muscles fall into logical groups on the basis of location, function, or both. The groups do not always have distinct anatomical boundaries. For example, a function such as extension of the vertebral column involves muscles along its entire length and movement at each of the intervertebral joints. We will discuss the axial muscles in four groups:

- *The Muscles of the Head and Neck.* This group includes muscles that move the face, tongue, and larynx. They are responsible for verbal and nonverbal communication—laughing, talking, frowning, smiling, whistling, and so on. You also use these muscles while eating—especially in sucking and chewing—and even while looking for food, as some of them control your eye movements. This group does not include muscles of the neck that are involved with movements of the vertebral column.
- *The Muscles of the Vertebral Column.* This group includes numerous flexors, extensors, and rotators of the vertebral column.
- *The Oblique and Rectus Muscles.* This group forms the muscular walls of the trunk between the first thoracic vertebra and the pelvis. In the thoracic area the ribs separate these muscles, but over the abdominal surface the muscles form broad muscular sheets. The neck also has oblique and rectus muscles. They do not form a complete muscular wall, but they share a common developmental origin with the oblique and rectus muscles of the trunk.
- *The Muscles of the Pelvic Floor.* These muscles extend between the sacrum and pelvic girdle. This group forms the *perineum* (per-ih-NĒ-um), a region anterior to the sacrum and coccyx between the inner thighs.

Muscles of the Head and Neck

We can divide the muscles of the head and neck into several functional groups. The *muscles of facial expression*, the *muscles of mastication* (chewing), the *muscles of the tongue*, and the *muscles of the pharynx* originate on the skull or hyoid bone.

11

Muscles involved with sight and hearing also are based on the skull. Here, we will consider the *extrinsic eye muscles*—those associated with movements of the eye. In Chapter 17 we discuss the intrinsic eye muscles, which control the diameter of the pupil and the shape of the lens, and the tiny skeletal muscles associated with the auditory ossicles.

Among the *muscles of the anterior neck*, the extrinsic muscles of the larynx adjust the position of the hyoid bone and larynx. We examine the intrinsic laryngeal muscles, including those of the vocal cords, in Chapter 23.

Muscles of Facial Expression

The muscles of facial expression originate on the surface of the skull (Figure 11–5). At their insertions, the fibers of the epimysium are woven into those of the superficial fascia and the dermis of the skin. For this reason, when these muscles contract, the skin moves.

The largest group of facial muscles is associated with the mouth. The **orbicularis oris** constricts the opening, and other muscles move the lips or the corners of the mouth. The **buccinator** has two functions related to eating (in addition to its importance to musicians). During chewing, it cooperates with the masticatory muscles by moving food back across the teeth from the *vestibule*, the space inside the cheeks. In infants, the buccinator provides suction for suckling at the breast.

Smaller groups of muscles control movements of the eyebrows and eyelids, the scalp, the nose, and the external ear. The **epicranium** (ep-i-KRĀ-nē-um; *epi-*, on + *kranion*, skull), or scalp, contains the **temporoparietalis** and the **occipitofrontalis** (ok-sip-ih-tō-fron-TAL-is), which has a *frontal belly* and an *occipital belly*. The two bellies are separated by the **epicranial aponeurosis**, a thick, collagenous sheet.

The **platysma** (pla-TIZ-muh; *platys*, flat) covers the anterior surface of the neck. This muscle extends from the base of the neck to the periosteum of the mandible and the fascia at the corner of the mouth. One of the effects of aging is the loss of muscle tone in the platysma, resulting in a looseness of the skin of the anterior throat.

Now take another look at Figure 11–5 with **Table 11–2**, which summarizes the muscles of facial expression by region. Focus on identifying the location and function of each of the major muscles named in this discussion.

Figure 11–5 **Muscles of Facial Expression.** ATLAS: Plate 3a–d

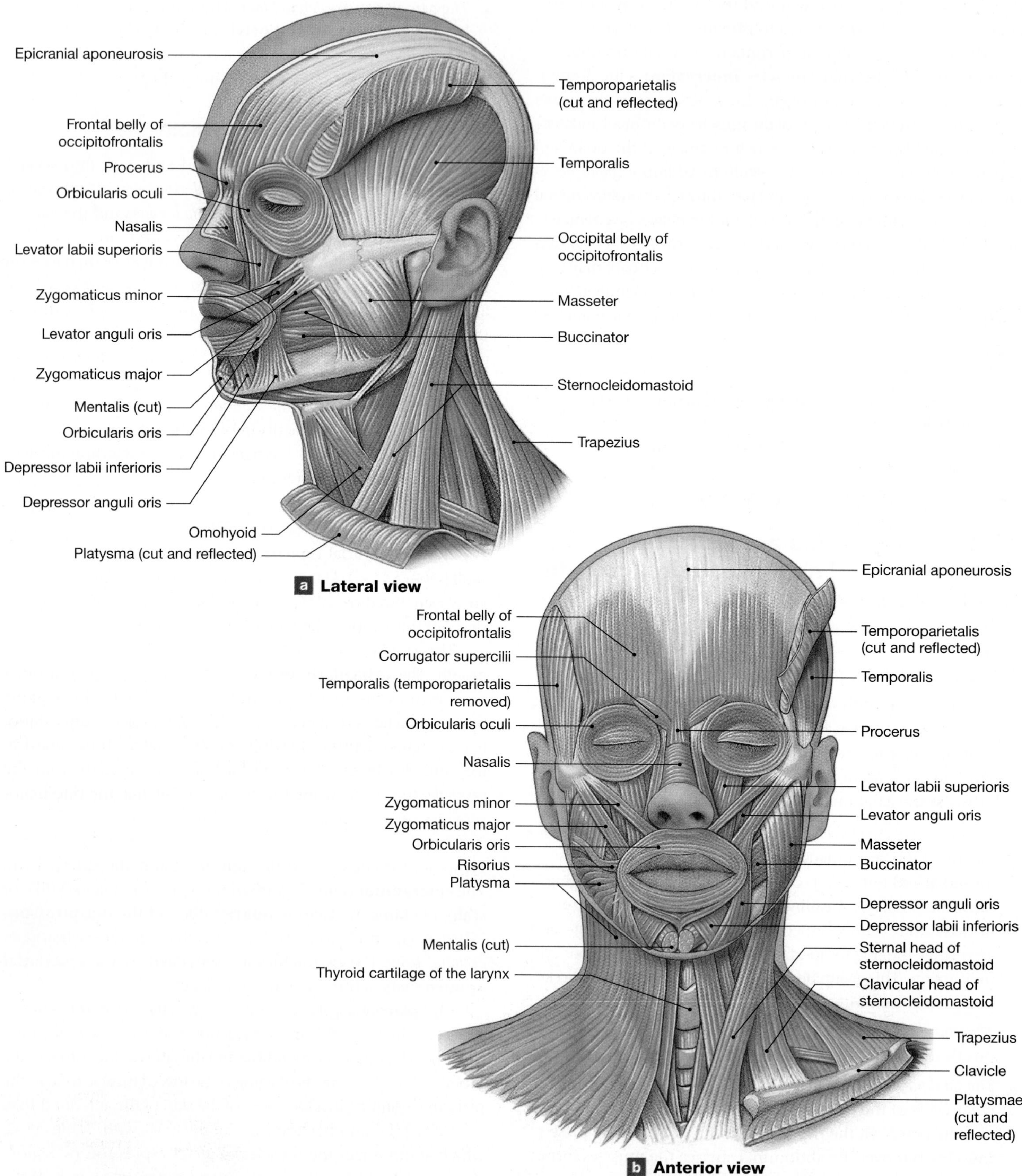

Epicranial aponeurosis
Temporoparietalis (cut and reflected)
Frontal belly of occipitofrontalis
Temporalis
Procerus
Orbicularis oculi
Nasalis
Occipital belly of occipitofrontalis
Levator labii superioris
Zygomaticus minor
Masseter
Levator anguli oris
Buccinator
Zygomaticus major
Sternocleidomastoid
Mentalis (cut)
Orbicularis oris
Trapezius
Depressor labii inferioris
Depressor anguli oris
Omohyoid
Platysma (cut and reflected)
Epicranial aponeurosis
Frontal belly of occipitofrontalis
Temporoparietalis (cut and reflected)
Corrugator supercilii
Temporalis (temporoparietalis removed)
Temporalis
Orbicularis oculi
Procerus
Nasalis
Levator labii superioris
Zygomaticus minor
Levator anguli oris
Zygomaticus major
Orbicularis oris
Masseter
Risorius
Buccinator
Platysma
Depressor anguli oris
Depressor labii inferioris
Mentalis (cut)
Sternal head of sternocleidomastoid
Thyroid cartilage of the larynx
Clavicular head of sternocleidomastoid
Trapezius
Clavicle
Platysmae (cut and reflected)

a Lateral view

b Anterior view

? Given its name, what does the levator anguli oris do?

Table 11–2 Muscles of Facial Expression (Figure 11–5)

Region and Muscle	Origin	Insertion	Action	Innervation
MOUTH				
Buccinator	Alveolar process of maxilla and alveolar part of the mandible	Blends into fibers of orbicularis oris	Compresses cheeks	Facial nerve (VII)*
Depressor labii inferioris	Mandible between the anterior midline and the mental foramen	Skin of lower lip	Depresses lower lip	Facial nerve (VII)
Levator labii superioris	Inferior margin of orbit, superior to the infra-orbital foramen	Orbicularis oris	Elevates upper lip	Facial nerve (VII)
Levator anguli oris	Maxilla below the infra-orbital foramen	Corner of mouth	Elevates corner of mouth	Facial nerve (VII)
Mentalis	Incisive fossa of mandible	Skin of chin	Elevates and protrudes lower lip	Facial nerve (VII)
Orbicularis oris	Maxilla and mandible	Lips	Compresses, purses lips	Facial nerve (VII)
Risorius	Fascia surrounding parotid salivary gland	Angle of mouth	Draws corner of mouth to the side	Facial nerve (VII)
Depressor anguli oris	Anterolateral surface of mandibular body	Skin at angle of mouth	Depresses corner of mouth	Facial nerve (VII)
Zygomaticus major	Zygomatic bone near zygomaticomaxillary suture	Angle of mouth	Retracts and elevates corner of mouth	Facial nerve (VII)
Zygomaticus minor	Zygomatic bone posterior to zygomaticotemporal suture	Upper lip	Retracts and elevates upper lip	Facial nerve (VII)
EYE				
Corrugator supercilii	Orbital rim of frontal bone near nasal suture	Eyebrow	Pulls skin inferiorly and anteriorly; wrinkles brow	Facial nerve (VII)
Levator palpebrae superioris (Figure 11–6a,b)	Tendinous band around optic foramen	Upper eyelid	Elevates upper eyelid	Oculomotor nerve (III)**
Orbicularis oculi	Medial margin of orbit	Skin around eyelids	Closes eye	Facial nerve (VII)
NOSE				
Procerus	Nasal bones and lateral nasal cartilages	Aponeurosis at bridge of nose and skin of forehead	Moves nose, changes position and shape of nostrils	Facial nerve (VII)
Nasalis	Maxilla and nasal cartilages	Dorsum of nose	Compresses bridge, depresses tip of nose; elevates corners of nostrils	Facial nerve (VII)
EAR				
Temporoparietalis	Fascia around external ear	Epicranial aponeurosis	Tenses scalp, moves auricle of ear	Facial nerve (VII)
SCALP (EPICRANIUM)				
Occipitofrontalis **Frontal belly**	Epicranial aponeurosis	Skin of eyebrow and bridge of nose	Raises eyebrows, wrinkles forehead	Facial nerve (VII)
Occipital belly	Occipital bone and mastoid region of temporal bones	Epicranial aponeurosis	Tenses and retracts scalp	Facial nerve (VII)
NECK				
Platysma	Superior thorax between cartilage of 2nd rib and acromion of scapula	Mandible and skin of cheek	Tenses skin of neck; depresses mandible; pulls lower lip down	Facial nerve (VII)

*A Roman numeral refers to a cranial nerve.

**This muscle originates in association with the extrinsic eye muscles, so its innervation is unusual.

Extrinsic Eye Muscles

Six **extrinsic eye muscles**, also known as the *oculomotor muscles*, originate on the surface of the orbit and control the position of each eye. These muscles, shown in Figure 11–6, are the **inferior rectus**, **medial rectus**, **superior rectus**, **lateral rectus**, **inferior oblique**, and **superior oblique**. Compare the illustrations in Figure 11–6 with the details on these muscles in Table 11–3.

11

Figure 11–6 Extrinsic Eye Muscles. ATLAS: Plates 12a; 16a,b

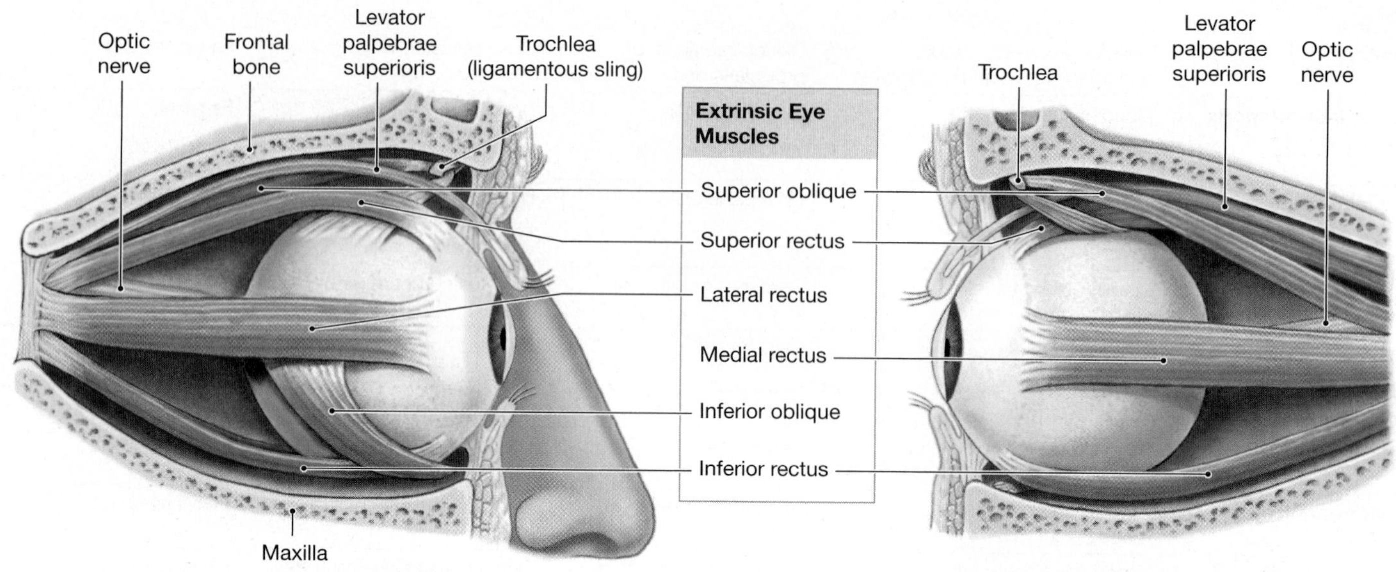
Optic nerve
Frontal bone
Levator palpebrae superioris
Trochlea (ligamentous sling)
Extrinsic Eye Muscles
Superior oblique
Superior rectus
Lateral rectus
Medial rectus
Inferior oblique
Inferior rectus
Maxilla
Trochlea
Levator palpebrae superioris
Optic nerve

a Lateral surface, right eye

b Medial surface, right eye

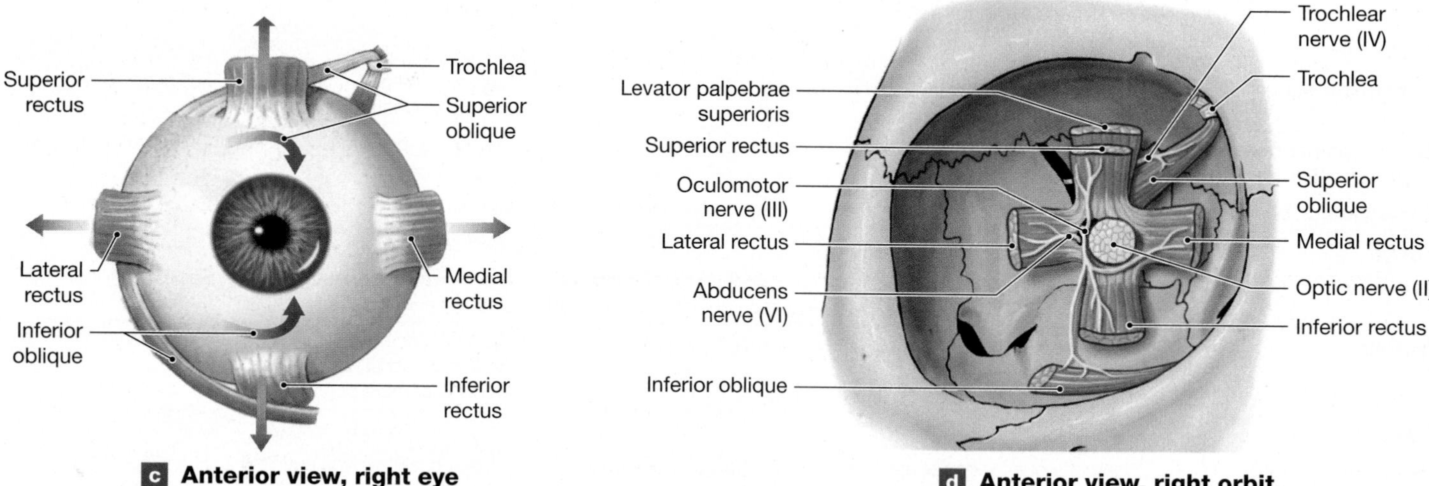
Superior rectus
Trochlea
Superior oblique
Lateral rectus
Medial rectus
Inferior oblique
Inferior rectus
Trochlear nerve (IV)
Trochlea
Levator palpebrae superioris
Superior rectus
Oculomotor nerve (III)
Lateral rectus
Abducens nerve (VI)
Inferior oblique
Superior oblique
Medial rectus
Optic nerve (II)
Inferior rectus

c Anterior view, right eye

d Anterior view, right orbit

? Six muscles control eye movement. How many of these muscles can you see from the lateral surface and how many can you see from the medial surface, respectively?

Table 11–3 Extrinsic Eye Muscles (Figure 11–6)

Muscle	Origin	Insertion	Action	Innervation
Inferior rectus	Sphenoid around optic canal	Inferior, medial surface of eyeball	Eye looks inferiorly	Oculomotor nerve (III)
Medial rectus	Sphenoid around optic canal	Medial surface of eyeball	Eye looks medially	Oculomotor nerve (III)
Superior rectus	Sphenoid around optic canal	Superior surface of eyeball	Eye looks superiorly	Oculomotor nerve (III)
Lateral rectus	Sphenoid around optic canal	Lateral surface of eyeball	Eye looks laterally	Abducens nerve (VI)
Inferior oblique	Maxilla at anterior portion of orbit	Inferior, lateral surface of eyeball	Eye rolls, looks superiorly and laterally	Oculomotor nerve (III)
Superior oblique	Sphenoid around optic canal	Superior, lateral surface of eyeball	Eye rolls, looks inferiorly and laterally	Trochlear nerve (IV)

Figure 11–7 Muscles of Mastication. ATLAS: Plate 3c,d

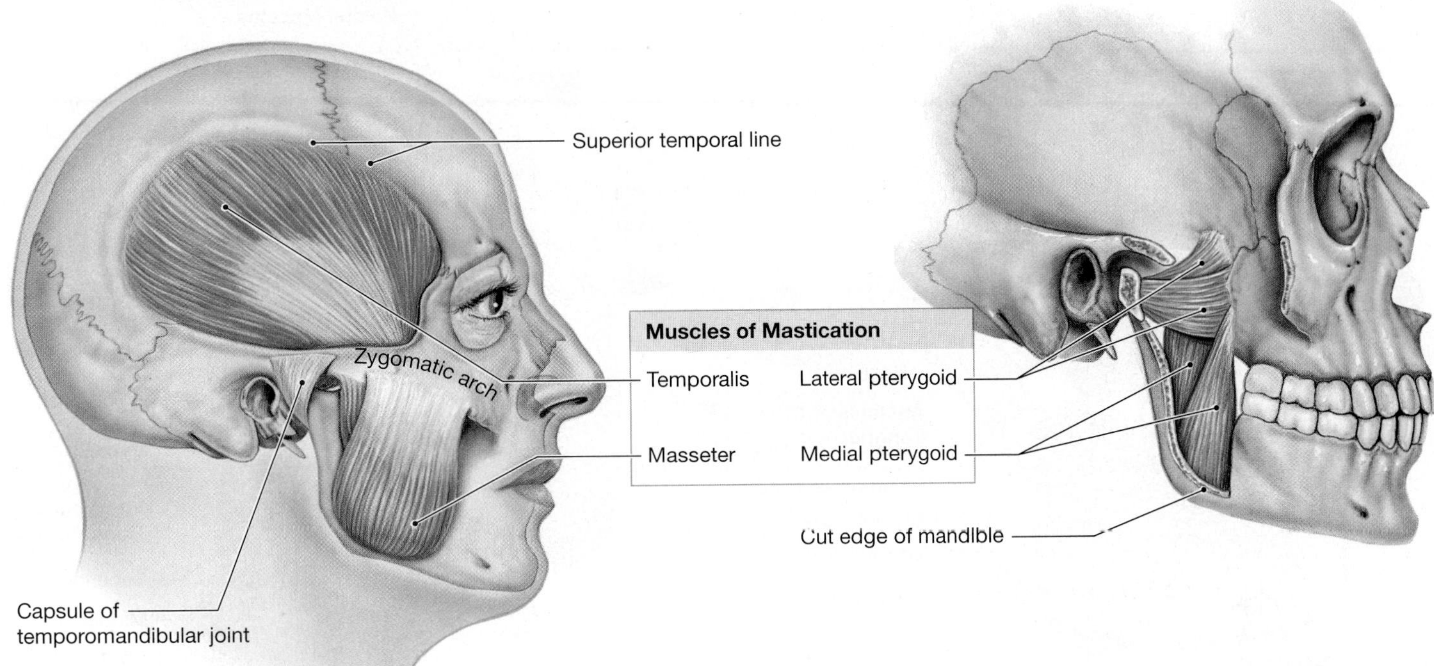
Superior temporal line
Zygomatic arch
Muscles of Mastication
Temporalis
Lateral pterygoid
Masseter
Medial pterygoid
Cut edge of mandible
Capsule of temporomandibular joint

a Lateral view. The temporalis passes medial to the zygomatic arch to insert on the coronoid process of the mandible. The masseter inserts on the angle and lateral surface of the mandible.

b Lateral view, pterygoid muscles exposed. The location and orientation of the pterygoid muscles are seen after the overlying muscles and a portion of the mandible are removed.

Table 11–4 Muscles of Mastication (Figure 11–7)

Muscle	Origin	Insertion	Action	Innervation
Masseter	Zygomatic arch	Lateral surface of mandibular ramus	Elevates mandible and closes the jaws	Trigeminal nerve (V), mandibular division
Temporalis	Along temporal lines of skull	Coronoid process of mandible	Elevates mandible	Trigeminal nerve (V), mandibular division
Pterygoids (medial and lateral)	Lateral pterygoid plate	Medial surface of mandibular ramus	*Medial:* Elevates the mandible and closes the jaws, or slides the mandible from side to side (lateral excursion)	Trigeminal nerve (V), mandibular division
			Lateral: Opens jaws, protrudes mandible, or performs lateral excursion	Trigeminal nerve (V), mandibular division

Muscles of Mastication

The muscles of mastication move the mandible at the temporomandibular joint (TMJ) (Figure 11–7). The large **masseter** is the strongest jaw muscle. The **temporalis** assists in elevating the mandible. You can feel these muscles in action by clenching your teeth while resting your hand on the side of your face below and then above the zygomatic arch. The **pterygoid** (TER-ih-goyd) muscles, used in various combinations, can elevate, depress, or protract the mandible or slide it from side to side, a movement called *lateral excursion.* These movements are important in making efficient use of your teeth while you chew foods of various consistencies. Now go ahead and compare Figure 11–7 with the summary information on these muscles in Table 11–4.

Tips & Tools

The medial and lateral *ptery*goid muscles are named for their origin on the pterygoid ("winged") portion of the sphenoid bone. The *ptero*dactyl was a prehistoric winged reptile.

11

Figure 11–8 Muscles of the Tongue.

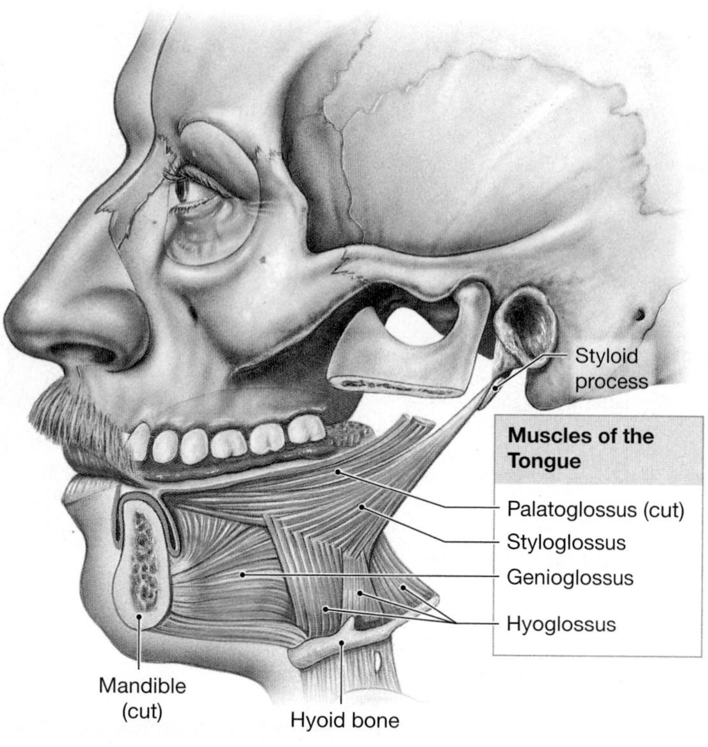
Styloid process
Muscles of the Tongue
Palatoglossus (cut)
Styloglossus
Genioglossus
Hyoglossus
Mandible (cut)
Hyoid bone

Figure 11–9 Muscles of the Pharynx.

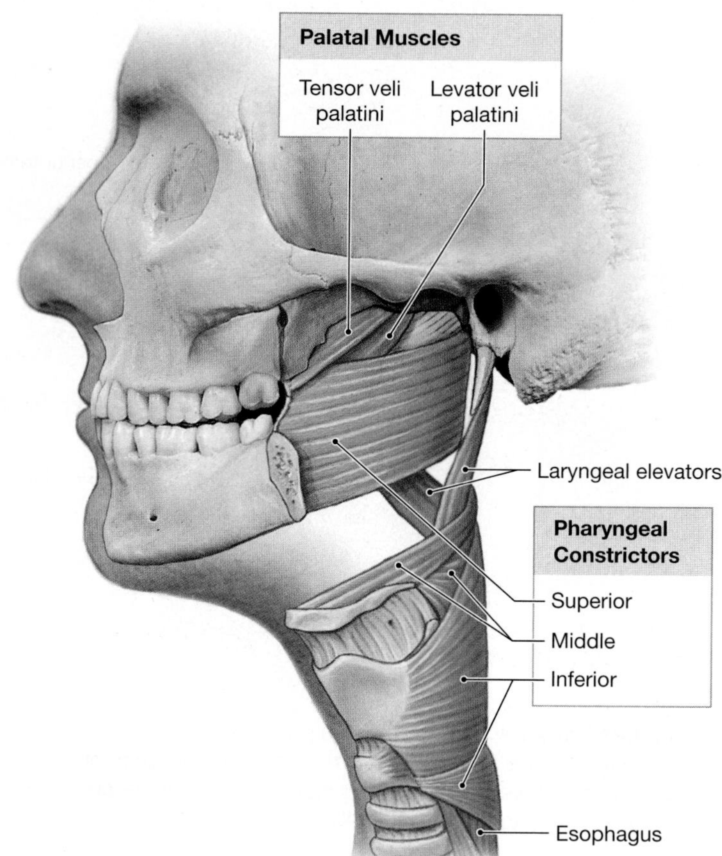
Palatal Muscles
Tensor veli palatini
Levator veli palatini
Laryngeal elevators
Pharyngeal Constrictors
Superior
Middle
Inferior
Esophagus

Muscles of the Tongue

The four muscles of the tongue have names ending in *glossus*, the Greek word for "tongue." The **palatoglossus** originates at the palate, the **styloglossus** at the styloid process of the temporal bone, the **genioglossus** at the chin, and the **hyoglossus** (hī-ō-GLOS-us) at the hyoid bone (Figure 11–8). These muscles, used in various combinations, move the tongue in the delicate and complex patterns necessary for speech. They also maneuver food within the mouth in preparation for swallowing. Now please compare the illustrations of these muscles in Figure 11–8 with the summary in Table 11–5.

Muscles of the Pharynx

The muscles of the pharynx are responsible for initiating the swallowing process (Figure 11–9). The **pharyngeal** (fa-RIN-jē-al) **constrictor** muscles (*superior*, *middle*, and *inferior*) move food (in the form of a compacted mass called a *bolus*) into the esophagus by constricting the pharyngeal walls. The two **palatal** muscles—the *tensor veli palatini* and the *levator veli palatini*—elevate the soft palate and adjacent portions of the pharyngeal wall and also pull open the entrance to the auditory tube. The **laryngeal elevators** raise the larynx. As a result, swallowing repeatedly can help you adjust to pressure changes when you fly or dive by opening the entrance to the auditory tube. Used together, the illustrations in Figure 11–9 and the summary in Table 11–6 will give you a more complete understanding of these muscles.

Table 11–5 Muscles of the Tongue (Figure 11–8)

Muscle	Origin	Insertion	Action	Innervation
Genioglossus	Medial surface of mandible around chin	Body of tongue, hyoid bone	Depresses and protracts tongue	Hypoglossal nerve (XII)
Hyoglossus	Body and greater horn of hyoid bone	Side of tongue	Depresses and retracts tongue	Hypoglossal nerve (XII)
Palatoglossus	Anterior surface of soft palate	Side of tongue	Elevates tongue, depresses soft palate	Internal branch of accessory nerve (XI)
Styloglossus	Styloid process of temporal bone	Along the side to tip and base of tongue	Retracts tongue, elevates side of tongue	Hypoglossal nerve (XII)

Table 11–6 Muscles of the Pharynx (Figure 11–9)

Muscle	Origin	Insertion	Action	Innervation
PHARYNGEAL CONSTRICTORS				
Superior constrictor	Pterygoid process of sphenoid, medial surfaces of mandible	Median raphe attached to occipital bone	Constricts pharynx to propel bolus into esophagus	Branches of pharyngeal plexus (from CN X)
Middle constrictor	Horns of hyoid bone	Median raphe	Constricts pharynx to propel bolus into esophagus	Branches of pharyngeal plexus (from CN X)
Inferior constrictor	Cricoid and thyroid cartilages of larynx	Median raphe	Constricts pharynx to propel bolus into esophagus	Branches of pharyngeal plexus (from CN X)
PALATAL MUSCLES				
Levator veli palatini	Petrous part of temporal bone; tissues around the auditory tube	Soft palate	Elevates soft palate	Branches of pharyngeal plexus (from CN X)
Tensor veli palatini	Sphenoidal spine; tissues around the auditory tube	Soft palate	Elevates soft palate	Trigeminal nerve (V)
LARYNGEAL ELEVATORS*				
	Ranges from soft palate, to cartilage around inferior portion of auditory tube, to styloid process of temporal bone	Thyroid cartilage	Elevate larynx	Branches of pharyngeal plexus (from CN IX and CN X)

*Refers to the palatopharyngeus, salpingopharyngeus, and stylopharyngeus, assisted by the thyrohyoid, geniohyoid, stylohyoid, and hyoglossus, discussed in Tables 11–5 and 11–7.

Muscles of the Anterior Neck

The anterior muscles of the neck include (1) muscles that control the position of the larynx, (2) muscles that depress the mandible and tense the floor of the mouth, and (3) muscles that provide a stable foundation for muscles of the tongue and pharynx (Figure 11–10).

The **digastric** (dī-GAS-trik) is one of the main muscles that controls the position of the larynx. This muscle has two bellies, as the name implies (*di-*, two + *gaster*, stomach). The anterior belly extends from the chin to the hyoid bone. The posterior belly continues from the hyoid bone to the mastoid portion of the temporal bone. Depending on which belly contracts and whether fixator muscles are stabilizing the position of the hyoid bone, the digastric can open the mouth by depressing the mandible, or it can elevate the larynx by raising the hyoid bone.

The digastric covers the broad, flat **mylohyoid**, which elevates the floor of the mouth or depresses the jaw when the

+ Clinical Note Intramuscular Injections

Drugs are commonly injected into muscle or adipose tissues rather than directly into the bloodstream. (Accessing blood vessels is more complicated.) An **intramuscular (IM) injection** introduces a drug into the mass of a large skeletal muscle. Depending on the size of the muscle, up to 5 mL of fluid may be injected at one time. This fairly large volume of drug then enters the circulation gradually. Uptake is generally faster and accompanied by less tissue irritation than when drugs are administered *intradermally* (injected into the dermis) or *subcutaneously* (injected into the subcutaneous layer). A decision on the injection technique and the injection site is based on the type of drug and its concentration.

For IM injections, the most common complications involve accidental injection into a blood vessel or nerve. The sudden entry of massive quantities of drug into the bloodstream can have fatal consequences. Damage to a nerve can cause motor paralysis or sensory loss. For these reasons, the site of the injection must be selected with care. Bulky muscles that contain few large vessels or nerves are ideal sites. The gluteus medius or the posterior, lateral, superior part of the gluteus maximus is commonly selected. The deltoid of the arm, about 2.5 cm (1 in.) distal to the acromion, is another effective site. From a technical point of view, the vastus lateralis of the thigh is a good site. Injections into this thick muscle will not encounter vessels or nerves, but may cause pain later when the muscle is used in walking. This is the preferred injection site in infants before they start walking, as their gluteal and deltoid muscles are relatively small. The site is also used in elderly patients or others with atrophied gluteal and deltoid muscles.

hyoid bone is fixed. It is aided by the deeper **geniohyoid** (jē-nē-ō-HĪ-oyd) that extends between the hyoid bone and the chin. The **stylohyoid** forms a muscular connection between the hyoid bone and the styloid process of the skull.

The **sternocleidomastoid** (ster-nō-klī-dō-MAS-toyd) extends from the clavicle and the sternum to the mastoid region of the skull and turns the head obliquely to the opposite side (see Figure 11–10). The **omohyoid** attaches to the scapula, the clavicle and first rib, and the hyoid bone. The other members of this group are strap-like muscles that extend between the sternum and larynx (*sternothyroid*) or hyoid bone (*sternohyoid*), and between the larynx and hyoid bone (*thyrohyoid*).

11

Now take a look at the muscles discussed here and depicted in Figure 11–10 with the summary details in Table 11–7.

Muscles of the Vertebral Column

The muscles of the vertebral column are covered by more superficial back muscles, such as the trapezius and latissimus dorsi (look back at Figure 11–4b). The main muscles of this group are the **erector spinae** (SPĪ-nē) muscles, which include superficial and deep layers. The superficial layer can be divided into **spinalis**, **longissimus** (lon-JIS-ih-mus), and **iliocostalis** (ē-lē-ō-kos-TAL-is) groups (Figure 11–11). In the inferior lumbar and sacral regions, the boundary between the longissimus and iliocostalis is indistinct. When contracting together, the erector spinae extend the vertebral column. When the muscles on only one side contract, the result is lateral flexion of the vertebral column.

Figure 11–10 Muscles of the Anterior Neck. ATLAS: Plates 3a–d; 17; 18a–c; 25

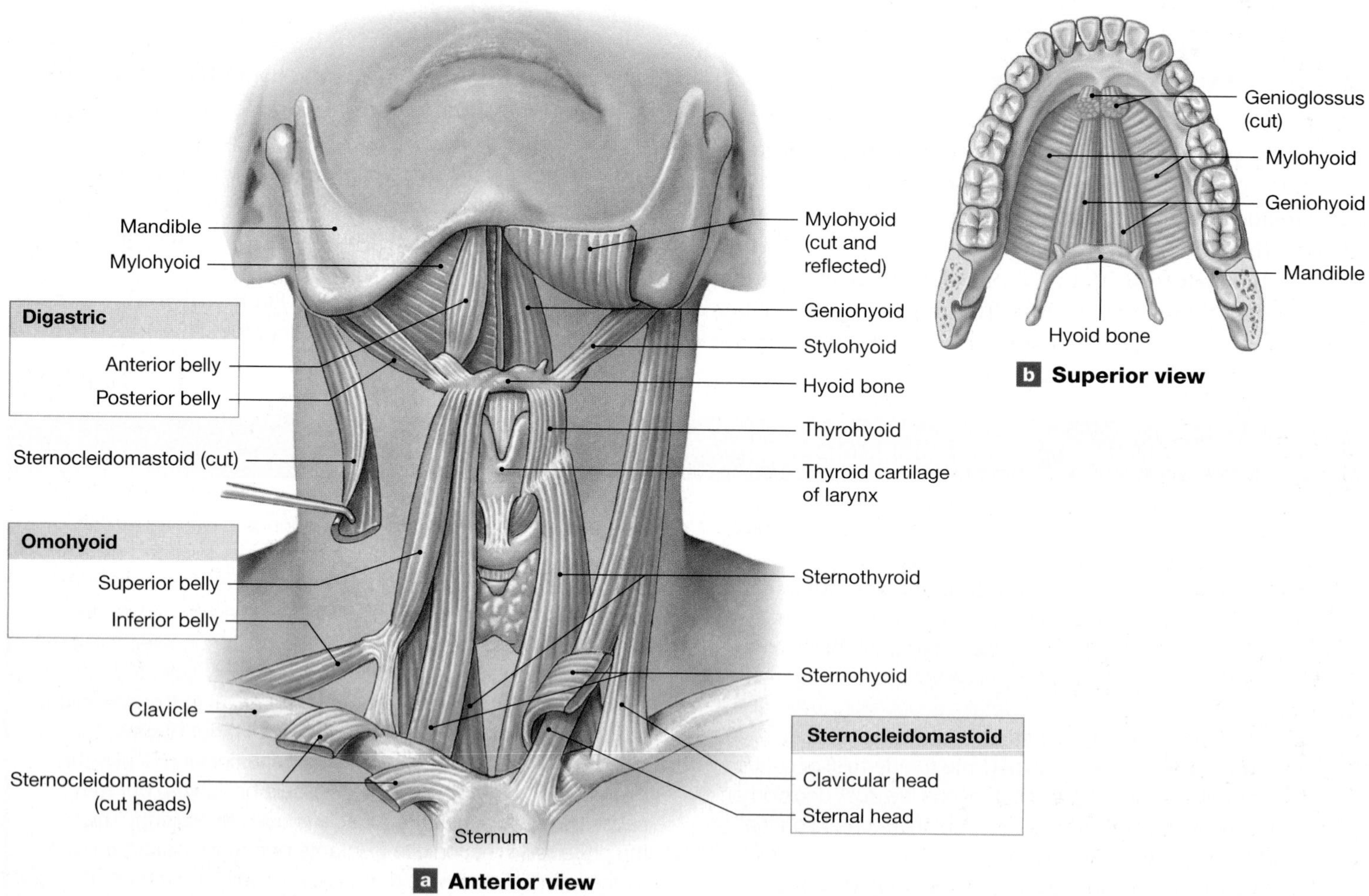

Mandible
Mylohyoid
Digastric
Anterior belly
Posterior belly
Sternocleidomastoid (cut)
Omohyoid
Superior belly
Inferior belly
Clavicle
Sternocleidomastoid (cut heads)
Sternum
Mylohyoid (cut and reflected)
Geniohyoid
Stylohyoid
Hyoid bone
Thyrohyoid
Thyroid cartilage of larynx
Sternothyroid
Sternohyoid
Sternocleidomastoid
Clavicular head
Sternal head
a Anterior view
Genioglossus (cut)
Mylohyoid
Geniohyoid
Mandible
Hyoid bone
b Superior view

? Which muscles insert on the hyoid bone?

Table 11–7 Muscles of the Anterior Neck (Figure 11–10)

Muscle	Origin	Insertion	Action	Innervation
Digastric	Two bellies: *anterior* from inferior surface of mandible at chin; *posterior* from mastoid region of temporal bone	Hyoid bone	Depresses mandible or elevates larynx	*Anterior belly:* Trigeminal nerve (V), mandibular division *Posterior belly:* Facial nerve (VII)
Geniohyoid	Medial surface of mandible at chin	Hyoid bone	As above and pulls hyoid bone anteriorly	Cervical nerve C_1 by hypoglossal nerve (XII)
Mylohyoid	Mylohyoid line of mandible	Median connective tissue band (raphe) that runs to hyoid bone	Elevates floor of mouth and hyoid bone or depresses mandible	Trigeminal nerve (V), mandibular division
Omohyoid (superior and inferior bellies united at central tendon anchored to clavicle and first rib)	Superior border of scapula near scapular notch	Hyoid bone	Depresses hyoid bone and larynx	Cervical spinal nerves C_2–C_3
Sternohyoid	Clavicle and manubrium	Hyoid bone	Depresses hyoid bone and larynx	Cervical spinal nerves C_1–C_3
Sternothyroid	Dorsal surface of manubrium and first costal cartilage	Thyroid cartilage of larynx	Depresses hyoid bone and larynx	Cervical spinal nerves C_1–C_3
Stylohyoid	Styloid process of temporal bone	Hyoid bone	Elevates larynx	Facial nerve (VII)
Thyrohyoid	Thyroid cartilage of larynx	Hyoid bone	Elevates thyroid, depresses hyoid bone	Cervical spinal nerves C_1–C_2 by hypoglossal nerve (XII)
Sternocleidomastoid	Two bellies: *clavicular head* attaches to sternal end of clavicle; *sternal head* attaches to manubrium	Mastoid region of skull and lateral portion of superior nuchal line	Together, they flex the neck; alone, one side flexes head toward shoulder and rotates face to opposite side	Accessory nerve (XI) and cervical spinal nerves (C_2–C_3) of cervical plexus

Deep to the spinalis muscles, smaller muscles interconnect and stabilize the vertebrae. These muscles include the **semispinalis** group; the **multifidus** (mul-TIF-ih-dus); and the **interspinales**, **intertransversarii** (in-ter-tranz-ver-SAR-ē-ī), and **rotatores** (rō-teh-TOR-ēz) (see Figure 11–11). In various combinations, they produce slight extension or rotation of the vertebral column. They are also important in making delicate adjustments in the positions of individual vertebrae, and they stabilize adjacent vertebrae. If injured, these muscles can start a cycle of pain → muscle stimulation → contraction → pain. Resultant pressure on adjacent spinal nerves can lead to sensory losses and mobility limitations. Many of the warm-up and stretching exercises recommended before athletic activity are intended to prepare these small but very important muscles for their supporting role.

The muscles of the vertebral column include many posterior extensors, but few anterior flexors. Why doesn't the vertebral column have massive **flexor** muscles? One reason is that many of the large trunk muscles flex the vertebral column when they contract. A second reason is that most of the body weight lies anterior to the vertebral column, so gravity tends to flex the spine. However, a few spinal flexors are associated with the anterior surface of the vertebral column. In the neck, the **longus capitis** (KAP-ih-tus) and the **longus colli** rotate or flex the neck, depending on whether the muscles of one or both sides are contracting (see Figure 11–11). In the lumbar region, the large **quadratus** (kwad-RĀ-tus) **lumborum** flexes the vertebral column and depresses the ribs.

Now use the illustrations in Figure 11–11 together with the details in Table 11–8 for a further understanding of these muscles.

+ Clinical Note Signs of Stroke

A *stroke*, or *cerebrovascular accident*, may manifest in the muscular system. This happens, for example, when a clot in a blood vessel interrupts blood flow to the brain, and a muscle does not receive the signal to contract. Early recognition and treatment of a stroke can make all the difference in the outcome of this threatening disorder. Public service announcements by the National Stroke Association alert the public about stroke by means of a catchy phrase—"Act FAST"—where

- **F** stands for **F**ace: Ask the person to smile and look for a droop in the facial muscles on the affected side.
- **A** stands for **A**rm: Ask the person to raise both arms and see if he or she has the strength to do this.
- **S** stands for **S**peech: Ask the person to repeat a few words and listen for slurred speech to determine if pharyngeal and laryngeal muscles are affected.
- **T** stands for **T**ime: If you observe any of these signs, it's time to call 9-1-1 for help.

Figure 11–11 Muscles of the Vertebral Column.

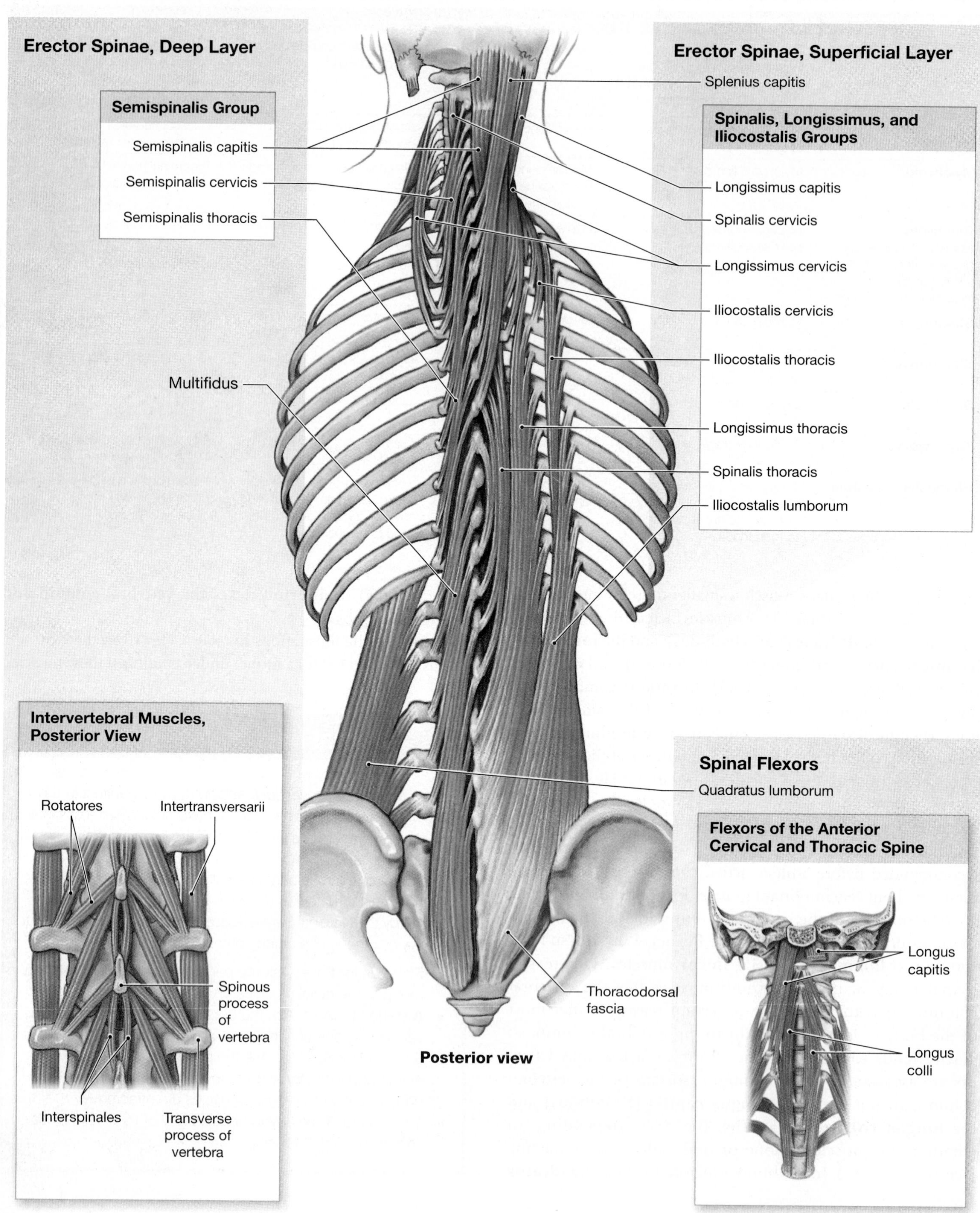
Erector Spinae, Deep Layer
Semispinalis Group
Semispinalis capitis
Semispinalis cervicis
Semispinalis thoracis
Multifidus
Erector Spinae, Superficial Layer
Splenius capitis
Spinalis, Longissimus, and Iliocostalis Groups
Longissimus capitis
Spinalis cervicis
Longissimus cervicis
Iliocostalis cervicis
Iliocostalis thoracis
Longissimus thoracis
Spinalis thoracis
Iliocostalis lumborum
Intervertebral Muscles, Posterior View
Rotatores
Intertransversarii
Spinous process of vertebra
Interspinales
Transverse process of vertebra
Spinal Flexors
Quadratus lumborum
Flexors of the Anterior Cervical and Thoracic Spine
Longus capitis
Longus colli
Thoracodorsal fascia
Posterior view

Table 11–8 Muscles of the Vertebral Column (Figure 11–11)

Group and Muscles	Origin	Insertion	Action	Innervation
SUPERFICIAL LAYER				
Splenius (splenius capitis, splenius cervicis)	Spinous processes and ligaments connecting inferior cervical and superior thoracic vertebrae	Mastoid process, occipital bone of skull, and superior cervical vertebrae	Together, the two sides extend neck; alone, each rotates and laterally flexes neck to that side	Cervical spinal nerves
Erector spinae				
Spinalis group				
Spinalis cervicis	Inferior portion of ligamentum nuchae and spinous process of C_7	Spinous process of axis	Extends neck	Cervical spinal nerves
Spinalis thoracis	Spinous processes of inferior thoracic and superior lumbar vertebrae	Spinous processes of superior thoracic vertebrae	Extends vertebral column	Thoracic and lumbar spinal nerves
Longissimus group				
Longissimus capitis	Transverse processes of inferior cervical and superior thoracic vertebrae	Mastoid process of temporal bone	Together, the two sides extend head; alone, each rotates and laterally flexes neck to that side	Cervical and thoracic spinal nerves
Longissimus cervicis	Transverse processes of superior thoracic vertebrae	Transverse processes of middle and superior cervical vertebrae	Together, the two sides extend head; alone, each rotates and laterally flexes neck to that side	Cervical and thoracic spinal nerves
Longissimus thoracis	Broad aponeurosis and transverse processes of inferior thoracic and superior lumbar vertebrae; joins iliocostalis	Transverse processes of superior vertebrae and inferior surfaces of ribs	Together, the two sides extend vertebral column; alone, each produces lateral flexion to that side	Thoracic and lumbar spinal nerves
Iliocostalis group				
Iliocostalis cervicis	Superior borders of vertebrosternal ribs near the angles	Transverse processes of middle and inferior cervical vertebrae	Extends or laterally flexes neck, elevates ribs	Cervical and superior thoracic spinal nerves
Iliocostalis thoracis	Superior borders of inferior seven ribs medial to the angles	Upper ribs and transverse process of last cervical vertebra	Stabilizes thoracic vertebrae in extension	Thoracic spinal nerves
Iliocostalis lumborum	Iliac crest, sacral crests, and spinous processes	Inferior surfaces of inferior seven ribs near their angles	Extends vertebral column, depresses ribs	Inferior thoracic and lumbar spinal nerves
DEEP LAYER				
Semispinalis group				
Semispinalis capitis	Articular processes of inferior cervical and transverse processes of superior thoracic vertebrae	Occipital bone, between nuchal lines	Together, the two sides extend head; alone, each extends and laterally flexes neck	Cervical spinal nerves
Semispinalis cervicis	Transverse processes of T_1–T_5 or T_6	Spinous processes of C_2–C_5	Extends vertebral column and rotates toward opposite side	Cervical spinal nerves
Semispinalis thoracis	Transverse processes of T_6–T_{10}	Spinous processes of C_5–T_4	Extends vertebral column and rotates toward opposite side	Thoracic spinal nerves
Multifidus	Sacrum and transverse processes of each vertebra	Spinous processes of the third or fourth more superior vertebrae	Extends vertebral column and rotates toward opposite side	Cervical, thoracic, and lumbar spinal nerves
Rotatores	Transverse processes of each vertebra	Spinous processes of adjacent, more superior vertebra	Extends vertebral column and rotates toward opposite side	Cervical, thoracic, and lumbar spinal nerves
Interspinales	Spinous processes of each vertebra	Spinous processes of more superior vertebra	Extends vertebral column	Cervical, thoracic, and lumbar spinal nerves
Intertransversarii	Transverse processes of each vertebra	Transverse process of more superior vertebra	Laterally flexes the vertebral column	Cervical, thoracic, and lumbar spinal nerves
SPINAL FLEXORS				
Longus capitis	Transverse processes of cervical vertebrae	Base of the occipital bone	Together, the two sides flex the neck; alone, each rotates head to that side	Cervical spinal nerves
Longus colli	Anterior surfaces of cervical and superior thoracic vertebrae	Transverse processes of superior cervical vertebrae	Flexes or rotates neck; limits hyperextension	Cervical spinal nerves
Quadratus lumborum	Iliac crest and iliolumbar ligament	Last rib and transverse processes of lumbar vertebrae	Together, they depress ribs; alone, each side laterally flexes vertebral column	Thoracic and lumbar spinal nerves

Oblique and Rectus Muscles and the Diaphragm

The oblique and rectus muscle groups of the anterior body wall share developmental origins. We include the diaphragm here because it develops in association with the other muscles of the chest wall.

Oblique and Rectus Muscles

The oblique and rectus muscle groups lie within the body wall, between the spinous processes of vertebrae and the ventral midline (look back at Figure 11–4a for an overview, and then take a first look at Figure 11–12). The oblique muscles compress underlying structures or rotate the vertebral column, depending

Figure 11–12 **Oblique and Rectus Muscles and the Diaphragm.** ATLAS: Plates 39b–d; 41a,b; 46

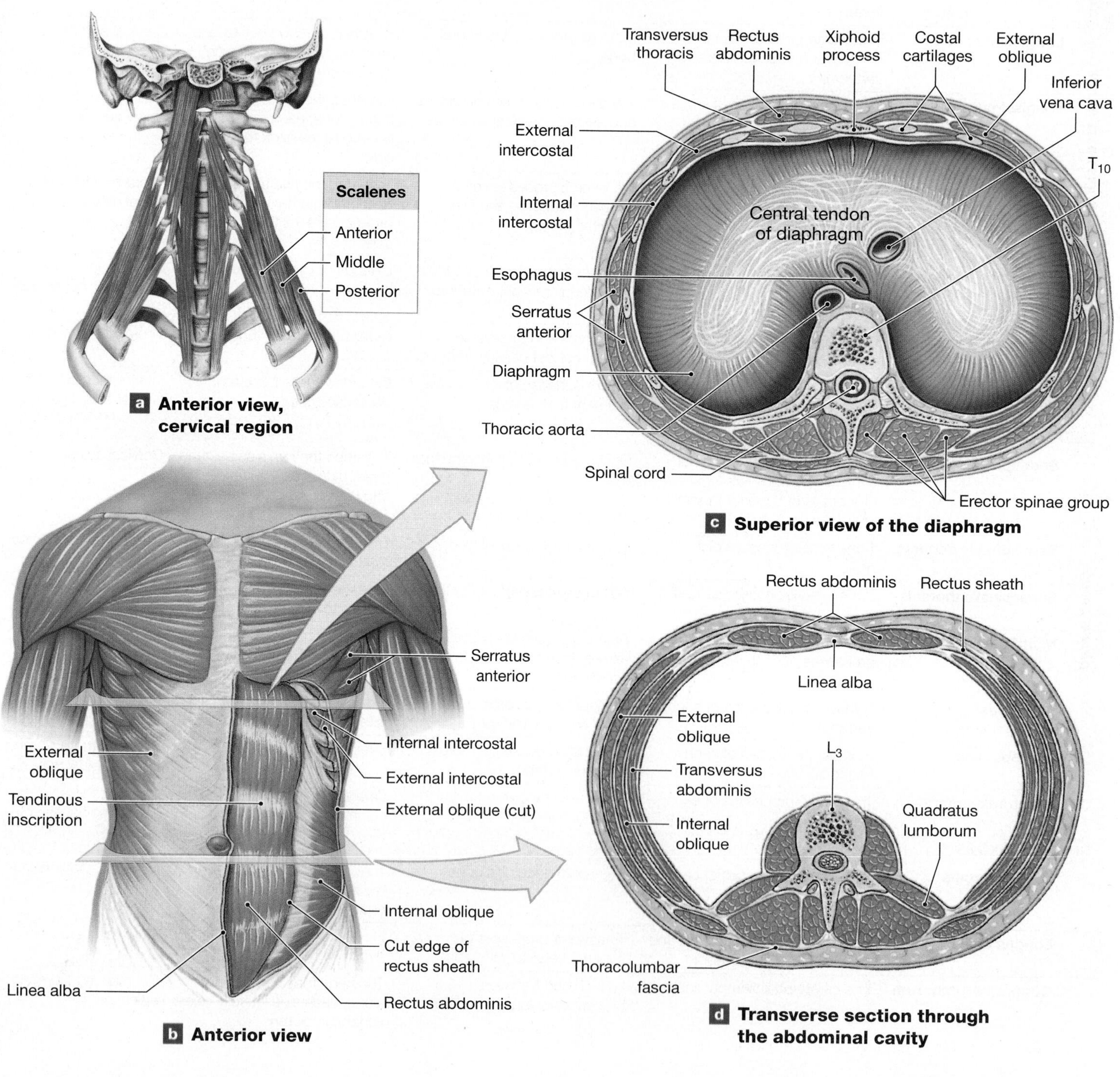

Scalenes
Anterior
Middle
Posterior
a Anterior view, cervical region
Transversus thoracis
Rectus abdominis
Xiphoid process
Costal cartilages
External oblique
Inferior vena cava
External intercostal
T10
Internal intercostal
Central tendon of diaphragm
Esophagus
Serratus anterior
Diaphragm
Thoracic aorta
Spinal cord
Erector spinae group
c Superior view of the diaphragm
Serratus anterior
External oblique
Internal intercostal
External intercostal
Tendinous inscription
External oblique (cut)
Internal oblique
Cut edge of rectus sheath
Linea alba
Rectus abdominis
b Anterior view
Rectus abdominis
Rectus sheath
Linea alba
External oblique
L3
Transversus abdominis
Quadratus lumborum
Internal oblique
Thoracolumbar fascia
d Transverse section through the abdominal cavity

on whether one or both sides contract. The rectus muscles are important flexors of the vertebral column, acting in opposition to the erector spinae. We can subdivide each of these groups into cervical, thoracic, and abdominal regions (Table 11–9).

The oblique group includes the **scalene** muscles of the neck (Figure 11–12a) and the **intercostal** and **transversus** muscles of the thorax (Figure 11–12b,c). The scalene muscles (*anterior, middle,* and *posterior*) elevate the first two ribs and assist in flexion of the neck. In the thorax, the oblique muscles extend between the ribs, with the **external intercostal muscles** covering the **internal intercostal muscles**. Both groups of intercostal muscles aid in breathing movements of the ribs. A small **transversus thoracis** (THOR-ah-sis) crosses the posterior surface of the sternum and is separated from the pleural cavity by the parietal pleura, a *serous membrane.* p. 140 The sternum occupies the place where we might otherwise expect thoracic rectus muscles to be.

The same basic pattern of musculature extends unbroken across the abdominopelvic surface (Figure 11–12b, d). Here, the muscles are the **external oblique**, **internal oblique**, **transversus abdominis**, and **rectus abdominis** (commonly called the "abs"). The rectus abdominis inserts at the xiphoid process and originates near the pubic symphysis. This muscle is longitudinally divided by the **linea alba** (white line), a median collagenous partition (see Figure 11–12b). The rectus abdominis is separated into segments by transverse

Table 11–9 Oblique and Rectus Muscle Groups (Figure 11–12)

Group and Muscles	Origin	Insertion	Action	Innervation*
OBLIQUE GROUP				
Cervical region				
Scalenes (anterior, middle, and posterior)	Transverse and costal processes of cervical vertebrae	Superior surfaces of first two ribs	Elevate ribs or flex neck	Cervical nerves
Thoracic region				
External intercostals	Inferior border of each rib	Superior border of more inferior rib	Elevate ribs	Intercostal nerves (branches of thoracic nerves)
Internal intercostals	Superior border of each rib	Inferior border of the preceding rib	Depress ribs	Intercostal nerves (branches of thoracic nerves)
Transversus thoracis	Posterior surface of sternum	Cartilages of ribs	Depress ribs	Intercostal nerves (branches of thoracic nerves)
Serratus posterior superior (*Figure 11–14b*)	Spinous processes of C_7–T_3 and ligamentum nuchae	Superior borders of ribs 2–5 near angles	Elevates ribs, enlarges thoracic cavity	Thoracic nerves (T_1–T_4)
Serratus posterior inferior (*Figure 11–14b*)	Aponeurosis from spinous processes of T_{10}–L_3	Inferior borders of ribs 8–12	Pulls ribs inferiorly; also pulls outward, opposing diaphragm	Thoracic nerves (T_9–T_{12})
Abdominal region				
External oblique	External and inferior borders of ribs 5–12	Linea alba and iliac crest	Compresses abdomen, depresses ribs, flexes or bends spine	Intercostal, iliohypogastric, and ilioinguinal nerves
Internal oblique	Thoracolumbar fascia and iliac crest	Inferior ribs, xiphoid process, and linea alba	Compresses abdomen, depresses ribs, flexes or bends spine	Intercostal, iliohypogastric, and ilioinguinal nerves
Transversus abdominis	Cartilages of ribs 6–12, iliac crest, and thoracolumbar fascia	Linea alba and pubis	Compresses abdomen	Intercostal, iliohypogastric, and ilioinguinal nerves
RECTUS GROUP				
Cervical region	*See muscles in Table 11–6*			
Thoracic region				
Diaphragm	Xiphoid process, cartilages of ribs 4–10, and anterior surfaces of lumbar vertebrae	Central tendinous sheet	Contraction expands thoracic cavity and compresses abdominopelvic cavity	Phrenic nerve (C_3–C_5)
Abdominal region				
Rectus abdominis	Superior surface of pubis around symphysis	Inferior surfaces of costal cartilages (ribs 5–7) and xiphoid process	Depresses ribs, flexes vertebral column, compresses abdomen	Intercostal nerves (T_7–T_{12})

*Where appropriate, spinal nerves involved are given in parentheses.

Figure 11–13 Muscles of the Pelvic Floor.

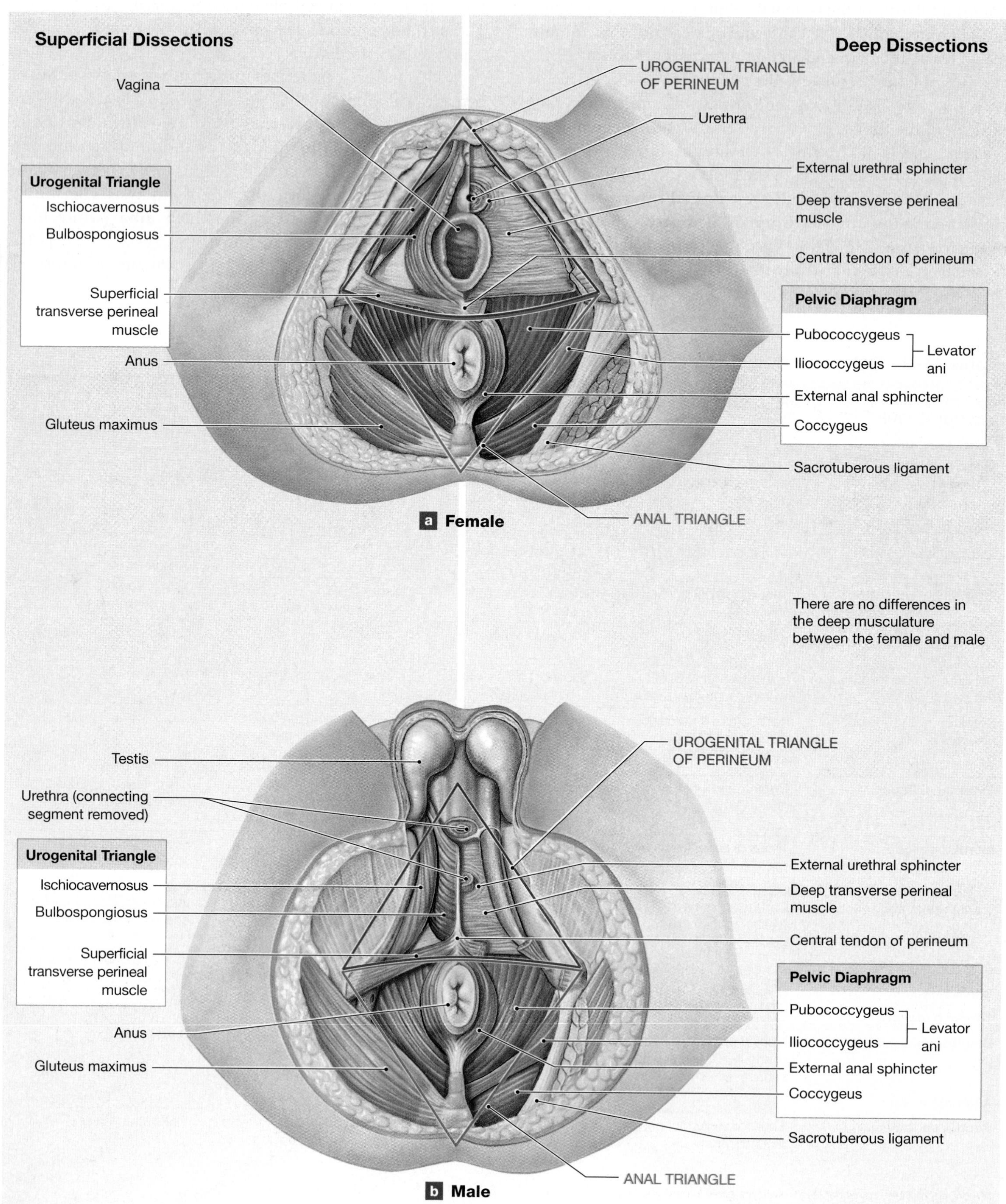
Superficial Dissections
Deep Dissections
Vagina
UROGENITAL TRIANGLE OF PERINEUM
Urethra
Urogenital Triangle
Ischiocavernosus
Bulbospongiosus
Superficial transverse perineal muscle
External urethral sphincter
Deep transverse perineal muscle
Central tendon of perineum
Pelvic Diaphragm
Pubococcygeus
Iliococcygeus
Levator ani
Anus
External anal sphincter
Gluteus maximus
Coccygeus
Sacrotuberous ligament
ANAL TRIANGLE
a Female
There are no differences in the deep musculature between the female and male
Testis
Urethra (connecting segment removed)
UROGENITAL TRIANGLE OF PERINEUM
Urogenital Triangle
Ischiocavernosus
Bulbospongiosus
Superficial transverse perineal muscle
External urethral sphincter
Deep transverse perineal muscle
Central tendon of perineum
Pelvic Diaphragm
Pubococcygeus
Iliococcygeus
Levator ani
Anus
Gluteus maximus
External anal sphincter
Coccygeus
Sacrotuberous ligament
ANAL TRIANGLE
b Male

bands of collagen fibers called **tendinous inscriptions**. Each segment contains muscle fibers that extend longitudinally, originating and inserting on the tendinous inscriptions. The bulging of enlarged muscle fibers of the rectus abdominis between these tendinous inscriptions produces what we call a "six-pack."

The Diaphragm

The term *diaphragm* refers to any muscular sheet that forms a wall. When used without a modifier, however, **diaphragm** specifies the muscular partition that separates the abdominopelvic and thoracic cavities (see Figure 11–12c). The diaphragm is a major muscle used in breathing and is further discussed in Chapter 23.

Now that you have been introduced to Figure 11–12, please study it together with the summary details presented in Table 11–9.

Muscles of the Pelvic Floor

The muscles of the pelvic floor extend from the sacrum and coccyx to the ischium and pubis (Figure 11–13). These muscles (1) support the organs of the pelvic cavity, (2) flex the sacrum and coccyx, and (3) control the movement of materials through the urethra and anus. They are summarized in Table 11–10.

The boundaries of the **perineum** are formed by the inferior margins of the pelvis. A line drawn between the ischial

Table 11–10 Muscles of the Pelvic Floor (Figure 11–13)

	Group and Muscle	Origin	Insertion	Action	Innervation*
	UROGENITAL TRIANGLE				
Superficial muscles	**Bulbospongiosus**				
	Males	Collagen sheath at base of penis; fibers cross over urethra	Median raphe and central tendon of perineum	Compresses base and stiffens penis; ejects urine or semen	Pudendal nerve, perineal branch (S_2–S_4)
	Females	Collagen sheath at base of clitoris; fibers run on either side of urethral and vaginal opening	Central tendon of perineum	Compresses and stiffens clitoris; narrows vaginal opening	Pudendal nerve, perineal branch (S_2–S_4)
	Ischiocavernosus	Ischial ramus and tuberosity	Pubic symphysis anterior to base of penis or clitoris	Compresses and stiffens penis or clitoris	Pudendal nerve, perineal branch (S_2–S_4)
	Superficial transverse perineal muscle	Ischial ramus	Central tendon of perineum	Stabilizes central tendon of perineum	Pudendal nerve, perineal branch (S_2–S_4)
Deep muscles	**Deep transverse perineal muscle**	Ischial ramus	Central tendon of perineum	Stabilizes central tendon of perineum	Pudendal nerve, perineal branch (S_2–S_4)
	External urethral sphincter				
	Males	Ischial and pubic rami	To median raphe at base of penis; inner fibers encircle urethra	Closes urethra; compresses prostate and bulbourethral glands	Pudendal nerve, perineal branch (S_2–S_4)
	Females	Ischial and pubic rami	To median raphe; inner fibers encircle urethra	Closes urethra; compresses vagina and greater vestibular glands	Pudendal nerve, perineal branch (S_2–S_4)
	ANAL TRIANGLE				
	Pelvic diaphragm				
	Coccygeus	Ischial spine	Lateral, inferior borders of sacrum and coccyx	Flexes coccygeal joints; tenses and supports pelvic floor	Inferior sacral nerves (S_4–S_5)
	Levator ani **Iliococcygeus**	Ischial spine, pubis	Coccyx and median raphe	Tenses floor of pelvis; flexes coccygeal joints; elevates and retracts anus	Pudendal nerve (S_2–S_4)
	Pubococcygeus	Inner margins of pubis	Coccyx and median raphe	Tenses floor of pelvis; flexes coccygeal joints; elevates and retracts anus	Pudendal nerve (S_2–S_4)
	External anal sphincter	By tendon from coccyx	Encircles anal opening	Closes anal opening	Pudendal nerve, hemorrhoidal branch (S_2–S_4)

*Where appropriate, spinal nerves involved are given in parentheses.

tuberosities divides the perineum into two triangles: an anterior **urogenital triangle** and a posterior **anal triangle** (see Figure 11–13). The superficial muscles of the urogenital triangle are the muscles of the external genitalia. They cover deeper muscles that strengthen the pelvic floor and encircle the urethra. An even more extensive muscular sheet, the **pelvic diaphragm**, forms the muscular foundation of the anal triangle. This layer extends as far as the pubic symphysis.

The urogenital and pelvic diaphragms do not completely close the pelvic outlet. The urethra and anus (in males and females), as well as the vagina in females, pass through them to open on the exterior. Muscular sphincters surround the passageways, and the external sphincters permit voluntary control of urination and defecation. Muscles, nerves, and blood vessels also pass through the pelvic outlet as they travel to or from the lower limbs.

Once again, please compare the illustrations in Figure 11–13 with the specific details in Table 11–10.

11

Checkpoint

13. **If you were contracting and relaxing your masseter muscle, what would you probably be doing?**
14. **Which facial muscle is well developed in a professional trumpet player?**
15. **Why can swallowing help alleviate the pressure sensations at the eardrum when you are in an airplane that is changing altitude?**
16. **Damage to the external intercostal muscles would interfere with what important process?**
17. **If someone hit you in your rectus abdominis, how would your body position change?**
18. **After spending an afternoon carrying heavy boxes from his basement to his attic, Joe complains that the muscles in his back hurt. Which muscles are most likely sore?**

See the blue Answers tab at the back of the book.

11-7 Appendicular muscles are muscles of the shoulders, upper limbs, pelvis, and lower limbs

Learning Outcome Identify the principal appendicular muscles of the body, plus their origins, insertions, actions, and innervation, and compare the major functional differences between the upper and lower limbs. They are summarized in Table 11–10.

The appendicular musculature positions and stabilizes the pectoral and pelvic girdles and moves the upper and lower limbs. There are two major groups of appendicular muscles: (1) *the muscles of the shoulders (pectoral girdles) and upper limbs* and (2) *the muscles of the pelvis (pelvic girdle) and lower limbs.*

The functions and required ranges of motion are very different between these groups. The muscular connections between the pectoral girdles and the axial skeleton increase the mobility of the arms and also act as shock absorbers. For example, while you jog, you can still perform delicate hand movements, because the muscular connections between the axial and appendicular components of the skeleton smooth out the bounces in your stride. In contrast, the pelvic girdle has evolved to transfer weight from the axial to the appendicular skeleton. Rigid, bony articulations are essential, because the emphasis is on strength rather than versatility, and a muscular connection would reduce the efficiency of the transfer.

Figure 11–14 and Table 11–11 provide an introduction to the organization of the appendicular muscles of the trunk.

Table 11–11 Muscles That Position the Pectoral Girdle (Figures 11–14 and 11–15)

Muscle	Origin	Insertion	Action	Innervation*
Levator scapulae	Transverse processes of first four cervical vertebrae	Vertebral border of scapula near superior angle	Elevates scapula	Cervical nerves C_3–C_4 and dorsal scapular nerve (C_5)
Pectoralis minor	Anterior–superior surfaces of ribs 2–4, 2–5, or 3–5 depending on anatomical variation	Coracoid process of scapula	Depresses and protracts shoulder; rotates scapula so glenoid cavity moves inferiorly (downward rotation); elevates ribs if scapula is stationary	Medial pectoral nerve (C_8, T_1)
Rhomboid major	Spinous processes of superior thoracic vertebrae	Vertebral border of scapula from spine to inferior angle	Adducts scapula and performs downward rotation	Dorsal scapular nerve (C_5)
Rhomboid minor	Spinous processes of vertebrae C_7–T_1	Vertebral border of scapula near spine	Adducts scapula and performs downward rotation	Dorsal scapular nerve (C_5)
Serratus anterior	Anterior and superior margins of ribs 1–8 or 1–9	Anterior surface of vertebral border of scapula	Protracts shoulder; rotates scapula so glenoid cavity moves superiorly (upward rotation)	Long thoracic nerve (C_5–C_7)
Subclavius	First rib	Clavicle (inferior border)	Depresses and protracts shoulder	Nerve to subclavius (C_5–C_6)
Trapezius	Occipital bone, ligamentum nuchae, and spinous processes of thoracic vertebrae	Clavicle and scapula (acromion and scapular spine)	Depends on active region and state of other muscles; may (1) elevate, retract, depress, or rotate scapula upward, (2) elevate clavicle, or (3) extend neck	Accessory nerve (XI) and cervical nerves (C_3–C_4)

*Where appropriate, spinal nerves involved are given in parentheses.

Go to MasteringA&P™ > Study Area > Menu > Animations & Videos > A&PFlix > Origins, Insertions, Actions, Innervations: Rectus abdominis

Figure 11–14 An Overview of the Appendicular Muscles of the Trunk.

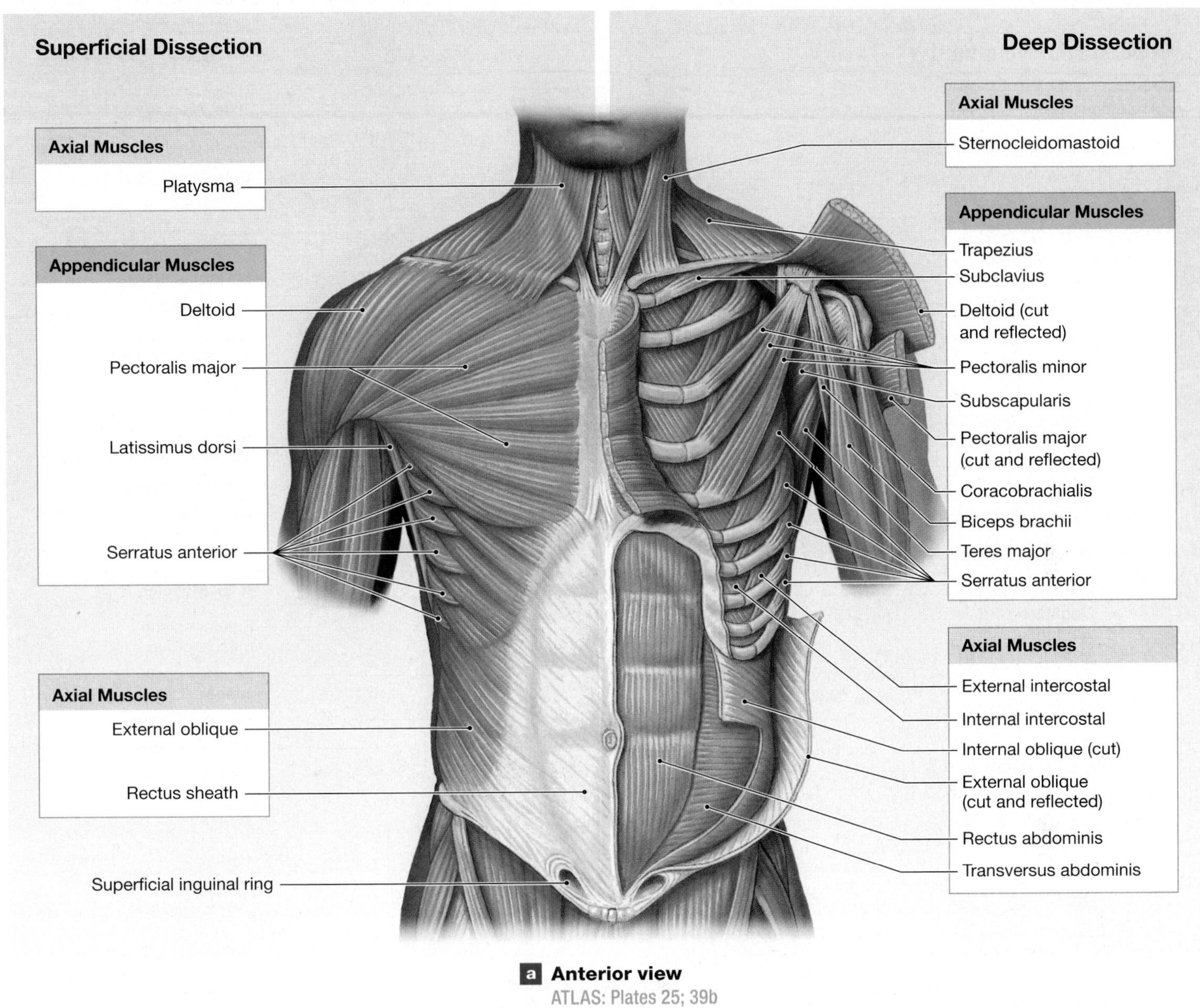
Superficial Dissection
Axial Muscles
Platysma
Appendicular Muscles
Deltoid
Pectoralis major
Latissimus dorsi
Serratus anterior
Axial Muscles
External oblique
Rectus sheath
Superficial inguinal ring
Deep Dissection
Axial Muscles
Sternocleidomastoid
Appendicular Muscles
Trapezius
Subclavius
Deltoid (cut and reflected)
Pectoralis minor
Subscapularis
Pectoralis major (cut and reflected)
Coracobrachialis
Biceps brachii
Teres major
Serratus anterior
Axial Muscles
External intercostal
Internal intercostal
Internal oblique (cut)
External oblique (cut and reflected)
Rectus abdominis
Transversus abdominis

a **Anterior view**
ATLAS: Plates 25; 39b

Muscles of the Shoulders and Upper Limbs

We can divide the muscles associated with the shoulders and upper limbs into four groups: (1) *muscles that position the pectoral girdle*, (2) *muscles that move the arm*, (3) *muscles that move the forearm and hand*, and (4) *muscles that move the fingers*.

Muscles That Position the Pectoral Girdle

The large, superficial **trapezius**, commonly called the "traps," cover the back and portions of the neck, reaching to the base of the skull. This muscle originates along the midline of the neck and back and inserts on the clavicles and the scapular spines (Figure 11–15). The trapezius is innervated by more than one nerve. For this reason, specific regions can be made to contract independently, and their actions are quite varied.

On the chest, the **serratus** (seh-RĀ-tus) **anterior** originates along the anterior surfaces of several ribs (see Figure 11–15a,b). This fan-shaped muscle inserts along the anterior margin of the vertebral border of the scapula. When the serratus anterior contracts, it abducts (protracts) the scapula and swings the shoulder anteriorly.

Two other deep chest muscles arise along the anterior surfaces of the ribs on either side. The **subclavius** (sub-KLĀ-vē-us; *sub-*, below + *clavius*, clavicle) inserts on the inferior border of the

Figure 11–14 An Overview of the Appendicular Muscles of the Trunk. (*continued*)

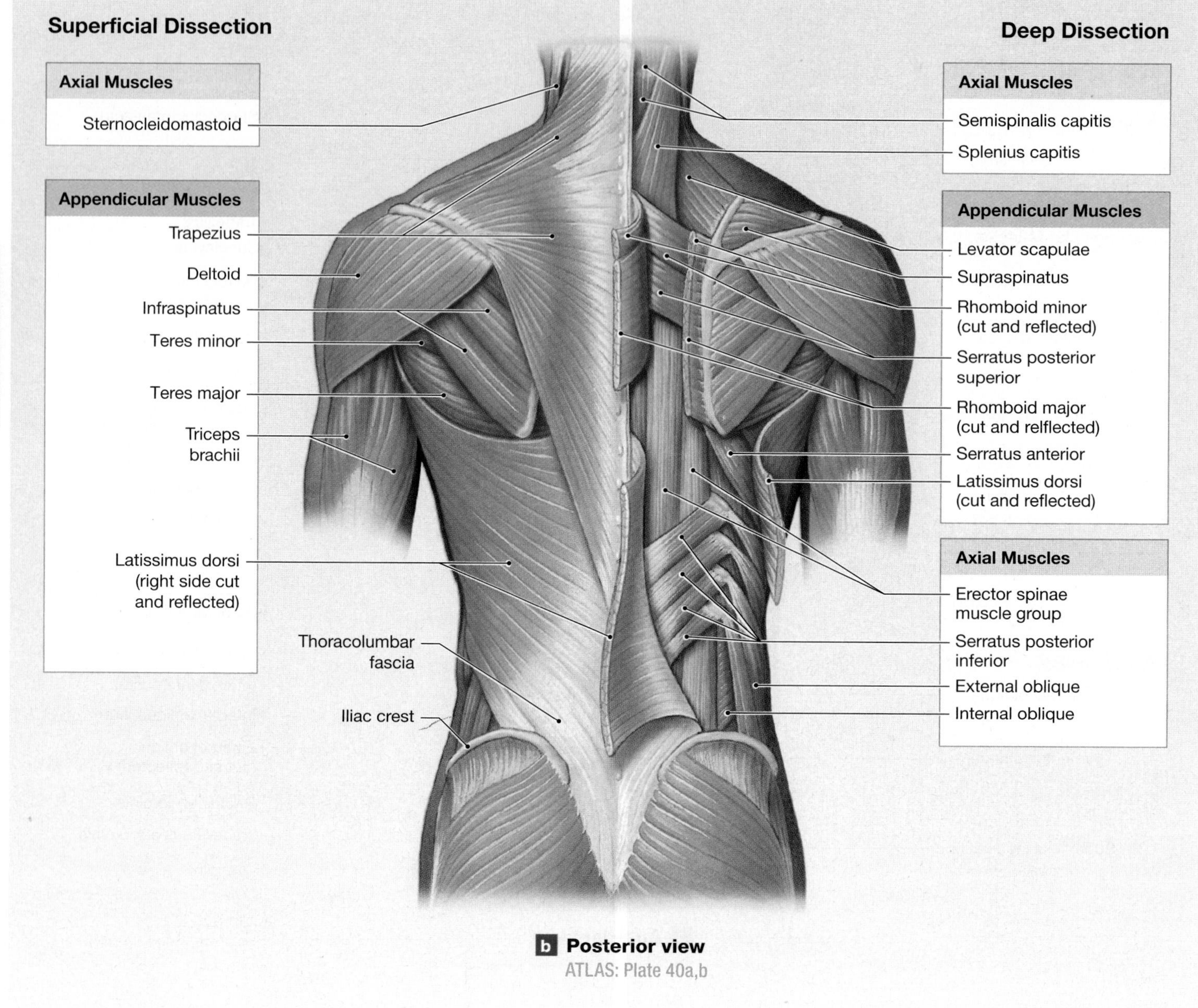
Superficial Dissection
Axial Muscles
Sternocleidomastoid
Appendicular Muscles
Trapezius
Deltoid
Infraspinatus
Teres minor
Teres major
Triceps brachii
Latissimus dorsi (right side cut and reflected)
Thoracolumbar fascia
Iliac crest
Deep Dissection
Axial Muscles
Semispinalis capitis
Splenius capitis
Appendicular Muscles
Levator scapulae
Supraspinatus
Rhomboid minor (cut and reflected)
Serratus posterior superior
Rhomboid major (cut and relflected)
Serratus anterior
Latissimus dorsi (cut and reflected)
Axial Muscles
Erector spinae muscle group
Serratus posterior inferior
External oblique
Internal oblique

b **Posterior view**
ATLAS: Plate 40a,b

clavicle (see Figure 11–15a). When it contracts, it depresses and protracts the scapular end of the clavicle. Because ligaments connect this end to the shoulder joint and scapula, those structures move as well. The **pectoralis** (pek-tō-RA-lis) **minor** attaches to the coracoid process of the scapula. The contraction of this muscle generally acts as a synergist to the subclavius.

Removing the trapezius reveals the **rhomboid major**, **rhomboid minor**, and **levator scapulae** (see Figure 11–15b). These muscles are attached to the posterior surfaces of the cervical and thoracic vertebrae. They insert along the vertebral border of each scapula, between the superior and inferior angles. Contraction of a rhomboid adducts (retracts) the scapula on that side. The levator scapulae, as its name implies, elevates the scapula.

Now take a look at how Figure 11–15 compares with the summary in Table 11–11.

Muscles That Move the Arm

Muscles that move the arm have their actions at the shoulder joint (Figure 11–16 and Table 11–12). The **deltoid** is the major abductor, but the **supraspinatus** (sū-pra-spī-NĀ-tus) acts as a synergist at the start of this movement. The **subscapularis** and **teres major** produce medial rotation at the shoulder, whereas the **infraspinatus** and the **teres minor** produce

Figure 11–15 Muscles That Position the Pectoral Girdle.

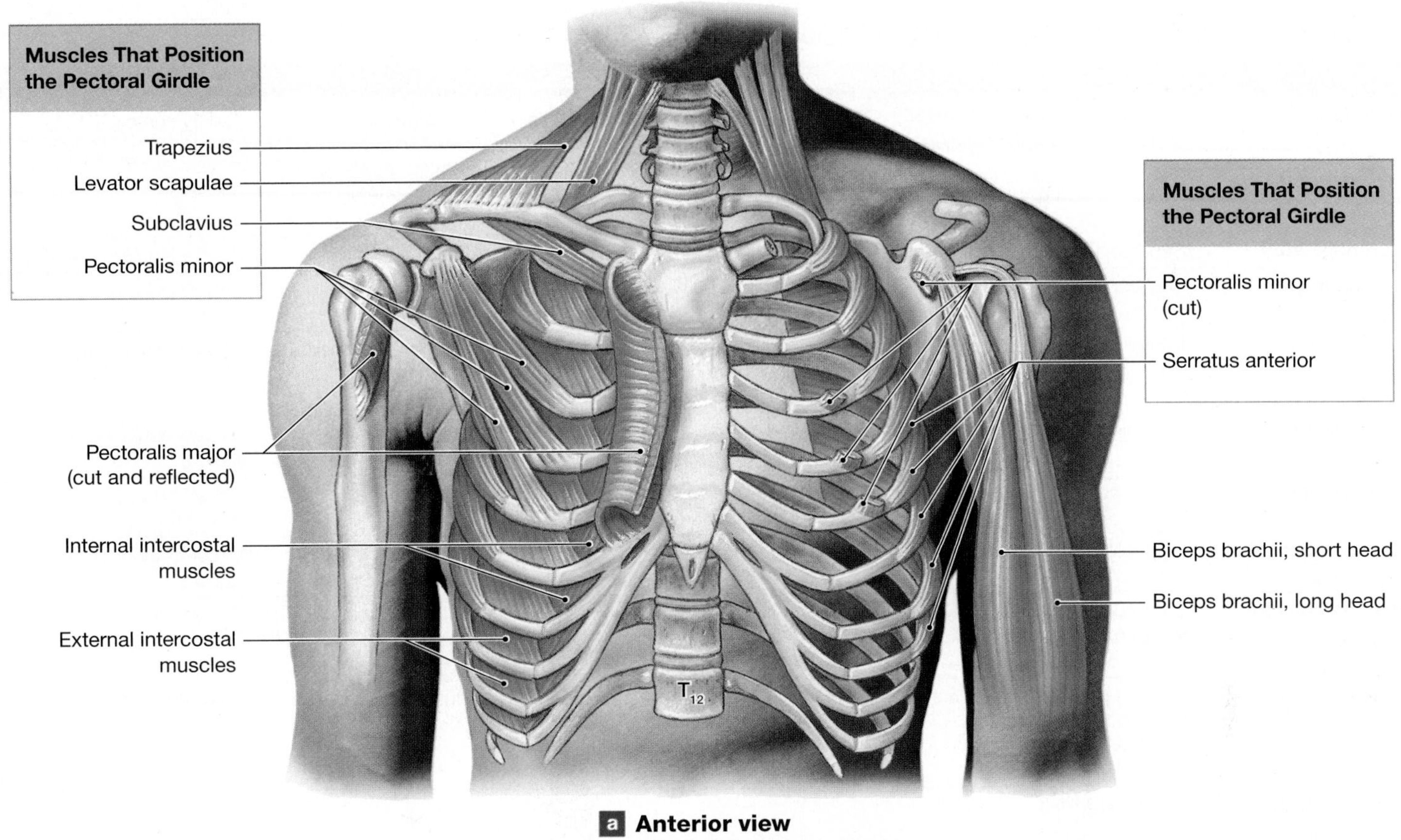

Muscles That Position the Pectoral Girdle
Trapezius
Levator scapulae
Subclavius
Pectoralis minor
Pectoralis major (cut and reflected)
Internal intercostal muscles
External intercostal muscles
T_{12}
Muscles That Position the Pectoral Girdle
Pectoralis minor (cut)
Serratus anterior
Biceps brachii, short head
Biceps brachii, long head

a **Anterior view**
ATLAS: Plates 39a–d; 40a–b

Superficial Dissection

Deep Dissection

Muscles That Position the Pectoral Girdle

Trapezius

Muscles That Position the Pectoral Girdle

Levator scapulae

Rhomboid minor

Rhomboid major

Serratus anterior

Scapula

Triceps brachii

T_{12} vertebra

b **Posterior view**
ATLAS: Plates 27b; 40a–b

Go to MasteringA&P™ > Study Area > Menu > Animations & Videos > *A&PFlix* > Group Muscle Actions & Joints: Muscles of the pectoral girdle

Table 11–12 Muscles That Move the Arm (Figures 11–14 to 11–16)

Muscle	Origin	Insertion	Action	Innervation*
Deltoid	Clavicle and scapula (acromion and adjacent scapular spine)	Deltoid tuberosity of humerus	*Whole muscle:* abduction at shoulder; *anterior part:* flexion and medial rotation; *posterior part:* extension and lateral rotation	Axillary nerve (C_5–C_6)
Supraspinatus	Supraspinous fossa of scapula	Greater tubercle of humerus	Abduction at the shoulder	Suprascapular nerve (C_5)
Subscapularis	Subscapular fossa of scapula	Lesser tubercle of humerus	Medial rotation at shoulder	Subscapular nerves (C_5–C_6)
Teres major	Inferior angle of scapula	Passes medially to reach the medial lip of intertubercular sulcus of humerus	Extension, adduction, and medial rotation at shoulder	Lower subscapular nerve (C_5–C_6)
Infraspinatus	Infraspinous fossa of scapula	Greater tubercle of humerus	Lateral rotation at shoulder	Suprascapular nerve (C_5–C_6)
Teres minor	Lateral border of scapula	Passes laterally to reach the greater tubercle of humerus	Lateral rotation at shoulder	Axillary nerve (C_5)
Coracobrachialis	Coracoid process	Medial margin of shaft of humerus	Adduction and flexion at shoulder	Musculocutaneous nerve (C_5–C_7)
Pectoralis major	Cartilages of ribs 2–6, body of sternum, and inferior, medial portion of clavicle	Crest of greater tubercle and lateral lip of intertubercular sulcus of humerus	Flexion, adduction, and medial rotation at shoulder	Pectoral nerves (C_5–T_1)
Latissimus dorsi	Spinous processes of inferior thoracic and all lumbar vertebrae, ribs 8–12, and thoracolumbar fascia	Floor of intertubercular sulcus of the humerus	Extension, adduction, and medial rotation at shoulder	Thoracodorsal nerve (C_6–C_8)
Triceps brachii (long head)	*See Table 11–13*			

*Where appropriate, spinal nerves involved are given in parentheses.

lateral rotation. All these muscles originate on the scapula. The small **coracobrachialis** (KOR-uh-kō-brā-kē-AH-lis) is the only muscle attached to the scapula that produces flexion and adduction at the shoulder (see Figure 11–16a).

Tips & Tools

The supraspinatus and infraspinatus are named for their origins above and below the spine of the scapula, respectively, not because they are located on the spinal column.

The **pectoralis major** extends between the anterior portion of the chest and the crest of the greater tubercle of the humerus (see Figure 11–16a). The **latissimus dorsi** (la-TIS-i-mus DOR-sē) extends between the thoracic vertebrae at the posterior midline and the intertubercular sulcus of the humerus (see Figure 11–16b). The pectoralis major produces flexion at the shoulder joint, and the latissimus dorsi produces extension. These muscles, commonly known as the "pecs" and the "lats," can also work together to produce adduction and medial rotation of the humerus at the shoulder.

Collectively, the supraspinatus, infraspinatus, teres minor, and subscapularis and their associated tendons form the **rotator cuff**. Sports that involve throwing a ball, such as baseball or football, place considerable strain on the rotator cuff, and rotator cuff injuries are common.

Tips & Tools

The acronym SITS is useful in remembering the four muscles of the rotator cuff.

See how the illustrations in Figure 11–16 relate to the summary information in Table 11–12.

Muscles That Move the Forearm and Hand

The forearm is the region of the upper limb between the elbow and the hand, and the hand is the region distal to the radiocarpal joint, comprising the wrist, palm, and fingers. The muscles of this region can be grouped according to their actions at the elbow and hand, and include flexors, extensors, pronators, and supinators.

Action at the Elbow. Most of the muscles that move the forearm and hand originate on the humerus and insert on the forearm and wrist. There are two exceptions: the biceps brachii and triceps brachii. The **biceps brachii** and the *long head* of the **triceps brachii** originate on the scapula and insert on the bones of the forearm (Figure 11–17).

Figure 11–16 Muscles That Move the Arm. ATLAS: Plates 39a–d; 40a–b

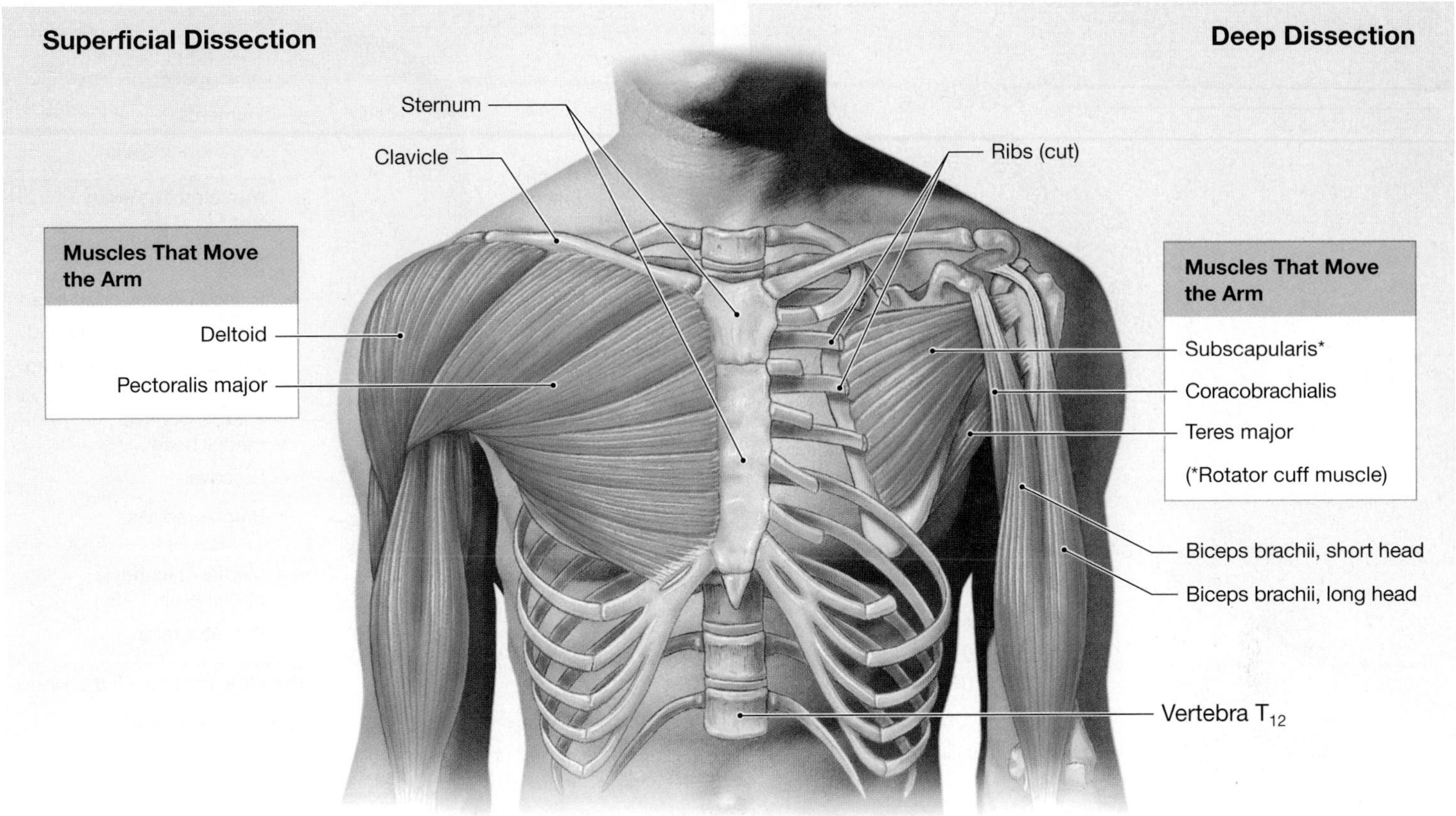

Superficial Dissection

Deep Dissection

Sternum

Clavicle

Ribs (cut)

Muscles That Move the Arm

Deltoid

Pectoralis major

Muscles That Move the Arm

Subscapularis*

Coracobrachialis

Teres major

(*Rotator cuff muscle)

Biceps brachii, short head

Biceps brachii, long head

Vertebra T_{12}

a Anterior view

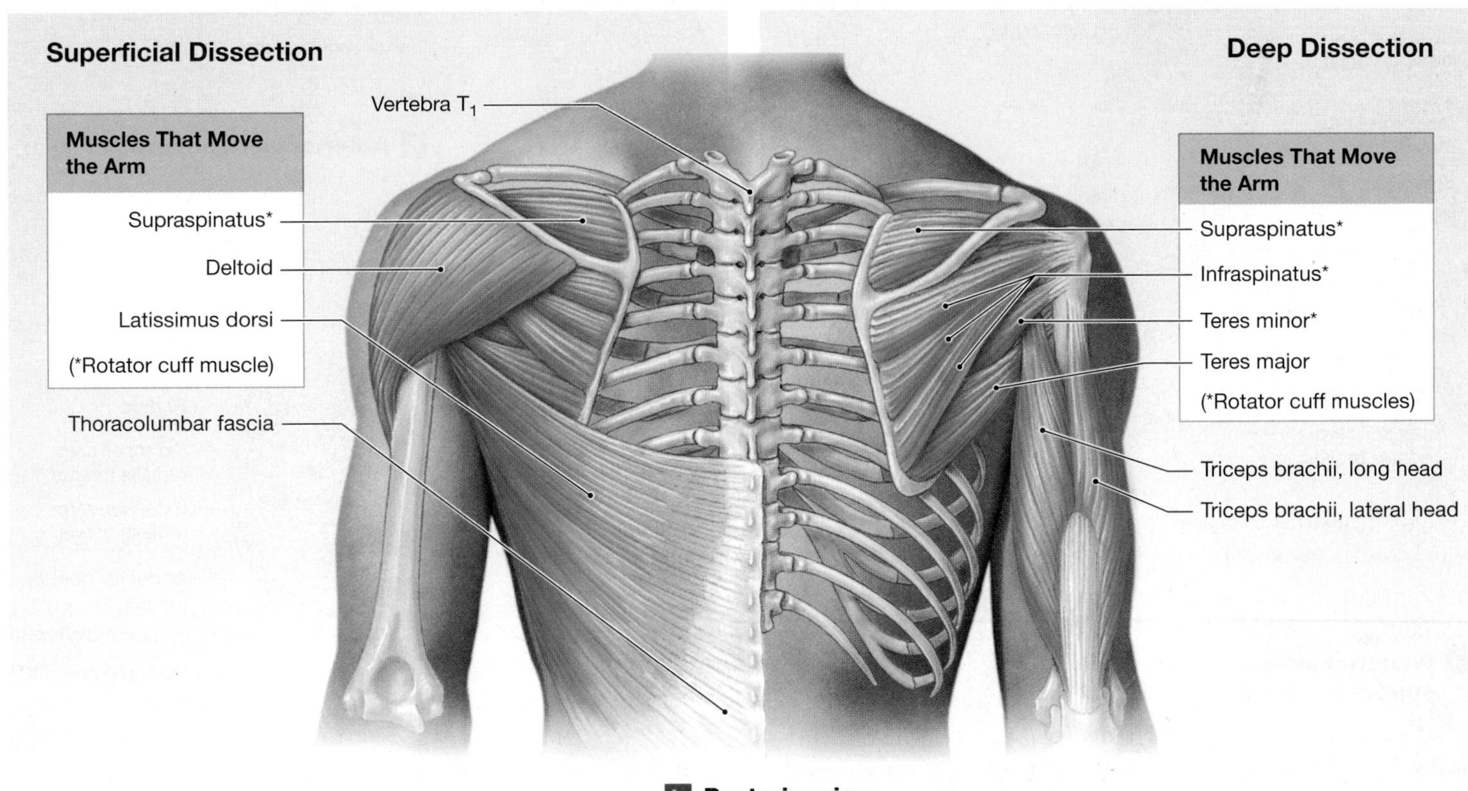

Superficial Dissection
Deep Dissection
Vertebra T_1
Muscles That Move the Arm
Supraspinatus*
Deltoid
Latissimus dorsi
(*Rotator cuff muscle)
Thoracolumbar fascia
Muscles That Move the Arm
Supraspinatus*
Infraspinatus*
Teres minor*
Teres major
(*Rotator cuff muscles)
Triceps brachii, long head
Triceps brachii, lateral head

b Posterior view

> Go to MasteringA&P™ > Study Area > Menu > Animations & Videos > A&PFlix > Origins, Insertions, Actions, Innervations: Biceps Brachii

11

Figure 11–17 **Muscles That Move the Forearm and Hand.** ATLAS: Plates 27a–c; 29a; 30; 33a–d; 37a,b

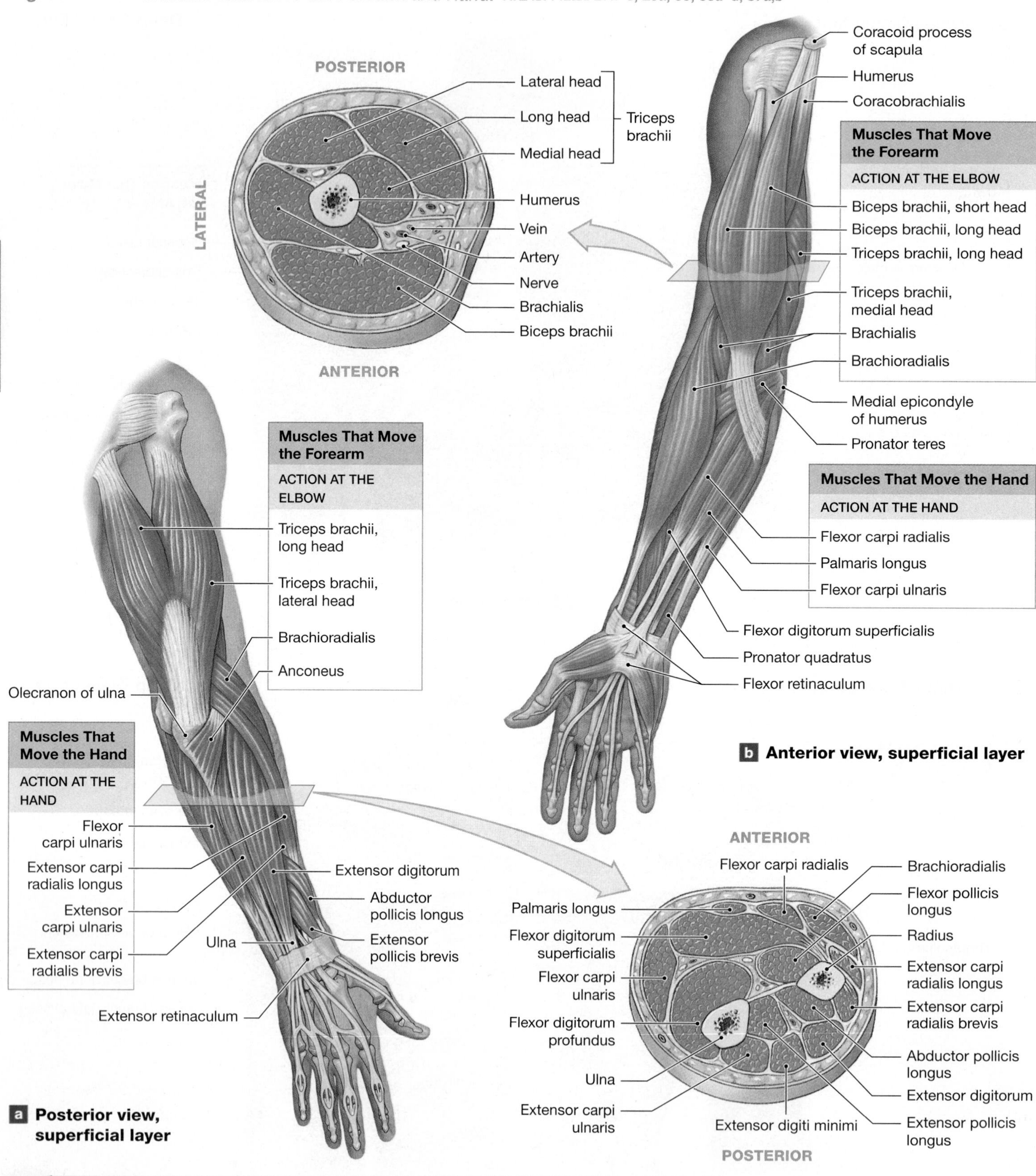

POSTERIOR
Lateral head
Long head
Medial head
Triceps brachii
Humerus
Vein
Artery
Nerve
Brachialis
Biceps brachii
LATERAL
ANTERIOR
Coracoid process of scapula
Humerus
Coracobrachialis
Muscles That Move the Forearm
ACTION AT THE ELBOW
Biceps brachii, short head
Biceps brachii, long head
Triceps brachii, long head
Triceps brachii, medial head
Brachialis
Brachioradialis
Medial epicondyle of humerus
Pronator teres
Muscles That Move the Hand
ACTION AT THE HAND
Flexor carpi radialis
Palmaris longus
Flexor carpi ulnaris
Flexor digitorum superficialis
Pronator quadratus
Flexor retinaculum
Muscles That Move the Forearm
ACTION AT THE ELBOW
Triceps brachii, long head
Triceps brachii, lateral head
Brachioradialis
Anconeus
Olecranon of ulna
Muscles That Move the Hand
ACTION AT THE HAND
Flexor carpi ulnaris
Extensor carpi radialis longus
Extensor carpi ulnaris
Extensor carpi radialis brevis
Ulna
Extensor digitorum
Abductor pollicis longus
Extensor pollicis brevis
Extensor retinaculum
ANTERIOR
Flexor carpi radialis
Brachioradialis
Palmaris longus
Flexor pollicis longus
Flexor digitorum superficialis
Radius
Flexor carpi ulnaris
Extensor carpi radialis longus
Extensor carpi radialis brevis
Flexor digitorum profundus
Abductor pollicis longus
Ulna
Extensor digitorum
Extensor carpi ulnaris
Extensor digiti minimi
Extensor pollicis longus
POSTERIOR

a **Posterior view, superficial layer**

b **Anterior view, superficial layer**

? Give the origins of the heads of the triceps brachii.

> Go to MasteringA&P™ > Study Area > Menu > Animations & Videos > *A&PFlix* > Group Muscle Actions & Joints: Muscles of the elbow joint

Table 11–13 Muscles That Move the Forearm and Hand (Figure 11–17 and Figure 11–18e)

	Muscle	Origin	Insertion	Action	Innervation*
ACTION AT THE ELBOW					
Flexors	**Biceps brachii**	*Short head* from the coracoid process; *long head* from the supraglenoid tubercle (both on the scapula)	Tuberosity of radius	Flexion at elbow and shoulder; supination	Musculocutaneous nerve (C_5–C_6)
Flexors	**Brachialis**	Anterior, distal surface of humerus	Tuberosity of ulna	Flexion at elbow	Musculocutaneous nerve (C_5–C_6) and radial nerve (C_7–C_8)
Flexors	**Brachioradialis**	Ridge superior to the lateral epicondyle of humerus	Lateral aspect of styloid process of radius	Flexion at elbow	Radial nerve (C_5–C_6)
Extensors	**Anconeus**	Posterior, inferior surface of lateral epicondyle of humerus	Lateral margin of olecranon on ulna	Extension at elbow	Radial nerve (C_7–C_8)
Extensors	**Triceps brachii**				
Extensors	**Lateral head**	Superior, lateral margin of humerus	Olecranon of ulna	Extension at elbow	Radial nerve (C_6–C_8)
Extensors	**Long head**	Infraglenoid tubercle of scapula	Olecranon of ulna	As above, plus extension and adduction at the shoulder	Radial nerve (C_6–C_8)
Extensors	**Medial head**	Posterior surface of humerus inferior to radial groove	Olecranon of ulna	Extension at elbow	Radial nerve (C_6–C_8)
PRONATORS/SUPINATORS					
	Pronator quadratus	Anterior and medial surfaces of distal portion of ulna	Anterolateral surface of distal portion of radius	Pronation	Median nerve (C_8–T_1)
	Pronator teres	Medial epicondyle of humerus and coronoid process of ulna	Midlateral surface of radius	Pronation	Median nerve (C_6–C_7)
	Supinator	Lateral epicondyle of humerus, annular ligament, and ridge near radial notch of ulna	Anterolateral surface of radius distal to the radial tuberosity	Supination	Deep radial nerve (C_6–C_8)
ACTION AT THE HAND					
Flexors	**Flexor carpi radialis**	Medial epicondyle of humerus	Bases of second and third metacarpal bones	Flexion and abduction at wrist	Median nerve (C_6–C_7)
Flexors	**Flexor carpi ulnaris**	Medial epicondyle of humerus; adjacent medial surface of olecranon and anteromedial portion of ulna	Pisiform, hamate, and base of fifth metacarpal bone	Flexion and adduction at wrist	Ulnar nerve (C_8–T_1)
Flexors	**Palmaris longus**	Medial epicondyle of humerus	Palmar aponeurosis and flexor retinaculum	Flexion at wrist	Median nerve (C_6–C_7)
Extensors	**Extensor carpi radialis longus**	Lateral supracondylar ridge of humerus	Base of second metacarpal bone	Extension and abduction at wrist	Radial nerve (C_6–C_7)
Extensors	**Extensor carpi radialis brevis**	Lateral epicondyle of humerus	Base of third metacarpal bone	Extension and abduction at wrist	Radial nerve (C_6–C_7)
Extensors	**Extensor carpi ulnaris**	Lateral epicondyle of humerus; adjacent dorsal surface of ulna	Base of fifth metacarpal bone	Extension and adduction at wrist	Deep radial nerve (C_6–C_8)

*Where appropriate, spinal nerves involved are given in parentheses.

The triceps brachii inserts on the olecranon of the ulna. Contraction of the triceps brachii extends the elbow, as when you do push-ups. The biceps brachii inserts on the radial tuberosity, a roughened bump on the anterior surface of the radius. p. 250 Contraction of the biceps brachii flexes the elbow and supinates the forearm. With the forearm pronated (palm facing posteriorly), the biceps brachii cannot function effectively. As a result, you are strongest when you flex your elbow with a supinated forearm. The biceps brachii then makes a prominent bulge.

The biceps brachii is important in stabilizing the shoulder joint. The short head originates on the coracoid process and supports the posterior surface of the joint capsule. The long head originates at the supraglenoid tubercle, inside the shoulder joint. p. 247 After crossing the head of the humerus, it passes along the intertubercular sulcus. In this position, the tendon helps to hold the head of the humerus within the glenoid cavity while arm movements are under way.

More muscles than these are shown in **Figure 11–17** and summarized in **Table 11–13**. As you study these muscles, notice

Figure 11–18 Muscles That Move the Hand and Fingers.

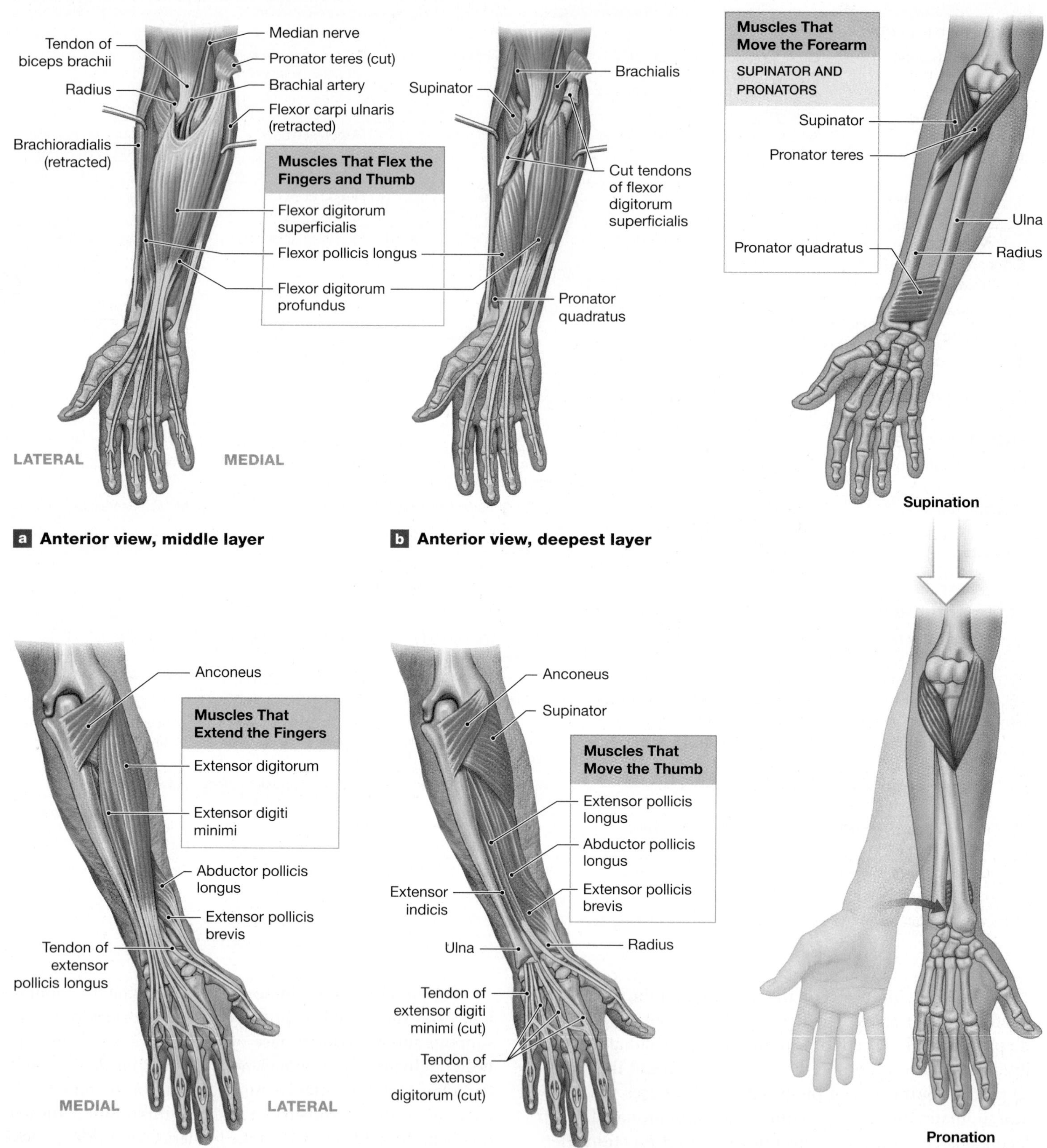
Tendon of biceps brachii
Radius
Brachioradialis (retracted)
Median nerve
Pronator teres (cut)
Brachial artery
Flexor carpi ulnaris (retracted)
Muscles That Flex the Fingers and Thumb
Flexor digitorum superficialis
Flexor pollicis longus
Flexor digitorum profundus
LATERAL
MEDIAL
Supinator
Brachialis
Cut tendons of flexor digitorum superficialis
Pronator quadratus
Muscles That Move the Forearm
SUPINATOR AND PRONATORS
Supinator
Pronator teres
Pronator quadratus
Ulna
Radius
Supination
Anconeus
Muscles That Extend the Fingers
Extensor digitorum
Extensor digiti minimi
Abductor pollicis longus
Extensor pollicis brevis
Tendon of extensor pollicis longus
MEDIAL
LATERAL
Anconeus
Supinator
Muscles That Move the Thumb
Extensor pollicis longus
Abductor pollicis longus
Extensor pollicis brevis
Extensor indicis
Ulna
Radius
Tendon of extensor digiti minimi (cut)
Tendon of extensor digitorum (cut)
Pronation

a Anterior view, middle layer

b Anterior view, deepest layer

c Posterior view, middle layer

d Posterior view, deepest layer

e Supination and pronation

Table 11–14 Muscles That Move the Hand and Fingers (Figure 11–18)

Muscle	Origin	Insertion	Action	Innervation*
Abductor pollicis longus	Proximal dorsal surfaces of ulna and radius	Lateral margin of first metacarpal bone	Abduction at joints of thumb and wrist	Deep radial nerve (C_6–C_7)
Extensor digitorum	Lateral epicondyle of humerus	Posterior surfaces of the phalanges, fingers 2–5	Extension at finger joints and wrist	Deep radial nerve (C_6–C_8)
Extensor pollicis brevis	Shaft of radius distal to origin of abductor pollicis longus	Base of proximal phalanx of thumb	Extension at joints of thumb; abduction at wrist	Deep radial nerve (C_6–C_7)
Extensor pollicis longus	Posterior and lateral surfaces of ulna and interosseous membrane	Base of distal phalanx of thumb	Extension at joints of thumb; abduction at wrist	Deep radial nerve (C_6–C_8)
Extensor indicis	Posterior surface of ulna and interosseous membrane	Posterior surface of phalanges of index finger (2), with tendon of extensor digitorum	Extension and adduction at joints of index finger	Deep radial nerve (C_6–C_8)
Extensor digiti minimi	By extensor tendon to lateral epicondyle of humerus and from intermuscular septa	Posterior surface of proximal phalanx of little finger (5)	Extension at joints of little finger	Deep radial nerve (C_6–C_8)
Flexor digitorum superficialis	Medial epicondyle of humerus; adjacent anterior surfaces of ulna and radius	Midlateral surfaces of middle phalanges of fingers 2–5	Flexion at proximal interphalangeal, metacarpophalangeal, and wrist joints	Median nerve (C_7–T_1)
Flexor digitorum profundus	Medial and posterior surfaces of ulna, medial surface of coronoid process, and interosseus membrane	Bases of distal phalanges of fingers 2–5	Flexion at distal interphalangeal joints and, to a lesser degree, proximal interphalangeal joints and wrist	Palmar interosseous nerve, from median nerve, and ulnar nerve (C_8–T_1)
Flexor pollicis longus	Anterior shaft of radius, interosseous membrane	Base of distal phalanx of thumb	Flexion at joints of thumb	Median nerve (C_8–T_1)

*Where appropriate, spinal nerves involved are given in parentheses.

that, in general, the extensor muscles lie along the posterior and lateral surfaces of the arm, whereas the flexors are on the anterior and medial surfaces. Connective tissue partitions separate major muscle groups, dividing the muscles into *compartments* formed by dense collagenous sheets.

The **brachialis** and **brachioradialis** (BRĀ-kē-ō-rā-dē-AH-lis) flex the elbow and are opposed by the **anconeus** and the triceps brachii, respectively.

Supination and Pronation. Supination and pronation involve changes in the relationship between the bones of the forearm, the radius and ulna. Such changes also result in movement of the palm of the hand. For this reason, the muscles involved in supination and pronation are shown in Figure 11–18e. The **supinator** and **pronator teres** originate on both the humerus and ulna. These muscles rotate the radius without either flexing or extending the elbow. The square-shaped **pronator quadratus** originates on the ulna and assists the pronator teres in opposing the actions of the supinator or biceps brachii. During pronation, the tendon of the biceps brachii rotates with the radius. As a result, this muscle cannot assist in flexion of the elbow when the forearm is pronated.

Action at the Hand. The **flexor carpi ulnaris**, **flexor carpi radialis**, and **palmaris longus** are superficial muscles that work together to produce flexion of the wrist. The flexor carpi radialis flexes and *ab*ducts, and the flexor carpi ulnaris flexes and *ad*ducts. *Pitcher's arm* is an inflammation at the origins of the flexor carpi muscles at the medial epicondyle of the humerus. This condition results from forcibly flexing the wrist just before releasing a baseball.

The **extensor carpi radialis** and the **extensor carpi ulnaris** have a similar relationship to that between the flexor carpi muscles. That is, the extensor carpi radialis produces extension and *ab*duction, whereas the extensor carpi ulnaris produces extension and *ad*duction.

Please take a look at the detailed summary listed in Table 11–13, and compare it with the illustrations in Figure 11–17 and Figure 11–18e.

Several superficial and deep muscles of the forearm flex and extend the finger joints (Figure 11–18 and Table 11–14). These large muscles end before reaching the wrist, and only their tendons cross the articulation. This arrangement ensures maximum mobility at both the wrist and hand. The tendons that cross the posterior and anterior surfaces of the wrist pass through **synovial tendon sheaths**, elongated bursae that reduce friction. ⮌ p. 269

The fascia of the forearm thickens on the posterior surface of the wrist, forming the **extensor retinaculum**

Figure 11–19 **Intrinsic Muscles of the Hand.** ATLAS: Plates 37b; 38c–f

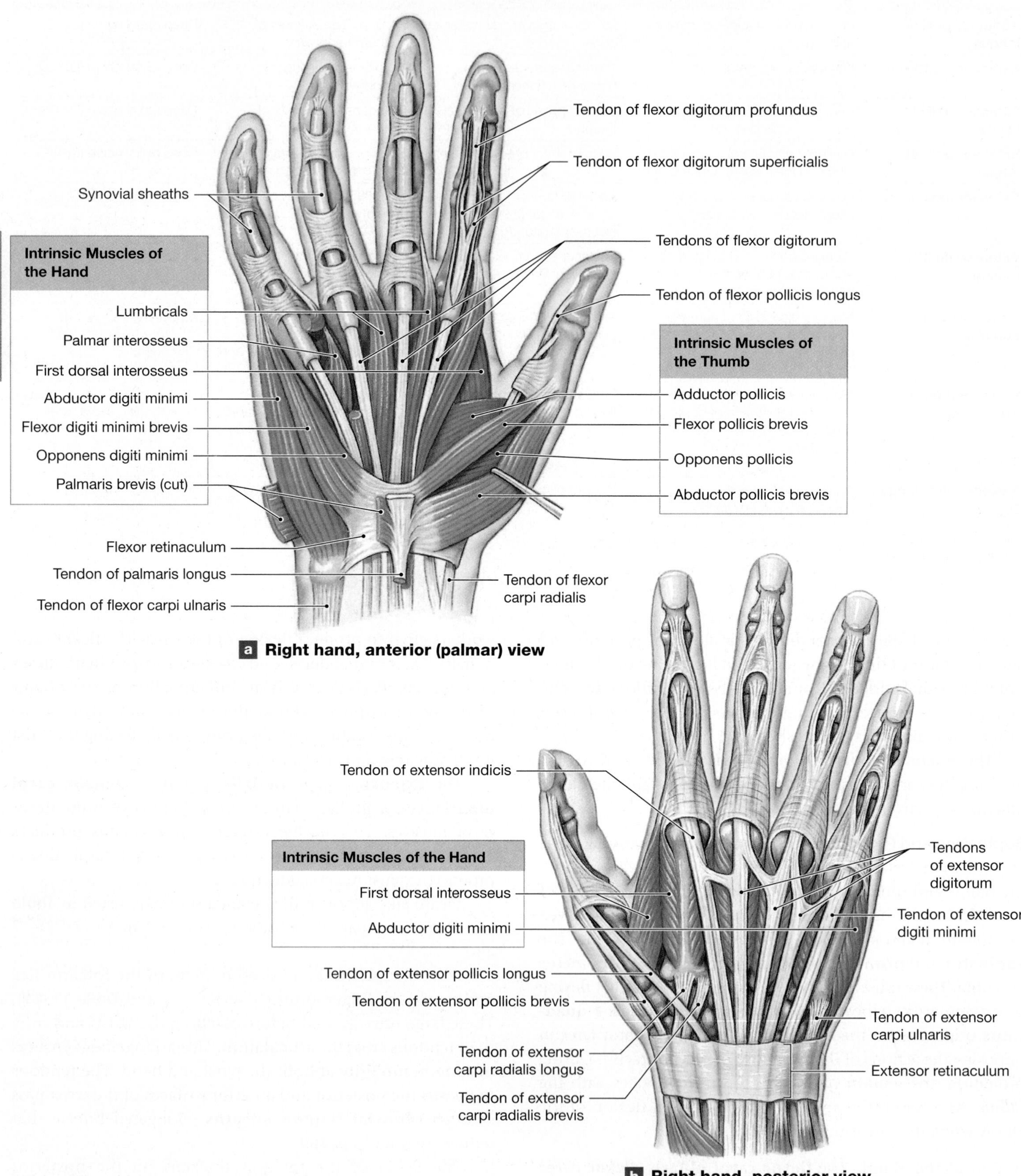

Tendon of flexor digitorum profundus
Tendon of flexor digitorum superficialis
Synovial sheaths
Tendons of flexor digitorum
Tendon of flexor pollicis longus
Intrinsic Muscles of the Hand
Lumbricals
Palmar interosseus
First dorsal interosseus
Abductor digiti minimi
Flexor digiti minimi brevis
Opponens digiti minimi
Palmaris brevis (cut)
Intrinsic Muscles of the Thumb
Adductor pollicis
Flexor pollicis brevis
Opponens pollicis
Abductor pollicis brevis
Flexor retinaculum
Tendon of palmaris longus
Tendon of flexor carpi ulnaris
Tendon of flexor carpi radialis
Tendon of extensor indicis
Intrinsic Muscles of the Hand
First dorsal interosseus
Abductor digiti minimi
Tendons of extensor digitorum
Tendon of extensor digiti minimi
Tendon of extensor pollicis longus
Tendon of extensor pollicis brevis
Tendon of extensor carpi ulnaris
Tendon of extensor carpi radialis longus
Extensor retinaculum
Tendon of extensor carpi radialis brevis

a Right hand, anterior (palmar) view

b Right hand, posterior view

? Which two intrinsic thumb muscles are an agonist-antagonist pair?

Table 11–15 Intrinsic Muscles of the Hand (Figure 11–19)

Muscle	Origin	Insertion	Action	Innervation*
Palmaris brevis	Palmar aponeurosis	Skin of medial border of hand	Moves skin on medial border toward midline of palm	Ulnar nerve, superficial branch (C_8)
ADDUCTION/ABDUCTION				
Adductor pollicis	Metacarpal and carpal bones	Proximal phalanx of thumb	Adduction of thumb	Ulnar nerve, deep branch (C_8–T_1)
Palmar interosseus ** **(3–4)**	Sides of metacarpal bones II, IV, and V	Bases of proximal phalanges of fingers 2, 4, and 5	Adduction at metacarpophalangeal joints of fingers 2, 4, and 5; flexion at metacarpophalangeal joints; extension at interphalangeal joints	Ulnar nerve, deep branch (C_8–T_1)
Abductor pollicis brevis	Transverse carpal ligament, scaphoid, and trapezium	Radial side of base of proximal phalanx of thumb	Abduction of thumb	Median nerve (C_6–C_7)
Dorsal interosseus (4)	Each originates from opposing faces of two metacarpal bones (I and II, II and III, III and IV, IV and V)	Bases of proximal phalanges of fingers 2–4	Abduction at metacarpophalangeal joints of fingers 2 and 4; flexion at metacarpophalangeal joints; extension at interphalangeal joints	Ulnar nerve, deep branch (C_8–T_1)
Abductor digiti minimi	Pisiform	Proximal phalanx of little finger	Abduction of little finger and flexion at its metacarpophalangeal joint	Ulnar nerve, deep branch (C_8–T_1)
FLEXION				
Flexor pollicis brevis	Flexor retinaculum, trapezium, capitate, and ulnar side of first metacarpal bone	Radial and ulnar sides of proximal phalanx of thumb	Flexion and adduction of thumb	Branches of median and ulnar nerves
Lumbricals (4)	Tendons of flexor digitorum profundus	Tendons of extensor digitorum to digits 2–5	Flexion at metacarpophalangeal joints 2–5; extension at proximal and distal interphalangeal joints, digits 2–5	Median nerve (lumbricals 1 and 2); ulnar nerve, deep branch (lumbricals 3 and 4)
Flexor digiti minimi brevis	Hamate	Proximal phalanx of little finger	Flexion at joints of little finger	Ulnar nerve, deep branch (C_8–T_1)
OPPOSITION				
Opponens pollicis	Trapezium and flexor retinaculum	First metacarpal bone	Opposition of thumb	Median nerve (C_6–C_7)
Opponens digiti minimi	Trapezium and flexor retinaculum	Fifth metacarpal bone	Opposition of fifth metacarpal bone	Ulnar nerve, deep branch (C_8–T_1)

*Where appropriate, spinal nerves involved are given in parentheses.

**The deep, medial portion of the flexor pollicis brevis originating on the first metacarpal bone is sometimes called the *first palmar interosseus muscle;* it inserts on the ulnar side of the phalanx and is innervated by the ulnar nerve.

(ret-i-NAK-ū-lum; plural, *retinacula*), a wide band of connective tissue. The extensor retinaculum holds the tendons of the extensor muscles in place. On the anterior surface, the fascia also thickens to form another wide band of connective tissue, the **flexor retinaculum**, which stabilizes the tendons of the flexor muscles.

Inflammation of the retinacula and synovial tendon sheaths can restrict movement and put pressure on the distal portions of the *median nerve*, a mixed (sensory and motor) nerve that innervates the hand. This condition, known as *carpal tunnel syndrome*, causes tingling, numbness, weakness, and chronic pain. A common cause is repetitive hand or wrist movements. p. 251

Please compare **Figure 11–18** with the information summarized in **Table 11–14** before you go on.

Muscles That Move the Fingers

The muscles of the forearm provide strength and gross motor movement of the hand and fingers. These muscles are known as the *extrinsic muscles of the hand.* Fine motor movement of the hand involves small *intrinsic muscles*, which originate on the carpal and metacarpal bones. No muscles originate on the phalanges, and only tendons extend across the distal joints of the fingers. The intrinsic muscles of the hand are detailed in **Figure 11–19** and **Table 11–15**.

Muscles of the Pelvis and Lower Limbs

The pelvic girdle is tightly bound to the axial skeleton, permitting little movement. Few muscles of the axial musculature can influence the position of the pelvis. However, the muscles

Go to MasteringA&P™ > Study Area > Menu > Animations & Videos > A&PFlix > Group Muscle Actions & Joints: Anterior muscles of the wrist and fingers

11

Table 11–16 Muscles That Move the Thigh (Figure 11–20)

Group and Muscle	Origin	Insertion	Action	Innervation*
GLUTEAL GROUP				
Gluteus maximus	Iliac crest, posterior gluteal line, and lateral surface of ilium; sacrum, coccyx, and thoracolumbar fascia	Iliotibial tract and gluteal tuberosity of femur	Extension and lateral rotation at hip	Inferior gluteal nerve (L_5–S_2)
Gluteus medius	Anterior iliac crest of ilium, lateral surface between posterior and anterior gluteal lines	Greater trochanter of femur	Abduction and medial rotation at hip	Superior gluteal nerve (L_4–S_1)
Gluteus minimus	Lateral surface of ilium between inferior and anterior gluteal lines	Greater trochanter of femur	Abduction and medial rotation at hip	Superior gluteal nerve (L_4–S_1)
Tensor fasciae latae	Iliac crest and lateral surface of anterior superior iliac spine	Iliotibial tract	Extension of the knee and lateral rotation of the leg acting through the iliotibial tract; abduction*** and medial rotation of the thigh	Superior gluteal nerve (L_4–S_1)
LATERAL ROTATOR GROUP				
Obturators (externus and internus)	Lateral and medial margins of obturator foramen	Trochanteric fossa of femur (externus); medial surface of greater trochanter (internus)	Lateral rotation at hip	Obturator nerve (externus: L_3–L_4) and special nerve from sacral plexus (internus: L_5–S_2)
Piriformis	Anterolateral surface of sacrum	Greater trochanter of femur	Lateral rotation and abduction at hip	Branches of sacral nerves (S_1–S_2)
Gemellus (superior and inferior)	Ischial spine and tuberosity	Medial surface of greater trochanter with tendon of obturator internus	Lateral rotation at hip	Nerves to obturator internus and quadratus femoris
Quadratus femoris	Lateral border of ischial tuberosity	Intertrochanteric crest of femur	Lateral rotation at hip	Special nerve from sacral plexus (L_4–S_1)
ADDUCTOR GROUP				
Adductor brevis	Inferior ramus of pubis	Linea aspera of femur	Adduction, flexion, and medial rotation at hip	Obturator nerve (L_3–L_4)
Adductor longus	Inferior ramus of pubis anterior to adductor brevis	Linea aspera of femur	Adduction, flexion, and medial rotation at hip	Obturator nerve (L_3–L_4)
Adductor magnus	Inferior ramus of pubis posterior to adductor brevis and ischial tuberosity	Linea aspera and adductor tubercle of femur	Adduction at hip; superior part produces flexion and medial rotation; inferior part produces extension and lateral rotation	Obturator and sciatic nerves
Pectineus	Superior ramus of pubis	Pectineal line inferior to lesser trochanter of femur	Flexion, medial rotation, and adduction at hip	Femoral nerve (L_2–L_4)
Gracilis	Inferior ramus of pubis	Medial surface of tibia inferior to medial condyle	Flexion at knee; adduction and medial rotation at hip	Obturator nerve (L_3–L_4)
ILIOPSOAS GROUP**				
Iliacus	Iliac fossa of ilium	Femur distal to lesser trochanter; tendon fused with that of psoas major	Flexion at hip	Femoral nerve (L_2–L_3)
Psoas major	Anterior surfaces and transverse processes of vertebrae (T_{12}–L_5)	Lesser trochanter in company with iliacus	Flexion at hip or lumbar intervertebral joints	Branches of the lumbar plexus (L_2–L_3)

*Where appropriate, spinal nerves involved are given in parentheses.

**The psoas major and iliacus are often considered collectively as the iliopsoas.

***Role in abduction is debatable.

that position the lower limbs provide a range of movements in these limbs. These muscles can be divided into three functional groups: (1) *muscles that move the thigh,* (2) *muscles that move the leg,* and (3) *muscles that move the foot and toes.*

Muscles That Move the Thigh

Gluteal muscles cover the lateral surfaces of the ilia (**Figure 11–20a–c**). The **gluteus maximus** is the largest and most posterior of the gluteal muscles. Its origin includes parts of the ilium; the sacrum, and coccyx; and the thoracolumbar fascia. Acting alone, this massive muscle produces extension and lateral rotation at the hip joint (see **Table 11–16**).

The gluteus maximus shares an insertion with the **tensor fasciae latae** (FASH-ē-ē LAH-tā), which originates on the iliac crest and the anterior superior iliac spine. Together, these muscles pull on the **iliotibial** (il-ē-ō-TIB-ē-ul) **tract**, a band of collagen

Figure 11–20 **Muscles That Move the Thigh.** ATLAS: Plates 68a–c; 72a,b; 73a,b

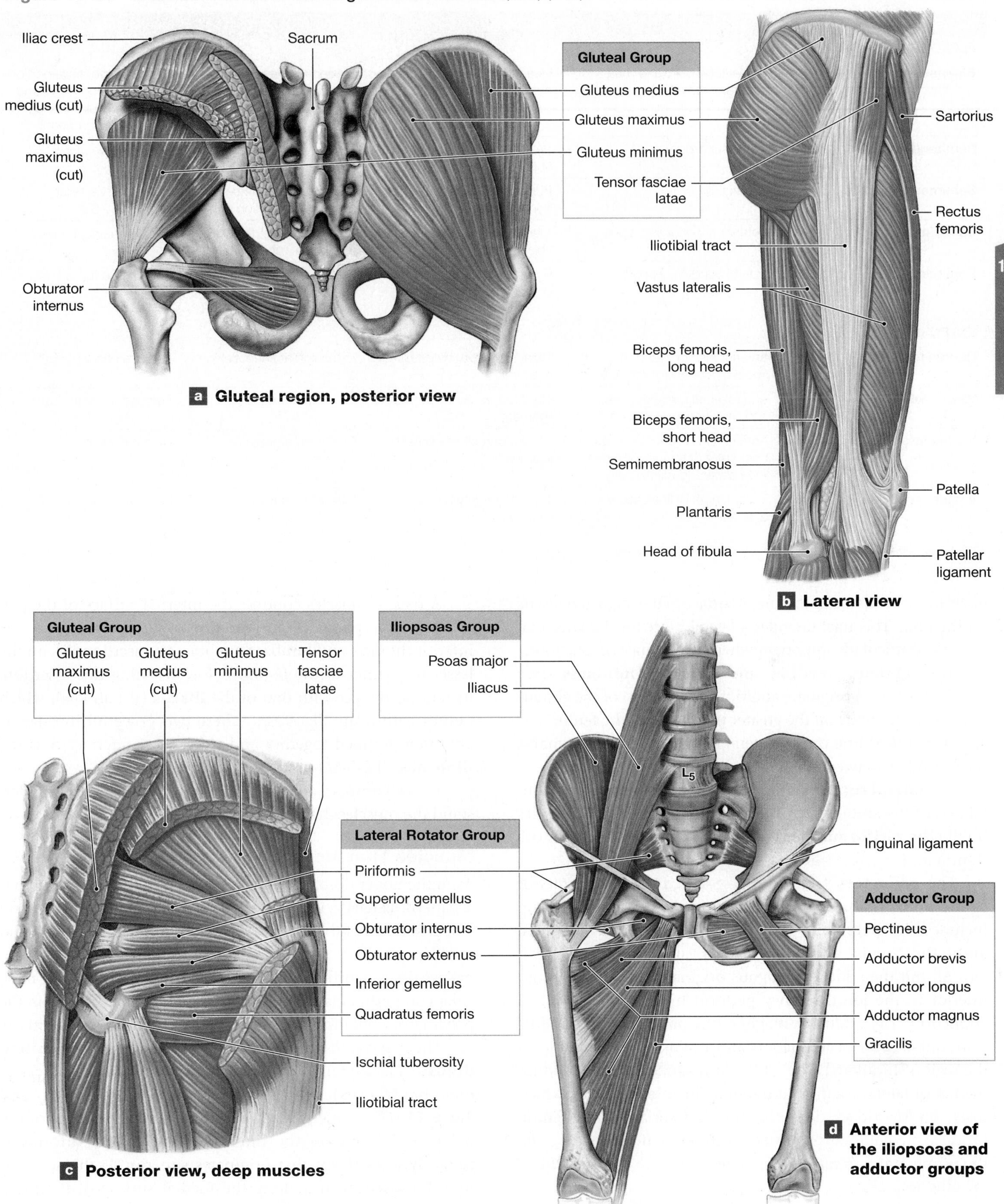
Iliac crest
Sacrum
Gluteus medius (cut)
Gluteus maximus (cut)
Obturator internus
Gluteal Group
Gluteus medius
Gluteus maximus
Gluteus minimus
Tensor fasciae latae
Sartorius
Rectus femoris
Iliotibial tract
Vastus lateralis
Biceps femoris, long head
Biceps femoris, short head
Semimembranosus
Patella
Plantaris
Head of fibula
Patellar ligament
Gluteal Group
Gluteus maximus (cut)
Gluteus medius (cut)
Gluteus minimus
Tensor fasciae latae
Iliopsoas Group
Psoas major
Iliacus
L_5
Lateral Rotator Group
Piriformis
Superior gemellus
Obturator internus
Obturator externus
Inferior gemellus
Quadratus femoris
Ischial tuberosity
Iliotibial tract
Inguinal ligament
Adductor Group
Pectineus
Adductor brevis
Adductor longus
Adductor magnus
Gracilis

a **Gluteal region, posterior view**

b **Lateral view**

c **Posterior view, deep muscles**

d **Anterior view of the iliopsoas and adductor groups**

11

Table 11–17 Muscles That Move the Leg (Figure 11–21)

Muscle	Origin	Insertion	Action	Innervation*
FLEXORS OF THE KNEE				
Biceps femoris	Ischial tuberosity and linea aspera of femur	Head of fibula, lateral condyle of tibia	Flexion at knee; extension and lateral rotation at hip	Sciatic nerve; tibial portion (S_1–S_3; to long head) and common fibular branch (L_5–S_2; to short head)
Semitendinosus	Ischial tuberosity	Proximal, medial surface of tibia near insertion of gracilis	Flexion at knee; extension and medial rotation at hip	Sciatic nerve (tibial portion; L_5–S_2)
Semimembranosus	Ischial tuberosity	Posterior surface of medial condyle of tibia	Flexion at knee; extension and medial rotation at hip	Sciatic nerve (tibial portion; L_5–S_2)
Sartorius	Anterior superior iliac spine	Medial surface of tibia near tibial tuberosity	Flexion at knee; flexion and lateral rotation at hip	Femoral nerve (L_2–L_3)
Popliteus	Lateral condyle of femur	Posterior surface of proximal tibial shaft	Medial rotation of tibia (or lateral rotation of femur); flexion at knee	Tibial nerve (L_4–S_1)
EXTENSORS OF THE KNEE				
Rectus femoris	Anterior inferior iliac spine and superior acetabular rim of ilium	Tibial tuberosity by patellar ligament	Extension at knee; flexion at hip	Femoral nerve (L_2–L_4)
Vastus intermedius	Anterolateral surface of femur and linea aspera (distal half)	Tibial tuberosity by patellar ligament	Extension at knee	Femoral nerve (L_2–L_4)
Vastus lateralis	Anterior and inferior to greater trochanter of femur and along linea aspera (proximal half)	Tibial tuberosity by patellar ligament	Extension at knee	Femoral nerve (L_2–L_4)
Vastus medialis	Entire length of linea aspera of femur	Tibial tuberosity by patellar ligament	Extension at knee	Femoral nerve (L_2–L_4)

*Where appropriate, spinal nerves involved are given in parentheses.

fibers that extends along the lateral surface of the thigh and inserts on the tibia. This tract provides a lateral brace for the knee that becomes particularly important when you balance on one foot.

The **gluteus medius** and **gluteus minimus** (see Figure 11–20a–c) originate anterior to the origin of the gluteus maximus and insert on the greater trochanter of the femur. The anterior gluteal line on the lateral surface of the ilium marks the boundary between these muscles.

The **lateral rotators** originate at or inferior to the horizontal axis of the acetabulum. There are six lateral rotator muscles in all. The **piriformis** (pir-i-FOR-mis) and the **obturator** are dominant (Figure 11–20c,d).

The **adductors** (see Figure 11–20c,d) originate inferior to the horizontal axis of the acetabulum. This muscle group includes the **pectineus** (pek-ti-NĒ-us), **adductor magnus**, **adductor brevis**, **adductor longus**, and **gracilis** (GRAS-i-lis). All but the adductor magnus originate both anterior and inferior to the joint, so they perform hip flexion as well as adduction. The adductor magnus can produce either adduction and flexion or adduction and extension, depending on the region stimulated. The adductor magnus can also produce medial or lateral rotation at the hip. The other muscles, which insert on low ridges along the posterior surface of the femur, produce medial rotation. When an athlete suffers a *pulled groin*, the problem is a *strain*—a muscle tear—in one of these adductor muscles.

A pair of muscles controls the internal surface of the pelvis. The large **psoas** (SŌ-us) **major** originates alongside the inferior thoracic and lumbar vertebrae. Its insertion lies on the lesser trochanter of the femur. Before reaching this insertion, its tendon merges with that of the **iliacus** (il-Ī-ah-kus), which nestles within the iliac fossa. These two powerful hip flexors are often grouped together and collectively referred to as the **iliopsoas** (il-ē-ō-SŌ-us) (see Figure 11–20d).

Please compare Figure 11–20 with Table 11–16 to understand the muscles that move the thigh.

Muscles That Move the Leg

As in the upper limb, muscle distribution in the lower limb has a pattern: Extensor muscles are located along the anterior and lateral surfaces of the leg, and flexors lie along the posterior and medial surfaces (Figure 11–21). As in the upper limb, sturdy connective tissue partitions divide the lower limb into separate muscular compartments. The flexors and adductors originate on the pelvic girdle, but most extensors originate on the femoral surface.

The *flexors of the knee* include the **biceps femoris**, **semitendinosus** (sem-ē-ten-di-NŌ-sus), **semimembranosus** (sem-ē-mem-bra-NŌ-sus), and **sartorius** (see Figure 11–21a and Table 11–17). These muscles originate along the edges of the pelvis and insert on the tibia and fibula. The sartorius is the only knee flexor that originates superior to the acetabulum. Its insertion lies along the medial surface of the tibia.

Figure 11–21 **Muscles That Move the Leg.** ATLAS: Plates 69a,b; 70b; 72a,b; 74; 76a,b; 78a–g

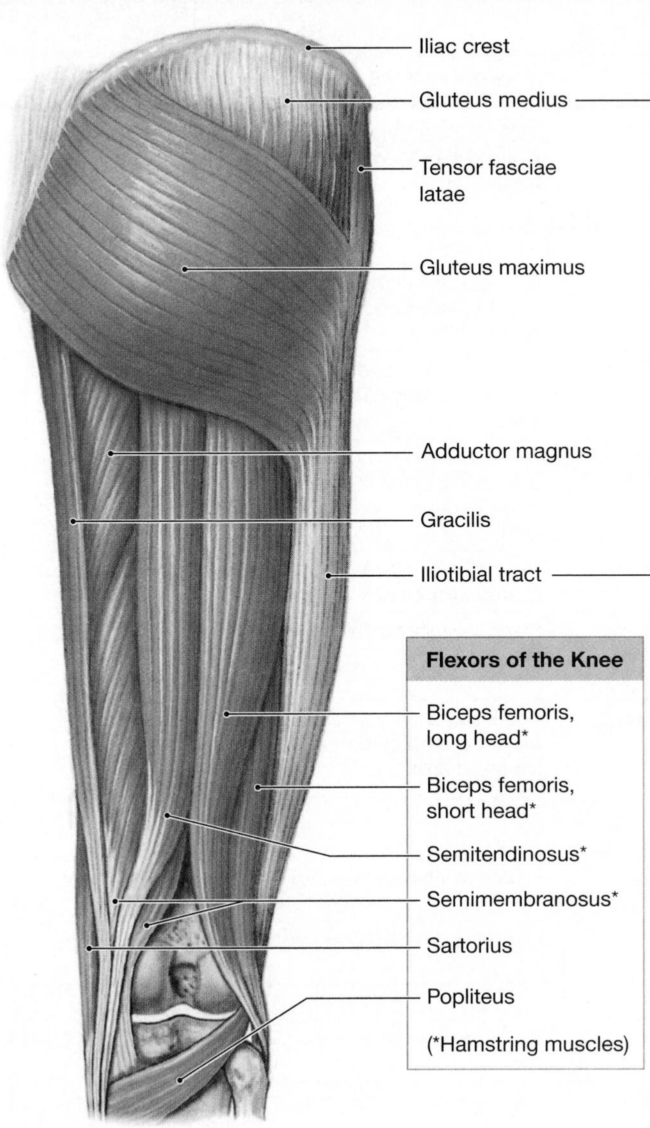

Iliac crest
Gluteus medius
Tensor fasciae latae
Gluteus maximus
Adductor magnus
Gracilis
Iliotibial tract
Flexors of the Knee
Biceps femoris, long head*
Biceps femoris, short head*
Semitendinosus*
Semimembranosus*
Sartorius
Popliteus
(*Hamstring muscles)

a Hip and thigh, posterior view

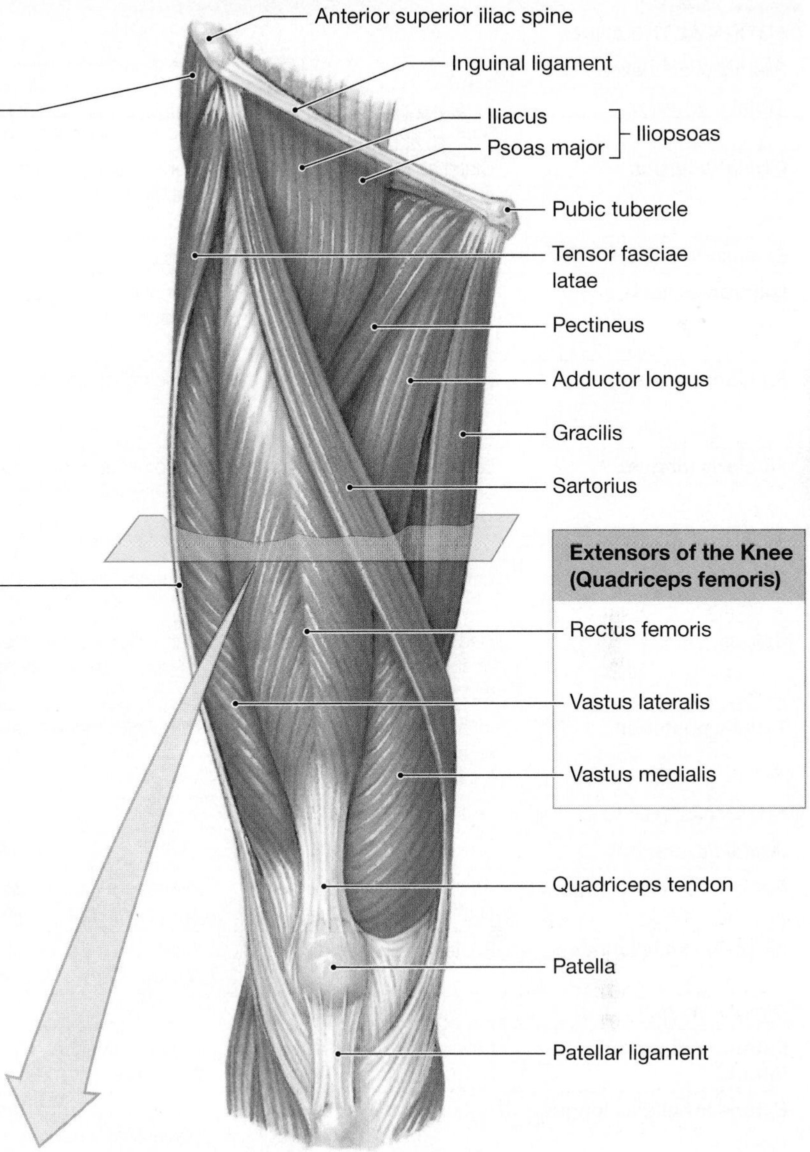

Anterior superior iliac spine
Inguinal ligament
Iliacus
Psoas major
Iliopsoas
Pubic tubercle
Tensor fasciae latae
Pectineus
Adductor longus
Gracilis
Sartorius
Extensors of the Knee (Quadriceps femoris)
Rectus femoris
Vastus lateralis
Vastus medialis
Quadriceps tendon
Patella
Patellar ligament

b Quadriceps femoris and thigh muscles, anterior view

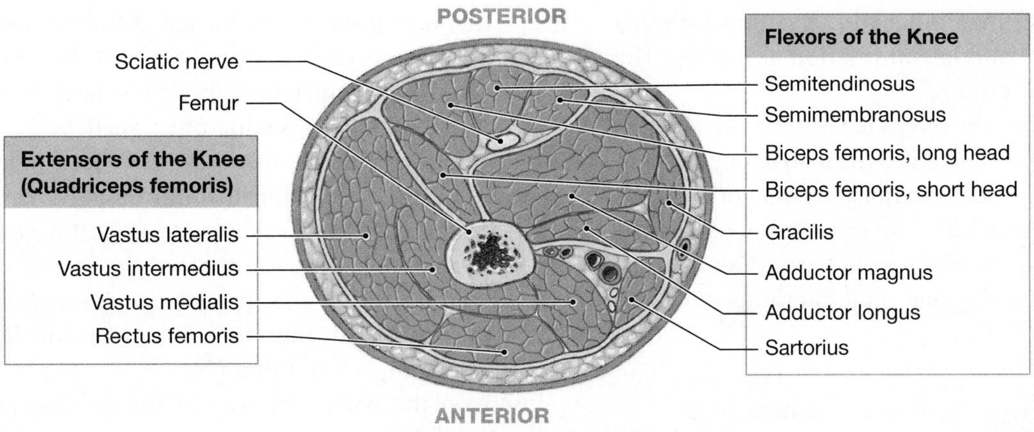

POSTERIOR
Sciatic nerve
Femur
Extensors of the Knee (Quadriceps femoris)
Vastus lateralis
Vastus intermedius
Vastus medialis
Rectus femoris
Flexors of the Knee
Semitendinosus
Semimembranosus
Biceps femoris, long head
Biceps femoris, short head
Gracilis
Adductor magnus
Adductor longus
Sartorius
ANTERIOR

c Sectional view

? Which quadriceps femoris muscle is not visible from the superficial anterior thigh?

Table 11–18 Extrinsic Muscles That Move the Foot and Toes (Figure 11–22)

Muscle	Origin	Insertion	Action	Innervation*
ACTION AT THE ANKLE				
Flexors (dorsiflexors)				
Tibialis anterior	Lateral condyle and proximal shaft of tibia	Base of first metatarsal bone and medial cuneiform bone	Flexion (dorsiflexion) at ankle; inversion of foot	Deep fibular nerve (L_4–S_1)
Fibularis tertius	Distal anterior surface of fibula and interosseous membrane	Dorsal surface of fifth metatarsal bone	Flexion (dorsiflexion) at ankle; eversion of foot	Deep fibular nerve (L_5–S_1)
Extensors (plantar flexors)				
Gastrocnemius	Femoral condyles	Calcaneus by calcaneal tendon	Extension (plantar flexion) at ankle; inversion of foot; flexion at knee	Tibial nerve (S_1–S_2)
Fibularis brevis	Midlateral margin of fibula	Base of fifth metatarsal bone	Eversion of foot and extension (plantar flexion) at ankle	Superficial fibular nerve (L_4–S_1)
Fibularis longus	Lateral condyle of tibia, head and proximal shaft of fibula	Base of first metatarsal bone and medial cuneiform bone	Eversion of foot and extension (plantar flexion) at ankle; supports longitudinal arch	Superficial fibular nerve (L_4–S_1)
Plantaris	Lateral supracondylar ridge	Posterior portion of calcaneus	Extension (plantar flexion) at ankle; flexion at knee	Tibial nerve (L_4–S_1)
Soleus	Head and proximal shaft of fibula and adjacent posteromedial shaft of tibia	Calcaneus by calcaneal tendon (with gastrocnemius)	Extension (plantar flexion) at ankle	Sciatic nerve, tibial branch (S_1–S_2)
Tibialis posterior	Interosseous membrane and adjacent shafts of tibia and fibula	Tarsal and metatarsal bones	Adduction and inversion of foot; extension (plantar flexion) at ankle	Sciatic nerve, tibial branch (S_1–S_2)
ACTION AT THE TOES				
Digital flexors				
Flexor digitorum longus	Posteromedial surface of tibia	Inferior surfaces of distal phalanges, toes 2–5	Flexion at joints of toes 2–5	Sciatic nerve, tibial branch (L_5–S_1)
Flexor hallucis longus	Posterior surface of fibula	Inferior surface, distal phalanx of great toe	Flexion at joints of great toe	Sciatic nerve, tibial branch (L_5–S_1)
Digital extensors				
Extensor digitorum longus	Lateral condyle of tibia, anterior surface of fibula	Superior surfaces of phalanges, toes 2–5	Extension at joints of toes 2–5	Deep fibular nerve (L_4–S_1)
Extensor hallucis longus	Anterior surface of fibula	Superior surface, distal phalanx of great toe	Extension at joints of great toe	Deep fibular nerve (L_4–S_1)

*Where appropriate, spinal nerves involved are given in parentheses.

The sartorius crosses over two joints and its contraction produces flexion at the knee and lateral rotation at the hip. This happens when you cross your legs.

Three of these muscles—the biceps femoris, semitendinosus, and semimembranous—are often called the **hamstrings**. Because they originate on the pelvic surface inferior and posterior to the acetabulum, their contractions produce not only flexion at the knee, but also extension at the hip. A *pulled hamstring* is a common sports injury caused by a strain affecting one of the hamstring muscles.

Tips & Tools

To remember that three muscles make up the **ham**strings, think "the three little pigs." These three muscles are portions of the cut of meat known as ham.

The knee joint can be locked at full extension by a slight lateral rotation of the tibia. p. 281 The small **popliteus** (pop-LI-tē-us) originates on the femur near the lateral condyle and inserts on the posterior tibial shaft (see Figure 11–22a). When flexion is started, this muscle contracts to produce a slight medial rotation of the tibia that unlocks the knee joint.

Collectively, four *knee extensors*—the three **vastus muscles**, which originate along the shaft of the femur, and the **rectus femoris**—make up the **quadriceps femoris** (the "quads"). Together, the vastus muscles cradle the rectus femoris the way a bun surrounds a hot dog (Figure 11–21c). All four muscles insert on the patella by way of the quadriceps tendon. The force of their contraction is relayed to the tibial tuberosity by way of the patellar ligament. The rectus femoris originates on

Figure 11–22 **Extrinsic Muscles That Move the Foot and Toes.** ATLAS: Plates 81a,b; 82a,b; 84a,b

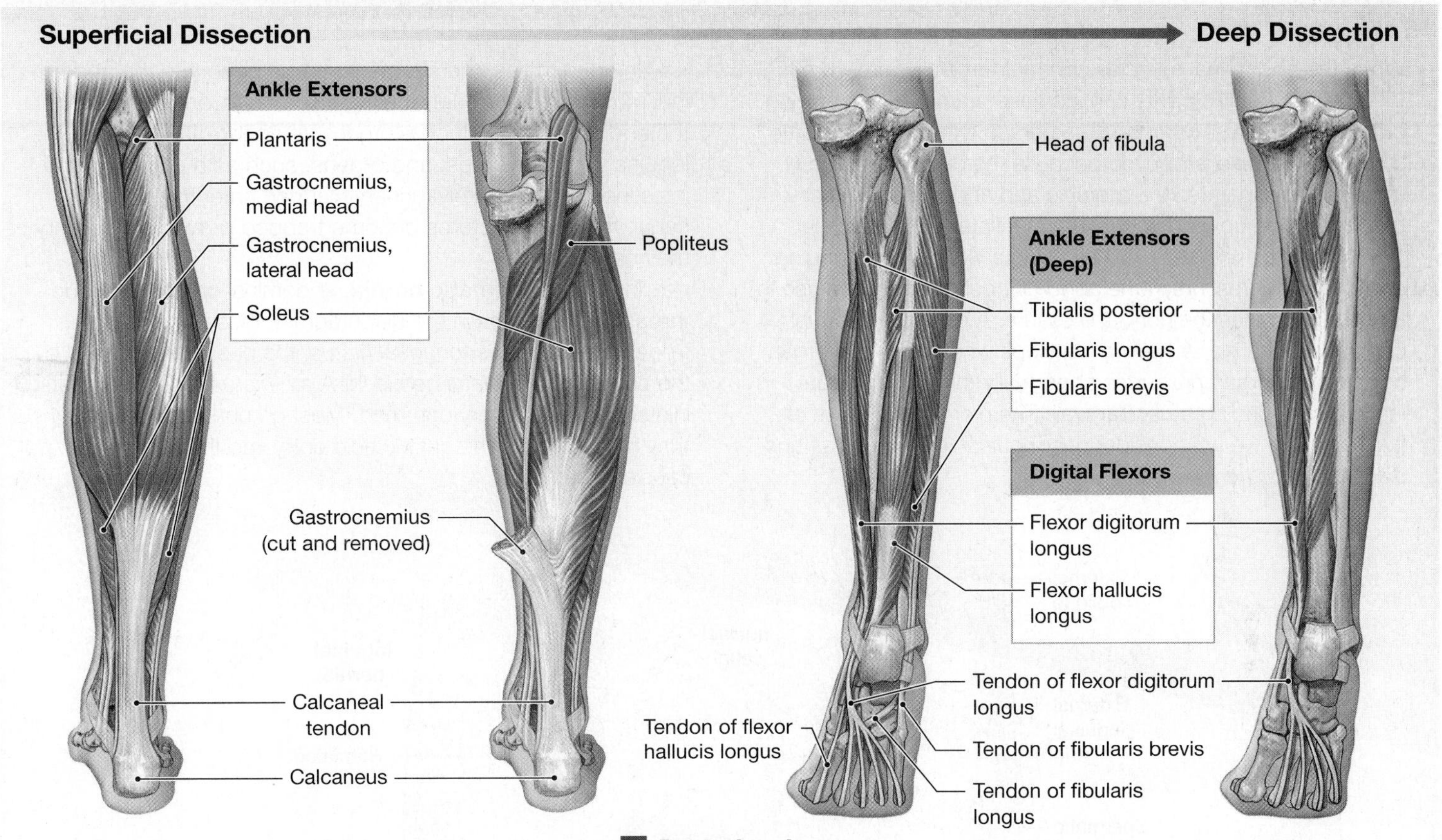

Superficial Dissection → **Deep Dissection**

Ankle Extensors
Plantaris
Gastrocnemius, medial head
Gastrocnemius, lateral head
Soleus

Popliteus

Gastrocnemius (cut and removed)

Calcaneal tendon

Calcaneus

Head of fibula

Ankle Extensors (Deep)
Tibialis posterior
Fibularis longus
Fibularis brevis

Digital Flexors
Flexor digitorum longus
Flexor hallucis longus

Tendon of flexor digitorum longus
Tendon of flexor hallucis longus
Tendon of fibularis brevis
Tendon of fibularis longus

a Posterior views

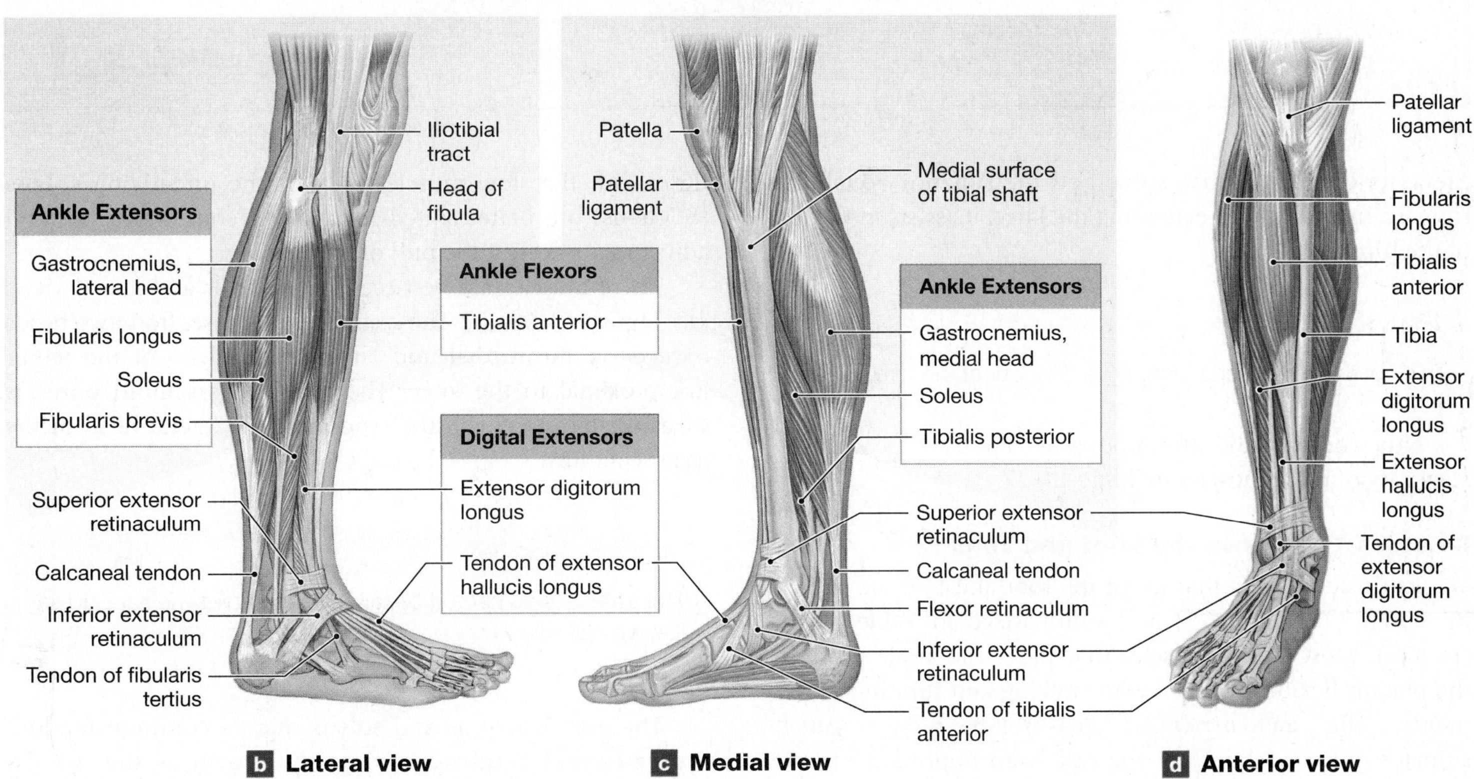
Iliotibial tract
Head of fibula
Ankle Extensors
Gastrocnemius, lateral head
Fibularis longus
Soleus
Fibularis brevis
Superior extensor retinaculum
Calcaneal tendon
Inferior extensor retinaculum
Tendon of fibularis tertius
Ankle Flexors
Tibialis anterior
Digital Extensors
Extensor digitorum longus
Tendon of extensor hallucis longus
Patella
Patellar ligament
Medial surface of tibial shaft
Ankle Extensors
Gastrocnemius, medial head
Soleus
Tibialis posterior
Superior extensor retinaculum
Calcaneal tendon
Flexor retinaculum
Inferior extensor retinaculum
Tendon of tibialis anterior
Patellar ligament
Fibularis longus
Tibialis anterior
Tibia
Extensor digitorum longus
Extensor hallucis longus
Tendon of extensor digitorum longus

b Lateral view **c Medial view** **d Anterior view**

+ Clinical Note Hernia

When the abdominal muscles contract forcefully, pressure in the abdominopelvic cavity can increase dramatically, applying pressure to internal organs. If the person exhales at the same time, the pressure is relieved because the diaphragm, the muscle that separates the thoracic and abdominopelvic cavities, can move upward as the lungs deflate. But during vigorous isometric exercises or when lifting a weight while holding one's breath, pressure in the abdominopelvic cavity can rise to roughly 100 times the normal pressure. A pressure that high can cause a hernia. A **hernia** develops when a visceral organ or part of an organ protrudes abnormally through an opening in a surrounding muscular wall. There are many types of hernias. Here we will consider only *inguinal* (groin) *hernias* and *diaphragmatic hernias.*

In adult males, the spermatic cords penetrate the abdominal musculature through the inguinal canals on their way to the abdominal reproductive organs. In an **inguinal hernia**, the inguinal canal enlarges, and organs, such as a portion of the intestine, or (more rarely) urinary bladder, enter the inguinal canal. If these structures become trapped or twisted, surgery may be required.

In a **diaphragmatic hernia**, abdominal organs protrude through a weakness in the diaphragm. If these organs protrude through the *esophageal hiatus*, the passageway used by the esophagus, a *hiatal hernia* (hī-Ā-tal; *hiatus*, opening) exists. Hiatal hernias are very common. Most go unnoticed, although they may increase the gastric acid entry into the esophagus, causing *heartburn*.

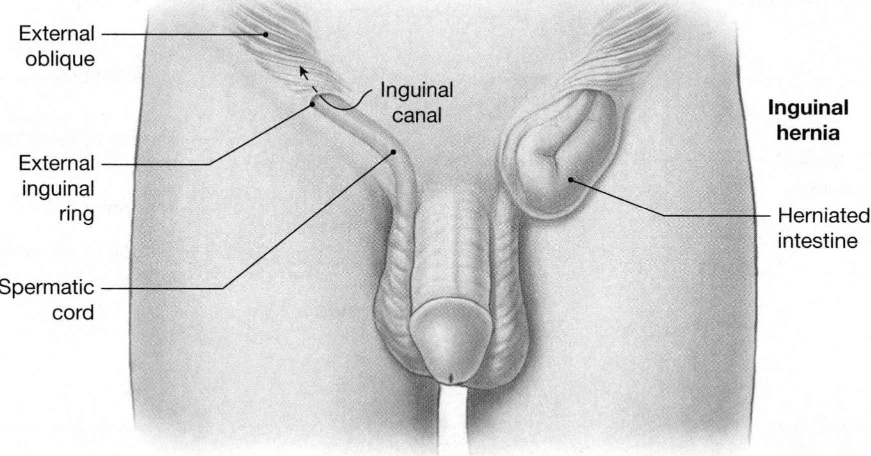
External oblique
Inguinal canal
External inguinal ring
Spermatic cord
Inguinal hernia
Herniated intestine

the anterior inferior iliac spine and the superior acetabular rim—so in addition to extending the knee, it assists in flexion of the hip.

Tips & Tools

Think of the quadriceps femoris as "the four at the fore."

Now compare the illustrations in Figure 11–21 with the summary of these muscles in Table 11–17.

Muscles That Move the Foot and Toes

The extrinsic muscles that move the foot and toes are shown in Figure 11–22 (p. 379) and summarized in Table 11–18 (p. 378). Most of the muscles that move the ankle produce the plantar flexion involved with walking and running movements. The **gastrocnemius** (gas-trok-NĒ-mē-us; *gaster*, stomach + *kneme*, knee) of the calf is an important plantar flexor, but the slow muscle fibers of the underlying **soleus** (SŌ-lē-us) are better suited for making continuous postural adjustments against the pull of gravity.

These muscles are best seen in posterior and lateral views (see Figure 11–22a,b). The gastrocnemius arises from two heads located on the medial and lateral epicondyles of the femur just proximal to the knee. The *fabella*, a sesamoid bone, is occasionally present in the tendon of the lateral head of the gastrocnemius.

Tips & Tools

The **sole**us is so named because it resembles the flat-bodied fish we call sole.

The gastrocnemius and soleus share a common tendon, the **calcaneal tendon**, commonly known as the *Achilles*

Table 11–19 Intrinsic Muscles of the Foot (Figure 11–23)

Muscle	Origin	Insertion	Action	Innervation*
FLEXION/EXTENSION				
Flexor hallucis brevis	Cuboid and lateral cuneiform bones	Proximal phalanx of great toe	Flexion at metatarsophalangeal joint of great toe	Medial plantar nerve (L_4–L_5)
Flexor digitorum brevis	Calcaneus (tuberosity on inferior surface)	Sides of middle phalanges, toes 2–5	Flexion at proximal interphalangeal joints of toes 2–5	Medial plantar nerve (L_4–L_5)
Quadratus plantae	Calcaneus (medial, inferior surfaces)	Tendon of flexor digitorum longus	Flexion at joints of toes 2–5	Lateral plantar nerve (L_4–L_5)
Lumbricals (4)	Tendons of flexor digitorum longus	Tendons of extensor digitorum longus	Flexion at metatarsophalangeal joints; extension at proximal interphalangeal joints of toes 2–5	Medial plantar nerve (lumbrical 1), lateral plantar nerve (lumbricals, 2, 3, and 4)
Flexor digiti minimi brevis	Base of metatarsal bone V	Lateral side of proximal phalanx of toe 5	Flexion at metatarsophalangeal joint of toe 5	Lateral plantar nerve (S_1–S_2)
Extensor digitorum brevis	Calcaneus (superior and lateral surfaces)	Dorsal surfaces of toes 1–4	Extension at metatarsophalangeal joints of toes 1–4	Deep fibular nerve (L_5–S_1)
Extensor hallucis brevis	Superior surface of anterior calcaneus	Dorsal surface of the base of proximal phalanx of great toe	Extension of great toe	Deep fibular nerve (L_5–S_1)
ADDUCTION/ABDUCTION				
Adductor hallucis	Bases of metatarsal bones II–IV and plantar ligaments	Proximal phalanx of great toe	Adduction at metatarsophalangeal joint of great toe	Lateral plantar nerve (S_1–S_2)
Abductor hallucis	Calcaneus (tuberosity on inferior surface)	Medial side of proximal phalanx of great toe	Abduction at metatarsophalangeal joint of great toe	Medial plantar nerve (L_4–L_5)
Plantar interosseus (3)	Bases and medial sides of metatarsal bones	Medial sides of toes 3–5	Adduction at metatarsophalangeal joints of toes 3–5	Lateral plantar nerve (S_1–S_2)
Dorsal interosseus (4)	Sides of metatarsal bones	Medial and lateral sides of toe 2; lateral sides of toes 3 and 4	Abduction at metatarsophalangeal joints of toes 3 and 4	Lateral plantar nerve (S_1–S_2)
Abductor digiti minimi	Inferior surface of calcaneus	Lateral side of proximal phalanx, toe 5	Abduction at metatarsophalangeal joint of toe 5	Lateral plantar nerve (L_4–L_5)

*Where appropriate, spinal nerves involved are given in parentheses.

tendon. This term comes from Greek mythology, as Achilles was a warrior who was invincible but for one vulnerable spot: the calcaneal tendon. Outside mythology, damage to the calcaneal tendon isn't a fatal problem. Although it is among the largest, strongest tendons in the body, its rupture is common. The applied forces increase markedly during rapid acceleration or deceleration. Sprinters can rupture the calcaneal tendon pushing off from starting blocks, and the elderly often snap this tendon during a stumble or fall. Surgery may be necessary to reposition and reconnect the torn ends of the tendon to promote healing.

Deep to the gastrocnemius and soleus are a pair of **fibularis** muscles, longus, and brevis, or *peroneus* (see Figure 11–22a,b). The fibularis produces eversion and extension (plantar flexion) at the ankle. Inversion is caused by the contraction of the **tibialis** (tib-ē-AH-lis) muscles. The large **tibialis anterior** flexes the ankle and opposes the gastrocnemius (see Figure 11–22b,c).

Important digital muscles originate on the surface of the tibia, the fibula, or both (see Figure 11–22c,d). Large synovial tendon sheaths surround the tendons of the tibialis anterior, **extensor digitorum longus**, and **extensor hallucis longus**, where they cross the ankle joint. The positions of these sheaths are stabilized by superior and inferior **extensor retinacula**.

The summary listed in Table 11–18 can be compared with Figure 11–22 for a more complete understanding of these muscles.

Intrinsic muscles of the foot originate on the tarsal and metatarsal bones (Figure 11–23 and Table 11–19). Their contractions move the toes and maintain the longitudinal arch of the foot. ⮌ p. 259

Figure 11–23 Intrinsic Muscles of the Foot. ATLAS: Plates 84a; 85a,b; 86c; 87a–c; 89

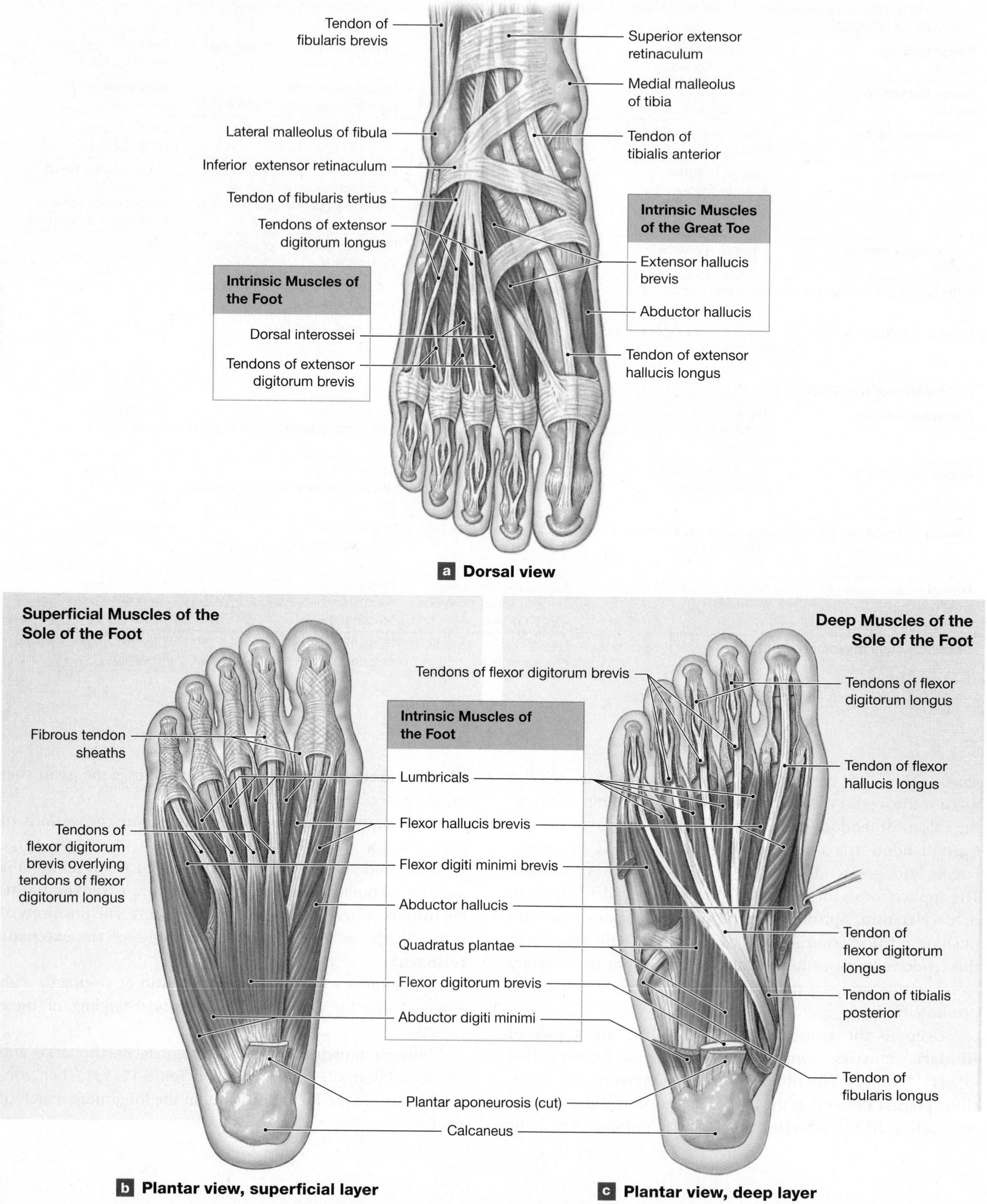

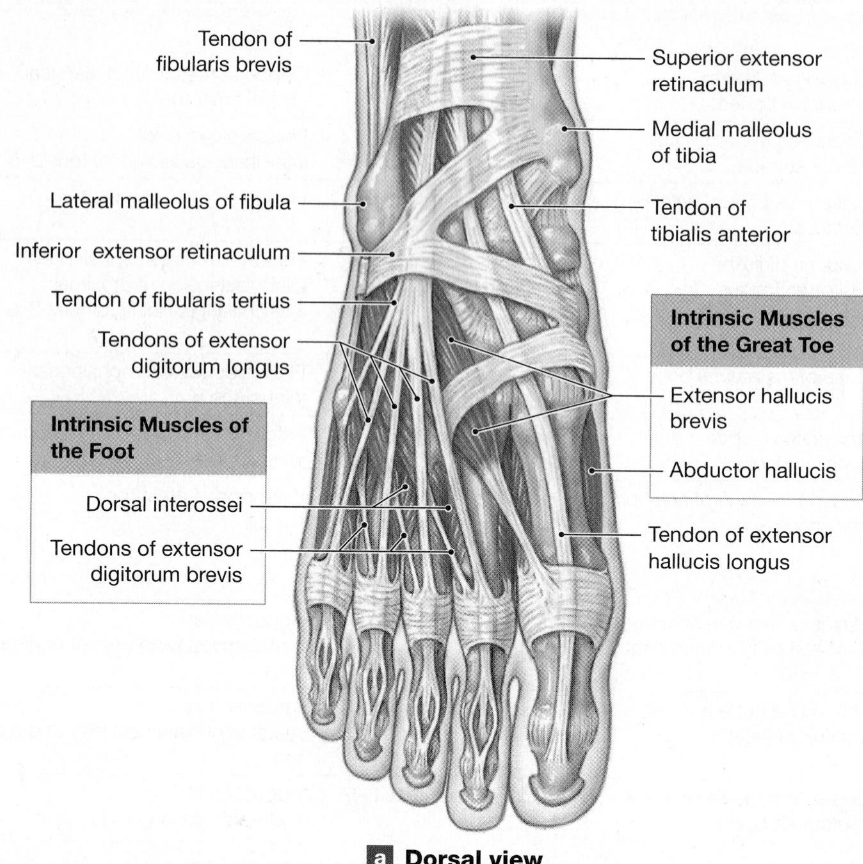
Tendon of fibularis brevis
Superior extensor retinaculum
Medial malleolus of tibia
Lateral malleolus of fibula
Tendon of tibialis anterior
Inferior extensor retinaculum
Tendon of fibularis tertius
Intrinsic Muscles of the Great Toe
Tendons of extensor digitorum longus
Extensor hallucis brevis
Intrinsic Muscles of the Foot
Abductor hallucis
Dorsal interossei
Tendons of extensor digitorum brevis
Tendon of extensor hallucis longus

a **Dorsal view**

Superficial Muscles of the Sole of the Foot
Deep Muscles of the Sole of the Foot
Tendons of flexor digitorum brevis
Tendons of flexor digitorum longus
Fibrous tendon sheaths
Intrinsic Muscles of the Foot
Lumbricals
Tendon of flexor hallucis longus
Flexor hallucis brevis
Tendons of flexor digitorum brevis overlying tendons of flexor digitorum longus
Flexor digiti minimi brevis
Abductor hallucis
Tendon of flexor digitorum longus
Quadratus plantae
Flexor digitorum brevis
Tendon of tibialis posterior
Abductor digiti minimi
Tendon of fibularis longus
Plantar aponeurosis (cut)
Calcaneus

b **Plantar view, superficial layer**

c **Plantar view, deep layer**

Checkpoint

19. **Shrugging your shoulders uses which muscles?**
20. **Baseball pitchers sometimes suffer from rotator cuff injuries. Which muscles are involved in this type of injury?**
21. **An injury to the flexor carpi ulnaris would impair which two movements?**
22. **Which leg movement would be impaired by injury to the obturator?**
23. **To what does a "pulled hamstring" refer?**
24. **How would a torn calcaneal tendon affect movement of the foot?**

See the blue Answers tab at the back of the book.

11-8 Exercise of the muscular system produces responses in multiple body systems

Learning Outcome Explain the functional relationship between the muscular system and other body systems, and explain the role of exercise in producing various responses in other body systems.

To operate at maximum efficiency, the muscular system must be supported by many other systems. The changes that take place during exercise provide a good example of such interaction. As noted earlier, active muscles consume oxygen and generate carbon dioxide and heat. The immediate effects of exercise on various body systems include the following:

- *Cardiovascular System:* Blood vessels in active muscles and the skin dilate, and heart rate increases. These adjustments speed up oxygen and nutrient delivery to and carbon dioxide removal from the muscle. They also bring heat to the skin for radiation into the environment.
- *Respiratory System:* Respiratory rate and depth of respiration increase. Air moves into and out of the lungs more quickly, keeping pace with the increased rate of blood flow through the lungs.
- *Integumentary System:* Blood vessels dilate, and sweat gland secretion increases. This combination increases evaporation at the skin surface and removes the excess heat generated by muscular activity.
- *Nervous and Endocrine Systems:* The responses of other systems are directed and coordinated through neural and endocrine (hormonal) adjustments in heart rate, respiratory rate, sweat gland activity, and mobilization of stored nutrient reserves.

Even when the body is at rest, the muscular system has extensive interactions with other systems. Build Your Knowledge Figure 11–24 summarizes the range of interactions between the muscular system and other body systems studied so far.

Checkpoint

25. **What major function does the muscular system perform for the body as a whole?**
26. **Identify the physiological effects of exercise on the cardiovascular, respiratory, and integumentary systems, and indicate the relationship between those physiological effects and the nervous and endocrine systems.**

See the blue Answers tab at the back of the book.

Build Your Knowledge

Figure 11–24 Integration of the MUSCULAR system with the other body systems presented so far.

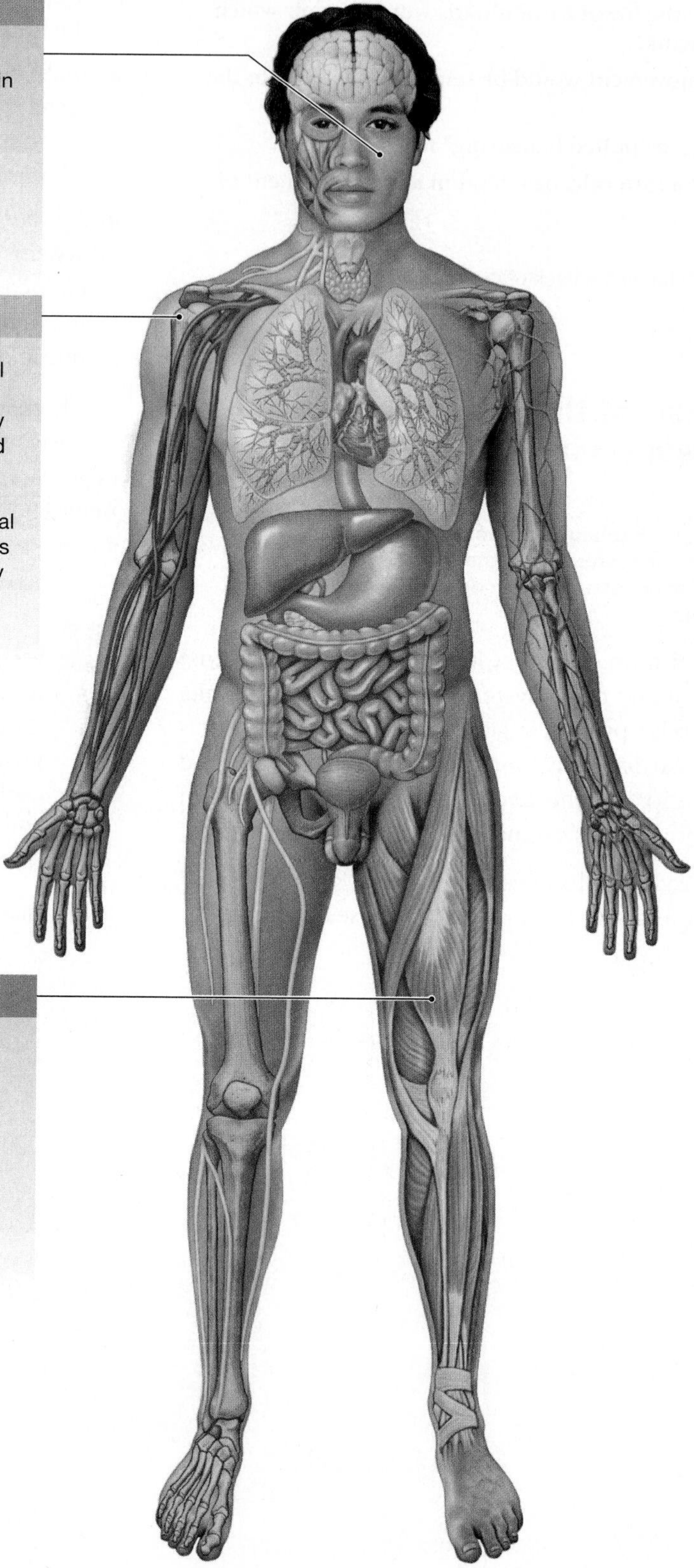

Integumentary System

- The Integumentary System removes excess body heat, synthesizes vitamin D_3 for calcium and phosphate absorption, and protects underlying muscles.
- The muscular system includes facial muscles that pull on the skin of the face to produce facial expressions

Skeletal System

- The Skeletal System provides mineral reserves for maintaining normal calcium and phosphate levels in body fluids, supports skeletal muscles, and provides sites of muscle attachment.
- The muscular system provides skeletal movement and support, and stabilizes bones and joints. Stresses exerted by tendons maintain normal bone structure and bone mass.

Muscular System

The muscular system performs these primary functions for the human body:

- It produces skeletal movement
- It helps maintain posture and body position
- It supports soft tissues
- It guards entrances and exits to the body
- It helps maintain body temperature

11 Chapter Review

Study Outline

An Introduction to the Muscular System p. 337

1. Structural variations among skeletal muscles affect their power, range, and speed of movement.

11-1 Fascicle arrangement is correlated with muscle power and range of motion p. 337

2. A muscle can be classified as a **parallel muscle**, **convergent muscle**, **pennate muscle**, or **circular muscle (sphincter)** according to its arrangement of fascicles. A pennate muscle may be *unipennate, bipennate,* or *multipennate. (Figure 11–1)*

11-2 The use of bones as levers increases muscle efficiency p. 339

3. A **lever** is a rigid structure that moves around a fixed point called the **fulcrum**. Levers can change the direction and effective strength of an applied force, and the distance and speed of the movement such a force produces.
4. Levers are classified as **first-class**, **second-class**, or **third-class levers** based on the relative position of three elements: applied force (AF), fulcrum (F), and load (L). Third-class levers are the most common levers in the body. *(Figure 11–2)*

11-3 The origins and insertions of muscles determine their actions p. 339

5. Each muscle can be identified by its *origin*, *insertion*, and *action*.
6. The site of attachment of the fixed end of a muscle is called the **origin**; the site where the movable end of the muscle attaches to another structure is called the **insertion**.
7. The movement produced when a muscle contracts is its **action**. *(Spotlight Figure 11–3)*
8. According to the function of its action, a muscle can be classified as an **agonist**, or **prime mover**; an **antagonist**; a **synergist**; or a **fixator**.

11-4 Descriptive terms are used to name skeletal muscles p. 343

9. The names of muscles commonly provide clues to their body region, origin and insertion, fascicle arrangement, position, structural characteristics, and action. *(Table 11–1)*

11-5 Axial muscles position the axial skeleton, and appendicular muscles support and move the appendicular skeleton p. 344

10. The **axial musculature** arises on the axial skeleton; it positions the head and vertebral column and moves the rib cage. The **appendicular musculature** stabilizes or moves the limbs of the appendicular skeleton. *(Figure 11–4)*
11. **Innervation** refers to the distribution of nerves that control a region or organ, including a muscle.

11-6 Axial muscles are muscles of the head and neck, vertebral column, trunk, and pelvic floor p. 347

12. The axial muscles fall into logical groups on the basis of location, function, or both.
13. Important muscles of facial expression include the **orbicularis oris**, **buccinator**, **occipitofrontalis**, and **platysma**. *(Figure 11–5; Table 11–2)*
14. Six extrinsic eye muscles (*oculomotor muscles*) control eye movements: the **inferior** and **superior rectus**, the **lateral** and **medial rectus**, and the **inferior** and **superior oblique**. *(Figure 11–6; Table 11–3)*
15. The muscles of mastication (chewing) are the **masseter**, **temporalis**, and **pterygoid** muscles. *(Figure 11–7; Table 11–4)*
16. The four muscles of the tongue are necessary for speech and swallowing and assist in mastication. They are the **palatoglossus**, **styloglossus**, **genioglossus**, and **hyoglossus**. *(Figure 11–8; Table 11–5)*
17. The muscles of the pharynx constrict the pharyngeal walls (**pharyngeal constrictors**), raise the larynx (**laryngeal elevators**), or raise the soft palate (**palatal muscles**). *(Figure 11–9; Table 11–6)*
18. The anterior muscles of the neck control the position of the larynx, depress the mandible, and provide a foundation for the muscles of the tongue and pharynx. The neck muscles include the **digastric** and **sternocleidomastoid** and seven muscles that originate or insert on the hyoid bone. *(Figure 11–10; Table 11–7)*
19. The superficial muscles of the spine can be classified into the **spinalis**, **longissimus**, and **iliocostalis** groups. *(Figure 11–11; Table 11–8)*
20. Other muscles of the spine include the **longus capitis** and **longus colli** of the neck, the small intervertebral muscles of the deep layer, and the **quadratus lumborum** of the lumbar region. *(Figure 11–11; Table 11–8)*
21. The oblique muscles include the **scalene** muscles and the **intercostal** and **transversus** muscles. The **external** and **internal intercostal muscles** are important in respiratory (breathing) movements of the ribs. Also important to respiration is the **diaphragm**. *(Figures 11–11, 11–12; Table 11–9)*
22. The **perineum** can be divided into an anterior **urogenital triangle** and a posterior **anal triangle**. The pelvic floor contains the muscles comprising the **pelvic diaphragm**. *(Figure 11–13; Table 11–10)*

11-7 Appendicular muscles are muscles of the shoulders, upper limbs, pelvis, and lower limbs p. 362

23. The **trapezius** affects the positions of the shoulder girdle, head, and neck. Other muscles inserting on the scapula include the **rhomboid**, **levator scapulae**, **serratus anterior**, **subclavius**, and **pectoralis minor**. *(Figures 11–14, 11–16; Table 11–11)*
24. Among muscles that move the arm, the **deltoid** and the **supraspinatus** are important abductors. The **subscapularis** and **teres major** produce medial rotation at the shoulder; the **infraspinatus** and **teres minor** produce lateral rotation; and the **coracobrachialis** produces flexion and adduction at the shoulder. *(Figures 11–14, 11–15, 11–16; Table 11–12)*
25. The **pectoralis major** flexes the shoulder joint, and the **latissimus dorsi** extends it. *(Figures 11–14, 11–15, 11–16; Table 11–12)*
26. Among muscles that move the forearm and hand, the actions of the **biceps brachii** and the **triceps brachii** (long head) affect the elbow joint. The **brachialis** and **brachioradialis** flex the elbow, opposed by the **anconeus**. The **flexor carpi ulnaris**, **flexor carpi radialis**, and **palmaris longus** cooperate to flex the wrist. The **extensor carpi radialis** and the **extensor carpi ulnaris** oppose them. The **pronator teres** and **pronator quadratus** pronate the forearm and are opposed by the **supinator**. *(Figures 11–16 to 11–19; Tables 11–13 to 11–15)*
27. Among muscles that move the thigh, **gluteal muscles** cover the lateral surfaces of the ilia. The largest is the **gluteus maximus**, which shares an insertion with the **tensor fasciae latae**. Together, these muscles pull on the **iliotibial tract**. *(Figure 11–20; Table 11–16)*
28. The **piriformis** and the **obturator** are the most important **lateral rotators**. The **adductors** can produce a variety of movements. *(Figure 11–20; Table 11–16)*
29. The **psoas major** and **iliacus** merge to form the **iliopsoas**, a powerful flexor of the hip. *(Figures 11–20, 11–21; Table 11–16)*
30. Among muscles that move the leg, the flexors of the knee include the **biceps femoris**, **semimembranosus**, and **semitendinosus** (the three **hamstrings**) and the **sartorius**. The **popliteus** unlocks the knee joint. *(Figures 11–21, 11–22a; Table 11–17)*
31. Collectively, the knee extensors are known as the **quadriceps femoris**. This group consists of the three **vastus** muscles (intermedius, lateralis, medialis) and the **rectus femoris**. *(Figure 11–21; Table 11–17)*
32. Among muscles that move the foot and toes, the **gastrocnemius** and **soleus** produce plantar flexion (ankle extension). The **fibularis brevis** and **longus** produce eversion as well as extension (plantar flexion) at the ankle. The fibularis tertius produces eversion and flexion (dorsiflexion) at the ankle. *(Figure 11–22; Table 11–18)*
33. Smaller muscles of the calf and shin (tibia) position the foot and move the toes. Muscles originating at the tarsal and metatarsal bones provide precise control of the phalanges. *(Figure 11–23; Table 11–19)*

11-8 Exercise of the muscular system produces responses in multiple body systems p. 382

34. Exercise illustrates the integration of the muscular system with the cardiovascular, respiratory, integumentary, nervous, and endocrine systems. *(Figure 11–24)*

11

Review Questions

See the blue Answers tab at the back of the book.

LEVEL 1 Reviewing Facts and Terms

1. Name the three pennate muscles in the following figure, and for each muscle indicate the type of pennate muscle based on the relationship of muscle fascicles to the tendon.

(a) ________
____
(b) ________
____
(c) ________
____

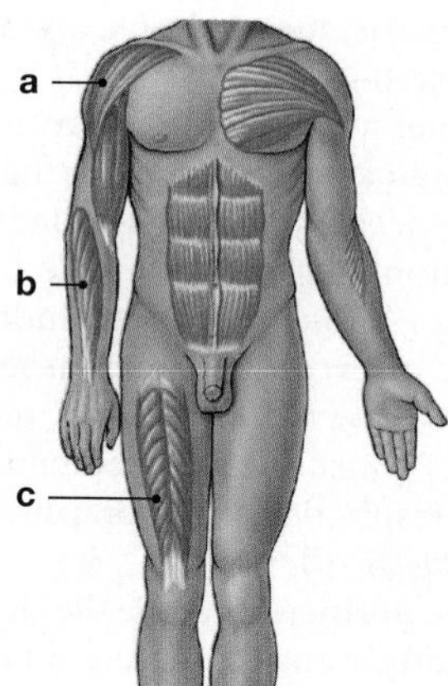
a
b
c

2. Label the three visible muscles of the rotator cuff in the following posterior view of the deep muscles that move the arm.

(a) ________
(b) ________
(c) ________

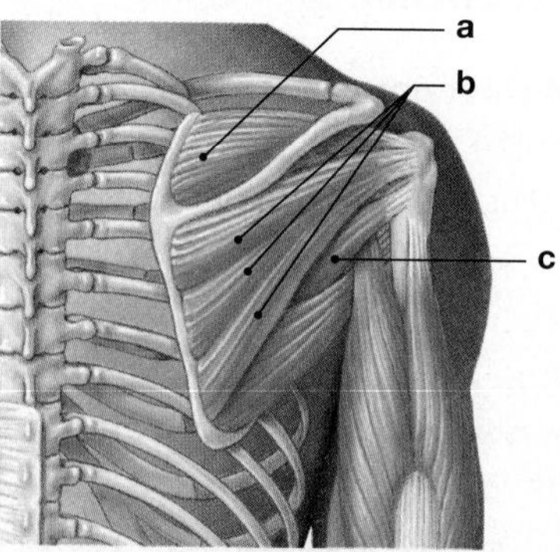
a
b
c

3. The bundles of muscle fibers within a skeletal muscle are called **(a)** muscles, **(b)** fascicles, **(c)** fibers, **(d)** myofilaments, **(e)** groups.

4. Levers make action more versatile by all of the following, *except* **(a)** changing the location of the muscle's insertion, **(b)** changing the speed of movement produced by an applied force, **(c)** changing the distance of movement produced by an applied force, **(d)** changing the strength of an applied force, **(e)** changing the direction of an applied force.
5. The more movable end of a muscle is the **(a)** insertion, **(b)** belly, **(c)** origin, **(d)** proximal end, **(e)** distal end.
6. The muscles of facial expression are innervated by cranial nerve **(a)** VII, **(b)** V, **(c)** IV, **(d)** VI.
7. The strongest masticatory muscle is the **(a)** pterygoid, **(b)** masseter, **(c)** temporalis, **(d)** mandible.
8. The muscle that rotates the eye medially is the **(a)** superior oblique, **(b)** inferior rectus, **(c)** medial rectus, **(d)** lateral rectus.
9. Important flexors of the vertebral column that act in opposition to the erector spinae are the **(a)** rectus abdominis, **(b)** longus capitis, **(c)** longus colli, **(d)** scalene.
10. The major extensor of the elbow is the **(a)** triceps brachii, **(b)** biceps brachii, **(c)** deltoid, **(d)** subscapularis.
11. The muscles that rotate the radius without producing either flexion or extension of the elbow are the **(a)** brachialis and brachioradialis, **(b)** pronator teres and supinator, **(c)** biceps brachii and triceps brachii, **(d)** a, b, and c.
12. The powerful flexor of the hip is the **(a)** piriformis, **(b)** obturator, **(c)** pectineus, **(d)** iliopsoas.
13. Knee extensors known as the quadriceps femoris consist of the **(a)** three vastus muscles and the rectus femoris, **(b)** biceps femoris, gracilis, and sartorius, **(c)** popliteus, iliopsoas, and gracilis, **(d)** gastrocnemius, tibialis, and peroneus.
14. List the four patterns of fascicle arrangement used to classify the different types of skeletal muscles.
15. What is an aponeurosis? Give two examples.
16. Which four muscle groups make up the axial musculature?
17. What three functions are accomplished by the muscles of the pelvic floor?
18. On which bones do the four rotator cuff muscles originate and insert?
19. What three functional groups make up the muscles of the lower limbs?

LEVEL 2 Reviewing Concepts

20. Of the following actions, the one that illustrates that of a second-class lever is **(a)** knee extension, **(b)** ankle extension (plantar flexion), **(c)** flexion at the elbow, **(d)** none of these.
21. Compartment syndrome can result from all of the following *except* **(a)** compressing a nerve in the wrist, **(b)** compartments swelling with blood due to an injury involving blood vessels, **(c)** torn ligaments in a given compartment, **(d)** pulled tendons in the muscles of a given compartment, **(e)** torn muscles in a particular compartment.
22. A(n) ____ develops when an organ protrudes through an abnormal opening.
23. Elongated bursae that reduce friction and surround the tendons that cross the posterior and anterior surfaces of the wrist form ____.
24. The muscles of the vertebral column include many posterior extensors but few anterior flexors. Why?
25. Why does a convergent muscle exhibit more versatility when contracting than does a parallel muscle?
26. Why can a pennate muscle generate more tension than can a parallel muscle of the same size?
27. Why is it difficult to lift a heavy object when the elbow is at full extension?
28. Which types of movements are affected when the hamstrings are injured?

LEVEL 3 Critical Thinking and Clinical Applications

29. Mary sees Jill coming toward her and immediately contracts her frontalis and procerus. She also contracts her right levator labii. Is Mary glad to see Jill? How can you tell?
30. Mary's newborn is having trouble suckling. The doctor suggests that it may be a problem with a particular muscle. What muscle is the doctor probably referring to? **(a)** orbicularis oris, **(b)** buccinator, **(c)** masseter, **(d)** risorius, **(e)** zygomaticus.
31. While unloading her car trunk, Amy strains a muscle and as a result has difficulty moving her arm. The doctor in the emergency room tells her that she strained her pectoralis major. Amy tells you that she thought the pectoralis major was a chest muscle and doesn't understand what that has to do with her arm. What should you tell her?

11

+ CLINICAL CASE Wrap-Up Downward-Facing Dog

At his 3-month visit, the nurse practitioner evaluates Rick's gains following the regular, low-intensity yoga exercise program. She conducts a review of systems, with a focus on his musculoskeletal system. She is looking for improved muscular strength and increased range of motion at the joints. She'll also inquire about Rick's arthritis pain level and how successfully he is able to perform his ADLs, or activities of daily living.

She observes that his muscle tone has improved overall. He can flex and extend more freely at joints in his arms and legs. He can get up from a chair and sit down with smoother motion than he could at the last visit. He is still working on balance issues, but he can reach up and around, as in putting on his coat. "I was a gymnast for my college team, back in the day," Rick tells her. "This is the best I've felt in years!"

1. Look at the photo of the person executing Downward-Facing Dog. Which muscles are contracting in this pose?
2. Which of Rick's muscles will help him maintain better posture as he goes about his day?

See the blue Answers tab at the back of the book.

Related Clinical Terms

charley horse: Common name for a muscle spasm, especially in the leg.

compartment syndrome: A condition in which increased pressure within the muscle compartment of a limb produces ischemia or "blood starvation."

fibromyositis: Chronic illness characterized by widespread musculoskeletal aches, pains, and stiffness, and soft tissue tenderness.

groin pull: An injury that is due to a strain of the muscles of the inner thigh.

impingement syndrome: Pain on elevation of the shoulder due to an injured or inflamed tendon or bursa coming into contact with the overlying acromial process.

physical therapist: Healthcare professional who uses specially designed exercises and equipment to help patients regain or improve their physical abilities.

plantar fasciitis: Inflammation of the plantar fascia causing foot or heel pain.

shin splints: Pain along the shinbone (tibia) caused by an overload on the tibia and connective tissues that connect muscle to the bone.

tenosynovitis: Inflammation of a tendon and the sheath that covers it.

torticollis: A shortening or contraction of the muscles of the neck resulting in the head being tipped to one side with the chin turned to the other side.

12 Nervous Tissue

Learning Outcomes

These Learning Outcomes correspond by number to this chapter's sections and indicate what you should be able to do after completing the chapter.

- 12-1 ■ **Describe the anatomical and functional divisions of the nervous system.** p. 390
- 12-2 ■ **Sketch and label the structure of a typical neuron, describe the functions of each component, and classify neurons on the basis of their structure and function.** p. 392
- 12-3 ■ **Describe the locations and functions of the various types of neuroglia.** p. 395
- 12-4 ■ **Explain how the resting membrane potential is established and maintained and how the membrane potential can change.** p. 402
- 12-5 ■ **Describe the events involved in the generation and propagation of an action potential and the factors involved in determining the speed of action potential propagation.** p. 409
- 12-6 ■ **Describe the structure of a synapse, and explain the mechanism involved in synaptic activity.** p. 416
- 12-7 ■ **Describe the major types of neurotransmitters and neuromodulators, and discuss their effects on postsynaptic membranes.** p. 420
- 12-8 ■ **Discuss the interactions that enable information processing to occur in nervous tissue.** p. 424

CLINICAL CASE Did President Franklin D. Roosevelt Really Have Polio?

On August 10, 1921, at age 39, Franklin D. Roosevelt went to bed early feeling tired and complaining of back pain, fever, and chills. The next day, he experienced progressive weakness: first in one leg, then the other, in an ascending pattern. By August 12, both of his legs were paralyzed and numb, though he still felt pain when lightly touched. Over the following month, the signs and symptoms progressed to involve weakness and abnormal sensations of his upper extremities and his face, and he developed an inability to voluntarily urinate or defecate. His diagnosis of *poliomyelitis*, or *polio*, included him in one of the most feared epidemics of the early 20th century.

In the following months, Mr. Roosevelt gradually recovered from the severe weakness and pain. He also regained control of his bodily functions. His legs and buttocks, however, were permanently weakened.

Franklin Roosevelt went on to become the 32nd and longest-serving president of the United States. He assisted in the founding of the March of Dimes, an organization that contributed to the elimination of polio from most of the world. **But looking back on his clinical course, did President Franklin D. Roosevelt really suffer from poliomyelitis, a viral disease of the motor neurons of the spinal cord? To find out, turn to the Clinical Case Wrap-Up on p. 431.**

12

An Introduction to the Nervous System and Nervous Tissue

The **nervous system** includes the various nervous organs throughout the body. ⮌ p. 144 These include the *brain*; the *spinal cord*; the *receptors* in complex sense organs, such as the eye and ear; and the *nerves* that link the nervous system with other body systems. Together these organs receive information from both external and internal stimuli, process that information, and send signals to initiate any necessary response. The organs of the nervous system are made up of **nervous tissue**, as well as supporting blood vessels and connective tissues.

The basic functional units of the nervous system are cells called *neurons* (NŪ-ronz), which are cells specialized for intercellular communication. Nervous tissue also includes supporting cells called *neuroglia* (nū-RŌG-lē-uh; *glia*, glue), or *glial cells.* Neuroglia have functions essential to the survival and functionality of neurons and to preserving the physical and biochemical structure of nervous tissue.

After providing a brief introduction to the nervous system as a whole, this chapter explores the structure of nervous tissue and the functional principles that govern neural activities. Later chapters then cover the various parts of the nervous system.

12-1 The nervous system has anatomical and functional divisions

Learning Outcome Describe the anatomical and functional divisions of the nervous system.

We can look at the nervous system from both anatomical and functional perspectives.

The Anatomical Divisions of the Nervous System

Viewed anatomically, the nervous system has three divisions: the *central nervous system* the *peripheral nervous system*, and the *enteric nervous system* (Figure 12–1). The **central nervous system (CNS)** consists of the brain and spinal cord. These complex organs include not only nervous tissue, but also blood vessels and the various connective tissues that physically protect and support them.

What are the general functions of the CNS? It is responsible for integrating, processing, and coordinating sensory data and motor commands. Sensory data convey information about conditions inside or outside the body. Motor commands control or adjust peripheral organs, such as skeletal muscles. When you stumble, for example, the CNS integrates information about your balance and the position of your limbs and then coordinates your recovery by sending motor commands to appropriate skeletal muscles—all in a split second and without your conscious effort. Additionally, the brain is the seat of higher functions, such as intelligence, memory, learning, and emotion.

The **peripheral nervous system (PNS)** includes all the nervous tissue outside the CNS and the ENS. There are two subdivisions of the PNS: the *somatic nervous system* and *autonomic nervous system*. The PNS delivers sensory information to the CNS and carries motor commands to peripheral tissues and systems, except those of the enteric nervous system. Bundles of axons, called *nerve fibers*, carry sensory information and motor commands in the PNS. ⮌ p. 144 Such bundles, with associated blood vessels and connective tissues, are called *peripheral nerves*, or simply **nerves**. Nerves connected to the brain are called **cranial nerves**, and those attached to the spinal cord are called **spinal nerves**.

Figure 12–1 An Overview of the Nervous System.

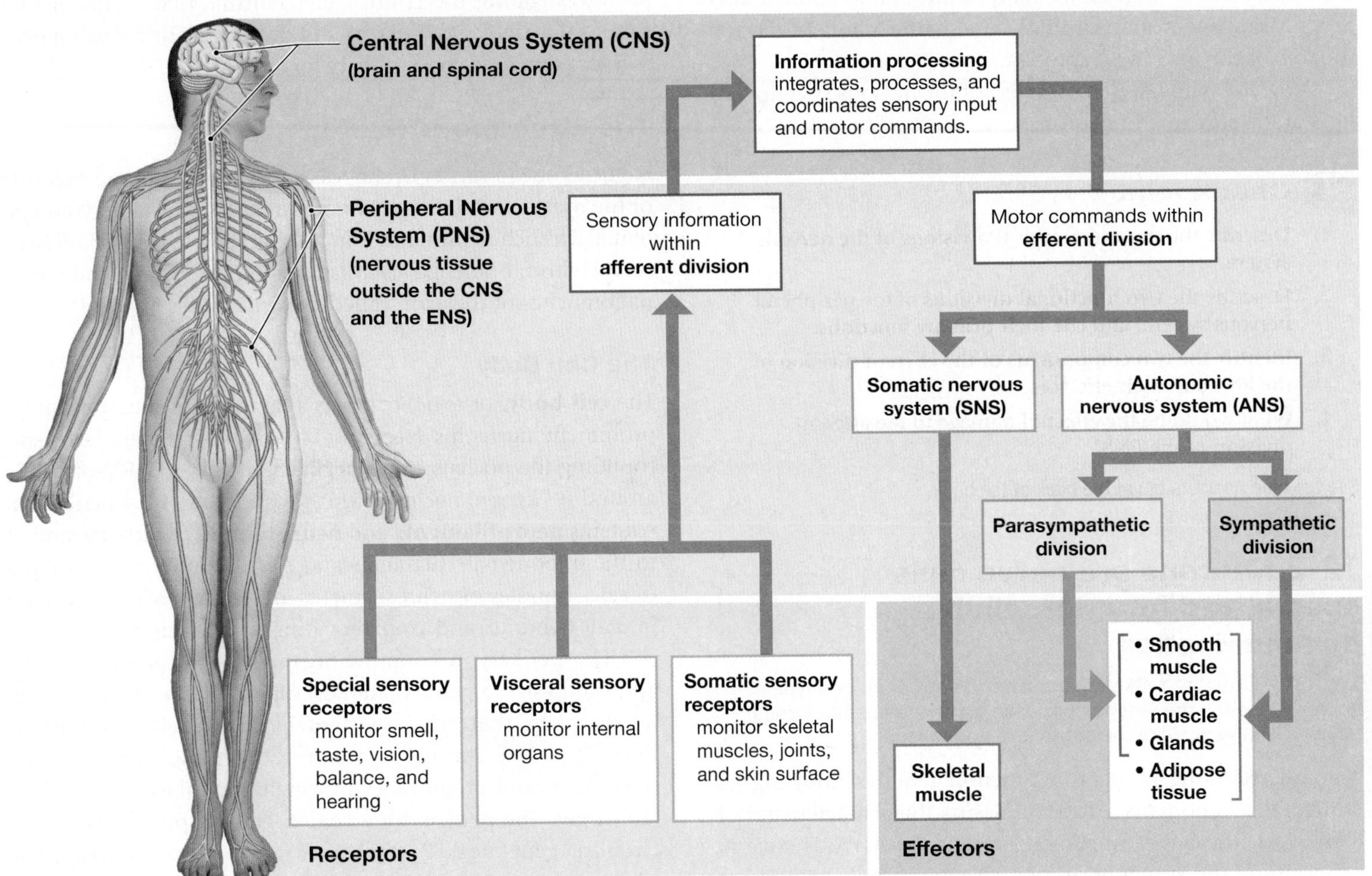
Central Nervous System (CNS)
(brain and spinal cord)
Information processing
integrates, processes, and coordinates sensory input and motor commands.
Peripheral Nervous System (PNS)
(nervous tissue outside the CNS and the ENS)
Sensory information within afferent division
Motor commands within efferent division
Somatic nervous system (SNS)
Autonomic nervous system (ANS)
Parasympathetic division
Sympathetic division
• Smooth muscle
• Cardiac muscle
• Glands
• Adipose tissue
Special sensory receptors monitor smell, taste, vision, balance, and hearing
Visceral sensory receptors monitor internal organs
Somatic sensory receptors monitor skeletal muscles, joints, and skin surface
Skeletal muscle
Receptors
Effectors

The Functional Divisions of the Nervous System

The CNS does not have separated functional divisions like the PNS, which is divided into afferent and efferent divisions (see Figure 12–1). The **afferent division** (*afferens*, to bring to) of the PNS brings sensory information *to* the CNS from receptors in peripheral tissues and organs. **Receptors** are sensory structures that either detect changes in the environment (internal or external) or respond to specific stimuli. Our receptors range from the slender cytoplasmic extensions of single cells to complex receptor organs, such as the eye and ear. Sensory receptors may be neurons or specialized cells of other tissues.

The **efferent division** (*effero*, to bring out) of the PNS carries motor commands *from* the CNS to muscles, glands, and adipose tissue. These target organs, which respond by *doing* something, are called **effectors**. The efferent division itself has two divisions, *somatic* and *autonomic*:

- The **somatic nervous system (SNS)** controls skeletal muscle contractions. *Voluntary* contractions are under conscious control. For example, you exert conscious control over your arm as you raise a full glass of water to your lips. *Involuntary* contractions may be simple, automatic responses or complex movements, but they are controlled at the subconscious level, outside your awareness. For instance, if you accidentally place your hand on a hot stove, you will withdraw it immediately, usually before you even notice any pain. This type of automatic response is called a **reflex**.
- The **autonomic nervous system (ANS)**, or *visceral motor system*, automatically regulates smooth muscle, cardiac muscle, glandular secretions, and adipose tissue at the subconscious level. The ANS includes a *parasympathetic division* and a *sympathetic division*, which commonly have antagonistic (opposite) effects. For example, parasympathetic activity slows the heart rate, whereas sympathetic activity accelerates the heart rate.

The third nervous system division is the **enteric nervous system (ENS)**. The **enteric nervous system** is an extensive network of neurons and nerve networks in the walls of the digestive tract. Although the sympathetic and parasympathetic divisions both influence ENS activities, the ENS initiates and coordinates many complex visceral reflexes locally and without instructions from the CNS. Altogether, the ENS has about 100 million neurons—at least as many as the spinal cord. It also uses the same neurotransmitters found in the brain. We discuss the role of the ENS in Chapter 24.

Now that you have an overview of the nervous system, let's look at the structure of neurons.

Checkpoint

1. **Describe the three anatomical divisions of the nervous system.**
2. **Describe the two functional divisions of the peripheral nervous system, and cite their primary functions.**
3. **Identify the two components of the efferent division of the PNS and their effectors.**
4. **What would be the effect of damage to the afferent division of the PNS?**

See the blue Answers tab at the back of the book.

12

12-2 Neurons are nerve cells specialized for intercellular communication

Learning Outcome Sketch and label the structure of a typical neuron, describe the functions of each component, and classify neurons on the basis of their structure and function.

An understanding of neuron function requires knowing its structural components. After we discuss the characteristics of a **neuron**, the basic functional unit of the nervous system, we examine the structure of a representative neuron and see how it is specialized for intercellular communication. Then we will consider the structural and functional classifications of neurons.

Functional Characteristics of Neurons

Neurons are generally long-lived and have a high metabolic rate. Like skeletal muscle cells, neurons have an excitable plasma membrane. The generation and propagation of action potentials places a substantial demand on a neuron's energy resources. Energy is also expended in synthesizing and secreting the chemical compounds necessary for sending information from neuron to neuron. Mitochondria in various parts of a neuron generate the energy needed by an active neuron.

Most neurons lack centrioles, important organelles that help to organize the cytoskeleton and the microtubules that move chromosomes during mitosis. ⮌ p. 105 As a result, typical CNS neurons cannot divide. For this reason, they cannot be replaced if lost to injury or disease. Neural stem cells persist in the adult nervous system, but they are typically inactive except in the nose, where the regeneration of olfactory (smell) receptors maintains our sense of smell, and in the *hippocampus*, a part of the brain involved in storing memories. Researchers are investigating the control mechanisms that trigger neural stem cell activity, with the goal of preventing or reversing neuron loss due to trauma, disease, or aging.

The Structure of Neurons

Neurons have a variety of shapes. Figure 12–2 shows an example of the most common type of neuron in the central nervous system. Each such neuron has four general regions: a large *cell body*; several short, branched *dendrites*; a single, long *axon*; and terminal branches of the axon called *telodendria* (see Figure 12–2a).

The Cell Body

The **cell body**, or *soma*, contains a large, round nucleus with a prominent nucleolus (see Figure 12–2b). The cytoplasm surrounding the nucleus is the **perikaryon** (per-i-KAR-ē-on; *peri-*, around + *karyon*, nucleus). The cytoskeleton of the perikaryon contains **neurofilaments** and **neurotubules**, which are similar to the intermediate filaments and microtubules of other types of cells. Bundles of neurofilaments, called **neurofibrils**, extend into the dendrites and axon, providing internal support for them.

The perikaryon contains organelles that provide energy and synthesize organic materials, especially the chemical *neurotransmitters* that are important in cell-to-cell communication. ⮌ p. 304 The numerous mitochondria, free and fixed ribosomes, and membranes of rough endoplasmic reticulum (RER) give the perikaryon a coarse, grainy appearance. Mitochondria generate ATP to meet neuronal energy demand. The ribosomes and RER synthesize proteins.

Some areas of the perikaryon contain clusters of RER and free ribosomes. These regions, which stain darkly with cresyl violet, are called *Nissl bodies* (after the German neurologist Franz Nissl, who first described them). Nissl bodies give a gray color to areas containing neuron cell bodies—the *gray matter* seen in gross dissection of the brain and spinal cord.

Dendrites, Axons, and Telodendria

A variable number of slender, sensitive processes (extensions) known as **dendrites** extend and branch out from the cell body (see Figure 12–2a). Dendrites play key roles in intercellular communication. Typical dendrites are highly branched, and some branches are studded with fine 0.5 to 1 μm long projections called *dendritic spines*, which participate in synapses. In the CNS, a neuron receives information from other neurons primarily at the dendritic spines, which may represent 80–90 percent of the neuron's total surface area.

An **axon** is a long cytoplasmic process, in fact up to a meter in length, which is capable of propagating an electrical impulse known as an *action potential*. ⮌ p. 304 The **axoplasm** (AK-sō-plazm), or cytoplasm of the axon, contains neurofibrils, neurotubules, small vesicles, lysosomes, mitochondria, and various

Figure 12–2 The Anatomy of a Typical Neuron.

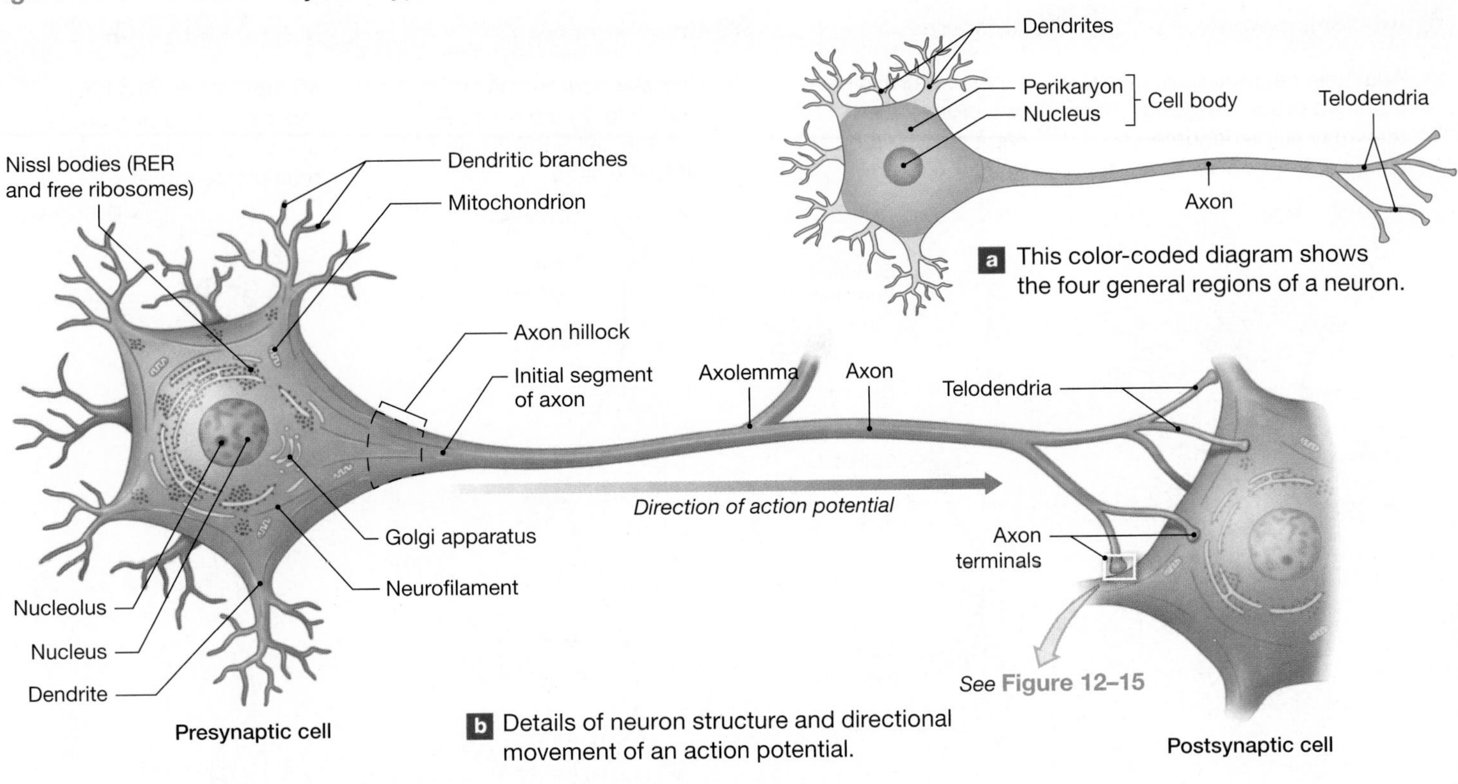
Dendrites
Perikaryon
Nucleus
Cell body
Telodendria
Axon
Nissl bodies (RER and free ribosomes)
Dendritic branches
Mitochondrion
Axon hillock
Initial segment of axon
Axolemma
Axon
Telodendria
Direction of action potential
Golgi apparatus
Neurofilament
Axon terminals
Nucleolus
Nucleus
Dendrite
Presynaptic cell
See Figure 12–15
Postsynaptic cell

a This color-coded diagram shows the four general regions of a neuron.

b Details of neuron structure and directional movement of an action potential.

? The axon hillock is a region between what two general regions of a neuron?

enzymes. The **axolemma** (*lemma*, husk), the plasma membrane of the axon, surrounds the axoplasm. In the CNS, the axolemma may be exposed to the interstitial fluid or, as we'll see, it may be covered by the cellular processes of neuroglia. The base, or **initial segment**, of the axon in a typical neuron joins the cell body at a thickened region known as the **axon hillock** (see Figure 12–2b).

An axon may branch along its length, producing side branches known as **collaterals**. Collaterals enable a single neuron to communicate with several other cells. The main axon trunk and any collaterals end in a series of fine extensions called **telodendria** (tel-ō-DEN-drē-uh; *telo-*, end + *dendron*, tree), or *terminal branches*. The telodendria, in turn, end at **axon terminals** (also called *synaptic terminals*), which play a role in communication with another cell. A *synapse* is where a neuron communicates with another cell, possibly another neuron (covered in Section 12-6).

The movement of materials between the cell body and axon terminals is called **axonal (axoplasmic) transport**. These materials travel the length of the axon on neurotubules in the axoplasm, pulled along by protein "molecular motors," called *kinesin* and *dynein*, which use ATP. Some materials travel slowly, at rates of a few millimeters per day. This transport mechanism is known as the "slow stream." Some vesicles move much more rapidly, traveling in the "fast stream" at a rate as high as 1000 mm per day, but the average is around 300 mm per day.

Axonal transport occurs simultaneously in both directions. The flow of materials from the cell body to the axon terminal is *anterograde* (AN-ter-ō-grād; *antero-*, forward) *flow*, carried by kinesin. At the same time, other substances are transported from the axon terminal toward the cell body in *retrograde* (RET-rō-grād) *flow* (*retro*, backward), carried by dynein. If debris or unusual chemicals appear in the axon terminal, retrograde flow soon delivers them to the cell body. There they may then alter the activity of the cell by turning certain genes on or off.

+ Clinical Note Rabies

Rabies is an example of a clinical condition involving retrograde flow. When a bite from a rabid animal injects the rabies virus into peripheral tissues, the virus particles quickly enter axon terminals and peripheral axons. Retrograde flow then carries the virus into the CNS, with fatal results. Many toxins (including heavy metals), some pathogenic bacteria, and other viruses also bypass CNS defenses through axonal transport.

12

Figure 12–3 Structural Classifications of Neurons.

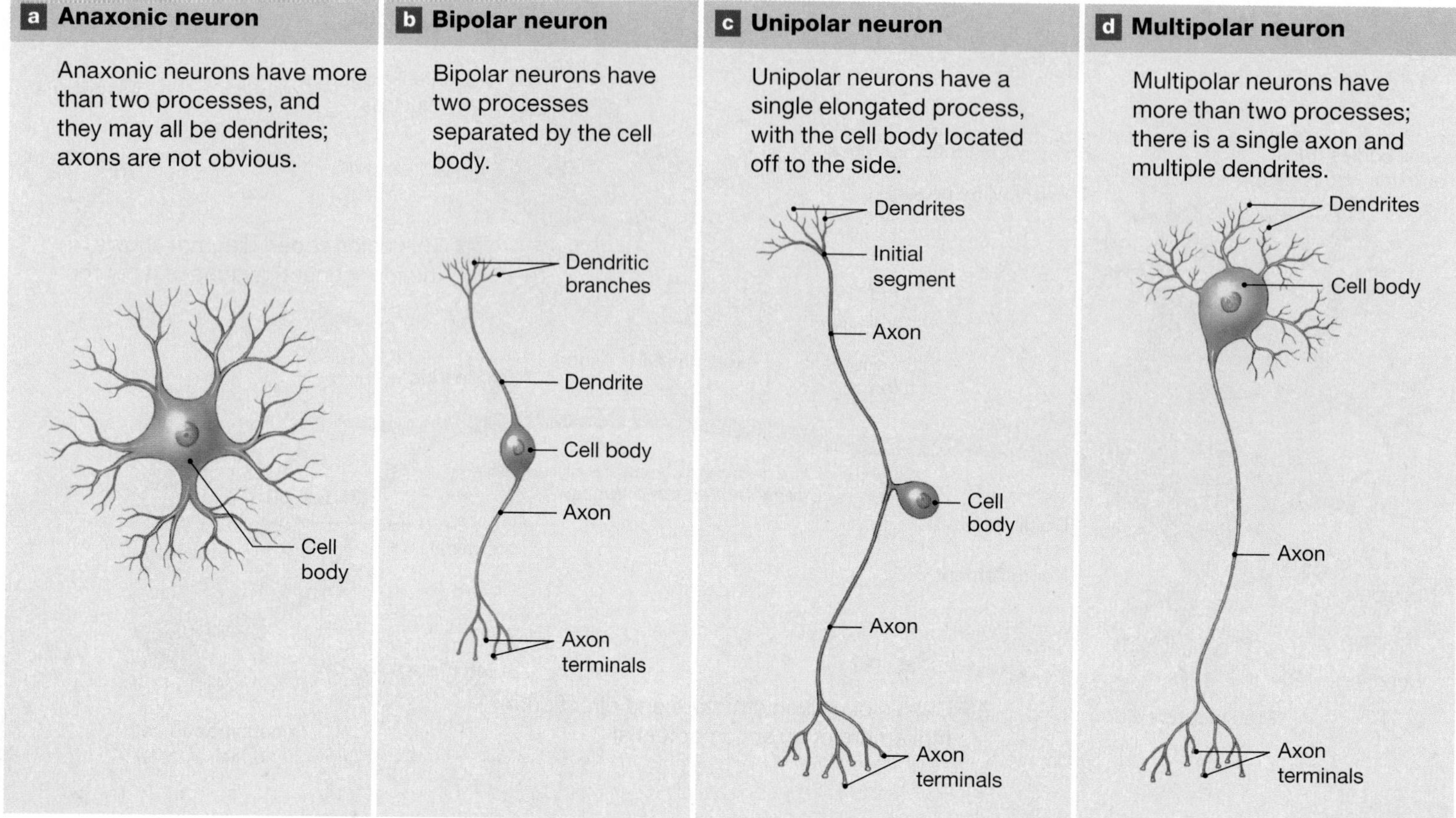
a Anaxonic neuron
Anaxonic neurons have more than two processes, and they may all be dendrites; axons are not obvious.
Cell body
b Bipolar neuron
Bipolar neurons have two processes separated by the cell body.
Dendritic branches
Dendrite
Cell body
Axon
Axon terminals
c Unipolar neuron
Unipolar neurons have a single elongated process, with the cell body located off to the side.
Dendrites
Initial segment
Axon
Cell body
Axon
Axon terminals
d Multipolar neuron
Multipolar neurons have more than two processes; there is a single axon and multiple dendrites.
Dendrites
Cell body
Axon
Axon terminals

The Classification of Neurons

We can group neurons by structure or by function.

Structural Classification of Neurons

Neurons are classified as *anaxonic, bipolar, unipolar,* or *multipolar* on the basis of the relationship of the dendrites to the cell body and the axon (Figure 12–3):

- **Anaxonic** (an-aks-ON-ik; *an-*, without) **neurons** are small and have numerous dendrites, but no obvious axons. They do have axons, yet they are not readily visible, even when viewed with a microscope. Anaxonic neurons are located in the brain and in special sense organs. Their functions are poorly understood.
- **Bipolar neurons** have two distinct processes—one dendrite (that branches extensively into dendritic branches at its distal tip) and one axon—with the cell body between the two. Bipolar neurons are rare. They occur in special sense organs, where they relay information about sight, smell, or hearing from receptor cells to other neurons. Bipolar neurons are relatively small: The largest measure less than 30 μm from end to end.
- In a **unipolar neuron**, or *pseudounipolar neuron*, the dendrites and axon are continuous—basically, fused—and the cell body lies off to one side. In such a neuron, the initial segment lies where the dendrites converge. The rest of the process, which carries action potentials, is usually considered an axon. Most sensory neurons of the peripheral nervous system are unipolar. Their axons may extend a meter or more, ending at synapses in the central nervous system. The longest axons of unipolar neurons carry sensations from the tips of the toes to the spinal cord.
- **Multipolar neurons** have two or more dendrites and a single axon. They are the most common neurons in the CNS. All the motor neurons that control skeletal muscles, for example, are multipolar neurons. The longest axons of multipolar neurons carry motor commands from the spinal cord to small muscles that move the toes.

Functional Classification of Neurons

Alternatively, we can categorize neurons by function as (1) *sensory neurons,* (2) *motor neurons,* or (3) *interneurons.*

Sensory Neurons. Sensory neurons, or *afferent neurons,* form the afferent division of the PNS. The cell bodies of sensory neurons are located in peripheral *sensory ganglia.* (A **ganglion** is a collection of neuron cell bodies in the PNS.) Sensory neurons are unipolar neurons whose processes, known as **afferent**

fibers, extend between a sensory receptor and the CNS (they deliver information from sensory receptors to the spinal cord or brain). The human body's 10 million or so sensory neurons collect information about the external or internal environment. **Somatic sensory neurons** monitor the outside world and our position within it. **Visceral sensory neurons** monitor internal conditions and the status of other organ systems (see Figure 12–1).

Sensory receptors are either the processes of specialized sensory neurons or cells monitored by sensory neurons. We can broadly categorize these receptors into three groups:

- **Interoceptors** (*intero-*, inside) monitor the digestive, respiratory, cardiovascular, urinary, and reproductive systems, and provide sensations of distension (stretch), deep pressure, and pain.
- **Exteroceptors** (*extero-*, outside) provide information about the external environment in the form of touch, temperature, or pressure sensations and the more complex senses of taste, smell, sight, equilibrium (balance), and hearing.
- **Proprioceptors** (prō-prē-ō-SEP-terz) monitor the position and movement of skeletal muscles and joints.

Motor Neurons. Motor neurons, or *efferent neurons*, form the efferent division of the PNS. The half a million motor neurons carry instructions from the CNS to peripheral effectors in a peripheral tissue, organ, or organ system. Axons of these neurons that travel away from the CNS are called **efferent fibers**. As noted earlier, the two major efferent systems are the somatic nervous system (SNS) and the autonomic (visceral) nervous system (ANS) (see Figure 12–1).

The somatic nervous system includes all the **somatic motor neurons** that innervate skeletal muscles. You have conscious control over the activity of the SNS. The cell body of a somatic motor neuron lies in the CNS, and its axon runs within a peripheral nerve to innervate skeletal muscle fibers at neuromuscular junctions.

You do not have conscious control over the activities of the ANS. **Visceral motor neurons** innervate all peripheral effectors other than skeletal muscles—that is, smooth muscle, cardiac muscle, glands, and adipose tissue throughout the body. The axons of visceral motor neurons in the CNS innervate a second set of visceral motor neurons in peripheral *autonomic ganglia*. The neurons whose cell bodies are located in those ganglia innervate and control peripheral effectors.

Tips & Tools

To distinguish between *afferent* and *efferent*, think of the **SAME** principle: **S** is for **s**ensory, **A** is for **a**fferent, **M** is for **m**otor, and **E** is for **e**fferent. This way, you associate the **S** and **A** together and the **M** and **E** together.

To get from the CNS to a visceral effector such as a smooth muscle cell, a signal must travel along one axon, be relayed across a synapse, and then travel along a second axon to its final destination. The axons extending from the CNS to an autonomic ganglion are called *preganglionic fibers*, and axons connecting the ganglion cells with the peripheral effectors are known as *postganglionic fibers*.

Interneurons. Interneurons are located between sensory and motor neurons. The 20 billion or so interneurons (*association neurons*) outnumber all other types of neurons combined. Most are located within the brain and spinal cord, but some are in autonomic ganglia. The main function of interneurons is integration—they distribute sensory information and coordinate motor activity. The more complex the response to a given stimulus, the more interneurons are involved. Interneurons also play a part in all higher functions, such as memory, planning, and learning.

We now turn our attention to the neuroglia, cells that support and protect the neurons.

✓ Checkpoint

5. **Name the structural components of a typical neuron.**
6. **Classify neurons according to their structure.**
7. **Classify neurons according to their function.**
8. **Are unipolar neurons in a tissue sample more likely to function as sensory neurons or motor neurons?**

See the blue Answers tab at the back of the book.

12-3 CNS and PNS neuroglia support and protect neurons

Learning Outcome Describe the locations and functions of the various types of neuroglia.

Neuroglia separate and protect neurons, provide a supportive framework for nervous tissue, act as phagocytes, and help regulate the composition of the interstitial fluid. Neuroglia are abundant and diverse; in fact, they make up about half the volume of the nervous system, far outnumbering the neurons.

The organization of nervous tissue in the CNS differs from that in the PNS, primarily because the CNS has a greater variety of neuroglia. Histological descriptions have been available for quite some time, but the technical problems involved in isolating and manipulating individual glial cells have provided obstacles to understanding their functions. Figure 12–4 summarizes what we know about the major neuroglial populations in the CNS and PNS.

Neuroglia of the Central Nervous System

The central nervous system has four types of neuroglia: (1) *astrocytes*, (2) *ependymal cells*, (3) *oligodendrocytes*, and (4) *microglia* (see Figure 12–4).

Figure 12–4 An Introduction to Neuroglia.

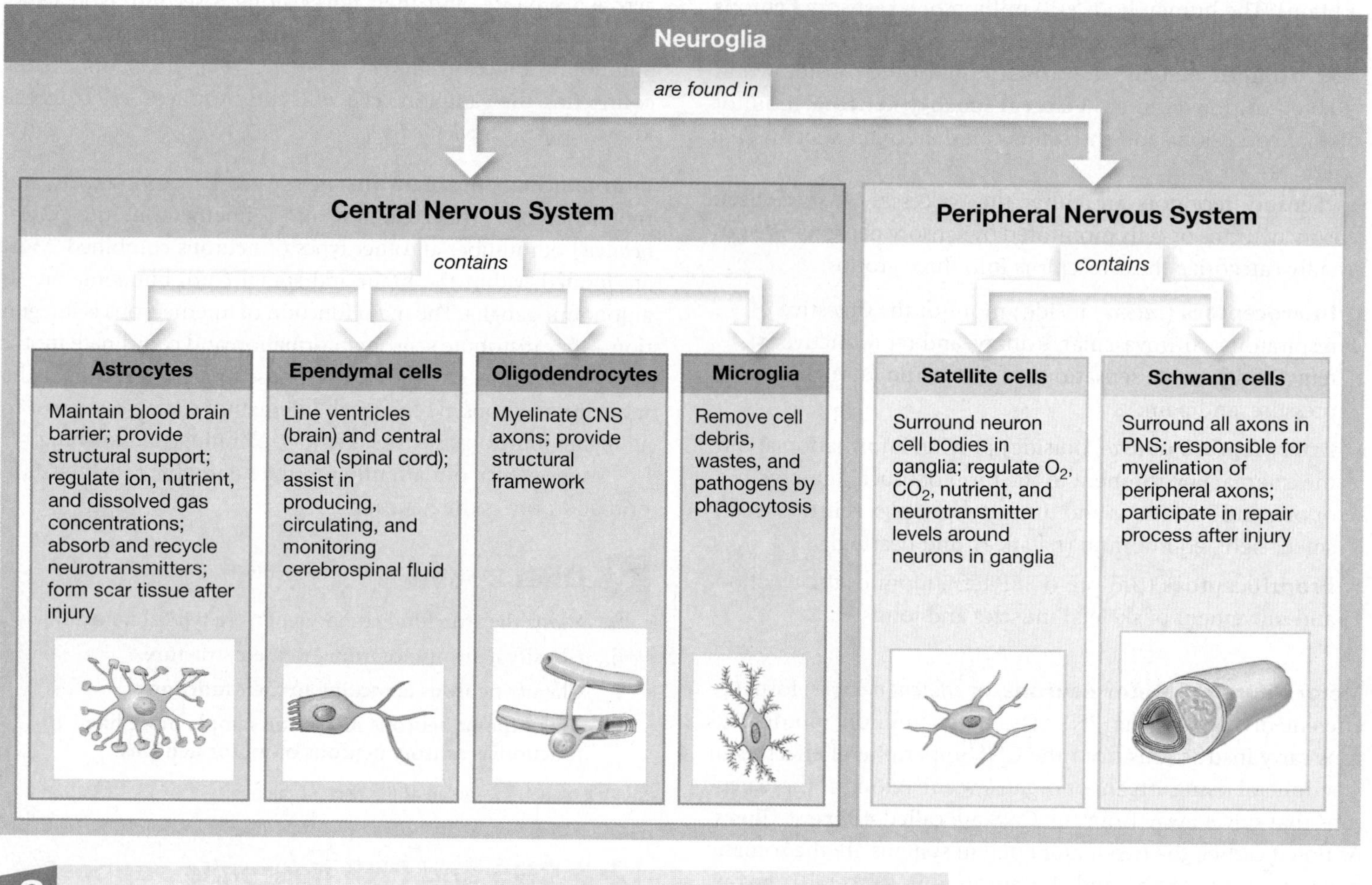
Neuroglia
are found in
Central Nervous System
contains
Astrocytes
Maintain blood brain barrier; provide structural support; regulate ion, nutrient, and dissolved gas concentrations; absorb and recycle neurotransmitters; form scar tissue after injury
Ependymal cells
Line ventricles (brain) and central canal (spinal cord); assist in producing, circulating, and monitoring cerebrospinal fluid
Oligodendrocytes
Myelinate CNS axons; provide structural framework
Microglia
Remove cell debris, wastes, and pathogens by phagocytosis
Peripheral Nervous System
contains
Satellite cells
Surround neuron cell bodies in ganglia; regulate O_2, CO_2, nutrient, and neurotransmitter levels around neurons in ganglia
Schwann cells
Surround all axons in PNS; responsible for myelination of peripheral axons; participate in repair process after injury

12

? Which neuroglia myelinate CNS axons and which neuroglia myelinate PNS axons?

Astrocytes

Astrocytes (AS-trō-sīts; *astro-*, star + *-cyte*, cell) are the largest and most numerous neuroglia in the CNS (Figure 12–5). The star-shaped astrocytes (hence their name) have many slender cytoplasmic processes that end in expanded "feet." Astrocytes are packed with microfilaments that extend across the breadth of the cell and its processes. These cells are connected to each other by gap junctions.

Astrocytes have a variety of functions:

- *Maintaining the Blood Brain Barrier.* The **blood brain barrier (BBB)** is a filtering mechanism of the capillaries that carry blood to the brain and spinal cord. This selectively permeable barrier allows the passage of some substances, but limits the passage of others. Astrocytes create this BBB, isolating the CNS from the general circulation.

  The expanded "feet" of astrocytes form a complete blanket around the capillaries, interrupted only where other neuroglia come in contact with the capillary walls. Astrocytes secrete chemicals that are responsible for maintaining the special permeability characteristics of the capillary endothelial cells.(We discuss the blood brain barrier further in Chapter 14.)
- *Creating a Three-Dimensional Framework for the CNS.* Their extensive cytoskeleton helps astrocytes to provide a structural framework for the neurons of the brain and spinal cord.
- *Repairing Damaged Nervous Tissue.* In the CNS, damaged nervous tissue seldom regains normal function. However, astrocytes that move into an injury site can make structural repairs that stabilize the tissue and prevent further injury. We discuss neural damage and repair later in this section.
- *Guiding Neuron Development.* Astrocytes in the embryonic brain appear to be involved in directing both the growth and interconnection of developing neurons.

Figure 12–5 Neuroglia in the CNS.

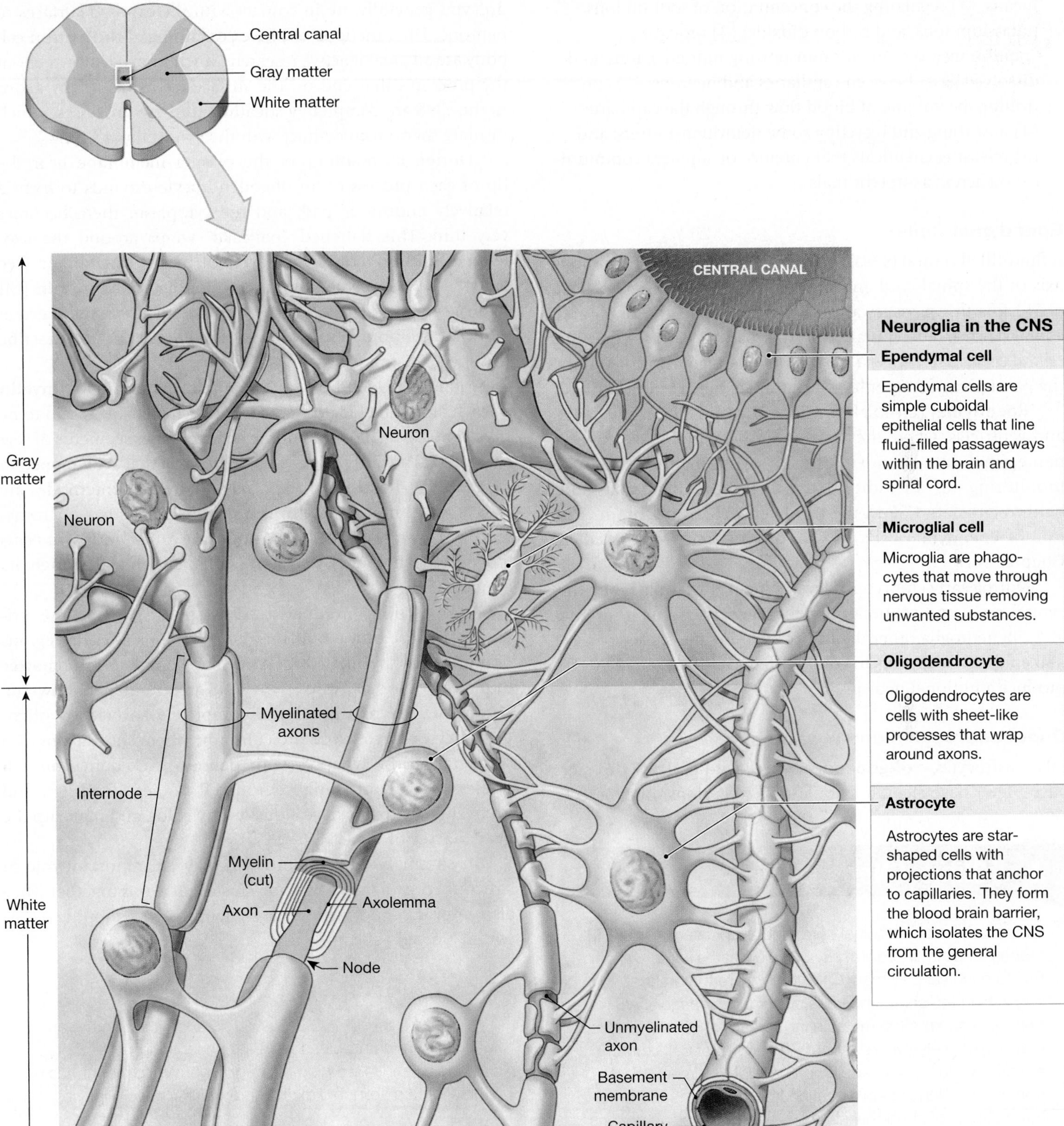
Central canal
Gray matter
White matter
CENTRAL CANAL
Neuron
Gray matter
Neuron
Myelinated axons
Internode
Myelin (cut)
Axon
Axolemma
Node
White matter
Unmyelinated axon
Basement membrane
Capillary
Neuroglia in the CNS
Ependymal cell
Ependymal cells are simple cuboidal epithelial cells that line fluid-filled passageways within the brain and spinal cord.
Microglial cell
Microglia are phagocytes that move through nervous tissue removing unwanted substances.
Oligodendrocyte
Oligodendrocytes are cells with sheet-like processes that wrap around axons.
Astrocyte
Astrocytes are star-shaped cells with projections that anchor to capillaries. They form the blood brain barrier, which isolates the CNS from the general circulation.

A diagram of nervous tissue in the CNS showing relationships between neuroglia and neurons

12

- *Controlling the Interstitial Environment.* Astrocytes appear to adjust the composition of interstitial fluid by several means: (1) regulating the concentration of sodium ions, potassium ions, and carbon dioxide; (2) providing a "rapid-transit system" for transporting nutrients, ions, and dissolved gases between capillaries and neurons; (3) controlling the volume of blood flow through the capillaries; (4) absorbing and recycling some neurotransmitters; and (5) releasing chemicals that enhance or suppress communication across axon terminals.

Ependymal Cells

A fluid-filled central passageway extends along the longitudinal axis of the spinal cord and brain. Protective *cerebrospinal fluid (CSF)* fills this passageway and also surrounds the brain and spinal cord. In the spinal cord, this passageway is called the *central canal* (see Figure 12–5). In several regions of the brain, the passageway forms enlarged chambers called *ventricles.*

Ependymal (eh-PEN-di-mul) **cells** line the central canal and ventricles, where they form a simple cuboidal to columnar epithelium. Ependymal cells assist in producing and monitoring CSF. In addition, the cilia of the ependymal cells help to circulate CSF in the third ventricle of the brain. The role of ependymal cells in CSF production is discussed in Chapter 14.

During early embryonic development, stem cells lining the central canal and ventricles divide to give rise to neurons and all CNS neuroglia other than microglia. In adults, the ependymal epithelium appears to contain stem cells that can divide to produce additional neurons.

Oligodendrocytes and Myelination

Like astrocytes, **oligodendrocytes** (ol-i-gō-DEN-drō-sīts; *oligo-*, few) have slender cytoplasmic extensions, but the cell bodies of oligodendrocytes are smaller, with fewer processes, than astrocytes (see Figure 12–5). The processes of oligodendrocytes generally are in contact with the exposed surfaces of neurons. The functions of processes ending at the neuron cell body are an area of active research. Much more is known about the processes that end on the surfaces of axons. Many axons in the CNS are completely sheathed in these processes, which insulate them from contact with the extracellular fluid.

During its maturation, the plasma membrane near the tip of each process of an oligodendrocyte expands to form a relatively enormous pad, and the cytoplasm there becomes very thin. This flattened "pancake" wraps around the axolemma, forming concentric layers of plasma membrane (see Figure 12–5). This membranous wrapping, called **myelin** (MĪ-e-lin), serves as electrical insulation and increases the speed at which an action potential travels along the axon. (We describe this mechanism in Section 12-5.)

Many oligodendrocytes cooperate in forming a **myelin sheath** along the length of an axon. Such an axon is said to be **myelinated**. Each oligodendrocyte myelinates segments of several axons. The fairly large areas of the axon that are wrapped in myelin are called **internodes** (*inter*, between). Internodes are typically 1–2 mm in length. The small gaps of a few micrometers that separate adjacent internodes are called **nodes**, or *nodes of Ranvier* (rahn-vē-Ā). An axon's collateral branches originate at nodes.

In dissection, myelinated axons appear glossy white, primarily because of the lipids in the myelin. As a result, regions dominated by myelinated axons are known as the **white matter** of the CNS. Not all axons in the CNS are myelinated, however. **Unmyelinated axons** are not completely covered by oligodendrocytes. Such axons are common where short axons and collaterals form synapses with densely packed neuron cell bodies. Areas containing neuron cell bodies, dendrites, and unmyelinated axons have a dusky gray color, and make up the **gray matter** of the CNS (see Figure 12–5).

In summary, oligodendrocytes play an important role in structural organization by tying clusters of axons together. They also improve the functioning of neurons by wrapping axons within a myelin sheath.

Tips & Tools

The overall color of CNS tissue is related to its structure and function. **Gr**ay matter has a **gr**eat concentration of neuron cell bodies and is a region of inte**gr**ation. **Wh**ite matter has a **wh**ole lot of myelinated axons and **wh**isks nerve impulses.

Microglia

The least numerous and smallest neuroglia in the CNS are phagocytic cells called **microglia** (mī-KRŌG-lē-uh). pp. 98, 101 Their slender processes have many fine branches (see Figure 12–5).

+ Clinical Note CNS Tumors

Tumors of the brain, spinal cord, and associated membranes result in approximately 90,000 deaths in the United States each year. *Primary CNS tumors* are tumors that originate in the central nervous system. *Secondary CNS tumors* arise from the metastasis (spread) of cancer cells that originate elsewhere. About 75 percent of CNS tumors are primary tumors. In adults, primary CNS tumors result from the divisions of abnormal neuroglia rather than from the divisions of abnormal neurons, because typical neurons in adults cannot divide. However, through the divisions of stem cells, neurons increase in number until children reach age 4. For this reason, primary CNS tumors involving abnormal neurons can occur in young children.

These cells can migrate through nervous tissue, acting as a wandering janitorial service and police force by engulfing cellular debris, wastes, and pathogens.

Microglia appear early in embryonic development, originating from mesodermal stem cells related to stem cells that produce monocytes and macrophages. pp. 128, 136 Microglia migrate into the CNS as the nervous system forms.

Neuroglia of the Peripheral Nervous System

Recall that the cell bodies of neurons in the PNS are clustered in masses called ganglia. The processes of neuroglia completely insulate neuronal cell bodies and most axons in the PNS from their surroundings. The two types of neuroglia in the PNS are *satellite cells* and *Schwann cells* (look back at Figure 12–4).

Satellite cells surround neuron cell bodies in ganglia. They regulate the interstitial fluid around the neurons, much as astrocytes do in the CNS.

Schwann cells, or *neurolemmocytes*, either form a thick, myelin sheath or indented folds of plasma membrane around peripheral axons in all parts of the peripheral nervous system. Wherever a Schwann cell covers an axon, the outer surface of the Schwann cell is called the **neurolemma**. It shields the axon from contact with interstitial fluids. Gaps between Schwann cells are called *nodes*.

A myelinating Schwann cell myelinates only one axon (Figure 12–6a), whereas an oligodendrocyte in the CNS may myelinate several axons (look back at Figure 12–5). Nonmyelinating Schwann cells can *enclose* segments of several unmyelinated axons. A series of Schwann cells is required to enclose an axon along its entire length (Figure 12–6b). Stages in the formation of a myelin sheath by a myelinating Schwann cell around the axon of a peripheral neuron are shown in Figure 12–6c.

Neural Responses to Injuries

What happens when a neuron is injured? It responds to injury in a very limited way, although in general more repair is possible in PNS neurons than in those of the CNS.

12

+ Clinical Note Demyelination

Demyelination is the progressive destruction of myelin sheaths, both in the CNS and in the PNS. The result is a loss of sensation and motor control that leaves affected regions numb and paralyzed. Many unrelated conditions that result in the destruction of myelin can cause symptoms of demyelination.

Chronic exposure to **heavy-metal ions**, such as arsenic, lead, or mercury, can cause heavy-metal poisoning, leading to neuroglial damage and demyelination.

Diphtheria (dif-THER-ē-uh) is a bacterial disease that damages Schwann cells and destroys myelin sheaths in the PNS. The resulting demyelination leads to sensory and motor problems that can ultimately produce a fatal paralysis. Due to an effective vaccine (the "D" in your "DPT" immunization record), cases are relatively rare.

Multiple sclerosis (skler-Ō-sis; *sklerosis*, hardness), or **MS**, is the most common demyelinating disease. It is characterized by recurrent demyelinating episodes that affect axons in the optic nerve, brain, and spinal cord. Although the precise cause of MS is yet to be discovered, myelin becomes the target of an autoimmune reaction. Inflammation occurs at multiple sites where white matter is located. Healing produces multiple scars, which accounts for the name of the disease. Common signs include partial loss of vision and problems with speech, balance, and general motor coordination, including bowel and urinary bladder control. The course of MS varies from case to case. Most often, the person experiences a cycle of attacks and recovery. The first attack typically occurs in people 20–40 years old. The incidence in women is twice that of men. Curiously, MS is more common in cold climates and in persons of Northern European descent, regardless of where they live. Although no cure currently exists, corticosteroid or interferon injections have slowed the progression of the disease in some patients. MS is the most common disabling neurological disease of young adults.

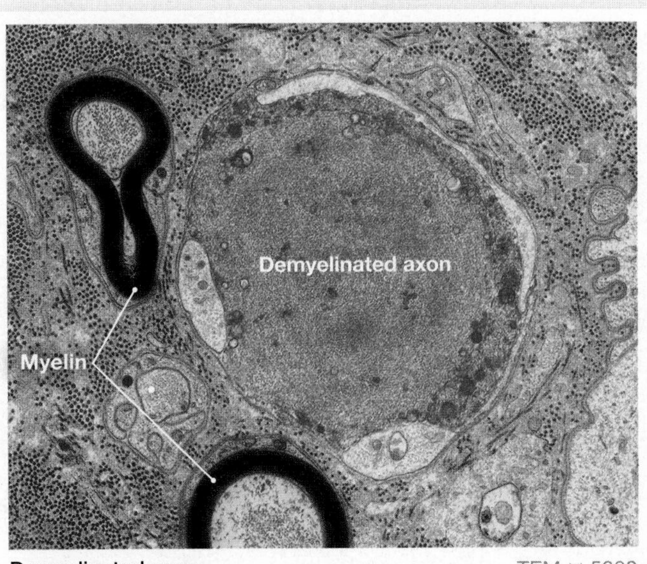
Demyelinated axon
Myelin

Demyelinated axon TEM × 5000

Figure 12–6 Schwann Cells, Peripheral Axons, and Formation of the Myelin Sheath in the PNS.

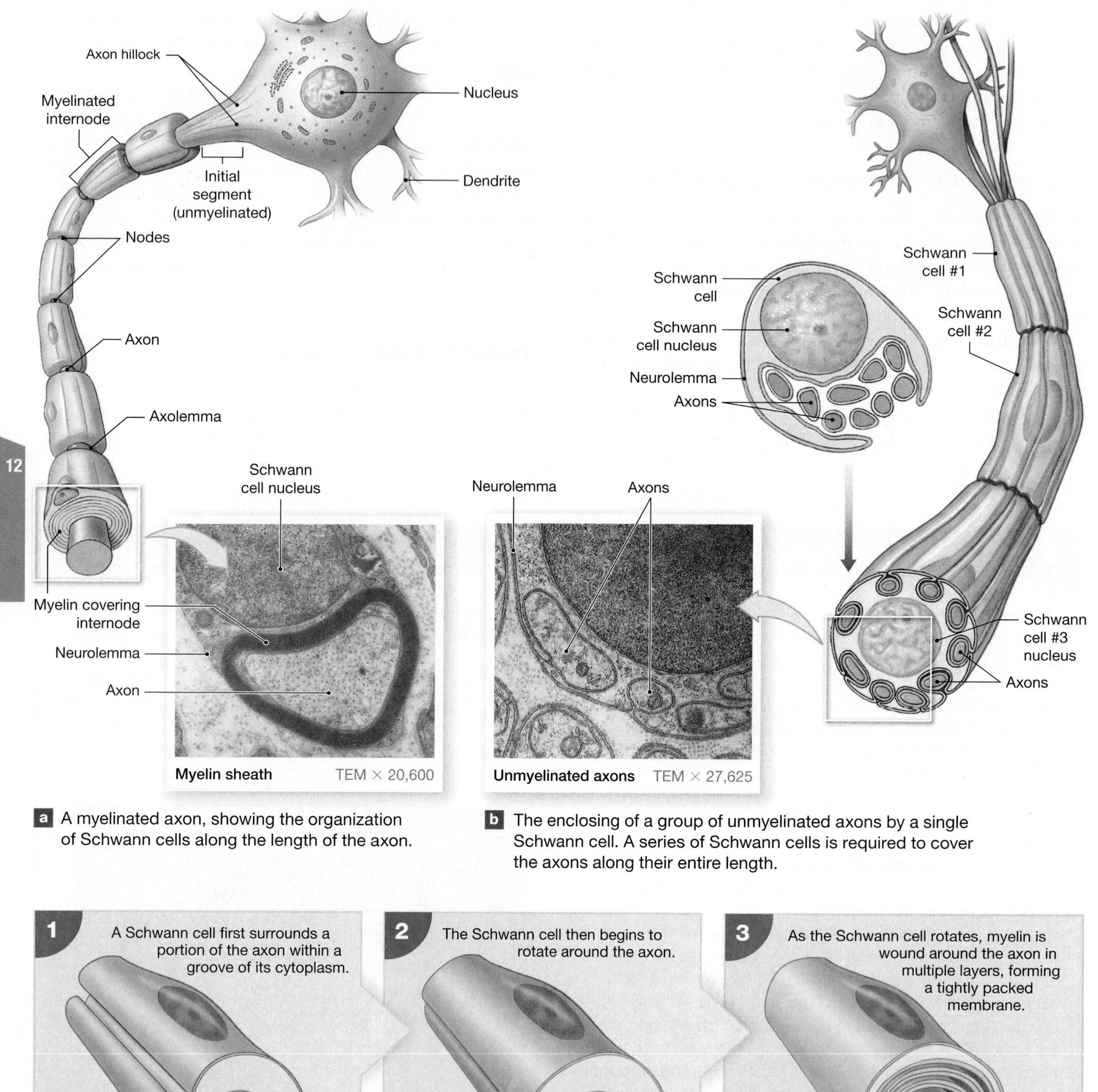

Axon hillock
Myelinated internode
Nucleus
Initial segment (unmyelinated)
Dendrite
Nodes
Axon
Axolemma
Schwann cell nucleus
Myelin covering internode
Neurolemma
Axon
Myelin sheath
TEM × 20,600
Neurolemma
Axons
Unmyelinated axons
TEM × 27,625
Schwann cell
Schwann cell nucleus
Neurolemma
Axons
Schwann cell #1
Schwann cell #2
Schwann cell #3 nucleus
Axons
1
A Schwann cell first surrounds a portion of the axon within a groove of its cytoplasm.
Schwann cell
Axon
2
The Schwann cell then begins to rotate around the axon.
3
As the Schwann cell rotates, myelin is wound around the axon in multiple layers, forming a tightly packed membrane.
Myelin
Schwann cell cytoplasm

a A myelinated axon, showing the organization of Schwann cells along the length of the axon.

b The enclosing of a group of unmyelinated axons by a single Schwann cell. A series of Schwann cells is required to cover the axons along their entire length.

c Stages in the formation of a myelin sheath by a single Schwann cell along a portion of a single axon.

After an injury, the Nissl bodies in the cell body disperse and the nucleus moves away from its centralized location as the cell increases its rate of protein synthesis. If the neuron recovers and regains function, it will regain its normal appearance.

The key to recovery appears to be events in the axon. If, for example, the pressure applied during a crushing injury produces a local decrease in blood flow and oxygen, the affected axolemma becomes unexcitable. If the pressure is alleviated after an hour or two, the neuron will recover within a few weeks. More severe or prolonged pressure produces effects similar to those caused by cutting the axon.

In the PNS, Schwann cells play a part in repairing damaged nerves. In the process known as **Wallerian** (vah-LEHR-ē-an) **degeneration**, the axon distal to the injury site degenerates, and macrophages migrate into the area to clean up the debris (Figure 12–7). The Schwann cells do not degenerate. Instead, they proliferate and form a solid cellular cord that follows the path of the original axon. As the neuron recovers, its axon grows into the site of injury, and the Schwann cells wrap around the axon.

If the axon grows alongside the appropriate cord of Schwann cells, it may eventually reestablish its normal synaptic contacts. However, if it stops growing or wanders off in some new direction, normal function will not return. The growing axon is most likely to arrive at its appropriate destination if the cut edges of the original nerve bundle remain in contact.

Limited regeneration can occur in the CNS, but the situation is more complicated because (1) many more axons are likely to be involved, (2) astrocytes produce scar tissue that can prevent axon growth across the damaged area, and (3) astrocytes release chemicals that block the regrowth of axons.

Checkpoint

9. List the neuroglia of the central nervous system.
10. Identify the neuroglia of the peripheral nervous system.
11. Which type of neuroglia would increase in number in the brain tissue of a person with a CNS infection?

See the blue Answers tab at the back of the book.

Figure 12–7 **Peripheral Nerve Regeneration after Injury.**

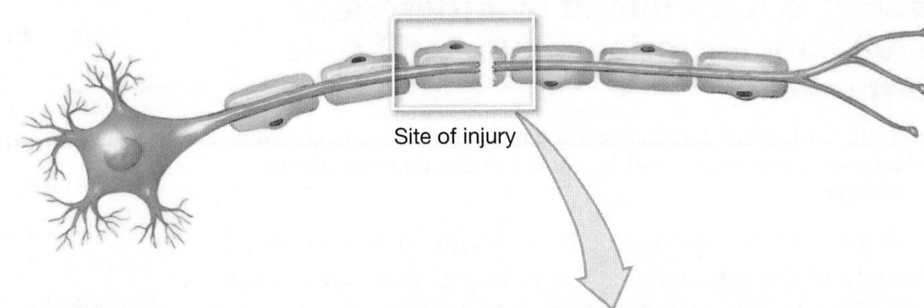
Site of injury

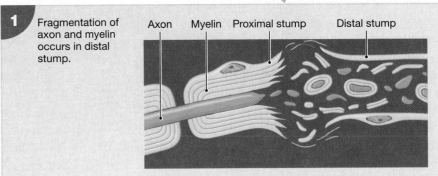

1 Fragmentation of axon and myelin occurs in distal stump.

Axon
Myelin
Proximal stump
Distal stump

2 Schwann cells form cord, grow into cut, and unite stumps. Macrophages engulf degenerating axon and myelin.

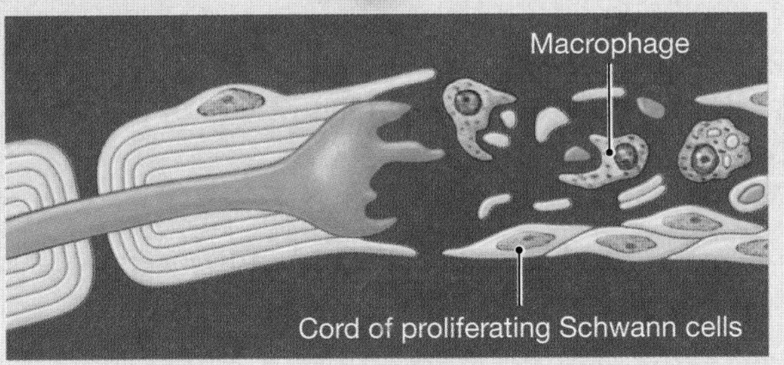
Macrophage
Cord of proliferating Schwann cells

3 Axon sends buds into network of Schwann cells and then starts growing along cord of Schwann cells.

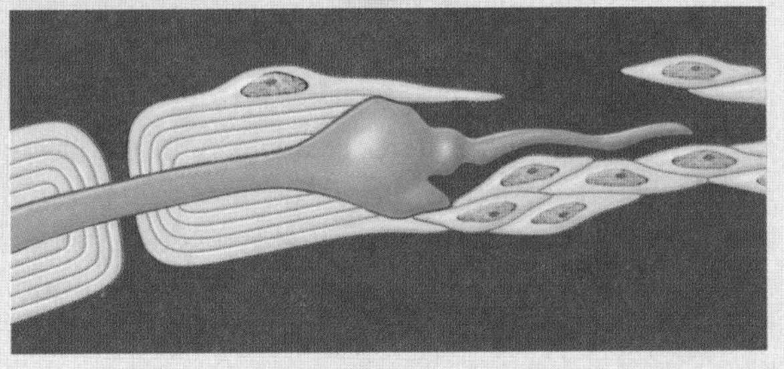

4 Axon continues to grow into distal stump and is enclosed by Schwann cells.

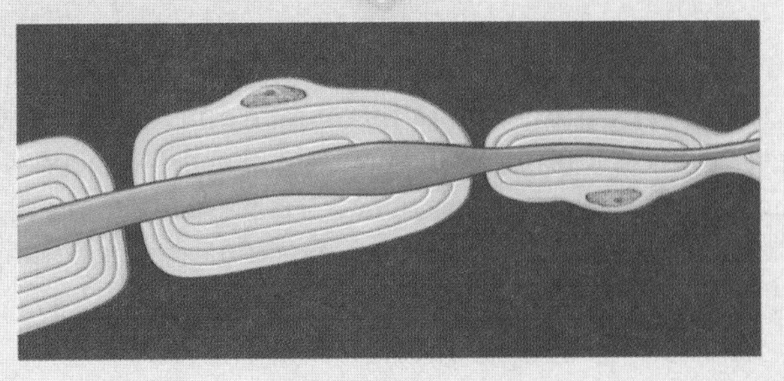

12-4 The membrane potential of a neuron is determined by differences in ion concentrations and membrane permeability

Learning Outcome Explain how the resting membrane potential is established and maintained and how the membrane potential can change.

In Chapter 3, we introduced the concepts of the *membrane potential* and the resting membrane potential, two characteristic physiological features of all cells. p. 99 In this discussion, we focus on the membranes of neurons, but many of the principles discussed apply to other types of cells as well.

All living cells have a membrane potential that varies from moment to moment depending on the activities of the cell. The *resting membrane potential* is the membrane potential of an unstimulated, resting cell. All neural activities begin with a change in the resting membrane potential of a neuron. A typical stimulus produces a temporary, localized change in the resting membrane potential. The effect, which decreases with distance from the stimulus, is called a *graded potential.* If the graded potential is large enough, it triggers an *action potential* in the membrane of the axon, a process that we cover in the next section. Here we look at the important membrane processes of the resting membrane potential and graded potentials.

The Resting Membrane Potential

Chapter 3 introduced three important concepts regarding the membrane potential:

- *The extracellular fluid (ECF) and intracellular fluid (cytosol) differ greatly in ionic composition.* The extracellular fluid contains high concentrations of sodium ions (Na^+) and chloride ions (Cl^-), whereas the cytosol contains high concentrations of potassium ions (K^+) and negatively charged proteins.
- *Cells have selectively permeable membranes.* If the plasma membrane were freely permeable, diffusion would continue until all the ions were evenly distributed across the membrane and a state of equilibrium existed. But an even distribution does not occur, because cells have selectively permeable membranes. p. 90 Ions cannot freely cross the lipid portions of the plasma membrane. They can enter or leave the cell only through membrane channels. Many kinds of membrane channels exist, each with its own properties. At the **resting membrane potential**, or membrane potential of an undisturbed cell, ions move through *leak channels*—membrane channels that are always open. p. 92 Active transport mechanisms, such as the sodium–potassium exchange pump, also move specific ions into or out of the cell. p. 97
- *Membrane permeability varies by ion.* The cell's passive and active transport mechanisms do not ensure an equal distribution of charges across its plasma membrane, because membrane permeability varies by ion. For example, it is easier for K^+ to diffuse out of the cell through a potassium leak channel than it is for Na^+ to enter the cell through a sodium leak channel. Additionally, negatively charged proteins inside the cell are too large to cross the membrane. As a result, the membrane's inner surface has an excess of negative charges with respect to the outer surface.

Both passive and active forces act across the plasma membrane to determine the membrane potential at any moment. **Spotlight Figure 12–8** shows the processes that lead to the normal resting membrane potential.

Passive Processes Acting across the Plasma Membrane: The Electrochemical Gradient

The passive processes acting across the plasma membrane involve both chemical and electrical gradients (see **Spotlight Figure 12–8**). Positive and negative charges attract one another. If nothing separates them, oppositely charged ions will move together and eliminate the potential difference between them. A movement of charges to eliminate a potential difference is called a **current**. If a barrier (such as a plasma membrane) separates the oppositely charged ions, the amount of current depends on how easily the ions can cross the membrane. The **resistance** of the membrane is a measure of how much the membrane restricts ion movement. If the resistance is high, the current is very small, because few ions can cross the membrane. If the resistance is low, the current is very large, because ions flood across the membrane. The resistance of a plasma membrane can change as ion channels open or close. The changes result in currents carrying ions into or out of the cytosol.

Electrical gradients can either reinforce or oppose the chemical gradient for each ion. The **electrochemical gradient** for a specific ion is the sum of the chemical and electrical forces acting on that ion across the plasma membrane. The electrochemical gradients for K^+ and Na^+ are the primary factors affecting the resting membrane potential of most cells, including neurons. Let's consider the forces acting on each ion independently.

The intracellular concentration of potassium ions is relatively high, whereas the extracellular concentration is very low. Therefore, the chemical gradient for potassium ions tends to drive them out of the cell, as indicated by the thicker orange arrow in Figure 12–9a. However, the electrical gradient opposes this movement, because K^+ inside and outside the cell are attracted to the negative charges on the inside of the plasma membrane, and repelled by the positive charges on the outside of the plasma membrane. The white arrow in Figure 12–9a indicates the size and direction of this electrical gradient. The chemical gradient is strong enough to overpower the electrical gradient, but the electrical gradient weakens the force driving K^+ out of the cell. The thinner orange arrow represents the net driving force of the potassium ion electrochemical gradient.

SPOTLIGHT

Figure 12–8

Go to MasteringA&P™ > Study Area > Menu > Animations & Videos > A&PFlix > A&P > Resting Membrane Potential

Processes That Produce the Resting Membrane Potential

All living cells have a membrane potential that varies from moment to moment depending on the activities of the cell. In an unstimulated, resting cell, the *potential difference* between positive and negative ions on either side of the membrane is called the **resting membrane potential**. Passive and active forces determine the resting membrane potential.

Passive Chemical Gradients

The intracellular concentration of potassium ions (K^+) is relatively high, so these ions tend to move out of the cell through potassium leak channels. Similarly, the extracellular concentration of sodium ions (Na^+) is relatively high, so these ions move into the cell through sodium leak channels. Each ion's movement is driven by a concentration gradient, or **chemical gradient**.

Active Na^+/K^+ Pumps

Sodium–potassium (Na^+/K^+) exchange pumps maintain the concentration gradients of sodium and potassium ions across the plasma membrane.

Passive Electrical Gradients

Potassium ions leave the cytosol more rapidly than sodium ions enter because the plasma membrane is much more permeable to potassium than to sodium. As a result, there are more positive charges outside the plasma membrane. Negatively charged protein molecules within the cytosol cannot cross the plasma membrane, so there are more negative charges on the cytosol side of the plasma membrane. This results in an **electrical gradient** across the plasma membrane.

Resting Membrane Potential

Whenever positive and negative ions are held apart, a potential difference arises. We measure the size of that potential difference in millivolts (mV). The resting membrane potential for most neurons is about –70 mV. The minus sign shows that the inner surface of the plasma membrane is negatively charged with respect to the exterior.

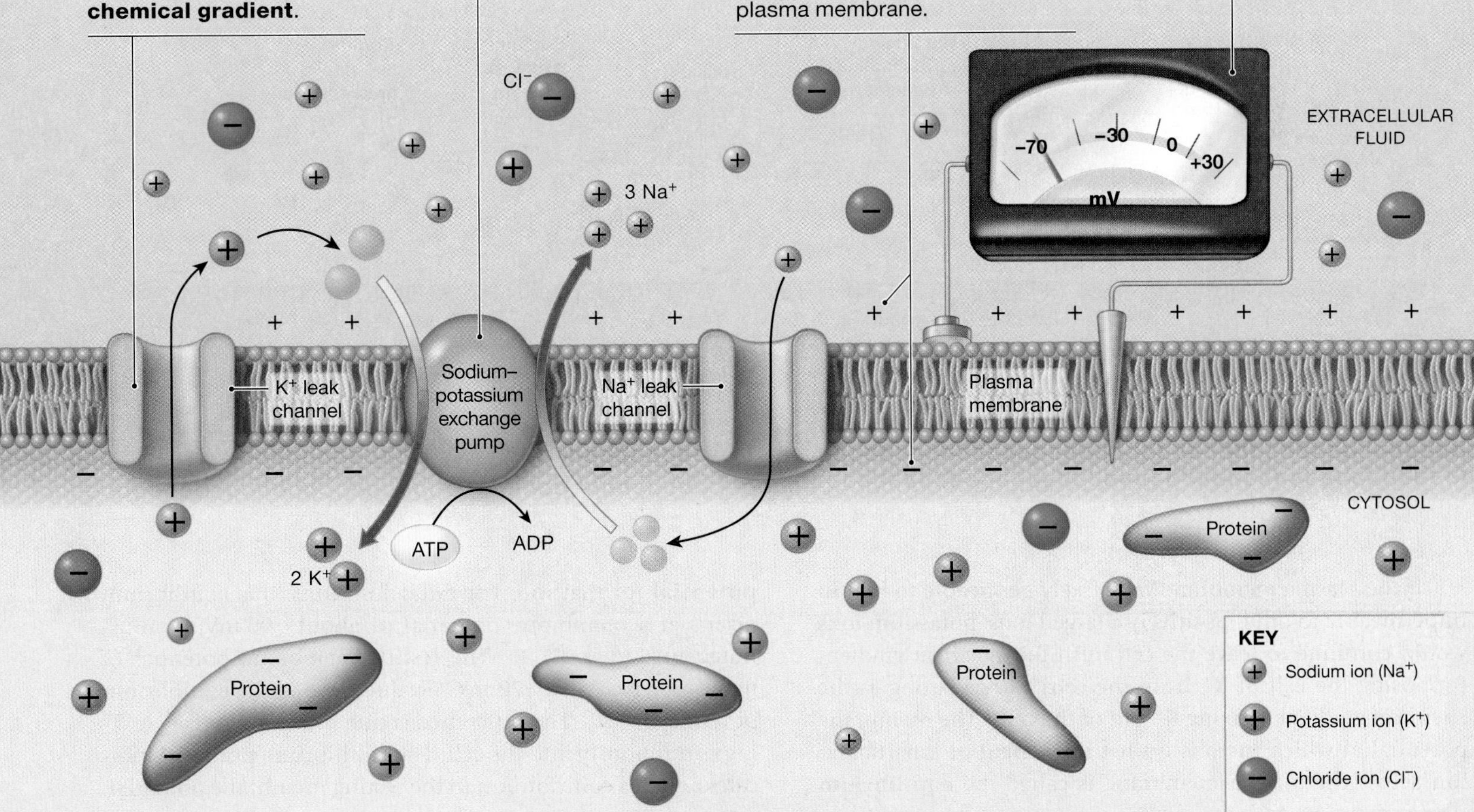

Cl⁻
3 Na⁺
–70
–30
0
+30
mV
EXTRACELLULAR FLUID
K⁺ leak channel
Sodium–potassium exchange pump
Na⁺ leak channel
Plasma membrane
ATP
ADP
2 K⁺
CYTOSOL
Protein
Protein
Protein
Protein
KEY
Sodium ion (Na⁺)
Potassium ion (K⁺)
Chloride ion (Cl⁻)

Figure 12–9 **Electrochemical Gradients for Potassium and Sodium Ions.**

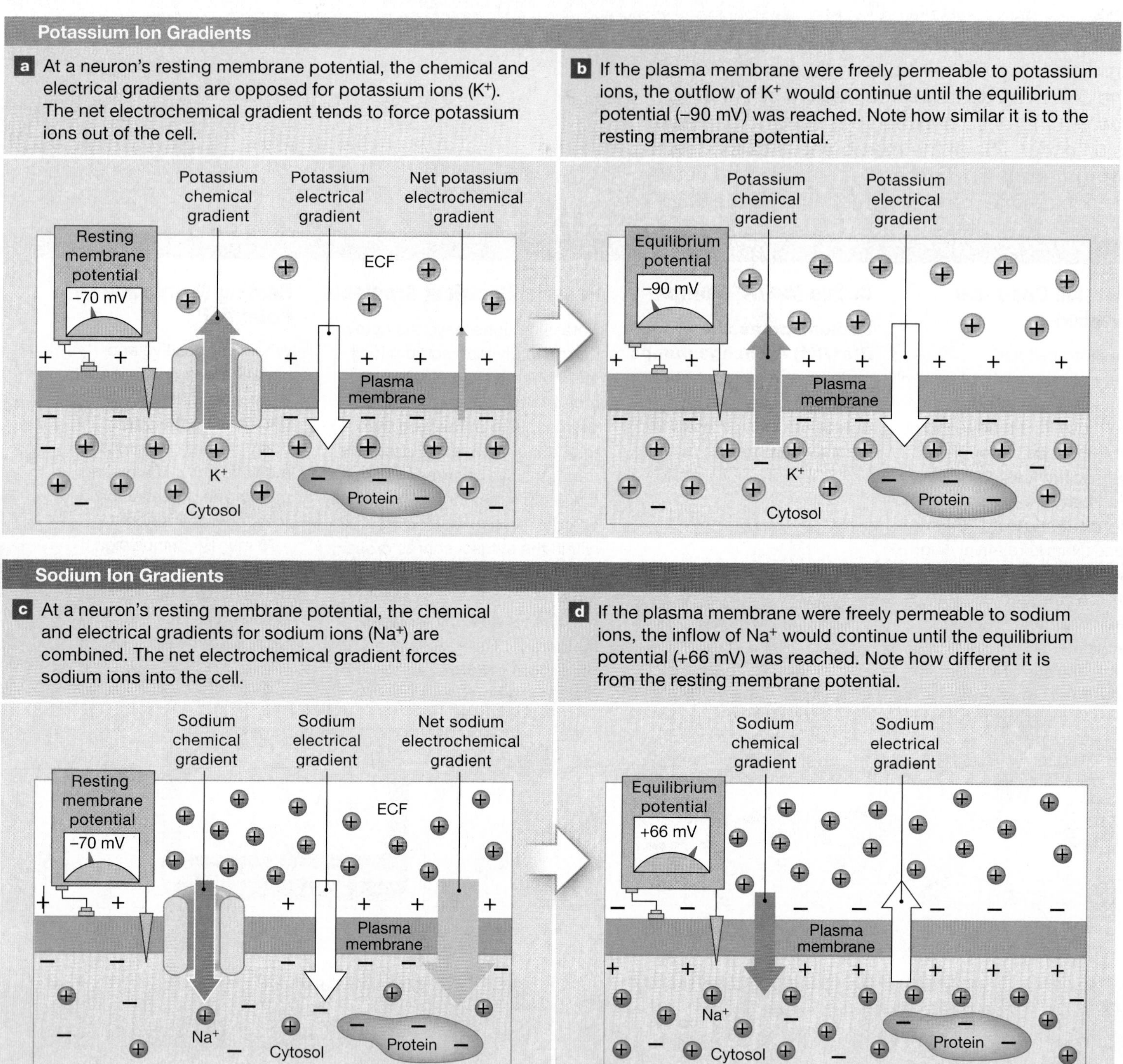
Potassium Ion Gradients
a At a neuron's resting membrane potential, the chemical and electrical gradients are opposed for potassium ions (K+). The net electrochemical gradient tends to force potassium ions out of the cell.
b If the plasma membrane were freely permeable to potassium ions, the outflow of K+ would continue until the equilibrium potential (–90 mV) was reached. Note how similar it is to the resting membrane potential.
Potassium chemical gradient
Potassium electrical gradient
Net potassium electrochemical gradient
Resting membrane potential
–70 mV
ECF
Plasma membrane
K+
Protein
Cytosol
Equilibrium potential
–90 mV
Sodium Ion Gradients
c At a neuron's resting membrane potential, the chemical and electrical gradients for sodium ions (Na+) are combined. The net electrochemical gradient forces sodium ions into the cell.
d If the plasma membrane were freely permeable to sodium ions, the inflow of Na+ would continue until the equilibrium potential (+66 mV) was reached. Note how different it is from the resting membrane potential.
Sodium chemical gradient
Sodium electrical gradient
Net sodium electrochemical gradient
Na+
+66 mV

If the plasma membrane were freely permeable to K^+ but impermeable to other positively charged ions, potassium ions would continue to leave the cell until the electrical gradient (opposing the exit of K^+ from the cell) was as strong as the chemical gradient (driving K^+ out of the cell). The membrane potential at which there is no net movement of a particular ion across the plasma membrane is called the **equilibrium potential** for that ion. For potassium ions, this equilibrium occurs at a membrane potential of about –90 mV, as illustrated in Figure 12–9b. The resting membrane potential of neurons is typically −70 mV, a value close to the equilibrium potential for K^+. The difference is due primarily to Na^+ leaking continuously into the cell. The equilibrium potential indicates an ion's contribution to the resting membrane potential.

Tips & Tools

To remember the relative distribution of ions across the resting cell's plasma membrane, associate **N**egative with the i**N**side and p**O**sitive with the **O**utside.

The sodium ion concentration is relatively high in the extracellular fluid, but extremely low inside the cell. As a result, there is a strong chemical gradient driving Na^+ into the cell (the thinner purple arrow in Figure 12–9c). In addition, the extracellular sodium ions are attracted by the excess of negative charges on the inner surface of the plasma membrane. The white arrow in Figure 12–9c shows the relative size and direction of this electrical gradient. Both electrical forces and chemical forces drive Na^+ into the cell, and the larger lavender arrow represents the net driving force.

If the plasma membrane were freely permeable to Na^+, these ions would continue to cross it until the interior of the cell contained enough excess positive charges to reverse the electrical gradient. In other words, ion movement would continue until the interior developed such a strongly positive charge that repulsion between the positive charges would prevent any further net movement of Na^+ into the cell. The equilibrium potential for Na^+ is approximately +66 mV, as shown in Figure 12–9d. The resting membrane potential is nowhere near that value, because the resting membrane permeability to Na^+ is very low, and because ion pumps in the plasma membrane eject sodium ions as fast as they cross the membrane.

An electrochemical gradient is a form of *potential energy.* ↺ p. 38 Potential energy is stored energy—the energy of position, as exists in a stretched spring, a charged battery, or water behind a dam. Without a plasma membrane, diffusion would eliminate all electrochemical gradients. In effect, the plasma membrane acts like a dam across a river. Without the dam, water would simply respond to gravity and flow downstream, gradually losing energy. With the dam in place, even a small opening releases water under tremendous pressure. Similarly, any stimulus that increases the permeability of the plasma membrane to sodium or potassium ions produces sudden and dramatic ion movement. For example, a stimulus that opens sodium ion channels triggers a rush of Na^+ into the cell. Note that the nature of the stimulus does not determine the amount of ion movement: If the stimulus opens the door, the electrochemical gradient does the rest.

Active Processes across the Membrane: The Sodium–Potassium Exchange Pump

We can compare a cell to a leaky fishing boat loaded with tiny fish floating in the sea. The hull represents the plasma membrane; the fish, K^+; and the ocean water, Na^+. As the boat pitches and rolls, water comes in through the cracks, and fish swim out. If the boat is to stay afloat and the catch kept, we must pump the water out and recapture the lost fish.

Table 12–1 The Resting Membrane Potential

- Because the plasma membrane is highly permeable to potassium ions, the resting membrane potential of approximately −70 mV is fairly close to −90 mV, the equilibrium potential for K^+.
- The electrochemical gradient for sodium ions is very large, but the membrane's permeability to these ions is very low. Consequently, Na^+ has only a small effect on the normal resting membrane potential, making it just slightly less negative than the equilibrium potential for K^+.
- The sodium–potassium exchange pump ejects 3 Na^+ ions for every 2 K^+ ions that it brings into the cell. It serves to stabilize the resting membrane potential when the ratio of Na^+ entry to K^+ loss through passive channels is 3:2.
- At the normal resting membrane potential, these passive and active mechanisms are in balance. The resting membrane potential varies widely with the type of cell. A typical neuron has a resting membrane potential of approximately −70 mV.

Similarly, at the normal resting membrane potential, the cell must bail out sodium ions that leak in and recapture potassium ions that leak out. The "bailing" takes place through the activity of the sodium–potassium exchange pump powered by ATP (look back at **Spotlight Figure 12–8**). This pump exchanges three intracellular sodium ions for two extracellular potassium ions. At the normal resting membrane potential, this pump ejects sodium ions as quickly as they enter the cell. In this way, the activity of the exchange pump exactly balances the passive forces of diffusion, and the resting membrane potential remains stable because the ionic concentration gradients are maintained.

Table 12–1 summarizes the important features that lead to the resting membrane potential.

Changes in the Resting Membrane Potential: Membrane Channels

As noted previously, the resting membrane potential is the membrane potential of an "undisturbed" cell. Recall that it exists because (1) the cytosol differs from extracellular fluid in chemical and ionic composition, and (2) the plasma membrane is selectively permeable. Yet cells are dynamic structures that continually modify their activities, either in response to external stimuli or to perform specific functions. The membrane potential is equally dynamic, rising or falling in response to temporary changes in membrane permeability. Those changes result from the opening or closing of specific membrane channels.

Membrane channels control the movement of ions across the plasma membrane. We will focus on the permeability of the membrane to sodium and potassium ions. These ions are the primary determinants of the membrane potential of many cell types, including neurons. Sodium and potassium ion channels are either passive or active.

Passive Ion Channels

Passive ion channels, called **leak channels**, are always open. However, their permeability can vary from moment to moment as the proteins that make up the channel change shape in response to local conditions. As noted earlier, leak channels are important in establishing the normal resting membrane potential of the cell (look back at **Spotlight Figure 12–8**).

Active Ion Channels

Plasma membranes also contain active channels, called **gated ion channels**, that open or close in response to specific stimuli. Three classes of gated channels exist: chemically gated, or ligand-gated channels; voltage-gated channels; and mechanically gated channels.

- **Chemically gated ion channels**, or **ligand-gated ion channels**, open or close when they bind specific chemicals, or ligands (Figure 12–10a). For example, the receptors that bind acetylcholine (ACh) at the neuromuscular junction are chemically gated ion channels. Chemically gated ion channels are most abundant on the dendrites and cell body of a neuron, the areas where most synaptic communication occurs.
- **Voltage-gated ion channels** open or close in response to changes in the membrane potential. They are characteristic of areas of *excitable cell membrane*, a membrane capable of

12

Figure 12–10 Types of Gated Ion Channels.

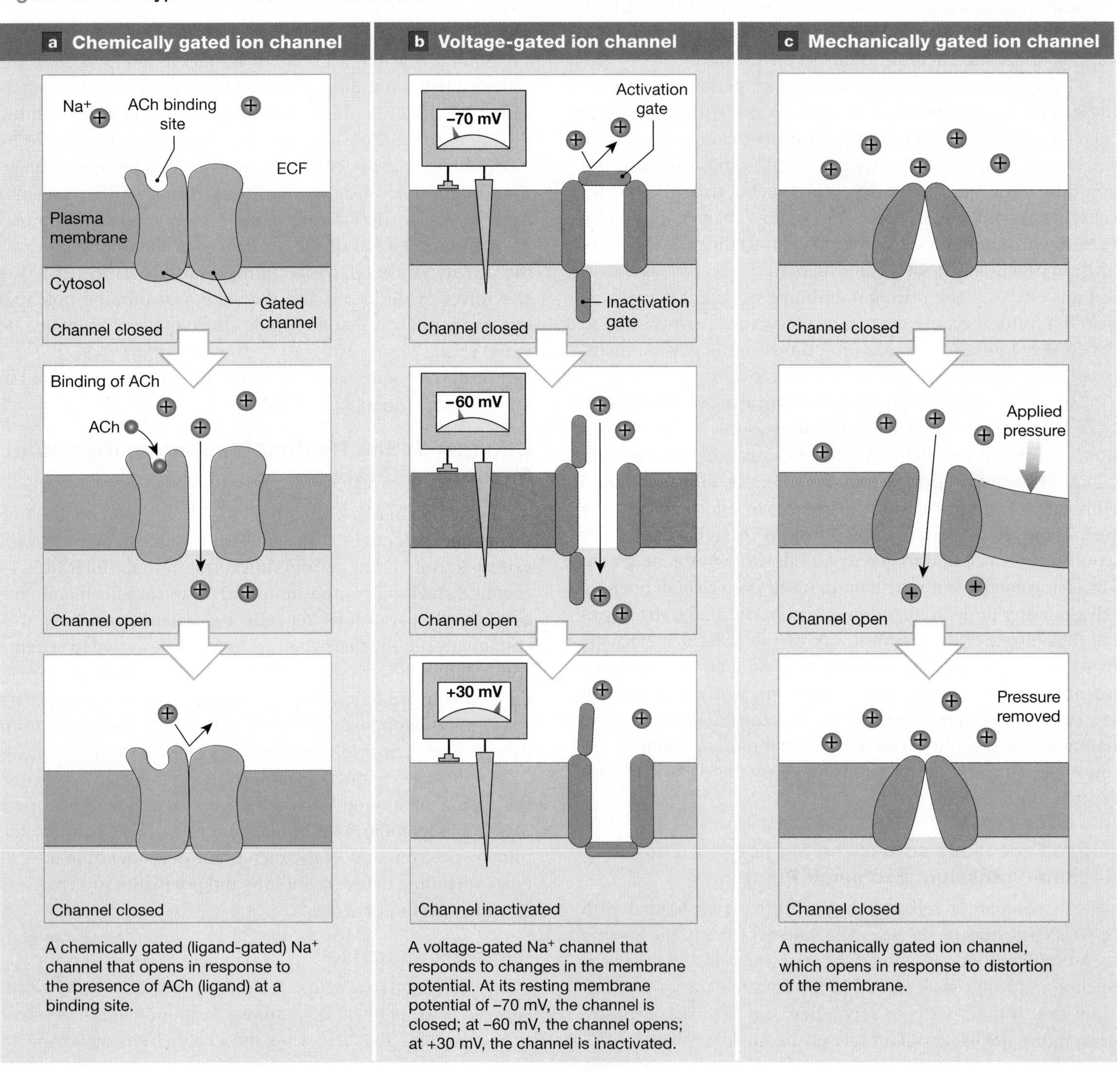
a Chemically gated ion channel
Na+
ACh binding site
ECF
Plasma membrane
Cytosol
Gated channel
Channel closed
Binding of ACh
ACh
Channel open
Channel closed
b Voltage-gated ion channel
Activation gate
–70 mV
Inactivation gate
Channel closed
–60 mV
Channel open
+30 mV
Channel inactivated
c Mechanically gated ion channel
Channel closed
Applied pressure
Channel open
Pressure removed
Channel closed

A chemically gated (ligand-gated) Na^+ channel that opens in response to the presence of ACh (ligand) at a binding site.

A voltage-gated Na^+ channel that responds to changes in the membrane potential. At its resting membrane potential of –70 mV, the channel is closed; at –60 mV, the channel opens; at +30 mV, the channel is inactivated.

A mechanically gated ion channel, which opens in response to distortion of the membrane.

generating and propagating an action potential. Examples of excitable membranes are the axons of unipolar and multipolar neurons, and the sarcolemma (including T tubules) of skeletal muscle fibers and cardiac muscle cells. The most important voltage-gated ion channels, for our purposes, are voltage-gated sodium ion channels, potassium ion channels, and calcium ion channels. These sodium ion channels have two gates that function independently: an *activation gate* that opens on stimulation, letting sodium ions into the cell, and an *inactivation gate* that closes to stop the entry of sodium ions (Figure 12–10b). Therefore each channel can be in one of three states: (1) closed but capable of opening, (2) open (**activated**), or (3) closed and incapable of opening (**inactivated**).

- **Mechanically gated ion channels** open or close in response to physical distortion of the membrane surface, such as when pressure is applied due to the touch of a hand (Figure 12–10c). Such channels are important in sensory receptors that respond to touch, pressure, or vibration. We discuss these receptors in more detail in Chapter 15.

The fact that the distribution of membrane channels varies from one region of the plasma membrane to another affects how and where a cell responds to specific stimuli. For example, in a neuron, chemically gated ion channels are located on the cell body and dendrites. Voltage-gated Na^+ and K^+ channels are found along the axon. Additionally, voltage-gated Ca^{2+} channels occur at axon terminals. Later sections will show how these differences in distribution affect the way that neurons function.

At the resting membrane potential, most gated channels are closed. When gated channels open, the rate of ion movement across the plasma membrane increases, changing the membrane potential. Let's explore more about how this works.

Graded Potentials

Any stimulus that opens a gated channel produces a graded potential. **Graded potentials**, or *local potentials*, are changes in the membrane potential that cannot spread far from the site of stimulation.

Figure 12–11 shows what happens when the plasma membrane of a resting cell is exposed to a chemical that opens chemically gated sodium ion channels:

1. Sodium ions enter the cell and are attracted to the negative charges along the inner surface of the membrane. As these additional positive charges spread out, the membrane potential shifts toward 0 mV. Any shift from the resting membrane potential toward a less negative potential is called a **depolarization**. Note that this term applies to both changes in potential from −70 mV to lesser negative values (−65 mV, −45 mV, −10 mV), as well as to membrane potentials above 0 mV (+10 mV, +30 mV). In all these changes, the membrane potential becomes more positive.

2. As the plasma membrane depolarizes, sodium ions are released from its outer surface. These ions, along with other extracellular sodium ions, then move toward the open channels, replacing ions that have already entered the cell. This movement of positive charges parallel to the inner and outer surfaces of a membrane that spreads the depolarization is called a **local current**.

In a graded potential, the degree of depolarization decreases with distance away from the stimulation site; in other words, the local current dissipates. Why? Depolarization lessens with distance because the cytosol offers considerable resistance to ion movement, and because some of the sodium ions entering the cell then move back out across the membrane through sodium leak channels. At some distance from the entry point, the effects on the membrane potential are undetectable.

The maximum change in the membrane potential is proportional to the size of the stimulus, which determines the number of open sodium ion channels. The more open channels, the more sodium ions enter the cell, the greater the membrane area affected, and the greater the degree of depolarization.

When the chemical stimulus is removed and normal membrane permeability is restored, the membrane potential soon returns to the resting level. The process of restoring the normal resting membrane potential after depolarization is called **repolarization** (Figure 12–12). Repolarization typically involves a combination of ion movement through membrane channels and the activities of ion pumps, especially the sodium–potassium exchange pump.

However, if a gated potassium ion channel opens in response to a stimulus, the opposite effect occurs. The rate of K^+ outflow increases, and the interior of the cell loses positive ions. In other words, the inside of the cell becomes more negative. The loss of positive ions produces **hyperpolarization**, an increase in the negativity of the resting membrane potential, for example, from −70 mV to perhaps −80 mV or more (see Figure 12–12). Again, a local current distributes the effect to adjacent portions of the plasma membrane, and the effect decreases with distance from the open channels. Table 12–2 summarizes the basic characteristics of graded potentials.

Table 12–2 Graded Potentials

Graded potentials, whether depolarizing or hyperpolarizing, share four basic characteristics:

- The membrane potential is most changed at the site of stimulation, and the effect decreases with distance.
- The effect spreads passively, due to local currents.
- The graded change in membrane potential may involve either depolarization or hyperpolarization. The properties and distribution of the membrane channels involved determine the nature of the change. For example, in a resting membrane, the opening of sodium ion channels causes depolarization, whereas the opening of potassium ion channels causes hyperpolarization. That is, the change in membrane potential reflects whether positive charges enter or leave the cell.
- The stronger the stimulus, the greater the change in the membrane potential and the larger the area affected.

12

Figure 12–11 Graded Potentials. The depolarization radiates in all directions away from the source of stimulation. For clarity, only gated ion channels are shown. Leak channels are present, but are not responsible for the production of graded potentials. Color changes in the plasma membrane indicate that the resting membrane potential has been disturbed and that the membrane potential is no longer −70 mV.

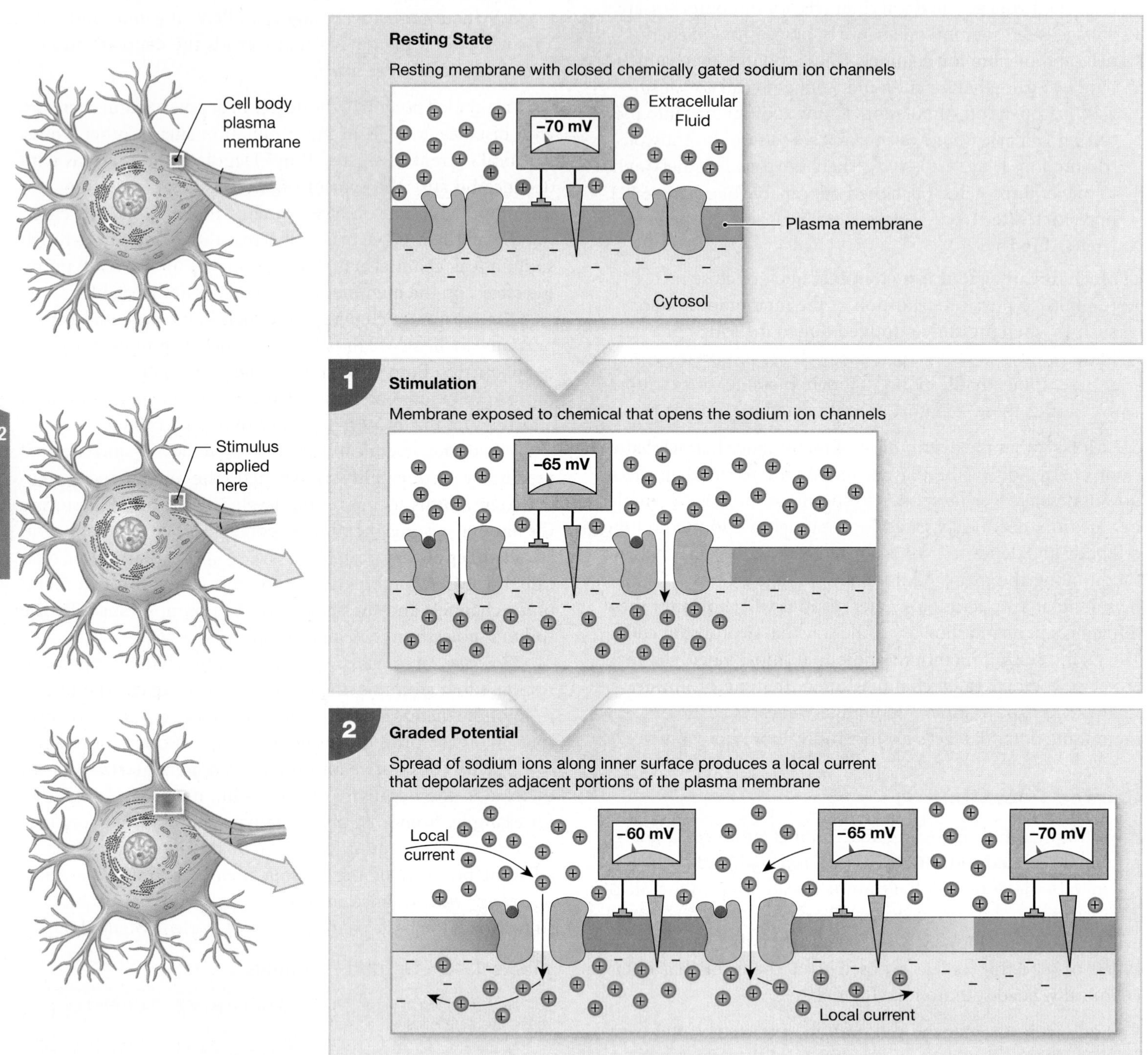
Cell body plasma membrane
Resting State
Resting membrane with closed chemically gated sodium ion channels
Extracellular Fluid
–70 mV
Plasma membrane
Cytosol
Stimulus applied here
1
Stimulation
Membrane exposed to chemical that opens the sodium ion channels
–65 mV
2
Graded Potential
Spread of sodium ions along inner surface produces a local current that depolarizes adjacent portions of the plasma membrane
Local current
–60 mV
–65 mV
–70 mV
Local current

Graded potentials occur in the membranes of many types of cells—not just nerve and muscle cells, but epithelial cells, gland cells, adipocytes, and a variety of sensory receptors. Graded potentials often trigger specific cell functions. For example, a graded potential at the surface of a gland cell may trigger the exocytosis of secretory vesicles. Similarly, at a neuromuscular junction, ACh may stimulate the depolarization of the motor end plate by a graded potential, which in turn may trigger an action potential in adjacent portions of the sarcolemma. The motor end plate supports graded potentials, but the rest of the sarcolemma consists

Figure 12–12 **Depolarization, Repolarization, and Hyperpolarization.**

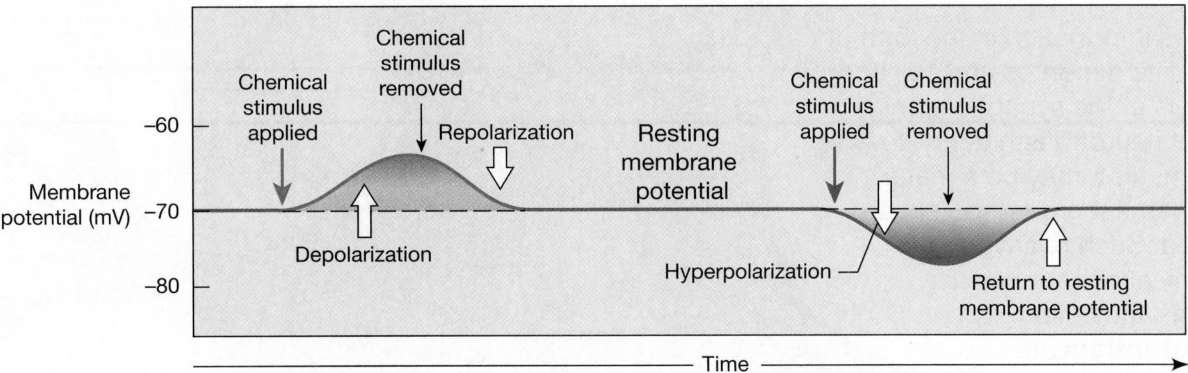
Chemical stimulus applied
Chemical stimulus removed
Repolarization
Resting membrane potential
Chemical stimulus applied
Chemical stimulus removed
-60
Membrane potential (mV)
-70
-80
Depolarization
Hyperpolarization
Return to resting membrane potential
Time

of excitable membrane. These areas of membrane are different because they contain voltage-gated ion channels.

Let's look at how graded potentials can lead to action potentials in a neuron. If a graded potential causes hyperpolarization, an action potential becomes less likely, but if the graded potential causes depolarization, an action potential becomes more likely to occur.

Checkpoint

12. **Define *resting membrane potential*.**
13. **What effect would a chemical that blocks the voltage-gated sodium ion channels in the plasma membrane of a neuron have on its ability to depolarize?**
14. **What effect would decreasing the concentration of extracellular potassium ions have on the membrane potential of a neuron?**

See the blue Answers tab at the back of the book.

12-5 An action potential is an all-or-none electrical event used for long-distance communication

Learning Outcome Describe the events involved in the generation and propagation of an action potential and the factors involved in determining the speed of action potential propagation.

Action potentials, also known as *nerve impulses*, are changes in the membrane potential that, once initiated, affect an entire excitable membrane. That is, action potentials are not graded potentials. An action potential is *propagated* (spread) along the surface of an axon and does not diminish as it moves away from its source. This impulse travels along the axon to the axon terminals.

Now let's take a closer look at the nature of action potentials, and how they are generated and propagated. Generation and propagation are closely related concepts, in terms of both time and space: An action potential must be generated at one site before it can be propagated away from that site.

Threshold and the All-or-None Principle

A graded potential must depolarize the axolemma to a particular level in order to stimulate an action potential. The membrane potential at which an action potential begins is called the **threshold**. Threshold for an axon is typically between −60 mV and −55 mV, corresponding to a depolarization of 10 to 15 mV. A stimulus that shifts the resting membrane potential from −70 mV to −62 mV will not produce an action potential, only a graded depolarization. When such a stimulus is removed, the membrane potential returns to the resting level. Local currents resulting from the graded depolarization of the axon hillock cause the depolarization of the initial segment of the axon.

The initial depolarization acts like pressure on the trigger of a gun. If you apply a slight pressure, the gun does not fire. It fires only when you apply a certain minimum pressure to the trigger. Once the pressure on the trigger reaches this threshold, the firing pin drops and the gun discharges. At that point, it no longer matters whether you applied the pressure gradually or suddenly or whether you moved just one finger or clenched your entire hand. The speed and range of the bullet that leaves the gun do not change, regardless of the forces that you applied to the trigger.

In the case of an axon or another area of excitable membrane, a graded depolarization is similar to the pressure on the trigger, and the action potential is like the firing of the gun. All stimuli that bring the membrane to threshold generate identical action potentials. In other words, the properties of the action potential are independent of the relative strength of the depolarizing stimulus, as long as that stimulus exceeds the threshold. This concept is called the **all-or-none principle**, because a given stimulus either triggers a typical action potential, or none at all. The all-or-none principle applies to all excitable membranes.

Generation of Action Potentials

Spotlight Figure 12–13 diagrams the steps involved in generating an action potential from the resting state. Recall that voltage-gated sodium ion channels are abundant on the axon,

Go to MasteringA&P™ > Study Area > Menu > iP2 > Nervous System I: Resting Membrane Potential

SPOTLIGHT

Figure 12–13

Generation of an Action Potential

Each neuron receives information in the form of graded potentials on its dendrites and cell body, and graded potentials at the synaptic terminals trigger the release of neurotransmitters. However, the two ends of the neuron may be a meter apart, and even the largest graded potentials affect only a tiny area. Such relatively long-range communication requires a different mechanism—the action potential. **Action potentials** are propagated changes in the membrane potential that, once initiated, affect an entire excitable membrane. Action potentials are dependent on the presence of both voltage-gated sodium and potassium ion channels.

Axon hillock

Initial segment

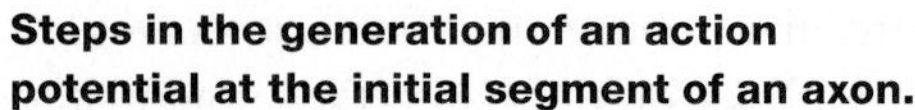

Steps in the generation of an action potential at the initial segment of an axon. The first step is a graded depolarizaton caused by the opening of chemically gated sodium ion channels, usually at the axon hillock. Note that when illustrating action potentials, we can ignore both the leak channels and the chemically gated channels, because their properties do not change. The membrane colors in steps 1–4 match the colors of the line graph showing membrane potential changes.

RESTING MEMBRANE POTENTIAL

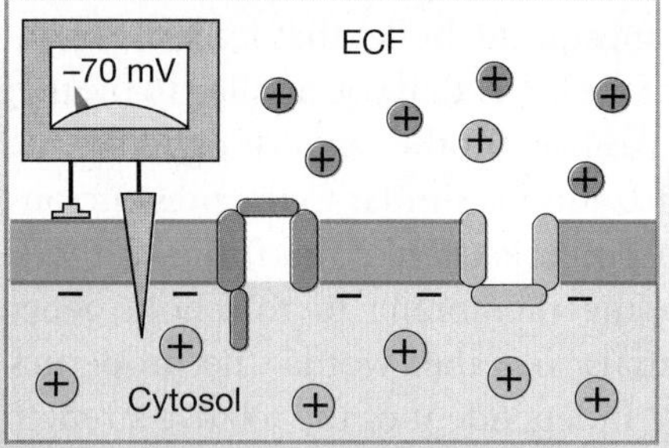

–70 mV
ECF
Cytosol

The axolemma contains both voltage-gated sodium channels and voltage-gated potassium channels that are closed when the membrane is at the resting membrane potential.

KEY

⊕ = Sodium ion

⊕ = Potassium ion

1 Depolarization to Threshold

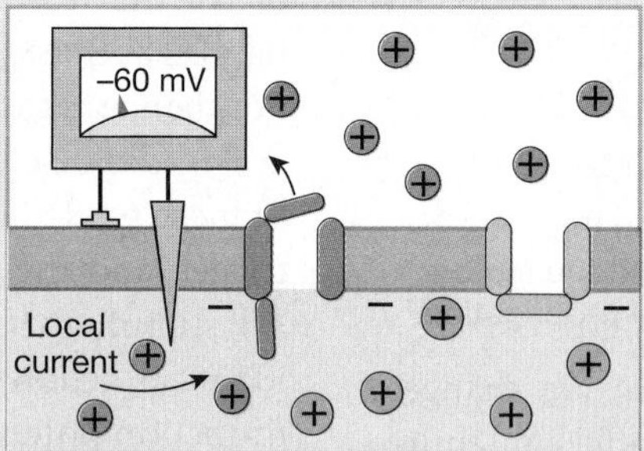

–60 mV
Local current

The stimulus that initiates an action potential is a graded depolarization large enough to open voltage-gated sodium channels. The opening of the channels occurs at a membrane potential known as the threshold.

2 Activation of Sodium Ion Channels and Rapid Depolarization

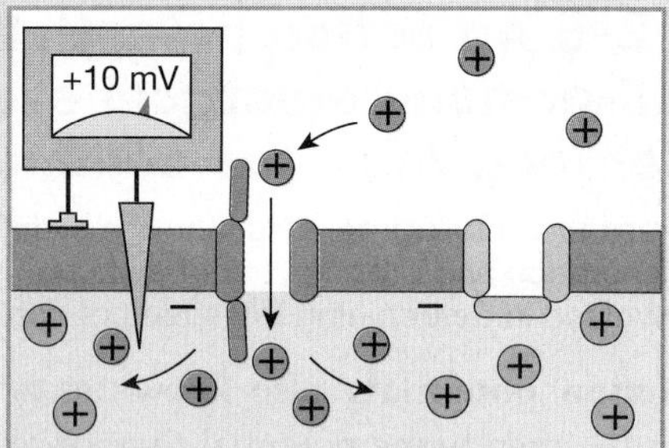

+10 mV

When the sodium channel activation gates open, the plasma membrane becomes much more permeable to Na^+. Driven by the large electrochemical gradient, sodium ions rush into the cytosol, and rapid depolarization occurs. The inner membrane surface now has more positive ions than negative ones, and the membrane potential has changed from –60 mV to a positive value.

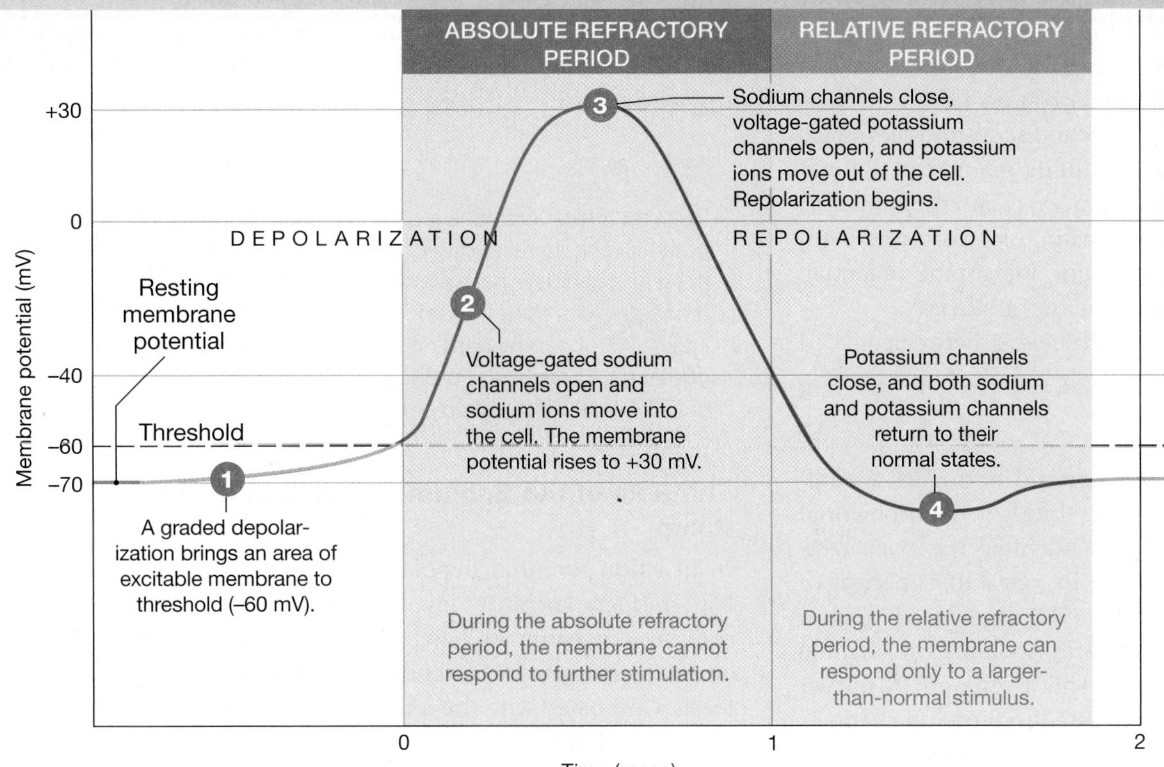
ABSOLUTE REFRACTORY PERIOD
RELATIVE REFRACTORY PERIOD
+30
0
-40
-60
-70
Membrane potential (mV)
DEPOLARIZATION
REPOLARIZATION
Resting membrane potential
Threshold
1
A graded depolarization brings an area of excitable membrane to threshold (–60 mV).
2
Voltage-gated sodium channels open and sodium ions move into the cell. The membrane potential rises to +30 mV.
3
Sodium channels close, voltage-gated potassium channels open, and potassium ions move out of the cell. Repolarization begins.
4
Potassium channels close, and both sodium and potassium channels return to their normal states.
During the absolute refractory period, the membrane cannot respond to further stimulation.
During the relative refractory period, the membrane can respond only to a larger-than-normal stimulus.
0
1
2
Time (msec)

Changes in the membrane potential at one location during the generation of an action potential. The circled numbers in the graph correspond to the steps illustrated below.

3 Inactivation of Sodium Ion Channels and Activation of Potassium Ion Channels Starts Repolarization

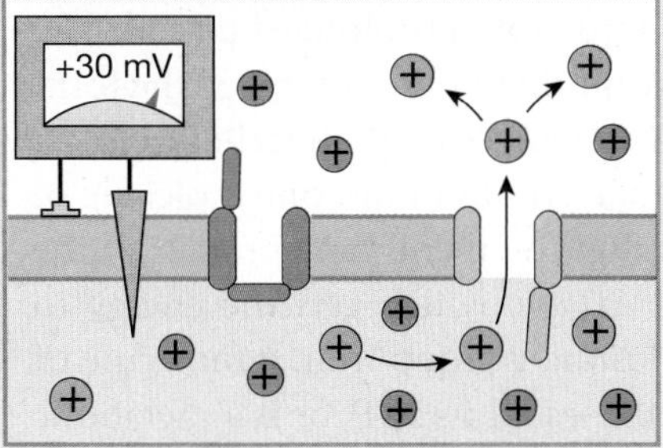
+30 mV

As the membrane potential approaches +30 mV, the inactivation gates of the voltage-gated sodium channels close. This step is known as **sodium channel inactivation**, and it coincides with the opening of voltage-gated potassium channels. Positively charged potassium ions move out of the cytosol, shifting the membrane potential back toward the resting level. Repolarization now begins.

4 Time Lag in Closing All Potassium Ion Channels Leads to Temporary Hyperpolarization

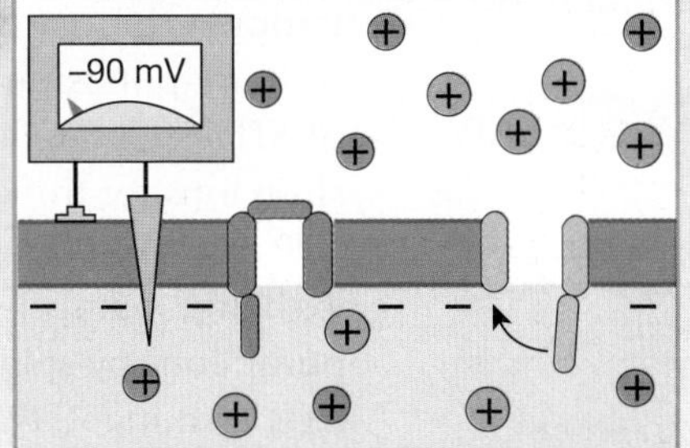
–90 mV

The voltage-gated sodium channels remain inactivated until the membrane has repolarized to near threshold level. At this time, they regain their normal status: closed but capable of opening. The voltage-gated potassium channels begin closing as the membrane reaches the normal resting membrane potential (about –70 mV). Until all of these potassium channels have closed, potassium ions continue to leave the cell. This produces a brief hyperpolarization.

RESTING MEMBRANE POTENTIAL

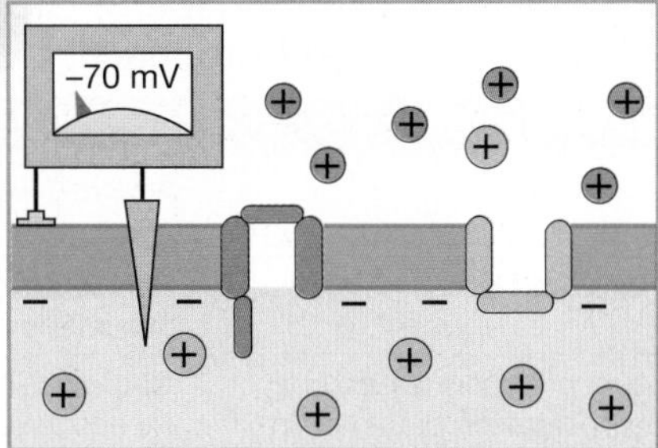
–70 mV

After all the voltage-gated potassium channels close, the membrane potential returns to the normal resting level. The action potential is now over, and the membrane is once again at the resting membrane potential.

its branches, and its axon terminals. At the resting membrane potential, the activation gates of the voltage-gated sodium ion channels are closed. Once the membrane reaches threshold, the first step in generating an action potential is the opening of voltage-gated sodium ion channels at one site, usually the initial segment of the axon. Then the steps proceed as follows: (1) depolarization to threshold; (2) activation of voltage-gated sodium ion channels and rapid depolarization; (3) inactivation of voltage-gated sodium ion channels and activation of voltage-gated potassium ion channels, beginning repolarization; and (4) closing of voltage-gated potassium ion channels, producing a brief hyperpolarization (due to a lag in total closing of channels), and then a return to the resting membrane potential. Please study this figure thoroughly before going on.

Table 12–3 reviews the key differences between graded potentials and action potentials.

The Refractory Period

12

The plasma membrane does not respond normally to additional depolarizing stimuli from the time an action potential begins until the resting membrane potential has been reestablished. This period is known as the **refractory period** of the membrane. From the moment the voltage-gated sodium ion channels open at threshold until sodium ion channel inactivation ends, the membrane cannot respond to further stimulation because all the voltage-gated sodium ion channels either are already open or are inactivated. This first part of the refractory period, called the **absolute refractory period**, lasts 0.4–1.0 msec.

The **relative refractory period** begins when the sodium ion channels regain their normal resting condition, and continues until the membrane potential stabilizes at the resting level. Another action potential can occur during this period if the membrane is sufficiently depolarized. That depolarization, however, requires a larger-than-normal stimulus, because (1) the local current must deliver enough Na^+ to counteract the exit of positively charged K^+ through open voltage-gated K^+ channels, and (2) the membrane is hyperpolarized to some degree through most of the relative refractory period.

Table 12–3 A Comparison of Graded Potentials and Action Potentials

Graded Potentials	Action Potentials
Depolarizing or hyperpolarizing	Always depolarizing
No threshold level	Depolarization to threshold level must occur before action potential begins
Amount of depolarization or hyperpolarization depends on intensity of stimulus; summation occurs	All-or-none event; all stimuli that exceed threshold produce identical action potentials; no summation occurs
Passive spread outward from site of stimulation	Action potential at one site depolarizes adjacent sites to threshold level
Effect on membrane potential decreases with distance from stimulation site	Propagated along entire membrane surface without decrease in strength
No refractory period	Refractory period occurs
Occur in most plasma membranes	Occur only in excitable membranes of specialized cells such as neurons and muscle cells

Tips & Tools

Flushing a toilet provides a useful analogy for an action potential and its refractory period. Nothing happens while you press the handle, until the water starts to flow (threshold is reached). After that, the amount of water that is released is independent of how hard or quickly you pressed the handle (all-or-none principle). Finally, you cannot flush the toilet again until the tank refills (refractory period).

The Role of the Sodium–Potassium Exchange Pump

In an action potential, depolarization results from the influx of Na^+, and repolarization involves the loss of K^+. Over time, the sodium–potassium exchange pump returns the intracellular and extracellular concentrations of these ions to prestimulation levels. Compared with the total number of ions inside and outside the cell, however, the number involved in a single action potential is insignificant. Tens of thousands of action potentials can occur before intracellular ion concentrations change enough to disrupt the entire mechanism. For this reason, the exchange pump is not essential to any single action potential.

However, a maximally stimulated neuron can generate action potentials at a rate of 1000 per second. Under these circumstances, the exchange pump is needed to keep ion concentrations within acceptable limits over a prolonged period. The sodium–potassium exchange pump requires energy in the form of ATP. Each time the pump exchanges two extracellular potassium ions for three intracellular sodium ions, one molecule of ATP is broken down. Recall that the membrane protein of the exchange pump is *Na^+/K^+ ATPase*, which gets the energy to pump ions by splitting a phosphate group from a molecule of ATP, forming ADP. If the cell uses all its ATP, or if a metabolic poison inactivates Na^+/K^+ ATPase, the neuron will soon stop functioning.

Propagation of Action Potentials

The events that generate an action potential take place in a small portion of the total plasma membrane surface. But unlike graded potentials, action potentials spread along the entire excitable membrane. To understand how this happens, imagine that you are standing by the doors of a movie theater at the start of a long line. Everyone is waiting for the doors to open. The manager steps outside and says to you, "Let everyone know that we're opening in 15 minutes." How would you spread the news?

If you treated the line as an unexcitable membrane, you would shout, "The doors open in 15 minutes!" as loudly as you could. The closest people in the line would hear the news very clearly, but those farther away might not hear the entire message, and those at the end of the line might not hear you at all.

If, however, you treated the crowd as an excitable membrane, you would tell the message to the next person in line, with instructions to pass it on. In that way, the message would travel along the line undiminished, until everyone had heard the news. Such a message "moves" as each person repeats it to someone else. Distance is not a factor, and the line can contain 50 people or 5000.

Having each person repeat the message is comparable to the way an action potential spreads along an excitable membrane. An action potential (message) is relayed from one location to another in a series of steps. At each step, the message is repeated. Because the same events take place over and over, the term **propagation** is preferable to the term *conduction*, which suggests a flow of charge similar to that in a conductor such as a copper wire. (In fact, compared to wires, axons are poor conductors of electricity.)

Types of Propagation

When sodium ions move into an axon and depolarize adjacent sites, this triggers the opening of additional voltage-gated sodium ion channels. The result is a chain reaction that spreads across the surface of the membrane like a line of falling dominoes. In this way, the action potential is propagated along the length of the axon, ultimately reaching the axon terminals.

Action potentials may travel along an axon in two ways: by *continuous propagation* (unmyelinated axons) or by *saltatory propagation* (myelinated axons). Let's explore the similarities and differences between these two types of propagation.

Continuous Propagation. In an unmyelinated axon, an action potential moves along by **continuous propagation** (Spotlight Figure 12–14a). As you learn the steps of this process, for convenience, think of the axon (axolemma) membrane as a series of adjacent segments.

As mentioned earlier, the action potential begins at the axon's initial segment. For a brief moment at the peak of the action potential, the membrane potential becomes positive rather than negative (1). A local current then develops as sodium ions begin moving in the cytosol and the extracellular fluid (2). The local current spreads in all directions, depolarizing adjacent portions of the membrane. (The axon hillock cannot respond with an action potential because, like the rest of the cell body, it lacks voltage-gated sodium ion channels.) The process then continues in a chain reaction (3 and 4).

Each time a local current develops, the action potential moves forward, but not backward, because the previous segment of the axon is still in the absolute refractory period. As a result, an action potential always proceeds away from the site of generation and cannot reverse direction. Eventually, the most distant portions of the plasma membrane are affected.

As in our "movie line" model, the message is relayed from one location to another. At each step along the way, the message is retold, so distance has no effect on the process. The action potential reaching the axon terminal is identical to the one generated at the initial segment. The net effect is the same as if a single action potential had traveled across the surface of the membrane.

In continuous propagation, an action potential appears to move across the surface of the membrane in a series of tiny steps. Even though the events at any one location take only about a millisecond, they must be repeated at each step along the way. Continuous propagation along unmyelinated axons occurs at a speed of about 1 meter per second (approximately 2 mph). For a second action potential to occur at the same site, a second stimulus must be applied.

12

Tips & Tools

The "wave" performed by fans in a football stadium illustrates the continuous propagation of an action potential. The "wave" moves, but the people remain in place.

Saltatory Propagation. An action potential in a myelinated axon moves by **saltatory propagation** (*saltare*, leaping) (Spotlight Figure 12–14b). Saltatory propagation in the CNS and PNS carries action potentials along an axon much more rapidly than does continuous propagation. To get the general idea, let's return to the line in front of the movie theater, and assume that it takes 1 second to relay the message to another person. In a model of continuous propagation, the people are jammed together. In 4 seconds, four people would hear the news, and the message would move perhaps 2 meters along the line. In a model of saltatory propagation, in contrast, the people in the line are spaced 5 meters apart. So after 4 seconds the same message would move 20 meters.

In a myelinated axon, the "people" are the nodes, and the spaces between them are the internodes wrapped in myelin (look back at Figure 12–6a). Continuous propagation cannot occur along a myelinated axon, because myelin increases resistance to the flow of ions across the membrane. Ions can readily cross the axolemma only at the nodes. As a result, only the nodes with their voltage-gated ion channels can respond to a depolarizing stimulus.

When an action potential appears at the initial segment of a myelinated axon, the local current skips the internodes and depolarizes the closest node to threshold. Because the nodes may be 1–2 mm apart in a large myelinated axon, the action potential "jumps" from node to node rather than moving along

SPOTLIGHT

Figure 12–14

Propagation of an Action Potential

a Continuous Propagation along an Unmyelinated Axon

In an unmyelinated axon, an action potential moves along by continuous propagation. The action potential spreads by depolarizing the adjacent region of the axon membrane. This process continues to spread as a chain reaction down the axon.

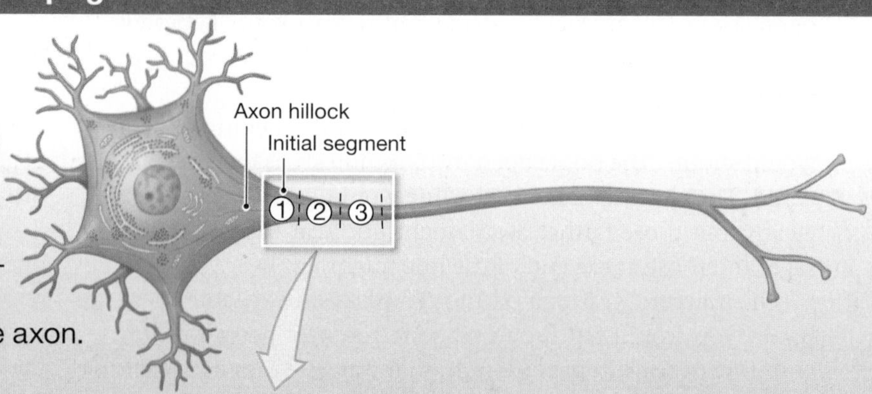

Axon hillock
Initial segment
1
2
3

1 As an action potential develops at the initial segment ①, the membrane potential at this site depolarizes to +30 mV.

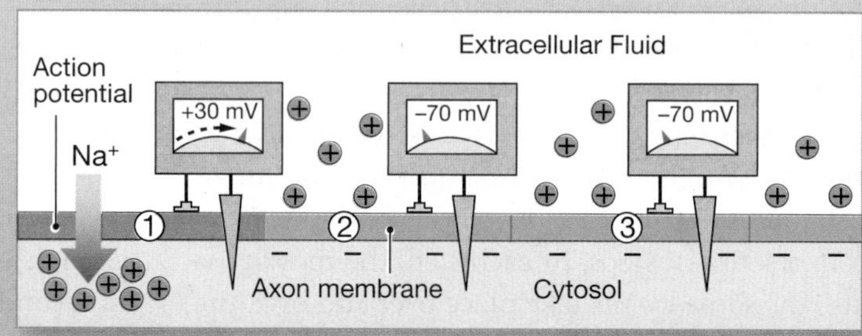

Extracellular Fluid
Action potential
Na+
+30 mV
−70 mV
−70 mV
1
2
3
Axon membrane
Cytosol

2 A local current then develops as the sodium ions entering at ① spread away from the open voltage-gated channels. A graded depolarization quickly brings the axon membrane (axolemma) in segment ② to threshold.

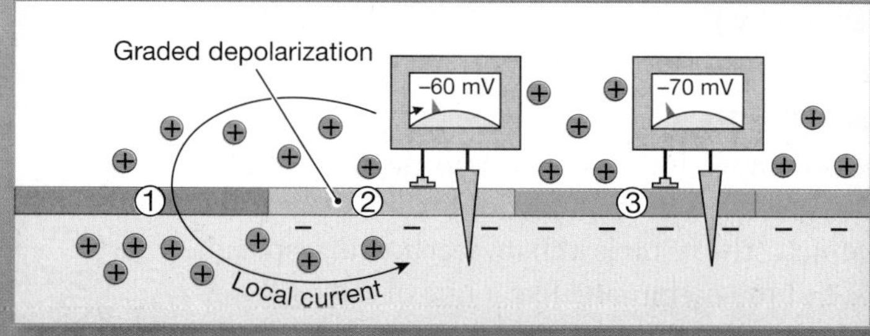

Graded depolarization
−60 mV
−70 mV
1
2
3
Local current

3 An action potential now occurs in segment ① while segment ② begins repolarization.

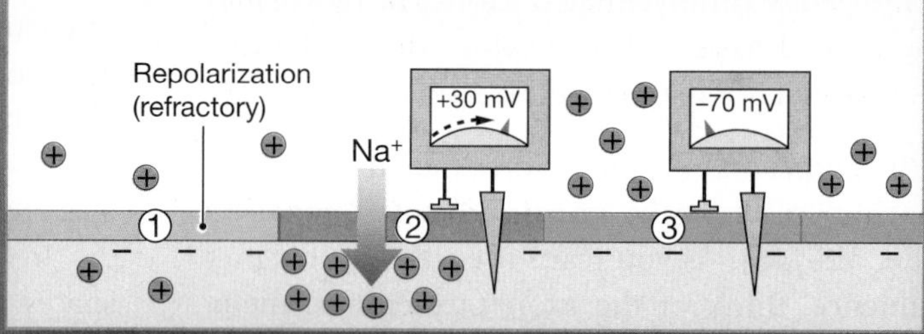

Repolarization (refractory)
Na+
+30 mV
−70 mV
1
2
3

4 As the sodium ions entering at segment ② spread laterally, a graded depolarization quickly brings the membrane in segment ③ to threshold, and the cycle is repeated.

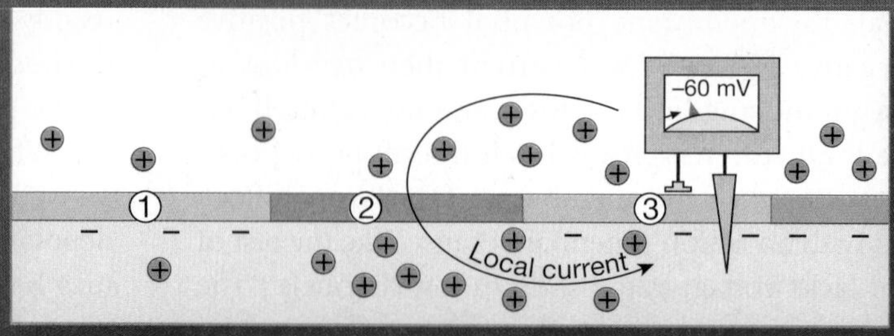

−60 mV
1
2
3
Local current

Go to MasteringA&P™ > Study Area > Menu > Animations & Videos > A&PFlix > A&P> Propagation of an Action Potential

b Saltatory Propagation along a Myelinated Axon

Because myelin limits the movement of ions across the axon membrane, the action potential must "jump" from node to node during propagation. This results in much faster propagation along the axon.

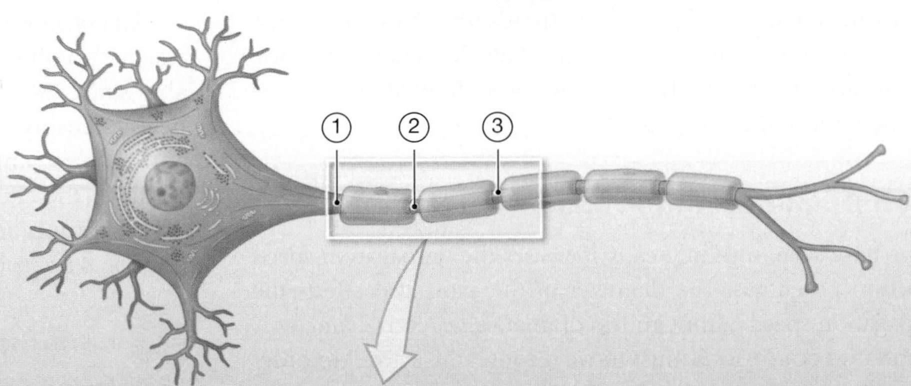
1
2
3

1 An action potential has occurred at the initial segment ①.

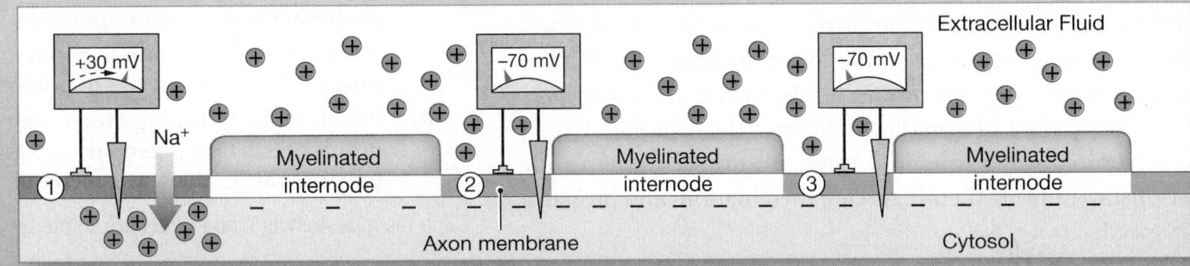
Extracellular Fluid
+30 mV
–70 mV
–70 mV
Na+
Myelinated internode
Myelinated internode
Myelinated internode
1
2
3
Axon membrane
Cytosol

2 A local current produces a graded depolarization that brings the axon membrane (axolemma) at the next node to threshold.

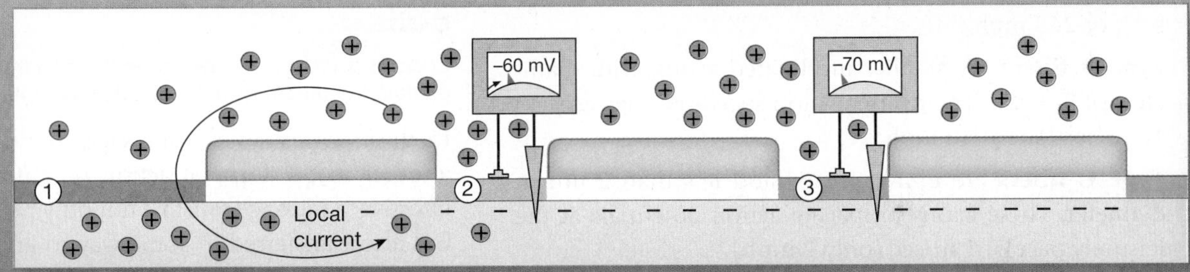
–60 mV
–70 mV
1
2
3
Local current

3 An action potential develops at node ②.

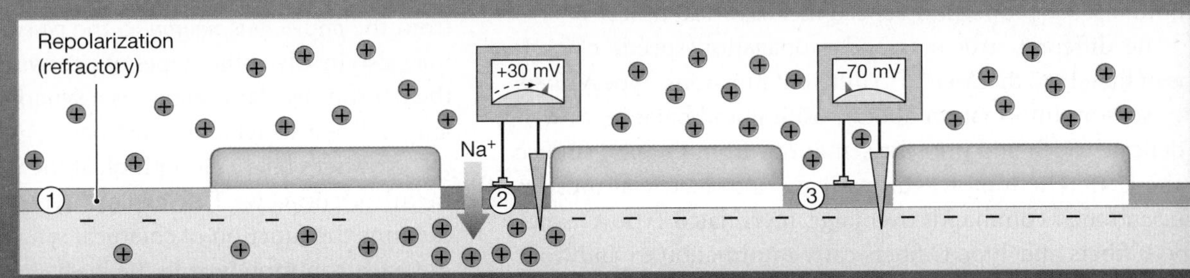
Repolarization (refractory)
+30 mV
–70 mV
Na+
1
2
3

4 A local current produces a graded depolarization that brings the axolemma at node ③ to threshold.

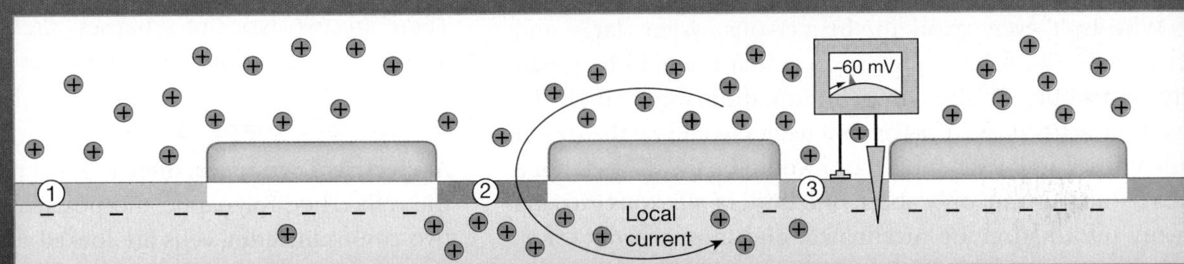
–60 mV
1
2
3
Local current

the axon in a series of tiny steps. In addition to being faster, saltatory propagation uses proportionately less energy, because less surface area is involved and fewer sodium ions must be pumped out of the cytoplasm.

Axon Diameter and Propagation Speed

As we have seen, myelin greatly increases the propagation speed of action potentials. The diameter of the axon also affects the propagation speed, although less dramatically. Axon diameter is important because ions must move through the cytosol in order to depolarize adjacent portions of the plasma membrane. Cytosol offers much less resistance to ion movement than does the axon membrane. In this instance, an axon behaves like an electrical cable: The larger the diameter, the lower the resistance. (This is why motors with large current demands, such as the starter on a car, an electric stove, or a big air conditioner, use such thick wires.)

We can classify axons into three groups according to the relationships among their diameter, myelination, and propagation speed:

1. **Type A fibers** are the largest myelinated axons, with diameters ranging from 4 to 20 μm. These fibers carry action potentials at speeds of up to 120 meters per second (m/sec), or 268 mph.
2. **Type B fibers** are smaller myelinated axons, with diameters of 2–4 μm. Their propagation speeds average around 18 m/sec (about 40 mph).
3. **Type C fibers** are unmyelinated and less than 2 µm in diameter. These axons propagate action potentials at the leisurely pace of 1 m/sec (only 2 mph).

The advantage of myelin becomes clear when you compare Type C to Type A fibers and note that the diameter increases tenfold but the propagation speed increases 120 times.

The different structures and propagation speeds of each type of fiber lend themselves to different functions. Type A fibers carry sensory information about position and balance, as well as delicate touch and pressure sensations from the skin surface to the CNS. The motor neurons that control skeletal muscles also send their commands over large, myelinated Type A axons. Type B fibers and Type C fibers carry information to and from the CNS. They deliver temperature, pain, and general touch and pressure sensations. They also carry instructions to smooth muscle, cardiac muscle, glands, and other peripheral effectors.

Why isn't every axon in the nervous system large and myelinated? The most likely reason is that it would be physically impossible. If all sensory information were carried by large Type A fibers, your peripheral nerves would be the size of garden hoses, and your spinal cord would be the diameter of a garbage can. Instead, only about one-third of all axons carrying sensory information are myelinated, and most sensory information arrives over slender Type C fibers.

In essence, information transfer in the nervous system represents a compromise between propagation time and available space. Messages are routed according to priority: Urgent news—sensory information about things that threaten survival and motor commands that prevent injury—travels over Type A fibers (the equivalent of instant messaging, or texting). Less urgent sensory information and motor commands are relayed by Type B fibers (e-mail) or Type C fibers (regular "snail mail").

Checkpoint

15. **Define *action potential*.**
16. **Identify the steps involved in the generation and propagation of an action potential.**
17. **What is the relationship between myelin and the propagation speed of action potentials?**
18. **Which of the following axons is myelinated: one that propagates action potentials at 50 meters per second, or one that carries them at 1 meter per second?**

See the blue Answers tab at the back of the book.

12-6 Synapses transmit signals among neurons or between neurons and other cells

Learning Outcome Describe the structure of a synapse, and explain the mechanism involved in synaptic activity.

In the nervous system, messages move from one location to another in the form of action potentials along axons. To be effective, a message must be not only propagated along an axon but also transferred in some way to another cell. This transfer takes place at a **synapse**, a specialized site where the neuron communicates with another cell.

At a synapse between two neurons, information passes from the *presynaptic neuron* to the *postsynaptic neuron*. Synapses may also involve other types of postsynaptic cells. For example, the neuromuscular junction is a synapse in which the postsynaptic cell is a skeletal muscle fiber.

Now let's take a closer look at the ways that synapses work. In this section, we first explore types of synapses, and then examine the function of chemical synapses with an example of the most common type in the body.

Types of Synapses

There are two types of synapses, *electrical* and *chemical*. Let's look at the similarities and differences between these types.

Electrical Synapses

At **electrical synapses**, there is direct physical contact between the cells. The presynaptic and postsynaptic membranes of the two communicating cells are locked together at gap junctions (look back at Figure 4–3, p. 119). The lipid portions of adjacent membranes, separated by only 2 nm, are held in position by binding between integral membrane proteins in these junctions called *connexons.* These proteins form pores that permit ions to pass between the cells. Because the two cells are linked in this way, changes in the membrane potential of one cell produce local currents that affect the other cell as if the two shared a common membrane. As a result, an electrical synapse propagates action potentials quickly and efficiently from one cell to the next.

Electrical synapses are extremely rare in both the adult CNS and PNS. They occur in some areas of the brain, including the vestibular nuclei (involved in balance), the eye, and in at least one pair of PNS ganglia (the ciliary ganglia). They do occur in some embryonic structures as well.

Chemical Synapses

A **chemical synapse** is where one neuron sends chemical signals to another cell, often a second neuron. Every chemical synapse involves two cells: (1) the axon terminal of the **presynaptic cell**, which sends a message, and (2) the **postsynaptic cell**, which receives the message. A narrow space called the **synaptic cleft** separates the two cells (Figure 12–15).

The presynaptic cell is usually a neuron. (Specialized receptor cells may form synaptic connections with the dendrites of neurons, a topic we describe in Chapter 15.) The postsynaptic cell can be either a neuron or another type of cell. One neuron may communicate with another at a synapse on a dendrite, on the cell body, or along the length of the axon of the receiving cell. At an *axoaxonic synapse*, a synapse occurs between the axons of two neurons. At an *axosomatic synapse*, the junction occurs at the axon terminal of one neuron with the cell body of another; at an *axodendritic synapse*, the synaptic contact is between the axon terminal of one neuron and the dendrite of another neuron.

Figure 12–15 **The Structure of a Typical Chemical Synapse.**

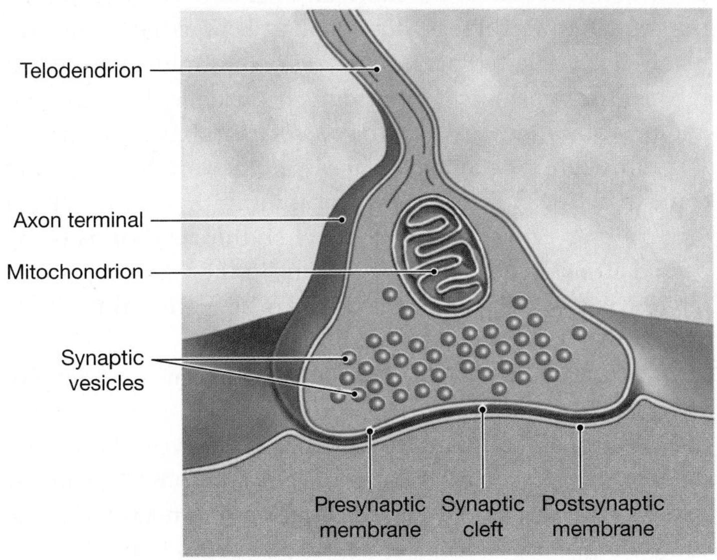
Telodendrion
Axon terminal
Mitochondrion
Synaptic vesicles
Presynaptic membrane
Synaptic cleft
Postsynaptic membrane

A synapse between a neuron and a skeletal muscle cell is called a **neuromuscular junction**. p. 303 At a **neuroglandular junction**, a neuron controls or regulates the activity of a secretory (gland) cell. Neurons also *innervate* (are distributed to) a variety of other cell types, such as adipocytes (fat cells). We consider the nature of that innervation in later chapters.

The axon terminal of the presynaptic cell releases chemicals called **neurotransmitters** into the synaptic cleft. Inside the axon terminal, neurotransmitters are contained in *synaptic vesicles.*

The structure of the axon terminal varies with the type of postsynaptic cell. A relatively simple, round axon terminal occurs where the postsynaptic cell is another neuron. At a synapse, the narrow synaptic cleft separates the presynaptic membrane, where neurotransmitters are released, from the postsynaptic membrane, which bears receptors for neurotransmitters (see Figure 12–15). The axon terminal at a neuromuscular junction is much more structurally complex.

Each axon terminal contains mitochondria and thousands of vesicles filled with neurotransmitter molecules. The axon terminal reabsorbs breakdown products of neurotransmitters formed at the synapse and reassembles them. It also receives a continuous supply of neurotransmitters synthesized in the cell body, along with enzymes and lysosomes, by anterograde flow.

Chemical synapses are by far the most abundant type of synapse. Most synapses among neurons, and all communications between neurons and other types of cells, involve chemical synapses. Normally, communication across a chemical synapse takes place in only one direction: from the presynaptic membrane to the postsynaptic membrane.

Function of Chemical Synapses

As we saw in Chapter 10, neurotransmitter release is triggered by electrical events, such as the arrival of an action potential. The presynaptic cell typically releases neurotransmitters. The neurotransmitters then flood the synaptic cleft and bind to receptors on the postsynaptic plasma membrane, changing its permeability and producing graded potentials. The mechanism is comparable to that of the neuromuscular junction, described in Chapter 10. pp. 304–305

The situation at a chemical synapse is far more variable than that at an electrical synapse, because the cells are not directly coupled. For example, an action potential that reaches an electrical synapse is *always* propagated to the next cell. But at a chemical synapse, an arriving action potential *may or may not* release enough neurotransmitter to bring the postsynaptic neuron to threshold. In addition, other factors may intervene and make the postsynaptic cell more or less sensitive to arriving stimuli. In essence, the postsynaptic cell at a chemical synapse is not a slave to the presynaptic neuron, and its activity can be adjusted, or "tuned," by a variety of factors.

Let's continue our discussion of chemical synapses with a look at a synapse that releases the neurotransmitter **acetylcholine (ACh)**. Section 12-7 introduces the neurotransmitters that you will encounter in later chapters.

Example of Chemical Synaptic Function: Cholinergic Synapses

Synapses that release ACh are known as **cholinergic** (kol-in-ER-jik) **synapses**. ACh is the most widespread (and best-studied) neurotransmitter. It is released at (1) all neuromuscular junctions involving skeletal muscle fibers, (2) many synapses in the CNS, (3) all neuron-to-neuron synapses in the PNS, and (4) all neuromuscular and neuroglandular junctions in the parasympathetic division of the ANS.

Tips & Tools

Cholinergic synapses are so named because the neurotransmitter involved is acetyl**cholin**e.

12

At a cholinergic synapse between two neurons, the presynaptic and postsynaptic membranes are separated by a synaptic cleft that averages 20 nm (0.02 μm) in width. Most of the ACh in the axon terminal is packaged in synaptic vesicles, each containing several thousand molecules of the neurotransmitter. A single axon terminal may contain a million such vesicles. Let's look at events at the cholinergic synapse, and then talk about how the neurotransmitter is released, diffuses across the synapse and binds to receptors in a process known as *synaptic delay*. We conclude with *synaptic fatigue*, the short-term inability of a neuron to generate an action potential.

Figure 12–16 diagrams the events that take place at a cholinergic synapse between neurons after an action potential arrives at an axon terminal. For convenience, we will assume that this synapse is adjacent to the initial segment of the axon, an arrangement that is easy to illustrate.

1. ***An Action Potential Arrives at the Presynaptic Axon Terminal and Depolarizes the Membrane.***

2. ***Extracellular Calcium Ions Enter the Axon Terminal, Triggering the Exocytosis of ACh in Synaptic Vesicles.*** The depolarization of the axon terminal briefly opens its voltage-gated calcium ion channels, allowing calcium ions to rush in. Their arrival triggers exocytosis of synaptic vesicles containing ACh into the synaptic cleft.

   The ACh is released in packets of roughly 3000 molecules, the average number of ACh molecules in a single vesicle. ACh release stops very soon, because active transport mechanisms rapidly remove the calcium ions from the cytosol in the axon terminal. These ions are either pumped out of the cell or transferred into mitochondria.

3. ***ACh Binds to Receptors on the Postsynaptic Membrane and Depolarizes the Membrane.*** ACh diffuses across the synaptic cleft toward receptors on the postsynaptic membrane. These ACh receptors are chemically gated cation (Na^+, K^+) channels. The primary response is an increased permeability to Na^+, producing a depolarization in the postsynaptic membrane that lasts about 20 msec. (These cation channels also let potassium ions out of the cell, but because sodium ions are driven by a much stronger electrochemical gradient, the net effect is a slight depolarization of the postsynaptic membrane.)

   This depolarization is a graded potential. The greater the amount of ACh released at the presynaptic membrane, the greater the number of open cation channels in the postsynaptic membrane, and so the larger the depolarization. If the depolarization brings an adjacent area of excitable membrane (such as the initial segment of an axon) to threshold, an action potential occurs in the postsynaptic neuron.

4. ***ACh Is Removed from the Synaptic Cleft by AChE.*** The neurotransmitter's effects on the postsynaptic membrane are temporary, because the synaptic cleft and the postsynaptic membrane contain the enzyme *acetylcholinesterase* (*AChE*, or *cholinesterase*). Roughly half of the ACh released at the presynaptic membrane is broken down before it reaches receptors on the postsynaptic membrane. ACh molecules that bind to receptor sites are generally broken down within 20 msec of their arrival.

AChE breaks down molecules of ACh (by hydrolysis) into **acetate** and **choline**. The choline is actively absorbed by the axon terminal and is used to synthesize more ACh, using acetate provided by *coenzyme A (CoA)*. (Recall from Chapter 2 that coenzymes derived from vitamins are required in many enzymatic reactions. p. 56) Acetate diffusing away from the synapse can be absorbed and metabolized by the postsynaptic cell or by other cells and tissues.

Synaptic Delay

A **synaptic delay** of 0.2–0.5 msec occurs between the arrival of the action potential at the axon terminal and the effect on the postsynaptic membrane. Most of that delay reflects the time involved in calcium ion influx and neurotransmitter release, not in the neurotransmitter's diffusion—the synaptic cleft is narrow, and neurotransmitters can diffuse across it in very little time.

Although a delay of 0.5 msec is not very long, in that time an action potential may travel more than 7 cm (about 3 in.) along a myelinated axon. When information is being passed along a chain of interneurons in the CNS, the cumulative synaptic delay may exceed the propagation time along the axons.

This is why reflexes are important for survival—they involve only a few synapses and thus provide rapid and automatic responses to stimuli. The fewer synapses involved, the shorter the total synaptic delay and the faster the response. The fastest reflexes have just one synapse, with a sensory neuron directly controlling a motor neuron (see Chapter 13 for examples).

Figure 12–16 Events in the Functioning of a Cholinergic Synapse.

1 An arriving action potential depolarizes the axon terminal membrane of a presynaptic neuron.

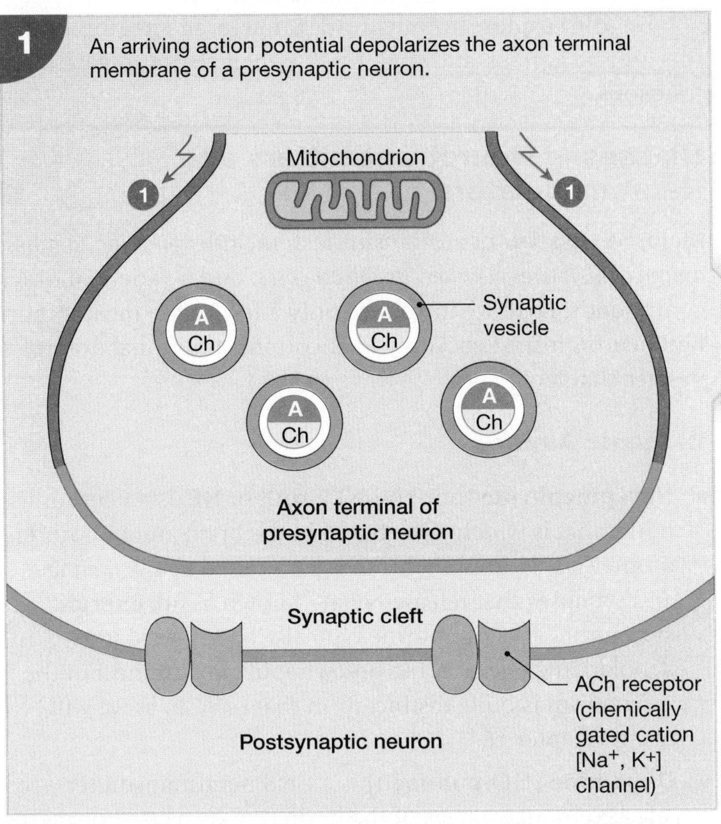
Mitochondrion
1
1
A
Ch
Synaptic vesicle
Axon terminal of presynaptic neuron
Synaptic cleft
Postsynaptic neuron
ACh receptor (chemically gated cation [Na+, K+] channel)

2 Depolarization of the axon terminal membrane opens voltage-gated calcium ion channels, and calcium ions (Ca^{2+}) enter the cytosol of the axon terminal. This results in ACh release from the synaptic vesicles by exocytosis.

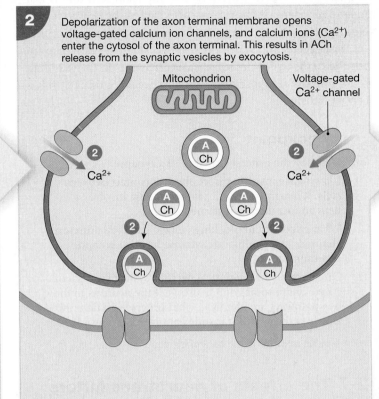

Mitochondrion

Voltage-gated Ca^{2+} channel

Ca^{2+}

Ca^{2+}

3 ACh diffuses across the synaptic cleft and binds to receptors on the postsynaptic membrane. Cation (Na^+, K^+) channels open, producing a graded depolarization due to Na^+ inflow.

Na^+

Chemically gated cation channels

Na^+

Initiation of a graded potential, or action potential, if threshold is reached at the initial segment

4 Depolarization ends as ACh is broken down into acetate and choline by AChE. The axon terminal reabsorbs choline from the synaptic cleft and uses it to resynthesize ACh.

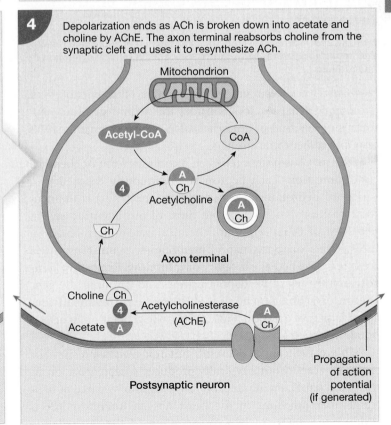

Mitochondrion

Acetyl-CoA

CoA

Acetylcholine

Axon terminal

Choline

Acetylcholinesterase (AChE)

Acetate

Propagation of action potential (if generated)

Postsynaptic neuron

? Which part of an acetylcholine (ACh) molecule undergoes reuptake by the axon terminal to be reused in another acetylcholine molecule?

12

Synaptic Fatigue

Because ACh molecules are recycled, the axon terminal is not totally dependent on the ACh synthesized in the cell body and delivered by axonal transport. But under intensive stimulation, resynthesis and transport mechanisms may not keep up with the demand for neurotransmitter. **Synaptic fatigue** then occurs, and the response of the synapse weakens until ACh has been replenished.

Checkpoint

19. **Describe the general structure of a synapse.**
20. **If a synapse involves direct physical contact between cells, it is termed ______; if the synapse involves a neurotransmitter, it is termed ______.**
21. **What effect would blocking voltage-gated calcium ion channels at a cholinergic synapse have on synaptic communication?**
22. **One pathway in the central nervous system consists of 3 neurons, another of 5 neurons. If the neurons in the two pathways are identical, what process will determine which pathway will transmit impulses more rapidly?**

See the blue Answers tab at the back of the book.

12

12-7 The effects of neurotransmitters and neuromodulators depend on their receptors

Learning Outcome Describe the major types of neurotransmitters and neuromodulators, and discuss their effects on postsynaptic membranes.

Now that we have examined the actions of acetylcholine at cholinergic synapses, let's consider the actions of other neurotransmitters, and of *neuromodulators*, which change the cell's response to neurotransmitters.

The nervous system relies on a complex form of chemical communication. Each neuron is continuously exposed to a variety of neurotransmitters. Acetylcholine (ACh) is the neurotransmitter that has received most of our attention so far, but there are other important chemical transmitters. Based on their effects on postsynaptic membranes, neurotransmitters are often classified as excitatory or inhibitory. **Excitatory neurotransmitters** cause depolarization and promote the generation of action potentials. **Inhibitory neurotransmitters** cause hyperpolarization and suppress the generation of action potentials.

This classification is useful, but not always precise. For example, acetylcholine typically produces a depolarization in the postsynaptic membrane, but acetylcholine released at neuromuscular junctions in the heart has an inhibitory effect, producing a transient hyperpolarization of the postsynaptic membrane. This situation highlights an important aspect of neurotransmitter function: *The effect of a neurotransmitter on the postsynaptic membrane depends on the properties of the receptor, not on the nature of the neurotransmitter.*

Let's first explore how neurotransmitters and neuromodulators are categorized and then discuss their receptor-dependent functions.

Classes of Neurotransmitters and Neuromodulators

Major classes of neurotransmitters include *biogenic amines, amino acids, neuropeptides, dissolved gases,* and a variety of other compounds. Here we introduce only a few of the most important neurotransmitters.You will encounter additional examples in later chapters.

Biogenic Amines

- **Norepinephrine** (nor-ep-i-NEF-rin), or **NE**, is a neurotransmitter that is widely distributed in the brain and in portions of the ANS. Norepinephrine is also called *noradrenaline,* and synapses that release NE are known as **adrenergic synapses**. Norepinephrine typically has an excitatory, depolarizing effect on the postsynaptic membrane, but the mechanism is quite distinct from that of ACh, as we will see in Chapter 16.
- **Dopamine** (DŌ-puh-mēn) is a CNS neurotransmitter released in many areas of the brain. It may have either inhibitory or excitatory effects. Inhibitory effects play an important role in our precise control of movements. For example, dopamine release in one portion of the brain prevents the overstimulation of neurons that control skeletal muscle tone. If the neurons that produce dopamine are damaged or destroyed, the result can be the characteristic rigidity and stiffness of *Parkinson's disease,* a condition we describe in Chapter 14. At other sites, dopamine release has excitatory effects. Cocaine inhibits the removal of dopamine from synapses in specific areas of the brain. The resulting rise in dopamine concentration at these synapses is responsible for the "high" experienced by cocaine users.
- **Serotonin** (ser-ō-TŌ-nin) is another important CNS neurotransmitter. Inadequate serotonin production can have widespread effects on a person's attention and emotional states and may be responsible for many cases of severe chronic depression. *Fluoxetine (Prozac), paroxetine (Paxil), sertraline (Zoloft),* and related antidepressant drugs inhibit the reabsorption of serotonin by axon terminals (hence their classification as *s*elective *s*erotonin *r*euptake *i*nhibitors, or *SSRIs*). This inhibition leads to increased serotonin concentration at synapses, and over time, the increase may relieve the symptoms of depression. Interactions among serotonin, norepinephrine, and other neurotransmitters are thought to be involved in the regulation of sleep and wake cycles.

Go to MasteringA&P™ > Study Area > Video Tutor > Events at a Cholinergic Synapse

Amino Acids

- **Gamma-aminobutyric** (a-MĒ-nō-bū-TĒR-ik) **acid**, or **GABA**, generally has an inhibitory effect. Roughly 20 percent of the synapses in the brain release GABA, but its functions remain incompletely understood. In the CNS, GABA release appears to reduce anxiety, and some antianxiety drugs work by enhancing this effect.

Neuropeptides

It is convenient to discuss each synapse as if it were releasing only one neurotransmitter. However, axon terminals may release a mixture of active compounds, either through diffusion across the membrane or by exocytosis, along with neurotransmitter molecules. These compounds may have a variety of functions. Those that alter the rate of neurotransmitter release by the presynaptic neuron or change the postsynaptic cell's response to neurotransmitters are called **neuromodulators** (nū-rō-MOD-ū-lā-torz). These substances are typically **neuropeptides**, small peptide chains synthesized and released by the axon terminal. Most neuromodulators act by binding to receptors in the presynaptic or postsynaptic membranes and activating cytoplasmic enzymes. In general, neuromodulators (1) have long-term effects that are relatively slow to appear; (2) trigger responses that involve a number of steps and intermediary compounds; (3) may affect the presynaptic membrane, the postsynaptic membrane, or both; and (4) can be released alone or along with a neurotransmitter.

Neuromodulators called **opioids** (Ō-pē-oydz) have effects similar to those of the drugs *opium* and *morphine*, because they bind to the same group of postsynaptic receptors. Three main classes of opioids in the CNS are (1) **enkephalins** (en-KEF-a-linz), (2) **endorphins** (en-DOR-finz), and (3) **dynorphins** (DĪ-nor-finz). The primary function of opioids is probably to relieve pain. They inhibit the release of the neurotransmitter *substance P* at synapses that relay pain sensations. Dynorphins have far more powerful pain-relieving effects than morphine or the other opioids.

Tips & Tools

Endorphins are so named because they act like **endo**genous (coming from within the body) m**orphin**e.

Dissolved Gases

- In addition, two lipid-soluble gases, nitric oxide and carbon monoxide, are known to be important neurotransmitters. **Nitric oxide (NO)** is generated by axon terminals that innervate smooth muscle in the walls of blood vessels in the PNS, and at synapses in several regions of the brain. **Carbon monoxide (CO)**, best known as a component of automobile exhaust, is also generated by specialized axon terminals in the brain, where it functions as a neurotransmitter.

The functions of many neurotransmitters are not well understood. In a clear demonstration of the principle "the more you look, the more you see," over 100 neurotransmitters have been identified, including certain amino acids, peptides, polypeptides, prostaglandins, and ATP.

Table 12–4 lists major neurotransmitters and neuromodulators of the brain and spinal cord, and their primary effects (if known). In practice, it can be very difficult to distinguish neurotransmitters from neuromodulators on either biochemical or functional grounds: A neuropeptide may function in one site as a neuromodulator and in another as a neurotransmitter. For this reason, Table 12–4 does not distinguish between neurotransmitters and neuromodulators.

The Functions of Neurotransmitters and Neuromodulators and Their Receptors

Functionally, neurotransmitters and neuromodulators fall into one of three groups: (1) *compounds that have a direct effect on membrane potential*, (2) *compounds that have an indirect effect on membrane potential through G proteins*, or (3) *lipid-soluble gases that have an indirect effect on membrane potential through intracellular enzymes*.

12

Direct Effects

Compounds that have direct effects on membrane potential open or close chemically gated ion channels (Figure 12–17a). Examples include ACh and the amino acids *glutamate* and *aspartate*. Because these neurotransmitters alter ion movement across the membrane, they are said to have *ionotropic* (ī-ō-nō-TRŌ-pik) *effects*.

A few neurotransmitters, notably glutamate, GABA, norepinephrine, and serotonin, have both direct and indirect effects, because these compounds target two different classes of receptors. The direct effects are ionotropic. The indirect effects, which involve changes in the metabolic activity of the postsynaptic cell, are called *metabotropic*.

Indirect Effects by Second Messengers

Compounds that have an indirect effect on membrane potential work through intermediaries known as *second messengers*. The neurotransmitter represents a *first messenger*, because it delivers the message to receptors on the plasma membrane or within the cell. Second messengers are ions or molecules that are produced or released inside the cell when a first messenger binds to one of these receptors.

Many neurotransmitters—including epinephrine, norepinephrine, dopamine, serotonin, histamine, and GABA—and many neuromodulators bind to receptors in the plasma membrane called *G-protein-coupled receptors (GPCRs)*. In these instances, the link between the first messenger and the second messenger is a **G protein**, an enzyme complex coupled to a

Table 12–4 Representative Neurotransmitters and Neuromodulators

Class and Neurotransmitter	Chemical Structure	Mechanism of Action	Location(s)	Comments
Acetylcholine	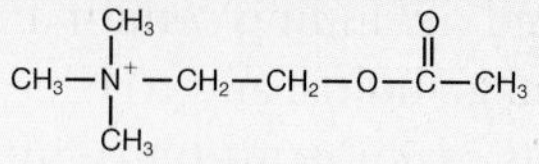 $CH_3-N^+(CH_3)(CH_3)-CH_2-CH_2-O-C(=O)-CH_3$	Primarily direct, through binding to chemically gated ion channels	CNS: synapses throughout brain and spinal cord PNS: neuromuscular junctions; preganglionic synapses of ANS; neuroglandular junctions of parasympathetic division and (rarely) sympathetic division of ANS; amacrine cells of retina	Widespread in CNS and PNS; best known and most studied of the neurotransmitters
BIOGENIC AMINES				
Norepinephrine	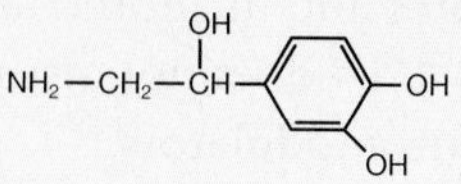 $NH_2-CH_2-CH(OH)-$[benzene ring with two OH]	Indirect: G proteins and second messengers	CNS: cerebral cortex, hypothalamus, brainstem, cerebellum, spinal cord PNS: most neuromuscular and neuroglandular junctions of sympathetic division of ANS	Involved in attention and consciousness, control of body temperature, and regulation of pituitary gland secretion
Epinephrine	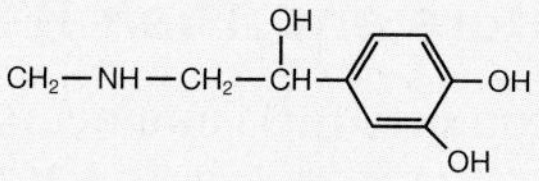 $CH_2-NH-CH_2-CH(OH)-$[benzene ring with two OH]	Indirect: G proteins and second messengers	CNS: thalamus, hypothalamus, midbrain, spinal cord	Uncertain functions
Dopamine	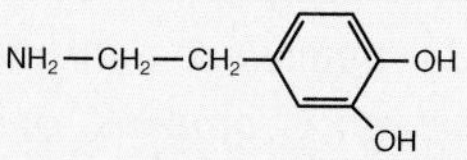 $NH_2-CH_2-CH_2-$[benzene ring with two OH]	Indirect: G proteins and second messengers	CNS: hypothalamus, midbrain, limbic system, cerebral cortex, retina	Regulation of subconscious motor function; receptor abnormalities have been linked to development of schizophrenia
Serotonin	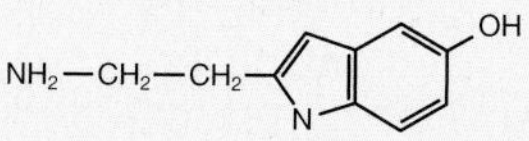 $NH_2-CH_2-CH_2-$[indole ring (N) with OH]	Primarily indirect: G proteins and second messengers	CNS: hypothalamus, limbic system, cerebellum, spinal cord, retina	Important in emotional states, moods, and body temperature; several illegal hallucinogenic drugs, such as Ecstasy, target serotonin receptors
Histamine	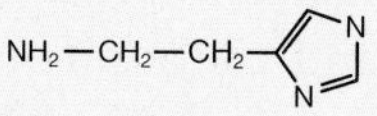 $NH_2-CH_2-CH_2-$[imidazole ring (N, N)]	Indirect: G proteins and second messengers	CNS: neurons in hypothalamus, with axons projecting throughout the brain	Receptors are primarily on presynaptic membranes; functions in sexual arousal, pain threshold, pituitary hormone secretion, thirst, and blood pressure control
AMINO ACIDS				
Excitatory: **Glutamate**	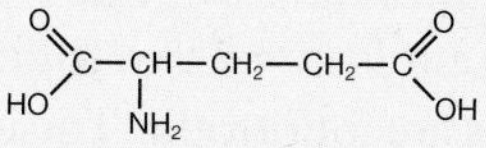 $HO(O=)C-CH(NH_2)-CH_2-CH_2-C(=O)OH$	Indirect: G proteins and second messengers Direct: opens calcium/sodium ion channels on pre- and postsynaptic membranes	CNS: cerebral cortex and brainstem	Important in memory and learning; most important excitatory neurotransmitter in the brain
Aspartate	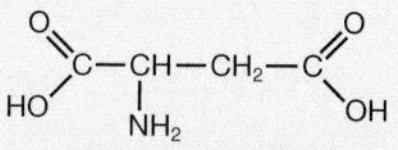 $HO(O=)C-CH(NH_2)-CH_2-C(=O)OH$	Direct or indirect (G proteins), depending on type of receptor	CNS: cerebral cortex, retina, and spinal cord	Used by pyramidal cells in the brain that provide voluntary motor control over skeletal muscles
Inhibitory: **Gamma-aminobutyric acid (GABA)**	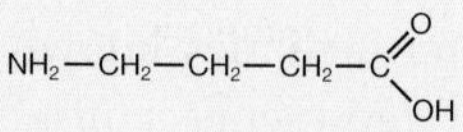 $NH_2-CH_2-CH_2-CH_2-C(=O)OH$	Direct or indirect (G proteins), depending on type of receptor	CNS: cerebral cortex, cerebellum, interneurons throughout brain and spinal cord	Direct effects: open Cl^- channels; indirect effects: open K^+ channels and block entry of Ca^{2+}
Glycine	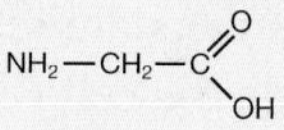 $NH_2-CH_2-C(=O)OH$	Direct: opens Cl^- channels	CNS: interneurons in brainstem, spinal cord, and retina	Produces postsynaptic inhibition; the poison *strychnine* produces fatal convulsions by blocking glycine receptors

(Continued)

Table 12–4 (continued)

Class and Neurotransmitter	Chemical Structure	Mechanism of Action	Location(s)	Comments
NEUROPEPTIDES				
Substance P	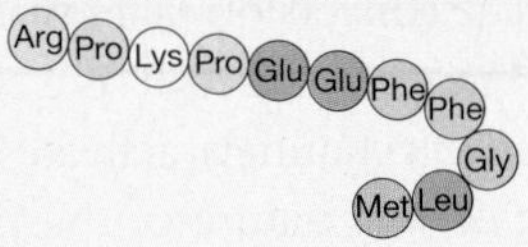 Arg Pro Lys Pro Glu Glu Phe Phe Gly Leu Met	Indirect: G proteins and second messengers	CNS: synapses of pain receptors within spinal cord, hypothalamus, and other areas of the brain PNS: enteric nervous system (network of neurons along the digestive tract)	Important in pain pathway, regulation of pituitary gland function, control of digestive tract reflexes
Neuropeptide Y	*36-amino-acid peptide*	Indirect: G proteins and second messengers	CNS: hypothalamus PNS: sympathetic neurons	Stimulates appetite and food intake
Opioids	*31-amino-acid peptide*	Indirect: G proteins and second messengers	CNS: thalamus, hypothalamus, brainstem, retina	Pain control; emotional and behavioral effects poorly understood
Enkephalins	 Tyr Gly Gly Phe Met	Indirect: G proteins and second messengers	CNS: basal nuclei, hypothalamus, midbrain, pons, medulla oblongata, spinal cord	Pain control; emotional and behavioral effects poorly understood
Endorphin	*9- or 10-amino-acid peptide*	Indirect: G proteins and second messengers	CNS: thalamus, hypothalamus, basal nuclei	Pain control; emotional and behavioral effects poorly understood
Dynorphin	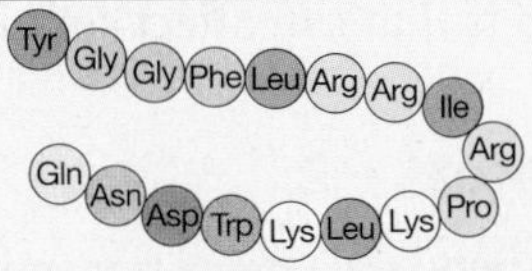 Tyr Gly Gly Phe Leu Arg Arg Ile Arg Pro Lys Leu Lys Trp Asp Asn Gln	Indirect: G proteins and second messengers	CNS: hypothalamus, midbrain, medulla oblongata	Pain control; emotional and behavioral effects poorly understood
PURINES				
ATP, GTP	*(see Figure 2–25)*	Direct or indirect (G proteins), depending on type of receptor	CNS: spinal cord PNS: autonomic ganglia	
Adenosine	*(see Figure 2–26)*	Indirect: G proteins and second messengers	CNS: cerebral cortex, hippocampus, cerebellum	Produces drowsiness; stimulatory effect of caffeine is due to inhibition of adenosine activity
HORMONES				
ADH, oxytocin, insulin, glucagon, secretin, CCK, GIP, VIP, inhibins, ANP, BNP, and many others	Peptide containing fewer than 200 amino acids	Typically indirect: G proteins and second messengers	CNS: brain (widespread)	Numerous, complex, and incompletely understood
GASES				
Carbon monoxide (CO)	C≡O	Indirect: by diffusion to enzymes activating second messengers	CNS: brain PNS: some neuromuscular and neuroglandular junctions	Localization and function poorly understood
Nitric oxide (NO)	N≡O	Indirect: by diffusion to enzymes activating second messengers	CNS: brain, especially at blood vessels PNS: some sympathetic neuromuscular and neuroglandular junctions	
LIPIDS				
Anandamide	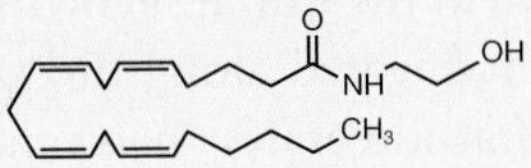 O, NH, OH, CH_3	Indirect: G proteins and second messengers	CNS: cerebral cortex, hippocampus, cerebellum	Euphoria, drowsiness, appetite; receptors are targeted by the active compound in marijuana

12

Figure 12–17 Mechanisms of Neurotransmitter and Receptor Function.

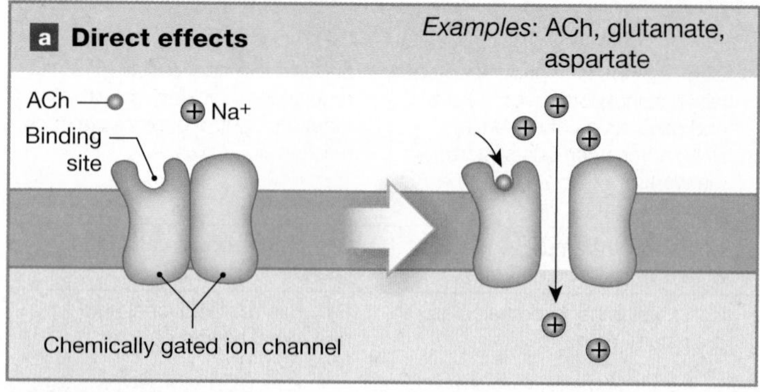
a Direct effects
Examples: ACh, glutamate, aspartate
ACh
Na+
Binding site
Chemically gated ion channel

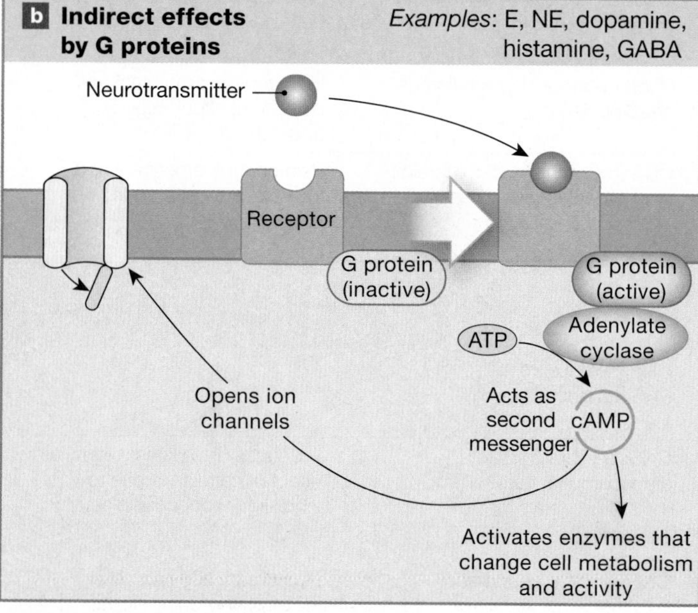
b Indirect effects by G proteins
Examples: E, NE, dopamine, histamine, GABA
Neurotransmitter
Receptor
G protein (inactive)
G protein (active)
Adenylate cyclase
ATP
Opens ion channels
Acts as second messenger
cAMP
Activates enzymes that change cell metabolism and activity

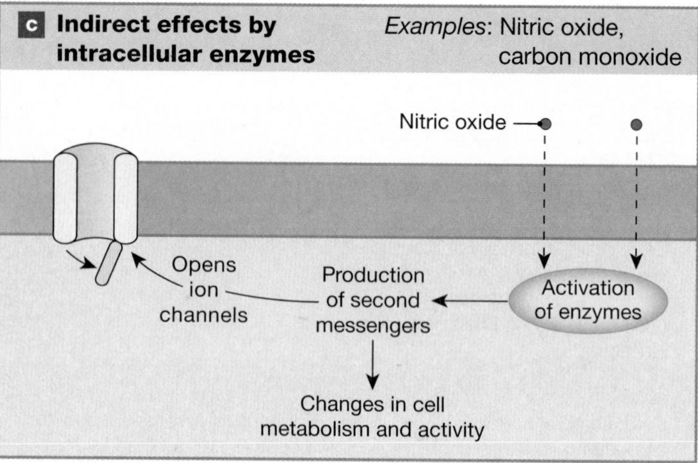
c Indirect effects by intracellular enzymes
Examples: Nitric oxide, carbon monoxide
Nitric oxide
Opens ion channels
Production of second messengers
Activation of enzymes
Changes in cell metabolism and activity

membrane receptor. The name *G protein* refers to the fact that these proteins bind GTP, a high-energy compound introduced in Chapter 2. p. 60 Several types of G protein exist, but each type includes an enzyme that is "turned on" when an extracellular compound binds to its associated receptor at the cell surface.

Figure 12–17b shows one possible result of this binding: the activation of the enzyme **adenylate cyclase**. This enzyme converts ATP, the energy currency of the cell, to *cyclic-AMP* (cAMP), a ring-shaped form of the compound AMP that was introduced in Chapter 2. p. 60 The conversion takes place at the inner surface of the plasma membrane. Cyclic-AMP is a second messenger that may open membrane ion channels, activate intracellular enzymes, or both, depending on the nature of the postsynaptic cell. This is only an overview of the function of one type of G protein. We examine several types of G proteins more closely in later chapters.

Indirect Effects by Intracellular Enzymes

The two lipid-soluble gases nitric oxide (NO) and carbon monoxide (CO) function as neurotransmitters in specific regions of the brain. Because they can diffuse through lipid membranes, these gases can enter the cell and bind to enzymes on the inner surface of the plasma membrane or elsewhere in the cytoplasm (Figure 12–17c). These enzymes then promote the appearance of second messengers that can affect cellular activity.

Checkpoint

23. **Differentiate between a neurotransmitter and a neuromodulator.**
24. **Identify the three functional groups of neurotransmitters and neuromodulators.**

See the blue Answers tab at the back of the book.

12-8 Individual neurons process information by integrating excitatory and inhibitory stimuli

Learning Outcome Discuss the interactions that enable information processing to occur in nervous tissue.

A single neuron may receive information across thousands of synapses. As we have seen, some of the neurotransmitters arriving at the postsynaptic cell at any moment may be excitatory, and others may be inhibitory. So how does the neuron respond? The net effect on the membrane potential of the cell body—specifically, in the area of the axon hillock—determines how the neuron responds from moment to moment. If the net effect is a depolarization at the axon hillock, that depolarization affects the membrane potential at the initial segment. If threshold is reached at the initial segment, an action potential is generated and propagated along the axon. Thus it is really the axon hillock that integrates the excitatory and inhibitory stimuli affecting the cell body and dendrites at any given moment. This integration process, which determines the rate of action potential generation at the initial segment, is the simplest level of **information processing** in the nervous system.

Once the signal gets to the postsynaptic cell, its response ultimately depends on what the stimulated receptors do and what other stimuli are influencing the cell at the same time. The excitatory and inhibitory stimuli are integrated through interactions between *postsynaptic potentials*, which we discuss in this section. Presynaptic inhibition, presynaptic facilitation, and the rate of action potential generation are also discussed. Higher levels of information processing involve interactions among neurons and among groups of neurons. We address these topics in later chapters.

Postsynaptic Potentials

Postsynaptic potentials are graded potentials that develop in the postsynaptic membrane in response to a neurotransmitter. (Figure 12–12 illustrated graded depolarizations and hyperpolarizations.) Two major types of postsynaptic potentials develop at neuron-to-neuron synapses: excitatory postsynaptic potentials and inhibitory postsynaptic potentials.

An **excitatory postsynaptic potential**, or **EPSP**, is a graded depolarization caused by the arrival of a neurotransmitter at the postsynaptic membrane. An EPSP results from the opening of chemically gated ion channels in the plasma membrane that lead to membrane depolarization. For example, the graded depolarization produced by the binding of ACh is an EPSP. Because it is a graded potential, an EPSP affects only the area immediately surrounding the synapse, as shown earlier in Figure 12–12.

We have already noted that not all neurotransmitters have an excitatory (depolarizing) effect. An **inhibitory postsynaptic potential**, or **IPSP**, is a graded hyperpolarization of the postsynaptic membrane. For example, an IPSP may result from the opening of chemically gated potassium ion channels. While the hyperpolarization continues, the neuron is said to be **inhibited**, because a larger-than-usual depolarizing stimulus is needed to bring the membrane potential to threshold. A stimulus that shifts the membrane potential by 10 mV (from −70 mV to −60 mV) would normally produce an action potential, but if the membrane potential were reset at −85 mV by an IPSP, the same stimulus would depolarize it to only −75 mV, which is below threshold.

Integrating Postsynaptic Potentials: Summation

An individual EPSP has a small effect on the membrane potential, typically producing a depolarization of about 0.5 mV at the postsynaptic membrane. Before an action potential will arise in the initial segment, local currents must depolarize that region by at least 10 mV. Therefore, a single EPSP will not result in an action potential, even if the synapse is on the axon hillock. But individual EPSPs combine through the process of **summation**, which integrates the effects of all the graded potentials that affect one portion of the plasma membrane. The graded potentials may be EPSPs, IPSPs, or both. We will consider EPSPs in our discussion. Two forms of summation exist: temporal summation and spatial summation (Figure 12–18).

Temporal Summation. Temporal summation (*tempus*, time) is the addition of stimuli occurring in rapid succession at a *single synapse* that is active *repeatedly*. This form of summation can be likened to using a bucket to fill up a bathtub: You can't fill the tub with a single bucket of water, but you will fill it eventually if you keep repeating the process. In the case of temporal summation, the water in a bucket corresponds to the sodium ions that enter the cytosol during an EPSP. A typical EPSP lasts about 20 msec, but under maximum stimulation an action potential can reach the axon terminal each millisecond. Figure 12–18a shows what happens when a second EPSP arrives before the effects of the first EPSP have disappeared: The effects of the two are combined. Every time an action potential arrives, a group of vesicles discharges ACh into the synaptic cleft, and every time more ACh molecules arrive at the postsynaptic membrane, more chemically gated ion channels open, and the degree of depolarization increases. In this way, a series of small steps can eventually bring the initial segment to threshold.

Spatial Summation. Spatial summation occurs when simultaneous stimuli applied at different locations have a cumulative effect on the membrane potential. In other words, spatial summation involves *multiple synapses* that are active *simultaneously*. In terms of our bucket analogy, you could fill the bathtub immediately if 10 friends emptied their buckets into it at the same time.

12

In spatial summation, more than one synapse is active at the same time (Figure 12–18b), and each "pours" sodium ions across the postsynaptic membrane, producing a graded potential with localized effects. At each active synapse, the sodium ions that produce the EPSP spread out along the inner surface of the membrane and mingle with those entering at other synapses. As a result, the effects on the initial segment are cumulative. The degree of depolarization depends on how many synapses are active at any moment, and on their distance from the initial segment. As in temporal summation, an action potential results when the membrane potential at the initial segment reaches threshold.

Facilitation of Neurons

Now consider a situation in which summation of EPSPs is under way, but the initial segment has not been depolarized to threshold. The closer the initial segment gets to threshold, the easier it will be for the *next* depolarizing stimulus to trigger an action potential. A neuron whose membrane potential shifts closer to threshold is said to be **facilitated**. The larger the degree of facilitation, the smaller the additional stimulus needed to trigger an action potential. In a highly facilitated neuron, even a small depolarizing stimulus produces an action potential.

Facilitation can result from the summation of EPSPs or from the exposure of a neuron to certain drugs in the extracellular fluid. For example, the nicotine in cigarettes stimulates postsynaptic ACh receptors, producing prolonged EPSPs that facilitate CNS neurons. Nicotine also increases the release of another neurotransmitter, dopamine, producing feelings of pleasure and reward, thereby leading to addiction.

Figure 12–18 **Temporal and Spatial Summation.**

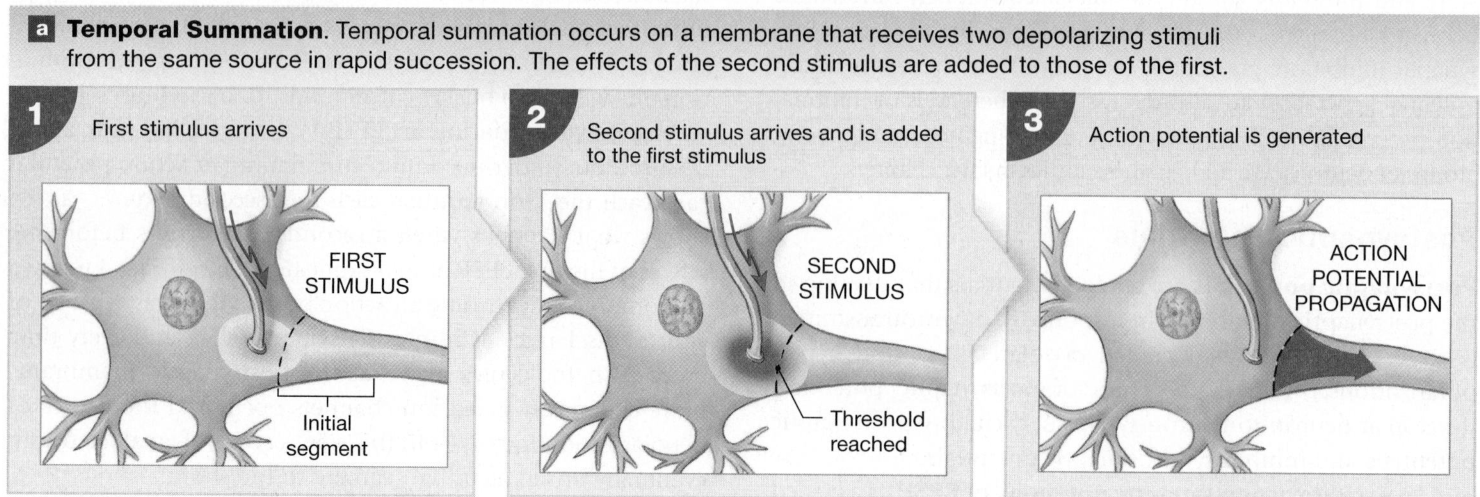
a **Temporal Summation**. Temporal summation occurs on a membrane that receives two depolarizing stimuli from the same source in rapid succession. The effects of the second stimulus are added to those of the first.
1 First stimulus arrives
FIRST STIMULUS
Initial segment
2 Second stimulus arrives and is added to the first stimulus
SECOND STIMULUS
Threshold reached
3 Action potential is generated
ACTION POTENTIAL PROPAGATION

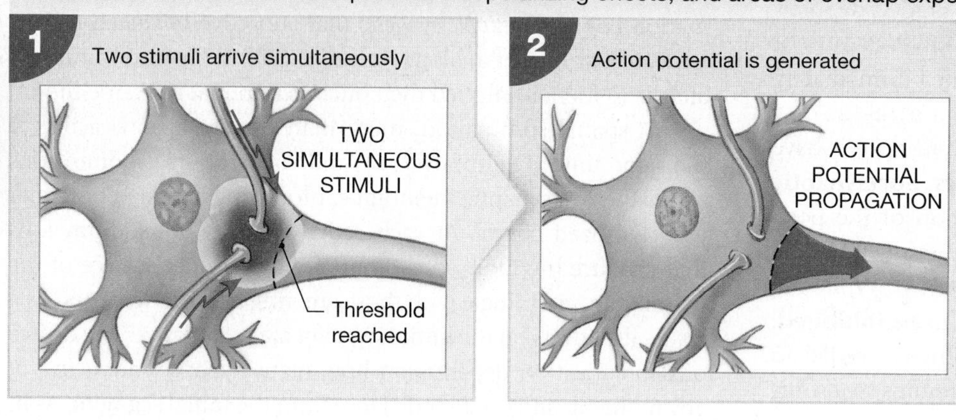
b **Spatial Summation**. Spatial summation occurs when two stimuli arrive at the same time, but at different locations. Local currents spread the depolarizing effects, and areas of overlap experience the combined effects.
1 Two stimuli arrive simultaneously
TWO SIMULTANEOUS STIMULI
Threshold reached
2 Action potential is generated
ACTION POTENTIAL PROPAGATION

12

Figure 12–19 **Interactions between EPSPs and IPSPs.** At time 1, a small depolarizing stimulus produces an EPSP. At time 2, a small hyperpolarizing stimulus produces an IPSP of comparable magnitude. If the two stimuli are applied simultaneously, as they are at time 3, summation occurs. Because the two are equal in size but have opposite effects, the membrane potential remains at the resting level. If the EPSP were larger, a net depolarization would result; if the IPSP were larger, a net hyperpolarization would result instead.

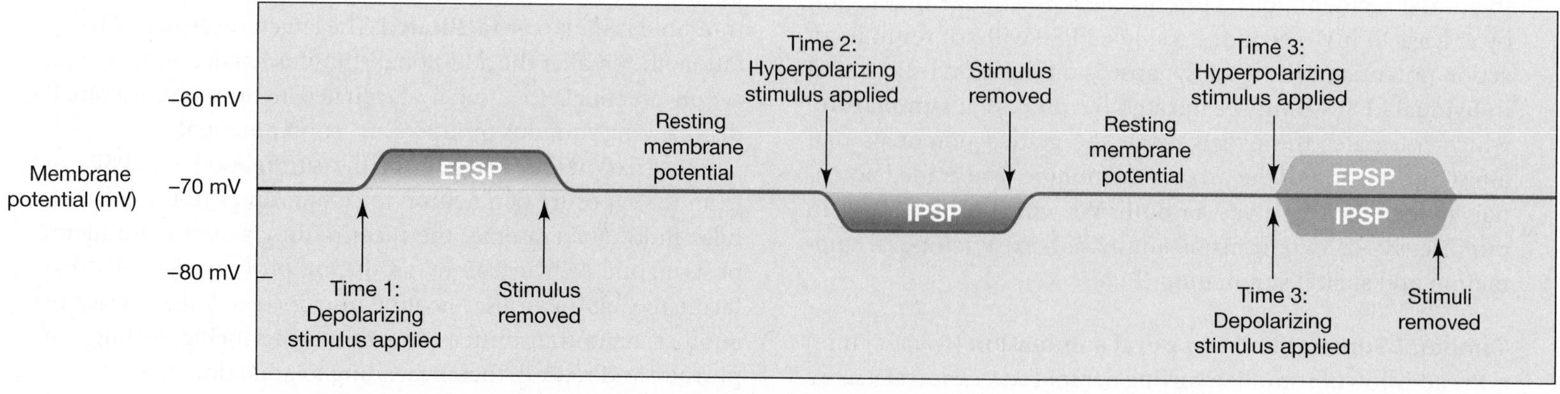
Membrane potential (mV)
−60 mV
−70 mV
−80 mV
EPSP
Time 1: Depolarizing stimulus applied
Stimulus removed
Resting membrane potential
Time 2: Hyperpolarizing stimulus applied
IPSP
Stimulus removed
Resting membrane potential
Time 3: Hyperpolarizing stimulus applied
EPSP
IPSP
Time 3: Depolarizing stimulus applied
Stimuli removed
Time

Summation of EPSPs and IPSPs

Like EPSPs, IPSPs combine spatially and temporally. EPSPs and IPSPs reflect the activation of different types of chemically gated ion channels, producing opposing effects on the membrane potential. The antagonism between IPSPs and EPSPs is important in cellular information processing. Using our bucket analogy, EPSPs put water into the bathtub, and IPSPs take water out. If more buckets add water than remove water, the water level in the tub rises. If more buckets remove water, the level falls. If a bucket of water is removed every time another bucket is dumped in, the level remains stable. Comparable interactions between EPSPs and IPSPs determine the membrane potential at the boundary between the axon hillock and the initial segment (Figure 12–19).

Neuromodulators, hormones, or both can change the postsynaptic membrane's sensitivity to excitatory or inhibitory neurotransmitters. By shifting the balance between EPSPs and IPSPs, these compounds promote facilitation or inhibition of CNS and PNS neurons.

Presynaptic Regulation: Inhibition and Facilitation

Inhibitory or excitatory responses may occur not only at postsynaptic neurons but also at presynaptic neurons. An axoaxonic (axon to axon) synapse at the axon terminal can either decrease (inhibit) or increase (facilitate) the rate of neurotransmitter release at the presynaptic membrane. In one form of **presynaptic inhibition**, the release of GABA inhibits the opening of voltage-gated calcium ion channels in the axon terminal (Figure 12–20a). This inhibition reduces the amount of neurotransmitter released when an action potential arrives there, and thus reduces the effects of synaptic activity on the postsynaptic membrane.

In **presynaptic facilitation**, activity at an axoaxonic synapse increases the amount of neurotransmitter released when an action potential arrives at the axon terminal (Figure 12–20b). This increase enhances and prolongs the neurotransmitter's effects on the postsynaptic membrane. The neurotransmitter *serotonin* is involved in presynaptic facilitation. In the presence of serotonin released at an axoaxonic synapse, voltage-gated calcium ion channels remain open longer.

The Rate of Action Potential Generation

In the nervous system, complex information is translated into action potentials that are propagated along axons. On arrival, the message is often interpreted solely on the basis of the frequency of action potentials. For example, action potentials arriving at a neuromuscular junction at the rate of 1 per second may produce a series of isolated twitches in the associated

12

Figure 12–20 Presynaptic Inhibition and Presynaptic Facilitation.

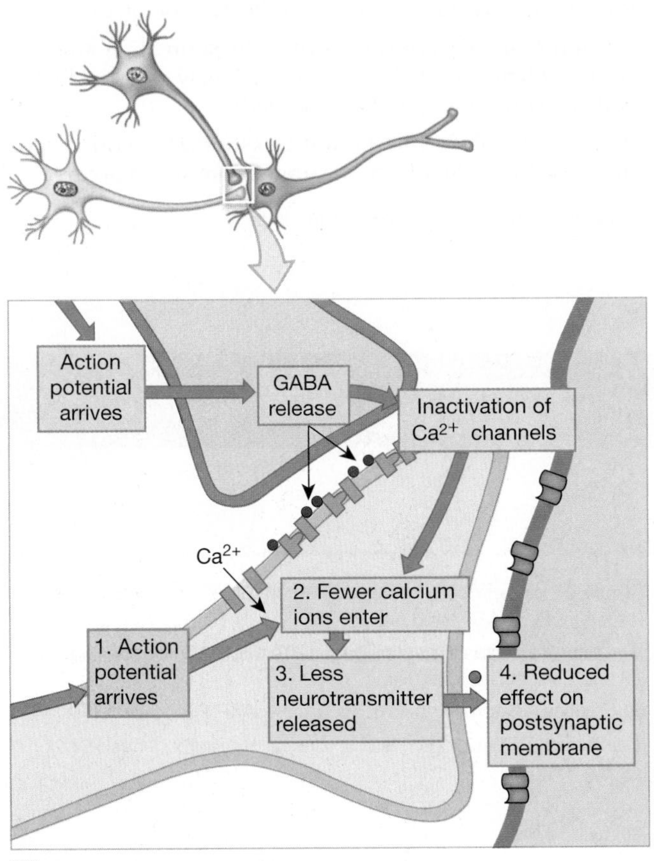
Action potential arrives
GABA release
Inactivation of Ca^{2+} channels
Ca^{2+}
2. Fewer calcium ions enter
1. Action potential arrives
3. Less neurotransmitter released
4. Reduced effect on postsynaptic membrane

a **Presynaptic inhibition**

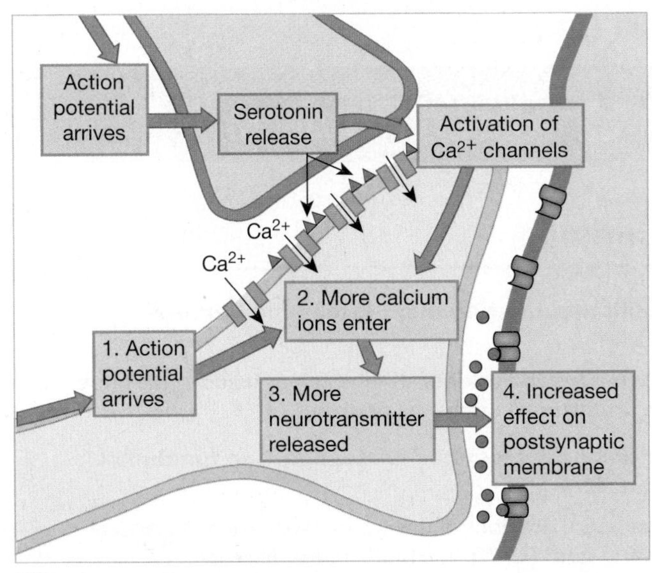
Action potential arrives
Serotonin release
Activation of Ca^{2+} channels
Ca^{2+}
Ca^{2+}
2. More calcium ions enter
1. Action potential arrives
3. More neurotransmitter released
4. Increased effect on postsynaptic membrane

b **Presynaptic facilitation**

skeletal muscle fiber, but at the rate of 100 per second they cause a sustained tetanic contraction. Similarly, you may perceive a few action potentials per second along a sensory fiber as a feather-light touch, but you would perceive hundreds of action potentials per second along that same axon as unbearable pressure. In general, the degree of sensory stimulation or the strength of the motor response is proportional to the frequency of action potentials. In this section, we examine factors that vary the rate of generation of action potentials.

If a graded potential briefly depolarizes the axon hillock such that the initial segment reaches its threshold, an action potential is generated. But what happens when the axon hillock *remains* depolarized past threshold for an extended period? The longer the initial segment remains above threshold, the more action potentials it produces. The *frequency* of action potentials depends on the degree of depolarization above threshold: The greater the degree of depolarization, the higher the frequency of action potentials. The membrane can respond to a second stimulus as soon as the absolute refractory period ends. Holding the membrane potential above threshold has the same effect as applying a second, larger-than-normal stimulus.

Action potentials can be generated at a maximum rate when the relative refractory period has been completely eliminated. For this reason, the maximum theoretical frequency of action potentials is established by the duration of the absolute refractory period. The absolute refractory period is shortest in large-diameter axons, in which the *theoretical* maximum frequency of action potentials is 2500 per second. However, the highest frequencies recorded from axons in the body range between 500 and 1000 per second.

Table 12–5 summarizes the basic principles of information processing.

You are now familiar with the basic components of nervous tissue, and the origin and significance of action potentials.

Table 12–5 Information Processing

- Information is relayed in the form of action potentials. In general, the degree of sensory stimulation or the strength of the motor response is proportional to the frequency of action potentials.
- The neurotransmitters released at a synapse may have either excitatory or inhibitory effects. The effect on the axon's initial segment reflects a summation of the stimuli that arrive at any moment. The frequency of generation of action potentials is an indication of the degree of sustained depolarization at the axon hillock.
- Neuromodulators can alter either the rate of neurotransmitter release or the response of a postsynaptic neuron to specific neurotransmitters.
- Neurons may be facilitated or inhibited by extracellular chemicals other than neurotransmitters or neuromodulators.
- The response of a postsynaptic neuron to the activation of a presynaptic neuron can be altered by (1) the presence of neuromodulators or other chemicals that cause facilitation or inhibition at the synapse, (2) activity under way at other synapses affecting the postsynaptic cell, and (3) modification of the rate of neurotransmitter release through presynaptic facilitation or presynaptic inhibition.

In the following four chapters, we consider higher levels of anatomical and functional organization within the nervous system, examine information processing at these levels, and see how a single process—the generation of action potentials—can be responsible for the incredible diversity of sensations and movements that we experience each day.

✓ Checkpoint

25. **One EPSP depolarizes the initial segment from a resting membrane potential of −70 mV to −65 mV, and threshold is at −60 mV. Will an action potential be generated?**
26. **Given the situation in Checkpoint Question 25, if a second, identical EPSP occurs immediately after the first, will an action potential be generated?**
27. **If the two EPSPs in Checkpoint Question 26 occurred at the same time, what form of summation would occur?**

See the blue Answers tab at the back of the book.

12 Chapter Review

Study Outline

An Introduction to the Nervous System and Nervous Tissue p. 390

1. The nervous system includes all the nervous tissue in the body. The basic functional unit is the **neuron**.

12-1 The nervous system has anatomical and functional divisions p. 390

2. The anatomical divisions of the nervous system are the **central nervous system (CNS)** (the brain and spinal cord), the

MasteringA&P™ Access more chapter study tools online in the MasteringA&P Study Area:

- Chapter Quizzes, Chapter Practice Test, MP3 Tutor Sessions, and Clinical Case Studies
- Practice Anatomy Lab PAL 3.0
- A&P Flix A&PFlix
- Interactive Physiology iP2
- PhysioEx PhysioEx 9.1

peripheral nervous system (PNS) (all the nervous tissue outside the CNS and ENS), and the enteric nervous system (ENS) (neurons and nerve networks in the walls of the digestive tract). The two subdivisions of the PNS are the *somatic nervous system* and *autonomic nervous system*. Bundles of **axons** (*nerve fibers*) in the PNS are called **nerves**. *(Figure 12–1)*

3. Functionally, the PNS can be divided into an **afferent division**, which brings sensory information from **receptors** to the CNS, and an **efferent division**, which carries motor commands to muscles and glands called **effectors**. *(Figure 12–1)*
4. The efferent division of the PNS includes the **somatic nervous system (SNS)**, which controls skeletal muscle contractions, and the **autonomic nervous system (ANS)**, which controls smooth muscle, cardiac muscle, glandular activity, and adipose tissue. *(Figure 12–1)*

12-2 Neurons are nerve cells specialized for intercellular communication p. 392

5. The **perikaryon** of a multipolar neuron surrounds the nucleus and contains organelles, including **neurofilaments**, **neurotubules**, and **neurofibrils**. The **axon hillock** connects the **initial segment** of the **axon** to the **cell body**, or *soma*. The **axoplasm** contains numerous organelles. *(Figure 12–2)*
6. **Collaterals** may branch from an axon, with **telodendria** branching from the axon's tip.
7. Neurons are structurally classified as **anaxonic**, **bipolar**, **unipolar**, or **multipolar**. *(Figure 12–3)*
8. The three functional categories of neurons are sensory neurons, motor neurons, and interneurons.
9. **Sensory neurons**, which form the afferent division of the PNS, deliver information received from **interoceptors**, **exteroceptors**, and **proprioceptors** to the CNS.
10. **Motor neurons**, which form the efferent division of the PNS, stimulate or modify the activity of a peripheral tissue, organ, or organ system.
11. **Interneurons** (*association neurons*) are always located in the CNS and may be situated between sensory and motor neurons. They distribute sensory inputs and coordinate motor outputs.

12-3 CNS and PNS neuroglia support and protect neurons p. 395

12. The four types of **neuroglia**, or *glial cells*, in the CNS are (1) **astrocytes**, the largest and most numerous neuroglia; (2) **ependymal cells**, with functions related to the **cerebrospinal fluid (CSF)**; (3) **oligodendrocytes**, which are responsible for the **myelination** of CNS axons; and (4) **microglia**, or phagocytic cells. *(Figures 12–4, 12–5)*
13. Neuron cell bodies in the PNS are clustered into **ganglia**.
14. **Satellite cells** surround neuron cell bodies within ganglia. **Schwann cells** cover axons in the PNS. A single Schwann cell may myelinate one segment of an axon or enclose segments of several unmyelinated axons. *(Figure 12–6)*
15. In the PNS, functional repair of axons may follow **Wallerian degeneration**. In the CNS, many factors complicate the repair process and reduce the chances of functional recovery. *(Figure 12–7)*

12-4 The membrane potential of a neuron is determined by differences in ion concentrations and membrane permeability p. 402

16. All normal neural signaling depends on events that occur at the plasma membrane. *(Spotlight Figure 12–8)*
17. The **electrochemical gradient** is the sum of all chemical and electrical forces acting across the plasma membrane. *(Figures 12–9)*
18. The sodium–potassium exchange pump stabilizes the resting membrane potential at approximately −70 mV. *(Table 12–1)*
19. The plasma membrane contains **passive (leak) ion channels**, which are always open, and **active (gated) ion channels**, which open or close in response to specific stimuli. *(Figures 12–9, 12–10)*
20. The three types of gated channels are **chemically gated ion channels (ligand-gated ion channels)**, **voltage-gated ion channels**, and **mechanically gated ion channels**. *(Figure 12–10)*
21. A localized **depolarization** or **hyperpolarization** is a **graded potential** (a change in potential that decreases with distance). *(Figures 12–11, 12–12; Table 12–2)*

12-5 An action potential is an all-or-none electrical event used for long-distance communication p. 409

22. An **action potential** arises when a region of excitable membrane depolarizes to its **threshold**. The steps involved, in order, are membrane depolarization to threshold, activation of voltage-gated sodium ion channels and rapid depolarization, inactivation of voltage-gated sodium ion channels and activation of voltage-gated potassium ion channels, and the return to normal permeability. *(Spotlight Figure 12–13; Table 12–3)*
23. The generation of an action potential follows the **all-or-none principle**. The **refractory period** lasts from the time an action potential begins until the normal resting membrane potential has returned. *(Spotlight Figure 12–13; Table 12–3)*
24. In **continuous propagation**, an action potential spreads across the entire excitable membrane surface in a series of small steps. *(Spotlight Figure 12–14)*
25. In **saltatory propagation**, an action potential appears to leap from node to node, skipping the intervening myelinated membrane surface. Saltatory propagation carries nerve impulses many times more rapidly than does continuous propagation. *(Spotlight Figure 12–14)*
26. Axons are classified as **Type A fibers**, **Type B fibers**, or **Type C fibers** on the basis of their diameter, myelination, and propagation speed.

12-6 Synapses transmit signals among neurons or between neurons and other cells p. 416

27. A **synapse** is a site of intercellular communication. At a synapse between two neurons, information passes from the **presynaptic neuron** to the **postsynaptic neuron**.
28. **Neurotransmitters** released from the axon terminals of the presynaptic cell affect the postsynaptic cell, which may be a neuron or another type of cell. *(Figures 12–2, 12–15)*
29. A synapse is either *electrical* (with direct physical contact between cells) or *chemical* (involving a neurotransmitter).
30. **Electrical synapses** occur in the CNS and PNS, but they are rare. At an electrical synapse, the presynaptic and postsynaptic plasma membranes are bound by interlocking membrane proteins at a gap junction. Pores formed by these proteins permit the passage of local currents, and the two neurons act as if they share a common plasma membrane.
31. **Chemical synapses** are far more common than electrical synapses. **Cholinergic synapses** release the neurotransmitter **acetylcholine (ACh)**. Communication moves from the presynaptic neuron to the postsynaptic neuron across a synaptic

cleft. A **synaptic delay** occurs because calcium ion influx and the release of the neurotransmitter takes an appreciable length of time. *(Figure 12–16)*

32. **Choline** released during the breakdown of ACh in the synaptic cleft is reabsorbed and recycled by the axon terminal. If stores of ACh are exhausted, **synaptic fatigue** can occur.

12-7 The effects of neurotransmitters and neuromodulators depend on their receptors p. 420

33. **Excitatory neurotransmitters** cause depolarization and promote the generation of action potentials, whereas **inhibitory neurotransmitters** cause hyperpolarization and suppress the generation of action potentials.
34. The effect of a neurotransmitter on the postsynaptic membrane depends on the properties of the receptor, not on the nature of the neurotransmitter.
35. **Adrenergic synapses** release **norepinephrine (NE)**, also called *noradrenaline.* Other important neurotransmitters include **dopamine**, **serotonin**, and **gamma aminobutyric acid (GABA)**.
36. **Neuromodulators** influence the postsynaptic cell's response to neurotransmitters. *(Table 12–4)*
37. Neurotransmitters can have a direct effect on membrane potential, indirect effect on membrane potential through G proteins, or, if lipid-soluble gases, an indirect effect through intracellular enzymes *(Figure 12–17)*

12

12-8 Individual neurons process information by integrating excitatory and inhibitory stimuli p. 424

38. Excitatory and inhibitory stimuli are integrated through interactions between **postsynaptic potentials**. This interaction is the simplest level of **information processing** in the nervous system.
39. A depolarization caused by a neurotransmitter is an **excitatory postsynaptic potential (EPSP)**. Individual EPSPs can combine through **summation**, which can be either **temporal** (occurring at a single synapse when a second EPSP arrives before the effects of the first have disappeared) or **spatial** (resulting from the cumulative effects of multiple synapses at various locations). *(Figure 12–18)*
40. Hyperpolarization of the postsynaptic membrane is an **inhibitory postsynaptic potential (IPSP)**.
41. The most important determinants of neural activity are EPSP–IPSP interactions. *(Figure 12–19)*
42. In **presynaptic inhibition**, GABA release at an *axoaxonic synapse* inhibits the opening of voltage-gated calcium ion channels in the axon terminal. This inhibition reduces the amount of neurotransmitter released when an action potential arrives at the axon terminal. *(Figure 12–20a)*
43. In **presynaptic facilitation**, activity at an axoaxonic synapse increases the amount of neurotransmitter released when an action potential arrives at the axon terminal. This increase enhances and prolongs the effects of the neurotransmitter on the postsynaptic membrane. *(Figure 12–20b)*
44. The neurotransmitters released at a synapse have excitatory or inhibitory effects. The effect on the initial segment reflects an integration of the stimuli arriving at any moment. The frequency of generation of action potentials depends on the degree of depolarization above threshold at the axon hillock. *(Table 12–5)*
45. Neuromodulators can alter either the rate of neurotransmitter release or the response of a postsynaptic neuron to specific neurotransmitters. Neurons may be facilitated or inhibited by extracellular chemicals other than neurotransmitters or neuromodulators. *(Table 12–5)*
46. The effect of a presynaptic neuron's activation on a postsynaptic neuron may be altered by other neurons. *(Table 12–5)*
47. The greater the degree of sustained depolarization at the axon hillock, the higher the frequency of generation of action potentials. At a frequency of about 1000 per second, the relative refractory period has been eliminated, and further depolarization will have no effect. *(Table 12–5)*

Review Questions

See the blue Answers tab at the back of the book.

LEVEL 1 Reviewing Facts and Terms

1. Label the structures in the following diagram of a neuron.

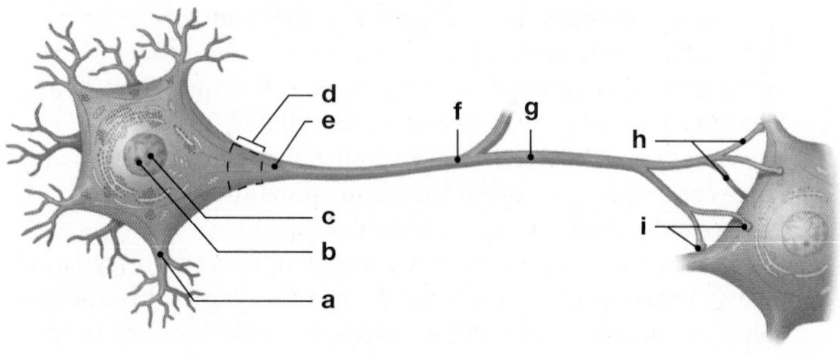
d
e
f
g
h
c
b
a
i

(a) ________ (b) ________
(c) ________ (d) ________
(e) ________ (f) ________
(g) ________ (h) ________
(i) ________

2. Regulation by the nervous system provides **(a)** relatively slow, but long-lasting, responses to stimuli, **(b)** swift, long-lasting responses to stimuli, **(c)** swift, but brief, responses to stimuli, **(d)** relatively slow, short-lived responses to stimuli.
3. In the CNS, a neuron typically receives information from other neurons at its **(a)** axon, **(b)** Nissl bodies, **(c)** dendrites, **(d)** nucleus.
4. Phagocytic cells in nervous tissue of the CNS are **(a)** astrocytes, **(b)** ependymal cells, **(c)** oligodendrocytes, **(d)** microglia.
5. The neural cells responsible for the analysis of sensory inputs and coordination of motor outputs are **(a)** neuroglia, **(b)** interneurons, **(c)** sensory neurons, **(d)** motor neurons.
6. Depolarization of a neuron plasma membrane will shift the membrane potential toward **(a)** 0 mV, **(b)** −70 mV, **(c)** −90 mV, **(d)** all of these.
7. What factor determines the direction that ions will move through an open membrane channel? **(a)** the membrane permeability to sodium ions, **(b)** the electrochemical gradient, **(c)** intracellular negatively charged proteins, **(d)** negatively charged chloride ions in the ECF.

8. Receptors that bind acetylcholine at the postsynaptic membrane are **(a)** chemically gated channels, **(b)** voltage-gated channels, **(c)** passive channels, **(d)** mechanically gated channels.
9. What are the major components of **(a)** the central nervous system? **(b)** the peripheral nervous system? **(c)** the enteric nervous system?
10. Which two types of neuroglia insulate neuron cell bodies and axons in the PNS from their surroundings?
11. What three *functional* groups of neurons are found in the nervous system? What is the function of each type of neuron?

LEVEL 2 Reviewing Concepts

12. If the resting membrane potential of a neuron is −70 mV and the threshold is −55 mV, a membrane potential of −60 mV will **(a)** produce an action potential, **(b)** make it easier to produce an action potential, **(c)** make it harder to produce an action potential, **(d)** hyperpolarize the membrane.
13. Why can't most neurons in the CNS be replaced when they are lost to injury or disease?
14. What is the difference between anterograde flow and retrograde flow?
15. What is the *functional* difference among chemically gated (ligand-gated), voltage-gated, and mechanically gated ion channels?
16. State the all-or-none principle of action potentials.
17. Describe the steps involved in the generation of an action potential.
18. What is meant by saltatory propagation? How does it differ from continuous propagation?
19. What are the structural and functional differences among type A, B, and C fibers?
20. Describe the events that occur during nerve impulse transmission at a typical cholinergic synapse.
21. What is the difference between temporal summation and spatial summation?

LEVEL 3 Critical Thinking and Clinical Applications

22. Harry has a kidney condition that causes changes in his body's electrolyte levels (concentration of ions in the extracellular fluid). As a result, he is exhibiting tachycardia, an abnormally fast heart rate. Which ion is involved, and how does a change in its concentration cause Harry's symptoms?
23. Twenty neurons synapse with a single receptor neuron. Fifteen of the 20 neurons release neurotransmitters that produce EPSPs at the postsynaptic membrane, and the other 5 release neurotransmitters that produce IPSPs. Each time one of the neurons is stimulated, it releases enough neurotransmitter to produce a 2-mV change in potential at the postsynaptic membrane. If the threshold of the postsynaptic neuron is 10 mV, how many of the excitatory neurons must be stimulated to produce an action potential in the receptor neuron if all 5 inhibitory neurons are stimulated? (Assume that spatial summation occurs.)
24. In multiple sclerosis, there is intermittent and progressive damage to the myelin sheath of peripheral nerves. This results in poor motor control of the affected area. Why does destruction of the myelin sheath affect motor control?
25. What factor determines the maximum frequency of action potentials that could be propagated by an axon?

12

+ CLINICAL CASE Wrap-Up Did President Franklin D. Roosevelt Really Have Polio?

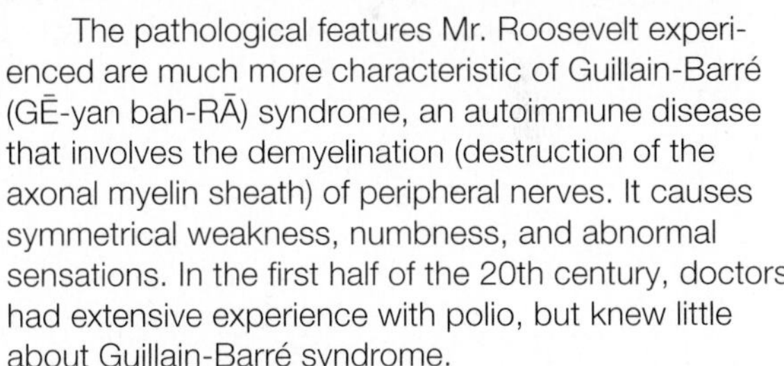

Poliomyelitis is caused by a virus that spreads easily in human populations. Most people with polio infections show no signs of disease. However, in a small percentage of victims, the virus enters the central nervous system, attacks the motor neurons of the spinal cord, and produces a flaccid paralysis (characterized by weakness and loss of muscle tone).

Certain details about Mr. Roosevelt's case call into question whether he really suffered from polio. Polio is a disease primarily of children. By age 30, most adults have naturally acquired immunity, a resistance to infection developed after a previous exposure. In addition, paralysis caused by polio is usually asymmetric, does not appear in an ascending pattern, and progresses for only 3 to 5 days—not for nearly a month, as in Mr. Roosevelt's case. Polio almost never involves facial muscles or disturbs urinary bladder and bowel functions. Polio also does not cause pain upon light touch, other pain, or numbness—only motor weakness. Mr. Roosevelt recovered from his paralysis in a descending pattern, opposite to its onset.

The pathological features Mr. Roosevelt experienced are much more characteristic of Guillain-Barré (GĒ-yan bah-RĀ) syndrome, an autoimmune disease that involves the demyelination (destruction of the axonal myelin sheath) of peripheral nerves. It causes symmetrical weakness, numbness, and abnormal sensations. In the first half of the 20th century, doctors had extensive experience with polio, but knew little about Guillain-Barré syndrome.

While it remains inconclusive which condition led to the lifetime impairment of our 32nd president, Franklin D. Roosevelt's polio diagnosis ultimately contributed to ending polio. In fact, a current goal of the Bill & Melinda Gates Foundation is to eradicate polio worldwide.

1. Which cells of the CNS are targeted by the polio virus?
2. If Mr. Roosevelt suffered from Guillain-Barré syndrome, where in his nervous system was the pathology taking place?

See the blue Answers tab at the back of the book.

Related Clinical Terms

anesthetic: An agent that produces a local or general loss of sensation including feelings of pain.

anticholinesterase drug: A drug that blocks the breakdown of ACh by AChE.

atropine: A drug that prevents ACh from binding to the postsynaptic membrane of cardiac muscle cells and smooth muscle cells.

d-tubocurarine: A drug, derived from curare, a South American rain forest vine. It produces paralysis by preventing ACh from binding to the postsynaptic membrane (motor end plate) of skeletal muscle fibers.

dysthymia: A form of clinical depression with a depressed mood for most of the time for at least two years along with at least two of the following signs and symptoms: poor appetite or overeating; insomnia or excessive sleep; low energy or fatigue; low self-esteem; poor concentration or indecisiveness; and hopelessness.

excitotoxicity: Continuous and exaggerated stimulation by a neurotransmitter, especially for the excitatory neurotransmitter glutamate. The nerve cells can become damaged and killed by the overactivation of receptors.

neuroblastoma: A malignant tumor composed of neuroblasts, most commonly in the adrenal gland; it is the most common cancer in infancy.

neuropathy: Condition that causes tingling, numbness, and/or pain in parts of the body, notably the hands and feet.

neurotoxin: A compound that disrupts normal nervous system function by interfering with the generation or propagation of action potentials. Examples include *tetrodotoxin (TTX)*, *saxitoxin (STX)*, and *ciguatoxin (CTX)*.

Tay–Sachs disease: A genetic abnormality involving the metabolism of gangliosides, important lipid components of neuron plasma membranes. The result is a gradual deterioration of neurons due to the buildup of metabolic by-products and the release of lysosomal enzymes.

13 The Spinal Cord, Spinal Nerves, and Spinal Reflexes

Learning Outcomes

These Learning Outcomes correspond by number to this chapter's sections and indicate what you should be able to do after completing the chapter.

- 13-1 ■ Describe the basic structural and organizational characteristics of the nervous system. p. 434
- 13-2 ■ Discuss the structure and functions of the spinal cord, and describe the three meningeal layers that surround the central nervous system. p. 435
- 13-3 ■ Explain the roles of white matter and gray matter in processing and relaying sensory information and motor commands. p. 440
- 13-4 ■ Describe the major components of a spinal nerve, describe a nerve plexus, and relate the distribution pattern of spinal nerves to the regions they innervate. p. 442
- 13-5 ■ Discuss the significance of neuronal pools, and describe the major patterns of interaction among neurons within and among these pools. p. 452
- 13-6 ■ Describe the steps in a neural reflex, and classify the types of reflexes. p. 453
- 13-7 ■ Distinguish among the types of motor responses produced by various reflexes, and explain how reflexes interact to produce complex behaviors. p. 457
- 13-8 ■ Explain how higher brain centers control and modify reflex responses. p. 460

CLINICAL CASE Prom Night

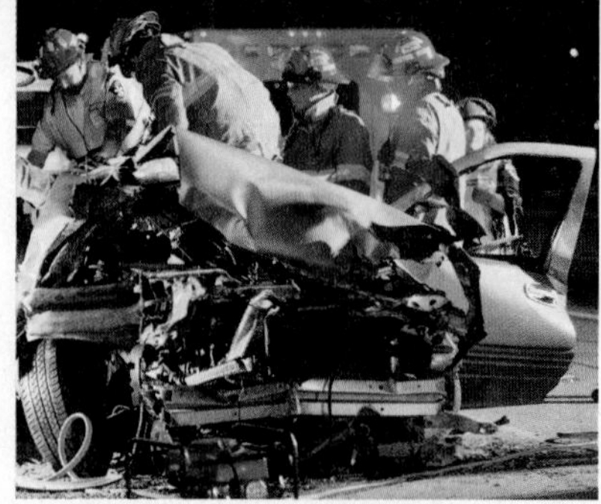

Joe is still buzzing from champagne as he kisses his girlfriend goodnight and gets into his car to head home. It's been a fun prom. The food was great, the band was awesome, and the afterparty at his friend's house had gone until 3 a.m. Joe pulls the car onto the road, turns the stereo to full blast, and steps on the gas. Then suddenly tired, he leans back and notices the moon through the trees. That is the last thing he remembers.

The emergency medical technicians (EMTs) at the motor vehicle accident scene note that Joe's car had left the road at a high rate of speed and rolled over at least three times. Joe is still unconscious when he is extricated from the car. He is breathing on his own but has obvious abrasions and bruising on his back.

The EMTs splint his entire axial skeleton, using a cervical collar and sandbags on a rigid backboard. They start an IV in the field and transport Joe to the closest Level I trauma center.

In the emergency department Joe can hear shouting in his ear as he is moved onto a cart. "Wiggle your toes," someone commands, but Joe can move nothing below his waist. "Can you feel this?" a voice asks. Joe can feel someone poking him with a pin from his chest down to his abdomen, just inferior to his navel, but nothing below that. "Can you feel me put this catheter in?" a nurse asks as she inserts a small tube into Joe's bladder. Joe cannot feel the catheter. **What has happened to Joe?** **To find out, turn to the Clinical Case Wrap-Up on p. 464.**

13

An Introduction to the Spinal Cord, Spinal Nerves, and Spinal Reflexes

Organization is usually the key to success in any complex environment. A large corporation, for example, has both a system to distribute messages on specific topics and executive assistants who decide whether an issue can be ignored or easily responded to. Only the most complex and important problems reach the desk of the president.

The nervous system works in much the same way. It has input pathways that route sensations, and processing centers that prioritize and distribute information. There are also several levels that issue motor responses. Your conscious mind (the president) gets involved only in a fraction of the day-to-day activities. The other decisions are handled at lower "management" levels, such as the spinal cord, that operate outside your awareness. This very efficient system works only because it is so highly organized.

In this chapter, we discuss the functional anatomy and organization of the spinal cord and spinal nerves. We also describe simple spinal reflexes.

13-1 This text's coverage of the nervous system parallels its simple-to-complex levels of organization

Learning Outcome **Describe the basic structural and organizational characteristics of the nervous system.**

The nervous system has so many components and does so much that we will cover it in several chapters. If our primary interest were the anatomy of this system, we would probably start with an examination of the **central nervous system** (brain and spinal cord) and then move on to the **peripheral nervous system** (cranial nerves and spinal nerves). However, our primary interest is how the nervous system *functions*, so we will consider the system from a functional perspective. Our basic approach for Chapters 13 and 14 is diagrammed in Figure 13–1.

Note that Chapter 12 considered the function of individual neurons. In this chapter, we consider the spinal cord and spinal nerves, the basics of neural reflexes, and simple spinal reflexes. *Reflexes* are quick, automatic responses triggered by specific stimuli. *Spinal reflexes* are controlled in the spinal cord. They function without any input from the brain. For example, a reflex controlled in the spinal cord makes you drop a frying pan you didn't realize was sizzling hot. Before the information reaches your brain and you become aware of the pain, you've already released the pan. More complex spinal reflexes exist, but this functional pattern still applies.

Your spinal cord is structurally and functionally integrated with your brain. In Chapter 14 we give an overview of the major structures and functions of the brain and cranial nerves. We also discuss *cranial reflexes*—localized reflex responses comparable to those of the spinal cord.

In Chapters 15 and 16 we consider the nervous system as an integrated functional unit. In Chapter 15 we deal with the interplay between the brain and spinal cord that occurs when processing sensory information. We then examine the conscious and subconscious control of skeletal muscles by the *somatic nervous system (SNS)*.

In Chapter 16 we continue with a discussion of visceral functions by the *autonomic nervous system (ANS)*. The ANS has processing centers in the brain, spinal cord, and peripheral nervous system. It controls visceral effectors, such as smooth muscles, cardiac muscle, glands, and adipose tissue. We then conclude the chapter by examining *higher-order functions*: memory, learning, consciousness, and personality. These fascinating topics are

Figure 13–1 An Overview of Chapters 13 and 14.

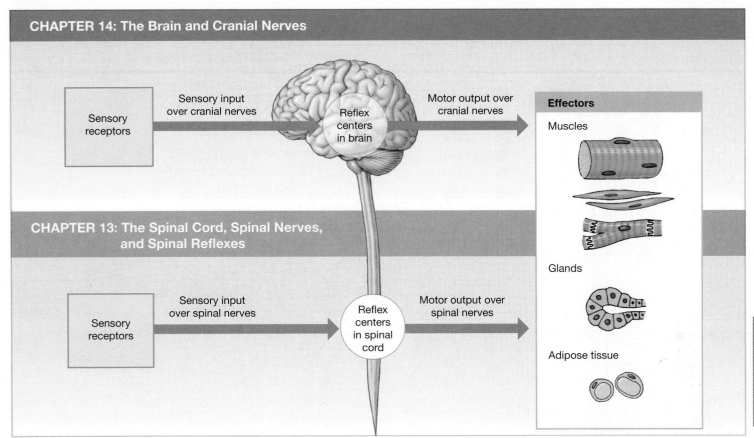

CHAPTER 14: The Brain and Cranial Nerves

Sensory receptors

Sensory input over cranial nerves

Reflex centers in brain

Motor output over cranial nerves

Effectors

Muscles

CHAPTER 13: The Spinal Cord, Spinal Nerves, and Spinal Reflexes

Glands

Sensory receptors

Sensory input over spinal nerves

Reflex centers in spinal cord

Motor output over spinal nerves

Adipose tissue

13

difficult to investigate, but they affect activity along the sensory and motor pathways and alter our perception of those activities.

With these basic principles, definitions, and strategies in mind, let's begin our examination of the functional organization in the nervous system.

Checkpoint

1. Name the structures of the central nervous system (CNS) and of the peripheral nervous system (PNS).
2. Define *spinal reflex*.

See the blue Answers tab at the back of the book.

13-2 The spinal cord is surrounded by three meninges and has spinal nerve roots

Learning Outcome Discuss the structure and functions of the spinal cord, and describe the three meningeal layers that surround the central nervous system.

The **spinal cord** is an organ primarily made up of nervous tissue that is found within protective membranes and the bony vertebral column. This organ carries sensory and motor information between the brain and most other parts of the body. To carry out this function, the spinal cord is the origin of many *spinal nerves*, which allow communication with the peripheral organs. This section focuses on the anatomy of the spinal cord and its protection, while we cover its function in Section 13–3. We begin this section with the gross and cross-sectional anatomy of the spinal cord. Then we examine the three membranous layers that surround the spinal cord: the *spinal meninges* (meh-NIN-jēz).

Gross Anatomy of the Spinal Cord

The adult spinal cord (Figure 13–2a) is approximately 45 cm (18 in.) long, with a maximum width of about 14 mm (0.55 in.). Note that the cord itself is not as long as the vertebral column. Instead, the adult spinal cord extends inferiorly from the brain and ends between vertebrae L_1 and L_2. Like the vertebrae, there are four spinal cord regions: *cervical, thoracic, lumbar*, and *sacral*. The spinal cord contains both gray matter (primarily made up of neuron cell bodies) and white matter (primarily myelinated axons). Figure 13–2b shows a lateral view, and Figure 13–2c is a series of cross-sectional views that illustrate the variations in the proportions of gray matter and white matter in the various regions of the spinal cord. The entire spinal cord is divided into 31 segments on the basis of the origins of the spinal nerves.

Figure 13–2 **Gross Anatomy of the Adult Spinal Cord.** ATLAS: Plates 2a; 20a,b; 24a–c

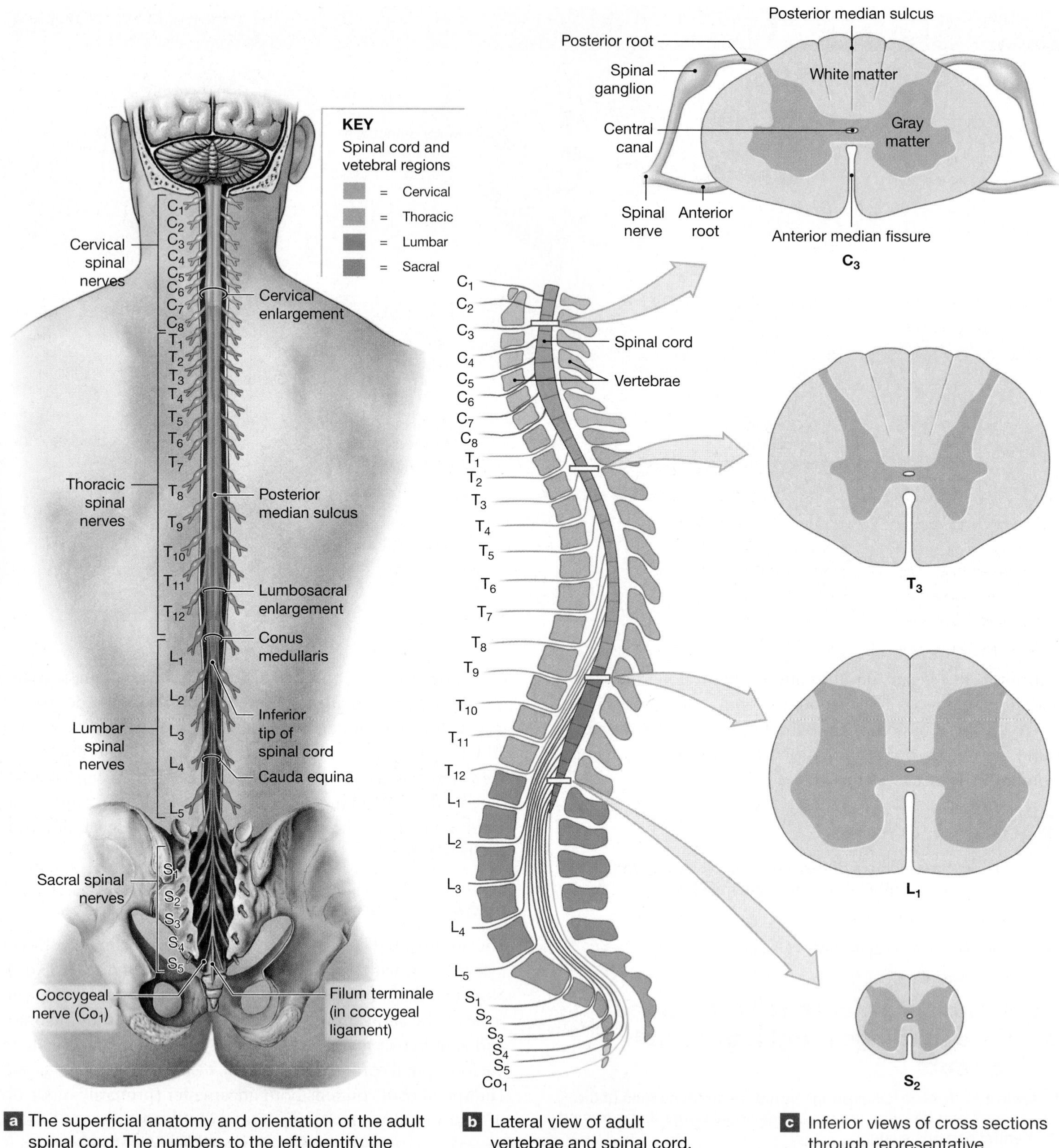

KEY
Spinal cord and vetebral regions
= Cervical
= Thoracic
= Lumbar
= Sacral
Cervical spinal nerves
Thoracic spinal nerves
Lumbar spinal nerves
Sacral spinal nerves
Coccygeal nerve (Co1)
C1 C2 C3 C4 C5 C6 C7 C8
T1 T2 T3 T4 T5 T6 T7 T8 T9 T10 T11 T12
L1 L2 L3 L4 L5
S1 S2 S3 S4 S5
Co1
Cervical enlargement
Posterior median sulcus
Lumbosacral enlargement
Conus medullaris
Inferior tip of spinal cord
Cauda equina
Filum terminale (in coccygeal ligament)
Spinal cord
Vertebrae
Posterior median sulcus
Posterior root
Spinal ganglion
White matter
Central canal
Gray matter
Spinal nerve
Anterior root
Anterior median fissure
C3
T3
L1
S2

a The superficial anatomy and orientation of the adult spinal cord. The numbers to the left identify the spinal nerves and indicate where the nerve roots leave the vertebral canal. The adult spinal cord extends from the brain only to the level of vertebrae L_1–L_2; the spinal segments found at representative locations are indicated in the cross sections.

b Lateral view of adult vertebrae and spinal cord. Note that the spinal cord segments for S_1–S_5 are level with the T_{12}–L_1 vertebrae.

c Inferior views of cross sections through representative segments of the spinal cord, showing the arrangement of gray matter and white matter.

A letter and number designation, the same method used to identify vertebrae, identifies each segment. For example, C_3, the spinal segment in the uppermost section of Figure 13–2c, is the third cervical segment.

The posterior (dorsal) surface of the spinal cord has a shallow longitudinal groove, the **posterior median sulcus**. The **anterior median fissure** is a deeper groove along the anterior (ventral) surface. In addition, the spinal cord has an internal passageway called the **central canal**. As noted in Chapter 12, this space is filled with **cerebrospinal fluid (CSF)**, which acts as a shock absorber and a diffusion medium for dissolved gases, nutrients, chemical messengers, and wastes. p. 398

The amount of gray matter is greatest in segments of the spinal cord dedicated to the sensory and motor control of the limbs. These segments are expanded, forming two **enlargements** of the spinal cord. The **cervical enlargement** supplies nerves to the shoulder and upper limbs. The **lumbosacral enlargement** innervates structures of the pelvis and lower limbs.

Inferior to the lumbosacral enlargement, the spinal cord becomes tapered and conical. This region is the **conus medullaris** (med-yū-LAR-us). The **filum terminale** (FĪ-lum ter-mi-NAH-lā; "terminal thread"), a slender strand of fibrous tissue, extends from the inferior tip of the conus medullaris. It continues along the length of the vertebral canal as far as the second sacral vertebra, where it longitudinally supports the spinal cord as a part of the *coccygeal ligament*.

Spinal Roots and Ganglia

Every spinal segment is associated with a pair of **spinal ganglia** (*dorsal root ganglia*) located near the spinal cord (see Figure 13–2c). These ganglia contain the cell bodies of sensory neurons. The axons of the neurons form the **posterior roots** or *dorsal roots*, which bring sensory information into the spinal cord. A pair of **anterior roots**, or *ventral roots* contains the axons of motor neurons that extend into the periphery to control somatic and visceral effectors. The roots of all spinal nerves divide in fanlike fashion, forming **rootlets**, before entering or leaving the spinal cord. A spinal posterior root may divide into 8–12 rootlets (Figure 13–3a).

On both sides, the posterior and anterior roots of each segment pass between the vertebral canal and the periphery at the *intervertebral foramen* between successive vertebrae. The spinal ganglion lies between the pedicles of the adjacent vertebrae. (You can review vertebral anatomy in Chapter 7. pp. 229–230)

The spinal cord continues to enlarge and elongate until approximately 4 years of age. Up to that time, enlargement of the spinal cord keeps pace with the growth of the vertebral column. Throughout this time, the posterior and anterior roots are very short, and they enter the intervertebral foramina immediately adjacent to their spinal segment. After age 4, the vertebral column continues to elongate, but the spinal cord does not.

Figure 13–3 **The Spinal Cord and Spinal Meninges.**

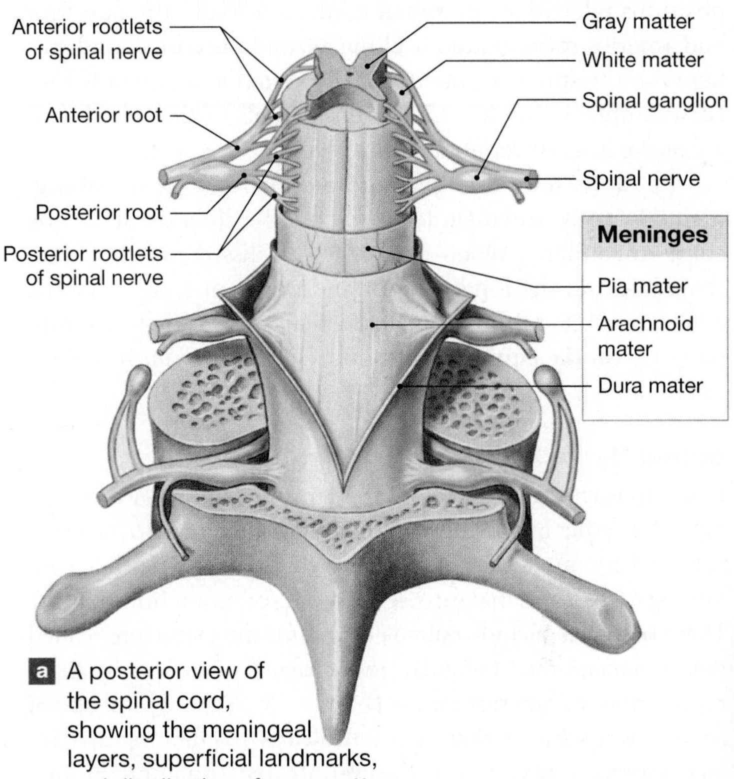
Anterior rootlets of spinal nerve
Anterior root
Posterior root
Posterior rootlets of spinal nerve
Gray matter
White matter
Spinal ganglion
Spinal nerve
Meninges
Pia mater
Arachnoid mater
Dura mater

a A posterior view of the spinal cord, showing the meningeal layers, superficial landmarks, and distribution of gray matter and white matter

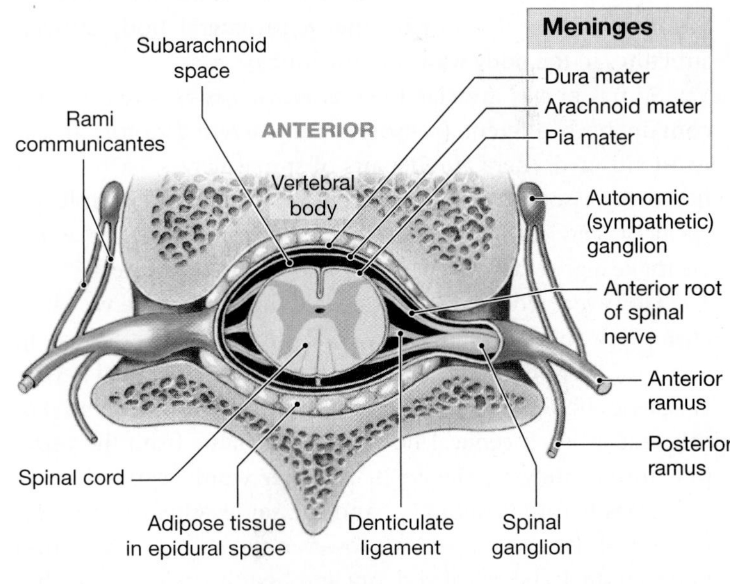
Subarachnoid space
Rami communicantes
ANTERIOR
Vertebral body
Meninges
Dura mater
Arachnoid mater
Pia mater
Autonomic (sympathetic) ganglion
Anterior root of spinal nerve
Anterior ramus
Posterior ramus
Spinal cord
Adipose tissue in epidural space
Denticulate ligament
Spinal ganglion
POSTERIOR

b A sectional view through the spinal cord and meninges, showing the relationship of the meninges, spinal cord, and spinal nerves

? What is the area between the arachnoid mater and the pia mater?

This vertebral growth moves the intervertebral foramina and thus the spinal nerves farther and farther from their original positions relative to the spinal cord. As a result, the posterior and anterior roots gradually elongate, and the correspondence between the spinal segment and the vertebral segment is lost. For example, in adults, the sacral segments of the spinal cord are at the level of vertebrae L_1–L_2 (see Figure 13–2).

In adults the posterior and anterior roots of spinal segments L_2 to S_5 extend inferiorly, past the inferior tip of the conus medullaris. When seen in gross dissection, the filum terminale and the long anterior and posterior roots resemble a horse's tail. For this reason, early anatomists called this complex the **cauda equina** (KAW-duh ek-WĪ-nuh; *cauda*, tail + *equus*, horse).

Spinal Nerves and Rami (Branches)

Distal to each spinal ganglion, the sensory and motor roots are bound together into a single **spinal nerve** just lateral to the intervertebral foramen. The spinal nerve then forms branches containing nerve fibers that carry sensory and/or motor information.
13 These branches include paired communicating structures called *rami communicantes* (singular, *ramus communicans*), a posterior ramus, and an anterior ramus (Figure 13–3b). A typical spinal nerve has a **white ramus communicans** (containing myelinated axons), a **gray ramus communicans** (containing unmyelinated fibers that innervate glands and smooth muscles in the body wall or limbs), a **posterior ramus** (providing sensory and motor innervation to the skin and muscles of the back), and an **anterior ramus** (supplying the ventrolateral body surface, structures in the body wall, and the limbs).

Spinal nerves are classified as *mixed nerves*—that is, they contain both afferent (sensory) and efferent (motor) fibers. As mentioned, there are 31 pairs of spinal nerves, each identified by its association with adjacent vertebrae. For example, we speak of "cervical spinal nerves" or even "cervical nerves" when we make a general reference to spinal nerves of the neck.

However, when we indicate specific spinal nerves, it is customary to give them a regional number, as indicated in Figure 13–2. In the cervical region, the first pair of spinal nerves, C_1, passes between the skull and the first cervical vertebra. For this reason, each cervical nerve takes its name from the vertebra immediately *inferior* to it. In other words, cervical nerve C_2 comes before vertebra C_2, and the same system is used for the rest of the cervical series. However, the spinal nerve found between the last cervical vertebra and first thoracic vertebra has been designated C_8. Therefore, although there are only seven cervical vertebrae, there are *eight* pairs of cervical nerves. Then each spinal nerve inferior to the first thoracic vertebra (down through the lumbar and sacral vertebrae) takes its name from the vertebra immediately *superior* to it. For example, spinal nerve T_1 emerges immediately inferior to vertebra T_1, spinal nerve T_2 follows vertebra T_2, and so forth.

Protection of the Spinal Cord: Spinal Meninges

The vertebral column and its surrounding ligaments, tendons, and muscles isolate the spinal cord from the rest of the body. These structures also protect against bumps and shocks to the back. The delicate neural tissues must also be protected from damaging contacts with the surrounding bony walls of the vertebral canal. The **spinal meninges** (me-NIN-jēz), a series of specialized membranes surrounding the spinal cord, provide the necessary physical stability and shock absorption. In addition, blood vessels branching within these layers deliver oxygen and nutrients to the spinal cord.

The spinal meninges consist of three layers: (1) the *dura mater*, (2) the *arachnoid mater*, and (3) the *pia mater*. The relationships among these membranes are shown in Figure 13–3. At the foramen magnum of the skull, the spinal meninges are continuous with the **cranial meninges**, which surround the brain. (The cranial meninges have the same three layers. We discuss them in Chapter 14.)

The Dura Mater

The tough, fibrous **dura mater** (DŪ-ruh MĀ-ter; *durus*, hard + *mater*, mother) is the layer that forms the outermost covering of the spinal cord (see Figure 13–3a). This layer contains dense collagen fibers that are oriented along the longitudinal axis of the cord. Between the dura mater and the walls of the vertebral canal lies the **epidural space**, a region that contains areolar tissue, blood vessels, and a protective padding of adipose tissue (see Figure 13–3b).

The connections of the spinal dura mater to its bony surroundings vary. The outer layer of the spinal dura mater fuses with the periosteum of the occipital bone of the cranium around the margins of the foramen magnum. There, the spinal dura mater becomes continuous with the cranial dura mater. However, the spinal dura mater does not have extensive, firm connections to the surrounding vertebrae. Attachment sites at either end of the vertebral canal provide longitudinal stability. In addition, the dura mater extends between adjacent vertebrae at each intervertebral foramen, fusing with the connective tissues that surround the spinal nerves. Within the sacral canal on the posterior surface of the sacrum, the spinal dura mater tapers from a sheath to a dense cord of collagen fibers that blends with components of the filum terminale to form the **coccygeal ligament** (look back at Figure 13–2a). The coccygeal ligament continues along the sacral canal, ultimately blending into the periosteum of the coccyx.

The Arachnoid Mater

In most anatomical and tissue specimens, a narrow **subdural space** separates the dura mater from deeper meningeal layers. In a healthy living person no such space exists. Instead, the inner surface of the dura mater is in contact with the outer surface of the **arachnoid** (a-RAK-noyd; *arachne*, spider) **mater**, the middle meningeal layer (see Figure 13–3b). The inner surface of the dura mater and the outer surface of the arachnoid mater are covered by simple squamous epithelia. The arachnoid mater includes this epithelium, called the *arachnoid membrane*, and the *arachnoid trabeculae*, a delicate network of collagen and elastic fibers that extends between the arachnoid membrane and the outer surface of the pia mater. The region between is called the **subarachnoid space**, and it is filled with CSF.

The spinal arachnoid mater extends inferiorly as far as the filum terminale, and the posterior and anterior roots of the cauda equina lie within the fluid-filled subarachnoid space. In adults, the withdrawal of CSF involves the insertion of a needle into the subarachnoid space in the inferior lumbar region. This procedure is known as a **lumbar puncture**, or **spinal tap**.

The Pia Mater

The subarachnoid space extends between the arachnoid epithelium and the innermost meningeal layer, the **pia mater** (PĒ-ah; *pia*, delicate + *mater*, mother). The pia mater consists of a meshwork of elastic and collagen fibers that is firmly bound to the underlying neural tissue (see Figure 13–3). These connective tissue fibers are interwoven with those that span the subarachnoid space, firmly binding the arachnoid to the pia mater. The blood vessels servicing the spinal cord run along the surface of the spinal pia mater, within the subarachnoid space (Figure 13–4).

Along the length of the spinal cord, paired **denticulate ligaments** extend from the pia mater through the arachnoid mater to the dura mater (Figure 13–3b, 13–4). Denticulate ligaments, which originate along either side of the spinal cord, prevent lateral (side-to-side) movement. The dural connections at the foramen magnum and the coccygeal ligament prevent longitudinal (superior–inferior) movement.

The spinal meninges accompany the posterior and anterior roots as these roots pass through the intervertebral foramina.

Figure 13–4 The Spinal Cord and Associated Structures. An anterior view of the cervical spinal cord and spinal nerve roots in the vertebral canal. The dura mater and arachnoid mater have been cut and reflected. Notice the blood vessels that run in the subarachnoid space, bound to the outer surface of the delicate pia mater.

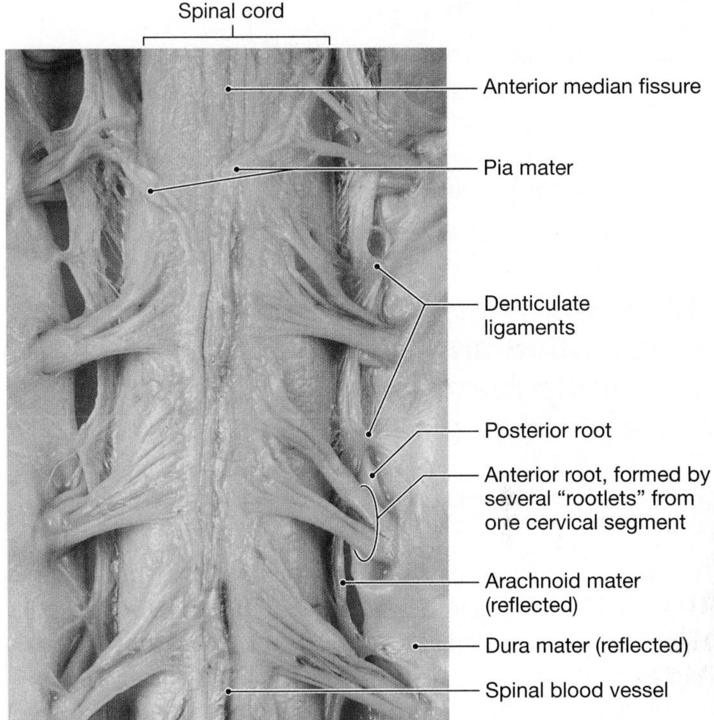
Spinal cord
Anterior median fissure
Pia mater
Denticulate ligaments
Posterior root
Anterior root, formed by several "rootlets" from one cervical segment
Arachnoid mater (reflected)
Dura mater (reflected)
Spinal blood vessel

As the sectional view in Figure 13–3b shows, the meningeal membranes are continuous with the connective tissues that surround the spinal nerves and their peripheral branches.

Bacterial or viral infection can cause **meningitis**, or inflammation of the meningeal membranes. Meningitis is dangerous because it can disrupt the normal circulation of CSF, damaging

+ Clinical Note Anesthesia

Anesthetics are often injected into the epidural space. Introduced in this way, a drug should affect only the spinal nerves in the immediate area of the injection. The result is a regional pain reliever (analgesic) called *epidural anesthesia*, which blocks pain sensations.

This procedure can be done either peripherally, as when skin lacerations are sewn up, or at sites around the spinal cord for more widespread anesthetic effects, as during labor and delivery. An epidural anesthetic has at least two advantages: (1) It affects only the spinal nerves in the immediate area of the injection; and (2) it provides mainly sensory anesthesia.

Local anesthetics can also be introduced as a single dose into the subarachnoid space of the spinal cord. This procedure is commonly called **spinal anesthesia**. The effects include both temporary muscle paralysis and sensory loss. These effects tend to spread as the movement of cerebrospinal fluid distributes the anesthetic along the spinal cord. Problems with overdosing are seldom serious, because controlling the patient's position during administration can limit the distribution of the drug. Breathing continues even if all thoracic and abdominal segments have been paralyzed because the upper cervical spinal nerves control the diaphragm.

In the United States, anesthesia is often administered by a certified registered nurse anesthetist (CRNA). This select vocation entails master's-level training following receipt of a bachelor of science in nursing (BSN).

or killing neurons and neuroglia in the affected areas. An initial diagnosis may locate the infection in the meninges of the spinal cord (*spinal meningitis*) or brain (*cerebral meningitis*), but in later stages the entire meningeal system is usually affected.

Checkpoint

3. **Identify the three spinal meninges.**
4. **Damage to which root of a spinal nerve would interfere with motor function?**
5. **Where is the cerebrospinal fluid that surrounds the spinal cord located?**

See the blue Answers tab at the back of the book.

13-3 Spinal cord gray matter integrates information and initiates commands, and white matter carries information from place to place

13

Learning Outcome Explain the roles of white matter and gray matter in processing and relaying sensory information and motor commands.

We have seen that the patterns of gray and white matter found in cross sections of the spinal cord vary with the spinal cord regions (see Figure 13–2c). Recall that in nervous tissue **gray matter** is dominated by the cell bodies of neurons, neuroglia, and unmyelinated axons, while the **white matter** contains large numbers of myelinated and unmyelinated axons. In this section we relate this cross-sectional organization to the functional organization of both gray and white matter in the spinal cord (Figure 13–5).

Functional Organization of Gray Matter

The left side of Figure 13–5a shows the structural pattern of gray matter in the spinal cord. In cross section, this gray matter forms an H, or butterfly, shape and surrounds the narrow central canal. The areas of gray matter on each side of the spinal cord (to the left or right halves defined by the anterior median fissure and posterior median sulcus) are called **horns**. There are **posterior**, **lateral**, and **anterior horns**, depending on their positions (Figure 13–5b). The narrow bridges of gray matter posterior and anterior to the central canal are **gray commissures** (*commissura*, a joining together).

Now let's think about what this organization of gray matter can tell us about its function. Masses of gray matter within the CNS are called **nuclei**; there are both sensory and motor nuclei. **Sensory nuclei** receive and relay sensory information from peripheral receptors. **Motor nuclei** issue motor commands to peripheral effectors. These motor and sensory nuclei are organized together in the spinal cord. If you took a frontal section along the length of the central canal of the spinal cord, it would separate the sensory (posterior, or dorsal) nuclei from the motor (anterior, or ventral) nuclei. The posterior horns contain somatic and visceral sensory nuclei, and the anterior horns contain somatic motor nuclei. The lateral horns, located only in the thoracic and lumbar segments, contain visceral motor nuclei. The gray commissures contain axons that cross from one side of the cord to the other before they reach an area in the gray matter.

The right side of Figure 13–5a shows the relationship between the function of a particular nucleus (sensory or motor) and its position in the gray matter of the spinal cord. Although sensory and motor nuclei appear small in transverse section, nuclei are also spatially organized—they may extend for a considerable distance along the length of the spinal cord. In the cervical enlargement, for example, the anterior horns contain nuclei whose motor neurons control the muscles of the upper limbs. On each side of the spinal cord, in medial to lateral sequence, are somatic motor nuclei that control (1) muscles that position the pectoral girdle, (2) muscles that move the arm, (3) muscles that move the forearm and hand, and (4) muscles that move the hand and fingers. Within each of these regions, the motor neurons that control flexor muscles are grouped separately from those that control extensor muscles. Because the spinal cord is so highly organized, we can predict which muscles will be affected by damage to a specific area of gray matter.

Functional Organization of White Matter

How is the white matter functionally organized? Let's look at the structural organization of white matter first (look back at the left side of Figure 13–5a). The white matter on each side of the spinal cord can be divided into three regions called **columns**). The **posterior white columns** lie between the posterior horns and the posterior median sulcus. The **anterior white columns** lie between the anterior horns and the anterior median fissure. The anterior white columns are interconnected by the **anterior white commissure**, a region where axons cross from one side of the spinal cord to the other. The white matter between the anterior and posterior columns on each side makes up the **lateral white column**.

Now let's connect the structural organization with the function of the columns. Each column contains *tracts* whose axons share functional and structural characteristics. A **tract** is a bundle of axons in the CNS that is somewhat uniform in diameter, myelination, and propagation speed. All the axons within a tract relay the same type of information (sensory or motor) in the same direction. Short tracts carry sensory or motor signals between segments of the spinal cord, and longer tracts connect the spinal cord with the brain. **Ascending tracts** carry *sensory* information toward the brain, and **descending tracts** convey *motor* commands to the spinal cord. We describe the major tracts and their functions in Chapters 15 and 16. Because spinal tracts have very specific functions, damage to one produces a characteristic loss of sensation or motor control.

Figure 13–5 The Sectional Organization of the Spinal Cord.

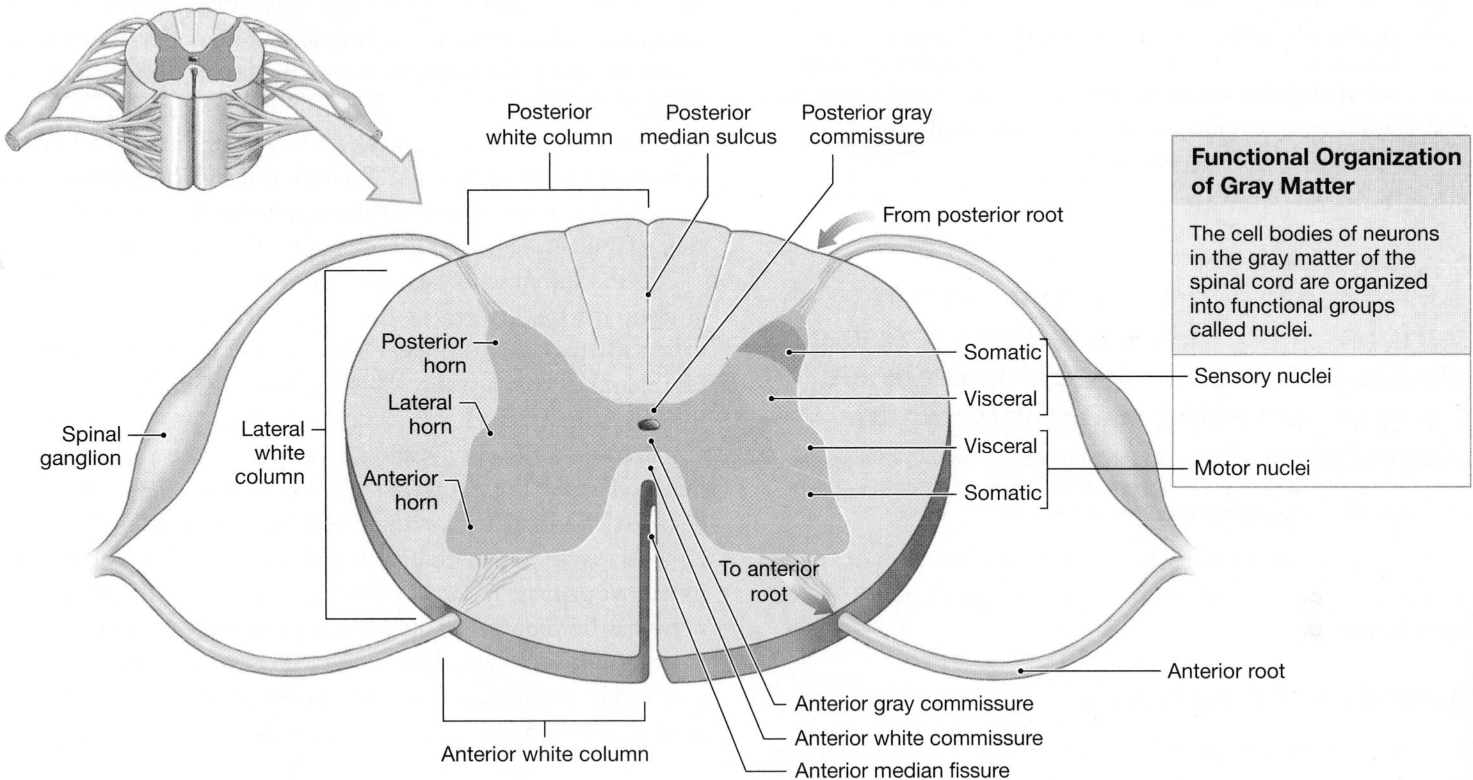
Posterior white column
Posterior median sulcus
Posterior gray commissure
From posterior root
Functional Organization of Gray Matter
The cell bodies of neurons in the gray matter of the spinal cord are organized into functional groups called nuclei.
Sensory nuclei
Motor nuclei
Somatic
Visceral
Visceral
Somatic
Posterior horn
Lateral horn
Anterior horn
Spinal ganglion
Lateral white column
To anterior root
Anterior root
Anterior gray commissure
Anterior white commissure
Anterior median fissure
Anterior white column

a The left half of this sectional view shows important anatomical landmarks, including the three columns of white matter. The right half indicates the functional organization of the nuclei in the anterior, lateral, and posterior horns. The red arrows represent sensory input from the posterior root and motor output to the anterior root.

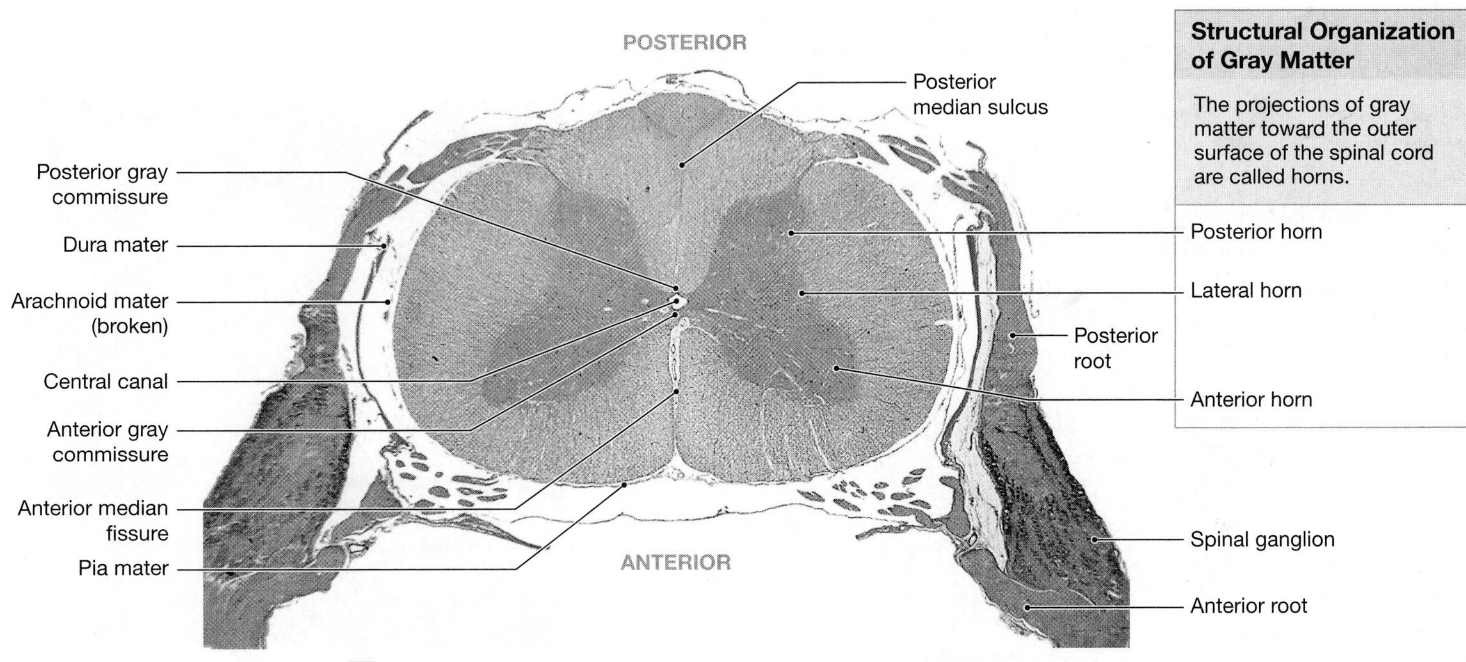
POSTERIOR
ANTERIOR
Posterior median sulcus
Structural Organization of Gray Matter
The projections of gray matter toward the outer surface of the spinal cord are called horns.
Posterior horn
Lateral horn
Anterior horn
Posterior root
Spinal ganglion
Anterior root
Posterior gray commissure
Dura mater
Arachnoid mater (broken)
Central canal
Anterior gray commissure
Anterior median fissure
Pia mater

b A micrograph of a transverse section through the spinal cord, showing major landmarks in and surrounding the cord.

? In which horn are somatic motor nuclei located?

13

Checkpoint

6. **Differentiate between sensory nuclei and motor nuclei.**
7. **A man with polio has lost the use of his leg muscles. In which area of his spinal cord would you expect the virus-infected motor neurons to be?**
8. **A disease that damages myelin sheaths would affect which part of the spinal cord?**

See the blue Answers tab at the back of the book.

13-4 Spinal nerves extend to form peripheral nerves, sometimes forming plexuses along the way; these nerves carry sensory and motor information

Learning Outcome Describe the major components of a spinal nerve, describe a nerve plexus, and relate the distribution pattern of spinal nerves to the regions they innervate.

In this section we consider the structure and function of spinal nerves and the interwoven networks of spinal nerves called *nerve plexuses*.

13

Anatomy of Spinal Nerves

Every segment of the spinal cord is connected to a pair of spinal nerves. Surrounding each spinal nerve is a series of connective tissue layers continuous with those of the associated peripheral nerves (Figure 13–6). These wrapped layers, best seen in sectional view, are comparable to those associated with skeletal muscles. p. 294 The **epineurium** (ep-ih-NŪR-ē-um), or outermost layer, consists of a dense network of collagen fibers. The fibers of the **perineurium**, the middle layer, extend inward from the epineurium. These connective tissue partitions divide the nerve into a series of compartments, called *fascicles*, that contain bundles of axons. Delicate connective tissue fibers of the **endoneurium**, the innermost layer, extend from the perineurium and surround individual axons.

Arteries and veins penetrate the epineurium and branch within the perineurium. Capillaries leaving the perineurium branch in the endoneurium. They supply the axons and Schwann cells of the nerve and the fibroblasts of the connective tissues.

As the spinal nerves extend into the periphery, they branch and interconnect, forming the **peripheral nerves** that innervate body tissues and organs. The connective tissue sheaths of peripheral nerves are the same as, and are continuous with, those of spinal nerves.

Recovery from an injury to a peripheral nerve may depend on the extent of the injury. If a peripheral axon is severed but not displaced, normal function may eventually return as the cut stump grows across the site of injury, away from the cell body and along its former path. p. 401 Repairs made after an entire peripheral nerve has been damaged are generally incomplete, primarily due to problems with axon alignment and regrowth. Various technologically sophisticated procedures designed to improve nerve regeneration and repair are currently under evaluation.

Peripheral Distribution and Function of Spinal Nerves

A **dermatome** is the specific bilateral region of the skin surface monitored by a single pair of spinal nerves. Each pair of

Figure 13–6 A Peripheral Nerve. A diagrammatic view and an electron micrograph of a typical spinal nerve. Note the connective tissue layers that are continuous with the associated spinal nerve.

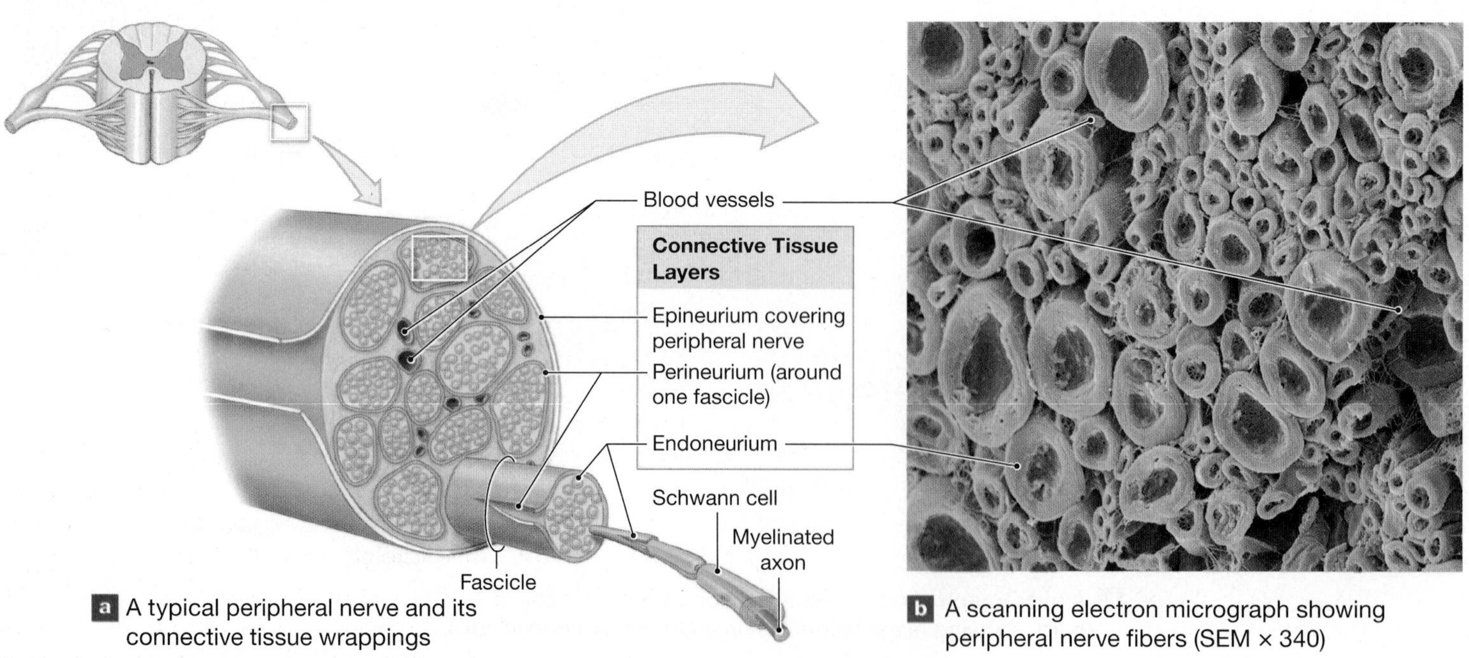
Blood vessels
Connective Tissue Layers
Epineurium covering peripheral nerve
Perineurium (around one fascicle)
Endoneurium
Schwann cell
Myelinated axon
Fascicle

a A typical peripheral nerve and its connective tissue wrappings

b A scanning electron micrograph showing peripheral nerve fibers (SEM × 340)

Go to MasteringA&P™ > Study Area > Menu > Lab Tools > PAL 3.0 > Histology > Nervous Tissue

Figure 13–7 Dermatomes. Anterior and posterior distributions of dermatomes on the surface of the skin. N V = cranial nerve five (trigeminal nerve).

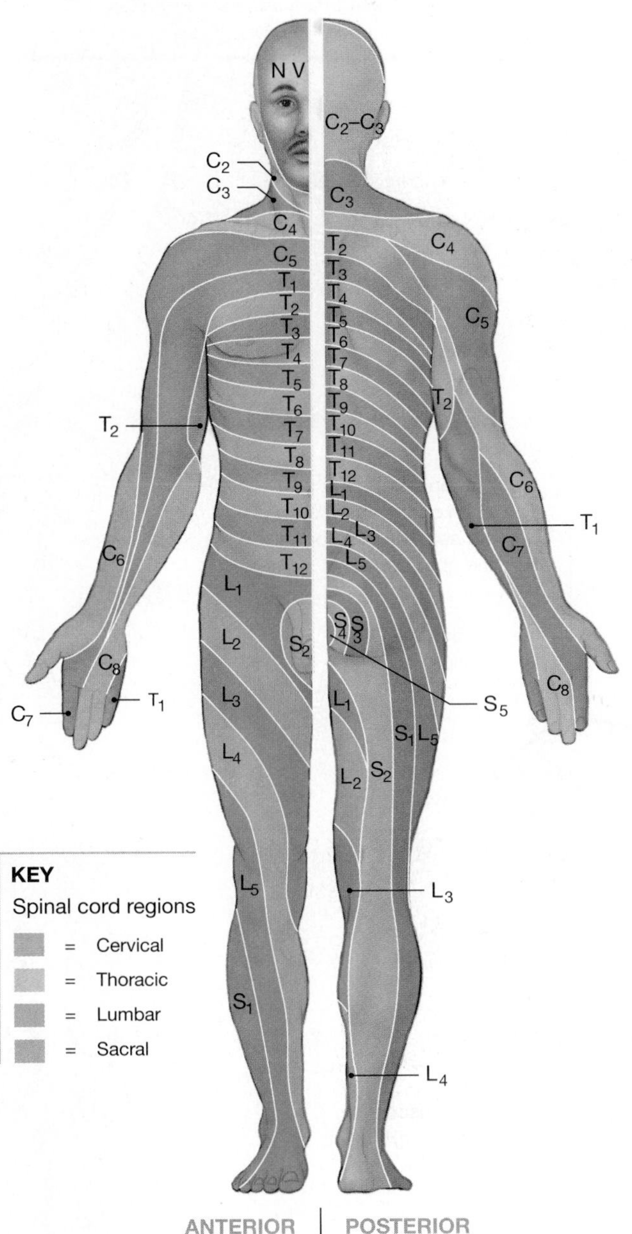
N V
C_2–C_3
C_2
C_3
C_4
C_5
T_1
T_2
T_3
T_4
T_5
T_6
T_7
T_8
T_9
T_{10}
T_{11}
T_{12}
L_1
L_2
L_3
L_4
L_5
C_6
C_7
C_8
S_1
S_2
S_3
S_4
S_5
KEY
Spinal cord regions
= Cervical
= Thoracic
= Lumbar
= Sacral
ANTERIOR
POSTERIOR

+ Clinical Note Shingles

The same virus that causes chickenpox, the varicella-zoster virus (VZV), also causes **shingles** (derived from the Latin *cingulum*, girdle). This herpesvirus attacks neurons within the posterior roots of spinal nerves and sensory ganglia of cranial nerves. Shingles produces a painful rash and blisters whose distribution corresponds to that of the affected sensory nerves and follows its dermatome, as shown in the photo. Treatment for shingles typically involves large doses of antiviral drugs. Fortunately for those affected, attacks of shingles usually heal, often without treatment, and leave behind only unpleasant memories.

Anyone who has had chickenpox is at risk of developing shingles. After the initial encounter, the virus remains dormant within neurons of the anterior horns of the spinal cord. It is not known what triggers the reactivation of this pathogen.

Many people who contract shingles suffer just a single episode in their adult lives. Some experience a painful condition known as *postherpetic neuralgia* after a bout of shingles. Moreover, the problem can recur in people with weakened immune systems, including the elderly, pregnant women, and individuals with AIDS or some forms of cancer. A VZV vaccine is available for people 60 years or older.

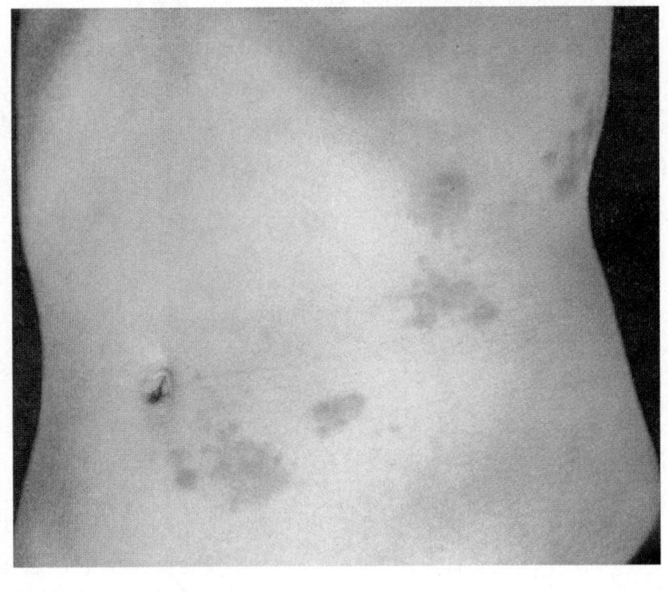

spinal nerves supplies its own dermatome (**Figure 13–7**). Note that the boundaries of adjacent dermatomes overlap to some degree. **Spotlight Figure 13–8** shows the distribution, or functional pathway, of a typical spinal nerve that originates from the thoracic or superior lumbar segments (T_1–L_2) of the spinal cord. This figure considers the pathways of both the sensory information and the motor commands.

Dermatomes are clinically important because damage or infection of a spinal nerve or spinal ganglion produces a loss of sensation in the corresponding region of the skin. Additionally, characteristic signs may appear on the skin supplied by that specific nerve. **Peripheral neuropathies** are regional losses of sensory and motor function most often resulting from nerve trauma or compression. (If your arm or leg has ever "fallen asleep" after you leaned or sat in an uncomfortable position, you have experienced a mild, temporary version of this disorder.) The location of the affected dermatomes provides clues to the location of injuries along

SPOTLIGHT

Figure 13–8

Structure, Function, and the Peripheral Distribution of Spinal Nerves (T_1–L_2)

SENSORY INFORMATION

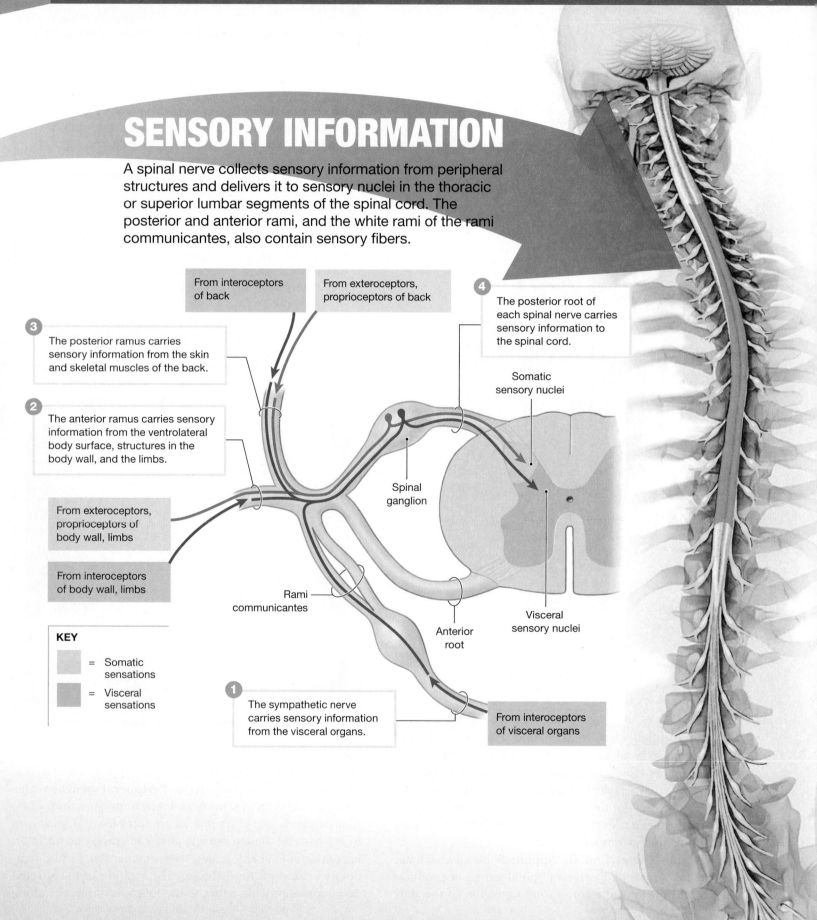

A spinal nerve collects sensory information from peripheral structures and delivers it to sensory nuclei in the thoracic or superior lumbar segments of the spinal cord. The posterior and anterior rami, and the white rami of the rami communicantes, also contain sensory fibers.

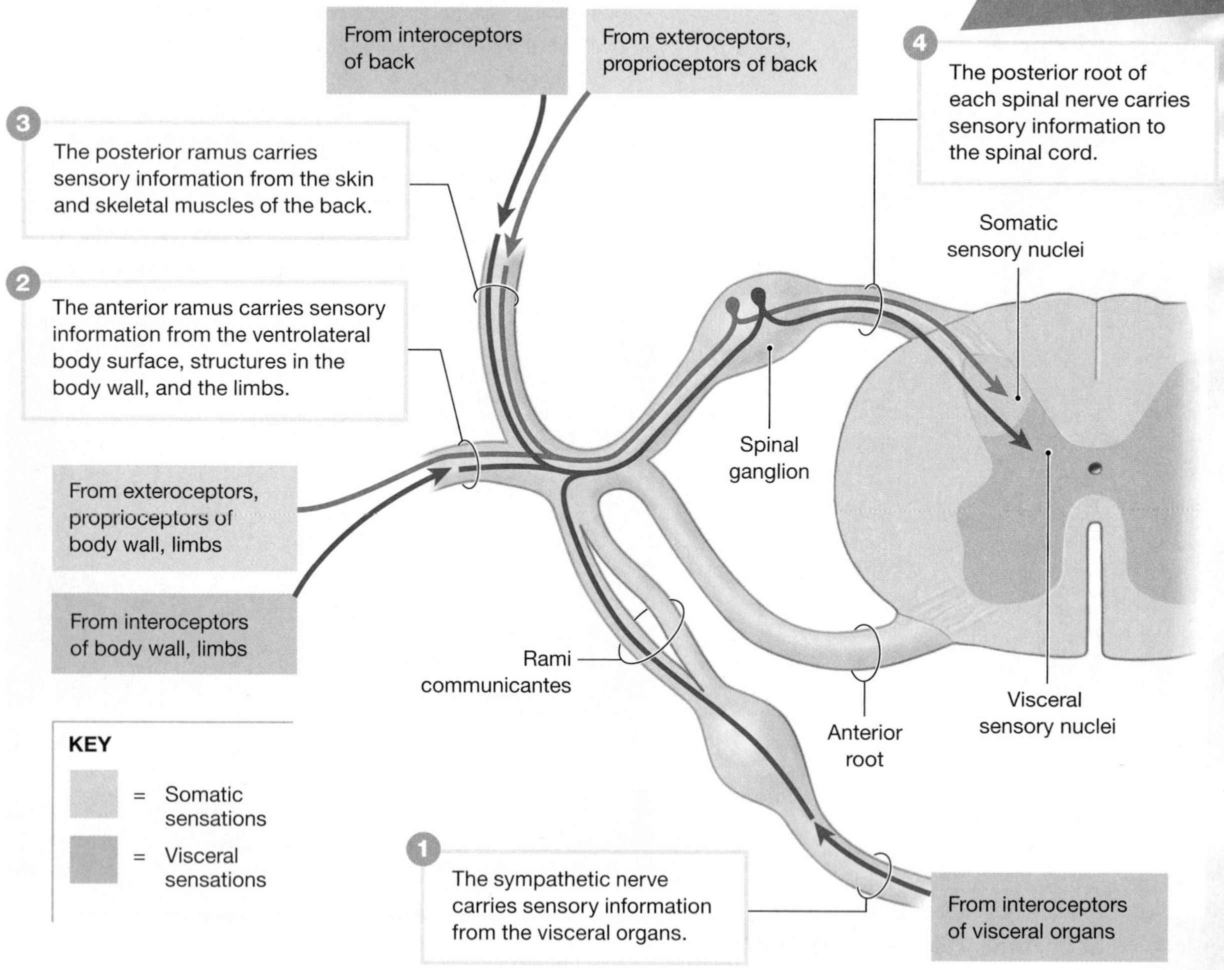

From interoceptors of back

From exteroceptors, proprioceptors of back

4 The posterior root of each spinal nerve carries sensory information to the spinal cord.

3 The posterior ramus carries sensory information from the skin and skeletal muscles of the back.

2 The anterior ramus carries sensory information from the ventrolateral body surface, structures in the body wall, and the limbs.

Somatic sensory nuclei

Spinal ganglion

From exteroceptors, proprioceptors of body wall, limbs

From interoceptors of body wall, limbs

Rami communicantes

Anterior root

Visceral sensory nuclei

KEY

- = Somatic sensations
- = Visceral sensations

1 The sympathetic nerve carries sensory information from the visceral organs.

From interoceptors of visceral organs

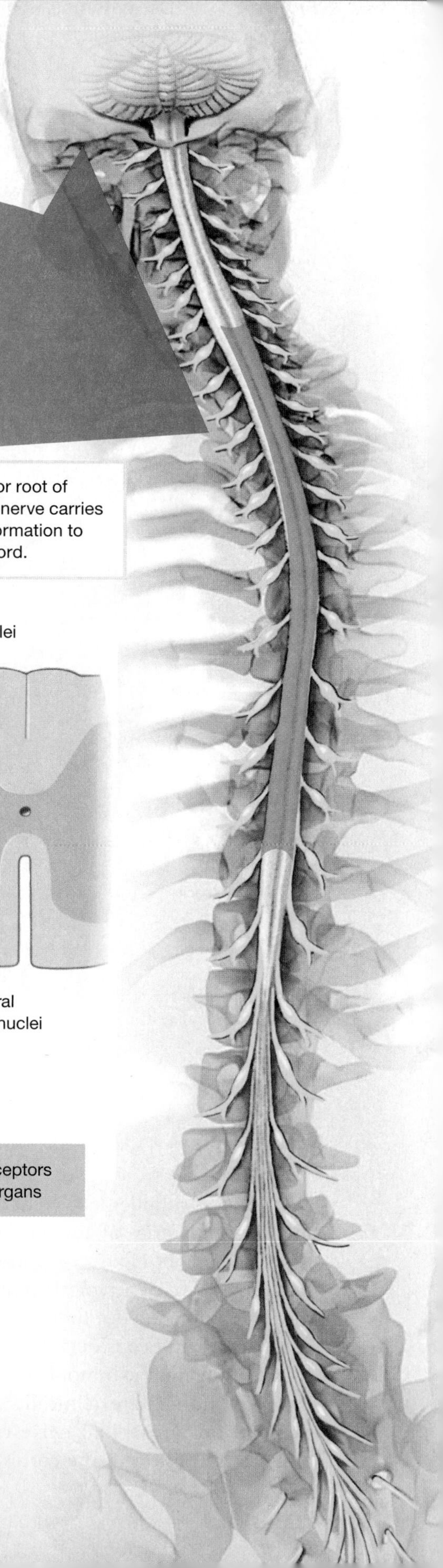

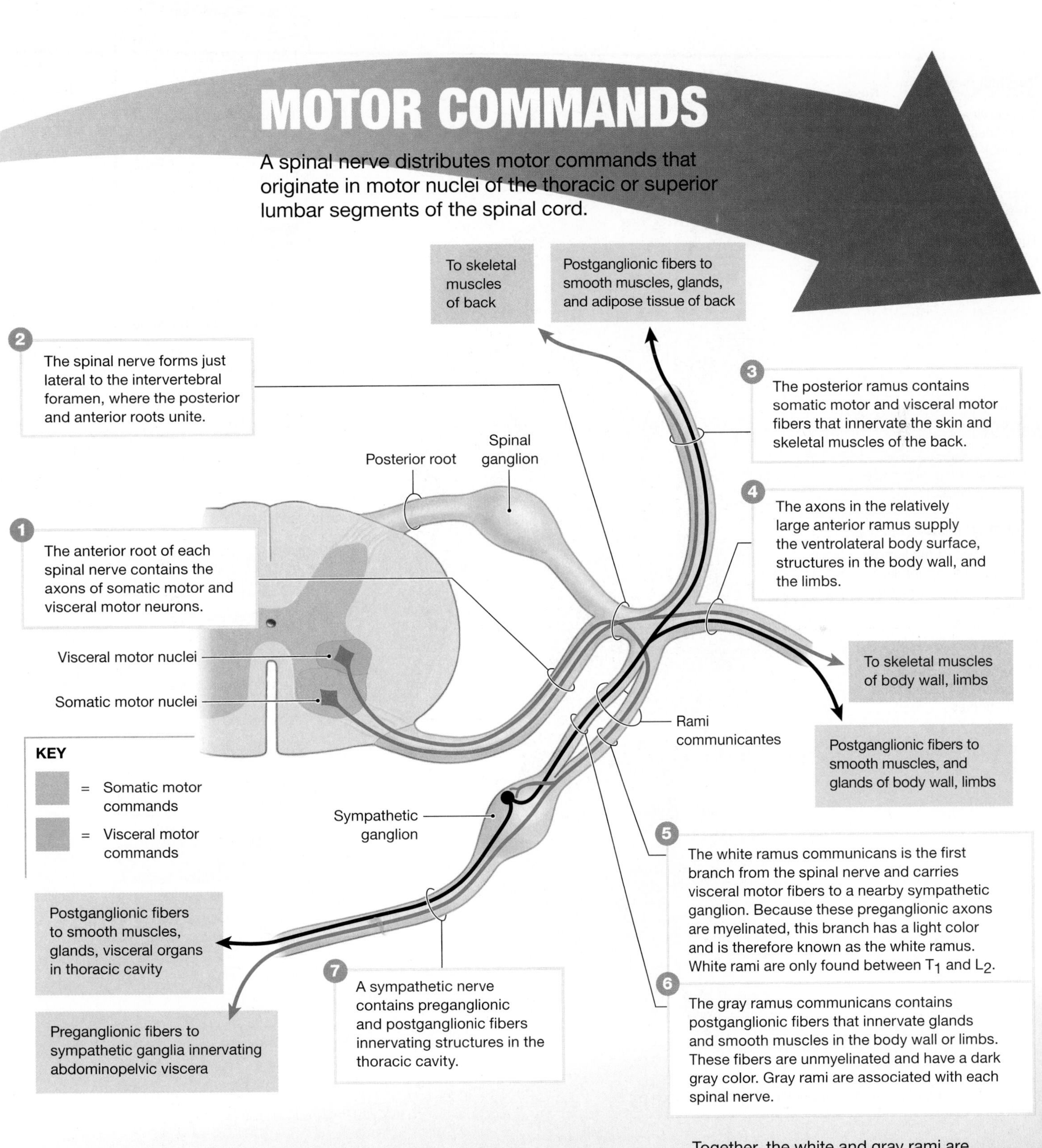

MOTOR COMMANDS
A spinal nerve distributes motor commands that originate in motor nuclei of the thoracic or superior lumbar segments of the spinal cord.
To skeletal muscles of back
Postganglionic fibers to smooth muscles, glands, and adipose tissue of back
2 The spinal nerve forms just lateral to the intervertebral foramen, where the posterior and anterior roots unite.
3 The posterior ramus contains somatic motor and visceral motor fibers that innervate the skin and skeletal muscles of the back.
Posterior root
Spinal ganglion
4 The axons in the relatively large anterior ramus supply the ventrolateral body surface, structures in the body wall, and the limbs.
1 The anterior root of each spinal nerve contains the axons of somatic motor and visceral motor neurons.
Visceral motor nuclei
Somatic motor nuclei
To skeletal muscles of body wall, limbs
Rami communicantes
Postganglionic fibers to smooth muscles, and glands of body wall, limbs
KEY
= Somatic motor commands
= Visceral motor commands
Sympathetic ganglion
5 The white ramus communicans is the first branch from the spinal nerve and carries visceral motor fibers to a nearby sympathetic ganglion. Because these preganglionic axons are myelinated, this branch has a light color and is therefore known as the white ramus. White rami are only found between T1 and L2.
Postganglionic fibers to smooth muscles, glands, visceral organs in thoracic cavity
7 A sympathetic nerve contains preganglionic and postganglionic fibers innervating structures in the thoracic cavity.
6 The gray ramus communicans contains postganglionic fibers that innervate glands and smooth muscles in the body wall or limbs. These fibers are unmyelinated and have a dark gray color. Gray rami are associated with each spinal nerve.
Preganglionic fibers to sympathetic ganglia innervating abdominopelvic viscera

Together, the white and gray rami are known as the **rami communicantes** (RĀ-mī ko-mū-ni-KAN-tēz), or communicating branches (singular, *ramus communicans*).

Figure 13–9 **Nerve Plexuses and Peripheral Nerves.** ATLAS: Plate 2a

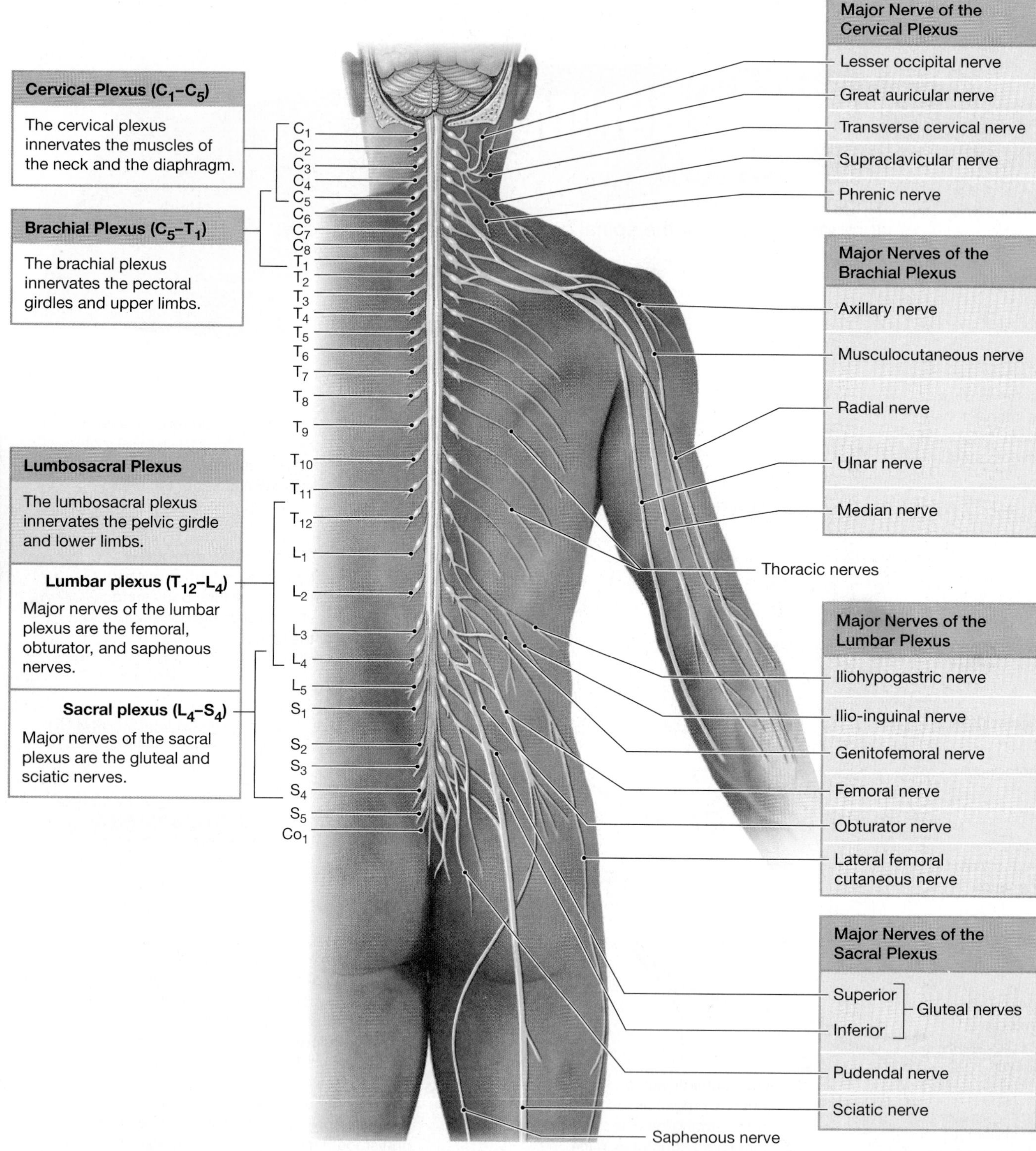
Cervical Plexus (C_1–C_5)
The cervical plexus innervates the muscles of the neck and the diaphragm.
Brachial Plexus (C_5–T_1)
The brachial plexus innervates the pectoral girdles and upper limbs.
Lumbosacral Plexus
The lumbosacral plexus innervates the pelvic girdle and lower limbs.
Lumbar plexus (T_{12}–L_4)
Major nerves of the lumbar plexus are the femoral, obturator, and saphenous nerves.
Sacral plexus (L_4–S_4)
Major nerves of the sacral plexus are the gluteal and sciatic nerves.
C_1 C_2 C_3 C_4 C_5 C_6 C_7 C_8
T_1 T_2 T_3 T_4 T_5 T_6 T_7 T_8 T_9 T_{10} T_{11} T_{12}
L_1 L_2 L_3 L_4 L_5
S_1 S_2 S_3 S_4 S_5
Co_1
Major Nerve of the Cervical Plexus
Lesser occipital nerve
Great auricular nerve
Transverse cervical nerve
Supraclavicular nerve
Phrenic nerve
Major Nerves of the Brachial Plexus
Axillary nerve
Musculocutaneous nerve
Radial nerve
Ulnar nerve
Median nerve
Thoracic nerves
Major Nerves of the Lumbar Plexus
Iliohypogastric nerve
Ilio-inguinal nerve
Genitofemoral nerve
Femoral nerve
Obturator nerve
Lateral femoral cutaneous nerve
Major Nerves of the Sacral Plexus
Superior
Inferior
Gluteal nerves
Pudendal nerve
Sciatic nerve
Saphenous nerve

? The axillary, radial, and ulnar nerves branch from which plexus?

the spinal cord, but the information is not precise. More exact conclusions can be drawn if there is loss of motor control, based on the origin and distribution of the peripheral nerves originating at nerve plexuses.

Nerve Plexuses

The simple distribution pattern of posterior and anterior rami in **Spotlight Figure 13–8** applies to spinal nerves T_1–L_2. But in segments controlling the skeletal muscles of the neck, upper limbs, or lower limbs, the situation is more complicated. During development, small skeletal muscles innervated by different anterior rami typically fuse to form larger muscles with compound origins. The anatomical distinctions between the component muscles may disappear, but separate anterior rami continue to provide sensory innervation and motor control to each part of the compound muscle. As they converge, the anterior rami of adjacent spinal nerves blend their fibers, producing a series of compound nerve trunks. Such a complex interwoven network of nerves is called a **nerve plexus** (PLEK-sus; ***plexus***, braid).

Only the anterior rami form plexuses. These four major plexuses are the (1) *cervical plexus,* (2) *brachial plexus,* (3) *lumbar plexus,* and (4) *sacral plexus* (**Figure 13–9**). The nerves arising at these plexuses contain sensory as well as motor fibers because these nerves form from the fusion of anterior rami (see **Spotlight Figure 13–8**).

Recall that Chapter 11 includes tables that list the nerves that control the major axial and appendicular muscles. As we proceed, you may find it helpful to refer to Tables 11–2 through 11–19 in that chapter. ↺ pp. 349–381

The Cervical Plexus

The **cervical plexus** consists of the anterior rami of spinal nerves C_1–C_5 (**Figures 13–9, 13–10**). The branches of the cervical

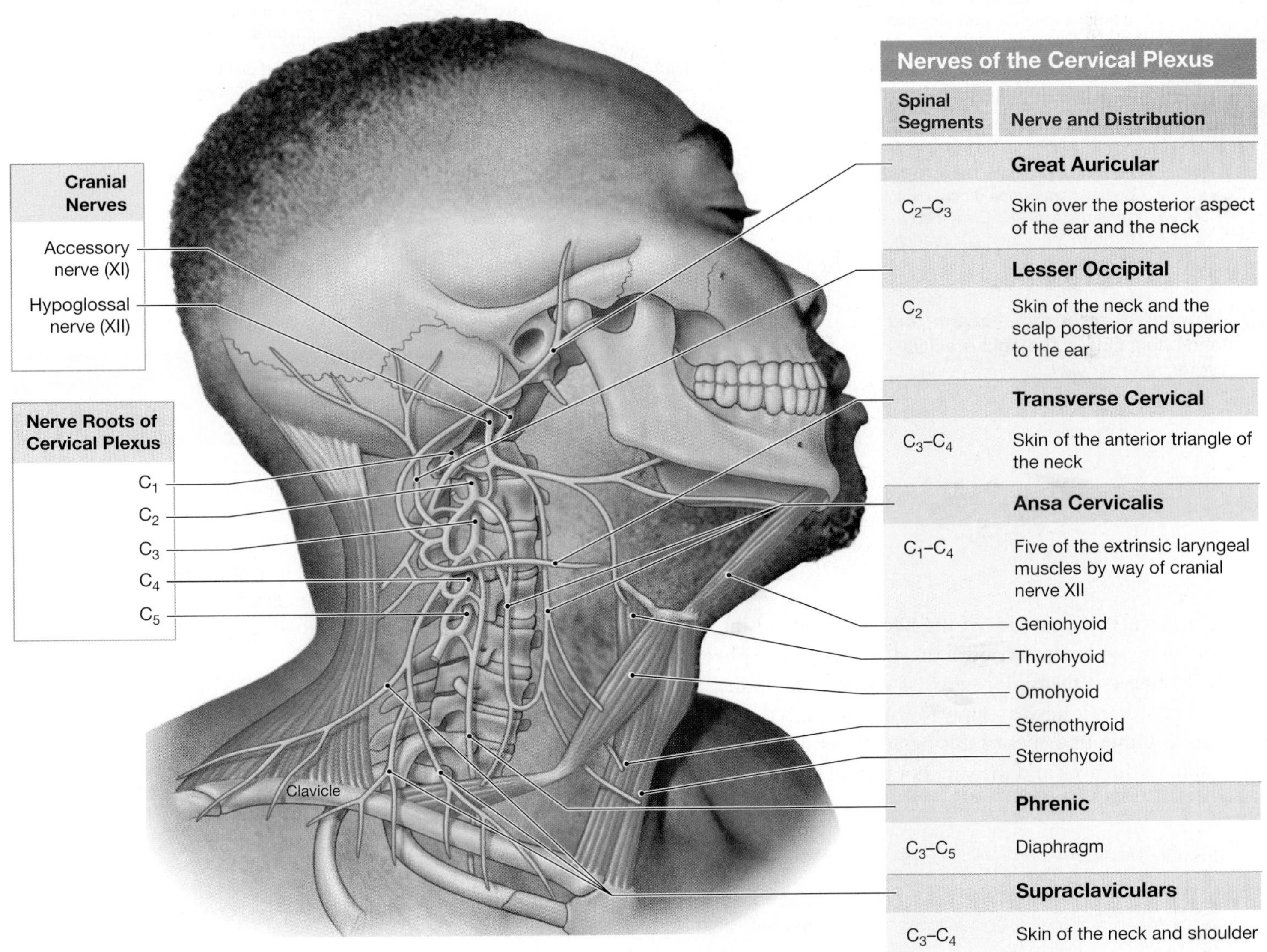
Cranial Nerves
Accessory nerve (XI)
Hypoglossal nerve (XII)
Nerve Roots of Cervical Plexus
C1
C2
C3
C4
C5
Clavicle
Geniohyoid
Thyrohyoid
Omohyoid
Sternothyroid
Sternohyoid

Nerves of the Cervical Plexus

Spinal Segments	Nerve and Distribution
	Great Auricular
C_2–C_3	Skin over the posterior aspect of the ear and the neck
	Lesser Occipital
C_2	Skin of the neck and the scalp posterior and superior to the ear
	Transverse Cervical
C_3–C_4	Skin of the anterior triangle of the neck
	Ansa Cervicalis
C_1–C_4	Five of the extrinsic laryngeal muscles by way of cranial nerve XII
	Phrenic
C_3–C_5	Diaphragm
	Supraclaviculars
C_3–C_4	Skin of the neck and shoulder

Figure 13–10 **The Cervical Plexus.** ATLAS: Plates 3c,d; 18a–c

Figure 13–11 **The Brachial Plexus.** ATLAS: Plates 27a–c; 29b,c; 30

Brachial Plexus: Major Nerves of the Arm and Forearm

Spinal Segments	Nerve and Distribution
	Musculocutaneous
C_5–T_1	Flexor muscles on the arm (biceps brachii, brachialis, and coracobrachialis); sensory from skin over the lateral surface of the forearm through the lateral antebrachial cutaneous nerve
	Radial
C_5–T_1	Many extensor muscles on the arm and forearm (triceps brachii, anconeus, extensor carpi radialis, extensor carpi ulnaris, and brachioradialis); and abductor pollicis by the deep branch; sensory from skin over the posterolateral surface of the limb through the posterior brachial cutaneus nerve, posterior antebrachial cutaneous nerve, and the superficial branch (radial half of the hand)
	Median
C_6–T_1	Flexor muscles on the forearm (flexor carpi radialis and palmaris longus); pronator quadratus and pronator teres; digital flexors (through the anterior interosseous nerve); sensory from skin over the anterolateral surface of the hand
	Ulnar
C_8–T_1	Flexor carpi ulnaris, flexor digitorum profundus, adductor pollicis, and small digital muscles by the deep branch; sensory from skin over medial surface of the hand through the superficial branch

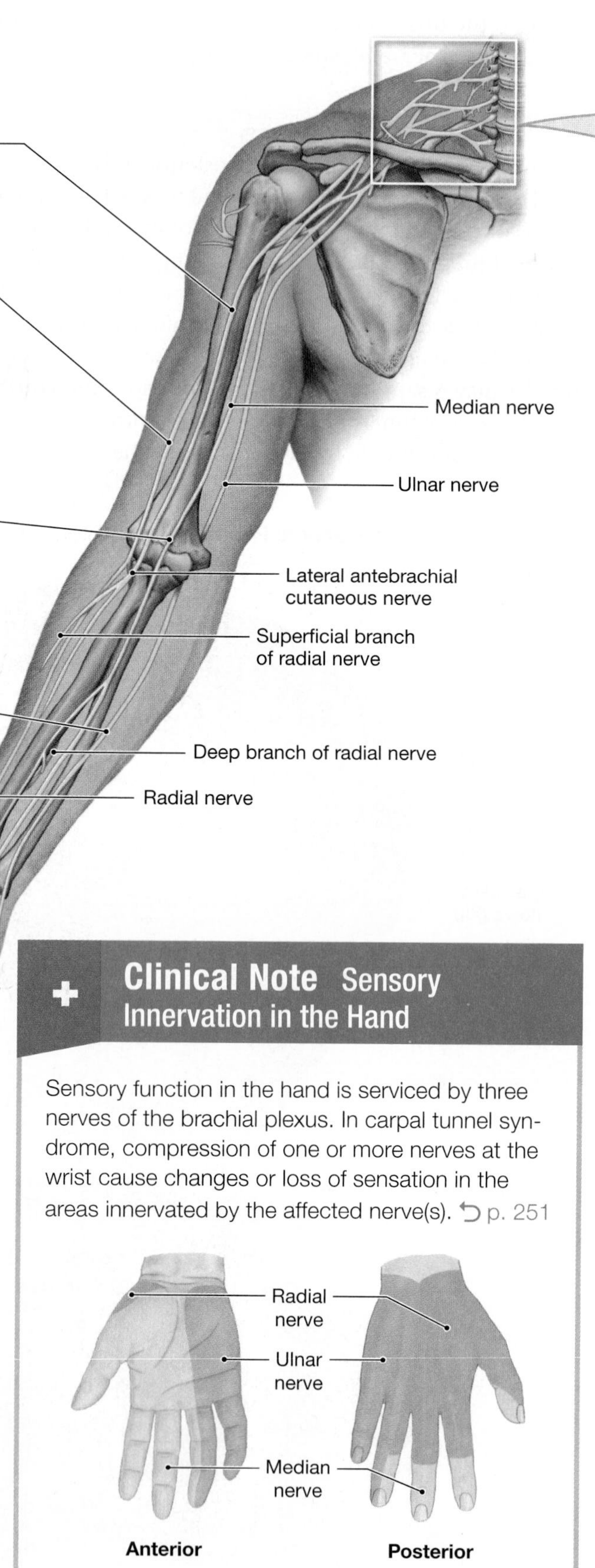
Median nerve
Ulnar nerve
Lateral antebrachial cutaneous nerve
Superficial branch of radial nerve
Deep branch of radial nerve
Radial nerve

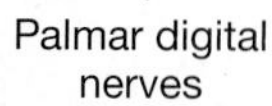
Palmar digital nerves

a Major nerves of the arm, forearm, and hand originating at the right brachial plexus, anterior view

plexus innervate the muscles of the neck and extend into the thoracic cavity, where they control the diaphragm. The **phrenic** (FREN-ik) **nerve** is the major nerve of the cervical plexus. The left and right phrenic nerves supply the diaphragm, a key breathing muscle. Other branches of this nerve plexus are distributed to the skin of the neck and the superior part of the chest.

Tips & Tools

A useful mnemonic for remembering a function of three important cervical nerves is "C3, C4, C5 keep the diaphragm alive."

+ Clinical Note Sensory Innervation in the Hand

Sensory function in the hand is serviced by three nerves of the brachial plexus. In carpal tunnel syndrome, compression of one or more nerves at the wrist cause changes or loss of sensation in the areas innervated by the affected nerve(s). ⮌ p. 251

Radial nerve
Ulnar nerve
Median nerve

Anterior **Posterior**

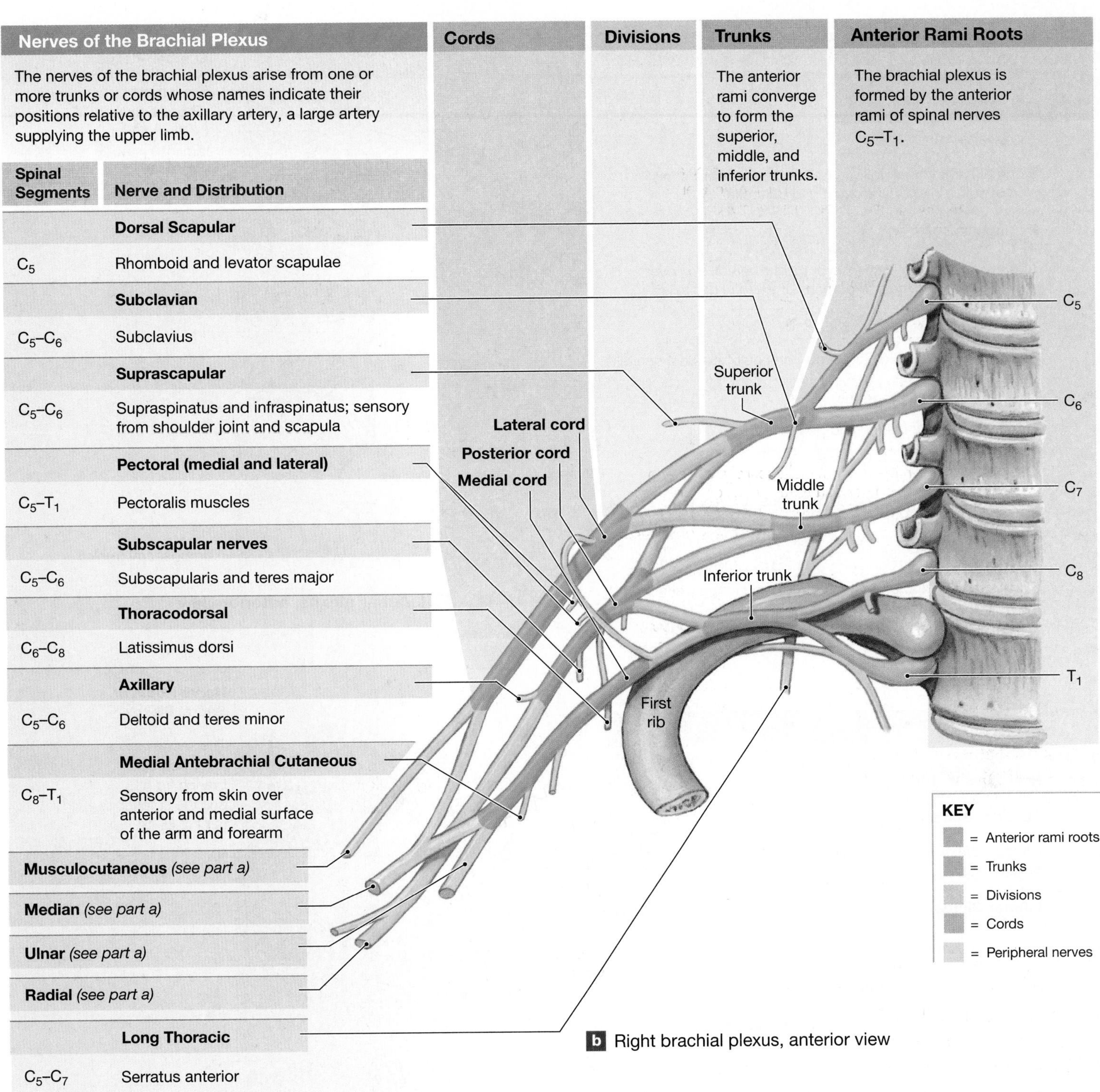
Nerves of the Brachial Plexus
The nerves of the brachial plexus arise from one or more trunks or cords whose names indicate their positions relative to the axillary artery, a large artery supplying the upper limb.
Spinal Segments
Nerve and Distribution
Dorsal Scapular
C5
Rhomboid and levator scapulae
Subclavian
C5–C6
Subclavius
Suprascapular
C5–C6
Supraspinatus and infraspinatus; sensory from shoulder joint and scapula
Pectoral (medial and lateral)
C5–T1
Pectoralis muscles
Subscapular nerves
C5–C6
Subscapularis and teres major
Thoracodorsal
C6–C8
Latissimus dorsi
Axillary
C5–C6
Deltoid and teres minor
Medial Antebrachial Cutaneous
C8–T1
Sensory from skin over anterior and medial surface of the arm and forearm
Musculocutaneous (see part a)
Median (see part a)
Ulnar (see part a)
Radial (see part a)
Long Thoracic
C5–C7
Serratus anterior
Cords
Divisions
Trunks
The anterior rami converge to form the superior, middle, and inferior trunks.
Anterior Rami Roots
The brachial plexus is formed by the anterior rami of spinal nerves C5–T1.
Superior trunk
Lateral cord
Posterior cord
Medial cord
Middle trunk
Inferior trunk
First rib
C5
C6
C7
C8
T1
KEY
= Anterior rami roots
= Trunks
= Divisions
= Cords
= Peripheral nerves

b Right brachial plexus, anterior view

13

The Brachial Plexus

The **brachial plexus** innervates the pectoral girdle and upper limb, with contributions from the anterior rami (also called roots) of spinal nerves C_5–T_1 (Figures 13–9, 13–11). The brachial plexus can also have fibers from C_4, T_2, or both.

The rami of these spinal nerves form trunks and cords before they become peripheral nerves. **Trunks** are large bundles of spinal nerve axons, which become the smaller branches called **cords**. We name both trunks and cords according to their location relative to the *axillary artery*, a large artery supplying the upper limb. Hence, we have *superior, middle,* and *inferior trunks*, and *lateral, medial,* and *posterior cords*. The lateral cord forms the **musculocutaneous nerve** exclusively and, together with the medial cord, contributes to the **median nerve**.

Figure 13–12 The Lumbar and Sacral Plexuses. ATLAS: Plates 70b; 76b; 82b

Lumbar Plexus

Spinal Segments	Nerve and Distribution
	Iliohypogastric
T_{12}–L_1	External and internal oblique and transverse abdominis; skin over the inferior abdomen and buttocks
	Ilio-inguinal
L_1	Abdominal muscles (with iliohypogastric nerve); skin over superior, medial thigh and portions of external genitalia
	Genitofemoral
L_1–L_2	Skin over anteromedial thigh and portions of external genitalia
	Lateral Femoral Cutaneous
L_2–L_3	Skin over anterior, lateral, and posterior thigh
	Femoral
L_2–L_4	Quadriceps femoris, sartorius, pectineus, and iliopsoas; skin of the anteromedial thigh, and medial surface of the leg and foot
	Obturator
L_2–L_4	Gracilis, and adductor magnus, brevis and longus; skin from the medial surface of the thigh

T_{12} subcostal nerve

Nerve Roots of Lumbar Plexus

The lumbar plexus is formed by the anterior rami of T_{12} – L_4.

T_{12}
L_1
L_2
L_3
L_4
L_5
Lumbosacral trunk

a Lumbar plexus, anterior view

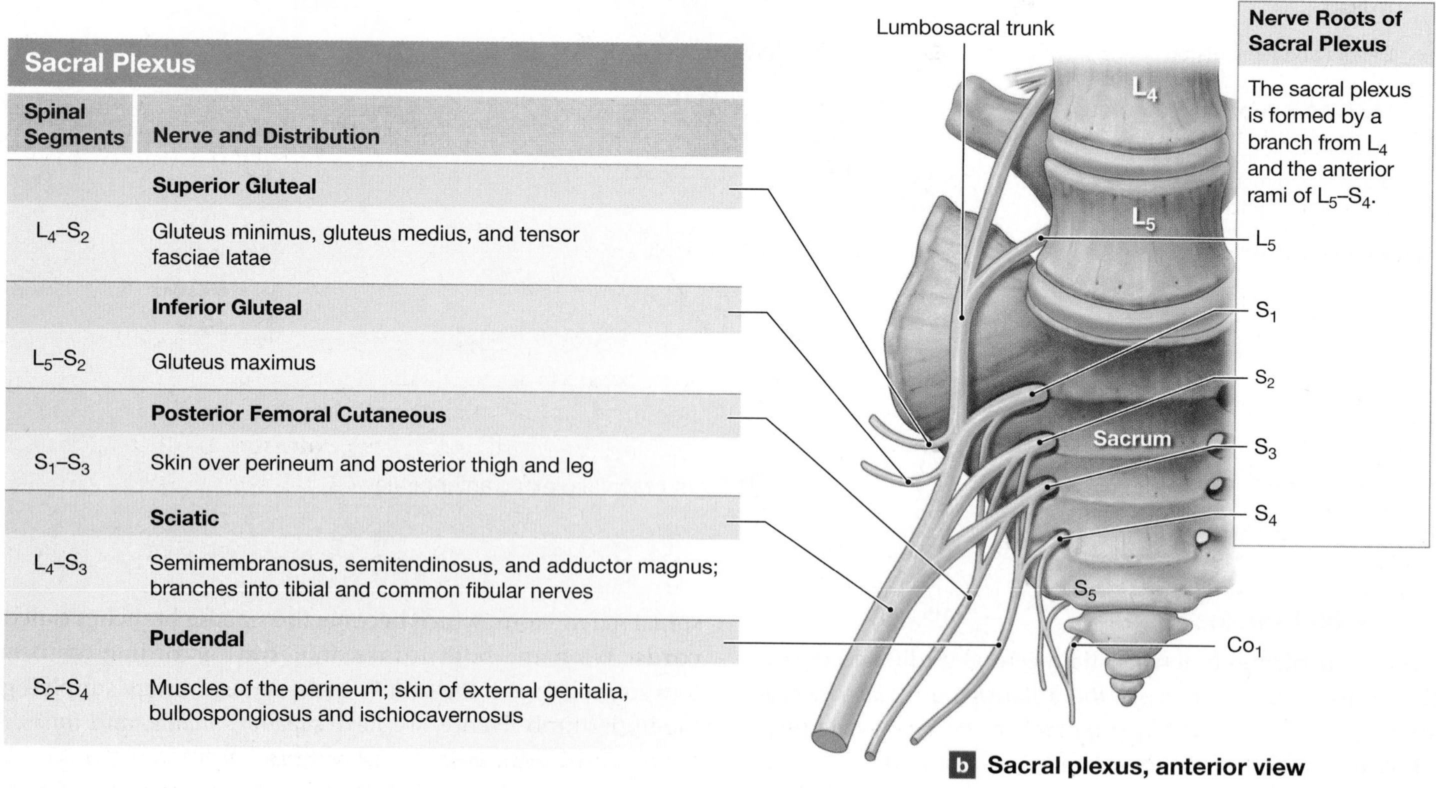

Sacral Plexus

Spinal Segments	Nerve and Distribution
	Superior Gluteal
L_4–S_2	Gluteus minimus, gluteus medius, and tensor fasciae latae
	Inferior Gluteal
L_5–S_2	Gluteus maximus
	Posterior Femoral Cutaneous
S_1–S_3	Skin over perineum and posterior thigh and leg
	Sciatic
L_4–S_3	Semimembranosus, semitendinosus, and adductor magnus; branches into tibial and common fibular nerves
	Pudendal
S_2–S_4	Muscles of the perineum; skin of external genitalia, bulbospongiosus and ischiocavernosus

Lumbosacral trunk
Nerve Roots of Sacral Plexus
The sacral plexus is formed by a branch from L_4 and the anterior rami of L_5–S_4.
L_4
L_5
S_1
S_2
Sacrum
S_3
S_4
S_5
Co_1

b Sacral plexus, anterior view

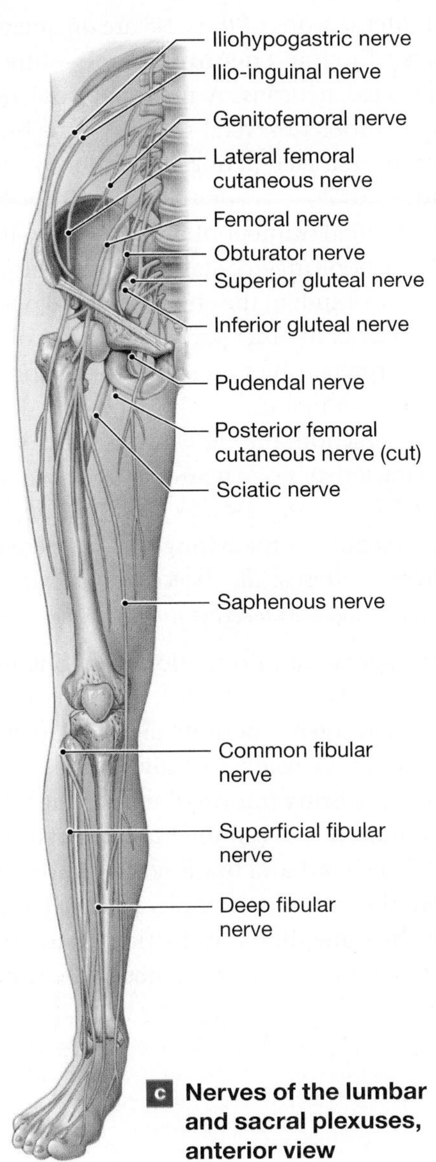
Iliohypogastric nerve
Ilio-inguinal nerve
Genitofemoral nerve
Lateral femoral cutaneous nerve
Femoral nerve
Obturator nerve
Superior gluteal nerve
Inferior gluteal nerve
Pudendal nerve
Posterior femoral cutaneous nerve (cut)
Sciatic nerve
Saphenous nerve
Common fibular nerve
Superficial fibular nerve
Deep fibular nerve

c Nerves of the lumbar and sacral plexuses, anterior view

Clinical Note
Sensory Innervation in the Ankle and Foot

This is the normal distribution of sensory nerves innervating the ankle and foot. Mapping of touch and pain perception can be combined with observations of muscle function to detect nerve damage affecting specific peripheral nerves.

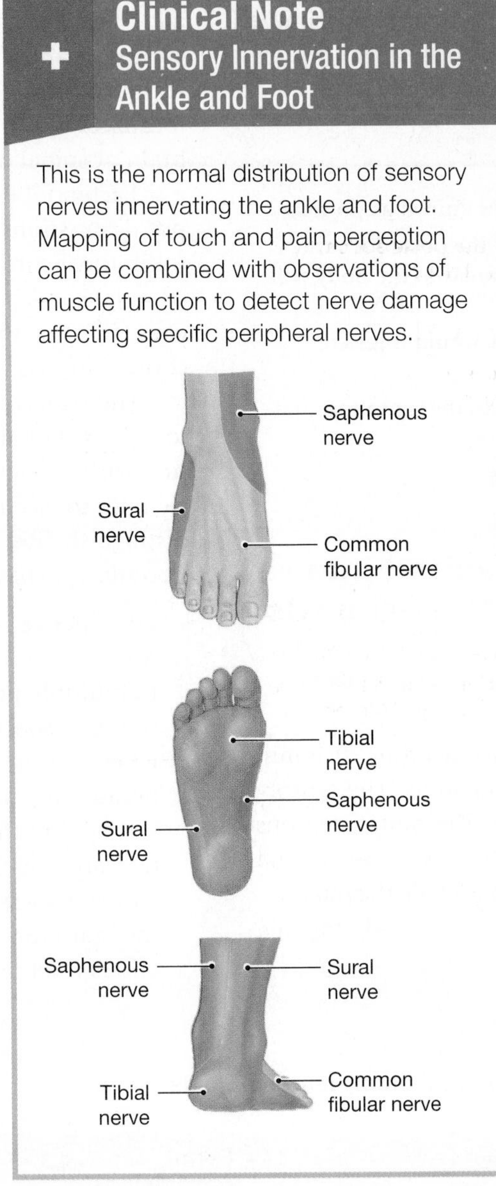
Saphenous nerve
Sural nerve
Common fibular nerve
Tibial nerve
Saphenous nerve
Sural nerve
Saphenous nerve
Sural nerve
Tibial nerve
Common fibular nerve

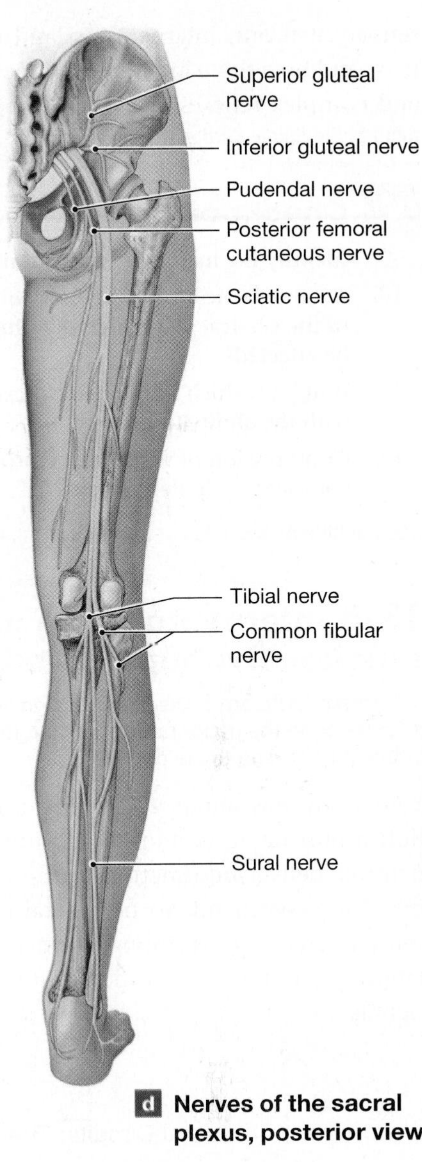
Superior gluteal nerve
Inferior gluteal nerve
Pudendal nerve
Posterior femoral cutaneous nerve
Sciatic nerve
Tibial nerve
Common fibular nerve
Sural nerve

d Nerves of the sacral plexus, posterior view

The **ulnar nerve** is the other major nerve of the medial cord. The posterior cord gives rise to the **axillary nerve** and the **radial nerve**.

The Lumbar and Sacral Plexuses

The **lumbar plexus** and the **sacral plexus** arise from the lumbar and sacral segments of the spinal cord, respectively. The nerves arising at these plexuses innervate the pelvic girdle and lower limbs (Figures 13–9, 13–12).

The lumbar plexus contains axons from the anterior rami of spinal nerves T_{12}–L_4. The major nerves of this plexus are the **genitofemoral nerve**, the **lateral femoral cutaneous nerve**, and the **femoral nerve**.

The sacral plexus contains axons from the anterior rami of spinal nerves L_4–S_4. Two major nerves arise at this plexus: the **sciatic** (sī-AT-ik) **nerve** and the **pudendal nerve**. The sciatic nerve passes posterior to the femur, deep to the long head of the biceps femoris. As it approaches the knee, the sciatic nerve divides into two branches: the **common fibular nerve** (or *common peroneal nerve*) and the **tibial nerve**. The *sural nerve*, formed by branches of the tibial nerve, is a sensory nerve innervating the lateral portion of the foot. A section of this nerve is often removed for use in nerve grafts.

Discussions of motor performance usually make a distinction between the conscious ability to control motor function—something that requires communication and feedback between the brain and spinal cord—and automatic motor responses coordinated entirely within the spinal cord. These automatic responses, called reflexes, are motor responses to specific stimuli. In the rest of this chapter, we look at how sensory neurons, interneurons, and motor neurons interconnect, and how these interconnections can produce both simple and complex reflexes. ATLAS: Embryology Summary 11: The Development of the Spinal Cord and Spinal Nerves

Checkpoint

9. Identify the major networks of nerves known as plexuses.
10. An anesthetic blocks the function of the posterior rami of the cervical spinal nerves. Which areas of the body will be affected?
11. Injury to which of the nerve plexuses would interfere with the ability to breathe?
12. Compression of which nerve produces the sensation that your leg has "fallen asleep"?

See the blue Answers tab at the back of the book.

13-5 Interneurons are organized into functional groups called neuronal pools

Learning Outcome Discuss the significance of neuronal pools, and describe the major patterns of interaction among neurons within and among these pools.

Your body has about 10 million sensory neurons, 0.5 million motor neurons, and 20 *billion* interneurons. The sensory neurons deliver information to the CNS. The motor neurons distribute commands to peripheral effectors, such as skeletal muscles. Interactions among interneurons provide the interpretation, planning, and coordination of incoming and outgoing signals.

The billions of interneurons of the CNS are organized into a few hundred to a few thousand **neuronal pools**—functional groups of interconnected neurons. A neuronal pool may be scattered, involving neurons in several regions of the brain, or localized, with neurons restricted to one specific location in the brain or spinal cord.

Each pool has a limited number of input sources and output destinations. Each pool may contain both excitatory and inhibitory neurons. The output of the entire neuronal pool may stimulate or depress activity in other parts of the brain or spinal cord, affecting the interpretation of sensory information or the coordination of motor commands.

The pattern of interaction among neurons provides clues to the functional characteristics of a neuronal pool. We refer to the "wiring diagrams" in Figure 13–13 as *neural circuits*, just as we refer to electrical circuits in the wiring of a house. We can distinguish five circuit patterns: *divergence*, *convergence*, *serial processing*, *parallel processing*, and *reverberation*.

- **Divergence** is the spread of information from one neuron to several neurons (Figure 13–13a), or from one pool to multiple pools. Divergence permits the broad distribution of a specific input. Considerable divergence occurs when sensory neurons bring information into the CNS, because the information is distributed to neuronal pools throughout the spinal cord and brain. Visual information arriving from the eyes, for example, reaches your consciousness at the same time it is distributed to areas of the brain that control posture and balance at the subconscious level.

Figure 13–13 Neural Circuits: The Organization of Neuronal Pools.

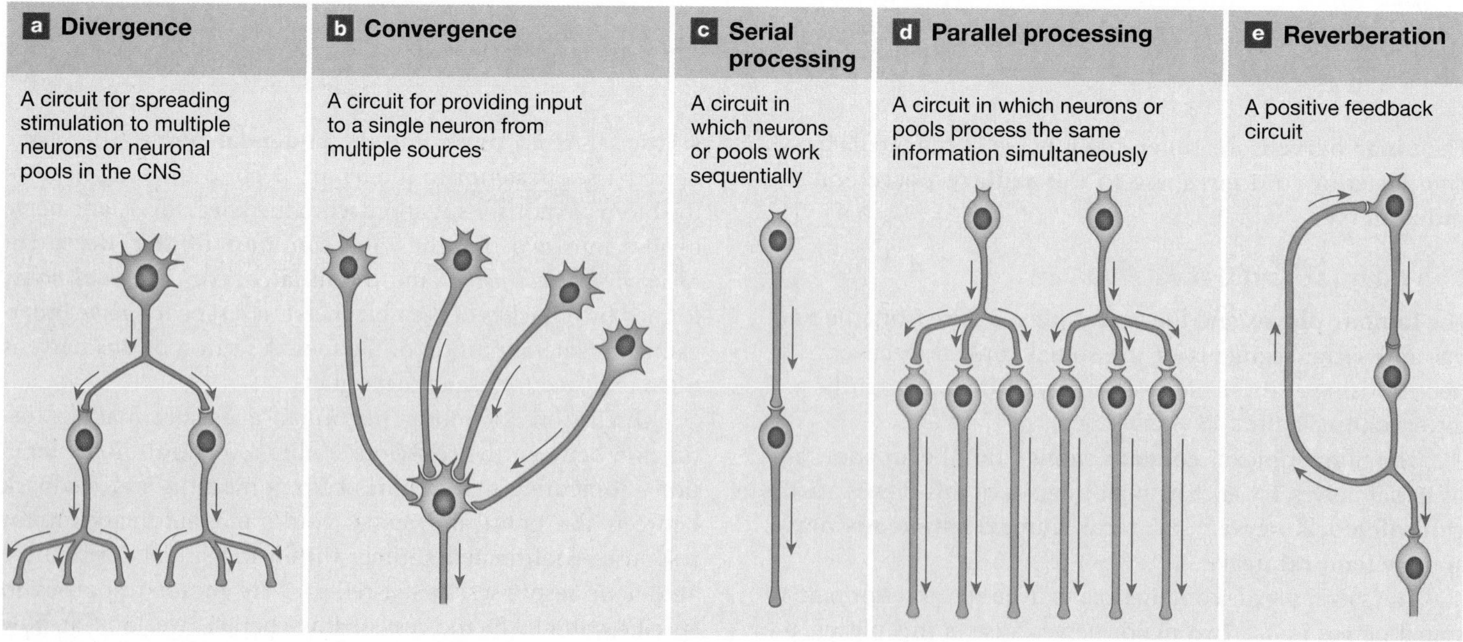
a Divergence
A circuit for spreading stimulation to multiple neurons or neuronal pools in the CNS
b Convergence
A circuit for providing input to a single neuron from multiple sources
c Serial processing
A circuit in which neurons or pools work sequentially
d Parallel processing
A circuit in which neurons or pools process the same information simultaneously
e Reverberation
A positive feedback circuit

- In **convergence**, several neurons synapse on a single postsynaptic neuron (Figure 13–13b). Several patterns of activity in the presynaptic neurons can therefore have the same effect on the postsynaptic neuron. Through convergence, the same motor neurons can be subject to both conscious and subconscious control. For example, the movements of your diaphragm and ribs are now being controlled by your brain at the subconscious level. But you can also consciously control the same motor neurons, as when you take a deep breath and hold it. Two neuronal pools are involved, both synapsing on the same motor neurons.
- In **serial processing**, information is relayed in a stepwise fashion, from one neuron to another or from one neuronal pool to the next (Figure 13–13c). This pattern occurs as sensory information is relayed from one part of the brain to another. For example, pain sensations on their way to your consciousness make stops at two neuronal pools along the pain pathway.
- **Parallel processing** occurs when several neurons or neuronal pools process the same information simultaneously (Figure 13–13d). Divergence must take place before parallel processing can occur. As a result, many responses can occur simultaneously. For example, stepping on a sharp object stimulates sensory neurons that distribute the information to several neuronal pools. As a result of parallel processing, you might withdraw your foot, shift your weight, move your arms, feel the pain, and shout "Ouch!" at about the same time.
- In **reverberation**, collateral branches of axons somewhere along the circuit extend back toward the source of an impulse and further stimulate the presynaptic neurons (Figure 13–13e). Reverberation is like a positive feedback loop involving neurons: Once a reverberating circuit has been activated, it will continue to function until synaptic fatigue or inhibitory stimuli break the cycle. Reverberation can take place within a single neuronal pool, or it may involve a series of interconnected pools. Highly complicated examples of reverberation among neuronal pools in the brain may help maintain consciousness, muscular coordination, and normal breathing.

The functions of the nervous system depend on the interactions among neurons organized in neuronal pools. The most complex neural processing steps take place between the spinal cord and brain. The simplest steps occur between the PNS and the spinal cord. These steps control reflexes that are a bit like Legos: Individually, the reflexes are quite simple, but they can be combined in a great variety of ways to create very complex responses. For this reason, reflexes are the basic building blocks of neural function, as you will see in the next section.

13. **Define** ***neuronal pool.***
14. **List the five neuronal pool circuit patterns.**

See the blue Answers tab at the back of the book.

13-6 The different types of neural reflexes are all rapid, automatic responses to stimuli

Learning Outcome **Describe the steps in a neural reflex, and classify the types of reflexes.**

Conditions inside or outside the body can change quickly and unexpectedly. **Reflexes**, which are rapid and automatic responses to specific stimuli, preserve homeostasis by making quick adjustments in the function of organs or organ systems. The response shows little variability: Each time a particular reflex is activated, it normally produces the same motor response.

In Chapter 1, we introduced the basic functional components involved in all types of homeostatic regulation: a *receptor*, a *control center* (or *integration center*), and an *effector*. ⮌ p. 19 Here we consider *neural reflexes*, in which sensory fibers deliver information from peripheral receptors to an integration center in the CNS, and motor fibers carry motor commands to peripheral effectors. This includes the events that happen when a reflex is triggered, as well as the types of neural reflexes. In Chapter 18, we examine a *neuroendocrine response reflex*, in which hormones in the bloodstream carry the commands to peripheral tissues and organs.

The Reflex Arc

The route followed by nerve impulses to produce a reflex is called a **reflex arc**. Spotlight Figure 13–14a diagrams the five steps in a simple neural reflex.

1 Arrival of a Stimulus and Activation of a Receptor. Reflex arcs begin with activation of a **receptor**, which may be a specialized cell or the dendrites of a sensory neuron. Receptors are sensitive to physical or chemical changes in the body or to changes in the external environment. We introduced the general categories of sensory receptors in Chapter 12. ⮌ p. 395 If you lean on a tack, for example, pain receptors in the palm of your hand are activated. These receptors, the dendrites of sensory neurons, respond to stimuli that cause or accompany tissue damage. (We discuss the link between receptor stimulation and sensory neuron activation further in Chapter 15.)

2 Activation of a Sensory Neuron. When the dendrites are stretched, this causes a graded depolarization that leads to the formation and propagation of action potentials along the axons of the sensory neurons. This information reaches the spinal cord by way of a posterior root. In our example, 1 and 2 involve the same cell. However, the two steps may involve different

SPOTLIGHT

Figure 13–14

Spinal Reflexes

Reflexes are rapid, automatic responses to specific stimuli. They preserve homeostasis by making rapid adjustments in the function of organs or organ systems.

SmartArt

a EVENTS IN A REFLEX ARC

The "wiring" of a single reflex is called a **reflex arc**. It begins at a receptor and ends at a peripheral effector such as a muscle fiber or a gland cell. A simple reflex arc, such as this withdrawal reflex, consists of five steps.

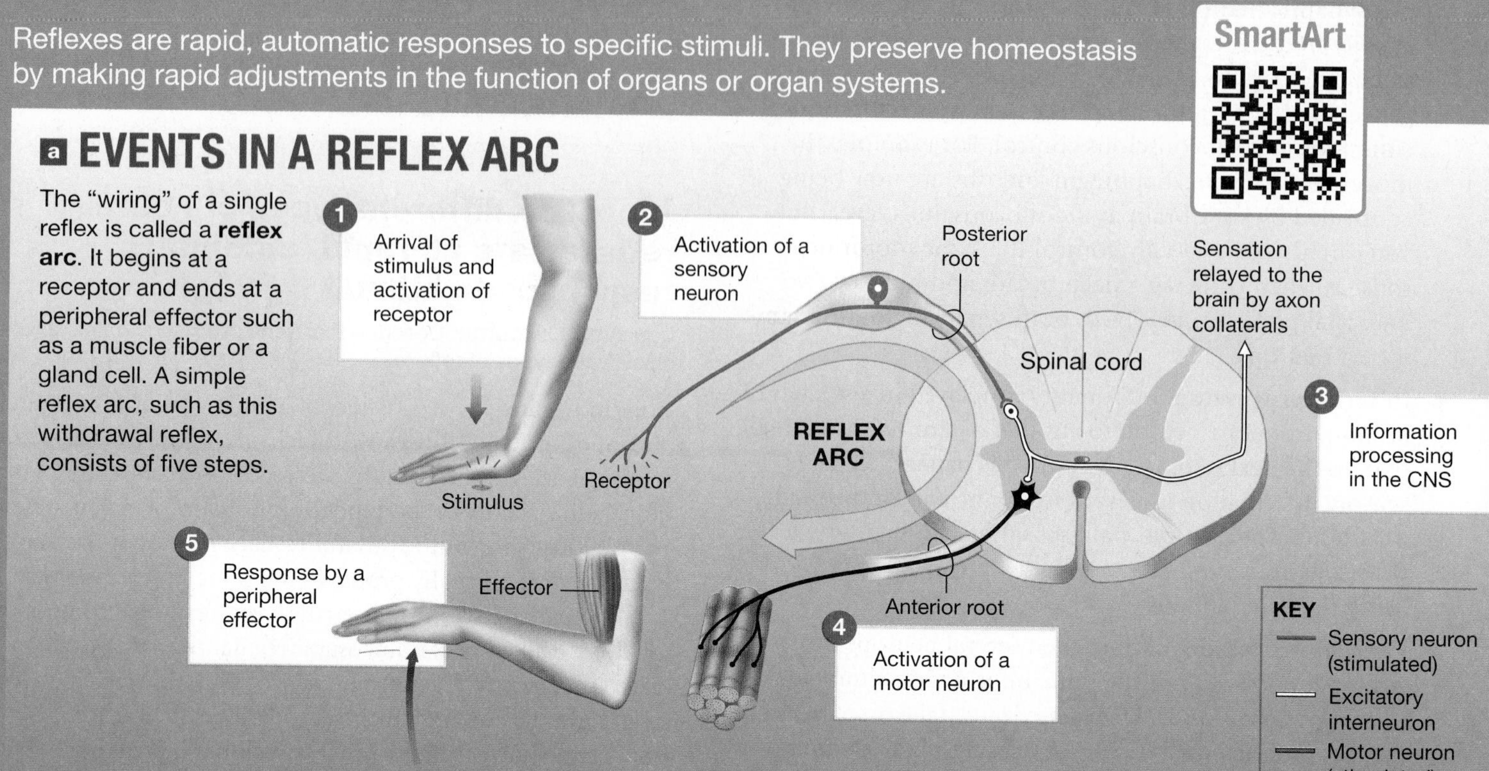

1 Arrival of stimulus and activation of receptor
Stimulus
Receptor
2 Activation of a sensory neuron
Posterior root
Spinal cord
Sensation relayed to the brain by axon collaterals
REFLEX ARC
3 Information processing in the CNS
Anterior root
4 Activation of a motor neuron
5 Response by a peripheral effector
Effector
KEY
Sensory neuron (stimulated)
Excitatory interneuron
Motor neuron (stimulated)

MONOSYNAPTIC REFLEXES

In the simplest reflex arc, a sensory neuron synapses directly on a motor neuron. Because there is only one synapse, it is called a **monosynaptic reflex**. Transmission across a chemical synapse always involves a synaptic delay, but with only one synapse, the delay between the stimulus and the response is minimized.

b Stretch Reflex

The **stretch reflex** is an example of a monosynaptic reflex. Because there is only one synapse, there is little delay between sensory input and motor output. These reflexes control the most rapid motor responses of the nervous system. A stretch reflex provides automatic regulation of skeletal muscle length.

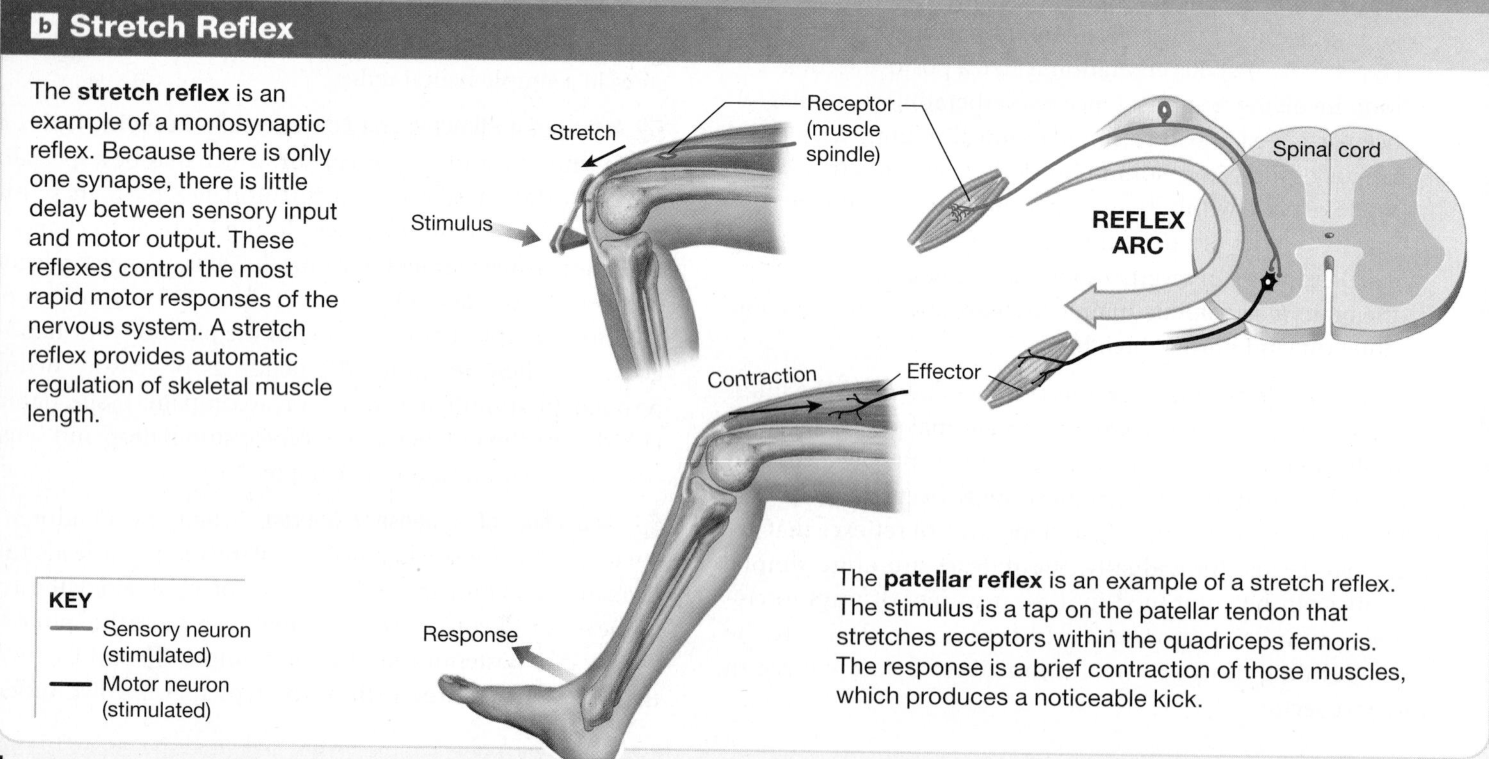

Stretch
Receptor (muscle spindle)
Spinal cord
Stimulus
REFLEX ARC
Contraction
Effector
Response
KEY
Sensory neuron (stimulated)
Motor neuron (stimulated)

The **patellar reflex** is an example of a stretch reflex. The stimulus is a tap on the patellar tendon that stretches receptors within the quadriceps femoris. The response is a brief contraction of those muscles, which produces a noticeable kick.

> Go to MasteringA&P™ > Study Area > Menu > Animations & Videos > A&PFlix A&P > The Stretch Reflex

POLYSYNAPTIC REFLEXES

Polysynaptic reflexes can produce far more complicated responses than monosynaptic reflexes, because the interneurons can control motor neurons that activate several muscle groups simultaneously.

c Withdrawal Reflex

A **withdrawal reflex** moves affected parts of the body away from a stimulus. A **flexor reflex** is an example of a withdrawal reflex that affects the muscles of a limb. In this example, the stimulus of a hot frying pan causes the contraction of the flexor muscles of the arm, yanking the forearm and hand away from the pan. This response occurs while pain sensations are simultaneously ascending to the brain.

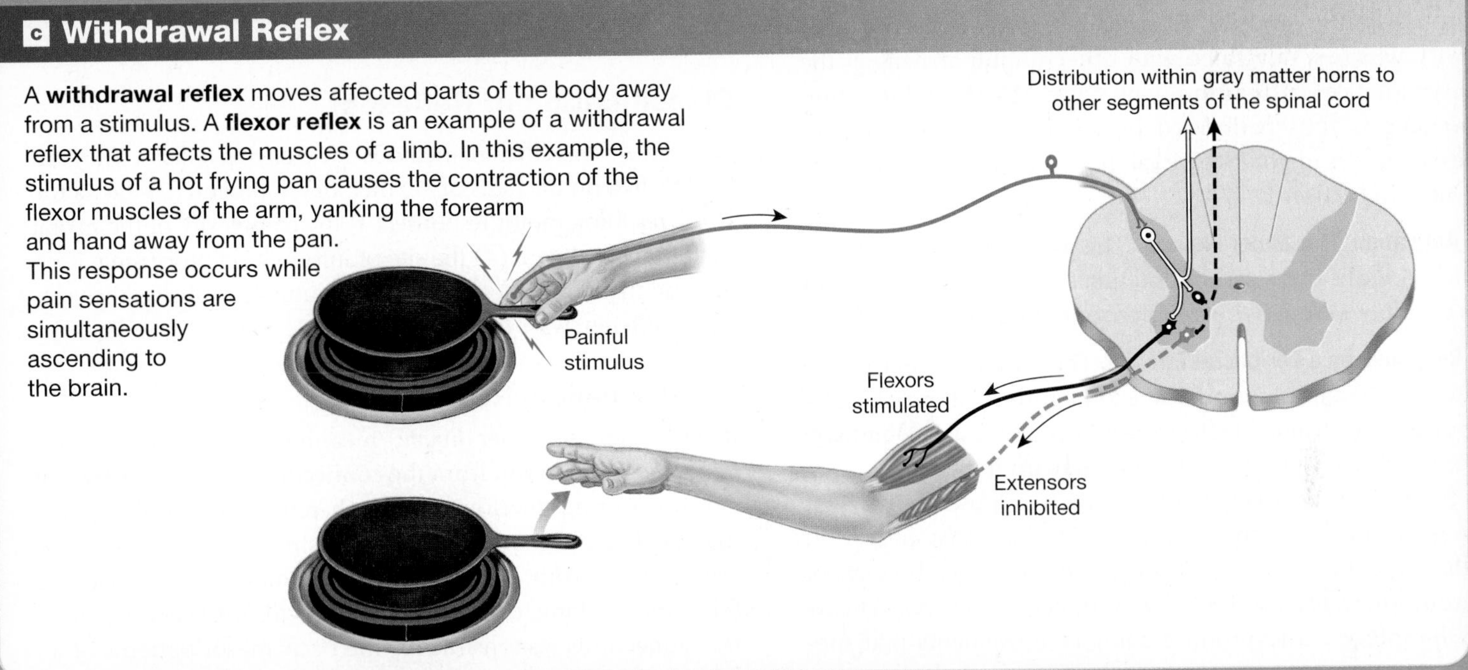
Distribution within gray matter horns to other segments of the spinal cord
Painful stimulus
Flexors stimulated
Extensors inhibited

d Crossed Extensor Reflex

A **crossed extensor reflex** involves a contralateral reflex arc (*contra*, opposite). In other words, a motor response to the stimulus also occurs on a side opposite the stimulus. The crossed extensor reflex complements the flexor reflex, and the two occur simultaneously. When you step on a tack, the flexor reflex pulls the affected foot away from the ground, while the crossed extensor reflex straightens the other leg to support your body weight. In the crossed extensor reflex, interneurons responding to the pain have axons that cross to the other side of the spinal cord. There they stimulate motor neurons that control the extensor muscles of the uninjured leg. As a result, your opposite leg straightens to support the shifting weight. Reverberating circuits use positive feedback to ensure that the movement lasts long enough to be effective. All of this happens without motor commands from higher centers of the brain.

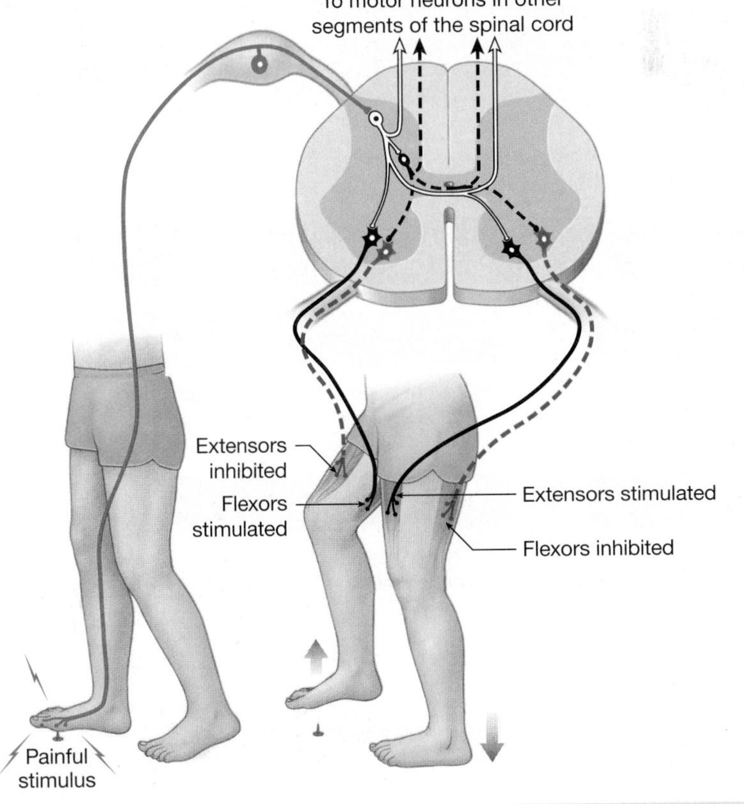
To motor neurons in other segments of the spinal cord
Extensors inhibited
Flexors stimulated
Extensors stimulated
Flexors inhibited
Painful stimulus

KEY

Sensory neuron (stimulated)	Motor neuron (inhibited)
Excitatory interneuron	Inhibitory interneuron
Motor neuron (stimulated)	

cells. For example, reflexes triggered by loud sounds begin when receptor cells in the internal ear release neurotransmitters that stimulate sensory neurons.

3 Information Processing in the CNS. In our example, information processing begins when excitatory neurotransmitter molecules, released by the axon terminal of a sensory neuron, arrive at the postsynaptic membrane of an interneuron. The neurotransmitter produces an excitatory postsynaptic potential (EPSP), which is integrated with other stimuli arriving at the postsynaptic cell at that moment. ⮌ p. 425 The information processing is thus performed by the interneuron. Different types of reflexes might involve varying numbers of interneurons (covered in Section 13-7).

4 Activation of a Motor Neuron. The axons of the stimulated motor neurons carry action potentials into the periphery—in this example, through the anterior root of a spinal nerve.

5 Response by a Peripheral Effector. The release of neurotransmitters by the motor neurons at axon terminals then leads to a response by a peripheral effector—in this case, a skeletal muscle
13 whose contraction pulls your hand away from the tack.

A reflex response generally removes or opposes the original stimulus, and this is an example of *negative feedback*. ⮌ p. 20 In the simplest reflexes, such as the *stretch reflex*, the sensory neuron innervates a motor neuron directly. However, reflexes may be integrated to produce complex movements that support or enhance the primary response. Two examples are given in **Spotlight Figure 13–14**. By opposing potentially harmful changes in the internal or external environment, reflexes play an important role in homeostasis. However, the immediate reflex response is typically not the only response to a stimulus. Reflexes may be integrated to produce complex movements that support or enhance the primary response. The other responses, which are directed by your brain, involve multiple synapses and take longer to organize and coordinate.

Classification of Reflexes

There are several different methods of classifying reflexes. They can be classified on the basis of (1) their development, (2) the nature of the resulting motor response, (3) the complexity of the neural circuit involved, and (4) the site of information processing. Keep in mind that these categories are not mutually exclusive; they represent different ways of describing a single reflex (**Figure 13–15**).

Development of Reflexes

Reflexes can be either *innate*, meaning inborn, or *acquired*. **Innate reflexes** result from the connections that form between neurons during development. Such reflexes generally appear in a predictable sequence, from the simplest reflex responses (withdrawal from pain) to more complex motor patterns (chewing, sucking, or tracking objects with the eyes). The neural connections responsible for the basic motor patterns of an innate reflex are genetically programmed. Examples include the

Figure 13–15 The Classification of Reflexes.

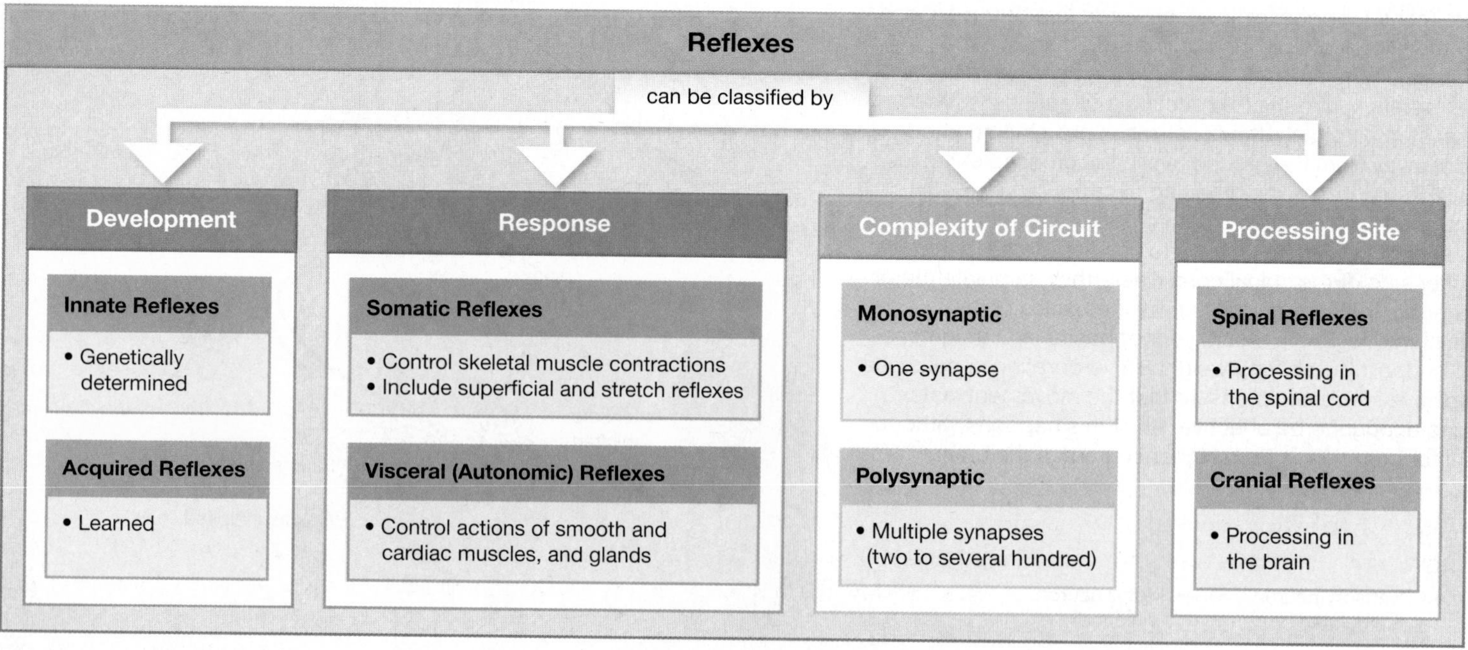
Reflexes
can be classified by
Development
Innate Reflexes
• Genetically determined
Acquired Reflexes
• Learned
Response
Somatic Reflexes
• Control skeletal muscle contractions
• Include superficial and stretch reflexes
Visceral (Autonomic) Reflexes
• Control actions of smooth and cardiac muscles, and glands
Complexity of Circuit
Monosynaptic
• One synapse
Polysynaptic
• Multiple synapses (two to several hundred)
Processing Site
Spinal Reflexes
• Processing in the spinal cord
Cranial Reflexes
• Processing in the brain

? What are the four ways reflexes can be classified?

reflexive removal of your hand from a hot stovetop and blinking when your eyelashes are touched.

More complex, learned motor patterns are called **acquired reflexes**. For example, an experienced driver steps on the brake when trouble appears ahead, or a professional skier makes equally quick adjustments in body position while racing. These motor responses are rapid and automatic, but they were learned rather than preestablished. Repetition enhances such reflexes.

The distinction between innate and acquired reflexes is not absolute. Some people can learn motor patterns more quickly than others. The differences probably have a genetic basis.

Most reflexes, whether innate or acquired, can be modified over time or suppressed through conscious effort. For example, while walking a tightrope over the Grand Canyon, you might ignore a bee sting on your hand. Under other circumstances you would probably withdraw your hand immediately, while shouting and thrashing about as well.

Nature of the Motor Response

Reflexes can be either *somatic* or *visceral*, depending on what type of motor responses they involve. **Somatic reflexes** provide a mechanism for the involuntary control of the muscular system. There are many types of somatic reflexes. For example, *superficial reflexes* are triggered by stimuli at the skin or mucous membranes. *Stretch reflexes* are triggered by the sudden elongation of a tendon, and thus of the muscle to which it attaches. A familiar example is the *patellar*, or *"knee-jerk," reflex* that is usually tested during physical exams. These reflexes are also known as *deep tendon reflexes*. **Visceral reflexes**, or *autonomic reflexes*, control the activities of other systems. We consider somatic reflexes in detail in Section 13-7 and visceral reflexes in Chapter 16.

The movements directed by somatic reflexes are neither delicate nor precise. You might wonder why they exist at all, because we have voluntary control over the same muscles. In fact, somatic reflexes are absolutely vital, primarily because they are *immediate*. Making decisions and coordinating voluntary responses take time. In an emergency—when you slip while walking down a flight of stairs, or accidentally press your hand against a knife edge—any delay increases the likelihood of severe injury. Thus, somatic reflexes provide a rapid response that can be modified later, if necessary, by voluntary motor commands.

Complexity of the Neural Circuit

Reflexes can also be defined as either *monosynaptic* or *polysynaptic* by how many neurons their circuits involve. **Monosynaptic reflexes** involve the simplest reflex arc. In these reflexes, the sensory neuron innervates a motor neuron directly. In that case, the motor neuron performs the information processing. Most reflexes are more complicated, however, and have at least one interneuron between the sensory neuron and the motor neuron. These types are called **polysynaptic reflexes**. Polysynaptic reflexes have a longer delay between stimulus and response. The length of the delay is proportional to the number of synapses involved. Because this classification method for reflexes focuses on functional organization, this is the system that we will use in the rest of the chapter. We discuss details and examples of these types of reflexes in the next section. **Spotlight Figure 13–14** highlights these reflexes.

Site of Information Processing

There is one final classification system for reflexes we should mention briefly, which is based on where they are processed. In **spinal reflexes**, the important interconnections and processing events occur in the spinal cord. We consider reflexes processed in the brain, called **cranial reflexes**, in Chapters 14, 16, and 17.

Checkpoint

15. **Define *reflex*.**
16. **What is the minimum number of neurons in a reflex arc?**
17. **One of the first somatic reflexes to develop is the suck reflex. Which type of reflex is this?**

See the blue Answers tab at the back of the book.

13

13-7 Monosynaptic reflexes produce simple responses, while polysynaptic reflexes can produce complex behaviors

Learning Outcome **Distinguish among the types of motor responses produced by various reflexes, and explain how reflexes interact to produce complex behaviors.**

As we discussed, spinal reflexes range in complexity from simple monosynaptic reflexes involving a single segment of the spinal cord to polysynaptic reflexes that involve many segments. In the most complicated polysynaptic reflexes, called **intersegmental reflex arcs**, many segments interact to produce a coordinated, highly variable motor response. Let's take a closer look at both of these types of reflexes.

Monosynaptic Reflexes

In monosynaptic reflexes, there is little delay between sensory input and motor output. These reflexes control the most rapid, *stereotyped* (preexisting, mechanically repetitive) *motor responses* of the nervous system to specific stimuli. Let's look at how a common example of these reflexes works.

The Stretch Reflex and Muscle Spindles

The best-known monosynaptic reflex is the **stretch reflex**, which automatically regulates skeletal muscle length. An example is the knee-jerk, or **patellar, reflex**. When a physician taps your patellar tendon with a reflex hammer, receptors in the quadriceps femoris are stretched (see **Spotlight Figure 13–14b**). The distortion of the receptors in turn stimulates sensory neurons that extend into the spinal cord, where they synapse on motor neurons that control the motor units in the stretched muscle. This leads to a reflexive contraction of the stretched muscle that extends the knee in a brief kick.

To summarize: The stimulus (increasing muscle length) activates a sensory neuron, which triggers an immediate motor response (contraction of the stretched muscle) that counteracts the stimulus. The entire reflex is completed within 20–40 msec because the action potentials traveling toward and away from the spinal cord are propagated along large, myelinated Type A fibers.

Stretch reflexes are found all over the body, in places where we need automatic responses in order to maintain our upright stance and posture. Let's look in detail at the function of these reflexes, including the sensory receptors involved in the stretch reflex, which are called **muscle spindles**.

Muscle Spindle Structure. Each muscle spindle consists of a bundle of small, specialized skeletal muscle fibers called **intrafusal muscle fibers** (Figure 13–16). The muscle spindle is surrounded by larger skeletal muscle fibers, called **extrafusal muscle fibers**. These fibers are responsible for the resting muscle tone and, at greater levels of stimulation, for the contraction of the entire muscle.

Both sensory and motor neurons innervate each intrafusal fiber. The dendrites of the sensory neuron spiral around the intrafusal fiber in a central sensory region. Axons from spinal motor neurons form neuromuscular junctions on either end of this fiber. Motor neurons innervating intrafusal fibers are called **gamma motor neurons**. Their axons are called **gamma efferents**.

Figure 13–16 A Muscle Spindle. The location, structure, and innervation of a muscle spindle.

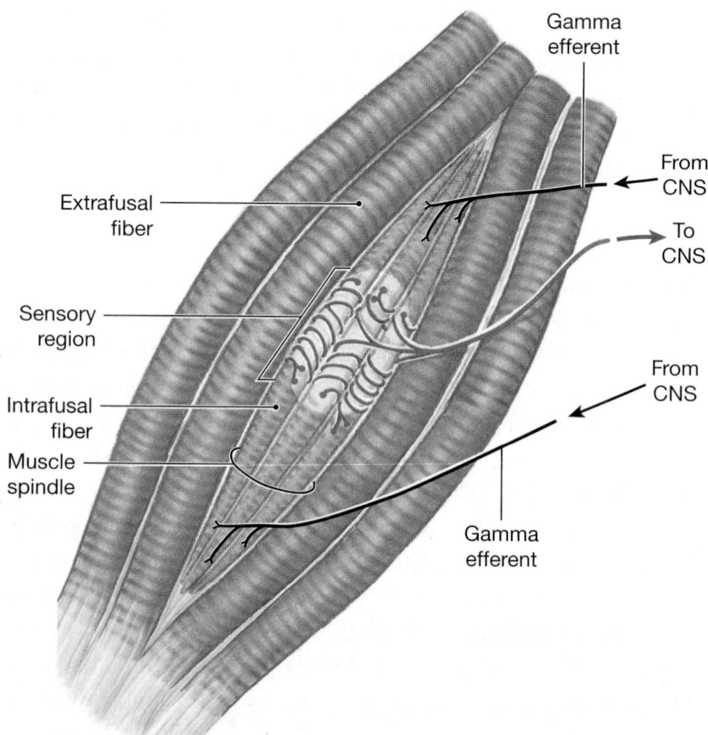
Gamma efferent
From CNS
Extrafusal fiber
To CNS
Sensory region
From CNS
Intrafusal fiber
Muscle spindle
Gamma efferent

An intrafusal fiber has one set of myofibrils at each end. These myofibrils run from the end of the intrafusal fiber only to the sarcolemma in the central region that is closely monitored by the sensory neuron.

Function of Muscle Spindles. The stretching of muscle spindles produces a sudden burst of activity in the sensory neurons that monitor them. This in turn leads to stimulation of motor neurons that control the motor units in the stretched muscle. The result is rapid muscle shortening, which returns the muscle spindles to their resting length. The rate of action potential generation in the sensory neurons then decreases, causing a drop in muscle tone to the resting level.

Now let's see how the muscle spindle normally functions as a sensory receptor and look at its effects on the surrounding extrafusal fibers. The sensory neuron is always active, conducting impulses to the CNS. The axon enters the CNS in a posterior root and synapses on motor neurons in the anterior horn of the spinal cord. Axon collaterals distribute the information to the brain, providing information about the muscle spindle. Stretching the central portion of the intrafusal fiber distorts the dendrites and stimulates the sensory neuron, increasing the frequency of action potential generation. Compressing the central portion inhibits the sensory neuron, decreasing the frequency of action potential generation.

The axon of the sensory neuron synapses on CNS motor neurons that control the extrafusal muscle fibers of the same muscle. An increase in sensory neuron stimulation, caused by stretching of the intrafusal fiber, increases stimulation to the motor neuron controlling the surrounding extrafusal fibers, so muscle tone increases. This increase provides automatic resistance that reduces the chance of muscle damage due to overstretching. The patellar reflex and similar reflexes serve this function.

A decrease in the stimulation of the sensory neuron, due to compression of the intrafusal fiber, leads to a decrease in the stimulation of the motor neuron controlling the surrounding extrafusal fibers, so muscle tone decreases. This decrease reduces resistance to the movement under way. For example, if your elbow is flexed and you let gravity extend it, the triceps brachii, which is compressed by this movement, relaxes.

Now let's turn to the role of the gamma efferents, which enable the CNS to adjust the sensitivity of the muscle spindle. Gamma efferents play a vital role whenever voluntary contractions change the length of a muscle. Impulses arriving over gamma efferents cause the contraction of myofibrils in the intrafusal fibers as the biceps brachii shortens. The myofibrils pull on the sarcolemma in the central portion of the intrafusal fiber—the region monitored by the sensory neuron—until that membrane is stretched to its normal resting length. As a result, the muscle spindles remain sensitive to any externally imposed changes in muscle length. For this reason, if someone drops a ball into your palm when your elbow is partially flexed, the muscle spindles automatically adjust the muscle tone to compensate for the increased load.

Postural Reflexes

Many of our body's reflexes, including both stretch reflexes and complex polysynaptic reflexes, are **postural reflexes**—reflexes that help us maintain a normal upright posture. Standing, for example, involves a cooperative effort by many muscle groups. Some of these muscles work in opposition to one another, exerting forces that keep the body's weight balanced over the feet. If your body leans forward, stretch receptors in your calf muscles are stimulated. Those muscles then respond by contracting, returning your body to an upright position. If the muscles overcompensate and your body begins to lean back, your calf muscles relax. But then stretch receptors in muscles of your shins and thighs are stimulated, and the problem is corrected immediately.

Postural muscles generally maintain a firm muscle tone and have extremely sensitive stretch receptors. As a result, very fine adjustments are continually being made, and you are not aware of the cycles of contraction and relaxation that occur.

Polysynaptic Reflexes

Polysynaptic reflexes can produce far more complicated responses than can monosynaptic reflexes. One reason is that the interneurons involved can control several muscle groups. Moreover, some of these interneurons release excitatory neurotransmitters (*excitatory interneurons*), and others release inhibitory neurotransmitters (*inhibitory interneurons*). This allows these interneurons to produce either excitatory or inhibitory postsynaptic potentials (EPSPs or IPSPs) at CNS motor nuclei. As a result, the response can involve the stimulation of some muscles and the inhibition of others.

Let's explore some of the different examples of polysynaptic reflexes.

The Tendon Reflex

The **tendon reflex** monitors the external tension produced during a muscular contraction and prevents tearing of the tendons. The sensory receptors for this reflex are *Golgi tendon organs*. They are stimulated when the collagen fibers are stretched to a dangerous degree. These receptors activate sensory neurons that stimulate inhibitory interneurons in the spinal cord. These interneurons in turn innervate the motor neurons controlling the skeletal muscle. The greater the tension in the tendon, the greater the inhibitory effect on the motor neurons. As a result, a skeletal muscle generally cannot develop enough tension to tear its tendons.

Withdrawal Reflexes

Withdrawal reflexes move affected parts of the body away from a stimulus. Painful stimuli trigger the strongest withdrawal reflexes, but these reflexes are also sometimes initiated by the stimulation of touch receptors or pressure receptors.

The **flexor reflex**, a representative withdrawal reflex, affects the muscles of a limb (see **Spotlight Figure 13–14c**). Recall from Chapters 9 and 11 that flexion reduces the angle between two articulating bones, and that the contractions of flexor muscles perform this movement. ⮌ pp. 274, 342 If you grab an unexpectedly hot pan on the stove, a dramatic flexor reflex will occur. When the pain receptors in your hand are stimulated, the sensory neurons activate interneurons in the spinal cord that stimulate motor neurons in the anterior horns. The result is a contraction of flexor muscles that yanks your hand away from the stove.

When a specific muscle contracts, opposing muscles must relax to permit the movement. For example, flexor muscles that bend the elbow (such as the biceps brachii) are opposed by extensor muscles (such as the triceps brachii) that straighten it out. A potential conflict exists: In theory, the contraction of a flexor muscle should trigger a stretch reflex in the extensors that would cause them to contract, opposing the movement. Interneurons in the spinal cord prevent such competition through **reciprocal inhibition**. When one set of motor neurons is stimulated, those neurons that control antagonistic muscles are inhibited. The term *reciprocal* refers to the fact that the system works both ways. When the flexors contract, the extensors relax, and when the extensors contract, the flexors relax.

Withdrawal reflexes are much more complex than any monosynaptic reflex. They also show tremendous versatility, because the sensory neurons activate many pools of interneurons. If the stimuli are strong, interneurons will carry excitatory and inhibitory impulses up and down the spinal cord, affecting motor neurons in many segments. The end result is always the same: a coordinated movement away from the stimulus. But the distribution of the effects and the strength and character of the motor responses depend on the intensity and location of the stimulus.

Mild discomfort might provoke a brief contraction in muscles of your hand and wrist. More powerful stimuli would produce coordinated muscular contractions affecting the positions of your hand, wrist, forearm, and arm. Severe pain would also stimulate contractions of your arm, shoulder, and trunk muscles. These contractions could last for several seconds, due to the activation of reverberating circuits. In contrast, monosynaptic reflexes are invariable and brief.

Crossed Extensor Reflexes

The stretch, tendon, and withdrawal reflexes involve *ipsilateral* (*ipsi*, same + *lateral*, side) *reflex arcs*: The sensory stimulus and the motor response occur on the same side of the body. The crossed extensor reflex involves a *contralateral reflex arc* (*contra*, opposite), because the motor response occurs on the side opposite the stimulus. Please read the information on this type of reflex in **Spotlight Figure 13–14d** before moving on.

General Characteristics of Polysynaptic Reflexes

Polysynaptic reflexes range in complexity from a simple tendon reflex to the complex and variable reflexes associated with standing, walking, and running. Yet all polysynaptic reflexes share the following basic characteristics:

- *They Involve Pools of Interneurons.* Processing takes place in pools of interneurons before motor neurons are activated.

The result may be excitation or inhibition. The tendon reflex inhibits motor neurons, and the flexor and crossed extensor reflexes direct specific muscle contractions.

- *They Involve More Than One Spinal Segment.* The interneuron pools extend across spinal segments and may activate muscle groups in many parts of the body.
- *They Involve Reciprocal Inhibition.* Reciprocal inhibition coordinates muscular contractions and reduces resistance to movement. In the flexor and crossed extensor reflexes, the contraction of one muscle group is associated with the inhibition of opposing muscles.
- *They Have Reverberating Circuits That Prolong the Reflexive Motor Response.* Positive feedback between interneurons that innervate motor neurons and the processing pool maintains the stimulation even after the initial stimulus has faded.
- *Several Reflexes May Cooperate to Produce a Coordinated, Controlled Response.* As a reflex movement gets under way, antagonistic reflexes are inhibited. For example, during the stretch reflex, antagonistic muscles are inhibited. In contrast, in the tendon reflex, antagonistic muscles are stimulated. In complex polysynaptic reflexes, commands may be distributed along the length of the spinal cord, producing a well-coordinated response.

✓ Checkpoint

18. **Identify the basic characteristics of polysynaptic reflexes.**
19. **For the patellar (knee-jerk) reflex, how would the stimulation of the muscle spindle by gamma motor neurons affect the speed of the reflex?**
20. **A weight lifter is straining to lift a 200-kg barbell above his head. Shortly after he lifts it to chest height, his muscles appear to relax and he drops the barbell. Which reflex has occurred?**
21. **During a withdrawal reflex of the foot, what happens to the limb on the side opposite the stimulus? What is this response called?**

See the blue Answers tab at the back of the book.

13-8 The brain can affect spinal cord–based reflexes

Learning Outcome Explain how higher brain centers control and modify reflex responses.

Reflex motor behaviors happen automatically, without instructions from higher brain centers. However, higher centers can have a profound effect on the performance of a reflex. Processing centers in the brain can facilitate or inhibit reflex motor patterns based in the spinal cord. Descending tracts originating in the brain synapse on interneurons and motor neurons throughout the spinal cord. These synapses are continuously active, producing EPSPs or IPSPs at the postsynaptic membrane. Let's find out how this process works.

Voluntary Movements and Reflex Motor Patterns

Spinal reflexes produce consistent, stereotyped motor patterns that are triggered by specific external stimuli. However, the same motor patterns can also be activated as needed by centers in the brain. By making use of these preexisting patterns, relatively few descending fibers can control complex motor functions.

For example, neuronal pools in the spinal cord direct the motor patterns for walking, running, and jumping. The descending pathways from the brain provide appropriate facilitation, inhibition, or "fine-tuning" of the established patterns. This is a very efficient system that is similar to an order given in a military drill: A single command triggers a complex, predetermined sequence of events.

Motor control involves a series of interacting levels. At the lowest level are monosynaptic reflexes that are rapid, but stereotyped and relatively inflexible. At the highest level, centers in the brain can modulate or build on reflexive motor patterns.

Reinforcement and Inhibition

A single EPSP may not depolarize the postsynaptic neuron sufficiently to generate an action potential, but it does make that neuron more sensitive to other excitatory stimuli. We introduced this process of *facilitation* in Chapter 12. Alternatively, an IPSP makes the neuron less responsive to excitatory stimulation, through the process of *inhibition.* p. 425 By stimulating excitatory or inhibitory interneurons within the brainstem or spinal cord, higher centers can adjust the sensitivity of reflexes by creating EPSPs or IPSPs at the motor neurons involved in reflex responses.

When many of the excitatory synapses are chronically active, the postsynaptic neuron can enter a state of generalized facilitation. This facilitation of reflexes can result in **reinforcement**, an enhancement of spinal reflexes. Reinforcement techniques are useful in clinical situations. For example, a stimulus may fail to elicit a particular reflex response during a clinical exam. There can be many reasons for the failure. The person may be consciously suppressing the response, the nerves involved may be damaged, or there may be underlying problems inside the CNS.

The clinician may then attempt the reflex procedure again, this time using reinforcement. For example, a method used to overemphasize the patellar reflex is called the Jendrassik maneuver. To do this, the subject hooks the hands together by interlocking the fingers and then tries to pull the hands apart while a light tap is applied to the patellar tendon. This reinforcement produces a big kick rather than a twitch.

Reinforced reflexes are usually too strong to suppress consciously. If the reflex still fails to appear, the likelihood of nerve or CNS damage is increased. The clinician may then order more sophisticated tests, such as nerve conduction studies or scans.

Other descending fibers have an inhibitory effect on spinal reflexes. In adults, stroking the lateral sole of the foot produces a curling of the toes, called a **plantar reflex**, after about a 1-second delay (Figure 13–17a). In contrast, stroking an infant's foot on the lateral sole produces a fanning of the toes known as the **Babinski reflex** (Figure 13–17b). This response disappears around age 2 as descending motor pathways develop. If either the higher centers or the descending tracts are damaged, the Babinski sign reappears in an adult. As a result, clinicians often test this reflex when CNS injury is suspected.

Figure 13–17 **The Plantar Reflex and Babinski Reflex.**

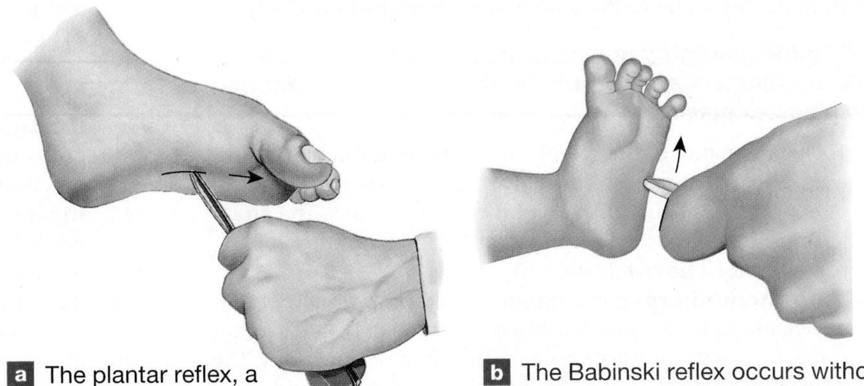

a The plantar reflex, a curling of the toes, is seen in healthy adults.

b The Babinski reflex occurs without descending inhibition. It is normal in infants, but pathological in adults.

Tips & Tools

Facilitation and inhibition are similar to what happens when a symphony conductor raises or lowers one hand to control the music's volume while keeping the rhythm going with the baton hand: The basic pattern of beats doesn't change, but the loudness does.

Checkpoint

22. **Define *reinforcement* as it pertains to spinal reflexes.**
23. **After injuring her back, Tina exhibits a Babinski reflex. What does this imply about Tina's injury?**

See the blue Answers tab at the back of the book.

13 Chapter Review

Study Outline

An Introduction to the Spinal Cord, Spinal Nerves, and Spinal Reflexes p. 434

1. The nervous system has input pathways that route sensations, and processing centers that prioritize and distribute information.

13-1 This text's coverage of the nervous system parallels its simple-to-complex levels of organization p. 434

2. The CNS consists of the brain and spinal cord; the PNS includes cranial nerves and spinal nerves. *(Figure 13–1)*

13-2 The spinal cord is surrounded by three meninges and has spinal nerve roots p. 435

3. The adult spinal cord includes two localized **enlargements**, which provide innervation to the limbs. The spinal cord has 31 segments, each associated with a pair of **posterior roots** or *dorsal roots* and a pair of **anterior roots** or *ventral roots*. *(Figure 13–2)*
4. The **filum terminale** (a strand of fibrous tissue) originates at the **conus medullaris** and ultimately becomes part of the **coccygeal ligament**. *(Figure 13–2)*
5. **Spinal nerves** are **mixed nerves**: They contain both afferent (sensory) and efferent (motor) fibers.

6. The three **spinal meninges** provide physical stability and shock absorption for nervous tissues of the spinal cord; the **cranial meninges** surround the brain. *(Figure 13–3)*
7. The **dura mater**, the outer meningeal layer, covers the spinal cord; inferiorly, it tapers into the **coccygeal ligament**. The **epidural space** separates the dura mater from the walls of the vertebral canal. *(Figures 13–3, 13–4)*
8. Interior to the inner surface of the dura mater are the **subdural space**, the **arachnoid mater** (the second meningeal layer), and the **subarachnoid space**. The subarachnoid space contains **cerebrospinal fluid (CSF)**, which acts as a shock absorber and a diffusion medium for dissolved gases, nutrients, chemical messengers, and waste products. *(Figures 13–3, 13–4)*

9. The **pia mater**, a meshwork of elastic and collagen fibers, is the innermost meningeal layer. **Denticulate ligaments** extend from the pia mater to the dura mater. *(Figures 13–3, 13–4)*

13-3 Spinal cord gray matter integrates information and initiates commands, and white matter carries information from place to place p. 440

10. The white matter of the spinal cord contains myelinated and unmyelinated axons, and the gray matter contains cell bodies of neurons and neuroglia and unmyelinated axons. The projections of gray matter toward the outer surface of the spinal cord are called **horns**. *(Figure 13–5)*
11. The **posterior horns** contain somatic and visceral sensory nuclei; nuclei in the **anterior horns** function in somatic motor control. The **lateral horns** contain visceral motor neurons. The **gray commissures** contain axons that cross from one side of the spinal cord to the other. *(Figure 13–5)*
12. The white matter can be divided into six **columns**, each of which contains **tracts**. **Ascending tracts** relay information from the spinal cord to the brain, and **descending tracts** carry information from the brain to the spinal cord. *(Figure 13–5)*

13-4 Spinal nerves extend to form peripheral nerves, sometimes forming plexuses along the way; these nerves carry sensory and motor information p. 442

13. There are 31 pairs of spinal nerves. Each has an **epineurium** (outermost layer), a **perineurium**, and an **endoneurium** (innermost layer). *(Figure 13–6)*
14. Each pair of spinal nerves monitors a region of the body surface called a **dermatome**. *(Figure 13–7)*
15. A typical spinal nerve has a **white ramus communicans** (containing myelinated axons), a **gray ramus communicans** (containing unmyelinated fibers that innervate glands and smooth muscles in the body wall or limbs), a **posterior ramus** (providing sensory and motor innervation to the skin and muscles of the back), and an **anterior ramus** (supplying the ventrolateral body surface, structures in the body wall, and the limbs). *(Spotlight Figure 13–8)*
16. A complex, interwoven network of nerves is a **nerve plexus**. The four large plexuses are the **cervical plexus**, the **brachial plexus**, the **lumbar plexus**, and the **sacral plexus**. *(Figures 13–9 to 13–12)*

13-5 Interneurons are organized into functional groups called neuronal pools p. 452

17. The body has sensory neurons, which deliver information to the CNS; motor neurons, which distribute commands to peripheral effectors; and interneurons, which interpret information and coordinate responses.
18. An organized, functional group of interconnected interneurons is a **neuronal pool**.
19. The neural circuit patterns are **divergence**, **convergence**, **serial processing**, **parallel processing**, and **reverberation**. *(Figure 13–13)*

13-6 The different types of neural reflexes are all rapid, automatic responses to stimuli p. 453

20. A *neural reflex* involves sensory fibers delivering information to the CNS, and motor fibers carrying commands to the effectors by the PNS.
21. A **reflex arc** is the neural "wiring" of a single reflex. *(Spotlight Figure 13–14)*
22. The five steps involved in a neural reflex are (1) the arrival of a stimulus and activation of a receptor, (2) the activation of a sensory neuron, (3) information processing in the CNS, (4) the activation of a motor neuron, and (5) a response by an effector. *(Spotlight Figure 13–14)*
23. Reflexes are classified according to (1) their development, (2) the nature of the resulting motor response, (3) the complexity of the neural circuit involved, and (4) the site of information processing. *(Figure 13–15)*
24. **Innate reflexes** result from the genetically determined connections that form between neurons during development. **Acquired reflexes** are learned and typically are more complex.
25. **Somatic reflexes** control skeletal muscles; **visceral reflexes** (*autonomic reflexes*) control the activities of other systems.
26. In a **monosynaptic reflex**—the simplest reflex arc—a sensory neuron synapses directly on a motor neuron, which acts as the processing center. A **polysynaptic reflex** has at least one interneuron between the sensory afferent and the motor efferent, and there is a longer delay between stimulus and response. *(Spotlight Figure 13–14)*
27. Reflexes processed in the brain are **cranial reflexes**. In a **spinal reflex**, the important interconnections and processing events occur in the spinal cord.

13-7 Monosynaptic reflexes produce simple responses, while polysynaptic reflexes can produce complex behaviors p. 457

28. Spinal reflexes range from simple monosynaptic reflexes to more complex polysynaptic and **intersegmental reflexes**, in which many spinal segments interact to produce a coordinated motor response.
29. The **stretch reflex** (such as the **patellar**, or **knee-jerk**, **reflex**) is a monosynaptic reflex that automatically regulates skeletal muscle length and muscle tone. The sensory receptors involved are **muscle spindles**. *(Spotlight Figure 13–14; Figure 13–16)*
30. A **postural reflex** maintains a person's normal upright posture.
31. Polysynaptic reflexes can produce more complicated responses than can monosynaptic reflexes. Examples include the **tendon reflex** (which monitors the tension produced during muscular contractions and prevents damage to tendons) and **withdrawal reflexes** (which move affected portions of the body away from a source of stimulation). The **flexor reflex** is a withdrawal reflex affecting the muscles of a limb. The **crossed extensor reflex** complements withdrawal reflexes. *(Spotlight Figure 13–14)*
32. All polysynaptic reflexes (1) involve pools of interneurons, (2) are intersegmental in distribution, (3) involve reciprocal inhibition, (4) have reverberating circuits, which prolong the reflexive motor response, and (5) can cooperate to produce a coordinated response.

13-8 The brain can affect spinal cord–based reflexes p. 460

33. The brain can facilitate or inhibit reflex motor patterns based in the spinal cord.
34. Motor control involves a series of interacting levels. Monosynaptic reflexes form the lowest level; at the highest level are the centers in the brain that can modulate or build on reflexive motor patterns.
35. Facilitation can produce an enhancement of spinal reflexes known as **reinforcement**. Spinal reflexes may also be inhibited, as when the **plantar reflex** in adults replaces the **Babinski reflex** in infants. *(Figure 13–17)*

Review Questions

See the blue Answers tab at the back of the book.

LEVEL 1 Reviewing Facts and Terms

1. Label the anatomical structures of the spinal cord in the following figure.

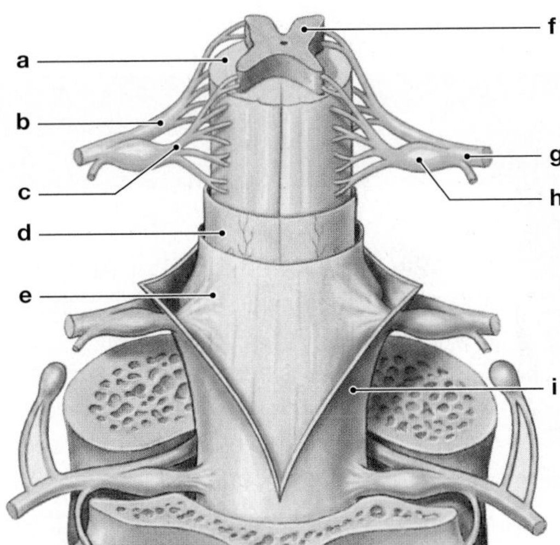

a
b
c
d
e
f
g
h
i

(a) ________ (b) ________
(c) ________ (d) ________
(e) ________ (f) ________
(g) ________ (h) ________
(i) ________

2. The anterior roots of each spinal segment **(a)** bring sensory information into the spinal cord, **(b)** control peripheral effectors, **(c)** contain the axons of somatic motor and visceral motor neurons, **(d)** do both b and c.
3. Spinal nerves are called mixed nerves because they **(a)** contain sensory and motor fibers, **(b)** exit at intervertebral foramina, **(c)** are associated with a pair of spinal ganglia, **(d)** are associated with posterior and anterior roots.
4. The adult spinal cord extends only to **(a)** the coccyx, **(b)** the sacrum, **(c)** the third or fourth lumbar vertebra, **(d)** the first or second lumbar vertebra, **(e)** the last thoracic vertebra.
5. Which of the following statements is *false* concerning the gray matter of the spinal cord? **(a)** It is located in the interior of the spinal cord around the central canal. **(b)** It functions in processing neural information. **(c)** It is primarily involved in relaying information to the brain. **(d)** It contains motor neurons. **(e)** It is divided into regions called horns.
6. The following are the steps involved in a neural reflex.
   (1) activation of a sensory neuron
   (2) activation of a motor neuron
   (3) response by an effector
   (4) arrival of a stimulus and activation of a receptor
   (5) information processing

   The proper sequence of these steps is
   (a) 1, 3, 4, 5, 2. **(b)** 4, 5, 3, 1, 2.
   (c) 4, 1, 5, 2, 3. **(d)** 4, 3, 1, 5, 2.
   (e) 3, 1, 4, 5, 2.
7. A sensory region monitored by the posterior rami of a single spinal segment is **(a)** a ganglion, **(b)** a fascicle, **(c)** a dermatome, **(d)** a ramus.
8. The major nerve of the cervical plexus that innervates the diaphragm is the **(a)** median nerve, **(b)** axillary nerve, **(c)** phrenic nerve, **(d)** common fibular nerve.
9. The genitofemoral, femoral, and lateral femoral cutaneous nerves are major nerves of the **(a)** lumbar plexus, **(b)** sacral plexus, **(c)** brachial plexus, **(d)** cervical plexus.
10. The synapsing of several neurons on the same postsynaptic neuron is called **(a)** serial processing, **(b)** reverberation, **(c)** divergence, **(d)** convergence.
11. The reflexes that control the most rapid, stereotyped motor responses to stimuli are **(a)** monosynaptic reflexes, **(b)** polysynaptic reflexes, **(c)** tendon reflexes, **(d)** extensor reflexes.
12. An example of a stretch reflex triggered by passive muscle movement is the **(a)** tendon reflex, **(b)** patellar reflex, **(c)** flexor reflex, **(d)** ipsilateral reflex.
13. The contraction of flexor muscles and the relaxation of extensor muscles illustrate the principle of **(a)** reverberating circuitry, **(b)** generalized facilitation, **(c)** reciprocal inhibition, **(d)** reinforcement.
14. Reflex arcs in which the sensory stimulus and the motor response occur on the same side of the body are **(a)** contralateral, **(b)** paraesthetic, **(c)** ipsilateral, **(d)** monosynaptic.
15. Proceeding deep from the most superficial structure, number the following in the correct sequence:
   (a) ____ walls of vertebral canal **(b)** ____ dura mater
   (c) ____ subdural space **(d)** ____ epidural space
   (e) ____ pia mater **(f)** ____ arachnoid membrane
   (g) ____ subarachnoid space **(h)** ____ spinal cord

LEVEL 2 Reviewing Concepts

16. Explain the anatomical significance of the fact that spinal cord growth stops at age 4.
17. List, in sequence, the five steps involved in a neural reflex.
18. Polysynaptic reflexes can produce far more complicated responses than can monosynaptic reflexes because **(a)** the response time is quicker, **(b)** the response is initiated by highly sensitive receptors, **(c)** motor neurons carry impulses at a faster rate than do sensory neurons, **(d)** the interneurons involved can control several muscle groups.
19. Why do cervical nerves outnumber cervical vertebrae?
20. If the anterior horns of the spinal cord were damaged, what type of control would be affected?
21. In which meningeal space is cerebrospinal fluid (CSF) found? What are the functions of CSF?
22. What five characteristics are common to all polysynaptic reflexes?
23. Predict the effects on the body of a spinal cord transection at segment C_7. How would these effects differ from those of a spinal cord transection at segment T_{10}?
24. The subarachnoid space contains **(a)** cerebrospinal fluid, **(b)** lymph, **(c)** air, **(d)** connective tissue and blood vessels, **(e)** denticulate ligaments.
25. Side-to-side movements of the spinal cord are prevented by the **(a)** filum terminale, **(b)** denticulate ligaments, **(c)** dura mater, **(d)** pia mater, **(e)** arachnoid mater.
26. Ascending tracts **(a)** carry sensory information to the brain, **(b)** carry motor information to the brain, **(c)** carry sensory information from the brain, **(d)** carry motor information from the brain, **(e)** connect perceptive areas with the brain.
27. What effect does the stimulation of a sensory neuron that innervates an intrafusal muscle fiber have on muscle tone?

LEVEL 3 Critical Thinking and Clinical Applications

28. Mary complains that when she wakes up in the morning, her thumb and forefinger are always "asleep." She mentions this condition to her physician, who asks Mary whether she sleeps with her wrists flexed. She replies that she does. The physician tells Mary that sleeping in that position may compress a portion of one of her peripheral nerves, producing her symptoms. Which nerve is involved?

29. The improper use of crutches can produce a condition known as "crutch paralysis," characterized by a lack of response by the extensor muscles of the arm, and a condition known as "wrist drop," consisting of an inability to extend the fingers and wrist. Which nerve is involved?
30. Bowel and urinary bladder control involve spinal reflex arcs that are located in the sacral region of the spinal cord. In both instances, two sphincter muscles—an inner sphincter of smooth muscle and an outer sphincter of skeletal muscle—control the passage of wastes (feces and urine) out of the body. How would a transection of the spinal cord at the L_1 segment level affect an individual's bowel and bladder control?
31. Karen falls down a flight of stairs and suffers lumbar and sacral spinal cord damage due to hyperextension of her back. The injury resulted in edema around the central canal that compressed the anterior horn of the lumbar region. What signs would you expect to observe as a result of this injury?

+ CLINICAL CASE Wrap-Up Prom Night

The emergency room team "log rolls" Joe over onto his side while maintaining his spinal alignment. The abrasions and bruising of his back are obvious. The doctor can feel that Joe's spinous processes, from the 7th cervical to the 11th thoracic vertebrae are in good alignment. There is an obvious gap, however, between the 11th and 12th thoracic spinous processes. The 12th thoracic spinous process and lumbar spine are displaced lateral to the midline. Palpating in this area causes Joe a great deal of discomfort. Joe has no sensation inferior to his 11th thoracic dermatome, just inferior to the level of the navel. Joe also has no voluntary muscle control of any muscles of his hips or inferior limbs.

X-rays taken in the trauma room show a fracture dislocation of the thoracic spine between the 11th thoracic vertebra (T_{11}) and the 12th thoracic vertebra (T_{12}). An MRI confirms the clinical impression that Joe's spinal cord is completely disrupted at this level.

Joe's spine is surgically realigned and stabilized (fixed in place) with internal fixation (surgical wires and pins). This enables him to begin sitting within the first week after his accident. Within 2 weeks he is transferred to a spinal rehabilitation facility to begin working on independence as a *paraplegic*—a person with loss of motor control and sensation to the lower limbs (but not the upper limbs).

1. Will Joe regain motor or sensory activity in his lower extremities? Why or why not?
2. Would Joe exhibit plantar reflexes or Babinski reflexes? Why?

See the blue Answers tab at the back of the book.

Related Clinical Terms

areflexia: Absence of reflexes.

Brown-Sequard syndrome: Loss of sensation and motor function that results from unilateral spinal cord lesions. Proprioception loss and weakness occur ipsilateral to the lesion while pain and temperature loss occur contralateral.

equinovarus: The foot is plantar flexed, inverted, and adducted; also called *talipes equinovalgus*.

Erb palsy (Erb-Duchenne palsy): Obstetric condition characterized by paralysis or weakness of a newborn's upper arm muscles caused by a stretch injury to the brachial plexus.

hemiparesis: Slight paralysis or weakness affecting one side of the body.

Kernig sign: Symptom of meningitis where patient cannot extend the leg at the knee due to stiffness in the hamstring muscles.

myelography: A diagnostic procedure in which a radiopaque dye is introduced into the cerebrospinal fluid to obtain an x-ray image of the spinal cord and cauda equina.

nerve conduction study: Test often performed along with electromyography (EMG); the test stimulates certain nerves and records their ability to send an impulse to the muscle; it can indicate where any blockage of the nerve pathway exists.

nerve growth factor: A peptide that promotes the growth and maintenance of neurons. Other factors that are important to neuron growth and repair include BDNF, NT-3, NT-4, and GAP-43.

paraplegia: Paralysis involving a loss of motor control of the lower, but not the upper, limbs.

quadriplegia: Paralysis involving the loss of sensation and motor control of the upper and lower limbs.

spinal shock: Term applied to all phenomena surrounding physiological or anatomical transection of the spinal cord that results in temporary loss or depression of all or most spinal reflex activity inferior to the level of the injury.

tabes dorsalis: Slow progressive degeneration of the myelin layer of the sensory neurons of the spinal cord that occurs in the tertiary (third) phase of syphilis. Common signs and symptoms are pain, weakness, diminished reflexes, unsteady gait, and loss of coordination.

14 The Brain and Cranial Nerves

Learning Outcomes

These Learning Outcomes correspond by number to this chapter's sections and indicate what you should be able to do after completing the chapter.

14-1 ▪ Name the major brain regions, vesicles, and ventricles, and describe the locations and functions of each. p. 466

14-2 ▪ Explain how the brain is protected and supported, and discuss the formation, circulation, and function of cerebrospinal fluid. p. 469

14-3 ▪ Describe the anatomical differences between the medulla oblongata and the spinal cord, and identify the main components and functions of the medulla oblongata. p. 474

14-4 ▪ List the main components of the pons, and specify the functions of each. p. 477

14-5 ▪ List the main components of the midbrain, and specify the functions of each. p. 478

14-6 ▪ List the main components of the cerebellum, and specify the functions of each. p. 478

14-7 ▪ List the main components of the diencephalon, and specify the functions of each. p. 481

14-8 ▪ Identify the main components of the limbic system, and specify the locations and functions of each. p. 484

14-9 ▪ Identify the major anatomical subdivisions and functions of the cerebrum, and discuss the origin and significance of the major types of brain waves seen in an electroencephalogram. p. 486

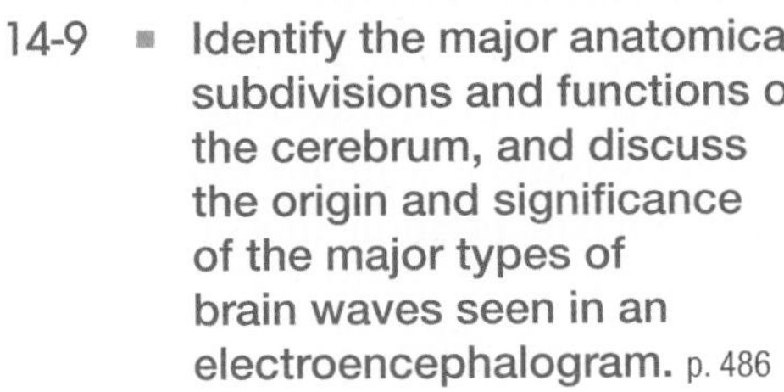

14-10 ▪ Describe representative examples of cranial reflexes that produce somatic responses or visceral responses to specific stimuli. p. 506

CLINICAL CASE The Neuroanatomist's Stroke

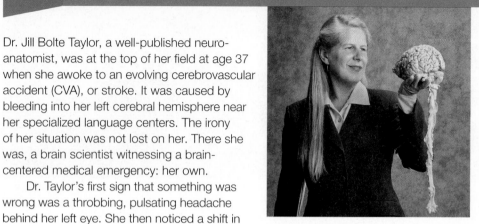

Dr. Jill Bolte Taylor, a well-published neuroanatomist, was at the top of her field at age 37 when she awoke to an evolving cerebrovascular accident (CVA), or stroke. It was caused by bleeding into her left cerebral hemisphere near her specialized language centers. The irony of her situation was not lost on her. There she was, a brain scientist witnessing a brain-centered medical emergency: her own.

Dr. Taylor's first sign that something was wrong was a throbbing, pulsating headache behind her left eye. She then noticed a shift in her perceptions and consciousness, and experienced a slowing of her thoughts and movements. By the time she was aware that her right arm was paralyzed, she was barely able to call for help. When she arrived at the hospital, she could not walk, talk, read, write, or recall anything about her life. She felt her spirit surrender and braced for death.

Dr. Taylor awoke later that day, shocked to be alive. She was unable to speak or understand speech. Sounds irritated her. She could not recognize or use numbers. Less than 3 weeks later, doctors performed open brain surgery to remove a large blood clot that was pressing on the left side of her brain. **Would Dr. Taylor ever fully recover? To find out,** **turn to the Clinical Case Wrap-Up on p. 510.**

An Introduction to the Brain and Cranial Nerves

14

In this chapter we introduce the functional organization of the *brain* and *cranial nerves*, and describe simple *cranial reflexes*. The adult human brain contains almost 97 percent of the body's nervous tissue. As we discussed in Chapter 12, the **brain** is responsible for integrating and processing sensory and motor information, as well as being the seat of higher mental functions such as intelligence, memory, and emotions.

A "typical" brain has a volume of 1200 mL (71 in.^{3}). However, brain size varies considerably among individuals. The brains of males are, on average, about 10 percent larger than those of females, due to differences in average body size. No correlation exists between brain size and intelligence. Individuals with the smallest brains (750 mL) and the largest brains (2100 mL) are functionally normal.

14-1 The brain develops four major regions: the cerebrum, cerebellum, diencephalon, and brainstem

Learning Outcome **Name the major brain regions, vesicles, and ventricles, and describe the locations and functions of each.**

In this section we introduce the anatomical organization of the brain. We begin with an overview of the brain's major regions and landmarks. Then we discuss the brain's embryonic origins and some prominent internal cavities: the *ventricles* of the brain.

Major Brain Regions and Landmarks

There are four major brain regions: the *cerebrum, cerebellum, diencephalon,* and *brainstem.* Figure 14–1 shows the boundaries and general functions of these regions.

The Cerebrum

The largest portion of the adult brain is the **cerebrum** (se-RĒ-brum; see Figure 14–1a). Conscious thoughts, sensations, intellect, memory, and complex movements all originate in the cerebrum. It is made up of large, paired left and right **cerebral hemispheres**. The surfaces of the cerebral hemispheres are highly folded and covered by a collection of neurons that form a thin superficial layer of gray matter called the **cerebral cortex** (*cortex,* bark). This cerebral cortex forms a series of rounded elevations called **gyri** (JĪ-rī; singular, *gyrus*) that increase its surface area. The gyri are separated by shallow grooves called **sulci** (SUL-sī; singular, *sulcus*) or by deeper grooves called **fissures**. Fissures separate larger brain regions.

The Cerebellum

The **cerebellum** (ser-e-BEL-um) is the second-largest part of the brain (see Figure 14–1b). It is partially hidden by the cerebral hemispheres. Like the cerebrum, the cerebellum has hemispheres covered by a sheet of gray matter called the *cerebellar cortex*. The cerebellum adjusts ongoing movements by comparing arriving sensations with previously experienced sensations, allowing you to perform the same movements over and over.

The Diencephalon

The other major anatomical regions of the brain can best be examined after the cerebral and cerebellar hemispheres have been removed. The walls of the **diencephalon** (dī-en-SEF-a-lon; *dia,* through + *encephalos,* brain) are composed of the *thalamus* (THAL-a-mus) and *hypothalamus* (see Figure 14–1c). The thalamus contains relay and processing centers for sensory information. The *hypothalamus* (*hypo-,* below) is the floor of the diencephalon. It contains centers involved with emotions,

Figure 14–1 An Introduction to Brain Structures and Functions.

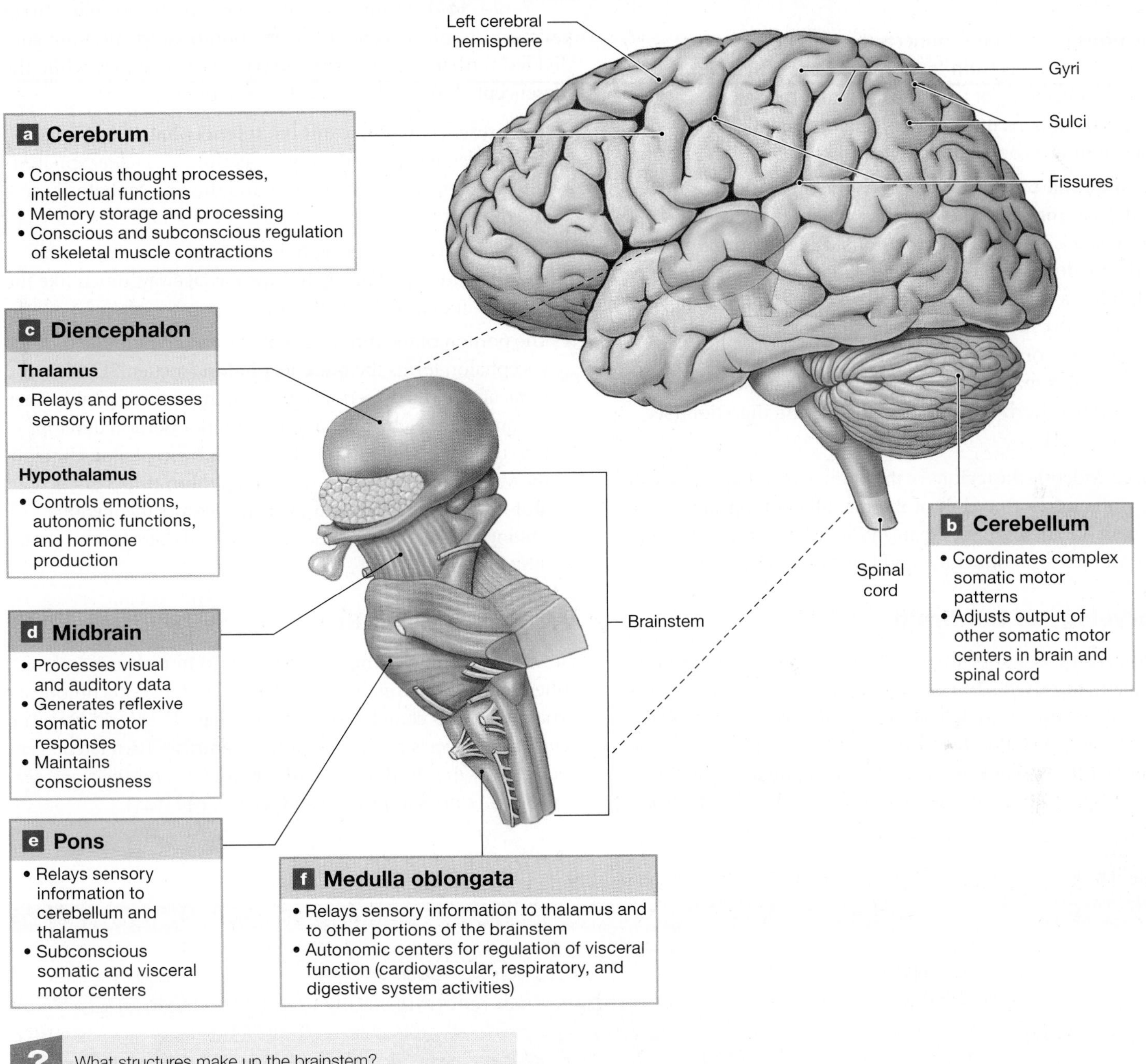
Left cerebral hemisphere
Gyri
Sulci
Fissures
a Cerebrum
• Conscious thought processes, intellectual functions
• Memory storage and processing
• Conscious and subconscious regulation of skeletal muscle contractions
c Diencephalon
Thalamus
• Relays and processes sensory information
Hypothalamus
• Controls emotions, autonomic functions, and hormone production
d Midbrain
• Processes visual and auditory data
• Generates reflexive somatic motor responses
• Maintains consciousness
e Pons
• Relays sensory information to cerebellum and thalamus
• Subconscious somatic and visceral motor centers
f Medulla oblongata
• Relays sensory information to thalamus and to other portions of the brainstem
• Autonomic centers for regulation of visceral function (cardiovascular, respiratory, and digestive system activities)
Brainstem
Spinal cord
b Cerebellum
• Coordinates complex somatic motor patterns
• Adjusts output of other somatic motor centers in brain and spinal cord

? What structures make up the brainstem?

autonomic function, and hormone production. The *infundibulum*, a narrow stalk, connects the hypothalamus to the **pituitary gland**, a part of the endocrine system. This connection integrates the nervous and endocrine systems.

The Brainstem

The diencephalon is a structural and functional link between the cerebral hemispheres and the brainstem. The **brainstem** contains a variety of important processing centers and nuclei (clearly distinguishable masses of brain neurons) that relay information headed to or from the cerebrum or cerebellum. As Figure 14–1d–f shows, the brainstem includes the *midbrain*, *pons*, and *medulla oblongata*.

- The **midbrain** contains nuclei that process visual and auditory information and control reflexes triggered by these stimuli. For example, your immediate, reflexive responses to a loud, unexpected noise (eye movements and head turning) are directed by nuclei in the midbrain. This region also contains centers (groups of nerve cells governing a specific function) that help maintain consciousness (see Figure 14–1d).

- The **pons** of the brain connects the cerebellum to the brainstem (*pons* is Latin for "bridge"). In addition to **tracts** (collections of CNS axons) and relay centers, the pons contains nuclei involved with somatic and visceral motor control (see Figure 14–1e).
- The **medulla oblongata** connects the brain to the spinal cord. Near the pons, the posterior wall of the medulla oblongata is thin and membranous. The inferior portion of the medulla oblongata resembles the spinal cord in that it has a narrow **central canal**. The medulla oblongata relays sensory information to the thalamus and to centers in other portions of the brainstem. The medulla oblongata also contains major centers that regulate autonomic function, such as heart rate, blood pressure, and digestion (see Figure 14–1f).

In considering the regions of the brain in this chapter, we will begin at the inferior portion of the medulla oblongata because this region has the simplest organization. We will then ascend to regions of increasing structural and functional complexity.

Embryology of the Brain

The embryonic origin of the brain can help us understand its internal organization. The central nervous system (CNS) begins as a hollow cylinder known as the *neural tube*. At the cephalic (rostral) portion of the neural tube, three swellings called the **primary brain vesicles** develop. They are named for their relative positions: the *prosencephalon* (prōz-en-SEF-ah-lon; *proso*, forward), or "forebrain"; the *mesencephalon*, or "midbrain"; and the *rhombencephalon* (rom-ben-SEF-ah-lon), or "hindbrain."

Table 14–1 summarizes the fates of the primary brain vesicles. The prosencephalon and rhombencephalon are subdivided further, forming **secondary brain vesicles**, while the mesencephalon develops but does not divide:

- The prosencephalon forms the **telencephalon** (tel-en-SEF-a-lon; *telos*, end) and the diencephalon. The telencephalon ultimately forms the cerebrum and the diencephalon forms the thalamus.
- The walls of the mesencephalon thicken, and the neural tube becomes a relatively narrow passageway, much like the central canal of the spinal cord.
- The portion of the rhombencephalon adjacent to the mesencephalon forms the **metencephalon** (met-en-SEF-ah-lon; *meta*, after). The posterior portion of the metencephalon later becomes the cerebellum, and the anterior portion develops into the pons. The portion of the rhombencephalon closer to the spinal cord forms the **myelencephalon** (mī-el-en-SEF-ah-lon; *myelon*, spinal cord), which becomes the medulla oblongata. ATLAS: Embryology Summary 12: The Development of the Brain and Cranial Nerves

Ventricles of the Brain

During development, the neural tube within the cerebral hemispheres, diencephalon, metencephalon, and medulla oblongata expands to form chambers called **ventricles** (VEN-trih-klz). The *ventricular system* is composed of four ventricles (two lateral ventricles, then the third and fourth ventricles) and their passageways (*interventricular foramen* and *cerebral aqueduct*).

Table 14–1 Development of the Brain

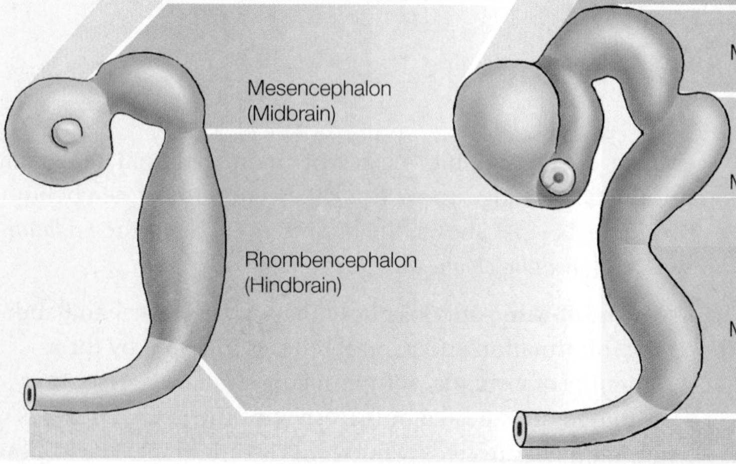

Primary Brain Vesicles (3 weeks)	Secondary Brain Vesicles (6 weeks)	Brain Regions at Birth	Ventricular System
Prosencephalon (Forebrain)	Telencephalon	Cerebrum	Lateral ventricles
	Diencephalon	Diencephalon	Third ventricle
Mesencephalon (Midbrain)	Mesencephalon	Midbrain	Cerebral aqueduct
Rhombencephalon (Hindbrain)	Metencephalon	Cerebellum and Pons	Fourth ventricle
	Myelencephalon	Medulla oblongata	Fourth ventricle

Figure 14–2 Ventricular System. The orientation and extent of the ventricles as they would appear if the brain were transparent. ATLAS: Plates 10; 12a–c; 13a–e

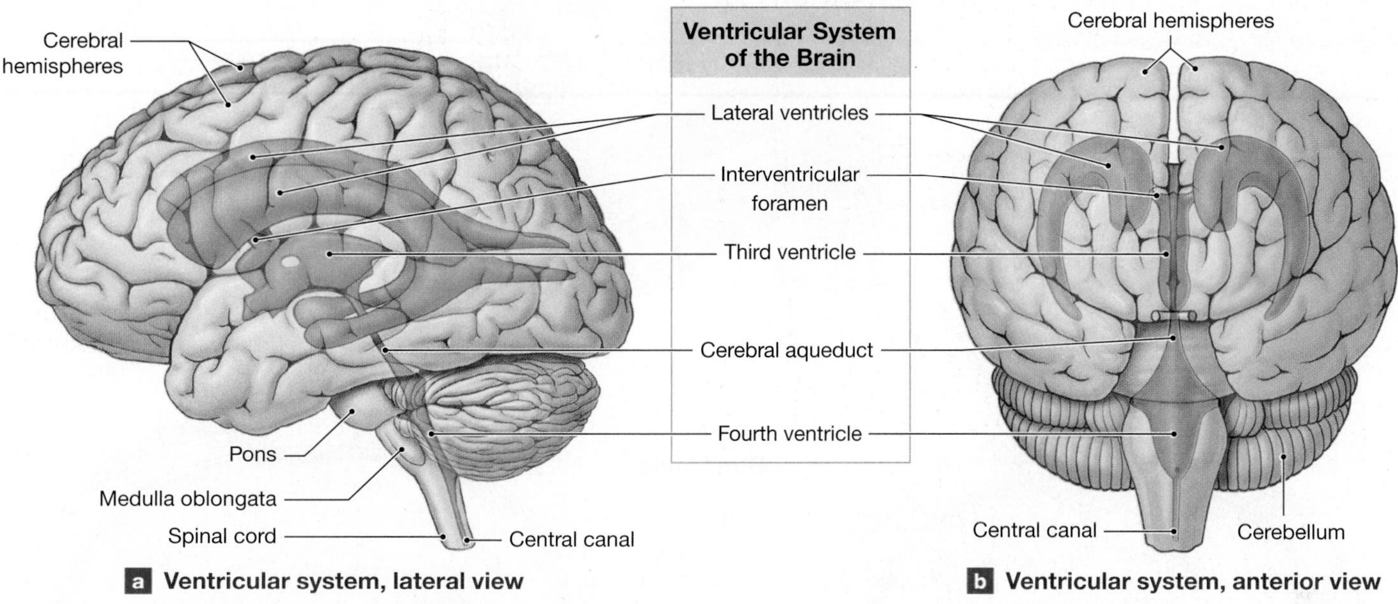
Cerebral hemispheres
Ventricular System of the Brain
Lateral ventricles
Interventricular foramen
Third ventricle
Cerebral aqueduct
Fourth ventricle
Pons
Medulla oblongata
Spinal cord
Central canal
Cerebral hemispheres
Central canal
Cerebellum

a Ventricular system, lateral view

b Ventricular system, anterior view

Each cerebral hemisphere contains a large **lateral ventricle** (Figure 14–2). The *septum pellucidum*, a thin plate of brain tissue, separates the two lateral ventricles. Because there are *two* lateral ventricles, the ventricle in the diencephalon is called the **third ventricle**. The two lateral ventricles are not directly connected, but each communicates with the third ventricle of the diencephalon through an **interventricular foramen**.

The midbrain has a slender canal known as the **cerebral aqueduct**. This passageway connects the third ventricle with the **fourth ventricle**. The superior portion of the fourth ventricle lies between the posterior surface of the pons and the anterior surface of the cerebellum. The fourth ventricle extends into the superior portion of the medulla oblongata. There this ventricle narrows and becomes continuous with the central canal of the spinal cord.

Ependymal cells, a type of neuroglia, line the ventricles. ⮌ p. 396 These cells produce *cerebrospinal fluid (CSF)* that fills the ventricles and continuously circulates in the CNS. The CSF passes between the interior and exterior of the CNS through three foramina in the roof of the fourth ventricle. We describe these foramina in Section 14-7.

✓ Checkpoint

1. **Name the four major regions of the brain.**
2. **What brain regions make up the brainstem?**
3. **Which primary brain vesicle is destined to form the cerebellum, pons, and medulla oblongata?**

See the blue Answers tab at the back of the book.

14-2 The brain is protected and supported by the cranial meninges, cerebrospinal fluid, and the blood brain barrier

Learning Outcome **Explain how the brain is protected and supported, and discuss the formation, circulation, and function of cerebrospinal fluid.**

The delicate tissues of the brain are protected from physical forces by the bones of the cranium, membranes called the *cranial meninges*, and cerebrospinal fluid. In addition, the nervous tissue of the brain is biochemically isolated from the general circulation by the *blood brain barrier*. Refer back to Figures 7–3 and 7–4 (pp. 212–215) for a review of the bones of the cranium. We discuss the other protective factors here.

The Cranial Meninges

The layers that make up the **cranial meninges**—the cranial *dura mater, arachnoid mater*, and *pia mater*—are continuous with the spinal meninges. ⮌ p. 437 However, the cranial meninges have distinctive anatomical and functional characteristics (Figure 14–3a):

Dura Mater and Dural Folds

The **dura mater** consists of outer and inner fibrous layers. The outer layer is fused to the periosteum of the cranial bones. As a result, there is no epidural space superficial to the dura mater, as occurs along the spinal cord. The outer *periosteal cranial dura* and inner *meningeal cranial dura* are typically fused together.

Figure 14–3 The Relationships among the Brain, Cranium, and Cranial Meninges. ATLAS: Plates 7a–d

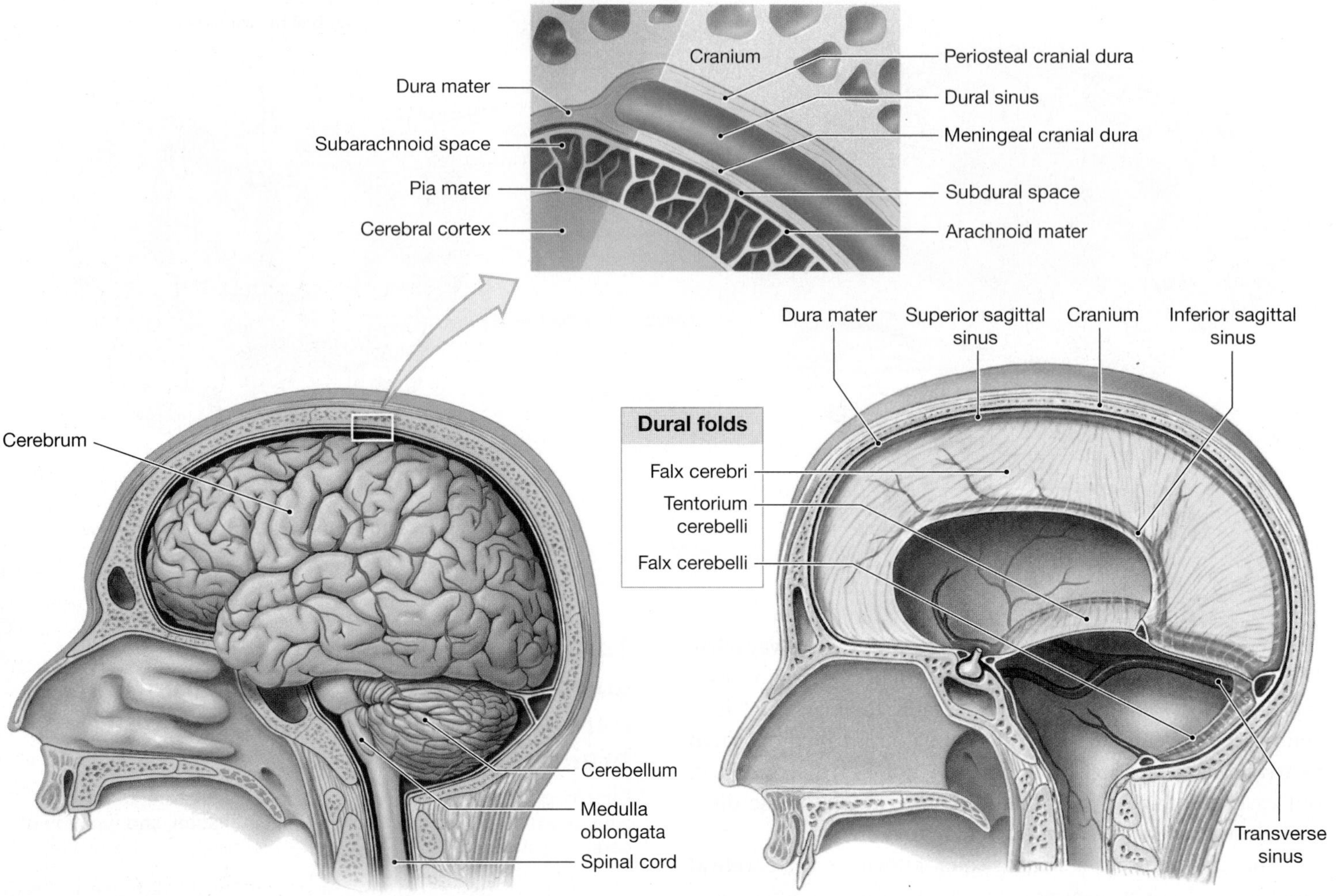

Dura mater
Subarachnoid space
Pia mater
Cerebral cortex
Cranium
Periosteal cranial dura
Dural sinus
Meningeal cranial dura
Subdural space
Arachnoid mater
Cerebrum
Cerebellum
Medulla oblongata
Spinal cord

a A lateral view of the brain, showing its position in the cranium and the organization of the meninges

Dura mater
Superior sagittal sinus
Cranium
Inferior sagittal sinus
Dural folds
Falx cerebri
Tentorium cerebelli
Falx cerebelli
Transverse sinus

b A diagrammatic view, showing the orientation of the three largest dural folds: the falx cerebri, tentorium cerebelli, and falx cerebelli

List the maters surrounding the brain from the deepest layer to the most superficial layer.

In several locations, the meningeal cranial dura extends into the cranial cavity, forming a sheet that dips inward and then returns. These inward projections, known as **dural folds**, provide additional stabilization and support to the brain. **Dural venous sinuses** are large collecting veins located within the dural folds. The veins of the brain open into these sinuses, which deliver the venous blood to the veins of the neck. The three largest dural folds are called the *falx cerebri*, the *tentorium cerebelli*, and the *falx cerebelli* (Figure 14–3b):

- The **falx cerebri** (FALKS SER-e-brī; *falx*, sickle shaped) is a fold of dura mater that projects between the cerebral hemispheres in the longitudinal cerebral fissure. Its inferior portions attach anteriorly to the crista galli of the ethmoid and posteriorly to the *internal occipital crest*, a ridge along the inner surface of the occipital bone. The **superior sagittal sinus** and the **inferior sagittal sinus**, two large dural venous sinuses, lie within this dural fold. The posterior margin of the falx cerebri intersects the tentorium cerebelli.
- The **tentorium cerebelli** (ten-TŌ-rē-um ser-e-BEL-ē; *tentorium*, tent) protects the cerebellum and separates the cerebral hemispheres from the cerebellum. It extends across the cranium at right angles to the falx cerebri. The **transverse sinus** is a paired dural venous sinus that runs along the occipital bone. It allows blood to drain from the back of the head.
- The **falx cerebelli** divides the two cerebellar hemispheres along the midsagittal line inferior to the tentorium cerebelli.

The dura mater may be separated from the next layer of membrane, the arachnoid mater, by a narrow gap called the **subdural space**.

Arachnoid Mater

The **arachnoid mater** is a membrane that resembles a spider web. In life, its arachnoid barrier cell layer is attached to the dural border cell layer of the dura mater, and no natural space occurs at this interface. (However, a subdural space may form as a result of trauma, disease, or lack of cerebrospinal fluid, as in a cadaver.) The cranial arachnoid mater covers the brain, providing a smooth surface that does not follow the brain's underlying folds. The **subarachnoid space**, containing cells and fibers of the arachnoid trabeculae, lies between the arachnoid mater and the pia mater.

Pia Mater

The **pia mater** sticks closely to the surface of the brain, anchored by the processes of astrocytes. It also accompanies the branches of cerebral blood vessels as they penetrate the surface of the brain to reach internal structures.

Cerebrospinal Fluid

Cerebrospinal fluid (CSF) completely surrounds and bathes the exposed surfaces of the CNS. The CSF has several important functions, including:

- *Supporting the Brain.* In essence, the brain is suspended inside the cranium and floats in the CSF. A human brain weighs about 1400 g (3.09 lb) in air, but only about 50 g (1.8 oz) when supported by CSF.
- *Cushioning Delicate Neural Structures.* CSF cushions the brain and spinal cord against physical trauma.
- *Transporting Nutrients, Chemical Messengers, and Wastes.* Except in the areas where CSF is produced, the ependymal cell lining is freely permeable. The CSF is in constant chemical communication with the interstitial fluid that surrounds the neurons and neuroglia of the CNS.

The Formation and Circulation of CSF

A **choroid** (KOR-oyd) **plexus** (*choroeides*, like a membrane; *plexus*, network) is an area within each ventricle that produces CSF. Specialized ependymal cells bearing microvilli and interconnected by tight junctions surround the capillaries of a choroid plexus. The process by which these cells form CSF is shown in step 1 of **Spotlight Figure 14–4**. CSF is a filtrate produced by the fluid leaking out of the capillaries in the choroid plexuses. The ependymal cells then secrete CSF into the ventricles. They also remove wastes from the CSF and adjust its composition over time.

Because free exchange occurs between the interstitial fluid of the brain and the CSF in this way, changes in CNS function can produce changes in the composition of the CSF and vice versa. The differences in composition between CSF and blood plasma (blood with the cellular elements removed) are quite noticeable. For example, blood plasma contains high concentrations of soluble proteins, but the CSF does not. The concentrations of individual ions and the levels of amino acids, lipids, and waste products are also different. As noted in Chapter 13, a *spinal tap* procedure that is done to collect a sample of CSF can provide useful clinical information about CNS injury, infection, or disease. ⮌ p. 439

Steps 2–5 of **Spotlight Figure 14–4** show how the 150 mL of CSF is circulated in the CNS; please study these steps before going on. If the normal circulation or resorption of CSF is

+ Clinical Note Epidural and Subdural Hemorrhages

A severe head injury may damage meningeal blood vessels and cause bleeding into the cranial cavity. The most serious cases involve an arterial break, because arterial blood pressure is relatively high. If blood is forced between the dura mater and the skull, the condition is known as an **epidural hemorrhage** (HEM-ō-raj). The elevated fluid pressure then distorts the underlying tissues of the brain. The individual loses consciousness for a period lasting from minutes to hours after the injury, and death follows in untreated cases. An epidural hemorrhage involving a damaged vein does not produce massive symptoms immediately. The individual may not have neurological problems for hours, days, or even weeks after the original injury. As a result, the problem may not be noticed until the nervous tissue has been severely damaged. Epidural hemorrhages are rare, occurring in less than 1 percent of head injuries. However, the mortality rate is 100 percent in untreated cases and over 50 percent even after the blood pool has been removed and the damaged vessels have been closed.

A **subdural hemorrhage** is bleeding between the dura mater and the arachnoid mater. Subdural hemorrhages are twice as common as epidural hemorrhages. Small veins between the arachnoid mater and dura mater or one of the dural venous sinuses cause the bleeding. Because the venous blood pressure in a subdural hemorrhage is lower than that in an arterial epidural hemorrhage, the distortion produced is gradual. The effects on brain function can be quite variable and difficult to diagnose.

SPOTLIGHT

Figure 14–4

Formation and Circulation of Cerebrospinal Fluid

Cerebrospinal fluid (CSF) completely surrounds and bathes the exposed surfaces of the CNS. The CSF has three important functions:

- Supporting the brain.
- Cushioning the brain and spinal cord against physical trauma.
- Transporting nutrients, chemical messengers, and wastes.

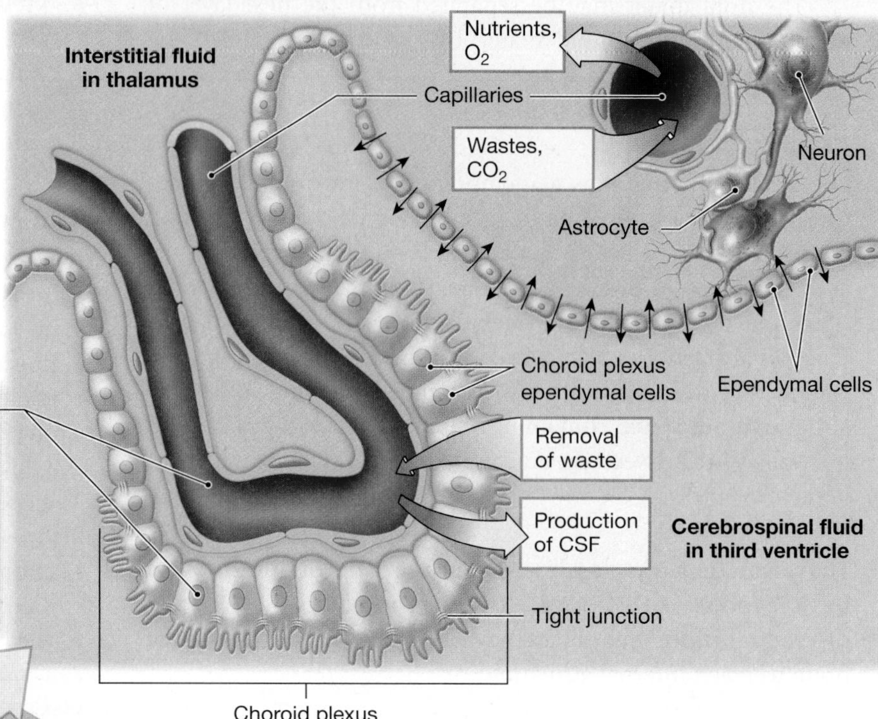

Interstitial fluid in thalamus
Nutrients, O_2
Capillaries
Wastes, CO_2
Neuron
Astrocyte
Choroid plexus ependymal cells
Ependymal cells
Removal of waste
Production of CSF
Cerebrospinal fluid in third ventricle
Tight junction
Choroid plexus

1 The choroid plexus produces and maintains CSF. Two extensive folds of the choroid plexus originate in the roof of the third ventricle and extend through the interventricular foramina. These folds cover the floors of the lateral ventricles. In the inferior brainstem, a region of the choroid plexus in the roof of the fourth ventricle projects between the cerebellum and the pons.

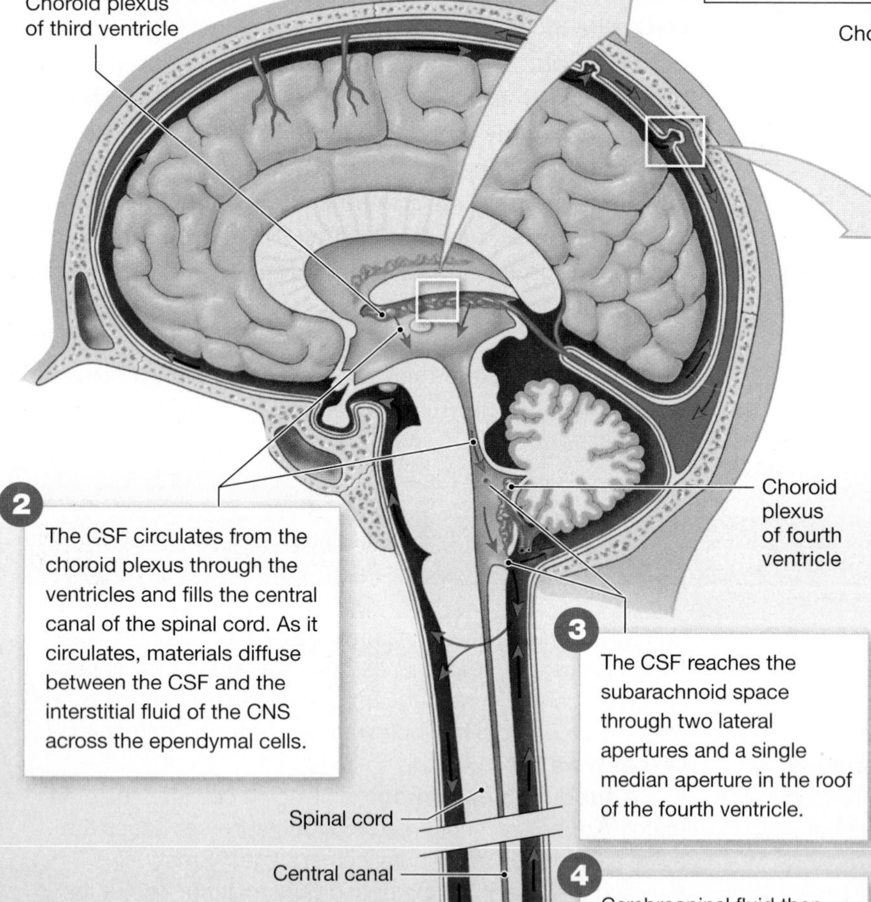

Choroid plexus of third ventricle
Choroid plexus of fourth ventricle
Spinal cord
Central canal
Dura mater
Conus medullaris
Arachnoid mater
Cauda equina
Filum terminale

Dura mater (periosteal layer)
Arachnoid granulation
Arachnoid trabecula
Cranium
Superior sagittal sinus
CSF fluid movement
Dura mater (meningeal layer)
Subdural space
Arachnoid mater
Subarachnoid space
Cerebral cortex
Pia mater

2 The CSF circulates from the choroid plexus through the ventricles and fills the central canal of the spinal cord. As it circulates, materials diffuse between the CSF and the interstitial fluid of the CNS across the ependymal cells.

3 The CSF reaches the subarachnoid space through two lateral apertures and a single median aperture in the roof of the fourth ventricle.

4 Cerebrospinal fluid then flows through the subarachnoid space surrounding the brain, spinal cord, and cauda equina.

5 Fingerlike extensions of the arachnoid membrane, called the arachnoid villi, penetrate the meningeal layer of the dura mater and extend into the superior sagittal sinus. In adults, these extensions form large arachnoid granulations. CSF is absorbed into the venous circulation at the arachnoid granulations.

The choroid plexus produces CSF at a rate of about 500 mL, or 2.1 cups, per day. The total volume of CSF at any moment is approximately 150 mL. The entire volume of CSF is replaced about every 8 hours. Despite this rapid turnover, the composition of CSF is closely regulated, and the rate of removal normally keeps pace with the rate of production.

interrupted, a variety of clinical problems may appear. For example, a problem with the resorption of CSF in infancy causes *hydrocephalus*, or "water on the brain." Infants with this condition have enormously expanded skulls due to the presence of an abnormally large volume of CSF. In adults, a failure of resorption or a blockage of CSF circulation can distort and damage the brain.

The Protective Function of the Cranial Meninges and CSF

The overall shape of the brain corresponds to that of the cranial cavity (look back at Figure 14–3a). The cranial bones physically protect the brain by cradling it, but they also pose a threat to its safety that is countered by the cranial meninges and the CSF. The brain and cranium are like a person driving a car: If the car hits a tree, the car protects the driver from contact with the tree, but serious injury will occur unless a seat belt or airbag protects the driver from contact with the car. The tough, fibrous dural folds act like seat belts that hold the brain in position and restrain it from being damaged by contact with the hard cranium. The cerebrospinal fluid in the subarachnoid space acts like an airbag by cushioning the brain against sudden jolts.

Cranial trauma is a head injury resulting from impact with another object. Each year in the United States, about 8 million cases of cranial trauma occur, but only 1 case in 8 results in serious brain damage. The percentage is relatively low because the cranial meninges and CSF are so effective in protecting the brain.

The Blood Supply to the Brain

Your brain, with its billions of neurons, is an extremely active organ with a continuous demand for nutrients and oxygen. However, as noted in Chapter 12, neurons have no energy reserves in the form of carbohydrates or lipids. Neurons also have no equivalent to myoglobin to store oxygen. Therefore, an extensive network of blood vessels meets these demands.

Arterial blood reaches the brain through the *internal carotid arteries* and the *vertebral arteries*. Most of the venous blood from the brain leaves the cranium in the *internal jugular veins*, which drain the dural venous sinuses.

Cerebrovascular diseases are cardiovascular disorders that interfere with the normal blood supply to the brain. The particular distribution of the vessel involved determines the signs and symptoms. The degree of oxygen or nutrient starvation determines their severity. A **cerebrovascular accident (CVA)**, or *stroke*, occurs when the blood supply to a portion of the brain is shut off. Affected neurons begin to die in a matter of minutes.

The Blood Brain Barrier

Nervous tissue in the CNS is isolated from the general circulation by the **blood brain barrier (BBB)**. This barrier is formed by capillary endothelial cells that are extensively interconnected by tight junctions. These junctions prevent materials from diffusing between endothelial cells. In general, only lipid-soluble compounds (including carbon dioxide; oxygen; ammonia; lipids, such as steroids or prostaglandins; and small alcohols) can diffuse across the membranes of endothelial cells into the interstitial fluid of the brain and spinal cord. Water and ions must pass through channels in the apical and basal surfaces of plasma membranes. Larger, water-soluble compounds can cross the capillary walls only by active or passive transport.

Astrocytes are critical components of the blood brain barrier. p. 396 These neuroglia play a key supporting role in the blood brain barrier because they release chemicals that lead to the formation of strong tight junctions and control the permeability of the endothelium to various substances, such as nutrients and ions. Astrocytes are in close contact with CNS capillaries. The processes of astrocytes cover the outer surfaces of the endothelial cells. If the astrocytes are damaged or stop stimulating the endothelial cells, the blood brain barrier disappears.

The choroid plexus is not part of the nervous tissue of the brain, so no astrocytes are in contact with the endothelial cells there. Substances do not have free access to the CNS, however, because specialized ependymal cells create a **blood CSF barrier**. These cells, also interconnected by tight junctions, surround the capillaries of the choroid plexus.

Transport across the blood brain and blood CSF barriers is selective and directional. Even the passage of small ions, such as sodium, hydrogen, potassium, or chloride, is controlled. As a result, the pH and concentrations of sodium, potassium, calcium, and magnesium ions in the blood and CSF are different. Some organic compounds are readily transported, and others cross only in very small amounts.

Neurons have a constant need for glucose that must be met regardless of the glucose concentrations in the blood and interstitial fluid. Even when the circulating glucose level is low, endothelial cells continue to transport glucose from the blood to the interstitial fluid of the brain. In contrast, only trace amounts of circulating *neurohormones* (chemical messengers released by neurons into the bloodstream), such as norepinephrine, epinephrine, dopamine, and serotonin, pass into the interstitial fluid or CSF of the brain. This limitation is important, because the entry of these neurotransmitters from the bloodstream (where their concentrations can be relatively high) could result in the uncontrolled stimulation of neurons throughout the brain.

The blood brain barrier remains intact throughout the CNS except in small areas known as *circumventricular organs (CVOs)*, which have fenestrated capillaries (blood vessels with ultramicroscopic pores). Because they are outside the BBB, they provide a direct link between the CNS and the peripheral blood and they play a role in neuroendocrine function. In summary, these noteworthy exceptions include:

- Portions of the hypothalamus, where the capillary endothelium is extremely permeable. This permeability exposes hypothalamic nuclei to circulating hormones and permits hypothalamic hormones to diffuse into the circulation.
- Capillaries in the posterior lobe of the pituitary gland, which is continuous with the floor of the hypothalamus. At this site, the hormones antidiuretic hormone and oxytocin, produced by hypothalamic neurons, are released into the circulation.
- Capillaries in the *pineal gland*. This endocrine gland is located on the posterior, superior surface of the diencephalon. The capillary permeability allows pineal secretions to enter the general circulation.
- Capillaries at a choroid plexus. The capillary characteristics of the blood brain barrier are lost there, but the transport activities of specialized ependymal cells in the choroid plexuses maintain the blood CSF barrier.

Physicians must sometimes get specific compounds into the interstitial fluid of the brain to fight CNS infections or to treat other neurological disorders. To do this, they must understand the limitations of the blood brain barrier and blood CSF barrier. For example, when considering possible treatments, the antibiotic *tetracycline* is not used to treat meningitis or other CNS infections because this drug cannot cross the barriers to reach the brain. In contrast, *sulfisoxazole* and *sulfadiazine* enter the CNS very rapidly. Sometimes, chemotherapeutic drugs or imaging dyes may be injected directly into the CSF by lumbar puncture.

Checkpoint

4. **From superficial to deep, name the layers that make up the cranial meninges.**
5. **What would happen if the normal circulation or resorption of CSF were blocked?**
6. **How would decreased diffusion across the arachnoid granulations affect the volume of cerebrospinal fluid in the ventricles?**
7. **Many water-soluble molecules that are abundant in the blood occur in small amounts or not at all in the cerebrospinal fluid (CSF) of the brain. Why?**

See the blue Answers tab at the back of the book.

14-3 Brainstem: The medulla oblongata relays signals between the rest of the brain and the spinal cord

Learning Outcome Describe the anatomical differences between the medulla oblongata and the spinal cord, and identify the main components and functions of the medulla oblongata.

As mentioned, the **medulla oblongata** is the most inferior of the brainstem regions. Figure 14–5 shows the position of the medulla oblongata in relation to the other components of the brainstem and the diencephalon. It also illustrates the attachment sites for 11 of the 12 pairs of cranial nerves. The individual cranial nerves are identified by capital letters CN followed by a Roman numeral (I–XII). (We introduce the full names and functions of these nerves in a later section.)

In sectional view, the inferior portion of the medulla oblongata resembles the spinal cord, with a small central canal. However, the gray matter and white matter organization is more complex. As the central canal travels superiorly through the medulla oblongata, the canal opens to become the fourth ventricle, and the similarity to the spinal cord all but disappears.

The medulla oblongata includes three groups of nuclei that we introduce here and discuss in later chapters (Figure 14–6). First are nuclei and processing centers that control visceral functions. The medulla oblongata is a center for the coordination of complex autonomic reflexes. Second are sensory and motor nuclei of the CNS. Third are relay stations that move all communication between the brain and spinal cord by tracts that ascend or descend through the medulla oblongata.

- *Reflex Centers: Autonomic and Reflex Activity.* The **reticular formation** is a closely intermingled mass of gray and white matter that contains embedded nuclei. It extends throughout the central core of the medulla oblongata and into the diencephalon and plays a role in autonomic (breathing, blood pressure, and thermoregulation) functions, endocrine functions, as well as body posture, skeletomuscular reflex activity, alertness, and sleep.

  There are two major groups of reflex centers in the medulla oblongata, cardiovascular and respiratory. The **cardiovascular centers** adjust the heart rate, the strength of cardiac contractions, and the flow of blood through peripheral tissues. (In terms of function, the cardiovascular centers are subdivided into *cardiac* and *vasomotor centers*, but their anatomical boundaries are difficult to determine.) The **respiratory rhythmicity centers** set the basic pace for respiratory movements. Their activity is regulated by inputs from the respiratory centers of the pons.
- *Sensory and Motor Nuclei of Cranial Nerves.* The medulla oblongata contains sensory and motor nuclei associated with five of the cranial nerves (VIII, IX, X, XI, and XII). These cranial nerves provide motor commands to muscles of the pharynx, neck, and back as well as to the visceral organs of the thoracic and peritoneal cavities. Cranial nerve VIII carries sensory information from receptors in the internal ear to the vestibular (involved with balance) and cochlear (involved with hearing) nuclei, which extend from the pons into the medulla oblongata.
- *Relay Stations along Sensory and Motor Pathways.* The **gracile nucleus** and the **cuneate nucleus** pass somatic sensory

Figure 14–5 The Diencephalon and Brainstem. The brainstem is made up of the medulla oblongata, pons, and midbrain.

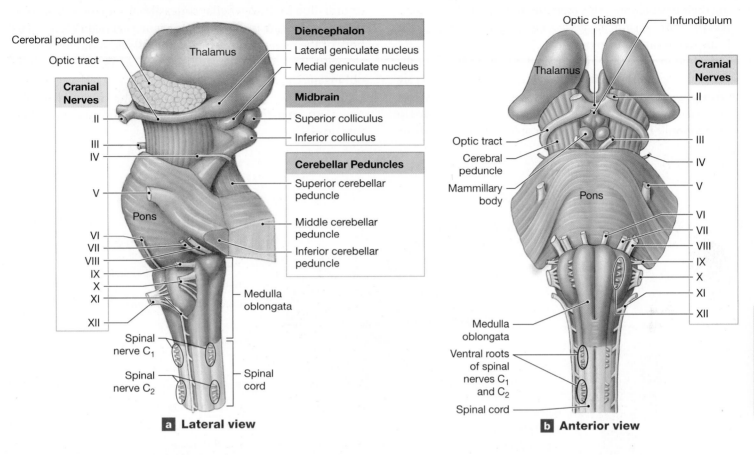

Cerebral peduncle
Optic tract
Thalamus

Cranial Nerves
II
III
IV
V
VI
VII
VIII
IX
X
XI
XII

Pons

Diencephalon
Lateral geniculate nucleus
Medial geniculate nucleus

Midbrain
Superior colliculus
Inferior colliculus

Cerebellar Peduncles
Superior cerebellar peduncle
Middle cerebellar peduncle
Inferior cerebellar peduncle

Medulla oblongata
Spinal nerve C_1
Spinal nerve C_2
Spinal cord

a Lateral view

Optic chiasm
Infundibulum
Thalamus

Cranial Nerves
II
III
IV
V
VI
VII
VIII
IX
X
XI
XII

Optic tract
Cerebral peduncle
Mammillary body
Pons
Medulla oblongata
Ventral roots of spinal nerves C_1 and C_2
Spinal cord

b Anterior view

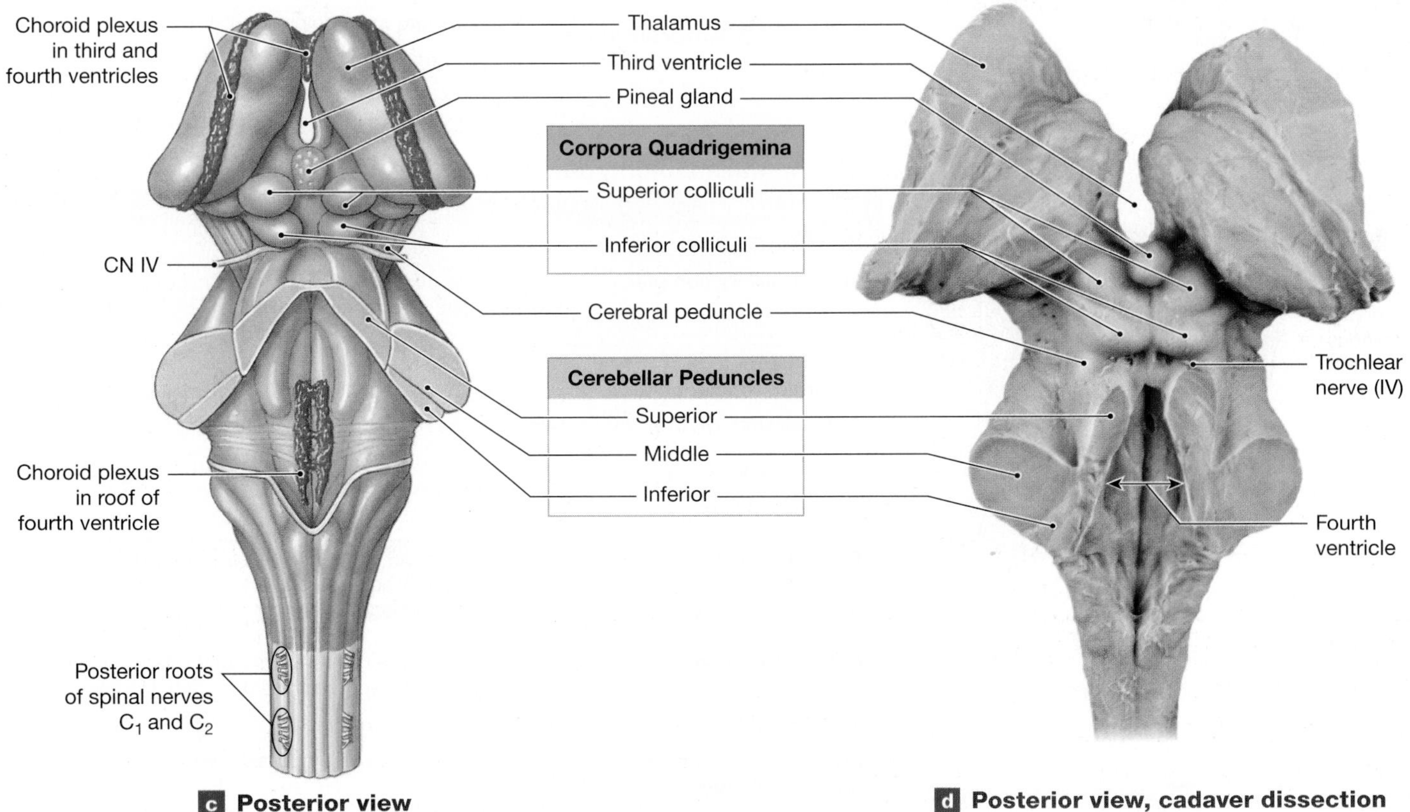

Choroid plexus in third and fourth ventricles
Thalamus
Third ventricle
Pineal gland
Corpora Quadrigemina
Superior colliculi
Inferior colliculi
CN IV
Cerebral peduncle
Cerebellar Peduncles
Superior
Middle
Inferior
Trochlear nerve (IV)
Choroid plexus in roof of fourth ventricle
Fourth ventricle
Posterior roots of spinal nerves C_1 and C_2

c Posterior view

d Posterior view, cadaver dissection

information to the thalamus. Tracts leaving these brainstem nuclei cross to the opposite side of the brain before reaching their destinations. This crossing over is called a *decussation* (dē-kuh-SĀ-shun; *decussatio*, crossing over), and the site is the **decussation of pyramids**.

In addition, a few important paired nuclei are found in the medulla oblongata. The **solitary nuclei** (*nuclei of solitary tract*) are the visceral sensory nuclei, receiving information from the spinal and cranial nerves. This information is integrated and forwarded to other autonomic centers in the medulla oblongata and elsewhere.

The **inferior olivary complex** consists of three nuclei that collectively form the inferior olivary nucleus. They relay information to the cerebellar cortex about somatic motor commands as they are issued by motor centers at higher levels. The bulk of the olivary nuclei creates the **olives**, prominent olive-shaped bulges along the ventrolateral surface of the medulla oblongata.

Figure 14–6 also contains a table that summarizes the major components of the medulla oblongata and their functions.

Figure 14–6 **The Medulla Oblongata.** ATLAS: Plates 9a–c; 11c

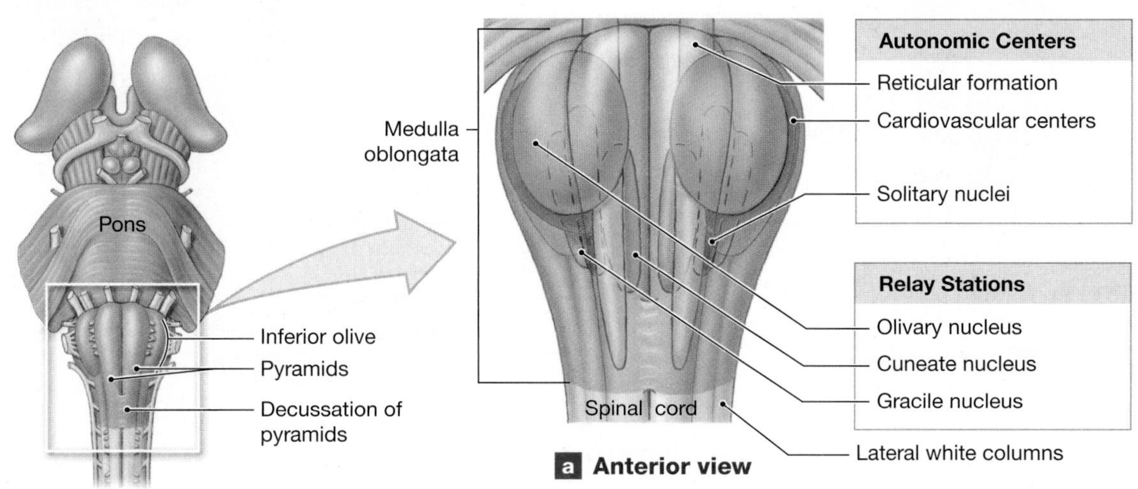

Pons
Medulla oblongata
Inferior olive
Pyramids
Decussation of pyramids
Autonomic Centers
Reticular formation
Cardiovascular centers
Solitary nuclei
Relay Stations
Olivary nucleus
Cuneate nucleus
Gracile nucleus
Spinal cord
Lateral white columns
a Anterior view

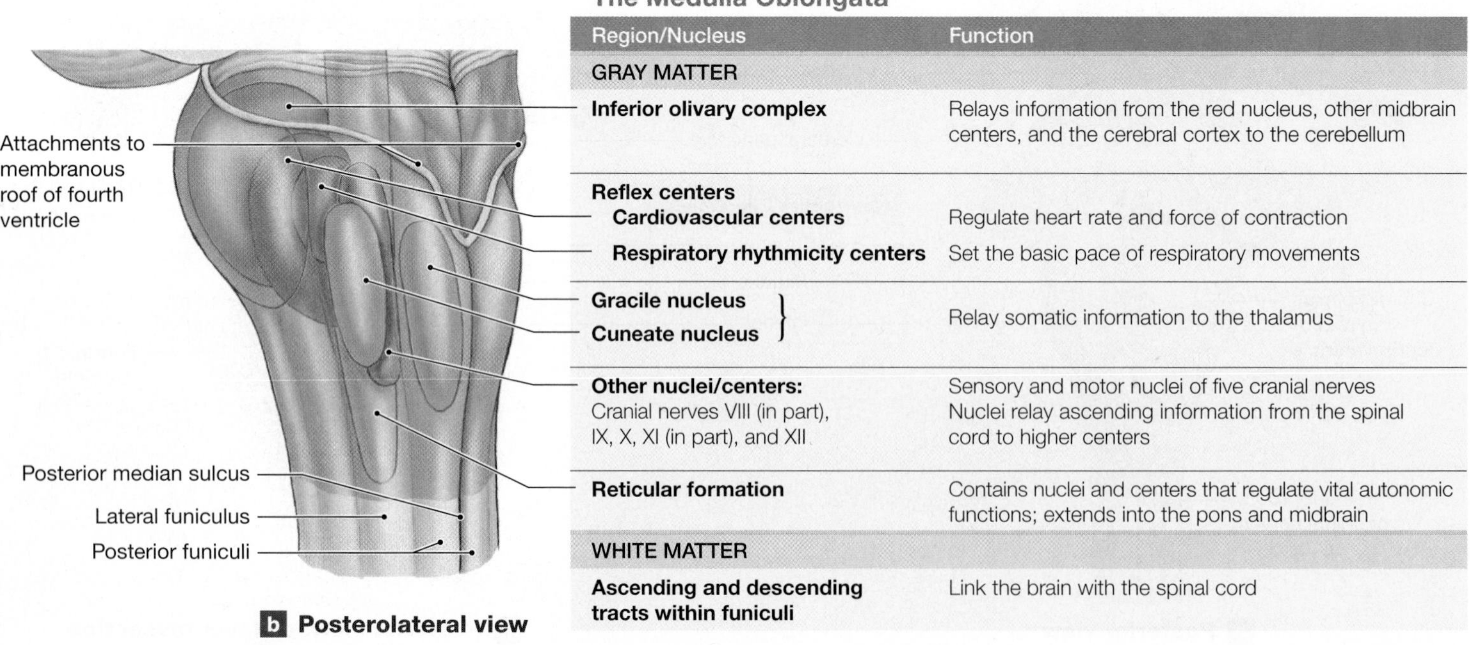

Attachments to membranous roof of fourth ventricle
Posterior median sulcus
Lateral funiculus
Posterior funiculi
b Posterolateral view

The Medulla Oblongata

Region/Nucleus	Function
GRAY MATTER	
Inferior olivary complex	Relays information from the red nucleus, other midbrain centers, and the cerebral cortex to the cerebellum
Reflex centers	
Cardiovascular centers	Regulate heart rate and force of contraction
Respiratory rhythmicity centers	Set the basic pace of respiratory movements
Gracile nucleus **Cuneate nucleus**	Relay somatic information to the thalamus
Other nuclei/centers: Cranial nerves VIII (in part), IX, X, XI (in part), and XII	Sensory and motor nuclei of five cranial nerves Nuclei relay ascending information from the spinal cord to higher centers
Reticular formation	Contains nuclei and centers that regulate vital autonomic functions; extends into the pons and midbrain
WHITE MATTER	
Ascending and descending tracts within funiculi	Link the brain with the spinal cord

Checkpoint

8. **Identify the components of the medulla oblongata that are responsible for relaying somatic sensory information to the thalamus.**
9. **The medulla oblongata is one of the smallest sections of the brain, yet damage there can cause death, whereas similar damage in the cerebrum might go unnoticed. Why?**

See the blue Answers tab at the back of the book.

14-4 Brainstem: The pons contains nuclei that process and tracts that relay sensory and motor information

Learning Outcome List the main components of the pons, and specify the functions of each.

The **pons** links the cerebellum with the midbrain, diencephalon, cerebrum, and spinal cord. Look back at the relationship of the pons with these structures in Figure 14–5.

Important features and regions of the pons are shown in Figure 14–7, along with a summary of its major components and their functions. The pons contains four groups of components:

- *Sensory and Motor Nuclei of Cranial Nerves.* These cranial nerves (V, VI, VII, and VIII) innervate the jaw muscles, the anterior surface of the face, one of the extrinsic eye muscles (the lateral rectus), and the sense organs of the internal ear (the vestibular and cochlear nuclei).
- *Nuclei Involved with the Control of Respiration.* Research on laboratory animals has identified regions in the medulla and pons, known as *centers*, that appear to modify breathing activities. Two pontine centers, the *apneustic* (ap-NŪS-tik) *center* located in the middle or lower pons, and the *pneumotaxic* (nū-mō-TAK-sik) *center* located in the rostral pons, process information originating in the respiratory rhythmicity centers of the medulla oblongata. These centers are further discussed in Chapter 23.
- *Nuclei and Tracts That Process and Relay Information Sent to or from the Cerebellum.* The pons links the cerebellum with the brainstem, cerebrum, and spinal cord.
- *Ascending, Descending, and Transverse Pontine Fibers.* Longitudinal tracts interconnect other portions of the CNS. Tracts of the cerebellum (middle cerebellar peduncles) are connected to the **transverse pontine fibers**, which cross the anterior surface of the pons. These fibers are axons that link nuclei of the pons (*pontine nuclei*) with the cerebellum of the opposite side.

Checkpoint

10. **Name the four groups of components found in the pons.**
11. **If the respiratory centers of the pons were damaged, what respiratory controls might be lost?**

See the blue Answers tab at the back of the book.

Figure 14–7 The Pons. ATLAS: Plates 9a–c; 11c

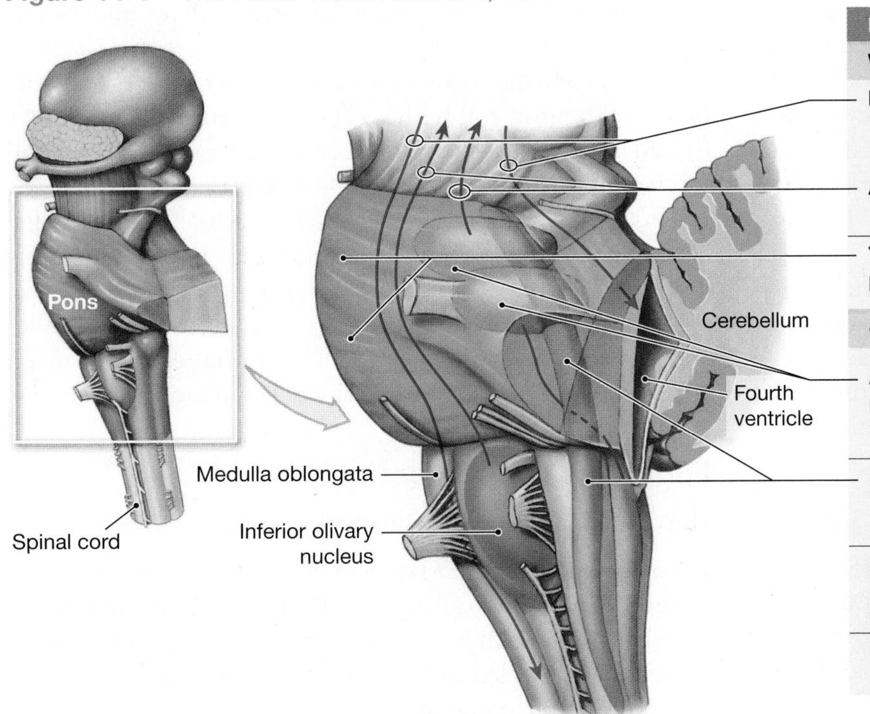
Pons
Cerebellum
Fourth ventricle
Medulla oblongata
Spinal cord
Inferior olivary nucleus

The Pons

Region/Nucleus	Function
WHITE MATTER	
Descending tracts	Carry motor commands from higher centers to nuclei of cranial or spinal nerves
Ascending tracts	Carry sensory information from brainstem nuclei to the thalamus
Transverse pontine fibers	Interconnect cerebellar hemispheres
GRAY MATTER	
Apneustic and pneumotaxic centers	Adjust activities of the respiratory rhythmicity centers in the medulla oblongata
Reticular formation	Automatic processing of incoming sensations and outgoing motor commands
Nuclei of cranial nerves V, VI, VII, and VIII (in part)	Relay sensory information and issue somatic motor commands
Other nuclei/relay centers	Relay sensory and motor information to the cerebellum

14-5 Brainstem: The midbrain regulates visual and auditory reflexes and controls alertness

Learning Outcome **List the main components of the midbrain, and specify the functions of each.**

Figure 14–8a shows the external anatomy of the midbrain, and lists its major nuclei. The **tectum**, or roof of the midbrain, is the region posterior (dorsal) to the cerebral aqueduct. It contains two pairs of sensory nuclei known collectively as the **corpora quadrigemina** (KOR-pōr-uh kwa-dri-JEM-ih-nuh; Figure 14–8b). These nuclei, the *superior* and *inferior colliculi*, process visual and auditory sensations. Each **superior colliculus** (ko-LIK-ū-lus; *colliculus*, hill) receives visual inputs from the lateral geniculate body of the thalamus on that side. Each **inferior colliculus** receives auditory input from nuclei in the medulla oblongata and pons. Some of this information may be forwarded to the medial geniculate body of the thalamus on the same side. The superior colliculi control the reflex movements of the eyes, head, and neck in response to visual stimuli, such as a bright light. The inferior colliculi control reflex movements of the head, neck, and trunk in response to auditory stimuli, such as a loud noise.

14

The area anterior to the cerebral aqueduct is called the **tegmentum**. On each side, the tegmentum contains a *red nucleus* and the *substantia nigra* (Figure 14–8c). The **red nucleus** contains numerous blood vessels, which give it a rich red color. This nucleus receives information from the cerebrum and cerebellum, and issues subconscious motor commands that affect upper limb position and background muscle tone. The **substantia nigra** (NĪ-gruh; *nigra*, black) is the largest midbrain nucleus. It lies lateral to the red nucleus. The neurons of the substantia nigra release the neurotransmitter *dopamine*. ⮌ p. 420 The gray matter in this region contains darkly pigmented cells, giving it a black color. The pigment is melanin and here it is a by-product of dopamine synthesis. The substantia nigra inhibits activity of the basal nuclei in the cerebrum. The *basal nuclei* are involved in the subconscious control of muscle tone and learned movements. As discussed in Section 14-9, *Parkinson's disease* is characterized by the loss of neuronal activity in the substantia nigra.

The nerve fiber bundles on the ventrolateral surfaces of the midbrain (see Figure 14–8b,c) are the **cerebral peduncles** (PEH-dung-kelz; *peduncles*, little feet). They contain (1) descending fibers that go to the cerebellum by way of the pons and (2) descending fibers that carry voluntary motor commands issued by the cerebral hemispheres.

Although the reticular formation extends the length of the brainstem, its headquarters resides in the midbrain. The midbrain also contains the *reticular activating system (RAS)*, a specialized component of the reticular formation. Stimulation of the RAS makes you more alert and attentive; damage to the RAS produces unconsciousness. We will consider the role of the RAS in maintaining consciousness in Chapter 16.

✓ Checkpoint

12. **Identify the sensory nuclei within the corpora quadrigemina**
13. **Which area of the midbrain controls reflexive movements of the eyes, head, and neck in response to visual stimuli?**

See the blue Answers tab at the back of the book.

14-6 The cerebellum coordinates reflexive and learned patterns of muscular activity at the subconscious level

Learning Outcome **List the main components of the cerebellum, and specify the functions of each.**

The cerebellum is an automatic processing and coordination center for patterns of muscular activity. Let's look at its structure and function.

Structure of the Cerebellum

The cerebellum has a complex, highly convoluted surface composed of gray matter called the **cerebellar cortex** (Figure 14–9a). The **folia** (FŌ-lē-uh; leaves), or folds of the cerebellum surface, are less prominent than the folds in the surfaces of the cerebral hemispheres. The **primary fissure** separates the **anterior** and **posterior lobes**. Along the midline, a narrow band of cortex known as the **vermis** separates the **cerebellar hemispheres**. The slender **flocculonodular** (flok-yū-lō-NOD-yū-lar) **lobe** lies between the roof of the fourth ventricle and the cerebellar hemispheres and vermis (Figure 14–9b).

The cerebellar cortex contains a layer of large, highly branched neuron cell bodies called the **Purkinje** (pur-KIN-jē) **cell layer**. The extensive dendrites of each Purkinje cell receive input from up to 200,000 synapses—more than any other type of brain cell. The internal white matter of the cerebellum forms a branching array that in sectional view resembles a tree. Anatomists call it the **arbor vitae** (VĒ-tā), or "tree of life" (see Figure 14–9b). It connects the cerebellar cortex and nuclei with the tracts of white matter that form the **cerebellar peduncles**, which we discuss in more detail soon.

Functions of the Cerebellum

The cerebellum has two primary functions:

- *Adjusting the Postural Muscles of the Body.* The cerebellum coordinates rapid, automatic adjustments that maintain balance and equilibrium. It makes these alterations in muscle tone and position by modifying the activities of motor centers in the brainstem.

Go to MasteringA&P™ > Study Area > Menu > Lab Tools > Bone & Dissection Videos > Sheep Brain: Brainstem

Figure 14–8 The Midbrain. ATLAS: Plates 7b; 9a–c; 11c; 12a,c

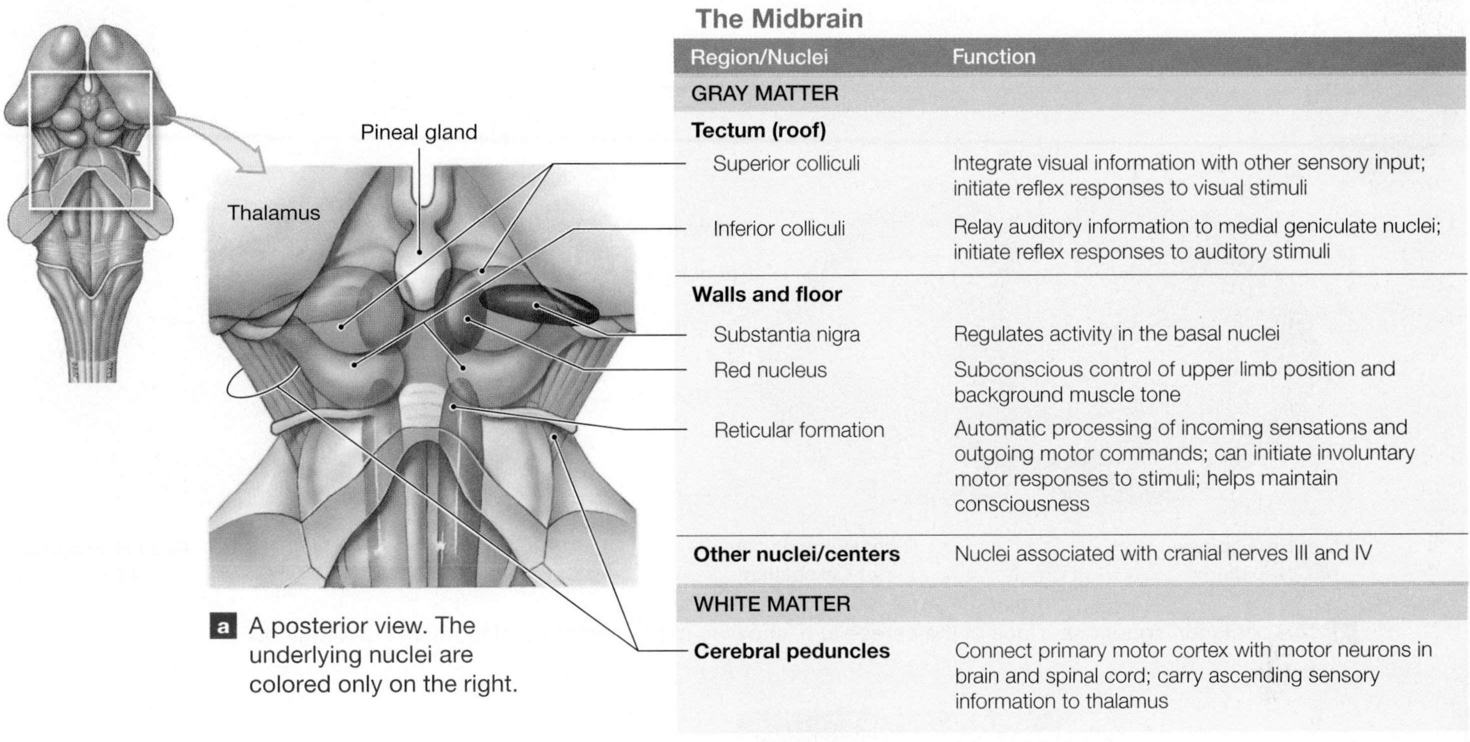
Pineal gland
Thalamus

The Midbrain

Region/Nuclei	Function
GRAY MATTER	
Tectum (roof)	
Superior colliculi	Integrate visual information with other sensory input; initiate reflex responses to visual stimuli
Inferior colliculi	Relay auditory information to medial geniculate nuclei; initiate reflex responses to auditory stimuli
Walls and floor	
Substantia nigra	Regulates activity in the basal nuclei
Red nucleus	Subconscious control of upper limb position and background muscle tone
Reticular formation	Automatic processing of incoming sensations and outgoing motor commands; can initiate involuntary motor responses to stimuli; helps maintain consciousness
Other nuclei/centers	Nuclei associated with cranial nerves III and IV
WHITE MATTER	
Cerebral peduncles	Connect primary motor cortex with motor neurons in brain and spinal cord; carry ascending sensory information to thalamus

a A posterior view. The underlying nuclei are colored only on the right.

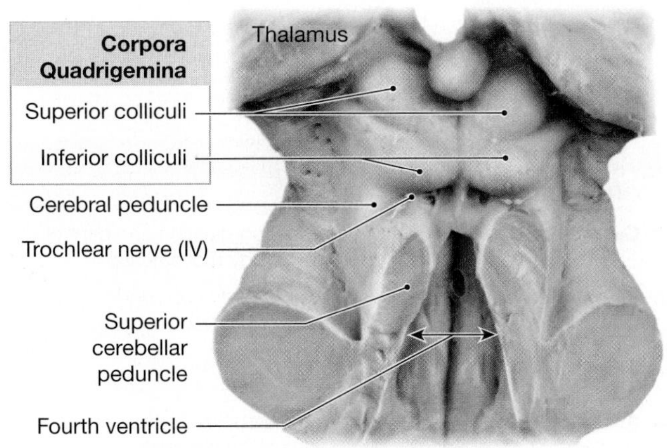
Thalamus
Corpora Quadrigemina
Superior colliculi
Inferior colliculi
Cerebral peduncle
Trochlear nerve (IV)
Superior cerebellar peduncle
Fourth ventricle

b Posterior view of a cadaver dissection of the diencephalon and brainstem.

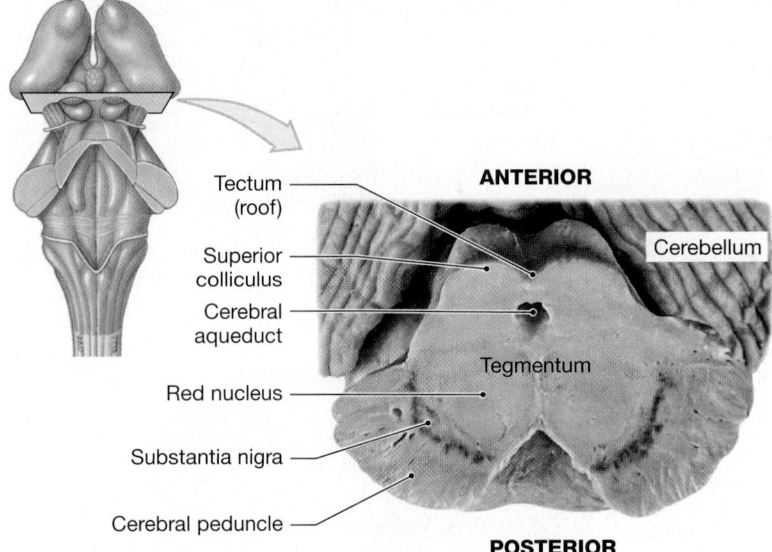
ANTERIOR
Tectum (roof)
Superior colliculus
Cerebral aqueduct
Cerebellum
Tegmentum
Red nucleus
Substantia nigra
Cerebral peduncle
POSTERIOR

c A superior view of a transverse section at the level of the midbrain.

- *Programming and Fine-Tuning Movements Controlled at the Conscious and Subconscious Levels.* The cerebellum refines learned movement patterns, such as riding a bicycle or playing the piano. It performs this function indirectly by regulating activity along motor pathways at the cerebral cortex, basal nuclei, and motor centers in the brainstem. The cerebellum compares the motor commands with *proprioceptive* information (position sense) and stimulates any adjustments needed to make the movement smooth.

The cerebellum receives proprioceptive information from the spinal cord and monitors all proprioceptive, visual, tactile, balance, and auditory sensations received by the brain. Most axons that carry sensory information do not synapse in the cerebellar nuclei but pass through the deeper layers of the cerebellum on their way to the Purkinje cells of the cerebellar cortex. Information about the motor commands issued at the conscious and subconscious levels reaches the Purkinje cells indirectly, after being relayed by nuclei in the pons or by the **cerebellar nuclei** embedded within the arbor vitae.

14

Figure 14–9 The Cerebellum.

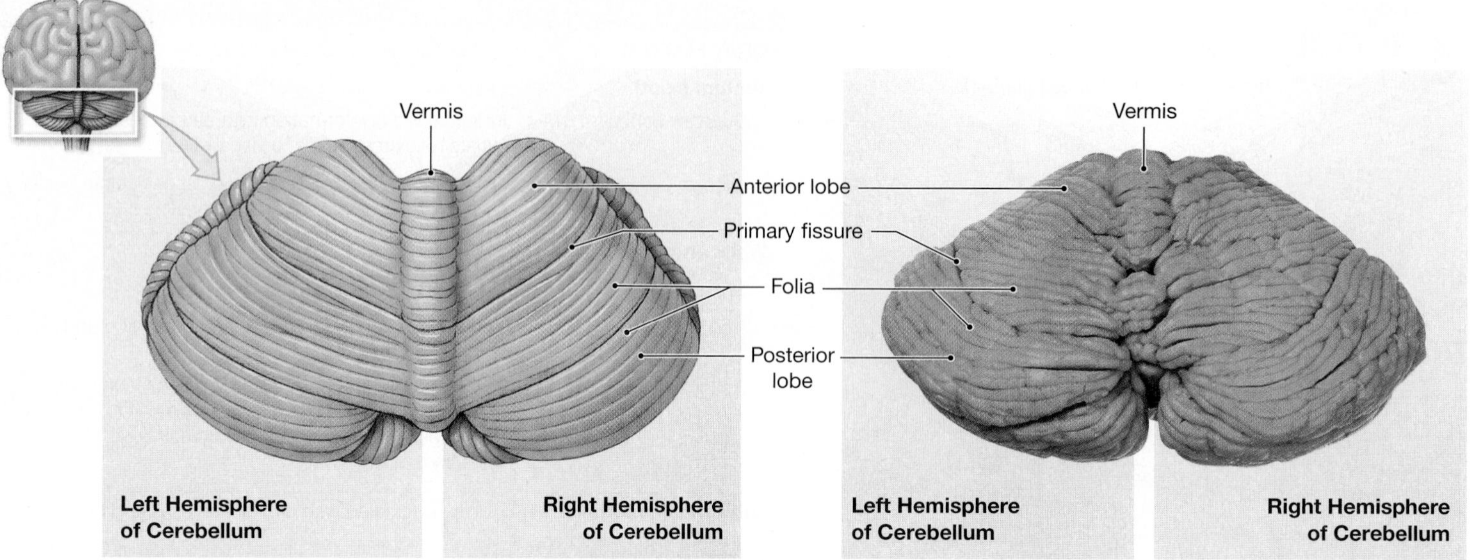
Vermis
Anterior lobe
Primary fissure
Folia
Posterior lobe
Left Hemisphere of Cerebellum
Right Hemisphere of Cerebellum
Vermis
Left Hemisphere of Cerebellum
Right Hemisphere of Cerebellum

a The posterior, superior surface of the cerebellum, showing major anatomical landmarks and regions

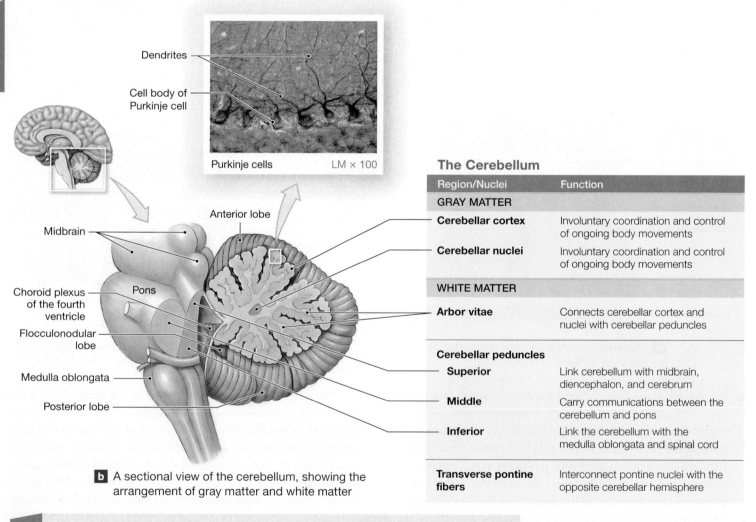

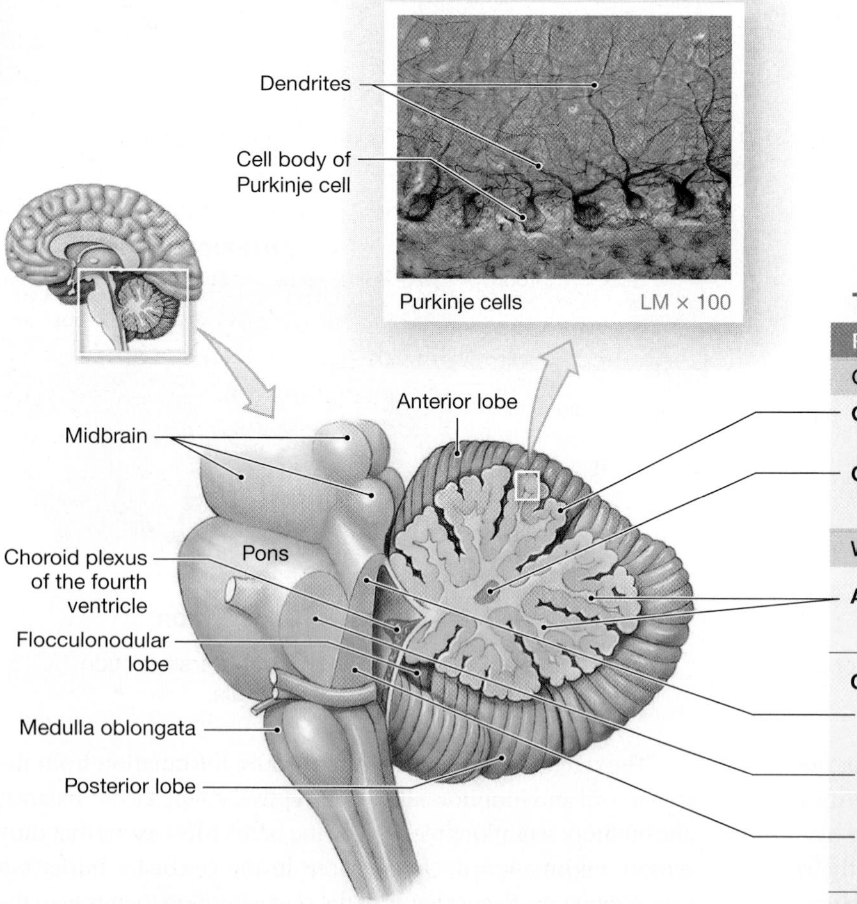

Dendrites

Cell body of Purkinje cell

Purkinje cells LM × 100

Anterior lobe

Midbrain

Choroid plexus of the fourth ventricle

Pons

Flocculonodular lobe

Medulla oblongata

Posterior lobe

b A sectional view of the cerebellum, showing the arrangement of gray matter and white matter

The Cerebellum

Region/Nuclei	Function
GRAY MATTER	
Cerebellar cortex	Involuntary coordination and control of ongoing body movements
Cerebellar nuclei	Involuntary coordination and control of ongoing body movements
WHITE MATTER	
Arbor vitae	Connects cerebellar cortex and nuclei with cerebellar peduncles
Cerebellar peduncles	
Superior	Link cerebellum with midbrain, diencephalon, and cerebrum
Middle	Carry communications between the cerebellum and pons
Inferior	Link the cerebellum with the medulla oblongata and spinal cord
Transverse pontine fibers	Interconnect pontine nuclei with the opposite cerebellar hemisphere

? What makes up the gray matter in the cerebellum?

Tracts that link the cerebellum with the brainstem, cerebrum, and spinal cord leave the cerebellar hemispheres as the *superior, middle,* and *inferior cerebellar peduncles*:

- The **superior cerebellar peduncles** link the cerebellum with nuclei in the midbrain, diencephalon, and cerebrum.
- The **middle cerebellar peduncles** are connected to the transverse pontine fibers. The middle cerebellar peduncles also connect the cerebellar hemispheres with sensory and motor nuclei in the pons.
- The **inferior cerebellar peduncles** communicate between the cerebellum and nuclei in the medulla oblongata and carry ascending and descending cerebellar tracts from the spinal cord.

The cerebellum can be permanently damaged by trauma or stroke. Drugs such as alcohol can also temporarily affect it. The result is **ataxia** (ah-TAK-sē-uh; *ataxia,* lack of order), a disturbance in muscular coordination. In severe ataxia, the individual cannot sit or stand without assistance.

Checkpoint

14. **Identify the components of the cerebellar gray matter.**
15. **What part of the brain has the arbor vitae? What is its function?**

See the blue Answers tab at the back of the book.

14-7 The diencephalon integrates sensory information with motor output at the subconscious level

Learning Outcome List the main components of the diencephalon, and specify the functions of each.

The diencephalon is a division of the brain that consists of the epithalamus, thalamus, and hypothalamus. Look back at Figure 14–5 to see its position and its relationship to landmarks on the brainstem.

The *epithalamus* is the roof of the diencephalon superior to the third ventricle. The anterior portion of the epithalamus contains an extensive area of choroid plexus that extends through the interventricular foramina into the lateral ventricles. The posterior portion of the epithalamus contains the **pineal** (PIN-ē-ul) **gland** (look back at Figure 14–5c,d), an endocrine structure that secretes the hormone **melatonin**. Melatonin is important in regulating day–night cycles and reproductive functions. (We describe the role of melatonin in Chapter 18.)

Most of the nervous tissue in the diencephalon is concentrated in the *thalamus,* which forms the lateral walls, and the *hypothalamus,* which forms the floor. The rest of this section looks at the structure and critical functions of the thalamus and hypothalamus.

The Thalamus

On each side of the diencephalon, the **thalamus** is the final relay point for sensory information ascending to the cerebral cortex. Ascending sensory information from the spinal cord and cranial nerves (other than the olfactory tract) synapses in a nucleus in the left side or right side of the thalamus before reaching the cerebral cortex and our conscious awareness. It acts as a filter, passing on only a small portion of the arriving sensory information. The thalamus also coordinates the activities of the basal nuclei (which we discuss shortly) and the cerebral cortex by relaying information between them.

Structure of the Thalamus

The third ventricle separates the thalamus into left and right sides. Each side consists of a rounded mass of groups of *thalamic nuclei* (Figure 14–10). Viewed in a midsagittal section through the brain (Figure 14–11a,b), each side of the thalamus extends from the anterior commissure to the inferior base of the pineal gland. A projection of gray matter called an **interthalamic adhesion** extends into the ventricle from the thalamus on each side, but no fibers cross the midline (see Figure 14–11b).

Functions of Thalamic Nuclei

The thalamic nuclei deal primarily with the relay of sensory information to the basal nuclei and cerebral cortex. Five major groups of thalamic nuclei are shown in Figure 14–10b and listed in Table 14–2. They are the anterior, medial, ventral, and dorsal nuclei, and the lateral and medial geniculate bodies.

- The **anterior nuclei of thalamus** are part of the *limbic system*. This system affects emotional states and integrates sensory information. We discuss it in Section 14-8.
- The **medial nuclei of thalamus** provide an awareness of emotional states by connecting emotional centers in the hypothalamus with the *frontal lobes* of the cerebral hemispheres. The medial nuclei also receive and relay sensory information from other portions of the thalamus.
- The **ventral nuclei of thalamus** relay information from the basal nuclei of the cerebrum and the cerebellum to somatic motor areas of the cerebral cortex. Ventral group nuclei also relay sensory information about touch, pressure, pain, temperature, and proprioception (position) to the sensory areas of the cerebral cortex.
- The **dorsal nuclei of thalamus** include the lateral dorsal nucleus and pulvinar nuclei. The **pulvinar** is the expanded region of the thalamus overlying the geniculate bodies. These dorsal nuclei integrate sensory information for sending signals (projecting) to the cerebral cortex.

Figure 14–10 The Thalamus.

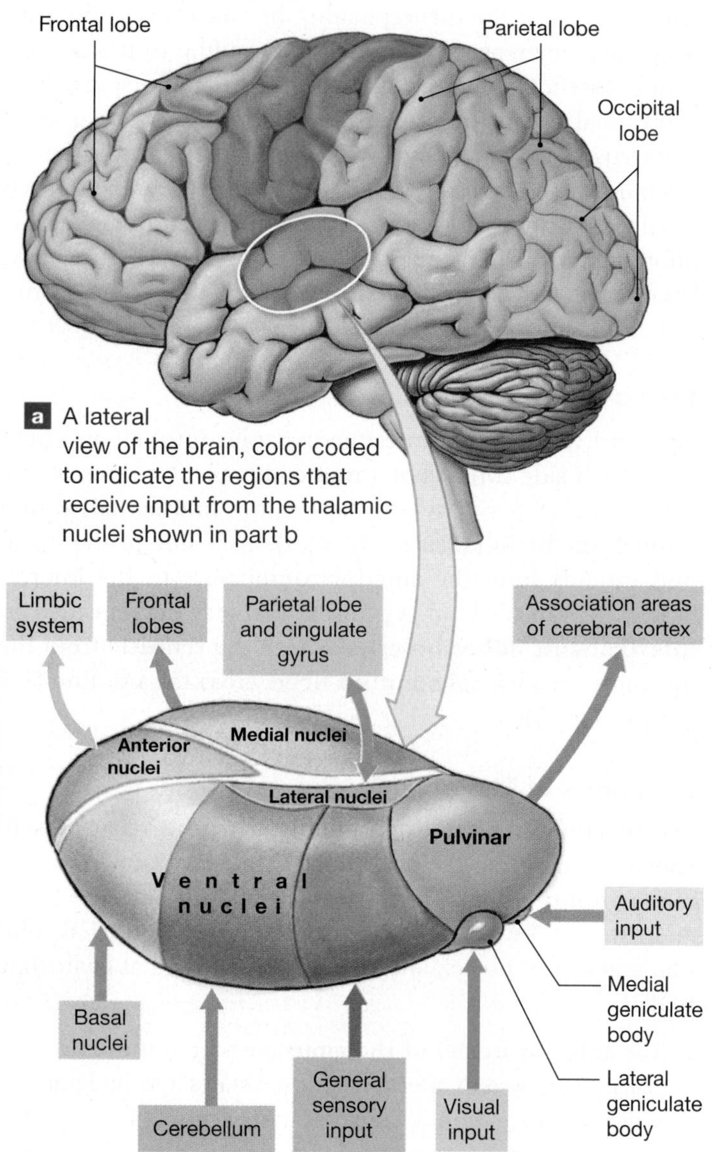
Frontal lobe
Parietal lobe
Occipital lobe
Limbic system
Frontal lobes
Parietal lobe and cingulate gyrus
Association areas of cerebral cortex
Anterior nuclei
Medial nuclei
Lateral nuclei
Pulvinar
Ventral nuclei
Auditory input
Medial geniculate body
Lateral geniculate body
Basal nuclei
Cerebellum
General sensory input
Visual input

a A lateral view of the brain, color coded to indicate the regions that receive input from the thalamic nuclei shown in part b

b An enlarged view of the thalamic nuclei of the left side

- The **lateral geniculate** (je-NIK-yū-lāt) **body** receives visual information over the *optic tract,* which originates at the eyes. The output of the **lateral geniculate body** goes to the *occipital lobes* of the cerebral hemispheres and to the midbrain. The **medial geniculate body** relays auditory information to the appropriate area of the cerebral cortex from specialized receptors of the internal ear. This body forms feedback loops with the limbic system and the *parietal lobes* of the cerebral hemispheres.

The Hypothalamus

The hypothalamus contains centers involved with emotions and visceral processes that affect the cerebrum as well as components of the brainstem. It also controls a variety of autonomic functions and forms the link between the nervous and endocrine systems. Let's look at its structure and main functions.

Table 14–2 The Thalamus

Nuclei/Body	Function
Anterior nuclei	Part of the limbic system
Medial nuclei	Integrate sensory information for projection to the frontal lobes
Ventral nuclei	Project sensory information to the primary sensory cortex; relay information from cerebellum and basal nuclei to motor area of cerebral cortex
Dorsal nuclei	
Lateral dorsal nucleus	Projects information to parietal, occipitoparietal, and temporal cortex; may play a role in memory
Pulvinar nuclei	Integrate sensory information for projection to association areas of cerebral cortex
Lateral geniculate body	Projects visual information to the visual cortex; integrates sensory information and influences emotional states
Medial geniculate body	Projects auditory information to the auditory cortex; integrates sensory information and influences emotional states

Structure of the Hypothalamus

The **hypothalamus** extends from the area superior to the *optic chiasm* (the location where axons within the optic nerves arrive at the brain and cross over) to the posterior margins of the **mammillary** (*mammilla,* nipple) **bodies** (see Figure 14–11a). Immediately posterior to the optic chiasm, a narrow stalk called the **infundibulum** (in-fun-DIB-ū-lum; *infundibulum,* funnel) extends inferiorly, connecting the floor of the hypothalamus to the pituitary gland (see Figure 14–11b). The floor of the hypothalamus between the infundibulum and the mammillary bodies is the **tuber cinereum** (TŪ-ber si-nir-Ē-um; *tuber,* swelling).

Functions of the Hypothalamus

The hypothalamus contains important nuclei that function as control and integrative centers in addition to those associated with the limbic system. These centers are shown in Figure 14–11b, along with a summary of their functions.

Hypothalamic centers may be stimulated by (1) sensory information from the cerebrum, brainstem, and spinal cord; (2) changes in the compositions of the CSF and interstitial fluid; or (3) chemical stimuli in the circulating blood that move rapidly across highly permeable capillaries to enter the hypothalamus (where there is no blood brain barrier).

Figure 14–11 The Hypothalamus in Sagittal Section.

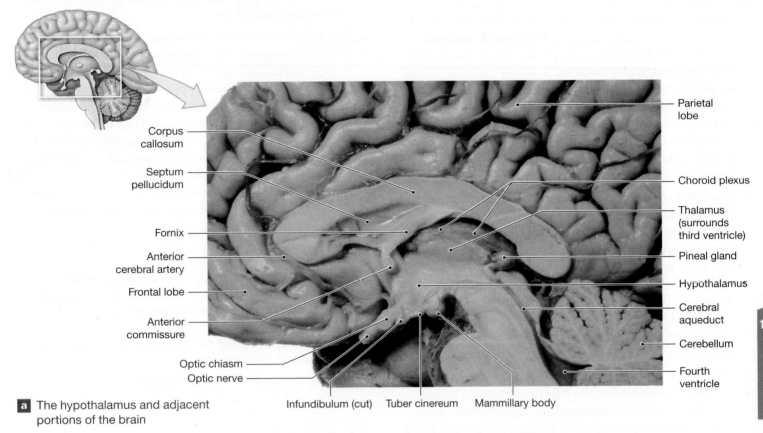

Corpus callosum
Septum pellucidum
Fornix
Anterior cerebral artery
Frontal lobe
Anterior commissure
Optic chiasm
Optic nerve
Infundibulum (cut)
Tuber cinereum
Mammillary body
Parietal lobe
Choroid plexus
Thalamus (surrounds third ventricle)
Pineal gland
Hypothalamus
Cerebral aqueduct
Cerebellum
Fourth ventricle

a The hypothalamus and adjacent portions of the brain

14

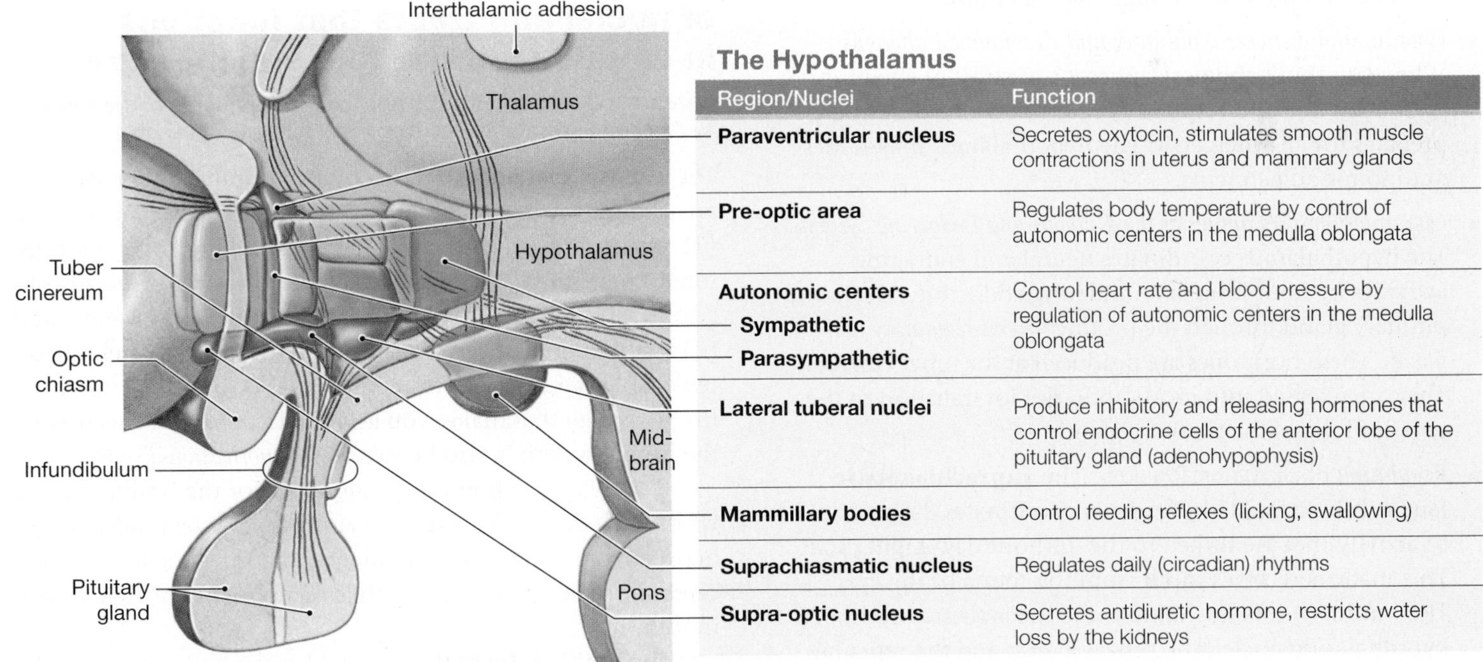
Interthalamic adhesion
Thalamus
Hypothalamus
Tuber cinereum
Optic chiasm
Infundibulum
Pituitary gland
Mid-brain
Pons

The Hypothalamus

Region/Nuclei	Function
Paraventricular nucleus	Secretes oxytocin, stimulates smooth muscle contractions in uterus and mammary glands
Pre-optic area	Regulates body temperature by control of autonomic centers in the medulla oblongata
Autonomic centers **Sympathetic** **Parasympathetic**	Control heart rate and blood pressure by regulation of autonomic centers in the medulla oblongata
Lateral tuberal nuclei	Produce inhibitory and releasing hormones that control endocrine cells of the anterior lobe of the pituitary gland (adenohypophysis)
Mammillary bodies	Control feeding reflexes (licking, swallowing)
Suprachiasmatic nucleus	Regulates daily (circadian) rhythms
Supra-optic nucleus	Secretes antidiuretic hormone, restricts water loss by the kidneys

b A diagram of the hypothalamus, showing the locations and functions of major nuclei and centers

The hypothalamus performs the following functions:

- *Secretion of Two Hormones.* The hypothalamus secretes *antidiuretic hormone* (*ADH*, also called vasopressin) and *oxytocin* (ok-sih-TOH-sin). The **supra-optic nucleus** produces ADH, which restricts water loss by the kidneys. The **paraventricular nucleus** produces oxytocin, which stimulates smooth muscle contractions in the uterus and mammary glands of females and the prostate of males. These hormones are transported along axons that pass through the infundibulum to the posterior lobe of the pituitary gland. There the hormones are released into the blood for distribution throughout the body.
- *Regulation of Body Temperature.* The **pre-optic area** of the hypothalamus is responsible for thermoregulation. It coordinates the activities of other CNS centers and regulates other physiological systems to maintain normal body temperature. If body temperature falls, the preoptic area communicates with the vasomotor center, an autonomic center in the medulla oblongata that controls blood flow by regulating the diameter of peripheral blood vessels. In response, the vasomotor center decreases the blood supply to the skin, reducing the rate of heat loss.
- *Control of Autonomic Function.* The hypothalamus adjusts and coordinates the activities of autonomic centers in the pons and medulla oblongata that regulate heart rate, blood pressure, respiration, and digestive functions.
- *Coordination between Voluntary and Autonomic Functions.* When you think about a dangerous or stressful situation, your heart rate and respiratory rate go up and your body prepares for an emergency. The hypothalamus makes these autonomic adjustments.
- *Coordination of Activities of the Nervous and Endocrine Systems.* The hypothalamus coordinates neural and endocrine activities by inhibiting or stimulating endocrine cells in the pituitary gland through the production of *regulatory hormones*. These hormones are produced at the tuber cinereum. They are released into local capillaries for transport to the anterior lobe of the pituitary gland.
- *Regulation of Circadian Rhythms.* The **suprachiasmatic** (sup-rah-kī-az-MAT-ik) **nucleus** coordinates daily cycles of activity that are linked to the 24-hour day–night cycle. This nucleus receives input from the retina of the eye. The output of this nucleus adjusts the activities of other hypothalamic nuclei, the pineal gland, and the reticular formation.
- *Subconscious Control of Skeletal Muscle Contractions.* The hypothalamus directs somatic motor patterns associated with rage, pleasure, pain, and sexual arousal by stimulating centers in other portions of the brain. For example, hypothalamic centers control the changes in facial expression that accompany rage and the basic movements associated with sexual activity.
- *Production of Emotions and Behavioral Drives.* Specific hypothalamic centers produce sensations that lead to conscious or subconscious changes in behavior. For example, stimulation of the **feeding center** produces the sensation of hunger. Stimulation of the **thirst center** produces the sensation of thirst. The **satiety** (sah-TĪ-eh-tē) **center** is the region that regulates food intake. These unfocused "impressions" originating in the hypothalamus are called **drives**. The conscious sensations are only part of the hypothalamic response. For instance, the thirst center also orders the release of ADH by neurons in the supra-optic nucleus.

14

Checkpoint

16. **Name the main components of the diencephalon.**
17. **Damage to the lateral geniculate body of the thalamus would interfere with the functions of which special sense?**
18. **Which component of the diencephalon is stimulated by changes in body temperature?**

See the blue Answers tab at the back of the book.

14-8 The limbic system is a group of nuclei and tracts that functions in emotion, motivation, and memory

Learning Outcome **Identify the main components of the limbic system, and specify the locations and functions of each.**

The **limbic system** is a functional grouping rather than an anatomical one. Functions of the limbic system include (1) establishing emotional states; (2) linking the conscious, intellectual functions of the cerebral cortex with the unconscious and autonomic functions of the brainstem; and (3) facilitating memory storage and retrieval. The cerebral cortex enables you to perform complex tasks, but it is largely the limbic system that makes you *want* to do them. For this reason, the limbic system is also known as the *motivational system*.

Figure 14–12 shows the major parts of the limbic system, which includes nuclei and tracts along the border (*limbus*, edge) between the cerebrum and diencephalon. Let's look at the cerebral, diencephalic, and other components of this system in turn.

The **limbic lobe** of the cerebral hemisphere consists of the superficial folds, or gyri, and underlying structures adjacent to the diencephalon. The gyri curve along the *corpus callosum*, a fiber tract that links the two cerebral hemispheres. The gyri then continue on to the medial surface of the cerebrum lateral to the diencephalon (Figure 14–12a). There are three gyri in the limbic lobe. The **cingulate** (SIN-gyū-lāt) **gyrus** (*cingulum*, girdle or belt) sits superior to the corpus callosum. The **dentate gyrus** and the **parahippocampal** (pah-rah-hip-ō-KAM-pal) **gyrus** form the posterior and inferior portions of the limbic lobe (Figure 14–12b).

These gyri conceal the **hippocampus**, a nucleus inferior to the floor of the lateral ventricle. To early anatomists, this structure resembled a sea horse (*hippocampus*). This nucleus is important in learning, especially in the storage and retrieval of new long-term memories.

Another cerebral nucleus is the **amygdaloid** (ah-MIG-dah-loyd; *amygdale*, almond) **body**, commonly referred to as the *amygdala* (Figure 14–12b). It appears to act as an interface between the limbic system, the cerebrum, and various sensory systems. It plays a role in regulating heart rate, in responding to fear and anxiety as well as controlling the "fight or flight" response by the sympathetic division of the autonomic nervous system (ANS), and in linking emotions with specific memories.

The **fornix** (FOR-niks, arch) is a tract of cerebral white matter that connects the hippocampus with the hypothalamus

Figure 14–12 The Limbic System.

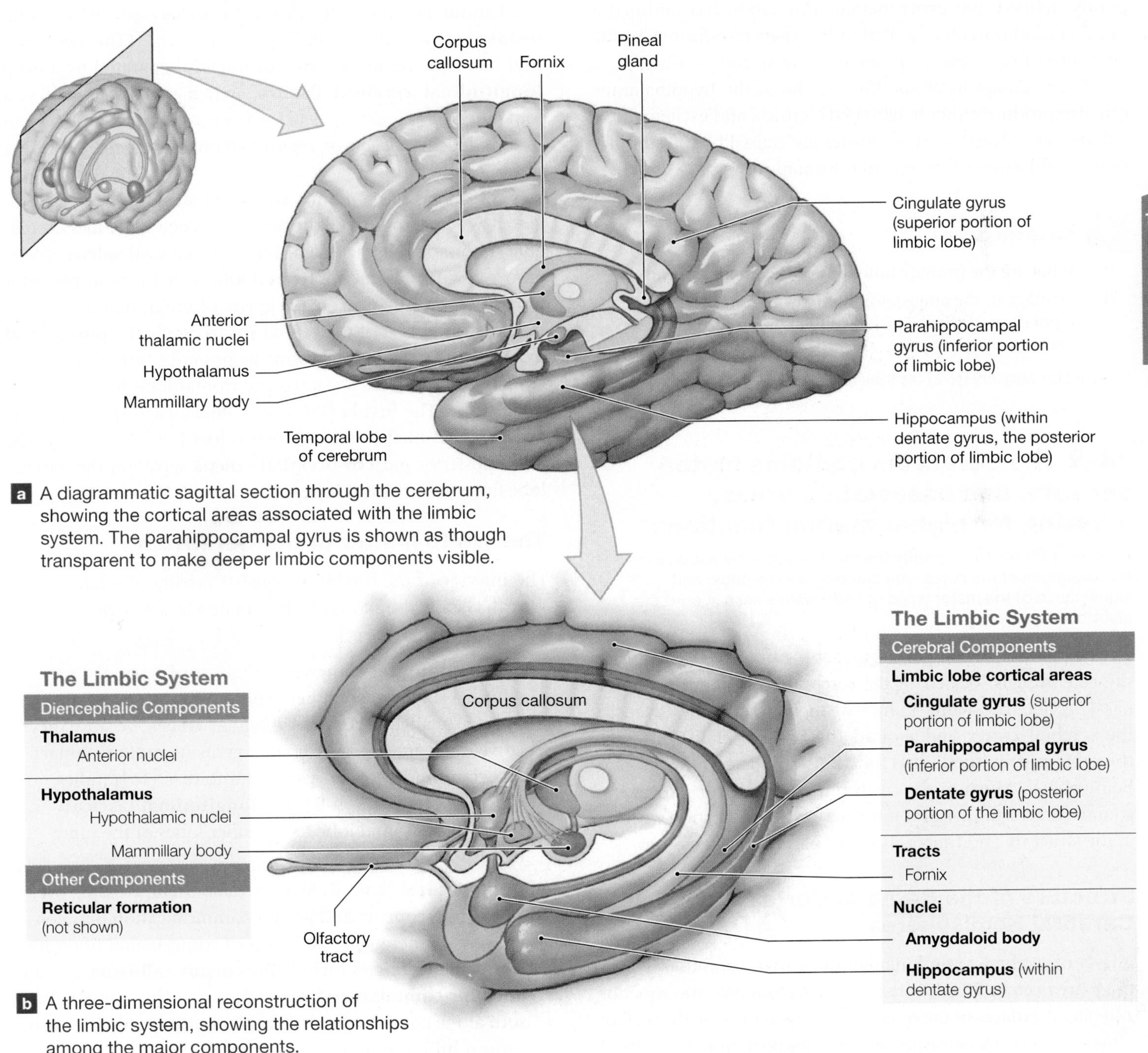

Corpus callosum
Fornix
Pineal gland
Cingulate gyrus (superior portion of limbic lobe)
Anterior thalamic nuclei
Hypothalamus
Mammillary body
Parahippocampal gyrus (inferior portion of limbic lobe)
Temporal lobe of cerebrum
Hippocampus (within dentate gyrus, the posterior portion of limbic lobe)
The Limbic System
Diencephalic Components
Thalamus
Anterior nuclei
Hypothalamus
Hypothalamic nuclei
Mammillary body
Other Components
Reticular formation (not shown)
Corpus callosum
Olfactory tract
The Limbic System
Cerebral Components
Limbic lobe cortical areas
Cingulate gyrus (superior portion of limbic lobe)
Parahippocampal gyrus (inferior portion of limbic lobe)
Dentate gyrus (posterior portion of the limbic lobe)
Tracts
Fornix
Nuclei
Amygdaloid body
Hippocampus (within dentate gyrus)

a A diagrammatic sagittal section through the cerebrum, showing the cortical areas associated with the limbic system. The parahippocampal gyrus is shown as though transparent to make deeper limbic components visible.

b A three-dimensional reconstruction of the limbic system, showing the relationships among the major components.

14

(see Figure 14–12b). From the hippocampus, the fornix curves medially, meeting its counterpart from the opposing hemisphere. The fornix proceeds anteriorly, inferior to the corpus callosum, before curving toward the hypothalamus. Many fibers of the fornix end in the mammillary bodies of the hypothalamus.

In the diencephalon, several other nuclei in the wall (thalamus) and floor (hypothalamus) are components of the limbic system. The anterior nuclei of thalamus (see Figure 14–12b) relay information from the mammillary body (of the hypothalamus) to the cingulate gyrus on that side. The boundaries between the hypothalamic nuclei of the limbic system are often poorly defined, but experimental stimulation has outlined a number of important hypothalamic centers responsible for the emotions of rage, fear, pain, sexual arousal, and pleasure.

The stimulation of specific regions of the hypothalamus can also produce either heightened alertness and excitement or lethargy and sleep. These responses are caused by the stimulation or inhibition of the reticular formation.

Checkpoint

19. **What are the primary functions of the limbic system?**
20. **Damage to the amygdaloid body would interfere with regulation of what division of the autonomic nervous system (ANS)?**

See the blue Answers tab at the back of the book.

14-9 The cerebrum contains motor, sensory, and association areas, allowing for higher mental functions

Learning Outcome **Identify the major anatomical subdivisions and functions of the cerebrum, and discuss the origin and significance of the major types of brain waves seen in an electroencephalogram.**

The cerebrum is the largest region of the brain. Gray matter in the cerebrum is located in the cerebral cortex and in deeper basal nuclei. The white matter of the cerebrum lies deep to the cerebral cortex and around the basal nuclei. Conscious thoughts and all intellectual functions originate in the cerebral hemispheres. Much of the cerebrum is involved in processing somatic sensory and motor information. In this section we look at the structure and functions of the parts of the cerebrum.

Structure of the Cerebral Cortex and Cerebral Hemispheres

A layer of cerebral cortex ranging from 1 to 4.5 mm thick covers the paired cerebral hemispheres, which dominate the superior and lateral surfaces of the cerebrum. The gyri increase the surface area of the cerebral hemispheres, and thus the number of cortical neurons they contain. The total surface area of the cerebral hemispheres is roughly equivalent to 2200 cm^2 (2.5 ft^2) of flat surface.

The entire brain has enlarged over the course of human evolution, but the cerebral hemispheres have enlarged at a much faster rate than has the rest of the brain. This enlargement reflects the large numbers of neurons needed for complex analytical and integrative functions. Since the neurons involved are in the superficial layer of cortex, it is there that the expansion has been most pronounced. The only solution available, other than an enlargement of the entire skull, was for the cortical layer to fold like a crumpled piece of paper into the sulci and gyri.

Landmarks and features on the surface of one cerebral hemisphere are shown in Figure 14–13a,b. (The two cerebral hemispheres are almost completely separated by a deep **longitudinal cerebral fissure**, which you can see if you look ahead to Figure 14–14b). Each cerebral hemisphere can be divided into *lobes*, or regions, named after the overlying bones of the skull.

Your brain has a unique pattern of sulci and gyri, as individual as a fingerprint, but the boundaries between lobes are reliable landmarks. On each hemisphere, the **central sulcus**, a deep groove, divides the anterior **frontal lobe** from the more posterior **parietal lobe**. The **precentral gyrus** of the frontal lobe forms the anterior border of the central sulcus, while the **postcentral gyrus** of the parietal lobe forms its posterior border. The horizontal **lateral sulcus** separates the frontal lobe from the **temporal lobe**. The **insula** (IN-sū-luh; *insula*, island), an "island" of cortex, lies medial to the lateral sulcus (Figure 14–13c). The more posterior **parieto-occipital sulcus** separates the parietal lobe from the **occipital lobe** (Figure 14–13d).

The White Matter of the Cerebrum

The interior of the cerebrum consists mostly of white matter. We can classify the axons of white matter as *association*, *commissural*, or *projection fibers* (Figure 14–14).

- **Association fibers** interconnect areas of cerebral cortex within a single cerebral hemisphere. Shorter association fibers are called **arcuate** (AR-kyū-āt) **fibers**, because they curve in an arc to pass from one gyrus to another. Longer association fibers are organized into discrete bundles, or *fasciculi* (fah-SIK-yū-lī). The **longitudinal fasciculi** connect the frontal lobe to the other lobes of the same hemisphere (Figure 14–14a).
- **Commissural** (kom-ih-SŪR-ul; *commissura*, crossing over) **fibers** interconnect and permit communication between the cerebral hemispheres. Bands of commissural fibers linking the hemispheres include the **corpus callosum** and the **anterior commissure** (Figure 14–14b). The corpus callosum alone contains more than 200 million axons, carrying some 4 billion impulses per second!

Figure 14–13 **The Brain in Lateral View.** ATLAS: Plates 11a,b; 13b–e; 14a–d

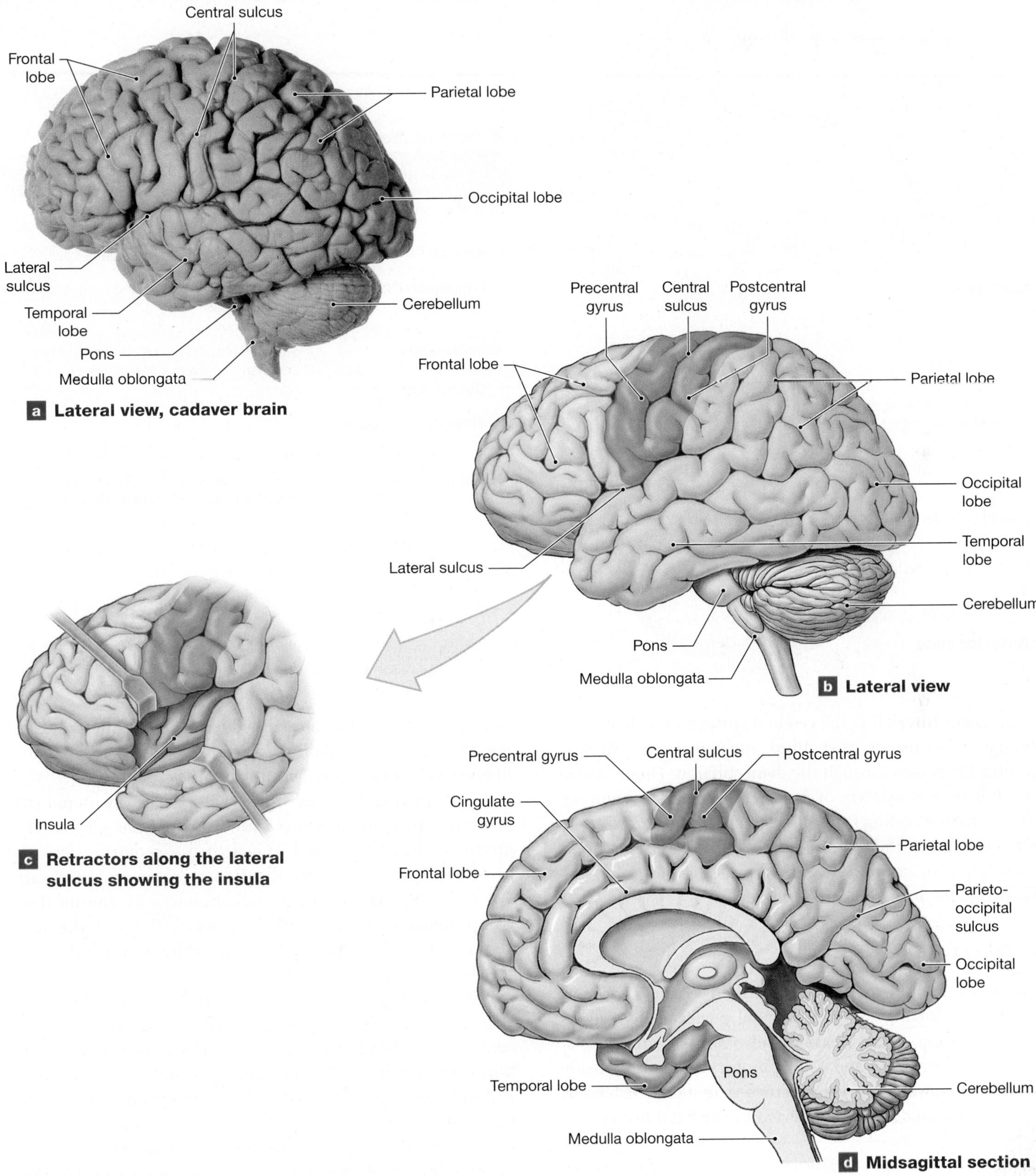
Central sulcus
Frontal lobe
Parietal lobe
Occipital lobe
Lateral sulcus
Temporal lobe
Cerebellum
Pons
Medulla oblongata
a Lateral view, cadaver brain
Precentral gyrus
Central sulcus
Postcentral gyrus
Frontal lobe
Parietal lobe
Occipital lobe
Temporal lobe
Lateral sulcus
Cerebellum
Pons
Medulla oblongata
b Lateral view
Insula
c Retractors along the lateral sulcus showing the insula
Precentral gyrus
Central sulcus
Postcentral gyrus
Cingulate gyrus
Parietal lobe
Frontal lobe
Parieto-occipital sulcus
Occipital lobe
Pons
Temporal lobe
Cerebellum
Medulla oblongata
d Midsagittal section

? The central sulcus divides which two lobes?

Figure 14–14 **Fibers of the White Matter of the Cerebrum.**

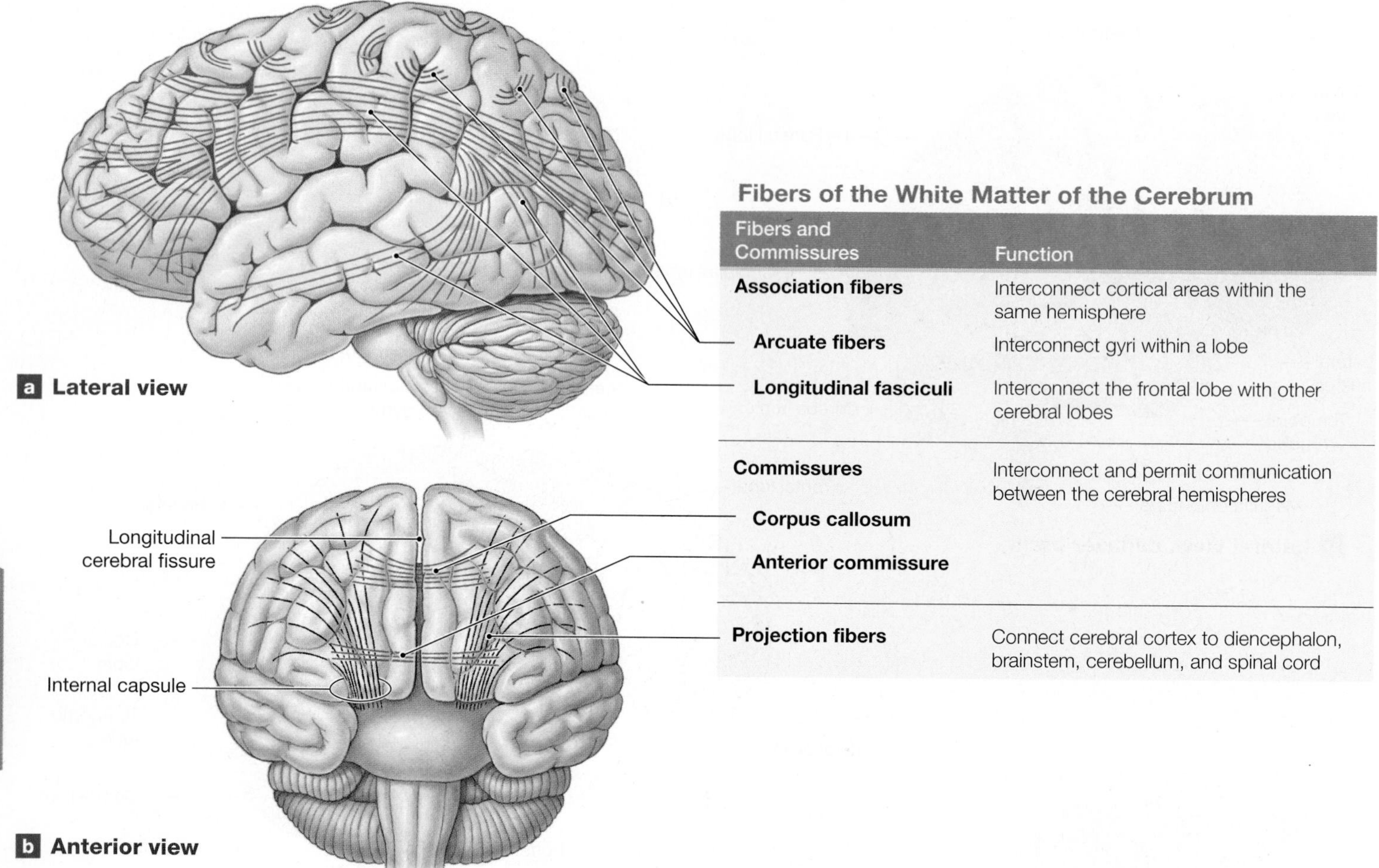
a Lateral view
Longitudinal cerebral fissure
Internal capsule
b Anterior view

Fibers of the White Matter of the Cerebrum

Fibers and Commissures	Function
Association fibers	Interconnect cortical areas within the same hemisphere
Arcuate fibers	Interconnect gyri within a lobe
Longitudinal fasciculi	Interconnect the frontal lobe with other cerebral lobes
Commissures	Interconnect and permit communication between the cerebral hemispheres
Corpus callosum	
Anterior commissure	
Projection fibers	Connect cerebral cortex to diencephalon, brainstem, cerebellum, and spinal cord

- **Projection fibers** link the cerebral cortex to the diencephalon, brainstem, cerebellum, and spinal cord. All projection fibers pass through the diencephalon. There, axons heading to sensory areas of the cerebral cortex pass among the axons descending from motor areas of the cortex. In gross dissection, the ascending fibers and descending fibers look alike. The entire collection of projection fibers is known as the **internal capsule** (see Figure 14–14b).

The Basal Nuclei

While your cerebral cortex is consciously directing a complex movement or solving some intellectual puzzle, other centers of your cerebrum, diencephalon, and brainstem are processing sensory information and issuing motor commands outside your conscious awareness. Many of these activities, which occur at the subconscious level, are directed by the basal nuclei.

Structure of the Basal Nuclei

The **basal nuclei** are masses of gray matter that lie within each hemisphere deep to the floor of the lateral ventricle (Figure 14–15a). They are embedded in the white matter of the cerebrum. The radiating projection fibers and commissural fibers travel around or between these nuclei.

Historically, the basal nuclei have been considered part of a larger functional group known as the *basal ganglia*. This group included the basal nuclei of the cerebrum and the associated motor nuclei in the diencephalon and midbrain. We will consider the functional interactions among these components in Chapter 15, but we will avoid the term "basal ganglia" because ganglia are otherwise restricted to the PNS.

The **caudate** (KOW-dāt) **nucleus** has a large head and a slender, curving tail that follows the curve of the lateral ventricle. The head of the caudate nucleus lies anterior to the **lentiform nucleus**. The lentiform nucleus consists of a lateral **putamen** (pū-TĀ-men; Figure 14–15b) and a medial **globus pallidus** (GLŌ-bus PAL-ih-dus; pale globe). The term *corpus striatum* (striated body) has been used to refer to the caudate and lentiform nuclei, or to the caudate nucleus and putamen. The name refers to the striated (striped) appearance of the internal capsule as its fibers pass among these nuclei.

Figure 14–15 The Basal Nuclei. ATLAS: Plates 11c; 13a,f

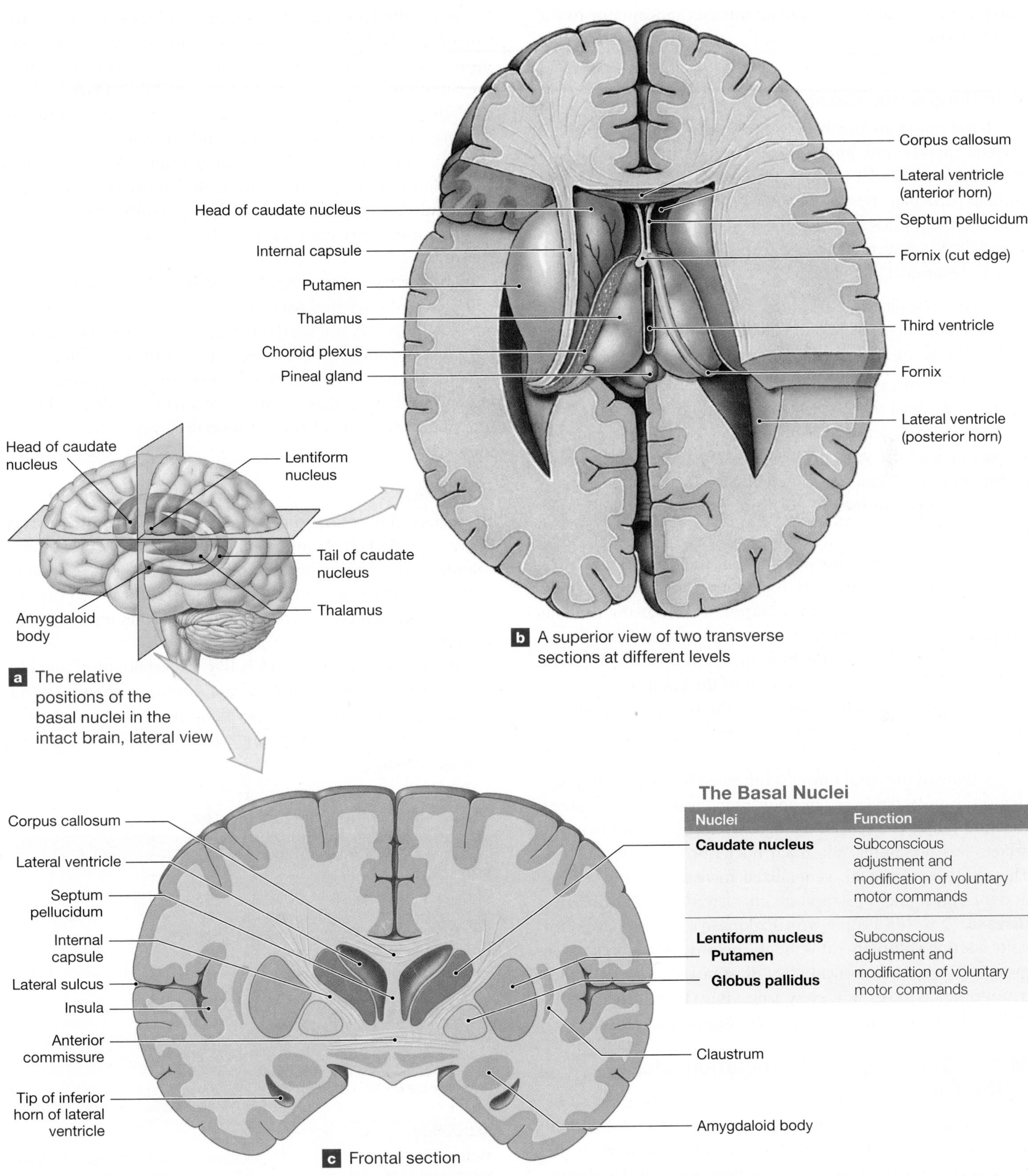
Head of caudate nucleus
Lentiform nucleus
Tail of caudate nucleus
Thalamus
Amygdaloid body
Corpus callosum
Lateral ventricle (anterior horn)
Septum pellucidum
Fornix (cut edge)
Third ventricle
Fornix
Lateral ventricle (posterior horn)
Head of caudate nucleus
Internal capsule
Putamen
Thalamus
Choroid plexus
Pineal gland
Corpus callosum
Lateral ventricle
Septum pellucidum
Internal capsule
Lateral sulcus
Insula
Anterior commissure
Tip of inferior horn of lateral ventricle
Claustrum
Amygdaloid body

a The relative positions of the basal nuclei in the intact brain, lateral view

b A superior view of two transverse sections at different levels

c Frontal section

The Basal Nuclei

Nuclei	Function
Caudate nucleus	Subconscious adjustment and modification of voluntary motor commands
Lentiform nucleus **Putamen** **Globus pallidus**	Subconscious adjustment and modification of voluntary motor commands

? Which portion of the lentiform nucleus is more medial and which portion is more lateral?

The **claustrum** is a thin layer of gray matter lying close to the putamen. The amygdaloid body, part of the limbic system, lies anterior to the tail of the caudate nucleus and inferior to the lentiform nucleus (Figure 14–15c).

Functions of the Basal Nuclei

The basal nuclei are involved with the subconscious control of skeletal muscle tone and the coordination of learned movement patterns. Under normal conditions, these nuclei do not initiate particular movements. But once a movement is under way, the basal nuclei provide the general pattern and rhythm, especially for movements of the trunk and proximal limb muscles. For example:

- When you walk, the basal nuclei control the cycles of arm and thigh movements that occur between the time you decide to "start" walking and the time you give the "stop" order.
- As you begin a voluntary movement, the basal nuclei control and adjust muscle tone, particularly in the appendicular muscles, to set your body position. When you decide to pick up a pencil, you consciously reach and grasp with your forearm, wrist, and hand while the basal nuclei operate at the subconscious level to position your shoulder and stabilize your arm.

14

How do the basal nuclei carry out their functions? They alter the motor commands issued by the cerebral cortex through a feedback loop. Information arrives at the caudate nucleus and putamen from the cerebral cortex. Lateral processing then takes place in these nuclei and in the adjacent globus pallidus. Most of the output of the basal nuclei leaves the globus pallidus and synapses in the thalamus. Nuclei in the thalamus then project the information to appropriate areas of the cerebral cortex.

Activity of the basal nuclei is inhibited by the dopamine (a neurotransmitter) released by neurons in the substantia nigra of the midbrain. If the substantia nigra is damaged or the neurons secrete less dopamine, the basal nuclei become more active. The result is a gradual, generalized increase in muscle tone and the appearance of symptoms characteristic of **Parkinson's disease.** ↺ p. 478 People with Parkinson's disease have difficulty starting voluntary movements, because opposing muscle groups do not relax. Instead, they must be overpowered. Once a movement is under way, every aspect must be voluntarily controlled through intense effort and concentration.

Motor, Sensory, and Association Areas of the Cortex

The cortex of the cerebral hemispheres contains functional areas whose boundaries are less clearly defined than are those of the cerebral lobes. Some of these areas deal with sensory information, called *sensory areas*, and others with motor commands, called *motor areas*. The central sulcus separates the motor and sensory areas of the cortex. There are also cortical areas called *association areas* that coordinate incoming and outgoing data from the sensory and motor areas. The major motor, sensory, and association areas of the cerebral cortex are shown in Figure 14–16 and listed in **Table 14–3**.

The cortex of each cerebral hemisphere receives somatosensory information from, and sends motor commands to, the opposite side of the body. For example, the motor areas of the left cerebral hemisphere control muscles on the right side, and the right cerebral hemisphere controls muscles on the left side. This crossing over has no known functional significance.

The correspondence between a specific function and a specific area of the cerebral cortex is imprecise. Because the boundaries are indistinct and have considerable overlap, one area may have several functions. Some aspects of cortical function, such as consciousness, cannot easily be assigned to any single area. However, we know that normal individuals use all portions of the brain. Let's look at these important areas of the cortex in more detail.

Motor Areas

The surface of the precentral gyrus is the **primary motor cortex** (Figure 14–16a). Neurons of the primary motor cortex direct voluntary movements by controlling somatic motor neurons in the brainstem and spinal cord. These cortical neurons are called **pyramidal cells**, because their cell bodies resemble little pyramids.

The primary motor cortex is like the keyboard of a piano. If you strike a specific piano key, you produce a specific sound. Similarly, if you stimulate a specific motor neuron in the primary motor cortex, you generate a contraction in a specific skeletal muscle.

Table 14–3 The Cerebral Cortex

Lobe/Area	Function
FRONTAL LOBE	
Primary motor cortex	Voluntary control of skeletal muscles
PARIETAL LOBE	
Primary somatosensory cortex	Conscious perception of touch, pressure, pain, vibration, taste, and temperature
OCCIPITAL LOBE	
Visual cortex	Conscious perception of visual stimuli
TEMPORAL LOBE	
Auditory cortex and olfactory cortex	Conscious perception of auditory (hearing) and olfactory (smell) stimuli
ALL LOBES	
Motor, sensory, somatosensory, and association areas	Integration and processing of sensory data; processing and initiation of motor activities

Figure 14–16 Motor, Sensory, and Association Areas of the Cerebral Cortex.

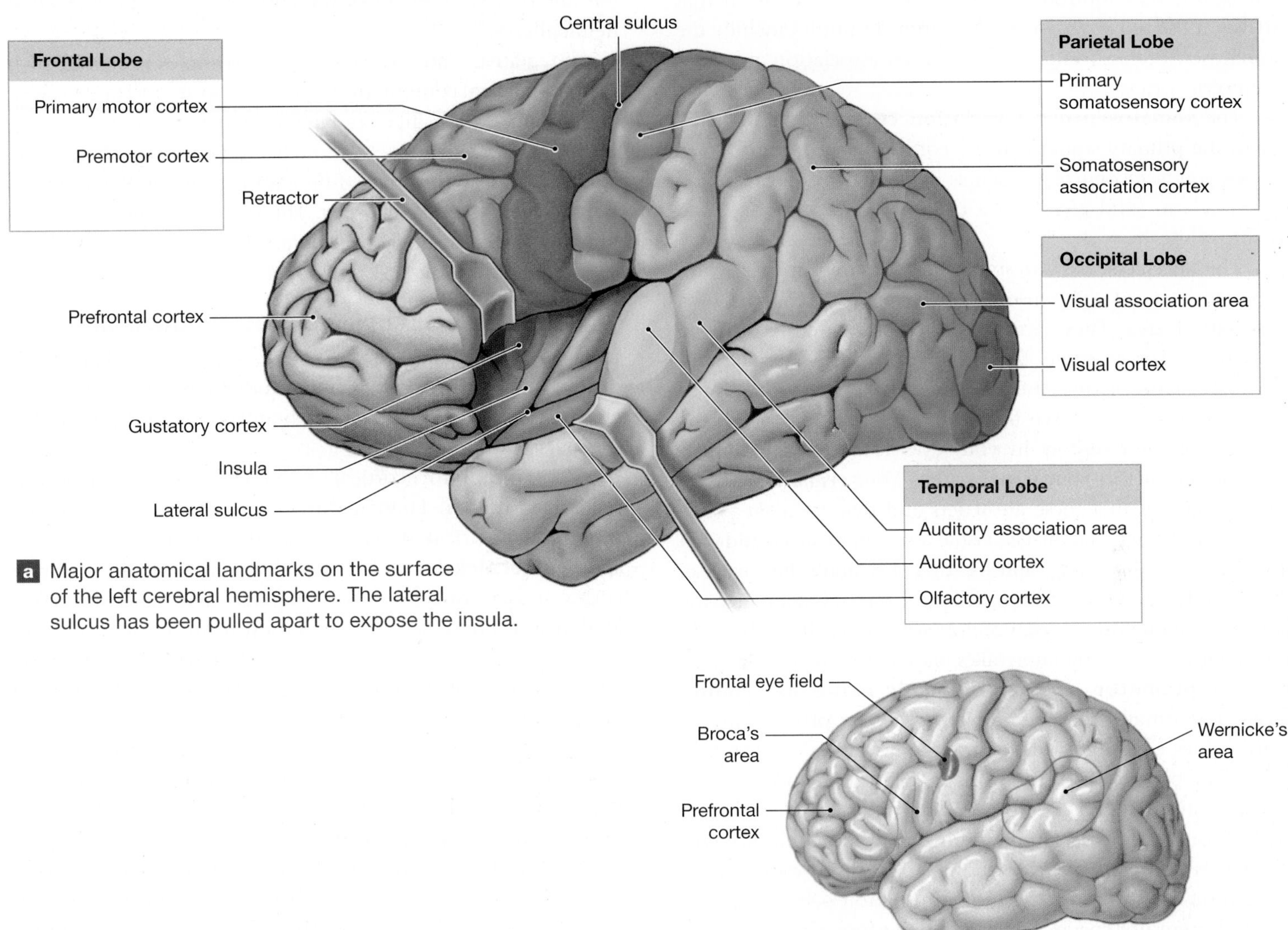
Central sulcus
Frontal Lobe
Primary motor cortex
Premotor cortex
Retractor
Prefrontal cortex
Gustatory cortex
Insula
Lateral sulcus
Parietal Lobe
Primary somatosensory cortex
Somatosensory association cortex
Occipital Lobe
Visual association area
Visual cortex
Temporal Lobe
Auditory association area
Auditory cortex
Olfactory cortex
Frontal eye field
Broca's area
Prefrontal cortex
Wernicke's area

a Major anatomical landmarks on the surface of the left cerebral hemisphere. The lateral sulcus has been pulled apart to expose the insula.

b The left hemisphere generally contains the specialized language areas. The prefrontal cortex of each hemisphere is involved with conscious intellectual functions.

14

Sensory Areas

Like the monitoring gauges in the dashboard of a car, the sensory areas of the cerebral cortex report key information. At each location, sensory information is reported in the pattern of neuron activity in the cortex. The surface of the postcentral gyrus contains the **primary somatosensory cortex**. Neurons in this region receive general somatic sensory information from receptors for touch, pressure, pain, vibration, or temperature. You are aware of these sensations only when nuclei in the thalamus relay the information to the primary somatosensory cortex.

Sight, sound, smell, and taste sensations arrive at other areas of the cerebral cortex (Figure 14–16a). The **visual cortex** of the occipital lobe receives visual information. The **auditory cortex** and **olfactory cortex** of the temporal lobe receive information about hearing and smell, respectively. The **gustatory cortex**, which receives information from taste receptors of the tongue and pharynx, lies in the anterior portion of the insula and adjacent portions of the frontal lobe.

Association Areas

The sensory and motor regions of the cortex are connected to nearby **association areas**, regions of the cortex that interpret incoming data or coordinate a motor response (Figure 14–16a). Like the information provided by the gauges in a car, the arriving data must be noticed and interpreted before the driver can take appropriate action. *Sensory association areas* are cortical regions that monitor and interpret the information that arrives at the sensory areas of the cortex. Examples include the somatosensory association cortex, visual association area, and premotor cortex.

The **somatosensory association cortex** monitors activity in the primary somatosensory cortex. This association area allows you to recognize a touch as light as a mosquito landing on your arm (and gives you a chance to swat the mosquito before it bites).

The special senses of smell, sight, and hearing involve separate areas of the sensory cortex, and each has its own association area. These areas monitor and interpret arriving sensations. For example, the **visual association area** monitors the patterns of activity in the visual cortex and interprets the results. You see the symbols *c*, *a*, and *r* when the stimulation of receptors in your eyes leads to the stimulation of neurons in your visual cortex. Your visual association area recognizes that these are letters and that $c + a + r = car$. An individual with a damaged visual association area could scan the lines of a printed page and see rows of symbols that are clear, but would perceive no meaning from the symbols. Similarly, the **auditory association area** monitors sensory activity in the auditory cortex. Word recognition takes place in this association area.

The **premotor cortex**, or **somatic motor association area,** coordinates learned movements. The primary motor cortex does nothing on its own, any more than a piano keyboard can play itself. The neurons in the primary motor cortex must be stimulated by neurons in other parts of the cerebrum. When you perform a voluntary movement, the premotor cortex relays the instructions to the primary motor cortex. With repetition, the proper pattern of stimulation becomes stored in your premotor cortex. You can then perform the movement smoothly and easily by triggering the *pattern* rather than by controlling the individual neurons. This principle applies to any learned movement, from something as simple as picking up a glass to something as complex as playing the piano. One area of the premotor cortex, the *frontal eye field*, controls learned eye movements, such as when you scan these lines of type (see Figure 14–16b). Individuals with damage to the frontal eye field can understand written letters and words but cannot read, because their eyes cannot follow the lines on a printed page.

Integrative Centers and Higher Mental Functions

Integrative centers receive information from many association areas and direct extremely complex motor activities. They also perform analytical functions. For example, the *prefrontal cortex* of the frontal lobe integrates information from sensory association areas and performs intellectual functions, such as predicting the consequences of possible responses. These centers are located in the lobes and cortical areas of both cerebral hemispheres.

Integrative centers concerned with complex processes, such as mathematical computation, speech, writing, and understanding spatial relationships are largely restricted to either the left or the right hemisphere. Although the two hemispheres look almost identical, they have different functions, a phenomenon called *hemispheric lateralization*. The corresponding regions on the opposite hemisphere are also active, but their functions are less well defined.

The Specialized Language Areas in the Brain

Language processing is much more complicated than previously thought, and brain imaging studies have shown that language processing occurs in both hemispheres and varies from one person to the next.

Two important cortical areas with varying functions related to human language are *Wernicke's area* and *Broca's area*. Both are primarily associated with the left cerebral hemisphere. **Wernicke's** (VER-nih-kēz) area is near the auditory cortex and is associated with language comprehension. This analytical center receives information from the sensory association areas and plays an important role in your personality by integrating sensory information and coordinating access to visual and auditory memories.

Broca's area (*motor speech area*) is near the motor cortex and is associated with speech production. Broca's area regulates the patterns of breathing and vocalization needed for normal speech. This regulation involves coordinating the activities of the respiratory muscles, the laryngeal and pharyngeal muscles, and the muscles of the tongue, cheeks, lips, and jaws. A person with damage to Broca's area can make sounds but not words.

The motor commands issued by the motor speech area are adjusted by feedback from the auditory association area, also called the *receptive speech area*. Damage to the related sensory areas can cause a variety of speech-related problems. (See the discussion of *aphasia* at right.) Some affected individuals have difficulty speaking although they know exactly which words to use. Others talk constantly but use all the wrong words.

The Prefrontal Cortex

The **prefrontal cortex** of the frontal lobe coordinates information relayed from all of the cortical association areas (see Figure 14–16b). In doing so, it performs such abstract intellectual functions as predicting the consequences of events or actions. The prefrontal cortex does not fully develop until the early 20s. Damage to the prefrontal cortex leads to difficulties in estimating temporal relationships between events. Questions such as "How long ago did this happen?" or "What happened first?" become difficult to answer.

+ Clinical Note Aphasia and Dyslexia

Speech therapists help children and adults struggling with speaking and swallowing disorders. **Aphasia** (ah-FĀ-zē-ah; *a*-, without + *phasia*, speech) is a disorder affecting the ability to speak or read. *Global aphasia* results from extensive damage to the specialized language areas of the brain or to the associated sensory tracts. Affected individuals are unable to speak, read, or understand speech. Global aphasia often accompanies a stroke that is severe enough to affect a large area of cortex, including the speech and language areas. Recovery is possible, but the process often takes months or even years.

Dyslexia (*dys*-, difficult, faulty + *lexis*, diction) is a disorder affecting the comprehension and use of written words. Up to 15 percent of children in the United States have some degree of dyslexia. Children with dyslexia have difficulty reading and writing, although their other intellectual functions may be normal or above normal. Their writing looks uneven and disorganized. They typically write letters in the wrong order (*dig* becomes *gid*) or reverse them (*E* becomes Ǝ). Recent evidence suggests that at least some forms of dyslexia result from problems in processing, sorting, and integrating visual or auditory information.

The prefrontal cortex has extensive connections with other cortical areas and with other portions of the brain. Feelings of frustration, tension, and anxiety are generated as the prefrontal cortex interprets ongoing events and makes predictions about future situations or consequences. If the connections between the prefrontal cortex and other brain regions are physically severed, the frustrations, tensions, and anxieties are removed. During the middle of the 20th century, this rather drastic procedure, called **prefrontal lobotomy**, was used to "cure" a variety of mental illnesses, especially those associated with violent or antisocial behavior. After a lobotomy, the patient would no longer be concerned about what had previously been a major problem, whether psychological (hallucinations) or physical (severe pain). However, the individual was often equally unconcerned about tact, decorum, and toilet training. Since then, drugs that target specific pathways and regions of the CNS have been developed, so lobotomies are no longer used to change behavior.

Hemispheric Lateralization

As mentioned, each of the two cerebral hemispheres is responsible for specific functions that are not ordinarily performed by the opposite hemisphere, a type of specialization known as **hemispheric lateralization**. In most people, the left hemisphere contains the specialized language areas of the brain and is responsible for language-based skills. For example, reading, writing, and speaking depend on processing done in the left cerebral hemisphere. In addition, the premotor cortex involved with hand movements is larger on the left side for right-handed people than for left-handed people. The left hemisphere is also important in performing analytical tasks, such as mathematical calculations and logical decision making.

The right cerebral hemisphere analyzes sensory information and relates the body to the sensory environment. Interpretive centers in this hemisphere permit you to identify familiar objects by touch, smell, sight, or taste. For example, the right hemisphere plays a dominant role in recognizing faces and in understanding three-dimensional relationships. It is also important in analyzing the emotional context of a conversation—for instance, distinguishing between the threat "Get lost!" and the question "Get lost?" Individuals with a damaged right hemisphere may be unable to add emotional inflections to their own words.

Left-handed people represent about 9 percent of the human population. In most cases, the primary motor cortex of the right hemisphere controls motor function for the dominant (left) hand. However, the centers involved with speech and analytical function are in the left hemisphere, just as they are in right-handed people. Interestingly, an unusually high percentage of musicians and artists are left-handed. Additionally, the more a person favored one hand over the other, the stronger the connection was with the other side of the brain. This suggests that as a species, we became left- or right-handed when we started developing language about 100,000 years ago.

Monitoring Brain Activity: The Electroencephalogram

Various methods have led to maps of brain activity and brain function. The primary somatosensory cortex and the primary motor cortex have been mapped by direct stimulation in patients undergoing brain surgery. The functions of other regions of the cerebrum can be revealed by the behavioral changes that follow localized injuries or strokes. The activities of specific regions can be visualized by a *PET scan* or through *functional magnetic resonance imaging* (*fMRI*). ⮌ p. 17

Clinicians commonly monitor the electrical activity of the brain to assess brain activity. Neural function depends on electrical events of the plasma membrane of neurons. The brain contains billions of neurons, and their activity generates an electrical field that can be measured by electrodes placed on the brain or on the outer surface of the skull. The electrical activity changes constantly, as nuclei and cortical areas are stimulated or quiet down. A printed recording of the electrical activity of the brain is called an **electroencephalogram (EEG)**. The electrical patterns observed are called *brain waves*.

Figure 14–17 **Brain Waves.** The four electrical patterns revealed by electroencephalograms (EEGs). The heights (amplitudes) of the four waves are not drawn to the same scale.

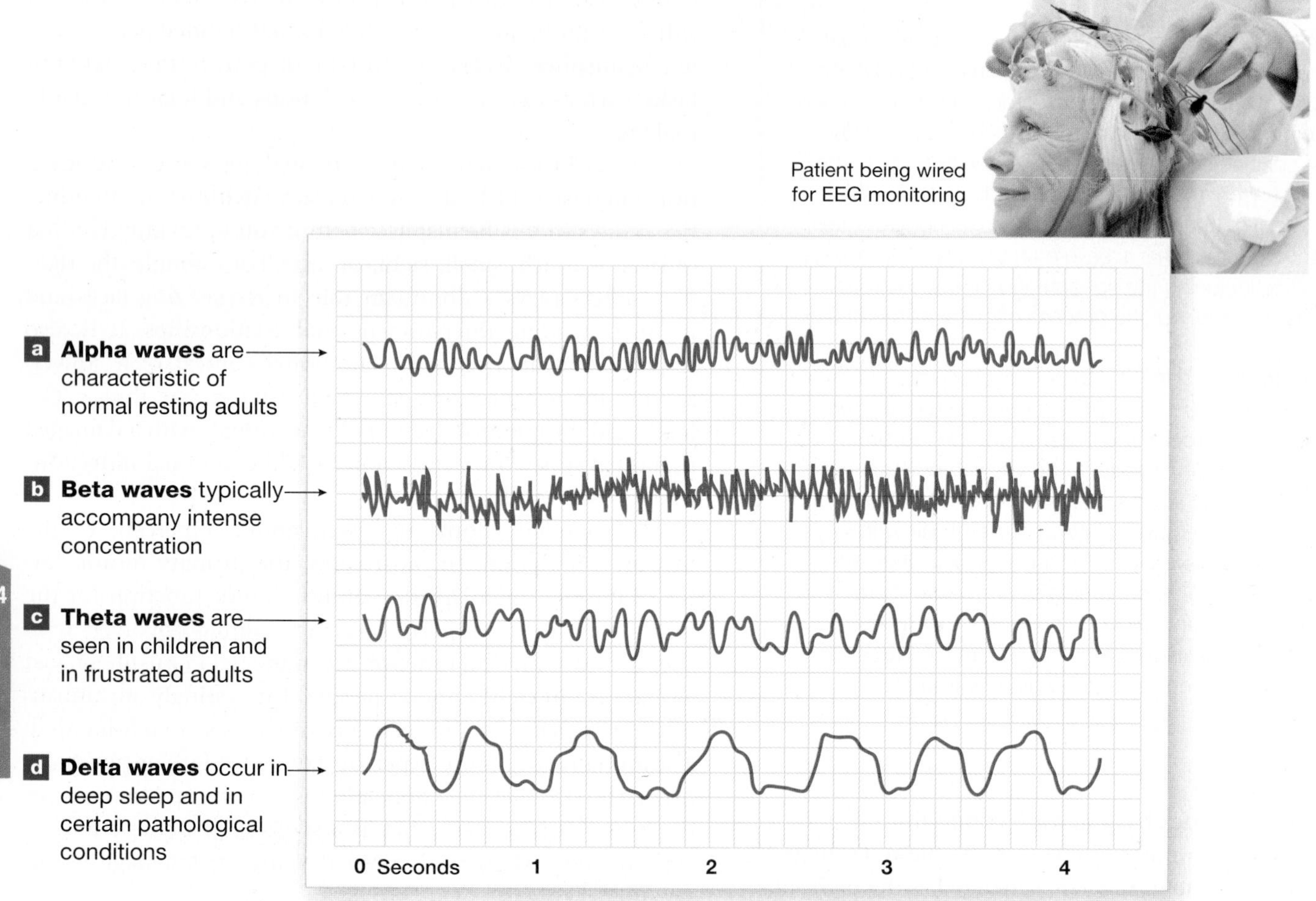
Patient being wired for EEG monitoring
a Alpha waves are characteristic of normal resting adults
b Beta waves typically accompany intense concentration
c Theta waves are seen in children and in frustrated adults
d Delta waves occur in deep sleep and in certain pathological conditions
0 Seconds
1
2
3
4

The four types of typical brain waves are shown in Figure 14–17. **Alpha waves** occur in the brains of healthy, awake adults who are resting with their eyes closed. Alpha waves disappear during sleep, but they also vanish when the individual begins to concentrate on some specific task. During attention to stimuli or tasks, higher-frequency **beta waves** replace alpha waves. Beta waves are typical of individuals who are either concentrating on a task, under stress, or in a state of psychological tension. **Theta waves** may appear transiently during sleep in normal adults but are most often observed in children and in intensely frustrated adults. The presence of theta waves under other circumstances may indicate the presence of a brain disorder, such as a tumor. **Delta waves** are very-large-amplitude, low-frequency waves. They are normally seen during deep sleep in individuals of all ages. Delta waves are also seen in the brains of infants (whose brains are still developing) and in awake adults when a tumor, vascular blockage, or inflammation has damaged portions of the brain.

Electrical activity in the two hemispheres is generally synchronized by a "pacemaker" mechanism that appears to involve the thalamus. Lack of synchronization between the hemispheres can indicate localized damage or other cerebral abnormalities. For example, a tumor or injury affecting one hemisphere typically changes the pattern in that hemisphere, and the patterns of the two hemispheres are no longer aligned.

A **seizure** is a temporary cerebral disorder accompanied by abnormal movements, unusual sensations, inappropriate behavior, or some combination of these symptoms. Clinical conditions characterized by seizures are known as seizure disorders, or *epilepsies*. The nature of the signs and symptoms produced depends on the region of the cortex involved. If a seizure affects the primary motor cortex, movements will occur. If it affects the auditory cortex, the individual will hear strange sounds. Seizures of all kinds are accompanied by a marked change in the pattern of electrical activity recorded in an electroencephalogram. The change begins in one portion of the cerebral cortex but may then spread across the entire cortical surface, like a wave on the surface of a pond.

Now we move on to the cranial nerves, so please study the following Focus: Cranial Nerves section thoroughly. Then we discuss the cranial reflexes that rely on these nerves.

FOCUS Cranial Nerves

Cranial nerves are PNS components that connect directly to the brain. The 12 pairs of cranial nerves are visible on the ventral surface of the brain (Figure 14–18). Each has a name related to its distribution or its function.

The number assigned to a cranial nerve corresponds to the nerve's position along the longitudinal axis of the brain, beginning at the cerebrum. In scientific writing, the abbreviation for cranial nerve, CN, is followed by a Roman numeral. For example, CN II refers to cranial nerve II or the optic nerve. If the full name of the cranial nerve is given, then only the Roman numeral is necessary, such as optic nerve (II).

Each cranial nerve attaches to the brain near the associated sensory or motor nuclei. The sensory nuclei act as switching centers, with the postsynaptic neurons relaying the information to other nuclei or to processing centers in the cerebral or cerebellar cortex. In a similar way, the motor nuclei receive convergent inputs from higher centers or from other nuclei along the brainstem.

In this section, we classify cranial nerves as primarily sensory, special sensory, motor, or mixed (sensory and motor). In this classification, sensory nerves carry somatic sensory information, including touch, pressure, vibration, temperature, or pain. Special sensory nerves carry the sensations of smell, sight, hearing, or balance. Motor nerves are dominated by the axons of somatic motor neurons. Mixed nerves have a mixture of sensory and motor fibers. This classification scheme is useful, but it is based on the primary function, and a cranial nerve can have important secondary functions. Here are three examples:

Figure 14–18 Origins of the Cranial Nerves. An inferior view of the brain.

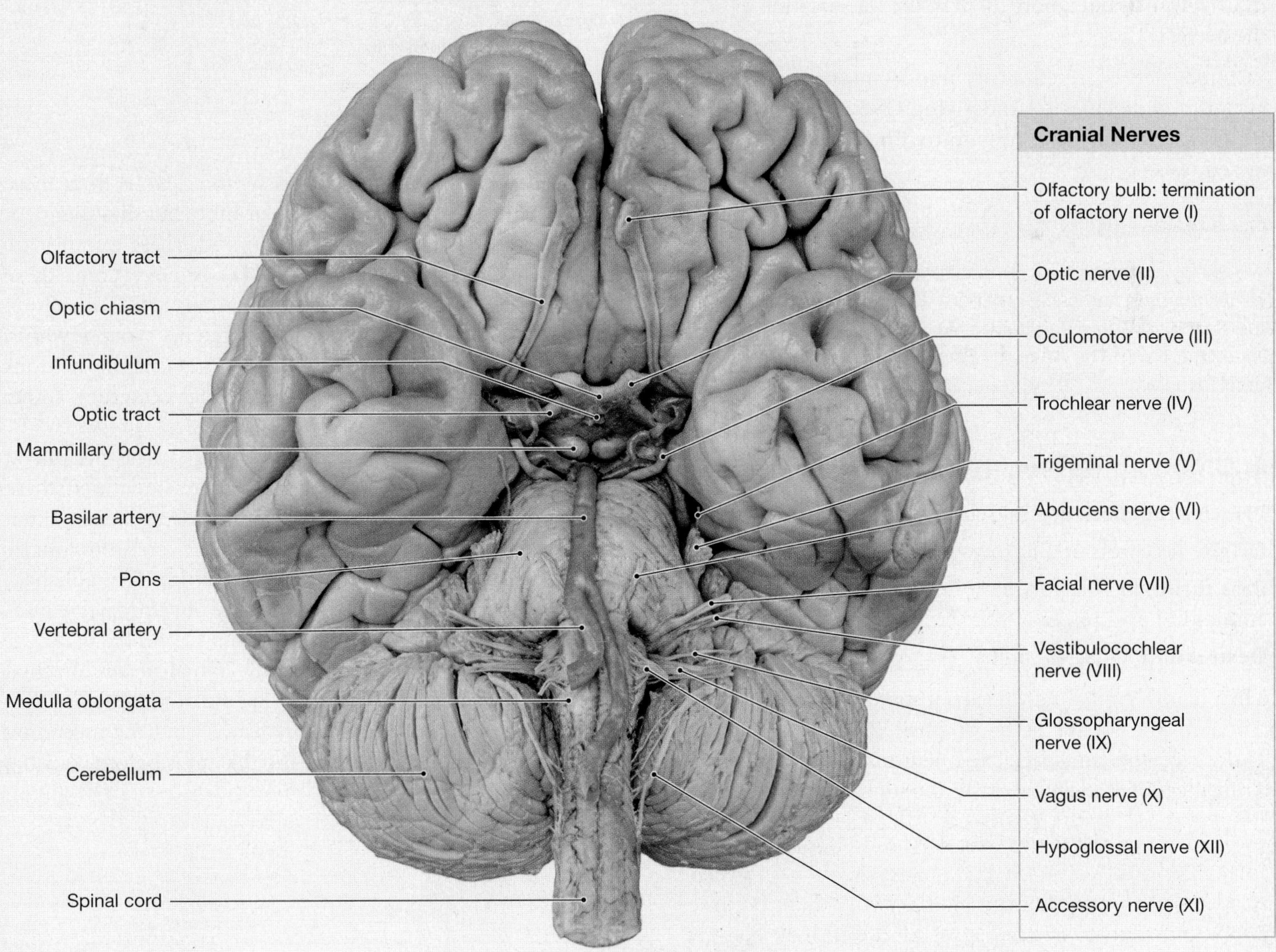

Olfactory tract
Optic chiasm
Infundibulum
Optic tract
Mammillary body
Basilar artery
Pons
Vertebral artery
Medulla oblongata
Cerebellum
Spinal cord
Cranial Nerves
Olfactory bulb: termination of olfactory nerve (I)
Optic nerve (II)
Oculomotor nerve (III)
Trochlear nerve (IV)
Trigeminal nerve (V)
Abducens nerve (VI)
Facial nerve (VII)
Vestibulocochlear nerve (VIII)
Glossopharyngeal nerve (IX)
Vagus nerve (X)
Hypoglossal nerve (XII)
Accessory nerve (XI)

- The olfactory receptors, the visual receptors, and the receptors of the internal ear are innervated by cranial nerves that are dedicated almost entirely to carrying special sensory information. In contrast, the sensation of taste, considered to be one of the special senses, is carried by axons that form only a small part of large cranial nerves that have other primary functions.
- As elsewhere in the PNS, a nerve containing tens of thousands of motor fibers that lead to a skeletal muscle also contains sensory fibers from muscle spindles and Golgi tendon organs (proprioceptive sensory nerve endings) in that muscle. We assume that these sensory fibers are present but ignore them in classifying the nerve.
- Regardless of their other functions, several cranial nerves (III, VII, IX, and X) distribute autonomic fibers to peripheral ganglia, just as spinal nerves deliver them to ganglia along the spinal cord. We will note the presence of small numbers of autonomic fibers (and discuss them further in Chapter 16) but ignore them in the classification of the nerve.

Cranial nerves are clinically important, in part because they can provide clues to underlying CNS problems. Clinicians assess cranial nerve function with a number of standardized tests.

Tips & Tools

Two useful mnemonics for remembering the names of the cranial nerves in numerical order are "**O**h **O**h **O**h, **T**o **T**ouch **A**nd **F**eel **V**ery **G**reen **V**egetables, **A**h **H**eaven!" and "**O**h, **O**nce **O**ne **T**akes **T**he **A**natomy **F**inal, **V**ery **G**ood **V**acations **A**re **H**eavenly!"

The Olfactory Nerves (I)

- **Primary function:** Special sensory (smell)
- **Origin:** Receptors of olfactory epithelium
- **Pass through:** Olfactory foramina in cribriform plate of ethmoid pp. 215, 220
- **Destination:** Olfactory bulbs

The first pair of cranial nerves carries special sensory information responsible for the sense of smell (Figure 14–19). The olfactory receptors are specialized neurons in the epithelium covering the roof of the nasal cavity, the superior nasal conchae, and the superior parts of the nasal septum. Axons from these sensory neurons collect to form 20 or more bundles that penetrate the cribriform plate of the ethmoid. These bundles are components of the **olfactory nerves (I)**. Almost at once these bundles enter the **olfactory bulbs**, neural masses on either side of the crista galli. The olfactory afferents synapse within the olfactory bulbs. The axons of the postsynaptic neurons proceed to the cerebrum along the slender **olfactory tracts** (Figures 14–19, 14–20).

Figure 14–19 **The Olfactory Nerve.**

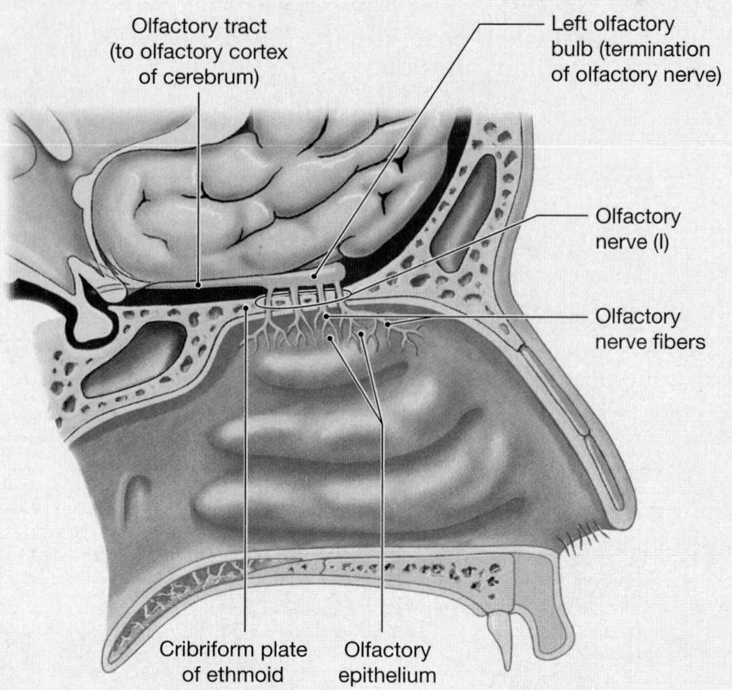
Olfactory tract
(to olfactory cortex
of cerebrum)
Left olfactory
bulb (termination
of olfactory nerve)
Olfactory
nerve (I)
Olfactory
nerve fibers
Cribriform plate
of ethmoid
Olfactory
epithelium

Because the olfactory tracts look like typical peripheral nerves, anatomists about a century ago misidentified these tracts as the first cranial nerve. Later studies demonstrated that the olfactory tracts and bulbs are part of the cerebrum, but by then the numbering system was already firmly established. Anatomists were left with a forest of tiny olfactory nerve bundles lumped together as cranial nerve I.

The olfactory nerves are the only cranial nerves attached directly to the cerebrum. The rest originate or terminate within nuclei of the diencephalon or brainstem, and the ascending sensory information synapses in the thalamus before reaching the cerebrum.

Figure 14–20 The Optic Nerve.

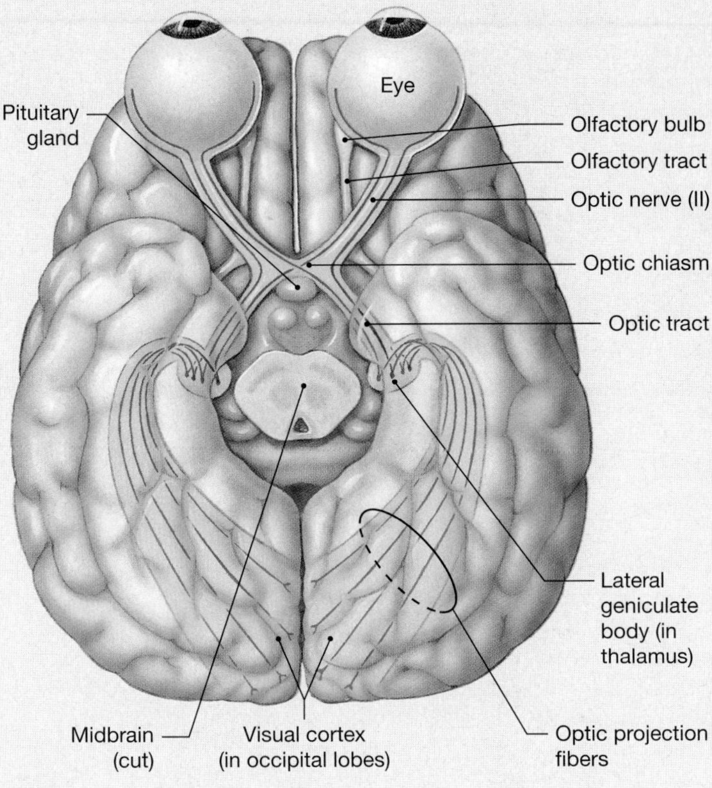

Eye
Pituitary gland
Olfactory bulb
Olfactory tract
Optic nerve (II)
Optic chiasm
Optic tract
Lateral geniculate body (in thalamus)
Midbrain (cut)
Visual cortex (in occipital lobes)
Optic projection fibers

? Which structure receives visual information before it reaches the visual cortex?

The Optic Nerves (II)

- **Primary function:** Special sensory (vision)
- **Origin:** Retina of eye
- **Pass through:** Optic canals of sphenoid p. 219
- **Destination:** Diencephalon by way of the optic chiasm

The **optic nerves (II)** carry visual information from special sensory ganglia in the eyes (Figure 14–20). These nerves contain about 1 million sensory nerve fibers. The optic nerves pass through the optic canals of the sphenoid. Then they converge at the ventral, anterior margin of the diencephalon, at the **optic chiasm** (*chiasma*, a crossing). At the optic chiasm, fibers from the medial half of each retina cross over to the opposite side of the brain.

The reorganized axons continue toward the lateral geniculate body of the thalamus as the **optic tracts** (Figures 14–19, 14–21). After synapsing in the lateral geniculate body, projection fibers deliver the information to the visual cortex of the occipital lobes. With this arrangement, each cerebral hemisphere receives visual information from the lateral half of the retina of the eye on that side and from the medial half of the retina of the eye of the opposite side. A few axons in the optic tracts bypass the lateral geniculate body and synapse in the superior colliculi of the midbrain. We consider that pathway in Chapter 17.

The Oculomotor Nerves (III)

- **Primary function:** Motor (eye movements)
- **Origin:** Midbrain
- **Pass through:** Superior orbital fissures of sphenoid pp. 213, 220, 222, 225
- **Destination:** *Somatic motor:* superior, inferior, and medial rectus muscles; inferior oblique; levator palpebrae superioris. *Visceral motor*: intrinsic eye muscles

The midbrain contains the motor nuclei controlling the third and fourth cranial nerves. Each **oculomotor nerve (III)** innervates four of the six extrinsic muscles that move the eye, and the levator palpebrae superioris, which raises the upper eyelid (see Figure 14–21). On each side of the brain, CN III emerges from the ventral surface of the midbrain and penetrates the posterior wall of the orbit at the superior orbital fissure. Individuals with damage to this nerve often complain of pain over the eye, droopy eyelids, and double vision, because the movements of the left and right eyes cannot be coordinated properly.

The oculomotor nerve also delivers preganglionic autonomic fibers to neurons of the **ciliary ganglion**. These neurons control intrinsic eye muscles. These muscles change the diameter of the pupil, adjusting the amount of light entering the eye. They also change the shape of the lens to focus images on the retina.

Figure 14–21 Cranial Nerves Controlling the Extrinsic Eye Muscles. ATLAS: Plates 16a,b

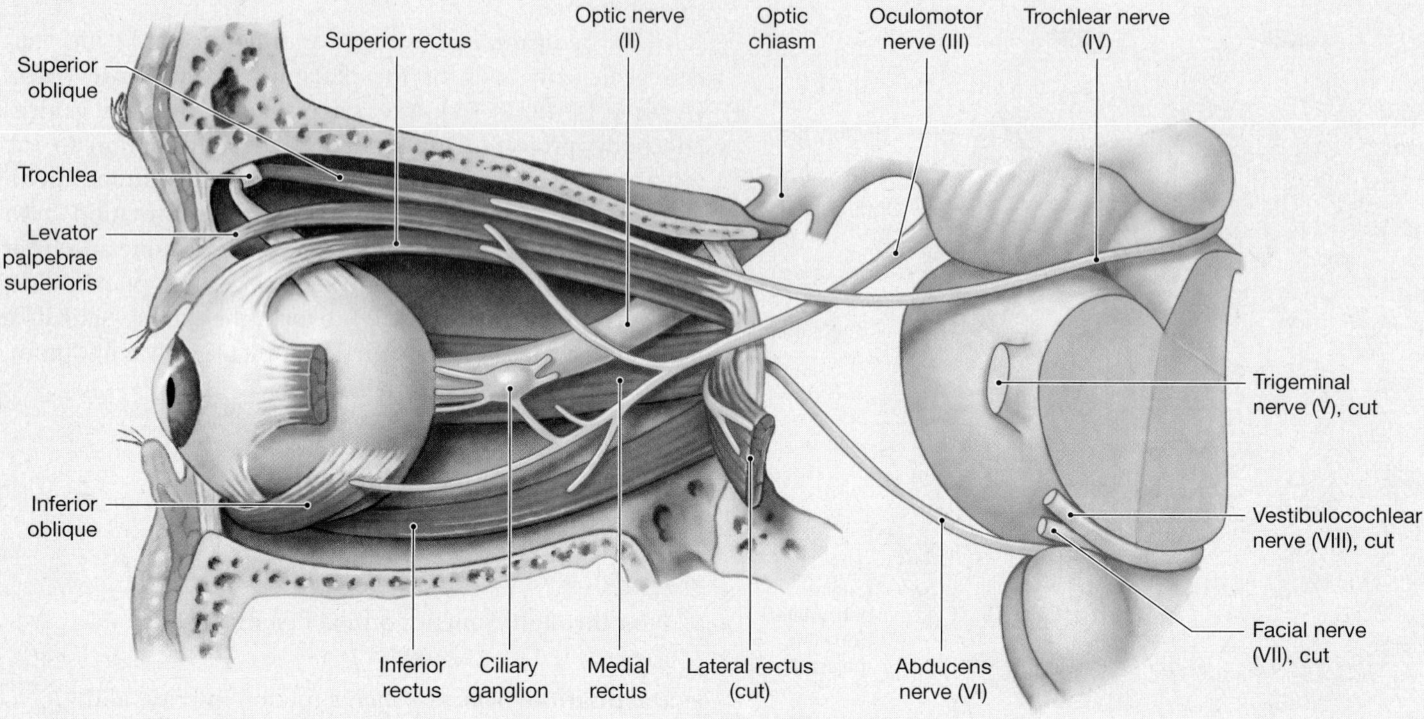
Superior oblique
Trochlea
Levator palpebrae superioris
Inferior oblique
Superior rectus
Optic nerve (II)
Optic chiasm
Oculomotor nerve (III)
Trochlear nerve (IV)
Trigeminal nerve (V), cut
Vestibulocochlear nerve (VIII), cut
Facial nerve (VII), cut
Inferior rectus
Ciliary ganglion
Medial rectus
Lateral rectus (cut)
Abducens nerve (VI)

The Trochlear Nerves (IV)

- **Primary function:** Motor (eye movements)
- **Origin:** Midbrain
- **Pass through:** Superior orbital fissures of sphenoid ⮌ pp. 213, 219, 222, 225
- **Destination:** Superior oblique

A **trochlear** (TRŌK-lē-ar; *trochlea*, a pulley) **nerve** (IV), the smallest cranial nerve, innervates the superior oblique of each eye (**Figure 14–21**). The trochlea is a pulley-shaped, ligamentous sling. Each superior oblique passes through a trochlea on its way to its insertion on the surface of the eye. An individual with damage to cranial nerve IV or to its nucleus has difficulty looking down and to the side.

The Abducens Nerves (VI)

- **Primary function:** Motor (eye movements)
- **Origin:** Pons
- **Pass through:** Superior orbital fissures of sphenoid ⮌ pp. 213, 219, 222, 225
- **Destination:** Lateral rectus

The **abducens** (ab-DŪ-senz) **nerves (VI)** innervate the lateral rectus, the sixth pair of extrinsic eye muscles. Contraction of the lateral rectus makes the eye look to the side. In essence, the *abducens* causes *abduction* of the eye. Each abducens nerve emerges from the inferior surface of the brainstem at the border between the pons and the medulla oblongata (**Figure 14–22**). Along with the oculomotor and trochlear nerves from that side, it reaches the orbit through the superior orbital fissure.

Figure 14–22 The Trigeminal Nerve.

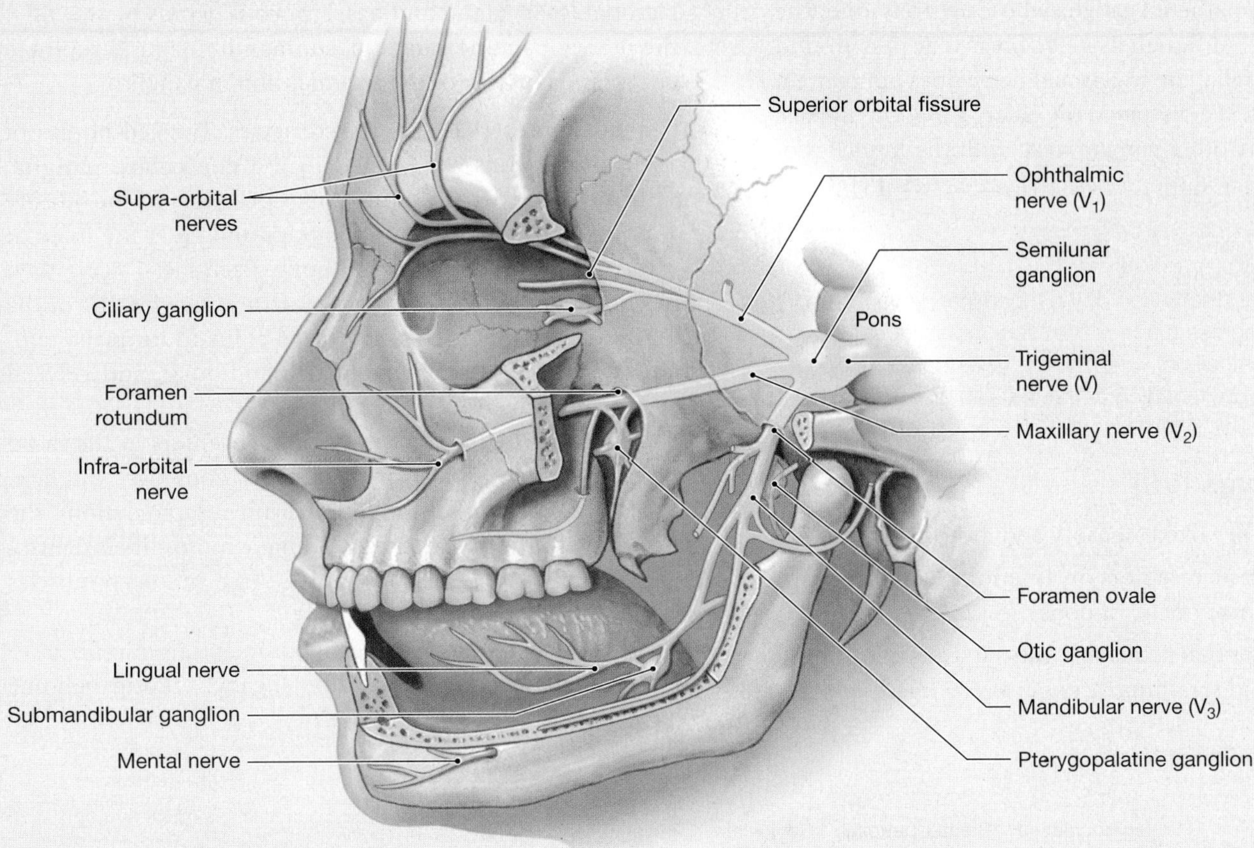

Superior orbital fissure
Supra-orbital nerves
Ophthalmic nerve (V_1)
Semilunar ganglion
Ciliary ganglion
Pons
Foramen rotundum
Trigeminal nerve (V)
Maxillary nerve (V_2)
Infra-orbital nerve
Foramen ovale
Otic ganglion
Lingual nerve
Submandibular ganglion
Mandibular nerve (V_3)
Mental nerve
Pterygopalatine ganglion

? What are the three divisions of the trigeminal nerve?

The Trigeminal Nerves (V)

- **Primary function:** Mixed (sensory and motor) to face
- **Origin:** *Ophthalmic nerve (V_1)* (sensory): orbital structures, cornea, nasal cavity, skin of forehead, upper eyelid, eyebrow, nose (part). *Maxillary nerve (V_2)* (sensory): lower eyelid, upper lip, gums, and teeth; cheek; nose, palate, and pharynx (part). *Mandibular nerve (V_3)* (mixed): sensory from lower gums, teeth, and lips; palate and tongue (part); motor from motor nuclei of pons
- **Pass through (on each side):** Ophthalmic nerve through superior orbital fissure, maxillary nerve through foramen rotundum, mandibular nerve through foramen ovale pp. 213, 215, 219, 225
- **Destination:** Ophthalmic, maxillary, and mandibular nerves to sensory nuclei in pons; mandibular nerve also innervates muscles of mastication p. 351

The pons contains the nuclei associated with three cranial nerves (V, VI, and VII) and contributes to a fourth (VIII). The **trigeminal** (trī-JEM-i-nal) **nerves (V)**, the largest cranial nerves, are mixed nerves. Each provides both somatic sensory information from the head and face, and motor control over the muscles of mastication. Sensory (posterior) and motor (anterior) roots originate on the lateral surface of the pons (Figure 14–22). The sensory branch is larger, and the enormous **semilunar ganglion** contains the cell bodies of the sensory neurons.

As the name implies, the trigeminal has three major divisions. The small motor root contributes to only one of the three. **Tic douloureux** (dū-luh-ROO; *douloureux*, painful), or *trigeminal neuralgia*, is a painful condition affecting the area innervated by the maxillary and mandibular divisions of the trigeminal nerve. Sufferers complain of debilitating pain triggered by contact with the lip, tongue, or gums. The cause of the condition is unknown.

The trigeminal nerve divisions are associated with the *ciliary, sphenopalatine, submandibular,* and *otic ganglia.* These are autonomic (parasympathetic) ganglia whose neurons innervate structures of the face. Although its nerve fibers may pass around or through these ganglia, the trigeminal nerve does not contain visceral motor fibers. We discussed the ciliary ganglion on page 497 and describe the other ganglia next, with the branches of the *facial nerves* (VII) and the *glossopharyngeal nerves* (IX).

Tips & Tools

To learn the sensory distribution of the trigeminal nerve, place your first 3 fingers across the face with your index finger just below your eye (ophthalmic), your middle finger on your upper lip (maxillary), and your third finger (mandibular) on your chin.

The Facial Nerves (VII)

- **Primary function:** Mixed (sensory and motor) to face
- **Origin:** *Sensory*: taste receptors on anterior two-thirds of tongue. *Motor*: motor nuclei of pons
- **Pass through:** Internal acoustic meatus to canals leading to the stylomastoid foramina pp. 213, 215, 219
- **Destination:** *Sensory*: sensory nuclei of pons. *Somatic motor*: muscles of facial expression. p. 347 *Visceral motor*: lacrimal (tear) gland and nasal mucous glands by way of the pterygopalatine ganglion; submandibular and sublingual glands by way of the submandibular ganglion

The **facial nerves (VII)** are mixed nerves. The cell bodies of the sensory neurons are located in the **geniculate ganglia**, and the motor nuclei are in the pons. On each side, the sensory and motor roots emerge from the pons and enter the internal acoustic meatus of the temporal bone. Each facial nerve then passes through the facial canal to reach the face by way of the stylomastoid foramen. The nerve then splits to form the temporal, zygomatic, buccal, marginal mandibular, and cervical branches (Figure 14–23).

The sensory neurons monitor proprioceptors in the facial muscles and provide deep pressure sensations over the face. They also receive taste information from receptors along the anterior two-thirds of the tongue. Somatic motor fibers control the superficial muscles of the scalp and face and deep muscles near the ear.

The facial nerves carry preganglionic autonomic fibers to the pterygopalatine and submandibular ganglia. Postganglionic

Figure 14–23 The Facial Nerve.

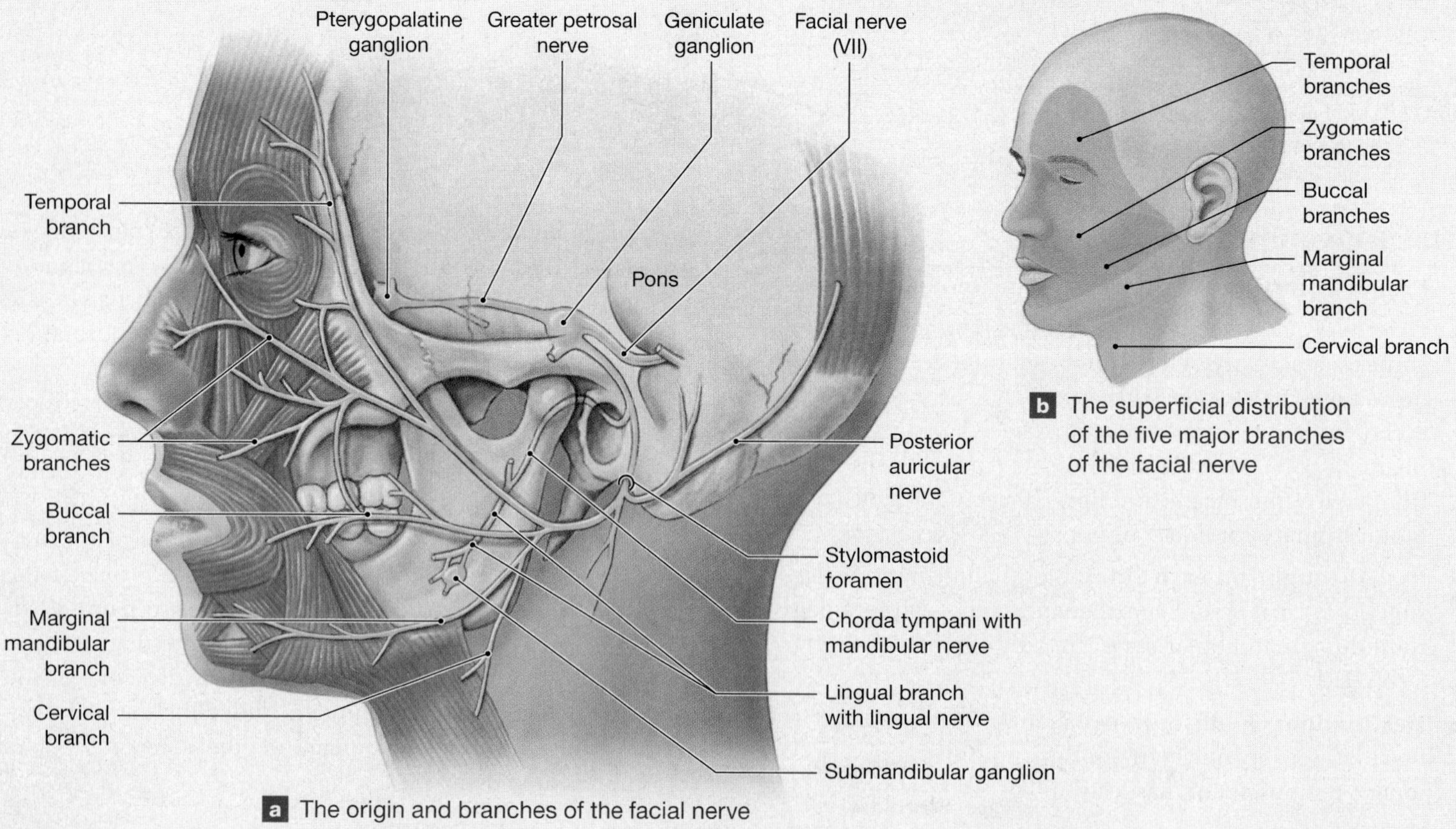

Pterygopalatine ganglion
Greater petrosal nerve
Geniculate ganglion
Facial nerve (VII)
Temporal branch
Pons
Zygomatic branches
Buccal branch
Marginal mandibular branch
Cervical branch
Posterior auricular nerve
Stylomastoid foramen
Chorda tympani with mandibular nerve
Lingual branch with lingual nerve
Submandibular ganglion
Temporal branches
Zygomatic branches
Buccal branches
Marginal mandibular branch
Cervical branch

a The origin and branches of the facial nerve

b The superficial distribution of the five major branches of the facial nerve fibers from the **pterygopalatine ganglia** innervate the lacrimal glands and small glands of the nasal cavity and pharynx. The **submandibular ganglia** innervate the *submandibular* and *sublingual* (*sub-*, under + *lingual*, pertaining to the tongue) *glands*.

Bell's palsy is a cranial nerve disorder that results from an inflammation of a facial nerve. The condition is probably due to a viral infection. Signs and symptoms include paralysis of facial muscles on the affected side and loss of taste sensations from the anterior two-thirds of the tongue. The condition is usually painless. In most cases the symptoms fade after a few weeks or months.

The Vestibulocochlear Nerves (VIII)

- **Primary function:** Special sensory: balance and equilibrium (vestibular nerve) and hearing (cochlear nerve)
- **Origin:** Monitor receptors of the internal ear (vestibule and cochlea)
- **Pass through:** Internal acoustic meatus of temporal bones ⮌ pp. 215, 218, 219
- **Destination:** Vestibular and cochlear nuclei of pons and medulla oblongata

The **vestibulocochlear nerves (VIII)** are also known as the *acoustic nerves*, the *auditory nerves*, and the *stato-acoustic nerves*. We will use *vestibulocochlear*, because this term indicates the names of its two major nerve branches: the *vestibular nerve* and the *cochlear nerve*. Each vestibulocochlear nerve lies posterior to the origin of the facial nerve, straddling the boundary between the pons and the medulla oblongata (Figure 14–24). This nerve reaches the sensory receptors of the internal ear by entering the internal acoustic meatus in company with the facial nerve.

Each vestibulocochlear nerve has two distinct bundles of sensory fibers. The **vestibular** (*vestibulum*, cavity) **nerve** originates at the receptors of the *vestibule*, the portion of the internal ear concerned with balance sensations. The sensory neurons are located in an adjacent sensory ganglion, and their axons target the **vestibular nuclei** of the pons and medulla oblongata. These afferents convey information about the orientation and movement of the head. The **cochlear** (*cochlea*, snail shell) **nerve** monitors the receptors in the *cochlea*, the portion of the internal ear that provides the sense of hearing. The cell bodies of the sensory neurons are located within a peripheral ganglion (the *spiral ganglion*), and their axons synapse within the **cochlear nuclei** of the pons and medulla oblongata. Axons leaving the vestibular and cochlear nuclei relay the sensory information to other centers or initiate reflexive motor responses.We discuss balance and the sense of hearing in Chapter 17.

Figure 14–24 The Vestibulocochlear Nerve.

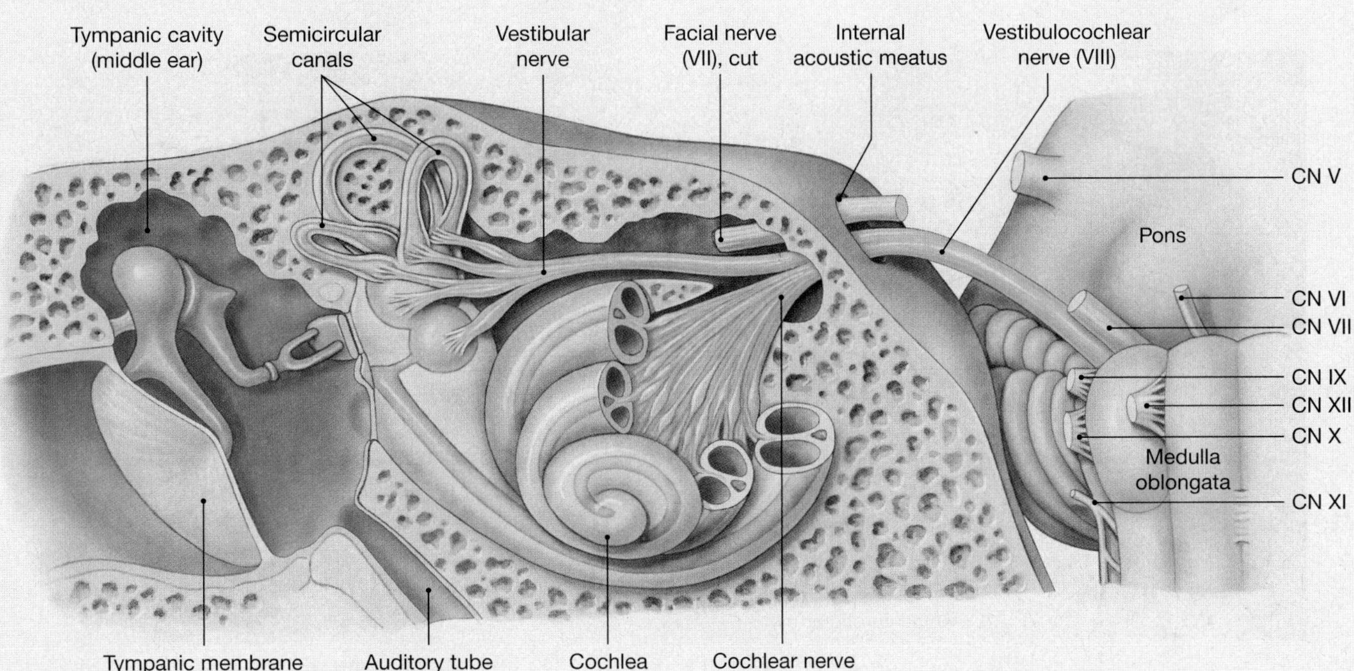

Tympanic cavity (middle ear)
Semicircular canals
Vestibular nerve
Facial nerve (VII), cut
Internal acoustic meatus
Vestibulocochlear nerve (VIII)
CN V
Pons
CN VI
CN VII
CN IX
CN XII
CN X
Medulla oblongata
CN XI
Tympanic membrane
Auditory tube
Cochlea
Cochlear nerve

The Glossopharyngeal Nerves (IX)

- **Primary function:** Mixed (sensory and motor) to head and neck
- **Origin:** *Sensory*: posterior one-third of the tongue, part of the pharynx and palate, carotid arteries of the neck. *Motor*: motor nuclei of medulla oblongata
- **Pass through:** Jugular foramina between the occipital bone and the temporal bones ⮌ pp. 213, 215
- **Destination:** *Sensory*: sensory nuclei of medulla oblongata. *Somatic motor*: pharyngeal muscles involved in swallowing. *Visceral motor*: parotid gland by way of the otic ganglion

The medulla oblongata contains the sensory and motor nuclei of cranial nerves IX, X, XI, and XII, in addition to the vestibular nucleus of nerve VIII. The **glossopharyngeal** (glos-ō-fah-RIN-jē-al; *glossum*, tongue) **nerves (IX)** innervate the tongue and pharynx. Each glossopharyngeal nerve penetrates the cranium within the jugular foramen, with cranial nerves X and XI.

The glossopharyngeal nerves are mixed nerves, but sensory fibers are most abundant. The sensory neurons on each side are in the **superior** (*jugular*) **ganglion** and **inferior** (*petrosal*) **ganglion** (Figure 14–25). The sensory fibers carry general sensory information from the lining of the pharynx and the soft palate to a nucleus in the medulla oblongata. These nerves also provide taste sensations from the posterior third of the tongue. Additionally, they have special receptors that monitor the blood pressure and dissolved gas concentrations in the carotid arteries, major blood vessels in the neck.

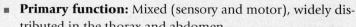

The somatic motor fibers control the pharyngeal muscles involved in swallowing. Visceral motor fibers synapse in the **otic ganglion**. Postganglionic fibers innervate the parotid gland of the cheek.

Figure 14–25 The Glossopharyngeal Nerve.

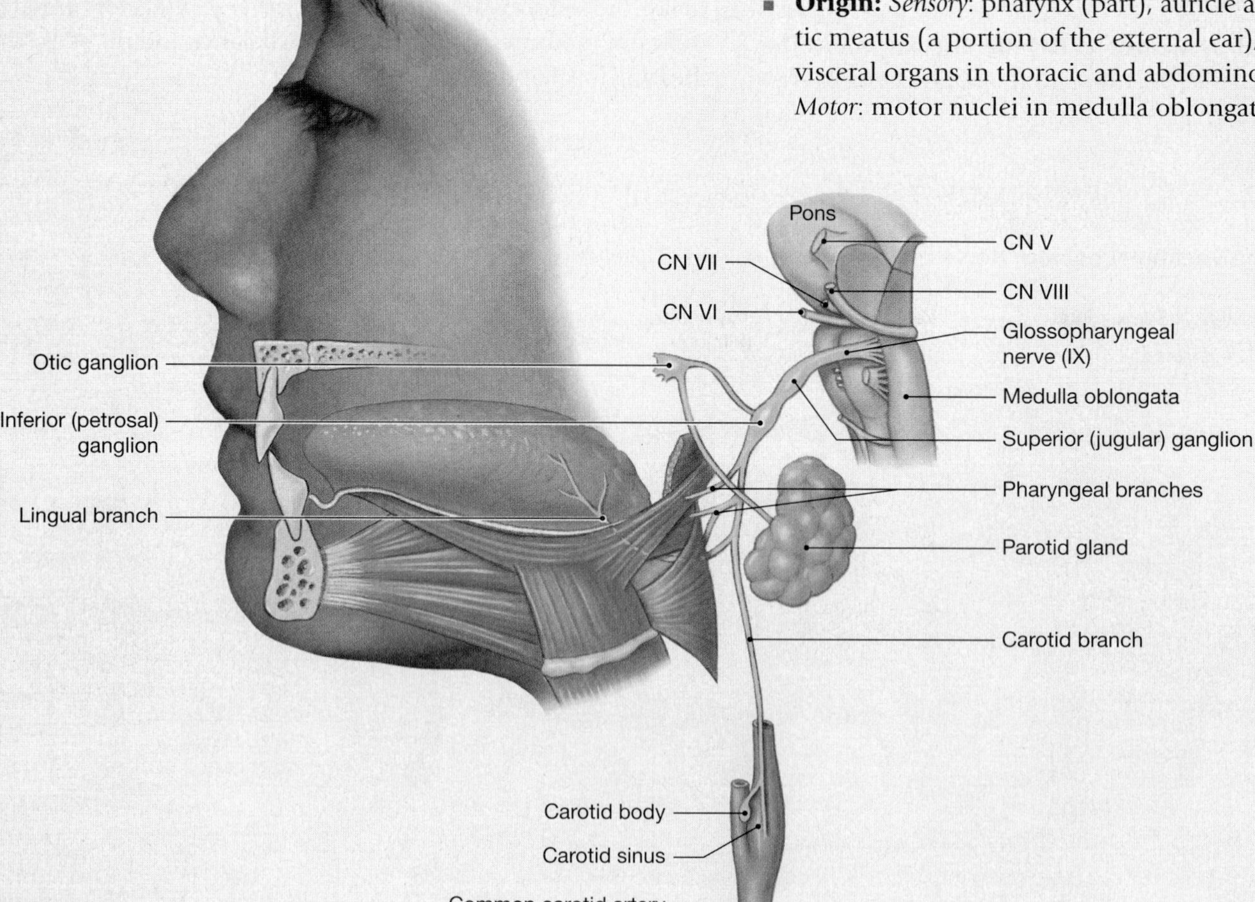

Pons
CN V
CN VII
CN VIII
CN VI
Glossopharyngeal nerve (IX)
Otic ganglion
Medulla oblongata
Inferior (petrosal) ganglion
Superior (jugular) ganglion
Pharyngeal branches
Lingual branch
Parotid gland
Carotid branch
Carotid body
Carotid sinus
Common carotid artery

The Vagus Nerves (X)

- **Primary function:** Mixed (sensory and motor), widely distributed in the thorax and abdomen
- **Origin:** *Sensory*: pharynx (part), auricle and external acoustic meatus (a portion of the external ear), diaphragm, and visceral organs in thoracic and abdominopelvic cavities. *Motor*: motor nuclei in medulla oblongata

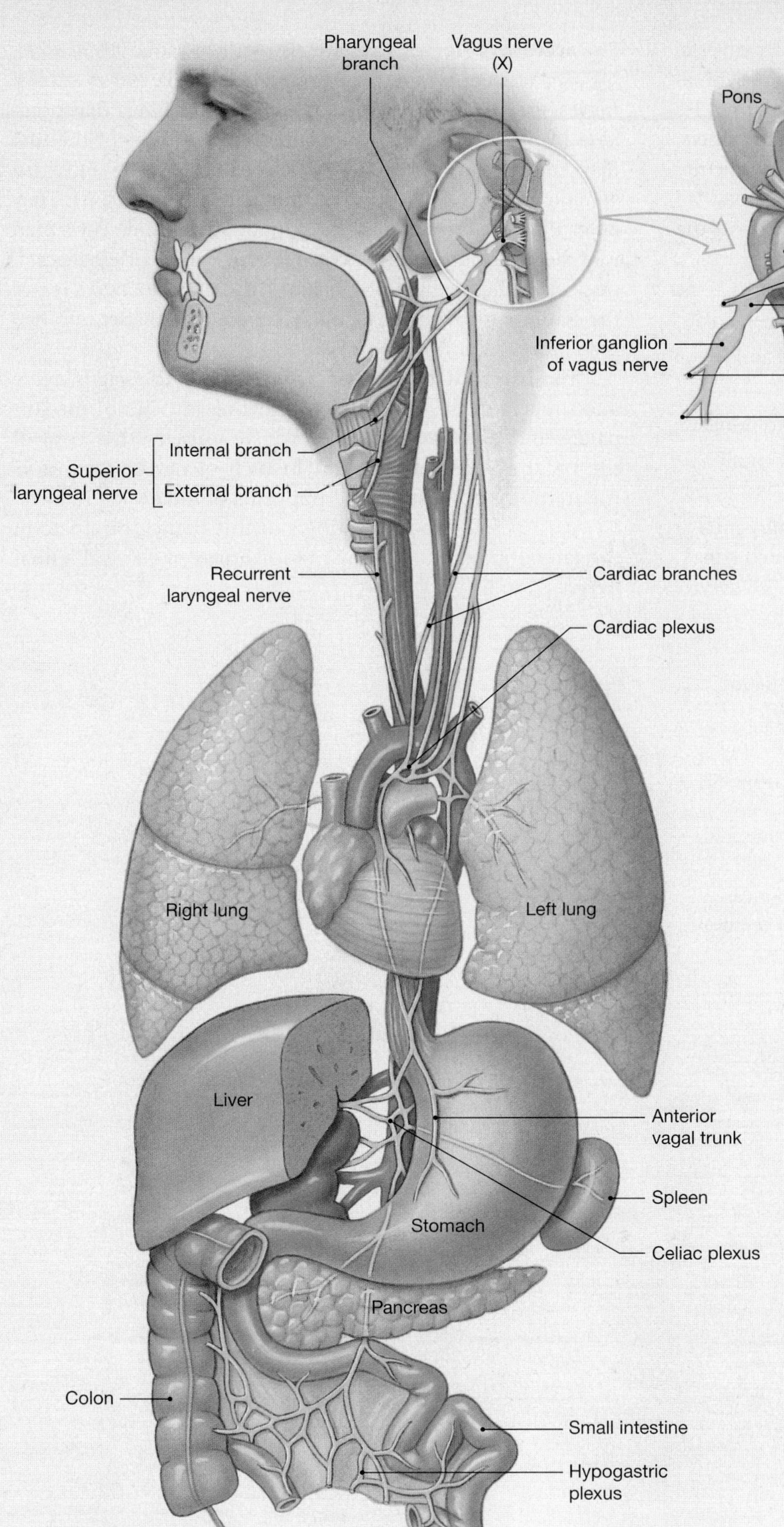

Pharyngeal branch
Vagus nerve (X)
Pons
Medulla oblongata
Auricular branch to external ear
Superior ganglion of vagus nerve
Inferior ganglion of vagus nerve
Pharyngeal branch
Superior laryngeal nerve
Superior laryngeal nerve
Internal branch
External branch
Recurrent laryngeal nerve
Cardiac branches
Cardiac plexus
Right lung
Left lung
Liver
Anterior vagal trunk
Spleen
Stomach
Celiac plexus
Pancreas
Colon
Small intestine
Hypogastric plexus

Figure 14–26 The Vagus Nerve.

- **Pass through:** Jugular foramina between the occipital bone and the temporal bones ⮌ pp. 213, 215
- **Destination:** *Sensory*: sensory nuclei and autonomic centers of medulla oblongata. *Visceral motor*: muscles of the palate, pharynx, digestive, respiratory, and cardiovascular systems in the thoracic and abdominal cavities

The **vagus** (VĀ-gus) **nerves (X)** arise immediately posterior to the attachment of the glossopharyngeal nerves. Many small rootlets contribute to their formation. Developmental studies indicate that these nerves probably represent the fusion of several smaller cranial nerves during our evolutionary history. *Vagus* means wandering, and the vagus nerves branch and radiate extensively. Figure 14–26 shows only the general pattern of distribution.

Sensory neurons are located in the **superior** (*jugular*) **ganglion** and the **inferior** (*nodose*; NŌ-dōs) **ganglion**. Each vagus nerve provides somatic sensory information about the external acoustic meatus and the diaphragm, and special sensory information from pharyngeal taste receptors. But most of the vagal afferents carry visceral sensory information from receptors along the esophagus, respiratory tract, and abdominal viscera as distant as the last portions of the large intestine. This visceral sensory information is vital to the autonomic control of visceral function.

The motor components of the vagus are equally diverse. Each vagus nerve carries preganglionic autonomic (parasympathetic) fibers that affect the heart and control smooth muscles and glands within the areas monitored by its sensory fibers, including the stomach, intestines, and gallbladder. Difficulty in swallowing is one of the most common signs of damage to either cranial nerve IX or X. Damage to either one prevents the coordination of the swallowing reflex.

The Accessory Nerves (XI)

- **Primary function:** Motor to muscles of the neck and upper back
- **Origin:** Motor nuclei of spinal cord and medulla oblongata
- **Pass through:** Jugular foramina between the occipital bone and the temporal bones pp. 213, 215
- **Destination:** Internal branch innervates voluntary muscles of palate, pharynx, and larynx; external branch controls sternocleidomastoid and trapezius

The **accessory nerves (XI)** are also known as the *spinal accessory nerves* or the *spino-accessory nerves*. Unlike other cranial nerves, each accessory nerve has some motor fibers that originate in the lateral part of the anterior gray horns of the first five cervical segments of the spinal cord (Figure 14–27). These somatic motor fibers form the **spinal root** of nerve XI. They enter the cranium through the foramen magnum. They then join the motor fibers of the **cranial root**, which originates at a nucleus in the medulla oblongata. The composite nerve leaves the cranium through the jugular foramen and divides into two branches.

The **internal branch** of nerve XI joins the vagus nerve and innervates the voluntary swallowing muscles of the soft palate and pharynx and the intrinsic muscles that control the vocal cords. The **external branch** of nerve XI controls the sternocleidomastoid and trapezius of the neck and back. pp. 355, 362 The motor fibers of this branch originate in the lateral gray part of the anterior horns of cervical spinal nerves C_1 to C_5.

Figure 14–27 The Accessory and Hypoglossal Nerves. ATLAS: Plates 18a–c; 25

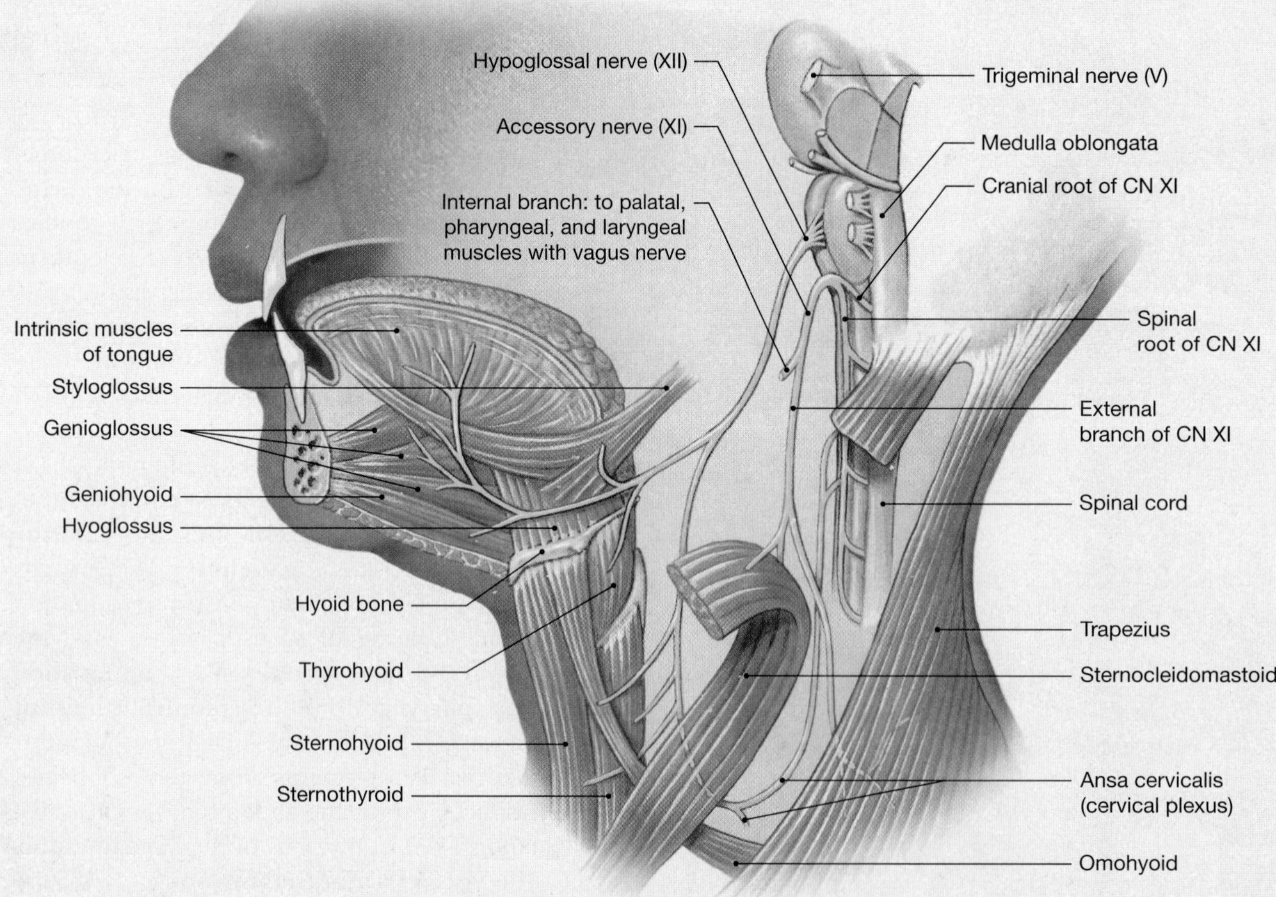
Hypoglossal nerve (XII)
Accessory nerve (XI)
Internal branch: to palatal, pharyngeal, and laryngeal muscles with vagus nerve
Intrinsic muscles of tongue
Styloglossus
Genioglossus
Geniohyoid
Hyoglossus
Hyoid bone
Thyrohyoid
Sternohyoid
Sternothyroid
Trigeminal nerve (V)
Medulla oblongata
Cranial root of CN XI
Spinal root of CN XI
External branch of CN XI
Spinal cord
Trapezius
Sternocleidomastoid
Ansa cervicalis (cervical plexus)
Omohyoid

The Hypoglossal Nerves (XII)

- **Primary function:** Motor (tongue movements)
- **Origin:** Motor nuclei of medulla oblongata
- **Pass through:** Hypoglossal canals of occipital bone ⮌ pp. 214, 215, 216
- **Destination:** Muscles of the tongue ⮌ p. 352

Each **hypoglossal** (hī-pō-GLOS-al) **nerve (XII)** leaves the cranium through the hypoglossal canal. The nerve then curves to reach the skeletal muscles of the tongue (see Figure 14–27). This cranial nerve provides voluntary motor control over movements of the tongue. Its condition is checked by having you stick out your tongue. Damage to one hypoglossal nerve or to its associated nuclei causes the tongue to veer toward the affected side.

Table 14–4 summarizes the basic distribution and function of each cranial nerve.

Table 14–4 Cranial Nerve Branches and Functions

Cranial Nerve (Number)	Sensory Ganglion	Branch	Primary Function	Foramen	Innervation
Olfactory (I)			Special sensory	Olfactory foramina of ethmoid	Olfactory epithelium
Optic (II)			Special sensory	Optic canal	Retina of eye
Oculomotor (III)			Motor	Superior orbital fissure	Inferior, medial, superior rectus, inferior oblique and levator palpebrae superioris; intrinsic eye muscles
Trochlear (IV)			Motor	Superior orbital fissure	Superior oblique
Trigeminal (V)	Semilunar		Mixed	Superior orbital fissure	Areas associated with the jaws
		Ophthalmic nerve (V_1)	Sensory	Superior orbital fissure	Orbital structures, nasal cavity, skin of forehead, upper eyelid, eyebrows, nose (part)
		Maxillary nerve (V_2)	Sensory	Foramen rotundum	Lower eyelid; superior lip, gums, and teeth; cheek, nose (part), palate, and pharynx (part)
		Mandibular nerve (V_3)	Mixed	Foramen ovale	*Sensory:* inferior gums, teeth, lips, palate (part), and tongue (part) *Motor:* muscles of mastication
Abducens (VI)			Motor	Superior orbital fissure	Lateral rectus
Facial (VII)	Geniculate		Mixed	Internal acoustic meatus to facial canal; exits at stylomastoid foramen	*Sensory:* taste receptors on anterior 2/3 of tongue *Motor:* muscles of facial expression, lacrimal gland, submandibular gland, sublingual glands
Vestibulocochlear (VIII)		Cochlear nerve Vestibular nerve	Special sensory	Internal acoustic meatus	Cochlea (receptors for hearing) Vestibule (receptors for motion and balance)
Glossopharyngeal (IX)	Superior (jugular) and inferior (petrosal)		Mixed	Jugular foramen	*Sensory:* posterior 1/3 of tongue; pharynx and palate (part); receptors for blood pressure, pH, oxygen, and carbon dioxide concentrations *Motor:* pharyngeal muscles and parotid gland
Vagus (X)	Superior (jugular) and inferior (nodose)		Mixed	Jugular foramen	*Sensory:* pharynx; auricle and external acoustic meatus; diaphragm; visceral organs in thoracic and abdominopelvic cavities *Motor:* palatal and pharyngeal muscles and visceral organs in thoracic and abdominopelvic cavities
Accessory (XI)		Internal	Motor	Jugular foramen	Skeletal muscles of palate, pharynx, and larynx (with vagus nerve)
		External	Motor	Jugular foramen	Sternocleidomastoid and trapezius
Hypoglossal (XII)			Motor	Hypoglossal canal	Tongue musculature

Checkpoint

21. What name is given to fibers carrying information between the brain and spinal cord, and through which brain regions do they pass?
22. What symptoms would you expect to observe in an individual who has damage to the basal nuclei?
23. A patient suffers a head injury that damages her primary motor cortex. Where is this area located?
24. Which senses would be affected by damage to the temporal lobes of the cerebrum?
25. After suffering a stroke, a patient is unable to speak. He can understand what is said to him, and he can understand written messages, but he cannot express himself verbally. Which part of his brain has been affected by the stroke?
26. A patient is having a difficult time remembering facts and recalling long-term memories. Which part of his cerebrum is probably involved?

See the blue Answers tab at the back of the book.

14

14-10 Cranial reflexes are rapid, automatic responses involving sensory and motor fibers of cranial nerves

Learning Outcome **Describe representative examples of cranial reflexes that produce somatic responses or visceral responses to specific stimuli.**

Cranial reflexes are automatic responses to stimuli that involve the sensory and motor fibers of cranial nerves. They include both monosynaptic and polysynaptic reflex arcs. Later chapters contain numerous examples of cranial reflexes. In this section we will simply provide an overview and general introduction.

Table 14–5 lists representative examples of cranial reflexes and their functions. There are two main types of cranial reflexes, somatic and visceral. These reflexes are clinically important because they provide a quick and easy method for observing the condition of cranial nerves and specific nuclei and tracts in the brain. The somatic reflexes mediated by the cranial nerves are seldom more complex than the somatic reflexes of the spinal cord that we discussed in Chapter 13. ⮌ p. 457 These cranial reflexes are often used to check for damage to the cranial nerves or the associated processing centers in the brain.

The brainstem contains many reflex centers that control visceral motor activity. The motor output of these visceral reflexes is distributed by the autonomic nervous system. As you will see in Chapter 16, the cranial nerves carry most of the commands issued by the parasympathetic division of the ANS. Spinal nerves T_1–L_2 carry the sympathetic commands. Many of the centers that coordinate autonomic reflexes are located in the medulla oblongata. These centers can direct very complex visceral motor responses that are essential to the control of respiratory, digestive, and cardiovascular functions.

Checkpoint

27. What are cranial reflexes?
28. Which reflex and type are you testing if you shine a light in a person's eye?

See the blue Answers tab at the back of the book.

+ Clinical Note Concussion and Beyond

Contact sports can threaten the athlete's brain—think of boxers taking multiple punches to the head, or soccer players heading the ball. The soft tissue of the brain accelerates and decelerates inside the hard bony box of the cranium.

A *concussion* is the most common type of traumatic brain injury that disturbs brain function. Effects, such as headaches and light sensitivity, are usually temporary. An acutely concussed brain often manifests with headaches that worsen, confusion, and slurred speech. However, the key factor in chronic brain injury is **repetition**. Football players have left the sport through retirement or injury only to experience progressive signs of diminished concentration, mood disorders (especially depression), personality changes (anger and aggression), cognitive decline (*cognition*, the ability to think), and memory loss. This phenomenon was first identified in the 1920s by the expression "punch drunk." The contemporary term is *chronic traumatic encephalopathy*, or *CTE*.

No specific biomarkers (cellular indicators) have yet been identified for CTE. On autopsy, the cerebral hemispheres show diffuse atrophy (loss of mass) and enlarged ventricles. Both neurons and astrocytes appear abnormal. As a result, axonal transmission is impaired.

Why do some athletes develop CTE and others do not? Risk factors remain to be specified, such as the severity and frequency of injury. Prevention strategies include wearing protective headgear. If an athlete sustains a concussion, it is crucial to observe strict "return to play" guidelines. The brain is especially vulnerable to a second injury if the rest period is curtailed.

Table 14–5 Cranial Reflexes

Reflex	Stimulus	Afferents	Central Synapse	Efferents	Response
SOMATIC REFLEXES					
Corneal reflex	Contact with corneal surface	Trigeminal (V)	Motor nucleus for facial (VII)	Facial (VII)	Blinking of eyelids
Tympanic reflex	Loud noise	Vestibulocochlear (VIII)	Inferior colliculus	Facial (VII)	Reduced movement of auditory ossicles
Auditory reflexes	Loud noise	Vestibulocochlear (VIII)	Motor nuclei of brainstem and spinal cord	CN III, IV, VI, VII, X, and cervical nerves	Eye and/or head movements triggered by sudden sounds
Vestibulo-ocular reflexes	Rotation of head	Vestibulocochlear (VIII)	Motor nuclei controlling eye muscles	CN III, IV, VI	Opposite movement of eyes to stabilize field of vision
VISCERAL REFLEXES					
Pupillary reflex	Light striking photoreceptors in one eye	Optic (II)	Superior colliculus	Oculomotor (III)	Pupil constricts
Consensual light reflex	Light striking photoreceptors in one eye	Optic (II)	Superior colliculus	Oculomotor (III)	Pupil constricts in both eyes

14 Chapter Review

Study Outline

MasteringA&P™ Access more chapter study tools online in the MasteringA&P Study Area:

- Chapter Quizzes, Chapter Practice Test, MP3 Tutor Sessions, and Clinical Case Studies
- Practice Anatomy Lab PAL 3.0
- A&P Flix A&PFlix
- Interactive Physiology iP2
- PhysioEx PhysioEx 9.1

An Introduction to the Brain and Cranial Nerves p. 466

1. The adult human brain contains almost 97 percent of the body's nervous tissue and averages 1.4 kg (3 lb) in weight and 1200 mL (71 in.^{3}) in volume.

14-1 The brain develops four main regions: the cerebrum, cerebellum, diencephalon, and brainstem p. 466

2. The four main regions in the adult brain are the **cerebrum**, **cerebellum**, **diencephalon**, and **brainstem**. The brainstem is made up of the **midbrain** (mesencephalon), **pons**, and **medulla oblongata**. *(Figure 14–1)*
3. The brain contains extensive areas of **cortex**, a layer of gray matter on the surfaces of the cerebrum (**cerebral cortex**) and cerebellum.
4. The brain forms from three swellings at the superior tip of the developing *neural tube*: the *prosencephalon*, the *mesencephalon*, and the *rhombencephalon*. The prosencephalon ("forebrain") forms the **telencephalon** (which becomes the cerebrum) and diencephalon; the mesencephalon becomes the midbrain; the rhombencephalon ("hindbrain") forms the **metencephalon** (cerebellum and pons) and **myelencephalon** (medulla oblongata). *(Table 14–1)*
5. During development, the central passageway of the brain expands to form chambers called **ventricles**, which contain cerebrospinal fluid (CSF). *(Figure 14–2)*

14-2 The brain is protected and supported by the cranial meninges, cerebrospinal fluid, and the blood brain barrier p. 469

6. The cranial meninges (the **dura mater**, **arachnoid mater**, and **pia mater**) are continuous with the meninges of the spinal cord.
7. Folds of dura mater, including the **falx cerebri**, **tentorium cerebelli**, and **falx cerebelli**, stabilize the position of the brain. *(Figure 14–3)*
8. Cerebrospinal fluid (CSF) (1) protects delicate neural structures, (2) supports the brain, and (3) transports nutrients, chemical messengers, and waste products.
9. Cerebrospinal fluid is produced at the **choroid plexuses**, reaches the subarachnoid space through the **lateral** and **median aperture**, and diffuses across the **arachnoid granulations** into the **superior sagittal sinus**. *(Spotlight Figure 14–4)*
10. The **blood brain barrier (BBB)** isolates nervous tissue from the general circulation.
11. The blood brain barrier is incomplete in parts of the **hypothalamus**, the **pituitary gland**, the **pineal gland**, and the **choroid plexus**.

14-3 Brainstem: The medulla oblongata relays signals between the rest of the brain and the spinal cord p. 474

12. The medulla oblongata connects the brain and spinal cord. It contains relay stations such as the **olivary nuclei**, and **reflex centers**, including the **cardiovascular** and **respiratory**

rhythmicity centers. The **reticular formation** begins in the medulla oblongata and extends into more superior portions of the brainstem. *(Figures 14–5, 14–6)*

14-4 Brainstem: The pons contains nuclei that process and tracts that relay sensory and motor information p. 477

13. The pons contains (1) sensory and motor nuclei for four cranial nerves; (2) nuclei that help control respiration; (3) nuclei and tracts linking the cerebellum with the brainstem, cerebrum, and spinal cord; and (4) ascending, descending, and transverse tracts. *(Figure 14–7)*

14-5 Brainstem: The midbrain regulates visual and auditory reflexes and controls alertness p. 478

14. The **tectum** (roof of the midbrain) contains the **corpora quadrigemina** (**superior colliculi** and **inferior colliculi**). The tegmentum contains the **red nucleus**, the **substantia nigra**, the **cerebral peduncles**, and the headquarters of the reticular activating system (RAS). *(Figure 14–8)*

14

14-6 The cerebellum coordinates reflexive and learned patterns of muscular activity at the subconscious level p. 478

15. The cerebellum adjusts postural muscles and programs and tunes ongoing movements. The two **cerebellar hemispheres** consist of the **anterior** and **posterior lobes**, the **vermis**, and the **flocculonodular lobe**. *(Figure 14–9)*
16. The **superior**, **middle**, and **inferior cerebellar peduncles** link the cerebellum with the brainstem, diencephalon, cerebrum, and spinal cord and interconnect the two cerebellar hemispheres.

14-7 The diencephalon integrates sensory information with motor output at the subconscious level p. 481

17. The diencephalon is composed of the epithalamus, the thalamus, and the hypothalamus. *(Figures 14–10, 14–11)*
18. The thalamus is the final relay point for ascending sensory information and coordinates the activities of the basal nuclei and cerebral cortex. *(Figures 14–10, 14–11; Table 14–2)*
19. The hypothalamus can (1) secrete certain hormones, (2) regulate body temperature, (3) control autonomic function, (4) coordinate voluntary and autonomic functions, (5) coordinate activities of the nervous and endocrine systems, (6) regulate circadian cycles of activity, (7) control skeletal muscle contractions at the subconscious level, and (8) produce emotions and behavioral **drives**. *(Figure 14–11)*

14-8 The limbic system is a group of nuclei and tracts that functions in emotion, motivation, and memory p. 484

20. The **limbic system**, or *motivational system*, includes the **amygdaloid body**, **cingulate gyrus**, **dentate gyrus**, **parahippocampal gyrus**, **hippocampus**, and **fornix**. The functions of the limbic system involve emotional states, behavioral drives, and memory. *(Figure 14–12)*

14-9 The cerebrum contains motor, sensory, and association areas, allowing for higher mental functions p. 486

21. The cortical surface contains **gyri** (elevated ridges) separated by **sulci** (shallow depressions) or **fissures** (deeper grooves). The **longitudinal cerebral fissure** separates the two **cerebral hemispheres**. The **central sulcus** separates the **frontal** and **parietal lobes**. Other sulci form the boundaries of the **temporal** and **occipital lobes**. *(Figure 14–13)*
22. The white matter of the cerebrum contains **association fibers**, **commissural fibers**, and **projection fibers**. *(Figure 14–14)*
23. The **basal nuclei** include the **caudate nucleus**, **globus pallidus**, and **putamen**; they control muscle tone and coordinate learned movement patterns and other somatic motor activities. *(Figure 14–15)*
24. The **primary motor cortex** of the **precentral gyrus** directs voluntary movements. The **primary somatosensory cortex** of the **postcentral gyrus** receives somatic sensory information from touch, pressure, pain, vibration, taste, and temperature receptors. *(Figure 14–16; Table 14–3)*
25. **Association areas**, such as the **somatic sensory association area**, **visual association area**, and **premotor cortex** (**somatic motor association area**), control our ability to understand sensory information and coordinate a motor response. *(Figure 14–16)*
26. The **specialized language areas of the brain** receive information from all the sensory association areas. It is usually dominant in only one hemisphere—generally the left. *(Figures 14–16b)*
27. The **prefrontal cortex** coordinates information from the secondary and special association areas of the entire cortex and performs abstract intellectual functions. *(Figure 14–16)*
28. The left hemisphere typically contains the **Broca's area** and **Wernicke's area** and is responsible for language-based skills. The right hemisphere is typically responsible for spatial relationships and analyses.
29. Brain activity is measured using an **electroencephalogram**. **Alpha waves** appear in healthy resting adults; **beta waves** occur when adults are concentrating; **theta waves** appear in children; and **delta waves** are normal during sleep. *(Figure 14–17)*

Focus: Cranial Nerves p. 495

30. We have 12 pairs of cranial nerves. Except for CN I and CN II, each nerve attaches to the ventrolateral surface of the brainstem near the associated sensory or motor nuclei. *(Figure 14–18)*
31. The **olfactory nerves (I)** carry sensory information responsible for the sense of smell. The olfactory afferents synapse within the **olfactory bulbs**. *(Figures 14–18, 14–19)*
32. The **optic nerves (II)** carry visual information from special sensory receptors in the eyes. *(Figures 14–18, 14–20)*
33. The **oculomotor nerves (III)** are the primary source of innervation for four of the extrinsic eye muscles. *(Figure 14–21)*
34. The **trochlear nerves (IV)**, the smallest cranial nerves, innervate the superior oblique muscles of the eyes. *(Figure 14–21)*
35. The **trigeminal nerves (V)**, the largest cranial nerves, are mixed nerves with three divisions: the *ophthalmic nerve*, *maxillary nerve*, and *mandibular nerve*. *(Figure 14–22)*

36. The **abducens nerves (VI)** innervate the lateral rectus muscles of the eyes. *(Figure 14–21)*
37. The **facial nerves (VII)** are mixed nerves that control muscles of the scalp and face. They provide pressure sensations over the face and receive taste information from the tongue. *(Figure 14–23)*
38. The **vestibulocochlear nerves (VIII)** contain the **vestibular nerve**, which monitors sensations of balance, position, and movement, and the **cochlear nerve**, which monitors hearing receptors. *(Figure 14–24)*
39. The **glossopharyngeal nerves (IX)** are mixed nerves that innervate the tongue and pharynx and control the action of swallowing. *(Figure 14–25)*
40. The **vagus nerves (X)** are mixed nerves that are vital to the autonomic control of visceral function in the thorax and abdomen. *(Figure 14–26)*
41. The **accessory nerves (XI)** have **internal branches**, which innervate voluntary swallowing muscles of the soft palate and pharynx, and **external branches**, which control muscles associated with the pectoral girdle. *(Figure 14–27)*
42. The **hypoglossal nerves (XII)** provide voluntary motor control over tongue movements. *(Figure 14–27)*
43. The branches and functions of the cranial nerves are summarized in *Table 14–4*.

14-10 Cranial reflexes are rapid, automatic responses involving sensory and motor fibers of cranial nerves

p. 506

44. **Cranial reflexes** are monosynaptic and polysynaptic reflex arcs that involve sensory and motor fibers of cranial nerves. *(Table 14–5)*

Review Questions

See the blue Answers tab at the back of the book.

LEVEL 1 Reviewing Facts and Terms

1. Identify the major regions of the brain in the following diagram.

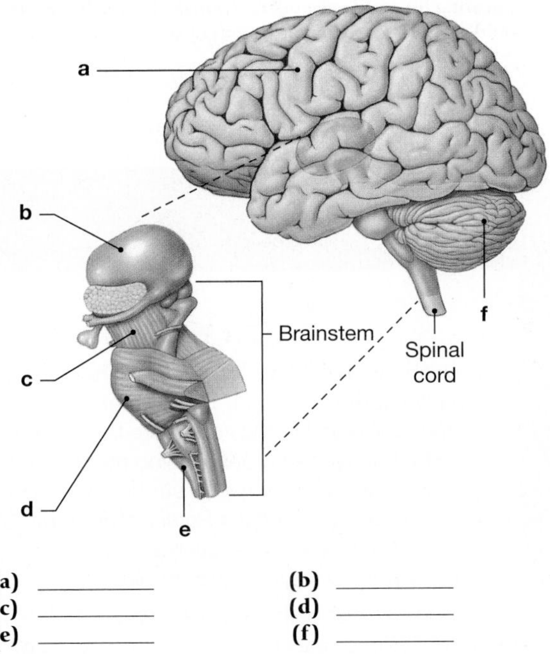
a
b
c
d
e
f
Brainstem
Spinal cord

(a) ________ (b) ________
(c) ________ (d) ________
(e) ________ (f) ________

2. Identify the layers of the cranial meninges in the following diagram.

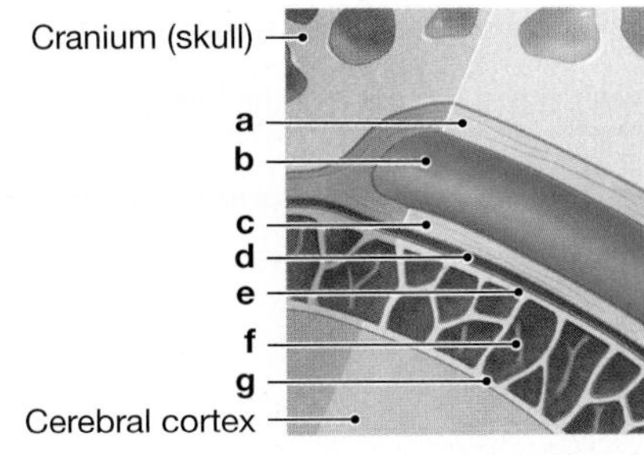
Cranium (skull)
a
b
c
d
e
f
g
Cerebral cortex

(a) ________ (b) ________
(c) ________ (d) ________
(e) ________ (f) ________
(g) ________

3. The term *higher brain centers* refers to those areas of the brain involved in higher-order functions. These centers would probably include nuclei, centers, and cortical areas of **(a)** the cerebrum, **(b)** the cerebellum, **(c)** the diencephalon, **(d)** all of these, **(e)** a and c only.
4. Which of the following is the site of cerebrospinal fluid production? **(a)** dural venous sinus, **(b)** choroid plexus, **(c)** falx cerebri, **(d)** tentorium cerebelli, **(e)** insula.
5. The pons contains **(a)** sensory and motor nuclei for six cranial nerves, **(b)** nuclei concerned with control of blood pressure, **(c)** tracts that link the cerebellum with the brainstem, **(d)** no ascending or descending tracts, **(e)** both a and b.
6. The dural fold that divides the two cerebellar hemispheres is the **(a)** transverse sinus, **(b)** falx cerebri, **(c)** tentorium cerebelli, **(d)** falx cerebelli.
7. Cerebrospinal fluid is produced and secreted by **(a)** neurons, **(b)** ependymal cells, **(c)** Purkinje cells, **(d)** basal nuclei.
8. The primary purpose of the blood brain barrier (BBB) is to **(a)** provide the brain with oxygenated blood, **(b)** drain venous blood by the internal jugular veins, **(c)** isolate neural tissue in the CNS from the general circulation, **(d)** do all of these.
9. The centers in the pons that modify the activity of the respiratory rhythmicity centers in the medulla oblongata are the **(a)** apneustic and pneumotaxic centers, **(b)** inferior and superior peduncles, **(c)** cardiac and vasomotor centers, **(d)** gracile nucleus and cuneate nucleus.
10. The final relay point for ascending sensory information that will be projected to the primary sensory cortex is the **(a)** hypothalamus, **(b)** thalamus, **(c)** spinal cord, **(d)** medulla oblongata.
11. The establishment of emotional states is a function of the **(a)** limbic system, **(b)** tectum, **(c)** mammillary bodies, **(d)** thalamus.
12. Coordination of learned movement patterns at the subconscious level is performed by **(a)** the cerebellum, **(b)** the substantia nigra, **(c)** association fibers, **(d)** the hypothalamus.
13. The two cerebral hemispheres are functionally different, even though anatomically they appear the same. **(a)** true, **(b)** false.
14. What are the three important functions of the CSF?

15. Which three areas in the brain are not isolated from the general circulation by the blood–brain barrier?
16. List the names of the 12 pairs of cranial nerves. (Hint: use the mnemonic device "Oh, Once One Takes The Anatomy Final, Very Good Vacations Are Heavenly.")

LEVEL 2 Reviewing Concepts

17. Why can the brain respond to stimuli with greater versatility than the spinal cord?
18. Briefly summarize the overall function of the cerebellum.
19. The only cranial nerves that are attached to the cerebrum are the _______ nerves. **(a)** optic, **(b)** oculomotor, **(c)** trochlear, **(d)** olfactory, **(e)** abducens.
20. If symptoms characteristic of Parkinson's disease appear, which part of the midbrain is inhibited from secreting a neurotransmitter? Which neurotransmitter is it?
21. What varied roles does the hypothalamus play in the body?
22. Stimulation of which part of the brain would produce sensations of hunger and thirst?
23. Which structure in the brain is your A&P instructor referring to when talking about a nucleus that resembles a sea horse and that is important in the storage and retrieval of long-term memories? In which functional system of the brain is it located?
24. What are the principal functional differences between the right and left cerebral hemispheres?
25. Damage to the vestibular nucleus would lead to **(a)** loss of sight, **(b)** loss of hearing, **(c)** inability to sense pain, **(d)** difficulty in maintaining balance, **(e)** inability to swallow.
26. A cerebrovascular accident occurs when **(a)** the reticular activating system fails to function, **(b)** the prefrontal lobe is damaged, **(c)** the blood supply to a portion of the brain is cut off, **(d)** a descending tract in the spinal cord is severed, **(e)** brainstem nuclei hypersecrete dopamine.
27. What kinds of problems are associated with the presence of lesions in the Wernicke's area and the Broca's area?

LEVEL 3 Critical Thinking and Clinical Applications

28. Smelling salts can sometimes help restore consciousness after a person has fainted. The active ingredient of smelling salts is ammonia, and it acts by irritating the lining of the nasal cavity. Propose a mechanism by which smelling salts would raise a person from the unconscious state to the conscious state.
29. A police officer has just stopped Bill on suspicion of driving while intoxicated. The officer asks Bill to walk the yellow line on the road and then to place the tip of his index finger on the tip of his nose. How would these activities indicate Bill's level of sobriety? Which part of the brain is being tested by these activities?
30. Colleen falls down a flight of stairs and bumps her head several times. Soon after, she develops a headache and blurred vision. Diagnostic tests at the hospital reveal an epidural hematoma in the temporoparietal area. The hematoma is pressing against the brainstem. What other signs and symptoms might she experience as a result of the injury?
31. Cerebral meningitis is a condition in which the meninges of the brain become inflamed as the result of viral or bacterial infection. This condition can be life threatening. Why?
32. Infants have little to no control of the movements of their head. One of the consequences of this is that they are susceptible to shaken baby syndrome, caused by vigorous shaking of an infant or young child by the arms, legs, chest, or shoulders. Forceful shaking can cause brain damage leading to mental retardation, speech and learning disabilities, paralysis, seizures, hearing loss, and even death. Damage to which areas of the brain would account for the clinical signs observed in this syndrome?

14

+ CLINICAL CASE Wrap-Up The Neuroanatomist's Stroke

Dr. Taylor's full recovery took about 8 years. While the left side of her brain was injured and shut down, the right side continued functioning. Because language and thoughts are controlled in the left hemisphere, Dr. Taylor "sat in an absolutely silent mind" for the first month after her stroke. Similarly, since the general interpretive center for mathematical calculation is in the left hemisphere, she had to learn to use numbers all over again. And because the primary motor cortex governing the right side of the body resides in the precentral gyrus of the left hemisphere, she had to learn to use her right arm again.

The stroke had injured some cells in her brain beyond repair. But some cells were dormant, capable of recovery and of making new neuronal connections. This ability of nerve cells to make these new connections, called *neuroplasticity*, permits healing as the brain reorganizes itself after an injury.

Dr. Taylor wants anatomy and physiology students to know two things. First, "if you study the brain, you will never be bored." Second, "if you treat stroke patients like they will recover, they are more likely to recover."

She has written a best-selling memoir about her personal experience with stroke, *My Stroke of Insight: A Brain Scientist's Personal Journey*. She offers suggestions for the stroke survivor, caregiver, and everyone who deals with stroke.

1. How would you know, just based on signs and symptoms, which side of Dr. Taylor's brain has been injured by a stroke?
2. What is neuroplasticity and why was it important in Dr. Taylor's recovery?

See the blue Answers tab at the back of the book.

Related Clinical Terms

attention deficit hyperactivity disorder (ADHD): Disorder occurring mainly in children characterized by hyperactivity, inability to concentrate, and impulsive or inappropriate behavior.

autism: Any of a range of behavioral disorders occurring primarily in children, including such symptoms as poor concentration, hyperactivity, and impulsivity.

Creutzfeldt-Jakob disease (CJD): A rare, degenerative, invariably fatal brain disorder that is marked by rapid mental deterioration. The disease, which is caused by a prion (an infectious protein particle), typically starts by causing mental and emotional problems, then progresses to affect motor skills, such as walking and talking.

delirium: An acutely disturbed state of mind that occurs in fever, intoxication, and other disorders and is characterized by restlessness, hallucinations, and incoherence of thought and speech.

dementia: A chronic or persistent disorder of the mental processes caused by brain disease or injury and marked by memory disorders, personality changes, and impaired reasoning.

Glasgow coma scale: The most widely used scoring system to quantify the level of consciousness of a victim of a traumatic brain injury. It rates three functions: eye opening, verbal response, and motor response.

migraine: A type of headache marked by severe debilitating head pain lasting several hours or longer.

myoclonus: A quick, involuntary muscle jerk or contraction; persistent myoclonus usually indicates a nervous system disorder.

pallidectomy: The destruction of all or part of the globus pallidus by chemicals or freezing; used in the treatment of Parkinson's disease.

prosopagnosia: The inability to recognize other humans by their faces.

psychosis: A severe mental disorder in which thought and emotions are so impaired that contact with reality is lost.

stupor: A state of near-unconsciousness or insensibility.

transient ischemic attack (TIA): An episode in which a person has stroke-like symptoms that last less than 24 hours and result in no permanent injury to the brain, but may be a warning sign of the potential for a major stroke.

15 Sensory Pathways and the Somatic Nervous System

Learning Outcomes

These Learning Outcomes correspond by number to this chapter's sections and indicate what you should be able to do after completing the chapter.

- 15-1 ■ **Specify the components of the afferent and efferent divisions of the nervous system, and explain what is meant by the somatic nervous system.** p. 513
- 15-2 ■ **Explain why receptors respond to specific stimuli, and how the organization of a receptor affects its sensitivity.** p. 514
- 15-3 ■ **Identify the receptors for the general senses, and describe how they function.** p. 517
- 15-4 ■ **Identify the major sensory pathways, and explain how it is possible to distinguish among sensations that originate in different areas of the body.** p. 522
- 15-5 ■ **Describe the components, processes, and functions of the somatic motor pathways, and the levels of information processing involved in motor control.** p. 527

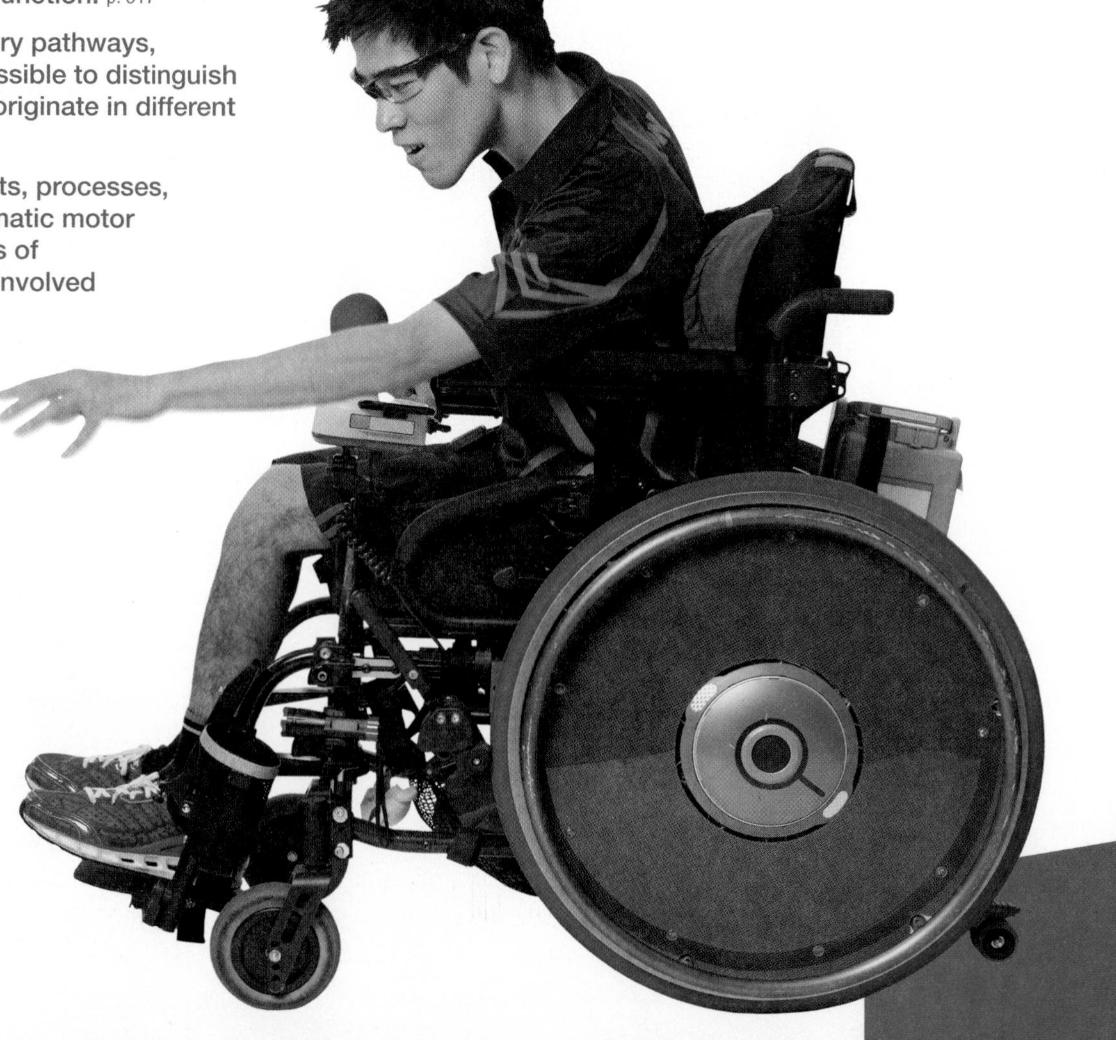

CLINICAL CASE Living with Cerebral Palsy

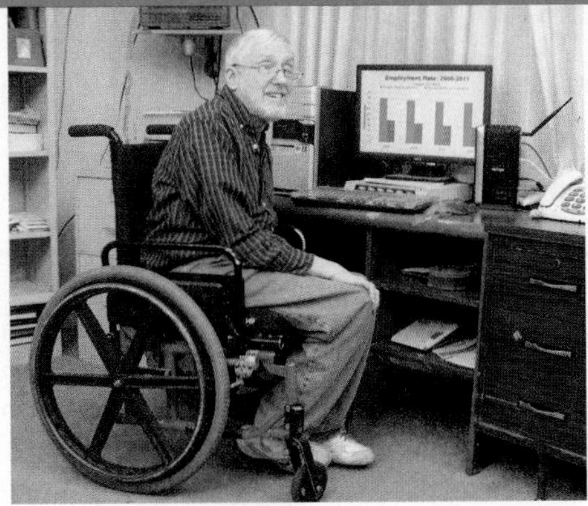

Michael Ward had a challenging start in life. His mother had a difficult and prolonged labor in the days when cesarean sections were rarely performed. Michael was a large baby and suffered brain damage from inadequate oxygen during birth, requiring an extended resuscitation. The lack of oxygen to his brain caused a condition known as *cerebral palsy (CP)*.

Michael was critically ill as a newborn and remained hospitalized for weeks. Because of swallowing difficulties, he had aspiration pneumonia (lung inflammation from inhaling food) three times in his first year of life. He was delayed in reaching gross and fine motor developmental milestones, and although he had plenty to say, his speech was difficult to understand. When he was finally able to walk, by age 4, he used crutches or held a parent's hand. His gait was slow, unsteady, and crouched. He fell often. He has used a wheelchair regularly since high school.

Cerebral palsy is the most common congenital neurological condition, affecting as many as 1 million Americans. In spite of advances in obstetric and neonatal care, the incidence of CP has not changed in the past half century. Cerebral palsy is caused by an injury to the developing brain, and can occur before, during, or within a few years after birth. The specific symptoms of the person with CP relate to the areas of the brain affected by the injury. It is not contagious. It is not progressive. It is a condition, not a disease. **How did Michael's life turn out? See the Clinical Case Wrap-Up on p. 534.**

An Introduction to Sensory Pathways and the Somatic Nervous System

In this chapter we examine how the nervous system works as an integrated unit. We first consider the sensory receptors and sensory processing centers in the brain that make up *sensory pathways*. In our discussion, we will focus primarily on the *general senses* that provide information about the body and its environment, rather than the *special senses* such as sight and hearing. Most of the coverage of the special senses is in Chapter 17.

Then we look at the *motor pathways* that provide for the voluntary and involuntary motor functions of the nervous system. Motor commands issued by the CNS are distributed to the PNS by the *somatic nervous system* and the *autonomic nervous system*. Here we cover the somatic nervous system, which controls the contractions of skeletal muscles. The output of the SNS is under voluntary control. The autonomic nervous system, or *visceral motor system*, controls visceral effectors, such as smooth muscle, cardiac muscle, glands, and adipocytes. We examine the organization and functions of the ANS in Chapter 16.

15-1 Sensory stimuli cause signals to be sent along sensory pathways, and in response motor commands are sent along motor pathways

Learning Outcome Specify the components of the afferent and efferent divisions of the nervous system, and explain what is meant by the somatic nervous system.

The basic events that occur along sensory and motor pathways are shown in Figure 15–1. Along **sensory pathways**, a series of neurons relays sensory information from the initiating point (the sensory receptor) to its final destination inside the CNS. *Sensory receptors* are specialized cells or cell processes that monitor specific conditions in the body or the external environment. When stimulated, a receptor generates action potentials that are propagated along the axon of a sensory neuron and then relayed to other neurons within the pathway. This signal is finally delivered to the CNS, where it can be processed.

Taken together, the somatic and visceral sensory pathways make up the *afferent division* of the nervous system. p. 391 Somatic and visceral sensory information often travels along the same pathway. Somatic sensory information is distributed to *sensory processing centers* in the brain—either the primary somatosensory cortex of the cerebral hemispheres or appropriate areas of the cerebellar hemispheres. Visceral sensory information is distributed primarily to reflex centers in the brainstem and diencephalon.

Once sensory information is integrated by the CNS, somatic (and visceral) motor commands may be issued. Somatic motor commands are carried out by the somatic motor portion of the *efferent division*—the *somatic motor pathways* that control peripheral effectors such as skeletal muscles (see Figure 15–1). Whether somatic motor commands arise at the conscious (voluntary) or subconscious (involuntary) levels, they travel from motor centers in the brain along these somatic motor pathways. The motor neurons and pathways that control skeletal muscles form the somatic nervous system.

Motor commands may be modified by planning, memories, and learning—the so-called *higher-order functions* of the brain. We also consider these higher-order functions in Chapter 16.

Figure 15–1 An Overview of Events Occurring Along the Sensory and Motor Pathways.

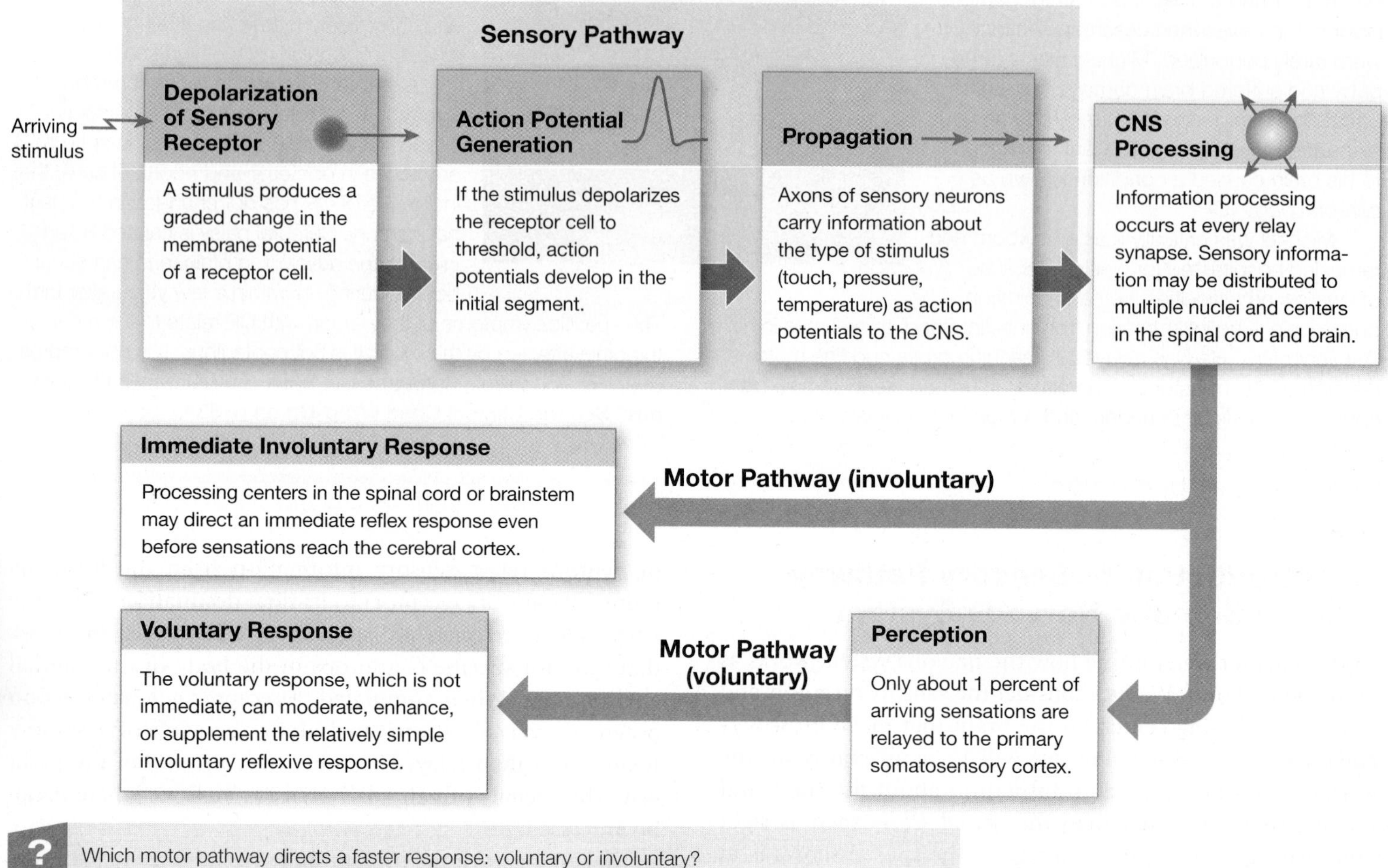
Sensory Pathway
Arriving stimulus
Depolarization of Sensory Receptor
A stimulus produces a graded change in the membrane potential of a receptor cell.
Action Potential Generation
If the stimulus depolarizes the receptor cell to threshold, action potentials develop in the initial segment.
Propagation
Axons of sensory neurons carry information about the type of stimulus (touch, pressure, temperature) as action potentials to the CNS.
CNS Processing
Information processing occurs at every relay synapse. Sensory information may be distributed to multiple nuclei and centers in the spinal cord and brain.
Immediate Involuntary Response
Processing centers in the spinal cord or brainstem may direct an immediate reflex response even before sensations reach the cerebral cortex.
Motor Pathway (involuntary)
Voluntary Response
The voluntary response, which is not immediate, can moderate, enhance, or supplement the relatively simple involuntary reflexive response.
Motor Pathway (voluntary)
Perception
Only about 1 percent of arriving sensations are relayed to the primary somatosensory cortex.

? Which motor pathway directs a faster response: voluntary or involuntary?

15

Checkpoint

1. **What do we call the body's specialized cells that monitor specific internal or external conditions?**
2. **Is it possible for somatic motor commands to occur at the subconscious level?**

See the blue Answers tab at the back of the book.

15-2 Sensory receptors connect our internal and external environments with the nervous system

Learning Outcome **Explain why receptors respond to specific stimuli and how the organization of a receptor affects its sensitivity.**

As we noted, **sensory receptors** are either the processes of specialized sensory neurons or cells monitored by sensory neurons. These receptors provide your central nervous system with information about conditions inside or outside the body. You have general sensory receptors scattered throughout your body. Some of the information they send to the CNS reaches the primary somatosensory cortex and so your awareness. The arriving information is called a **sensation**. The conscious awareness of a sensation is called a **perception**.

Our senses can be divided into general and special, as mentioned. We use the term **general senses** to describe our sensitivity to temperature, pain, touch, pressure, vibration, and *proprioception* (body position). The **special senses** are *olfaction* (smell), *gustation* (taste), *vision* (sight), *equilibrium* (balance), and *hearing*. These sensations are provided by receptors that are structurally more complex than those of the general senses. *Special sensory receptors* are located in **sense organs** such as the eye or ear, where surrounding tissues protect the receptors. The information these receptors provide is distributed to specific areas of the cerebral cortex (the olfactory cortex, the visual cortex, and so forth) and to centers throughout the brainstem. We cover the common functions of both general and special sensory receptors here, and then go on to cover the specific function of each type of special sensory receptor in Chapter 17.

Sensory receptors represent the interface between your nervous system and your internal and external environments. When a sensory receptor detects an arriving stimulus and converts it into an action potential that can be propagated to the CNS, this conversion process is called *transduction*. If transduction does not take place, then as far as you are concerned, the stimulus doesn't exist. For example, bees can see ultraviolet light that you can't see. Dogs can respond to sounds you can't hear. In each case the stimuli are there—but your receptors cannot detect them. Now let's examine the basic concepts of receptor function and sensory processing.

The Detection of Stimuli

How do receptors detect stimuli? Each receptor has a characteristic sensitivity. For example, a touch receptor is very sensitive to pressure but relatively insensitive to chemical stimuli. A taste receptor is sensitive to dissolved chemicals but insensitive to pressure. This feature is called *receptor specificity*.

Specificity may result from the structure of the receptor cell, or from the presence of accessory cells or structures that shield the receptor cell from other stimuli. The simplest receptors are the dendrites of sensory neurons. The branching tips of these dendrites, called *free nerve endings*, are not protected by accessory structures. Free nerve endings extend through a tissue the way grass roots extend into the soil. Many different stimuli can stimulate them. For this reason, free nerve endings show little receptor specificity. For example, chemicals, pressure, temperature changes, or trauma may stimulate free nerve endings that respond to tissue damage by providing pain sensations.

Complex receptors, such as the eye's visual receptors, are protected by accessory cells and connective tissue layers. These cells are seldom exposed to any stimulus other than light and so provide very specific information.

Receptive Fields

The area monitored by a single receptor cell is its *receptive field* (Figure 15–2). Whenever a sufficiently strong stimulus arrives in the receptive field, the CNS receives the information "stimulus arriving at receptor X." The larger the receptive field, the poorer your ability to localize a stimulus. A touch receptor on the general body surface, for example, may have a receptive field 7 cm (2.5 in.) in diameter. As a result, you can describe a light touch there as affecting only a general area, not an exact spot. On your tongue or fingertips, the receptive fields are less than a millimeter in diameter. In these areas, you can be very precise about the location of a stimulus.

The Process of Transduction

An arriving stimulus can take many forms. It may be a physical force (such as pressure), a dissolved chemical, a sound, or light. Regardless of the nature of the stimulus, however, sensory information is sent to the CNS only in the form of action potentials, which are electrical events.

As noted earlier, **transduction** is the conversion of an arriving stimulus into an action potential by a sensory receptor.

Figure 15–2 Receptors and Receptive Fields. Each receptor cell monitors a specific area known as the receptive field.

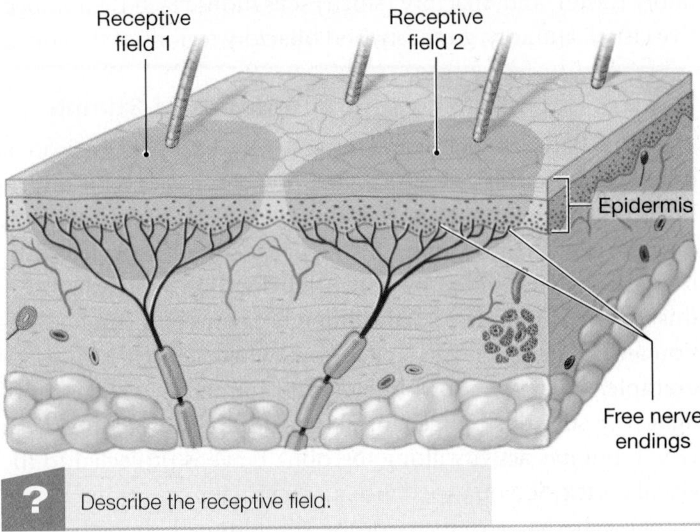
Receptive field 1
Receptive field 2
Epidermis
Free nerve endings

? Describe the receptive field.

Transduction begins when a stimulus changes the membrane potential of the receptor cell. This change, called a *receptor potential*, is either a graded depolarization or a graded hyperpolarization. The stronger the stimulus, the larger the receptor potential.

The typical receptors for the general senses are the free nerve endings (dendrites) of sensory neurons. In these cases, the sensory neuron is the receptor cell. Any receptor potential that depolarizes the plasma membrane will bring the membrane closer to threshold. A depolarizing receptor potential in a neural receptor is called a *generator potential*.

Specialized receptor cells that communicate with sensory neurons across chemical synapses give us sensations of taste, hearing, equilibrium, and vision. In these cases, the receptor potential and the generator potential occur in different cells. The receptor potential develops in the receptor cell, and the generator potential appears later, in the sensory neuron. The receptor cells develop graded receptor potentials in response to stimulation, and the change in membrane potential alters the rate of neurotransmitter release at the synapse. The result is a depolarization or hyperpolarization of the sensory neuron. If sufficient depolarization occurs, an action potential appears in the sensory neuron.

Whenever a sufficiently large generator potential appears, action potentials develop in the axon of a sensory neuron. For reasons discussed in Chapter 12, the greater the degree of sustained depolarization at the axon hillock, the higher the frequency of action potentials in the afferent fiber. p. 414

The Interpretation of Sensory Information

Sensory information that arrives at the CNS is routed according to the location and nature of the stimulus. In previous chapters we emphasized the fact that axons in the CNS are organized in tracts with specific origins and destinations. Once a receptor is stimulated, the series of neurons in that sensory pathway relay that signal to a neuron at a specific site in the cerebral cortex. For example, sensations of touch, pressure, pain, and temperature arrive at the primary somatosensory cortex. Visual, auditory, gustatory (taste), and olfactory (smell) sensations reach their respective visual, auditory, gustatory, and olfactory regions of the cortex.

Determination of the Characteristics of Stimuli

The link between peripheral receptor and cortical neuron is called a **labeled line**. Each labeled line consists of axons carrying information about one *modality*, or type of stimulus (touch, pressure, light, and sound). The type of stimulus perceived by the CNS is determined by which labeled line is activated. For this reason, you cannot tell the difference between a "true" sensation and a "false" one generated somewhere along the line. For example, when you rub your eyes, you commonly see flashes of light. The stimulus is mechanical (physical pressure) rather than visual, but *any* activity along the optic nerve is projected to the visual cortex. So you experience it as a visual perception.

All other characteristics of the stimulus—its strength, duration, and variation—are conveyed by the frequency and pattern of action potentials. As noted in Chapter 12, the CNS interprets sensory information on the basis of the frequency of arriving action potentials. ⮌ p. 427 For example, when pressure sensations are arriving, the greater the pressure, the higher the frequency of action potentials. The conversion of complex sensory information into meaningful patterns of action potentials is called *sensory coding*.

Transduction in Tonic and Phasic Receptors

Receptors can be categorized by the nature of their response to stimulation (Figure 15–3). Some sensory neurons, called **tonic receptors**, are always active (Figure 15–3a). A *tonic response* is an increase or decrease in the frequency of action potentials that directly reflects an increase or decrease in stimulation. Other receptors, called **phasic receptors**, are usually not active. Phasic receptors provide information about the intensity and rate of change of a stimulus (Figure 15–3b). A *phasic response* to stimulation begins with a burst of action potentials that ends when the stimulus stops or does not change in intensity. Receptors that combine phasic and tonic coding can convey extremely complicated sensory information.

Adaptation of Sensory Receptors

Adaptation is a reduction of receptor sensitivity in the presence of a constant stimulus. You seldom notice the rumble of the tires when you ride in a car or the background noise of an air conditioner. The reason is that your nervous system quickly adapts to stimuli that are painless and constant.

There are two types of sensory receptor adaptation: peripheral and central. *Peripheral adaptation* occurs in the PNS when the level of receptor activity changes. The receptor responds strongly at first, but then its activity gradually decreases, in part because the size of the generator potential gradually decreases. This response is characteristic of phasic receptors, which are also called **fast-adapting receptors**. For example, temperature receptors are phasic receptors, so you seldom notice the room temperature unless it changes suddenly.

Tonic receptors show little peripheral adaptation and so are called **slow-adapting receptors**. Pain receptors are slow-adapting receptors. This is one reason why pain sensations remind you of an injury long after the initial damage has taken place.

Adaptation also occurs along sensory pathways inside the CNS. For example, a few seconds after you have been exposed to a new smell, awareness of the stimulus virtually disappears, although the sensory neurons are still quite active. This process

Figure 15–3 Tonic and Phasic Sensory Receptors.

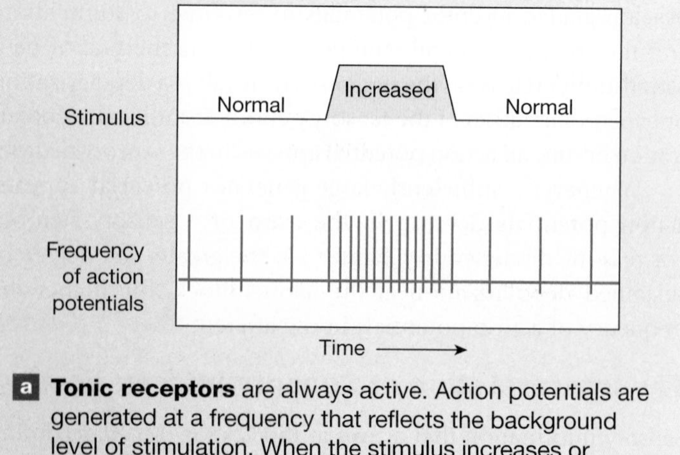

Stimulus
Normal
Increased
Normal
Frequency of action potentials
Time

a **Tonic receptors** are always active. Action potentials are generated at a frequency that reflects the background level of stimulation. When the stimulus increases or decreases, the rate of action potential generation changes accordingly.

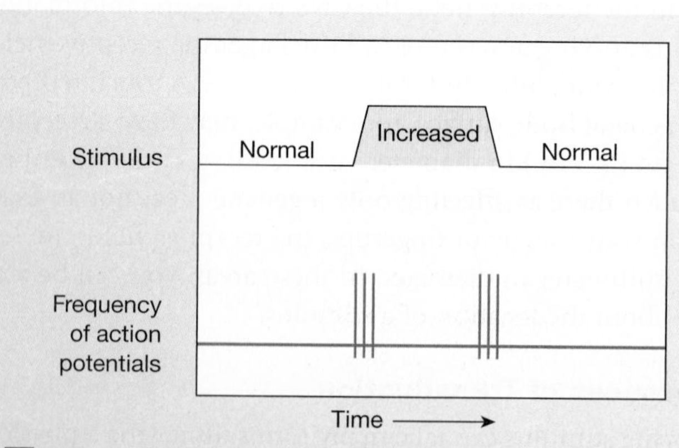

Stimulus
Normal
Increased
Normal
Frequency of action potentials
Time

b **Phasic receptors** are normally inactive. Action potentials are generated only for a short time in response to a change in the conditions they are monitoring.

is known as *central adaptation*. Central adaptation generally involves the inhibition of nuclei along a sensory pathway.

Peripheral adaptation reduces the amount of information that reaches the CNS. Central adaptation at the subconscious level further restricts the amount of detail that arrives at the cerebral cortex. Most of the incoming sensory information is processed in centers along the spinal cord or brainstem at the subconscious level. This processing can produce reflexive motor responses, but you are seldom consciously aware of either the stimuli or the responses.

The output from higher centers can increase receptor sensitivity or facilitate transmission along a sensory pathway. The reticular activating system in the midbrain helps focus our attention and heightens or reduces our awareness of arriving sensations. ⮌ p. 478 This adjustment of sensitivity can take place under conscious or subconscious direction. When you "listen carefully," your sensitivity and awareness of auditory stimuli increase. Output from higher centers can also inhibit transmission along a sensory pathway. Such inhibition occurs when you enter a noisy factory or walk along a crowded city street. You automatically tune out the high level of background noise. Next let's consider how these basic concepts of receptor function and sensory processing apply to the general senses.

✓ Checkpoint

3. **Receptor A has a circular receptive field on the skin surface with a diameter of 2.5 cm. Receptor B has a circular receptive field 7.0 cm in diameter. Which receptor provides more precise sensory information?**
4. **Define *adaptation*.**

See the blue Answers tab at the back of the book.

15-3 General sensory receptors can be classified by the type of stimulus that excites them

Learning Outcome **Identify the receptors for the general senses, and describe how they function.**

Using the classification scheme introduced in Chapter 12, we can divide the receptors for the general senses into three types by location of the stimulus. ⮌ p. 395 *Exteroceptors* provide information about the external environment. *Proprioceptors* report the positions and movements of skeletal muscles and joints. *Interoceptors* monitor visceral organs and functions.

However, other classification schemes for the general sensory receptors are also used. One useful scheme is to divide them into four types by the nature of the stimulus that excites them: *nociceptors* (pain), *thermoreceptors* (temperature), *mechanoreceptors* (physical distortion such as touch), and *chemoreceptors* (chemical concentration). Each class of receptors has distinct structural and functional characteristics, although all general sensory receptors are simple in structure.

There are both somatic and visceral versions of these receptors; the difference between them has to do with location, not structure. A pain receptor in the gut looks and acts like a pain receptor in the skin, but the two sensations are delivered to separate locations in the CNS. The visceral organs also have fewer pain, temperature, and touch receptors than we find elsewhere in the body. The sensory information you receive from these receptors is poorly localized because their receptive fields are very large and may be widely separated.

The processing of general sensations is widely distributed in the CNS, but most processing takes place in centers along the sensory pathways in the spinal cord or brainstem. Only about 1 percent of the information provided by afferent fibers reaches the cerebral cortex and our awareness. For example, you usually do not feel the clothes you wear.

Let's take a closer look at the structure and function of the receptors as classified by the nature of their stimuli.

Nociceptors and Pain

Nociceptors (*noxa*, harm), or pain receptors, are especially common in the superficial portions of the skin, in joint capsules, within the periostea of bones, and around the walls of blood vessels. Other deep tissues and most visceral organs have few nociceptors. Pain receptors are free nerve endings with large receptive fields (see Figure 15–2). As a result, it is often difficult to determine the exact source of a painful sensation.

Different nociceptors may be sensitive to (1) temperature extremes, (2) mechanical damage, and (3) dissolved chemicals, such as chemicals released by injured cells. Very strong stimuli, however, will excite all three receptor types. For that reason, people describing very painful sensations—whether caused by acids, heat, or a deep cut—use similar descriptive terms, such as "burning."

Stimulation of the dendrites of a nociceptor causes depolarization. When the initial segment of the axon reaches threshold, an action potential heads toward the CNS.

Types of Pain

Two types of axons—Type A and Type C fibers—carry painful sensations. ⮌ p. 416 Myelinated Type A fibers carry sensations of **fast pain**, or *prickling pain*. An injection or a deep cut produces this type of pain. These sensations reach the CNS quickly, where they often trigger somatic reflexes. They are also relayed to the primary somatosensory cortex and so receive conscious attention. In most cases, the arriving information allows you to localize the stimulus to an area several inches in diameter.

Unmyelinated Type C fibers carry sensations of **slow pain**, or *burning and aching pain*. These sensations cause a generalized activation of the reticular formation and thalamus. You become aware of the pain but only have a general idea of the area affected.

Pain Perception

Nociceptors are tonic receptors (see Figure 15–3a). Significant peripheral adaptation does not occur in these receptors, so they continue to respond as long as the painful stimulus remains. Painful sensations stop only after tissue damage has ended. However, central adaptation may reduce the *perception* of the pain while nociceptors remain stimulated. This effect involves the inhibition of centers in the thalamus, reticular formation, lower brainstem, and spinal cord.

An understanding of the origins of pain sensations and an ability to control or reduce pain levels have always been among the most important aspects of medical treatment. After all, it is usually pain that causes us to seek treatment. We usually tolerate or ignore conditions that are not painful. Although we often use the term *pain pathways* to refer to the route on which pain signals travel to the brain, it is clear that pain distribution and perception are extremely complex.

The sensory neurons that bring pain sensations into the CNS release *glutamate* and/or *substance P* as neurotransmitters. These neurotransmitters facilitate neurons along the pain pathways. As a result, the level of pain experienced (especially chronic pain) can be out of proportion to the amount of painful stimuli or the apparent tissue damage. This effect may be one reason why people differ so widely in their perception of the pain associated with headaches, back pain, or childbirth. This facilitation is also presumed to play a role in *phantom limb syndrome*, when pain is still felt in an amputated limb (discussed in Section 15-4): The sensory neurons may be inactive, but the hyperexcitable interneurons may continue to generate pain sensations.

The release of the neuromodulators called endorphins and enkephalins within the CNS can reduce the level of pain that a person feels. As noted in Chapter 12, endorphins and enkephalins, which are structurally similar to morphine, inhibit activity along pain pathways in the brain. p. 421 They are found in the limbic system, hypothalamus, and reticular formation.

Tips & Tools

The ***P*** in substance ***P*** stands for ***p***eptide and is involved with ***p***ain, which it transmits ***p***eripherally.

Thermoreceptors

Thermoreceptors, or temperature receptors, are free nerve endings located in the dermis, skeletal muscles, liver, and hypothalamus. Cold receptors are three or four times more numerous than warm receptors. There are no structural differences between warm and cold thermoreceptors. Temperature sensations are conducted along the same pathways that carry pain sensations. They are sent to the reticular formation, the thalamus, and (to a lesser extent) the primary somatosensory cortex.

Thermoreceptors are phasic receptors (see Figure 15–3b). They are very active when the temperature is changing, but they quickly adapt to a stable temperature. This is why when you enter an air-conditioned classroom on a hot summer day, the temperature change seems extreme at first, but you quickly become comfortable as adaptation occurs.

Mechanoreceptors

Mechanoreceptors (MEK-ah-nō-rē-sep-terz) are sensitive to physical stimuli that distort their plasma membranes. These membranes contain *mechanically gated ion channels*. The gates open or close in response to stretching, compression, twisting, or other distortions of the membrane.

There are three classes of mechanoreceptors, tactile receptors, baroreceptors, and proprioceptors:

- **Tactile receptors** provide the closely related sensations of touch, pressure, and vibration. Touch sensations provide information about shape or texture, whereas pressure sensations indicate the degree of mechanical distortion. Vibration sensations indicate a pulsing pressure. The receptors involved may be specialized in some way. For example, rapidly adapting tactile receptors are best suited for detecting vibration. But your interpretation of a sensation as touch rather than pressure is typically a matter of the degree of stimulation, and not of differences in the type of receptor stimulated.
- **Baroreceptors** (bar-ō-rē-SEP-terz; *baro-*, pressure) detect pressure changes in the walls of blood vessels and in portions of the digestive, respiratory, and urinary tracts.
- **Proprioceptors** monitor the positions of joints and skeletal muscles. They are the most structurally and functionally complex of the general sensory receptors.

Tactile Receptors

Fine touch and pressure receptors provide detailed information about a source of stimulation, including its exact location, shape, size, texture, and direction of movement. These receptors are extremely sensitive and have narrow receptive fields. In contrast, *crude touch and pressure receptors* provide poor localization. They give little additional information about the stimulus because they have large receptive fields.

Tactile receptors range in complexity from free nerve endings to specialized sensory complexes with accessory cells and supporting structures. Figure 15–4 shows six types of tactile receptors in the skin: free nerve endings, root hair plexuses, tactile discs, bulbous corpuscles, lamellar corpuscles, and tactile corpuscles.

- *Free nerve endings* sensitive to touch and pressure are located between epidermal cells (Figure 15–4a). Free nerve endings that provide touch sensations are tonic receptors with small receptive fields. There appear to be no structural differences between these receptors and the free nerve

Figure 15–4 Tactile Receptors in the Skin.

a Free nerve endings

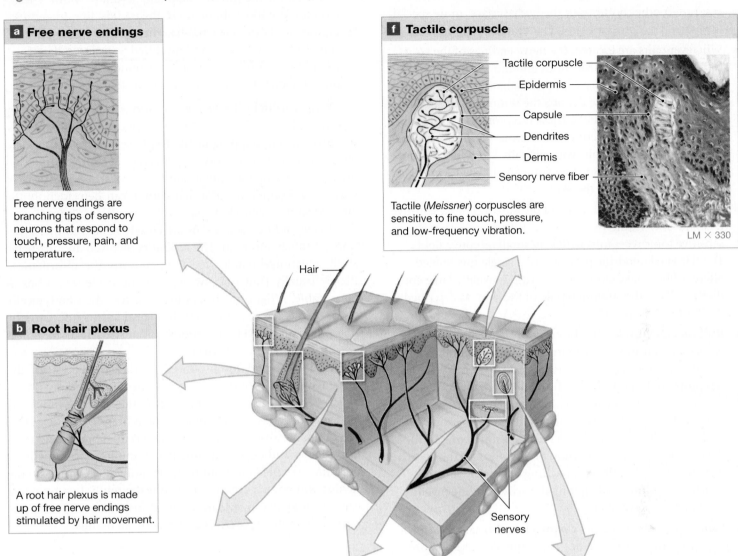

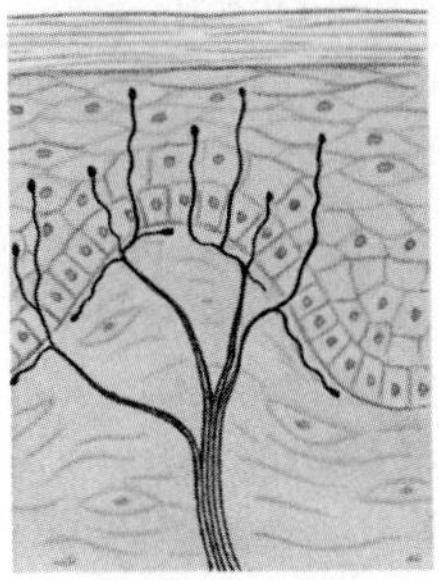

Free nerve endings are branching tips of sensory neurons that respond to touch, pressure, pain, and temperature.

f Tactile corpuscle

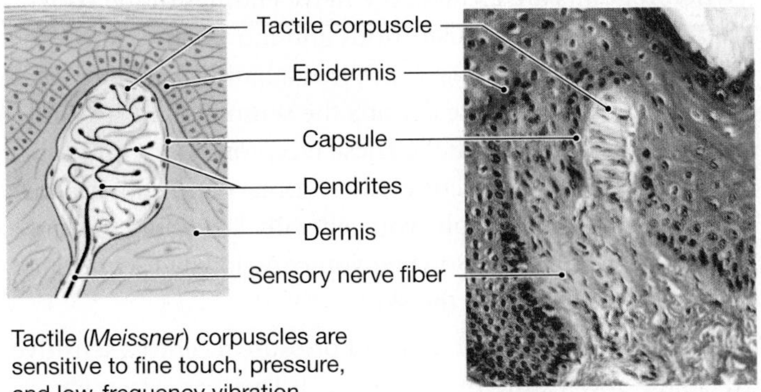

Tactile corpuscle

Epidermis

Capsule

Dendrites

Dermis

Sensory nerve fiber

LM × 330

Tactile (*Meissner*) corpuscles are sensitive to fine touch, pressure, and low-frequency vibration.

Hair

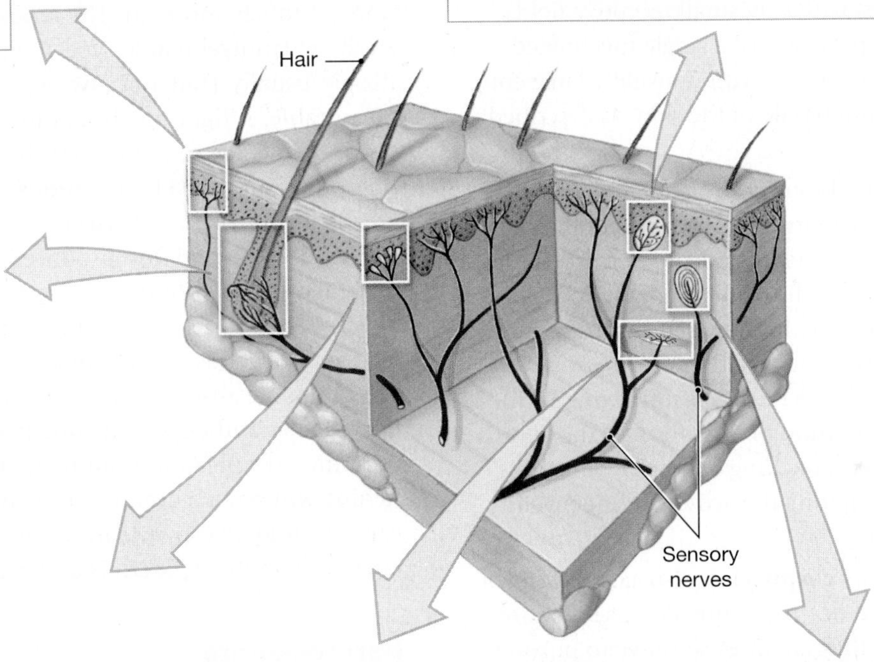

b Root hair plexus

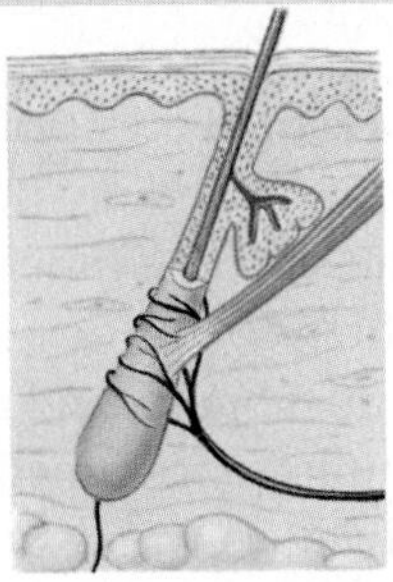

A root hair plexus is made up of free nerve endings stimulated by hair movement.

Sensory nerves

15

c Tactile discs

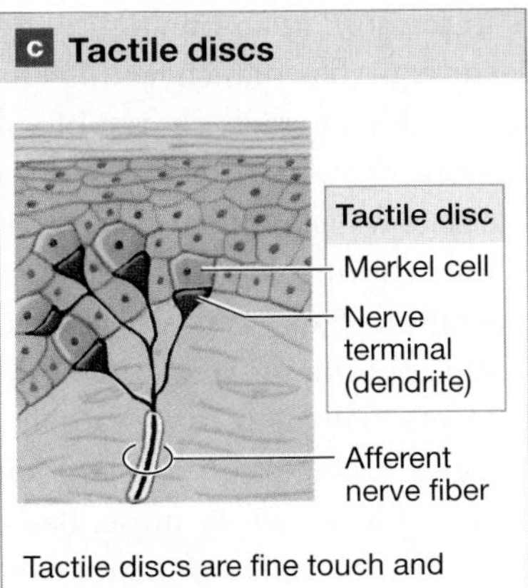
Tactile disc
Merkel cell
Nerve terminal (dendrite)
Afferent nerve fiber

Tactile discs are fine touch and pressure receptors sensitive to shape and texture.

d Bulbous corpuscle

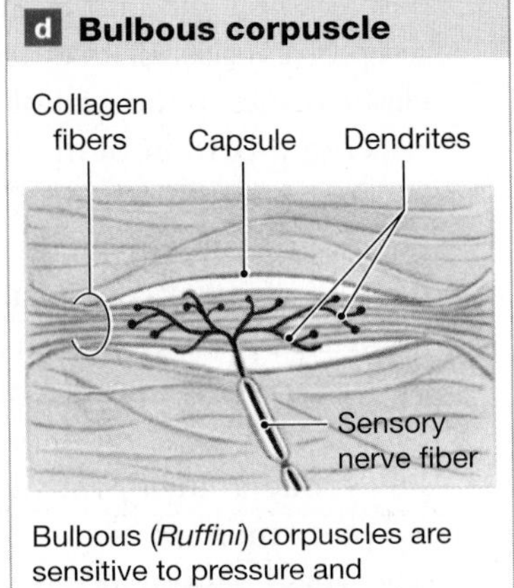
Collagen fibers
Capsule
Dendrites
Sensory nerve fiber

Bulbous (*Ruffini*) corpuscles are sensitive to pressure and distortion of the deep dermis.

e Lamellar corpuscle

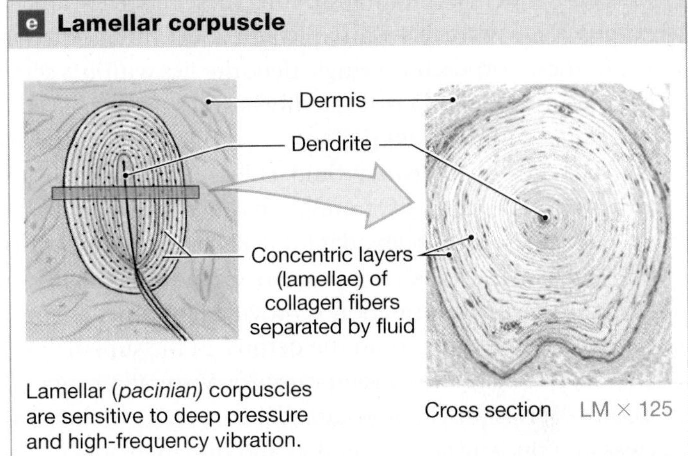
Dermis
Dendrite
Concentric layers (lamellae) of collagen fibers separated by fluid
Cross section
LM × 125

Lamellar (*pacinian*) corpuscles are sensitive to deep pressure and high-frequency vibration.

? Which tactile receptors are located in the dermis?

endings that provide temperature or pain sensations. These are the only sensory receptors on the corneal surface of the eye, but in other portions of the body surface, more specialized tactile receptors are probably more important.

- Wherever hairs are located, the nerve endings of the **root hair plexus** monitor distortions and movements across the body surface (Figure 15–4b). When a hair is displaced, the movement of its follicle distorts the sensory dendrites and produces action potentials. These receptors adapt rapidly, so they are best at detecting initial contact and subsequent movements. For example, you generally feel your clothing only when you move or when you consciously focus on tactile sensations from the skin.
- **Tactile discs** are fine-touch and pressure receptors sensitive to shape and texture (Figure 15–4c). They are extremely sensitive tonic receptors, with very small receptive fields. The flattened dendritic processes of a single myelinated afferent fiber make close contact with unusually large epithelial cells in the stratum basale of the skin, as described in Chapter 5. ⮌ p. 155
- **Bulbous corpuscles** (KOR-pus-elz), also called *Ruffini corpuscles*, are sensitive to pressure and distortion of the skin, and they are located in the reticular (deep) dermis. These receptors are tonic and show little if any adaptation. The capsule surrounds a core of collagen fibers that are continuous with those of the surrounding dermis (Figure 15–4d). In the capsule, a network of dendrites is intertwined with the collagen fibers. Any tension or distortion of the dermis tugs or twists these fibers, stretching or compressing the attached dendrites and altering the activity in the myelinated afferent nerve fiber.
- **Lamellar** (LAM-eh-lāt-ar) **corpuscles**, also called *pacinian corpuscles*, are sensitive to deep pressure. Because they are fast-adapting receptors, they are most sensitive to pulsing or high-frequency vibrating stimuli. These relatively large corpuscles may reach 4 mm in length and 1 mm in diameter. In these corpuscles, a single dendrite lies within a series of concentric layers of collagen fibers and supporting cells (specialized fibroblasts) (Figure 15–4e). The concentric layers, separated by interstitial fluid, shield the dendrite from virtually every source of stimulation other than direct pressure. Lamellar corpuscles adapt quickly because distortion of the capsule soon relieves pressure on the sensory process. Somatic sensory information is provided by lamellar corpuscles located throughout the dermis, in the superficial and deep fasciae, and in joint capsules. Visceral sensory information is provided by lamellar corpuscles in the pancreas and the walls of the urethra and urinary bladder.
- **Tactile corpuscles**, also called *Meissner corpuscles*, give us sensations of fine touch and pressure and low-frequency vibration. They adapt to stimulation within a second after contact. Tactile corpuscles are fairly large structures, measuring about 100 μm in length and 50 μm in width. These receptors are most abundant in the eyelids, lips, fingertips, nipples, and external genitalia. The dendrites are highly coiled and interwoven, and modified Schwann cells surround them. A fibrous capsule surrounds the entire complex and anchors it within the dermis (Figure 15–4f).

Your sensitivity to tactile sensations may be altered by infection, disease, or damage to sensory neurons or pathways. For this reason, mapping tactile responses can sometimes aid clinical assessment. Sensory losses with clear regional boundaries indicate trauma to spinal nerves. For example, sensory loss within the boundaries of a dermatome can help identify the affected spinal nerve (look back at Figure 13–7). ⮌ p. 443

Tickle and itch sensations are closely related to the sensations of touch and pain. The receptors involved are free nerve endings. Unmyelinated Type C fibers carry the information. People usually (but not always) describe tickle sensations as pleasurable. A light touch that moves across the skin typically produces a tickle sensation. Psychological factors influence the interpretation of tickle sensations, and tickle sensitivity differs greatly from person to person.

Itching is probably produced by the stimulation of the same receptors, although its precise mechanism is unknown. Specific "itch spots" can be mapped in the skin, the inner surfaces of the eyelids, and the mucous membrane of the nose. Itch sensations are absent from other mucous membranes and from deep tissues and viscera. Itching is very unpleasant, sometimes even more unpleasant than pain, as individuals with extreme itching will scratch even when pain is the result. Itch receptors can be stimulated by the injection of histamine or proteolytic enzymes into the epidermis and superficial dermis.

Baroreceptors

Baroreceptors monitor changes in pressure in an organ. A baroreceptor consists of free nerve endings that branch within the elastic tissues in the wall of a distensible organ, such as a blood vessel or a portion of the respiratory, digestive, or urinary tract. When the pressure changes, the elastic walls of the tract recoil or expand. This movement distorts the dendritic branches and alters the rate of action potential generation. Baroreceptors respond immediately to a change in pressure, but they adapt rapidly, and the output along the afferent fibers gradually returns to normal.

Baroreceptors monitor pressure changes in organs that stretch due to their contents, such as blood, air, or urine. Baroreceptors monitor blood pressure in the walls of major vessels. These vessels include the carotid arteries (at the *carotid sinuses*) and the aorta (at the *aortic sinuses*). The information plays a major role in regulating cardiac function and adjusting blood flow to vital tissues. Baroreceptors in the lungs monitor lung expansion. This information is relayed to the respiratory rhythmicity centers, which set the pace of respiration. Stretch receptors at various sites in the digestive and urinary tracts trigger a variety of visceral reflexes, including defecation and urination reflexes (Figure 15–5). We describe baroreceptor reflexes in the digestive and urinary system chapters.

Proprioceptors

Proprioception is a somatic sensation—there are no proprioceptors in visceral organs of the thoracic and abdominopelvic cavities. (Therefore, you cannot tell, for example, where your spleen, appendix, or pancreas is at the moment.) Proprioceptors are found in joints, tendons, and muscles, however. These receptors monitor the position of joints, the tension in tendons and ligaments, and the state of muscular contraction.

There are three major groups of proprioceptors:

- *Muscle Spindles.* As we discussed in Chapter 13, **muscle spindles** monitor skeletal muscle length and trigger stretch reflexes. p. 457
- *Golgi Tendon Organs.* **Golgi tendon organs** are similar in function to bulbous corpuscles but are located at the junction between a skeletal muscle and its tendon. In a Golgi tendon organ, dendrites branch repeatedly and wind around the densely packed collagen fibers of the tendon. These receptors are stimulated by tension in the tendon. They monitor the external tension developed during muscle contraction.
- *Receptors in Joint Capsules.* Joint capsules are richly innervated by free nerve endings that detect pressure, tension, and movement at the joint.

Your sense of body position results from the integration of information from joint capsule receptors, muscle spindles, Golgi tendon organs, and the receptors of the internal ear.

Proprioceptors do not adapt to constant stimulation, and each receptor continuously sends information to the CNS. Most proprioceptive information is processed at subconscious levels. Only a small proportion of the arriving proprioceptive information reaches your awareness.

Figure 15–5 Baroreceptors and the Regulation of Autonomic Functions.

Baroreceptors in the Body
Baroreceptors of Carotid Sinus and Aortic Sinus
Provide information on blood pressure to cardiovascular and respiratory control centers
Baroreceptors of Lung
Provide information on lung expansion to respiratory rhythmicity centers for control of respiratory rate
Baroreceptors of Digestive Tract
Provide information on volume of tract segments, trigger reflex movement of materials along tract
Baroreceptors of Colon
Provide information on volume of fecal material in colon, trigger defecation reflex
Baroreceptors of Bladder Wall
Provide information on volume of urinary bladder, trigger urination reflex

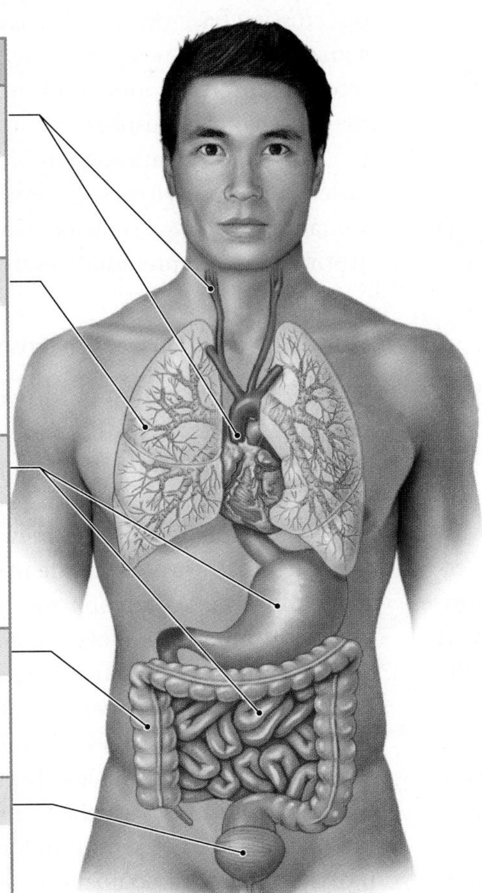

Chemoreceptors

Chemoreceptors (KĒ-mō-rē-sep-terz) are specialized nerve cells that detect small changes in the concentration of specific chemicals or compounds. In general, chemoreceptors respond only to water-soluble and lipid-soluble substances that are dissolved in body fluids [interstitial fluid, blood plasma, and cerebrospinal fluid (CSF)]. These receptors exhibit peripheral adaptation over a period of seconds, although they may also show central adaptation.

The chemoreceptors included in the general senses do not send information to the primary somatosensory cortex, so you are not consciously aware of the sensations they provide. The arriving sensory information is routed to brainstem centers that deal with the autonomic control of respiratory and cardiovascular functions. Neurons in the respiratory centers of the brain respond to the concentrations of hydrogen ions (pH) and carbon dioxide molecules in the CSF.

Chemoreceptors are also located in the **carotid bodies**, near the origin of the internal carotid arteries on each side of the neck, and in the **aortic bodies**, between the major branches of the aortic arch. (The carotid bodies and aortic bodies are small tissue masses of receptors and supporting cells.) These receptors monitor the pH and the carbon dioxide and oxygen levels in arterial blood and play an important role in the reflexive control of breathing and cardiovascular function. The afferent fibers leaving the carotid or aortic bodies reach the respiratory centers by traveling within cranial nerves IX (glossopharyngeal) and X (vagus) (Figure 15–6).

Figure 15–6 Locations and Functions of Chemoreceptors.

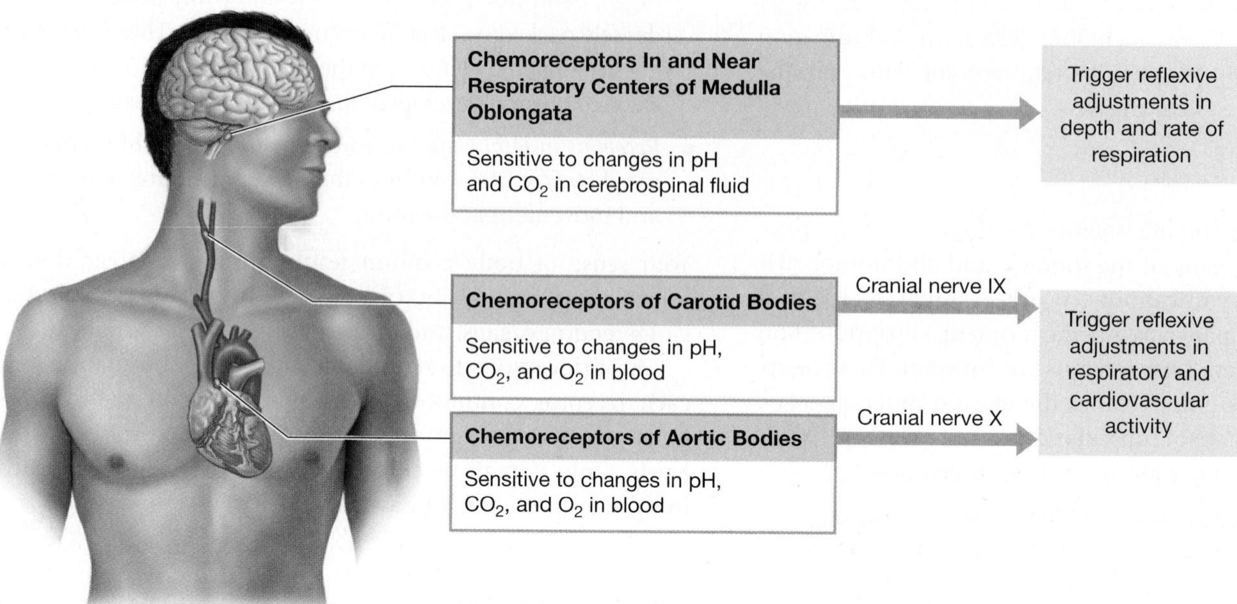
Chemoreceptors In and Near Respiratory Centers of Medulla Oblongata
Sensitive to changes in pH and CO_2 in cerebrospinal fluid
Trigger reflexive adjustments in depth and rate of respiration
Chemoreceptors of Carotid Bodies
Sensitive to changes in pH, CO_2, and O_2 in blood
Cranial nerve IX
Trigger reflexive adjustments in respiratory and cardiovascular activity
Chemoreceptors of Aortic Bodies
Sensitive to changes in pH, CO_2, and O_2 in blood
Cranial nerve X

Checkpoint

5. **List the major types of general sensory receptors, and identify the nature of the stimulus that excites each type.**
6. **Identify the three classes of mechanoreceptors.**
7. **What would happen to you if the information from proprioceptors in your legs were blocked from reaching the CNS?**

See the blue Answers tab at the back of the book.

15-4 The afferent division is made up of separate somatic sensory and visceral sensory pathways that deliver sensory information to the CNS

Learning Outcome Identify the major sensory pathways, and explain how it is possible to distinguish among sensations that originate in different areas of the body.

When a sensory signal is being sent to the CNS along a sensory pathway, it goes through a series of neurons that can be distinguished by their locations. A sensory neuron that delivers sensations directly to the CNS is called a **first-order neuron**. The cell body of a first-order general sensory neuron is located in a spinal ganglion or cranial nerve ganglion. In the CNS, the axon of that sensory neuron synapses (passes signals to) an interneuron known as a **second-order neuron**, which may be located in the spinal cord or brainstem. For us to become aware of the sensation, the second-order neuron in turn must pass on the signal to a **third-order neuron** in the thalamus.

Somewhere along its length, the axon of the second-order neuron crosses over to the opposite side of the CNS in a process called **decussation**. As a result, the right side of the thalamus receives sensory information from the left side of the body, and the left side of the thalamus receives sensory information from the right side. The axon of the third-order neuron then ascends (without crossing back over) and synapses on neurons of the primary somatosensory cortex. Therefore, the right cerebral hemisphere receives sensory information from the left side of the body, and the left cerebral hemisphere receives sensations from the right side.

Our sensory pathways all include first-, second-, and third-order neurons. However, several different pathways bring somatic and visceral sensory information to the CNS. Let's look at some major representatives of each type.

Somatic Sensory Pathways

Somatic sensory pathways carry sensory information from the skin and muscles of the body wall, head, neck, and limbs to the CNS. We will consider three major somatic sensory pathways: (1) the *spinothalamic pathway*, (2) the *posterior column pathway*, and (3) the *spinocerebellar pathway*.

These pathways are made up of pairs of spinal tracts, symmetrically arranged on opposite sides of the spinal cord. All the axons within a tract share a common origin and destination. Figure 15–7 shows the relative positions of the spinal tracts in each of these somatic sensory pathways.

Note that tract names often give clues to their function. For example, if the name of a tract begins with *spino-*, the tract *starts* in the spinal cord and *ends* in the brain. It must therefore be an ascending tract that carries sensory information. The rest of the name indicates the tract's destination. Thus, a *spinothalamic tract* begins in the spinal cord and carries sensory information to the thalamus.

+ Clinical Note Assessment of Tactile Sensitivities

Regional sensitivity to light touch can be checked by gentle contact with a fingertip or a slender wisp of cotton. The **two-point discrimination test** provides a more detailed sensory map of tactile receptors. Two fine points of a bent paper clip or another object are applied to the skin surface simultaneously. The person then describes the contact. When the points fall within a single receptive field, the person will report only one point of contact. A person usually loses two-point discrimination at 1 mm (0.04 in.) on the surface of the tongue, at 2–3 mm (0.08–0.12 in.) on the lips, at 3–5 mm (0.12–0.20 in.) on the backs of the hands and feet, and at 4–7 cm (1.6–2.75 in.) over the general body surface.

Applying the base of a tuning fork to the skin tests vibration receptors. Damage to an individual spinal nerve produces insensitivity to vibration along the paths of the related sensory nerves. If the sensory loss results from spinal cord damage, the injury site can typically be located by walking the tuning fork down the spinal column, resting its base on the vertebral spinous processes.

Various descriptive terms indicate the degree of sensitivity in an area. *Anesthesia* implies a total loss of sensation. The person cannot perceive touch, pressure, pain, or temperature sensations in that area. *Paresthesia* is the presence of abnormal sensations such as the prickling sensation when an arm or leg "falls asleep" as a result of pressure on a peripheral nerve.

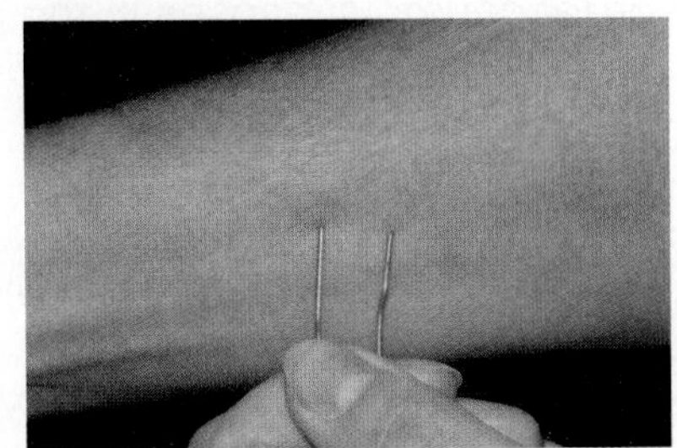

Figure 15–7 Sensory Pathways and Ascending Tracts in the Spinal Cord. A cross-sectional view of the spinal cord showing the locations of the major ascending (sensory) tracts. For information about these tracts, see **Table 15–1**.

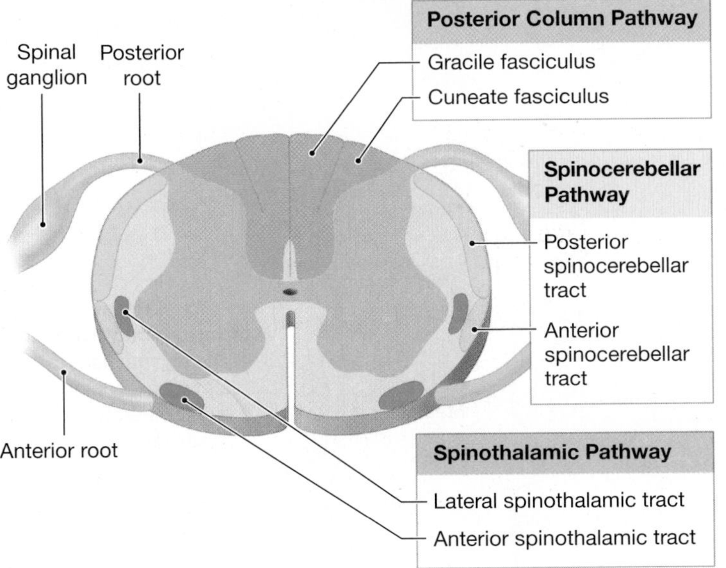
Spinal ganglion
Posterior root
Anterior root
Posterior Column Pathway
Gracile fasciculus
Cuneate fasciculus
Spinocerebellar Pathway
Posterior spinocerebellar tract
Anterior spinocerebellar tract
Spinothalamic Pathway
Lateral spinothalamic tract
Anterior spinothalamic tract

? Which pathway or tract detects sensations of pain and temperature?

The Spinothalamic Pathway

The **spinothalamic pathway** carries sensations of crude touch, pressure, pain, and temperature. This pathway includes larger tracts that carry sensations destined for the cerebral cortex as well as small tracts that deliver sensations to reflex centers in the brainstem. We will ignore the smaller tracts in this discussion.

The larger tracts are the **anterior spinothalamic tract** and the **lateral spinothalamic tract**. Sensations bound for the cerebral cortex ascend within these tracts, which end at third-order neurons in the ventral nuclei of the thalamus. After the sensations have been sorted and processed, they are relayed to the primary somatosensory cortex (**Spotlight Figure 15–8a**).

15

How we perceive an arriving stimulus, whether as painful, cold, hot, or vibrating, depends on which second-order and third-order neurons are stimulated. The ability to connect that stimulus with a specific location in the body—to *localize* the sensation—depends on the projection of information from the thalamus to an appropriate area of the primary somatosensory cortex.

Any abnormality along the pathway can result in inappropriate sensations or inaccurate localization of the source. Consider these examples:

- A person can experience painful sensations that are not produced where they are perceived to originate. For example, a person may continue to experience pain in an amputated limb. This **phantom limb syndrome** is caused by activity in the sensory neurons or interneurons along the spinothalamic pathway. The neurons involved were once part of the labeled line that monitored conditions in the intact limb. These labeled lines and pathways are developmentally programmed. For this reason, even people born without limbs can have phantom limb syndrome.
- A person can feel pain in an uninjured part of the body when the pain actually originates at another location, a phenomenon called **referred pain**. For example, strong visceral pain sensations arriving at a segment of the spinal cord can stimulate interneurons that are part of the spinothalamic pathway. Activity in these interneurons leads to the stimulation of the primary somatosensory cortex, so the individual feels referred

SPOTLIGHT

Figure 15–8
Somatic Sensory Pathways

a SPINOTHALAMIC PATHWAY

The **spinothalamic pathway** provides conscious sensations of poorly localized ("crude") touch, pressure, pain, and temperature. In this pathway, the axons of first-order neurons enter the spinal cord and synapse on second-order neurons within the posterior horns. The axons of these interneurons cross to the opposite side of the spinal cord before ascending to the thalamus. The third-order neuron synapses in the primary somatosensory cortex.

KEY

- Axon of first-order neuron
- Second-order neuron
- Third-order neuron

A Sensory Homunculus

The functional mapping of the somatosensory cortex is called **somatotopy**. Such a somatotopic map is called a *homunculus*, meaning "little man" – as in, the little man in the brain. The map looks like a caricature, showing the relative size of cortex devoted to a particular body region is proportional to the density of sensory input received from that body region.

Torso
Upper limb
Lower limb
Genitalia
Primary Somatosensory Cortex
Face and lips
Jaw
Tongue
Midbrain
Medulla oblongata
Spinal cord

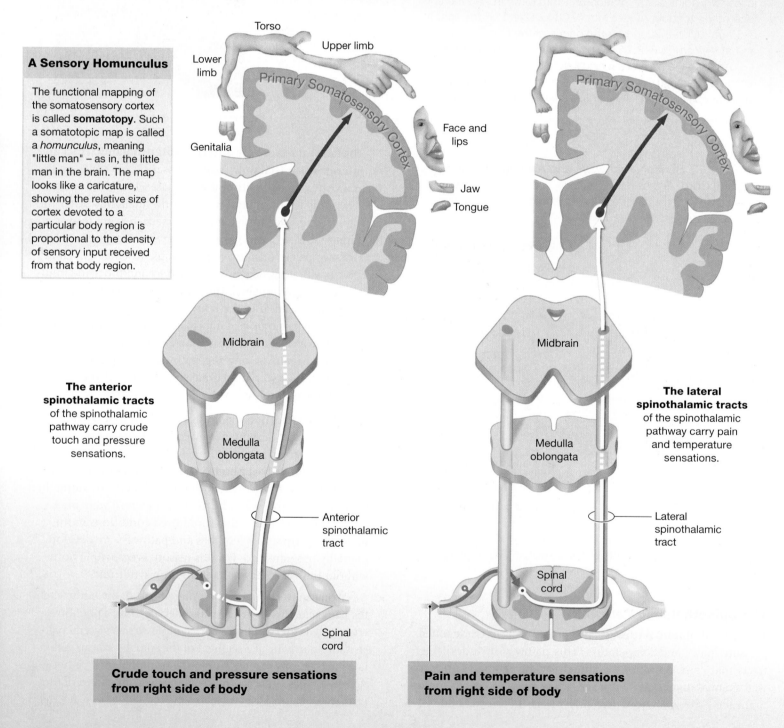

The anterior spinothalamic tracts of the spinothalamic pathway carry crude touch and pressure sensations.

Anterior spinothalamic tract

Crude touch and pressure sensations from right side of body

The lateral spinothalamic tracts of the spinothalamic pathway carry pain and temperature sensations.

Lateral spinothalamic tract

Pain and temperature sensations from right side of body

b POSTERIOR COLUMN PATHWAY

The **posterior column pathway** carries sensations of highly localized ("fine") touch, pressure, vibration, and proprioception. This pathway is also known as the posterior column–medial lemniscus pathway. It begins at a peripheral receptor and ends at the primary somatosensory cortex of the cerebral hemispheres.

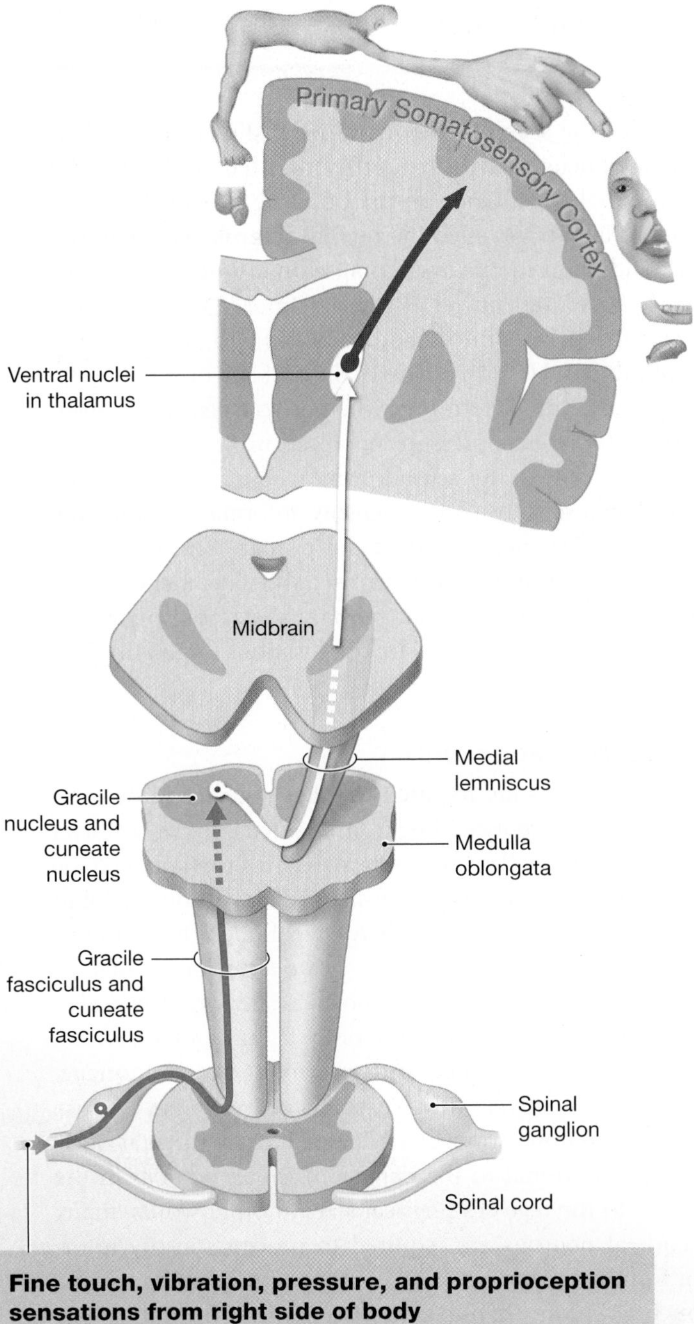

Primary Somatosensory Cortex
Ventral nuclei in thalamus
Midbrain
Medial lemniscus
Gracile nucleus and cuneate nucleus
Medulla oblongata
Gracile fasciculus and cuneate fasciculus
Spinal ganglion
Spinal cord
Fine touch, vibration, pressure, and proprioception sensations from right side of body

c SPINOCEREBELLAR PATHWAY

The cerebellum receives proprioceptive information about the position of skeletal muscles, tendons, and joints along the **spinocerebellar pathway**. The **posterior spinocerebellar tracts** contain axons that do not cross over to the opposite side of the spinal cord. These axons reach the cerebellar cortex by the inferior cerebellar peduncle of that side. The **anterior spinocerebellar tracts** are dominated by axons that have crossed over to the opposite side of the spinal cord.

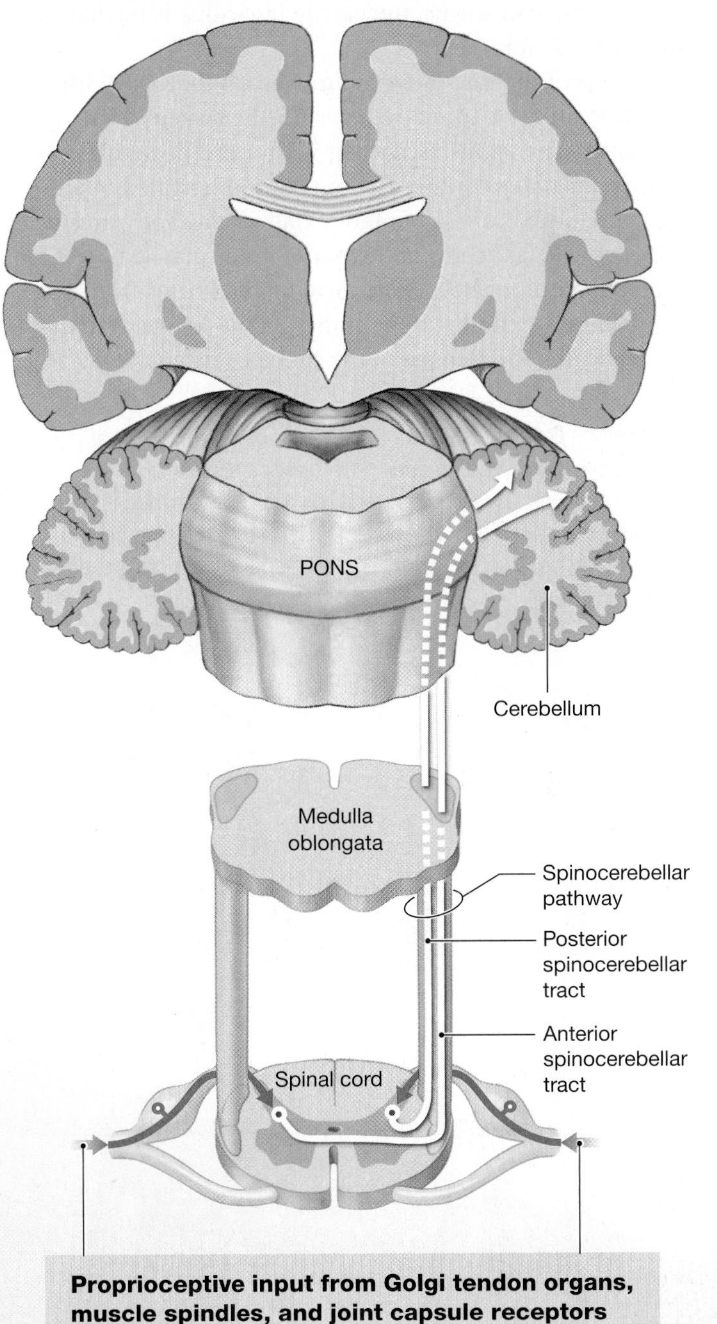

PONS
Cerebellum
Medulla oblongata
Spinocerebellar pathway
Posterior spinocerebellar tract
Anterior spinocerebellar tract
Spinal cord
Proprioceptive input from Golgi tendon organs, muscle spindles, and joint capsule receptors

pain in a specific part of the body surface. One familiar example is the pain of a heart attack, which is frequently felt in the left arm. Women are more likely than men to experience shortness of breath, nausea, vomiting, back pain, or jaw pain. Additional examples for both anterior and posterior sides of the body are shown in Figure 15–9.

The Posterior Column Pathway

The **posterior column pathway**, shown in Spotlight Figure 15–8b, carries sensations of fine touch, vibration, pressure, and proprioception. The spinal tracts involved are the left and right **gracile fasciculus** (*gracilis*, slender) and the left and right **cuneate fasciculus** (*cuneus*, wedge shaped). On each side of the posterior median sulcus, the gracile fasciculus is medial to the cuneate fasciculus (see Figure 15–7).

The axons of the first-order neurons reach the CNS within the posterior roots of spinal nerves and the sensory roots of cranial nerves. The axons ascending within the posterior column are organized according to the region innervated. Axons carrying sensations from the inferior half of the body ascend within the gracile fasciculus and synapse in the gracile nucleus of the medulla oblongata. Axons carrying sensations from the superior half of the trunk, upper limbs, and neck ascend in the cuneate fasciculus and synapse in the cuneate nucleus. ↺ p. 474

Figure 15–9 Referred Pain. Pain sensations from visceral organs are often perceived as arising from areas distant from the actual origin (arrows). This occurs because the body surface is innervated by the same spinal segments. Each region of perceived pain is labeled according to the organ at which the pain originates.

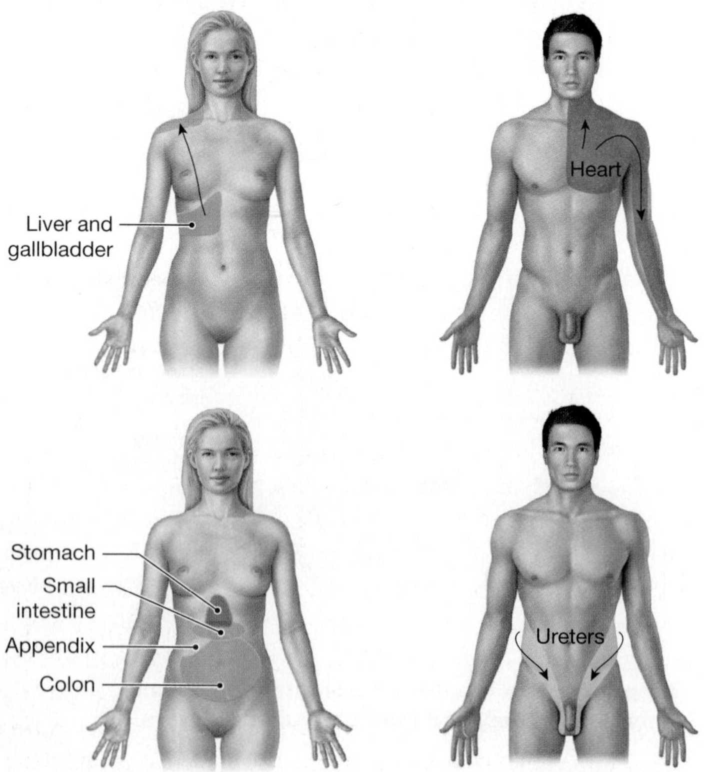
Liver and gallbladder
Heart
Stomach
Small intestine
Appendix
Colon
Ureters

Clinical Note Phantom Limb Syndrome

A soldier in the Afghan War unwittingly steps on a landmine. Although he survives, he loses his leg. During his extended rehabilitation he is at risk for a curious phenomenon called *phantom limb syndrome*. His leg—the one he lost—keeps him up at night with pain. Sometimes the leg burns; other times it feels like pins and needles. He says this leg feels touch, pressure, and vibration. It even itches! Despite the amputation, the sensory pathway for this leg persists and continues to convey messages to the thalamus. If the veteran is fit for a prosthesis (an artificial limb replacement) and wears it early and regularly, the device will ease his phantom limb pain by applying a real physical stimulus to the limb stump.

Axons of the second-order neurons of the gracile nucleus and cuneate nucleus decussate in the brainstem as they ascend toward the thalamus. Once on the opposite side of the brain, the axons enter a tract called the **medial lemniscus** (*lemniskos*, ribbon). The axons in this tract synapse on third-order neurons in one of the ventral nuclei of the thalamus. ↺ p. 481 These nuclei sort the arriving information according to (1) the nature of the stimulus and (2) the region of the body involved. Processing in the thalamus determines whether you perceive a given sensation as fine touch, pressure, or vibration.

We then localize the sensation by where it arrives in the primary somatosensory cortex. Sensory information from the toes arrives at the medial end of the primary sensory cortex, and information from the head arrives at the lateral end. When neurons in one portion of your primary somatosensory cortex are stimulated, you become aware of sensations originating at a specific location.

The Sensory Homunculus

The same sensations are reported whether the cortical neurons are activated by axons ascending from the thalamus or by direct electrical stimulation. Researchers have electrically stimulated the primary somatosensory cortex in awake individuals during brain surgery and asked the subjects where they thought the stimulus originated. The results were used to create a functional map of the primary somatosensory cortex known as a *sensory homunculus*.

In a sensory homunculus, the body features are distorted. For example, the face is huge, with enormous lips and tongue, but the back is relatively tiny. These distortions arise because the area of somatosensory cortex devoted to a particular body region is proportional to the density *of sensory neurons* in the region, not to the region's physical size. In other words, many more cortical neurons are required to process sensory information from the tongue, which has tens of thousands of taste and touch receptors, than to process sensations from the back, where touch receptors are few and far between.

The Spinocerebellar Pathway

The **spinocerebellar pathway** conveys information about muscle, tendon, and joint positions from the spine to the cerebellum (**Spotlight Figure 15–8c**). This information does not reach our awareness. The axons of first-order sensory neurons synapse on interneurons in the posterior horns of the spinal cord. The axons of these second-order neurons ascend in one of the spinocerebellar tracts: the **posterior spinocerebellar tracts** or the **anterior spinocerebellar tracts**.

The sensations carried by the anterior spinocerebellar tracts reach the cerebellar cortex by the superior cerebellar peduncle. Interestingly, many of the axons that cross over and ascend to the cerebellum then cross over again within the cerebellum, synapsing on the same side as the original stimulus. The functional significance of this "double cross" is unknown.

The information carried by the spinocerebellar pathway ultimately arrives at the *Purkinje cells* of the cerebellar cortex. ⮌ p. 478 Proprioceptive information from each part of the body is relayed to a specific portion of the cerebellar cortex. We will consider the integration of proprioceptive information and the role of the cerebellum in somatic motor control in Section 15-5.

Visceral Sensory Pathways

Visceral sensory information is collected by interoceptors monitoring visceral tissues and organs, primarily within the thoracic and abdominopelvic cavities. These interoceptors include nociceptors, thermoreceptors, tactile receptors, baroreceptors, and chemoreceptors, although none of them is as numerous as they are in somatic tissues.

Sensory information from these interoceptors is carried from the mouth, palate, pharynx, larynx, trachea, esophagus, and associated vessels and glands by cranial nerves V, VII, IX, and X. The posterior roots of spinal nerves T_1–L_2 carry visceral sensory information provided by receptors in organs located between the diaphragm and the pelvic cavity. The posterior roots of spinal nerves S_2–S_4 carry visceral sensory information from organs in the inferior pelvic cavity. These organs include the last segment of the large intestine, the urethra and base of the urinary bladder, and the prostate gland (males) or the cervix of the uterus and portions of the vagina (females).

The axons of first-order neurons that carry visceral sensory information usually travel in company with autonomic motor fibers innervating the same visceral structures. These first-order neurons deliver the sensory information to second-order interneurons whose axons ascend within the spinothalamic pathway. Most of this information is delivered to the **solitary nucleus**, a large nucleus on each side of the medulla oblongata that has extensive connections with the various cardiovascular and respiratory centers as well as with the reticular formation. These connections allow the solitary nuclei to function as major processing and sorting centers for visceral sensory information. Because there is no third-order neuron in this pathway, this sensory information never reaches the primary sensory cortex, and you remain unaware of these sensations.

Table 15–1 summarizes the somatic sensory pathways discussed in this section.

Checkpoint

8. **As a result of pressure on her spinal cord, Jill cannot feel fine touch or pressure on her lower limbs. Which spinal tract is being compressed?**
9. **Which spinal tract carries action potentials generated by nociceptors?**
10. **Which cerebral hemisphere receives impulses carried by the right gracile fasciculus?**

See the blue Answers tab at the back of the book.

15-5 The somatic nervous system is an efferent division made up of somatic motor pathways that control skeletal muscles

Learning Outcome Describe the components, processes, and functions of the somatic motor pathways, and the levels of information processing involved in motor control.

Let's take a look now at the structure of the **somatic nervous system (SNS)**, which allows us to control skeletal muscles. The SNS is made up of **somatic motor pathways** that consist of motor nuclei, tracts, and nerves. Throughout this discussion we will use the terms *motor neuron* and *motor control* to refer specifically to these somatic motor neurons and pathways.

Somatic motor pathways always involve at least two motor neurons: an **upper motor neuron**, whose cell body lies in a CNS processing center (the premotor and/or primary motor cortexes), and a **lower motor neuron**, whose cell body lies in a nucleus of the brainstem or spinal cord. Only the axon of the lower motor neuron extends outside the CNS. The upper motor neuron synapses on the lower motor neuron, which in turn innervates a single motor unit in a skeletal muscle. ⮌ p. 315

Activity in the upper motor neuron may facilitate or inhibit the lower motor neuron. Activation of the lower motor neuron triggers a contraction in the innervated muscle. Destruction of or damage to a lower motor neuron eliminates voluntary and reflex control over the innervated motor unit.

Figure 15–10 indicates the positions of the associated motor (descending) tracts in the spinal cord. Conscious and subconscious motor commands control skeletal muscles by traveling over three integrated motor pathways: the *corticospinal pathway*, the *medial pathway*, and the *lateral pathway*. Note that because the name of the corticospinal tract ends in *-spinal*, this indicates that the tract ends in the spinal cord. The first part of the name indicates the nucleus or cortical area of the brain where the tract originates. Therefore, the name of the corticospinal tract tells you that it carries motor commands from the cerebral cortex to the spinal cord.

Table 15–1 Principal Ascending (Sensory) Pathways

		Location of Neuron Cell Bodies			Location of Neuron Cell Bodies	
Pathway/Tract	**Sensation(s)**	**First-Order**	**Second-Order**	**Third-Order**	**Final Destination**	**Site of Decussation**
SPINOTHALAMIC PATHWAY						
Lateral spinothalamic tracts	Pain and temperature	Spinal ganglia; axons enter CNS in posterior roots	Interneurons in posterior horn; axons enter lateral spinothalamic tract on opposite side	Ventral nuclei of thalamus	Primary sensory cortex on side opposite stimulus	Axons of second-order neurons at level of entry
Anterior spinothalamic tracts	Crude touch and pressure	Spinal ganglia; axons enter CNS in posterior roots	Interneurons in posterior horn; axons enter anterior spinothalamic tract on opposite side	Ventral nuclei of thalamus	Primary sensory cortex on side opposite stimulus	Axons of second-order neurons at level of entry
POSTERIOR COLUMN PATHWAY						
Gracile fasciculus	Proprioception and fine touch, anterior pressure, and vibration from inferior half of body	Spinal ganglia of inferior half of body; axons enter CNS in posterior roots and join gracile fasciculus	Gracile nucleus of medulla oblongata; axons cross over before entering medial lemniscus	Ventral nuclei of thalamus	Primary sensory cortex on side opposite stimulus	Axons of second-order neurons before entering the medial lemniscus
Cuneate fasciculus	Proprioception and fine touch, anterior pressure, and vibration from superior half of body	Spinal ganglia of superior half of body; axons enter CNS in posterior roots and join cuneate fasciculus	Cuneate nucleus of medulla oblongata; axons cross over before entering medial lemniscus	Ventral nuclei of thalamus	Primary sensory cortex on side opposite stimulus	Axons of second-order neurons before entering the medial lemniscus
SPINOCEREBELLAR PATHWAY						
Posterior spinocerebellar tracts	Proprioception	Spinal ganglia; axons enter CNS in posterior roots	Interneurons in posterior horn; axons enter posterior spinothalamic tract on same side	Not present	Cerebellar cortex on side of stimulus	None
Anterior spinocerebellar tracts	Proprioception	Spinal ganglia; axons enter CNS in posterior roots	Interneurons in same spinal section; axons enter anterior spinocerebellar tract on the same or opposite side	Not present	Cerebellar cortex on side opposite (and side of) stimulus	Axons of most second-order neurons cross over before entering tract; many recross at cerebellum

The basal nuclei and cerebellum monitor and adjust activity within these motor pathways. The output of these centers stimulates or inhibits the activity of either (1) motor nuclei or (2) the primary motor cortex. Let's look at these motor pathways and how they are monitored.

The Corticospinal Pathway

The **corticospinal pathway** (Figure 15–11), sometimes called the *pyramidal system*, provides voluntary control over skeletal muscles. This system begins at the *pyramidal cells* of the primary motor cortex. ↺ p. 490 The axons of these upper motor neurons descend into the brainstem and spinal cord to synapse on lower motor neurons that control skeletal muscles. In general, the corticospinal pathway is direct: The upper motor neurons synapse directly on the lower motor neurons. However, the corticospinal pathway also works indirectly, as it innervates centers of the medial and lateral pathways.

The corticospinal pathway contains three pairs of descending tracts: (1) the *corticobulbar tracts*, (2) the *lateral corticospinal tracts*, and (3) the *anterior corticospinal tracts*. These tracts enter the white matter of the internal capsule, descend into the brainstem, and emerge on either side of the midbrain as the *cerebral peduncles*. ↺ p. 478

The Corticobulbar Tracts

Axons in the **corticobulbar** (kor-ti-kō-BUL-bar; *bulbar*, brainstem) **tracts** synapse on lower motor neurons in the motor nuclei of cranial nerves III, IV, V, VI, VII, IX, XI, and XII. The corticobulbar tracts provide conscious control over skeletal muscles that move the eyes, jaw, and face, and some muscles of the neck and pharynx. The corticobulbar tracts also innervate the motor centers of the medial and lateral pathways.

The Corticospinal Tracts

Axons in the **corticospinal tracts** synapse on lower motor neurons in the anterior horns of the spinal cord. As they descend, the

Figure 15–10 **Descending (Motor) Tracts in the Spinal Cord.** A cross-sectional view showing the locations of the major descending (motor) tracts that contain the axons of upper motor neurons. The origins and destinations of these tracts are listed in **Table 15–2**.

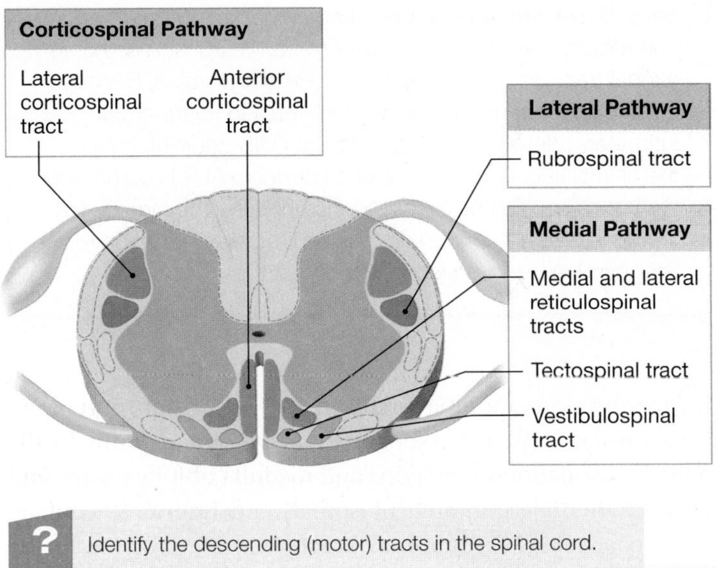

Corticospinal Pathway
Lateral corticospinal tract
Anterior corticospinal tract
Lateral Pathway
Rubrospinal tract
Medial Pathway
Medial and lateral reticulospinal tracts
Tectospinal tract
Vestibulospinal tract

? Identify the descending (motor) tracts in the spinal cord.

corticospinal tracts are visible along the anterior surface of the medulla oblongata as a pair of thick bands, the **pyramids** (which is why these tracts are sometimes called the *pyramidal system*).

Along the length of the pyramids, roughly 85 percent of the axons decussate to enter the descending **lateral corticospinal tracts** on the opposite side of the spinal cord. The other 15 percent continue uncrossed along the spinal cord as the **anterior corticospinal tracts**. Then, at the spinal segment it targets, an axon in the anterior corticospinal tract decussates in the anterior white commissure. The axon synapses on lower motor neurons in the anterior horns.

The Motor Homunculus

The activity of pyramidal cells in a specific portion of the primary motor cortex results in the contraction of specific peripheral muscles. Which muscles are stimulated depends on the region of motor cortex that is active. As in the primary somatosensory cortex, the primary motor cortex corresponds point by point with specific regions of the body. The cortical areas have been mapped out in diagrammatic form, creating a **motor homunculus**. Figure 15–11 shows the motor homunculus of the left cerebral hemisphere and the corticospinal pathway controlling skeletal muscles on the right side of the body.

The proportions of the motor homunculus are quite different from those of the actual body, because the area of primary motor cortex devoted to a specific body region is proportional to the density of motor units involved in the region's control, not to the region's physical size. For this reason, the homunculus indicates the degree of fine motor control available. For example, the hands, face, and tongue are all capable of varied and complex movements, so they appear very large. In contrast, the trunk is relatively small. These proportions are similar to those of the sensory homunculus (see **Spotlight Figure 15–8**).

The sensory and motor homunculi differ, however, with respect to some other areas of the body. For example, some highly sensitive regions, such as the sole of the foot, contain few motor units. Also, some areas with an abundance of motor units, such as the eye muscles, are not particularly sensitive.

Figure 15–11 **The Corticospinal Pathway.** The corticospinal pathway originates at the primary motor cortex. The corticobulbar tracts end at the motor nuclei of cranial nerves on the opposite side of the brain. Most fibers in this pathway cross over in the medulla oblongata and enter the lateral corticospinal tracts; the rest descend in the anterior corticospinal tracts and cross over after reaching target segments in the spinal cord.

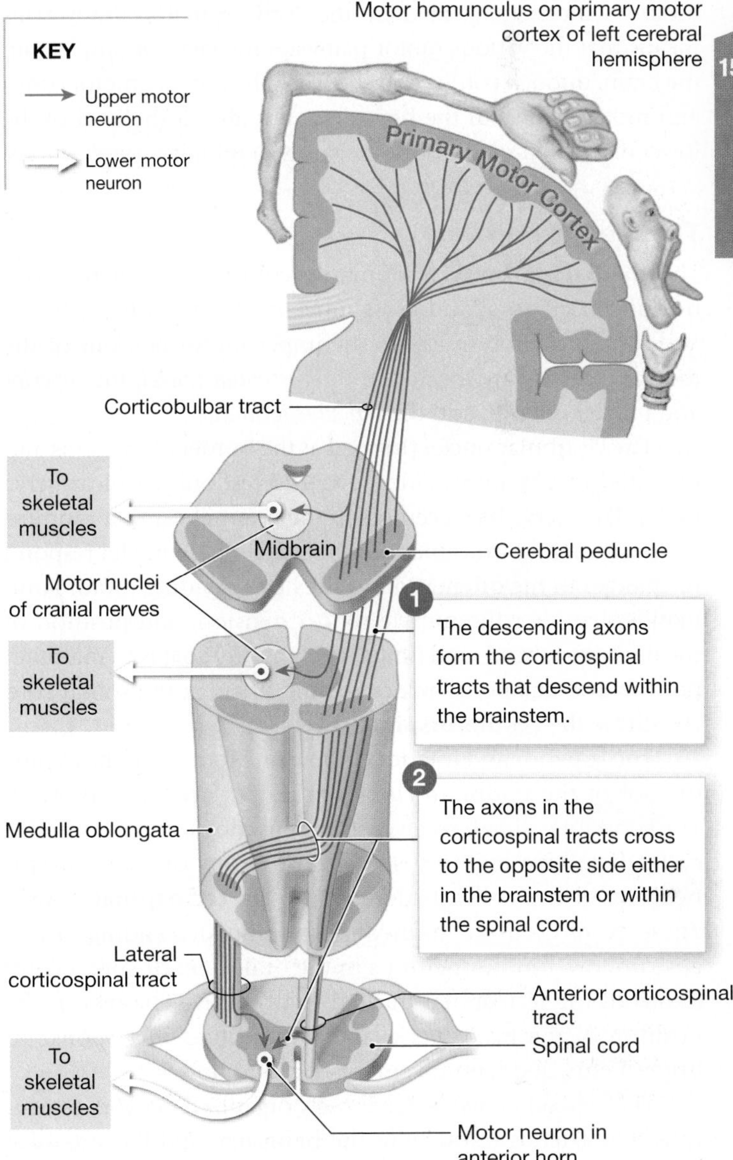

KEY
Upper motor neuron
Lower motor neuron
Motor homunculus on primary motor cortex of left cerebral hemisphere
Primary Motor Cortex
Corticobulbar tract
To skeletal muscles
Midbrain
Cerebral peduncle
Motor nuclei of cranial nerves
1 The descending axons form the corticospinal tracts that descend within the brainstem.
To skeletal muscles
2 The axons in the corticospinal tracts cross to the opposite side either in the brainstem or within the spinal cord.
Medulla oblongata
Lateral corticospinal tract
Anterior corticospinal tract
Spinal cord
To skeletal muscles
Motor neuron in anterior horn

15

The Medial and Lateral Pathways

Several centers in the cerebrum, diencephalon, and brainstem may issue somatic motor commands as a result of processing performed at a subconscious level. These centers and their associated tracts were long known as the *extrapyramidal system (EPS)* because they were thought to operate independently of, and in parallel with, the corticospinal pathway (pyramidal system). However, this classification scheme is both inaccurate and misleading, because motor control is integrated at all levels through extensive feedback loops and interconnections. We now group these nuclei and tracts in terms of their primary functions: The components of the *medial pathway* help control gross movements of the trunk and proximal limb muscles, and those of the *lateral pathway* help control the distal limb muscles that perform more precise movements.

The medial and lateral pathways can modify or direct skeletal muscle contractions by stimulating, facilitating, or inhibiting lower motor neurons. Note that the axons of upper motor neurons in the medial and lateral pathways synapse on the same lower motor neurons innervated by the corticospinal pathway. This means that the various motor pathways interact not only within the brain, through connections between the primary motor cortex and motor centers in the brainstem, but also at the level of the lower motor neuron, through excitatory or inhibitory interactions.

The Medial Pathway

The **medial pathway** is primarily concerned with the control of muscle tone and gross movements of the neck, trunk, and proximal limb muscles. The upper motor neurons of the medial pathway are located in the *vestibular nuclei*, the *superior* and *inferior colliculi*, and the *reticular formation*.

The vestibular nuclei (located at the border of the pons and medulla) receive information over the vestibulocochlear nerve (VIII). This nerve has receptors in the internal ear that monitor the position and movement of the head. These nuclei respond to changes in the orientation of the head, sending motor commands that alter the muscle tone, extension, and position of the neck, eyes, head, and limbs. The primary goal is to maintain posture and balance. The descending fibers in the spinal cord constitute the **vestibulospinal tracts** (see Figure 15–10).

The superior and inferior colliculi are located in the *tectum*, or roof of the midbrain (look back at Figure 14–8, p. 479). The superior colliculi receive visual sensations, and the inferior colliculi receive auditory sensations. Axons of upper motor neurons in the colliculi descend in the **tectospinal tracts**. These axons decussate immediately, before descending to synapse on lower motor neurons in the brainstem or spinal cord. Axons in the tectospinal tracts direct reflexive changes in the position of the head, neck, and upper limbs in response to bright lights, sudden movements, or loud noises.

The *reticular formation* is a loosely organized network of neurons that extends throughout the brainstem and has extensive interconnections with the cerebrum, the cerebellum, and brainstem nuclei. p. 474 Axons of upper motor neurons in the reticular formation of the pons and medulla oblongata descend into the **medial** (pontoreticulospinal) and **lateral** (medullary reticulospinal) **reticulospinal tracts** without crossing to the opposite side. The reticular formation receives input from almost every ascending and descending pathway. The region stimulated determines the effects of reticular formation stimulation. For example, the stimulation of upper motor neurons in one portion of the reticular formation produces eye movements. The stimulation of another portion activates respiratory muscles.

Clinical Note Amyotrophic Lateral Sclerosis

Amyotrophic lateral sclerosis (ALS) is a progressive, degenerative disorder that affects motor neurons in the spinal cord, brainstem, and cerebral hemispheres. The degeneration affects both upper and lower motor neurons. A defect in axonal transport is thought to underlie the disease. Because a motor neuron and its dependent muscle fibers are so intimately related, the destruction of CNS neurons causes the associated skeletal muscles to atrophy. ALS is commonly known as Lou Gehrig's disease, named after the famous New York Yankees player who died of the disorder. Noted physicist Stephen Hawking is also afflicted with this condition.

The Lateral Pathway

The **lateral pathway** is primarily concerned with the control of muscle tone and the more precise movements of the distal parts of the limbs. The upper motor neurons of the lateral pathway lie within the red nuclei of the midbrain. p. 478 Axons of upper motor neurons in the red nuclei decussate in the brain and descend into the spinal cord in the **rubrospinal tracts** (*ruber*, red). In humans, the rubrospinal tracts are small and extend only to the cervical spinal cord. There they provide motor control over distal muscles of the upper limbs. Normally, their role is insignificant as compared with that of the lateral corticospinal tracts. However, the rubrospinal tracts can be important in maintaining motor control and muscle tone in the upper limbs if the lateral corticospinal tracts are damaged.

Table 15–2 summarizes the descending (motor) tracts discussed in this section.

The Monitoring Role of the Basal Nuclei and Cerebellum

The *basal nuclei* and *cerebellum* are responsible for coordination and feedback control over muscle contractions. Those contractions can be consciously or subconsciously directed.

Table 15–2 Principal Descending (Motor) Pathways

Tract	Location of Upper Motor Neurons	Destination	Site of Decussation	Action
CORTICOSPINAL PATHWAY				
Corticobulbar tracts	Primary motor cortex (cerebral hemisphere)	Lower motor neurons of cranial nerve nuclei in brainstem	Brainstem	Conscious motor control of skeletal muscles
Lateral corticospinal tracts	Primary motor cortex (cerebral hemisphere)	Lower motor neurons of anterior horns of spinal cord	Pyramids of medulla oblongata	Conscious motor control of skeletal muscles
Anterior corticospinal tracts	Primary motor cortex (cerebral hemisphere)	Lower motor neurons of anterior horns of spinal cord	Level of lower motor neuron	Conscious motor control of skeletal muscles
MEDIAL PATHWAY				
Vestibulospinal tracts	Vestibular nuclei (at border of pons and medulla oblongata)	Lower motor neurons of anterior horns of spinal cord	None (uncrossed)	Subconscious regulation of balance and muscle tone
Tectospinal tracts	Tectum (midbrain: superior and inferior colliculi)	Lower motor neurons of anterior horns (cervical spinal cord only)	Brainstem (midbrain)	Subconscious regulation of eye, head, neck, and upper limb position in response to visual and auditory stimuli
Reticulospinal tracts	Reticular formation (network of nuclei in brainstem)	Lower motor neurons of anterior horns of spinal cord	None (uncrossed)	Subconscious regulation of reflex activity
LATERAL PATHWAY				
Rubrospinal tracts	Red nuclei of midbrain	Lower motor neurons of anterior horns of spinal cord	Brainstem (midbrain)	Subconscious regulation of upper limb muscle tone and movement

15

The Basal Nuclei and Patterns of Movement

The **basal nuclei** of the cerebrum provide the background patterns of movement involved in voluntary motor activities. For example, they may control muscles that determine the background position of the trunk or limbs, or they may direct rhythmic cycles of movement, as in walking or running. These nuclei do not exert direct control over lower motor neurons. Instead, they adjust the activities of upper motor neurons in the various motor pathways based on input from all portions of the cerebral cortex, as well as from the substantia nigra.

The basal nuclei adjust or establish patterns of movement by two major pathways:

- One group of axons synapses on thalamic neurons, whose axons extend to the *premotor cortex*, the motor association area that directs activities of the primary motor cortex. This arrangement creates a feedback loop that changes the sensitivity of the pyramidal cells and alters the pattern of instructions carried by the corticospinal tracts.
- A second group of axons synapses in the reticular formation, altering the excitatory or inhibitory output of the reticulospinal tracts.

Two distinct populations of interneurons exist in the basal nuclei: one that stimulates neurons by releasing acetylcholine (ACh), and another that inhibits neurons by releasing gamma aminobutyric acid (GABA). Under normal conditions, the excitatory interneurons are kept inactive, and the axons leaving the basal nuclei have an inhibitory effect on upper motor neurons. In *Parkinson's disease*, the excitatory neurons become more active, leading to problems with the voluntary control of movement. ⮌ p. 109

If the primary motor cortex is damaged, the person loses fine motor control over skeletal muscles. However, the basal nuclei can still control some voluntary movements. In effect, the medial and lateral pathways function as they usually do, but the corticospinal pathway cannot fine-tune the movements. For example, after damage to the primary motor cortex, the basal nuclei can still receive information about planned movements from the prefrontal cortex and can perform preparatory movements of the trunk and limbs. But because the corticospinal pathway is inoperative, precise movements of the forearms, wrists, and hands cannot occur. An individual in this condition can stand, maintain balance, and even walk, but all movements are hesitant, awkward, and poorly controlled.

The Cerebellum and Movement Adjustments

Whenever movement is occurring, the cerebellum monitors proprioceptive (position) sensations, visual information from the eyes, and vestibular (balance) sensations from the internal ear. Axons within the spinocerebellar tracts deliver proprioceptive information to the cerebellar cortex. Visual information is relayed from the superior colliculi, and balance information is relayed from the vestibular nuclei. During this process the cerebellum is comparing the arriving sensations with those experienced during previous movements.

When motor commands are issued, all motor pathways also send information to the cerebellum. In general, any voluntary movement begins with the activation of far more motor units than are required. The cerebellum then acts like a brake, adjusting the activities of the upper motor neurons involved to inhibit the number of motor commands used for that movement. The output of the cerebellum affects upper motor neuron activity in the corticospinal, medial, and lateral pathways. The pattern and degree of inhibition changes from moment to moment, making the movement efficient, smooth, and precisely controlled.

The patterns of cerebellar activity are learned by trial and error, over many repetitions. Many of the basic patterns are established early in life. Examples include the fine balancing adjustments you make while standing and walking. The ability to fine-tune a complex pattern of movement improves with practice, until the movements become fluid and automatic. Consider the relaxed, smooth movements of acrobats, golfers, and sushi chefs. These people move without thinking about the details of their movements. Conversely, think about what happens when you concentrate on voluntary control of various movements. The rhythm and pattern of the movement usually fall apart as your primary motor cortex starts overriding the commands of the basal nuclei and cerebellum.

✓ Checkpoint

11. **What is the anatomical basis for the fact that the left side of the brain controls motor function on the right side of the body?**
12. **An injury involving the superior portion of the motor cortex affects which region of the body?**
13. **What effect would increased stimulation of the motor neurons of the red nucleus have on muscle tone?**

See the blue Answers tab at the back of the book.

15 Chapter Review

Study Outline

MasteringA&P™ Access more chapter study tools online in the MasteringA&P Study Area:

- Chapter Quizzes, Chapter Practice Test, MP3 Tutor Sessions, and Clinical Case Studies
- Practice Anatomy Lab PAL 3.0
- A&P Flix A&PFlix
- Interactive Physiology iP2
- PhysioEx PhysioEx 9.1

An Introduction to Sensory Pathways and the Somatic Nervous System p. 513

1. The nervous system works as an integrated unit. This chapter considers sensory pathways (sensory receptors and sensory processing centers in the brain) and motor pathways (including both conscious and subconscious motor functions). *(Figure 15–1)*

15-1 Sensory stimuli cause signals to be sent along sensory pathways, and in response motor commands are sent along motor pathways p. 513

2. The brain, spinal cord, and peripheral nerves continuously communicate with each other and with the internal and external environments. Information arrives by sensory receptors and ascends within the afferent division, while motor commands descend and are distributed by the efferent division.

15-2 Sensory receptors connect our internal and external environments with the nervous system p. 514

3. A sensory receptor is a specialized cell or cell process that monitors specific conditions within the body or in the external environment. Arriving information is a **sensation**; awareness of a sensation is a **perception**.
4. The **general senses** are our sensitivity to temperature, pain, touch, pressure, vibration, and proprioception. Receptors for these senses are distributed throughout the body. **Special senses**, located in specific **sense organs**, are structurally more complex.
5. Each receptor cell monitors a specific *receptive field*. *Transduction* begins when a large enough stimulus depolarizes the *receptor potential* or *generator potential* to the point where action potentials are produced. *(Figure 15–2)*
6. A **labeled line** is a link between a peripheral receptor and a cortical neuron. **Tonic receptors** are always active. **Phasic receptors** provide information about the intensity and rate of change of a stimulus. **Adaptation** is a reduction in sensitivity in the presence of a constant stimulus. Phasic receptors are **fast-adapting receptors**, while tonic receptors are **slow-adapting receptors**. *(Figure 15–3)*

15-3 General sensory receptors can be classified by the type of stimulus that excites them p. 517

7. Three types of **nociceptors** in the body provide information on pain as related to extremes of temperature, mechanical damage, and dissolved chemicals. Myelinated Type A fibers carry **fast pain**. Unmyelinated Type C fibers carry **slow pain**.
8. **Thermoreceptors** are found in the dermis. **Mechanoreceptors** are sensitive to distortion of their membranes, and include **tactile receptors**, **baroreceptors**, and **proprioceptors**. There are six types of tactile receptors in the skin, and three kinds of proprioceptors. **Chemoreceptors** include **carotid bodies** and **aortic bodies**. *(Figures 15–4, 15–5, 15–6)*

15-4 The afferent division is made up of separate somatic sensory and visceral sensory pathways that deliver sensory information to the CNS p. 522

9. Sensory neurons that deliver sensation to the CNS are referred to as **first-order neurons**. These synapse on **second-order neurons** in the brainstem or spinal cord. The next neuron in this chain is a **third-order neuron**, found in the thalamus.
10. Three major somatic sensory pathways carry sensory information from the skin and musculature of the body wall, head, neck, and limbs: the *spinothalamic pathway*, the *posterior column pathway*, and the *spinocerebellar pathway*. *(Figure 15–7)*

11. The **spinothalamic pathway** carries poorly localized ("crude") sensations of touch, pressure, pain, and temperature. The axons involved decussate in the spinal cord and ascend within the **anterior** and **lateral spinothalamic tracts** to the ventral nuclei of the thalamus. Abnormalities along the spinothalamic pathway can lead to **phantom limb syndrome**, painful sensations that are perceived despite a missing limb, and **referred pain**, inaccurate localizations of the source of pain. *(Spotlight Figure 15–8, Figure 15–9; Table 15–1)*
12. The **posterior column pathway** carries fine touch, pressure, and proprioceptive sensations. The axons ascend within the **gracile fasciculus** and **cuneate fasciculus** and relay information to the thalamus by the **medial lemniscus**. Before the axons enter the medial lemniscus, they cross over to the opposite side of the brainstem. This crossing over is called **decussation**. *(Spotlight Figure 15–8; Table 15–1)*
13. The **spinocerebellar pathway**, including the **posterior** and **anterior spinocerebellar tracts**, carries sensations to the cerebellum concerning the position of muscles, tendons, and joints. *(Spotlight Figure 15–8; Table 15–1)*
14. Visceral sensory pathways carry information collected by interoceptors. Sensory information from cranial nerves V, VII, IX, and X is delivered to the **solitary nucleus** in the medulla oblongata. Posterior roots of spinal nerves T_1–L_2 carry visceral sensory information from organs between the diaphragm and the pelvic cavity. Posterior roots of spinal nerves S_2–S_4 carry sensory information from more inferior structures.

15-5 The somatic nervous system is an efferent division made up of somatic motor pathways that control skeletal muscles p. 527

15. Somatic motor (descending) pathways always involve an **upper motor neuron** (whose cell body lies in a CNS processing center) and a **lower motor neuron** (whose cell body is located in a nucleus of the brainstem or spinal cord). *(Figure 15–10)*
16. The neurons of the primary motor cortex are *pyramidal cells*. The **corticospinal pathway** provides voluntary skeletal muscle control. The **corticobulbar tracts** end at the cranial nerve nuclei. The **corticospinal tracts** synapse on lower motor neurons in the anterior horns of the spinal cord. The corticospinal tracts are visible along the medulla oblongata as a pair of thick bands, the **pyramids**, where most of the axons decussate to enter the descending **lateral corticospinal tracts**. Those that do not cross over enter the **anterior corticospinal tracts**. The corticospinal pathway provides a rapid, direct mechanism for controlling skeletal muscles. *(Figure 15–11; Table 15–2)*
17. The **medial** and **lateral pathways** include several other centers that issue motor commands as a result of processing performed at a subconscious level. *(Table 15–2)*
18. The medial pathway primarily controls gross movements of the neck, trunk, and proximal limbs. It includes the vestibulospinal, tectospinal, and medial and lateral reticulospinal tracts. The **vestibulospinal tracts** carry information related to maintaining balance and posture. Commands carried by the **tectospinal tracts** change the position of the head, neck, and upper limbs in response to bright lights, sudden movements, or loud noises. Motor commands carried by the **reticulospinal tracts** vary according to the region stimulated. *(Table 15–2)*
19. The lateral pathway consists of the **rubrospinal tracts**, which primarily control muscle tone and movements of the distal muscles of the upper limbs. *(Table 15–2)*
20. The basal nuclei adjust the motor commands issued in other processing centers and provide background patterns of movement involved in voluntary motor activities.
21. The cerebellum monitors proprioceptive sensations, visual information, and vestibular (balance) sensations. The integrative activities performed by neurons in the cortex and nuclei of the cerebellum are essential for the precise control of movements.

Review Questions

See the blue Answers tab at the back of the book.

LEVEL 1 Reviewing Facts and Terms

1. The larger the receptive field, the **(a)** larger the stimulus needed to stimulate a sensory receptor, **(b)** fewer sensory receptors there are, **(c)** harder it is to locate the exact point of stimulation, **(d)** larger the area of the somatosensory cortex in the brain that deals with the area, **(e)** closer together the receptor cells.
2. ___________ receptors are normally inactive, but become active for a short time whenever there is a change in the modality that they monitor.
3. The CNS interprets information entirely on the basis of the **(a)** number of action potentials that it receives, **(b)** kind of action potentials that it receives, **(c)** line over which sensory information arrives, **(d)** intensity of the sensory stimulus, **(e)** number of sensory receptors that are stimulated.
4. The area of primary somatosensory cortex devoted to a body region is relative to the **(a)** size of the body area, **(b)** distance of the body area from the brain, **(c)** density of motor units in the area of the body, **(d)** density of sensory receptors in the area of the body, **(e)** size of the nerves that serve the area of the body.
5. Identify and shade in the locations of all the ascending sensory tracts in the following diagram of the spinal cord.

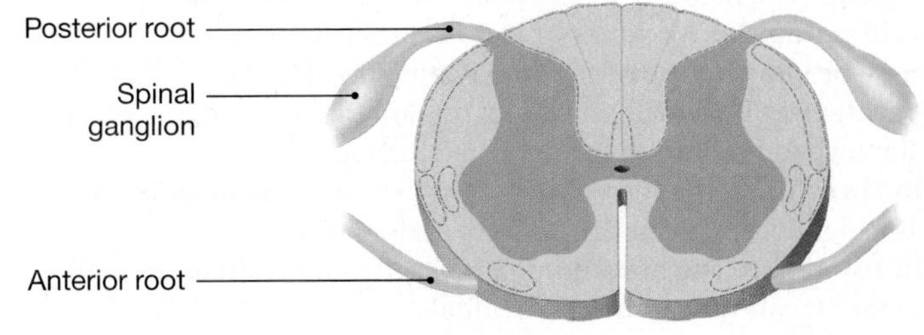
Posterior root
Spinal ganglion
Anterior root

6. Identify six types of tactile receptors located in the skin, and describe their sensitivities.
7. What three types of mechanoreceptors respond to stretching, compression, twisting, or other distortions of their plasma membrane?
8. What are the three major somatic sensory pathways and their functions?
9. Which three pairs of descending tracts make up the corticospinal pathway?
10. Which three motor tracts make up the medial pathway?
11. What are the two primary functional roles of the cerebellum?
12. The corticospinal tract **(a)** carries motor commands from the cerebral cortex to the spinal cord, **(b)** carries sensory information from the spinal cord to the brain, **(c)** starts in the spinal cord and ends in the brain, **(d)** does all of these.
13. What three steps are necessary for transduction to occur?

LEVEL 2 Reviewing Concepts

14. Differentiate between a tonic receptor and a phasic receptor.
15. What is a motor homunculus? How does it differ from a sensory homunculus?
16. Describe the relationship among first-order, second-order, and third-order neurons in a sensory pathway.
17. Damage to the posterior spinocerebellar tract on the left side of the spinal cord at the L_1 level would interfere with the coordinated movement of which limb(s)?
18. What effect does injury to the primary motor cortex have on peripheral muscles?
19. By which structures and in which part of the brain is the level of muscle tone in the body's skeletal muscles controlled? How is this control exerted?
20. Explain the phenomenon of *referred pain* in terms of labeled lines and organization of sensory tracts and pathways.

LEVEL 3 Critical Thinking and Clinical Applications

21. Kayla is having difficulty controlling her eye movements and has lost some control of her facial muscles. After an examination and testing, Kayla's physician tells her that her cranial nerves are perfectly normal but that a small tumor is putting pressure on certain fiber tracts in her brain. This pressure is the cause of Kayla's symptoms. Where is the tumor most likely located?
22. Harry, a construction worker, suffers a fractured skull when a beam falls on his head. Diagnostic tests indicate severe damage to the primary motor cortex. His wife is anxious to know if he will ever be able to move or walk again. What would you tell her?
23. Denzel had to have his arm amputated at the elbow after an accident. He tells you that he can sometimes still feel pain in his fingers even though the hand is gone. He says this is especially true when he bumps the stub. How can this be?

15

+ CLINICAL CASE Wrap-Up Living with Cerebral Palsy

Communication between the periphery and the central nervous system follows neural pathways. Somatic sensory tracts are ascending, ending in the brain. Somatic motor tracts are descending, beginning in the brain. Between the input and the output, nerve impulses are analyzed, influenced, and modified in various parts of the brain, including the cerebrum, cerebellum, basal nuclei, hippocampus, and thalamus. These were the very areas of Michael's brain that were injured during birth.

The damage affected the sensory and motor processing centers in the brain. As a result, Michael's voluntary movements, muscle tone, postural muscle control, and balance were all affected. These deficits were stable and did not get worse. There was no limit to the development of the rest of Michael's brain, or how brilliant he could become.

In spite of his disabilities, Michael chose to define himself in terms of his abilities. He earned his Ph.D. in special education. He was a pioneer and became a catalyst in introducing the concept of self-determination to the field of education. He has been an educator, writer, and motivator his entire life. He is not just living, but thriving despite his cerebral palsy.

1. Name the principal descending motor pathways that carry information that would be affected by the brain damage causing Michael's cerebral palsy.
2. Besides aspiration pneumonia, what other difficulties could result from poor motor control of the muscles used in swallowing?

See the blue Answers tab at the back of the book.

Related Clinical Terms

analgesia: The inability to feel pain while still conscious.

flaccid paralysis: Weakness or loss of muscle tone as a result of disease or injury to the nerves innervating the muscles.

hyperalgesia: Increased sensitivity to pain that may be caused by damage to nociceptors or peripheral nerves.

pain threshold: The lowest intensity of stimulation at which pain is experienced. It can vary among individuals.

pain tolerance: The maximum pain level an individual can withstand. It can vary among individuals.

spastic paralysis: The chronic pathological condition in which muscles are affected by persistent spasms and exaggerated tendon reflexes due to damage to motor nerves of the central nervous system.

syphilis: A sexually transmitted disease (STD), also called a sexually transmitted infection (STI), caused by the bacterium *Treponema pallidum*. Many of the signs and symptoms are indistinguishable from those of other diseases. Late-stage untreated syphilis can cause progressive degeneration of the posterior column pathway, spinocerebellar pathway, and posterior roots of the spinal cord.

16 The Autonomic Nervous System and Higher-Order Functions

Learning Outcomes

These Learning Outcomes correspond by number to this chapter's sections and indicate what you should be able to do after completing the chapter.

CLINICAL CASE Remember Me?

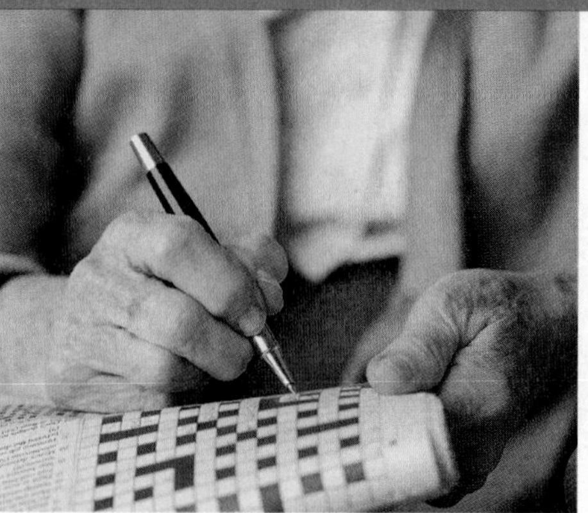

It started with the little things. Helen kept misplacing her keys. I'd try and get her to think, "When did you last have them?" I could tell it upset her. She was always so organized when she was raising our four kids. She'd be fine for a while. Then she'd get turned around in the neighborhood grocery store. The day she turned 60, she forgot we were having her big birthday celebration. She was anxious; I was worried. We made an appointment with the doctor.

Our family doctor has known us for years. She listened carefully to Helen and me and so was able to trace the start of the problem back to a year ago. Her physical exam turned out normal. Then Doc performed a little test. She told Helen a list of three things and had her repeat them back. Later in the appointment, Doc asked her again to repeat the items. You could see Helen try to concentrate, but she could only remember one.

Well, it went downhill from there, slowly but steadily. Doc gave us a medication to try and it helped for a while. I'm retired and I could do a lot for Helen. We used the adult day care at the senior center so she'd have something fun to do in the mornings and I could get a few errands done. Then Helen needed more and more attention. She was restless. She didn't talk anymore. When our son moved back home to help, she was afraid of him. This is hardest on Helen, I guess, but it is hard on all of us, really. We've lost the Helen we knew. **Why can't Helen remember? To find out, turn to the Clinical Case Wrap-Up on p. 564.**

16

An Introduction to the Autonomic Nervous System and Higher-Order Functions

In this chapter we focus on the *autonomic nervous system*, or *visceral motor system*, which adjusts our basic life support systems without our conscious control. We also consider aspects of higher-order functions such as consciousness, learning, and intelligence. Finally, we look at the effects of aging on the nervous system and conclude with an overview of interactions between the nervous system and other body systems.

16-1 The autonomic nervous system, which has sympathetic and parasympathetic divisions, is involved in the unconscious regulation of visceral functions

Learning Outcome **Compare the organization of the autonomic nervous system with that of the somatic nervous system, and name the divisions and major functions of the ANS.**

Your conscious thoughts, plans, and actions are a tiny fraction of what the nervous system does. If all consciousness were eliminated, your vital physiological processes would continue virtually unchanged. After all, a night's sleep is not a life-threatening event. Longer, deeper states of unconsciousness are not necessarily more dangerous, as long as you get nourishment and other basic care. People with severe brain injuries can survive in a coma for decades.

How do people survive under these conditions? Their survival is possible because the **autonomic nervous system (ANS)** makes routine homeostatic adjustments in physiological systems without instructions or interference from the conscious mind. By itself the ANS coordinates cardiovascular, respiratory, digestive, urinary, and reproductive functions. It also adjusts internal water, electrolyte, nutrient, and dissolved gas concentrations in body fluids.

Somatic or visceral sensory information can trigger *visceral reflexes* (responses to stimuli in visceral organs), and the ANS distributes the motor commands of those reflexes. Sometimes those motor commands control the activities of target organs. For example, in cold weather, the ANS stimulates the arrector pili muscles that give you "goosebumps." ⮌ p. 166 In other cases, the motor commands may alter some ongoing activity. A sudden, loud noise can startle you and make you jump, but thanks to the ANS, that sound can also increase your heart rate dramatically and temporarily stop all digestive gland secretion. These changes in visceral activity take place in response to *neurotransmitters* released by neurons of the ANS. As noted in Chapter 12, a specific neurotransmitter may stimulate or inhibit activity, depending on the response of particular plasma membrane receptors. We consider the major types of receptors in Section 16-3. Let's begin our examination of the ANS by comparing its organization with that of the somatic nervous system, and then go on to look at the ANS divisions and their overall functions.

Comparison of the Somatic and Autonomic Nervous Systems

Figure 16–1 compares the functional organization of the somatic and autonomic nervous systems. Both are efferent divisions that carry motor commands. The *somatic nervous system (SNS)* allows for voluntary control of skeletal muscles, while the ANS

Figure 16–1 Comparison of the Functional Organizations of the Somatic and Autonomic Nervous Systems.

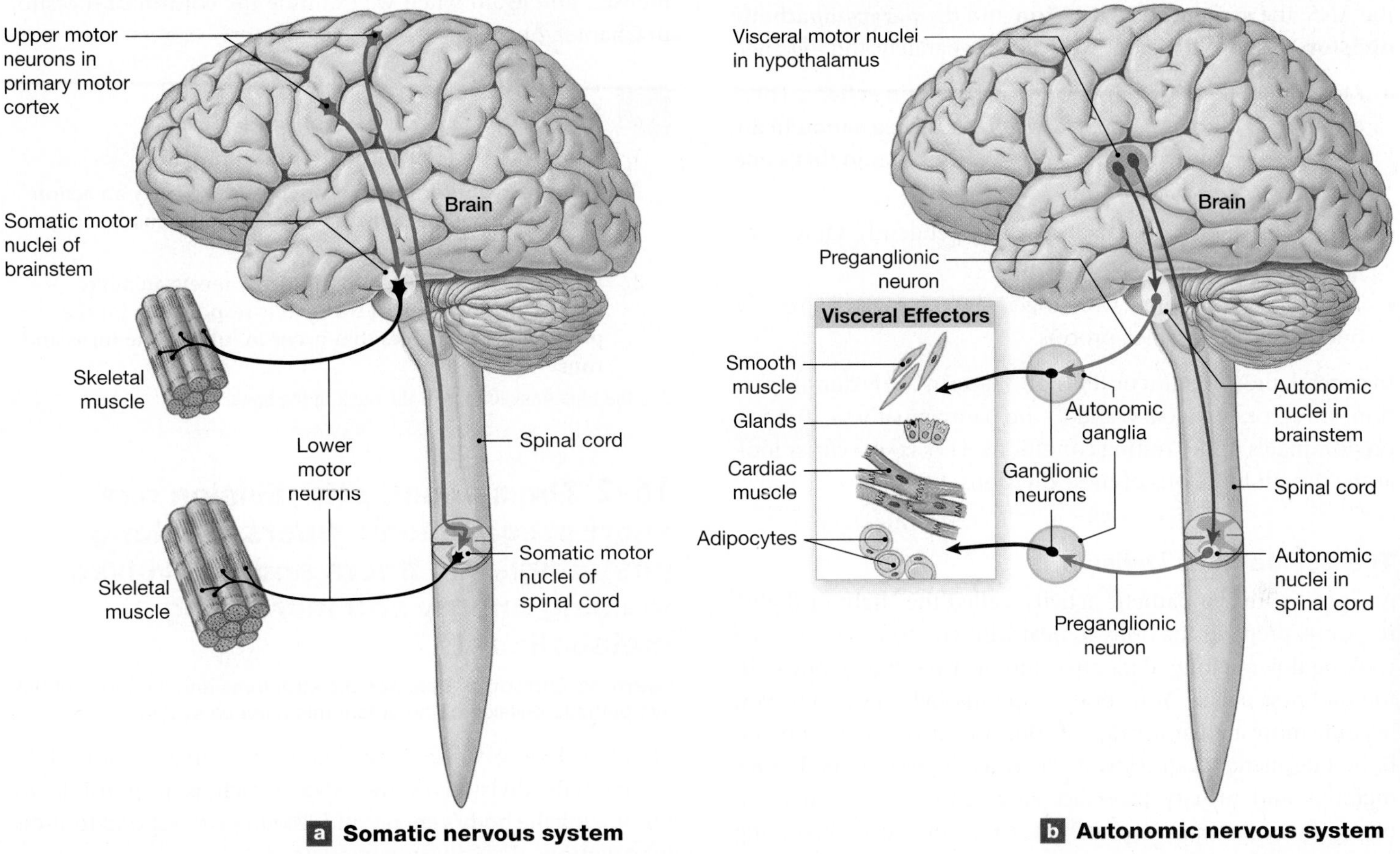

a Somatic nervous system

b Autonomic nervous system

? The somatic nervous system affects skeletal muscles. Name effectors of the autonomic nervous system.

is responsible for involuntary control of visceral effectors such as smooth muscle, glands, cardiac muscle, and adipocytes. The primary structural difference between the two is that in the SNS, motor neurons of the central nervous system (CNS) exert direct control over skeletal muscles (Figure 16–1a). In the ANS, by contrast, motor neurons of the CNS synapse on visceral motor neurons in autonomic ganglia (clusters of nerve cell bodies in the ANS), and these ganglionic neurons control visceral effectors (Figure 16–1b).

Organization of the ANS

The hypothalamus contains the integrative centers for autonomic activity. The neurons in these centers are comparable to the upper motor neurons in the SNS. Visceral motor neurons in the brainstem and spinal cord are known as **preganglionic neurons** because they extend to ganglia. These neurons are part of *visceral reflex arcs*. Most of their activities represent direct reflex responses, rather than responses to commands from the hypothalamus. The axons of preganglionic neurons are called **preganglionic fibers**.

Preganglionic fibers leave the CNS and synapse on **ganglionic neurons**, or **postganglionic neurons**—visceral motor neurons in peripheral ganglia. These ganglia, which contain hundreds to thousands of ganglionic neurons, are called **autonomic ganglia**. Ganglionic neurons innervate visceral effectors such as smooth muscle, glands, cardiac muscle, and adipocytes. The axons of ganglionic neurons are called **postganglionic fibers**, because they begin at the autonomic ganglia and extend to the peripheral target organs.

Tips & Tools

Each autonomic ganglion functions somewhat like a baton handoff zone in a relay race. Within the ganglion, one runner (the preganglionic fiber) hands off the baton (a neurotransmitter) to the next runner (the postganglionic fiber), who then continues on toward the finish line (the target effector).

Divisions of the ANS

In Chapter 12 we introduced the two main subdivisions of the ANS: the **sympathetic division** and the **parasympathetic division**. The two divisions work both separately and together:

- Most often, these two divisions have opposing effects. For example, if the sympathetic division causes excitation in an effector, the parasympathetic causes inhibition in that same effector.
- The two divisions may also work independently. Only one division innervates some structures.
- The two divisions may work together, with each controlling one stage of a complex process.

In general, the sympathetic division "kicks in" only during exertion, stress, or emergency, and t he parasympathetic division predominates under resting conditions. Let's take a closer look at the overall functions of these divisions.

The Sympathetic Division

An increase in sympathetic activity, called the "fight or flight" response, prepares the body to deal with emergencies. Imagine walking down a long, dark alley and hearing strange noises in the darkness ahead. Your body responds right away, and you become more alert and aware of your surroundings. Your metabolic rate increases quickly, up to twice its resting level. Your digestive and urinary processes stop temporarily, and more blood flows to your skeletal muscles. You begin breathing more quickly and more deeply. Both your heart rate and blood pressure increase, circulating your blood more rapidly. You feel warm and begin to perspire. We can summarize this general pattern of responses to an increased level of sympathetic activity as follows: (1) heightened mental alertness, (2) increased metabolic rate, (3) decreased digestive and urinary functions, (4) activation of energy reserves, (5) increased respiratory rate and dilation of respiratory passageways, (6) increased heart rate and blood pressure, and (7) activation of sweat glands.

The Parasympathetic Division

The parasympathetic division stimulates visceral activity. For example, it brings about the state of "rest and digest" after you eat a big dinner. Your body relaxes, energy demands are minimal, and both your heart rate and blood pressure are relatively low. Meanwhile, your digestive organs are highly stimulated. The overall pattern of responses to increased parasympathetic activity is as follows: (1) decreased metabolic rate, (2) decreased heart rate and blood pressure, (3) increased secretion by salivary and digestive glands, (4) increased motility and blood flow in the digestive tract, and (5) stimulation of urination and defecation.

In this chapter, we focus primarily on the sympathetic and parasympathetic divisions, which integrate and coordinate visceral functions throughout the body. We consider the enteric nervous system in Section 16-8, when we discuss visceral reflexes, and again when we examine the control of digestion in Chapter 24.

Checkpoint

1. **Identify the two major divisions of the ANS.**
2. **How many motor neurons are needed to carry an action potential from the spinal cord to smooth muscles in the wall of the intestine?**
3. **While out for a walk, Julie suddenly meets an angry dog. Which division of the ANS is responsible for the physiological changes that occur in Julie as she turns and runs?**

See the blue Answers tab at the back of the book.

16-2 The sympathetic division has short preganglionic fibers and long postganglionic fibers and is involved in using energy and increasing metabolic rate

Learning Outcome **Describe the structures and functions of the sympathetic division of the autonomic nervous system.**

Spotlight Figure 16–2a shows the overall organization of the sympathetic division of the ANS, which is responsible for mobilizing the body's energy and resources to respond to stressful situations. This division consists of:

- *Short Preganglionic Fibers in Thoracic and Lumbar Segments of the Spinal Cord.* This division is also called the *thoracolumbar division* because its preganglionic neurons are located between segments T_1 and L_2 of the spinal cord. The cell bodies of the preganglionic neurons are in the lateral horns, and their axons enter the anterior roots of these segments.
- *Ganglionic Neurons in Ganglia near the Spinal Cord.* The preganglionic fibers are relatively short because the ganglia are located near the spinal cord. They release acetylcholine (ACh), stimulating ganglionic neurons.
- *Long Postganglionic Fibers to Target Organs.* The postganglionic fibers are relatively long. (An exception occurs at the adrenal medullae, which we discuss later in this section.)

Functional Organization of the Sympathetic Division

Spotlight Figure 16–2a provides a detailed view of the sympathetic division as a whole. The left side of the figure shows the somatic distribution to the skin, skeletal muscles, and other tissues of the body wall. The right side shows the innervation of visceral organs.

Sympathetic preganglionic fibers originate from the lateral horns and become part of the spinal anterior roots. We described the basic pattern of sympathetic innervation in these regions in **Spotlight Figure 13–8**, pp. 444–445. After passing through the intervertebral foramen, each anterior root gives rise to a myelinated white ramus communicans, which carries myelinated preganglionic fibers into a nearby *sympathetic chain ganglion* found on either side of the vertebral column. These fibers diverge extensively, with one preganglionic fiber synapsing on two dozen or more ganglionic neurons. Preganglionic fibers running between the sympathetic chain ganglia interconnect them, making the chain look like a string of pearls. Each sympathetic chain ganglion innervates a particular body organ or group of organs.

Where ganglionic neurons synapse, or which pathway they follow, determines which target organs they will affect. These neurons may synapse in three locations: within the sympathetic chain ganglia, and/or at one of the collateral ganglia, or in the adrenal medullae (Figure 16–3).

- *Sympathetic Chain Ganglia.* **Sympathetic chain ganglia**, also called *paravertebral ganglia*, lie on each side of the vertebral column (Figure 16–3a). Neurons in these ganglia control effectors in the body wall, the thoracic cavity, head, neck, and limbs.
- *Collateral Ganglia.* **Collateral ganglia**, also known as *prevertebral ganglia*, are anterior to the vertebral column (Figure 16–3b). Collateral ganglia contain ganglionic neurons that innervate abdominopelvic tissues and viscera.
- *Adrenal Medullae.* The center of each adrenal (*ad-*, near + *renal*, kidney) gland, the **adrenal medulla**, is a modified sympathetic ganglion (Figure 16–3c). The ganglionic neurons of the adrenal medullae have very short axons. When these medullary cells are stimulated, they secrete neurotransmitters directly into the bloodstream, not at a synapse. This mode of release allows the neurotransmitters to function as *hormones*, affecting target cells throughout the body.

Sympathetic Chain Ganglia Pathway: Upper Body

If a preganglionic fiber carries motor commands targeting structures in the body wall, thoracic cavity, head, neck, or limbs, it synapses in one or more sympathetic chain ganglia. Each sympathetic chain contains 3 cervical, 10–12 thoracic, 4–5 lumbar, and 4–5 sacral ganglia, plus 1 coccygeal ganglion. (The numbers vary because adjacent ganglia sometimes fuse.) Preganglionic neurons are limited to spinal cord segments T_1–L_2, and these spinal nerves have both *white rami communicantes* (myelinated preganglionic fibers) and *gray rami communicantes* (unmyelinated postganglionic fibers). The neurons in the cervical, inferior lumbar, and sacral sympathetic chain ganglia are innervated by preganglionic fibers that run along the axis of the chain. In turn, these chain ganglia provide postganglionic fibers, through gray rami communicantes, to the cervical, lumbar, and sacral spinal nerves. As a result, only spinal nerves T_1–L_2 have white rami communicantes, and every spinal nerve has a gray ramus communicans that carries sympathetic postganglionic fibers for distribution in the body wall.

The unmyelinated postganglionic fibers then follow one of two different paths, depending on where their targets lie:

- Postganglionic fibers that control visceral effectors in the body wall, head, neck, or limbs enter the gray ramus communicans and return to the spinal nerve for subsequent distribution (see Figure 16–3a, right). These postganglionic fibers innervate sweat glands, smooth muscles in superficial blood vessels, and arrector pili muscles in the skin (see **Spotlight Figure 16–2a**).
- Postganglionic fibers innervating visceral organs in the thoracic cavity, such as the heart and lungs, form bundles known as **sympathetic nerves** (see Figure 16–3a, left).

16

For the sake of clarity, Figure 16–3a shows sympathetic nerves on the left side and spinal nerve distribution on the right but in reality *both* innervation patterns occur on *each* side of the body.

The spinal nerves provide somatic motor innervation to skeletal muscles of the body wall and limbs, but they also distribute sympathetic postganglionic fibers (see Figure 16–3a and **Spotlight Figure 16–2a**). About 8 percent of the axons in each spinal nerve are sympathetic postganglionic fibers. In the head and neck, sympathetic postganglionic fibers leaving the superior cervical sympathetic ganglia supply the regions and structures innervated by cranial nerves III, VII, IX, and X. ⮌ pp. 497, 500, 502

In summary:

- The cervical, inferior lumbar, and sacral chain ganglia receive preganglionic fibers from spinal segments T_1–L_2.
- Only the thoracic and superior lumbar ganglia (T_1–L_2) receive preganglionic fibers from white rami communicantes.
- Every spinal nerve receives a gray ramus communicans from a ganglion of the sympathetic chain.

Collateral Ganglia Pathway: Abdominopelvic Cavity

The abdominopelvic viscera receive sympathetic innervation by sympathetic preganglionic fibers that synapse in separate collateral ganglia (Figure 16–3b). These fibers pass through the sympathetic chain without synapsing. They form the *splanchnic* (SPLANK-nik) *nerves*, which lie in the posterior wall of the abdominal cavity. In adults the collateral ganglia are typically

SPOTLIGHT

Figure 16–2
The Autonomic Nervous System

AUTONOMIC NERVOUS SYSTEM

Consists of two divisions

a Sympathetic Division (Thoracolumbar Division)

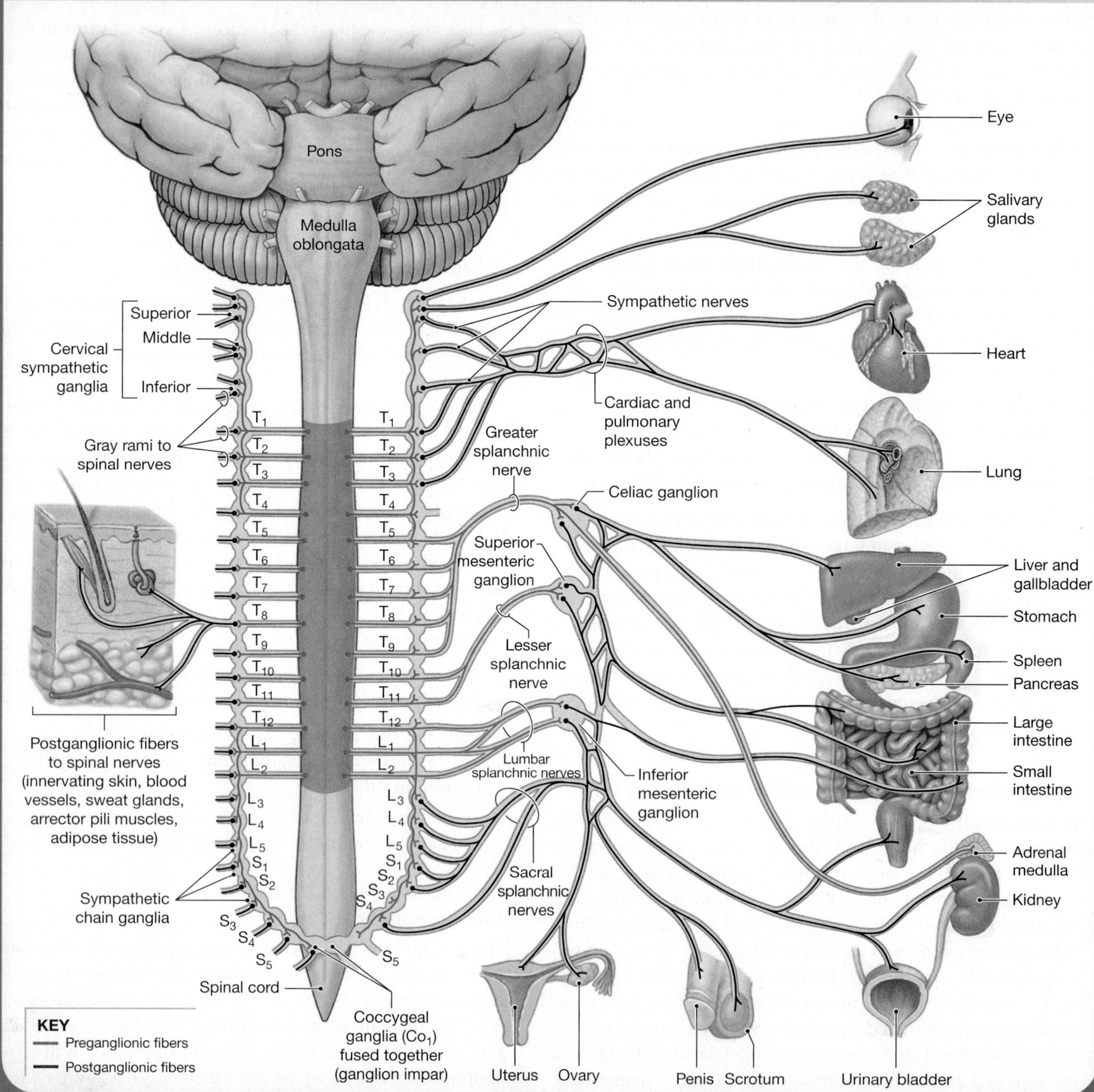

Pons
Medulla oblongata
Eye
Salivary glands
Sympathetic nerves
Heart
Superior
Middle
Cervical sympathetic ganglia
Inferior
Cardiac and pulmonary plexuses
Gray rami to spinal nerves
Greater splanchnic nerve
Lung
Celiac ganglion
Superior mesenteric ganglion
Liver and gallbladder
Stomach
Lesser splanchnic nerve
Spleen
Pancreas
Large intestine
Lumbar splanchnic nerves
Small intestine
Postganglionic fibers to spinal nerves (innervating skin, blood vessels, sweat glands, arrector pili muscles, adipose tissue)
Inferior mesenteric ganglion
Adrenal medulla
Sacral splanchnic nerves
Kidney
Sympathetic chain ganglia
Spinal cord
Coccygeal ganglia (Co_1) fused together (ganglion impar)
Uterus
Ovary
Penis
Scrotum
Urinary bladder
KEY
Preganglionic fibers
Postganglionic fibers

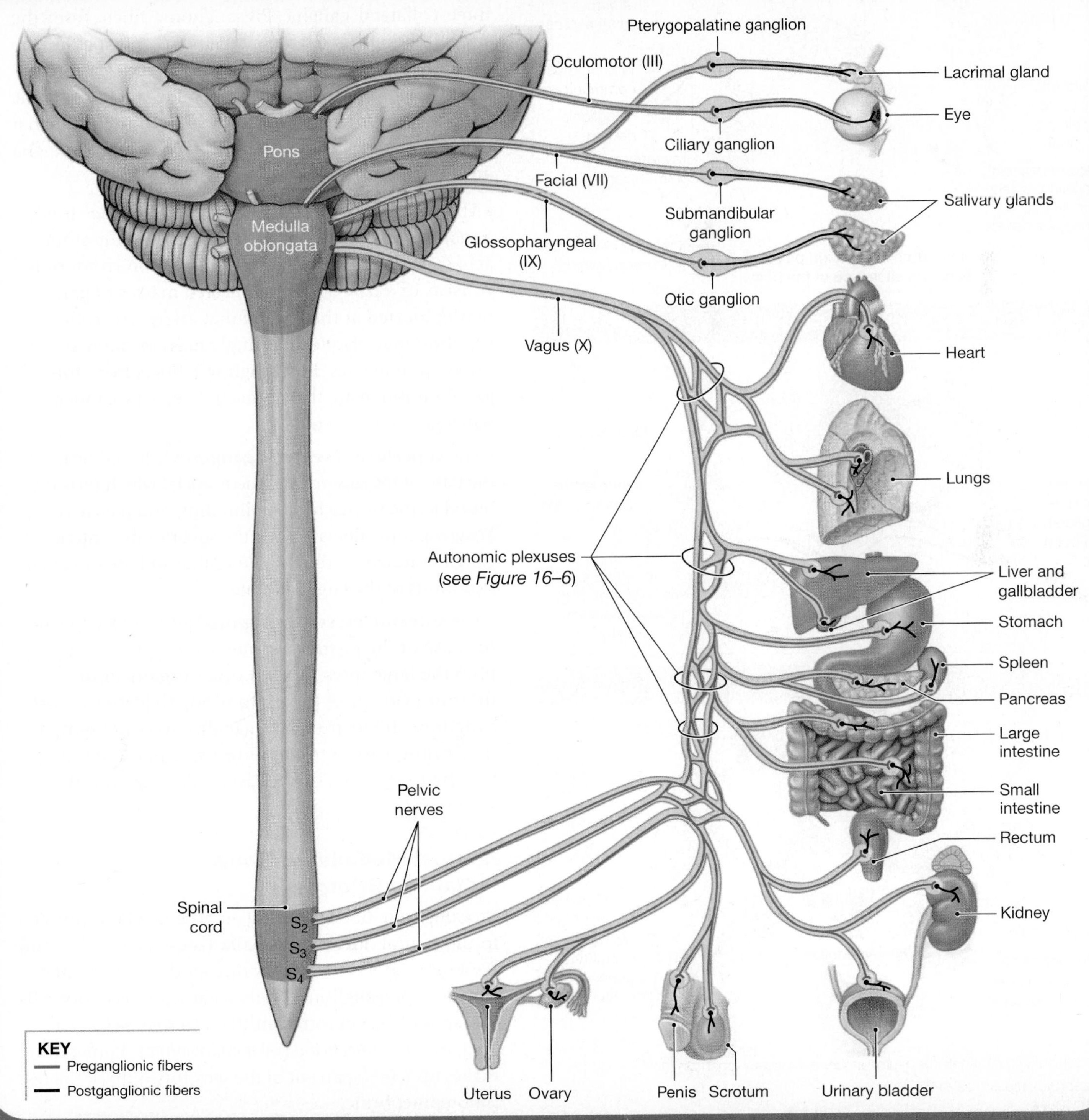

b Parasympathetic Division (Craniosacral Division)
Pterygopalatine ganglion
Oculomotor (III)
Lacrimal gland
Eye
Pons
Ciliary ganglion
Facial (VII)
Salivary glands
Medulla oblongata
Submandibular ganglion
Glossopharyngeal (IX)
Otic ganglion
Vagus (X)
Heart
Lungs
Autonomic plexuses (see Figure 16–6)
Liver and gallbladder
Stomach
Spleen
Pancreas
Large intestine
Small intestine
Rectum
Pelvic nerves
Kidney
Spinal cord
S_2
S_3
S_4
KEY
Preganglionic fibers
Postganglionic fibers
Uterus
Ovary
Penis
Scrotum
Urinary bladder

Figure 16–3 Sites of Ganglia in Sympathetic Pathways. Inferior views of sections through the thoracic spinal cord, showing the three major patterns of distribution for preganglionic and postganglionic fibers.

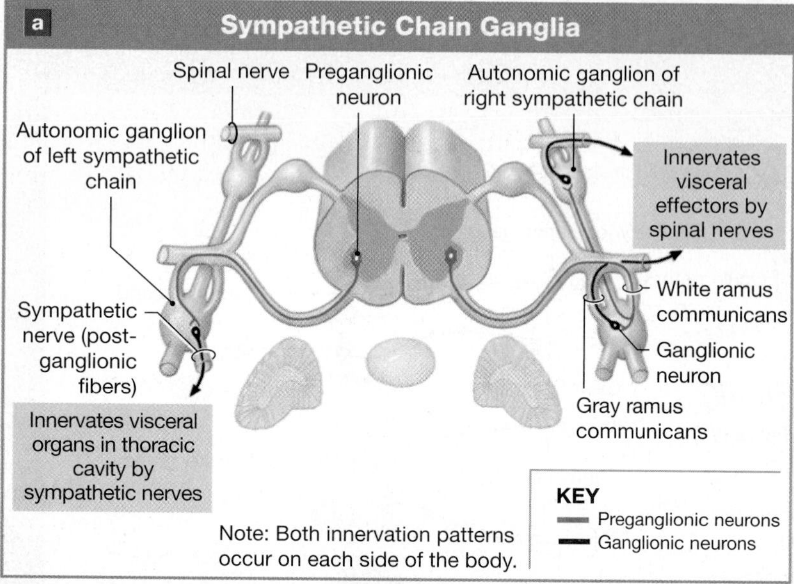

a Sympathetic Chain Ganglia
Spinal nerve
Preganglionic neuron
Autonomic ganglion of right sympathetic chain
Autonomic ganglion of left sympathetic chain
Innervates visceral effectors by spinal nerves
White ramus communicans
Sympathetic nerve (postganglionic fibers)
Ganglionic neuron
Gray ramus communicans
Innervates visceral organs in thoracic cavity by sympathetic nerves
KEY
Preganglionic neurons
Ganglionic neurons
Note: Both innervation patterns occur on each side of the body.

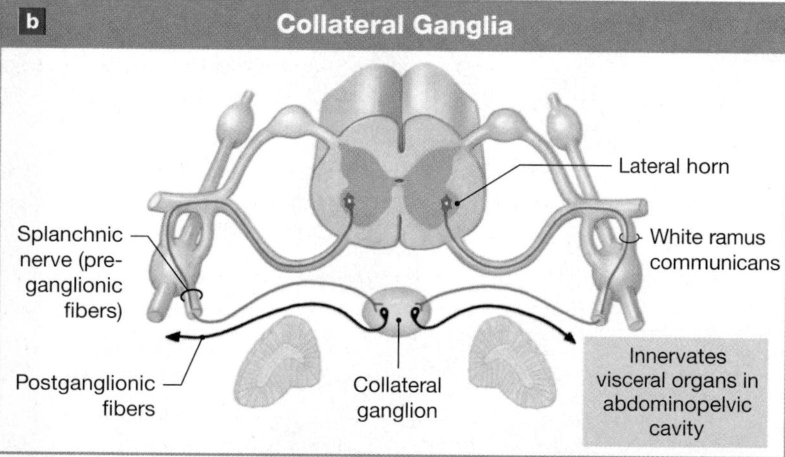

b Collateral Ganglia
Lateral horn
Splanchnic nerve (preganglionic fibers)
White ramus communicans
Postganglionic fibers
Collateral ganglion
Innervates visceral organs in abdominopelvic cavity

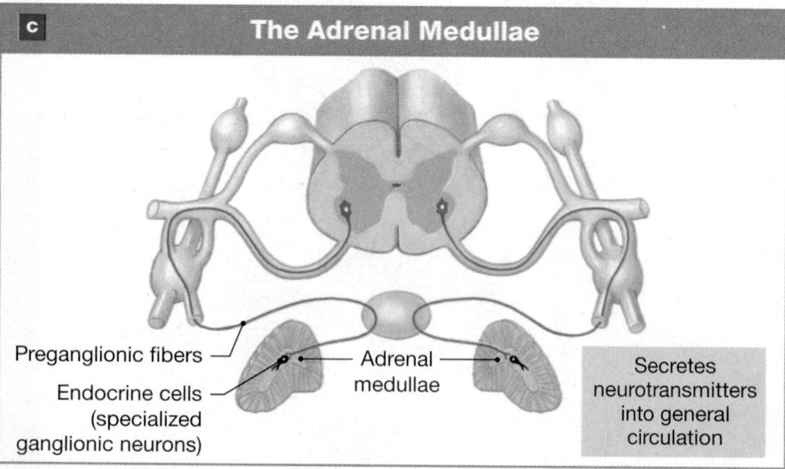

c The Adrenal Medullae
Preganglionic fibers
Adrenal medullae
Endocrine cells (specialized ganglionic neurons)
Secretes neurotransmitters into general circulation

? Which type of ganglionic neuron innervates visceral organs in the abdominopelvic cavity?

single rather than paired. They originate as paired ganglia (left and right), but the two usually fuse.

Postganglionic fibers leaving the collateral ganglia extend throughout the abdominopelvic cavity, innervating a variety of visceral tissues and organs. Their general function is to (1) reduce blood flow and energy use by organs that are not important to immediate survival (such as the digestive tract) and (2) release stored energy.

The preganglionic **splanchnic nerves** innervate three collateral ganglia. Preganglionic fibers from the seven inferior thoracic spinal segments end at either the *celiac* (SĒ-lē-ak) *ganglion* or the *superior mesenteric ganglion* (see **Spotlight Figure 16–2a**). Preganglionic fibers from the lumbar segments form splanchnic nerves that end at the *inferior mesenteric ganglion*. All three ganglia are named after nearby arteries:

- The **celiac ganglion** is named after the *celiac trunk*, a major artery and its branches supplying the stomach, spleen, and liver. The celiac ganglion commonly consists of a pair of interconnected masses of gray matter located at the base of that artery. The celiac ganglion may also form a single mass or many small, interwoven masses. Postganglionic fibers from this ganglion innervate the stomach, liver, gallbladder, pancreas, and spleen.
- The **superior mesenteric ganglion** is found near the base of the *superior mesenteric artery*, which provides blood to the stomach, small intestine, and pancreas. Postganglionic fibers leaving the superior mesenteric ganglion innervate the small intestine and the proximal two-thirds of the large intestine.
- The **inferior mesenteric ganglion** is located near the base of the *inferior mesenteric artery*, which supplies the large intestine and other organs in the inferior portion of the abdominopelvic cavity. Postganglionic fibers from this ganglion provide sympathetic innervation to the kidneys, urinary bladder, terminal segments of the large intestine, and the sex organs.

Adrenal Medulla Pathway: Metabolic Effects

Preganglionic fibers entering an adrenal gland proceed to the central **adrenal medulla** (see **Figure 16–3c** and **Spotlight Figure 16–2a**). In this modified sympathetic ganglion, preganglionic fibers synapse on secretory cells that secrete the neurotransmitters *epinephrine* (or *adrenaline*) and *norepinephrine* (or *noradrenaline*). Epinephrine makes up 75–80 percent of the secretory output; the rest is norepinephrine.

The bloodstream carries the neurotransmitters throughout the body, where they cause changes in the metabolic activities of many different cells. These effects resemble those produced by the stimulation of sympathetic postganglionic fibers. They differ, however, in two respects: (1) Cells not innervated by sympathetic postganglionic fibers are affected; and (2) the effects last much longer than those produced by direct sympathetic innervation, because the chemicals continue to diffuse out of the bloodstream for an extended period.

Sympathetic Activation

The sympathetic division can change the activities of tissues and organs by releasing norepinephrine at peripheral synapses, and by distributing epinephrine and norepinephrine throughout the body in the bloodstream. The visceral motor fibers that target specific effectors, such as smooth muscle fibers in blood vessels of the skin, can be activated in reflexes that do not involve other visceral effectors.

In a crisis, however, the entire division responds. This event, called **sympathetic activation**, is controlled by sympathetic centers in the hypothalamus. The effects are not limited to peripheral tissues, because sympathetic activation also alters CNS activity. When sympathetic activation occurs, a person experiences the following changes:

- Increased alertness by stimulation of the reticular activating system, causing the individual to feel "on edge."
- A feeling of energy and euphoria, often associated with a disregard for danger and a temporary insensitivity to painful stimuli.
- Increased activity in the cardiovascular and respiratory centers of the pons and medulla oblongata, which increases blood pressure, heart rate, breathing rate, and depth of respiration.
- A general elevation in muscle tone through stimulation of the medial and lateral pathways, so the person *looks* tense and may begin to shiver.
- The mobilization of energy reserves, through the accelerated breakdown of glycogen in muscle and liver cells and the release of lipids by adipose tissues.

These changes, plus the peripheral changes already noted, prepare the person to cope with a stressful situation.

✓ Checkpoint

4. **Describe the positions of the sympathetic chain ganglia and collateral ganglia relative to the vertebral column.**
5. **Where do the nerves that synapse in collateral ganglia originate?**

See the blue Answers tab at the back of the book.

16-3 Different types of neurotransmitters and receptors lead to different sympathetic effects

Learning Outcome **Describe the types of neurotransmitters and receptors and explain their mechanisms of action.**

We have examined the structure of the sympathetic division of the ANS and the general effects of sympathetic activation. Now let's consider the cellular basis of these effects on peripheral organs.

When stimulated, sympathetic preganglionic neurons release *acetylcholine (ACh)* at synapses with ganglionic neurons. Synapses that use ACh as a transmitter are called *cholinergic.* ⮌ p. 418 The effect on the ganglionic neurons is always excitatory.

These ganglionic neurons then release neurotransmitters at specific target organs. The axon terminals differ from those at neuromuscular (nerve to muscle cell) junctions of the somatic nervous system. In the sympathetic nervous system, the telodendria form a branching network, with each branch resembling a string of pearls. Each "pearl" is a swollen segment called a **varicosity** that is packed with neurotransmitter vesicles (Figure 16–4). Chains of varicosities are found along or near the surfaces of effector cells such as smooth muscle cells. These cells have no specialized postsynaptic membranes, such as the motor end plates in skeletal muscle cells. Instead, membrane receptors are scattered across the surfaces of the target cells.

Most sympathetic ganglionic neurons release norepinephrine (NE) at their varicosities. Neurons that release NE are called *adrenergic.* ⮌ p. 420 A small but significant number of ganglionic neurons in the sympathetic division release ACh rather than NE. Varicosities releasing ACh are located in the body wall, the skin, the brain, and skeletal muscles.

The NE released by varicosities affects its targets until it is reabsorbed or inactivated by enzymes. Fifty to 80 percent of the NE is reabsorbed by varicosities and is either reused or broken down by the enzyme *monoamine oxidase (MAO).* The rest of the NE diffuses out of the area or is broken down by the enzyme *catechol-O-methyltransferase (COMT)* in surrounding tissues.

In general, the effects of NE on the postsynaptic membrane last for a few seconds, significantly longer than the 20-msec duration of ACh effects. (As usual, the responses of the target cells vary with the nature of their receptors.) When the adrenal medullae release norepinephrine (NE) and epinephrine (E) into the bloodstream, the effects last even longer because (1) the bloodstream does not contain MAO or COMT, and (2) most tissues contain relatively low concentrations of those enzymes. After the adrenal medullae are stimulated, tissue concentrations of norepinephrine and epinephrine throughout the body may remain elevated for as long as 30 seconds, and the effects may persist for several minutes. Let's look at the nature of the receptors that these neurotransmitters stimulate, and the effects of such stimulation.

16

> Go to MasteringA&P™ > Study Area > Menu > Lab Tools > PAL 3.0 > Human Cadaver > Nervous System > Autonomic Nervous System

Figure 16–4 Sympathetic Varicosities.

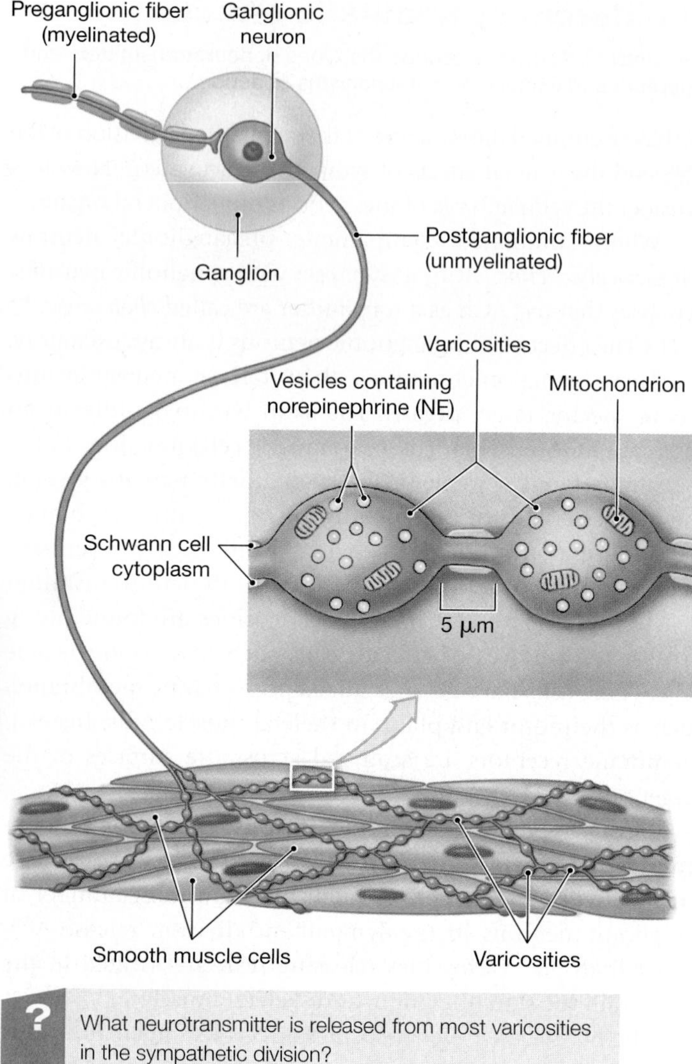
Preganglionic fiber (myelinated)
Ganglionic neuron
Postganglionic fiber (unmyelinated)
Ganglion
Varicosities
Vesicles containing norepinephrine (NE)
Mitochondrion
Schwann cell cytoplasm
5 μm
Smooth muscle cells
Varicosities

? What neurotransmitter is released from most varicosities in the sympathetic division?

Effects of Sympathetic Stimulation of Adrenergic Synapses and Receptors

The effects of sympathetic stimulation result primarily from the interactions of NE and E with adrenergic membrane receptors. There are two classes of these receptors: *alpha receptors* and *beta receptors.* In general, NE stimulates alpha receptors to a greater degree than it does beta receptors, and E stimulates both classes of receptors. For this reason, localized sympathetic activity, involving the release of NE at varicosities, primarily affects nearby alpha receptors. By contrast, generalized sympathetic activation and the release of E by the adrenal medullae affect both alpha and beta receptors throughout the body.

Alpha receptors and beta receptors are *G-protein–coupled receptors.* As we saw in Chapter 12, the effects of stimulating such a receptor depend on the production of *second messengers,* intracellular intermediaries with varied functions. p. 421

Alpha Receptors

The stimulation of **alpha (α) receptors** activates their associated G proteins on the cytoplasmic surface of the plasma membrane. There are two types of alpha receptors: *alpha-1* (α_1) and *alpha-2* (α_2).

- Alpha-1 receptors, the more common type of alpha receptor, are found primarily in smooth muscle cells of many different organs. The stimulation of this receptor results in G protein activation that brings about the release of intracellular calcium ions into the cytosol from reserves in the endoplasmic reticulum. This action generally has an excitatory effect on the target cell. For example, the stimulation of α_1 receptors on smooth muscle cells causes peripheral blood vessels to constrict and sphincters along the digestive tract and urinary bladder to close.
- Alpha-2 receptors are found on preganglionic or postganglionic sympathetic neurons. Stimulation of α_2 receptors and G protein activation result in a lowering of the cyclic-AMP (cAMP) level in the cytoplasm. Cyclic-AMP is an important second messenger that can activate or inactivate key enzymes. p. 424 This decrease generally has an inhibitory effect on the target cell. The presence of α_2 receptors in the parasympathetic division helps coordinate sympathetic and parasympathetic activities. When the sympathetic division is active, the NE released binds to α_2 receptors at parasympathetic neuromuscular and neuroglandular junctions and inhibits their activity.

Beta Receptors

Beta (β) receptors are located on the plasma membranes of cells in many organs, including skeletal muscles, the lungs, the heart, and the liver. The stimulation of beta receptors and G protein activation trigger changes in the metabolic activity of the target cell. These changes occur indirectly, as G protein activation results in an increase in intracellular cAMP levels. There are three major types of beta receptors: *beta-1* (β_1), *beta-2* (β_2), and *beta-3* (β_3).

- The stimulation of β_1 receptors leads to an increase in metabolic activity. For example, the stimulation of β_1 receptors in skeletal muscles accelerates the metabolic activities of the muscles. The stimulation of β_1 receptors in the heart increases heart rate and force of contraction.
- The stimulation of β_2 receptors causes inhibition, triggering the relaxation of smooth muscles along the respiratory tract. As a result, respiratory passageways dilate, making breathing easier. The inhalers used to treat asthma trigger this response.
- The β_3 receptor is found in adipose tissue. Stimulation of β_3 receptors leads to *lipolysis*, the breakdown of triglycerides stored within adipocytes. The fatty acids generated through lipolysis are released into the bloodstream for use by other tissues.

Effects of Sympathetic Stimulation on Other Types of Synapses

The vast majority of sympathetic postganglionic fibers are adrenergic (release NE), but as we noted, a few are cholinergic (release ACh). These postganglionic fibers innervate skin sweat glands and the blood vessels to skeletal muscles and the brain. The activation of these sympathetic fibers stimulates sweat gland secretion and dilates the blood vessels.

In some areas of the body, the sympathetic division requires the use of two types of synapses. For example, in the body wall and skeletal muscles, both ACh and NE are needed to regulate visceral functions with precision. ACh causes most small peripheral arteries to dilate (*vasodilation*), and NE causes them to constrict (*vasoconstriction*). This means that the sympathetic division can increase blood flow to skeletal muscles by releasing ACh to activate cholinergic synapses. At the same time, it can reduce the blood flow to other tissues in the body wall, by releasing NE to stimulate adrenergic synapses.

The sympathetic division also includes *nitroxidergic synapses*, which release *nitric oxide (NO)* as a neurotransmitter. Such synapses occur where neurons innervate smooth muscles in the walls of blood vessels in many regions, notably in skeletal muscles and the brain. The activity of these synapses produces vasodilation and increased blood flow through these regions.

Checkpoint

6. **How would a drug that stimulates cholinergic receptors affect the sympathetic nervous system?**
7. **A person with high blood pressure is given a medication that blocks beta receptors. How could this medication help correct that person's condition?**

See the blue Answers tab at the back of the book.

16-4 The parasympathetic division has long preganglionic fibers and short postganglionic fibers and is involved in conserving energy and lowering metabolic rate

Learning Outcome **Describe the structures and functions of the parasympathetic division of the autonomic nervous system.**

The parasympathetic division of the ANS is responsible for conserving the body's energy and maintaining the resting metabolic rate. This division (**Spotlight Figure 16–2b**) consists of:

- *Long Preganglionic Fibers in the Brainstem and in Sacral Segments of the Spinal Cord.* All parts of the brainstem, the midbrain, pons, and medulla oblongata, contain autonomic nuclei associated with several cranial nerves. In sacral segments of the spinal cord, the parasympathetic nuclei lie in the lateral horns of spinal segments S_2–S_4. This divided pattern of neuron locations explains why the parasympathetic division is also called the *craniosacral division*.
- *Ganglionic Neurons in Peripheral Ganglia within or Adjacent to Target Organs.* Preganglionic fibers of the parasympathetic division do not diverge as extensively as do those of the sympathetic division. A typical preganglionic fiber synapses on six to eight ganglionic neurons, all in the same ganglion. The ganglion may be a **terminal ganglion**, located near the target organ, or an **intramural** (*murus*, wall) **ganglion**, embedded in the tissues of the target organ. Terminal ganglia are usually paired. Intramural ganglia typically consist of interconnected masses and clusters of ganglion cells.
- *Short Postganglionic Fibers in or near Target Organs.* The postganglionic fibers then extend from the terminal or intramural ganglion to the target organ. Having the ganglia and postganglionic neurons near the target organs makes the effects of parasympathetic stimulation very specific and localized.

Let's look at the organization of this division more closely, and at what happens when the division is activated.

Functional Organization of the Parasympathetic Division

Parasympathetic fibers innervate organs of three main regions: the cranial, trunk, and pelvic regions.

Parasympathetic preganglionic fibers leave the brain in cranial nerves III (oculomotor), VII (facial), IX (glossopharyngeal), and X (vagus) (**Spotlight Figure 16–2b**). These fibers carry the cranial parasympathetic output to visceral structures in the head. These fibers synapse in the *ciliary*, *pterygopalatine*, *submandibular*, and *otic ganglia*. ↺ pp. 497, 499, 500 Short postganglionic fibers continue to their peripheral targets.

The vagus nerve alone provides roughly 75 percent of all parasympathetic outflow (output from nerves leaving the CNS). It supplies preganglionic parasympathetic innervation to structures in the neck and in the thoracic and abdominopelvic cavities as distant as the distal portion of the large intestine. The many branches of the vagus nerve mingle with preganglionic and postganglionic fibers of the sympathetic division, forming plexuses comparable to those formed by spinal nerves innervating the limbs. We consider these plexuses in Section 16-7.

The preganglionic fibers in the sacral segments of the spinal cord carry the sacral parasympathetic output. These fibers do not join the anterior roots of the spinal nerves. Instead, they form distinct **pelvic nerves**, which innervate intramural ganglia in the walls of the kidneys, urinary bladder, terminal portions of the large intestine, and sex organs.

Parasympathetic Activation

The functions of the parasympathetic division center on relaxation, food processing, and nutrient absorption. This division has been called the *anabolic system* (*anabole*, a raising up), because its stimulation leads to a general increase in the nutrient content of the blood. In response to this increase, cells throughout the body absorb nutrients and use them to support growth and cell division and to create energy reserves in the form of lipids or glycogen.

The major effects of the parasympathetic division include the following:

- Constriction of the pupils (to restrict the amount of light that enters the eyes) and focusing of the lenses of the eyes on nearby objects.
- Secretion by digestive glands, including salivary glands, gastric glands, duodenal glands, intestinal glands, the pancreas (exocrine and endocrine), and the liver.
- The secretion of hormones that promote the absorption and use of nutrients by peripheral cells.
- Changes in blood flow and glandular activity associated with sexual arousal.
- An increase in smooth muscle activity along the digestive tract.
- The stimulation and coordination of defecation.
- Contraction of the urinary bladder during urination.
- Constriction of the respiratory passageways.
- A reduction in heart rate and force of contraction.

16

Checkpoint

8. **Which nerve is responsible for the parasympathetic innervation of the lungs, heart, stomach, liver, pancreas, and parts of the small and large intestines?**
9. **Why is the parasympathetic division sometimes referred to as the anabolic system?**

See the blue Answers tab at the back of the book.

16-5 Different types of receptors lead to different parasympathetic effects

Learning Outcome Describe the mechanisms of parasympathetic neurotransmitter release and their effects on target organs and tissues.

All parasympathetic neurons release ACh as a neurotransmitter. The effects on the postsynaptic cell can vary widely, however, due to different types of receptors or to the nature of the second messenger involved.

Effects of Parasympathetic Stimulation of Cholinergic Receptors

When parasympathetic neurons release ACh, the neurotransmitter may affect neurons, muscle fibers, or effector cells. The effects of stimulation are both localized and short lived. Most of the ACh released is inactivated at the synapse by *acetylcholinesterase (AChE)*. Any ACh diffusing into the surrounding tissues is inactivated by the enzyme *tissue cholinesterase.*

The effects of ACh release depend to a large degree on what type of cholinergic receptor is stimulated. Two types of cholinergic receptors occur on the postsynaptic membranes: *nicotinic* and *muscarinic*.

Nicotinic Receptors

Nicotinic (nik-ō-TIN-ik) **receptors** occur on ganglion cells of both the parasympathetic and sympathetic divisions. (They also occur at neuromuscular junctions of the somatic nervous system.) When ACh activates nicotinic receptors, it opens chemically gated Na^+ channels in the postsynaptic membrane of the ganglionic neuron or motor end plate of the muscle fiber, which always causes excitation of that cell.

Muscarinic Receptors

Muscarinic (mus-kah-RIN-ik) **receptors** occur at cholinergic neuromuscular or neuroglandular junctions in the parasympathetic division. They also occur at the few cholinergic junctions in the sympathetic division. Muscarinic receptors are G protein–coupled receptors. Their stimulation and G protein activation produce longer-lasting effects than does the stimulation of nicotinic receptors. The response can be excitatory or inhibitory, depending on the activation or inactivation of specific enzymes.

Effects of Toxins on Cholinergic Receptors

The names *nicotinic* and *muscarinic* originated with researchers who found that dangerous environmental toxins bind to these receptor sites. Nicotinic receptors bind *nicotine*, a powerful toxin obtained from a variety of sources, including tobacco leaves. Muscarinic receptors are stimulated by *muscarine*, a toxin produced by some poisonous mushrooms.

These toxins have completely separate actions. Nicotine targets the autonomic ganglia and skeletal neuromuscular junctions. Muscarine acts at the parasympathetic neuromuscular or neuroglandular junctions. They produce dangerously exaggerated, uncontrolled responses due to abnormal stimulation of cholinergic or adrenergic receptors.

Nicotine poisoning occurs if as little as 50 mg of the compound is ingested or absorbed through the skin. (The highest nicotine level in tobacco is about 3 mg per gram.) The signs reflect widespread autonomic activation. They include vomiting, diarrhea, high blood pressure, rapid heart rate, sweating, and profuse salivation. Convulsions may occur because the neuromuscular junctions of the somatic nervous system are also stimulated. In severe cases, the stimulation of nicotinic receptors inside the CNS can lead to coma and death within minutes.

The signs and symptoms of muscarine poisoning are almost entirely restricted to the parasympathetic division. They include salivation, nausea, vomiting, diarrhea, constriction

Table 16–1 Adrenergic and Cholinergic Receptors of the ANS

Receptor	Location	Response	Mechanism
ADRENERGIC			
α_1	Widespread, found in most tissues	Excitation, stimulation of metabolism	Enzyme activation; intracellular release of Ca^{2+}
α_2	Sympathetic neuromuscular or neuroglandular junctions	Inhibition of effector cell	Reduction of cAMP concentration
α_2	Parasympathetic neuromuscular or neuroglandular junctions	Inhibition of neurotransmitter release	Reduction of cAMP concentration
β_1	Heart, kidneys, liver, adipose tissue*	Stimulation, increased energy consumption	Enzyme activation
β_2	Smooth muscle in vessels of heart and skeletal muscle; smooth muscle layers in intestines, lungs, bronchi	Inhibition, relaxation	Enzyme activation
CHOLINERGIC			
Nicotinic	All autonomic synapses between preganglionic and ganglionic neurons; neuromuscular junctions of SNS	Stimulation, excitation; muscular contraction	Opening of chemically gated Na^+ channels
Muscarinic	All parasympathetic and cholinergic sympathetic neuromuscular or neuroglandular junctions	Variable	Enzyme activation causing changes in membrane permeability to K^+

*Adipocytes also contain an additional receptor type, β_3, not found in other tissues. Stimulation of β_3 receptors causes lipolysis.

of respiratory passages, low blood pressure, and an abnormally slow heart rate (bradycardia). Sweating, a sympathetic response, is also prominent. For this reason, *Amanita muscaria*, commonly known as the fly-agaric mushroom, is used in very small amounts in some Native American sweat lodge rituals.

Table 16–1 summarizes details about the adrenergic and cholinergic receptors of the ANS.

✓ Checkpoint

10. **What neurotransmitter is released by all parasympathetic neurons?**
11. **Name the two types of cholinergic receptors on the postsynaptic membranes of parasympathetic neurons.**
12. **How would the stimulation of muscarinic receptors in cardiac muscle affect the heart?**

See the blue Answers tab at the back of the book.

16-6 The differences in the organization of sympathetic and parasympathetic structures lead to widespread sympathetic effects and specific parasympathetic effects

Learning Outcome Compare and contrast the sympathetic and parasympathetic nervous systems.

Figure 16–5 and **Table 16–2** compare key structural features of the sympathetic and parasympathetic divisions of the ANS. The differences in structure lead to differences in functions. The sympathetic division has widespread impact, reaching organs and tissues throughout the body. The parasympathetic division innervates only visceral structures that are supplied by the cranial nerves or that lie within the abdominopelvic cavity.

Figure 16–5 A Summary Comparison of the Sympathetic and Parasympathetic Divisions.

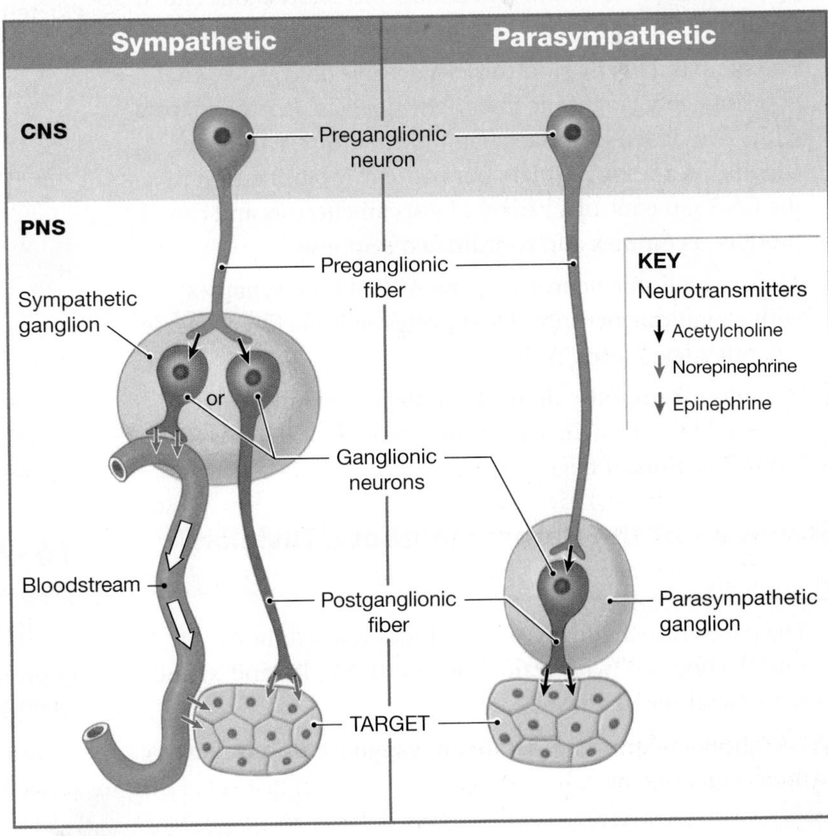
Sympathetic
Parasympathetic
CNS
PNS
Preganglionic neuron
Preganglionic fiber
Sympathetic ganglion
or
Ganglionic neurons
Bloodstream
Postganglionic fiber
TARGET
Parasympathetic ganglion
KEY
Neurotransmitters
Acetylcholine
Norepinephrine
Epinephrine

16

Table 16–2 A Structural Comparison of the Sympathetic and Parasympathetic Divisions of the ANS

Characteristic	Sympathetic Division	Parasympathetic Division
Location of CNS visceral motor neurons	Lateral horns of spinal segments T_1–L_2	Brainstem and spinal segments S_2–S_4
Location of PNS ganglia	Near vertebral column	Typically intramural
Preganglionic fibers		
Length	Relatively short	Relatively long
Neurotransmitter released	Acetylcholine	Acetylcholine
Postganglionic fibers		
Length	Relatively long	Relatively short
Neurotransmitter released	Normally NE; sometimes NO or ACh	Acetylcholine
Neuromuscular or neuroglandular junction	Varicosities and enlarged axonic terminals that release transmitter near target cells	Junctions that release transmitter to special receptor surface
Degree of divergence from CNS to ganglion cells	Approximately 1:32	Approximately 1:6
General function	Stimulates metabolism; increases alertness; prepares for emergency ("fight or flight")	Promotes relaxation, nutrient uptake, energy storage ("rest and digest")

16

Summary of the Sympathetic Division

To summarize:

- The sympathetic division of the ANS includes two sets of sympathetic chain ganglia, one on each side of the vertebral column; three collateral ganglia anterior to the vertebral column; and two adrenal medullae.
- The preganglionic fibers are short, because the ganglia are close to the spinal cord. The postganglionic fibers are longer and extend a considerable distance to their target organs. (In the case of the adrenal medullae, very short axons end at capillaries that carry their secretions to the bloodstream.)
- The sympathetic division shows extensive divergence. That is, a single preganglionic fiber may innervate (form synapses with) two dozen or more ganglionic neurons in different ganglia. As a result, a single sympathetic motor neuron in the CNS can control a variety of visceral effectors and can produce a complex and coordinated response.
- All preganglionic neurons release ACh at their synapses with ganglionic neurons. Most postganglionic fibers release NE, but a few release ACh or NO.
- The effector response depends on the second messengers triggered by G protein activation when NE or E binds to alpha receptors or beta receptors.

Summary of the Parasympathetic Division

In summary:

- The parasympathetic division includes visceral motor nuclei associated with cranial nerves III, VII, IX, and X, and with sacral segments S_2–S_4.
- Ganglionic neurons are located in ganglia within or next to their target organs.
- The parasympathetic division innervates regions serviced by the cranial nerves and organs enclosed by the thoracic and abdominopelvic cavities.
- The parasympathetic division has approximately one-fifth as much divergence as the sympathetic division.
- All parasympathetic neurons are cholinergic. Ganglionic neurons have nicotinic receptors, which are excited by ACh. Muscarinic receptors at neuromuscular or neuroglandular junctions produce either excitation or inhibition, depending on the action of enzymes triggered by G protein activation when ACh binds to the receptor.
- The effects of parasympathetic stimulation are generally brief and restricted to specific organs and sites.

Let's explore next how these different functions provide coordinated responses to changing bodily environments.

✓ Checkpoint

13. **Compare the anatomy of the sympathetic division with that of the parasympathetic division.**
14. **Compare the amount of divergence of the sympathetic division with the parasympathetic division and explain the significance of divergence.**

See the blue Answers tab at the back of the book.

16-7 Dual innervation of organs allows the sympathetic and parasympathetic divisions to coordinate vital functions

Learning Outcome **Discuss the functional significance of dual innervation and autonomic tone.**

Some organs are innervated by either the sympathetic or parasympathetic division, but most vital organs receive **dual innervation,** so they receive instructions from both divisions. Where dual innervation exists, the two divisions commonly have opposing effects. Dual innervation with opposing effects is most obvious in the digestive tract, heart, and lungs. At other sites, the responses may be separate or complementary. **Table 16–3** provides a functional comparison of the two divisions, noting the effects of sympathetic or parasympathetic activity on specific organs and systems. We also discuss *autonomic tone* in this section, the background level of ANS activation, both with and without dual innervation.

Anatomy of Dual Innervation

In the head, parasympathetic postganglionic fibers from the ciliary, pterygopalatine, submandibular, and otic ganglia travel by the cranial nerves to their peripheral destinations. Sympathetic innervation reaches the same structures by traveling directly from the superior cervical ganglia of the sympathetic chain.

In the thoracic and abdominopelvic cavities, the sympathetic postganglionic fibers mingle with parasympathetic preganglionic fibers, forming a series of nerve networks collectively called *autonomic plexuses.* Nerves leaving these networks travel with the blood vessels and lymphatic vessels that supply visceral organs.

Autonomic fibers entering the thoracic cavity intersect at the **cardiac plexus** and the **pulmonary plexus** (**Figure 16–6**). These plexuses contain sympathetic and parasympathetic fibers to the heart and lungs, respectively, as well as the parasympathetic ganglia whose output affects those organs. The **esophageal plexus** contains descending branches of the vagus nerves and splanchnic nerves leaving the sympathetic chain on either side.

Figure 16–6 The Autonomic Plexuses and Ganglia.

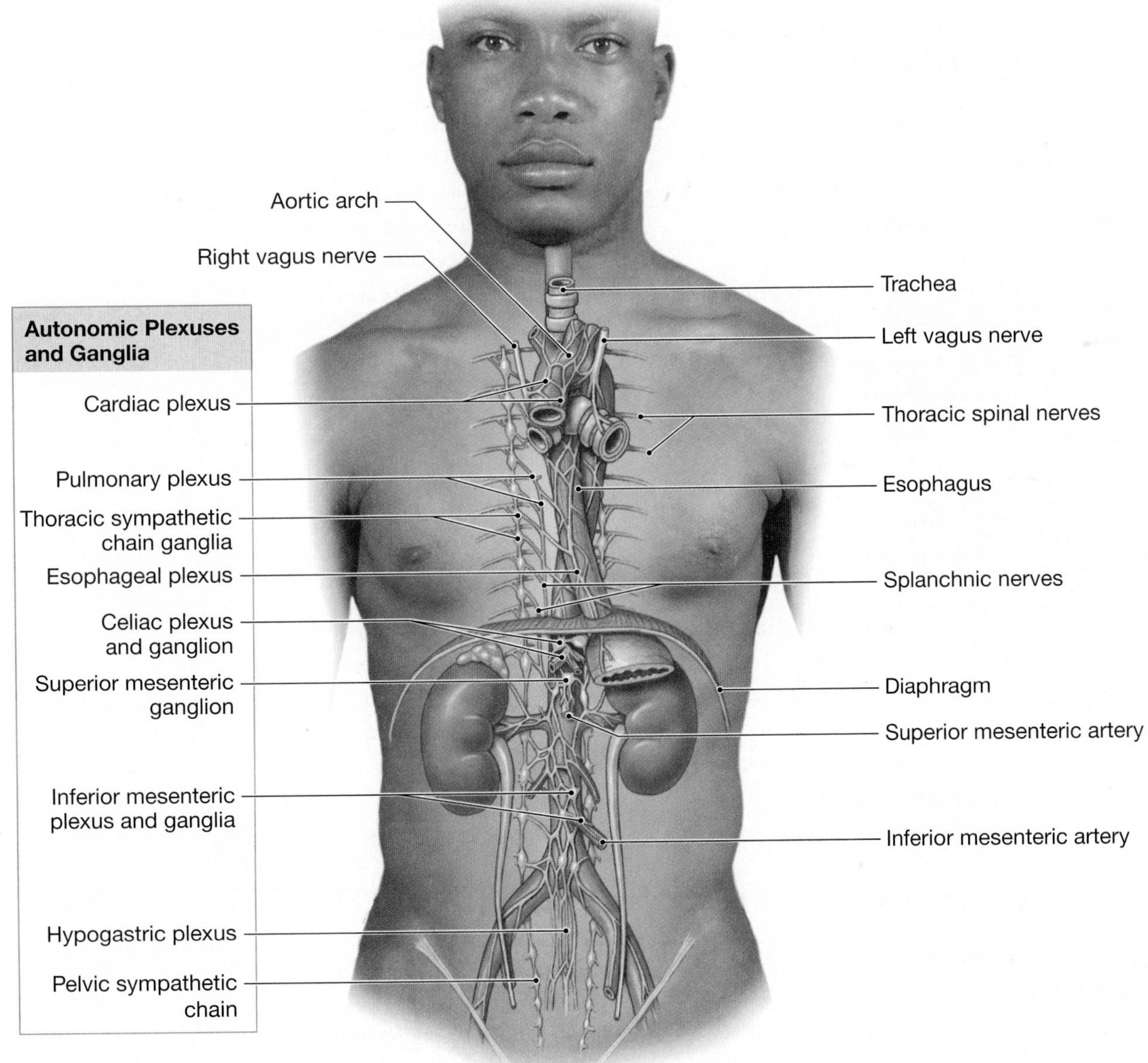
Aortic arch
Right vagus nerve
Autonomic Plexuses and Ganglia
Cardiac plexus
Pulmonary plexus
Thoracic sympathetic chain ganglia
Esophageal plexus
Celiac plexus and ganglion
Superior mesenteric ganglion
Inferior mesenteric plexus and ganglia
Hypogastric plexus
Pelvic sympathetic chain
Trachea
Left vagus nerve
Thoracic spinal nerves
Esophagus
Splanchnic nerves
Diaphragm
Superior mesenteric artery
Inferior mesenteric artery

Table 16–3 A Functional Comparison of the Sympathetic and Parasympathetic Divisions of the ANS

Structure	Sympathetic Effects (receptor or synapse type)	Parasympathetic Effects (all muscarinic receptors)
EYE	Dilation of pupil (α_1); accommodation for distance vision (β_2)	Constriction of pupil; accommodation for close vision
Lacrimal glands	None (not innervated)	Secretion
SKIN		
Sweat glands	Increased secretion, palms and soles (α_1); generalized increase in secretion (cholinergic)	None (not innervated)
Arrector pili muscles	Contraction; erection of hairs (α_1)	None (not innervated)
CARDIOVASCULAR SYSTEM		
Blood vessels		None (not innervated)
To skin	Dilation (β_2 and cholinergic); constriction (α_1)	None (not innervated)
To skeletal muscles	Dilation (β_2 and cholinergic; nitroxidergic)	None (not innervated)
To heart	Dilation (β_2); constriction (α_1, α_2)	None (not innervated)
To lungs	Dilation (α_2); constriction (α_1)	None (not innervated)
To digestive viscera	Constriction (β_2); dilation (α_2)	None (not innervated)
To kidneys	Constriction, decreased urine production (α_1, α_2); dilation, increased urine production (β_1, β_2)	None (not innervated)
To brain	Dilation (cholinergic and nitroxidergic)	None (not innervated)
Veins	Constriction (α_1, β_2)	None (not innervated)
Heart	Increased heart rate, force of contraction, and blood pressure (α_1, β_1)	Decreased heart rate, force of contraction, and blood pressure
ENDOCRINE SYSTEM		
Adrenal gland	Secretion of epinephrine, norepinephrine by adrenal medulla	None (not innervated)
Posterior lobe of pituitary gland	Secretion of ADH (β_1, β_2)	None (not innervated)
Pancreas	Decreased insulin secretion (α_2)	Increased insulin secretion
Pineal gland	Increased melatonin secretion (β_1, β_2)	Inhibition of melatonin synthesis
RESPIRATORY SYSTEM		
Airways	Increased airway diameter (β_2)	Decreased airway diameter
Secretory glands	Mucus secretion (α_1)	None (not innervated)
DIGESTIVE SYSTEM		
Salivary glands	Production of viscous secretion (α_1, β_1) containing mucins and enzymes	Production of copious, watery secretion
Sphincters	Constriction (α_1)	Dilation
General level of activity	Decreased (α_2, β_2)	Increased
Secretory glands	Inhibition (α_2)	Stimulation
Liver	Glycogen breakdown, glucose synthesis and release (α_1, β_2)	Glycogen synthesis
Pancreas	Decreased exocrine secretion (α_1)	Increased exocrine secretion
SKELETAL MUSCLES		
	Increased force of contraction, glycogen breakdown (β_2)	None (not innervated)
	Facilitation of ACh release at neuromuscular junction (α_2)	None (not innervated)
ADIPOSE TISSUE	Lipolysis, fatty acid release (α_1, β_1, β_3)	None (not innervated)
URINARY SYSTEM		
Kidneys	Secretion of renin (β_1)	Uncertain effects on urine production
Urinary bladder	Constriction of internal sphincter; relaxation of urinary bladder (α_1, β_2)	Tensing of urinary bladder, relaxation of internal sphincter to eliminate urine
MALE REPRODUCTIVE SYSTEM	Increased glandular secretion and ejaculation (α_1)	Erection
FEMALE REPRODUCTIVE SYSTEM		
	Increased glandular secretion; contraction of pregnant uterus (α_1)	Variable (depending on hormones present)
	Relaxation of nonpregnant uterus (β_2)	Variable (depending on hormones present)

16

Parasympathetic preganglionic fibers of the vagus nerves enter the abdominopelvic cavity with the esophagus. There the fibers enter the **celiac plexus**, also known as the *solar plexus*. The celiac plexus and associated smaller plexuses, such as the **inferior mesenteric plexus**, innervate viscera within the abdominal cavity (see Figure 16–6). The **hypogastric plexus** innervates the digestive, urinary, and reproductive organs of the pelvic cavity. This plexus contains the parasympathetic outflow of the pelvic nerves, sympathetic postganglionic fibers from the inferior mesenteric ganglion, and splanchnic nerves from the sacral sympathetic chain.

Autonomic Tone

Even without stimuli, autonomic motor neurons show a resting level of spontaneous activity. This background level of activity determines an individual's **autonomic tone**. Autonomic tone is an important aspect of ANS function, just as muscle tone is a key aspect of SNS function. If a nerve is absolutely inactive under normal conditions, then all it can do is increase its activity on demand. But if the nerve maintains a background level of activity, then it can increase or decrease its activity, providing a greater range of control options.

Autonomic tone is significant where dual innervation occurs and the two ANS divisions have opposing effects. It is even more important where dual innervation does not occur. To demonstrate how autonomic tone affects ANS function, let's consider one example of each situation.

The heart receives dual innervation. Recall that the heart consists of cardiac muscle tissue, and that specialized pacemaker cells trigger its contractions. p. 144 The two autonomic divisions have opposing effects on heart function. Acetylcholine, released by postganglionic fibers of the parasympathetic division, causes a reduction in heart rate. Norepinephrine, released by varicosities of the sympathetic division, accelerates heart rate. Small amounts of both of these neurotransmitters are released continuously, creating autonomic tone. However, the parasympathetic division dominates under resting conditions. Heart rate can be controlled very precisely to meet the demands of active tissues through small adjustments in the balance between parasympathetic and sympathetic stimulation. In a crisis, stimulation of the sympathetic innervation and inhibition of the parasympathetic innervation accelerate the heart rate to the maximum extent possible.

The sympathetic control of blood vessel diameter demonstrates how autonomic tone allows fine adjustment of peripheral activities in target organs that are innervated by only one ANS division. Blood flow to specific organs must be controlled to meet the tissue demands for oxygen and nutrients. When a blood vessel dilates, blood flow through it increases; when it constricts, blood flow is reduced. Sympathetic postganglionic fibers release NE at the smooth muscle cells in the walls of peripheral vessels. This background sympathetic tone keeps these muscles partially contracted, so the blood vessels are ordinarily at about half their maximum diameter. When increased blood flow is needed, the rate of NE release decreases and sympathetic cholinergic fibers are stimulated. As a result, the smooth muscle cells relax, the vessels dilate, and blood flow increases. By adjusting sympathetic tone and the activity of cholinergic fibers, the sympathetic division can exert precise control of vessel diameter over its entire range.

Checkpoint

15. **List general responses to increased sympathetic activity and to parasympathetic activity.**
16. **What effect would the loss of sympathetic tone have on blood flow to a tissue?**
17. **What physiological changes would you expect in a patient who is about to undergo a dental root canal and is quite anxious about the procedure?**

See the blue Answers tab at the back of the book.

16-8 Various levels of autonomic regulation allow for the integration and control of autonomic functions

Learning Outcome Describe the hierarchy of interacting levels of control in the autonomic nervous system, including the significance of visceral reflexes.

Recall that centers involved in somatic motor control are found in all portions of the CNS, from lower motor neurons involved in cranial and spinal reflex arcs to the pyramidal motor neurons of the primary motor cortex. Similarly, the ANS is also organized into a series of interacting levels. At the lowest level are visceral motor neurons in the lower brainstem and spinal cord that are involved in cranial and spinal visceral reflexes. **Visceral reflexes** are autonomic reflexes initiated in the viscera. As in the SNS, ANS simple reflexes based in the spinal cord respond rapidly and automatically to stimuli. At the highest level, centers in the brainstem that regulate specific visceral functions control the levels of activity of the sympathetic and parasympathetic divisions.

Visceral Reflexes

Visceral reflexes provide automatic motor responses that can be modified, facilitated, or inhibited by higher centers, especially those of the hypothalamus. For example, when a light is shone in one of your eyes, a visceral reflex constricts the pupils of *both* eyes (the *consensual light reflex*). The visceral motor commands are distributed by parasympathetic fibers. In darkness, your pupils dilate, but this *pupillary reflex* is directed by sympathetic fibers. However, the motor nuclei directing pupillary constriction or dilation are also controlled by hypothalamic centers concerned with emotional states. For example, when you are queasy or nauseated, your pupils constrict. When you are sexually aroused, your pupils dilate.

Figure 16–7 **Visceral Reflexes.** Visceral reflexes have the same basic components as somatic reflexes, but all visceral reflexes are polysynaptic. Note that short visceral reflexes bypass the CNS altogether.

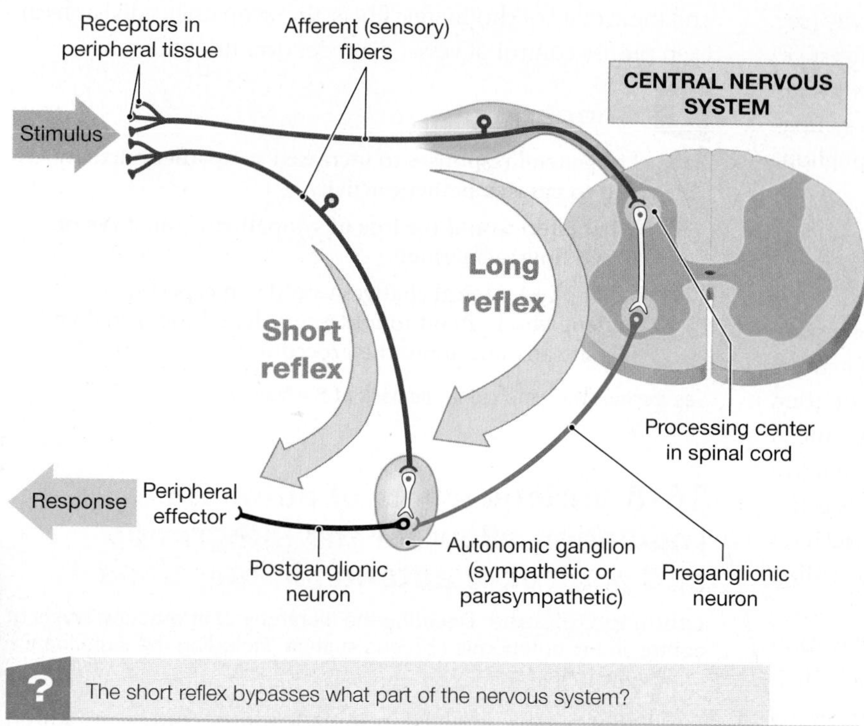
Receptors in peripheral tissue
Afferent (sensory) fibers
CENTRAL NERVOUS SYSTEM
Stimulus
Long reflex
Short reflex
Processing center in spinal cord
Response
Peripheral effector
Autonomic ganglion (sympathetic or parasympathetic)
Postganglionic neuron
Preganglionic neuron

? The short reflex bypasses what part of the nervous system?

Each **visceral reflex arc** consists of a receptor, a sensory neuron, a processing center (one or more interneurons), and two visceral motor neurons (Figure 16–7). All visceral reflexes are polysynaptic. They can be *long reflexes* or *short reflexes*.

Long reflexes are the autonomic equivalents of the polysynaptic reflexes introduced in Chapter 13. p. 459 Visceral sensory neurons deliver information to the CNS. It travels along the posterior roots of spinal nerves, within the sensory branches of cranial nerves, and within the autonomic nerves that innervate visceral effectors. The processing steps involve interneurons within the CNS. The ANS then carries the motor commands to the appropriate visceral effectors. Long reflexes typically coordinate the activities of an entire organ.

Short reflexes bypass the CNS entirely. They involve sensory neurons and interneurons whose cell bodies lie in autonomic ganglia. These interneurons synapse on postganglionic neurons, and their postganglionic fibers distribute the motor commands. Short reflexes control very simple motor responses with localized effects. In general, short reflexes may control patterns of activity in one small part of a target organ.

In most organs, long reflexes are most important in regulating visceral activities, but this is not the case with the digestive tract and its associated glands. In these areas, short reflexes provide most of the control and coordination for normal functioning. The neurons involved form the enteric nervous system, introduced previously. p. 391 The ganglia in the walls of the digestive tract contain the cell bodies of visceral sensory neurons, interneurons, and visceral motor neurons, and all their axons form extensive nerve nets. Parasympathetic innervation by visceral motor neurons can stimulate and coordinate various digestive activities, but the enteric nervous system is capable of controlling digestive functions independent of the CNS. We consider the functions of the enteric nervous system further in Chapter 24.

As we examine other body systems in later chapters, you will come across many examples of autonomic reflexes involved in respiration, cardiovascular function, and other visceral activities. Look at Table 16–4 to preview some of the most important ones. Notice that the parasympathetic division participates in a variety of reflexes that affect individual organs and systems. This specialization reflects its relatively restricted pattern of innervation. In contrast, fewer sympathetic reflexes exist. The sympathetic division is typically activated as a whole. One reason is that it has such a high degree of divergence. Another reason is that the release of hormones by the adrenal medullae produces widespread peripheral effects.

Higher Levels of Autonomic Control

Processing centers in the medulla oblongata of the brainstem coordinate more complex sympathetic and parasympathetic reflexes. In addition to the cardiovascular and respiratory centers, the medulla oblongata contains centers and nuclei involved with salivation, swallowing, digestive secretions, peristalsis, and urinary function. These centers are in turn subject to regulation by the hypothalamus. p. 482

The term *autonomic* was originally applied because the regulatory centers involved with the control of visceral function were thought to operate autonomously—that is, independent of other CNS activities. We now know that because the hypothalamus interacts with all other areas of the brain, activity in the limbic system, thalamus, or cerebral cortex can have dramatic effects on autonomic function. For example, what happens when you become angry? Your heart rate accelerates, your blood pressure rises, and your respiratory rate increases. What happens when you think about your next meal? Your stomach "growls" and your mouth waters.

The Integration of ANS and SNS Activities

Figure 16–8 and Table 16–5 show how the activities of the autonomic nervous system and the somatic nervous system are integrated. We have considered visceral and somatic motor

Table 16–4 Representative Visceral Reflexes

Reflex	Stimulus	Response	Comments
PARASYMPATHETIC REFLEXES			
Gastric and intestinal reflexes *(Chapter 24)*	Pressure and physical contact	Smooth muscle contractions that propel food materials and mix with secretions	Initiated by vagus nerves and controlled locally by ENS
Defecation *(Chapter 24)*	Distention of rectum	Relaxation of internal anal sphincter	Requires voluntary relaxation of external anal sphincter
Urination *(Chapter 26)*	Distention of urinary bladder	Contraction of walls of urinary bladder; relaxation of internal urethral sphincter	Requires voluntary relaxation of external urethral sphincter
Consensual light response *(Chapter 14)*	Bright light shining in eye(s)	Constriction of pupils of both eyes	
Swallowing reflex *(Chapter 24)*	Movement of food and liquids into pharynx	Smooth muscle and skeletal muscle contractions	Coordinated by medullary swallowing center
Coughing reflex *(Chapter 23)*	Irritation of respiratory tract	Sudden explosive ejection of air	Coordinated by medullary coughing center
Baroreceptor reflex *(Chapters 17, 20, 21)*	Sudden rise in carotid blood pressure	Reduction in heart rate and force of contraction	Coordinated in cardiac centers of medulla oblongata
Sexual arousal *(Chapter 28)*	Erotic stimuli (visual or tactilo)	Increased glandular secretions, sensitivity, erection	
SYMPATHETIC REFLEXES			
Cardioacceleratory reflex *(Chapter 21)*	Sudden decline in blood pressure in carotid artery	Increase in heart rate and force of contraction	Coordinated in cardiac centers of medulla oblongata
Vasomotor reflexes *(Chapter 21)*	Changes in blood pressure in major arteries	Changes in diameter of peripheral vessels	Coordinated in vasomotor center in medulla oblongata
Pupillary light reflex *(Chapter 17)*	Light striking photoreceptors in one eye	Dilation of pupil	
Ejaculation (in males) *(Chapter 28)*	Erotic stimuli (tactile)	Skeletal muscle contractions ejecting semen	

Table 16–5 A Comparison of the ANS and SNS

Characteristic	ANS	SNS
Innervation	Visceral effectors, including cardiac muscle, smooth muscle, glands, adipocytes	Skeletal muscles
Activation	In response to sensory stimuli or from commands of higher centers	In response to sensory stimuli or from commands of higher centers
Relay and processing centers	Brainstem	Brainstem and thalamus
Headquarters	Hypothalamus	Cerebral cortex
Feedback received from	Limbic system and thalamus	Cerebellum and basal nuclei
Control method	Adjustment of activity in brainstem processing centers that innervate preganglionic neurons	Direct (corticospinal) and indirect (medial and lateral) pathways that innervate lower motor neurons
Reflexes	Polysynaptic (short and long)	Monosynaptic and polysynaptic (always long)

pathways separately, but the two have many parallels, in terms of both organization and function. Integration takes place at the level of the brainstem, and both systems are under the influence of higher centers.

✓ Checkpoint

18. **Define *visceral reflex*.**
19. **Luke has a brain tumor that is interfering with the function of his hypothalamus. Explain why this tumor would interfere with autonomic function.**

See the blue Answers tab at the back of the book.

16-9 Higher-order functions include memory and states of consciousness, and neurotransmitters influence behavior

Learning Outcome **Explain how memories are created, stored, and recalled; distinguish among the levels of consciousness and unconsciousness; and describe how neurotransmitters influence brain function.**

Higher-order functions share three characteristics:

- The cerebral cortex is required for their performance. They involve complex interactions among areas of the cortex and between the cortex and other areas of the brain.

Figure 16–8 **A Comparison of Somatic and Autonomic Function.** The SNS and ANS have parallel organization and are integrated at the level of the brainstem. Blue arrows indicate ascending sensory information; red arrows, descending motor commands; dashed lines indicate pathways of communication and feedback among higher centers.

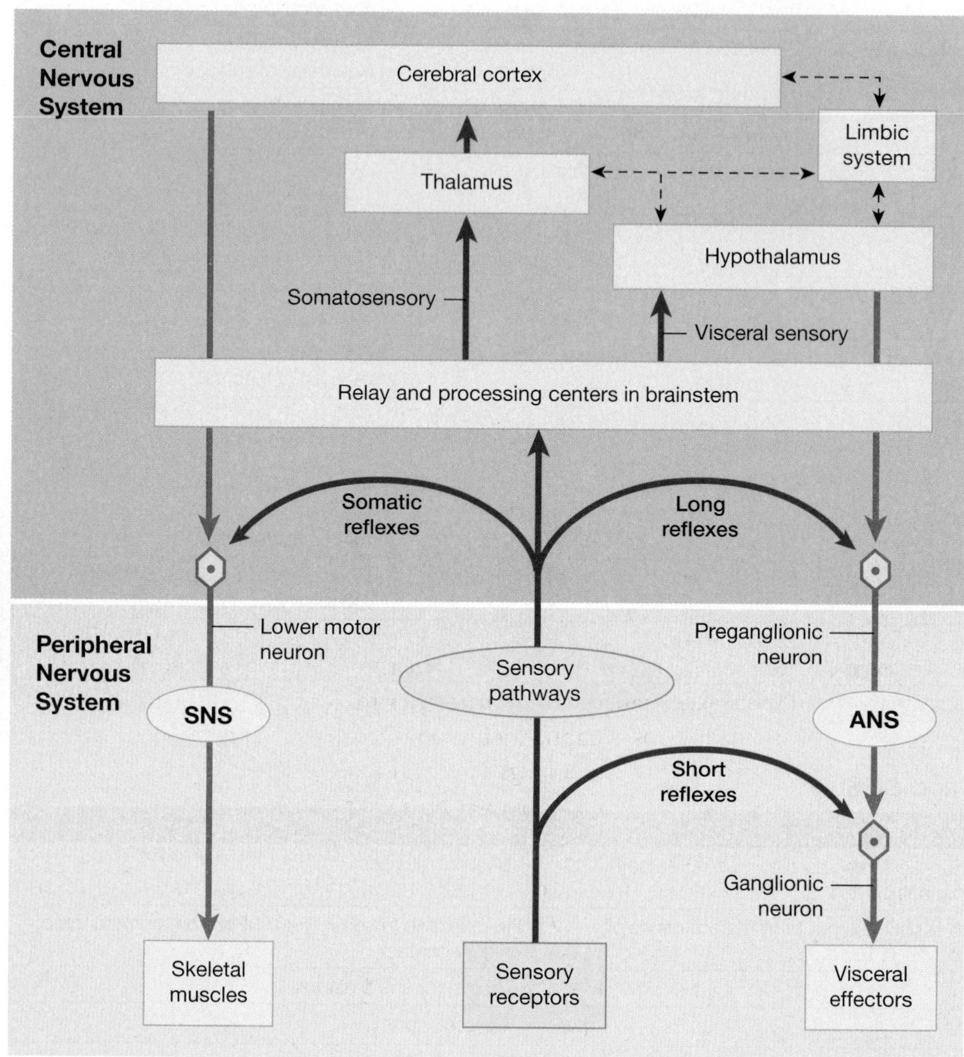

Central Nervous System
Cerebral cortex
Limbic system
Thalamus
Hypothalamus
Somatosensory
Visceral sensory
Relay and processing centers in brainstem
Somatic reflexes
Long reflexes
Peripheral Nervous System
Lower motor neuron
Preganglionic neuron
Sensory pathways
SNS
ANS
Short reflexes
Ganglionic neuron
Skeletal muscles
Sensory receptors
Visceral effectors

- They involve both conscious and unconscious information processing.
- Higher-order functions are subject to adjustment over time. They are not innate (inborn), fixed reflexive behaviors.

In Chapter 14, we considered functional areas of the cerebral cortex and also hemispheric lateralization. pp. 490–492 In this section, we consider the mechanisms of memory and learning. We also describe the neural interactions responsible for consciousness, sleep, and arousal.

Memory

What was the topic of the last sentence you read? What is your Social Security number? What does a hot dog taste like? To answer these questions, you access **memories**, stored bits of information gathered through experience. **Fact memories** are specific bits of information, such as the color of a stop sign or the smell of a perfume. **Skill memories** are learned motor behaviors. You can probably remember how to light a match or open a screw-top jar, for example. With repetition, skill memories become incorporated at the unconscious level. Skill memories related to innate behaviors, such as eating, are stored in appropriate portions of the brainstem. Complex skill memories, such as how to ski or play the violin, involve the integration of motor patterns in the basal nuclei, cerebral cortex, and cerebellum.

Two classes of memories are recognized, short term and long term (Figure 16–9). **Short-term memories** do not last long, but while they persist the information can be recalled immediately. Short-term memories contain small bits of information, such as a person's name or a telephone number. Repeating a phone number or other bit of information reinforces the original short-term memory and allows it to be converted to a long-term memory.

Long-term memories last much longer, in some cases for an entire lifetime. The conversion from short-term to long-term memory is called **memory consolidation**. There are two types of long-term memory: (1) *Secondary memories* are long-term memories that fade with time and may require considerable effort to recall. (2) *Tertiary memories* are long-term memories that are with you for a lifetime, such as your name or the contours of your own body.

Brain Regions Involved in Memory Consolidation and Access

The amygdaloid body (amygdala) and the hippocampus, two components of the limbic system (look back at Figure 14–12, p. 485), are essential to memory consolidation. Damage to the hippocampus leads to an inability to convert short-term memories to new long-term memories, although existing long-term memories remain intact and accessible. Tracts leading from the amygdaloid body to the hypothalamus may link memories to specific emotions.

The **nucleus basalis**, a cerebral nucleus near the diencephalon, plays an uncertain role in memory storage and retrieval. Tracts connect this nucleus with the hippocampus,

Figure 16–9 **Memory Storage.** Steps in the storage of memories and the conversion from short-term memory to long-term memory.

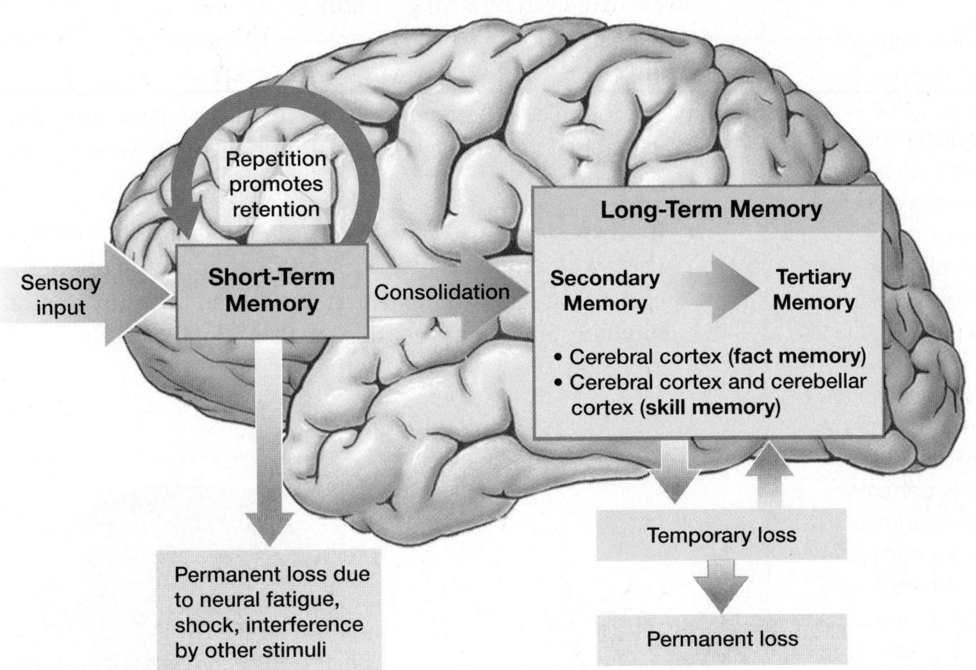
Repetition promotes retention
Sensory input
Short-Term Memory
Consolidation
Long-Term Memory
Secondary Memory
Tertiary Memory
• Cerebral cortex (fact memory)
• Cerebral cortex and cerebellar cortex (skill memory)
Permanent loss due to neural fatigue, shock, interference by other stimuli
Temporary loss
Permanent loss

amygdaloid body, and all areas of the cerebral cortex. Damage to this nucleus is associated with changes in emotional states, memory, and intellectual function (as we see in the discussion of Alzheimer's disease later in this chapter).

Most long-term memories are stored in the cerebral cortex. Conscious motor and sensory memories are referred to the appropriate association areas. For example, visual memories are stored in the visual association area, and memories of voluntary motor activity are stored in the premotor cortex. Special portions of the occipital and temporal lobes are crucial to the memories of faces, voices, and words.

In at least some cases, a specific memory probably depends on the activity of a single neuron. For example, in one portion of the temporal lobe an individual neuron responds to the sound of one word and ignores others. A specific neuron may also be activated by the proper combination of sensory stimuli associated with a particular individual, such as your grandmother. As a result, these neurons are called "grandmother cells."

Information on one subject is parceled out to many different regions of the brain. Your memories of cows are stored in the visual association area (what a cow looks like, that the letters *c-o-w* mean "cow"), the auditory association area (the "moo" sound and how the word *cow* sounds), the speech center (how to say the word *cow*), and the frontal lobes (how big cows are, what they eat). Related information, such as how you feel about cows and what milk tastes like, is stored in other locations. If one of those storage areas is damaged, your memory will be incomplete in some way. How these memories are accessed and assembled on demand remains an area of active research.

Cellular Mechanisms of Memory Formation and Storage

Memory consolidation at the cellular level involves anatomical and physiological changes in neurons and synapses. For legal, ethical, and practical reasons, scientists do not conduct much research on these mechanisms with human subjects. Research on other animals, commonly those with relatively simple nervous systems, has indicated that the following mechanisms may be involved:

- *Increased Neurotransmitter Release.* A synapse that is frequently active increases the amount of neurotransmitter it stores, and it releases more with each stimulation. The more neurotransmitter released, the greater the effect on the postsynaptic neuron.
- *Facilitation at Synapses.* When a neural circuit is repeatedly activated, the axon terminals begin continuously releasing neurotransmitter in small quantities. The neurotransmitter binds to receptors on the postsynaptic membrane, producing a graded depolarization that brings the membrane closer to threshold. The facilitation that results affects all neurons in the circuit.
- *The Formation of Additional Synaptic Connections.* When one neuron repeatedly communicates with another, the axon tip branches and forms additional synapses on the postsynaptic neuron. As a result, stimulation of the presynaptic neuron has a greater effect on the membrane potential of the postsynaptic neuron.

Such processes create anatomical changes that facilitate communication along a specific neural circuit. This facilitated communication is thought to be the basis of memory storage. A single circuit that corresponds to a single memory has been called a **memory engram**. This definition is based on function rather than structure. We know too little about the organization and storage of memories to be able to describe the neural circuits involved. Memory engrams form as the result of experience and repetition. Repetition is crucial—that's why you probably need to read these chapters more than once before an exam.

Efficient conversion of a short-term memory into a memory engram takes time, usually at least an hour. Whether that conversion will occur depends on several factors. They include the nature, intensity, and frequency of the original stimulus. Very strong, repeated, or exceedingly pleasant or unpleasant events are most likely to be converted to long-term memories. Drugs that stimulate the CNS, such as caffeine and nicotine, may enhance

+ Clinical Note Insomnia

Insomnia (literally, "no sleep") includes trouble falling asleep, frequent awakenings with trouble falling back to sleep, and poor sleep quality. One in ten adults in the United States (more commonly females and older adults) suffers from chronic insomnia. This condition is more than a mere nuisance. Insomnia results in physical and mental disorders, job and relationship dysfunction, and diminished quality of life. The insomniac also suffers from constant hyperarousal, including an increased heart rate, elevated blood pressure, increased muscle tone, and disturbed sleep pattern, due to an overactive sympathetic nervous system.

Insomnia can be treated both medically and psychologically. An overnight sleep study can be used to collect objective data about sleep habits and behaviors. In such a study, sleep technicians record brain activity patterns. Psychologists use the "3P" model to evaluate insomnia: **p**redisposing factors (such as genetics and anxious personality type), **p**recipitating factors (such as a stressful life event), and **p**erpetuating factors (such as substance abuse and poor preparation for sleep). All of this information is then incorporated into a treatment plan.

memory consolidation through facilitation. We discussed the membrane effects of those drugs in Chapter 12. pp. 423, 425

The hippocampus plays a key role in consolidating memories. The mechanism remains unknown, but it is linked to the presence of *NMDA* (N-methyl D-aspartate) *receptors,* which are chemically gated calcium ion channels. When activated by the neurotransmitter *glutamate,* the gates open and calcium ions enter the cell. Blocking NMDA receptors in the hippocampus prevents long-term memory formation.

States of Consciousness

The difference between a conscious individual and an unconscious one might seem obvious. A conscious individual is alert and attentive, and an unconscious individual is not. Yet there are many degrees of both states. *Conscious* implies an awareness of and attention to external events and stimuli, but a healthy conscious person can be nearly asleep, wide awake, or high-strung and jumpy. *Unconscious* can refer to conditions ranging from the deep, unresponsive state induced by anesthesia before major surgery, to deep sleep, to the light, drifting "nod" that occasionally plagues students who are reading anatomy and physiology textbooks.

A person's degree of wakefulness at any moment indicates the level of ongoing CNS activity. When you are asleep, you are unconscious but can still be awakened by normal sensory stimuli. Healthy individuals cycle between the alert, conscious state and sleep each day. When CNS function becomes abnormal or depressed, the state of wakefulness can be affected. An individual in a *coma,* for example, is unconscious and cannot be awakened, even by strong stimuli.

Sleep

We recognize two general levels of sleep, *deep* and *REM sleep,* each with characteristic patterns of brain wave activity (**Figure 16–10a**):

- In **deep sleep**, also called *slow wave* or *non-REM (NREM) sleep,* your entire body relaxes, and activity at the cerebral cortex is at a minimum. Heart rate, blood pressure, respiratory rate, and energy use decline by up to 30 percent.

Figure 16–10 States of Sleep.

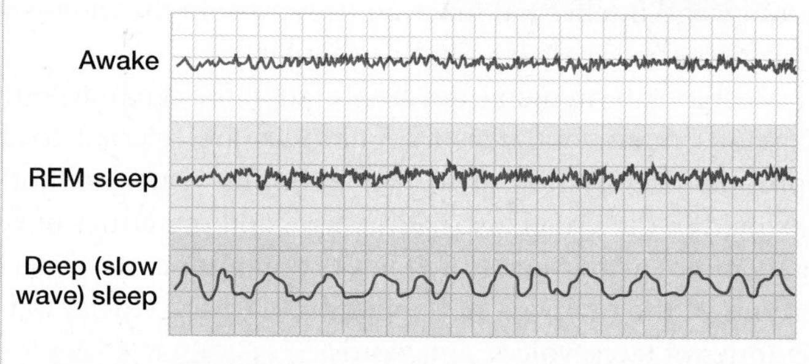
Awake
REM sleep
Deep (slow wave) sleep

a An EEG of the various sleep states. The EEG pattern during REM sleep resembles the alpha waves typical of awake adults.

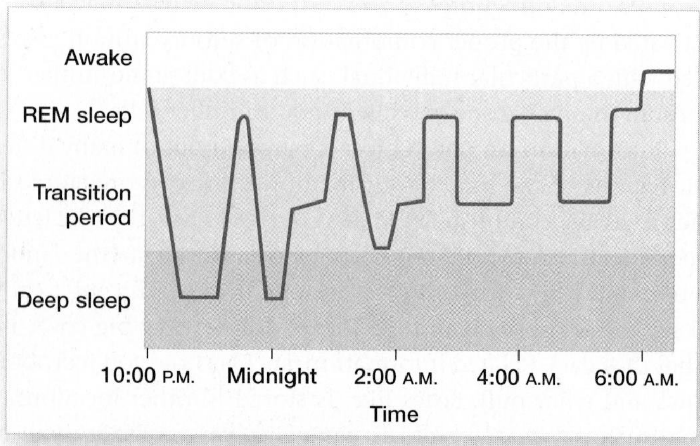
Awake
REM sleep
Transition period
Deep sleep
10:00 P.M.
Midnight
2:00 A.M.
4:00 A.M.
6:00 A.M.
Time

b Typical pattern of sleep stages in a healthy young adult during a single night's sleep.

- During **rapid eye movement (REM) sleep**, you dream actively and your blood pressure and respiratory rate change. Although an EEG taken during REM sleep resembles that of the awake state, you become even less receptive to outside stimuli than in deep sleep, and your muscle tone decreases markedly. p. 493 Intense inhibition of somatic motor neurons probably prevents you from acting out the responses you envision while dreaming. The neurons controlling your eye muscles escape this inhibitory influence, and your eyes move rapidly as dream events unfold.

Periods of REM and deep sleep alternate throughout the night, beginning with deep sleep for about an hour and a half (**Figure 16–10b**). REM periods initially average about 5 minutes, but over an 8-hour night they gradually increase to about 20 minutes. Each night we probably spend less than 2 hours dreaming, but variation among individuals is significant. For example, children spend more time in REM sleep than do adults, and extremely tired individuals have very short and infrequent REM periods.

Sleep produces only minor changes in the physiological activities of other organs and systems, and none of these changes appear to be essential to normal function. The significance of sleep must lie in its impact on the CNS, but the physiological or biochemical basis remains to be determined. We do know that protein synthesis in neurons increases during sleep.

Extended periods without sleep lead to a variety of disturbances in mental function. Roughly 25 percent of the U.S. population experiences some form of *sleep disorder*. Examples include abnormal patterns or duration of REM sleep or unusual behaviors during sleep, such as sleepwalking. In some cases, these problems affect the individual's conscious activities. Slowed reaction times, irritability, and behavioral changes may result. Memory loss has also been linked to sleep disorders.

Arousal and the Reticular Activating System

Arousal, or awakening from sleep, appears to be one of the functions of the reticular formation. The reticular formation is especially well suited for providing "watchdog" services, because it has extensive interconnections with the sensory, motor, and integrative nuclei and pathways all along the brainstem.

Your state of consciousness results from complex interactions between the reticular formation and the cerebral cortex. One of the most important brainstem components is the **reticular activating system (RAS)**. p. 478 This diffuse network in the reticular formation extends from the medulla oblongata to the midbrain (**Figure 16–11**). The output of the RAS projects to thalamic nuclei that influence large areas of the cerebral cortex. When the RAS is inactive, so is the cerebral cortex, and stimulation of the RAS produces a widespread activation of the cerebral cortex.

The midbrain portion of the RAS appears to be the "headquarters" of the system. Stimulating this area produces the most pronounced and long-lasting effects on the cerebral cortex. Stimulating other portions of the RAS seems to have an effect only to the degree that it changes the activity of the midbrain region. The greater the stimulation to the midbrain region of the RAS, the more alert and attentive the individual will be to incoming sensory information. The thalamic nuclei associated with the RAS may also play an important role in focusing attention on specific mental processes.

Sleep may be ended by any stimulus sufficient to activate the reticular formation and RAS. Arousal occurs rapidly, but the effects of a single stimulation of the RAS last less than a minute. After that, consciousness can be maintained by positive feedback, because activity in the cerebral cortex, basal nuclei, and sensory and motor pathways will continue to stimulate the RAS.

After many hours of activity, the reticular formation becomes less responsive to stimulation. You become less alert and more lethargic. The precise mechanism remains unknown, but neural fatigue probably plays a relatively minor role in the decreasing RAS activity. Evidence suggests that the regulation of sleep–wake cycles involves an interplay between brainstem nuclei that use different neurotransmitters. One group of nuclei stimulates the RAS with norepinephrine and maintains the awake, alert state. Another group depresses RAS activity with serotonin, promoting deep sleep. These "dueling" nuclei are located in the brainstem. 16

Figure 16–11 The Reticular Activating System (RAS). The midbrain "headquarters" of the reticular formation receives collateral inputs from a variety of sensory pathways. Stimulation of this region produces arousal and heightened states of attentiveness.

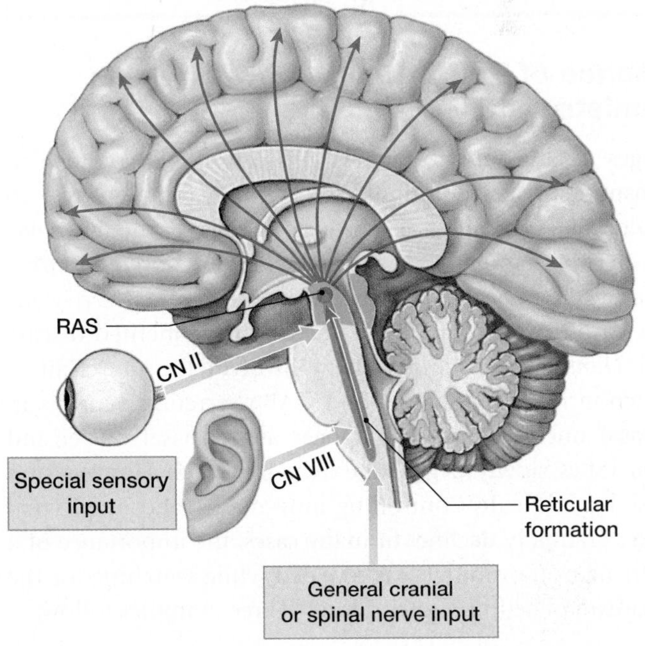
RAS
CN II
CN VIII
Special sensory input
Reticular formation
General cranial or spinal nerve input

? Special sensory inputs arrive by which cranial nerves?

Clinical Note Summary of Nervous System Disorders

Nervous tissue is extremely delicate. The characteristics of its extracellular environment must be kept within narrow homeostatic limits. Signs and symptoms of neurological disorders appear when homeostatic regulatory mechanisms break down. This may happen under the stress of genetic or environmental factors, infection, or trauma.

Literally hundreds of disorders affect the nervous system. They can be categorized as:

- *Infections*, which include diseases such as rabies, meningitis, and polio
- *Congenital disorders*, such as spina bifida and hydrocephalus
- *Degenerative disorders*, such as Parkinson's disease and Alzheimer's disease
- *Tumors* of neural origin
- *Trauma*, such as spinal cord injuries and concussions
- *Toxins*, such as heavy metals and the neurotoxins found in certain seafoods
- *Secondary disorders*, which are problems resulting from dysfunction in other systems, for example, strokes and several demyelination disorders.

A standard physical examination includes a neurological component. The physician checks the general status of the CNS and PNS. In a *neurological examination*, a physician attempts to trace the location of a specific problem by evaluating the sensory, motor, behavioral, and cognitive functions of the nervous system.

Influence of Neurotransmitters on Brain Chemistry and Behavior

Changes in the normal balance between two or more neurotransmitters can profoundly affect brain function and so behavior. For example, as we just discussed, the interplay between populations of neurons releasing norepinephrine and serotonin appears to be involved in regulating sleep–wake cycles. Another example is provided by the inherited disease called *Huntington's disease,* which involves the unexplained destruction of ACh-secreting and GABA-secreting neurons in the basal nuclei. Symptoms appear as the basal nuclei and frontal lobes slowly degenerate. A person with Huntington's disease has difficulty controlling movements, and intellectual abilities gradually decline. In many cases, the importance of a specific neurotransmitter was revealed while searching for the mechanism of action of other drugs. Three examples follow.

- *Serotonin.* An extensive network of tracts delivers serotonin to nuclei and higher centers throughout the brain, and variations in the serotonin level affect sensory interpretation and emotional states. Compounds that enhance the effects of serotonin produce hallucinations. For instance, *lysergic acid diethylamide (LSD)* is a powerful hallucinogenic drug that activates serotonin receptors in the brainstem, hypothalamus, and limbic system. Compounds that merely enhance the effects of serotonin also produce hallucinations, whereas compounds that inhibit serotonin production or block its action cause severe depression and anxiety. An effective antidepressive drug now in widespread use, *fluoxetine (Prozac)*, slows the removal of serotonin at synapses, causing an increase in the serotonin concentration at the postsynaptic membrane. Such drugs are classified as selective serotonin reuptake inhibitors (SSRIs). Other important SSRIs include *Celexa, Luvox, Paxil,* and *Zoloft.*
- *Norepinephrine.* Norepinephrine (NE) has pathways throughout the brain. Drugs that stimulate NE release cause exhilaration, while those that depress its release cause depression. One inherited form of depression has been linked to a defective enzyme involved in NE synthesis.
- *Dopamine.* Disturbances in dopamine transmission have been linked to several neurological disorders. We have already seen that inadequate dopamine causes the motor problems of Parkinson's disease. p. 478 Excessive production of dopamine may be associated with *schizophrenia*, a psychological disorder marked by pronounced disturbances of mood, thought patterns, and behavior. Amphetamines, or "speed," stimulate dopamine secretion and, in large doses, can produce symptoms resembling those of schizophrenia. Dopamine is thus important not only in the nuclei involved in the control of intentional movements, but in many other centers of the diencephalon and cerebrum.

Checkpoint

20. **List three characteristics of higher-order functions.**
21. **As you recall facts while you take your A&P test, which type of memory are you using?**
22. **What would happen if your RAS were suddenly stimulated while you were sleeping?**
23. **What would be an effect of a drug that substantially increases the amount of serotonin released in the brain?**
24. **Amphetamines stimulate the secretion of which neurotransmitter?**

See the blue Answers tab at the back of the book.

16-10 Aging produces various structural and functional changes in the nervous system

Learning Outcome **Summarize the effects of aging on the nervous system and give examples of interactions between the nervous system and other organ systems.**

The aging process affects all body systems, and the nervous system is no exception. Anatomical and physiological changes probably begin by age 30 and accumulate over time. An estimated 85 percent of people above age 65 lead relatively normal lives, but they exhibit noticeable changes in mental performance and in CNS function. Common age-related anatomical changes in the nervous system include the following:

- *Reduction in Brain Size and Weight.* This reduction results primarily from a decrease in the volume of the cerebral cortex. The brains of elderly individuals have narrower gyri and wider sulci than do those of young people, and the subarachnoid space is larger.
- *Reduction in the Number of Neurons.* Brain shrinkage has been linked to a loss of cortical neurons, although evidence indicates that neurons are not lost (at least to the same degree) in brainstem nuclei.
- *Decrease in Blood Flow to the Brain.* With age, fatty deposits gradually build up in the walls of blood vessels. Just as a clog in a drain reduces water flow, these deposits reduce the rate of blood flow through arteries. (This process, called *arteriosclerosis,* affects arteries throughout the body, as we discuss further in Chapter 21.) Even if the reduction in blood flow is not large enough to damage neurons, arteriosclerosis increases the chances that the affected vessel wall will rupture, damaging the surrounding nervous tissue and producing signs and symptoms of a *cerebrovascular accident (CVA)*, also called a *stroke*.
- *Changes in the Synaptic Organization of the Brain.* In many areas, the number of dendritic branches, spines, and interconnections appears to decrease. Synaptic connections are lost, and the rate of neurotransmitter production declines.
- *Intracellular and Extracellular Changes in CNS Neurons.* Many neurons in the brain accumulate abnormal intracellular deposits, including lipofuscin and neurofibrillary tangles. **Lipofuscin** is a granular pigment with no known function. **Neurofibrillary tangles** are masses of neurofibrils that form dense mats inside the cell body and axon. **Plaques** are extracellular accumulations of fibrillar proteins, surrounded by abnormal dendrites and axons. Both plaques and tangles contain deposits of several peptides—primarily two forms of **amyloid (A) protein**, fibrillar and soluble. They appear in brain regions such as the hippocampus, specifically associated with memory processing. Their significance is unknown. Evidence indicates that they appear in all aging brains. In excess, they seem to be associated with clinical abnormalities.

The anatomical changes associated with aging of the nervous system affect all neural functions. For example, memory consolidation typically becomes more difficult, and secondary memories, especially those of the recent past, become harder to access. The sensory systems of elderly people—notably, hearing, balance, vision, smell, and taste—become less sensitive. Lights must be brighter, sounds louder, and smells stronger before they are perceived. Reaction rates are slowed, and reflexes—even some withdrawal reflexes—weaken or disappear. The precision of motor control decreases, and it takes longer for an elderly person to perform a given motor pattern than it did 20 years earlier.

For roughly 85 percent of the elderly population, these changes do not interfere with their abilities to function in society. But for reasons yet unknown, some elderly individuals become incapacitated by progressive CNS changes. These degenerative changes, which can include memory loss, anterograde amnesia, and emotional disturbances, are often lumped together as **senile dementia**, or *senility*. By far the most common and disabling form of senile dementia is **Alzheimer's disease**.

The relationships between the nervous system and the integumentary, skeletal, and muscular systems are shown in Build Your Knowledge Figure 16–12.

Checkpoint

25. **What are some possible reasons for the slower recall and loss of memory that occur with age?**
26. **Identify several common age-related anatomical changes in the nervous system.**
27. **Name the most common form of senile dementia.**
28. **Identify the relationships between the nervous system and the body systems studied so far.**

See the blue Answers tab at the back of the book.

16

Clinical Note Fainting

A stressful or emotional event may trigger a sudden and overwhelming loss of consciousness, commonly known as *fainting*. The technical term for fainting is the *vasovagal response* (*vaso*, vascular; *vagal*, vagus nerve).

This response is the result of a momentary malfunction of the sympathetic division, causing the parasympathetic division (sometimes called the "faint and freeze" division) to go into overdrive. In this process, parasympathetic stimulation of the vagus nerve causes it to release ACh at the cardiac plexus. However, without any counteracting stimulation from the sympathetic nervous system, the heart rate slows, the force of contractions decreases, and blood pressure drops. As a result, blood flow to the brain decreases, causing the fainting episode.

Rapid recovery begins as soon as the sympathetic division takes over, releasing norepinephrine at the cardiac plexus and throughout the body. Heart rate and blood pressure increase, as does level of consciousness.

To prevent a vasovagal response, a person beginning to feel faint should lie down before he or she falls down. This action will deliver more blood flow to the brain and help prevent an injury.

Build Your Knowledge

Figure 16–12 Integration of the NERVOUS system with the other body systems presented so far.

Integumentary System

- The Integumentary System provides sensations of touch, pressure, pain, vibration, and temperature; hair provides some protection and insulation for skull and brain; protects peripheral nerves
- The nervous system controls the contraction of arrector pili muscles and the secretion of sweat glands

Nervous System

The nervous system is your most complex organ system. It:

- monitors the body's internal and external environments
- integrates sensory information
- directs immediate responses to stimuli by coordinating voluntary and involuntary responses of many other organ systems

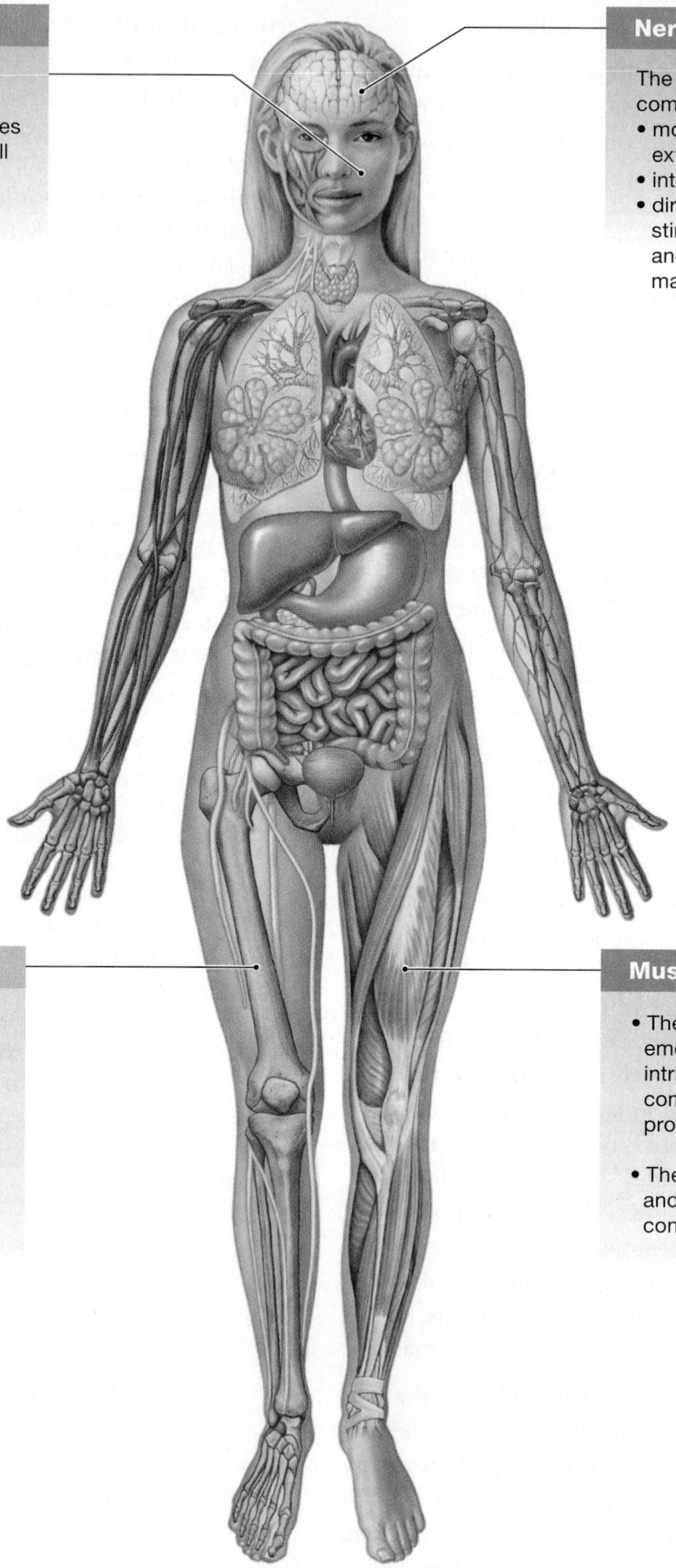

Skeletal System

- The Skeletal System provides calcium for neural function and protects the brain and spinal cord
- The nervous system affects bone thickening and maintenance by controlling muscle contractions

Muscular System

- The Muscular System expresses emotional states with facial muscles; intrinsic laryngeal muscles permit communication; muscle spindles provide proprioceptive sensations
- The nervous system controls voluntary and involuntary skeletal muscle contractions

16 Chapter Review

Study Outline

MasteringA&P™ Access more chapter study tools online in the MasteringA&P Study Area:
- Chapter Quizzes, Chapter Practice Test, MP3 Tutor Sessions, and Clinical Case Studies
- Practice Anatomy Lab PAL 3.0
- A&P Flix A&PFlix
- Interactive Physiology iP2
- PhysioEx PhysioEx 9.1

An Introduction to the Autonomic Nervous System and Higher-Order Functions p. 536

1. The autonomic nervous system adjusts our basic life support systems without conscious control.

16-1 The autonomic nervous system, which has sympathetic and parasympathetic divisions, is involved in the unconscious regulation of visceral functions p. 536

2. The **autonomic nervous system (ANS)** coordinates cardiovascular, respiratory, digestive, urinary, and reproductive functions.
3. **Preganglionic neurons** in the CNS send axons to synapse on **ganglionic neurons** in **autonomic ganglia** outside the CNS. *(Figure 16–1)*
4. Preganglionic fibers from the thoracic and lumbar segments form the **sympathetic division**, or *thoracolumbar division* ("fight or flight" system), of the ANS. Preganglionic fibers leaving the brain and sacral segments form the **parasympathetic division**, or *craniosacral division* ("rest and digest" system).

16-2 The sympathetic division has short preganglionic fibers and long postganglionic fibers and is involved in using energy and increasing metabolic rate p. 538

5. The sympathetic division consists of preganglionic neurons between segments T_1 and L_2, ganglionic neurons in ganglia near the vertebral column, and specialized neurons in the adrenal glands. *(Spotlight Figure 16–2a)*
6. The two types of sympathetic ganglia are **sympathetic chain ganglia** (*paravertebral ganglia*) and **collateral ganglia** (*prevertebral ganglia*). *(Figure 16–3)*
7. In spinal segments T_1–L_2, anterior roots give rise to the myelinated white ramus communicans, which, in turn, leads to the sympathetic chain ganglia. *(Spotlight Figure 16–2a, Figure 16–3)*
8. **Postganglionic fibers** targeting structures in the body wall and limbs rejoin the spinal nerves and reach their destinations by way of the posterior and anterior rami. *(Spotlight Figure 16–2a, Figure 16–3)*
9. Postganglionic fibers targeting structures in the thoracic cavity form **sympathetic nerves**, which go directly to their visceral destinations. Preganglionic fibers run between the sympathetic chain ganglia and interconnect them. *(Spotlight Figure 16–2a, Figure 16–3)*
10. The abdominopelvic viscera receive sympathetic innervation by preganglionic fibers that synapse within collateral ganglia. The preganglionic fibers that innervate the collateral ganglia form the **splanchnic nerves**. *(Spotlight Figure 16–2a, Figure 16–3)*
11. The **celiac ganglion** innervates the stomach, liver, gallbladder, pancreas, and spleen; the **superior mesenteric ganglion** innervates the small intestine and initial segments of the large intestine; and the **inferior mesenteric ganglion** innervates the kidneys, urinary bladder, the terminal portions of the large intestine, and the sex organs. *(Spotlight Figure 16–2a)*
12. Preganglionic fibers entering an adrenal gland synapse within the **adrenal medulla**. *(Spotlight Figure 16–2a, Figure 16–3)*
13. In a crisis, the entire sympathetic division responds—an event called **sympathetic activation**. Its effects include increased alertness, a feeling of energy and euphoria, increased cardiovascular and respiratory activities, a general elevation in muscle tone, and a mobilization of energy reserves.

16-3 Different types of neurotransmitters and receptors lead to different sympathetic effects p. 543

14. The stimulation of the sympathetic division has two distinctive results: the release of either ACh or *norepinephrine (NE)* at specific locations, and the secretion of *epinephrine (E)* and NE into the general circulation.
15. Sympathetic ganglionic neurons end in telodendria studded with **varicosities** containing neurotransmitters. *(Figure 16–4)*
16. The two types of sympathetic receptors are **alpha receptors** and **beta receptors**.
17. Most postganglionic fibers are *adrenergic*; a few are *cholinergic* or *nitroxidergic*. *(Table 16–1)*
18. The sympathetic division includes two sympathetic chain ganglia, three collateral ganglia, and two adrenal medullae. *(Figure 16–5; Tables 16–2, 16–3)*

16-4 The parasympathetic division has long preganglionic fibers and short postganglionic fibers and is involved in conserving energy and lowering metabolic rate p. 545

19. The parasympathetic division includes preganglionic neurons in the brainstem and sacral segments of the spinal cord, and ganglionic neurons in peripheral ganglia located within (**intramural**) or next to (**terminal**) target organs.
20. Preganglionic fibers leave the brain as components of cranial nerves III, VII, IX, and X. Those leaving the sacral segments form **pelvic nerves**. *(Spotlight Figure 16–2b)*
21. The effects produced by the parasympathetic division center on relaxation, food processing, and energy absorption.

16-5 Different types of receptors lead to different parasympathetic effects p. 546

22. All parasympathetic preganglionic and postganglionic fibers release ACh. The effects are short lived because ACh is inactivated by *acetylcholinesterase (AChE)* and by *tissue cholinesterase. (Figure 16–5)*
23. Postsynaptic membranes have two types of ACh receptors. The stimulation of **muscarinic receptors** produces a longer-lasting effect than does the stimulation of **nicotinic receptors**. *(Table 16–1)*

16-6 The differences in the organization of sympathetic and parasympathetic structures lead to widespread sympathetic effects and specific parasympathetic effects p. 547

24. The sympathetic division has widespread influence on visceral and somatic structures. *(Figure 16–5; Table 16–2)*
25. The parasympathetic division innervates areas serviced by cranial nerves and organs in the thoracic and abdominopelvic cavities. *(Figure 16–5; Table 16–2)*

16-7 Dual innervation of organs allows the sympathetic and parasympathetic divisions to coordinate vital functions p. 548

16

26. The parasympathetic division innervates only visceral structures that are serviced by cranial nerves or enclosed by the thoracic and abdominopelvic cavities. Organs with **dual innervation** receive input from both divisions. *(Tables 16–2, 16–3)*
27. The parasympathetic and sympathetic nerves intermingle in the thoracic and abdominopelvic cavities to form a series of characteristic *autonomic plexuses* (nerve networks): the **cardiac**, **pulmonary**, **esophageal**, **celiac**, **inferior mesenteric**, and **hypogastric plexuses**. *(Figure 16–6)*
28. Important physiological and functional differences exist between the sympathetic and parasympathetic divisions. *(Figure 16–5; Tables 16–2, 16–3)*
29. Even when stimuli are absent, autonomic motor neurons show a resting level of activation, the **autonomic tone**.

16-8 Various levels of autonomic regulation allow for the integration and control of autonomic functions p. 551

30. **Visceral reflex arcs** perform the simplest function of the ANS, and can be either **long reflexes** (with interneurons) or **short reflexes** (bypassing the CNS). *(Figure 16–7)*
31. Parasympathetic reflexes govern respiration, cardiovascular functions, and other visceral activities. *(Table 16–4)*
32. Levels of activity in the sympathetic and parasympathetic divisions of the ANS are controlled by centers in the brainstem that regulate specific visceral functions.
33. The SNS and ANS are organized in parallel. Integration occurs at the level of the brainstem and higher centers. *(Figure 16–8; Table 16–5)*

16-9 Higher-order functions include memory and states of consciousness, and neurotransmitters influence behavior p. 553

34. Higher-order functions (1) are performed by the cerebral cortex and involve complex interactions among areas of the cerebral cortex and between the cortex and other areas of the brain, (2) involve conscious and unconscious information processing, and (3) are subject to modification and adjustment over time.
35. Memories can be classified as **short term** or **long term**.
36. The conversion from short-term to long-term memory is **memory consolidation**. *(Figure 16–9)*
37. **Amnesia** is the loss of memory as a result of disease or trauma.
38. In **deep sleep** (*slow wave* or *non-REM sleep*), the body relaxes and cerebral cortex activity is low. In **rapid eye movement (REM) sleep**, active dreaming occurs. *(Figure 16–10)*
39. The **reticular activating system (RAS)**, a network in the reticular formation, is most important to arousal and the maintenance of consciousness. *(Figure 16–11)*
40. Changes in the normal balance between two or more neurotransmitters can profoundly affect brain function.

16-10 Aging produces various structural and functional changes in the nervous system p. 558

41. Age-related changes in the nervous system include a reduction in brain size and weight, a reduction in the number of neurons, a decrease in blood flow to the brain, changes in the synaptic organization of the brain, and intracellular and extracellular changes in CNS neurons.
42. The nervous system monitors all other systems and issues commands that adjust their activities. The efficiency of these activities typically decreases with aging. *(Figure 16–12)*

Review Questions

See the blue Answers tab at the back of the book.

LEVEL 1 Reviewing Facts and Terms

1. The preganglionic and ganglionic neurons are missing from the diagram below showing the distribution of sympathetic innervation. Indicate their distribution using red for the preganglionic neurons and black for the ganglionic neurons.

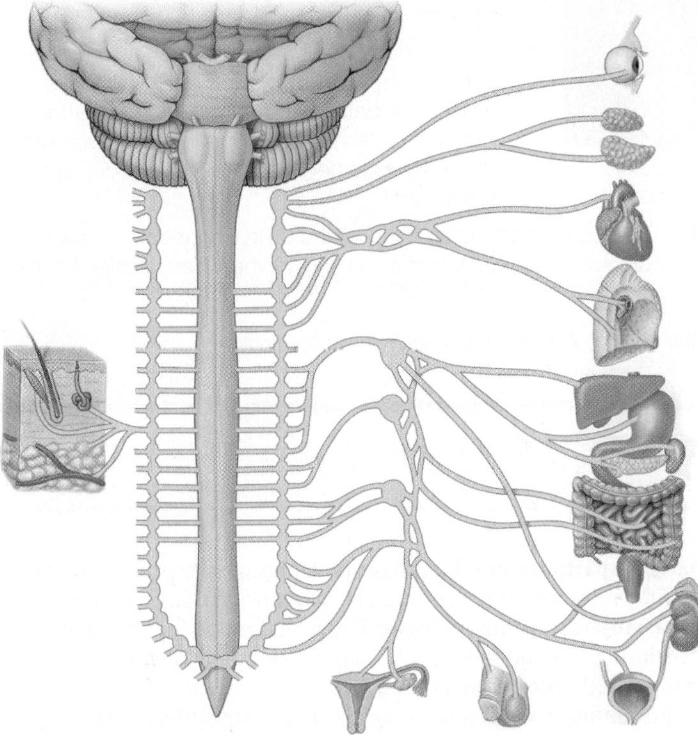

2. The autonomic division of the nervous system directs **(a)** voluntary motor activity, **(b)** conscious control of skeletal muscles, **(c)** unconscious control of skeletal muscles, **(d)** processes that maintain homeostasis, **(e)** sensory input from the skin.
3. The division of the ANS that prepares the body for activity and stress is the ____ division. **(a)** sympathetic, **(b)** parasympathetic, **(c)** craniosacral, **(d)** intramural, **(e)** somatomotor.
4. Effects produced by the parasympathetic branch of the ANS include **(a)** dilation of the pupils, **(b)** increased secretion by digestive glands, **(c)** dilation of respiratory passages, **(d)** increased heart rate, **(e)** increased breakdown of glycogen by the liver.
5. A progressive disorder characterized by the loss of higher-order cerebral functions is **(a)** Parkinson's disease, **(b)** parasomnia, **(c)** Huntington's disease, **(d)** Alzheimer's disease.
6. Starting in the spinal cord, trace an action potential through the sympathetic division of the ANS until it reaches a target organ in the abdominopelvic region.
7. Which four ganglia serve as origins for postganglionic fibers involved in control of visceral structures in the head?
8. What are the components of a visceral reflex arc?
9. What cellular mechanisms identified in animal studies are thought to be involved in memory formation and storage?
10. What physiological activities distinguish non-REM sleep from REM sleep?
11. What anatomical and functional changes in the brain are linked to alterations that occur with aging?
12. All preganglionic autonomic fibers release ____ at their axon terminals, and the effects are always ____. **(a)** norepinephrine, inhibitory, **(b)** norepinephrine, excitatory, **(c)** acetylcholine, excitatory, **(d)** acetylcholine, inhibitory.
13. The neurotransmitter at all synapses and neuromuscular or neuroglandular junctions in the parasympathetic division of the ANS is **(a)** epinephrine, **(b)** norepinephrine, **(c)** cyclic-AMP, **(d)** acetylcholine.
14. How does the emergence of sympathetic fibers from the spinal cord differ from the emergence of parasympathetic fibers?
15. Which three collateral ganglia serve as origins for ganglionic neurons that innervate organs or tissues in the abdominopelvic region?
16. What two distinctive results are produced by the stimulation of sympathetic ganglionic neurons?
17. Which four pairs of cranial nerves are associated with the cranial segment of the parasympathetic division of the ANS?
18. Which six plexuses in the thoracic and abdominopelvic cavities innervate visceral organs, and what are the effects of sympathetic versus parasympathetic stimulation?
19. What three characteristics are shared by higher-order functions?

LEVEL 2 Reviewing Concepts

20. Dual innervation refers to situations in which **(a)** vital organs receive instructions from both sympathetic and parasympathetic fibers, **(b)** the atria and ventricles of the heart receive autonomic stimulation from the same nerves, **(c)** sympathetic and parasympathetic fibers have similar effects, **(d)** all of these are correct.
21. Damage to the hippocampus, a component of the limbic system, leads to **(a)** a loss of emotion due to forgetfulness, **(b)** a loss of consciousness, **(c)** a loss of long-term memory, **(d)** an immediate loss of short-term memory.
22. Why does sympathetic function remain intact even when the anterior roots of the cervical spinal nerves are damaged?
23. During sympathetic stimulation, a person may begin to feel "on edge"; this is the result of **(a)** increased energy metabolism by muscle tissue, **(b)** increased cardiovascular activity, **(c)** stimulation of the reticular activating system, **(d)** temporary insensitivity to painful stimuli, **(e)** decreased levels of epinephrine in the blood.
24. Under which of the following circumstances would the diameter of peripheral blood vessels be greatest? **(a)** increased sympathetic stimulation, **(b)** decreased sympathetic stimulation, **(c)** increased parasympathetic stimulation, **(d)** decreased parasympathetic stimulation, **(e)** both increased parasympathetic and sympathetic stimulation.
25. A possible side effect of a drug used to open the airways of someone suffering from an asthma attack is **(a)** decreased activity of the digestive system, **(b)** diarrhea, **(c)** profuse urination, **(d)** increased blood pressure, **(e)** decreased heart rate.
26. You are home alone at night when you hear what sounds like breaking glass. What physiological effects would this experience probably produce, and what would be their cause?
27. Why is autonomic tone a significant part of ANS function?
28. Nicotine stimulates cholinergic receptors of the ANS. Based on this information, how would cigarette smoking affect the cardiovascular system?
29. The condition known as shock is characterized in part by a decreased return of venous blood to the heart. How could an upsetting situation, such as the sight of a tragic accident or very bad news, produce some temporary symptoms of shock?

LEVEL 3 Critical Thinking and Clinical Applications

30. Phil is stung on his cheek by a wasp. Because Phil is allergic to wasp venom, his throat begins to swell and his respiratory passages constrict. Would acetylcholine or epinephrine be more helpful in relieving his condition? Why?
31. While studying the activity of smooth muscle in blood vessels, Shelly discovers that, when applied to a muscle plasma membrane, a molecule chemically similar to a neurotransmitter triggers an increase in intracellular calcium ions. Which neurotransmitter is the molecule mimicking, and to which receptors is it binding?

+ CLINICAL CASE Wrap-Up Remember Me?

Helen has the progressive neurological disorder Alzheimer's disease. The hallmark of Alzheimer's is loss of memory. But patients also experience a gradual loss of cognitive functioning involving orientation, concentration, problem solving, judgment, and language. In the end stage, the person can no longer perform any activities of daily living and needs total care. This incurable disease can run for 5 or more years before the patient dies.

During life, the brain with Alzheimer's is deficient in the neurotransmitter acetylcholine (ACh). Diagnosis is established by ruling out other causes of memory loss. On autopsy Alzheimer's can be confirmed by an atrophied brain with wide sulci and shrunken gyri; under the microscope, we see abnormal plaques and neurofibrillary tangles that impair nerve transmission.

1. If memory loss is her first symptom, what part of Helen's brain is first affected? As her disease progresses, which additional brain regions become involved?
2. Helen's doctor will check her mental orientation. A normal result is "oriented x3," which means the patient recognizes the three attributes of person, place, and time. In which order did Helen lose her orientation as her disease worsened?
3. The doctor gave Helen a trial of the drug donepezil, a cholinesterase inhibitor. Why did this help? Where specifically did the drug act?

See the blue Answers tab at the back of the book.

Related Clinical Terms

alpha$_1$-receptor agonists: Drugs used to treat hypotension (low-blood pressure) by stimulating α_1 receptors to cause vasoconstriction of blood vessels.

alpha$_2$-receptor agonists: Drugs used to treat hypertension (high blood pressure) by stimulating α_2-adrenergic receptors to inhibit sympathetic vasomotor centers.

beta-adrenergic blockers: Drugs that decrease heart rate and force of contraction, lowering peripheral blood pressure by acting on beta-adrenergic receptors to diminish the effects of epinephrine.

parasympathetic blocking agents: Drugs that target the muscarinic receptors at neuromuscular or neuroglandular junctions.

parasympathomimetic drugs: Drugs that mimic parasympathetic stimulation and increase the activity along the digestive tract.

sympathetic blocking agents: Drugs that bind to receptor sites, preventing a normal response to neurotransmitters or sympathomimetic drugs.

sympathomimetic drugs: Drugs that mimic the effects of sympathetic stimulation.

17 The Special Senses

Learning Outcomes

These Learning Outcomes correspond by number to this chapter's sections and indicate what you should be able to do after completing the chapter.

- 17-1 ■ Describe the sensory organs of smell, trace the olfactory pathways to their destinations in the brain, and explain the physiological basis of olfactory discrimination. p. 566
- 17-2 ■ Describe the sensory organs of taste, trace the gustatory pathways to their destinations in the brain, and explain the physiological basis of gustatory discrimination. p. 570
- 17-3 ■ Identify the internal and accessory structures of the eye, and explain the functions of each. p. 572
- 17-4 ■ Describe how refraction and the focusing of light on the retina lead to vision. p. 581
- 17-5 ■ Explain color and depth perception, describe how light stimulates the production of nerve impulses, and trace the visual pathways to their destinations in the brain. p. 583
- 17-6 ■ Describe the structures of the external, middle, and internal ear, explain their roles in equilibrium and hearing, and trace the pathways for equilibrium and hearing to their destinations in the brain. p. 592

CLINICAL CASE A Chance to See

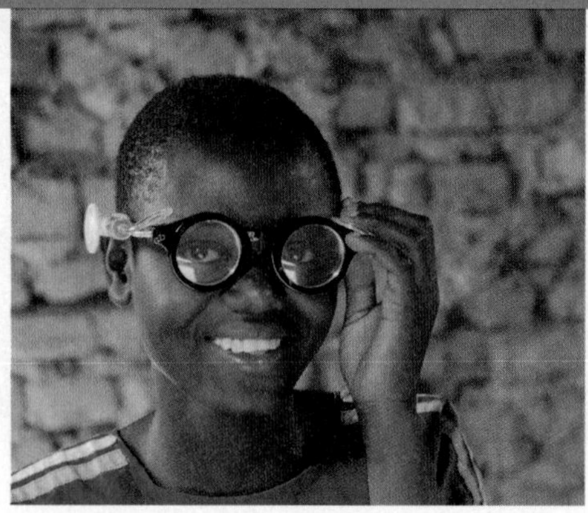

Makena is a 12-year-old Kenyan girl living with her family in a small village in sub-Saharan Africa. Their village is miles away from roads, electricity, and health care. Her family is very poor, and lives on less than $1.25 per day. Makena's daily chores include walking 2 hours to collect water, every morning before school and again after school. Water is so precious in this part of the world that nobody in Makena's village has ever taken a shower.

Makena is a bright girl, but she does not do well in school. She can't see the chalkboard in her classroom. She can't see the soccer ball when she tries to play with her classmates. She even has trouble seeing the path that leads to the dry riverbed where she collects water from a deep hole. She has no idea that other people can see these things better.

No one in Makena's village wears eyeglasses, and most do not even know what they are. Makena's teacher, however, studied at the University of Nairobi and knows that glasses can correct vision problems. The teacher thinks perhaps nearsightedness is a problem for Makena. **Can a desperately poor child, many miles away from health care of any kind, obtain glasses to correct her poor vision? To find out, turn to the Clinical Case Wrap-Up on p. 609.**

An Introduction to the Special Senses

17

Our knowledge of the world around us is limited to those characteristics that stimulate our sensory receptors. We may not realize it, but our picture of the environment is incomplete. Colors we cannot see guide insects to flowers. Sounds we cannot hear and smells we cannot detect give dolphins, dogs, and cats key information about their surroundings.

What we *do* perceive varies considerably with the state of our nervous system. For example, during sympathetic activation, you experience a heightened awareness of sensory information. You hear sounds that normally you would not notice. Yet, when concentrating on a difficult problem, you may be unaware of fairly loud noises.

Finally, our perception of any stimulus reflects activity in the cerebral cortex, but that activity can be generated by the nervous system itself. In cases of phantom pain syndrome, for example, a person feels pain in a missing limb. During an epileptic seizure, a person may experience sights, sounds, or smells that have no physical basis.

In our discussion of the general senses and sensory pathways in Chapter 15, we introduced basic principles of receptor function and sensory processing. We now turn to the five *special senses: olfaction* (smell), *gustation* (taste), *vision*, *equilibrium* (balance), and *hearing*. The sense organs involved are structurally more complex than those of the general senses, but the same basic principles of receptor function apply. That is, all of the special sense receptors also transduce an arriving stimulus into action potentials that are then sent to the CNS for interpretation and possible response. ⮌ p. 515 ATLAS: Embryology Summary 13: The Development of Special Sense Organs

17-1 Olfaction, the sense of smell, involves olfactory receptors responding to airborne chemical stimuli

Learning Outcome **Describe the sensory organs of smell, trace the olfactory pathways to their destinations in the brain, and explain the physiological basis of olfactory discrimination.**

The sense of smell, called **olfaction**, is made possible by paired **olfactory organs**. These olfactory organs, which are located in the **nasal cavity** on either side of the nasal septum, contain *olfactory sensory neurons*. What happens when you inhale through your nose? The air swirls within your nasal cavity. This turbulence brings airborne substances, including water-soluble and lipid-soluble substances called *odorants*, to your olfactory organs. A normal, relaxed inhalation carries a small sample (about 2 percent) of the inhaled air to the olfactory organs. If you sniff repeatedly, you increase the flow of air and so odorants, increasing the stimulation of the olfactory receptors. Once these receptors are stimulated, they send signals to the *olfactory cortex*, which interprets them. Let's look more closely at this process.

Anatomy of the Olfactory Organs

The olfactory organs are made up of two layers: the *olfactory epithelium* and the *lamina propria* (Figure 17–1a). The **olfactory epithelium** contains the olfactory sensory neurons, supporting cells, and regenerative *basal epithelial cells* (*stem cells*) (Figure 17–1b). This epithelium covers the inferior surface of the cribriform plate, the superior portion of the perpendicular plate, and the superior nasal conchae of the ethmoid. ⮌ p. 220 The second layer, the underlying lamina propria, consists of areolar tissue, numerous blood vessels, and nerves. This layer

Figure 17–1 The Olfactory Organs.

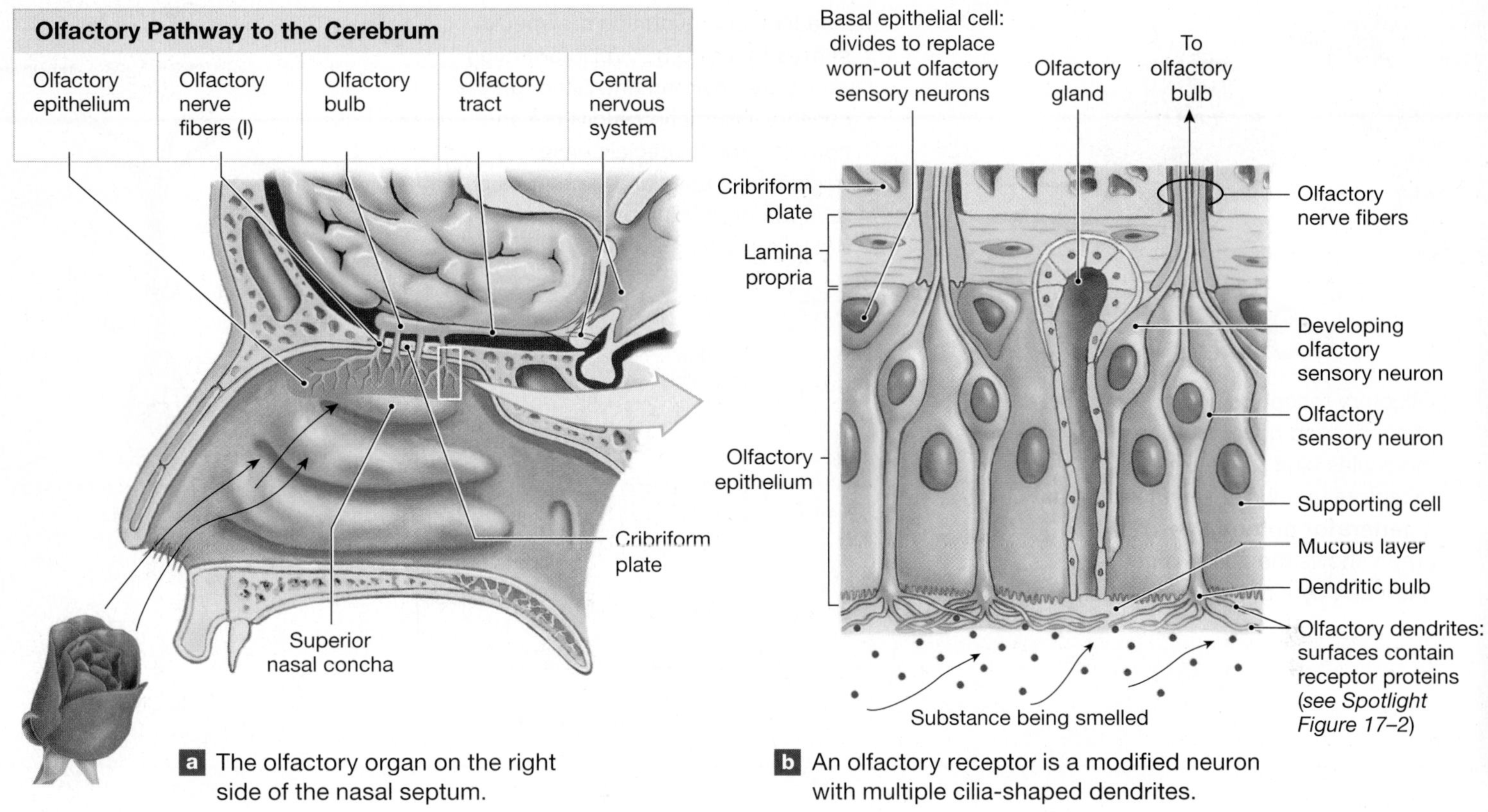

Olfactory Pathway to the Cerebrum
Olfactory epithelium
Olfactory nerve fibers (I)
Olfactory bulb
Olfactory tract
Central nervous system
Cribriform plate
Superior nasal concha
Basal epithelial cell: divides to replace worn-out olfactory sensory neurons
Olfactory gland
To olfactory bulb
Cribriform plate
Lamina propria
Olfactory epithelium
Olfactory nerve fibers
Developing olfactory sensory neuron
Olfactory sensory neuron
Supporting cell
Mucous layer
Dendritic bulb
Olfactory dendrites: surfaces contain receptor proteins (see Spotlight Figure 17–2)
Substance being smelled

a The olfactory organ on the right side of the nasal septum.

b An olfactory receptor is a modified neuron with multiple cilia-shaped dendrites.

also contains **olfactory glands**. Their secretions absorb water and form thick, pigmented *mucus*.

Olfactory sensory neurons are highly modified nerve cells. The exposed tip of each sensory neuron forms a prominent dendritic bulb that projects beyond the epithelial surface (see Figure 17–1b). The dendritic bulb is a base for up to 20 cilia-shaped dendrites that extend into the surrounding mucus. These dendrites lie parallel to the epithelial surface, exposing their considerable surface area to odorants.

Between 10 and 20 million olfactory receptors fill an area of roughly 5 cm^2 (0.8 $in.^2$). If we take into account the exposed dendritic surfaces, the actual sensory area probably approaches that of our entire body surface.

Olfactory Receptors and the Physiology of Olfaction

Olfactory reception begins with the binding of an odorant to a G protein–coupled receptor in the plasma membrane of an olfactory dendrite. This creates a depolarization called a **generator potential**. This potential leads to the generation of action potentials, which are then carried to the CNS by sensory afferent fibers. Please study this process in **Spotlight Figure 17–2a** before reading on.

Olfactory Pathways

The olfactory pathway begins with afferent fibers leaving the olfactory epithelium that collect into 20 or more bundles. These bundles penetrate the cribriform plate of the ethmoid bone to reach the *olfactory bulbs* of the cerebrum, where the first synapse occurs (see Figure 17–1a). Efferent fibers from nuclei elsewhere in the brain also innervate neurons of the olfactory bulbs. Axons leaving the olfactory bulb travel along the olfactory tract to the **olfactory cortex** of the cerebral hemispheres, the hypothalamus, and portions of the limbic system.

Olfactory stimulation is the only type of sensory information that reaches the cerebral cortex directly. All other sensations are relayed from processing centers in the thalamus. Certain smells can trigger profound emotional and behavioral responses, as well as memories, due to the fact that olfactory information is also distributed to the limbic system and hypothalamus.

The activation of an afferent fiber does not guarantee an awareness of the stimulus. Considerable convergence takes place along the olfactory pathway, and inhibition at the intervening synapses can prevent the sensations from reaching the *olfactory cortex*. p. 491 As seen in **Spotlight Figure 17–2a**, the olfactory receptors themselves adapt very little to an ongoing stimulus.

SPOTLIGHT

Figure 17–2
Olfaction and Gustation

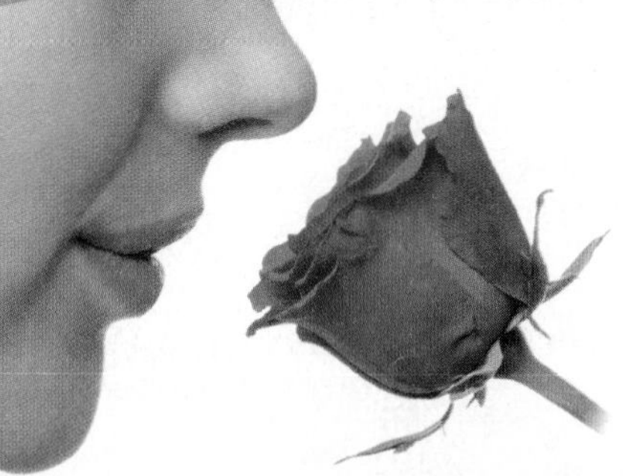

Olfaction and gustation are special senses that give us vital information about our environment. Although the sensory information is diverse and complex, each special sense originates at specialized neurons or receptor cells that communicate with other sensory neurons.

a OLFACTION

Olfactory receptors are the dendrites of specialized neurons. When odorant molecules bind to the olfactory receptors, a depolarization known as a **generator potential** results. This graph shows the action potentials produced by a generator potential.

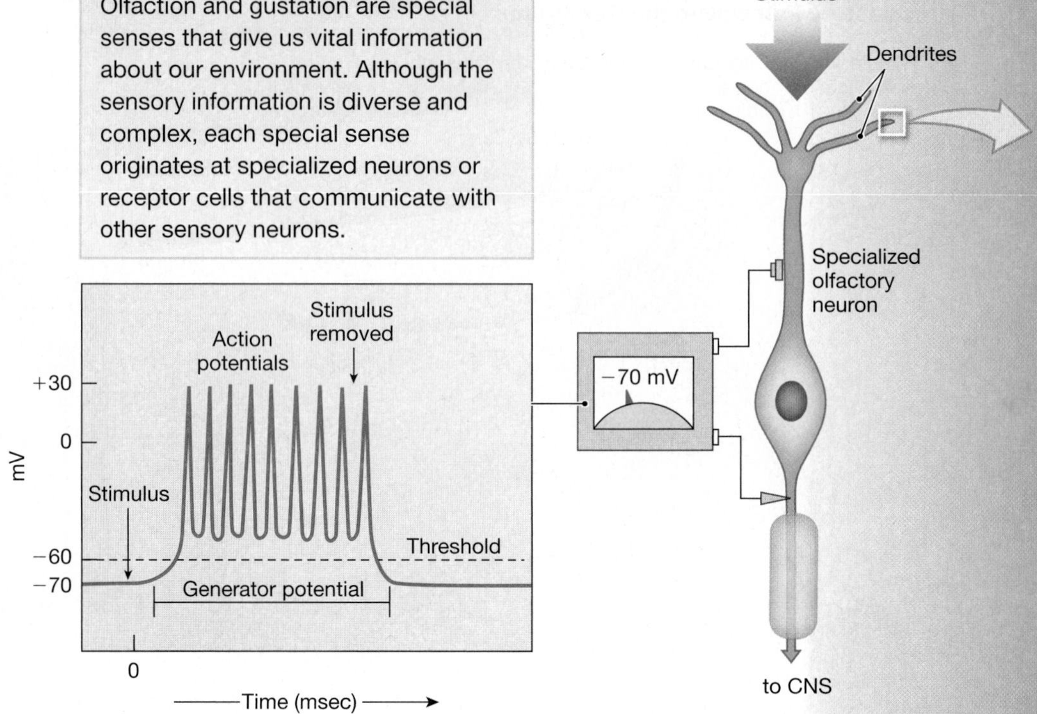

Stimulus
Dendrites
Specialized olfactory neuron
Action potentials
Stimulus removed
+30
0
mV
Stimulus
−60
−70
Threshold
Generator potential
0
Time (msec)
−70 mV
to CNS

b GUSTATION

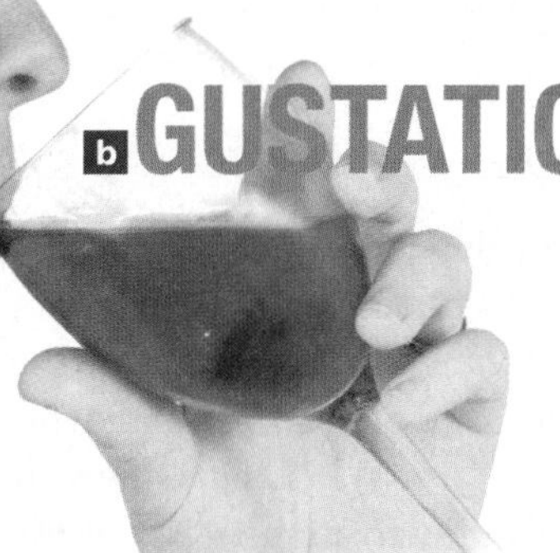

The receptors for the senses of taste, vision, equilibrium, and hearing are specialized cells that have unexcitable membranes and form synapses with the processes of sensory neurons. As this upper graph shows, the membrane of the stimulated receptor cell undergoes a graded depolarization that triggers the release of chemical transmitters at the synapse. These transmitters then depolarize the sensory neuron, creating a generator potential and action potentials that are propagated to the CNS. Because a synapse is involved, there is a slight synaptic delay. However, this arrangement permits modification of the sensitivity of the receptor cell by presynaptic facilitation or inhibition.

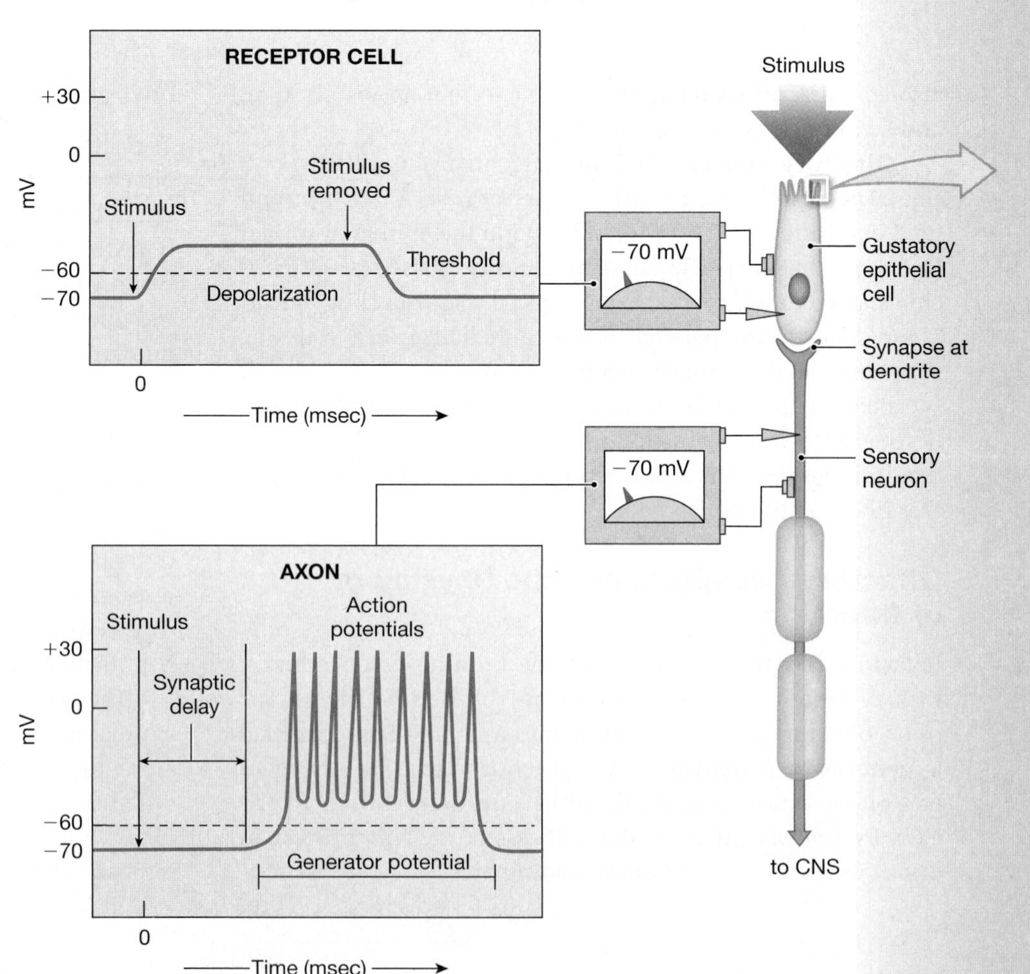

RECEPTOR CELL
+30
0
mV
Stimulus
Stimulus removed
−60
−70
Threshold
Depolarization
0
Time (msec)
Stimulus
−70 mV
Gustatory epithelial cell
Synapse at dendrite
Sensory neuron
−70 mV
AXON
Stimulus
Action potentials
+30
0
mV
Synaptic delay
−60
−70
Generator potential
0
Time (msec)
to CNS

Olfactory reception occurs on the surface membranes of the olfactory dendrites. **Odorants**—substances that stimulate olfactory receptors—interact with receptors called odorant-binding proteins on the membrane surface.

In general, odorants are small organic molecules. The strongest smells are associated with molecules of either high water or high lipid solubilities. As few as four odorant molecules can activate an olfactory receptor.

1 The binding of an **odorant** to its receptor protein leads to the activation of adenylate cyclase, the enzyme that converts ATP to cyclic AMP (cAMP).

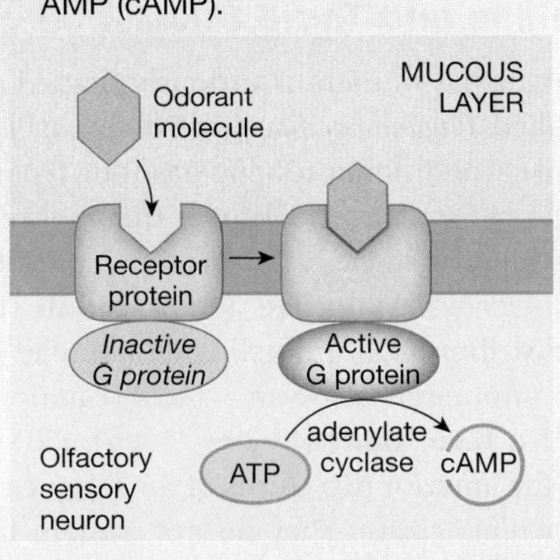

Odorant molecule
MUCOUS LAYER
Receptor protein
Inactive G protein
Active G protein
adenylate cyclase
Olfactory sensory neuron
ATP
cAMP

2 The cAMP opens sodium ion channels in the plasma membrane, which then begins to depolarize.

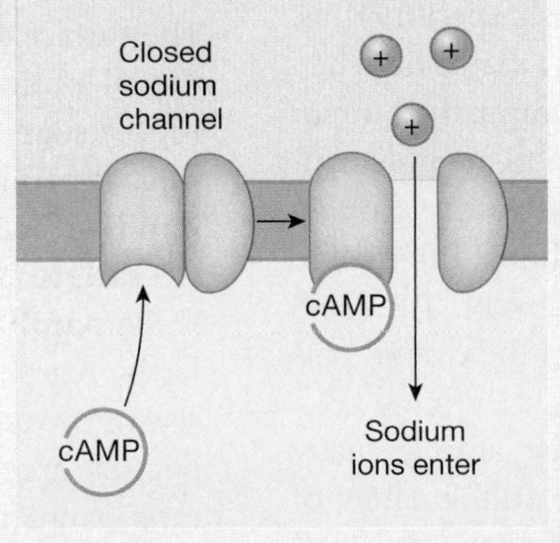

Closed sodium channel
cAMP
cAMP
Sodium ions enter

3 If sufficient depolarization occurs, an action potential is triggered in the axon, and the information is relayed to the CNS.

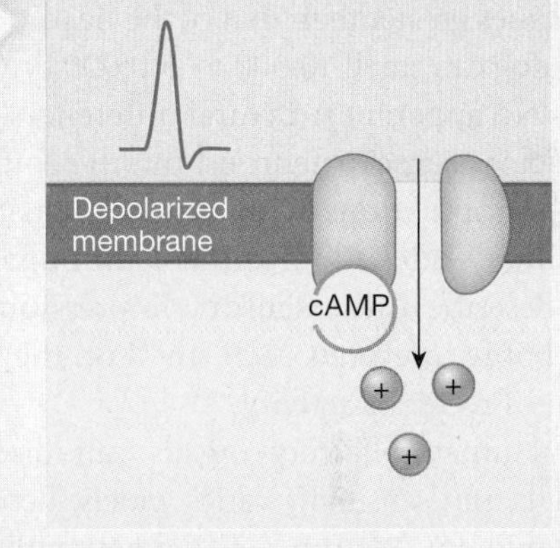

Depolarized membrane
cAMP

Some 90 percent of the gustatory epithelial cells respond to two or more different taste stimuli. The different tastes involve different receptor mechanisms. A salty stimulus involves the diffusion of Na^+ ions through a sodium ion leak channel common in epithelial cells. Stimuli for a sour or acidic taste include H^+ ions that diffuse through the same epithelial Na^+ channel. The intracellular increase in cations leads to depolarization and neurotransmitter release. Sweet, bitter, and umami stimuli bind to specific G protein–coupled receptors. The resulting multiple chemical pathways lead to depolarization and neurotransmitter release.

Salt and Sour Channels

The diffusion of sodium ions from salt solutions or hydrogen ions from acids or sour solutions into the gustatory epithelial cell leads to depolarization.

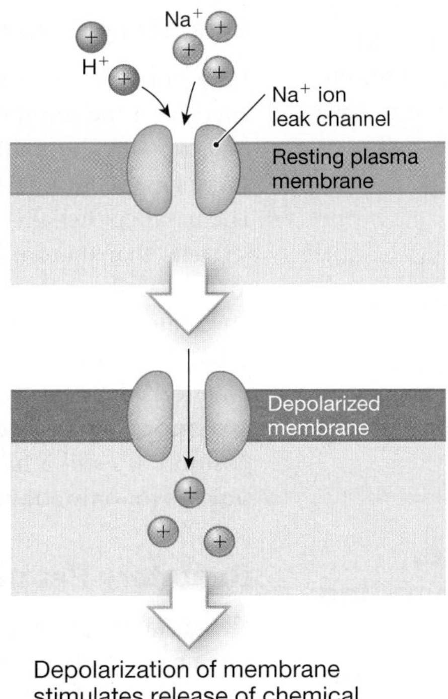

Na^+
H^+
Na^+ ion leak channel
Resting plasma membrane
Depolarized membrane

Depolarization of membrane stimulates release of chemical neurotransmitters.

Sweet, Bitter, and Umami Receptors

Receptors responding to stimuli that produce sweet, bitter, and umami sensations are linked to G proteins called **gustducins** (GUST-doos- inz)—protein complexes that use second messengers to produce their effects.

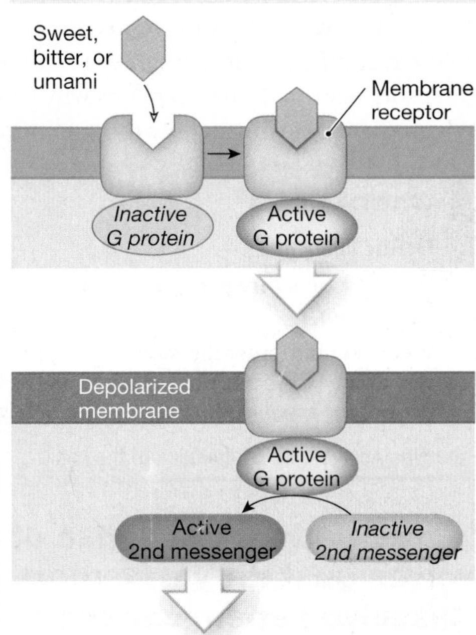

Sweet, bitter, or umami
Membrane receptor
Inactive G protein
Active G protein
Depolarized membrane
Active G protein
Active 2nd messenger
Inactive 2nd messenger

Activation of second messengers stimulates release of chemical neurotransmitters.

Rather, central adaptation (provided by the innervation of the olfactory bulbs by other brain nuclei) ensures that you quickly lose awareness of a new smell but remain sensitive to others.

Olfactory Discrimination

The human olfactory system can discriminate between, or make subtle distinctions among, 2000–4000 chemical stimuli. Yet, our olfactory sensitivities cannot compare with those of other vertebrates such as dogs, cats, or fishes. A German shepherd dog sniffing for smuggled drugs or explosives has an olfactory receptor surface 72 times greater than that of the nearby customs inspector! Thus, the dog can smell 10,000 to 100,000 times better than a human.

No apparent structural differences exist among the human olfactory sensory neurons, but the epithelium as a whole contains populations with distinct sensitivities. Upwards of 50 "primary smells" are known, although it is almost impossible to describe these sensory impressions effectively. The CNS probably interprets each smell on the basis of the overall pattern of receptor activity.

17

Human olfactory organs can discriminate among many smells, but sensitivity varies widely, depending on the nature of the odorant. We can detect many odorants in amazingly small concentrations. One example is mercaptan, an odorant commonly added to natural gas, which is otherwise odorless. Because we can smell mercaptan at an extremely low concentration (a few parts per billion), its addition enables us to detect a gas leak almost at once.

The olfactory receptor population is replaced frequently. Basal epithelial cells in the epithelium divide and differentiate to produce new sensory neurons. This turnover is one of the few examples of neuronal replacement in adult humans. Despite this process, the total number of neurons declines with age, and those that remain become less sensitive. As a result, elderly people have difficulty detecting odors in low concentrations. This sensory neuron decline explains why Grandpa's aftershave smells so strong: He must apply more to be able to smell it.

✓ Checkpoint

1. **Define *olfaction*.**
2. **Trace the olfactory pathway, beginning at the olfactory epithelium.**
3. **When you first enter the A&P lab for dissection, you are very aware of the odor of preservatives. By the end of the lab period, the smell doesn't seem to be nearly as strong. Why?**

See the blue Answers tab at the back of the book.

17-2 Gustation, the sense of taste, involves gustatory receptors responding to dissolved chemical stimuli

Learning Outcome **Describe the sensory organs of taste, trace the gustatory pathways to their destinations in the brain, and explain the physiological basis of gustatory discrimination.**

Gustation, or *taste*, provides information about the foods and liquids we eat and drink. *Gustatory* (GUS-ta-tor-ē) *epithelial cells*, or *taste receptors* are found in *taste buds* that are distributed over the superior surface of the tongue and adjacent portions of the pharynx and larynx. These receptors are stimulated by dissolved food molecules. This stimulation leads to action potentials that are sent to the *gustatory cortex* for interpretation. Let's look further into this process.

Anatomy of Papillae and Taste Buds

The surface of the tongue has numerous variously shaped epithelial projections called *lingual papillae* (pa-PIL-ē; *papilla*, a nipple-shaped mound). The human tongue has four types of lingual papillae (Figure 17–3a,b): (1) **filiform** (*filum*, thread) **papillae**, (2) **fungiform** (*fungus*, mushroom) **papillae**, (3) **vallate** (VAL-āt; *vallum*, wall) **papillae**, and (4) **foliate** (FŌ-lē-āt) **papillae**. The distribution of these lingual papillae varies by region. Their components also vary—most contain the sensory structures called **taste buds** (Figure 17–3c). Filiform papillae are found in the anterior two-thirds of the tongue running parallel to the *midline groove*. They do not contain taste buds, but they do provide an abrasive coat that creates friction to help move food around the mouth. Fungiform papillae are scattered around the tongue with concentration along the tip and sides. Each small fungiform papilla contains about 5 taste buds. Vallate papillae appear as an inverted "V" near the posterior margin of the tongue. There are up to 12 vallate papillae, and each contains as many as 100 taste buds. The foliate papillae are found as a series of folds along the lateral margins with taste buds embedded in their surfaces.

Gustatory Receptors

Taste buds are recessed into the surrounding epithelium, isolated from the unprocessed contents of the mouth. Each taste bud contains about 40–100 gustatory epithelial cells and many small stem cells called basal epithelial cells (see Figure 17–3b,c). The basal epithelial cells continually divide to produce daughter cells that mature in three stages—basal, transitional, and mature. Cells at all stages are innervated by sensory neurons. The mature cells of the last stage are the **gustatory epithelial cells**. Each gustatory epithelial cell extends microvilli, sometimes called *taste hairs*, into the surrounding fluids through the **taste pore**, a narrow opening. Despite this relatively protected position, it's still a hard life: A typical gustatory epithelial cell survives for only about 10 days before it is replaced.

Gustatory Pathways

The gustatory pathway starts with taste buds, which are innervated by cranial nerves VII (facial), IX (glossopharyngeal), and X (vagus). The facial nerve innervates all the taste buds located on the anterior two-thirds of the tongue, from the tip to the

Figure 17–3 Papillae, Taste Buds, and Gustatory Epithelial Cells.

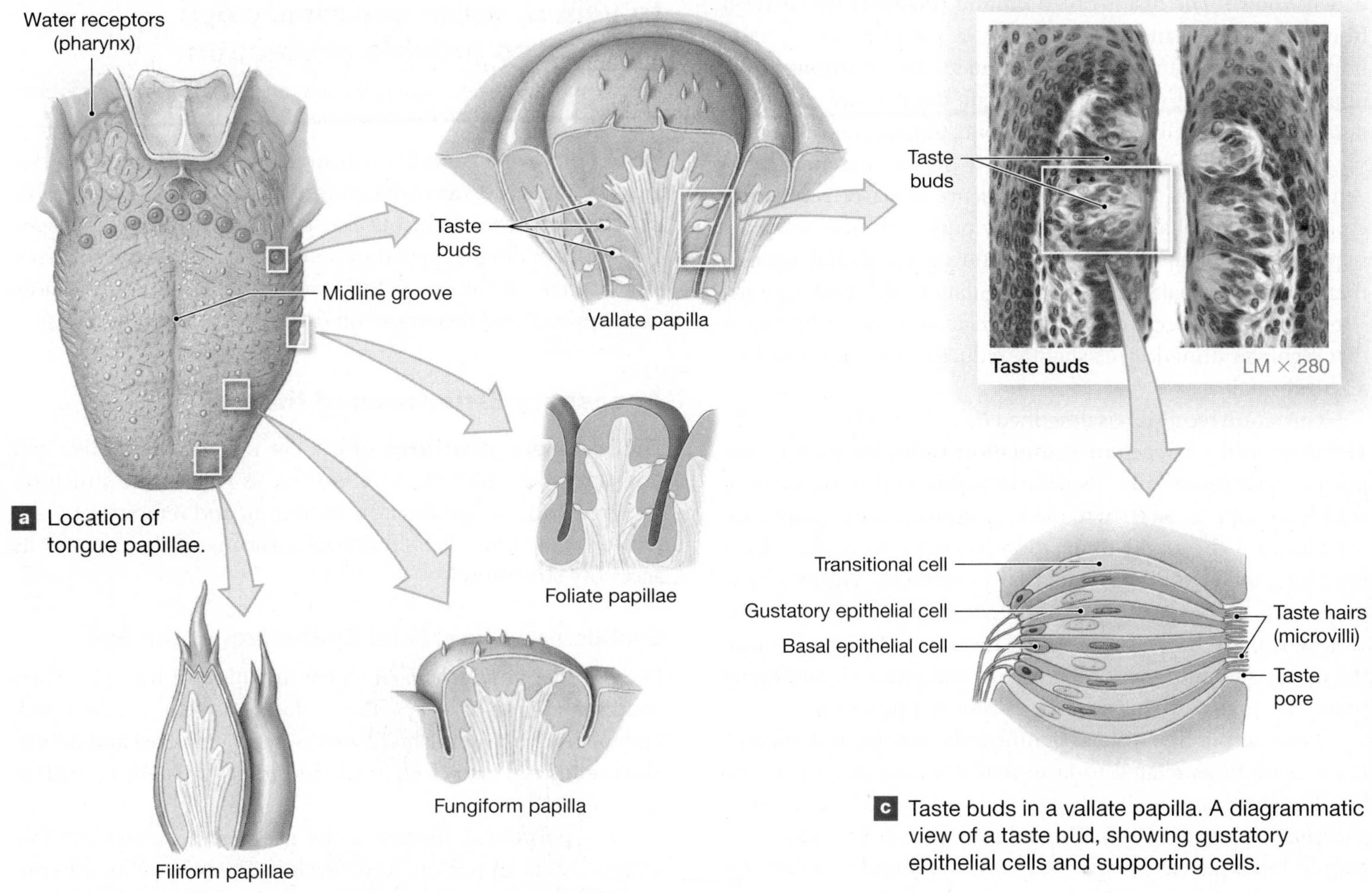
Water receptors (pharynx)
Taste buds
Midline groove
Vallate papilla
Taste buds
Taste buds
LM × 280
Foliate papillae
Fungiform papilla
Filiform papillae
Transitional cell
Gustatory epithelial cell
Basal epithelial cell
Taste hairs (microvilli)
Taste pore

a Location of tongue papillae.

b The structure and representative locations of the four types of lingual papillae. Taste receptors are located in taste buds, which form pockets in the epithelium of fungiform, foliate, and vallate papillae.

c Taste buds in a vallate papilla. A diagrammatic view of a taste bud, showing gustatory epithelial cells and supporting cells.

line of vallate papillae. The glossopharyngeal nerve innervates the vallate papillae and the posterior one-third of the tongue. The vagus nerve innervates taste buds scattered on the surface of the epiglottis.

The sensory afferent fibers carried by these cranial nerves synapse in the solitary nucleus of the medulla oblongata. The axons of the postsynaptic neurons then enter the medial lemniscus. There, the neurons join axons that carry somatic sensory information on touch, pressure, and proprioception. After another synapse in the thalamus, the information is sent to the appropriate portions of the gustatory cortex of the insula.

You have a conscious perception of taste as the brain correlates information received from the taste buds with other sensory data. Information about the texture of food, along with taste-related sensations such as "peppery," comes from sensory afferent fibers in the trigeminal cranial nerve (V).

In addition, the level of stimulation from the olfactory receptors plays a major role in taste perception. The combination of taste and smell is what provides the *flavor,* or distinctive quality of a particular food or drink. You are several thousand times more sensitive to "tastes" when your olfactory organs are fully functional. By contrast, when you have a cold and your nose is stuffed up, airborne molecules cannot reach your olfactory receptors, so meals taste dull and unappealing. The reduction in flavor happens even though the taste buds may be responding normally. Without accompanying olfactory sensations, you are now limited to the basic taste sensations provided by your gustatory receptors.

Gustatory Discrimination and Physiology of Gustation

You are probably already familiar with the four **primary taste sensations: sweet, salty, sour,** and **bitter**. There is some evidence for differences in sensitivity to tastes along the axis of the tongue, with greatest sensitivity to salty–sweet anteriorly and sour–bitter posteriorly. However, there are no differences in the structure of the taste buds. Taste buds in all portions of the tongue provide all four primary taste sensations.

Humans have two additional taste sensations that are less widely known. The first is called **umami** (oo-MAH-mē, derived from Japanese meaning "delicious"), a pleasant, savory taste imparted by the amino acid glutamate. The distribution of umami receptors is not known in detail, but they are present in taste buds of the vallate papillae. Second, although most people say that water has no flavor, research on humans and other vertebrates has demonstrated the presence of **water receptors**, especially in the pharynx. The sensory output of these receptors is processed in the hypothalamus and affects several systems that affect water balance and the regulation of blood volume. For example, the secretion of *antidiuretic hormone* (a hormone that regulates urination) is slightly reduced each time you take a long drink.

Gustation reception is described in **Spotlight Figure 17–2b**. The threshold for receptor stimulation varies for each of the primary taste sensations. Also, the taste receptors respond more readily to unpleasant than to pleasant stimuli. For example, we are almost a thousand times more sensitive to acids, which taste sour, than to either sweet or salty chemicals. We are a hundred times more sensitive to bitter compounds than to acids. This sensitivity has survival value, because acids can damage the mucous membranes of the mouth and pharynx, and many potent biological toxins have an extremely bitter taste.

Taste sensitivity differs significantly among individuals. Many conditions related to taste sensitivity are inherited. The best-known example involves sensitivity to the compound *phenylthiocarbamide* (PTC). This substance tastes bitter to some people, but is tasteless to others. PTC is not found in foods, but compounds related to PTC are found in Brussels sprouts, broccoli, cabbage, and cauliflower.

Our tasting abilities change with age. Many elderly people find that their food tastes bland and unappetizing, while children tend to find the same foods too spicy. What accounts for these differences? We begin life with more than 10,000 taste buds, but by the time we reach adulthood, the taste receptors on the pharynx, larynx, and epiglottis have decreased in number. By age 50, the number begins declining dramatically. The sensory loss becomes especially significant because, as we have already noted, older individuals also experience a decline in the number of olfactory receptors.

✓ Checkpoint

4. **Define *gustation*.**
5. **If you completely dry the surface of your tongue and then place salt or sugar on it, you can't taste the substance. Why not?**
6. **Your grandfather can't understand why foods he used to enjoy just don't taste the same anymore. How would you explain this to him?**

See the blue Answers tab at the back of the book.

17-3 Internal eye structures contribute to vision, while accessory eye structures provide protection

Learning Outcome Identify the internal and accessory structures of the eye, and explain the functions of each.

We rely more on **vision** than on any other special sense. Our eyes are elaborate structures containing our visual receptors that enable us not only to detect light, but also to create detailed visual images. We begin our discussion of these fascinating organs with the *accessory structures* of the eye (which provide protection, lubrication, and support), and then move on to the structures of the *eyeball*.

Accessory Structures of the Eye

The **accessory structures** of the eye include the eyelids and the superficial epithelium of the eye, as well as the structures involved with the production, secretion, and removal of tears. Figure 17–4 shows the superficial anatomy of the eye and its accessory structures.

Eyelids and Superficial Epithelium of the Eye

The **eyelids**, or *palpebrae*, are a continuation of the skin. Their continual blinking keeps the surface of the eye lubricated. They also act like windshield wipers, removing dust and debris. The eyelids can also close firmly to protect the delicate surface of the eye.

The **palpebral fissure** is the gap that separates the free margins of the upper and lower eyelids. The two eyelids are connected, however, at the **medial angle** (*medial canthus*) and the **lateral angle** (*lateral canthus*) (Figure 17–4a). The **eyelashes**, along the margins of the eyelids, are very robust hairs. They help prevent foreign matter (including insects) from reaching the surface of the eye.

The eyelashes are associated with unusually large sebaceous glands. Along the inner margin of the lid are small, modified sebaceous glands called **tarsal glands** (not shown in the figure), or *Meibomian* (mī-BŌ-mē-an) *glands.* They secrete a lipid-rich substance that helps keep the eyelids from sticking together. At the medial angle of eye, the **lacrimal caruncle** (KAR-ung-kul), a mass of soft tissue, contains glands producing the thick secretions that contribute to the gritty deposits that sometimes appear after a good night's sleep.

Bacteria occasionally invade and infect these various glands. A *chalazion* (kah-LĀ-zē-on; small lump) is a cyst that results from an infection of a tarsal gland. An infection in a sebaceous gland associated with an eyelash produces a painful localized swelling known as a *sty.*

The skin covering the visible surface of the eyelid is very thin. Deep to the skin lie the muscle fibers of the *orbicularis oculi* and *levator palpebrae superioris.* ⮌ p. 350 These skeletal muscles close the eyelids and raise the upper eyelid, respectively.

Figure 17–4 **External Features and Accessory Structures of the Eye.** ATLAS: Plates 3c; 12a; 16a,b

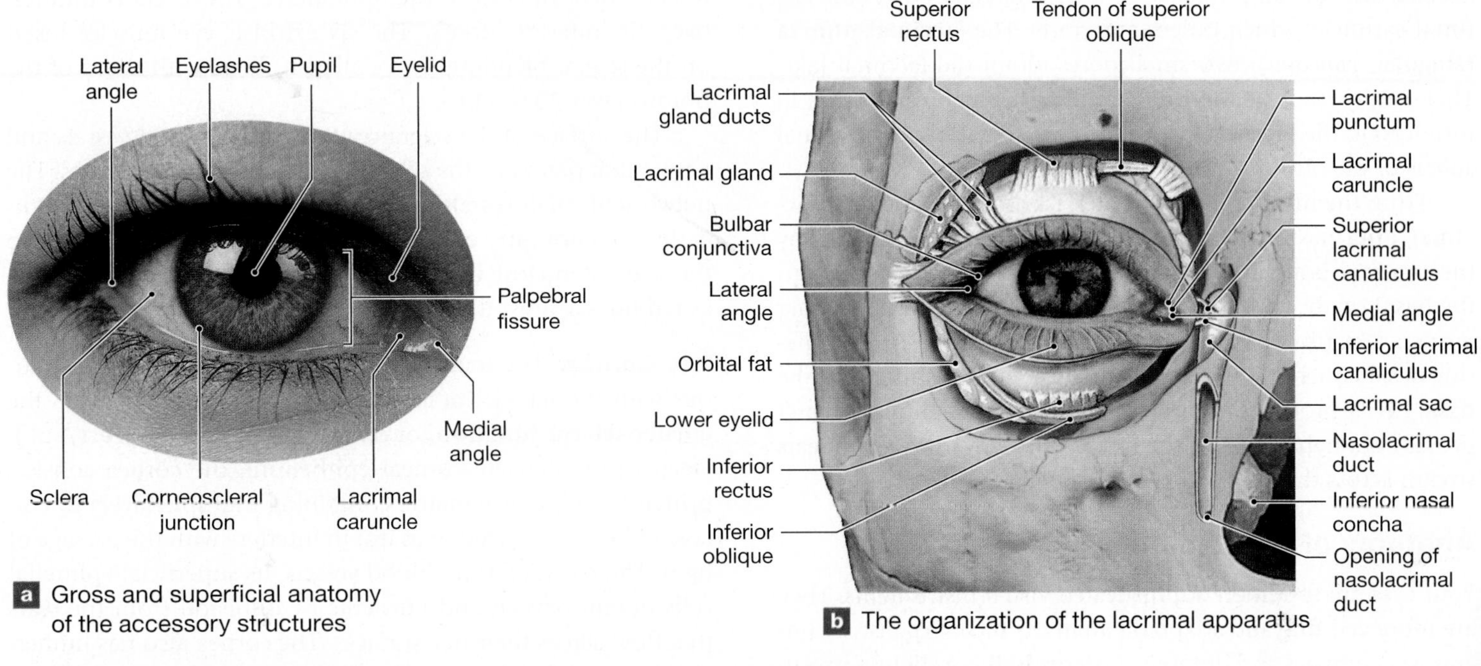
Lateral angle
Eyelashes
Pupil
Eyelid
Palpebral fissure
Medial angle
Sclera
Corneoscleral junction
Lacrimal caruncle
Superior rectus
Tendon of superior oblique
Lacrimal gland ducts
Lacrimal gland
Bulbar conjunctiva
Lateral angle
Orbital fat
Lower eyelid
Inferior rectus
Inferior oblique
Lacrimal punctum
Lacrimal caruncle
Superior lacrimal canaliculus
Medial angle
Inferior lacrimal canaliculus
Lacrimal sac
Nasolacrimal duct
Inferior nasal concha
Opening of nasolacrimal duct

a Gross and superficial anatomy of the accessory structures

b The organization of the lacrimal apparatus

? Choose the correct word from each pairing: The lacrimal gland is located (inferior, superior) and (lateral, medial) to the eye.

17

The epithelium covering the inner surfaces of the eyelids and the outer surface of the eyeball is called the **conjunctiva** (kon-junk-TĪ-vuh). It is a mucous membrane covered by a specialized stratified squamous epithelium. The **palpebral conjunctiva** covers the inner surface of the eyelids. The **bulbar conjunctiva**, or *ocular conjunctiva,* covers the anterior surface of the eye (Figure 17–4b).

The bulbar conjunctiva extends to the edges of the **cornea** (KOR-nē-uh), a transparent part of the outer fibrous layer of the eye. The cornea is covered by a very delicate squamous *corneal epithelium,* five to seven cells thick, that is continuous with the bulbar conjunctiva. A constant supply of fluid washes over the surface of the eyeball, keeping the bulbar conjunctiva and cornea moist and clean. Mucous cells in the epithelium assist the accessory glands in lubricating the surfaces of the conjunctiva to prevent drying out and friction.

Conjunctivitis, or *pinkeye,* is an inflammation of the conjunctiva. The most obvious sign, redness, is due to the dilation of blood vessels deep to the conjunctival epithelium. This condition may be caused by pathogenic infection or by physical, allergic, or chemical irritation of the conjunctival surface.

The Lacrimal Apparatus

A constant flow of tears keeps conjunctival surfaces moist and clean. Tears reduce friction, remove debris, prevent bacterial infection, and provide nutrients and oxygen to portions of the conjunctival epithelium. The **lacrimal apparatus** produces, distributes, and removes tears. The lacrimal apparatus of each eye consists of (1) a *lacrimal gland* with associated ducts, (2) paired *lacrimal canaliculi,* (3) a *lacrimal sac,* and (4) a *nasolacrimal duct* (see Figure 17–4b).

The **fornix** of the eye is a pocket created where the palpebral conjunctiva becomes continuous with the bulbar conjunctiva. The lateral portion of the superior fornix receives 10–12 ducts from the **lacrimal gland**, or tear gland (see Figure 17–4b). This gland is about the size and shape of an almond, measuring roughly 12–20 mm (0.5–0.75 in.). It nestles within a depression in the frontal bone, just inside the orbit and superior and lateral to the eyeball. p. 217

The lacrimal gland normally provides the key ingredients and most of the volume of the tears that bathe the conjunctival surfaces. The lacrimal secretions supply nutrients and oxygen to the corneal cells by diffusion. The lacrimal secretions are watery and slightly alkaline. They contain the antibacterial enzyme **lysozyme** and antibodies that attack pathogens before they enter the body.

Each lacrimal gland produces about 1 mL of tears each day. Once the lacrimal secretions have reached the ocular surface, they mix with the products of accessory glands and the oily secretions of the tarsal glands. The result is a superficial "oil slick" that aids lubrication and slows evaporation.

Blinking sweeps the tears across the ocular surface. Tears accumulate at the medial angle of eye in an area known as the *lacrimal lake,* or "lake of tears." The lacrimal lake covers the lacrimal caruncle, which bulges anteriorly. The **lacrimal puncta** (singular, *punctum*), two small pores, drain the lacrimal lake. They empty into the **lacrimal canaliculi**, small canals that in turn lead to the **lacrimal sac**, which nestles within the lacrimal sulcus of the orbit (see Figure 17–4b). ⮌ p. 223

From the inferior portion of the lacrimal sac, the **nasolacrimal duct** passes through the *nasolacrimal canal,* formed by the lacrimal bone and the maxilla. This duct delivers tears to the nasal cavity on that side. The duct empties into the *inferior meatus,* a narrow passageway inferior and lateral to the inferior nasal concha. When a person cries, tears rushing into the nasal cavity produce a runny nose. If the lacrimal puncta can't provide enough drainage, the lacrimal lake overflows and tears stream across the face.

Anatomy of the Eyeball

Your eyes are extremely sophisticated visual instruments. They are more versatile and adaptable than the most expensive cameras, yet compact and durable. Each **eyeball** is a slightly irregular spheroid (Figure 17–5a), a little smaller than a Ping-Pong ball, with an average diameter of 24 mm (almost 1 inch) and also a weight of about 8 g (0.28 oz). Within the orbit, the eyeball shares space with the extrinsic eye muscles, the lacrimal gland, and the cranial nerves and blood vessels that supply the eye and adjacent portions of the orbit and face. **Orbital fat** cushions and insulates the eye.

The wall of the eyeball has three distinct layers, formerly called tunics (Figure 17–5b): (1) an outer *fibrous layer* (2) an intermediate *vascular layer (uvea),* and (3) a deep *inner layer (retina).* The visual receptors, or *photoreceptors,* are located in the inner layer.

The eyeball itself is hollow and filled with fluid. Its interior can be divided into two cavities, anterior and posterior (Figure 17–5c). The **anterior cavity** has two chambers, the *anterior chamber* (between the cornea and the iris) and the *posterior chamber* (between the iris and the transparent *lens*). A clear, watery fluid called *aqueous (Ā-kwē-us) humor* fills the entire anterior cavity. The **posterior cavity**, or *vitreous chamber,* contains a gelatinous substance called the *vitreous body.*

The Fibrous Layer

The **fibrous layer**, the outermost layer of the eyeball, consists of the whitish *sclera* (SKLER-uh) and the transparent *cornea.* The fibrous layer (1) supports and protects the eye, (2) serves as an attachment site for the extrinsic eye muscles, and (3) contains structures that assist in focusing.

The Sclera. The **sclera**, or "white of the eye," covers most of the ocular surface (see Figure 17–5c). The sclera consists of a dense fibrous connective tissue containing both collagen and elastic fibers. This layer is thickest over the posterior surface of the eye, near the exit of the optic nerve. The sclera is thinnest over the anterior surface. The six extrinsic eye muscles insert on the sclera, blending their collagen fibers with those of the fibrous layer. ⮌ p. 349

The surface of the sclera contains small blood vessels and nerves that penetrate the sclera to reach internal structures. The network of small vessels interior to the bulbar conjunctiva generally does not carry enough blood to lend an obvious color to the sclera. On close inspection, however, the vessels are visible as red lines against the white background of collagen fibers.

The Cornea. The transparent **cornea** is structurally continuous with the sclera. The border between the two is called the **corneoscleral junction**, or *corneal limbus* (see Figure 17–5c). Deep to the delicate corneal epithelium, the cornea consists primarily of a dense matrix containing multiple layers of collagen fibers, organized so as not to interfere with the passage of light. The cornea has no blood vessels. Its superficial epithelial cells obtain oxygen and nutrients by diffusion from the tears that flow across their free surfaces. The cornea also has numerous free nerve endings, making it the most sensitive portion of the eye.

Damage to the cornea may cause blindness even though the functional components of the eye—including the photoreceptors—are perfectly normal. The cornea has a very restricted ability to repair itself. For this reason, corneal injuries must be treated immediately to prevent serious vision losses.

Restoring vision after corneal scarring generally requires replacing the cornea through a *corneal transplant.* Corneal replacement is probably the most common form of transplant surgery. Such transplants can be performed between unrelated individuals, because there are no blood vessels to carry white blood cells, which attack foreign tissues, into the area. Corneal grafts are obtained from donor eyes. For best results, the tissues must be removed within 5 hours after the donor's death.

The Vascular Layer

The **vascular layer**, or *uvea* (YŪ-vē-uh), is a pigmented region that includes the *iris, ciliary body,* and *choroid* (see Figure 17–5b,c). It contains numerous blood vessels, lymphatic vessels, and the intrinsic (smooth) muscles of the eye. This middle layer (1) provides a route for blood vessels and lymphatics that supply the tissues of the eye; (2) regulates the amount of light that enters the eye; (3) secretes and reabsorbs the aqueous humor that circulates within the chambers of the eye; and (4) controls the shape of the lens, an essential part of the focusing process.

The Iris. The **iris**, a pigmented, flattened ring structure, is visible through the transparent corneal surface. The iris contains blood vessels, pigment cells (melanocytes), and two layers

Figure 17–5 The Sectional Anatomy of the Eye. ATLAS: Plates 12a; 16a,b

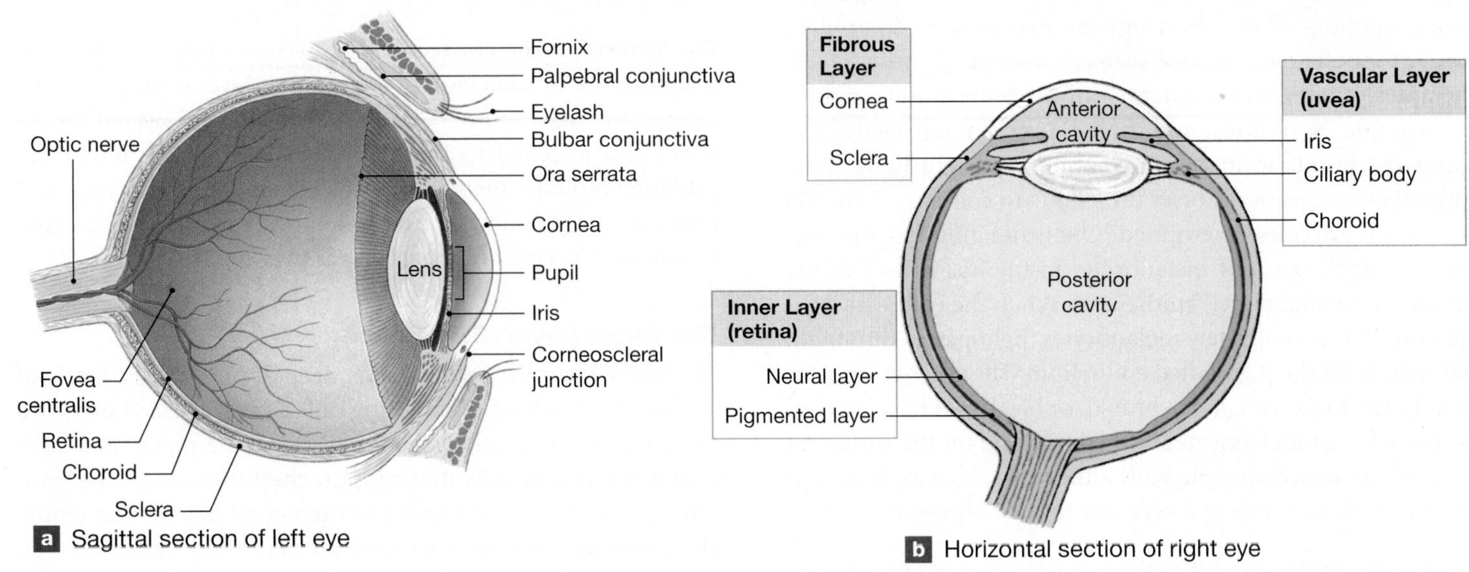
Fornix
Palpebral conjunctiva
Eyelash
Optic nerve
Bulbar conjunctiva
Ora serrata
Cornea
Lens
Pupil
Iris
Corneoscleral junction
Fovea centralis
Retina
Choroid
Sclera
Fibrous Layer
Cornea
Sclera
Anterior cavity
Vascular Layer (uvea)
Iris
Ciliary body
Choroid
Posterior cavity
Inner Layer (retina)
Neural layer
Pigmented layer

a Sagittal section of left eye

b Horizontal section of right eye

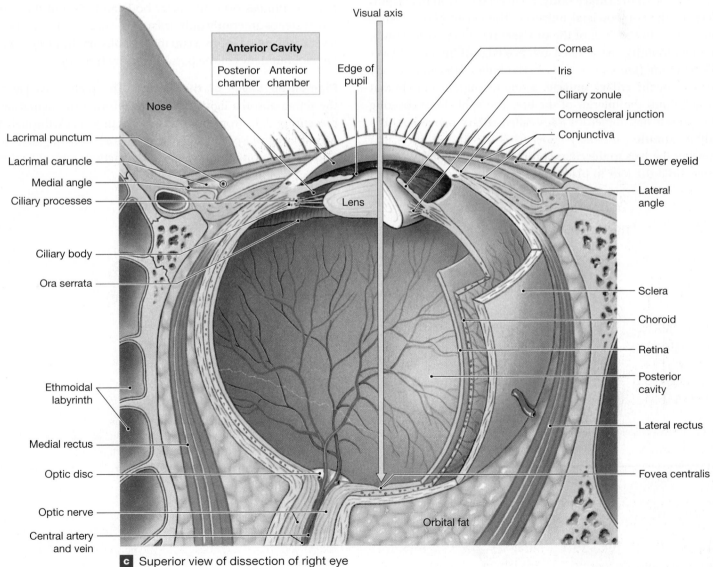

Visual axis
Anterior Cavity
Posterior chamber
Anterior chamber
Edge of pupil
Nose
Cornea
Iris
Ciliary zonule
Corneoscleral junction
Conjunctiva
Lacrimal punctum
Lacrimal caruncle
Medial angle
Ciliary processes
Lens
Lower eyelid
Lateral angle
Ciliary body
Ora serrata
Sclera
Choroid
Retina
Posterior cavity
Ethmoidal labyrinth
Medial rectus
Lateral rectus
Optic disc
Fovea centralis
Optic nerve
Orbital fat
Central artery and vein

c Superior view of dissection of right eye

of smooth muscle fibers called *pupillary muscles.* When these muscles contract, they change the diameter of the **pupil**, the central opening of the iris. There are two groups of pupillary muscles: *dilator pupillae* and *sphincter pupillae* (Figure 17–6). The autonomic nervous system controls both muscle groups. For example, in response to dim light, sympathetic activation causes the pupils to dilate. In response to bright light, parasympathetic activation causes the pupils to constrict. ⮌ p. 550

How is eye color determined? Our genes influence the density and distribution of melanocytes in the iris, as well as the density of the pigmented epithelium. When the connective tissue of the iris contains few melanocytes, light passes through it and reflects off the pigmented epithelium. The eye then appears blue. Individuals with green, brown, or black eyes have increasing numbers of melanocytes in the body and on the surface of the iris. The eyes of people with albinism, whose cells do not produce melanin, appear a very pale gray or blue-gray.

The Ciliary Body. At its periphery, the iris attaches to the anterior portion of the **ciliary body**, a thickened region that begins deep to the corneoscleral junction. The ciliary body extends posteriorly to the level of the **ora serrata** (Ō-ra ser-RAH-tuh; serrated mouth), the serrated anterior edge of the neural layer of the retina (see Figure 17–5c). The bulk of the ciliary body consists of the **ciliary muscle**, a ring of smooth muscle that projects into the interior of the eye. The epithelium covering this muscle has numerous folds called **ciliary processes**. The **ciliary zonule** (*suspensory ligament*) is the ring of fibers that attaches the lens to the ciliary processes. The connective tissue fibers hold the lens in place posterior to the iris and centered on the pupil. As a result, any light passing through the pupil also passes through the lens.

The Choroid. The **choroid** (KOR-oyd) is a vascular layer that separates the fibrous layer and the inner layer posterior to the ora serrata (see Figure 17–5c). The choroid is covered by the sclera and attached to the outermost layer of the retina. An extensive capillary network in the choroid delivers oxygen and nutrients to the retina. The choroid also contains melanocytes, which are especially numerous near the sclera.

The Inner Layer: The Retina

The **inner layer**, or **retina**, is the deepest of the three layers of the eye. It consists of a thin lining called the *pigmented layer,* and a thicker, covering called the *neural layer.* The pigmented layer contains pigment cells that support the functions of the photoreceptors, which are located in the neural layer of the retina. These two layers of the retina are normally very close together, but not tightly interconnected. The pigmented layer of the retina continues over the ciliary body and iris, but the neural layer extends anteriorly only as far as the ora serrata. The neural layer of the retina forms a cup that establishes the posterior and lateral boundaries of the posterior cavity (see Figure 17–5b,c).

Pigmented Layer of the Retina. The pigmented layer of the retina absorbs light that passes through the neural layer, preventing light from bouncing (reflecting) back through the neural layer and producing visual "echoes." The pigment cells also have important biochemical interactions with the retina's photoreceptors.

Figure 17–6 **The Pupillary Muscles.**

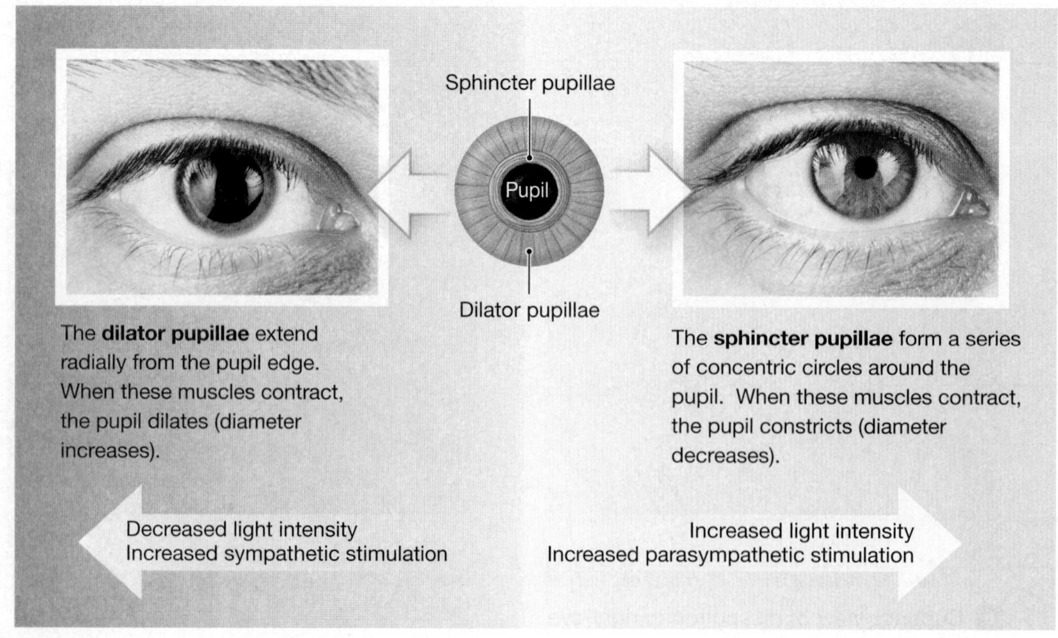

Sphincter pupillae
Pupil
Dilator pupillae
The **dilator pupillae** extend radially from the pupil edge. When these muscles contract, the pupil dilates (diameter increases).
The **sphincter pupillae** form a series of concentric circles around the pupil. When these muscles contract, the pupil constricts (diameter decreases).
Decreased light intensity
Increased sympathetic stimulation
Increased light intensity
Increased parasympathetic stimulation

Neural Layer: Cellular Organization. In sectional view, the neural layer of the retina contains several layers of cells (Figure 17–7a). The inner layers of the retina contain supporting cells and neurons that do preliminary processing and integration of visual information. The outermost part, closest to the pigmented layer of the retina, contains the **photoreceptors**, the cells that detect light.

The eye has two main types of photoreceptors: rods and cones. **Rods** do not discriminate among colors of light. Highly sensitive to light, they enable us to see in dimly lit rooms, at twilight, and in pale moonlight. **Cones** give us color vision. Cones give us sharper, clearer images than rods do, but cones require more intense light. If you sit outside at sunset with your textbook open to a colorful illustration, you can detect the gradual shift in your visual system from cone-based vision (a clear image in full color) to rod-based vision (a relatively grainy image in black and white). A third type of photoreceptor is the *intrinsically photosensitive retinal ganglion cell* (ipRGC). The photopigment in the ipRGC is melanopsin. These cells are known to respond to different levels of brightness and influence the body's 24-hour circadian rhythm (biological clock).

Tips & Tools

Associate the **r** in **r**od with the **r** in da**r**k, and associate the **c** in **c**ones with the **c** in **c**olor and in a**c**uity (sharpness).

Rods and cones are not evenly distributed across the retina. Approximately 125 million rods form a broad band around the periphery of the retina. As you move toward the center of the retina, the density of rods gradually decreases. In contrast, most of the roughly 6 million cones are concentrated in the area where a visual image arrives after it passes through the cornea and lens. This region, known as the **macula** (MAK-yū-luh; spot), has no rods. The highest concentration of cones occurs at the center of the macula, an area called the **fovea centralis** (FŌ-vē-uh; shallow depression), or simply the **fovea** (Figure 17–7b). The fovea centralis is the site of sharpest color vision. When you look directly at an object, its image falls on this portion of the retina. An imaginary line drawn from the center of that object through the center of the lens to the fovea centralis establishes the **visual axis** of the eye (look back at Figure 17–5c).

What are some of the visual consequences of this distribution of photoreceptors? When you look directly at an object, you are placing its image on the fovea centralis. You see a very good image as long as there is enough light to stimulate the cones. But in very dim light, cones cannot function. That is why you can't see a dim star if you stare directly at it, but you can see it if you shift your gaze to one side or the other. Shifting your gaze moves the image of the star from the fovea centralis, where it does not provide enough light to stimulate the cones, to the periphery, where it can affect the more sensitive rods.

Rods and cones synapse with about 6 million neurons called **bipolar cells** (see Figure 17–7a). These cells in turn synapse within the layer of the 1 million neurons called **ganglion cells**, which lie adjacent to the posterior cavity. A network of **horizontal cells** extends across the neural layer of the retina at the level of the synapses between photoreceptors and bipolar cells. A comparable network of **amacrine** (AM-ah-krin) **cells** occurs where bipolar cells synapse with ganglion cells. Horizontal and amacrine cells can facilitate or inhibit communication between photoreceptors and ganglion cells, altering the sensitivity of the retina. These cells play an important role in the eye's adjustment to dim or brightly lit environments.

Neural Layer: The Optic Disc. Axons from an estimated 1 million ganglion cells converge on the **optic disc**, a circular region just medial to the fovea centralis. The optic disc is the origin of the optic nerve (II). From this point, the axons turn, penetrate the wall of the eye, and proceed toward the diencephalon (Figure 17–7c). The *central retinal artery* and *central retinal vein*, which supply the retina, pass through the center of the optic nerve and emerge on the surface of the optic disc (see Figure 17–7c).

The optic disc has no photoreceptors or other structures typical of the rest of the retina. Because light striking the optic disc goes unnoticed, this area is commonly called the **blind spot**. Why don't you see a blank spot in your field of vision?

+ Clinical Note Diabetic Retinopathy

A *retinopathy* is a disease of the retina. **Diabetic retinopathy** develops in many people with *diabetes mellitus*, an endocrine disorder that interferes with glucose metabolism. Diabetes affects many systems and is a risk factor for heart disease. Diabetic retinopathy develops over years. It results from the blockage of small retinal blood vessels, followed by excessive growth of abnormal blood vessels. These blood vessels invade the retina and extend into the space between its two parts—the pigmented layer and the neural layer. Visual acuity is gradually lost through damage to photoreceptors (which are deprived of oxygen and nutrients), leakage of blood into the posterior cavity, and the overgrowth of blood vessels. Laser therapy can seal leaking vessels and block new vessel growth. The posterior cavity can be drained and the cloudy fluid replaced by a suitably clear substitute. However, these are only temporary, imperfect fixes. Diabetic retinopathy is the primary cause of blindness in the United States.

Figure 17–7 The Organization of the Retina.

Horizontal cell
Cone
Rod
Amacrine cell
Choroid
Pigmented layer of retina
Rods and cones
Bipolar cells
Ganglion cells
Posterior cavity
LIGHT
Retina
LM × 350
Nuclei of ganglion cells
Nuclei of rods and cones
Nuclei of bipolar cells

a The cellular organization of the retina. The photoreceptors are closer to the choroid than they are to the posterior cavity (vitreous chamber).

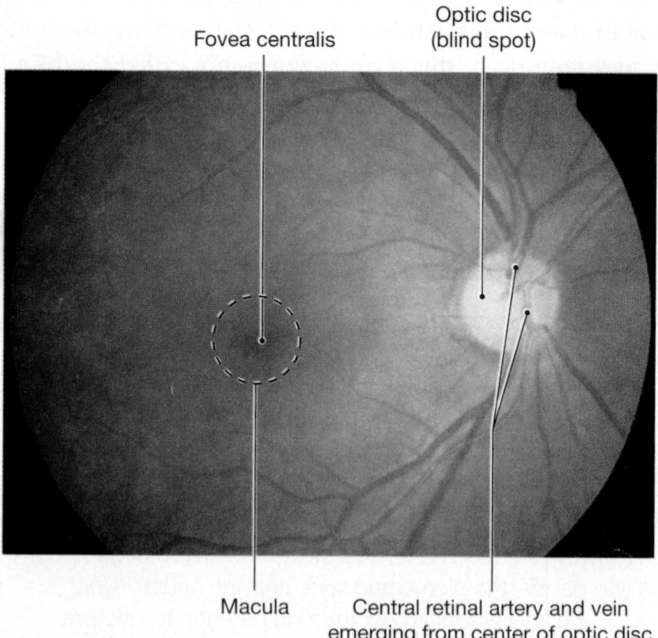
Fovea centralis
Optic disc (blind spot)
Macula
Central retinal artery and vein emerging from center of optic disc

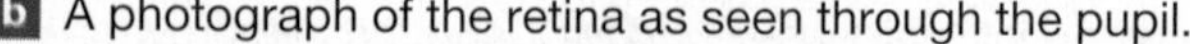

b A photograph of the retina as seen through the pupil.

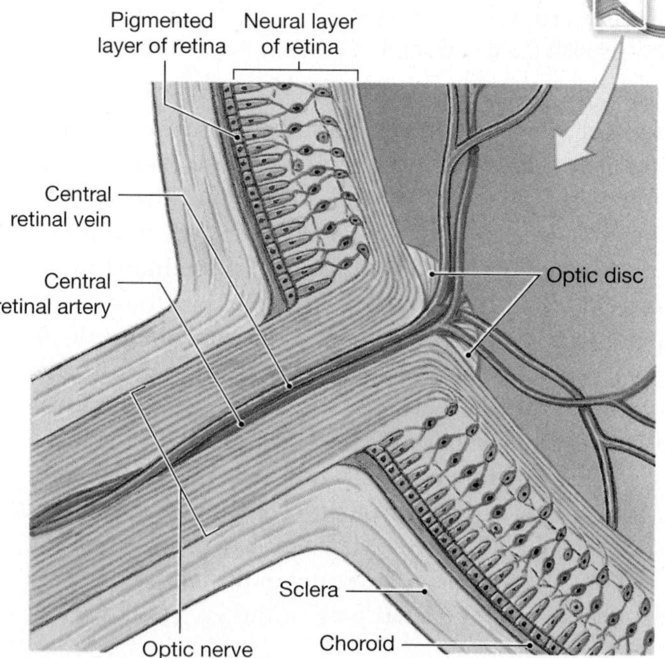
Pigmented layer of retina
Neural layer of retina
Central retinal vein
Central retinal artery
Optic disc
Sclera
Optic nerve
Choroid

c The optic disc in diagrammatic sagittal section.

? Amacrine cells are found where what other types of cells synapse? Horizontal cells are found where what other types of cells synapse?

The reason is that involuntary eye movements keep the visual image moving and allow your brain to fill in the missing information. However, try the simple activity in Figure 17–8 to prove that a blind spot really exists in your field of vision.

The Chambers of the Eye

Recall that the ciliary body and lens divide the interior of the eye into a large posterior cavity and a smaller anterior cavity made up of anterior and posterior chambers (look back at Figure 17–5c). Both chambers are filled with aqueous humor. The posterior cavity also contains aqueous humor, but the gelatinous *vitreous body* takes up most of its volume.

Aqueous Humor. Aqueous humor is a fluid that circulates within the anterior cavity. It passes from the posterior chamber to the anterior chamber through the pupil (Figure 17–9). It also freely diffuses through the posterior cavity and across the surface of the retina. Its composition resembles that of cerebrospinal fluid.

Aqueous humor forms through active secretion by epithelial cells of the ciliary body's ciliary processes. The epithelial

Figure 17–8 A Demonstration of the Presence of a Blind Spot. Close your left eye and stare at the plus sign with your right eye, keeping the plus sign in the center of your field of vision. Begin with the page a few inches away from your eye, and gradually increase the distance. The dot will disappear when its image falls on the blind spot, at your optic disc. To check the blind spot in your left eye, close your right eye and repeat the sequence while you stare at the dot.

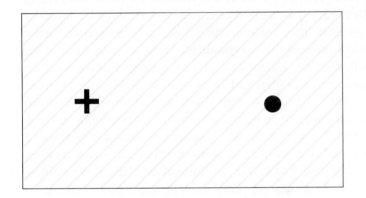

> **+ Clinical Note** Detached Retina
>
> Photoreceptors depend entirely on the diffusion of oxygen and nutrients from blood vessels in the choroid. In a **detached retina**, the neural layer of the retina becomes separated from the pigmented layer. This condition can result from a variety of factors, including a sudden hard impact to the eye. This is a medical emergency—unless the two parts of the retina are reattached, the photoreceptors will degenerate and vision will be lost. Reattachment involves "welding" the two parts together using laser beams focused through the cornea. These beams heat the two parts, fusing them together at several points around the retina. However, the procedure destroys the photoreceptors at the "welds," producing permanent blind spots.

Figure 17–9 The Circulation of Aqueous Humor. Aqueous humor, which is secreted at the ciliary body, circulates through the posterior and anterior chambers before it is reabsorbed through the scleral venous sinus.

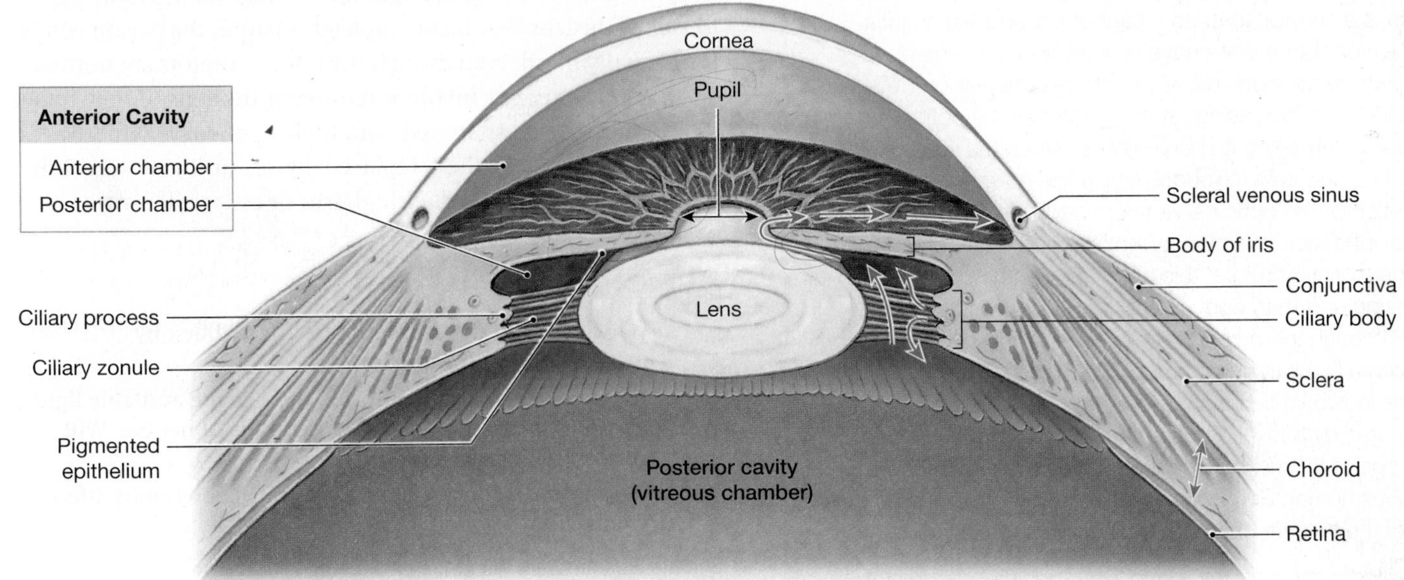
Cornea
Pupil
Anterior Cavity
Anterior chamber
Posterior chamber
Scleral venous sinus
Body of iris
Conjunctiva
Lens
Ciliary body
Ciliary process
Ciliary zonule
Sclera
Pigmented epithelium
Posterior cavity (vitreous chamber)
Choroid
Retina

cells regulate its composition. Because aqueous humor circulates, it provides an important route for nutrient and waste transport. It also serves as a fluid cushion.

Because the eyeball is filled with aqueous humor, pressure provided by this fluid helps retain the eye's shape. Fluid pressure also stabilizes the position of the retina, pressing the neural layer against the pigmented layer. In effect, the aqueous humor acts like the air inside a balloon.

The eye's **intra-ocular pressure** can be measured in the anterior chamber, where the fluid pushes against the inner surface of the cornea. Intra-ocular pressure is usually checked by applanation tonometry whereby a small, flat disk is placed on the anesthetized cornea to measure the tension. Normal intraocular pressure ranges from 12 to 21 mm Hg.

Aqueous humor is secreted into the posterior chamber at a rate of 1–2 mL per minute. It leaves the anterior chamber at the same rate. After filtering through a network of connective tissues located near the base of the iris, aqueous humor enters the **scleral venous sinus** (*canal of Schlemm*), a passageway that extends completely around the eye at the level of the corneoscleral junction. Collecting channels then deliver the aqueous humor to veins in the sclera.

17

Aqueous humor is removed and recycled within a few hours of its formation. The rate of removal normally keeps pace with the rate of generation at the ciliary processes.

The Vitreous Body. The posterior cavity of the eye contains the **vitreous body**, a gelatinous mass. Its fluid portion is called **vitreous humor**. The vitreous body helps stabilize the shape of the eye. Otherwise, the eye might distort as the extrinsic eye muscles change its position within the orbit. Specialized cells embedded in the vitreous body produce the collagen fibers and proteoglycans that account for the gelatinous consistency of this mass. Unlike the aqueous humor, the vitreous body is formed during development and is not replaced.

+ Clinical Note Glaucoma

If aqueous humor cannot drain into the scleral venous sinus, intra-ocular pressure rises due to the continued production of aqueous humor, and **glaucoma** results. The sclera is a fibrous coat, so it cannot expand like an inflating balloon, but it does have one weak point—the optic disc, where the optic nerve penetrates the wall of the eye. Gradually the increasing pressure pushes the optic nerve outward, damaging its nerve fibers. When the intra-ocular pressure has risen to roughly twice the normal level, the distortion of the optic nerve fibers begins to interfere with the propagation of action potentials, and peripheral vision begins to deteriorate. If this condition is not corrected, tunnel vision and then complete blindness may result.

Although the primary factors responsible are not known, glaucoma is relatively common. For this reason, most eye exams include a test of intra-ocular pressure. Glaucoma may be treated by topical drugs that constrict the pupil and tense the edge of the iris, making the surface more permeable to aqueous humor. Surgical correction involves perforating the wall of the anterior chamber to encourage drainage.

The Lens

The **lens** is a transparent, biconvex (outwardly curving) flexible disc that lies posterior to the cornea and is held in place by the ciliary zonule that originates on the ciliary body of the choroid (see Figure 17–9). The primary function of the lens is to focus the visual image on the photoreceptors. The lens does so by changing its shape.

The lens consists of concentric layers of cells. A dense fibrous capsule covers the entire lens. Many of the capsular fibers are elastic. Unless an outside force is applied, they will contract and make the lens spherical. Around the edges of the lens, these capsular fibers intermingle with those of the ciliary zonule.

The cells in the interior of the lens are called **lens fibers**. These highly specialized cells have lost their nuclei and other organelles. They are long and slender and are filled with transparent proteins called **crystallins**. These proteins give the lens both its clarity and its focusing power. Crystallins are extremely stable proteins—they remain intact and functional for a lifetime.

The transparency of the lens depends on the crystallin proteins maintaining a precise combination of structural and biochemical characteristics. Because these proteins are not renewed, any modifications they undergo over time can accumulate and result in the lens losing its transparency. This abnormality is known as a **cataract**. Cataracts can result from injuries, UV radiation, or reaction to drugs. **Senile cataracts**, however, are a natural consequence of aging and are the most common form.

Over time, the lens turns yellowish and eventually begins to lose its transparency. As the lens becomes "cloudy," the individual needs brighter and brighter light for reading. Visual clarity begins to fade. If the lens becomes completely opaque, the person will be functionally blind, even though the photoreceptors are normal.

Cataract surgery involves removing the lens, either intact or after it has been shattered with high-frequency sound waves. The missing lens is then replaced by an artificial substitute. Vision is then fine-tuned with glasses or contact lenses.

✓ Checkpoint

7. **Which layer of the eye would be affected first by inadequate tear production?**
8. **As Sue enters a dimly lit room, most of the available light becomes focused on the fovea centralis of her eye. Will she be able to see very clearly?**
9. **How would a blockage of the scleral venous sinus affect your vision?**

See the blue Answers tab at the back of the book.

17-4 The focusing of light on the retina leads to the formation of a visual image

Learning Outcome Describe how refraction and the focusing of light on the retina lead to vision.

Each of the millions of photoreceptors monitors the light striking a specific site on the retina. The processing of information from all the receptors produces a visual image. To understand how this happens, we need to look into the properties of light itself, and how light interacts with the structures of our eyes.

An Introduction to Light

Light energy is a form of *radiant energy* that travels in waves with characteristic *wavelengths* (distances between wave peaks). Let's look at the relationship between wavelengths and color, and learn what happens when light rays encounter an object such as our eyes.

Wavelength and Color

Our eyes are sensitive to wavelengths of 700–400 nm, the spectrum of visible light. Remember this spectrum, as seen in a rainbow, with the acronym ROY G. BIV (**R**ed, **O**range, **Y**ellow, **G**reen, **B**lue, **I**ndigo, **V**iolet). Visible light is also described as being made up of photons, small energy packets with characteristic wavelengths. Photons of red light carry the least energy and have the longest wavelength. Photons from the violet portion of the spectrum contain the most energy and have the shortest wavelength.

Refraction and Focusing of Light

The eye is often compared to a camera. To provide useful information, the lens of the eye, like a camera lens, must *focus* the arriving image. To say that an image is "in focus" means that the light rays arriving from an object strike the sensitive surface of the retina (or the semiconductor device that records light electronically in digital cameras) in precise order so as to form a miniature image of the object. If the rays are not perfectly focused, the image is blurry.

Focusing normally occurs in two steps, as light passes first through the cornea and then the lens. To focus the light on the retina, the lens must change shape, or undergo *accommodation*.

Refraction. Light is **refracted**, or bent, when it passes from one medium to another medium with a different density. You can see this effect if you stick a pencil into a glass of water. Notice that the shaft of the pencil appears to bend sharply at the air–water interface. This effect occurs when light is refracted as it passes into the air from the much denser water.

In the human eye, the greatest amount of refraction occurs when light passes from the air into the corneal tissues, which have a density close to that of water. When you open your eyes under water, you cannot see clearly because refraction at the corneal surface has been largely eliminated. Light passes unbent from one watery medium to another.

Additional refraction takes place when the light passes from the aqueous humor in the anterior chamber into the relatively dense lens. The lens provides the extra refraction needed to focus the light rays from an object toward a **focal point**—a specific point of intersection on the retina.

The distance between the center of the lens and its focal point is the **focal distance** of the lens. Whether in the eye or in a camera, the focal distance is determined by two factors:

- *The Distance of the Object from the Lens.* The closer an object is to the lens, the greater the focal distance (Figure 17–10a,b).
- *The Shape of the Lens.* The rounder the lens, the more refraction occurs. So, a very round lens has a shorter focal distance than a flatter one (Figure 17–10b,c).

If light passing through the cornea and lens is not refracted properly, the visual image will be distorted. In the condition called **astigmatism**, the degree of curvature in the cornea or lens varies from one axis to another. Minor astigmatism is very common. The image distortion may be so minimal that people are unaware of the condition.

Figure 17–10 Factors Affecting Focal Distance. Light rays from a source are refracted when they reach the lens of the eye. The rays are then focused onto a single focal point.

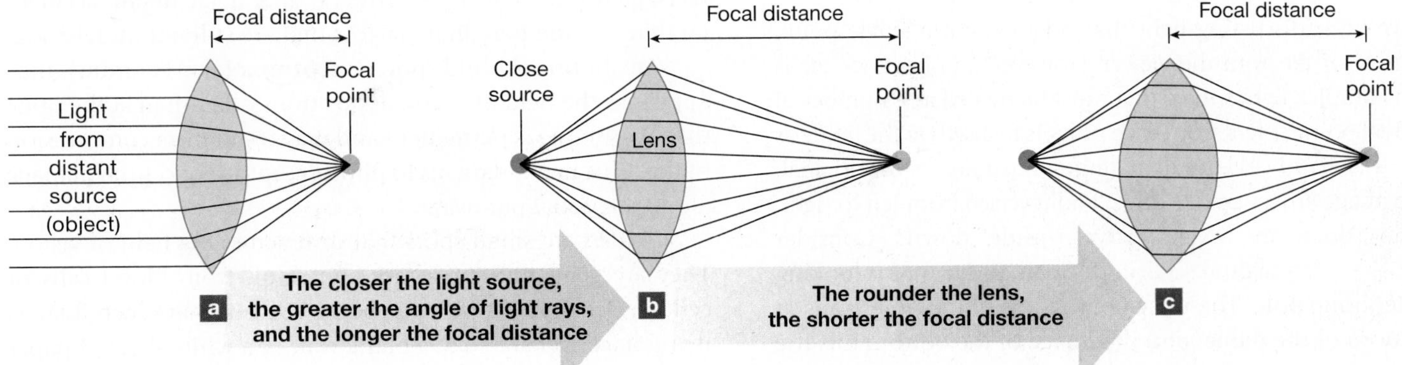
Focal distance
Focal point
Light from distant source (object)
a
The closer the light source, the greater the angle of light rays, and the longer the focal distance
Focal distance
Close source
Lens
Focal point
b
The rounder the lens, the shorter the focal distance
Focal distance
Focal point
c

Figure 17–11 **Accommodation.** For the eye to form a sharp image, the lens becomes rounder for close objects and flatter for distant objects.

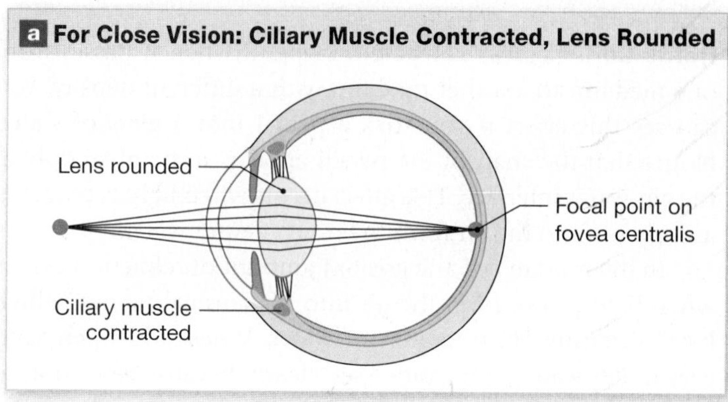
a For Close Vision: Ciliary Muscle Contracted, Lens Rounded
Lens rounded
Focal point on fovea centralis
Ciliary muscle contracted

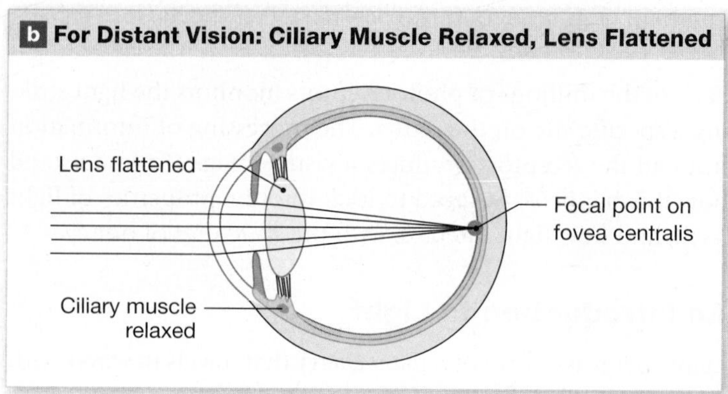
b For Distant Vision: Ciliary Muscle Relaxed, Lens Flattened
Lens flattened
Focal point on fovea centralis
Ciliary muscle relaxed

17

Accommodation. **Accommodation** is the automatic adjustment of the eye to give us clear vision (Figure 17–11). During accommodation, the lens becomes rounder to focus the image of a nearby object on the retina. The lens becomes flatter to focus the image of a distant object on the retina.

How does the lens change shape? The lens is held in place by the ciliary zonule that originates at the ciliary body. Smooth muscle fibers in the ciliary body act like sphincter muscles. When the ciliary muscle contracts, the ciliary body moves toward the lens, thereby reducing the tension in the ciliary zonule. The elastic capsule then pulls the lens into a rounder shape that increases the refractive (bending) power of the lens. This enables it to bring light from nearby objects into focus on the retina (see Figure 17–11a). When the ciliary muscle relaxes, the ciliary zonule pulls at the circumference of the lens, making the lens flatter (see Figure 17–11b).

The greatest amount of refraction is required to view objects that are very close to the lens. The inner limit of clear vision, known as the *near point of vision,* is determined by the degree of elasticity in the lens. Children can usually focus on something 7–9 cm (3–4 in.) from the eye, but over time the lens tends to become stiffer and less responsive. A young adult can usually focus on objects 15–20 cm (6–8 in.) away. As we age, this distance gradually increases. The near point at age 60 is typically about 83 cm (33 in.).

Image Formation and Reversal

We have been discussing light that originates at a single point, either near or far from the viewer. However, an object we see is really a complex light source that must be treated as a number of individual points. Light from each point is focused on the retina as in Figure 17–12a,b. The result is a miniature image of the original, but the image arrives upside down and reversed from left to right.

Why does an image arrive upside down? Consider Figure 17–12c, a sagittal section through an eye that is looking at a telephone pole. The image of the top of the pole lands at the bottom of the retina, and the image of the bottom hits the top of the retina. Now consider Figure 17–12d, a horizontal section through an eye that is looking at a picket fence. The image of the left edge of the fence falls on the right side of the retina, and the image of the right edge falls on the left side of the retina. The brain compensates for this image reversal, so you are not aware of any difference between the orientation of the image on the retina and that of the object.

Visual Acuity

How well you see, or your **visual acuity**, is rated by comparison to a "normal" standard. The standard vision rating of 20/20 is defined as the level of detail seen at a distance of 20 feet by a person with normal vision. That is, a person with a visual acuity of 20/20 sees clearly at 20 feet what should normally be seen at 20 feet. Vision rated as 20/15 is better than average, because at 20 feet the person is able to see details that would be clear to a normal eye only at a distance of 15 feet. Conversely, a person with 20/30 vision must be 20 feet from an object to discern details that a person with normal vision could make out at a distance of 30 feet.

When visual acuity falls below 20/200, even with the help of glasses or contact lenses, the individual is considered to be legally blind. It is estimated that there are 1.3 million legally blind people in the United States, and more than half of those are over 65 years old. The term *blindness* implies a total absence of vision due to damage to the eyes or to the optic pathways. Common causes of blindness include diabetes mellitus, cataracts, glaucoma, corneal scarring, retinal detachment, accidental injuries, and hereditary factors that are still not understood.

An abnormal blind spot, or **scotoma** (skō-TŌ-muh), may appear in the field of vision at positions other than at the optic disc. A scotoma is permanent and may result from compression of the optic nerve, damage to photoreceptors, or central damage along the visual pathway.

Floaters are small spots that drift across the field of vision. They are generally temporary and result from blood cells or cellular debris in the vitreous body. You may have seen floaters if you have ever stared at a blank wall or a white sheet of paper and noticed these little spots. Spotlight Figure 17–13 describes common visual abnormalities.

Figure 17–12 **Image Formation.** These illustrations are not drawn to scale because the fovea centralis occupies a small area of the retina, and the projected images are very tiny. As a result, the crossover of light rays is shown in the lens, but it actually occurs very close to the fovea centralis.

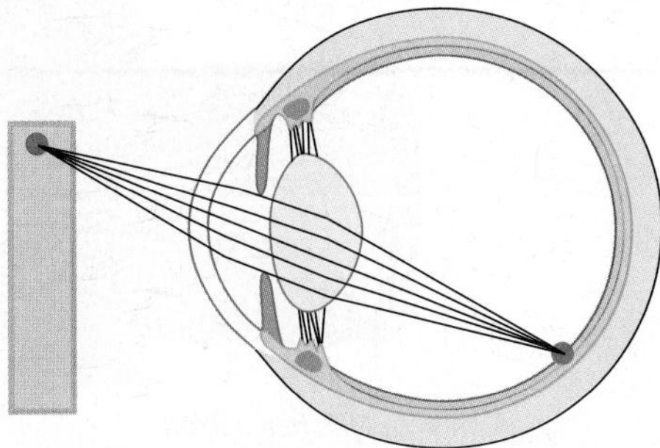

a Light from a point at the top of an object is focused on the lower retinal surface.

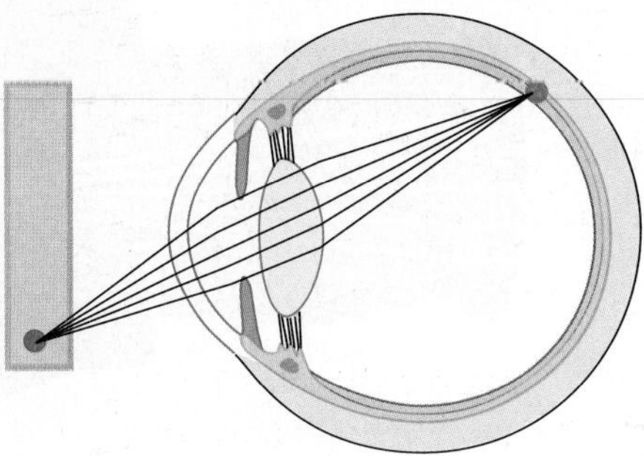

b Light from a point at the bottom of an object is focused on the upper retinal surface.

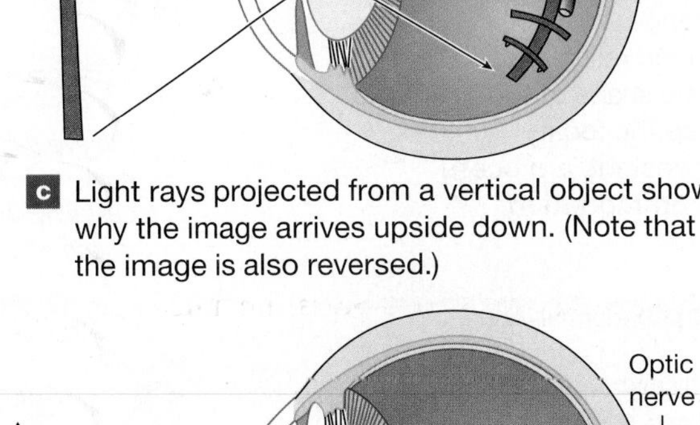
Optic
nerve

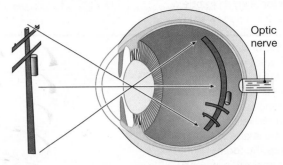

c Light rays projected from a vertical object show why the image arrives upside down. (Note that the image is also reversed.)

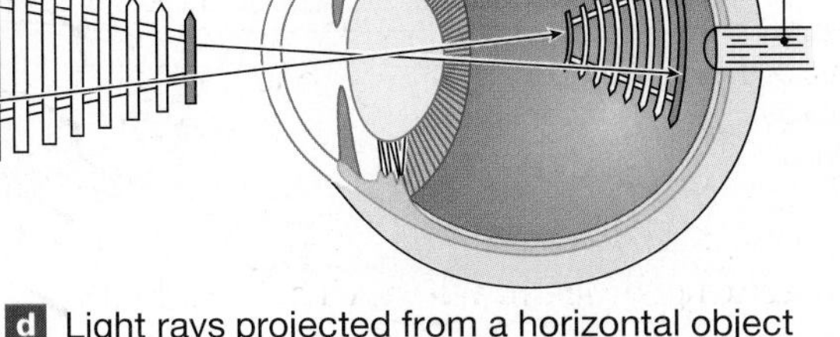
Optic
nerve

d Light rays projected from a horizontal object show why the image arrives with a left and right reversal. The image also arrives upside down.

Checkpoint

10. **Define *focal point*.**
11. **When the lens of your eye is more rounded, are you looking at an object that is close to you or far from you?**

See the blue Answers tab at the back of the book.

17-5 Photoreceptors transduce light into electrical signals that are then processed in the visual cortex

Learning Outcome Explain color and depth perception, describe how light stimulates the production of nerve impulses, and trace the visual pathways to their destinations in the brain.

How does our special sense of vision work? Let's begin to answer this question by examining the structure of photoreceptors, and then the physiology of vision, the way in which photoreceptors function. Finally, we will consider the structure and function of the visual pathways.

Physiology of Vision

The rods and cones of the retina are called photoreceptors because they detect *photons*, basic units of visible light. Rods provide the central nervous system with information about the presence or absence of photons, with little regard to their wavelength. Cones provide information about the wavelength of arriving photons, giving us the perception of color.

Anatomy of Photoreceptors: Rods and Cones

Figure 17–14a compares the structures of **photoreceptors**, rods and cones. The names *rod* and *cone* refer to the shape of each photoreceptor's outer segment. Both rods and cones contain special organic compounds called *visual pigments.* These pigments are located in each photoreceptor's outer segment, in flattened membranous plates called *discs.* Please study this figure before reading on.

Physiology of Photoreceptors

Visual pigments are able to absorb photons; this is the first key step in the process of **photoreception** —the detection of light. This section looks at the function of photoreceptors and the visual pigments they contain.

Visual Pigments. Visual pigments are derivatives of the compound **rhodopsin** (rō-DOP-sin), or *visual purple,* the visual pigment found in rods (Figure 17–14b). Rhodopsin consists of a protein, **opsin**, bound to the pigment **retinal** (RET-ih-nal),

SPOTLIGHT

Figure 17–13 Refractive Problems

A camera focuses an image by moving the lens toward or away from the film or semiconductor device. This method cannot work in our eyes, because the distance from the lens to the macula cannot change. We focus images on the retina by changing the shape of the lens to keep the focal distance constant, a process called **accommodation**.

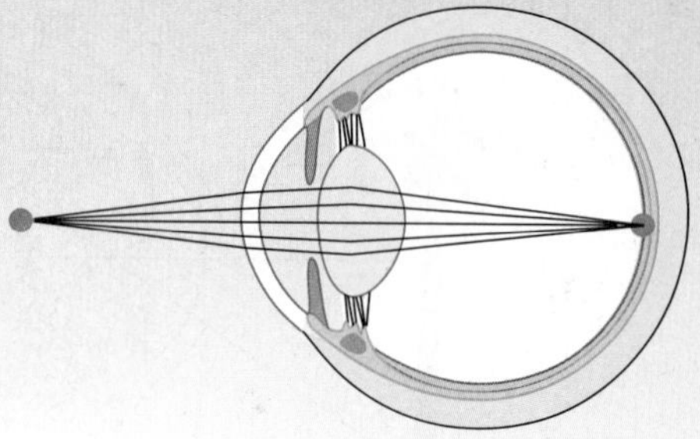

The eye has a fixed focal distance and focuses by varying the shape of the lens.

A camera lens has a fixed size and shape and focuses by varying the distance to the film or semiconductor device.

Emmetropia (normal vision)

In the healthy eye, when the ciliary muscle is relaxed and the lens is flattened, a distant image will be focused on the retina's surface. This condition is called **emmetropia** (*emmetro-*, proper + *opia,* vision).

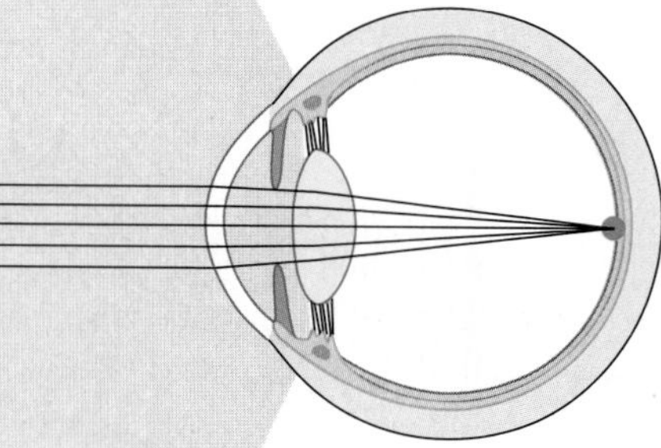

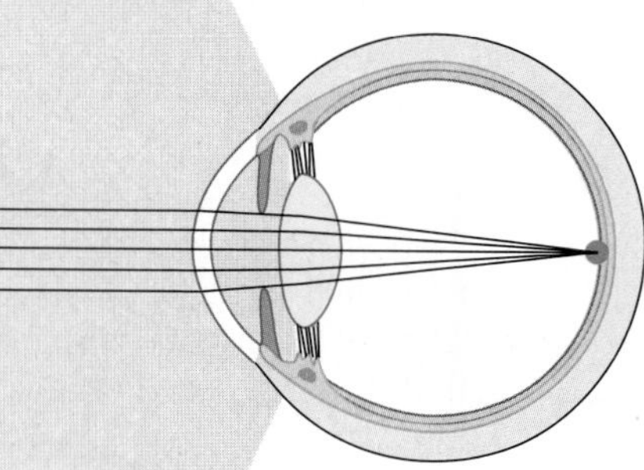

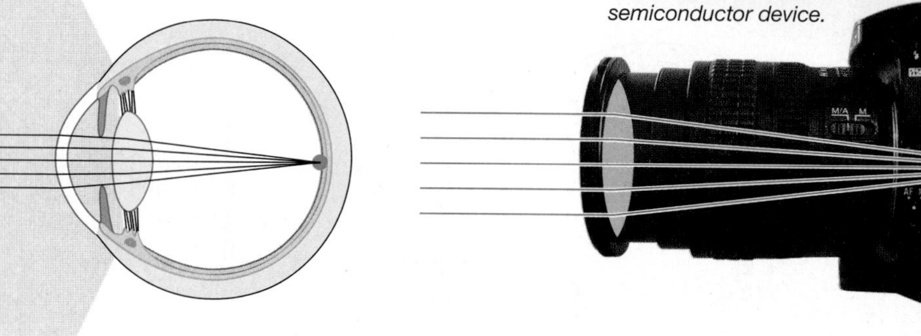

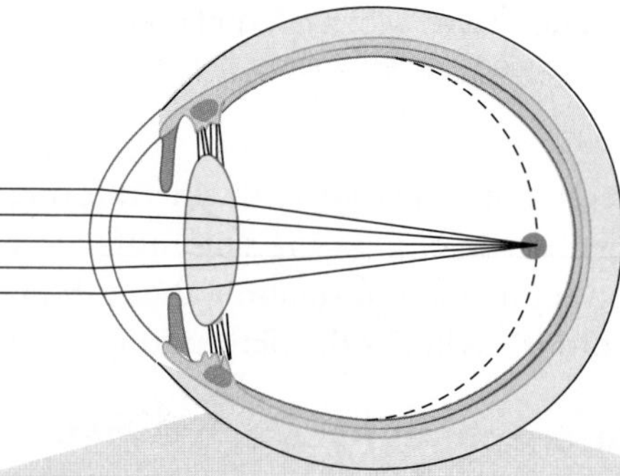

Myopia (nearsightedness)

If the eyeball is too deep or the resting curvature of the lens is too great, the image of a distant object is projected in front of the retina. The person will see distant objects as blurry and out of focus. Vision at close range will be normal because the lens is able to round as needed to focus the image on the retina.

Myopia corrected with a diverging, concave lens

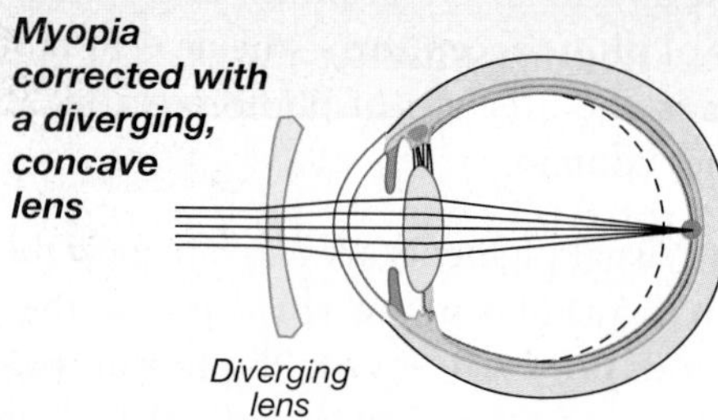
Diverging lens

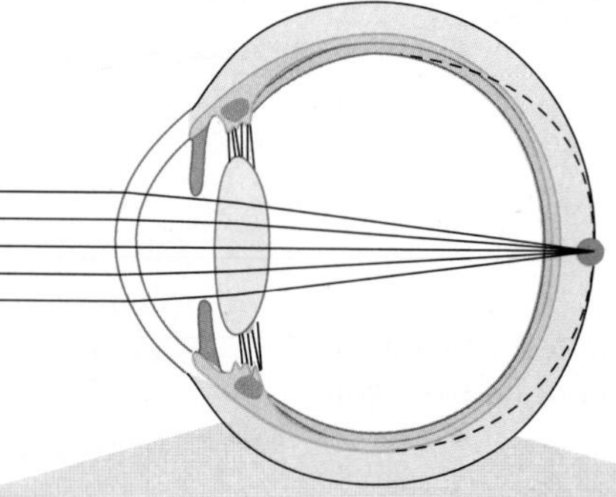

Hyperopia (farsightedness)

If the eyeball is too shallow or the lens is too flat, hyperopia results. The ciliary muscle must contract to focus even a distant object on the retina. And at close range the lens cannot provide enough refraction to focus an image on the retina. Older people become farsighted as their lenses lose elasticity, a form of hyperopia called **presbyopia** (*presbys*, old man).

Hyperopia corrected with a converging, convex lens

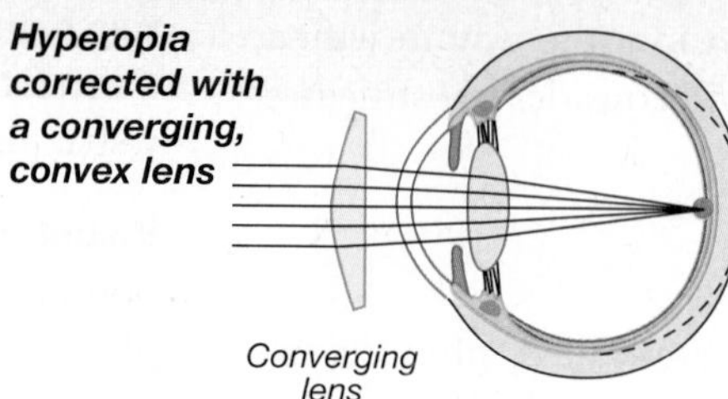
Converging lens

Surgical Correction

Variable success at correcting myopia and hyperopia has been achieved by surgery that reshapes the cornea. In **photorefractive keratectomy** (PRK) a computer-guided laser shapes the cornea to exact specifications. The entire procedure can be done in less than a minute. A variation on PRK is called *LASIK* (*Laser-Assisted in-Situ Keratomileusis*). In this procedure the interior layers of the cornea are reshaped and then re-covered by the flap of original outer corneal epithelium. Roughly 70 percent of LASIK patients achieve normal vision, and LASIK has become the most common form of refractive surgery.

Even after surgery, many patients still need reading glasses, and both immediate and long-term visual problems can occur.

Figure 17–14 Structure of Rods, Cones, and the Rhodopsin Molecule.

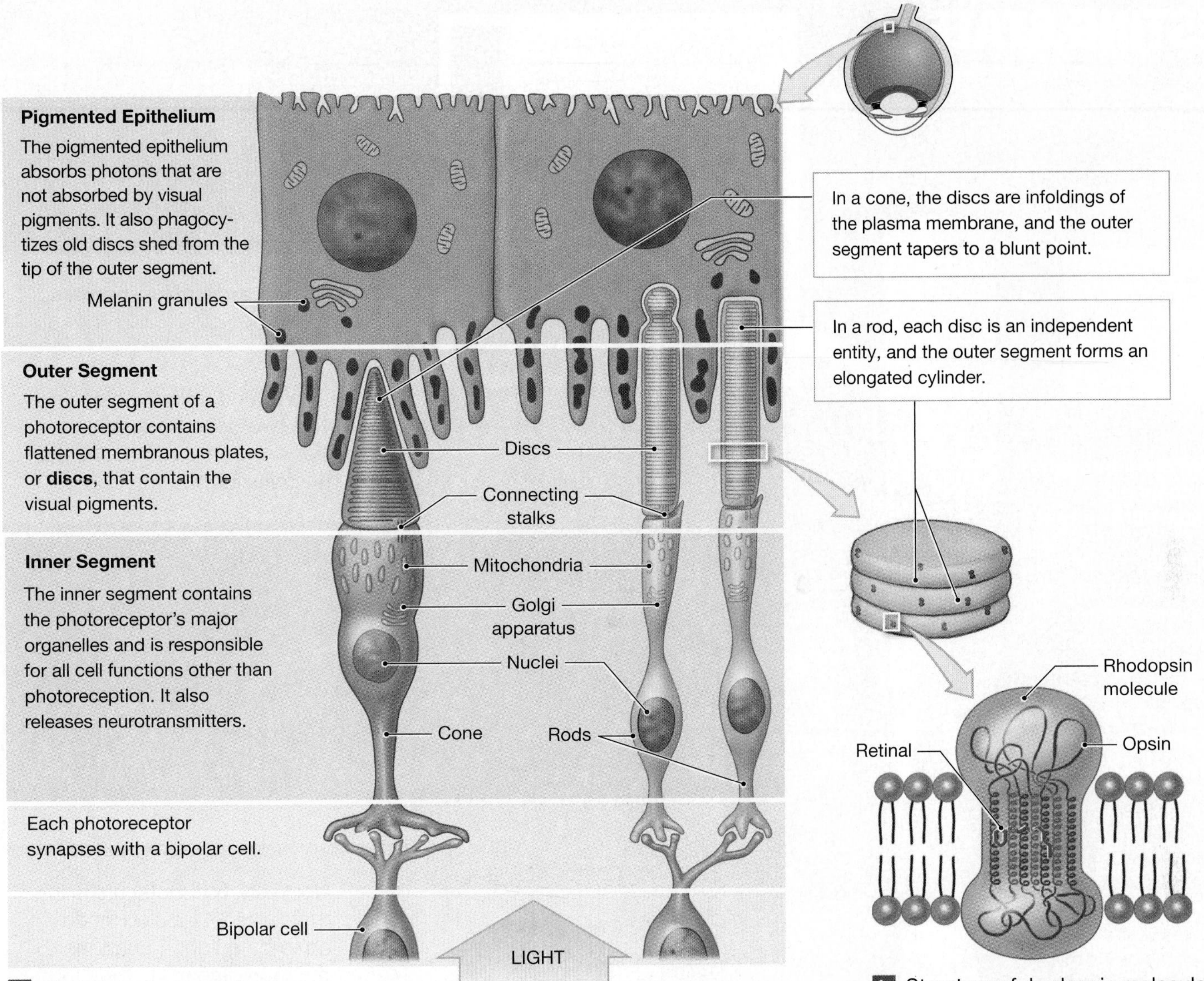

Pigmented Epithelium

The pigmented epithelium absorbs photons that are not absorbed by visual pigments. It also phagocytizes old discs shed from the tip of the outer segment.

Melanin granules

Outer Segment

The outer segment of a photoreceptor contains flattened membranous plates, or **discs**, that contain the visual pigments.

Inner Segment

The inner segment contains the photoreceptor's major organelles and is responsible for all cell functions other than photoreception. It also releases neurotransmitters.

Each photoreceptor synapses with a bipolar cell.

Bipolar cell

Discs

Connecting stalks

Mitochondria

Golgi apparatus

Nuclei

Cone

Rods

LIGHT

In a cone, the discs are infoldings of the plasma membrane, and the outer segment tapers to a blunt point.

In a rod, each disc is an independent entity, and the outer segment forms an elongated cylinder.

a Structure of rods and cones

Rhodopsin molecule

Retinal

Opsin

b Structure of rhodospin molecule

which is synthesized from **vitamin A**. All rods contain the same form of opsin. Cones contain the same retinal pigment that rods do, but the retinal is attached to different forms of opsin.

We have three types of cones—**blue cones, green cones,** and **red cones**. Each type has a different form of opsin and is sensitive to a different range of wavelengths. Their sensitivities overlap, but each type is most sensitive to a specific portion of the visual spectrum (Figure 17–15). Their stimulation in various combinations is the basis for color vision. In an individual with normal vision, the cone population consists of 16 percent blue cones, 10 percent green cones, and 74 percent red cones.

The process of photoreception by visual pigments is described in Spotlight Figure 17–16. Please study this figure before moving on.

Figure 17–15 Cone Types and Sensitivity to Color. This graph compares the absorptive characteristics of blue, green, and red cones with those of typical rods.

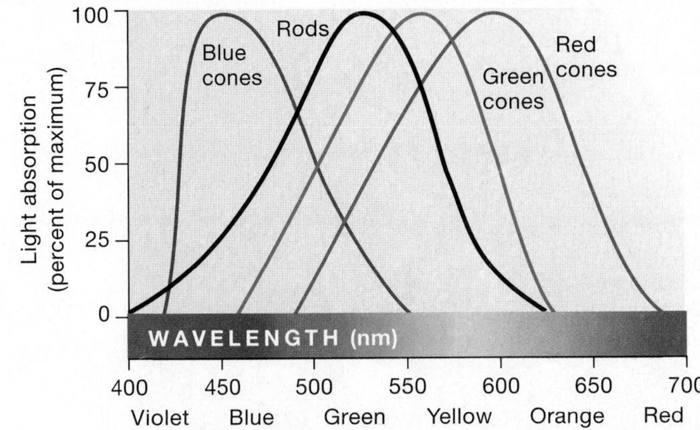
Light absorption (percent of maximum)
100
75
50
25
0
Blue cones
Rods
Green cones
Red cones
WAVELENGTH (nm)
400
450
500
550
600
650
700
Violet
Blue
Green
Yellow
Orange
Red

RESTING STATE

DARKNESS

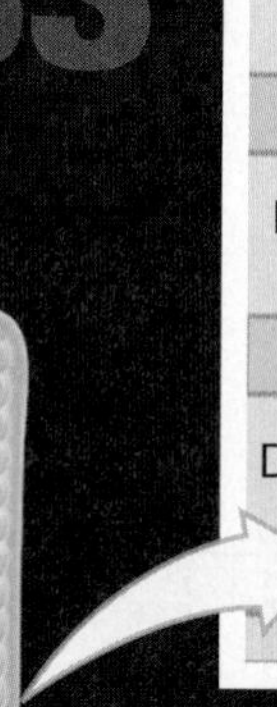

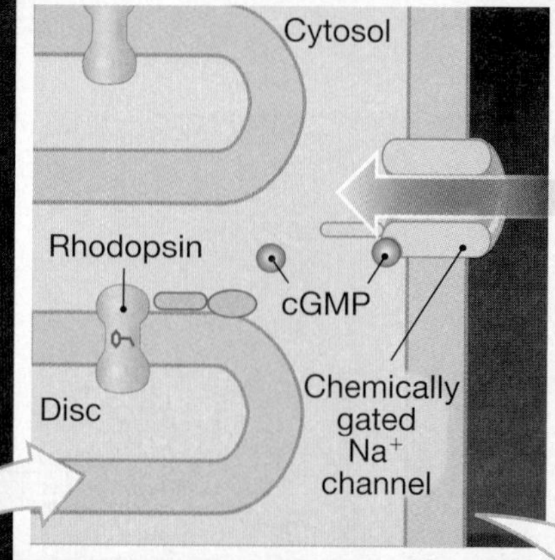

Cytosol
Rhodopsin
cGMP
Disc
Chemically gated Na^+ channel

−40 mV

Sodium entry through gated channels produces dark current

The plasma membrane in the outer segment of the photoreceptor contains chemically gated sodium ion channels. In darkness, these gated channels are kept open in the presence of *cGMP (cyclic guanosine monophosphate)*, a derivative of the high-energy compound *guanosine triphosphate (GTP)*. Because the channels are open, the membrane potential is approximately –40 mV, rather than the –70 mV typical of resting neurons. At the –40 mV membrane potential, the photoreceptor is continuously releasing neurotransmitters (in this case, glutamate) across synapses at the inner segment. The inner segment also continuously pumps sodium ions out of the cytosol.

Na^+

The movement of ions into the outer segment, on to the inner segment, and out of the cell is known as the **dark current**.

Na^+

Rod

Neurotransmitter Release

Bipolar cell

1 Opsin activation occurs

The bound retinal molecule has two possible configurations: the 11-*cis* form and the 11-*trans* form.

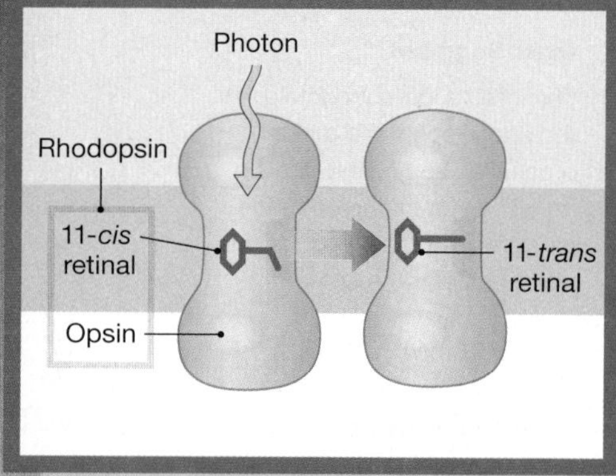

Photon
Rhodopsin
11-*cis* retinal
11-*trans* retinal
Opsin

Normally, the molecule is in the curved 11-*cis* form; on absorbing light it changes to the more linear 11-*trans* form. This change activates the opsin molecule.

ACTIVE STATE IN LIGHT

2 Opsin activates transducin, which in turn activates phosphodiesterase (PDE)

Transducin is a G protein—a membrane-bound enzyme complex

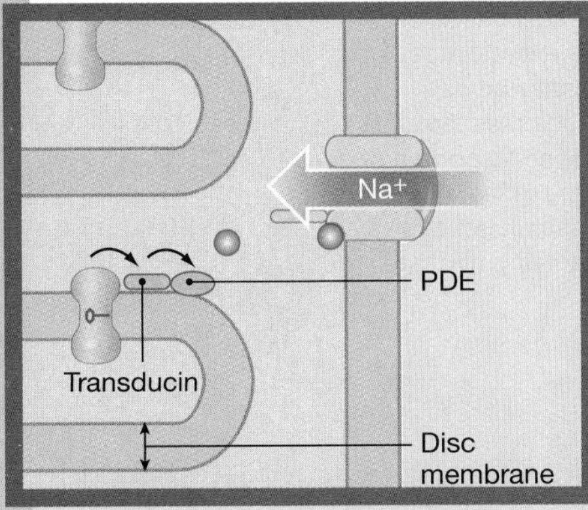

Na+
PDE
Transducin
Disc membrane

In this case, opsin activates transducin, and transducin in turn activates **phosphodiesterase** (**PDE**).

3 Cyclic GMP levels decline and chemically gated sodium ion channels close

Phosphodiesterase is an enzyme that breaks down cGMP.

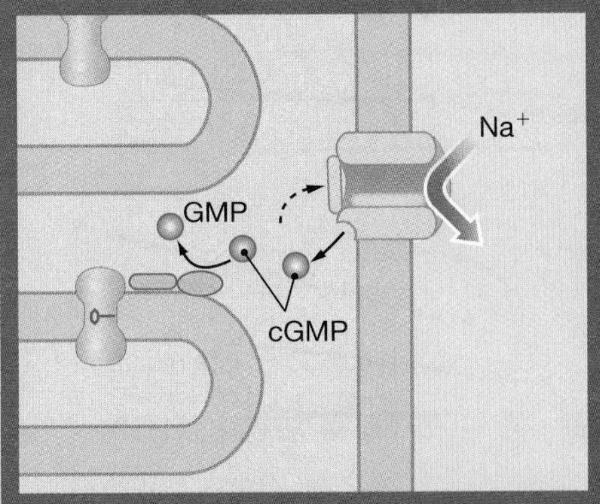

Na+
GMP
cGMP

The removal of cGMP from the chemically gated sodium ion channels results in their inactivation. The rate of Na^+ entry into the cytosol then decreases.

4 Dark current is reduced and rate of neurotransmitter release declines

The reduction in the rate of Na^+ entry reduces the dark current. At the same time, active transport continues to export Na^+ from the cytosol. When the sodium ion channels close, the membrane potential drops toward −70 mV. As the plasma membrane hyperpolarizes, the rate of neurotransmitter release decreases. This decrease signals the adjacent bipolar cell that the photoreceptor has absorbed a photon. After absorbing a photon, retinal does not spontaneously revert to the 11-*cis* form. Instead, the entire rhodopsin molecule must be broken down into retinal and opsin, in a process called *bleaching*. It is then reassembled.

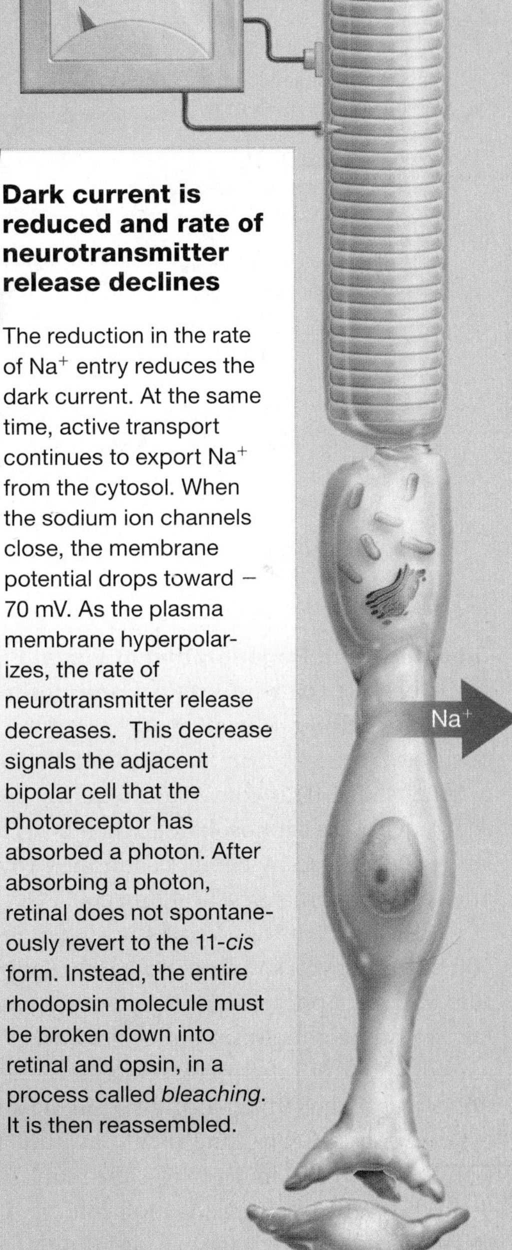

−70 mV
Na+

Figure 17–17 Bleaching and Regeneration of Visual Pigments.

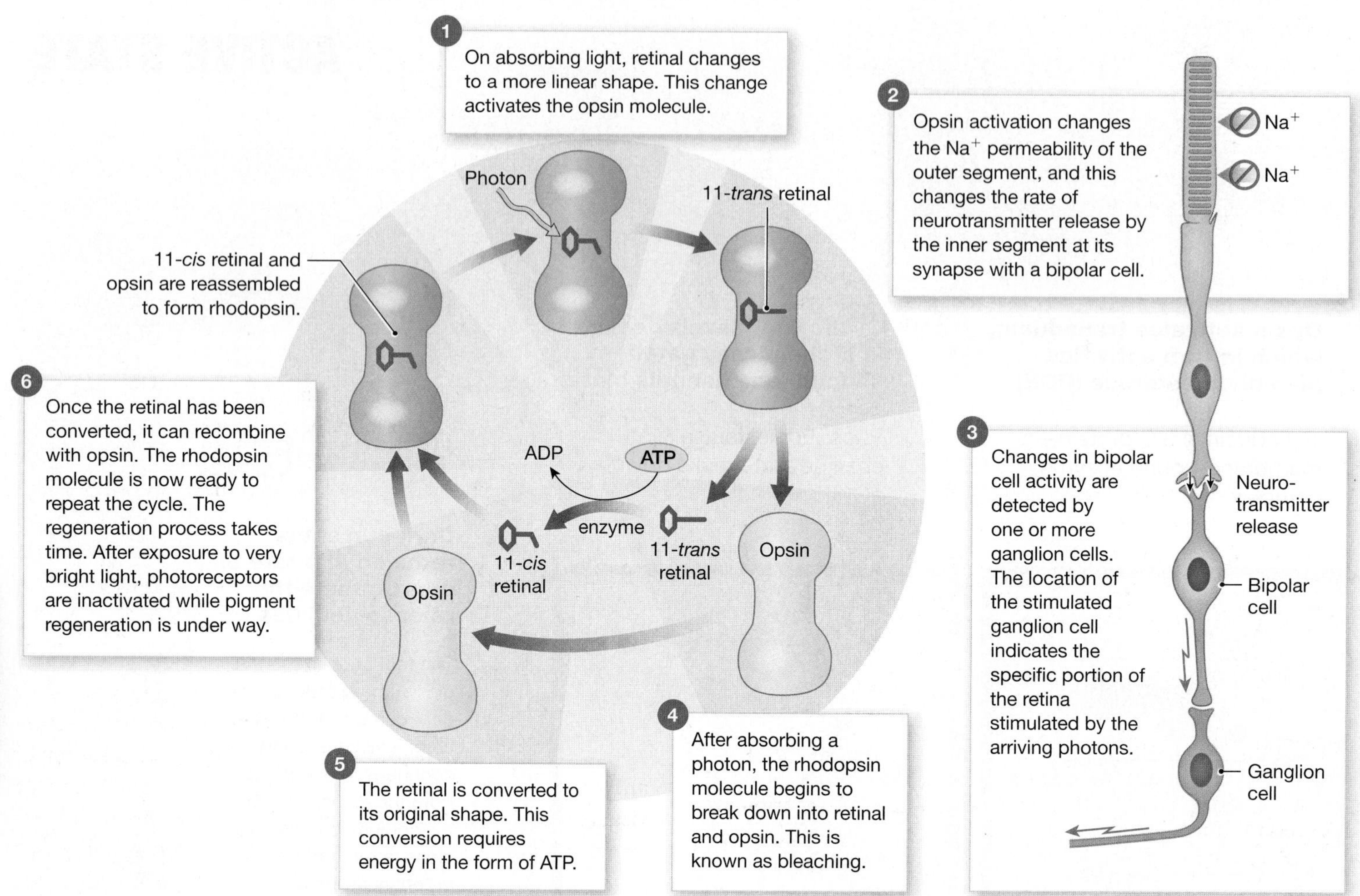
1
On absorbing light, retinal changes to a more linear shape. This change activates the opsin molecule.
2
Opsin activation changes the Na+ permeability of the outer segment, and this changes the rate of neurotransmitter release by the inner segment at its synapse with a bipolar cell.
Na+
Na+
Photon
11-trans retinal
11-cis retinal and opsin are reassembled to form rhodopsin.
6
Once the retinal has been converted, it can recombine with opsin. The rhodopsin molecule is now ready to repeat the cycle. The regeneration process takes time. After exposure to very bright light, photoreceptors are inactivated while pigment regeneration is under way.
ADP
ATP
enzyme
11-cis retinal
11-trans retinal
Opsin
Opsin
3
Changes in bipolar cell activity are detected by one or more ganglion cells. The location of the stimulated ganglion cell indicates the specific portion of the retina stimulated by the arriving photons.
Neuro-transmitter release
Bipolar cell
Ganglion cell
4
After absorbing a photon, the rhodopsin molecule begins to break down into retinal and opsin. This is known as bleaching.
5
The retinal is converted to its original shape. This conversion requires energy in the form of ATP.

17

Bleaching and Regeneration of Visual Pigments. Rhodopsin is broken apart into retinal and opsin through a process called **bleaching** (Figure 17–17). Before recombining with opsin, the retinal must be enzymatically converted back to its original shape. This conversion requires energy in the form of ATP (adenosine triphosphate), and it takes time. Then rhodopsin is regenerated by being recombined with opsin. Bleaching and regeneration is a cyclical process.

Bleaching contributes to the lingering visual impression you have after you see a camera's flash. Following intense exposure to light, a photoreceptor cannot respond to further stimulation until its rhodopsin molecules have been regenerated. As a result, a "ghost" afterimage remains on the retina. We seldom notice bleaching under ordinary circumstances, because our eyes are constantly making small, involuntary changes in position that move the image across the retina's surface.

While the rhodopsin molecule is being reassembled (regenerated), membrane permeability of the outer segment is returning to normal. Opsin is inactivated when bleaching occurs, and the breakdown of cGMP halts as a result. As other enzymes generate cGMP in the cytosol, the chemically gated sodium ion channels reopen.

Synthesis and Recycling of Visual Pigments. The body contains vitamin A reserves sufficient to synthesize visual pigments for several months. A significant amount is stored in the cells of the pigmented layer of the retina.

New discs containing visual pigment are continuously assembled at the base of the outer segment of both rods and cones. A completed disc then moves toward the tip of the segment. After about 10 days, the disc is shed in a small droplet of cytoplasm. The pigment cells absorb droplets with shed discs, break down the disc membrane's components, and reconvert the retinal to vitamin A. The pigment cells then store the vitamin A for later transfer to the photoreceptors.

If dietary sources are inadequate, these reserves are gradually used up and the amount of visual pigment in the photoreceptors begins to decline. Daylight vision is affected, but in daytime the light is usually bright enough to stimulate any visual pigments that remain within the densely packed cone population of the fovea centralis. As a result, the problem first becomes apparent at night, when the dim light proves insufficient to activate the rods. This condition, known as **night blindness**, or **nyctalopia**, can be treated by eating a diet rich in vitamin A. The body can convert the carotene pigments in many vegetables to vitamin A. Carrots are a particularly good source of carotene, leading to the old adage that carrots are good for your eyes.

Light and Dark Adaptation of Visual Pigments

The sensitivity of your visual system varies with the intensity of illumination. After 30 minutes or more in the dark, almost all visual pigments will have recovered from photobleaching. This state in which the pigments are fully receptive to stimulation is called the **dark-adapted state**. When dark adapted, the visual system is extremely sensitive. For example, a single rod will hyperpolarize in response to a single photon of light. Even more remarkable, if as few as seven rods absorb photons at one time, you will see a flash of light.

When the lights come on, at first they seem almost unbearably bright, but over the next few minutes your sensitivity decreases as bleaching occurs. Eventually, the rate of bleaching is balanced by the rate at which the visual pigments reassemble. This condition is the **light-adapted state**. If you moved from the depths of a cave to the full sunlight of midday, your receptor sensitivity would decrease by a factor of 25,000.

The autonomic nervous system adjusts to incoming light. The pupils constrict, reducing the amount of light entering your eye to 1/30th the maximum dark-adapted level. Conversely, dilating the pupil fully can produce a 30-fold increase in the amount of light entering the eye. In addition, facilitating some of the synapses along the visual pathway can perhaps triple its sensitivity. In these ways, the sensitivity of the entire system may increase by a factor of more than 1 million.

Retinitis pigmentosa (RP) is an inherited disease characterized by progressive retinal degeneration. As the visual receptors gradually deteriorate, blindness eventually results. The mutations that are responsible change the structure of the photoreceptors—specifically, the visual pigments of the membrane discs. We do not know how the altered pigments lead to the destruction of photoreceptors.

Color Vision

An ordinary lightbulb or the sun emits photons of all wavelengths, which stimulate both rods and cones. Your eyes also detect photons that reach your retina after they bounce off objects around you. If photons of all visible wavelengths bounce off an object, the object will appear white to you. If the object absorbs all the photons (so that none reaches the retina), the object will appear black.

An object appears to have a particular color when it reflects (or transmits) photons from one portion of the visible spectrum and absorbs the rest. Color discrimination takes place through the integration of information arriving from all three types of cones: blue, green, and red. A person who sees these three primary colors has normal color vision. For example, you perceive yellow from a combination of inputs from highly stimulated green cones, less strongly stimulated red cones, and relatively unaffected blue cones (look back at Figure 17–15, p. 585). If all three cone populations are stimulated, we perceive the color as white. We also perceive white if rods, rather than cones, are stimulated. For this reason, everything appears black and white when you enter dimly lit surroundings or walk by starlight.

Persons who are unable to distinguish certain colors have a form of **color blindness**. The standard tests for color vision involve picking numbers or letters out of a complex colored picture such as the one shown in Figure 17–18. Color blindness occurs when one or more types of cones are nonfunctional. The cones may be absent, or they may be present but unable to manufacture the necessary visual pigments. In the most common type of color blindness (red–green color blindness), the red cones are missing, so the individual cannot distinguish red light from green light. Inherited color blindness involving one or two cone pigments is not unusual. Ten percent of all males show some color blindness, but only about 0.67 percent of all females have color blindness. Total color blindness is extremely rare. Only 1 person in 300,000 does not manufacture any cone pigments. We consider the inheritance of color blindness in Chapter 29.

The Visual Pathways

The visual pathways begin at the photoreceptors and end at the *visual cortex* of the cerebral hemispheres. In other sensory pathways we have examined, only one synapse lies between a receptor and a sensory neuron that delivers information to the CNS. In the visual pathways, the message must cross two synapses (photoreceptor to bipolar cell, and bipolar cell to ganglion cell). Only then does it move toward the brain. The extra synapse increases the synaptic delay, but it provides an opportunity for the processing and integration of visual information before it leaves the retina. Let's look at both retinal and cortical processing of visual signals.

Figure 17–18 A Standard Test for Color Vision. People who lack one or more types of cones cannot see the number 12 in this pattern.

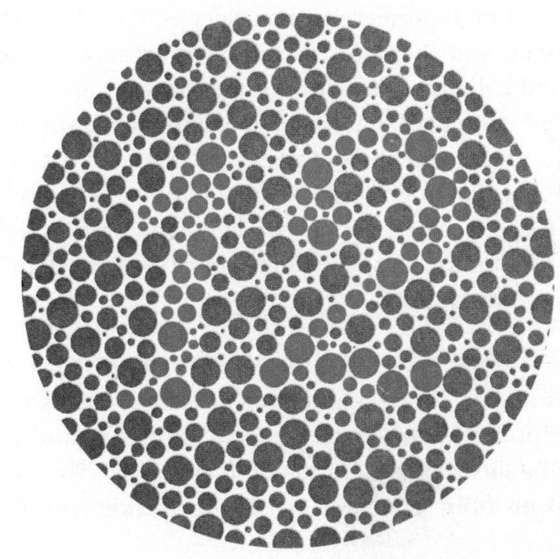

Processing by the Retina

Each of the millions of photoreceptors in the retina monitors a specific receptive field. Given that there are also millions of bipolar and ganglion cells, a considerable amount of convergence must take place at the start of the visual pathway. The degree of convergence differs between rods and cones; rods have a large degree of convergence, whereas cones typically show very little. Regardless of the amount of convergence, each ganglion cell monitors a specific portion of the field of vision, its *receptive field*.

As many as a thousand rods may pass information by their bipolar cells to a single ganglion cell. The fairly large ganglion cells that monitor rods are called **M cells** (*magnocells; magnus,* great). They provide information about the general form of an object, motion, and shadows in dim lighting. Because so much convergence occurs, the activation of an M cell indicates that light has arrived in a general area rather than at a specific location.

17 The loss of specificity due to convergence from rod stimuli is partially overcome by the fact that the activity of ganglion cells varies according to the pattern of activity in their receptive field, which is usually shaped like a circle. Typically, a ganglion cell responds differently to stimuli that arrive in the center of its receptive field than to stimuli that arrive at the edges (Figure 17–19). Some ganglion cells, called **on-center neurons**, are excited by light arriving in the center of their receptive field and are inhibited when light strikes its edges. Others, known as **off-center neurons**, are inhibited by light in the central zone, but are stimulated by light at the edges. On-center and off-center neurons provide information about which portion of their receptive field is illuminated. This kind of retinal processing within ganglion receptive fields improves the detection of the edges of objects within the field of vision.

The processing of visual stimuli from cones is different because of their lack of convergence. In the fovea centralis, the ratio of cones to ganglion cells is 1:1. The ganglion cells that monitor cones, called **P cells** (*parvo cells; parvus,* small), are smaller and more numerous than M cells. P cells are active in bright light, and they provide information about edges, fine detail, and color. Because little convergence occurs, the activation of a P cell means that light has arrived at one specific location. As a result, cones provide more precise information about a visual image than do rods. We could say that images formed by rods have a coarse, grainy, pixelated appearance that blurs details. By contrast, images produced by cones are sharp, clear, and of high resolution.

Figure 17–19 Convergence and Ganglion Cell Function. Photoreceptors are organized in groups within a receptive field. Each ganglion cell monitors a well-defined portion of that field. Some ganglion cells (on-center neurons, labeled A) respond strongly to light arriving at the center of their receptive field. Others (off-center neurons, labeled B) respond most strongly to illumination of the edges of their receptive field.

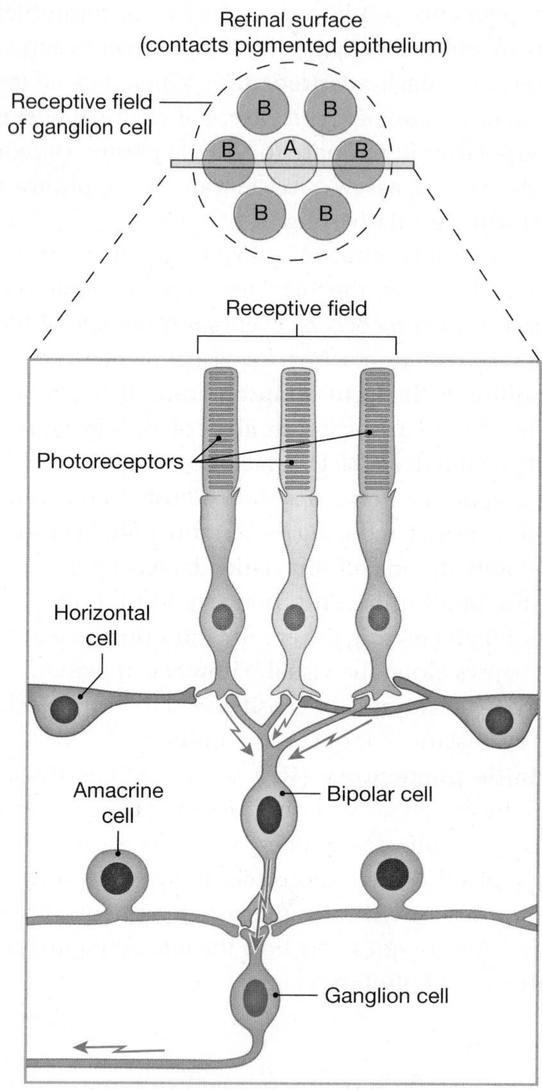
Retinal surface
(contacts pigmented epithelium)
Receptive field of ganglion cell
B
B
B
A
B
B
B
Receptive field
Photoreceptors
Horizontal cell
Amacrine cell
Bipolar cell
Ganglion cell

Central Processing of Visual Information

Axons from the entire population of ganglion cells converge on the optic disc, penetrate the wall of the eye, and proceed toward the diencephalon as the optic nerve (II). The two optic nerves, one from each eye, reach the diencephalon after a partial crossover at the **optic chiasm** (Figure 17–20). From that point, approximately half the fibers proceed toward the lateral geniculate body of the same side of the brain, whereas the other half cross over to reach the lateral geniculate body of the opposite side. p. 482 From each lateral geniculate body, visual information travels to the **visual cortex** in the occipital lobe of the cerebral hemisphere on that side. The bundle of projection fibers linking the lateral geniculates with the visual cortex is known as the **optic radiation**. Collaterals from the fibers synapsing in the lateral geniculate continue to subconscious processing centers in the diencephalon and brainstem.

Go to MasteringA&P™ > Study Area > Menu > Lab Tools > Bone & Dissection Videos > Cow Eye

Figure 17–20 The Visual Pathways. The crossover of some nerve fibers occurs at the optic chiasm. As a result, each hemisphere receives visual information from the medial half of the field of vision of the eye on that side, and from the lateral half of the field of vision of the eye on the opposite side. Visual association areas integrate this information to develop a composite picture of the entire field of vision.

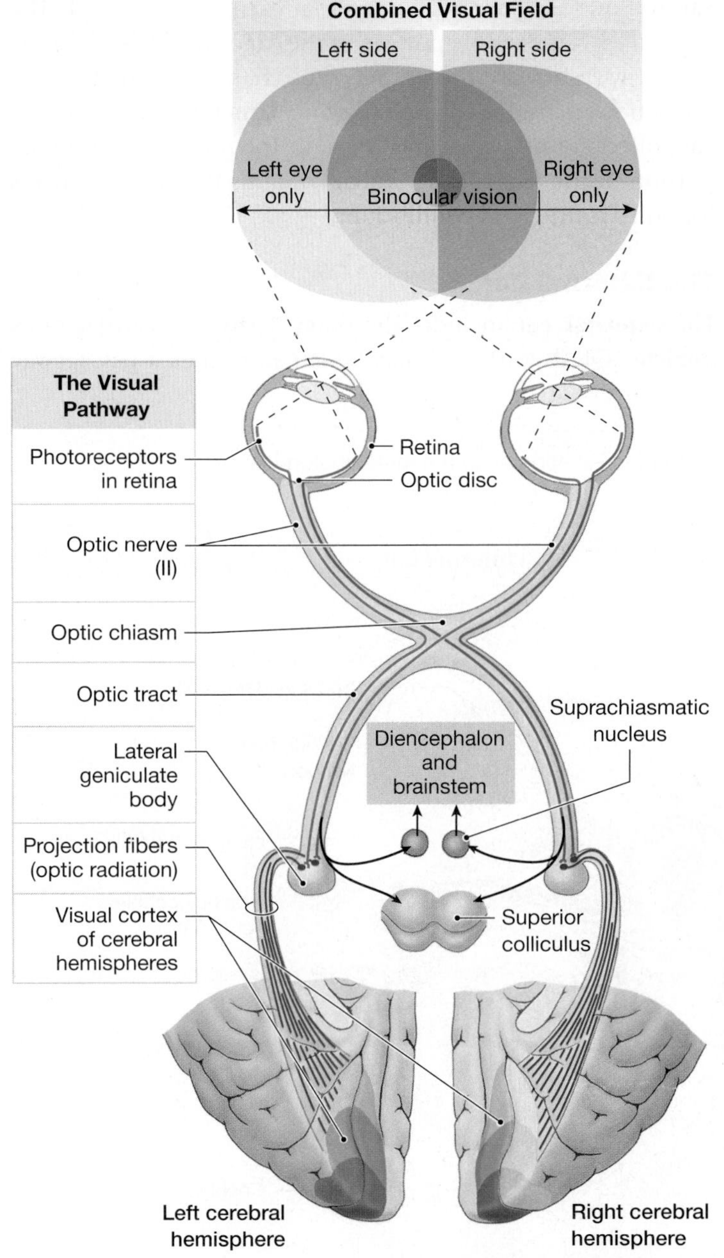

Combined Visual Field
Left side
Right side
Left eye only
Binocular vision
Right eye only
The Visual Pathway
Photoreceptors in retina
Optic nerve (II)
Optic chiasm
Optic tract
Lateral geniculate body
Projection fibers (optic radiation)
Visual cortex of cerebral hemispheres
Retina
Optic disc
Suprachiasmatic nucleus
Diencephalon and brainstem
Superior colliculus
Left cerebral hemisphere
Right cerebral hemisphere

The Field of Vision. You perceive a visual image due to the integration of information that arrives at the visual cortex of the occipital lobes. Each eye receives a slightly different visual image. One reason is that the foveae in your two eyes are 5–7.5 cm (2–3.0 in.) apart. Another reason is that your nose and eye socket block the view of the opposite side. **Depth perception** is the ability to judge depth or distance by interpreting the three-dimensional relationships among objects in view. Your brain resolves it by comparing the relative positions of objects within the images received by your two eyes.

When you look straight ahead, the visual images from your left and right eyes overlap (see Figure 17–20); the combined areas are your **field of vision**. The image received by the fovea centralis of each eye lies in the center of the region of overlap. A vertical line drawn through the center of this region marks the division of visual information at the optic chiasm. Visual information from the left half of the combined field of vision reaches the visual cortex of your right occipital lobe. Visual information from the right half of the combined field of vision arrives at the visual cortex of your left occipital lobe.

The cerebral hemispheres thus contain a map of the entire field of vision. As in the case of the somatosensory cortex, the map does not faithfully duplicate the areas within the sensory field. For example, the area assigned to the macula and fovea centralis covers about 35 times the surface it would cover if the map were proportionally accurate. The map is also upside down and reversed, duplicating the orientation of the visual image at the retina.

The Brainstem and Visual Processing. Many centers in the brainstem receive visual information, either from the lateral geniculate bodies or through collaterals from the optic tracts. Collaterals that bypass the lateral geniculates synapse in the superior colliculi or in the hypothalamus. The superior colliculi of the midbrain issue motor commands that control unconscious eye, head, or neck movements in response to visual stimuli. For example, the pupillary reflexes and reflexes that control eye movement are triggered by collaterals carrying information to the superior colliculi.

Visual inputs to the suprachiasmatic nucleus of the hypothalamus affect the function of other brainstem nuclei. p. 484 The suprachiasmatic nucleus and the *pineal gland* of the epithalamus receive visual information and use it to establish a **circadian** (ser-KĀ-dē-an) **rhythm** (*circa,* about + *dies,* day), which is a daily pattern of visceral activity that is tied to the day–night cycle. This circadian rhythm affects your metabolic rate, endocrine function, blood pressure, digestive activities, sleep–wake cycle, and other physiological and behavioral processes.

Checkpoint

12. **If you had been born without cones in your eyes, would you still be able to see? Explain.**
13. **How could a diet deficient in vitamin A affect vision?**
14. **What effect would a decrease in phosphodiesterase activity in photoreceptors have on vision?**

See the blue Answers tab at the back of the book.

Go to MasteringA&P™ > Study Area > Menu > Lab Tools > PAL 3.0 > Histology > Special Senses
Go to MasteringA&P™ > Study Area > Menu > Lab Tools > PAL 3.0 > Anatomical Models > Nervous System > Special Senses

17-6 Equilibrium sensations monitor head position and movement, while hearing involves the detection and interpretation of sound waves

Learning Outcome Describe the structures of the external, middle, and internal ear, explain their roles in equilibrium and hearing, and trace the pathways for equilibrium and hearing to their destinations in the brain.

Equilibrium sensations inform us of the position of the head in space by monitoring gravity, linear acceleration, and rotation. *Hearing* enables us to detect and interpret sound waves. Both of these senses are provided by the *internal ear,* a receptor complex located in the temporal bone of the skull.

The basic receptor mechanism for both senses is the same. The receptors, called *hair cells,* are mechanoreceptors. The complex structure of the internal ear and the different arrangement of accessory structures enable hair cells to respond to different stimuli. Once the hair cells have transduced these stimuli, equilibrium input is sent to the brainstem, while hearing input is sent to the brainstem and/or *auditory cortex* for processing and possible response. In this section, we look at the anatomy of the ear before going into the processes of equilibrium and hearing.

Anatomy of the Ear

The **ear** is divided into three anatomical regions: the external ear, the middle ear, and the internal ear (Figure 17–21). The *external ear*—the visible portion of the ear—collects and directs sound waves toward the *middle ear,* a chamber located within the petrous part of the temporal bone. Structures of the middle ear collect sound waves and transmit them to an appropriate portion of the *internal ear,* which contains the sensory organs for both hearing and equilibrium.

The External Ear

The **external ear** includes the outer fleshy and cartilaginous **auricle** (OR-ih-kul), or *pinna,* which surrounds a passageway

17

Figure 17–21 The Anatomy of the Ear. The dashed lines indicate the boundaries separating the three anatomical regions of the ear (external, middle, and internal).

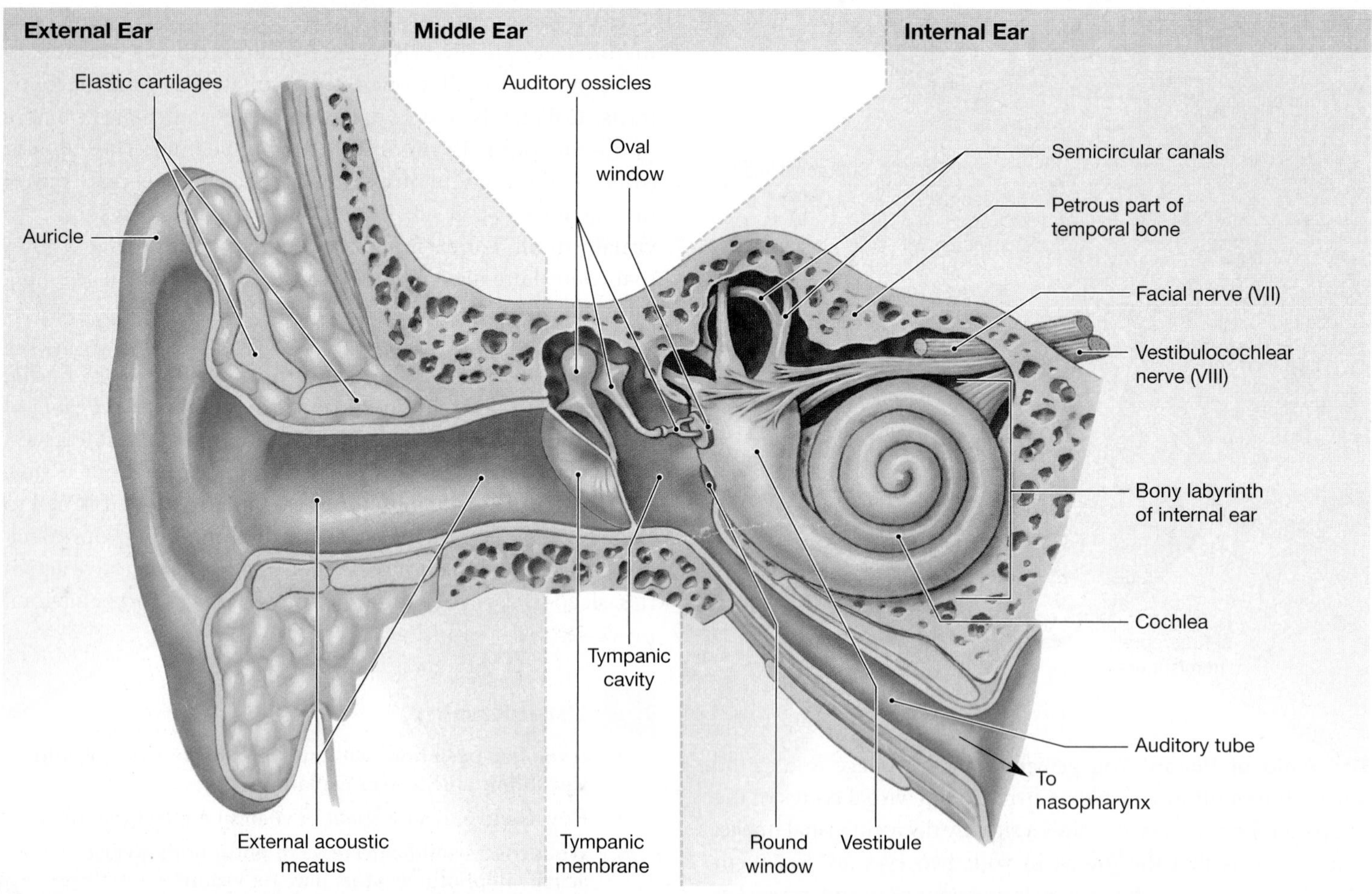
External Ear
Middle Ear
Internal Ear
Elastic cartilages
Auricle
Auditory ossicles
Oval window
Semicircular canals
Petrous part of temporal bone
Facial nerve (VII)
Vestibulocochlear nerve (VIII)
Bony labyrinth of internal ear
Cochlea
Auditory tube
To nasopharynx
Tympanic cavity
External acoustic meatus
Tympanic membrane
Round window
Vestibule

? The tympanic cavity and auditory ossicles are found in which anatomical region of the ear?

called the **external acoustic meatus**, or *auditory canal.* The auricle protects the opening of the canal and provides directional sensitivity. Sounds coming from behind the head are blocked by the auricle, but sounds coming from the side or front are collected and channeled into the external acoustic meatus. (When you "cup" your ear with your hand to hear a faint sound more clearly, you are exaggerating this effect.) The external acoustic meatus ends at the **tympanic membrane** (*tympanon,* drum), or *eardrum.* This thin, semitransparent sheet separates the external ear from the middle ear.

The tympanic membrane is very delicate. The auricle and the narrow external acoustic meatus provide some protection for it from accidental injury. In addition, **ceruminous glands**—integumentary glands along the external acoustic meatus—secrete a waxy material called **cerumen** (or *earwax*) that helps keep out foreign objects or small insects. Cerumen also slows the growth of microorganisms and so reduces the chances of infection. In addition, the canal is lined with many small, outwardly projecting hairs. These hairs trap debris and also provide increased tactile sensitivity through their root hair plexuses.

The Middle Ear

The **middle ear**, or **tympanic cavity**, is an air-filled chamber separated from the external acoustic meatus by the tympanic membrane (see Figure 17–21). The middle ear communicates both with the *nasopharynx* (the superior portion of the pharynx), through the *auditory tube,* and with the mastoid air cells, through a number of small connections (Figure 17–22a).

Figure 17–22 **The Middle Ear.**

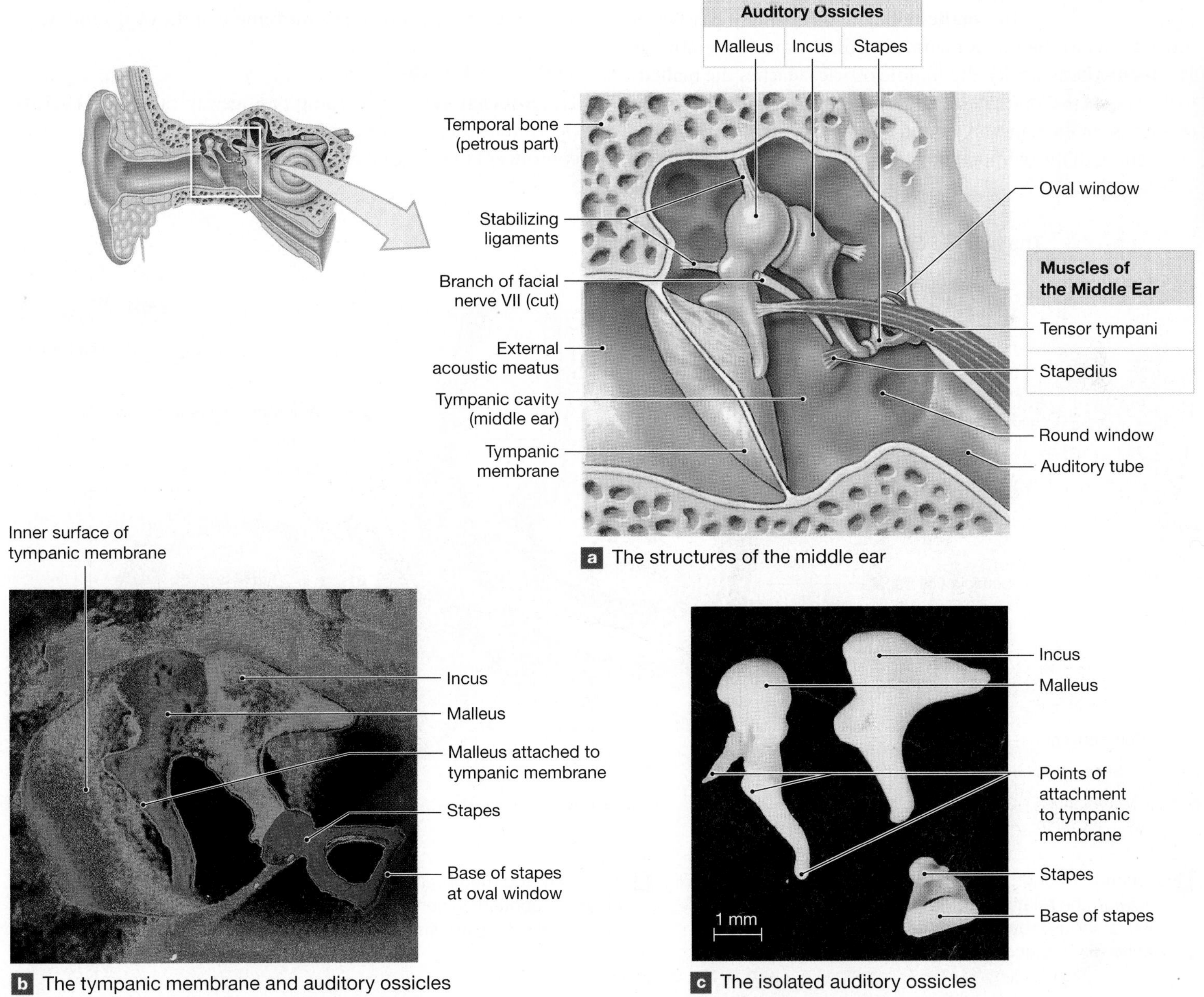

Auditory Ossicles
Malleus
Incus
Stapes
Temporal bone (petrous part)
Stabilizing ligaments
Branch of facial nerve VII (cut)
External acoustic meatus
Tympanic cavity (middle ear)
Tympanic membrane
Oval window
Muscles of the Middle Ear
Tensor tympani
Stapedius
Round window
Auditory tube
Inner surface of tympanic membrane
Incus
Malleus
Malleus attached to tympanic membrane
Stapes
Base of stapes at oval window
Incus
Malleus
Points of attachment to tympanic membrane
Stapes
Base of stapes
1 mm

a The structures of the middle ear

b The tympanic membrane and auditory ossicles

c The isolated auditory ossicles

17

The **auditory tube** (also called the *pharyngotympanic tube* or the *Eustachian tube) is* about 4 cm (1.6 in.) long and consists of two portions. The portion near the connection to the middle ear is narrow and is supported by elastic cartilage. The portion near the opening into the nasopharynx is broad and funnel shaped. The auditory tube equalizes pressure on either side of the tympanic membrane. Unfortunately, the auditory tube can also allow microorganisms to travel from the nasopharynx into the middle ear. Invasion by microorganisms can lead to an unpleasant middle ear infection known as *otitis media.*

The Auditory Ossicles. The middle ear contains three tiny ear bones, collectively called **auditory ossicles** (OS-ih-kulz). These ear bones connect the tympanic membrane with the internal ear (see Figure 17–21). The articulations between the auditory ossicles are the smallest synovial joints in the body. Each has a tiny joint capsule and supporting extracapsular ligaments. The three auditory ossicles are the malleus, the incus, and the stapes (Figure 17–22b,c). The **malleus** (*malleus,* hammer) attaches at three points to the interior surface of the tympanic membrane. The **incus** (*incus,* anvil), the middle ossicle, attaches the malleus to the **stapes** (STĀ-pēz; *stapes,* stirrup), the inner ossicle. The edges of the base of the stapes are bound to the edges of the *oval window,* an opening in the temporal bone that surrounds the internal ear.

17

Muscles of the Middle Ear. When sound waves cause the tympanic membrane to vibrate, the ossicles conduct the vibrations to the internal ear. When we are exposed to very loud noises, two small muscles in the middle ear (see Figure 17–22a) protect the tympanic membrane and ossicles from violent movements:

- The **tensor tympani** (TEN-sor tim-PAN-ē) is a short ribbon of muscle originating on the petrous part of the temporal bone and the auditory tube, and inserting on the "handle" of the malleus. When the tensor tympani contracts, it pulls the malleus medially, stiffening the tympanic membrane. This increased stiffness reduces the amount of movement possible. The tensor tympani is innervated by motor fibers of the mandibular nerve (V_3), a division of the trigeminal nerve (V).
- The stapedius (sta-PĒ-dē-us), innervated by the facial nerve (VII), originates from the posterior wall of the middle ear and inserts on the stapes. Contraction of the stapedius pulls on the stapes, reducing its movement at the oval window.

The Internal Ear

The **internal ear** is a winding passageway called a **labyrinth** (*labyrinthos,* network of canals) (Figure 17–23). The superficial contours of the internal ear are formed by a layer of dense bone

Figure 17–23 The Internal Ear.

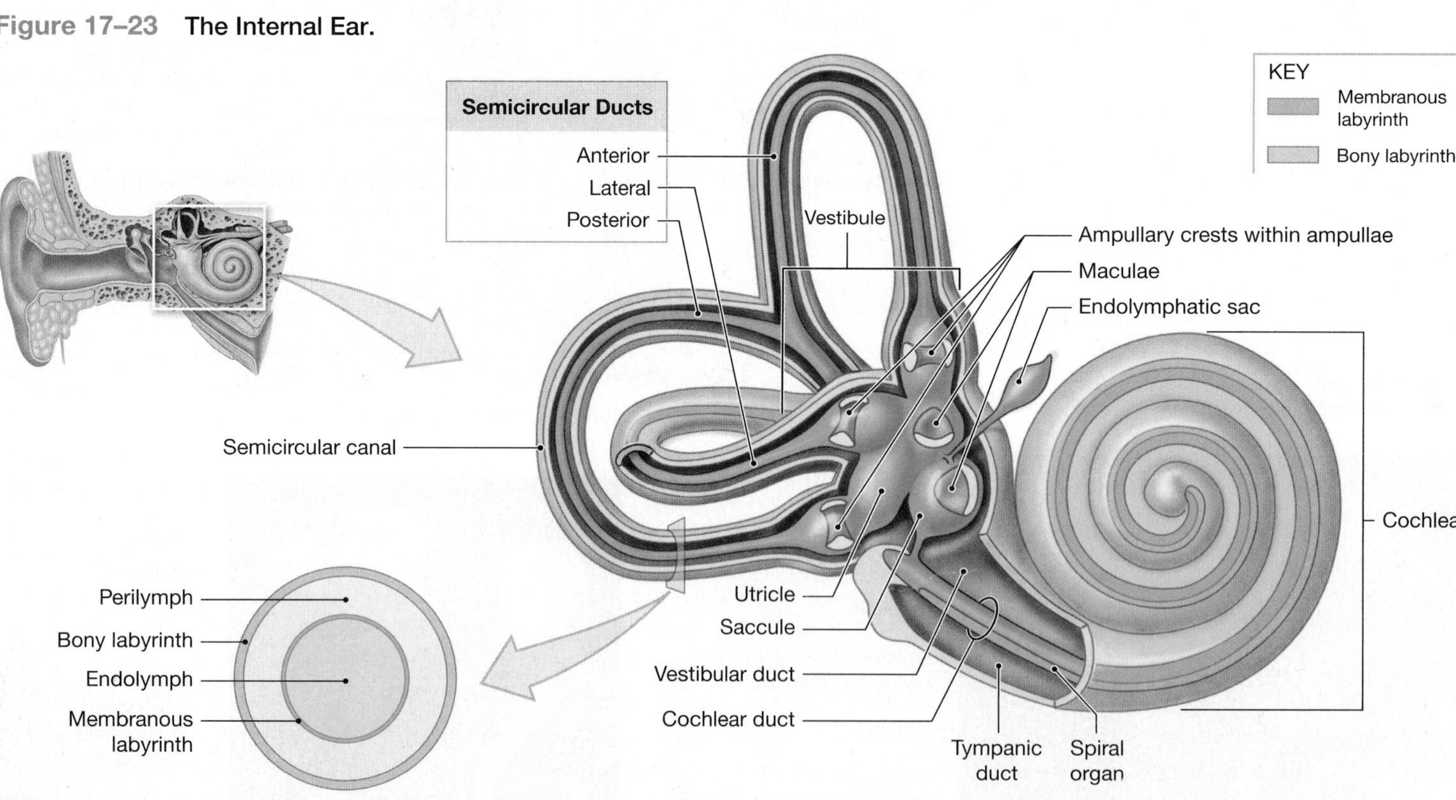
Semicircular Ducts
Anterior
Lateral
Posterior
Vestibule
KEY
Membranous labyrinth
Bony labyrinth
Ampullary crests within ampullae
Maculae
Endolymphatic sac
Semicircular canal
Cochlea
Perilymph
Bony labyrinth
Endolymph
Membranous labyrinth
Utricle
Saccule
Vestibular duct
Cochlear duct
Tympanic duct
Spiral organ

a A section through one of the semicircular canals, showing the relationship between the bony and membranous labyrinths, and the boundaries of perilymph and endolymph.

b The bony and membranous labyrinths. Areas of the membranous labyrinth containing sensory receptors (ampullary crests, maculae, and spiral organ) are shown in purple.

known as the **bony labyrinth**. The walls of the bony labyrinth are continuous with the surrounding temporal bone. The inner contours of the bony labyrinth are closely followed by the contours of the **membranous labyrinth**, a delicate, interconnected network of fluid-filled tubes. The receptors of the internal ear are found within these tubes.

Between the bony and membranous labyrinths flows **perilymph** (PEHR-ih-limf), while the membranous labyrinth contains **endolymph** (EN-dō-limf) (Figure 17–23a). Perilymph closely resembles cerebrospinal fluid. In contrast, endolymph has electrolyte concentrations that differ from those of typical body fluids. (See the Appendix for a chemical analysis of perilymph, endolymph, and other body fluids.)

We can subdivide the bony labyrinth into the *vestibule,* the *semicircular canals,* and the *cochlea* (KOK-lē-ah) (Figure 17–23b). The **vestibule** (VES-tih-byūl) consists of a pair of membranous sacs: the *saccule* (SAK-yūl) and the *utricle* (YŪ-trih-kul). Receptors in these sacs provide equilibrium sensations. The three **semicircular canals** enclose three slender *semicircular ducts.* The combination of vestibule and semicircular canals is called the *vestibular complex*. The fluid-filled chambers within the vestibule are continuous with those of the semicircular canals.

The **cochlea** (*cochlea,* a snail shell) is a spiral-shaped, bony chamber that contains the *cochlear duct* of the membranous labyrinth. Receptors within the cochlear duct provide the sense of hearing. The duct is sandwiched between a pair of perilymph-filled chambers. The entire complex spirals around a central bony hub, much like a snail shell.

The walls of the bony labyrinth consist of dense bone everywhere except at two small areas near the base of the cochlear spiral (see Figure 17–21). The **round window** is a thin, membranous partition that separates the perilymph of the cochlear chambers from the air-filled middle ear. Collagen fibers connect the bony margins of the opening known as the **oval window** to the base of the stapes.

Equilibrium

Equilibrium is the state of physical balance. We'll explore the structures within the internal ear that enable us to maintain this balance.

The Vestibular Complex and Physiology of Equilibrium

As just noted, receptors of the **vestibular complex** provide you with equilibrium sensations. In both the semicircular ducts and the vestibule, these receptors are **hair cells**. The hair cells of the semicircular ducts convey information about rotational movements of the head. The hair cells in the saccule and the utricle of the vestibule convey information about your position with respect to gravity, and tell you if you are accelerating or decelerating.

The Semicircular Ducts: Rotational Movements. The hair cells of the semicircular ducts are active during a rotational movement, but are quiet when the body is motionless. For example, when you turn your head to the left, receptors stimulated in the semicircular ducts tell you how rapid the movement is, and in which direction. The **anterior**, **posterior**, and **lateral semicircular ducts** are continuous with the utricle (Figure 17–24a). Each semicircular duct contains an **ampulla**, an expanded region that contains the receptors. The region in the wall of the ampulla that contains the hair cells is known as the **ampullary crest** (Figure 17–24b). Each ampullary crest is bound to an **ampullary cupula** (KŪ-pū-luh), a gelatinous structure that extends the full width of the ampulla.

Hair cells are always surrounded by supporting cells and monitored by the dendrites of sensory neurons. The free surface of each hair cell supports 80–100 long **stereocilia**, which resemble very long microvilli (Figure 17–24d). Each hair cell in the vestibular complex also contains a single large cilium called a **kinocilium** (kī-nō-SIL-ē-um). At an ampullary crest, the kinocilia and stereocilia of the hair cells are embedded in the ampullary cupula (see Figure 17–24b).

Hair cells do not actively move their kinocilia or stereocilia. However, when an external force pushes against these processes, the distortion of the plasma membrane alters the rate at which the hair cell releases neurotransmitters. The action potentials generated in response to these neurotransmitters allow the hair cells to provide information about the direction and strength of mechanical stimuli. The stimuli involved, however, depend on the hair cell's location: gravity or acceleration in the vestibule and rotation in the semicircular canals (plus sound in the cochlea). The sensitivities of the hair cells differ, because each of these regions has different accessory structures that determine which stimulus will provide the force to deflect the kinocilia and stereocilia.

The ampullary cupula with its hair cells has a density very close to that of the surrounding endolymph, so it essentially floats above the receptor surface. When your head rotates in the plane of a semicircular duct, the movement of endolymph along the length of the duct pushes the ampullary cupula to the side and distorts the receptor processes (Figure 17–24c). Movement of fluid in one direction stimulates the hair cells, and movement in the opposite direction inhibits them. When the endolymph stops moving, the elastic nature of the ampullary cupula makes it return to its normal position.

Even the most complex movement can be analyzed in terms of motion in three rotational planes. Each semicircular duct responds to one of these rotational movements. A horizontal rotation, as in shaking your head "no," stimulates the hair cells of the lateral semicircular duct. Nodding "yes" excites the anterior duct, and tilting your head from side to side activates receptors in the posterior duct.

Figure 17–24 The Semicircular Ducts.

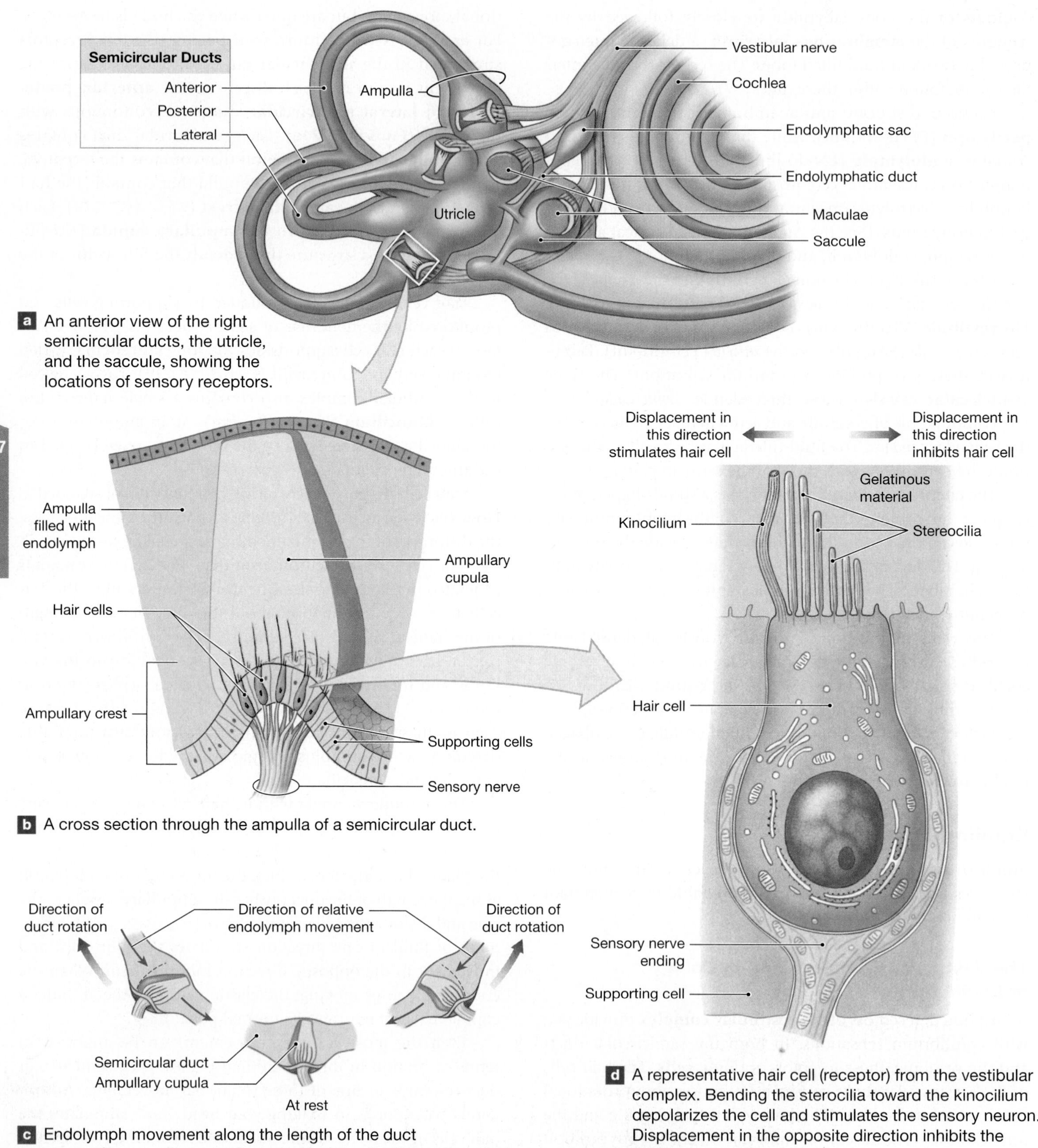
Semicircular Ducts
Anterior
Posterior
Lateral
Ampulla
Utricle
Vestibular nerve
Cochlea
Endolymphatic sac
Endolymphatic duct
Maculae
Saccule
Ampulla filled with endolymph
Ampullary cupula
Hair cells
Ampullary crest
Supporting cells
Sensory nerve
Displacement in this direction stimulates hair cell
Displacement in this direction inhibits hair cell
Gelatinous material
Kinocilium
Stereocilia
Hair cell
Sensory nerve ending
Supporting cell
Direction of duct rotation
Direction of relative endolymph movement
Direction of duct rotation
Semicircular duct
Ampullary cupula
At rest

a An anterior view of the right semicircular ducts, the utricle, and the saccule, showing the locations of sensory receptors.

b A cross section through the ampulla of a semicircular duct.

c Endolymph movement along the length of the duct moves the ampullary cupula and stimulates the hair cells.

d A representative hair cell (receptor) from the vestibular complex. Bending the sterocilia toward the kinocilium depolarizes the cell and stimulates the sensory neuron. Displacement in the opposite direction inhibits the sensory neuron.

17

The Utricle and Saccule: Position and Acceleration. The hair cells of the **utricle** and **saccule** provide position and linear movement sensations, whether the body is moving or stationary. For example, if you stand with your head tilted to one side, these receptors report the angle involved and whether your head tilts forward or backward. The two chambers are connected by a slender passageway that is continuous with the narrow **endolymphatic duct** (see Figure 17–24a). This duct ends in a closed cavity called the **endolymphatic sac**. This sac projects through the dura mater that lines the temporal bone and into the subarachnoid space, where a capillary network surrounds it.

Portions of the endolymphatic duct secrete endolymph continuously, and excess fluid returns to the general circulation at the endolymphatic sac. There the capillaries absorb endolymph removed by a combination of active transport and vesicular transport.

The hair cells of the utricle and saccule are clustered in oval structures called **maculae** (MAK-yū-lē; *macula,* spot) (Figure 17–25a). The **macula of utricle** is sensitive to changes in horizontal movement. The **macula of saccule** is sensitive to changes in vertical movement. As in the ampullae, the hair cell processes are embedded in a gelatinous structure, here the *otolithic membrane.* This membrane's surface contains densely packed calcium carbonate crystals called **otoliths** ("ear stones").

The macula of utricle is diagrammed in Figure 17–25b. Its functioning is shown in Figure 17–25c. When your head is in the normal, upright position, the otoliths sit atop the otolithic membrane of macula of utricle ❶. Their weight presses on the macular surface, pushing the hair cell processes down rather than to one side or another. When your head is tilted, the pull of gravity on the otoliths shifts them to the side, distorting the hair cell processes and stimulating the macular receptors ❷. The change in receptor activity tells the CNS that your head is no longer level.

A similar mechanism accounts for your perception of linear acceleration when you are in a car that speeds up suddenly. The otoliths lag behind due to their inertia, and the effect on the hair cells is comparable to tilting your head back. Under normal circumstances, your nervous system distinguishes between the sensations of tilting and linear acceleration by integrating vestibular sensations with visual information. Many amusement

Figure 17–25 The Saccule and Utricle.

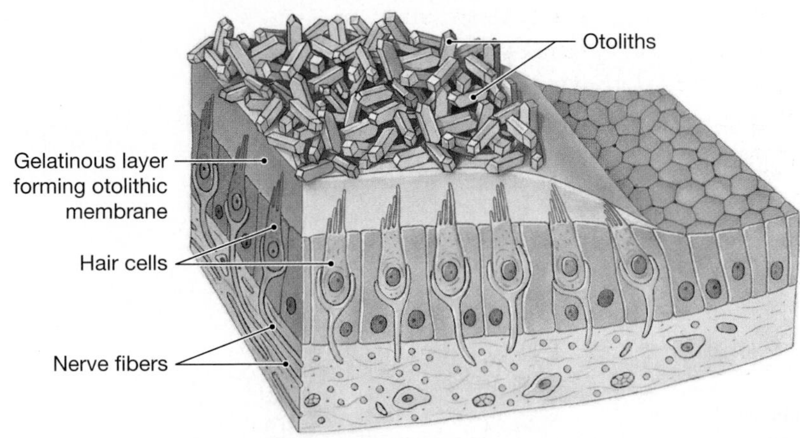

Otoliths
Gelatinous layer forming otolithic membrane
Hair cells
Nerve fibers

b The structure of an individual macula of utricle

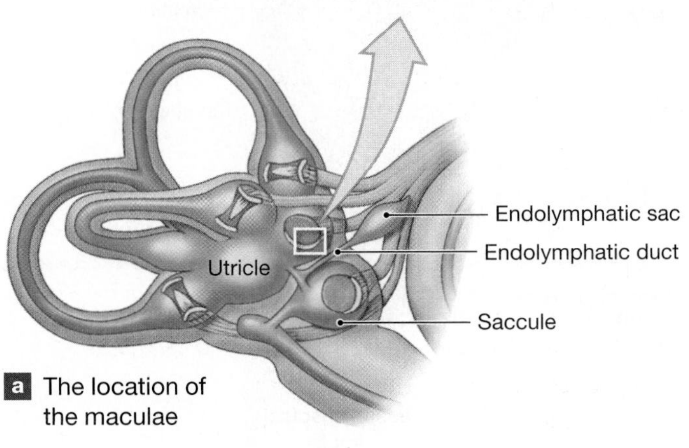

Endolymphatic sac
Endolymphatic duct
Utricle
Saccule

a The location of the maculae

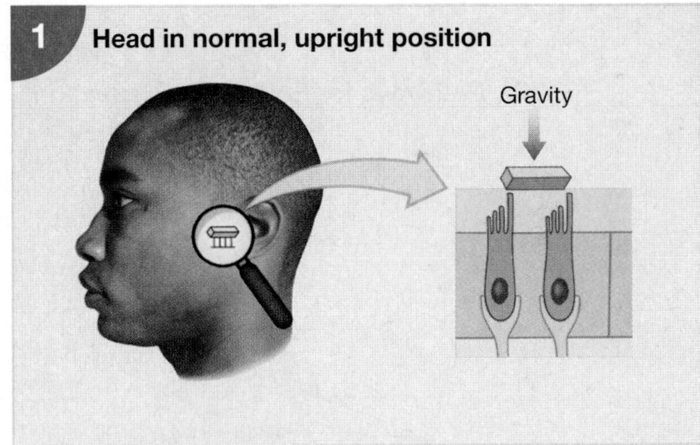

1
Head in normal, upright position
Gravity

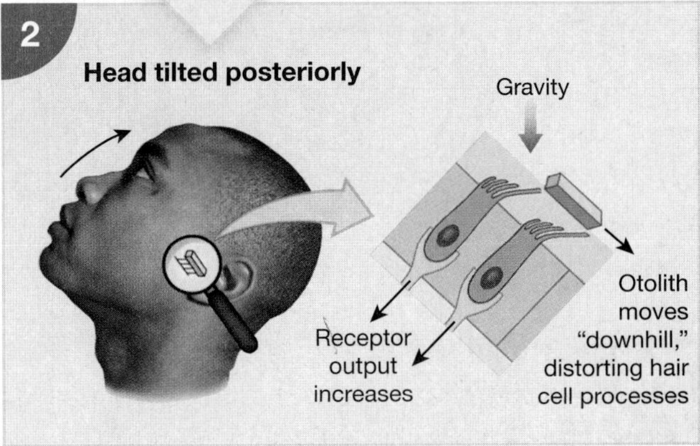

2
Head tilted posteriorly
Gravity
Receptor output increases
Otolith moves "downhill," distorting hair cell processes

c A diagram of the functioning of the macula of utricle when the head is held normally ❶ and then tilted back ❷ park rides confuse your sense of equilibrium by combining rapid rotation with changes in position and acceleration while providing restricted or misleading visual information.

Pathways for Equilibrium Sensations

Hair cells of the vestibule and semicircular ducts are monitored by sensory neurons located in adjacent **vestibular ganglia**. Sensory fibers from these ganglia form the **vestibular nerve**, a division of the vestibulocochlear nerve (VIII). p. 501 These fibers innervate neurons within the pair of **vestibular nuclei** at the boundary between the pons and the medulla oblongata.

The vestibular nuclei have four functions:

- Integrating sensory information about balance and equilibrium that arrives from both sides of the head.
- Relaying information from the vestibular complex to the cerebellum.
- Relaying information from the vestibular complex to the cerebral cortex, providing a conscious sense of head position and movement.
- Sending commands to motor nuclei in the brainstem and spinal cord.

The reflexive motor commands issued by the vestibular nuclei are distributed to the motor nuclei for cranial nerves involved with eye, head, and neck movements (CN III, IV, VI, and XI). Instructions descending in the *vestibulospinal tracts* of the spinal cord adjust peripheral muscle tone and complement the reflexive movements of the head or neck. p. 530 These pathways are indicated in Figure 17–26.

The automatic movements of the eyes in response to sensations of motion are directed by the *superior colliculi* of the midbrain. p. 478 These movements attempt to keep your gaze focused on a specific point in space, despite changes in body position and orientation. If your body is turning or spinning rapidly, your eyes will fix on one point for a moment and then jump ahead to another in a series of short, jerky movements.

If either the brainstem or the internal ear is damaged, this type of eye movement can occur even when the body is stationary. Individuals with this condition, which is called **nystagmus** (nis-TAG-mus), have trouble controlling their eye movements. Physicians commonly check for nystagmus by asking patients to watch a small penlight as it is moved across the field of vision.

Hearing

The receptors of the cochlear duct provide a sense of **hearing** that enables us to detect the quietest whisper, yet remain

Figure 17–26 Pathways for Equilibrium Sensations.

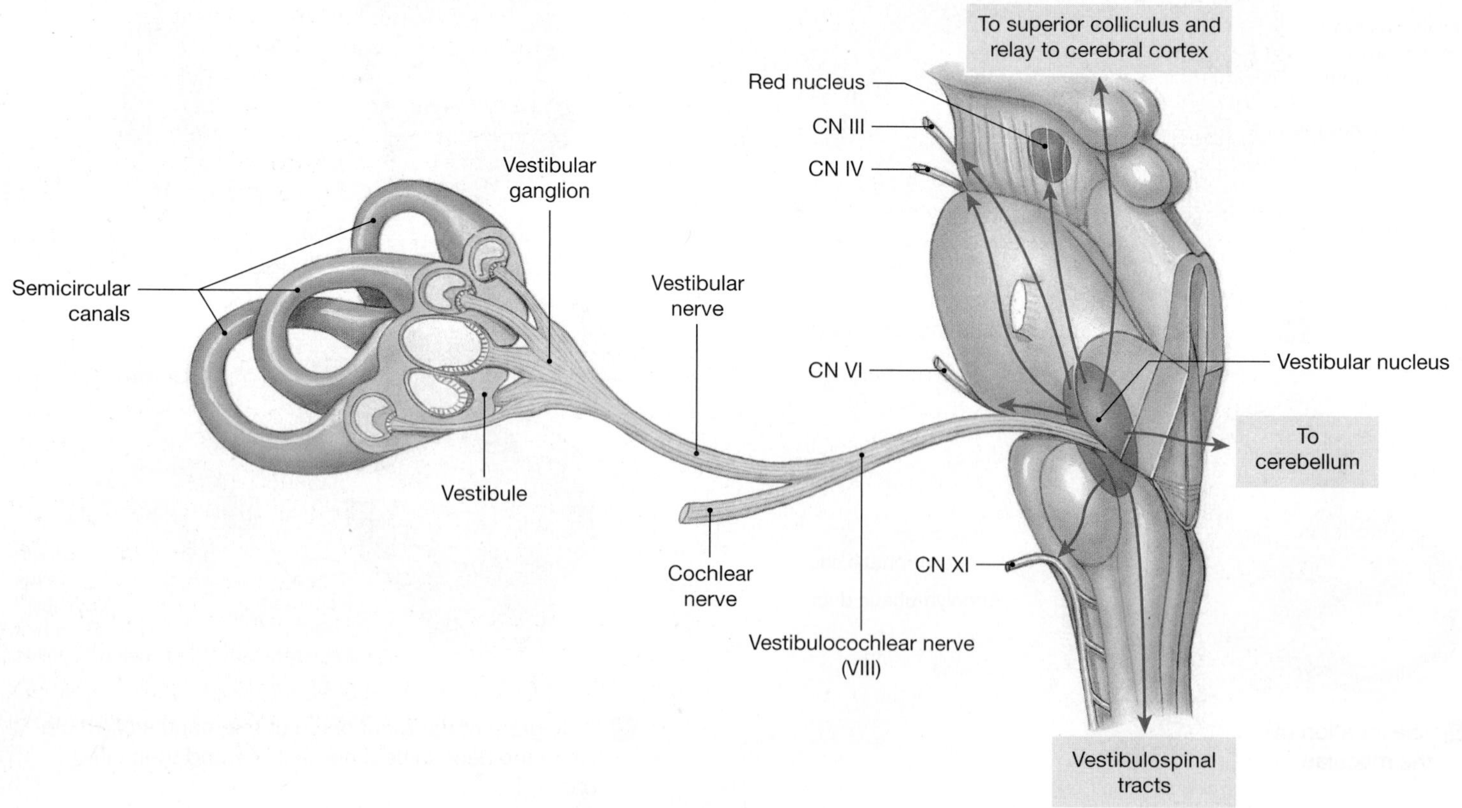
To superior colliculus and relay to cerebral cortex
Red nucleus
CN III
CN IV
Vestibular ganglion
Semicircular canals
Vestibular nerve
CN VI
Vestibular nucleus
To cerebellum
Vestibule
Cochlear nerve
CN XI
Vestibulocochlear nerve (VIII)
Vestibulospinal tracts

> **+ Clinical Note** Motion Sickness
>
> The signs and symptoms of **motion sickness** are exceedingly unpleasant. They include headache, sweating, facial flushing, nausea, vomiting, and various changes in mental perspective. Motion sickness may result when central processing stations, such as the tectum of the midbrain, receive conflicting sensory information. Why and how these conflicting reports result in nausea, vomiting, and other signs and symptoms are not known. Sitting below decks on a moving boat or reading in a car or airplane tends to provide the necessary conditions. Your eyes (which are tracking lines on a page) report that your position in space is not changing, but your semicircular ducts report that your body is lurching and turning. To counter this effect, seasick sailors watch the horizon rather than their immediate surroundings, so that their eyes will provide visual confirmation of the movements detected by their internal ears. It is not clear why some individuals are almost immune to motion sickness, but others find travel by boat or plane almost impossible.
>
> Drugs commonly used to prevent motion sickness include dimenhydrinate (*Dramamine*), scopolamine (*Transderm Scop*), and promethazine. These drugs depress activity at the vestibular nuclei. Sedatives, such as prochlorperazine (*Compazine*), may also be effective.

functional in a noisy room. The receptors responsible for auditory sensations are hair cells similar to those of the vestibular complex. However, their placement within the cochlear duct and the organization of the surrounding accessory structures shield them from stimuli other than sound.

In the process of hearing, arriving sound waves are converted into mechanical movements by vibration of the tympanic membrane. These vibrations are then conducted to the internal ear by the auditory ossicles. In the internal ear, the vibrations are converted to pressure waves in fluid, which eventually are detected by the hair cells in the cochlear duct. This sensory information is sent to the *auditory cortex* of the brain to be interpreted. Before discussing the mechanics of this remarkably elegant process, let's first look at the basic properties of sound, which are key to understanding this process.

An Introduction to Sound

Hearing is the perception of sound, but what is sound? It consists of waves of pressure conducted through a medium such as air or water. In air, each *pressure wave* consists of a region where the air molecules are crowded together and an adjacent zone where they are farther apart (Figure 17–27a). These waves are sine waves—that is, S-shaped curves that repeat in a regular pattern. They travel through the air at about 1235 km/h (768 mph).

The *wavelength* of sound is the distance between two adjacent wave crests (peaks), or the distance between two adjacent wave troughs (Figure 17–27b). The **frequency** of a sound is the number of waves that pass a fixed reference point in a given time. Physicists use the term **cycles** rather than *waves.* We measure the frequency of a sound in terms of the number of cycles per second (cps), a unit called **hertz (Hz)**. Wavelength and frequency are inversely related.

Figure 17–27 The Nature of Sound.

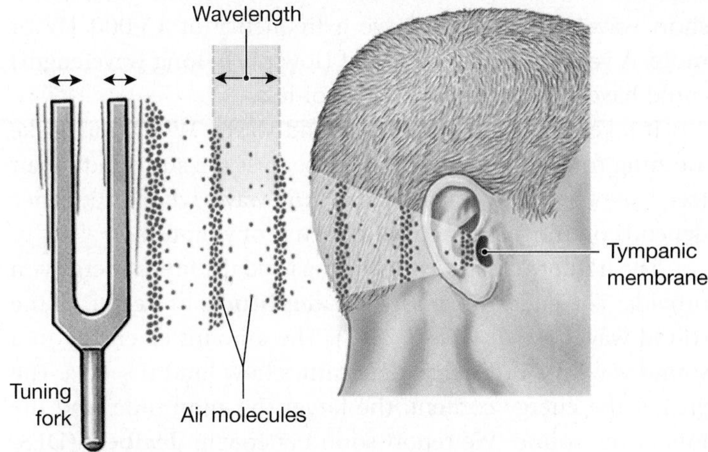
Wavelength
Tympanic membrane
Tuning fork
Air molecules

a Sound waves (here, generated by a tuning fork) travel through the air as pressure waves.

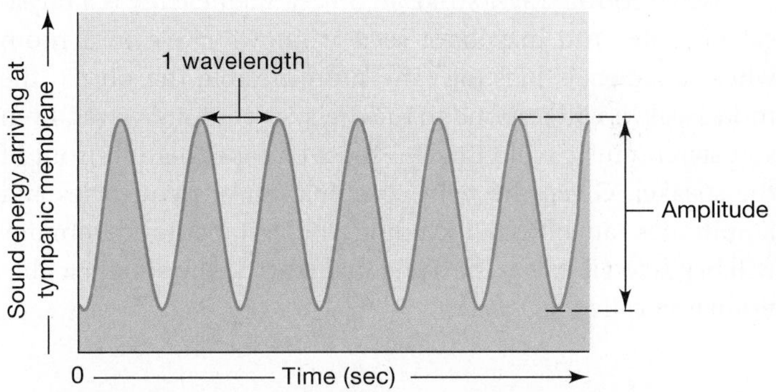
Sound energy arriving at tympanic membrane
1 wavelength
Amplitude
0
Time (sec)

b A graph showing the sound energy arriving at the tympanic membrane. The distance between wave peaks is the wavelength. The number of waves arriving each second is the frequency, which we perceive as pitch. Frequencies are reported in cycles per second (cps), or hertz (Hz). The amount of energy carried by the wave is its amplitude. The greater the amplitude, the louder the sound.

What we perceive as the **pitch** of a sound is our sensory response to its frequency. A *high-frequency* sound (high pitch, short wavelength) might have a frequency of 15,000 Hz or more. A very *low-frequency* sound (low pitch, long wavelength) could have a frequency of 100 Hz or less.

It takes energy to produce sound waves. When you strike a tuning fork, it vibrates, pushing against the surrounding air (see Figure 17–27a). It produces sound waves whose frequency depends on the instrument's frequency of vibration.

The harder you strike the tuning fork, the more energy you provide. The energy increases the **amplitude**, or height, of the sound wave (see Figure 17–27b). The amount of energy in a sound wave, its **intensity**, determines how loud it seems. The greater the energy content, the larger the amplitude, and the louder the sound. We report sound energy in **decibels** (DES-ih-belz, dB). Table 17–1 shows the decibel levels of familiar sounds.

17 When sound waves strike an object, their energy is a physical pressure. You may have seen windows move in a room where a stereo is blasting. The more flexible the object, the more easily it will respond to the pressure of sound waves. Even soft stereo music will vibrate a sheet of paper held in front of the speaker. Given the right combination of frequencies and amplitudes, an object, including your tympanic membrane, will begin to vibrate at the same frequency as the sound, a phenomenon called *resonance.*

Table 17–1 Intensity of Representative Sounds

Typical Decibel Level	Example	Dangerous Time Exposure
0	Lowest audible sound	
30	Quiet library; soft whisper	
40	Quiet office; living room; bedroom away from traffic	
50	Light traffic at a distance; refrigerator; gentle breeze	
60	Air conditioner at 20 feet; conversation; sewing machine in operation	
70	Busy traffic; noisy restaurant	Some damage if continuous
80	Subway; heavy city traffic; alarm clock at 2 feet; factory noise	More than 8 hours
90	Truck traffic; noisy home appliances; shop tools; gas lawn mower	Less than 8 hours
100	Chain saw; boiler shop; pneumatic drill	2 hours
120	"Heavy metal" rock concert; sandblasting; thunderclap nearby	Immediate danger
140	Gunshot; jet plane	Immediate danger
160	Rocket launching pad	Hearing loss inevitable

Anatomy of the Cochlear Duct

In sectional view (Figure 17–28), the **cochlear duct**, or **scala media**, lies between a pair of perilymphatic chambers, or *scalae:* the **scala vestibuli** (SKĀ-luh ves-TIB-yū-lē), or **vestibular duct**, and the **scala tympani** (TIM-pa-nē), or **tympanic duct**. The outer surfaces of all three ducts are encased by the bony labyrinth everywhere except at the oval window (the base of the scala vestibuli) and the round window (the base of the scala tympani). These scalae really form one long and continuous perilymphatic chamber because they are interconnected at the tip of the spiral-shaped cochlea. This chamber begins at the oval window; extends through the scala vestibuli, around the top of the cochlea, and along the scala tympani; and ends at the round window.

The hair cells of the cochlear duct are located in a structure called the **spiral organ** (**organ of Corti**) (Figures 17–28b and Figure 17–29). This sensory structure sits on the **basilar membrane**, a membrane that separates the cochlear duct from the scala tympani. The hair cells are arranged in a series of longitudinal rows. They lack kinocilia, and their stereocilia are in contact with the overlying **tectorial** (tek-TOR-ē-al; *tectum*, roof) **membrane**. This membrane is firmly attached to the inner wall of the cochlear duct.

Auditory Discrimination

Our hearing abilities are remarkable, but it is difficult to assess the degree of auditory discrimination. The range from the softest audible sound to the loudest tolerable blast represents a trillion-fold increase in power. The receptor mechanism is so sensitive that, if we were to remove the stapes, we could, in theory, hear air molecules bouncing off the oval window. We never use the full potential of this system, because body movements and our internal organs produce squeaks, groans, thumps, and other sounds that are tuned out by central and peripheral adaptation. When other environmental noises fade away, the level of adaptation drops and the system becomes increasingly sensitive. For example, when you relax in a quiet room, your heartbeat seems to get louder and louder as the auditory system adjusts to the level of background noise.

Young children have the greatest hearing range: They can detect sounds ranging from a 20-Hz buzz to a 20,000-Hz whine. With age, damage due to loud noises or other injuries accumulates. The tympanic membrane gets less flexible, the articulations between the ossicles stiffen, and the round window may begin to ossify. As a result, older individuals show some degree of hearing loss.

Probably more than 6 million people in the United States have at least a partial hearing deficit, either congenital (inborn) or acquired. Such deficits are due to problems with the transfer of vibrations by the auditory ossicles, or damage to the receptors or the auditory pathways.

Figure 17–28 The Cochlea.

Round window
Stapes at oval window
KEY
From oval window to tip of spiral
From tip of spiral to round window
Scala vestibuli
Cochlear duct
Scala tympani
Semicircular canals
Cochlear nerve
Vestibular nerve
Vestibulocochlear Nerve (VIII)

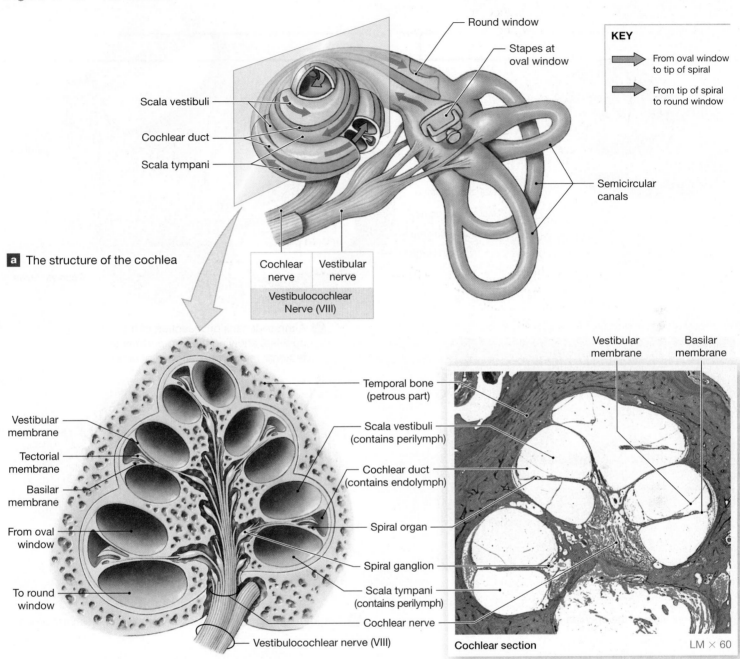

a The structure of the cochlea

Vestibular membrane
Tectorial membrane
Basilar membrane
From oval window
To round window
Temporal bone (petrous part)
Scala vestibuli (contains perilymph)
Cochlear duct (contains endolymph)
Spiral organ
Spiral ganglion
Scala tympani (contains perilymph)
Cochlear nerve
Vestibulocochlear nerve (VIII)
Vestibular membrane
Basilar membrane
Cochlear section
LM × 60

b Diagrammatic and sectional views of the spiral-shaped cochlea

The scala vestibuli and scala tympani contain a different fluid than the cochlear duct. What type of fluid does each chamber contain?

The Physiology of Hearing

We divide the process of hearing into six basic steps (Figure 17–30):

1 Sound waves arrive at the tympanic membrane. Sound waves enter the external acoustic meatus and travel toward the tympanic membrane. The orientation of this meatus, or canal, provides some directional sensitivity. Sound waves approaching a particular side of the head have direct access to the tympanic membrane on that side. Sounds arriving from another direction must bend around corners or pass through the auricle or other body tissues.

Figure 17–29 The Spiral Organ.

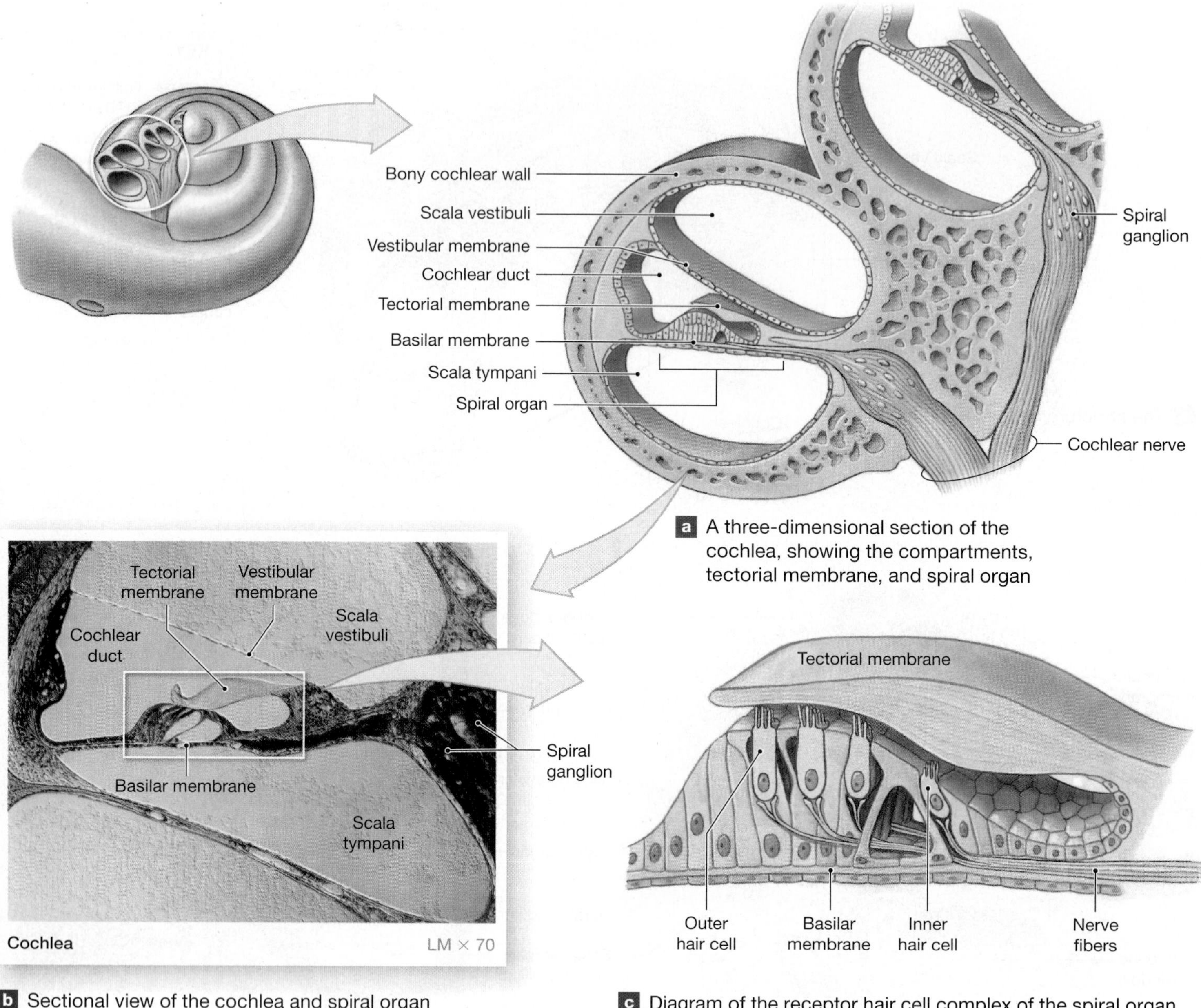
Bony cochlear wall
Scala vestibuli
Vestibular membrane
Cochlear duct
Tectorial membrane
Basilar membrane
Scala tympani
Spiral organ
Spiral ganglion
Cochlear nerve
Tectorial membrane
Vestibular membrane
Cochlear duct
Scala vestibuli
Basilar membrane
Spiral ganglion
Scala tympani
Cochlea
LM × 70
Tectorial membrane
Outer hair cell
Basilar membrane
Inner hair cell
Nerve fibers

a A three-dimensional section of the cochlea, showing the compartments, tectorial membrane, and spiral organ

b Sectional view of the cochlea and spiral organ

c Diagram of the receptor hair cell complex of the spiral organ

2 **Movement of the tympanic membrane displaces the auditory ossicles.** The tympanic membrane provides a surface for the collection of sound. The membrane vibrates in resonance to sound waves with frequencies between approximately 20 and 20,000 Hz. When the tympanic membrane vibrates, so do the malleus, incus, and stapes. The ossicles are connected in such a way that an in–out movement of the tympanic membrane produces a rocking motion of the stapes. In this way, the sound is amplified.

3 **Movement of the stapes at the oval window produces pressure waves in the perilymph of the scala vestibuli.** Liquids are not compressible. If you squeeze one part of a water-filled balloon, it bulges somewhere else. Because the rest of the cochlea is sheathed in bone, pressure applied at the oval window can be relieved only at the round window. The movement of the stapes can be understood by focusing on just its in–out component. Basically, when the stapes moves inward, the round window bulges outward, into the middle ear cavity. As the stapes moves in and out,

Figure 17–30 **Sound and Hearing.** Steps in the reception and transduction of sound energy.

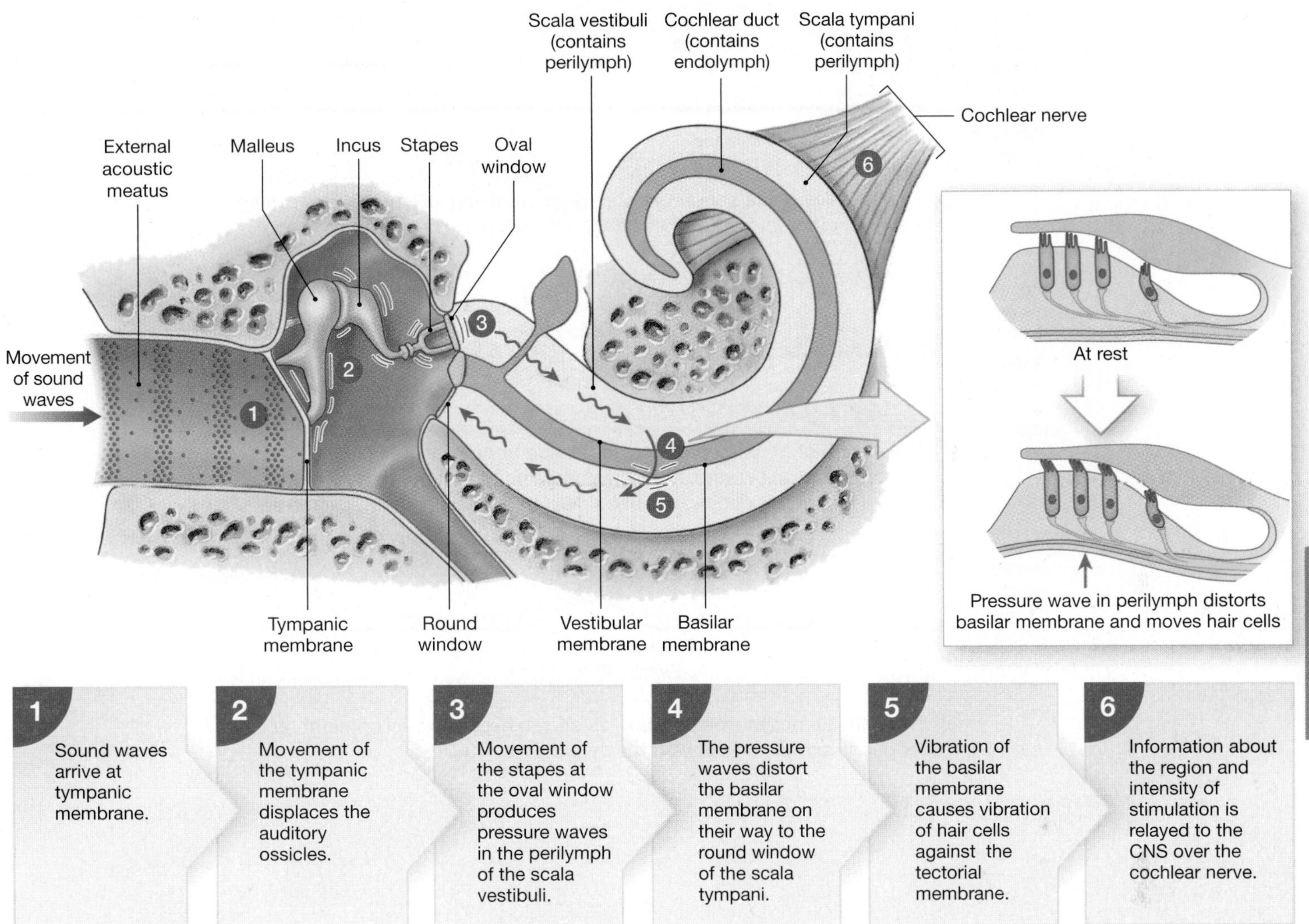

Scala vestibuli (contains perilymph)
Cochlear duct (contains endolymph)
Scala tympani (contains perilymph)
Cochlear nerve
External acoustic meatus
Malleus
Incus
Stapes
Oval window
Movement of sound waves
At rest
Pressure wave in perilymph distorts basilar membrane and moves hair cells
Tympanic membrane
Round window
Vestibular membrane
Basilar membrane

1 Sound waves arrive at tympanic membrane.

2 Movement of the tympanic membrane displaces the auditory ossicles.

3 Movement of the stapes at the oval window produces pressure waves in the perilymph of the scala vestibuli.

4 The pressure waves distort the basilar membrane on their way to the round window of the scala tympani.

5 Vibration of the basilar membrane causes vibration of hair cells against the tectorial membrane.

6 Information about the region and intensity of stimulation is relayed to the CNS over the cochlear nerve.

vibrating at the frequency of the sound arriving at the tympanic membrane, it produces pressure waves within the perilymph.

4 The pressure waves distort the basilar membrane on their way to the round window of the scala tympani. These pressure waves travel through the perilymph of the scala vestibuli and scala tympani to reach the round window. In doing so, the waves distort the basilar membrane. The location of maximum distortion varies with the frequency of the sound, due to regional differences in the width and flexibility of the basilar membrane along its length. High-frequency sounds, which have a very short wavelength, vibrate the basilar membrane near the oval window. The lower the frequency of the sound and the longer the wavelength, the farther the area of maximum distortion is from the oval window (Figure 17–31). In this way, information about frequency is translated into information about *position* along the basilar membrane.

The *amount* of movement at a given location depends on the amount of force applied by the stapes, which in turn depends on the energy content of the sound. The louder the sound, the more the basilar membrane moves.

5 Vibration of the basilar membrane causes hair cells to vibrate against the tectorial membrane. Vibration of the affected region of the basilar membrane moves hair cells against the tectorial membrane. This movement leads to the displacement of the stereocilia, which in turn opens ion channels in the plasma membranes of the hair cells. The resulting inrush of ions depolarizes the hair cells, leading to the release of neurotransmitters that stimulate sensory neurons.

The hair cells of the spiral organ are arranged in several rows. A very soft sound may stimulate only a few hair cells in a portion of one row. As the intensity of a sound increases, not only do these hair cells become more active, but additional hair

Figure 17–31 Frequency Discrimination.

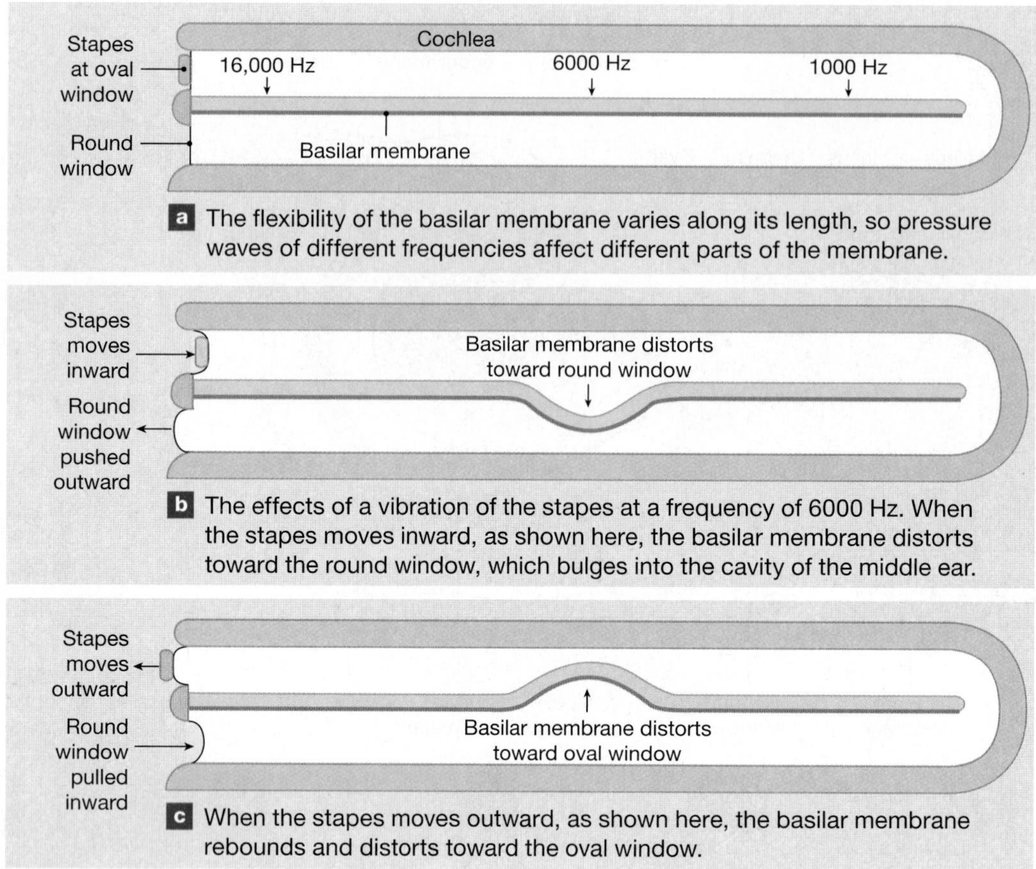
Stapes at oval window
Round window
Cochlea
16,000 Hz
6000 Hz
1000 Hz
Basilar membrane
Stapes moves inward
Round window pushed outward
Basilar membrane distorts toward round window
Stapes moves outward
Round window pulled inward
Basilar membrane distorts toward oval window

a The flexibility of the basilar membrane varies along its length, so pressure waves of different frequencies affect different parts of the membrane.

b The effects of a vibration of the stapes at a frequency of 6000 Hz. When the stapes moves inward, as shown here, the basilar membrane distorts toward the round window, which bulges into the cavity of the middle ear.

c When the stapes moves outward, as shown here, the basilar membrane rebounds and distorts toward the oval window.

cells—first in the same row and then in adjacent rows—are stimulated as well. The number of hair cells responding in a given region of the spiral organ provides information on the intensity of the sound.

6 Information about the region and the intensity of stimulation is relayed to the CNS over the cochlear nerve, a division of cranial nerve VIII. The cell bodies of the bipolar sensory neurons that monitor the cochlear hair cells are located at the center of the bony cochlea, in the **spiral ganglion** (see Figure 17–29a). From there, the information is carried by the cochlear nerve to the cochlear nuclei of the medulla oblongata for distribution to other brain centers.

Auditory Pathways

The ascending pathways for auditory sensations are shown in Figure 17–32. Stimulation of hair cells along the basilar membrane activates sensory neurons whose cell bodies are in the adjacent spiral ganglion 1. The afferent fibers of those neurons form the **cochlear nerve**. These axons enter the medulla oblongata, where they synapse at the **cochlear nucleus** on that side 2. From there, information ascends to both **superior olivary nuclei** of the pons and both inferior colliculi of the midbrain 3. This midbrain processing center coordinates a variety of unconscious motor responses to acoustic stimuli, including auditory reflexes that involve skeletal muscles of the head, face, and trunk 4. These reflexes automatically change the position of your head in response to a sudden loud noise. You usually turn your head and your eyes toward the source of the sound.

Before reaching the cerebral cortex and your awareness, ascending auditory sensations synapse in the medial geniculate body of the thalamus 5. Projection fibers then deliver the information to the **auditory cortex** of the temporal lobe 6. Information travels to the cortex over labeled lines: High-frequency sounds activate one portion of the cortex, low-frequency sounds another. In effect, the auditory cortex contains a map of the spiral organ. So, information about frequency, translated into information about position on the basilar membrane, is projected in that form onto the auditory cortex. There it is interpreted to produce your subjective sensation of pitch.

Figure 17–32 **Pathways for Auditory Sensations.**

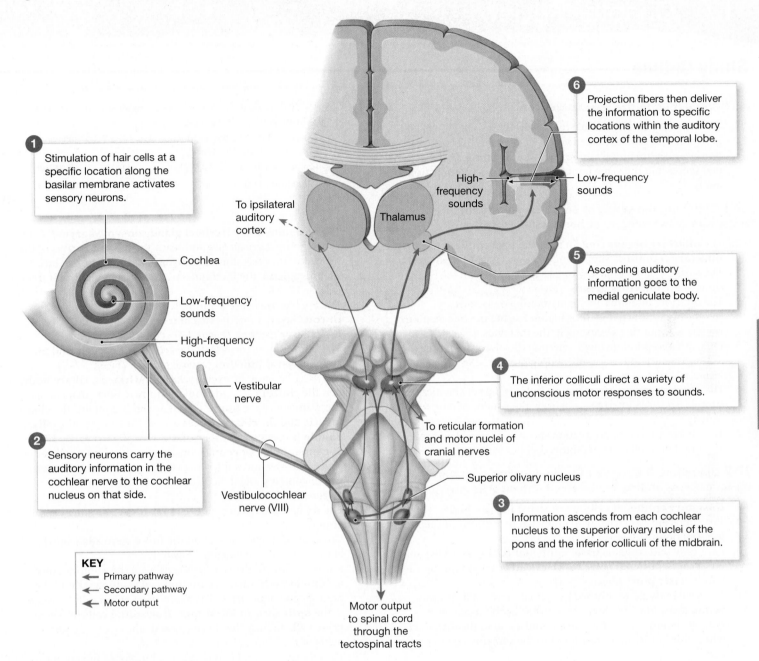

1 Stimulation of hair cells at a specific location along the basilar membrane activates sensory neurons.

Cochlea

Low-frequency sounds

High-frequency sounds

Vestibular nerve

2 Sensory neurons carry the auditory information in the cochlear nerve to the cochlear nucleus on that side.

Vestibulocochlear nerve (VIII)

To ipsilateral auditory cortex

Thalamus

High-frequency sounds

Low-frequency sounds

6 Projection fibers then deliver the information to specific locations within the auditory cortex of the temporal lobe.

5 Ascending auditory information goes to the medial geniculate body.

4 The inferior colliculi direct a variety of unconscious motor responses to sounds.

To reticular formation and motor nuclei of cranial nerves

Superior olivary nucleus

3 Information ascends from each cochlear nucleus to the superior olivary nuclei of the pons and the inferior colliculi of the midbrain.

Motor output to spinal cord through the tectospinal tracts

KEY
- Primary pathway
- Secondary pathway
- Motor output

Most of the auditory information from one cochlea is projected to the auditory complex of the cerebral hemisphere on the opposite side of the brain. However, each auditory cortex also receives information from the cochlea on that side. These interconnections play a role in localizing left/right sounds. They can also reduce the functional impact of damage to a cochlea or ascending pathway.

An individual whose auditory cortex is damaged will respond to sounds and have normal acoustic reflexes, but will find it difficult or impossible to interpret the sounds and recognize a pattern in them. Damage to the adjacent auditory association area leaves the ability to detect the tones and patterns intact, but produces an inability to comprehend their meaning.

Checkpoint

15. **If the round window were not able to bulge out with increased pressure in the perilymph, how would the perception of sound be affected?**
16. **How would the loss of stereocilia from hair cells of the spiral organ affect hearing?**
17. **Why would blockage of the auditory tube produce an earache?**

See the blue Answers tab at the back of the book.

17 Chapter Review

Study Outline

MasteringA&P™ Access more chapter study tools online in the MasteringA&P Study Area:
- Chapter Quizzes, Chapter Practice Test, MP3 Tutor Sessions, and Clinical Case Studies
- Practice Anatomy Lab PAL 3.0
- A&P Flix A&PFlix
- Interactive Physiology iP2
- PhysioEx PhysioEx 9.1

An Introduction to the Special Senses p. 566

1. The sense organs responsible for the five *special senses—olfaction* (smell), *gustation* (taste), *vision, equilibrium* (balance), and *hearing*—are structurally more complex than those of the general senses, but the same basic principles of receptor function apply.

17-1 Olfaction, the sense of smell, involves olfactory receptors responding to airborne chemical stimuli p. 566

2. The **olfactory organs** contain the **olfactory epithelium** with **olfactory receptors**, supporting cells, and **basal (stem) cells**. The surfaces of the olfactory organs are coated with the secretions of the **olfactory glands**. *(Figure 17–1)*
3. The olfactory receptors are highly modified neurons.
4. In olfaction, the arriving information reaches the information centers without first synapsing in the thalamus. *(Figure 17–1)*
5. Olfactory reception involves detecting dissolved chemicals as they interact with odorant-binding proteins. *(Spotlight Figure 17–2)*
6. The olfactory system can distinguish thousands of chemical stimuli. The CNS interprets smells by the pattern of receptor activity.
7. The olfactory receptor population shows considerable turnover. The number of olfactory receptors declines with age.

17-2 Gustation, the sense of taste, involves gustatory receptors responding to dissolved chemical stimuli p. 570

8. **Gustatory epithelial cells** are clustered in **taste buds**.
9. Taste buds are associated with epithelial projections (*lingual papillae*) on the tongue. *(Figure 17–3)*
10. Each taste bud contains **basal epithelial cells** (stem cells) and **gustatory epithelial cells**, which extend *taste hairs* through a narrow **taste pore**. *(Figure 17–3)*
11. The taste buds are monitored by cranial nerves that synapse within the solitary nucleus of the medulla oblongata. Postsynaptic neurons carry the nerve impulses on to the thalamus, where third-order neurons project to the somatosensory cortex.
12. The **primary taste sensations** are sweet, salty, sour, and bitter. Receptors also exist for **umami** and **water**. *(Spotlight Figure 17–2)*
13. Taste sensitivity exhibits significant individual differences, some of which are inherited.
14. The number of taste buds declines with age.

17-3 Internal eye structures contribute to vision, while accessory eye structures provide protection p. 572

15. The **accessory structures** of the eye include the **eyelids** (*palpebrae*), separated by the **palpebral fissure**, the **eyelashes**, and the **tarsal glands**. *(Figures 17–4, 17–5)*
16. An epithelium called the **conjunctiva** covers the inner surfaces of the eyelids and most of the exposed surface of the eye. The **cornea** is transparent. *(Figure 17–4, 17–5)*
17. The secretions of the **lacrimal gland** contain **lysozyme**. Tears collect in the *lacrimal lake* and reach the inferior meatus of the nose after they pass through the **lacrimal puncta**, the **lacrimal canaliculi**, the **lacrimal sac**, and the **nasolacrimal duct**. *(Figure 17–4)*
18. The eye has three layers, formerly called tunics: an outer **fibrous layer**, a middle **vascular layer** (uvea), and a deeper, **inner layer** (retina). *(Figure 17–5)*
19. The fibrous layer consists of the **sclera**, the **cornea**, and the **corneoscleral junction** (*corneal limbus*). *(Figure 17–5)*
20. The vascular layer, or **uvea**, includes the **iris**, the **ciliary body**, and the **choroid**. The iris contains muscle fibers that change the diameter of the **pupil**. The ciliary body contains the **ciliary muscle** and the **ciliary processes**, which attach to the **ciliary zonule** (*suspensory ligament*) of the lens. *(Figures 17–5, 17–6)*
21. The **inner layer**, or **retina**, consists of a thin lining called the *pigmented layer* and a thicker covering called the *neural layer*, which contains visual receptors and associated neurons. *(Figures 17–5, 17–7)*
22. The retina contains two types of **photoreceptors: rods** and **cones**.
23. Cones are densely clustered in the **fovea centralis** (*fovea*), at the center of the **macula**. *(Figure 17–7)*
24. The direct line to the CNS proceeds from the photoreceptors to **bipolar cells**, then to **ganglion cells**, and, finally, to the brain by the optic nerve. The axons of ganglion cells converge at the **optic disc**, or **blind spot**. **Horizontal cells** and **amacrine cells** modify the signals passed among other components of the retina. *(Figures 17–7, 17–8)*
25. The ciliary body and lens divide the interior of the eye into a large **posterior cavity**, or *vitreous chamber*, and a smaller **anterior cavity**. The anterior cavity is subdivided into the **anterior chamber**, which extends from the cornea to the iris, and a **posterior chamber**, between the iris and the ciliary body and lens. *(Figure 17–9)*
26. The fluid **aqueous humor** circulates within the eye and reenters the circulation after diffusing through the walls of the anterior chamber and into the **scleral venous sinus** (*canal of Schlemm*). *(Figure 17–9)*
27. The **lens** lies posterior to the cornea and forms the anterior boundary of the posterior cavity. This cavity contains the **vitreous body**, a clear, gelatinous mass that helps stabilize the shape of the eye and support the retina. *(Figure 17–9)*
28. The lens focuses a visual image on the photoreceptors. A **cataract** is the condition in which a lens has lost its transparency.

17-4 The focusing of light on the retina leads to the formation of a visual image p. 581

29. Light is **refracted** (bent) when it passes through the cornea and lens. During **accommodation**, the shape of the lens changes to focus an image on the retina. "Normal" **visual acuity** is rated 20/20. *(Figures 17–10 to Spotlight Figure 17–13)*

17-5 Photoreceptors transduce light into electrical signals that are then processed in the visual cortex p. 583

30. The two types of photoreceptors are rods, which respond to almost any photon, regardless of its energy content, and cones, which have characteristic ranges of sensitivity. *(Figure 17–14)*
31. Each photoreceptor contains an **outer segment** with membranous **discs**. A narrow stalk connects the outer segment to the **inner segment**. Light absorption occurs in the **visual pigments**, which are derivatives of **rhodopsin** (opsin plus the pigment retinal, which is synthesized from vitamin A). *(Figure 17–14)*
32. Color sensitivity depends on the integration of information from **red**, **green**, and **blue** cones. **Color blindness** is the inability to detect certain colors. *(Figures 17–15, 17–18)*
33. In the absence of photons, a photoreceptor produces a dark current and constantly releases neurotransmitter. Photon absorption reduces the dark current and decreases the release of neurotransmitter to the bipolar cell. *(Spotlight Figure 17–16)*
34. During the process of **bleaching**, rhodopsin molecules are broken apart to regenerate retinal back to its photon-absorbing 11-*cis* form. *(Spotlight Figure 17–16, Figure 17–17)*
35. In the **dark-adapted state**, most visual pigments are fully receptive to stimulation. In the **light-adapted state**, the pupil constricts and bleaching of the visual pigments takes place.
36. The ganglion cells that monitor rods, called **M cells** (*magnocells*), are relatively large. The ganglion cells that monitor cones, called **P cells** *(parvo cells)*, are smaller and more numerous. *(Figure 17–19)*
37. Visual data from the left half of the combined field of vision arrive at the visual cortex of the right occipital lobe. Data from the right half of the combined field of vision arrive at the visual cortex of the left occipital lobe. *(Figure 17–20)*
38. **Depth perception** is resolved by comparing relative positions of objects between the left- and right-eye images. *(Figure 17–20)*
39. Visual inputs to the suprachiasmatic nucleus of the hypothalamus affect the function of other brainstem nuclei. This nucleus establishes a visceral **circadian rhythm**, which is tied to the day–night cycle and affects other metabolic processes.

17-6 Equilibrium sensations monitor head position and movement, while hearing involves the detection and interpretation of sound waves p. 592

40. The receptors of the internal ear provide the senses of equilibrium and hearing.
41. The ear is divided into the **external ear**, the **middle ear**, and the **internal ear**. *(Figure 17–21)*
42. The external ear includes the **auricle**, or *pinna*, which surrounds the entrance to the **external acoustic meatus**, which ends at the **tympanic membrane** (*eardrum*). *(Figure 17–21)*
43. The middle ear communicates with the nasopharynx by the **auditory** (*pharyngotympanic*) **tube**. The middle ear encloses and protects the **auditory ossicles**. *(Figures 17–21, 17–22)*
44. The **membranous labyrinth** (the chambers and tubes) of the internal ear contains the fluid **endolymph**. The **bony labyrinth** surrounds and protects the membranous labyrinth and can be subdivided into the **vestibule**, the **semicircular canals**, and the **cochlea**. *(Figures 17–21, 17–23)*
45. The vestibule of the internal ear encloses the **saccule** and **utricle**. The semicircular canals contain the **semicircular ducts**. The cochlea contains the **cochlear duct**, an elongated portion of the membranous labyrinth. *(Figure 17–23)*
46. The **round window** separates the **perilymph** from the air spaces of the middle ear. The **oval window** is connected to the base of the stapes. *(Figure 17–21)*
47. The basic sensory receptors of the internal ear are **hair cells**, which provide information about the direction and strength of mechanical stimuli. *(Figure 17–24)*
48. The **anterior**, **posterior**, and **lateral semicircular ducts** are continuous with the utricle. Each duct contains an **ampullary crest** with a gelatinous **ampullary cupula** and associated sensory receptors. *(Figures 17–23, 17–24)*
49. The saccule and utricle are connected by a passageway that is continuous with the **endolymphatic duct**. This duct terminates in the **endolymphatic sac**. In the saccule and utricle, hair cells cluster within **maculae**, where their hair cells contact the gelatinous *otolithic membrane*. The surface of the membrane contains calcium carbonate crystals called **otoliths**. *(Figures 17–24, 17–25)*
50. The vestibular receptors activate sensory neurons of the **vestibular ganglia**. The axons form the **vestibular nerve**, synapsing within the **vestibular nuclei**. *(Figure 17–26)*
51. The energy content of a sound determines its *intensity*, or loudness, measured in **decibels**. Sound waves travel toward the tympanic membrane, which vibrates, and the auditory ossicles then conduct these vibrations to the internal ear. Movement at the oval window applies pressure to the perilymph of the scala vestibuli. *(Figures 17–27, 17–30; Table 17–1)*
52. The **cochlear duct** (**scala media**) lies between the **scala vestibuli** (*vestibular duct*) and the **scala tympani** (*tympanic duct*). The hair cells of the cochlear duct lie within the **spiral organ** (**organ of Corti**). *(Figures 17–28, 17–29)*
53. Pressure waves distort the **basilar membrane** and push the hair cells of the spiral organ against the **tectorial membrane**. Two muscles, the **tensor tympani** and **stapedius**, contract to reduce the amount of motion when very loud sounds arrive. *(Figures 17–30, 17–31)*
54. The sensory neurons are located in the **spiral ganglion** of the cochlea. The afferent fibers of these neurons form the **cochlear nerve**, synapsing at their respective left or right **cochlear nucleus**. Ascending auditory information from each ear reaches the auditory cortex on both sides of the brain. *(Figure 17–32)*

Review Questions

See the blue Answers tab at the back of the book.

LEVEL 1 Reviewing Facts and Terms

1. Identify the structures in the following horizontal section of the eye.

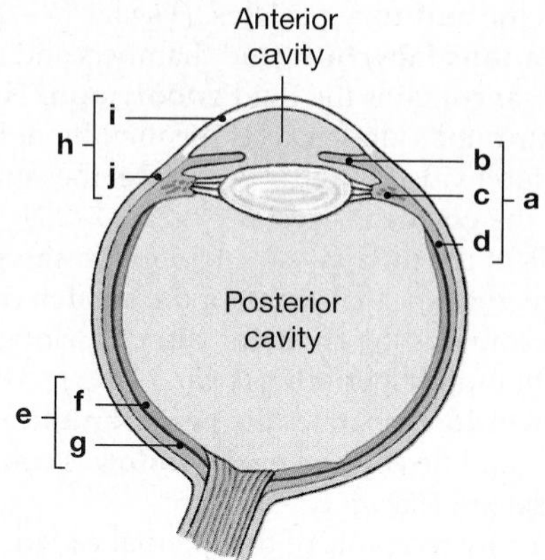

Anterior cavity
Posterior cavity
a
b
c
d
e
f
g
h
i
j

(a) ________ (b) ________
(c) ________ (d) ________
(e) ________ (f) ________
(g) ________ (h) ________
(i) ________ (j) ________

2. A reduction in sensitivity in the presence of a constant stimulus is **(a)** transduction, **(b)** sensory coding, **(c)** line labeling, **(d)** adaptation.
3. A blind spot occurs in the retina where **(a)** the fovea centralis is located, **(b)** ganglion cells synapse with bipolar cells, **(c)** the optic nerve attaches to the retina, **(d)** rod cells are clustered to form the macula, **(e)** amacrine cells are located.
4. Sound waves are converted into mechanical movements by the **(a)** auditory ossicles, **(b)** cochlea, **(c)** oval window, **(d)** round window, **(e)** tympanic membrane.
5. The basic receptors in the internal ear are the **(a)** utricles, **(b)** saccules, **(c)** hair cells, **(d)** supporting cells, **(e)** ampullae.
6. The retina is also called **(a)** the vascular layer, **(b)** the fibrous layer, **(c)** the inner layer, **(d)** all of these.
7. At sunset, your visual system adapts to **(a)** fovea vision, **(b)** rod-based vision, **(c)** macular vision, **(d)** cone-based vision.
8. A better-than-average visual acuity rating is **(a)** 20/20, **(b)** 20/30, **(c)** 15/20, **(d)** 20/15.
9. The malleus, incus, and stapes are the tiny bones located in the **(a)** external ear, **(b)** middle ear, **(c)** internal ear, **(d)** membranous labyrinth.
10. Identify the structures of the external, middle, and internal ear in the following figure.

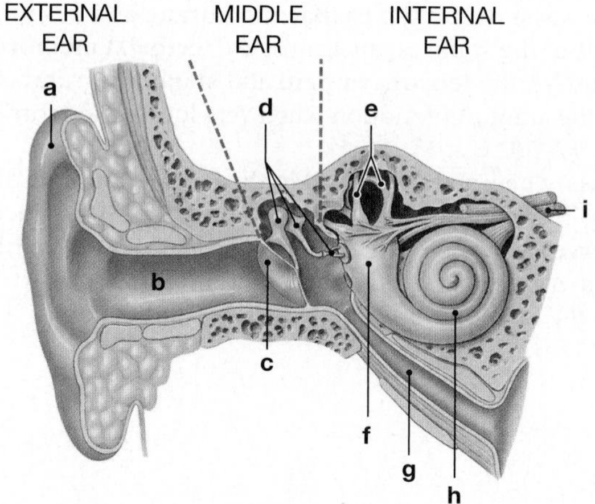

EXTERNAL EAR
MIDDLE EAR
INTERNAL EAR
a
b
c
d
e
f
g
h
i

(a) ________ (b) ________
(c) ________ (d) ________
(e) ________ (f) ________
(g) ________ (h) ________
(i) ________

11. Receptors in the saccule and utricle provide sensations of **(a)** angular acceleration, **(b)** hearing, **(c)** vibration, **(d)** gravity and linear acceleration and deceleration.
12. The spiral organ is located in the ____ of the internal ear. **(a)** utricle, **(b)** bony labyrinth, **(c)** vestibule, **(d)** cochlea.
13. Auditory information about the frequency and intensity of stimulation is relayed to the CNS over the cochlear nerve, a division of cranial nerve **(a)** IV, **(b)** VI, **(c)** VIII, **(d)** X.
14. What are the four types of papillae on the human tongue?
15. **(a)** What structures make up the fibrous layer of the eye? **(b)** What are the functions of the fibrous layer?
16. What structures make up the vascular layer of the eye?
17. What are the three auditory ossicles in the middle ear, and what are their functions?

LEVEL 2 Reviewing Concepts

18. Trace the olfactory pathway from the time an odor reaches the olfactory epithelium until nerve impulses reach their final destination in the brain.
19. Why are olfactory sensations long lasting and an important part of our memories and emotions?
20. What is the usual result if a sebaceous gland of an eyelash or a tarsal gland becomes infected?
21. Displacement of stereocilia toward the kinocilium of a hair cell **(a)** produces a depolarization of the membrane, **(b)** produces a hyperpolarization of the membrane, **(c)** decreases the membrane permeability to sodium ions, **(d)** increases the membrane permeability to potassium ions, **(e)** does not affect the membrane potential of the cell.
22. Damage to the ampullary cupula of the lateral semicircular duct would interfere with the perception of **(a)** the direction of gravitational pull, **(b)** linear acceleration, **(c)** horizontal rotation of the head, **(d)** vertical rotation of the head, **(e)** angular rotation of the head.
23. When viewing an object *close* to you, your lens should be more ____. **(a)** rounded, **(b)** flattened, **(c)** concave, **(d)** lateral, **(e)** medial.

LEVEL 3 Critical Thinking and Clinical Applications

24. You are at a park watching some deer 35 feet away from you. A friend taps you on the shoulder to ask a question. As you turn to look at your friend, who is standing just 2 feet away, what changes would your eyes undergo?
25. Your friend Shelly suffers from myopia (nearsightedness). You remember from your physics class that concave lenses cause light waves to spread or diverge and that convex lenses cause light waves to converge. What type of corrective lenses would you suggest to your friend? **(a)** concave lenses, **(b)** convex lenses.
26. Tom has surgery to remove polyps (growths) from his sinuses. After he heals from the surgery, he notices that his sense of smell is not as keen as it was before the surgery. Can you suggest a reason for this?
27. For a few seconds after you ride the express elevator from the 20th floor to the ground floor, you still feel as if you are descending, even though you have come to a stop. Why?
28. Juan tells his physician that he has been feeling dizzy, especially when he closes his eyes. He is asked to stand with his feet together and arms extended forward. As long as he keeps his eyes open, he exhibits very little movement. But when he closes his eyes, his body begins to sway a great deal, and his arms tend to drift together toward the left side of his body. Why does this occur?

+ CLINICAL CASE Wrap-Up A Chance to See

Makena suffers from *myopia*, or nearsightedness. Due to the elongated shape of Makena's eyeballs, distant images are projected in front of her retina and appear blurry and out of focus to her. Glasses can correct this easily, but nobody in Makena's village has ever met an eye doctor of any kind. The closest optical shop is far out of this family's reach, both in distance and cost.

Makena's teacher sets up a makeshift eye clinic with the help of volunteers from the Centre for Vision in the Developing World, a group dedicated to correcting vision for poor children. To test her vision, Makena attempts to read from an eye chart taped to the wall of the school building 20 feet away. Makena's visual acuity tests at 20/200.

Next, Makena puts on a simple pair of universal glasses, made with a clear membrane filled with silicone oil, held between two plastic discs on each side. A removable syringe filled with silicone oil is attached to each side. With her right eye covered, Makena uses the syringe to inject more silicone into the left lens until her left eye can read the eye chart perfectly. Then she repeats the process for her right eye. The final step is to remove the syringes. Now she can see things she never knew she had been missing—and she gets to keep the glasses.

1. What does Makena's initial eye test finding of 20/200 mean?
2. If Makena's vision is corrected to a state of emmetropia, what does this mean?

See the blue Answers tab at the back of the book.

17

Related Clinical Terms

age-related macular degeneration: A disease associated with aging that gradually destroys sharp, central vision by affecting the macula, the part of the eye that allows a person to see fine detail. It is painless and in some cases can slowly worsen over time, causing little concern to the person, or it can rapidly progress and may cause blindness in both eyes. It is the leading cause of blindness in persons over the age of 60. It has two forms, wet and dry.

ageusia: A rare inability to taste. More common is hypogeusia, a disorder in which the person affected has trouble distinguishing between tastes.

anosmia: The complete loss of smell, which can be temporary (caused by an obstruction such as a polyp) or permanent (perhaps due to aging or a brain tumor).

blepharitis: Common and persistent inflammation of the eyelid caused by poor hygiene, excessive oil production by the glands of the eyelid, or a bacterial infection. Signs and symptoms include itching, flakes on the eyelashes, and a gritty, sandy feeling.

conductive hearing loss: Deafness resulting from conditions in the external or middle ear that block the transfer of vibrations from the tympanic membrane to the oval window.

hyposmia: A lessened sensitivity to odors.

mydriasis: Dilation of the pupils of the eye induced by medical eye drops or caused by disease.

ophthalmologist: A physician who specializes in ophthalmology, which is the branch of medicine dealing with the diseases and surgery of the visual pathways, including the eye, brain, and areas surrounding the eye.

optometrist: A primary eye care doctor who diagnoses, manages, and treats disorders of the visual system and eye diseases and who also measures vision for the purpose of correcting one's vision problems with specifically prescribed lenses.

otalgia: Pain in the ear; an earache.

photophobia: An oversensitivity to light possibly leading to tearing, discomfort, or pain. Causes include abrasions to the corneal area, inflammation, disease, and some medications.

sensorineural hearing loss: Deafness resulting from problems within the cochlea or along the auditory nerve pathway.

Snellen chart: A printed chart of block letters in graduated type sizes used to measure visual acuity.

strabismus: The abnormal alignment of one or both eyes that prevents the person from gazing on the same point with both eyes.

synesthesia: Abnormal condition in which sensory nerve messages connect to the wrong centers of the brain. For example, touching an object may produce the perception of a sound, while hearing a tone may produce the visualization of a color.

tinnitus: A buzzing, whistling, or ringing sound heard in the absence of an external stimulus. Causes include injury, disease, inflammation, or some drugs.

vertigo: A feeling that you are dizzily spinning or that things are dizzily turning about you. Vertigo is usually caused by a problem with the internal ear, but can also be due to vision problems.

18 The Endocrine System

Learning Outcomes

These Learning Outcomes correspond by number to this chapter's sections and indicate what you should be able to do after completing this chapter.

- 18-1 ▪ **Explain the importance of intercellular communication, describe the mechanisms involved, and compare the modes of intercellular communication that occur in the endocrine and nervous systems.** p. 611
- 18-2 ▪ **Compare the cellular components of the endocrine system with those of other systems, contrast the major structural classes of hormones, and explain the general mechanisms of hormonal action on target organs.** p. 613
- 18-3 ▪ **Describe the location, hormones, and functions of the pituitary gland, and discuss the effects of abnormal pituitary hormone production.** p. 619
- 18-4 ▪ **Describe the location, hormones, and functions of the thyroid gland, and discuss the effects of abnormal thyroid hormone production.** p. 627
- 18-5 ▪ **Describe the location, hormone, and functions of the parathyroid glands, and discuss the effects of abnormal parathyroid hormone production.** p. 632
- 18-6 ▪ **Describe the location, structure, hormones, and general functions of the adrenal glands, and discuss the effects of abnormal adrenal hormone production.** p. 634
- 18-7 ▪ **Describe the location of the pineal gland, and discuss the functions of the hormone it produces.** p. 637
- 18-8 ▪ **Describe the location, structure, hormones, and functions of the pancreas, and discuss the effects of abnormal pancreatic hormone production.** p. 637
- 18-9 ▪ **Describe the functions of the hormones produced by the kidneys, heart, thymus, testes, ovaries, and adipose tissue.** p. 641
- 18-10 ▪ **Explain how hormones interact to produce coordinated physiological responses and influence behavior, describe the role of hormones in the general adaptation syndrome, and discuss how aging affects hormone production and give examples of interactions between the endocrine system and other organ systems.** p. 646

CLINICAL CASE Stones, Bones, and Groans

Monica is back in her doctor's office, following up on her lab results. She feels a little crazy, seeing a new family practitioner with so many vague complaints. She has a history of hypertension (high blood pressure) and heart palpitations (irregular heartbeats). She has had several kidney stones. She has osteopenia and suffers from nagging bone pain. Recently she has experienced vague abdominal pain with loss of appetite and constipation. Furthermore, she has been feeling chronic fatigue, some of which she attributes to having

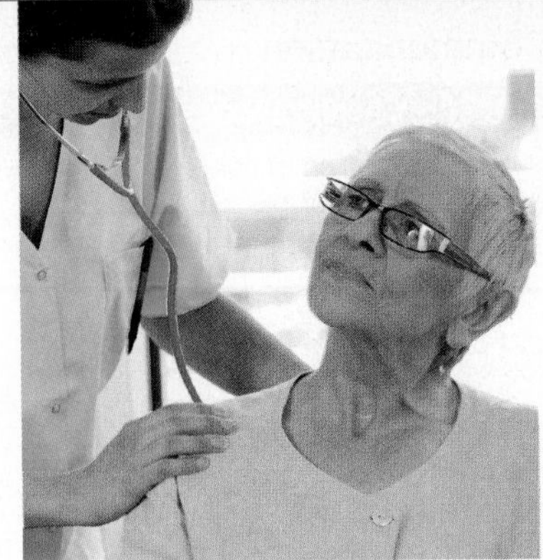

trouble falling asleep and regularly awakening in the middle of the night. Monica has also noticed that her memory is getting worse and that she feels increasingly depressed. Worst of all, her husband of 40 years has complained that she is getting more irritable and cranky.

"Monica," says her doctor, "I think I know what is causing your symptoms. The answer is right here in your lab results." **What could be causing Monica's many complaints? To find out, turn to the Clinical Case Wrap-Up on p. 654.**

An Introduction to the Endocrine System

About 30 chemical messengers known as *hormones* regulate human activities such as sleep, body temperature, hunger, and stress management. These hormones are products of the *endocrine system*, which along with the nervous system, controls and coordinates our body processes.

In this chapter we examine the structural and functional organization of the endocrine system and compare it to the nervous system. After an overview of the endocrine system and the characteristics of the hormones it produces, we consider the structure and function of the body's various *endocrine glands*. Finally, we look at the ways in which hormones modify metabolic operations, and the interactions between the endocrine system and other body systems. Let's begin by considering the role of intercellular communication in maintaining homeostasis.

18-1 Homeostasis is preserved through intercellular communication by the nervous and endocrine systems

Learning Outcome **Explain the importance of intercellular communication, describe the mechanisms involved, and compare the modes of intercellular communication that occur in the endocrine and nervous systems.**

To preserve homeostasis, cellular activities must be coordinated throughout the body. Cells coordinate their activities by sending and receiving chemical messages. Both the nervous system, which we have explored in Chapters 12–17, and the endocrine system accomplish this coordination. However, they provide different types of coordination because of how their cells communicate. To understand how cellular messages are generated and interpreted, let's take a closer look at how cells communicate with one another.

Mechanisms of Intercellular Communication

In a few specialized cases, adjacent cells coordinate cellular activities by exchanging ions and molecules across gap junctions. This **direct communication** occurs between two cells of the same type, and the cells must be in extensive physical contact. The two cells communicate so closely that they function as a single entity. Gap junctions (1) coordinate ciliary movement among epithelial cells, (2) coordinate the contractions of cardiac muscle cells, and (3) facilitate the propagation of action potentials from one neuron to the next at electrical synapses.

Most communication between cells involves the release and receipt of chemical messages. When these messages occur between cells within a single tissue, this is called **paracrine communication**, and the chemicals involved are *paracrines*. Each cell continuously "talks" to its neighbors by releasing chemicals into the extracellular fluid. These chemicals tell cells what their neighbors are doing at any moment. The result is the coordination of tissue function locally (within the same tissue). An example of a paracrine is somatostatin, which is released by some pancreatic cells to inhibit the release of insulin by other pancreatic cells. **Autocrine communication** occurs when the messages affect the same cells that secrete them, and the chemicals involved are *autocrines*. Examples of autocrines include prostaglandins, which are secreted by smooth muscle cells to cause contraction of those cells.

Endocrine communication occurs when the endocrine system uses chemical messengers called **hormones** to relay information and instructions between cells in distant portions of the body. Endocrine cells release these hormones in one tissue, and they are then transported in the bloodstream and distributed throughout the body. Each hormone has **target cells**, specific cells in other tissues that have the *receptors* needed to bind and "read" the hormonal message when it arrives. These messages alter the operations of target cells by changing the types, quantities, or activities of important enzymes and structural proteins. A hormone can modify the physical structure or biochemical properties of its target cells. Because the target cells can be anywhere in the body, a single hormone can alter the metabolic activities of multiple tissues and organs at the same time.

Table 18–1 Mechanisms of Intercellular Communication

Mechanism	Transmission	Chemical Mediators	Distribution of Effects
Direct communication	Through gap junctions	Ions, small solutes, lipid-soluble materials	Usually limited to adjacent cells of the same type that are interconnected by connexons
Paracrine communication	Through extracellular fluid	Paracrines	Primarily limited to a local area, where paracrine concentrations are relatively high; target cells must have appropriate receptors
Autocrine communication	Through extracellular fluid	Autocrines	Limited to the cell that secretes the hormone
Endocrine communication	Through the bloodstream	Hormones	Target cells are primarily in other tissues and organs and must have appropriate receptors
Synaptic communication	Across synapses	Neurotransmitters	Limited to very specific area; target cells must have appropriate receptors

18

Synaptic Communication

The nervous system, too, relies primarily on chemical communication, but it does not send messages through the bloodstream. Instead, as we have seen, in **synaptic communication** neurons release a neurotransmitter at a synapse very close to target cells that have the appropriate receptors. The signal then travels rapidly from one location to another in the form of action potentials propagated along axons. The nervous system can thus carry high-speed "messages" to specific destinations throughout the body.

Table 18–1 summarizes the five ways cells and tissues communicate with one another.

Comparison of Endocrine and Nervous Communication

How are the cellular communication mechanisms of the endocrine and nervous systems similar and different? Viewed from a general perspective, the differences between these systems seem relatively clear. In the nervous system, neurons use action potentials and neurotransmitters to control specific cells or groups of cells. This arrangement allows synaptic communication to provide crisis management in situations requiring split-second responses: If you are in danger of being hit by a speeding bus, the nervous system can coordinate and direct your leap to safety. Only a small fraction of all the cells in the body are innervated, however, and the commands from the nervous system are very specific and relatively short-lived. However, in the endocrine system, the hormones produced by endocrine cells reach almost every cell in the body. The effects of hormones on their target cells may be slow to appear, but they typically last for days. Consequently, endocrine communication is effective in coordinating cell, tissue, and organ activities such as growth and development or affecting metabolic activities on a sustained, long-term basis. These broad organizational and functional distinctions are the basis for treating the nervous and endocrine systems as separate systems.

Yet when we consider these two systems in detail, we see that they also have many similarities. For example:

- Both systems rely on the release of chemicals that bind to specific receptors on their target cells.

- The two systems share many chemical messengers. For example, norepinephrine and epinephrine are called hormones when released into the bloodstream, but neurotransmitters when released across synapses.
- Both systems are regulated mainly by negative feedback control mechanisms.
- The two systems share a common goal: to preserve homeostasis by coordinating and regulating the activities of other cells, tissues, organs, and systems.

Next we introduce the components and functions of the endocrine system.

Checkpoint

1. **Define *hormone*.**
2. **Describe paracrine communication.**
3. **Identify five mechanisms of intercellular communication.**

See the blue Answers tab at the back of the book.

18-2 The endocrine system regulates physiological processes by releasing bloodborne hormones that bind to receptors on remote target organs

Learning Outcome Compare the cellular components of the endocrine system with those of other systems, contrast the major structural classes of hormones, and explain the general mechanisms of hormonal action on target organs.

Because of their wide-reaching and long-term effects, the hormones of the endocrine system can produce complex changes in the body's physical structure and physiological capabilities. The major processes that are affected by hormone actions include:

- Growth and development
- Reproduction
- Regulation of cell metabolism and energy balance
- Regulation of body water content and levels of electrolytes and organic nutrients
- Mobilization of body defenses.

Overview of Endocrine Organs and Tissues

The **endocrine system** includes all the endocrine cells and tissues of the body that produce hormones or paracrines with effects beyond their tissues of origin. As noted in Chapter 4, *endocrine cells* are glandular secretory cells that release their secretions into the extracellular fluid. This characteristic distinguishes them from *exocrine cells,* which secrete their products onto epithelial surfaces, generally by way of ducts. ↺ p. 122

Figure 18–1 introduces the tissues, organs, and hormones of the endocrine system. Notice that *endocrine organs* are scattered throughout the body rather than being clustered together like those of many systems. Some of these organs, such as the *pituitary gland,* have endocrine secretion as a primary function. Others, such as the *pancreas,* have many other functions in addition to endocrine secretion. We consider specific endocrine organs in later sections, and we cover them in more detail in chapters on other systems.

Classes of Hormones

We can divide hormones into three classes on the basis of their chemical structure: (1) *amino acid derivatives,* (2) *peptide hormones,* and (3) *lipid derivatives.*

Amino acid derivatives, sometimes known as *biogenic amines,* are relatively small molecules that are structurally related to amino acids, the building blocks of proteins. ↺ p. 53 These hormones are synthesized from the amino acids tyrosine and tryptophan. Those made from tyrosine include (1) *thyroid hormones,* produced by the *thyroid gland,* and (2) the compounds epinephrine (E), norepinephrine (NE), and dopamine, which are sometimes called the *catecholamines* (kat-eh-KŌ-lah-mēnz). The primary hormone made from tryptophan is *melatonin* (mel-ah-TŌ-nin), produced by the *pineal gland.*

We can divide peptide hormones into two groups. One group consists of glycoproteins, while the other is made up of short polypeptides and small proteins.

There are two groups of lipid derivatives: (1) *eicosanoids* and (2) *steroid hormones.* Eicosanoids are signaling molecules and include leukotrienes, **prostaglandins**, thromboxanes, and prostacyclins. We consider eicosanoids in chapters that discuss individual tissues and organs, including Chapters 19 (the blood), 22 (the lymphatic system), and 28 (the reproductive system). Steroid hormones are lipids structurally similar to cholesterol (look back at Figure 2–18, p. 50). The individual hormones differ in the side chains attached to the basic ring structure.

Spotlight Figure 18–2 summarizes the structural classification of hormones.

Transport and Inactivation of Hormones

Hormones are typically released where capillaries are abundant, and the hormones quickly enter the bloodstream for distribution throughout the body. Within the blood, hormones may circulate freely or travel bound to special carrier proteins. A freely circulating hormone remains functional for less than 1 hour, and sometimes for as little as 2 minutes. It is inactivated when (1) it diffuses out of the bloodstream and binds to receptors on target cells, (2) it is absorbed and broken down by cells of the liver or kidneys, or (3) it is broken down by enzymes in the blood or interstitial fluids.

Figure 18–1 Organs and Tissues of the Endocrine System.

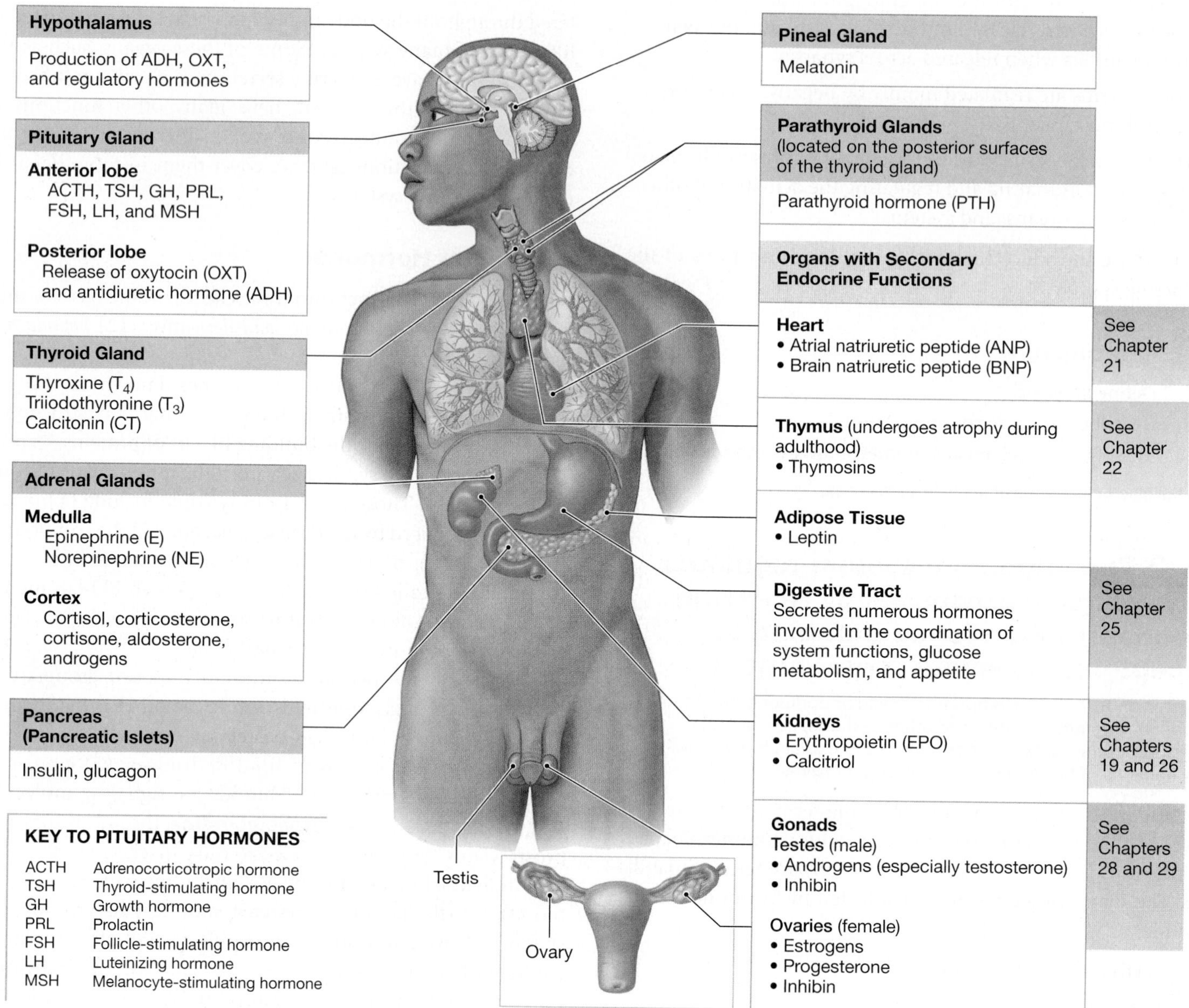
Hypothalamus
Production of ADH, OXT, and regulatory hormones
Pituitary Gland
Anterior lobe
ACTH, TSH, GH, PRL, FSH, LH, and MSH
Posterior lobe
Release of oxytocin (OXT) and antidiuretic hormone (ADH)
Thyroid Gland
Thyroxine (T_4)
Triiodothyronine (T_3)
Calcitonin (CT)
Adrenal Glands
Medulla
Epinephrine (E)
Norepinephrine (NE)
Cortex
Cortisol, corticosterone, cortisone, aldosterone, androgens
Pancreas (Pancreatic Islets)
Insulin, glucagon
KEY TO PITUITARY HORMONES
ACTH Adrenocorticotropic hormone
TSH Thyroid-stimulating hormone
GH Growth hormone
PRL Prolactin
FSH Follicle-stimulating hormone
LH Luteinizing hormone
MSH Melanocyte-stimulating hormone
Testis
Ovary
Pineal Gland
Melatonin
Parathyroid Glands (located on the posterior surfaces of the thyroid gland)
Parathyroid hormone (PTH)
Organs with Secondary Endocrine Functions
Heart
• Atrial natriuretic peptide (ANP)
• Brain natriuretic peptide (BNP)
See Chapter 21
Thymus (undergoes atrophy during adulthood)
• Thymosins
See Chapter 22
Adipose Tissue
• Leptin
Digestive Tract
Secretes numerous hormones involved in the coordination of system functions, glucose metabolism, and appetite
See Chapter 25
Kidneys
• Erythropoietin (EPO)
• Calcitriol
See Chapters 19 and 26
Gonads
Testes (male)
• Androgens (especially testosterone)
• Inhibin
Ovaries (female)
• Estrogens
• Progesterone
• Inhibin
See Chapters 28 and 29

Thyroid hormones and steroid hormones remain functional much longer, because when these hormones enter the bloodstream, more than 99 percent of them become attached to special transport proteins. For each hormone, an equilibrium state exists between its free and bound forms. As the free hormones are removed and inactivated, bound hormones are released to replace them. At any given time, the bloodstream contains a substantial reserve (several weeks' supply) of bound hormones.

Mechanisms of Hormone Action

In this chapter we focus on circulating hormones that function primarily to coordinate activities in many tissues and organs on a sustained basis. When such hormones bind to their specific receptors, this modifies cellular activities. Specifically, the binding of a hormone may:

- *Alter Genetic Activity.* The synthesis of an enzyme or a structural protein not already present in the cytoplasm may be stimulated by activating appropriate genes in the cell nucleus.
- *Alter the Rate of Protein Synthesis.* The rate of synthesis of a particular enzyme or other protein may be increased or decreased by changing the rate of transcription or translation.
- *Change Membrane Permeability.* An existing enzyme or membrane channel may be turned "on" or "off" by changing its shape or structure.

SPOTLIGHT

Figure 18–2

Structural Classification of Hormones

HORMONES

The hormones of the body can be divided into three classes on the basis of their chemical structure.

Amino Acid Derivatives

Amino acid derivatives are small molecules that are structurally related to amino acids, the building blocks of proteins.

Derivatives of Tyrosine

Thyroid Hormones

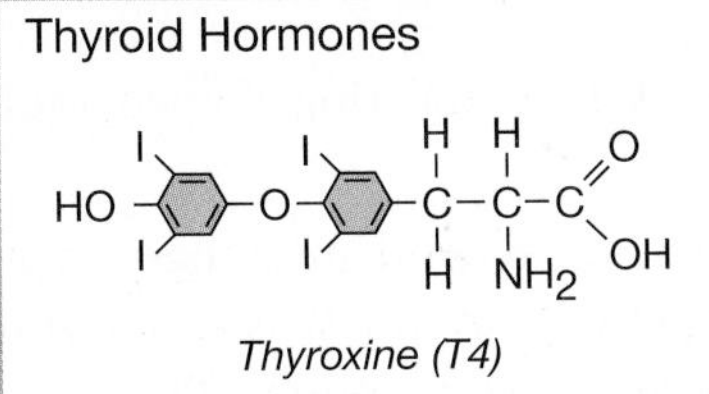

Thyroxine (T4)

Catecholamines

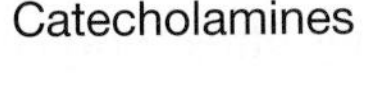

Epinephrine

Sources of tyrosine include meat, dairy, and fish.

Derivative of Tryptophan

Melatonin

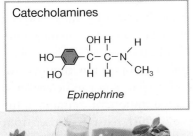

Turkey is a well-known source of tryptophan. Other sources include chocolate, oats, bananas, dried dates, milk, cottage cheese, and peanuts.

Peptide Hormones

Peptide hormones are chains of amino acids. Most peptide hormones are synthesized as **prohormones**—inactive molecules that are converted to active hormones before or after they are secreted.

Glycoproteins

These proteins are more than 200 amino acids long and have carbohydrate side chains. The glycoproteins include thyroid-stimulating hormone (TSH), luteinizing hormone (LH), and follicle-stimulating hormone (FSH) from the anterior lobe of the pituitary gland, as well as several hormones produced in other organs.

Short Polypeptides/ Small Proteins

This group of peptide hormones is large and diverse. It includes hormones that range from **short-chain polypeptides**, such as antidiuretic hormone (ADH) and oxytocin (OXT) (each 9 amino acids long), to **small proteins**, such as insulin (51 amino acids), growth hormone (GH; 191 amino acids), and prolactin (PRL; 198 amino acids). This group includes all the hormones secreted by the hypothalamus, heart, thymus, digestive tract, pancreas, and posterior lobe of the pituitary gland, as well as several hormones produced in other organs.

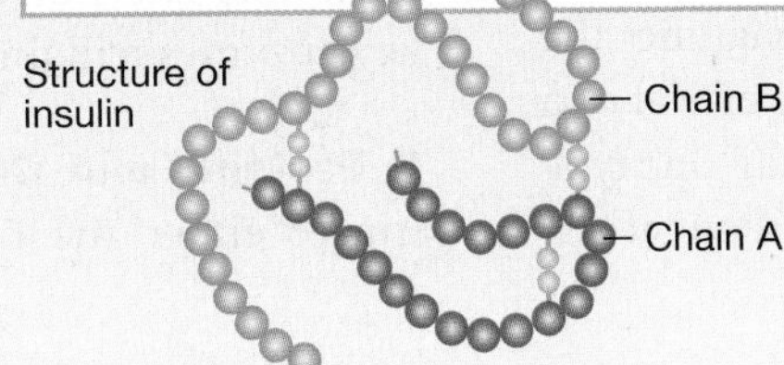
Chain B
Chain A

Structure of insulin

Lipid Derivatives

There are two classes of lipid derivatives: **eicosanoids**, derived from arachidonic acid, a 20-carbon fatty acid; and **steroid hormones**, derived from cholesterol.

Eicosanoids

Eicosanoids (Ī-kō-sa-noydz) are important paracrines that coordinate cellular activities and affect enzymatic processes (such as blood clotting in extracellular fluids). Some eicosanoids, such as **leukotrienes** (lu-kō-TRĪ-ēns), have secondary roles as hormones. A second group of eicosanoids—**prostaglandins**—are involved primarily in coordinating local cellular activities. In some tissues, prostaglandins are converted to **thromboxanes** (throm-BOX-ānz) and **prostacyclins** (pros-ta-SĪ-klinz), which also have strong paracrine effects.

Prostaglandin E

Aspirin suppresses the production of prostaglandins.

Steroid Hormones

Steroid hormones are released by the reproductive organs (androgens by the testes in males, estrogens and progesterone by the ovaries in females), by the cortex of the adrenal glands (corticosteroids), and by the kidneys (calcitriol). Because circulating steroid hormones are bound to specific transport proteins in the plasma, they remain in circulation longer than do secreted peptide hormones.

Estrogen

To affect a target cell, a hormone must first interact with an appropriate receptor. A **hormone receptor**, like a neurotransmitter receptor, is a protein molecule to which a particular molecule binds strongly. Each cell has receptors for several different hormones, but cells in different tissues have different combinations of receptors. This arrangement is one reason hormones have different effects on different tissues. For every cell, the presence or absence of a specific receptor determines the cell's hormonal sensitivities. Cells throughout the body are exposed to hormones whether or not they have the necessary receptors. If a cell has a receptor that can bind a particular hormone, that cell responds to the hormone. If a cell lacks that receptor, the hormone has no effect on that cell.

The presence or absence of a hormone can also affect the nature and number of hormone receptor proteins. **Down-regulation** is a process in which the presence of a hormone triggers a decrease in the number of hormone receptors. In down-regulation, cells become *less* sensitive to high levels of a particular hormone. In contrast, **up-regulation** is a process in which the absence of a hormone triggers an increase in the number of hormone receptors. In up-regulation, cells become *more* sensitive to low levels of a particular hormone.

Hormone receptors are located either on plasma membranes (*extracellular receptors*) or within target cells (*intracellular receptors*). These different types of receptors respond with different mechanisms. Using a few specific examples, let's consider the basic mechanisms involved.

Hormones and Extracellular Receptors: Second Messengers

The receptors for catecholamines (E, NE, and dopamine) and peptide hormones are extracellular receptors. Catecholamines and peptide hormones cannot penetrate a plasma membrane because they are not lipid soluble. Instead, these hormones bind to receptor proteins at the *outer* surface of the plasma membrane of their target cells.

Communication between the hormone and the cell uses a mechanism involving first and second messengers. A *first messenger* is a hormone that binds to an extracellular receptor. A **second messenger** is an intermediary molecule that appears due to a hormone–receptor interaction. Important second messengers are (1) *cyclic AMP (cAMP)*, a derivative of ATP; (2) *cyclic GMP (cGMP)*, a derivative of GTP, another high-energy compound; and (3) calcium ions (Ca^{2+}).

When a small number of hormone molecules binds to extracellular receptors, thousands of second messengers may appear in a cell. This process, called *amplification*, magnifies the effect of a hormone on the target cell. Moreover, the arrival of a single hormone may promote the release of more than one type of second messenger, or the production of a linked sequence of enzymatic reactions known as a *receptor cascade*. Through such mechanisms, the hormone can alter many aspects of cell function at the same time.

The link between the first messenger and the second messenger generally involves a **G protein**, an enzyme complex coupled to a membrane receptor. (Roughly 80 percent of prescription drugs target receptors coupled to G proteins.) The name *G protein* refers to the fact that these proteins bind GTP. ⮌ p. 421 A G protein is activated when a hormone binds to its receptor at the membrane surface. What happens next depends on the nature of the G protein and its effects on second messengers in the cytoplasm. Spotlight Figure 18–3, which focuses on cAMP and Ca^{2+}, diagrams three major patterns of response to G protein activation.

G Proteins and cAMP. Spotlight Figure 18–3a (left panel) shows the steps involved in *increasing* the cAMP level:

1. The activated G protein activates the enzyme **adenylate cyclase**.
2. Adenylate cyclase converts ATP to the ring-shaped molecule *cyclic AMP*.
3. Cyclic AMP then functions as a second messenger, typically by activating a *kinase* (KĪ-nās). A kinase is an enzyme that attaches a high-energy phosphate group ($-PO_4^{3-}$) to another molecule in a process called *phosphorylation*.
4. Generally, cyclic AMP activates kinases that phosphorylate proteins. The effect on the target cell depends on the nature of these proteins. The phosphorylation of plasma membrane proteins, for example, can open ion channels. In the cytoplasm, many important enzymes can be activated only by phosphorylation. One important example is the enzyme *glycogen phosphorylase*, which releases glucose from glycogen reserves in skeletal muscles and the liver.

Many hormones, including epinephrine, calcitonin, parathyroid hormone, ADH, ACTH, FSH, LH, TSH, and glucagon, produce their effects by this mechanism. The increase in the cAMP level is usually short-lived, because the cytoplasm contains another enzyme, **phosphodiesterase (PDE)**, which inactivates cyclic AMP by converting it to AMP (adenosine monophosphate).

Spotlight Figure 18–3a (center panel) depicts one way the activation of a G protein can *decrease* the concentration of cAMP within the cell. In this case, the activated G protein stimulates PDE activity and inhibits adenylate cyclase activity. The cAMP level then declines, because cAMP breakdown accelerates while cAMP synthesis is prevented. The decline has an inhibitory effect on the cell, because without phosphorylation, key enzymes remain inactive.

G Proteins and Calcium Ions. An activated G protein can trigger either the opening of calcium ion channels in the

SPOTLIGHT

Figure 18–3
G Proteins and Second Messengers

MECHANISM OF HORMONE ACTION

First and Second Messengers

A hormone that binds to receptors in the plasma membrane cannot directly affect the activities inside the target cell. For example, it cannot begin building a protein or catalyzing a specific reaction. Instead, the hormone uses an intracellular intermediary to bring about its effects. The hormone, or **first messenger**, does something that leads to the appearance of a **second messenger** in the cytoplasm. The second messenger may act as an enzyme activator, inhibitor, or cofactor. The net result is a change in the rates of various metabolic reactions. The two most important second messengers are cyclic AMP (cAMP), a derivative of ATP, and calcium ions (Ca^{2+}).

The first messenger (a peptide hormone, catecholamine, or eicosanoid) binds to a membrane receptor and activates a G protein.

A **G protein** is an enzyme complex coupled to a membrane receptor that serves as a link between the first and second messenger.

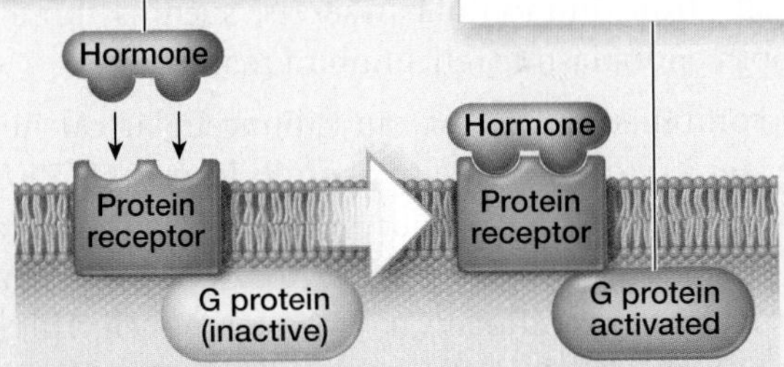
Hormone
Protein receptor
G protein (inactive)
Hormone
Protein receptor
G protein activated

a Effects on cAMP Level

Many G proteins, once activated, exert their effects by changing the concentration of cyclic AMP, which acts as the second messenger within the cell.

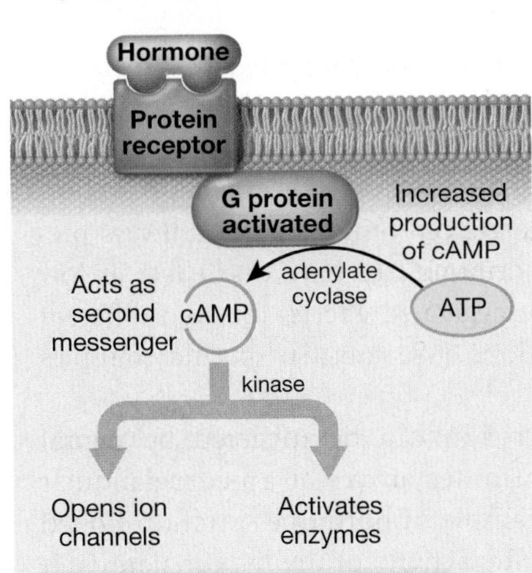
Hormone
Protein receptor
G protein activated
Increased production of cAMP
adenylate cyclase
ATP
Acts as second messenger
cAMP
kinase
Opens ion channels
Activates enzymes

If levels of cAMP increase, enzymes may be activated or ion channels may be opened, accelerating the metabolic activity of the cell.

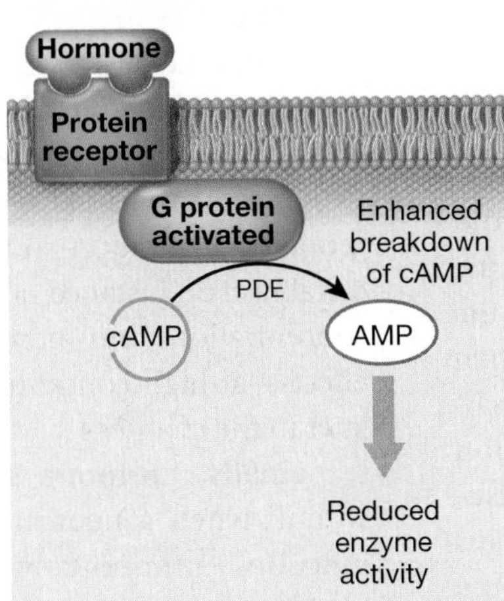
Hormone
Protein receptor
G protein activated
Enhanced breakdown of cAMP
PDE
cAMP
AMP
Reduced enzyme activity

In some instances, G protein activation results in decreased levels of cAMP in the cytoplasm. This decrease has an inhibitory effect on the cell.

First Messenger Examples
- Epinephrine and norepinephrine
- Calcitonin
- Parathyroid hormone
- ADH, ACTH, FSH, LH, TSH

First Messenger Examples
- Epinephrine and norepinephrine

b Effects on Ca^{2+} Level

Some G proteins use Ca^{2+} as a second messenger within the cell.

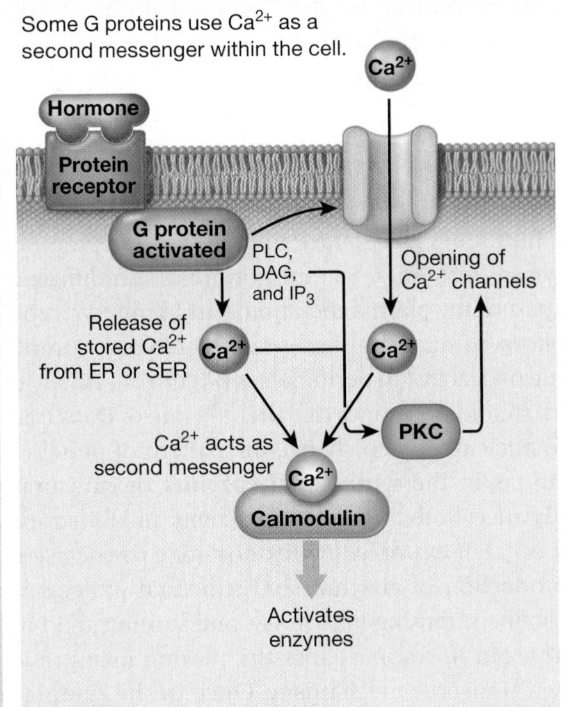
Ca2+
Hormone
Protein receptor
G protein activated
PLC, DAG, and IP3
Opening of Ca2+ channels
Release of stored Ca2+ from ER or SER
Ca2+
Ca2+
PKC
Ca2+ acts as second messenger
Ca2+
Calmodulin
Activates enzymes

First Messenger Examples
- Epinephrine and norepinephrine
- Oxytocin
- Regulatory hormones of hypothalamus
- Several eicosanoids

plasma membrane or the release of calcium ions from intracellular compartments. Spotlight Figure 18–3b (right panel) diagrams the steps involved:

1. The G protein first activates the enzyme **phospholipase C (PLC)**.
2. This enzyme triggers a receptor cascade that begins with the production of **diacylglycerol (DAG)** and **inositol trisphosphate (IP_3)** from membrane phospholipids.
3. IP_3 diffuses into the cytoplasm and triggers the release of Ca^{2+} from intracellular reserves, such as those in the smooth endoplasmic reticulum of many cells.
4. The combination of DAG and intracellular calcium ions activates a cytosolic protein: **protein kinase C (PKC)**. The activation of PKC leads to the phosphorylation of calcium ion channel proteins, a process that opens the channels and permits extracellular Ca^{2+} to enter the cell. This sets up a positive feedback loop that rapidly elevates the intracellular calcium ion concentration.
5. The calcium ions themselves serve as messengers, generally in combination with an intracellular protein called **calmodulin**. Once it has bound calcium ions, calmodulin can activate specific cytoplasmic enzymes. Calmodulin activation is also involved in the responses to oxytocin and to several regulatory hormones secreted by the hypothalamus.

18

Hormones and Intracellular Receptors: Gene Activation

Most hormones that target intracellular receptors are steroid or thyroid hormones. Steroid hormones can diffuse across the lipid part of the plasma membrane and bind to receptors in the cytoplasm or nucleus. The hormone–receptor complexes then activate or deactivate specific genes (Figure 18–4a). By this mechanism, steroid hormones can alter the rate of DNA transcription in the nucleus, and so change the pattern of protein synthesis. Alterations in the synthesis of enzymes or structural proteins directly affect both the metabolic activity and the structure of the target cell. For example, the sex hormone *testosterone* stimulates the production of enzymes and structural proteins in skeletal muscle fibers, causing muscle size and so strength to increase.

Thyroid hormones cross the plasma membrane primarily by a transport mechanism. Once in the cytoplasm, these hormones bind to receptors within the nucleus and on mitochondria (Figure 18–4b). The hormone–receptor complexes in the nucleus activate specific genes or change the rate of transcription. The change in transcription rate affects the metabolic activities of the cell by increasing or decreasing the concentration of specific enzymes. Thyroid hormones bound to mitochondria increase the mitochondrial rates of ATP production.

Control of Hormone Secretion

Hormone secretion, or endocrine activity, is mainly controlled by negative feedback. A stimulus triggers the production of a hormone whose direct or indirect effects reduce the intensity of the stimulus.

Hormone secretion can be triggered by (1) *humoral stimuli* (changes in the composition of the extracellular fluid), (2) *hormonal stimuli* (the arrival or removal of a specific hormone), or (3) *neural stimuli* (the arrival of neurotransmitters at neuroglandular junctions).

In the simplest case, only one hormone is involved. The endocrine cells involved respond directly to changes in the composition of the extracellular fluid by continuously releasing hormones. The secreted hormone adjusts the activities of target cells and restores homeostasis.

Hormone secretion by the heart, pancreas, parathyroid glands, and digestive tract are controlled by humoral stimuli. For example, when the blood glucose level increases, the pancreas increases its secretion of *insulin*, a hormone that stimulates glucose uptake and use. As the insulin level increases, the glucose level decreases, reducing the stimulation of the insulin-secreting cells. As the glucose level returns to normal, the rate of insulin secretion returns to resting levels. (We introduced this regulatory pattern, called *negative feedback*, in Chapter 1 when we considered the control of body temperature. ↺ p. 19)

Hormone secretion may also be controlled by hormonal stimuli, such as changes in the levels of circulating hormones. Such control may involve one or more intermediary steps and two or more hormones. However, the relationship between hormone concentration and target cell response is not always predictable. For instance, a hormone can have one effect at low concentrations and more exaggerated effects—or even different effects—at high concentrations. (We consider specific examples later in the chapter.)

Finally, hormone secretion can be triggered by neural stimuli when a neurotransmitter arrives at a neuroglandular junction. An important example of hormone secretion linked to neural stimuli involves the activity of the hypothalamus. It provides the highest level of endocrine control, as we discuss in the next section.

✓ Checkpoint

4. **What primary factor determines each cell's hormonal sensitivities?**
5. **How would the presence of a substance that inhibits the enzyme adenylate cyclase affect the activity of a hormone that produces its cellular effects by way of the second messenger cAMP?**
6. **Describe negative feedback as it relates to humoral stimuli.**

See the blue Answers tab at the back of the book.

Figure 18–4 **Effects of Intracellular Hormone Binding.**

a Steroid hormones diffuse through the plasma membrane and bind to receptors in the cytoplasm or nucleus. The complex then binds to DNA in the nucleus, activating specific genes.

1 Diffusion through membrane lipids
CYTOPLASM
Target cell response
Alteration of cellular structure or activity
6 Translation and protein synthesis
Receptor
2 Binding of hormone to cytoplasmic or nuclear receptors
5 Transcription and mRNA production
Receptor
4 Gene activation
Nuclear pore
Nuclear envelope
3 Binding of hormone–receptor complex to DNA

b Thyroid hormones enter the cytoplasm and bind to receptors in the nucleus to activate specific genes. They also bind to receptors on mitochondria and accelerate ATP production.

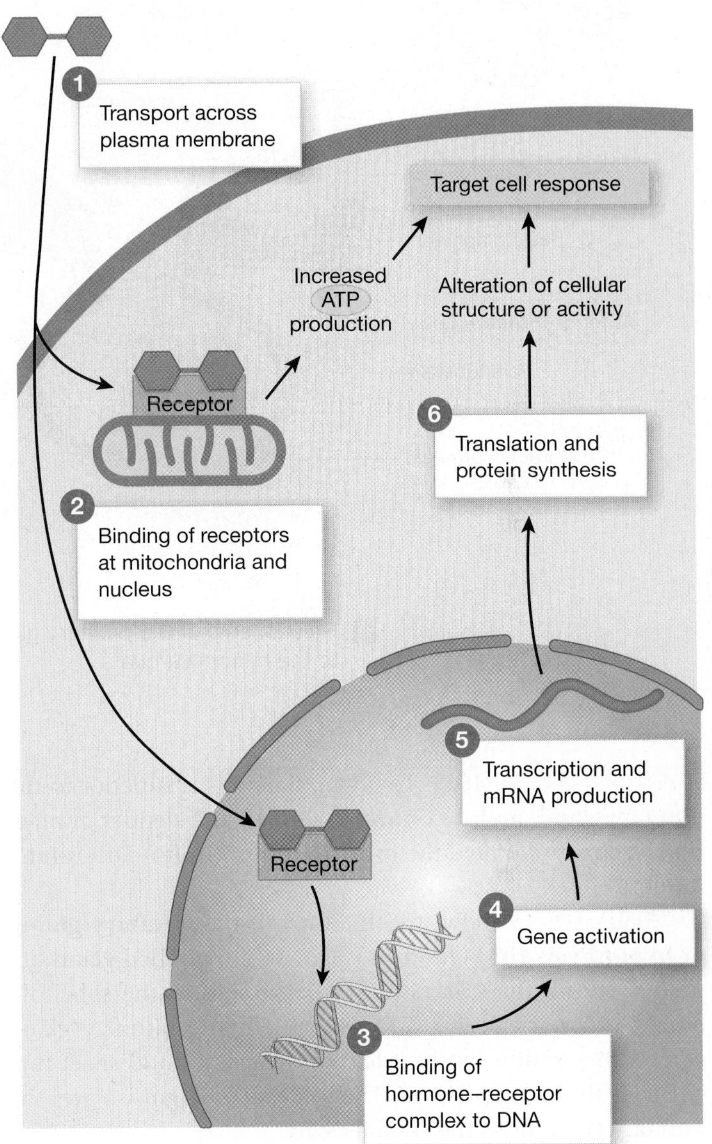

1 Transport across plasma membrane
Target cell response
Increased ATP production
Alteration of cellular structure or activity
Receptor
6 Translation and protein synthesis
2 Binding of receptors at mitochondria and nucleus
5 Transcription and mRNA production
Receptor
4 Gene activation
3 Binding of hormone–receptor complex to DNA

18-3 The anterior lobe of the pituitary gland produces and releases hormones under hypothalamic control, while the posterior lobe releases hypothalamic hormones

Learning Outcome Describe the location, hormones, and functions of the pituitary gland, and discuss the effects of abnormal pituitary hormone production.

The *pituitary gland* has *anterior* and *posterior lobes* that differ in function and developmental anatomy. It releases nine important peptide hormones; seven come from the anterior lobe and two from the posterior lobe. The *hypothalamus* has a close anatomical and functional relationship with both lobes of the pituitary gland. Let's start our exploration with the anatomy of these critical organs. ATLAS: Embryology Summary 14: The Development of the Endocrine System

Anatomy of the Hypothalamus and Pituitary Gland

The **hypothalamus**, which contains both brain centers and endocrine tissue, is found inferior to the thalamus (look back

Figure 18–5 The Orientation and Anatomy of the Pituitary Gland.

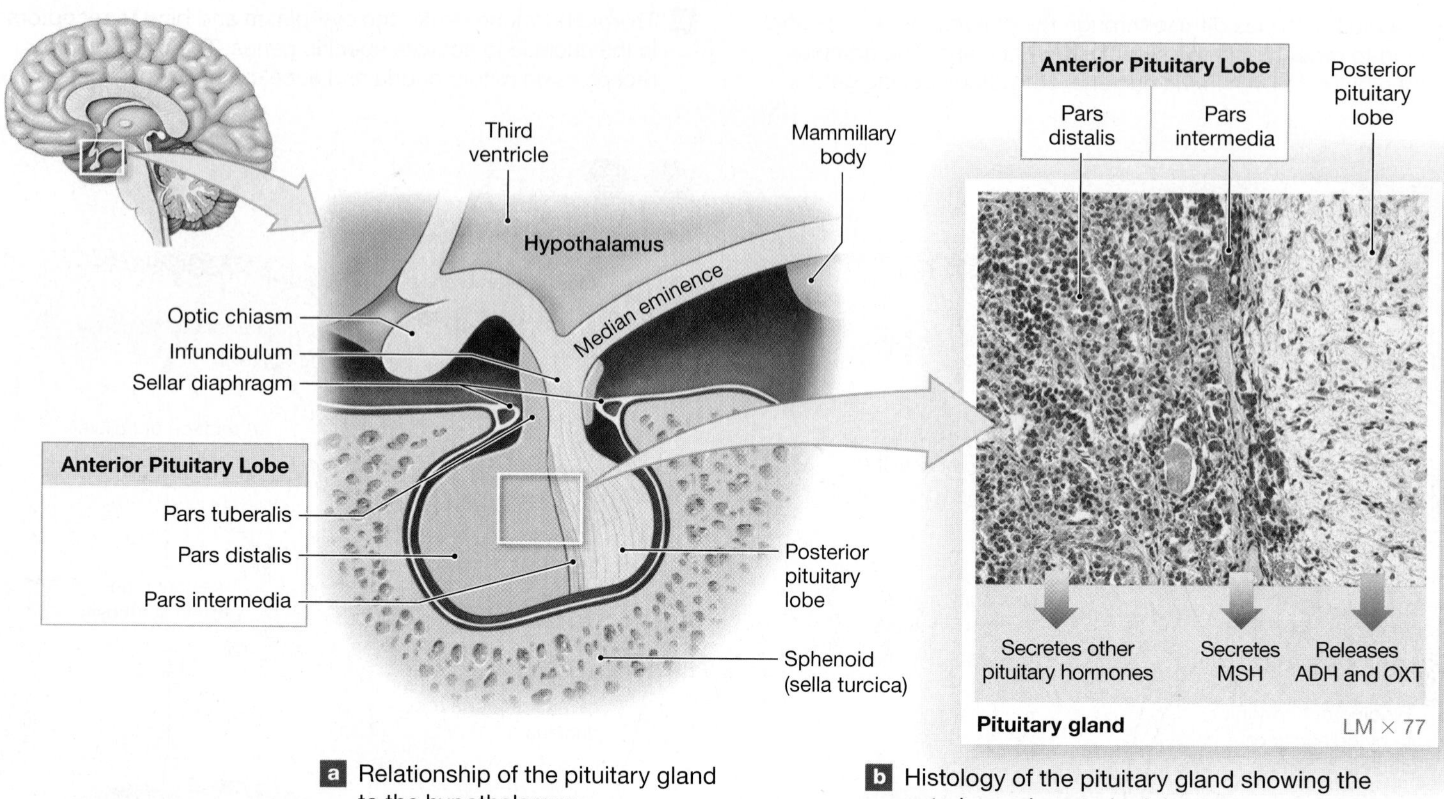
Third ventricle
Mammillary body
Hypothalamus
Median eminence
Optic chiasm
Infundibulum
Sellar diaphragm
Anterior Pituitary Lobe
Pars tuberalis
Pars distalis
Pars intermedia
Posterior pituitary lobe
Sphenoid (sella turcica)
Anterior Pituitary Lobe
Pars distalis
Pars intermedia
Posterior pituitary lobe
Secretes other pituitary hormones
Secretes MSH
Releases ADH and OXT
Pituitary gland
LM × 77

a Relationship of the pituitary gland to the hypothalamus

b Histology of the pituitary gland showing the anterior and posterior lobes

at Figure 14–11, p. 483). The hypothalamus is superior to the pituitary gland, and is connected to it by the slender, funnel-shaped structure called the **infundibulum** (in-fun-DIB-ū-lum; funnel).

Figure 18–5a shows the anatomy of the **pituitary gland**, or **hypophysis** (hī-POF-ih-sis). This small, bilobed gland lies nestled within the *sella turcica*, a depression in the sphenoid bone (look back at Figure 7–8, p. 219). The pituitary gland is held in position by the *sellar diaphragm*, a dural sheet that encircles the infundibulum. The sellar diaphragm isolates the pituitary gland from the cranial cavity.

Control of Pituitary Activity by the Hypothalamus

The hypothalamus regulates the functions of both the anterior and posterior lobes of the pituitary gland. It also integrates the activities of the nervous and endocrine systems in three ways (Figure 18–6):

- The hypothalamus itself acts as an endocrine organ. Hypothalamic neurons synthesize hormones and transport them along axons to the posterior lobe of the pituitary gland, where they are released into the circulation. We introduced these hormones, antidiuretic hormone (ADH) and oxytocin (ok-sih-TŌ-sin; OXT), in Chapter 14. p. 474 These hormones then go on to cause effects in their target organs.
- The hypothalamus secretes **regulatory hormones**. The hypothalamic regulatory hormones control the secretory activities of endocrine cells in the anterior lobe of the pituitary gland. In turn, the hormones from the anterior lobe control the activities of endocrine cells in target organs.
- The hypothalamus contains autonomic centers that exert direct neural control over the endocrine cells of the adrenal medulla. When the sympathetic division is activated, the adrenal medulla releases the hormones epinephrine (E) and norepinephrine (NE) into the bloodstream.

The hypothalamus secretes regulatory hormones and ADH in response to changes in the composition of the circulating blood. The secretion of OXT, E, and NE involves both neural and hormonal mechanisms. For example, the adrenal medulla secretes E and NE in response to action potentials rather than to circulating hormones. Such pathways are called *neuroendocrine responses*, because they include both neural and endocrine components.

Several hypothalamic and pituitary hormones are released in sudden bursts called *pulses*, rather than continuously. When hormones arrive in pulses, target cells (such as the cells of the

Figure 18–6 Three Mechanisms of Hypothalamic Control over Endocrine Function.

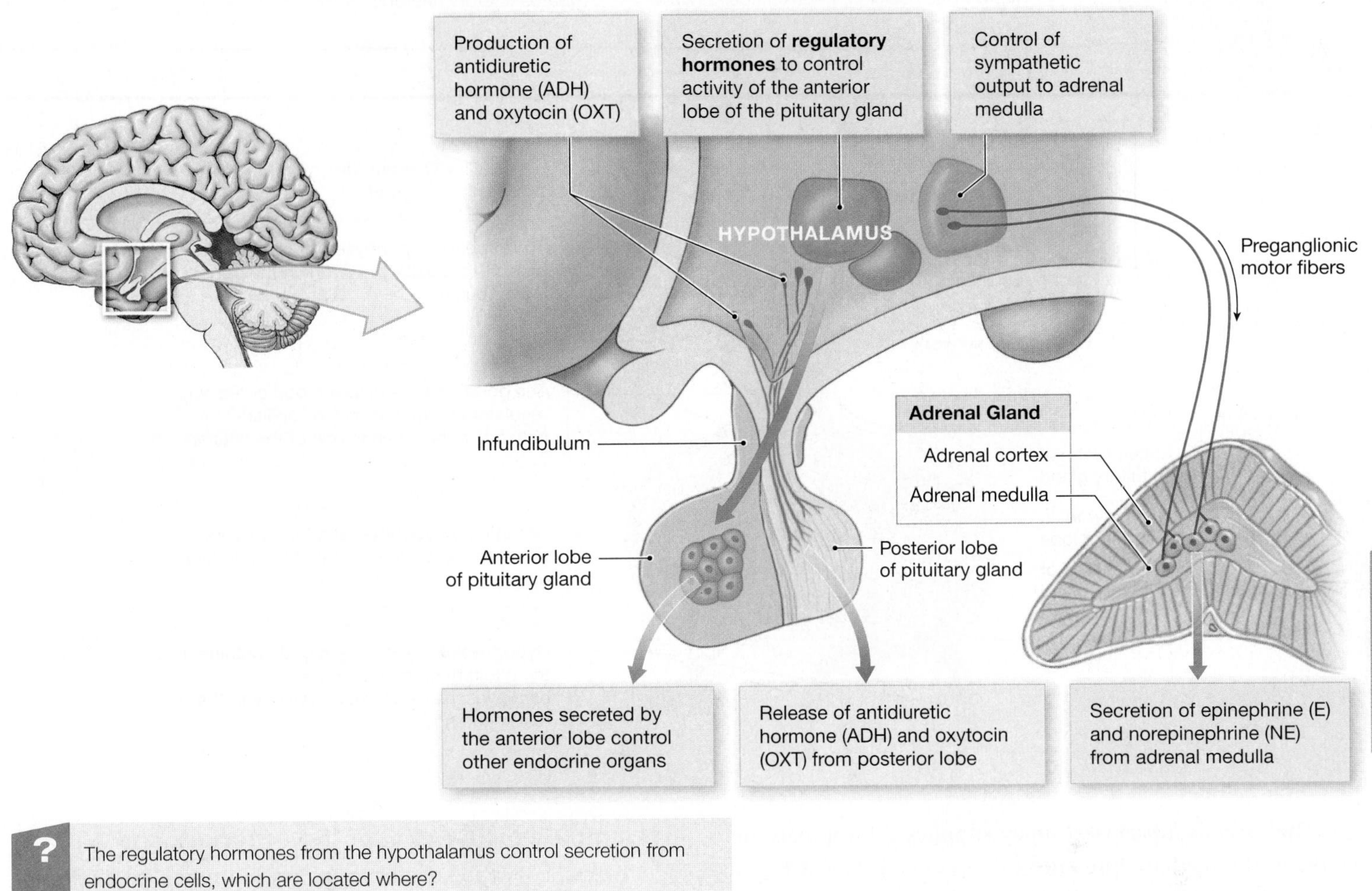
Production of antidiuretic hormone (ADH) and oxytocin (OXT)
Secretion of **regulatory hormones** to control activity of the anterior lobe of the pituitary gland
Control of sympathetic output to adrenal medulla
HYPOTHALAMUS
Preganglionic motor fibers
Adrenal Gland
Adrenal cortex
Adrenal medulla
Infundibulum
Anterior lobe of pituitary gland
Posterior lobe of pituitary gland
Hormones secreted by the anterior lobe control other endocrine organs
Release of antidiuretic hormone (ADH) and oxytocin (OXT) from posterior lobe
Secretion of epinephrine (E) and norepinephrine (NE) from adrenal medulla

? The regulatory hormones from the hypothalamus control secretion from endocrine cells, which are located where?

18

anterior lobe of the pituitary gland in response to hypothalamic regulatory hormones) may vary their response with the frequency of the pulses. For example, the target cell response to one pulse every 3 hours can differ from the response when pulses arrive every 30 minutes. The most complicated hormonal instructions from the hypothalamus involve changes in the frequency of pulses *and* in the amount secreted in each pulse.

The Anterior Lobe of the Pituitary Gland

Let's look at how the anatomy of the anterior lobe and its associated blood vessels facilitates its function. We then describe how the hypothalamus controls the secretion of anterior lobe hormones, and the actions of those secreted hormones.

Histology of the Anterior Lobe

The **anterior lobe** of the pituitary gland, also called the **adenohypophysis** (ad-eh-nō-hī-POF-ih-sis), contains a variety of endocrine cells. The anterior lobe has three regions: (1) the **pars distalis** (dis-TAL-is; distal part), the largest and most anterior portion of the pituitary gland; (2) an extension called the **pars tuberalis**, which wraps around the adjacent portion of the infundibulum; and (3) the slender **pars intermedia**, a narrow band bordering the posterior lobe (Figure 18–5b). An extensive capillary network radiates through these regions, giving every endocrine cell immediate access to the bloodstream.

Hormone Secretion and the Hypophyseal Portal System

The hypothalamus controls the production of hormones in the anterior lobe of the pituitary gland by secreting specific regulatory hormones. At the *median eminence,* a swelling near the attachment of the infundibulum, hypothalamic neurons release regulatory hormones into the surrounding interstitial fluids. Their secretions enter the bloodstream quite easily, because the endothelial cells lining the capillaries in this region are unusually permeable. These **fenestrated** (FEN-es-trā-ted; **fenestra**, window) **capillaries** allow relatively large molecules to enter or leave the bloodstream. p. 473

Figure 18–7 **The Hypophyseal Portal System and the Blood Supply to the Pituitary Gland.**

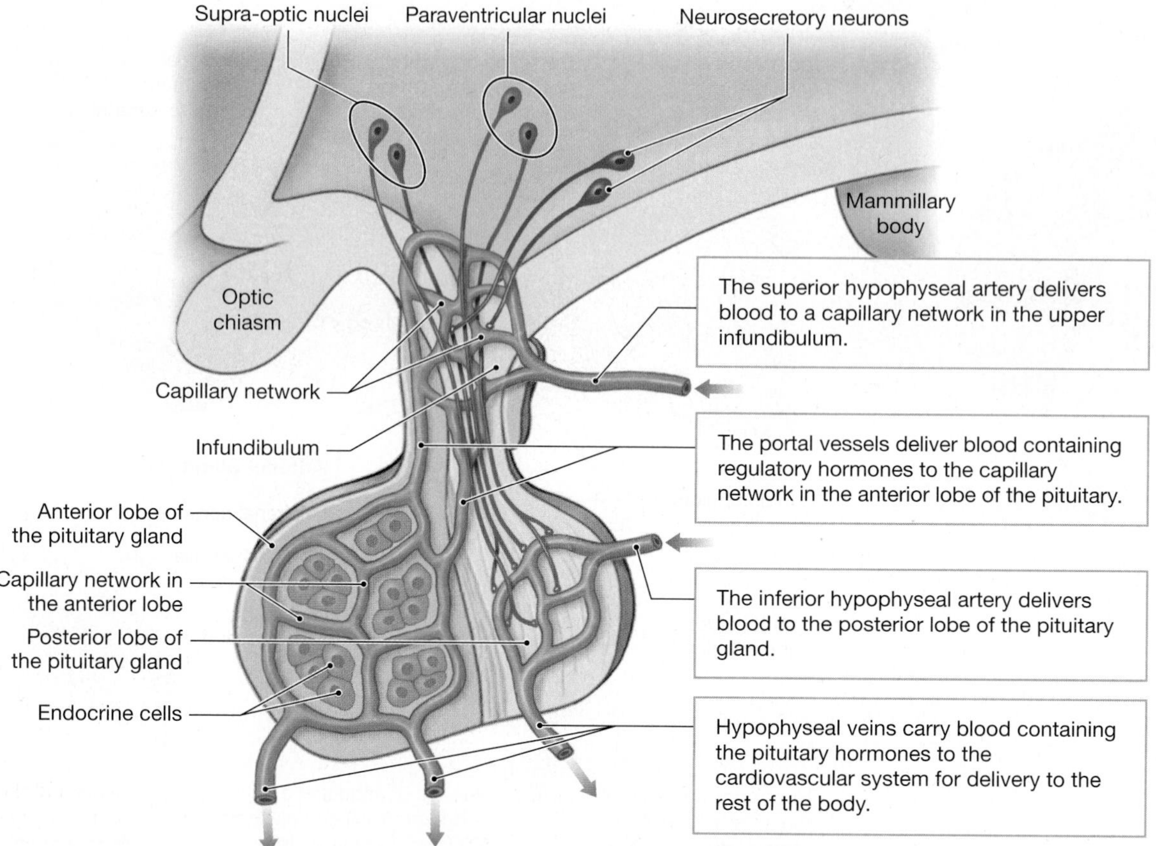
Supra-optic nuclei
Paraventricular nuclei
Neurosecretory neurons
Mammillary body
Optic chiasm
The superior hypophyseal artery delivers blood to a capillary network in the upper infundibulum.
Capillary network
Infundibulum
The portal vessels deliver blood containing regulatory hormones to the capillary network in the anterior lobe of the pituitary.
Anterior lobe of the pituitary gland
Capillary network in the anterior lobe
The inferior hypophyseal artery delivers blood to the posterior lobe of the pituitary gland.
Posterior lobe of the pituitary gland
Endocrine cells
Hypophyseal veins carry blood containing the pituitary hormones to the cardiovascular system for delivery to the rest of the body.

18

The *superior hypophyseal artery* supplies the capillary networks in the median eminence (Figure 18–7). Before leaving the hypothalamus, the capillary networks unite to form a series of larger vessels that spiral around the infundibulum to reach the anterior lobe. In the anterior lobe, these vessels form a second capillary network that branches among the endocrine cells.

This vascular arrangement is unusual because it carries blood from one capillary network to another. In contrast, a typical artery conducts blood from the heart to a capillary network, and a typical vein carries blood from a capillary network back to the heart. Blood vessels that link two capillary networks are called **portal vessels**. In this case, they have the histological structure of veins, so they are also called *portal veins*. The entire complex is a **portal system**. Portal systems are named for their destinations. This particular network is the **hypophyseal** (hī-pō-FIZ-ē-al) **portal system**.

Portal systems provide an efficient means of chemical communication. This one ensures that all the hypothalamic hormones entering the portal vessels reach the target cells in the anterior lobe before being diluted through mixing with the general circulation. The communication is strictly one way, however. Any chemicals released by the cells "downstream" must do a complete circuit of the cardiovascular system before they reach the capillaries of the portal system.

Hypothalamic Control of the Anterior Lobe

Two classes of hypothalamic regulatory hormones exist: releasing hormones and inhibiting hormones. A **releasing hormone (RH)** stimulates the synthesis and secretion of one or more hormones at the anterior lobe. In contrast, an **inhibiting hormone (IH)** prevents the synthesis and secretion of hormones from the anterior lobe. Releasing hormones, inhibiting hormones, or some combination of the two may control an endocrine cell in the anterior lobe.

Once secreted, the anterior pituitary hormones then stimulate the cells of their target organs. Many of these organs are endocrine glands themselves, which then secrete hormones in response. In fact, the hormones of the anterior lobe are sometimes called *tropic hormones* (*trope*, a turning), because they "turn on" endocrine glands or support the functions of other organs. Then negative feedback from these target organ hormones controls the rate at which the hypothalamus secretes regulatory hormones. The primary regulatory patterns are diagrammed in Figure 18–8a–c, and we will refer to them as we examine specific pituitary hormones.

Hormones of the Anterior Lobe

The control mechanisms and functions of seven hormones from the anterior lobe are reasonably well understood. The regulation of *thyroid-stimulating hormone, adrenocorticotropic hormone,* and two gonadotropins called *follicle-stimulating hormone*

Figure 18–8 Feedback Control of Endocrine Secretion.

Typical pattern of regulation when multiple endocrine organs are involved

a The hypothalamus produces a releasing hormone (RH) to stimulate hormone production by other glands. Homeostatic control occurs by negative feedback.

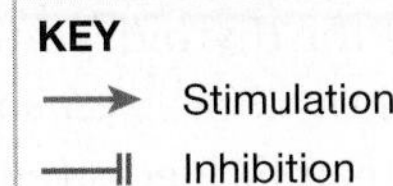
KEY
Stimulation
Inhibition

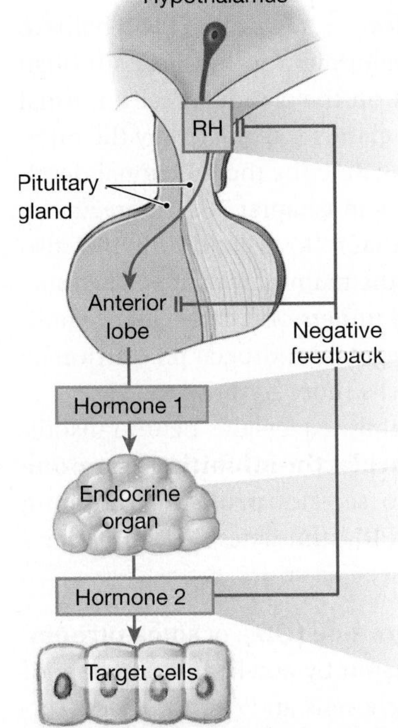
Hypothalamus
RH
Pituitary gland
Anterior lobe
Negative feedback
Hormone 1
Endocrine organ
Hormone 2
Target cells

The effects of hypothalamic releasing hormones that follow the typical pattern of regulation

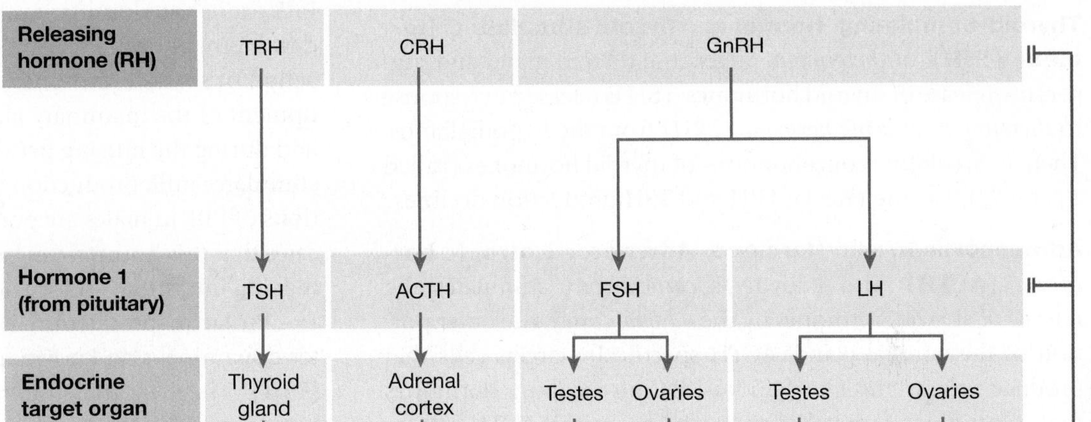

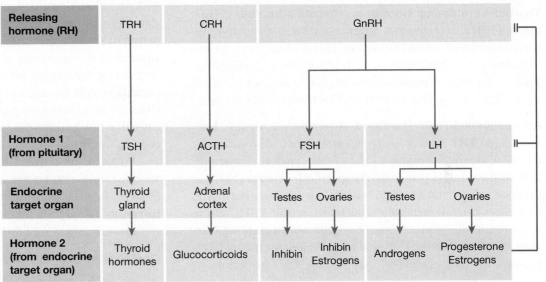

Releasing hormone (RH)	TRH	CRH	GnRH			
Hormone 1 (from pituitary)	TSH	ACTH	FSH		LH	
Endocrine target organ	Thyroid gland	Adrenal cortex	Testes	Ovaries	Testes	Ovaries
Hormone 2 (from endocrine target organ)	Thyroid hormones	Glucocorticoids	Inhibin	Inhibin Estrogens	Androgens	Progesterone Estrogens

Variations on the typical pattern of regulation of endocrine organs by the hypothalamus and anterior pituitary lobe

b The regulation of prolactin (PRL) production by the anterior lobe. In this case, the hypothalamus produces both a releasing hormone (PRH) and an inhibiting hormone (PIH). When one is stimulated, the other is inhibited.

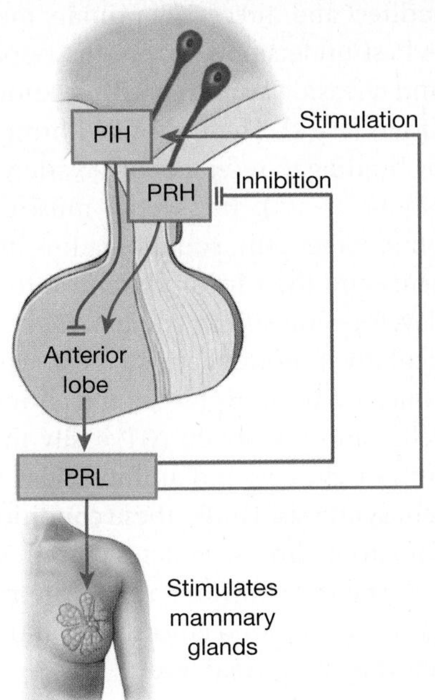
PIH
Stimulation
PRH
Inhibition
Anterior lobe
PRL
Stimulates mammary glands

c The regulation of growth hormone (GH) production by the anterior lobe. When GH–RH release is inhibited, GH–IH release is stimulated.

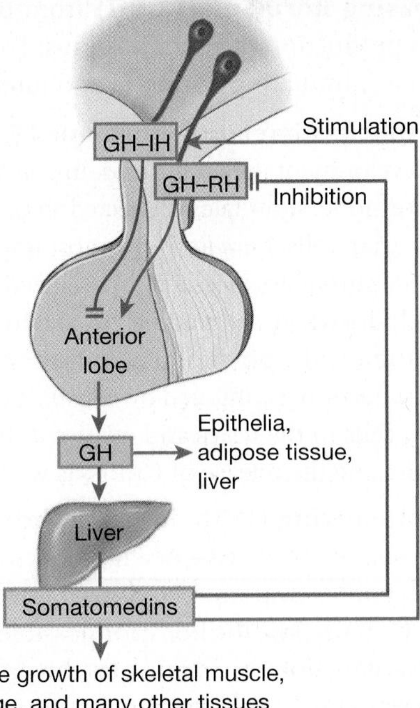
GH–IH
Stimulation
GH–RH
Inhibition
Anterior lobe
GH
Epithelia, adipose tissue, liver
Liver
Somatomedins
Stimulate growth of skeletal muscle, cartilage, and many other tissues

? In a typical regulation pattern of endocrine secretion, which hormone is responsible for negative feedback?

18

and *luteinizing hormone* follows the typical regulatory pattern shown in Figure 18–8a. Two others, *prolactin* and *growth hormone,* follow the atypical patterns shown in Figures 18–8b and c, respectively. Of these six hormones, which are all produced by the pars distalis, four regulate the production of hormones by other endocrine glands such as the thyroid, adrenal cortex, and reproductive organs. A final hormone, *melanocyte-stimulating hormone,* is produced by the pars intermedia (and is not shown in the figure). All seven hormones bind to extracellular receptors, and all use cAMP as a second messenger.

Thyroid-Stimulating Hormone. Thyroid-stimulating hormone (TSH), or *thyrotropin,* targets the thyroid gland and triggers the release of thyroid hormones. TSH is released in response to *thyrotropin-releasing hormone (TRH)* from the hypothalamus. Then as circulating concentrations of thyroid hormones (called T_3 and T_4) rise, the rates of TRH and TSH production decline.

Adrenocorticotropic Hormone. Adrenocorticotropic hormone (ACTH), also known as *corticotropin,* stimulates the release of steroid hormones by the *adrenal cortex,* the outer portion of the adrenal gland. ACTH specifically targets cells that produce *glucocorticoids* (glū-kō-KOR-tih-koydz), hormones that affect glucose metabolism such as cortisol. ACTH release occurs under the stimulation of **corticotropin-releasing hormone (CRH)** from the hypothalamus. As glucocorticoid levels increase, the rates of CRH release and ACTH release decline.

Gonadotropins. The hormones called **gonadotropins** (gō-nad-ō-TRŌ-pinz) regulate the activities of the *gonads.* (These organs—the *testes* in males and the *ovaries* in females—produce reproductive cells as well as hormones.) **Gonadotropin-releasing hormone (GnRH)** from the hypothalamus stimulates production of gonadotropins. The two gonadotropins are follicle-stimulating hormone and luteinizing hormone:

- **Follicle-stimulating hormone (FSH)** promotes follicle development in females and, in combination with luteinizing hormone, stimulates the secretion of *estrogens* (ES-trō-jenz) by ovarian cells. *Estradiol* is the most important estrogen. In males, FSH stimulates *nurse cells,* specialized cells in the seminiferous tubules where sperm differentiate. In response, the nurse cells promote the physical maturation of developing sperm. FSH production is inhibited by *inhibin,* a peptide hormone released by cells in the testes and ovaries. (The role of inhibin in suppressing the release of GnRH as well as FSH is under debate.)
- **Luteinizing** (LŪ-tē-in-ī-zing) **hormone (LH)** induces *ovulation,* the release of a reproductive cell in females. It also promotes ovarian secretion of estrogens and *progesterone,* which prepare the body for possible pregnancy. In males, this gonadotropin stimulates the production of sex hormones by the *interstitial endocrine cells* of the testes. These male sex hormones are called **androgens** (AN-drō-jenz; *andros,* man). The most important one is *testosterone.*

  LH production, like FSH production, is stimulated by GnRH from the hypothalamus. Estrogens, progesterone, and androgens inhibit GnRH production.

An abnormally low production of gonadotropins results in **hypogonadism**. Children with this condition do not mature sexually, and adults with hypogonadism cannot produce functional sperm (males) or oocytes (females).

Prolactin. Prolactin (*pro-,* before + *lac,* milk) **(PRL)** helps to stimulate mammary gland development in females. Although PRL exerts the dominant effect on the glandular cells, normal development of the mammary glands is regulated by the interaction of several hormones. (We describe the functional development of the mammary glands in Chapter 28.) In pregnancy and during the nursing period that follows childbirth, PRL also stimulates milk production by the mammary glands. The functions of PRL in males are poorly understood, but evidence indicates that this hormone helps regulate androgen production by making interstitial endocrine cells more sensitive to LH.

Prolactin production is inhibited by the neurotransmitter dopamine, also known as **prolactin-inhibiting hormone (PIH)**. The hypothalamus also secretes **prolactin-releasing hormone (PRH)**. Circulating PRL stimulates PIH release and inhibits the secretion of PRH (see Figure 18–8b).

Growth Hormone. Growth hormone (GH), or **somatotropin**, stimulates cell growth and division by accelerating the rate of protein synthesis. Skeletal muscle cells and chondrocytes (cartilage cells) are particularly sensitive to GH, although virtually every tissue responds to some degree.

The stimulation of growth by GH involves two mechanisms, indirect and direct. The primary mechanism, which is indirect, is best understood. Liver cells respond to GH by synthesizing and releasing **somatomedins**, known also as *insulin-like growth factors (IGFs)*. These peptide hormones stimulate tissue growth by binding to receptors on a variety of plasma membranes (see Figure 18–8c). In skeletal muscle fibers, cartilage cells, and other target cells, somatomedins increase the uptake of amino acids and their incorporation into new proteins. These effects develop almost immediately after GH is released. They are particularly important after a meal, when the blood contains high concentrations of glucose and amino acids. In functional terms, cells can now obtain ATP easily through the aerobic metabolism of glucose, and amino acids are readily available for protein synthesis. Under these conditions, GH, acting through the somatomedins, stimulates protein synthesis and cell growth.

The direct actions of GH are more selective. They tend to appear after blood glucose and amino acid concentrations have returned to normal levels:

- In epithelia and connective tissues, GH stimulates stem cell divisions and the differentiation of daughter cells. (Somatomedins then stimulate the growth of these daughter cells.)

- In adipose tissue, GH stimulates the breakdown of stored triglycerides by adipocytes (fat cells), which then release fatty acids into the blood. As the level of circulating fatty acids rises, many tissues stop breaking down glucose to generate ATP and instead start breaking down fatty acids. This process is termed a **glucose-sparing effect**.
- In the liver, GH stimulates the breakdown of glycogen reserves by liver cells, which then release glucose into the bloodstream. Because most other tissues are now metabolizing fatty acids rather than glucose, the blood glucose concentration begins to climb to a level significantly higher than normal. This elevation of the blood glucose level by GH has been called a **diabetogenic effect**, because *diabetes mellitus*, an endocrine disorder we consider later in the chapter, is characterized by an abnormally high blood glucose concentration.

The production of GH is regulated by **growth hormone–releasing hormone (GH–RH)** and **growth hormone–inhibiting hormone (GH–IH)**, or *somatostatin*, from the hypothalamus. Somatomedins inhibit GH–RH and stimulate GH–IH.

Melanocyte-Stimulating Hormone. The pars intermedia may secrete two forms of **melanocyte-stimulating hormone (MSH)**. As the name indicates, MSH stimulates the melanocytes of the skin, increasing their production of melanin, a brown, black, or yellow-brown pigment. p. 161 Dopamine inhibits the release of MSH.

Melanocyte-stimulating hormone from the pituitary gland is important in the control of skin pigmentation in fishes, amphibians, reptiles, and many mammals other than primates. In humans, MSH is produced locally, within sun-exposed skin. The pars intermedia in adult humans is virtually nonfunctional, and the circulating blood usually does not contain MSH. However, the human pars intermedia secretes MSH (1) during fetal development, (2) in very young children, (3) in pregnant women, and (4) in the course of some diseases. The functional significance of MSH secretion under these circumstances is not known.

The Posterior Lobe of the Pituitary Gland

The **posterior lobe** of the pituitary gland, also called the **neurohypophysis** (nū-rō-hī-POF-ih-sis), contains the axons of hypothalamic neurons. Neurons of the **supra-optic** and **paraventricular nuclei** manufacture antidiuretic hormone (ADH) and oxytocin (OXT), respectively (see Figure 18–7). These hormones move along axons in the infundibulum to axon terminals, which end on the basement membranes of capillaries in the posterior lobe. They travel by means of axoplasmic transport. p. 393 The posterior lobe does not have a portal system. The *inferior hypophyseal artery* delivers blood to it, and the *hypophyseal veins* carry the blood and hormones away. At their target organs, both hormones bind to extracellular receptors, and both use cAMP as a second messenger.

Antidiuretic Hormone

Antidiuretic hormone (ADH), also known as *vasopressin (VP)*, is released in response to a variety of stimuli, most notably an increase in the solute concentration in the blood or a decrease in blood volume or blood pressure. An increase in the solute concentration stimulates specialized neurons in the hypothalamus. These neurons are called *osmoreceptors* because they respond to a change in the osmotic concentration of body fluids. The osmoreceptors then stimulate the neurosecretory neurons that release ADH.

The primary function of ADH is to act on the kidneys to retain water and decrease urination. With losses minimized, any water absorbed from the digestive tract will be retained, reducing the concentrations of electrolytes in the extracellular fluid. At a high concentration, ADH also causes *vasoconstriction*, a narrowing of peripheral blood vessels that helps elevate blood pressure. Alcohol inhibits ADH release, which explains why people find themselves making frequent trips to the bathroom after consuming alcoholic beverages.

Oxytocin

In women, **oxytocin** (*oxy-*, quick + *tokos*, childbirth) **(OXT)** stimulates smooth muscle contraction in the wall of the uterus, promoting labor and delivery. After delivery, oxytocin promotes the ejection of milk by stimulating the contraction of myoepithelial cells around the secretory alveoli and the ducts of the mammary glands. 18

Until the last stages of pregnancy, the uterine smooth muscles are relatively insensitive to oxytocin, but they become more sensitive as delivery approaches. The trigger for normal labor

+ Clinical Note Diabetes Insipidus

Diabetes occurs in several forms. All are characterized by excessive urine production (*polyuria*; pol-ē-YŪR-ē-ah). Most forms result from endocrine abnormalities, but physical damage to the kidneys can also cause diabetes. The two most prevalent forms are *diabetes insipidus* and *diabetes mellitus*. (We describe diabetes mellitus in the Spotlight figure on page 642.) **Diabetes insipidus** generally develops because the posterior lobe of the pituitary gland no longer releases adequate amounts of antidiuretic hormone (ADH). Water conservation by the kidneys is impaired, and excessive amounts of water are lost in the urine. As a result, the person is constantly thirsty, but the body does not retain the fluids consumed.

Mild cases of diabetes insipidus may not require treatment if fluid and electrolyte intake keep pace with urinary losses. In severe cases, the fluid losses can reach 10 liters per day (normal urinary output is about 1–2 liters per day), and dehydration and electrolyte imbalances are fatal without treatment. This condition can be effectively treated with desmopressin, a synthetic form of ADH.

and delivery is probably a sudden rise in the oxytocin level at the uterus. There is good evidence, however, that the uterus and fetus secrete most of the oxytocin involved. Oxytocin from the posterior lobe plays only a supporting role.

Oxytocin secretion and milk ejection are part of a neuroendocrine response reflex called the *milk ejection reflex (milk let-down reflex)*. Milk ejection occurs only in response to oxytocin release at the posterior lobe of the pituitary gland. The normal stimulus is an infant suckling at the breast, and sensory nerves innervating the nipples relay the information to the hypothalamus. Oxytocin is then released into the circulation at the posterior lobe, and the myoepithelial cells respond by squeezing milk from the secretory alveoli into large collecting ducts. Any factor that affects the hypothalamus can modify this reflex. For example, anxiety, stress, and other factors can prevent the flow of milk, even when the mammary glands are fully functional. In contrast, nursing mothers can become conditioned to associate a baby's crying with suckling. In these women, milk ejection may begin as soon as they hear a baby cry.

The functions of oxytocin in sexual activity remain uncertain, but it is known that the circulating concentration of oxytocin rises during sexual arousal and peaks at orgasm in both sexes. Evidence indicates that in men, oxytocin stimulates smooth muscle contractions in the walls of the *ductus deferens* (sperm duct) and prostate gland. These actions may be important in *emission*—the ejection of sperm, secretions of the prostate gland, and the secretions of other glands into the male reproductive tract before ejaculation. Studies suggest that the oxytocin released in females during intercourse may stimulate smooth muscle contractions in the uterus and vagina that promote the transport of sperm toward the uterine tubes.

Summary: The Hormones of the Pituitary Gland

Table 18–2 and **Figure 18–9** summarize important information about the hormones of the pituitary gland. Review them carefully before considering the structure and function of other endocrine organs.

Table 18–2 The Pituitary Hormones

	Hormone	Target	Hormonal Effect	Hypothalamic Regulatory Hormone
ANTERIOR LOBE				
Pars distalis	**Thyroid-stimulating hormone (TSH)**	Thyroid gland	Secretion of thyroid hormones	Thyrotropin-releasing hormone (TRH)
	Adrenocorticotropic hormone (ACTH)	Adrenal cortex (zona fasciculata)	Secretion of glucocorticoids (cortisol, corticosterone, and cortisone)	Corticotropin-releasing hormone (CRH)
	Gonadotropins:			
	Follicle-stimulating hormone (FSH)	Follicle cells of ovaries	Secretion of estrogen, follicle development	Gonadotropin-releasing hormone (GnRH)
		Nurse cells of testes	Stimulation of sperm maturation	Gonadotropin-releasing hormone (GnRH)
	Luteinizing hormone (LH)	Follicle cells of ovaries	Ovulation, formation of corpus luteum, secretion of progesterone	Gonadotropin-releasing hormone (GnRH)
		Interstitial endocrine cells of testes	Secretion of testosterone	Gonadotropin-releasing hormone (GnRH)
	Prolactin (PRL)	Mammary glands	Production of milk	Prolactin-releasing factor (PRF)
				Prolactin-inhibiting hormone (PIH)
	Growth hormone (GH)	All cells	Growth, protein synthesis, lipid mobilization and catabolism	Growth hormone–releasing hormone (GH–RH)
				Growth hormone–inhibiting hormone (GH–IH)
Pars intermedia (not active in normal adults)	**Melanocyte-stimulating hormone (MSH)**	Melanocytes	Increased melanin synthesis in epidermis	Melanocyte-stimulating hormone–inhibiting hormone (MSH–IH)
POSTERIOR LOBE				
	Antidiuretic hormone (ADH)	Kidneys	Reabsorption of water, elevation of blood volume and pressure	None; transported along axons from supra-optic nucleus to the posterior lobe of the pituitary gland
	Oxytocin (OXT)	Uterus, mammary glands (females)	Labor contractions, milk ejection	None; transported along axons from paraventricular nucleus to the posterior lobe of the pituitary gland
		Ductus deferens and prostate gland (males)	Contractions of ductus deferens and prostate gland	

Figure 18–9 Pituitary Hormones and Their Targets.

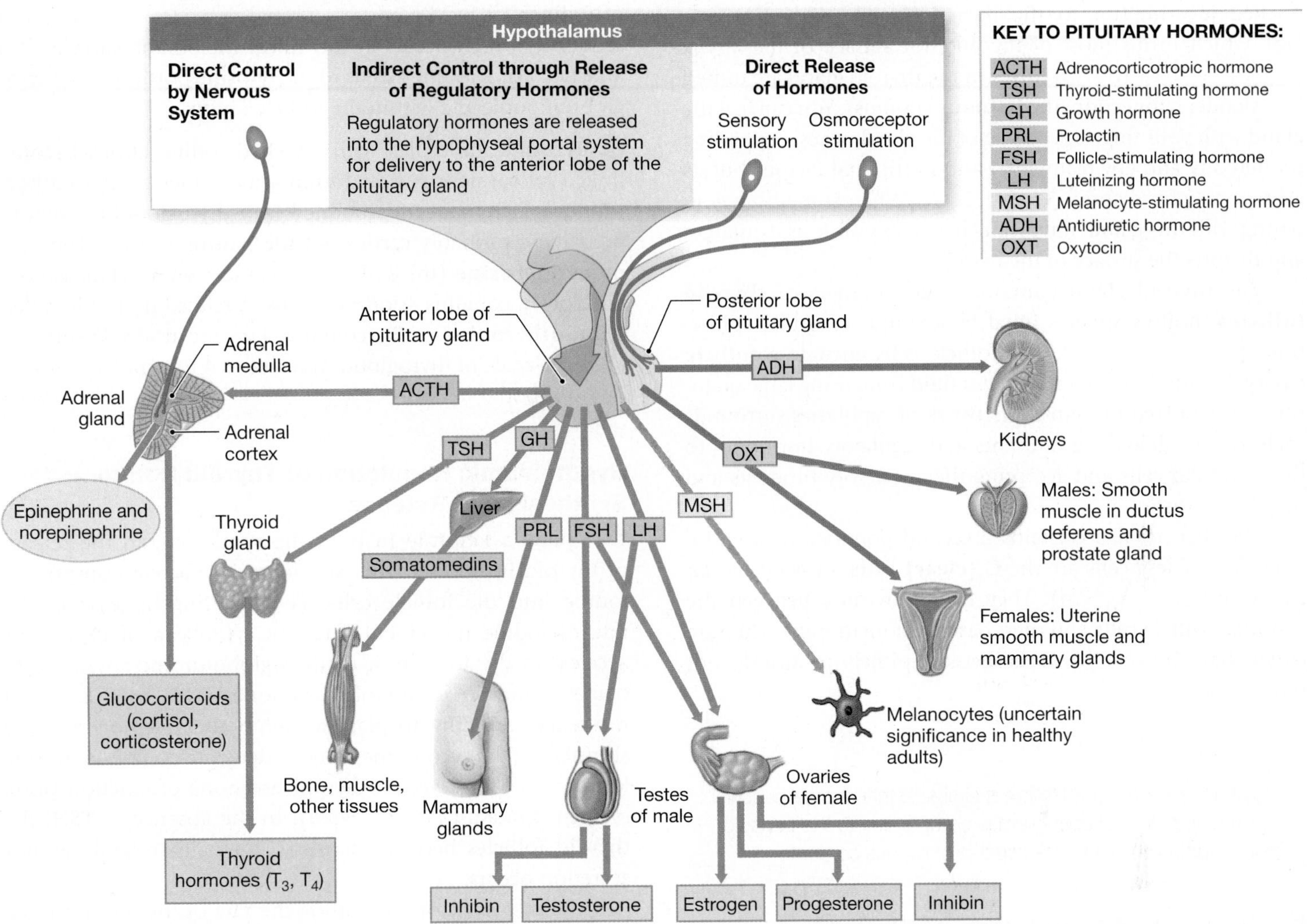
Hypothalamus
Direct Control by Nervous System
Indirect Control through Release of Regulatory Hormones
Regulatory hormones are released into the hypophyseal portal system for delivery to the anterior lobe of the pituitary gland
Direct Release of Hormones
Sensory stimulation
Osmoreceptor stimulation
KEY TO PITUITARY HORMONES:
ACTH Adrenocorticotropic hormone
TSH Thyroid-stimulating hormone
GH Growth hormone
PRL Prolactin
FSH Follicle-stimulating hormone
LH Luteinizing hormone
MSH Melanocyte-stimulating hormone
ADH Antidiuretic hormone
OXT Oxytocin
Anterior lobe of pituitary gland
Posterior lobe of pituitary gland
Adrenal medulla
Adrenal gland
Adrenal cortex
ACTH
ADH
Kidneys
TSH
GH
OXT
Epinephrine and norepinephrine
Thyroid gland
Liver
MSH
PRL
FSH
LH
Somatomedins
Males: Smooth muscle in ductus deferens and prostate gland
Females: Uterine smooth muscle and mammary glands
Melanocytes (uncertain significance in healthy adults)
Glucocorticoids (cortisol, corticosterone)
Bone, muscle, other tissues
Mammary glands
Testes of male
Ovaries of female
Thyroid hormones (T_3, T_4)
Inhibin
Testosterone
Estrogen
Progesterone
Inhibin

? What hormones are released from the posterior lobe of the pituitary gland and how does their release differ from hormones released by the anterior lobe of the pituitary gland?

Checkpoint

7. Identify the two lobes of the pituitary gland.
8. If a person were dehydrated, how would the amount of ADH released by the posterior pituitary lobe change?
9. The release of which pituitary hormone would lead to an increased level of somatomedins in the blood?
10. What effect would an elevated circulating level of cortisol, a steroid hormone from the adrenal cortex, have on the pituitary secretion of ACTH?

See the blue Answers tab at the back of the book.

18-4 The thyroid gland synthesizes thyroid hormones that affect the rate of metabolism

Learning Outcome Describe the location, hormones, and functions of the thyroid gland, and discuss the effects of abnormal thyroid hormone production.

The thyroid gland is a ductless gland in the neck that secretes hormones regulating growth, development, and metabolism. Cells of this gland extract iodine from the blood to synthesize thyroid hormones. Let's look first at the anatomy.

Anatomy of the Thyroid Gland

The **thyroid gland** curves across the anterior surface of the trachea just inferior to the *thyroid* ("shield-shaped") *cartilage*, which forms most of the anterior surface of the larynx (Figure 18–10a). The two **lobes** of the thyroid gland are united by a slender connection, the **isthmus** (IS-mus). You can feel the gland with your fingers. The size of the gland varies, depending on heredity and environmental and nutritional factors, but its average weight is about 34 g (1.2 oz). When something goes wrong with this gland, it typically becomes visible as it enlarges and distorts the surface of the neck.

The thyroid gland contains large numbers of **thyroid follicles**, hollow spheres lined by a simple cuboidal epithelium (Figure 18–10b,c). The follicle cells encircle a **follicle cavity** that holds a viscous *colloid*, a fluid containing large quantities of dissolved proteins. A network of capillaries surrounds each follicle, delivering nutrients and regulatory hormones to the glandular cells and accepting their secretory products and metabolic wastes.

18

The thyroid also contains a second population of endocrine cells. These cells are the **C (clear) cells**, or *parafollicular cells* (see Figure 18–10c). They lie sandwiched between the cuboidal follicle cells and their basement membrane. They are larger than the cells of the follicular epithelium and do not stain as clearly.

Tips & Tools

The structure of a thyroid follicle is similar to that of a gel capsule: The simple cuboidal epithelium is comparable to the capsule itself, and the colloid to the capsule's viscous contents.

Synthesis and Regulation of Thyroid Hormones

The thyroid gland produces two types of **thyroid hormones**, called T_3 and T_4. *Thyroid-stimulating hormone* (*TSH*) regulates both the synthesis and release of thyroid hormones. The synthesis of these hormones requires iodine, which we obtain from the food we eat.

Thyroid Hormone Synthesis

Follicle cells synthesize a globular protein called **thyroglobulin** (thī-rō-GLOB-yū-lin) and secrete it into the colloid of the thyroid follicles. Thyroglobulin molecules contain the amino acid tyrosine, the building block of thyroid hormones. The formation of thyroid hormones involves the following basic steps (Figure 18–11a, p. 630):

1. *Iodide ions* (I^-) are absorbed from the diet by the digestive tract, and the bloodstream delivers them to the thyroid gland. TSH-sensitive carrier proteins in the basement membrane of the follicle cells actively transport iodide ions into the cytoplasm.

2. The iodide ions diffuse to the apical surface of each follicle cell, where they lose an electron and are converted to an atom of iodine (I^0) by the enzyme *thyroid peroxidase*. This reaction sequence, which occurs at the apical membrane surface, also attaches 1 or 2 iodine atoms to the tyrosine portions of a thyroglobulin molecule within the follicle cavity.

3. Tyrosine molecules with attached iodine atoms become linked by covalent bonds, forming molecules of thyroid hormones that remain incorporated into thyroglobulin. Thyroid peroxidase probably carries out the pairing process. The hormone **thyroxine** (thī-ROK-sēn), also known as *tetraiodothyronine*, or $\mathbf{T_4}$, contains 4 iodine atoms. A related molecule called **triiodothyronine**, or $\mathbf{T_3}$, contains 3 iodine atoms. Eventually, each molecule of thyroglobulin contains 4 to 8 molecules of T_3 or T_4, or both.

Hypothalamic Regulation of Thyroid Hormone Synthesis and Release

TSH plays a key role in both the synthesis and the release of thyroid hormones. TSH stimulates the active transport of iodide into the follicle cells. The resulting increase in the rate of iodide movement into the cytoplasm of these cells accelerates the formation of thyroglobulin and thyroid peroxidase, and so of thyroid hormones. TSH acts as a first messenger, binding to plasma membrane receptors and, by stimulating the second messenger adenylate cyclase, activating key enzymes involved in thyroid hormone production (look back at **Spotlight Figure 18–3a**). In the absence of TSH, the thyroid follicles become inactive, and neither synthesis nor secretion occurs.

The major factor controlling the rate of thyroid hormone release is the concentration of TSH in the circulating blood (Figure 18–11b). Under the influence of TSH, the following steps occur (see Figure 18–11a):

4. Follicle cells remove thyroglobulin from the follicles by endocytosis.

5. Lysosomal enzymes break down the thyroglobulin, and the amino acids and thyroid hormones enter the cytoplasm. The amino acids are then recycled and used to synthesize more thyroglobulin.

6. The released molecules of T_3 and T_4 diffuse across the basement membrane and enter the bloodstream. About 90 percent of all thyroid secretions is T_4, and T_3 is secreted in comparatively small amounts.

7. About 75 percent of the T_4 molecules and 70 percent of the T_3 molecules entering the bloodstream become attached to transport proteins called **thyroid-binding globulins (TBGs)**. Most of the rest of the T_4 and T_3 in circulation is attached to **transthyretin**, or to *albumin*, one of the plasma proteins.

Figure 18–10 Anatomy of the Thyroid Gland.

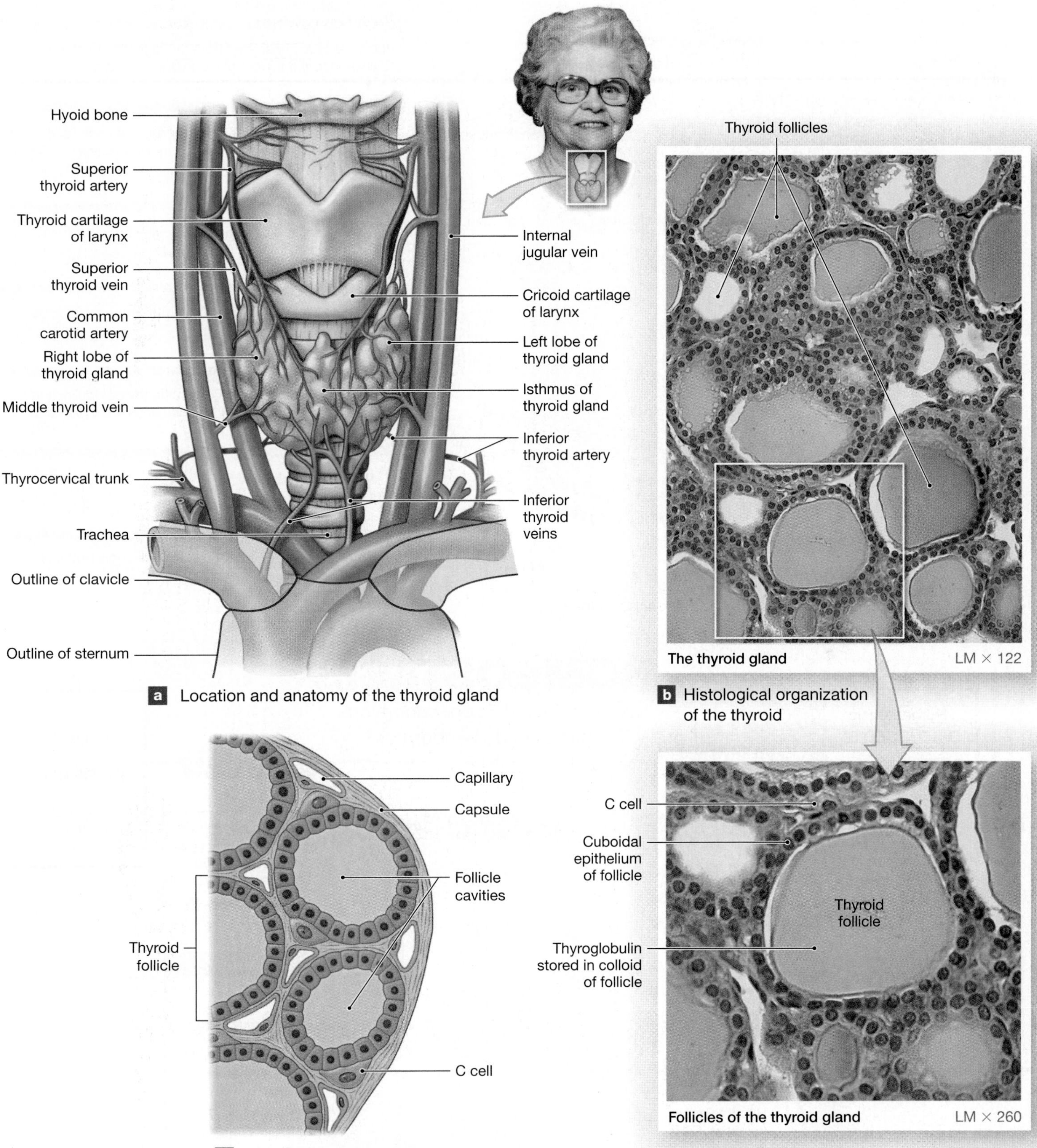
Hyoid bone
Superior thyroid artery
Thyroid cartilage of larynx
Superior thyroid vein
Common carotid artery
Right lobe of thyroid gland
Middle thyroid vein
Thyrocervical trunk
Trachea
Outline of clavicle
Outline of sternum
Internal jugular vein
Cricoid cartilage of larynx
Left lobe of thyroid gland
Isthmus of thyroid gland
Inferior thyroid artery
Inferior thyroid veins
Thyroid follicles
The thyroid gland
LM × 122
Capillary
Capsule
Follicle cavities
Thyroid follicle
C cell
C cell
Cuboidal epithelium of follicle
Thyroid follicle
Thyroglobulin stored in colloid of follicle
Follicles of the thyroid gland
LM × 260
18

a Location and anatomy of the thyroid gland

b Histological organization of the thyroid

c Histological details of the thyroid gland showing thyroid follicles and both cell types in the follicular epithelium ATLAS: Plate 18c

Figure 18–11 **Synthesis and Regulation of Thyroid Hormones.**

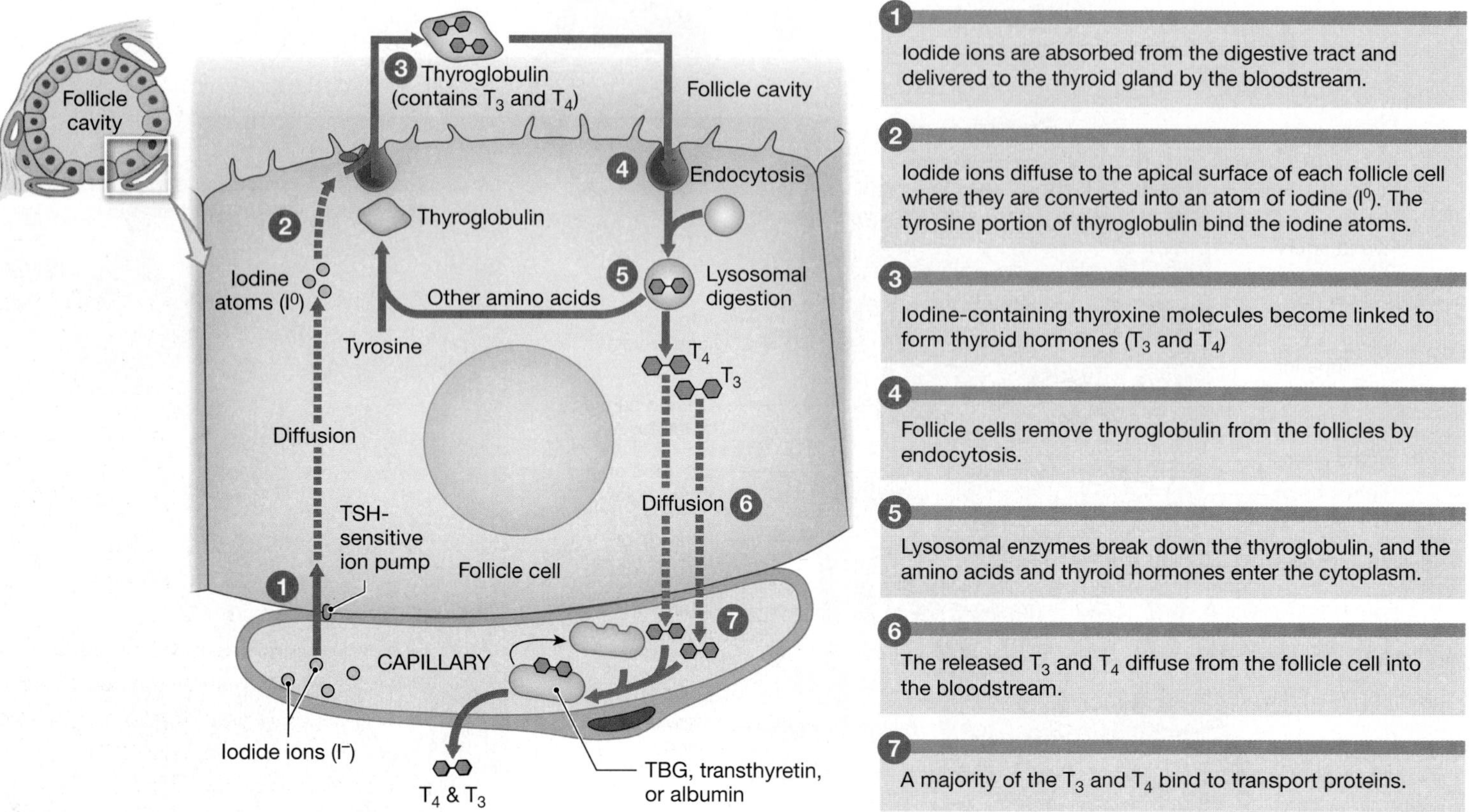
Follicle cavity
3 Thyroglobulin (contains T_3 and T_4)
Follicle cavity
4 Endocytosis
Thyroglobulin
2
Iodine atoms (I^0)
5 Lysosomal digestion
Other amino acids
Tyrosine
T_4
T_3
Diffusion
Diffusion 6
TSH-sensitive ion pump
Follicle cell
1
7
CAPILLARY
Iodide ions (I^-)
T_4 & T_3
TBG, transthyretin, or albumin
1 Iodide ions are absorbed from the digestive tract and delivered to the thyroid gland by the bloodstream.
2 Iodide ions diffuse to the apical surface of each follicle cell where they are converted into an atom of iodine (I^0). The tyrosine portion of thyroglobulin bind the iodine atoms.
3 Iodine-containing thyroxine molecules become linked to form thyroid hormones (T_3 and T_4)
4 Follicle cells remove thyroglobulin from the follicles by endocytosis.
5 Lysosomal enzymes break down the thyroglobulin, and the amino acids and thyroid hormones enter the cytoplasm.
6 The released T_3 and T_4 diffuse from the follicle cell into the bloodstream.
7 A majority of the T_3 and T_4 bind to transport proteins.

a The synthesis, storage, and secretion of thyroid hormones.

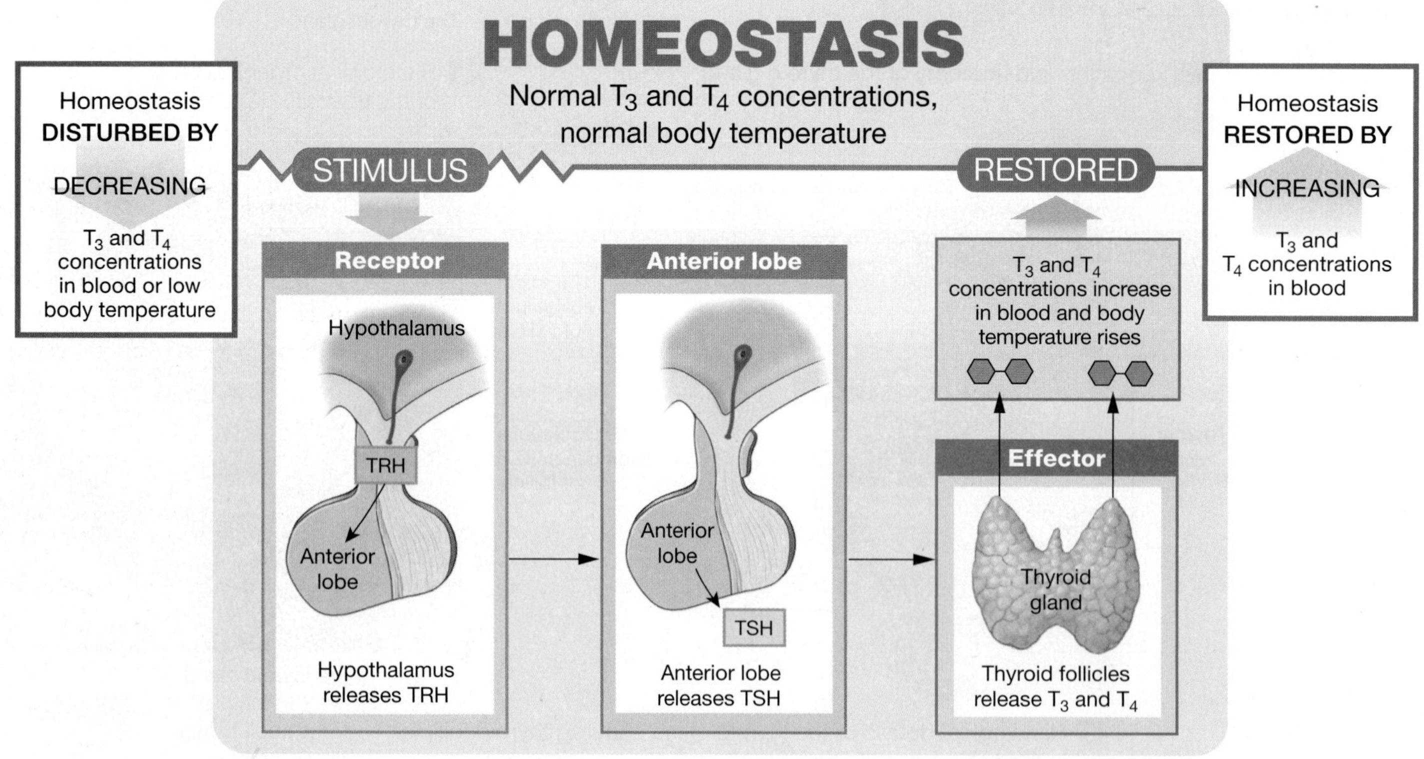
HOMEOSTASIS
Normal T_3 and T_4 concentrations, normal body temperature
Homeostasis DISTURBED BY
DECREASING
T_3 and T_4 concentrations in blood or low body temperature
STIMULUS
RESTORED
Homeostasis RESTORED BY
INCREASING
T_3 and T_4 concentrations in blood
Receptor
Hypothalamus
TRH
Anterior lobe
Hypothalamus releases TRH
Anterior lobe
Anterior lobe
TSH
Anterior lobe releases TSH
T_3 and T_4 concentrations increase in blood and body temperature rises
Effector
Thyroid gland
Thyroid follicles release T_3 and T_4

b The regulation of thyroid secretion.

Only the relatively small quantities of thyroid hormones that remain unbound—roughly 0.3 percent of the circulating T_3 and 0.03 percent of the circulating T_4—are free to diffuse into peripheral tissues.

Equilibrium exists between the amount of bound and unbound thyroid hormones. At any moment, free thyroid hormones are being bound to transport proteins at the same rate at which bound hormones are being released. When unbound thyroid hormones diffuse out of the bloodstream and into other tissues, the equilibrium is disturbed. The transport proteins then release additional thyroid hormones until a new equilibrium is reached. The bound thyroid hormones make up a substantial reserve: The bloodstream normally has more than a week's supply of thyroid hormones.

Iodine and Thyroid Hormones

Each day the follicle cells in the thyroid gland absorb 120–150 μg of **iodine ions** (I^-), the minimum dietary amount needed to maintain normal thyroid function. As a result, the concentration of I^- inside thyroid follicle cells is generally about 30 times higher than that in the blood. If the blood I^- level rises, so does its level inside the follicle cells. The thyroid follicles contain most of the iodide reserves in the body.

The typical diet in the United States provides approximately 500 μg of iodide per day, about three times the minimum daily requirement. Much of the excess is due to the I^- added to table salt (iodized salt), which is consumed in large quantities in processed foods. For this reason, iodide deficiency is seldom responsible for limiting the rate of thyroid hormone production. (This is not necessarily the case in other countries.) The kidneys remove excess I^- from the blood and eliminate them in the urine. Each day the liver excretes a small amount of I^- (about 20 μg) into the *bile*, an exocrine product stored in the gallbladder. The I^- excreted in bile is then eliminated in the feces. Because the losses in the bile continue even if the diet contains less than the minimum iodide requirement, this process can gradually deplete the iodide reserves in the thyroid. Thyroid hormone production then declines, regardless of the circulating TSH level.

Functions of Thyroid Hormones

Thyroid hormones affect almost every cell in the body; they produce a strong, immediate, and short-lived increase in the rate of cellular metabolism. They enter target cells by means of an energy-dependent transport system. Inside a target cell, they bind to receptors in three locations: (1) in the cytoplasm, (2) on the surfaces of mitochondria, and (3) in the nucleus.

- Thyroid hormones bound to cytoplasmic receptors are essentially held in storage. If intracellular levels of thyroid hormones decline, the bound thyroid hormones are released into the cytoplasm.
- The thyroid hormones binding to mitochondria increase the rates of mitochondrial ATP production.
- The binding to receptors in the nucleus activates genes that control the synthesis of enzymes involved in energy transformation and use. One specific effect is the accelerated production of *sodium–potassium ATPase*. Recall that this membrane protein ejects intracellular sodium ions and recovers extracellular potassium ions. As noted in Chapter 3, this exchange pump uses large amounts of ATP. p. 97

Thyroid hormones also activate genes that code for enzymes involved in glycolysis and ATP production. Coupled with the direct effect of thyroid hormones on mitochondria, this effect increases the metabolic rate of the cell. The effect is called the **calorigenic effect** (*calor*, heat) of thyroid hormones because the cell consumes more energy and generates more heat. In young children, TSH production increases in cold weather, and the calorigenic effect may help them adapt to cold climates. (This response does not occur in adults.) In growing children, thyroid hormones are also essential to normal development of the skeletal, muscular, and nervous systems.

The thyroid gland produces large amounts of T_4, but T_3 is primarily responsible for the observed effects of thyroid hormones. At any moment, T_3 released from the thyroid gland accounts for only 10–15 percent of the T_3 in peripheral tissues. However, enzymes in the liver, kidneys, and other tissues convert T_4 to T_3. Roughly 85–90 percent of the T_3 that reaches the target cells comes from the conversion of T_4 within peripheral tissues. Table 18–3 summarizes the effects of thyroid hormones on major organs and systems.

Synthesis and Functions of Calcitonin

The C cells of the thyroid gland produce the hormone **calcitonin (CT)**, which helps to regulate the Ca^{2+} concentration in body fluids. We introduced the functions of this hormone in Chapter 6. p. 199 Calcitonin stimulates Ca^{2+} excretion by the kidneys and prevents Ca^{2+} absorption by the digestive tract.

Calcitonin might be most important during childhood, when it stimulates bone growth and skeletal mineral deposition.

Table 18–3 Effects of Thyroid Hormones on Major Organs and Systems

- Elevate rates of oxygen consumption and energy consumption; in children, may cause a rise in body temperature
- Increase heart rate and force of contraction; generally results in a rise in blood pressure
- Increase sensitivity to sympathetic stimulation
- Maintain normal sensitivity of respiratory centers to changes in oxygen and carbon dioxide concentrations
- Stimulate red blood cell formation and thus enhances oxygen delivery
- Stimulate activity in other endocrine tissues
- Accelerate turnover of minerals in bone

It also appears to be important in reducing the loss of bone mass (1) during prolonged starvation and (2) in the late stages of pregnancy. Late in pregnancy, the maternal skeleton competes with the developing fetus for calcium ions from the diet. The role of calcitonin on the bones of the skeleton in the healthy nonpregnant adult is unclear.

In several chapters you have seen the importance of Ca^{2+} in controlling muscle cell and neuron activities. The calcium ion concentration also affects the sodium ion permeabilities of excitable membranes. At a high Ca^{2+} concentration, sodium ion permeability decreases and membranes become less responsive. Such problems are relatively rare.

Lower-than-normal Ca^{2+} concentrations are dangerous. When the calcium ion concentration declines, sodium ion permeabilities increase. As a result, cells become extremely excitable. If the calcium ion level falls too far, convulsions or muscular spasms can result. The *parathyroid glands* and *parathyroid hormone* are largely responsible for maintaining an adequate blood calcium level.

✓ Checkpoint

11. **Identify the hormones of the thyroid gland.**
12. **What signs and symptoms would you expect to see in a person whose diet lacks iodine?**
13. **When a person's thyroid gland is removed, signs of decreased thyroid hormone concentration do not appear until about 1 week later. Why?**

See the blue Answers tab at the back of the book.

18-5 The four parathyroid glands secrete parathyroid hormone, which increases the blood calcium ion level

Learning Outcome Describe the location, hormone, and functions of the parathyroid glands, and discuss the effects of abnormal parathyroid hormone production.

There are normally two pairs of **parathyroid glands** embedded in the posterior surfaces of the thyroid gland (Figure 18–12a). A dense capsule surrounding each parathyroid gland separates

18

Figure 18–12 Anatomy of the Parathyroid Glands.

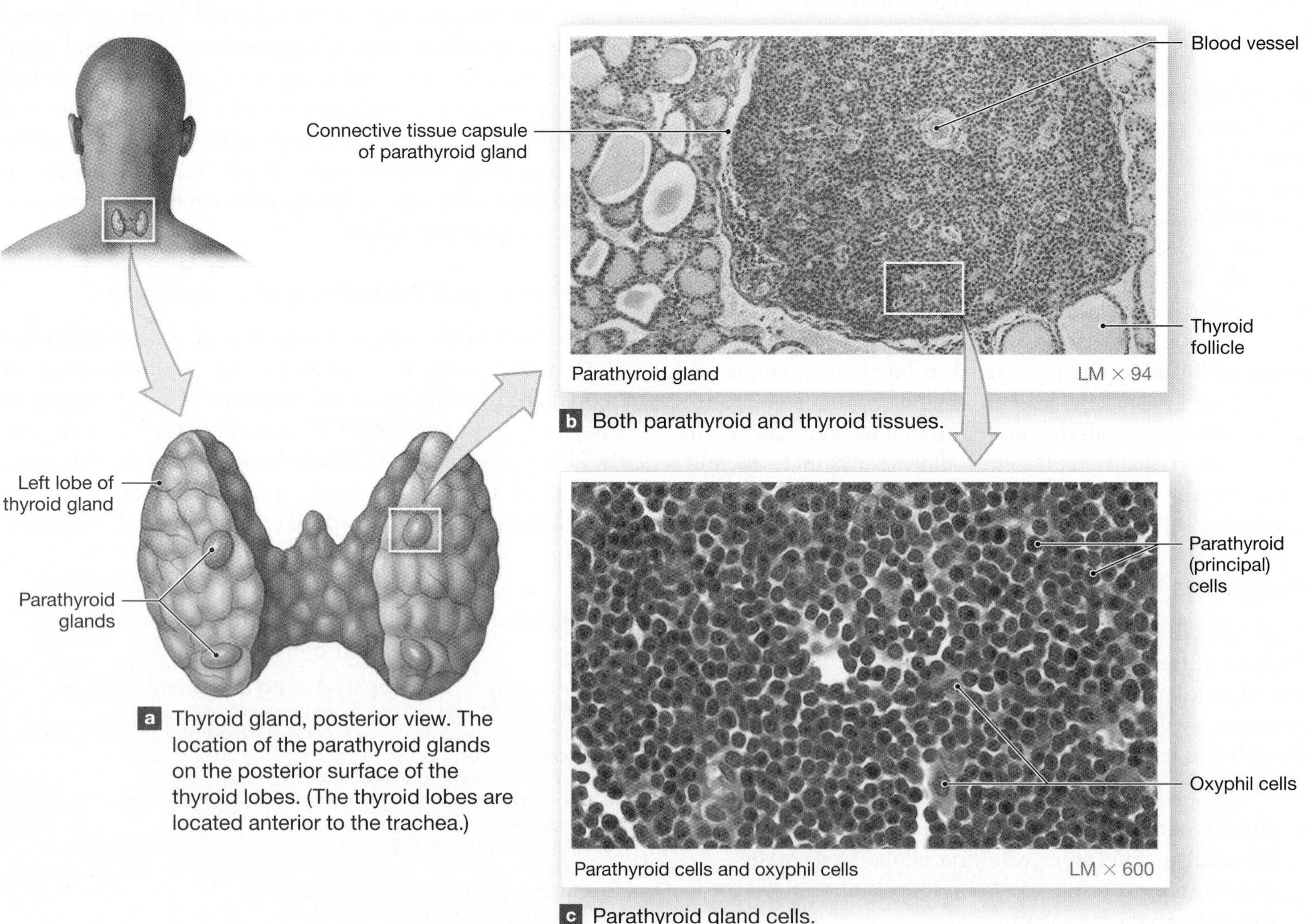
Connective tissue capsule of parathyroid gland
Blood vessel
Thyroid follicle
Parathyroid gland
LM × 94
Left lobe of thyroid gland
Parathyroid glands
Parathyroid (principal) cells
Oxyphil cells
Parathyroid cells and oxyphil cells
LM × 600

a Thyroid gland, posterior view. The location of the parathyroid glands on the posterior surface of the thyroid lobes. (The thyroid lobes are located anterior to the trachea.)

b Both parathyroid and thyroid tissues.

c Parathyroid gland cells.

it from the cells of the thyroid gland. Altogether, the four parathyroid glands weigh a mere 1.6 g (0.06 oz). The histology of a single parathyroid gland is shown in Figure 18–12b,c. The parathyroid glands have at least two cell populations. The **parathyroid (principal) cells** produce parathyroid hormone. The functions of the other cells, called *parathyroid oxyphil cells*, are unknown. They do not appear until after puberty, and then their numbers increase with age.

Like the C cells of the thyroid gland, the parathyroid cells monitor the circulating concentration of calcium ions. When the Ca^{2+} concentration of the blood falls below normal, the parathyroid cells secrete **parathyroid hormone (PTH)**, or *parathormone*. The net result of PTH secretion is an increase in Ca^{2+} concentration in body fluids. Parathyroid hormone has three major effects:

- It mobilizes calcium from bone by affecting osteoblast and osteoclast activity. PTH stimulates osteoblasts to secrete a growth factor known as RANKL (receptor activator of nuclear factor-kappa-B ligand). Osteoclasts have no PTH receptors, but both precursor and mature osteoclasts have RANKL receptors. This growth factor results in an increase in osteoclasts and osteoclast activity. With more active osteoclasts, the rates of mineral turnover and Ca^{2+} release accelerate. As bone matrix erodes, the blood Ca^{2+} level rises.
- It enhances the resorption of Ca^{2+} by the kidneys, reducing urinary losses.
- It stimulates the formation and secretion of *calcitriol* by the kidneys. In general, the effects of calcitriol complement or enhance those of PTH, but calcitriol also enhances Ca^{2+} and PO_4^{3-} absorption by the digestive tract. p. 197

Figure 18–13 illustrates the role of PTH in regulating the blood Ca^{2+} concentration. PTH, aided by calcitriol, is probably the primary regulator of the circulating calcium ion concentration in healthy adults.

Information about the hormones of the thyroid gland and parathyroid glands is summarized in Table 18–4.

18

Figure 18–13 Homeostatic Regulation of the Blood Calcium Ion Concentration.

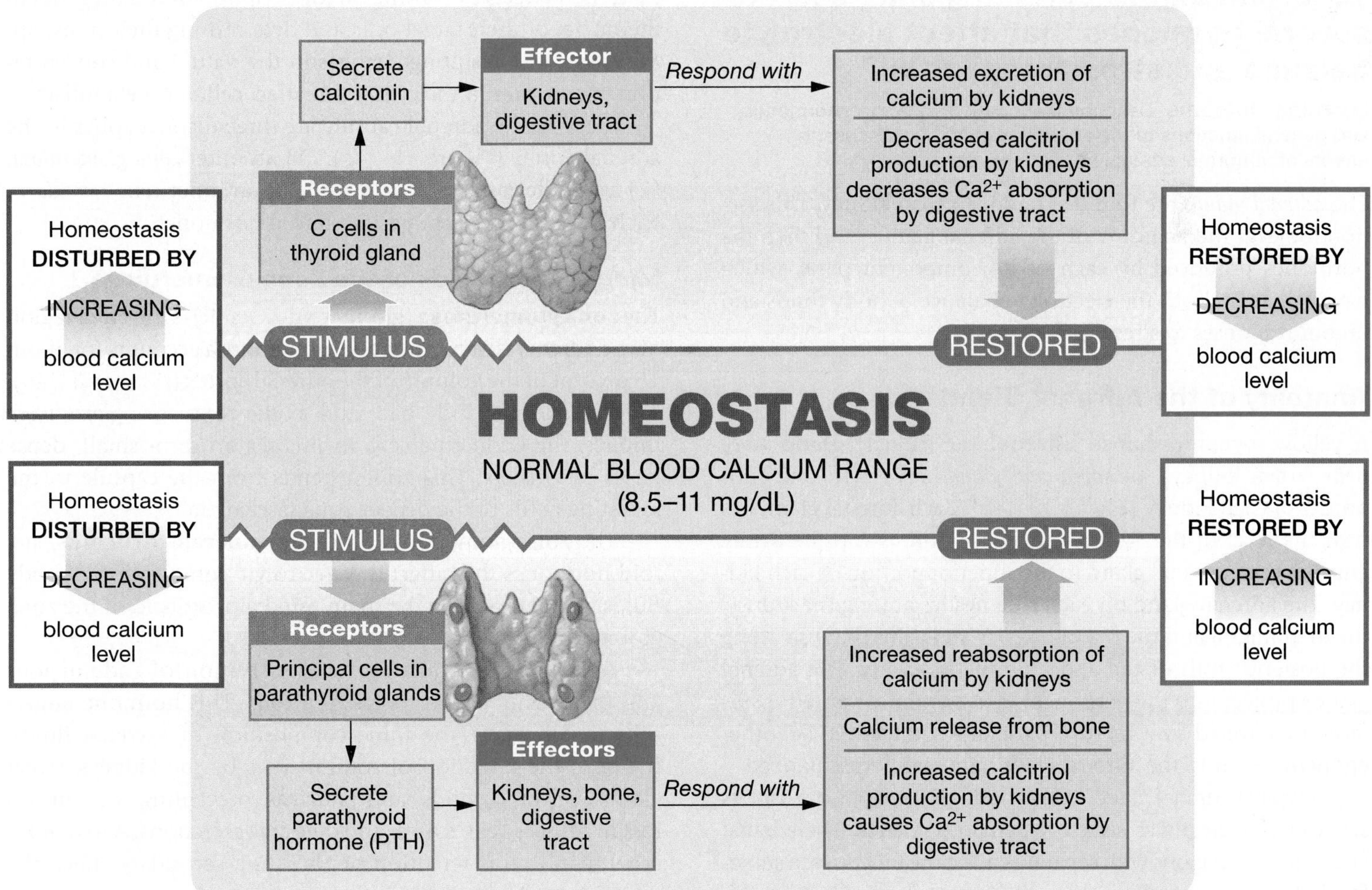

Secrete calcitonin
Effector
Kidneys, digestive tract
Respond with
Increased excretion of calcium by kidneys
Decreased calcitriol production by kidneys decreases Ca2+ absorption by digestive tract
Receptors
C cells in thyroid gland
Homeostasis DISTURBED BY INCREASING blood calcium level
STIMULUS
RESTORED
Homeostasis RESTORED BY DECREASING blood calcium level
HOMEOSTASIS
NORMAL BLOOD CALCIUM RANGE
(8.5–11 mg/dL)
Homeostasis DISTURBED BY DECREASING blood calcium level
STIMULUS
RESTORED
Homeostasis RESTORED BY INCREASING blood calcium level
Receptors
Principal cells in parathyroid glands
Increased reabsorption of calcium by kidneys
Calcium release from bone
Increased calcitriol production by kidneys causes Ca2+ absorption by digestive tract
Effectors
Kidneys, bone, digestive tract
Respond with
Secrete parathyroid hormone (PTH)

Table 18–4 Hormones of the Thyroid Gland and Parathyroid Glands

Gland/Cells	Hormone	Target	Hormonal Effect	Regulatory Control
THYROID GLAND				
Follicular epithelium	Thyroxine (T_4) Triiodothyronine (T_3)	Most cells	Increases energy utilization, oxygen consumption, growth, and development	Stimulated by TSH from the anterior lobe of the pituitary gland
C cells	Calcitonin (CT)	Kidneys, bone (mechanism unclear)	Decreases Ca^{2+} concentrations in body fluids	Stimulated by elevated blood Ca^{2+} levels; actions opposed by PTH
PARATHYROID GLANDS				
Parathyroid (principal) cells	Parathyroid hormone (PTH)	Bone, kidneys	Increases Ca^{2+} concentrations in body fluids	Stimulated by low blood Ca^{2+} levels; PTH effects enhanced by calcitriol and opposed by calcitonin

✓ Checkpoint

14. **Describe the location of the parathyroid glands.**
15. **Identify the hormone secreted by the parathyroid glands.**
16. **The removal of the parathyroid glands would result in a decrease in the blood concentration of which important mineral?**

See the blue Answers tab at the back of the book.

18

18-6 The paired adrenal glands secrete several hormones that affect electrolyte balance and stress responses

Learning Outcome Describe the location, structure, hormones, and general functions of the adrenal glands, and discuss the effects of abnormal adrenal hormone production.

The *adrenal glands* are found superior to each kidney. This section looks at the anatomy of the adrenal glands, and then the hormones produced by each of its component parts, which have effects on both the electrolyte balance of body fluids and on our responses to stress.

Anatomy of the Adrenal Glands

A yellow, pyramid-shaped **adrenal** (ad-RĒ-nal) **gland** (*ad-*, near + *ren*, kidney), or *suprarenal gland*, sits on the superior border of each kidney (Figure 18–14a). Each adrenal gland lies near the level of the 12th rib. A dense fibrous capsule firmly attaches each adrenal gland to the superior portion of each kidney. The adrenal gland on each side nestles among the kidney, the diaphragm, and the major arteries and veins that run along the posterior wall of the abdominopelvic cavity. The adrenal glands project into the peritoneal cavity, and their anterior surfaces are covered by a layer of parietal peritoneum. Like other endocrine glands, the adrenal glands are highly vascularized.

A typical adrenal gland weighs about 5.0 g (0.18 oz), but its size can vary greatly as secretory demands change. The adrenal gland has two regions with separate endocrine functions: a superficial *adrenal cortex* and an inner *adrenal medulla* (Figure 18–14b).

Corticosteroids of the Adrenal Cortex

The yellowish color of the adrenal cortex is due to stored lipids, especially cholesterol and various fatty acids. The **adrenal cortex** produces more than two dozen steroid hormones that are collectively called **corticosteroids**. In the bloodstream, these hormones are bound to transport proteins called *transcortins*.

Corticosteroids are vital. If the adrenal glands are destroyed or removed, the person will die unless corticosteroids are administered. Like other steroid hormones, corticosteroids exert their effects by turning on transcription of certain genes in the nuclei of their target cells and determining their transcription rates. The resulting changes in the nature and concentration of enzymes in the cytoplasm affect cellular metabolism.

Deep to the adrenal capsule are three distinct zones in the adrenal cortex (Figure 18–14c): (1) an outer *zona glomerulosa;* (2) a middle *zona fasciculata;* and (3) an inner *zona reticularis*. Each zone synthesizes specific steroid hormones.

Mineralocorticoids of the Zona Glomerulosa

The **zona glomerulosa** (glō-mer-yū-LŌ-suh) is the outer region of the adrenal cortex. The zona glomerulosa accounts for about 15 percent of the volume of the adrenal cortex (Figure 18–14c). A *glomerulus* is a little ball, and as the term *zona glomerulosa* implies, the endocrine cells in this region form small, dense knots or clusters. This zone extends from the capsule to the radiating cords of the deeper zona fasciculata.

The zona glomerulosa produces **mineralocorticoids**, steroid hormones that affect the electrolyte composition of body fluids. **Aldosterone** is the main mineralocorticoid of the zona glomerulosa.

Aldosterone stimulates the conservation of sodium ions and the elimination of potassium ions. This hormone targets cells that regulate the ionic composition of excreted fluids. It causes the retention of sodium ions by the kidneys, sweat glands, salivary glands, and pancreas, preventing sodium ion loss in urine, sweat, saliva, and digestive secretions. A loss of K^+ accompanies this retention of Na^+. As a secondary effect, the retention of Na^+ enhances the osmotic reabsorption of water

Figure 18–14 The Adrenal Gland and Adrenal Hormones. ATLAS: Plates 61a,b; 62b

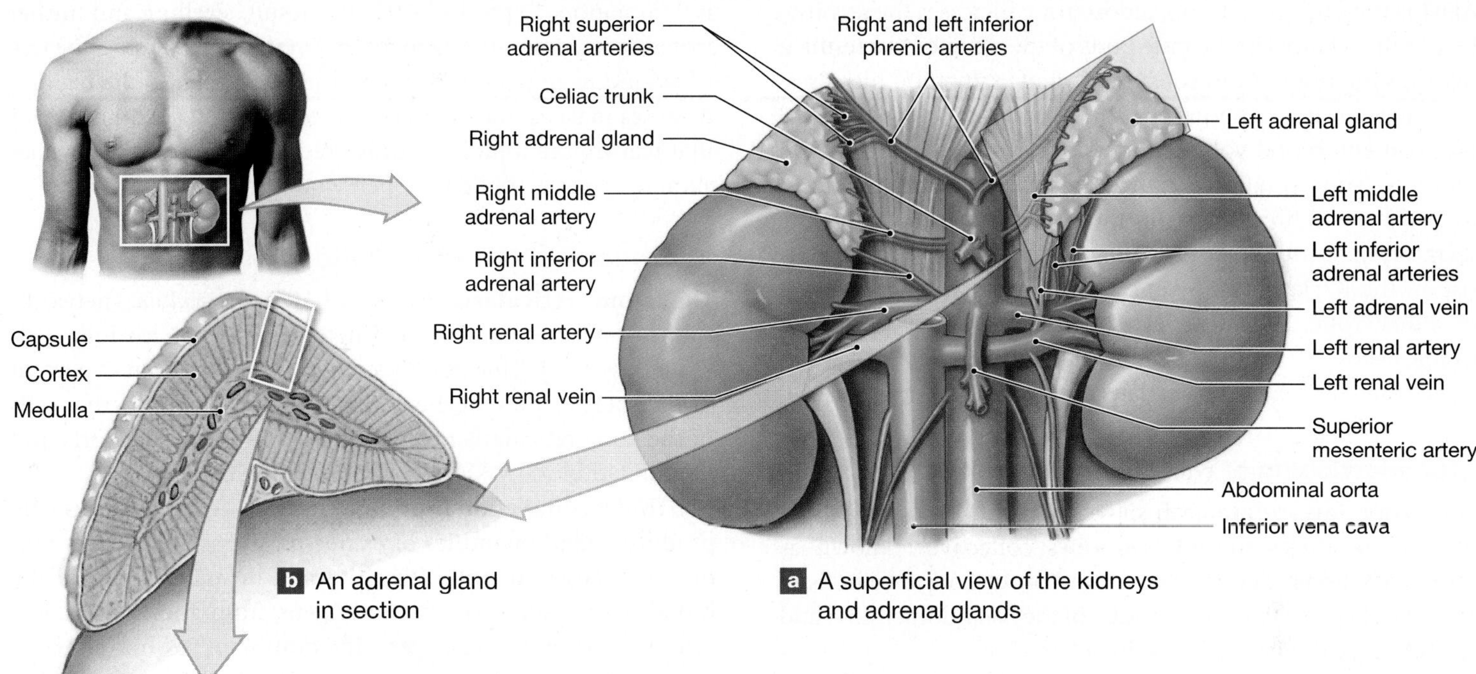
Right superior adrenal arteries
Right and left inferior phrenic arteries
Celiac trunk
Right adrenal gland
Left adrenal gland
Right middle adrenal artery
Left middle adrenal artery
Right inferior adrenal artery
Left inferior adrenal arteries
Left adrenal vein
Right renal artery
Left renal artery
Left renal vein
Right renal vein
Superior mesenteric artery
Abdominal aorta
Inferior vena cava
Capsule
Cortex
Medulla

b An adrenal gland in section

a A superficial view of the kidneys and adrenal glands

The Adrenal Hormones

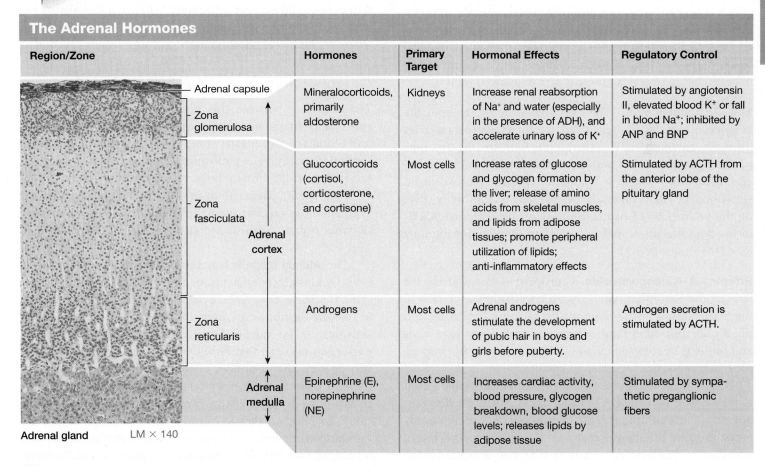

Region/Zone		Hormones	Primary Target	Hormonal Effects	Regulatory Control
Adrenal capsule; Zona glomerulosa	Adrenal cortex	Mineralocorticoids, primarily aldosterone	Kidneys	Increase renal reabsorption of Na^+ and water (especially in the presence of ADH), and accelerate urinary loss of K^+	Stimulated by angiotensin II, elevated blood K^+ or fall in blood Na^+; inhibited by ANP and BNP
Zona fasciculata	Adrenal cortex	Glucocorticoids (cortisol, corticosterone, and cortisone)	Most cells	Increase rates of glucose and glycogen formation by the liver; release of amino acids from skeletal muscles, and lipids from adipose tissues; promote peripheral utilization of lipids; anti-inflammatory effects	Stimulated by ACTH from the anterior lobe of the pituitary gland
Zona reticularis	Adrenal cortex	Androgens	Most cells	Adrenal androgens stimulate the development of pubic hair in boys and girls before puberty.	Androgen secretion is stimulated by ACTH.
	Adrenal medulla	Epinephrine (E), norepinephrine (NE)	Most cells	Increases cardiac activity, blood pressure, glycogen breakdown, blood glucose levels; releases lipids by adipose tissue	Stimulated by sympathetic preganglionic fibers

Adrenal gland LM × 140

c The major regions and zones of an adrenal gland and the hormones they produce

? Which zone of the adrenal cortex produces mineralocorticoids and what is the primary mineralocorticoid?

by the kidneys, sweat glands, salivary glands, and pancreas. The effect at the kidneys is most dramatic when a normal level of ADH is present. In addition, aldosterone increases the sensitivity of salt receptors in the taste buds of the tongue. As a result, a person's interest in (and consumption of) salty food increases.

Aldosterone secretion occurs in response to a drop in blood Na^+ content, blood volume, or blood pressure, or to a rise in blood K^+ concentration. Changes in either Na^+ or K^+ concentration have a direct effect on the zona glomerulosa, but the secretory cells are most sensitive to changes in the K^+ level. A rise in the K^+ level is very effective in stimulating the release of aldosterone. Aldosterone release also occurs in response to *angiotensin II*. We discuss this hormone, part of the *renin-angiotensin-aldosterone system*, later in this chapter.

Glucocorticoids of the Zona Fasciculata

The **zona fasciculata** (fah-sik-yū-LAH-tuh; *fasciculus*, little bundle) produces steroid hormones collectively known as **glucocorticoids**, due to their effects on glucose metabolism. This zone begins at the inner border of the zona glomerulosa and extends toward the adrenal medulla (see Figure 18–14c). It contributes about 78 percent of the cortical volume. The endocrine cells are larger and contain more lipids than those of the zona glomerulosa, and the lipid droplets give the cytoplasm a pale, foamy appearance. The cells of the zona fasciculata form individual cords composed of stacks of cells. Adjacent cords are separated by flattened blood vessels (sinusoids) with fenestrated walls.

The Glucocorticoids. When stimulated by ACTH from the anterior lobe of the pituitary gland, the zona fasciculata secretes primarily **cortisol** (KOR-tih-sol) along with smaller amounts of the related steroid **corticosterone** (kor-tih-KOS-tehr-ōn) and **cortisone**. Negative feedback regulates glucocorticoid secretion: The glucocorticoids released have an inhibitory effect on the production of corticotropin-releasing hormone (CRH) in the hypothalamus, and on ACTH in the anterior pituitary lobe (see Figure 18–8a).

Effects of Glucocorticoids. Glucocorticoids speed up the rates of glucose synthesis and glycogen formation, especially in the liver. Adipose tissue responds by releasing fatty acids into the blood, and other tissues begin to break down fatty acids and proteins instead of glucose. This process is another example of a glucose-sparing effect. p. 625

Glucocorticoids also show **anti-inflammatory** effects. That is, they inhibit the activities of white blood cells and other components of the immune system. "Steroid creams" are commonly used to control irritating allergic rashes, such as poison ivy rash, and injections of glucocorticoids may be used to control more severe allergic reactions. How do these treatments work? Glucocorticoids slow the migration of phagocytic cells into an injury site and cause phagocytic cells already in the area to become less active. In addition, mast cells exposed to these steroids are less likely to release histamine and other chemicals that promote inflammation. pp. 145–147 As a result, swelling and further irritation are dramatically reduced. On the negative side, the rate of wound healing decreases, and the weakening of the region's defenses makes it more susceptible to infectious organisms. For that reason, the topical steroids used to treat superficial rashes should never be applied to open wounds.

Androgens of the Zona Reticularis

The **zona reticularis** (re-tik-ū-LAR-is; *reticulum*, network) forms a narrow band bordering each adrenal medulla (see Figure 18–14c). This zone accounts for only about 7 percent of the total volume of the adrenal cortex. The endocrine cells of the zona reticularis form a folded, branching network, and fenestrated blood vessels wind among the cells.

Under stimulation by ACTH, the zona reticularis normally produces small quantities of *androgens*, the sex hormones produced in large quantities by the testes in males. Once in the bloodstream, some of the androgens from the zona reticularis are converted to *estrogens*, the dominant sex hormones in females. Adrenal androgens stimulate the development of pubic hair in boys and girls before puberty. Adrenal androgens are not important in adult men, but in adult women they promote muscle mass and blood cell formation, and support the sex drive.

Catecholamines of the Adrenal Medulla

The boundary between the adrenal cortex and the adrenal medulla is irregular, and the supporting connective tissues and blood vessels are extensively interconnected. The **adrenal medulla** is pale gray or pink, due in part to the many blood vessels in the area. It contains large, rounded cells—similar to cells in sympathetic ganglia that are innervated by preganglionic sympathetic fibers. The sympathetic division of the autonomic nervous system controls secretion by the adrenal medulla. p. 539

The adrenal medulla contains two populations of secretory cells: One produces epinephrine (E), the other norepinephrine (NE), both catecholamines. Evidence suggests that the two types of cells are distributed in different areas and that their secretory activities can be independently controlled. The secretions are packaged in vesicles that form dense clusters just inside plasma membranes. The hormones in these vesicles are continuously released at low levels by exocytosis. Sympathetic stimulation dramatically accelerates the rate of exocytosis and hormone release.

Epinephrine makes up 75–80 percent of the secretions from the adrenal medulla. The rest is norepinephrine. These hormones interact with alpha and beta receptors on plasma membranes, as we described in Chapter 16. p. 544 Stimulation of α_1 and β_1 receptors, the most common types, speeds up the use of cellular energy and the mobilization of energy reserves.

Activation of the adrenal medulla has the following effects:

- In skeletal muscles, E and NE trigger mobilization of glycogen reserves and accelerate the breakdown of glucose to provide ATP. This combination increases both muscular strength and endurance.
- In adipose tissue, stored fats are broken down into fatty acids, which are released into the bloodstream for other tissues to use for ATP production.
- In the liver, glycogen molecules are broken down. The resulting glucose molecules are released into the bloodstream, primarily for use by nervous tissue, which cannot shift to fatty acid metabolism.
- In the heart, the stimulation of β_1 receptors triggers an increase in the rate and force of cardiac muscle contraction.

The metabolic changes that follow the release of E and NE reach their peak about 30 seconds after adrenal stimulation, and they persist for several minutes. As a result, the effects of stimulating the adrenal medulla outlast the other signs of sympathetic activation.

✓ Checkpoint

17. **Identify the two regions of the adrenal gland, and cite the hormones secreted by each.**
18. **List the three zones of the adrenal cortex.**
19. **What effect would an elevated cortisol level have on the blood glucose level?**

See the blue Answers tab at the back of the book.

18-7 The pineal gland secretes melatonin, which affects the circadian rhythm

Learning Outcome Describe the location of the pineal gland, and discuss the functions of the hormone it produces.

The **pineal gland**, the main part of the epithalamus, lies in the posterior portion of the roof of the third ventricle (Figure 18–15). The pineal gland contains neurons, neuroglia, and special secretory cells called **pinealocytes** (pin-Ē-al-ō-sīts). These cells synthesize the hormone **melatonin** from molecules of the neurotransmitter *serotonin*. Collaterals from the visual pathways enter the pineal gland and affect the rate of melatonin production. This rate is lowest during daylight hours and highest at night.

Among the functions suggested for melatonin in humans are the following:

- *Influencing Circadian Rhythms.* Because its activity is cyclical, the pineal gland may be involved in maintaining basic *circadian rhythms*—daily changes in physiological processes that follow a regular day–night pattern. p. 484 Increased melatonin secretion in darkness has been suggested as a primary cause of *seasonal affective disorder (SAD)*. This condition can develop during the winter in people who live at high latitudes, where sunlight is scarce or lacking. It is characterized by changes in mood, eating habits, and sleeping patterns.
- *Inhibiting Reproductive Functions.* In some mammals, melatonin slows the maturation of sperm, oocytes, and reproductive organs by reducing the rate of GnRH secretion. The significance of this effect in humans remains unclear. Circumstantial evidence suggests that melatonin may play a role in the timing of human sexual maturation. The melatonin level in the blood declines at puberty, and pineal tumors that eliminate melatonin production cause premature puberty in young children.
- *Protecting against Damage by Free Radicals.* Melatonin is a very effective *antioxidant*. It may protect CNS neurons from free radicals, such as nitric oxide (NO) or hydrogen peroxide (H_2O_2), that may form in active nervous tissue.

Figure 18–15 Anatomy of the Pineal Gland.

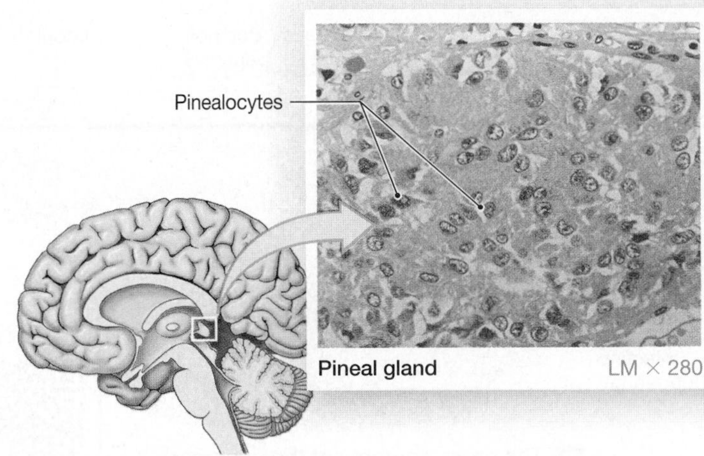
Pinealocytes
Pineal gland
LM × 280

✓ Checkpoint

20. **Identify the hormone-secreting cells of the pineal gland.**
21. **Increased amounts of light would inhibit the production of which hormone?**
22. **List three possible functions of melatonin.**

18-8 The pancreas is both an exocrine organ and an endocrine gland that produces hormones affecting the blood glucose level

Learning Outcome Describe the location, structure, hormones, and functions of the pancreas, and discuss the effects of abnormal pancreatic hormone production.

The pancreas is a large gland located within the first fold of the small intestine (duodenum), close to the stomach. As an exocrine organ, it secretes digestive enzymes into the duodenum.

Figure 18–16 Anatomy of the Pancreas. ATLAS: Plate 49e

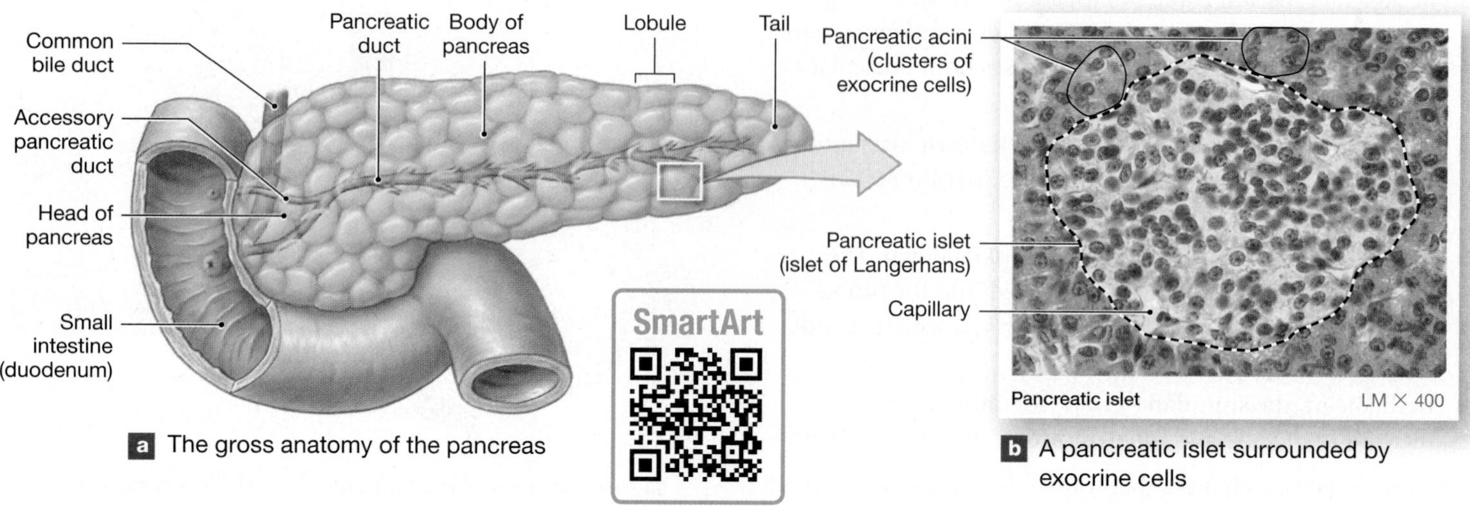
Common bile duct
Accessory pancreatic duct
Head of pancreas
Small intestine (duodenum)
Pancreatic duct
Body of pancreas
Lobule
Tail
SmartArt
Pancreatic acini (clusters of exocrine cells)
Pancreatic islet (islet of Langerhans)
Capillary
Pancreatic islet
LM × 400

a The gross anatomy of the pancreas

b A pancreatic islet surrounded by exocrine cells

18

Embedded in the pancreas are endocrine cells that secrete two hormones, insulin and glucagon, into the bloodstream. Our discussion begins with the anatomy of the pancreas.

Anatomy of the Pancreas

The **pancreas** is mostly retroperitoneal and lies in the loop between the inferior border of the stomach and the proximal portion of the small intestine (see Figure 18–1). It is a slender, pale organ with a nodular (lumpy) texture (Figure 18–16a). The pancreas is 20–25 cm (8–10 in.) long and weighs about 80 g (2.8 oz) in adults.

The pancreas is made up of both exocrine and endocrine cells. The **exocrine pancreas** consists of clusters of gland cells, called *pancreatic acini,* and their attached ducts. The exocrine pancreas takes up roughly 99 percent of the pancreatic volume. Together the gland and duct cells secrete large quantities of an alkaline, enzyme-rich fluid that reaches the lumen of the digestive tract through a network of secretory ducts. We consider its anatomy further in Chapter 24.

The **endocrine pancreas** consists of small groups of cells scattered among the exocrine cells. The endocrine clusters are known as **pancreatic islets**, or the *islets of Langerhans* (LAHNG-er-hanz) (Figure 18–16b). Pancreatic islets account for only about 1 percent of all cells in the pancreas. Nevertheless, a typical pancreas contains approximately 2 million pancreatic islets. The pancreatic islets are surrounded by an extensive, fenestrated capillary network.

Functions of Pancreatic Islet Cells

The hormones secreted by the pancreatic islet cells into the bloodstream are vital to our survival. Each islet contains four types of cells:

- **Alpha (α) cells** produce the hormone *glucagon* (GLŪ-kah-gon). Glucagon raises the blood glucose level by increasing the rates of glycogen breakdown and glucose release by the liver.
- **Beta (β) cells** produce the hormone *insulin* (IN-suh-lin). Insulin lowers the blood glucose level by increasing the rate of glucose uptake and use by most body cells, and by increasing glycogen synthesis in skeletal muscles and the liver. Beta cells also secrete *amylin,* a peptide hormone whose level rises after a meal.
- **Delta (δ) cells** produce a peptide hormone identical to growth hormone–inhibiting hormone (GH–IH), a hypothalamic regulatory hormone. GH–IH suppresses the release of glucagon and insulin by other islet cells and slows the rates of food absorption and enzyme secretion along the digestive tract.
- **Pancreatic polypeptide cells (PP cells)** produce the hormone pancreatic polypeptide (PP). PP inhibits gallbladder contractions and regulates the production of some pancreatic enzymes. It may also help control the rate of nutrient absorption by the digestive tract.

Hormones That Regulate the Blood Glucose Level

The hormones that regulate the blood glucose level are insulin and glucagon (Figure 18–17). When the blood glucose level increases, beta cells secrete insulin, which then stimulates the transport of glucose across plasma membranes and into target cells. When the blood glucose level decreases, alpha cells secrete glucagon, which stimulates glycogen breakdown and glucose release by the liver.

Figure 18–17 Homeostatic Regulation of the Blood Glucose Concentration.

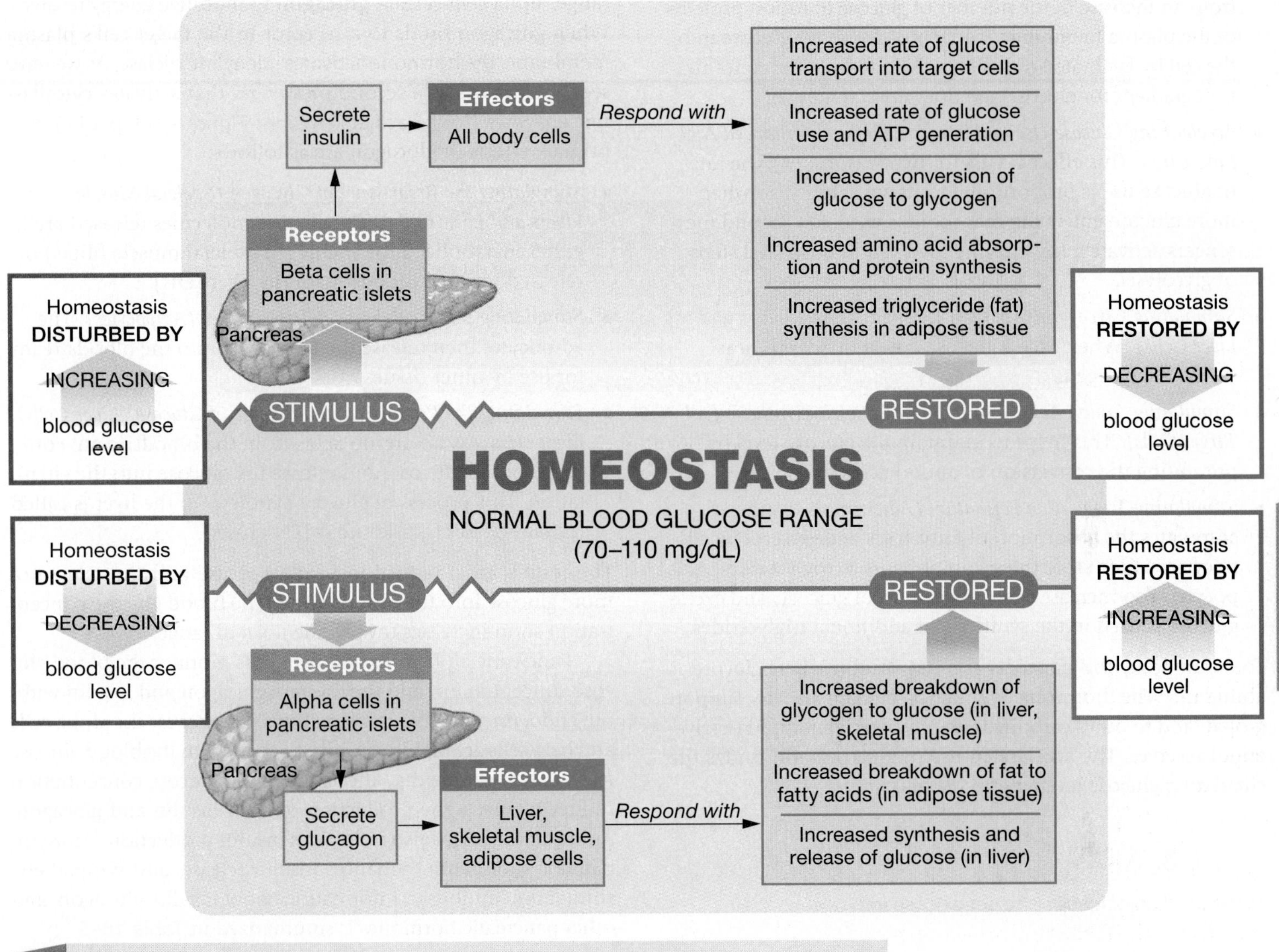
Secrete insulin
Effectors
All body cells
Respond with
Increased rate of glucose transport into target cells
Increased rate of glucose use and ATP generation
Increased conversion of glucose to glycogen
Increased amino acid absorption and protein synthesis
Increased triglyceride (fat) synthesis in adipose tissue
Receptors
Beta cells in pancreatic islets
Pancreas
Homeostasis DISTURBED BY INCREASING blood glucose level
STIMULUS
RESTORED
Homeostasis RESTORED BY DECREASING blood glucose level
HOMEOSTASIS
NORMAL BLOOD GLUCOSE RANGE (70–110 mg/dL)
Homeostasis DISTURBED BY DECREASING blood glucose level
STIMULUS
RESTORED
Homeostasis RESTORED BY INCREASING blood glucose level
Receptors
Alpha cells in pancreatic islets
Pancreas
Secrete glucagon
Effectors
Liver, skeletal muscle, adipose cells
Respond with
Increased breakdown of glycogen to glucose (in liver, skeletal muscle)
Increased breakdown of fat to fatty acids (in adipose tissue)
Increased synthesis and release of glucose (in liver)
18

? Which pancreatic cells secrete glucagon? List the effects of glucagon.

Tips & Tools

To help in differentiating between the polysaccharide storage molecule "glycogen" and the pancreatic hormone "glucagon," remember that glyco**gen** literally means "**gen**erates sugar"; and associate glucag**on** with the Pentag**on**, both of which "issue orders."

Insulin

Insulin is a peptide hormone released by beta cells when the blood glucose concentration exceeds the normal range (70–110 mg/dL). Elevated levels of some amino acids, including arginine and leucine, also stimulate secretion of insulin. This hormone affects cellular metabolism in a series of steps that begins when insulin binds to receptor proteins on the plasma membrane of a target cell. Binding activates the receptor, which functions as a kinase, attaching phosphate groups to intracellular enzymes. These enzymes then produce primary and secondary effects in the cell. The biochemical details of these effects remain unresolved.

One of the most important effects of insulin is the enhancement of glucose absorption and use. Insulin receptors are present in most plasma membranes, and cells that have them are called *insulin dependent*. However, cells in the brain and kidneys, cells in the lining of the digestive tract, and red blood cells lack insulin receptors. These cells are called *insulin independent*, because they can absorb and use glucose without insulin stimulation.

These are the effects of insulin on its target cells:

- *Accelerating Glucose Uptake (All Target Cells)*. This effect results from an increase in the number of glucose transport proteins in the plasma membrane. These proteins move glucose into the cell by facilitated diffusion, which follows the concentration gradient for glucose and does not require ATP.
- *Accelerating Glucose Use (All Target Cells) and Enhanced ATP Production.* This effect occurs for two reasons: (1) The rate of glucose use is proportional to its availability, so when more glucose enters the cell, more is used. (2) Second messengers activate a key enzyme involved in the initial steps of glycolysis.
- *Stimulating Glycogen Formation (Skeletal Muscle Fibers and Liver Cells).* When excess glucose enters these cells, it is stored as glycogen.
- *Stimulating Amino Acid Absorption and Protein Synthesis (All Target Cells).* This helps to maintain the glucose level by preventing the conversion of amino acids to glucose.
- *Stimulating Triglyceride Formation (Adipocytes).* Insulin stimulates the absorption of fatty acids and glycerol by adipocytes, which store these components as triglycerides. Adipocytes also increase their absorption of glucose, and excess glucose is used in the synthesis of additional triglycerides.

18

To summarize, the pancreas secretes insulin when glucose is abundant. The hormone stimulates glucose use to support growth and to build carbohydrate (glycogen) and lipid (triglyceride) reserves. The accelerated use of glucose soon brings the circulating glucose level within normal limits.

Tips & Tools

The function of **in**sulin is to get glucose **in**to cells.

Glucagon

When the glucose concentration decreases below the normal range, alpha cells release **glucagon** to mobilize energy reserves. When glucagon binds to a receptor in the target cell's plasma membrane, the hormone activates adenylate cyclase. As we have seen, cAMP acts as a second messenger that activates cytoplasmic enzymes (look back at **Spotlight Figure 18–3**, p. 617). The primary effects of glucagon are as follows:

- *Stimulating the Breakdown of Glycogen (Skeletal Muscle Fibers and Liver Cells).* The glucose molecules released are either metabolized for energy (in skeletal muscle fibers) or released into the bloodstream (by liver cells).
- *Stimulating the Breakdown of Triglycerides (Adipocytes).* The adipocytes then release the fatty acids into the bloodstream for use by other tissues.
- *Stimulating the Production and Release of Glucose (Liver Cells).* Liver cells absorb amino acids from the bloodstream, convert them to glucose, and release the glucose into the circulation. This process of glucose synthesis in the liver is called *gluconeogenesis* (glū-kō-nē-ō-JEN-eh-sis).

The results are a reduction in glucose use and the release of more glucose into the bloodstream. The blood glucose concentration soon increases toward the normal range.

Pancreatic alpha cells and beta cells monitor the blood glucose concentration, and they secrete glucagon and insulin without endocrine or nervous instructions. Yet because the alpha cells and beta cells are highly sensitive to changes in the blood glucose level, any hormone that affects the blood glucose concentration indirectly affects the production of both insulin and glucagon. Autonomic activity also influences insulin production: Parasympathetic stimulation enhances insulin release, and sympathetic stimulation inhibits it. Information about insulin, glucagon, and other pancreatic hormones is summarized in **Table 18–5**.

Table 18–5 Hormones Produced by the Pancreatic Islets

Structure/Cells	Hormone	Primary Targets	Hormonal Effect	Regulatory Control
PANCREATIC ISLETS				
Alpha (α) cells	Glucagon	Liver, adipose tissue	Mobilizes lipid reserves; promotes glucose synthesis and glycogen breakdown in liver; elevates blood glucose concentration	Stimulated by low blood glucose concentrations; inhibited by GH–IH from delta cells
Beta (β) cells	Insulin	Most cells	Facilitates uptake of glucose by target cells; stimulates formation and storage of lipids and glycogen	Stimulated by high blood glucose concentrations, parasympathetic stimulation, and high levels of some amino acids; inhibited by GH–IH from delta cells and by sympathetic activation
Delta (δ) cells	Somatostatin (GH–IH)	Other islet cells, digestive epithelium	Inhibits insulin and glucagon secretion; slows rates of nutrient absorption and enzyme secretion along digestive tract	Stimulated by protein-rich meal; mechanism unclear
Pancreatic polypeptide cells	Pancreatic polypeptide (PP)	Digestive organs	Inhibits gallbladder contraction; regulates production of pancreatic enzymes; influences rate of nutrient absorption by digestive tract	Stimulated by protein-rich meal and by parasympathetic stimulation

Diabetes Mellitus

Whether glucose is absorbed by the digestive tract or manufactured and released by the liver, very little glucose leaves the body once it has entered the bloodstream. The kidneys reabsorb virtually all glucose, so glucose does not appear in the urine. However, in **diabetes mellitus**, glucose accumulates in the blood and urine as a result of faulty glucose metabolism.

Diabetes mellitus can be caused by genetic abnormalities, and some of the genes responsible have been identified. Mutations that result in inadequate insulin production, the synthesis of abnormal insulin molecules, or the production of defective receptor proteins produce comparable symptoms. Under these conditions, obesity accelerates the onset and severity of the disease. Diabetes mellitus can also result from other pathological conditions, injuries, immune disorders, or hormonal imbalances.

The two major types of diabetes mellitus are Type 1 diabetes (previously known as juvenile diabetes) and Type 2 diabetes. This disorder is described in **Spotlight Figure 18–18**.

Checkpoint

23. **Identify the types of cells in the pancreatic islets and the hormones produced by each.**
24. **Why does a person with Type 1 or Type 2 diabetes urinate frequently and have increased thirst?**
25. **What effect would an increased blood glucagon level have on the amount of glycogen stored in the liver?**

See the blue Answers tab at the back of the book.

18-9 Many organs have secondary endocrine functions

Learning Outcome Describe the functions of the hormones produced by the kidneys, heart, thymus, testes, ovaries, and adipose tissue.

As we noted earlier, many organs of other body systems have secondary endocrine functions. Examples are the intestines (digestive system), the kidneys (urinary system), the heart (cardiovascular system), the thymus (lymphatic system), and the *gonads*—the testes in males and the ovaries in females (reproductive system).

Several new hormones from these endocrine tissues have been identified. In many cases, their structures and modes of action remain uncertain, and we have not described them in this chapter. However, in one instance, researchers traced a significant new hormone to an unexpected site of origin, leading to the realization that the body's adipose tissue has important endocrine functions. We include the endocrine functions of adipose tissue in this section, although all the details have yet to be worked out. **Table 18–6** provides an overview of some of the hormones those organs of other systems produce.

The Intestines

The intestines process and absorb nutrients. They release a variety of hormones that coordinate the activities of the digestive system. Most digestive processes are hormonally controlled locally, although the autonomic nervous system can affect the pace of digestive activities. We describe these hormones in Chapter 24.

18

The Kidneys

The kidneys release the steroid hormone *calcitriol*, the peptide hormone *erythropoietin*, and the enzyme *renin*. Calcitriol is important for calcium ion homeostasis. Erythropoietin and renin are involved in the regulation of blood volume and blood pressure.

Calcitriol

Calcitriol is a steroid hormone secreted by the kidneys in response to parathyroid hormone (PTH) (**Figure 18–19a**). *Cholecalciferol* (vitamin D_3) is a related steroid that is synthesized in the skin or absorbed from the diet. Cholecalciferol is converted to calcitriol, although not directly. The term *vitamin D* applies

Table 18–6 Representative Hormones Produced by Organs of Other Systems

Organ	Hormone	Primary Target	Hormonal Effect
Intestines	Many (secretin, gastrin, cholecystokinin, etc.)	Other regions and organs of the digestive system	Coordinate digestive activities
Kidneys	Erythropoietin (EPO) Calcitriol	Red bone marrow Intestinal lining, bone, kidneys	Stimulates red blood cell production Stimulates calcium and phosphate absorption; stimulates Ca^{2+} release from bone; inhibits PTH secretion
Heart	Natriuretic peptides (ANP and BNP)	Kidneys, hypothalamus, adrenal gland	Increase water and salt loss at kidneys; decrease thirst; suppress secretion of ADH and aldosterone
Thymus	Thymosins (many)	Lymphocytes and other cells of the immune response	Coordinate and regulate immune response
Gonads	*See Table 18–7*		
Adipose tissue	Leptin	Hypothalamus	Suppression of appetite; permissive effects on GnRH and gonadotropin synthesis

Go to MasteringA&P™ > Study Area > Menu > Lab Tools > PhysioEx 9.1 > Exercise 4: Endocrine System Physiology > Activity 2: Plasma Glucose, Insulin, and Diabetes Mellitus

SPOTLIGHT

Figure 18–18
Diabetes Mellitus

Untreated diabetes mellitus disrupts metabolic activities throughout the body. Clinical problems arise because the tissues involved are experiencing an energy crisis—in essence, most of the tissues are responding as they would during chronic starvation, breaking down lipids and even proteins because they are unable to absorb glucose from their surroundings. Problems involving abnormal changes in blood vessel structure are particularly dangerous. An estimated 25.8 million people in the United States have some form of diabetes.

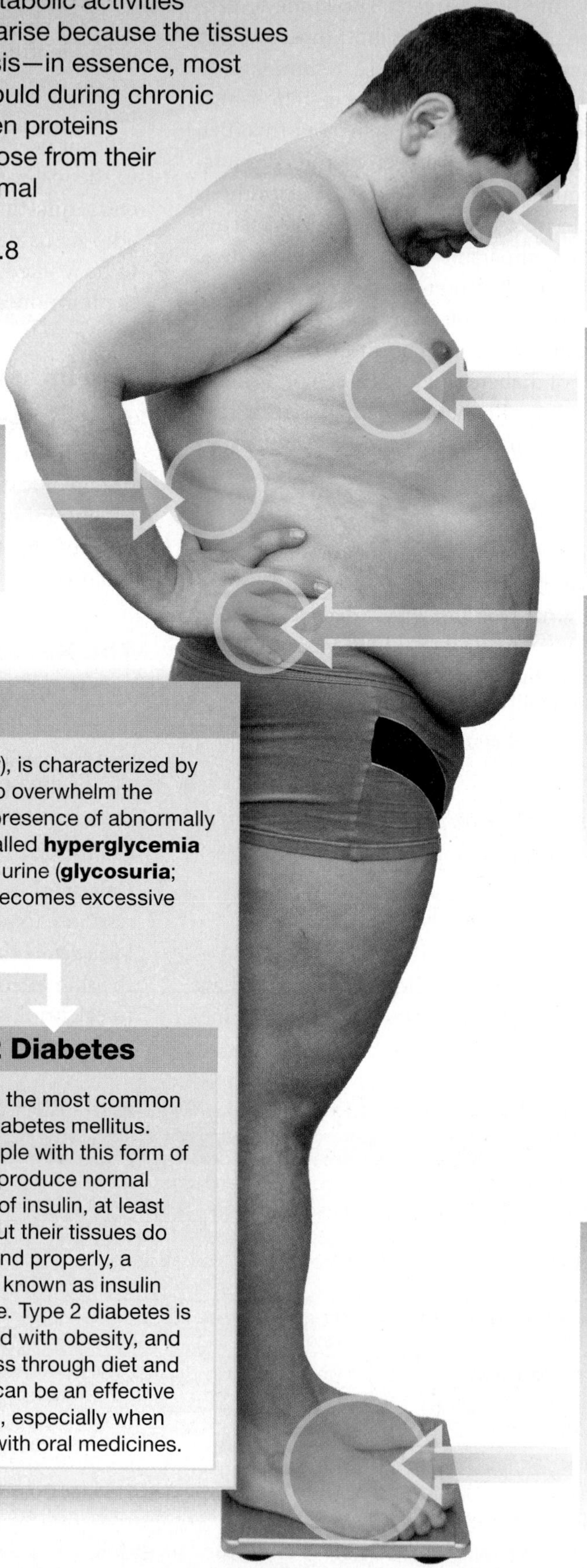

Retinal Damage

The proliferation of capillaries and hemorrhaging at the retina may cause partial or complete blindness. This condition is called **diabetic retinopathy**.

Early Heart Attacks

Degenerative blockages in cardiac circulation can lead to early heart attacks. For a given age group, heart attacks are three to five times more likely in people with diabetes than in nondiabetic people.

Kidney Degeneration

Degenerative changes in the kidneys, a condition called **diabetic nephropathy**, can lead to kidney failure.

Peripheral Nerve Problems

Abnormal blood flow to neural tissues is probably responsible for a variety of neural problems with peripheral nerves, including abnormal autonomic function. These disorders are collectively termed **diabetic neuropathies**.

Diabetes Mellitus

Diabetes mellitus (mel-Ī-tus; *mellitum*, honey), is characterized by glucose concentrations that are high enough to overwhelm the reabsorption capabilities of the kidneys. (The presence of abnormally high glucose levels in the blood in general is called **hyperglycemia** [hī-per-glī-SĒ-mē-ah].) Glucose appears in the urine (**glycosuria**; glī-kō-SYŪ-rē-a), and urine volume generally becomes excessive (**polyuria**).

subdivided into

Type 1 Diabetes

Type 1 is characterized by inadequate insulin production by the pancreatic beta cells. Persons with Type 1 diabetes require insulin to live and usually require multiple injections daily, or continuous infusion through an insulin pump or other device. This form of diabetes accounts for approximately 5% of cases. It usually develops in children and young adults.

Type 2 Diabetes

Type 2 is the most common form of diabetes mellitus. Most people with this form of diabetes produce normal amounts of insulin, at least initially, but their tissues do not respond properly, a condition known as insulin resistance. Type 2 diabetes is associated with obesity, and weight loss through diet and exercise can be an effective treatment, especially when coupled with oral medicines.

Peripheral Tissue Damage

Blood flow to the distal portions of the limbs is reduced, and peripheral tissues may suffer as a result. For example, a reduction in blood flow to the feet can lead to tissue death, ulceration, infection, and loss of toes or a major portion of one or both feet.

Figure 18–19 **Endocrine Functions of the Kidneys.**

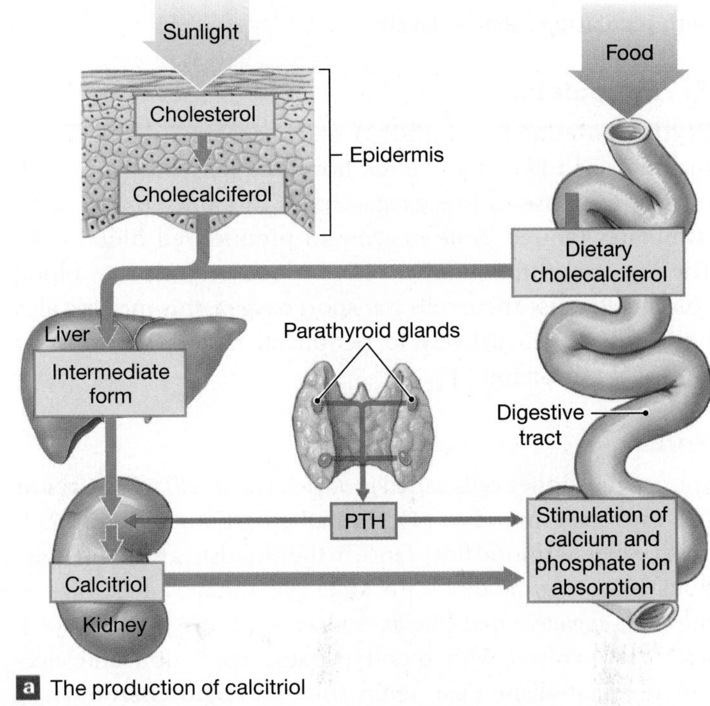
Sunlight
Food
Cholesterol
Epidermis
Cholecalciferol
Dietary cholecalciferol
Liver
Intermediate form
Parathyroid glands
Digestive tract
PTH
Stimulation of calcium and phosphate ion absorption
Calcitriol
Kidney

a The production of calcitriol

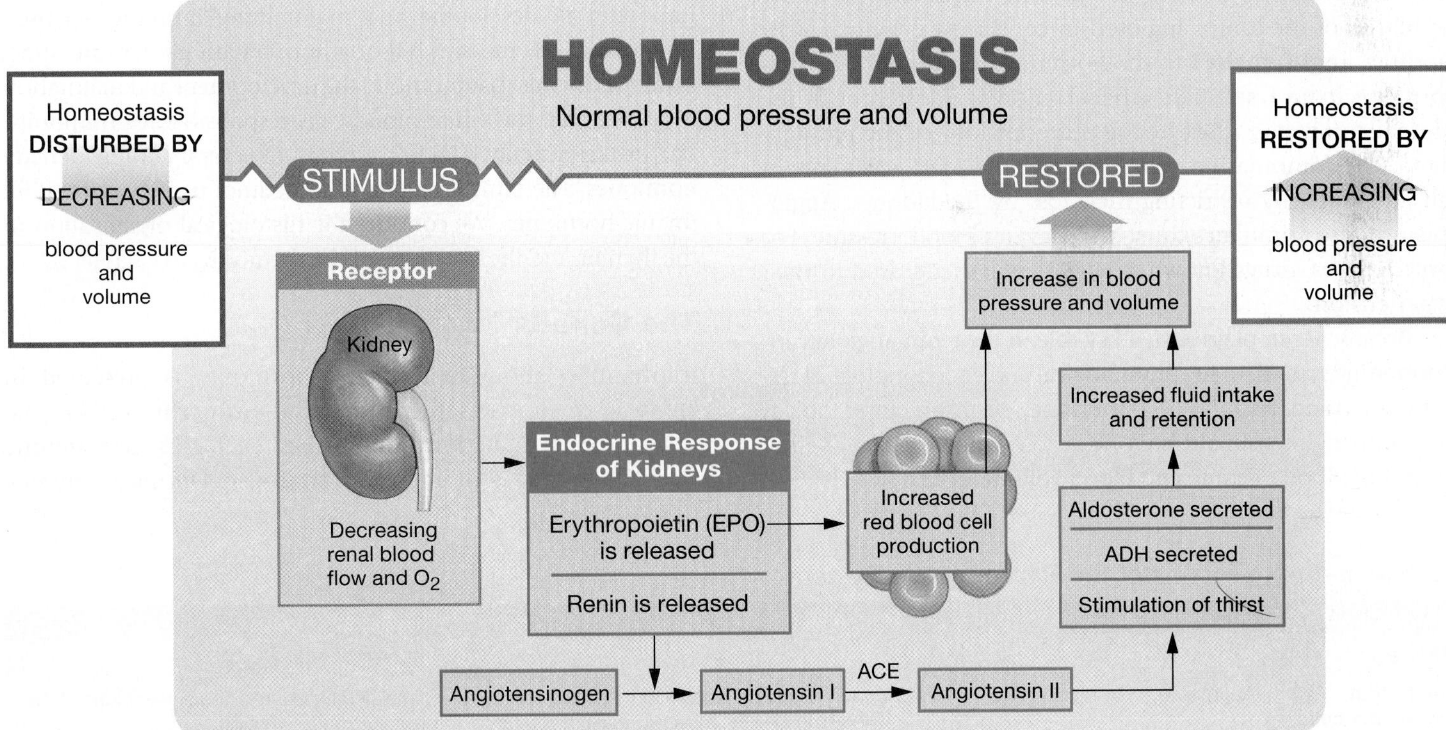
HOMEOSTASIS
Normal blood pressure and volume
Homeostasis DISTURBED BY DECREASING blood pressure and volume
STIMULUS
RESTORED
Homeostasis RESTORED BY INCREASING blood pressure and volume
Receptor
Kidney
Decreasing renal blood flow and O2
Endocrine Response of Kidneys
Erythropoietin (EPO) is released
Renin is released
Increased red blood cell production
Increase in blood pressure and volume
Increased fluid intake and retention
Aldosterone secreted
ADH secreted
Stimulation of thirst
Angiotensinogen
Angiotensin I
ACE
Angiotensin II

b The release of renin and erythropoietin, and an overview of the renin-angiotensin-aldosterone system beginning with the activation of angiotensinogen by renin

to the entire group of related steroids, including calcitriol, cholecalciferol, and various intermediate products.

The best-known function of calcitriol is to stimulate calcium and phosphate ion absorption along the digestive tract. The effects of PTH on Ca^{2+} absorption result primarily from stimulation of calcitriol release. Calcitriol's other effects on calcium metabolism include (1) stimulating the formation and differentiation of osteogenic cells and osteoclasts, (2) stimulating bone resorption by osteoclasts, (3) stimulating Ca^{2+} reabsorption by the kidneys, and (4) suppressing PTH production.

Evidence indicates that calcitriol also affects lymphocytes and keratinocytes in the skin, but these effects have nothing to do with regulating calcium levels.

Erythropoietin

Erythropoietin (e-rith-rō-POY-e-tin; *erythros,* red + *poiesis,* making), or **EPO**, is a peptide hormone released by the kidneys in response to low oxygen levels in kidney tissues. EPO stimulates the red bone marrow to produce red blood cells. The increase in the number of red blood cells elevates blood volume. Because these cells transport oxygen, this increase also improves oxygen delivery to peripheral tissues. We consider EPO again in Chapter 19.

Renin

Specialized kidney cells called *juxtaglomerular cells* release **renin** (RĒ-nin) in response to (1) sympathetic stimulation or (2) a decline in renal blood flow. Once in the bloodstream, renin functions as an enzyme that starts an enzymatic cascade known as the *renin-angiotensin-aldosterone system (RAAS)* (Figure 18–19b). RAAS is involved with blood pressure regulation and electrolyte metabolism. First, renin converts **angiotensinogen**, a plasma protein produced by the liver, to angiotensin I. In the capillaries of the lungs, angiotensin-converting enzyme (ACE) modifies **angiotensin I** to the hormone **angiotensin II**. This hormone, in turn, stimulates the secretion of aldosterone by the adrenal cortex and ADH by the posterior lobe of the pituitary gland. The combination of aldosterone and ADH acts to retain salt and water by restricting their loss by the kidneys. Angiotensin II also stimulates thirst and elevates blood pressure. For these reasons, drugs known as *ACE inhibitors* are used to treat hypertension.

Because renin plays such a key role in the renin-angiotensin-aldosterone system, many physiological and endocrinological references consider renin to be a hormone. We take a closer look at the renin-angiotensin-aldosterone system when we examine the control of blood pressure and blood volume in Chapter 21.

The Heart

The endocrine cells in the heart are cardiac muscle cells in the walls of the *atria* (chambers that receive blood from the veins) and the *ventricles* (chambers that pump blood to the rest of the body). If blood volume becomes too great, these cells are stretched excessively, to the point at which they begin to secrete **natriuretic peptides** (nā-trē-ū-RET-ik; *natrium,* sodium + *ouresis,* urination). In general, the effects of natriuretic peptides oppose those of angiotensin II: Natriuretic peptides promote the loss of Na^+ and water by the kidneys, and inhibit renin release and the secretion of ADH and aldosterone. They also suppress thirst and prevent angiotensin II and norepinephrine from elevating blood pressure. The net result is a reduction in both blood volume and blood pressure, thereby reducing the stretching of the cardiac muscle cells in the heart walls. We discuss two natriuretic peptides—*ANP* (atrial natriuretic peptide) and *BNP* (brain natriuretic peptide)—when we consider the control of blood pressure and volume in Chapters 21 and 26.

The Thymus

The **thymus** is located in the mediastinum, generally just deep to the sternum. This gland produces several hormones that are important in developing and maintaining immune defenses. **Thymosin** (THĪ-mō-sin) is the name originally given to an extract from the thymus that promotes the development and maturation of *lymphocytes,* the white blood cells responsible for immunity. The extract actually contains a blend of several complementary hormones. The term *thymosins* is sometimes used to refer to all thymic hormones. We consider the histological organization of the thymus and the functions of the thymosins in Chapter 22.

The Gonads

Information about reproductive hormones is presented in Table 18–7. In males, the **interstitial endocrine cells** of the testes produce male hormones called *androgens.* Testosterone (tes-TOS-ter-ōn) is an important androgen. During embryonic

Table 18–7 Hormones of the Reproductive System

Structure/Cells	Hormone	Primary Target	Hormonal Effect	Regulatory Control
TESTES				
Interstitial endocrine cells	Androgens	Most cells	Support functional maturation of sperm, protein synthesis in skeletal muscles, male secondary sex characteristics, and associated behaviors	Stimulated by LH from the anterior lobe of the pituitary gland
Nurse cells	Inhibin	Pituitary gland	Inhibits secretion of FSH	Stimulated by FSH from the anterior lobe
OVARIES				
Follicular cells	Estrogens	Most cells	Support follicle maturation, female secondary sex characteristics, and associated behaviors	Stimulated by FSH and LH from the anterior lobe of the pituitary gland
	Inhibin	Pituitary gland	Inhibits secretion of FSH	Stimulated by FSH from anterior lobe
Corpus luteum	Progesterone	Uterus, mammary glands	Prepares uterus for implantation; prepares mammary glands for secretory activity	Stimulated by LH from the anterior lobe of the pituitary gland

Clinical Note Endocrine Disorders

Regulation of hormone levels often involves negative feedback control mechanisms. These mechanisms involve the endocrine organ, neural regulatory factors, and the target tissues. Hormone overproduction (*hypersecretion*) or underproduction (*hyposecretion*) may cause problems. Abnormal cellular sensitivity to the hormone can also lead to problems.

Primary disorders arise from problems within the endocrine organ. The underlying cause may be a metabolic factor. Hypothyroidism due to a lack of dietary iodine is an example. An endocrine organ may also malfunction due to physical damage that destroys cells or disrupts the normal blood supply. Congenital problems may also affect the regulation, production, or release of hormones by endocrine cells.

Secondary disorders are due to problems in other organs or target tissues. Such disorders often involve the hypothalamus or pituitary gland. For example, secondary hypothyroidism occurs when the hypothalamus or pituitary gland doesn't produce enough TRH and/or TSH.

Abnormalities in target cells can affect their sensitivity or responsiveness to a particular hormone. For example, Type 2 diabetes is due to the target cells' decreased sensitivity to insulin.

Endocrine disorders often reflect either abnormal hormone production or abnormal cellular sensitivity to hormones. The signs and symptoms highlight the significance of normally "silent" hormonal contributions. The characteristics of these disorders are summarized in **Table 18–8**.

Table 18–8 Clinical Implications of Endocrine Malfunctions

Hormone	Underproduction or Tissue Insensitivity	Principal Signs and Symptoms	Overproduction or Tissue Hypersensitivity	Principal Signs and Symptoms
Growth hormone (GH)	Pituitary growth failure	Retarded growth, abnormal fat distribution, low blood glucose hours after a meal	Gigantism, acromegaly	Excessive growth
Antidiuretic hormone (ADH) or vasopressin (VP)	Diabetes insipidus	Polyuria, dehydration, thirst	SIADH (syndrome of inappropriate ADH secretion)	Increased body weight and water content
Thyroxine (T_4), triiodothyronine (T_3)	Hypothyroidism, infantile hypothyroidism, myxedema	Low metabolic rate; low body temperature; impaired physical and mental development	Hyperthyroidism, Graves disease	High metabolic rate and body temperature
Parathyroid hormone (PTH)	Hypoparathyroidism	Muscular weakness, neurological problems, formation of dense bones, tetany due to low blood Ca^{2+} concentration	Hyperparathyroidism	Neurological, mental, muscular problems due to high blood Ca^{2+} concentration; weak and brittle bones
Insulin	Diabetes mellitus (Type 1)	High blood glucose, impaired glucose utilization, dependence on lipids for energy; glycosuria	Excess insulin production or administration	Low blood glucose levels, possibly causing coma
Mineralocorticoids (MCs)	Hypoaldosteronism	Polyuria, low blood volume, high blood K^+, low blood Na^+ concentration	Aldosteronism	Increased body weight due to Na^+ and water retention; low blood K^+ concentration
Glucocorticoids (GCs)	Addison disease	Inability to tolerate stress, mobilize energy reserves, or maintain normal blood glucose concentration	Cushing disease	Excessive breakdown of tissue proteins and lipid reserves; impaired glucose metabolism
Epinephrine (E), norepinephrine (NE)	None identified		Pheochromocytoma	High metabolic rate, body temperature, and heart rate; elevated blood glucose level
Estrogens (females)	Hypogonadism	Sterility, lack of secondary sex characteristics	Adrenogenital syndrome	Overproduction of androgens by zona reticularis of adrenal cortex leads to masculinization
			Precocious puberty	Premature sexual maturation and related behavioral changes
Androgens (males)	Hypogonadism	Sterility, lack of secondary sex characteristics	Adrenogenital syndrome (gynecomastia)	Abnormal production of estrogen, due to adrenal or interstitial endocrine cell tumors; leads to breast enlargement
			Precocious puberty	Premature sexual maturation and related behavioral changes

18

development, the production of testosterone affects the development of CNS structures, including hypothalamic nuclei, which will later influence sexual behaviors. **Nurse cells** (*Sertoli cells*) in the testes support the differentiation and physical maturation of sperm. Under FSH stimulation, these cells secrete the hormone **inhibin**. It inhibits the secretion of FSH at the anterior lobe of the pituitary gland and perhaps suppresses GnRH release at the hypothalamus.

In females, steroid hormones called **estrogens** are produced in the ovaries under FSH and LH stimulation. **Estradiol** is the main estrogen. Circulating FSH stimulates the secretion of inhibin by ovarian cells, and inhibin suppresses FSH release through a feedback mechanism comparable to that in males.

At ovulation, follicles in the ovary release an immature gamete, or oocyte. The remaining follicle cells then reorganize into a *corpus luteum* (LŪ-tē-um; "yellow body") that releases a mixture of estrogens and **progesterone** (prō-JES-ter-ōn). During pregnancy, the placenta and uterus produce additional hormones that interact with those produced by the ovaries and the pituitary gland to promote normal fetal development and delivery. We consider the hormonal aspects of pregnancy in Chapter 29.

Adipose Tissue

Recall from Chapter 4 that adipose tissue is a type of loose connective tissue. p. 130 Adipose tissue produces a peptide hormone, called **leptin**, that has several functions. Its best known function is feedback control of appetite. When we eat, adipose tissue absorbs glucose and lipids and synthesizes triglycerides for storage. At the same time, it releases leptin into the bloodstream. Leptin binds to hypothalamic neurons involved with emotion and appetite control. The result is a sense of fullness (satiation) and the suppression of appetite.

Leptin was first discovered in a strain of obese mice that had a defective leptin gene. When treated with leptin, these overweight mice quickly turned into slim, athletic animals. The initial hope that leptin could be used to treat human obesity was soon dashed, however. Most obese people appear to have defective leptin receptors (or leptin pathways) in the appetite centers of the CNS. Their circulating leptin levels are already several times higher than those in individuals of normal body weight. Additional leptin would have no effect. Researchers are now investigating the structure of the receptor protein and the biochemistry of the pathway triggered by leptin binding.

Leptin must be present for normal levels of GnRH and gonadotropin synthesis to take place. This explains why (1) thin girls commonly enter puberty relatively late, (2) an increase in body fat can improve fertility, and (3) women stop menstruating when their body fat content becomes very low.

✓ Checkpoint

26. **Identify two hormones secreted by the kidneys.**
27. **Identify a hormone released by adipose tissue.**
28. **Describe the action of renin in the bloodstream.**

See the blue Answers tab at the back of the book.

18-10 Hormones interact over our lifetime to produce coordinated physiological responses

Learning Outcome Explain how hormones interact to produce coordinated physiological responses and influence behavior, describe the role of hormones in the general adaptation syndrome, and discuss how aging affects hormone production and give examples of interactions between the endocrine system and other organ systems.

We usually study hormones individually, but the extracellular fluids contain a mixture of hormones whose concentrations change daily or even hourly. As a result, cells never respond to only one hormone. Instead, they respond to multiple hormones. When a cell receives instructions from two hormones at the same time, four outcomes are possible:

- The two hormones may have opposing, or **antagonistic, effects**, as in the case of insulin and glucagon. The net result depends on the balance between the two hormones. In general, when two antagonistic hormones are present, the observed effects are weaker than those produced by either hormone acting unopposed.
- The two hormones may have additive effects, so that the net result is greater than the effect that each would produce acting alone. In some cases, the net result is greater than the *sum* of the hormones' individual effects. This interaction is a **synergistic effect** (sin-er-JIS-tik; *sunergos*, working together). An example is the glucose-sparing action of GH and glucocorticoids.
- One hormone can have a **permissive effect** on another. In such cases, the first hormone is needed for the second to produce its effect. For example, epinephrine does not change energy consumption unless thyroid hormones are also present in normal concentrations.
- Finally, hormones may produce different, but complementary, results in specific tissues and organs. These **integrative effects** are important in coordinating the activities of diverse physiological systems. The differing effects of calcitriol and parathyroid hormone on tissues involved in calcium metabolism are an example.

When multiple hormones regulate a complex process, it is very difficult to determine whether a hormone has synergistic, permissive, or integrative effects. Next we consider three examples of processes regulated by complex hormonal interactions: growth, the response to stress, and behavior.

Role of Hormones in Growth

Several endocrine organs work together to bring about normal growth. Several hormones—GH, thyroid hormones, insulin, PTH, calcitriol, and reproductive hormones—are especially important. Many others have secondary effects on growth. The circulating concentrations of these hormones are regulated independently. Every time the hormonal mixture changes, metabolic operations are modified to some degree. The modifications vary in duration and intensity, producing unique individual growth patterns.

- *Growth Hormone (GH).* The effects of GH on protein synthesis and cellular growth are most apparent in children. GH supports their muscular and skeletal development. In adults, growth hormone helps to maintain normal blood glucose concentrations and to mobilize lipid reserves in adipose tissue. GH is not the primary hormone involved, however. An adult with a GH deficiency but normal levels of thyroxine (T_4), insulin, and glucocorticoids will have no physiological problems.
- *Thyroid Hormones.* Normal growth also requires appropriate levels of thyroid hormones. If these hormones are absent during fetal development or for the first year after birth, the nervous system fails to develop normally, and developmental delay results. If T_4 concentrations decline later in life but before puberty, normal skeletal development does not continue.
- *Insulin.* Growing cells need adequate supplies of energy and nutrients. Without insulin, the passage of glucose and amino acids across plasma membranes stops or is drastically reduced.
- *Parathyroid Hormone (PTH) and Calcitriol.* Parathyroid hormone and calcitriol promote the absorption of calcium salts from the bloodstream for deposition in bone. Without adequate levels of both hormones, bones can still enlarge, but are poorly mineralized, weak, and flexible. For example, *rickets* is a condition typically caused by inadequate calcitriol production due to vitamin D deficiency in growing children. As a result, the lower limb bones are so weak that they bend under the body's weight. ⮌ p. 164
- *Reproductive Hormones.* The presence or absence of reproductive hormones (androgens in males, estrogens in females) affects the activity of osteoblasts in key locations and the growth of specific cell populations. Androgens and estrogens stimulate cell growth and differentiation in their target tissues, but their targets differ. The differential growth induced by each accounts for gender-related differences in skeletal proportions and secondary sex characteristics.

The Hormonal Responses to Stress

Any condition—physical or emotional—that threatens homeostasis is a form of **stress**. Specific homeostatic adjustments oppose many stresses. For example, a decrease in body temperature leads to shivering or changes in the pattern of blood flow, which can restore normal body temperature.

In addition, the body has a *general* response to stress that can occur while other, more specific responses are under way. A wide variety of stress-causing factors produce the same general pattern of hormonal and physiological adjustments. These responses are part of the **general adaptation syndrome (GAS)**, also known as the **stress response**. GAS has three phases: the *alarm phase*, the *resistance phase*, and the *exhaustion phase* (Spotlight Figure 18–20).

The Effects of Hormones on Behavior

As we have seen, the hypothalamus regulates many endocrine functions, and hypothalamic neurons monitor the levels of many circulating hormones. Other portions of the CNS are also quite sensitive to hormonal stimulation.

We can see the behavioral effects of specific hormones most clearly in individuals whose endocrine glands are oversecreting or undersecreting. But even normal changes in circulating hormone levels can cause behavioral changes. For example, in *precocious* (premature) *puberty*, sex hormones are produced at an inappropriate time, perhaps as early as age 5 or 6. Not only does an affected child begin to develop adult secondary sex characteristics, but the child's behavior also changes. The "nice little kid" disappears, and the child becomes aggressive and assertive due to the effects of sex hormones on CNS function. In normal teenagers, these behaviors are usually attributed to environmental stimuli, such as peer pressure, but here we can see that they have a physiological basis as well. In adults, changes in the mixture of hormones reaching the CNS can affect intellectual capabilities, memory, learning, and emotional states.

We now briefly turn to the effects of aging on hormone production.

Aging and Hormone Production

The endocrine system undergoes few functional changes with age. An exception is the decline in the concentrations of reproductive hormones. We noted the effects of these hormonal changes on the skeletal system in Chapter 6 ⮌ p. 204, and we continue the discussion in Chapter 29.

Blood and tissue concentrations of many other hormones, including TSH, thyroid hormones, ADH, PTH, prolactin, and glucocorticoids, do not change with advancing age. Circulating hormone levels may remain within normal limits, but some endocrine tissues become less responsive to stimulation. For example, in elderly people, smaller amounts of GH and insulin are secreted after a carbohydrate-rich meal. The reduced levels of GH and other tropic hormones affect tissues throughout the body. These hormonal effects involve the reductions in bone density and muscle mass noted in earlier chapters.

Finally, age-related changes in peripheral tissues may make them less responsive to some hormones. This loss of sensitivity has been documented in the case of glucocorticoids and ADH.

SPOTLIGHT

Figure 18–20
The General Adaptation Syndrome

ALARM

Alarm Phase ("Fight or Flight")

During the **alarm phase**, an immediate response to the stress occurs. The sympathetic division of the autonomic nervous system directs this response. In the alarm phase, (1) energy reserves are mobilized, mainly in the form of glucose, and (2) the body prepares to deal with the stress-causing factor by "fight or flight" responses. Epinephrine is the dominant hormone of the alarm phase. Its secretion is part of a generalized sympathetic activation.

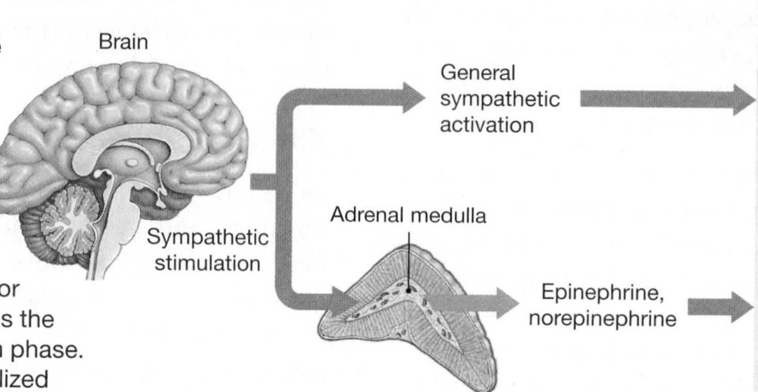

Brain
General sympathetic activation
Adrenal medulla
Sympathetic stimulation
Epinephrine, norepinephrine

Immediate Short-Term Responses to Crises

- Increased mental alertness
- Increased energy use by all cells
- Mobilization of glycogen and lipid reserves
- Changes in circulation
- Decreased digestive activity and urine production
- Increased sweat gland secretion
- Increased heart rate and respiratory rate

RESISTANCE

Resistance Phase

If a stress lasts longer than a few hours, the person enters the **resistance phase** of GAS. Glucocorticoids are the dominant hormones of the resistance phase. Epinephrine, GH, and thyroid hormones are also involved. Energy demands in the resistance phase remain higher than normal, due to the combined effects of these hormones. Nervous tissue has a high demand for energy, and requires a reliable supply of glucose. If blood glucose level falls too low, neural function deteriorates. Glycogen reserves can meet neural demand during the alarm phase, but become depleted after several hours. Hormones of the resistance phase mobilize lipids and amino acids as energy sources to conserve glucose for use by nervous tissue.

Sympathetic stimulation → Growth hormone

Pancreas → Glucagon

ACTH → Adrenal cortex → Glucocorticoids

Kidney → Renin-angiotensin-aldosterone system → Mineralocorticoids (with ADH)

Long-Term Metabolic Adjustments

- Mobilization of remaining energy reserves: Lipids are released by adipose tissue; amino acids are released by skeletal muscle
- Conservation of glucose: Peripheral tissues (except nervous) break down lipids to obtain energy
- Elevation of blood glucose concentration: Liver synthesizes glucose from other carbohydrates, amino acids, and lipids
- Conservation of salts and water, loss of K^+ and H^+

EXHAUSTION

Exhaustion Phase

The body's lipid reserves are sufficient to maintain the resistance phase for weeks or even months. But when the resistance phase ends, homeostatic regulation breaks down and the **exhaustion phase** begins. Unless corrective actions are taken almost immediately, the failure of one or more organ systems will prove fatal. The production of aldosterone throughout the resistance phase results in a conservation of Na^+ at the expense of K^+. As the body's K^+ content decreases, a variety of cells begin to malfunction. The underlying problem of the exhaustion phase is the body's inability to sustain the endocrine and metabolic adjustments of the resistance phase.

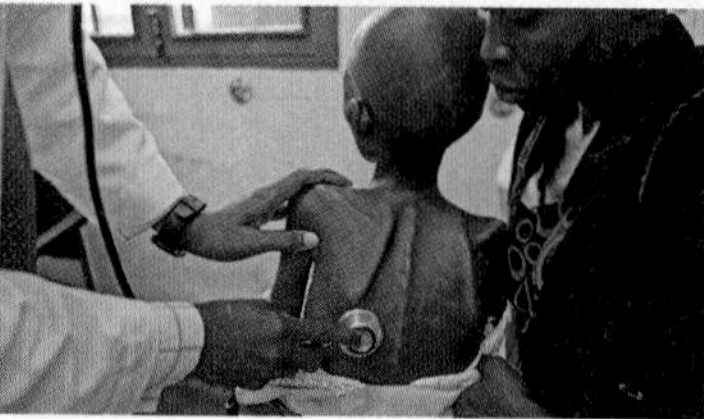

Collapse of Vital Systems

- Exhaustion of lipid reserves
- Cumulative structural or functional damage to vital organs
- Inability to produce glucocorticoids
- Failure of electrolyte balance

Clinical Note Hormones and Athletic Performance

The use of hormones to improve athletic performance is widely banned. The International Olympic Committee, the U.S. Olympic Committee, the National Collegiate Athletic Association, Major League Baseball, and the National Football League all ban it. The American Medical Association and the American College of Sports Medicine condemn the practice. Yet athletes such as Lance Armstrong, the seven-time winner of the Tour de France cycling event, admit to "doping." Athletes most often use anabolic steroids, growth hormone, erythropoietin, and a variety of synthetic hormones.

Androgen Abuse

The use of *anabolic steroids* (synthetic androgens) has become popular with many amateur and professional athletes. The goal of steroid use is to increase muscle mass, endurance, and "competitive spirit." One supposed justification for steroid use is the unfounded opinion that compounds made in the body are not only safe, but good for you. In reality, the administration of natural or synthetic androgens in abnormal amounts carries unacceptable health risks. Androgens are known to produce several complications. These problems include (1) premature closure of epiphyseal cartilages, (2) various liver dysfunctions (such as jaundice and liver tumors), (3) prostate gland enlargement and urinary tract obstruction, and (4) testicular atrophy and infertility. Links to heart attacks, impaired cardiac function, and strokes have also been suggested.

The normal regulation of androgen production involves a feedback mechanism comparable to that described for adrenal steroids earlier in this chapter. GnRH stimulates the production of LH, and LH stimulates the secretion of testosterone and other androgens by the interstitial endocrine cells of the testes. Circulating androgens, in turn, inhibit the production of both GnRH and LH. For this reason, high doses of synthetic androgens can suppress the normal production of testosterone and depress the manufacture of GnRH by the hypothalamus. *This suppression of GnRH release can be permanent*.

The use of androgens as "bulking agents" by female bodybuilders may add muscle mass, but it can also alter muscular proportions and secondary sex characteristics. For example, women taking steroids can develop irregular menstrual periods and changes in body hair distribution (such as baldness). Finally, androgen abuse can depress the immune system.

Erythropoietin Abuse

Erythropoietin (EPO) is readily available because it is now synthesized by recombinant DNA techniques. Endurance athletes, such as cyclists and marathon runners, sometimes use it to boost the number of oxygen-carrying red blood cells in the bloodstream. This effect increases the oxygen content of the blood, but it also makes the blood more viscous. For this reason, the heart must work harder to push the "thicker" blood through the blood vessels. This effort can result in death due to heart failure or stroke in young and otherwise healthy individuals. Lance Armstrong admitted publicly to repeated use of performance-enhancing drugs including EPO, testosterone, and growth hormone. He was stripped of all seven cycling titles and has been banned from the sport for life.

Androgens and EPO are hormones with reasonably well-understood effects. Drug testing is now widespread in amateur and professional sports, but athletes interested in "getting an edge" are experimenting with drugs not easily detected by standard tests. The long-term and short-term effects of these drugs are difficult to predict.

Gamma-hydroxybutyrate Use

Another drug used by amateur athletes is *gamma-hydroxybutyrate (GHB)*. It was tested for use as an anesthetic in the 1960s, but rejected in part because it was linked to seizures. In 1990, the drug appeared in health-food stores. It was sold as an anabolic agent and diet aid. It has also been used as a "date rape" drug. According to the FDA, GHB and related compounds—sold or distributed under the names Renewtrient, Revivarant, Blue Nitro, Firewater, and Serenity—have recently been responsible for serious illnesses and even deaths. Signs and symptoms include reduced heart rate, lowered body temperature, confusion, hallucinations, seizures, and coma at doses from 0.25 teaspoon to 4 tablespoons.

Extensive integration occurs between the endocrine system and other body systems. For all systems, the endocrine system adjusts metabolic rates and use of substrates such as glucose, triglycerides, and amino acids. It also regulates growth and development. Build Your Knowledge **Figure 18–21** shows the functional relationships between the endocrine system and other body systems we have studied so far.

Checkpoint

29. **Insulin decreases the blood glucose level and glucagon increases the blood glucose level. This is an example of which type of hormonal interaction?**
30. **The lack of which hormones would inhibit skeletal formation?**
31. **Why do levels of GH–RH and CRH rise during the resistance phase of the general adaptation syndrome?**
32. **Discuss the general role of the endocrine system in the functioning of other body systems.**
33. **Discuss the functional relationship between the endocrine system and the muscular system.**

See the blue Answers tab at the back of the book.

Go to MasteringA&P™ > Study Area > Menu > Lab Tools > PhysioEx 9.1 > Exercise 4: Endocrine System Physiology > Activity 4: Measuring Cortisol and Adrenocorticotropic Hormone

Go to MasteringA&P™ > Study Area > Menu > Lab Tools > PAL 3.0 > Histology > Endocrine System

18

Build Your Knowledge

Figure 18–21 Integration of the ENDOCRINE system with the other body systems presented so far.

Integumentary System

- The Integumentary System protects superficial endocrine organs; epidermis synthesizes vitamin D_3 (precursor to calcitriol production)
- The endocrine system secretes sex hormones that stimulate sebaceous gland activity, influence hair growth, fat distribution, and apocrine sweat gland activity; PRL that stimulates development of mammary glands; adrenal hormones that alter dermal blood flow; MSH that stimulates melanocyte activity

Skeletal System

- The Skeletal System protects endocrine organs, especially in brain, chest, and pelvic cavity
- The endocrine system regulates skeletal growth with several hormones; calcium homeostasis regulated primarily by parathyroid hormone; sex hormones speed growth and closure of epiphyseal cartilages at puberty and help maintain bone mass in adults

Muscular System

- The Muscular System provides protection for some endocrine organs
- The endocrine system secretes hormones that adjust muscle metabolism, energy production, and growth; regulate calcium and phosphate levels in body fluids; speed skeletal muscle growth

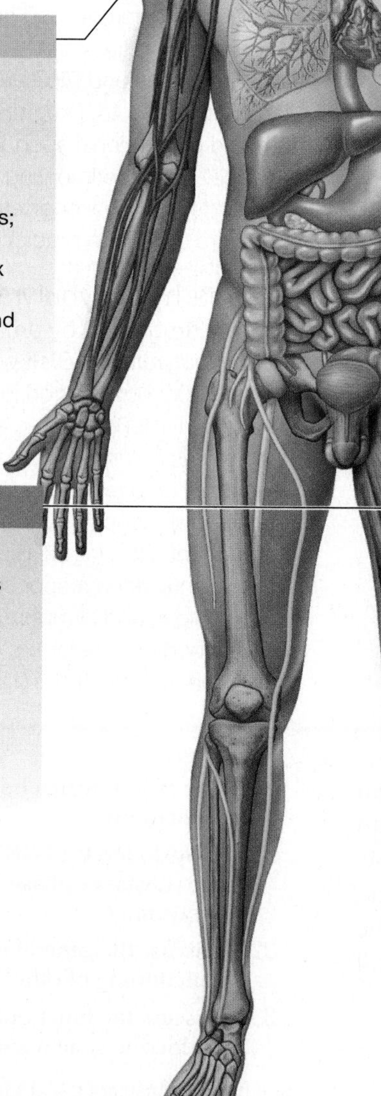

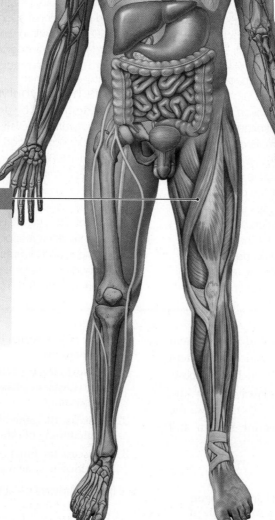

Nervous System

- The Nervous System produces hypothalamic hormones that directly control pituitary secretions and indirectly control secretions of other endocrine organs; controls adrenal medulla; secretes ADH and oxytocin
- The endocrine system secretes several hormones that affect neural metabolism and brain development; hormones that help regulate fluid and electrolyte balance; reproductive hormones that influence CNS development and behaviors

Endocrine System

The endocrine system provides long-term regulation and adjustments of homeostatic mechanisms that affect many body functions. It:

- regulates fluid and electrolyte balance
- regulates cell and tissue metabolism
- regulates growth and development
- regulates reproductive functions
- responds to stressful stimuli through the general adaptation syndrome (GAS)

18 Chapter Review

Study Outline

MasteringA&P™ Access more chapter study tools online in the MasteringA&P Study Area:
- Chapter Quizzes, Chapter Practice Test, MP3 Tutor Sessions, and Clinical Case Studies
- Practice Anatomy Lab PAL 3.0
- A&P Flix A&PFlix
- Interactive Physiology iP2
- PhysioEx PhysioEx 9.1

An Introduction to the Endocrine System p. 611

1. Chemical messengers known as *hormones* regulate human activities such as sleep, body temperature, hunger, and stress management. They are products of the *endocrine system*, which along with the nervous system, controls and coordinates our body processes.

18-1 Homeostasis is preserved through intercellular communication by the nervous and endocrine systems p. 611

2. In general, the nervous system performs short-term "crisis management," whereas the endocrine system regulates longer-term, ongoing metabolic processes.
3. **Paracrine communication** involves the use of chemical signals to transfer information from cell to cell within a single tissue. **Autocrine communication** involves chemicals affecting the same cell that secreted them.
4. **Endocrine communication** results when chemicals, called **hormones**, are released into the circulation by *endocrine cells*. The hormones alter the metabolic activities of many tissues and organs simultaneously by modifying the activities of **target cells**. *(Table 18–1)*

18-2 The endocrine system regulates physiological processes by releasing bloodborne hormones that bind to receptors on remote target organs p. 613

5. The endocrine system includes all the cells and endocrine tissues of the body that produce hormones or paracrine factors. *(Figure 18–1)*
6. Hormones can be divided into three classes according to their chemical structure: *amino acid derivatives; peptide hormones;* and *lipid derivatives*, including **steroid hormones** and **eicosanoids**. *(Spotlight Figure 18–2)*
7. Hormones may circulate freely or bound to transport proteins. Free hormones are rapidly removed from the bloodstream.
8. Receptors for *catecholamines*, peptide hormones, and eicosanoids are in the plasma membranes of target cells. Thyroid and steroid hormones cross the plasma membrane and bind to receptors in the cytoplasm or nucleus, activating or inactivating specific genes. *(Spotlight Figure 18–3, Figure 18–4)*
9. Hormone secretion can be triggered by (1) *humoral stimuli* (changes in the composition of the extracellular fluid), (2) *hormonal stimuli* (the arrival or removal of a specific hormone), or (3) *neural stimuli* (the arrival of neurotransmitters at neuroglandular junctions).

18-3 The anterior lobe of the pituitary gland produces and releases hormones under hypothalamic control, while the posterior lobe releases hypothalamic hormones p. 619

10. The **hypothalamus** contains both brain centers and endocrine tissue. It is connected to the **pituitary gland**, or **hypophysis**, by the **infundibulum**. *Figure 18–5)*
11. The **anterior lobe** of the pituitary gland, or **adenohypophysis**, can be subdivided into the **pars distalis**, the **pars intermedia**, and the **pars tuberalis**. *(Figure 18–5)*
12. The hypothalamus regulates the activities of the nervous and endocrine systems. It acts as an endocrine organ by releasing hormones into the bloodstream at the posterior lobe of the pituitary gland; secreting **regulatory hormones**, which control the activities of endocrine cells in the anterior lobe of the pituitary gland; and exerting direct neural control over the endocrine cells of the adrenal medulla. *(Figure 18–6)*
13. At the median eminence of the hypothalamus, neurons release regulatory hormones (either **releasing hormones**, **RH**, or **inhibiting hormones**, **IH**) into the surrounding interstitial fluids through **fenestrated capillaries**. *(Figure 18–7)*
14. The **hypophyseal portal system** ensures that these regulatory hormones reach the intended target cells in the pituitary before they enter the general circulation. *(Figure 18–7)*
15. The pituitary gland releases nine important peptide hormones; seven come from the anterior lobe and two from the posterior lobe. The tropic hormones of the anterior lobe "turn on" endocrine glands or support the functions of other organs. *(Figure 18–8; Table 18–2)*
16. **Thyroid-stimulating hormone (TSH)** triggers the release of thyroid hormones. *Thyrotropin-releasing hormone (TRH)* from the hypothalamus promotes the pituitary's secretion of TSH. *(Figure 18–9)*
17. **Adrenocorticotropic hormone (ACTH)** stimulates the release of *glucocorticoids* by the adrenal cortex. Corticotropin-releasing hormone (CRH) from the hypothalamus causes the pituitary to secrete ACTH. *(Figure 18–9)*
18. **Follicle-stimulating hormone (FSH)** stimulates follicle development and estrogen secretion in females and sperm production in males. **Luteinizing hormone (LH)** causes *ovulation* and *progesterone* production in females, and androgen production in males. Gonadotropin-releasing hormone (GnRH) from the hypothalamus promotes the pituitary's secretion of both FSH and LH. *(Figure 18–9)*
19. **Prolactin (PRL)** from the pituitary, together with other hormones, stimulates both the development of the mammary glands and milk production. *(Figure 18–9)*
20. **Growth hormone** (**GH**, or **somatotropin**) from the pituitary stimulates cell growth and replication through the release of **somatomedins**, also called **insulin-like growth factors (IGFs)** from liver cells. The production of GH is regulated

by **growth hormone–releasing hormone (GH–RH)** and **growth hormone–inhibiting hormone (GH–IH)** from the hypothalamus. *(Figure 18–9)*

21. **Melanocyte-stimulating hormone (MSH)** may be secreted by the pars intermedia of the pituitary during fetal development, early childhood, pregnancy, or certain diseases. This hormone stimulates melanocytes to produce melanin.
22. The **posterior lobe** of the pituitary gland, or **neurohypophysis**, contains the unmyelinated axons of hypothalamic neurons. Neurons of the **supra-optic** and **paraventricular nuclei** manufacture **antidiuretic hormone (ADH)** and **oxytocin (OXT)**, respectively. ADH decreases the amount of water lost at the kidneys and, in higher concentrations, elevates blood pressure. In women, oxytocin stimulates contractile cells in the mammary glands and has a stimulatory effect on smooth muscles in the uterus. *(Figure 18–9; Table 18–2)*

18-4 The thyroid gland synthesizes thyroid hormones that affect the rate of metabolism p. 627

23. The thyroid gland lies anterior to the *thyroid cartilage* of the larynx and consists of two **lobes** connected by a narrow **isthmus**. *(Figure 18–10)*
24. The thyroid gland contains numerous **thyroid follicles**. Thyroid follicles release several hormones, including **thyroxine** and **triiodothyronine**. *(Figures 18–10, 18–11; Table 18–4)*
25. Most of the thyroid hormones entering the bloodstream are attached to special **thyroid-binding globulins (TBGs)**; the rest are attached to either **transthyretin** or albumin. *(Figure 18–11)*
26. In target cells, thyroid hormones are held in storage in the cytoplasm, bound to mitochondria (where they increase ATP production), or bound to receptors activating genes that control energy utilization. They also exert a **calorigenic effect**. *(Table 18–3)*
27. The **C cells** of the thyroid follicles produce **calcitonin (CT)**, which possibly helps regulate Ca^{2+} concentrations in body fluids, especially during childhood and pregnancy. *(Figure 18–10; Table 18–4)*

18-5 The four parathyroid glands secrete parathyroid hormone, which increases the blood calcium ion level p. 632

28. Four **parathyroid glands** are embedded in the posterior surfaces of the thyroid gland. **Parathyroid (principal) cells** produce **parathyroid hormone (PTH)** in response to lower-than-normal concentrations of Ca^{2+}. The parathyroid glands, aided by *calcitriol*, are the primary regulators of blood calcium levels in healthy adults. *(Figures 18–12, 18–13; Table 18–4)*

18-6 The paired adrenal glands secrete several hormones that affect electrolyte balance and stress responses p. 634

29. One **adrenal** *(suprarenal)* **gland** lies along the superior border of each kidney. The gland is subdivided into the superficial **adrenal cortex** and the inner **adrenal medulla**. *(Figure 18–14)*
30. The adrenal cortex manufactures steroid hormones called **corticosteroids**. The cortex can be subdivided into three areas: (1) the **zona glomerulosa**, which releases **mineralocorticoids**, principally **aldosterone**; (2) the **zona fasciculata**, which produces **glucocorticoids**, notably **cortisol, corticosterone, and cortisone**; and (3) the **zona reticularis**, which produces androgens under ACTH stimulation. *(Figure 18–14)*
31. The adrenal medulla produces epinephrine (75–80 percent of medullary secretion) and norepinephrine (20–25 percent). *(Figure 18–14)*

18-7 The pineal gland secretes melatonin, which affects the circadian rhythm p. 637

32. The **pineal gland** contains **pinealocytes**, which synthesize **melatonin**. Suggested functions include inhibiting reproductive functions, protecting against damage by free radicals, and setting circadian rhythms. *(Figure 18–15)*

18-8 The pancreas is both an exocrine organ and an endocrine gland that produces hormones affecting the blood glucose level p. 637

33. The pancreas contains both exocrine and endocrine cells. Cells of the endocrine pancreas form clusters called **pancreatic islets** *(islets of Langerhans)*. These islets contain **alpha (α) cells**, which secrete the hormone **glucagon; beta (β) cells,** which secrete **insulin; delta (δ) cells**, which secrete **somatostatin (GH–IH)**; and **pancreatic polypeptide (PP) cells**, which secrete **pancreatic polypeptide**. *(Figure 18–16; Table 18–5)*
34. Insulin lowers blood glucose by increasing the rate of glucose uptake and utilization by most body cells; glucagon raises blood glucose by increasing the rates of glycogen breakdown and glucose manufacture in the liver. *(Figure 18–17; Table 18–5)*
35. **Diabetes mellitus** is an endocrine disorder characterized by high blood glucose levels. *(Spotlight Figure 18–18)*

18-9 Many organs have secondary endocrine functions p. 641

36. The intestines produce hormones important in coordinating digestive activities. *(Table 18–6)*
37. Endocrine cells in the kidneys produce the hormones *calcitriol* and *erythropoietin* and the enzyme *renin*. *(Table 18–6)*
38. **Calcitriol** stimulates calcium and phosphate ion absorption along the digestive tract. *(Figure 18–19)*
39. **Erythropoietin (EPO)** stimulates red blood cell production by the red bone marrow. *(Figure 18–19)*
40. In the *renin-angiotensin-aldosterone system (RAAS)*, **renin** converts **angiotensinogen** to **angiotensin I**. In the capillaries of the lungs, angiotensin-converting enzyme (ACE) converts angiotensin I to **angiotensin II**, a hormone that (1) stimulates the adrenal production of aldosterone, (2) stimulates the pituitary release of ADH, (3) promotes thirst, and (4) elevates blood pressure. *(Figure 18–19)*
41. Specialized muscle cells in the heart produce **natriuretic peptides** (*ANP* and *BNP*) when blood volume becomes excessive. In general, their actions oppose those of angiotensin II. *(Table 18–6)*
42. The thymus produces several hormones, collectively known as **thymosins**, which play a role in developing and maintaining normal immune defenses. *(Table 18–6)*

43. The **interstitial endocrine cells** of the testes produce androgens. **Testosterone** is the most important sex hormone in males. *(Table 18–7)*
44. In females, *oocytes* develop in follicles, and follicle cells produce **estrogens**, especially **estradiol**. After ovulation, the remaining follicle cells reorganize into a *corpus luteum*. Those cells release a mixture of estrogens and **progesterone**. *(Table 18–7)*
45. Adipose tissue secretes **leptin** (a feedback control for appetite). *(Table 18–6)*

18-10 Hormones interact over our lifetime to produce coordinated physiological responses p. 646

46. Endocrine system hormones often interact, producing **antagonistic** (opposing) **effects**; **synergistic** (additive) **effects**; **permissive effects**, in which one hormone is necessary for another to produce its effect; or **integrative effects**, in which hormones produce different, but complementary, results.
47. Normal growth requires the cooperation of several endocrine organs. Several hormones are especially important: GH, thyroid hormones, insulin, PTH, calcitriol, and reproductive hormones.
48. Any condition that threatens homeostasis is a **stress**. Our bodies respond to a variety of stress-causing factors through the **general adaptation syndrome (GAS)**, or **stress response**.
49. The GAS can be divided into three phases: (1) the **alarm phase** (an immediate, "fight or flight" response, under the direction of the sympathetic division of the ANS); (2) the **resistance phase**, dominated by glucocorticoids; and (3) the **exhaustion phase**, the eventual breakdown of homeostatic regulation and failure of one or more organ systems. *(Spotlight Figure 18–20)*
50. Many hormones affect the CNS. Changes in the normal mixture of hormones can significantly alter intellectual capabilities, memory, learning, and emotional states.
51. The endocrine system undergoes few functional changes with advanced age. The major changes include a decline in the concentration of growth hormone and reproductive hormones.
52. The endocrine system provides long-term regulation and homeostatic adjustments that affect many body systems. *(Figure 18–21)*

Review Questions

See the blue Answers tab at the back of the book.

LEVEL 1 Reviewing Facts and Terms

1. The use of a chemical messenger to transfer information from cell to cell within a single tissue is referred to as _____ communication. **(a)** direct, **(b)** paracrine, **(c)** hormonal, **(d)** endocrine.
2. Cyclic AMP functions as a second messenger to **(a)** build proteins and catalyze specific reactions, **(b)** activate adenylate cyclase, **(c)** open ion channels and activate key enzymes in the cytoplasm, **(d)** bind the hormone–receptor complex to DNA segments.
3. Adrenocorticotropic hormone (ACTH) stimulates the release of **(a)** thyroid hormones by the hypothalamus, **(b)** gonadotropins by the adrenal glands, **(c)** growth hormones by the hypothalamus, **(d)** steroid hormones by the adrenal glands.
4. FSH production in males supports **(a)** the maturation of sperm by stimulating nurse cells, **(b)** the development of muscles and strength, **(c)** the production of male sex hormones, **(d)** an increased desire for sexual activity.
5. The two hormones released by the posterior lobe of the pituitary gland are **(a)** GH and gonadotropin, **(b)** estrogen and progesterone, **(c)** GH and prolactin, **(d)** ADH and oxytocin.
6. All of the following are true of the endocrine system, *except* that it **(a)** releases chemicals into the bloodstream for distribution throughout the body, **(b)** releases hormones that simultaneously alter the metabolic activities of many different tissues and organs, **(c)** produces effects that can last for hours, days, and even longer, **(d)** produces rapid, local, brief-duration responses to specific stimuli, **(e)** functions to control ongoing metabolic processes.
7. A cell's hormonal sensitivities are determined by the **(a)** chemical nature of the hormone, **(b)** quantity of circulating hormone, **(c)** shape of the hormone molecules, **(d)** presence or absence of appropriate receptors, **(e)** thickness of its plasma membrane.
8. Identify the endocrine glands and tissues in the following diagram.

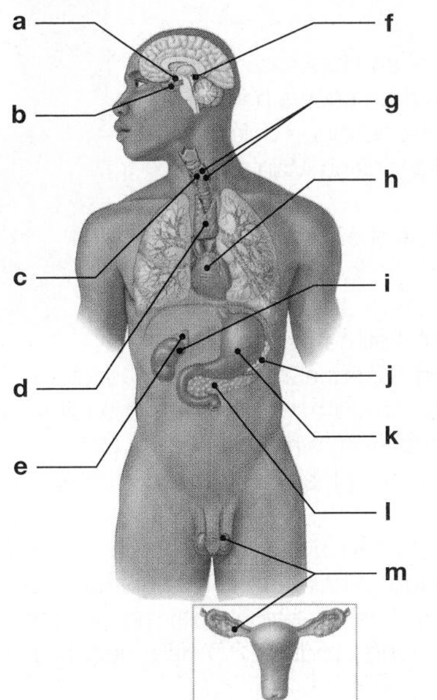
a
b
c
d
e
f
g
h
i
j
k
l
m

(a) __________
(b) __________
(c) __________
(d) __________
(e) __________
(f) __________
(g) __________
(h) __________
(i) __________
(j) __________
(k) __________
(l) __________
(m) __________

9. In what three ways is the hypothalamus involved in integrating the activities of the nervous and endocrine systems?
10. Which seven hormones are released by the anterior lobe of the pituitary gland?

11. What six hormones primarily affect growth?
12. What five primary effects result from the action of thyroid hormones?
13. What effects does parathyroid hormone have on blood calcium level?
14. What three zones make up the adrenal cortex, and what kind of hormones does each zone produce?
15. Which two hormones are released by the kidneys, and what is the importance of each hormone?
16. What are the four opposing effects of natriuretic peptides and angiotensin II?
17. What four cell populations make up the endocrine pancreas? Which hormone does each type of cell produce?

LEVEL 2 Reviewing Concepts

18. What is the primary difference in the way the nervous and endocrine systems communicate with their target cells?
19. In what ways can a hormone modify the activities of its target cells?
20. How would blocking the activity of phosphodiesterase (PDE) affect a cell that responds to hormonal stimulation by the cAMP second messenger system?
21. How does control of the adrenal medulla differ from control of the adrenal cortex?
22. A researcher observes that stimulation by a particular hormone induces a marked increase in the activity of G proteins in the target plasma membrane. The hormone being studied is probably **(a)** a steroid, **(b)** a peptide, **(c)** testosterone, **(d)** estrogen, **(e)** aldosterone.
23. Decreased blood calcium level would result in *increased* **(a)** secretion of calcitonin, **(b)** secretion of PTH, **(c)** elimination of calcium by the kidneys, **(d)** osteoclast activity, **(e)** excitability of neural membranes.
24. In Type 2 diabetes, insulin levels are frequently normal, yet the target cells are less sensitive to the effects of insulin. This suggests that the target cells **(a)** are impermeable to insulin, **(b)** may lack enough insulin receptors, **(c)** cannot convert insulin to an active form, **(d)** have adequate internal supplies of glucose, **(e)** both b and c.

LEVEL 3 Critical Thinking and Clinical Applications

25. Roger has been extremely thirsty. He drinks numerous glasses of water every day and urinates a great deal. Name two disorders that could produce these signs and symptoms. What test could a clinician perform to determine which disorder Roger has?
26. Julie is pregnant but is not receiving prenatal care. She has a poor diet consisting mostly of fast food. She drinks no milk, preferring colas instead. How would this situation affect Julie's parathyroid hormone level?
27. Sherry tells her physician that she has been restless and irritable lately. She has a hard time sleeping and complains of diarrhea and weight loss. During the examination, her physician notices a higher-than-normal heart rate and a fine tremor in her outstretched fingers. What tests could the physician perform to make a positive diagnosis of Sherry's condition?
28. What are two benefits of having a portal system connect the median eminence of the hypothalamus with the anterior lobe of the pituitary gland?
29. Pamela and her teammates are considering taking anabolic steroids (synthetic hormones derived from testosterone) to enhance their competitive skills. What natural effects of testosterone are they hoping to gain? What additional side effects might these women expect should they begin an anabolic steroid regime?

18

+ CLINICAL CASE Wrap-Up Stones, Bones, and Groans

The doctor jokes with Monica that she has a case of "stones, bones, and groans." He then shows her the lab results, which indicate an elevated blood calcium (Ca^{2+}) level—a pivotal key to diagnosing Monica's condition.

Calcium is critical in the control of many systems. It provides the means for cholinergic synapses to work, including those in the CNS, all neuron-to-neuron synapses in the PNS, and all neuromuscular junctions. Calcium also plays an essential role in the muscular contraction cycles of all skeletal, smooth, and cardiac muscles. In addition, it strengthens the skeleton, which acts as a Ca^{2+} storage system. Calcium is the most closely regulated element in our bodies.

The parathyroid glands function to increase the blood calcium ion concentration. Normally the four small parathyroid glands produce parathyroid hormone (PTH). Their principal cells monitor the circulating concentration of Ca^{2+} and produce PTH in a negative feedback control system.

It turns out that one of Monica's parathyroid glands has a benign tumor that is producing PTH without feedback control. The excess PTH causes the elevated blood Ca^{2+} level, osteopenia, and bone pain. The elevated blood Ca^{2+} is responsible for Monica's heart palpitations, kidney stones, muscle fatigue, memory loss, depression, and irritability.

Outpatient surgery is performed to remove the tumor. The remaining three parathyroid glands take over normal Ca^{2+} regulation. Monica soon begins to feel like herself again.

1. In addition to mobilizing calcium from bone, how else does PTH increase the blood calcium ion level?
2. What hormone counteracts PTH and decreases the blood Ca^{2+} level?

See the blue Answers tab at the back of the book. Note also that calcium homeostasis was introduced in Chapter 6 and can be reviewed in Figure 6–16.

Related Clinical Terms

adrenalectomy: Surgical removal of an adrenal gland.

empty sella syndrome: Condition in which the pituitary gland becomes shrunken or flattened.

exophthalmos: Abnormal protrusion of the eyeballs.

galactorrhea: A milky discharge from the nipple unrelated to normal breast feeding.

Hashimoto disease: Disorder that affects the thyroid gland, causing the immune system to attack the thyroid gland. It is the most common cause of hypothyroidism in the United States; also known as *chronic lymphocytic thyroiditis*.

hirsutism: Excessive growth of facial or body hair in a woman. Hirsutism is a sign of hyperandrogenism, or the presence of abnormally high levels of androgens. It may be a sign of polycystic ovarian syndrome, congenital adrenal hyperplasia (CAH), or androgen-secreting tumors, all of which may cause infertility in women.

hypocalcemic tetany: Muscle spasms affecting the face and upper extremities; caused by low Ca^{2+} concentrations in body fluids.

hypophysectomy: Surgical removal of the pituitary gland.

posttraumatic stress disorder (PTSD): A common anxiety disorder that develops after being exposed to a life-threatening situation or terrifying event.

prolactinoma: Noncancerous pituitary tumor that produces prolactin, resulting in too much prolactin in the blood.

psychosocial dwarfism: Growth disorder occurring between the ages of 2 and 15, caused by extreme emotional deprivation or stress; also called *deprivation dwarfism*.

thyroid function tests: Blood and radionuclide tests to determine thyroid gland activity.

thyroidectomy: Surgical removal of all or part of the thyroid gland.

thyrotoxicosis: A condition caused by the oversecretion of thyroid hormones (hyperthyroidism). Signs and symptoms include increases in metabolic rate, blood pressure, and heart rate; excitability and emotional instability; and lowered energy reserves.

virilism: A disorder of females in which there is development of secondary male sexual characteristics such as hirsutism and lowered voice caused by a number of conditions that affect hormone regulation.

19 Blood

Learning Outcomes

These Learning Outcomes correspond by number to this chapter's sections and indicate what you should be able to do after completing the chapter.

- 19-1 ■ **Describe the components and major functions of blood, identify blood collection sites, list the physical characteristics of blood, and specify the composition and functions of plasma.** p. 657
- 19-2 ■ **List the characteristics and functions of red blood cells, describe the structure and functions of hemoglobin, describe how red blood cell components are recycled, explain erythropoiesis, and discuss respiratory gas transport.** p. 659
- 19-3 ■ **Explain the importance of blood typing and the basis for ABO and Rh incompatibilities.** p. 666
- 19-4 ■ **Categorize white blood cell types based on their structures and functions, and discuss the factors that regulate the production of each type.** p. 670
- 19-5 ■ **Describe the structure, function, and production of platelets.** p. 678
- 19-6 ■ **Discuss the mechanisms that control blood loss after an injury, and describe the reaction sequences responsible for blood clotting.** p. 678

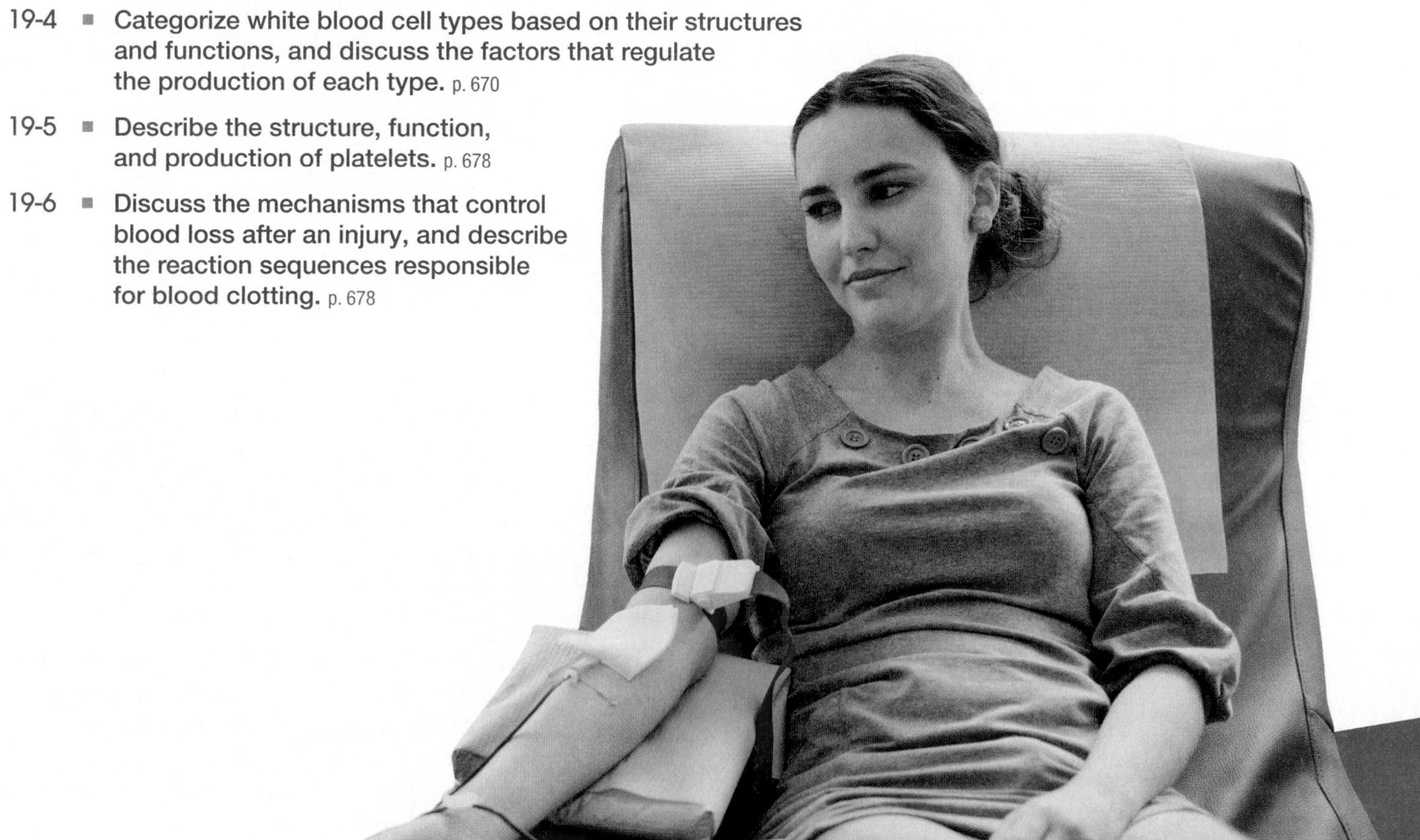

+ CLINICAL CASE Crisis in the Blood

She feels that same pain she had as a child: This time it comes from her legs, sharp and desperate. Clare has let things get out of control and she knows better. It was just hard to manage with the pressures of her classes in the pre-nursing program and her part-time job. She is exhausted, isn't eating well, and has badly neglected her fluid intake. Clare looks down at her legs and sees how red and swollen they look. Her feet feel icy. She winces as the bone-deep pain surges along her shins. Now that she is studying anatomy and physiology, Clare understands better than ever what is happening in her own bloodstream. Clare knows she is in for another trip to the hospital.

TEST EARLY FOR
SICKLE CELL
USA
37
2004

Her roommate gives her a ride to the local emergency room. They don't know her here like they did at home. The nurse practitioner listens carefully and learns about Clare's medical history. On physical examination, Clare is running a temperature, and her heart rate and blood pressure are elevated. Her sunken eyes and loss of skin elasticity reveal significant dehydration. The nurse practitioner sends a sample of blood to the lab to confirm Clare's blood condition. Clare's leg pain has become so severe that she is given a dose of morphine (an opioid painkiller) and is admitted for stabilization. **What crisis is occurring in Clare's blood? To find out, turn to the Clinical Case Wrap-Up on p. 686.**

An Introduction to Blood and the Cardiovascular System

In this chapter we discuss the nature of *blood,* the fluid connective tissue of the **cardiovascular system**. This body system also includes a pump (the heart) that circulates the fluid and a series of conducting hoses (the blood vessels) that carry it throughout the body. In this chapter we take a close look at the structure and functions of blood. We discuss the heart in Chapter 20 and the blood vessels in Chapter 21.

19-1 Blood, composed of plasma and formed elements, provides transport, regulation, and protective services to the body

Learning Outcome **Describe the components and major functions of blood, identify blood collection sites, list the physical characteristics of blood, and specify the composition and functions of plasma.**

Blood is a specialized connective tissue that contains cells suspended in a fluid matrix. In Chapter 4 we introduced the components and properties of this tissue. ⮌ p. 135 In this chapter we cover the components of blood, and the processes of these components that allow this fluid tissue to perform its functions.

Functions of Blood

The blood of the cardiovascular system performs many vital roles. These services are essential to each of the body's roughly 75 trillion cells—so much so that any cell deprived of circulation may die in a matter of minutes. Let's look a bit closer at these functions of blood:

- *Transporting Dissolved Gases, Nutrients, Hormones, and Metabolic Wastes.* Circulating blood contains *red blood cells* that carry oxygen from the lungs to peripheral tissues, and carbon dioxide from those tissues back to the lungs. Blood distributes nutrients absorbed by the digestive tract or released from storage in adipose tissue or in the liver. It carries hormones from endocrine glands toward their target cells, as noted in Chapter 18. It also absorbs the wastes produced by tissue cells and carries them to the kidneys for excretion.
- *Regulating the pH and Ion Composition of Interstitial Fluids.* Diffusion between interstitial fluids and blood eliminates local deficiencies or excesses of ions, such as calcium or potassium. Blood also absorbs and neutralizes acids, such as lactic acid, generated by active skeletal muscles.
- *Restricting Fluid Losses at Injury Sites.* Blood contains enzymes and other substances that respond to breaks in vessel walls by initiating the *clotting* process. A blood clot acts as a temporary patch that prevents further blood loss.
- *Defending against Toxins and Pathogens.* Blood transports *white blood cells,* specialized cells that migrate into other tissues to fight infections or remove debris. Blood also delivers *antibodies,* proteins that specifically attack invading pathogens.
- *Stabilizing Body Temperature.* Blood absorbs the heat generated by active skeletal muscles and redistributes it to other tissues. If body temperature is already high, that heat will be lost across the skin surface. If body temperature is too low, the warm blood is directed to the brain and to other temperature-sensitive organs.

Characteristics of Blood

Whole blood from any source—veins, capillaries, or arteries—has the same basic physical characteristics:

- Blood temperature is about 38°C (100.4°F), slightly above normal body temperature.
- Blood is five times as viscous as water—that is, five times as sticky, five times as cohesive, and five times as resistant to flow as water. The high viscosity results from interactions among dissolved proteins, formed elements, and water molecules in plasma.
- Blood is slightly alkaline, with a pH between 7.35 and 7.45.

Adult males typically have more blood than do adult females. We can estimate blood volume in liters for a person of either sex by calculating 7 percent of the body weight in kilograms. For example, a 75-kg (165-lb) person would have a blood volume of approximately 5.25 liters (5.4 quarts).

Components of Blood

Spotlight Figure 19–1 (p. 660) describes the composition of **whole blood**, which is made up of fluid *plasma* and *formed elements* (cells and cell fragments). The components of whole blood can be **fractionated**, or separated, for analytical or clinical purposes.

Plasma

19

As shown in the top half of Spotlight Figure 19–1, **plasma** makes up about 55 percent of the volume of whole blood. The components of plasma include plasma proteins, other solutes (primarily nutrients, electrolytes, and wastes), and water.

Plasma Proteins. Plasma contains significant quantities of dissolved proteins, namely *albumins*, *globulins*, and *fibrinogen*. These three types make up more than 99 percent of the plasma proteins. (The remaining 1 percent of plasma proteins is composed of enzymes and hormones whose levels vary widely.)

- **Albumins** (al-BYŪ-minz) make up the majority of the plasma proteins. Thus, they are the major contributors to plasma osmolarity and osmotic pressure. Albumins are also important for transporting fatty acids, thyroid hormones, some steroid hormones, and other substances.
- **Globulins** are the second most-abundant proteins in plasma. Examples of plasma globulins include antibodies (*immunoglobulins*) that aid in body defense and *transport globulins*. We cover antibodies in depth with immunity in Chapter 22. Transport globulins bind small ions, hormones, and substances that might otherwise be removed by the kidneys or that have very low water solubility. Transport globulins include *hormone-binding proteins*, which provide a reserve of hormones in the bloodstream. Examples include thyroid-binding globulin and transthyretin, which transport thyroid hormones, and transcortin, which transports ACTH. There are also *metalloproteins* that transport metal ions. Later we will see an example of this in transferrin, which transports iron (Fe^{2+}). In addition, *apolipoproteins* (ap-ō-lip-ō-PRŌ-tēnz) carry triglycerides and other lipids in blood. When bound to lipids, an apolipoprotein becomes a *lipoprotein* (LĪ-pō-prō-tēn). Finally, *steroid-binding proteins* transport steroid hormones in blood. For example, testosterone-binding globulin (TeBG) binds and transports testosterone.
- The third major type of plasma protein, **fibrinogen**, functions in clotting. If steps are not taken to prevent clotting in a blood sample, fibrinogen (a soluble protein) will be converted to *fibrin* (an insoluble protein). This conversion removes the clotting proteins, leaving a fluid known as **serum**. The clotting process also removes calcium ions and other materials from solution, so plasma and serum differ in several significant ways. (See Appendix.) Therefore, the results of a blood test generally indicate whether a sample was plasma or serum.

+ Clinical Note Plasma Expanders

Plasma expanders can be used to increase blood volume temporarily, over a period of hours. They are often used to buy time for lab work to determine a person's blood type. (Transfusion of the wrong blood type can kill the recipient.) Isotonic electrolyte solutions such as normal (physiological) saline can be used as plasma expanders, but their effects are short-lived due to diffusion into interstitial fluid and cells. This fluid loss is slowed by the addition of solutes that cannot freely diffuse across plasma membranes. One example is lactated *Ringer's solution*, an isotonic saline also containing lactate, potassium chloride, and calcium chloride ions. The effects of Ringer's solution fade gradually as the liver, skeletal muscles, and other tissues absorb and metabolize the lactate ions. Another option is isotonic saline solution containing purified human albumin.

However, the plasma expanders in clinical use often contain large carbohydrate molecules, rather than proteins, to maintain proper osmotic concentration. These carbohydrates are not metabolized, but phagocytes gradually remove them from the bloodstream, and blood volume slowly declines. Plasma expanders are easily stored. Their sterile preparation avoids viral or bacterial contamination, which can be a problem with donated plasma. Note that plasma expanders provide a temporary solution to low blood volume, but they do not increase the amount of oxygen carried by the blood. Red blood cells are needed for that function.

Origins of the Plasma Proteins. The liver synthesizes and releases more than 90 percent of the plasma proteins, including all albumins and fibrinogen, most globulins, and various proenzymes. Hence, liver disorders can alter the composition and functional properties of blood. For example, some forms of liver disease can lead to uncontrolled bleeding due to the inadequate synthesis of fibrinogen and other proteins involved in clotting.

Formed Elements

The **formed elements** of blood are made up primarily of red blood cells, white blood cells, and cell fragments known as *platelets*. Formed elements make up about 45 percent of whole blood volume. The percentage of a blood sample that consists of formed elements (most of which are red blood cells) is known as the *hematocrit* (hē-MAH-tō-krit). Formed elements are produced in the process of **hemopoiesis** (hē-mō-POY-eh-sis). Please study the bottom half of **Spotlight Figure 19–1** now for an introduction to these elements and hemopoiesis. Most of the rest of this chapter discusses the functions of the formed elements, starting with red blood cells.

Checkpoint

1. **List five major functions of blood.**
2. **Identify the three types of formed elements in blood.**
3. **List the three major types of plasma proteins.**
4. **What would be the effects of a decrease in the amount of plasma proteins?**

See the blue Answers tab at the back of the book.

19-2 Red blood cells, formed by erythropoiesis, contain hemoglobin that transports respiratory gases

Learning Outcome **List the characteristics and functions of red blood cells, describe the structure and functions of hemoglobin, describe how red blood cell components are recycled, explain erythropoiesis, and discuss respiratory gas transport.**

The most abundant blood cells are the **red blood cells (RBCs)**, also called **erythrocytes** (eh-RITH-rō-sītz), which account for 99.9 percent of the formed elements. These cells give whole blood its deep red color because they contain the red pigment *hemoglobin* (HĒ-mō-glō-bin), which binds and transports the respiratory gases oxygen and carbon dioxide.

Abundance of RBCs: The Hematocrit

RBCs make up roughly one-third of all cells in the human body. A standard blood test reports the number of RBCs per microliter (μL) of whole blood as the *red blood cell count*. In adult males, 1 microliter, or 1 *cubic millimeter* (mm^3), of whole blood contains 4.5–6.3 million RBCs. In adult females, 1 microliter contains 4.2–5.5 million. A single drop of whole blood contains approximately 260 million RBCs. The blood of an average adult has 25 trillion RBCs.

The **hematocrit** is determined by spinning a blood sample in a centrifuge so that all the formed elements come out of suspension. Whole blood contains about 1000 red blood cells for each white blood cell. After centrifugation, the white blood cells and platelets form a very thin *buffy coat* above a thick layer of RBCs.

Many conditions can affect the hematocrit. For example, the hematocrit increases during dehydration, due to a decrease in plasma volume, or after stimulation with the hormone *erythropoietin* (eh-rith-rō-POY-eh-tin). p. 644 The hematocrit decreases as a result of internal bleeding or problems with RBC formation. So, the hematocrit alone does not provide specific diagnostic information. Still, an abnormal hematocrit is an indication that other, more specific tests are needed. (Examples of such tests are described in **Table 19–1** on p. 665.)

Relationship of RBC Structure to RBC Function

Red blood cells are among the most specialized cells of the body. A red blood cell is very different from the "typical cell" discussed in Chapter 3. Each RBC is a biconcave disc with a thin

> **+ Clinical Note** Collecting Blood for Analysis
>
> Fresh whole blood is generally collected from a superficial vein, such as the *median cubital vein* on the anterior surface of the elbow. The procedure is called **venipuncture** (VĒN-i-punk-chur; *vena*, vein + *punctura*, a piercing). It is a common sampling technique because (1) superficial veins are easy to locate, (2) the walls of veins are thinner than those of comparably sized arteries, and (3) blood pressure in the venous system is relatively low, so the puncture wound seals quickly. The most common clinical procedures examine venous blood.
>
> A small drop of blood can be used to prepare a *blood smear*, a thin film of blood on a microscope slide. The blood smear is then stained with special dyes to show each type of formed element. Blood from peripheral capillaries can be obtained by puncturing the tip of a finger, an earlobe, or (in infants) the great toe or heel. Small amounts of capillary blood can also be used to test glucose, cholesterol, and hemoglobin levels. This method is valuable when veins are difficult to access.
>
> An **arterial puncture**, or "arterial stick," can be used for blood gas analysis, an indicator of lung function. Samples are generally drawn from the *radial artery* at the wrist or the *brachial artery* at the elbow.

> Go to MasteringA&P™ > Study Area > Menu > Lab Tools > PAL 3.0 > Histology > Cardiovascular System

19

SPOTLIGHT

Figure 19–1
The Composition of Whole Blood

A Fluid Connective Tissue

Blood is a fluid connective tissue with a unique composition. It consists of a matrix called **plasma** (PLAZ-muh) and formed elements (cells and cell fragments). The term **whole blood** refers to the combination of plasma and the formed elements. The cardiovascular system of an adult male contains 5–6 liters (5.3–6.4 quarts) of whole blood. An adult female contains 4–5 liters (4.2–5.3 quarts). The sex differences in blood volume primarily reflect differences in average body size.

The **hematocrit** (he-MAT-ō-krit) is the percentage of formed elements in a sample of blood. The normal hematocrit, or **packed cell volume (PCV)**, in adult males is 46 and in adult females is 42. The sex difference in hematocrit primarily reflects the fact that androgens (male hormones) stimulate red blood cell production, whereas estrogens (female hormones) do not.

PLASMA

Plasma, the fluid matrix of blood, makes up 46–63% of the volume of whole blood. In many respects, the composition of plasma resembles that of interstitial fluid. This similarity exists because water, ions, and small solutes are continuously exchanged between plasma and interstitial fluids across the walls of capillaries. The primary differences between plasma and interstitial fluid involve (1) the levels of respiratory gases (oxygen and carbon dioxide, due to the respiratory activities of tissue cells), and (2) the concentrations and types of dissolved proteins (because plasma proteins cannot cross capillary walls).

Z Serum Clot Activator

consists of

Plasma
55%
(Range 46-63%)

Plasma Proteins 7%
Other Solutes 1%
Water 92%

+

Formed Elements
45%
(Range 37–54%)

Platelets < .1%
White Blood Cells < .1%
Red Blood Cells 99.9%

Formed elements are blood cells and cell fragments that are suspended in plasma. These elements account for 37–54% of the volume of whole blood. Three types of formed elements exist: platelets, white blood cells, and red blood cells. Formed elements are produced through the process of **hemopoiesis** (hē-mō-poy-Ē-sis), also called **hematopoiesis**. Two populations of stem cells—myeloid stem cells and lymphoid stem cells—are responsible for the production of formed elements.

FORMED ELEMENTS

Plasma Proteins

Plasma proteins are in solution rather than forming insoluble fibers like those in other connective tissues, such as loose connective tissue or cartilage. On average, each 100 mL of plasma contains 7.6 g of protein, almost five times the concentration in interstitial fluid. The large size and globular shapes of most blood proteins prevent them from crossing capillary walls, so they remain trapped within the bloodstream. The liver synthesizes and releases more than 90% of the plasma proteins, including all albumins and fibrinogen, most globulins, and various proenzymes.

Albumins (al-BŪ-minz) constitute about 60% of the plasma proteins. As the most abundant plasma proteins, they are major contributors to the osmotic pressure of plasma.

Globulins (GLOB-ū-linz) account for approximately 35% of the proteins in plasma. Important plasma globulins include antibodies and transport globulins. **Antibodies**, also called **immunoglobulins** (i-mū-nō-GLOB-ū-linz), attack foreign proteins and pathogens. **Transport globulins** bind small ions, hormones, and other compounds.

Fibrinogen (fī-BRIN-ō-jen) functions in clotting, and normally accounts for roughly 4% of plasma proteins. Under certain conditions, fibrinogen molecules interact, forming large, insoluble strands of **fibrin** (FĪ-brin) that form the basic framework for a blood clot.

Plasma also contains enzymes and hormones whose concentrations vary widely.

Other Solutes

Other solutes are generally present in concentrations similar to those in the interstitial fluids. However, because blood is a transport medium there may be differences in nutrient and waste product concentrations between arterial blood and venous blood.

Organic Nutrients: Organic nutrients are used for ATP production, growth, and cell maintenance. This category includes lipids (fatty acids, cholesterol, glycerides), carbohydrates (primarily glucose), and amino acids.

Electrolytes: Normal extracellular ion composition is essential for vital cellular activities. The major plasma electrolytes are Na^+, K^+, Ca^{2+}, Mg^{2+}, Cl^-, HCO_3^-, HPO_4^-, and SO_4^{2-}.

Organic Wastes: Wastes are carried to sites for breakdown or excretion. Examples of organic wastes include urea, uric acid, creatinine, bilirubin, and ammonium ions.

Platelets

Platelets (PLĀT-lets) are small, membrane-bound cell fragments that contain enzymes and other substances important for clotting.

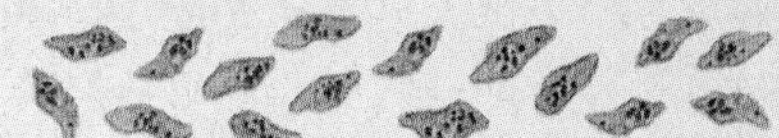

White Blood Cells

White blood cells (**WBCs**), or **leukocytes** (LŪ-kō-sīts; *leukos*, white + *-cyte*, cell), play a role in the body's defense mechanisms. There are five classes of leukocytes, each with slightly different functions that will be explored later in the chapter.

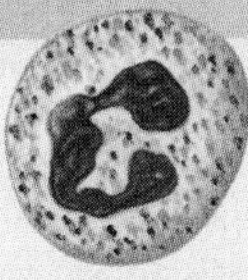

Neutrophils

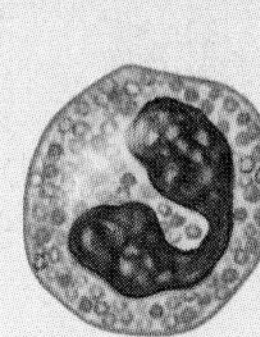

Eosinophils

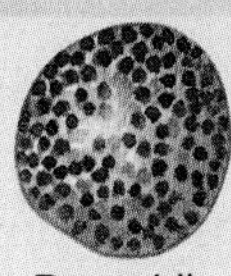

Basophils

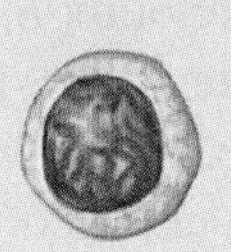

Lymphocytes

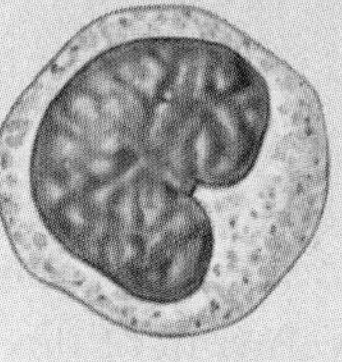

Monocytes

Red Blood Cells

Red blood cells (**RBCs**), or **erythrocytes** (e-RITH-rō-sīts; *erythros*, red + *-cyte*, cell), are the most abundant blood cells. These specialized cells are essential for the transport of oxygen in the blood.

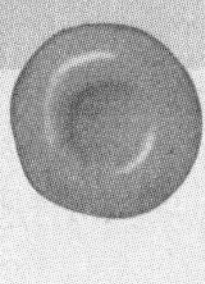
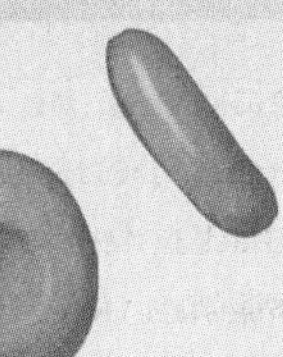

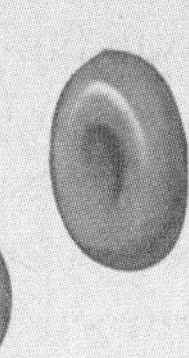

Figure 19–2 The Structure of Red Blood Cells.

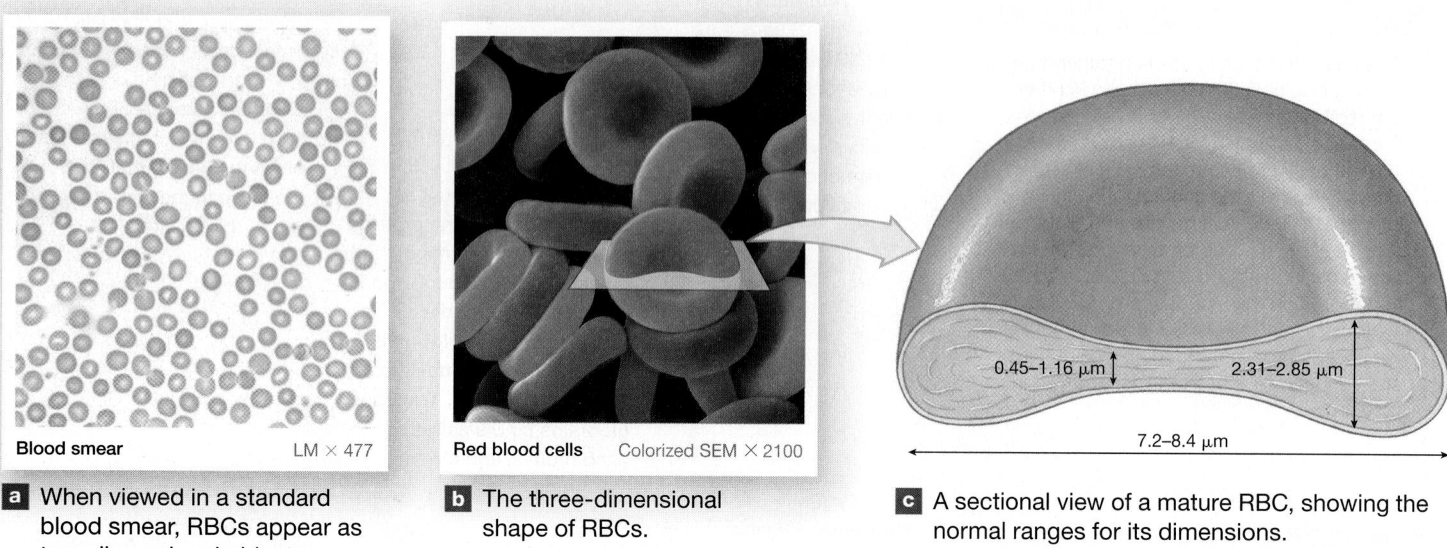
Blood smear
LM × 477
Red blood cells
Colorized SEM × 2100
0.45–1.16 μm
2.31–2.85 μm
7.2–8.4 μm

a When viewed in a standard blood smear, RBCs appear as two-dimensional objects, because they are flattened against the surface of the slide.

b The three-dimensional shape of RBCs.

c A sectional view of a mature RBC, showing the normal ranges for its dimensions.

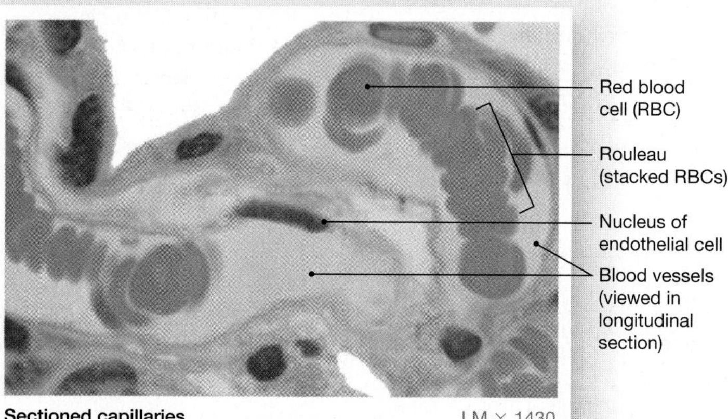
Red blood cell (RBC)
Rouleau (stacked RBCs)
Nucleus of endothelial cell
Blood vessels (viewed in longitudinal section)
Sectioned capillaries
LM × 1430

d When traveling through relatively narrow capillaries, RBCs may stack like dinner plates.

central region and a thicker outer margin (Figure 19–2a–c). An average RBC has a diameter of 7.8 μm and a maximum thickness of 2.85 μm, although the center thins to about 0.8 μm. In addition, the cytoplasmic surface of an RBC plasma membrane is a meshwork of flexible proteins.

The biconcave shape and flexible plasma membrane have three important effects on RBC function:

- *Gives Each RBC a Large Surface-Area-to-Volume Ratio.* Each RBC carries oxygen bound to intracellular proteins. That oxygen must be absorbed or released quickly as the RBC passes through the capillaries of the lungs or peripheral tissues. The greater the surface area per unit volume, the faster the exchange between the RBC's interior and the surrounding plasma. The total surface area of all the RBCs in the blood of a typical adult is about 3800 square meters (nearly 4600 square yards), some 2000 times the total surface area of the body.
- *Enables RBCs to Form Stacks That Smooth Blood Flow through Narrow Blood Vessels.* These stacks of RBCs are like stacks of dinner plates (Figure 19–2d). Known as *rouleaux* (rū-LŌ; singular, *rouleau,*), the stacks form and dissociate repeatedly without affecting the cells involved. An entire stack can pass along a blood vessel that is only slightly larger than the diameter of a single RBC. In contrast, individual cells would bump the walls, bang together, and form logjams that could restrict or prevent blood flow.
- *Enables RBCs to Bend and Flex When Entering Small Capillaries.* Red blood cells are very flexible. By changing shape, individual RBCs can squeeze through capillaries as narrow as 4 μm.

When a developing red blood cell differentiates, it loses any organelle not directly associated with its primary function: the transport of respiratory gases. Mature RBCs are **anucleate**, without nuclei; they retain only the cytoskeleton. Without nuclei and ribosomes, circulating RBCs cannot divide or synthesize structural proteins or enzymes. As a result, the RBCs cannot repair themselves, so their life span is relatively short—normally less than 120 days. With few organelles and no ability to synthesize proteins, their energy demands are low. Without mitochondria, they obtain the energy they need through the anaerobic metabolism of glucose that is absorbed from the surrounding plasma. The lack of mitochondria ensures that absorbed oxygen will be carried to peripheral tissues, not "stolen" by mitochondria in the RBC.

Hemoglobin

Hemoglobin (Hb or **Hgb)** is responsible for the red blood cell's ability to transport oxygen and carbon dioxide. Molecules of hemoglobin make up more than 95 percent of its intracellular proteins. The hemoglobin content of whole blood is reported in grams of Hb per deciliter (100 mL) of whole blood (g/dL). Normal ranges are 14–18 g/dL in males and 12–16 g/dL in females.

Hemoglobin Structure

Hemoglobin molecules have a complex quaternary structure. p. 55 Each Hb molecule has two *alpha* (α) *chains* and two *beta* (β) *chains* of polypeptides (Figure 19–3). Each chain is a globular protein subunit that resembles the myoglobin in skeletal and cardiac muscle cells. Like myoglobin, each Hb chain contains a single molecule of **heme**, a nonprotein pigment complex that forms a ring. Each heme unit holds an iron ion in such a way that the iron can interact with an oxygen molecule, forming **oxyhemoglobin**, **HbO_2**. Blood that contains RBCs filled with oxyhemoglobin is bright red.

The binding of an oxygen molecule to the iron in a heme unit is reversible. The iron–oxygen interaction is very weak. The iron and oxygen can easily dissociate without damaging the heme unit or the oxygen molecule. A hemoglobin molecule whose iron is not bound to oxygen is called **deoxyhemoglobin**. Blood containing RBCs filled with deoxyhemoglobin is dark red.

The RBCs of an embryo or a fetus contain a different form of hemoglobin, known as *fetal hemoglobin*. It binds oxygen more readily than does adult hemoglobin. For this reason, a developing fetus can "steal" oxygen from the maternal bloodstream at the placenta. The conversion from fetal hemoglobin to the adult form begins shortly before birth and continues over the next year. The production of fetal hemoglobin can be stimulated in adults by the administration of drugs such as *hydroxyurea* or *butyrate*. This is one method of treatment for *sickle cell disease* (SCD), a group of conditions that result from the production of abnormal forms of adult hemoglobin.

Hemoglobin Function: Binding and Releasing Oxygen and Carbon Dioxide

Each RBC contains about 280 million Hb molecules. Because 1 Hb molecule contains 4 heme units, each RBC can potentially carry more than a billion molecules of oxygen at a time. Roughly 98.5 percent of the oxygen carried by the blood is bound to Hb molecules inside RBCs.

The amount of oxygen bound to hemoglobin depends mostly on the oxygen content of the plasma. When the plasma oxygen level is low, hemoglobin releases oxygen. Under these conditions, typical of peripheral capillaries, the plasma carbon dioxide level is elevated. The alpha and beta chains of hemoglobin then bind carbon dioxide, forming **carbaminohemoglobin**. In the lung capillaries, the plasma oxygen level is high and the carbon dioxide level is low. Upon reaching these capillaries, RBCs absorb oxygen (which is then bound to hemoglobin) and release carbon dioxide. We revisit these processes in Chapter 23.

Figure 19–3 **The Structure of Hemoglobin.** Hemoglobin consists of four globular protein subunits. Each subunit contains a single molecule of heme—a nonprotein ring surrounding a single ion of iron.

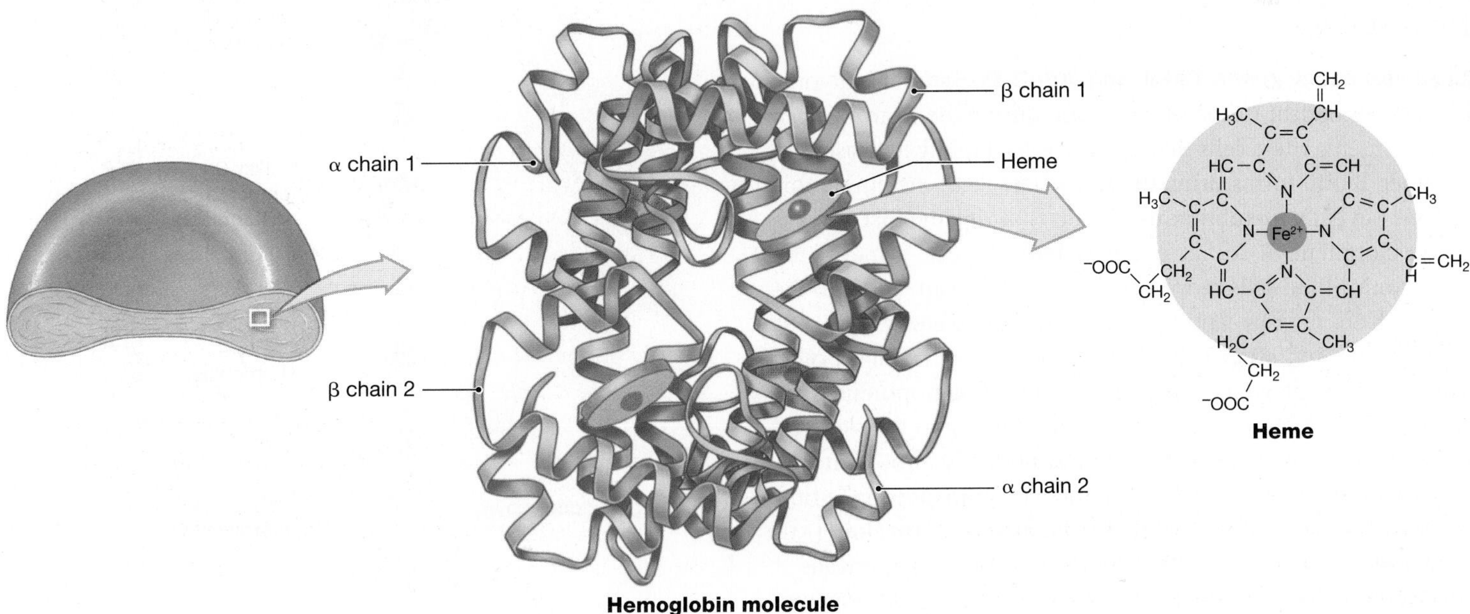
β chain 1
α chain 1
Heme
β chain 2
α chain 2
Hemoglobin molecule
CH2
CH
H3C
C=C
HC–C
C=CH
N
H3C
C
CH3
N–Fe2+–N
C=CH2
H
-OOC
CH2
CH2
HC–C
C=CH
C=C
H2C
CH3
CH2
-OOC
Heme

? Each molecule of heme contains a single ion of what element?

Normal activity levels can be sustained only when the tissue oxygen level is kept within normal limits. If the hematocrit is low or the Hb content of the RBCs is reduced, a condition known as **anemia** results. Anemia interferes with oxygen delivery to peripheral tissues. Every system is affected as organ function deteriorates due to oxygen starvation. Anemic individuals become weak and lethargic. They may also become confused, because the brain is affected as well.

RBC Formation and Turnover

An RBC is exposed to severe physical stresses. A single round-trip from the heart, through the peripheral tissues, and back to the heart usually takes less than a minute. In that time, the RBC gets pumped out of the heart and forced along vessels, where it bounces off the walls and collides with other RBCs. It forms stacks, contorts, and squeezes through tiny capillaries. Then it is rushed back to the heart to make another round trip.

With all this wear and tear and no repair mechanisms, a typical RBC has a short life span. After it travels about 700 miles in 120 days, either its plasma membrane ruptures or some other damage is detected by phagocytes of the spleen, liver, and red bone marrow, which engulf it. The continuous elimination of RBCs usually goes unnoticed, because new ones enter the bloodstream at a comparable rate. About 1 percent of the circulating RBCs are replaced each day, and in the process approximately 3 million new RBCs enter the bloodstream *each second*!

RBC Production: Erythropoiesis

19

Erythropoiesis is the formation of RBCs and it occurs throughout life. Many factors stimulate this process, such as a decreased oxygen level or increased erythropoietin. Let's look at erythropoiesis in detail.

Erythropoiesis in the Fetal and Adult Periods. Embryonic blood cells appear in the bloodstream during the third week of development. These cells divide repeatedly, rapidly increasing in number. Blood forms primarily in the vessels of the embryonic *yolk sac* (an early membranous structure) during the first 8 weeks of development. As other organ systems appear, some of the embryonic blood cells move out of the bloodstream and into the liver, spleen, thymus, and bone marrow. These embryonic cells differentiate into **stem cells** that divide to produce blood cells. The liver and spleen are the primary sites of hemopoiesis from the second to fifth months of development, but as the skeleton enlarges, the bone marrow becomes increasingly important.

In adults, red blood cell formation, or **erythropoiesis** (e-rith-rō-poy-Ē-sis), takes place only in *red bone marrow*, or **myeloid** (MĪ-e-loyd; *myelos*, marrow) **tissue**. This tissue is located in portions of the vertebrae, sternum, ribs, skull, scapulae, pelvis, and proximal limb bones. Other marrow areas contain a fatty tissue known as *yellow bone marrow*. p. 188 Under extreme stimulation, such as severe and sustained blood loss, areas of yellow marrow can convert to red marrow, increasing the rate of RBC formation.

Stages of RBC Maturation. For RBCs to be produced, first cells called **hemocytoblasts** (*hemo-*, blood + *cyte*, cell + *blastos*, precursor), or **hematopoietic stem cells (HSCs)**, in the red bone marrow must divide. These stem cells produce two types of cells: (1) **myeloid stem cells**, which in turn divide to produce red blood cells and several classes of white blood cells, and (2) **lymphoid stem cells**, which divide to produce white blood cells called lymphocytes.

In the red bone marrow, a red blood cell matures in a series of stages. Blood specialists, known as **hematologists** (hē-mah-TOL-ah-jists), have named six key developmental stages that occur over about a week (Figure 19–4). On day 1, cells destined to become RBCs first differentiate into **proerythroblasts**, and then on days 2 and 3 proceed through various **erythroblast**

Figure 19–4 Stages of RBC Maturation: Erythropoiesis. Red blood cells are produced in the red bone marrow. The color density in the cytoplasm indicates the abundance of hemoglobin. Note the reductions in the sizes of the cell and nucleus leading up to the formation of a reticulocyte.

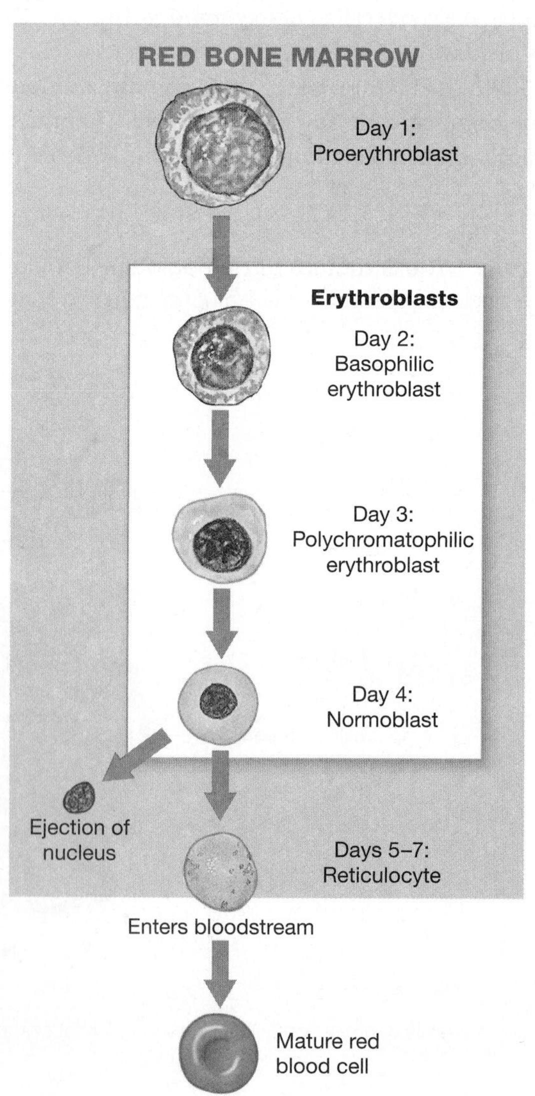
RED BONE MARROW
Day 1:
Proerythroblast
Erythroblasts
Day 2:
Basophilic
erythroblast
Day 3:
Polychromatophilic
erythroblast
Day 4:
Normoblast
Ejection of
nucleus
Days 5–7:
Reticulocyte
Enters bloodstream
Mature red
blood cell

stages. Erythroblasts, which actively synthesize hemoglobin, are named based on total size, amount of hemoglobin, and size and appearance of the nucleus.

After day 4, the erythroblast, now called a *normoblast*, sheds its nucleus and becomes a **reticulocyte** (re-TIK-yū-lō-sīt), which contains 80 percent of the Hb of a mature RBC. Hemoglobin synthesis then continues for days 5–7. During this period, while the cell is synthesizing Hb and other proteins, its cytoplasm still contains RNA, which can be seen under the microscope with certain stains. After 2 days in the red bone marrow, the reticulocyte enters the bloodstream. At this time, reticulocytes normally account for about 0.8 percent of the RBC population in the blood and can still be detected by staining. After a final day in circulation, the reticulocytes complete their maturation and become indistinguishable from other mature RBCs.

Regulation of Erythropoiesis

Erythropoiesis is stimulated directly by the hormone **erythropoietin (EPO)** (⮌ p. 644) and indirectly by several hormones, including thyroxine, androgens, and growth hormone. As noted earlier in **Spotlight Figure 19–1** on pp. 660–661, estrogens do not stimulate erythropoiesis. This fact accounts for the differences in hematocrit values between males and females.

Erythropoietin is a glycoprotein formed by the kidneys and liver. EPO appears in the plasma when peripheral tissues, especially the kidneys, are exposed to a low concentration of oxygen. A low oxygen level in tissues is called **hypoxia** (hī-POKS-ē-uh; *hypo-*, below + *oxy-*, presence of oxygen). Erythropoietin is released (1) during anemia; (2) when blood flow to the kidneys declines; (3) when the oxygen content of air in the lungs declines, due to disease or high altitude; and (4) when the respiratory surfaces of the lungs are damaged.

Once in the bloodstream, EPO travels to the red bone marrow, where it stimulates stem cells and developing RBCs. Erythropoietin has two major effects: (1) It stimulates cell division rates in erythroblasts and in the stem cells that produce erythroblasts, and (2) it speeds up the maturation of RBCs, mainly by accelerating Hb synthesis. Under maximum EPO stimulation, red bone marrow can increase RBC formation 10-fold, to about 30 million cells per second.

The ability to increase the rate of blood formation quickly and dramatically is important to a person recovering from a severe blood loss. But if EPO is administered to a healthy person, the hematocrit may rise to 65 or more. Such an increase can place an intolerable strain on the heart. Similar problems can occur after **blood doping**, a practice in which athletes elevate their hematocrits by re-infusing packed RBCs that were removed and stored at an earlier date. The goal is to improve oxygen delivery to muscles, thereby enhancing performance. The strategy can be dangerous, however, because it makes blood more viscous and increases the workload on the heart.

One more factor may limit red blood cell formation. For erythropoiesis to proceed normally, the red bone marrow needs adequate supplies of amino acids, iron, and vitamins (including B_{12}, B_6, and folic acid) for protein synthesis. We obtain **vitamin B_{12}** from dairy products and meat. To absorb vitamin B_{12}, we need *intrinsic factor*, which is produced in the stomach. Without vitamin B_{12} from the diet, normal stem cell divisions cannot occur and *pernicious anemia* results. Thus, pernicious anemia may be caused by a vitamin B_{12} deficiency, a problem with the production of intrinsic factor, or a problem with the absorption of vitamin B_{12} bound to intrinsic factor.

Blood tests provide information about the general health of an individual, usually with a minimum of trouble and expense. Several common blood tests, called *RBC tests*, assess the number, size, shape, and maturity of circulating RBCs, indicating the erythropoietic activities under way. The tests can also be useful in detecting problems, such as internal bleeding, that may not produce other obvious signs or symptoms. **Table 19–1** lists examples of important red blood cell tests and related terms. (These tests are performed as part of a *complete blood count*, or *CBC*).

Table 19–1 RBC Tests and Related Terminology

		Terms Associated with Abnormal Values	
Test	Determines	Elevated	Depressed
Hematocrit (Hct)	Percentage of formed elements in whole blood Normal = 37–54%	Polycythemia (may reflect erythrocytosis or leukocytosis)	Anemia
Reticulocyte count (Retic.)	Percentage of circulating reticulocytes Normal = 0.8%	Reticulocytosis	Diminished erythropoiesis
Hemoglobin concentration (Hb; Hgb)	Concentration of hemoglobin in blood Normal = 12–18 g/dL	Polycythemia	Anemia
RBC count	Number of RBCs per μL of whole blood Normal = 4.2–6.3 million cells/μL	Erythrocytosis/polycythemia	Anemia
Mean corpuscular volume (MCV)	Average volume of one RBC Normal = 82–101 μm^3/cell (normocytic)	Macrocytic	Microcytic
Mean corpuscular hemoglobin (MCH)	Average weight of Hb in one RBC Normal = 27–34 pg/RBC (normochromic)	Hyperchromic	Hypochromic

Hemoglobin Recycling

Macrophages of the spleen, liver, and red bone marrow play a central role in recycling red blood cell components. These phagocytes engulf aged red blood cells and also detect and remove Hb molecules from **hemolyzed**, or ruptured, RBCs (Figure 19–5). Hemoglobin remains intact only inside RBCs. If the Hb released by hemolysis is not phagocytized, its components will not be recycled.

Once a phagocytic cell has engulfed and broken down an RBC, each part of the Hb molecule has a different fate (see Figure 19–5). The alpha and beta chains of Hb are filtered by the kidneys and eliminated in urine. The globular proteins are broken apart into their component amino acids, which are then either metabolized by the cell or released into the bloodstream for use by other cells. Only the iron of each heme unit is recycled. The remaining portion is processed separately.

When abnormally large numbers of RBCs break down in the bloodstream, urine may turn red or brown. This condition is called **hemoglobinuria**. The presence of intact RBCs in urine—a sign called **hematuria** (hē-mah-TYŪ-rē-uh)—occurs only after kidney damage or damage to vessels along the urinary tract.

Fate of Heme. Each heme unit is stripped of its iron and converted to **biliverdin** (bil-ih-VER-din), an organic compound with a green color. (Bad bruises commonly develop a greenish tint when biliverdin forms in the blood-filled tissues.) Biliverdin is then converted to **bilirubin** (bil-ih-RŪ-bin), an orange-yellow pigment, and released into the bloodstream. There, the bilirubin binds to albumin and is transported to the liver for excretion in bile.

19

If the bile ducts are blocked or the liver cannot absorb or excrete bilirubin, the circulating level of the compound climbs rapidly. Bilirubin then diffuses into peripheral tissues, giving them a yellow color that is most apparent in the skin and over the sclera of the eyes. This combination of yellow skin and eyes is **jaundice** (JAWN-dis).

In the large intestine, bacteria convert bilirubin to related pigments called *urobilinogens* (yūr-ō-bī-LIN-ō-jens) and *stercobilinogens* (ster-kō-bī-LIN-ō-jens). Some of the urobilinogens are absorbed into the bloodstream and are then excreted into urine. When exposed to oxygen, some of the urobilinogens and stercobilinogens are converted to **urobilins** (ūr-ō-BĪ-lins) and **stercobilins** (ster-kō-BĪ-lins).

Recycling of Iron. Large quantities of free iron are toxic to cells, so in the body iron is generally bound to transport or storage proteins. Iron ions (Fe^{2+}) extracted from heme molecules may be bound and stored in a phagocytic cell or released into the bloodstream, where they bind to **transferrin** (trans-FER-in), a plasma protein. Red blood cells developing in the red bone marrow absorb the amino acids and transferrins from the bloodstream and use them to synthesize new Hb molecules. Excess transferrins are removed in the liver and spleen, and the iron is stored in two special iron–protein complexes: **ferritin** (FER-ih-tin) and **hemosiderin** (hē-mō-SID-eh-rin).

This recycling system is remarkably efficient. Although roughly 26 mg of iron is incorporated into new Hb molecules each day, a dietary supply of 1–2 mg can keep pace with the incidental losses that occur at the kidneys and digestive tract.

Any impairment in iron uptake or metabolism can cause serious clinical problems, because RBC formation will be affected. For example, *iron-deficiency anemia* results from a lack of iron in the diet or from problems with iron absorption. Too much iron can also cause problems, due to excessive buildup in secondary storage sites, such as the liver and cardiac muscle tissue. Excessive iron deposition in cardiac muscle cells has been linked to heart disease.

✓ Checkpoint

5. **Describe the structure of hemoglobin.**
6. **How would a person's hematocrit change after a significant blood loss?**
7. **What effect would obstruction to the renal arteries have on a person's hematocrit?**
8. **How would liver disease affect the level of bilirubin in the blood?**

See the blue Answers tab at the back of the book.

19-3 The ABO and Rh blood groups are based on antigen–antibody responses

Learning Outcome Explain the importance of blood typing and the basis for ABO and Rh incompatibilities.

Antigens are substances that can trigger a protective defense mechanism called an *immune response.* Most antigens are proteins, but some other types of organic molecules are antigens as well. Your plasma membranes contain **surface antigens**, substances that your immune system recognizes as "normal," or "self." In other words, your immune system ignores these substances rather than attacking them as "foreign." Let's look at how surface antigens on red blood cells determine *blood type.* Blood type then determines how blood reacts during *transfusions.*

ABO and Rh Blood Groups

Your **blood type** is determined by the presence or absence of specific surface antigens in RBC plasma membranes. The surface antigens involved are integral membrane glycoproteins whose characteristics are genetically determined. Red blood cells have at least 50 kinds of surface antigens, but three surface antigens are of particular importance: **A**, **B**, and **Rh** (or **D**).

Go to MasteringA&P™ > Study Area > Menu > Lab Tools > PhysioEx 9.1 > Exercise 11: Blood Analysis > Activity 1: Hematocrit Determination

Figure 19–5 Recycling of Red Blood Cell Components.

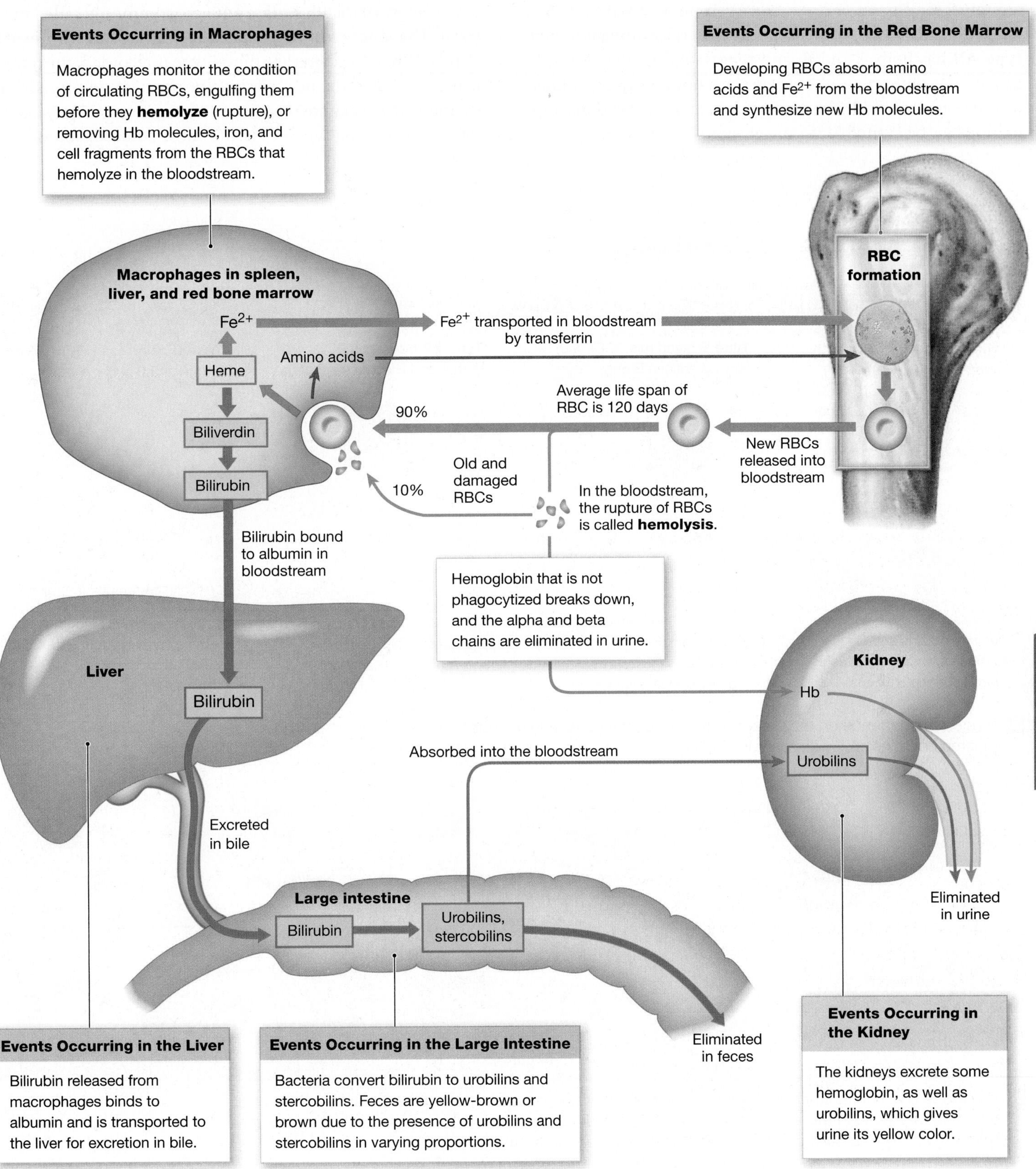
Events Occurring in Macrophages
Macrophages monitor the condition of circulating RBCs, engulfing them before they **hemolyze** (rupture), or removing Hb molecules, iron, and cell fragments from the RBCs that hemolyze in the bloodstream.
Events Occurring in the Red Bone Marrow
Developing RBCs absorb amino acids and Fe^{2+} from the bloodstream and synthesize new Hb molecules.
Macrophages in spleen, liver, and red bone marrow
Fe^{2+}
Fe^{2+} transported in bloodstream by transferrin
Heme
Amino acids
Biliverdin
Bilirubin
RBC formation
Average life span of RBC is 120 days
90%
New RBCs released into bloodstream
Old and damaged RBCs
10%
In the bloodstream, the rupture of RBCs is called **hemolysis**.
Bilirubin bound to albumin in bloodstream
Hemoglobin that is not phagocytized breaks down, and the alpha and beta chains are eliminated in urine.
Liver
Bilirubin
Kidney
Hb
Absorbed into the bloodstream
Urobilins
Excreted in bile
Large intestine
Bilirubin
Urobilins, stercobilins
Eliminated in urine
Eliminated in feces
Events Occurring in the Liver
Bilirubin released from macrophages binds to albumin and is transported to the liver for excretion in bile.
Events Occurring in the Large Intestine
Bacteria convert bilirubin to urobilins and stercobilins. Feces are yellow-brown or brown due to the presence of urobilins and stercobilins in varying proportions.
Events Occurring in the Kidney
The kidneys excrete some hemoglobin, as well as urobilins, which gives urine its yellow color.

? Where are red blood cells produced?

19

The *ABO blood group* is based on the presence or absence of the A and B surface antigens. According to this group, there are four blood types (**Figure 19–6a**): **Type A** blood has RBCs with surface antigen A only, **type B** has surface antigen B only, **type AB** has both A and B, and **type O** has neither A nor B. Individuals with these blood types are not evenly distributed throughout the world. The average percentages for various populations in the United States are given in **Table 19–2**.

The *Rh blood group* is based on the presence or absence of the Rh surface antigen. The term **Rh positive (Rh^+)** indicates the presence of the Rh surface antigen, commonly called the *Rh factor*. The absence of this antigen is indicated as **Rh negative (Rh^-)**. When the complete blood type is recorded, the term *Rh* is usually omitted, and a positive or negative sign is used. For example, the types are reported as O negative (O^-), A positive (A^+), and so on. Type O^+ is the most common blood type.

Figure 19–6 Blood Types and Cross-Reactions.

Type A	Type B	Type AB	Type O
Type A blood has RBCs with surface antigen A only.	**Type B** blood has RBCs with surface antigen B only.	**Type AB** blood has RBCs with both A and B surface antigens.	**Type O** blood has RBCs lacking both A and B surface antigens.
Surface antigen A	Surface antigen B		
If you have Type A blood, your plasma contains anti-B antibodies, which will attack Type B surface antigens.	If you have Type B blood, your plasma contains anti-A antibodies, which will attack Type A surface antigens.	If you have Type AB blood, your plasma has neither anti-A nor anti-B antibodies.	If you have Type O blood, your plasma contains both anti-A and anti-B antibodies.

19

a Blood type depends on the presence of surface antigens (agglutinogens) on RBC surfaces. The plasma contains antibodies (agglutinins) that will react with foreign surface antigens.

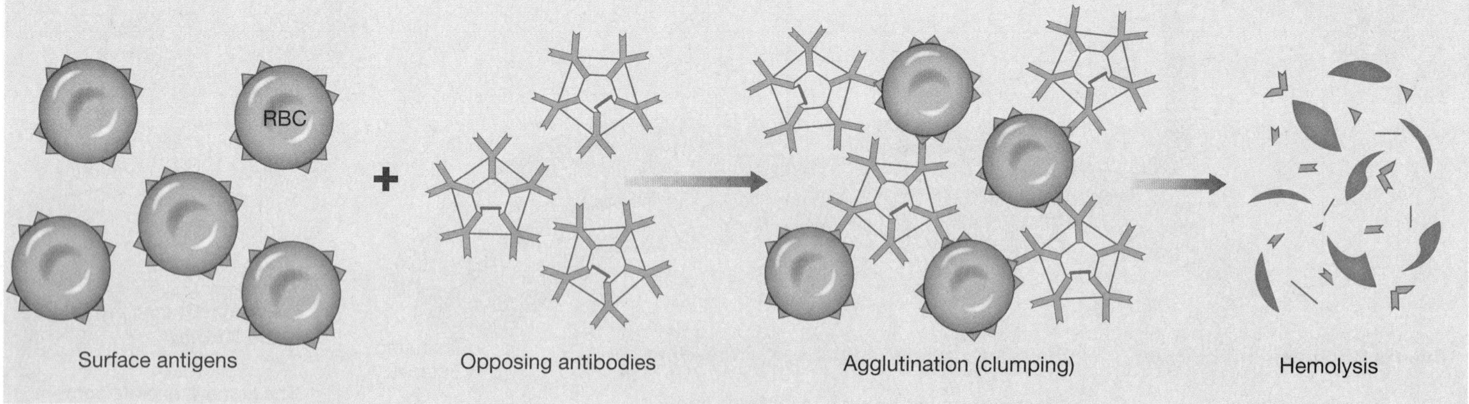
RBC
+
Surface antigens
Opposing antibodies
Agglutination (clumping)
Hemolysis

b In a cross-reaction, antibodies react with their target antigens causing agglutination and hemolysis of the affected RBCs. In this example, anti-B antibodies encounter B surface antigens, which cause the RBCs bearing the B surface antigens to clump together and break up.

? Type A blood has what type of RBC surface antigen and what type of opposing antibodies?

Table 19–2 Differences in Blood Group Distribution

Population	Percentage with Each Blood Type				
	O	A	B	AB	Rh^+
UNITED STATES					
Black American	49	27	20	4	95
Caucasian	45	40	11	4	85
Chinese American	42	27	25	6	100
Filipino American	44	22	29	6	100
Hawaiian	46	46	5	3	100
Hispanic	57	31	10	2	92
Japanese American	31	39	21	10	100
Korean American	32	28	30	10	100
NATIVE NORTH AMERICAN	79	16	4	1	100
NATIVE SOUTH AMERICAN	100	0	0	0	100
AUSTRALIAN ABORIGINE	44	56	0	0	100

As in the distribution of A and B surface antigens, Rh type differs by ethnic group and by region (see Table 19–2).

Your immune system ignores these surface antigens—called **agglutinogens** (ah-glū-TIN-ō-jenz)—on your own RBCs. However, your plasma contains antibodies, sometimes called **agglutinins** (ah-GLŪ-tih-ninz), that will attack the antigens on "foreign" RBCs. When these antibodies attack, the foreign cells *agglutinate*, or clump together; this process is called **agglutination**. If you have type A blood, your plasma contains anti-B antibodies, which will attack type B surface antigens. If you have type B blood, your plasma contains anti-A antibodies. The RBCs of an individual with type O blood have neither A nor B surface antigens, and that person's plasma contains both anti-A and anti-B antibodies. A type AB individual has RBCs with both A and B surface antigens, and the plasma does not contain anti-A or anti-B antibodies. The presence of anti-A and/or anti-B antibodies is genetically determined. These antibodies are present throughout life, regardless of whether the individual has ever been exposed to foreign RBCs.

In contrast, the plasma of an Rh-negative individual does not contain anti-Rh antibodies. These antibodies are present only if the individual has been **sensitized** by previous exposure to Rh^+ RBCs. Such exposure can occur accidentally during a transfusion, but it can also accompany a seemingly normal pregnancy involving an Rh^- mother and an Rh^+ fetus. (We cover this process in more detail later in this section.)

Transfusions

A **transfusion** is the process of transferring whole blood or blood plasma from one person to another. If the donor's cells and the recipient's plasma are not compatible, a potentially life-threatening *cross-reaction* can occur. Such cross-reactions can be prevented by testing to make sure that the blood types of the donor and the recipient are **compatible**—that is, that the donor's blood cells and the recipient's plasma will not cross-react.

Cross-Reactions

During a transfusion, if an antibody meets its specific surface antigen, the RBCs agglutinate and may also hemolyze. This is called a **cross-reaction**, or *transfusion reaction* (Figure 19–6b). For instance, an anti-A antibody that encounters A surface antigens will cause the RBCs bearing the A surface antigens to clump or even break up. Clumps and fragments of RBCs under attack form drifting masses that can plug small blood vessels in the kidneys, lungs, heart, or brain, damaging or destroying affected tissues.

In practice, the surface antigens on the donor's cells are more important in determining compatibility than are the antibodies in the donor's plasma. Why? Unless large volumes of whole blood or plasma are transferred, cross-reactions between the donor's plasma and the recipient's blood cells will not produce significant agglutination. This is because the donated plasma is diluted quickly through mixing with the larger plasma volume of the recipient. (One unit of whole blood, 500 mL, contains roughly 275 mL of plasma, only about 10 percent of normal plasma volume.) When the goal is to increase the blood's oxygen-carrying capacity rather than its plasma volume, packed RBCs, with a minimal amount of plasma, are often transfused. This practice minimizes the risk of a cross-reaction.

Compatibility Testing

To avoid cross-reactions, a compatibility test is usually performed in advance of any transfusion. This process normally involves two steps: (1) a determination of blood type and (2) a cross-match test.

Figure 19–7 Blood Type Testing. Test results for blood samples from four individuals. Drops of blood are mixed with solutions containing antibodies to the surface antigens A, B, AB, and Rh. Anti-A antibodies agglutinate A antigens. Anti-B antigens agglutinate B antigens. The tested person's blood types are shown at right.

Anti-A	Anti-B	Anti-Rh	Blood type
Clumping	No clumping	Clumping	A^+
No clumping	Clumping	Clumping	B^+
Clumping	Clumping	Clumping	AB^+
No clumping	No clumping	No clumping	O^-

The standard test for blood type considers only the three surface antigens most likely to produce dangerous cross-reactions: A, B, and Rh (Figure 19–7). The test involves taking drops of blood and mixing them separately with solutions containing anti-A, anti-B, and anti-Rh (anti-D) antibodies. Any cross-reactions are then recorded. For example, if an individual's RBCs clump together when exposed to anti-A and to anti-B antibodies, the individual has type AB blood. If no reactions occur after exposure, that person must have type O blood. The presence or absence of the Rh surface antigen is also noted, and the individual is classified as Rh positive or Rh negative on that basis.

Standard blood typing of both donor and recipient can be completed in a matter of minutes. However, in an emergency, there may not be time for preliminary testing. For example, a person with a severe gunshot wound may require 5 *liters* or more of blood before the damage can be repaired. Under these circumstances, type O blood (preferably O^-) will be administered. Because the donated RBCs lack both A and B surface antigens, the recipient's blood can have anti-A antibodies, anti-B antibodies, or both and still not cross-react with the donor's blood. Because cross-reactions with type O blood are very unlikely, type O individuals are sometimes called *universal donors*. Type AB individuals were once called *universal recipients*, because they lack anti-A and anti-B antibodies that would attack donated RBCs, and so can safely receive blood of any type. However, now that blood supplies are usually adequate and compatibility testing is regularly done, the term has largely been dropped. If the recipient's blood type is known to be AB, type AB blood will be administered.

Enzymes can now be used to strip off the A or B surface antigens from RBCs and create type O blood in the laboratory. The procedure is expensive and time consuming and has limited use in emergency treatment. Still, cross-reactions can occur, even to type O^- blood, because at least 48 other surface antigens are present. For this reason, whenever time and facilities permit, further testing is performed to ensure complete compatibility between donor blood and recipient blood. **Cross-match testing** involves exposing the donor's RBCs to a sample of the recipient's plasma under controlled conditions. This procedure reveals significant cross-reactions involving surface antigens other than A, B, or Rh. Another way to avoid compatibility problems is to replace lost blood with synthetic blood substitutes, which do not contain surface antigens that can trigger a cross-reaction.

Because blood groups are inherited, blood tests are also used as paternity tests and in crime detection. The blood collected cannot prove that a particular individual *is* a certain child's father or *is* guilty of a specific crime, but it can prove that the individual is *not* involved. For example, an adult with type AB blood cannot be the parent of an infant with type O blood. Testing for additional surface antigens, other than the standard ABO groups, can increase the accuracy of the conclusions. When available, DNA identity testing, which has nearly 100 percent accuracy, has replaced blood type identity testing.

Spotlight Figure 19–8 describes a serious condition known as *hemolytic disease of the newborn* (*HDN*). Hemolytic disease of the newborn does not involve a transfusion; rather it involves Rh incompatibility between a pregnant mother and her developing fetus.

Checkpoint

9. **What is the function of surface antigens on RBCs?**
10. **Which blood type can be safely transfused into a person with type O blood?**
11. **Why can't a person with type A blood safely receive blood from a person with type B blood?**

See the blue Answers tab at the back of the book.

19-4 The various types of white blood cells contribute to the body's defenses

Learning Outcome Categorize white blood cell types based on their structures and functions, and discuss the factors that regulate the production of each type.

Unlike red blood cells, **white blood cells (WBCs)**, or **leukocytes**, have nuclei and other organelles, and they lack hemoglobin.

Go to MasteringA&P™ > Study Area > Menu > Lab Tools > PhysioEx 9.1 > Exercise 11: Blood Analysis > Activity 4: Blood Typing

White blood cells help defend the body against invasion by pathogens. They also remove toxins, wastes, and abnormal or damaged cells.

A typical microliter of blood contains 5000 to 10,000 WBCs, compared with 4.2 to 6.3 million RBCs. Most of the WBCs in the body at any moment are in connective tissue proper or in organs of the lymphatic system. Circulating WBCs are only a small fraction of the total WBC population.

WBC Characteristics and Functions

Unlike RBCs, WBCs circulate for only a short time during their life span. White blood cells migrate through the loose and dense connective tissues of the body. They use the bloodstream to travel from one organ to another and for rapid transportation to areas of infection or injury. As they travel along the miles of capillaries, WBCs can detect the chemical signs of damage to surrounding tissues. When these cells detect problems, they leave the bloodstream and enter the damaged area.

Circulating WBCs have the following functional characteristics:

- *All Can Migrate Out of the Bloodstream.* When WBCs in the bloodstream are activated, they contact and adhere to the vessel walls in a process called *margination*. After further interaction with endothelial cells, the activated WBCs squeeze between adjacent endothelial cells and enter the surrounding tissue. This process is called *emigration*, or *diapedesis* (dī-ah-peh-DĒ-sis; *dia*, through + *pedesis*, a leaping).
- *All Are Capable of Amoeboid Movement. Amoeboid movement* is a gliding motion made possible by the flow of cytoplasm into slender cellular processes extended in the direction of movement. (The movement is so named because it is similar to that of an *amoeba*, a type of protozoan.) The mechanism is not fully understood, but it involves the continuous rearrangement of bonds between actin filaments in the cytoskeleton. It also requires calcium ions and ATP. This mobility allows WBCs to move through the endothelial lining and into peripheral tissues.
- *All Are Attracted to Specific Chemical Stimuli.* This characteristic, called **positive chemotaxis** (kē-mō-TAK-sis), guides WBCs to invading pathogens, damaged tissues, and other active WBCs.

In addition, keep in mind that some types of WBCs are capable of phagocytosis. These WBCs may engulf pathogens, cell debris, or other materials.

Types of WBCs

Several types of WBCs can be distinguished microscopically in a blood smear by using either of two standard stains: *Wright's stain* or *Giemsa* (JEM-suh) *stain*. Traditionally, WBCs have been divided into two groups based on their appearance after staining: (1) *granular leukocytes*, or *granulocytes* (with abundant stained granules)—the *neutrophils, eosinophils*, and *basophils*; and (2) *agranular leukocytes*, or *agranulocytes* (with few, if any, stained granules)—the *monocytes* and *lymphocytes*. This categorization is convenient but somewhat misleading. In reality, the granules in "granular leukocytes" are secretory vesicles and lysosomes, and the "agranular leukocytes" also contain vesicles and lysosomes, but they are just smaller and difficult to see with a light microscope.

Neutrophils, eosinophils, basophils, and monocytes are part of the body's *nonspecific defenses*. A variety of stimuli activate such defenses, which do not discriminate between one type of threat and another. Neutrophils and eosinophils are sometimes called *microphages*, to distinguish them from the larger macrophages in connective tissues. Macrophages are monocytes that have moved out of the bloodstream and have become actively phagocytic. ⮌ p. 129 Lymphocytes, in contrast, are responsible for *specific defenses*, which mount a counterattack against specific types of invading pathogens or foreign proteins. We discuss the interactions among WBCs and the relationships between specific and nonspecific defenses in Chapter 22.

Neutrophils

Fifty to 70 percent of the circulating WBCs are **neutrophils** (NŪ-trō-filz). This name reflects the fact that the granules of these WBCs are chemically neutral and thus are difficult to stain with either acidic or basic (alkaline) dyes. A mature neutrophil has a very dense, segmented nucleus with two to five lobes resembling beads on a string (Figure 19–9a). This structure has given neutrophils another name: **polymorphonuclear** (pol-ē-mor-fō-NŪ-klē-ar) **leukocytes** (*poly*, many + *morphe*, form), or *PMNs*. "Polymorphs," or "polys," as they are often called, are roughly 12 μm in diameter. Their cytoplasm is packed with pale granules containing lysosomal enzymes and bactericidal (bacteria-killing) compounds.

Neutrophils are highly mobile. They are generally the first of the WBCs to arrive at the site of an injury. These very active phagocytic cells specialize in attacking and digesting bacteria that have been "marked" with antibodies or with *complement proteins*—plasma proteins involved in tissue defenses. (We discuss the complement system in Chapter 22.)

When a neutrophil encounters a bacterium, it quickly engulfs the invader, and the metabolic rate of the neutrophil increases dramatically. Along with this *respiratory burst*, the neutrophil produces highly reactive, destructive chemical agents, including hydrogen peroxide (H_2O_2) and superoxide anions (O_2^-), which can kill bacteria.

Meanwhile, the vesicle containing the engulfed pathogen fuses with lysosomes that contain digestive enzymes and small peptides called *defensins*. This process is called **degranulation** because it reduces the number of granules in the cytoplasm.

SPOTLIGHT

Figure 19–8

Hemolytic Disease of the Newborn

Hemolytic disease of the newborn is an RBC-related disorder caused by a cross-reaction between fetal and maternal blood types. Genes controlling the presence or absence of any surface antigen in the plasma membrane of a red blood cell are provided by both parents, so a child can have a blood type different from that of either parent. During pregnancy, when fetal and maternal vascular systems are closely intertwined, the mother's antibodies may cross the placenta, attacking and destroying fetal RBCs. The resulting condition, called **hemolytic disease of the newborn (HDN)**, has many forms, some quite dangerous and others so mild as to remain undetected.

Rh^- mother

Rh^+ fetus

The most common form of hemolytic disease of the newborn develops after an Rh^- woman has carried an Rh^+ fetus.

FIRST PREGNANCY

1 Problems seldom develop during a first pregnancy, because very few fetal cells enter the maternal bloodstream then, and thus the mother's immune system is not stimulated to produce anti-Rh antibodies.

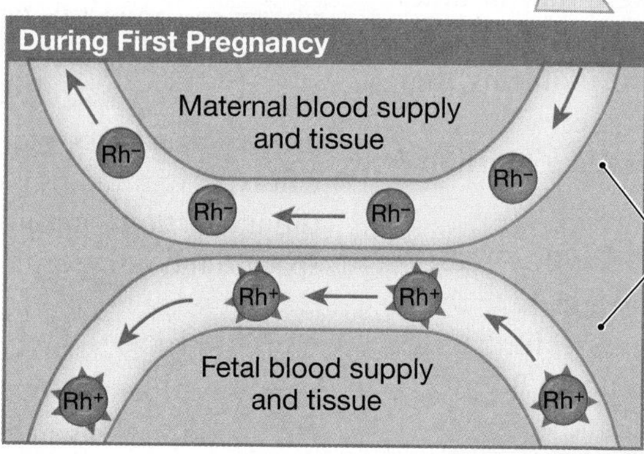
During First Pregnancy
Maternal blood supply and tissue
Rh−
Rh+
Fetal blood supply and tissue
Placenta

2 Exposure to fetal red blood cell antigens generally occurs during delivery, when bleeding takes place at the placenta and uterus. Such mixing of fetal and maternal blood can stimulate the mother's immune system to produce anti-Rh antibodies, leading to sensitization.

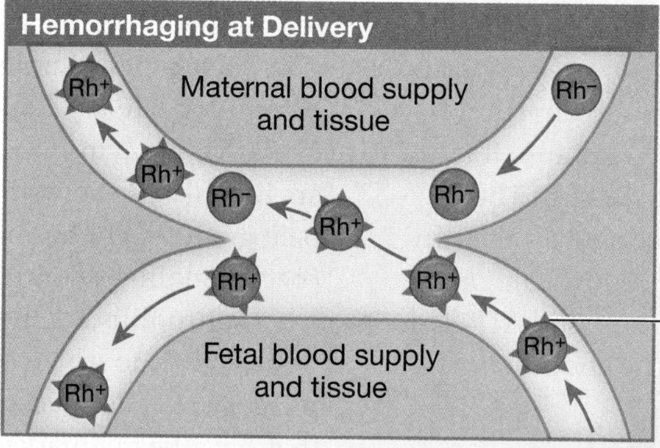
Hemorrhaging at Delivery
Maternal blood supply and tissue
Rh+
Rh−
Fetal blood supply and tissue
Rh antigen on fetal red blood cells

3 About 20% of Rh^- mothers who carried Rh^+ children become sensitized within 6 months of delivery. Because the anti-Rh antibodies are not produced in significant amounts until after delivery, a woman's first infant is not affected.

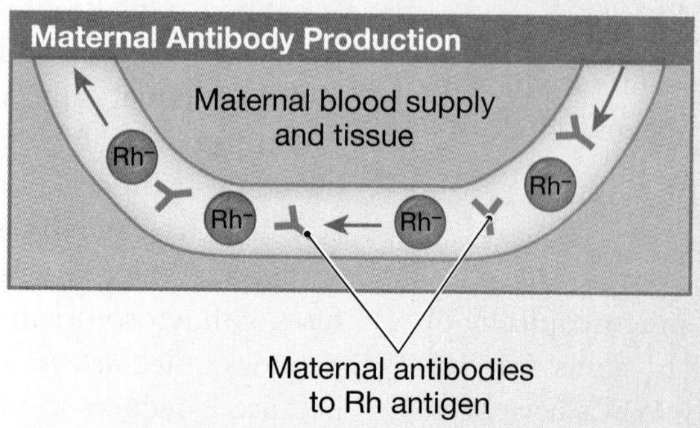
Maternal Antibody Production
Maternal blood supply and tissue
Rh−
Maternal antibodies to Rh antigen

Rh⁻ mother

Rh⁺ fetus

SECOND PREGNANCY

4

If a future pregnancy involves an Rh⁺ fetus, maternal anti-Rh antibodies produced after the first delivery cross the placenta and enter the fetal bloodstream. These antibodies destroy fetal RBCs, producing a dangerous anemia. The fetal demand for blood cells increases, and they leave the bone marrow and enter the bloodstream before completing their development. Because these immature RBCs are erythroblasts, HDN is also known as **erythroblastosis fetalis** (e-rith-rō-blas-TŌ-sis fē-TAL-is). Without treatment, the fetus may die before delivery or shortly thereafter. A newborn with severe HDN is anemic, and the high concentration of circulating bilirubin produces jaundice. Because the maternal antibodies remain active in the newborn for 1 to 2 months after delivery, the infant's entire blood volume may require replacement to remove the maternal anti-Rh antibodies, as well as the damaged RBCs. Fortunately, the mother's anti-Rh antibody production can be prevented if such antibodies (available under the name RhoGAM) are administered to the mother in weeks 26–28 of pregnancy and during and after delivery. These antibodies destroy any fetal RBCs that cross the placenta before they can stimulate a maternal immune response. Because maternal sensitization does not occur, no anti-Rh antibodies are produced. In the United States, this relatively simple procedure has almost entirely eliminated HDN mortality caused by Rh incompatibilities.

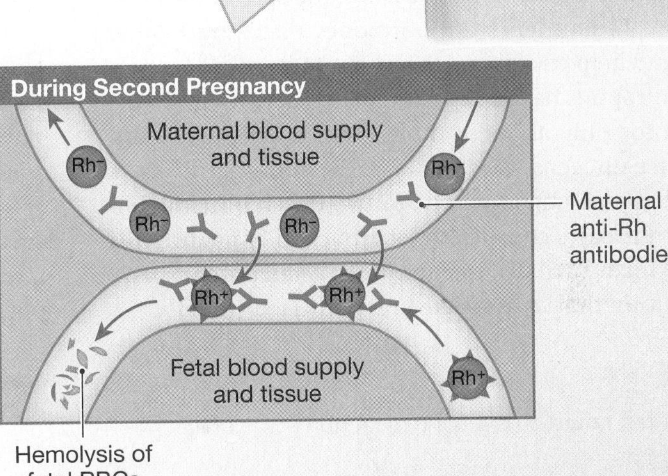
During Second Pregnancy
Maternal blood supply and tissue
Rh⁻
Rh⁻
Rh⁻
Rh⁻
Maternal anti-Rh antibodies
Rh⁺
Rh⁺
Rh⁺
Fetal blood supply and tissue
Hemolysis of fetal RBCs

Figure 19–9 Types of White Blood Cells.

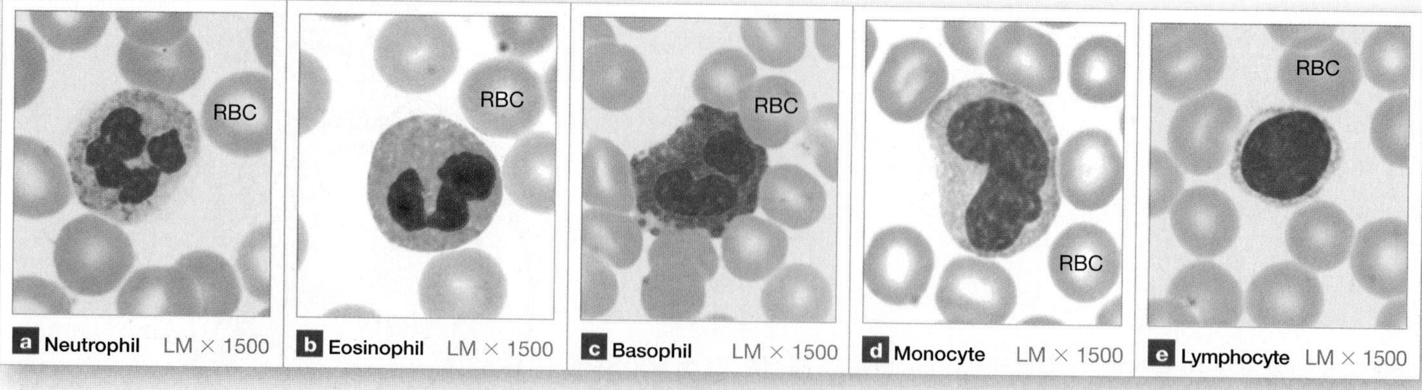
RBC
RBC
RBC
RBC
RBC
RBC

a Neutrophil LM × 1500 b Eosinophil LM × 1500 c Basophil LM × 1500 d Monocyte LM × 1500 e Lymphocyte LM × 1500

Defensins kill a variety of pathogens, including bacteria, fungi, and enveloped viruses, by combining to form large channels in their plasma membranes. The digestive enzymes then break down the bacterial remains.

While actively attacking bacteria, a neutrophil releases prostaglandins and leukotrienes. p. 615 The prostaglandins increase capillary permeability in the affected region, contributing to local inflammation and restricting the spread of injury and infection. Leukotrienes are hormones that attract other phagocytes and help coordinate the immune response.

Most neutrophils have a short life span. They survive in the bloodstream for only about 10 hours. When actively engulfing debris or pathogens, they may last 30 minutes or less. A neutrophil dies after engulfing one to two dozen bacteria, but its breakdown releases chemicals that attract other neutrophils to the site. A mixture of dead neutrophils, cellular debris, and other wastes form the *pus* associated with infected wounds.

Tips & Tools

Remember that **neu**trophils are the most **nu**merous of the white blood cells.

Eosinophils

Eosinophils (ē-ō-SIN-ō-filz) have granules that stain darkly with *eosin*, a red dye. The granules also stain with other acid dyes, so the name **acidophils** (ah-SID-ō-filz) applies as well. Eosinophils generally represent 2–4 percent of the circulating WBCs. They are similar in size to neutrophils. However, the combination of deep red granules and a typically bilobed (two-lobed) nucleus makes eosinophils easy to identify (Figure 19–9b).

Eosinophils attack objects that are coated with antibodies. These phagocytic cells will engulf antibody-marked bacteria, protozoa, or cellular debris, but their primary mode of attack is the exocytosis of toxic compounds, including nitric oxide and cytotoxic enzymes. This is particularly effective against multicellular parasites that are too big to engulf, such as flukes or parasitic roundworms. The number of circulating eosinophils increases dramatically during a parasitic infection.

Eosinophils increase in number during allergic reactions as well, because they are sensitive to circulating *allergens* (materials that trigger allergies). Eosinophils are also attracted to sites of injury, where they release enzymes that reduce inflammation produced by mast cells and neutrophils. This controls the spread of inflammation to adjacent tissues.

Basophils

Basophils (BĀ-sō-filz; *baso-*, base + *phileo*, to love) have numerous granules that stain darkly with basic dyes. In a standard blood smear, the inclusions are deep purple or blue (Figure 19–9c). Measuring 8–10 μm in diameter, basophils are smaller than neutrophils or eosinophils. They are also relatively rare, accounting for less than 1 percent of the circulating WBC population.

Basophils migrate to injury sites and cross the capillary **endothelium** to accumulate in the damaged tissues. There they discharge their granules into the interstitial fluids. The granules contain *histamine*, which dilates blood vessels, and *heparin*, a compound that prevents blood clotting. Stimulated basophils release these chemicals into the interstitial fluids to enhance the local inflammation initiated by mast cells. p. 129 Although mast cells release the same compounds in damaged connective tissues, mast cells and basophils are distinct populations. Stimulated basophils also release other chemicals that attract eosinophils and other basophils to the area.

Monocytes

Monocytes (MON-ō-sīts) are spherical cells that may exceed 15 μm in diameter, nearly twice the diameter of a typical RBC. When flattened in a blood smear, they look even larger, so monocytes are fairly easy to identify. The nucleus is large and tends to be oval or kidney bean–shaped rather than lobed (Figure 19–9d). Monocytes normally account for 2–8 percent of circulating WBCs.

An individual monocyte is transported in the bloodstream. It remains in circulation for only about 24 hours before entering peripheral tissues to become a tissue macrophage. Macrophages are aggressive phagocytes. They often attempt to engulf items as large as or larger than themselves. When active, they release chemicals that attract and stimulate neutrophils, monocytes, and other phagocytic cells. Active macrophages also secrete substances that draw fibroblasts into the region. The fibroblasts then begin producing scar tissue to wall off the injured area. Macrophages are discussed further in Chapter 22.

Tips & Tools

A **mon**ocyte is the **mon**ster cell that engulfs debris and pathogens.

Lymphocytes

Typical **lymphocytes** (LIM-fō-sīts) are slightly larger than RBCs and lack abundant, deeply staining granules. In blood smears, lymphocytes typically have a large, round nucleus surrounded by a thin halo of cytoplasm (Figure 19–9e).

Lymphocytes account for 20–40 percent of the circulating WBC population. Lymphocytes continuously migrate from the bloodstream, through peripheral tissues, and back to the bloodstream. Circulating lymphocytes represent only a small fraction of all lymphocytes. At any moment, most of your body's lymphocytes are in other connective tissues and in organs of the lymphatic system.

The circulating blood contains three functional classes of lymphocytes. These classes cannot be distinguished using just a light microscope. They are as follows:

- **T cells** (T lymphocytes) are responsible for *cell-mediated immunity*, a specific defense mechanism against invading foreign cells, and for the coordination of the immune response. T cells either enter peripheral tissues and attack foreign cells directly, or control the activities of other lymphocytes.
- **B cells** (B lymphocytes) are responsible for *humoral immunity*, a specific defense mechanism that involves the production of antibodies. These antibodies are distributed by blood, lymph, and interstitial fluid and are capable of attacking foreign antigens throughout the body. Activated B cells differentiate into **plasma cells**, which are specialized to synthesize and secrete antibodies. The T cells responsible for cellular immunity must migrate to their targets, but the antibodies produced by plasma cells in one location can destroy antigens almost anywhere in the body.
- **Natural killer (NK) cells** carry out *immune surveillance*—the detection and subsequent destruction of abnormal cells. NK cells, sometimes known as *large granular lymphocytes*, are important in preventing cancer.

Tips & Tools

To remember the various white blood cell populations, think "Never let monkeys eat bananas" for neutrophils, lymphocytes, monocytes, eosinophils, and basophils.

The Differential Count and Changes in WBC Profiles

A variety of conditions, including infection, inflammation, and allergic reactions, cause characteristic changes in circulating populations of WBCs. By examining a stained blood smear, we can obtain a **differential count** of the WBC population. The values reported provide a *WBC profile* indicating the number of each type of cell in a sample of 100 WBCs. The normal range of abundance for each type of WBC is shown in Table 19–3.

The term **leukopenia** (lū-kō-PĒ-nē-uh; *penia*, poverty) indicates inadequate numbers of WBCs. **Leukocytosis** (loo-kō-sī-TŌ-sis) refers to excessive numbers of WBCs. A modest leukocytosis is normal during an infection. Extreme leukocytosis (100,000/µL or more) generally indicates the presence of some form of **leukemia** (lū-KĒ-mē-uh), a cancer of the WBCs.

WBC Production: Leukopoiesis

The process of WBC production is called **leukopoiesis** (lū-kō-POY-eh-sis). Stem cells that produce WBCs originate in the red bone marrow, with the divisions of hemocytoblasts (Figure 19–10). As we have noted, hemocytoblast divisions produce myeloid stem cells and lymphoid stem cells. Myeloid stem cells divide to create **progenitor cells**, which give rise to all the formed elements except lymphocytes. One type of progenitor cell produces daughter cells that mature into RBCs. A second type produces cells that manufacture platelets. A third type produces daughter cells that develop into neutrophils, eosinophils, basophils, and monocytes.

Granulocytes (basophils, eosinophils, and neutrophils) complete their development in the red bone marrow. These WBCs go through a series of maturational stages, proceeding from *blast cells* to *myelocytes* to *band cells* before becoming mature WBCs. For example, a cell differentiating into a neutrophil goes from a *myeloblast* to a *neutrophilic myelocyte* and then becomes a *neutrophilic band cell*. Some band cells enter the bloodstream before completing their maturation. Normally, 3–5 percent of all circulating WBCs are band cells.

Monocytes begin to differentiate in the red bone marrow. Then they enter the bloodstream and complete development when they become free macrophages in peripheral tissues.

The process of lymphocyte production is called **lymphocytopoiesis**. Some lymphocytes are derived from lymphoid stem cells that remain in red bone marrow. These lymphocytes differentiate into either B cells or natural killer cells.

Table 19–3 Formed Elements of the Blood

Cell	Abundance (average number per µL)	Appearance in a Stained Blood Smear	Functions	Remarks
RED BLOOD CELLS	5.2 million (range: 4.2–6.3 million)	Flattened, circular cell; no nucleus, mitochondria, or ribosomes; red	Transport oxygen from lungs to tissues and carbon dioxide from tissues to lungs	Remain in bloodstream; 120-day life expectancy; amino acids and iron recycled; produced in red bone marrow
WHITE BLOOD CELLS	7000 (range: 5000–10,000)			
Neutrophils	4150 (range: 1800–7300) Differential count: 50–70%	Round cell; nucleus lobed and may resemble a string of beads; cytoplasm contains large, pale inclusions	Phagocytic: Engulf pathogens or debris in tissues, release cytotoxic enzymes and chemicals	Move into tissues after several hours; may survive minutes to days, depending on tissue activity; produced in red bone marrow
Eosinophils	165 (range: 0–700) Differential count: 2–4%	Round cell; nucleus generally in two lobes; cytoplasm contains large granules that generally stain bright red	Phagocytic: Engulf antibody-labeled materials, release cytotoxic enzymes, reduce inflammation; increase in allergic and parasitic situations	Move into tissues after several hours; survive minutes to days, depending on tissue activity; produced in red bone marrow
Basophils	44 (range: 0–150) Differential count: <1%	Round cell; nucleus generally cannot be seen through dense, blue-stained granules in cytoplasm	Enter damaged tissues and release histamine and other chemicals that promote inflammation	Survival time unknown; assist mast cells of tissues in producing inflammation; produced in red bone marrow
Monocytes	456 (range: 200–950) Differential count: 2–8%	Very large cell; kidney bean–shaped nucleus; abundant pale cytoplasm	Enter tissues to become macrophages; engulf pathogens or debris	Move into tissues after 1–2 days; survive for months or longer; produced primarily in red bone marrow
Lymphocytes	2185 (range: 1500–4000) Differential count: 20–40%	Generally round cell, slightly larger than RBC; round nucleus; very little cytoplasm	Cells of lymphatic system, providing defense against specific pathogens or toxins	Survive for months to decades; circulate from blood to tissues and back; produced in red bone marrow and lymphatic tissues
PLATELETS	350,000 (range: 150,000–500,000)	Round to disc-shaped cytoplasmic fragment; contain enzymes, proenzymes, actin, and myosin; no nucleus	Hemostasis: Clump together and stick to vessel wall (platelet phase); activate intrinsic pathway of coagulation phase	Remain in bloodstream or in vascular organs; remain intact for 7–12 days; produced by megakaryocytes in red bone marrow

Many of the lymphoid stem cells that produce lymphocytes migrate from the red bone marrow to peripheral **lymphatic tissues**, including the thymus, spleen, and lymph nodes. As a result, lymphocytes are produced in these organs as well as in the red bone marrow. Lymphoid stem cells migrating to the thymus mature into T cells.

Regulation of WBC Production

Factors that regulate lymphocyte maturation are incompletely understood. Until adulthood, hormones produced by the thymus promote the differentiation and maintenance of T cell populations. The importance of the thymus in adults, especially with respect to aging, remains unclear. In adults, the production of B and T lymphocytes is regulated primarily by exposure to antigens (foreign proteins, cells, or toxins). When antigens appear, lymphocyte production escalates. We describe the control mechanisms in Chapter 22.

Several hormones, called **colony-stimulating factors (CSFs)**, are involved in regulating other WBC populations. Four CSFs have been identified. Each stimulates the formation of WBCs or both WBCs and RBCs. The designation for each factor indicates its target, as shown in the blue ovals in Figure 19–10:

- **Multi-CSF** accelerates the production of granulocytes, monocytes, platelets, and RBCs.

Figure 19–10 The Origins and Differentiation of Formed Elements. Hemocytoblast divisions give rise to myeloid stem cells or lymphoid stem cells. Myeloid stem cells produce progenitor cells that divide to produce red blood cells, platelets, and white blood cells except for lymphocytes. The targets of erythropoietin (EPO) and the four colony-stimulating factors (CSFs) are indicated. Lymphoid stem cells produce the various lymphocytes.

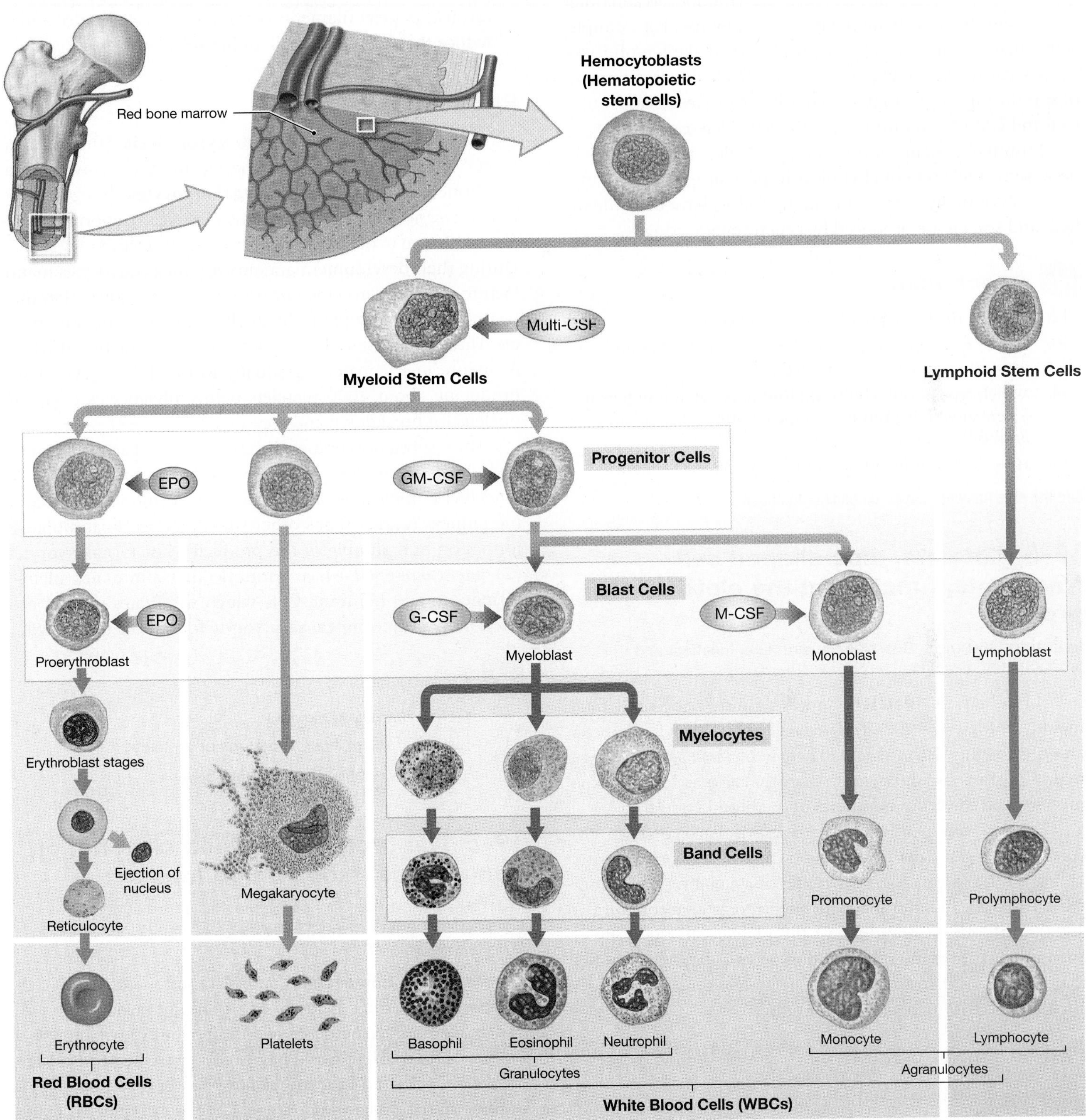

Red bone marrow
Hemocytoblasts (Hematopoietic stem cells)
Multi-CSF
Myeloid Stem Cells
Lymphoid Stem Cells
Progenitor Cells
EPO
GM-CSF
Blast Cells
EPO
G-CSF
M-CSF
Proerythroblast
Myeloblast
Monoblast
Lymphoblast
Myelocytes
Erythroblast stages
Band Cells
Ejection of nucleus
Megakaryocyte
Reticulocyte
Promonocyte
Prolymphocyte
Erythrocyte
Platelets
Basophil
Eosinophil
Neutrophil
Monocyte
Lymphocyte
Granulocytes
Agranulocytes
Red Blood Cells (RBCs)
White Blood Cells (WBCs)

? Granulocytes consist of what three types of mature cells and from what blast cell do granulocytes differentiate?

- **GM-CSF** stimulates the production of both granulocytes and monocytes. **G-CSF** stimulates the production of granulocytes (neutrophils, eosinophils, and basophils)
- **M-CSF** stimulates the production of monocytes.

Chemical communication between lymphocytes and other WBCs helps to coordinate the immune response. For example, active macrophages release chemicals that make lymphocytes more sensitive to antigens and also accelerate the development of specific immunity. In turn, active lymphocytes release multi-CSF and GM-CSF, reinforcing nonspecific defenses.

Immune system hormones are under intensive study because of their potential clinical importance. The molecular structures of many of the stimulating factors have been identified, and several are produced by genetic engineering.

Checkpoint

12. **Identify the five types of white blood cells.**
13. **Which type of white blood cell would you find in the greatest numbers in an infected cut?**
14. **Which type of cell would you find in elevated numbers in a person who is producing large amounts of circulating antibodies to combat a virus?**
15. **How do basophils respond to an injury?**

See the blue Answers tab at the back of the book.

19

19-5 Platelets, disc-shaped cell fragments, function in the clotting process

Learning Outcome Describe the structure, function, and production of platelets.

In a blood smear, **platelets** appear as disc-shaped cell fragments (see Figure 19–10). They average about 3 μm in diameter and are roughly 1 μm thick. Platelets, or *thrombocytes*, play a major role in a vascular *clotting system* that also includes plasma proteins and the cells and tissues of the blood vessels.

Platelets are continuously replaced. Each platelet circulates for 9–12 days before being removed by phagocytes, mainly in the spleen. Each microliter of circulating blood contains 150,000–500,000 platelets; the average concentration is 350,000/μL. About one-third of the platelets in the body at any moment are in the spleen and other vascular organs, rather than in the bloodstream. These reserves are mobilized during a circulatory crisis, such as severe bleeding.

Platelet Functions

The functions of platelets include:

- *Releasing Chemicals Important to the Clotting Process.* By releasing enzymes and other factors at the appropriate times, platelets help initiate and control the clotting process.
- *Forming a Temporary Patch in the Walls of Damaged Blood Vessels.* Platelets clump together at an injury site, forming a *platelet plug*, which can slow blood loss while clotting takes place.
- *Reducing the Size of a Break in a Vessel Wall.* Platelets contain filaments of actin and myosin. After a blood clot has formed, platelet filaments contract to shrink the clot and reduce the size of the break in the vessel wall.

Platelet Production

Platelet production, or **thrombocytopoiesis** (throm-bō-sī-tō-POY-eh-sis), takes place in the red bone marrow. Normal red bone marrow contains **megakaryocytes** (meg-ah-KAR-ē-ō-sīts; *mega-*, big + *karyon*, nucleus + *-cyte*, cell), enormous cells (up to 160 μm in diameter) with large nuclei (see Figure 19–10). During their development and growth, megakaryocytes manufacture structural proteins, enzymes, and membranes. They then begin shedding cytoplasm in small membrane-enclosed packets. These packets are the platelets that enter the bloodstream. A mature megakaryocyte gradually loses all of its cytoplasm, producing about 4000 platelets before phagocytes engulf its nucleus for breakdown and recycling.

Three substances influence the rate of megakaryocyte activity and platelet formation. They are (1) *thrombopoietin* (*TPO*), or *thrombocyte-stimulating factor*, a peptide hormone produced in the kidneys (and perhaps other sites) that accelerates platelet formation and stimulates the production of megakaryocytes; (2) *interleukin-6* (*IL-6*), a hormone that stimulates platelet formation; and (3) multi-CSF, which stimulates platelet production by promoting megakaryocyte formation and growth.

Checkpoint

16. **Define *thrombocytopoiesis*.**
17. **List the three primary functions of platelets.**

See the blue Answers tab at the back of the book.

19-6 The process of blood clotting, or hemostasis, stops blood loss

Learning Outcome Discuss the mechanisms that control blood loss after an injury, and describe the reaction sequences responsible for blood clotting.

The process of **hemostasis** (*haima*, blood + *stasis*, halt), the stopping of bleeding, halts the loss of blood through the walls of damaged vessels. At the same time, it establishes a framework for tissue repairs. Hemostasis has three phases: the *vascular phase*, the *platelet phase*, and the *coagulation phase* (Figure 19–11). The boundaries of these phases are somewhat arbitrary. In reality, hemostasis is a complex *cascade*, or chain reaction, in which many things happen at once, and all of them interact to some degree. The end of the process is *clot retraction*.

The Vascular Phase

Cutting the wall of a blood vessel triggers a contraction in the smooth muscle fibers of the vessel wall (see Figure 19–11 ①). This local contraction of the vessel is a **vascular spasm**, which decreases the diameter of the vessel at the site of injury. Such a constriction can slow or even stop blood loss through the wall of a small vessel. The vascular spasm lasts about 30 minutes, a period called the **vascular phase** of hemostasis.

During the vascular phase, changes take place in the endothelium of the vessel at the injury site:

- *The Endothelial Cells Contract and Expose the Underlying Basement Membrane to the Bloodstream.*
- *The Endothelial Cells Begin Releasing Chemical Factors and Local Hormones.* We will discuss several of these factors, including *ADP*, *tissue factor*, and *prostacyclin*, later in this section. Endothelial cells also release peptide hormones called **endothelins**. These local hormones (1) stimulate smooth muscle contraction and promote vascular spasms, and (2) stimulate the division of endothelial cells, smooth muscle cells, and fibroblasts to accelerate the repair process.
- *The Endothelial Plasma Membranes Become "Sticky."* A tear in the wall of a small artery or vein may be partially sealed off by the attachment of endothelial cells on either side of the break. In small capillaries, endothelial cells on opposite sides of the vessel may stick together and prevent blood flow along the damaged vessel. The stickiness is also important because it helps platelets to attach as the platelet phase gets under way.

The Platelet Phase

The attachment of platelets to sticky endothelial surfaces, to the basement membrane, and to exposed collagen fibers marks the start of the **platelet phase** of hemostasis (see Figure 19–11 ②). The attachment of platelets to exposed surfaces is called **platelet adhesion**. As more and more platelets arrive, they begin sticking to one another as well. This process, called **platelet aggregation**, forms a **platelet plug** that may close the break in the vessel wall if the damage is not severe or the vessel is relatively small. Platelet aggregation begins within 15 seconds after an injury occurs.

As platelets arrive at the injury site, they become activated. The first sign of this activation is that the platelets become more spherical and develop cytoplasmic processes that extend toward nearby platelets. At this time, the platelets begin releasing a wide variety of compounds. These compounds include (1) *adenosine diphosphate (ADP)*, which stimulates platelet aggregation and secretion; (2) *thromboxane* A_2 and *serotonin*, which stimulate vascular spasms; (3) *clotting factors*, proteins that play a role in blood clotting; (4) *platelet-derived growth factor (PDGF)*, a peptide that promotes vessel repair; and (5) calcium ions, which are required for platelet aggregation and in several steps in the clotting process.

The platelet phase proceeds quickly, as ADP, thromboxane, and calcium ions released from each arriving platelet stimulate further aggregation. This positive feedback loop ultimately produces a platelet plug that will be reinforced as clotting occurs. ⮌ p. 22 However, platelet aggregation must be controlled and restricted to the injury site.

Several key factors limit the growth of the platelet plug: (1) **prostacyclin**, a prostaglandin that inhibits platelet aggregation and is released by endothelial cells; (2) inhibitory compounds released by WBCs entering the area; (3) circulating plasma enzymes that break down ADP near the plug; (4) compounds that, when abundant, inhibit plug formation (for example, serotonin, which at a high concentration blocks the action of ADP); and (5) the development of a blood clot, which reinforces the platelet plug but isolates it from the general circulation.

The Coagulation Phase

The vascular and platelet phases begin within a few seconds after the injury. The **coagulation** (kō-ag-yū-LĀ-shun) **phase**, or **blood clotting**, does not start until 30 seconds or more after the vessel has been damaged. Figure 19–11 ③) shows the sequence of steps involved in the formation and structure of a blood clot.

Clotting Factors

Normal blood clotting depends on the presence of **clotting factors**, or **procoagulants**, in the plasma. Important clotting factors include Ca^{2+} and 11 different proteins (Table 19–4).

Many of the clotting proteins are **proenzymes**, or inactive enzymes. When converted to active enzymes, these proteins direct essential reactions in the **clotting response**. During the coagulation phase, enzymes and proenzymes interact successively. The activation of one proenzyme commonly creates an enzyme that activates a second proenzyme, and so on in a cascade.

The Coagulation Process

The process of coagulation involves the chain reactions of the *extrinsic*, *intrinsic*, and *common pathways* (see Figure 19–11 ③). The extrinsic pathway begins outside the bloodstream, in the vessel wall. The intrinsic pathway begins inside the bloodstream, with the activation of a circulating proenzyme. These two pathways converge at the common pathway.

The Extrinsic Pathway. The **extrinsic pathway** begins when damaged endothelial cells or peripheral tissues release **Factor III**, also termed *tissue factor* or *thromboplastin*. The greater the damage, the more tissue factor is released and the faster clotting takes place. Factor III then combines with Ca^{2+} and another

FIGURE 19–11 The Phases of Hemostasis (Vascular, Platelet, and Coagulation) and Clot Retraction.

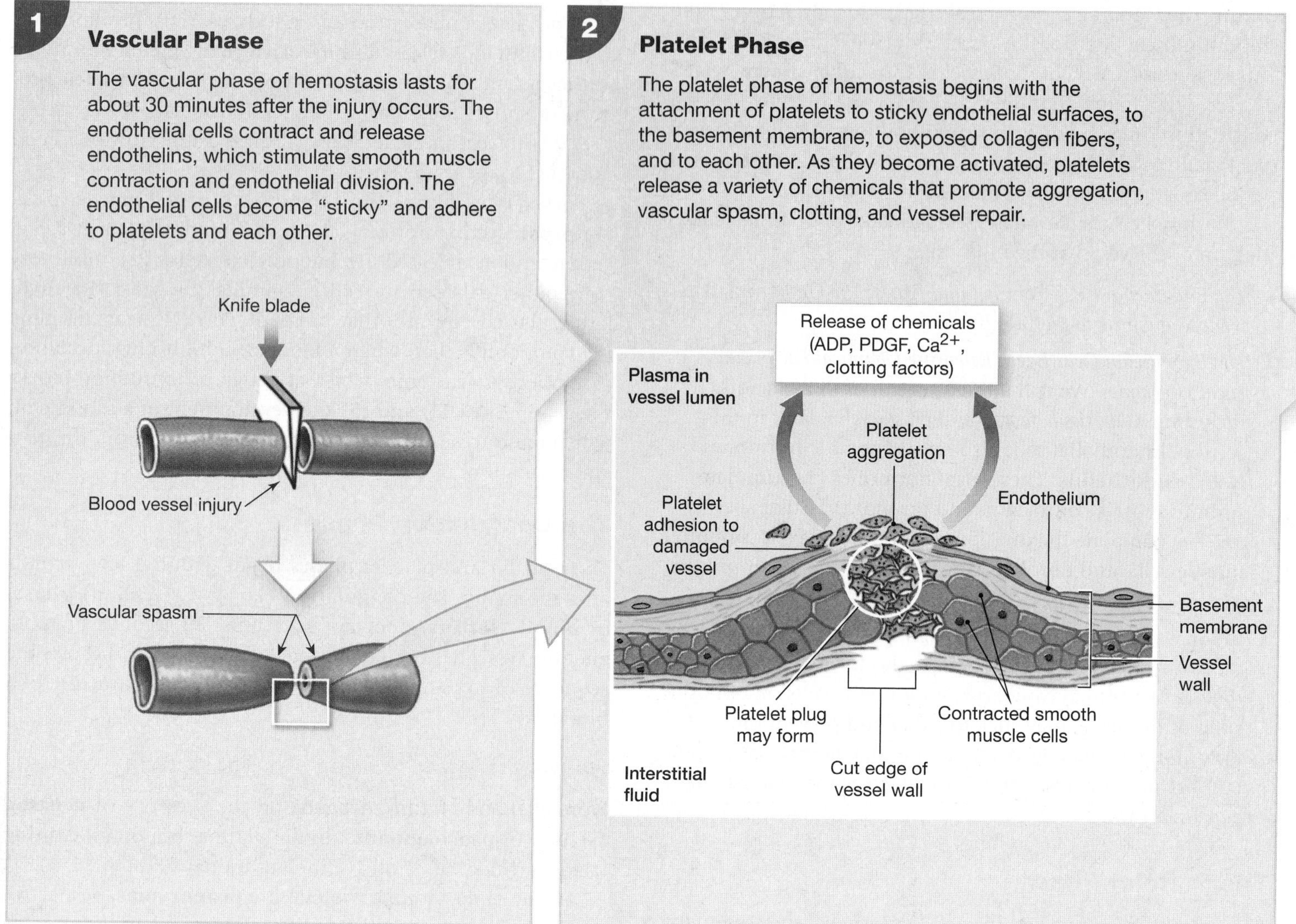
1
Vascular Phase
The vascular phase of hemostasis lasts for about 30 minutes after the injury occurs. The endothelial cells contract and release endothelins, which stimulate smooth muscle contraction and endothelial division. The endothelial cells become "sticky" and adhere to platelets and each other.
Knife blade
Blood vessel injury
Vascular spasm
2
Platelet Phase
The platelet phase of hemostasis begins with the attachment of platelets to sticky endothelial surfaces, to the basement membrane, to exposed collagen fibers, and to each other. As they become activated, platelets release a variety of chemicals that promote aggregation, vascular spasm, clotting, and vessel repair.
Release of chemicals (ADP, PDGF, Ca^{2+}, clotting factors)
Plasma in vessel lumen
Platelet aggregation
Platelet adhesion to damaged vessel
Endothelium
Basement membrane
Vessel wall
Platelet plug may form
Contracted smooth muscle cells
Interstitial fluid
Cut edge of vessel wall

19

clotting factor (Factor VII) to form an enzyme complex capable of activating Factor X, the first step in the common pathway.

The Intrinsic Pathway. The **intrinsic pathway** begins with the activation of proenzymes (usually Factor XII) exposed to collagen fibers at the injury site (or to a glass surface of a slide or collection tube). This pathway proceeds with the assistance of **PF-3**, a platelet factor released by aggregating platelets. Platelets also release a variety of other factors that accelerate the reactions of the intrinsic pathway. After a series of linked reactions, activated Factors VIII and IX combine to form an enzyme complex capable of activating Factor X.

The Common Pathway. The **common pathway** begins when enzyme complexes from either pathway activate Factor X. Activated Factor X activates a complex called **prothrombin activator**. Prothrombin activator converts the proenzyme prothrombin into the enzyme **thrombin**. Thrombin then completes the clotting process by converting fibrinogen, a soluble plasma protein, to insoluble strands of **fibrin**, producing a **blood clot** (see the SEM in Figure 19–11 3).

Interactions among the Pathways

When a blood vessel is damaged, both the extrinsic and intrinsic pathways respond. The extrinsic pathway is shorter and faster than the intrinsic pathway, and it is usually the first to initiate clotting by the common pathway. In essence, the extrinsic pathway stimulates the common pathway to produce a small amount of thrombin very quickly. This quick patch is reinforced by the intrinsic pathway, which later stimulates the common pathway to produce more thrombin.

Then the activity of the common pathway in turn stimulates both the intrinsic and extrinsic pathways. Thrombin

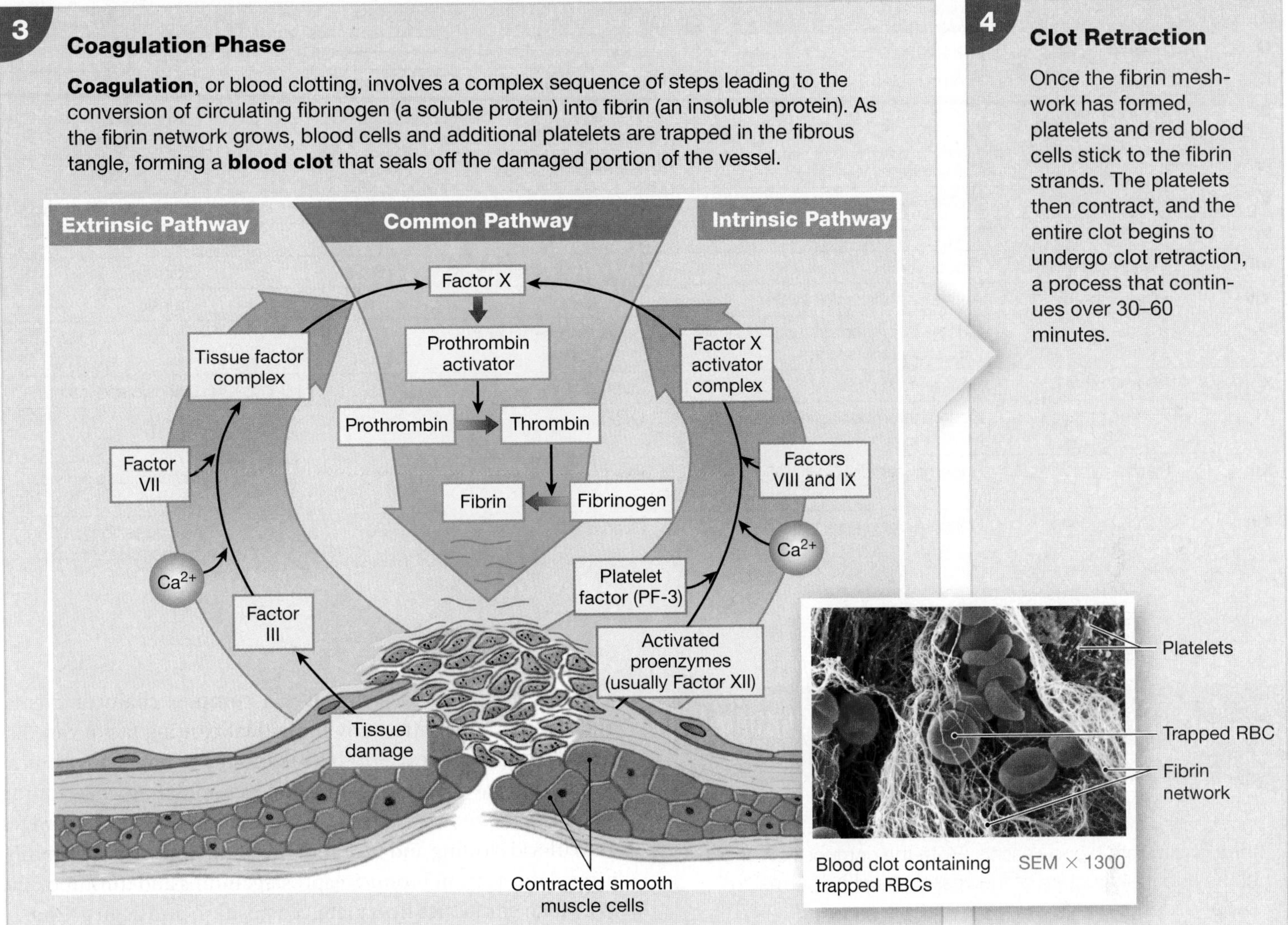
3
Coagulation Phase
Coagulation, or blood clotting, involves a complex sequence of steps leading to the conversion of circulating fibrinogen (a soluble protein) into fibrin (an insoluble protein). As the fibrin network grows, blood cells and additional platelets are trapped in the fibrous tangle, forming a blood clot that seals off the damaged portion of the vessel.
Extrinsic Pathway
Common Pathway
Intrinsic Pathway
Factor X
Prothrombin activator
Tissue factor complex
Factor X activator complex
Prothrombin
Thrombin
Factor VII
Factors VIII and IX
Fibrin
Fibrinogen
Ca2+
Ca2+
Platelet factor (PF-3)
Factor III
Activated proenzymes (usually Factor XII)
Tissue damage
Contracted smooth muscle cells
Platelets
Trapped RBC
Fibrin network
Blood clot containing trapped RBCs
SEM × 1300
4
Clot Retraction
Once the fibrin meshwork has formed, platelets and red blood cells stick to the fibrin strands. The platelets then contract, and the entire clot begins to undergo clot retraction, a process that continues over 30–60 minutes.

generated in the common pathway stimulates blood clotting by (1) stimulating the formation of tissue factor and (2) stimulating the release of PF-3 by platelets (see Figure 19–11 ③). This positive feedback loop accelerates the clotting process, and speed can be very important in reducing blood loss after a severe injury.

The time required to complete clot formation varies with the site and the nature of the injury. In tests of the clotting system, blood held in fine glass tubes normally clots in 8–18 minutes (the *coagulation time*). A small puncture wound typically stops bleeding in 1–4 minutes (the *bleeding time*).

Feedback Control of Blood Clotting

Blood clotting is restricted by substances that either deactivate or remove clotting factors and other stimulatory agents from the blood. Examples include the following:

- Normal plasma contains several **anticoagulants**—enzymes that inhibit clotting. One, **antithrombin-III**, inhibits several clotting factors, including thrombin.
- **Heparin**, a compound released by basophils and mast cells, is a cofactor that accelerates the activation of antithrombin-III. Heparin is also used clinically to impede or prevent clotting.
- **Thrombomodulin** is released by endothelial cells. This protein binds to thrombin and converts it to an enzyme that activates protein C. **Protein C** is a plasma protein that inactivates several clotting factors and stimulates the formation of *plasmin*, an enzyme that gradually breaks down fibrin strands.
- Prostacyclin released during the platelet phase inhibits platelet aggregation and opposes the stimulatory action of thrombin, ADP, and other factors.

Table 19–4 Clotting Factors

Factor	Structure	Name	Source	Concentration in Plasma (μg/mL)	Pathway
I	Plasma protein	Fibrinogen	Liver	2500–3500	Common
II	Plasma protein	Prothrombin	Liver, requires vitamin K	100	Common
III	Glycoprotein	Tissue factor (TF)	Damaged tissue, activated platelets	0	Extrinsic
IV	Ion	Calcium ions (Ca^{2+})	Bone, diet, platelets	100	Entire process
V	Protein	Pro-accelerin	Liver, platelets	10	Extrinsic and intrinsic
VI	(No longer used)				
VII	Plasma protein	Proconvertin	Liver, requires vitamin K	0.5	Extrinsic
VIII	Plasma protein	Antihemophilic factor (AHF)	Liver, lung endothelial cells	15	Intrinsic
IX	Plasma protein	Plasma thromboplastin component (PTC), Christmas factor	Liver, requires vitamin K	3	Intrinsic
X	Plasma protein	Stuart–Prower factor	Liver, requires vitamin K	10	Extrinsic and intrinsic
XI	Plasma protein	Plasma thromboplastin antecedent (PTA)	Liver	<5	Intrinsic
XII	Plasma protein	Hageman factor	Liver	<5	Intrinsic; also activates plasmin
XIII	Plasma protein	Fibrin-stabilizing factor (FSF)	Liver, platelets	20	Stabilizes fibrin, slows fibrinolysis

19

+ Clinical Note Bleeding and Clotting Extremes

In a healthy circulation pattern, blood moves relatively freely through smooth blood vessels. Both the composition of the blood and the structure of the vessels must be normal to support ideal blood flow.

Bleeding problems occur when the blood lacks platelets, or when clotting proteins are deficient or defective. The most common reason for bleeding is **thrombocytopenia**, a condition in which the platelet count is too low. This may result from drug toxicity, underlying bone marrow disease, or for unknown reasons. **Hemophilia** is a genetic disorder seen in males and characterized by the lack of clotting factors such as Factor VIII or IX. When hemostasis is activated (for example, due to trauma), the coagulation pathway cannot proceed, and so the fibrin clot cannot form. The person experiences bleeding that may prove fatal depending on the severity of the injury.

Clotting problems are called **thrombophilia**. When clots form in the venous system, the serious condition **deep vein thrombosis (DVT)** can result. These clots can loosen, travel through the circulation, and end up wedged in blood vessels of the lungs. The lungs are then hampered in oxygenating the blood. This condition, called *pulmonary embolism*, is fatal in 30 percent of cases.

The clotting process involves a complex chain of events, and disorders that affect any individual clotting factor can disrupt the entire process. For this reason, managing many clinical conditions involves controlling or manipulating the clotting response. *Aspirin*, for example, is a common drug that interferes with blood clotting and prolongs bleeding time. It acts by inhibiting platelets from producing prostaglandins and thromboxane A_2, which stimulates new platelet activation and aggregation.

Calcium Ions, Vitamin K, and Blood Clotting

Calcium ions and **vitamin K** affect almost every aspect of the clotting process. All three pathways (intrinsic, extrinsic, and common) require Ca^{2+}, so any disorder that lowers the plasma Ca^{2+} concentration will impair blood clotting.

Adequate amounts of vitamin K must be present for the liver to synthesize four of the clotting factors, including prothrombin. Vitamin K is a fat-soluble vitamin, present in green vegetables, grain, and organ meats, that is absorbed with dietary lipids. We obtain roughly half of our daily requirement from the diet. Bacteria in the large intestine manufacture the other half. A vitamin K deficiency may arise from a diet inadequate in fats or in vitamin K, or a disorder that affects fat digestion and absorption (such as problems with bile production), or prolonged use of antibiotics that kill normal intestinal bacteria. Vitamin K deficiency will cause the eventual breakdown of the common pathway due to a lack of clotting factors. Ultimately, the deficiency can deactivate the entire clotting system.

Clot Retraction

Clot retraction pulls the torn edges of the vessel closer together (see Figure 19–11 ④). This process reduces residual bleeding and stabilizes the injury site. The process also reduces the size of the damaged area, making it easier for fibroblasts, smooth muscle cells, and endothelial cells to complete repairs.

Fibrinolysis

As the repairs proceed, the clot gradually dissolves in a process called **fibrinolysis** (fī-brih-NOL-uh-sis). The process begins when two enzymes activate the proenzyme **plasminogen** (plaz-MIN-ō-jen). These enzymes are thrombin, produced by the common pathway, and **tissue plasminogen activator (t-PA)**, released by damaged tissues at the site of injury. The activation of plasminogen produces the enzyme **plasmin**, which begins digesting the fibrin strands and eroding the clot.

To perform its vital functions, blood must be kept in motion. On average, an RBC completes two circuits around the cardiovascular system each minute. Blood begins to circulate in the third week of embryonic development and continues throughout life. If the blood supply is cut off, dependent tissues may die in a matter of minutes. In Chapter 20, we examine the structure and function of the heart—the pump that maintains this vital blood flow.

Checkpoint

18. **A decreased number of megakaryocytes would interfere with what body process?**
19. **How could a fat-free diet affect blood clotting?**
20. **Unless chemically treated, whole blood will coagulate in a test tube. This clotting process begins when Factor XII becomes activated. Which clotting pathway is involved in this process?**

See the blue Answers tab at the back of the book.

19 Chapter Review

Study Outline

MasteringA&P™ Access more chapter study tools online in the MasteringA&P Study Area:

- Chapter Quizzes, Chapter Practice Test, MP3 Tutor Sessions, and Clinical Case Studies
- Practice Anatomy Lab PAL 3.0
- A&P Flix A&PFlix
- Interactive Physiology iP2
- PhysioEx PhysioEx 9.1

An Introduction to Blood and the Cardiovascular System p. 657

1. The **cardiovascular system** enables the rapid transport of nutrients, respiratory gases, waste products, and cells within the body.

19-1 Blood, composed of plasma and formed elements, provides transport, regulation, and protective services to the body p. 657

2. **Blood** is a specialized fluid connective tissue. Its functions include (1) transporting dissolved gases, nutrients, hormones, and metabolic wastes; (2) regulating the pH and ion composition of interstitial fluids; (3) restricting fluid losses at injury sites; (4) defending the body against toxins and pathogens; and (5) regulating body temperature by absorbing and redistributing heat.
3. Blood contains **plasma** and **formed elements—red blood cells (RBCs)**, **white blood cells (WBCs)**, and **platelets**. The plasma and formed elements make up **whole blood**, which can be **fractionated** for analytical or clinical purposes. *(Spotlight Figure 19–1)*
4. **Hemopoiesis** is the process of blood cell formation. Myeloid and lymphoid stem cells divide to form all types of blood cells. *(Spotlight Figure 19–1)*
5. Whole blood from any region of the body has roughly the same temperature, viscosity, and pH.
6. Plasma accounts for 55 percent of the volume of blood; roughly 92 percent of plasma is water. *(Spotlight Figure 19–1)*
7. Plasma differs from interstitial fluid in terms of its oxygen and carbon dioxide levels and the concentrations and types of dissolved proteins.
8. The three major types of plasma proteins are *albumins*, *globulins*, and *fibrinogen*.

Go to MasteringA&P™ > Study Area > Menu > Lab Tools > PhysioEx 9.1 > Exercise 11: Blood Analysis > Activity 5: Blood Cholesterol

9. **Albumins** make up about 60 percent of plasma proteins. **Globulins** constitute roughly 35 percent of plasma proteins; they include **antibodies (immunoglobulins)**, which attack foreign proteins and pathogens, and **transport globulins**, which bind ions, hormones, and other compounds. **Fibrinogen** molecules are converted to **fibrin** in the clotting process. **Serum** is plasma without fibrinogen.
10. The liver synthesizes and releases more than 90 percent of the plasma proteins.

19-2 Red blood cells, formed by erythropoiesis, contain hemoglobin that transports respiratory gases p. 659

11. Red blood cells make up slightly less than half of the blood volume and 99.9 percent of the formed elements. The **hematocrit** value indicates the percentage of formed elements within whole blood. It is commonly reported as the *packed cell volume (PCV)*. *(Spotlight Figure 19–1; Table 19–1)*
12. Each RBC is a biconcave disc, providing a large surface-to-volume ratio. This shape and a flexible plasma membrane allow RBCs to stack, bend, and flex. *(Figure 19–2)*
13. Red blood cells lack most organelles, including mitochondria and nuclei, retaining only the cytoskeleton. They typically degenerate after about 120 days in the bloodstream.
14. Molecules of **hemoglobin (Hb; Hgb)** make up more than 95 percent of the proteins in RBCs. Hemoglobin is a globular protein formed from two pairs of polypeptide subunits. Each subunit contains a single molecule of **heme**, which also has an iron atom that can reversibly bind an oxygen molecule.
15. **Erythropoiesis**, the formation of red blood cells, occurs only in *red bone marrow* (**myeloid tissue**). The process speeds up under stimulation by erythropoietin (EPO or **erythropoiesis-stimulating hormone**). Stages in RBC development include **erythroblasts** and **reticulocytes**. *(Figure 19–4)*
16. Phagocytes recycle RBCs. Damaged RBCs are continuously replaced at a rate of approximately 3 million new RBCs entering the bloodstream per second. They are replaced before they **hemolyze**. *(Figure 19–5)*
17. The components of hemoglobin are individually recycled. The heme is stripped of its iron and converted to **biliverdin**, which is converted to **bilirubin**. If bile ducts of the liver are blocked, bilirubin builds up in skin and eyes, resulting in **jaundice**. *(Figure 19–5)*
18. Iron is recycled by being stored in phagocytic cells or transported through the bloodstream, bound to **transferrin**.

19-3 The ABO and Rh blood groups are based on antigen–antibody responses p. 666

19. **Blood type** is determined by the presence or absence of specific **surface antigens** (*agglutinogens*) in the RBC plasma membranes: antigens **A**, **B**, and **Rh (D)**. Antibodies (*agglutinins*) in the plasma will react with RBCs that have different surface antigens. When an antibody meets its specific surface antigen, a **cross-reaction** results. *(Figures 19–6 to Spotlight Figure 19–8; Table 19–2)*

19-4 The various types of white blood cells contribute to the body's defenses p. 670

20. White blood cells, or **leukocytes**, have nuclei and other organelles. They defend the body against pathogens and remove toxins, wastes, and abnormal or damaged cells.
21. White blood cells are capable of *margination*, amoeboid movement, and **positive chemotaxis**. Some WBCs are also capable of *phagocytosis*.
22. *Granular leukocytes (granulocytes)* are subdivided into **neutrophils**, **eosinophils**, and **basophils**. Fifty to 70 percent of circulating WBCs are neutrophils, which are highly mobile phagocytes. The much less common eosinophils are phagocytes attracted to foreign compounds that have reacted with circulating antibodies. The fairly rare basophils migrate to damaged tissues and release *histamine* and *heparin*, aiding the inflammatory response. *(Figure 19–9)*
23. *Agranular leukocytes (agranulocytes)* include **monocytes** and **lymphocytes**. Monocytes that migrate into peripheral tissues become tissue macrophages. Lymphocytes, the primary cells of the lymphatic system, include **T cells** (which enter peripheral tissues and attack foreign cells directly, or affect the activities of other lymphocytes); **B cells** (which produce antibodies); and **natural killer (NK) cells** (which destroy abnormal cells). *(Figure 19–9; Table 19–3)*
24. A **differential count** of the WBC population can indicate a variety of disorders. **Leukemia** is indicated by extreme **leukocytosis**—that is, excessive numbers of WBCs. *(Table 19–3)*
25. Granulocytes and monocytes are produced by myeloid stem cells in the red bone marrow that divide to create **progenitor cells**. Lymphoid stem cells also originate in the red bone marrow, but many migrate to peripheral **lymphatic tissues**. *(Figure 19–10)*
26. Factors that regulate lymphocyte maturation are not completely understood. Several **colony-stimulating factors (CSFs)** are involved in regulating other WBC populations and in coordinating RBC and WBC production. *(Figure 19–10)*

19-5 Platelets, disc-shaped cell fragments, function in the clotting process p. 678

27. Platelets are disc-shaped cell fragments. They circulate for 9–12 days before being removed by phagocytes. *(Figure 19–11)*
28. The functions of platelets include (1) releasing chemicals important to the clotting process, (2) forming a temporary patch in the walls of damaged blood vessels, and (3) reducing the size of a break in a vessel wall.
29. During **thrombocytopoiesis**, **megakaryocytes** in the red bone marrow release packets of cytoplasm (platelets) into the circulating blood. The rate of platelet formation is stimulated by thrombopoietin or thrombocyte-stimulating factor, interleukin-6, and multi-CSF.

19-6 The process of blood clotting, or hemostasis, stops blood loss p. 678

30. **Hemostasis** halts the loss of blood through the walls of damaged vessels. It consists of three phases: the *vascular phase*, the *platelet phase*, and the *coagulation phase*.

31. The **vascular phase** is a period of local blood vessel constriction, or **vascular spasm**, at the injury site. *(Figure 19–11)*
32. The **platelet phase** follows as platelets are activated, aggregate at the site, and adhere to the damaged surfaces. *(Figure 19–11)*
33. The **coagulation phase** occurs as factors released by platelets and endothelial cells interact with **clotting factors** (through either the **extrinsic pathway**, the **intrinsic pathway**, or the **common pathway**) to form a **blood clot**. In this reaction sequence, soluble fibrinogen is converted to large, insoluble fibers of fibrin. *(Figure 19–11; Table 19–4)*
34. During **clot retraction**, platelets contract and pull the torn edges of the damaged vessel closer together. *(Figure 19–11)*
35. During **fibrinolysis**, the clot gradually dissolves through the action of **plasmin**, the activated form of circulating **plasminogen**.

Review Questions

See the blue Answers tab at the back of the book.

LEVEL 1 Reviewing Facts and Terms

1. Identify the five types of white blood cells in the following photographs,

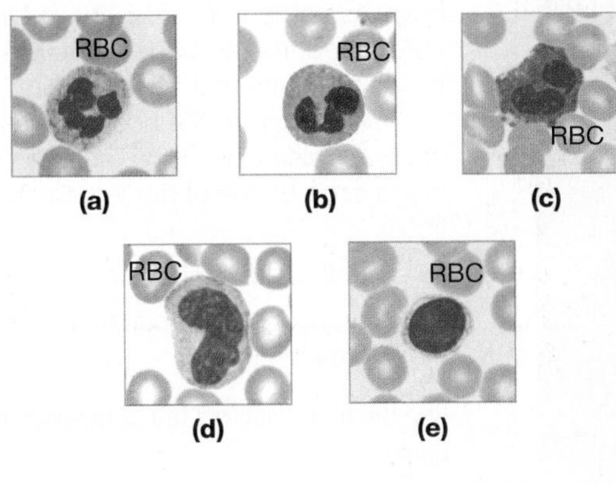
RBC
RBC
RBC
(a)
(b)
(c)
RBC
RBC
(d)
(e)

(a) ________ (b) ________
(c) ________ (d) ________
(e) ________

2. The formed elements of the blood include **(a)** plasma, fibrin, and serum, **(b)** albumins, globulins, and fibrinogen, **(c)** WBCs, RBCs, and platelets, **(d)** all of these.
3. Blood temperature is approximately ______, and blood pH averages ______. **(a)** 36°C, 7.0, **(b)** 39°C, 7.8, **(c)** 38°C, 7.4, **(d)** 37°C, 7.0.
4. Plasma contributes approximately ______ percent of the volume of whole blood, and water accounts for ______ percent of the plasma volume. **(a)** 55, 92, **(b)** 25, 55, **(c)** 92, 55, **(d)** 35, 72.
5. Serum is **(a)** the same as blood plasma, **(b)** plasma minus the formed elements, **(c)** plasma minus the proteins, **(d)** plasma minus fibrinogen, **(e)** plasma minus the electrolytes.
6. A hemoglobin molecule is composed of **(a)** two protein chains, **(b)** three protein chains, **(c)** four protein chains and nothing else, **(d)** four protein chains and four heme groups, **(e)** four heme groups but no protein.
7. The following is a list of the phases involved in the process of hemostasis.
   (1) coagulation
   (2) fibrinolysis
   (3) vascular spasm
   (4) retraction
   (5) platelet phase
   The correct sequence of these phases is **(a)** 5, 1, 4, 2, 3, **(b)** 3, 5, 1, 4, 2, **(c)** 2, 3, 5, 1, 4, **(d)** 3, 5, 4, 1, 2, **(e)** 4, 3, 5, 2, 1.
8. Stem cells responsible for lymphocytopoiesis are located in **(a)** the thymus and spleen, **(b)** the lymph nodes, **(c)** the red bone marrow, **(d)** all of these structures.
9. ______ and ______ affect almost every aspect of the clotting process. **(a)** Calcium, vitamin K, **(b)** Calcium, vitamin B_{12}, **(c)** Sodium, vitamin K, **(d)** Sodium, vitamin B_{12}.
10. What five major functions are performed by blood?
11. Name the three major types of plasma proteins and identify their functions,
12. Which type of antibodies does plasma contain for each of the following blood types? **(a)** type A, **(b)** type B, **(c)** type AB, **(d)** type O.
13. What four characteristics of WBCs are important to their response to tissue invasion or injury?
14. Which kinds of WBCs contribute to the body's nonspecific defenses?
15. Name the three types of lymphocytes and identify their functions,
16. What is the difference between prothrombin and thrombin?
17. What four conditions cause the release of erythropoietin?
18. What contribution from the intrinsic and the extrinsic pathways is necessary for the common pathway to begin?

LEVEL 2 Reviewing Concepts

19. Dehydration would **(a)** cause an increase in the hematocrit, **(b)** cause a decrease in the hematocrit, **(c)** have no effect on the hematocrit, **(d)** cause an increase in plasma volume.
20. Erythropoietin directly stimulates RBC formation by **(a)** increasing rates of mitotic divisions in erythroblasts, **(b)** speeding up the maturation of red blood cells, **(c)** accelerating the rate of hemoglobin synthesis, **(d)** all of these.
21. The waste product bilirubin is formed from **(a)** transferrin, **(b)** globin, **(c)** heme, **(d)** hemosiderin, **(e)** ferritin.
22. A difference between the A, B, and O blood types and the Rh factor is **(a)** Rh agglutinogens are not found on the surface of red blood cells, **(b)** Rh agglutinogens do not produce a cross-reaction, **(c)** individuals who are Rh^- do not carry agglutinins to Rh factor unless they have been previously sensitized, **(d)** Rh agglutinogens are found free in the plasma, **(e)** Rh agglutinogens are found bound to plasma proteins.
23. How do red blood cells differ from white blood cells in both form and function?
24. How do elements of blood defend against toxins and pathogens in the body?
25. What is the role of blood in the stabilization and maintenance of body temperature?
26. Relate the structure of hemoglobin to its function,
27. Why is aspirin sometimes prescribed for the prevention of vascular problems?

LEVEL 3 Critical Thinking and Clinical Applications

28. A test for prothrombin time is used to identify deficiencies in the extrinsic clotting pathway; prothrombin time is prolonged if any of the factors are deficient. A test for activated partial thromboplastin time is used in a similar fashion to detect deficiencies in the intrinsic clotting pathway. Which factor would be deficient if a person had a prolonged prothrombin time but a normal partial thromboplastin time?
29. In the disease mononucleosis ("mono"), the spleen enlarges because of increased numbers of phagocytes and other cells. Common signs and symptoms of this disease include pale complexion, a tired feeling, and a lack of energy sometimes to the point of not being able to get out of bed. What might cause these signs and symptoms?
30. Almost half of our vitamin K is synthesized by bacteria that inhabit the large intestine. Based on this information, how could taking a broad-spectrum antibiotic for a long time cause frequent nosebleeds?
31. After Randy was diagnosed with stomach cancer, nearly all of his stomach had to be removed. Postoperative treatment included regular injections of vitamin B_{12}. Why was this vitamin prescribed, and why were injections specified?

+ CLINICAL CASE Wrap-Up Crisis in the Blood

Clare knows what her blood test will show. Instead of looking like tiny, double-dimpled Frisbees, many of her RBCs look sharp and crescent shaped like the "sickles" for which they are named. Inside her RBCs the hemoglobin molecule has a malformation in the globular protein subunit called the beta chain. The quaternary structure of the molecule is disrupted and is responsible for the distorted shape of the red blood cell. Clare has **sickle cell disease (SCD)**.

19

In **SCD**, two copies of a gene called hemoglobin-S are inherited, one from each parent. The gene mutation for SCD evolved in Africa where the sickle shape prevented the malarial parasite from continuing its life cycle in the blood, and so conferred a survival advantage against malaria. Anemia frequently accompanies SCD because sickled red blood cells live only 10–20 days (as compared to 120 days for normal RBCs) and the red bone marrow is not able to keep up with their replacement.

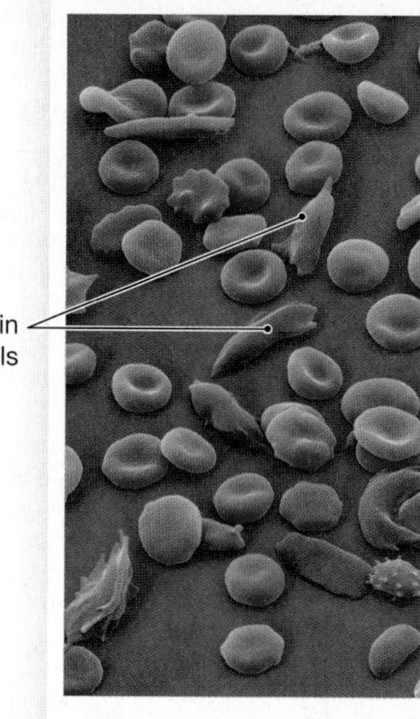
Sickling in red blood cells

Clare's trip to the hospital was brought on by a sickle cell crisis. The crisis occurred in the large bones of her leg, the tibia and fibula.

1. Why is Clare in such pain?
2. Why must she be careful to stay hydrated?

See the blue Answers tab at the back of the book.

Related Clinical Terms

arterial stick: The taking of a blood sample from an artery rather than a vein. It is usually more painful due to arteries being deeper, having more nerves, and having thicker walls.

blood bank: Place where blood is collected, typed, separated into components, stored, and prepared for transfusion to recipients.

bone marrow biopsy: The removal of a small piece of bone marrow for either laboratory analysis, to diagnose and stage some forms of cancer, to diagnose other blood disorders, to find the source of unexplained fever, or to diagnose fibrosis of bone marrow or myeloma, a tumor composed of cells normally found in the bone marrow.

disseminated intravascular coagulation: A serious disorder in which the proteins that control blood clotting become abnormally active, causing small blood clots to form, which can prevent blood from reaching vital organs.

dyscrasia: An abnormal condition, especially of the blood.

ecchymosis: Skin discoloration caused by the escape of blood into tissues from ruptured blood vessels.

embolism: A condition in which a drifting blood clot (an embolus) becomes stuck in a blood vessel, blocking circulation to the area downstream.

hematology: The science concerned with the medical study of blood and blood-producing organs.

hemochromatosis: A rare metabolic disorder wherein the skin has a bronze coloration; accompanied by cirrhosis and severe diabetes mellitus; caused by the deposit of the iron–protein complex hemosiderin in tissues.

hemophilia: Inherited disorders characterized by the inadequate production of clotting factors.

hemopoietic growth factor: A group of proteins that cause blood cells to grow and mature.

hemosiderosis: An increase in tissue iron stores without any associated damage.

hypervolemic: Having an excessive blood volume.

hypovolemic: Having a low blood volume.

iron overload: Pathology in which iron accumulates in the tissues; characterized by bronzed skin, enlarged liver, diabetes mellitus, and abnormalities of the pancreas.

myeloproliferative disorder: A group of slow-growing blood cancers, including chronic myelogenous leukemia, characterized by large numbers of abnormal RBCs, WBCs, or platelets growing and spreading in the bone marrow and the peripheral blood.

normovolemic: Referring to a normal blood volume.

phlebotomist: Medical technician who extracts blood by venipuncture for treatment or for laboratory analysis.

plaque: An abnormal accumulation of large quantities of lipids within a blood vessel wall.

plasmapheresis: A procedure consisting of the removal of blood from a person, separating the blood cells from plasma, and returning these blood cells to the person's circulation, diluted with fresh plasma or a substitute. Used to treat autoimmune disorders.

Schilling test: The test to determine whether the body absorbs vitamin B_{12} normally.

septicemia: Systemic toxic illness due to bacterial invasion of the bloodstream from a local infection. Signs and symptoms include chills, fever, and exhaustion. The disorder is treated with massive doses of antibiotics. Also known as *blood poisoning*.

thrombolytic: An agent that causes the breakup of a thrombus (clot).

thrombus: A blood clot attached to the luminal (inner) surface of a blood vessel.

20 The Heart

Learning Outcomes

These Learning Outcomes correspond by number to this chapter's sections and indicate what you should be able to do after completing the chapter.

20-1 ■ Describe the anatomy of the heart, including vascular supply and pericardium structure, and trace the flow of blood through the heart, identifying the major blood vessels, chambers, and heart valves. p. 689

20-2 ■ Explain the events of an action potential in cardiac muscle, indicate the importance of calcium ions to the contractile process, describe the conducting system of the heart, and identify the electrical events associated with a normal electrocardiogram. p. 702

20-3 ■ Explain the events of the cardiac cycle, including atrial and ventricular systole and diastole, and relate the heart sounds to specific events in the cycle. p. 711

20-4 ■ Define cardiac output, describe the factors that influence heart rate and stroke volume, and explain how adjustments in stroke volume and cardiac output are coordinated at different levels of physical activity. p. 716

CLINICAL CASE A Needle to the Chest

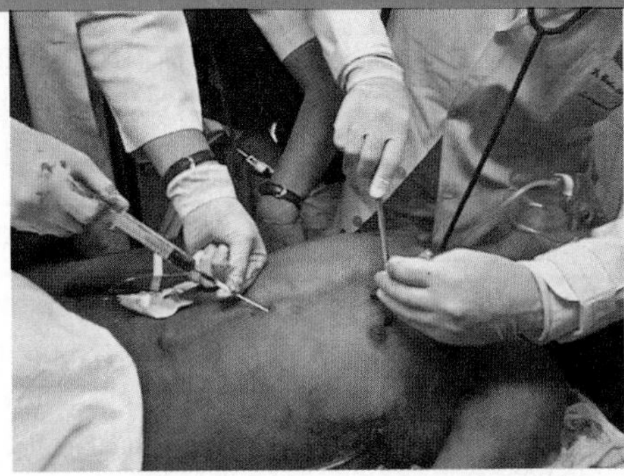

It has been a slow night in the emergency room. Dr. Jim is reading an emergency medicine journal and nodding off. He perks up when he hears the speakerphone crackle.

"We have a male, mid-40s, driver of a vehicle involved in a head-on collision with a truck," reports an EMT. "He was unconscious at the scene, but is now semi-alert, complaining of severe chest pain. His respirations are 30 and shallow. His blood pressure is 80/58. Pulse 130. Carotid pulses in the neck are present, but barely palpable (able to be felt). He has massive jugular venous distension (the veins in his neck are backed up with blood). The seat belt was broken and he has a steering wheel contusion on his anterior chest."

Dr. Jim instructs the EMTs to take the victim directly into Trauma Room 1. The patient is loaded supine on the trauma cart and Dr. Jim quickly introduces himself to the barely conscious man. A chest x-ray shows a fractured sternum and several fractured ribs, but the heart shadow (outline on the x-ray) appears normal. The heart sounds are distant and muffled through a stethoscope. After a rapid bedside ultrasound, Dr. Jim puts on sterile gloves, does a skin prep of the patient's anterior chest, and sticks a large-bore needle with a 100-mL syringe attached into the man's sub-xiphoid space, directing the needle toward his left shoulder. The syringe fills with dark blood twice. **What is Dr. Jim doing?** **To find out, turn to the Clinical Case Wrap-Up on p. 726.**

An Introduction to the Heart as Part of the Cardiovascular System

The *heart,* along with the *blood* and *blood vessels,* is part of the **cardiovascular system**, as we introduced in Chapter 19. This system functions to distribute oxygen and nutrients to the cells of the body, and to take away carbon dioxide and other wastes.

In this chapter we consider the structure and function of the heart, a small organ about the size of your clenched fist. This extraordinary organ beats approximately 100,000 times each day. Each day the heart pumps about 8000 liters of blood—enough to fill forty 55-gallon drums, or 8800 quart-sized milk cartons. Try transferring a gallon of water with a cooking baster, and you'll appreciate just how hard the heart has to work to keep you alive. Let's look first at the structure of this organ, and then at several important aspects of its function.

20-1 The heart is a four-chambered organ that pumps blood through the systemic and pulmonary circuits

Learning Outcome **Describe the anatomy of the heart, including vascular supply and pericardium structure, and trace the flow of blood through the heart, identifying the major blood vessels, chambers, and heart valves.**

The heart is a hollow muscular organ that pumps oxygen-poor blood to the lungs within the *pulmonary circuit* and oxygen-rich blood to the rest of the body within the *systemic circuit.* The pulmonary circuit carries oxygen-poor blood from the heart's lower right chamber (right ventricle), through the pulmonary arteries, to the lungs, and oxygen-rich blood back through the pulmonary veins to the heart's upper left chamber (left atrium). The systemic circuit carries oxygen-rich blood from the heart's lower left chamber (left ventricle), through the systemic arteries, and oxygen-poor blood through systemic veins back to the heart's upper right chamber (right atrium). Let's start with an overview of heart function.

Overview of Heart Function: The Pulmonary and Systemic Circuits

In the cardiovascular system, blood flows through a network of blood vessels that extend between the heart and peripheral tissues. Those blood vessels make up a **pulmonary circuit**, which carries blood to and from the gas exchange surfaces of the lungs, and a **systemic circuit**, which transports blood to and from the rest of the body (Figure 20–1). Each circuit begins and ends at the heart, and blood travels through these circuits in sequence. Thus, blood returning to the heart from the systemic circuit must complete the pulmonary circuit before reentering the systemic circuit.

The blood vessels of both circuits include arteries, capillaries, and veins. *Arteries* carry blood away from the heart, and *veins* return blood to the heart. The *great vessels* are the largest veins and arteries in the body, those connected to the heart. Microscopic thin-walled vessels called *capillaries* interconnect the smallest arteries and the smallest veins. Capillaries are called *exchange vessels,* because their thin walls permit the exchange of nutrients, dissolved gases (called *gas exchange*), and wastes between the blood and surrounding tissues.

The **heart** has four muscular chambers, two associated with each circuit—in essence, the heart is two side-by-side pumps. The *right atrium* (Ā-trē-um; entry chamber; plural, *atria*) receives blood from the systemic circuit and passes it to the *right ventricle* (VEN-tri-kul; little belly), which then pumps blood into the pulmonary circuit. The *left atrium* collects blood from the pulmonary circuit and empties it into the *left ventricle,* which pumps blood into the systemic circuit. When the heart beats,

Figure 20–1 An Overview of the Cardiovascular System. Driven by the pumping of the heart, blood flows through the pulmonary and systemic circuits in sequence. Each circuit begins and ends at the heart and contains arteries, capillaries, and veins.

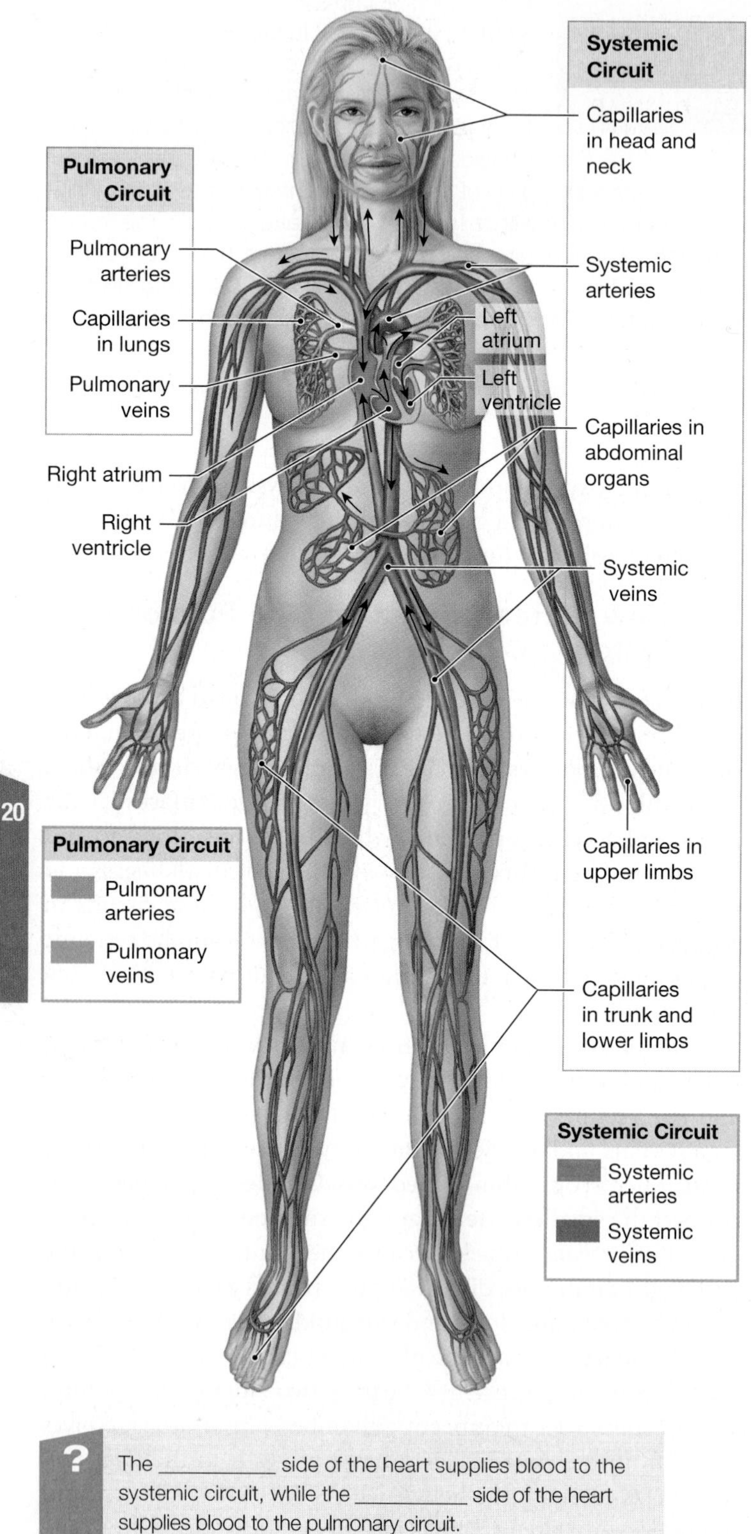
Systemic Circuit
Capillaries in head and neck
Pulmonary Circuit
Pulmonary arteries
Capillaries in lungs
Pulmonary veins
Systemic arteries
Left atrium
Left ventricle
Capillaries in abdominal organs
Right atrium
Right ventricle
Systemic veins
Capillaries in upper limbs
Pulmonary Circuit
Pulmonary arteries
Pulmonary veins
Capillaries in trunk and lower limbs
Systemic Circuit
Systemic arteries
Systemic veins
?
The __________ side of the heart supplies blood to the systemic circuit, while the __________ side of the heart supplies blood to the pulmonary circuit.

first the atria contract, and then the ventricles contract. The two ventricles contract at the same time and eject equal volumes of blood into the pulmonary and systemic circuits.

Heart Location and Position

The heart is located in the thoracic cavity near the anterior chest wall, directly posterior to the sternum (Figure 20–2a,b). The great vessels, both veins and arteries, are connected to the superior end of the heart at its base. The **base** sits posterior to the sternum at the level of the third costal cartilage, centered about 1.2 cm (0.5 in.) to the left side (Figure 20–3a). The inferior, pointed tip of the heart is the **apex** (Ā-peks). Standard measurements for determining heart size take into account age, height, weight, and sex. A midsagittal section through the trunk does not divide the heart into two equal halves. Note that the center of the base lies slightly to the left of the midline.

The heart sits in the anterior portion of the **mediastinum**, the region between the two pleural cavities. The mediastinum contains the great vessels, which are attached at the base of the heart, as well as the thymus, esophagus, and trachea. Figure 20–2b is a superior view that shows the position of the heart relative to other structures in the mediastinum.

Heart Superficial Anatomy, Heart Wall, and Cardiac Skeleton

The walls of the heart, from deep to superficial, are the *endocardium* (inner layer whose simple squamous epithelium is continuous with the endothelial lining of blood vessels), *myocardium* (spiral bundles of cardiac muscle cells), and *pericardium* (fibrous pericardium and serous pericardium that protect, anchor, and prevent overfilling). The two-layered serous pericardium is made up of a parietal layer and a visceral layer (epicardium). These layers are separated by a fluid-filled pericardial cavity. Another structure, the *cardiac skeleton*, is a crisscrossing, interlacing layer of dense connective tissue that anchors muscle fibers, supports the great vessels and heart valves, and limits the spread of action potentials. Our discussion begins with the pericardium.

Outer Covering of the Heart: The Pericardium

The **pericardium** (pehr-ih-KAR-dē-um) surrounds the heart and consists of an outer *fibrous pericardium* and an inner *serous pericardium*. The fibrous pericardium contains a dense network of collagen fibers that stabilize the position of the heart and associated vessels within the mediastinum.

The inner **serous pericardium** is a two-layered membrane composed of an outer **parietal layer** and an inner **visceral layer**. The visceral layer is also known as the *epicardium*. The potential, fluid-filled space between these two serous layers is the **pericardial cavity**.

The pericardial cavity normally contains 15–50 mL of **pericardial fluid**, secreted by the pericardial membranes. This fluid acts as a lubricant, reducing friction between the opposing visceral and parietal surfaces as the heart beats. To visualize

Figure 20–2 The Location of the Heart in the Thoracic Cavity. ATLAS: Plates 47a,b

Trachea
Thyroid gland
First rib (cut)
Base of heart
Right lung
Left lung
Apex of heart
Diaphragm

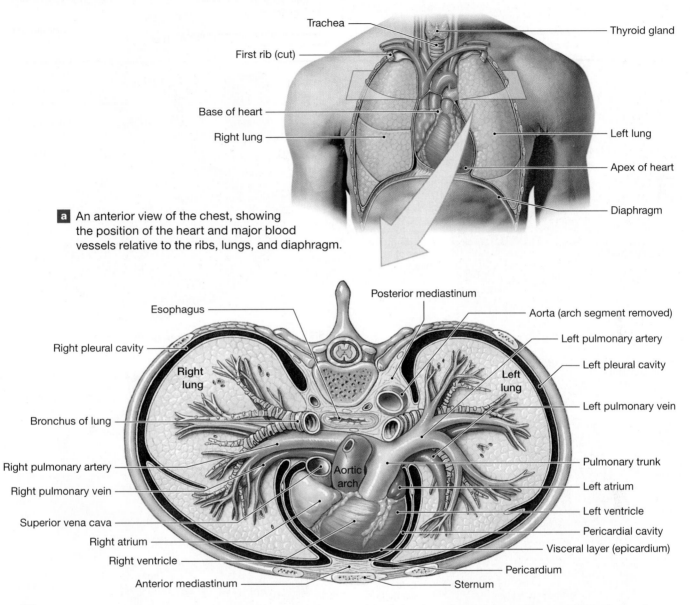

a An anterior view of the chest, showing the position of the heart and major blood vessels relative to the ribs, lungs, and diaphragm.

Posterior mediastinum
Esophagus
Aorta (arch segment removed)
Left pulmonary artery
Right pleural cavity
Left pleural cavity
Right lung
Left lung
Left pulmonary vein
Bronchus of lung
Pulmonary trunk
Right pulmonary artery
Aortic arch
Right pulmonary vein
Left atrium
Left ventricle
Superior vena cava
Pericardial cavity
Right atrium
Visceral layer (epicardium)
Right ventricle
Pericardium
Anterior mediastinum
Sternum

b A superior view of the organs in the mediastinum; portions of the lungs have been removed to reveal blood vessels and airways. The heart is located in the anterior part of the mediastinum, immediately posterior to the sternum.

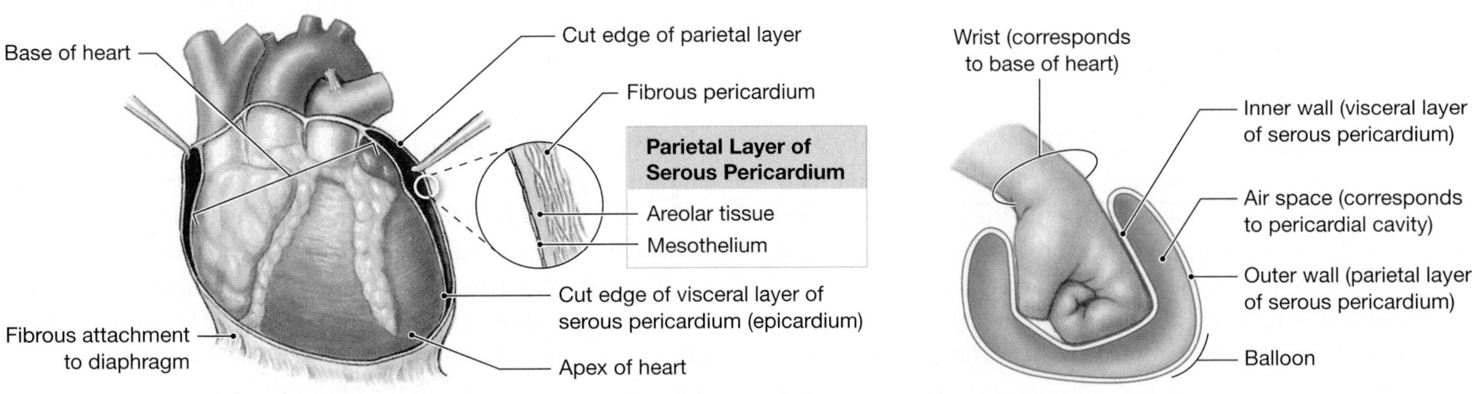
Base of heart
Cut edge of parietal layer
Fibrous pericardium
Parietal Layer of Serous Pericardium
Areolar tissue
Mesothelium
Cut edge of visceral layer of serous pericardium (epicardium)
Fibrous attachment to diaphragm
Apex of heart
Wrist (corresponds to base of heart)
Inner wall (visceral layer of serous pericardium)
Air space (corresponds to pericardial cavity)
Outer wall (parietal layer of serous pericardium)
Balloon

c The relationship between the heart and the pericardial cavity; compare with the fist-and-balloon example.

Figure 20–3 The Position and Superficial Anatomy of the Heart.

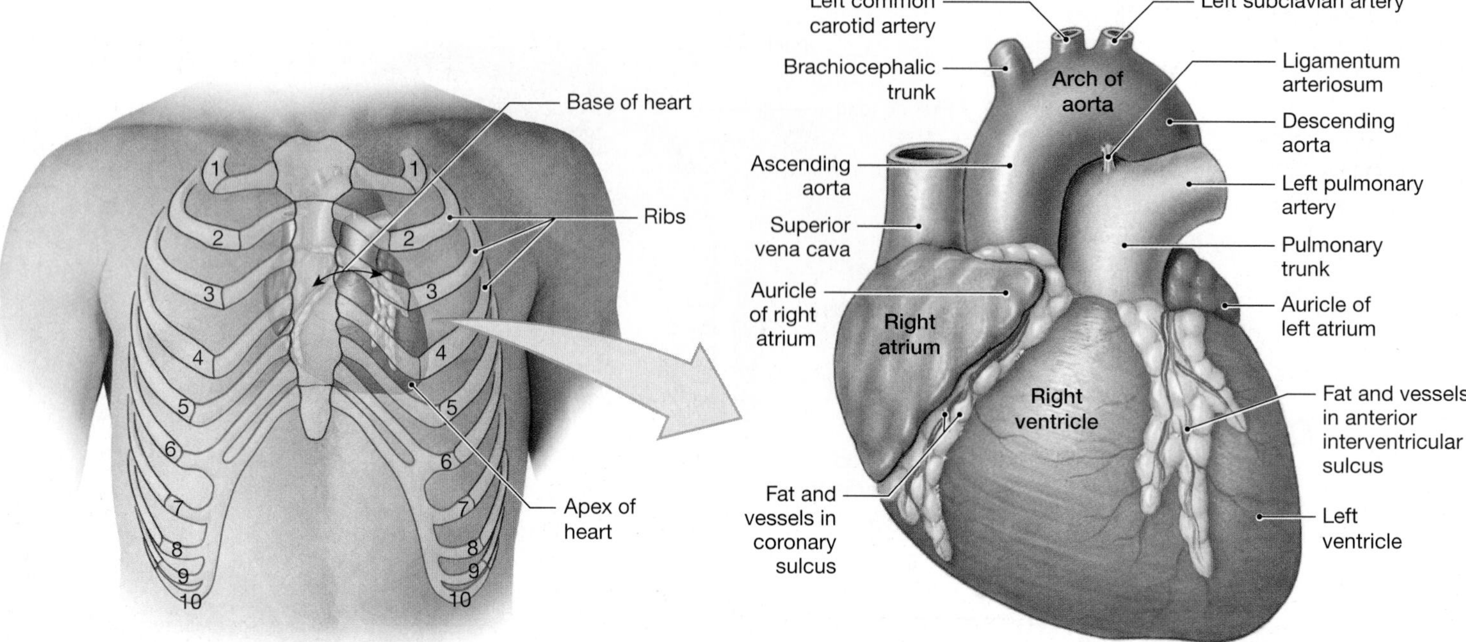
Base of heart
Ribs
Apex of heart
1
2
3
4
5
6
7
8
9
10
Left common carotid artery
Left subclavian artery
Brachiocephalic trunk
Arch of aorta
Ligamentum arteriosum
Descending aorta
Ascending aorta
Left pulmonary artery
Superior vena cava
Pulmonary trunk
Auricle of right atrium
Auricle of left atrium
Right atrium
Right ventricle
Fat and vessels in anterior interventricular sulcus
Fat and vessels in coronary sulcus
Left ventricle

a Heart position relative to the rib cage.

b Major anatomical features on the anterior surface.

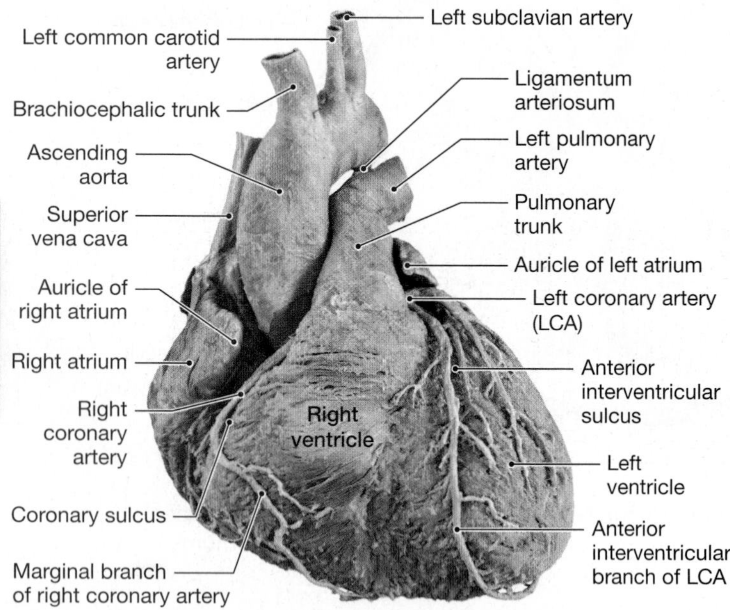
Left subclavian artery
Left common carotid artery
Brachiocephalic trunk
Ligamentum arteriosum
Ascending aorta
Left pulmonary artery
Superior vena cava
Pulmonary trunk
Auricle of left atrium
Auricle of right atrium
Left coronary artery (LCA)
Right atrium
Anterior interventricular sulcus
Right coronary artery
Right ventricle
Left ventricle
Coronary sulcus
Anterior interventricular branch of LCA
Marginal branch of right coronary artery

c Anterior surface of the heart, cadaver dissection.

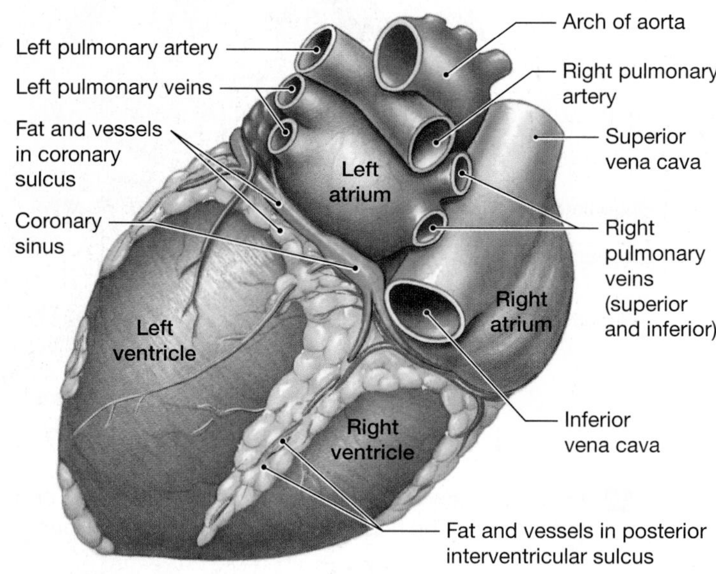
Arch of aorta
Left pulmonary artery
Left pulmonary veins
Right pulmonary artery
Fat and vessels in coronary sulcus
Superior vena cava
Left atrium
Coronary sinus
Right pulmonary veins (superior and inferior)
Right atrium
Left ventricle
Inferior vena cava
Right ventricle
Fat and vessels in posterior interventricular sulcus

d Major anatomical features on the posterior surface. Coronary arteries (which supply the heart itself) are shown in red; coronary veins are shown in blue.

the relationship between the heart and the pericardial cavity, imagine pushing your fist toward the center of a large, partially inflated balloon. The balloon represents the pericardium, and your fist is the heart. Your wrist, where the balloon folds back on itself, corresponds to the base of the heart (**Figure 20–2c**).

Pathogens can infect the pericardium, producing inflammation and the condition **pericarditis**. The inflamed pericardial surfaces rub against one another, making a distinctive scratching sound (called a *friction rub*) that can be heard through an instrument called a *stethoscope* that is placed on the chest. The pericardial inflammation also commonly results in increased production of pericardial fluid. Fluid then collects in the pericardial cavity, restricting the movement of the heart. This condition, called **cardiac tamponade** (tam-pō-NĀD; *tampon*, plug), can also result from traumatic injuries (such as stab wounds) that produce bleeding into the pericardial cavity.

> Go to MasteringA&P™ > Study Area > Menu > Lab Tools > PAL 3.0 > Anatomical Models > Cardiovascular System > Heart

20

Superficial Anatomy of the Heart

The four chambers of the heart can be identified in a superficial view (see Figure 20–3). The two atria have relatively thin muscular walls and are highly expandable. When not filled with blood, the outer portion of each atrium deflates and becomes a lumpy, wrinkled flap. This expandable extension of an atrium is called an **auricle** (AW-rih-kul; *auris*, ear), because it reminded early anatomists of the external ear (Figure 20–3b).

The **coronary sulcus**, a deep groove, marks the border between the atria and the ventricles. The **anterior interventricular sulcus** and the **posterior interventricular sulcus** are shallower depressions that mark the boundary between the left and right ventricles (Figure 20–3b–d). Substantial amounts of fat generally lie in the coronary and interventricular sulci. In fresh or preserved hearts, this fat must be stripped away to expose the underlying grooves. These sulci also contain the arteries and veins that carry blood to and from the cardiac muscle.

The Heart Wall

A section through the wall of the heart reveals three distinct layers: the outer epicardium, a middle myocardium, and an inner endocardium (Figure 20–4a). The details of these three layers, from superficial to deep, are covered in the following list.

1. The **visceral layer of serous pericardium (epicardium)** covers the surface of the heart. This serous membrane consists of an exposed mesothelium and an underlying layer of areolar connective tissue that is attached to the myocardium. The **parietal layer of serous pericardium** consists of an outer dense fibrous layer, an areolar layer, and an inner mesothelium.
2. The **myocardium** is cardiac muscle tissue that forms the atria and ventricles. This muscular layer contains cardiac muscle cells, connective tissues, blood vessels, and nerves. The atrial myocardium contains muscle bundles that wrap around the atria and form figure eights that encircle the great vessels (Figure 20–4b). Superficial ventricular muscles wrap around both ventricles, and deeper muscle layers spiral around and between the ventricles toward the apex in a figure-eight pattern.
3. The **endocardium** covers the inner surfaces of the heart, including those of the heart valves. It is made up of a simple squamous epithelium and underlying areolar tissue. This simple squamous epithelium, or endothelium, is continuous with the endothelium of the attached great vessels.

Figure 20–4 The Heart Wall.

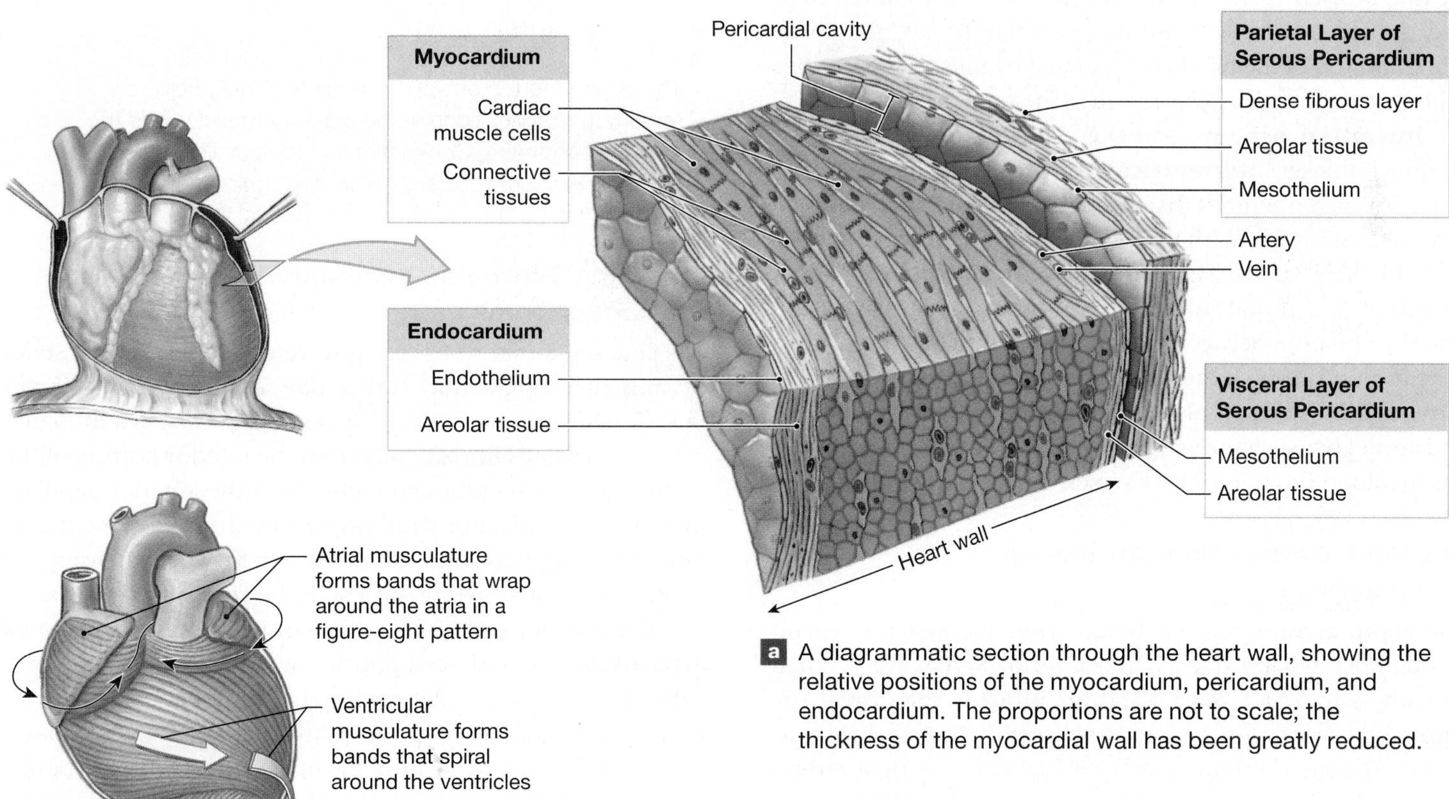
Myocardium
Cardiac muscle cells
Connective tissues
Endocardium
Endothelium
Areolar tissue
Pericardial cavity
Parietal Layer of Serous Pericardium
Dense fibrous layer
Areolar tissue
Mesothelium
Artery
Vein
Visceral Layer of Serous Pericardium
Mesothelium
Areolar tissue
Heart wall
Atrial musculature forms bands that wrap around the atria in a figure-eight pattern
Ventricular musculature forms bands that spiral around the ventricles

a A diagrammatic section through the heart wall, showing the relative positions of the myocardium, pericardium, and endocardium. The proportions are not to scale; the thickness of the myocardial wall has been greatly reduced.

b Cardiac muscle tissue forms concentric layers that wrap around the atria or spiral within the walls of the ventricles.

Connective Tissues and the Cardiac Skeleton

The connective tissues of the heart include large numbers of collagen and elastic fibers. Each cardiac muscle cell is wrapped in a strong, but elastic, sheath. Adjacent cells are tied together by fibrous cross-links, or "struts." These fibers are, in turn, interwoven into sheets that separate the superficial and deep muscle layers. The connective tissue fibers (1) provide physical support for the cardiac muscle fibers, blood vessels, and nerves of the myocardium; (2) help distribute the forces of contraction; (3) add strength and prevent overexpansion of the heart; and (4) provide elasticity that helps return the heart to its original size and shape after a contraction.

The **cardiac skeleton** (sometimes called the *fibrous skeleton*) of the heart consists of four dense bands of tough elastic tissue that encircle the heart valves and the bases of the pulmonary trunk and aorta (look ahead to Figure 20–7). These bands stabilize the positions of the heart valves and ventricular muscle cells. They also electrically insulate the ventricular cells from the atrial cells.

Heart Chambers, Valves, and Great Vessels

Next let's examine the major landmarks and structures visible on the interior surface of the heart. In a sectional view, you can see that the right atrium communicates with the right ventricle, and the left atrium with the left ventricle (Figure 20–5a,b). The chambers of the heart are separated by muscular partitions called *septa* (singular *septum,* wall). The atria are separated by the **interatrial septum**, and the ventricles are separated by the much thicker **interventricular septum**. The heart also has *valves*, covered openings that direct the flow of blood between chambers and vessels. The cardiac skeleton stabilizes the positions of these valves. The two **atrioventricular (AV) valves** (tricuspid and mitral) are folds of fibrous tissue that extend into the openings between the atria and ventricles. These valves permit blood to flow only in one direction: from the atria to the ventricles. There are also two **semilunar valves** (pulmonary and aortic) between the ventricles and their great vessels, which ensure blood flows into these vessels.

The Vena Cavae, Right Atrium, and Tricuspid Valve

The **right atrium** receives blood from the systemic circuit through the two great veins: the **superior vena cava** (VĒ-nah KĀ-vuh; *venae cavae,* plural) and the **inferior vena cava**. The superior vena cava opens into the posterior and superior portion of the right atrium. It delivers blood to the right atrium from the head, neck, upper limbs, and chest. The inferior vena cava opens into the posterior and inferior portion of the right atrium. It carries blood to the right atrium from the rest of the trunk, the viscera, and the lower limbs. Note there are no valves between the venae cavae and the right atrium.

From the fifth week of embryonic development until birth, an oval opening called the **foramen ovale** penetrates the interatrial septum and connects the two atria of the fetal heart. Before birth, the foramen ovale permits blood to flow from the right atrium to the left atrium while the lungs are developing. At birth, the foramen ovale closes, and the opening is permanently sealed off within 3 months of delivery. (If the foramen ovale does not close, serious cardiovascular problems may result. We consider these in Chapter 21.) A small, shallow depression called the **fossa ovalis** remains at this site in the adult heart (see Figure 20–5a). ATLAS: Embryology Summary 15: The Development of the Heart

The posterior walls of the right atrium and the interatrial septum have smooth surfaces. In contrast, the anterior atrial wall and the inner surface of the auricle contain prominent muscular ridges called the **pectinate muscles** (*pectin,* comb) (see Figure 20–5a).

Blood travels from the right atrium into the right ventricle through a broad opening bordered by three fibrous flaps. These flaps, called **cusps**, are part of the **tricuspid** (trī-KUS-pid; *tri,* three) **valve**, also known as the *right atrioventricular (AV) valve*. The free edge of each cusp is attached to connective tissue fibers called the **chordae tendineae** (KOR-dē TEN-dih-nē-ē; tendinous cords). The fibers originate at the **papillary** (PAP-ih-lehr-ē) **muscles**, conical muscular projections that arise from the inner surface of the right ventricle (Figure 20–5c).

Tips & Tools

The saying "To tug on your heartstrings" may help you remember the functions of the papillary muscles and the chordae tendineae: Contractions of the papillary muscles pull on the chordae tendineae, which "tug" on your heart's valves.

The Right Ventricle, Pulmonary Valve, and Pulmonary Trunk

The internal surface of the **right ventricle** contains a series of muscular ridges: the **trabeculae carneae** (trah-BEK-yū-lē KAR-nē-ē; *carneus,* fleshy). The *moderator band* is a muscular ridge that extends horizontally from the inferior portion of the interventricular septum and connects to the anterior papillary muscle. The moderator band delivers the stimulus for contraction to the papillary muscles. As a result, they begin tensing the chordae tendineae before the rest of the ventricle contracts.

The superior end of the right ventricle tapers to the **conus arteriosus**, a cone-shaped pouch that ends at the **pulmonary valve**, or *pulmonary semilunar valve*. The pulmonary valve consists of three semilunar (half-moon shaped) cusps of thick connective tissue. Blood flowing from the right ventricle passes through this valve into the **pulmonary trunk**, the start of the pulmonary circuit. Once in the pulmonary trunk, blood flows into the **left pulmonary arteries** and the **right pulmonary arteries**. These vessels branch repeatedly within the lungs before supplying the capillaries, where gas exchange occurs.

Figure 20–5 The Sectional Anatomy of the Heart.

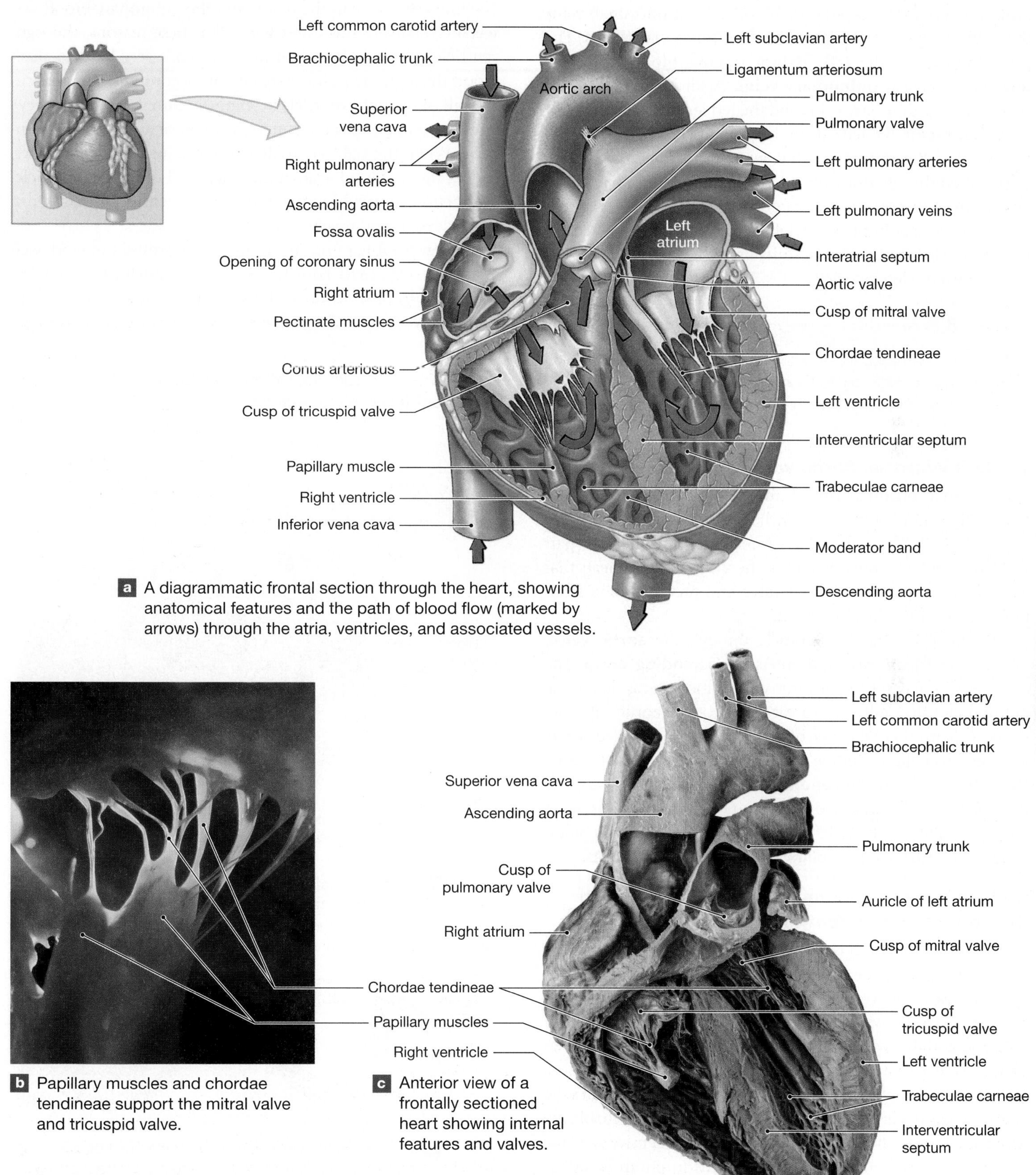
Left common carotid artery
Brachiocephalic trunk
Aortic arch
Left subclavian artery
Ligamentum arteriosum
Pulmonary trunk
Pulmonary valve
Left pulmonary arteries
Left pulmonary veins
Superior vena cava
Right pulmonary arteries
Ascending aorta
Fossa ovalis
Opening of coronary sinus
Right atrium
Pectinate muscles
Conus arteriosus
Cusp of tricuspid valve
Papillary muscle
Right ventricle
Inferior vena cava
Left atrium
Interatrial septum
Aortic valve
Cusp of mitral valve
Chordae tendineae
Left ventricle
Interventricular septum
Trabeculae carneae
Moderator band
Descending aorta
Left subclavian artery
Left common carotid artery
Brachiocephalic trunk
Superior vena cava
Ascending aorta
Pulmonary trunk
Cusp of pulmonary valve
Auricle of left atrium
Right atrium
Cusp of mitral valve
Chordae tendineae
Papillary muscles
Right ventricle
Cusp of tricuspid valve
Left ventricle
Trabeculae carneae
Interventricular septum
20

a A diagrammatic frontal section through the heart, showing anatomical features and the path of blood flow (marked by arrows) through the atria, ventricles, and associated vessels.

b Papillary muscles and chordae tendineae support the mitral valve and tricuspid valve.

c Anterior view of a frontally sectioned heart showing internal features and valves.

? Beginning with the right atrium, what is the order of the valves through which blood will pass?

The Pulmonary Veins, Left Atrium, and Mitral Valve

From the respiratory capillaries, blood collects into small veins that ultimately unite to form the four pulmonary veins. The posterior wall of the **left atrium** receives blood from two **left** and two **right pulmonary veins**. Again there is no valve between the pulmonary veins and the left atrium. Like the right atrium, however, the left atrium has an auricle.

A valve, the **mitral** (MĪ-tral; *mitre,* a bishop's hat) **valve** (also called the *left atrioventricular [AV] valve,* or *bicuspid valve*) guards the entrance to the left ventricle (see Figure 20–5a,c). As the name *bicuspid* implies, the left AV valve contains two cusps, not three. The mitral valve permits blood to flow from the left atrium into the left ventricle.

Tips & Tools

To remember the locations of the tricuspid and bicuspid (mitral) valves, think "***try*** to be ***right***" for the ***tri***cuspid, and associate the ***l*** in mitral with the ***l*** in ***l***eft.

The Left Ventricle, Aortic Valve, and Ascending Aorta

The internal organization of the **left ventricle** resembles that of the right ventricle, but it has no moderator band (see Figure 20–5a). The trabeculae carneae are prominent. A pair of large papillary muscles tenses the chordae tendineae that anchor the cusps of the mitral valve and prevent blood from flowing back into the left atrium.

20

Blood leaves the left ventricle through the **aortic valve,** or *aortic semilunar valve,* and enters the **ascending aorta**. The arrangement of cusps in the aortic valve is the same as that in the pulmonary valve. Adjacent to each cusp of the aortic valve are saclike expansions of the base of the ascending aorta called **aortic sinuses**. From the ascending aorta, blood flows through the **aortic arch** and into the **descending aorta** (see Figure 20–5a). The pulmonary trunk is attached to the aortic arch by the *ligamentum arteriosum,* a fibrous band left over from an important fetal blood vessel that once linked the pulmonary and systemic circuits.

Structural and Functional Differences between the Left and Right Ventricles

The function of the atria is to collect blood that is returning to the heart and to convey it to the ventricles. The demands on the right and left atria are similar, and the two chambers look almost identical. The demands on the right and left ventricles, however, are very different, and the two have significant structural differences. Even though the two ventricles hold and pump equal amounts of blood, the left ventricle is much larger than the right ventricle. What's the reason? It has thicker walls. These thick, muscular walls enable the left ventricle to push blood through the body's extensive systemic circuit. In contrast, the right ventricle needs to pump blood, at lower pressure, only about 15 cm (6 in.) to and from the lungs.

Anatomical differences between the left and right ventricles are easiest to see in a three-dimensional view (Figure 20–6a). The lungs are close to the heart, and the pulmonary blood vessels are relatively short and wide. For these reasons, the right ventricle normally does not need to work very hard to push blood through the pulmonary circuit. Accordingly, the muscular wall of the right ventricle is relatively thin. In sectional view, it resembles a pouch attached to the massive wall of the left ventricle. When the right ventricle contracts, it acts like a bellows, squeezing the blood against the thick wall of the left ventricle. This action moves blood very efficiently with minimal effort, but it develops relatively low pressures.

A comparable pumping arrangement would not work well for the left ventricle. Four to six times as much pressure must be exerted to push through the systemic circuit as through the pulmonary circuit. The left ventricle has an extremely thick

Figure 20–6 Structural Differences between the Left and Right Ventricles. ATLAS: Plate 45d

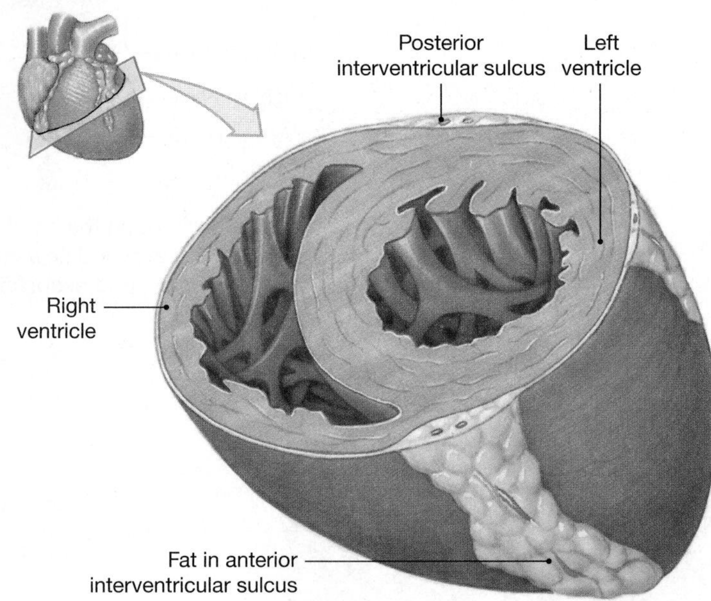
Posterior interventricular sulcus
Left ventricle
Right ventricle
Fat in anterior interventricular sulcus

a A diagrammatic sectional view through the heart, showing the relative thicknesses of the two ventricular walls. Note the pouchlike shape of the right ventricle and the greater thickness of the left ventricular muscle.

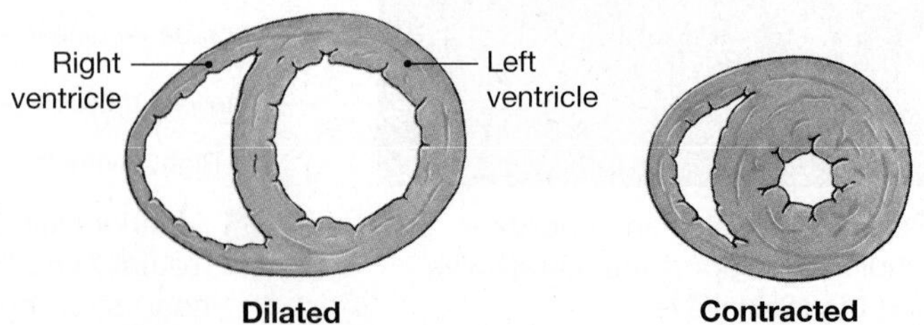
Right ventricle
Left ventricle
Dilated
Contracted

b Diagrammatic views of the ventricles just before a contraction (dilated) and just after a contraction (contracted).

muscular wall and is round in cross section (see Figure 20–6a). When this ventricle contracts, it shortens and narrows. In other words, (1) the distance between the base and apex decreases, and (2) the diameter of the ventricular chamber decreases. The effect is similar to simultaneously squeezing and rolling up the end of a toothpaste tube. The pressure generated is more than enough to open the aortic valve and eject blood into the ascending aorta.

As the powerful left ventricle contracts, it bulges into the right ventricular cavity (Figure 20–6b). This action increases the pumping efficiency of the right ventricle. Individuals with severe damage to the right ventricle may survive, because the contraction of the left ventricle helps push blood into the pulmonary circuit. We return to this topic in Chapter 21, where we consider the integrated functioning of the cardiovascular system.

Blood Flow through the Heart Valves

As we have seen, the heart has two pairs of one-way valves. These valves prevent the backflow of blood as the chambers contract. Let's look at the function of these heart valves, and the path of blood flow through each half of the heart.

The AV Valves: Atria to Ventricles

The atrioventricular (AV) valves prevent the backflow of blood from the ventricles to the atria when the ventricles are contracting. The chordae tendineae and papillary muscles play important roles in the normal function of the AV valves. When the ventricles are relaxed, the chordae tendineae are loose, and the AV valves offer no resistance as blood flows from the atria into the ventricles (Figure 20–7a). When the ventricles contract, blood moving back toward the atria swings the cusps together, closing the valves (Figure 20–7b). At the same time, the contraction of the papillary muscles tenses the chordae tendineae, stopping the cusps before they swing into the atria. If the chordae tendineae were cut or the papillary muscles were damaged, backflow, called **regurgitation**, of blood into the atria would occur each time the ventricles contracted.

+ Clinical Note Faulty Heart Valves

Each of the four heart valves must open and close crisply and precisely to permit the proper flow of blood through the heart.

The most common valve to falter is the mitral valve. One scenario responsible for mitral malfunction is an untreated bacterial or viral infection that infiltrates the valve cusps. The cusps become inflamed and later scar, resulting in a faulty valve.

A valve can malfunction in one of three ways: (1) It can become rigid (a *stenotic* valve) so that it does not open fully, (2) it can fail to close properly (a *regurgitant* valve), or (3) it can actually flop backwards (a *prolapsed* valve). Faulty valves are heard as *heart murmurs* with a stethoscope.

The Semilunar Valves: Ventricles to Great Vessels

The pulmonary and aortic semilunar valves prevent the backflow of blood from the pulmonary trunk and aorta into the right and left ventricles, respectively. Unlike the AV valves, the semilunar valves do not need muscular braces, because the arterial walls do not contract and the relative positions of the cusps are stable. When the semilunar valves close, the three symmetrical cusps support one another like the legs of a tripod (see Figure 20–7a). When the aortic valve opens, the aortic sinuses prevent the individual cusps from sticking to the wall of the aorta (see Figure 20–7b).

Serious valve problems can interfere with the working of the heart. If valve function deteriorates to the point at which the heart cannot maintain adequate circulatory flow, symptoms of **valvular heart disease (VHD)** appear. Congenital malformations may be responsible, but in many cases the condition develops after **carditis**, an inflammation of the heart, occurs. One important cause of carditis is **rheumatic** (rū-MAT-ik) **fever**, an inflammatory autoimmune response to an infection by streptococcal bacteria. It most often occurs in children.

The Blood Supply to the Heart

The heart works continuously, so cardiac muscle cells need reliable supplies of oxygen and nutrients. A great volume of blood flows through the chambers of the heart, but the myocardium has its own, separate blood supply. The **coronary circulation** supplies blood to the muscle tissue of the heart. During maximum exertion, the heart's demand for oxygen rises considerably. The blood flow to the myocardium may then increase to nine times that of the resting level. The coronary circulation includes an extensive network of coronary blood vessels, both arteries and veins (Figure 20–8).

The Coronary Arteries

The left and right **coronary arteries** originate at the base of the ascending aorta, at the aortic sinuses (Figure 20–8a). Blood pressure here is the highest in the systemic circuit. When the left ventricle contracts and forces blood into the aorta, the high pressure of this blood stretches the elastic walls of the aorta. When the left ventricle relaxes, blood no longer flows into the aorta, pressure declines, and the walls of the aorta recoil. This recoil, called *elastic rebound*, pushes blood both forward, into the systemic circuit, and backward, through the left and right aortic sinuses and then into the respective coronary arteries. In this way, the combination of elevated blood pressure and elastic rebound ensures a continuous flow of blood to meet the demands of active cardiac muscle tissue.

The *right* and *left coronary arteries*, which deliver blood to the myocardium, originate at the right and left aortic sinuses. The **right coronary artery** follows the coronary sulcus around the heart (see Figure 20–8). It supplies blood to (1) the right atrium, (2) portions of both ventricles, and (3) portions of the

Figure 20–7 Valves of the Heart and Blood Flow. Red (oxygenated) and blue (deoxygenated) arrows indicate blood flow into or out of a ventricle, smaller red arrows show blood flow into an atrium, and green arrows indicate contraction of the ventricles.

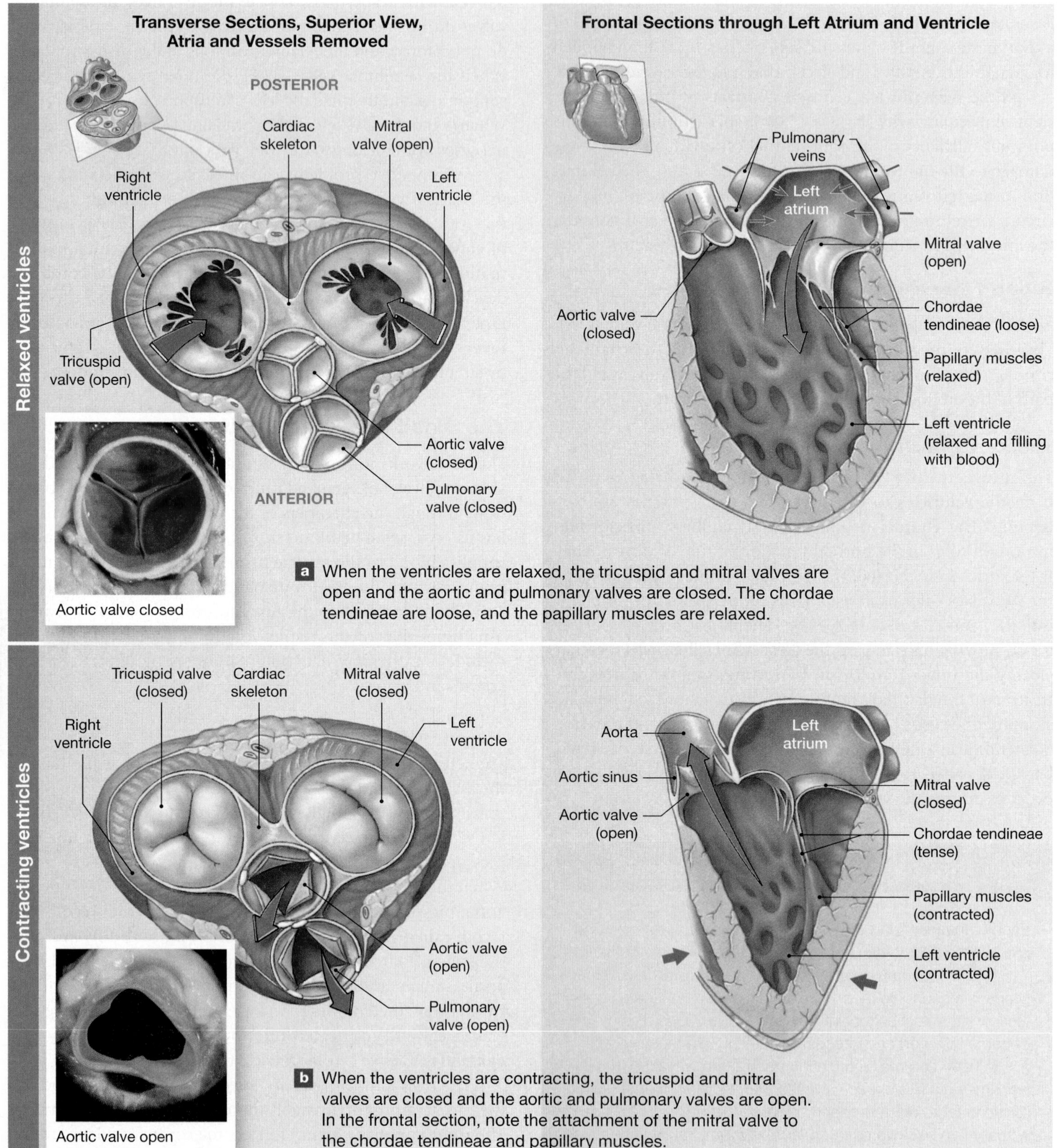
Relaxed ventricles
Transverse Sections, Superior View, Atria and Vessels Removed
POSTERIOR
Cardiac skeleton
Mitral valve (open)
Right ventricle
Left ventricle
Tricuspid valve (open)
Aortic valve (closed)
Pulmonary valve (closed)
ANTERIOR
Aortic valve closed
Frontal Sections through Left Atrium and Ventricle
Pulmonary veins
Left atrium
Mitral valve (open)
Aortic valve (closed)
Chordae tendineae (loose)
Papillary muscles (relaxed)
Left ventricle (relaxed and filling with blood)
Contracting ventricles
Tricuspid valve (closed)
Cardiac skeleton
Mitral valve (closed)
Right ventricle
Left ventricle
Aortic valve (open)
Pulmonary valve (open)
Aortic valve open
Aorta
Aortic sinus
Aortic valve (open)
Left atrium
Mitral valve (closed)
Chordae tendineae (tense)
Papillary muscles (contracted)
Left ventricle (contracted)

a When the ventricles are relaxed, the tricuspid and mitral valves are open and the aortic and pulmonary valves are closed. The chordae tendineae are loose, and the papillary muscles are relaxed.

b When the ventricles are contracting, the tricuspid and mitral valves are closed and the aortic and pulmonary valves are open. In the frontal section, note the attachment of the mitral valve to the chordae tendineae and papillary muscles.

Figure 20–8 **The Coronary Circulation.** ATLAS: Plates 45b,c

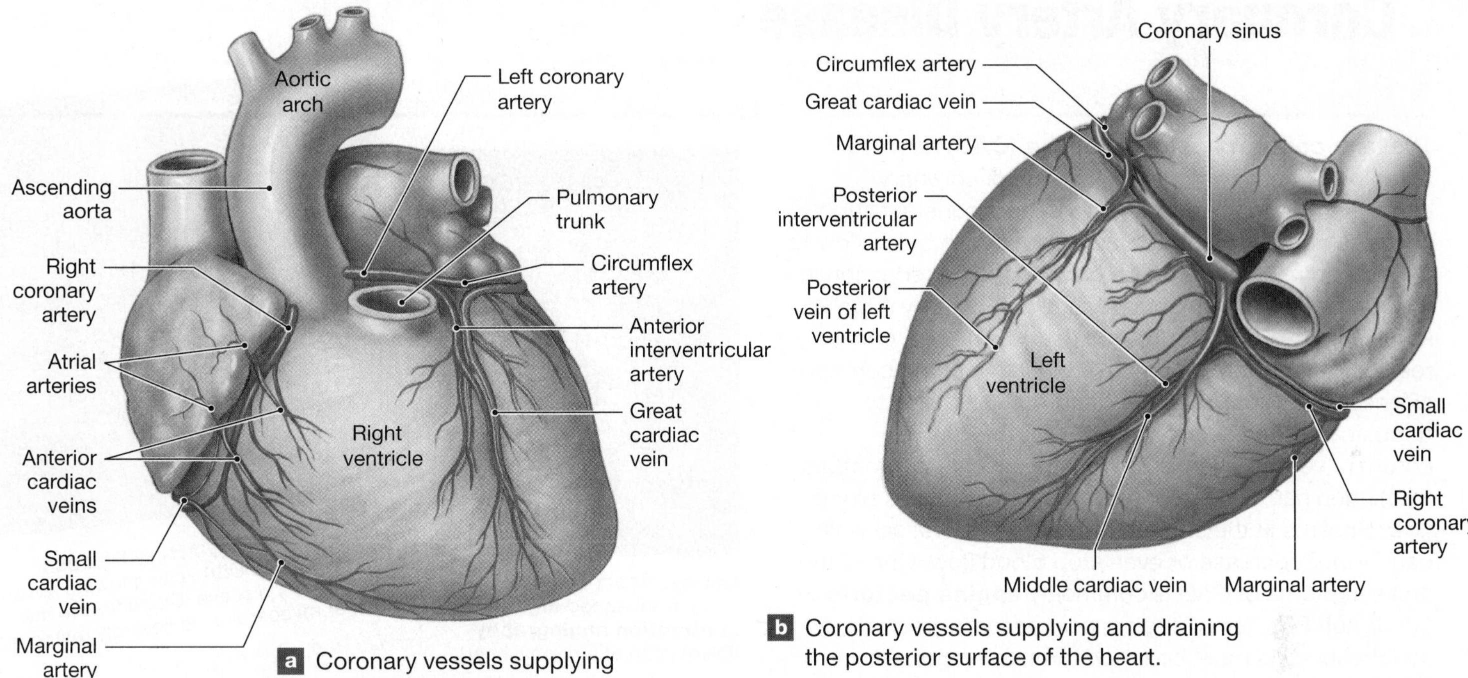
Aortic arch
Left coronary artery
Ascending aorta
Pulmonary trunk
Right coronary artery
Circumflex artery
Anterior interventricular artery
Atrial arteries
Great cardiac vein
Right ventricle
Anterior cardiac veins
Small cardiac vein
Marginal artery
Coronary sinus
Circumflex artery
Great cardiac vein
Marginal artery
Posterior interventricular artery
Posterior vein of left ventricle
Left ventricle
Small cardiac vein
Right coronary artery
Middle cardiac vein
Marginal artery

a Coronary vessels supplying and draining the anterior surface of the heart.

b Coronary vessels supplying and draining the posterior surface of the heart.

electrical conducting system of the heart. Inferior to the right atrium, the right coronary artery generally gives rise to one or more **marginal arteries**, which extend across the surface of the right ventricle. The right coronary artery then continues across the posterior surface of the heart. It supplies the **posterior interventricular artery**, or *posterior descending artery*, which runs toward the apex within the posterior interventricular sulcus. The posterior interventricular artery supplies blood to the interventricular septum and adjacent portions of the ventricles.

The **left coronary artery** supplies blood to the left ventricle, left atrium, and interventricular septum (see Figure 20–8). As it reaches the anterior surface of the heart, it gives rise to a circumflex branch and an anterior interventricular branch. The **circumflex artery** curves to the left around the coronary sulcus. It eventually meets and fuses with small branches of the right coronary artery.

The much larger **anterior interventricular artery**, or *left anterior descending artery (LAD)*, swings around the pulmonary trunk and runs along the surface within the anterior interventricular sulcus. The anterior interventricular artery supplies small tributaries continuous with those of the posterior interventricular artery. Such interconnections between arteries are called **arterial anastomoses** (ah-nas-tō-MŌ-sēz; *anastomosis*, outlet). Because the arteries are interconnected in this way, the blood supply to the cardiac muscle remains relatively constant despite pressure fluctuations in the left and right coronary arteries as the heart beats.

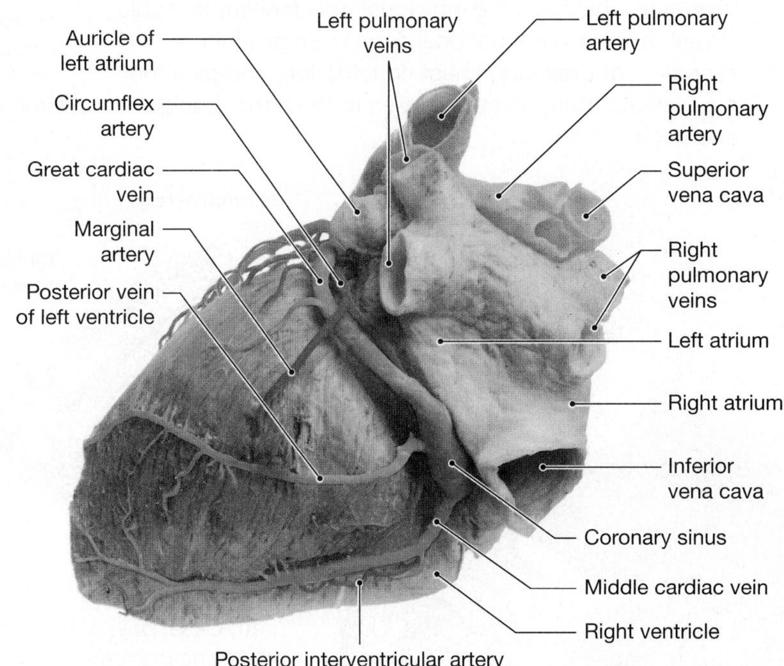
Auricle of left atrium
Left pulmonary veins
Left pulmonary artery
Circumflex artery
Right pulmonary artery
Great cardiac vein
Superior vena cava
Marginal artery
Right pulmonary veins
Posterior vein of left ventricle
Left atrium
Right atrium
Inferior vena cava
Coronary sinus
Middle cardiac vein
Right ventricle
Posterior interventricular artery

c A posterior view of the heart; the vessels have been injected with colored latex (liquid rubber).

Coronary artery disease is characterized by interrupted blood flow to the myocardium. **Spotlight Figure 20–9** describes this condition, along with myocardial infarction.

The Cardiac Veins

The various cardiac veins are also shown in Figure 20–8. The **great cardiac vein** begins on the anterior surface of the

20

Coronary Artery Disease

HEART DISEASE

The term **coronary artery disease (CAD)** refers to areas of partial or complete blockage of coronary circulation. Cardiac muscle cells need a constant supply of oxygen and nutrients, so any reduction in blood flow to the heart muscle produces a corresponding reduction in cardiac performance. Such reduced circulatory supply, known as **coronary ischemia** (iss-KĒ-mē-uh), generally results from partial or complete blockage of the coronary arteries. The usual cause is the formation of a fatty deposit, or *atherosclerotic plaque*, in the wall of a coronary vessel. The plaque, or an associated *thrombus* (clot), then narrows the passageway and reduces blood flow. Spasms in the smooth muscles of the vessel wall can further decrease or even stop blood flow. One of the first symptoms of CAD is commonly **angina pectoris** (an-JĪ-nuh PEK-tor-is; *angina*, pain spasm + *pectoris*, of the chest). In its most common form, a temporary ischemia develops when the workload of the heart increases. Although the individual may feel comfortable at rest, exertion or emotional stress can produce a sensation of pressure, chest constriction, and pain that may radiate from the sternal area to the arms, back, and neck.

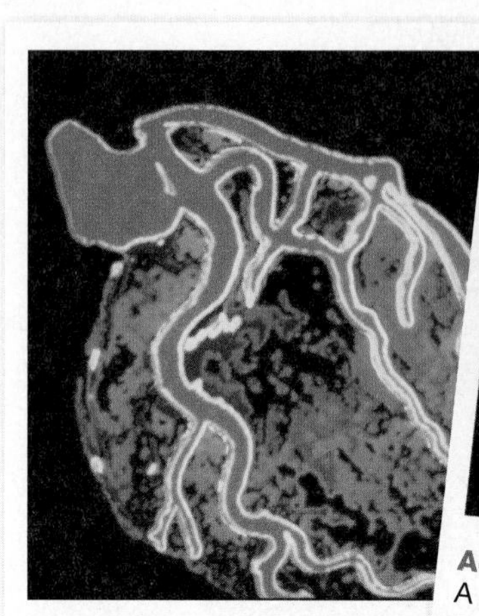

Normal Heart
A color-enhanced **digital subtraction angiography (DSA)** scan of a normal heart.

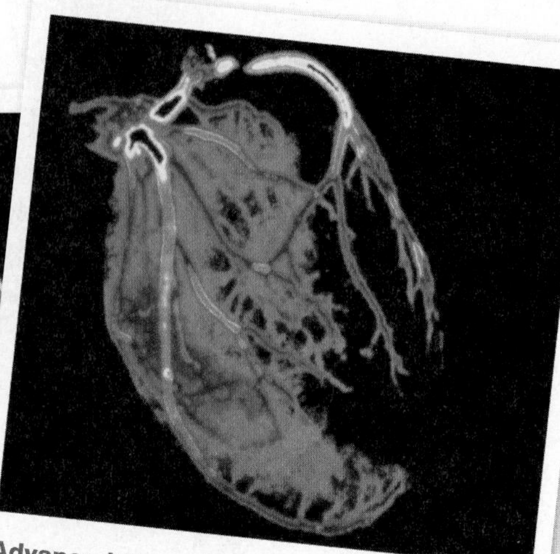

Advanced Coronary Artery Disease
A color-enhanced DSA scan showing advanced coronary artery disease. Blood flow to the ventricular myocardium is severely restricted.

Plaques may be visible by ***angiography*** *or high-resolution ultrasound, and the effects on coronary blood flow can be detected in digital subtraction angiography (DSA) scans of the heart as shown above.*

Normal Artery

Tunica externa

Tunica media

Cross section

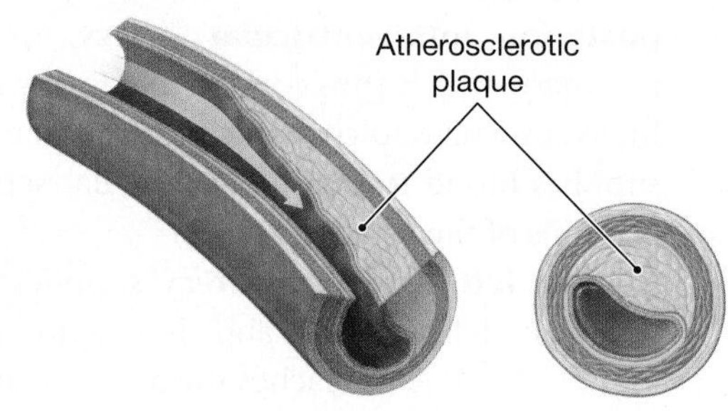
Narrowing of Artery
Atherosclerotic plaque
Cross section

Left: *Chest pain or discomfort is a common symptom of angina pectoris. Although similar symptoms may accompany a heart attack, during a heart attack, pain often radiates down the left arm.*

Below: *The presence of any two of these coronary risk factors more than doubles the risk of heart attack.*

Risk Factors for CAD and Myocardial Infarction

SMOKING • HIGH BLOOD PRESSURE • HIGH CHOLESTEROL •

• MALE OVER AGE 70 • SEVERE EMOTIONAL STRESS •

Occluded coronary artery
Damaged heart muscle

Myocardial Infarction

HEART ATTACK

In a **myocardial** (mī-ō-KAR-dē-al) **infarction (MI)**, or *heart attack*, part of the coronary circulation becomes blocked, and cardiac muscle cells die from lack of oxygen. The death of affected tissue creates a nonfunctional area known as an *infarct*. Heart attacks most commonly result from severe coronary artery disease (CAD). The consequences depend on the site and nature of the circulatory blockage. If it occurs near the start of one of the coronary arteries, the damage will be widespread and the heart may stop beating. If the blockage involves one of the smaller arterial branches, the individual may survive the immediate crisis but may have many complications such as reduced contractility and cardiac arrhythmias.

A crisis often develops as a result of thrombus formation at a plaque (the most common cause of an MI), a condition called **coronary thrombosis**. A vessel already narrowed by plaque formation may also become blocked by a sudden spasm in the smooth muscles of the vascular wall.

Individuals having an MI experience intense pain, similar to that felt in angina, but persisting even at rest. However, pain does not always accompany a heart attack, and silent heart attacks may be even more dangerous than more apparent attacks, because the condition may go undiagnosed and may not be treated before a fatal MI occurs. A myocardial infarction can usually be diagnosed with an electrocardiogram (ECG) and blood studies. Damaged myocardial cells release enzymes into the circulation, and these elevated enzymes can be measured in diagnostic blood tests. The enzymes include **cardiac troponin T**, **cardiac troponin I**, and a special form of creatinine phosphokinase, **CK-MB**.

Treatment of CAD and Myocardial Infarction

About 25% of MI patients die before obtaining medical assistance, and 65% of MI deaths among those under age 50 occur within an hour after the initial infarction.

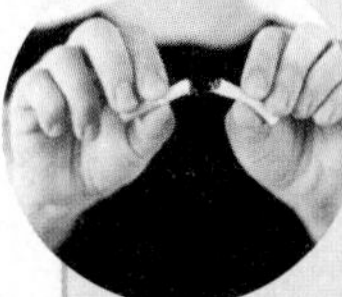

Risk Factor Modification

Stop smoking, treat high blood pressure, adjust diet to lower cholesterol and promote weight loss, reduce stress, and increase physical activity.

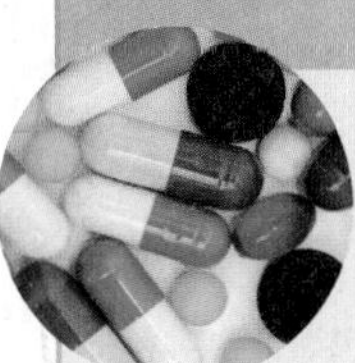

Drug Treatment

- Drugs that reduce coagulation and therefore the risk of thrombosis, such as aspirin and coumadin
- Drugs that block sympathetic stimulation (propranolol or metoprolol)
- Drugs that cause vasodilation, such as **nitroglycerin** (nī-trō-GLIS-er-in)
- Drugs that block calcium ion movement into the cardiac and vascular smooth muscle cells (calcium ion channel blockers)
- In a myocardial infarction, drugs to relieve pain, fibrinolytic agents to help dissolve clots, and oxygen

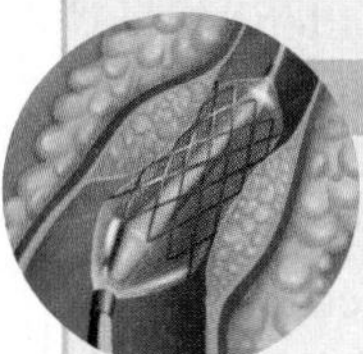

Noninvasive Surgery

- **Atherectomy**. Blockage by a single, soft plaque may be reduced with the aid of a long, slender **catheter** (KATH-e-ter) inserted into a coronary artery to the plaque. A variety of surgical tools can be slid into the catheter, and the plaque can then be removed.
- **Balloon angioplasty** (AN-jē-ō-plas-tē; *angeion*, vessel). In balloon angioplasty, the tip of the catheter contains an inflatable balloon. Once in position, the balloon is inflated, pressing the plaque against the vessel walls. Because plaques commonly redevelop after angioplasty, a fine tubular wire mesh called a *stent* may be inserted into the vessel, holding it open.

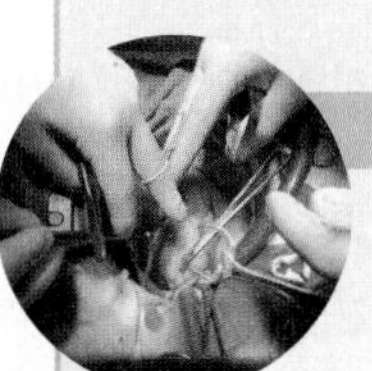

Coronary Artery Bypass Graft (CABG)

In a **coronary artery bypass graft**, a small section is removed from either a small artery or a peripheral vein and is used to create a detour around the obstructed portion of a coronary artery. As many as four coronary arteries can be rerouted this way during a single operation. The procedures are named according to the number of vessels repaired, so we speak of single, double, triple, or quadruple coronary bypasses.

IABETES • HIGH CIRCULATING LOW-DENSITY LIPOPROTEI

SITY • SEDENTARY LIFESTYLE • GENETIC PREDISPOSITI

Clinical Note Broken-Heart Syndrome

The popular notion that grief can "break your heart" finds support in the novel diagnosis of *broken-heart syndrome*, also called *stress-induced cardiomyopathy*. First noted in Japan in the early 2000s, elderly women were presenting with apparent heart attacks. The precipitating event was an acute emotional stress, such as the death of a loved one. The women's ECGs were abnormal; their left ventricles were significantly dilated. But their coronary arteries—the site of the lesions responsible for heart attacks—were clear! All of these changes reverted to normal within a month of supportive care.

ventricles, along the interventricular sulcus. This vein drains blood from the region supplied by the anterior interventricular artery, a branch of the left coronary artery. The great cardiac vein reaches the level of the atria and then curves around the left side of the heart within the coronary sulcus.

The cardiac veins draining the myocardium return blood to the **coronary sinus**, a large, thin-walled vein. The coronary sinus opens into the right atrium near the base of the inferior vena cava.

Other cardiac veins empty into the great cardiac vein or the coronary sinus. These veins include (1) the **posterior vein of left ventricle**, draining the area served by the circumflex artery; (2) the **middle cardiac vein**, draining the area supplied by the posterior interventricular artery; and (3) the **small cardiac vein**, which receives blood from the posterior surfaces of the right atrium and ventricle. The **anterior cardiac veins** (also called the *anterior veins of right ventricle*), which drain the anterior surface of the right ventricle, empty directly into the right atrium.

20

Checkpoint

1. **Damage to the semilunar valve of the right ventricle would affect blood flow into which vessel?**
2. **What prevents the AV valves from swinging into the atria?**
3. **Why is the left ventricle more muscular than the right ventricle?**

See the blue Answers tab at the back of the book.

20-2 The cells of the conducting system distribute electrical impulses through the heart, causing cardiac contractile cells to contract

Learning Outcome **Explain the events of an action potential in cardiac muscle, indicate the importance of calcium ions to the contractile process, describe the conducting system of the heart, and identify the electrical events associated with a normal electrocardiogram.**

Next we start to look at how the heart works—cardiac physiology. Then we examine the structure and function of the different types of cardiac muscle cells. Some form the *conducting system*, an internal network that uses electrical signals to coordinate the contractions of cardiac muscle cells. The electrical events of the heart can then be recorded in an *electrocardiogram (ECG)*. Finally we discuss the contraction of the heart at the cellular level.

Cardiac Physiology: Electrical Impulses Leading to the Contractions Making Up a Heartbeat

In a single cardiac contraction, or **heartbeat**, all of the chambers of the heart contract in series—first the atria and then the ventricles. Two types of cardiac muscle cells are involved in a normal heartbeat: (1) Specialized *autorhythmic cells* (pacemaker and conducting) of the conducting system control and coordinate the heartbeat, and (2) *contractile cells* produce the powerful contractions that propel blood. The electrical impulses of the conducting system initiate the contraction of the heart chambers.

Each heartbeat begins with an action potential generated by cells of the conducting system. Other cells of this system then propagate and distribute this electrical impulse to stimulate contractile cells to push blood in the right direction at the proper time. The arrival of an electrical impulse at a cardiac contractile cell's plasma membrane produces an action potential that is comparable to an action potential in a skeletal muscle fiber. As in a skeletal muscle fiber, this action potential triggers the contraction of the cardiac contractile cell. The actual contraction lags behind the beginning of the action potential. Because of the coordination provided by the conducting system, the atria contract first, driving blood into the ventricles through the AV valves, and the ventricles contract next, driving blood out of the heart through the semilunar valves.

A heartbeat lasts only about 370 milliseconds (msec). Although brief, it is a very busy period! Let's follow the steps that produce a single heartbeat.

The Conducting System: Pacemaker and Conducting Cells

Unlike skeletal muscle, cardiac muscle tissue contracts on its own, without neural or hormonal stimulation; this property is called **autorhythmicity**. The cells that initiate and distribute the stimulus to contract are part of the heart's **conducting system.** This system is a network of specialized cardiac muscle cells, called *pacemaker cells* and *conducting cells*, that initiates and distributes electrical impulses. Let's look at these types of cells, and how they function in the conducting system.

Go to MasteringA&P™ > Study Area > Menu > Lab Tools > Bone and Dissection Videos > Sheep Heart: External

Figure 20–10 The Conducting System of the Heart and the Pacemaker Potential

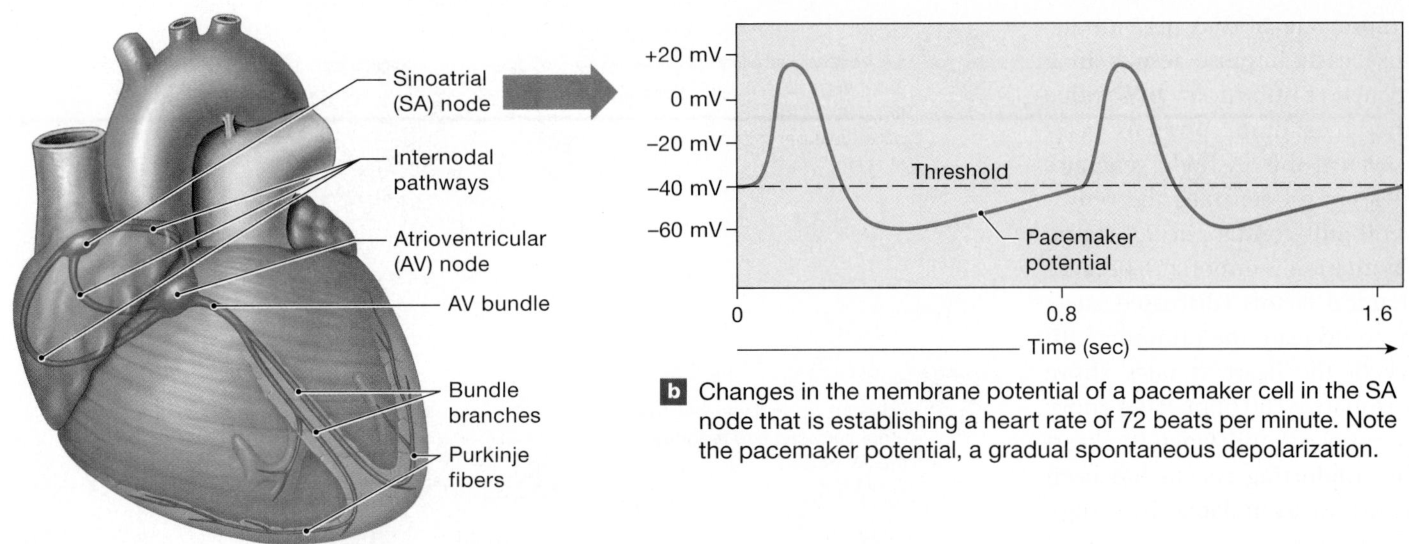
Sinoatrial (SA) node
Internodal pathways
Atrioventricular (AV) node
AV bundle
Bundle branches
Purkinje fibers
+20 mV
0 mV
–20 mV
–40 mV
–60 mV
Threshold
Pacemaker potential
0
0.8
1.6
Time (sec)

a Components of the conducting system.

b Changes in the membrane potential of a pacemaker cell in the SA node that is establishing a heart rate of 72 beats per minute. Note the pacemaker potential, a gradual spontaneous depolarization.

Components of the Conducting System

The specialized cardiac cells of the conducting system are collected in various locations. This section introduces the components of this system (Figure 20–10a).

Pacemaker Cells in the Sinoatrial and Atrioventricular Nodes. Pacemaker cells are essential to establishing the normal *heart rate*. They are also called *nodal cells* because they are found in two nodes, or locations:

- The **sinoatrial** (sī-nō-Ā-trē-al) **(SA) node** is embedded in the posterior wall of the right atrium, near the entrance of the superior vena cava. Because the SA node is the primary driver of the heart rate, it is also known as the *cardiac pacemaker*.
- The relatively large **atrioventricular (AV) node** is located at the junction between the atria and ventricles, near the opening of the coronary sinus. The pacemaker cells of this node send on signals from the cells of the SA node, and act as backup to the SA node pacemaker cells.

Conducting Cells in the Internodal Pathways, AV Bundles, Bundle Branches, and Purkinje Fibers. Conducting cells interconnect the SA and AV nodes, and distribute the contractile stimulus throughout the myocardium. These cells are found in two main locations:

- In the atria, conducting cells are found in **internodal pathways** in the atrial walls. These pathways distribute the contractile stimulus to atrial muscle cells as this electrical impulse travels from the SA node to the AV node.
- In the ventricles, conducting cells include those in the **atrioventricular (AV) bundle** (also called the *bundle of His*) and the **bundle branches** that run between the ventricles, as well as the **Purkinje** (pur-KIN-jē) **fibers**, which distribute the stimulus to the ventricular myocardium.

Most of the cells of the conducting system are smaller than the contractile cells of the myocardium and contain very few myofibrils. Purkinje fibers, however, are much larger in diameter than the contractile cells.

20

Pacemaker Cells and Heart Rate

Pacemaker cells of the SA and AV nodes share a special characteristic: Their excitable membranes do not have a stable resting membrane potential. Each time a pacemaker cell repolarizes, its membrane potential drifts toward threshold. This gradual depolarization is called a **pacemaker potential** (Figure 20–10b). The pacemaker potential results from a slow inflow of Na^+ without a compensating outflow of K^+.

The rate of spontaneous depolarization differs in various parts of the conducting system. It is fastest at the SA node. Without neural or hormonal stimulation, the SA node generates action potentials at a rate of 60–100 per minute. Isolated cells of the AV node depolarize more slowly, generating 40–60 action potentials per minute. Because the SA node reaches threshold first, it establishes the basic *heart rhythm*, or **sinus rhythm**. (While heart rhythm is essential to heart rate, the heart rate is the number of beats per minute.) In other words, the impulse generated by the SA node brings the AV pacemaker cells to threshold faster than does the pacemaker potential of the AV pacemaker cells. The normal resting heart rate is somewhat slower than 80–100 beats per minute (bpm), however, due to the effects of parasympathetic innervation. (We discuss the influence of autonomic innervation on heart rate in Section 20–4.)

The pacemaker cells of the AV node can conduct impulses at a maximum rate of 230 per minute. Because each impulse results in a ventricular contraction, this value is the maximum normal heart rate. Even if the SA node generates impulses at a faster rate, the ventricles will still contract at 230 bpm. This limitation is important, because mechanical factors (discussed later) begin to decrease the pumping efficiency of the heart at rates above approximately 180 bpm. Rates above 230 bpm occur only when the heart or the conducting system has been damaged or stimulated by drugs. As ventricular rates increase toward their theoretical maximum limit of 300–400 bpm, pumping effectiveness becomes dangerously, if not fatally, reduced.

Impulse Conduction through the Heart

Now let's trace the path of an impulse from its initiation at the SA node, examining its effects on the surrounding myocardium as we proceed.

Functions of the Pacemaker Cells. An action potential produced by the pacemaker cells of the SA node (Figure 20–11 ①) takes approximately 50 msec to travel to the AV node along the internodal pathways. Along the way, the conducting cells pass the stimulus to contractile cells of both atria. The action potential then spreads across the atrial surfaces by cell-to-cell contact (Figure 20–11 ②). The stimulus affects only the atria, because the cardiac skeleton isolates the atrial myocardium from the ventricular myocardium.

The impulse slows as it leaves the internodal pathways and enters the AV node, because the nodal cells are smaller in diameter than the conducting cells. (Chapter 12 discussed

Figure 20–11 Impulse Conduction through the Heart and Accompanying ECG Tracings. The ECG tracings are explained in the next section and in Figure 20–12 on page 706. After reading about the electrocardiogram, come back to this figure to integrate the electrical activity of the heart and atrial and ventricular contractions.

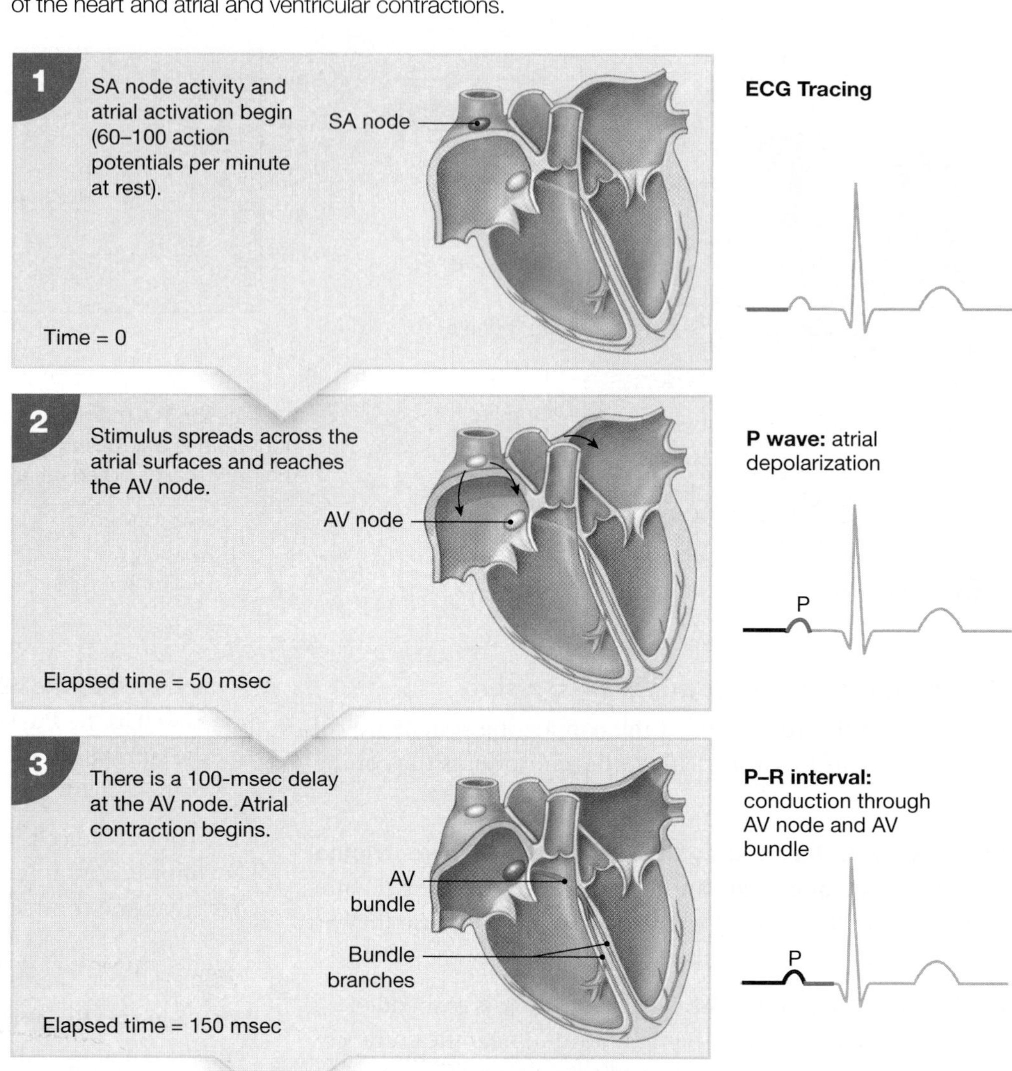
1
SA node activity and atrial activation begin (60–100 action potentials per minute at rest).
SA node
Time = 0
ECG Tracing
2
Stimulus spreads across the atrial surfaces and reaches the AV node.
AV node
Elapsed time = 50 msec
P wave: atrial depolarization
P
3
There is a 100-msec delay at the AV node. Atrial contraction begins.
AV bundle
Bundle branches
Elapsed time = 150 msec
P–R interval: conduction through AV node and AV bundle
P

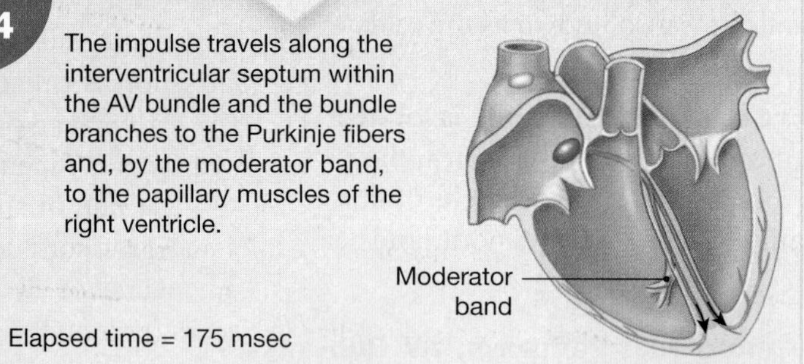
4
The impulse travels along the interventricular septum within the AV bundle and the bundle branches to the Purkinje fibers and, by the moderator band, to the papillary muscles of the right ventricle.
Moderator band
Elapsed time = 175 msec

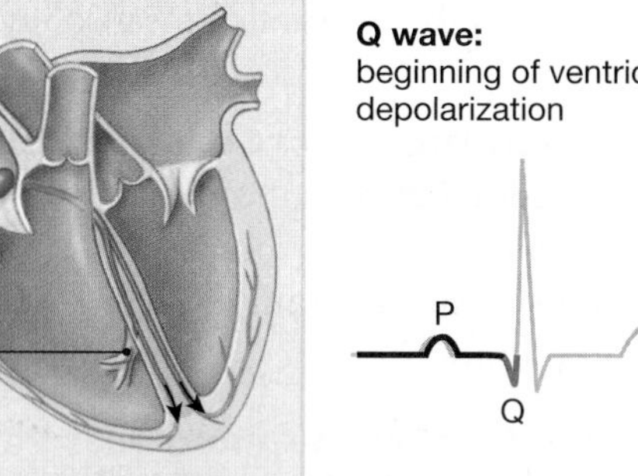
Q wave: beginning of ventricular depolarization
P
Q

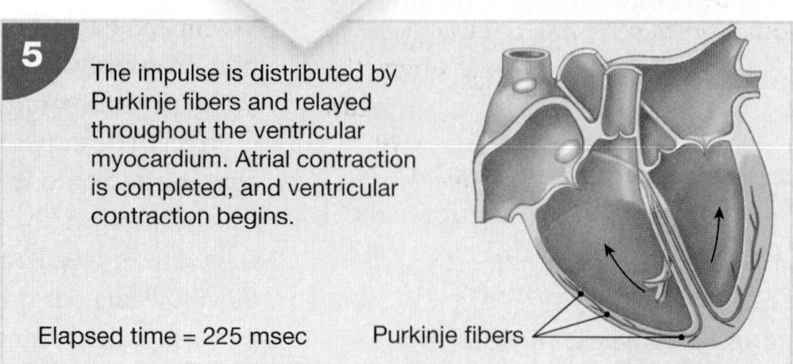
5
The impulse is distributed by Purkinje fibers and relayed throughout the ventricular myocardium. Atrial contraction is completed, and ventricular contraction begins.
Elapsed time = 225 msec
Purkinje fibers

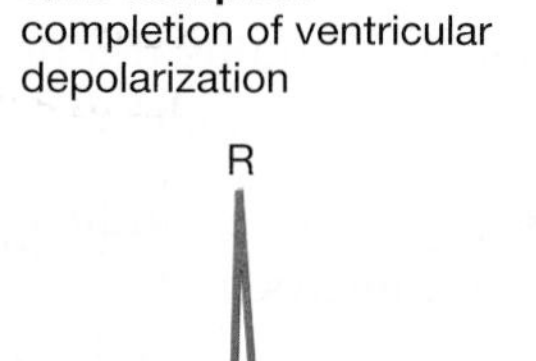
QRS complex: completion of ventricular depolarization
R
P
Q S

? What happens to the electrical impulse when it reaches the AV node?

the relationship between axon diameter and propagation speed. ⮌ p. 416) In addition, the connections between pacemaker cells are less efficient than those between conducting cells at relaying the impulse from one cell to another. As a result, the impulse takes about 100 msec to pass through the AV node (Figure 20–11 ③). This delay is important because it allows the atria to contract before the ventricles do. (The delay is a result of the time it takes for calcium ions to enter the sarcoplasm and activate the contraction process, as described in Chapter 10. ⮌ p. 306) Otherwise, contraction of the powerful ventricles would close the AV valves and prevent blood flow from the atria into the ventricles.

Functions of the Conducting Cells. After the brief delay at the AV node, the impulse is conducted along the AV bundle and the bundle branches to the Purkinje fibers and the papillary muscles (Figure 20–11 ④). The connection between the AV node and the AV bundle is normally the only electrical connection between the atria and the ventricles. Once an impulse enters the AV bundle, it travels to the interventricular septum and enters the **right** and **left bundle branches**. The left bundle branch, which supplies the massive left ventricle, is much larger than the right bundle branch. Both branches extend toward the apex of the heart, turn, and fan out deep to the endocardial surface.

As the bundle branches diverge, they conduct the impulse to both the Purkinje fibers and, through the moderator band, to the papillary muscles of the right ventricle. Because the impulse is delivered to the papillary muscles directly, these muscles begin contracting before the rest of the ventricular musculature does. Contraction of the papillary muscles applies tension to the chordae tendineae, bracing the AV valves. This tension limits the movement of the cusps, preventing the backflow of blood into the atria when the ventricles contract.

The Purkinje fibers then distribute the impulse to the ventricular myocardium (Figure 20–11 ⑤). Purkinje fibers, which radiate from the apex toward the base of the heart, conduct action potentials very rapidly—as fast as small myelinated axons. Within about 75 msec, the signal to begin a contraction has reached all the ventricular cardiac contractile cells, and ventricular contraction begins. By this time, the atria have completed their contractions and ventricular contraction can safely occur. Because of the location of the Purkinje fibers, the ventricles contract in a wave that begins at the apex of the heart and spreads toward the base. The contraction pushes blood toward the base, into the aorta and pulmonary trunk.

In summary, each time the heart beats, a wave of depolarization spreads through the atria, pauses at the AV node, then travels down the interventricular septum to the apex, turns, and spreads through the ventricular myocardium toward the base (see Figure 20–11). The entire process, from the generation of an impulse at the SA node to the complete depolarization of the ventricular myocardium, normally takes around 225 msec.

Disturbances in Heart Rhythm

A number of clinical problems result from abnormal pacemaker function. **Bradycardia** (brād-ē-KAR-dē-uh; *bradys*, slow) is a condition in which the heart rate is slower than normal. **Tachycardia** (tak-ē-KAR-dē-uh; *tachys*, swift) is a faster-than-normal heart rate. These terms are relative, and in clinical practice the definitions vary with the normal resting heart rate of the individual.

Damage to the conducting pathways disturbs the normal rhythm of the heart. The resulting problems are called *conduction deficits*. If the SA node or any of the internodal pathways becomes damaged, the heart continues to beat, but at a slower rate, usually 40–60 bpm, as dictated by the AV node. Certain cells in the Purkinje fiber network depolarize spontaneously at an even slower rate. If the rest of the conducting system is damaged, these cells can stimulate a heart rate of 20–40 bpm. Under normal conditions, cells of the AV bundle, the bundle branches, and most Purkinje fibers do not depolarize spontaneously. If, due to damage or disease, these cells *do* begin depolarizing spontaneously, the heart may no longer pump blood effectively. Death can result if the problem persists.

If an abnormal conducting cell or ventricular contractile cell begins generating action potentials at a higher rate, the impulses can override those of the SA or AV node. The origin of these abnormal signals is called an **ectopic** (ek-TOP-ik; out of place) **pacemaker**. The activity of an ectopic pacemaker partially or completely bypasses the conducting system, disrupting the timing of ventricular contraction. The result may be a dangerous reduction in the pumping efficiency of the heart. Such conditions are commonly diagnosed with the aid of an electrocardiogram.

The Electrocardiogram (ECG)

The electrical events in the heart are powerful enough to be detected by electrodes on the body. A procedure known as *electrocardiography* can monitor the electrical events of the conducting system. A recording of these events is an **electrocardiogram** (ē-lek-trō-KAR-dē-ō-gram), also called an **ECG**, or **EKG** (from the German term). Clinicians can use an ECG to assess the functions of specific pacemaker, conducting, and contractile cells. When a portion of the heart has been damaged by a heart attack, for example, the ECG reveals an abnormal pattern of impulse conduction.

The monitoring electrodes for an ECG are placed at different locations on the body surface. The appearance of the ECG varies with the placement of the monitoring electrodes, or *leads*. Figure 20–12a shows the leads in one of the standard configurations. Figure 20–12b depicts the important features of

Figure 20–12 An Electrocardiogram (ECG).

a Electrode placement for recording a standard ECG.

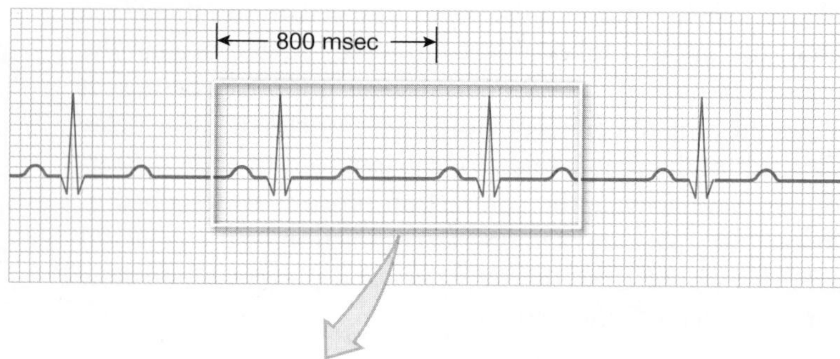
800 msec

SmartArt

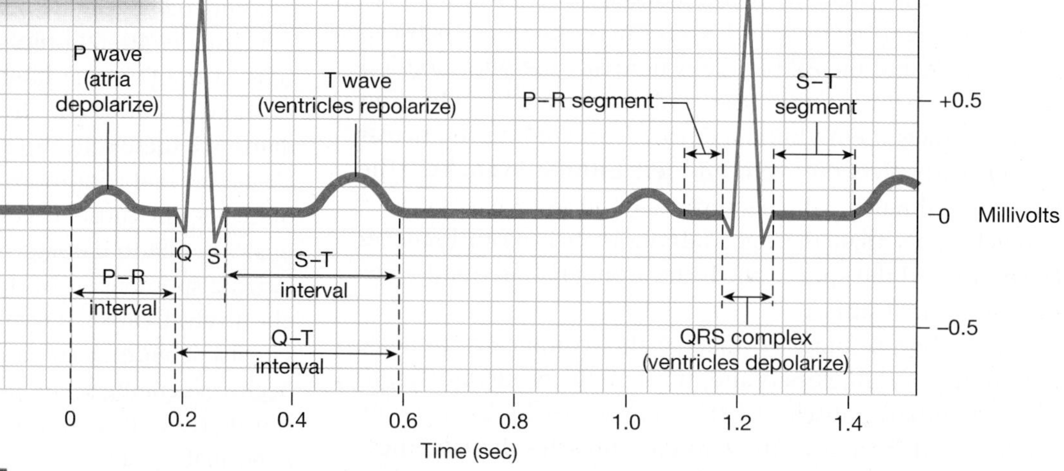
R
R
+1
P wave
(atria
depolarize)
T wave
(ventricles repolarize)
P–R segment
S–T
segment
+0.5
Q
S
S–T
interval
P–R
interval
Q–T
interval
QRS complex
(ventricles depolarize)
–0
Millivolts
–0.5
0
0.2
0.4
0.6
0.8
1.0
1.2
1.4
Time (sec)

b An ECG printout is a strip of graph paper containing a record of the electrical events monitored by the electrodes. The placement of electrodes on the body surface affects the size and shape of the waves recorded. The example is a normal ECG; the enlarged section indicates the major components of the ECG and the measurements most often taken during clinical analysis.

an ECG recorded with that configuration. Note the following ECG features:

- The small **P wave**, which accompanies the depolarization of the atrial contractile cells. Depolarization of these cells causes atrial contraction. (Also see Figure 20–11 ❷.)
- The **QRS complex**, which appears as the ventricle contractile cells depolarize. This electrical signal is relatively strong, because the ventricular muscle is much more massive than that of the atria. It is also a complex signal, largely because of the complex pathway that the spread of depolarization takes through the ventricles. The ventricles begin contracting shortly after the peak of the **R wave**. Atrial repolarization takes place while the ventricles are depolarizing and it is obscured by the more powerful QRS complex. (Also see Figure 20–11 ❹ and ❺.)
- The smaller **T wave**, which indicates repolarization of the ventricular contractile cells.

To analyze an ECG, you must measure the size of the voltage changes and determine the durations and temporal (time) relationships of the various components. The amount of depolarization during the P wave and the QRS complex is particularly important in making a diagnosis. For example, an excessively large QRS complex often indicates that the heart has become enlarged. A smaller-than-normal electrical signal may mean that the mass of the heart muscle has decreased (although monitoring problems are more often responsible). The size and shape of the T wave may also be affected by any condition that slows ventricular repolarization. For example, starvation and low cardiac energy reserves, coronary ischemia (inadequate blood flow to cardiac cells), or abnormal ion concentrations reduce the size of the T wave.

The times between waves are reported as *segments* and *intervals*. Segments generally extend from the end of one wave to the start of another. Intervals are more variable, but always include at least one entire wave. Commonly used segments and intervals are labeled in Figure 20–12b. The names, however, can be somewhat misleading. For example:

- The **P–R interval** extends from the start of atrial depolarization to the start of the QRS complex (ventricular depolarization) rather than to R, because in abnormal ECGs the peak at R can be difficult to determine. Extension of the P–R interval to more than 200 msec can indicate damage to the AV node or conducting pathways.
- The **Q–T interval** indicates the time required for the ventricles to undergo a single cycle of depolarization and repolarization. It is usually measured from the end of the P–R interval rather than from the bottom of the Q wave. The Q–T interval can be lengthened by electrolyte disturbances, some medications, conduction problems, coronary ischemia, or myocardial damage. A congenital heart defect that can cause sudden death without warning may be detectable as a prolonged Q–T interval.

An *arrhythmia* (ā-RITH-mē-uh) is an irregularity in the normal rhythm or force of the heartbeat. Serious arrhythmias may indicate damage to the myocardium, injuries to the pacemaker cells or conducting cell pathways, exposure to drugs, or abnormalities in the electrolyte composition of extracellular fluids. Spotlight Figure 20–13 describes cardiac arrhythmias.

Cardiac Contractions: Contractile Cells

The Purkinje fibers distribute the stimulus to the **cardiac contractile cells**, which form the bulk of the atrial and ventricular walls. These cells account for about 99 percent of the muscle cells in the heart. As noted in Chapter 10, cardiac contractile cells are interconnected by **intercalated** (in-TER-kah-lā-ted) **discs** (Figure 20–14a). Intercalated discs transfer the force of contraction from cell to cell and propagate action potentials. At an intercalated disc, the interlocking membranes of adjacent cells are held together by desmosomes and linked by gap junctions (Figure 20–14b,c). The desmosomes prevent cells from separating during contraction, while the gap junctions allow ions to pass and electrically couple adjacent cells. This allows the heart muscle to behave as a *functional syncytium* (sin-SI-shē-um), a mechanically, chemically, and electrically coupled, multinucleate tissue.

Tips & Tools

The term *intercalated* means "inserted between other elements." So, intercalated discs appear to have been inserted between cardiac muscle cells.

Cardiac Contractile Cells Compared with Skeletal Muscle Fibers

Table 20–1 provides a review of the structural and functional differences between cardiac contractile cells and skeletal muscle fibers. Histological characteristics that distinguish cardiac contractile cells from skeletal muscle fibers include (1) small size; (2) a single, centrally located nucleus; (3) branching interconnections between cells; and (4) the presence of intercalated discs.

Table 20–1 Structural and Functional Differences between Cardiac Contractile Cells and Skeletal Muscle Fibers

Feature	Cardiac Contractile Cells	Skeletal Muscle Fibers
Size	10–20 μm × 50–100 μm	100 μm × up to 40 cm
Nuclei	Typically 1 (rarely 2–5)	Multiple (hundreds)
Contractile proteins	Sarcomeres along myofibrils	Sarcomeres along myofibrils
Internal membranes	Short T tubules; no triads formed with sarcoplasmic reticulum	Long T tubules form triads with cisternae of the sarcoplasmic reticulum
Mitochondria	Abundant (25% of cell volume)	Much less abundant
Inclusions	Myoglobin, lipids, glycogen	Little myoglobin, few lipids, but extensive glycogen reserves
Blood supply	Very extensive	More extensive than in most connective tissues, but sparse compared with supply to cardiac muscle cells
Metabolism (resting)	Not applicable	Aerobic, primarily lipid based
Metabolism (active)	Aerobic, primarily using lipids and carbohydrates	Anaerobic, through breakdown of glycogen reserves
Contractions	Twitches with brief relaxation periods; long refractory period prevents tetanic contractions	Usually sustained contractions
Stimulus for contraction	Autorhythmicity of pacemaker cells generates action potentials	Activity of somatic motor neuron generates action potentials in sarcolemma
Trigger for contraction	Calcium ion entry from the extracellular fluid and calcium ion release from the sarcoplasmic reticulum	Calcium ion release from the sarcoplasmic reticulum
Intercellular connections	Branching network with plasma membranes locked together at intercalated discs; connective tissue fibers tie adjacent layers together	Adjacent fibers tied together by connective tissue fibers

SPOTLIGHT

Figure 20–13

Cardiac Arrhythmias

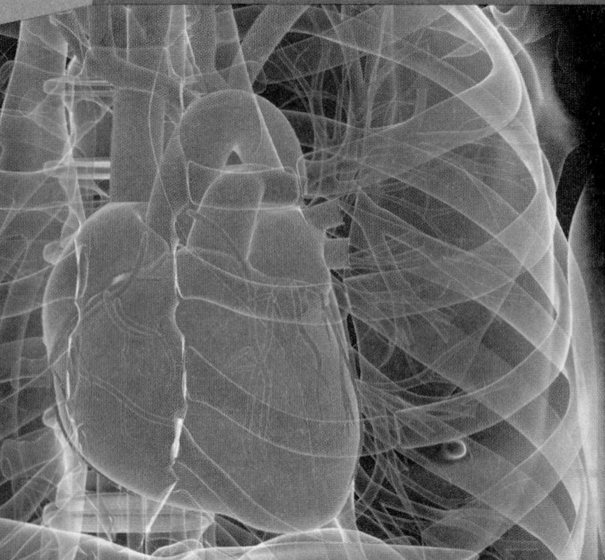

Despite the variety of sophisticated equipment available to assess or visualize cardiac function, in the majority of cases the ECG provides the most important diagnostic information. ECG analysis is especially useful in detecting and diagnosing **cardiac arrhythmias** (ā-RITH-mē-uz)—abnormal patterns of cardiac electrical activity. Momentary arrhythmias are not inherently dangerous, but clinical problems appear when arrhythmias reduce the pumping efficiency of the heart.

Premature Atrial Contractions (PACs)

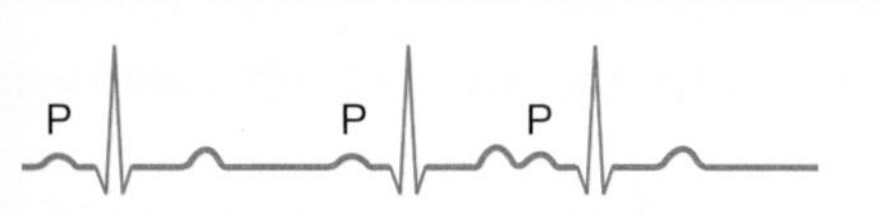
P
P
P

Premature atrial contractions (PACs) often occur in healthy individuals. In a PAC, the normal atrial rhythm is momentarily interrupted by a "surprise" atrial contraction. Stress, caffeine, and various drugs may increase the incidence of PACs, presumably by increasing the permeabilities of the SA pacemakers. The impulse spreads along the conduction pathway, and a normal ventricular contraction follows the atrial beat.

Paroxysmal Atrial Tachycardia (PAT)

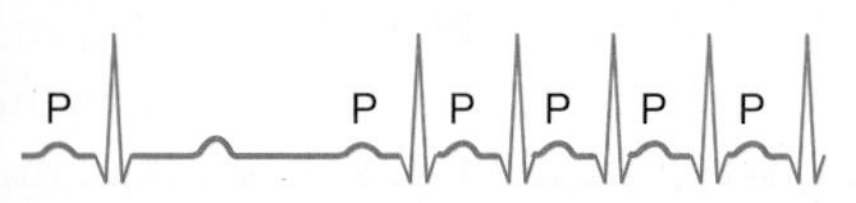
P
P
P
P
P
P

In **paroxysmal** (par-ok-SIZ-mal) **atrial tachycardia**, or **PAT**, a premature atrial contraction triggers a flurry of atrial activity. The ventricles are still able to keep pace, and the heart rate jumps to about 180 beats per minute.

Atrial Fibrillation (AF)

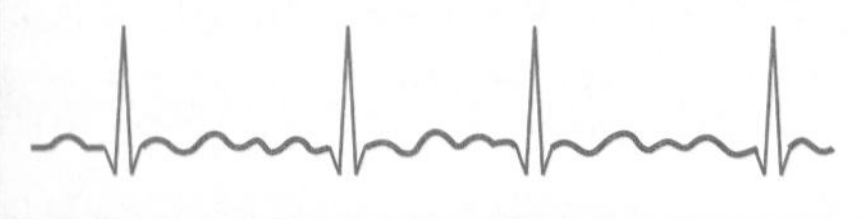

During **atrial fibrillation** (fib-ri-LĀ-shun), the impulses move over the atrial surface at rates of perhaps 500 beats per minute. The atrial wall quivers instead of producing an organized contraction. The ventricular rate cannot follow the atrial rate and may remain within normal limits. Even though the atria are now nonfunctional, their contribution to ventricular end-diastolic volume (the maximum amount of blood the ventricles can hold at the end of atrial contraction) is so small that the condition may go unnoticed in older individuals.

Premature Ventricular Contractions (PVCs)

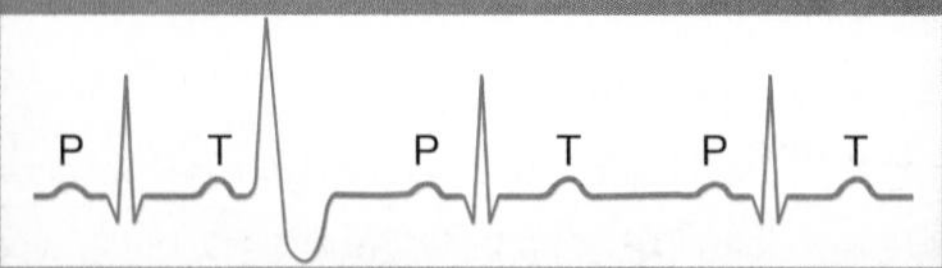
P
T
P
T
P
T

Premature ventricular contractions (PVCs) occur when a Purkinje cell or ventricular myocardial cell depolarizes to threshold and triggers a premature contraction. Single PVCs are common and not dangerous. The cell responsible is called an **ectopic pacemaker**. The frequency of PVCs can be increased by exposure to epinephrine, to other stimulatory drugs, or to ionic changes that depolarize cardiac muscle plasma membranes.

Ventricular Tachycardia (VT)

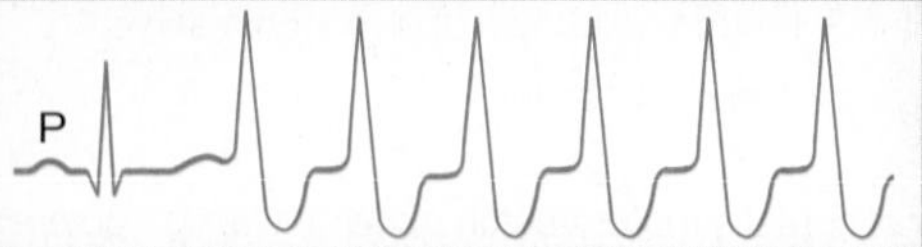
P

Ventricular tachycardia is defined as four or more PVCs without intervening normal beats. It is also known as **VT** or **V-tach**. Multiple PVCs and VT may indicate that serious cardiac problems exist.

Ventricular Fibrillation (VF)

Ventricular fibrillation (VF) is responsible for the condition known as **cardiac arrest**. VF is rapidly fatal, because the ventricles quiver and stop pumping blood.

Figure 20–14 Cardiac Contractile Cells.

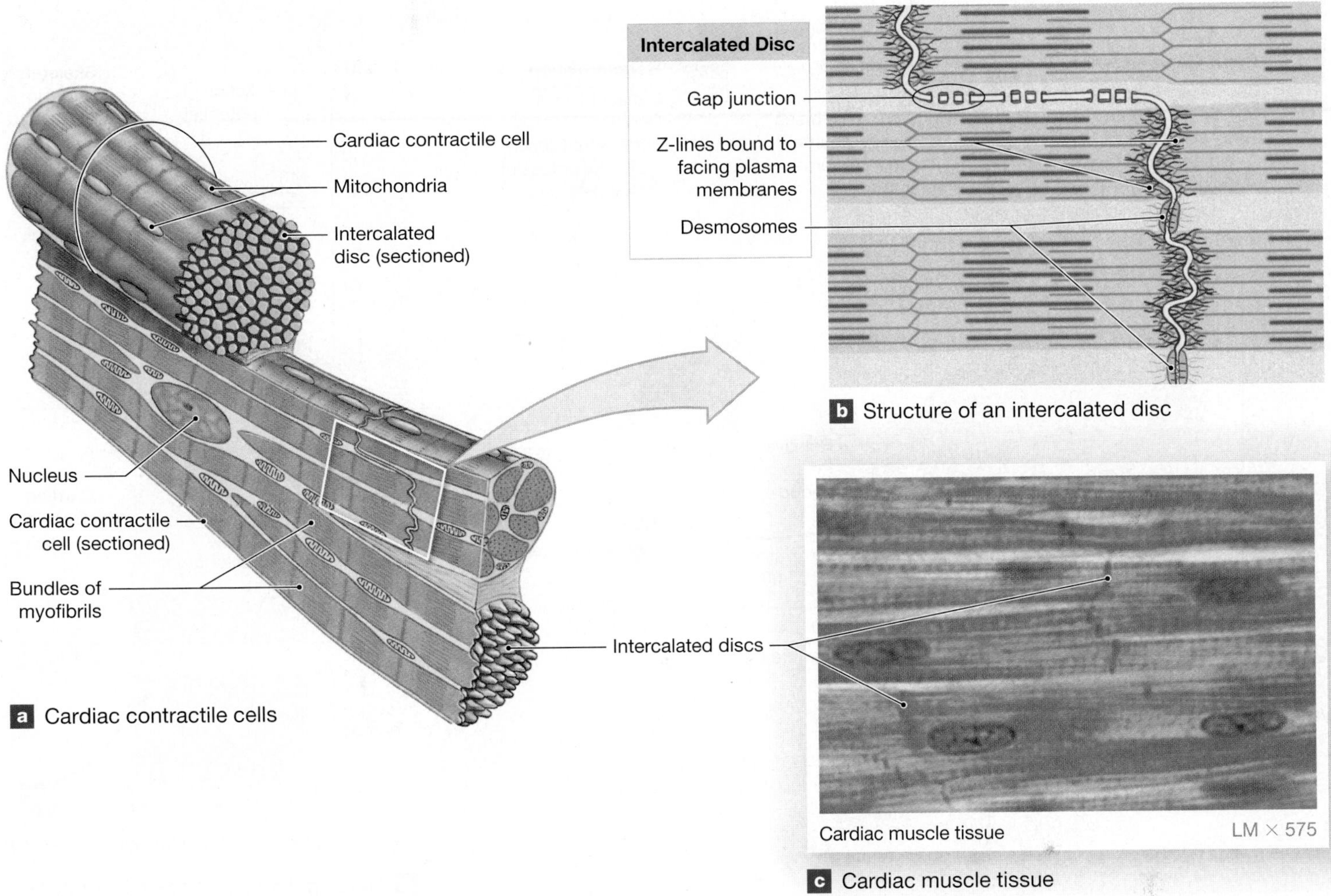
Cardiac contractile cell
Mitochondria
Intercalated disc (sectioned)
Nucleus
Cardiac contractile cell (sectioned)
Bundles of myofibrils
Intercalated discs
Intercalated Disc
Gap junction
Z-lines bound to facing plasma membranes
Desmosomes
Cardiac muscle tissue
LM × 575

a Cardiac contractile cells

b Structure of an intercalated disc

c Cardiac muscle tissue

The functions of cardiac contractile cells and skeletal muscle cells are similar but have some differentiating characteristics. In both types of cells, (1) an action potential leads to the appearance of Ca^{2+} among the myofibrils, and (2) the binding of Ca^{2+} to troponin on the thin filaments initiates the contraction. But cardiac contractile cells and skeletal muscle fibers differ in terms of the nature of the action potential, the source of the Ca^{2+}, and the duration of the resulting contraction. p. 331

The Action Potential in Cardiac Contractile Cells

The resting membrane potential of a ventricular contractile cell is approximately −90 mV, comparable to that of a resting skeletal muscle fiber (−85 mV). (The resting membrane potential of an atrial contractile cell is about −80 mV, but the basic principles described here apply to atrial cells as well.)

Process of the Action Potential. An action potential begins when the membrane of the ventricular contractile cell reaches threshold, usually at about −75 mV. Threshold is normally reached in a portion of the membrane next to an intercalated disc. The typical stimulus is the excitation of an adjacent contractile cell. Once threshold has been reached, the action potential proceeds in three basic steps (Figure 20–15a):

1 **Rapid Depolarization.** The stage of *rapid depolarization* in a cardiac contractile cell resembles that in a skeletal muscle fiber. At threshold, voltage-gated sodium ion channels open, and the membrane suddenly becomes permeable to Na^+. A massive influx of sodium ions rapidly depolarizes the sarcolemma. The channels involved are called **fast sodium channels**, because they open quickly and remain open for only a few milliseconds.

2 **The Plateau.** A partial repolarization takes place as some potassium ions (K^+) leave the cell before most K^+ channels close. Voltage-gated calcium channels then open and *extracellular* calcium ions (Ca^{2+}) enter the cytosol. The **slow calcium channels** remain open for a relatively long period—roughly 175 msec. The inflow of positive charges (Ca^{2+}) and reduced outflow of K^+ delay repolarization, causing the membrane

Figure 20–15 Action Potentials in Cardiac Contractile Cells and Skeletal Muscle Fibers.

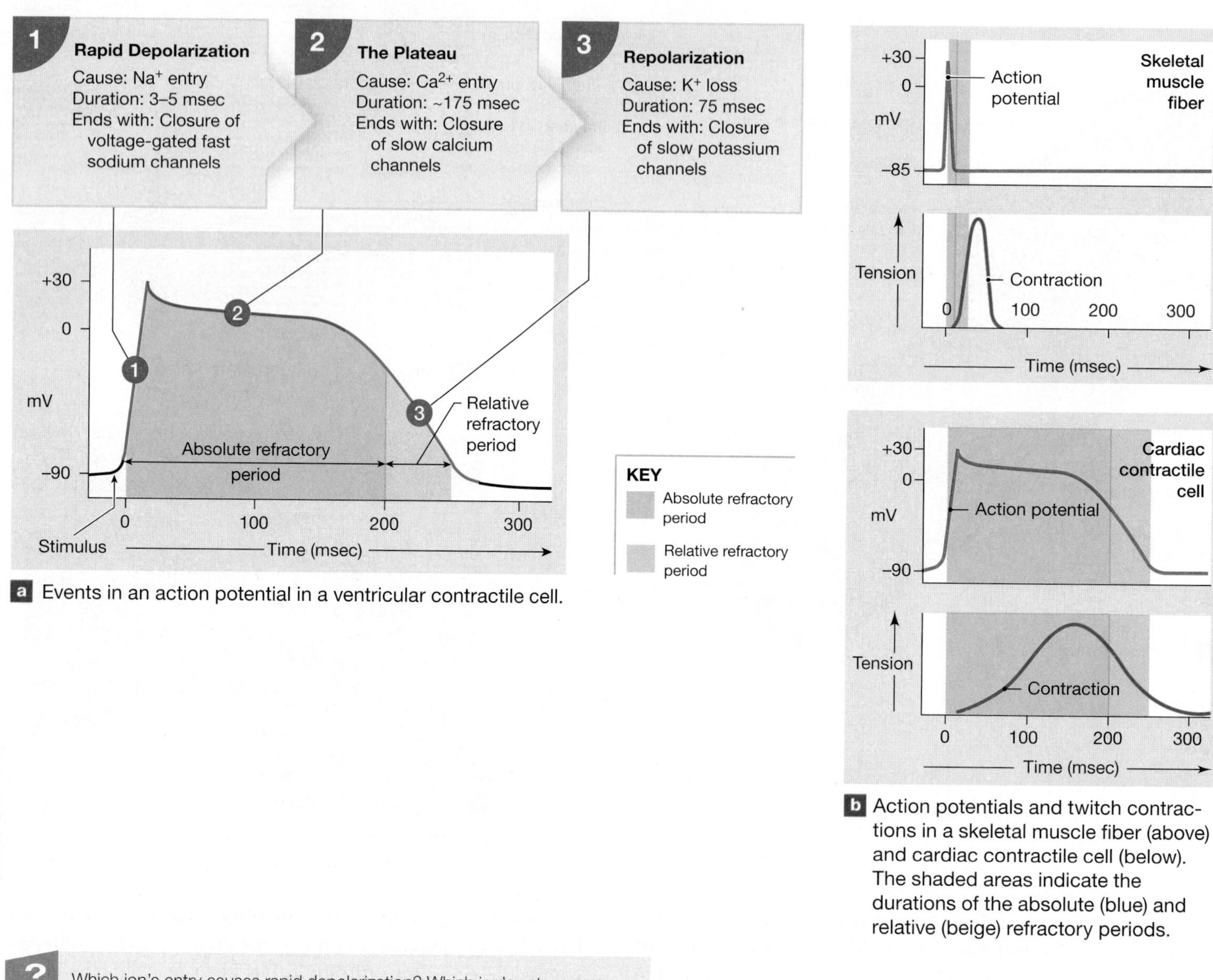
1 Rapid Depolarization
Cause: Na^+ entry
Duration: 3–5 msec
Ends with: Closure of voltage-gated fast sodium channels
2 The Plateau
Cause: Ca^{2+} entry
Duration: ~175 msec
Ends with: Closure of slow calcium channels
3 Repolarization
Cause: K^+ loss
Duration: 75 msec
Ends with: Closure of slow potassium channels
+30
0
mV
−90
Absolute refractory period
Relative refractory period
Stimulus
0
100
200
300
Time (msec)
KEY
Absolute refractory period
Relative refractory period
Skeletal muscle fiber
Action potential
−85
Tension
Contraction
Cardiac contractile cell

a Events in an action potential in a ventricular contractile cell.

b Action potentials and twitch contractions in a skeletal muscle fiber (above) and cardiac contractile cell (below). The shaded areas indicate the durations of the absolute (blue) and relative (beige) refractory periods.

? Which ion's entry causes rapid depolarization? Which ion's entry causes the plateau? Which ion's exit causes repolarization?

potential to remain near 0 mV for an extended period. This portion of the action potential is called the *plateau*. The extracellular calcium ions initiate contraction and also delay repolarization. Their increased concentration within the cell also triggers the release of Ca^{2+} from reserves in the sarcoplasmic reticulum (SR), which continues the contraction.

3 Repolarization. As the plateau continues, slow calcium channels begin closing, and **slow potassium channels** begin opening. As these channels open, potassium ions (K^+) rush out of the cell, and the net result is a period of rapid repolarization that restores the resting membrane potential.

The Refractory Period. As with skeletal muscle fiber contractions, the membrane of a cardiac contractile cell will not respond normally to a second stimulus for some time after an action potential begins. This time is called the **refractory period**. Initially, in the *absolute refractory period*, the membrane cannot respond at all, because the sodium ion channels are either already open or closed and inactivated. In a ventricular contractile cell, the absolute refractory period lasts approximately 200 msec. It includes the plateau and the initial period of rapid repolarization.

The absolute refractory period is followed by a shorter (50-msec) *relative refractory period*. During this period, the voltage-gated sodium ion channels are closed, but can open. The membrane will respond to a stronger-than-normal stimulus by initiating another action potential. In total, an action potential in a ventricular contractile cell lasts 250–300 msec, about 30 times as long as a typical action potential in a skeletal muscle fiber.

The Role of Calcium Ions in Cardiac or Skeletal Muscle Contractions

The appearance of an action potential in the cardiac contractile cell plasma membrane produces a contraction by causing an increase in the concentration of Ca^{2+} around the myofibrils. This process takes place in two steps:

1. Extracellular calcium ions crossing the plasma membrane during the plateau phase of the action potential provide roughly 20 percent of the Ca^{2+} required for a contraction.
2. The arrival of extracellular Ca^{2+} triggers the release of additional Ca^{2+} from reserves in the sarcoplasmic reticulum (SR).

Extracellular calcium ions thus have both direct and indirect effects on cardiac contractile cell contraction. For this reason, cardiac muscle tissue is highly sensitive to changes in the Ca^{2+} concentration of the extracellular fluid.

In a skeletal muscle fiber, the action potential is relatively brief and ends as the resulting twitch contraction begins (Figure 20–15b). The twitch contraction is short and ends as the SR reclaims the Ca^{2+} it released. In a cardiac contractile cell, as we have seen, the action potential is prolonged, and calcium ions continue to enter the cell throughout the plateau. As a result, the cell actively contracts until the plateau ends. As the slow calcium channels close, the intracellular calcium ions are pumped back into the SR or out of the cell, and the contractile cell relaxes.

In skeletal muscle fibers, the refractory period ends before peak tension develops. As a result, twitches can summate and tetanus can occur. In cardiac contractile cells, the absolute refractory period continues until relaxation is under way. Because summation is not possible, tetanic contractions cannot occur in a normal cardiac contractile cell, regardless of the frequency or intensity of stimulation. This feature is vital: A heart in tetany could not pump blood. With a single twitch lasting 250 msec or longer, a normal cardiac contractile cell could reach 300–400 contractions per minute under maximum stimulation. This rate is not reached in a normal heart, due to limitations imposed by the conducting system.

The Energy for Cardiac Contractions

When a normal heart is beating, it gets energy as the mitochondria break down fatty acids (stored as lipid droplets) and glucose (stored as glycogen). These aerobic reactions can occur only when oxygen is readily available. ⮌ p. 320

In addition to obtaining oxygen from the coronary circulation, cardiac contractile cells maintain their own sizable reserves of oxygen. In these cells, oxygen molecules are bound to the heme units of *myoglobin* molecules. (We discussed this globular protein, which reversibly binds oxygen molecules, and its function in skeletal muscle fibers in Chapter 10.) ⮌ p. 323 Normally, the combination of circulatory supplies plus myoglobin reserves is enough to meet the oxygen demands of the heart, even when it is working at maximum capacity.

Checkpoint

4. **Define *autorhythmicity*.**
5. **Which structure of the heart is known as the cardiac pacemaker?**
6. **If the cells of the SA node did not function, how would the heart rhythm change and the heart rate be affected?**
7. **Why is it important for impulses from the atria to be delayed at the AV node before they pass into the ventricles?**

See the blue Answers tab at the back of the book.

20-3 The contraction–relaxation events that occur during a complete heartbeat make up a cardiac cycle

Learning Outcome **Explain the events of the cardiac cycle, including atrial and ventricular systole and diastole, and relate the heart sounds to specific events in the cycle.**

The electrical events of the conducting system initiate the mechanical events of heart contraction. A brief resting phase follows each heartbeat, allowing time for the chambers to relax and prepare for the next heartbeat. The period between the start of one heartbeat and the beginning of the next is a single **cardiac cycle**. It includes alternating periods of contraction and relaxation. For any one chamber in the heart, the cardiac cycle can be divided into two phases: systole and diastole. During **systole** (SIS-tō-lē), or contraction, the chamber contracts and pushes blood into an adjacent chamber or into an arterial trunk. Systole is followed by **diastole** (dī-AS-tō-lē), or relaxation. During diastole, the chamber fills with blood and prepares for the next cardiac cycle.

An Introduction to Pressure and Flow in the Heart

Fluids flow from an area of higher pressure to an area of lower pressure. In the cardiac cycle, the blood pressure within each chamber rises during systole and falls during diastole. Valves between adjacent chambers help ensure that blood flows in the required direction, but blood flows from one chamber to another only if the pressure in the first chamber exceeds that in the second. This basic principle governs the movement of blood between atria and ventricles, between ventricles and arterial trunks, and between major veins and atria.

Figure 20–16 Phases of the Cardiac Cycle. Thin black arrows indicate blood flow, green arrows indicate contractions, and the red-hued areas indicate which heart chambers are in systole.

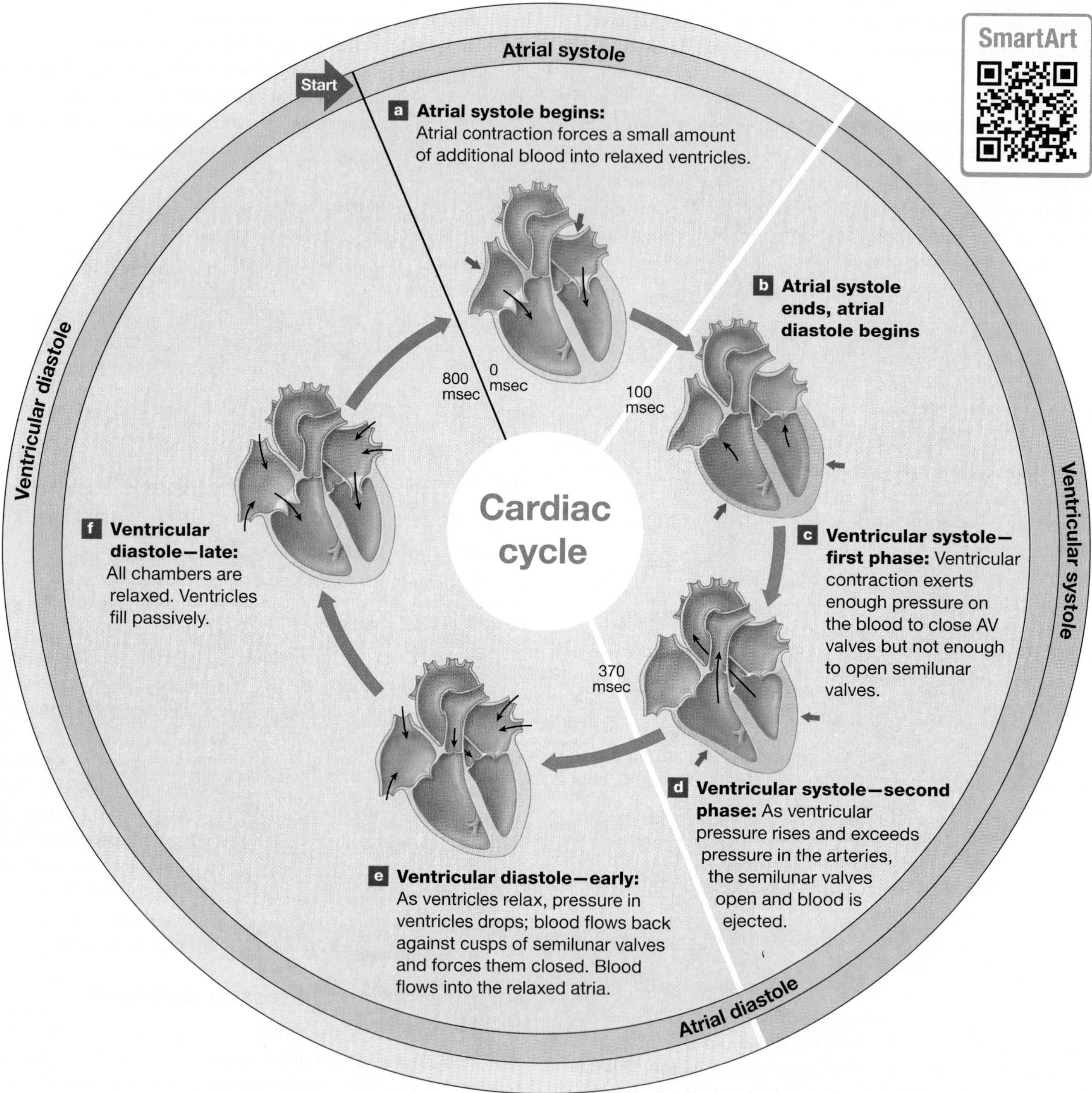
SmartArt
Atrial systole
Start
a Atrial systole begins: Atrial contraction forces a small amount of additional blood into relaxed ventricles.
0 msec
800 msec
100 msec
b Atrial systole ends, atrial diastole begins
Ventricular diastole
Ventricular systole
Cardiac cycle
c Ventricular systole—first phase: Ventricular contraction exerts enough pressure on the blood to close AV valves but not enough to open semilunar valves.
f Ventricular diastole—late: All chambers are relaxed. Ventricles fill passively.
370 msec
d Ventricular systole—second phase: As ventricular pressure rises and exceeds pressure in the arteries, the semilunar valves open and blood is ejected.
e Ventricular diastole—early: As ventricles relax, pressure in ventricles drops; blood flows back against cusps of semilunar valves and forces them closed. Blood flows into the relaxed atria.
Atrial diastole

The correct pressure relationships depend on the careful timing of contractions. For example, blood could not move in the proper direction if an atrium and its attached ventricle contracted at precisely the same moment. The heart's elaborate pacemaking and conducting systems normally provide the required spacing between atrial systole and ventricular systole.

Phases of the Cardiac Cycle

The phases of the cardiac cycle—*atrial systole, atrial diastole, ventricular systole,* and *ventricular diastole*—are diagrammed in Figure 20–16 for a representative heart rate of 75 bpm. At this heart rate, a sequence of systole and diastole in either the atria or the ventricles lasts 800 msec. When the cardiac cycle begins, all four chambers are relaxed, and the ventricles are partially filled with blood.

Let's start by focusing on the atria. During **atrial systole**, the atria contract, filling the ventricles completely with blood (Figure 20–16a,b). Atrial systole lasts 100 msec. The atria next enter **atrial diastole**, which continues until the start of the next cardiac cycle.

Ventricular systole begins at the same time as atrial diastole. Ventricular systole lasts 270 msec. During this period, the ventricles push blood through the systemic and pulmonary circuits and toward the atria (Figure 20–16c,d). The heart then enters **ventricular diastole** (Figure 20–16e,f), which lasts 530 msec (the 430 msec remaining in this cardiac cycle, plus the first 100 msec of the next, when the atria are again contracting). For the rest of this cycle, filling occurs passively, and both the atria and the ventricles are relaxed. The next cardiac cycle begins with atrial systole, which completes the filling of the ventricles.

When the heart rate increases, all the phases of the cardiac cycle are shortened. The greatest reduction occurs in the length of time spent in diastole. When the heart rate climbs from 75 bpm to 200 bpm, the time spent in systole drops by less than 40 percent, but the duration of diastole is reduced by almost 75 percent.

Pressure and Volume Changes in the Cardiac Cycle

Figure 20–17 plots the pressure and volume changes during the cardiac cycle. It also shows an ECG for the cardiac cycle. The circled numbers in the figure correspond to numbered paragraphs in the text. The figure shows pressure and volume within the left atrium and left ventricle, but our discussion applies to both sides of the heart. Although pressures are lower in the right atrium and right ventricle, both sides of the heart contract at the same time, and they eject equal volumes of blood.

Atrial Systole

In atrial systole:

1 **Atrial contraction begins.** At the start of atrial systole, the ventricles are already filled to about 70 percent of their normal capacity, due to passive blood flow during the end of the previous cardiac cycle.

2 **Atria eject blood into the ventricles.** As the atria contract, rising atrial pressures provide the remaining 30 percent by pushing blood into the ventricles through the open right and left AV valves. Atrial systole essentially "tops off" the ventricles. Over this period, blood cannot flow into the atria from the veins because atrial pressure exceeds venous pressure. Yet there is very little backflow into the veins, even though the connections with the venous system lack valves. The reason is that blood takes the path of least resistance. Resistance to blood flow through the wide opening between the atria and ventricles is less than that through the smaller, angled openings of the large veins.

At the end of atrial systole, each ventricle contains the maximum amount of blood that it will hold in this cardiac cycle. That quantity is called the *end-diastolic volume (EDV)*. In an adult who is standing at rest, the end-diastolic volume is typically about 130 mL (about 4.4 oz).

Ventricular Systole and Atrial Diastole

In ventricular systole:

3 **Atrial systole ends; AV valves close.** As atrial systole ends, ventricular systole begins. As the pressures in the ventricles rise above those in the atria, the AV valves are pushed closed.

4 **Isovolumetric ventricular contraction occurs.** During the early stage of ventricular systole, the ventricles are contracting, but blood flow has yet to occur. Ventricular pressures are not yet high enough to force open the semilunar valves and push blood into the pulmonary or aortic trunk. Over this period, the ventricles contract isometrically; in other words, they generate tension and pressures rise inside them, but blood does not flow out. The ventricles are in **isovolumetric contraction**: All the heart valves are closed, the volumes of the ventricles do not change, and ventricular pressures are rising.

5 **Ventricular ejection occurs.** Once pressure in the ventricles exceeds that in the arterial trunks, the semilunar valves are pushed open and blood flows into the pulmonary and aortic trunks. This point marks the beginning of **ventricular ejection**. The ventricles now contract isotonically: The muscle cells shorten, and tension production remains relatively constant. (To review isotonic versus isometric contractions, look back at Figure 10–18, p. 317.)

During ventricular ejection, each ventricle ejects 70–80 mL of blood, the *stroke volume (SV)* of the heart. The stroke volume at rest is roughly 60 percent of the end-diastolic volume.

After reaching a peak, ventricular pressures gradually decline near the end of ventricular systole. Figure 20–17 shows values for the left ventricle and aorta. The right ventricle also undergoes periods of isovolumetric contraction and ventricular ejection.

6 **Semilunar valves close.** As the end of ventricular systole approaches, ventricular pressures fall rapidly. Blood in the aorta and pulmonary trunk now starts to flow back toward the ventricles, and this movement closes the semilunar valves. As the backflow begins, pressure decreases in the aorta. When the semilunar valves close, pressure rises again as the elastic arterial walls recoil. This small, temporary rise produces a valley in the aortic pressure tracing, called a *dicrotic* (dī-KROT-ik; *dikrotos*, double beating) *notch*.

The amount of blood remaining in the ventricle when the semilunar valve closes is the *end-systolic volume (ESV)*. At rest, the end-systolic volume is 50 mL, about 40 percent of the end-diastolic volume.

Figure 20–17 Pressure and Volume Relationships in the Cardiac Cycle. Major features of the cardiac cycle are shown for a heart rate of 75 bpm. The circled numbers correspond to those in the associated box. For further details, see the numbered list in the text (pp. 713–715).

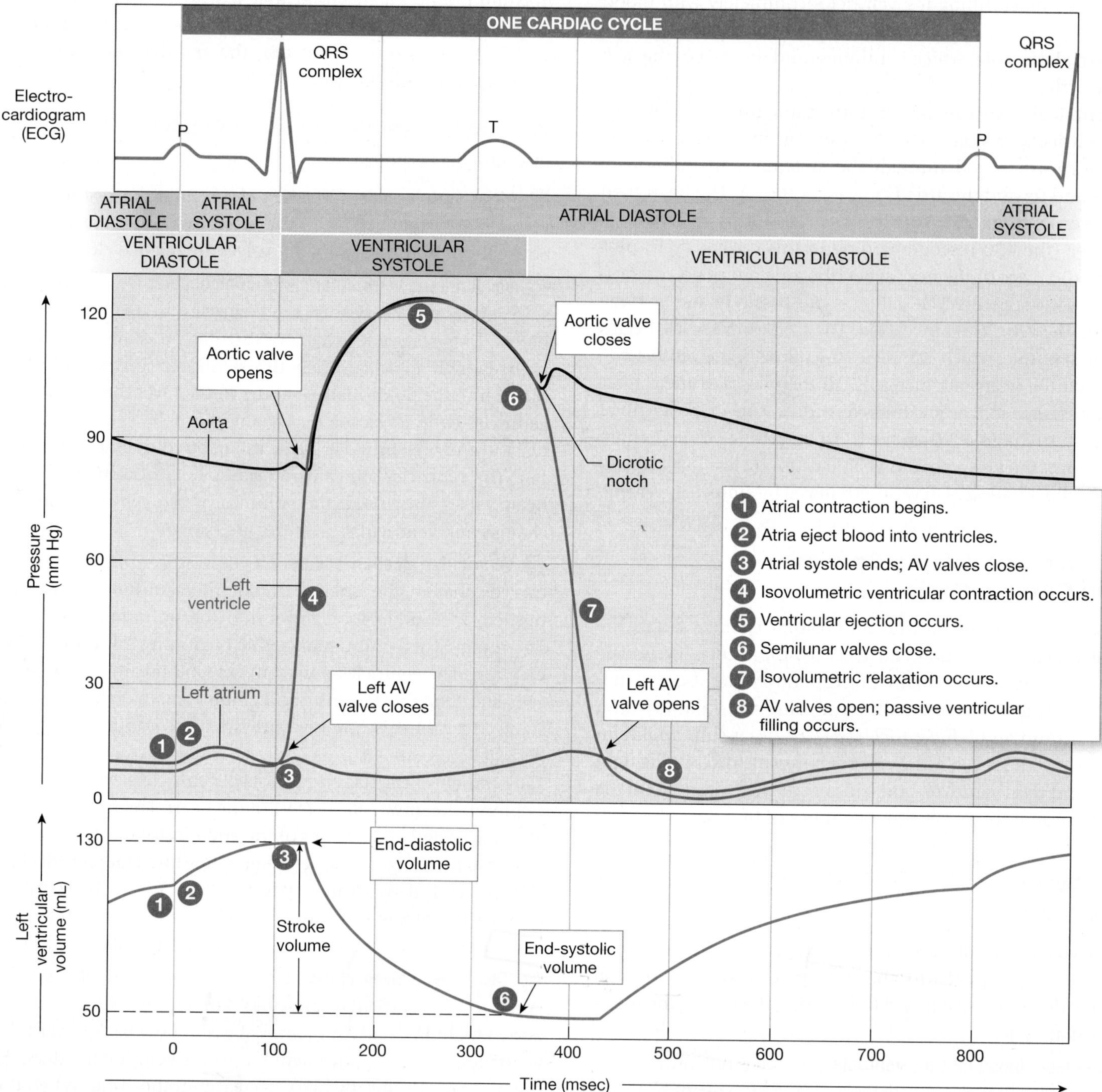

ONE CARDIAC CYCLE
QRS complex
QRS complex
Electro-cardiogram (ECG)
P
T
P
ATRIAL DIASTOLE
ATRIAL SYSTOLE
ATRIAL DIASTOLE
ATRIAL SYSTOLE
VENTRICULAR DIASTOLE
VENTRICULAR SYSTOLE
VENTRICULAR DIASTOLE
Pressure (mm Hg)
120
90
60
30
0
Aortic valve opens
Aorta
Aortic valve closes
Dicrotic notch
Left ventricle
Left atrium
Left AV valve closes
Left AV valve opens
1 Atrial contraction begins.
2 Atria eject blood into ventricles.
3 Atrial systole ends; AV valves close.
4 Isovolumetric ventricular contraction occurs.
5 Ventricular ejection occurs.
6 Semilunar valves close.
7 Isovolumetric relaxation occurs.
8 AV valves open; passive ventricular filling occurs.
Left ventricular volume (mL)
130
50
End-diastolic volume
Stroke volume
End-systolic volume
0
100
200
300
400
500
600
700
800
Time (msec)

Ventricular Diastole

The period of ventricular diastole lasts for the remainder of the current cardiac cycle and continues through atrial systole in the next cycle.

7 Isovolumetric relaxation occurs. All the heart valves are now closed, and the ventricular myocardium is relaxing. Because ventricular pressures are still higher than atrial pressures, blood cannot flow into the ventricles. This is **isovolumetric relaxation**. Ventricular pressures drop rapidly over this period, because the elasticity of the connective tissues of the heart and cardiac skeleton helps re-expand the ventricles toward their resting dimensions.

8 AV valves open; passive ventricular filling occurs. When ventricular pressures fall below those of the atria, the atrial pressures force the AV valves open. Blood now flows from the atria into the ventricles. Both the atria and the ventricles are in diastole, but the ventricular pressures continue to fall as the ventricular chambers expand. Throughout this period, pressures in the ventricles are so far below those in the major veins that blood pours through the relaxed atria and on through the open AV valves into the ventricles. This passive flow of blood is the primary method of ventricular filling. The ventricles become nearly three-quarters full before the cardiac cycle ends.

The relatively minor contribution that atrial systole makes to ventricular volume explains why individuals can survive quite normally when their atria have been so severely damaged that they can no longer function. In contrast, damage to one or both ventricles can leave the heart unable to pump enough blood through peripheral tissues and organs. A condition of **heart failure** then exists.

Heart Sounds

Listening to the heart, a technique called *auscultation* (aws-kul-TĀ-shun), is a simple and effective method of cardiac assessment. Clinicians use an instrument called a **stethoscope** to listen for normal and abnormal heart sounds. Where the stethoscope is placed depends on which valve is under examination (Figure 20–18a). Valve sounds must pass through the pericardium, surrounding tissues, and the chest wall, and some tissues muffle sounds more than others. For this reason, the placement of the stethoscope differs somewhat from the position of the valve under review.

There are four heart sounds, named S_1 through S_4 (Figure 20–18b). If you listen to your own heart with a stethoscope, you will clearly hear the *first* and *second heart sounds.* These sounds accompany the closing of your heart valves. The first heart sound (S_1), known as "lubb," lasts a little longer than the second (S_2), called "dupp." S_1 marks the start of ventricular contraction, when the AV valves close and the semilunar valves open. S_2 occurs at the beginning of ventricular filling, when the semilunar valves close and the AV valves open. The *third* and *fourth heart sounds* are usually very faint and are seldom audible in healthy adults. These sounds are associated with blood flowing into the ventricles (S_3) and atrial contraction (S_4), rather than with valve action.

If the valve cusps are malformed or there are problems with the papillary muscles or chordae tendineae, the heart valves may not close properly. AV valve regurgitation then occurs during ventricular systole. The surges, swirls, and eddies of turbulent blood flow that accompany regurgitation create a rushing, gurgling sound known as a **heart murmur**. Minor heart murmurs are common and inconsequential.

Figure 20–18 Heart Sounds.

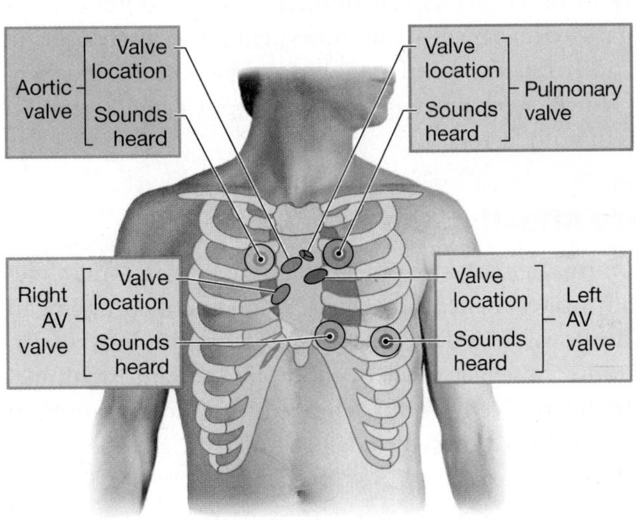
Aortic valve
Valve location
Sounds heard
Pulmonary valve
Valve location
Sounds heard
Right AV valve
Valve location
Sounds heard
Left AV valve
Valve location
Sounds heard

a Placements of a stethoscope for listening to the different sounds produced by individual valves

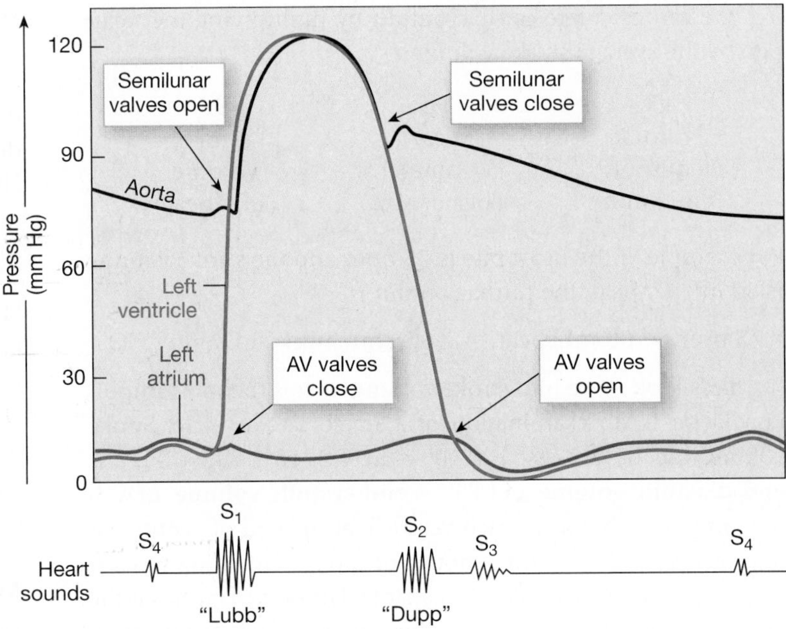
120
90
60
30
0
Pressure (mm Hg)
Semilunar valves open
Semilunar valves close
Aorta
Left ventricle
Left atrium
AV valves close
AV valves open
Heart sounds
S4
S1
S2
S3
S4
"Lubb"
"Dupp"

b The relationship between heart sounds and key events in the cardiac cycle

Checkpoint

8. Give the technical terms for heart contraction and heart relaxation.
9. List the phases of the cardiac cycle.
10. Is the heart always pumping blood when pressure in the left ventricle is rising? Explain.
11. What could cause an increase in the size of the QRS complex in an electrocardiogram?

See the blue Answers tab at the back of the book.

20-4 Cardiac output is determined by heart rate and stroke volume

Learning Outcome Define cardiac output, describe the factors that influence heart rate and stroke volume, and explain how adjustments in stroke volume and cardiac output are coordinated at different levels of physical activity.

When considering cardiac function over time, physicians generally are most interested in the **cardiac output (CO)**, the amount of blood pumped by the left ventricle in 1 minute. In essence, cardiac output is an indication of the blood flow through peripheral tissues—and without adequate blood flow, homeostasis cannot be maintained. The cardiac output provides a useful indication of ventricular efficiency over time.

Figure 20–19 summarizes the factors involved in the normal regulation of cardiac output. Cardiac output can be adjusted by changes in either heart rate or stroke volume. The **heart rate (HR)**, the number of heartbeats per minute, can be adjusted by the activities of the autonomic nervous system or by circulating hormones. **Stroke volume (SV)** is the amount of blood pumped out of a ventricle during each contraction.

20

We can calculate cardiac output by multiplying the heart rate by the average stroke volume:

$$\underset{\substack{\text{cardiac}\\\text{output}\\\text{(mL/min)}}}{\text{CO}} = \underset{\substack{\text{heart}\\\text{rate}\\\text{(beats/min)}}}{\text{HR}} \times \underset{\substack{\text{stroke}\\\text{volume}\\\text{(mL/beat)}}}{\text{SV}}$$

For example, if the heart rate is 75 bpm and the stroke volume is 80 mL per beat, the cardiac output is

$$75 \text{ bpm} \times 80 \text{ mL/beat} = 6000 \text{ mL/min (6 L/min) CO}$$

Let's look more into stroke volume, as it is the most important factor in an examination of a single cardiac cycle. Stroke volume can be expressed as SV = EDV − ESV, where EDV is **end-diastolic volume** and ESV is **end-systolic volume**. EDV is the amount of blood in each ventricle at the end of ventricular diastole, while ESV is the amount of blood remaining in each ventricle at the end of ventricular systole. The *ejection fraction* is the percentage of EDV that is ejected during a ventricular contraction.

If the heart were a simple hand pump, the stroke volume would be the amount of water pumped in one up–down cycle of the handle (Figure 20–20). Where you stop when you lift the handle determines how much water the pump contains—the EDV. How far down you push the handle determines how much water remains in the pump at the end of the cycle—the ESV. You pump the maximum amount of water when you pull the handle all the way to the top and then push it all the way to the bottom. In other words, you get the largest stroke volume when the EDV is as large as it can be and the ESV is as small as it can be.

Figure 20–19 Factors Affecting Cardiac Output.

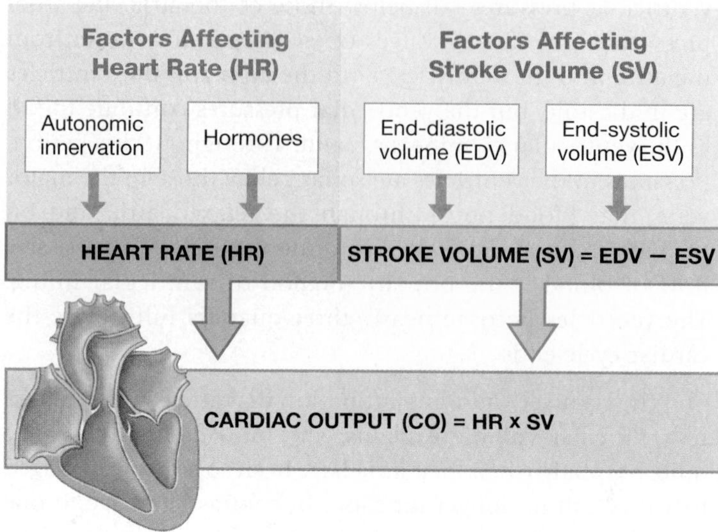
Factors Affecting Heart Rate (HR)
Factors Affecting Stroke Volume (SV)
Autonomic innervation
Hormones
End-diastolic volume (EDV)
End-systolic volume (ESV)
HEART RATE (HR)
STROKE VOLUME (SV) = EDV − ESV
CARDIAC OUTPUT (CO) = HR x SV

The body precisely adjusts cardiac output to supply peripheral tissues with enough blood as conditions change. When necessary, the heart rate can increase by 250 percent, and stroke volume in a normal heart can almost double. (In addition, a variety of other factors can influence cardiac output under abnormal circumstances.) For convenience, the discussions in this section consider heart rate and stroke volume independently. However, changes in cardiac output generally reflect changes in both factors.

Factors Affecting the Heart Rate

Under normal circumstances, autonomic activity and circulating hormones make homeostatic adjustments to the heart rate as cardiovascular demands change, thereby adjusting cardiac output. These factors act by modifying the autorhythmicity of the heart. Even a heart removed for a heart transplant continues to beat unless it is kept chilled in a preservation solution.

Autonomic Innervation

The sympathetic and parasympathetic divisions of the autonomic nervous system innervate the heart by means of the nerve network known as the *cardiac plexus* (Figure 16–6,

Go to MasteringA&P™ > Study Area > iP2™ > Cardiovascular System: Cardiac Cycle

Figure 20–20 A Simple Model of Stroke Volume. The stroke volume of the heart can be compared to the amount of water ejected from a simple pump. The amount of water ejected varies with the amount of movement of the pump handle.

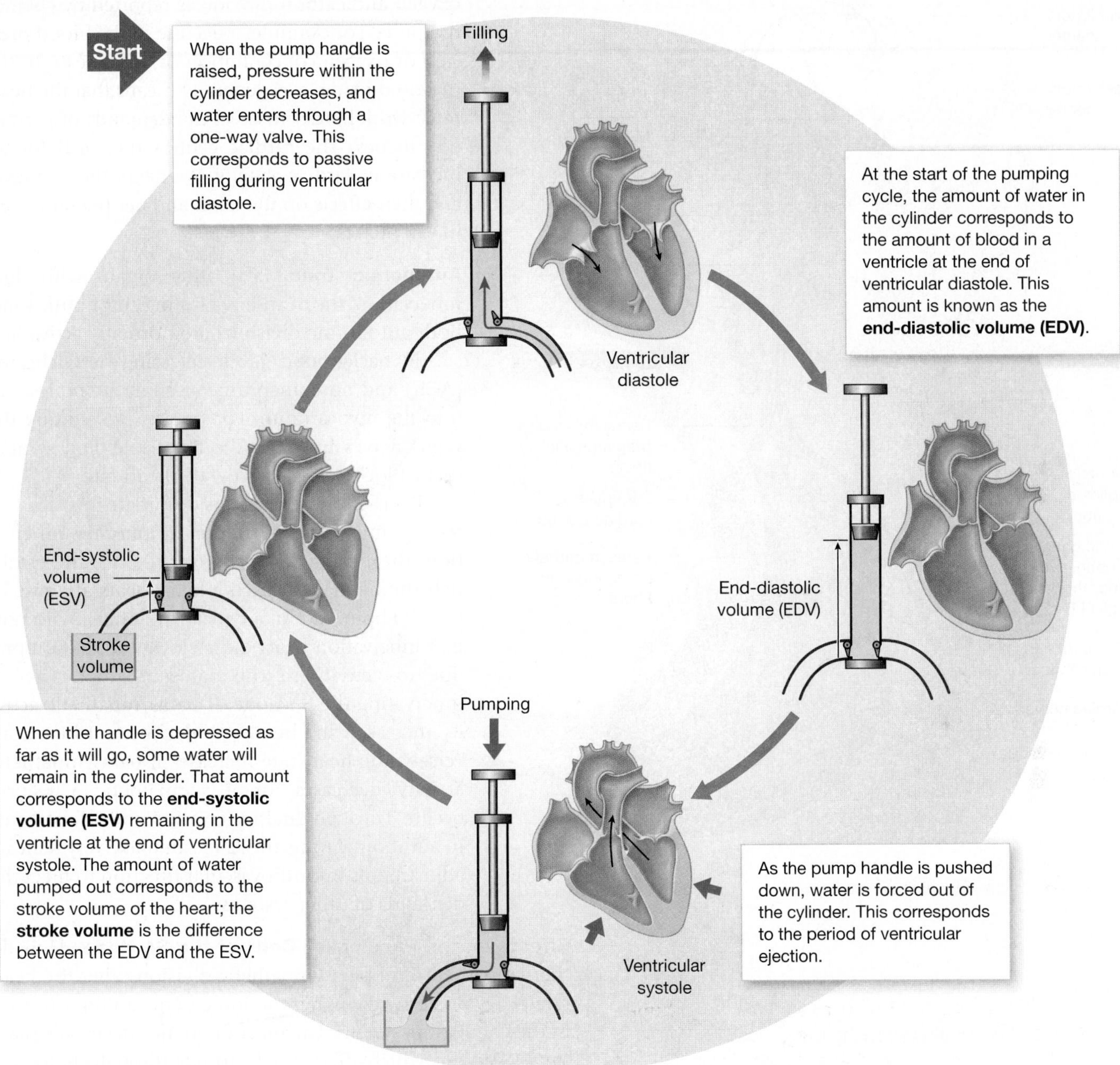

Start
When the pump handle is raised, pressure within the cylinder decreases, and water enters through a one-way valve. This corresponds to passive filling during ventricular diastole.
Filling
Ventricular diastole
At the start of the pumping cycle, the amount of water in the cylinder corresponds to the amount of blood in a ventricle at the end of ventricular diastole. This amount is known as the **end-diastolic volume (EDV)**.
End-diastolic volume (EDV)
End-systolic volume (ESV)
Stroke volume
Pumping
When the handle is depressed as far as it will go, some water will remain in the cylinder. That amount corresponds to the **end-systolic volume (ESV)** remaining in the ventricle at the end of ventricular systole. The amount of water pumped out corresponds to the stroke volume of the heart; the **stroke volume** is the difference between the EDV and the ESV.
As the pump handle is pushed down, water is forced out of the cylinder. This corresponds to the period of ventricular ejection.
Ventricular systole

p. 549, and **Figure 20–21**). Postganglionic sympathetic neurons are located in the cervical and upper thoracic ganglia. The vagus nerves (CN X) carry parasympathetic preganglionic fibers to small ganglia in the cardiac plexus. Both ANS divisions innervate the SA and AV nodes and the atrial contractile cells. Both divisions also innervate ventricular contractile cells, but sympathetic fibers far outnumber parasympathetic fibers there.

The *cardiac centers* of the medulla oblongata contain the autonomic headquarters for cardiac control. ⮌ p. 474 The **cardioacceleratory center** controls sympathetic neurons that increase the heart rate. The adjacent **cardioinhibitory center** controls the parasympathetic neurons that slow the heart rate. Reflex pathways regulate the cardiac centers. They also receive input from higher centers, especially from the parasympathetic and sympathetic centers in the hypothalamus.

Figure 20–21 Autonomic Innervation of the Heart.

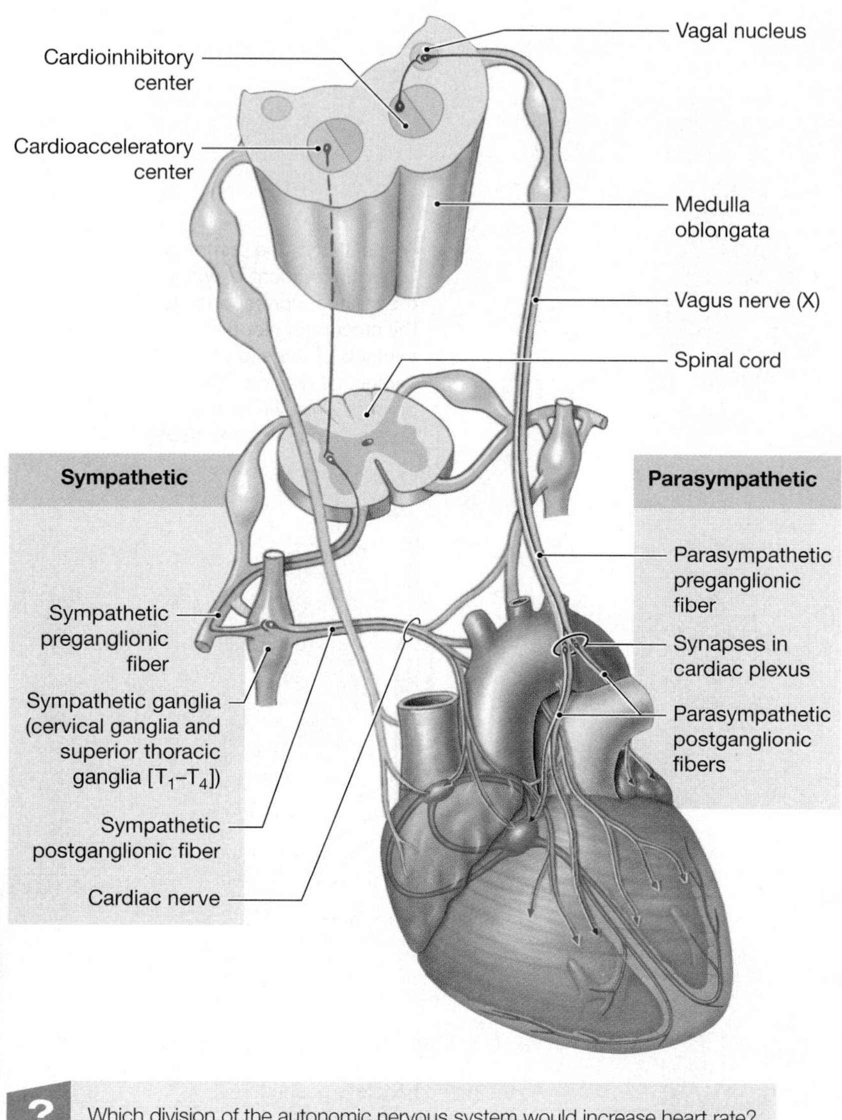
Cardioinhibitory center
Cardioacceleratory center
Vagal nucleus
Medulla oblongata
Vagus nerve (X)
Spinal cord
Sympathetic
Parasympathetic
Parasympathetic preganglionic fiber
Synapses in cardiac plexus
Parasympathetic postganglionic fibers
Sympathetic preganglionic fiber
Sympathetic ganglia (cervical ganglia and superior thoracic ganglia [T1–T4])
Sympathetic postganglionic fiber
Cardiac nerve

? Which division of the autonomic nervous system would increase heart rate?

The cardiac centers respond to changes in blood pressure as reported by baroreceptors, and to changes in arterial concentrations of dissolved oxygen and carbon dioxide as reported by chemoreceptors. For example, a decline in the blood pressure or oxygen concentration or an increase in the carbon dioxide level generally means that the heart must work harder to meet the demands of peripheral tissues. The cardiac centers then call for an increase in cardiac activity. We detail these reflexes and their effects on the heart and peripheral vessels in Chapter 21.

Autonomic Tone. Like other organs with dual innervation, the heart has a resting autonomic tone. Both autonomic divisions are normally active at a steady background level, releasing acetylcholine (ACh) and norepinephrine (NE) at the nodes and into the myocardium. For this reason, cutting the vagus nerves increases the heart rate, and sympathetic blocking agents slow the heart rate.

Parasympathetic effects dominate in a healthy, resting individual. Without autonomic innervation, the pacemaker cells of the SA node establish the heart rate. Such a heart beats at a rate of 80–100 bpm. At rest, a typical adult heart with normal innervation beats more slowly, at 70–80 bpm, due to activity in the parasympathetic nerves innervating the SA node. If parasympathetic activity increases, the heart rate declines further. Conversely, the heart rate increases if parasympathetic activity decreases, or if sympathetic activation occurs. Through dual innervation and adjustments in autonomic tone, the ANS can make very delicate adjustments in cardiovascular function to meet the demands of other systems.

Tips & Tools

To remember the effect of the sympathetic nervous system on cardiac performance, remember that **s**ympathetic input **s**peeds and **s**trengthens the heartbeat.

Cardiac Reflexes. Information about the status of the cardiovascular system arrives over visceral sensory fibers accompanying the vagus nerve and the sympathetic nerves of the cardiac plexus. The cardiac centers monitor baroreceptors and chemoreceptors innervated by the glossopharyngeal (IX) and vagus (X) nerves. ↺ pp. 502, 503 On the basis of the information received, the cardiac centers adjust the heart's activity to maintain adequate circulation to vital organs, such as the brain.

Effects on the Pacemaker Cells of the SA Node. How do the sympathetic and parasympathetic divisions alter the heart rate? They do so by changing the ionic permeabilities of cells in the conducting system. The most dramatic effects take place at the SA node, which affects the heart rate through changes in the rate at which impulses are generated.

Consider the SA node of a resting individual whose heart is beating at 75 bpm (Figure 20–22a). Any factor that changes the rate of spontaneous depolarization or the duration of repolarization in pacemaker cells will alter the heart rate by changing the time required for these cells to reach threshold. Acetylcholine released by parasympathetic neurons opens chemically gated K^+ channels in the plasma membrane. Then K^+ leaves the pacemaker cells, dramatically slowing their rate of spontaneous depolarization and also slightly extending their duration of repolarization (Figure 20–22b). As a result, heart rate declines.

Figure 20–22 Autonomic Regulation of Pacemaker Cell Function.

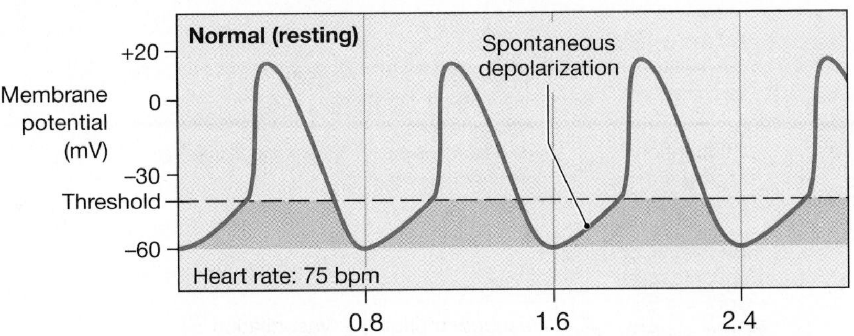
Normal (resting)
Spontaneous depolarization
+20
0
-30
-60
Membrane potential (mV)
Threshold
Heart rate: 75 bpm
0.8
1.6
2.4

a Pacemaker cells have membrane potentials closer to threshold than those of cardiac contractile cells (–60 mV versus –90 mV). Their plasma membranes spontaneously depolarize to threshold, producing action potentials at a frequency determined by (1) the membrane potential and (2) the rate of depolarization.

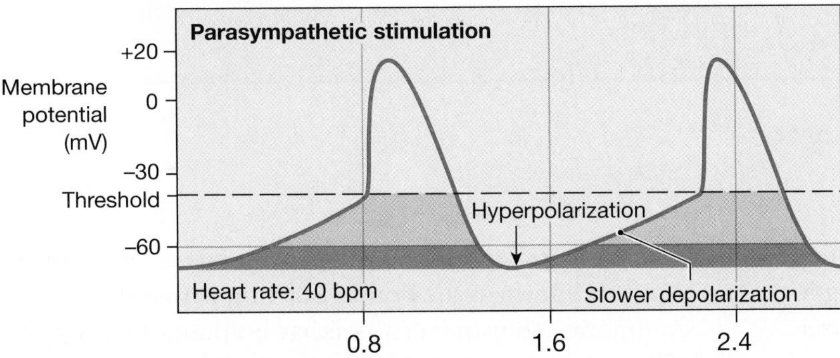
Parasympathetic stimulation
+20
0
-30
-60
Membrane potential (mV)
Threshold
Hyperpolarization
Heart rate: 40 bpm
Slower depolarization
0.8
1.6
2.4

b Parasympathetic stimulation releases ACh, which extends repolarization and decreases the rate of spontaneous depolarization. The heart rate slows.

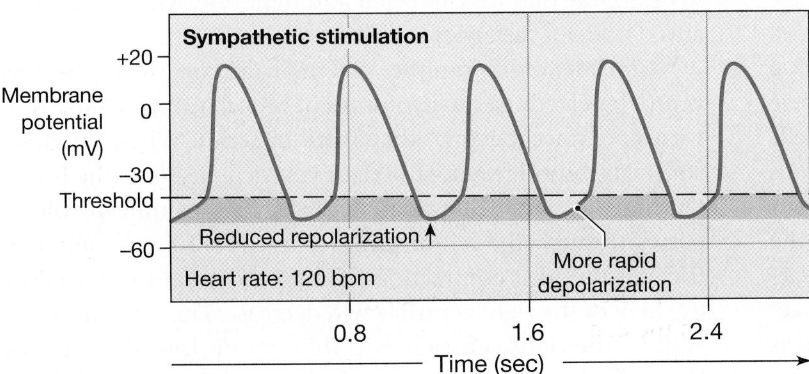
Sympathetic stimulation
+20
0
-30
-60
Membrane potential (mV)
Threshold
Reduced repolarization
More rapid depolarization
Heart rate: 120 bpm
0.8
1.6
2.4
Time (sec)

c Sympathetic stimulation releases NE, which shortens repolarization and accelerates the rate of spontaneous depolarization. As a result, the heart rate increases.

Norepinephrine released by sympathetic neurons binds to beta-1 receptors, leading to the opening of sodium ion and calcium ion channels. Then an influx of positively charged ions increases the rate of depolarization and shortens the period of repolarization. The pacemaker cells reach threshold more quickly, and the heart rate increases (Figure 20–22c).

Venous Return and the Bainbridge Reflex

Venous return is the amount of blood returning to the heart through veins. Venous return directly affects pacemaker cells. When venous return increases, the atria receive more blood and the walls are stretched. Stretching of the cardiac pacemaker cells of the SA node leads to more rapid depolarization and an increase in the heart rate.

Venous return also has an indirect effect on heart rate. The **Bainbridge reflex**, or *atrial reflex*, involves adjustments in heart rate in response to an increase in the venous return. When the walls of the right atrium are stretched, stretch receptors there trigger a reflexive increase in heart rate by stimulating sympathetic activity (see Figure 20–22). Thus, when the rate of venous return to the heart increases, so does the heart rate, and for this reason the cardiac output increases as well.

Hormones and Factors Affecting Heart Rate

Epinephrine (E), norepinephrine (NE), and thyroid hormone (T_3,) increase heart rate by their effects on the SA node. The effects of E on the pacemaker cells of the SA node are similar to those of NE. Epinephrine also affects the cardiac contractile cells. After massive sympathetic stimulation of the adrenal medullae, the myocardium may become so excitable that abnormal contractions occur. Factors that increase the heart rate are *positively chronotropic* (*chrono-*, time) and factors that decrease the heart rate are *negatively chronotropic*.

20

Factors Affecting the Stroke Volume

Changes in either EDV or ESV can change the stroke volume, and thus cardiac output. The factors involved in the regulation of stroke volume are indicated in Figure 20–23.

The End-Diastolic Volume (EDV)

Two factors affect the EDV, the amount of blood in a ventricle at the end of diastole: the filling time and the venous return. **Filling time** is the duration of ventricular diastole. It depends entirely on the heart rate: The faster the heart rate, the shorter the filling time. Venous return is variable over this period. It varies in response to changes in cardiac output, blood volume, peripheral circulation, skeletal muscle activity, and other factors that affect the rate of blood flow back to the heart. (We explore these factors in Chapter 21.)

Preload. The degree of stretching in ventricular muscle cells during ventricular diastole is called the **preload**. The preload is directly proportional to the EDV: The greater the EDV,

Figure 20–23 **Factors Affecting Stroke Volume.**

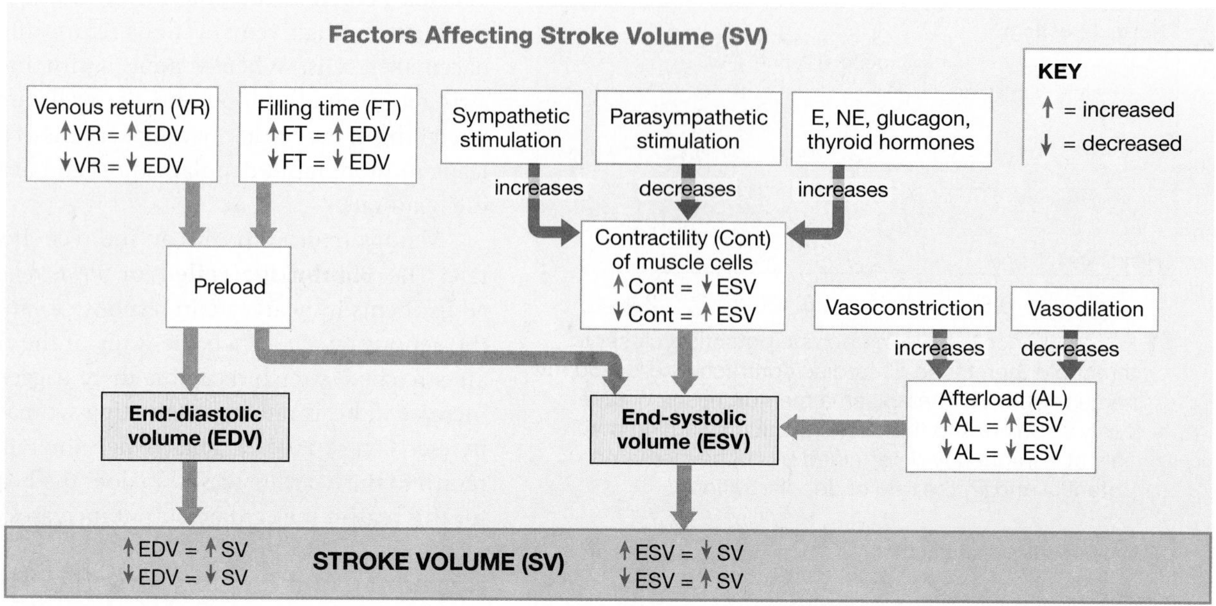
Factors Affecting Stroke Volume (SV)
KEY
↑ = increased
↓ = decreased
Venous return (VR)
↑VR = ↑EDV
↓VR = ↓EDV
Filling time (FT)
↑FT = ↑EDV
↓FT = ↓EDV
Sympathetic stimulation
Parasympathetic stimulation
E, NE, glucagon, thyroid hormones
increases
decreases
increases
Contractility (Cont) of muscle cells
↑Cont = ↓ESV
↓Cont = ↑ESV
Preload
Vasoconstriction
Vasodilation
increases
decreases
Afterload (AL)
↑AL = ↑ESV
↓AL = ↓ESV
End-diastolic volume (EDV)
End-systolic volume (ESV)
↑EDV = ↑SV
↓EDV = ↓SV
STROKE VOLUME (SV)
↑ESV = ↓SV
↓ESV = ↑SV

the larger the preload. Preload matters because it affects the ability of muscle cells to produce tension. As sarcomere length increases past resting length, the amount of force produced during systole increases.

The amount of preload, and hence the degree of myocardial stretching, varies with the demands on the heart. When you are standing at rest, your EDV is low. The ventricular muscle is stretched very little, and the sarcomeres are relatively short. During ventricular systole, the cardiac contractile cells develop little power, and the ESV (the amount of blood in the ventricle after contraction) is relatively high because the contractile cells contracted only a short distance. If you begin exercising, venous return increases and more blood flows into your heart. Your EDV increases, and the myocardium stretches further. As the sarcomeres approach optimal lengths, the ventricular contractile cells can contract more efficiently and produce more forceful contractions. They also shorten more, and more blood is pumped out of your heart.

The EDV and Stroke Volume: The Frank–Starling Principle. In general, the greater the EDV, the larger the stroke volume. Stretching the cardiac contractile cells *past* their optimal length would *reduce* the force of contraction, but this degree of stretching does not normally take place. Myocardial connective tissues, the cardiac skeleton, and the pericardium all limit the expansion of the ventricles.

The relationship between the amount of ventricular stretching and the contractile force means that, within normal physiological limits, increasing the EDV results in a corresponding increase in the stroke volume. This general rule of "more in = more out" is known as the **Frank–Starling principle**.

Autonomic adjustments to cardiac output normally make the effects of the Frank–Starling principle difficult to see. However, we can see the effects more clearly in heart transplant patients, because the implanted heart is not innervated by the ANS. So, the outputs of the left and right ventricles remain balanced under a variety of conditions.

Consider, for example, a person at rest, with the two ventricles ejecting equal volumes of blood. Although the ventricles contract together, they work in series: When the heart contracts, blood leaving the right ventricle heads to the lungs. During the next ventricular diastole, that volume of blood passes through the left atrium, to be ejected by the left ventricle at the next contraction. If the venous return decreases, the EDV of the right ventricle will decrease. During ventricular systole, the right ventricle will then pump less blood to the lungs. In the next cardiac cycle, the EDV of the left ventricle will be reduced, and that ventricle will eject a smaller volume of blood. The output of the two ventricles will again be in balance, but both will have smaller stroke volumes than they did initially.

The End-Systolic Volume (ESV)

After the ventricle has contracted and ejected the stroke volume, the amount of blood that remains in the ventricle at the end of ventricular systole is the ESV. Three factors that influence the ESV are preload, ventricular contractility, and afterload.

Contractility. **Contractility** is the amount of force produced during a contraction, at a given preload. Under normal circumstances, autonomic innervation or circulating hormones can alter contractility. Under special circumstances, drugs or abnormal ion concentrations in the extracellular fluid can alter contractility.

Factors that strengthen heart contraction are *positively inotropic* (ī-nō-trō-pik) (*ino-*, fiber). Factors that weaken heart contraction are *negatively inotropic*. Positive inotropic agents typically stimulate Ca^{2+} entry into cardiac contractile cells, thus increasing the force and duration of ventricular contractions. Negative inotropic agents may block Ca^{2+} movement or depress cardiac muscle metabolism. Positive and negative inotropic factors include ANS activity, hormones, and changes in extracellular ion concentrations.

Effects of Autonomic Activity on Contractility. Autonomic activity alters the degree of contraction and changes the ESV in the following ways:

- Sympathetic stimulation has a positive inotropic effect. It causes the release of norepinephrine (NE) by postganglionic fibers of the cardiac nerves and the secretion of epinephrine (E) and NE by the adrenal medullae. These hormones affect heart rate. They also stimulate alpha and beta receptors in cardiac contractile cell plasma membranes. This stimulation increases the metabolism of these cells. The net effect is that the ventricles contract more forcefully, increasing the ejection fraction and decreasing the ESV.
- Parasympathetic stimulation from the vagus nerves has a negative inotropic effect. The primary effect of acetylcholine (ACh) is at the cardiac contractile cell membrane surface, where it produces hyperpolarization and inhibition. As a result, the force of cardiac contractions is reduced. The atria show the greatest changes in contractile force because the ventricles are not extensively innervated by the parasympathetic division. However, under strong parasympathetic stimulation or with drugs that mimic ACh, the ventricles contract less forcefully, the ejection fraction decreases, and the ESV becomes larger.

Hormones Affecting Contractility. Many hormones affect heart contractility. For example, epinephrine, norepinephrine, glucagon, and thyroid hormones are positively inotropic. Before synthetic inotropic drugs were available, glucagon was used to stimulate cardiac function. It is still used in cardiac emergencies and to treat some forms of heart disease.

The drugs *isoproterenol, dopamine,* and *dobutamine* mimic E and NE by stimulating beta-1 receptors on cardiac contractile cells. Dopamine (at high doses) and dobutamine also stimulate Ca^{2+} entry through alpha-1 receptor stimulation. *Digitalis* elevates the intracellular Ca^{2+} concentration by interfering with Ca^{2+} removal from the cytosol of cardiac contractile cells.

Many of the drugs used to treat *hypertension* (high blood pressure) are negatively inotropic. Beta-blocking drugs such as *propranolol, metoprolol, atenolol,* and *labetalol* block beta receptors, alpha receptors, or both, and prevent sympathetic stimulation of the heart. Calcium-channel blockers such as *nifedipine* or *verapamil* are negatively inotropic.

Afterload. The **afterload** is the amount of tension that the contracting ventricle must produce to force open the semilunar valve and eject blood. Afterload increases with increased resistance to blood flow out of the ventricle. The greater the afterload, the longer the period of isovolumetric contraction, the shorter the duration of ventricular ejection, and the larger the ESV. In other words, as the afterload increases, the stroke volume decreases.

Any factor that restricts blood flow through the arterial system increases afterload. For example, either the constriction of peripheral blood vessels or a circulatory blockage elevates arterial blood pressure and increases the afterload. If the afterload is too great, the ventricles cannot eject blood. Such a high afterload is rare in a normal heart. However, damage to the heart muscle can weaken the myocardium enough that even a modest rise in arterial blood pressure can decrease the stroke volume to a dangerously low level, producing heart failure. Damage to the semilunar valve that restricts blood flow will also increase afterload.

Summary: The Control of Cardiac Output

Figure 20–24 summarizes the factors that regulate heart rate and stroke volume, which interact to determine cardiac output. The heart rate is influenced by the autonomic nervous system, circulating hormones, and venous return.

- Sympathetic stimulation increases the heart rate, and parasympathetic stimulation decreases it. Under resting conditions, parasympathetic tone dominates, and the heart rate is slightly slower than the intrinsic heart rate. When the activity level rises, venous return increases and triggers the Bainbridge reflex. The result is an increase in sympathetic tone and an increase in heart rate.
- Circulating hormones, specifically E, NE, and T_3, accelerate heart rate.
- An increase in venous return stretches the pacemaker cells and increases heart rate.
- The stroke volume (SV) is the difference between the end-diastolic volume (EDV) and the end-systolic volume (ESV).
- The EDV is determined by the available filling time and the rate of venous return.
- The ESV is determined by the amount of preload (the degree of myocardial stretching), the degree of contractility (adjusted by hormones and autonomic innervation), and the afterload (the arterial resistance to blood flow out of the heart).

Figure 20–24 A Summary of the Factors Affecting Cardiac Output.

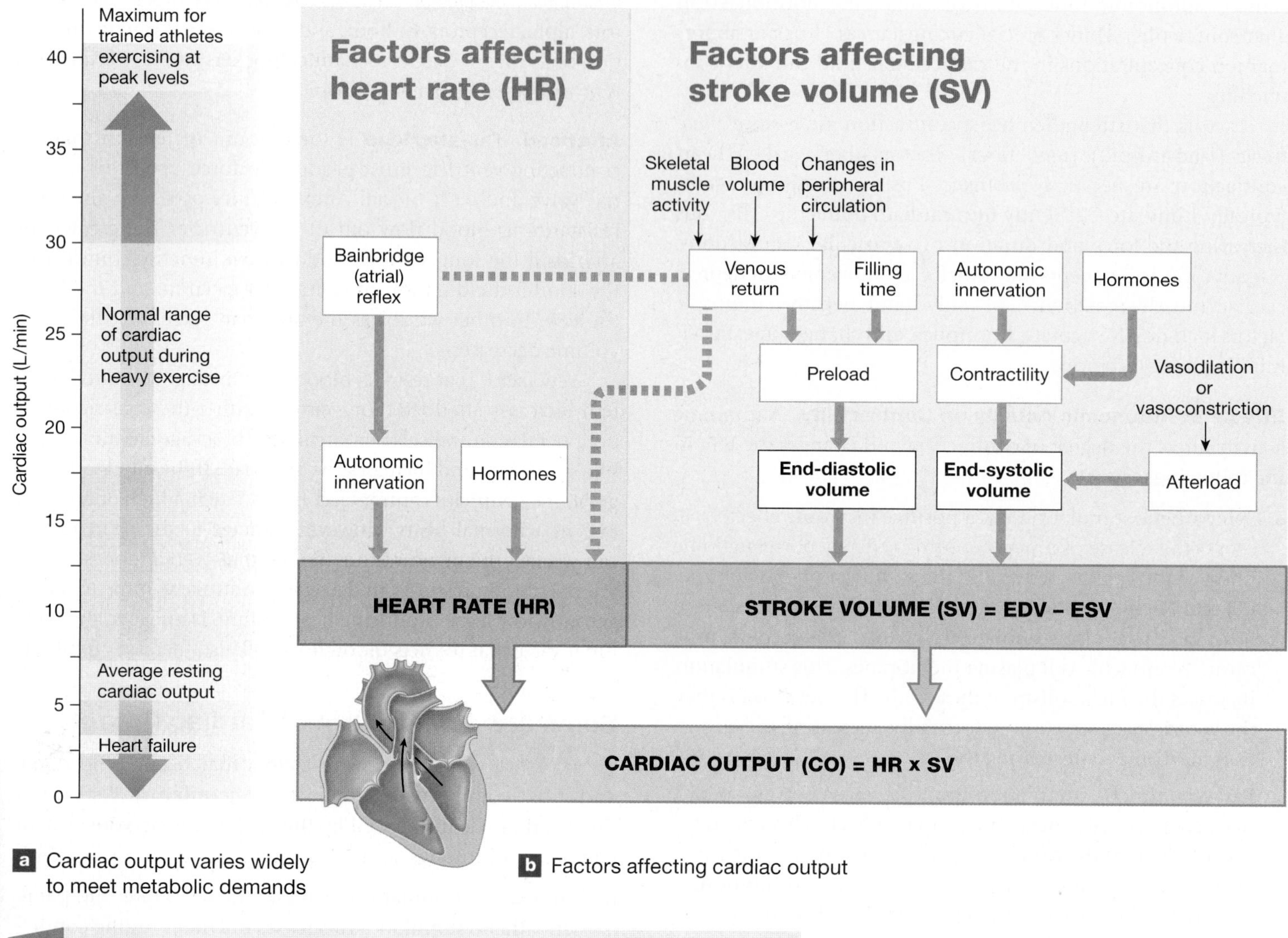
Cardiac output (L/min)
40
35
30
25
20
15
10
5
0
Maximum for trained athletes exercising at peak levels
Normal range of cardiac output during heavy exercise
Average resting cardiac output
Heart failure
Factors affecting heart rate (HR)
Factors affecting stroke volume (SV)
Skeletal muscle activity
Blood volume
Changes in peripheral circulation
Bainbridge (atrial) reflex
Venous return
Filling time
Autonomic innervation
Hormones
Preload
Contractility
Vasodilation or vasoconstriction
Autonomic innervation
Hormones
End-diastolic volume
End-systolic volume
Afterload
HEART RATE (HR)
STROKE VOLUME (SV) = EDV – ESV
CARDIAC OUTPUT (CO) = HR x SV

a Cardiac output varies widely to meet metabolic demands

b Factors affecting cardiac output

? What is considered a "normal" cardiac output when resting? Please provide units.

During exercise, increasing both the stroke volume and the heart rate can increase the cardiac output by 300–500 percent, to 18–30 L/min. The difference between resting and maximal cardiac outputs is the **cardiac reserve**. Trained athletes exercising at maximal levels may increase cardiac output by nearly 700 percent, to 40 L/min.

Cardiac output cannot increase indefinitely, primarily because the available filling time shortens as the heart rate increases. At heart rates up to 160–180 bpm, the combination of increased venous return rate and increased contractility compensates for the reduced filling time. Over this range, cardiac output and heart rate increase together. But if the heart rate continues to climb, the stroke volume begins to drop. Cardiac output first plateaus and then declines.

The Heart and the Vessels of the Cardiovascular System

The purpose of cardiovascular regulation is to maintain adequate blood flow to all body tissues. The heart cannot accomplish this by itself, and it does not work in isolation. For example, when blood pressure changes, the cardiac centers adjust not only the heart rate but also the diameters of peripheral blood vessels. These adjustments work together to keep the blood pressure within normal limits and to maintain circulation to vital tissues and organs. In Chapter 21 we complete this story by detailing the cardiovascular responses to changing activities and circulatory emergencies. We then conclude our discussion of the cardiovascular system by examining the anatomy of the pulmonary and systemic circuits.

> Go to MasteringA&P™ > Study Area > Menu > **iP2** > Cardiovascular System: Cardiac Output

Checkpoint

12. Define *cardiac output*.
13. Caffeine has effects on cardiac conducting cells and contractile cells that are similar to those of NE. What effect would drinking large amounts of caffeinated drinks have on the heart?
14. Why is it a potential problem if the heart beats too rapidly?
15. What effect would stimulating the acetylcholine receptors of the heart have on cardiac output?
16. What effect would an increase in venous return have on the stroke volume?
17. Joe's end-systolic volume is 40 mL, and his end-diastolic volume is 125 mL. What is Joe's stroke volume?

See the blue Answers tab at the back of the book.

20 Chapter Review

Study Outline

MasteringA&P™ Access more chapter study tools online in the MasteringA&P Study Area:

- Chapter Quizzes, Chapter Practice Test, MP3 Tutor Sessions, and Clinical Case Studies
- Practice Anatomy Lab PAL 3.0
- A&P Flix A&PFlix
- Interactive Physiology iP2
- PhysioEx PhysioEx 9.1

An Introduction to the Heart as Part of the Cardiovascular System p. 689

1. The blood vessels can be subdivided into the **pulmonary circuit** (which carries blood to and from the lungs) and the **systemic circuit** (which transports blood to and from the rest of the body).
2. **Arteries** carry blood away from the heart; **veins** return blood to the heart. **Capillaries**, or *exchange vessels*, are thin-walled, narrow-diameter vessels that connect the smallest arteries and veins. *(Figure 20–1)*
3. The heart has four chambers: the **right atrium** and **right ventricle**, and the **left atrium** and **left ventricle**.

20-1 The heart is a four-chambered organ that pumps blood through the pulmonary and systemic circuits p. 689

4. The heart is surrounded by the **pericardium** and lies within the anterior portion of the **mediastinum**, which separates the two pleural cavities. *(Figure 20–2)*
5. The pericardial cavity is lined by the **pericardium**. The **visceral layer of serous pericardium (epicardium)** covers the heart's outer surface, and the **parietal layer of serous pericardium** lines the inner surface. *(Figure 20–2)*
6. The **coronary sulcus**, a deep groove, marks the boundary between the atria and the ventricles. Other surface markings also provide useful reference points in describing the heart and associated structures. *(Figure 20–3)*
7. The bulk of the heart consists of the muscular **myocardium**. The **endocardium** lines the inner surfaces of the heart, and the **visceral layer of serous pericardium** covers the outer surface. The endothelium of the endocardium is continuous with the endothelium of the attached great vessels. *(Figure 20–4)*
8. The atria are separated by the **interatrial septum**, and the ventricles are divided by the **interventricular septum**. The right atrium receives blood from the systemic circuit via two large veins, the **superior vena cava** and the **inferior vena cava**. (The atrial walls contain the **pectinate muscles**, prominent muscular ridges.) *(Figure 20–5)*
9. Blood flows from the right atrium into the right ventricle by the **tricuspid valve (right atrioventricular valve)**. This opening is bounded by three **cusps** of fibrous tissue braced by the **chordae tendineae**, which are connected to **papillary muscles**. *(Figure 20–5)*
10. Blood leaving the right ventricle enters the **pulmonary trunk** after passing through the **pulmonary valve**. The pulmonary trunk divides to form the **left** and **right pulmonary arteries**. The **left** and **right pulmonary veins** return blood from the lungs to the left atrium. Blood leaving the left atrium flows into the left ventricle through the **mitral valve (left atrioventricular valve** or *bicuspid valve*). Blood leaving the left ventricle passes through the **aortic valve** and into the systemic circuit by the **ascending aorta**. *(Figure 20–5)*
11. Anatomical differences between the ventricles reflect the functional demands placed on them. The wall of the right ventricle is relatively thin, whereas the left ventricle has a massive muscular wall. *(Figure 20–6)*
12. The connective tissues of the heart (mainly collagen and elastic fibers) and the **cardiac skeleton** support the heart's valves and ventricular muscle cells. *(Figure 20–7)*
13. Valves normally permit blood flow in only one direction, preventing the **regurgitation** (backflow) of blood. *(Figure 20–7)*
14. The **coronary circulation** meets the high oxygen and nutrient demands of cardiac muscle cells. The **coronary arteries** originate at the base of the ascending aorta. Interconnections between arteries, called **arterial anastomoses**, ensure a constant blood supply. The **great**, **posterior**, **small**, **anterior**, and **middle cardiac veins** are epicardial vessels that carry blood from the coronary capillaries to the **coronary sinus**. *(Figure 20–8)*
15. In **coronary artery disease (CAD)**, portions of the coronary circulation undergo partial or complete blockage. A **myocardial infarction (MI)**, or heart attack, occurs when part of the coronary circulation becomes blocked and muscle tissue dies when it cannot be oxygenated. *(Spotlight Figure 20–9)*

20-2 The cells of the conducting system distribute electrical impulses through the heart, causing cardiac contractile cells to contract p. 702

16. Two general classes of cardiac muscle cells are involved in the normal **heartbeat**: **autorhythmic cells** and **contractile cells**.
17. The **conducting system** is composed of the *sinoatrial node*, the *atrioventricular node*, and *conducting cells*. The conducting system initiates and distributes electrical impulses within the heart. Nodal cells establish the rate of cardiac contraction, and conducting cells distribute the contractile stimulus from the SA node to the atrial myocardium and the AV node (along *internodal pathways*), and from the AV node to the ventricular myocardium. *(Figure 20–10)*
18. Unlike skeletal muscle, cardiac muscle contracts without neural or hormonal stimulation. **Pacemaker cells** in the **sinoatrial (SA) node** (*cardiac pacemaker*) normally establish the rate of contraction. From the SA node, the stimulus travels to the **atrioventricular (AV) node**, and then to the **AV bundle**, which divides into **bundle branches**. From there, **Purkinje fibers** convey the impulses to the ventricular myocardium. *(Figures 20–10, 20–11)*
19. A recording of electrical activities in the heart is an **electrocardiogram (ECG** or **EKG)**. Important landmarks of an ECG include the **P wave** (atrial depolarization), the **QRS complex** (ventricular depolarization), and the **T wave** (ventricular repolarization). *(Figure 20–12)*
20. **Cardiac arrhythmias** are abnormal patterns of electrical activity in the heart. *(Spotlight Figure 20–13)*
21. **Cardiac contractile cells** are interconnected by **intercalated discs**, which convey the force of contraction from cell to cell and conduct action potentials. *(Figure 20–14;* Table 20–1*)*
22. **Contractile cells** form the bulk of the atrial and ventricular walls. Cardiac contractile cells have a long refractory period, so rapid stimulation produces twitches rather than tetanic contractions. *(Figure 20–15)*

20

20-3 The contraction–relaxation events that occur during a complete heartbeat make up a cardiac cycle p. 711

23. The **cardiac cycle** includes periods of **atrial** and **ventricular systole** (contraction) and **atrial** and **ventricular diastole** (relaxation). *(Figure 20–16)*
24. When the heart beats, the two ventricles eject equal volumes of blood. *(Figure 20–17)*
25. The closing of valves and rushing of blood through the heart cause characteristic heart sounds, which can be heard during *auscultation*. *(Figure 20–18)*

20-4 Cardiac output is determined by heart rate and stroke volume p. 716

26. Cardiac output can be adjusted by changes in either heart rate or stroke volume. *(Figure 20–19)*
27. The amount of blood ejected by a ventricle during a single beat is the **stroke volume (SV)**. The amount of blood pumped by a ventricle each minute is the **cardiac output (CO)**. *(Figure 20–20)*
28. The **cardioacceleratory center** in the medulla oblongata activates sympathetic neurons; the **cardioinhibitory center** controls the parasympathetic neurons that slow the heart rate. These cardiac centers receive inputs from higher centers and from receptors monitoring blood pressure and the concentrations of dissolved gases. *(Figure 20–21)*
29. The basic heart rate is established by the pacemaker cells of the SA node, but it can be modified by the autonomic nervous system. The **Bainbridge reflex** *(atrial reflex)* accelerates the heart rate when the walls of the right atrium are stretched. *(Figure 20–22)*
30. Sympathetic activity increases the force of contraction (contractility), which reduces the ESV. Parasympathetic stimulation slows the heart rate, reduces contractility, and raises the ESV.
31. Cardiac output is affected by various factors, including autonomic innervation and hormones. *(Figure 20–22)*
32. The stroke volume is the difference between the **end-diastolic volume (EDV)** and the **end-systolic volume (ESV)**. The **filling time** and **venous return** interact to determine the EDV. Normally, the greater the EDV, the more powerful the succeeding contraction (the **Frank–Starling principle**). *(Figure 20–23)*
33. The difference between resting and maximal cardiac outputs is the **cardiac reserve**. *(Figure 20–24)*
34. The heart does not work in isolation in maintaining adequate blood flow to all tissues.

Review Questions

See the blue Answers tab at the back of the book.

LEVEL 1 Reviewing Facts and Terms

1. The great cardiac vein drains blood from the heart muscle to the **(a)** left ventricle, **(b)** right ventricle, **(c)** right atrium, **(d)** left atrium.
2. The autonomic centers for cardiac function are located in **(a)** the myocardial tissue of the heart, **(b)** the cardiac centers of the medulla oblongata, **(c)** the cerebral cortex, **(d)** all of these structures.
3. The serous membrane covering the outer surface of the heart is the **(a)** parietal layer of the serous pericardium, **(b)** endocardium, **(c)** myocardium, **(d)** visceral layer of the serous pericardium.
4. The simple squamous epithelium covering the heart valves is the **(a)** epicardium, **(b)** endocardium, **(c)** myocardium, **(d)** endothelium.
5. The heart is surrounded by the **(a)** pleural cavity, **(b)** peritoneal cavity, **(c)** abdominopelvic cavity, **(d)** mediastinum, **(e)** abdominal cavity.
6. The cardiac skeleton of the heart has which *two* of the following functions? **(a)** It physically isolates the muscle fibers of the atria from those of the ventricles. **(b)** It maintains the normal shape of the heart. **(c)** It helps distribute the forces of cardiac contraction. **(d)** It allows more rapid contraction of the ventricles. **(e)** It strengthens and helps prevent overexpansion of the heart.
7. Cardiac output is equal to the **(a)** difference between the end-diastolic volume and the end-systolic volume, **(b)** product of heart rate and

stroke volume, **(c)** difference between the stroke volume at rest and the stroke volume during exercise, **(d)** stroke volume less the end-systolic volume, **(e)** product of heart rate and blood pressure.

8. Identify the superficial structures in the following diagram of the heart.

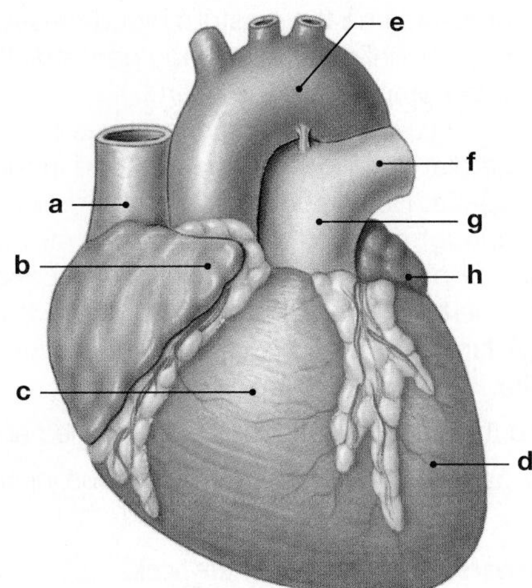
a
b
c
d
e
f
g
h

(a) __________ **(b)** __________
(c) __________ **(d)** __________
(e) __________ **(f)** __________
(g) __________ **(h)** __________

9. Identify the structures in the following diagram of a sectional view of the heart.

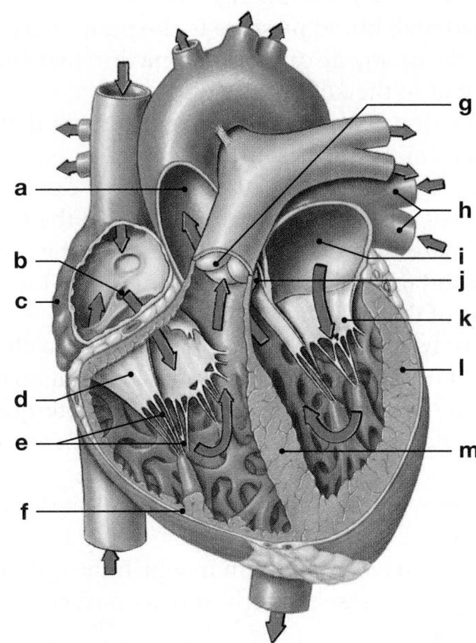
a
b
c
d
e
f
g
h
i
j
k
l
m

(a) __________ **(b)** __________ **(c)** __________
(d) __________ **(e)** __________ **(f)** __________
(g) __________ **(h)** __________ **(i)** __________
(j) __________ **(k)** __________ **(l)** __________
(m) __________

10. During diastole, a chamber of the heart **(a)** relaxes and fills with blood, **(b)** contracts and pushes blood into an adjacent chamber, **(c)** experiences a sharp increase in pressure, **(d)** reaches a pressure of approximately 120 mm Hg.
11. During the cardiac cycle, the amount of blood ejected from the left ventricle when the semilunar valve opens is the **(a)** stroke volume (SV), **(b)** end-diastolic volume (EDV), **(c)** end-systolic volume (ESV), **(d)** cardiac output (CO).
12. What role do the chordae tendineae and papillary muscles play in the normal function of the AV valves?
13. Describe the three distinct layers that make up the heart wall.
14. What are the valves in the heart, and what is the function of each?
15. Trace the normal pathway of an electrical impulse through the conducting system of the heart.
16. What is the cardiac cycle? What phases and events are necessary to complete a cardiac cycle?
17. What three factors regulate stroke volume to ensure that the left and right ventricles pump equal volumes of blood?

LEVEL 2 Reviewing Concepts

18. The cells of the conducting system differ from the contractile cells of the heart in that **(a)** conducting cells are larger and contain more myofibrils, **(b)** contractile cells exhibit pacemaker potentials, **(c)** contractile cells do not normally exhibit autorhythmicity, **(d)** both a and b are correct.
19. Which of the following is *longer*? **(a)** the refractory period of a cardiac contractile cell, **(b)** the refractory period of skeletal muscle fiber.
20. If the papillary muscles fail to contract, **(a)** the ventricles will not pump blood, **(b)** the atria will not pump blood, **(c)** the semilunar valves will not open, **(d)** the AV valves will not close properly, **(e)** none of these happen.
21. Cardiac output cannot increase indefinitely because **(a)** the available filling time becomes shorter as the heart rate increases, **(b)** the cardiovascular centers adjust the heart rate, **(c)** the rate of spontaneous depolarization decreases, **(d)** the ion concentrations of pacemaker plasma membranes decrease.
22. Describe the function of the SA node in the cardiac cycle. How does this function differ from that of the AV node?
23. What are the sources and significance of the four heart sounds?
24. Differentiate between stroke volume and cardiac output. How is cardiac output calculated?
25. What factors influence cardiac output?
26. What effect does sympathetic stimulation have on the heart? What effect does parasympathetic stimulation have on the heart?
27. Describe the effects of epinephrine, norepinephrine, glucagon, and thyroid hormones on the contractility of the heart.

LEVEL 3 Critical Thinking and Clinical Applications

28. Vern is suffering from cardiac arrhythmias and is brought into the emergency room of a hospital. In the emergency room he begins to exhibit tachycardia and as a result loses consciousness. Explain why Vern lost consciousness.
29. Harvey has a heart murmur in his left ventricle that produces a loud "gurgling" sound at the beginning of systole. Which valve is probably faulty?
30. The following measurements were made on two individuals (the values recorded remained stable for 1 hour):
Person 1: heart rate, 75 bpm; stroke volume, 60 mL
Person 2: heart rate, 90 bpm; stroke volume, 95 mL
Which person has the greater venous return? Which person has the longer ventricular filling time?
31. Karen is taking the medication verapamil, a drug that blocks the calcium channels in cardiac muscle cells. What effect should this medication have on Karen's stroke volume?

CLINICAL CASE Wrap-Up A Needle to the Chest

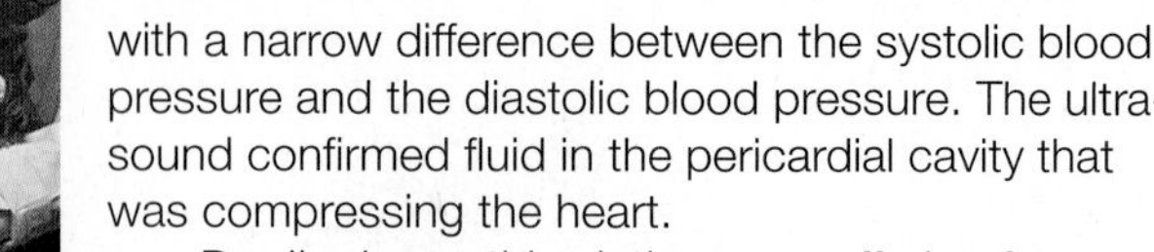

With the force of the head-on collision, the patient's chest first broke the seat belt and then collided with the steering wheel. This caused fractures of the anterior chest wall, including the sternum and ribs.

The heart, located just deep to the sternum, then hit the steering wheel hard enough to cause bleeding into the pericardial cavity. As the delicate parietal and visceral pericardial linings bled into the pericardial cavity, the tough, fibrous connective tissue of the parietal layer itself did not expand.

The blood in the pericardial cavity compressed the heart, squeezing the atria and ventricles. During atrial diastole, the atria could not fill. This caused a backup of the venous system that was obvious in the veins of the patient's neck. The compressed ventricles could not fill with blood during ventricular diastole. Because the end-diastolic volume (EDV) was so low, the stroke volume (SV) was also very low. This was evident in the blood pressure of 80/58, with a narrow difference between the systolic blood pressure and the diastolic blood pressure. The ultrasound confirmed fluid in the pericardial cavity that was compressing the heart.

Dr. Jim knew this victim was suffering from cardiac tamponade due to blood in the pericardial space. He also knew the problem was severe enough that with the decreasing stroke volume, the heart could not pump adequate blood to the patient's brain. So he removed the blood from the pericardial space with a needle, a procedure known as *pericardiocentesis*, and saved this patient's life.

1. Why do you think the heart sounds were muffled and distant?
2. What other cardiac structures were likely to be injured by this collision?

See the blue Answers tab at the back of the book.

Related Clinical Terms

20

artificial pacemaker: A small, battery-operated device that keeps one's heart beating in a regular rhythm. It may be permanently implanted or temporarily placed externally.

asystole: The absence of cardiac activity with no contraction and no output.

automated external defibrillator (AED): A device that, when applied, automatically checks the function of the heart. Upon detecting a condition that may respond to an electric shock, it delivers a shock to restore normal heartbeat rhythm.

automatic implantable cardioverter defibrillator (AICD): A surgically implanted battery-operated device that monitors the function of the heart. Upon detecting a condition that may respond to an electric shock, such as a disorganized heartbeat, the device delivers a shock to restore normal heartbeat rhythm.

cardiac arrest: Sudden stopping of the pumping action of the heart causing the loss of arterial blood pressure.

cardiology: The branch of medicine dealing with the diagnosis and treatment of heart disorders and related conditions.

cardiomegaly: An enlarged heart, which is a sign of some other condition such as stress, weakening of the heart muscle, coronary artery disease, heart valve problems, or abnormal heart rhythms.

cardiomyoplasty: A surgical procedure that uses stimulated latissimus dorsi muscle to assist with cardiac function. The latissimus dorsi muscle is relocated and wrapped around the left and right ventricles and stimulated to contract during cardiac systole by means of an implanted burst-stimulator.

commotio cordis: Sudden cardiac arrest as the result of a blunt hit or impact to the chest.

congestive heart failure: The heart condition of weakness, edema, and shortness of breath caused by the inability of the heart to maintain adequate blood circulation in the peripheral tissues and the lungs.

cor pulmonale: Weakness of the right ventricle of the heart due to prolonged high blood pressure in the pulmonary artery and right ventricle; or any disease or malfunction that affects the pulmonary circuit in the lungs.

echocardiography: A noninvasive diagnostic test that uses ultrasound to make images of the heart chambers, valves, and surrounding structures. This diagnostic tool can also measure cardiac output, detect inflammation around the heart, identify abnormal anatomy, and detect infections of the heart valves.

endocarditis: Inflammation or infection of the endocardium, the inner lining of the heart muscle.

fibrillation: Fast twitching of the heart muscle fibers with little or no movement of the muscle as a whole. Atrial fibrillation occurs in the atria of the heart and is characterized by chaotic quivers and irregular ventricular beating with both atria and ventricles being out of sync.

heart block: Delay in the normal electrical pulses that cause the heart to beat.

mitral valve prolapse: A condition in which the mitral (bicuspid) valve cusps do not close properly and are pushed back toward the left atrium.

myocarditis: Inflammation of the myocardium, the middle layer of the heart wall tissue.

palpitation: Irregular and rapid beating of the heart.

percutaneous transluminal coronary angioplasty (PTCA): The surgical use of a balloon-tipped catheter to enlarge a narrowed artery.

sick sinus syndrome: A group of heart rhythm disorders or problems in which the sinoatrial node does not work properly to regulate the heart rhythms.

21 Blood Vessels and Circulation

Learning Outcomes

These Learning Outcomes correspond by number to this chapter's sections and indicate what you should be able to do after completing the chapter.

CLINICAL CASE Did Ancient Mummies Have Atherosclerosis?

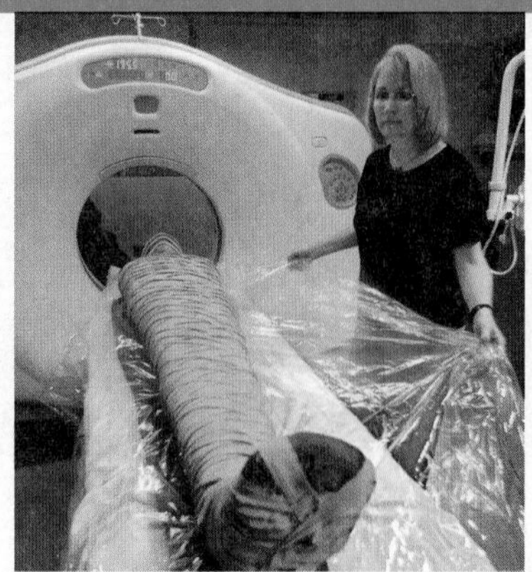

Dr. Iacullo is part of a research team studying ancient mummies from Egypt, Peru, New Mexico, and the Aleutian Islands in Alaska. The study spans 4000 years of human history and cultures from three different continents. Some of the mummies were once members of Egypt's upper class, while others were likely common people from the Americas. Their diets varied greatly and included nuts, berries, farmed corn, domesticated animals, wild game, sea otters, and whales. Their activity patterns varied—the wealthy Egyptians, for instance, were much more sedentary than the hunter-gatherers of the Americas.

Dr. Iacullo is studying the incidence of atherosclerosis (the buildup of arterial plaque) in mummies. Her team uses computerized tomography (CT) scans to look for calcifications in the walls of the mummies' arteries. They discover that one-third of the mummies have CT evidence of plaque buildup. This finding spans all four geographic populations, all diets, and all lifestyle activity levels. **Why do you think these ancient people developed atherosclerosis, which today we usually associate with the unhealthy diet and inactivity of a modern lifestyle?** **To find out, turn to the Clinical Case Wrap-Up on p. 783.**

An Introduction to Blood Vessels and Circulation

Blood circulates throughout the body, moving from the heart through the tissues and back to the heart, in tubular structures called *blood vessels*. In this chapter we examine the organization of blood vessels and consider the integrated functions of the cardiovascular system as a whole. We begin with the histology of *arteries*, *capillaries*, and *veins*. Then we explore the functions of these vessels, the basic principles of cardiovascular regulation, and the distribution of major blood vessels in the body.

21

21-1 Arteries, which are elastic or muscular, and veins, which contain valves, have three-layered walls; capillaries have thin walls with only one layer

Learning Outcome Distinguish among the types of blood vessels based on their structure and function.

The cardiovascular system has five general classes of blood vessels: arteries, arterioles, capillaries, venules, and veins. **Arteries** carry blood away from the heart. As they enter peripheral tissues, they branch repeatedly, and the branches decrease in diameter. The smallest arterial branches are called **arterioles** (ar-TĒR-ē-ōlz). From the arterioles, blood moves into **capillaries**, where diffusion between blood and interstitial fluid takes place. From the capillaries, blood enters small **venules** (VEN-yūlz), which unite to form larger **veins** that return blood to the heart.

Blood leaves the heart through the pulmonary trunk (which originates at the right ventricle) and the aorta (which originates at the left ventricle). The pulmonary arteries that branch from the pulmonary trunk carry blood to the lungs. The systemic arteries that branch from the aorta distribute blood to all other organs. Within these organs, the vessels branch into several hundred million tiny arterioles that supply blood to more than 10 billion capillaries. These capillaries are barely the diameter of a single red blood cell. If all the capillaries in your body were placed end to end, their combined length would be more than 25,000 miles—that's enough to circle the planet!

Blood vessels must be resilient enough to withstand changes in pressure, and flexible enough to move with underlying tissues and organs. The pressures inside vessels vary with distance from the heart, and the structures of different vessels reflect this fact. The functional differences between arteries, veins, and capillaries are associated with distinctive anatomical features.

Vessel Wall Structure in Arteries and Veins

The walls of arteries and veins have three distinct layers, from deep to superficial—the *tunica intima, tunica media*, and *tunica externa* (**Figure 21–1**):

1. The **tunica intima** (IN-ti-muh) is the inner layer of a blood vessel. It includes the endothelial lining and a surrounding layer of connective tissue with a variable number of elastic fibers. In arteries, the outer margin of the tunica intima contains a thick layer of elastic fibers called the **internal elastic membrane**.
2. The **tunica media** is the middle layer of a blood vessel. It contains concentric sheets of smooth muscle tissue in a framework of loose connective tissue. The collagen fibers bind the tunica media to the tunica intima and tunica externa. The tunica media is commonly the thickest layer in a small artery. It is separated from the surrounding tunica externa by a thin band of elastic fibers called the **external elastic membrane**. The smooth muscle cells of the tunica media encircle the endothelium that lines the **lumen** (interior space) of the blood vessel. Large arteries also contain layers of longitudinally arranged smooth muscle cells.

Figure 21–1 **Comparison of a Typical Artery and a Typical Vein.**

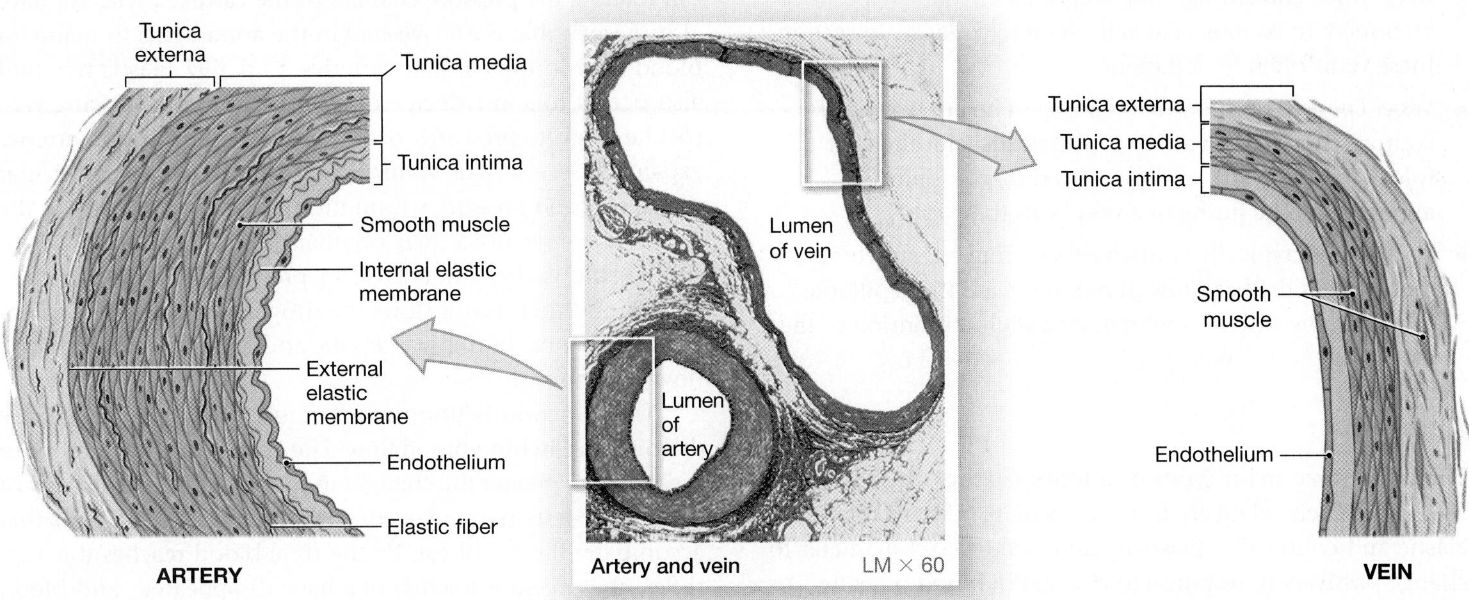
Tunica externa
Tunica media
Tunica intima
Smooth muscle
Internal elastic membrane
External elastic membrane
Endothelium
Elastic fiber
ARTERY
Lumen of vein
Lumen of artery
Artery and vein
LM × 60
Tunica externa
Tunica media
Tunica intima
Smooth muscle
Endothelium
VEIN

Feature	Typical Artery	Typical Vein
GENERAL APPEARANCE IN SECTIONAL VIEW	Usually round, with relatively thick wall	Usually flattened or collapsed, with relatively thin wall
TUNICA INTIMA		
Endothelium	Usually rippled, due to vessel constriction	Often smooth
Internal elastic membrane	Present	Absent
TUNICA MEDIA	Thick, dominated by smooth muscle cells and elastic fibers	Thin, dominated by smooth muscle cells and collagen fibers
External elastic membrane	Present	Absent
TUNICA EXTERNA	Collagen and elastic fibers	Collagen and elastic fibers and smooth muscle cells

3. The **tunica externa** (eks-TER-nuh), or *tunica adventitia* (ad-ven-TISH-uh), is the outer layer of a blood vessel. It is a connective tissue sheath. In arteries, it contains collagen fibers with scattered bands of elastic fibers. In veins, it is generally thicker than the tunica media and contains networks of elastic fibers and bundles of smooth muscle cells. The connective tissue fibers of the tunica externa typically blend into those of adjacent tissues, stabilizing and anchoring the blood vessel.

Their layered walls give arteries and veins considerable strength. The muscular and elastic components also permit controlled changes in diameter as blood pressure or blood volume changes. When the smooth muscle of the tunica media contracts, the vessel decreases in diameter, and when it relaxes, the diameter increases. However, the walls of arteries and veins are too thick to allow diffusion between the bloodstream and surrounding tissues, or even between the blood and the tissues of the vessel itself. For this reason, the walls of large vessels contain small arteries and veins that supply the smooth muscle cells and fibroblasts of the tunica media and tunica externa. These blood vessels are called the **vasa vasorum** ("vessels of vessels").

Differences between Arteries and Veins

Arteries and veins supplying the same region lie side by side (see the light micrograph in Figure 21–1). In sectional view, you can distinguish arteries and veins by the following features:

- *Vessel Walls.* In general, the walls of arteries are thicker than the walls of veins. The tunica media of an artery contains more smooth muscle and elastic fibers than a vein does. These components help resist the arterial pressure generated by the heart as it pumps blood into the pulmonary trunk and aorta.
- *Vessel Lumen.* When not opposed by blood pressure, the elastic fibers in the arterial walls recoil, which constricts the lumen. Thus, seen on dissection or in sectional view, the lumen of an artery often looks smaller than that of the

corresponding vein. Because the walls of arteries are relatively thick and strong, they keep their circular shape when sectioned. In contrast, cut veins tend to collapse. In section, these veins often look flattened.

- *Vessel Lining.* The endothelial lining of an artery cannot contract, so when an artery constricts, its endothelium folds. Therefore, the sectioned arteries have a pleated appearance. The lining of a vein lacks these folds.
- *Valves.* Veins typically contain *valves*—internal structures that prevent the backflow of blood toward the capillaries. In a vein, the valve is apparent by a slight distension of the vessel wall.

Arteries

There are three main types of arteries: elastic, muscular, and arterioles. Their relatively thick, muscular walls make arteries elastic and contractile. Elasticity allows the vessel diameter to change passively in response to changes in blood pressure. For example, it allows arteries to absorb the surging pressure waves that accompany the contractions of the ventricles. Arteries carry blood under great pressure, and their walls are adapted to handle that pressure.

When arterial walls actively change diameter, this change takes place by actions of the sympathetic nervous system. When stimulated, arterial smooth muscles contract, constricting the artery—a process called **vasoconstriction**. When these smooth muscles relax, the diameter of the lumen increases—a process called **vasodilation**. Vasoconstriction and vasodilation affect (1) the afterload on the heart, (2) peripheral blood pressure, and (3) capillary blood flow. We explore these effects in Section 21-3. Vessel contractility is also important during the vascular phase of hemostasis, when the contraction of a damaged vessel wall helps reduce bleeding. p. 679

21

In traveling from the heart to peripheral capillaries, blood passes through *elastic arteries, muscular arteries*, and *arterioles* (Figure 21–2). Keep in mind that vessel characteristics change gradually with distance from the heart. Each type of vessel described here actually represents the midpoint in a portion of a continuum. For example, the largest muscular arteries contain a considerable amount of elastic tissue, and the smallest resemble heavily muscled arterioles.

Elastic Arteries

Elastic arteries carry large volumes of blood away from the heart. They are large vessels with an internal, or *luminal*, diameter up to 2.5 cm (1 in.) (see Figure 21–2). The pulmonary trunk and aorta, as well as their major branches (the *pulmonary, common carotid, subclavian,* and *common iliac arteries*), are elastic arteries.

The walls of elastic arteries are extremely resilient because the tunica media contains a high density of elastic fibers and relatively few smooth muscle cells. As a result, elastic arteries can tolerate the pressure changes of the cardiac cycle. We have already seen that *elastic rebound* in the aorta helps to maintain blood flow in the coronary arteries. p. 697 Elastic rebound also occurs to some degree in all elastic arteries. During ventricular systole, pressures rise rapidly and the elastic arteries expand as the stroke volume is ejected. During ventricular diastole, blood pressure within the arterial system falls and the elastic fibers recoil to their original dimensions. Their expansion cushions the sudden rise in pressure during ventricular systole, and their recoil slows the drop in pressure during ventricular diastole. In this way, elastic arteries help to make blood flow continuous.

This function is important because blood pressure is the driving force behind blood flow: The greater the fluctuations in pressure, the greater the changes in blood flow. The elasticity of the arterial system dampens the pressure peaks and valleys that accompany the heartbeat. By the time blood reaches the arterioles, the pressure fluctuations have disappeared, and blood flow is continuous.

Muscular Arteries

Muscular arteries, or *medium-sized arteries*, distribute blood to the body's skeletal muscles and internal organs. Most of the vessels of the arterial system are muscular arteries. They are characterized by a thick tunica media, which contains more smooth muscle cells than the tunica media of elastic arteries (see Figure 21–2). A typical muscular artery has a luminal diameter of approximately 4 mm (0.16 in.), but some have diameters as small as 0.5 mm. The *external carotid arteries* of the neck, the *brachial arteries* of the arms, the *mesenteric arteries* of the abdomen, and the *femoral arteries* of the thighs are examples of muscular arteries.

Superficial muscular arteries are important as *pressure points*—places in the body where muscular arteries can be pressed against deeper bones to reduce blood flow and control severe bleeding. Major arterial pressure points are the common carotid, radial, brachial, femoral, popliteal, posterior tibial, and dorsal pedal.

Arterioles

Arterioles, with an internal diameter of 30 μm or less, are considerably smaller than muscular arteries. Arterioles have a poorly defined tunica externa. In the larger arterioles, the tunica media consists of one or two layers of smooth muscle cells (see Figure 21–2). The wall of an arteriole with a luminal diameter of 30 μm can have a 20-μm-thick layer of smooth muscle. In the smallest arterioles, the tunica media contains scattered smooth muscle cells that do not form a complete layer.

The luminal diameters of smaller muscular arteries and arterioles change in response to local conditions or to sympathetic or endocrine stimulation. For example, arterioles in most

Figure 21–2 **Histological Structure of Blood Vessels.** Representative diagrammatic views of arteries, veins, and capillaries. The relative proportions of wall thickness to luminal diameter are not drawn to scale.

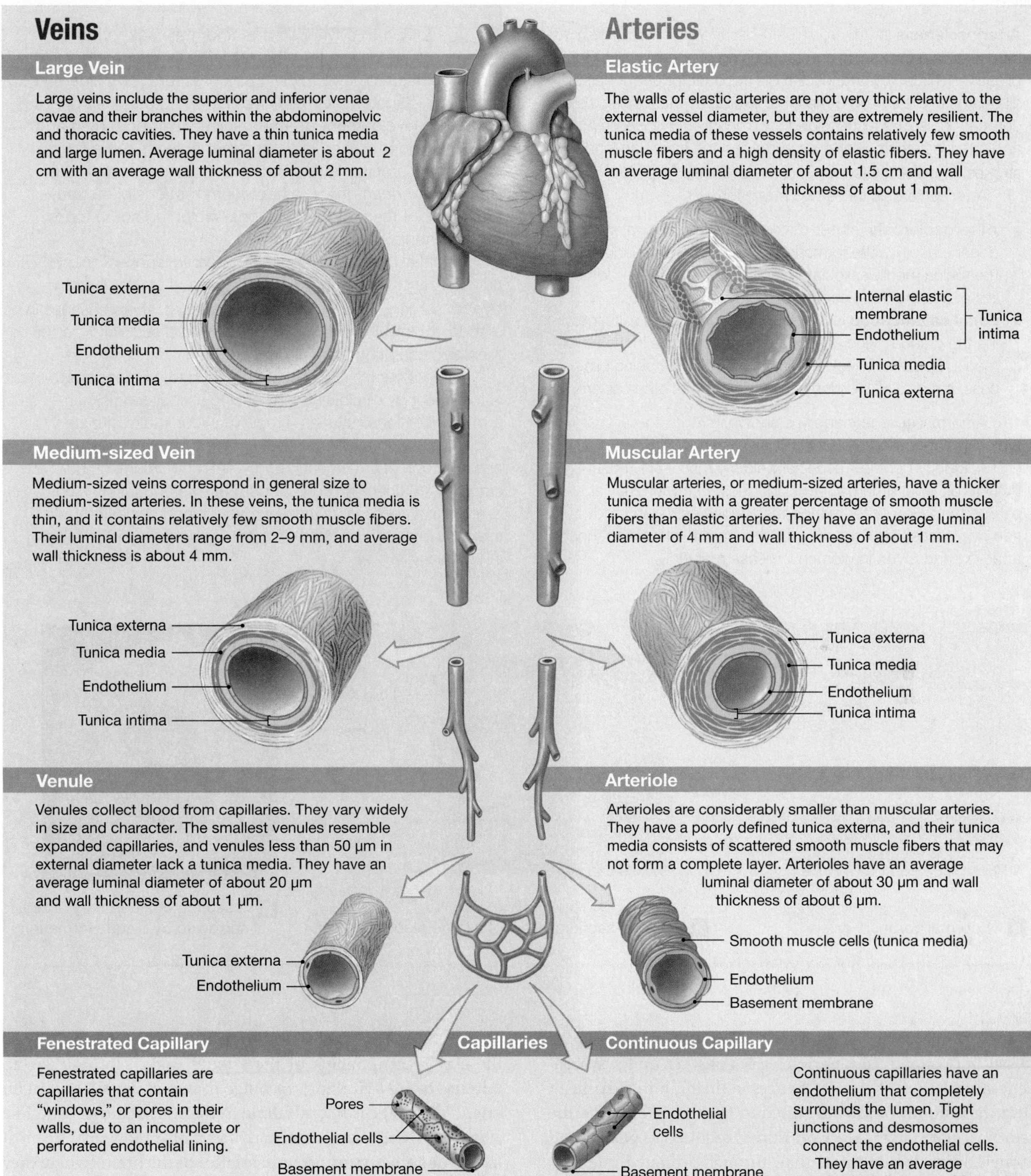
Veins
Large Vein
Large veins include the superior and inferior venae cavae and their branches within the abdominopelvic and thoracic cavities. They have a thin tunica media and large lumen. Average luminal diameter is about 2 cm with an average wall thickness of about 2 mm.
Tunica externa
Tunica media
Endothelium
Tunica intima
Medium-sized Vein
Medium-sized veins correspond in general size to medium-sized arteries. In these veins, the tunica media is thin, and it contains relatively few smooth muscle fibers. Their luminal diameters range from 2–9 mm, and average wall thickness is about 4 mm.
Tunica externa
Tunica media
Endothelium
Tunica intima
Venule
Venules collect blood from capillaries. They vary widely in size and character. The smallest venules resemble expanded capillaries, and venules less than 50 μm in external diameter lack a tunica media. They have an average luminal diameter of about 20 μm and wall thickness of about 1 μm.
Tunica externa
Endothelium
Arteries
Elastic Artery
The walls of elastic arteries are not very thick relative to the external vessel diameter, but they are extremely resilient. The tunica media of these vessels contains relatively few smooth muscle fibers and a high density of elastic fibers. They have an average luminal diameter of about 1.5 cm and wall thickness of about 1 mm.
Internal elastic membrane
Endothelium
Tunica intima
Tunica media
Tunica externa
Muscular Artery
Muscular arteries, or medium-sized arteries, have a thicker tunica media with a greater percentage of smooth muscle fibers than elastic arteries. They have an average luminal diameter of 4 mm and wall thickness of about 1 mm.
Tunica externa
Tunica media
Endothelium
Tunica intima
Arteriole
Arterioles are considerably smaller than muscular arteries. They have a poorly defined tunica externa, and their tunica media consists of scattered smooth muscle fibers that may not form a complete layer. Arterioles have an average luminal diameter of about 30 μm and wall thickness of about 6 μm.
Smooth muscle cells (tunica media)
Endothelium
Basement membrane
Fenestrated Capillary
Fenestrated capillaries are capillaries that contain "windows," or pores in their walls, due to an incomplete or perforated endothelial lining.
Pores
Endothelial cells
Basement membrane
Capillaries
Continuous Capillary
Endothelial cells
Basement membrane
Continuous capillaries have an endothelium that completely surrounds the lumen. Tight junctions and desmosomes connect the endothelial cells. They have an average luminal diameter of 8 μm.
21

? Which type of blood vessel has (a) the largest lumen and (b) the thickest tunica media?

Clinical Note Arteriosclerosis

Arteriosclerosis (ar-tēr-ē-ō-skler-Ō-sis; *arterio-*, artery + *sklerosis*, hardness) is a thickening and toughening of arterial walls. Complications related to arteriosclerosis account for about half of all deaths in the United States. For example, arteriosclerosis of coronary vessels is responsible for *coronary artery disease (CAD)*, and arteriosclerosis of arteries supplying the brain can lead to strokes, or *cerebrovascular accidents* (*CVAs*). p. 700

Arteriosclerosis takes two major forms:

- **Atherosclerosis** (ath-er-ō-skler-Ō-sis; *athero-*, fatty degeneration) is the formation of lipid deposits (plaque) in the tunica media associated with damage to the endothelial lining. It is the most common form of arteriosclerosis.
- **Focal calcification** is the deposition of calcium salts following the gradual degeneration of smooth muscle in the tunica media. Some focal calcification is a part of the aging process. It may also develop in association with atherosclerosis.

A normal coronary artery is shown in **Figure a**, and a typical lesion, called a *plaque* (PLAK), is shown in **Figures b** and **c**. Elderly people—especially elderly men—are most likely to develop atherosclerotic plaques. Estrogen may slow plaque formation. After menopause, when estrogen production decreases, the risks of CAD, myocardial infarctions (heart attacks), and CVAs in women increase markedly.

In addition to advanced age and male sex, other important risk factors for atherosclerosis include high blood cholesterol level, high blood pressure, and cigarette smoking. Diabetes mellitus, obesity, and stress can promote the development of atherosclerosis in both men and women. Evidence also indicates that at least some forms of atherosclerosis may be linked to chronic infection with *Chlamydia pneumoniae*, a bacterium responsible for several types of respiratory infections, including some forms of pneumonia.

Potential treatments for atherosclerotic plaques include catheterization with balloon angioplasty and stenting, and bypass surgery. p. 701 In the many cases where changes in diet do not lower circulating cholesterol level sufficiently, drug therapies can bring them under control.

Without question, the best approach to atherosclerosis is to avoid it by eliminating or reducing associated risk factors. Suggestions include (1) reducing your intake of dietary cholesterol, saturated fats, and trans fatty acids by restricting consumption of fatty meats, egg yolks, and cream; (2) not smoking; (3) checking your blood pressure and taking steps to lower it if necessary; (4) having your blood cholesterol level checked annually; (5) controlling your weight; and (6) exercising regularly.

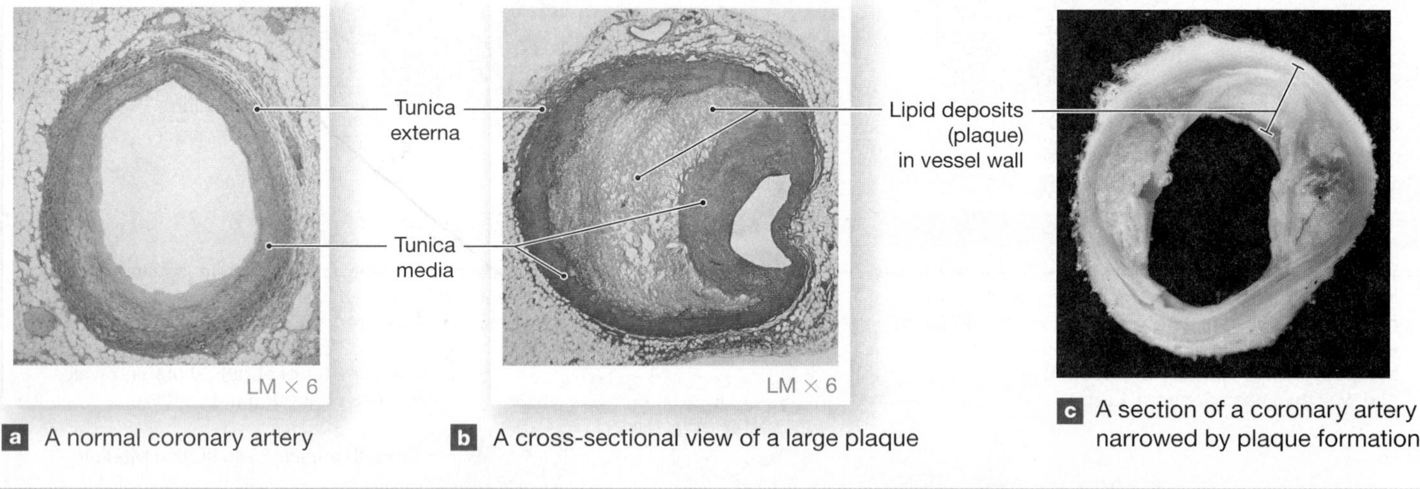
Tunica
externa
Tunica
media
Lipid deposits
(plaque)
in vessel wall
LM × 6
LM × 6

a A normal coronary artery

b A cross-sectional view of a large plaque

c A section of a coronary artery narrowed by plaque formation

tissues vasodilate when the blood oxygen level is low. Also, as we saw in Chapter 16, arterioles vasoconstrict under sympathetic stimulation. p. 544 Changes in their luminal diameter affect the amount of force required to push blood around the cardiovascular system: More pressure is required to push blood through a constricted vessel than through a dilated one. The force opposing blood flow is called *resistance*, so arterioles are also called **resistance vessels**.

Occasionally, local blood pressure exceeds the capacity of the elastic components of the arterial walls. The result is an **aneurysm** (AN-yū-rizm), or bulge in the weakened wall of an artery. The bulge is like a bubble in the wall of a tire—and like a bad tire, the artery can suffer a catastrophic blowout. The most dangerous aneurysms occur in arteries of the brain (where they cause strokes) or in the aorta (where a rupture will cause fatal bleeding in a matter of minutes).

Capillaries

When we think of the cardiovascular system, we think first of the heart or the great blood vessels connected to it. But the microscopic capillaries that permeate most tissues do the real work of the cardiovascular system. These delicate vessels weave throughout active tissues, forming intricate networks that surround muscle fibers. Capillaries radiate through connective tissues, and branch beneath the basement membrane of epithelia.

A typical capillary consists of an endothelial tube inside a thin basement membrane. Neither a tunica media nor a tunica externa is present (see Figure 21–2). The average luminal diameter of a capillary is a mere 8 μm, very close to that of a single red blood cell.

Types of Capillaries

The two major types of capillaries are *continuous capillaries* and *fenestrated capillaries*. *Sinusoids* are a third type of capillary, related in structure and function to fenestrated capillaries.

Continuous Capillaries. Continuous capillaries supply most regions of the body. In a **continuous capillary**, the endothelium is a complete lining. A cross section through a large continuous capillary cuts across several endothelial cells (Figure 21–3a). In a small continuous capillary, a single endothelial cell may completely encircle the lumen.

Continuous capillaries are located in all tissues except epithelia and cartilage. Continuous capillaries permit water, small solutes, and lipid-soluble substances to diffuse into the interstitial fluid. At the same time, they prevent the loss of blood cells and plasma proteins. In addition, some exchange may occur between blood and interstitial fluid by *bulk transport*—the movement of materials by endocytosis or exocytosis at the inner endothelial surface. ⮌ p. 97

In specialized continuous capillaries in most of the central nervous system and in the thymus, the endothelial cells are bound together by tight junctions. These capillaries have very restricted permeability. We discussed one example—the capillaries responsible for the *blood brain barrier*—in Chapters 12 and 14. ⮌ pp. 396, 473

Fenestrated Capillaries. Fenestrated (FEN-es-trā-ted; *fenestra*, window) **capillaries** contain "windows," or pores, that penetrate the endothelial lining (Figure 21–3b). The pores allow

Figure 21–3 **Capillary Structure.**

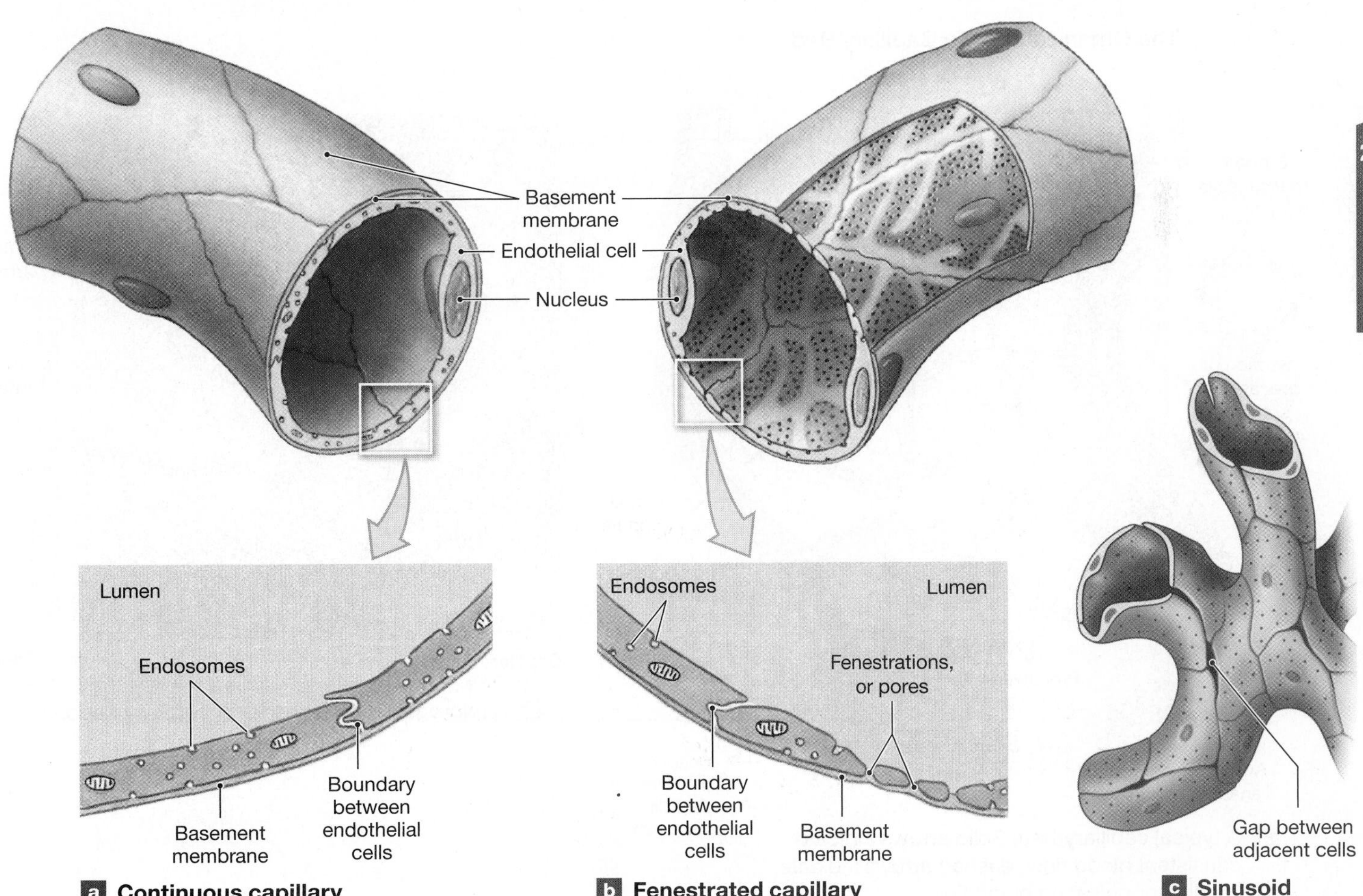
Basement membrane
Endothelial cell
Nucleus
Lumen
Endosomes
Boundary between endothelial cells
Basement membrane
a Continuous capillary
Endosomes
Lumen
Fenestrations, or pores
Boundary between endothelial cells
Basement membrane
b Fenestrated capillary
Gap between adjacent cells
c Sinusoid

21

rapid exchange of water and solutes between blood and interstitial fluid. Examples of fenestrated capillaries include the *choroid plexus* of the brain and the blood vessels in a variety of endocrine organs, such as the hypothalamus and the pituitary, pineal, and thyroid glands. Fenestrated capillaries are also found along absorptive areas of the intestinal tract and at filtration sites in the kidneys. Both the number of pores and their permeability vary from one region of the capillary to another.

Sinusoids. Sinusoids (SĪ-nuh-soydz), or **sinusoidal capillaries**, resemble fenestrated capillaries that are flattened and irregularly shaped (Figure 21–3c). In contrast to continuous and fenestrated capillaries, they have a discontinuous endothelium. In addition to being fenestrated, sinusoids commonly have gaps between adjacent endothelial cells, and the basement membrane is either thinner or absent. As a result, sinusoids permit the free exchange of water and solutes, such as plasma proteins, between blood and interstitial fluid.

Blood moves through sinusoids relatively slowly, maximizing the time available for exchange across the sinusoidal walls. Sinusoids occur in the liver, bone marrow, spleen, and many endocrine organs, including the pituitary and adrenal glands. At liver sinusoids, plasma proteins secreted by liver cells enter the bloodstream. Along sinusoids of the liver, spleen, and bone marrow, phagocytic cells monitor the passing blood, engulfing damaged red blood cells, pathogens, and cellular debris.

Capillary Beds

Capillaries function as part of an interconnected collective network called a **capillary bed** (Figure 21–4). A single arteriole generally gives rise to dozens of capillaries, which empty into several venules. A **precapillary sphincter** surrounds the entrance of a capillary and controls arterial blood flow to the tissues. Contraction of the smooth muscle cells of this sphincter narrows the capillary entrance, reducing or stopping the flow of blood. When one precapillary sphincter constricts, blood is diverted into other branches of the network. When a precapillary sphincter relaxes, the entrance dilates, and blood flows into the capillary. The central passageway in the arteriole system is the **thoroughfare channel**. Blood flow through this channel is controlled by the sphincter effect of the arteriole.

Figure 21–4 **The Organization of a Capillary Bed.**

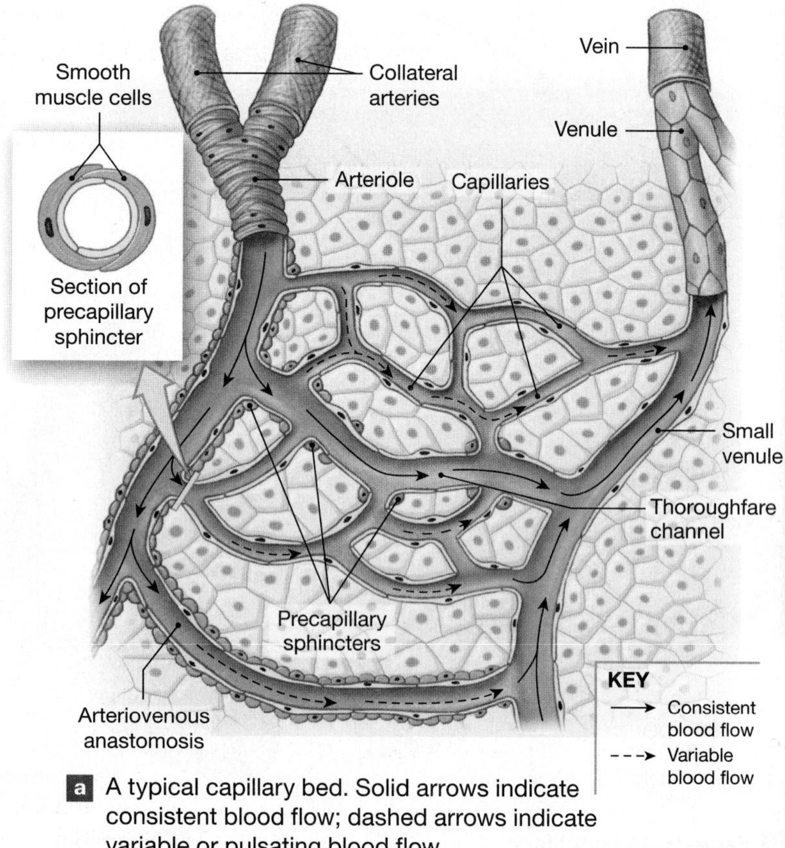
Smooth muscle cells
Collateral arteries
Vein
Venule
Arteriole
Capillaries
Section of precapillary sphincter
Small venule
Thoroughfare channel
Precapillary sphincters
Arteriovenous anastomosis
KEY
Consistent blood flow
Variable blood flow

a A typical capillary bed. Solid arrows indicate consistent blood flow; dashed arrows indicate variable or pulsating blood flow.

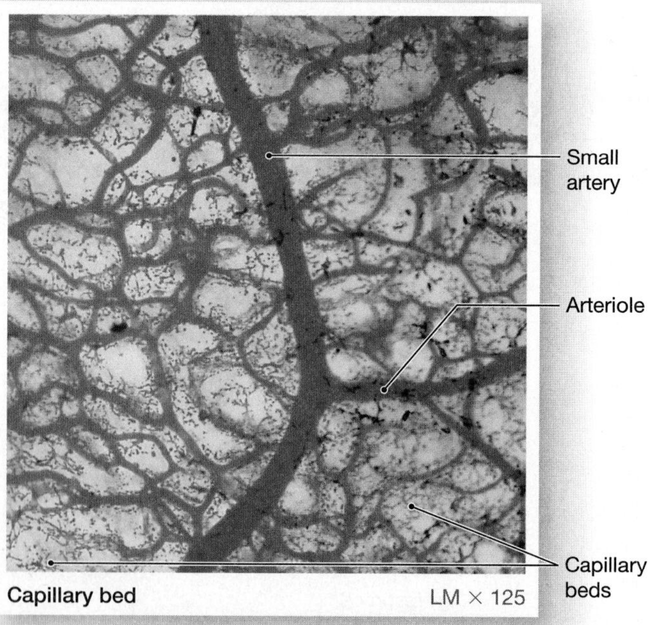
Small artery
Arteriole
Capillary beds
Capillary bed
LM × 125

b A micrograph of a number of capillary beds.

Anastomoses and Angiogenesis

More than one artery may supply blood to a capillary bed. The multiple arteries are called **collaterals**. They fuse before giving rise to arterioles. The fusion of two collateral arteries that supply a capillary bed is an example of an **arterial anastomosis**. (An *anastomosis* is the joining of blood vessels.) The interconnections between the *anterior* and *posterior interventricular arteries* of the heart are arterial anastomoses. ↺ p. 699 An arterial anastomosis acts like an insurance policy: If one artery is blocked, capillary circulation will continue.

Arteriovenous (ar-tēr-ē-o-VĒ-nus) **anastomoses** are direct connections between arterioles and venules. When an arteriovenous anastomosis is dilated, blood bypasses the capillary bed and flows directly into the venous circulation. The pattern of blood flow through an arteriovenous anastomosis is regulated primarily by sympathetic innervation under the control of the cardiovascular center of the medulla oblongata.

Angiogenesis (an-jē-ō-JEN-e-sis; *angio-*, blood vessel + *genesis*, production) is the formation of new blood vessels from preexisting vessels and occurs under the direction of **vascular endothelial growth factor (VEGF)**. Angiogenesis occurs during embryonic and fetal development as tissues and organs develop. It may also occur at other times in any body tissue in response to factors released by oxygen-starved cells. Clinically, angiogenesis is probably most important in cardiac muscle, where it takes place in response to a chronically constricted or occluded vessel.

Veins

Veins collect blood from all tissues and organs and return it to the heart. The walls of veins can be thinner than those of corresponding arteries because the blood pressure in veins is lower than that of arteries. We classify veins on the basis of their size. Even though their walls are thinner, in general veins have larger luminal diameters than their corresponding arteries. (Review Figure 21–2 to compare arteries and veins.)

Types of Veins

Types of veins include venules, medium-sized veins, and large veins.

Venules. Venules, which are the smallest veins, collect blood from capillary beds. They vary widely in size and structure. An average venule has a luminal diameter of roughly 20 μm. Venules smaller than 50 μm lack a tunica media, and the smallest venules resemble expanded capillaries.

Medium-Sized and Large Veins. Medium-sized veins are comparable in size to muscular arteries. They range from 2 to 9 mm in luminal diameter. Their tunica media is thin and contains relatively few smooth muscle cells. The thickest layer of a medium-sized vein is the tunica externa, which contains longitudinal bundles of elastic and collagen fibers.

Large veins include the superior and inferior venae cavae and their branches within the abdominopelvic and thoracic cavities. All large veins have all three layers. The slender tunica media is surrounded by a thick tunica externa composed of a mixture of elastic and collagen fibers.

Venous Valves

The arterial system is a high-pressure system: Almost all the force developed by the heart is required to push blood along the network of arteries and through miles of capillaries. However, blood pressure in a peripheral venule is only about 10 percent of that in the ascending aorta, and pressures continue to fall along the venous system.

The blood pressure in venules and medium-sized veins is so low that it cannot overcome the force of gravity. In the limbs, veins of this size contain **valves**, folds of the tunica intima that project from the vessel wall into the lumen and point in the direction of blood flow. These valves, like those in the heart, permit blood flow in one direction only. Venous valves prevent blood from moving back toward the capillaries (Figure 21–5).

As long as the valves function normally, any movement that compresses a vein pushes blood toward the heart. This effect improves *venous return*, the amount of blood per unit of time returning to the heart. ↺ p. 719 Valves compartmentalize the blood within the veins, dividing the weight of the blood among the compartments.

Figure 21–5 **The Function of Valves in the Venous System.**

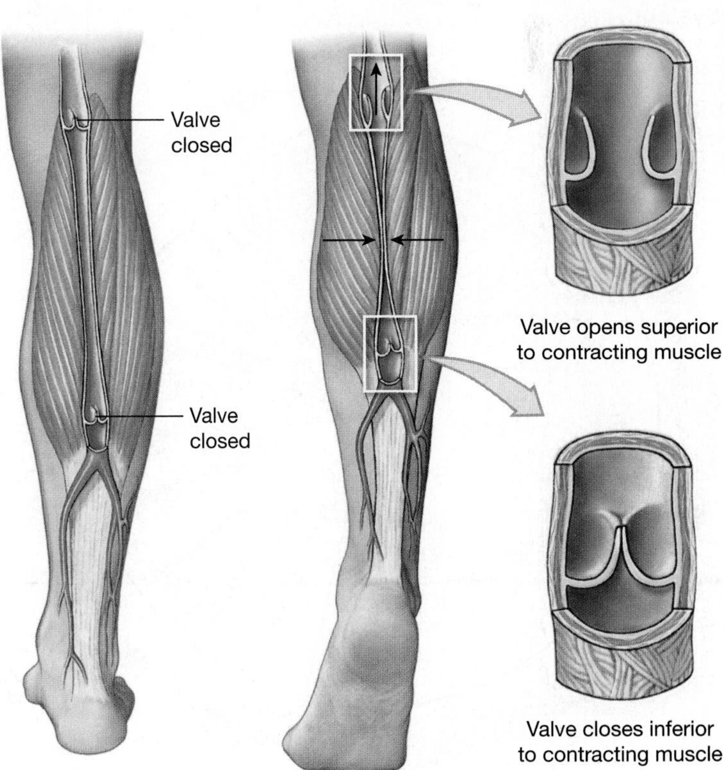
Valve closed
Valve closed
Valve opens superior to contracting muscle
Valve closes inferior to contracting muscle

If the walls of the veins near the valves weaken or become stretched, the valves may not work properly. Blood then pools in the veins, and the vessels become grossly distended. The effects range from mild discomfort and a cosmetic problem, as in superficial *varicose veins* in the thighs and legs, to painful distortion of adjacent tissues, as in **hemorrhoids** (HEM-uh-roydz).

The Distribution of Blood

Our total blood volume is unevenly distributed among arteries, veins, and capillaries (Figure 21–6). The heart, arteries, and capillaries in the pulmonary and systemic circuits normally contain 30–35 percent of the blood volume (roughly 1.5 liters of whole blood). The venous system contains the rest (65–70 percent, or about 3.5 liters). About one-third of the blood in the venous system (about a liter) is circulating within the liver, bone marrow, and skin. These organs have extensive venous networks that at any moment contain large volumes of blood.

Veins are much more *distensible*, or expandable, than arteries because their walls are thinner, with less smooth muscle. For a given rise in blood pressure, a typical vein stretches about eight times as much as a corresponding artery. The *capacitance* of a blood vessel is the relationship between the volume of blood it contains and the blood pressure. Veins behave like a balloon, expanding easily at low pressures, and they have high capacitance. Arteries behave more like a truck tire, expanding only at high pressures, and they have low capacitance. Veins are called **capacitance vessels**. Because veins have high capacitance, they act as **blood reservoirs**, which can accommodate large changes in blood volume. If the blood volume rises or falls, the elastic walls stretch or recoil, changing the volume of blood in the venous system.

If serious hemorrhaging (profuse blood loss) occurs, sympathetic nerves stimulate smooth muscle cells in the walls of medium-sized veins in the systemic system to constrict. This process, called **venoconstriction** (vē-nō-kon-STRIK-shun), decreases the amount of blood within the venous system, increasing the volume within the arterial system and capillaries. Venoconstriction can keep the blood volume within the arteries and capillaries at near-normal levels despite a significant blood loss. As a result, blood flow to active skeletal muscles and delicate organs, such as the brain, can be increased or maintained.

Figure 21–6 The Distribution of Blood in the Cardiovascular System.

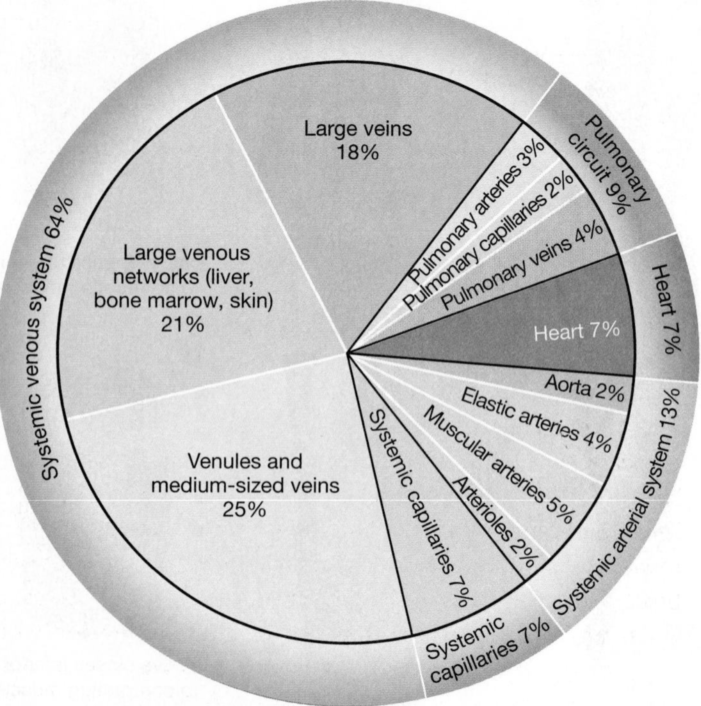

Large veins 18%
Large venous networks (liver, bone marrow, skin) 21%
Venules and medium-sized veins 25%
Systemic venous system 64%
Pulmonary arteries 3%
Pulmonary capillaries 2%
Pulmonary veins 4%
Pulmonary circuit 9%
Heart 7%
Heart 7%
Aorta 2%
Elastic arteries 4%
Muscular arteries 5%
Arterioles 2%
Systemic arterial system 13%
Systemic capillaries 7%
Systemic capillaries 7%

✓ Checkpoint

1. **List the five general classes of blood vessels.**
2. **Which type of vessel is characterized by thin walls with very little smooth muscle tissue in the tunica media?**
3. **Why are valves located in veins but not in arteries?**
4. **Where in the body would you find fenestrated capillaries?**

See the blue Answers tab at the back of the book.

+ Clinical Note Varicose Veins

Distended veins run like thick, blue ropes up the lower leg and popliteal region in some people. These **varicose veins** have a familial link and are more commonly found in people who stand for long time periods. The veins are unsightly, but superficial varicose veins are only a cosmetic issue. Minimally invasive techniques such as *sclerotherapy* (the injection of an irritant chemical) can shut down the dilated veins and improve the appearance of the legs.

However, varicosity of deep leg veins poses a threat. Here venous blood flow slows to a crawl and triggers coagulation. The result is **deep vein thrombosis (DVT)**. The *thrombus*, or clot, can impair venous return to the right atrium. Worse still, multiple DVTs can dislodge and become *emboli*, or traveling clots. The clots may end up in the lungs, producing a potentially fatal condition called **pulmonary embolism**.

21-2 Pressure and resistance determine blood flow and affect rates of capillary exchange

Learning Outcome Describe the factors that influence blood pressure, and explain how and where fluid and dissolved materials enter and leave the cardiovascular system.

Adequate amounts of blood must flow to the capillary beds in organs and peripheral tissues. Under normal circumstances, blood flow is equal to cardiac output. When cardiac output increases, blood flow through capillary beds increases, and when cardiac output decreases, capillary blood flow decreases.

Introduction to Pressure and Flow in Blood Vessels

Capillary blood flow is determined by the interplay between **pressure (*P*)** and **resistance (*R*)** in the cardiovascular network. To keep blood moving, the heart must generate enough pressure to overcome the resistance to blood flow in the pulmonary and systemic circuits. In general terms, **flow (*F*)** is directly proportional to the pressure (increased pressure → increased flow), and inversely proportional to resistance (increased resistance → decreased flow). However, the absolute pressure is less important than the pressure *gradient*—the difference in pressure from one end of the vessel to the other. This relationship can be summarized as

$$F \propto \frac{\Delta P}{R}$$

where the symbol ∝ means "is proportional to" and Δ means "the difference in." The largest pressure gradient is found between the base of the aorta and the proximal ends of peripheral capillary beds. The cardiovascular center can alter this pressure gradient, and change the rate of capillary blood flow, by adjusting cardiac output and peripheral resistance.

Blood leaving the peripheral capillaries enters the venous system. The pressure gradient across the venous system is relatively small, but venous resistance is very low. The low venous blood pressure—aided by valves, skeletal muscle contraction, gravity, and other factors—is enough to return the blood to the heart. When necessary, cardiovascular control centers can raise venous pressure (through venoconstriction) to improve venous return and maintain adequate cardiac output.

We begin this section by examining blood pressure and resistance more closely. We then consider the mechanisms of *capillary exchange*, the transfer of liquid and solutes between the blood and interstitial fluid. Capillary exchange provides tissues with oxygen and nutrients and removes the carbon dioxide and wastes generated by active cells.

Active tissues need more blood flow than inactive ones. Even something as simple as a change in position—going from sitting to standing, for instance—triggers a number of cardiovascular changes. We end this section with a discussion of what those changes are and how they are coordinated.

Pressures Affecting Blood Flow

When talking about cardiovascular pressures, three values are important:

- *Blood Pressure.* The term **blood pressure (BP)** refers to arterial pressure, usually reported in millimeters of mercury (mm Hg). Systemic arterial pressures in adults average 120 mm Hg at the entrance to the aorta to roughly 35 mm Hg at the start of a capillary network.
- *Capillary Hydrostatic Pressure. Hydrostatic pressure* is the force exerted by a fluid pressing against a wall. **Capillary hydrostatic pressure (CHP)**, or *capillary pressure,* is the pressure of blood within capillary walls. Along the length of a typical capillary, pressures decline from roughly 35 mm Hg to about 18 mm Hg.
- *Venous Pressure.* **Venous pressure** is the pressure of the blood within the venous system. Venous pressure is quite low: The pressure gradient from the venules to the right atrium is only about 18 mm Hg.

The difference in pressure (ΔP) across the entire systemic circuit, sometimes called the *circulatory pressure,* averages about 100 mm Hg.

Total Peripheral Resistance

For circulation to occur, the circulatory pressure must overcome the **total peripheral resistance**—the resistance of the entire cardiovascular system. The arterial network has by far the largest pressure gradient (85 mm Hg), and this primarily reflects the relatively high resistance of the arterioles. The total peripheral resistance of the cardiovascular system reflects a combination of factors: *vascular resistance, blood viscosity,* and *turbulence.*

Vascular Resistance

Vascular resistance, the forces that oppose blood flow in the blood vessels, is the largest component. The most important factor in vascular resistance is **friction** between blood and the vessel walls. The amount of friction depends on two factors: vessel length and internal vessel diameter (Figure 21–7).

Vessel Length. Increasing the length of a blood vessel increases friction: The longer the vessel, the greater the surface area in contact with blood. You can easily blow the water out of a snorkel that is 2.5 cm (1 in.) in diameter and 25 cm (10 in.) long, but you cannot blow the water out of a 15-m (16-yard)-long garden hose, because the total friction is too great. The most dramatic changes in blood vessel length occur between birth and maturity, as individuals grow to adult size. In adults, vessel length can increase or decrease gradually when individuals gain or lose weight, but on a day-to-day basis this component of vascular resistance can be considered constant.

Figure 21–7 Factors Affecting Friction and Vascular Resistance.

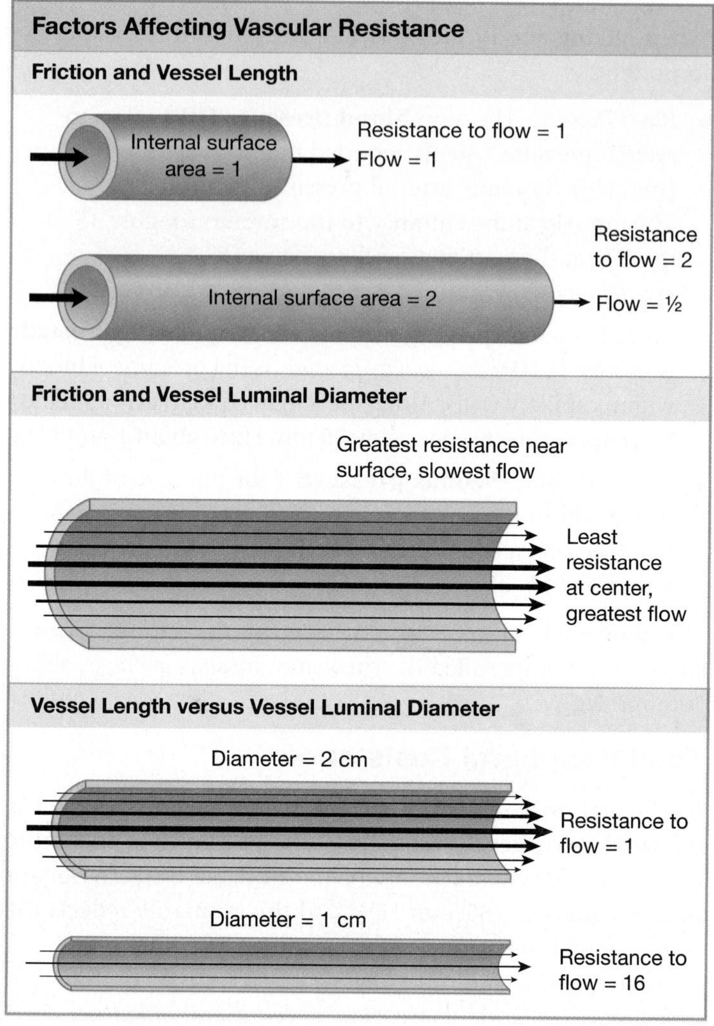
Factors Affecting Vascular Resistance
Friction and Vessel Length
Internal surface area = 1
Resistance to flow = 1
Flow = 1
Internal surface area = 2
Resistance to flow = 2
Flow = ½
Friction and Vessel Luminal Diameter
Greatest resistance near surface, slowest flow
Least resistance at center, greatest flow
Vessel Length versus Vessel Luminal Diameter
Diameter = 2 cm
Resistance to flow = 1
Diameter = 1 cm
Resistance to flow = 16

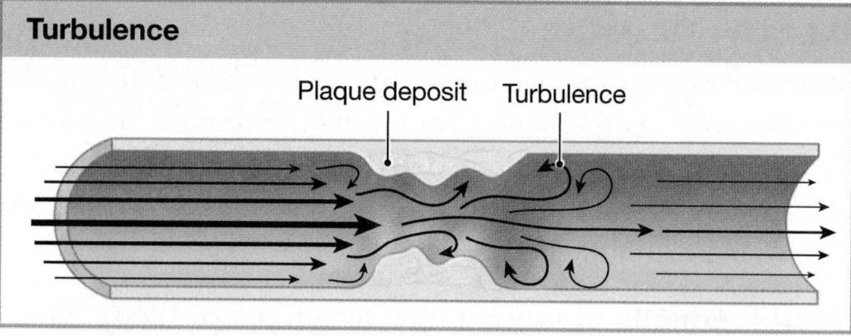
Turbulence
Plaque deposit
Turbulence

? Which of the following would *increase* vascular resistance? increased vessel length, increased vessel luminal diameter, turbulence.

Vessel Luminal Diameter. The effects of friction on blood act in a narrow zone closest to the internal vessel wall. In a small-diameter vessel, friction with the luminal wall slows nearly all the blood. Resistance is therefore relatively high. Blood near the center of a large-diameter vessel does not encounter friction with the luminal wall, so the resistance in large vessels is fairly low.

Differences in diameter have much more significant effects on resistance than do differences in length. If two vessels are equal in radius (one-half the diameter) but one is twice as long as the other, the longer vessel offers twice as much resistance to blood flow. But for two vessels of equal length, one twice the radius of the other, the narrower one offers 16 times as much resistance to blood flow. This relationship, expressed in terms of the vessel radius *r* and resistance *R*, can be summarized as $R \propto 1/r^4$.

More significantly, there is no way to control vessel length, but vessel luminal diameter can change quickly through vasoconstriction or vasodilation. Most of the peripheral resistance occurs in arterioles. When these smooth muscles contract or relax, peripheral resistance increases or decreases, respectively. Because a small change in luminal diameter produces a large change in resistance, mechanisms that alter the diameters of arterioles provide control over peripheral resistance and blood flow.

Blood Viscosity

Viscosity is the resistance to flow caused by interactions among molecules and suspended materials in a liquid. Liquids of low viscosity, such as water (viscosity 1.0), flow at low pressures. Syrupy fluids, such as molasses (viscosity 300), flow only under higher pressures. Whole blood has a viscosity about five times that of water, due to its plasma proteins and blood cells. Under normal conditions, the viscosity of blood remains stable. Anemia, polycythemia, and other disorders that affect the hematocrit also change blood viscosity, and thus peripheral resistance.

Turbulence

High flow rates, irregular surfaces, and sudden changes in vessel luminal diameter upset the smooth flow of blood, creating eddies and swirls. This phenomenon, called **turbulence**, increases resistance and slows blood flow.

Turbulence normally occurs when blood flows between the atria and the ventricles, and between the ventricles and the aortic and pulmonary trunks. It also develops in large arteries, such as the aorta, when cardiac output and arterial flow rates are very high. However, turbulence seldom occurs in smaller vessels unless their walls are damaged. For example, an atherosclerotic plaque creates abnormal turbulence and restricts blood flow (Figure 21–7). Because turbulence makes a distinctive sound, or *bruit* (brū-Ē), plaques in large blood vessels can often be detected with a stethoscope.

Table 21–1 provides a review of the terms and relationships discussed in this section.

Table 21–1 Key Terms and Relationships Pertaining to Blood Circulation

Blood flow (*F*)	The volume of blood flowing per unit of time through a vessel or a group of vessels; may refer to circulation through a capillary, a tissue, an organ, or the entire vascular network. Total blood flow is equal to cardiac output.
Blood pressure (BP)	The hydrostatic pressure in the arterial system that pushes blood through capillary beds.
Circulatory pressure	The pressure difference between the base of the ascending aorta and the entrance to the right atrium.
Hydrostatic pressure	A pressure exerted by a liquid in response to an applied force.
Peripheral resistance (PR)	The resistance of the arterial system; affected by such factors as vascular resistance, viscosity, and turbulence.
Resistance (*R*)	A force that opposes movement (in this case, blood flow).
Total peripheral resistance	The resistance of the entire cardiovascular system.
Turbulence	A resistance due to the irregular, swirling movement of blood at high flow rates or exposure to irregular surfaces.
Vascular resistance	A resistance due to friction within a blood vessel, primarily between the blood and the vessel walls. Increases with increasing length or decreasing diameter; vessel length is constant, but vessel diameter can change.
Venous pressure	The hydrostatic pressure in the venous system.
Viscosity	A resistance to flow due to interactions among molecules within a liquid.
RELATIONSHIPS AMONG THE PRECEDING TERMS	
$F \propto \Delta P$	Flow is proportional to the pressure gradient.
$F \propto 1/R$	Flow is inversely proportional to resistance.
$F \propto \Delta P/R$	Flow is directly proportional to the pressure gradient, and inversely proportional to resistance.
$F \propto BP/PR$	Flow is directly proportional to blood pressure, and inversely proportional to peripheral resistance.
$R \propto 1/r^4$	Resistance is inversely proportional to the fourth power of the vessel radius.

An Overview of Cardiovascular Pressures

Look at the graphs in Figure 21–8 for an overview of the vessel luminal diameters, cross-sectional area of vessel lumen, pressures, and velocity of blood flow in the systemic circuit.

- *Vessel Luminal Diameters.* As blood flows from the aorta toward the capillaries, vessels diverge. The arteries branch repeatedly, and each branch is smaller in luminal diameter than the preceding one (Figure 21–8a). As blood flows from the capillaries toward the venae cavae, vessels converge. Vessel luminal diameters increase as venules combine to form small and medium-sized veins.
- *Total Cross-Sectional Areas.* Although arterioles, capillaries, and venules are small in luminal diameter, the body has large numbers of them. All the blood flowing through the aorta also flows through peripheral capillaries. Blood pressure and the speed of blood flow are proportional to the cross-sectional area of the vessels involved. What is important is not the cross-sectional area of each individual vessel, but the *combined* cross-sectional area of *all* the vessel lumens (Figure 21–8b). In effect, your blood moves from one big pipe (the aorta, with a cross-sectional area of 4.5 cm^2) into countless tiny ones (the peripheral capillaries, with a total cross-sectional area of 5000 cm^2), and then blood travels back to the heart through two large pipes (the venae cavae).
- *Pressures.* As arteries branch, their total cross-sectional area increases, and blood pressure falls rapidly (Figure 21–8c). Most of the decline takes place in the small arteries and arterioles. Venous pressures are relatively low.
- *Velocity of Blood Flow.* As the total cross-sectional area of the vessel lumens increases from the aorta toward the capillaries, the velocity of blood flow decreases (Figure 21–8d). Blood flow velocity then increases as the total cross-sectional area decreases from the capillaries toward the venae cavae.

Figure 21–9 graphs the blood pressure throughout the cardiovascular system. Systemic pressures are highest in the aorta, peaking at about 120 mm Hg. Pressures reach a minimum of 2 mm Hg at the entrance to the right atrium. Pressures in the pulmonary circuit are much lower than those in the systemic circuit. The right ventricle does not ordinarily develop high pressures because the pulmonary vessels are much shorter and more distensible than the systemic vessels, thus providing less resistance to blood flow. Let's look at pressures throughout the systemic circuit.

Arterial Blood Pressure

Arterial pressure is important because it maintains blood flow through capillary beds. To do this, it must always be high enough to overcome the peripheral resistance. Arterial pressure is not constant. Rather, it rises during ventricular systole and falls during ventricular diastole. The peak blood pressure measured during ventricular systole is called **systolic pressure**, and the minimum blood pressure at the end of ventricular diastole is called **diastolic pressure**. In recording blood pressure, we separate systolic and diastolic pressures by a slash, as in "120/80" ("one-twenty over eighty") or "110/75."

Figure 21–8 **Relationships among Vessel Luminal Diameter, Cross-Sectional Area, Blood Pressure, and Blood Velocity within the Systemic Circuit.**

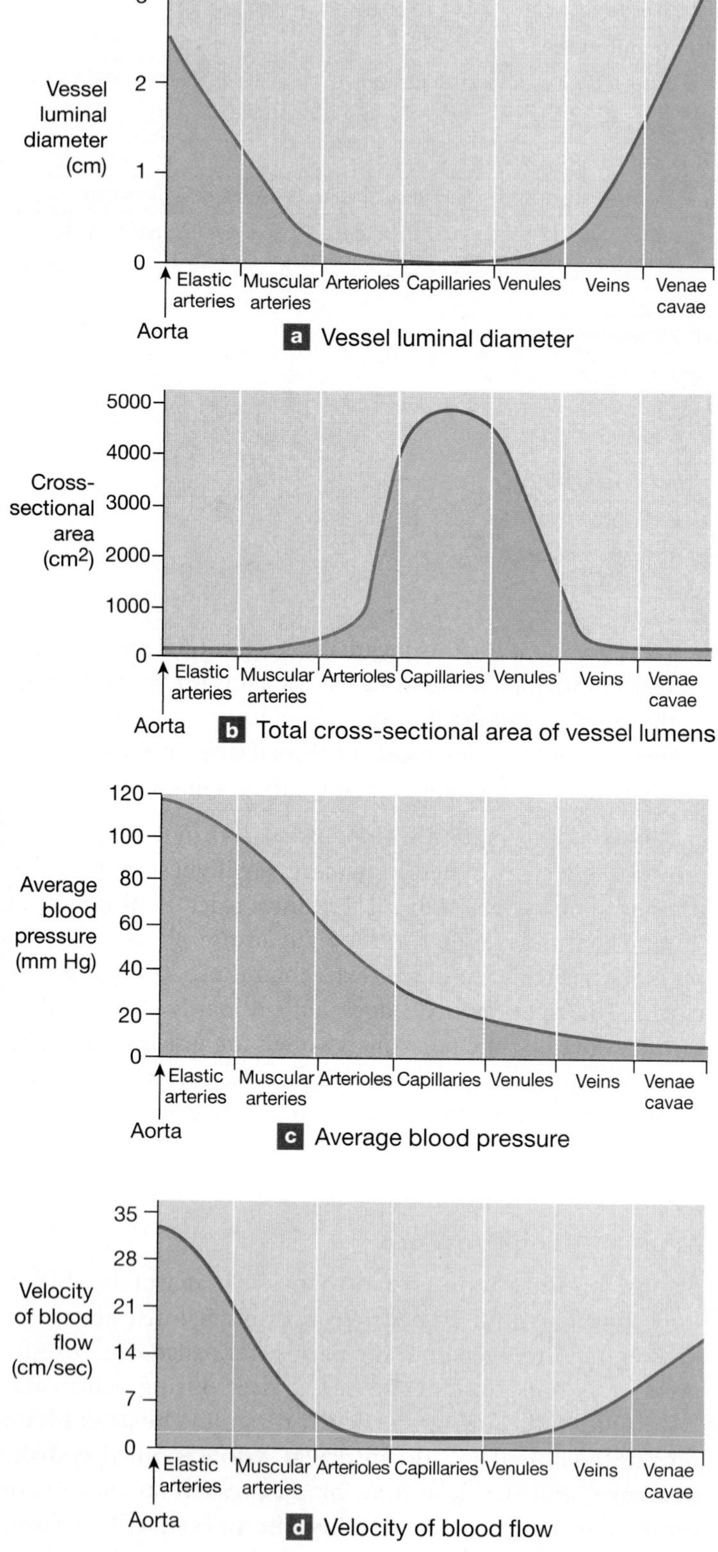
Vessel luminal diameter (cm)
Elastic arteries
Muscular arteries
Arterioles
Capillaries
Venules
Veins
Venae cavae
Aorta
a Vessel luminal diameter
Cross-sectional area (cm2)
b Total cross-sectional area of vessel lumens
Average blood pressure (mm Hg)
c Average blood pressure
Velocity of blood flow (cm/sec)
d Velocity of blood flow

Figure 21–9 **Pressures within the Systemic Circuit.** Note the general reduction in circulatory pressure within the systemic circuit and the elimination of the pulse pressure within the arterioles.

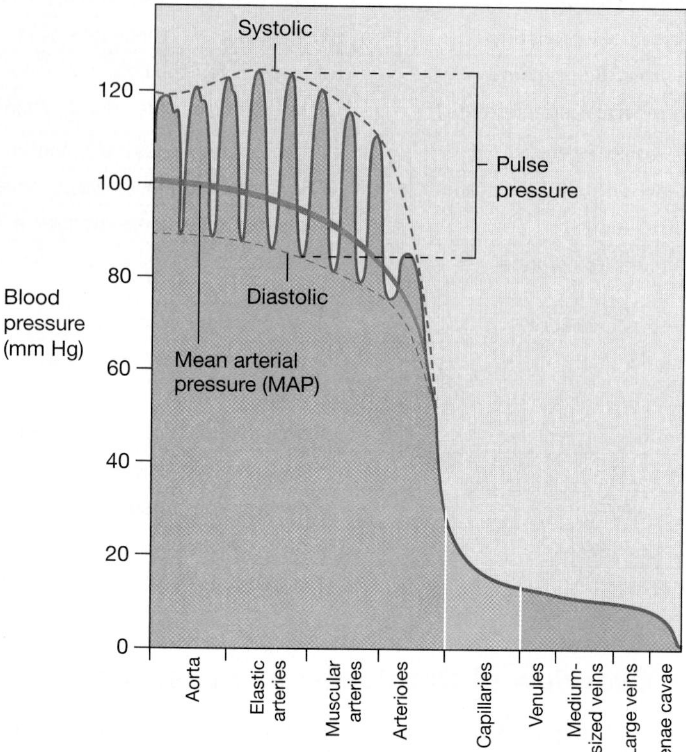
Systolic
Pulse pressure
Diastolic
Mean arterial pressure (MAP)
Blood pressure (mm Hg)
Aorta
Elastic arteries
Muscular arteries
Arterioles
Capillaries
Venules
Medium-sized veins
Large veins
Venae cavae

21

A *pulse* is a rhythmic fluctuation in pressure that accompanies each heartbeat. The difference between the systolic and diastolic pressures is the **pulse pressure** (Figure 21–9). To report a single blood pressure value, we use the **mean arterial pressure (MAP)**. It is calculated for someone at rest by adding one-third of the pulse pressure to the diastolic pressure:

$$\text{MAP} = \text{diastolic pressure} + \frac{\text{pulse pressure}}{3}$$

For a systolic pressure of 120 mm Hg and a diastolic pressure of 90 mm Hg, we calculate MAP as follows:

$$\text{MAP} = 90 + \frac{(120 - 90)}{3} = 90 + 10 = 100\,\text{mm Hg}$$

A normal range of systolic and diastolic pressures occurs in healthy individuals. When pressures shift outside the normal range, clinical problems develop. Abnormally high blood pressure is termed **hypertension**. Abnormally low blood pressure is **hypotension**. Hypertension is much more common. In fact, many cases of hypotension result from overly aggressive drug treatment for hypertension.

The usual criterion established by the American Heart Association for stage 1 hypertension in adults is a systolic blood pressure range of 140–159 and a diastolic range of 90–99.

Values from 120–139 for systolic pressure and 80–89 for diastolic pressure indicate *pre-hypertension*. Blood pressure less than 120/80 is normal. Cardiologists often recommend some combination of diet modification and drug therapy for people whose blood pressures are consistently pre-hypertensive.

Cardiac workload is how hard the heart has to work to pump blood throughout the body. Hypertension significantly increases the workload on the heart, and the left ventricle gradually enlarges. More muscle mass means a greater demand for oxygen. When the coronary circulation cannot keep pace, signs and symptoms of *coronary ischemia* appear. p. 700 Increased arterial pressures also place a physical stress on the walls of blood vessels throughout the body. This stress promotes or accelerates the development of arteriosclerosis. It also increases the risk of aneurysms, heart attacks, and strokes.

Elastic Rebound in Arteries

As systolic pressure climbs, the arterial walls stretch, just as an extra puff of air expands a partially inflated balloon. This expansion allows the arterial system to accommodate some of the blood ejected during ventricular systole. When diastole begins and blood pressures fall, the arteries recoil to their original dimensions. This phenomenon is called **elastic rebound**. Some blood is forced back toward the left ventricle, closing the aortic valve and helping to drive additional blood into the coronary arteries. However, most of the push from elastic rebound forces blood toward the capillaries. This maintains blood flow along the arterial network while the left ventricle is in diastole.

Pressures in Muscular Arteries and Arterioles

The mean arterial pressure and the pulse pressure become smaller as the distance from the heart increases (see Figure 21–9):

- The mean arterial pressure declines as the arterial branches become smaller and more numerous. In essence, blood pressure decreases as it overcomes friction and produces blood flow.
- The pulse pressure lessens due to the cumulative effects of elastic rebound along the arterial system. The effect can be likened to a series of ever-softer echoes following a loud shout. Each time an echo is produced, the reflecting surface absorbs some of the sound energy. Eventually, the echo disappears. The pressure surge accompanying ventricular ejection is like the shout, and it is reflected by the wall of the aorta, echoing down the arterial system until it finally disappears at the level of the small arterioles. By the time blood reaches a precapillary sphincter, no pressure fluctuations remain, and the blood pressure is steady at approximately 35 mm Hg.

Venous Pressure and Venous Return

Venous pressure, although low, determines **venous return**—the amount of blood arriving at the right atrium each minute. Venous return has a direct impact on cardiac output. p. 718 Blood pressure at the start of the venous system is only about one-tenth that at the start of the arterial system, but the blood must still travel through a vascular network as complex as the arterial system before returning to the heart.

Pressures at the entrance to the right atrium fluctuate, but they average about 2 mm Hg. Thus, the effective pressure in the venous system is roughly 16 mm Hg (from 18 mm Hg in the venules to 2 mm Hg in the venae cavae). This venous pressure gradient compares with 65 mm Hg in the arterial system (from 100 mm Hg at the aorta to 35 mm Hg at the capillaries). Yet, although venous pressures are low, veins offer comparatively little resistance, so pressure declines very slowly as blood moves through the venous system. As blood moves toward the heart, the veins become larger, resistance drops, and the velocity of blood flow increases (look back at Figure 21–8).

When you stand, the venous blood returning from your body inferior to the heart must overcome gravity as it travels up the inferior vena cava. Two factors assist the low venous pressures in propelling blood toward your heart: *muscular compression* of peripheral veins and the *respiratory pump* during inhalation.

Muscular Compression. The contractions of skeletal muscles near a vein compress it, helping to push blood toward the heart. The valves in small and medium-sized veins ensure that blood flows in one direction only (look back at Figure 21–5). When you are standing and walking, the cycles of contraction and relaxation that accompany your normal movements assist venous return. The mechanism is particularly important when you are standing, because blood returning from your feet must overcome gravity to ascend to the heart. Although you are probably not aware of it, when you stand, your leg muscles undergo cycles of rapid contraction and relaxation, helping to push blood toward the trunk. When you lie down, venous valves play a smaller part in venous return, because your heart and major vessels are at the same level.

If you stand at attention, with knees locked and leg muscles immobilized, that assistance is lost. The reduction in venous return then leads to a decline in cardiac output, which reduces the blood supply to the brain. This decline is sometimes enough to cause **fainting**, a temporary loss of consciousness. You would then collapse, but while you were in the horizontal position, both venous return and cardiac output would return to normal.

The Respiratory Pump. As you inhale, your thoracic cavity expands, reducing the pressure within the thoracic cavity. When the pressure in the thoracic cavity drops below atmospheric pressure, air will enter the lungs. At the same time, the drop in venous pressure in the chest makes it easier to force blood into the inferior vena cava and right atrium from the smaller veins of your abdominal cavity and lower body. The effect on venous return through the superior vena cava is less noticeable, because blood in that vessel is normally assisted by gravity. As you exhale, your thoracic cavity decreases in size. Internal pressure then rises, forcing air out of your lungs and pushing venous blood into the right atrium. This mechanism is called the **respiratory pump**. Such pumping action becomes more important during heavy exercise, when respirations are deep and frequent.

Capillary Exchange and Capillary Pressures

The vital functions of the cardiovascular system depend entirely on events at the capillary level: All chemical and gaseous exchange between blood and interstitial fluid takes place across capillary walls. This process, called **capillary exchange**, plays a key role in homeostasis. Cells rely on capillary exchange to obtain nutrients and oxygen and to remove metabolic wastes, such as carbon dioxide and urea.

Exchange takes place very rapidly, because the distances involved are very short. Few cells lie more than 125 μm (0.005 in.) from a capillary. In addition, blood flows through capillaries fairly slowly, allowing time for the diffusion or active transport of materials across the capillary walls. In this way, the histological structure of capillaries permits a two-way exchange of substances between blood and interstitial fluid. Capillaries are the *only* blood vessels whose walls permit such exchange. The most important processes that move materials across typical capillary walls are *diffusion*, *filtration*, and *reabsorption*.

Diffusion

21

As we saw in Chapter 3, *diffusion* is the net movement of ions or molecules from an area where their concentration is higher to an area where their concentration is lower. p. 90 The difference between the high and low concentrations represents a *concentration gradient*. Diffusion tends to eliminate that gradient. Diffusion occurs most rapidly when (1) the distances involved are short, (2) the concentration gradient is steep, and (3) the ions or molecules involved are small.

Different substances diffuse across capillary walls by different routes:

- *Water, ions, and small organic molecules, such as glucose, amino acids, and urea,* can usually enter or leave the bloodstream by diffusion between adjacent endothelial cells or through the pores of fenestrated capillaries.
- *Many ions, including sodium, potassium, calcium, and chloride,* can diffuse across endothelial cells by passing through channels in plasma membranes.
- *Large water-soluble compounds* are unable to enter or leave the bloodstream except at fenestrated capillaries, such as those of the hypothalamus, the kidneys, many endocrine organs, and the intestinal tract.
- *Lipids, such as fatty acids and steroids, and lipid-soluble materials, including soluble gases such as oxygen and carbon dioxide,* can cross capillary walls by diffusion through the endothelial plasma membranes.
- *Plasma proteins* are normally unable to cross the endothelial lining anywhere except in sinusoids, such as those of the liver, where plasma proteins enter the bloodstream.

Filtration

Filtration is the removal of solutes as a solution flows across a porous membrane. Solutes too large to pass through the pores are filtered out of the solution. The driving force for filtration is hydrostatic pressure. As we saw earlier, it pushes water from an area of higher pressure to an area of lower pressure.

In *capillary filtration*, water and small solutes are forced across a capillary wall, leaving larger solutes and suspended proteins in the bloodstream (Figure 21–10). The solute molecules that leave the bloodstream are small enough to pass between adjacent endothelial cells or through the pores in a fenestrated capillary. Filtration takes place primarily at the arterial end of a capillary, where capillary hydrostatic pressure (CHP) is highest.

Reabsorption

Reabsorption occurs as the result of osmosis. *Osmosis* is the diffusion of water across a selectively permeable membrane that

Figure 21–10 Capillary Filtration. Capillary hydrostatic pressure (CHP) forces water and solutes through the gaps between adjacent endothelial cells in continuous capillaries. The sizes of solutes that move across the capillary wall are determined by the dimensions of the gaps.

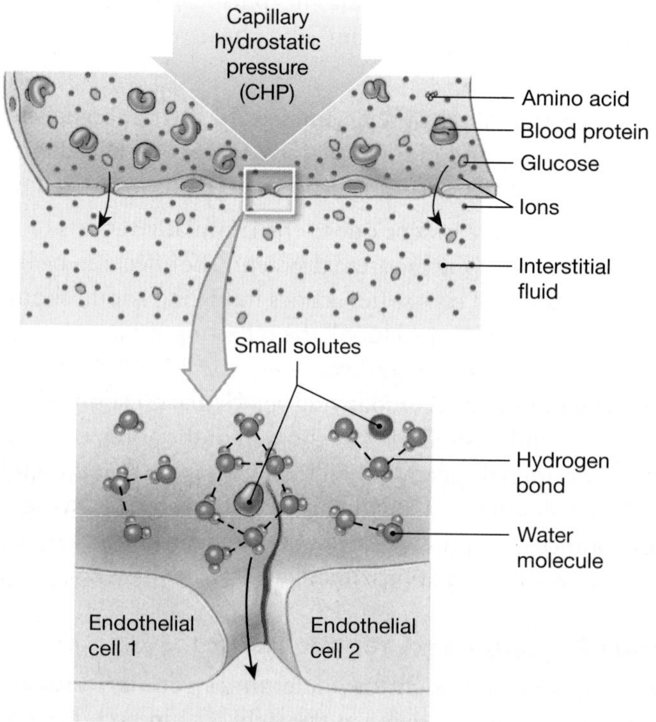

Capillary hydrostatic pressure (CHP)
Amino acid
Blood protein
Glucose
Ions
Interstitial fluid
Small solutes
Hydrogen bond
Water molecule
Endothelial cell 1
Endothelial cell 2

separates two solutions of differing solute concentrations. Water molecules tend to diffuse across a membrane *toward* the solution containing the higher solute concentration (look back at Figure 3–16, p. 93).

The **osmotic pressure (OP)** of a solution represents the pressure that must be applied to prevent osmotic movement across a membrane. The higher the solute concentration of a solution, the greater the solution's osmotic pressure. Suspended proteins that cannot cross capillary walls creates an osmotic pressure called *blood colloid osmotic pressure (BCOP)*. Clinicians often use the term *oncotic pressure* (*onkos*, a swelling) when referring to the colloid osmotic pressure of body fluids. The two terms are equivalent. Osmotic water movement continues until either the solute concentrations are equalized or an opposing hydrostatic pressure prevents the movement.

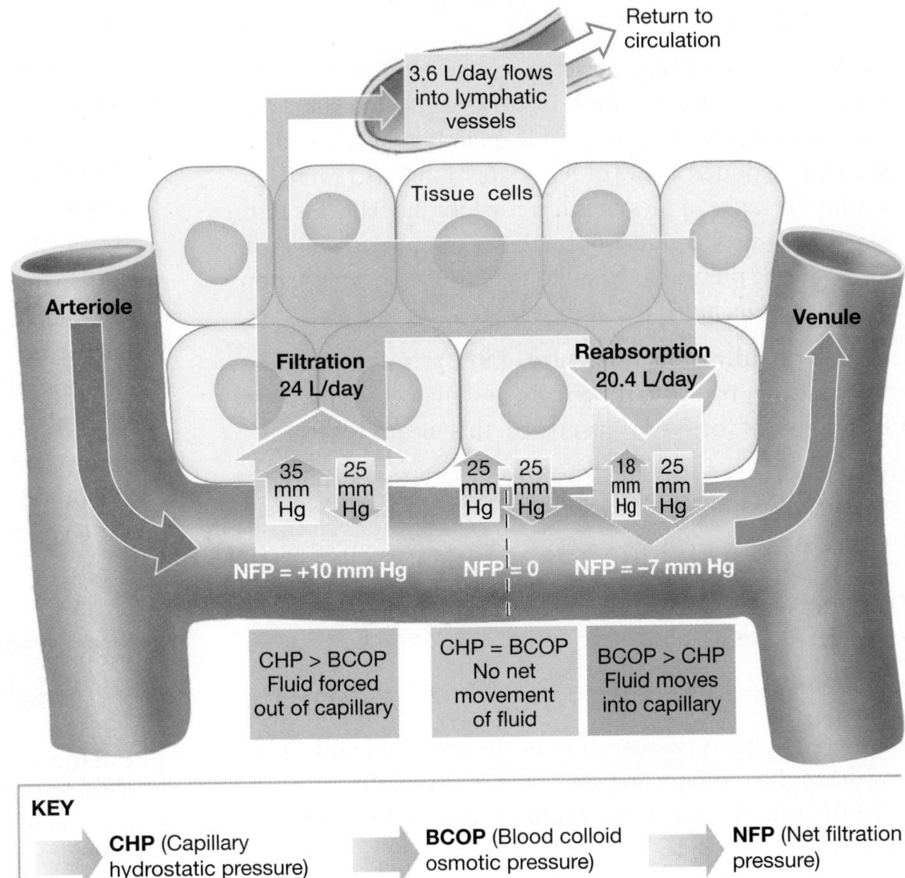
Return to circulation
3.6 L/day flows into lymphatic vessels
Tissue cells
Arteriole
Venule
Filtration 24 L/day
Reabsorption 20.4 L/day
35 mm Hg
25 mm Hg
25 mm Hg
25 mm Hg
18 mm Hg
25 mm Hg
NFP = +10 mm Hg
NFP = 0
NFP = −7 mm Hg
CHP > BCOP Fluid forced out of capillary
CHP = BCOP No net movement of fluid
BCOP > CHP Fluid moves into capillary
KEY
CHP (Capillary hydrostatic pressure)
BCOP (Blood colloid osmotic pressure)
NFP (Net filtration pressure)

Figure 21–11 **Forces Acting across Capillary Walls.** At the arterial end of a capillary, capillary hydrostatic pressure (CHP) is greater than blood colloid osmotic pressure (BCOP), so fluid moves out of the capillary (filtration). Near the venule, CHP is lower than BCOP, so fluid moves into the capillary (reabsorption). In this model, interstitial fluid colloid osmotic pressure (ICOP) and interstitial fluid hydrostatic pressure (IHP) are assumed to be 0 mm Hg and so are not shown.

Bulk Flow: The Interplay between Filtration and Reabsorption

Now let's look at how filtration and reabsorption function together along the length of a typical capillary. The continuous net movement of water out of the capillaries, through peripheral tissues, and then back to the bloodstream by way of the lymphatic system is known as *bulk flow*. This process has four important functions:

- It ensures that plasma and interstitial fluid, two major components of extracellular fluid, communicate and exchange materials constantly.
- It accelerates the distribution of nutrients, hormones, and dissolved gases throughout tissues.
- It assists in the transport of insoluble lipids and tissue proteins that cannot cross the capillary walls and enter the bloodstream.
- It flushes out bacterial toxins and other chemicals, carrying them to lymphatic tissues and organs responsible for immunity to disease.

Capillary blood pressure declines as blood flows from the arterial end to the venous end of a capillary. As a result, the rates of filtration and reabsorption gradually change as blood passes along the length of a capillary. The factors involved are diagrammed in Figure 21–11.

The *net capillary hydrostatic pressure* is the difference between the hydrostatic pressure inside the capillary and the hydrostatic pressure outside the capillary. This pressure tends to push water and solutes out of capillaries and into the interstitial fluid. Factors that contribute to the net hydrostatic pressure include

- the capillary hydrostatic pressure (CHP), which ranges from 35 mm Hg at the arterial end of a capillary to 18 mm Hg at the venous end, and
- the *interstitial fluid hydrostatic pressure (IHP)*. Measurements of IHP have yielded very small values that differ from tissue to tissue—from +6 mm Hg in the brain to −6 mm Hg in subcutaneous tissues. A positive IHP opposes CHP, so in this situation the tissue hydrostatic pressure must be overcome before fluid can move out of a capillary. A negative IHP assists CHP, and additional fluid will leave the capillary. However, under normal circumstances the average IHP is 0 mm Hg, and we can assume that the net hydrostatic pressure is equal to CHP. (For this reason, IHP is not included in Figure 21–11.)

Plasma proteins in capillary blood create capillary colloid osmotic pressure. The *net capillary colloid osmotic pressure* tends to cause water and solutes, by bulk flow, to enter the capillary from the interstitial fluid. The net colloid osmotic pressure is the difference between

- the *blood colloid osmotic pressure (BCOP)*, which is about 25 mm Hg, and
- the *interstitial fluid colloid osmotic pressure (ICOP)*. The ICOP is as variable and low as the IHP, because the interstitial fluid in most tissues contains negligible quantities of suspended proteins. Reported values of ICOP are from 0 to 5 mm Hg, within the range of pressures recorded for the IHP. It is thus safe to assume that under normal circumstances the net colloid osmotic pressure is equal to the BCOP. (For this reason, ICOP is not included in Figure 21–11.)

The **net filtration pressure (NFP)** is the difference between the net hydrostatic pressure and the net osmotic pressure. In terms of the factors just listed, this means that

$$\text{net filtration pressure} = \text{net hydrostatic pressure} - \text{net colloid osmotic pressure}$$

$$\text{NFP} = (\text{CHP} - \text{IHP}) - (\text{BCOP} - \text{ICOP})$$

At the arterial end of a capillary, the net filtration pressure can be calculated as follows:

$$\text{NFP} = (35 - 0) - (25 - 0) = 35 - 25 = 10 \text{ mm Hg}$$

Because this value is positive, it indicates that fluid will tend to move *out of* the capillary and into the interstitial fluid.

21

At the venous end of the capillary, the net filtration pressure will be

$$\text{NFP} = (18 - 0) - (25 - 0) = 18 - 25 = -7 \text{ mm Hg}$$

The minus sign indicates that fluid tends to move *into* the capillary; that is, reabsorption is occurring.

The transition between filtration and reabsorption occurs where the CHP is 25 mm Hg, because at that point the hydrostatic and osmotic forces are equal—that is, the NFP is 0 mm Hg. If the maximum filtration pressure at the arterial end of the capillary were equal to the maximum reabsorption pressure at the venous end, this transition point would lie midway along the length of the capillary. Under these circumstances, filtration would occur along the first half of the capillary, and an identical amount of reabsorption would occur along the second half. However, the maximum filtration pressure is higher than the maximum reabsorption pressure, so the transition point between filtration and reabsorption normally lies closer to the venous end of the capillary than to the arterial end. As a result, more filtration than reabsorption occurs along the capillary. Of the nearly 24 liters of fluid that move out of the plasma and into the interstitial fluid each day, 20.4 liters (85 percent) are reabsorbed. The remainder (3.6 liters) flows through the tissues and into lymphatic vessels, for eventual return to the venous system (see Figure 21–11).

Any condition that affects hydrostatic or osmotic pressures in the blood or tissues will shift the balance between hydrostatic and osmotic forces. We can predict the effects of these changes on the basis of an understanding of capillary dynamics. For example:

- If hemorrhaging occurs, both blood volume and blood pressure decrease. This reduction in CHP lowers the NFP and increases the amount of reabsorption. The result is a decrease in the volume of interstitial fluid and an increase in the circulating plasma volume. This process is known as a *recall of fluids*.
- If dehydration occurs, the plasma volume decreases due to water loss, and the concentration of plasma proteins increases. The increase in BCOP accelerates reabsorption and a recall of fluids that delays the onset and severity of clinical signs and symptoms.
- If the CHP rises or the BCOP declines, fluid moves out of the blood and builds up in peripheral tissues, a condition called *edema*.

+ Clinical Note Edema

Edema (eh-DĒ-muh) is an abnormal accumulation of interstitial fluid. The underlying problem in all types of edema is a disturbance in the normal balance between hydrostatic and osmotic forces at the capillary level. For instance:

- You usually have swelling at a bruise. When a capillary is damaged, plasma proteins cross the capillary wall and enter the interstitial fluid. The resulting rise in the interstitial fluid colloid osmotic pressure (ICOP) reduces the rate of capillary reabsorption and produces a localized edema.
- In starvation, the liver cannot synthesize enough plasma proteins to maintain normal concentrations in the blood, so the blood colloid osmotic pressure (BCOP) declines. Fluids then begin moving from the blood into peripheral tissues. In children, fluid builds up in the abdominopelvic cavity, producing the swollen bellies typical of starvation victims. BCOP is also reduced after severe burns and in several types of liver and kidney diseases.
- In the U.S. population, most serious cases of edema result from increases in arterial blood pressure, venous pressure, or total circulatory pressure. The increase may be due to heart problems such as heart failure or venous blood clots that elevate venous pressures. The net result is an increase in capillary hydrostatic pressure (CHP) that accelerates fluid movement into the tissues.

Checkpoint

5. **Identify the factors that contribute to total peripheral resistance.**
6. **In a healthy person, where is blood pressure greater: at the aorta or at the inferior vena cava? Explain.**
7. **While standing in the hot sun, Lailah begins to feel light-headed and faints. Explain what happened.**
8. **Mike's blood pressure is 125/70. What is his mean arterial pressure (MAP)?**

See the blue Answers tab at the back of the book.

21-3 Blood flow and pressure in tissues are controlled by both autoregulation and central regulation

Learning Outcome Describe the control mechanisms that regulate blood flow and pressure in tissues, and explain how the activities of the cardiac, vasomotor, and respiratory centers are coordinated to control blood flow through the tissues.

Homeostatic mechanisms regulate cardiovascular activity to ensure that the amount of blood flow through tissues, called **tissue perfusion**, is adequate to meet the demand for oxygen and nutrients. The factors that affect tissue perfusion are (1) cardiac output, (2) peripheral resistance, and (3) blood pressure. We discussed cardiac output in Chapter 20. p. 716 We considered peripheral resistance and blood pressure earlier in this chapter.

Most cells are relatively close to capillaries. When a group of cells becomes active, the circulation to that region must increase to bring the necessary oxygen and nutrients, and to carry away the wastes and carbon dioxide they generate. The purpose of cardiovascular regulation is to ensure that these blood flow changes occur (1) at an appropriate time, (2) in the right area, and (3) without drastically changing blood pressure and blood flow to vital organs. Let's look first at how blood is distributed to capillaries, and then at the regulatory mechanisms that control how that distribution occurs.

Vasomotion

Although blood normally flows from arterioles to venules at a constant rate, the flow within each capillary is quite variable. Each precapillary sphincter alternately contracts and relaxes, perhaps a dozen times per minute. These rhythmic changes in vessel diameter are called **vasomotion**. As a result, the blood flow within any capillary occurs in pulses rather than as a steady and constant stream. The net effect is that blood may reach the venules by one route now and by a different route later.

When you are at rest, blood flows through about 25 percent of the vessels within a typical capillary bed in your body. Your cardiovascular system does not contain enough blood to maintain adequate blood flow to all the capillaries in all the capillary beds in your body at the same time. As a result, when many tissues become active, the blood flow through capillary beds must be coordinated.

Overview of Autoregulation and Central Regulation

Homeostatic mechanisms operating at the local, regional, and systemic levels adjust blood flow through the capillaries to meet the demands of peripheral tissues. We can group these mechanisms as follows:

- *Autoregulation.* Local factors change the pattern of blood flow within capillary beds as precapillary sphincters open and close in response to chemical changes in interstitial fluids. This is an example of autoregulation at the tissue level. Autoregulation causes immediate, localized homeostatic adjustments.

  If autoregulation fails to normalize conditions at the tissue level, *central regulation*, involving both neural and endocrine mechanisms, is activated. These central regulation mechanisms focus on controlling cardiac output and blood pressure to restore adequate blood flow after blood pressure decreases.
- *Neural Mechanisms.* Neural mechanisms respond to changes in arterial pressure or blood gas levels sensed at specific sites. When those changes occur, the cardiovascular center of the autonomic nervous system makes short-term adjustments to cardiac output and peripheral resistance to maintain blood pressure and ensure adequate blood flow.
- *Endocrine Mechanisms.* The endocrine system releases hormones that enhance short-term adjustments and that direct long-term changes in cardiovascular performance.

Now let's see how each of these regulatory mechanisms responds to inadequate perfusion of skeletal muscles. The regulatory relationships are diagrammed in Figure 21–12.

Autoregulation of Blood Flow within Tissues

Under normal resting conditions, cardiac output remains stable, and peripheral resistance within individual tissues is adjusted to control local blood flow. Vasomotion is controlled locally by changes in the concentrations of chemicals and dissolved gases in the interstitial fluids. This process is an example of capillary **autoregulation**.

Factors that promote the dilation of precapillary sphincters are called **vasodilators**. For example, when the dissolved oxygen concentration decreases within a tissue, **local vasodilators** cause the capillary sphincters to relax, so blood flow to the area increases. Local vasodilators include the following:

- Decreased oxygen level or increased CO_2 level in tissue.
- Lactate generated by tissue cells.
- Nitric oxide (NO) released from endothelial cells.

Go to MasteringA&P™ > Study Area > Lab Tools > PhysioEx 9.1 > Exercise 5: Cardiovascular Dynamics

Figure 21–12 **Short-Term and Long-Term Cardiovascular Responses.** This diagram indicates general mechanisms that compensate for a decrease in blood flow, blood volume, and blood pressure.

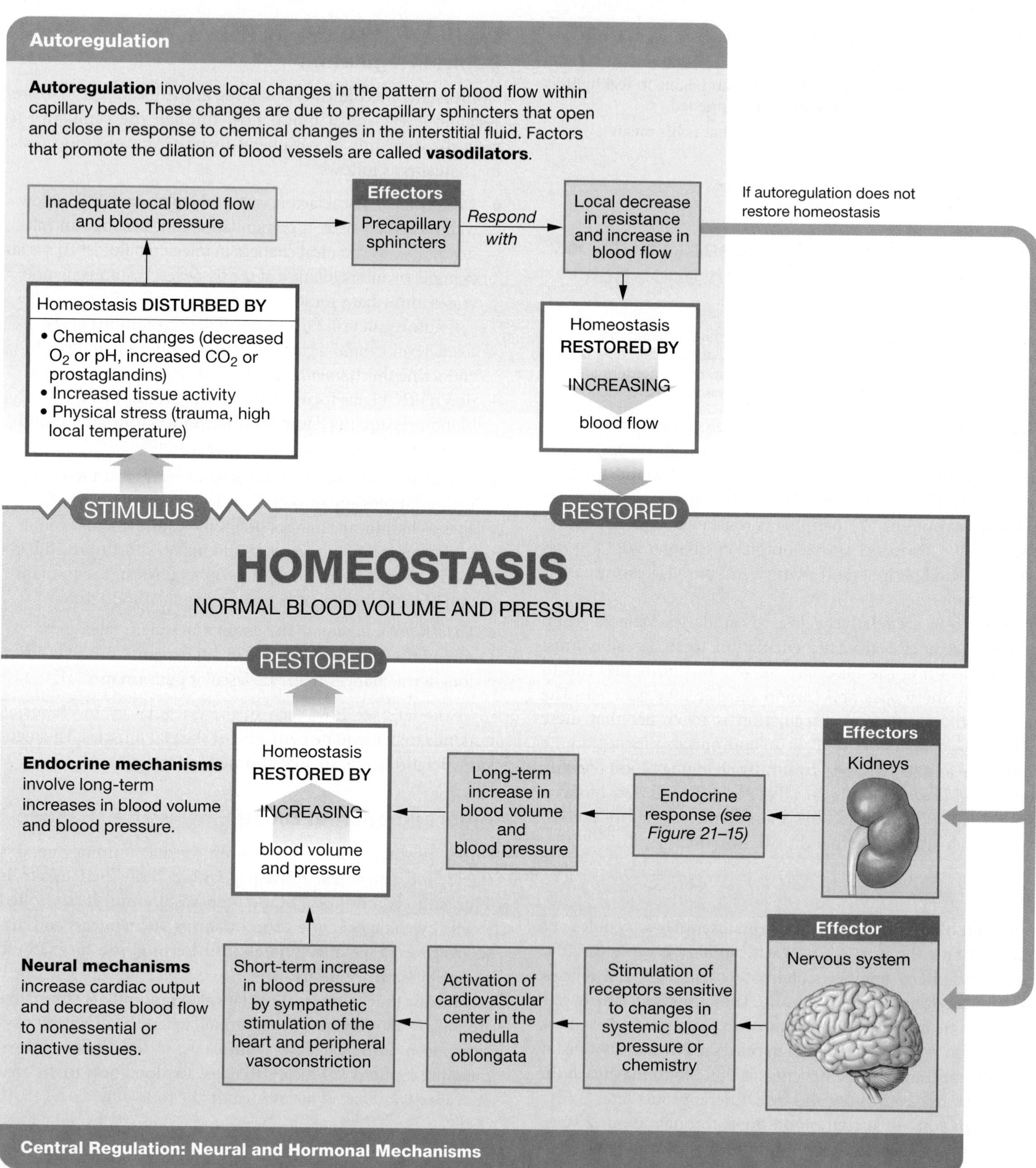
Autoregulation
Autoregulation involves local changes in the pattern of blood flow within capillary beds. These changes are due to precapillary sphincters that open and close in response to chemical changes in the interstitial fluid. Factors that promote the dilation of blood vessels are called **vasodilators**.
Inadequate local blood flow and blood pressure
Effectors
Precapillary sphincters
Respond with
Local decrease in resistance and increase in blood flow
If autoregulation does not restore homeostasis
Homeostasis DISTURBED BY
• Chemical changes (decreased O_2 or pH, increased CO_2 or prostaglandins)
• Increased tissue activity
• Physical stress (trauma, high local temperature)
Homeostasis RESTORED BY
INCREASING
blood flow
STIMULUS
RESTORED
HOMEOSTASIS
NORMAL BLOOD VOLUME AND PRESSURE
RESTORED
Endocrine mechanisms involve long-term increases in blood volume and blood pressure.
Homeostasis RESTORED BY
INCREASING
blood volume and pressure
Long-term increase in blood volume and blood pressure
Endocrine response (see Figure 21–15)
Effectors
Kidneys
Neural mechanisms increase cardiac output and decrease blood flow to nonessential or inactive tissues.
Short-term increase in blood pressure by sympathetic stimulation of the heart and peripheral vasoconstriction
Activation of cardiovascular center in the medulla oblongata
Stimulation of receptors sensitive to changes in systemic blood pressure or chemistry
Effector
Nervous system
Central Regulation: Neural and Hormonal Mechanisms

? For blood flow to be autoregulated, what types of stimuli signal inadequate local blood flow and blood pressure?

21

- Rising concentration of potassium ions or hydrogen ions in the interstitial fluid.
- Chemicals released during local inflammation, including histamine. p. 147
- Elevated local temperature.

The presence of any of these vasodilators indicates that tissue conditions are in some way abnormal. The increase in blood flow, which brings oxygen, nutrients, and buffers, may be sufficient to restore homeostasis.

As noted in Chapter 19, aggregating platelets and damaged tissues produce compounds that stimulate precapillary sphincters to constrict. These compounds are **local vasoconstrictors**. Examples include prostaglandins and thromboxanes released by activated platelets and white blood cells, and the endothelins released by damaged endothelial cells.

Local vasodilators and vasoconstrictors control blood flow within a single capillary bed (look back at Figure 21–4). In high concentrations, these factors also affect arterioles, increasing or decreasing blood flow to all the capillary beds in a given area.

Central Regulation: Neural Mechanisms

The nervous system adjusts cardiac output and peripheral resistance to maintain adequate blood flow to vital tissues and organs. The center primarily responsible for these regulatory activities is the **cardiovascular (CV) center** of the medulla oblongata. The CV center is made up of the *cardiac centers* and the *vasomotor center*. Although these centers are difficult to distinguish anatomically (p. 474), in functional terms, they often act independently.

The cardiac centers are composed of the cardioinhibitory center and the cardioacceleratory center. As noted in Chapter 20, the *cardioacceleratory center* increases cardiac output through sympathetic innervation. The *cardioinhibitory center* decreases cardiac output through parasympathetic innervation. p. 717 The vasomotor center directs vasomotor responses in blood vessels. Let's look at the functions of these centers.

The Vasomotor Center and Vasomotor Tone

The vasomotor center contains two populations of neurons: (1) a very large group responsible for widespread vasoconstriction and (2) a smaller group responsible for the vasodilation of arterioles in skeletal muscles and the brain. The vasomotor center controls the activity of sympathetic motor neurons:

- *Control of Vasoconstriction.* The neurons innervating peripheral blood vessels in most tissues are *adrenergic*; that is, they release the neurotransmitter norepinephrine (NE). NE stimulates smooth muscles in the walls of arterioles, producing vasoconstriction.
- *Control of Vasodilation.* Vasodilator neurons innervate blood vessels in skeletal muscles and in the brain. The stimulation of these neurons relaxes smooth muscle cells in the walls of arterioles, producing vasodilation. This relaxation is triggered by the appearance of NO in the surroundings. The vasomotor center may control NO release indirectly or directly. The most common vasodilator synapses are *cholinergic*—their axon terminals release ACh. In turn, ACh stimulates endothelial cells in the area to release NO, causing local vasodilation. Other vasodilator synapses are *nitroxidergic*—their axon terminals release NO as a neurotransmitter. Nitric oxide has an immediate and direct relaxing effect on the vascular smooth muscle cells in the area.

In Chapter 16, we saw how autonomic tone sets a background level of neural activity that can increase or decrease on demand. p. 551 The sympathetic vasoconstrictor nerves are always active, producing a significant **vasomotor tone**. This vasoconstrictor activity normally keeps the arterioles partially constricted. Under maximal stimulation, arterioles constrict to about half their resting diameter. (To dilate fully, an arteriole increases its resting diameter by about 1.5 times.)

Constriction has a large effect on resistance, because, as we saw earlier, resistance increases sharply as the luminal diameter of a vessel decreases. The resistance of a maximally constricted arteriole is roughly 80 *times* that of a fully dilated arteriole. Because blood pressure varies directly with peripheral resistance, the vasomotor center can control arterial blood pressure very effectively by making modest adjustments in vessel diameters. Extreme stimulation of the vasomotor center also produces venoconstriction and mobilizes the venous blood reservoirs.

21

Reflex Control of Cardiovascular Function by the Cardiovascular Center

The cardiovascular (CV) center detects changes in tissue demand by monitoring arterial blood, especially its blood pressure, pH, and concentrations of dissolved gases. The *baroreceptor reflexes* (*baro-*, pressure) respond to changes in blood pressure, and the *chemoreceptor reflexes* (*chemo-*, chemical) monitor changes in the chemical composition of arterial blood. These reflexes are regulated through a negative feedback loop: The stimulation of a receptor by an abnormal condition leads to a short-term response that counteracts the stimulus and restores normal conditions.

Baroreceptor Reflexes. Baroreceptors are specialized receptors that monitor the degree of stretch in the walls of expandable organs. p. 520 The baroreceptors involved in cardiovascular regulation are found in the walls of (1) the **carotid sinuses**, expanded chambers near the bases of the *internal carotid arteries* of the neck (look ahead to Figure 21–21); (2) the **aortic arch**, which curves over the superior surface of the heart (look back at Figure 20–8, p. 699); and (3) the wall of the right atrium. These receptors are part of the **baroreceptor reflexes**, which adjust cardiac output and peripheral resistance to maintain normal arterial pressures.

Aortic baroreceptors monitor blood pressure within the ascending aorta. Any changes trigger the **aortic reflex**, which adjusts blood pressure to maintain adequate blood pressure and blood flow through the systemic circuit. Carotid sinus baroreceptors trigger reflexes that maintain adequate blood flow to the brain. The carotid sinus receptors are extremely sensitive because blood flow to the brain must remain constant. Figure 21–13 presents the basic organization of the baroreceptor reflexes triggered by changes in blood pressure at the carotid sinuses and aortic arch.

When blood pressure rises, the increased output from the baroreceptors alters activity in the CV center and produces two major effects (see the top half of Figure 21–13):

- *A decrease in cardiac output*, due to parasympathetic stimulation and the inhibition of sympathetic activity. This results from the stimulation of the cardioinhibitory center and the inhibition of the cardioacceleratory center.
- *Widespread peripheral vasodilation*, due to the inhibition of excitatory neurons in the vasomotor center.

Figure 21–13 Baroreceptor Reflexes of the Carotid Sinuses and Aortic Arch.

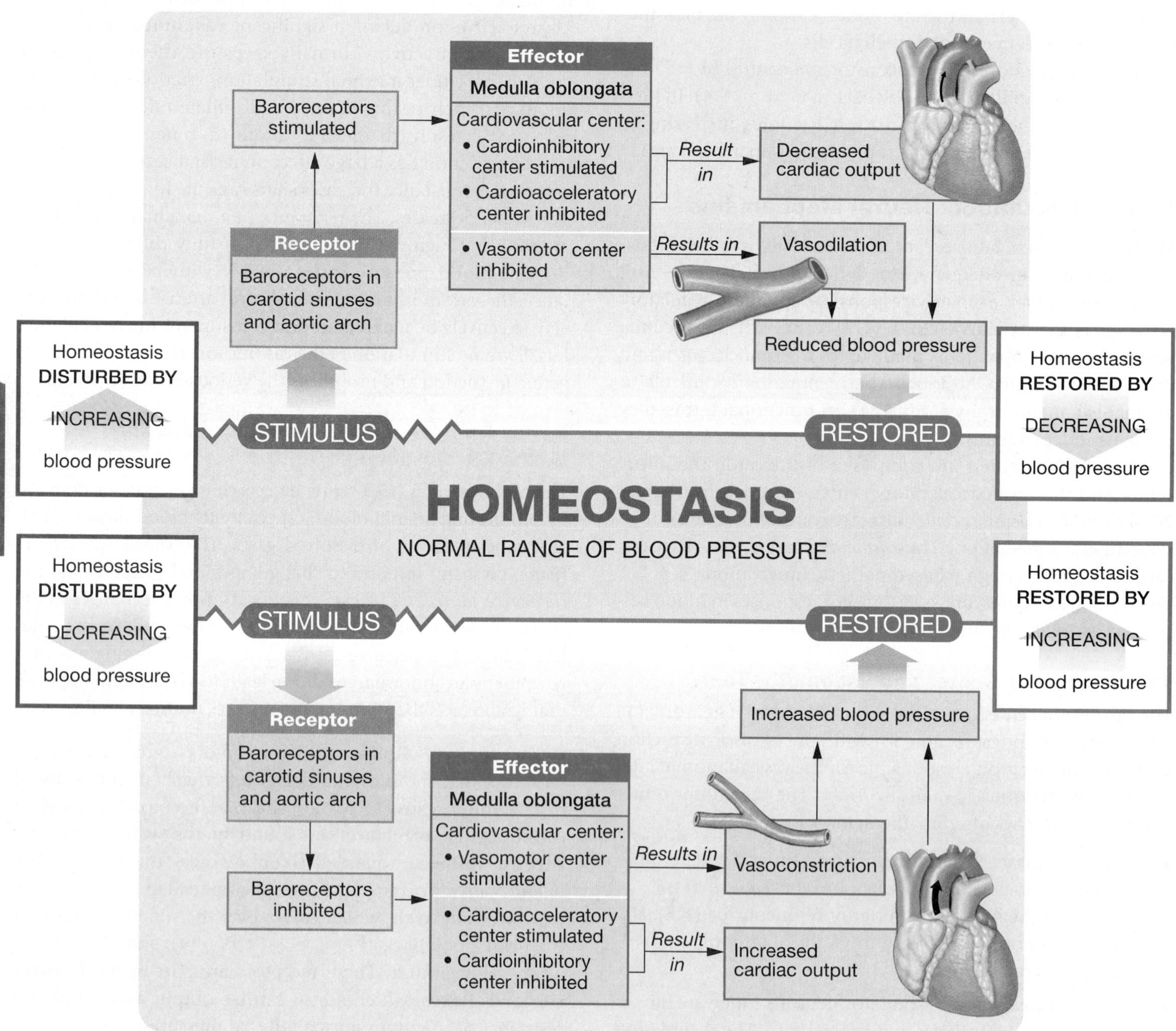
Effector
Medulla oblongata
Baroreceptors stimulated
Cardiovascular center:
• Cardioinhibitory center stimulated
• Cardioacceleratory center inhibited
• Vasomotor center inhibited
Result in
Decreased cardiac output
Results in
Vasodilation
Receptor
Baroreceptors in carotid sinuses and aortic arch
Reduced blood pressure
Homeostasis DISTURBED BY INCREASING blood pressure
STIMULUS
RESTORED
Homeostasis RESTORED BY DECREASING blood pressure
HOMEOSTASIS
NORMAL RANGE OF BLOOD PRESSURE
Homeostasis DISTURBED BY DECREASING blood pressure
STIMULUS
RESTORED
Homeostasis RESTORED BY INCREASING blood pressure
Increased blood pressure
Receptor
Baroreceptors in carotid sinuses and aortic arch
Effector
Medulla oblongata
Cardiovascular center:
• Vasomotor center stimulated
• Cardioacceleratory center stimulated
• Cardioinhibitory center inhibited
Baroreceptors inhibited
Results in
Vasoconstriction
Result in
Increased cardiac output

21

The decrease in cardiac output reflects primarily a reduction in heart rate due to the release of acetylcholine at the sinoatrial (SA) node. p. 718 The widespread vasodilation lowers peripheral resistance, and this effect, combined with a reduction in cardiac output, leads to a fall in blood pressure to a normal level.

When blood pressure falls below normal, baroreceptor output is reduced accordingly (see the lower half of Figure 21–13). This change has two major effects working together to raise blood pressure:

- *An increase in cardiac output,* through the stimulation of sympathetic innervation to the heart. This results from stimulation of the cardioacceleratory center that is accompanied by inhibition of the cardioinhibitory center.
- *Widespread peripheral vasoconstriction,* caused by the stimulation of sympathetic vasoconstrictor neurons by the vasomotor center.

The effects on the heart result from the release of NE by sympathetic neurons innervating the SA node, the atrioventricular (AV) node, and the general myocardium. In a crisis, such as significant blood loss, sympathetic activation occurs, and its effects are enhanced by the release of both NE and epinephrine (E) from the adrenal medullae. The net effect is an immediate increase in heart rate and stroke volume, and a corresponding increase in cardiac output.

The vasoconstriction, which also results from the release of NE by sympathetic neurons, increases peripheral resistance. These adjustments—increased cardiac output and increased peripheral resistance—work together to raise blood pressure.

Atrial baroreceptors monitor blood pressure at the end of the systemic circuit—at the venae cavae and the right atrium. Recall from Chapter 20 that the **Bainbridge reflex** responds to a stretching of the wall of the right atrium. p. 719

Under normal circumstances, the heart pumps blood into the aorta at the same rate at which blood arrives at the right atrium. When blood pressure rises at the right atrium, blood is arriving at the heart faster than it is being pumped out. The atrial baroreceptors correct the situation by stimulating the CV center to increase cardiac output until the backlog of venous blood is removed. Atrial pressure then returns to normal.

A procedure known as the Valsalva maneuver is a simple way to check for normal cardiovascular responses to changes in arterial pressure and venous return. The *Valsalva maneuver* involves trying to exhale forcefully with closed lips and nostrils so that no air can leave the lungs and pressure in the thoracic cavity rises sharply. This action causes reflexive changes in blood pressure and cardiac output due to increased intrathoracic pressure, which impedes venous return to the right atrium. When internal pressures rise, the venae cavae collapse, and venous return decreases. The resulting drop in cardiac output and blood pressure stimulates the aortic and carotid baroreceptors, causing a reflexive increase in heart rate and peripheral vasoconstriction. When the glottis opens and pressures return to normal, venous return increases suddenly and so does cardiac output. Because vasoconstriction has occurred, blood pressure rises sharply, and this inhibits the baroreceptors. As a result, cardiac output, heart rate, and blood pressure quickly return to normal levels.

Chemoreceptor Reflexes. The **chemoreceptor reflexes** respond to changes in carbon dioxide, oxygen, or pH levels in blood and cerebrospinal fluid (CSF) (Figure 21–14). The chemoreceptors involved are sensory neurons. They are located in the **carotid bodies**, situated in the neck near the carotid sinus, and the **aortic bodies**, near the arch of the aorta. p. 521 These receptors monitor the composition of arterial blood. Additional chemoreceptors located on the ventrolateral surfaces of the medulla oblongata monitor the composition of CSF.

When chemoreceptors in the carotid bodies or aortic bodies detect either an increase in the carbon dioxide concentration or a decrease in the pH of arterial blood, the cardioacceleratory and vasomotor centers are stimulated. At the same time, the cardioinhibitory center is inhibited. This dual effect causes an increase in cardiac output, peripheral vasoconstriction, and an increase in blood pressure. A decrease in the oxygen level at the aortic bodies has the same effects. Strong stimulation of the carotid or aortic chemoreceptors causes widespread sympathetic activation, with more dramatic increases in heart rate and cardiac output.

The chemoreceptors of the medulla oblongata are involved primarily with the control of respiratory function, and secondarily with regulating blood flow to the brain. For example, a steep rise in the CSF carbon dioxide level triggers the vasodilation of cerebral vessels, but produces vasoconstriction in most other organs. The result is increased blood flow—and increased oxygen delivery—to the brain.

Coordination of cardiovascular and respiratory activities is vital, because accelerating blood flow in the tissues is useful only if the circulating blood contains an adequate amount of oxygen. The arterial CO_2 level can be decreased and the O_2 level increased most effectively by coordinating cardiovascular and respiratory activities. Chemoreceptor stimulation also stimulates the respiratory centers, and the increase in cardiac output and blood pressure is associated with an increased respiratory rate. In addition, an increase in the respiratory rate accelerates venous return through the action of the respiratory pump. (We consider other aspects of chemoreceptor activity and respiratory control in Chapter 23.)

The Cardiovascular Center and CNS Activities

The output of the cardiovascular center can also be influenced by activities in other areas of the brain. For example,

Figure 21–14 **The Chemoreceptor Reflexes.**

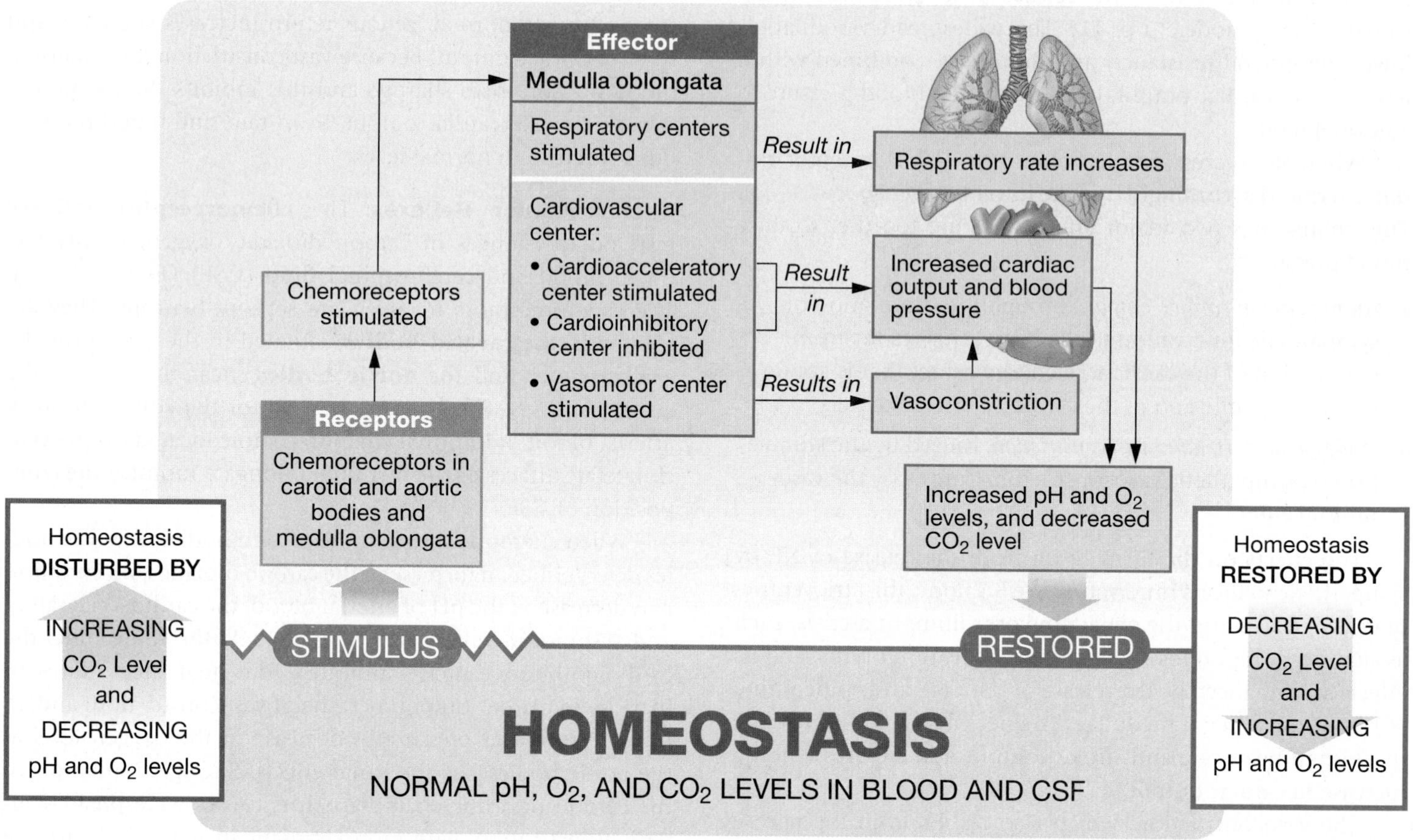

Effector
Medulla oblongata
Respiratory centers stimulated
Result in
Respiratory rate increases
Cardiovascular center:
• Cardioacceleratory center stimulated
• Cardioinhibitory center inhibited
Result in
Increased cardiac output and blood pressure
• Vasomotor center stimulated
Results in
Vasoconstriction
Chemoreceptors stimulated
Receptors
Chemoreceptors in carotid and aortic bodies and medulla oblongata
Increased pH and O_2 levels, and decreased CO_2 level
Homeostasis DISTURBED BY
INCREASING CO_2 Level and DECREASING pH and O_2 levels
STIMULUS
RESTORED
Homeostasis RESTORED BY
DECREASING CO_2 Level and INCREASING pH and O_2 levels
HOMEOSTASIS
NORMAL pH, O_2, AND CO_2 LEVELS IN BLOOD AND CSF

the activation of either division of the autonomic nervous system affects output from the cardiovascular center. A general sympathetic activation stimulates the cardioacceleratory and vasomotor centers. As a result, cardiac output and blood pressure increase. In contrast, when the parasympathetic division is activated, the cardioinhibitory center is stimulated. The result is a decrease in cardiac output. Parasympathetic activity does not directly affect the vasomotor center, but vasodilation takes place as sympathetic activity declines.

The higher brain centers can also affect blood pressure. Our thought processes and emotional states can produce significant changes in blood pressure by influencing cardiac output and vasomotor tone. For example, strong emotions of anxiety, fear, and rage are accompanied by an increase in blood pressure, due to cardiac stimulation and vasoconstriction.

Central Regulation: Endocrine Mechanisms

The endocrine system regulates cardiovascular performance in both the short term and the long term. As we have seen, E and NE from the adrenal medullae stimulate cardiac output and peripheral vasoconstriction. Here we will look at other hormones important in regulating cardiovascular function: (1) antidiuretic hormone (ADH), (2) angiotensin II, (3) erythropoietin (EPO), and (4) the natriuretic peptides (ANP and BNP). p. 644 All four are concerned primarily with the long-term regulation of blood volume (Figure 21–15). ADH and angiotensin II also affect blood pressure.

Antidiuretic Hormone

Antidiuretic hormone (ADH) is released at the posterior lobe of the pituitary gland in response to a decrease in blood volume, to an increase in the osmotic concentration of the plasma, or (secondarily) to circulating angiotensin II. It brings about a peripheral vasoconstriction that elevates blood pressure. This hormone also stimulates water conservation by the kidneys, thus preventing a reduction in blood volume that would further reduce blood pressure (see Figure 21–15).

Angiotensin II

Angiotensin II appears in the blood when specialized cells in kidney arterioles, called *juxtaglomerular cells,* release the enzyme *renin* (RĒ-nin) in response to a decrease in renal blood pressure (see Figure 21–15). Once in the bloodstream, renin starts an enzymatic chain reaction. In the first step, renin converts *angiotensinogen,* a plasma protein produced by the liver, to *angiotensin I.*

Figure 21–15 **The Hormonal Regulation of Blood Pressure and Blood Volume.**

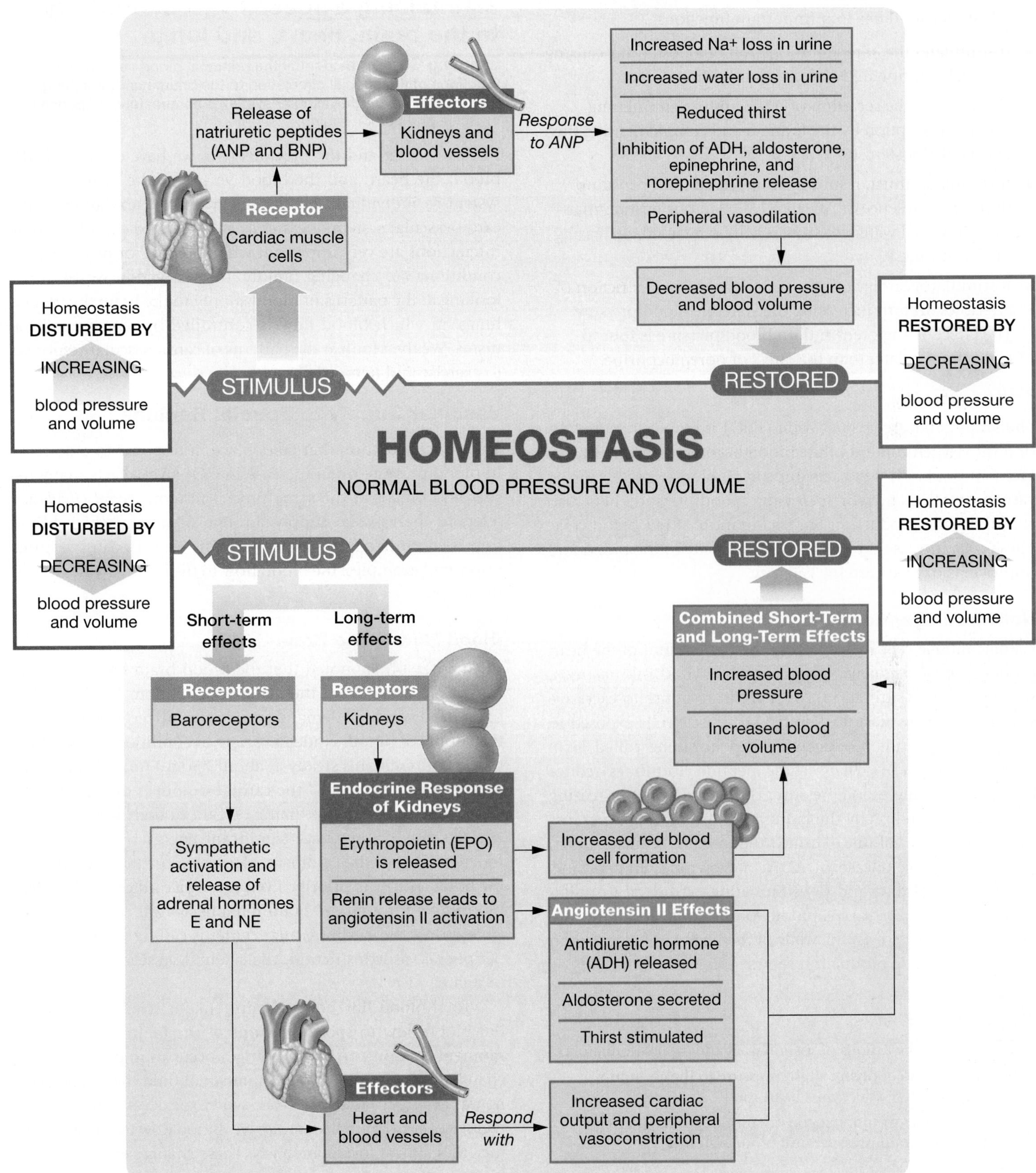
Release of natriuretic peptides (ANP and BNP)
Effectors
Kidneys and blood vessels
Response to ANP
Increased Na+ loss in urine
Increased water loss in urine
Reduced thirst
Inhibition of ADH, aldosterone, epinephrine, and norepinephrine release
Peripheral vasodilation
Receptor
Cardiac muscle cells
Decreased blood pressure and blood volume
Homeostasis DISTURBED BY INCREASING blood pressure and volume
STIMULUS
RESTORED
Homeostasis RESTORED BY DECREASING blood pressure and volume
HOMEOSTASIS
NORMAL BLOOD PRESSURE AND VOLUME
Homeostasis DISTURBED BY DECREASING blood pressure and volume
STIMULUS
RESTORED
Homeostasis RESTORED BY INCREASING blood pressure and volume
Short-term effects
Long-term effects
Combined Short-Term and Long-Term Effects
Increased blood pressure
Increased blood volume
Receptors
Baroreceptors
Receptors
Kidneys
Endocrine Response of Kidneys
Erythropoietin (EPO) is released
Renin release leads to angiotensin II activation
Increased red blood cell formation
Sympathetic activation and release of adrenal hormones E and NE
Angiotensin II Effects
Antidiuretic hormone (ADH) released
Aldosterone secreted
Thirst stimulated
Effectors
Heart and blood vessels
Respond with
Increased cardiac output and peripheral vasoconstriction

? What hormones are released as a short-term response to a decrease in blood pressure and blood volume?

In the capillaries of the lungs, *angiotensin-converting enzyme (ACE)* then modifies angiotensin I to angiotensin II, an active hormone with diverse effects.

Angiotensin II has four important functions:

- It stimulates the adrenal production of aldosterone, causing Na^+ reabsorption and K^+ loss by the kidneys.
- It stimulates the secretion of ADH, in turn stimulating water reabsorption by the kidneys and complementing the effects of aldosterone.
- It stimulates thirst, resulting in increased fluid consumption. (The presence of ADH and aldosterone ensures that the additional water consumed will be retained, increasing blood volume.)
- It stimulates cardiac output and triggers the constriction of arterioles, in turn increasing the systemic blood pressure. The effect of angiotensin II on blood pressure is four to eight times greater than the effect of norepinephrine.

Erythropoietin

The kidneys release *erythropoietin (EPO)* if blood pressure falls or if the oxygen content of the blood becomes abnormally low (see **Figure 21–15**). EPO acts directly on blood vessels, causing vasoconstriction, thereby increasing blood pressure. EPO also stimulates the production and maturation of red blood cells. These cells increase the volume and viscosity of the blood and improve its oxygen-carrying capacity.

21

Natriuretic Peptides

Cardiac muscle cells in the wall of the right atrium of the heart produce *atrial natriuretic peptide* (nā-trē-yū-RET-ik; *natrium,* sodium + *ouresis,* urination), or *ANP,* in response to excessive stretching during diastole. Ventricular muscle cells exposed to comparable stimuli produce a related hormone called *brain natriuretic peptide,* or *BNP.* These peptide hormones reduce blood volume and blood pressure. They do so by (1) increasing sodium ion excretion by the kidneys; (2) promoting water loss by increasing the volume of urine produced; (3) reducing thirst; (4) blocking the release of ADH, aldosterone, epinephrine, and norepinephrine; and (5) stimulating peripheral vasodilation (see **Figure 21–15**). As blood volume and blood pressure decrease, the stresses on the walls of the heart are reduced, and natriuretic peptide production ceases.

Checkpoint

9. **Describe the actions of vasodilators and local vasodilators.**
10. **How would applying slight pressure to the common carotid artery affect your heart rate?**
11. **What effect would vasoconstriction of the renal artery have on blood pressure and blood volume?**

See the blue Answers tab at the back of the book.

21-4 The cardiovascular system adapts to physiological stress while maintaining a special vascular supply to the brain, heart, and lungs

Learning Outcome Identify the principal blood vessels and functions of the special circulation to the brain, heart, and lungs, and explain the cardiovascular system's homeostatic response to exercise and hemorrhaging.

In this chapter and the previous two, we have considered the blood, the heart, and the blood vessels of the cardiovascular system as individual entities. Yet in our day-to-day lives, the cardiovascular system operates as an integrated complex. These interactions are very important when physical or physiological conditions are changing rapidly. In this section we begin by looking at the patterns of blood supply to the brain, heart, and lungs, in which blood flow is controlled by separate mechanisms. We then look at the patterns of cardiovascular responses to exercise and hemorrhaging.

Vascular Supply to Special Regions

The vasoconstriction that takes place in response to a decrease in blood pressure or an increase in CO_2 level affects many tissues and organs at the same time. The term *special circulation* refers to the vascular supply through organs in which blood flow is controlled by separate mechanisms. Let's consider three important examples: the blood flow to the brain, the heart, and the lungs.

Blood Flow to the Brain

In Chapter 14, we noted that the blood brain barrier isolates most CNS tissue from the general circulation. p. 473 The brain has a very high demand for oxygen and receives a substantial supply of blood. Under a variety of conditions, blood flow to the brain remains steady at about 750 mL/min. That means that about 12 percent of the cardiac output is delivered to an organ that makes up less than 2 percent of body weight.

Neurons do not have significant energy reserves, and in functional terms the cardiovascular system treats blood flow to the brain as the top priority. Even during a cardiovascular crisis, blood flow through the brain remains as near normal as possible: While the cardiovascular center is calling for widespread peripheral vasoconstriction, the cerebral vessels are instructed to dilate.

Total blood flow to the brain remains relatively constant, but blood flow to specific regions of the brain changes from moment to moment. These changes occur in response to local changes in the composition of interstitial fluid that accompany neural activity. When you read, write, speak, or walk, specific regions of your brain become active. Blood flow to those regions increases almost instantaneously. These changes ensure that the active neurons will receive the oxygen and nutrients they require.

The brain receives arterial blood through four arteries. A slowly developing interruption of flow in any one of these large vessels may not significantly reduce blood flow to the brain as a whole, because these arteries form anastomoses inside the cranium and other vessels may enlarge or establish alternative routes. However, a plaque or a blood clot may still block a small artery, and weakened arteries may rupture. Such incidents temporarily or permanently shut off blood flow to a localized area of the brain, damaging or killing the dependent neurons. Signs and symptoms of a *stroke*, or *cerebrovascular accident (CVA)*, then appear.

Blood Flow to the Heart

The coronary arteries arise at the base of the ascending aorta, where systemic pressures are highest (see Chapter 20 p. 697). Each time the heart contracts, it squeezes the coronary vessels, so blood flow is reduced. In the left ventricle, systolic pressures are high enough that blood can flow into the myocardium only during diastole. Over this period, elastic rebound helps drive blood along the coronary vessels. Cardiac contractile cells can tolerate these brief circulatory interruptions because they have substantial oxygen reserves.

When you are at rest, coronary blood flow is about 250 mL/min. When the workload on your heart increases, local factors, such as reduced O_2 and increased lactate levels, dilate the coronary vessels and increase blood flow. Epinephrine released during sympathetic stimulation promotes the vasodilation of coronary vessels. It also increases heart rate and the strength of cardiac contractions. As a result, coronary blood flow increases while vasoconstriction occurs in other tissues.

For reasons that are not clear, some people have *coronary spasms*. These spasms can temporarily restrict coronary circulation and produce symptoms of angina. The heart's ability to increase its output, even under maximal stimulation, can be limited by certain conditions. These conditions include a permanent restriction or blockage of coronary vessels (as in coronary artery disease) and tissue damage (as caused by a myocardial infarction). When the cardiac workload increases much above the resting level, individuals with these conditions experience signs and symptoms of *heart failure*.

Blood Flow to the Lungs

The lungs have about 300 million *alveoli* (al-VĒ-ō-lī; *alveolus*, sac), delicate epithelial pockets where gas exchange takes place. An extensive capillary network surrounds each alveolus. Local responses to the oxygen level within each alveolus regulate blood flow through the lungs. How does this local regulation work? When an alveolus contains plenty of oxygen, its blood vessels dilate. Blood flow then increases, promoting the absorption of oxygen from air inside the alveolus. When the oxygen content of the air is very low, the vessels constrict. They shunt blood to other alveoli that contain more oxygen. This local mechanism makes the respiratory system very efficient. There is no benefit in circulating blood through the capillaries of an alveolus that contains little oxygen.

This mechanism is precisely the opposite of that in other tissues, where a decline in the oxygen level causes local vasodilation rather than vasoconstriction. The difference makes functional sense, but its physiological basis remains unclear.

Blood pressure in pulmonary capillaries (average: 10 mm Hg) is lower than that in systemic capillaries. The BCOP (25 mm Hg) is the same as elsewhere in the bloodstream. As a result, reabsorption exceeds filtration in pulmonary capillaries. Fluid moves continuously into the pulmonary capillaries across the alveolar surfaces. This flow prevents fluid from building up in the alveoli and interfering with the diffusion of respiratory gases. If the blood pressure in pulmonary capillaries rises above 25 mm Hg, fluid enters the alveoli, causing *pulmonary edema*.

The Cardiovascular Response to Exercise

At rest, cardiac output averages about 5800 mL/min (5.8 L/min). That amount changes dramatically during exercise. In addition, the pattern of blood distribution changes markedly, as detailed in **Table 21–2**.

Light Exercise

Before you begin to exercise, your heart rate increases slightly due to a general rise in sympathetic activity as you think about the workout ahead. As you begin light exercise, three interrelated changes take place:

- *Extensive vasodilation occurs* as skeletal muscles consume oxygen more quickly. Peripheral resistance decreases, blood flow through the capillaries increases, and blood enters the venous system at a faster rate.
- *The venous return increases* as skeletal muscle contractions squeeze blood along the peripheral veins and faster breathing pulls blood into the venae cavae by the respiratory pump.
- *Cardiac output rises,* primarily in response to (1) increased venous return (the Frank–Starling principle) and (2) atrial

21

Table 21–2 Changes in Blood Distribution at Rest and during Exercise

	Tissue Blood Flow (mL/min)		
Organ	Rest	Light Exercise	Strenuous Exercise
Skeletal muscles	1200	4500	12,500
Heart	250	350	750
Brain	750	750	750
Skin	500	1500	1900
Kidney	1100	900	600
Abdominal viscera	1400	1100	600
Miscellaneous	600	400	400
Total cardiac output	5800	9500	17,500

stretching (the Bainbridge reflex). p. 719 Some sympathetic stimulation occurs, leading to increases in heart rate and contractility, but the sympathetic nervous system is not greatly activated. The increased cardiac output keeps pace with the elevated demand, and arterial pressures are maintained despite the drop in peripheral resistance.

This regulation by venous feedback produces a gradual increase in cardiac output to about double resting levels. The increase supports accelerated blood flow to skeletal muscles, cardiac muscle, and the skin. The flow to skeletal and cardiac muscles increases as arterioles and precapillary sphincters dilate in response to local factors. The flow to the skin increases in response to the rise in body temperature.

Heavy Exercise

At higher levels of exertion, other physiological adjustments take place as the cardiac and vasomotor centers activate the sympathetic nervous system. Cardiac output increases toward maximal levels. Major changes in the peripheral distribution of blood improve blood flow to active skeletal muscles.

Under intense sympathetic stimulation, the cardioacceleratory center can increase cardiac output to levels as high as 20–25 L/min. But that is still not enough to meet the demands of active skeletal muscles unless the vasomotor center severely restricts the blood flow to "nonessential" organs, such as those of the digestive system. During exercise at maximal levels, your blood essentially races between the skeletal muscles and the lungs and heart. Although blood flow to most tissues is diminished, skin perfusion increases further, to cool the body as body temperature continues to climb. Only the blood supply to the brain remains unaffected.

Exercise, Cardiovascular Fitness, and Health

Cardiovascular performance improves significantly with training. **Table 21–3** compares the cardiac performance of athletes with that of nonathletes. Trained athletes have larger hearts and greater stroke volumes than do nonathletes. These functional differences are important—recall that cardiac output is equal to the stroke volume times the heart rate. For the same cardiac output, the person with a greater stroke volume has a slower heart rate. An athlete at rest can maintain normal blood flow to peripheral tissues at a heart rate as low as 32 bpm (beats per minute). When necessary, the athlete's cardiac output can increase to levels 50 percent higher than those of nonathletes. Thus, a trained athlete can tolerate sustained levels of activity that are well beyond the capabilities of nonathletes.

Exercise and Cardiovascular Disease

Regular exercise can lower the risk of cardiovascular disease. Even a moderate exercise routine (jogging 5 miles a week, for example) can lower total blood cholesterol levels. Exercise lowers cholesterol by stimulating enzymes that help move *low-density lipoproteins* (*LDLs*, or so-called "bad cholesterol") from the blood to the liver. In the liver, the cholesterol is converted to bile and excreted from the body. Exercise also increases the size of the lipoprotein particles that carry cholesterol, making it harder for small proteins to lodge in the vessel walls. A high cholesterol level is one of the major risk factors for atherosclerosis, which leads to cardiovascular disease and strokes. In addition, a healthy lifestyle that includes regular exercise, a balanced diet, weight control, and not smoking reduces stress, lowers blood pressure, and slows the formation of plaques.

Regular moderate exercise (30 minutes most days of the week) may cut the incidence of heart attack, or *myocardial infarction*, almost in half. However, only an estimated 8 percent of adults in the United States currently exercise at the recommended level. Exercise also speeds recovery after a heart attack. Regular light-to-moderate exercise (such as walking, jogging, or bicycling), coupled with a low-fat diet and a low-stress lifestyle, reduces symptoms of *coronary artery disease* (such as *angina*, an-JĪ-nuh; chest pain). Such exercise also improves a person's mood and overall quality of life. However, exercise does not remove underlying medical problems. For example, atherosclerotic plaques, described on p. 732, do not disappear and seldom grow smaller with exercise.

There is no evidence that *intense* athletic training lowers the incidence of cardiovascular disease. On the contrary, the strains placed on all body systems—including the cardiovascular system—during an ultramarathon, Ironman Triathlon, or other extreme athletic event can be severe. People with congenital aneurysms, cardiomyopathy (chronic disease of heart muscle), or cardiovascular disease risk fatal cardiovascular problems, such as an arrhythmia or heart attack, during severe exercise. Even healthy individuals can develop acute physiological disorders, such as kidney failure, after extreme exercise.

Table 21–3 Effects of Training on Cardiovascular Performance

Subject	Heart Weight (g)	Stroke Volume (mL)	Heart Rate (bpm)	Cardiac Output (L/min)	Blood Pressure (systolic/diastolic)
Nonathlete (rest)	300	60	83	5.0	120/80
Nonathlete (maximum)		104	192	19.9	187/75
Trained athlete (rest)	500	100	53	5.3	120/80
Trained athlete (maximum)		167	182	30.4	200/90*

* Diastolic pressures in athletes during maximal activity have not been accurately measured.

The Cardiovascular Response to Hemorrhaging and Shock

In Chapter 19, we considered the local cardiovascular reaction to a break in the wall of a blood vessel. p. 679 When hemostasis fails to prevent significant blood loss, the entire cardiovascular system makes adjustments (Figure 21–16). The immediate problem is to maintain adequate blood pressure and peripheral blood flow. The long-term problem is to restore normal blood volume.

Short-Term Elevation of Blood Pressure

Almost as soon as the pressure starts to decrease, several short-term responses appear:

- In the initial neural response, carotid and aortic reflexes increase cardiac output and cause peripheral vasoconstriction. With blood volume reduced, an increase in heart rate, typically up to 180–200 bpm, maintains cardiac output.
- The combination of stress and anxiety stimulates the sympathetic nervous system headquarters in the hypothalamus, which in turn triggers a further increase in vasomotor tone, constricting the arterioles and raising blood pressure. At the same time, venoconstriction mobilizes the venous blood reservoirs and quickly improves venous return.
- Short-term hormonal effects also occur. For instance, sympathetic activation causes the adrenal medullae to secrete E and NE. These hormones increase cardiac output and extend peripheral vasoconstriction.

This combination of short-term responses elevates blood pressure and improves peripheral blood flow. It often restores normal arterial pressures and peripheral circulation after blood losses of up to 20 percent of total blood volume. Such adjustments are more than enough to compensate for the blood loss when you donate blood. (Most blood banks collect 500 mL of

Figure 21–16 Cardiovascular Responses to Blood Loss. These mechanisms cope with blood losses equivalent to approximately 30 percent of total blood volume.

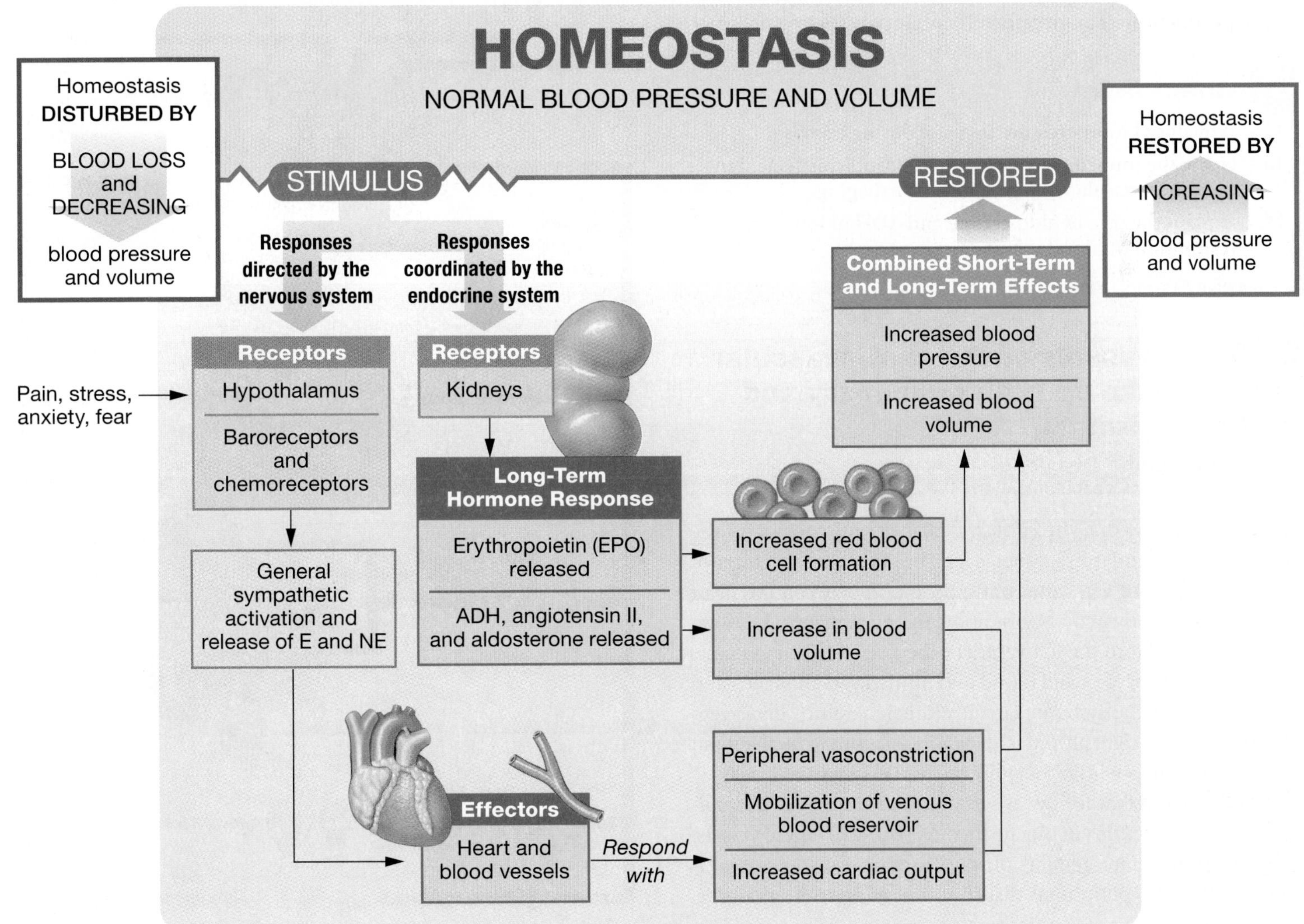
HOMEOSTASIS
NORMAL BLOOD PRESSURE AND VOLUME
Homeostasis
DISTURBED BY
BLOOD LOSS
and
DECREASING
blood pressure
and volume
STIMULUS
RESTORED
Homeostasis
RESTORED BY
INCREASING
blood pressure
and volume
Responses directed by the nervous system
Responses coordinated by the endocrine system
Combined Short-Term and Long-Term Effects
Increased blood pressure
Increased blood volume
Receptors
Hypothalamus
Baroreceptors and chemoreceptors
Pain, stress, anxiety, fear
Receptors
Kidneys
Long-Term Hormone Response
Erythropoietin (EPO) released
ADH, angiotensin II, and aldosterone released
Increased red blood cell formation
Increase in blood volume
General sympathetic activation and release of E and NE
Effectors
Heart and blood vessels
Respond with
Peripheral vasoconstriction
Mobilization of venous blood reservoir
Increased cardiac output

whole blood, roughly 10 percent of your total blood volume.) If compensatory mechanisms fail, the individual develops signs of *shock*.

Long-Term Restoration of Blood Volume

Short-term responses temporarily compensate for a reduction in blood volume. Long-term responses are geared to restoring normal blood volume. This process can take several days after a serious hemorrhage. The responses include the following:

- The decrease in capillary blood pressure triggers a recall of fluids from the interstitial spaces.
- Aldosterone and ADH promote fluid retention and reabsorption at the kidneys, preventing a further reduction in blood volume.
- Thirst increases, and the digestive tract absorbs additional water. This intake of fluid increases the blood volume and ultimately replaces the interstitial fluids "borrowed" at the capillaries.
- Erythropoietin targets the red bone marrow. It stimulates the maturation of red blood cells, which increases blood volume and improves oxygen delivery to peripheral tissues.
- ADH release by the posterior lobe of the pituitary gland and the production of angiotensin II enhance vasoconstriction.

Checkpoint

12. **Why does blood pressure increase during exercise?**
13. **Name the immediate and long-term problems related to the cardiovascular response to hemorrhaging.**
14. **Explain the role of aldosterone and ADH in long-term restoration of blood volume.**

See the blue Answers tab at the back of the book.

21

21-5 The vessels of the cardiovascular system make up both pulmonary and systemic circuits

Learning Outcome Describe the pulmonary and systemic circuits of the cardiovascular system.

You already know that the cardiovascular system consists of the *pulmonary circuit* and the *systemic circuit*. The pulmonary circuit consists of arteries and veins that transport blood between the heart and the lungs. This circuit begins at the right ventricle and ends at the left atrium. From the left ventricle, the arteries of the systemic circuit transport oxygenated blood and nutrients to all organs and tissues. Veins of the systemic circuit ultimately return deoxygenated blood to the right atrium. Figure 21–17 summarizes the main distribution routes within the pulmonary and systemic circuits.

In the pages that follow, we examine the vessels of the pulmonary and systemic circuits further. As you study these vessels, keep in mind some general observations about their organization. First, the peripheral distributions of arteries and veins on the body's left and right sides are generally identical, except near the heart, where the largest vessels connect to the atria or ventricles. Corresponding arteries and veins usually follow the same path. For example, the left and right subclavian *arteries* parallel the left and right subclavian *veins*. In addition, a single

Figure 21–17 A Schematic Overview of the Pattern of Circulation. RA stands for right atrium, LA for left atrium.

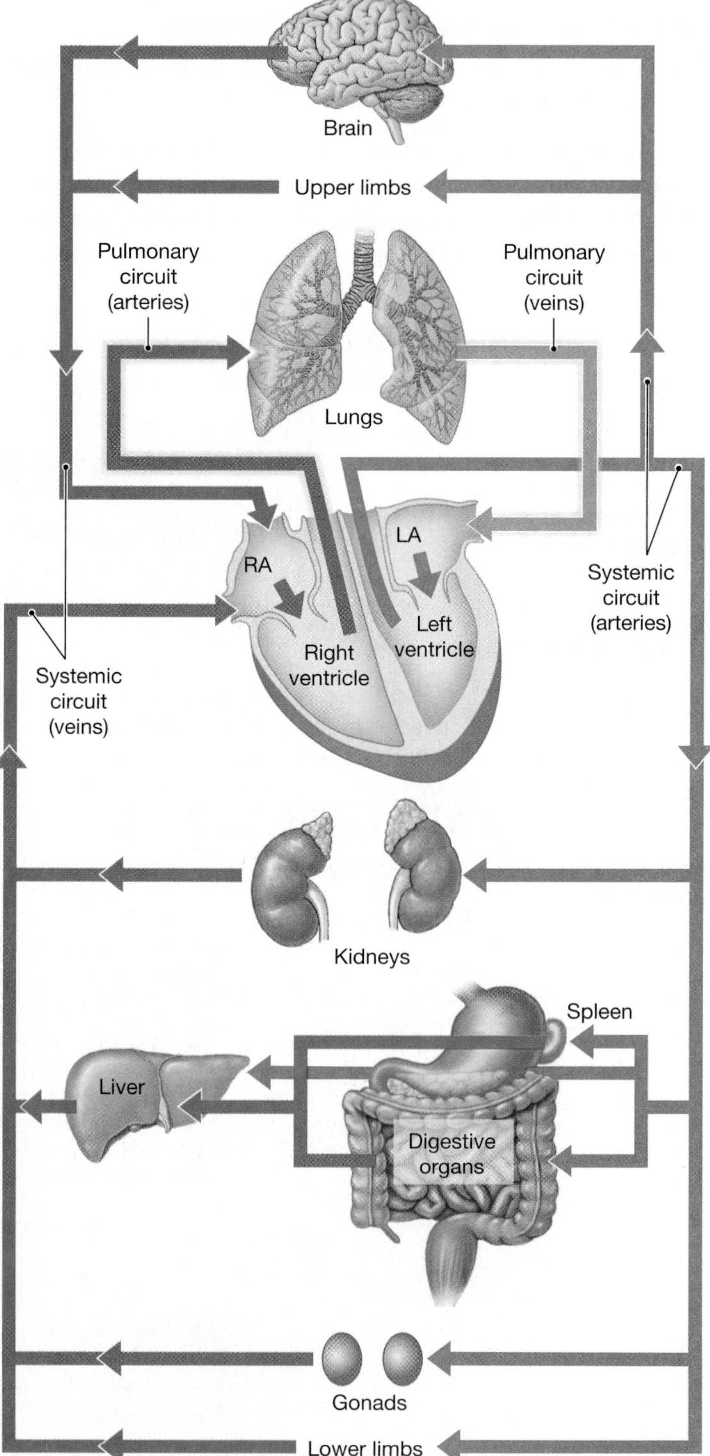
Brain
Upper limbs
Pulmonary circuit (arteries)
Pulmonary circuit (veins)
Lungs
RA
LA
Right ventricle
Left ventricle
Systemic circuit (arteries)
Systemic circuit (veins)
Kidneys
Spleen
Liver
Digestive organs
Gonads
Lower limbs

vessel may have several names as it crosses specific anatomical boundaries. These names make accurate anatomical descriptions possible when the vessel extends far into the periphery. For example, the *external iliac artery* becomes the *femoral artery* as it leaves the trunk and enters the lower limb. Finally, several arteries and veins usually service tissues and organs. Often, anastomoses between adjacent arteries or veins reduce the impact of a temporary or even permanent blockage, or *occlusion*, in a single blood vessel.

✓ Checkpoint

15. **Identify the two circuits of the cardiovascular system.**
16. **Identify the major patterns of blood vessel organization seen in the pulmonary and systemic circuits of the cardiovascular system.**

See the blue Answers tab at the back of the book.

21-6 In the pulmonary circuit, deoxygenated blood enters the lungs in arteries, and oxygenated blood leaves the lungs by veins

Learning Outcome **Identify the major arteries and veins of the pulmonary circuit.**

Blood entering the right atrium has just returned from the peripheral capillary beds, where it released oxygen and absorbed carbon dioxide. After traveling through the right atrium and ventricle, this deoxygenated blood enters the pulmonary trunk, the start of the pulmonary circuit (Figure 21–18). At the lungs, oxygen is replenished, and carbon dioxide is released. The oxygenated blood returns to the heart for distribution by the systemic circuit.

Compared with the systemic circuit, the pulmonary circuit is short: The base of the pulmonary trunk and the lungs are only about 15 cm (6 in.) apart.

Figure 21–18 **The Pulmonary Circuit.** The pulmonary circuit consists of pulmonary arteries, which deliver deoxygenated blood from the right ventricle to the lungs; pulmonary capillaries, where gas exchange occurs; and pulmonary veins, which deliver oxygenated blood to the left atrium. As the enlarged view shows, diffusion across the capillary walls at alveoli removes carbon dioxide and provides oxygen to the blood. ATLAS: Plates 42a; 44c; 47b

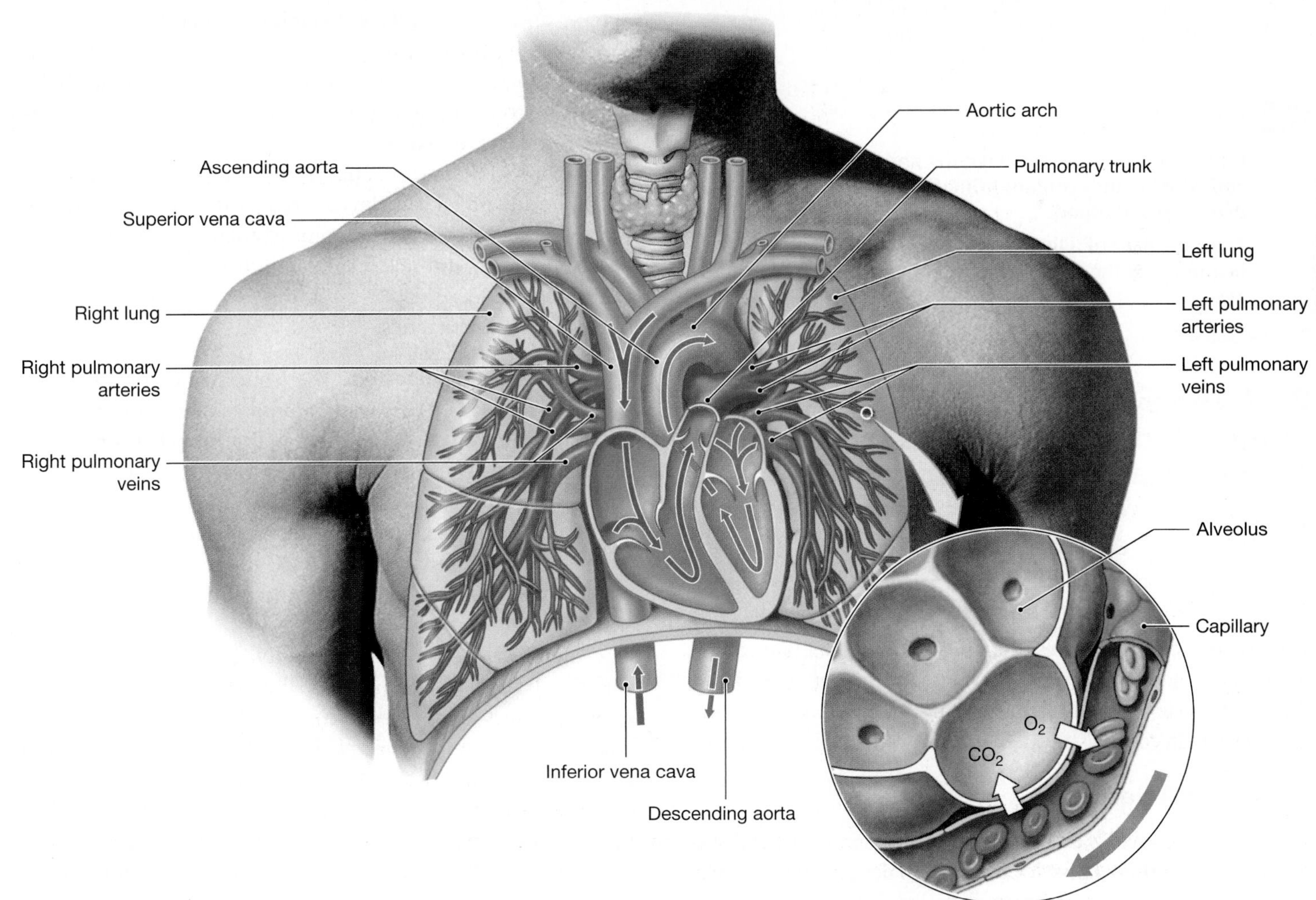
Aortic arch
Ascending aorta
Pulmonary trunk
Superior vena cava
Left lung
Right lung
Left pulmonary arteries
Right pulmonary arteries
Left pulmonary veins
Right pulmonary veins
Alveolus
Capillary
O_2
CO_2
Inferior vena cava
Descending aorta

The arteries of the pulmonary circuit differ from those of the systemic circuit in that they carry deoxygenated blood. (For this reason, these color-coded diagrams show the pulmonary arteries in blue, the same color as systemic veins.) As the pulmonary trunk curves over the superior border of the heart, it gives rise to the **left** and **right pulmonary arteries**. These large arteries enter the lungs before branching repeatedly, giving rise to smaller and smaller arteries. The smallest branches, the *pulmonary arterioles*, provide blood to capillary networks that surround *alveoli*. The walls of these small air pockets are thin enough for gas to be exchanged between the capillary blood and inspired air. As oxygenated blood leaves the alveolar capillary network, it enters venules that in turn unite to form larger vessels carrying blood toward the **pulmonary veins**. These four veins, two from each lung, empty into the left atrium, to complete the pulmonary circuit.

Tips & Tools

Arteries and veins are defined by the direction of blood flow relative to the heart, not by the oxygen content of the blood they carry. So if you remember that ***a***rteries carry blood ***a***way from the heart, and veins carry blood toward the heart, you can remember that the pulmonary ***a***rteries carry oxygen-poor blood ***a***way from the heart to the lungs, and the pulmonary veins deliver oxygen-rich blood to the heart.

✓ Checkpoint

21

17. **Name the blood vessels that enter and exit the lungs, and whether they contain primarily oxygenated or deoxygenated blood.**
18. **Trace the path of a drop of blood through the lungs, beginning at the right ventricle and ending at the left atrium.**

See the blue Answers tab at the back of the book.

21-7 The systemic circuit carries oxygenated blood from the left ventricle to tissues and organs other than the lungs, and returns deoxygenated blood to the right atrium

Learning Outcome **Identify the major arteries and veins of the systemic circuit.**

The systemic circuit supplies the capillary beds in all parts of the body not serviced by the pulmonary circuit. This circuit begins at the left ventricle and ends at the right atrium. At any moment the systemic circuit contains about 84 percent of total blood volume.

Systemic Arteries

Figure 21–19 provides an overview of the systemic arterial system and shows the relative locations of major systemic arteries. Figures 21–20 to 21–25 show the detailed distribution of these vessels and their branches. By convention, several large arteries are called *trunks*. Examples are the *pulmonary*, *brachiocephalic*, *thyrocervical*, and *celiac trunks*. Most of the major arteries are paired, with one artery of each pair on each side of the body. For this reason, the terms *right* and *left* appear in figures only when the arteries on both sides are labeled.

The Ascending Aorta

The **ascending aorta** (Figure 21–20) begins at the aortic valve of the left ventricle. The left and right coronary arteries originate in the aortic sinus at the base of the ascending aorta, just superior to the aortic valve. The distribution of coronary vessels was described in Chapter 20 and illustrated in Figure 20–8, p. 699.

The Aortic Arch

The **aortic arch** curves like the handle of a cane across the superior surface of the heart, connecting the ascending aorta with the *descending aorta* (see Figure 21–19). Three elastic arteries originate along the aortic arch and deliver blood to the head, neck, shoulders, and upper limbs: (1) the **brachiocephalic** (brā-kē-ō-se-FAL-ik) **trunk**, (2) the **left common carotid artery**, and (3) the **left subclavian artery** (Figure 21–21). The brachiocephalic trunk, also called the *innominate artery* (i-NOM-i-nat; unnamed), ascends for a short distance before branching to form the **right subclavian artery** and the **right common carotid artery**.

We have only one brachiocephalic trunk, with the left common carotid and left subclavian arteries arising separately from the aortic arch. However, in terms of their peripheral distribution, the vessels on the left side are mirror images of those on the right side. Figures 21–20 and 21–22 illustrate the major branches of these arteries.

The Subclavian Arteries. The subclavian arteries supply blood to the arms, chest wall, shoulders, back, and CNS (see Figures 21–19 and 21–20). Three major branches arise before a subclavian artery leaves the thoracic cavity: (1) the **internal thoracic artery**, supplying the pericardium and anterior wall of the chest; (2) the **vertebral artery**, which provides blood to the brain and spinal cord; and (3) the **thyrocervical trunk**, which provides blood to muscles and other tissues of the neck, shoulder, and upper back.

After leaving the thoracic cavity and passing across the superior border of the first rib, the subclavian is called the **axillary artery**. This artery crosses the axilla to enter the arm, where it gives rise to *humeral circumflex arteries*, which supply structures near the head of the humerus. Distally, it becomes the **brachial artery**, which supplies blood to the rest of the upper limb. The brachial artery gives rise to the *deep brachial artery*, which supplies deep structures on the posterior aspect of the arm, and the *ulnar collateral arteries*, which supply the area around the elbow.

Figure 21–19 An Overview of the Major Systemic Arteries.

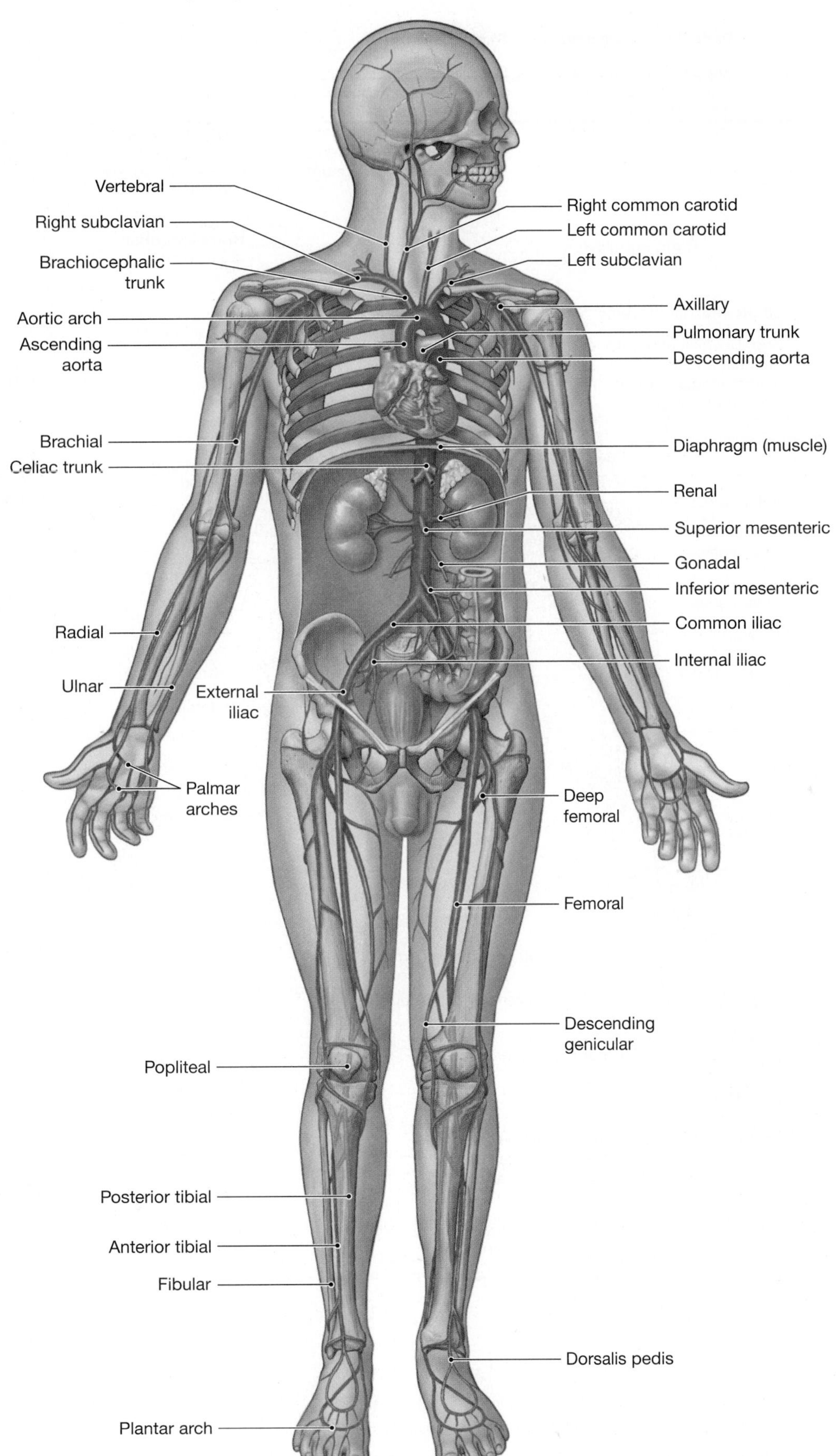
Vertebral
Right subclavian
Brachiocephalic trunk
Aortic arch
Ascending aorta
Brachial
Celiac trunk
Radial
Ulnar
External iliac
Palmar arches
Popliteal
Posterior tibial
Anterior tibial
Fibular
Plantar arch
Right common carotid
Left common carotid
Left subclavian
Axillary
Pulmonary trunk
Descending aorta
Diaphragm (muscle)
Renal
Superior mesenteric
Gonadal
Inferior mesenteric
Common iliac
Internal iliac
Deep femoral
Femoral
Descending genicular
Dorsalis pedis

Figure 21–20 Arteries of the Chest and Upper Limb. ATLAS: Plates 27a–c; 29c; 30; 45a

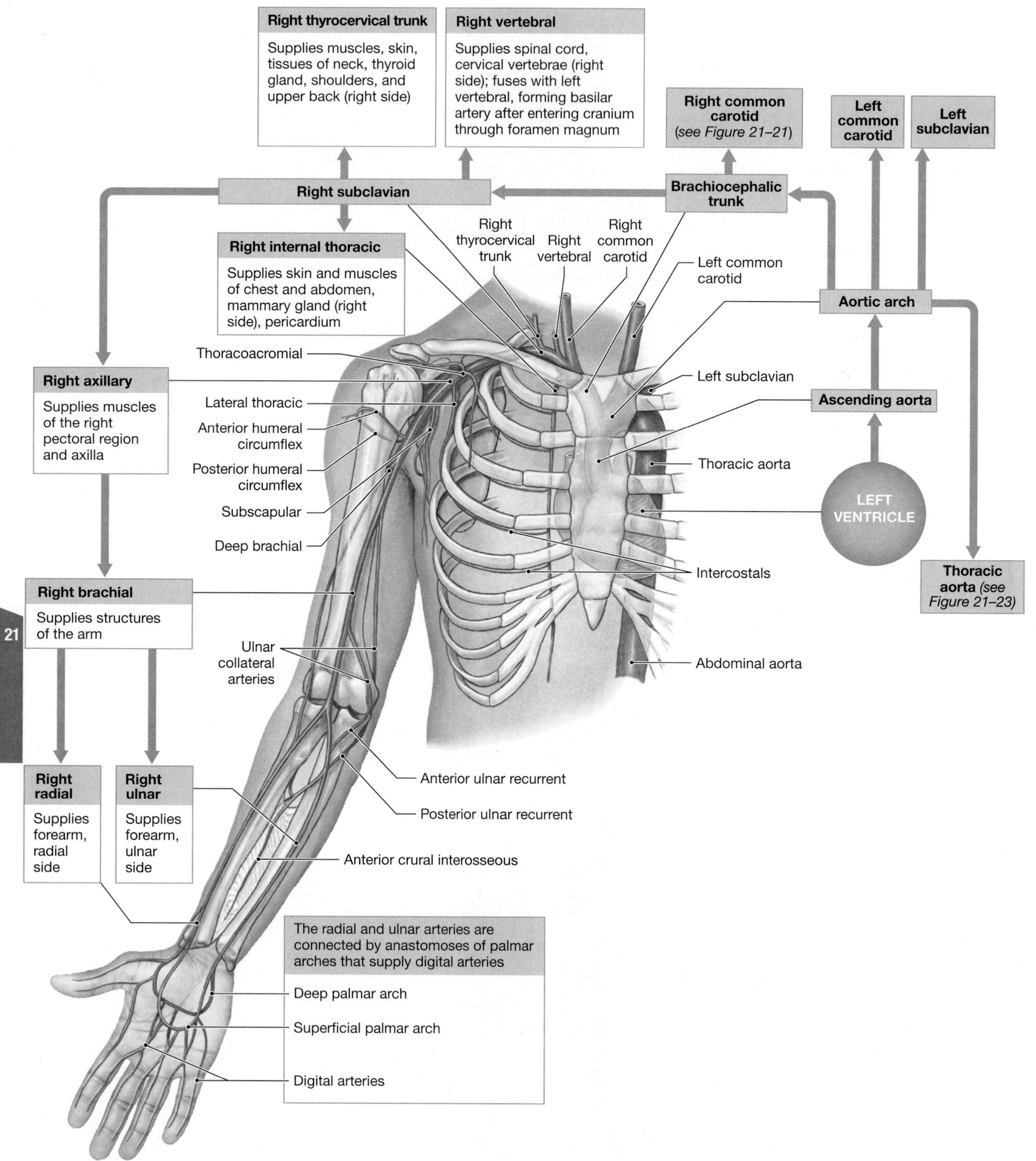
Right thyrocervical trunk
Supplies muscles, skin, tissues of neck, thyroid gland, shoulders, and upper back (right side)
Right vertebral
Supplies spinal cord, cervical vertebrae (right side); fuses with left vertebral, forming basilar artery after entering cranium through foramen magnum
Right common carotid (see Figure 21–21)
Left common carotid
Left subclavian
Right subclavian
Brachiocephalic trunk
Right internal thoracic
Supplies skin and muscles of chest and abdomen, mammary gland (right side), pericardium
Right thyrocervical trunk
Right vertebral
Right common carotid
Left common carotid
Aortic arch
Thoracoacromial
Right axillary
Supplies muscles of the right pectoral region and axilla
Lateral thoracic
Anterior humeral circumflex
Posterior humeral circumflex
Subscapular
Deep brachial
Left subclavian
Ascending aorta
Thoracic aorta
LEFT VENTRICLE
Intercostals
Thoracic aorta (see Figure 21–23)
Right brachial
Supplies structures of the arm
Ulnar collateral arteries
Abdominal aorta
Right radial
Supplies forearm, radial side
Right ulnar
Supplies forearm, ulnar side
Anterior ulnar recurrent
Posterior ulnar recurrent
Anterior crural interosseous
The radial and ulnar arteries are connected by anastomoses of palmar arches that supply digital arteries
Deep palmar arch
Superficial palmar arch
Digital arteries

Figure 21–21 Arteries of the Neck and Head. Shown as seen from the right side. ATLAS: Plates 3c,d; 15b; 18a–c; 45a

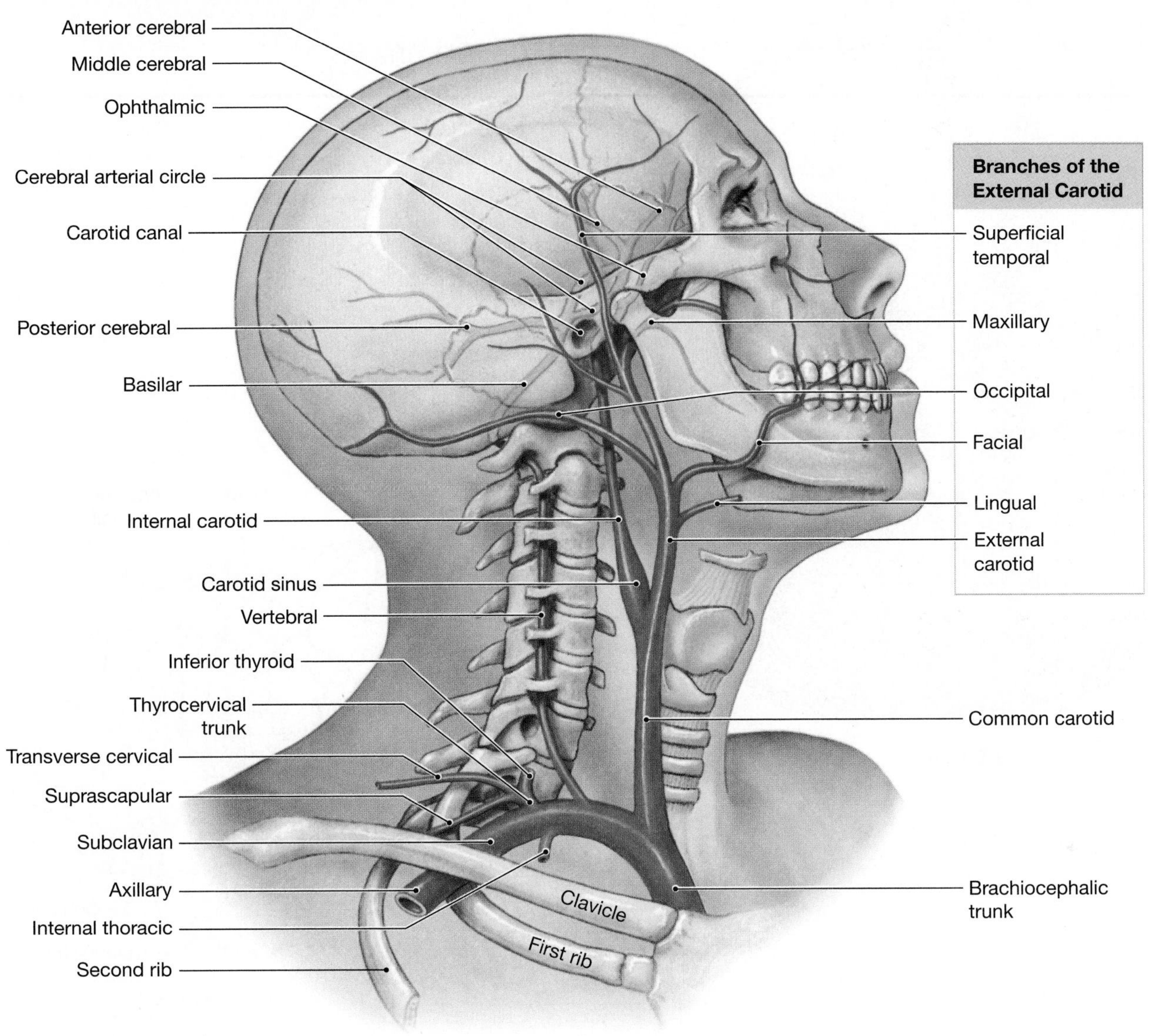
Anterior cerebral
Middle cerebral
Ophthalmic
Cerebral arterial circle
Carotid canal
Posterior cerebral
Basilar
Internal carotid
Carotid sinus
Vertebral
Inferior thyroid
Thyrocervical trunk
Transverse cervical
Suprascapular
Subclavian
Axillary
Internal thoracic
Second rib
Clavicle
First rib
Branches of the External Carotid
Superficial temporal
Maxillary
Occipital
Facial
Lingual
External carotid
Common carotid
Brachiocephalic trunk

? The superficial temporal, maxillary, occipital, facial, and lingual arteries are all branches of what carotid artery?

As it approaches the coronoid fossa of the humerus, the brachial artery divides into the **radial artery**, which follows the radius, and the **ulnar artery**, which follows the ulna to the wrist. These arteries supply blood to the forearm and, through the *ulnar recurrent arteries*, the region around the elbow. At the wrist, the radial and ulnar arteries fuse to form the **superficial** and **deep palmar arches**, which supply blood to the hand and to the **digital arteries** of the thumb and fingers.

The Carotid Artery and the Blood Supply to the Brain. The common carotid arteries ascend deep in the tissues of the neck. You can usually locate the carotid artery by pressing gently along either side of the windpipe (trachea) until you feel a strong pulse.

Each common carotid artery divides into an **external carotid artery** and an **internal carotid artery** (Figure 21–21). The *carotid sinus*, located at the base of the internal carotid artery, may extend along a portion of the common carotid. The external carotid arteries supply blood to the structures of the neck, esophagus, pharynx, larynx, lower jaw, and face. The internal carotid arteries enter the skull through the carotid canals of the temporal bones, delivering blood to the brain (see Figure 7–3 and Spotlight Figure 7–4, pp. 212–215).

21

Figure 21–22 Arteries of the Brain. ATLAS: Plates 15a–c

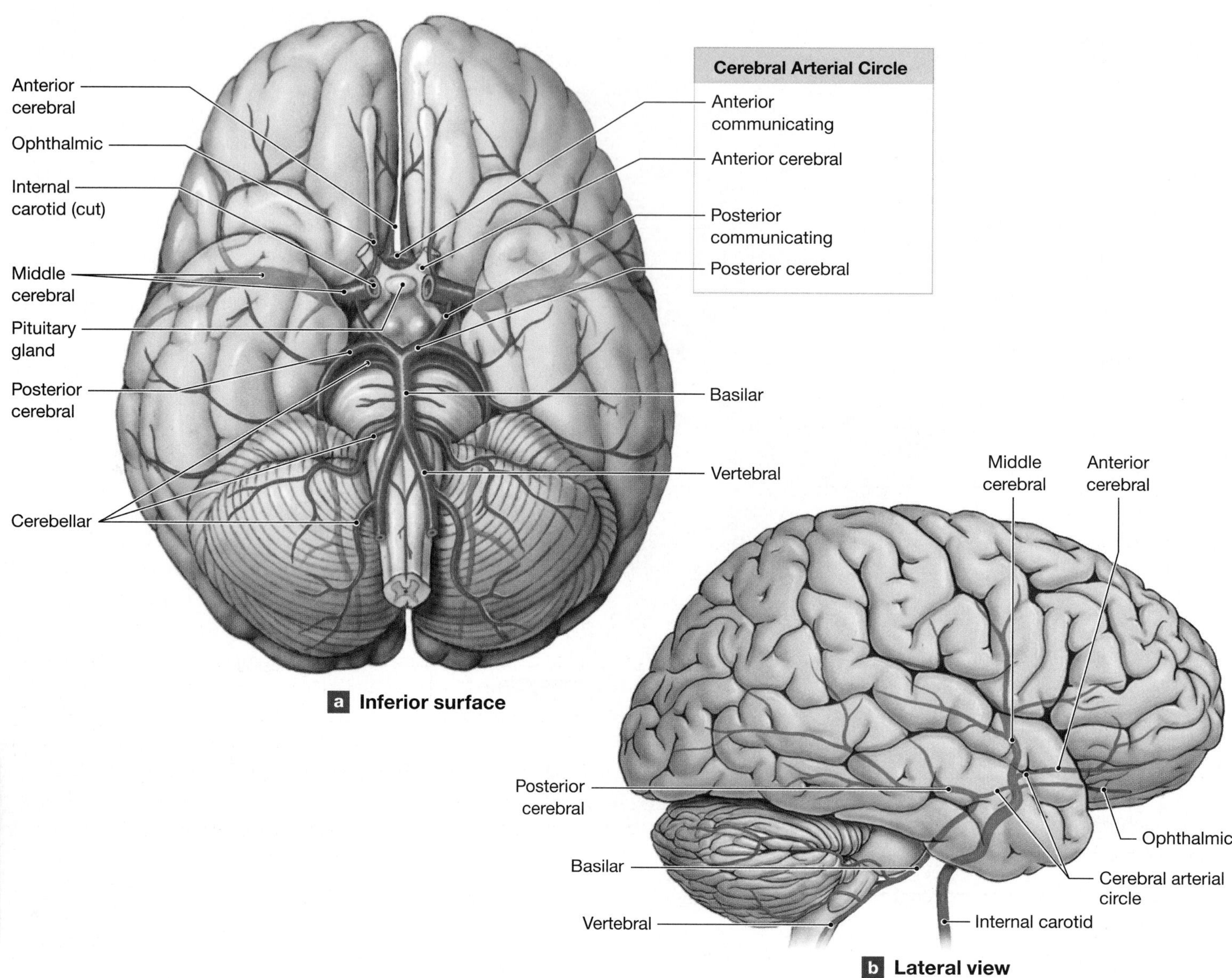
Anterior cerebral
Ophthalmic
Internal carotid (cut)
Middle cerebral
Pituitary gland
Posterior cerebral
Cerebellar
Cerebral Arterial Circle
Anterior communicating
Anterior cerebral
Posterior communicating
Posterior cerebral
Basilar
Vertebral
a Inferior surface
Middle cerebral
Anterior cerebral
Posterior cerebral
Basilar
Vertebral
Ophthalmic
Cerebral arterial circle
Internal carotid
b Lateral view

21

Tips & Tools

Remember that the *external* carotid artery supplies the face (which is external), and that the *internal* carotid supplies the brain (which is internal).

The internal carotid arteries ascend to the level of the optic nerves, where each artery divides into three branches: (1) an **ophthalmic artery**, which supplies the eyes; (2) an **anterior cerebral artery**, which supplies the frontal and parietal lobes of the brain; and (3) a **middle cerebral artery**, which supplies the midbrain and lateral surfaces of the cerebral hemispheres (see Figures 21–21 and 21–22).

The brain is extremely sensitive to changes in blood supply. An interruption of blood flow for several seconds produces unconsciousness. After 4 minutes some permanent neural damage can occur. Such crises are rare, because blood reaches the brain through the vertebral arteries as well as by way of the internal carotid arteries. The left and right vertebral arteries arise from the subclavian arteries and ascend within the transverse foramina of the cervical vertebrae (look back at Figure 7–19b,c, p. 231). The vertebral arteries enter the cranium at the foramen magnum, where they fuse along the ventral surface of the medulla oblongata to form the **basilar artery**. The vertebral arteries and the basilar artery supply blood to the spinal cord, medulla oblongata, pons, and cerebellum. They then divide into the **posterior cerebral arteries**, which in turn branch off into the **posterior communicating arteries** (Figure 21–22).

The internal carotid arteries normally supply the arteries of the anterior half of the cerebrum, and the rest of the brain receives blood from the vertebral arteries. But this pattern of blood flow can easily change, because the internal carotid arteries and the basilar artery are interconnected. They form a ring-shaped anastomosis called the **cerebral arterial circle**, or *circle of Willis*, which encircles the optic chiasm and the infundibulum of the pituitary gland (see Figure 21–22). With this arrangement, the brain can receive blood from either the carotid or the vertebral arteries, reducing the likelihood of a serious interruption of circulation.

Strokes, or *cerebrovascular accidents (CVAs)*, are interruptions of the vascular supply to a portion of the brain. The *middle cerebral artery*, a major branch of the cerebral arterial circle, is the most common site of a stroke. Superficial branches deliver blood to the temporal lobe and large portions of the frontal and parietal lobes. Deep branches supply the basal nuclei and portions of the thalamus. If a stroke blocks the middle cerebral artery on the left side of the brain, aphasia and a sensory and motor paralysis of the right side of the body result. In a stroke affecting the middle cerebral artery on the right side, the individual experiences a loss of sensation and motor control over the left side of the body and has difficulty drawing or interpreting spatial relationships. Strokes affecting vessels that supply the brainstem also produce distinctive symptoms. Strokes affecting the lower brainstem are commonly fatal.

The Descending Aorta

The **descending aorta** is continuous with the aortic arch. The diaphragm divides the descending aorta into a superior **thoracic aorta** and an inferior **abdominal aorta** (look ahead to Figures 21–23 and 21–24).

The Thoracic Aorta. The thoracic aorta begins at the level of vertebra T_5 and penetrates the diaphragm at the level of vertebra T_{12}. It travels within the mediastinum, on the posterior thoracic wall, slightly to the left of the vertebral column. This vessel supplies blood to branches that service the tissues and organs of the mediastinum, the muscles of the chest and the diaphragm, and the thoracic spinal cord.

We group the branches of the thoracic aorta anatomically as either visceral or somatic (Figure 21–23):

- *Visceral branches* supply the organs of the chest. The **bronchial arteries** supply the tissues of the lungs not involved in gas exchange. The **pericardial arteries** supply the pericardium. The **esophageal arteries** supply the esophagus, and the **mediastinal arteries** supply the tissues of the mediastinum.
- *Somatic branches* supply the chest wall. The **intercostal arteries** supply the chest muscles and the vertebral column area. The **superior phrenic** (FREN-ik) **arteries** deliver blood to the superior surface of the diaphragm, which separates the thoracic and abdominopelvic cavities.

+ Clinical Note Aortic Aneurysm

Over a period of many years, atherosclerosis and hypertension may take their toll on the aorta. The wall of the aorta becomes damaged and stretches out to form a pouch, or aneurysm. This structural change can weaken the aorta so much that it tears. The ensuing pain can come on without warning. Picture the path of the aorta as it plunges deep through the thorax and abdomen, and you will appreciate how daunting it is to reach a ruptured aortic aneurysm in order to stop the loss of blood and save the life of a patient.

The Abdominal Aorta. The abdominal aorta is a continuation of the thoracic aorta (see Figure 21–23). It begins immediately inferior to the diaphragm and descends slightly to the left of the vertebral column but posterior to the peritoneal cavity. A cushion of adipose tissue commonly surrounds the abdominal aorta. At the level of vertebra L_4, the abdominal aorta splits into two major arteries—the *left* and *right common iliac arteries*—that supply deep pelvic structures and the lower limbs. The region where the abdominal aorta splits is called the *terminal segment of the aorta*.

The abdominal aorta delivers blood to all the abdominopelvic organs and structures. The major branches to visceral organs are unpaired. They arise on the anterior surface of the abdominal aorta and extend into the mesenteries. By contrast, branches to the body wall, the kidneys, the urinary bladder, and other structures outside the peritoneal cavity are paired. They originate along the lateral surfaces of the abdominal aorta. Figure 21–23 shows the major arteries of the trunk after most thoracic and abdominal organs have been removed. Figure 21–24 shows the distribution of those arteries to abdominopelvic organs.

Tips & Tools

The aorta resembles a walking cane: the ascending aorta, aortic arch, and the start of the descending aorta form the cane's handle, while the thoracic and abdominal segments of the descending aorta form the cane's shaft.

The abdominal aorta gives rise to three unpaired arteries (see Figures 21–23 and 21–24).

1. The **celiac** (SĒ-lē-ak) **trunk** delivers blood to the liver, stomach, and spleen. The celiac trunk divides into three branches: (a) The **left gastric artery** supplies the stomach and the inferior portion of the esophagus. (b) The **splenic artery** supplies the spleen and arteries to the stomach (*left gastro-epiploic artery*) and pancreas (*pancreatic arteries*). (c) The **common hepatic artery** supplies arteries to the liver (*hepatic artery proper*), stomach (*right gastric artery*), gallbladder

Figure 21–23 Major Arteries of the Trunk. ATLAS: Plates 47d; 53c,e; 62a,b

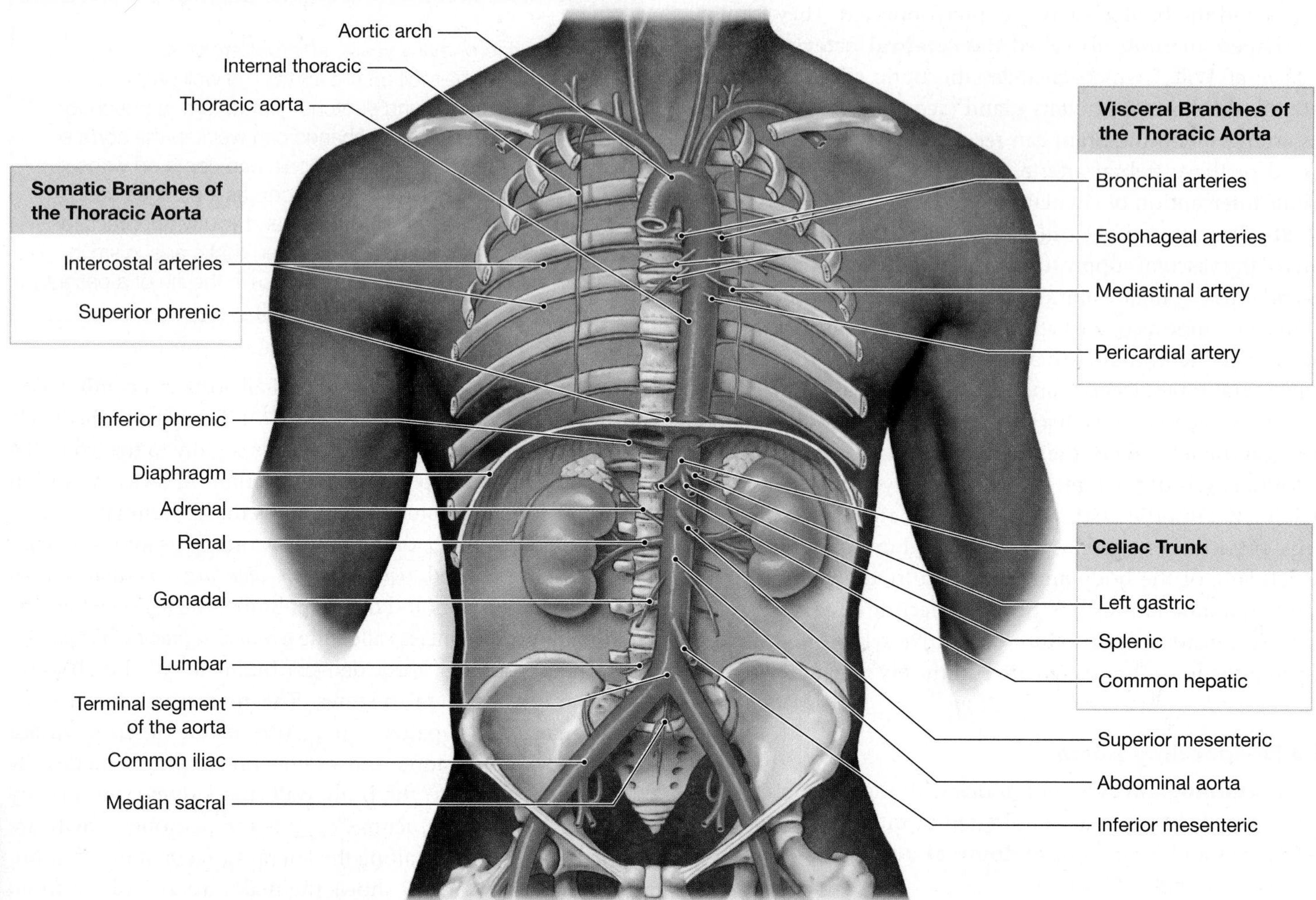
Aortic arch
Internal thoracic
Thoracic aorta
Somatic Branches of the Thoracic Aorta
Intercostal arteries
Superior phrenic
Visceral Branches of the Thoracic Aorta
Bronchial arteries
Esophageal arteries
Mediastinal artery
Pericardial artery
Inferior phrenic
Diaphragm
Adrenal
Renal
Gonadal
Lumbar
Terminal segment of the aorta
Common iliac
Median sacral
Celiac Trunk
Left gastric
Splenic
Common hepatic
Superior mesenteric
Abdominal aorta
Inferior mesenteric

a A diagrammatic view, with most of the thoracic and abdominal organs removed

21

(*cystic artery*), and duodenal area (*gastroduodenal, right gastroepiploic,* and *superior pancreaticoduodenal arteries*).

2. The **superior mesenteric** (mez-en-TER-ik) **artery** arises about 2.5 cm (1 in.) inferior to the celiac trunk. It supplies arteries to the pancreas and duodenum (*inferior pancreaticoduodenal artery*), small intestine (*intestinal arteries*), and most of the large intestine (*right* and *middle colic* and the *ileocolic arteries*).
3. The **inferior mesenteric artery** arises about 5 cm (2 in.) superior to the terminal aorta. It delivers blood to the terminal portions of the colon (*left colic* and *sigmoid arteries*) and the rectum (*rectal arteries*).

The abdominal aorta also gives rise to five paired arteries (Figure 21–23):

1. The **inferior phrenic arteries** supply the inferior surface of the diaphragm and the inferior portion of the esophagus.
2. The **adrenal arteries** originate on either side of the aorta near the base of the superior mesenteric artery. Each adrenal artery supplies one adrenal gland, which caps the superior part of a kidney.
3. The short (about 7.5 cm) **renal arteries** arise along the posterolateral surface of the abdominal aorta, about 2.5 cm (1 in.) inferior to the superior mesenteric artery. They travel posterior to the peritoneal lining to reach the adrenal glands and kidneys.
4. The **gonadal** (gō-NAD-al) **arteries** originate between the superior and inferior mesenteric arteries. In males, they are called *testicular arteries* and are long, thin arteries that supply blood to the testes and scrotum. In females, they are termed *ovarian arteries* and supply blood to the ovaries, uterine tubes, and uterus. The distribution of gonadal vessels (both arteries and veins) differs by sex.
5. Small **lumbar arteries** arise on the posterior surface of the aorta. They supply the vertebrae, spinal cord, and abdominal wall.

Figure 21–23 Major Arteries of the Trunk. (*Continued*)

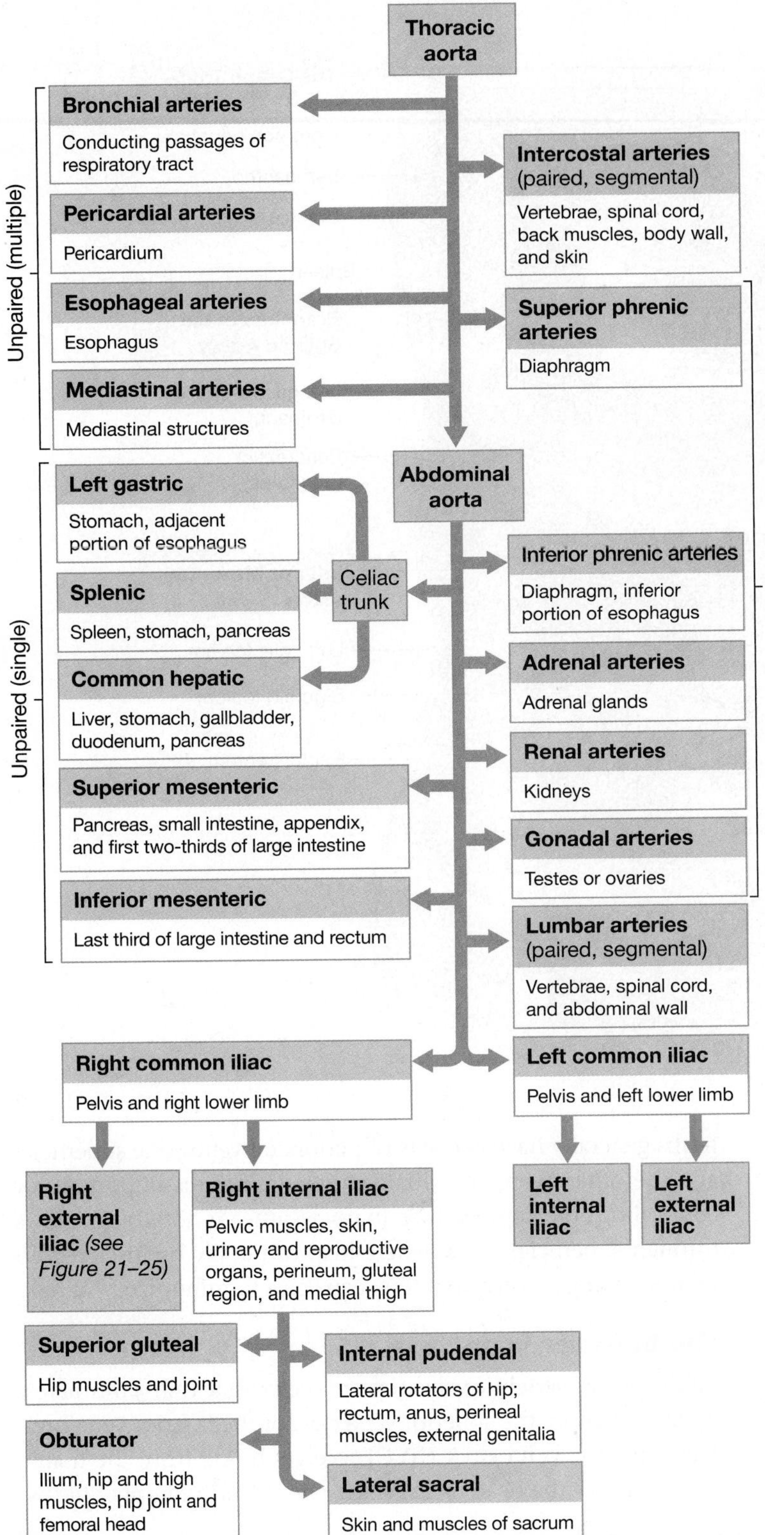
Thoracic aorta
Unpaired (multiple)
Bronchial arteries
Conducting passages of respiratory tract
Pericardial arteries
Pericardium
Esophageal arteries
Esophagus
Mediastinal arteries
Mediastinal structures
Intercostal arteries (paired, segmental)
Vertebrae, spinal cord, back muscles, body wall, and skin
Superior phrenic arteries
Diaphragm
Abdominal aorta
Unpaired (single)
Left gastric
Stomach, adjacent portion of esophagus
Celiac trunk
Splenic
Spleen, stomach, pancreas
Common hepatic
Liver, stomach, gallbladder, duodenum, pancreas
Superior mesenteric
Pancreas, small intestine, appendix, and first two-thirds of large intestine
Inferior mesenteric
Last third of large intestine and rectum
Paired
Inferior phrenic arteries
Diaphragm, inferior portion of esophagus
Adrenal arteries
Adrenal glands
Renal arteries
Kidneys
Gonadal arteries
Testes or ovaries
Lumbar arteries (paired, segmental)
Vertebrae, spinal cord, and abdominal wall
Right common iliac
Pelvis and right lower limb
Left common iliac
Pelvis and left lower limb
Right external iliac (see Figure 21–25)
Right internal iliac
Pelvic muscles, skin, urinary and reproductive organs, perineum, gluteal region, and medial thigh
Left internal iliac
Left external iliac
Superior gluteal
Hip muscles and joint
Internal pudendal
Lateral rotators of hip; rectum, anus, perineal muscles, external genitalia
Obturator
Ilium, hip and thigh muscles, hip joint and femoral head
Lateral sacral
Skin and muscles of sacrum

b A flowchart showing major arteries of the trunk

Arteries of the Pelvis and Lower Limbs

Near the level of vertebra L_4, the terminal segment of the abdominal aorta divides to form a pair of elastic arteries—the **right** and **left common iliac** (IL-ē-ak) **arteries**—plus the small **median sacral artery** (Figure 21–23a). The common iliac arteries carry blood to the pelvis and lower limbs. They descend posterior to the cecum and sigmoid colon along the inner surface of the ilium. At the level of the lumbosacral joint, each common iliac divides to form an **internal iliac artery** and an **external iliac artery** (Figure 21–23b). The internal iliac arteries enter the pelvic cavity to supply the urinary bladder, the internal and external walls of the pelvis, the external genitalia, the medial side of the thigh, and, in females, the uterus and vagina. The major branches of the internal iliac artery are the *gluteal, internal pudendal, obturator,* and *lateral sacral arteries.* The external iliac arteries supply blood to the lower limbs. They are much larger in diameter than the internal iliac arteries.

Arteries of the Thigh and Leg. Each external iliac artery crosses the surface of an iliopsoas muscle and penetrates the abdominal wall midway between the anterior superior iliac spine and the pubic symphysis on that side. It emerges on the anterior, medial surface of the thigh as the **femoral artery** (Figure 21–25a,b). Roughly 5 cm (2 in.) distal to the emergence of the femoral artery, the **deep femoral artery** branches off its lateral surface. The deep femoral artery supplies blood to the ventral and lateral regions of the skin and deep muscles of the thigh. It gives rise to the *femoral circumflex arteries.*

21

The femoral artery continues inferiorly and posterior to the femur. As it approaches the knee, it gives rise to the *descending genicular artery,* which supplies the area around the knee. At the popliteal fossa, posterior to the knee joint, the femoral artery becomes the **popliteal** (pop-LIT-ē-al) **artery**, which then branches to form the **posterior** and **anterior tibial arteries**. The posterior tibial artery gives rise to the **fibular artery**, or *peroneal* (*perone,* fibula) *artery,* and then continues inferiorly along the posterior surface of the tibia. The anterior tibial artery passes between the tibia and fibula. It emerges on the anterior surface of the tibia. As it descends toward the foot, the anterior tibial artery provides blood to the skin and muscles of the anterior portion of the leg.

Arteries of the Foot. At the ankle, the anterior tibial artery becomes the **dorsalis pedis artery**. It then branches repeatedly, supplying the ankle and dorsal portion of the foot (see Figure 21–25a,b). Figure 21–25c charts the flow of blood from the external iliac artery to the lower limbs.

At the ankle, the posterior tibial artery divides to form the **medial** and **lateral plantar arteries**. They supply blood to the plantar surface of the foot. These arteries are connected to the dorsalis pedis artery through a pair of

Figure 21–24 **Arteries Supplying the Abdominopelvic Organs.** *(See also Figure 24–24, p. 922.)* ATLAS: Plates 53a–e; 54c; 55

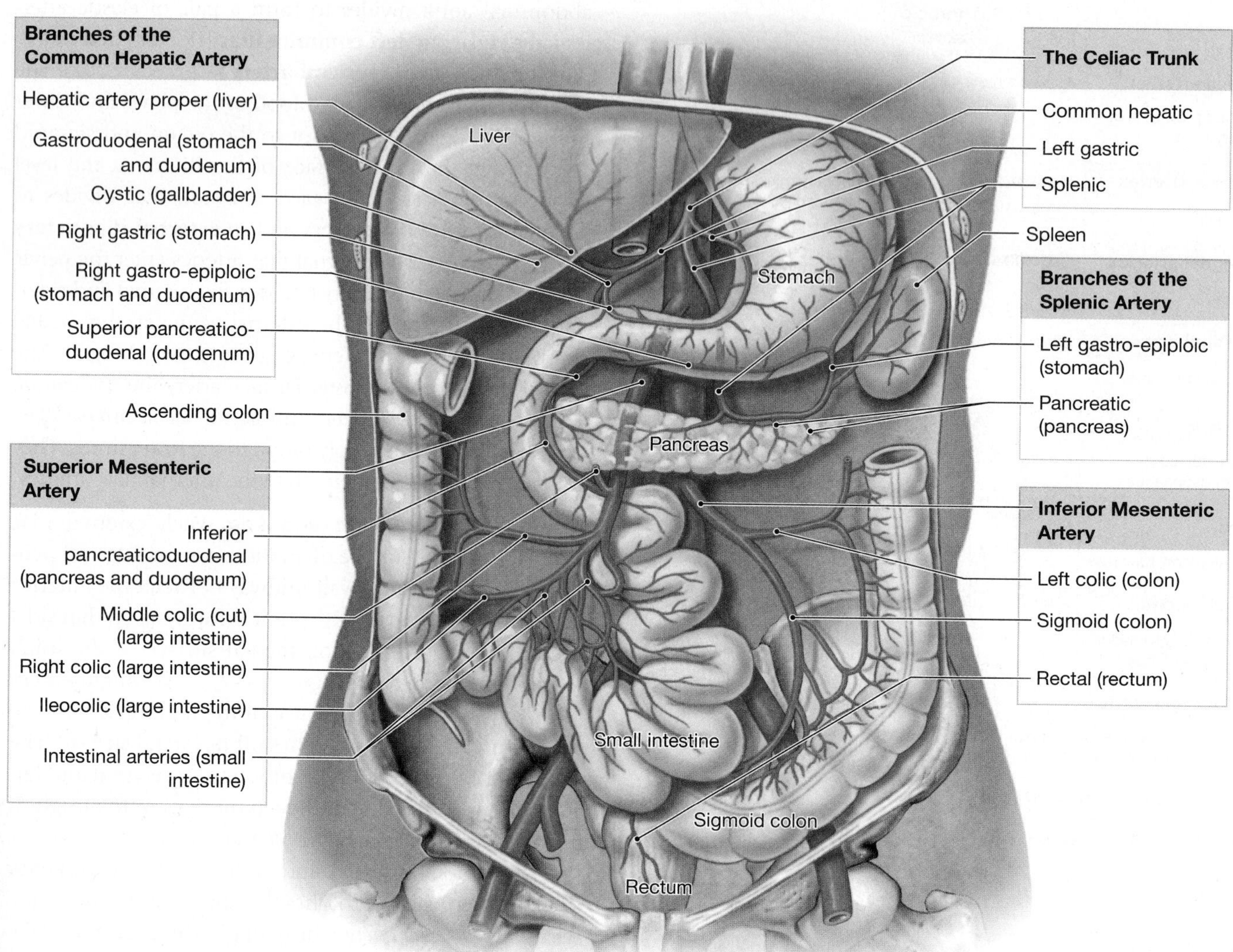
Branches of the Common Hepatic Artery
Hepatic artery proper (liver)
Gastroduodenal (stomach and duodenum)
Cystic (gallbladder)
Right gastric (stomach)
Right gastro-epiploic (stomach and duodenum)
Superior pancreatico-duodenal (duodenum)
Ascending colon
Superior Mesenteric Artery
Inferior pancreaticoduodenal (pancreas and duodenum)
Middle colic (cut) (large intestine)
Right colic (large intestine)
Ileocolic (large intestine)
Intestinal arteries (small intestine)
Liver
Stomach
Pancreas
Small intestine
Sigmoid colon
Rectum
The Celiac Trunk
Common hepatic
Left gastric
Splenic
Spleen
Branches of the Splenic Artery
Left gastro-epiploic (stomach)
Pancreatic (pancreas)
Inferior Mesenteric Artery
Left colic (colon)
Sigmoid (colon)
Rectal (rectum)

anastomoses. The arrangement produces a **dorsal arch** (*arcuate arch*) and a **plantar arch**. Small arteries branching off these arches supply the distal portions of the foot and the toes.

Systemic Veins

Veins collect blood from the tissues and organs of the body by means of an elaborate venous network that drains into the right atrium of the heart by the superior and inferior venae cavae (Figure 21–26). The branching pattern of peripheral veins is much more variable than is the branching pattern of arteries. We base the discussion that follows on the most common arrangement of veins. Complementary arteries and veins commonly run side by side. In many cases they have comparable names.

One significant difference between the arterial and venous systems involves the distribution of major veins in the neck and limbs. Arteries in these areas are deep beneath the skin, protected by bones and surrounding soft tissues. In contrast, the neck and limbs generally have two sets of peripheral veins, one superficial and the other deep. This dual venous drainage is important for controlling body temperature. In hot weather, venous blood flows through superficial veins, where heat can easily be lost. In cold weather, blood is routed to the deep veins to minimize heat loss.

The Superior Vena Cava

All the body's systemic veins (except the cardiac veins) ultimately drain into either the superior vena cava or the inferior vena cava. The **superior vena cava (SVC)** receives blood from the tissues and organs of the head, neck, chest, shoulders, and upper limbs.

Venous Return from the Cranium. Numerous veins drain the cerebral hemispheres. The *superficial cerebral veins* and small veins of the brainstem empty into a network of dural sinuses (Figure 21–27a,b). These sinuses include the *superior* and *inferior sagittal sinuses*, the *petrosal sinuses*, the *occipital sinus*, the *left* and *right transverse sinuses*, and the *straight sinus* (Figure 21–27c).

Figure 21–25 **Arteries of the Lower Limb.** ATLAS: Plates 68c; 70b; 78b–g

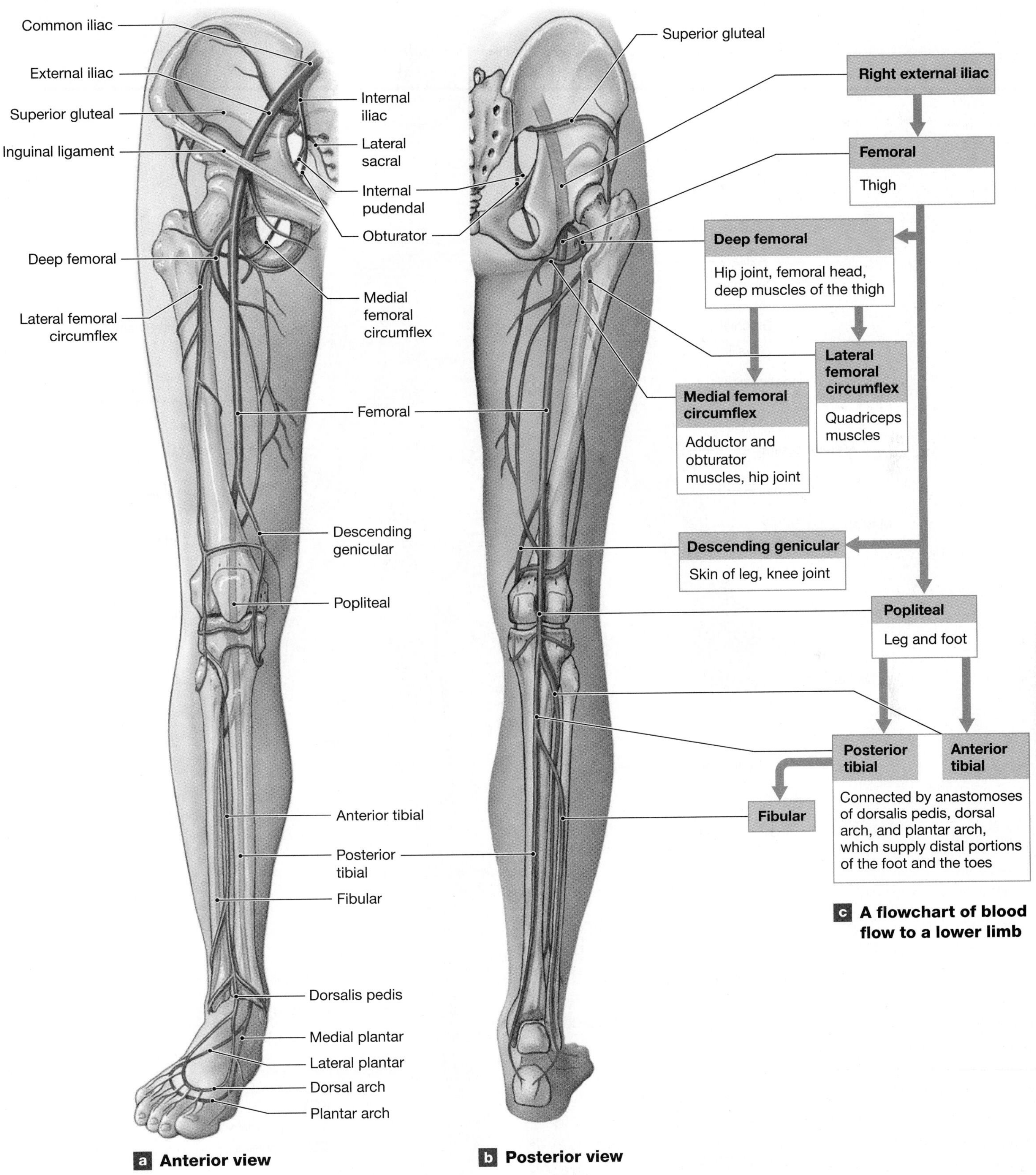

Common iliac
External iliac
Superior gluteal
Inguinal ligament
Internal iliac
Lateral sacral
Internal pudendal
Obturator
Deep femoral
Lateral femoral circumflex
Medial femoral circumflex
Femoral
Descending genicular
Popliteal
Anterior tibial
Posterior tibial
Fibular
Dorsalis pedis
Medial plantar
Lateral plantar
Dorsal arch
Plantar arch
Superior gluteal
Right external iliac
Femoral
Thigh
Deep femoral
Hip joint, femoral head, deep muscles of the thigh
Lateral femoral circumflex
Quadriceps muscles
Medial femoral circumflex
Adductor and obturator muscles, hip joint
Descending genicular
Skin of leg, knee joint
Popliteal
Leg and foot
Posterior tibial
Anterior tibial
Fibular
Connected by anastomoses of dorsalis pedis, dorsal arch, and plantar arch, which supply distal portions of the foot and the toes
21

a Anterior view

b Posterior view

c A flowchart of blood flow to a lower limb

Figure 21–26 **An Overview of the Major Systemic Veins.**

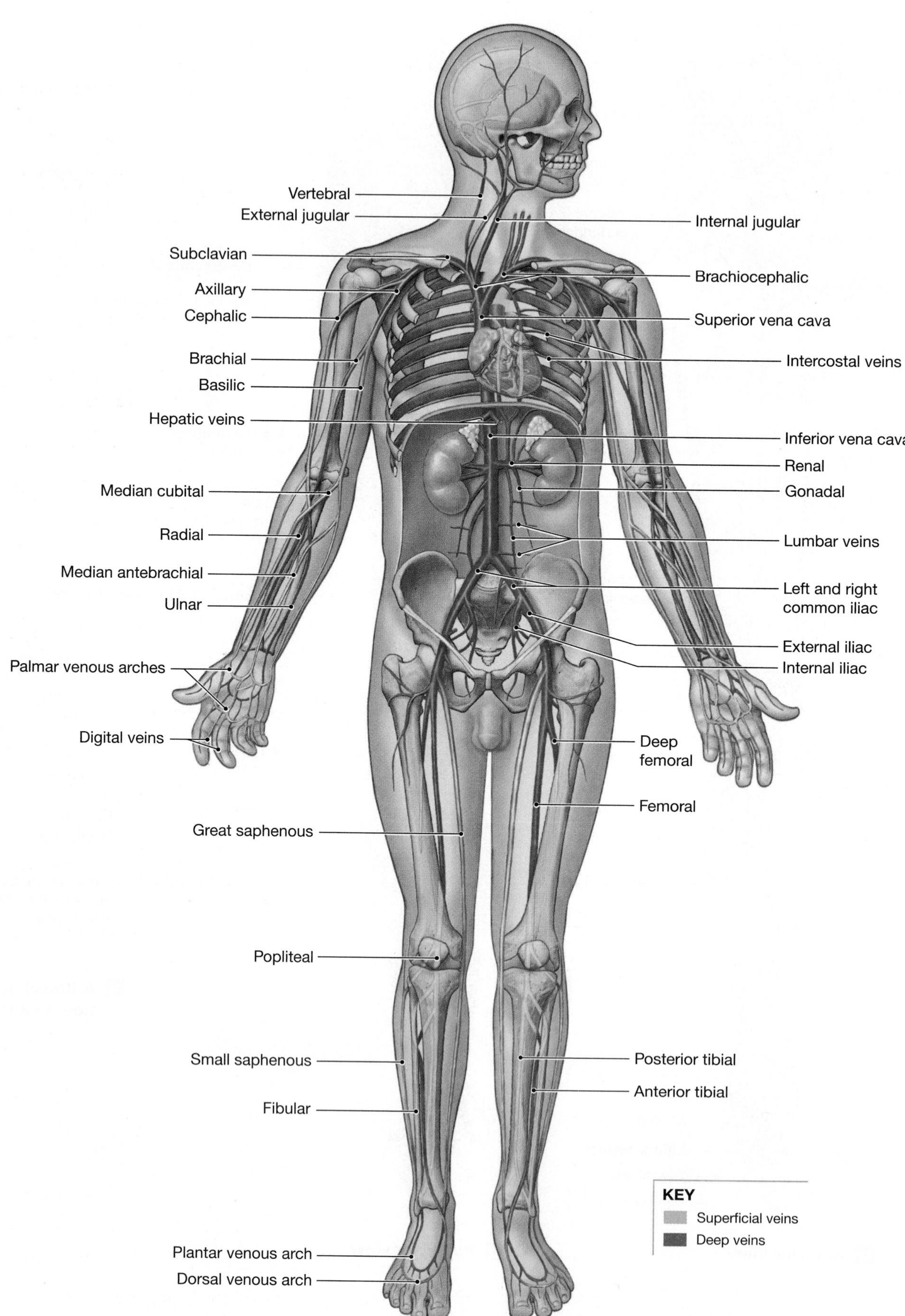
Vertebral
External jugular
Internal jugular
Subclavian
Brachiocephalic
Axillary
Cephalic
Superior vena cava
Brachial
Intercostal veins
Basilic
Hepatic veins
Inferior vena cava
Renal
Median cubital
Gonadal
Radial
Lumbar veins
Median antebrachial
Left and right common iliac
Ulnar
External iliac
Internal iliac
Palmar venous arches
Digital veins
Deep femoral
Femoral
Great saphenous
Popliteal
Small saphenous
Posterior tibial
Anterior tibial
Fibular
KEY
Superficial veins
Deep veins
Plantar venous arch
Dorsal venous arch

21

Figure 21–27 **Major Veins of the Head, Neck, and Brain.** ATLAS: Plates 3c,d; 18a–c

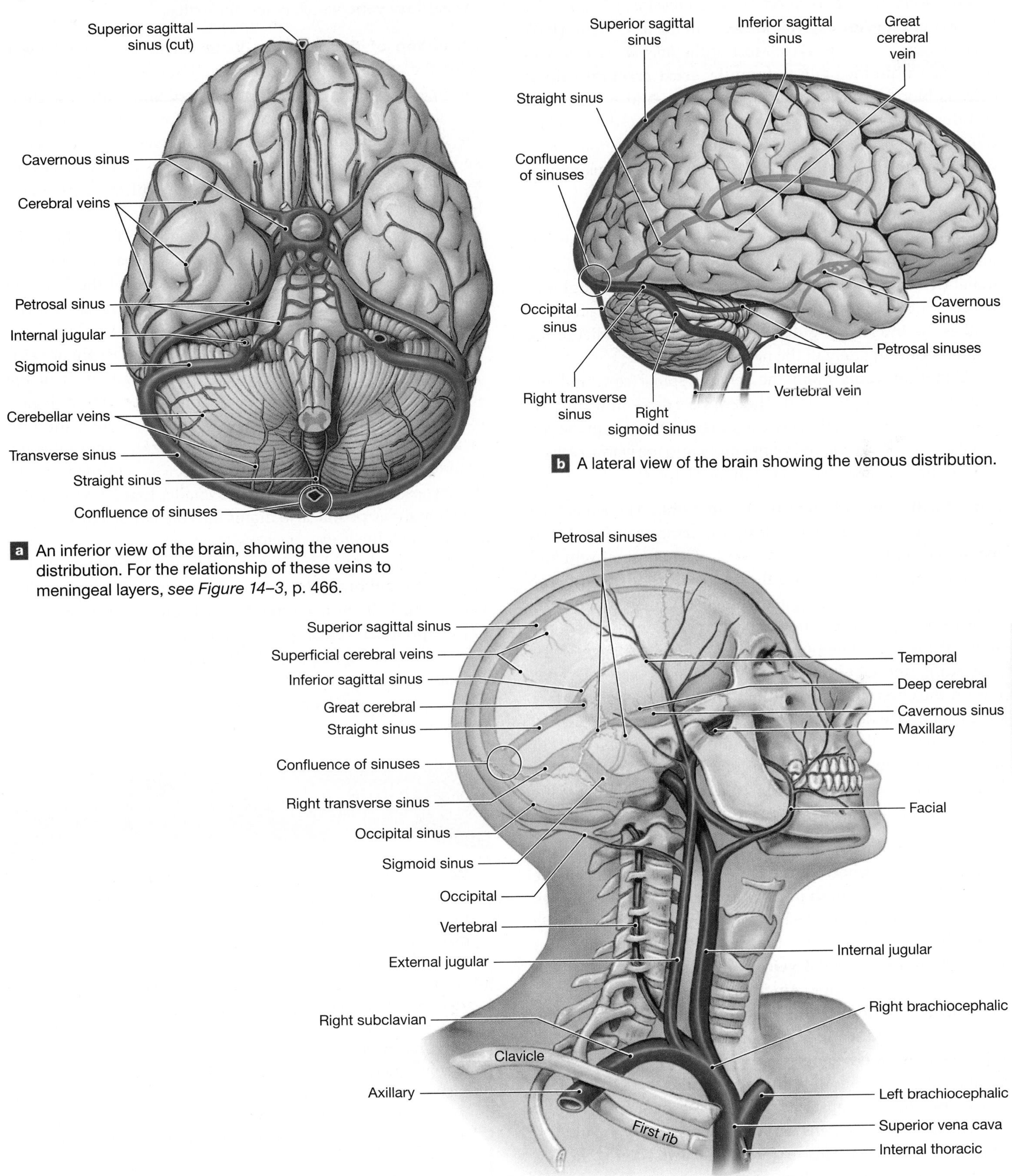
Superior sagittal sinus (cut)
Cavernous sinus
Cerebral veins
Petrosal sinus
Internal jugular
Sigmoid sinus
Cerebellar veins
Transverse sinus
Straight sinus
Confluence of sinuses
Superior sagittal sinus
Inferior sagittal sinus
Great cerebral vein
Straight sinus
Confluence of sinuses
Occipital sinus
Right transverse sinus
Right sigmoid sinus
Cavernous sinus
Petrosal sinuses
Internal jugular
Vertebral vein
Petrosal sinuses
Superior sagittal sinus
Superficial cerebral veins
Inferior sagittal sinus
Great cerebral
Straight sinus
Confluence of sinuses
Right transverse sinus
Occipital sinus
Sigmoid sinus
Occipital
Vertebral
External jugular
Right subclavian
Clavicle
Axillary
First rib
Temporal
Deep cerebral
Cavernous sinus
Maxillary
Facial
Internal jugular
Right brachiocephalic
Left brachiocephalic
Superior vena cava
Internal thoracic
21

a An inferior view of the brain, showing the venous distribution. For the relationship of these veins to meningeal layers, *see Figure 14–3*, p. 466.

b A lateral view of the brain showing the venous distribution.

c Veins draining the brain and the superficial and deep portions of the head and neck.

The superior sagittal sinus, occipital sinus, straight sinus, and two transverse sinuses connect at the *confluence of sinuses*. The largest, the **superior sagittal sinus**, is in the falx cerebri (look back at Figure 14–3, p. 470). Most of the *inferior cerebral veins* converge within the brain to form the **great cerebral vein**. It delivers blood from the interior of the cerebral hemispheres and the choroid plexus to the **straight sinus**. Other cerebral veins drain into the **cavernous sinus** with numerous small veins from the orbit. Blood from the cavernous sinus reaches the internal jugular vein through the petrosal sinuses.

The venous sinuses converge within the dura mater in the region of the lambdoid suture. The left and right transverse sinuses begin at the confluence of the occipital, sagittal, and straight sinuses. Each transverse sinus drains into a **sigmoid sinus**, which penetrates the jugular foramen and leaves the skull as the **internal jugular vein**. It descends parallel to the common carotid artery in the neck. p. 761

Vertebral veins drain the cervical spinal cord and the posterior surface of the skull. These vessels descend within the transverse foramina of the cervical vertebrae, along with the vertebral arteries. The vertebral veins empty into the *brachiocephalic veins* of the chest (discussed later in the chapter).

Superficial Veins of the Head and Neck. The superficial veins of the head converge to form the **temporal**, **facial**, and **maxillary veins** (see Figure 21–27c). The temporal vein and the maxillary vein drain into the **external jugular vein**. The facial vein drains into the internal jugular vein. A broad anastomosis between the external and internal jugular veins at the angle of the mandible provides dual venous drainage of the face, scalp, and cranium. The external jugular vein descends toward the chest just deep to the skin on the anterior surface of the sternocleidomastoid muscle. Posterior to the clavicle, the external jugular vein empties into the *subclavian vein*. In healthy individuals, the external jugular vein is easily palpable. A *jugular venous pulse (JVP)* is sometimes detectable at the base of the neck.

Venous Return from the Upper Limbs. The **digital veins** empty into the **superficial** and **deep palmar veins** of the hand, which interconnect to form the **palmar venous arches** (Figure 21–28). The superficial arch empties into the **cephalic vein**, which ascends along the radial side of the forearm; the **median antebrachial vein**; and the **basilic vein**, which ascends on the ulnar side. Anterior to the elbow is the superficial **median cubital vein**, which passes from the cephalic vein, medially and at an oblique angle, to connect to the basilic vein. (Venous blood samples are typically collected from the median cubital.) From the elbow, the basilic vein passes superiorly along the medial surface of the biceps brachii muscle.

The deep palmar veins drain into the **radial vein** and the **ulnar vein**. These veins fuse to form the **brachial vein**, running parallel to the brachial artery. As the brachial vein continues toward the trunk, it merges with the basilic vein and becomes the **axillary vein**, which enters the axilla.

Formation of the Superior Vena Cava. The cephalic vein joins the axillary vein on the lateral surface of the first rib, forming the **subclavian vein**, which continues into the chest. The subclavian vein passes superior to the first rib and along the superior margin of the clavicle. It merges with the external and internal jugular veins of that side. This fusion creates the **brachiocephalic vein**, or *innominate vein*, which penetrates the body wall and enters the thoracic cavity.

Each brachiocephalic vein receives blood from the vertebral vein of the same side, which drains the back of the skull and spinal cord. Near the heart, at the level of the first and second ribs, the left and right brachiocephalic veins join, creating the superior vena cava. Close to the point of fusion, the **internal thoracic vein** empties into the brachiocephalic vein.

The **azygos** (AZ-ī-gos) **vein** is the major branch of the superior vena cava. This vein ascends from the lumbar region over the right side of the vertebral column to enter the thoracic cavity through the diaphragm. The azygos vein joins the superior vena cava at the level of vertebra T_2. On the left side, the azygos receives blood from the smaller **hemi-azygos vein**, which in many people also drains into the left brachiocephalic vein through the *highest intercostal vein*.

The azygos and hemi-azygos veins are the chief collecting vessels of the thorax. They receive blood from (1) **intercostal veins**, which in turn receive blood from the chest muscles; (2) **esophageal veins**, which drain blood from the inferior portion of the esophagus; and (3) smaller veins draining other mediastinal structures.

Figure 21–29 diagrams the venous branches of the superior vena cava.

+ Clinical Note Preparing the Circulation for Dialysis

When a patient's kidneys fail from chronic disease (diabetes mellitus, for example), the patient requires **dialysis** to cleanse the blood as the kidneys would normally do. A minor outpatient surgical procedure called an **arteriovenous fistula** prepares the circulation for successful dialysis. The surgeon connects a large vein, usually in the arm, to the nearby artery. Soon the higher pressure in the artery acts on the vein to enlarge its diameter. In a few weeks, the fistula has healed and can accept a dialysis needle. The patient's "dirty" blood (blood containing nitrogenous waste, such as urea) is withdrawn from his or her artery, "cleaned" in the dialysis machine, and returned easily to their enlarged vein. The fistula can last for several years before a new one needs to be created at another site.

Figure 21–28 The Venous Drainage of the Abdomen and Chest. ATLAS: Plates 27c; 29c; 47b,d; 61a; 62a,b

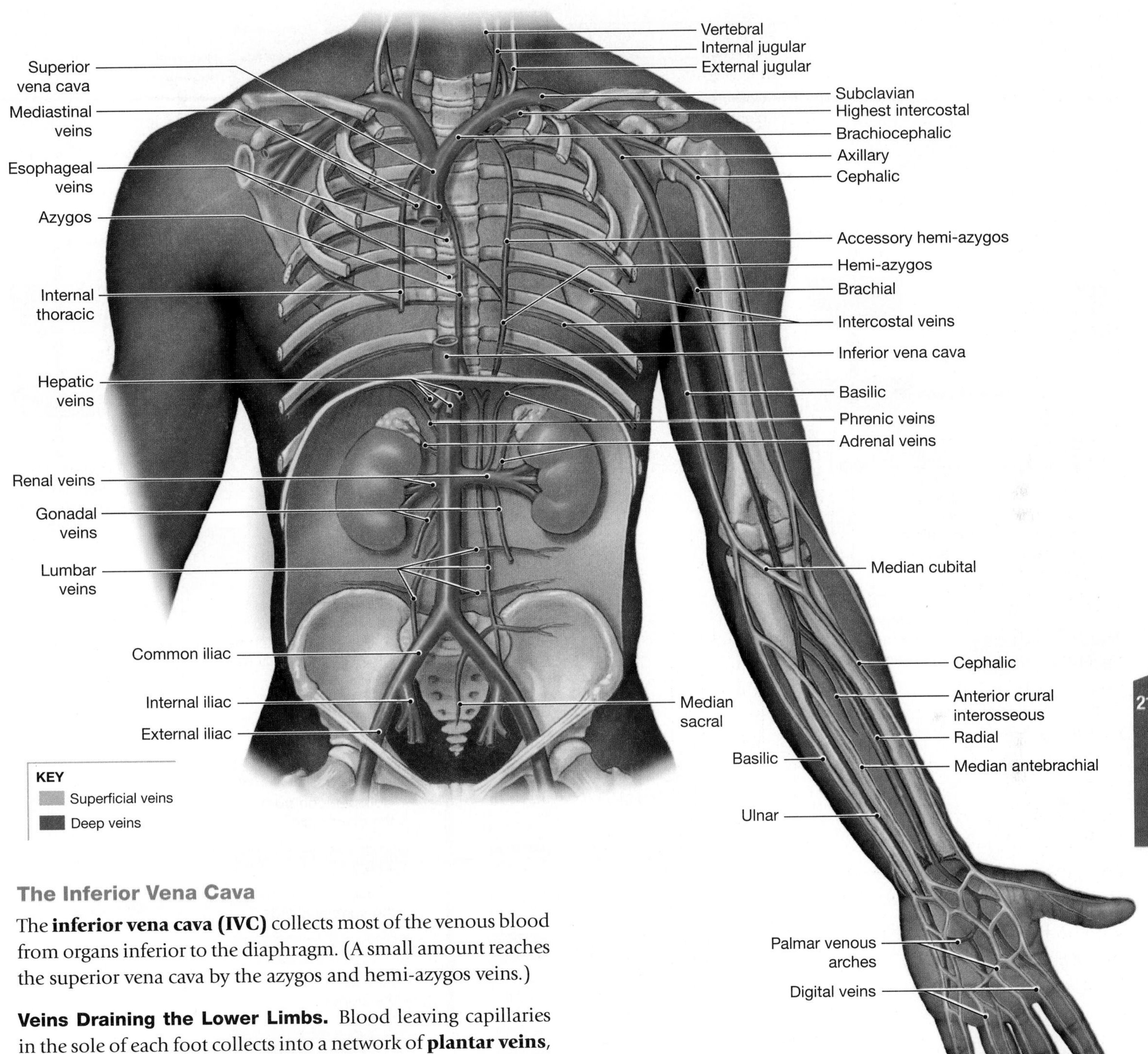
Vertebral
Internal jugular
External jugular
Superior vena cava
Mediastinal veins
Esophageal veins
Azygos
Subclavian
Highest intercostal
Brachiocephalic
Axillary
Cephalic
Accessory hemi-azygos
Hemi-azygos
Brachial
Internal thoracic
Intercostal veins
Inferior vena cava
Hepatic veins
Basilic
Phrenic veins
Adrenal veins
Renal veins
Gonadal veins
Lumbar veins
Median cubital
Common iliac
Cephalic
Internal iliac
Median sacral
Anterior crural interosseous
External iliac
Radial
Basilic
Median antebrachial
KEY
Superficial veins
Deep veins
Ulnar
Palmar venous arches
Digital veins

21

The Inferior Vena Cava

The **inferior vena cava (IVC)** collects most of the venous blood from organs inferior to the diaphragm. (A small amount reaches the superior vena cava by the azygos and hemi-azygos veins.)

Veins Draining the Lower Limbs. Blood leaving capillaries in the sole of each foot collects into a network of **plantar veins**, which supply the **plantar venous arch** (Figure 21–30a). The plantar network sends blood to the deep veins of the leg: the **anterior tibial vein**, the **posterior tibial vein**, and the **fibular vein**. The **dorsal venous arch** collects blood from capillaries on the superior surface of the foot and the **digital veins** of the toes. The plantar arch and the dorsal arch are extensively interconnected, and the path of blood flow can easily shift from superficial to deep veins.

Two superficial veins drain the dorsal venous arch: the **great saphenous** (sa-FĒ-nus; *saphenes*, prominent) **vein** and the **small saphenous vein**. The great saphenous vein ascends along the medial aspect of the leg and thigh, draining into the *femoral vein* near the hip joint. The small saphenous vein arises from the dorsal venous arch and ascends along the posterior and lateral aspect of the calf. This vein then enters the popliteal fossa, where it meets the **popliteal vein**, formed by the union

Figure 21–29 **Flowchart of Circulation to the Superior and Inferior Venae Cavae.**

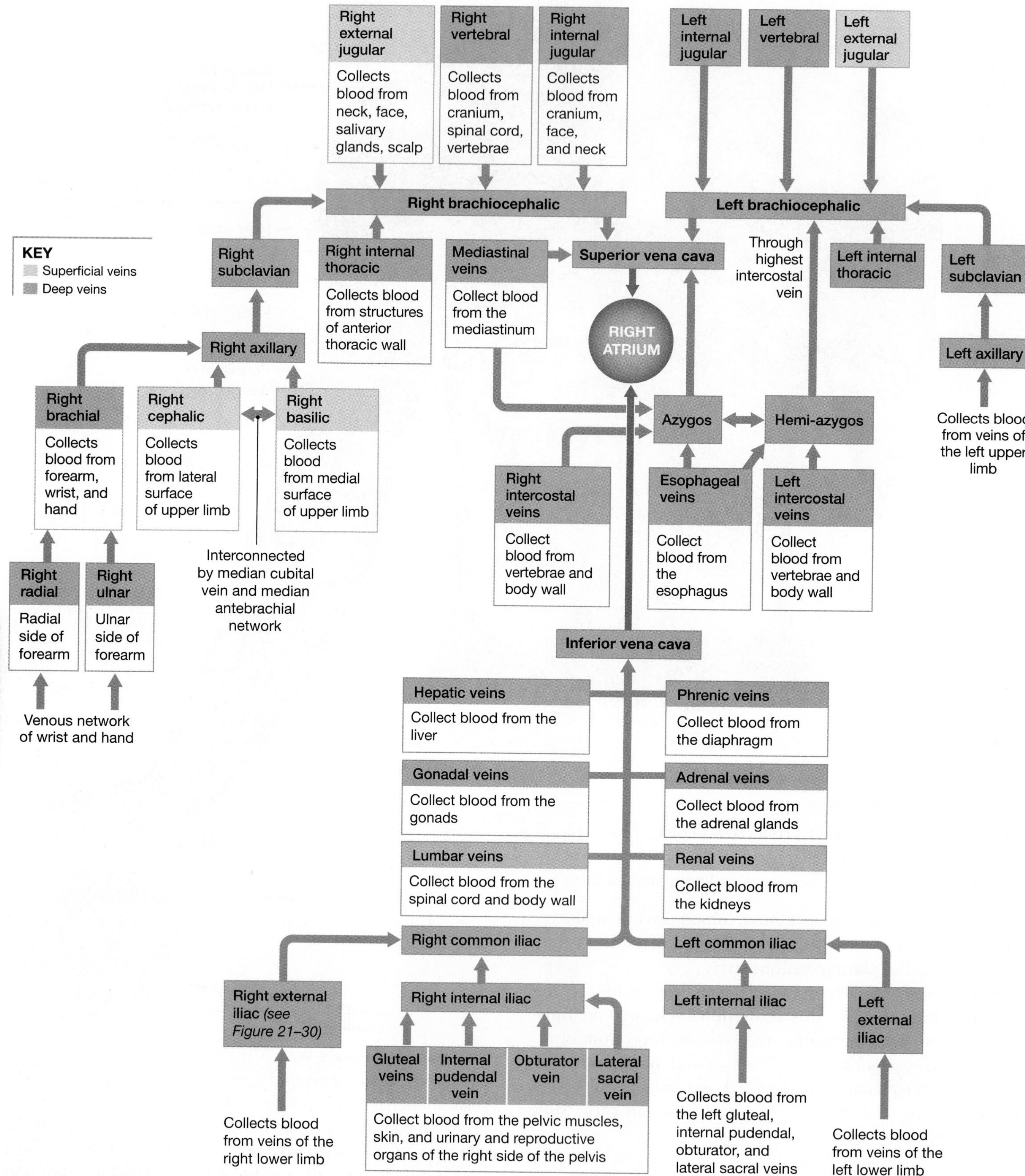
Right external jugular
Collects blood from neck, face, salivary glands, scalp
Right vertebral
Collects blood from cranium, spinal cord, vertebrae
Right internal jugular
Collects blood from cranium, face, and neck
Left internal jugular
Left vertebral
Left external jugular
Right brachiocephalic
Left brachiocephalic
KEY
Superficial veins
Deep veins
Right subclavian
Right internal thoracic
Collects blood from structures of anterior thoracic wall
Mediastinal veins
Collect blood from the mediastinum
Superior vena cava
Through highest intercostal vein
Left internal thoracic
Left subclavian
RIGHT ATRIUM
Right axillary
Left axillary
Right brachial
Collects blood from forearm, wrist, and hand
Right cephalic
Collects blood from lateral surface of upper limb
Right basilic
Collects blood from medial surface of upper limb
Azygos
Hemi-azygos
Collects blood from veins of the left upper limb
Right intercostal veins
Collect blood from vertebrae and body wall
Esophageal veins
Collect blood from the esophagus
Left intercostal veins
Collect blood from vertebrae and body wall
Right radial
Radial side of forearm
Right ulnar
Ulnar side of forearm
Interconnected by median cubital vein and median antebrachial network
Inferior vena cava
Venous network of wrist and hand
Hepatic veins
Collect blood from the liver
Phrenic veins
Collect blood from the diaphragm
Gonadal veins
Collect blood from the gonads
Adrenal veins
Collect blood from the adrenal glands
Lumbar veins
Collect blood from the spinal cord and body wall
Renal veins
Collect blood from the kidneys
Right common iliac
Left common iliac
Right external iliac (see Figure 21–30)
Right internal iliac
Left internal iliac
Left external iliac
Gluteal veins
Internal pudendal vein
Obturator vein
Lateral sacral vein
Collect blood from the pelvic muscles, skin, and urinary and reproductive organs of the right side of the pelvis
Collects blood from veins of the right lower limb
Collects blood from the left gluteal, internal pudendal, obturator, and lateral sacral veins
Collects blood from veins of the left lower limb

21

? What veins are shown draining directly into the superior vena cava?

Figure 21–30 Venous Drainage from the Lower Limb. ATLAS: Plates 70b; 74; 78a–g

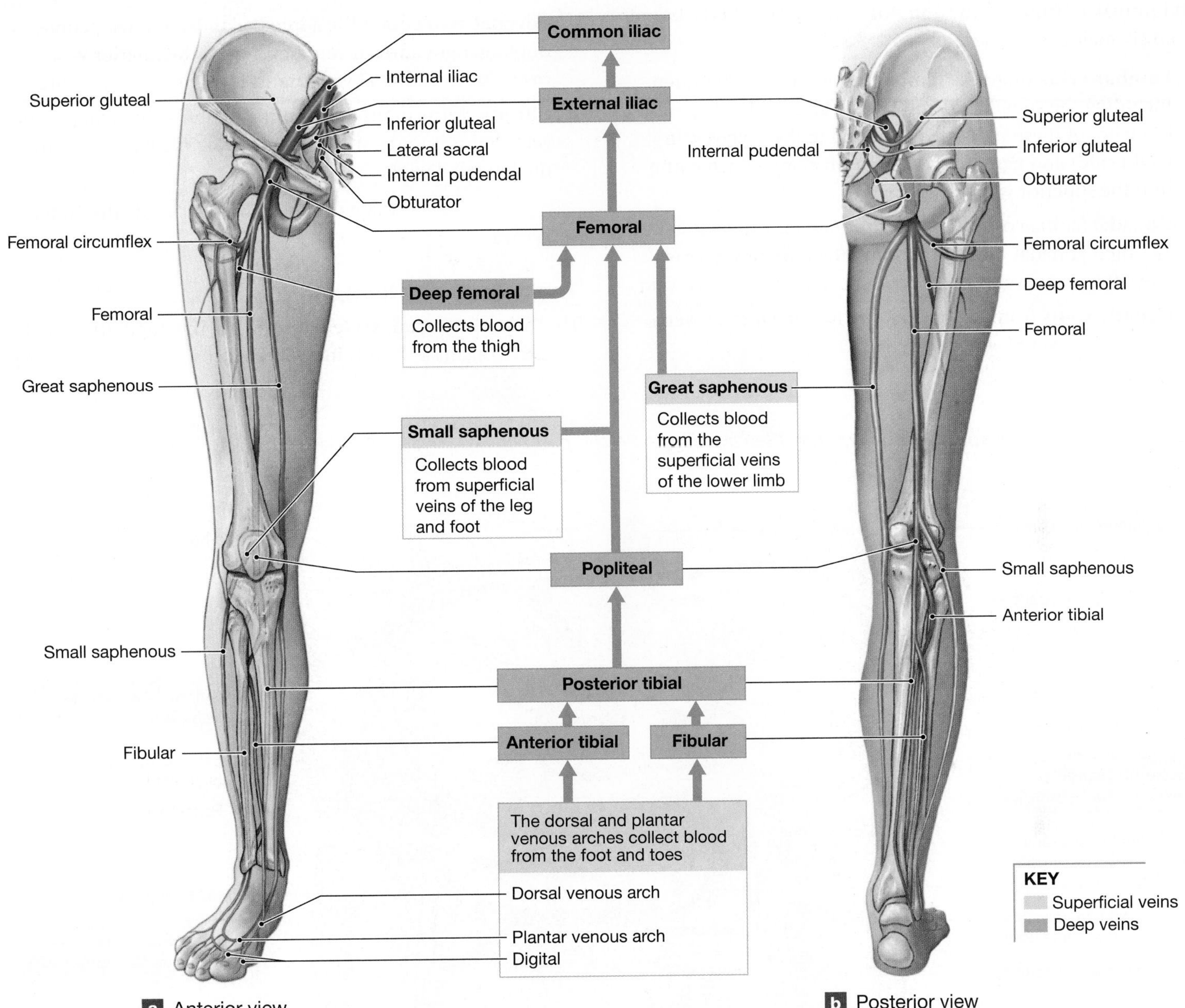
Common iliac
Internal iliac
External iliac
Superior gluteal
Inferior gluteal
Lateral sacral
Internal pudendal
Obturator
Femoral circumflex
Femoral
Deep femoral
Collects blood from the thigh
Great saphenous
Collects blood from the superficial veins of the lower limb
Small saphenous
Collects blood from superficial veins of the leg and foot
Popliteal
Posterior tibial
Anterior tibial
Fibular
The dorsal and plantar venous arches collect blood from the foot and toes
Dorsal venous arch
Plantar venous arch
Digital
a Anterior view
b Posterior view
KEY
Superficial veins
Deep veins

21

of the fibular and both tibial veins (Figure 21–30a,b). The popliteal vein is easily palpated in the popliteal fossa adjacent to the adductor magnus muscle. At the femur, the popliteal vein becomes the **femoral vein**, which ascends along the thigh, next to the femoral artery. Immediately before penetrating the abdominal wall, the femoral vein receives blood from (1) the great saphenous vein; (2) the **deep femoral vein**, which collects blood from deeper structures in the thigh; and (3) the **femoral circumflex vein**, which drains the region around the neck and head of the femur. The femoral vein penetrates the body wall and emerges in the pelvic cavity as the **external iliac vein**.

Veins Draining the Pelvis. The external iliac veins receive blood from the lower limbs, the pelvis, and the lower abdomen. As the left and right external iliac veins cross the inner surface of the ilium, they are joined by the **internal iliac veins**, which drain the pelvic organs (see Figure 21–29). The internal iliac veins are formed by the fusion of the *gluteal, internal pudendal, obturator,* and *lateral sacral veins* (Figure 21–30a).

The union of external and internal iliac veins forms the **common iliac vein**. Its right and left branches ascend at an oblique angle. The left common iliac vein receives blood from the *median sacral vein*, which drains the area supplied by the median sacral artery (see Figure 21–28). Anterior to vertebra L_5, the common iliac veins unite to form the inferior vena cava.

Veins Draining the Abdomen. The inferior vena cava ascends posterior to the peritoneal cavity, parallel to the aorta. The abdominal portion of the inferior vena cava collects blood from six major veins (see Figures 21–28 and 21–29):

- **Lumbar veins** drain the lumbar portion of the abdomen, including the spinal cord and body wall muscles. Superior branches of these veins are connected to the azygos vein (right side) and hemi-azygos vein (left side), which empty into the superior vena cava.
- **Gonadal** (*ovarian* or *testicular*) **veins** drain the ovaries or testes. The right gonadal vein empties into the inferior vena cava. The left gonadal vein generally drains into the left renal vein.
- **Hepatic veins** from the liver empty into the inferior vena cava at the level of vertebra T_{10}.
- **Renal veins**, the largest branches of the inferior vena cava, collect blood from the kidneys.
- **Adrenal veins** drain the adrenal glands. In most people, only the right adrenal vein drains into the inferior vena cava. The left adrenal vein drains into the left renal vein.
- **Phrenic veins** drain the diaphragm. Only the right phrenic vein drains into the inferior vena cava. The left drains into the left renal vein.

Figure 21–29 diagrams the branches of the inferior vena cava.

The Hepatic Portal System

The **hepatic portal system** begins in the capillaries of the digestive organs and ends in the liver sinusoids (Figure 21–31).

Figure 21–31 The Hepatic Portal System. *(See also Figure 24–24, p. 922.)* ATLAS: Plates 53b; 54a–c; 55; 57a,b

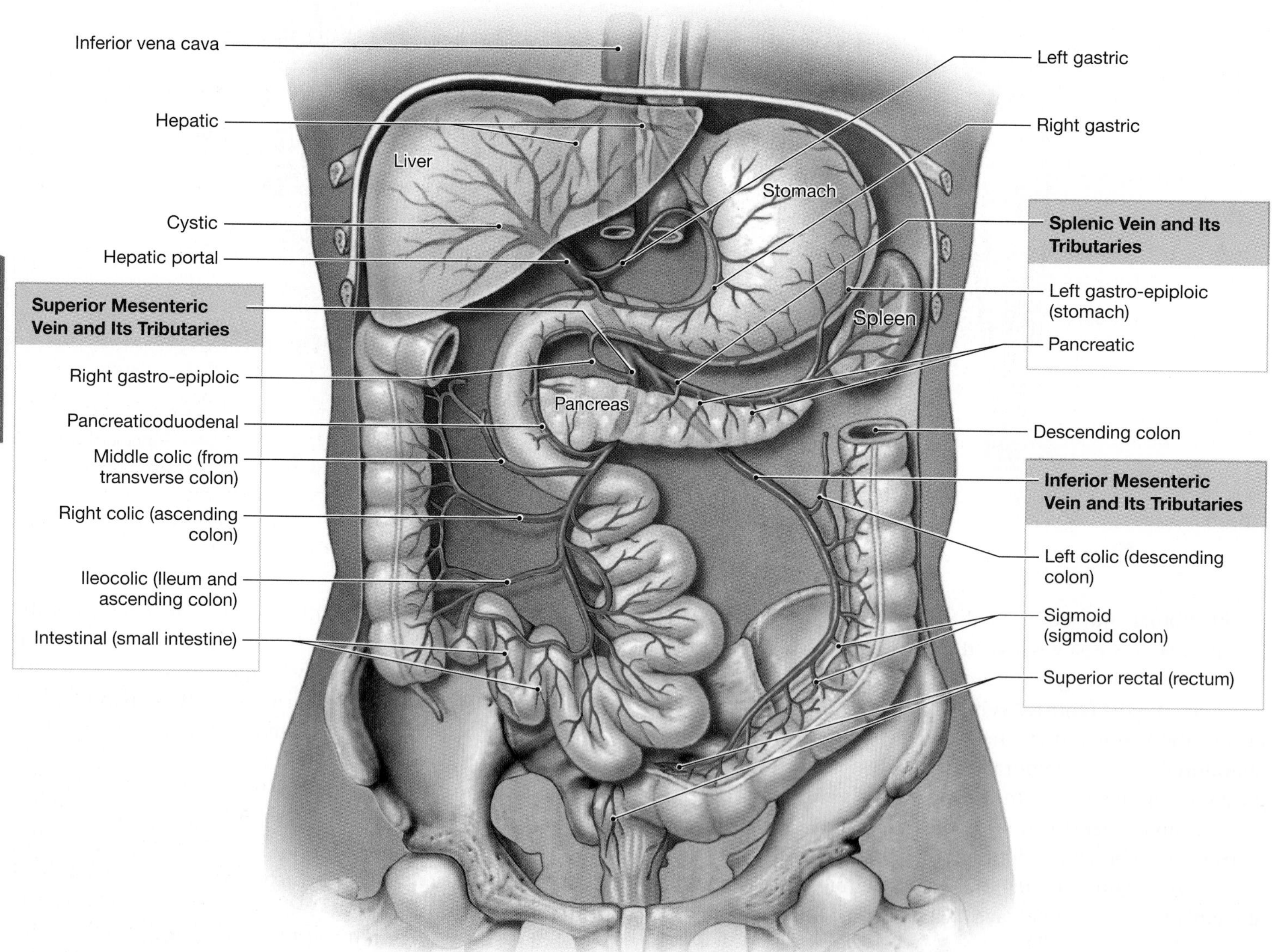
Inferior vena cava
Hepatic
Liver
Cystic
Hepatic portal
Superior Mesenteric Vein and Its Tributaries
Right gastro-epiploic
Pancreaticoduodenal
Middle colic (from transverse colon)
Right colic (ascending colon)
Ileocolic (ileum and ascending colon)
Intestinal (small intestine)
Stomach
Spleen
Pancreas
Left gastric
Right gastric
Splenic Vein and Its Tributaries
Left gastro-epiploic (stomach)
Pancreatic
Descending colon
Inferior Mesenteric Vein and Its Tributaries
Left colic (descending colon)
Sigmoid (sigmoid colon)
Superior rectal (rectum)

21

(As you may recall from Chapter 18, a blood vessel connecting two capillary beds is called a *portal vessel*. The network is a *portal system*.) Blood flowing in the hepatic portal system is quite different from blood in other systemic veins, because it contains substances absorbed from the stomach and intestines. For example, levels of blood glucose and amino acids in the hepatic portal vein often exceed those found anywhere else in the cardiovascular system. The hepatic portal system delivers these and other absorbed compounds directly to the liver for storage, metabolic conversion, or excretion.

The largest vessel of the hepatic portal system is the **hepatic portal vein** (see Figure 21–31). It delivers venous blood to the liver. It receives blood from three large veins draining organs within the peritoneal cavity:

- The **inferior mesenteric vein** collects blood from capillaries along the inferior portion of the large intestine. Its branches include the *left colic vein*, *sigmoid vein*, and the *superior rectal veins*. They drain the descending colon, sigmoid colon, and rectum.
- The **splenic vein** is formed by the union of the inferior mesenteric vein and veins from the spleen, the lateral border of the stomach (*left gastro-epiploic vein*), and the pancreas (*pancreatic veins*).
- The **superior mesenteric vein** collects blood from veins draining the stomach (*right gastro-epiploic vein*), the small intestine (*intestinal* and *pancreaticoduodenal veins*), and two-thirds of the large intestine (*ileocolic*, *right colic*, and *middle colic veins*).

The hepatic portal vein forms through the fusion of the superior mesenteric, inferior mesenteric, and splenic veins. The superior mesenteric vein normally contributes the greater volume of blood and most of the nutrients. As it proceeds, the hepatic portal vein receives blood from the left and right **gastric veins**, which drain the medial border of the stomach, and from the **cystic vein** of the gallbladder.

After passing through liver sinusoids, blood collects in the hepatic veins, which empty into the inferior vena cava.

The composition of the blood in the systemic circuit is relatively stable despite changes in diet and digestive activity. The reason for this stability is that blood from the intestines goes to the liver first, and the liver regulates the nutrient content of the blood before it enters the inferior vena cava.

✓ Checkpoint

19. **A blockage of which branch from the aortic arch would interfere with blood flow to the left arm?**
20. **Why would compression of the common carotid arteries cause a person to lose consciousness?**
21. **Isabella is in an automobile accident, and her celiac trunk is ruptured. Which organs will be affected most directly by this injury?**
22. **Whenever Noah gets angry, a large vein bulges in the lateral region of his neck. Which vein is this?**
23. **A thrombus that blocks the popliteal vein would interfere with blood flow in which other veins?**

See the blue Answers tab at the back of the book.

21-8 Modifications of fetal and maternal cardiovascular systems promote the exchange of materials; the fetal cardiovascular system changes to function independently after birth

Learning Outcome **Identify the differences between fetal and adult circulation patterns, and describe the changes in the patterns of blood flow that occur at birth.**

The fetal cardiovascular system differs from the adult cardiovascular system because the fetus and the adult have different sources of respiratory and nutritional support. Most strikingly, the fetal lungs are collapsed and nonfunctional, and the digestive tract has nothing to digest. Instead, diffusion across the placenta provides for the respiratory and nutritional needs of the fetus. This situation changes dramatically at birth, when the newborn's system is separated from that of the mother.

Fetal Circulatory Route and Placental Blood Supply

Fetal patterns of blood flow are diagrammed in Figure 21–32a. Blood flows to the placenta through a pair of **umbilical arteries**. They arise from the internal iliac arteries and enter the umbilical cord. Blood returns from the placenta in the single **umbilical vein**, bringing oxygen and nutrients to the developing fetus. The umbilical vein drains into the **ductus venosus**, a vascular connection to an intricate network of veins within the developing liver. The ductus venosus collects blood from the veins of the liver and from the umbilical vein, and empties into the inferior vena cava.

Fetal Heart and Great Vessels

One of the most interesting aspects of cardiovascular development is that of the heart, which reflects the differences between the life of an embryo or fetus and that of an infant. The interatrial and interventricular septa develop early in fetal life, but the interatrial partition remains functionally incomplete until birth. The **foramen ovale**, or *interatrial opening*, is associated with a long flap that acts as a valve. Blood can flow freely from the right atrium to the left atrium, but any backflow closes the valve and isolates the two chambers from one another. Thus, blood entering the heart at the right atrium can bypass the pulmonary circuit. A second short-circuit exists between the pulmonary and

Go to MasteringA&P™ > Study Area > Lab Tools > PAL 3.0 > Human Cadaver > Cardiovascular System > Blood Vessels
Go to MasteringA&P™ > Study Area > Lab Tools > PAL 3.0 > Anatomical Models > Cardiovascular System

Figure 21–32 **Fetal Circulation.**

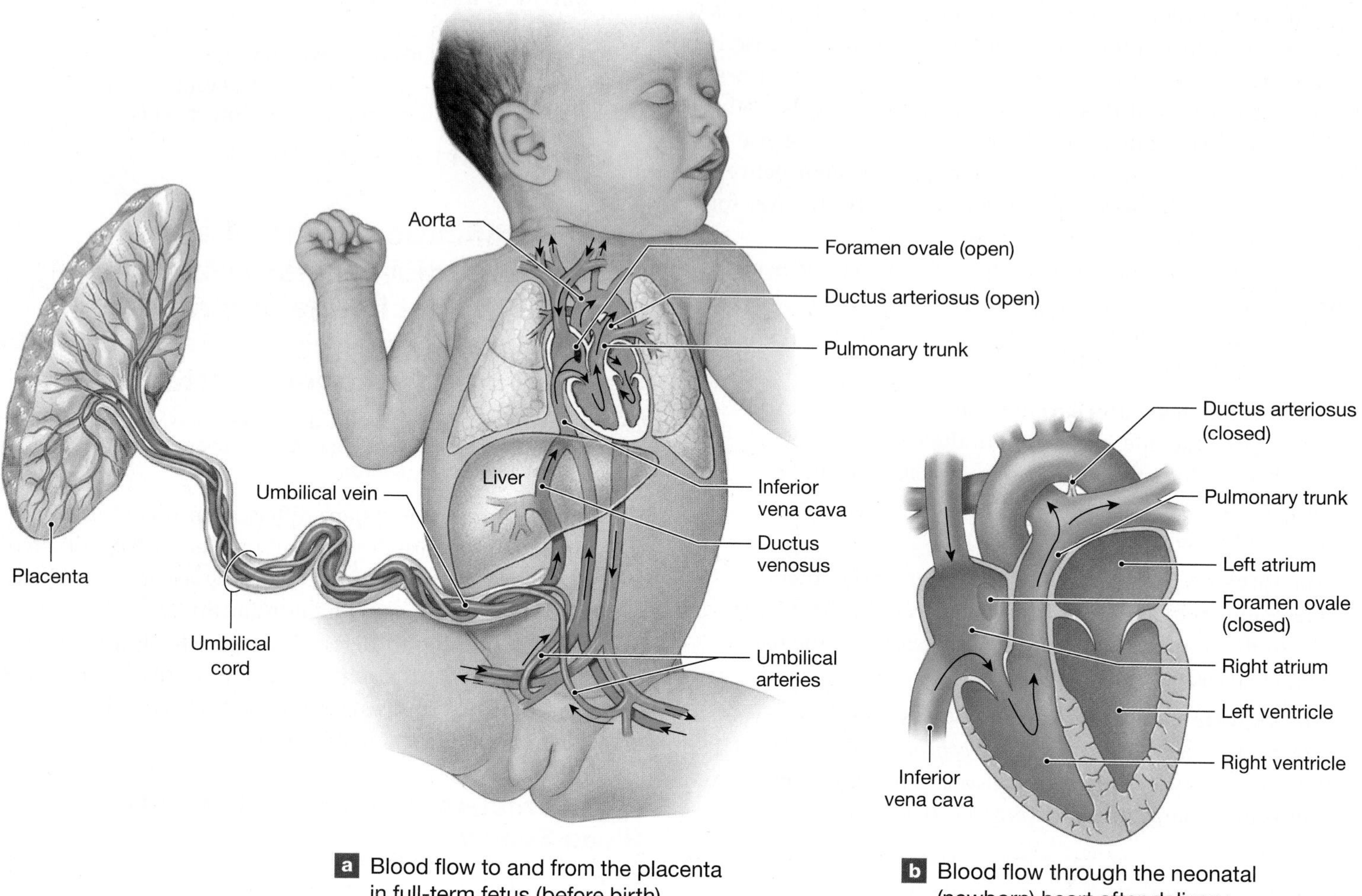
Aorta
Foramen ovale (open)
Ductus arteriosus (open)
Pulmonary trunk
Liver
Umbilical vein
Placenta
Umbilical cord
Inferior vena cava
Ductus venosus
Umbilical arteries
Ductus arteriosus (closed)
Pulmonary trunk
Left atrium
Foramen ovale (closed)
Right atrium
Left ventricle
Right ventricle
Inferior vena cava

a Blood flow to and from the placenta in full-term fetus (before birth)

b Blood flow through the neonatal (newborn) heart after delivery

aortic trunks. This connection, the **ductus arteriosus**, consists of a short, muscular vessel. ATLAS: Embryology Summary 16: The Development of the Cardiovascular System

With the lungs collapsed, the capillaries are compressed and little blood flows through the lungs. During diastole, blood enters the right atrium and flows into the right ventricle, but it also passes into the left atrium through the foramen ovale. About 25 percent of the blood arriving at the right atrium bypasses the pulmonary circuit in this way. In addition, more than 90 percent of the blood leaving the right ventricle passes through the ductus arteriosus and enters the systemic circuit rather than continuing to the lungs.

Cardiovascular Changes at Birth

At birth, dramatic changes take place. When the placental connection is broken at birth, blood stops flowing in the umbilical vessels, and they soon degenerate. Remnants of these vessels persist throughout life as fibrous cords. When an infant takes the first breath, the lungs expand, and so do the pulmonary vessels. The resistance in the pulmonary circuit declines suddenly, and blood rushes into the pulmonary vessels. Within a few seconds, rising O_2 levels stimulate the constriction of the ductus arteriosus, isolating the pulmonary and aortic trunks from one another. As pressures rise in the left atrium, the valvular flap closes the foramen ovale. In adults, the interatrial septum bears the *fossa ovalis*, a shallow depression that marks the site of the foramen ovale (look back at Figure 20–5a, p. 695). The remnants of the ductus arteriosus persist throughout life as the *ligamentum arteriosum*, a fibrous cord.

If the proper cardiovascular changes do not take place at birth or shortly after, problems eventually develop. Their severity depends on which connection remains open and on the size of the opening. Treatment may involve surgery to close the foramen ovale, the ductus arteriosus, or both. Other congenital heart defects result from abnormal cardiac development or inappropriate connections between the heart and major arteries and veins. Spotlight Figure 21–33 focuses on congenital heart problems.

SPOTLIGHT

Figure 21–33

Congenital Heart Problems

Although minor individual variations in the vascular network are quite common, congenital heart problems serious enough to threaten homeostasis are relatively rare.

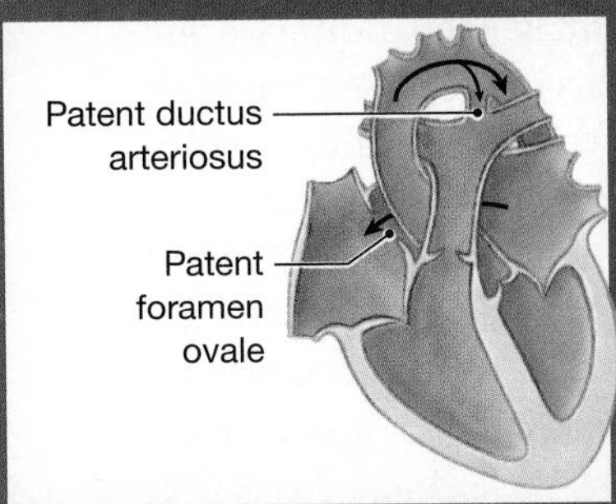

Patent ductus arteriosus
Patent foramen ovale

Patent Foramen Ovale and Patent Ductus Arteriosus

If the foramen ovale remains open, or *patent*, blood recirculates through the pulmonary circuit instead of entering the left ventricle. The movement, driven by the relatively high systemic pressure, is called a "left-to-right shunt." Arterial oxygen content is normal, but the left ventricle must work much harder than usual to provide adequate blood flow through the systemic circuit. Hence, pressures rise in the pulmonary circuit. If the pulmonary pressures rise enough, they may force blood into the systemic circuit through the ductus arteriosus. This condition—a patent ductus arteriosus—creates a "right-to-left shunt." Because the circulating blood is not adequately oxygenated, it develops a deep red color. The skin then develops the blue tones typical of cyanosis and the infant is known as a "blue baby."

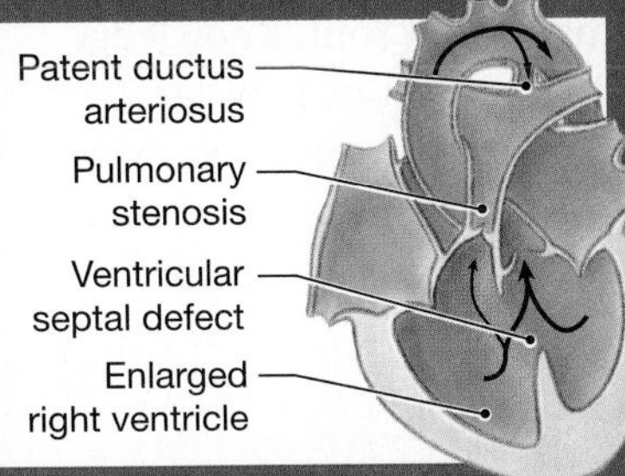

Patent ductus arteriosus
Pulmonary stenosis
Ventricular septal defect
Enlarged right ventricle

Tetralogy of Fallot

The tetralogy of Fallot (fa-LŌ) is a complex group of heart and circulatory defects that affect 0.10% of newborn infants. In this condition, (1) the pulmonary trunk is abnormally narrow (pulmonary stenosis), (2) the interventricular septum is incomplete, (3) the aorta originates where the interventricular septum normally ends, and (4) the right ventricle is enlarged and both ventricles thicken in response to the increased workload.

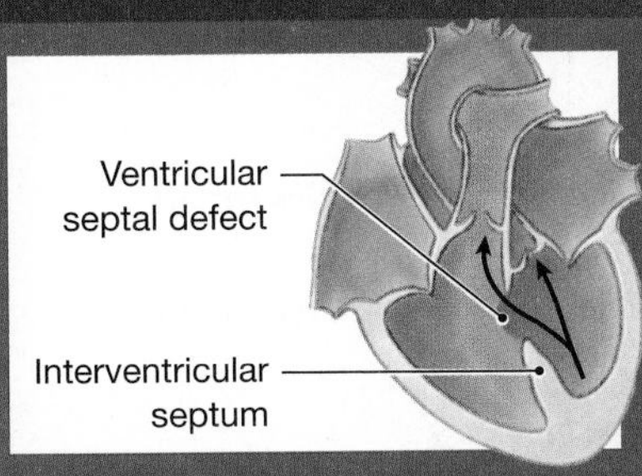

Ventricular septal defect
Interventricular septum

Ventricular Septal Defect

A ventricular septal defect is an abnormal opening in the wall (septum) between the left and right ventricles. It affects 0.12% of newborns. The opening between the two ventricles has an effect similar to a connection between the atria: When the more powerful left ventricle beats, it ejects blood into the right ventricle and pulmonary circuit.

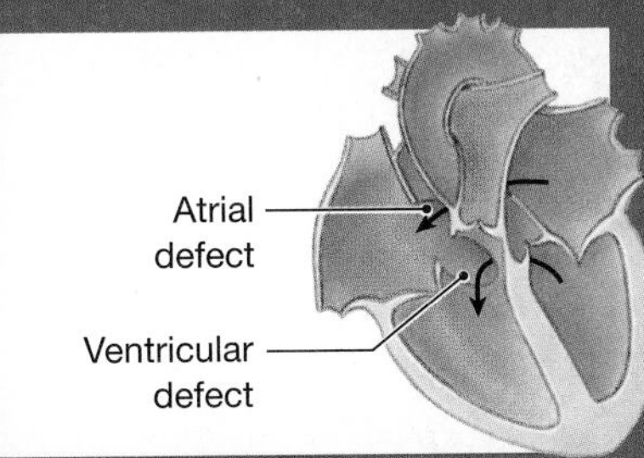

Atrial defect
Ventricular defect

Atrioventricular Septal Defect

In an atrioventricular septal defect, both the atria and ventricles are incompletely separated. The results are quite variable, depending on the extent of the defect and the effects on the atrioventricular valves. This type of defect most commonly affects infants with Down syndrome, a disorder caused by the presence of an extra copy of chromosome 21.

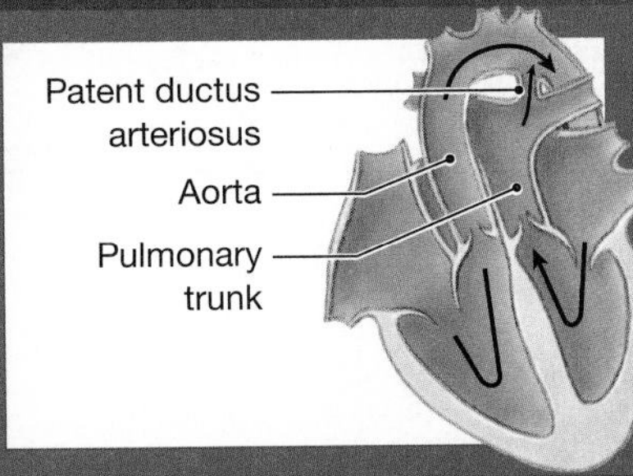

Patent ductus arteriosus
Aorta
Pulmonary trunk

Transposition of the Great Vessels

In the transposition of the great vessels, the aorta is connected to the right ventricle instead of to the left ventricle, and the pulmonary artery is connected to the left ventricle instead of the right ventricle. This malformation affects 0.05% of newborn infants.

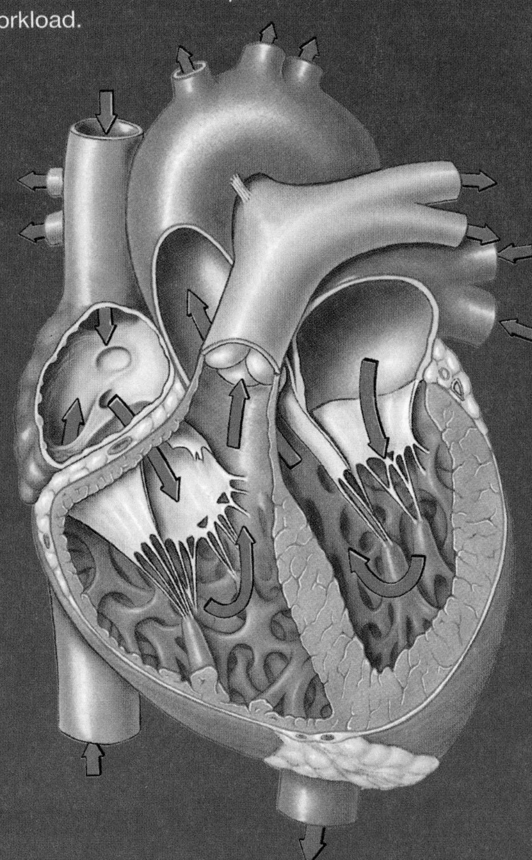

Normal Heart Structure
Most congenital heart problems result from abnormal formation of the heart or problems with the connections between the heart and the great vessels.

Checkpoint

24. **Name the three vessels that carry blood to and from the placenta.**
25. **A blood sample taken from an umbilical cord contains high levels of oxygen and nutrients, and low levels of carbon dioxide and waste products. Is this sample from an umbilical artery or the umbilical vein? Explain.**
26. **Name the structures in the fetal circulation that stop functioning at birth. What becomes of these structures?**

See the blue Answers tab at the back of the book.

21-9 Aging affects the blood, heart, and blood vessels

Learning Outcome **Discuss the effects of aging on the cardiovascular system, and give examples of interactions between the cardiovascular system and other organ systems.**

The capabilities of the cardiovascular system gradually decline. As you age, your cardiovascular system undergoes the following major changes:

- **Age-related changes in blood** may include (1) a decreased hematocrit; (2) constriction or blockage of peripheral veins by a stationary blood clot called a *thrombus*, which can become detached, pass through the heart, and become wedged in a small artery (commonly in the lungs), causing *pulmonary embolism*; and (3) pooling of blood in the veins of the legs because valves are not working effectively.
- **Age-related changes in the heart** include (1) a reduction in maximum cardiac output, (2) changes in the activities of pacemaker and conducting cells, (3) a reduction in the elasticity of the cardiac (fibrous) skeleton, (4) progressive atherosclerosis that can restrict coronary circulation, and (5) replacement of damaged cardiac contractile cells by scar tissue.
- **Age-related changes in blood vessels** may be linked to arteriosclerosis: (1) The inelastic walls of arteries become less tolerant of sudden pressure increases, which can lead to an aneurysm, whose rupture may (depending on the vessel) cause a cerebrovascular accident (CVA), myocardial infarction, or massive blood loss; (2) calcium salts can be deposited on weakened vascular walls, increasing the risk of a CVA or myocardial infarction; (3) lipid deposits in the tunica media associated with a damaged endothelial lining and calcium salts can form atherosclerotic plaques; and (4) thrombi can form at atherosclerotic plaques.

The cardiovascular system is both anatomically and functionally linked to all other systems. Build Your Knowledge Figure 21–34 shows the functional relationships between the cardiovascular system and the other body systems we have studied so far. We are now ready to consider the organization and function of the lymphatic system, the focus of Chapter 22.

Checkpoint

27. **Identify components of the cardiovascular system that are affected by age.**
28. **Define *thrombus*.**
29. **Define *aneurysm*.**
30. **Describe what the cardiovascular system provides for all other body systems.**
31. **What is the relationship between the skeletal system and the cardiovascular system?**

See the blue Answers tab at the back of the book.

Build Your Knowledge

Figure 21–34 Integration of the CARDIOVASCULAR system with the other body systems presented so far.

Integumentary System

- The Integumentary System has mast cells that trigger localized changes in blood flow and capillary permeability
- The cardiovascular system delivers immune system cells to injury sites; clotting response seals breaks in skin surface; carries away toxins from sites of infection; provides heat

Skeletal System

- The Skeletal System provides calcium needed for normal cardiac muscle contraction; protects blood cells developing in red bone marrow
- The cardiovascular system transports calcium and phosphate for bone deposition; delivers EPO to red bone marrow, parathyroid hormone and calcitonin to osteoblasts and osteoclasts

Muscular System

- The Muscular System assists venous circulation through skeletal muscle contractions; protects superficial blood vessels, especially in neck and limbs
- The cardiovascular system delivers oxygen and nutrients, removes carbon dioxide, lactate, and heat during skeletal muscle activity

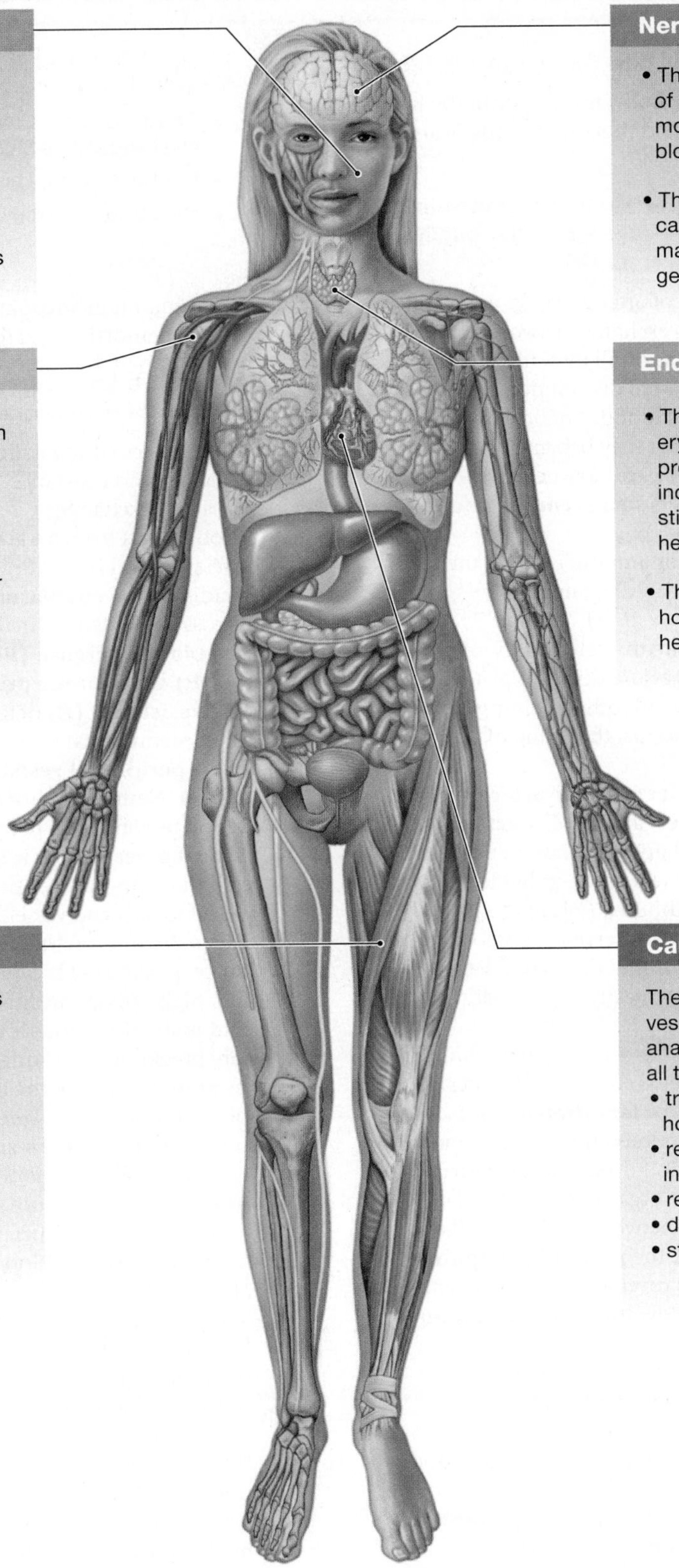

Nervous System

- The Nervous System controls patterns of circulation in peripheral tissues; modifies heart rate and regulates blood pressure; releases ADH
- The cardiovascular system has capillaries whose endothelial cells maintain the blood brain barrier; helps generate CSF

Endocrine System

- The Endocrine System produces erythropoietin (EPO), which regulates production of RBCs; several hormones increase blood pressure; epinephrine stimulates cardiac muscle, increasing heart rate and force of contraction
- The cardiovascular system distributes hormones throughout the body; the heart secretes ANP and BNP

Cardiovascular System

The cardiovascular system has blood vessels that provide extensive anatomical connections between it and all the other organ systems. It:

- transports dissolved gases, nutrients, hormones, and metabolic wastes
- regulates pH and ion composition of interstitial fluid
- restricts fluid losses at injury sites
- defends against toxins and pathogens
- stabilizes body temperature

21 Chapter Review

Study Outline

An Introduction to Blood Vessels and Circulation p. 728

1. Blood circulates throughout the body, moving from the heart, then through the body tissues, and then back to the heart, in tubular structures called *blood vessels.*

21-1 Arteries, which are elastic or muscular, and veins, which contain valves, have three-layered walls; capillaries have thin walls with only one layer p. 728

2. Blood flows through a network of arteries, veins, and capillaries. All chemical and gaseous exchange between blood and interstitial fluid takes place across capillary walls.
3. **Arteries** and **veins** form an internal distribution system through which the heart propels blood. Arteries branch repeatedly, decreasing in size until they become **arterioles**. From the arterioles, blood enters **capillary** networks. Blood flowing from the capillaries enters small **venules** before entering larger veins.
4. The walls of arteries and veins contain three layers: the innermost **tunica intima**, the **tunica media**, and the outermost **tunica externa**. *(Figure 21–1)*
5. In general, the walls of arteries are thicker than those of veins. Arteries constrict when blood pressure does not distend them, but veins constrict very little. The endothelial lining cannot contract, so when constriction occurs, the lining of an artery becomes folded. *(Figure 21–1)*
6. The arterial system includes the large **elastic arteries**, medium-sized **muscular arteries**, and smaller **arterioles**. As blood proceeds toward the capillaries, the number of vessels increases, but the luminal diameters of the individual vessels decrease and the walls become thinner. *(Figure 21–2)*
7. **Atherosclerosis**, a type of **arteriosclerosis**, is associated with changes in the endothelial lining of arteries. Fatty masses of tissue called **plaques** typically develop during atherosclerosis.
8. Capillaries are the only blood vessels whose walls are thin enough to permit an exchange between blood and interstitial fluid. Capillaries are **continuous** or **fenestrated**. **Sinusoids** have fenestrated walls, a discontinuous endothelium, and form networks that allow very slow blood flow. Sinusoids occur in the liver and in various endocrine organs. *(Figure 21–3)*
9. Capillaries form interconnected networks called **capillary beds.** A band of smooth muscle, the **precapillary sphincter**, adjusts the blood flow into each capillary. The entire capillary bed may be bypassed by blood flow through **arteriovenous anastomoses**. *(Figure 21–4)*
10. Venules collect blood from the capillaries and merge into **medium-sized veins** and then **large veins**. The arterial system is a high-pressure system; blood pressure in veins is much lower. **Valves** in veins prevent the backflow of blood. *(Figures 21–1, 21–2, 21–5)*
11. Veins expand easily at low pressures and have high **capacitance**. Veins act as **blood reservoirs** and can accommodate large changes in blood volume. Peripheral **venoconstriction** helps maintain adequate blood volume in the arterial system after a hemorrhage. *(Figure 21–6)*

> MasteringA&P™ Access more chapter study tools online in the MasteringA&P Study Area:
> - Chapter Quizzes, Chapter Practice Test, MP3 Tutor Sessions, and Clinical Case Studies
> - Practice Anatomy Lab PAL 3.0
> - Interactive Physiology iP2
> - A&P Flix A&PFlix
> - PhysioEx PhysioEx 9.1

21-2 Pressure and resistance determine blood flow and affect rates of capillary exchange p. 737

12. Cardiovascular regulation involves the manipulation of blood pressure and resistance to control the rates of blood flow and capillary exchange.
13. Blood flows from an area of higher pressure to one of lower pressure, and blood flow is proportional to the pressure gradient. The **circulatory pressure** is the pressure gradient across the systemic circuit. It is reported as three values: arterial **blood pressure (BP)**, **capillary hydrostatic pressure (CHP)**, and **venous pressure**.
14. The **resistance (*R*)** determines the rate of blood flow through the systemic circuit. The major determinant of blood flow rate is the **peripheral resistance**—the resistance of the arterial system. Neural and hormonal control mechanisms regulate blood pressure and peripheral resistance.
15. **Vascular resistance** is the resistance of blood vessels. It is the largest component of peripheral resistance and depends on vessel length and vessel luminal diameter. *(Figure 21–7)*
16. **Viscosity** and **turbulence** also contribute to peripheral resistance. *(Table 21–1)*
17. The high arterial pressures overcome peripheral resistance and maintain blood flow through peripheral tissues. Capillary pressures are normally low, and small changes in capillary pressure determine the rate of movement of fluid into or out of the bloodstream. Venous pressure, normally low, determines **venous return** and affects cardiac output and peripheral blood flow. *(Figures 21–8, 21–9; Table 21–1)*
18. Arterial blood pressure rises during ventricular systole and falls during ventricular diastole. The difference between these two blood pressures is the **pulse pressure**. *(Figures 21–8, 21–9)*
19. Valves, muscular compression, and the **respiratory pump** help the relatively low venous pressures propel blood toward the heart. *(Figures 21–5, 21–8)*
20. At the capillaries, hydrostatic pressure (blood pressure) forces water and solutes out of the plasma, across capillary walls. Water moves out of the capillaries, through the peripheral tissues, and back to the bloodstream by way of the lymphatic system. Water movement across capillary walls is determined by the interplay between hydrostatic pressures and osmotic pressures. *(Figure 21–10)*

21. **Osmotic pressure (OP)** is a measure of the pressure that must be applied to prevent osmotic movement across a membrane. Osmotic water movement continues until either solute concentrations are equalized or the movement is prevented by an opposing hydrostatic pressure.
22. The rates of filtration and reabsorption gradually change as blood passes along the length of a capillary, as determined by the **net filtration pressure** (the difference between the net hydrostatic pressure and the net osmotic pressure). *(Figure 21–11)*

21-3 Blood flow and pressure in tissues are controlled by both autoregulation and central regulation p. 745

23. Homeostatic mechanisms ensure that **tissue perfusion** (blood flow) delivers adequate oxygen and nutrients.
24. Autoregulation, neural mechanisms, and endocrine mechanisms influence the coordinated regulation of cardiovascular function. Autoregulation involves local factors changing the pattern of blood flow within capillary beds in response to chemical changes in interstitial fluids. Neural mechanisms respond to changes in arterial pressure or blood gas levels. Hormones can assist in short-term adjustments (changes in cardiac output and peripheral resistance) and long-term adjustments (changes in blood volume that affect cardiac output and gas transport). *(Figure 21–12)*
25. Peripheral resistance is adjusted at the tissues by local factors that result in the dilation or constriction of precapillary sphincters. *(Figure 21–4)*
26. **The cardiovascular (CV) center** of the medulla oblongata is responsible for adjusting cardiac output and peripheral resistance to maintain adequate blood flow. Its vasomotor center contains one group of neurons responsible for controlling vasoconstriction, and another group responsible for controlling vasodilation. Its cardiac centers include a *cardioacceleratory center* (which increases cardiac output through sympathetic innervation) and *cardioinhibitory center* (which decreases cardiac output through parasympathetic innervation).
27. **Baroreceptor reflexes** monitor the degree of stretch within expandable organs. Baroreceptors are located in the **carotid sinuses**, the **aortic arch**, and the right atrium. *(Figure 21–13)*
28. **Chemoreceptor reflexes** respond to changes in the oxygen or CO_2 levels in the blood. They are triggered by sensory neurons located in the **carotid bodies** and the **aortic bodies**. *(Figure 21–14)*
29. The endocrine system provides short-term regulation of cardiac output and peripheral resistance with epinephrine and norepinephrine from the adrenal medullae. Hormones involved in the long-term regulation of blood pressure and volume are *antidiuretic hormone (ADH)*, *angiotensin II*, *erythropoietin (EPO)*, and *natriuretic peptides (ANP* and *BNP)*. *(Figure 21–15)*

21-4 The cardiovascular system adapts to physiological stress while maintaining a special vascular supply to the brain, heart, and lungs p. 752

30. The blood brain barrier, the coronary circulation, and the capillary network in the lungs are examples of special circulations, in which cardiovascular dynamics and regulatory mechanisms differ from those in other tissues.
31. During exercise, blood flow to skeletal muscles increases at the expense of blood flow to nonessential organs, and cardiac output rises. Cardiovascular performance improves with training. Athletes have larger stroke volumes, slower resting heart rates, and larger cardiac reserves than do nonathletes. *(Tables 21–2, 21–3)*
32. Blood loss lowers blood volume and venous return and decreases cardiac output. Compensatory mechanisms include an increase in cardiac output, mobilization of venous blood reservoirs, peripheral vasoconstriction, and the release of hormones that promote the retention of fluids and the manufacture of erythrocytes. *(Figure 21–16)*

21-5 The vessels of the cardiovascular system make up both pulmonary and systemic circuits p. 756

33. The peripheral distributions of arteries and veins are generally identical on both sides of the body, except near the heart. *(Figure 21–17)*

21-6 In the pulmonary circuit, deoxygenated blood enters the lungs in arteries, and oxygenated blood leaves the lungs by veins p. 757

34. The pulmonary circuit includes the pulmonary trunk, the **left** and **right pulmonary arteries**, and the **pulmonary veins**, which empty into the left atrium. *(Figure 21–18)*

21-7 The systemic circuit carries oxygenated blood from the left ventricle to tissues and organs other than the lungs, and returns deoxygenated blood to the right atrium p. 758

35. The **ascending aorta** gives rise to the coronary circulation. The **aortic arch** communicates with the **descending aorta**. *(Figures 21–19 to 21–25)*
36. Three elastic arteries originate along the aortic arch: the **left common carotid artery**, the **left subclavian artery**, and the **brachiocephalic trunk**. *(Figures 21–20 to 21–22)*
37. The remaining major arteries of the body originate from the **descending aorta**. *(Figures 21–23 to 21–25)*
38. Arteries in the neck and limbs are deep beneath the skin; in contrast, there are generally two sets of peripheral veins, one superficial and one deep. This dual venous drainage is important for controlling body temperature. *(Figure 21–26)*
39. The **superior vena cava** receives blood from the head, neck, chest, shoulders, and arms. *(Figures 21–26 to 21–29)*
40. The **inferior vena cava** collects most of the venous blood from organs inferior to the diaphragm. *(Figures 21–28 to 21–30)*
41. The **hepatic portal system** directs blood from the other digestive organs to the liver before the blood returns to the heart. *(Figure 21–31)*

21-8 Modifications of fetal and maternal cardiovascular systems promote the exchange of materials; the fetal cardiovascular system changes to function independently after birth p. 775

42. Blood flows to the placenta in a pair of **umbilical arteries** and from the placenta in a single **umbilical vein**. *(Figure 21–32)*
43. The interatrial partition remains functionally incomplete until birth. The **foramen ovale** allows blood to flow freely from the right to the left atrium, and the **ductus arteriosus** short-circuits the pulmonary trunk.
44. At birth, the foramen ovale closes, leaving the fossa ovalis. The ductus arteriosus constricts, leaving the ligamentum arteriosum. *(Figure 21–32)*
45. Congenital cardiovascular problems generally reflect abnormalities of the heart or of interconnections between the heart and great vessels. *(Spotlight Figure 21–33)*

21-9 Aging affects the blood, heart, and blood vessels p. 778

46. Age-related changes in the blood include (1) a decreased hematocrit, (2) constriction or blockage of peripheral veins by a *thrombus* (stationary blood clot), and (3) pooling of blood in the veins of the legs because valves are not working effectively.
47. Age-related changes in the heart include (1) a reduction in the maximum cardiac output, (2) changes in the activities of nodal and conducting cells, (3) a reduction in the elasticity of the cardiac skeleton, (4) progressive atherosclerosis that can restrict coronary circulation, and (5) the replacement of damaged cardiac muscle cells by scar tissue.
48. Age-related changes in blood vessels, commonly related to arteriosclerosis, include (1) a weakening in the walls of arteries, potentially leading to the formation of an *aneurysm;* (2) deposition of calcium salts on weakened vascular walls, increasing the risk of a stroke or myocardial infarction; (3) lipid deposits in the tunica media associated with damaged endothelial lining and calcium salts forming atherosclerotic plaques; and (4) the formation of a thrombus at atherosclerotic plaques.
49. The cardiovascular system is anatomically and functionally connected to all other body systems. *(Figure 21–34)*

Review Questions

See the blue Answers tab at the back of the book.

LEVEL 1 Reviewing Facts and Terms

1. Identify the major arteries in the following diagram.

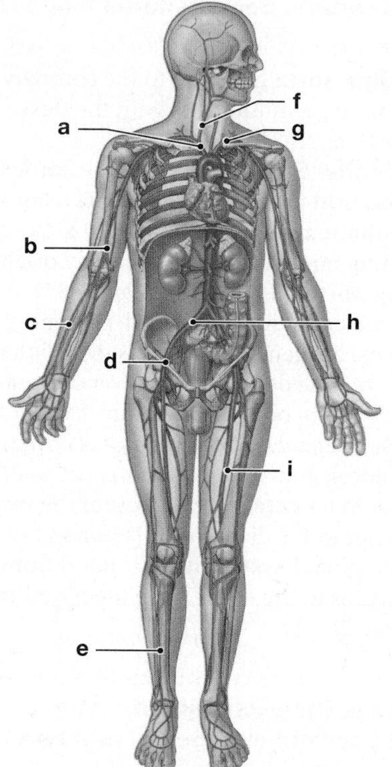

a
b
c
d
e
f
g
h
i

(a) ________ (b) ________
(c) ________ (d) ________
(e) ________ (f) ________
(g) ________ (h) ________
(i) ________

2. The blood vessels that play the most important role in regulating blood pressure and blood flow to a tissue are the **(a)** arteries, **(b)** arterioles, **(c)** veins, **(d)** venules, **(e)** capillaries.
3. Cardiovascular function is regulated by all of the following *except* **(a)** local factors, **(b)** neural factors, **(c)** endocrine factors, **(d)** venous return, **(e)** conscious control.
4. Baroreceptors that function in the regulation of blood pressure are located in the **(a)** left ventricle, **(b)** brainstem, **(c)** carotid sinus, **(d)** common iliac artery, **(e)** pulmonary trunk.
5. The two-way exchange of substances between blood and body cells occurs only through **(a)** arterioles, **(b)** capillaries, **(c)** venules, **(d)** all of these.
6. Large molecules such as peptides and proteins move into and out of the bloodstream by way of **(a)** continuous capillaries, **(b)** fenestrated capillaries, **(c)** thoroughfare channels, **(d)** venules.
7. The local control of blood flow due to the action of precapillary sphincters is **(a)** vasomotion, **(b)** autoregulation, **(c)** selective resistance, **(d)** turbulence.
8. Blood is transported through the venous system by means of **(a)** skeletal muscle contractions, **(b)** decreasing blood pressure, **(c)** the respiratory pump, **(d)** a and c.
9. Identify the major veins in the following diagram.

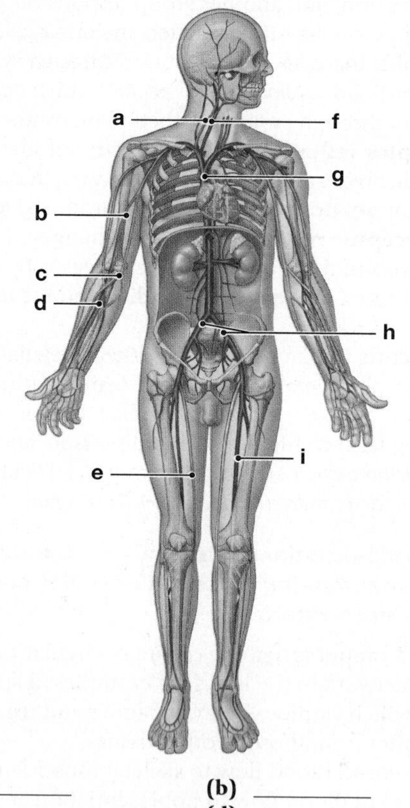

a
b
c
d
e
f
g
h
i

(a) ________ (b) ________
(c) ________ (d) ________
(e) ________ (f) ________
(g) ________ (h) ________
(i) ________

10. The most important factor in vascular resistance is **(a)** the viscosity of the blood, **(b)** the diameter of the lumen of blood vessels, **(c)** turbulence due to irregular surfaces of blood vessels, **(d)** the length of the blood vessels.
11. Net hydrostatic pressure forces water ________ a capillary; net osmotic pressure reabsorbs water ________ a capillary. **(a)** into, out of, **(b)** out of, into, **(c)** out of, out of, **(d)** into, into.
12. The two arteries formed by the division of the brachiocephalic trunk are the **(a)** aorta and internal carotid, **(b)** axillary and brachial, **(c)** external and internal carotid, **(d)** common carotid and subclavian.
13. The unpaired arteries supplying blood to the visceral organs include **(a)** the adrenal, renal, and lumbar arteries, **(b)** the iliac, gonadal, and femoral arteries, **(c)** the celiac and superior and inferior mesenteric arteries, **(d)** all of these.
14. The paired arteries supplying blood to the body wall and other structures outside the abdominopelvic cavity include the **(a)** left gastric, hepatic, splenic, and phrenic arteries, **(b)** adrenal, colic, lumbar, and gonadal arteries, **(c)** iliac, femoral, and lumbar arteries, **(d)** celiac, left gastric, and superior and inferior mesenteric arteries.
15. The vein that drains the dural sinuses of the brain is the **(a)** cephalic vein, **(b)** great saphenous vein, **(c)** internal jugular vein, **(d)** superior vena cava.
16. The vein that collects most of the venous blood inferior to the diaphragm is the **(a)** superior vena cava, **(b)** great saphenous vein, **(c)** inferior vena cava, **(d)** azygos vein.
17. What are the primary forces that cause fluid to move **(a)** out of a capillary at its arterial end and into the interstitial fluid, **(b)** into a capillary at its venous end from the interstitial fluid?
18. What cardiovascular changes occur at birth?

LEVEL 2 Reviewing Concepts

19. A major difference between the arterial and venous systems is that **(a)** arteries are usually more superficial than veins, **(b)** in the limbs there is dual venous drainage, **(c)** veins are usually less branched compared to arteries, **(d)** veins exhibit a much more orderly pattern of branching in the limbs, **(e)** veins are not found in the abdominal cavity.
20. Which of the following conditions would have the *greatest* effect on peripheral resistance? **(a)** doubling the length of a vessel, **(b)** doubling the diameter of a vessel, **(c)** doubling the viscosity of the blood, **(d)** doubling the turbulence of the blood, **(e)** doubling the number of white cells in the blood.
21. Which of the following is *greater*? **(a)** the osmotic pressure of the interstitial fluid during inflammation, **(b)** the osmotic pressure of the interstitial fluid during normal conditions, **(c)** neither is greater.
22. Relate the anatomical differences between arteries and veins to their functions.
23. Why do capillaries permit the diffusion of materials, whereas arteries and veins do not?
24. How is blood pressure maintained in veins to counter the force of gravity?
25. How do pressure and resistance affect cardiac output and peripheral blood flow?
26. Why is blood flow to the brain relatively continuous and constant?
27. Compare the effects of the cardioacceleratory and cardioinhibitory centers on cardiac output and blood pressure.

LEVEL 3 Critical Thinking and Clinical Applications

28. Bob is sitting outside on a warm day and is sweating profusely. Mary wants to practice taking blood pressures, and he agrees to play the patient. Mary finds that Bob's blood pressure is elevated, even though he is resting and has lost fluid from sweating. (She reasons that fluid loss should lower blood volume and, thus, blood pressure.) Why is Bob's blood pressure high instead of low?
29. People with allergies commonly take antihistamines with decongestants to relieve their symptoms. The container warns that individuals who are being treated for high blood pressure should not take the medication. Why not?
30. Jolene awakens suddenly to the sound of her alarm clock. Realizing that she is late for class, she jumps to her feet, feels light-headed, and falls back on her bed. What probably caused this reaction? Why doesn't this happen all the time?

+ CLINICAL CASE Wrap-Up Did Ancient Mummies Have Atherosclerosis?

The majority of American men and women today have atherosclerotic plaques in their arteries by the age of 50. The disease is usually thought of as a modern one, the result of too many cheeseburgers and hours spent in front of computers and flat-screen TVs. But what could account for its incidence in the ancient mummies?

The degree of atherosclerosis discovered in the mummies turns out to be correlated with their age at death. The mean age at death for mummies without visible atherosclerosis is only 32. Among the mummies of people over age 40 that were examined, 50 percent had atherosclerosis.

Atherosclerosis is a vascular disease that progresses over many years. It is influenced by both genetics and environment. The most important risk factor, however, seems to be increasing age. Sex, diet, high levels of cholesterol in the blood, smoking, hypertension, and obesity are additional risk factors.

The prevalence of atherosclerosis in varied, preindustrial populations raises the possibility of a basic, human predisposition to the disease. Atherosclerosis appears to have been stalking humans for millennia.

1. Why does the finding of calcium, which is easily seen on a CT scan, within systemic arteries indicate the presence of atherosclerosis?
2. Why do you think atherosclerosis is never seen in veins?

See the blue Answers tab at the back of the book.

Related Clinical Terms

angiogram: An x-ray of a blood vessel that becomes visible due to a prior injection of dye into the subject's bloodstream.
carotid sinus massage: A procedure that involves rubbing the large part of the arterial wall at the point where the common carotid artery divides into its two main branches.
intermittent claudication: A limp that results from cramping leg pain that is typically caused by obstruction of the arteries.
Korotkoff sounds: Distinctive sounds, caused by turbulent arterial blood flow, heard through the stethoscope while measuring blood pressure .
normotensive: Having normal blood pressure.
orthostatic hypotension: A form of low blood pressure that occurs when you stand up from sitting or lying down. It can cause dizziness or a light-headed feeling.
phlebitis: Inflammation of a vein.
Raynaud phenomenon: A condition resulting in the discoloration of the fingers and/or the toes when a person is subjected to changes in temperature or to emotional stress.
sclerotherapy: The treatment of varicose veins in which an irritant is injected to cause inflammation, coagulation of blood, and a narrowing of the blood vessel wall.
sphygmomanometer: A device that measures blood pressure using an inflatable cuff placed around a limb.
syncope: A temporary loss of consciousness due to a sudden drop in blood pressure.
thrill: A vibration felt in a blood vessel that usually occurs due to abnormal blood flow. It is also often noticed at the fistula of a hemodialysis patient.
thrombophlebitis: An inflammation in a vein associated with the formation of a thrombus (clot).
vascular murmur: Periodic abnormal sounds heard upon auscultation that are produced as a result of turbulent blood flow.
white coat hypertension: A short-term increase in blood pressure triggered by the sight of medical personnel in white coats or other medical attire.

22 The Lymphatic System and Immunity

Learning Outcomes

These Learning Outcomes correspond by number to this chapter's sections and indicate what you should be able to do after completing the chapter.

CLINICAL CASE Isn't There a Vaccine for That?

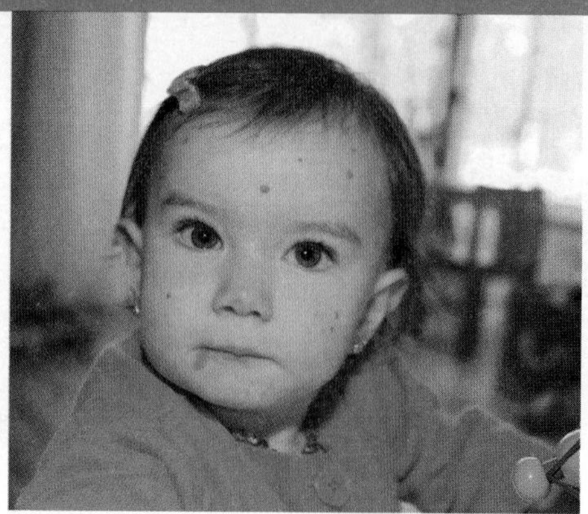

Baby Ruthie is standing and holding on to a play table when her mother arrives to pick her up from day care. Ruthie is 11 months old and not yet walking. She hasn't had an appetite since yesterday, and last night she had a fever. When her mother gives Ruthie a hug, she notices a small blister near the corner of Ruthie's mouth. She also notices that Ruthie seems feverish again.

"Ruthie didn't do much today," reports the day care attendant. "Did you notice this little blister on her face?" asks Ruthie's mother. "No. It must have just appeared," the attendant replies. "Tommy, one of our older kids, had a little blister like that on his cheek just last week. He hasn't been back since."

That night, Ruthie's fever worsens. She won't eat and seems to have a sore throat when trying to swallow. She is listless. When her mother undresses her for bed, she notices more small blisters on Ruthie's face, abdomen, and back. She takes her straight to the Urgent Care Center.

"This looks like chickenpox," says the doctor. "Thanks to vaccines, we don't see this disease very often any more." **Ruthie's mother is certain that Ruthie is current on all her vaccinations. How could Ruthie have come down with chickenpox? To find out, turn to the Clinical Case Wrap-Up on p. 833.**

An Introduction to the Lymphatic System and Immunity

The world is not always kind to the human body. Accidental bumps, cuts, and scrapes; chemical and thermal burns; extreme cold; and ultraviolet radiation are just a few of the hazards in our physical environment. Making matters worse, the world around us contains an assortment of viruses, bacteria, fungi, and parasites capable of not only surviving but thriving inside our bodies—and potentially causing us great harm. These disease-causing organisms, called **pathogens**, are responsible for many diseases. Each pathogen has a different lifestyle and attacks the body in a specific way. For example, viruses spend most of their time hidden within cells, which they often eventually destroy. Many bacteria multiply in interstitial fluids, where they release foreign proteins—enzymes or toxins—that can damage cells, tissues, even entire organ systems. Some large parasites, such as roundworms, actually burrow through internal organs. And as if that were not enough, we are constantly at risk from renegade body cells that have the potential to produce lethal cancers. ⮌ p. 107

Many organs and systems work together to keep us alive and healthy. One system that plays a central role in the ongoing struggle to maintain health is the lymphatic system. The **lymphatic system** (also called the *lymphoid system*) includes the cells, tissues, and organs responsible for defending the body. This system acts both against environmental hazards, such as various pathogens, and against internal threats, such as cancer cells. We introduced *lymphocytes*, the primary cells of the lymphatic system, in Chapters 4 and 19. ⮌ pp. 136, 675 These cells are vital to the body's ability to resist or overcome infection and disease.

The ability to resist infection and disease is called **immunity**. All the cells and tissues involved in producing immunity are part of the **immune system**. The immune system is a *functional* system, and it includes parts of the integumentary, skeletal, lymphatic, cardiovascular, respiratory, and digestive systems.

We begin by examining the organization of the lymphatic system, lymphoid tissues, and lymphoid organs, and then we look at how this system interacts with other systems to defend the body against infection and disease.

22-1 The vessels, tissues, and organs of the lymphatic system maintain fluid volume and function in body defenses

Learning Outcome Identify the major components of the lymphatic system, describe the structure and functions of each component, and discuss the importance of lymphocytes.

The lymphatic system consists of (1) *lymph*, a fluid that resembles plasma but contains a much lower concentration of suspended proteins; (2) a network of *lymphatic vessels*, which begin in peripheral tissues and connect to veins; (3) an array of **lymphoid tissues** and **lymphoid organs** scattered throughout the body; and (4) *lymphoid cells*, including lymphocytes and smaller numbers of phagocytes and other cells.

Primary lymphoid tissues and organs are sites where lymphocytes are formed and mature. They include *red bone marrow* and the *thymus*. Recall that red bone marrow, covered in Chapter 19, is where other defense cells, the monocytes and macrophages, are also formed. ⮌ p. 674 *Secondary lymphoid tissues and organs* are where lymphocytes are activated. These structures include the *tonsils*, *MALT* (mucosa-associated

Figure 22–1 The Components of the Lymphatic System.

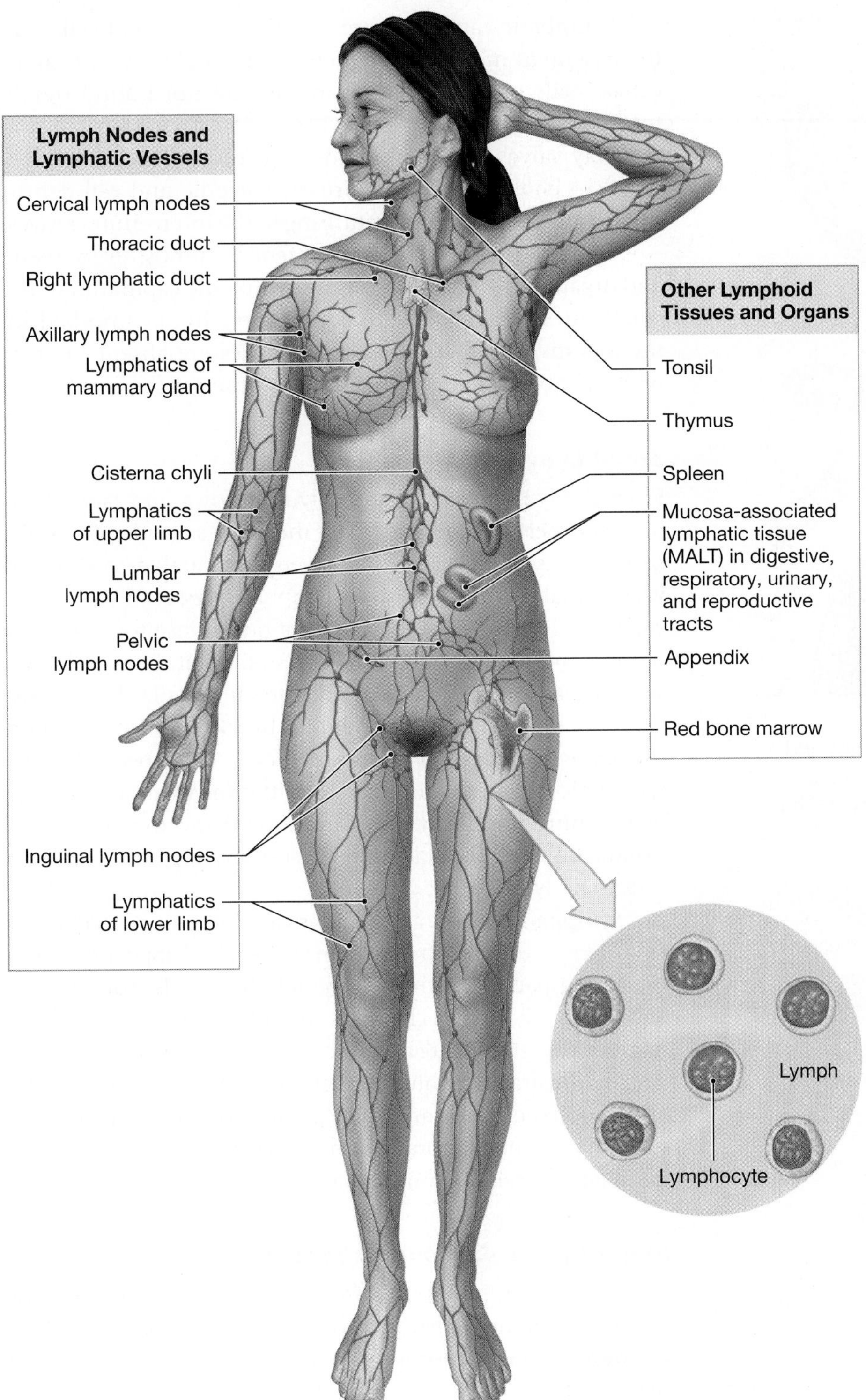
Lymph Nodes and Lymphatic Vessels
Cervical lymph nodes
Thoracic duct
Right lymphatic duct
Axillary lymph nodes
Lymphatics of mammary gland
Cisterna chyli
Lymphatics of upper limb
Lumbar lymph nodes
Pelvic lymph nodes
Inguinal lymph nodes
Lymphatics of lower limb
Other Lymphoid Tissues and Organs
Tonsil
Thymus
Spleen
Mucosa-associated lymphatic tissue (MALT) in digestive, respiratory, urinary, and reproductive tracts
Appendix
Red bone marrow
Lymph
Lymphocyte

lymphatic tissue), *lymph nodes,* and the *spleen.* Figure 22–1 provides a general overview of the components of the lymphatic system.

Functions of the Lymphatic System

The primary function of the lymphatic system is to produce, maintain, and distribute lymphocytes and other lymphoid cells that provide defense against infections and foreign substances. To provide an effective defense, lymphocytes must detect problems, and they must be able to reach the site of injury or infection. Lymphocytes and other cells circulate within the blood. They are able to enter or leave the capillaries that supply most of the body's tissues.

As noted in Chapter 21, capillaries normally deliver more fluid to peripheral tissues than they carry away. ⮌ p. 743 The excess fluid returns to the bloodstream through lymphatic vessels. This continuous circulation of extracellular fluid helps transport lymphocytes and white blood cells from one organ to another. In the process, it maintains normal blood volume. It also eliminates local variations in the composition of the interstitial fluid by distributing hormones, nutrients, and wastes from their tissues of origin to the general circulation.

Lymphatic Vessels and Circulation of Lymph

Lymphatic vessels, often called *lymphatics,* carry lymph from peripheral tissues to the venous system. **Lymph** is interstitial fluid that has entered lymphatic vessels, which vary in size. Lymph first enters lymphatic capillaries and then drains into larger *major lymph-collecting vessels*, known as *trunks* and *ducts*.

Lymphatic Capillaries

The lymphatic network begins with **lymphatic capillaries**, which branch through peripheral tissues. Lymphatic capillaries differ from blood capillaries in several ways. They (1) are closed at one end rather than forming continuous tubes, (2) have larger luminal diameters, (3) have thinner

22

Figure 22–2 Lymphatic Capillaries.

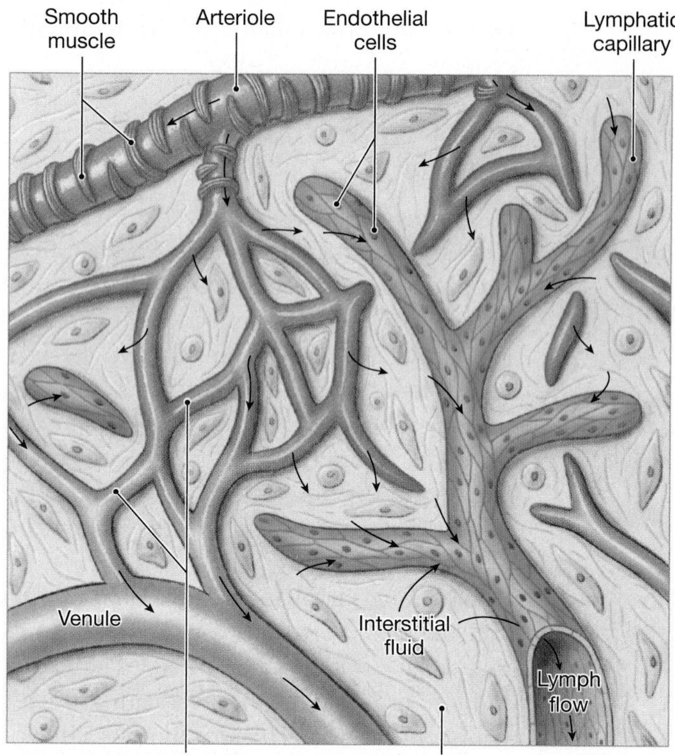
Smooth muscle
Arteriole
Endothelial cells
Lymphatic capillary
Venule
Interstitial fluid
Lymph flow
Blood capillaries
Areolar tissue

a The interwoven network formed by blood capillaries and lymphatic capillaries. Arrows indicate the movement of fluid out of blood capillaries and the net flow of interstitial fluid and lymph.

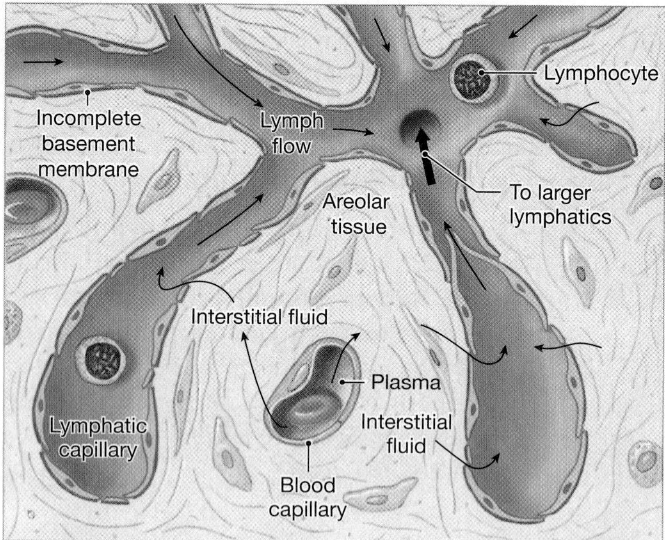
Lymphocyte
Incomplete basement membrane
Lymph flow
Areolar tissue
To larger lymphatics
Interstitial fluid
Plasma
Lymphatic capillary
Interstitial fluid
Blood capillary

b A sectional view indicating the movement of fluid from the plasma, through the tissues as interstitial fluid, and into the lymphatic system as lymph.

walls, and (4) typically have a flattened or irregular outline in sectional view (Figure 22–2).

Lymphatic capillaries are lined by endothelial cells, but the basement membrane is incomplete or absent. The endothelial cells of a lymphatic capillary are not bound tightly together, but they do overlap. The region of overlap acts as a one-way valve. It permits fluids and solutes (including proteins) to enter, along with viruses, bacteria, and cell debris, but it prevents them from returning to the intercellular spaces.

Lymphatic capillaries are present in almost every tissue and organ in the body. Important lymphatic capillaries in the small intestine, called *lacteals*, transport lipids absorbed by the digestive tract. Lymphatic capillaries are absent in areas without a blood supply, such as the cornea of the eye.

Small Lymphatic Vessels

From the lymphatic capillaries, lymph flows into larger lymphatic vessels that lead toward the body's trunk. The walls of these vessels contain three layers comparable to those of veins. Small to medium-sized lymphatic vessels also contain valves (Figure 22–3). The valves are quite close together, and produce noticeable bulges. As a result, some lymphatic vessels have a beaded appearance due to an irregular lumen (see Figure 22–3a). The valves prevent the backflow of lymph within lymphatic vessels, especially in the limbs. Pressures within the lymphatic system are minimal, and the valves are essential to maintaining normal lymph flow toward the thoracic cavity. Contractions of skeletal muscles surrounding the lymphatic vessels aid lymph flow.

Lymphatic vessels are often associated with blood vessels (see Figure 22–3a). Differences in size, general appearance, and branching pattern distinguish lymphatic vessels from arteries and veins. We can see characteristic color differences in living tissues. Most arteries are bright red, veins are dark red (although usually illustrated as blue to distinguish them from arteries), and lymphatic vessels are a pale golden color. In general, a tissue contains many more lymphatic vessels than veins, but the lymphatic vessels are much smaller.

Major Lymph-Collecting Vessels

Two sets of lymphatic vessels collect lymph from the lymphatic capillaries: superficial lymphatics and deep lymphatics. **Superficial lymphatics** are located in the subcutaneous layer deep to the skin; in the areolar tissues of the mucous membranes lining the digestive, respiratory, urinary, and reproductive tracts; and in the areolar tissues of the serous membranes lining the pleural, pericardial, and peritoneal cavities. **Deep lymphatics** are larger lymphatic vessels that accompany deep arteries and veins supplying skeletal muscles and other organs of the neck, limbs, and trunk, and the walls of visceral organs.

Figure 22–3 Lymphatic Vessels and Valves.

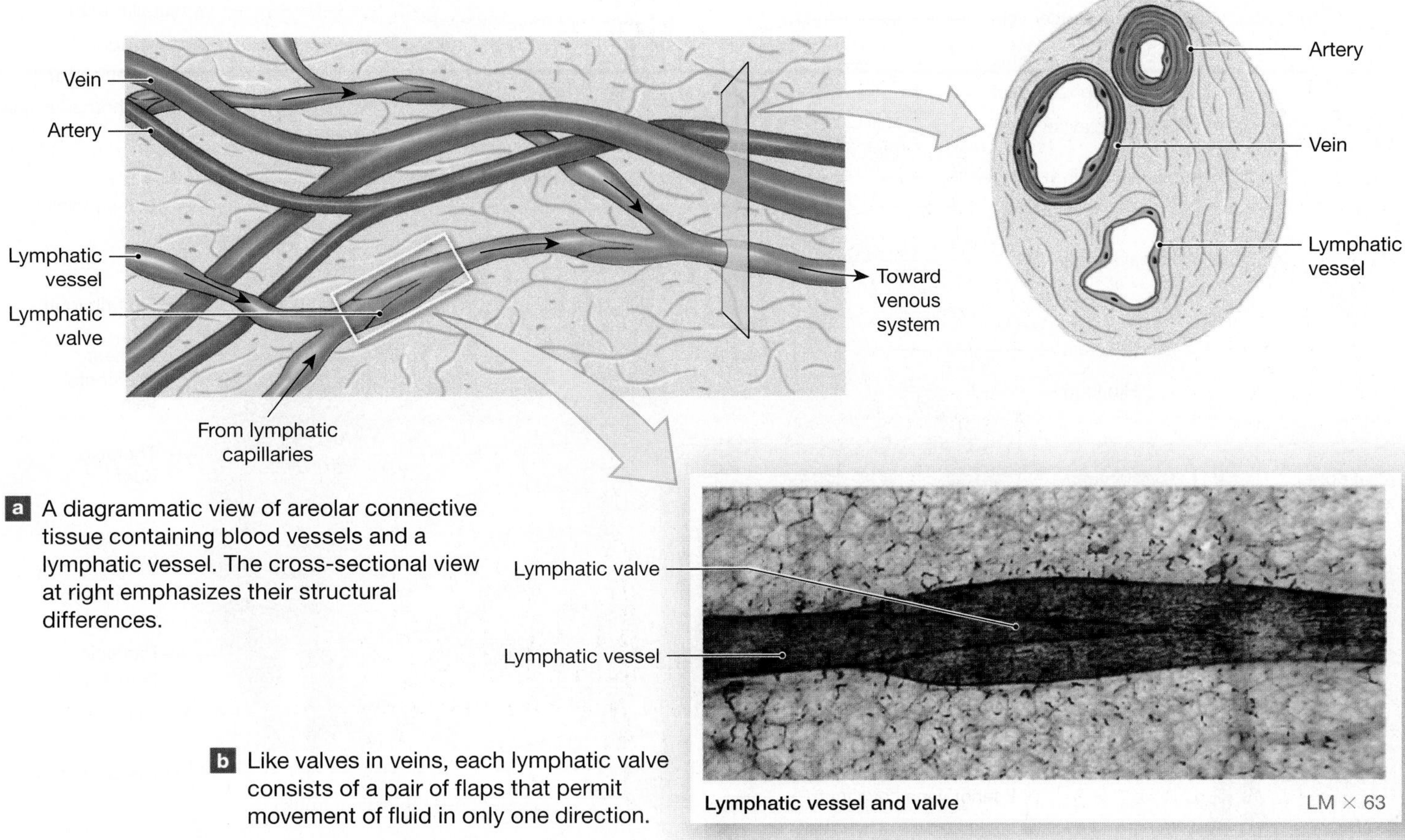
Vein
Artery
Lymphatic vessel
Lymphatic valve
From lymphatic capillaries
Toward venous system
Artery
Vein
Lymphatic vessel
Lymphatic valve
Lymphatic vessel
Lymphatic vessel and valve
LM × 63

a A diagrammatic view of areolar connective tissue containing blood vessels and a lymphatic vessel. The cross-sectional view at right emphasizes their structural differences.

b Like valves in veins, each lymphatic valve consists of a pair of flaps that permit movement of fluid in only one direction.

? What feature do lymphatic vessels share with veins that allows lymph to flow in only one direction?

Superficial and deep lymphatics converge to form even larger vessels called **lymphatic trunks**. The trunks in turn empty into two large collecting vessels: the thoracic duct and the right lymphatic duct. The **thoracic duct** collects lymph from the body inferior to the diaphragm and from the left side of the body superior to the diaphragm. The smaller **right lymphatic duct** collects lymph from the right side of the body superior to the diaphragm (Figure 22–4a).

The thoracic duct begins inferior to the diaphragm at the level of vertebra L_2 (Figure 22–4b). The base of the thoracic duct is an expanded, saclike chamber called the **cisterna chyli** (KĪ-lī; *chylos*, juice). The cisterna chyli receives lymph from the inferior part of the abdomen, the pelvis, and the lower limbs by way of the *right* and *left lumbar trunks* and the *intestinal trunk*.

The inferior segment of the thoracic duct lies anterior to the vertebral column. From the second lumbar vertebra, it passes posterior to the diaphragm alongside the aorta. It then ascends along the left side of the vertebral column to the level of the left clavicle. It collects lymph from the *left bronchomediastinal trunk*, the *left subclavian trunk*, and the *left jugular trunk*, and then empties into the left subclavian vein near the left internal jugular vein (see Figure 22–4b). In this way, lymph reenters the venous circulation from the left side of the head, neck, and thorax, as well as from the entire body inferior to the diaphragm.

The *right lymphatic duct* is formed by the merging of the *right jugular*, *right subclavian*, and *right bronchomediastinal trunks* in the area near the right clavicle. This duct empties into the right subclavian vein, delivering lymph from the right side of the body superior to the diaphragm.

Obstruction of lymphatic vessels produces **lymphedema** (limf-e-DĒ-muh), as large amounts of lymph accumulate in the affected region. This causes swelling, especially in subcutaneous tissues. If the condition persists, the connective tissues lose their elasticity and the swelling becomes permanent. Lymphedema by itself does not pose a major threat to life. The danger comes from

Figure 22–4 **The Relationship between the Lymphatic Ducts and the Venous System.** ATLAS: Plates 48a,b

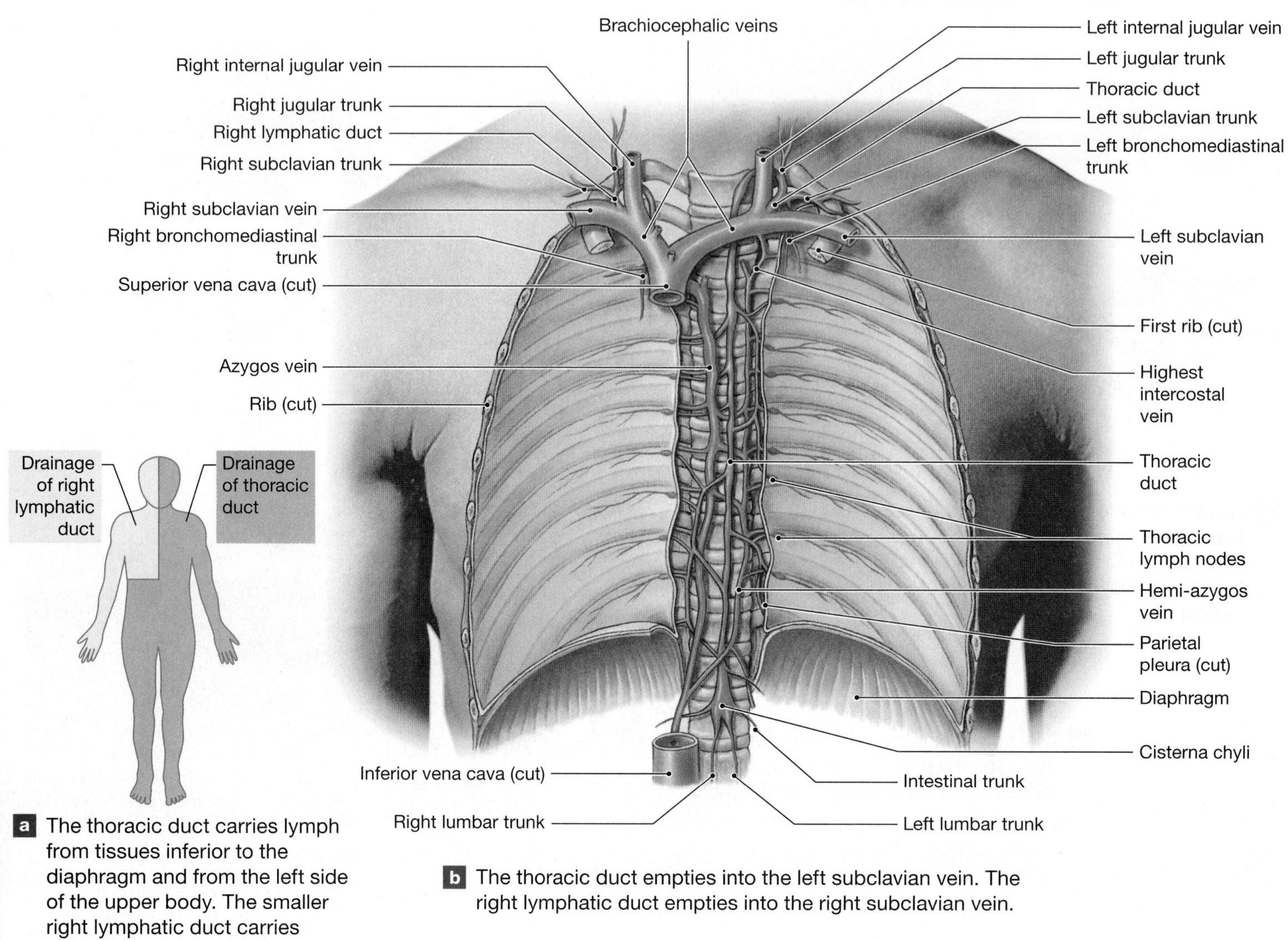
Brachiocephalic veins
Right internal jugular vein
Right jugular trunk
Right lymphatic duct
Right subclavian trunk
Right subclavian vein
Right bronchomediastinal trunk
Superior vena cava (cut)
Azygos vein
Rib (cut)
Drainage of right lymphatic duct
Drainage of thoracic duct
Left internal jugular vein
Left jugular trunk
Thoracic duct
Left subclavian trunk
Left bronchomediastinal trunk
Left subclavian vein
First rib (cut)
Highest intercostal vein
Thoracic duct
Thoracic lymph nodes
Hemi-azygos vein
Parietal pleura (cut)
Diaphragm
Cisterna chyli
Inferior vena cava (cut)
Intestinal trunk
Right lumbar trunk
Left lumbar trunk

a The thoracic duct carries lymph from tissues inferior to the diaphragm and from the left side of the upper body. The smaller right lymphatic duct carries lymph from the rest of the body.

b The thoracic duct empties into the left subclavian vein. The right lymphatic duct empties into the right subclavian vein.

the constant risk that an uncontrolled infection will develop in the affected area. Because the interstitial fluids are essentially stagnant, toxins and pathogens can accumulate and overwhelm local defenses without fully activating the immune system.

Lymphoid Cells

Lymphoid cells consist of immune system cells found in lymphoid tissues and the cells that support those tissues. Immune system cells that function in defense include phagocytes and lymphocytes. Phagocytes include cell types such as *macrophages* and *microphages*. These cells function as a general first line of defense against pathogens (such as bacteria or viruses).

Lymphocytes respond to specific invading pathogens, as well as to abnormal body cells (such as virus-infected cells or cancer cells) and foreign proteins (such as the toxins released by some bacteria). They eliminate these threats or render them harmless through a combination of physical and chemical attacks. Lymphocytes account for 20–40 percent of circulating leukocytes. Three classes of lymphocytes circulate in blood: (1) *T* (thymus-dependent) *cells*, (2) *B* (bone marrow–derived) *cells*, and (3) *NK* (natural killer) *cells*. Each type has distinctive biochemical and functional properties. However, circulating lymphocytes are only a small fraction of the total lymphocyte population. The body contains some 10^{12} lymphocytes, with a combined weight of more than a kilogram (2.2 lb).

Lymphoid Tissues

Lymphoid tissues are connective tissues dominated by lymphocytes. In a **lymphoid nodule**, or *lymphatic nodule*, the lymphocytes are densely packed in an area of areolar tissue. In many areas, lymphoid nodules form large clusters. Lymphoid nodules occur in the connective tissue deep to the epithelia

Figure 22–5 Lymphoid Nodules.

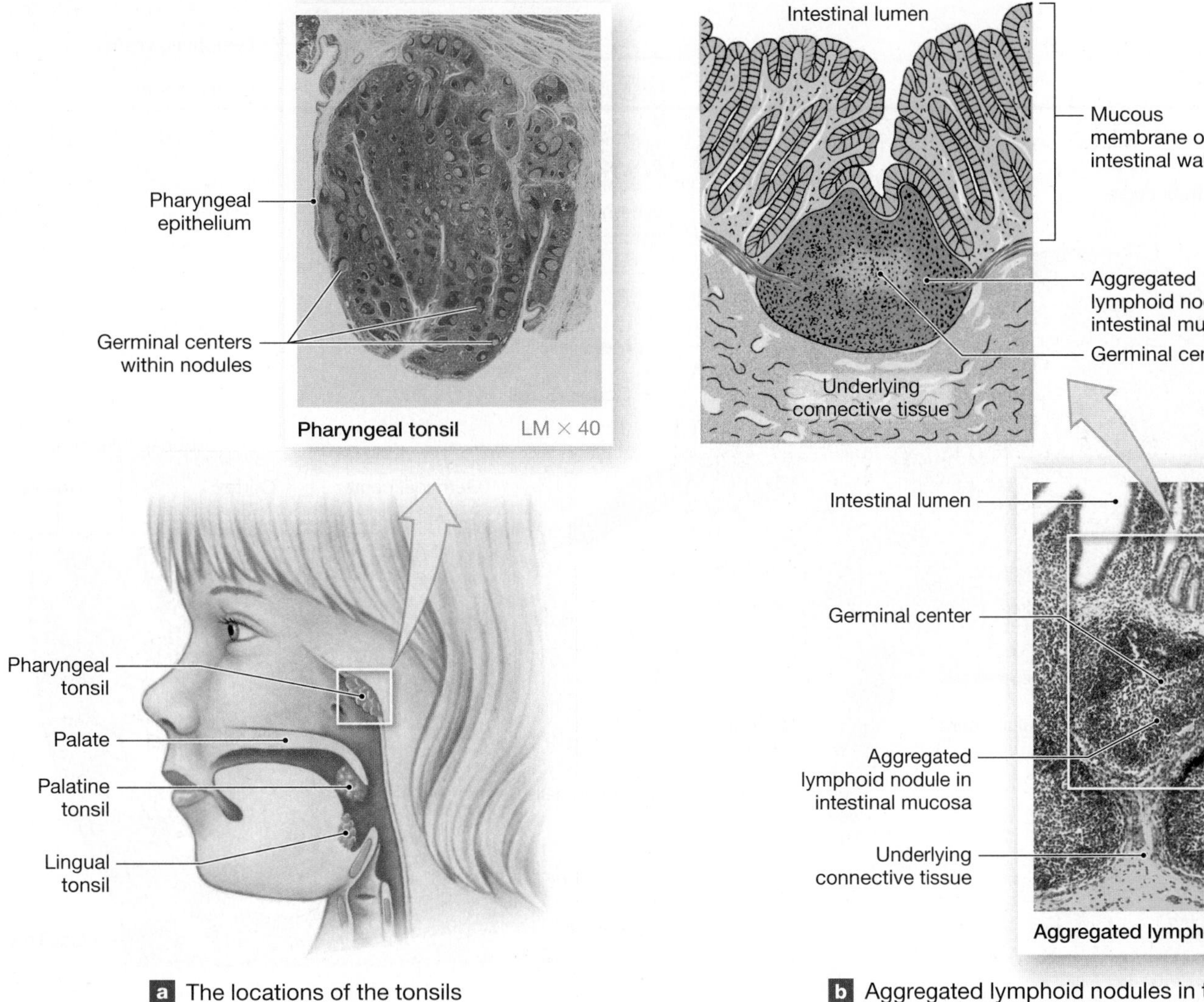
Pharyngeal epithelium
Germinal centers within nodules
Pharyngeal tonsil
LM × 40
Pharyngeal tonsil
Palate
Palatine tonsil
Lingual tonsil
Intestinal lumen
Mucous membrane of intestinal wall
Aggregated lymphoid nodule in intestinal mucosa
Germinal center
Underlying connective tissue
Intestinal lumen
Germinal center
Aggregated lymphoid nodule in intestinal mucosa
Underlying connective tissue
Aggregated lymphoid nodules LM × 20

a The locations of the tonsils

b Aggregated lymphoid nodules in the intestine

22

lining the respiratory tract, where they are known as *tonsils*, and along the digestive, respiratory, urinary, and reproductive tracts (Figure 22–5). They are also found within more complex lymphoid organs, such as lymph nodes or the spleen.

A single nodule averages about a millimeter in diameter. Its boundaries are not distinct, because no fibrous capsule surrounds it. Each nodule often has a central zone called a **germinal center**, which contains dividing lymphocytes (see Figure 22–5).

Tonsils

The **tonsils** are large lymphoid nodules in the walls of the pharynx (Figure 22–5a). Most people have five tonsils. A single **pharyngeal tonsil**, often called the *adenoid*, lies in the posterior superior wall of the nasopharynx. Left and right **palatine tonsils** are located at the posterior, inferior margin of the oral cavity, along the boundary of the pharynx. A pair of **lingual tonsils** lies deep to the mucous epithelium covering the base (pharyngeal portion) of the tongue. Because of their location, the lingual tonsils are usually not visible unless they become infected and swollen. **Tonsillitis** is an inflammation of the tonsils (especially the palatine tonsils) although the other tonsils may also be affected. Tonsils reach their largest size by puberty and then begin to atrophy.

MALT

The collection of lymphoid tissues that protect the epithelia of the digestive, respiratory, urinary, and reproductive systems is called the **mucosa-associated lymphoid tissue (MALT)**. Clusters of lymphoid nodules deep to the epithelial lining of the intestine are known as **aggregated lymphoid nodules**, or *Peyer's patches* (Figure 22–5b). Other examples of MALT include the appendix and the tonsils.

The *appendix*, or *vermiform* ("worm-shaped") *appendix*, is a tube-shaped sac opening into the junction between the small and large intestines. Its walls contain a mass of fused lymphoid nodules.

Figure 22–6 **The Structure of a Lymph Node.** ATLAS: Plate 70a

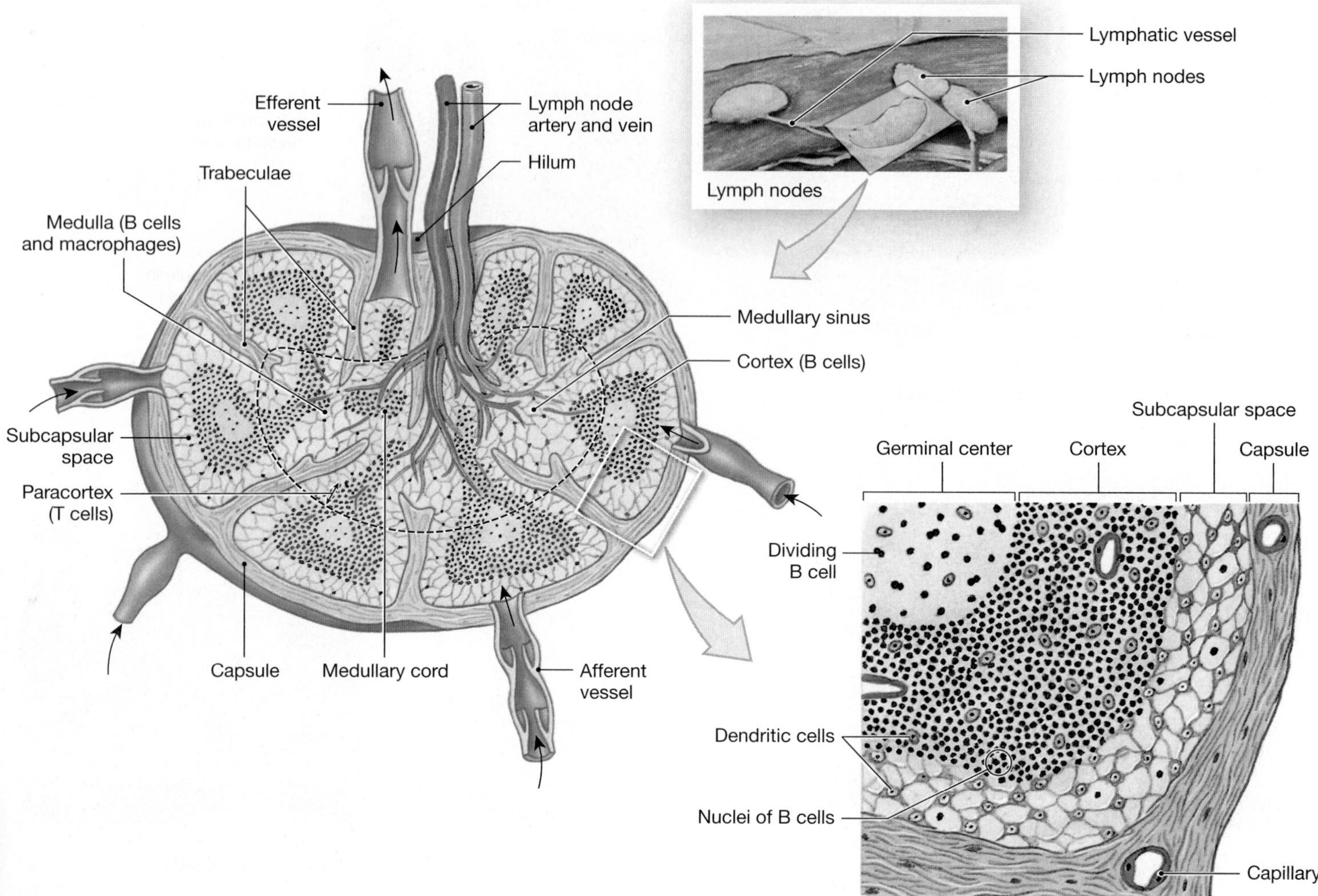
Lymphatic vessel
Lymph nodes
Lymph nodes
Efferent vessel
Lymph node artery and vein
Hilum
Trabeculae
Medulla (B cells and macrophages)
Medullary sinus
Cortex (B cells)
Subcapsular space
Paracortex (T cells)
Capsule
Medullary cord
Afferent vessel
Subcapsular space
Germinal center
Cortex
Capsule
Dividing B cell
Dendritic cells
Nuclei of B cells
Capillary

Lymphoid Organs

Lymphoid organs include the lymph nodes, thymus, and spleen. A fibrous connective tissue capsule separates these organs from surrounding tissues.

Lymph Nodes

Lymph nodes are small lymphoid organs ranging in diameter from 1 mm to 25 mm (to about 1 in.). The greatest number of lymph nodes is located in the neck, axillae, and groin, where they defend us against bacteria and other invaders. Look back at Figure 22–1 to see the general pattern of lymph node distribution in the body.

A dense connective tissue capsule covers each lymph node (Figure 22–6). Bundles of collagen fibers extend from the capsule into the interior of the node. These fibrous partitions are called **trabeculae** (*trabecula*, a beam).

The typical lymph node is shaped like a kidney bean (see Figure 22–6). Blood vessels and nerves reach the lymph node at a shallow indentation called the **hilum**. Two sets of lymphatic vessels, afferent lymphatics and efferent lymphatics, are connected to each lymph node. **Afferent** (*afferens*, to bring to) **lymphatics** bring lymph to the lymph node from peripheral tissues. The afferent lymphatics penetrate the capsule of the lymph node on the side opposite the hilum. **Efferent** (*efferens*, to bring out) **lymphatics** leave the lymph node at the hilum. These vessels carry lymph away from the lymph node and toward the venous circulation.

Lymph Flow through Nodes. Lymph from the afferent lymphatics flows through the lymph node within a network of sinuses, open passageways with incomplete walls. The lymph node interior is divided into an outer cortex, an inner medulla, and a region between the two called the paracortex (see Figure 22–6). Lymph first enters a *subcapsular space* that contains a meshwork of branching reticular fibers, macrophages, and dendritic cells. *Dendritic cells* are involved in starting an immune response. Lymph passes through the subcapsular space and then flows through the

cortex of the node. The periphery of the cortex contains B cells within germinal centers similar to those of lymphoid nodules.

Lymph then continues through lymph sinuses toward an ill-defined area known as the **paracortex**, which is dominated by T cells. Here lymphocytes leave the bloodstream and enter the lymph node by crossing the walls of blood vessels.

After flowing through the sinuses of the paracortex, lymph continues into the core, or **medulla**. The medulla contains B cells and macrophages. Elongated masses of dense lymphoid tissue between the sinuses in the medulla are known as **medullary cords**. Lymph passes through a network of sinuses in the medulla and then enters the efferent lymphatics at the hilum.

Lymph Node Function. A lymph node functions like a kitchen water filter: It purifies lymph before it reaches the veins. As lymph flows through a lymph node, at least 99 percent of the antigens (foreign substances) in the lymph are removed. Fixed macrophages in the walls of the lymphatic sinuses engulf debris or pathogens in lymph as it flows past.

In addition to filtering, lymph nodes function as an early-warning defense system. Any infection or other abnormality in a peripheral tissue puts antigens into the interstitial fluid, and thus into the lymph leaving the area. These antigens then stimulate macrophages and lymphocytes in nearby lymph nodes. Antigens are also carried by dendritic cells from peripheral tissues to lymph nodes.

To protect a house against intruders, you might guard all the entrances and exits or place locks on the windows and doors. The distribution of lymphoid tissues and lymph nodes follows such a pattern. The largest lymph nodes are located where peripheral lymphatics connect with the trunk, such as in the groin, the axillae, and the base of the neck. These nodes are often called *lymph glands*. Because lymph is monitored in these nodes, potential problems can be detected and dealt with before they affect the vital organs of the trunk. The mesenteries of the gut also have aggregations of lymph nodes, located near the trachea and passageways leading to the lungs, and in association with the thoracic duct. These lymph nodes protect against pathogens and other antigens within the digestive and respiratory systems.

A minor infection commonly produces a slight enlargement of the nodes along the lymphatic vessels draining the region. This sign is often called "swollen glands." It typically indicates inflammation in peripheral structures. The enlargement generally results from an increase in the number of lymphocytes and phagocytes in the node. Inflammation of lymph nodes is termed *lymphadenitis*.

The Thymus

The thymus is a primary lymphoid organ. It is necessary in early life for immunity, but later in life it atrophies and becomes inactive. We begin this section by describing its anatomy.

+ Clinical Note Lymphadenopathy

We have approximately 600 lymph nodes scattered throughout our bodies. Some are superficial, but many are deep; few are *palpable* (able to be felt). Abnormal lymph nodes that are palpable may come to our attention during a physical examination. Any disease that affects lymph nodes is termed **lymphadenopathy** (lim-fad-e-NOP-a thē).

Most enlarged lymph nodes are benign and reflect an underlying localized infection. Think of *pharyngitis* (a sore throat) draining to nodes in the cervical region, or a sinus infection infiltrating submandibular nodes. An infected cut on the hand will show up in the axillary nodes, whereas a sexually transmitted disease will invade the inguinal nodes.

Generalized lymphadenopathy should always be evaluated. It may be evidence of something more serious, such as *lymphoma* (a primary cancer of the lymph nodes) or metastatic cancer (cancerous nodes caused by cancer spreading from another site).

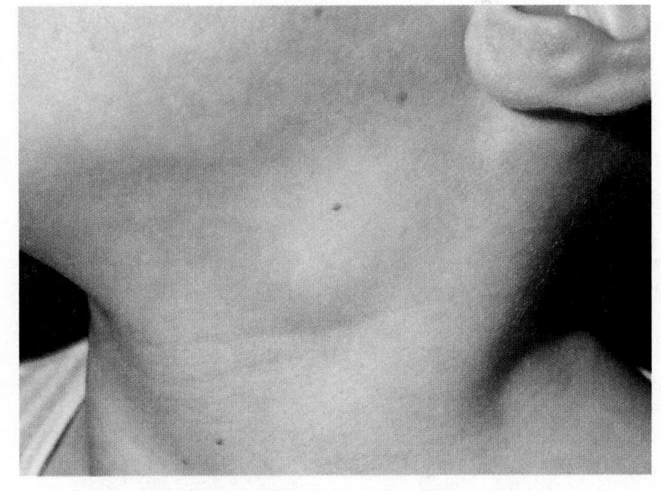

Anatomy of the Thymus. The **thymus** is a pink, grainy organ located in the mediastinum, generally just posterior to the sternum (Figure 22–7a,b). In newborn infants and young children, the thymus is relatively large. It commonly extends from the base of the neck to the superior border of the heart. The thymus reaches its greatest size relative to body size in the first year or two after birth. (The organ continues to increase in mass throughout childhood, but the body as a whole grows even faster, so the size of the thymus relative to that of the other organs in the mediastinum gradually decreases.)

The thymus reaches its maximum absolute size, at a weight of about 40 g (1.4 oz), just before puberty. After puberty, it gradually diminishes in size and becomes increasingly fibrous and fatty, a process called *involution*. By the time a person reaches age 50, the thymus may weigh less than 12 g (0.3 oz).

Figure 22–7 The Thymus. ATLAS: Plate 47a

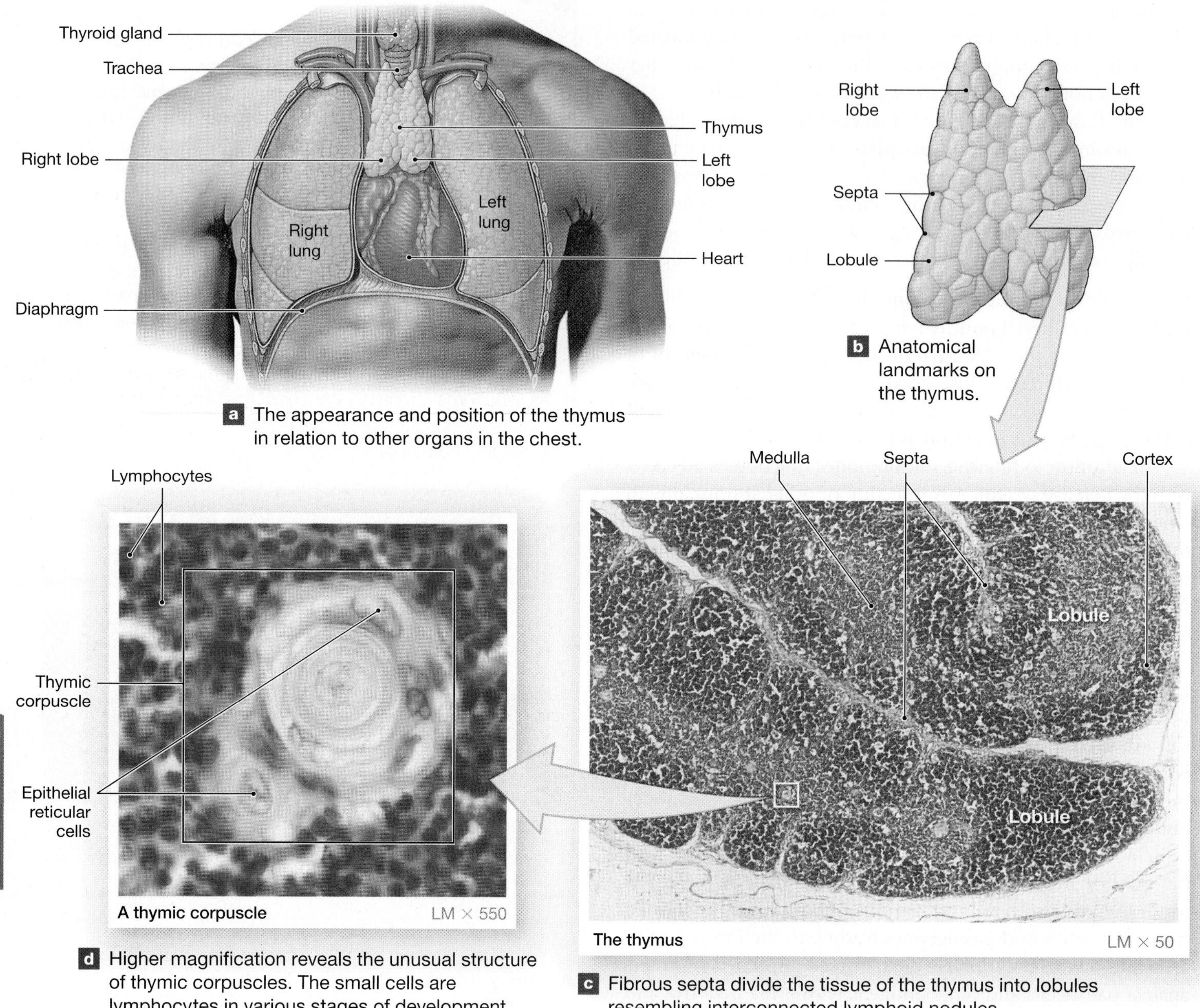
Thyroid gland
Trachea
Right lobe
Thymus
Left lobe
Left lung
Right lung
Heart
Diaphragm
Right lobe
Left lobe
Septa
Lobule
Medulla
Septa
Cortex
Lobule
Lobule
Lymphocytes
Thymic corpuscle
Epithelial reticular cells
A thymic corpuscle
LM × 550
The thymus
LM × 50

a The appearance and position of the thymus in relation to other organs in the chest.

b Anatomical landmarks on the thymus.

c Fibrous septa divide the tissue of the thymus into lobules resembling interconnected lymphoid nodules.

d Higher magnification reveals the unusual structure of thymic corpuscles. The small cells are lymphocytes in various stages of development.

22

The capsule that covers the thymus divides it into two thymic **lobes** (Figure 22–7b). Fibrous partitions called **septa** (singular, *septum*) originate at the capsule and divide the lobes into **lobules** averaging 2 mm in diameter (Figure 22–7b,c). Each lobule consists of an outer **cortex** densely packed with lymphocytes and a paler, central **medulla**.

Lymphocytes in the cortex are arranged in clusters that are completely surrounded by **epithelial reticular cells**. The epithelial reticular cells in the medulla cluster together in concentric layers, forming distinctive structures known as **thymic (Hassall's) corpuscles** (Figure 22–7d). Epithelial reticular cells also encircle the blood vessels of the cortex.

Function of the Thymus. The cortex contains actively dividing T cell lymphocytes (T lymphocytes). Epithelial reticular cells maintain the **blood thymus barrier** around the blood vessels of the cortex, which separates developing T cells from the general circulation. These epithelial cells also regulate T cell development and function. Maturing T cells leave the cortex and enter the medulla of the thymus. The medulla has no blood thymus barrier. After about 3 weeks, these T cells leave the thymus by entering one of the medullary blood vessels. T cells in the medulla can enter or leave the bloodstream across the walls of blood vessels in this region or within one of the efferent lymphatics that collect lymph from the thymus.

Figure 22–8 **The Spleen.** ATLAS: Plates 49e; 55; 56e,f; 57a,b

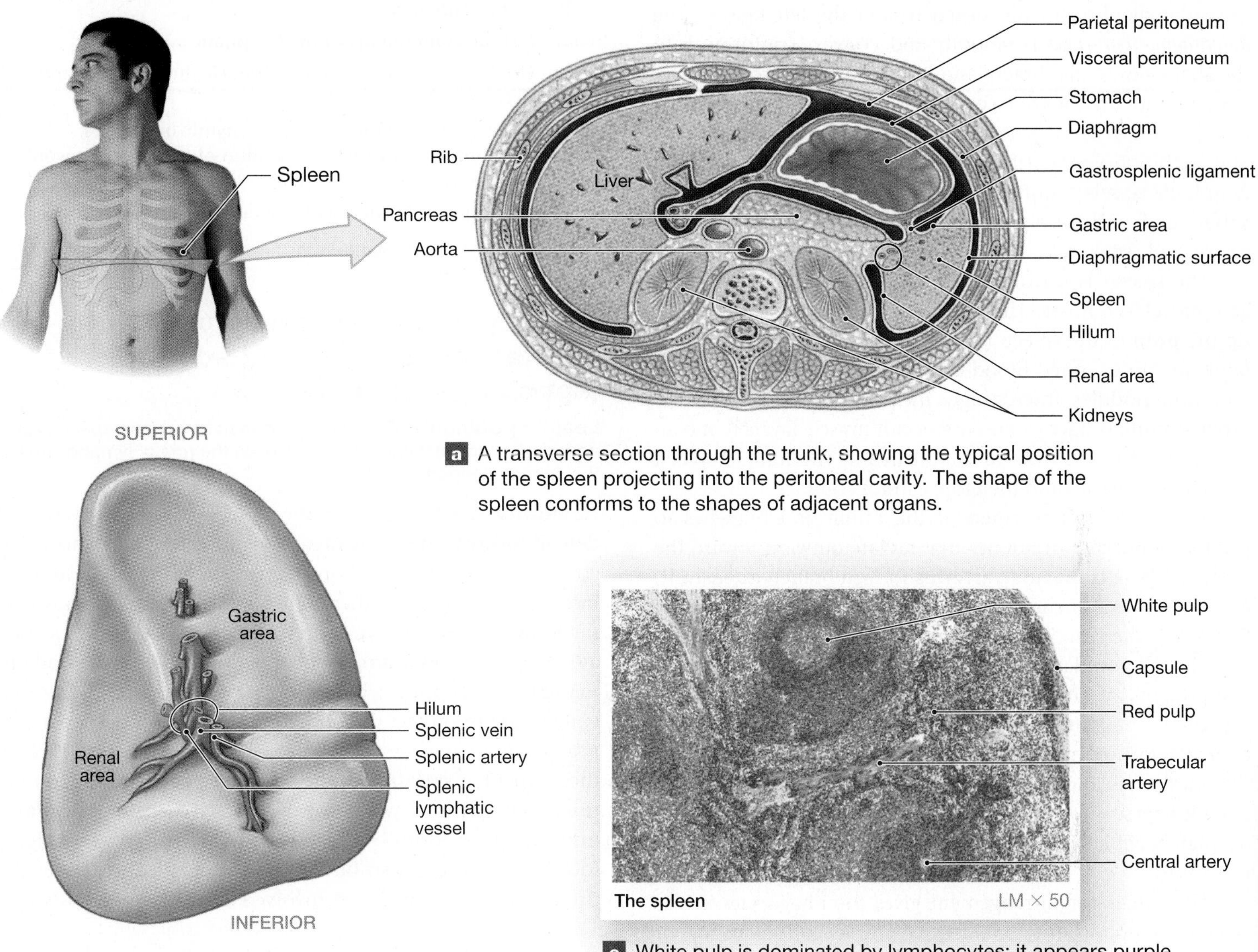

Spleen
Rib
Liver
Pancreas
Aorta
Parietal peritoneum
Visceral peritoneum
Stomach
Diaphragm
Gastrosplenic ligament
Gastric area
Diaphragmatic surface
Spleen
Hilum
Renal area
Kidneys
SUPERIOR
Gastric area
Renal area
Hilum
Splenic vein
Splenic artery
Splenic lymphatic vessel
INFERIOR
White pulp
Capsule
Red pulp
Trabecular artery
Central artery
The spleen
LM × 50

a A transverse section through the trunk, showing the typical position of the spleen projecting into the peritoneal cavity. The shape of the spleen conforms to the shapes of adjacent organs.

b A posterior view of the surface of an intact spleen, showing major anatomical landmarks.

c White pulp is dominated by lymphocytes; it appears purple because the nuclei of lymphocytes stain very darkly. Red pulp contains large numbers of red blood cells.

? What types of cells are found in the white pulp of the spleen and in the red pulp of the spleen?

22

Thymic Hormones. The thymus produces several hormones that are important to the development and maintenance of T cells for normal immunological defenses. *Thymosin* (THĪ-mō-sin) is the name originally given to an extract from the thymus that promotes the development and maturation of T cells. This extract actually contains several complementary hormones. They include *thymosin-a, thymosin-b, thymosin V, thymopoietin,* and *thymulin*. The plural term *thymosins* is sometimes used to refer to all thymic hormones.

The Spleen

The adult **spleen** contains the largest collection of lymphoid tissue in the body. In essence, the spleen performs the same functions for blood that lymph nodes perform for lymph. Functions of the spleen can be summarized as (1) removing abnormal blood cells and other blood components by phagocytosis, (2) storing iron recycled from red blood cells, and (3) initiating immune responses by B cells and T cells in response to antigens in circulating blood.

Anatomy and Histology of the Spleen. The spleen is about 12 cm (5 in.) long and weighs, on average, nearly 160 g (5.6 oz). In gross dissection, the spleen is deep red, due to the blood it contains. The spleen lies along the curving lateral border of the stomach, extending between the 9th and 11th ribs on the left side. It is attached to the lateral border of the stomach by the **gastrosplenic ligament**, a broad band of mesentery (Figure 22–8a).

The spleen has a soft texture, so its shape reflects the shapes of the structures around it. The spleen is in contact with the muscular diaphragm, the stomach, and the left kidney. The *diaphragmatic surface* is smooth and convex, conforming to the shape of the diaphragm and body wall. The *visceral surface* contains indentations that conform to the shape of the stomach (the *gastric area*) and the kidney (the *renal area*) (Figure 22–8b). Splenic blood vessels (the *splenic artery* and *splenic vein*) and lymphatic vessels communicate with the spleen on the visceral surface at the **hilum**, a groove marking the border between the gastric and renal areas.

The spleen is surrounded by a capsule containing collagen and elastic fibers. The cellular components within make up the **pulp** of the spleen (Figure 22–8c). **Red pulp** contains large quantities of red blood cells, and **white pulp** resembles lymphoid nodules. Interestingly, the spleens of dogs, cats, and other mammals have extensive smooth muscle layers that contract to eject blood into the bloodstream. The human spleen lacks these muscles and therefore cannot contract.

The splenic artery enters at the hilum and branches to produce a number of arteries that radiate outward toward the capsule. These **trabecular arteries** in turn branch extensively, and their finer branches are surrounded by areas of white pulp. Capillaries then discharge the blood into the red pulp.

The cell population of the red pulp includes all the normal components of circulating blood, plus fixed and free macrophages. The structural framework of the red pulp consists of a network of reticular fibers. The blood passes through this meshwork and enters large sinusoids, also lined by fixed macrophages. The sinusoids empty into small veins, which ultimately collect into **trabecular veins** that continue toward the hilum.

22

This circulatory arrangement gives the phagocytes in the spleen an opportunity to identify and engulf damaged or infected cells in circulating blood. Macrophages are scattered throughout the red pulp, and the area surrounding the white pulp has a high concentration of lymphocytes and dendritic cells. For this reason, any microorganism or other antigen in the blood quickly triggers an immune response.

Rupture of the Spleen. The spleen tears so easily that a seemingly minor hit to the left side of the abdomen can rupture the capsule. The result is serious internal bleeding and eventual circulatory shock. Such an injury is a known risk of contact sports (such as football, hockey, and rugby) and of more individual athletic activities, such as skiing and sledding.

Because the spleen is so fragile, it is very difficult to repair surgically. (Sutures typically tear out before they have been tensed enough to stop the bleeding.) A severely ruptured spleen is removed, a process called a **splenectomy** (splē-NEK-tō-mē). A person can survive without a spleen but lives with an increased risk of bacterial infection.

✓ Checkpoint

1. **Define *pathogen*.**
2. **List the components of the lymphatic system.**
3. **How would blockage of the thoracic duct affect lymph circulation?**
4. **If the thymus failed to produce thymic hormones, development of which population of lymphocytes would be affected?**
5. **Why do lymph nodes enlarge during some infections?**

See the blue Answers tab at the back of the book.

22-2 Lymphocytes are important to innate (nonspecific) and adaptive (specific) immunity

Learning Outcome Distinguish between innate (nonspecific) and adaptive (specific) immunity, and explain the role of lymphocytes in the immune response.

Immunity (i-MYŪ-ni-tē) is the body's ability to resist and defend against infectious organisms or other substances that could damage tissues and organs. The body's reaction to these infectious agents and other abnormal substances is known as the **immune** (*immune*, protected) **response**. The closely associated process is **resistance**, which is the ability of the body to maintain its immunity. We have two types of immunity that collaborate to protect the human body. These types are *innate (nonspecific) immunity* and *adaptive (specific) immunity*. Each of these types has specific lymphocytes associated with their function, although as we will discover in this section, innate immunity uses other forms of defense as well. We also take a closer look at the origin, distribution, and general functions of the three classes of lymphocytes involved in the immune response.

Types of Immunity

Humans have two types of immunity: innate (natural and "one-size-fits all") and adaptive (acquired and specific). Both function to defend us against foreign organisms or substances.

1. *Innate (Nonspecific) Immunity.* The body has several physical barriers and internal defense processes that either prevent or slow the entry of infectious organisms, or attack them if they do enter. For example, the skin serves as a physical barrier, and phagocytes attack invading bacteria. This type of immunity is called *innate* because you are born with it, and *nonspecific* because it does not distinguish one potential threat from another. Therefore, this category of immunity is called **innate (nonspecific) immunity** and it depends on a class of lymphocytes called NK cells.
2. *Adaptive (Specific) Immunity.* In contrast, two classes of lymphocytes, T cells and B cells, respond to specific antigens. If a bacterial pathogen invades peripheral tissues,

Go to MasteringA&P™ > Study Area > Menu > Lab Tools > PAL 3.0 > Human Cadaver and Anatomical Models > Lymphatic System these lymphocytes organize a defense against that particular type of bacterium, but not other bacteria and viruses. More importantly, this type of immunity also protects us against further attacks by the same type of pathogen. For this reason, we say that T cells and B cells provide an *adaptive* defense. Many specific defenses develop after birth as a result of accidental or deliberate exposure to antigens. **Adaptive (specific) immunity** depends on the activities of specific lymphocytes—B cells and T cells.

Lymphocytes

In the previous section, we introduced the main classes of lymphocytes that circulate in blood. These are **B cells**, **T cells**, and **NK cells**. Most lymphocytes are T cells. These cells recognize various abnormal and foreign antigens, and produce a defensive, immune response. Let's look at the circulation and life span of these lymphocytes, as well as their production.

Circulation and Life Span of Lymphocytes

The various types of lymphocytes are not evenly distributed in the blood, red bone marrow, spleen, thymus, and peripheral lymphoid tissues. The ratio of B cells to T cells varies among tissues and organs. For example, B cells are seldom found in the thymus, and T cells outnumber B cells in blood by a ratio of 8:1.

The lymphocytes in these organs are visitors, not residents. All types of lymphocytes move throughout the body. They wander through tissues and then enter blood vessels or lymphatic vessels for transport.

T cells move quickly. For example, a wandering T cell may spend about 30 minutes in the blood, 5–6 hours in the spleen, and 15–20 hours in a lymph node. B cells move more slowly. A typical B cell spends about 30 hours in a lymph node before moving on.

Lymphocytes have relatively long life spans. About 80 percent survive 4 years, and some last 20 years or more. Throughout your life, you maintain normal lymphocyte populations by producing new lymphocytes in your red bone marrow and lymphoid tissues.

Lymphocyte Production

In Chapter 19, we discussed *hemopoiesis*—the formation of the formed elements of blood. ⮌ pp. 660, 664, 675 In adults, red blood cell formation, or *erythropoiesis*, is normally confined to red bone marrow. In contrast, lymphocyte formation, or **lymphocytopoiesis** (lim-fō-SĪ-tō-poy-Ē-sis), involves the red bone marrow, thymus, and peripheral lymphoid tissues (Figure 22–9).

Red bone marrow plays the primary role in maintaining normal lymphocyte populations. Hemocytoblasts divide in the red bone marrow of adults to generate the lymphoid stem cells that produce all types of lymphocytes. The red bone marrow produces two distinct populations of lymphoid stem cells.

One group of lymphoid stem cells remains in the red bone marrow and the other group migrates to the thymus (see Figure 22–9). Lymphoid stem cells in the red bone marrow divide to produce immature B cells and NK cells. B cell development involves intimate contact with large *stromal cells* (not shown) in the red bone marrow. Bone marrow stromal cells provide a specialized microenvironment to control hemopoiesis. The cytoplasmic extensions of stromal cells contact or even wrap around the developing B cells. Stromal cells of the red bone marrow produce a cytokine called *interleukin-7*, which promotes the differentiation of B cells.

As they mature, B cells and NK cells enter the bloodstream and migrate to peripheral tissues (see Figure 22–9). Most of the B cells move into lymph nodes, the spleen, or other lymphoid tissues. The NK cells patrol the body, moving through peripheral tissues in search of abnormal cells.

The second group of lymphoid stem cells migrates to the thymus (see Figure 22–9). These stem cells and their descendants develop and mature in an environment that is isolated from the general circulation by the blood thymus barrier. Under the influence of thymic hormones, the lymphoid stem cells divide repeatedly, producing the various kinds of T cells. When the development of the T cells nears completion, they reenter the bloodstream and travel to peripheral tissues, including lymphoid tissues and organs, such as the spleen (see Figure 22–9).

The T cells and B cells that migrate from their sites of origin retain the ability to divide, producing daughter cells of the same type. For example, a dividing B cell produces other B cells, not T cells or NK cells. As we will see, the ability of specific types of lymphocytes to increase in number is crucial to the success of the immune response.

Checkpoint

6. **Compare immunity with resistance.**
7. **Explain the difference between innate immunity and adaptive immunity.**

See the blue Answers tab at the back of the book.

22-3 Innate defenses respond the same regardless of the invader

Learning Outcome List the body's innate defenses, and describe the components, mechanisms, and functions of each.

Innate (nonspecific) defenses make up the first line of defense in innate immunity. They include physical barriers, phagocytes, immune surveillance, interferons, complement, inflammation, and fever. These innate defenses are summarized in Figure 22–10.

Physical Barriers

To cause trouble, a foreign substance or pathogen must enter body tissues. In other words, it must cross an epithelium—either

Figure 22–9 The Origin and Distribution of Lymphocytes.

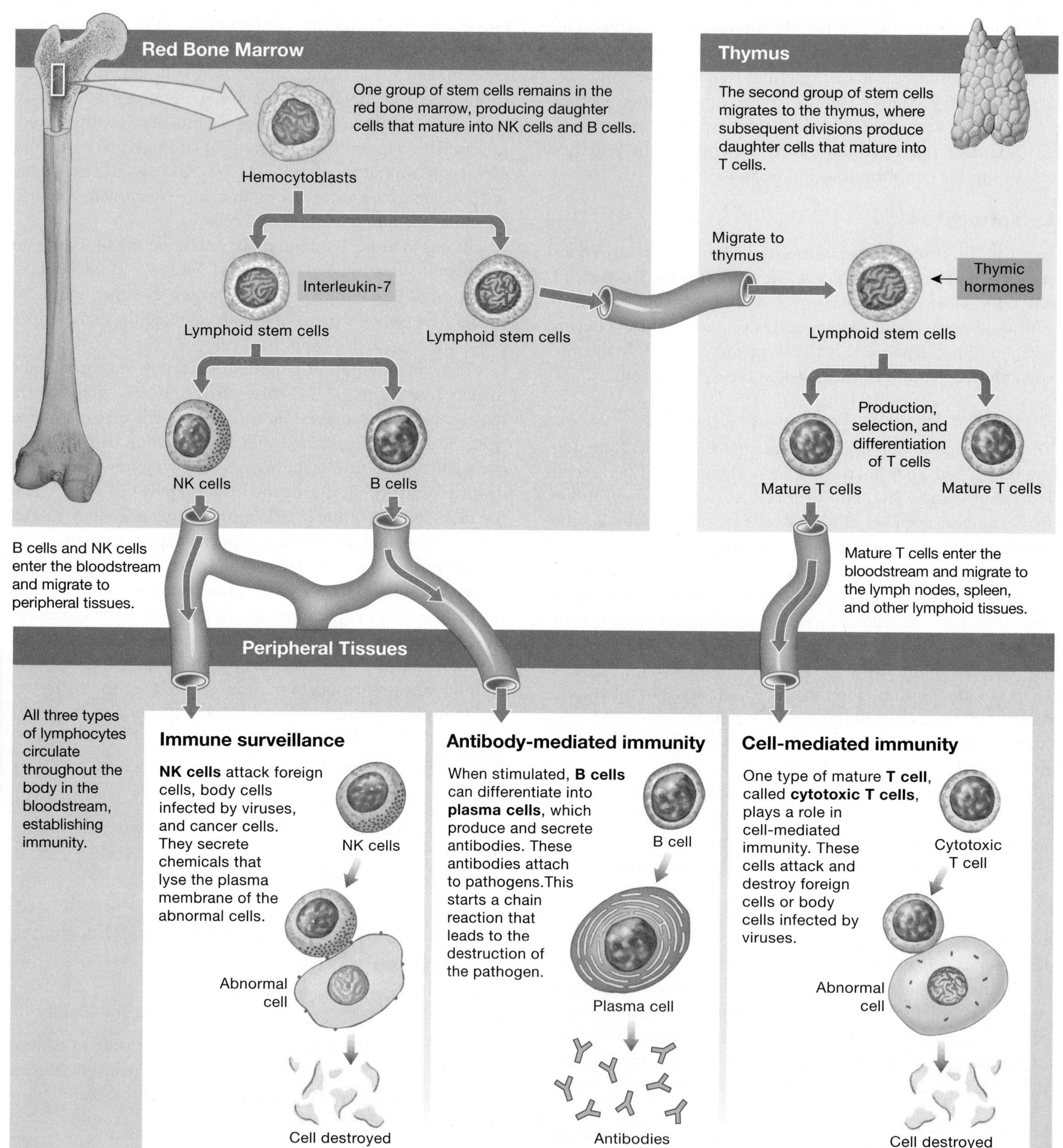
Red Bone Marrow
One group of stem cells remains in the red bone marrow, producing daughter cells that mature into NK cells and B cells.
Hemocytoblasts
Interleukin-7
Lymphoid stem cells
Lymphoid stem cells
NK cells
B cells
Thymus
The second group of stem cells migrates to the thymus, where subsequent divisions produce daughter cells that mature into T cells.
Migrate to thymus
Thymic hormones
Lymphoid stem cells
Production, selection, and differentiation of T cells
Mature T cells
Mature T cells
B cells and NK cells enter the bloodstream and migrate to peripheral tissues.
Mature T cells enter the bloodstream and migrate to the lymph nodes, spleen, and other lymphoid tissues.
Peripheral Tissues
All three types of lymphocytes circulate throughout the body in the bloodstream, establishing immunity.
Immune surveillance
NK cells attack foreign cells, body cells infected by viruses, and cancer cells. They secrete chemicals that lyse the plasma membrane of the abnormal cells.
NK cells
Abnormal cell
Cell destroyed
Antibody-mediated immunity
When stimulated, B cells can differentiate into plasma cells, which produce and secrete antibodies. These antibodies attach to pathogens.This starts a chain reaction that leads to the destruction of the pathogen.
B cell
Plasma cell
Antibodies
Cell-mediated immunity
One type of mature T cell, called cytotoxic T cells, plays a role in cell-mediated immunity. These cells attack and destroy foreign cells or body cells infected by viruses.
Cytotoxic T cell
Abnormal cell
Cell destroyed

22

Figure 22–10 **Innate Defenses.** Innate (nonspecific) defenses deny pathogens access to the body or destroy them without distinguishing among specific types.

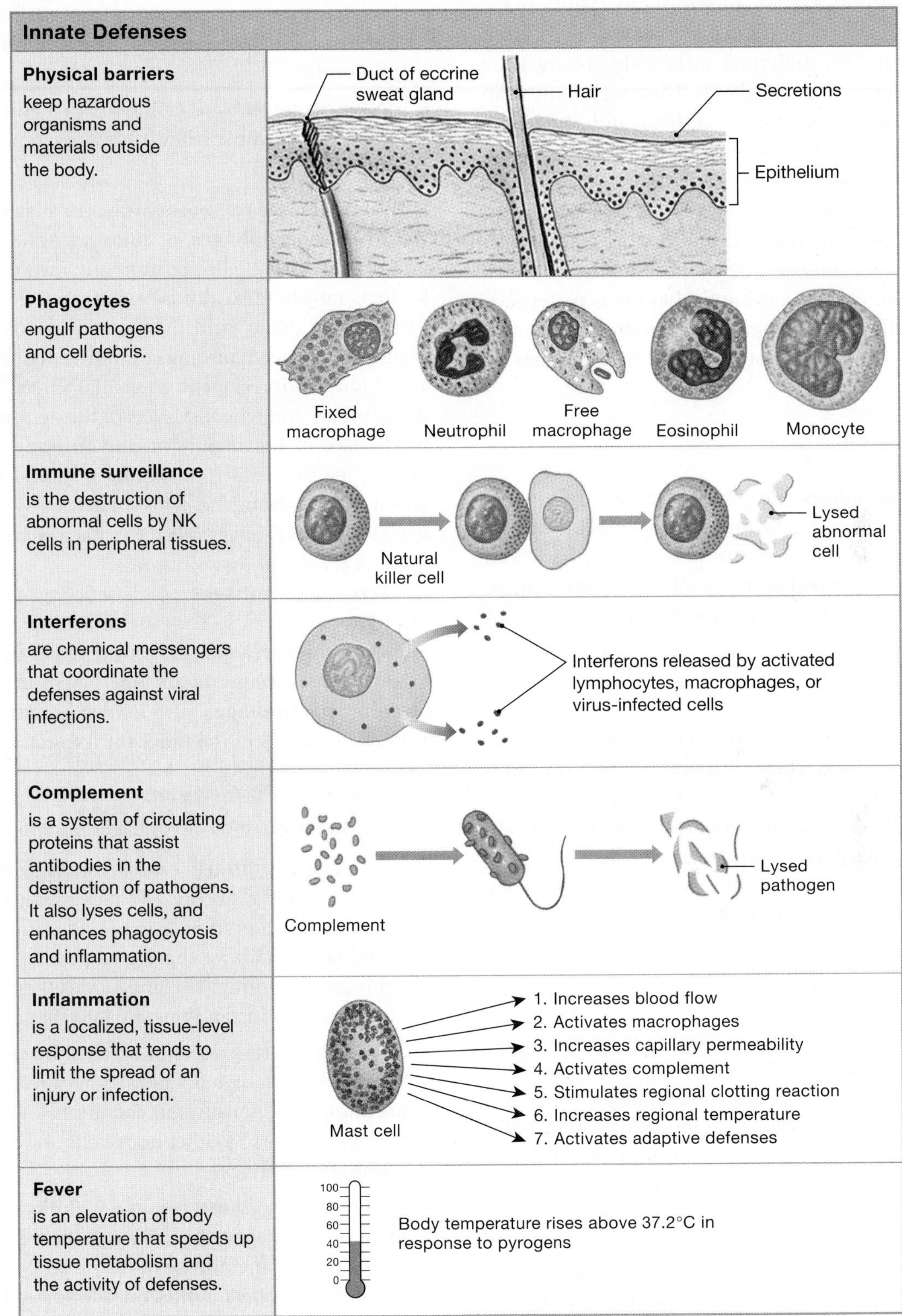

Innate Defenses
Physical barriers
keep hazardous organisms and materials outside the body.
Duct of eccrine sweat gland
Hair
Secretions
Epithelium
Phagocytes
engulf pathogens and cell debris.
Fixed macrophage
Neutrophil
Free macrophage
Eosinophil
Monocyte
Immune surveillance
is the destruction of abnormal cells by NK cells in peripheral tissues.
Natural killer cell
Lysed abnormal cell
Interferons
are chemical messengers that coordinate the defenses against viral infections.
Interferons released by activated lymphocytes, macrophages, or virus-infected cells
Complement
is a system of circulating proteins that assist antibodies in the destruction of pathogens. It also lyses cells, and enhances phagocytosis and inflammation.
Complement
Lysed pathogen
Inflammation
is a localized, tissue-level response that tends to limit the spread of an injury or infection.
Mast cell
1. Increases blood flow
2. Activates macrophages
3. Increases capillary permeability
4. Activates complement
5. Stimulates regional clotting reaction
6. Increases regional temperature
7. Activates adaptive defenses
Fever
is an elevation of body temperature that speeds up tissue metabolism and the activity of defenses.
100
80
60
40
20
0
Body temperature rises above 37.2°C in response to pyrogens

at the *skin* or across a *mucous membrane*. The epithelial covering of the skin has multiple layers, a keratin coating, and a network of desmosomes that lock adjacent cells together. pp. 118–119, 153–157 These physical barriers are very effective in protecting underlying tissues. Even along the more delicate internal passageways of the respiratory, digestive, urinary, and reproductive tracts, epithelial cells are tied together by tight junctions. These cells generally are supported by a dense and fibrous basement membrane.

In addition, specialized accessory structures and secretions protect most epithelia. The hairs on most areas of your body provide some protection against abrasion, especially on the scalp. They often prevent hazardous materials or insects from contacting your skin. The epidermal surface also receives the secretions of sebaceous and sweat glands. These secretions flush the surface and wash away microorganisms and chemicals. Such secretions may also contain chemicals that kill bacteria, destructive enzymes called *lysozymes*, and antibodies.

The epithelia lining the digestive, respiratory, urinary, and reproductive tracts are more delicate, but they are equally well defended. For example, mucus bathes most surfaces of your digestive and respiratory tracts, urine flushes the urinary tract, and glandular secretions do the same for the reproductive tract. Your stomach also contains a powerful acid that can destroy many pathogens.

Phagocytes

Phagocytes serve as janitors and police in peripheral tissues. They remove cellular debris and respond to invasion by foreign substances or pathogens. Many phagocytes attack and remove microorganisms even before lymphocytes detect them. The human body has two general classes of phagocytes: microphages and macrophages.

Microphages

22

Microphages are the neutrophils and eosinophils that normally circulate in the blood. These phagocytes leave the bloodstream and enter peripheral tissues that have been subjected to injury or infection. As noted in Chapter 19, neutrophils are abundant, mobile, and quick to phagocytize cellular debris or invading bacteria. ↺ p. 671 Eosinophils are less abundant. They target foreign substances or pathogens that have been coated with antibodies.

Macrophages

Macrophages are large, active phagocytes. Your body has several types of macrophages. Most are derived from the monocytes in circulating blood. Typically, macrophages are either fixed in position or freely mobile. As a result they are usually classified as fixed macrophages or free macrophages, but the distinction is not absolute. During an infection, fixed macrophages may lose their attachments and begin roaming around the damaged tissue.

No organs or tissues are dominated by phagocytes, but almost every tissue in the body shelters resident or visiting macrophages. This relatively diffuse collection of phagocytes is the **monocyte–macrophage system**, or the *reticuloendothelial system*.

How does an activated macrophage respond to a pathogen? It may:

- engulf a pathogen or other foreign object and destroy it with lysosomal enzymes.
- bind to or remove a pathogen from the interstitial fluid, but be unable to destroy the invader until assisted by other cells.
- destroy its target by releasing toxic chemicals, such as *tumor necrosis factor*, nitric oxide, or hydrogen peroxide, into the interstitial fluid.

We consider these responses further in a later section.

Fixed macrophages, or *histiocytes*, reside in specific tissues and organs. These cells are normally incapable of movement, so their targets must diffuse or otherwise move through the surrounding tissue until they are within range. Fixed macrophages are scattered among connective tissues, usually in close association with collagen or reticular fibers. They are found in the papillary and reticular layers of the dermis, in the subarachnoid space of the meninges, and in bone marrow. In some organs, the fixed macrophages have special names. **Microglia** are macrophages in the central nervous system, and **stellate macrophages** *(Kupffer cells)* are macrophages located on the luminal surface of liver sinusoids.

Free macrophages, or *wandering macrophages*, travel throughout the body. They arrive at the site of an injury by migrating through adjacent tissues or by leaving the circulating blood. Some tissues contain free macrophages. For example, **alveolar macrophages**, also known as *dust cells*, monitor the epithelial surfaces of the lungs for respiratory pathogens.

Functions of Phagocytes

Free macrophages and microphages function in similar ways:

- Both can move through capillary walls by squeezing between adjacent endothelial cells. This process is known as *emigration*. ↺ p. 671 The endothelial cells in an injured area develop membrane "markers" that signal passing blood cells that something is wrong. The phagocytes then attach to the endothelial lining and migrate into the surrounding tissues.
- Both may be attracted to or repelled by chemicals in the surrounding fluids, a phenomenon called **chemotaxis**. They are particularly sensitive to chemicals known as *cytokines* that are released by other body cells and to chemicals released by pathogens.
- For both, phagocytosis begins with **adhesion**, the attachment of the phagocyte to its target. In adhesion, receptors on the plasma membrane of the phagocyte bind to the surface of the target. Adhesion is followed by the formation of a vesicle containing the bound target (look back at Figure 3–22, p. 101). The contents of the vesicle are digested once the vesicle fuses with lysosomes or peroxisomes.

All phagocytes function in much the same way, although their targets may differ. The life span of a phagocyte can be

Figure 22–11 **How Natural Killer Cells Kill Cellular Targets.**

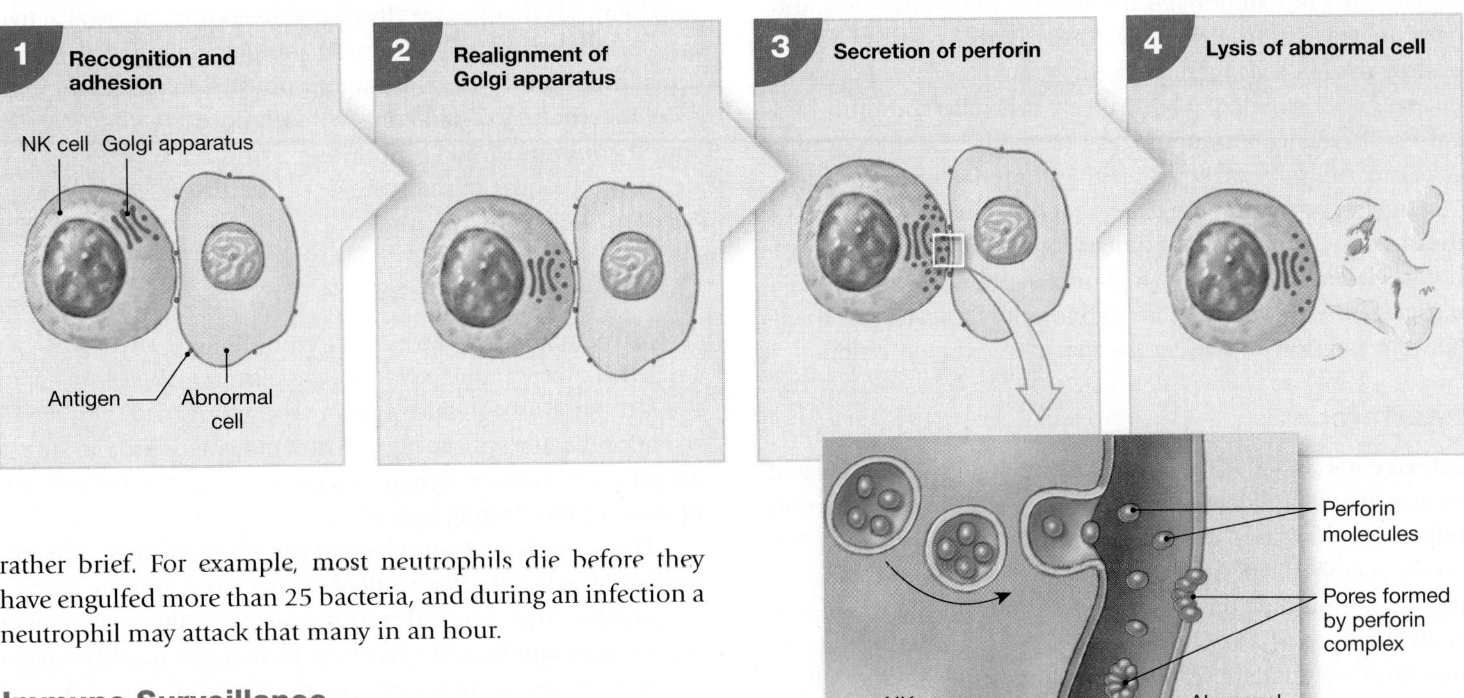
1 Recognition and adhesion
NK cell
Golgi apparatus
Antigen
Abnormal cell
2 Realignment of Golgi apparatus
3 Secretion of perforin
4 Lysis of abnormal cell
Perforin molecules
Pores formed by perforin complex
NK cell
Abnormal cell

rather brief. For example, most neutrophils die before they have engulfed more than 25 bacteria, and during an infection a neutrophil may attack that many in an hour.

Immune Surveillance

The immune system generally ignores the body's own cells unless they become abnormal in some way. Natural killer (NK) cells, also known as *large granular lymphocytes*, are responsible for recognizing and destroying abnormal cells when they appear in peripheral tissues. Their continuous "policing" of peripheral tissues has been called **immune surveillance**.

The plasma membrane of an abnormal cell generally contains antigens that are not found on the membranes of normal cells. NK cells recognize an abnormal cell by detecting those antigens. NK cells are much less selective about their targets than are other lymphocytes: They respond to a *variety* of abnormal antigens that may appear anywhere on a plasma membrane. They also attack *any* membrane containing abnormal antigens. As a result, NK cells are highly versatile defenders. A single NK cell can attack bacteria in the interstitial fluid, body cells infected with viruses, or cancer cells.

NK cells respond much more rapidly than T cells or B cells. The activation of T cells and B cells involves a complex and time-consuming sequence of events. NK cells respond immediately on contact with an abnormal cell.

Activated NK cells react in a predictable way (Figure 22–11):

1 **Recognition and Adhesion.** If a cell has unusual components in its plasma membrane, an NK cell recognizes that cell as abnormal. This recognition activates the NK cell, which then adheres to its target cell.

2 **Realignment of Golgi Apparatus.** The Golgi apparatus moves around the nucleus until the maturing (*trans*) face points directly toward the abnormal cell. The process might be compared to the rotation of a tank turret to point the cannon toward the enemy. The Golgi apparatus then produces a flood of secretory vesicles that contain proteins called **perforins**. These vesicles move through the cytosol toward the cell surface.

3 **Secretion of Perforin.** The perforins are released at the NK cell surface by exocytosis and diffuse across the narrow gap between the NK cell and its target.

4 **Lysis of Abnormal Cell.** When perforin molecules reach the opposing plasma membrane, the perforins interact with one another and create a network of pores in it. These pores are large enough to allow the free passage of ions, proteins, and other intracellular materials. As a result, the target cell can no longer maintain its internal environment, and it quickly disintegrates.

Tips & Tools

Perforin gets its name because it ***perfor***ates the target cell.

Why doesn't perforin affect the plasma membrane of the NK cell? The answer is not clear, but NK cell membranes contain a second protein, called *protectin*. It may bind and inactivate perforin.

NK cells attack cancer cells and body cells infected with viruses. Cancer cells probably appear throughout life, but their plasma membranes generally contain unusual proteins called **tumor-specific antigens**, which NK cells recognize as abnormal. The NK cells then destroy the cancer cells, preserving the tissue.

22

Unfortunately, some cancer cells avoid detection, perhaps because they lack tumor-specific antigens or because these antigens are covered in some way. Other cancer cells are able to destroy the NK cells that detect them. This process of avoiding detection or neutralizing body defenses is called **immunological escape**. Once immunological escape has occurred, cancer cells can multiply and spread without interference by NK cells.

In viral infections, the viruses replicate inside cells, beyond the reach of circulating antibodies. However, infected cells incorporate viral antigens into their plasma membranes, and NK cells recognize these infected cells as abnormal. By destroying them, NK cells can slow or prevent the spread of a viral infection.

Interferons

Interferons (in-ter-FĒR-onz) **(IFNs)** are small proteins released by activated lymphocytes and macrophages, and by tissue cells infected with viruses. An interferon binds to surface receptors on the membrane of a normal cell and, by second messengers, triggers the production of antiviral proteins in the cytoplasm. Antiviral proteins do not prevent viruses from entering the cell. Instead, they interfere with viral replication inside the cell. In addition to slowing the spread of viral infections, IFNs stimulate the activities of macrophages and NK cells.

At least three types of interferons exist: **interferon alpha** (α), **interferon beta** (β), and **interferon gamma** (γ) (Figure 22–12). Most cells other than lymphocytes and macrophages respond to viral infection by secreting interferon beta.

Figure 22–12 Interferons.

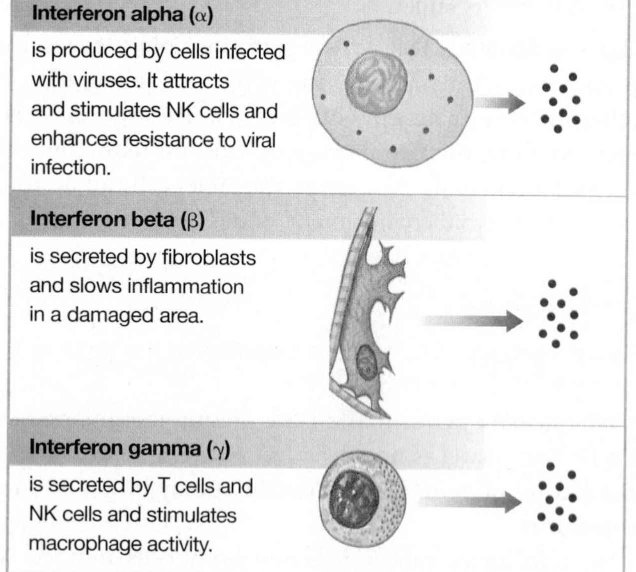
Interferon alpha (α)
is produced by cells infected with viruses. It attracts and stimulates NK cells and enhances resistance to viral infection.
Interferon beta (β)
is secreted by fibroblasts and slows inflammation in a damaged area.
Interferon gamma (γ)
is secreted by T cells and NK cells and stimulates macrophage activity.

? Which type of interferon attracts and stimulates NK cells as a way to enhance viral resistance to viral infection?

Interferons are examples of cytokines. **Cytokines** (SĪ-tō-kīnz) are chemicals that tissue cells release to coordinate local activities. Cytokines include hormones and paracrine-like glycoproteins important to the immune response. Most cells produce cytokines only for paracrine communication—that is, cell-to-cell communication within one tissue. p. 611 However, defense cells also release cytokines; we discuss their role in Section 22–5.

Complement System

Plasma contains over 30 special **complement proteins** that form the **complement system**. The term *complement* refers to the fact that this system complements, or enhances, the action of antibodies and phagocytes. The complement proteins interact with one another in chain reactions, or *cascades*, reminiscent of those of the clotting system.

The activation of complement can occur by three different routes: the *classical pathway*, the *lectin pathway*, and the *alternative pathway*. The **classical pathway** activates the complement system most rapidly and effectively. Its proteins are abbreviated "C" with a number that corresponds to its discovery. It begins with the binding of complement protein C1 to two nearby antibodies already attached to its specific antigen, such as a bacterial cell wall. The **lectin pathway** activates the complement system by the binding of the protein mannose-binding lectin (MBL) to carbohydrates on pathogen surfaces, such as bacterial cell walls. The **alternative pathway** also activates the complement system without antibody interactions. Its proteins are given specific names or called *factors*. The alternative pathway begins when several complement proteins—including **properdin** (*factor P*), *factor B*, and *factor D*—interact in the plasma. This interaction is triggered by exposure to foreign materials, such as the cell wall of a bacterium.

All three pathways involve the splitting of inactive C3 protein to activated C3b and C3a proteins. Complement activation brings about the following effects: killing of the pathogen by cell lysis, enhanced phagocytosis (**opsonization**), and enhanced inflammation (histamine release). Figure 22–13 shows the three pathways of complement activation and its effects.

Inflammation

Inflammation is a localized tissue response to injury. p. 145 Inflammation produces local *redness*, *swelling*, *heat*, and *pain*. These are known as the so-called *cardinal signs and symptoms of inflammation*. Sometimes a fifth is included: *lost function*.

Many stimuli can produce inflammation. They include impact, abrasion, distortion, chemical irritation, infection by pathogens, and extreme temperatures (hot or cold). Each stimulus kills cells, damages connective tissue fibers, or injures the tissue in some other way. The changes alter the chemical composition of the interstitial fluid. Damaged cells release

Figure 22–13 Pathways of Complement Activation.

Classical Pathway

The most rapid and effective activation of the complement system occurs through the classical pathway.

Antibodies

Bacterial cell wall

Antibodies bind to bacterial cell wall

C1

Attachment of C1

C1 must attach to two antibodies for its activation.

C2 C4 C3 C3a C3b

Activation and Cascade

The attached C1 protein acts as an enzyme, catalyzing a series of reactions involving other complement proteins that split C3 into C3a and C3b. C3a diffuses away and activates an inflammatory response.

C3b

C3b Attachment (classical pathway)

C3b binds to the bacterial cell wall and enhances phagocytosis.

Lectin Pathway

The lectin pathway is activated by the protein mannose-binding lectin (MBL). MBL binds to carbohydrates on bacterial surfaces.

Bacterial surface

Lectin

Lectin binds to bacterial cell wall

MBL binds to carbohydrates on the bacterial surface.

C3 C3a C3b

C3 Activation

Bound MBL forms a complex that splits C3 into C3a and C3b. C3a activates an inflammatory response.

C3b

C3b Attachment (lectin pathway)

C3b binds to bacterial surface and enhances phagocytosis.

Alternative Pathway

The alternative pathway is important in the defense against bacteria, some parasites, and virus-infected cells.

Complement proteins interact in plasma

Properdin
Factor B
Factor D
C3 C3a C3b

Bacterial cell wall

The alternative pathway begins when several complement proteins, notably properdin, interact in the plasma causing C3 to split into C3a and C3b. This interaction can be triggered by exposure to foreign materials, such as the capsule of a bacterium. C3a activates an inflammatory response.

C3b Attachment (alternative pathway)

C3b

C3b protein binds to the bacterial cell wall.

Complement activation causes

Killing of Pathogen (Cell Lysis)

Once an activated C3b protein has attached to the cell wall, additional complement proteins may form a **membrane attack complex (MAC)** in the membrane that destroys the integrity of the target cell.

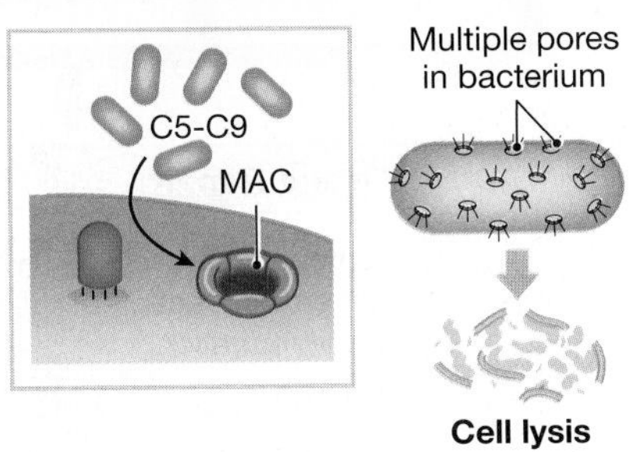
C5-C9
MAC

Multiple pores in bacterium

Cell lysis

Enhanced Phagocytosis (Opsonization)

The attached C3b may also act to enhance phagocytosis. This enhancement of phagocytosis, a process called **opsonization**, occurs because macrophage membranes contain receptors that detect and bind to complement proteins and bound antibodies.

Inflammation (Histamine Release)

Release of histamine by mast cells and basophils increases the degree of local inflammation, attracts and activates phagocytes, and accelerates blood flow to the region.

22

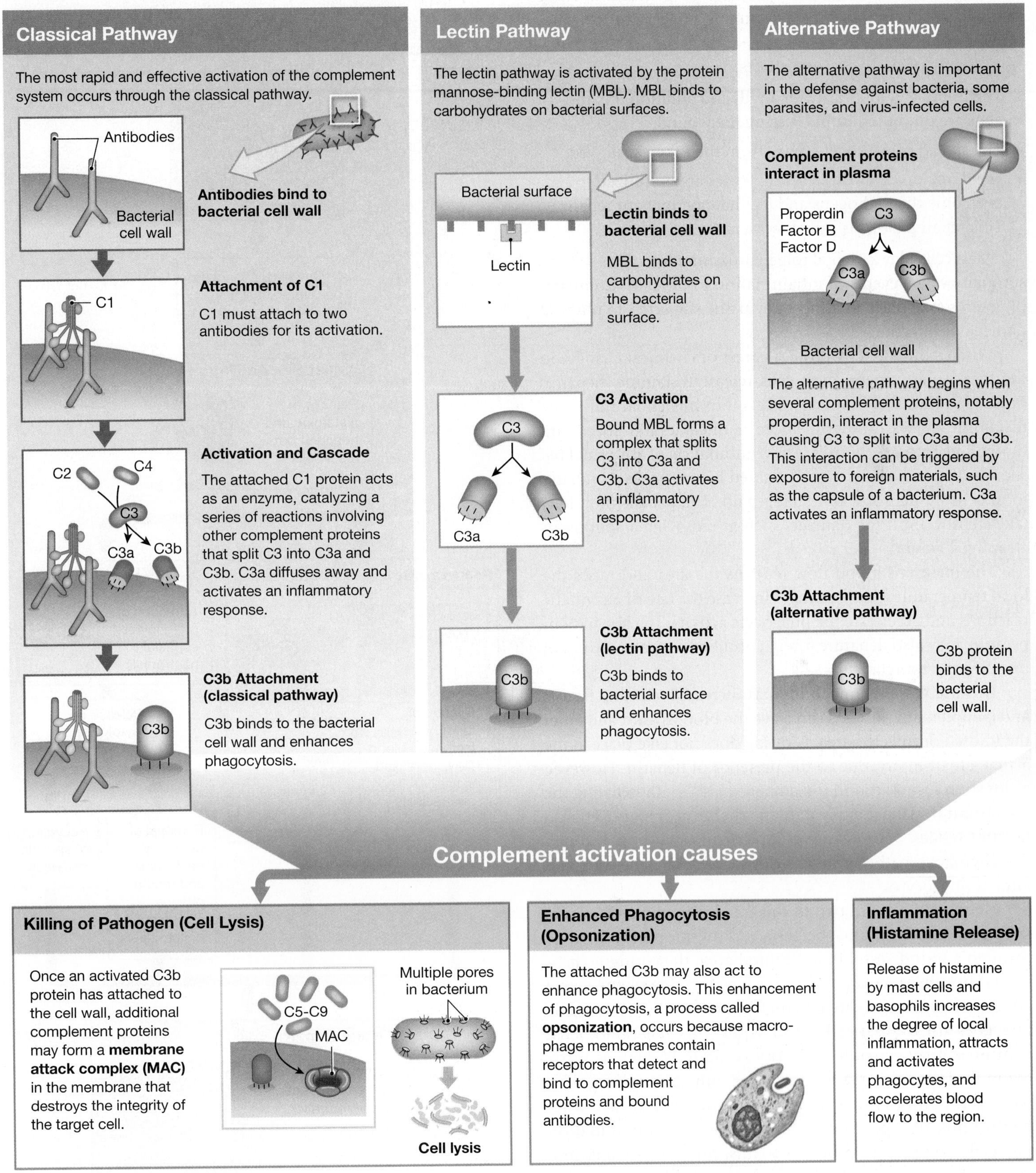

prostaglandins, proteins, and potassium ions. The injury itself may have introduced foreign proteins or pathogens. These changes in the interstitial environment trigger the complex process of inflammation.

Inflammation has several effects:

- The injury is temporarily repaired, and additional pathogens are prevented from entering the wound.
- The spread of pathogens away from the injury is slowed.
- Local, regional, and systemic defenses are mobilized to overcome the pathogens and facilitate permanent repairs. This repair process is called *regeneration*.

Mast cells play a pivotal role in inflammation. Figure 22–14 summarizes the events of inflammation in the skin. Comparable events take place in almost any tissue subjected to physical damage or infection.

When stimulated by physical stress or chemical changes in the local environment, mast cells release histamine, heparin, prostaglandins, and other chemicals into interstitial fluid. The histamine makes capillaries more permeable and speeds up blood flow through the area. The combination of abnormal tissue conditions and chemicals released by mast cells stimulates local sensory neurons, producing pain. The person then may take steps to limit the damage, such as removing a splinter or cleaning a wound.

The increased blood flow reddens the area and raises the local temperature. These changes increase the rate of enzymatic reactions and accelerate the phagocyte activity. The rise in temperature may also denature foreign proteins or vital enzymes of invading microorganisms.

Because vessel permeability has increased, clotting factors and complement proteins can leave the bloodstream and enter the injured or infected area. Clotting does not take place at the actual site of injury, due to the presence of heparin. However, a clot soon forms around the damaged area, both isolating the region and slowing the spread of the chemical or pathogen into healthy tissues. Meanwhile, complement activation through the alternative pathway breaks down bacterial cell walls and attracts phagocytes.

Neutrophils are drawn to the area by chemotaxis, where they attack debris and bacteria. As neutrophils circulate through a blood vessel in an injured area, they undergo *activation*. In this process (1) they stick to the side of the vessel and move into the tissue by emigration; (2) their metabolic rate goes up dramatically, and while this *respiratory burst* continues, they generate reactive compounds, such as nitric oxide and hydrogen peroxide, that can destroy engulfed pathogens; and (3) they secrete cytokines that attract other neutrophils and macrophages to the area. As inflammation proceeds, the foreign proteins, toxins, microorganisms, and active phagocytes in the area activate the body's specific defenses.

Figure 22–14 Inflammation and the Steps in Tissue Repair.

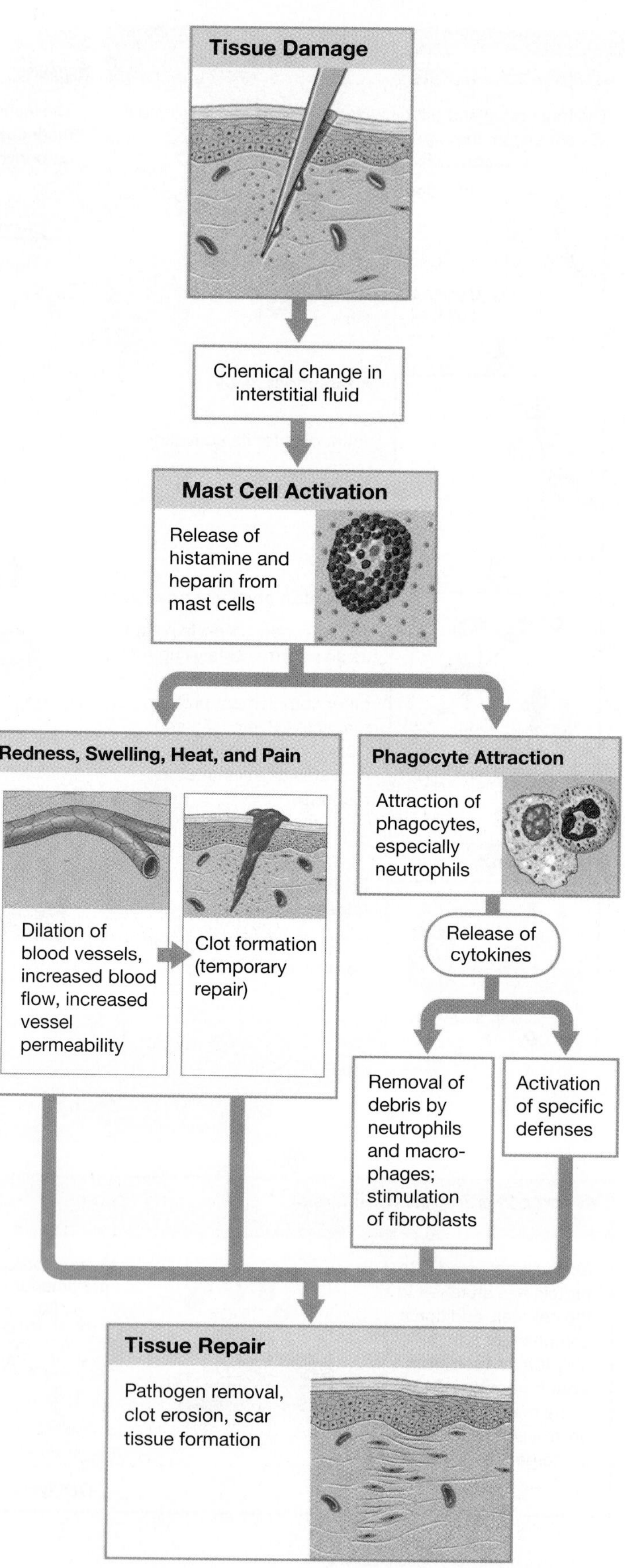
Tissue Damage
Chemical change in interstitial fluid
Mast Cell Activation
Release of histamine and heparin from mast cells
Redness, Swelling, Heat, and Pain
Dilation of blood vessels, increased blood flow, increased vessel permeability
Clot formation (temporary repair)
Phagocyte Attraction
Attraction of phagocytes, especially neutrophils
Release of cytokines
Removal of debris by neutrophils and macrophages; stimulation of fibroblasts
Activation of specific defenses
Tissue Repair
Pathogen removal, clot erosion, scar tissue formation

Fixed and free macrophages engulf pathogens and cell debris. At first, neutrophils outnumber these cells. Then, as the macrophages and neutrophils continue to secrete cytokines, the number of macrophages increases rapidly. Eosinophils may get involved if antibodies coat the foreign materials.

The cytokines released by active phagocytes stimulate fibroblasts in the area. The fibroblasts then begin forming scar tissue that reinforces the clot and slows the invasion of adjacent tissues. Over time, the clot is broken down and the injured tissues are either repaired or replaced by scar tissue. The process is essentially complete, although subsequent remodeling may take place over a period of years.

After an injury, tissue conditions generally become even more abnormal before they begin to improve. The tissue destruction that occurs after cells have been injured or destroyed is called **necrosis** (ne-KRŌ-sis). The destruction begins several hours after the initial event, and the damage is due to lysosomal enzymes. Lysosomes break down by autolysis, releasing digestive enzymes that first destroy the injured cells and then attack surrounding tissues. ⮌ p. 80 As local inflammation continues, debris, fluid, dead and dying cells, and necrotic tissue components accumulate at the injury site. This viscous fluid mixture is known as **pus**. An accumulation of pus in an enclosed tissue space is called an **abscess**.

Fever

Fever is a body temperature greater than 37.2°C (99°F). Recall from Chapter 14 that the pre-optic area of the hypothalamus contains the temperature-regulating center. ⮌ p. 484 Fever-inducing agents called **pyrogens** (PĪ-rō-jenz; *pyro-*, fever) can reset this thermostat and raise body temperature.

Bacteria, molds, viruses, and yeasts produce pyrogens and release them into the bloodstream. Several endogenous, or internally produced, pyrogens have been identified including interleukin-1, interferons, and tumor necrosis factor.

Within limits, a fever can be beneficial. High body temperatures may inhibit some viruses and bacteria. The most likely benefit is on body metabolism. For each 1°C rise in body temperature, metabolic rate increases by 10 percent. Cells can move faster, and enzymatic reactions take place more quickly. As a result, tissue defenses can be mobilized more rapidly and the repair process speeds up.

✓ Checkpoint

8. **What types of cells would be affected by a decrease in the number of monocyte-forming cells in red bone marrow?**
9. **A rise in the level of interferon in the body suggests what kind of infection?**
10. **What effects do pyrogens have in the body?**

See the blue Answers tab at the back of the book.

22-4 Adaptive (specific) defenses respond to particular threats and are either cell mediated or antibody mediated

Learning Outcome Define adaptive (specific) defenses, identify the forms and properties of immunity, and distinguish between cell-mediated (cellular) immunity and antibody-mediated (humoral) immunity.

Lymphocytes of Adaptive Immunity

Adaptive (specific) defenses result from the coordinated activities of T lymphocytes and B lymphocytes, known as T cells and B cells. Under proper stimulation, T cells differentiate into several types of cells, which attack antigens and help to increase the immune response.

The primary types of T cells are as follows (Figure 22–15):

- **Cytotoxic T cells** are involved in direct cellular attack. These cells enter peripheral tissues and attack antigens physically and chemically.
- **Helper T cells** stimulate the responses of both T cells and B cells. Helper T cells are absolutely vital to the immune response, because they must activate B cells before the B cells can produce antibodies. The reduction in the helper T cell population that occurs in AIDS is largely responsible for the loss of immunity. (We discuss AIDS on p. 827.)
- **Regulatory T cells** are a subset of T cells that moderate the immune response.
- **Memory T cells** respond to antigens they have already encountered by the cloning (producing identical cellular copies) of more lymphocytes to ward off the invader.

The interplay between helper T cells and regulatory T cells helps establish and control the sensitivity of the immune response. Other types of regulatory T cells also take part in the immune response. For example, *inflammatory T cells* stimulate regional inflammation and local defenses in an injured tissue. *Suppressor/inducer T cells* suppress B cell activity but stimulate other T cells.

When stimulated, B cells differentiate into *plasma cells* that secrete soluble proteins called *antibodies*. ⮌ p. 661 The binding of an antibody to its target antigen leads to the destruction of the target invader.

Types of Adaptive Immunity

Adaptive (specific) defenses result from the coordinated activities of T cells and B cells. Cytotoxic T cells provide **cell-mediated immunity**, or *cellular immunity*, which defends against abnormal cells and pathogens inside cells. B cells provide **antibody-mediated immunity**, which defends against antigens and pathogens in body fluids.

Figure 22–15 Classes of Lymphocytes.

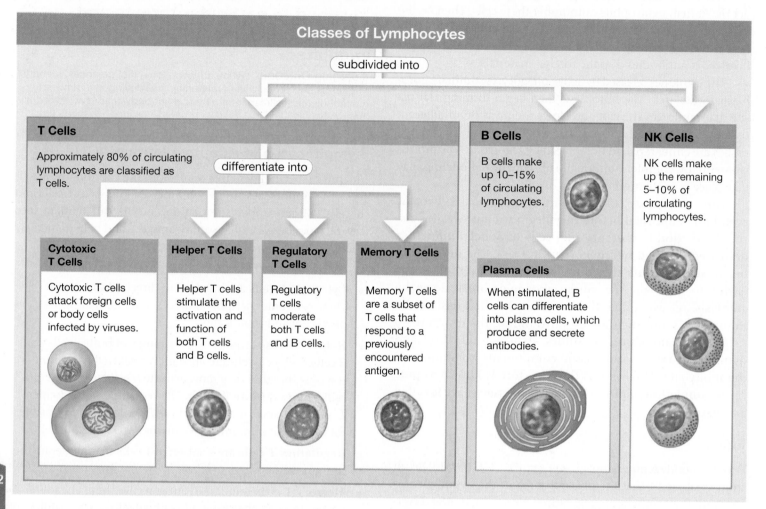

Classes of Lymphocytes

subdivided into

T Cells

Approximately 80% of circulating lymphocytes are classified as T cells.

differentiate into

Cytotoxic T Cells

Cytotoxic T cells attack foreign cells or body cells infected by viruses.

Helper T Cells

Helper T cells stimulate the activation and function of both T cells and B cells.

Regulatory T Cells

Regulatory T cells moderate both T cells and B cells.

Memory T Cells

Memory T cells are a subset of T cells that respond to a previously encountered antigen.

B Cells

B cells make up 10–15% of circulating lymphocytes.

Plasma Cells

When stimulated, B cells can differentiate into plasma cells, which produce and secrete antibodies.

NK Cells

NK cells make up the remaining 5–10% of circulating lymphocytes.

22

Antibody-mediated immunity is also known as *humoral* ("liquid") *immunity* because antibodies are found in body fluids.

Both kinds of immunity are important because they come into play under different circumstances. Both T and B cells respond to the presence of specific antigens. However, activated T cells do not respond to antigens in solution, and antibodies (produced by activated B cells) cannot cross plasma membranes. Moreover, helper T cells play a crucial role in antibody-mediated immunity by stimulating the activity of B cells.

An Introduction to Adaptive Immunity

Figure 22–16 provides an overview of adaptive immunity. When an antigen triggers an immune response, it usually activates both T cells and B cells. The activation of T cells generally occurs first, but only after phagocytes have been exposed to the antigen. Once activated, T cells attack the antigen and stimulate the activation of B cells. Activated B cells mature into cells that produce antibodies. Antibodies in the bloodstream then bind to and attack the antigen.

Antigens

Specific chemical targets called **antigens** stimulate the immune response. Most antigens are pathogens, parts or products of pathogens, or other foreign substances. Antigens are usually proteins, but some lipids, polysaccharides, and nucleic acids are also antigens.

Lymphocyte Activation and Clonal Selection

A lymphocyte becomes **activated** when it has contact with an appropriate antigen. When activated, a lymphocyte begins to divide, producing more lymphocytes with the same specificity. All the identical cells produced by these divisions make up a **clone**, and all the members of that clone are sensitive to the same specific antigen.

To understand how this system works, think about running a commercial kitchen with only samples on display. You can display a wide selection because the samples don't take up much space, and you don't have to expend energy now preparing food that might never be eaten. When a customer selects

Figure 22–16 An Overview of Adaptive Immunity.

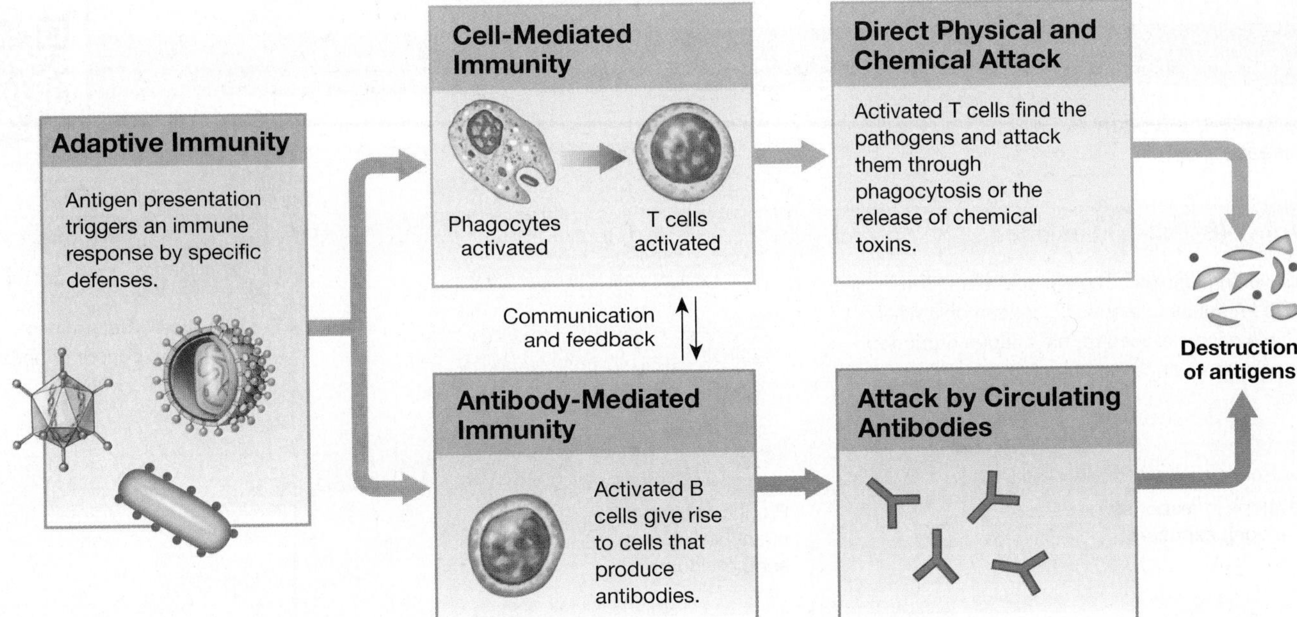
Adaptive Immunity
Antigen presentation triggers an immune response by specific defenses.
Cell-Mediated Immunity
Phagocytes activated
T cells activated
Communication and feedback
Antibody-Mediated Immunity
Activated B cells give rise to cells that produce antibodies.
Direct Physical and Chemical Attack
Activated T cells find the pathogens and attack them through phagocytosis or the release of chemical toxins.
Attack by Circulating Antibodies
Destruction of antigens

one of your samples and places an order for several dozen, you prepare them on the spot.

The same principle applies to the lymphatic system. Your body contains a small number of many different kinds of lymphocytes. When an antigen arrives, it "selects" only those lymphocytes sensitive to it. These lymphocytes become activated and divide to generate a large number of additional lymphocytes of the same type. This process of an antigen selecting particular lymphocytes for cloning is called **clonal selection**.

Forms of Adaptive Immunity

Adaptive immunity can be *active* or *passive* (Figure 22–17). These forms of immunity can be either naturally acquired or artificially acquired.

Active immunity develops after exposure to an antigen. It is a consequence of the immune response. In active immunity, the body responds to an antigen by making its own antibody. The immune system is *capable* of defending against a huge number of antigens. However, the appropriate defenses are mounted only after you encounter a particular antigen. Active immunity can result from natural exposure to an antigen in the environment (*naturally acquired active immunity*) or from deliberate exposure to an antigen (*artificially acquired active immunity*).

- **Naturally acquired active immunity** normally begins to develop after birth. It continues to build as you encounter "new" pathogens or other antigens. You might compare this process to the development of a child's vocabulary: The child begins with a few basic common words and learns new ones as they are encountered.
- **Artificially acquired active immunity** stimulates the body to produce antibodies under controlled conditions so that you will be able to overcome natural exposure to the pathogen in the future. This is the basic principle behind *immunization*, or *vaccination*, to prevent disease. A **vaccine** is a preparation designed to induce an immune response. It contains either a dead or an inactive pathogen, or antigens derived from that pathogen.

Passive immunity is produced by transferring antibodies from another source.

- In **naturally acquired passive immunity**, a baby receives antibodies from the mother, either during gestation (by crossing the placenta) or in early infancy (through breast milk).
- In **artificially acquired passive immunity**, a person receives antibodies to fight infection or prevent disease. For example, someone who has been bitten by a rabid animal gets injections containing antibodies against the rabies virus.

Properties of Adaptive Immunity

Regardless of the form, adaptive immunity has four general properties: (1) *specificity*, (2) *versatility*, (3) *memory*, and (4) *tolerance*.

Specificity

A specific defense is activated by a specific antigen, and the immune response targets that particular antigen and no others. **Specificity** results from the activation of appropriate lymphocytes and the production of antibodies with targeted effects. Specificity occurs because T cells and B cells respond to the

Figure 22–17 Forms of Immunity.

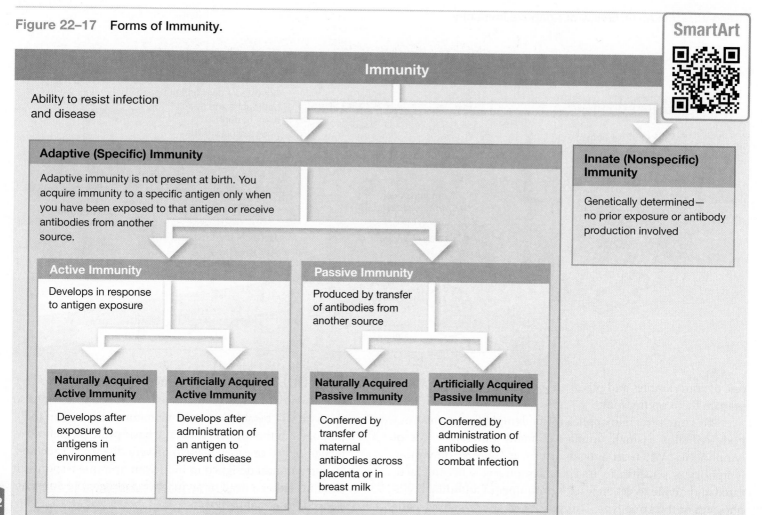

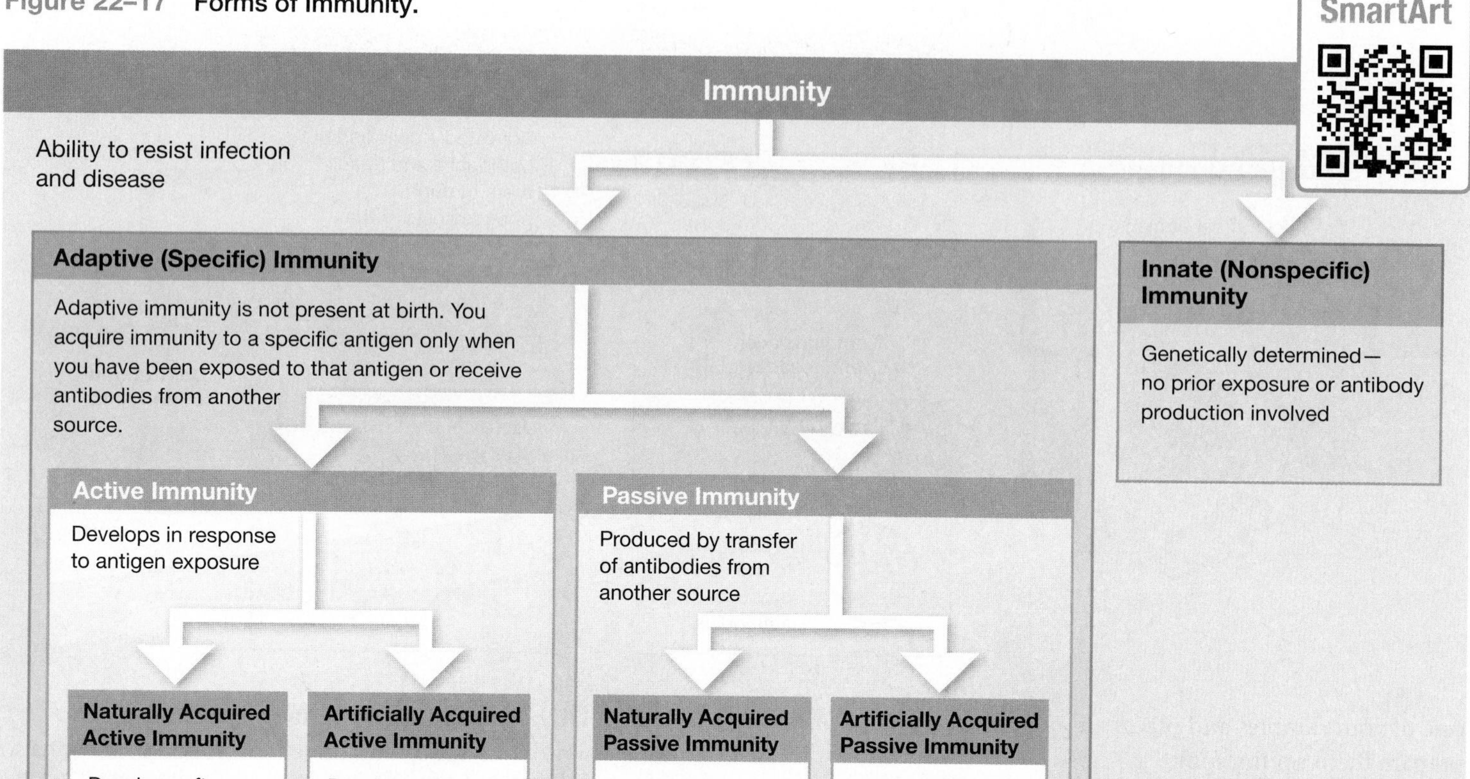
SmartArt
Immunity
Ability to resist infection and disease
Adaptive (Specific) Immunity
Adaptive immunity is not present at birth. You acquire immunity to a specific antigen only when you have been exposed to that antigen or receive antibodies from another source.
Innate (Nonspecific) Immunity
Genetically determined—no prior exposure or antibody production involved
Active Immunity
Develops in response to antigen exposure
Passive Immunity
Produced by transfer of antibodies from another source
Naturally Acquired Active Immunity
Develops after exposure to antigens in environment
Artificially Acquired Active Immunity
Develops after administration of an antigen to prevent disease
Naturally Acquired Passive Immunity
Conferred by transfer of maternal antibodies across placenta or in breast milk
Artificially Acquired Passive Immunity
Conferred by administration of antibodies to combat infection

? What type of immunity develops after receiving a vaccine?

molecular structure of an antigen. The shape and size of the antigen determine which lymphocytes will respond to it. Each T cell or B cell has receptors that will bind to one specific antigen, ignoring all others. The response of an activated T cell or B cell is equally specific. Either lymphocyte will destroy or inactivate that antigen without affecting other antigens or normal tissues.

Versatility

Millions of antigens in the environment can pose a threat to health. Over a normal lifetime, you encounter only a fraction of that number—perhaps tens of thousands of antigens. Your immune system, however, has no way of anticipating which antigens it will encounter. It must be ready to confront *any* antigen at *any* time. **Versatility** results in part from the large diversity of lymphocytes present in the body, and in part from variability in the structure of synthesized antibodies.

During development, cells in the lymphatic system differentiate to produce a huge number of lymphocytes with varied antigen sensitivities. The trillion or more T cells and B cells in your body include millions of different lymphocyte populations, distributed throughout the body. Each population contains several thousand cells with receptors in their membranes that differ from the receptors of other lymphocyte populations. As a result, each population of lymphocytes responds to a different antigen.

Memory

As we just saw, during the initial response to an antigen, lymphocytes that are sensitive to it undergo repeated cycles of cell division. **Memory** exists because those cell divisions produce two groups of cells. One group attacks the invader immediately. Another group remains inactive unless it meets the same antigen at a later date. This inactive group is made up of *memory cells* that enable your immune system to "remember" an antigen it has previously encountered, and to launch a faster, stronger, and longer-lasting counterattack if such an antigen appears again.

Tolerance

The immune system does not respond to all antigens. All cells and tissues in the body, for example, contain antigens that normally do not stimulate an immune response. We say that the immune system exhibits **tolerance** toward such antigens, called *self-antigens*.

During their differentiation in the red bone marrow (B cells) and thymus (T cells), any cells that react to self-antigens are destroyed. During this process up to 98 percent of developing T cells are deselected and undergo *apoptosis* (the process of genetically programmed cell death introduced in Chapter 3). ⮌ p. 102 As a result, mature B cells and T cells ignore self-antigens, but attack foreign antigens, or *nonself-antigens*. Tolerance can also develop over time in response to chronic exposure to an antigen in the environment. Such tolerance lasts only as long as the exposure continues.

✓ Checkpoint

11. **Explain the difference between cell-mediated (cellular) immunity and antibody-mediated (humoral) immunity.**
12. **Identify the four major types of T cells.**
13. **Identify the two forms of active immunity and the two forms of passive immunity.**
14. **List the four general properties of adaptive immunity.**

See the blue Answers tab at the back of the book.

22-5 In cell-mediated adaptive immunity, presented antigens activate T cells, which respond by producing cytotoxic and helper T cells

Learning Outcome Discuss the types of T cells and their roles in the adaptive immune response, and describe the mechanisms of T cell activation and differentiation.

This section describes the mechanisms involved in antigen recognition by T cells and their subsequent activation, clonal selection, and mitotic divisions.

Activation and Clonal Selection of T Cells

Before an immune response can begin, T cells must be activated by exposure to an antigen. This activation seldom occurs through direct interaction between a T cell and the antigen. Foreign substances or pathogens entering a tissue commonly fail to stimulate an immediate immune response. T cells only recognize antigens when they are processed and "presented" by cells called *antigen-presenting cells*, most of which are phagocytes such as macrophages. This process—*antigen presentation*—is generally the first step in the activation of the immune response (look back at Figure 22–16). The next step is antigen recognition, and then differentiation and production of clones of the various defensive cells, such as cytotoxic T cells and helper T cells.

Antigen Presentation with MHC Proteins

For T cells to recognize an antigen, the antigen must be bound to glycoproteins in the plasma membrane of another cell, such as an antigen-presenting cell. Recall that glycoproteins are integral membrane components. ⮌ p. 57 The structure of these glycoproteins is genetically determined, and these proteins identify the cell as "self" rather than foreign. The genes controlling their synthesis are located along one portion of chromosome 6, in a region called the **major histocompatibility complex (MHC)**. These membrane glycoproteins are therefore called **MHC proteins**, or *human leukocyte antigens (HLAs)*. **Antigen presentation** occurs when an antigen–MHC protein combination capable of activating T cells appears in a plasma membrane.

The amino acid sequences and the shapes of MHC proteins differ among individuals. Each MHC molecule has a distinct three-dimensional shape with a relatively narrow central groove. An antigen that fits into this groove can be held in position by hydrogen bonding.

+ Clinical Note Lab Tests for Organ Donation

A successful match between an organ donor and a recipient depends on the results of three tests: (1) blood type, (2) tissue type, and (3) cross-matching. Each of these tests seeks evidence of an immune response to a foreign antigen. Immune rejection either disqualifies the match or requires the recipient to receive more medication to suppress his or her immune response.

Blood typing establishes whether an individual has type O, A, B, or AB blood. Type O blood is the universal donor (all blood types can receive type O); type AB is the universal recipient (type AB can receive all blood types).

Tissue typing is performed by specialized laboratories. Also called histocompatibility testing, this test assesses HLAs, human leukocyte antigens. Out of more than 100 existing antigens, 6 have been determined to be the most important in organ donation, 3 inherited from each parent. Although it is rare to have a perfect 6-antigen match between donor and recipient, the higher the match, the better the success rate of organ or tissue transplantation.

Cross-matching tests for incompatibility between donor and recipient blood to avoid hemolytic reactions. It is done by mixing a sample of donor red blood cells with plasma from the recipient and mixing recipient red blood cells with plasma from the donor. If clumping occurs, donation is not possible because the blood types are incompatible.

Two major classes of MHC proteins are known: *class I* and *class II*. An antigen bound to a class I MHC protein acts like a red flag that in effect tells the immune system "I'm an abnormal cell—kill me!" An antigen bound to a class II MHC protein tells the immune system "This antigen is dangerous—get rid of it!"

Class I MHC Proteins. **Class I MHC proteins** are in the plasma membranes of all nucleated cells. These proteins are continuously synthesized and exported to the plasma membrane in vesicles formed by the Golgi apparatus. As they form, class I proteins pick up small peptides from the surrounding cytoplasm and carry them to the cell surface. If the cell is healthy and the peptides are normal, T cells ignore them. If the cytoplasm contains abnormal (non-self) peptides or viral proteins, they soon appear in the plasma membrane, and T cells will recognize them as foreign and be activated (Figure 22–18a).

Figure 22–18 Antigens and MHC Proteins.

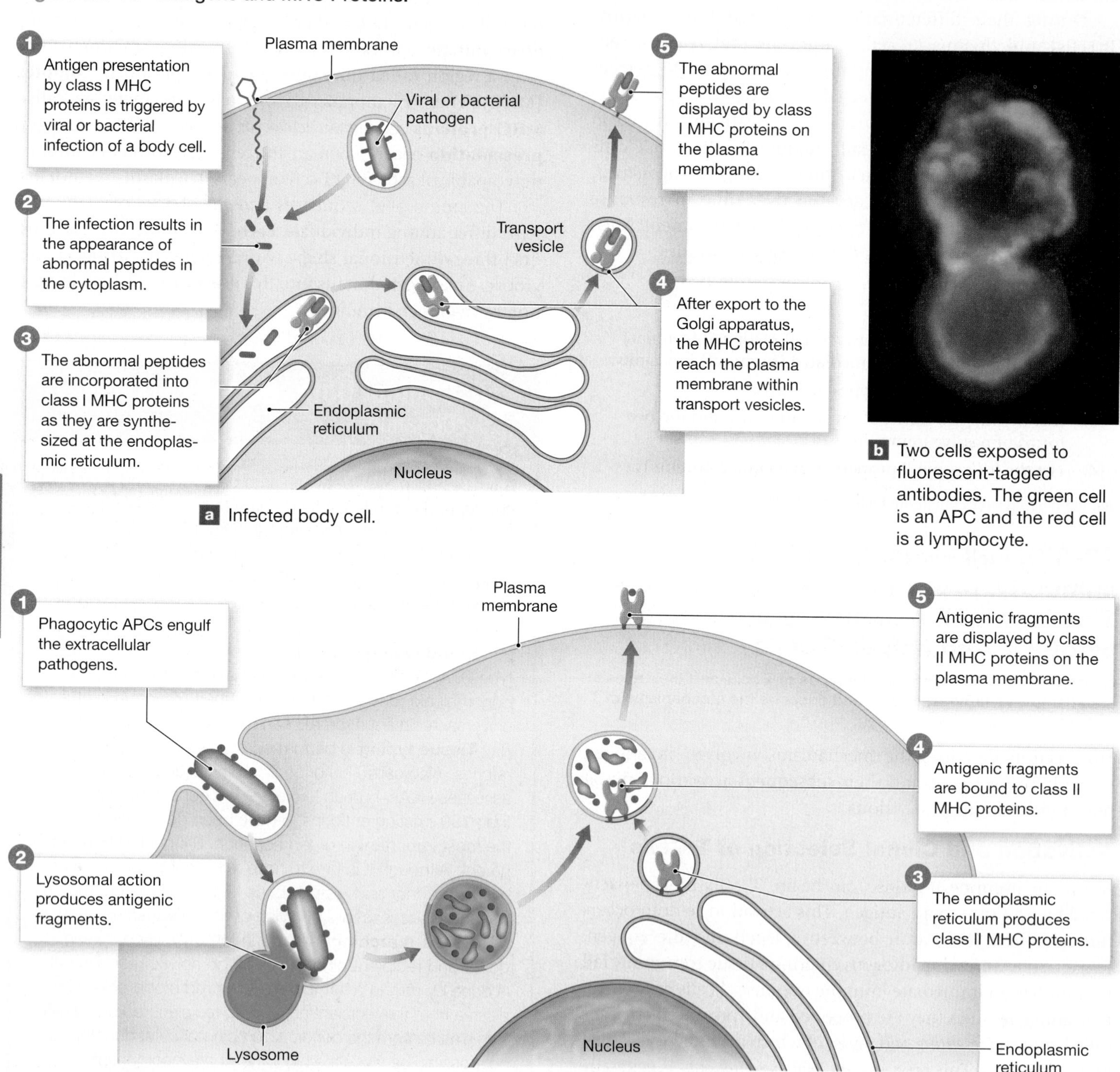
1 Antigen presentation by class I MHC proteins is triggered by viral or bacterial infection of a body cell.
Plasma membrane
Viral or bacterial pathogen
2 The infection results in the appearance of abnormal peptides in the cytoplasm.
3 The abnormal peptides are incorporated into class I MHC proteins as they are synthesized at the endoplasmic reticulum.
Transport vesicle
Endoplasmic reticulum
Nucleus
4 After export to the Golgi apparatus, the MHC proteins reach the plasma membrane within transport vesicles.
5 The abnormal peptides are displayed by class I MHC proteins on the plasma membrane.
1 Phagocytic APCs engulf the extracellular pathogens.
2 Lysosomal action produces antigenic fragments.
Lysosome
Plasma membrane
Nucleus
3 The endoplasmic reticulum produces class II MHC proteins.
4 Antigenic fragments are bound to class II MHC proteins.
5 Antigenic fragments are displayed by class II MHC proteins on the plasma membrane.
Endoplasmic reticulum

a Infected body cell.

b Two cells exposed to fluorescent-tagged antibodies. The green cell is an APC and the red cell is a lymphocyte.

c Antigen-presenting cell (APC).

22

+ Clinical Note Organ Donation

Each day in the United States, 79 people receive an organ donation and 22 people die waiting. Every 10 minutes, a new name is added to the current roster of 122,000 people on the National Transplant Waiting List. These names come from people with chronic disease, trauma, or birth defects. People need organs (kidneys, hearts, lungs, pancreases, livers) or tissues (such as bone marrow, bone, tendons, or corneas) to extend their life or improve its quality. Each donor organ or tissue must be scrupulously matched to the recipient to ensure the best chance of success.

Most living donors are female; most deceased donors are male, and most of these are accident victims. A single donor can save up to eight lives through the donation of multiple organs and tissues. In the case of a fatality, there is a window of 24 hours to initiate donation.

Individuals who are interested in becoming donors can sign up with a state registry or give their legal consent on their driver's license or in their living will. The U.S. Department of Health and Human Services (DHHS) sponsors the network for organ procurement and transplantation.

Ultimately, their activation leads to the destruction of the abnormal cells. This is the primary reason that donated organs are commonly rejected by the recipient. Despite preliminary cross-match testing, the recipient's T cells recognize the transplanted tissue as foreign.

Class II MHC Proteins and Antigen-Presenting Cells. **Class II MHC proteins** are present only in the plasma membranes of antigen-presenting cells and lymphocytes. **Antigen-presenting cells (APCs)** are specialized cells that activate T cell defenses against foreign cells (including bacteria) and foreign proteins. Antigen-presenting cells include all the phagocytes of the monocyte–macrophage group discussed in other chapters, including (1) free and fixed macrophages in connective tissues, (2) the stellate macrophages of the liver, and (3) the microglia in the central nervous system (see Chapter 12). ⮌ pp. 129, 398 The **dendritic cells** of the skin (where they are also known as *Langerhans* cells) and those of the lymph nodes and spleen are APCs that are not phagocytic. ⮌ p. 155

Antigen-presenting cells engulf and break down pathogens or foreign antigens. This **antigen processing** creates fragments of the antigen, which are then bound to class II MHC proteins and inserted into the plasma membrane (Figure 22–18b,c). Class II MHC proteins appear in the plasma membrane only when the cell is processing antigens. The APC then presents the combined class II MHC protein and processed antigen to appropriate T cells.

The dendritic cells remove antigenic materials from their surroundings by pinocytosis rather than phagocytosis. However, their plasma membranes still present antigens bound to class II MHC proteins. More important is that, unlike macrophages that do not migrate very far from an infection site, dendritic cells travel with their foreign antigens to lymphocyte-packed lymph nodes.

Antigen Recognition and CD Markers

How do T cells recognize antigens? Inactive T cells have receptors that can bind either class I or class II MHC proteins. These receptors also have binding sites for a specific target antigen. If an MHC protein contains any antigen other than the specific target of a particular kind of T cell, the T cell remains inactive. If the MHC protein contains the antigen that the T cell is programmed to detect, binding occurs. This process is called **antigen recognition**, because the T cell recognizes that it has found an appropriate target.

Some T cells can recognize antigens bound to class I MHC proteins, whereas others can recognize antigens bound to class II MHC proteins. Whether a T cell responds to antigens held by one class or the other depends on a type of protein in the T cell's own plasma membrane. The membrane proteins involved are members of a larger group of proteins called **CD** (*cluster of differentiation*) **markers**.

More than 350 types of CD markers exist, and each type is designated by a number. All T cells have a **CD3 receptor complex** in their plasma membranes, and this complex ultimately activates the T cell. Either of two other CD markers may be bound to the CD3 receptor complex, and these two CD markers are especially important in specific groups of T cells:

- **CD8** markers are found on cytotoxic T cells and regulatory T cells, which together are often called *CD8 T cells*. CD8 T cells respond to antigens presented by class I MHC proteins.
- **CD4** markers are found on helper T cells, often called *CD4 T cells*. CD4 T cells respond to antigens presented by class II MHC proteins.

T cells are not usually activated upon their first encounter with the antigen. Antigen recognition simply prepares the cell for activation.

Tips & Tools

You can remember that MHC class **I** and CD**8** go together, and MHC class *II* and CD*4* go together because both pairs multiplied are 8 ($1 \times 8 = 8$ and $2 \times 4 = 8$).

Costimulation and Clonal Selection

To proceed from recognition to activation, a T cell must also bind to the stimulating cell at a second site. This vital secondary binding process, called *costimulation*, essentially confirms the initial activation signal.

Costimulation is like the safety on a gun: It helps prevent T cells from mistakenly attacking normal (self) tissues. If a cell displays an unusual antigen but does not display the "I am an active phagocyte" or "I am infected" signal, T cell activation will not occur. Costimulation is important only in determining whether a T cell will become activated.

Once activation has occurred, the "safety" is off and the T cell will undergo mitosis, producing clones of identical cells. The types of cells that are produced depend on the type of T cell that was activated.

Functions of Activated CD8 T Cells

Two different types of CD8 T cells are activated by exposure to antigens bound to class I MHC proteins. One type responds quickly, giving rise to clones containing large numbers of active cytotoxic T cells and inactive *memory T cells* (Figure 22–19). The other type responds more slowly and produces relatively small numbers of regulatory T cells.

Cytotoxic T Cells (T_c)

Cytotoxic T (T_c) cells seek out and destroy abnormal and infected cells. They are highly mobile cells that roam throughout injured tissues. The constant monitoring of normal tissues by cytotoxic T cells and NK cells is called immune surveillance.

When a cytotoxic T cell encounters its target antigen bound to class I MHC proteins, it immediately destroys the target cell (see Figure 22–19). The T cell may (1) release perforins to destroy the target cell's plasma membrane, (2) release cytokines and activate genes in the target cell's nucleus telling that cell to undergo apoptosis, or (3) secrete a poisonous **lymphotoxin** (lim-fō-TOK-sin) to kill the target cell.

The entire sequence of events, from the appearance of the antigen in a tissue to cell destruction by cytotoxic T cells, takes a significant amount of time. After the first exposure to an antigen, 2 days or more may pass before the population of cytotoxic T cells reaches an effective level at the site of injury or infection. Over this period, the damage or infection may spread, making it more difficult to control.

Figure 22–19 Antigen Recognition and Activation of Cytotoxic T Cells.

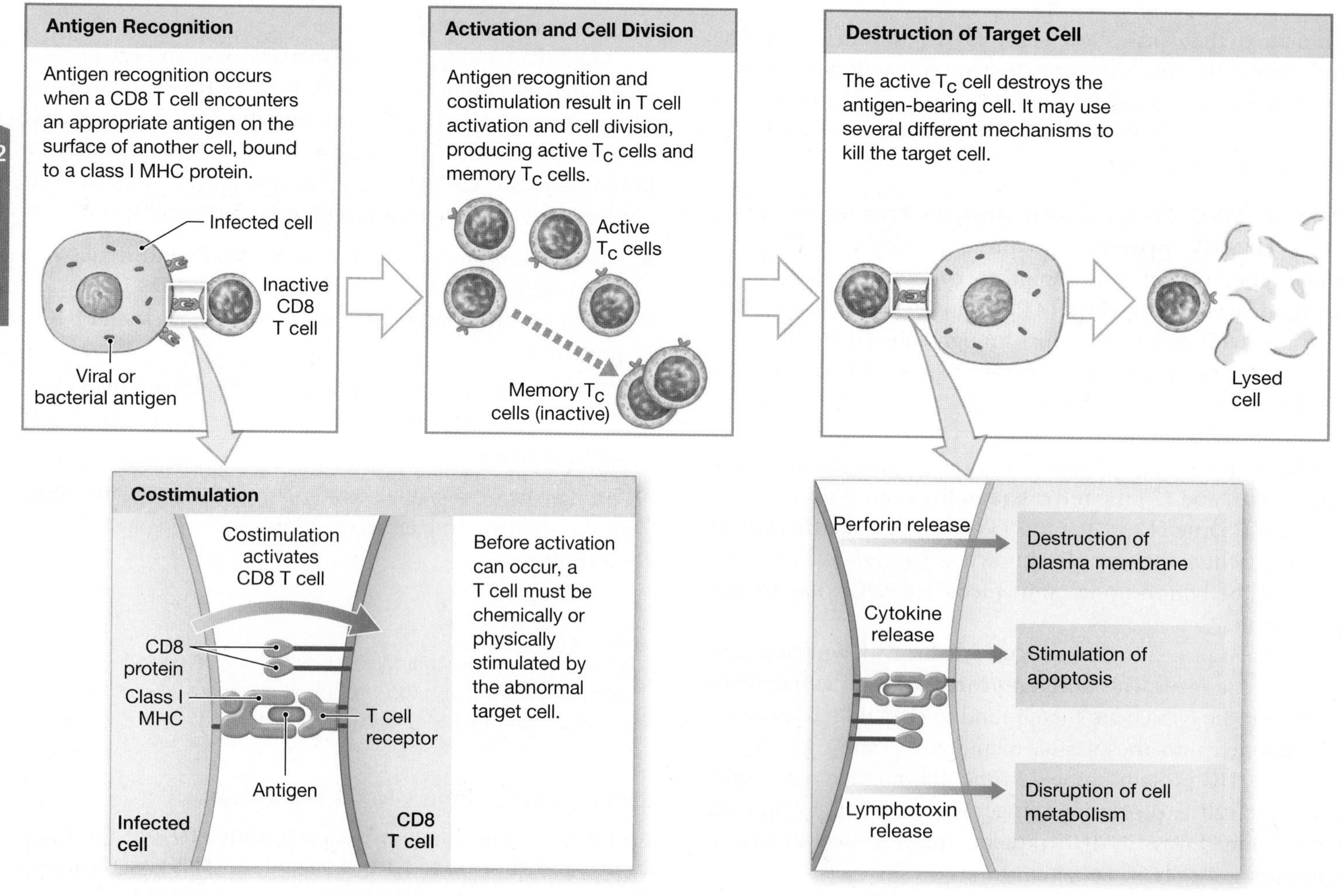
Antigen Recognition
Antigen recognition occurs when a CD8 T cell encounters an appropriate antigen on the surface of another cell, bound to a class I MHC protein.
Infected cell
Inactive CD8 T cell
Viral or bacterial antigen
Activation and Cell Division
Antigen recognition and costimulation result in T cell activation and cell division, producing active T_C cells and memory T_C cells.
Active T_C cells
Memory T_C cells (inactive)
Destruction of Target Cell
The active T_C cell destroys the antigen-bearing cell. It may use several different mechanisms to kill the target cell.
Lysed cell
Costimulation
Costimulation activates CD8 T cell
CD8 protein
Class I MHC
T cell receptor
Antigen
Infected cell
CD8 T cell
Before activation can occur, a T cell must be chemically or physically stimulated by the abnormal target cell.
Perforin release
Destruction of plasma membrane
Cytokine release
Stimulation of apoptosis
Lymphotoxin release
Disruption of cell metabolism

Memory T_C Cells

The same cell divisions that produce cytotoxic T cells also produce thousands of **memory T_C cells**. These cells do not differentiate further the first time the antigen triggers an immune response. However, if the same antigen appears a second time, memory T_C cells *immediately* differentiate into cytotoxic T cells and undergo repeated mitotic divisions. They produce a prompt, effective cellular response that can overwhelm the invader before it becomes well established in the tissues.

Regulatory T Cells (T_{regs})

Regulatory T cells, referred to as T_{regs}, moderate the responses of other T cells and of B cells by secreting inhibitory cytokines called *suppression factors*. Suppression does not take place right away, because activation takes much longer for regulatory T cells than for other types of T cells. In addition, upon activation, most of the CD8 T cells in the bloodstream produce cytotoxic T cells rather than regulatory T cells. As a result, regulatory T cells act *after* the initial immune response. In effect, these cells limit the degree of immune system activation from a single stimulus.

Functions of Activated CD4 T Cells: Helper T (T_H) and Memory T_H Cells

Upon activation, helper T (T_H) cells with CD4 markers undergo a series of divisions that produce both active helper T cells and **memory T_H cells** (Figure 22–20). The memory T_H cells remain in reserve. The active helper T cells secrete a variety of cytokines, our next topic.

Tips & Tools

Helper T cells "help" translate the message from the antigen-presenting cells of the nonspecific response to the cells of the specific immune responses.

Figure 22–20 Antigen Recognition and Activation of Helper T Cells. Inactive CD4 T cells (T_H cells) must be exposed to appropriate antigens bound to class II MHC proteins. The T_H cells then undergo activation, dividing to produce active T_H cells and memory T_H cells.

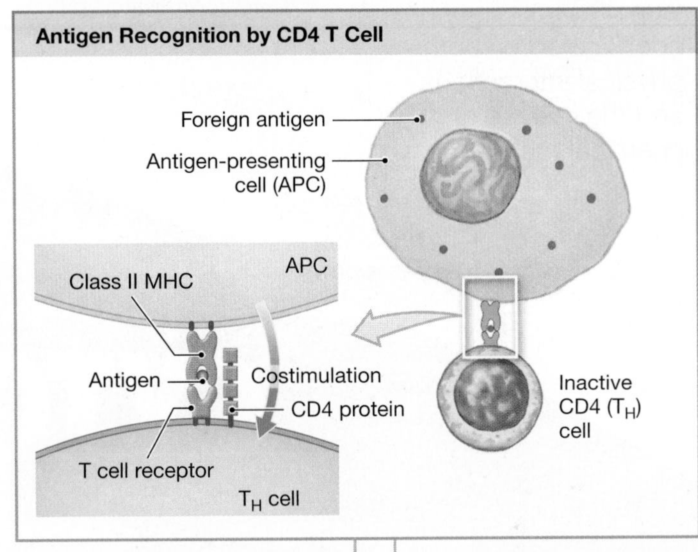
Antigen Recognition by CD4 T Cell
Foreign antigen
Antigen-presenting cell (APC)
APC
Class II MHC
Antigen
Costimulation
CD4 protein
T cell receptor
T_H cell
Inactive CD4 (T_H) cell

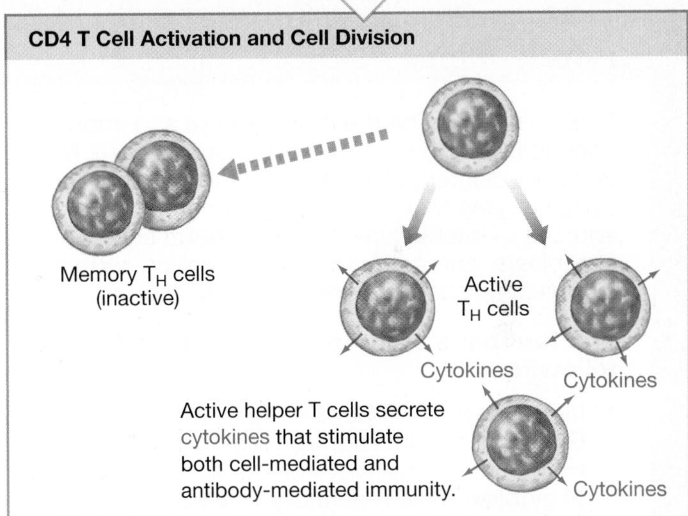
CD4 T Cell Activation and Cell Division
Memory T_H cells (inactive)
Active T_H cells
Cytokines
Cytokines
Active helper T cells secrete cytokines that stimulate both cell-mediated and antibody-mediated immunity.
Cytokines

22

Cytokines of Adaptive Defenses

Lymphocytes and other defensive cells produce cytokines that act as hormones, affecting cells and tissues throughout the body. Interleukins and tumor necrosis factors are examples of cytokines used in adaptive defenses (**Spotlight Figure 22–21**).

Cytokines Produced by Helper T Cells

Helper T cells produce cytokines that coordinate specific and nonspecific defenses and stimulate cell-mediated and antibody-mediated immunities. These cytokines do the following:

- Stimulate the T cell divisions that produce memory T cells and speed the maturation of cytotoxic T cells.
- Enhance nonspecific defenses by attracting macrophages to the affected area, preventing their departure, and stimulating their phagocytic activity and effectiveness.
- Attract and stimulate the activity of cytotoxic T cells, providing another means of destroying abnormal cells and pathogens.
- Promote the activation of B cells, leading ultimately to antibody production.

Summary of Cell-Mediated Adaptive Immunity

Figure 22–22 provides a review of the methods of antigen presentation and T cell stimulation. The plasma membranes of infected or otherwise abnormal cells trigger an immune response when CD8 T cells recognize antigens bound to class I MHC proteins. Extracellular pathogens or foreign proteins trigger an immune response when CD4 T cells recognize antigens displayed by

SPOTLIGHT

Figure 22–21

Cytokines of the Immune System

Innate (nonspecific) immunity and adaptive (specific) immunity are coordinated by both physical interactions and the release of chemical messengers.

The Immune System is coordinated by

Physical Interactions

Chemical Messengers

CYTOKINES

Many types of cells involved in the immune response secrete chemicals known as cytokines. Six major groups of cytokines are described here.

Interleukins

Interleukins may be the most diverse and important chemical messengers in the immune system. Nearly 20 types of interleukins have been identified. Lymphocytes and macrophages are the primary sources of interleukins, but endothelial cells, fibroblasts, and astrocytes also produce certain interleukins, such as interleukin-1 (IL-1).

The functions of interleukins include the following:

1. **Increasing T Cell Sensitivity to Antigens Exposed on Macrophage Membranes.** Heightened sensitivity speeds up the production of cytotoxic and regulatory T cells.
2. **Stimulating B Cell Activity, Plasma Cell Formation, and Antibody Production.** These events promote the production of antibodies and the development of antibody-mediated immunity.
3. **Enhancing Innate (Nonspecific) Immunity.** Known effects of interleukin production include (a) stimulation of inflammation, (b) formation of scar tissue by fibroblasts, (c) increasing body temperature by the pre-optic nucleus of the hypothalamus, (d) stimulation of mast cell formation, and (e) promotion of adrenocorticotropic hormone (ACTH) secretion by the anterior lobe of the pituitary gland.
4. **Moderating the Immune Response.** Some interleukins help suppress immune function and shorten the immune response.

Two interleukins, IL-1 and IL-2, are important in stimulating and maintaining the immune response. When released by activated macrophages and lymphocytes, these cytokines stimulate the activities of other immune cells and of the secreting cell. The result is a positive feedback loop that helps to recruit additional immune cells.

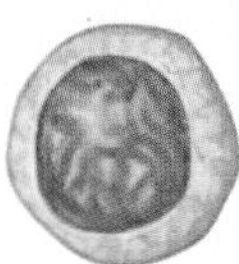
Lymphocyte

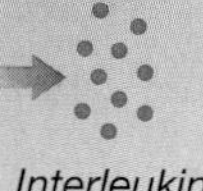
Interleukin

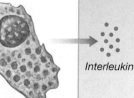

Fixed macrophage

Interleukin

Although mechanisms exist to control the degree of stimulation, the regulatory process sometimes breaks down. Massive production of interleukins can cause problems at least as severe as those of the primary infection. For example, in *Lyme disease*, activated macrophages release IL-1 in response to a localized bacterial infection. This release produces fever, pain, skin rash, and arthritis throughout the entire body.

Some interleukins enhance the immune response, while others suppress it. The quantities secreted at any moment affect the nature and intensity of the response to an antigen. During a typical infection, the pattern of interleukin secretion changes constantly. Whether stimulatory interleukins or suppressive interleukins predominate plays a part in determining whether the individual overcomes the infection. For this reason, interleukins and their interactions are now the focus of intensive research.

Interferons

Interferons (IFNs) make the cell that synthesizes them, and that cell's neighbors, resistant to viral infection. In this way, IFNs slow the spread of a virus. They may have other beneficial effects as well. For example, interferon alpha (α) and interferon gamma (γ) attract and stimulate NK cells. Interferon beta (β) slows the progress of inflammation associated with viral infection. Interferon gamma also stimulates macrophages, making them more effective at killing bacterial or fungal pathogens.

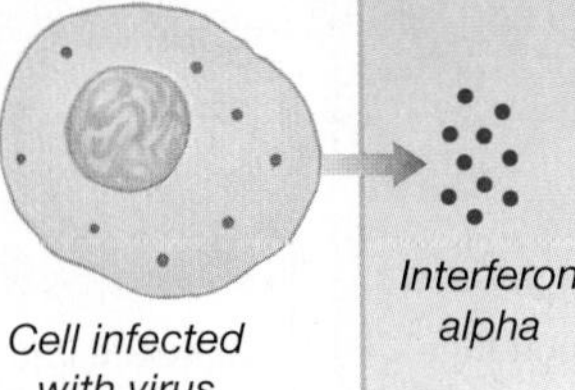
Cell infected with virus
Interferon alpha

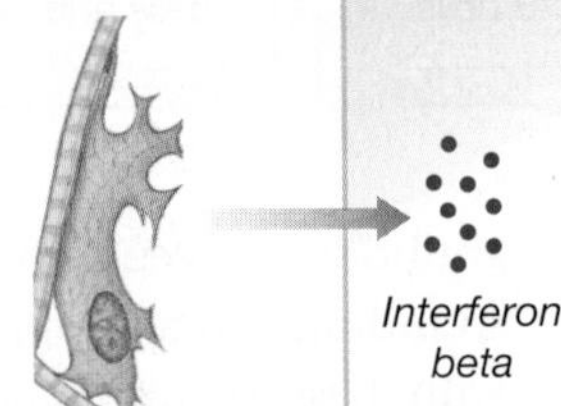
Fibroblast
Interferon beta

Because they stimulate NK cell activity, interferons can be used to fight some cancers. For example, interferon alpha has been used to treat malignant melanoma, bladder cancer, ovarian cancer, and some forms of leukemia. Interferon alpha or gamma may be used to treat Kaposi's sarcoma, a cancer that typically develops in patients with AIDS.

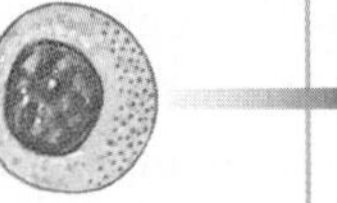
NK cell

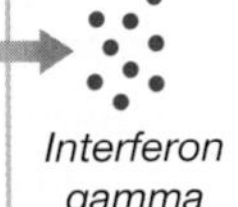
Interferon gamma

Tumor Necrosis Factors

Tumor necrosis factors (TNFs) slow the growth of a tumor and kill sensitive tumor cells. Activated macrophages secrete one type of TNF and carry the molecules in their plasma membranes. Cytotoxic T cells produce a different type of TNF. In addition to their effects on tumor cells, TNFs stimulate granular leukocyte production, promote eosinophil activity, act as pyrogens and cause fever, and increase T cell sensitivity to interleukins.

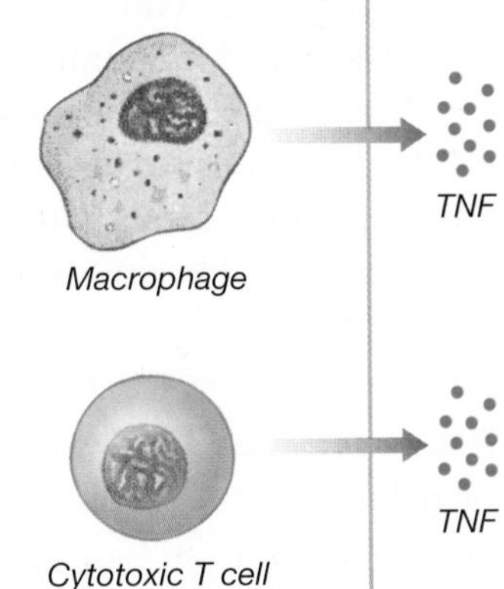
Macrophage
TNF
Cytotoxic T cell
TNF

Phagocyte-Activating Chemicals

Several cytokines coordinate immune defenses by adjusting the activities of phagocytic cells. These cytokines include factors that attract free macrophages and microphages and prevent their premature departure from the site of an injury.

Colony-Stimulating Factors

Colony-stimulating factors (CSFs) were introduced in Chapter 19. ⮌ p. 676 These factors are produced by active T cells, cells of the monocyte–macrophage system, endothelial cells, and fibroblasts. CSFs stimulate the production of blood cells in red bone marrow and lymphocytes in lymphoid tissues and organs.

Miscellaneous Cytokines

This general category includes many chemicals discussed in earlier chapters. Examples include leukotrienes, lymphotoxins, perforin, hemopoiesis-stimulating factor, and suppression factors.

NOTE: Cytokines are often classified according to their origins. For example, *lymphokines* are produced by lymphocytes. *Monokines* are secreted by active monocytes, macrophages, and other antigen-presenting cells. These terms are misleading, however, because lymphocytes and macrophages may secrete the same cytokines. In addition, cells involved in adaptive immunity and tissue repair can also secrete cytokines.

Figure 22–22 A Summary of the Pathways of T Cell Activation.

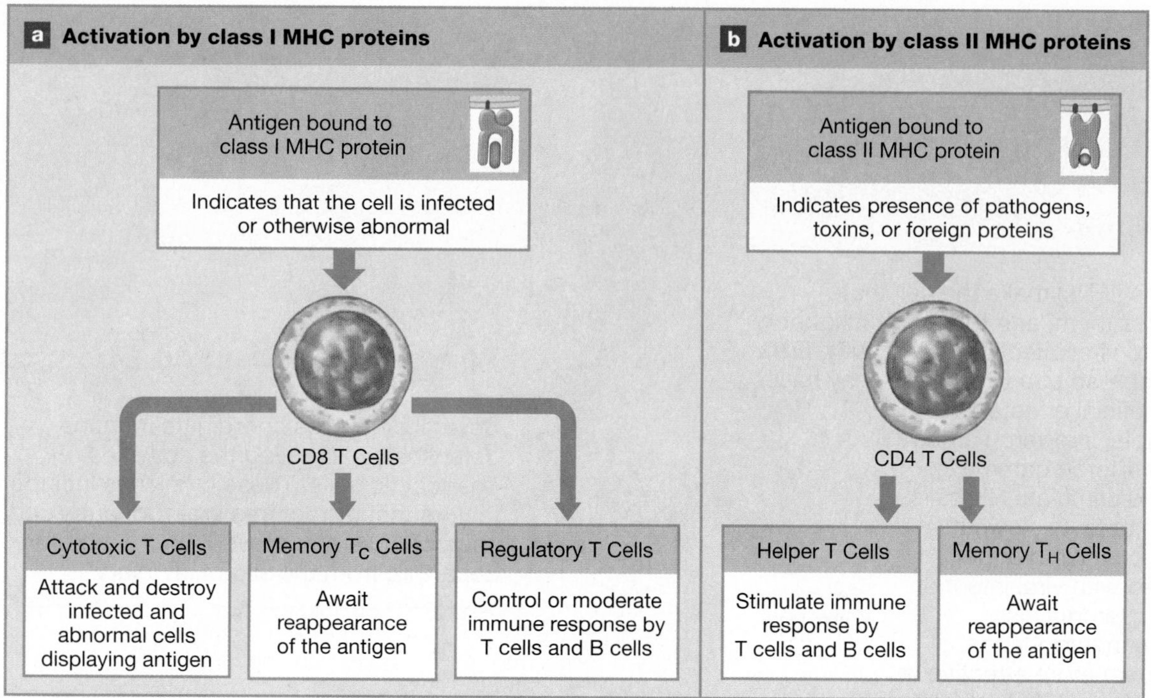
a Activation by class I MHC proteins
Antigen bound to class I MHC protein
Indicates that the cell is infected or otherwise abnormal
CD8 T Cells
Cytotoxic T Cells
Attack and destroy infected and abnormal cells displaying antigen
Memory T_C Cells
Await reappearance of the antigen
Regulatory T Cells
Control or moderate immune response by T cells and B cells
b Activation by class II MHC proteins
Antigen bound to class II MHC protein
Indicates presence of pathogens, toxins, or foreign proteins
CD4 T Cells
Helper T Cells
Stimulate immune response by T cells and B cells
Memory T_H Cells
Await reappearance of the antigen

class II MHC proteins. In the next section, we look at how the helper T cells derived from activated CD4 T cells in turn activate B cells that are sensitive to the same specific antigen.

Checkpoint

15. **How can the presence of an abnormal peptide in the cytoplasm of a cell initiate an immune response?**
16. **A decrease in the number of cytotoxic T cells would affect which type of immunity?**

See the blue Answers tab at the back of the book.

22-6 In antibody-mediated adaptive immunity, sensitized B cells respond to antigens by producing specific antibodies

Learning Outcome Discuss the mechanisms of B cell activation and differentiation, describe the structure and function of antibodies, and explain the primary and secondary responses to antigen exposure.

B cells are responsible for launching a chemical attack on antigens. They produce appropriate specific antibodies that eliminate the antigens. If the same antigens eliminated in the *primary response* reappear, a rapid *secondary response* is initiated by memory cells.

B Cell Sensitization and Activation

As noted earlier, the body has millions of B cell populations. Each B cell carries thousands of its own particular type of antibody molecules bound to its plasma membrane. If corresponding antigens appear in the interstitial fluid, they interact with these membrane-bound antibodies. When binding occurs, the B cell prepares to undergo activation. This preparatory process is called **sensitization** (Figure 22–23). Because B cells migrate throughout the body, pausing briefly in one lymphoid tissue or another, sensitization typically takes place within the lymph node nearest the site of infection or injury.

Recall that B cell plasma membranes contain class II MHC proteins. During sensitization, antigens are brought into the cell by endocytosis. The antigens subsequently appear on the surface of the B cell, bound to class II MHC proteins. The sensitized B cell is then on "standby" but generally does not undergo activation unless it receives the "OK" from a helper T cell. The need for activation by a helper T cell helps prevent inappropriate activation, the same way that costimulation acts as a "safety" in cell-mediated immunity.

What happens when a sensitized B cell meets a helper T cell that has already been activated by antigen presentation? The helper T cell binds to the B cell's MHC complex, recognizes the antigen, and begins secreting cytokines that promote B cell activation (see Figure 22–23).

Production of Plasma Cells, Antibodies, and Memory B Cells

The activated B cell typically divides several times, producing daughter cells that differentiate into **plasma cells** and *memory B cells* (see Figure 22–23). The plasma cells begin synthesizing

Figure 22–23 The Sensitization and Activation of B Cells. A B cell is sensitized by exposure to antigens. Once antigens are bound to antibodies in the B cell plasma membrane, the B cell displays those antigens on class II MHC proteins in its plasma membrane. Activated helper T cells encountering the antigens release cytokines that costimulate the sensitized B cell and trigger its activation. The activated B cell then divides, producing memory B cells and plasma cells that secrete antibodies.

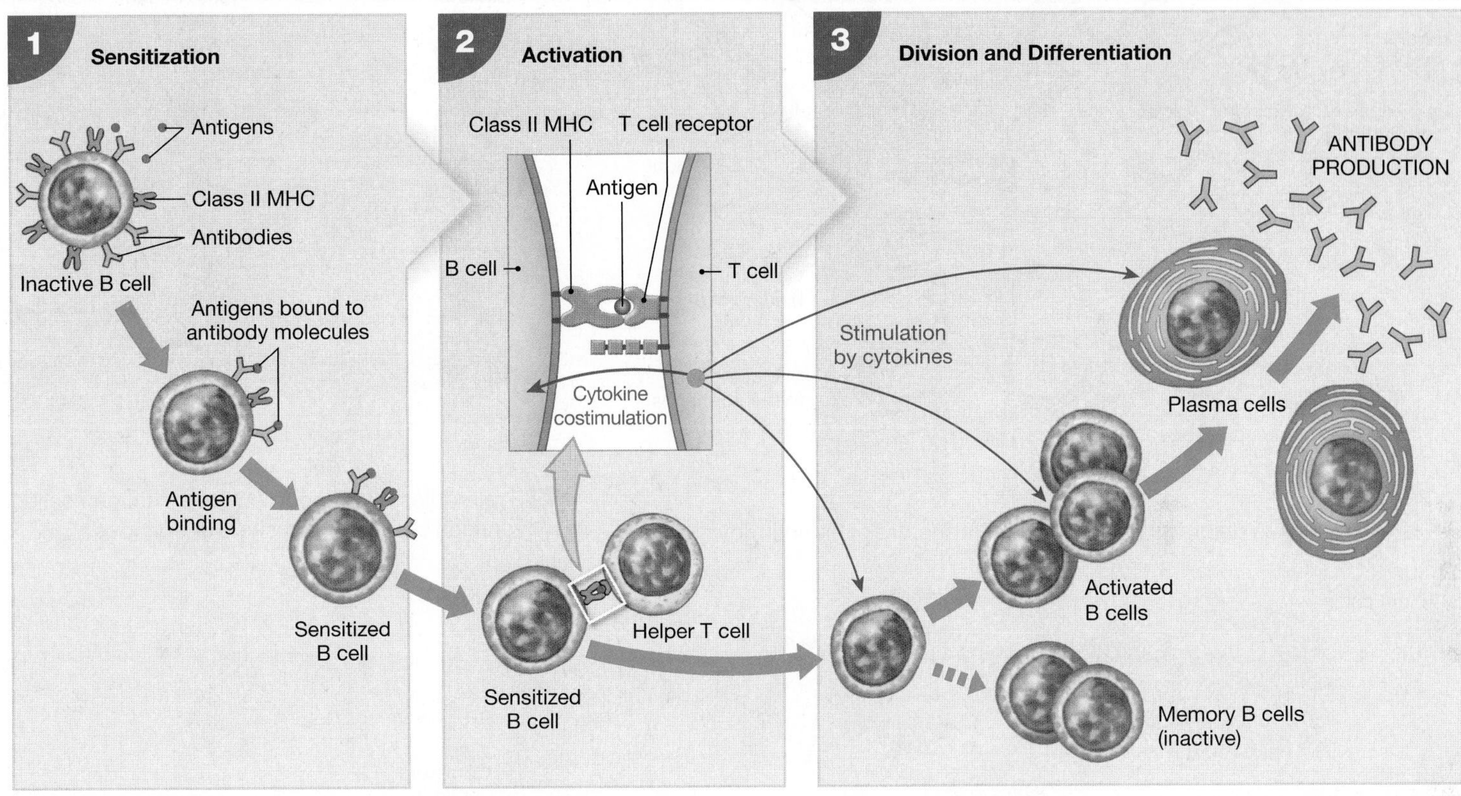
1 Sensitization
Antigens
Class II MHC
Antibodies
Inactive B cell
Antigens bound to antibody molecules
Antigen binding
Sensitized B cell
2 Activation
Class II MHC
T cell receptor
Antigen
B cell
T cell
Cytokine costimulation
Helper T cell
Sensitized B cell
3 Division and Differentiation
ANTIBODY PRODUCTION
Stimulation by cytokines
Plasma cells
Activated B cells
Memory B cells (inactive)

and secreting large quantities of antibodies into the interstitial fluid. These antibodies have the same target as the membrane-bound antibodies on the surface of the sensitized B cell. After activation of the B cell, helper T cells secrete cytokines that stimulate B cell division, speed plasma cell formation, and enhance antibody production. A plasma cell can secrete up to 100 million antibody molecules each hour.

Memory B cells perform the same role in antibody-mediated immunity that memory T cells perform in cell-mediated immunity. Memory B cells do not respond to a threat on first exposure. Instead, they remain in reserve to deal with subsequent injuries or infections that involve the same antigens. On subsequent exposure, the memory B cells divide and differentiate into plasma cells that secrete antibodies in massive quantities.

Antibody Structure and Function

We have discussed that **antibodies** are soluble proteins. ⮌ p. 661 The binding of an antibody to its target antigen starts a chain reaction that leads to the destruction of the target compound or organism.

Structure and Classes of Antibodies

A Y-shaped antibody molecule consists of two pairs of polypeptide chains: one pair of **heavy chains** and one pair of **light chains** (Figure 22–24). Each chain contains both *constant segments* and *variable segments.*

The constant segments of the heavy chains form the base of the antibody molecule (see Figure 22–24a,b). B cells produce only five types of constant segments. These are the basis of a scheme that identifies five classes of antibodies, or **immunoglobulins (Igs)**: *IgG, IgE, IgD, IgM,* and *IgA* (Table 22–1). The structure of the constant segments of the heavy chains determines the way the antibody is secreted and how it is distributed within the body. For example, antibodies in one class circulate in body fluids, whereas those of another class bind to the membranes of basophils and mast cells.

Tips & Tools

The classes of immunoglobulins—Ig***M***, Ig***A***, Ig***D***, Ig***G***, Ig***E***—spell ***MADGE***.

Figure 22–24 Antibody Structure and Function.

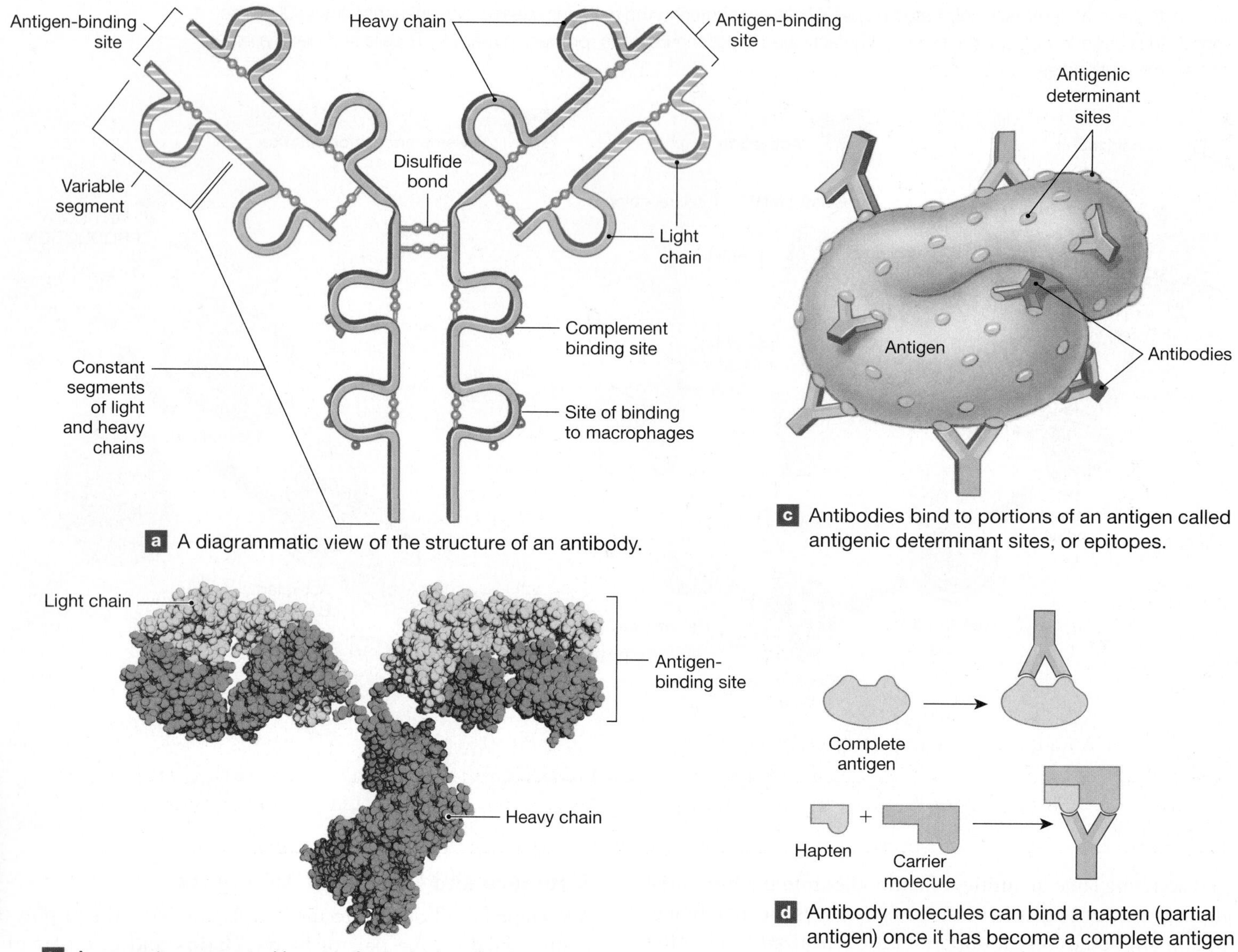
Antigen-binding site
Heavy chain
Antigen-binding site
Variable segment
Disulfide bond
Light chain
Complement binding site
Constant segments of light and heavy chains
Site of binding to macrophages
Antigenic determinant sites
Antigen
Antibodies
Light chain
Antigen-binding site
Heavy chain
Complete antigen
Hapten
Carrier molecule

a A diagrammatic view of the structure of an antibody.

b A computer-generated image of a typical antibody.

c Antibodies bind to portions of an antigen called antigenic determinant sites, or epitopes.

d Antibody molecules can bind a hapten (partial antigen) once it has become a complete antigen by combining with a carrier molecule.

The heavy-chain constant segments, which are bound to constant segments of the light chains, also contain binding sites that can activate the complement system. These binding sites are covered when the antibody is secreted but become exposed when the antibody binds to an antigen.

The specificity of an antibody molecule depends on the amino acid sequence of the variable segments of the light and heavy chains. The free tips of the two variable segments form the **antigen-binding sites** of the antibody molecule (see Figure 22–24a). The heavy-chain constant segments have no effect on the antibody's specificity, which is determined by the antigen-binding sites.

Small differences in the structure of the variable segments affect the precise shape of the antigen-binding site. These differences make antibodies specific for different antigens. The distinctions are the result of minor genetic variations that occur during the production, division, and differentiation of B cells. A normal adult body contains roughly 10 trillion B cells, which can produce an estimated 100 million types of antibodies, each with a different specificity.

The Antigen–Antibody Complex

When an antigen-binding site on an antibody molecule binds to its corresponding antigen molecule, an **antigen–antibody complex** is formed. Once the two molecules are in position, hydrogen bonding and other weak chemical forces lock them together. Antigen-binding sites interact with an antigen in the same way that the active site of an enzyme interacts with a substrate molecule. ⮌ p. 55

Table 22–1 Classes of Antibodies

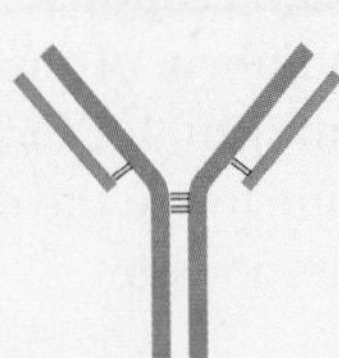	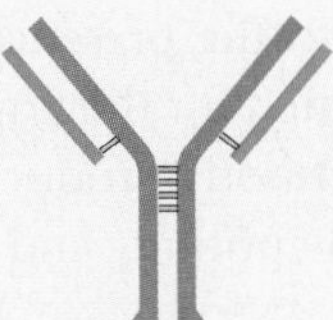	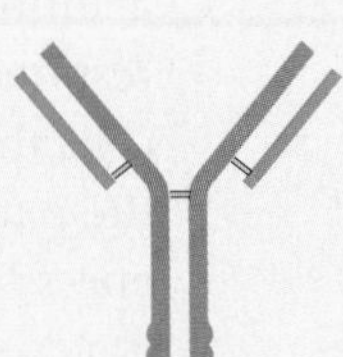	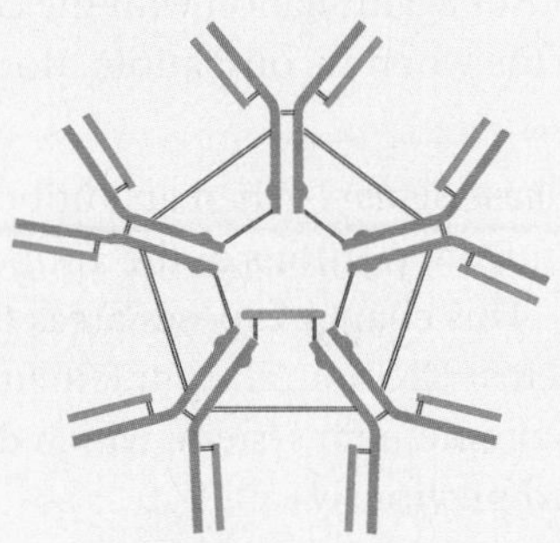	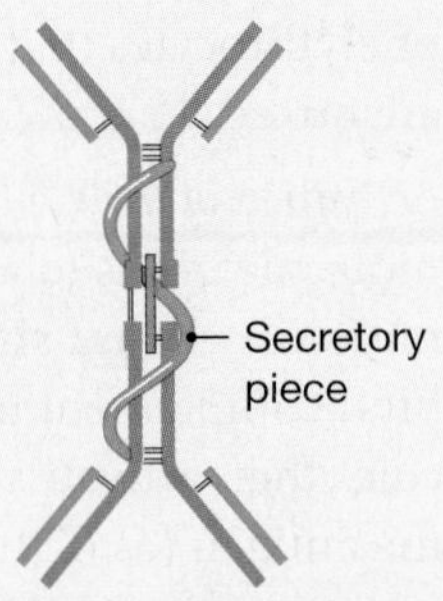 Secretory piece
IgG is the largest and most diverse class of antibodies. IgG antibodies account for 80 percent of all antibodies. They are responsible for resistance against many viruses, bacteria, and bacterial toxins. These antibodies can cross the placenta, and maternal IgG provides passive immunity to the fetus during embryonic development. However, the anti-Rh antibodies produced by Rh-negative mothers are also IgG antibodies and produce *hemolytic disease of the newborn*.	**IgE** attaches as an individual molecule to the exposed surfaces of basophils and mast cells. When a suitable antigen is bound by IgE molecules, the cell is stimulated to release histamine and other chemicals that accelerate inflammation in the immediate area. IgE is also important in the allergic response.	**IgD** is an individual molecule on the surfaces of B cells, where it can bind antigens in the extracellular fluid. This binding can play a role in the sensitization of the B cell involved.	**IgM** is the first class of antibody secreted after an antigen is encountered. IgM concentration declines as IgG production accelerates. Although plasma cells secrete individual IgM molecules, IgM circulates as a five-antibody starburst. The anti-A and anti-B antibodies responsible for the agglutination of incompatible blood types are IgM antibodies. IgM antibodies may also attack bacteria that are insensitive to IgG.	**IgA** is found primarily in glandular secretions such as mucus, tears, saliva, and semen. These antibodies attack pathogens before they gain access to internal tissues. IgA antibodies circulate in blood as individual molecules or in pairs. Epithelial cells absorb them from the blood and attach a *secretory piece*, which confers solubility, before secreting the IgA molecules onto the epithelial surface.

Antibodies do not bind to the entire antigen. Instead, they bind to specific portions of its exposed surface—regions called **antigenic determinant sites**, or **epitopes** (Figure 22–24c). The specificity of the binding depends initially on the three-dimensional "fit" between the variable segments of the antibody molecule and the corresponding antigenic determinant sites. A **complete antigen** has at least two antigenic determinant sites, one for each of the antigen-binding sites on an antibody molecule. Exposure to a complete antigen can lead to B cell sensitization and a subsequent immune response. Most environmental antigens have multiple antigenic determinant sites, and entire microorganisms may have thousands.

Haptens, or *partial antigens*, do not ordinarily cause B cell activation and antibody production. Haptens include short peptide chains, steroids and other lipids, and several drugs, including antibiotics such as *penicillin*. However, haptens may become attached to carrier molecules, forming combinations that can function as complete antigens (Figure 22–24d). In some cases, the carrier contributes an antigenic determinant site. Antibodies will then attack both the hapten and the carrier molecule. If the carrier molecule is normally present in the tissues, the antibodies may begin attacking and destroying normal cells. This process is the basis for several drug reactions, including allergies to penicillin.

Actions of Antibodies

The formation of an antigen–antibody complex may cause the elimination of the antigen in the following ways:

- *Neutralization.* Both viruses and bacterial toxins have specific sites that must bind to target regions on body cells before they can enter or injure those cells. Antibodies may bind to those sites, making the virus or toxin incapable of attaching itself to a cell. This mechanism is known as **neutralization**.
- *Precipitation and Agglutination.* Each antibody molecule has two antigen-binding sites, and most antigens have many antigenic determinant sites. If individual antigens (such as macromolecules or bacterial cells) are far apart, an antibody molecule will necessarily bind to two antigenic sites on the same antigen. However, if antigens are close together, an antibody can bind to antigenic determinant sites on two separate antigens. In this way, antibodies can link large numbers of antigens together. The three-dimensional structure created by such binding is known as an **immune complex**. When the antigen is a soluble molecule, such as a toxin, this process may create complexes that are too large to remain in solution. The formation of insoluble immune complexes is called **precipitation**. When the target antigen is on the surface of a cell or virus,

the formation of large complexes is called **agglutination**. We discussed an example of agglutination with the clumping of erythrocytes that occurs when incompatible blood types are mixed. p. 669

- *Activation of the Complement System.* When an antibody molecule binds to an antigen, portions of the antibody molecule change shape. This change exposes areas that bind complement proteins. The bound complement molecules then activate the complement system, which destroys the antigen (as discussed previously).
- *Attraction of Phagocytes.* Antigens covered with antibodies attract eosinophils, neutrophils, and macrophages. These cells phagocytize pathogens and destroy foreign or abnormal plasma membranes.
- *Opsonization.* A coating of antibodies and complement proteins increases the effectiveness of phagocytosis. This effect is called *opsonization*. Some bacteria have slick plasma membranes or capsules, but opsonization makes it easier for phagocytes to hold onto their prey before they engulf it. Phagocytes can bind more easily to antibodies and complement proteins than they can to the bare surface of a pathogen.
- *Stimulation of Inflammation.* Antibodies may promote inflammation by stimulating basophils and mast cells.
- *Prevention of Bacterial and Viral Adhesion.* Antibodies dissolved in saliva, mucus, and perspiration coat epithelia, adding an additional layer of defense. A covering of antibodies makes it difficult for pathogens to attach to and penetrate body surfaces.

Primary and Secondary Responses to Antigen Exposure

The initial immune response to an antigen is called the **primary response**. When the antigen appears again, it triggers a more extensive and prolonged **secondary response**. This response is due to the presence of large numbers of memory cells that are primed for the arrival of the antigen. Let's look at the pattern of antibody production over time to see the differences between the primary and secondary responses.

The Primary Response

The primary response takes time to develop because the antigen must activate the appropriate B cells. These cells must then differentiate into plasma cells. As plasma cells differentiate and begin secreting, the concentration of circulating antibodies makes a gradual, sustained rise (Figure 22–25a).

During the primary response, the **antibody titer**, or level of antibodies in the plasma, does not peak until 1 to 2 weeks after the initial exposure. If the person is no longer exposed to the antigen, the antibody concentration then declines. The antibody titer declines because (1) plasma cells have very high metabolic rates and survive for only a short time, and (2) regulatory T cells release suppression factors that inhibit further production of plasma cells. However, regulatory T cell activity does not begin immediately after exposure to the antigen. Also, under normal conditions helper T cells outnumber regulatory T cells by more than 3 to 1. As a result, many B cells are activated before regulatory T cell activity has a noticeable effect.

22

Figure 22–25 The Primary and Secondary Responses in Antibody-Mediated Immunity.

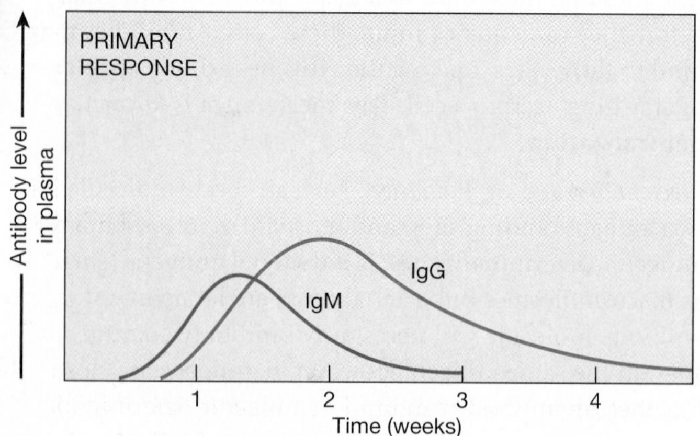
PRIMARY
RESPONSE
Antibody level in plasma
IgG
IgM
1
2
3
4
Time (weeks)

a The primary response takes about 2 weeks to develop peak antibody levels (titers). IgM and IgG antibody levels do not remain elevated.

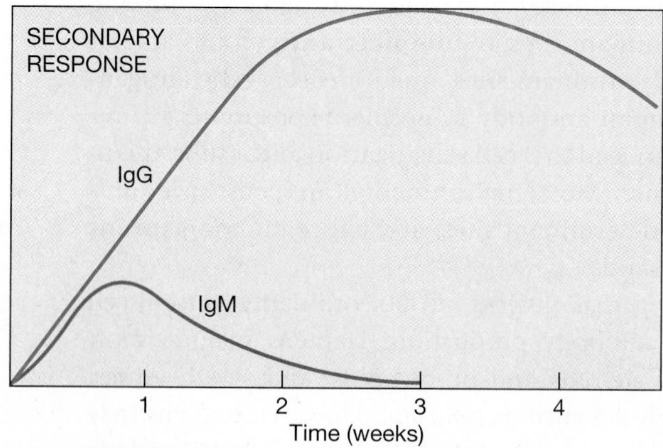
SECONDARY
RESPONSE
IgG
IgM
1
2
3
4
Time (weeks)

b The secondary response has a very rapid increase in IgG antibody concentration and rises to levels much higher than those of the primary response. Antibody levels remain elevated for an extended period after the second exposure to the antigen.

? During the primary response, which antibody peaks sooner? During the secondary response, which antibody level is higher?

Activated B cells start dividing immediately. At each cycle of division, some of the daughter cells differentiate into plasma cells, while others continue to divide. Molecules of *immunoglobulin M*, or *IgM*, are the first to appear in the bloodstream. The plasma cells that produce IgM differentiate after only a few cycles of B cell division. Levels of *immunoglobulin G*, or *IgG*, rise more slowly, because the plasma cells responsible differentiate only after repeated cell divisions that also generate large numbers of memory B cells. In general, IgM is less effective as a defense than IgG. However, IgM provides an immediate defense that can fight the infection until massive quantities of IgG can be produced.

The Secondary Response

Unless memory B cells are exposed to the same antigen a second time, they do not differentiate into plasma cells. If and when that exposure occurs, the memory B cells respond right away—faster than the B cells stimulated during the initial exposure. This response is immediate in part because memory B cells are activated at relatively low antigen concentrations. In addition, these cells synthesize more effective and destructive antibodies. Activated memory B cells divide and differentiate into plasma cells that secrete these antibodies in massive quantities. This secretion is the secondary response to antigen exposure.

During the secondary response, antibody concentrations and titers increase more rapidly and reach levels many times higher than they did in the primary response (Figure 22–25b). The secondary response appears even if the second exposure occurs years after the first. The reason is that memory cells may survive for 20 years or more.

The primary response develops slowly and does not produce antibodies in massive quantities. For these reasons, it may not prevent an infection the first time a pathogen appears in the body. However, a person who survives the first infection will probably be resistant to that pathogen in the future, thanks to a rapid and overwhelming secondary response. The effectiveness of the secondary response is one of the basic principles behind the use of immunization to prevent disease.

Tips & Tools

The antibody response is like ordering a custom suit. The first suit (the primary antibody response) takes time to make because the tailor (an activated B cell) must first make a pattern (a clone of memory cells). Subsequent suits (secondary responses) are made much more quickly because the pattern already exists.

Checkpoint

17. **How would a lack of helper T cells affect the antibody-mediated immune response?**
18. **Define *sensitization*.**
19. **A sample of lymph contains an elevated number of plasma cells. Would you expect the number of antibodies in the blood to be increasing or decreasing? Why?**
20. **Describe the structure of an antibody.**
21. **Which would be more negatively affected—the primary response or the secondary response—by a lack of memory B cells for a particular antigen?**

See the blue Answers tab at the back of the book.

22-7 Immunocompetence enables a normal immune response; abnormal responses result in immune disorders

Learning Outcome Describe the development of immunocompetence, discuss the effects of stress on immune function, and list and explain examples of immune disorders and allergies.

We begin this section by summarizing the normal immune response. Next, we go on to explore **immunocompetence**, the ability to produce an immune response after exposure to an antigen. Then we examine the role of stress in immune responses. Finally, we consider some of the disorders that result when immune responses go wrong.

Summary of Innate and Adaptive Immunity

We have now examined the basic cellular and chemical interactions that follow after a foreign antigen appears in the body. Table 22–2 reviews the cells that participate in tissue defenses and Figure 22–26 gives a timeline for their appearance at the site of a bacterial infection.

Figure 22–26 The Course of the Body's Response to a Bacterial Infection. The basic sequence of events begins with the appearance of bacteria in peripheral tissues at time 0.

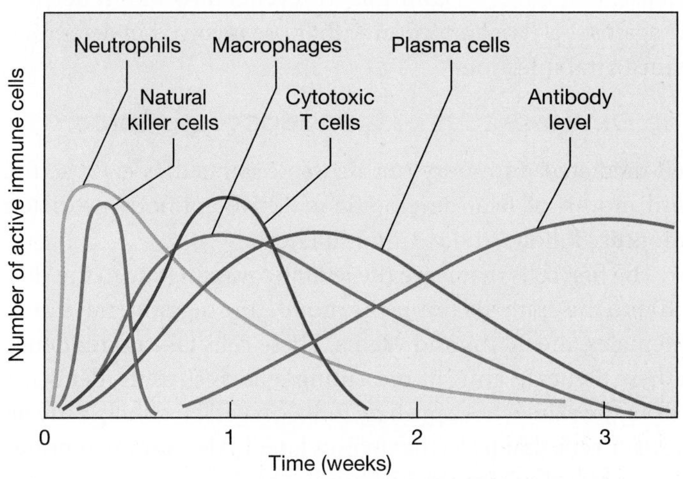
Neutrophils
Macrophages
Plasma cells
Natural killer cells
Cytotoxic T cells
Antibody level
Number of active immune cells
0
1
2
3
Time (weeks)

? After the appearance of bacteria in the body, which immune cell is present for the shortest amount of time?

Go to MasteringA&P™ > Study Area > Lab Tools > PhysioEx 9.1 > Exercise 12: Serological Testing

Table 22–2 Cells That Participate in Tissue Defenses

Cell	Functions
Neutrophils	Phagocytosis; stimulation of inflammation
Eosinophils	Phagocytosis of antigen–antibody complexes; suppression of inflammation; participation in allergic response
Mast cells and basophils	Stimulation and coordination of inflammation by release of histamine, heparin, leukotrienes, prostaglandins
ANTIGEN-PRESENTING CELLS	
Macrophages (free and fixed macrophages, stellate macrophages, and microglia)	Phagocytosis; antigen processing; antigen presentation with class II MHC proteins; secretion of cytokines, especially interleukins and interferons
Dendritic cells	Pinocytosis; antigen processing; antigen presentation bound to class II MHC proteins
LYMPHOCYTES	
NK cells	Destruction of plasma membranes containing abnormal antigens
Cytotoxic T cells (CD8 marker)	Lysis of plasma membranes containing antigens bound to class I MHC proteins; secretion of perforins, defensins, lymphotoxins, and other cytokines
Helper T cells (CD4 marker)	Secretion of cytokines that stimulate cell-mediated and antibody-mediated immunity; activation of sensitized B cells
B cells	Differentiation into plasma cells, which secrete antibodies and provide antibody-mediated immunity
Regulatory T cells (CD8 marker)	Secretion of suppression factors that inhibit the immune response
Memory cells (T, B)	Produced during the activation of T cells and B cells; remain in tissues awaiting antigen reappearance

In the early stages of infection, neutrophils and NK cells migrate into the threatened area and destroy bacteria. Over time, cytokines draw increasing numbers of phagocytes into the region. Cytotoxic T cells appear as arriving T cells are activated by antigen presentation. Last of all, the population of plasma cells rises as activated B cells differentiate. This rise is followed by a gradual, sustained increase in the level of circulating antibodies.

22

Figure 22–27 provides an integrated review of the relationship between innate and adaptive immunity. The basic sequence of events is similar when a viral infection occurs. The initial steps are different, however, because cytotoxic T cells provide an adaptive defense and NK cells provide an innate defense. Figure 22–28 contrasts the events involved in defending against bacterial infection with those involved in defending against viral infection.

The Development of Immunocompetence

Cell-mediated immunity can be demonstrated as early as the third month of fetal development. Active antibody-mediated immunity follows about 1 month later.

The first cells that leave the fetal thymus migrate to the skin and into the epithelia lining the mouth, the digestive tract, and, in females, the uterus and vagina. These cells take up residence in these tissues as antigen-presenting cells, such as the dendritic cells of the skin, whose primary function will be to help activate T cells. T cells that leave the thymus later in development populate lymphoid organs throughout the body.

The plasma membranes of the first B cells produced in the liver and bone marrow carry IgM antibodies. Sometime after the fourth month the fetus may produce IgM antibodies if exposed to specific pathogens. The developing fetus usually does not produce antibodies, however, because it has naturally acquired passive immunity due to IgG antibodies from the mother's bloodstream. These are the only antibodies that can cross the placenta. They include the antibodies responsible for the clinical problems due to Rh incompatibility, discussed in Chapter 19. ⮌ p. 672 Problems with incompatibilities involving the ABO blood groups rarely occur. The reason is that anti-A and anti-B antibodies are IgM antibodies, which cannot cross the placenta.

The natural immunity provided by maternal IgG may not be enough to protect the fetus if the maternal defenses are overwhelmed by a bacterial or viral infection. For example, the microorganisms responsible for syphilis and rubella ("German measles") can cross from the maternal to the fetal bloodstream, producing a congenital infection that leads to the production of fetal antibodies. IgM provides only a partial defense, and these infections can result in severe developmental problems for the fetus.

Delivery eliminates the maternal supply of IgG. Although the mother provides IgA antibodies in breast milk, the infant gradually loses its passive immunity. The amount of maternal IgG in the infant's bloodstream declines rapidly over the first 2 months after birth. During this period, the infant becomes vulnerable to infection by bacteria or viruses that were previously overcome by maternal antibodies. The infant's immune system begins to respond to infections, environmental antigens, and vaccinations. As a result, the infant begins producing its own IgG. From birth to age 12, children encounter a "new" antigen on average every 6 weeks. (This fact explains why most parents, who were exposed to the same antigens when they were children, remain healthy while their children develop runny noses and colds.) Over this period, the concentration of circulating

Figure 22–27 **An Integrated Summary of the Immune Response.**

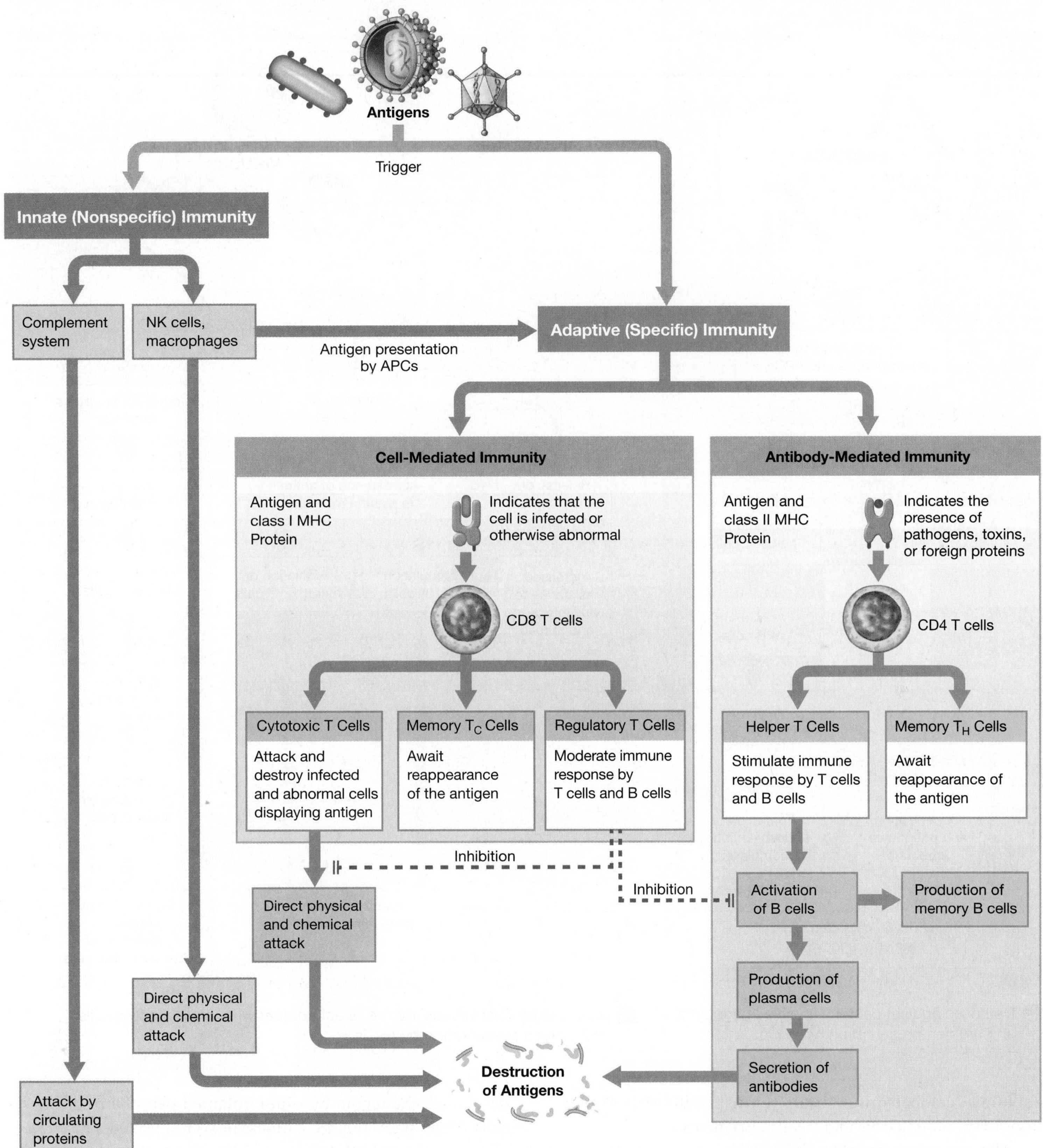
Antigens
Trigger
Innate (Nonspecific) Immunity
Complement system
NK cells, macrophages
Antigen presentation by APCs
Adaptive (Specific) Immunity
Cell-Mediated Immunity
Antigen and class I MHC Protein
Indicates that the cell is infected or otherwise abnormal
CD8 T cells
Cytotoxic T Cells
Attack and destroy infected and abnormal cells displaying antigen
Memory T_C Cells
Await reappearance of the antigen
Regulatory T Cells
Moderate immune response by T cells and B cells
Antibody-Mediated Immunity
Antigen and class II MHC Protein
Indicates the presence of pathogens, toxins, or foreign proteins
CD4 T cells
Helper T Cells
Stimulate immune response by T cells and B cells
Memory T_H Cells
Await reappearance of the antigen
Inhibition
Inhibition
Direct physical and chemical attack
Activation of B cells
Production of memory B cells
Production of plasma cells
Direct physical and chemical attack
Destruction of Antigens
Secretion of antibodies
Attack by circulating proteins

Figure 22–28 Defenses against Bacterial and Viral Pathogens.

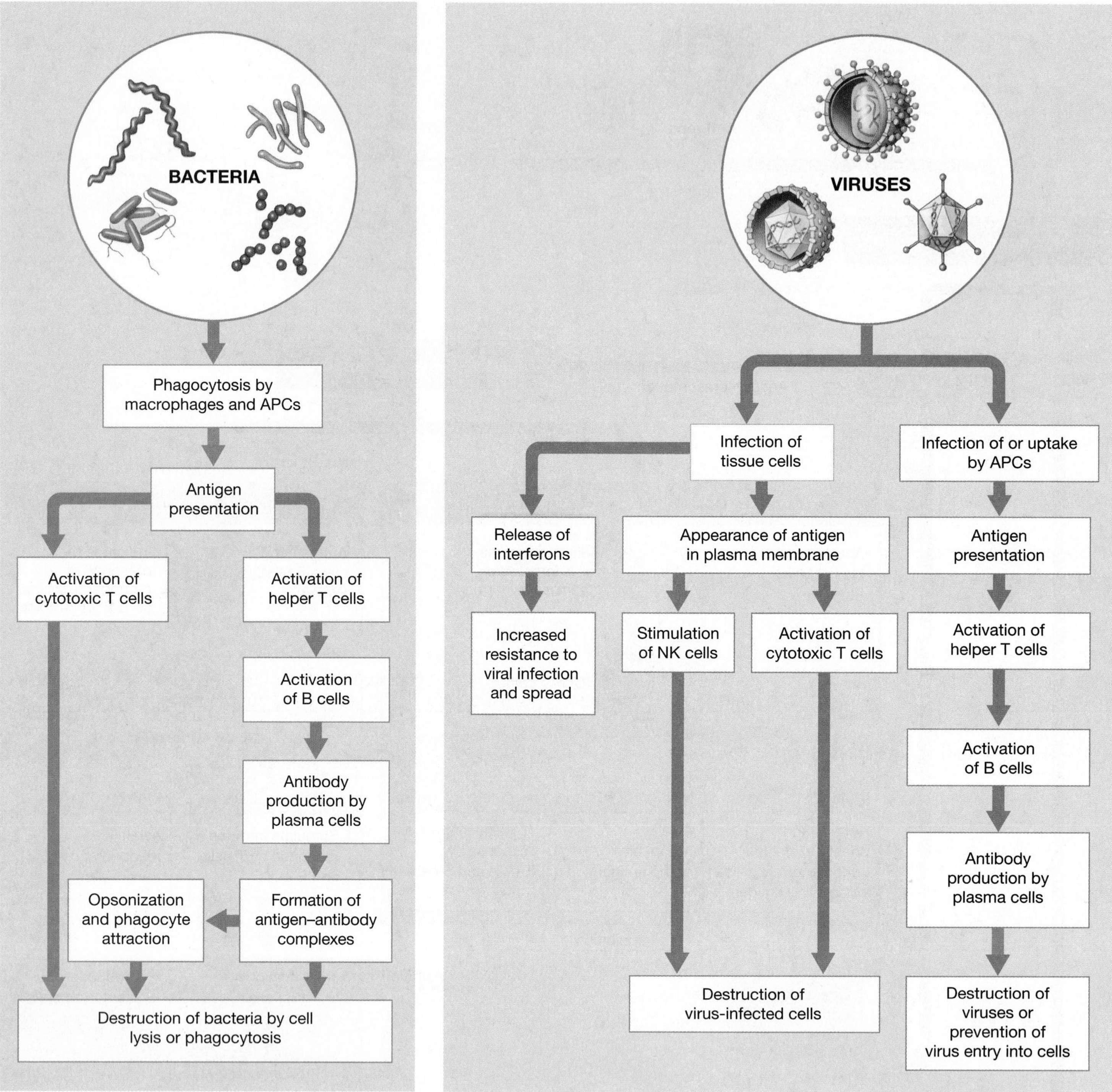
BACTERIA
Phagocytosis by macrophages and APCs
Antigen presentation
Activation of cytotoxic T cells
Activation of helper T cells
Activation of B cells
Antibody production by plasma cells
Opsonization and phagocyte attraction
Formation of antigen–antibody complexes
Destruction of bacteria by cell lysis or phagocytosis
VIRUSES
Infection of tissue cells
Infection of or uptake by APCs
Release of interferons
Appearance of antigen in plasma membrane
Antigen presentation
Increased resistance to viral infection and spread
Stimulation of NK cells
Activation of cytotoxic T cells
Activation of helper T cells
Activation of B cells
Antibody production by plasma cells
Destruction of virus-infected cells
Destruction of viruses or prevention of virus entry into cells

a Defenses against bacteria involve phagocytosis and antigen presentation by APCs.

b Defenses against viruses involve direct contact with virus-infected cells and antigen presentation by APCs.

antibodies gradually rises toward normal adult levels. Populations of memory B cells and T cells also increase.

Skin tests are sometimes used to determine whether an individual has been exposed to a particular antigen. In these procedures, small quantities of antigen are injected into the skin, generally on the anterior surface of the forearm. If resistance has developed, the region becomes inflamed over the next 2 to 4 days. Many states require a tuberculosis test, called a *tuberculin skin test*, before children enter public school. If the test is positive, further tests must then be performed to determine whether an infection is currently under way. Skin tests are also used to check for allergies to environmental antigens.

Stress and the Immune Response

One of the normal effects of interleukin-1 secretion is the stimulation of adrenocorticotropic hormone (ACTH) production by the anterior lobe of the pituitary gland. This hormone in turn prompts the secretion of glucocorticoids by the adrenal cortex. ⮌ p. 636 The anti-inflammatory effects of the glucocorticoids may help control the extent of the immune response.

However, chronic stress depresses the immune system and can be a serious threat to health. The long-term secretion of glucocorticoids, as in the resistance phase of the *stress response*, can inhibit the immune response and lower a person's resistance to disease. ⮌ p. 648 The effects of glucocorticoids that alter the effectiveness of innate and adaptive defenses include the following:

- *Depression of Inflammation.* Glucocorticoids inhibit mast cells and make capillaries less permeable. Inflammation becomes less likely. When it does occur, the reduced permeability of the capillaries slows the entry of fibrinogen, complement proteins, and cellular defenders into tissues.
- *Reduction in the Abundance and Activity of Phagocytes in Peripheral Tissues.* This reduction further impairs innate defense mechanisms. It also interferes with the processing and presentation of antigens to lymphocytes.
- *Inhibition of Interleukin Secretion.* A reduction in interleukin production depresses the response of lymphocytes, even to antigens bound to MHC proteins.

The mechanisms that bring about these changes are still under investigation.

Immune Disorders

Because the immune response is so complex, many opportunities exist for things to go wrong. A variety of clinical conditions result from disorders of the immune function. A common class of immune disorders is *allergies*, also called *hypersensitivities*. *Autoimmune disorders* develop when the immune response inappropriately targets normal body cells and tissues. In an *immunodeficiency disease*, either the immune system fails to develop normally or the immune response is blocked in some way. Autoimmune disorders and immunodeficiency diseases are fairly rare—clear evidence of the effectiveness of the immune system's control mechanisms.

Hypersensitivities

Hypersensitivities, commonly called allergies, are excessive immune responses to antigens. The sudden increase in cellular activity or antibody titers can have several unpleasant side effects. For example, neutrophils or cytotoxic T cells may destroy normal cells while they are attacking an antigen. Or the antigen–antibody complex may trigger massive inflammation. Antigens that set off allergic reactions are often called **allergens**.

There are several types of hypersensitivities. A complete classification has four categories: *immediate hypersensitivity (type I)*, *cytotoxic reactions (type II)*, *immune complex disorders (type III)*, and *delayed hypersensitivity (type IV)*. In Chapter 19 we discussed one example of a cytotoxic (type II) reaction: the cross-reaction that follows the transfusion of an incompatible blood type. ⮌ p. 669 Here we focus on immediate hypersensitivity (type I). It is probably the most common type of allergy.

Immediate hypersensitivity is a rapid and especially severe response to an antigen. One form, *allergic rhinitis*, includes hay fever and environmental allergies. This form may affect 15 percent of the U.S. population. Sensitization to an allergen during the initial exposure leads to the production of large quantities of IgE. Genes may determine a person's tendency to produce IgE in response to particular allergens.

Due to the lag time needed to activate B cells, produce plasma cells, and synthesize antibodies, the first exposure to an allergen does not produce signs and symptoms. Instead, it sets the stage for the next encounter. After sensitization, the IgE molecules become attached to basophils and mast cells throughout the body. When the individual meets the same allergen again, the bound antibodies stimulate these cells to release histamine, heparin, several cytokines, prostaglandins, and other chemicals into the surrounding tissues. This results in a sudden, massive inflammation of the affected tissues.

The cytokines and other mast cell secretions draw basophils, eosinophils, T cells, and macrophages into the area. These cells release their own chemicals, extending and intensifying the responses initiated by mast cells. The severity of the allergic reaction depends on the individual's sensitivity and on the location involved. If allergen exposure occurs at the body surface, the response may be restricted to that area. If the allergen enters the bloodstream, the response could be lethal.

In **anaphylaxis** (an-uh-fuh-LAK-sis; *ana-*, again + *phylaxis*, protection), a circulating allergen affects mast cells throughout the body (Figure 22–29). In drug reactions, such as allergies to penicillin, IgE antibodies are produced in response to a hapten (partial antigen) bound to a larger molecule that is widely distributed within the body because the combination acts as an allergen. A wide range of signs and symptoms can develop within minutes. Changes in capillary permeabilities produce swelling and edema in the dermis, and raised welts, or *hives*, appear on the skin. Smooth muscles along the respiratory passageways contract, and the narrowed passages make breathing extremely difficult. In severe cases, an extensive peripheral vasodilation occurs, producing a drop in blood pressure that can lead to a circulatory collapse. This response is **anaphylactic shock**.

The prompt administration of drugs that block the action of histamine, known as **antihistamines** (an-tē-HIS-ta-mēnz), can prevent many of the signs and symptoms of immediate hypersensitivity. *Benadryl (diphenhydramine hydrochloride)* is a popular antihistamine that is available over the counter. Severe anaphylaxis is treated with injections of antihistamines, corticosteroids, and epinephrine.

Figure 22–29 **The Mechanism of Anaphylaxis.**

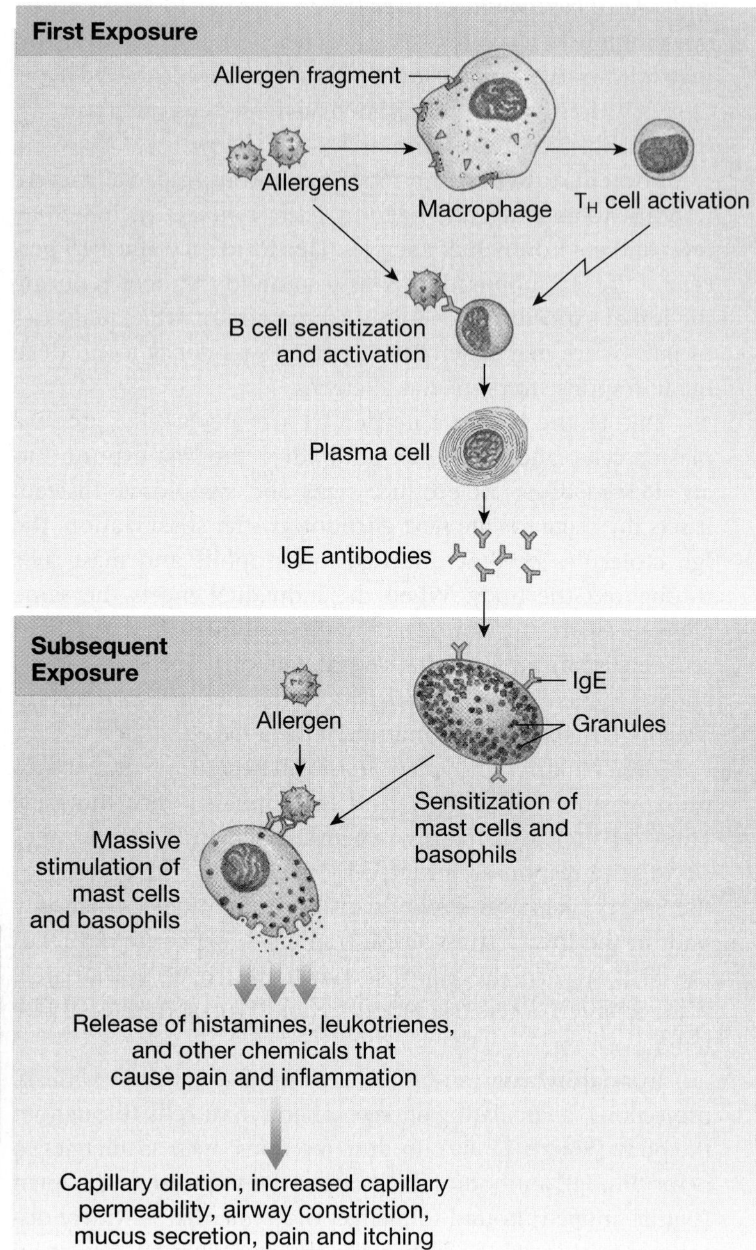
First Exposure
Allergen fragment
Allergens
Macrophage
T_H cell activation
B cell sensitization and activation
Plasma cell
IgE antibodies
Subsequent Exposure
Allergen
IgE
Granules
Sensitization of mast cells and basophils
Massive stimulation of mast cells and basophils
Release of histamines, leukotrienes, and other chemicals that cause pain and inflammation
Capillary dilation, increased capillary permeability, airway constriction, mucus secretion, pain and itching

Autoimmune Disorders

Autoimmune disorders affect an estimated 5 percent of adults in North America and Europe. In previous chapters we have cited many examples of the effects of autoimmune disorders on the function of major systems.

The immune system usually recognizes but ignores antigens normally found in the body—self-antigens. When the recognition system does not work correctly, however, activated B cells make antibodies against other body cells and tissues. These "misguided" antibodies are called **autoantibodies**. The trigger may be a reduction in suppressor T cell activity, the excessive stimulation of helper T cells, tissue damage that releases large quantities of antigenic fragments, haptens bound to compounds normally ignored, viral or bacterial toxins, or a combination of factors.

The resulting condition depends on the antigen that the autoantibodies attack. For example:

- The inflammation of *thyroiditis* is due to autoantibodies against thyroglobulin.
- *Rheumatoid arthritis* occurs when autoantibodies form immune complexes in connective tissues around the joints.
- Type 1 diabetes develops when autoantibodies attack cells in the pancreatic islets.

Many autoimmune disorders appear to be cases of mistaken identity. For example, proteins associated with the measles, Epstein–Barr, influenza, and other viruses contain amino acid sequences that are similar to those of myelin proteins. As a result, antibodies that target these viruses may also attack myelin sheaths. This accounts for the neurological complications that sometimes follow a vaccination or a viral infection. It also may be responsible for *multiple sclerosis*.

For unknown reasons, the risk of autoimmune problems increases if an individual has an unusual type of MHC protein. At least 50 clinical conditions have been linked to specific variations in MHC structure.

Immunodeficiency Diseases

Immunodeficiency diseases result from (1) problems with the embryonic development of lymphoid organs and tissues; (2) an infection with a virus, such as HIV, that depresses immune function; or (3) treatment with, or exposure to, immunosuppressive agents, such as radiation or drugs.

Individuals born with **severe combined immunodeficiency disease (SCID)** fail to develop either cell-mediated or antibody-mediated immunity. Their lymphocyte populations are low, and normal B cells and T cells are absent. Such infants cannot produce an immune response, so even a mild infection can prove fatal. Total isolation offers protection but at great cost—extreme restrictions on lifestyle. Bone marrow transplants from a compatible donor, normally a close relative, have been used to colonize lymphoid tissues with functional lymphocytes. Gene-splicing techniques have led to therapies that can treat at least one form of SCID.

AIDS is an immunodeficiency disease that results from a viral infection that targets helper T cells. We discuss AIDS in the Clinical Note on the next page.

Immunosuppressive drugs have been used for many years to prevent graft rejection after transplant surgery. Unfortunately, immunosuppressive agents can destroy stem cells and lymphocytes and lead to a complete failure of the immune response.

✓ Checkpoint

22. **Which kind of immunity protects a developing fetus, and how is that immunity produced?**
23. **How does increased stress reduce the effectiveness of the immune response?**
24. **Define *autoimmune disorder*.**

See the blue Answers tab at the back of the book.

+ Clinical Note AIDS

Acquired immunodeficiency syndrome (AIDS), or *late-stage HIV disease*, is caused by the **human immunodeficiency virus (HIV)**. This virus is a *retrovirus*: It carries its genetic information in RNA rather than in DNA. The virus enters human leukocytes by receptor-mediated endocytosis. p. 98 Specifically, the virus binds to CD4 T cells. HIV also infects several types of antigen-presenting cells, including those of the monocyte–macrophage line. It is the infection of helper T cells that leads to clinical problems.

The gradual destruction of helper T cells impairs the immune response, because these cells play a central role in coordinating cell-mediated and antibody-mediated responses to antigens. To make matters worse, regulatory T cells are relatively unaffected by the virus. Over time the excess of suppressing factors "turns off" the normal immune response. Circulating antibody levels decline and cell-mediated immunity is reduced. The body is left with impaired defenses against a wide variety of microbial invaders. Microorganisms that would ordinarily be harmless can now initiate lethal *opportunistic infections*. The risk of cancer also increases because immune surveillance is depressed.

Infection with HIV occurs through intimate contact with the body fluids of infected individuals. The major routes of transmission involve contact with blood, semen, or vaginal secretions, although all body fluids may contain the virus. Worldwide, most individuals with AIDS become infected through sexual contact with an HIV-infected person (who may *not* necessarily show clinical signs of AIDS). The next largest group of infected individuals is intravenous drug users who shared contaminated needles. Relatively few individuals have become infected with the virus after receiving a transfusion of contaminated blood or blood products. Finally, an increasing number of infants are born with AIDS acquired from infected mothers.

AIDS continues to be a public health problem of massive proportions. Current statistics from the Centers for Disease Control and Prevention (CDC) show that more than 1.2 million people in the United States are infected with HIV.

The most effective ways to prevent HIV infection include abstinence, mutual monogamy of uninfected partners, male circumcision (to reduce the risk of transmission), and the avoidance of needle sharing. Circumcised men are up to 70 percent less likely to become infected than uncircumcised men. They are also less likely to infect their female partners. The use of latex condoms labeled "for the prevention of disease" has been shown to prevent the passage of HIV as well as the hepatitis and herpes viruses. Preexposure prophylaxis (PrEP) is a new HIV prevention method aimed at people who are not infected with HIV but who have a high risk of becoming infected. PrEP involves taking a pill that contains a combination of two HIV medications, tenofovir and emtricitabine. These drugs prevent HIV infection, but they must be taken every day and the person must be under the supervision of a health care provider.

Clinical signs and symptoms of AIDS may not appear until 5–10 years or more after infection. When they do appear, they are commonly mild, consisting of swollen lymph nodes and chronic, but nonfatal, infections. So far as is known, however, AIDS is almost always fatal.

Despite intensive efforts, a vaccine has yet to be developed that prevents HIV infection in an uninfected person exposed to the virus. The survival rate for AIDS patients has been steadily increasing. That is, more people are living longer before dying of AIDS or its complications. New drugs and drug combinations slow the progression of the disease, and improved antibiotic therapies help combat secondary infections.

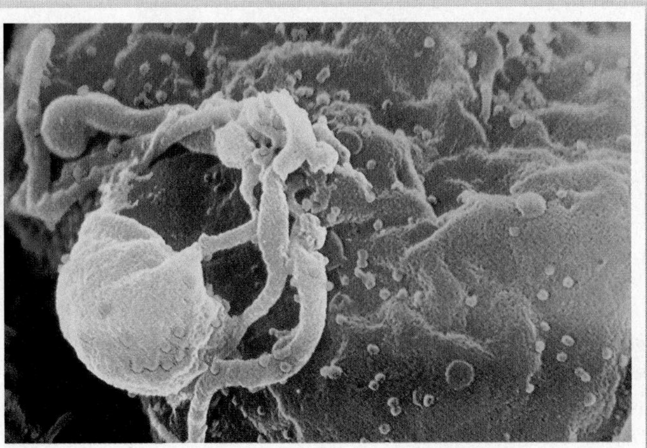
HIV (green) budding from an infected T_H cell (blue) SEM × 40,000

22-8 The immune response diminishes as we age

Learning Outcome **Describe the effects of aging on the lymphatic system and the immune response.**

As we age, the immune system becomes less effective at combating disease. T cells become less responsive to antigens, so fewer cytotoxic T cells respond to an infection. This effect may be due to the gradual decrease in the size of functional tissue and secretory abilities of the thymus and to lower circulating levels of thymic hormones. Because the number of helper T cells is also reduced, B cells are less responsive, so antibody levels do not rise as quickly after antigen exposure. The net result is an increased susceptibility to viral and bacterial infections. For this reason, vaccinations for acute viral diseases such as the flu (influenza), and for pneumococcal pneumonia, are strongly recommended for elderly individuals. The increased incidence of cancer in elderly people reflects the fact that immune surveillance declines, so tumor cells are not eliminated as effectively.

Build Your Knowledge

Figure 22–30 Integration of the LYMPHATIC system with the other body systems presented so far.

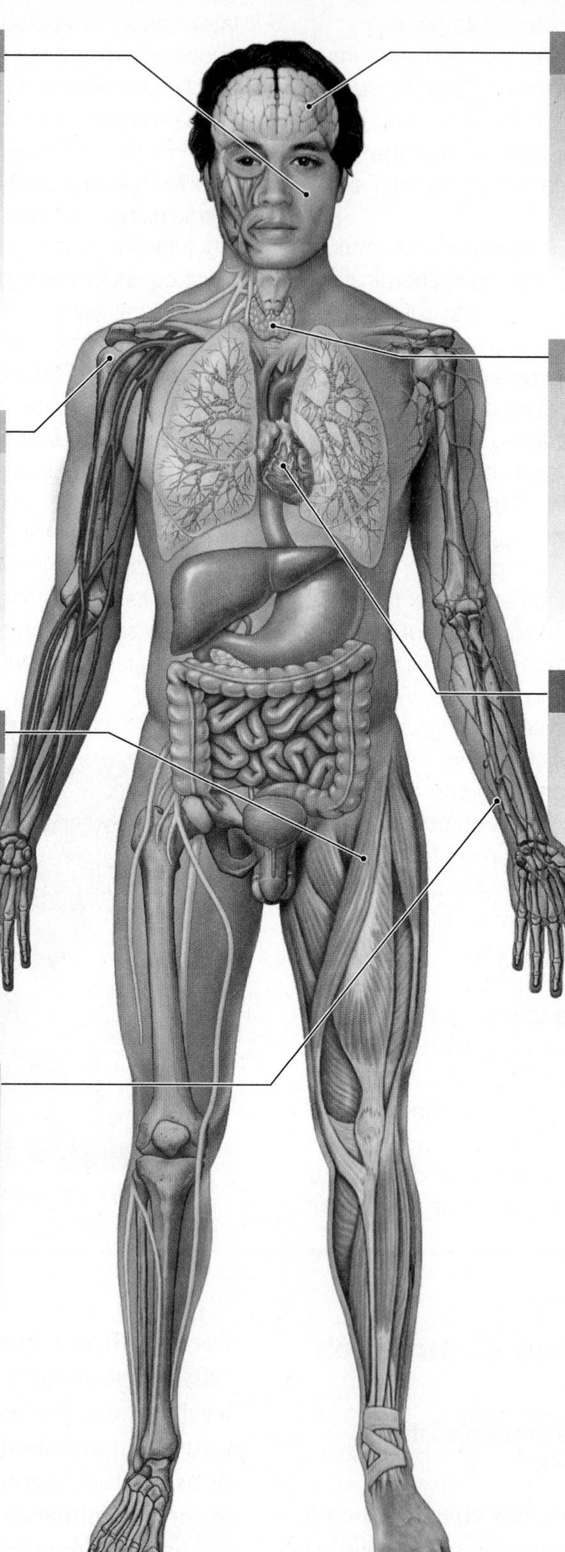

Integumentary System

- The Integumentary System provides physical barriers to pathogen entry; dendritic cells in epidermis and macrophages in dermis resist infection and present antigens to trigger the immune response; mast cells trigger inflammation, mobilize cells of lymphatic system
- The lymphatic system provides IgA antibodies for secretion onto integumentary surfaces

Skeletal System

- The Skeletal System supports the production of lymphocytes and other cells involved in the immune response in red bone marrow
- The lymphatic system assists in repair of bone after injuries; osteoclasts differentiate from monocyte–macrophage cell line

Muscular System

- The Muscular System protects superficial lymph nodes and the lymphatic vessels in abdominopelvic mesenteries; muscle contractions help propel lymph along lymphatic vessels
- The lymphatic system assists in repair after injuries

Lymphatic System

The lymphatic system provides adaptive (specific) defenses against infection for all body systems. It:

- produces, maintains, and distributes lymphocytes
- returns fluid and solutes from peripheral tissues to the blood
- distributes hormones, nutrients, and waste products from their tissues of origin to the general circulation

Nervous System

- The Nervous System has antigen-presenting microglia that stimulate adaptive defenses; glial cells secrete cytokines; innervates antigen-presenting cells
- The lymphatic system produces cytokines that affect production of CRH and TRH by hypothalamus

Endocrine System

- The Endocrine System produces glucocorticoids that have anti-inflammatory effects; thymosins that stimulate development and maturation of lymphocytes; many hormones that affect immune function
- The lymphatic system secretes thymosins from thymus; cytokines affect cells throughout the body

Cardiovascular System

- The Cardiovascular System distributes WBCs; carries antibodies that attack pathogens; clotting response helps restrict spread of pathogens; granulocytes and lymphocytes produced in red bone marrow
- The lymphatic system fights infections of cardiovascular organs; returns tissue fluid to circulation

Checkpoint

25. **Why are elderly people more susceptible to viral and bacterial infections?**
26. **What may account for the increased incidence of cancer among elderly people?**

See the blue Answers tab at the back of the book.

22-9 The nervous and endocrine systems influence the immune response

Learning Outcome Give examples of interactions between the lymphatic system and other organ systems we have studied so far and explain how the nervous and endocrine systems influence the immune response.

Build Your Knowledge Figure 22–30 summarizes the interactions between the lymphatic system and other body systems we have studied so far. Interactions among elements of the immune response and the nervous and endocrine systems are now the focus of intense research. For example:

- The thymus secretes oxytocin, ADH, and endorphins as well as thymic hormones. The effects on the CNS are not known, but removal of the thymus lowers brain endorphin levels.
- Both thymic hormones and cytokines help establish the normal levels of CRH and TRH produced by the hypothalamus.
- Other thymic hormones affect the anterior lobe of the pituitary gland directly, stimulating the secretion of prolactin and GH.

Conversely, the nervous system can apparently adjust the sensitivity of the immune response:

- The PNS innervates dendritic cells in the lymph nodes, spleen, skin, and other antigen-presenting cells. The nerve endings release neurotransmitters that heighten local immune responses. For this reason, some skin conditions, such as *psoriasis*, worsen when a person is under stress.
- Neuroglia in the CNS produce cytokines that promote an immune response.
- A sudden decline in immune function can occur after even a brief period of emotional distress.

Checkpoint

27. **What is the relationship between the endocrine system and the lymphatic system?**
28. **Identify the role of the lymphatic system for all body systems.**

See the blue Answers tab at the back of the book.

22 Chapter Review

Study Outline

An Introduction to the Lymphatic System and Immunity p. 786

1. The cells, tissues, and organs of the **lymphatic system** play a central role in the body's defenses against a variety of **pathogens**, or disease-causing organisms.
2. **Immunity** is the ability to resist infection and disease. All the cells and tissues involved in producing immunity make up a functional **immune system.**

22-1 The vessels, tissues, and organs of the lymphatic system maintain fluid volume and function in body defenses p. 786

3. An array of **lymphoid tissues** and **lymphoid organs** is connected to lymphatics. **Primary lymphoid tissues and organs** are sites where lymphocytes are formed and mature. **Secondary lymphoid tissues and organs** are where lymphocytes are activated and cloned. The lymphatic system includes a network of **lymphatic vessels**, or *lymphatics*, that carry **lymph** (interstitial fluid that has entered lymphatic vessels). *(Figure 22–1)*
4. The lymphatic system produces, maintains, and distributes lymphocytes (which attack invading organisms, abnormal cells, and foreign proteins); it also helps maintain blood volume and eliminate local variations in the composition of interstitial fluid.
5. Lymph flows along a network of lymphatic vessels, the smallest of which are the **lymphatic capillaries**. The lymphatic vessels empty into the **thoracic duct** and the **right lymphatic duct**. *(Figures 22–2 to 22–4)*
6. **Lymphoid cells** are immune system cells found in lymphoid tissues and the cells that support those tissues. Immune system cells that function in defense include phagocytes and

lymphocytes. **Lymphocytes** respond to specific invading pathogens, abnormal body cells, and foreign proteins. The three classes of lymphocytes are T (thymus-dependent) cells, B (bone marrow–derived) cells, and NK (natural killer) cells.

7. **Lymphoid tissues** are connective tissues dominated by lymphocytes. In a **lymphoid nodule**, the lymphocytes are densely packed in an area of areolar tissue. The lymphoid tissue that protects the epithelia of the digestive, respiratory, urinary, and reproductive tracts is called **mucosa-associated lymphoid tissue (MALT)**. *(Figure 22–5)*
8. Important lymphoid organs include the **lymph nodes**, the **thymus**, and the **spleen**. Lymphoid tissues and organs are distributed in areas that are especially vulnerable to injury or invasion.
9. Lymph nodes are encapsulated masses of lymphoid tissue. The **paracortex** is dominated by T cells; the **cortex** and **medulla** contain B cells. *(Figure 22–6)*
10. The thymus lies behind the sternum, in the anterior mediastinum. **Epithelial reticular cells** regulate T cell development and function. *(Figure 22–7)*
11. The adult spleen contains the largest mass of lymphoid tissue in the body. The cellular components form the **pulp** of the spleen. **Red pulp** contains large numbers of red blood cells, and **white pulp** resembles lymphoid nodules. *(Figure 22–8)*

22-2 Lymphocytes are important to innate (nonspecific) and adaptive (specific) immunity p. 796

12. The lymphatic system is a major component of the body's defenses, which are classified as either (1) **innate (nonspecific) immunity**, which protects without distinguishing one threat from another, or (2) **adaptive (specific) immunity**, which protects against particular threats only. The body's reaction to infectious agents and other abnormal substances is known as the **immune response**.
13. Lymphocytes continuously migrate into and out of the blood through the lymphoid tissues and organs. **Lymphocytopoiesis** (lymphocyte production) involves the red bone marrow, thymus, and peripheral lymphoid tissues. *(Figure 22–9)*

22

22-3 Innate defenses respond the same regardless of the invader p. 797

14. Innate (nonspecific) defenses prevent the approach, deny the entry, or limit the spread of living or nonliving hazards. *(Figure 22–10)*
15. Physical barriers include skin, mucous membranes, hair, epithelia, and various secretions of the integumentary and digestive systems.
16. The two types of phagocytes are **microphages** and **macrophages** (cells of the **monocyte–macrophage system**). Microphages are neutrophils and eosinophils in circulating blood.
17. **Phagocytes** leave the bloodstream by *emigration* (migration between adjacent endothelial cells), and exhibit **chemotaxis** (sensitivity and orientation to chemical stimuli).
18. NK cells (also called **large granular lymphocytes**) attack foreign cells, normal cells infected with viruses, and cancer cells. NK cells provide *immune surveillance*.
19. **Immune surveillance** involves constant monitoring of normal tissues by NK cells that are sensitive to abnormal antigens on the surfaces of otherwise normal cells. NK cells kill cancer cells that have **tumor-specific antigens** on their surfaces. *(Figure 22–11)*
20. **Interferons (IFNs)**—small proteins released by cells infected with viruses—trigger the production of **antiviral proteins**, which interfere with viral replication inside the cell. IFNs are **cytokines**, which are chemical messengers released by tissue cells to coordinate local activities. *(Figure 22–12)*
21. At least 30 **complement proteins** make up the **complement system**. These proteins interact with each other in cascades to destroy target cell walls, enhance phagocytosis (opsonization), or stimulate inflammation. The complement system can be activated by either the **classical pathway**, the **lectin pathway**, or the **alternative pathway**. *(Figure 22–13)*
22. **Inflammation** is a localized tissue response to injury. *(Figure 22–14)*
23. **Fever** is a body temperature greater than 37.2°C (99°F). Fevers can inhibit pathogens and accelerate metabolic processes. **Pyrogens** can reset the body's thermostat and raise the temperature.

22-4 Adaptive (specific) defenses respond to particular threats and are either cell mediated or antibody mediated p. 805

24. T cells are responsible for **cell-mediated (cellular) immunity**. The primary types of T cells are **cytotoxic T cells, helper T cells, regulatory T cells**, and **memory T cells.** *(Figure 22–15)*
25. B cells provide **antibody-mediated (humoral) immunity**. B cells differentiate into plasma cells, which secrete antibodies. *(Figure 22–15)*
26. The **immune response** is triggered by the presence of specific antigens and includes cell-mediated and antibody-mediated defenses. *(Figure 22–16)*
27. Forms of immunity include **innate immunity** (genetically determined and present at birth) or **adaptive immunity** (produced by prior exposure to an antigen or antibody production). The two types of adaptive immunity are **active immunity** (which appears after exposure to an antigen) and **passive immunity** (produced by the transfer of antibodies from another source). *(Figure 22–17)*
28. Immunity exhibits four general properties: **specificity**, **versatility**, **memory**, and **tolerance**. Specificity occurs because T cells and B cells respond to the molecular structure of specific antigens. Versatility is based on the large diversity of lymphocytes in the body and from variability in the structure of synthesized antibodies. *Memory cells* enable the immune system to "remember" previous target antigens. Tolerance is the ability of the immune system to ignore some antigens, such as those of normal body cells.

22-5 In cell-mediated adaptive immunity, presented antigens activate T cells, which respond by producing cytotoxic and helper T cells p. 809

29. **Antigen presentation** occurs when an antigen–glycoprotein combination appears in the plasma membrane of an antigen-presenting cell (typically of a macrophage or dendritic cell). T cells sensitive to this combination are activated if they contact the membrane of the antigen-presenting cell.
30. All body cells have plasma membrane glycoproteins. The genes controlling their synthesis make up a chromosomal region called the **major histocompatibility complex (MHC)**. The membrane glycoproteins are called **MHC proteins**. **APCs (antigen-presenting cells)** are involved in antigen stimulation.
31. Lymphocytes are not activated by free antigens, but respond instead to an antigen bound to either a **class I** or a **class II** MHC protein in a process called **antigen recognition**. *(Figure 22–18)*

32. Class I MHC proteins are in all nucleated body cells. class II MHC proteins are only in antigen-presenting cells (APCs) and lymphocytes.
33. Whether a T cell responds to antigens held in class I or class II MHC proteins depends on the structure of the T cell plasma membrane. T cell plasma membranes contain proteins called **CD** (*cluster of differentiation*) **markers**. **CD3 markers** are present on all T cells. **CD8 markers** are on cytotoxic and regulatory T cells. **CD4 markers** are on all helper T cells.
34. One type of CD8 cell responds quickly to a class I MHC–bound antigen, giving rise to large numbers of cytotoxic T cells and memory T cells. The other type of CD8 cell responds more slowly, giving rise to small numbers of regulatory T cells.
35. Cytotoxic T cells seek out and destroy abnormal and infected cells, using three different methods, including the secretion of **lymphotoxin**. *(Figure 22–19)*
36. Cell-mediated immunity (cellular immunity) results from the activation of CD8 cells by antigens bound to class I MHCs. When activated, most of these T cells divide to generate cytotoxic T cells and **memory T cells**, which remain in reserve to guard against future such attacks. Regulatory T cells moderate the responses of other T cells and of B cells. *(Figures 22–19, 22–22)*
37. Helper T cells, or CD4 cells, respond to antigens presented by class II MHC proteins. When activated, helper T cells secrete cytokines that aid in coordinating adaptive and innate defenses and also regulate cell-mediated and antibody-mediated immunity. *(Figures 22–20, Spotlight Figure 22–21)*
38. Cytokines are chemical messengers coordinated by the immune system. **Interleukins** increase T cell sensitivity to antigens exposed on macrophage membranes; stimulate B cell activity, plasma cell formation, and antibody production; enhance innate defenses; and moderate the immune response. *(Spotlight Figure 22–21)*
39. Interferons (IFNs) slow the spread of a virus by making the synthesizing cell and its neighbors resistant to viral infections. *(Spotlight Figure 22–21)*
40. **Tumor necrosis factors (TNFs)** slow tumor growth and kill tumor cells, and act as pyrogens. *(Spotlight Figure 22–21)*
41. Several cytokines adjust the activities of phagocytes to coordinate innate and adaptive defenses. *(Spotlight Figure 22–21)*
42. **Colony-stimulating factors (CSFs)** are factors produced by active T cells, cells of the monocyte–macrophage group, endothelial cells, and fibroblasts. *(Spotlight Figure 22–21)*

22-6 In antibody-mediated adaptive immunity, sensitized B cells respond to antigens by producing specific antibodies p. 816

43. B cells become **sensitized** when antibody molecules in their plasma membranes bind antigens. The antigens are then displayed on the class II MHC proteins of the B cells, which become activated by helper T cells activated by the same antigen. *(Figure 22–23)*
44. An active B cell may differentiate into a plasma cell or produce daughter cells that differentiate into plasma cells and memory B cells. Antibodies are produced by plasma cells. *(Figure 22–23)*
45. A Y-shaped antibody molecule consists of two parallel pairs of polypeptide chains containing *constant* and *variable segments*. *(Figure 22–24)*
46. The five classes of antibodies **(immunoglobulins, Ig)** in body fluids are (1) **IgG**, responsible for resistance against many viruses, bacteria, and bacterial toxins; (2) **IgE**, which releases chemicals that accelerate local inflammation; (3) **IgD**, located on the surfaces of B cells; (4) **IgM**, the first type of antibody secreted after an antigen arrives; and (5) **IgA**, found in glandular secretions. *(Table 22–1)*
47. When antibody molecules bind to an antigen, they form an **antigen–antibody complex**. Effects that appear after binding include **neutralization** (antibody binding that prevents viruses or bacterial toxins from binding to body cells); **precipitation** (formation of an insoluble **immune complex**) and **agglutination** (formation of large complexes); *opsonization* (coating of pathogens with antibodies and complement proteins to enhance phagocytosis); stimulation of inflammation; and prevention of bacterial or viral adhesion. *(Figure 22–24)*
48. In humoral immunity, the antibodies first produced by plasma cells are the agents of the **primary response**. The maximum antibody level appears during the **secondary response** to antigen exposure. *(Figure 22–25)*
49. The initial steps in the immune responses to viral and bacterial infections differ. *(Figures 22–26 to 22–28; Table 22–2)*

22-7 Immunocompetence enables a normal immune response; abnormal responses result in immune disorders p. 821

50. **Immunocompetence** is the ability to produce an immune response after exposure to an antigen. A developing fetus acquires passive immunity from antibodies in the maternal bloodstream. After delivery, the infant begins developing active immunity following exposure to environmental antigens.
51. During periods of stress, interleukin-1 released by active macrophages triggers the release of ACTH by the anterior lobe of the pituitary gland. Glucocorticoids produced by the adrenal cortex moderate the immune response, but their long-term secretion can lower a person's resistance to disease.
52. **Hypersensitivities** (allergies) are inappropriate or excessive immune responses to **allergens** (antigens that trigger allergic reactions). The four types of allergies are *immediate hypersensitivity (type I)*, *cytotoxic reactions (type II)*, *immune complex disorders (type III)*, and *delayed hypersensitivity (type IV)*.
53. In **anaphylaxis**, a circulating allergen affects mast cells throughout the body. *(Figure 22–29)*
54. **Autoimmune disorders** develop when the immune response inappropriately targets normal body cells and tissues.
55. In an **immunodeficiency disease**, either the immune system does not develop normally or the immune response is blocked.

22-8 The immune response diminishes as we age p. 827

56. With aging, the immune system becomes less effective at combating disease.

22-9 The nervous and endocrine systems influence the immune response p. 829

57. The lymphatic system has extensive interactions with the nervous and endocrine systems. *(Figure 22–30)*.

Review Questions

See the blue Answers tab at the back of the book.

LEVEL 1 Reviewing Facts and Terms

1. Identify the structures of the lymphatic system in the following diagram.

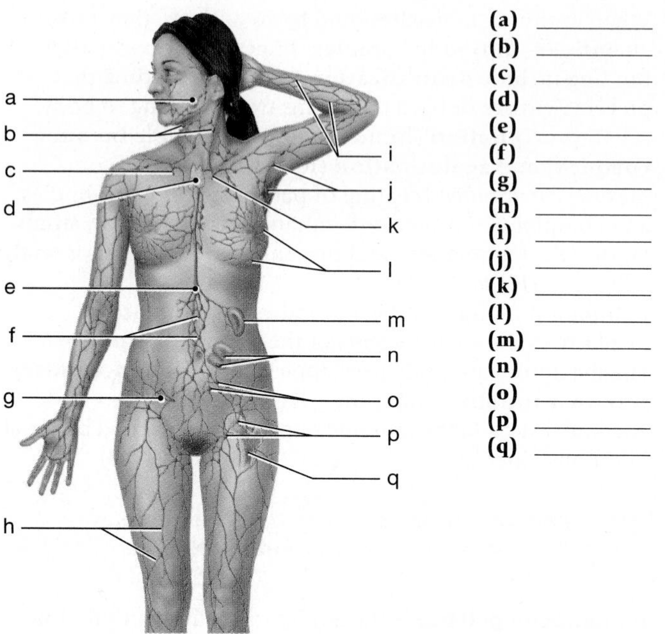
a
b
c
d
e
f
g
h
i
j
k
l
m
n
o
p
q

(a) ________
(b) ________
(c) ________
(d) ________
(e) ________
(f) ________
(g) ________
(h) ________
(i) ________
(j) ________
(k) ________
(l) ________
(m) ________
(n) ________
(o) ________
(p) ________
(q) ________

2. Lymph from the right arm, the right half of the head, and the right chest is received by the **(a)** cisterna chyli, **(b)** right lymphatic duct, **(c)** right thoracic duct, **(d)** aorta.
3. Anatomically, lymphatic vessels resemble **(a)** elastic arteries, **(b)** muscular arteries, **(c)** arterioles, **(d)** medium veins, **(e)** the venae cavae.
4. The specificity of an antibody is determined by the **(a)** fixed segment, **(b)** antigenic determinants, **(c)** variable region, **(d)** size of the antibody, **(e)** antibody class.
5. The major histocompatibility complex (MHC) **(a)** is responsible for forming lymphocytes, **(b)** produces antibodies in lymph glands, **(c)** is a group of genes that codes for human leukocyte antigens, **(d)** is a membrane protein that can recognize foreign antigens, **(e)** is the antigen found on bacteria that stimulates an immune response.
6. Red blood cells that are damaged or defective are removed from the bloodstream by the **(a)** thymus, **(b)** lymph nodes, **(c)** spleen, **(d)** tonsils.
7. Phagocytes move through capillary walls by squeezing between adjacent endothelial cells, a process known as **(a)** diapedesis, **(b)** chemotaxis, **(c)** adhesion, **(d)** perforation.
8. Perforins are proteins associated with the activity of **(a)** T cells, **(b)** B cells, **(c)** NK cells, **(d)** plasma cells.
9. Complement activation **(a)** stimulates inflammation, **(b)** attracts phagocytes, **(c)** enhances phagocytosis, **(d)** achieves a, b, and c.
10. The most beneficial effect of fever is that it **(a)** inhibits the spread of some bacteria and viruses, **(b)** increases the metabolic rate by up to 10 percent, **(c)** stimulates the release of pyrogens, **(d)** achieves a and b.
11. CD4 markers are associated with **(a)** cytotoxic T cells, **(b)** regulatory T cells, **(c)** helper T cells, **(d)** all of these.
12. List the specific functions of each of the body's primary and secondary lymphoid tissues and organs.
13. Give a function for each of the following: **(a)** cytotoxic T cells, **(b)** helper T cells, **(c)** regulatory T cells, **(d)** plasma cells, **(e)** NK cells, **(f)** stromal cells, **(g)** epithelial reticular cells, **(h)** interferons, **(i)** pyrogens, **(j)** T cells, **(k)** B cells, **(l)** interleukins, **(m)** tumor necrosis factor, **(n)** colony-stimulating factors.
14. What are the three classes of lymphocytes, and where does each class originate?
15. What seven defenses, present at birth, provide the body with the defensive capability known as innate (nonspecific) immunity?

LEVEL 2 Reviewing Concepts

16. Compared with innate defenses, adaptive defenses **(a)** do not distinguish between one threat and another, **(b)** are always present at birth, **(c)** protect against threats on an individual basis, **(d)** deny the entry of pathogens to the body.
17. Blocking the antigen receptors on the surface of lymphocytes would interfere with **(a)** phagocytosis of the antigen, **(b)** that lymphocyte's ability to produce antibodies, **(c)** antigen recognition, **(d)** the ability of the lymphocyte to present antigen, **(e)** opsonization of the antigen.
18. A *decrease* in which population of lymphocytes would impair all aspects of an immune response? **(a)** cytotoxic T cells, **(b)** helper T cells, **(c)** regulatory T cells, **(d)** B cells, **(e)** plasma cells.
19. Skin tests are used to determine if a person **(a)** has an active infection, **(b)** has been exposed to a particular antigen, **(c)** carries a particular antigen, **(d)** has measles, **(e)** can produce antibodies.
20. Compare and contrast the effects of complement with those of interferon.
21. How does a cytotoxic T cell destroy another cell displaying antigens bound to class I MHC proteins?
22. How does the formation of an antigen–antibody complex cause the elimination of an antigen?
23. Give one example of each type of immunity: innate immunity, naturally acquired active immunity, artificially acquired active immunity, artificially acquired passive immunity, and naturally acquired passive immunity.
24. An anesthesia technician is advised that she should be vaccinated against hepatitis B, which is caused by a virus. She is given one injection and is told to come back for a second injection in a month and a third injection after 6 months. Why is this series of injections necessary?

LEVEL 3 Critical Thinking and Clinical Applications

25. An investigator at a crime scene discovers some body fluid on the victim's clothing. The investigator carefully takes a sample and sends it to the crime lab for analysis. On the basis of the analysis of antibodies, could the crime lab determine whether the sample is blood plasma or semen? Explain.
26. Ted finds out that he has been exposed to measles. He is concerned that he might have contracted the disease, so he goes to see his physician. The physician takes a blood sample and sends it to a lab for antibody levels and titers. The results show an elevated level and activity of IgM antibodies to rubella (measles) virus but very few IgG antibodies to the virus. Has Ted contracted the disease?
27. While walking along the street, you and your friend see an elderly woman whose left arm appears to be swollen to several times its normal size. Your friend wonders aloud what might be its cause. You say that it may be likely that the woman had a radical mastectomy (the removal of a breast because of cancer). Explain the rationale behind your answer.
28. Paula's grandfather is diagnosed as having lung cancer. His physician takes biopsies of several lymph nodes from neighboring regions of the body, and Paula wonders why, since his cancer is in the lungs. What would you tell her?
29. Willy is allergic to ragweed pollen and tells you that he read about a medication that can help his condition by blocking certain antibodies. Do you think that this treatment could help Willy? Explain.

22

+ CLINICAL CASE Wrap-Up? Isn't There a Vaccine for That?

Vaccinations against infectious diseases are one of the most successful medical advancements in the history of public health. A vaccine, containing weakened or dead forms of the infecting microbe (antigen), is administered orally, intranasally, subcutaneously, or intramuscularly. The antigens are not virulent enough to cause the disease, but they provoke the adaptive immune response in the body. This immune response involves B cells that differentiate into plasma cells and make antibodies. It also involves cytotoxic T cells and helper T cells. Regulatory T cells moderate the immune response. Memory T cells and memory B cells help the body remember the antigen so when the protected person is exposed to the real disease, a strong secondary immune response occurs.

Vaccinations are as varied as the diseases they protect against. Some are given as live, attenuated (greatly weakened) forms of a virus. Others are given as inactivated antigens, made from small pieces of killed bacteria. Some vaccines require repeat doses (booster shots) to boost the secondary immune response and keep immune memory alive.

Baby Ruthie is due for her chickenpox vaccination next month, at her 12-month checkup. She is not yet immune to the varicella-zoster virus that causes chickenpox. Disease control in babies not yet immunized depends on the rest of the population being immunized. This is known as "herd" immunity. Ruthie's day care classmate, Tommy, had never been vaccinated and was spreading chickenpox to other unprotected children in the class.

1. Is immunity to chickenpox, developed after vaccination, innate immunity, or adaptive immunity? Why?
2. Does baby Ruthie still need her chickenpox vaccine at her 12-month checkup?

See the blue Answers tab at the back of the book.

Related Clinical Terms

adenitis: Inflammation of the adenoid (pharyngeal tonsil).

allograft: Transplant between compatible recipient and donor of the same species.

anamnestic response: An immune response that is initiated by memory cells.

autograft: A transplant of tissue that is taken from the same person.

Burkitt lymphoma: A malignant cancer of B lymphocytes.

chronic fatigue syndrome: A complicated disorder most often characterized by extreme fatigue that does not improve with rest, and which may worsen with physical activity.

congenital thymic aplasia: Congenital (present at birth) absence of the thymus and parathyroid glands and a deficiency of immunity.

Coombs test: A medical test to detect antibodies or complement in the blood.

dermatomyositis: An autoimmune disease characterized by inflammation of the skin and muscles.

eczema: A genetic inflammatory skin disorder, often with crusts, papules, and leaky eruptions.

Hodgkin lymphoma: A malignant lymphoma affecting lymph nodes and lymph organs.

host versus graft disease: A pathological condition in which cells from the transplanted tissue of a donor initiate an immune response, attacking the cells and tissue of the recipient.

hybridoma: A tissue culture composed of cancer cells fused to lymphocytes to mass produce a specific antibody.

immunology: Branch of biomedicine concerned with the structure and function of the immune system.

infectious mononucleosis: An acute disease caused by the Epstein–Barr virus, producing fever, swelling of the lymph nodes, sore throat, and increased lymphocytes in the bloodstream.

latex allergy: Hypersensitivity to products made of the sap of the rubber plant.

polymyositis: An autoimmune disease characterized by inflammation and atrophy of muscles.

sentinel node: The first lymph node to receive drainage from a tumor. It is used to determine if there is lymphatic metastasis in some types of cancer.

splenomegaly: Enlargement of the spleen.

systemic lupus erythematosus (SLE): An autoimmune disease in which a person's immune system attacks and injures its own organs and tissues in virtually every system of the body.

xenograft: A transplant that is made between two different species.

23 The Respiratory System

Learning Outcomes

These Learning Outcomes correspond by number to this chapter's sections and indicate what you should be able to do after completing the chapter.

CLINICAL CASE No Rest for the Weary

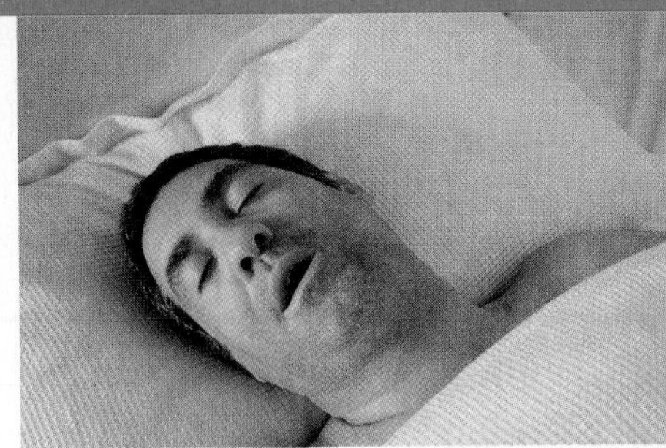

Doug reads the boss's e-mail for the third time. It's 3 p.m. at the office and he's had trouble concentrating all day. The commute home gets downright dangerous when he catches himself nodding off behind the wheel.

Doug is a big man and could stand to lose 30 pounds. He enjoys a few beers after dinner every night. When he finally hits the hay, he's exhausted. Every couple of hours his wife nudges him. "Doug, turn over. You're making a racket!" she says. She lies awake and notices that now he's quiet. Too quiet. Finally, he takes another rattling breath. This is their sleep pattern during most nights. In the morning, Doug wakes up with a headache, and neither one of them feels rested.

Doug's doctor refers him to an ENT (an ear, nose and throat specialist, also called an *otorhinolaryngologist*), who finds that Doug has a lengthened soft palate with a prominent uvula. During an overnight sleep study, it's noted that Doug sleeps on his back. The sleep technician documents several episodes of *apnea* (cessation of breathing). **What is causing Doug's nighttime drama?** **To find out, turn to the Case Study Wrap-Up on p. 883.**

An Introduction to the Respiratory System

When we think of the respiratory system, we generally think of breathing—pulling air into and out of our bodies. However, an efficient respiratory system must do more than merely move air. Cells need energy for maintenance, growth, defense, and division. Our cells obtain that energy mainly through aerobic mechanisms that require oxygen and produce carbon dioxide.

Our cells obtain oxygen and release carbon dioxide when these gases diffuse across the thin respiratory exchange surfaces inside of the *lungs*—a warm, moist, protected environment. In this environment, diffusion takes place between the air and the blood. Circulating blood then carries oxygen from the lungs to peripheral tissues. Your blood also transports the carbon dioxide generated by those tissues, delivering it to the lungs. Therefore, the cardiovascular system is the link between your interstitial fluids and the exchange surfaces of your lungs. In this chapter we describe how air enters and exits the lungs by the actions of respiratory muscles, and how oxygen and carbon dioxide are exchanged within the lungs.

23-1 The respiratory system, organized into an upper respiratory system and a lower respiratory system, functions primarily to aid gas exchange

Learning Outcome Describe the primary functions and organization of the respiratory system, and explain how the delicate respiratory exchange surfaces are protected from pathogens, debris, and other hazards.

The **respiratory system** is composed of structures involved in *ventilation* (providing air) and *gas exchange*. In this section we consider its functions and structures.

Functions of the Respiratory System

The respiratory system has the following basic functions:

- Providing an extensive surface area for gas exchange between air and circulating blood.
- Moving air to and from the exchange surfaces of the lungs along the respiratory passageways.
- Protecting respiratory surfaces from dehydration, temperature changes, and pathogens.
- Producing sounds for speaking, singing, and other forms of communication.
- Detecting odors by olfactory receptors in the superior portions of the nasal cavity.

Organization of the Respiratory System

We can organize the respiratory system from either an anatomical or a functional perspective. Anatomically, we can divide the system into an upper respiratory system and a lower respiratory system (Figure 23–1). The **upper respiratory system** consists of the *nose, nasal cavity, paranasal sinuses,* and *pharynx* (throat). The **lower respiratory system** includes the *larynx* (voice box), *trachea* (windpipe), *bronchi* and *bronchioles* (air-conducting passageways), and *alveoli* (air-filled pockets within the lungs).

Tips & Tools

To recall the boundary between the upper and lower respiratory systems, remember that the ***l***ower respiratory system begins at the ***l***arynx.

From a functional perspective, the term **respiratory tract** refers to the passageways that carry air to and from the lung exchange surfaces. The *conducting portion* of the respiratory tract begins at the entrance to the nasal cavity and extends through the pharynx, larynx, trachea, bronchi, and larger

Figure 23–1 **The Structures of the Respiratory System.** ATLAS: Plates 47a,b

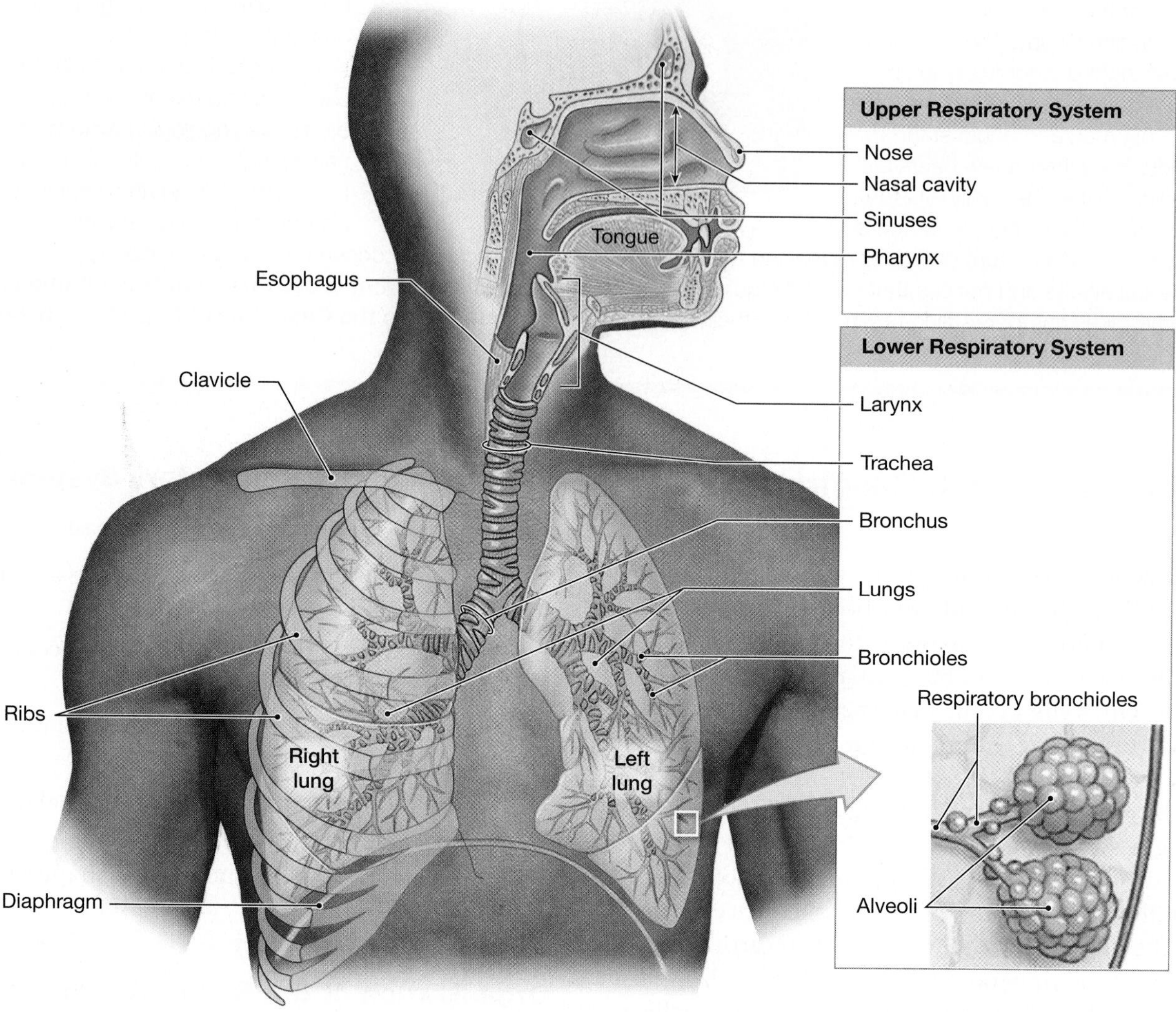
Upper Respiratory System
Nose
Nasal cavity
Sinuses
Pharynx
Tongue
Esophagus
Clavicle
Lower Respiratory System
Larynx
Trachea
Bronchus
Lungs
Bronchioles
Respiratory bronchioles
Ribs
Right lung
Left lung
Diaphragm
Alveoli

23

bronchioles. The *respiratory portion* of the tract includes the smallest, thinnest *respiratory bronchioles* and the associated **alveoli** (al-VĒ-ō-lī), where all gas exchange between air and blood takes place. The following sections combine these two methods of organization, so keep in mind that the upper respiratory system contains the upper part of the conducting portion, while the lower respiratory system includes the rest of the conducting portion and all of the respiratory portion of the respiratory system.

The passageways of the conducting portion of the upper respiratory system filter, warm, and humidify inhaled air, protecting the more delicate surfaces of the lower respiratory system. By the time air reaches the alveoli, most foreign particles and pathogens have been removed, and the humidity and temperature are within acceptable limits. The success of this "conditioning process" is due to the respiratory mucosa.

The Respiratory Mucosa and the Respiratory Defense System

The **respiratory mucosa** (myū-KŌ-suh) lines the conducting portion of the respiratory system. A *mucosa* is a *mucous membrane*, one of the four types of membranes introduced in Chapter 4. This mucosa, which changes along the length of the conducting portion, provides a series of filtration mechanisms that make up the **respiratory defense system**. This defense system reduces the amount of debris or pathogens in inhaled air, which can severely damage the exchange surfaces of the respiratory system.

The Respiratory Mucosa

The respiratory mucosa consists of an epithelium and an underlying layer of areolar tissue. ⮌ p. 140 The **lamina propria** (LAM-ih-nuh PRŌ-prē-uh) is the underlying layer of areolar

Figure 23–2 The Respiratory Epithelium of the Nasal Cavity and Conducting Portion of the Respiratory Tract.

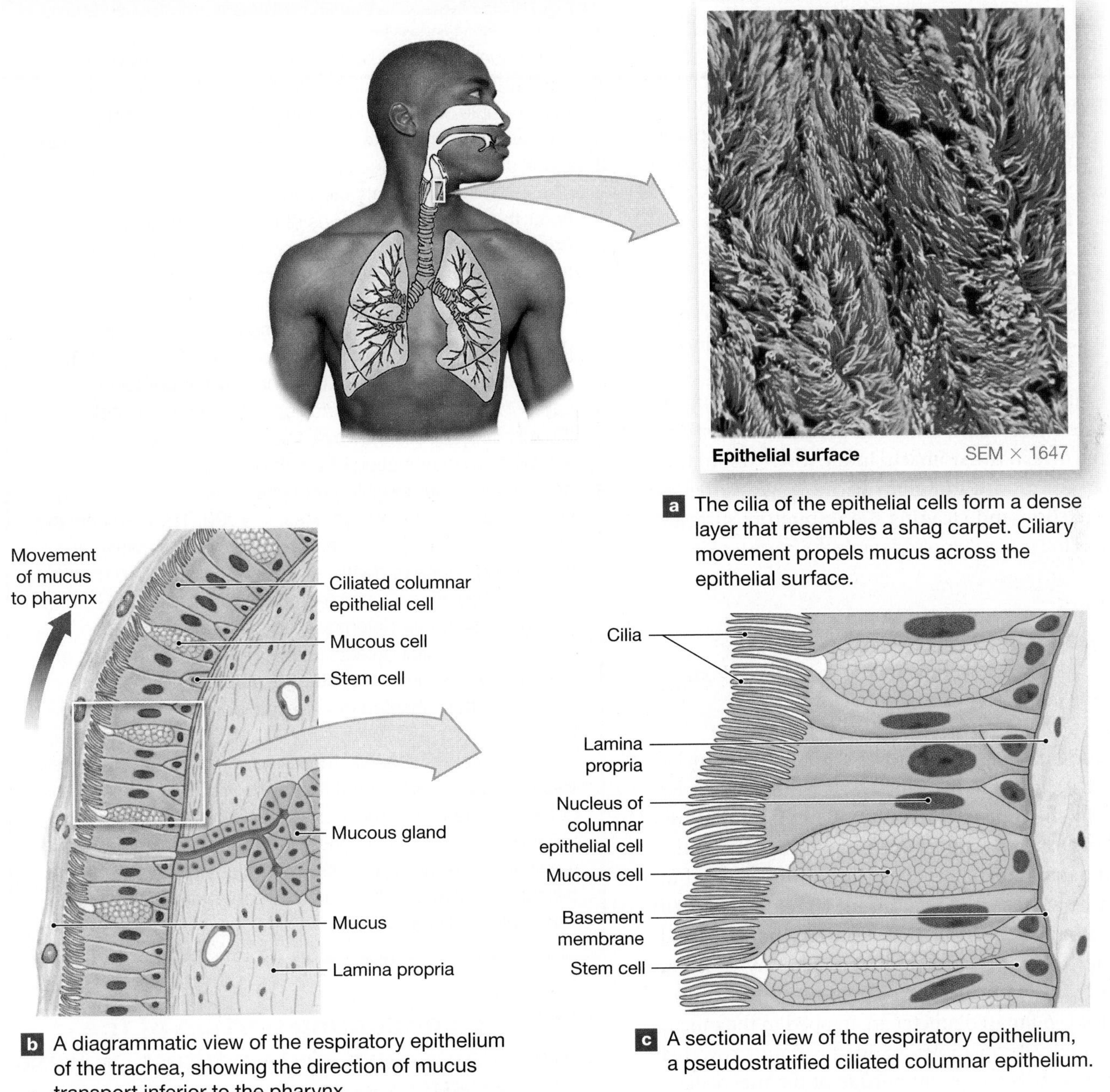
Epithelial surface
SEM × 1647
Movement of mucus to pharynx
Ciliated columnar epithelial cell
Mucous cell
Stem cell
Mucous gland
Mucus
Lamina propria
Cilia
Lamina propria
Nucleus of columnar epithelial cell
Mucous cell
Basement membrane
Stem cell

a The cilia of the epithelial cells form a dense layer that resembles a shag carpet. Ciliary movement propels mucus across the epithelial surface.

b A diagrammatic view of the respiratory epithelium of the trachea, showing the direction of mucus transport inferior to the pharynx.

c A sectional view of the respiratory epithelium, a pseudostratified ciliated columnar epithelium.

? What type of epithelium lines the conducting portion of the respiratory tract?

tissue that supports the respiratory epithelium. In the upper respiratory system, trachea, and bronchi, the lamina propria contains mucous glands that discharge their secretions onto the epithelial surface (Figure 23–2). The respiratory mucosa of the trachea consists of a mucosa, submucosa, hyaline cartilage, and an adventitial layer. The lamina propria in the conducting portions of the lower respiratory system contains bundles of smooth muscle cells. In the bronchioles, the smooth muscles form thick bands that encircle or spiral around the lumen.

A pseudostratified ciliated columnar epithelium with numerous mucous cells lines the nasal cavity and the superior portion of the pharynx. p. 122 The epithelium lining inferior portions of the pharynx is a stratified squamous epithelium similar to that of the oral cavity. These portions of the pharynx conduct air to the larynx and also carry food to the esophagus. The pharyngeal epithelium must therefore protect against abrasion and chemicals.

A pseudostratified ciliated columnar epithelium, comparable to that of the nasal cavity, lines the superior portion of the lower respiratory system. The smaller bronchioles have a cuboidal epithelium with scattered cilia. The exchange surfaces of the alveoli are made up of a simple squamous epithelium. Other, more specialized cells are scattered among the squamous cells and together they form the *alveolar epithelium*.

The Respiratory Defense System

Along much of the respiratory tract, mucous cells in the epithelium and mucous glands in the lamina propria produce a sticky mucus that bathes exposed surfaces. In the nasal cavity, cilia sweep that mucus and any trapped debris or microorganisms toward the pharynx. There it is swallowed and exposed to the acids and enzymes in the stomach. In the lower respiratory system, the cilia beat toward the pharynx, moving a carpet of mucus in that direction and cleaning the respiratory surfaces. This mechanism is the *mucociliary escalator* (see Figure 23–2b,c).

Tips & Tools

The mucus in the respiratory mucosa functions like sticky flypaper, but instead of flies, it traps particles and debris from the air that moves past it.

Filtration in the nasal cavity removes virtually all particles larger than about 10 μm from the inhaled air. Smaller particles may be trapped by the mucus of the nasopharynx or by secretions of the pharynx. The rate of mucus production in the nasal cavity and paranasal sinuses speeds up when exposed to unpleasant stimuli, such as noxious vapors, large quantities of dust and debris, allergens, or pathogens. (The familiar signs and symptoms of the "common cold" appear when any of more than 200 types of viruses invades the respiratory epithelium.)

Most particles 1–5 μm in diameter are trapped in the mucus coating the respiratory bronchioles or in the liquid covering the alveolar surfaces. These areas are outside of the mucociliary escalator, but the foreign particles can be engulfed by alveolar macrophages. Most particles smaller than about 0.5 μm remain suspended in the air.

Checkpoint

1. Identify several functions of the respiratory system.
2. List the two anatomical divisions of the respiratory system.
3. What membrane lines the conducting portion of the respiratory tract?

See the blue Answers tab at the back of the book.

Clinical Note Breakdown of the Respiratory Defense System

Large quantities of airborne particles may overload the respiratory defenses and produce a variety of illnesses. For example, irritants in the lining of the conducting passageways can provoke the formation of mucous plugs that block airflow and reduce pulmonary function. Damage to the epithelium in the affected area may allow irritants to enter the surrounding tissues of the lung. The irritants then produce local inflammation. Airborne irritants—such as those in cigarette smoke—are known to promote the development of lung cancer (p. 877).

Aggressive pathogens can also overwhelm respiratory defenses. **Tuberculosis** (tū-ber-kyū-LŌ-sis), or **TB**, results from an infection of the lungs by the bacterium called *Mycobacterium tuberculosis*. Bacteria may colonize the respiratory passageways, the interstitial spaces, the alveoli, or a combination of the three. Signs and symptoms generally include coughing and chest pain, plus fever, night sweats, fatigue, and weight loss. In 1900, TB, then known as "consumption" (because its victims wasted away), was the leading cause of death. An estimated one-third of the world's population is still infected with TB today.

The respiratory defense system can also fail due to inherited congenital defects affecting mucus production or transport. For example, **cystic fibrosis (CF)** is the most common lethal inherited disease in individuals of Northern European descent. It occurs in 1 in 3200 Caucasian births. The respiratory mucosa in these individuals produces dense, viscous mucus that cannot be transported by the respiratory defense system. The mucociliary escalator stops working, leading to frequent infections. Mucus also blocks the smaller respiratory passageways, making breathing difficult. The average life span for people with CF who live into adulthood is 37 years.

23-2 The conducting portion of the upper respiratory system filters, warms, and humidifies air

Learning Outcome **Identify the organs of the upper respiratory system, and describe their functions.**

As we have noted, the upper respiratory system consists of the nose, nasal cavity, paranasal sinuses, and pharynx (see Figure 23–1).

The Nose and Nasal Cavity

The nose provides an airway for breathing, moistens and warms entering air, filters and cleans inhaled air, serves as a resonating chamber for speech, and houses olfactory receptors. The nasal cavity filters coarse particles from inspired air and is lined with an olfactory mucosa.

The Nose

The **nose** is the primary passageway for air entering the respiratory system. Air normally enters through the paired **nostrils**, or **nares** (NĀ-res), which open into the *nasal cavity* (Figure 23–3a). The **nasal vestibule** is the space contained within the flexible tissues of the nose (Figure 23–3c). The epithelium of the vestibule contains coarse hairs that extend across the nostrils. Large airborne particles, such as sand, sawdust, or even insects, are trapped in these hairs and prevented from entering the nasal cavity.

The Nasal Cavity and Paranasal Sinuses

The **nasal cavity** is divided into left and right sides by the **nasal septum** (Figure 23–3b). The bony portion of the nasal septum is formed by the fusion of the perpendicular plate of the ethmoid and the vomer (look back at Figure 7–3d, p. 213). The anterior portion of the nasal septum is formed of hyaline cartilage. This cartilaginous plate supports the **dorsum of nose** (bridge) and the **apex of nose** (tip).

The maxillae, nasal bone, frontal bone, ethmoid, and sphenoid form the lateral and superior walls of the nasal cavity. The mucus produced in the *paranasal sinuses* (sinuses of the frontal bone, sphenoid, ethmoid, and paired maxillae and palatine bones) helps keep the surfaces of the nasal cavity moist and clean (look back at Figure 7–15, p. 226). The tears draining through the nasolacrimal ducts do so as well.

The *olfactory region* is the superior portion of the nasal cavity. It includes the areas lined by olfactory epithelium: (1) the inferior surface of the cribriform plate, (2) the superior portion of the nasal septum, and (3) the superior nasal conchae. Receptors in the olfactory epithelium provide your sense of smell. ⮌ p. 567 The *superior, middle,* and *inferior nasal conchae* project toward the nasal septum from the lateral walls of the nasal cavity. (Some references use the terms *superior, middle,* and *inferior turbinates.*) ⮌ pp. 220, 223 To pass from the vestibule to the **choanae** (KŌ-an-ē)—openings of the nasal cavity—air flows between adjacent conchae, through the **superior**, **middle**, and **inferior nasal meatuses** (mē-Ā-tus-ez;) (see Figure 23–3b). These are narrow grooves rather than open passageways. The incoming air bounces off the conchal surfaces and churns like a stream flowing over rocks. This turbulence serves several purposes. As the air swirls, small airborne particles are likely to come into contact with the mucus coating the nasal cavity. In addition, the turbulence provides extra time for warming and humidifying incoming air. It also creates circular air currents that bring olfactory stimuli to the olfactory receptors.

The bony **hard palate** is made up of portions of the paired maxillae and palatine bones. The hard palate forms the floor of the nasal cavity and separates it from the oral cavity. A fleshy **soft palate** extends posterior to the hard palate, marking the boundary between the superior *nasopharynx* (nā-zō-FAR-ingks) and the rest of the pharynx. The nasal cavity opens into the nasopharynx at the choanae.

Functions of the Nasal Mucosa

The mucosa of the nasal cavity prepares inhaled air for arrival at the lower respiratory system. Throughout much of the nasal cavity, the lamina propria contains an abundance of arteries, veins, and capillaries that bring nutrients and water to the secretory cells. The lamina propria of the nasal conchae also contains an extensive network of highly expandable veins. This vascularization warms and humidifies the incoming air (and dehumidifies and absorbs the heat of the outgoing air as well). As cool, dry air passes inward over the exposed surfaces of the nasal cavity, the warm epithelium radiates heat, and water in the mucus evaporates. In this way, air moving from your nasal cavity to your lungs is heated almost to body temperature. It is also nearly saturated with water vapor. These changes protect more delicate respiratory surfaces from chilling or drying out—two potentially disastrous events. If you breathe through your mouth, you eliminate much of these preliminary events. To avoid alveolar damage, patients breathing on a respirator (mechanical ventilator), which uses a tube to conduct air directly into the trachea, must receive air that has been externally filtered and humidified.

As air moves out of the respiratory tract, it again passes over the epithelium of the nasal cavity. This air is warmer and more humid than the air that enters. It warms the nasal mucosa, and moisture condenses on the epithelial surfaces. In this way, breathing through your nose helps prevent heat loss and water loss.

The extensive vascularization of the nasal cavity and the vulnerable position of the nose make a **nosebleed**, or *epistaxis* (ep-ih-STAK-sis), a fairly common event. (The leading cause of epistaxis in children is nose-picking.) This bleeding generally involves vessels of the mucosa covering the cartilaginous portion of the septum. Possible causes include trauma (such as a punch in the nose), drying, infections, allergies, or clotting disorders. Hypertension can also bring on a nosebleed by rupturing small vessels of the lamina propria.

The Pharynx

23

The **pharynx** (FAR-ingks), or throat, is a chamber shared by the digestive and respiratory systems. It extends between the choanae and the entrances to the larynx and esophagus. The curving superior and posterior walls of the pharynx are closely bound to the axial skeleton, but the lateral walls are flexible and muscular.

The pharynx is divided into three parts: the *nasopharynx*, the *oropharynx*, and the *laryngopharynx* (see Figure 23–3c).

1. The **nasopharynx** is the superior portion of the pharynx. It is connected to the posterior portion of the nasal cavity through the choanae. The soft palate separates it from the oral cavity. The nasopharynx is lined by the same pseudostratified ciliated columnar epithelium as in the nasal cavity. The *pharyngeal tonsil* is located on the posterior

Figure 23–3 **The Structures of the Upper Respiratory System.** ATLAS: Plate 19

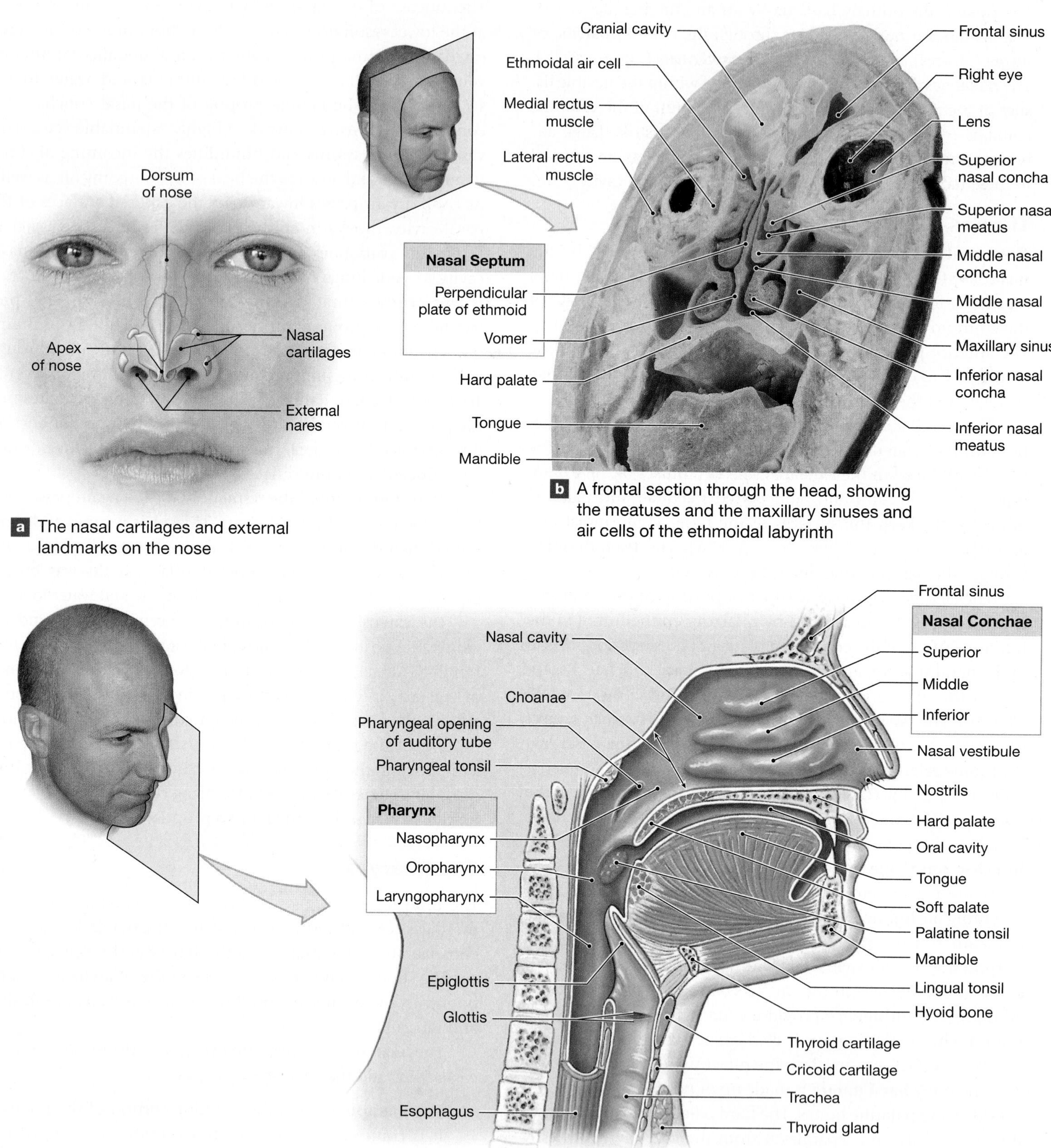
Dorsum of nose
Apex of nose
Nasal cartilages
External nares
Cranial cavity
Ethmoidal air cell
Medial rectus muscle
Lateral rectus muscle
Nasal Septum
Perpendicular plate of ethmoid
Vomer
Hard palate
Tongue
Mandible
Frontal sinus
Right eye
Lens
Superior nasal concha
Superior nasal meatus
Middle nasal concha
Middle nasal meatus
Maxillary sinus
Inferior nasal concha
Inferior nasal meatus
Nasal cavity
Choanae
Pharyngeal opening of auditory tube
Pharyngeal tonsil
Pharynx
Nasopharynx
Oropharynx
Laryngopharynx
Epiglottis
Glottis
Esophagus
Frontal sinus
Nasal Conchae
Superior
Middle
Inferior
Nasal vestibule
Nostrils
Hard palate
Oral cavity
Tongue
Soft palate
Palatine tonsil
Mandible
Lingual tonsil
Hyoid bone
Thyroid cartilage
Cricoid cartilage
Trachea
Thyroid gland

a The nasal cartilages and external landmarks on the nose

b A frontal section through the head, showing the meatuses and the maxillary sinuses and air cells of the ethmoidal labyrinth

c The nasal cavity and pharynx, as seen in sagittal section with the nasal septum removed wall of the nasopharynx. Each auditory tube opens into the nasopharynx at the **pharyngeal opening of auditory tube** on each side of the pharyngeal tonsil. pp. 593, 791

2. The **oropharynx** (*oris*, mouth) extends between the soft palate and the base of the tongue at the level of the hyoid bone. The posterior portion of the oral cavity connects directly with the oropharynx, as does the posterior inferior portion of the nasopharynx. At the boundary between the nasopharynx and the oropharynx, the epithelium changes from pseudostratified columnar epithelium to stratified squamous epithelium.
3. The narrow **laryngopharynx** (la-rin-gō-FAR-ingks) is the inferior part of the pharynx. It includes that portion of the pharynx between the hyoid bone and the entrance to the larynx and esophagus. Like the oropharynx, the laryngopharynx is lined with a stratified squamous epithelium that resists abrasion, chemicals, and pathogens.

Checkpoint

4. **Name the structures of the upper respiratory system.**
5. **Why is the vascularization of the nasal cavity important?**
6. **Why is the lining of the nasopharynx different from that of the oropharynx and the laryngopharynx?**

See the blue Answers tab at the back of the book.

23-3 The conducting portion of the lower respiratory system conducts air to the respiratory portion and produces sound

Learning Outcome **Describe the structure of the larynx, discuss its roles in normal breathing and in sound production, and identify structures of the airways.**

In this section we cover the conducting portion of the lower respiratory system. These structures include the larynx, which also allows us to produce sound. The trachea is a large air passageway that leads to the bronchi. The bronchi and their branching vessels form the *bronchial tree* that leads air into the lungs.

The Larynx

Inhaled air leaves the pharynx and enters the larynx through the *glottis*, a slit like opening between the vocal cords. The glottis is the "voice box," of the larynx. The **larynx** (LAR-ingks) is a cartilaginous tube that surrounds and protects the glottis. The larynx begins at the level of vertebra C_4 or C_5 and ends at the level of vertebra C_6. Essentially a cylinder, the larynx has incomplete cartilages that are stabilized by ligaments and laryngeal muscles (**Figure 23–4**).

Figure 23–4 The Anatomy of the Larynx.

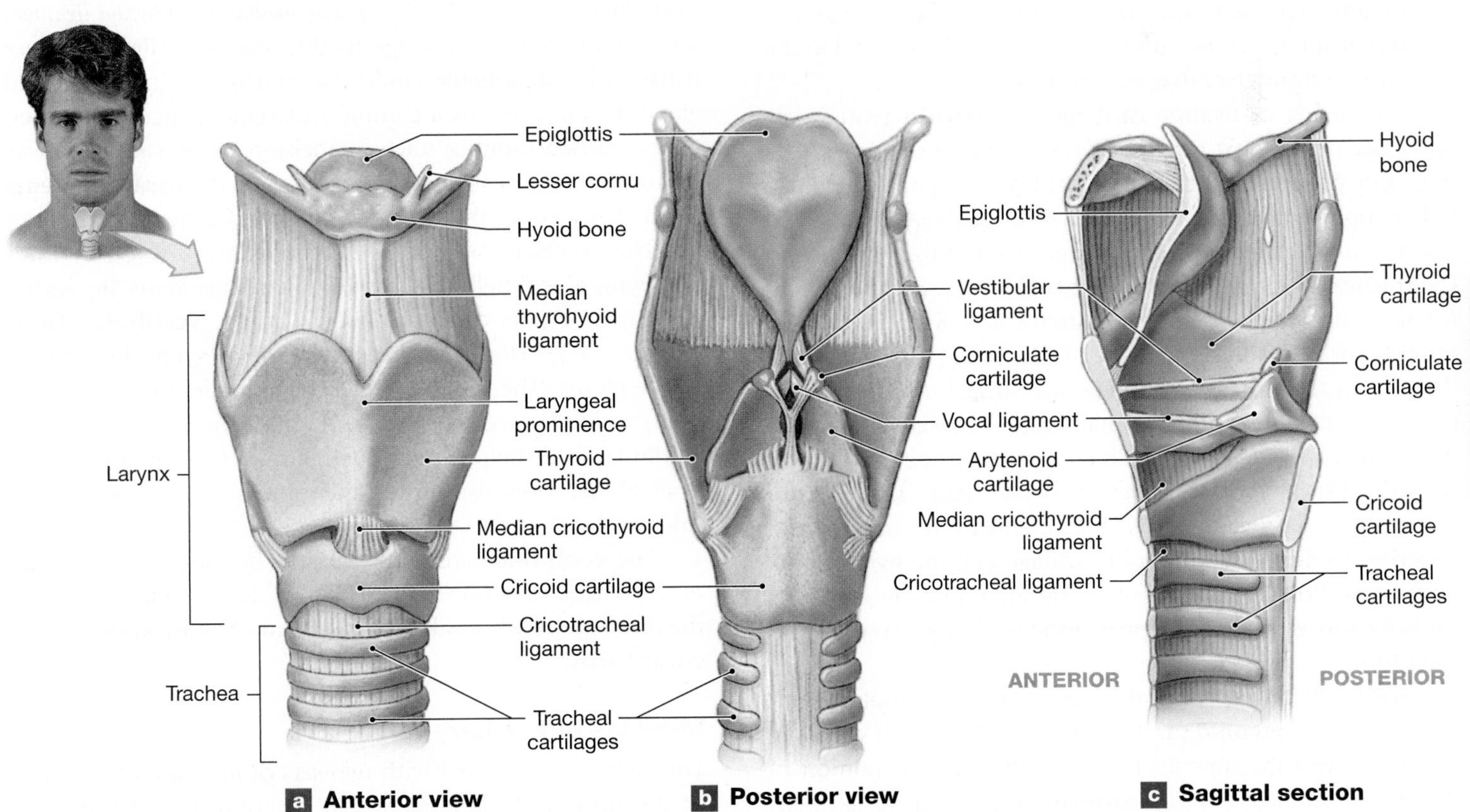

Epiglottis
Lesser cornu
Hyoid bone
Median thyrohyoid ligament
Laryngeal prominence
Thyroid cartilage
Median cricothyroid ligament
Cricoid cartilage
Cricotracheal ligament
Tracheal cartilages
Larynx
Trachea
a Anterior view
b Posterior view
Epiglottis
Vestibular ligament
Corniculate cartilage
Vocal ligament
Arytenoid cartilage
Median cricothyroid ligament
Cricotracheal ligament
Hyoid bone
Thyroid cartilage
Corniculate cartilage
Cricoid cartilage
Tracheal cartilages
ANTERIOR
POSTERIOR
c Sagittal section

Go to MasteringA&P™ > Study Area > Lab Tools > PAL 3.0 > Human Cadaver and Anatomical Models > Respiratory System

Figure 23–5 The Glottis and Surrounding Structures.

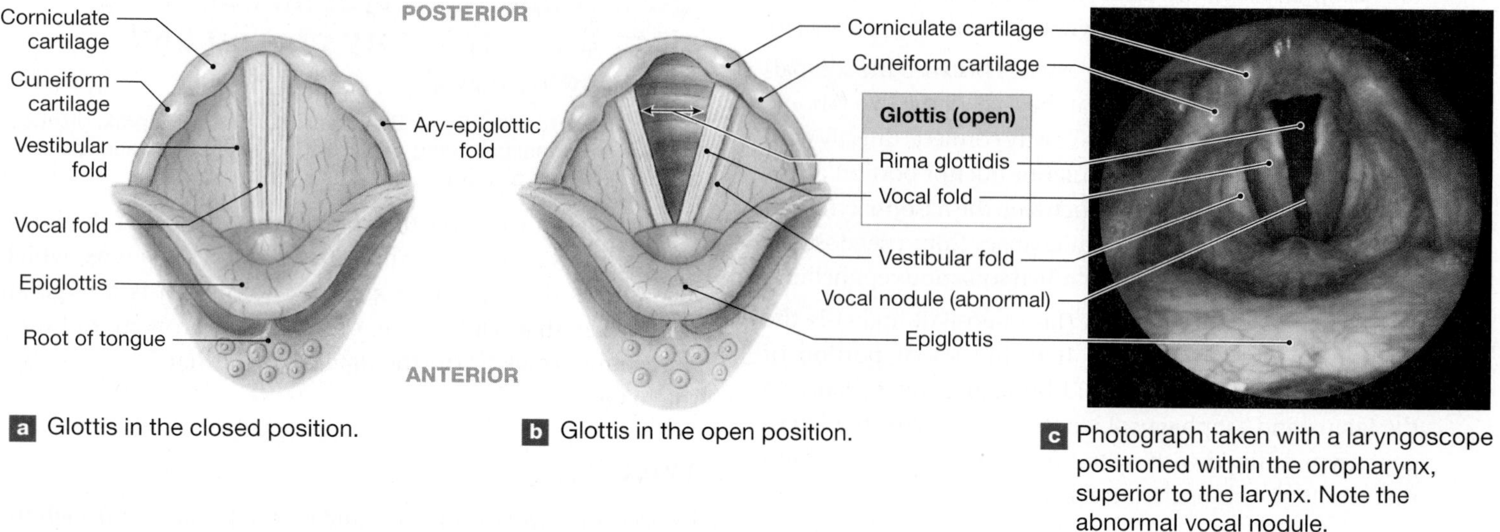
POSTERIOR
Corniculate cartilage
Cuneiform cartilage
Vestibular fold
Vocal fold
Epiglottis
Root of tongue
Ary-epiglottic fold
ANTERIOR
Corniculate cartilage
Cuneiform cartilage
Glottis (open)
Rima glottidis
Vocal fold
Vestibular fold
Vocal nodule (abnormal)
Epiglottis

a Glottis in the closed position.

b Glottis in the open position.

c Photograph taken with a laryngoscope positioned within the oropharynx, superior to the larynx. Note the abnormal vocal nodule.

Cartilages and Ligaments of the Larynx

Three large, unpaired cartilages form the larynx: (1) the thyroid cartilage, (2) the cricoid cartilage, and (3) the epiglottis (see Figure 23–4). Made of hyaline cartilage, the **thyroid cartilage** is the largest laryngeal cartilage. It forms most of the anterior and lateral walls of the larynx. In section, this cartilage is U-shaped, and posteriorly, it is incomplete. You can easily see and feel the prominent anterior surface of the thyroid cartilage, called the *laryngeal prominence*, or *Adam's apple*. The inferior surface articulates with the cricoid cartilage. The superior surface has ligamentous attachments to the hyoid bone and to the epiglottis and smaller laryngeal cartilages.

Also made of hyaline cartilage, the **cricoid** (KRĪ-koyd) **cartilage** is the most inferior of the laryngeal cartilages. The posterior portion of the cricoid cartilage is expanded to provide support where there is no thyroid cartilage. The cricoid and thyroid cartilages protect the glottis and the entrance to the trachea. Their broad surfaces provide sites for the attachment of important laryngeal muscles and ligaments. Ligaments attach the inferior surface of the cricoid cartilage to the first tracheal cartilage. The superior surface of the cricoid cartilage articulates with the small, paired *arytenoid cartilages*.

Made of elastic cartilage, the shoehorn-shaped **epiglottis** (ep-ih-GLOT-is) projects superior to the glottis and forms a lid over it. It has ligamentous attachments to the anterior and superior borders of the thyroid cartilage and the hyoid bone. During swallowing, the larynx is elevated and the epiglottis folds back over the glottis, preventing both liquids and solid food from entering the respiratory tract.

The larynx also contains three pairs of smaller hyaline cartilages: (1) The **arytenoid** (ar-ih-TĒ-noyd; ladle shaped) **cartilage** articulates with the superior border of the enlarged portion of the cricoid cartilage. (2) The **corniculate** (kor-NIK-yū-lāt; horn shaped) **cartilage** articulates with the arytenoid cartilage. The corniculate and arytenoid cartilages function in the opening and closing of the glottis and the production of sound. (3) An elongated, curving **cuneiform** (kyū-NĒ-ih-form; wedge shaped) **cartilage** is within the *ary-epiglottic folds*, tissue that extends between the lateral surface of each arytenoid cartilage and the epiglottis.

Ligaments bind together the various laryngeal cartilages. Additional ligaments attach the thyroid cartilage to the hyoid bone, and the cricoid cartilage to the trachea (cricotracheal ligament) (see Figure 23–4a,b). The *median cricothyroid ligament* attaches the thyroid cartilage to the cricoid cartilage. Because it has easily identifiable landmarks and is clear of the thyroid gland, this ligament is a common placement site for an emergency tracheostomy, a tracheal incision to bypass an airway obstruction. The **vestibular ligaments** and the **vocal ligaments** extend between the thyroid cartilage and the arytenoid cartilages.

The vestibular and vocal ligaments are covered by folds of laryngeal epithelium. The vestibular ligaments lie within the superior pair of folds, known as the **vestibular folds** (Figure 23–5). These folds are fairly inelastic and lie laterally to the glottis. The **glottis** (GLOT-is) is made up of the **vocal folds** and the space between them, called the *rima glottidis*. The vestibular folds help prevent foreign objects from entering the open glottis. They also protect the more inferior, delicate vocal folds of the glottis.

The vocal folds are highly elastic, because the vocal ligaments consist of elastic tissue. The vocal folds are involved with the production of sound. For this reason they are known as the **vocal cords**.

Muscles of the Larynx

The larynx is associated with two sets of muscles: (1) muscles of the neck and pharynx, which position and stabilize the larynx (pp. 352–354), and (2) smaller intrinsic muscles that control tension in the glottal vocal folds or that open and close the glottis. These smaller muscles insert on the thyroid, arytenoid, and corniculate cartilages. The opening or closing of the glottis involves rotational movements of the arytenoid cartilages.

When you swallow, both sets of muscles work together to prevent food or drink from entering the glottis. Food is crushed and chewed into a pasty mass, known as a *bolus*, before being swallowed. Muscles of the neck and pharynx then elevate the larynx, bending the epiglottis over the glottis, so that the bolus can glide across the epiglottis rather than falling into the larynx. While this movement is under way, the glottis is closed.

Foods or liquids that touch the vestibular folds or glottis trigger the *coughing reflex*. In a cough, the glottis is kept closed while the chest and abdominal muscles contract, compressing the lungs. When the glottis is opened suddenly, a blast of air from the trachea ejects material that blocks the entrance to the glottis.

Sound Production

How do you produce sounds? Air passing through your open glottis vibrates its vocal folds and produces sound waves. The pitch of the sound depends on the diameter, length, and tension in your vocal folds. The diameter and length are directly related to the size of your larynx. You control the tension by contracting voluntary muscles that reposition the arytenoid cartilages relative to the thyroid cartilage. When the distance increases, your vocal folds tense and the pitch rises. When the distance decreases, your vocal folds relax and the pitch falls.

Children have slender, short vocal folds, so their voices tend to be high pitched. At puberty, the larynx of males enlarges much more than that of females. The vocal cords of an adult male are thicker and longer, so they produce lower tones than those of an adult female.

Sound production at the larynx is called *phonation* (fō-NĀ-shun; *phone*, voice). Phonation is one part of speech production. Clear speech also requires *articulation*, the modification of those sounds by voluntary movements of other structures, such as the tongue, teeth, and lips to form words. In a stringed instrument, such as a guitar, the quality of the sound produced does not depend solely on the nature of the vibrating string. Rather, the entire instrument becomes involved as the walls vibrate and the composite sound echoes within the hollow body. Similar amplification and resonance take place within your pharynx, oral cavity, nasal cavity, and paranasal sinuses. The combination gives you the particular and distinctive sound of your voice. That sound changes when you have a sinus infection and your nasal cavity and paranasal sinuses are filled with mucus rather than air.

Tips & Tools

Intelligible sound requires both phonation and articulation. Saying "ahhhh" while your tongue is depressed during a tonsil examination is an example of phonation. Saying "hot" adds articulation to that sound.

An infection or inflammation of the larynx is known as *laryngitis* (lar-in-JĪ-tis). It commonly affects the vibrational qualities of the vocal folds. Hoarseness is the most familiar result. Mild cases are temporary and seldom serious. However, bacterial or viral infections of the epiglottis can be very dangerous. The resulting swelling may close the glottis and cause suffocation. This condition, *acute epiglottitis* (ep-ih-glot-TĪ-tis), can develop rapidly after a bacterial infection of the throat. Young children are most likely to be affected.

The Trachea

The **trachea** (TRĀ-kē-uh), or windpipe, is a tough, flexible tube with a diameter of about 2.5 cm (1 in.) and a length of about 11 cm (4.33 in.) (Figure 23–6). The trachea begins anterior to vertebra C_6 in a ligamentous attachment to the cricoid cartilage. It ends in the mediastinum, at the level of vertebra T_5, where it branches to form the *right* and *left main bronchi*.

The epithelium of the trachea is continuous with that of the larynx. The mucosa of the trachea resembles that of the nasal cavity and nasopharynx (look back at Figure 23–2a). The **submucosa** (sub-mū-KŌ-suh), a thick layer of connective tissue, surrounds the mucosa. The submucosa contains **tracheal glands** whose mucous secretions reach the tracheal lumen through a number of short ducts. The trachea contains 15–20 **tracheal cartilages** that stiffen the tracheal walls and protect the airway (see Figure 23–6a). They also prevent it from collapsing or overexpanding as pressure changes in the respiratory system.

Each tracheal cartilage is C-shaped. The closed portion of the C protects the anterior and lateral surfaces of the trachea. The open portion of the C faces posteriorly, toward the esophagus (see Figure 23–6b). Because these cartilages are not continuous, the posterior tracheal wall can easily distort when you swallow, allowing large masses of food to pass through the esophagus.

An elastic **anular ligament** and the **trachealis**, a band of smooth muscle, connect the ends of each tracheal cartilage (see Figure 23–6b). Contraction of the trachealis reduces the diameter of the trachea. This narrowing increases the tube's resistance to airflow. The normal diameter of the trachea changes from moment to moment, primarily under the control of the sympathetic division of the ANS. Sympathetic stimulation increases the diameter of the trachea and makes it easier to move large volumes of air along the respiratory passageways.

Figure 23–6 **The Anatomy of the Trachea.** ATLAS: Plates 42b,c

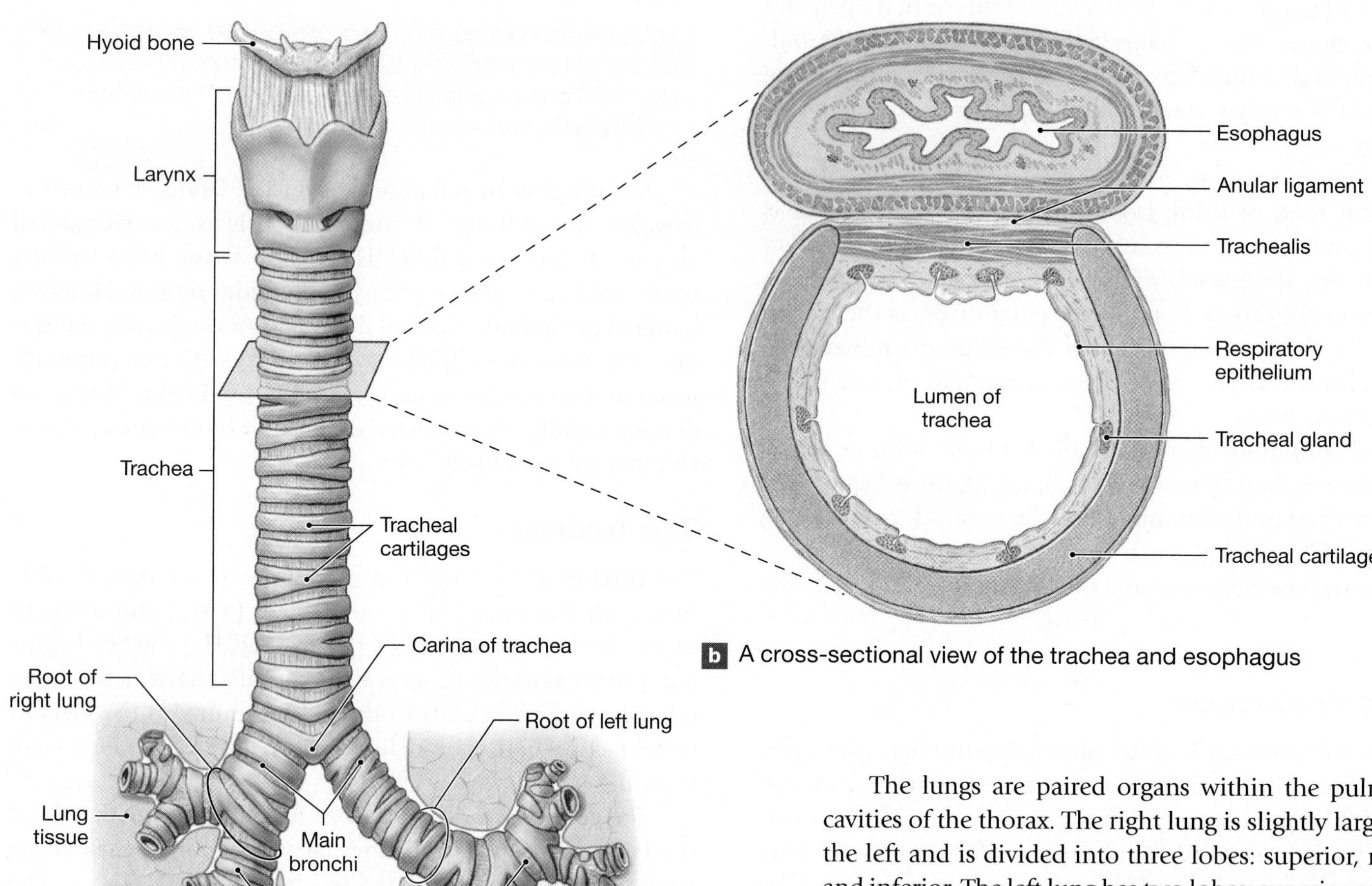

Hyoid bone
Larynx
Trachea
Tracheal cartilages
Carina of trachea
Root of right lung
Root of left lung
Lung tissue
Main bronchi
Lobar bronchi
Right lung
Left lung
Esophagus
Anular ligament
Trachealis
Respiratory epithelium
Lumen of trachea
Tracheal gland
Tracheal cartilage

b A cross-sectional view of the trachea and esophagus

a A diagrammatic anterior view showing the plane of section for part (b)

The Bronchial Tree

The trachea branches within the mediastinum into the **right main bronchus** and **left main bronchus** (BRONG-kus; plural, *bronchi*). These main bronchi enter the right and left lungs, respectively, and divide to form smaller passageways, the *lobar bronchi* (see Figure 23–6a). Together, the main and lobar bronchi and their branches form the **bronchial tree**.

The Bronchi

The **carina** (kah-RĪ-nuh) **of trachea** is a ridge that separates the openings of the right and left main bronchi at their junction with the trachea (see Figure 23–6a). Like the trachea, the main bronchi have C-shaped rings, but the ends of the C overlap. The right main bronchus supplies the right lung, and the left supplies the left lung. The right main bronchus is larger in diameter than the left, and descends toward the lung at a steeper angle. For these reasons, most foreign objects that enter the trachea find their way into the right bronchus rather than the left.

The lungs are paired organs within the pulmonary cavities of the thorax. The right lung is slightly larger than the left and is divided into three lobes: superior, middle, and inferior. The left lung has two lobes: superior and inferior. Each main bronchus divides to form **lobar bronchi**, which supply the lobes of the lungs. Because the right lung has three lobes, it has three lobar bronchi; because the left lung has two lobes, it has two lobar bronchi. Each lobe is further divided internally into pulmonary lobules.

Figure 23–7a,b depicts the branching pattern of the left main bronchus as it enters the lung. (The number of branches has been reduced for clarity.) In each lung, the lobar bronchi branch to form **segmental bronchi**. The branching pattern differs between the two lungs, but each segmental bronchus ultimately supplies air to a single **bronchopulmonary segment**, a specific region of one lung (see Figure 23–7a). The right lung has 10 bronchopulmonary segments. During development, the left lung also has 10 segments, but adjacent segmental bronchi fuse, generally reducing the number to 8 or 9.

The walls of the main, lobar, and segmental bronchi contain progressively less cartilage. In the lobar and segmental bronchi, the cartilages form plates arranged around the lumen. These cartilages serve the same structural purpose as the rings of cartilage in the trachea and primary bronchi. As the amount of cartilage decreases, the amount of smooth muscle increases. With less cartilaginous support, the degree of tension in those smooth muscles exerts a proportionally greater effect on bronchial diameter and the resistance to airflow. During a respiratory

Figure 23–7 **The Bronchi, Lobules, and Alveoli of the Lung.** ATLAS: Plates 42b,c; 47b–d

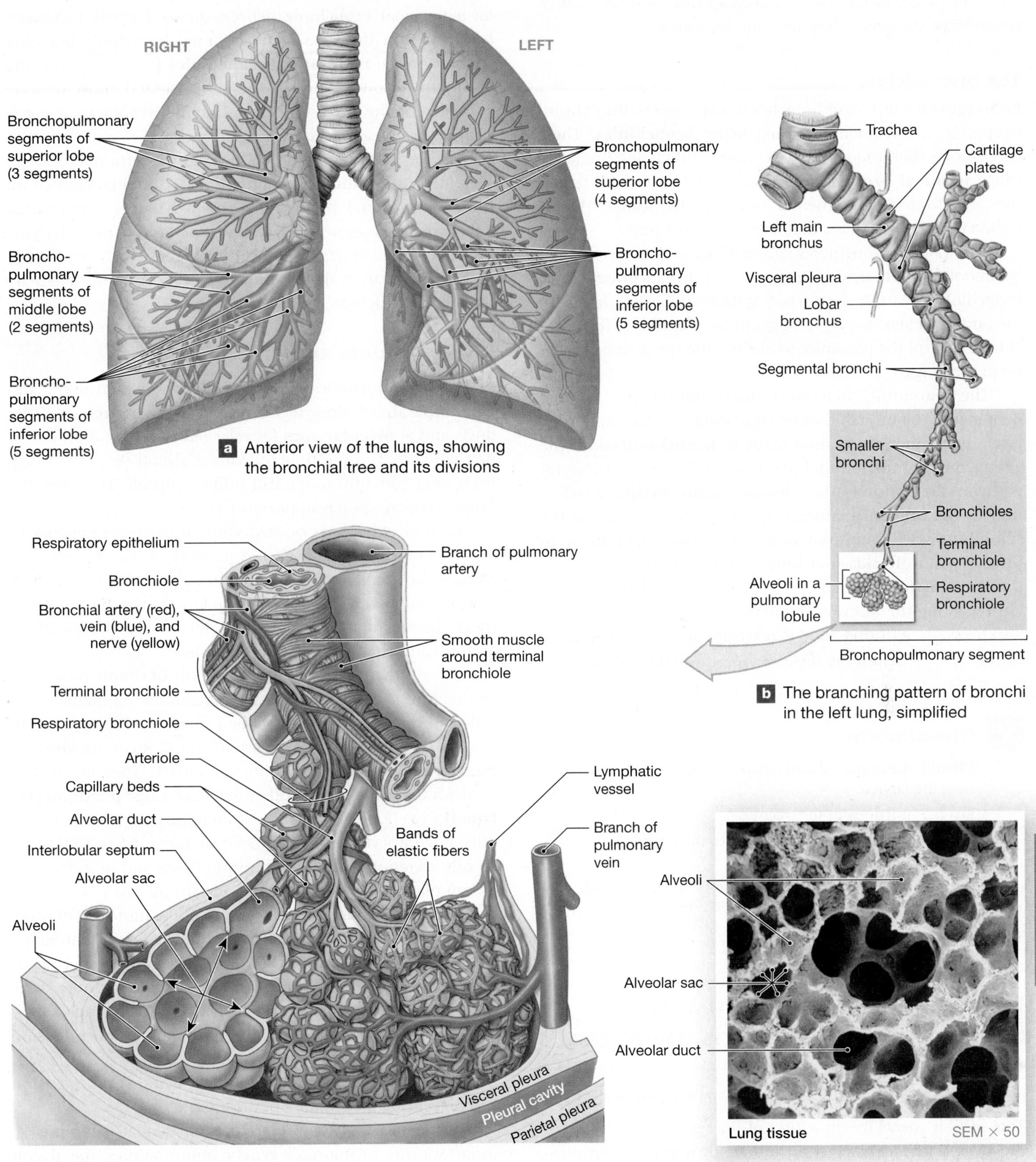
RIGHT
LEFT
Bronchopulmonary segments of superior lobe (3 segments)
Broncho-pulmonary segments of middle lobe (2 segments)
Broncho-pulmonary segments of inferior lobe (5 segments)
Bronchopulmonary segments of superior lobe (4 segments)
Broncho-pulmonary segments of inferior lobe (5 segments)
Trachea
Cartilage plates
Left main bronchus
Visceral pleura
Lobar bronchus
Segmental bronchi
Smaller bronchi
Bronchioles
Terminal bronchiole
Alveoli in a pulmonary lobule
Respiratory bronchiole
Bronchopulmonary segment
Respiratory epithelium
Bronchiole
Bronchial artery (red), vein (blue), and nerve (yellow)
Terminal bronchiole
Respiratory bronchiole
Arteriole
Capillary beds
Alveolar duct
Interlobular septum
Alveolar sac
Alveoli
Branch of pulmonary artery
Smooth muscle around terminal bronchiole
Bands of elastic fibers
Lymphatic vessel
Branch of pulmonary vein
Visceral pleura
Pleural cavity
Parietal pleura
Alveoli
Alveolar sac
Alveolar duct
Lung tissue
SEM × 50
23

a Anterior view of the lungs, showing the bronchial tree and its divisions

b The branching pattern of bronchi in the left lung, simplified

c The structure of a single pulmonary lobule, part of a bronchopulmonary segment

d SEM of lung tissue showing the appearance and organization of the alveoli infection, the bronchi and bronchioles can become inflamed and constricted, increasing resistance. In this condition, called **bronchitis**, the person has difficulty breathing.

The Bronchioles

Each segmental bronchus branches several times within a bronchopulmonary segment, forming many **bronchioles**. These bronchioles then branch into the finest conducting branches, called **terminal bronchioles**. Nearly 6500 terminal bronchioles arise from each segmental bronchus. The lumen of each terminal bronchiole has a diameter of 0.3–0.5 mm.

The walls of bronchioles lack cartilage and are dominated by smooth muscle tissue (Figure 23–7c). In functional terms, bronchioles are to the respiratory system what arterioles are to the cardiovascular system. Changes in the diameter of the bronchioles control the resistance to airflow and the distribution of air in the lungs.

The autonomic nervous system controls the luminal diameter of the bronchioles by regulating the smooth muscle layer. Sympathetic activation leads to **bronchodilation**, the enlargement of the luminal diameter of the airway. Parasympathetic stimulation leads to **bronchoconstriction**, a reduction in the luminal diameter of the airway. Tension in the smooth muscles commonly causes the bronchiole mucosa to form a series of folds that limits airflow. Excessive stimulation, as in **asthma** (AZ-muh), can almost completely prevent airflow along the terminal bronchioles. Bronchoconstriction also takes place during allergic reactions such as anaphylaxis, in response to histamine released by activated mast cells and basophils. ↺ p. 825

✓ Checkpoint

7. Identify the unpaired and paired cartilages associated with the larynx.
8. List the functions of the trachea.
9. Why are the cartilages that reinforce the trachea C-shaped?
10. If food accidentally enters the main bronchi, in which bronchus is it more likely to lodge? Why?

See the blue Answers tab at the back of the book.

23

23-4 The respiratory portion of the lower respiratory system is where gas exchange occurs

Learning Outcome Describe the functional anatomy of alveoli.

This section covers the structures of the respiratory portion of the lower respiratory system, including the respiratory bronchioles, the alveoli, and the blood air barrier. We cover much of the function of these structures in the following sections.

The Respiratory Bronchioles

Each terminal bronchiole delivers air to a single *pulmonary lobule*. Within the lobule, the terminal bronchiole branches to form several **respiratory bronchioles** (Figure 23–7c). The respiratory bronchioles are the thinnest and most delicate branches of the bronchial tree. They deliver air to the gas-exchange surfaces of the lungs.

Before incoming air moves beyond the terminal bronchioles, it has been filtered and humidified. Cuboidal epithelium lines the terminal bronchioles and respiratory bronchioles. There are only scattered cilia and no mucous cells or underlying mucous glands. If particulates or pathogens reach this part of the respiratory tract, there is little to prevent them from damaging the lung's delicate exchange surfaces.

Alveolar Ducts and Alveoli

Respiratory bronchioles are connected to individual alveoli and to multiple alveoli along regions called **alveolar ducts**. Alveolar ducts end at **alveolar sacs** (*alveolar saccules*), common chambers connected to multiple individual alveoli (Figure 23–8a). Each lung contains about 150 million alveoli. They give the lungs an open, spongy appearance (Figure 23–8b).

Each alveolus is associated with an extensive network of capillaries (see Figure 23–8a). A network of elastic fibers surrounds the capillaries. These fibers help maintain the relative positions of the alveoli and respiratory bronchioles. When these fibers recoil during exhalation, they reduce the size of the alveoli and help push air out of the lungs.

The alveolar cell layer consists mainly of simple squamous epithelium (Figure 23–8c). The squamous epithelial cells, called **pneumocytes type I** (*type I alveolar cells*), are unusually thin and are the sites of gas diffusion. Roaming **alveolar macrophages** patrol the epithelial surface. They engulf any particles that have eluded other defenses. Large **pneumocytes type II** (*type II alveolar cells*) are scattered among the squamous cells. Pneumocytes type II produce **surfactant** (sur-FAK-tant), an oily secretion containing phospholipids and proteins. Surfactant plays a key role in keeping the alveoli open.

Surfactant reduces surface tension in the thin layer of water coating the alveolar surface. Recall from Chapter 2 that *surface tension* results from the attraction between water molecules at an air–water boundary. ↺ p. 36 Surface tension not only keeps small objects from entering the water, but it also tends to collapse small air bubbles. Without surfactant, surface tension would collapse the alveoli in much the same way. Surfactant interacts with the water molecules, reducing surface tension and preventing the collapse of the alveoli.

If pneumocytes type II produce inadequate amounts of surfactant due to injury or genetic abnormalities, the alveoli collapse after each exhalation. Respiration then becomes difficult. With each breath, the inhalation must be forceful enough

Figure 23–8 Alveolar Organization and the Blood Air Barrier.

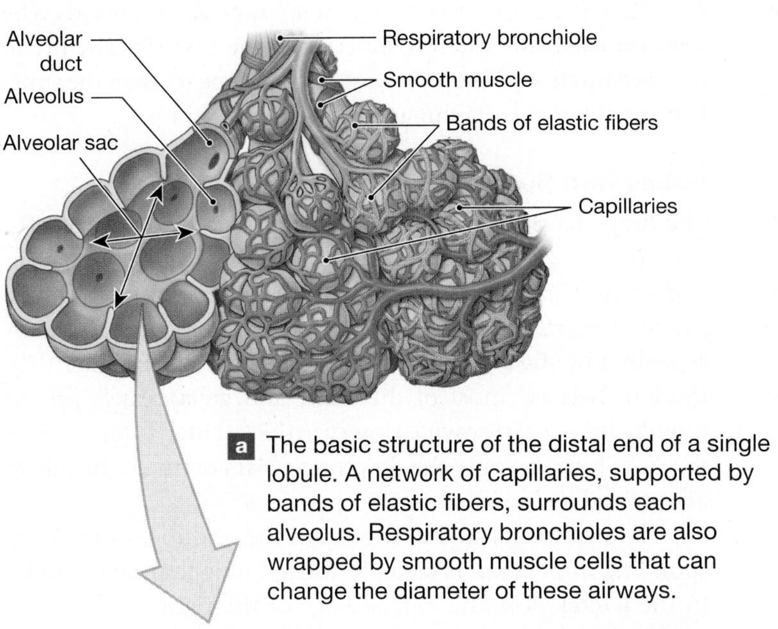
Alveolar duct
Alveolus
Alveolar sac
Respiratory bronchiole
Smooth muscle
Bands of elastic fibers
Capillaries

a The basic structure of the distal end of a single lobule. A network of capillaries, supported by bands of elastic fibers, surrounds each alveolus. Respiratory bronchioles are also wrapped by smooth muscle cells that can change the diameter of these airways.

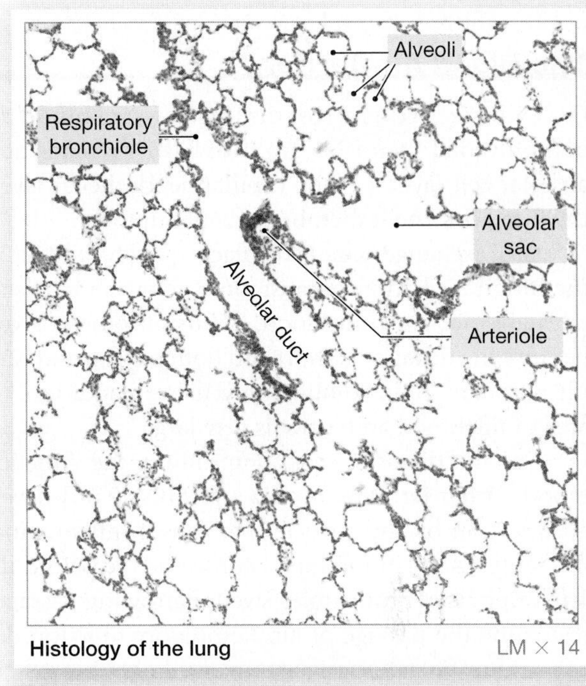
Alveoli
Respiratory bronchiole
Alveolar sac
Alveolar duct
Arteriole
Histology of the lung
LM × 14

b Low-power micrograph of lung tissue.

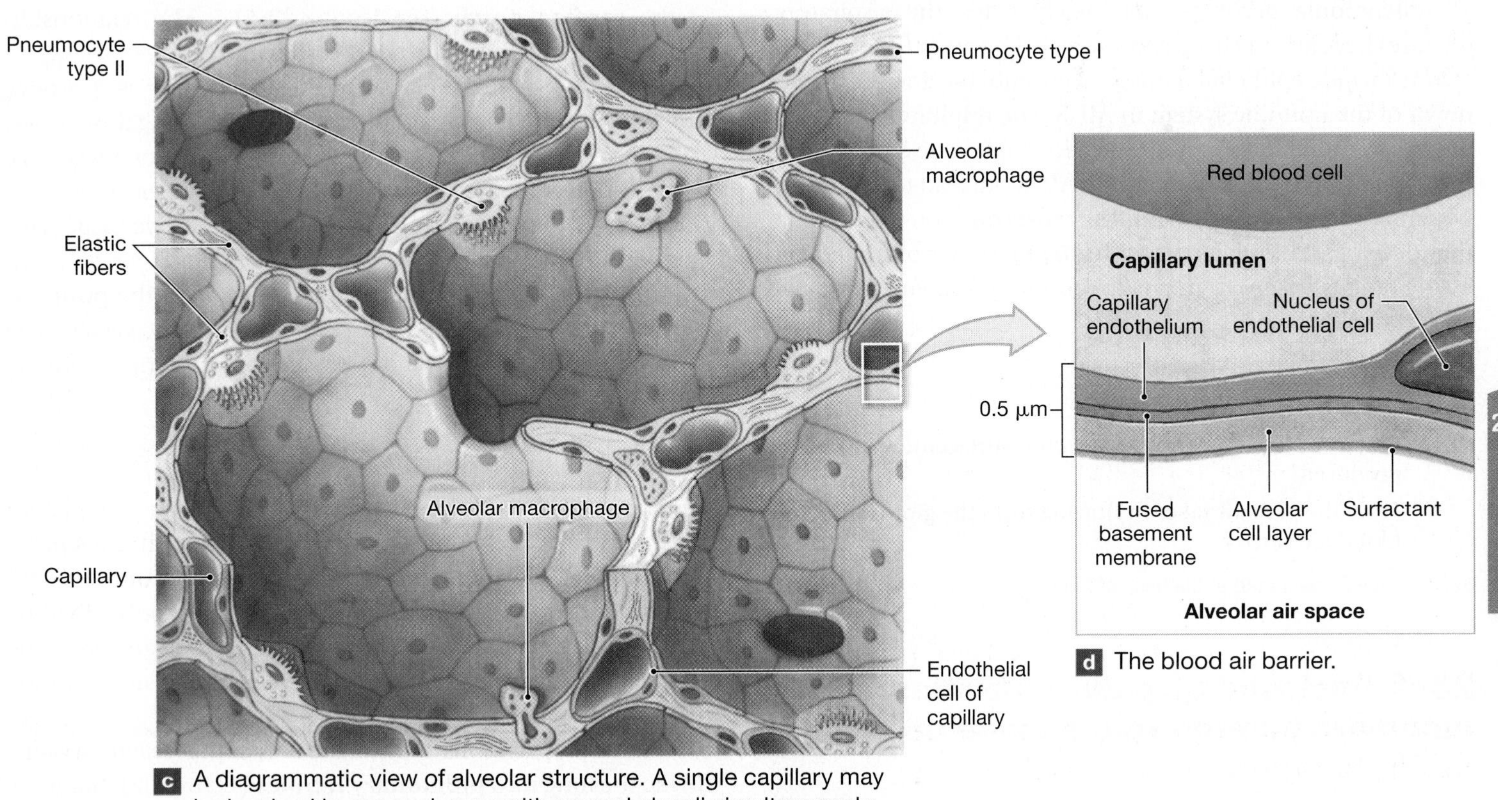
Pneumocyte type II
Pneumocyte type I
Alveolar macrophage
Elastic fibers
Alveolar macrophage
Capillary
Endothelial cell of capillary
Red blood cell
Capillary lumen
Capillary endothelium
Nucleus of endothelial cell
0.5 μm
Fused basement membrane
Alveolar cell layer
Surfactant
Alveolar air space

c A diagrammatic view of alveolar structure. A single capillary may be involved in gas exchange with several alveoli simultaneously.

d The blood air barrier.

? What type of muscle wraps around a respiratory bronchiole and can change the diameter of the airway?

to open the alveoli. A person without enough surfactant is soon exhausted by the effort of inflating and deflating the lungs. This condition is called *respiratory distress syndrome*.

The Blood Air Barrier

Gas exchange occurs across the three-layered **blood air barrier** of the alveoli (Figure 23–8d). This barrier is made up of (1) the alveolar cell layer, (2) the capillary endothelial layer, and (3) the fused basement membrane between them.

Gas exchange can take place quickly and efficiently at the blood air barrier because only a very short distance separates alveolar air from blood. Diffusion proceeds very rapidly because the distance is short and both oxygen and carbon dioxide are small, lipid-soluble molecules. In addition, the surface area of the blood air barrier is very large.

Certain diseases can compromise the function of the blood air barrier. **Pneumonia** (nū-MŌ-nē-uh) develops from an infection or any particle that causes inflammation within the lung. As inflammation occurs, fluids leak into the alveoli. The respiratory bronchioles swell, narrowing passageways and restricting the passage of air. Respiratory function deteriorates as a result. When bacteria are involved, they are generally types that normally inhabit the mouth and pharynx but have managed to evade the respiratory defenses.

Pneumonia becomes more likely when the respiratory defenses have already been compromised by other factors. Such factors include epithelial damage from smoking and the breakdown of the immune system in AIDS. The respiratory defenses of healthy individuals prevent infection and tissue damage, but the breakdown of those defenses in AIDS can lead to a massive, potentially fatal lung infection. The most common pneumonia that develops in individuals with AIDS results from infection by the fungus *Pneumocystis jiroveci* (formerly known as *Pneumocystis carinii*).

23

Checkpoint

11. **What would happen to the alveoli if surfactant were not produced?**
12. **Trace the path air takes in flowing from the glottis to the blood air barrier.**

See the blue Answers tab at the back of the book.

23-5 Enclosed by pleural cavities, the lungs are paired organs made up of multiple lobes

Learning Outcome **Describe the superficial anatomy of the lungs and the structure of a pulmonary lobule.**

The lungs contain both the bronchial tree and the respiratory portion of the lower respiratory system. The left and right **lungs** are surrounded by the left and right *pleural cavities*, respectively (Figure 23–9a). In this section we look at the anatomy of the lungs, including their blood supply.

Anatomy of the Lungs

Each lung is a blunt cone. Its tip, or apex, points superiorly. The apex on each side extends superior to the first rib. The broad concave inferior portion, or base, of each lung rests on the superior surface of the diaphragm.

Lobes and Surfaces of the Lungs

The lungs have distinct **lobes** that are separated by deep fissures (Figure 23–9b,c). The right lung has three lobes—*superior*, *middle*, and *inferior*—separated by the *horizontal* and *oblique fissures*. The left lung has only two lobes—*superior* and *inferior*—separated by the *oblique fissure*. The right lung is broader than the left, because most of the heart and great vessels project into the left thoracic cavity. However, the left lung is longer than the right lung, because the diaphragm rises on the right side to accommodate the liver.

The heart is located to the left of the midline, so its corresponding impression is larger in the left lung than in the right. In the lateral view, the medial edge of the right lung forms a vertical line, but the medial margin of the left lung is indented at the **cardiac notch** (see Figure 23–9b). The relationship between the lungs and the heart is shown in Figure 23–10.

Along the mediastinal surface of each lung is a groove called the **hilum**. Each main bronchus travels along this groove. The hilum also provides access for entry to pulmonary vessels, nerves, and lymphatics (see Figure 23–9c). The entire array is firmly anchored in a meshwork of dense connective tissue. This complex is the **root of the lung** (look back at Figure 23–6a). The root attaches to the mediastinum and fixes the positions of the major nerves, blood vessels, and lymphatic vessels. The roots of the lungs are anterior to vertebrae T_5 (right) and T_6 (left).

Pulmonary Lobules

The connective tissues of the root of each lung extend into the lung's *parenchyma*, or functional cells. These fibrous partitions, or *trabeculae*, contain elastic fibers, smooth muscles, and lymphatic vessels. The partitions branch repeatedly, dividing the lobes into ever-smaller compartments. The branches of the conducting passageways, pulmonary vessels, and nerves of the lungs follow these trabeculae.

The finest partitions, or **interlobular septa** (*septum*, a wall), divide the lung into **pulmonary lobules** (LOB-yūlz). Branches of the pulmonary arteries, pulmonary veins, and respiratory passageways supply each lobule. The connective tissues of the septa are, in turn, continuous with those of the *visceral pleura*, the serous membrane covering the lungs (see Figure 23–7).

Figure 23–9 The Gross Anatomy of the Lungs.
ATLAS: Plates 42–47

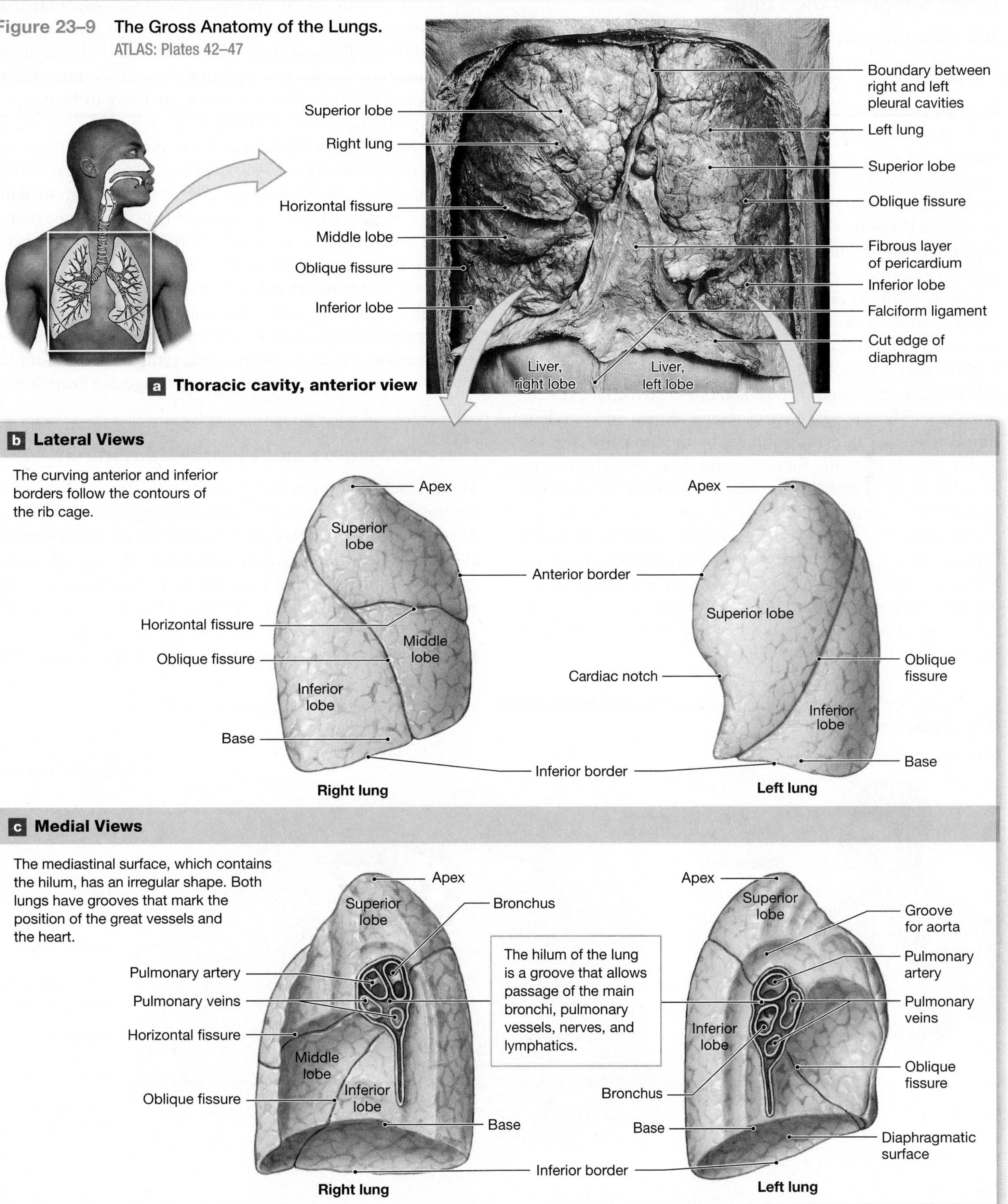
Superior lobe
Right lung
Horizontal fissure
Middle lobe
Oblique fissure
Inferior lobe
Boundary between right and left pleural cavities
Left lung
Superior lobe
Oblique fissure
Fibrous layer of pericardium
Inferior lobe
Falciform ligament
Cut edge of diaphragm
Liver, right lobe
Liver, left lobe
a Thoracic cavity, anterior view
b Lateral Views
The curving anterior and inferior borders follow the contours of the rib cage.
Apex
Superior lobe
Anterior border
Horizontal fissure
Oblique fissure
Middle lobe
Inferior lobe
Base
Inferior border
Right lung
Apex
Superior lobe
Cardiac notch
Oblique fissure
Inferior lobe
Base
Left lung
c Medial Views
The mediastinal surface, which contains the hilum, has an irregular shape. Both lungs have grooves that mark the position of the great vessels and the heart.
Apex
Superior lobe
Bronchus
Pulmonary artery
Pulmonary veins
Horizontal fissure
Middle lobe
Oblique fissure
Inferior lobe
Base
Inferior border
Right lung
The hilum of the lung is a groove that allows passage of the main bronchi, pulmonary vessels, nerves, and lymphatics.
Apex
Superior lobe
Groove for aorta
Pulmonary artery
Pulmonary veins
Inferior lobe
Oblique fissure
Bronchus
Base
Diaphragmatic surface
Left lung

? How many lobes does each lung have, and which lung has a cardiac notch?

23

Blood Supply to the Lungs

Two circuits nourish lung tissue. One supplies the *respiratory* portion of the lungs. The other perfuses the *conducting* portion.

The respiratory exchange surfaces receive blood from arteries of the pulmonary circuit. The pulmonary arteries carry deoxygenated blood. They enter the lungs at the hilum and branch with the bronchi as they approach the lobules. Each lobule receives an arteriole and a venule, and a network of capillaries surrounds each alveolus as part of the blood air barrier. Oxygen-rich blood from the alveolar capillaries passes through the pulmonary venules and then enters the pulmonary veins, which deliver the blood to the left atrium.

In addition to providing for gas exchange, the endothelial cells of the alveolar capillaries are the primary source of *angiotensin-converting enzyme (ACE)*, which converts circulating angiotensin I to angiotensin II. This enzyme plays an important role in regulating blood volume and blood pressure. ⮌ p. 750

The conducting passageways receive oxygen and nutrients from bronchial capillaries supplied by the bronchial arteries, which branch from the thoracic aorta. ⮌ p. 763 The venous blood from these bronchial capillaries may return to the heart by either the systemic circuit or pulmonary circuit. About one-eighth of the venous blood drains into bronchial veins, passes through the azygos, hemi-azygos, or highest intercostal vein, and enters the superior vena cava. The larger fraction of the venous blood enters anastomoses and drains into the pulmonary veins. This venous blood dilutes the oxygenated alveolar blood within the pulmonary veins.

Blood pressure in the pulmonary circuit is lower than that in the systemic circuit, so pulmonary vessels can easily become blocked by small blood clots, fat masses, or air bubbles in the pulmonary arteries. Because the lungs receive the entire cardiac output, any drifting objects in blood are likely to be trapped in the pulmonary arterial or capillary networks. Very small blood clots occasionally form in the venous system. These are usually trapped in the pulmonary (alveolar) capillary network, where they soon dissolve. Larger emboli are much more dangerous. Blocking a branch of a pulmonary artery stops blood flow to a group of lobules or alveoli. This condition is called a **pulmonary embolism**. If a pulmonary embolism is in place for several hours, the alveoli will permanently collapse. If the blockage occurs in a major pulmonary vessel rather than a minor branch, pulmonary resistance increases. The resistance places extra strain on the heart's right ventricle, which may be unable to maintain cardiac output, and congestive heart failure can result.

Pleural Cavities and Pleural Membranes

The thoracic cavity has the shape of a broad cone. Its walls are the rib cage, and the muscular diaphragm forms its floor. The two **pleural cavities** are separated by the mediastinum (Figure 23–10). Each lung is surrounded by a single pleural cavity, which is lined by a serous membrane called the **pleura** (PLOOR-uh; plural, *pleurae*). The pleura consists of two layers: the parietal pleura and the visceral pleura. The **parietal pleura** covers the inner surface of the thoracic wall

Figure 23–10 The Relationship between the Lungs and Heart. This transverse section was taken at the level of the cardiac notch.

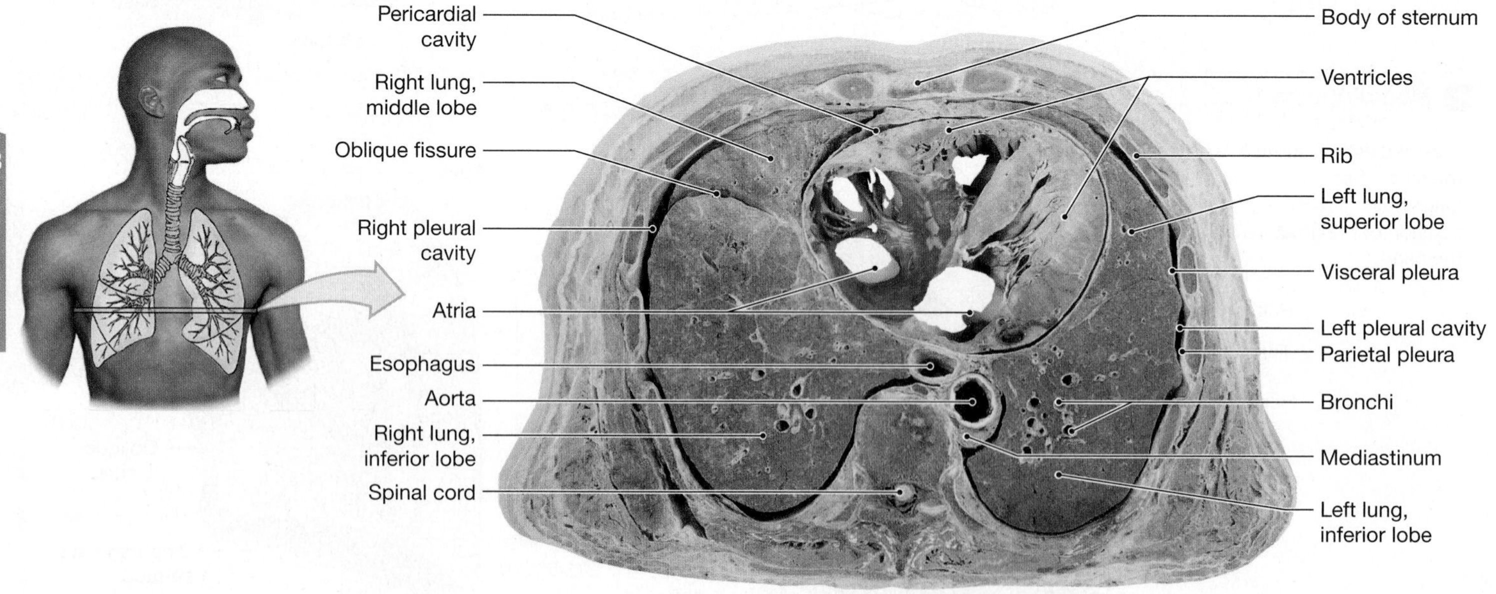
Pericardial cavity
Right lung, middle lobe
Oblique fissure
Right pleural cavity
Atria
Esophagus
Aorta
Right lung, inferior lobe
Spinal cord
Body of sternum
Ventricles
Rib
Left lung, superior lobe
Visceral pleura
Left pleural cavity
Parietal pleura
Bronchi
Mediastinum
Left lung, inferior lobe

23

and extends over the diaphragm and mediastinum. The **visceral pleura** covers the outer surfaces of the lungs, extending into the fissures between the lobes. Each pleural cavity actually represents a potential space rather than an open chamber, because the parietal and visceral pleurae are usually in close contact.

Both pleurae secrete a small amount of **pleural fluid**, a moist, slippery coating that lubricates. It reduces friction between the parietal and visceral surfaces as you breathe. Samples of pleural fluid, obtained through a long needle inserted between the ribs, are sometimes needed for diagnostic purposes. This sampling procedure is called *thoracentesis* (thōr-a-sen-TĒ-sis; *thora-*, thoracic + *centesis*, puncture). The extracted fluid is examined for bacteria, blood cells, or other abnormal components.

In some diseases, the normal coating of pleural fluid does not prevent friction between the pleural surfaces. The result is pain and pleural inflammation, a condition called *pleurisy*. When pleurisy develops, pleural fluid secretion may be excessive, or the inflamed pleurae may adhere to one another, limiting movement. In either case, breathing becomes difficult, and prompt medical attention is required.

Checkpoint

13. **Which arteries supply blood to the gas-exchange surfaces and conducting portions of the respiratory system?**
14. **List the functions of the pleura.**

See the blue Answers tab at the back of the book.

23-6 External respiration and internal respiration allow gas exchange within the body

Learning Outcome Define and compare the processes of external respiration and internal respiration.

The general term **respiration** includes two integrated processes: *external respiration* and *internal respiration*. The definitions of these terms vary among references. In our discussion, **external respiration** includes all the processes involved in the exchange of oxygen and carbon dioxide between the body's interstitial fluids and the external environment. The purpose of external respiration—and the primary function of the respiratory system—is to meet the respiratory demands of cells. **Internal respiration** is the absorption of oxygen and the release of carbon dioxide by those cells. We consider the biochemical pathways responsible for oxygen consumption and for the generation of carbon dioxide by mitochondria—pathways known collectively as *cellular respiration*—in Chapter 25.

Our discussion here focuses on three integrated steps in external respiration (Figure 23–11):

- *Pulmonary ventilation*, or breathing, which physically moves air into and out of the lungs
- *Gas diffusion* across the blood air barrier between alveolar air spaces and alveolar capillaries, and across capillary walls between blood and other tissues
- *Transport of oxygen and carbon dioxide* between alveolar capillaries and capillary beds in other tissues.

Figure 23–11 An Overview of the Key Steps in Respiration.

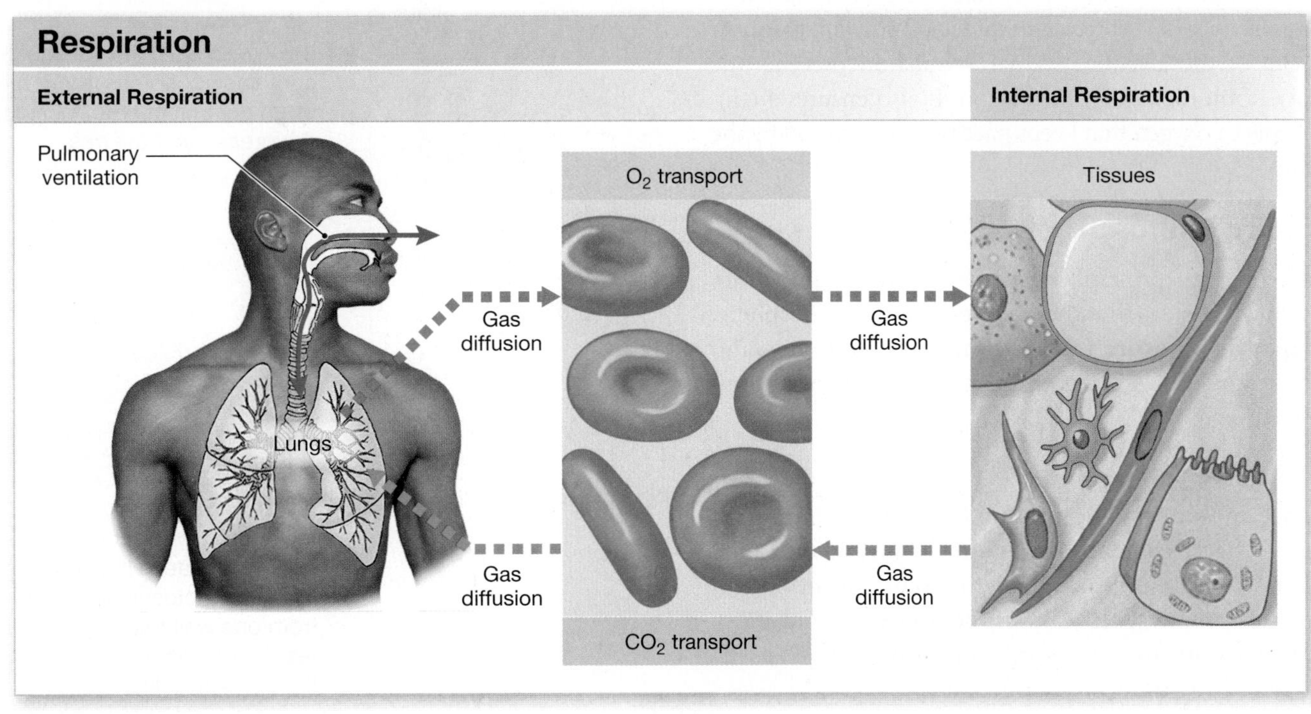
Respiration
External Respiration
Internal Respiration
Pulmonary ventilation
Lungs
Gas diffusion
Gas diffusion
O_2 transport
CO_2 transport
Gas diffusion
Gas diffusion
Tissues

Abnormalities affecting any of the steps involved in external respiration ultimately affect the concentration of gases in interstitial fluids, and thus cellular activities. If the oxygen level declines, the affected tissues will become starved for oxygen. **Hypoxia**, or a low tissue oxygen level, places severe limits on the metabolic activities of the affected area. For example, the effects of coronary ischemia result from chronic hypoxia affecting cardiac muscle cells. p. 700 If the oxygen supply is cut off completely, the condition called **anoxia** (an-OK-sē-uh; *an-*, without + *oxy-*, oxygen) results. Anoxia kills cells very quickly. Much of the damage from strokes and heart attacks results from local anoxia.

In the sections that follow, we examine each of the processes involved in external respiration in greater detail.

Checkpoint

15. **Define *external respiration* and *internal respiration*.**
16. **Name the integrated steps involved in external respiration.**

See the blue Answers tab at the back of the book.

23-7 Pulmonary ventilation—air exchange between the atmosphere and the lungs—involves muscle actions and volume changes that cause pressure changes

Learning Outcome **Summarize the physical principles controlling the movement of air into and out of the lungs, and describe the actions of the respiratory muscles.**

Pulmonary ventilation (breathing) is the physical movement of air into and out of the respiratory tract. Its primary function is to maintain adequate *alveolar ventilation*—movement of air into and out of the alveoli. Alveolar ventilation prevents the buildup of carbon dioxide in the alveoli. It also ensures a continuous supply of oxygen that keeps pace with absorption by the bloodstream.

An Introduction to Airflow

Some basic physical principles govern the movement of air. One of the most basic is that the weight of Earth's atmosphere, the **atmospheric pressure (atm)**, compresses our bodies and everything around us. Let's look at the role of atmospheric pressure on breathing.

Airflow and Pressure Gradients

The presence of atmospheric pressure has several important physiological effects. For example, air moves into and out of the respiratory tract due to pressure gradients, as the air pressure in the lungs cycles between below atmospheric pressure and above atmospheric pressure. In other words, air flows from an area of higher pressure to an area of lower pressure.

The Relationship between Gas Pressure and Volume (Boyle's Law)

The primary differences between liquids and gases reflect the interactions among individual molecules. The molecules in a liquid are in constant motion, but they are held closely together by weak interactions, such as the hydrogen bonding between adjacent water molecules. p. 36 Yet because the electrons of adjacent atoms tend to repel one another, liquids tend to resist compression. If you squeeze a balloon filled with water, it will change into a different shape, but the volumes of the two shapes will be the same.

In a gas, such as air, the molecules bounce around as independent objects. At normal atmospheric pressures, gas molecules are much farther apart than the molecules in a liquid. The forces acting between gas molecules are minimal because the molecules are too far apart for weak interactions to take place, so an applied pressure can push them closer together. Consider a sealed container of air at atmospheric pressure. The pressure exerted by the gas inside results from gas molecules bumping into the walls of the container. The greater the number of collisions, the higher the pressure.

You can change the gas pressure within a sealed container by changing the volume of the container, giving the gas molecules more or less room to bounce around. If you decrease the volume of the container, pressure increases (Figure 23–12a).

Figure 23–12 **The Relationship between Gas Pressure and Volume.**

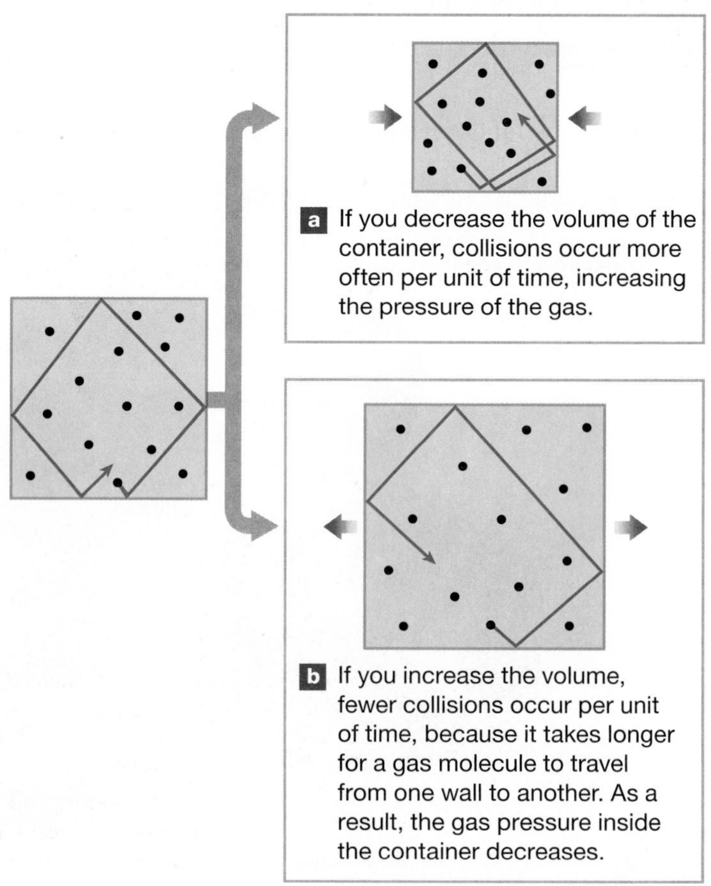

a If you decrease the volume of the container, collisions occur more often per unit of time, increasing the pressure of the gas.

b If you increase the volume, fewer collisions occur per unit of time, because it takes longer for a gas molecule to travel from one wall to another. As a result, the gas pressure inside the container decreases.

If you increase the volume of the container, pressure decreases (Figure 23–12b).

For a gas in a closed container and at a constant temperature, pressure (P) is inversely proportional to volume (V). That is, *if you decrease the volume of a gas, its pressure will increase. If you increase the volume of a gas, its pressure will decrease.* In particular, the relationship between pressure and volume is reciprocal: If you double the external pressure on a flexible container, its volume will drop by half, and if you reduce the external pressure by half, the volume of the container will double. This relationship, $P = 1/V$, is called **Boyle's law** because it was first recognized by Robert Boyle in the 1600s.

Overview of Pulmonary Ventilation: Volume Changes and Pressure Gradients

The tendency of air to flow in response to pressure gradients, plus the pressure–volume relationship of Boyle's law, provides the basis for pulmonary ventilation. A single **respiratory cycle** consists of an *inspiration*, or inhalation, and an *expiration*, or exhalation. Inhalation and exhalation involve changes in the volume of the lungs. These volume changes create pressure gradients that move air into or out of the respiratory tract.

Each lung is surrounded by a pleural cavity. The parietal and visceral pleurae are separated by only a thin film of pleural fluid. The two membranes can slide across one another, but they are held together by that fluid film. You can see the same principle when you set a wet glass on a smooth tabletop. You can slide the glass easily, but when you try to lift it, you feel considerable resistance. As you pull the glass away from the tabletop, you create a powerful suction. The only way to overcome it is to tilt the glass so that air is pulled between the glass and the table, breaking the fluid bond.

A comparable fluid bond exists between the parietal pleura and the visceral pleura covering the lungs. For this reason, the surface of each lung sticks to the inner wall of the chest and to the superior surface of the diaphragm. Movements of the diaphragm or rib cage that change the volume of the thoracic cavity also change the volume of the lungs (Spotlight Figure 23–13):

- The diaphragm forms the floor of the thoracic cavity. The relaxed diaphragm has the shape of a dome that projects superiorly into the thoracic cavity. When the diaphragm contracts, it tenses and moves inferiorly. This movement increases the volume of the thoracic cavity, decreasing the pressure within it. When the diaphragm relaxes, it returns to its original position, and the volume of the thoracic cavity decreases.
- Due to the nature of the articulations between the sternum and ribs and between the ribs and vertebrae, raising the rib cage involves both superior and anterior movements. These movements increase both the depth and width of the thoracic cavity, increasing its volume. Return of the rib cage to its original position reverses the process, decreasing the volume of the thoracic cavity.

At rest just prior to inhalation, pressures inside and outside the thoracic cavity are identical, and no air moves into or out of the lungs (see Spotlight Figure 23–13). When the thoracic cavity enlarges, the lungs expand to fill the additional space. This increase in volume decreases the pressure inside the lungs. Air then enters the respiratory passageways, because the pressure inside the lungs (P_{inside}) is lower than atmospheric pressure ($P_{outside}$). Air continues to enter the lungs until their volume stops increasing and the internal pressure is the same as that outside. When the thoracic cavity decreases in volume, pressures increase inside the lungs, forcing air out of the respiratory tract.

Actions of the Respiratory Muscles

The most important muscles involved in breathing are the *diaphragm* and the *external intercostals*. These muscles are the **primary respiratory muscles** and are active during normal breathing at rest. The **accessory respiratory muscles** become active when the depth and frequency of breathing must be greatly increased (Figure 23–14).

Muscles Used in Inhalation

Inhalation is an active process. It involves one or more of the following actions:

- Contraction of the diaphragm flattens the floor of the thoracic cavity, increasing its volume and drawing air into the lungs. Contraction of the diaphragm contributes about 75% of the volume of air in the lungs at rest.
- Contraction of the external intercostal muscles assists in inhalation by raising the ribs. This action contributes approximately 25% of the volume of air in the lungs at rest.
- Contraction of accessory muscles, including the sternocleidomastoid, scalenes, pectoralis minor, and serratus anterior, can assist the external intercostal muscles in elevating the ribs. These muscles increase the speed and amount of rib movement.

Muscles Used in Exhalation

Exhalation is either passive or active, depending on the level of respiratory activity. When exhalation is active, it may involve one or more of the following actions:

- The internal intercostal muscle and transversus thoracis depress the ribs. This action reduces the width and depth of the thoracic cavity.
- The abdominal muscles, including the external and internal oblique, transversus abdominis, and rectus abdominis, can assist the internal intercostal muscles in exhalation by compressing the abdomen. This action forces the diaphragm upward.

SPOTLIGHT

Figure 23–13

Pulmonary Ventilation

Pulmonary ventilation, or breathing, is the movement of air into and out of the respiratory system. It occurs by changing the volume of the lungs. Changes in lung volume take place through the contraction of skeletal muscles. The most important respiratory muscles are the diaphragm and the external intercostal muscles.

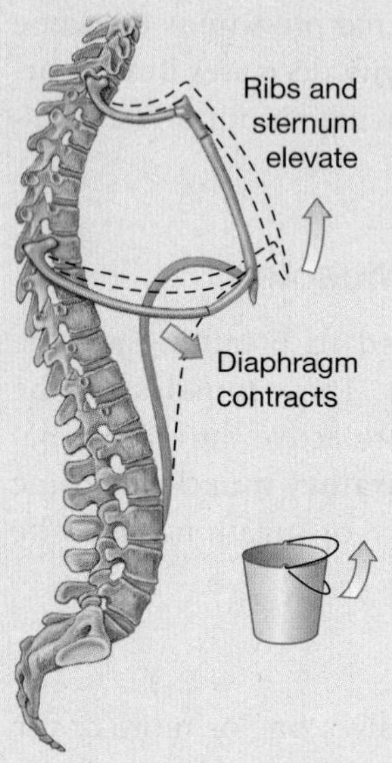

Ribs and sternum elevate
Diaphragm contracts

As the diaphragm is contracted or the rib cage (ribs and sternum) is elevated, the volume of the thoracic cavity increases and air moves into the lungs. The anterior movement of the ribs and sternum as they are elevated resembles the outward swing of a raised bucket handle.

AT REST

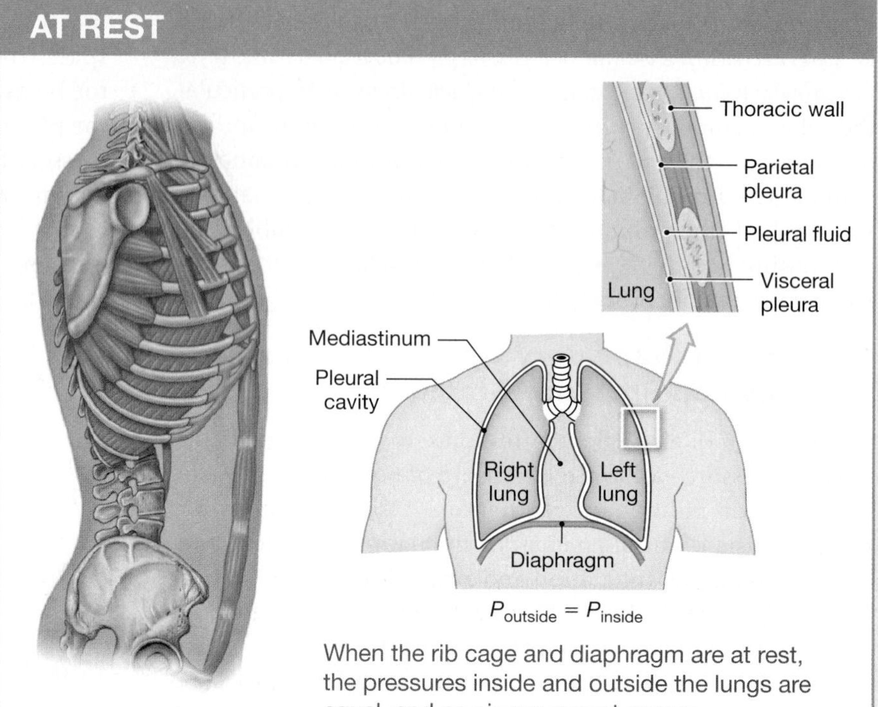

Thoracic wall
Parietal pleura
Pleural fluid
Visceral pleura
Lung
Mediastinum
Pleural cavity
Right lung
Left lung
Diaphragm

$P_{outside} = P_{inside}$

When the rib cage and diaphragm are at rest, the pressures inside and outside the lungs are equal, and no air movement occurs.

INHALATION

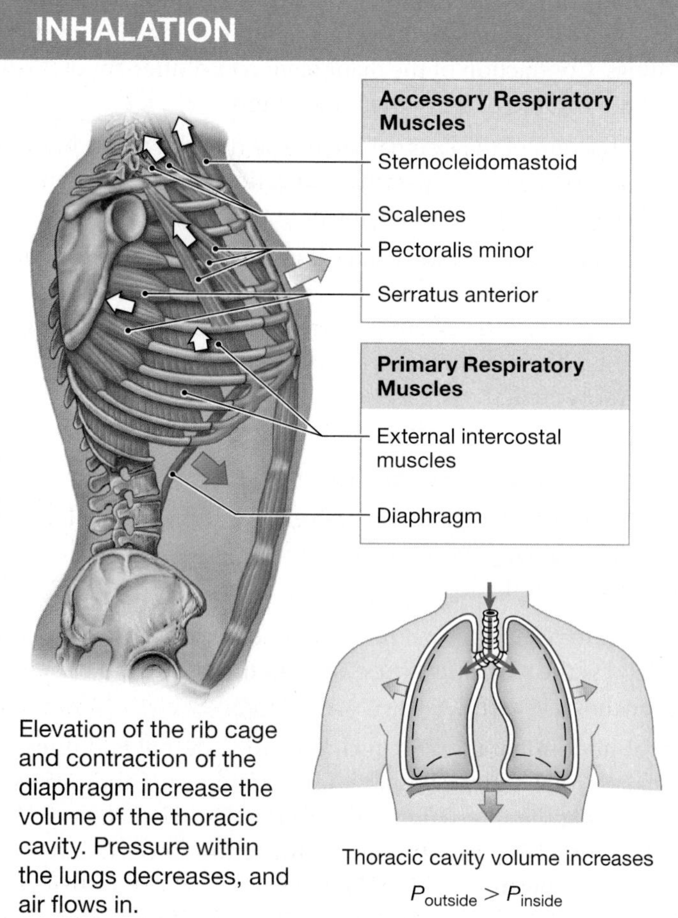

Accessory Respiratory Muscles
Sternocleidomastoid
Scalenes
Pectoralis minor
Serratus anterior
Primary Respiratory Muscles
External intercostal muscles
Diaphragm
Thoracic cavity volume increases

Elevation of the rib cage and contraction of the diaphragm increase the volume of the thoracic cavity. Pressure within the lungs decreases, and air flows in.

$P_{outside} > P_{inside}$

EXHALATION

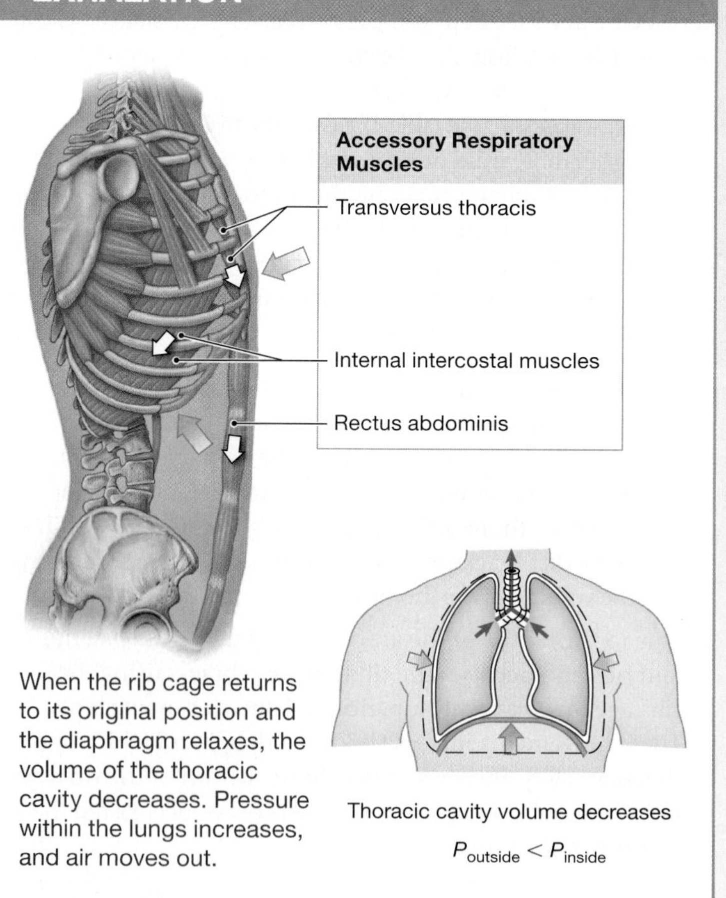

Accessory Respiratory Muscles
Transversus thoracis
Internal intercostal muscles
Rectus abdominis
Thoracic cavity volume decreases

When the rib cage returns to its original position and the diaphragm relaxes, the volume of the thoracic cavity decreases. Pressure within the lungs increases, and air moves out.

$P_{outside} < P_{inside}$

Figure 23–14 **Primary and Accessory Respiratory Muscles.**

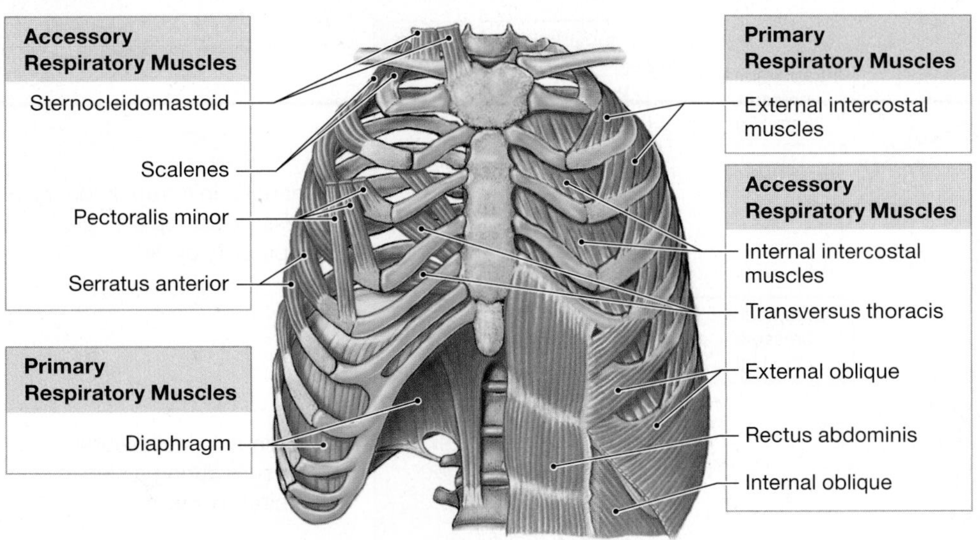
Accessory Respiratory Muscles
Sternocleidomastoid
Scalenes
Pectoralis minor
Serratus anterior
Primary Respiratory Muscles
Diaphragm
Primary Respiratory Muscles
External intercostal muscles
Accessory Respiratory Muscles
Internal intercostal muscles
Transversus thoracis
External oblique
Rectus abdominis
Internal oblique

? Name the primary muscles of respiration.

Volume Changes in Pulmonary Ventilation

During pulmonary ventilation, we use the respiratory muscles in various combinations, depending on the volume of air that must be moved into or out of the system. We usually classify respiratory movements as *quiet breathing* or *forced breathing* according to the pattern of muscle activity during a single respiratory cycle.

Volume Changes in Quiet Breathing

In **quiet breathing**, or **eupnea** (YŪP-nē-uh; *eu-*, normal + *-pnea*, breath), inhalation involves muscular contractions, but exhalation is a passive process. Inhalation usually involves contracting both the diaphragm and the external intercostal muscles. The relative contributions of these muscles can vary:

- During **diaphragmatic breathing**, or **deep breathing**, contraction of the diaphragm provides the necessary change in thoracic volume. Air is drawn into the lungs as the diaphragm contracts. Air is exhaled passively when the diaphragm relaxes.
- In **costal breathing**, or **shallow breathing**, the thoracic volume changes because the rib cage alters its shape. Inhalation takes place when contractions of the external intercostal muscles raise the ribs and enlarge the thoracic cavity. Exhalation takes place passively when these muscles relax.

During quiet breathing, expansion of the lungs stretches their elastic fibers. In addition, elevation of the rib cage stretches opposing skeletal muscles and elastic fibers in the connective tissues of the body wall. When the muscles of inhalation relax, these elastic components recoil, returning the diaphragm, the rib cage, or both to their original positions. This action is called **elastic rebound**.

We typically use diaphragmatic breathing at minimal levels of activity. As we need increased volumes of air, our inspiratory movements become larger and the contribution of rib movement increases. Even when we are at rest, costal breathing can predominate when abdominal pressures, fluids, or masses restrict diaphragmatic movements. For example, pregnant women rely more and more on costal breathing as the enlarging uterus pushes the abdominal organs against the diaphragm.

Volume Changes in Forced Breathing

Forced breathing, or **hyperpnea** (hī-PERP-nē-uh), involves active inspiratory and expiratory movements. In forced breathing, our accessory muscles assist with inhalation, and exhalation involves contraction of the internal intercostal muscles. At an absolute maximum level of forced breathing, our abdominal muscles take part in exhalation. Their contraction compresses the abdominal contents, pushing them up against the diaphragm. This action further reduces the volume of the thoracic cavity.

Pressure Gradients in Pulmonary Ventilation

Let's look more closely at the pressures inside and outside the respiratory tract. We know outside is made up only of atmospheric pressure, but the pressure inside the lungs has several different components. We use millimeters of mercury (mm Hg) to measure these pressures, as we did for blood pressure. Atmospheric pressure is also measured in *atmospheres*. At sea level, 1 atmosphere (atm) of pressure is equivalent to 760 mm Hg.

Intrapulmonary Pressure

The direction of airflow is determined by the relationship between atmospheric pressure and intrapulmonary pressure. **Intrapulmonary** (in-tra-PUL-mo-nār-ē) **pressure**, or **intra-alveolar** (in-tra-al-VĒ-ō-lar) **pressure**, is the pressure inside the respiratory tract, at the alveoli.

When you are relaxed and breathing quietly, the difference between atmospheric pressure and intrapulmonary pressure is relatively small. On inhalation, your lungs expand, and the intrapulmonary pressure drops to about 759 mm Hg. Because the intrapulmonary pressure is 1 mm Hg below atmospheric pressure, it is generally reported as −1 mm Hg. On exhalation, your lungs recoil, and intrapulmonary pressure rises to 761 mm Hg, or +1 mm Hg (Figure 23–15a).

The size of the pressure gradient increases when you breathe heavily. When a trained athlete breathes at maximum

FIGURE 23–15 **Pressure and Volume Changes during Inhalation and Exhalation.** One sequence of inhalation and exhalation makes up a single respiratory cycle.

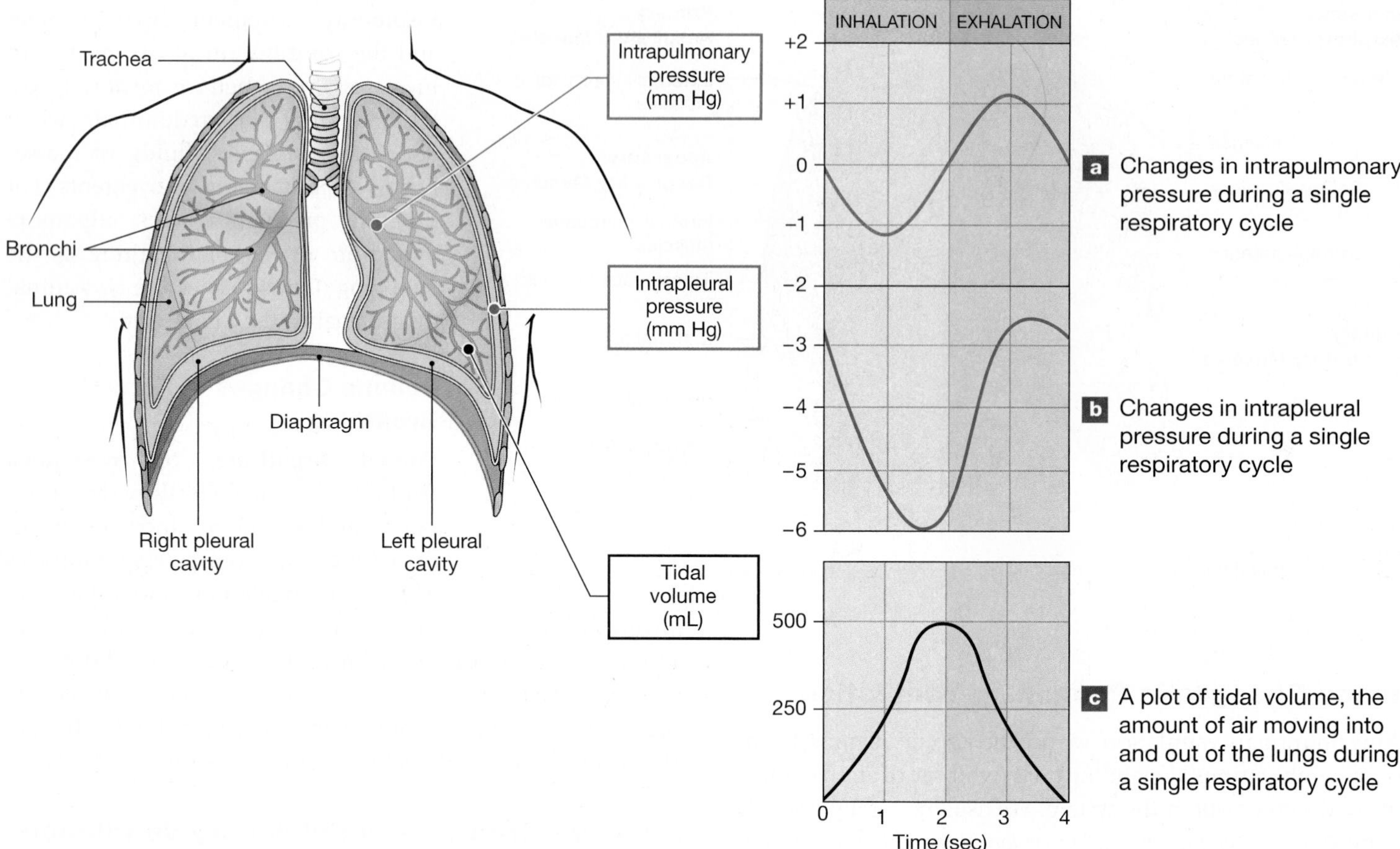
Trachea
Bronchi
Lung
Diaphragm
Right pleural cavity
Left pleural cavity
Intrapulmonary pressure (mm Hg)
Intrapleural pressure (mm Hg)
Tidal volume (mL)
INHALATION
EXHALATION
+2
+1
0
−1
−2
−3
−4
−5
−6
500
250
0
1
2
3
4
Time (sec)
a Changes in intrapulmonary pressure during a single respiratory cycle
b Changes in intrapleural pressure during a single respiratory cycle
c A plot of tidal volume, the amount of air moving into and out of the lungs during a single respiratory cycle

capacity, the pressure differentials can reach –30 mm Hg during inhalation and +100 mm Hg if the individual is straining with the glottis kept closed. This is one reason you are told to exhale while lifting weights. Exhaling keeps your intrapulmonary pressures and peritoneal pressure from increasing so high that an alveolar rupture or hernia could occur.

Intrapleural Pressure

Intrapleural pressure is the pressure in the pleural cavity, between the parietal and visceral pleurae. Intrapleural pressure averages about −4 mm Hg (Figure 23–15b). During a powerful inhalation, it can reach –18 mm Hg. This pressure is below atmospheric pressure, due to the relationship between the lungs and the body wall.

Earlier, we noted that the lungs are highly elastic. In fact, they would collapse to about 5% of their normal resting volume if their elastic fibers could recoil completely. The elastic fibers cannot recoil so much, however. The reason is that they are not strong enough to overcome the fluid bond between the parietal and visceral pleurae. The elastic fibers continuously oppose that fluid bond and pull the lungs away from the chest wall and diaphragm, lowering the intrapleural pressure slightly. The elastic fibers remain stretched even after a full exhalation. For this reason, intrapleural pressure remains below atmospheric pressure throughout normal cycles of inhalation and exhalation. The cyclical changes in the intrapleural pressure create the *respiratory pump* that assists the venous return to the heart. p. 741

+ Clinical Note Pneumothorax

Air can enter the pleural cavity due to an injury to the chest wall that penetrates the parietal pleura, or a rupture of the alveoli that breaks through the visceral pleura. This condition, called **pneumothorax** (nū-mō-THOR-aks; *pneumo*-, air), breaks the fluid bond between the pleurae and allows the elastic fibers to recoil, resulting in a "collapsed lung," or **atelectasis** (at-e-LEK-ta-sis; *ateles*, incomplete + *ektasis*, extension). The other lung is not affected because it is enclosed by a separate pleura. The treatment for a collapsed lung involves putting a needle into the cavity and drawing out the air. This treatment lowers the intrapleural pressure and reinflates the lung.

Summary of Volume Changes and Pressure Gradients during a Respiratory Cycle

As we have just seen, you move air into and out of the respiratory system by changing the volume of the lungs. Those changes alter the pressure relationships, producing air movement. The curves in Figure 23–15a,b show the intrapulmonary and intrapleural pressures during a single respiratory cycle of an individual at rest. The graph in Figure 23–15c plots the *tidal volume*, the amount of air you move into or out of your lungs during a single respiratory cycle.

Tips & Tools

Tidal volume floods and ebbs like the ocean *tides*.

At the start of the respiratory cycle, the intrapulmonary and atmospheric pressures are equal, and no air is moving. Inhalation begins when the thoracic cavity expands, causing a fall of intrapleural pressure. This pressure gradually falls to approximately –6 mm Hg. Over this period, intrapulmonary pressure drops to just under –1 mm Hg. It then begins to rise as air flows into the lungs.

When exhalation begins, intrapleural and intrapulmonary pressures rise rapidly, forcing air out of the lungs. At the end of exhalation, air again stops moving when the difference between intrapulmonary and atmospheric pressures has been eliminated. The amount of air moved into the lungs during inhalation, the tidal volume, equals the amount moved out of the lungs during exhalation.

Physical Factors Affecting Pulmonary Ventilation

Bronchodilation and bronchoconstriction are ways of adjusting the **resistance** to airflow. These actions direct airflow toward or away from specific portions of the respiratory exchange surfaces.

Recall that pneumocytes type II secrete surfactant onto the alveolar surfaces to reduce surface tension. Surface tension is the reason small air bubbles in water collapse. Without surfactant, alveoli collapse in much the same way. The thin surface layer of surfactant interacts with the water molecules, keeping the alveoli open.

Tips & Tools

The term ***surfactant*** is derived from its purpose as a ***surf***ace ***act***ive ***a***ge***nt***.

The **compliance** of the lungs is a measure of their expandability, or how easily the lungs expand in response to applied pressure. The lower the compliance, the greater the force required to fill the lungs. The greater the compliance, the easier it is to fill the lungs. Factors affecting compliance include the following:

- *The Connective Tissue of the Lungs.* The loss of supporting tissues due to alveolar damage, as in *emphysema*, increases compliance. This loss results in the alveoli increasing beyond their normal size, making the lungs easier to fill.
- *The Level of Surfactant Production.* On exhalation, the collapse of alveoli due to inadequate surfactant, as in respiratory distress syndrome, reduces compliance.
- *The Mobility of the Thoracic Cage.* Arthritis or other skeletal disorders that affect the articulations of the ribs or spinal column also reduce compliance.

At rest, the muscular activity involved in pulmonary ventilation accounts for 3–5% of the resting energy demand. If compliance is reduced, that figure climbs dramatically. An individual may become exhausted simply trying to continue breathing.

Measuring Respiratory Rates and Volumes

The respiratory system is extremely adaptable. You can be breathing slowly and quietly one moment, rapidly and deeply the next. The respiratory system adapts to meet the oxygen demands of the body by varying both the number of breaths per minute and the amount of air moved per breath. When you exercise at peak levels, the amount of air moving into and out of the respiratory tract can be 50 times the amount moved at rest.

Respiratory Rate

Your **respiratory rate** is the number of breaths you take each minute. As you read this, you are probably breathing quietly, with a low respiratory rate. The normal respiratory rate of a resting adult ranges from 12 to 18 breaths each minute, roughly one for every four heartbeats. Children breathe more rapidly, at rates of about 18–20 breaths per minute.

Respiratory Volumes

Recall that the **tidal volume**, symbolized V_T, is the amount of air moved into or out of the lungs during a single respiratory cycle. Let's look at the volumes that make up the tidal volume.

23

The Respiratory Minute Volume ($\dot{V}_E$). We can calculate the amount of air moved each minute, symbolized $\dot{V}_E$, by multiplying the respiratory rate f by the tidal volume V_T. This value is called the **respiratory minute volume**. The respiratory rate at rest averages 12 breaths per minute, and the tidal volume at rest averages around 500 mL per breath. On that basis, we calculate the respiratory minute volume as follows:

$$\begin{aligned}\dot{V}_E &= f \times V_T \\ \begin{pmatrix}\text{Volume of air moved}\\ \text{each minute}\end{pmatrix} &= \begin{pmatrix}\text{breaths per}\\ \text{minute}\end{pmatrix} \times \begin{pmatrix}\text{tidal}\\ \text{volume}\end{pmatrix} \\ &= 12 \text{ per minute} \times 500 \text{ mL} \\ &= 6000 \text{ mL per minute} \\ &= 6.0 \text{ liters per minute}\end{aligned}$$

In other words, the respiratory minute volume at rest is approximately 6 liters per minute.

Alveolar Ventilation (V_A). The respiratory minute volume measures pulmonary ventilation. It provides an indication of how much air is moving into and out of the respiratory tract. However, only some of the inhaled air reaches the alveolar exchange surfaces. A typical inhalation pulls about 500 mL of air into the respiratory system. The first 350 mL of inhaled air travels along the conducting passageways and enters the alveolar spaces. The last 150 mL of inhaled air never gets that far. It stays in the conducting passageways and thus does not participate in gas exchange with blood. This "lagging" volume of air in the conducting passages is known as the **anatomic dead space**, denoted $\mathbf{V_D}$. To calculate the anatomic dead space, multiply the tidal volume by 0.3. The example calculation below shows that with a tidal volume of 500 mL, anatomic dead space equals 150 mL.

$$V_D = V_T \times 0.3$$

$$V_D = 500 \text{ mL} \times 0.3 = 150 \text{ mL}$$

Alveolar ventilation, symbolized $\dot{V}_A$, is the amount of air reaching the alveoli each minute. The alveolar ventilation is less than the respiratory minute volume because some of the air never reaches the alveoli; it instead remains in the dead space of the lungs. We can calculate alveolar ventilation by subtracting the dead space from the tidal volume:

$$\dot{V}_A = f \times (V_T - V_D)$$

$$\begin{pmatrix}\text{Alveolar}\\ \text{ventilation}\end{pmatrix} = \begin{pmatrix}\text{breaths}\\ \text{per minute}\end{pmatrix} \times \begin{pmatrix}\text{tidal} & - & \text{anatomic}\\ \text{volume} & & \text{dead space}\end{pmatrix}$$

At rest, alveolar ventilation rates are approximately 4.2 liters per minute (12 per minute $\times$ 350 mL). However, the gas arriving in the alveoli is significantly different from the surrounding atmosphere. The reason is that inhaled air always mixes with "used" air in the conducting passageways on its way in. The air in alveoli thus contains less oxygen and more carbon dioxide than does atmospheric air.

Relationships among Tidal Volume (V_T), ($\dot{V}_E$), and ($\dot{V}_A$). The respiratory minute volume can be increased by increasing either the tidal volume or the respiratory rate. In other words, you can breathe more deeply or more quickly or both. Under maximum stimulation, the tidal volume can increase to roughly 4800 mL, or 4.8 liters. At peak respiratory rates of 40–50 breaths per minute and maximum cycles of inhalation and exhalation, the respiratory minute volume can approach 200 liters per minute.

In functional terms, the alveolar ventilation rate is more important than the respiratory minute volume, because it determines the rate of delivery of oxygen to the alveoli. The respiratory rate and the tidal volume together determine the alveolar ventilation rate:

- For a given respiratory rate, increasing the tidal volume (breathing more deeply) increases the alveolar ventilation rate.
- For a given tidal volume, increasing the respiratory rate (breathing more quickly) increases the alveolar ventilation rate.

The alveolar ventilation rate can change independently of the respiratory minute volume. In our previous example, the respiratory minute volume at rest was 6 liters and the alveolar ventilation rate was 4.2 L/min. If the respiratory rate rises to 20 breaths per minute, but the tidal volume drops to 300 mL, the respiratory minute volume remains the same ($20 \times 300 = 6000$). However, the alveolar ventilation rate drops to only 3 L/min ($20 \times [300 - 150] = 3000$). For this reason, whenever the demand for oxygen increases, both the tidal volume *and* the respiratory rate must be regulated closely. (We focus on the mechanisms involved in a later section.)

Respiratory Performance and Volume Relationships

Pulmonary function tests monitor several aspects of respiratory function by measuring rates and volumes of air movement. We exchange only a small proportion of the air in our lungs during a single quiet respiratory cycle. We can increase the tidal volume by inhaling more vigorously and exhaling more completely.

We can divide the total volume of the lungs into a series of *volumes* and *capacities* (each the sum of various volumes), as indicated in Figure 23–16. These measurements are obtained by an instrument called a **spirometer**. Spirometry values are useful in diagnosing problems with pulmonary ventilation. Adult females, on average, have smaller bodies and thus smaller lung volumes than do males. As a result, there are sex-related differences in respiratory volumes and capacities. The table in the figure shows representative values for both sexes.

Pulmonary volumes include the following:

- The **tidal volume (V_T)**, which we have already covered, is the amount of air you move into or out of your lungs during a single respiratory cycle. The tidal volume averages about 500 mL in both males and females.
- The **expiratory reserve volume (ERV)** is the amount of air that you can voluntarily expel after you have completed a normal, quiet respiratory cycle. As an example, with maximum use of the accessory muscles, males can expel an additional 1000 mL of air, on average. Female expiratory reserve volume averages 700 mL.
- The **residual volume** is the amount of air that remains in your lungs even after a maximal exhalation—typically about 1200 mL in males and 1100 mL in females. The **minimal volume**, a component of the residual volume, is the amount of air that would remain in your lungs if they were allowed to collapse. The minimal volume ranges from

Figure 23–16 Pulmonary Volumes and Capacities. The red line indicates the volume of air within the lungs during breathing.

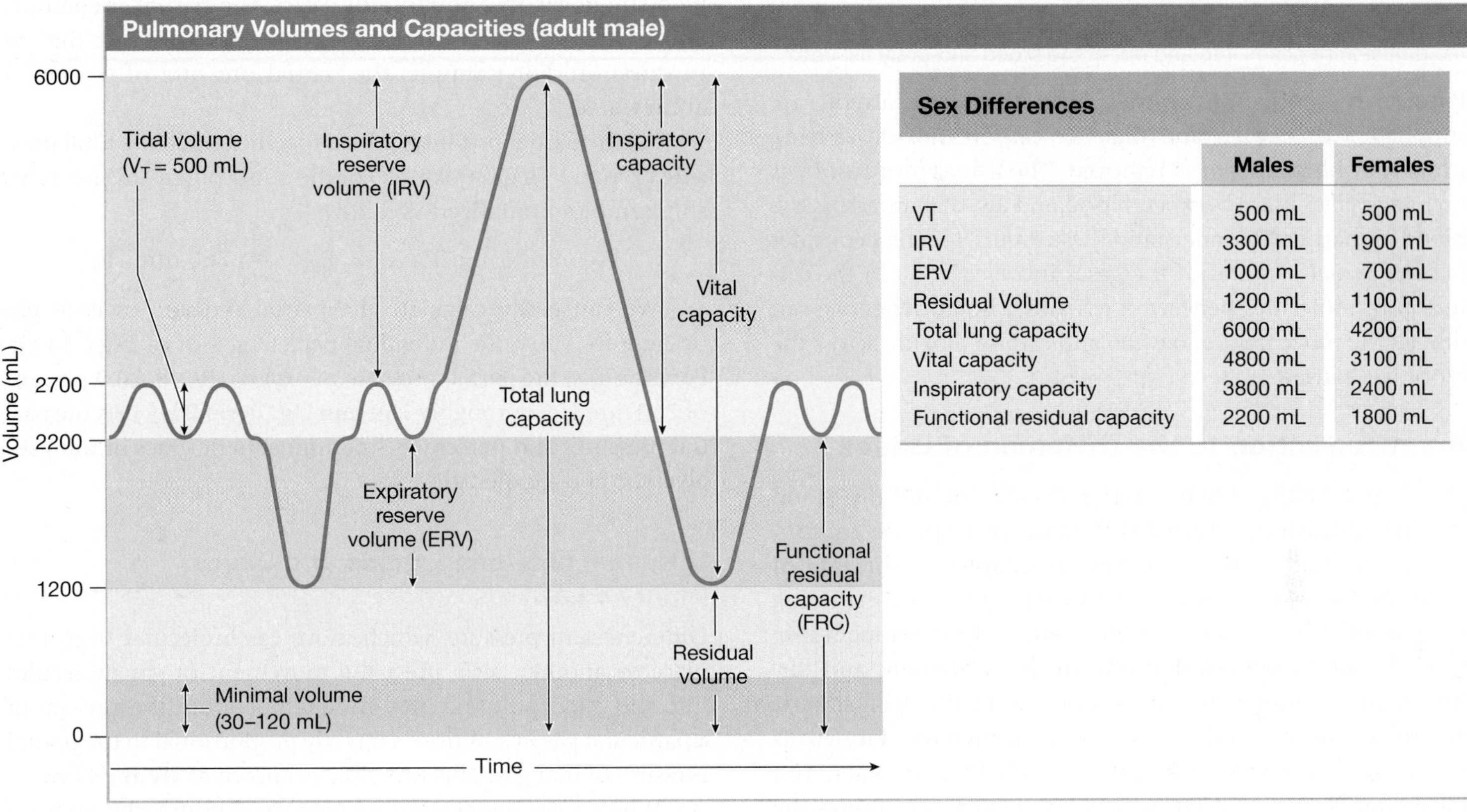
Pulmonary Volumes and Capacities (adult male)
Volume (mL)
6000
2700
2200
1200
0
Tidal volume
(V_T = 500 mL)
Inspiratory
reserve
volume (IRV)
Inspiratory
capacity
Vital
capacity
Total lung
capacity
Expiratory
reserve
volume (ERV)
Functional
residual
capacity
(FRC)
Residual
volume
Minimal volume
(30–120 mL)
Time
Sex Differences

	Males	Females
VT	500 mL	500 mL
IRV	3300 mL	1900 mL
ERV	1000 mL	700 mL
Residual Volume	1200 mL	1100 mL
Total lung capacity	6000 mL	4200 mL
Vital capacity	4800 mL	3100 mL
Inspiratory capacity	3800 mL	2400 mL
Functional residual capacity	2200 mL	1800 mL

? The inspiratory capacity is a total of what two pulmonary volumes?

30 to 120 mL. Unlike other volumes, it cannot be measured in a healthy person. The minimal volume and the residual volume are very different, because the fluid bond between the lungs and the chest wall normally prevents the recoil of the elastic fibers in the lungs. Some air remains in the lungs, even at minimal volume, because the surfactant coating the alveolar surfaces prevents their collapse.

- The **inspiratory reserve volume (IRV)** is the amount of air that you can take in over and above the tidal volume. On average, the lungs of males are larger than those of females. For this reason, the inspiratory reserve volume of males averages 3300 mL, compared with 1900 mL in females.

We can calculate respiratory capacities by adding the values of various volumes. Examples include the following:

- The **inspiratory capacity** is the amount of air that you can draw into your lungs after you have completed a quiet respiratory cycle. The inspiratory capacity is the sum of the tidal volume and the inspiratory reserve volume.
- The **functional residual capacity (FRC)** is the amount of air remaining in your lungs after you have completed a quiet respiratory cycle. The FRC is the sum of the expiratory reserve volume and the residual volume.
- The **vital capacity** is the maximum amount of air that you can move into or out of your lungs in a single respiratory cycle. The vital capacity is the sum of the expiratory reserve volume, the tidal volume, and the inspiratory reserve volume. It averages around 4800 mL in males and 3400 mL in females.
- The **total lung capacity** is the total volume of your lungs. We calculate it by adding the vital capacity and the residual volume. The total lung capacity averages around 6000 mL in males and 4200 mL in females.

✓ Checkpoint

17. **Define *compliance* and identify the factors that affect it.**
18. **Name the various measurable pulmonary volumes.**
19. **Mark breaks a rib that punctures the chest wall on his left side. What is likely to happen to his left lung as a result?**
20. **In pneumonia, fluid accumulates in the alveoli of the lungs. How would this accumulation affect vital capacity?**

See the blue Answers tab at the back of the book.

23-8 Gas exchange depends on the partial pressures of gases and the diffusion of gas molecules

Learning Outcome Summarize the physical principles governing the diffusion of gases into and out of the blood and body tissues.

Pulmonary ventilation ensures both that your alveoli are supplied with oxygen and that the carbon dioxide arriving from your bloodstream is removed. The actual process of gas exchange takes place between blood and alveolar air across the blood air barrier. To understand these events, let's first consider (1) the *partial pressures* of the gases involved and (2) the diffusion of molecules between a gas and a liquid. Then we can discuss the movement of oxygen and carbon dioxide across the blood air barrier.

An Introduction to the Diffusion of Gases

Gases are exchanged between the alveolar air and the blood through diffusion, which takes place in response to concentration gradients. As we saw in Chapter 3, the rate of diffusion varies in response to a variety of factors, including the size of the concentration gradient and the temperature. ⮌ p. 91 The principles that govern the movement and diffusion of gas molecules, such as those in the atmosphere, are relatively straightforward. These principles, known as *gas laws*, have been understood for about 250 years. You have already heard about Boyle's law, which determines the direction of air movement in pulmonary ventilation. In this section, you will learn about other gas laws and factors that determine the rate of oxygen and carbon dioxide diffusion across the blood air barrier.

Partial Pressures (Dalton's Law)

The air we breathe is a mixture of gases. Nitrogen molecules (N_2) are the most abundant, accounting for about 78.6% of atmospheric gas molecules. Oxygen molecules (O_2), the second most abundant, make up roughly 20.9% of air. Most of the remaining 0.5% consists of water molecules, with carbon dioxide (CO_2) contributing a mere 0.04%.

23

Atmospheric pressure, 760 mm Hg, represents the combined effects of collisions involving each type of molecule in air. At any moment, 78.6% of those collisions involve nitrogen molecules, 20.9% oxygen molecules, and so on. Thus, each of the gases contributes to the total pressure in proportion to its relative abundance. This relationship is known as **Dalton's law**.

The **partial pressure** of a gas is the pressure contributed by a single gas in a mixture of gases. The partial pressure is abbreviated by the symbol P or p. A subscript shows the gas in question. For example, the partial pressure of oxygen is abbreviated P_{O_2}.

All the partial pressures added together equal the total pressure exerted by a gas mixture. For the atmosphere, this relationship can be summarized as follows:

$$P_{N_2} + P_{O_2} + P_{H_2O} + P_{CO_2} = 760 \text{ mm Hg}$$

We can easily calculate the partial pressure of each gas because we know the individual percentages of each gas in air. For example, the partial pressure of oxygen, P_{O_2}, is 20.9 percent of 760 mm Hg, or roughly 159 mm Hg. Table 23–1 lists the partial pressures and percentages of atmospheric gases in inhaled, alveolar, and exhaled air.

Diffusion between Liquids and Gases (Henry's Law)

Differences in pressure, which move gas molecules from one place to another, also affect the movement of gas molecules into and out of solution. At a given temperature, the amount of a particular gas in solution is directly proportional to the partial pressure of that gas. This principle is known as **Henry's law**.

When a gas under pressure contacts a liquid, the pressure tends to force gas molecules into solution. At a given pressure, the number of dissolved gas molecules rises until an equilibrium is established. At equilibrium, gas molecules diffuse out of the liquid as quickly as they enter it, so the total number of gas molecules in solution remains constant. If the partial pressure increases, more gas molecules go into solution. If the partial pressure decreases, gas molecules come out of solution.

You see Henry's law in action whenever you open a can of soda. The soda was put into the can under pressure, and the gas (carbon dioxide) is in solution (Figure 23–17a). When you open the can, the pressure decreases and the gas molecules begin coming out of solution (Figure 23–17b). Theoretically, the process will continue until an equilibrium develops between the surrounding air and the gas in solution. In fact, the volume of the can is so small, and the volume of the atmosphere so great, that within a half hour or so virtually all the carbon dioxide comes out of solution. You are left with "flat" soda.

Table 23–1 Partial Pressures (mm Hg) and Normal Gas Concentrations (%) in Air

Source of Sample	Nitrogen (N_2)	Oxygen (O_2)	Carbon Dioxide (CO_2)	Water Vapor (H_2O)
Inhaled air (dry)	597 (78.6%)	159 (20.9%)	0.3 (0.04%)	3.7 (0.5%)
Alveolar air (saturated)	573 (75.4%)	100 (13.2%)	40 (5.2%)	47 (6.2%)
Exhaled air (saturated)	569 (74.8%)	116 (15.3%)	28 (3.7%)	47 (6.2%)

Figure 23–17 **Henry's Law and the Relationship between Solubility and Pressure.**

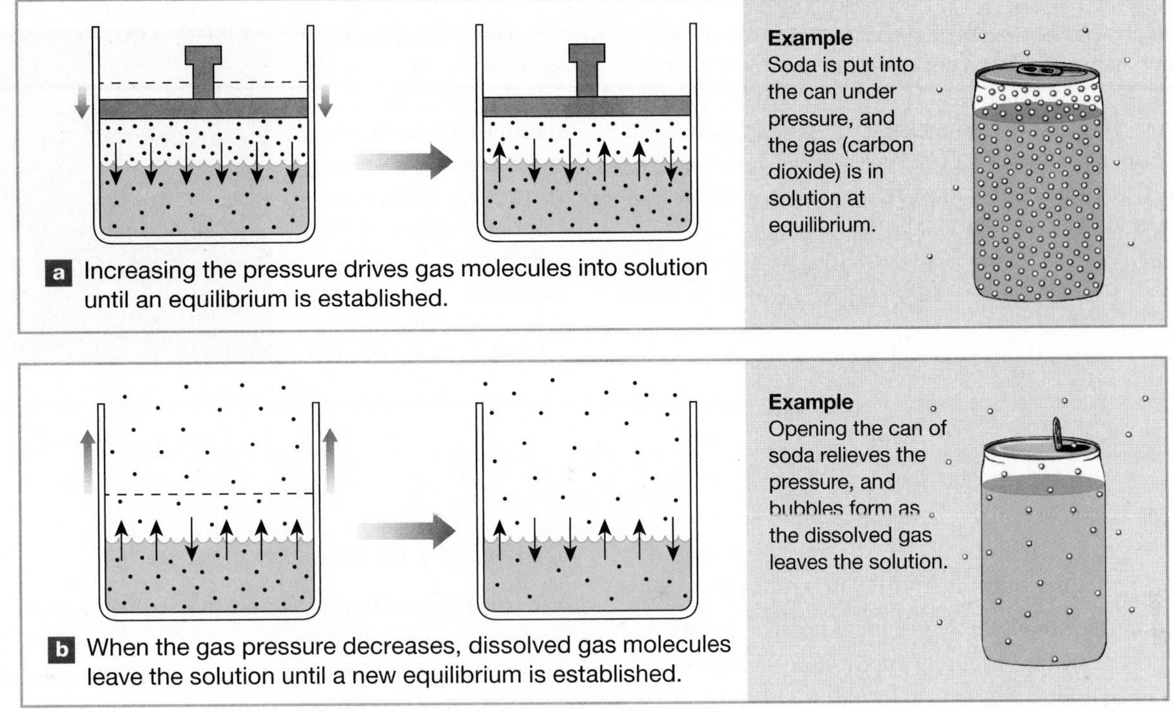
Example
Soda is put into the can under pressure, and the gas (carbon dioxide) is in solution at equilibrium.
a Increasing the pressure drives gas molecules into solution until an equilibrium is established.
Example
Opening the can of soda relieves the pressure, and bubbles form as the dissolved gas leaves the solution.
b When the gas pressure decreases, dissolved gas molecules leave the solution until a new equilibrium is established.

Solubility of Gases

The actual *amount* of a gas in solution at a given partial pressure and temperature depends on the solubility of the gas in that particular liquid. In body fluids, carbon dioxide is highly soluble, and oxygen is somewhat less soluble. Nitrogen has very limited solubility. The dissolved gas content is usually reported in milliliters of gas per 100 mL (1 dL) of solution. To see the differences in relative solubility of these three gases, we can compare the gas content of blood in the pulmonary veins with the partial pressure of each gas in the alveoli. In a pulmonary vein, plasma generally contains 2.62 mL/dL of dissolved CO_2 (P_{CO_2} = 40 mm Hg), 0.29 mL/dL of dissolved O_2 (P_{O_2} = 100 mm Hg), and 1.25 mL/dL of dissolved N_2 (P_{N_2} = 573 mm Hg). Thus even though the partial pressure of carbon dioxide is less than one-tenth the partial pressure of nitrogen, the plasma contains more than twice as much carbon dioxide as nitrogen in solution.

Diffusion of Gases across the Blood Air Barrier

Now, as part of our discussion of external respiration, let's consider how differing partial pressures and solubilities determine the direction and rate of gas diffusion in the lungs. This section covers the diffusion of oxygen and carbon dioxide across the blood air barrier that separates the air within the alveoli from the blood in alveolar capillaries.

The Composition of Alveolar Air

As we have noted, as soon as air enters the respiratory tract, its characteristics begin to change, being warmed, filtered, and humidified. When incoming air reaches the alveoli, it mixes with air remaining in the alveoli from the previous respiratory cycle. For this reason, alveolar air contains more carbon dioxide and less oxygen than does atmospheric air.

The last 150 mL of inhaled air never gets farther than the conducting passageways. It remains in the anatomic dead space of the lungs. During the next exhalation, the departing alveolar air mixes with air in the dead space, producing yet another mixture that differs from both atmospheric and alveolar samples. As mentioned, the differences in composition between atmospheric (inhaled) and alveolar air are given in **Table 23–1**.

Efficiency of Diffusion at the Blood Air Barrier

Gas exchange at the blood air barrier is efficient for the following reasons:

- *The differences in partial pressure across the blood air barrier are substantial.* This fact is important, because the greater the difference in partial pressure, the faster the rate of gas diffusion. Conversely, if P_{O_2} in alveoli decreases, the rate of oxygen diffusion into blood decreases. This is why many people feel light-headed at altitudes of 3000 m or more. The partial

+ Clinical Note Decompression Sickness

Decompression sickness is a painful condition that develops when a person is exposed to a sudden drop in atmospheric pressure. This condition is caused by nitrogen due to its high partial pressure in air. When the pressure drops, nitrogen comes out of solution, forming bubbles like those in a shaken can of soda. The bubbles may form in joint cavities, in the bloodstream, and in the cerebrospinal fluid. Individuals with decompression sickness typically curl up from the pain in affected joints. This reaction accounts for the condition's common name: *the bends*.

Decompression sickness most commonly affects scuba divers who return to the surface too quickly after breathing air under greater-than-normal pressure while submerged. It can also develop in airline passengers subject to sudden losses of cabin pressure.

pressure of oxygen in their alveoli has dropped so low that the rate of oxygen absorption is significantly reduced.

- *The distances involved in gas exchange are short.* The fusion of capillary and alveolar basement membranes reduces the distance for gas exchange to as little as 0.1 µm, with an average of 0.5 µm. Inflammation of the lung tissue or a buildup of fluid in alveoli increases the diffusion distance and impairs alveolar gas exchange.
- *The gases are lipid soluble.* Both oxygen and carbon dioxide diffuse readily through the surfactant layer and the alveolar and endothelial plasma membranes.
- *The total surface area is large.* To meet the metabolic requirements of peripheral tissues, the surface area for gas exchange in the lungs must be very large. Estimates of the surface area involved in gas exchange range from 70 m^2 to 140 m^2 (753 ft^2 to 1506 ft^2). It is about 35 times the surface area of your body, or slightly bigger than half a tennis court. Damage to alveolar surfaces, which occurs in emphysema, reduces the available surface area and the efficiency of gas transfer.
- *Blood flow and airflow are coordinated.* This close coordination makes both pulmonary ventilation and pulmonary circulation more efficient. For example, blood flow is greatest around alveoli with the highest P_{O_2} values, where oxygen uptake can proceed with maximum efficiency. This matching of alveolar ventilation with pulmonary blood perfusion is called the *ventilation-to-perfusion ratio* (or *V/Q ratio*). This ratio is the amount of air reaching the alveoli to the amount of blood reaching the alveoli. If the normal blood flow is impaired (as it is in *pulmonary embolism*), or if the normal airflow is interrupted (as it is in various forms of *pulmonary obstruction*), this coordination is lost and respiratory efficiency decreases.

Summary of Gas Exchange

We cover gas exchange as part of the processes of both external and internal respiration. Figure 23–18 illustrates gas exchange due to the partial pressures of oxygen and carbon dioxide in the pulmonary and systemic circuits.

External Respiration

Notice that blood arriving from the pulmonary arteries has a lower P_{O_2} and a higher P_{CO_2} than does alveolar air (see Figure 23–18a). Diffusion between the alveolar gas mixture and the alveolar capillaries then increases the P_{O_2} of blood and decreases its P_{CO_2}. By the time the blood enters the pulmonary venules, it has reached equilibrium with the alveolar air. Blood departs the alveoli with a P_{O_2} of about 100 mm Hg and a P_{CO_2} of roughly 40 mm Hg.

+ Clinical Note Blood Gas Analysis

Blood samples can be analyzed to determine their concentrations of dissolved gases. The usual tests include the determination of pH, P_{CO_2}, and P_{O_2} in an arterial sample. Such samples provide information about the degree of oxygenation in peripheral tissues. For example, if the arterial P_{CO_2} is very high and the P_{O_2} is very low, tissues are not receiving adequate oxygen. This problem may be solved by giving the patient a gas mixture that has a high P_{O_2} (or even pure oxygen, with a P_{O_2} of 760 mm Hg).

Blood gas measurements also provide information on the efficiency of gas exchange at the lungs. If the arterial P_{O_2} remains low despite the administration of oxygen, or if the P_{CO_2} is very high, pulmonary exchange problems may exist. Such problems may include pulmonary edema, asthma, or pneumonia.

Figure 23–18 A Summary of Respiratory Processes and Partial Pressures in Respiration.

SmartArt

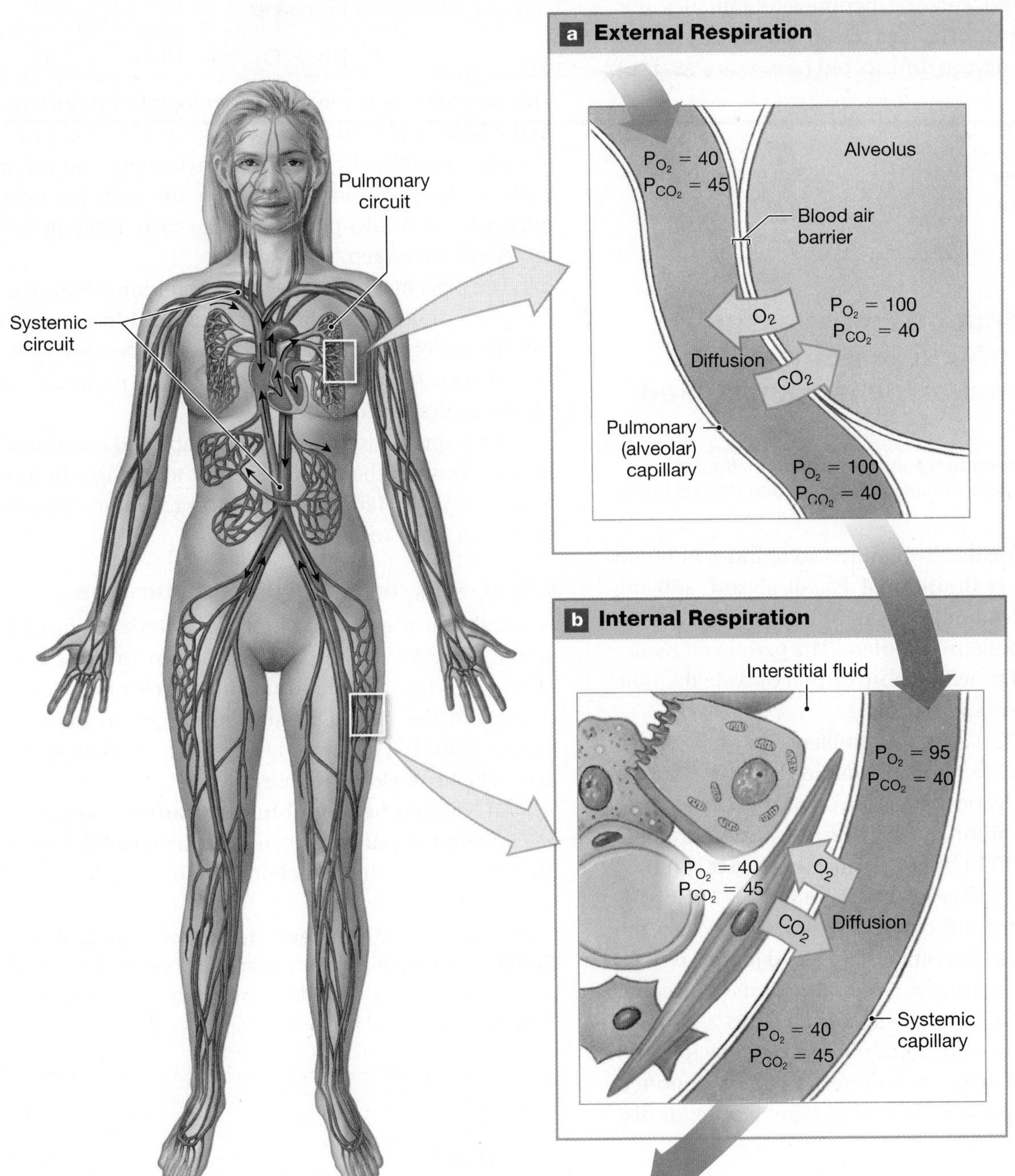
Pulmonary circuit
Systemic circuit
a External Respiration
P_{O_2} = 40
P_{CO_2} = 45
Alveolus
Blood air barrier
O_2
P_{O_2} = 100
P_{CO_2} = 40
Diffusion
CO_2
Pulmonary (alveolar) capillary
P_{O_2} = 100
P_{CO_2} = 40
b Internal Respiration
Interstitial fluid
P_{O_2} = 95
P_{CO_2} = 40
P_{O_2} = 40
P_{CO_2} = 45
O_2
CO_2
Diffusion
P_{O_2} = 40
P_{CO_2} = 45
Systemic capillary

Diffusion between alveolar air and blood in the alveolar capillaries occurs very rapidly. When you are at rest, a red blood cell moves through one of your alveolar capillaries in about 0.75 second. When you exercise, the passage takes less than 0.3 second. This amount of time is usually sufficient to reach an equilibrium between the alveolar air and the blood.

Internal Respiration

Oxygenated blood leaves the pulmonary (alveolar) capillaries to return to the heart, to be pumped into the systemic circuit. As this blood enters the pulmonary veins, it mixes with blood that flowed through capillaries around conducting passageways. Because gas exchange in the lungs occurs only at alveoli, the blood leaving the conducting passageways carries relatively little oxygen. The partial pressure of oxygen in the pulmonary veins therefore drops to about 95 mm Hg. This is the P_{O_2} in the blood that enters the systemic circuit. No further changes in partial pressure occur until the blood reaches the systemic capillaries (Figure 23–18b).

Normal interstitial fluid has a P_{O_2} of 40 mm Hg. As a result, oxygen diffuses out of the systemic capillaries until the capillary partial pressure is the same as that in the adjacent tissues.

Inactive peripheral tissues normally have a P_{CO_2} of about 45 mm Hg, whereas blood entering peripheral capillaries normally has a P_{CO_2} of 40 mm Hg. As a result, carbon dioxide diffuses into the blood as oxygen diffuses out (see Figure 23–18b).

✓ Checkpoint

21. Define *Dalton's law.*
22. Define *Henry's law.*

See the blue Answers tab at the back of the book.

23-9 In gas transport, most oxygen is transported bound to hemoglobin, whereas carbon dioxide is transported in three ways

Learning Outcome Describe the structure and function of hemoglobin, and the transport of oxygen and carbon dioxide in the blood.

Oxygen and carbon dioxide have limited solubilities in blood plasma. For example, at the normal P_{O_2} of alveoli, 100 mL of plasma absorbs only about 0.3 mL of oxygen. The limited solubilities of these gases are a problem. The peripheral tissues need more oxygen and generate more carbon dioxide than the plasma alone can absorb and transport.

Red blood cells (RBCs) solve this problem. They remove dissolved O_2 and CO_2 molecules from plasma and bind them (in the case of oxygen) or mostly use them to manufacture soluble compounds (in the case of carbon dioxide). As the reactions in RBCs remove dissolved gases from blood plasma, these gases continue to diffuse into the blood, but never reach equilibrium. These reactions are both temporary and completely reversible. When the oxygen or carbon dioxide concentration in blood plasma is high, RBCs remove the excess molecules. When the plasma concentrations are falling, the RBCs release their stored reserves.

In this section we cover the transport of oxygen by RBCs, as well as the transport of carbon dioxide. CO_2 can be transported in three ways: as carbonic acid, bound to hemoglobin, or dissolved in plasma.

Oxygen Transport

Each 100 mL of blood leaving the alveolar capillaries carries away about 20 mL of oxygen. Only 1.5% (about 0.3 mL) of this amount is oxygen molecules in solution. The rest of the oxygen molecules are bound to *hemoglobin (Hb) molecules*. This section covers how hemoglobin carries oxygen in the blood and the factors that affect this process.

Transport of O_2 on Hemoglobin

Recall that the hemoglobin molecule consists of four globular protein subunits, each containing a heme unit. The oxygen molecules bind to the iron ions in the center of heme units. p. 663 Thus, each hemoglobin molecule can bind four molecules of oxygen, forming **oxyhemoglobin (HbO_2)**. We can summarize this process as

$$Hb + O_2 \rightleftharpoons HbO_2$$

This reversible reaction allows hemoglobin to pick up oxygen in the lungs and then release it to body tissues elsewhere.

Each red blood cell has approximately 280 million molecules of hemoglobin. With four heme units per hemoglobin molecule, each RBC potentially can carry more than a billion molecules of oxygen.

The percentage of heme units containing bound oxygen at any given moment is called the **hemoglobin saturation**. If all the Hb molecules in the blood are fully loaded with oxygen, saturation is 100%. If, on average, each Hb molecule carries two O_2 molecules, saturation is 50%.

In Chapter 2, we saw that the shape and functional properties of a protein change in response to changes in its environment. p. 55 Hemoglobin is no exception. Any changes in shape can affect oxygen binding.

Effect of P_{O_2} on Hemoglobin Saturation

Normally, the most important factor affecting hemoglobin saturation is the P_{O_2} of blood. Hemoglobin in RBCs carries most of the oxygen in the bloodstream and releases it in response to changes in the partial pressure of oxygen in the surrounding plasma. If the P_{O_2} increases, hemoglobin binds oxygen. If the P_{O_2} decreases, hemoglobin releases oxygen.

An **oxygen-hemoglobin saturation curve**, or *oxygen-hemoglobin dissociation curve*, is a graph that relates the hemoglobin saturation to the partial pressure of oxygen (Figure 23–19).

Figure 23–19 An Oxygen-Hemoglobin Saturation Curve. The saturation characteristics of hemoglobin at various partial pressures of oxygen under normal conditions (body temperature of 37°C and blood pH of 7.4).

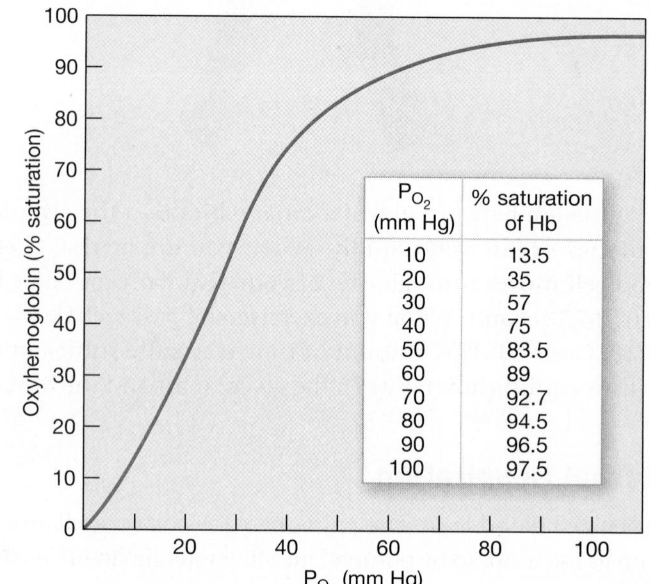
Oxyhemoglobin (% saturation)
0
10
20
30
40
50
60
70
80
90
100
20
40
60
80
100
P_{O_2} (mm Hg)

P_{O_2} (mm Hg)	% saturation of Hb
10	13.5
20	35
30	57
40	75
50	83.5
60	89
70	92.7
80	94.5
90	96.5
100	97.5

The binding and dissociation, or release, of oxygen to hemoglobin is a typical reversible reaction. At equilibrium, oxygen molecules bind to heme at the same rate at which other oxygen molecules are being released. If the P_{O_2} increases, then more oxygen molecules bind to hemoglobin, and fewer are released. Referring to the equation given in the previous section, we say that the reaction shifts to the right. If the P_{O_2} decreases, more oxygen molecules are released from hemoglobin, while fewer oxygen molecules bind. In this case, we say that the reaction shifts to the left.

Notice that the graph of this relationship between P_{O_2} and hemoglobin saturation is a curve rather than a straight line. It is a curve because the shape of the hemoglobin molecule changes slightly each time it binds an oxygen molecule. This change in shape enhances the ability of hemoglobin to bind *another* oxygen molecule. In other words, the binding of the first oxygen molecule makes it easier to bind the second; binding the second promotes binding of the third; and binding the third enhances binding of the fourth.

Because each arriving oxygen molecule makes it easier for hemoglobin to bind the *next* oxygen molecule, the saturation curve takes the form shown in Figure 23–19. Once the first oxygen molecule binds to the hemoglobin, the slope rises rapidly until it levels off near 100% saturation. Where the slope is steep, a very small change in plasma P_{O_2} results in a large change in the amount of oxygen bound to Hb or released from HbO_2. Notice that hemoglobin will be more than 90% saturated if exposed to an alveolar P_{O_2} above 60 mm Hg. Thus, near-normal oxygen transport can continue even when the oxygen content of alveolar air decreases below normal, or P_{O_2} = 100 mm Hg. Without this ability, you could not survive at high altitudes. Conditions that significantly reduce pulmonary ventilation would be immediately fatal.

At normal alveolar pressures (P_{O_2} = 100 mm Hg), the hemoglobin saturation is very high (97.5%), but complete saturation does not occur until the P_{O_2} reaches an excessively high level (about 250 mm Hg). In functional terms, the maximum saturation is not as important as the ability of hemoglobin to provide oxygen over the normal P_{O_2} range in body tissues. Over that range—from 100 mm Hg at the alveoli to perhaps 15 mm Hg in active tissues—the saturation drops from 97.5% to less than 20%, and a small change in P_{O_2} makes a big difference in terms of the amount of oxygen bound to hemoglobin.

Note that the relationship between P_{O_2} and hemoglobin saturation remains valid whether the P_{O_2} is rising or falling. If the P_{O_2} increases, the saturation goes up and hemoglobin stores oxygen. If the P_{O_2} decreases, hemoglobin releases oxygen into its surroundings. When oxygenated blood arrives in the peripheral capillaries, the blood P_{O_2} declines rapidly due to gas exchange with the interstitial fluid. As the P_{O_2} falls, hemoglobin gives up its oxygen.

The relationship between the P_{O_2} and hemoglobin saturation provides a mechanism for automatic regulation of oxygen delivery. Inactive tissues have little demand for oxygen, and the local P_{O_2} is usually about 40 mm Hg. Under these conditions, hemoglobin does not release much oxygen. As it passes through the capillaries, it goes from 97% saturation (P_{O_2} = 95 mm Hg) to 75% saturation (P_{O_2} = 40 mm Hg). Because hemoglobin still retains three-quarters of its oxygen, venous blood has a relatively large oxygen reserve. This reserve is important, because it can be mobilized if the tissue oxygen demand increases.

Active tissues use oxygen at an accelerated rate, so their P_{O_2} may drop to 15–20 mm Hg. Hemoglobin passing through these capillaries goes from 97% saturation to about 20% saturation. In practical terms, this means that as blood circulates through peripheral capillaries, active tissues receive 3.5 times as much oxygen as do inactive tissues.

Effect of Other Factors on Hemoglobin Saturation

At least three other factors may affect the oxygen saturation rate of hemoglobin. These factors are (1) blood pH, (2) temperature, and (3) the products of ongoing metabolic activity within RBCs.

Hemoglobin and pH. Hemoglobin helps to buffer acids by binding hydrogen ions (H^+). At a given P_{O_2}, hemoglobin releases additional oxygen if the blood pH decreases. The oxygen–hemoglobin saturation curve in Figure 23–19 was determined in normal blood, with a pH of 7.4 and a temperature of 37°C. In addition to consuming oxygen, active tissues generate acids that lower the pH of the interstitial fluid. When the pH decreases, the shape of hemoglobin molecules changes. As a result of this change, the molecules release their oxygen reserves more readily, so the slope of the hemoglobin saturation curve changes (Figure 23–20a). In other words, as pH decreases, the saturation declines. At a tissue P_{O_2} of 40 mm Hg, for example, you can see in Figure 23–20a that a pH decrease from 7.4 to 7.2 changes hemoglobin saturation from 75% to 60%. This means that hemoglobin molecules release 20% more oxygen in peripheral tissues at a pH of 7.2 than they do at a pH of 7.4. This effect of pH on the hemoglobin saturation curve is called the **Bohr effect**.

Carbon dioxide is the primary compound responsible for the Bohr effect. When CO_2 diffuses into the blood, it rapidly diffuses into red blood cells. There, an enzyme called **carbonic anhydrase** catalyzes the reaction of CO_2 with water molecules:

$$CO_2 + H_2O \xrightleftharpoons{\text{carbonic anhydrase}} H_2CO_3 \rightleftharpoons H^+ + HCO_3^-$$

The product of this enzymatic reaction, H_2CO_3, is called *carbonic acid*, because it dissociates into a hydrogen ion (H^+) and a bicarbonate ion (HCO_3^-). The rate of carbonic acid formation depends on the amount of carbon dioxide in solution, which, as noted earlier, depends on the P_{CO_2}. When the P_{CO_2} rises, the rate of carbonic acid formation accelerates and the reaction proceeds from left to right. The hydrogen ions that are generated diffuse out of the RBCs, and the pH of the plasma decreases. When the P_{CO_2} declines, hydrogen ions diffuse out of

Figure 23–20 The Effects of Blood pH and Temperature on Hemoglobin Saturation.

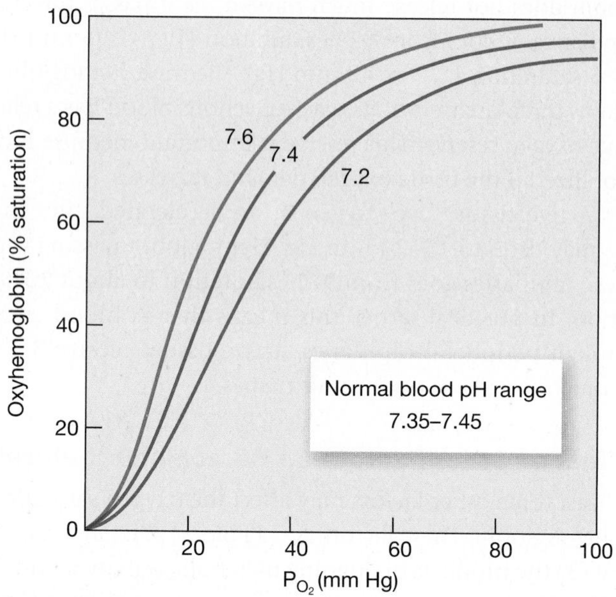
100
80
60
40
20
0
Oxyhemoglobin (% saturation)
7.6
7.4
7.2
Normal blood pH range
7.35–7.45
20
40
60
80
100
P_{O_2} (mm Hg)

a **Effect of pH.** When the blood pH decreases below the normal range, more oxygen is released; the oxygen-hemoglobin saturation curve shifts to the right. When the pH increases, less oxygen is released; the curve shifts to the left.

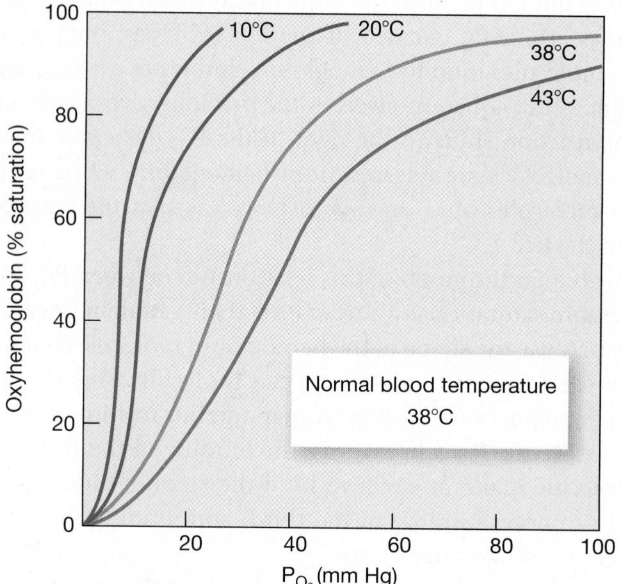
100
80
60
40
20
0
Oxyhemoglobin (% saturation)
10°C
20°C
38°C
43°C
Normal blood temperature
38°C
20
40
60
80
100
P_{O_2} (mm Hg)

b **Effect of temperature.** When the temperature increases, more oxygen is released; the oxygen-hemoglobin saturation curve shifts to the right. When the temperature decreases, the curve shifts to the left.

? According to these graphs, more oxygen is released when the pH is (higher or lower) than normal and when temperature is (higher or lower) than normal.

the plasma and into the RBCs. As a result, the pH of the plasma increases as the reaction proceeds from right to left.

Hemoglobin and Temperature. At a given P_{CO_2}, hemoglobin releases additional oxygen if the temperature of the blood increases. Changes in temperature affect the slope of the hemoglobin saturation curve (Figure 23–20b). As the temperature increases, hemoglobin releases more oxygen. As the temperature decreases, hemoglobin holds oxygen more tightly. Temperature effects are significant only in active tissues that are generating large amounts of heat. For example, active skeletal muscles generate heat, and the heat warms blood that flows through these organs. As the blood warms, the Hb molecules release more oxygen than the active muscle fibers can use.

23

Hemoglobin and BPG. Red blood cells lack mitochondria. These cells produce ATP only by glycolysis. As a result, lactate is formed, as we saw in Chapter 10. p. 322 The metabolic pathways involved in glycolysis in an RBC also generate the compound **2,3-bisphosphoglycerate** (biz-fos-fō-GLIS-er-āt), or **BPG**. This compound has a direct effect on oxygen binding and release. For any partial pressure of oxygen, the higher the concentration of BPG, the greater the release of oxygen by Hb molecules. Normal RBCs always contain BPG.

Both BPG synthesis and the Bohr effect improve oxygen delivery when the pH changes: The BPG level rises when the pH increases, and the Bohr effect appears when the pH decreases. The concentration of BPG can be increased by high blood pH, thyroid hormones, growth hormone, epinephrine, and androgens. When the plasma P_{O_2} level is low for an extended time, red blood cells generate more BPG. These factors improve oxygen delivery to the tissues, because when the BPG level is elevated, hemoglobin releases about 10% more oxygen at a given P_{O_2} than it would do otherwise.

The production of BPG decreases as RBCs age. For this reason, the level of BPG can determine how long a blood bank can store fresh whole blood. When the BPG level gets too low, hemoglobin becomes firmly bound to the available oxygen. The blood is then useless for transfusions, because the RBCs will no longer release oxygen to peripheral tissues, even at a disastrously low P_{O_2}.

Transport of O_2 by Fetal Hemoglobin

The RBCs of a developing fetus contain **fetal hemoglobin**. The structure of fetal hemoglobin differs from that of adult hemoglobin, giving it a much higher affinity for oxygen. At the same P_{O_2}, fetal hemoglobin binds more oxygen than does adult hemoglobin (Figure 23–21). This trait is key to transferring oxygen across the placenta.

Figure 23–21 A Functional Comparison of Fetal and Adult Hemoglobin.

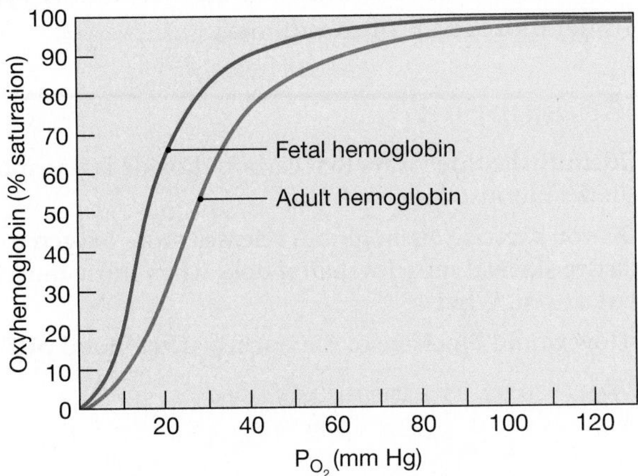
Oxyhemoglobin (% saturation)
100
90
80
70
60
50
40
30
20
10
0
Fetal hemoglobin
Adult hemoglobin
20
40
60
80
100
120
P_{O_2} (mm Hg)

A fetus obtains oxygen from the maternal bloodstream. At the placenta, maternal blood has a relatively low P_{O_2}, ranging from 35 to 50 mm Hg. If maternal blood arrives at the placenta with a P_{O_2} of 40 mm Hg, hemoglobin saturation is roughly 75%. The fetal blood arriving at the placenta has a P_{O_2} close to 20 mm Hg. However, because fetal hemoglobin has a higher affinity for oxygen, it is still 58% saturated.

As diffusion takes place between fetal blood and maternal blood, oxygen enters the fetal bloodstream until the P_{O_2} reaches equilibrium at 30 mm Hg. At this P_{O_2}, the maternal hemoglobin is less than 60% saturated, but the fetal hemoglobin is over 80% saturated, as you can see in Figure 23–21. The steeper slope of the saturation curve for fetal hemoglobin means that when fetal RBCs reach peripheral tissues of the fetus, the Hb molecules release a large amount of oxygen in response to a very small change in P_{O_2}.

+ Clinical Note Carbon Monoxide Poisoning

The exhaust from automobiles and other petroleum-burning engines, oil lamps, and fuel-fired space heaters contains *carbon monoxide (CO).* Each winter entire families die from **carbon monoxide poisoning**. Carbon monoxide competes with oxygen molecules for the binding sites on heme units. Unfortunately, the carbon monoxide usually wins the competition. The reason is that at very low partial pressures it has a much stronger affinity for hemoglobin than does oxygen. The bond formed between CO and heme is extremely strong. As a result, the attachment of a CO molecule essentially makes that heme unit inactive for respiratory purposes.

If CO molecules make up just 0.1% of inhaled air, enough hemoglobin is affected that human survival becomes impossible without medical assistance. Treatment includes (1) preventing further CO exposure; (2) administering pure oxygen, because at sufficiently high partial pressures, the oxygen molecules gradually replace CO at the hemoglobin molecules; and, if necessary, (3) transfusing compatible red blood cells.

Carbon Dioxide Transport

Carbon dioxide is generated by aerobic metabolism in peripheral tissues. Carbon dioxide travels in the bloodstream in three different ways. After entering the blood, a CO_2 molecule either (1) is converted to a molecule of carbonic acid, (2) binds to hemoglobin within red blood cells, or (3) dissolves in plasma. All three reactions are completely reversible, allowing carbon dioxide to be picked up from body tissues and then delivered to the alveoli. Let's consider the events that take place as blood enters peripheral tissues in which the P_{CO_2} is 45 mm Hg.

Transport of CO_2 as Bicarbonate Ions

Roughly 70% of the carbon dioxide absorbed by blood is transported as bicarbonate ions (HCO_3^-). Most of the carbon dioxide that enters the bloodstream diffuses into RBCs. Within the RBCs, carbon dioxide and water are combined and converted to carbonic acid through the activity of the enzyme *carbonic anhydrase* (Figure 23–22). The carbonic acid molecules immediately dissociate into hydrogen ions and bicarbonate ions, as

Figure 23–22 Carbon Dioxide Transport in Blood.

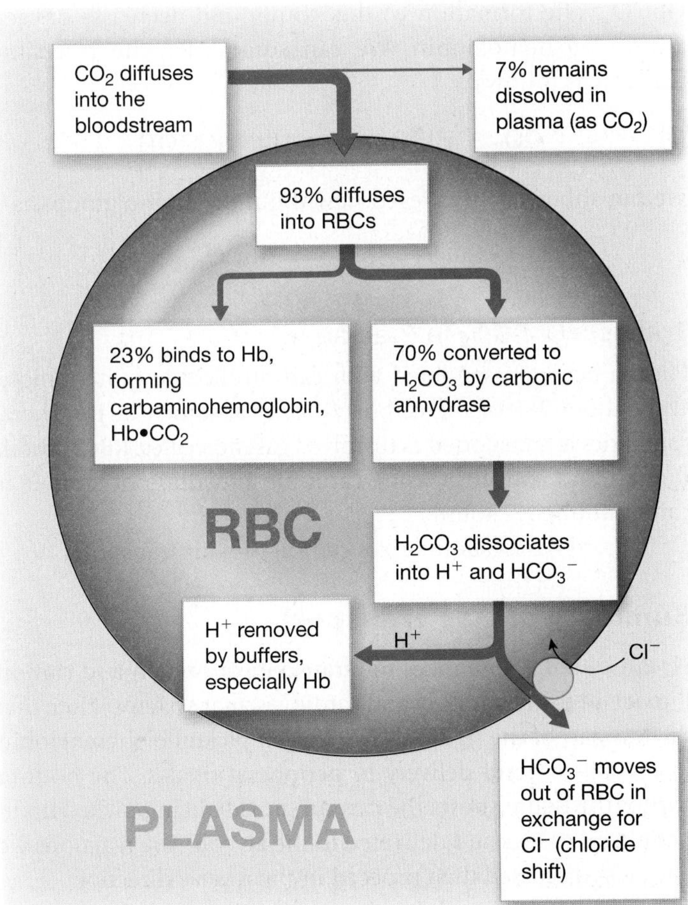
CO_2 diffuses into the bloodstream
7% remains dissolved in plasma (as CO_2)
93% diffuses into RBCs
23% binds to Hb, forming carbaminohemoglobin, Hb•CO_2
70% converted to H_2CO_3 by carbonic anhydrase
RBC
H_2CO_3 dissociates into H^+ and HCO_3^-
H^+ removed by buffers, especially Hb
H^+
Cl^-
HCO_3^- moves out of RBC in exchange for Cl^- (chloride shift)
PLASMA

described earlier (p. 865). We can ignore the intermediate steps in this sequence and summarize the reaction as

$$CO_2 + H_2O \underset{}{\overset{\text{carbonic anhydrase}}{\rightleftharpoons}} H^+ + HCO_3^-$$

This reaction is completely reversible. In peripheral capillaries, it proceeds vigorously, tying up large numbers of CO_2 molecules. The reaction continues as carbon dioxide diffuses out of the interstitial fluids.

The hydrogen ions and bicarbonate ions have different fates. Most of the hydrogen ions bind to hemoglobin molecules, forming HbH^+. The Hb molecules function as pH buffers, tying up the released H^+ before the ions can leave the RBCs and lower the pH of the plasma. The bicarbonate ions move into the plasma with the aid of a countertransport mechanism that exchanges intracellular bicarbonate ions (HCO_3^-) for extracellular chloride ions (Cl^-). This exchange trades one anion for another and does not require ATP. The result is a mass movement of chloride ions into the RBCs, an event known as the **chloride shift**.

Transport of CO_2 on Hemoglobin

About 23% of the carbon dioxide carried by blood is bound to the protein portions of Hb molecules inside RBCs. These CO_2 molecules are attached to exposed amino groups ($-NH_2$) of the Hb molecules. The resulting compound is called **carbaminohemoglobin** (kar-BAM-ih-nō-hē-mō-glō-bin), $HbCO_2$. The formation of this compound decreases oxygen affinity to hemoglobin. We can summarize the reversible reaction as follows:

$$CO_2 + HbNH_2 \rightleftharpoons HbNHCOOH$$

We can abbreviate this reaction without the amino groups as

$$CO_2 + Hb \rightleftharpoons HbCO_2$$

Transport of CO_2 in Plasma

23 Plasma becomes saturated with carbon dioxide quite rapidly. Only about 7% of the carbon dioxide absorbed at peripheral capillaries is transported as dissolved gas molecules. RBCs absorb the rest and convert it using carbonic anhydrase or store it as carbaminohemoglobin.

Figure 23–22 summarizes carbon dioxide transport.

Summary of Gas Transport

Figure 23–23 summarizes the transport of oxygen and carbon dioxide in the respiratory and cardiovascular systems. Note that the top part of the figure shows oxygen pickup by hemoglobin in the alveoli, and delivery to peripheral tissues. The bottom part of the figure shows the carbon dioxide being picked up in peripheral tissues and delivered to the alveoli. The reactions we have just discussed then proceed in the reverse direction.

Gas transport is a dynamic process. Its responses can be varied to meet changing circumstances. Some of the responses are automatic and result from the basic chemistry of the transport mechanisms. Other responses require coordinated adjustments in the activities of the cardiovascular and respiratory systems. We consider those levels of control next.

✓ Checkpoint

23. **Identify the three ways that carbon dioxide is transported in the bloodstream.**
24. **As you exercise, hemoglobin releases more oxygen to active skeletal muscles than it does when those muscles are at rest. Why?**
25. **How would blockage of the trachea affect blood pH?**

See the blue Answers tab at the back of the book.

23-10 Respiratory centers in the brainstem, along with respiratory reflexes, control respiration

Learning Outcome List the factors that influence respiration rate, and discuss reflex respiratory activity and the brain centers involved in the control of respiration.

Peripheral cells continuously absorb oxygen and generate carbon dioxide. Under normal conditions, the cellular rates of absorption and generation are matched by the capillary rates of delivery and removal. Both rates are identical to those of oxygen absorption and carbon dioxide excretion at the lungs.

If diffusion rates at the peripheral and alveolar capillaries become unbalanced, homeostatic mechanisms intervene to restore equilibrium. Such mechanisms involve (1) local changes in oxygen delivery to peripheral tissues and in the ventilation-to-perfusion ratio in the lungs, and (2) changes in the depth and rate of respiration under the control of the brain's *respiratory centers*. The activities of the respiratory centers are coordinated with changes in cardiovascular function, such as fluctuations in blood pressure and cardiac output.

Local Regulation of Oxygen Delivery and Ventilation-to-Perfusion Ratio

The rate of oxygen delivery in each tissue and the efficiency of oxygen pickup at the lungs are largely regulated at the local level. For example, when a peripheral tissue becomes more active, the interstitial P_{O_2} falls and the P_{CO_2} rises. These changes increase the difference between the partial pressures in the tissues and in the arriving blood. As a result, more oxygen is delivered and more carbon dioxide is carried away. In addition, the rising P_{CO_2} level causes the relaxation of smooth muscles in the walls of arterioles and capillaries in the area, increasing local blood flow.

At the lungs, local factors coordinate (1) *lung perfusion*, or blood flow to the alveoli, with (2) *alveolar ventilation*, or airflow. This local coordination, called the **ventilation-to-perfusion ratio**, or **V/Q ratio**, takes place over a wide range of conditions

Figure 23–23 **A Summary of the Primary Gas Transport Mechanisms.**

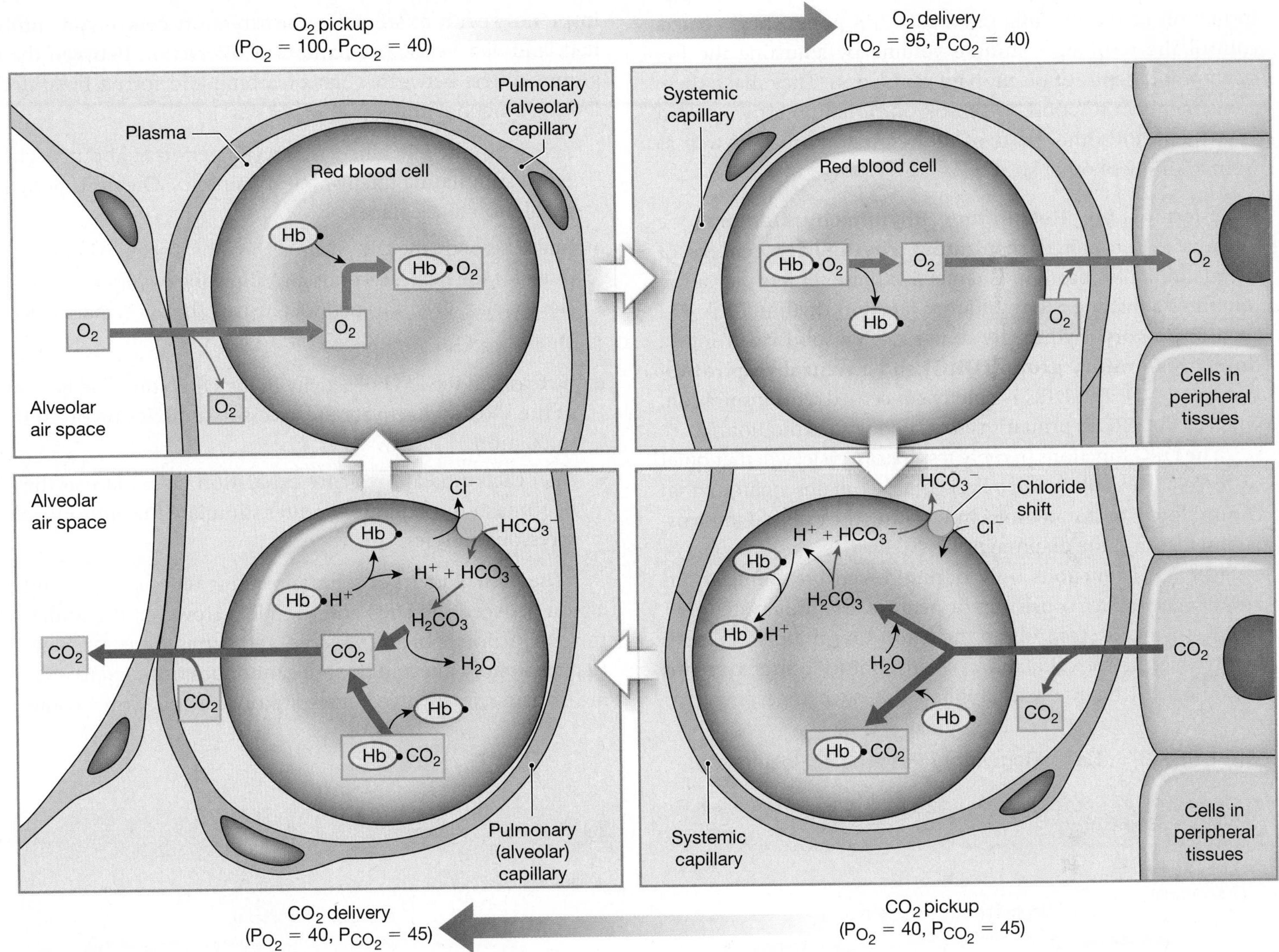

O_2 pickup
(P_{O_2} = 100, P_{CO_2} = 40)
O_2 delivery
(P_{O_2} = 95, P_{CO_2} = 40)
Plasma
Pulmonary (alveolar) capillary
Red blood cell
Hb
Hb • O_2
O_2
Alveolar air space
Systemic capillary
Cells in peripheral tissues
Cl⁻
HCO_3^-
$H^+ + HCO_3^-$
Hb • H^+
H_2CO_3
CO_2
H_2O
Hb • CO_2
Chloride shift
CO_2 delivery
(P_{O_2} = 40, P_{CO_2} = 45)
CO_2 pickup
(P_{O_2} = 40, P_{CO_2} = 45)

and activity levels. As blood flows toward the alveolar capillaries, it is directed toward lobules with a relatively high P_{O_2}. This movement takes place because alveolar capillaries constrict when the local P_{O_2} is low. (This response is the opposite of that seen in peripheral tissues, as we noted in Chapter 21. ⮌ p. 753)

Also in the lungs, smooth muscles in the walls of bronchioles are sensitive to the P_{CO_2} of the air they contain. When the P_{CO_2} goes up, the bronchioles increase in diameter (bronchodilation). When the P_{CO_2} declines, the bronchioles constrict (bronchoconstriction). Airflow is therefore directed to lobules with a high P_{CO_2}. These lobules get their carbon dioxide from blood and are actively engaged in gas exchange. The response of the bronchioles to P_{CO_2} is especially important, because improvements in airflow to functional alveoli can at least partially compensate for damage to pulmonary lobules.

Local adjustments improve the efficiency of gas transport by directing blood flow to alveoli with low CO_2 levels and increasing airflow to alveoli with high CO_2 levels. These adjustments in alveolar blood flow and bronchiole diameter take place automatically.

Neural Control of Respiration

When activity levels increase and the demand for oxygen rises, respiratory rates increase under neural control. Neural control of respiration has both involuntary and voluntary components. Your brain's involuntary *respiratory centers* set and adjust the respiratory rhythm by regulating the activities of the respiratory muscles.

The voluntary control of respiration reflects activity in the higher centers of the cerebral cortex. This activity affects the output of either the respiratory centers or motor neurons in the spinal cord that control respiratory muscles. These spinal motor neurons are generally controlled by *respiratory reflexes*, but they can also be controlled voluntarily through commands delivered by the corticospinal pathway. ⮌ p. 528

Involuntary Control of the Basic Respiratory Rhythm

The respiratory centers are three pairs of nuclei in the reticular formation of the medulla oblongata and pons. These centers control the respiratory minute volume by adjusting the frequency and depth of pulmonary ventilation. They make these adjustments in response to sensory information arriving from your lungs and other parts of the respiratory tract, as well as from a variety of other sites.

Function of the Respiratory Rhythmicity Centers. We introduced the paired **respiratory rhythmicity centers** of the medulla oblongata in Chapter 14; these centers play a key role in establishing the respiratory rate and rhythm. p. 474 Each respiratory rhythmicity center can be subdivided into a **dorsal respiratory group (DRG)** and a **ventral respiratory group (VRG)**. The DRG is chiefly concerned with inspiration, whereas the VRG is primarily associated with expiration.

The DRG functions in every respiratory cycle, whether quiet or forced. The DRG's *inspiratory center* contains neurons that control lower motor neurons innervating the external intercostal muscles and the diaphragm.

The VRG functions only during forced breathing. It has an *expiratory center* consisting of neurons that innervate lower motor neurons controlling accessory respiratory muscles involved in active exhalation. Its inspiratory center contains neurons involved in maximal inhalation, such as gasping.

Reciprocal inhibition takes place between the neurons involved with inhalation and exhalation. p. 459 When the inspiratory neurons are active, the expiratory neurons are inhibited, and vice versa. The pattern of interaction between these groups differs between quiet breathing and forced breathing. During quiet breathing (Figure 23–24a):

- Activity in the DRG increases over a period of about 2 seconds, stimulating the inspiratory muscles. Over this period, inhalation takes place.
- After 2 seconds, the DRG neurons become inactive. They remain quiet for the next 3 seconds and allow the inspiratory muscles to relax. Over this period, passive exhalation takes place.

During forced breathing (Figure 23–24b):

- Increases in the level of activity in the DRG stimulate neurons of the VRG that activate the accessory muscles involved in inhalation.
- After each inhalation, active exhalation takes place as the neurons of the expiratory center stimulate the appropriate accessory muscles.

The basic pattern of respiration thus reflects a cyclic interaction between the DRG and the VRG. However, the timing of this interaction is highly variable, and current research suggests that respiratory rate and rhythm cannot be attributed to a single area. The nuclei involved in respiratory control form a complex

Figure 23–24 **Basic Regulatory Patterns of Respiration.**

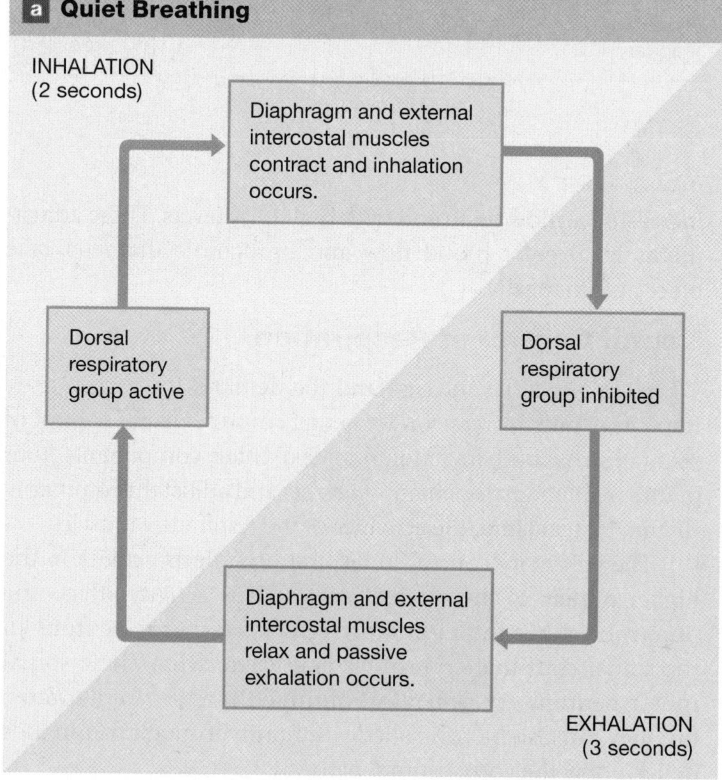
a Quiet Breathing
INHALATION (2 seconds)
Diaphragm and external intercostal muscles contract and inhalation occurs.
Dorsal respiratory group active
Dorsal respiratory group inhibited
Diaphragm and external intercostal muscles relax and passive exhalation occurs.
EXHALATION (3 seconds)

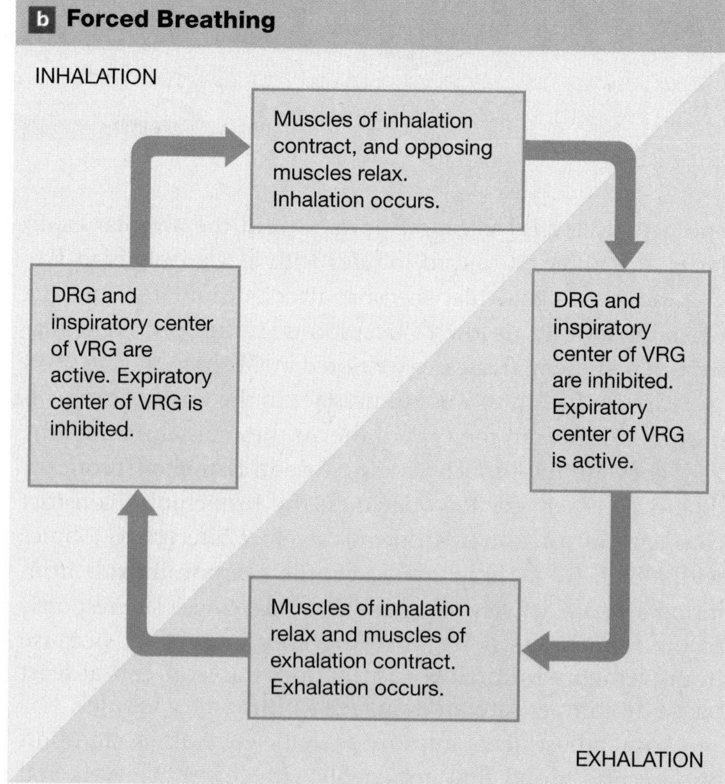
b Forced Breathing
INHALATION
Muscles of inhalation contract, and opposing muscles relax. Inhalation occurs.
DRG and inspiratory center of VRG are active. Expiratory center of VRG is inhibited.
DRG and inspiratory center of VRG are inhibited. Expiratory center of VRG is active.
Muscles of inhalation relax and muscles of exhalation contract. Exhalation occurs.
EXHALATION

23

network that includes (1) the rhythmicity centers, (2) centers in the pons that are modulated by peripheral and higher central inputs, and (3) an area in the ventrolateral medulla, known as the pre-Bötzinger complex (*central pattern generator*), that is essential to all forms of breathing. There are reciprocal pathways between all these nuclei, so deciphering their communication is extremely difficult, especially because the patterns change with different situations. Central nervous system stimulants, such as amphetamines or even caffeine, increase the respiratory rate by facilitating the respiratory centers. These actions are opposed by CNS depressants, such as barbiturates or opiates.

Respiratory control is an area of intense study and our understanding of the interactions and mechanisms remains incomplete. Due to the brainstem locations, it is nearly impossible to study this mechanism in humans, and much of the work has been done in cats and newborn rodents.

Function of the Apneustic and Pneumotaxic Centers. The *apneustic centers* and *pneumotaxic centers* (also called the *pontine respiratory group*) of the pons regulate the depth and rate of respiration in response to sensory stimuli or input from other centers in the brain. Each apneustic center provides continuous stimulation to the DRG on that side of the brainstem. During quiet breathing, stimulation from the apneustic center helps increase the intensity of inhalation over the next 2 seconds. Under normal conditions, after 2 seconds the apneustic center is inhibited by signals from the pneumotaxic center on that side. During forced breathing, the apneustic centers also respond to sensory input from the vagus nerves regarding the amount of lung inflation.

The pneumotaxic centers inhibit the apneustic centers and promote passive or active exhalation. Centers in the hypothalamus and cerebrum can alter the activity of the pneumotaxic centers, as well as the respiratory rate and depth. However, essentially normal respiratory cycles continue even if the brainstem superior to the pons has been severely damaged.

In some cases, the inhibitory output of the pneumotaxic centers is cut off by a stroke or other damage to the brainstem, or sensory innervation from the lungs is eliminated due to damage to the vagus nerves. In these cases, the person inhales to maximum capacity and maintains that state for 10–20 seconds at a time. Intervening exhalations are brief, and little pulmonary ventilation occurs.

The CNS regions involved with respiratory control are diagrammed in **Spotlight Figure 23–25**. Interactions between the DRG and the VRG establish the basic rate and depth of respiration. The pneumotaxic centers modify that rate: An increase in pneumotaxic output quickens the pace of respiration by shortening the duration of each inhalation. A decrease in pneumotaxic output slows the respiratory rate, but increases the depth of respiration, because the apneustic centers are more active.

Sudden infant death syndrome (SIDS), also known as *crib death*, is the leading cause of death for babies 1–12 months old and kills an estimated 2250 infants each year in the United States alone. Most crib deaths occur between midnight and 9:00 a.m., in the late fall or winter, and involve infants 2 to 4 months old. Eyewitness accounts indicate that the sleeping infant suddenly stops breathing, turns blue, and relaxes. Genetic factors appear to be involved, but controversy remains as to the relative importance of other factors. The age at the time of death corresponds with a period when the pacemaker complex and respiratory centers are establishing connections with other portions of the brain. It has been suggested that SIDS results from a problem in the interconnection process that disrupts the reflexive respiratory pattern.

Modification of Respiratory Rhythm by Respiratory Reflexes

The activities of the respiratory centers are modified by sensory information from several sources:

- Chemoreceptors sensitive to the P_{CO_2}, pH, or P_{O_2} of the blood or cerebrospinal fluid;
- Baroreceptors in the aortic or carotid sinuses sensitive to changes in blood pressure;
- Stretch receptors that respond to changes in the volume of the lungs (the *Hering–Breuer reflexes*);
- Irritating physical or chemical stimuli in the nasal cavity, larynx, or bronchial tree; and
- Other sensations, including pain, changes in body temperature, and abnormal visceral sensations.

Information from these receptors alters the pattern of respiration. The induced changes have been called *respiratory reflexes*.

Chemoreceptor Reflexes. The respiratory centers are strongly influenced by chemoreceptor inputs from cranial nerves IX and X, and from receptors that monitor the composition of the cerebrospinal fluid (CSF):

- The glossopharyngeal nerves (IX) carry chemoreceptive information from the carotid bodies, adjacent to the carotid sinus. pp. 521, 758 The carotid bodies are stimulated by a decrease in the pH or P_{O_2} of blood. Because changes in P_{CO_2} affect pH, an increase in the P_{CO_2} indirectly stimulates these receptors.
- The vagus nerves (X) monitor chemoreceptors in the aortic bodies, near the aortic arch. pp. 521, 758 These receptors are sensitive to the same stimuli as the carotid bodies. Carotid and aortic body receptors are often called *peripheral chemoreceptors*.
- Chemoreceptors are located on the ventrolateral surface of the medulla oblongata in a region known as the *chemosensitive area*. The neurons in that area respond only to the P_{CO_2} and pH of the CSF. They are often called *central chemoreceptors*.

SPOTLIGHT

Figure 23–25

Control of Respiration

Respiratory control involves multiple levels of regulation. Most of the regulatory activities occur outside of our awareness.

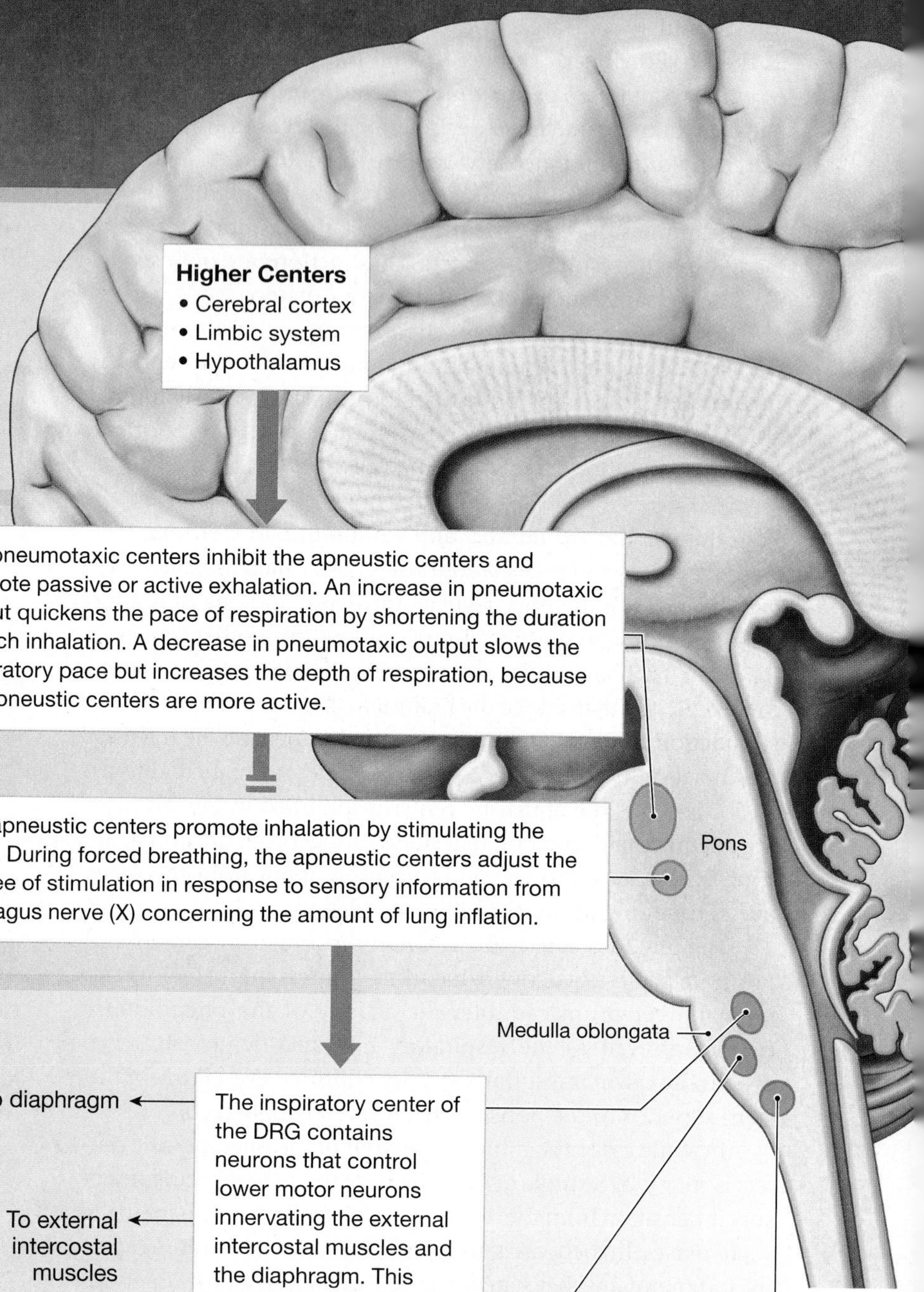

LEVEL 3

Higher Centers

Higher centers in the cerebral cortex, limbic system, and hypothalamus can alter the activity of the pneumotaxic centers, but essentially normal respiratory cycles continue even if the brainstem superior to the pons has been severely damaged.

Higher Centers
- Cerebral cortex
- Limbic system
- Hypothalamus

LEVEL 2

Apneustic and Pneumotaxic Centers in the Pons

The **apneustic** (ap-NŪ-stik) **centers** and the **pneumotaxic** (nū-mō-TAK-sik) **centers** of the pons are paired nuclei that adjust the output of the respiratory rhythmicity centers.

The pneumotaxic centers inhibit the apneustic centers and promote passive or active exhalation. An increase in pneumotaxic output quickens the pace of respiration by shortening the duration of each inhalation. A decrease in pneumotaxic output slows the respiratory pace but increases the depth of respiration, because the apneustic centers are more active.

The apneustic centers promote inhalation by stimulating the DRG. During forced breathing, the apneustic centers adjust the degree of stimulation in response to sensory information from the vagus nerve (X) concerning the amount of lung inflation.

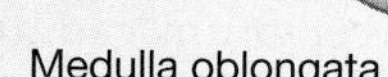

Pons

Medulla oblongata

LEVEL 1

Respiratory Rhythmicity Centers in the Medulla Oblongata

The most basic level of respiratory control involves pacemaker cells in the medulla oblongata. These neurons generate cycles of contraction and relaxation in the diaphragm. The **respiratory rhythmicity centers** set the pace of respiration by adjusting the activities of these pacemakers and coordinating the activities of additional respiratory muscles. Each rhythmicity center can be subdivided into a **dorsal respiratory group (DRG)** and a **ventral respiratory group (VRG)**. The DRG is mainly concerned with inspiration and the VRG is primarily associated with expiration.The DRG modifies its activities in response to input from chemoreceptors and baroreceptors that monitor O_2, CO_2, and pH in the blood and CSF, and from stretch receptors that monitor the degree of stretching in the walls of the lungs.

The inspiratory center of the DRG contains neurons that control lower motor neurons innervating the external intercostal muscles and the diaphragm. This center functions in every respiratory cycle.

To diaphragm

To external intercostal muscles

The VRG has inspiratory and expiratory centers that function only when breathing demands increase and accessory muscles become involved.

To accessory inspiratory muscles

To accessory expiratory muscles

In addition to the centers in the pons, the DRG, and the VRG, the pre-Bötzinger complex in the medulla is essential to all forms of breathing. Its mechanisms are poorly understood.

Respiratory Centers and Reflex Controls

This illustration shows the locations and relationships among the major respiratory centers in the pons and medulla oblongata and other factors important to the reflex control of respiration. Pathways for conscious control over respiratory muscles are not shown.

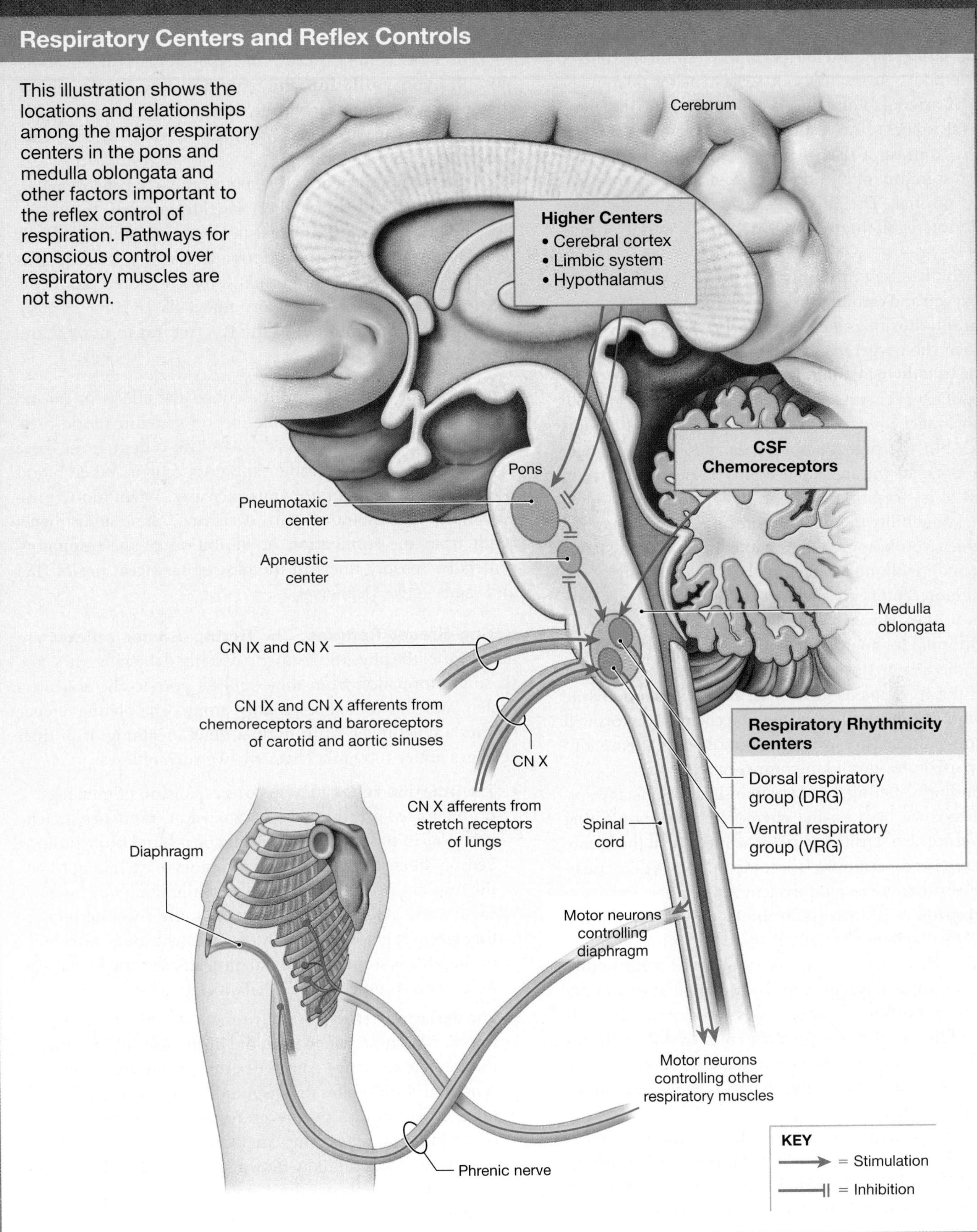

Cerebrum
Higher Centers
• Cerebral cortex
• Limbic system
• Hypothalamus
CSF Chemoreceptors
Pons
Pneumotaxic center
Apneustic center
Medulla oblongata
CN IX and CN X
CN IX and CN X afferents from chemoreceptors and baroreceptors of carotid and aortic sinuses
CN X
Respiratory Rhythmicity Centers
Dorsal respiratory group (DRG)
Ventral respiratory group (VRG)
CN X afferents from stretch receptors of lungs
Spinal cord
Diaphragm
Motor neurons controlling diaphragm
Motor neurons controlling other respiratory muscles
Phrenic nerve
KEY
= Stimulation
= Inhibition

We discussed chemoreceptors and their effects on cardiovascular function in Chapters 15 and 21. pp. 521, 749–750 Stimulation of these chemoreceptors leads to an increase in the rate and depth of respiration. Under normal conditions, a drop in arterial P_{O_2} has little effect on the respiratory centers, until the arterial P_{O_2} drops by about 40%, to below 60 mm Hg. If the P_{O_2} of arterial blood drops to 40 mm Hg (the level in peripheral tissues), the respiratory rate increases by only 50–70%. In contrast, a rise of just 10% in the arterial P_{CO_2} causes the respiratory rate to double, even if the P_{O_2} remains completely normal. The carbon dioxide level is therefore responsible for regulating respiratory activity under normal conditions.

Although the receptors monitoring the CO_2 level are more sensitive, oxygen and carbon dioxide receptors work together in a crisis. Carbon dioxide is generated during oxygen consumption, so when the oxygen concentration is falling rapidly, the CO_2 level is usually increasing. This cooperation breaks down only under unusual circumstances. For example, you can hold your breath longer than normal by taking deep, full breaths beforehand, but the practice is very dangerous. The danger lies in the fact that the increased ability is due not to extra oxygen, but to less carbon dioxide. If the P_{CO_2} is driven down far enough, your ability to hold your breath can increase to the point at which you become unconscious from oxygen starvation in the brain without ever feeling the urge to breathe.

The chemoreceptors are subject to adaptation—a decrease in sensitivity after chronic stimulation—if the P_{O_2} or P_{CO_2} remains abnormal for an extended period. This adaptation can complicate the treatment of chronic respiratory disorders. For example, if the P_{O_2} remains low for an extended period while the P_{CO_2} remains chronically elevated, the chemoreceptors will reset to those values. They will then oppose any attempts to return the partial pressures to the proper range.

Any condition altering the pH of blood or CSF affects respiratory performance because the chemoreceptors monitoring the CO_2 level are also sensitive to pH. The increase in the lactate level after exercise, for example, is accompanied by a decrease in pH that helps stimulate respiratory activity.

Hypercapnia is an increase in the P_{CO_2} of arterial blood. Figure 23–26a diagrams the central response to hypercapnia. Stimulation of chemoreceptors in the carotid and aortic bodies triggers this response. It is reinforced by the stimulation of CNS chemoreceptors. Carbon dioxide crosses the blood brain barrier quite rapidly. For this reason, a rise in arterial P_{CO_2} almost immediately increases the CO_2 level in the CSF, decreasing the pH of the CSF and stimulating the chemoreceptive neurons of the medulla oblongata.

These chemoreceptors stimulate the respiratory centers to increase the rate and depth of respiration. Your breathing becomes more rapid, and more air moves into and out of your lungs with each breath. Because more air moves into and out of the alveoli each minute, alveolar concentrations of carbon dioxide decline, accelerating the diffusion of carbon dioxide out of alveolar capillaries. In this way, homeostasis is restored.

The most common cause of hypercapnia is hypoventilation. In **hypoventilation**, the respiratory rate remains abnormally low and cannot meet the demands for normal oxygen delivery and carbon dioxide removal. Carbon dioxide then accumulates in the blood.

If the rate and depth of respiration exceed the demands for oxygen delivery and carbon dioxide removal, the condition called **hyperventilation** exists. Hyperventilation gradually leads to **hypocapnia**, an abnormally low P_{CO_2}. If the arterial P_{CO_2} drops below the normal level, chemoreceptor activity decreases and the respiratory rate falls (Figure 23–26b). This situation continues until the P_{CO_2} returns to normal and homeostasis is restored.

Baroreceptor Reflexes. We described the effects of carotid and aortic baroreceptor stimulation on systemic blood pressure in Chapter 21. pp. 747–748 The output from these baroreceptors also affects the respiratory centers. When blood pressure falls, the respiratory rate increases. When blood pressure rises, the respiratory rate decreases. These adjustments result from the stimulation or inhibition of the respiratory centers by sensory fibers in the glossopharyngeal nerves (IX) and vagus nerves (X) nerves.

Hering–Breuer Reflexes. The **Hering–Breuer reflexes** are named after the physiologists who described them in 1865. The sensory information from these reflexes goes to the apneustic centers and the ventral respiratory group. The Hering–Breuer reflexes are *not* involved in normal quiet breathing or in tidal volumes under 1000 mL. There are two such reflexes:

- The **inflation reflex** prevents overexpansion of the lungs during forced breathing. Lung expansion stimulates stretch receptors in the smooth muscle tissue around bronchioles. Sensory fibers from the stretch receptors of each lung reach the respiratory rhythmicity center on the same side by the vagus nerve. As lung volume increases, the dorsal respiratory group is gradually inhibited, and the expiratory center of the VRG is stimulated. Inhalation stops as the lungs near maximum volume. Active exhalation then begins.
- The **deflation reflex** normally functions only during forced exhalation, when both the inspiratory and expiratory centers are active. This reflex inhibits the expiratory centers and stimulates the inspiratory centers when the lungs are deflating. These receptors are distinct from those of the inflation reflex. They are located in the alveolar wall near the alveolar capillary network. The smaller the volume of the lungs, the greater the degree of inhibition. Finally, exhalation stops and inhalation begins.

Figure 23–26 The Chemoreceptor Response to Changes in P_{CO_2}.

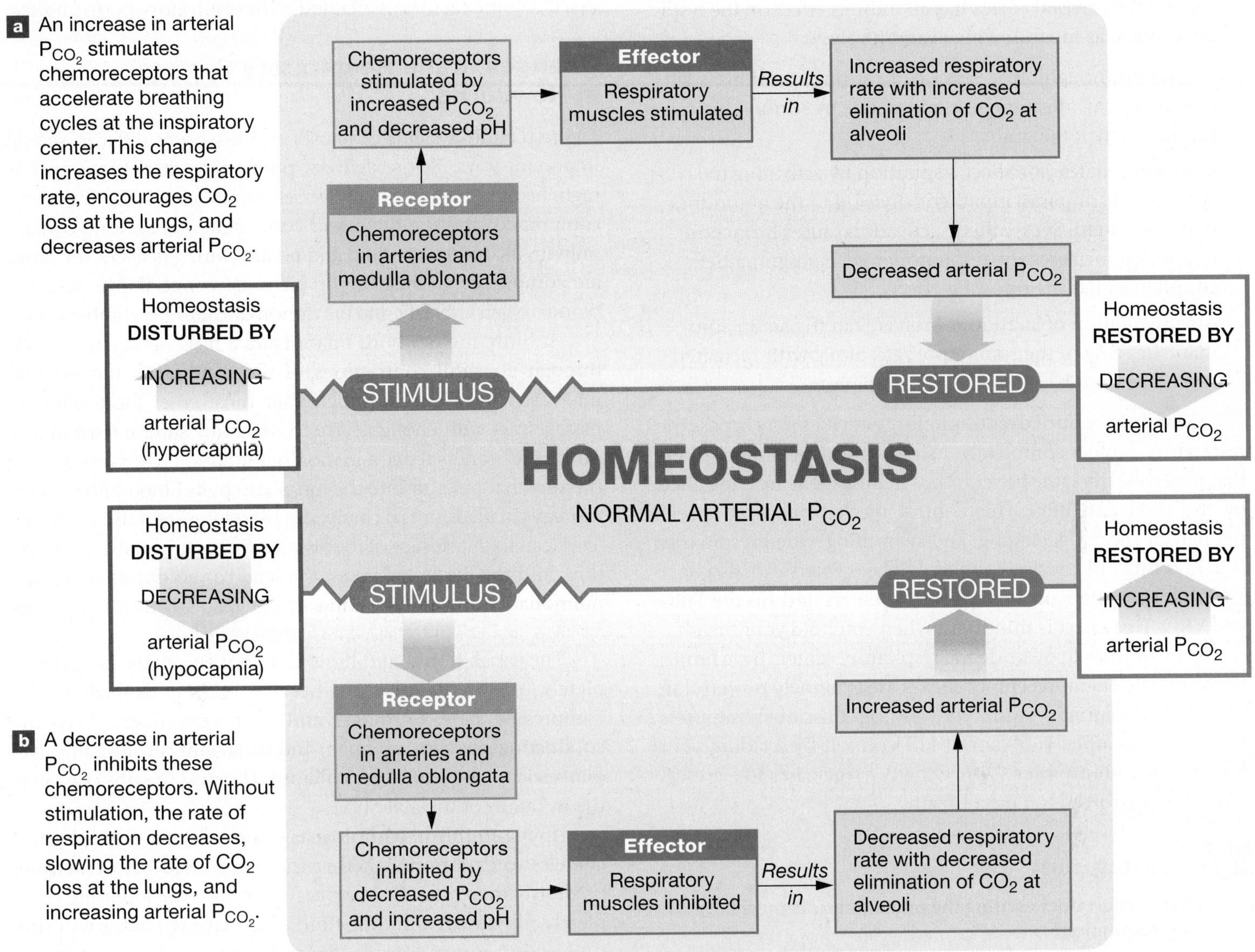
a An increase in arterial P_{CO_2} stimulates chemoreceptors that accelerate breathing cycles at the inspiratory center. This change increases the respiratory rate, encourages CO_2 loss at the lungs, and decreases arterial P_{CO_2}.
Chemoreceptors stimulated by increased P_{CO_2} and decreased pH
Effector
Respiratory muscles stimulated
Results in
Increased respiratory rate with increased elimination of CO_2 at alveoli
Receptor
Chemoreceptors in arteries and medulla oblongata
Decreased arterial P_{CO_2}
Homeostasis DISTURBED BY INCREASING arterial P_{CO_2} (hypercapnia)
STIMULUS
RESTORED
Homeostasis RESTORED BY DECREASING arterial P_{CO_2}
HOMEOSTASIS
NORMAL ARTERIAL P_{CO_2}
Homeostasis DISTURBED BY DECREASING arterial P_{CO_2} (hypocapnia)
STIMULUS
RESTORED
Homeostasis RESTORED BY INCREASING arterial P_{CO_2}
Receptor
Chemoreceptors in arteries and medulla oblongata
Increased arterial P_{CO_2}
b A decrease in arterial P_{CO_2} inhibits these chemoreceptors. Without stimulation, the rate of respiration decreases, slowing the rate of CO_2 loss at the lungs, and increasing arterial P_{CO_2}.
Chemoreceptors inhibited by decreased P_{CO_2} and increased pH
Effector
Respiratory muscles inhibited
Results in
Decreased respiratory rate with decreased elimination of CO_2 at alveoli

? Inhibition of medulla oblongata chemoreceptors and respiratory muscles has what effect on respiratory rate, elimination of CO_2 at alveoli, and arterial P_{CO_2}?

Protective Reflexes. Protective reflexes operate when you are exposed to toxic vapors, chemical irritants, or mechanical stimulation of the respiratory tract. The receptors involved are located in the epithelium of the respiratory tract. Examples of protective reflexes include sneezing, coughing, and laryngeal spasms.

Sneezing is triggered by an irritation of the nasal cavity wall. *Coughing* is triggered by an irritation of the larynx, trachea, or bronchi. Both reflexes involve **apnea** (AP-nē-uh), a period in which respiration is suspended. A forceful expulsion of air usually follows to remove the offending stimulus. The glottis is forcibly closed while the lungs are still relatively full. The abdominal and internal intercostal muscles then contract suddenly, creating pressures that blast air out of the respiratory passageways when the glottis reopens. Air leaving the larynx can travel at 99 mph (160 kph), carrying mucus, foreign particles, and irritating gases out of the respiratory tract by the nose or mouth.

Laryngeal spasms result when chemical irritants, foreign objects, or fluids enter the area around the glottis. This reflex generally closes the airway temporarily. A very strong stimulus, such as a toxic gas, could close the glottis so powerfully that you could lose consciousness and die without taking another breath. Fine chicken bones or fish bones that pierce the laryngeal walls can also stimulate laryngeal spasms, swelling, or both, restricting the airway.

Voluntary Control of Respiration by Higher Brain Centers

Activity of the cerebral cortex has an indirect effect on the respiratory centers, as the following examples show:

- Conscious thought processes tied to strong emotions, such as rage or fear, affect the respiratory rate by stimulating centers in the hypothalamus.
- Emotional states can affect respiration by activating the sympathetic or parasympathetic division of the autonomic nervous system. Sympathetic activation causes bronchodilation and increases the respiratory rate. Parasympathetic stimulation has the opposite effects.
- An anticipation of strenuous exercise can trigger an automatic increase in the respiratory rate, along with increased cardiac output, by sympathetic stimulation.

Conscious control over respiratory activities may bypass the respiratory centers completely, using cortical pyramidal fibers that innervate the same lower motor neurons that are controlled by the DRG and VRG. This control mechanism is an essential part of speaking, singing, and swimming, when respiratory activities must be precisely timed. Higher centers can also have an inhibitory effect on the apneustic centers and on the DRG and VRG. This effect is important when you hold your breath.

Your abilities to override the respiratory centers have limits, however. The chemoreceptor reflexes are extremely powerful in stimulating respiration, and you cannot consciously suppress them. For example, you cannot kill yourself by holding your breath "till you turn blue." Once the P_{CO_2} increases to a critical level, you are forced to take a breath.

Checkpoint

26. **What effect does exciting the pneumotaxic centers have on respiration?**
27. **Are peripheral chemoreceptors more or less sensitive to the level of carbon dioxide than they are to the level of oxygen? Why?**
28. **Little Johnny is angry with his mother, so he tells her that he will hold his breath until he turns blue and dies. Is Johnny likely to die if he holds his breath as long as he can?**

See the blue Answers tab at the back of the book.

23-11 Respiratory performance changes over the life span

Learning Outcome Describe age-related changes in the respiratory system.

From birth to death, our respiratory system undergoes age-related changes. In healthy people, many of the changes associated with age may not lead to symptoms. Some changes may decrease one's ability to engage in vigorous exercise, but heart function is more likely to affect pulmonary function. Here we discuss some factors associated with respiratory performance.

Changes in the Respiratory System in Newborns

The respiratory systems of fetuses and newborns differ in several important ways. Before delivery, pulmonary arterial resistance is high, because the pulmonary vessels are collapsed. The rib cage is compressed, and the lungs and conducting passageways contain only small amounts of fluid and no air. During delivery, the lungs are compressed further. As the placental connection is lost, the blood oxygen level falls and the carbon dioxide level climbs rapidly.

At birth, the newborn infant takes a truly heroic first breath through powerful contractions of the diaphragm and external intercostal muscles. The inhaled air must enter the respiratory passageways with enough force to overcome surface tension and inflate the bronchial tree and most of the alveoli. The same drop in pressure that pulls air into the lungs also pulls blood into the pulmonary circulation. The changes in blood flow and rise in oxygen level lead to the closure of the *foramen ovale*, an interatrial connection, and the *ductus arteriosus*, the fetal connection between the pulmonary trunk and the aorta. pp. 775–776 ATLAS: Embryology Summary 18: The Development of the Respiratory System

The exhalation that follows does not empty the lungs completely, because the rib cage does not return to its former, fully compressed state. Cartilages and connective tissues keep the conducting passageways open, and surfactant covering the alveolar surfaces prevents their collapse. The next breaths complete the inflation of the alveoli.

After a stillbirth, pathologists sometimes use these physical changes to determine whether the infant died before delivery or shortly thereafter. Before the first breath, the lungs are completely filled with amniotic fluid, and extracted lungs will sink if placed in water. After the infant's first breath, even the collapsed lungs contain enough air to keep them afloat.

Changes in the Respiratory System in Elderly Individuals

Many factors interact to reduce the efficiency of the respiratory system in elderly individuals. Here are three examples:

- With age, elastic tissue deteriorates throughout the body. These changes decrease lung compliance, lowering the vital capacity.
- Chest movements are restricted by arthritic changes in the rib articulations and by decreased flexibility at the costal cartilages. Along with elastic tissue deterioration, the stiffening and reduction in chest movement effectively limit the respiratory minute volume. This restriction contributes to the reduction in exercise performance and capabilities with increasing age.

Figure 23–27 Decline in Respiratory Performance with Age and Smoking. The relative respiratory performances of people who have never smoked, people who quit smoking at age 45, people who quit smoking at age 65, and lifelong smokers.

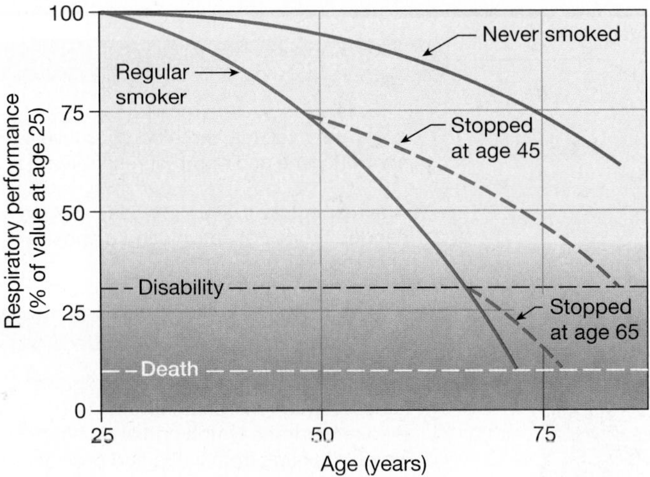

Respiratory performance (% of value at age 25)
100
75
50
25
0
Never smoked
Regular smoker
Stopped at age 45
Disability
Stopped at age 65
Death
25
50
75
Age (years)

- Some degree of emphysema is normal in individuals over age 50. However, the extent varies widely with the lifetime exposure to cigarette smoke and other respiratory irritants. Figure 23–27 compares the respiratory performance of individuals who have never smoked with individuals who have smoked for various periods of time. The message is quite clear: Some decrease in respiratory performance is inevitable, but you can prevent serious respiratory deterioration by stopping smoking or never starting.

✓ Checkpoint

29. **Name several age-related factors that affect the respiratory system.**

See the blue Answers tab at the back of the book.

23-12 The respiratory system provides oxygen to, and eliminates carbon dioxide from, other organ systems

Learning Outcome Give examples of interactions between the respiratory system and other organ systems studied so far.

The goal of respiratory activity is to maintain homeostatic oxygen and carbon dioxide levels in peripheral tissues. Changes in respiratory activity alone are seldom enough to accomplish this. Coordinated changes in cardiovascular activity must also take place.

These examples highlight the integration between the respiratory and cardiovascular systems:

- At the local level, changes in lung perfusion in response to changes in alveolar P_{O_2} improve the efficiency of gas exchange within or among lobules.

+ Clinical Note Smoking and the Lungs

Emphysema and lung cancer are two relatively common disorders associated with cigarette smoking. **Emphysema** (em-fih-ZĒ-muh) is a chronic, progressive condition. It is characterized by shortness of breath and an inability to tolerate physical exertion. The underlying problem is the destruction of alveolar surfaces and inadequate surface area for oxygen and carbon dioxide exchange. The alveoli gradually expand and merge to form larger air spaces supported by fibrous tissue without alveolar capillary networks. As connective tissues are eliminated, compliance increases, so air moves into and out of the lungs more easily than before. But the loss of respiratory surface area restricts oxygen absorption, so the individual becomes short of breath.

Emphysema has been linked to breathing air that contains fine particles or toxic vapors, such as those in cigarette smoke. Genetic factors also predispose individuals to the condition. Some degree of emphysema is a normal consequence of aging. An estimated 66% of adult males and 25% of adult females have detectable areas of emphysema in their lungs.

Lung cancer is an aggressive class of malignancies. These cancers affect the epithelial cells that line conducting passageways, mucous glands, or alveoli. Signs and symptoms generally do not appear until tumors restrict airflow or compress adjacent structures. Chest pain, shortness of breath, a cough or a wheeze, and weight loss are common. Treatment varies with the cellular organization of the tumor and whether *metastasis* (cancer cell migration) has occurred. Surgery, radiation, or chemotherapy may be involved. According to the CDC, more people die from lung cancer than any other type of cancer.

- Chemoreceptor stimulation not only increases the respiratory rate, but it also causes blood pressure to rise and cardiac output to increase.
- The stimulation of baroreceptors in the lungs has secondary effects on cardiovascular function. For example, the stimulation of airway stretch receptors not only triggers the inflation reflex, but also increases heart rate. Thus, as the lungs fill, cardiac output rises and more blood flows through the alveolar capillaries.

The adaptations that take place at high altitudes are an excellent example of the functional interplay between the respiratory and cardiovascular systems. Atmospheric pressure decreases with increasing altitude, and so do the partial pressures of the component gases, including oxygen. People living in Denver or Mexico City function normally with alveolar oxygen pressures in the 80–90 mm Hg range. At higher elevations, alveolar P_{O_2} is even lower. At 3300 meters (10,826 ft),

Build Your Knowledge

Figure 23–28 Integration of the RESPIRATORY system with the other body systems presented so far.

Integumentary System

- The Integumentary System protects portions of upper respiratory tract; hairs guard entry to external nares
- The respiratory system provides oxygen to tissues and removes carbon dioxide

Endocrine System

- The Endocrine System produces epinephrine and norepinephrine, which stimulate respiratory activity and dilate respiratory passageways
- The respiratory system produces angiotensin-converting enzyme (ACE) from capillaries in the lungs. ACE converts angiotensin I to angiotensin II

Skeletal System

- The Skeletal System protects the lungs with its axial division; movements of the ribs are important in breathing.
- The respiratory system provides oxygen to skeletal structures and disposes of carbon dioxide

Muscular System

- The Muscular System controls the entrances to respiratory tract; intrinsic laryngeal muscles control airflow through larynx and produce sounds; respiratory muscles change thoracic cavity volume so air moves into and out of lungs
- The respiratory system provides oxygen needed for muscle contractions and disposes of carbon dioxide generated by active muscles

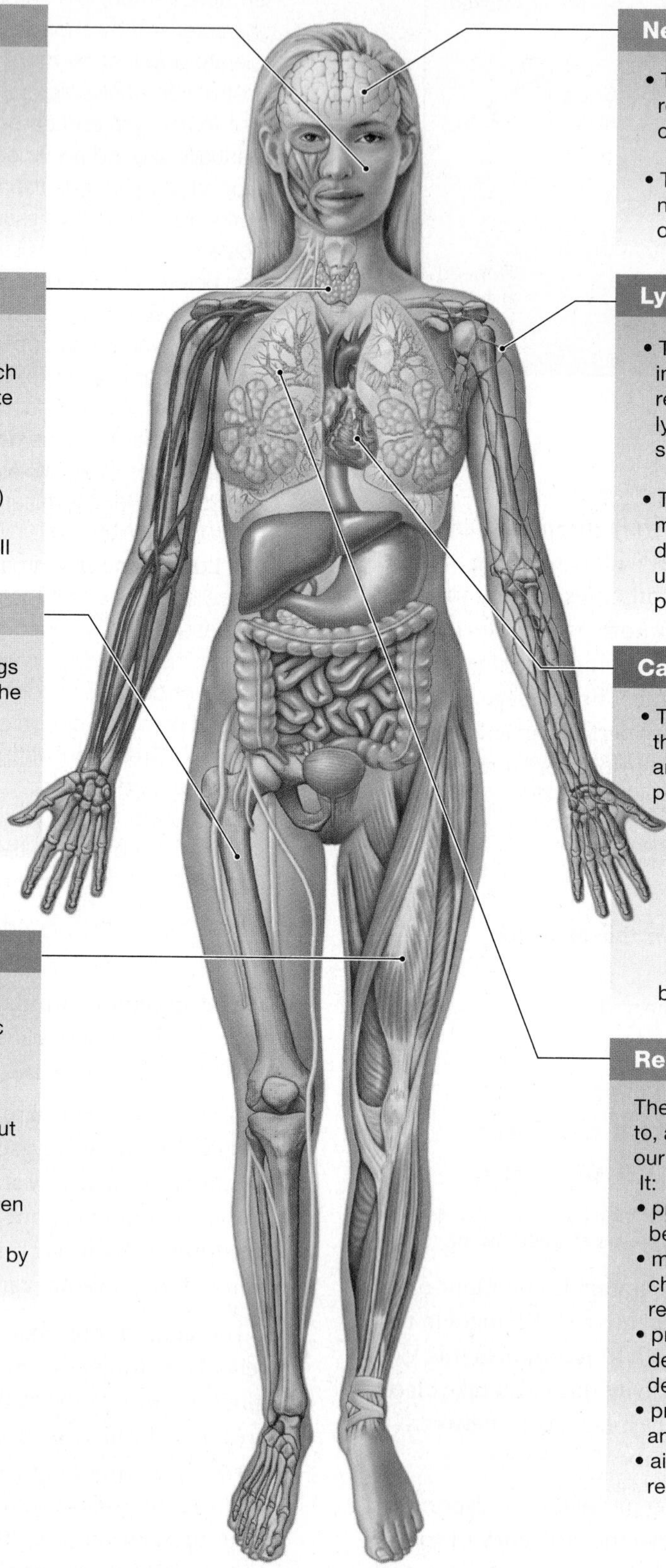

Nervous System

- The Nervous System monitors respiratory volume and blood gas levels; controls pace and depth of respiration
- The respiratory system provides oxygen needed for neural activity and disposes of carbon dioxide

Lymphatic System

- The Lymphatic System protects against infection with tonsils at entrance to respiratory tract; lymph nodes monitor lymph drainage from lungs and provide specific defenses when infection occurs
- The respiratory system has alveolar macrophages that provide nonspecific defenses; mucous membrane lining the upper respiratory system traps pathogens and protects deeper tissues

Cardiovascular System

- The Cardiovascular System circulates the red blood cells that transport oxygen and carbon dioxide between lungs and peripheral tissues
- The respiratory system provides oxygen needed for heart muscle contraction; bicarbonate ions contribute to the buffering capability of blood; activation of angiotensin II by ACE is important in regulation of blood pressure and volume

Respiratory System

The respiratory system provides oxygen to, and eliminates carbon dioxide from, our cells.

It:

- provides a large area for gas exchange between air and circulating blood
- moves air to and from the gas-exchange surfaces of the lungs along the respiratory passageways
- protects the respiratory surfaces from dehydration, temperature changes, and defends against invading pathogens
- produces sounds for speaking, singing, and other forms of communication
- aids the sense of smell by the olfactory receptors in the nasal cavity

an altitude many hikers and skiers have experienced, alveolar P_{O_2} is about 60 mm Hg.

Despite the low alveolar P_{O_2}, millions of people live and work at altitudes this high or higher. Important physiological adjustments include an increased respiratory rate, an increased heart rate, an increased BPG level, and an elevated hematocrit. Thus, even though the hemoglobin is not fully saturated, the bloodstream holds more of it, and the round-trip between the lungs and the peripheral tissues takes less time. However, most of these adaptations take days to weeks to develop. As a result, athletes planning to compete in events at high altitude must begin training under such conditions well in advance.

The respiratory system is functionally linked to all other body systems as well. Build Your Knowledge **Figure 23–28** illustrates these interrelationships with the other body systems studied so far.

Checkpoint

30. **Identify the functional relationship between the respiratory system and all other organ systems.**
31. **Describe the functional relationship between the respiratory system and the lymphatic system.**

See the blue Answers tab at the back of the book.

23 Chapter Review

Study Outline

MasteringA&P™ Access more chapter study tools online in the MasteringA&P Study Area:

- Chapter Quizzes, Chapter Practice Test, MP3 Tutor Sessions, and Clinical Case Studies
- Practice Anatomy Lab PAL 3.0
- A&P Flix A&PFlix
- Interactive Physiology iP2
- PhysioEx PhysioEx 9.1

An introduction to the respiratory system p. 835

1. Body cells must obtain oxygen and eliminate carbon dioxide. Gas exchange takes place at respiratory surfaces inside the lungs.

23-1 The respiratory system, organized into an upper respiratory system and a lower respiratory system, functions primarily to aid gas exchange p. 835

2. The functions of the **respiratory system** include (1) providing an area for gas exchange between air and circulating blood; (2) moving air to and from exchange surfaces; (3) protecting respiratory surfaces from environmental variations and defending the respiratory system and other tissues from invasion by pathogens; (4) producing sounds; and (5) facilitating the detection of olfactory stimuli.
3. The respiratory system includes the **upper respiratory system**, composed of the nose, nasal cavity, paranasal sinuses, and pharynx, and the **lower respiratory system**, which includes the larynx, trachea, bronchi, bronchioles, and alveoli of the lungs. *(Figure 23–1)*
4. The **respiratory tract** consists of the conducting airways that carry air to and from the **alveoli**. The passageways of the upper respiratory tract filter and humidify incoming air. The lower respiratory tract includes delicate conduction passages and the exchange surfaces of the alveoli.
5. The **respiratory mucosa** (respiratory epithelium and underlying connective tissue) lines the conducting portion of the respiratory tract.
6. The respiratory epithelium changes in structure along the respiratory tract. It is supported by the **lamina propria**, a layer of areolar tissue. *(Figure 23–2)*
7. Contamination of the respiratory system is prevented by the mucus and cilia of the **respiratory defense system**. *(Figure 23–2)*

23-2 The conducting portion of the upper respiratory system filters, warms, and humidifies air p. 838

8. The upper respiratory system consists of the nose, nasal cavity, paranasal sinuses, and pharynx. *(Figures 23–1, 23–3)*
9. Air normally enters the respiratory system through the **nostrils**, which open into the *nasal cavity*. The **nasal vestibule** (entryway) is guarded by hairs that screen out large particles. *(Figure 23–3)*
10. Incoming air flows through the **superior**, **middle**, and **inferior meatuses** (narrow grooves) and bounces off the conchal surfaces. *(Figure 23–3)*
11. The **hard palate** separates the oral and nasal cavities. The **soft palate** separates the superior nasopharynx from the rest of the pharynx. The **choanae** connect the nasal cavity and nasopharynx.
12. The nasal mucosa traps particles, warms and humidifies incoming air, and dehumidifies and absorbs heat of outgoing air.
13. The **pharynx**, or throat, is a chamber shared by the digestive and respiratory systems. The **nasopharynx** is the superior part of the pharynx. The **oropharynx** is continuous with the oral cavity. The **laryngopharynx** includes the narrow zone between the hyoid bone and the entrance to the esophagus. *(Figure 23–3)*

23-3 The conducting portion of the lower respiratory system conducts air to the respiratory portion and produces sound p. 841

14. Inhaled air passes through the glottis on the way to the lungs; the **larynx** surrounds and protects the glottis, or vocal apparatus. *(Figure 23–4)*
15. The cylindrical larynx is composed of three large cartilages (the **median thyroid cartilage**, median **cricoid cartilage**, and **epiglottis**) and three smaller pairs of cartilages (the **arytenoid**, **corniculate**, and **cuneiform cartilages**). The epiglottis projects into the pharynx. *(Figures 23–4, 23–5)*
16. A pair of inelastic **vestibular folds** surrounds the glottis. The **glottis** includes the elastic **vocal folds** and the space between them, the *rima glottidis*. *(Figure 23–5)*
17. Air passing through the open glottis vibrates its vocal folds, producing sound. The pitch of the sound depends on the diameter, length, and tension of the vocal folds.
18. The muscles of the neck and pharynx position and stabilize the larynx. The smaller intrinsic muscles regulate tension in the vocal folds or open and close the glottis. During swallowing, both sets of muscles help prevent particles from entering the glottis.
19. The **trachea** extends from the sixth cervical vertebra to the fifth thoracic vertebra. The **submucosa** contains C-shaped **tracheal cartilages**, which stiffen the tracheal walls and protect the airway. The posterior tracheal wall can distort to permit large masses of food to pass through the esophagus. *(Figure 23–6)*
20. The trachea branches within the mediastinum to form the **right** and **left main bronchi**. Each bronchus enters a lung at the **hilum** (a groove). The **root of the lung** is a connective tissue mass that includes the bronchus, pulmonary vessels, and nerves. *(Figures 23–6, 23–9)*
21. The main bronchi and their branches form the **bronchial tree**. The **lobar** and **segmental bronchi** are branches within the lungs. As they branch, the amount of cartilage in their walls decreases and the amount of smooth muscle increases. *(Figure 23–7)*
22. Each tertiary bronchus supplies air to a single **bronchopulmonary segment**. *(Figure 23–7)*
23. **Bronchioles** within the bronchopulmonary segments ultimately branch into **terminal bronchioles**.

23

23-4 The respiratory portion of the lower respiratory system is where gas exchange occurs p. 846

24. Each terminal bronchiole delivers air to a single **pulmonary lobule** in which the terminal bronchiole branches into **respiratory bronchioles**. The connective tissues of the root of the lung extend into the parenchyma of the lung as a series of *trabeculae* (partitions) that branch to form **interlobular septa**, which divide the lung into lobules. *(Figure 23–7)*
25. The respiratory bronchioles open into **alveolar ducts**, where many alveoli are interconnected. The respiratory exchange surfaces are extensively connected to the circulatory system by the capillaries of the pulmonary circuit. *(Figure 23–7)*
26. The **blood air barrier** consists of a simple squamous epithelium, the endothelial cell lining an adjacent capillary, and their fused basement membranes. **Pneumocytes type II** (type II alveolar cells) scattered in the blood air barrier produce **surfactant** that reduces surface tension and keeps the alveoli from collapsing. **Alveolar macrophages** patrol the epithelium and engulf foreign particles. *(Figure 23–8)*

23-5 Enclosed by pleural cavities, the lungs are paired organs made up of multiple lobes p. 848

27. The **lobes** of the lungs are separated by fissures. The right lung has three lobes, the left lung two. *(Figure 23–9)*
28. The anterior and lateral surfaces of the lungs follow the inner contours of the rib cage. The concavity of the medial surface of the left lung is the **cardiac notch**, which conforms to the shape of the pericardium. *(Figures 23–9, 23–10)*
29. The respiratory exchange surfaces receive blood from arteries of the pulmonary circuit. The conducting portions of the respiratory tract receive blood from the bronchial arteries. Most of the returning venous blood flows into the pulmonary veins, bypassing the rest of the systemic circuit and diluting the oxygenated blood leaving the alveoli.
30. Each lung is surrounded by a single **pleural cavity** lined by a **pleura** (serous membrane). The two types of pleurae are the **parietal pleura**, covering the inner surface of the thoracic wall, and the **visceral pleura**, covering the lungs.

23-6 External respiration and internal respiration allow gas exchange within the body p. 851

31. Respiratory physiology focuses on a series of integrated processes. **External respiration** is the exchange of oxygen and carbon dioxide between interstitial fluid and the external environment and includes *pulmonary ventilation* (breathing). **Internal respiration** is the exchange of oxygen and carbon dioxide between interstitial fluid and cells. If the oxygen content declines, the affected tissues suffer from **hypoxia**; if the oxygen supply is completely shut off, **anoxia** and tissue death result. *(Figure 23–11)*

23-7 Pulmonary ventilation—air exchange between the atmosphere and the lungs—involves muscle actions and volume changes that cause pressure changes p. 852

32. **Pulmonary ventilation** is the physical movement of air into and out of the respiratory tract.
33. Decreasing the volume of a gas increases its pressure. Increasing the volume of a gas decreases its pressure. This inverse relationship is **Boyle's law**. *(Figure 23–12)*
34. Movement of the diaphragm and ribs changes lung volume. **Primary respiratory muscles** are the diaphragm and external intercostal muscles. *(Spotlight Figure 23–13, Figure 23–14)*
35. The relationship between **intrapulmonary pressure** (the pressure inside the respiratory tract) and **atmospheric pressure(atm)** determines the direction of airflow. **Intrapleural pressure** is the pressure in the potential space between the parietal and visceral pleurae. *(Figure 23–15)*

36. A **respiratory cycle** is a single cycle of inhalation and exhalation. The amount of air moved in one respiratory cycle is the **tidal volume**. A **spirometer** is an instrument used to measure the capacity of the lungs. *(Figure 23–15)*
37. **Alveolar ventilation** is the amount of air reaching the alveoli each minute. The **vital capacity** includes the **tidal volume** plus the **expiratory** and **inspiratory reserve volumes**. The air left in the lungs at the end of maximum exhalation is the **residual volume**. *(Figure 23–16)*

23-8 Gas exchange depends on the partial pressures of gases and the diffusion of gas molecules p. 860

38. In a mixture of gases, the individual gases exert a pressure proportional to their abundance in the mixture (**Dalton's law**). The pressure contributed by a single gas is its **partial pressure**. *(Table 23–1)*
39. The amount of a gas in solution is directly proportional to the partial pressure of that gas (**Henry's law**). *(Figure 23–17)*
40. Alveolar air and atmospheric air differ in composition. Gas exchange across the blood air barrier is efficient due to differences in partial pressures, the short diffusion distance, lipid-soluble gases, the large surface area of all the alveoli combined, and the coordination of blood flow and airflow. *(Figure 23–18)*

23-9 In gas transport, most oxygen is transported bound to hemoglobin, whereas carbon dioxide is transported in three ways p. 864

41. Blood entering peripheral capillaries delivers oxygen and absorbs carbon dioxide. The transport of oxygen and carbon dioxide in blood involves reactions that are completely reversible.
42. Oxygen is carried mainly by RBCs, reversibly bound to hemoglobin. At alveolar P_{O_2}, hemoglobin is almost fully saturated; at the P_{O_2} of peripheral tissues, it releases oxygen but still retains a substantial oxygen reserve. The effect of pH on the hemoglobin saturation curve is called the **Bohr effect**. When low plasma P_{O_2} continues for extended periods, red blood cells generate more **2,3-bisphosphoglycerate (BPG)**, which reduces hemoglobin's affinity for oxygen. *(Figures 23–19, 23–20)*
43. **Fetal hemoglobin** has a stronger affinity for oxygen than does adult hemoglobin, aiding the removal of oxygen from maternal blood. *(Figure 23–21)*
44. Aerobic metabolism in peripheral tissues generates CO_2. Roughly 70% is transported as bicarbonate ions (HCO_3^-), 23% is bound as **carbaminohemoglobin**, and about 7% of the CO_2 in blood is dissolved in the plasma. *(Figure 23–22)*
45. Driven by differences in partial pressure, oxygen enters the blood at the lungs and leaves it in peripheral tissues; similar forces drive carbon dioxide into the blood at the tissues and into the alveoli at the lungs. *(Figure 23–23)*

23-10 Respiratory centers in the brainstem, along with respiratory reflexes, control respiration p. 868

46. Normally, the cellular rates of gas absorption and generation are matched by the capillary rates of delivery and removal and are identical to the rates of oxygen absorption and carbon dioxide removal at the lungs. When these rates are unbalanced, homeostatic mechanisms restore equilibrium.
47. Local factors regulate alveolar blood flow (*lung perfusion*) and airflow (*alveolar ventilation*). Alveolar capillaries constrict under conditions of low oxygen, and bronchioles dilate under conditions of high carbon dioxide.
48. The **respiratory centers** include three pairs of nuclei in the reticular formation of the pons and medulla oblongata. The *respiratory rhythmicity centers* set the pace for respiration; the **apneustic centers** cause strong, sustained inspiratory movements; and the **pneumotaxic centers** inhibit the apneustic centers and promote exhalation. *(Figure 23–24, Spotlight Figure 23–25)*
49. Stimulation of the chemoreceptor reflexes is based on the level of carbon dioxide in the blood and CSF. *(Spotlight Figure 23–25, Figure 23–26)*
50. The **inflation reflex** prevents overexpansion of the lungs during forced breathing. The **deflation reflex** stimulates inhalation when the lungs are collapsing.
51. Conscious and unconscious thought processes can affect respiration by affecting the respiratory centers.

23-11 Respiratory performance changes over the life span p. 876

52. Before delivery, the fetal lungs are filled with body fluids and are collapsed. At the first breath, the lungs inflate and do not collapse completely thereafter.
53. The respiratory system is generally less efficient in elderly individuals because (1) elastic tissue deteriorates, lowering compliance and the vital capacity of the lungs; (2) movements of the chest are restricted by arthritic changes and decreased flexibility of costal cartilages; and (3) some degree of emphysema is generally present. *(Figure 23–27)*

23-12 The respiratory system provides oxygen to, and eliminates carbon dioxide from, other organ systems p. 877

54. The respiratory system has extensive anatomical and physiological connections with the cardiovascular system. *(Figure 23–28)*

Review Questions

See the blue Answers tab at the back of the book.

LEVEL 1 Reviewing Facts and Terms

1. Surfactant **(a)** protects the surface of the lungs, **(b)** phagocytizes small particulates, **(c)** replaces mucus in the alveoli, **(d)** helps prevent the alveoli from collapsing, **(e)** is not found in healthy lung tissue.
2. The hard palate separates the **(a)** nasal cavity from the larynx, **(b)** left and right sides of the nasal cavity, **(c)** nasal cavity and the oral cavity, **(d)** nostrils from the choanae, **(e)** soft palate from the nasal cavity.
3. Air moves into the lungs because **(a)** the gas pressure in the lungs is less than atmospheric pressure, **(b)** the volume of the lungs decreases with inspiration, **(c)** the thorax is muscular, **(d)** contraction of the diaphragm decreases the volume of the thoracic cavity, **(e)** the respiratory control center initiates active expansion of the thorax.
4. The glottis closes partway through an exhalation. The abdominal and internal intercostal muscles then contract suddenly, creating pressure that blasts the air out of the respiratory passages. This describes a **(a)** sneeze, **(b)** hiccough, **(c)** cough, **(d)** laryngeal spasm, **(e)** gag.
5. When the diaphragm and external intercostal muscles contract, **(a)** exhalation occurs, **(b)** intrapulmonary pressure increases, **(c)** intrapleural pressure decreases, **(d)** the volume of the lungs decreases, **(e)** the size of the thoracic cavity increases.
6. Identify the structures of the respiratory system in the figure below:

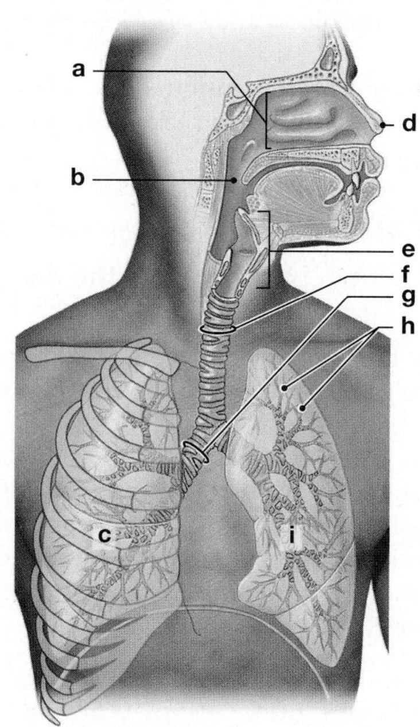
a
b
c
d
e
f
g
h
i

(a) ________
(b) ________
(c) ________
(d) ________
(e) ________
(f) ________
(g) ________
(h) ________
(i) ________

7. During the winter, Brad sleeps in a dorm room that lacks any humidifier for the heated air. In the mornings he notices that his nose is "stuffy" similar to when he has a cold, but after showering and drinking some water, the stuffiness disappears until the next morning. What might be the cause of Brad's nasal condition?
8. Distinguish the structures of the upper respiratory system from those of the lower respiratory system.
9. Name the three regions of the pharynx. Where is each region located?
10. List the cartilages of the larynx. What are the functions of each?
11. What three integrated steps are involved in external respiration?
12. What important physiological differences exist between fetal hemoglobin and maternal hemoglobin?
13. By what three ways is carbon dioxide transported in the bloodstream?

LEVEL 2 Reviewing Concepts

14. Which of the following does *not* occur in internal respiration? **(a)** Oxygen diffuses from the blood to the interstitial spaces. **(b)** Carbon dioxide diffuses from the interstitial spaces to the blood. **(c)** Hemoglobin binds more oxygen. **(d)** Bicarbonate ions are formed in red blood cells. **(e)** Chloride ions diffuse into red blood cells as bicarbonate ions diffuse out.
15. Gas exchange at the blood air barrier is efficient because **(a)** the differences in partial pressure are substantial, **(b)** the gases are lipid soluble, **(c)** the total surface area is large, **(d)** of all of these.
16. For any partial pressure of oxygen, if the concentration of 2,3-bisphosphoglycerate (BPG) increases, **(a)** the amount of oxygen released by hemoglobin will decrease, **(b)** the oxygen levels in hemoglobin will be unaffected, **(c)** the amount of oxygen released by hemoglobin will increase, **(d)** the amount of carbon dioxide carried by hemoglobin will increase.
17. An increase in the partial pressure of carbon dioxide in arterial blood causes chemoreceptors to stimulate the respiratory centers, resulting in **(a)** a decreased respiratory rate, **(b)** an increased respiratory rate, **(c)** hypocapnia, **(d)** hypercapnia.
18. Why is breathing through the nasal cavity more desirable than breathing through the mouth?
19. Correctly justify the statement "The bronchioles are to the respiratory system what the arterioles are to the cardiovascular system."
20. How are pneumocytes type II involved with keeping the alveoli from collapsing?
21. How does pulmonary ventilation differ from alveolar ventilation, and what is the function of each type of ventilation?
22. What is the significance of (a) Boyle's law, (b) Dalton's law, and (c) Henry's law to the process of respiration?
23. What happens to the process of respiration when a person is sneezing or coughing?
24. What are the differences between pulmonary volumes and respiratory capacities? How are pulmonary volumes and respiratory capacities determined?
25. What is the functional difference between the dorsal respiratory group (DRG) and the ventral respiratory group (VRG) of the medulla oblongata?

LEVEL 3 Critical Thinking and Clinical Applications

26. Billy's normal alveolar ventilation rate (AVR) during mild exercise is 6.0 L/min. While at the beach on a warm summer day, he goes snorkeling. The snorkel has a volume of 50 mL. Assuming that the water is not too cold and that snorkeling is mild exercise for Billy, what would his respiratory rate have to be for him to maintain an AVR of 6.0 L/min while snorkeling? (Assume a constant tidal volume of 500 mL and an anatomic dead space of 150 mL.)
27. Mr. B. has had chronic advanced emphysema for 15 years. While hospitalized with a respiratory infection, he goes into respiratory distress. Without thinking, his nurse immediately administers pure oxygen, which causes Mr. B. to stop breathing. Why?
28. Cary hyperventilates for several minutes before diving into a swimming pool. After he enters and begins swimming underwater, he blacks out and almost drowns. What caused this to happen?
29. Why do individuals who are anemic generally not exhibit an increase in respiratory rate or tidal volume, even though their blood is not carrying enough oxygen?
30. Doris has an obstruction of her right primary bronchus. As a result, how would the oxygen–hemoglobin saturation curve for her right lung compare with that for her left?

+ CLINICAL CASE Wrap-Up No Rest for the Weary

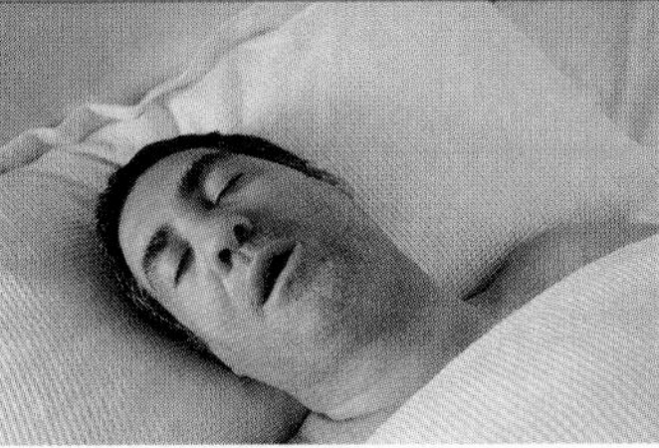

Doug has *obstructive sleep apnea*. During sleep his airway becomes blocked and he stops breathing. Several risk factors are promoting this condition. Doug's throat is tight due to the structure of his palate. Excess fat in his neck compresses his throat from the outside. When he falls asleep on his back, his pharyngeal muscles and tongue relax and drop back into his throat, helped by the muscle-relaxing effects of alcohol. As he breathes, especially on inhalation, the tissues in his throat vibrate and make the noise of snoring. The narrower a person's airway, the louder their snoring! Every time his throat collapses completely, Doug experiences apnea: He gets no oxygen for a period of multiple seconds to a minute, or more. Apnea puts strain on his heart and raises his blood pressure. His nighttime rest is disturbed and so is his daytime function. This is a serious problem.

The doctor wants Doug to reduce his risk factors by losing weight and avoiding alcohol at night. In the short term, the treatment of choice for sleep apnea is a CPAP (continuous positive airway pressure) machine. Doug wears the mask device at night. It keeps his airway open by blowing air into the back of his throat.

1. Which parts of the upper respiratory tract (URT) can contribute to obstructed breathing?
2. Children can snore too. A tonsillectomy might eliminate the snoring. Why?

See the blue Answers tab at the back of the book.

Related Clinical Terms

asbestosis: Pneumoconiosis, disease of the lungs, caused by the inhalation of asbestos particles over time.

asphyxia: Impaired oxygen–carbon dioxide exchange that results in suffocation.

aspirate: Drawing fluid from the body by suction; foreign material accidentally sucked into the lungs.

bronchography: A procedure in which radiopaque materials are introduced into the airways to improve x-ray imaging of the bronchial tree.

bronchoscope: A fiber-optic bundle small enough to be inserted into the trachea and finer airways; the procedure is called *bronchoscopy*.

cardiopulmonary resuscitation (CPR): The application of cycles of compression to the rib cage to restore cardiovascular and respiratory function.

Cheyne-Stokes breathing: Hyperpnea (deep, fast breathing) alternating with apnea (absence of breathing).

chronic obstructive pulmonary disease (COPD): A general term describing temporary or permanent lung disease of the bronchial tree.

dyspnea: The condition of labored breathing.

endotracheal tube: Tube that is passed through the mouth or nose to the trachea.

Heimlich maneuver or abdominal thrust: Sudden compression applied to the abdomen just inferior to the diaphragm, to force air out of the lungs to clear a blocked trachea or larynx.

hemoptysis: Coughing up blood or bloody mucus.

hemothorax: The condition of having blood in the pleural cavity.

hyperbaric oxygenation: Therapy to force more oxygen into the blood by use of a pressure chamber.

nasal polyps: Benign growths on the mucous lining of the nasal cavity.

orthopnea: Condition in which one has breathing difficulty except when in an upright position.

otorhinolaryngology: Branch of medicine dealing with disease and treatment of the ear, nose, and throat.

rales: Abnormal hissing or other respiratory sounds.

rhinitis: Inflammation of the nasal cavity.

rhinoplasty: Plastic surgery of the nose.

severe acute respiratory syndrome (SARS): A harsh viral respiratory illness caused by a coronavirus that typically progresses to pneumonia.

sputum: A mixture of saliva and mucus coughed up from the respiratory tract, often as the result of an infection.

stridor: Harsh, vibrating breathing sound caused by an obstruction in the windpipe or larynx.

stuttering: To speak with a continued involuntary repetition of sounds.

tachypnea: Rapid rate of breathing.

tussis: A cough.

wheeze: An audible whistling sound when breathing.

24 The Digestive System

Learning Outcomes

These Learning Outcomes correspond by number to this chapter's sections and indicate what you should be able to do after completing the chapter.

24-1 ■ Identify the organs of the digestive system, list their major functions, describe the histology of the digestive tract, and outline the mechanisms that regulate digestion. p. 885

24-2 ■ Discuss the anatomy of the oral cavity, and list the functions of its major structures and regions. p. 893

24-3 ■ Describe the structure and functions of the pharynx and the esophagus. p. 898

24-4 ■ Describe the anatomy of the stomach, including its histology, and discuss its roles in digestion and absorption. p. 901

24-5 ■ Describe the structure, functions, and regulation of the accessory digestive organs. p. 905

24-6 ■ Describe the anatomy and histology of the small intestine, and explain the functions and regulation of intestinal secretions. p. 915

24-7 ■ Describe the gross and histological structures of the large intestine, including its regional specializations and role in nutrient absorption. p. 921

24-8 ■ List the nutrients required by the body, describe the chemical events responsible for the digestion of organic nutrients, and describe the mechanisms involved in the absorption of organic and inorganic nutrients. p. 926

24-9 ■ Summarize the effects of aging on the digestive system. p. 931

24-10 ■ Give examples of interactions between the digestive system and other body systems studied so far. p. 932

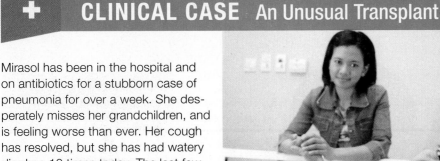

CLINICAL CASE An Unusual Transplant

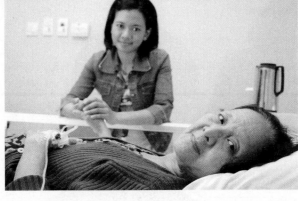

Mirasol has been in the hospital and on antibiotics for a stubborn case of pneumonia for over a week. She desperately misses her grandchildren, and is feeling worse than ever. Her cough has resolved, but she has had watery diarrhea 10 times today. The last few times, there was blood in her stool. She has lost her appetite and feels crampy abdominal pain. She is dehydrated and unable to drink enough liquids to keep up with her fluid loss from diarrhea. She also has a fever and has lost weight.

Her doctor enters the room with results of recent tests performed on her stool. "I'm afraid the normal, helpful bacteria that live in your colon have been killed off by the antibiotics used to cure your pneumonia. Now, harmful bacteria have taken over and are causing your bloody diarrhea," reports the doctor. "Your colon is overgrown with *Clostridium difficile* bacteria. These bacteria are producing toxins that are causing your colitis (inflammation of the colon). The usual treatment for this condition is more antibiotics."

"Doctor," cries Mirasol, "I don't think I can take any more antibiotics. I feel worse now than when I arrived."

"Well, I have an idea that may work," the doctor replies. "A transplant, from one of your healthy, adult grandchildren. It's a little unusual, however. You need to keep an open mind." **What kind of transplant can possibly help Mirasol? To find out, turn to the Clinical Case Wrap-Up on p. 938.**

An Introduction to the Digestive System

All living organisms must get nutrients from their environment to sustain life. These nutrients are used as raw materials for synthesizing essential compounds (anabolism). They are also broken down to provide the energy that cells need to continue functioning (catabolism). pp. 37–38, 320–322 The *digestive system,* working with the cardiovascular and lymphatic systems, provides the needed nutrients in the form of organic molecules (such as carbohydrates, fats, or proteins). In effect, the digestive system supplies both the fuel that keeps all the body's cells functioning and the building blocks needed for cell growth and repair.

In this chapter we discuss the structure and function of the major body passage known as the *digestive tract,* as well as several accessory organs. We look at the process of digestion and how it breaks down large and complex organic molecules into smaller fragments that can be absorbed by the digestive epithelium. We also see how some types of organic wastes are removed from the body.

24-1 The digestive system, consisting of the digestive tract and accessory organs, functions primarily to break down and absorb nutrients from food and to eliminate wastes

Learning Outcome **Identify the organs of the digestive system, list their major functions, describe the histology of the digestive tract, and outline the mechanisms that regulate digestion.**

The **digestive system** (*alimentary system*) consists of a muscular tube, the **digestive tract**, also called the *gastrointestinal (GI) tract* or *alimentary canal,* and various **accessory organs**. Figure 24–1 shows the major organs of the digestive system. The digestive tract begins at the *oral cavity* (mouth). It continues through the *pharynx* (throat), *esophagus, stomach, small intestine,* and *large intestine,* which opens to the exterior at the *anus.* Accessory digestive organs include the teeth, tongue, and various *glandular organs,* such as the *salivary glands, liver,* and *pancreas,* as well as the *gallbladder,* which only has a secretory function.

These structures have overlapping functions that all contribute to the main function of the digestive system, which is to break down food, absorb the nutrients that can be used by body cells, and eliminate the remaining waste. However, each of these structures also has certain areas of specialization and so shows distinctive histological characteristics. This section introduces the functions of the digestive system. We also look at several structural and functional characteristics of the system as a whole, including how its epithelial lining is specialized for absorption, how materials move through it, and how those movements are regulated.

Functions and Processes of the Digestive System

Food enters the digestive tract and passes along its length. Along the way, the digestive system performs six integrated processes. Let's take a look at these processes:

1. **Ingestion** takes place when food enters the oral cavity. Ingestion is an active process involving choices and decision making.
2. **Mechanical digestion and propulsion** involve the crushing and shearing of food and then propelling the food along the digestive tract. Mechanical digestion may not

Figure 24–1 **Organs of the Digestive System.**

Major Organs of the Digestive Tract	
Oral Cavity (Mouth)	Ingestion, mechanical digestion with accessory organs (teeth and tongue), moistening, mixing with salivary secretions
Pharynx	Muscular propulsion of materials into the esophagus
Esophagus	Transport of materials to the stomach
Stomach	Chemical digestion of materials by acid and enzymes; mechanical digestion through muscular contractions
Small Intestine	Enzymatic digestion and absorption of water, organic substrates, vitamins, and ions
Large Intestine	Dehydration and compaction of indigestible materials in preparation for elimination

Anus

Accessory Organs of the Digestive System	
Teeth	Mechanical digestion by chewing (mastication)
Tongue	Assists mechanical digestion with teeth, sensory analysis
Salivary Glands	Secretion of lubricating fluid containing enzymes that break down carbohydrates
Liver	Secretion of bile (important for lipid digestion), storage of nutrients, many other vital functions
Gallbladder	Storage and concentration of bile
Pancreas	Exocrine cells secrete buffers and digestive enzymes; endocrine cells secrete hormones

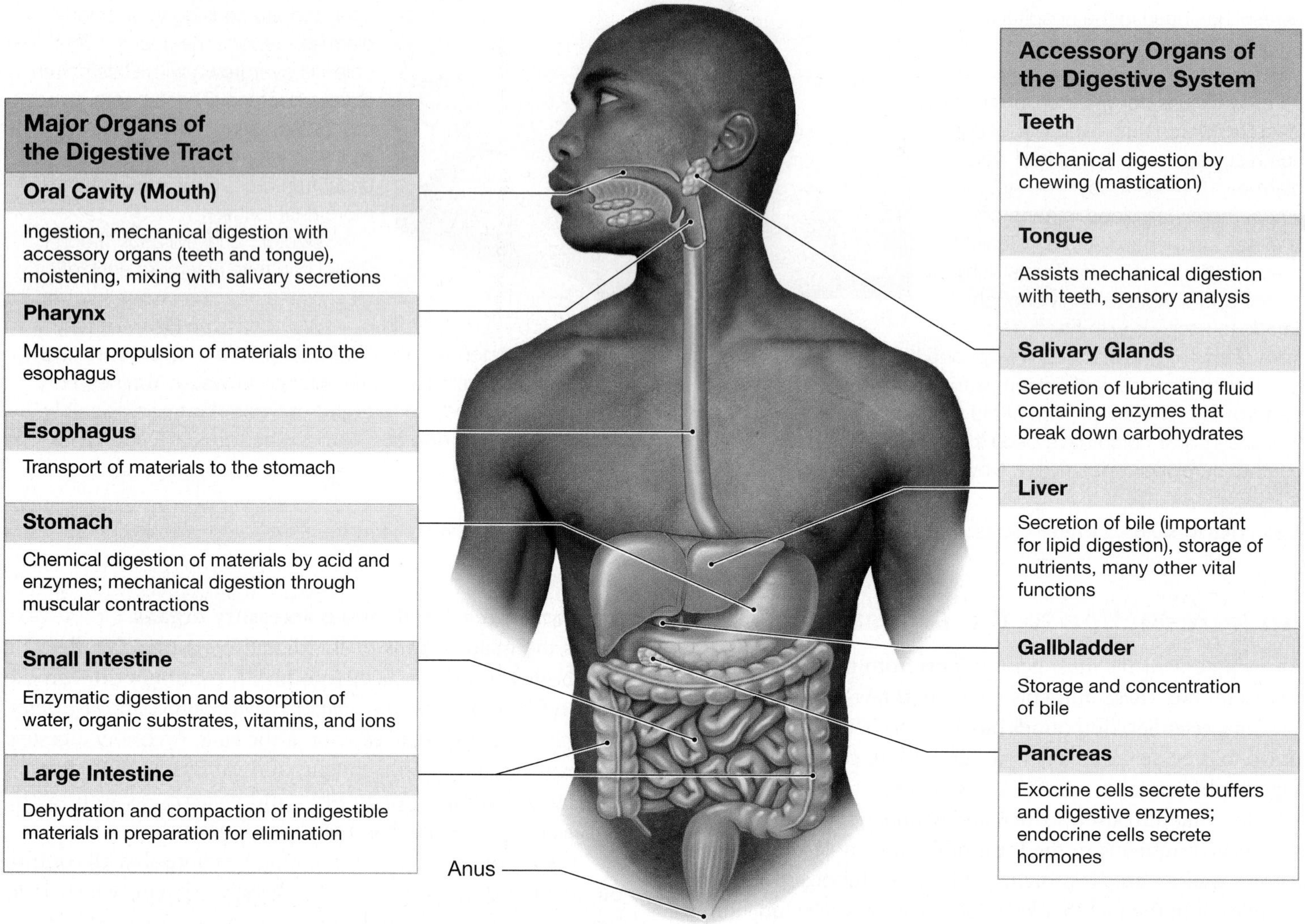

be required before ingestion. You can swallow liquids immediately, but you must chew most solids first. Tearing and mashing with the teeth, followed by squashing and compressing by the tongue, are examples of preliminary mechanical digestion. Swirling, mixing, and churning motions of the stomach and intestines provide mechanical digestion after ingestion. Mechanical digestion also increases the surface area of ingested materials, making them more susceptible to the actions of enzymes.

3. **Chemical digestion** refers to the chemical breakdown of food into small organic and inorganic molecules suitable for absorption by the digestive epithelium. Simple molecules in food, such as glucose, can be absorbed intact. However, epithelial cells have no way to absorb molecules with the size and complexity of proteins, polysaccharides, or triglycerides. Digestive enzymes must first disassemble these molecules. For example, the starches in a potato are of no nutritional value until enzymes have broken them down to simple sugars.

4. **Secretion** is the release of water, acids, enzymes, buffers, and salts by the epithelium of the digestive tract, glandular organs, and the gallbladder. The glandular organs and gallbladder secrete their products into ducts that empty into the digestive tract.

5. **Absorption** is the movement of organic molecules, electrolytes (inorganic ions), vitamins, minerals, and water across the digestive epithelium. These materials are absorbed into the interstitial fluid of the digestive tract for distribution to body cells.

6. **Defecation** (def-eh-KĀ-shun) is the elimination of wastes from the body. The digestive tract and glandular and secretory organs discharge wastes in secretions that enter the lumen of the tract. Most of these wastes mix with the indigestible food of the digestive process. These wastes are dehydrated and compacted into matter called *feces* and ejected from the digestive tract through the anus.

This chapter explores these specific processes in more detail as we consider the structure and function of individual regions of the digestive system.

In addition to its main function, the digestive system also plays a protective role. The lining of the digestive tract safeguards surrounding tissues against (1) the corrosive effects of digestive acids and enzymes; (2) mechanical stresses, such as abrasion; and (3) bacteria that are swallowed with food or live in the digestive tract. The digestive epithelium and its secretions provide a nonspecific defense against these bacteria. When bacteria reach the underlying layer of areolar tissue, the *lamina propria*, macrophages and other cells of the immune system attack them.

Relationship between the Digestive Organs and the Peritoneum: The Mesenteries

The abdominopelvic cavity contains the *peritoneal cavity*, which is lined by a serous membrane called the **peritoneum**. This membrane consists of a superficial mesothelium covering a layer of areolar tissue. pp. 18, 140 We divide the peritoneum into two parts. The serosa, or *visceral peritoneum*, covers organs that project into the peritoneal cavity. The *parietal peritoneum* lines the inner surfaces of the body wall.

The serous membrane lining the peritoneal cavity continuously produces peritoneal fluid, which lubricates the surfaces. Because a thin layer of peritoneal fluid separates them, the parietal and visceral surfaces can slide without friction and resulting irritation. The membrane secretes and reabsorbs about 7 liters (7.4 quarts) of fluid each day, but the volume within the peritoneal cavity at any one time is very small. Liver disease, kidney disease, and heart failure can increase the rate at which fluids move into the peritoneal cavity. The buildup of fluid creates a characteristic abdominal swelling called **ascites** (a-SĪ-tēz). This fluid can distort internal organs and cause symptoms such as heartburn, indigestion, and lower back pain.

Portions of the digestive tract are suspended within the peritoneal cavity by sheets of serous membrane that connect the parietal peritoneum with the visceral peritoneum. These **mesenteries** (MEZ-en-ter-ēz) are double sheets of peritoneal membrane. The areolar tissue between the mesothelial surfaces provides a route to and from the digestive tract for blood vessels, nerves, and lymphatic vessels. Mesenteries stabilize the positions of the attached organs. The mesenteries also prevent the intestines from becoming entangled during digestive movements or sudden changes in body position.

During embryonic development, the digestive tract and accessory organs are suspended within the peritoneal cavity by *dorsal* and *ventral mesenteries* (Figure 24–2a). The ventral mesentery later disappears along most of the digestive tract. It persists in adults in only two places: on the ventral surface of the stomach, between the stomach and the liver (the *lesser omentum*) (Figure 24–2b,d); and between the liver and the anterior abdominal wall (the *falciform ligament*) (Figure 24–2c,d). The **lesser omentum** (ō-MEN-tum; *omentum*, apron) stabilizes the position of the stomach and provides an access route for blood vessels and other structures entering or leaving the liver. The **falciform** (FAL-si-form; *falx*, sickle + *forma*, form) **ligament** helps stabilize the position of the liver relative to the diaphragm and abdominal wall. ATLAS: Embryology Summary 19: The Development of the Digestive System

As the digestive tract elongates, it twists and turns within the crowded peritoneal cavity. The dorsal mesentery of the stomach becomes greatly enlarged and forms an enormous pouch that extends inferiorly between the body wall and the anterior surface of the small intestine. This pouch is the **greater omentum** (Figure 24–2b,d). It hangs like an apron from the lateral and inferior borders of the stomach. Adipose tissue in the greater omentum conforms to the shapes of the surrounding organs, providing padding and protection across the anterior and lateral surfaces of the abdomen. When a person gains weight, this adipose tissue contributes to the characteristic "beer belly." The lipids in the adipose tissue are an important energy reserve. The greater omentum also insulates the body, reducing heat loss across the anterior abdominal wall.

All but the first 25 cm (10 in.) of the small intestine is suspended by the **mesentery proper**, a thick mesenterial sheet. It provides stability, but permits some independent movement. The mesentery associated with the initial portion of the small intestine (the *duodenum*) and the pancreas fuses with the posterior abdominal wall and locks those structures in place. Only their anterior surfaces remain covered by peritoneum. We describe these organs as **retroperitoneal** (*retro*, behind) because their mass lies posterior to, rather than surrounded by, the peritoneal cavity.

24

+ Clinical Note Peritonitis

An inflammation of the peritoneum, the serous membrane of the abdominopelvic cavity, is called **peritonitis** (pehr-ih-tō-NĪ-tis). This painful condition interferes with the normal functions of the affected organs. Physical damage, chemical irritation, and bacterial invasion of the peritoneum can lead to severe cases of peritonitis. Such cases can even be fatal. In untreated appendicitis, the appendix may rupture and release bacteria into the peritoneal cavity. This event may cause peritonitis. Peritonitis can also be a complication of any surgery that opens the peritoneal cavity. There is a danger of peritonitis in any disease or injury that perforates the stomach or intestines.

Figure 24–2 **The Mesenteries.**

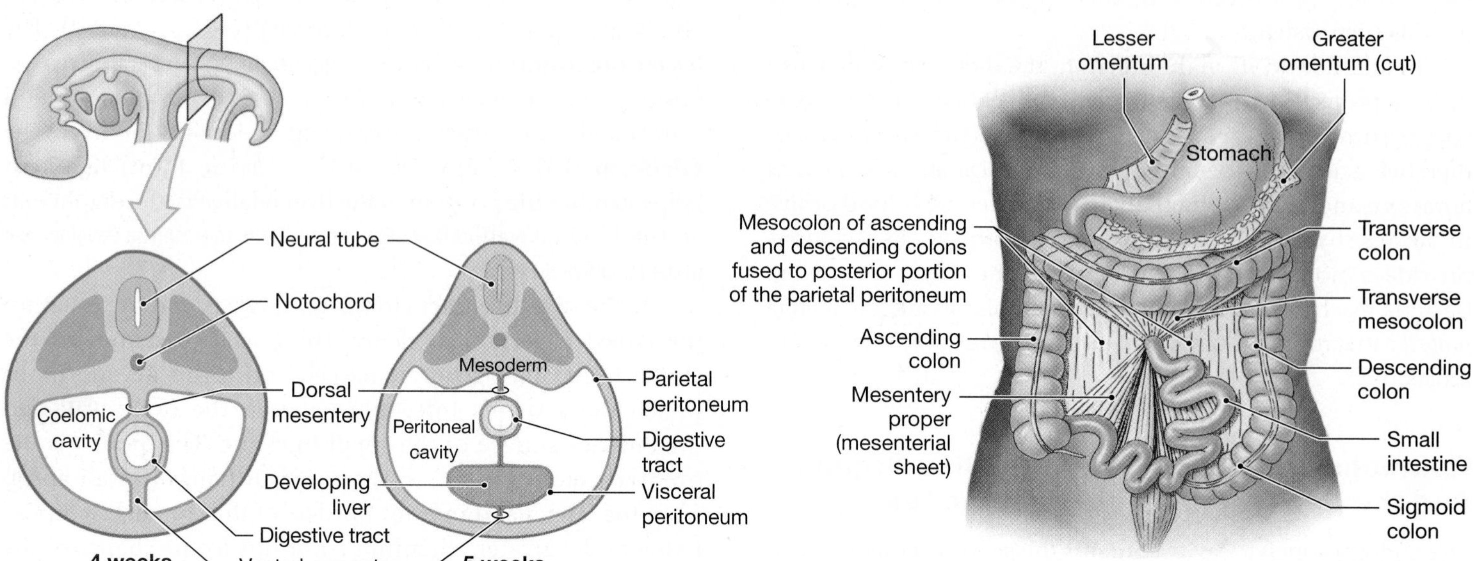
Neural tube
Notochord
Mesoderm
Dorsal mesentery
Coelomic cavity
Peritoneal cavity
Parietal peritoneum
Digestive tract
Developing liver
Visceral peritoneum
Digestive tract
Ventral mesentery
4 weeks
5 weeks
Lesser omentum
Greater omentum (cut)
Stomach
Mesocolon of ascending and descending colons fused to posterior portion of the parietal peritoneum
Transverse colon
Transverse mesocolon
Ascending colon
Descending colon
Mesentery proper (mesenterial sheet)
Small intestine
Sigmoid colon

a During embryonic development, the digestive tube is initially suspended by dorsal and ventral mesenteries. In adults, the ventral mesentery is lost except at the lesser omentum between the stomach and liver and the falciform ligament between the liver and diaphragm (see part d).

b A diagrammatic view of the organization of mesenteries in an adult. As the digestive tract enlarges, mesenteries associated with the proximal portion of the small intestine, the pancreas, and the ascending and descending portions of the colon fuse to the body wall.

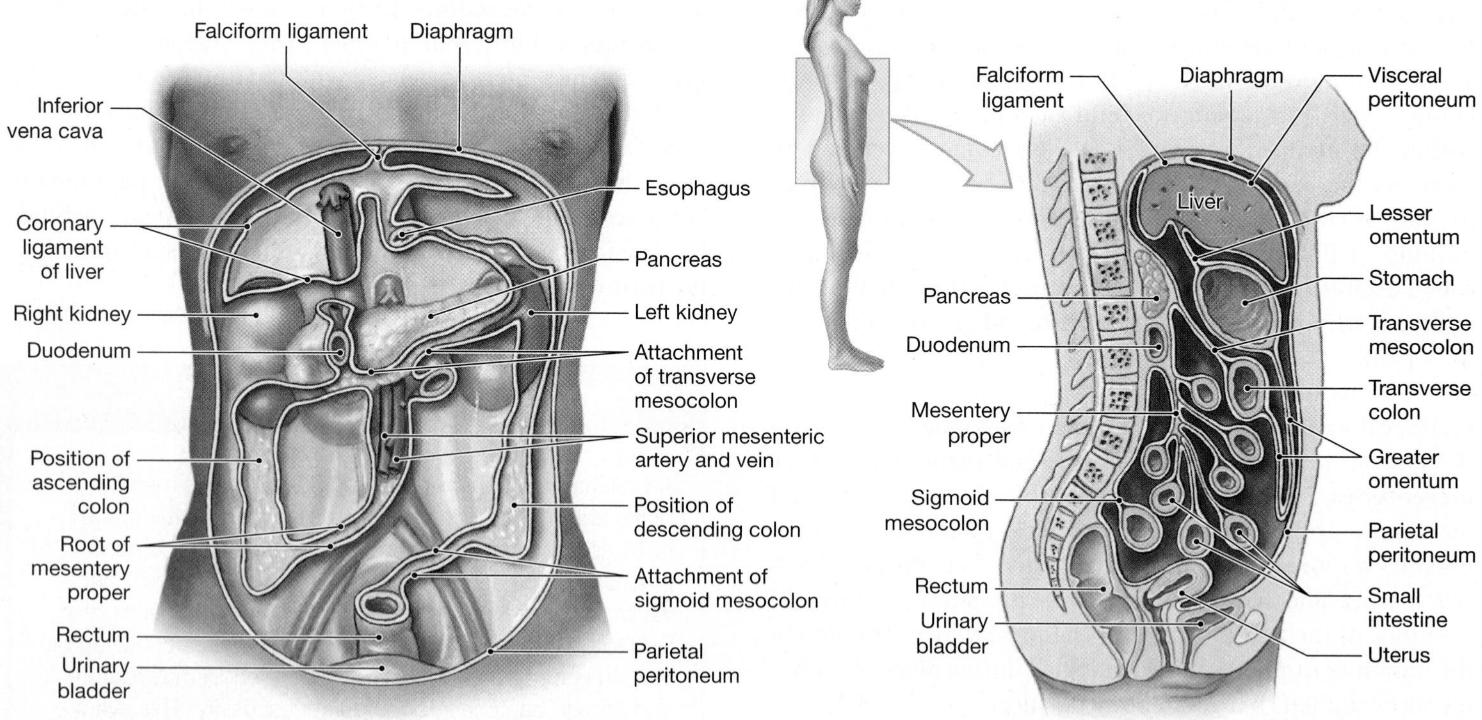
Falciform ligament
Diaphragm
Inferior vena cava
Esophagus
Coronary ligament of liver
Pancreas
Right kidney
Left kidney
Duodenum
Attachment of transverse mesocolon
Superior mesenteric artery and vein
Position of ascending colon
Position of descending colon
Root of mesentery proper
Attachment of sigmoid mesocolon
Rectum
Parietal peritoneum
Urinary bladder
Falciform ligament
Diaphragm
Visceral peritoneum
Liver
Lesser omentum
Stomach
Pancreas
Duodenum
Transverse mesocolon
Transverse colon
Mesentery proper
Greater omentum
Sigmoid mesocolon
Parietal peritoneum
Rectum
Small intestine
Urinary bladder
Uterus

c An anterior view of the empty peritoneal cavity, showing the attachment of mesenteries to the posterior body wall. Some visceral organs that were originally suspended within the peritoneal cavity are now retroperitoneal due to fusion of the serosa with the parietal peritoneum.

d A sagittal section showing the mesenteries of an adult. Notice that the pancreas, duodenum, and rectum are retroperitoneal.

24

A **mesocolon** is a mesentery associated with a portion of the large intestine. During normal development, the mesocolon of the *ascending colon*, the *descending colon*, and the *rectum* of the large intestine fuse to the posterior body wall. These regions become locked in place. Thereafter, these organs are retroperitoneal. The visceral peritoneum covers only their anterior surfaces and portions of their lateral surfaces (see Figure 24–2b,c,d).

Histology of the Digestive Tract

The four major layers of the digestive tract are (1) the *mucosa*, (2) the *submucosa*, (3) the *muscular layer*, and (4) the *serosa*. The structure of these layers varies by region. Figure 24–3 is a composite view. It most closely resembles the small intestine, the longest segment of the digestive tract.

The lining of the digestive tract varies by region. For example, the stomach lining has longitudinal folds, which disappear as the tract fills. The lining of the small intestine has permanent transverse folds called *circular folds*, or *plicae* (PLĪ-sē; folds) *circulares* (see Figure 24–3). This folding, along with small mucosal projections called *villi*, increases the surface area available for absorption. The secretions of gland cells in the mucosa and submucosa—or in accessory glandular organs and the gallbladder—are carried to the epithelial surfaces by ducts.

The Mucosa

The inner lining, or **mucosa**, of the digestive tract is a *mucous membrane*. It consists of an epithelium, moistened by glandular secretions, a *lamina propria* of areolar tissue, and a muscular *muscularis mucosae*.

The Digestive Epithelium. The mucosal epithelium is either simple or stratified, depending on its location and the stresses placed on it. Mechanical stresses are most severe in the oral cavity, pharynx, esophagus, and anal canal. These structures are lined by a stratified squamous epithelium. In contrast, the stomach, the small intestine, and almost the entire length of the large intestine have a simple columnar epithelium. In the stomach, this epithelium contains mucous cells, and in the intestines the epithelium contains goblet cells. Scattered among the columnar cells are **enteroendocrine cells**. They secrete hormones that coordinate the activities of the digestive tract and the accessory glands.

The Lamina Propria. The lamina propria is a layer of areolar tissue that also contains blood vessels, sensory nerve endings, lymphatic vessels, smooth muscle cells, and scattered lymphatic tissue. In the oral cavity, pharynx, esophagus, stomach, and the proximal portion of the small intestine, the lamina propria also contains the secretory cells of intestinal glands.

Figure 24–3 **Histological Organization of the Digestive Tract.** A diagrammatic view of a representative portion of the digestive tract. The features illustrated are typical of those of the small intestine.

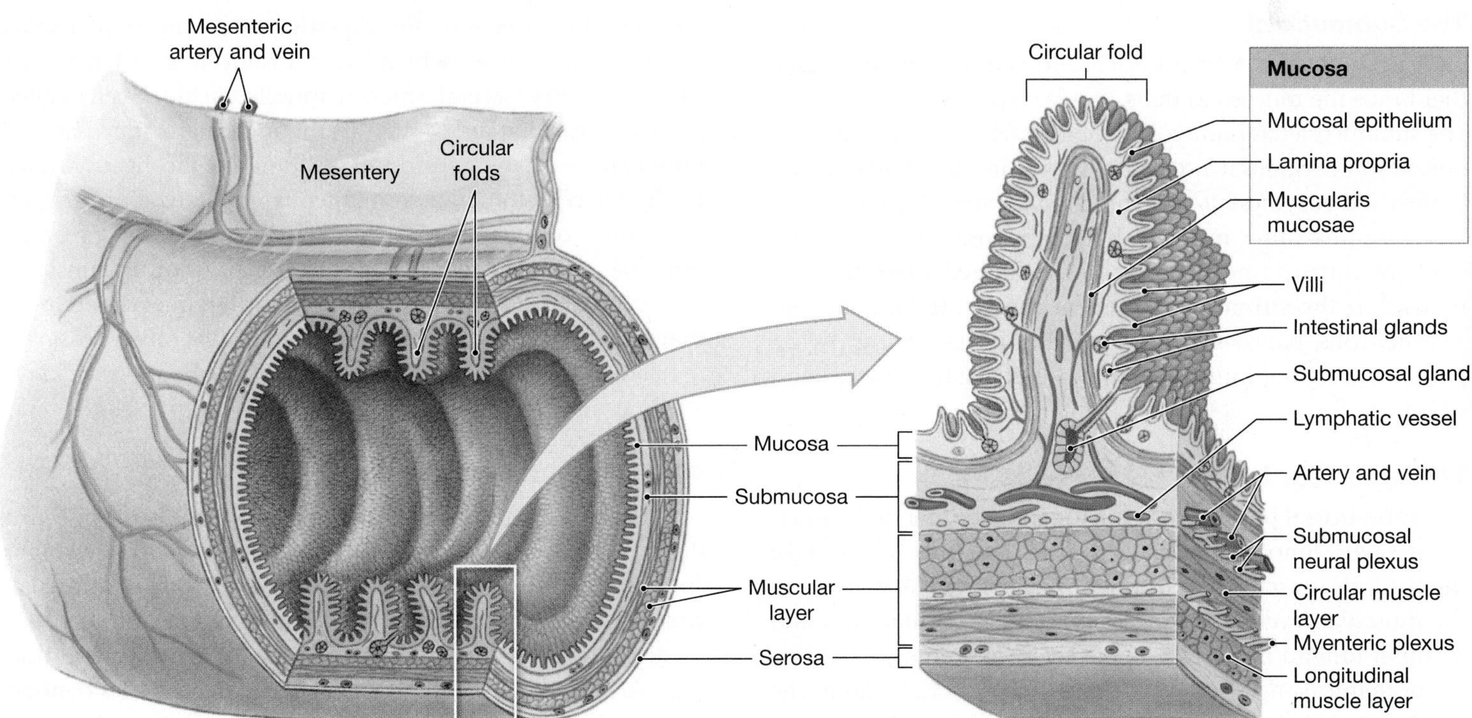
Mesenteric artery and vein
Mesentery
Circular folds
Mucosa
Submucosa
Muscular layer
Serosa
Circular fold
Mucosa
Mucosal epithelium
Lamina propria
Muscularis mucosae
Villi
Intestinal glands
Submucosal gland
Lymphatic vessel
Artery and vein
Submucosal neural plexus
Circular muscle layer
Myenteric plexus
Longitudinal muscle layer

24

Clinical Note Epithelial Renewal and Repair

The life span of a typical epithelial cell varies from 2 to 3 days in the esophagus to 6 days in the large intestine. Epithelial stem cells continuously divide to renew the lining of the entire digestive tract. These divisions normally keep pace with the rates of cell destruction and loss at epithelial surfaces. This high rate of cell division explains why radiation and anticancer drugs that inhibit mitosis have drastic effects on the digestive tract. Lost epithelial cells are no longer replaced. The cumulative damage to the epithelial lining quickly leads to problems in absorbing nutrients. In addition, the exposure of the lamina propria to digestive enzymes can cause internal bleeding and other serious problems.

The Muscularis Mucosae. In most areas of the digestive tract, deep to the lamina propria is a narrow sheet of smooth muscle and elastic fibers. This sheet is called the **muscularis** (mus-kyū-LAIR-is) **mucosae** (myū-KŌ-sē) (see Figure 24–3). The smooth muscle cells in the muscularis mucosae are arranged in two concentric layers. The inner layer encircles the lumen (the *circular muscle*), and the outer layer contains muscle cells arranged parallel to the long axis of the tract (the *longitudinal layer*). Contractions in these layers alter the shape of the lumen.

The Submucosa

The **submucosa** is a layer of dense irregular connective tissue that binds the mucosa to the muscular layer (see Figure 24–3). The submucosa has numerous blood vessels and lymphatic vessels. In some regions it also contains exocrine glands that secrete buffers and enzymes into the lumen of the digestive tract.

Along its outer margin, the submucosa contains a network of intrinsic nerve fibers and scattered neurons. This network is the **submucosal neural plexus**. It contains sensory neurons, parasympathetic ganglionic neurons, and sympathetic postganglionic fibers that innervate the mucosa and submucosa.

The Muscular Layer

The submucosal plexus lies along the inner border of the **muscular layer**. Smooth muscle cells dominate this region. Like the smooth muscle cells in the muscularis mucosae, those in the muscular layer are arranged in an inner circular layer and an outer longitudinal layer. These layers play an essential role in mechanical digestion and in moving materials along the digestive tract.

The movements of the digestive tract are coordinated primarily by the sensory neurons, interneurons, and motor neurons of the enteric nervous system (ENS). p. 552 The ENS is primarily innervated by the parasympathetic division of the ANS. Sympathetic postganglionic fibers also synapse here. Many of these fibers continue onward to innervate the mucosa and the **myenteric** (mī-en-TER-ik) **plexus** (*mys*, muscle + *enteron*, intestine). This plexus is a network of parasympathetic ganglia, sensory neurons, interneurons, and sympathetic postganglionic fibers. It lies sandwiched between the circular and longitudinal muscle layers. In general, parasympathetic stimulation *increases* muscle tone and activity. Sympathetic stimulation decreases muscle tone and activity.

Tips & Tools

Because the parasympathetic nervous system plays a dominant role in the digestive process, it is often referred to as the "rest and digest" division.

The Serosa

A serous membrane known as the **serosa** covers the muscular layer along most portions of the digestive tract enclosed by the peritoneal cavity (see Figure 24–3).

In areas where there is no serosa covering the muscular layer, a dense network of collagen fibers firmly attaches the digestive tract to adjacent structures. This fibrous sheath is called an ***adventitia*** (ad-ven-TISH-uh).

Motility of the Digestive Tract

The muscular layers of the digestive tract consist of *visceral smooth muscle tissue*, which we introduced in Chapter 10. p. 329 This visceral smooth muscle includes cells called *pacesetter cells* that are located in the muscularis mucosae and muscular layer, which surround the lumen of the digestive tract. Pacesetter cells undergo spontaneous depolarization, triggering a wave of contraction that spreads throughout the entire muscular sheet. The coordinated contractions of the muscular layer cause the digestive tract to rhythmically squeeze and move materials along the tract, a propulsion process called *peristalsis*. A related type of movement, or *motility*, called *segmentation*, also mechanically digests materials.

Propulsion: Peristalsis

The muscular layer propels materials from one portion of the digestive tract to another by contractions known as **peristalsis** (pehr-ih-STAL-sis). Peristalsis consists of waves of muscular contractions that move a *bolus* (BŌ-lus), or soft rounded ball of digestive contents, along the length of the digestive tract (Figure 24–4). During peristalsis, the circular muscles contract behind the bolus while circular muscles ahead of the bolus

Figure 24–4 Peristalsis. Peristalsis propels materials along the length of the digestive tract.

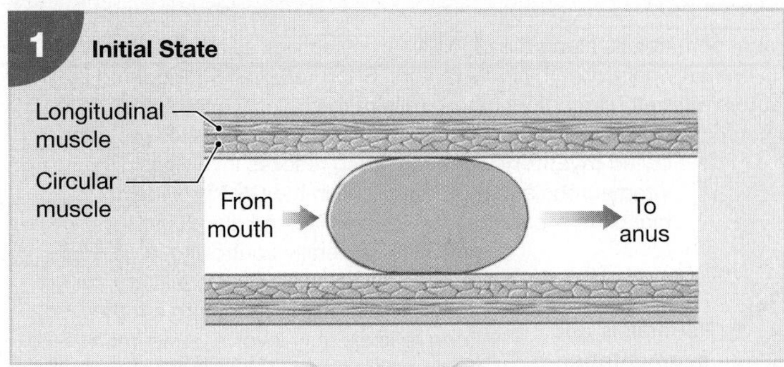
1 Initial State
Longitudinal muscle
Circular muscle
From mouth
To anus

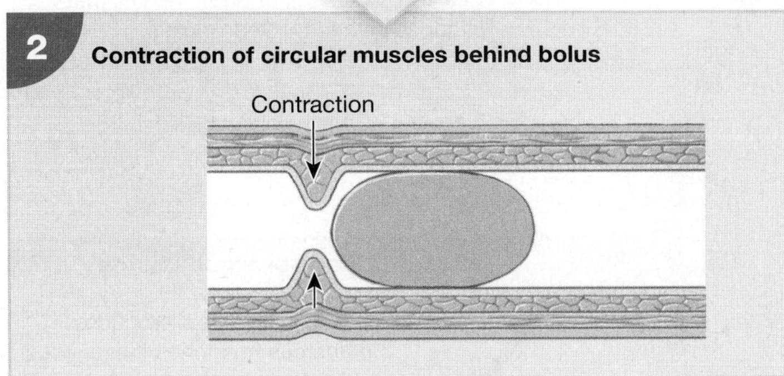
2 Contraction of circular muscles behind bolus
Contraction

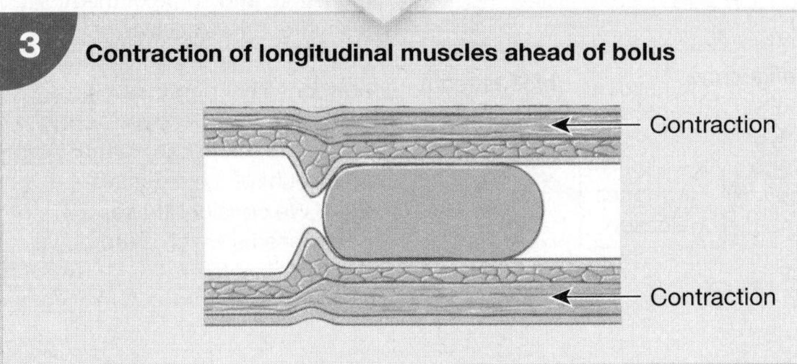
3 Contraction of longitudinal muscles ahead of bolus
Contraction
Contraction

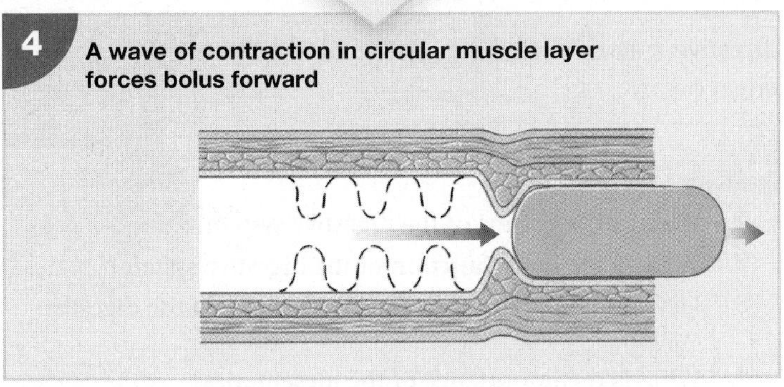
4 A wave of contraction in circular muscle layer forces bolus forward

? Why do circular muscles contract behind the bolus while longitudinal muscles contract ahead of the bolus?

relax. Longitudinal muscles ahead of the bolus then contract, shortening adjacent segments. A wave of contraction in the circular muscles then forces the bolus forward.

Tips & Tools

Squeezing toothpaste out of a tube is similar to peristalsis: Your squeezing hand (contracting circular muscles) forces toothpaste (the bolus) along and out of the tube (the digestive tract).

Mechanical Digestion: Segmentation

Most areas of the small intestine and some portions of the large intestine undergo cycles of contraction that churn and fragment the bolus, mixing the contents with intestinal secretions. This activity, called **segmentation**, does not follow a set pattern. For this reason, segmentation does not push materials along the tract in any one direction.

Regulation of Digestive Functions

Local factors interact with neural and hormonal mechanisms to regulate the activities of the digestive system (Figure 24–5, p. 892).

Local Factors

The initial regulation of digestive function occurs at the local level (1 Figure 24-5). Local environmental factors such as the pH, volume, or chemical composition of the intestinal contents can have a direct effect on digestive activity in that segment of the digestive tract. Some of these local factors have a direct effect on local digestive activities; for example, stretching of the intestinal wall can stimulate localized contractions of smooth muscles. In other cases, the local factors stimulate the release of chemicals. Prostaglandins, histamine, and other chemicals released into interstitial fluid may affect adjacent cells within a small segment of the tract. For example, the release of histamine in the lamina propria of the stomach stimulates the secretion of acid by cells in the adjacent epithelium.

24

Neural Mechanisms

Neural mechanisms control digestive tract movement (2 Figure 24-5). For example, sensory receptors in the walls of the digestive tract trigger peristaltic movements that are limited to a few centimeters. The visceral motor neurons that control smooth muscle contraction and glandular secretion are located in the myenteric plexus. These neurons are usually considered parasympathetic, because some of them synapse with parasympathetic preganglionic fibers. However, the plexus also contains sensory neurons, motor neurons, and interneurons responsible for local reflexes that operate entirely outside the control of the central nervous system. As we noted in Chapter 16, local reflexes are called *short reflexes*. ⮌ p. 552

Figure 24–5 The Regulation of Digestive Activities.

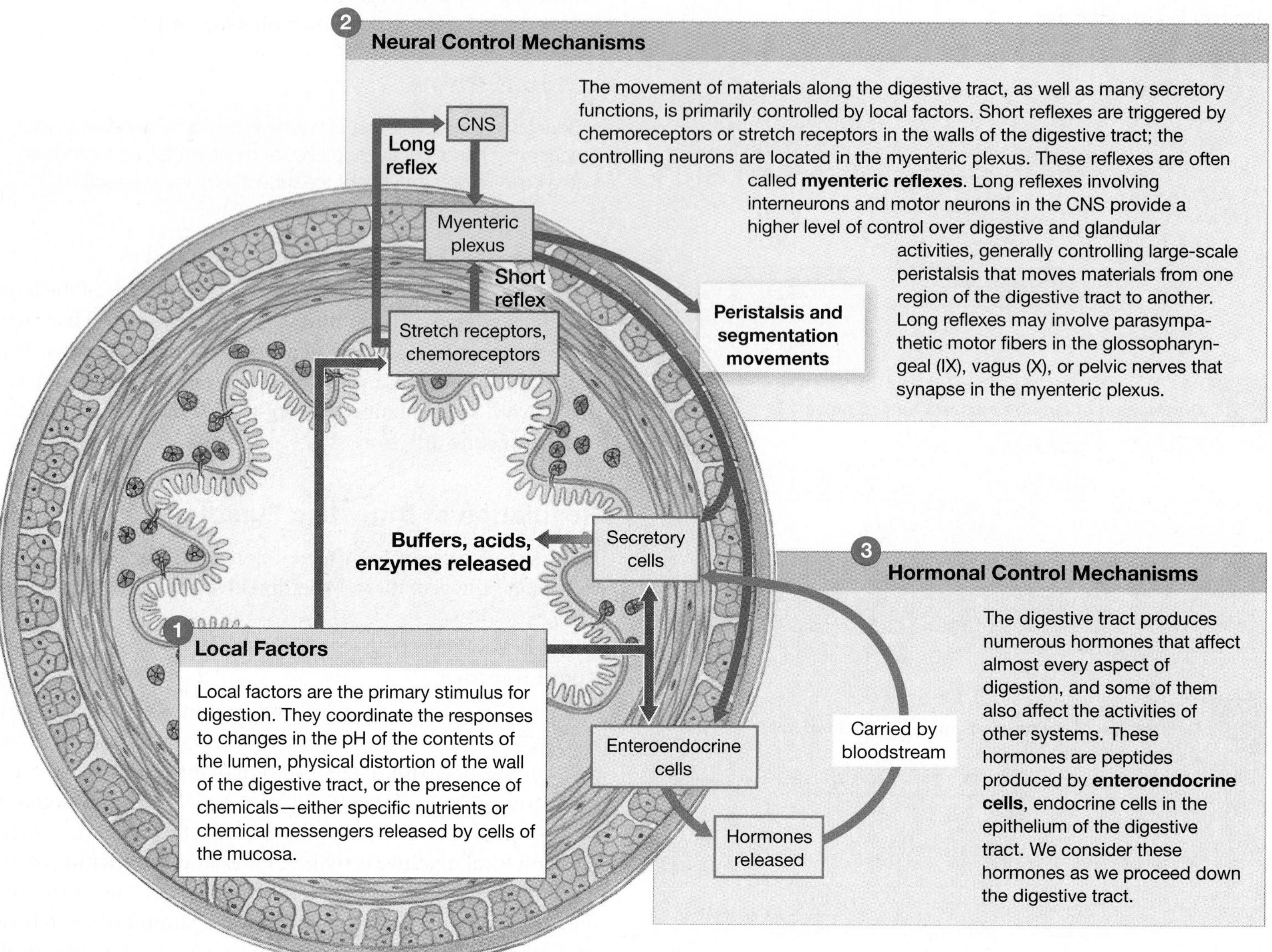
2 Neural Control Mechanisms
The movement of materials along the digestive tract, as well as many secretory functions, is primarily controlled by local factors. Short reflexes are triggered by chemoreceptors or stretch receptors in the walls of the digestive tract; the controlling neurons are located in the myenteric plexus. These reflexes are often called myenteric reflexes. Long reflexes involving interneurons and motor neurons in the CNS provide a higher level of control over digestive and glandular activities, generally controlling large-scale peristalsis that moves materials from one region of the digestive tract to another. Long reflexes may involve parasympathetic motor fibers in the glossopharyngeal (IX), vagus (X), or pelvic nerves that synapse in the myenteric plexus.
CNS
Long reflex
Myenteric plexus
Short reflex
Stretch receptors, chemoreceptors
Peristalsis and segmentation movements
Buffers, acids, enzymes released
Secretory cells
3 Hormonal Control Mechanisms
The digestive tract produces numerous hormones that affect almost every aspect of digestion, and some of them also affect the activities of other systems. These hormones are peptides produced by enteroendocrine cells, endocrine cells in the epithelium of the digestive tract. We consider these hormones as we proceed down the digestive tract.
1 Local Factors
Local factors are the primary stimulus for digestion. They coordinate the responses to changes in the pH of the contents of the lumen, physical distortion of the wall of the digestive tract, or the presence of chemicals—either specific nutrients or chemical messengers released by cells of the mucosa.
Enteroendocrine cells
Carried by bloodstream
Hormones released

In general, short reflexes control localized activities that involve small segments of the digestive tract. For example, they may coordinate local peristalsis and trigger secretion by digestive glands in response to the arrival of a bolus. Many neurons are involved. The enteric nervous system has about as many neurons as the spinal cord, and as many neurotransmitters as the brain. The specific functions and interactions of these neurotransmitters in the enteric nervous system are unknown.

Sensory information from receptors in the digestive tract also reaches the CNS. There it can trigger *long reflexes*, which involve interneurons and motor neurons in the CNS. p. 552

Hormonal Mechanisms

Digestive hormones can enhance or diminish the sensitivity of the smooth muscle cells to neural commands (3 Figure 24-5). These hormones, produced by enteroendocrine cells in the digestive tract, travel through the bloodstream to reach their target organs.

Checkpoint

1. **Identify the organs of the digestive system.**
2. **What is the main function of the digestive system?**
3. **List and define the six primary processes of the digestive system.**
4. **What is the importance of the mesenteries?**
5. **Name the layers of the gastrointestinal tract beginning at the surface of the lumen.**
6. **Which is more efficient in propelling intestinal contents from one place to another: peristalsis or segmentation?**
7. **What effect would a drug that blocks parasympathetic stimulation of the digestive tract have on peristalsis?**

See the blue Answers tab at the back of the book.

24-2 The oral cavity, which contains the tongue, teeth, and salivary glands, functions in the ingestion and mechanical digestion of food

Learning Outcome Discuss the anatomy of the oral cavity, and list the functions of its major structures and regions.

Let's start our exploration of the digestive tract with how ingested food is processed in the oral cavity. The process of chewing, or *mastication*, involves not only the tongue and teeth but also *saliva* produced by the salivary glands.

The Oral Cavity

We begin at the **oral cavity** (mouth), (Figure 24–6). The functions of the oral cavity include (1) *sensory analysis* of food before swallowing; (2) *mechanical digestion* through the actions of the teeth, tongue, and palatal surfaces; (3) *lubrication* by mixing with mucus and saliva; and (4) limited *chemical digestion* of carbohydrates and lipids.

The oral cavity is lined by the **oral mucosa**, which has a stratified squamous epithelium. A layer of keratinized cells covers regions exposed to severe abrasion, such as the superior surface of the tongue and the opposing surface of the hard palate (part of the roof of the mouth). The epithelial lining of the cheeks, lips, and inferior surface of the tongue is relatively thin and nonkeratinized. Nutrients are not absorbed in the oral cavity, but the mucosa inferior to the tongue is thin enough and vascular enough to permit the rapid absorption of lipid-soluble drugs. *Nitroglycerin* may be administered by this route to treat acute angina pectoris. p. 701

The mucosae of the **cheeks**, or lateral walls of the oral cavity, are supported by buccal fat pads and the buccinator muscles. Anteriorly, the mucosa of each cheek is continuous with that of the **lips**. The **oral vestibule** is the space between the cheeks (or lips) and the teeth. The **gingivae** (JIN-ji-vē), or *gums*, are ridges of oral mucosa that surround the base of each tooth on the alveolar processes of the maxillae and the alveolar part of the mandible. In most regions, the gingivae are firmly bound to the periostea of the underlying bones. The *frenulum of lower lip* and the *frenulum of upper lip* are folds of mucosa that extends from the gingiva to the lower lip and upper lip respectively, attaching the lips to the gum.

The hard and soft palates form the roof of the oral cavity, while the tongue dominates its floor. The floor of the mouth inferior to the tongue gains extra support from the geniohyoid and mylohyoid muscles. p. 354 The palatine processes of the maxillae and the horizontal plates of the palatine bones form the *hard palate*. A prominent central ridge, or *palatine raphe* (RĀ-fee), extends along the midline of the hard palate. The mucosa lateral and anterior to the raphe is thick, with complex ridges. When your tongue compresses food against the hard palate, these ridges provide traction. The *soft palate* lies posterior to the hard palate. A thinner and more delicate mucosa covers the posterior margin of the hard palate and extends onto the soft palate.

Figure 24–6 Anatomy of the Oral Cavity. ATLAS: Plates 11a; 19

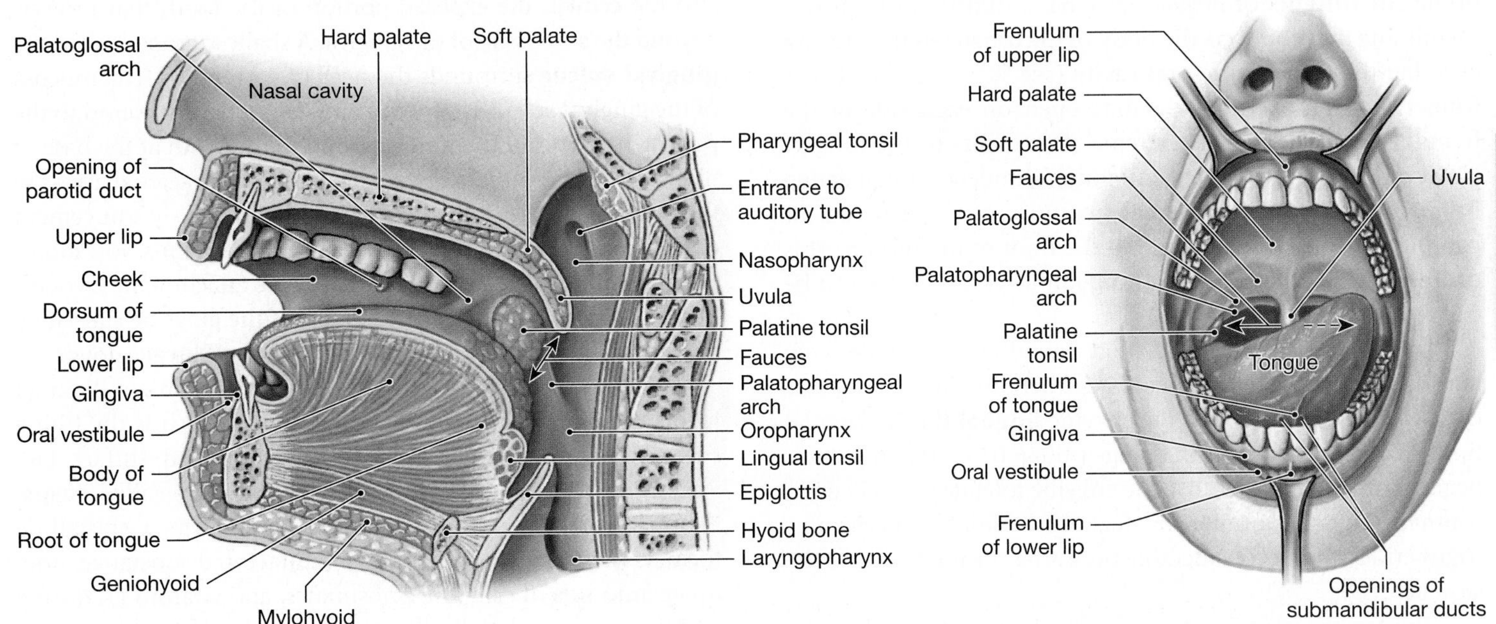
Palatoglossal arch
Hard palate
Soft palate
Nasal cavity
Opening of parotid duct
Upper lip
Cheek
Dorsum of tongue
Lower lip
Gingiva
Oral vestibule
Body of tongue
Root of tongue
Geniohyoid
Mylohyoid
Pharyngeal tonsil
Entrance to auditory tube
Nasopharynx
Uvula
Palatine tonsil
Fauces
Palatopharyngeal arch
Oropharynx
Lingual tonsil
Epiglottis
Hyoid bone
Laryngopharynx
Frenulum of upper lip
Hard palate
Soft palate
Fauces
Palatoglossal arch
Palatopharyngeal arch
Palatine tonsil
Frenulum of tongue
Gingiva
Oral vestibule
Frenulum of lower lip
Uvula
Tongue
Openings of submandibular ducts

a A sagittal section of the oral cavity

b An anterior view of the oral cavity

24

The posterior margin of the soft palate supports the **uvula** (YŪ-vyū-luh), a dangling process that helps prevent food from entering the pharynx too soon (see Figure 24–6a). On either side of the uvula are two pairs of muscular *palatal arches* (see Figure 24–6b). The more anterior **palatoglossal** (pal-a-tō-GLOS-al) **arch** extends between the soft palate and the base of the tongue. The space between the oral cavity and the pharynx, bounded by the soft palate and the base of the tongue, is called the **fauces** (FAW-sēz). The more posterior **palatopharyngeal** (pal-ah-tō-fah-RIN-jē-al) **arch** extends from the soft palate to the pharyngeal wall. A palatine tonsil lies between the palatoglossal and palatopharyngeal arches on each side.

The Tongue

The **tongue** has four primary functions (see Figure 24–6): (1) mechanical digestion by compression, abrasion, and distortion; (2) manipulation to assist in chewing and to prepare food for swallowing; (3) sensory analysis by touch, temperature, and taste receptors; and (4) secretion of mucins and the enzyme *lingual lipase*.

We can divide the tongue into an anterior **body**, or *oral portion*, and a posterior **root**, or *pharyngeal portion*. The superior surface, or *dorsum*, of the body contains a forest of fine projections, the *lingual papillae*. p. 570 The thickened epithelium covering each papilla assists the tongue in moving materials. A V-shaped line of vallate papillae roughly marks the boundary between the body and the root of the tongue, which is located in the oropharynx (see Figure 24–6a).

The epithelium covering the inferior surface of the tongue is thinner and more delicate than that of the dorsum. Along the inferior midline is the **frenulum** (FREN-yū-lum; *frenum*, bridle) **of tongue**, or *lingual frenulum*, a thin fold of mucous membrane that connects the body of the tongue to the mucosa covering the floor of the oral cavity (see Figure 24–6b. Ducts from two pairs of salivary glands open on each side of the frenulum, which prevents extreme movements of the tongue. However, an overly restrictive frenulum hinders normal eating or speech. A shortened lingual frenulum causes a partial or complete fusion of the tongue to the floor of the mouth. This condition, called *ankyloglossia* (ang-kih-lō-GLOS-ē-uh), can be corrected surgically.

Secretions of small glands that extend into the underlying lamina propria flush the tongue's epithelium. These secretions contain water, mucins, and the enzyme **lingual lipase**. Lingual lipase works over a broad pH range (3.0–6.0) so lipid digestion begins immediately. Because the enzyme tolerates an acid environment, it can continue to break down lipids—specifically, triglycerides—for a considerable time after the food reaches the stomach.

The tongue contains two groups of skeletal muscles. The large **extrinsic tongue muscles** perform all gross movements of the tongue. p. 352 The smaller **intrinsic tongue muscles** change the shape of the tongue and assist the extrinsic muscles during precise movements, as in speech. Both intrinsic and extrinsic tongue muscles are under the control of the hypoglossal cranial nerves (XII).

The Teeth

Tongue movements are important in passing food across the **teeth** during chewing. Let's look at the different types and sets of teeth.

Anatomy of a Tooth

Figure 24–7a is a sectional view through an adult tooth. The bulk of each tooth consists of a mineralized matrix similar to that of bone. This material, called **dentin**, differs from bone in that it does not contain cells. Instead, cytoplasmic processes extend into the dentin from cells in the central **pulp cavity**, an interior chamber. The pulp cavity receives blood vessels and nerves through the **root canal**, a narrow tunnel located at the **root**, or base, of the tooth. Blood vessels and nerves enter the root canal through an opening called the **apical foramen** to supply the pulp cavity.

The root of each tooth sits in a bony cavity or socket called the *tooth socket* or a *tooth alveolus*. Collagen fibers of the **periodontal ligament** extend from the dentin of the root to the alveolar bone, creating a strong articulation known as a ***gomphosis***. p. 267 A layer of **cement** (sē-MENT) covers the dentin of the root. Cement protects and firmly anchors the periodontal ligament. Cement is histologically similar to bone and is less resistant to erosion than is dentin.

The **neck** of the tooth marks the boundary between the root and the **crown**, the exposed portion of the tooth that projects beyond the soft tissue of the gingiva. A shallow groove called the **gingival sulcus** surrounds the neck of each tooth. The mucosa of the gingival sulcus is very thin and is not tightly bound to the periosteum. The epithelium is bound to the tooth at the base of the sulcus. This epithelial attachment prevents bacterial access to the lamina propria of the gingiva and the relatively soft cement of the root. When you brush and massage your gums, you stimulate the epithelial cells and strengthen the attachment. A condition called *gingivitis*, an inflammation of the gingivae caused by bacterial infection, can occur if the attachment breaks down.

A layer of **enamel** covers the dentin of the crown. Enamel forms the **occlusal surface** of each tooth, which is the biting surface that grinds food against the opposing tooth surface. Elevations or projections of the occlusal surface are called **cusps**. Enamel, which contains calcium phosphate in a crystalline form, is the hardest biologically manufactured substance. Adequate amounts of calcium, phosphates, and vitamin D_3 during childhood are essential if the enamel coating is to be complete and resistant to decay.

Figure 24–7 The Teeth.

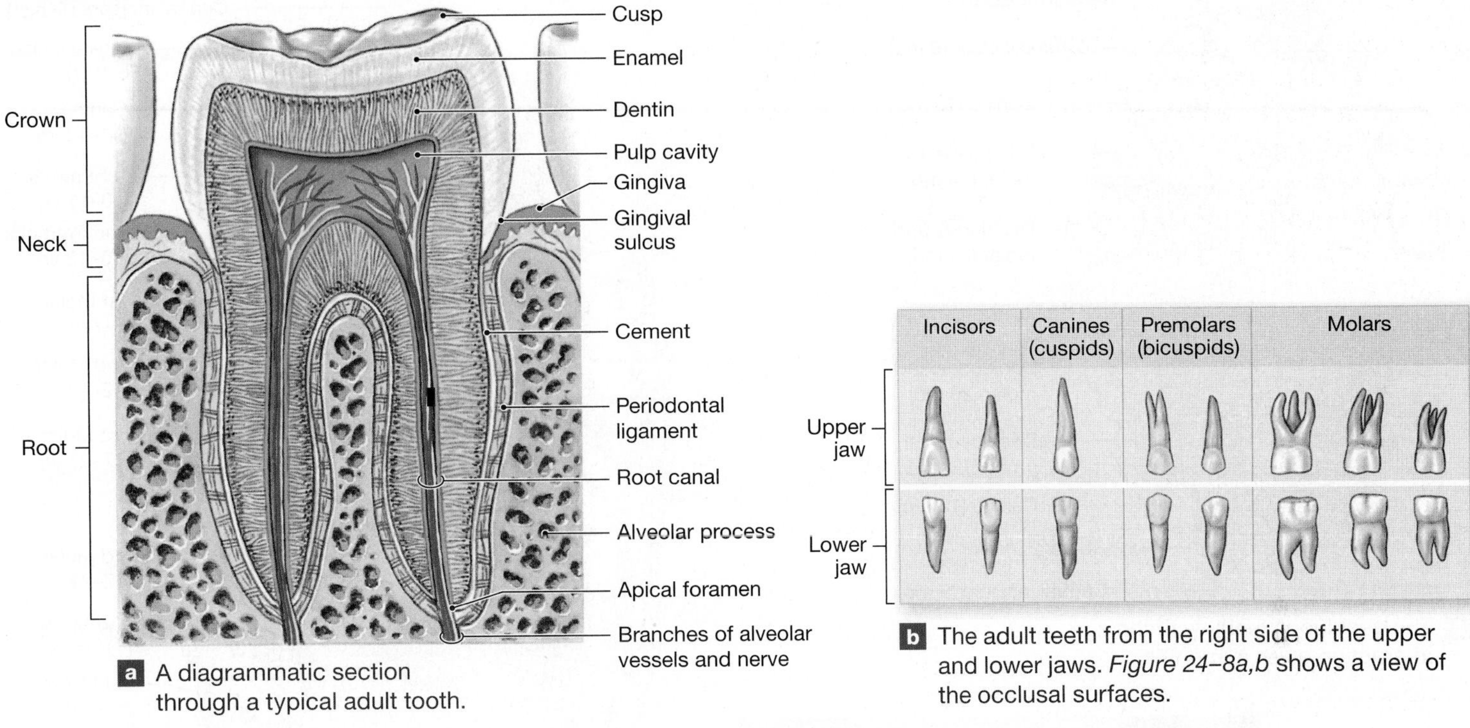
Cusp
Enamel
Dentin
Pulp cavity
Gingiva
Gingival sulcus
Cement
Periodontal ligament
Root canal
Alveolar process
Apical foramen
Branches of alveolar vessels and nerve
Crown
Neck
Root
Incisors
Canines (cuspids)
Premolars (bicuspids)
Molars
Upper jaw
Lower jaw

a A diagrammatic section through a typical adult tooth.

b The adult teeth from the right side of the upper and lower jaws. *Figure 24–8a,b* shows a view of the occlusal surfaces.

Tooth decay generally results from the action of bacteria that live in your mouth. Bacteria adhering to the surfaces of the teeth produce a sticky matrix that traps food particles and creates deposits known as *dental plaque*. Over time, this organic material can become calcified, forming a hard layer of *tartar*, or *dental calculus*, which can be difficult to remove. Tartar deposits most commonly develop at or near the gingival sulcus, where brushing cannot remove the soft plaque deposits.

Types of Teeth

The alveolar processes of the maxillae and the alveolar part of the mandible form the *maxillary dental arcade* and *mandibular dental arcade*, or upper and lower dental arcades, respectively. These arcades contain four types of teeth, each with specific functions (Figure 24–7b):

1. **Incisor** (in-SĪ-zer) **teeth** are blade-shaped teeth located at the front of the mouth. Incisors are useful for clipping or cutting, as when you nip off the tip of a carrot stick. These teeth have a single root.
2. **Canine** (KĀ-nīn) **teeth**, or *cuspids* (KUS-pidz), are conical with a single, pointed cusp. Also known as the "eyeteeth," because they lie directly under the eye, the canines are used for tearing or slashing. You might weaken a tough piece of celery using the clipping action of the incisors and then take advantage of the shearing action of the canines. Canines have a single root.
3. **Premolar teeth**, or *bicuspids* (bī-KUS-pidz), have flattened crowns with two prominent rounded cusps. They crush, mash, and grind. Bicuspids have one or two roots.
4. **Molar teeth** have very large, flattened crowns with four to five prominent rounded cusps adapted for crushing and grinding. You can usually shift a tough nut to your premolars and molars for successful crunching. Molars in the upper jaw typically have three roots while those in the lower jaw usually have two roots.

Sets of Teeth

The teeth are collectively called the *dentition*. Two such dentitions form during development: deciduous and permanent. The first to appear are the **deciduous teeth** (de-SID-yū-us; *deciduus*, falling off). Deciduous teeth are also called *primary teeth, milk teeth*, or *baby teeth*. Most children have 20 deciduous teeth (Figure 24–8a). On each side of the upper or lower jaw, the deciduous teeth consist of two incisors, one canine, and a pair of deciduous molars for a total of 20 teeth.

These teeth are later replaced by the **permanent teeth** of the larger adult jaws. As replacement proceeds, the periodontal ligaments and roots of the deciduous teeth erode. The deciduous teeth either fall out or are pushed aside by the **eruption**, or emergence, of the permanent teeth (Figure 24–8b). The permanent premolars take the place of the deciduous molars, while the permanent molars extend the tooth rows as the jaw

Figure 24–8 **Deciduous and Permanent Dentitions.**

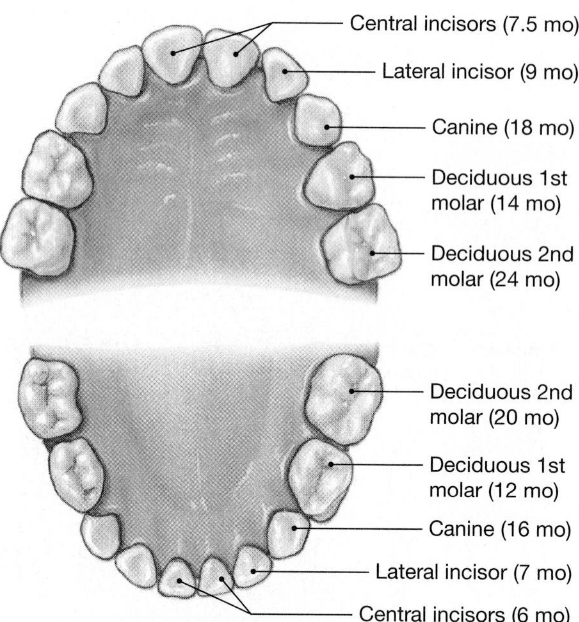
Central incisors (7.5 mo)
Lateral incisor (9 mo)
Canine (18 mo)
Deciduous 1st molar (14 mo)
Deciduous 2nd molar (24 mo)
Deciduous 2nd molar (20 mo)
Deciduous 1st molar (12 mo)
Canine (16 mo)
Lateral incisor (7 mo)
Central incisors (6 mo)

a The deciduous teeth, with the age at eruption given in months

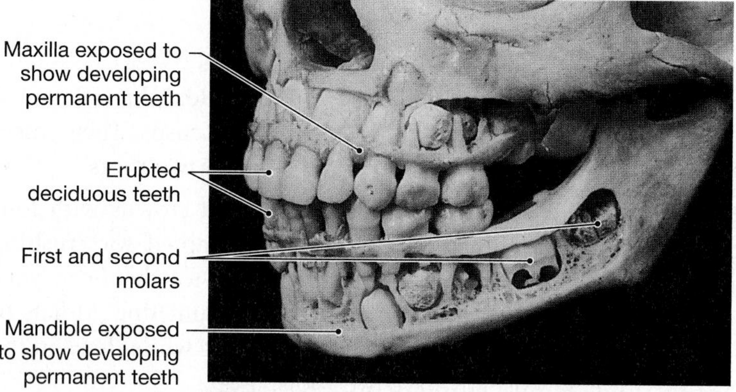
Maxilla exposed to show developing permanent teeth
Erupted deciduous teeth
First and second molars
Mandible exposed to show developing permanent teeth

b Maxilla and mandible with unerupted teeth exposed

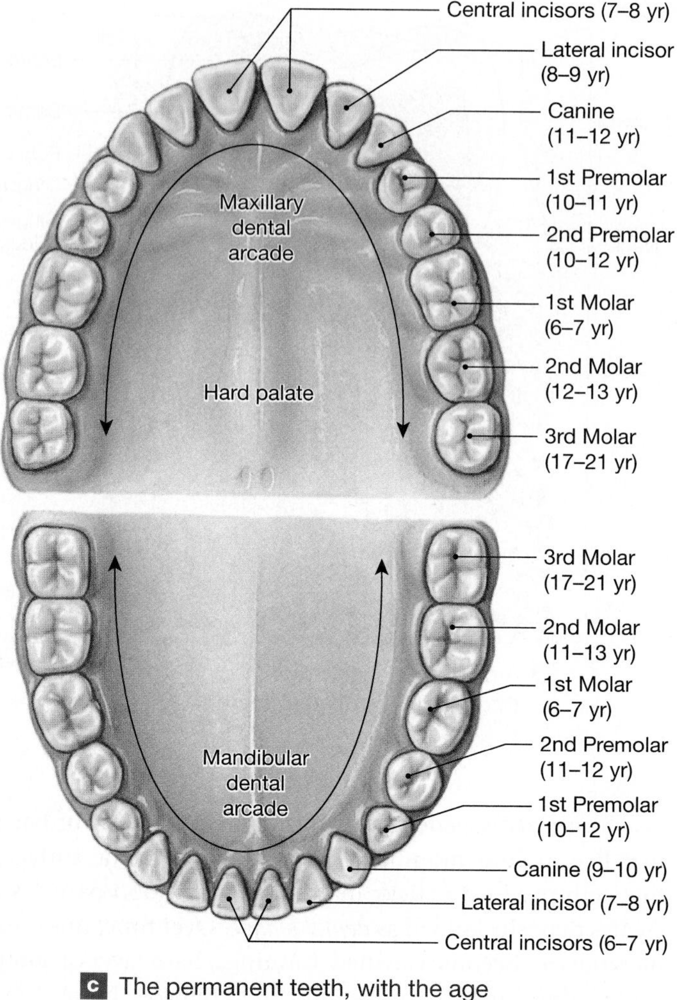
Central incisors (7–8 yr)
Lateral incisor (8–9 yr)
Canine (11–12 yr)
1st Premolar (10–11 yr)
2nd Premolar (10–12 yr)
1st Molar (6–7 yr)
2nd Molar (12–13 yr)
3rd Molar (17–21 yr)
Maxillary dental arcade
Hard palate
3rd Molar (17–21 yr)
2nd Molar (11–13 yr)
1st Molar (6–7 yr)
2nd Premolar (11–12 yr)
1st Premolar (10–12 yr)
Canine (9–10 yr)
Lateral incisor (7–8 yr)
Central incisors (6–7 yr)
Mandibular dental arcade

c The permanent teeth, with the age at eruption given in years

enlarges (Figure 24–8c). Three additional molars appear on each side of the upper and lower jaws as the person ages. These molars extend the rows of teeth posteriorly and bring the permanent tooth count to 32.

The third molars, or *wisdom teeth*, may not erupt before age 21. Wisdom teeth may not erupt because they develop in inappropriate positions or because there is not enough space. Any teeth that develop in locations that do not allow them to erupt are called *impacted teeth*. Impacted wisdom teeth can be surgically removed to prevent the formation of abscesses.

The Salivary Glands

Three major pairs of salivary glands secrete into the oral cavity (Figure 24–9a). Each pair has a distinctive cellular organization and produces *saliva*, a mixture of glandular secretions, with slightly different properties:

- The large **parotid** (pah-ROT-id) **glands** lie inferior to the zygomatic arch deep to the skin covering the lateral and posterior surface of the mandible. Each gland has an irregular shape. It extends from the mastoid process of the temporal bone across the outer surface of the masseter. The parotid glands produce a serous secretion containing large amounts of *salivary amylase*. This enzyme breaks down starches (complex carbohydrates). The secretions of each parotid gland are drained by a **parotid duct**, which empties into the vestibule at the second upper molar.
- The **sublingual** (sub-LING-gwal) **glands** are covered by the mucous membrane of the floor of the mouth. These

Figure 24–9 Anatomy of the Salivary Glands. ATLAS: Plates 3c,d; 18a,b

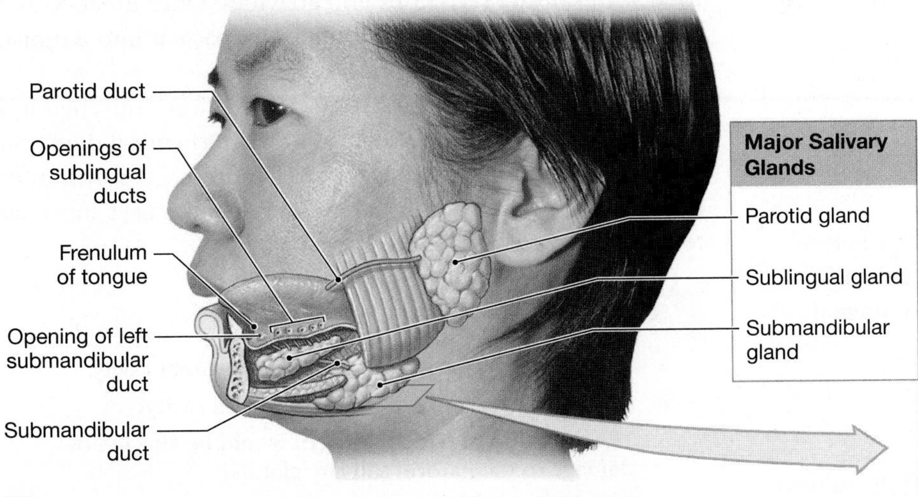
Parotid duct
Openings of sublingual ducts
Frenulum of tongue
Opening of left submandibular duct
Submandibular duct
Major Salivary Glands
Parotid gland
Sublingual gland
Submandibular gland

a A lateral view, showing the relative positions of the major salivary glands and ducts on the left side of the head. For clarity, the left ramus and body of the mandible have been removed. For the positions of the parotid and submandibular ducts in the oral cavity, see *Figure 24–6*.

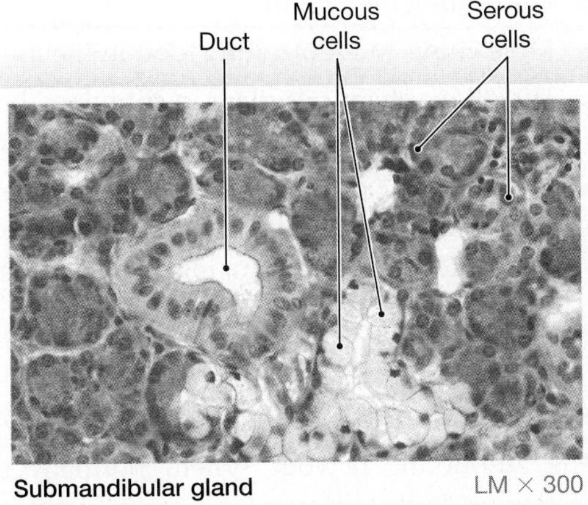
Duct
Mucous cells
Serous cells
Submandibular gland
LM × 300

b The submandibular gland secretes a mixture of mucins, produced by mucous cells, and enzymes, produced by serous cells.

? Which type of salivary glands produces glycoproteins called mucins?

glands produce mucus that acts as a buffer and lubricant. Numerous **sublingual ducts** open along either side of the lingual frenulum.

- The **submandibular glands** lie along the inner surfaces of the mandible within a depression called the *mandibular groove*. Cells of the submandibular glands (Figure 24–9b) secrete a mixture of buffers, glycoproteins called *mucins*, and salivary amylase. The **submandibular ducts** open into the mouth on each side of the frenulum of tongue immediately posterior to the teeth (see Figure 24–6b).

Minor salivary glands include some 800–1000 submucosal glands in the oral cavity. These small (1–2 mm in diameter) glands may share an excretory duct and mainly secrete mucus.

Composition and Functions of Saliva

The salivary glands produce 1.0–1.5 liters of **saliva** each day. About 70 percent of saliva comes from the submandibular glands. Another 25 percent comes from the parotid glands, and 5 percent from the sublingual glands.

Saliva is 99.4 percent water. The remaining 0.6 percent includes electrolytes (mainly Na^+, Cl^-, and HCO_3^-), buffers, glycoproteins, antibodies, enzymes, and wastes. Mucins give saliva its lubricating action.

Saliva continuously flushes the oral surfaces, helping to keep them clean. Buffers in the saliva keep the pH of your mouth near 7.0. They prevent the buildup of acids produced by bacteria. In addition, saliva contains antibodies (IgA) and *lysozyme*. Both help control populations of oral bacteria. A reduction in or elimination of salivary secretions—caused by radiation, emotional distress, certain drugs, sleep, or other factors—triggers a bacterial population explosion. Such a proliferation can

24

+ Clinical Note Mumps

The *mumps virus* most often targets the major salivary glands, especially the parotid glands, but other organs can also be infected. Infection typically occurs at 5–9 years of age. The first exposure stimulates the production of antibodies and, in most cases, confers permanent immunity. In an adult male, mumps may infect the testes and cause sterility. Infection of the pancreas by the mumps virus can produce temporary or permanent diabetes. Other organ systems, including the central nervous system, are affected in severe cases. A mumps vaccine effectively confers active immunity. Widespread administration of that vaccine has almost eliminated mumps in the United States.

rapidly lead to recurring infections and progressive erosion of the teeth and gums.

The saliva produced when you eat has a variety of functions, including the following:

- Lubricating the mouth.
- Moistening and lubricating food in the mouth.
- Dissolving chemicals that can stimulate the taste buds and provide sensory information about the food.
- Beginning the digestion of complex carbohydrates before the food is swallowed. The enzyme involved is **salivary amylase**. Saliva also contains a small amount of lingual lipase secreted by the glands of the tongue.

Regulation of Salivary Secretions

The autonomic nervous system normally controls salivary secretions. Each salivary gland has parasympathetic and sympathetic innervation. The parasympathetic efferents originate in the **superior salivatory nucleus** and the **inferior salivatory nucleus** of the medulla oblongata and synapses in the submandibular and otic ganglia. Any object in your mouth can trigger a salivary reflex. It stimulates receptors monitored by the trigeminal nerve (V) or taste buds innervated by cranial nerve VII, IX, or X. Parasympathetic stimulation speeds up secretion by all the salivary glands. As a result, you produce large amounts of saliva. The role of sympathetic innervation is unclear. Evidence suggests that it provokes the secretion of small amounts of very thick saliva.

The salivatory nuclei are also influenced by other brainstem nuclei, as well as by the activities of higher centers. For example, chewing with an empty mouth, smelling food, or even thinking about food increases salivation. That is why chewing gum keeps your mouth moist. Irritating stimuli in the esophagus, stomach, or intestines also speed up production of saliva, as does nausea. Increased saliva production in response to unpleasant stimuli helps reduce the stimulus by dilution, rinsing, or buffering strong acids or bases.

Mechanical Digestion: Mastication (Chewing)

During **mastication**, you force food from the oral cavity to the vestibule and back, crossing and recrossing the occlusal surfaces of the teeth. This movement results in part from the action of the *muscles of mastication*, which close your jaws and slide or rock your lower jaw from side to side. p. 351 The complicated process of mastication can involve any combination of mandibular elevation/depression, protraction/retraction, and medial/lateral movement (try classifying the movements involved the next time you eat). Control would be impossible, however, without help from the muscles of the cheeks, lips, and tongue.

Mastication breaks down tough connective tissues in meat and the plant fibers in vegetables. It also helps saturate the food with salivary secretions and enzymes. Once the food is a satisfactory consistency, your tongue compacts it into a moist, cohesive **bolus** that is fairly easy to swallow.

We have now discussed how mechanical and chemical digestion occur in the oral cavity. However, chemical digestion is not completed in the oral cavity because no absorption of nutrients takes place across the oral epithelium. Let's move on to more structures of the digestive tract.

✓ Checkpoint

8. **Name the structures associated with the oral cavity.**
9. **Which type of epithelium lines the oral cavity?**
10. **The digestion of which nutrient would be affected by damage to the parotid salivary glands?**
11. **Which type of tooth is most useful for chopping off bits of rigid food?**
12. **Where are the fauces located?**

See the blue Answers tab at the back of the book.

24-3 The pharynx and esophagus are passageways that transport the food bolus from the oral cavity to the stomach

Learning Outcome Describe the structure and functions of the pharynx and the esophagus.

The **pharynx** (FAR-ingks), or throat, is an anatomical space that serves as a common passageway for solid food, liquid, and air. Food normally passes through parts of the pharynx on its way to the esophagus. The **esophagus** is a hollow muscular tube that then conveys these ingested materials to the stomach.

The Pharynx

We described the regions of the pharynx—nasopharynx, oropharynx, and laryngopharynx—in Chapter 23. p. 839 These regions have a lining of stratified squamous epithelium similar to that of the oral cavity. The lamina propria contains scattered mucous glands and the lymphatic tissue of the pharyngeal, palatal, and lingual tonsils. Deep to the lamina propria lies a dense layer of elastic fibers, bound to the underlying skeletal muscles. Specific pharyngeal muscles work with muscles of the oral cavity to start swallowing, which pushes the bolus into the esophagus on its way to the stomach.

The Esophagus

Unlike the pharynx, the esophagus actively moves ingested materials down toward the stomach. Let's look at how its structure allows this function, and how this transport occurs.

Gross Anatomy of the Esophagus

The esophagus is approximately 25 cm (10 in.) in length and has a diameter of about 2 cm (0.80 in.) at its widest point. The esophagus begins posterior to the cricoid cartilage, at the level of vertebra C_6. It is narrowest at this point. The esophagus descends toward the thoracic cavity posterior to the trachea, continuing inferiorly along the posterior wall of the mediastinum. It then enters the abdominopelvic cavity through the **esophageal hiatus** (hī-Ā-tus), an opening in the diaphragm. The esophagus empties into the stomach anterior to vertebra T_7.

Histology of the Esophagus

The wall of the esophagus contains a mucosa, submucosa, and muscular layer comparable to those shown in Figure 24–3. Distinctive features of the esophageal wall include the following (Figure 24–10):

- The mucosa of the esophagus contains a nonkeratinized, stratified squamous epithelium similar to that of the pharynx and oral cavity.
- The mucosa and submucosa are packed into large folds that extend the length of the esophagus. These longitudinal folds allow for expansion during the passage of a large bolus. Muscle tone in the walls keeps the lumen closed, except when you swallow.
- The muscularis mucosae consists of an irregular layer of smooth muscle.
- The submucosa contains scattered *esophageal glands*. They produce mucus that reduces friction between the bolus and the esophageal lining.
- The muscular layer has the usual inner circular and outer longitudinal layers. However, in the superior third of the esophagus, these layers contain skeletal muscle fibers. The middle third contains a mixture of skeletal and smooth muscle tissue. Along the inferior third, only smooth muscle occurs.
- An adventitia of connective tissue outside the muscular layer anchors the esophagus to the posterior body wall. Over the 1–2 cm (0.4–0.8 in.) between the diaphragm and stomach, the esophagus is retroperitoneal. Peritoneum covers the anterior and left lateral surfaces.

Physiology of the Esophagus

The esophagus is innervated by parasympathetic and sympathetic fibers from the esophageal plexus. p. 549 Resting muscle tone in the circular muscle layer in the superior 3 cm (1.2 in.) of the esophagus normally prevents air from entering the esophagus. A comparable zone at the inferior end of the esophagus normally remains in a state of active contraction. This state prevents the backflow (reflux) of materials from the stomach into the esophagus. Neither region has a well-defined sphincter muscle. Nevertheless, we often use the terms *upper esophageal sphincter* and *lower esophageal sphincter (cardiac sphincter)* to describe these regions, which are similar in function to other sphincters.

Figure 24–10 Anatomy of the Esophagus.

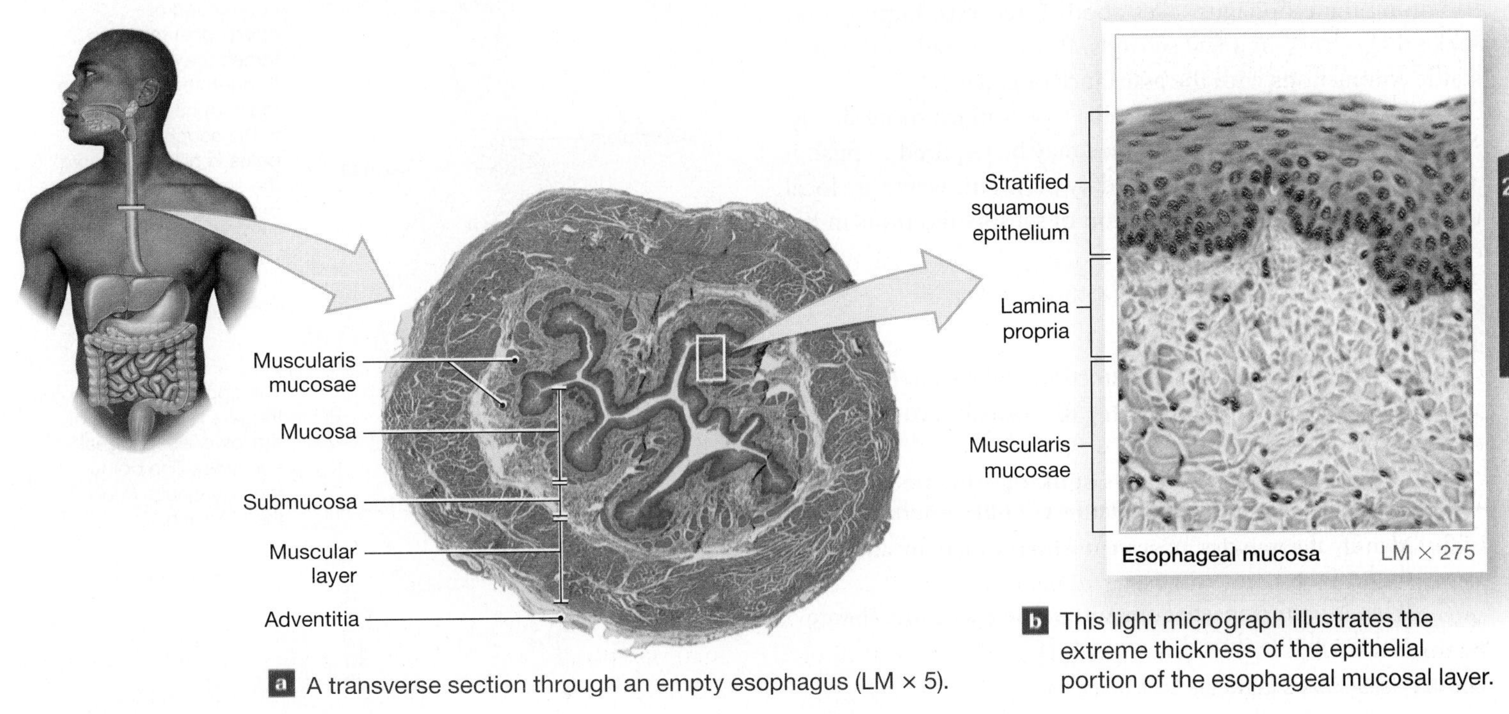
Muscularis mucosae
Mucosa
Submucosa
Muscular layer
Adventitia
Stratified squamous epithelium
Lamina propria
Muscularis mucosae
Esophageal mucosa
LM × 275

a A transverse section through an empty esophagus (LM × 5).

b This light micrograph illustrates the extreme thickness of the epithelial portion of the esophageal mucosal layer.

Ingestion: Deglutition (Swallowing)

Deglutition (dē-glū-TISH-un), or swallowing, is a complex process. We divide swallowing into buccal, pharyngeal, and esophageal phases, detailed in Figure 24–11. When you swallow, different pharyngeal muscles work together to prevent food or drink from entering the glottis. p. 352 Food is crushed and chewed into a bolus before being swallowed. When you are ready to swallow, the *palatal muscles* elevate the soft palate and adjacent portions of the pharyngeal wall. Muscles of the neck and pharynx, the *palatopharyngeus* and *stylopharyngeus*, then elevate the larynx. This bends the epiglottis over the glottis, so that the bolus can glide across the epiglottis rather than falling into the larynx. The *pharyngeal constrictor muscles* push the bolus toward and into the esophagus. While this movement is under way, the glottis is closed.

It takes less than a second for the pharyngeal muscles to propel the bolus into the esophagus. During this period, the respiratory centers are inhibited and breathing stops. Swallowing takes place at regular intervals as saliva collects at the back of the mouth. Each day you swallow approximately 2400 times.

Deglutition can be initiated voluntarily when you eat or drink, but proceeds automatically once it begins. The **swallowing reflex** begins when tactile receptors on the palatal arches and uvula are stimulated by the passage of the bolus. The information is relayed to the **swallowing center** of the medulla oblongata over the trigeminal nerves (V) and glossopharyngeal nerves (IX). Motor commands from this center then signal the pharyngeal musculature, producing a coordinated and stereotyped pattern of muscle contraction.

Primary peristaltic waves are peristaltic movements coordinated by afferent and efferent fibers in the glossopharyngeal nerves (IX) and vagus nerves (X). For a typical bolus, the entire trip along the esophagus takes about 9 seconds. Liquids may make the journey in a few seconds, flowing ahead of the peristaltic contractions with the assistance of gravity.

A dry or poorly lubricated bolus travels much more slowly. A series of *secondary peristaltic waves* may be required to push it all the way to the stomach. Secondary peristaltic waves are local reflexes triggered by the stimulation of sensory receptors in the esophageal walls.

24

✓ Checkpoint

13. Describe the structure and function of the pharynx.
14. Name the structure connecting the pharynx to the stomach.
15. Compared to other segments of the digestive tract, what is unusual about the muscular layer of the esophagus?
16. Identify the muscles associated with the pharynx and deglutition.
17. What is occurring when the soft palate and larynx elevate and the glottis closes?

See the blue Answers tab at the back of the book.

Figure 24–11 The Process of Deglutition (Swallowing). This sequence, based on a series of x-rays, shows the phases of swallowing and the movement of a bolus from the mouth to the stomach.

1 Buccal Phase

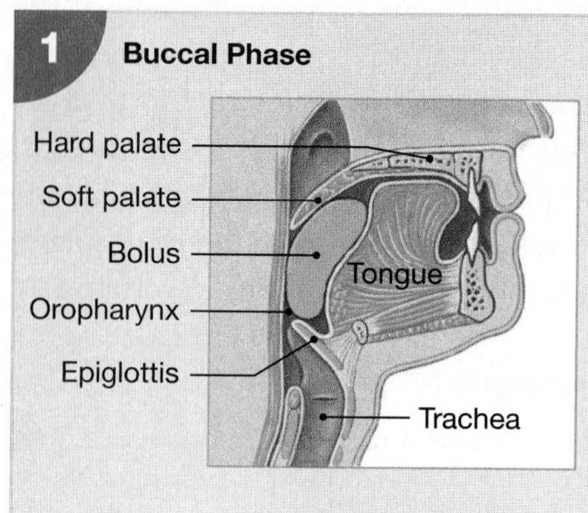

Hard palate
Soft palate
Bolus
Oropharynx
Epiglottis
Tongue
Trachea

The **buccal phase** begins with the compression of the bolus against the hard palate. Retraction of the tongue then forces the bolus into the oropharynx and assists in elevating the soft palate, thereby sealing off the nasopharynx. Once the bolus enters the oropharynx, reflex responses begin and the bolus is moved toward the stomach.

2 Pharyngeal Phase

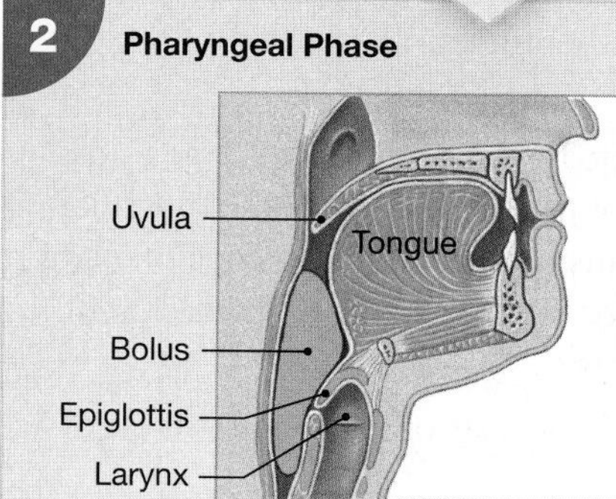

Uvula
Tongue
Bolus
Epiglottis
Larynx

The **pharyngeal phase** begins as the bolus comes into contact with the palatal arches and the posterior pharyngeal wall. Elevation of the larynx and folding of the epiglottis direct the bolus past the closed glottis. At the same time, the uvula and soft palate block passage back to the nasopharynx.

3 Esophageal Phase

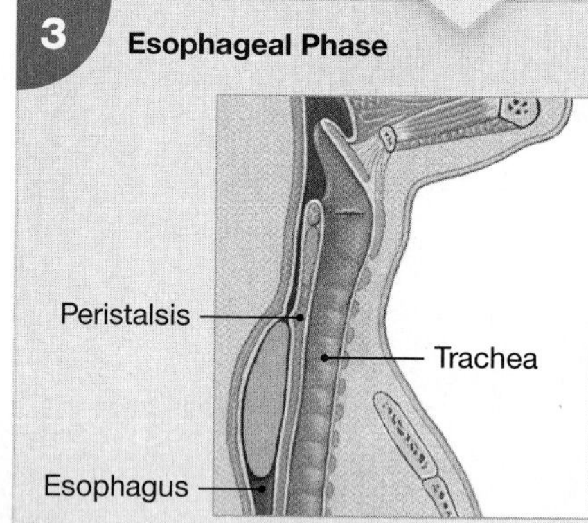

Peristalsis
Trachea
Esophagus

The **esophogeal phase** begins as the contraction of pharyngeal muscles forces the bolus through the entrance to the esophagus. Once in the esophagus, the bolus is pushed toward the stomach by a peristaltic wave.

4 Bolus Enters Stomach

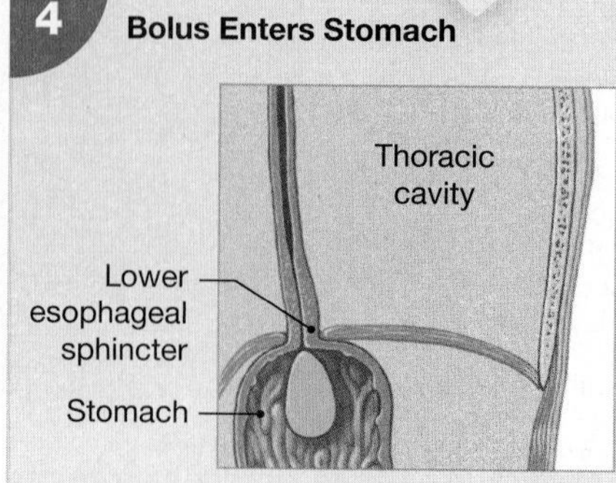

Thoracic cavity
Lower esophageal sphincter
Stomach

The approach of the bolus triggers the opening of the lower esophageal sphincter. The bolus then continues into the stomach.

Tips & Tools

Just as security gates control the passage of people by opening and closing, sphincters control the passage of material through them by dilating and constricting.

24-4 The stomach is a J-shaped organ that receives the bolus and aids in its chemical and mechanical digestion

Learning Outcome **Describe the anatomy of the stomach, including its histology, and discuss its roles in digestion and absorption.**

The **stomach** is an expandable tubelike organ found in the upper left quadrant of the peritoneal cavity. The major functions of the stomach are to (1) temporarily store ingested food received from the esophagus, (2) mechanically digest food through muscular contractions, and (3) chemically digest food through the action of acid and enzymes. Ingested substances combine with the secretions of the stomach glands, producing a viscous, acidic, soupy mixture of partially digested food called **chyme** (KĪM). This section examines the anatomy of the stomach and then details these important functions.

Gross Anatomy of the Stomach

The stomach has the shape of an expanded J (Figure 24–12). A short **lesser curvature** forms the medial surface of the organ, and a long **greater curvature** forms the lateral surface. The anterior and posterior surfaces are rounded. The shape and size of the stomach can vary greatly from person to person and even from one meal to the next. In an "average" stomach, the lesser curvature is approximately 10 cm (4 in.) long, and the greater curvature measures about 40 cm (16 in.). The stomach typically extends between the levels of vertebrae T_7 and L_3.

We divide the stomach into the following four regions (see Figure 24–12):

- *The Cardia.* The **cardia** (KAR-dē-uh) is the smallest region of the stomach. It consists of the superior, medial portion of the stomach within 3 cm (1.2 in.) of the junction between the stomach and the esophagus. The cardia contains abundant mucous glands. Their secretions coat the connection with the esophagus and help protect that tube from the acid and enzymes of the stomach.
- *The Fundus.* The **fundus** is the region of the stomach that is superior to the junction between the stomach and the esophagus. The fundus contacts the inferior, posterior surface of the diaphragm (see Figure 24–12a).
- *The Body.* The area of the stomach between the fundus and the curve of the J is the **body**, the largest region of the stomach. The body acts as a mixing tank for ingested food and secretions produced in the stomach.
- *The Pyloric Part.* The **pyloric part** (pī-LOR-ick; *pyloros*, gatekeeper) forms the portion of the stomach between the body and the duodenum (first segment of the small intestine). It is divided into a **pyloric antrum**, which is connected to the body, a **pyloric canal**, which empties into the duodenum, and the **pylorus** (pī-LOR-us), which is the muscular tissue surrounding the **pyloric orifice** (stomach outlet).

  During digestion, the shape of the pyloric part changes often. A thickening of the circular layer of muscle within the pylorus, called the **pyloric sphincter**, regulates the release of chyme into the duodenum.

The stomach's volume increases while you eat and then decreases as chyme enters the small intestine. When the stomach is relaxed (empty), the mucosa has prominent longitudinal folds called **rugae** (RŪ-gē; wrinkles). These temporary features let the gastric lumen expand (see Figure 24–12b). The stomach can stretch up to 50 times its empty size. As the stomach fills, the rugae gradually flatten out until, at maximum distension, they almost disappear. (The world record, set in 2016, for eating hot dogs with buns was 73 1/2 in 10 minutes!) When empty, the stomach resembles a muscular tube with a narrow, constricted lumen. When full, it can contain 1–1.5 liters of material.

Tips & Tools

Both circular folds and rugae allow a large surface area to fit within a small volume, just as crumpling a sheet of paper rearranges its surface area so that it occupies a smaller space.

The muscularis mucosae and muscular layer of the stomach contain extra layers of smooth muscle cells in addition to the usual circular and longitudinal layers. The muscularis mucosae generally contain an outer, circular layer of muscle cells. The muscular layer has an inner, **oblique layer** of smooth muscle (see Figure 24–12b). The extra layers of smooth muscle strengthen the stomach wall and assist in the mixing and churning essential to the formation of chyme.

Histology of the Stomach

Simple columnar epithelium lines all portions of the stomach (Figure 24–13a). The epithelium is a *secretory sheet*, which produces a carpet of mucus that covers the interior surface of the stomach. The alkaline mucous layer protects epithelial cells against the acid and enzymes in the gastric lumen.

Figure 24–12 Gross Anatomy of the Stomach. ATLAS: Plates 49a–c; 50a–c

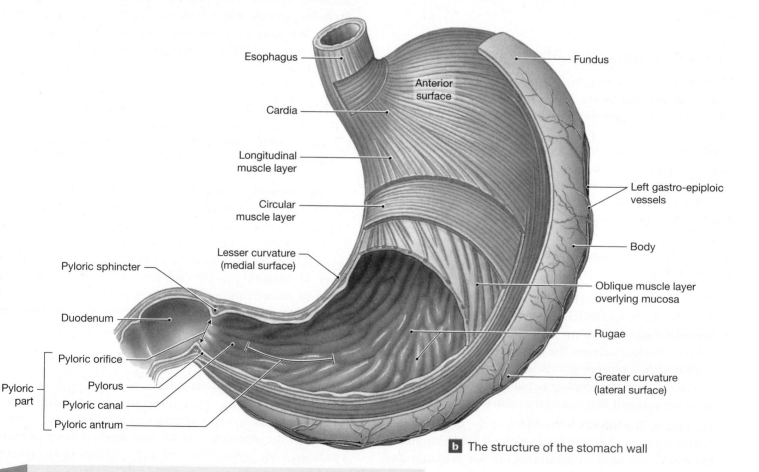

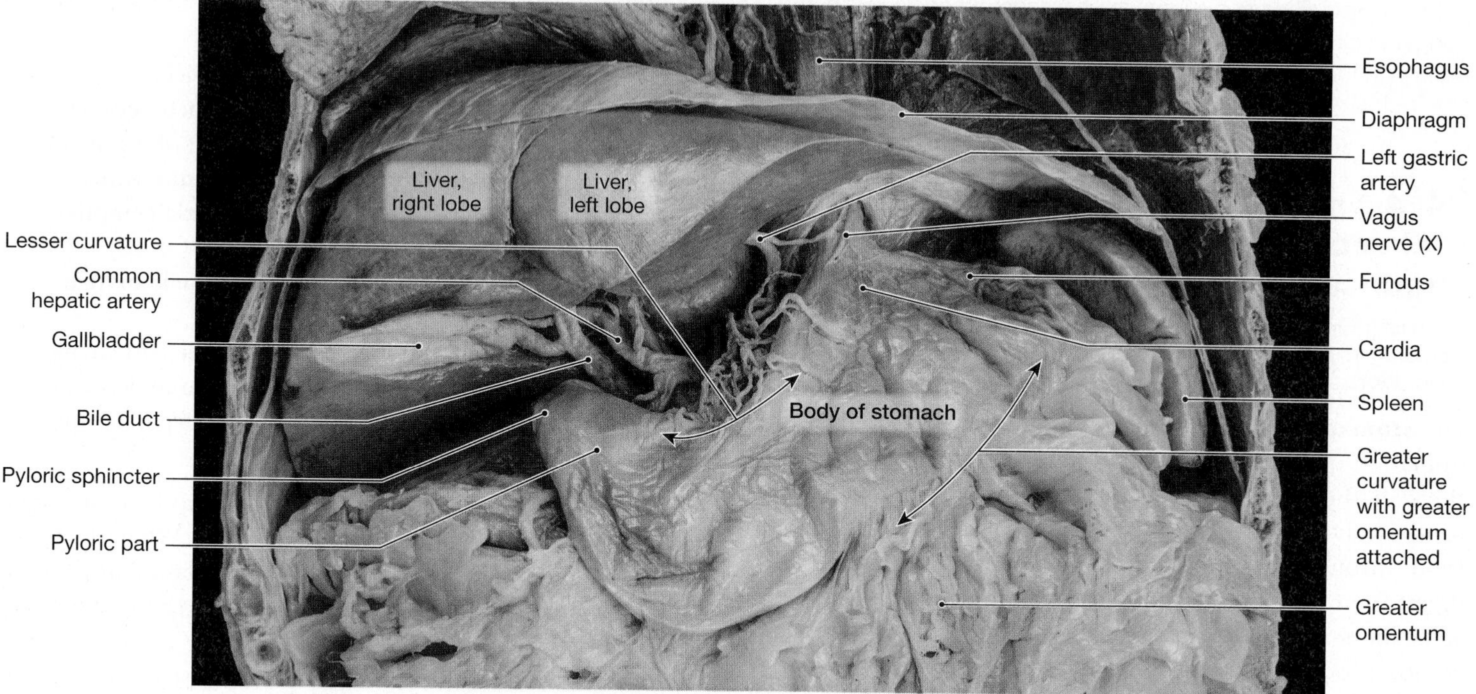
Esophagus
Diaphragm
Left gastric artery
Vagus nerve (X)
Fundus
Cardia
Spleen
Greater curvature with greater omentum attached
Greater omentum
Liver, right lobe
Liver, left lobe
Body of stomach
Lesser curvature
Common hepatic artery
Gallbladder
Bile duct
Pyloric sphincter
Pyloric part

a The position and external appearance of the stomach, showing superficial landmarks

Esophagus
Fundus
Anterior surface
Cardia
Longitudinal muscle layer
Circular muscle layer
Left gastro-epiploic vessels
Body
Lesser curvature (medial surface)
Pyloric sphincter
Oblique muscle layer overlying mucosa
Duodenum
Rugae
Pyloric part
Pyloric orifice
Pylorus
Pyloric canal
Pyloric antrum
Greater curvature (lateral surface)

b The structure of the stomach wall

? Name the region of the stomach that connects to the esophagus.

Figure 24–13 Histology of the Stomach Lining.

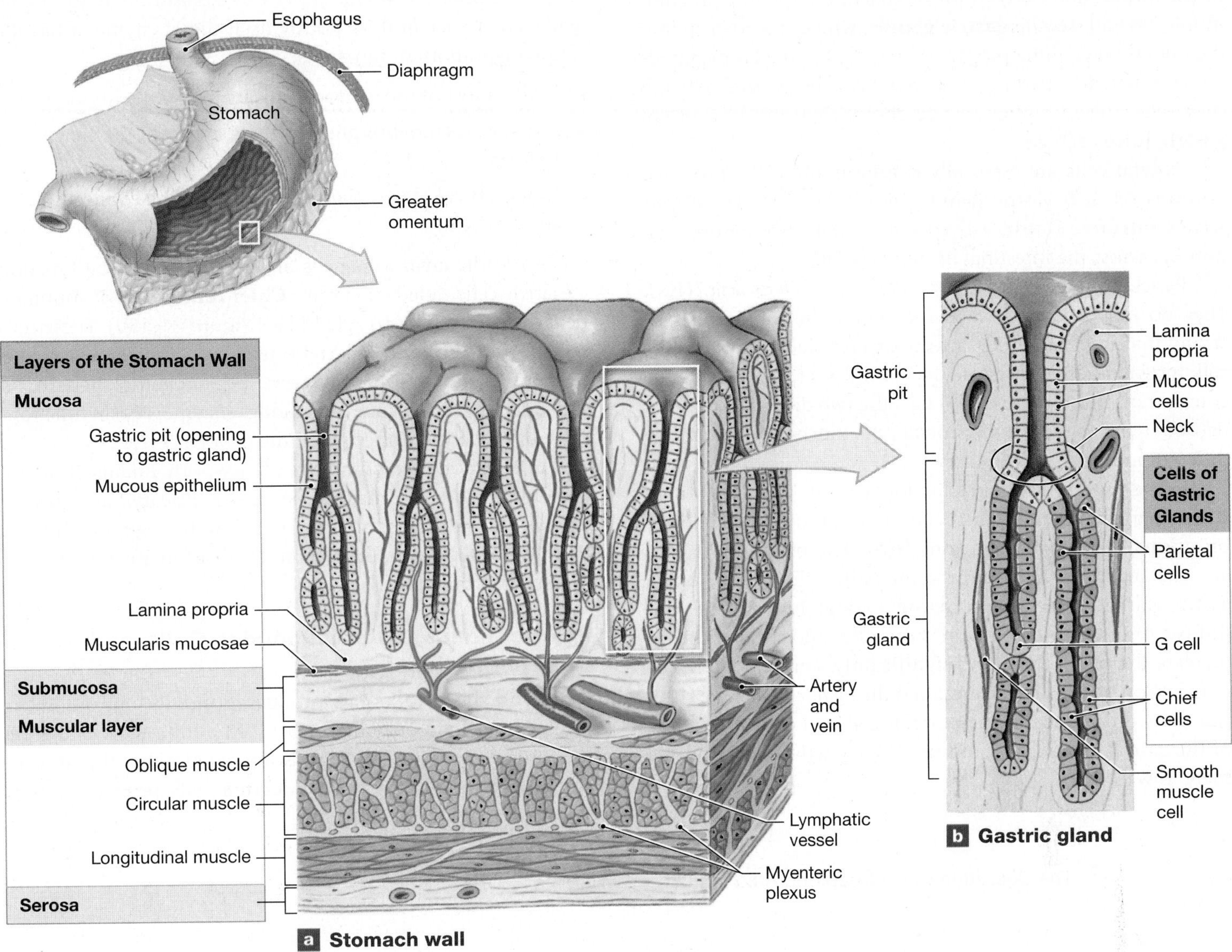
Esophagus
Diaphragm
Stomach
Greater omentum
Layers of the Stomach Wall
Mucosa
Gastric pit (opening to gastric gland)
Mucous epithelium
Lamina propria
Muscularis mucosae
Submucosa
Muscular layer
Oblique muscle
Circular muscle
Longitudinal muscle
Serosa
Artery and vein
Lymphatic vessel
Myenteric plexus
a Stomach wall
Gastric pit
Gastric gland
Lamina propria
Mucous cells
Neck
Cells of Gastric Glands
Parietal cells
G cell
Chief cells
Smooth muscle cell
b Gastric gland

Shallow depressions called **gastric pits** open onto the gastric surface (Figure 24–13b). The mucous cells at the base, or *neck*, of each gastric pit actively divide, replacing superficial cells that are shed into the chyme. A typical gastric epithelial cell has a life span of 3 to 7 days. Exposure to alcohol or other chemicals that damage or kill epithelial cells increases cell turnover.

Secretory Glands and Gastric Secretions

The gastric epithelium contains several types of glands that produce secretions needed for the function of the stomach. *Gastric* (*gaster*, stomach) *glands* in the fundus and body secrete most of the acid and enzymes involved in gastric digestion. Glands in the pyloric part secrete mucus and important digestive hormones, including *gastrin*, a hormone that stimulates gastric glands.

+ Clinical Note Gastritis and Peptic Ulcers

A superficial inflammation of the gastric mucosa is called *gastritis* (gas-TRĪ-tis). The condition can develop after a person has swallowed drugs, including alcohol and aspirin. Gastritis is also associated with smoking, severe emotional or physical stress, bacterial infection of the gastric wall, or ingestion of strongly acidic or alkaline chemicals. Over time, gastritis can lead to the erosion of the stomach or duodenal lining, and a *peptic ulcer* may form.

Gastric Glands and Gastric Juice

In the fundus and body of the stomach, each gastric pit communicates with several **gastric glands**, which extend deep into the underlying lamina propria (Figure 24–13b). Gastric glands are dominated by two types of secretory cells: *parietal cells* and *chief cells*. Together, they secrete about 1500 mL (1.6 qt) of **gastric juice** each day.

Parietal cells are especially common along the proximal portions of each gastric gland (Figure 24–13b). These cells secrete **intrinsic factor**, a glycoprotein that helps absorb vitamin B_{12} across the intestinal lining. p. 665

Parietal cells also indirectly secrete *hydrochloric acid (HCl)*. They do not produce HCl in the cytoplasm. This acid is so strong that it would erode a secretory vesicle and destroy the cell. Instead, these cells create the conditions for increasing the concentrations of both H^+ and Cl^-, the two dissociation products of HCl. Each ion exits parietal cells independently and by different mechanisms (Figure 24–14).

In the process of H^+ production, carbonic acid is formed within parietal cells. The dissociation of carbonic acid releases bicarbonate and hydrogen ions (1). The bicarbonate ions are exchanged for Cl^- from the interstitial fluid (2). When gastric glands are actively secreting, enough bicarbonate ions diffuse into the bloodstream from the interstitial fluid to increase the pH of the blood significantly. This sudden influx of bicarbonate ions has been called the *alkaline tide*. The chloride ions diffuse across the parietal cell and into the lumen of the gastric gland (3), and the H^+ are actively transported there as well (4).

The hydrochloric acid produced by these processes can keep the stomach contents at pH 1.5–2.0. Although this highly acidic environment does not by itself digest chyme, it has the following important functions:

- It kills most of the microorganisms ingested with food.
- It denatures proteins and inactivates most of the enzymes in food.
- It helps break down plant cell walls and the connective tissues in meat.

An acidic environment is also necessary for the function of gastric cells called chief cells. **Chief cells** are most abundant near the base of a gastric gland (see Figure 24–13b). These cells secrete **pepsinogen** (pep-SIN-ō-jen), an inactive proenzyme. Acid in the gastric lumen converts pepsinogen to **pepsin**, an active *proteolytic*, or protein-digesting, enzyme. Pepsin functions most effectively at a strongly acidic pH of 1.5–2.0.

In addition, the stomachs of newborn infants (but not of adults) produce **rennin**, also called *chymosin*, and **gastric lipase**. These enzymes are important for the digestion of milk. Rennin coagulates milk proteins. Gastric lipase initiates the digestion of milk fats.

Pyloric Glands, Enteroendocrine Cells, and Gastric Hormones

Glands in the pyloric part produce primarily mucous secretions, rather than enzymes or acid. In addition, several types of enteroendocrine cells are scattered among the mucus-secreting cells. These enteroendocrine cells produce at least

Figure 24–14 The Secretion of Hydrochloric Acid Ions.

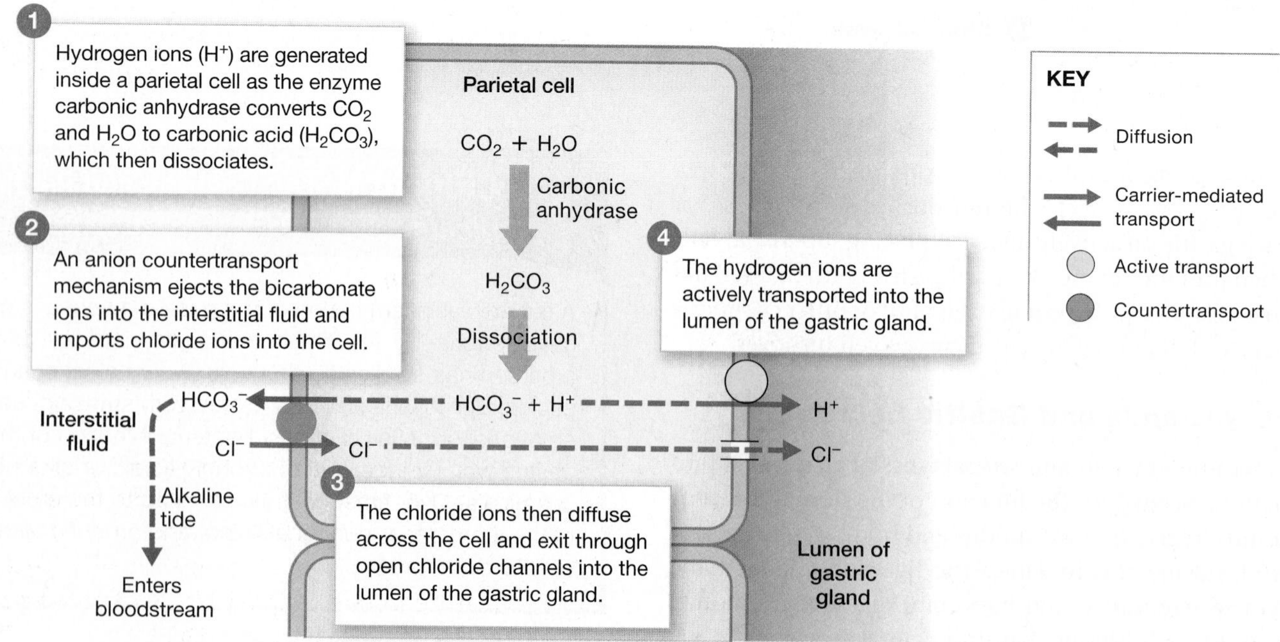
1
Hydrogen ions (H^+) are generated inside a parietal cell as the enzyme carbonic anhydrase converts CO_2 and H_2O to carbonic acid (H_2CO_3), which then dissociates.
2
An anion countertransport mechanism ejects the bicarbonate ions into the interstitial fluid and imports chloride ions into the cell.
Parietal cell
$CO_2 + H_2O$
Carbonic anhydrase
H_2CO_3
Dissociation
$HCO_3^- + H^+$
4
The hydrogen ions are actively transported into the lumen of the gastric gland.
Interstitial fluid
HCO_3^-
Cl^-
Cl^-
H^+
Cl^-
Alkaline tide
Enters bloodstream
3
The chloride ions then diffuse across the cell and exit through open chloride channels into the lumen of the gastric gland.
Lumen of gastric gland
KEY
Diffusion
Carrier-mediated transport
Active transport
Countertransport

seven hormones, most notably **gastrin**. Gastrin is produced by *G cells*, which are most abundant in the gastric pits of the pyloric antrum. Gastrin stimulates secretion by both parietal and chief cells, as well as contractions of the gastric wall that mix and stir the gastric contents.

The pyloric glands also contain *D cells*, which release **somatostatin**, a hormone that inhibits the release of gastrin. D cells continuously release their secretions into the interstitial fluid adjacent to the G cells. Neural and hormonal stimuli can override this inhibition of gastrin production when the stomach is preparing for digestion or is already engaged in digestion.

Several other hormones play a role in hunger and *satiety* (a feeling of not being hungry). The level of *ghrelin* (GREL-in), a hormone produced by *P/D1 cells* lining the fundic region of the stomach, increases before meals to initiate hunger. The ghrelin level decreases shortly after eating to curb appetite. Ghrelin is also antagonistic to *leptin*, a hormone derived from fat tissue that induces satiety. Another hormone from the stomach and small intestine, *obestatin*, is thought to decrease appetite and inhibit thirst. The same gene encodes both ghrelin and obestatin.

Tips & Tools

The stomach squeezes chyme into the small intestine just as you squeeze cake frosting out of a pastry bag.

Physiology of the Stomach: Chemical Digestion

The stomach is a holding tank in which food is saturated with gastric juices and exposed to stomach acid and the digestive effects of pepsin. When ingested food reaches the stomach, salivary amylase and lingual lipase continue the digestion of carbohydrates and lipids. These enzymes continue to work until the pH throughout the contents of the stomach falls below 4.5. They generally remain active 1 to 2 hours after a meal.

As the stomach contents become more fluid and the pH approaches 2.0, the preliminary digestion of proteins by pepsin increases. Protein digestion is not completed in the stomach, because time is limited and pepsin acts on only specific types of peptide bonds, not all of them. However, pepsin generally has enough time to break down complex proteins into smaller peptide and polypeptide chains before the chyme enters the duodenum.

Although partial or limited chemical digestion takes place in the stomach, nutrients are not absorbed there. This is because (1) the mucus covering the epithelial cells means that they are not directly exposed to chyme, (2) the epithelial cells lack the specialized transport mechanisms of cells that line the small intestine, (3) the gastric lining is somewhat impermeable to water, and (4) digestion has not been completed by the time chyme leaves the stomach. At this stage, most carbohydrates, lipids, and proteins are only partially broken down.

Some drugs can be absorbed in the stomach. For example, ethyl alcohol can diffuse through the mucous barrier and penetrate the lipid membranes of the epithelial cells. As a result, alcohol is absorbed in your stomach before any nutrients in a meal reach the bloodstream. Meals containing large amounts of fat slow the rate of alcohol absorption. Why? The reason is that alcohol is lipid soluble, and some of it will be dissolved in fat droplets in the chyme. Aspirin is another lipid-soluble drug that can enter the bloodstream across the gastric mucosa. Such drugs alter the properties of the mucous layer and can promote epithelial damage by stomach acid and enzymes. Prolonged use of aspirin can cause gastric bleeding, so individuals with stomach ulcers are usually advised to avoid aspirin.

Regulation of Gastric Activity in Phases of Digestion

The production of acid and enzymes by the gastric mucosa can be controlled by the CNS; regulated by short reflexes of the enteric nervous system, coordinated in the wall of the stomach; and regulated by hormones of the digestive tract. Gastric control proceeds in three overlapping phases. They are named according to the location of the control center: the *cephalic phase*, the *gastric phase*, and the *intestinal phase*. Please study **Spotlight Figure 24–15** carefully before moving on.

✓ Checkpoint

18. **Name the four major regions of the stomach.**
19. **Discuss the significance of the low pH in the stomach.**
20. **How does a large meal affect the pH of blood leaving the stomach?**
21. **When a person suffers from chronic gastric ulcers, the branches of the vagus nerves that supply the stomach are sometimes cut in an attempt to provide relief. Why might this be an effective treatment?**

See the blue Answers tab at the back of the book.

24

24-5 Accessory digestive organs, such as the pancreas and liver, produce secretions that aid in chemical digestion

Learning Outcome **Describe the structure, functions, and regulation of the accessory digestive organs.**

The pancreas provides digestive enzymes, as well as buffers that help neutralize chyme. The liver secretes *bile*, a solution stored in the gallbladder for discharge into the small intestine. Bile contains buffers and *bile salts*, compounds that facilitate the digestion and absorption of lipids. Let's look at these accessory organs and their functions in more detail.

> Go to MasteringA&P™ > Study Area > Menu > Lab Tools > PhysioEx 9.1 > Exercise 8: Chemical and Physical Processes of Digestion

SPOTLIGHT

Figure 24–15

The Regulation of Gastric Activity

The duodenum plays a key role in controlling digestive function because it monitors the contents of the chyme and adjusts the activities of the stomach and accessory glands to protect the delicate absorptive surfaces of the jejunum. This pivotal role of the duodenum is apparent when you consider the three phases of gastric secretion. The phases are named according to the location of the control center involved.

1 CEPHALIC PHASE

The **cephalic phase** of gastric secretion begins when you see, smell, taste, or think of food. This phase, which is directed by the CNS, prepares the stomach to receive food. The neural output proceeds by way of the parasympathetic division of the autonomic nervous system. The vagus nerves (X) innervate the submucosal plexus of the stomach. Next, postganglionic parasympathetic fibers innervate mucous cells, chief cells, parietal cells, and G cells of the stomach. In response to stimulation, the production of gastric juice speeds up, reaching rates of about 500 mL/h, or about 2 cups per hour. This phase generally lasts only minutes.

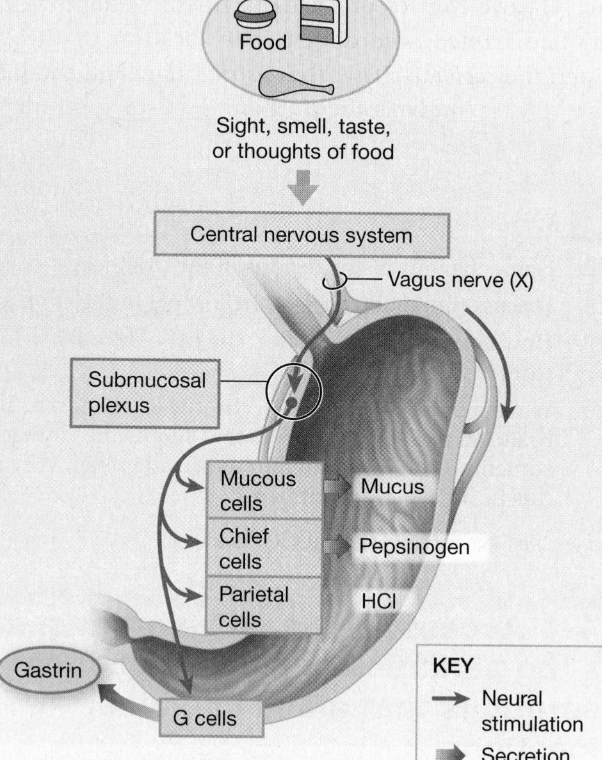

Food
Sight, smell, taste, or thoughts of food
Central nervous system
Vagus nerve (X)
Submucosal plexus
Mucous cells
Mucus
Chief cells
Pepsinogen
Parietal cells
HCl
Gastrin
G cells
KEY
Neural stimulation
Secretion

Emotional states can exaggerate or inhibit the cephalic phase. For example, anger or hostility leads to excessive gastric secretion. On the other hand, anxiety, stress, or fear decreases gastric secretion and gastric contractions, or *motility.*

2 GASTRIC PHASE

The **gastric phase** begins with the arrival of food in the stomach and builds on the stimulation provided during the cephalic phase. This phase may continue for three to four hours while the acid and enzymes process the ingested materials. The stimuli that initiate the gastric phase are (1) distension of the stomach, (2) an increase in the pH of the gastric contents, and (3) the presence of undigested materials in the stomach, especially proteins and peptides. The gastric phase consists of the following mechanisms:

Local Response

Distension of the gastric wall stimulates the release of histamine in the lamina propria, which binds to receptors on the parietal cells and stimulates acid secretion.

Neural Response

The stimulation of stretch receptors and chemoreceptors triggers short reflexes coordinated in the submucosal and myenteric plexuses. This in turn activates the stomach's secretory cells. The stimulation of the myenteric plexus produces powerful contractions called **mixing waves** in the muscularis externa.

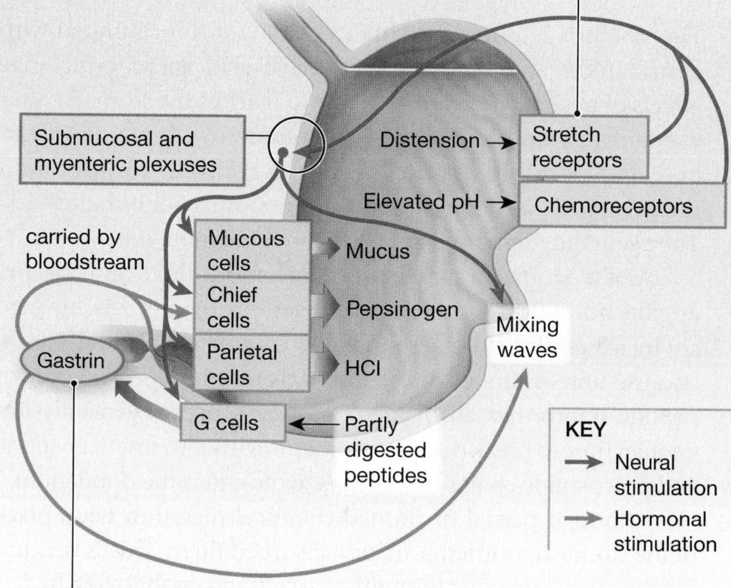

Submucosal and myenteric plexuses
Distension
Stretch receptors
Elevated pH
Chemoreceptors
carried by bloodstream
Mucous cells
Mucus
Chief cells
Pepsinogen
Mixing waves
Gastrin
Parietal cells
HCl
G cells
Partly digested peptides
KEY
Neural stimulation
Hormonal stimulation

Hormonal Response

Neural stimulation and the presence of peptides and amino acids in chyme stimulate the secretion of the hormone gastrin, primarily by G cells. Gastrin travels in the bloodstream to parietal and chief cells, whose increased secretions reduce the pH of the gastric juice. In addition, gastrin also stimulates gastric motility.

3 INTESTINAL PHASE

The **intestinal phase** of gastric secretion begins when chyme first enters the small intestine. The function of the intestinal phase is to control the rate of gastric emptying to ensure that the secretory, digestive, and absorptive functions of the small intestine can proceed with reasonable efficiency. Although here we consider the intestinal phase as it affects stomach activity, the arrival of chyme in the small intestine also triggers other neural and hormonal events that coordinate the activities of the intestinal tract, pancreas, liver, and gallbladder.

Neural Responses

Chyme leaving the stomach decreases the distension in the stomach, thereby reducing the stimulation of stretch receptors. Distension of the duodenum by chyme stimulates stretch receptors and chemoreceptors that trigger the **enterogastric reflex**. This reflex inhibits both gastrin production and gastric contractions and stimulates the contraction of the pyloric sphincter, which prevents further discharge of chyme. At the same time, local reflexes at the duodenum stimulate mucus production, which helps protect the duodenal lining from the arriving acid and enzymes.

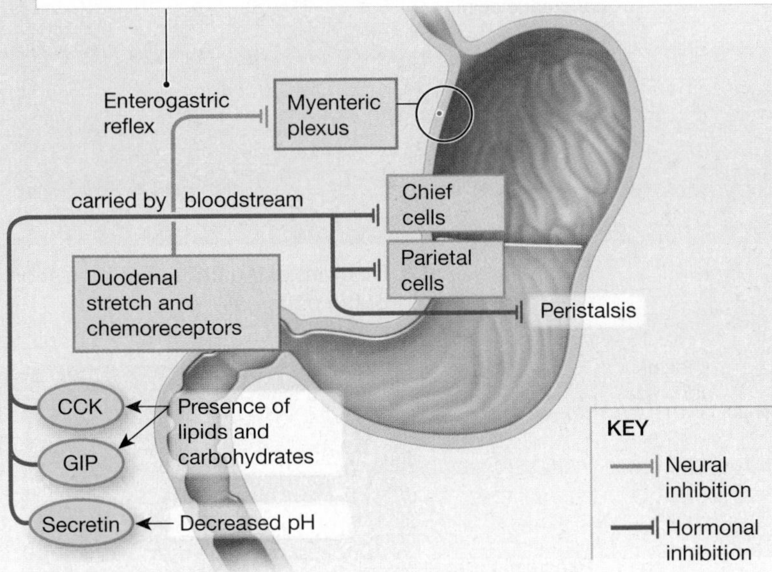

Enterogastric reflex
Myenteric plexus
carried by bloodstream
Chief cells
Parietal cells
Duodenal stretch and chemoreceptors
Peristalsis
CCK
GIP
Secretin
Presence of lipids and carbohydrates
Decreased pH
KEY
Neural inhibition
Hormonal inhibition

Hormonal Responses

The arrival of chyme in the duodenum triggers hormonal responses:

- Arrival of lipids and carbohydrates stimulates the secretion of cholecystokinin (CCK) and gastric inhibitory peptide (GIP).
- A decrease in pH below 4.5 stimulates the secretion of secretin.
- Partially digested proteins in the duodenum stimulate G cells that secrete gastrin, which circulates to the stomach and speeds gastric processing.

CENTRAL REFLEXES

Two central reflexes, the gastroenteric reflex and gastroileal reflex, are also triggered by the stimulation of stretch receptors in the stomach wall as it fills. These reflexes accelerate movement along the small intestine while the enterogastric reflex controls the rate of chyme entry into the duodenum.

Central Gastric Reflexes

The **gastroenteric reflex** stimulates motility and secretion along the entire small intestine.

The **gastroileal** (gas-trō-IL-ē-al) **reflex** triggers the opening of the ileocecal valve, allowing materials to pass from the small intestine into the large intestine.

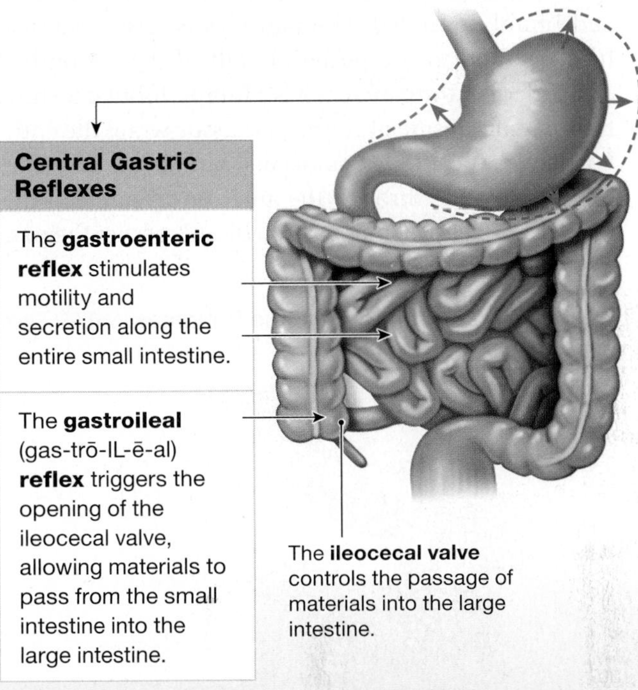

The **ileocecal valve** controls the passage of materials into the large intestine.

In general, the rate of movement of chyme into the small intestine is highest when the stomach is greatly distended and the meal contains little protein. A large meal containing small amounts of protein, large amounts of carbohydrates (such as rice or pasta), wine (alcohol), or after-dinner coffee (caffeine) will leave your stomach very quickly. One reason is that both alcohol and caffeine stimulate gastric secretion and motility.

The *vomiting* reflex occurs in response to irritation of the fauces, pharynx, esophagus, stomach, or proximal segment of the small intestine. These sensations are relayed to the vomiting center of the medulla oblongata, which coordinates motor responses. In preparation for vomiting, the pyloric sphincter relaxes and the contents of the duodenum are discharged back into the stomach by strong peristaltic waves that travel toward the stomach. Vomiting, or **emesis** (EM-e-sis), then occurs as the gastroesophageal sphincter relaxes and the stomach regurgitates its contents through the esophagus and pharynx and out through the mouth.

The Pancreas

The **pancreas** lies posterior to the stomach. It extends laterally from the duodenum toward the spleen. The pancreas is primarily an exocrine organ that produces digestive enzymes and buffers.

Gross Anatomy of the Pancreas

The pancreas is an elongate, pinkish-gray organ about 15 cm (6 in.) long and weighing about 80 g (3 oz) (Figure 24–16a). The broad **head** of the pancreas lies within the loop formed by the duodenum as it leaves the pyloric part. The slender **body** of the pancreas extends toward the spleen, and the **tail** is short and bluntly rounded. The pancreas is retroperitoneal and is firmly bound to the posterior wall of the abdominal cavity. The surface of the pancreas has a lumpy, lobular texture. A thin, transparent capsule of connective tissue wraps the entire organ. The pancreatic lobules, associated blood vessels, and excretory ducts are visible through the anterior capsule and the overlying layer of peritoneum. Arterial blood reaches the pancreas by way of branches of the splenic, superior mesenteric, and common hepatic arteries. The pancreatic arteries and pancreaticoduodenal arteries are the major branches from these vessels. The splenic vein and its branches drain the pancreas.

The large **pancreatic duct** delivers the secretions of the pancreas to the duodenum. (In 3–10 percent of the population, a small *accessory pancreatic duct* branches from the pancreatic duct.) The pancreatic duct extends within the attached mesentery to reach the duodenum, where it meets the *bile duct* from the liver and gallbladder (look ahead to Figure 24–19b). The two ducts then empty into the *duodenal ampulla*, a chamber located roughly halfway along the length of the duodenum. When present, the accessory pancreatic duct usually empties into the duodenum independently, outside the duodenal ampulla.

Histology of the Pancreas

Partitions of connective tissue divide the interior of the pancreas into distinct lobules. The blood vessels and tributaries of the pancreatic ducts are located within these connective tissue

Figure 24–16 Anatomy of the Pancreas. ATLAS: Plates 54d; 55; 57a

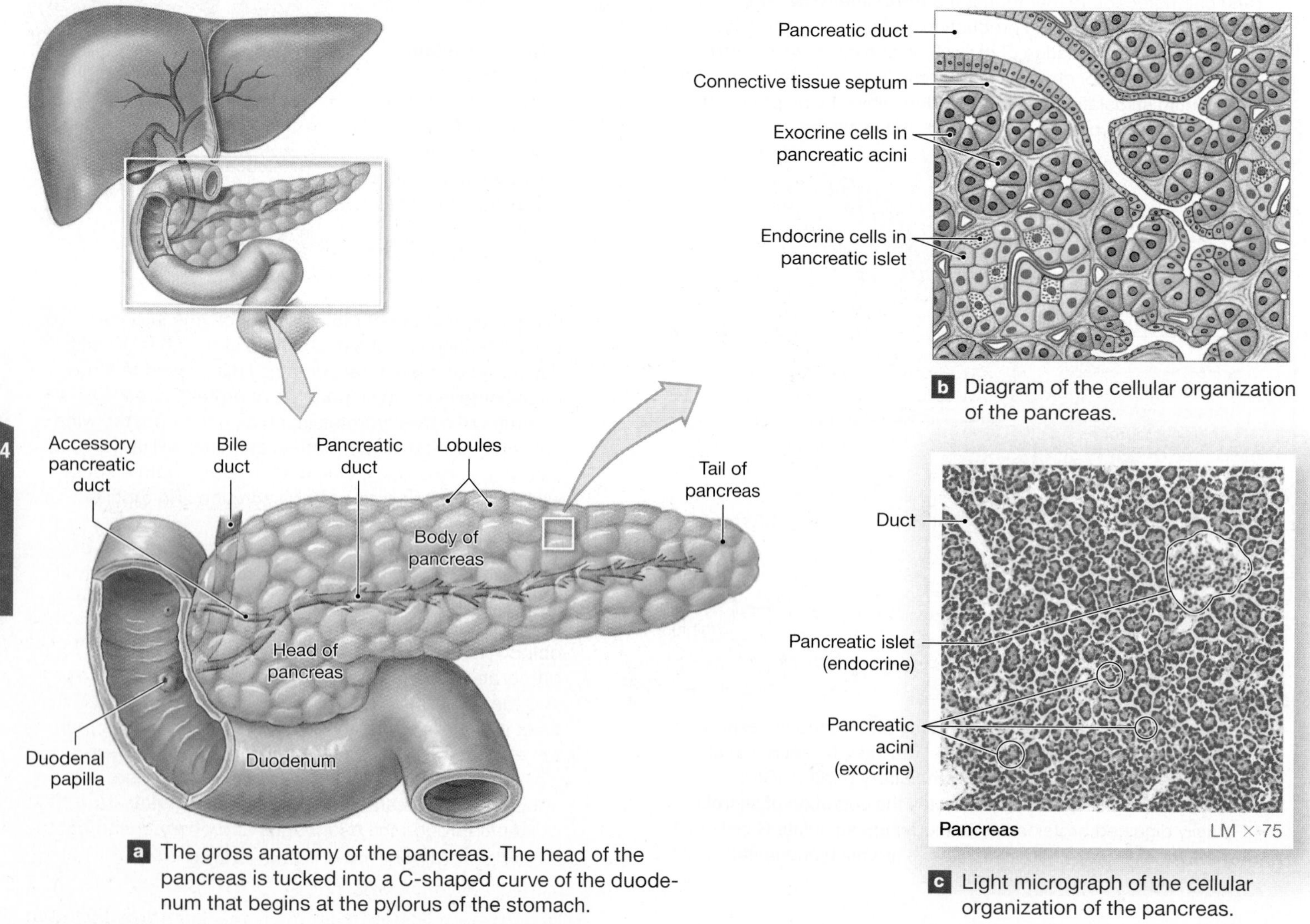
Accessory pancreatic duct
Bile duct
Pancreatic duct
Lobules
Tail of pancreas
Body of pancreas
Head of pancreas
Duodenal papilla
Duodenum
Pancreatic duct
Connective tissue septum
Exocrine cells in pancreatic acini
Endocrine cells in pancreatic islet
Duct
Pancreatic islet (endocrine)
Pancreatic acini (exocrine)
Pancreas
LM × 75

a The gross anatomy of the pancreas. The head of the pancreas is tucked into a C-shaped curve of the duodenum that begins at the pylorus of the stomach.

b Diagram of the cellular organization of the pancreas.

c Light micrograph of the cellular organization of the pancreas.

septa (Figure 24–16b). The pancreas is an example of a *compound tubulo-alveolar gland*, a structure described in Chapter 4. ⮌ p. 125 In each lobule, the ducts branch repeatedly before ending in pockets called **pancreatic acini** (AS-ih-nī). Each pancreatic acinus is lined with simple cuboidal epithelium. *Pancreatic islets*, the endocrine tissues of the pancreas, are scattered among the acini (Figure 24–16c). The islets account for only about 1 percent of the cell population of the pancreas.

Pancreatic Secretions

The pancreas has two distinct functions, one endocrine and the other exocrine. The endocrine cells of the pancreatic islets secrete insulin and glucagon into the bloodstream to control blood sugar. The exocrine cells include the acinar cells and the epithelial cells that line the duct system. Together, these exocrine cells secrete **pancreatic juice**—an alkaline mixture of digestive enzymes, water, and ions—into the small intestine. Acinar cells secrete pancreatic enzymes, which do most of the digestive work in the small intestine. Pancreatic enzymes break down ingested materials into small molecules suitable for absorption. The water and ions, secreted primarily by the cells lining the pancreatic ducts, help dilute and neutralize acid in the chyme.

Physiology of the Pancreas

Each day, the pancreas secretes about 1000 mL (about 1 qt) of pancreatic juice. Hormones from the duodenum control these secretory activities. When chyme arrives in the duodenum, it releases the hormone *secretin*. This hormone triggers the pancreatic secretion of a watery buffer solution with a pH of 7.5–8.8. Among its other components, the secretion contains bicarbonate and phosphate buffers that help raise the pH of the chyme.

Another duodenal hormone, *cholecystokinin (CCK)*, stimulates the production and secretion of pancreatic enzymes. Stimulation by the vagus nerves also increases the secretion of pancreatic enzymes. Recall that this stimulation takes place during the cephalic phase of gastric regulation, so the pancreas starts to synthesize enzymes before food even reaches the stomach. This head start is important, because enzyme synthesis takes much longer than the production of buffers. By starting early, the pancreatic cells are ready to meet the demand when chyme arrives in the duodenum.

The pancreatic enzymes include the following:

- **Pancreatic alpha-amylase**, a **carbohydrase** (kar-bō-HĪ-drās)—an enzyme that breaks down certain starches. Pancreatic alpha-amylase is almost identical to salivary amylase.
- **Pancreatic lipase**, which breaks down certain complex lipids, releasing products (such as fatty acids) that can be easily absorbed.
- **Nucleases**, which break down RNA or DNA.
- **Proteolytic enzymes**, which break apart proteins. These enzymes include **proteases** which break apart large protein complexes, and **peptidases**, which break small peptide chains into individual amino acids.

Proteolytic enzymes account for about 70 percent of total pancreatic enzyme production. These enzymes are secreted as inactive proenzymes. They are activated only after they reach the small intestine. Proenzymes discussed in previous chapters include pepsinogen, angiotensinogen, plasminogen, fibrinogen, and many of the clotting factors and enzymes of the complement system. ⮌ pp. 644, 661, 683, 802 As in the stomach, the release of a proenzyme rather than an active enzyme protects the secretory cells from the destructive effects of their own products. Among the proenzymes secreted by the pancreas are **trypsinogen** (trip-SIN-ō-jen), **chymotrypsinogen** (kī-mo-trip-SIN-ō-jen) **procarboxypeptidase** (prō-kar-bok-sē-PEP-ti-dās), and **proelastase** (pro-ē-LAS-tās).

Inside the duodenum, trypsinogen is converted to **trypsin**, an active protease. Trypsin then activates the other proenzymes, producing **chymotrypsin**, **carboxypeptidase**, and **elastase**. Each enzyme breaks peptide bonds linking specific amino acids and ignores others. Together, these enzymes break down proteins into a mixture of dipeptides, tripeptides, and amino acids.

+ Clinical Note Pancreatitis

Pancreatitis (pan-krē-a-TĪ-tis) is an inflammation of the pancreas. The condition is extremely painful. The factors that may produce it include a blockage of the excretory ducts, bacterial or viral infections, ischemia, and drug reactions—especially those involving alcohol. These stimuli provoke a crisis by injuring exocrine cells in at least a portion of the organ. Then lysosomes in the damaged cells release enzymes causing autolysis (self-digestion of cells), which spreads to surrounding normal cells. In most cases, only a portion of the pancreas is affected. The condition usually subsides in a few days. But in 10–15 percent of cases, the process continues and may ultimately destroy the pancreas. If the islet cells are damaged, diabetes mellitus may result. ⮌ p. 641

The Liver

The **liver** is the largest visceral organ, a firm, reddish-brown organ weighing about 1.5 kg (3.3 lb). Most of its mass lies in the right hypochondriac and epigastric regions, but it may extend into the left hypochondriac and umbilical regions as well. The versatile liver performs essential metabolic and synthetic functions.

Gross Anatomy of the Liver

Most of surface of the liver is wrapped in a tough fibrous capsule and is covered by a layer of visceral peritoneum. On the anterior surface, the **falciform ligament** marks the division between the organ's left and right lobes (Figure 24–17a,b). A thickening in the posterior margin of the falciform ligament

Figure 24–17 **Gross Anatomy of the Liver.** ATLAS: Plates 49a,b,e; 54a–c; 57a,b

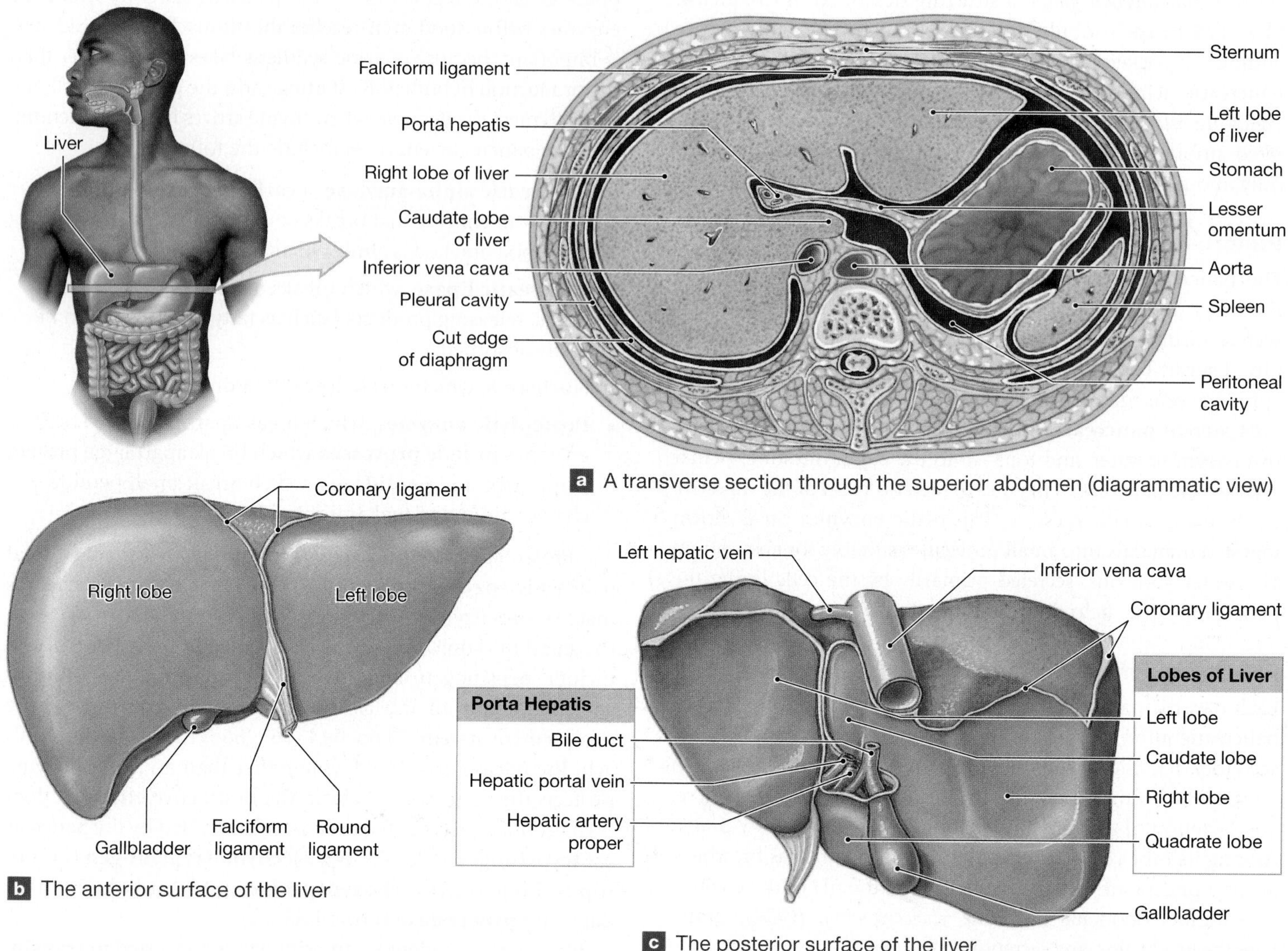
Liver
Falciform ligament
Porta hepatis
Right lobe of liver
Caudate lobe of liver
Inferior vena cava
Pleural cavity
Cut edge of diaphragm
Sternum
Left lobe of liver
Stomach
Lesser omentum
Aorta
Spleen
Peritoneal cavity
Coronary ligament
Right lobe
Left lobe
Gallbladder
Falciform ligament
Round ligament
Left hepatic vein
Inferior vena cava
Coronary ligament
Porta Hepatis
Bile duct
Hepatic portal vein
Hepatic artery proper
Lobes of Liver
Left lobe
Caudate lobe
Right lobe
Quadrate lobe
Gallbladder

a A transverse section through the superior abdomen (diagrammatic view)

b The anterior surface of the liver

c The posterior surface of the liver

is the **round ligament**, or *ligamentum teres*. This fibrous band marks the path of the fetal umbilical vein.

On the posterior surface of the liver, the impression left by the inferior vena cava marks the division between the right lobe and the small **caudate** (KAW-dāt) **lobe** (Figure 24–17c). Inferior to the caudate lobe lies the **quadrate lobe**, sandwiched between the left lobe and the gallbladder. Afferent blood vessels and other structures reach the liver by traveling within the connective tissue of the lesser omentum. They converge at a region called the **porta hepatis** ("doorway to the liver").

We discussed the circulation to the liver in Chapter 21 and summarized it in Figures 21-24 and 21-31, pp. 763, 774. Nearly one-third of the blood supply to the liver is arterial blood from the hepatic artery proper. The rest is venous blood from the hepatic portal vein. Liver cells, called **hepatocytes** (HEP-ah-tō-sīts), adjust circulating levels of nutrients through selective absorption and secretion. The blood leaving the liver returns to

+ Clinical Note Cirrhosis

Any condition that severely damages the liver is a serious threat to life. The liver has a high capacity to regenerate itself after injury. Even so, liver function does not fully recover unless the normal vascular pattern is restored. Examples of important types of liver disease include **cirrhosis** (sih-RŌ-sis), which is characterized by the replacement of lobules by fibrous tissue, and various forms of *hepatitis* caused by viral infections. In some cases, liver transplants are used to treat liver failure. Unfortunately, the supply of suitable donor tissue is limited. The success rate is highest in young, otherwise healthy individuals. A form of "liver dialysis" known as *ELAD* (extracorporeal liver assist device) is beneficial for individuals with liver failure or chronic liver disease.

the systemic circuit through the hepatic veins. These veins open into the inferior vena cava.

Histology of the Liver

Connective tissue divides each lobe of the liver into approximately 100,000 liver **lobules**, the basic functional units of the liver. The histological organization and structure of a typical liver lobule are shown in Figure 24–18.

Each lobule is roughly 1 mm in diameter. Adjacent lobules are separated by an *interlobular septum*. The hepatocytes in a liver lobule form a series of irregular plates arranged like the spokes of a wheel (see Figure 24–18a,b). The plates are only one cell thick. Exposed hepatocyte surfaces are covered with short microvilli. Within a lobule, sinusoids between adjacent plates empty into the **central vein**. (We introduced sinusoids in Chapter 21. p. 734) The liver sinusoids lack a basement membrane, so large openings between the endothelial cells allow solutes—even those as large as plasma proteins—to pass out of the bloodstream and into the spaces surrounding the hepatocytes.

Figure 24–18 Histology of the Liver.

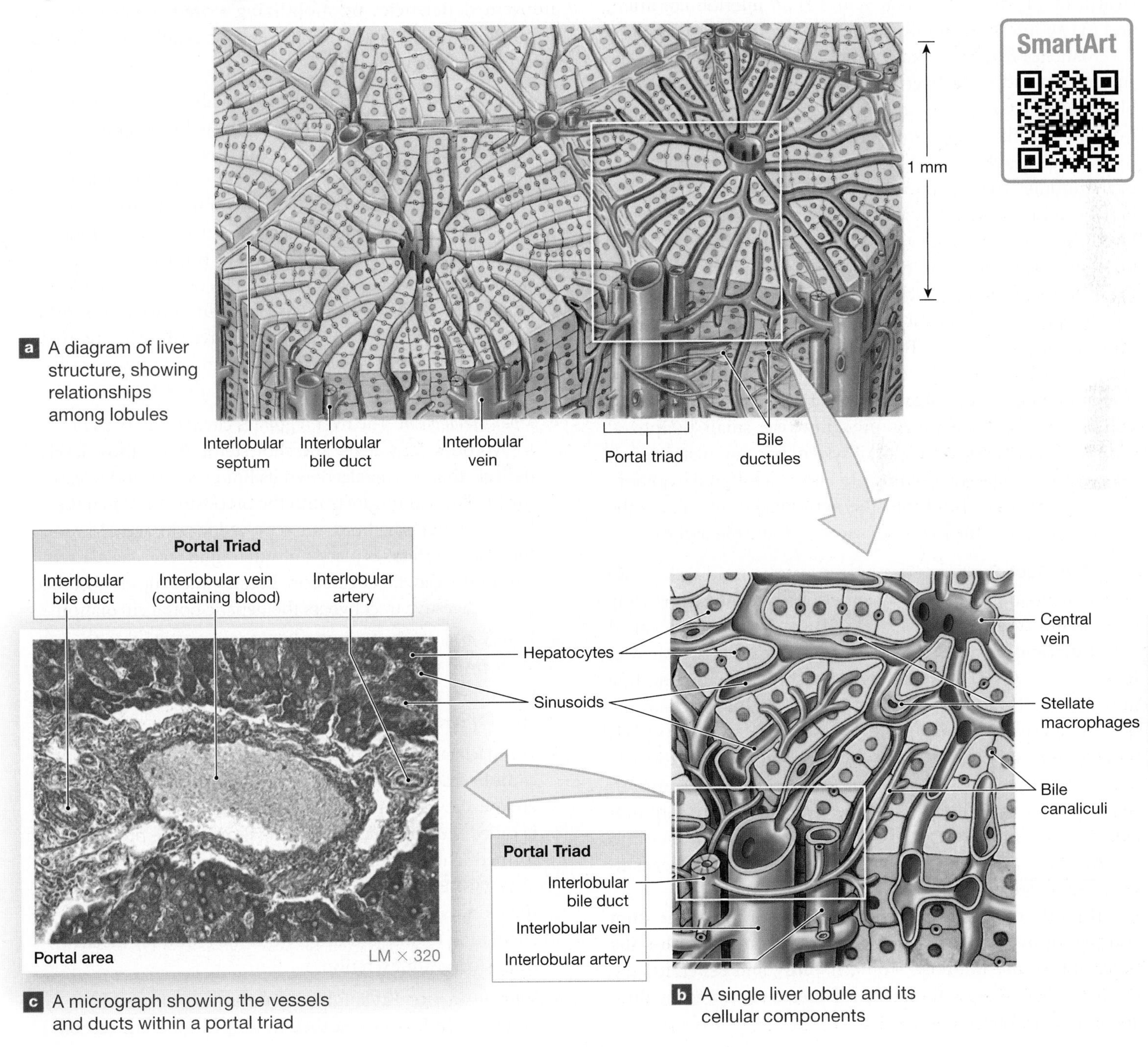

SmartArt
1 mm
Interlobular septum
Interlobular bile duct
Interlobular vein
Portal triad
Bile ductules
Portal Triad
Interlobular bile duct
Interlobular vein (containing blood)
Interlobular artery
Portal area
LM × 320
Hepatocytes
Sinusoids
Portal Triad
Interlobular bile duct
Interlobular vein
Interlobular artery
Central vein
Stellate macrophages
Bile canaliculi

a A diagram of liver structure, showing relationships among lobules

b A single liver lobule and its cellular components

c A micrograph showing the vessels and ducts within a portal triad

The lining of the sinusoids contains typical endothelial cells and a large number of **stellate macrophages** (*Kupffer cells*). These phagocytic cells are part of the monocyte–macrophage system. They engulf pathogens, cell debris, and damaged blood cells. Stellate macrophages also store iron, some lipids, and heavy metals (such as tin or mercury) that are absorbed by the digestive tract.

The Hepatic Portal System. Blood enters the liver sinusoids from small branches of the hepatic portal vein and hepatic artery proper. A typical liver lobule has a hexagonal shape in cross section (see Figure 24–18a). There are six **portal triads**, one at each corner of the lobule. A portal triad contains three structures: (1) an interlobular vein, (2) an interlobular artery, and (3) an interlobular bile duct (see Figure 24–18a-c).

Branches from the arteries and veins deliver blood to the sinusoids of adjacent liver lobules (see Figure 24–18a,b). As blood flows through the sinusoids, hepatocytes absorb solutes from the plasma and secrete materials such as plasma proteins. Blood then leaves the sinusoids and enters the central vein of the lobule. The central veins ultimately merge to form the hepatic veins, which then empty into the inferior vena cava. Liver diseases, such as the various forms of *hepatitis*, and conditions such as alcoholism, can lead to degenerative changes in the liver tissue and constriction of the circulatory supply.

Pressures in the hepatic portal system are usually low, averaging 10 mm Hg or less. This pressure can increase markedly, however, if blood flow through the liver is restricted by a blood clot or damage to the organ. Such a rise in portal pressure is called *portal hypertension*. As pressures rise, small peripheral veins and capillaries in the portal system become distended. If they rupture, extensive bleeding can take place. Portal hypertension can also force fluid into the peritoneal cavity across the serosal surfaces of the liver and viscera, producing ascites.

The Bile Duct System. The liver secretes a fluid called **bile** into a network of narrow channels between the opposing membranes of adjacent liver cells. These passageways, called **bile canaliculi**, extend outward, away from the central vein (see Figure 24–18b). Eventually, they connect with fine bile ductules (DUK-tūlz), which carry bile to the bile ducts in the nearest portal area (see Figure 24–18a). The **right** and **left hepatic ducts** (Figure 24–19a) collect bile from all the bile ducts of the liver lobes. These ducts unite to form the **common hepatic duct**, which leaves the liver. The bile in the common hepatic duct either flows into the *bile duct*, which empties into the duodenal ampulla, or enters the *cystic duct*, which leads to the gallbladder (see Figure 24–19a–c).

The **bile duct** is formed by the union of the **cystic duct** and the common hepatic duct. The bile duct passes within the lesser omentum toward the stomach, turns, and penetrates the wall of the duodenum to meet the pancreatic duct at the duodenal ampulla (see Figure 24–19b).

24

Physiology of the Liver

The liver carries out more than 200 functions. They fall into three general categories: (1) *metabolic regulation*, (2) *hematological regulation*, and (3) *bile production*. In this discussion we provide a general overview.

Metabolic Regulation. The liver is the primary organ involved in regulating the composition of circulating blood. All blood leaving the absorptive surfaces of the digestive tract enters the hepatic portal system and flows into the liver. Liver cells extract nutrients or toxins from the blood before it reaches the systemic circulation through the hepatic veins. The liver removes and stores excess nutrients. It corrects nutrient deficiencies by mobilizing stored reserves or performing synthetic activities. The liver's regulatory activities affect the following:

- *Carbohydrate Metabolism.* The liver stabilizes the blood glucose level at about 90 mg/dL. If the blood glucose level decreases, hepatocytes break down glycogen reserves and release glucose into the bloodstream. They also synthesize glucose from other carbohydrates or from available amino acids. The synthesis of glucose from other compounds is called *gluconeogenesis*. If the blood glucose level climbs, hepatocytes remove glucose from the bloodstream. They either store it as glycogen or use it to synthesize lipids that can be stored in the liver or other tissues. Circulating hormones, such as insulin and glucagon, regulate these metabolic activities. pp. 638–640
- *Lipid Metabolism.* The liver regulates circulating levels of triglycerides, fatty acids, and cholesterol. When those levels decline, the liver breaks down its lipid reserves and releases the breakdown products into the bloodstream. When the levels are high, the lipids are removed for storage. However, this regulation takes place only after lipid levels have risen within the general circulation, because most lipids absorbed by the digestive tract bypass the hepatic portal circulation.
- *Amino Acid Metabolism.* The liver removes excess amino acids from the bloodstream. These amino acids can be used to synthesize proteins or can be converted to lipids or glucose for energy storage.
- *Waste Removal.* When converting amino acids to lipids or carbohydrates, or when breaking down amino acids to get energy, the liver strips off the amino groups. This process is called *deamination*. Ammonia, a toxic waste product, is formed. The liver neutralizes ammonia by converting it to *urea* (yū-RĒ-ah), a fairly harmless compound excreted by the kidneys. The liver also removes other waste products, circulating toxins, and drugs from the blood for inactivation, storage, or excretion.
- *Vitamin Storage.* Fat-soluble vitamins (A, D, E, and K) and vitamin B_{12} are absorbed from the blood and stored

Figure 24–19 Anatomy and Physiology of the Gallbladder and Bile Ducts. ATLAS: Plates 49c,e; 51a; 54b–d

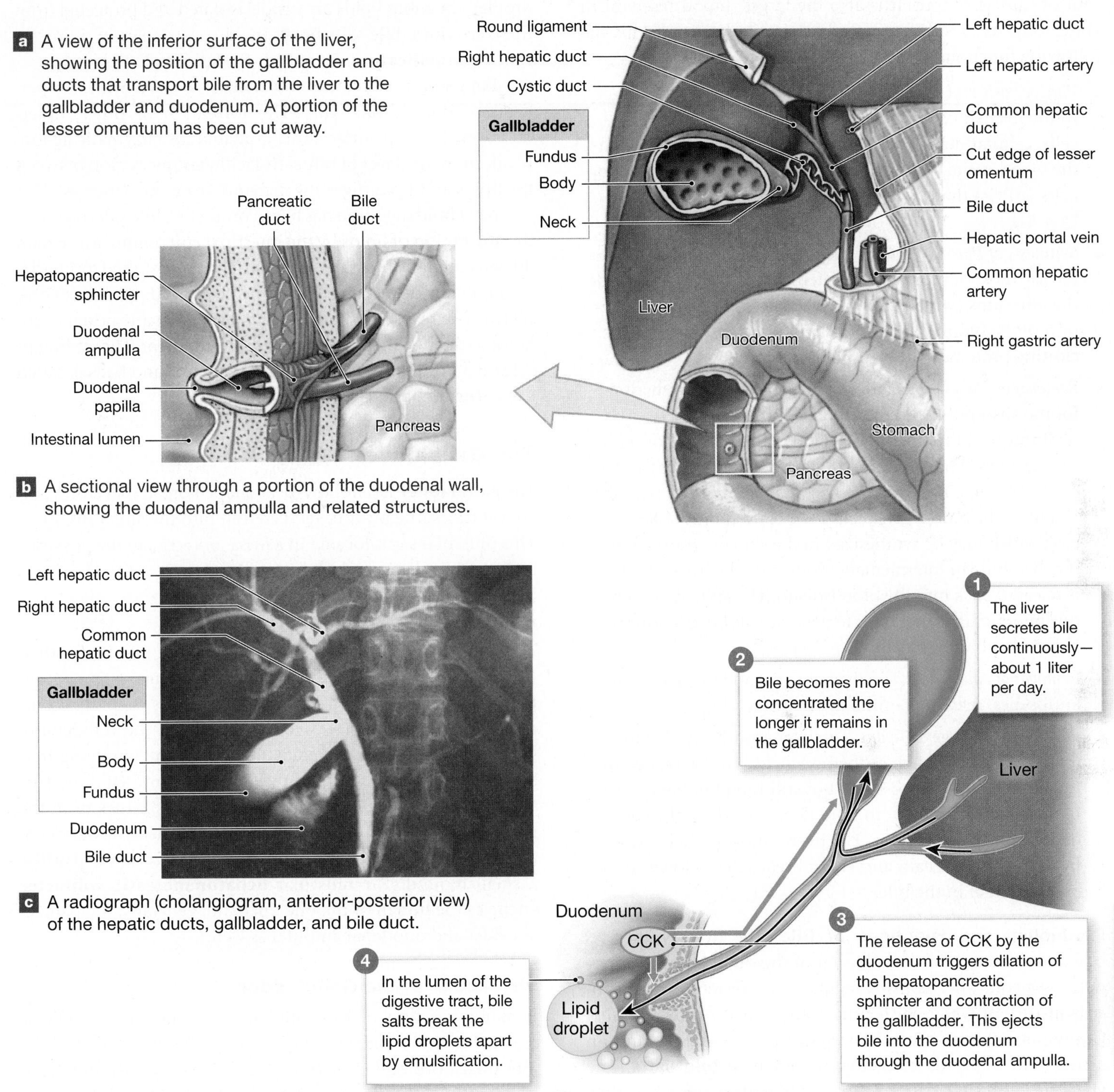

Round ligament
Right hepatic duct
Cystic duct
Gallbladder
Fundus
Body
Neck
Left hepatic duct
Left hepatic artery
Common hepatic duct
Cut edge of lesser omentum
Bile duct
Hepatic portal vein
Common hepatic artery
Right gastric artery
Liver
Duodenum
Stomach
Pancreas
Pancreatic duct
Bile duct
Hepatopancreatic sphincter
Duodenal ampulla
Duodenal papilla
Intestinal lumen
Pancreas
Left hepatic duct
Right hepatic duct
Common hepatic duct
Gallbladder
Neck
Body
Fundus
Duodenum
Bile duct
1 The liver secretes bile continuously—about 1 liter per day.
2 Bile becomes more concentrated the longer it remains in the gallbladder.
Liver
Duodenum
CCK
3 The release of CCK by the duodenum triggers dilation of the hepatopancreatic sphincter and contraction of the gallbladder. This ejects bile into the duodenum through the duodenal ampulla.
4 In the lumen of the digestive tract, bile salts break the lipid droplets apart by emulsification.
Lipid droplet
24

a A view of the inferior surface of the liver, showing the position of the gallbladder and ducts that transport bile from the liver to the gallbladder and duodenum. A portion of the lesser omentum has been cut away.

b A sectional view through a portion of the duodenal wall, showing the duodenal ampulla and related structures.

c A radiograph (cholangiogram, anterior-posterior view) of the hepatic ducts, gallbladder, and bile duct.

d Physiology of the gallbladder.

in the liver. These reserves are used when your diet contains inadequate amounts of those vitamins.

- *Mineral Storage.* The liver converts iron reserves to ferritin and stores this protein–iron complex. p. 666
- *Drug Inactivation.* The liver removes and breaks down circulating drugs, limiting the duration of their effects.

When physicians prescribe a particular drug, they must take into account the rate at which the liver removes that drug from the bloodstream. For example, a drug that is absorbed relatively quickly must be administered every few hours to keep the blood concentration at a therapeutic level.

Hematological Regulation. The liver receives about 25 percent of cardiac output. It is also the largest blood reservoir in your body. As blood passes through it, the liver performs the following functions:

- *Phagocytosis and Antigen Presentation.* Stellate macrophages in the liver sinusoids engulf old or damaged red blood cells, cellular debris, and pathogens, removing them from the bloodstream. Stellate macrophages are antigen-presenting cells (APCs) that can stimulate an immune response. p. 811
- *Synthesis of Plasma Proteins.* Hepatocytes synthesize and release most of the plasma proteins. These proteins include the albumins (which contribute to the osmotic concentration of the blood), the various types of transport proteins, clotting proteins, and complement proteins.
- *Removal of Circulating Hormones.* The liver is the primary site for the absorption and recycling of epinephrine, norepinephrine, insulin, thyroid hormones, and steroid hormones, such as the sex hormones (estrogens and androgens) and corticosteroids. The liver also absorbs *cholecalciferol* (vitamin D_3) from the blood. Liver cells then convert the cholecalciferol, which may be synthesized in the skin or absorbed in the diet, into an intermediary product, 25-hydroxy-D_3, that is released back into the bloodstream. The kidneys absorb this intermediary and use it to generate *calcitriol*, a hormone important to Ca^{2+} metabolism. p. 641
- *Removal of Antibodies.* The liver absorbs and breaks down antibodies, releasing amino acids for recycling.
- *Removal or Storage of Toxins.* The liver absorbs lipid-soluble toxins in the diet, such as the insecticide DDT (banned in the United States since 1972, but still found in the environment), and stores them in lipid deposits, where they do not disrupt cellular functions. The liver removes other toxins from the bloodstream and either breaks them down or excretes them in the bile.

Production and Functions of Bile. The liver synthesizes bile and secretes it into the lumen of the duodenum. Hormonal and neural mechanisms regulate bile secretion. Bile consists mostly of water, with minor amounts of ions, *bilirubin* (a pigment derived from hemoglobin), cholesterol, and an assortment of lipids collectively known as **bile salts**. The water and ions help dilute and buffer acids in chyme as they enter the small intestine. Bile salts are synthesized from cholesterol in the liver. Several related compounds are involved. The most abundant are derivatives of the steroids *cholate* and *chenodeoxycholate*.

Bile salts play a role in the digestion of lipids. Most dietary lipids are not water soluble, so mechanical digestion in the stomach creates large drops containing a variety of lipids. Pancreatic lipase is not lipid soluble, so the enzymes can interact with lipids only at the surface of a lipid droplet. The larger the droplet, the more lipids are inside, isolated and protected from these enzymes. Bile salts break the droplets apart in a process called **emulsification** (ē-mul-sih-fih-KĀ-shun).

Emulsification creates tiny *emulsion droplets* with a superficial coating of bile salts. The formation of tiny droplets increases the total surface area available for enzymatic action. In addition, the layer of bile salts facilitates interaction between the lipids and lipid-digesting enzymes from the pancreas.

After lipid digestion has been completed, bile salts promote the absorption of lipids by the intestinal epithelium. More than 90 percent of the bile salts are themselves reabsorbed, primarily in the ileum of the small intestine, as lipid digestion is completed. The reabsorbed bile salts enter the hepatic portal circulation. The liver then collects and recycles them. The cycling of bile salts from the liver to the small intestine and back is called the **enterohepatic circulation** of bile.

The Gallbladder

The **gallbladder** is a hollow, pear-shaped organ that stores and concentrates bile prior to its secretion into the small intestine. This muscular sac is located in a fossa, or recess, in the posterior surface of the liver's right lobe (see Figure 24–19a).

Anatomy of the Gallbladder

The gallbladder is divided into three regions: (1) the **fundus**, (2) the **body**, and (3) the **neck**. The cystic duct extends from the gallbladder to the point where it unites with the common hepatic duct to form the bile duct. At the duodenum, the bile duct meets the pancreatic duct before emptying into a chamber called the **duodenal ampulla** (am-PUL-luh) (see Figure 24–19b), which receives buffers and enzymes from the pancreas and bile from the liver and gallbladder. The duodenal ampulla opens into the duodenum at the **duodenal papilla**, a small mound. The muscular **hepatopancreatic sphincter** encircles the lumen of the bile duct and, generally, the pancreatic duct and duodenal ampulla as well.

Physiology of the Gallbladder

A major function of the gallbladder is bile storage. Bile is released into the duodenum only under the stimulation of the intestinal hormone CCK (see Figure 24–19d). Without CCK, the hepatopancreatic sphincter remains closed, so bile exiting the liver in the common hepatic duct cannot flow through the bile duct and into the duodenum. Instead, it enters the cystic duct and is stored within the expandable gallbladder. Whenever chyme enters the duodenum, CCK is released, relaxing the hepatopancreatic sphincter and stimulating contractions of the gallbladder that push bile into the small intestine. The amount of CCK secreted increases markedly when the chyme contains large amounts of lipids.

The gallbladder also functions in bile modification. When full, the gallbladder contains 40–70 mL of bile. The composition of bile gradually changes as it remains in the gallbladder: Much of the water is absorbed, and the bile salts and other components of bile become increasingly concentrated.

If bile becomes too concentrated, crystals of insoluble minerals and salts begin to form. These deposits are called *gallstones*. Small gallstones are not a problem as long as they can be flushed down the bile duct and excreted. In *cholecystitis* (kō-lē-sis-TĪ-tis; *chole*, bile + *kystis*, bladder + *-itis*, inflammation), the gallstones are so large that they can damage the wall of the gallbladder or block the cystic duct or bile duct. In that case, the gallbladder may need to be surgically removed. This does not seriously impair digestion, because bile production continues at the normal level. However, the bile is more dilute, and its entry into the small intestine is not as closely tied to the arrival of food in the duodenum.

 **Checkpoint**

22. **The digestion of which nutrient would be most impaired by damage to the exocrine pancreas?**
23. **Describe the portal triad.**
24. **What is the main function of the gallbladder?**

See the blue Answers tab at the back of the book.

24-6 The small intestine primarily functions in the chemical digestion and absorption of nutrients

Learning Outcome Describe the anatomy and histology of the small intestine, and explain the functions and regulation of intestinal secretions.

The **small intestine** is a long muscular tube where chemical digestion is completed and the products of digestion are absorbed. Ninety percent of nutrient absorption takes place in the small intestine (most of the rest occurs in the large intestine). This section examines the anatomy and physiology of this critical organ.

Gross Anatomy of the Small Intestine

The small intestine, also known as the *small bowel*, averages 6 m (19.7 ft) in length (range: 4.5–7.5 m; 14.8–24.6 ft). Its diameter ranges from 4 cm (1.6 in.) at the stomach to about 2.5 cm (1 in.) at the junction with the large intestine.

The small intestine fills much of the peritoneal cavity. It occupies all abdominal regions except the right and left hypochondriac and epigastric regions (look back at Figure 1–4b, p. 11). Its position is stabilized by the mesentery proper, a broad mesentery attached to the posterior body wall (look back at Figure 24–2c,d). The stomach, large intestine, abdominal wall, and pelvic girdle restrict movement of the small intestine during digestion. Blood vessels, lymphatic vessels, and nerves reach the small intestine within the connective tissue of the mesentery. The primary blood vessels involved are branches of the superior mesenteric artery and the superior mesenteric vein. ⮌ pp. 764, 775

The small intestine has three segments: the duodenum, the jejunum, and the ileum (Figure 24–20a). The **duodenum** (dū-ō-DĒ-num), 25 cm (10 in.) in length, is the segment closest to the stomach. This portion of the small intestine is a "mixing bowl." It receives chyme from the stomach and digestive secretions from the pancreas and liver. From its connection with the stomach, the duodenum curves in a C that encloses the pancreas. Except for the proximal 2.5 cm (1 in.), the duodenum is in a retroperitoneal position between vertebrae L_1 and L_4 (look back at Figure 24–2d).

A rather abrupt bend marks the boundary between the duodenum and the **jejunum** (jeh-JŪ-num). At this junction, the small intestine reenters the peritoneal cavity, supported by a sheet of mesentery. The jejunum is about 2.5 m (8.2 ft) long. The bulk of chemical digestion and nutrient absorption occurs there.

The (IL-ē-um), the final segment of the small intestine, is also the longest. It averages 3.5 m (11.5 ft) in length. The ileum ends at the **ileocecal** (il-ē-ō-SĒ-kal) **valve**. This sphincter controls the flow of material from the ileum into the *cecum* (SĒ-kum) of the large intestine.

Tips & Tools

To remember the order of the small intestine segments, beginning at the stomach, use this mnemonic: ***D***on't ***j***ump ***i***n—***d***uodenum, ***j***ejunum, and ***i***leum. Also, make sure not to confuse *ileum*, the last segment of the small intestine, with *ilium*, which is a bone.

Histology of the Small Intestine

The intestinal lining has a series of transverse folds called **circular folds** (see Figure 24–20b). Unlike the rugae in the stomach, the circular folds are permanent features—they do not disappear when the small intestine fills. The small intestine contains about 800 circular folds (roughly 2 per centimeter), which greatly increase the surface area available for absorption.

The Intestinal Villi

If the small intestine were a simple tube with smooth walls, it would have a total absorptive area of about 3300 cm^2 (3.6 ft^2). Instead, the mucosa contains circular folds. Each circular fold in turn has a series of fingerlike projections, the **intestinal villi** (Figure 24–21). The villi are covered by simple columnar epithelium that is carpeted with microvilli. The cells are said

Figure 24–20 Gross Anatomy and Segments of the Intestine. ATLAS: Plates 49a,b,d; 51a,b

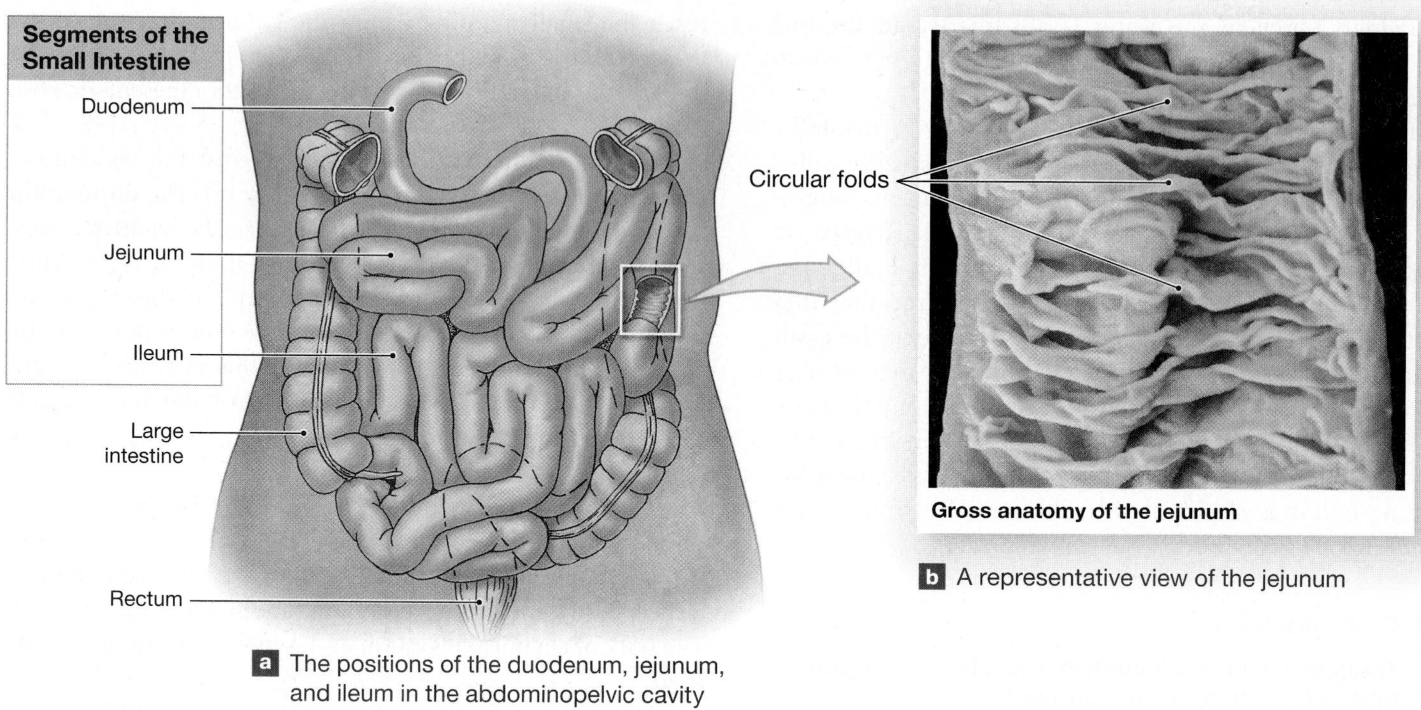
Segments of the Small Intestine
Duodenum
Jejunum
Ileum
Large intestine
Rectum
Circular folds
Gross anatomy of the jejunum

a The positions of the duodenum, jejunum, and ileum in the abdominopelvic cavity

b A representative view of the jejunum

to have a *brush border* because the microvilli project from the epithelium like the bristles on a brush (see Figure 24–21e). This arrangement increases the total area for absorption by a factor of more than 600, to approximately 2 million cm^2 (more than 2200 ft^2). This area is nearly the size of 10 lanes in a bowling alley.

The lamina propria of each villus contains an extensive network of capillaries that originate in a vascular network within the submucosa (see Figure 24–21d). These capillaries carry absorbed nutrients to the hepatic portal circulation for delivery to the liver.

24

In addition to capillaries and nerve endings, each villus contains a central lymphatic vessel called a **lacteal** (LAK-tē-ul; *lacteus*, milky) (see Figure 24–21b,d). Lacteals transport materials that cannot enter blood capillaries. For example, absorbed fatty acids are assembled into protein–lipid packages that are too large to diffuse into the bloodstream. These packets, called *chylomicrons*, reach the venous circulation through the thoracic duct, which delivers lymph into the left subclavian vein. The name *lacteal* refers to the pale, milky appearance of lymph that contains large quantities of lipids.

Intestinal Glands and Their Secretions

The duodenum has numerous goblet cells, both in the epithelium and deep to it. Goblet cells between the columnar epithelial cells eject mucins onto the intestinal surfaces (see Figure 24–21d,e). At the bases of the villi are the entrances to the **intestinal glands**, also called *intestinal crypts*. These glandular pockets extend deep into the underlying lamina propria. Near the base of each intestinal gland, *stem cells* divide and produce new generations of epithelial cells, which are continuously displaced toward the intestinal surface. In a few days the new cells reach the tip of a villus and are shed into the intestinal lumen. This ongoing process renews the epithelial surface. The disintegration of the shed cells adds enzymes to the lumen. **Paneth cells** at the base of the intestinal glands have a role in innate (nonspecific) immunity, and release defensins and lysozyme (Figure 24–21b,c). These secretions kill some bacteria and allow others to live, thereby establishing the flora of the intestinal lumen. Intestinal glands also contain enteroendocrine cells that produce several intestinal hormones, including gastrin, cholecystokinin, and secretin.

In addition to intestinal glands, the intestinal submucosa contains **duodenal submucosal glands**. Duodenal submucosal glands produce copious quantities of mucus when chyme arrives from the stomach. The mucus protects the epithelium from the acidity of chyme and also contains bicarbonate ions that help raise the pH of the chyme. The duodenal submucosal glands also secrete the hormone *urogastrone*, which inhibits gastric acid production and stimulates the division of epithelial stem cells along the digestive tract (look back at Figure 24–3).

Figure 24–21 Histology of the Intestinal Wall. ATLAS: Plates 51a–d

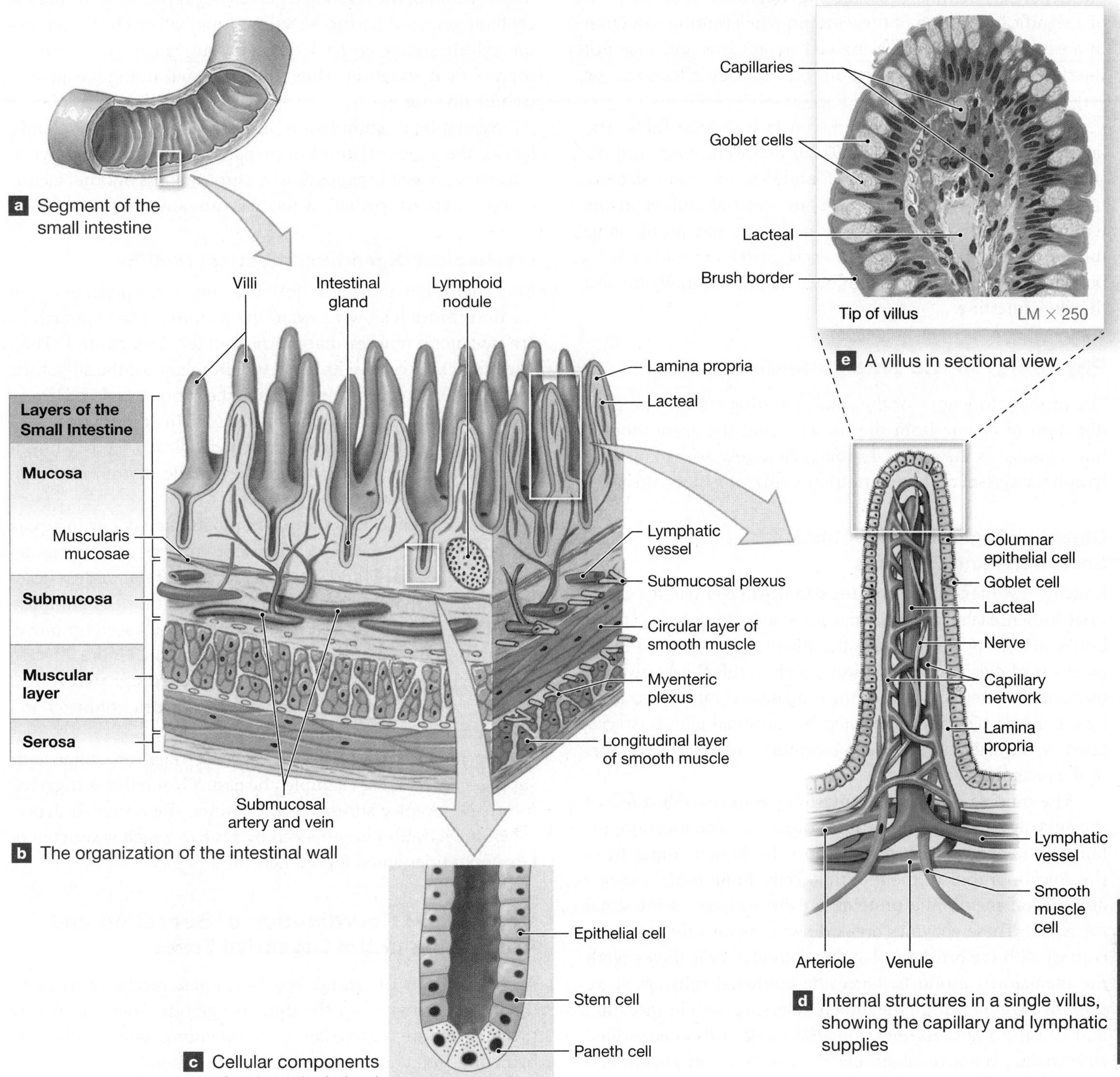

a Segment of the small intestine
Villi
Intestinal gland
Lymphoid nodule
Layers of the Small Intestine
Mucosa
Muscularis mucosae
Submucosa
Muscular layer
Serosa
Lamina propria
Lacteal
Lymphatic vessel
Submucosal plexus
Circular layer of smooth muscle
Myenteric plexus
Longitudinal layer of smooth muscle
Submucosal artery and vein
b The organization of the intestinal wall
Epithelial cell
Stem cell
Paneth cell
c Cellular components of an intestinal gland
Capillaries
Goblet cells
Lacteal
Brush border
Tip of villus
LM × 250
e A villus in sectional view
Columnar epithelial cell
Goblet cell
Lacteal
Nerve
Capillary network
Lamina propria
Lymphatic vessel
Smooth muscle cell
Arteriole
Venule
d Internal structures in a single villus, showing the capillary and lymphatic supplies

Specializations of the Segments

The segments of the small intestine—the duodenum, jejunum, and ileum—are distinguished by both histological specialization and primary function. The duodenum has few circular folds, and their villi are small. The primary function of the duodenum is to receive chyme from the stomach and neutralize its acids before they can damage the absorptive surfaces of the small intestine.

Over the proximal half of the jejunum, however, circular folds and villi are very prominent. Thereafter, the circular folds and villi gradually decrease in size. This reduction parallels a reduction in absorptive activity: Most nutrient absorption takes place before ingested materials reach the ileum. One rather drastic surgical method of promoting weight loss is the removal of a significant portion of the jejunum. The resulting reduction in absorptive area causes a marked weight loss and may not interfere with adequate nutrition, but the side effects can be very serious.

The distal portions of the ileum lack circular folds. The lamina propria there contains 20–30 masses of lymphoid tissue called aggregated lymphoid nodules, or *Peyer's patches*. p. 791 These lymphoid tissues are most abundant in the terminal portion of the ileum, near the entrance to the large intestine. The lymphocytes in the aggregated lymphoid nodules protect the small intestine from bacteria that normally inhabit the large intestine.

Physiology of the Small Intestine

The primary functions of the small intestine are the chemical digestion of chyme from the stomach and the absorption of nutrients across the mucosal surface. Fats are absorbed into the lymphatic system and other nutrients enter the bloodstream.

Chemical Digestion: Intestinal Secretions and Enzymes

Roughly 1.8 liters of watery **intestinal juice** enters the intestinal lumen each day. Intestinal juice moistens chyme, helps buffer acids, and keeps both the digestive enzymes and the products of digestion in solution. Much of this fluid arrives by osmosis, as water flows out of the mucosa and into the concentrated chyme. The rest is secreted by intestinal glands, stimulated by the activation of touch receptors and stretch receptors in the intestinal walls.

The mucosa of the small intestine produces only a few of the enzymes involved in chemical digestion. The most important of the enzymes introduced into the lumen comes from the apical portions of the intestinal cells. *Brush border enzymes* are integral membrane proteins on the surfaces of intestinal microvilli. These enzymes break down materials that come in contact with the brush border. The epithelial cells then absorb the breakdown products. Once the epithelial cells are shed, they disintegrate within the lumen, releasing both intracellular and brush border enzymes. *Enteropeptidase* (previously called *enterokinase*) is one brush border enzyme that enters the lumen in this way. It does not directly participate in digestion. Instead, it activates the key pancreatic proenzyme, trypsinogen.

The duodenal submucosal glands help protect the duodenal epithelium from gastric acids and enzymes. As chyme travels the length of the duodenum, its pH increases from 1–2 to 7–8. These glands increase their secretion in response to (1) local reflexes, (2) the release of the hormone *enterocrinin* by enteroendocrine cells of the duodenum, and (3) parasympathetic stimulation through the vagus nerves. The first two mechanisms operate only after chyme arrives in the duodenum. However, the duodenal glands begin secreting during the cephalic phase of gastric secretion, long before chyme reaches the pyloric sphincter. They do so because vagus nerve activity triggers their secretion. Thus, the duodenal lining has protection in advance.

Sympathetic stimulation inhibits the duodenal glands, leaving the duodenal lining unprepared for the arrival of chyme. This effect probably explains why chronic stress or other factors that promote sympathetic activation can cause duodenal ulcers.

Mechanical Digestion: Intestinal Motility

After chyme has arrived in the duodenum, weak peristaltic contractions move it slowly toward the jejunum. The contractions are myenteric reflexes that are not under CNS control. Their effects are limited to within a few centimeters of the site of the original stimulus. Motor neurons in the submucosal and myenteric plexuses control these short reflexes. In addition, some of the smooth muscle cells contract periodically, even without stimulation, establishing a basic contractile rhythm that then spreads from cell to cell.

The stimulation of the parasympathetic system increases the sensitivity of the weak myenteric reflexes and speeds up both local peristalsis and segmentation. More elaborate reflexes coordinate activities along the entire length of the small intestine. The gastroenteric and gastroileal reflexes speed up movement along the small intestine (see **Spotlight Figure 24–15**). This effect is the opposite from that of the **enterogastric reflex**, of which one result is constriction of the pyloric sphincter and the discharge of chyme.

Hormones released by the digestive tract can enhance or suppress reflexes. For example, the gastroileal reflex is triggered by stretch receptor stimulation. However, the degree of ileocecal valve relaxation is enhanced by gastrin, which is secreted in large quantities when food enters the stomach.

Regulation: Coordination of Secretion and Absorption in the Digestive Tract

A combination of neural and hormonal mechanisms coordinates the activities of the digestive glands. These regulatory mechanisms are centered on the duodenum, where acids must be neutralized and appropriate enzymes added.

Neural mechanisms involving the CNS prepare the digestive tract for activity (through parasympathetic innervation) or inhibit its activity (through sympathetic innervation). Neural mechanisms also coordinate the movement of materials along the length of the digestive tract (through the enterogastric, gastroenteric, and gastroileal reflexes).

In addition, motor neurons synapsing in the digestive tract release a variety of neurotransmitters. Many of these chemicals are also released in the CNS. In general, their functions are

Figure 24–22 The Secretion and Effects of Major Duodenal Hormones.

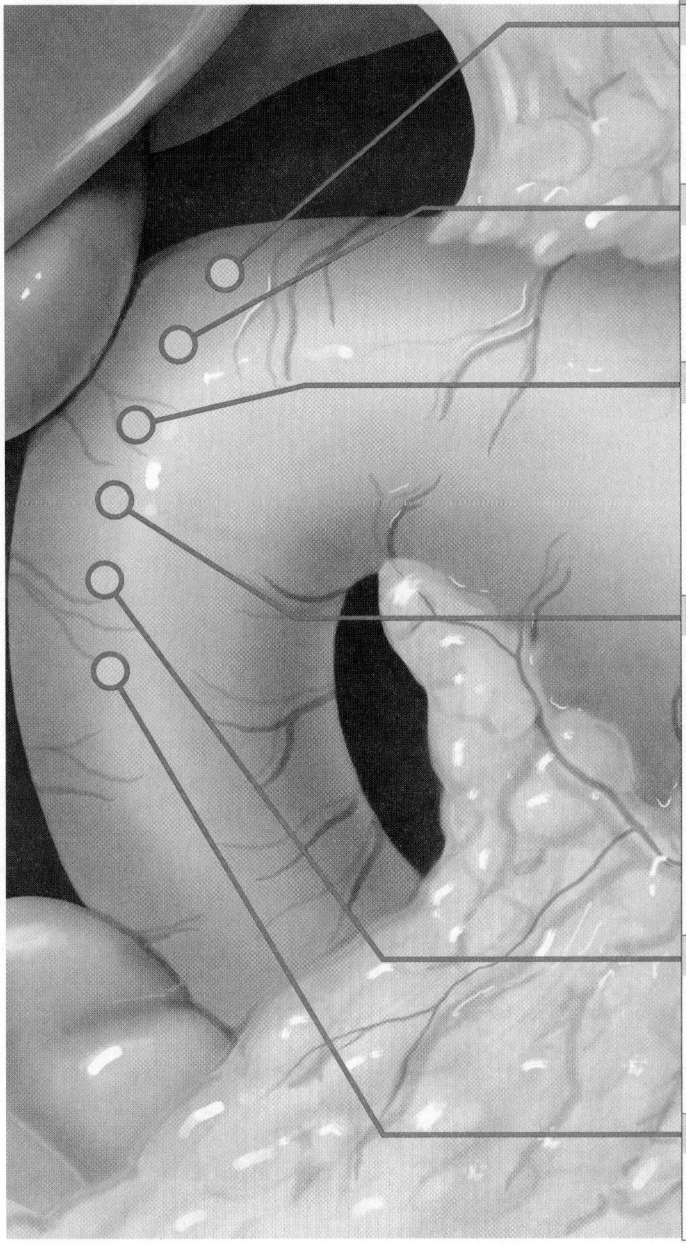

Major Hormones of the Duodenum	
Gastrin	Gastrin is secreted by G cells in the duodenum when they are exposed to large quantities of incompletely digested proteins. The functions of gastrin include promoting increased stomach motility and stimulating the production of gastric acids and enzymes. (Gastrin is also produced by the stomach.)
Secretin	Secretin is released when chyme arrives in the duodenum. Secretin's primary effect is an increase in the secretion of buffers by the pancreas, which increase the pH of the chyme. Secondarily, secretin also stimulates secretion of bile by the liver and reduces gastric motility and gastric secretory rates.
Gastric Inhibitory Peptide (GIP)	Gastric inhibitory peptide is secreted when fats and carbohydrates—especially glucose—enter the small intestine. The inhibition of gastric activity is accompanied by the stimulation of insulin release at the pancreatic islets. GIP has several secondary effects, including stimulating duodenal gland activity, stimulating lipid synthesis in adipose tissue, and increasing glucose use by skeletal muscles.
Cholecystokinin (CCK)	Cholecystokinin is secreted when chyme arrives in the duodenum, especially when the chyme contains lipids and partially digested proteins. In the pancreas, CCK accelerates the production and secretion of all types of digestive enzymes. It also causes a relaxation of the hepatopancreatic sphincter and contraction of the gallbladder, resulting in the ejection of bile and pancreatic juice into the duodenum. Thus, the net effects of CCK are to increase the secretion of pancreatic enzymes and to push pancreatic secretions and bile into the duodenum. The presence of CCK in high concentrations has two additional effects: It inhibits gastric activity, and it appears to have CNS effects that reduce the sensation of hunger.
Vasoactive Intestinal Peptide (VIP)	Vasoactive intestinal peptide stimulates the secretion of intestinal glands, dilates regional capillaries, and inhibits acid production in the stomach. By dilating capillaries in active areas of the intestinal tract, VIP provides an efficient mechanism for removing absorbed nutrients.
Enterocrinin	Enterocrinin is released when chyme enters the duodenum. It stimulates alkaline mucus production by the submucosal glands.

poorly understood. Examples of neurotransmitters that may be important include substance P, enkephalins, and endorphins.

We now summarize the information presented thus far on the regulation of intestinal and glandular function. We also consider some additional details about the regulatory mechanisms involved.

Secretion and Effects of Intestinal Hormones

The intestinal tract secretes a variety of peptide hormones with similar chemical structures. Many of these hormones have multiple effects in several regions of the digestive tract, and in the accessory glandular organs as well. The origins and primary effects of these important digestive hormones are shown in Figure 24–22.

Other intestinal hormones are produced in relatively small quantities. Examples include *motilin*, which stimulates intestinal contractions; *villikinin*, which promotes the movement of villi and the associated lymph flow; and *somatostatin*, which inhibits gastric secretion. Functional interactions among gastrin, GIP, secretin, CCK, and VIP are diagrammed in Figure 24–23.

Figure 24–23 The Secretion and Effects of Major Digestive Tract Hormones. The primary effects of gastrin, GIP, secretin, CCK, and VIP are shown.

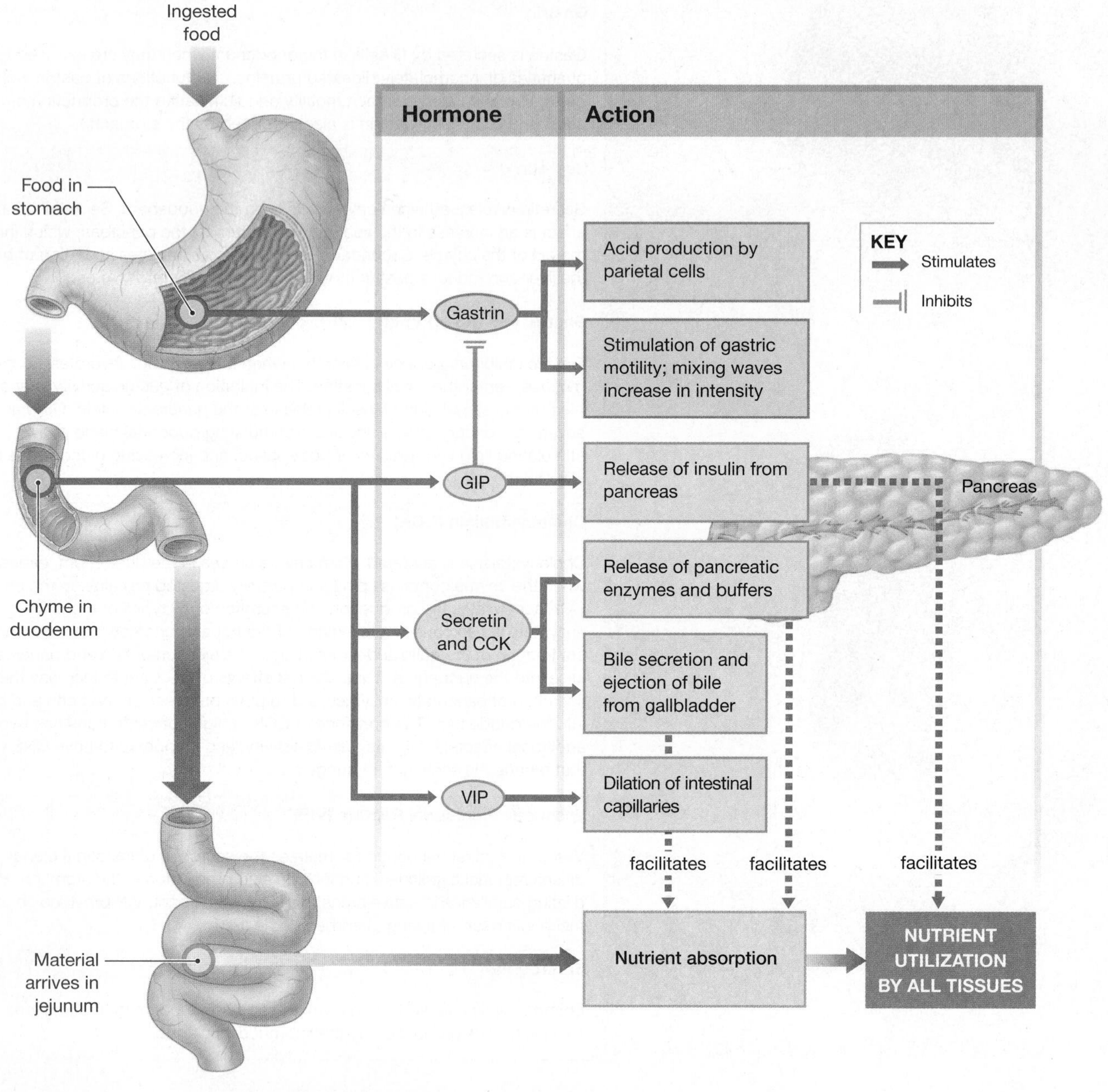
Ingested food
Food in stomach
Hormone
Action
Gastrin
Acid production by parietal cells
Stimulation of gastric motility; mixing waves increase in intensity
KEY
Stimulates
Inhibits
Chyme in duodenum
GIP
Release of insulin from pancreas
Pancreas
Secretin and CCK
Release of pancreatic enzymes and buffers
Bile secretion and ejection of bile from gallbladder
VIP
Dilation of intestinal capillaries
facilitates
facilitates
facilitates
Material arrives in jejunum
Nutrient absorption
NUTRIENT UTILIZATION BY ALL TISSUES

24

? Chyme in the duodenum stimulates the secretion of what hormones in the digestive tract?

Absorption in the Small Intestine

On average, it takes about 5 hours for materials to pass from the duodenum to the end of the ileum. The first of the materials to enter the duodenum after you eat breakfast may leave the small intestine at lunchtime. We discuss the absorption of nutrients later in **Spotlight Figure 24–27** on page 928.

Along the way, the organ's absorptive effectiveness is enhanced by the fact that so much of the mucosa is movable. The microvilli can be moved by their supporting microfilaments. The intestinal villi move back and forth, exposing the epithelial surfaces to the liquefied intestinal contents. Groups of villi are moved by the muscularis mucosae; and the muscularis mucosae and the muscular layer move the circular folds. These movements stir and mix the intestinal contents, making absorption more efficient by quickly eliminating local differences in nutrient concentration.

Checkpoint

25. Name the three regions of the small intestine from proximal to distal.
26. How is the small intestine adapted for nutrient absorption?
27. Does a high-fat meal raise or lower the level of cholecystokinin in the blood?
28. How would the pH of the intestinal contents be affected if the small intestine did not produce secretin?

See the blue Answers tab at the back of the book.

24-7 The large intestine, which is divided into three parts, absorbs water from digestive materials and eliminates the remaining waste as feces

Learning Outcome Describe the gross and histological structures of the large intestine, including its regional specializations and role in nutrient absorption.

The horseshoe-shaped **large intestine** begins at the end of the ileum and ends at the anus. The large intestine lies inferior to the stomach and liver and almost completely frames the small intestine (look back at Figure 24–1). The large intestine stores digestive wastes and reabsorbs water.

Gross Anatomy and Segments of the Large Intestine

The large intestine, also known as the *large bowel*, has an average length of about 1.5 meters (4.9 ft) and a width of 7.5 cm (3 in.). We can divide it into three segments: (1) the pouchlike *cecum*, the first portion of the large intestine; (2) the *colon*, the largest portion; and (3) the *rectum*, the last 15 cm (6 in.) of the large intestine and the end of the digestive tract (Figure 24–24a).

The Cecum and Appendix

Material arriving from the ileum first enters an expanded pouch called the **cecum**. The ileum attaches to the medial surface of the cecum and opens into the cecum at the *ileocecal valve* (see Figure 24–24a,b). The cecum collects and stores materials from the ileum and begins the process of compaction.

The slender, hollow **appendix**, or *vermiform appendix* (*vermis*, worm), is attached to the posteromedial surface of the cecum (see Figure 24–24a,b). The appendix is normally about 9 cm (3.6 in.) long, but its size and shape are quite variable. A small mesentery called the **meso-appendix** connects the appendix to the ileum and cecum. The primary function of the appendix is as an organ of the lymphatic system. Lymphoid nodules dominate the mucosa and submucosa of the appendix. Inflammation of the appendix is known as *appendicitis*.

The Colon

The **colon** has a larger diameter and a thinner wall than the small intestine. Distinctive features of the colon include the following (see Figure 24–24a):

- The wall of the colon forms a series of pouches, or **haustra** (HAWS-truh; singular, *haustrum*). The creases between the haustra affect the mucosal lining as well, producing a series of internal folds. Haustra permit the colon to expand and elongate.
- Three separate longitudinal bands of smooth muscle—called the **teniae coli** (TĒ-nē-ē KŌ-lē)—run along the outer surfaces of the colon just deep to the serosa. These bands correspond to the outer layer of the muscular layer in other portions of the digestive tract. Muscle tone within the teniae coli is what creates the haustra.
- The serosa of the colon contains numerous teardrop-shaped sacs of fat called **omental appendices** or the *fatty appendices of colon*.

We can subdivide the colon into four segments: the ascending colon, transverse colon, descending colon, and sigmoid colon (see Figure 24–24a).

1. The **ascending colon** begins at the superior border of the cecum and ascends along the right lateral and posterior wall of the peritoneal cavity to the inferior surface of the liver. There, the colon bends sharply to the left at the **right colic flexure**, or *hepatic flexure*. This bend marks the end of the ascending colon and the beginning of the transverse colon.
2. The **transverse colon** curves anteriorly from the right colic flexure and crosses the abdomen from right to left. The transverse colon is supported by the transverse mesocolon. It is separated from the anterior abdominal wall by the layers of the greater omentum. At the left side of the body, the transverse colon passes inferior to the greater curvature of the stomach. Near the spleen, the colon makes a 90° turn at the **left colic flexure**, or *splenic flexure*, and becomes the descending colon.
3. The **descending colon** proceeds inferiorly along the left side until reaching the iliac fossa formed by the inner surface of the left ilium. The descending colon is retroperitoneal and firmly attached to the abdominal wall. At the iliac fossa, the descending colon curves and becomes the sigmoid colon.
4. The **sigmoid** (SIG-moyd; *sigmeidos*, the Greek letter *S*) **colon** is an S-shaped segment that is only about 15 cm (6 in.) long. The sigmoid colon lies posterior to the urinary bladder, suspended from the sigmoid mesocolon. The sigmoid colon empties into the rectum.

The colon, and indeed the entire large intestine, receives blood from branches of the superior mesenteric and inferior mesenteric arteries. The superior mesenteric and inferior mesenteric veins collect venous blood from the large intestine. pp. 764, 775

Figure 24–24 **Anatomy of the Large Intestine.** ATLAS: Plates 49a–c; 58a–c; 59; 64; 65

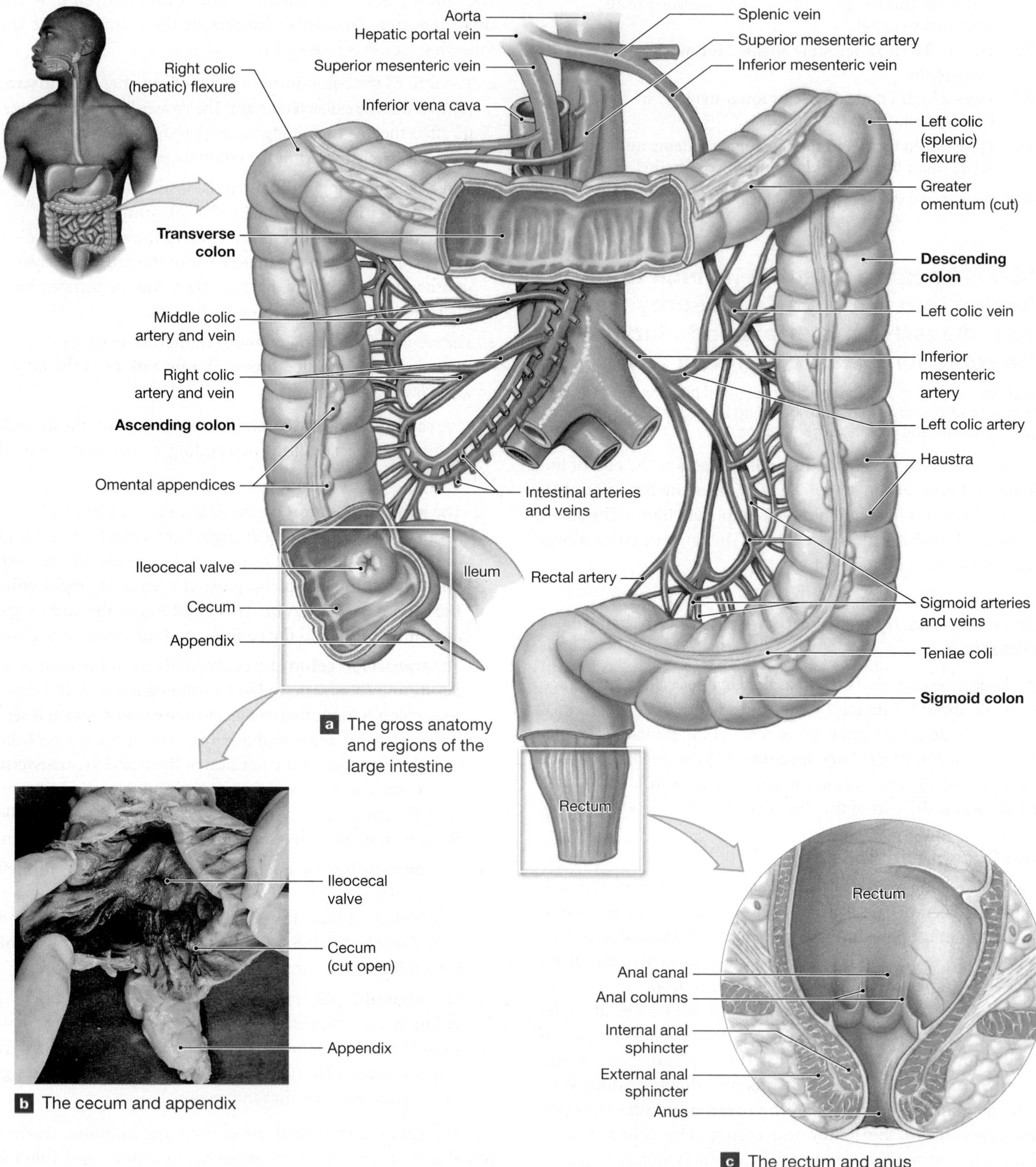

Aorta
Hepatic portal vein
Superior mesenteric vein
Inferior vena cava
Splenic vein
Superior mesenteric artery
Inferior mesenteric vein
Right colic (hepatic) flexure
Left colic (splenic) flexure
Greater omentum (cut)
Transverse colon
Descending colon
Middle colic artery and vein
Left colic vein
Right colic artery and vein
Inferior mesenteric artery
Ascending colon
Left colic artery
Haustra
Omental appendices
Intestinal arteries and veins
Ileocecal valve
Ileum
Rectal artery
Cecum
Sigmoid arteries and veins
Appendix
Teniae coli
Sigmoid colon
Rectum
Ileocecal valve
Cecum (cut open)
Appendix
Rectum
Anal canal
Anal columns
Internal anal sphincter
External anal sphincter
Anus

a The gross anatomy and regions of the large intestine

b The cecum and appendix

c The rectum and anus

24

+ Clinical Note Colorectal Cancer

Colorectal cancer is relatively common in the United States. Aside from skin cancers, colorectal cancer is the third most common cancer in the United States, affecting both men and women. The American Cancer Society estimates that 95,270 new cases of colon cancer and 39,220 new cases of rectal cancer will occur in 2016. The death rate has declined over the past 20 years for both men and women.

The best defense appears to be early detection and prompt treatment. Standard screening involves checking the feces for blood. This simple procedure can be performed easily on a stool (fecal) sample as part of a routine physical. For individuals at increased risk because of family history, associated disease, or older age, visual inspection of the intestinal lumen by fiber-optic colonoscopy is prudent. Such examination can find polyps before they develop into cancers. The 5-year survival rate for people whose cancer is found at an early stage and treated immediately is greater than 90 percent.

The Rectum

The **rectum** (REK-tum) forms the last 15 cm (6 in.) of the digestive tract (see Figure 24–24a,c). It is an expandable organ for the temporary storage of **feces**.

The last portion of the rectum, the **anal canal**, contains small longitudinal folds called **anal columns**. The distal margins of these columns are joined by transverse folds that mark the boundary between the columnar epithelium of the proximal rectum and a stratified squamous epithelium like that in the oral cavity. The **anus** is the exit of the anal canal. There, the epidermis becomes keratinized and identical to the surface of the skin.

The circular layer of muscle in this region forms the **internal anal sphincter** (see Figure 24–24c). The smooth muscle cells of this sphincter are not under voluntary control. The **external anal sphincter**, which guards the anus, consists of a ring of skeletal muscle fibers that encircles the distal portion of the anal canal. This sphincter is under voluntary control.

The lamina propria and submucosa of the anal canal contain a network of veins. If venous pressures there rise too high due to straining during defecation or pregnancy, the veins can become distended, producing *hemorrhoids*.

Histology of the Large Intestine

The luminal diameter of the large intestine is about three times that of the small intestine, but its wall is much thinner. Large lymphoid nodules are scattered throughout the lamina propria and submucosa. The muscular layer of the large intestine is unusual, because the longitudinal layer has been reduced to the muscular bands of the teniae coli.

The major characteristics of the colon are the lack of villi, the abundance of goblet cells, and the presence of distinctive intestinal glands (Figure 24–25). The glands in the large intestine are deeper than those of the small intestine and are dominated by goblet cells. The mucus provides lubrication as the fecal material becomes drier and more compact. Mucus is secreted as local stimuli, such as friction or exposure to harsh chemicals, trigger short reflexes involving local nerve plexuses.

Physiology of the Large Intestine

Any digestion that occurs in the large intestine results from enzymes introduced in the small intestine or from bacterial action. Although less than 10 percent of the nutrient absorption in the digestive tract occurs in the large intestine, the absorptive operations that do occur in this segment of the digestive tract are nevertheless important. The mixing and propulsive contractions of the colon resemble those of the small intestine. The large intestine also prepares fecal material for defecation.

Absorption and the Microbiome in the Large Intestine

The reabsorption of water is an important function of the large intestine. Roughly 1500 mL of material enters the colon each day, but only about 200 mL of feces is ejected. To appreciate how efficient our digestion is, consider the average composition of feces: 75 percent water, 5 percent bacteria, and the rest a mixture of indigestible materials, small quantities of inorganic matter, and the remains of epithelial cells.

In addition, the large intestine absorbs a number of other substances that remain in the feces or were secreted into the digestive tract along its length. Examples include useful compounds such as bile salts and vitamins; organic wastes; and various toxins generated by bacterial action. Most of the bile salts entering the large intestine are promptly reabsorbed in the cecum and transported in blood to the liver for secretion into bile.

Bacteria that inhabit the large intestine contribute to our *microbiome*. The human **microbiome** is made up of the microbes (bacteria, fungi, and viruses) and their genes that live in and on the human body. Most of the gut bacteria live in the large intestine where they break down nutrients not absorbed in the small intestine and produce vitamins.

Vitamins. *Vitamins* are organic molecules required in very small quantities. The normal bacterial residents of the colon generate three vitamins that are absorbed in the large intestine:

- *Vitamin K*, a fat-soluble vitamin the liver requires for synthesizing four clotting factors, including prothrombin
- *Biotin*, a water-soluble vitamin important in various reactions, notably those of glucose metabolism

Figure 24–25 Histology of the Colon.

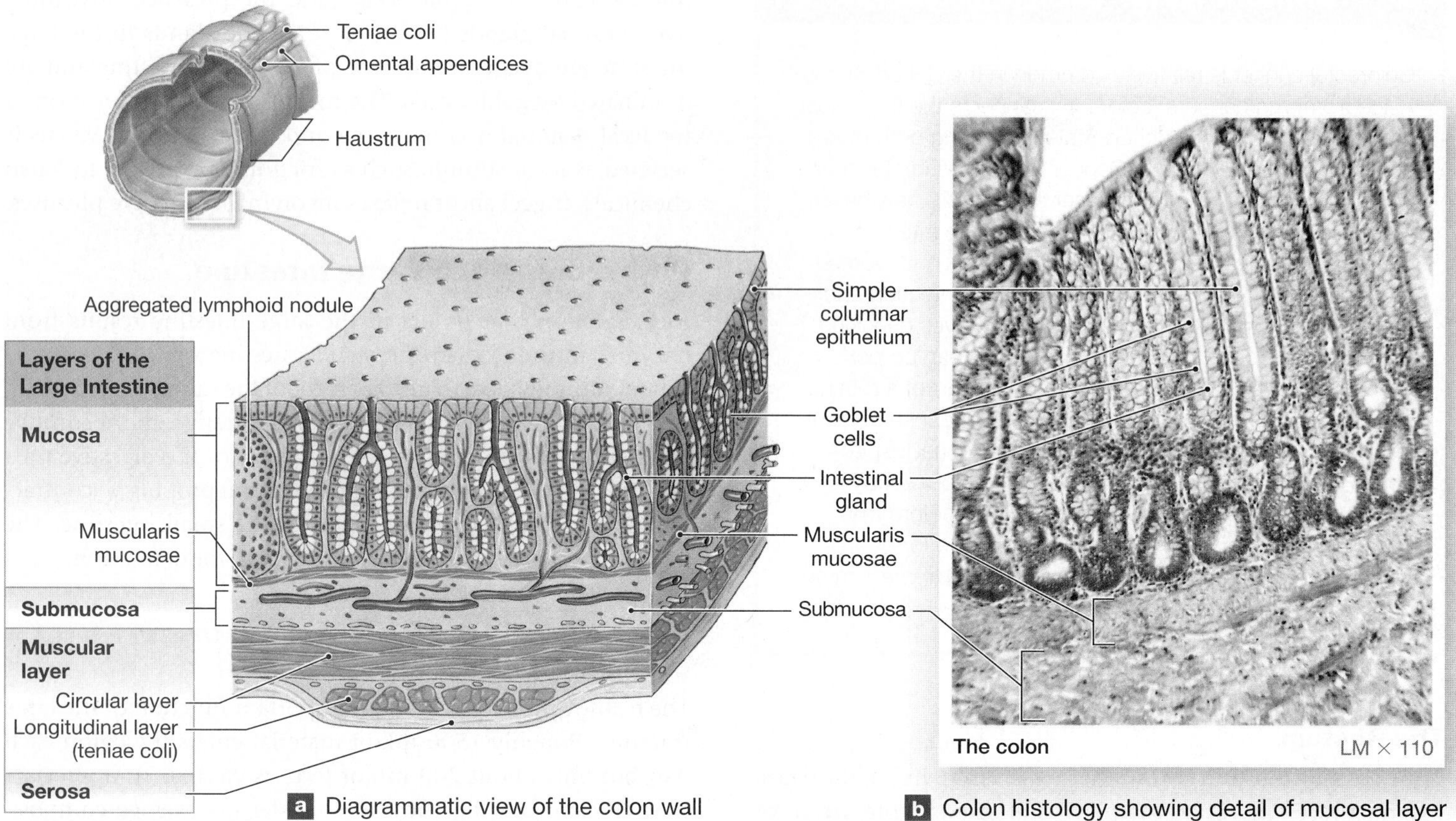
Teniae coli
Omental appendices
Haustrum
Aggregated lymphoid nodule
Layers of the Large Intestine
Mucosa
Muscularis mucosae
Submucosa
Muscular layer
Circular layer
Longitudinal layer (teniae coli)
Serosa
Simple columnar epithelium
Goblet cells
Intestinal gland
Muscularis mucosae
Submucosa
The colon
LM × 110

a Diagrammatic view of the colon wall

b Colon histology showing detail of mucosal layer

- *Vitamin* B_5 (pantothenic acid), a water-soluble vitamin required in the manufacture of steroid hormones and some neurotransmitters.

Intestinal bacteria produce about half of your daily vitamin K requirements. Vitamin K deficiencies lead to impaired blood clotting. They result from either (1) not enough lipids in the diet, which impairs the absorption of all fat-soluble vitamins; or (2) problems affecting lipid processing and absorption, such as inadequate bile production or chronic diarrhea (frequent, watery bowel movements). Because the intestinal bacteria generally produce sufficient amounts of biotin and vitamin B_5 to supplement any dietary shortage, disorders due to deficiencies of these vitamins are extremely rare after infancy.

Organic Wastes. We discussed the fate of bilirubin, a breakdown product of heme, in Chapter 19. p. 666 In the large intestine, bacteria convert bilirubin to *urobilinogens* and *stercobilinogens*. Some urobilinogens are absorbed into the bloodstream and then excreted in urine. The urobilinogens and stercobilinogens remaining within the colon are converted to **urobilins** and **stercobilins** by exposure to oxygen. These pigments in various proportions give feces a yellow-brown or brown color.

Bacterial action breaks down peptides that remain in the feces. This action generates (1) ammonia, in the form of soluble ammonium ions (NH_4^+); (2) *indole* and *skatole*, two nitrogen-containing compounds that are primarily responsible for the odor of feces; and (3) hydrogen sulfide (H_2S), a gas with a "rotten egg" odor. Significant amounts of ammonia and smaller amounts of other toxins cross the colonic epithelium and enter the hepatic portal circulation. The liver removes these toxins and converts them to relatively nontoxic compounds that can be released into the blood and excreted at the kidneys.

Intestinal enzymes do not alter indigestible carbohydrates, and the mucosa of the large intestine does not produce enzymes. These materials arrive in the colon virtually intact. These complex polysaccharides provide a reliable nutrient source for bacteria in the colon. The metabolic activities of these bacteria create small amounts of **flatus**, or intestinal gas. Foods with large amounts of indigestible carbohydrates (such as beans) stimulate bacterial gas production. Distension of the colon, cramps, and the frequent discharge of intestinal gases can result.

Motility of the Large Intestine and the Defecation Reflex

The gastroileal and gastroenteric reflexes move materials into the cecum while you eat. Movement from the cecum to the transverse colon is very slow, allowing hours for water absorption to convert the already thick material into a sludgy paste. Peristaltic waves move material along the length of the colon. Segmentation movements, called *haustral churning*, mix the contents of adjacent haustra. When the stomach and duodenum are distended, signals are relayed over the intestinal nerve plexuses. These signals cause

Figure 24–26 **The Defecation Reflex.**

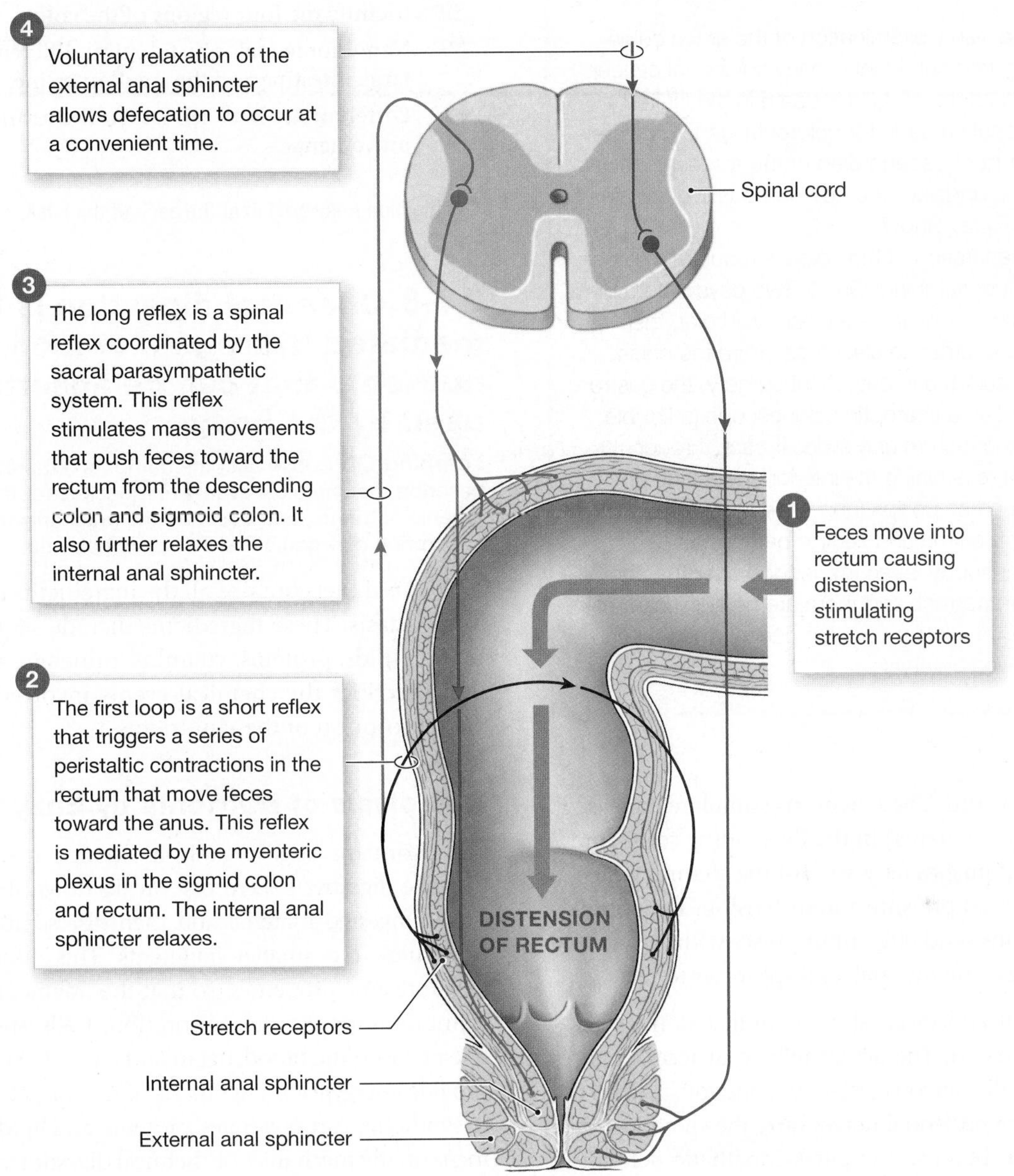
4
Voluntary relaxation of the external anal sphincter allows defecation to occur at a convenient time.
Spinal cord
3
The long reflex is a spinal reflex coordinated by the sacral parasympathetic system. This reflex stimulates mass movements that push feces toward the rectum from the descending colon and sigmoid colon. It also further relaxes the internal anal sphincter.
1
Feces move into rectum causing distension, stimulating stretch receptors
2
The first loop is a short reflex that triggers a series of peristaltic contractions in the rectum that move feces toward the anus. This reflex is mediated by the myenteric plexus in the sigmid colon and rectum. The internal anal sphincter relaxes.
DISTENSION OF RECTUM
Stretch receptors
Internal anal sphincter
External anal sphincter

? What stimulates stretch receptors in the rectum?

powerful peristaltic contractions called **mass movements** that move material from the transverse colon through the rest of the large intestine. The contractions force feces into the rectum. When feces move into the rectum, the distension on the rectal wall stimulates stretch receptors, initiating the *defecation reflex*.

The rectal chamber is usually empty, except when a mass movement forces feces out of the sigmoid colon into the rectum. The defecation reflex is an involuntary response to stimuli in the lower bowel that results in a bowel movement. This reflex involves two positive feedback loops, a short, *intrinsic myenteric defecation reflex* and a long, *parasympathetic defecation reflex* (Figure 24–26).

- *Intrinsic Myenteric Defecation Reflex.* In this short feedback loop, afferent fibers from the stretch receptors in the rectal walls stimulate the myenteric plexus to initiate a series of increased local peristaltic contractions in the sigmoid colon and rectum. The waves of contraction move feces toward the anus and further increase distension of the rectum. When the peristaltic contractions reach the anus, the internal anal sphincter relaxes.
- *Parasympathetic Defecation Reflex.* In this long feedback loop of the enteric nervous system, the stretch receptors in the rectal walls also stimulate parasympathetic motor neurons

+ Clinical Note Colonoscopy

Colonoscopy is the visual examination of the entire colon. This vital screening test can identify early colorectal cancer, the second leading cause of cancer death in the United States. On a personal crusade for colorectal cancer awareness in 2000 (after her husband died of the disease), news anchor Katie Couric underwent a now-famous live colonoscopy on the NBC Today Show.

A thorough preparation of the colon is required to properly evaluate the mucosal lining. One to two days before the procedure, the patient consumes a clear liquid diet, then takes laxatives and enemas to clear feces from the colon. After the patient is sedated in outpatient surgery, the gastroenterologist introduces a fiberoptic colonoscope (a flexible viewing tube) into the rectum and slides it back through the colon to the cecum, examining the interior and taking photos. Any lesions, such as polyps (abnormal fleshy growths), can be biopsied at the time and sent to pathology.

An initial colonoscopy is recommended after the age of 50. If the results are normal, and there are no significant risk factors, such as family history of colon cancer, the test can be repeated at 10-year intervals.

in the sacral spinal cord. These neurons stimulate increased peristalsis (mass movements) in the descending colon and sigmoid colon that pushes feces toward the rectum, which further increases rectal pressure and distension. The stimulated motor neurons send efferent impulses within the pelvic nerves that cause the internal anal sphincter to relax.

Both the internal and external anal sphincters must relax for feces to be eliminated. The actual release of feces requires activation of the somatic nervous system to consciously open the external sphincter. The pudendal nerves carry the somatic motor commands. However, because the parasympathetic defecation reflex magnifies the intrinsic myenteric defecation reflex, the force generated may be great enough to cause defecation, despite consciously controlling the voluntary external anal sphincter.

24

If the external anal sphincter remains constricted, the peristaltic contractions cease. As more feces enter the rectum, the expansion triggers further defecation reflexes. The urge to defecate usually develops when rectal pressure reaches about 15 mm Hg. If this pressure is more than 55 mm Hg, the external anal sphincter involuntarily relaxes and defecation takes place. This mechanism causes defecation in infants and in adults with spinal cord injuries.

In addition, other consciously directed activities can increase intra-abdominal pressure and help force fecal material out of the rectum. These activities include the *Valsalva maneuver*, a tensing of the abdominal muscles and an increase in intra-abdominal pressure by exhaling forcibly with a closed glottis.

✓ Checkpoint

29. **Identify the four regions of the colon.**
30. **Name some major histological differences between the large intestine and the small intestine.**
31. **Differentiate between haustral churning and mass movements.**

See the blue Answers tab at the back of the book.

24-8 Chemical digestion is the enzyme-mediated hydrolysis of food into nutrients that can be absorbed and used by the body

Learning Outcome List the nutrients required by the body, describe the chemical events responsible for the digestion of organic nutrients, and describe the mechanisms involved in the absorption of organic and inorganic nutrients.

A balanced diet contains all the ingredients needed to maintain homeostasis. These ingredients include six nutrients: carbohydrates, lipids, proteins, vitamins, minerals, and water. This section describes the chemical events involved in the processing and absorption of these nutrients.

Hydrolysis of Nutrients by Enzymes

Food contains large organic molecules, many of them insoluble. The digestive system first breaks down the physical structure of the ingested material and then disassembles the component molecules into smaller fragments. This disassembly eliminates any antigenic properties, so that the fragments do not trigger an immune response after absorption. Cells absorb the molecules released into the bloodstream and either (1) break them down to provide energy for the synthesis of ATP or (2) use these molecules to synthesize carbohydrates, proteins, and lipids. In this section we focus on the mechanics of chemical digestion and absorption. The fates of the compounds inside cells are the focus in Chapter 25.

Most ingested organic materials are complex chains, or polymers, of simpler molecules called monomers. In a typical dietary carbohydrate, the basic molecules are simple sugars. In a protein, the building blocks are amino acids. In lipids, they are generally fatty acids. And in nucleic acids, they are nucleotides. In a process called *hydrolysis*, digestive enzymes break the bonds between the component molecules of carbohydrates, proteins, lipids, and nucleic acids. pp. 38, 45

The classes of digestive enzymes differ with respect to their specific substrates. *Carbohydrases* break the bonds between simple sugars, *proteases* split the linkages between amino acids, and *lipases* separate fatty acids from glycerides. Some enzymes in each class are even more selective, breaking bonds between specific molecules. For example, a particular carbohydrase might break the bond between two glucose molecules, but not those between glucose and another simple sugar.

Digestive enzymes secreted by the salivary glands, tongue, stomach, and pancreas are mixed into the ingested material as it passes along the digestive tract. These enzymes break down large carbohydrates, proteins, lipids, and nucleic acids into smaller fragments. Typically these fragments in turn must be broken down even further before absorption can occur. The final enzymatic steps involve brush border enzymes, which are attached to the exposed surfaces of intestinal microvilli.

The digestive fates of carbohydrates, lipids, and proteins, the major dietary components, are shown in **Spotlight Figure 24–27**. The major digestive enzymes and their functions are summarized in **Table 24-1**. Next we take a closer look at the digestion and absorption of carbohydrates, lipids, proteins, and nucleic acids.

Carbohydrate Digestion and Absorption

Carbohydrates are a large group of organic compounds, such as sugars, starch, and cellulose, found in food. Our bodies break down carbohydrates for use in cellular metabolism. This section discusses how these processes occur.

Chemical Digestion of Carbohydrates

The digestion of complex carbohydrates (simple polysaccharides and starches) proceeds in two steps. The first step involves carbohydrases produced by the salivary glands and pancreas. The second step uses brush border enzymes.

Actions of Salivary and Pancreatic Enzymes. The digestion of complex carbohydrates involves two enzymes—salivary amylase and pancreatic alpha-amylase (see **Spotlight Figure 24–27**). Both function effectively at a pH of 6.7–7.5. Carbohydrate digestion

Table 24-1 Digestive Enzymes and Their Functions

Enzyme (proenzyme)	Source	Optimal pH	Substrate	Products	Remarks
CARBOHYDRASES					
Maltase, sucrase, lactase	Brush border of small intestine	7–8	Maltose, sucrose, lactose	Monosaccharides	Found in membrane surface of microvilli
Pancreatic alpha-amylase	Pancreas	6.7–7.5	Complex carbohydrates	Disaccharides and trisaccharides	Breaks bonds between sugars
Salivary amylase	Salivary glands	6.7–7.5	Complex carbohydrates	Disaccharides and trisaccharides	Breaks bonds between sugars
PROTEASES					
Carboxypeptidase (procarboxypeptidase)	Pancreas	7–8	Proteins, polypeptides, amino acids	Short-chain peptides	Activated by trypsin
Chymotrypsin (chymotrypsinogen)	Pancreas	7–8	Proteins, polypeptides	Short-chain peptides	Activated by trypsin
Dipeptidases, peptidases	Brush border of small intestine	7–8	Dipeptides, tripeptides	Amino acids	Found in membrane surface of brush border
Elastase (proelastase)	Pancreas	7–8	Elastin	Short-chain peptides	Activated by trypsin
Enteropeptidase	Brush border and lumen of small intestine	7–8	Trypsinogen	Trypsin	Reaches lumen through disintegration of shed epithelial cells
Pepsin (pepsinogen)	Chief cells of stomach	1.5–2.0	Proteins, polypeptides	Short-chain polypeptides	Secreted as proenzyme pepsinogen; activated by H^+ in stomach acid
Rennin	Stomach	3.5–4.0	Milk proteins		Secreted only in infants; causes protein coagulation
Trypsin (trypsinogen)	Pancreas	7–8	Proteins, polypeptides	Short-chain peptides	Proenzyme activated by enteropeptidase; activates other pancreatic proteases
LIPASES					
Lingual lipase	Glands of tongue	3.0–6.0	Triglycerides	Fatty acids and monoglycerides	Begins lipid digestion
Pancreatic lipase	Pancreas	7–8	Triglycerides	Fatty acids and monoglycerides	Bile salts must be present for efficient action
NUCLEASES					
	Pancreas	7–8	Nucleic acids	Nitrogenous bases and simple sugars	Includes ribonuclease for RNA and deoxyribonuclease for DNA

SPOTLIGHT

Figure 24–27

The Chemical Events of Digestion

A typical meal contains carbohydrates, proteins, lipids, water, minerals, electrolytes, and vitamins. The digestive system handles each component differently. Large organic molecules must be chemically broken down before they can be absorbed. Water, minerals, and vitamins can be absorbed without processing, but they may require special transport mechanisms.

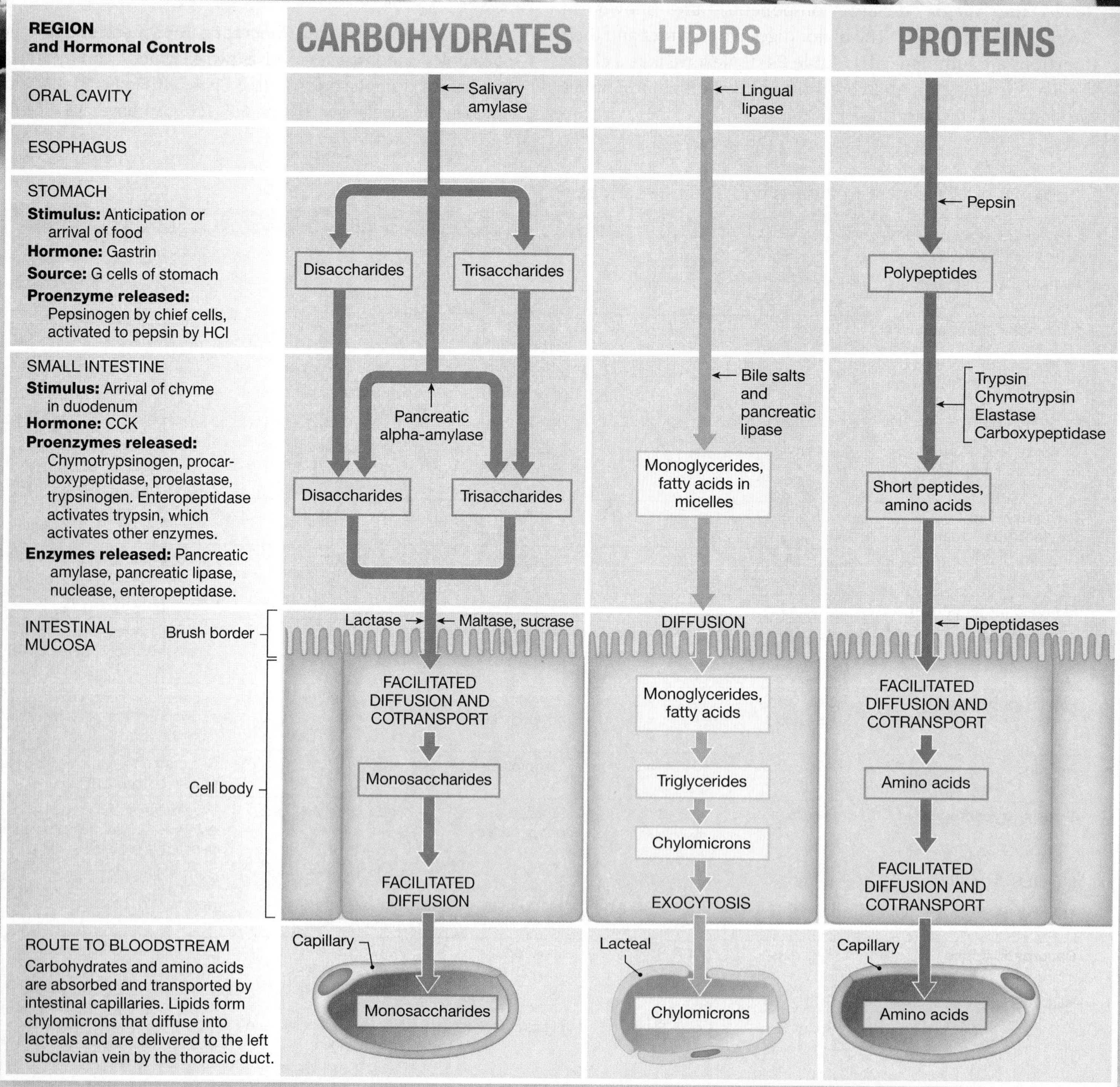

REGION and Hormonal Controls

ORAL CAVITY

ESOPHAGUS

STOMACH
Stimulus: Anticipation or arrival of food
Hormone: Gastrin
Source: G cells of stomach
Proenzyme released: Pepsinogen by chief cells, activated to pepsin by HCl

SMALL INTESTINE
Stimulus: Arrival of chyme in duodenum
Hormone: CCK
Proenzymes released: Chymotrypsinogen, procarboxypeptidase, proelastase, trypsinogen. Enteropeptidase activates trypsin, which activates other enzymes.
Enzymes released: Pancreatic amylase, pancreatic lipase, nuclease, enteropeptidase.

INTESTINAL MUCOSA
Brush border
Cell body

ROUTE TO BLOODSTREAM
Carbohydrates and amino acids are absorbed and transported by intestinal capillaries. Lipids form chylomicrons that diffuse into lacteals and are delivered to the left subclavian vein by the thoracic duct.

CARBOHYDRATES

Salivary amylase
Disaccharides
Trisaccharides
Pancreatic alpha-amylase
Disaccharides
Trisaccharides
Lactase
Maltase, sucrase
FACILITATED DIFFUSION AND COTRANSPORT
Monosaccharides
FACILITATED DIFFUSION
Capillary
Monosaccharides

LIPIDS

Lingual lipase
Bile salts and pancreatic lipase
Monoglycerides, fatty acids in micelles
DIFFUSION
Monoglycerides, fatty acids
Triglycerides
Chylomicrons
EXOCYTOSIS
Lacteal
Chylomicrons

PROTEINS

Pepsin
Polypeptides
Trypsin
Chymotrypsin
Elastase
Carboxypeptidase
Short peptides, amino acids
Dipeptidases
FACILITATED DIFFUSION AND COTRANSPORT
Amino acids
FACILITATED DIFFUSION AND COTRANSPORT
Capillary
Amino acids

begins in the mouth during mastication, through the action of salivary amylase from the parotid and submandibular salivary glands. Salivary amylase breaks down starches (complex carbohydrates) into a mixture composed mostly of *disaccharides* (two simple sugars, or *monosaccharides*, joined together) and *trisaccharides* (three monosaccharides). Salivary amylase continues to digest the starches and glycogen in food for 1 to 2 hours before stomach acids render the enzyme inactive. Only a small amount of digestion takes place over this period because the enzymatic content of saliva is not high.

In the duodenum, pancreatic alpha-amylase breaks down the remaining complex carbohydrates. Any disaccharides or trisaccharides produced, and any present in the food, are not broken down further until they contact the intestinal mucosa.

Actions of Brush Border Enzymes. Brush border enzymes of the intestinal microvilli break disaccharides and trisaccharides into monosaccharides prior to absorption. The enzyme **maltase** splits bonds between the two glucose molecules of the disaccharide **maltose**. **Sucrase** breaks the disaccharide **sucrose** into glucose and *fructose*, another six-carbon sugar. **Lactase** hydrolyzes the disaccharide **lactose** into a molecule of glucose and one of *galactose*.

Lactose is the main carbohydrate in milk, so lactase provides an essential function in infancy and early childhood by breaking down lactose. If the intestinal mucosa stops producing lactase, the person becomes **lactose intolerant**. After ingesting milk and other dairy products, people with lactose-intolerance can experience a variety of unpleasant digestive problems, including lower abdominal pain, gas, diarrhea, and vomiting.

Absorption of Monosaccharides

The intestinal epithelium then absorbs the monosaccharides by facilitated diffusion and cotransport mechanisms (look back at Figure 3–18, p. 96). Both methods involve a carrier protein. Facilitated diffusion and cotransport differ in the following major ways:

- *Number of Transported Substances.* Facilitated diffusion moves only one molecule or ion through the plasma membrane at a time. Cotransport moves more than one molecule or ion through the membrane at the same time. In cotransport, the transported substances move in the same direction.
- *Concentration Gradients.* Facilitated diffusion does not take place if there is an opposing concentration gradient for the particular molecule or ion. By contrast, cotransport can take place despite an opposing concentration gradient for one of the transported substances. For example, cells lining the small intestine continue to absorb glucose when glucose concentrations inside the cells are much higher than they are in the intestinal contents.
- *ATP Requirement.* Facilitated diffusion is a passive process and does not require ATP. Cotransport by itself does not consume ATP. However, the cell must often expend ATP to preserve homeostasis in a process called *secondary active transport.* p. 97 For example, in the cotransport of glucose and sodium ions, the sodium ions must later be pumped out of the cell.

Glucose cotransport is an example of sodium-linked cotransport. Comparable cotransport mechanisms exist for other simple sugars and for some amino acids. These mechanisms deliver valuable nutrients to the cytoplasm, but they also bring in sodium ions that must be ejected by the sodium–potassium exchange pump.

The simple sugars that are transported into the intestinal epithelial cell at its apical surface diffuse through the cytoplasm. They then reach the interstitial fluid by facilitated diffusion across the basolateral surfaces. These monosaccharides diffuse into the capillaries of the villus for eventual transport to the liver in the hepatic portal vein.

Lipid Digestion and Absorption

Lipid digestion involves lingual lipase from glands of the tongue, and pancreatic lipase from the pancreas (see **Spotlight Figure 24–27**). The most important and abundant dietary lipids are *triglycerides.* They consist of three fatty acids attached to a single molecule of glycerol (see Figure 2–17, p. 50). The lingual and pancreatic lipases break off two of the fatty acids, leaving *monoglycerides.*

Lipases are water-soluble enzymes, and lipids tend to form large drops that exclude water molecules. As a result, lipases can only interact with the exposed surfaces of the lipid drops. Lingual lipase begins breaking down triglycerides in the mouth and continues for a variable time within the stomach. The lipid drops are so large, however, and the available time so short, that only about 20 percent of the lipids have been digested by the time the chyme enters the duodenum.

Bile salts improve chemical digestion by emulsifying the lipid drops into tiny emulsion droplets, thereby providing better access for pancreatic lipase. The emulsification takes place only after the chyme has been mixed with bile in the duodenum. Pancreatic lipase then breaks apart the triglycerides to form a mixture of monoglycerides and fatty acids. As these molecules are released, they interact with bile salts in the surrounding chyme to form small lipid–bile salt complexes called **micelles** (mī-SELZ). A micelle is only about 2.5 nm (0.0025 μm) in diameter.

When a micelle contacts the intestinal epithelium, the lipids diffuse across the plasma membrane and enter the cytoplasm. The intestinal cells synthesize new triglycerides from the monoglycerides and fatty acids. These triglycerides, together with absorbed steroids, phospholipids, and fat-soluble vitamins, are then coated with proteins. The resulting complexes are known as **chylomicrons** (kī-lō-MĪ-kronz; *chylos*, juice + *mikros*, small).

The intestinal cells then secrete the chylomicrons into interstitial fluid by exocytosis. The protein coating keeps the chylomicrons suspended in the interstitial fluid, but they are generally too large to diffuse into capillaries. Most of the chylomicrons diffuse into the intestinal lacteals, which lack basement membranes and have large gaps between adjacent endothelial cells. From the lacteals, the chylomicrons proceed along the lymphatic vessels and through the thoracic duct. They finally enter the bloodstream at the left subclavian vein.

Most of the bile salts within micelles are reabsorbed by sodium-linked cotransport. Only about 5 percent of the bile salts secreted by the liver enters the colon. Only about 1 percent is lost in feces.

Protein Digestion and Absorption

Proteins have very complex structures, so protein digestion is both complex and time consuming. Proteases must access individual proteins, so the three-dimensional organization of food must first be disrupted. This step involves mechanical digestion in the oral cavity, through mastication, and chemical digestion in the stomach, through the action of hydrochloric acid. The strong acid in the stomach disrupts tertiary and secondary protein structure, exposing peptide bonds to enzymatic action.

The acidic content of the stomach also provides the proper environment for the activity of pepsin, the proteolytic enzyme secreted in an inactive form by chief cells of the stomach (see Spotlight Figure 24–27). Pepsin works effectively at a pH of 1.5–2.0. It breaks the peptide bonds within a polypeptide chain.

When chyme enters the duodenum, enteropeptidase from the small intestine triggers the conversion of trypsinogen to trypsin. Buffers increase the pH to 7–8. Pancreatic proteases can now begin working. Trypsin, chymotrypsin, and elastase are like pepsin in that they break specific peptide bonds within a polypeptide. For example, trypsin breaks peptide bonds involving the amino acids *arginine* or *lysine*. Chymotrypsin targets peptide bonds involving *tyrosine* or *phenylalanine*.

Carboxypeptidase also acts in the small intestine. This enzyme chops off the last amino acid of a polypeptide chain, no matter which amino acids are involved. Thus, while the other peptidases generate a variety of short peptides, carboxypeptidase produces free amino acids. The epithelial surfaces of the small intestine contain several peptidases, notably **dipeptidases**. These enzymes break short peptide chains into individual amino acids. (Dipeptidases break apart *dipeptides*.)

These amino acids, as well as those produced by the pancreatic enzymes, are absorbed through both facilitated diffusion and cotransport mechanisms. The amino acids diffuse through the cell to its basolateral surface. There they are released into interstitial fluid by facilitated diffusion and cotransport. Once in the interstitial fluid, the amino acids diffuse into intestinal capillaries for transport to the liver by means of the hepatic portal vein.

Nucleic Acid Digestion and Absorption

Nucleic acids are broken down into their component nucleotides. Brush border enzymes digest these nucleotides into sugars, phosphates, and nitrogenous bases that are absorbed by active transport. However, nucleic acids represent only a small fraction of all the nutrients absorbed each day.

Absorption of Water, Ions, and Vitamins

Water, electrolytes (inorganic ions), and vitamins are absorbed through various transport processes as they pass along the small and large intestines. Let's begin by looking at water absorption.

Water Absorption

Cells cannot actively absorb or secrete water. All movement of water across the lining of the digestive tract involves passive water flow down osmotic gradients. The production of glandular secretions also involves passive water flow down osmotic gradients. Recall that when a selectively permeable membrane separates two solutions, water tends to flow into the solution that has the higher concentration of solutes. p. 93 Osmotic movements are rapid, so interstitial fluid and the fluids in the intestinal lumen always have the same osmotic concentration of solutes, or *osmolarity*.

Intestinal epithelial cells continuously absorb nutrients and ions. These activities gradually lower the solute concentration in the lumen. As the solute concentration drops, water moves into the surrounding tissues, maintaining osmotic equilibrium.

Each day, about 2200 mL of water enters the digestive tract in the form of food and drink. Saliva, gastric secretions, bile, intestinal secretions, and mucus provide an additional 7200 mL. Of that total, only about 150 mL is lost in feces. The sites of secretion and absorption of water are shown in Figure 24–28.

Ion Absorption

Osmosis does not distinguish among solutes. All that matters is the total concentration of solutes. To maintain homeostasis, however, the concentrations of specific ions must be closely regulated. Thus, each ion must be handled individually, and the rate of intestinal absorption of each must be tightly controlled (Table 24-2). Many of the regulatory mechanisms controlling the rates of absorption are poorly understood.

Sodium ions (Na^+) are usually the most abundant cations in food. They may enter intestinal cells by diffusion, by cotransport with another nutrient, or by active transport. These ions are then pumped into interstitial fluid across the base of the cell.

The rate of Na^+ uptake from the lumen is generally proportional to the concentration of Na^+ in the intestinal contents. As a result, eating heavily salted foods leads to increased sodium ion absorption and an associated gain of water through osmosis. The rate of sodium ion absorption by the digestive tract is increased by *aldosterone*, a steroid hormone from the adrenal cortex. p. 634

Calcium ion (Ca^{2+}) absorption involves active transport at the epithelial surface. Calcitriol speeds up the rate of transport. p. 633

As other solutes move out of the lumen, the concentration of potassium ions (K^+) increases. These ions can diffuse into the epithelial cells, driven by the concentration gradient. The absorption of magnesium (Mg^{2+}), iron (Fe^{2+}), and other cations involves specific carrier proteins. The cell must use

Figure 24–28 **Digestive Secretion and Water Reabsorption in the Digestive Tract.** The purple arrows indicate secretion, and the blue arrows show water ingestion and reabsorption.

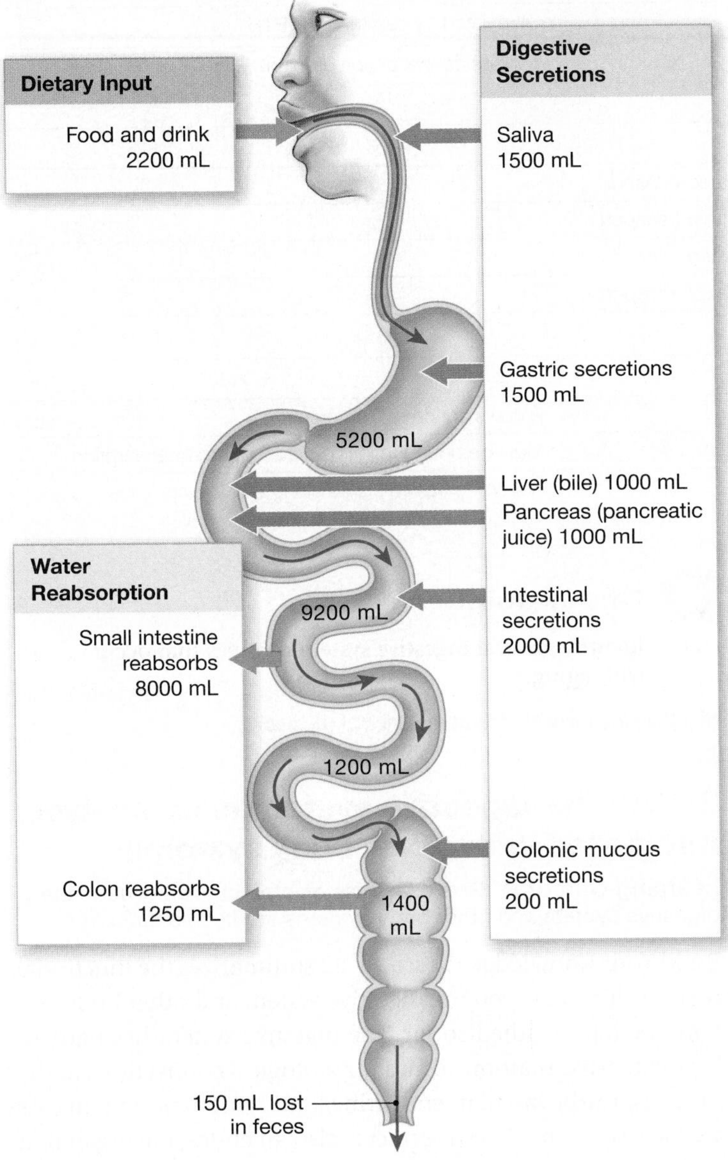
Dietary Input
Food and drink
2200 mL
Digestive Secretions
Saliva
1500 mL
Gastric secretions
1500 mL
5200 mL
Liver (bile) 1000 mL
Pancreas (pancreatic juice) 1000 mL
Water Reabsorption
Intestinal secretions
2000 mL
9200 mL
Small intestine reabsorbs
8000 mL
1200 mL
Colonic mucous secretions
200 mL
Colon reabsorbs
1250 mL
1400 mL
150 mL lost in feces

ATP to obtain and transport these ions to interstitial fluid. Regulatory factors controlling their absorption are poorly understood.

The anions chloride (Cl^-), iodide (I^-), bicarbonate (HCO_3^-), and nitrate (NO_3^-) are absorbed by diffusion or carrier-mediated transport. Phosphate (PO_4^{3-}) and sulfate (SO_4^{2-}) ions enter epithelial cells only by active transport.

Vitamin Absorption

As we have already discussed, **vitamins** are organic compounds required in very small quantities. However, their role in the body is crucial, as they are important cofactors or coenzymes in many metabolic pathways. We consider the functions of vitamins and associated nutritional problems in detail in Chapter 25.

There are two major groups of vitamins: water-soluble and fat-soluble vitamins. The nine **water-soluble vitamins** include the B vitamins, common in milk and meats, and vitamin C, found in citrus fruits. All but one of the water-soluble vitamins are easily absorbed by diffusion across the digestive epithelium. The exception is vitamin B_{12}, which you may recall from Chapter 19 is essential for normal erythropoiesis. By itself vitamin B_{12} cannot be absorbed by the intestinal mucosa in normal amounts. This vitamin must be bound to intrinsic factor, as we said a glycoprotein secreted by the parietal cells of the stomach. The combination is then absorbed through active transport.

Vitamins A, D, E, and K are **fat-soluble vitamins**, which as their name suggests can dissolve in lipids. Fat-soluble vitamins in the diet enter the duodenum in fat droplets, mixed with triglycerides. The vitamins remain in association with these lipids as they form emulsion droplets and, after further digestion, micelles. The fat-soluble vitamins are then absorbed from the micelles along with the fatty acids and monoglycerides. As discussed, vitamin K produced in the colon is absorbed with other lipids released through bacterial action. Taking supplements of fat-soluble vitamins while you have an empty stomach, are fasting, or are on a low-fat diet is relatively ineffective. The reason is that proper absorption of these vitamins requires the presence of other lipids.

✓ Checkpoint

32. **Name the nutrients the body requires.**
33. **What component of food would increase the number of chylomicrons in the lacteals?**
34. **The absorption of which vitamin would be impaired by the removal of the stomach?**
35. **Why is diarrhea potentially life threatening, but constipation (infrequent defecation) is not?**

See the blue Answers tab at the back of the book.

24-9 Many age-related changes affect digestion and absorption

Learning Outcome Summarize the effects of aging on the digestive system.

Normal digestion and absorption take place in elderly people. However, many changes in the digestive system parallel age-related changes we have already discussed in connection with other systems:

- *The Division Rate of Epithelial Stem Cells Declines.* The digestive epithelium becomes more susceptible to damage by abrasion, acids, or enzymes. Peptic ulcers therefore become more likely. Stem cells in the epithelium divide less frequently with age, so tissue repair is less efficient. In the mouth, esophagus, and anus, the stratified epithelium becomes thinner and more fragile.
- *Smooth Muscle Tone Decreases.* General motility decreases, and peristaltic contractions are weaker as a result of a decrease in smooth muscle tone. These changes slow the

Table 24-2 The Absorption of Ions and Vitamins

Ion or Vitamin	Transport Mechanism	Regulatory Factors
Na^+	Channel-mediated diffusion, cotransport, or active transport	Increased when sodium-linked cotransport is under way; stimulated by aldosterone
Ca^{2+}	Active transport	Stimulated by calcitriol and PTH
K^+	Channel-mediated diffusion	Follows concentration gradient
Mg^{2+}	Active transport	
Fe^{2+}	Active transport	
Cl^-	Channel-mediated diffusion or carrier-mediated transport	
I^-	Channel-mediated diffusion or carrier-mediated transport	
HCO_3^-	Channel-mediated diffusion or carrier-mediated transport	
NO_3^-	Channel-mediated diffusion or carrier-mediated transport	
PO_4^{3-}	Active transport	
SO_4^{2-}	Active transport	
Water-soluble vitamins (except B_{12})	Channel-mediated diffusion	Follows concentration gradient
Vitamin B_{12}	Active transport	Must be bound to intrinsic factor prior to absorption
Fat-soluble vitamins	Diffusion	Absorbed from micelles along with dietary lipids

rate of fecal movement and promote constipation. Sagging and inflammation of the haustra in the colon can occur. Straining to eliminate compacted feces can stress the less resilient walls of blood vessels, producing hemorrhoids. Problems are not restricted to the lower digestive tract. For example, weakening of muscular sphincters can lead to esophageal reflux and frequent bouts of "heartburn."

- *The Effects of Cumulative Damage Become Apparent.* A familiar example is the gradual loss of teeth due to *dental caries* (cavities) or gingivitis. Cumulative damage can involve internal organs as well. Toxins such as alcohol and other injurious chemicals that are absorbed by the digestive tract are transported to the liver for processing. The cells of the liver are not immune to these toxic compounds, and chronic exposure can lead to cirrhosis or other types of liver disease.
- *Cancer Rates Increase.* Cancers are most common in organs in which stem cells divide to maintain epithelial cell populations. Rates of colon cancer and stomach cancer rise with age. Oral, esophageal, and pharyngeal cancers are particularly common among elderly smokers.
- *Dehydration Is Common Among the Elderly.* One reason is that osmoreceptor sensitivity declines with age.
- *Changes in Other Systems Have Direct or Indirect Effects on the Digestive System.* For example, reductions in bone mass and calcium content in the skeleton are associated with erosion of the tooth sockets and eventual tooth loss. The decline in olfactory and gustatory sensitivities with age can lead to dietary changes that affect the entire body.

Checkpoint

36. **Identify general digestive system changes that occur with aging.**

See the blue Answers tab at the back of the book.

24-10 The digestive system is extensively integrated with other body systems

Learning Outcome **Give examples of interactions between the digestive system and other body systems studied so far.**

Build Your Knowledge **Figure 24–29** summarizes the functional relationships between the digestive system and other body systems we have studied so far. The digestive system has particularly extensive anatomical and physiological connections to the nervous, cardiovascular, endocrine, and lymphatic systems. As we have seen, the digestive tract is also an endocrine organ that produces a variety of hormones. Many of these hormones, and some of the neurotransmitters produced by the digestive system, can enter the circulation, cross the blood brain barrier, and alter CNS activity. In this way, a continual exchange of chemical information takes place among these systems.

Checkpoint

37. **Identify the functional relationships between the digestive system and other body systems.**
38. **What body systems may be affected by inadequate calcium ion absorption?**

See the blue Answers tab at the back of the book.

Build Your Knowledge

Figure 24–29 Integration of the DIGESTIVE system with the other body systems presented so far.

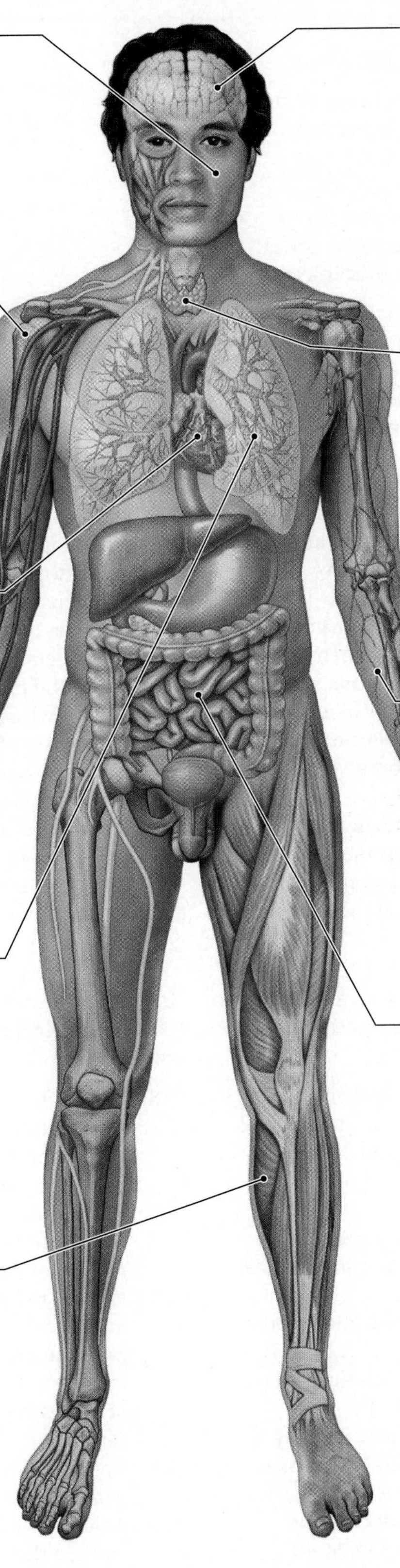

Integumentary System

- The Integumentary System provides vitamin D_3 needed for the absorption of calcium and phosphorus
- The digestive system provides lipids for storage by adipocytes in hypodermis

Skeletal System

- The Skeletal System (axial division and pelvic girdle) supports and protects parts of digestive tract; teeth are used in mechanical processing of food
- The digestive system absorbs calcium and phosphate ions for use in bone matrix; provides lipids for storage in yellow marrow

Cardiovascular System

- The Cardiovascular System distributes hormones of the digestive tract; carries nutrients, water, and ions from sites of absorption; delivers nutrients and toxins to liver
- The digestive system absorbs fluid to maintain normal blood volume; absorbs vitamin K; liver excretes heme (as bilirubin), synthesizes blood clotting proteins

Respiratory System

- The Respiratory System can assist in defecation by producing increased thoracic and abdominal pressure through contraction of respiratory muscles
- The digestive system produces pressure with digestive organs against the diaphragm that can assist in exhalation and limit inhalation

Muscular System

- The Muscular System protects and supports digestive organs in abdominal cavity; controls entrances and exits of digestive tract
- The digestive system regulates blood glucose and fatty acid levels and metabolizes lactate from active muscles with the liver

Nervous System

- The Nervous System regulates movement and secretion with the ANS and ENS; reflexes coordinate passage of materials along tract; control over skeletal muscles regulates ingestion and defecation; hypothalamic centers control hunger, satiation, and feeding
- The digestive system provides compounds essential for neurotransmitter synthesis

Endocrine System

- The Endocrine System produces epinephrine and norepinephrine that stimulate constriction of sphincters and depress digestive activity; hormones coordinate activity along digestive tract
- The digestive system provides nutrients and substrates to endocrine cells; endocrine cells of pancreas secrete insulin and glucagon; liver produces angiotensinogen

Lymphatic System

- The Lymphatic System defends against infection and toxins absorbed from the digestive tract; lymphatic vessels carry absorbed lipids to the general circulation
- The digestive system secretes acids and enzymes that provide innate (nonspecific) immunity against pathogens

Digestive System

The digestive system provides organic substrates, vitamins, minerals, electrolytes (inorganic ions), and water required by all cells. It:

- ingests solid and liquid materials into the oral cavity of the digestive tract
- mechanically digests solid materials to ease their movement and propels them down the digestive tract
- chemically digests food into smaller molecules in the digestive tract
- secretes water, acids, enzymes, and buffers into the digestive tract
- absorbs small organic molecules (organic substrates), ions, vitamins, and water across the digestive epithelium
- eliminates wastes and indigestible materials from the body by defecation

24 Chapter Review

Study Outline

An Introduction to the Digestive Systemd p. 885

1. The digestive system supplies both the fuel that keeps all the body's cells functioning and the building blocks needed for cell growth and repair.

24-1 The digestive system, consisting of the digestive tract and accessory organs, functions primarily to break down and absorb nutrients from food and to eliminate wastes p. 885

2. The **digestive system** consists of the muscular **digestive tract** and various **accessory organs**. *(Figure 24–1)*
3. The general functions of the digestive system are **ingestion, mechanical digestion and propulsion, chemical digestion, secretion, absorption, and defecation.**
4. Double sheets of peritoneal membrane called **mesenteries** suspend the digestive tract. The **greater omentum** lies anterior to the abdominal viscera. Its adipose tissue provides padding, protection, insulation, and an energy reserve. *(Figure 24–2)*
5. The epithelium, *lamina propria*, and *muscularis mucosae* form the **mucosa** (mucous membrane) that lines the lumen of the digestive tract. Deep to the mucosa is the **submucosa**, the **muscular layer**, and a fibrous layer of connective tissue called the *adventitia*. For viscera projecting into the peritoneal cavity, the muscular layer is covered by a serous membrane called the **serosa**. *(Figure 24–3)*
6. The muscular layer propels materials through the digestive tract by the contractions of **peristalsis**. **Segmentation** movements in the small intestine churn digestive materials. *(Figure 24–4)*
7. Local factors, neural mechanisms, and hormonal mechanisms control digestive tract activities. *(Figure 24–5)*

24-2 The oral cavity, which contains the tongue, teeth, and salivary glands, functions in the ingestion and mechanical digestion of food p. 893

8. The functions of the **oral cavity** are (1) *sensory analysis* of foods; (2) *mechanical digestion* by the teeth, tongue, and palatal surfaces; (3) *lubrication*, by mixing with mucus and salivary gland secretions; and (4) limited *chemical digestion* of carbohydrates and lipids.
9. The oral cavity is lined by the **oral mucosa**. The *hard* and *soft palates* form the roof of the oral cavity, and the *tongue* forms its floor. *(Figure 24–6)*
10. **Intrinsic** and **extrinsic tongue muscles** are controlled by the hypoglossal nerves. *(Figure 24–6)*
11. **Mastication** (chewing) of the **bolus** occurs through the contact of the **occlusal** (opposing) **surfaces** of the **teeth**. The **periodontal ligament** anchors each tooth in an *alveolus*, or bony socket. **Dentin** forms the basic structure of a tooth. The **crown** is coated with **enamel**, the **root** with **cement**. *(Figure 24–7)*
12. During childhood and early adulthood, the 20 **deciduous teeth**, are replaced by 32 **permanent teeth**. *(Figure 24–8)*
13. The **parotid**, **sublingual**, and **submandibular salivary glands** discharge their secretions into the oral cavity. *(Figure 24–9)*

> MasteringA&P™ Access more chapter study tools online in the MasteringA&P Study Area:
> - Chapter Quizzes, Chapter Practice Test, MP3 Tutor Sessions, and Clinical Case Studies
> - Practice Anatomy Lab PAL 3.0
> - Interactive Physiology iP2
> - A&P Flix A&PFlix
> - PhysioEx PhysioEx 9.1

24-3 The pharynx and esophagus are passageways that transport the food bolus from the oral cavity to the stomach p. 898

14. The **pharynx** serves as a common passageway for solid food, liquid, and air.
15. The **esophagus** carries solids and liquids from the pharynx to the stomach through the **esophageal hiatus**, an opening in the diaphragm. *(Figure 24–10)*
16. The esophageal mucosa consists of a stratified epithelium. Mucous secretion by esophageal glands of the submucosa reduces friction during the passage of foods. The proportions of skeletal and smooth muscle of the muscular layer change from the pharynx to the stomach. *(Figure 24–10)*
17. Swallowing, or **deglutition**, can be divided into **buccal**, **pharyngeal**, and **esophageal phases**. Swallowing begins with the compaction of a bolus and its movement through the pharynx, from contractions of the *pharyngeal constrictor* and the *palatal muscles*, followed by the elevation of the larynx, reflection of the *epiglottis*, and closure of the *glottis*. After the *upper esophageal sphincter* is opened, peristalsis moves the bolus down the esophagus to the *lower esophageal sphincter*. *(Figure 24–11)*

24-4 The stomach is a J-shaped organ that receives the bolus and aids in its chemical and mechanical digestion p. 901

18. The four regions of the stomach are the **cardia**, **fundus**, **body**, and **pyloric part**. The **pyloric sphincter** guards the exit from the stomach. In a relaxed state, the stomach lining contains numerous **rugae** (ridges and folds). *(Figure 24–12)*
19. The **stomach** functions include the following: (1) storage of ingested food, (2) mechanical breakdown of food, and (3) disruption of chemical bonds by acid and enzymes.
20. Within the **gastric glands**, *parietal cells* secrete *intrinsic factor* and the ions of *hydrochloric acid*. **Chief cells** secrete **pepsinogen**, which is converted by acids in the gastric lumen to the enzyme **pepsin**. **Enteroendocrine** cells of the stomach secrete several compounds, notably the hormone **gastrin**. *(Figures 24–13, 24–14)*
21. Gastric control involves (1) the **cephalic phase**, which prepares the stomach to receive ingested materials; (2) the **gastric phase**, which begins with the arrival of food in the stomach; and (3) the **intestinal phase**, which controls the rate of gastric emptying. Vomiting, or **emesis**, is reverse peristalsis. *(Spotlight Figure 24–15)*

24-5 Accessory digestive organs, such as the pancreas and liver, produce secretions that aid in chemical digestion p. 905

22. The **pancreas** has a slender **body** and a short tail. The **pancreatic duct** penetrates the wall of the duodenum, the first section of the small intestine. Within each lobule, ducts branch repeatedly before ending in the **pancreatic acini** (pockets). *(Figure 24–16)*
23. The pancreas has two functions: endocrine (secreting insulin and glucagon into the blood) and exocrine (secreting **pancreatic juice** into the small intestine). Pancreatic enzymes include **carbohydrases**, **lipases**, **nucleases**, and **proteases**.
24. The **liver** performs metabolic and hematological regulation and produces **bile**. The bile ducts from all the **liver lobules** unite to form the **common hepatic duct**. That duct meets the **cystic duct** to form the **bile duct**, which empties into the duodenum. *(Figures 24–17 to 24–19)*
25. The liver lobule is the organ's basic functional unit. **Hepatocytes** form irregular plates arranged in the form of spokes of a wheel. **Bile canaliculi** carry bile to the **bile ductules**, which lead to **portal triads**. *(Figure 24–18)*
26. In **emulsification**, **bile salts** break apart large drops of lipids, making the lipids accessible to lipases secreted by the pancreas.
27. The liver is the primary organ involved in regulating the composition of circulating blood. All the blood leaving the absorptive surfaces of the digestive tract flows into the liver before entering the systemic circulation. The liver regulates metabolism as it removes and stores excess nutrients, vitamins, and minerals from the blood; mobilizes stored reserves; synthesizes needed nutrients; and removes wastes.
28. The liver's hematological activities include the monitoring of circulating blood by phagocytes and antigen-presenting cells; the synthesis of plasma proteins; the removal of circulating hormones and antibodies; and the removal or storage of toxins.
29. The liver synthesizes bile, which is composed of water, ions, bilirubin, cholesterol, and bile salts (an assortment of lipids).
30. The **gallbladder** stores, modifies, and concentrates bile. *(Figure 24–19)*

24-6 The small intestine primarily functions in the chemical digestion and absorption of nutrients p. 915

31. Most of the important digestive and absorptive functions occur in the **small intestine** (*small bowel*). The pancreas, liver, and gallbladder provide digestive secretions and buffers.
32. The small intestine consists of the **duodenum**, the **jejunum**, and the **ileum**. A sphincter, the **ileocecal valve**, marks the transition between the small and large intestines. *(Figure 24–20)*
33. The intestinal mucosa has transverse folds called **circular folds** (*plicae circulares*) and small projections called **intestinal villi**. These folds and projections increase the surface area for absorption. Each villus contains a terminal lymphatic called a **lacteal**. Pockets called **intestinal glands** are lined by enteroendocrine, mucous, paneth, and stem cells. *(Figures 24–20, 24–21)*
34. **Intestinal juice** moistens chyme, helps buffer acids, and holds digestive enzymes and digestive products in solution.
35. The **duodenal submucosal glands** of the duodenum produce mucus, bicarbonate ions, and the hormone **urogastrone**. The ileum contains masses of lymphoid tissue called **aggregated lymphoid nodules**, or *Peyer's patches*, near the entrance to the large intestine.
36. The **gastroenteric reflex**, initiated by stretch receptors in the stomach, stimulates motility and secretion along the entire small intestine. The **gastroileal reflex** triggers the relaxation of the ileocecal valve.
37. Neural and hormonal mechanisms coordinate the activities of the digestive glands. Gastrointestinal activity is stimulated by parasympathetic innervation and inhibited by sympathetic innervation. The **enterogastric**, **gastroenteric**, and **gastroileal reflexes** coordinate movement from the stomach to the large intestine.
38. Intestinal hormones include **secretin**, **cholecystokinin (CCK)**, **gastric inhibitory peptide (GIP)**, **vasoactive intestinal peptide (VIP)**, **gastrin**, and **enterocrinin**. *(Figures 24–22, 24–23)*

24-7 The large intestine, which is divided into three parts, absorbs water from digestive materials and eliminates the remaining waste as feces p. 921

39. The main functions of the **large intestine** are to (1) reabsorb water and compact materials into feces, (2) absorb vitamins produced by bacteria, and (3) store fecal material prior to defecation. The large intestine (*large bowel*) consists of the *cecum*, *colon*, and *rectum*. *(Figure 24–24)*
40. The **cecum** collects and stores material from the ileum and begins the process of compaction. The **appendix** is attached to the cecum. *(Figure 24–24)*
41. The **colon** has a larger diameter and a thinner wall than the small intestine. The colon bears **haustra** (pouches), **teniae coli** (longitudinal bands of muscle), and sacs of fat called **omental appendices**. *(Figure 24–24)*
42. The four regions of the colon are the **ascending colon**, **transverse colon**, **descending colon**, and **sigmoid colon**. *(Figure 24–24)*
43. The **rectum** terminates in the **anal canal**, leading to the **anus**. *(Figure 24–24)*
44. Histological characteristics of the colon include the absence of villi and the presence of goblet cells and deep intestinal glands. *(Figure 24–25)*
45. The large intestine reabsorbs water and other substances such as vitamins, urobilinogen, bile salts, and toxins. Most of the gut bacteria of the human *microbiome* live in the large intestine. Bacteria are responsible for intestinal gas, or **flatus**.
46. The gastroileal reflex moves materials from the ileum into the cecum while you eat. Distension of the stomach and duodenum stimulates **mass movements** of materials from the transverse colon through the rest of the large intestine and into the rectum. Muscular sphincters control the passage of fecal material to the anus. Distension of the rectal wall triggers the **defecation reflex**. *(Figure 24–26)*

24-8 Chemical digestion is the enzyme-mediated hydrolysis of food into nutrients that can be absorbed and used by the body p. 926

47. The digestive system first breaks down the physical structure of the ingested material and then disassembles the component molecules into smaller fragments by *hydrolysis*. *(Spotlight Figure 24–27; Table 24–1)*
48. Salivary and pancreatic amylases break down complex carbohydrates into *disaccharides* and *trisaccharides*. These in turn are broken down into *monosaccharides* by enzymes at the epithelial surface. The monosaccharides are then absorbed by the intestinal epithelium by facilitated diffusion or cotransport. *(Spotlight Figure 24–27)*
49. *Triglycerides* are emulsified into lipid droplets that interact with bile salts to form **micelles**. The fatty acids and monoglycerides resulting from the action of pancreatic lipase diffuse from the micelles across the intestinal epithelium. Triglycerides are then synthesized and released into the interstitial fluid for transport to the general circulation by way of the lymphatic system. *(Spotlight Figure 24–27)*
50. Protein digestion involves a low pH (which breaks down tertiary and quaternary structure), the gastric enzyme pepsin, and various pancreatic proteases. Peptidases liberate amino acids

that are absorbed and exported to interstitial fluid. *(Spotlight Figure 24–27)*

51. About 2200 mL of water is ingested each day, and digestive secretions provide another 7200 mL. Nearly all is reabsorbed by osmosis. *(Figure 24–28)*
52. Various processes, including diffusion, cotransport, and carrier-mediated and active transport, are responsible for the movements of cations (sodium, calcium, potassium, and so on) and anions (chloride, iodide, bicarbonate, and so on) into epithelial cells. *(Table 24–2)*
53. The **water-soluble vitamins** (except B_{12}) diffuse easily across the digestive epithelium. **Fat-soluble vitamins** are enclosed within fat droplets and absorbed with the products of lipid digestion. *(Table 24–2)*

24-9 Many age-related changes affect digestion and absorption p. 931

54. Age-related changes include a thinner and more fragile epithelium due to a reduction in epithelial stem cell divisions, weaker peristaltic contractions as smooth muscle tone decreases, the effects of cumulative damage, increased cancer rates, and increased dehydration.

24-10 The digestive system is extensively integrated with other body systems p. 932

55. The digestive system has extensive anatomical and physiological connections to the nervous, endocrine, cardiovascular, and lymphatic systems. *(Figure 24–29)*

Review Questions

See the blue Answers tab at the back of the book.

LEVEL 1 Reviewing Facts and Terms

1. The enzymatic breakdown of large molecules into their basic building blocks is called **(a)** absorption, **(b)** secretion, **(c)** mechanical processing, **(d)** chemical digestion.
2. The outer layer of the digestive tract is known as the **(a)** serosa, **(b)** mucosa, **(c)** submucosa, **(d)** muscularis.
3. Double sheets of peritoneum that provide support and stability for the organs of the peritoneal cavity are the **(a)** mediastina, **(b)** mucous membranes, **(c)** omenta, **(d)** mesenteries.
4. A branch of the hepatic portal vein, hepatic artery proper, and branch of the bile duct form **(a)** a liver lobule, **(b)** the sinusoids, **(c)** a portal triad, **(d)** the hepatic duct, **(e)** the pancreatic duct.
5. Label the digestive system structures in the following figure.

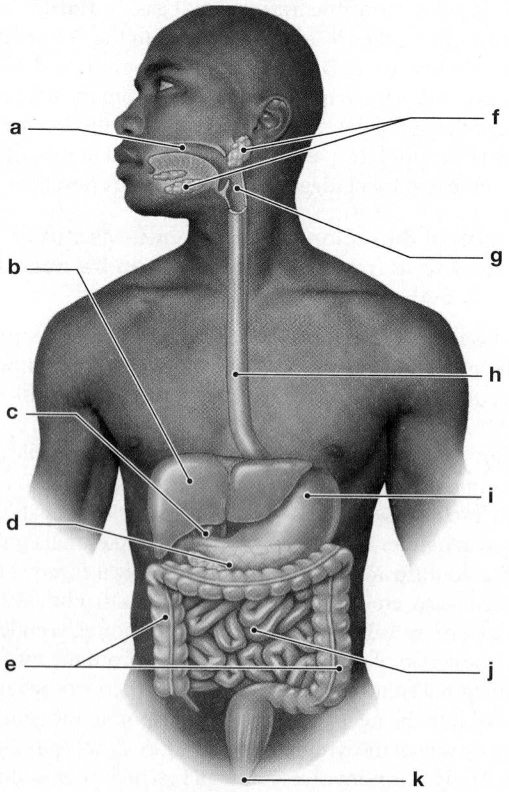
a
b
c
d
e
f
g
h
i
j
k

(a) ________ **(b)** ________
(c) ________ **(d)** ________
(e) ________ **(f)** ________
(g) ________ **(h)** ________
(i) ________ **(j)** ________
(k) ________

6. Label the four layers of the digestive tract in the following figure.

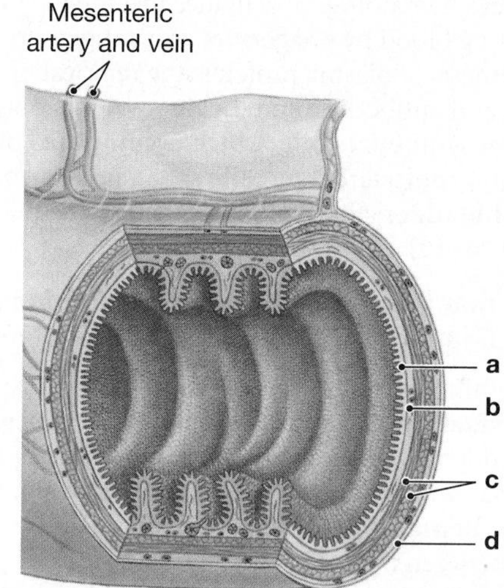
Mesenteric artery and vein
a
b
c
d

(a) ________ **(b)** ________
(c) ________ **(d)** ________

7. Most of the digestive tract is lined by ________ epithelium. **(a)** pseudostratified ciliated columnar, **(b)** cuboidal, **(c)** stratified squamous, **(d)** simple, **(e)** simple columnar.
8. Regional movements that occur in the small intestine and function to churn and fragment the digestive material are called **(a)** segmentation, **(b)** pendular movements, **(c)** peristalsis, **(d)** mass movements, **(e)** mastication.
9. Bile release from the gallbladder into the duodenum occurs only under the stimulation of **(a)** cholecystokinin, **(b)** secretin, **(c)** gastrin, **(d)** enteropeptidase.

10. Label the three segments of the small intestine in the following figure.

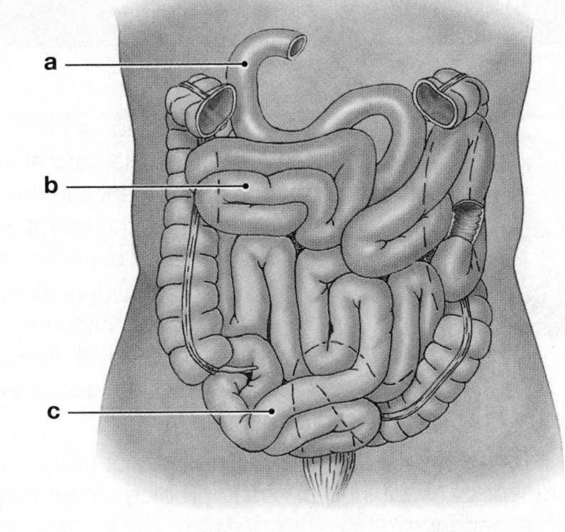
a
b
c

(a) ________ (b) ________ (c) ________

11. The major function(s) of the large intestine is (are) **(a)** reabsorption of water and compaction of feces, **(b)** absorption of vitamins liberated by bacterial action, **(c)** storage of fecal material prior to defecation, **(d)** all of these.
12. Vitamins generated by bacteria in the colon are **(a)** vitamins A, D, and E, **(b)** B complex vitamins and vitamin C, **(c)** vitamin K, biotin, and pantothenic acid, **(d)** niacin, thiamine, and riboflavin.
13. The final enzymatic steps in the digestive process are accomplished by **(a)** brush border enzymes of the intestinal microvilli, **(b)** enzymes secreted by the stomach, **(c)** enzymes secreted by the pancreas, **(d)** the action of bile from the gallbladder.
14. What are the six main processes of the digestive system?
15. Name and describe the layers of the digestive tract, proceeding from the innermost layer nearest the lumen to the outermost layer.
16. What three basic mechanisms regulate the activities of the digestive tract?
17. What are the three phases of swallowing, and how are they controlled?
18. What are the primary digestive functions of the pancreas, liver, and gallbladder?
19. Which hormones produced by duodenal enteroendocrine cells effectively coordinate digestive functions?
20. What are the three primary functions of the large intestine?
21. What two positive feedback loops are involved in the defecation reflex?

LEVEL 2 Reviewing Concepts

22. During defecation, **(a)** stretch receptors in the rectal wall initiate a series of peristaltic contractions in the colon and rectum, **(b)** stretch receptors in the rectal wall activate parasympathetic centers in the sacral region of the spinal cord, **(c)** the internal anal sphincter relaxes while the external anal sphincter contracts, **(d)** all of these occur, **(e)** only a and b occur.
23. *Increased* parasympathetic stimulation of the intestine would result in **(a)** decreased motility, **(b)** decreased secretion, **(c)** decreased sensitivity of local reflexes, **(d)** decreased segmentation, **(e)** none of these.
24. A drop in pH below 4.5 in the duodenum stimulates the secretion of **(a)** secretin, **(b)** cholecystokinin, **(c)** gastrin, **(d)** all of these.
25. Through which layers of a molar would an oral surgeon drill to perform a root canal (removal of the alveolar nerve in a severely damaged tooth)?
26. How is the stomach protected from digestion?
27. How does each of the three phases of gastric secretion promote and facilitate gastric control?
28. Nutritionists have found that after a heavy meal, the pH of blood increases slightly, especially in the veins that carry blood away from the stomach. What causes this increase in blood pH?

LEVEL 3 Critical Thinking and Clinical Applications

29. Some people with gallstones develop pancreatitis. How could this occur?
30. Harry is suffering from an obstruction in his colon. He notices that when he urinates, the color of his urine is much darker than normal, and he wonders if there is any relationship between the color of his urine and his intestinal obstruction. What would you tell him?
31. A condition known as lactose intolerance is characterized by painful abdominal cramping, gas, and diarrhea. The cause of the problem is an inability to digest the milk sugar lactose. How would this cause the observed signs and symptoms?
32. Recently, more people have turned to surgery to help them lose weight. One form of weight control surgery involves stapling a portion of the stomach shut, creating a smaller volume. How would such a surgery result in weight loss?

CLINICAL CASE Wrap-Up An Unusual Transplant

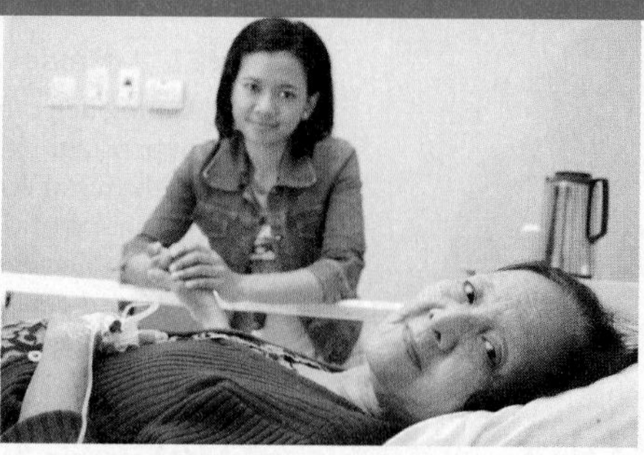

Normally the lining of the colon is slimy with mucus and covered with millions of beneficial bacteria. These friendly, "good" bacteria generate vitamins K, B_5, and biotin. They convert bilirubin (the breakdown product of blood) to urobilinogens and stercobilinogens for excretion. The bacteria also break down any remaining peptides in feces.

When these normal, beneficial intestinal bacteria have been killed by antibiotics, they are replaced by other bacteria. Often the replacement bacteria do harm by producing toxins, irritating the intestinal lining, and causing colitis. This is the case with *Clostridium difficile* infections.

One way of treating *C. difficile* infections is with a fecal transplant, as Mirasol's doctor suggests. Normal feces from a healthy donor contain populations of beneficial bacteria that usually colonize large intestines. With a minimal amount of processing, the donated feces can be infused through a nose tube (nasogastric tube). Then the normal intestinal bacteria can take up residence in the recipient's large intestine, crowd out the harmful *C. difficile* bacteria, and resume their usual work of protecting the lining of the large intestine, producing vitamins, and recycling nutrients.

After overcoming her initial disgust, Mirasol agrees to the transplant. She recovers completely and leaves the hospital 2 days after her treatment.

1. If Mirasol's *Clostridium difficile* colitis were not treated with a fecal transplant, what kind of health problems might she continue to experience?
2. If you could look inside Mirasol's large intestine with a colonoscope before the transplant, what would you likely see?

See the blue Answers tab at the back of the book.

Related Clinical Terms

borborygmus: A rumbling or gurgling sound made by the movement of fluids and gases in the intestines.
cathartics: Drugs that promote defecation.
cholelithiasis: The presence of gallstones in the gallbladder.
cholera: A bacterial infection of the digestive tract that causes massive fluid losses through diarrhea.
colitis: A general term for a condition characterized by inflammation of the colon.
Crohn's disease: An incurable chronic inflammatory bowel disease that can affect any part of the digestive tract, from the mouth to the anus. The presence of strictures, fistulas, and fissures is common.
diverticulitis: An infection and inflammation of mucosal pockets of the large intestine (diverticula).

diverticulosis: The formation of diverticula, generally along the sigmoid colon.
dysphagia: Difficulty or discomfort in swallowing due to disease.
esophageal varices: Swollen and fragile esophageal veins that result from portal hypertension.
fecal occult blood test: Test to check for hidden blood in feces.
gastrectomy: The surgical removal of the stomach, generally to treat advanced stomach cancer.
gastroesophageal reflux disease (GERD): Chronic condition in which the lower esophageal sphincter allows gastric acids to backflow into the esophagus, causing heartburn, acid indigestion, and possible injury to the esophageal lining.
gastroscope: A fiber-optic instrument inserted into the mouth and directed along the esophagus and into the stomach; used to examine the interior of the stomach and to perform minor surgical procedures.
halitosis: Bad breath that may be due to poor oral hygiene, an infection, diabetes, or other disease.
insoluble fiber: Indigestible plant carbohydrates that do not dissolve in water and pass through the GI tract unchanged. Found in many vegetables and the skins of fruits, insoluble fiber speeds up the passage of material in the GI tract. Individuals consuming diets rich in insoluble fiber decrease their risk for developing diabetes, atherosclerosis, and colorectal cancers, among other diseases.
irritable bowel syndrome (IBS): A common disorder affecting the large intestine, accompanied by cramping, abdominal pain, bloating, gas, diarrhea, and constipation.
pancreatic cancer: Malignancy of the pancreas that does not cause symptoms in its early stages, leading to late detection and a survival rate of only 4 percent.
periodontal disease: A loosening of the teeth within the alveolar sockets caused by erosion of the periodontal ligaments by acids produced through bacterial action.
polyps: Small growths with a stalk protruding from a mucous membrane that is usually benign.
pulpitis: An infection of the pulp of a tooth; treatment may involve a root canal procedure.
pyloric stenosis: Uncommon condition in which the muscle of the lower end of the stomach enlarges and prevents food from entering the small intestine.
pylorospasm: Spasm of the pyloric sphincter, accompanied by pain and vomiting.
root canal: Removal of the alveolar nerve in a severely damaged tooth.
soluble fiber: Indigestible plant carbohydrates found in beans, oats, and citrus fruits that dissolve in water when eaten, forming a gel within the digestive tract to slow the passage of material. Diets rich in soluble fiber lower blood cholesterol levels.

25 Metabolism, Nutrition, and Energetics

Learning Outcomes

These Learning Outcomes correspond by number to this chapter's sections and indicate what you should be able to do after completing the chapter.

25-1 ■ **Explain the relationship among metabolism, catabolism, and anabolism; and describe the energetics of an oxidation–reduction reaction.** p. 940

25-2 ■ **Describe the basic steps in glycolysis, the citric acid cycle, the electron transport chain, and chemiosmosis, and summarize the energy yield of cellular respiration (glycolysis and aerobic metabolism).** p. 943

25-3 ■ **Describe the pathways involved in lipid metabolism, and summarize the mechanisms of lipid transport and distribution.** p. 951

25-4 ■ **Summarize the main processes of protein metabolism, and discuss the use of protein as an energy source.** p. 956

25-5 ■ **Differentiate between the absorptive and postabsorptive metabolic states, and summarize the characteristics of each.** p. 957

25-6 ■ **Explain what makes up a balanced diet and why such a diet is important.** p. 959

25-7 ■ **Define metabolic rate, discuss the factors involved in determining an individual's BMR, and discuss the homeostatic mechanisms that maintain a constant body temperature.** p. 966

CLINICAL CASE The Miracle Supplement

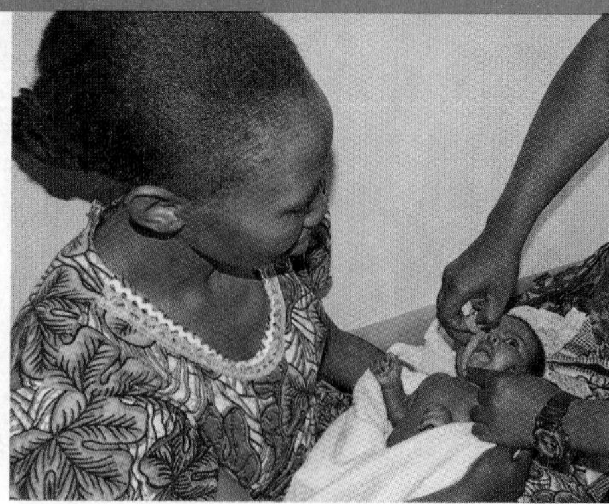

Justine, a pregnant 30-year-old Kenyan woman, is walking home from the clinic after her final prenatal visit. As the sun sets, she realizes she is suddenly no longer able to see. Terrified, she manages to walk home blind, in rural Kenya, after dark. The next morning, her vision returns.

Justine knows children who have experienced night blindness. Many of them died young; in fact, every family in Justine's village has experienced childhood death. Several of the village's children also have stunted growth and delayed skeletal development. They seem to have weak immunity, and commonly develop malarial, respiratory, and diarrheal infections.

Fortunately, Justine goes on to deliver a healthy baby at the clinic. Shortly after giving birth, doctors treat Justine with a single high-dose vitamin supplement. She is told this will help her night blindness and ensure that there is enough of this vitamin in her breast milk to keep her baby healthy, too.

When Justine's baby is 6 months old, Justine takes her to get her first dose of vitamin supplement. This dose will be repeated every 6 months until the child is 5 years old. Costing only 2¢ per dose and reducing overall childhood mortality (death rate) by 23 percent, this vitamin supplement has been called the best public health investment ever. **What could this critical vitamin be?** **To find out, turn to Clinical Case Wrap-Up on p. 975.**

An Introduction to Metabolism, Nutrition, and Energetics

The amount and type of *nutrients,* essential elements and molecules, you get from food can vary widely. Your body builds energy reserves when nutrients are abundant. It mobilizes these reserves when nutrients are in short supply. The nervous and endocrine systems control the storage and mobilization of these energy reserves. These systems adjust and coordinate these reactions as part of their regulation of all of the *metabolic* activities of the body's tissues. In this chapter we examine (1) how the body obtains energy from the breakdown of organic molecules, stores it in the form of the energy transfer molecule *ATP,* and uses it to support intracellular processes such as the construction of new organic molecules; (2) the role of diet in physiological function; and (3) the *energetics* of how the body balances heat gains and losses to maintain homeostasis.

25

25-1 Metabolism is the sum of all the catabolic and anabolic reactions in the body, and energetics is the flow and transformation of energy

Learning Outcome **Explain the relationship among metabolism, catabolism, and anabolism; and describe the energetics of an oxidation–reduction reaction.**

Cells are chemical factories that break down organic molecules to obtain energy. Our cells can then use this energy to generate ATP from ADP and P_i. Reactions within mitochondria provide most of the energy a typical cell needs. p. 81 To carry out these reactions, cells must have a reliable supply of oxygen and nutrients, including *water, vitamins, mineral ions,* and *organic substrates.* Substrates are the substances acted on by an enzyme. p. 55

Metabolism

Metabolism is the sum of all chemical and physical changes that occur in body tissues. It consists of **catabolism** (catabolic reactions) and **anabolism** (anabolic reactions). Catabolism converts large molecules into smaller ones, and anabolism converts small molecules into larger ones. Energy-rich macromolecules are obtained in our diet—namely, the fats, carbohydrates, and proteins we eat. The pathways making up these reactions extract energy from nutrients, transfer and use energy, and build and store excesses for later use. Nutrients, the components in food, such as water, vitamins, mineral ions, carbohydrates, fats, and proteins, are necessary for normal physiological function. All the available nutrient molecules distributed in the blood form a **nutrient pool** (Figure 25–1).

Catabolism

Catabolism proceeds in a series of steps. In general, the first steps take place in the cytosol. Fats are broken down into fatty acids and glycerol; carbohydrates are broken down into glucose; and proteins are catabolized into amino acids. Different processes are involved in the catabolism of each nutrient. For example, fat catabolism involves *lipolysis* (fat breakdown), and carbohydrate catabolism involves *glycogenolysis* (breakdown of glycogen into glucose) and *glycolysis* (breakdown of glucose into pyruvate). Every day the lungs absorb oxygen, and the digestive tract absorbs nutrients. The cardiovascular system then carries these substances throughout the body in the blood, where our cells can absorb and use them to generate ATP. Figure 25–1 shows the relationships of the various metabolic pathways, which we discuss in greater detail later in the chapter.

Figure 25–1 Metabolism of Organic Nutrients and Nutrient Pools. This flowchart illustrates the major organic macromolecules in the diet and the relationships among the metabolic pathways and nutrient pools of their building blocks (monomers)—free fatty acids, glucose, and amino acids.

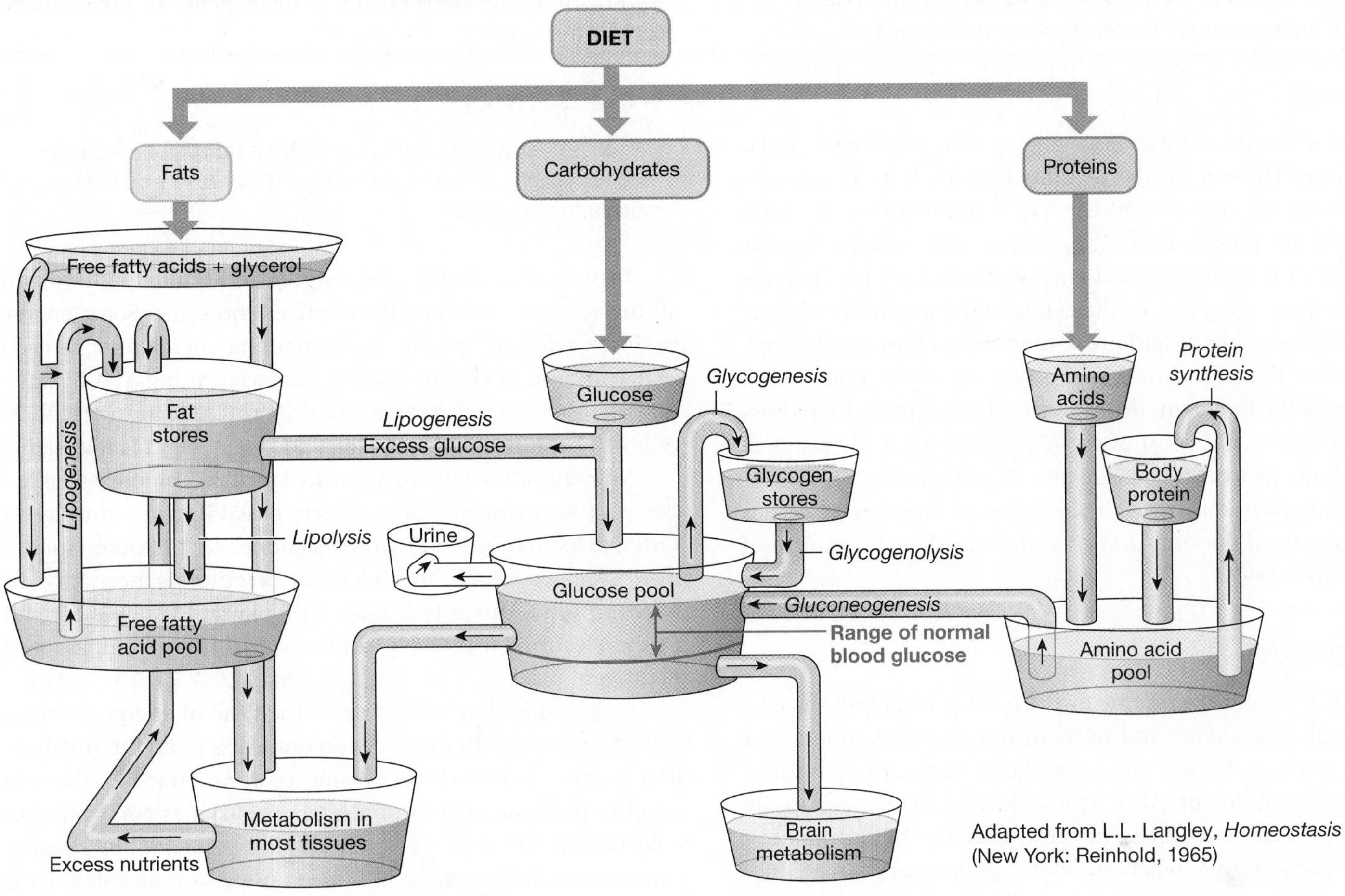
DIET
Fats
Carbohydrates
Proteins
Free fatty acids + glycerol
Fat stores
Lipogenesis
Lipogenesis
Excess glucose
Glucose
Glycogenesis
Amino acids
Protein synthesis
Glycogen stores
Body protein
Lipolysis
Urine
Glycogenolysis
Glucose pool
Gluconeogenesis
Range of normal blood glucose
Free fatty acid pool
Amino acid pool
Metabolism in most tissues
Excess nutrients
Brain metabolism
Adapted from L.L. Langley, *Homeostasis* (New York: Reinhold, 1965)

Anabolism

The ATP produced by mitochondria provides energy to support both *anabolism,* which is the synthesis of new organic molecules, and other cell functions. For example, muscle fibers need ATP to provide energy for contraction, but gland cells need ATP to synthesize and transport secretions. Here we focus on anabolism.

In terms of energy, anabolism is an "uphill" process that involves the formation of new chemical bonds. Cells synthesize new organic components for four basic reasons:

- *To Carry Out Structural Maintenance or Repairs.* All cells must expend energy for ongoing maintenance and repairs, because most structures in the cell are temporary rather than permanent. Cells continuously replace membranes, organelles, enzymes, and structural proteins in the process of *metabolic turnover* (see Appendix, Table 4).
- *To Support Growth.* Cells preparing to divide increase in size and synthesize extra proteins and organelles.
- *To Produce Secretions.* Secretory cells must synthesize their products and deliver them to the interstitial fluid.
- *To Store Nutrient Reserves.* Most cells "prepare for a rainy day"—a period of emergency, an interval of extreme activity, or a time when the supply of nutrients in the bloodstream is inadequate. Cells prepare for such times by building up nutrient reserves in a form that can be mobilized as needed. For example, in the process of *lipogenesis,* or fat formation, *triglycerides* (consisting primarily of *fatty acids*) are the most abundant storage form. In the process of *glycogenesis* (glycogen formation), carbohydrates are stored as *glycogen,* a branched chain of glucose molecules. If glucose is unavailable, it may be synthesized through the process of *gluconeogenesis.* Muscle cells and liver cells, for example, store glucose in the form of glycogen, but adipocytes and liver cells store triglycerides. Proteins are the most abundant organic components in the body. They perform a variety of vital functions for the cell. Anabolic activities require more amino acids than lipids, and few

25

carbohydrates. Therefore, when energy is available, cells synthesize additional proteins. When glucose and fatty acids are unavailable, cells can break down proteins into their component amino acids, and use the amino acids as an energy source. Proteins are so abundant and accessible that they represent an important "last-ditch" nutrient reserve (Figure 25–1).

Mitochondria provide the energy that supports cellular operations. The cell "feeds" its mitochondria from its nutrient pool. In return, the cell gets the ATP it needs. However, mitochondria are picky eaters: They accept only specific organic molecules for processing and energy production. For this reason, chemical reactions in the cytosol take whichever organic nutrients are available and break them down into smaller fragments that the mitochondria can process. The mitochondria then break the fragments down further, generating carbon dioxide, water, and ATP. Most of the ATP is formed as the result of a metabolic process called *oxidative phosphorylation*. This mitochondrial activity involves the *oxidation* of high-energy carrier molecules from the *citric acid cycle* and the phosphorylation of ADP to form ATP.

Energetics

Each tissue contains a unique mixture of various kinds of cells. As a result, the energy and nutrient requirements of any two tissues, such as loose connective tissue and cardiac muscle, can be quite different. Also, activity levels can change rapidly within a tissue, and such changes affect the body's requirements. For example, when skeletal muscles start contracting, oxygen demand increases dramatically. Thus, the energy and nutrient requirements vary from moment to moment (resting versus exercising), hour to hour (asleep versus awake), and year to year (growing child versus adult). Understanding energy requirements is one aspect of **energetics**, the study of the flow of energy and its transformation, or change, from one form to another.

25

Oxidation and Reduction

Most of the important energy changes in cells involve chemical reactions known as oxidation and reduction. Oxidation and reduction reactions may be described as transfers of oxygen, hydrogen, or electrons, but for our purposes the most important are those involving the transfer of electrons. A gain of oxygen, or a loss of hydrogen or electrons, is a form of **oxidation**, and the net result is a *decrease* in potential energy of the atom or molecule. A loss of oxygen, or a gain of hydrogen or electrons, is a form of **reduction**, and the net result is an *increase* in potential energy of the atom or molecule. The two reactions are always paired. When electrons pass from one molecule to another, the electron donor is *oxidized* and the electron recipient is *reduced*. Oxidation and reduction are important because electrons carry chemical energy. In a typical oxidation–reduction reaction, often called a *redox reaction*, the reduced atom or molecule gains energy at the expense of the oxidized atom or molecule.

Tips & Tools

To remember what occurs with electron transfer in oxidation and reduction reactions, think ***OIL RIG*** for **o**xidation **i**s **l**oss and **r**eduction **i**s **g**ain.

In such an exchange, the reduced molecule does not gain all the energy released by the oxidized molecule. Some energy is always released as heat. The remaining energy may be used to do physical or chemical work, such as forming ATP. By passing electrons through a series of oxidation–reduction reactions, cells can capture and use much of the energy that is released.

Mitochondria contain a series of protein complexes, called the *electron transport chain*, which pass electrons through a series of oxidation–reduction reactions. To maintain such a flow of electrons, a final electron acceptor is necessary. In mitochondria, this role is played by oxygen. Water is formed as the electrons ultimately combine with oxygen atoms and hydrogen ions.

Coenzymes play a key role in the flow of energy within a cell and its mitochondria. A **coenzyme** acts as an intermediary that accepts electrons from one molecule and transfers them to another molecule. For example, the coenzymes **NAD** (*nicotinamide adenine dinucleotide*) and **FAD** (*flavin adenine dinucleotide*) can remove hydrogen atoms from organic molecules. Each hydrogen atom consists of an electron (e^-) and a proton (a hydrogen ion, H^+). Thus, when a coenzyme accepts hydrogen atoms, the coenzyme is reduced and gains energy. The donor molecule loses electrons and energy as it gives up its hydrogen atoms.

The coenzyme FAD can accept 2 hydrogen atoms, gaining 2 electrons and forming $FADH_2$. The oxidized form of the coenzyme NAD has a positive charge (NAD^+). This coenzyme also gains 2 electrons as 2 hydrogen atoms are removed from the donor molecule, resulting in the formation of NADH and the release of a proton (H^+). For this reason, the reduced form of NAD is often described as "NADH + H^+."

In summary, the coenzymes NADH + H^+ and $FADH_2$ provide electrons to the electron transport chain. Oxygen accepts, or gains, those electrons. In other words, the coenzymes are oxidized and oxygen is reduced.

✓ Checkpoint

1. **Define *metabolism*, *catabolism*, and *anabolism*.**
2. **Compare oxidation and reduction.**

See the blue Answers tab at the back of the book.

25-2 Carbohydrate metabolism generates ATP by glucose catabolism and forms glucose by gluconeogenesis

Learning Outcome Describe the basic steps in glycolysis, the citric acid cycle, the electron transport chain, and chemiosmosis, and summarize the energy yield of cellular respiration (glycolysis and aerobic metabolism).

Carbohydrate metabolism is primarily concerned with glucose metabolism because most carbohydrates are converted to glucose. The steps involved in glucose catabolism and ATP production include glycolysis, the citric acid cycle, the electron transport chain, and chemiosmosis, whereas glucose anabolism involves gluconeogenesis and glycogenesis. We describe the important catabolic and anabolic cellular reactions for glucose in this section.

Overview of Glucose Catabolism

Most cells generate ATP and other high-energy compounds by breaking down carbohydrates, especially glucose. The complete oxidation of glucose to CO_2 and H_2O involves three major biochemical pathways: glycolysis, the citric acid cycle, and the electron transport chain. We can summarize the complete reaction as follows:

$$\underset{\text{glucose}}{C_6H_{12}O_6} + \underset{\text{oxygen}}{6O_2} \longrightarrow \underset{\text{carbon dioxide}}{6CO_2} + \underset{\text{water}}{6H_2O}$$

This overall reaction is called **cellular respiration**. The breakdown of glucose takes place in a series of small steps. Several of these steps release sufficient energy to convert ADP to ATP. The complete catabolism of 1 molecule of glucose provides a typical body cell a net gain of 30–32 molecules of ATP.

Most ATP production occurs inside the mitochondria, but the first steps take place in the cytosol. In Chapter 10 we introduced the process of *glycolysis*, which breaks down glucose in the cytosol into smaller molecules that mitochondria can absorb and use. ↩ p. 320 These reactions are *anaerobic* because glycolysis does not require oxygen. The subsequent reactions take place in mitochondria and are *aerobic*. The mitochondrial activity responsible for ATP production is called *aerobic metabolism*. (The terms *cellular respiration* and *aerobic metabolism* are often considered interchangeable, however cellular respiration includes both anaerobic and aerobic reactions.)

Glucose Catabolism: Glycolysis

Glycolysis (glī-KOL-ih-sis; *glykus*, sweet + *lysis*, a loosening) is the breakdown of glucose. In this process, a series of enzymatic steps breaks the 6-carbon glucose molecule ($C_6H_{12}O_6$) into two 3-carbon molecules of **pyruvic acid** (CH_3—CO—COOH). (At the normal pH inside cells, each pyruvic acid molecule loses a hydrogen ion and exists as the negatively charged anion CH_3—CO—COO^-. This ionized form is called **pyruvate**.)

Glycolysis requires (1) glucose molecules; (2) appropriate cytosolic enzymes; (3) ATP and ADP; (4) inorganic phosphate groups (P_i); and (5) NAD, a coenzyme that removes hydrogen atoms during one of the enzymatic reactions. Recall from Chapter 2 that coenzymes are organic molecules that are essential to enzyme function. ↩ p. 56 If any of these participants is missing, glycolysis cannot take place.

Figure 25–2 provides an overview of the steps in glycolysis. The beginning of glycolysis "costs" the cell 2 ATP molecules. Glycolysis begins when an enzyme **phosphorylates**—that is, attaches a phosphate group to—the last (sixth) carbon atom of a **glucose** molecule, creating **glucose-6-phosphate** and using 1 ATP molecule. It has two important results: (1) It traps the glucose molecule within the cell, because phosphorylated glucose cannot cross the plasma membrane; and (2) it prepares the glucose molecule for further biochemical reactions. A second phosphorylation, using one more ATP molecule, takes place in the cytosol before the 6-carbon chain is broken into two 3-carbon fragments.

Then energy benefits begin to appear as these fragments are converted to pyruvate. Two of the steps release enough energy to generate ATP from ADP and inorganic phosphate (P_i). In addition, 2 molecules of NAD are reduced to NADH + H^+. (For convenience, we will use NAD for NAD+ and NADH for NADH + H^+ in our discussion.) The net reaction of glycolysis looks like this:

$$\text{Glucose} + 2\ \text{NAD} + 2\ \text{ADP} + 2\ P_i \longrightarrow 2\ \text{Pyruvate} + 2\ \text{NADH} + 2\ \text{ATP}$$

This reaction sequence is anaerobic (without oxygen) and provides the cell a net gain of 2 molecules of ATP for each glucose molecule converted to 2 pyruvate molecules.

A few highly specialized cells, such as red blood cells, lack mitochondria and derive all their ATP through glycolysis. Skeletal muscle fibers rely on glycolysis for ATP production during periods of active contraction. Most cells can survive for brief periods using the ATP provided by glycolysis alone.

Here is a summary of the glycolytic pathway, which prepares glucose for the citric acid cycle:

1. One 6-carbon glucose molecule enters the pathway.
2. Two 3-carbon pyruvate molecules are produced.
3. Two molecules of ATP are produced from 2 ADP and 2 P_i.
4. Two pairs of high-energy electrons are passed to 2 NAD to form 2 NADH.
5. Specific enzymes catalyze each step of the process.

Glucose Catabolism: Fate of Pyruvate

A great deal of additional energy is still stored in the chemical bonds of pyruvate. The cell's ability to capture that energy depends on the availability of oxygen. If oxygen supplies are inadequate, pyruvate is reduced to form lactate and NAD is oxidized from NADH. Lactate formation from pyruvate does not

Figure 25–2 Glycolysis. Glycolysis breaks down a 6-carbon glucose molecule into two 3-carbon molecules of pyruvate through a series of enzymatic steps.

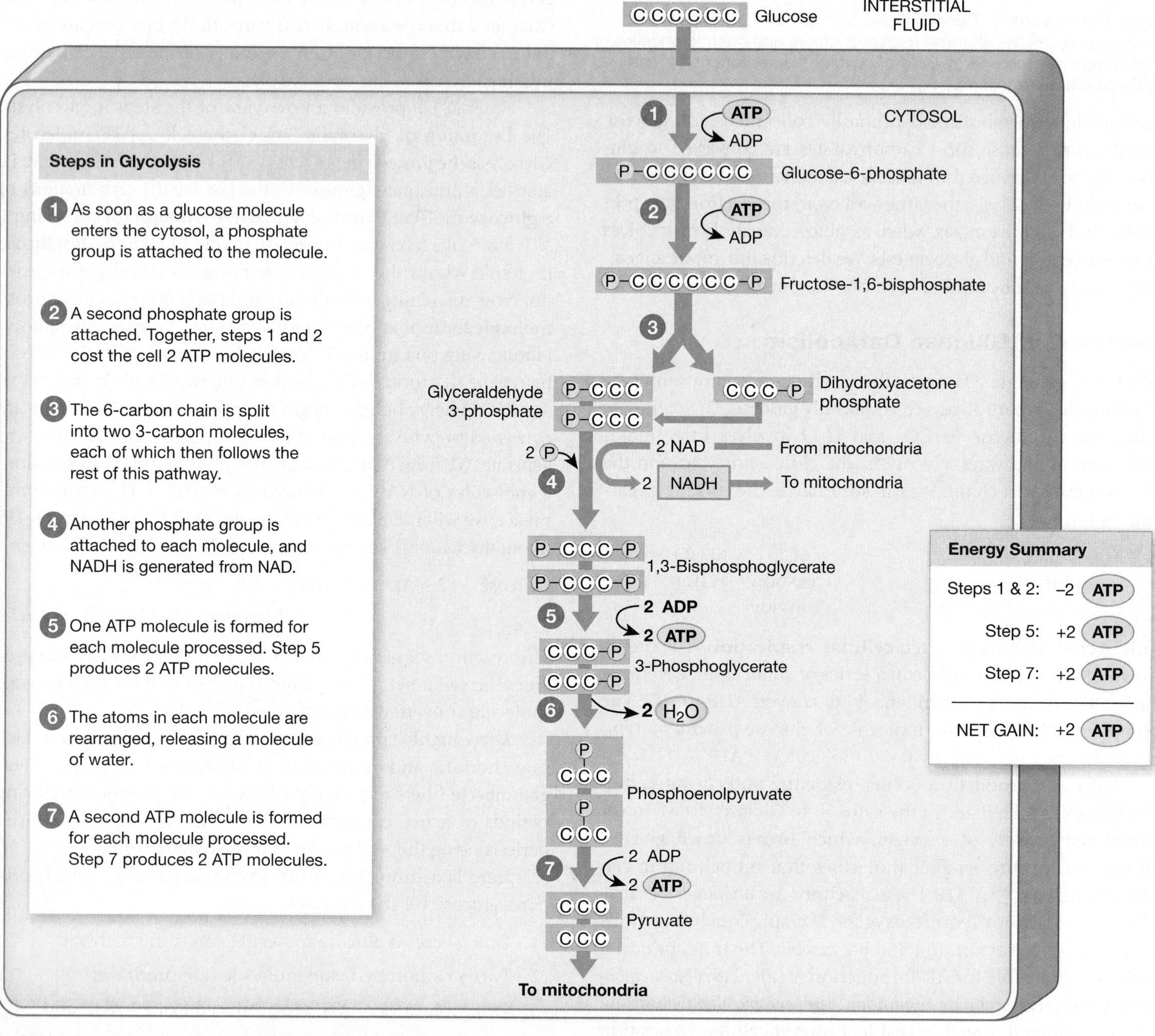

CCCCCC Glucose
INTERSTITIAL FLUID
CYTOSOL
Steps in Glycolysis
1 As soon as a glucose molecule enters the cytosol, a phosphate group is attached to the molecule.
2 A second phosphate group is attached. Together, steps 1 and 2 cost the cell 2 ATP molecules.
3 The 6-carbon chain is split into two 3-carbon molecules, each of which then follows the rest of this pathway.
4 Another phosphate group is attached to each molecule, and NADH is generated from NAD.
5 One ATP molecule is formed for each molecule processed. Step 5 produces 2 ATP molecules.
6 The atoms in each molecule are rearranged, releasing a molecule of water.
7 A second ATP molecule is formed for each molecule processed. Step 7 produces 2 ATP molecules.
ATP
ADP
P–CCCCCC Glucose-6-phosphate
P–CCCCCC–P Fructose-1,6-bisphosphate
Glyceraldehyde 3-phosphate
P–CCC
CCC–P Dihydroxyacetone phosphate
2 P
2 NAD From mitochondria
2 NADH To mitochondria
P–CCC–P 1,3-Bisphosphoglycerate
2 ADP
2 ATP
CCC–P 3-Phosphoglycerate
2 H_2O
Phosphoenolpyruvate
CCC Pyruvate
To mitochondria
Energy Summary
Steps 1 & 2: –2 ATP
Step 5: +2 ATP
Step 7: +2 ATP
NET GAIN: +2 ATP

? Glycolysis produces how many pyruvate molecules and how many ATP molecules?

occur without the presence of NAD. If oxygen is readily available, mitochondrial activity provides most of the ATP that cells need.

Glucose Catabolism: Aerobic Metabolism

When oxygen supplies are adequate, mitochondria absorb the pyruvate molecules. Recall that a double membrane surrounds each mitochondrion. The *outer membrane* contains large pores that are permeable to ions and small organic molecules such as pyruvate. Ions and molecules easily enter the *intermembrane space* separating the outer membrane from the *inner membrane*. The inner membrane contains a carrier protein that moves pyruvate into the mitochondrial matrix.

Within the mitochondria, in the process of **aerobic metabolism**, pyruvate is broken down and the energy released is used to produce a large amount of ATP. This process involves two pathways: the *citric acid cycle* and the *electron transport chain*.

Citric Acid Cycle

The citric acid cycle begins with pyruvate, which is where glycolysis left off, and it occurs in the mitochondria. This cycle involves oxidation, reduction, and decarboxylation reactions. Each pyruvate molecule ($CH_3—CO—COO^-$) is oxidized as its hydrogen atoms are removed by coenzymes, which are then reduced. These hydrogen atoms are ultimately the source of most of the cell's energy gain. The carbon and oxygen atoms are removed and released as carbon dioxide in a process called **decarboxylation** (dē-kar-boks-i-LĀ-shun). In total, each molecule of pyruvate in this cycle produces 3 molecules of CO_2, 3 molecules of H_2O, 4 molecules of NADH, 1 molecule of $FADH_2$, and 1 molecule of GTP (equivalent to ATP in terms of energy). Let's look at the cycle in detail.

Upon entering the mitochondrion, a pyruvate molecule participates in a complex reaction involving NAD and another coenzyme called **coenzyme A (CoA)**. This reaction yields 1 molecule of CO_2, 1 of NADH, and 1 of **acetyl-CoA** (a-SĒ-til KŌ-a)—a 2-carbon **acetyl group** (CH_3CO) bound to coenzyme A.

This sets the stage for a sequence of enzymatic reactions called the **citric acid cycle**. This cycle is also known as the *tricarboxylic* (trī-kar-bok-SIL-ik) *acid cycle* (TCA cycle) and the *Krebs cycle*. Because citric acid is the first substrate of the cycle, we will use citric acid cycle as the preferred term. The function of the citric acid cycle is to remove hydrogen atoms from organic molecules and transfer them to coenzymes. The overall pattern of the citric acid cycle is shown in Figure 25–3. (At physiological pH, the organic acids exist in their dissociated, ionized form.)

At the start of the citric acid cycle, the 2-carbon acetyl group carried by CoA is transferred to a 4-carbon oxaloacetate molecule to make the 6-carbon compound citric acid (citrate). Coenzyme A is released intact and can thus bind another acetyl group. A complete revolution or "turn" of the citric acid cycle removes 2 carbon atoms, regenerating the 4-carbon chain. (This is why the reaction sequence is called a *cycle*.) In the citric acid cycle, the coenzymes NAD and FAD remove hydrogen atoms from organic molecules, becoming reduced to NADH and $FADH_2$, respectively. We can summarize the fate of the atoms in the acetyl group as follows:

- The 2 carbon atoms are removed in enzymatic reactions that incorporate 4 oxygen atoms and form 2 molecules of carbon dioxide, a waste product.
- The hydrogen atoms are removed by the coenzymes NAD or FAD (see Figure 25–3).

Several of the steps in a turn of the citric acid cycle involve more than one reaction and require more than one enzyme. Water molecules are tied up in two of those steps. We can summarize the entire sequence as follows:

$$CH_3CO - CoA + 3NAD + FAD + GDP + P_i + 2H_2O \longrightarrow CoA + 2CO_2 + 3NADH + FADH_2 + 2H^+ + GTP$$

The only immediate energy benefit of one turn of the citric acid cycle is the formation of a single molecule of GTP (*guanosine triphosphate*) from GDP (*guanosine diphosphate*) and P_i. In practical terms, GTP is the equivalent of ATP, because GTP readily transfers a phosphate group to ADP, producing ATP:

$$GTP + ADP \longrightarrow GDP + ATP$$

The formation of GTP from GDP in the citric acid cycle is an example of **substrate-level phosphorylation**. In this process, an enzyme uses the energy released by a chemical reaction to transfer a phosphate group to a suitable acceptor molecule. GTP is formed in the citric acid cycle, but many reaction pathways in the cytosol phosphorylate ADP and form ATP directly. For example, the ATP produced during glycolysis is generated through substrate-level phosphorylation. Normally, however, substrate-level phosphorylation provides a relatively small amount of energy compared with *oxidative phosphorylation*, which we discuss next.

Oxidative Phosphorylation and the Generation of ATP

Oxidative phosphorylation is the generation of ATP as the result of the transfer of electrons from the coenzymes NADH and $FADH_2$ to oxygen by a sequence of electron carriers within mitochondria. The process produces more than 90 percent of the ATP used by body cells. The basis of oxidative phosphorylation is the formation of water, a very simple reaction:

$$2H_2 + O_2 \longrightarrow 2H_2O$$

Cells can easily obtain the ingredients for this reaction. Hydrogen is a component of all organic molecules, and oxygen is an atmospheric gas. The only problem is that the reaction releases a tremendous amount of energy all at once. In fact, this reaction releases so much energy that it is used to launch space shuttles into orbit. Cells cannot handle energy explosions. Instead, energy release must be gradual, as it is in oxidative phosphorylation. This powerful reaction proceeds in a series of small, enzymatically controlled oxidation–reduction reactions. These key reactions take place in the electron transport chain, a series of integral and peripheral protein complexes in the inner mitochondrial membrane. Under these controlled conditions, energy can be captured safely, and ATP generated.

Electron Transport Chain. The **electron transport chain (ETC)** is embedded in the inner mitochondrial membrane (Spotlight Figure 25–4). The ETC consists of four respiratory protein complexes, coenzyme Q, and electron carriers made up of a series of **cytochrome** (SĪ-tō-krōm; *cyto-*, cell + *chroma*, color) molecules (*b*, *c*, *a*, and a_3). Each cytochrome has two parts: a protein and a pigment. The protein is embedded in the inner membrane of a mitochondrion. The protein surrounds the pigment portion, which contains a metal ion, either iron (Fe^{3+}) or copper (Cu^{2+}).

Figure 25–3 The Citric Acid Cycle.

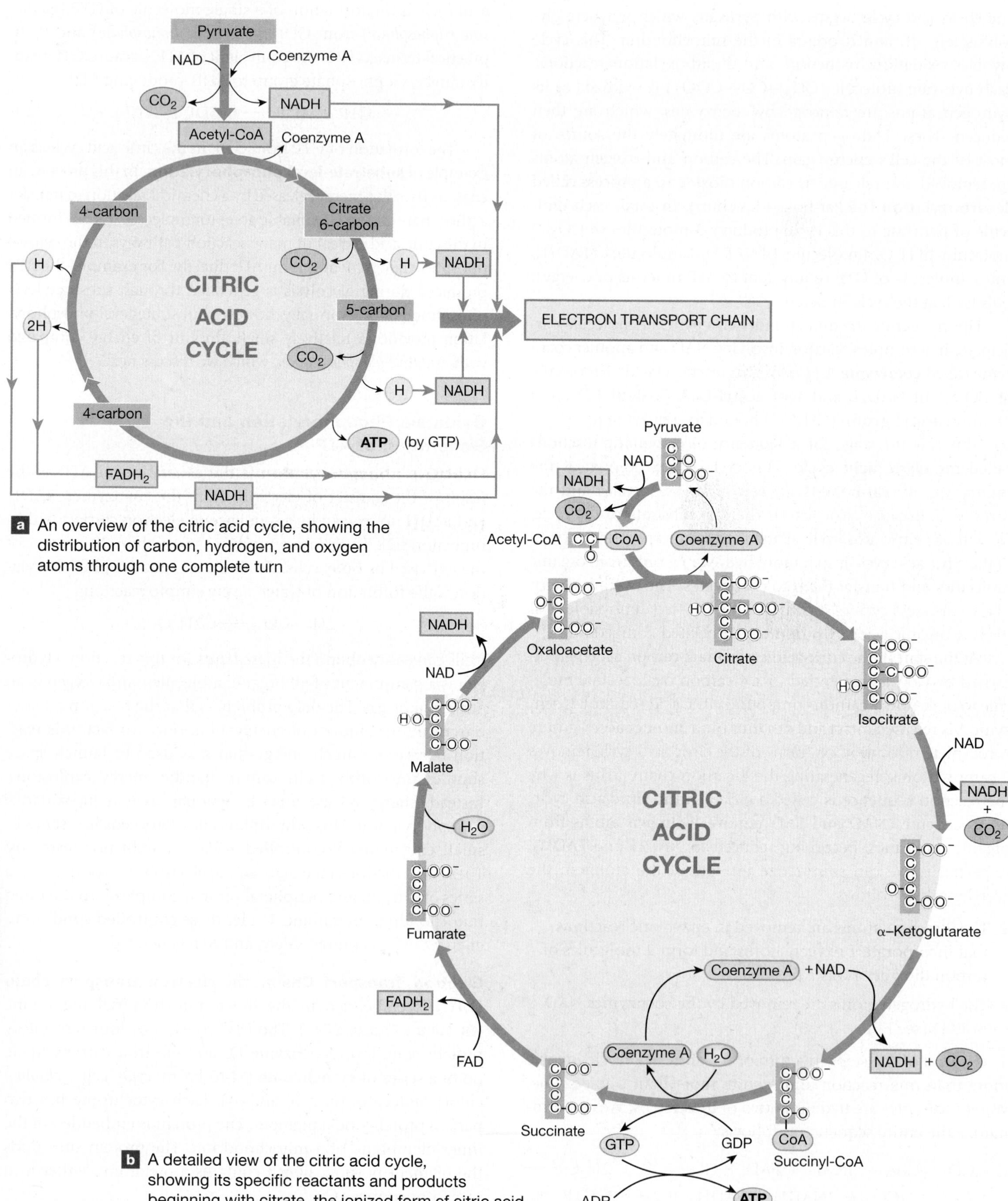
Pyruvate
NAD
Coenzyme A
CO2
NADH
Acetyl-CoA
Coenzyme A
4-carbon
Citrate 6-carbon
H
CO2
H
NADH
CITRIC ACID CYCLE
5-carbon
2H
ELECTRON TRANSPORT CHAIN
CO2
H
NADH
4-carbon
ATP (by GTP)
FADH2
NADH
Pyruvate
NAD
NADH
CO2
Acetyl-CoA
CoA
Coenzyme A
Oxaloacetate
Citrate
Isocitrate
NADH
NAD
Malate
NAD
NADH
CO2
H2O
CITRIC ACID CYCLE
Fumarate
α–Ketoglutarate
Coenzyme A
+NAD
FADH2
FAD
Coenzyme A
H2O
NADH
CO2
Succinate
GTP
GDP
CoA
Succinyl-CoA
ADP
ATP

a An overview of the citric acid cycle, showing the distribution of carbon, hydrogen, and oxygen atoms through one complete turn

b A detailed view of the citric acid cycle, showing its specific reactants and products beginning with citrate, the ionized form of citric acid

SPOTLIGHT

Figure 25–4

The Electron Transport Chain and ATP Formation

The final step in aerobic ATP production involves the indirect transfer of energy from the high-energy electrons carried by NADH and $FADH_2$ to ATP molecules. This energy transfer involves the electron transport chain and chemiosmosis along the inner mitochondrial membrane.

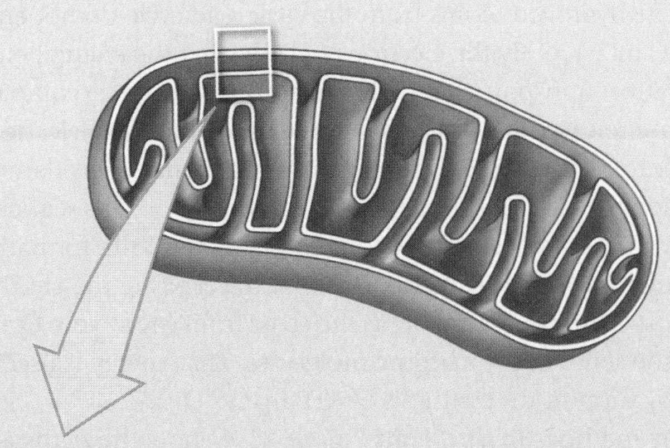

The electron transport chain consists of four respiratory complexes (I–IV), coenzyme Q, and cytochrome molecules (*b*, *c*, *a*, and a_3). Except for cytochrome c (*Cyt c*), these molecules are associated with respiratory complexes III and IV. The hydrogen atoms delivered by NADH and $FADH_2$ are first split into electrons (e^-) and protons (H^+). The high-energy electrons are passed along by redox reactions, such that their energy is released in a series of small steps. This energy is used to pump H^+ into the intermembrane space. The result is a proton (H^+) electrochemical gradient across the inner membrane. Chemiosmosis uses this proton (H^+) gradient to generate ATP through ATP synthase. The red line with the arrowhead indicates the paths and destination of the electrons.

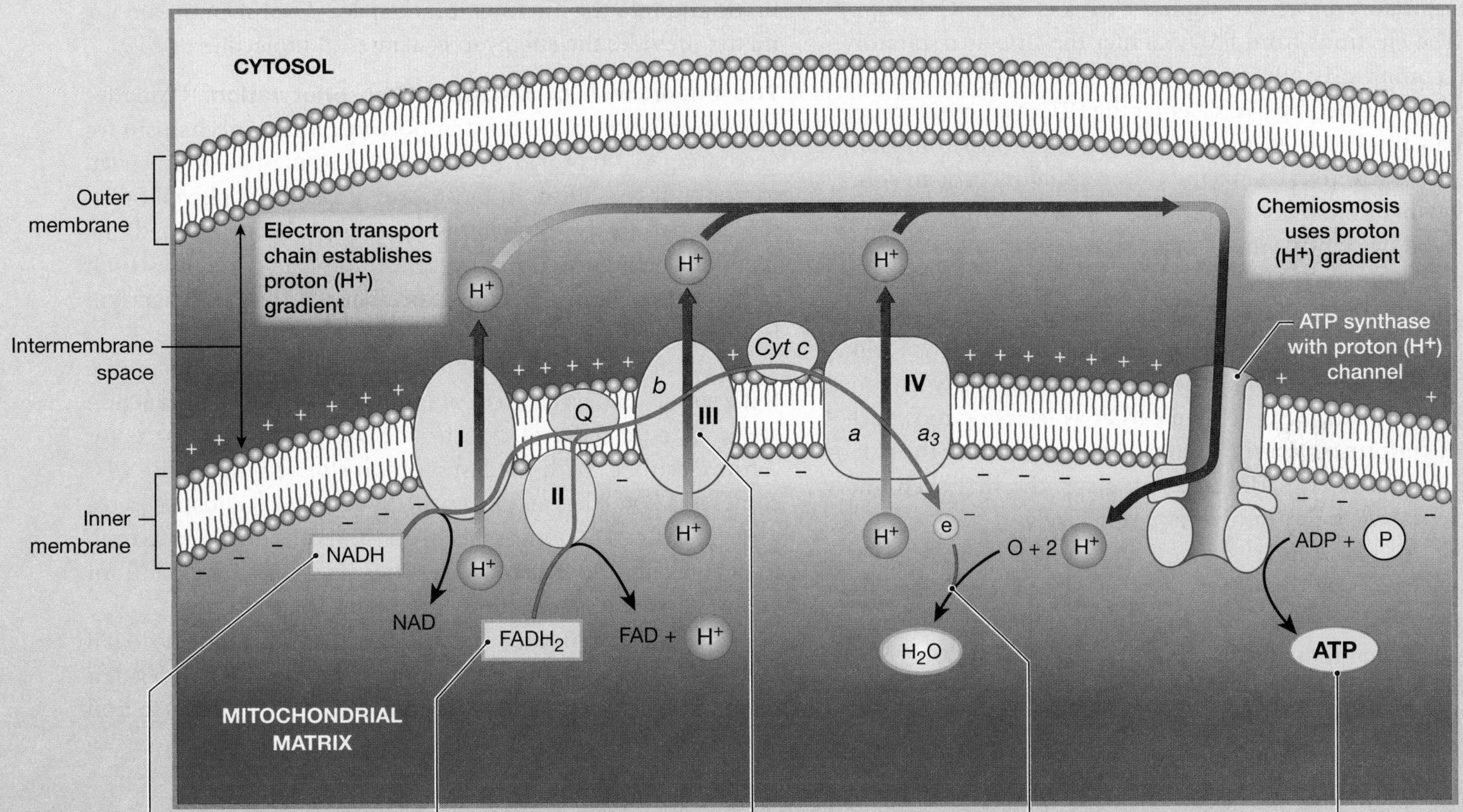

CYTOSOL

Outer membrane

Electron transport chain establishes proton (H^+) gradient

Intermembrane space

Inner membrane

Chemiosmosis uses proton (H^+) gradient

ATP synthase with proton (H^+) channel

H^+

I

II

III

IV

Q

b

Cyt c

a

a_3

NADH

NAD

$FADH_2$

FAD + H^+

e

O + 2 H^+

H_2O

ADP + P

ATP

MITOCHONDRIAL MATRIX

1 The electrons from NADH pass from respiratory complex I to coenzyme Q. The energy released by the electrons is used to pump protons (H^+) from the matrix into the intermembrane space.

2 The electrons from $FADH_2$ are transferred to respiratory complex II and then directly to coenzyme Q.

3 The electrons are transferred from coenzyme Q to respiratory complex III. More electron energy is released and used to pump H^+ into the intermembrane space.

4 The electrons transferred to respiratory complex IV release their remaining energy to pump H^+ into the intermembrane space. The electrons are released into the matrix and combine with oxygen and H^+ to form water molecules.

5 H^+ pass through hydrogen ion channels into the matrix. Their movement provides the energy for the ATP synthase complex to combine ADP and P into ATP.

The Process of Oxidative Phosphorylation. During the citric acid cycle, hydrogen atoms are passed to the coenzymes NAD and FAD. The hydrogen atoms from the citric acid cycle do not enter the ETC intact. Only the electrons (which carry the energy) enter the electron transport chain. The protons (H^+) that accompany them are released into the mitochondrial matrix. The electrons, which carry the chemical energy, enter a sequence of oxidation–reduction reactions in the electron transport chain. This sequence ends with the transfer of electrons to oxygen and the formation of a water molecule. The electrons that travel along the electron transport chain release energy as they pass from **coenzyme Q** and along the series of cytochrome molecules. That energy is used to *indirectly* support the synthesis of ATP from ADP.

The red lines in Spotlight Figure 25–4 show the paths and destinations of electrons. Those from NADH are transferred from respiratory protein complex I to coenzyme Q (*ubiquinone*). The electrons from $FADH_2$ enter the ETC at respiratory protein complex II and are passed to coenzyme Q. As a result, the electrons carried by NADH enter the ETC at a higher energy level than those carried by $FADH_2$. The electrons from both paths are passed from coenzyme Q to respiratory protein complex III, which contains cytochrome *b*. From there, the electrons are shuttled by cytochrome *c* to additional cytochromes within respiratory protein complex IV.

Notice that this reaction sequence starts with the removal of 2 hydrogen atoms from a substrate molecule and ends with the formation of water from 2 hydrogen ions and 1 oxygen ion. This is the reaction we described earlier as releasing too much energy if performed in a single step. However, because the reaction takes place in a series of small steps, the hydrogen and oxygen combine safely rather than explosively.

ATP Generation by Chemiosmosis. As we noted in Chapter 3, concentration gradients across membranes represent a form of potential energy that can be harnessed by the cell. The electron transport chain does not produce ATP directly. Instead, it creates the conditions necessary for ATP production. It creates a steep concentration gradient across the inner mitochondrial membrane. The electrons that travel along the electron transport chain release energy as they pass from coenzyme to cytochrome and from cytochrome to cytochrome. The energy released at each of several steps drives proton, or hydrogen ion, pumps. These pumps move hydrogen ions from the mitochondrial matrix into the intermembrane space between the inner and outer mitochondrial membranes. The pumps create a large electrochemical gradient for hydrogen ions across the inner membrane. This electrochemical gradient provides the energy to convert ADP to ATP.

Despite the electrochemical gradient, hydrogen ions cannot diffuse into the matrix because they are not lipid soluble. However, hydrogen ion channels in the inner membrane permit H^+ to diffuse into the matrix. These ion channels and their attached *coupling factors* make up a protein complex called **ATP synthase**. The kinetic energy of passing hydrogen ions generates ATP in a process known as **chemiosmosis** (kem-ē-oz-MŌ-sis), or *chemiosmotic phosphorylation*.

Spotlight Figure 25–4 diagrams the mechanism of ATP generation by chemiosmosis. Hydrogen ions are pumped into the intermembrane space at three places: (1) as respiratory complex I reduces coenzyme Q, (2) as cytochrome *b* reduces cytochrome *c*, and (3) as electrons are passed from cytochrome *a* to cytochrome a_3. For each pair of electrons removed from a substrate in the citric acid cycle by NAD, 6 hydrogen ions are pumped across the inner membrane of the mitochondrion and into the intermembrane space. Their reentry into the matrix provides the energy to generate 2.5 molecules of ATP. Alternatively, for each pair of electrons removed from a substrate in the citric acid cycle by FAD, 4 hydrogen ions are pumped across the inner membrane and into the intermembrane space. Their reentry into the matrix provides the energy to generate 1.5 molecules of ATP.

The Importance of Oxidative Phosphorylation. Oxidative phosphorylation is the most important cellular mechanism for generating ATP. It provides roughly 95 percent of the ATP needed to keep our cells alive. In most cases, if oxidative phosphorylation slows or stops, the cell dies. If many cells are affected, the individual may die. Oxidative phosphorylation requires both oxygen and electrons. For this reason, the rate of ATP generation is limited by the availability of either oxygen or electrons.

Cells obtain oxygen by diffusion from the extracellular fluid. If the supply of oxygen is cut off, mitochondrial ATP production stops, because reduced cytochrome a_3 will have no acceptor for its electrons. With the last reaction stopped, the entire ETC comes to a halt, like cars at a washed-out bridge. Oxidative phosphorylation can no longer take place, and cells quickly succumb to energy starvation. Because neurons have a high demand for energy, the brain is one of the first organs to be affected.

Hydrogen cyanide gas is sometimes used as a pesticide to kill rats or mice. In some states where capital punishment is legal, it is used to execute criminals. The cyanide ion (CN^-) binds to iron atoms in cytochrome a_3 and prevents the transfer of electrons to oxygen. As a result, cells die from energy starvation.

Glucose Catabolism: Energy Yield of Glycolysis and Aerobic Metabolism

For most cells, the main method of generating ATP is the complete reaction pathway that begins with glucose and ends with carbon dioxide and water. Figure 25–5 summarizes the process in terms of energy production:

- *Glycolysis.* During glycolysis, the cell gains a net 2 molecules of ATP for each glucose molecule broken down anaerobically to 2 molecules of pyruvate. Two molecules of NADH are also produced. In most cells, electrons are passed from NADH to FAD by an intermediate electron carrier in the intermembrane space, producing $FADH_2$, and then to the electron transport chain.

Figure 25–5 A Summary of the Energy Yield of Glycolysis and Aerobic Metabolism. For each glucose molecule broken down by glycolysis, only 2 molecules of ATP (net) are produced. However, glycolysis, the formation of acetyl-CoA, and the citric acid cycle all yield molecules of reduced coenzymes (NADH or $FADH_2$). Many additional ATP molecules are produced when electrons from these coenzymes pass through the electron transport chain. In most cells, the 2 NADH molecules produced in glycolysis provide another 3–5 ATP molecules. The 8 NADH molecules produced in the mitochondria yield 20 ATP molecules, for a total of 23–25. Another 3 ATP molecules are gained from the 2 $FADH_2$ molecules generated in the mitochondria. The citric acid cycle generates an additional 2 ATP molecules in the form of GTP.

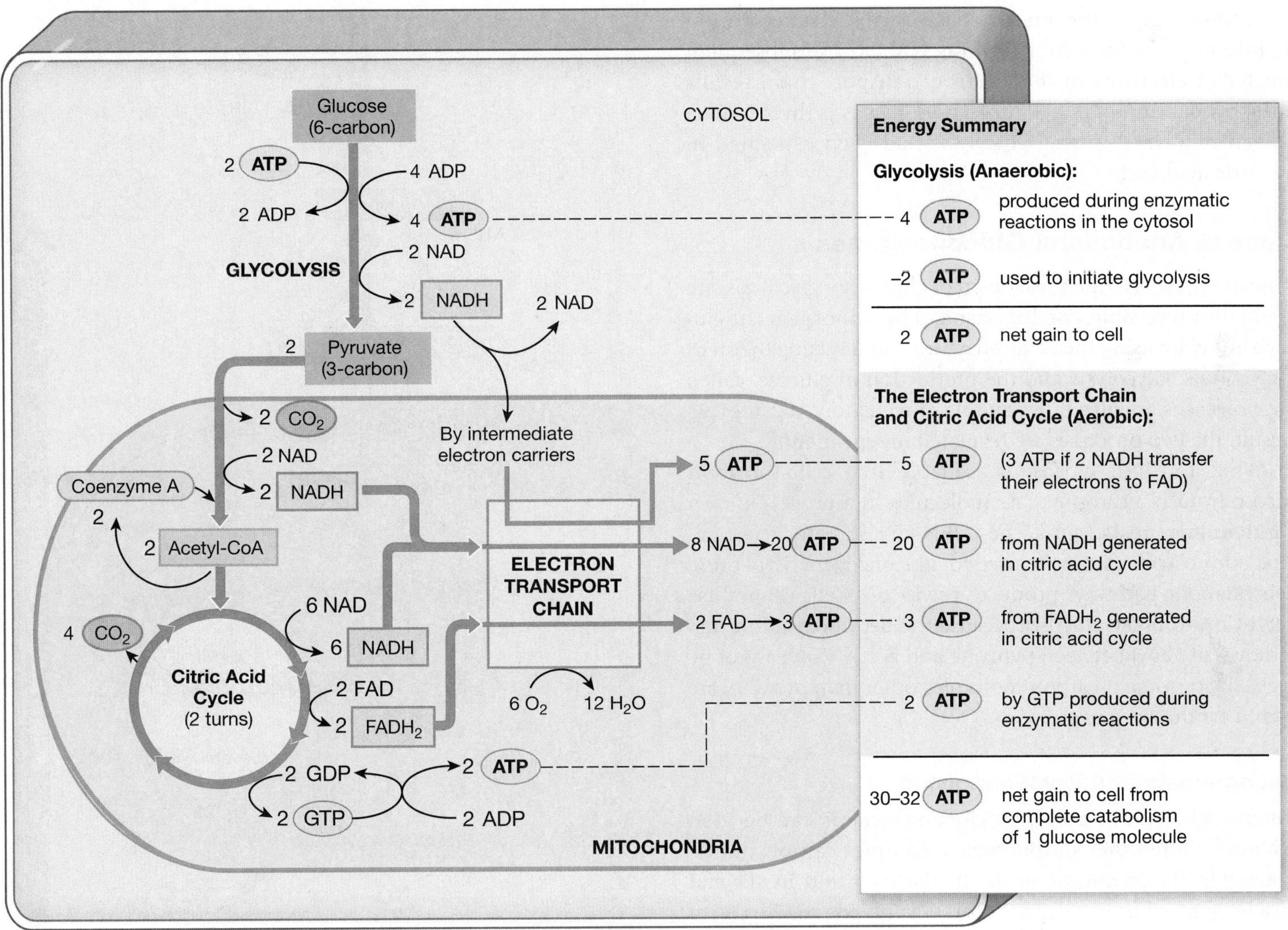

Glucose (6-carbon)
CYTOSOL
2 ATP
4 ADP
2 ADP
4 ATP
2 NAD
GLYCOLYSIS
2 NADH
2 NAD
2 Pyruvate (3-carbon)
2 CO2
By intermediate electron carriers
2 NAD
Coenzyme A
2 NADH
2
2 Acetyl-CoA
5 ATP
8 NAD→20 ATP
ELECTRON TRANSPORT CHAIN
2 FAD→3 ATP
6 NAD
4 CO2
6 NADH
Citric Acid Cycle (2 turns)
2 FAD
2 FADH2
6 O2
12 H2O
2 GDP
2 ATP
2 GTP
2 ADP
MITOCHONDRIA
Energy Summary
Glycolysis (Anaerobic):
4 ATP produced during enzymatic reactions in the cytosol
–2 ATP used to initiate glycolysis
2 ATP net gain to cell
The Electron Transport Chain and Citric Acid Cycle (Aerobic):
5 ATP (3 ATP if 2 NADH transfer their electrons to FAD)
20 ATP from NADH generated in citric acid cycle
3 ATP from FADH2 generated in citric acid cycle
2 ATP by GTP produced during enzymatic reactions
30–32 ATP net gain to cell from complete catabolism of 1 glucose molecule

? The energy produced from aerobic metabolism comes from what two sources?

- *The Citric Acid Cycle.* Each of the two revolutions of the citric acid cycle required to break down the 2 pyruvate molecules completely yields 1 molecule of ATP by way of GTP. This cycling provides an additional gain of 2 molecules of ATP. This cycle also transfers hydrogen atoms to NADH and $FADH_2$. These coenzymes provide electrons to the electron transport chain.
- *The Electron Transport Chain.* For each molecule of glucose broken down, a total of 10 NADH and 2 $FADH_2$ deliver their high-energy electrons to the electron transport chain. Each NADH yields 2.5 ATP and each $FADH_2$ yields 1.5 ATP. The 2 $FADH_2$ molecules from glycolysis yield 3 ATP molecules and 2 water molecules. Each of the 8 molecules of NADH from the citric acid cycle yields 2.5 molecules of ATP and 1 water molecule. Thus, the shuffling from the citric acid cycle to the ETC yields 23 molecules of ATP.

Summing up, for each glucose molecule processed, the cell gains 30–32 molecules of ATP: 2 from glycolysis, 3–5 from the NADH generated in glycolysis, 2 from the citric acid cycle (by means of GTP), and 23 from the electron transport chain. All but 2 of them are produced in the mitochondria.

Cardiac muscle cells and liver cells are able to gain an additional 2 ATP molecules for each glucose molecule broken down. They do so by increasing the energy yield from the 2 NADH generated during glycolysis. Although the NADH produced during glycolysis does not enter mitochondria in these cells, each NADH molecule passes its electrons to an intermediate molecule that can cross the mitochondrial membrane. Inside the matrix, NAD strips electrons from the intermediate molecule, forming NADH. The subsequent transfer of electrons to the electron transport chain results in the production of 5 ATP molecules, just as if the 2 NADH molecules formed during glycolysis had been generated in the citric acid cycle.

Glucose Anabolism: Gluconeogenesis

Some of the steps in the breakdown of glucose, or glycolysis, are essentially irreversible. For this reason, cells cannot generate glucose simply by using the same enzymes and reversing the steps of glycolysis. Glycolysis and the production of glucose, called *gluconeogenesis,* involve different sets of regulatory enzymes. As a result, the two processes are regulated independently.

When too little glucose is available, liver cells can form glucose from noncarbohydrate molecules in a process known as **gluconeogenesis** (glū-kō-nē-ō-JEN-eh-sis). Fatty acids and other amino acids cannot be used for gluconeogenesis, because their catabolic pathways produce acetyl-CoA. Cells cannot use acetyl-CoA to make glucose, because the decarboxylation step (removal of CO_2) between pyruvate and acetyl-CoA cannot be reversed. So, some 3-carbon molecules other than pyruvate are used to synthesize glucose.

Glycogenesis and Glycogenolysis

Glucose molecules formed by gluconeogenesis can be used to manufacture other simple sugars, complex carbohydrates, proteoglycans, or nucleic acids. In the liver and in skeletal muscle, glucose molecules are stored as **glycogen**. Glycogen is an important energy reserve that can be broken down when the cell cannot obtain enough glucose from interstitial fluid. Glycogen molecules are large, but glycogen reserves take up very little space because they form compact, insoluble granules.

When the amount of available glucose exceeds cellular needs, cells can form glycogen from excess glucose in a process known as **glycogenesis** (glī-kō-JEN-e-sis). It involves several steps and requires the high-energy compound *uridine triphosphate (UTP)*. The breakdown of glycogen to monomers of glucose is called **glycogenolysis** (glī-kō-je-NOL-i-sis). It takes place quickly and involves a single enzymatic step (Figure 25–6).

Figure 25–6 Glycolysis and Gluconeogenesis. Many of the reactions are freely reversible, but separate regulatory enzymes control the key steps, which are indicated by colored arrows. Some amino acids, carbohydrates, lactate, and glycerol can be converted to glucose. The enzymatic reaction that converts pyruvate to acetyl-CoA cannot be reversed.

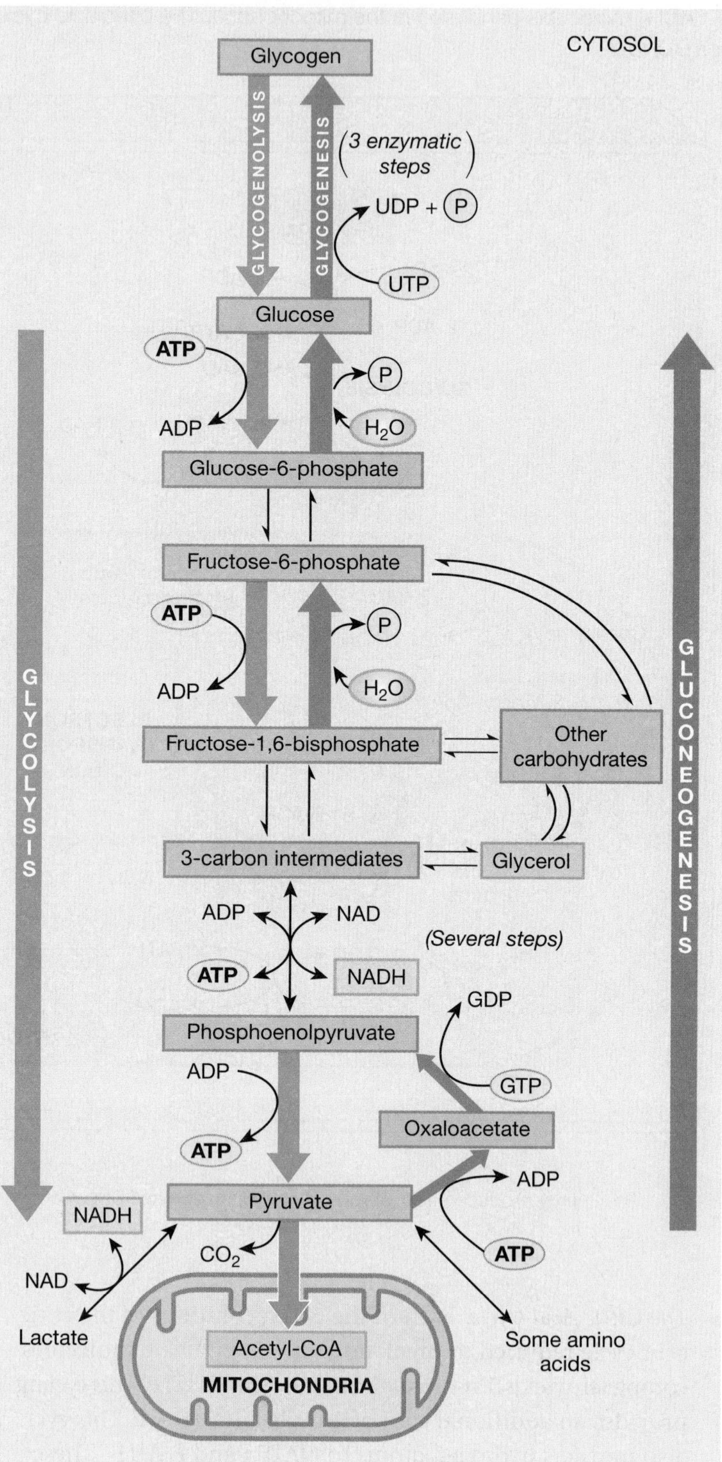
CYTOSOL
Glycogen
GLYCOGENOLYSIS
GLYCOGENESIS
(3 enzymatic steps)
UDP + P
UTP
Glucose
ATP
ADP
P
H2O
Glucose-6-phosphate
Fructose-6-phosphate
ATP
ADP
P
H2O
Fructose-1,6-bisphosphate
Other carbohydrates
3-carbon intermediates
Glycerol
ADP
NAD
ATP
NADH
(Several steps)
GDP
Phosphoenolpyruvate
ADP
GTP
ATP
Oxaloacetate
ADP
Pyruvate
NADH
ATP
NAD
CO2
Lactate
Acetyl-CoA
MITOCHONDRIA
Some amino acids
GLYCOLYSIS
GLUCONEOGENESIS

+ Clinical Note Carbohydrate Loading

Eating carbohydrates just before exercise does not improve your performance. In fact, it can decrease your endurance by slowing the mobilization of existing energy reserves. Runners or swimmers preparing for lengthy endurance events, such as a marathon or 5K swim, do not eat immediately before participating. For 2 hours before the event, they also limit their intake to just drinking water.

However, performance in endurance sports improves if muscles have large stores of glycogen. Endurance athletes try to build these stores by eating carbohydrate-rich diets for 3 days before competing. This practice is called *carbohydrate loading*. Studies in Sweden, Australia, and South Africa have recently shown that attempts to deplete stores by exercising to exhaustion before carbohydrate loading, a practice called *carbohydrate depletion/loading*, are less effective than 3 days of rest or minimal exercise during carbohydrate loading. The less intense approach improves mood and reduces the risks of muscle and kidney damage.

Tips & Tools

To make several similar-sounding terms easier to tell apart, learn these word parts: *genesis* means "the formation of," *lysis* means "a loosening," *glyco* and *gluco* refer to glucose, and *neo* means "new." So, ***gluconeogenesis*** is the formation of new glucose, ***glycogenesis*** is the formation of glycogen (the storage form of glucose), ***glycogenolysis*** is the breakdown of glycogen to glucose, and ***glycolysis*** is the breakdown of glucose to pyruvate.

✓ Checkpoint

3. **What is the primary role of the citric acid cycle in the production of ATP?**
4. **What process in the mitochondrion provides the conditions necessary for the synthesis of ATP by chemiosmosis?**
5. **How would a decrease in the level of NAD in the cytosol affect ATP production in mitochondria?**

See the blue Answers tab at the back of the book.

25-3 Lipid metabolism provides long-term storage and release of energy

Learning Outcome **Describe the pathways involved in lipid metabolism, and summarize the mechanisms of lipid transport and distribution.**

Like carbohydrates, lipid molecules contain carbon, hydrogen, and oxygen, but the atoms are present in different proportions. **Triglycerides** are the most abundant lipid in the body, so our discussion focuses on pathways for triglyceride breakdown and synthesis (see Figure 25–1).

Lipid Catabolism: Lipolysis

Lipid metabolism involves the breakdown of fats into glycerol, fatty acids, and monoglycerides. During lipid catabolism, or **lipolysis**, lipids are broken down into pieces that can be either converted to pyruvate or channeled directly into the citric acid cycle (Figure 25–7). A triglyceride is split into its component parts by hydrolysis, yielding 1 molecule of glycerol and 3 fatty acid molecules. This catabolism occurs under the influence of growth hormone and thyroid hormone.

Conversion of Glycerol

Enzymes in the cytosol convert glycerol to pyruvate. Assuming the presence of sufficient oxygen, the pyruvate is converted to acetyl-CoA and then enters the citric acid cycle.

Beta-Oxidation of Fatty Acids

The catabolism of fatty acids involves a process that generates acetyl-CoA directly. In the process of **beta-oxidation**, fatty acid molecules are broken down in a sequence of reactions into 2-carbon acetic acid fragments, and FAD and NAD^+ are reduced. Figure 25–7 diagrams one step in the process of beta-oxidation. Each step generates molecules of acetyl-CoA, NADH, and $FADH_2$, and leaves a shorter carbon chain bound to coenzyme A. This process takes place inside mitochondria, so the acetyl-CoA produced can enter the citric acid cycle immediately.

Beta-oxidation has substantial energy benefits. For example, 9 acetyl-CoA can be produced from one 18-carbon fatty acid. During the creation of an acetyl-CoA from a fatty acid, 1 NADH and 1 $FADH_2$ result. For an 18-carbon fatty acid, eight steps of beta-oxidation result in 9 acetyl-CoA and the formation of 8 NADH and 8 $FADH_2$. The breakdown of one acetyl-CoA yields 3 NADH, 1 GTP/ATP, and 1 $FADH_2$ in the citric acid cycle. So, the breakdown of 9 acetyl-CoA yields 27 NADH, 9 GTP/ATP, and 9 $FADH_2$. Recall that each NADH produces 2.5 ATP and each $FADH_2$ produces 1.5 ATP. The 35 NADH and 17 $FADH_2$ produce 113 ATP. Adding the 9 GTP/ATP makes a total of 122 ATP. Although one 1 ATP was required for activation, it counts as 2 ATP, because the result was an AMP, not an ADP. The cell can therefore gain 120 ATP molecules from the breakdown of one 18-carbon fatty acid molecule. By comparison, this yield of ATP molecules is almost 1.3 times the energy obtained by the complete breakdown of three 6-carbon glucose molecules (90–96 ATP). The catabolism of other lipids follows similar patterns, generally ending with the formation of acetyl-CoA.

Figure 25–7 Lipolysis and Beta-Oxidation. Triglycerides are broken down into 1 glycerol and 3 fatty acid molecules. During beta-oxidation, the carbon chains of fatty acids are broken down to yield molecules of acetyl-CoA, which can be used in the citric acid cycle. The reaction also donates hydrogen atoms to coenzymes, which deliver them to the electron transport chain.

Triglyceride
Absorption through endocytosis
CYTOSOL
Beta-oxidation
1 Lysosomal enzymes break down triglyceride molecules into 1 glycerol molecule and 3 fatty acids.
Fatty acid (18-carbon)
Glycerol
2 ATP
2 In the cytosol, the glycerol is converted to pyruvate through the glycolysis pathway.
MITOCHONDRIA
3 Fatty acids are absorbed into the mitochondria.
Fatty acid (18-carbon)
Pyruvate
Coenzyme A
CO_2
Coenzyme A
ATP
AMP + 2 P
Fatty acid (18-carbon) – CoA
Coenzyme A
NADH
$FADH_2$
Fatty acid (16-carbon) – CoA
Acetyl-CoA
Acetyl-CoA
13 ATP
Citric acid cycle
Coenzymes
Electron transport chain
O_2
H_2O
CO_2
4 An enzymatic reaction then breaks off the first 2 carbons as acetyl-CoA while leaving a shorter fatty acid bound to the second molecule of coenzyme A.

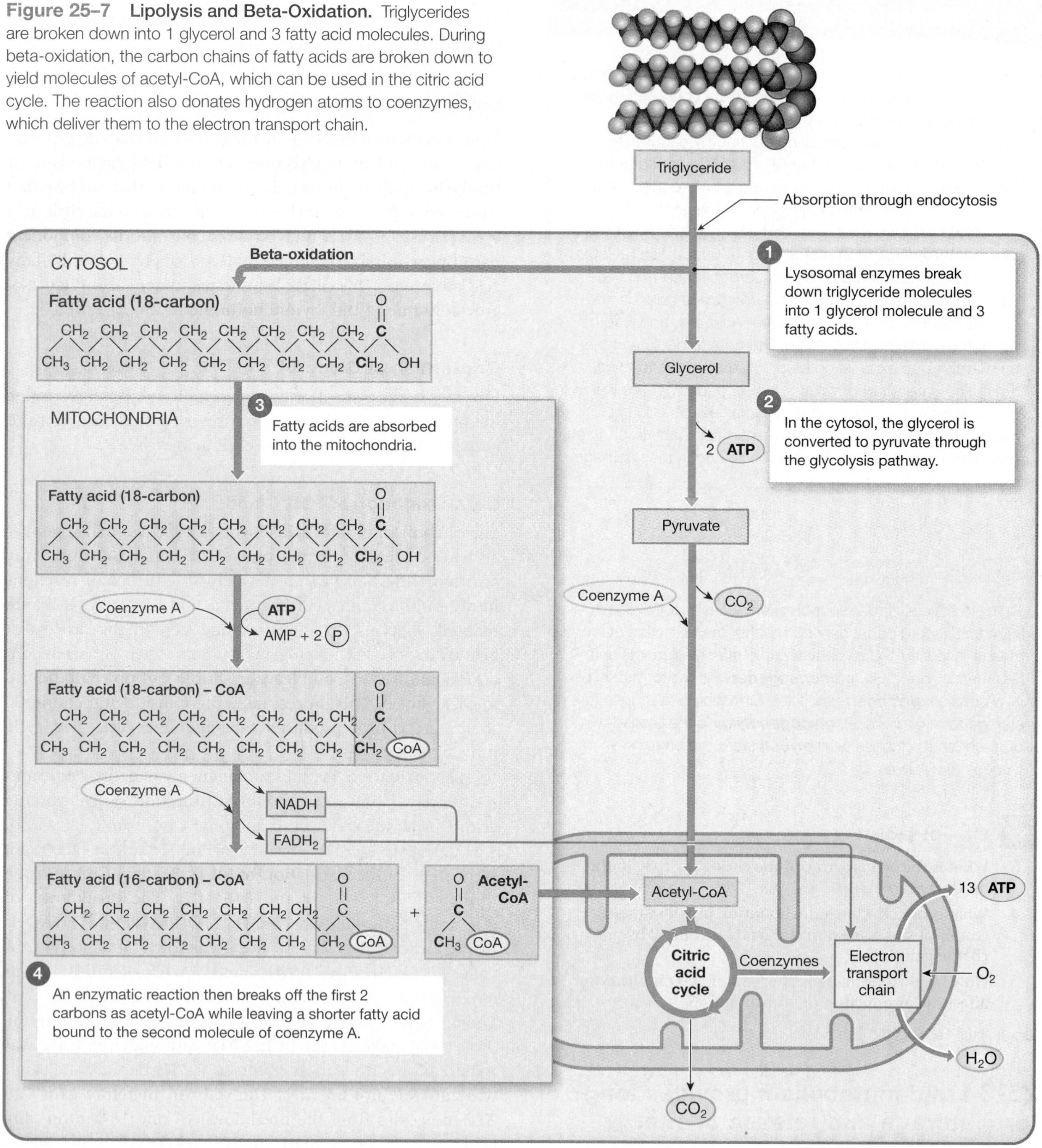

? Where does beta-oxidation take place?

Lipid Anabolism: Lipogenesis

Lipogenesis (lip-ō-JEN-e-sis) is the synthesis of lipids (see Figure 25–1). It is stimulated by a diet high in carbohydrates and several hormones, including insulin. Glycerol is synthesized from *dihydroxyacetone phosphate*, one of the 3-carbon intermediate products shared by the pathways of glycolysis and gluconeogenesis. The synthesis of most types of lipids, including nonessential fatty acids and steroids, begins with acetyl-CoA. Lipogenesis can use almost any organic substrate, because lipids, amino acids, and carbohydrates can be converted to acetyl-CoA. In other words, the body can turn just about anything we eat into fat.

Fatty acid synthesis involves a reaction sequence quite distinct from that of beta-oxidation. Body cells cannot *build* every fatty acid they can break down. For example, **linoleic acid** and **linolenic acid** are both 18-carbon unsaturated fatty acids synthesized by plants. They cannot be synthesized in the human body. They are considered **essential fatty acids**, because they must be included in your diet. These fatty acids are also needed to synthesize prostaglandins and some of the phospholipids in plasma membranes throughout the body.

Lipid Storage and Energy Release

Lipids are important energy reserves because their breakdown provides large amounts of ATP. Lipids can be stored in compact droplets in the cytosol because they are insoluble in water. This storage method saves space, but when the lipid droplets are large, it is difficult for water-soluble enzymes to get at them. For this reason, lipid reserves are more difficult to access than carbohydrate reserves. Also, most lipids are processed inside mitochondria, and mitochondrial activity is limited by the availability of oxygen.

The net result is that lipids cannot provide large amounts of ATP quickly. However, cells with modest energy demands can shift to lipid-based energy production when glucose supplies are limited. Skeletal muscle fibers normally cycle between lipid metabolism and carbohydrate metabolism. At rest (when energy demands are low), these cells break down fatty acids. During activity (when energy demands are high and immediate), skeletal muscle fibers shift to glucose metabolism.

Lipid Transport and Distribution

Like glucose, lipids are needed throughout the body. For example, all cells need lipids to maintain their plasma membranes, and important steroid hormones must reach target cells in many different tissues. However, free fatty acids make up only a small percentage of the total circulating lipids. Because most lipids are not soluble in water, special transport mechanisms carry them from one region of the body to another. Most lipids, especially *cholesterol*, circulate through the bloodstream as *lipoproteins* (Figure 25–8).

Free Fatty Acids

Free fatty acids (FFAs) are lipids that can diffuse easily across plasma membranes. In the blood, free fatty acids are generally bound to albumin, the most abundant plasma protein. Sources of free fatty acids in the blood include the following:

- Fatty acids that are not used in the synthesis of triglycerides, but diffuse out of the intestinal epithelium and into the blood
- Fatty acids that diffuse out of lipid reserves (such as those in the liver and adipose tissue) when triglycerides are broken down.

Liver cells, cardiac muscle cells, skeletal muscle fibers, and many other body cells can metabolize free fatty acids. These lipids are an important energy source during periods of starvation, when glucose supplies are limited.

Lipoproteins and Cholesterol

Lipoproteins are lipid–protein complexes that contain large insoluble glycerides and **cholesterol**. A superficial coating of phospholipids and proteins makes the entire complex soluble. Exposed proteins of the complexes bind to specific membrane receptors. For this reason, these membrane proteins determine which cells absorb the associated lipids.

Lipoproteins are usually classified into the following four major groups according to size and the relative proportions of lipid and protein:

- *Chylomicrons.* ***Chylomicrons*** (kī-lō-MĪ-kronz) are the largest lipoproteins, ranging in diameter from 0.03 to 0.5 μm. About 95 percent of the weight of a chylomicron consists of triglycerides. They are produced by intestinal epithelial cells from the fats in food. ⮌ p. 916 Chylomicrons carry absorbed lipids from the intestinal tract into the lymph and then to the bloodstream.

The liver is the primary source of all the other groups of lipoproteins, which shuttle lipids among various tissues:

- *Very Low Density Lipoproteins (VLDLs).* Very low density lipoproteins contain triglycerides manufactured by the liver, plus small amounts of phospholipids and cholesterol. The primary function of VLDLs is to transport these triglycerides to skeletal muscles and adipose tissues. The VLDLs range in diameter from 25 to 75 nm (0.025–0.075 μm).
- *Low-Density Lipoproteins (LDLs).* Low-density lipoproteins contain cholesterol, lesser amounts of phospholipids, and very few triglycerides. These lipoproteins are about 25 nm in diameter. They deliver cholesterol to peripheral tissues. LDL cholesterol is often called "bad cholesterol" because the cholesterol may wind up in arterial plaques.
- *High-Density Lipoproteins (HDLs).* **High-density lipoproteins** have roughly equal amounts of lipid and protein. The lipids

Figure 25–8 **Lipid Transport and Use.**

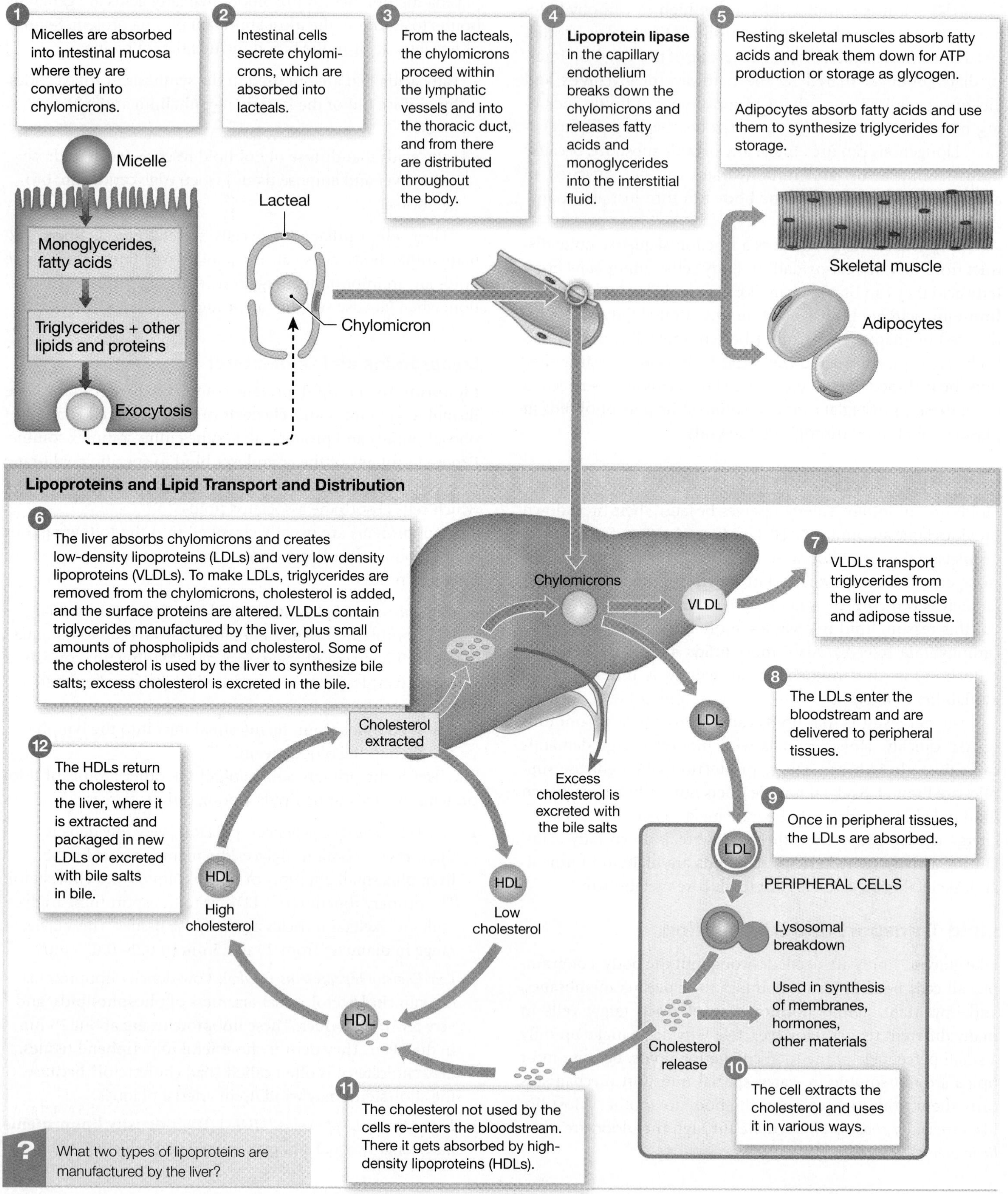

1 Micelles are absorbed into intestinal mucosa where they are converted into chylomicrons.
2 Intestinal cells secrete chylomicrons, which are absorbed into lacteals.
3 From the lacteals, the chylomicrons proceed within the lymphatic vessels and into the thoracic duct, and from there are distributed throughout the body.
4 **Lipoprotein lipase** in the capillary endothelium breaks down the chylomicrons and releases fatty acids and monoglycerides into the interstitial fluid.
5 Resting skeletal muscles absorb fatty acids and break them down for ATP production or storage as glycogen.
Adipocytes absorb fatty acids and use them to synthesize triglycerides for storage.
Micelle
Monoglycerides, fatty acids
Triglycerides + other lipids and proteins
Exocytosis
Lacteal
Chylomicron
Skeletal muscle
Adipocytes
Lipoproteins and Lipid Transport and Distribution
6 The liver absorbs chylomicrons and creates low-density lipoproteins (LDLs) and very low density lipoproteins (VLDLs). To make LDLs, triglycerides are removed from the chylomicrons, cholesterol is added, and the surface proteins are altered. VLDLs contain triglycerides manufactured by the liver, plus small amounts of phospholipids and cholesterol. Some of the cholesterol is used by the liver to synthesize bile salts; excess cholesterol is excreted in the bile.
Chylomicrons
VLDL
7 VLDLs transport triglycerides from the liver to muscle and adipose tissue.
LDL
8 The LDLs enter the bloodstream and are delivered to peripheral tissues.
Cholesterol extracted
Excess cholesterol is excreted with the bile salts
9 Once in peripheral tissues, the LDLs are absorbed.
LDL
PERIPHERAL CELLS
12 The HDLs return the cholesterol to the liver, where it is extracted and packaged in new LDLs or excreted with bile salts in bile.
HDL
High cholesterol
HDL
Low cholesterol
Lysosomal breakdown
Used in synthesis of membranes, hormones, other materials
HDL
Cholesterol release
10 The cell extracts the cholesterol and uses it in various ways.
11 The cholesterol not used by the cells re-enters the bloodstream. There it gets absorbed by high-density lipoproteins (HDLs).

? What two types of lipoproteins are manufactured by the liver?

25

Clinical Note Dietary Fats and Cholesterol

Due to concerns about cholesterol, such phrases as "low cholesterol," "contains no cholesterol," and "cholesterol free" are now widely used in food packaging. Cholesterol content alone, however, does not tell the entire story:

- *Cholesterol has many vital functions in the human body.* Cholesterol serves as a waterproofing for the epidermis, a lipid component of all plasma membranes, a key constituent of bile, and the precursor of several hormones (estrogen and testosterone, for example) and vitamin D.
- *The cholesterol content of the diet is not the only source of circulating cholesterol.* The liver can manufacture cholesterol, accounting for about 20 percent of the cholesterol in the bloodstream. The rest comes from metabolism of saturated fats in the diet. If the diet contains an abundance of saturated fats, cholesterol levels in the blood rise because excess lipids are broken down to acetyl-CoA and used to synthesize cholesterol. Consequently, individuals trying to lower serum cholesterol levels through diet must also restrict other lipids, especially saturated fats.
- *Genetic factors affect each individual's cholesterol level.* If you reduce the cholesterol in your diet, your body synthesizes more to maintain cholesterol concentrations in the blood. Because individuals have different genes, their cholesterol levels can vary, even on similar diets. In virtually everyone, however, dietary restrictions can lower blood cholesterol significantly.
- *Cholesterol levels vary with age.* At age 19, males have fasting cholesterol levels (measured 8–12 hours after a meal) below 170 mg/dL, whereas females have levels below 175 mg/dL. As age increases, the cholesterol levels gradually climb. Cholesterol levels are considered unhealthy if they are higher than those of 90 percent of the population in that age group. For males, this level is 185 mg/dL at age 19 and 250 mg/dL at age 70. For females, the comparable levels are 190 mg/dL and 275 mg/dL, respectively.

The *2015-2020 Dietary Guidelines for Americans* recommends that we eat as little dietary cholesterol as possible. Healthy diets typically contain 100–300 mg of cholesterol.

are largely cholesterol and phospholipids. HDLs are about 10 nm in diameter. Their primary function is to transport excess cholesterol from peripheral tissues back to the liver for storage or excretion in the bile. HDL cholesterol is called "good cholesterol" because it is returning from peripheral tissues and does not cause circulatory problems. Actually, applying the terms *good* and *bad* to cholesterol can be misleading, for cholesterol metabolism is complex and variable. (For more details, see the Clinical Note "Dietary Fats and Cholesterol".) Figure 25–8 also shows the relationship between LDL and HDL lipoproteins and the transport of cholesterol.

Exercise may lower cholesterol by several mechanisms. For example, exercise stimulates enzymes that move LDL from the blood into the liver and exercise increases the size of the protein particles that carry cholesterol in the bloodstream. Recent studies have also shown that vigorous exercise—the equivalent of 20 miles of jogging per week—lowered the LDL level even more than moderate exercise. However, the amount of exercise a person needs to experience the cholesterol-lowering benefit is still uncertain.

Checkpoint

6. **Define *beta-oxidation*.**
7. **Identify the four major groups of lipoproteins.**
8. **Why are high-density lipoproteins (HDLs) considered beneficial to health?**

See the blue Answers tab at the back of the book.

Clinical Note Blood Testing for Fat

When ordering a blood test for cholesterol, most physicians also request information about circulating triglycerides. In fasting individuals, triglycerides are usually found within the range of 40–150 mg/dL. (After a person has consumed a fatty meal, the triglyceride level may be temporarily elevated.)

When cholesterol levels are high, or when an individual has a family history of atherosclerosis or coronary artery disease (CAD), further tests and calculations may be performed. The **HDL** level is measured, and the **LDL** level is calculated as follows:

$$\text{LDL} = \text{cholesterol} - \text{HDL} - \frac{\text{triglycerides}}{5}$$

A high total cholesterol value linked to a high LDL level spells trouble. In effect, too much cholesterol is being exported to peripheral tissues. Problems can also exist in individuals with high total cholesterol—or even normal total cholesterol—but low HDL levels (below 35 mg/dL). In such cases, excess cholesterol delivered to the tissues cannot easily be returned to the liver for excretion. In either event, the amount of cholesterol in peripheral tissues—and especially in arterial walls—is likely to increase.

For years, LDL:HDL ratios were considered valid predictors of the risk of developing atherosclerosis. Risk-factor analysis and the LDL level are now thought to be more accurate indicators. For males with more than one risk factor, many clinicians recommend dietary changes and drug therapy if the LDL level exceeds 130 mg/dL, or even 100 mg/dL, regardless of the total cholesterol or HDL level.

25-4 Protein metabolism provides amino acids and synthesizes proteins

Learning Outcome Summarize the main processes of protein metabolism, and discuss the use of protein as an energy source.

The body can synthesize 100,000 to 140,000 different proteins. They have various functions and structures. Yet, each protein contains some combination of the same 20 **amino acids**. Amino acids enter body cells by active transport, and growth hormone influences the entry of amino acids into cells. The amino acids are synthesized into proteins that function as enzymes, hormones, structural elements, and neurotransmitters. Very little protein is used as an energy source. Let's begin with amino acid catabolism.

Amino Acid Catabolism

Before protein catabolism occurs, proteins must be converted to substances that can enter the citric acid cycle. This conversion involves transamination (removal of an amino group from an amino acid and its transfer to a keto acid), deamination (removal of an amino group), and the urea cycle (producing urea from ammonium ions derived from deamination reactions and carbon dioxide). Amino acids may be converted into glucose, fatty acids, and ketone bodies. The first step in amino acid catabolism is the removal of the amino group ($-NH_2$). This step requires a coenzyme derivative of **vitamin B_6** (*pyridoxine*).

Transamination

Transamination (tranz-am-i-NĀ-shun) is a reversible chemical reaction that transfers the amino group of an amino acid to a **keto acid**. Keto acids resemble amino acids except that the second carbon binds to an oxygen atom rather than to an amino group (Figure 25–9a). This transfer converts the keto acid into an amino acid that can leave the mitochondrion and enter the cytosol. There it can be used for protein synthesis. In the process, the original amino acid becomes a keto acid that can be broken down in the citric acid cycle.

Cells perform transamination in order to synthesize many of the amino acids needed for protein synthesis. For example, cells of the liver, skeletal muscles, heart, lung, kidney, and brain are particularly active in protein synthesis, so they carry out many transamination reactions.

Deamination

Deamination (dē-am-i-NĀ-shun) prepares an amino acid for breakdown in the citric acid cycle (Figure 25–9b). Deamination is the removal of an amino group and a hydrogen atom in a reaction that generates a toxic ammonium ion (NH_4^+). Liver cells are the primary sites of deamination. They also have enzymes that remove ammonium ions by synthesizing **urea**, a fairly harmless water-soluble compound excreted in urine. The **urea cycle** is the reaction sequence that produces urea (Figure 25–9c).

Several inherited metabolic disorders result from an inability to produce specific enzymes involved in amino acid metabolism. For example, individuals with *phenylketonuria* (fen-il-kē-tō-NŪ-rē-uh), or *PKU*, have a defect in the enzyme *phenylalanine hydroxylase*. They cannot convert phenylalanine to tyrosine. This reaction is an essential step in the synthesis of norepinephrine, epinephrine, dopamine, and melanin. If the enzyme is absent, phenylalanine undergoes a transamination and makes phenylpyruvic acid, which can be detected in urine. If PKU is not detected in infancy, phenylalanine and its deaminated products build up and central nervous system development is inhibited. Severe brain damage results. The condition is common enough that a warning is printed on the packaging of products that contain phenylalanine, such as diet drinks.

Amino Acids and ATP Production

When glucose supplies are low and lipid reserves are inadequate, mitochondria can generate ATP by breaking down amino acids in the citric acid cycle. Liver cells break down internal proteins and absorb additional amino acids from the blood (see Figure 25–1). The amino acids are deaminated, and their carbon chains are sent to the mitochondria. Not all amino acids enter the cycle at the same point, however, so the ATP benefits vary. Nonetheless, the average ATP yield per gram is comparable to that of carbohydrate catabolism.

However, three factors make protein catabolism an impractical source of quick energy:

- Proteins are more difficult to break apart than are complex carbohydrates or lipids.
- One of the by-products, ammonium ions, is toxic to cells.
- Proteins form the most important structural and functional components of any cell. Extensive protein catabolism threatens homeostasis at both the cellular and system levels.

Protein Synthesis

We discussed the basic mechanism for protein synthesis in Chapter 3 (look back at Figures 3–12 and 3–13, (⮌ pp. 86, 88–89). Your body can synthesize about half of the various amino acids needed to build proteins. There are 10 **essential amino acids**, which must come from the diet. Your body cannot synthesize eight of them (*isoleucine, leucine, lysine, threonine, tryptophan, phenylalanine, valine,* and *methionine*). The other two (*arginine* and *histidine*) can be synthesized, but not in the amounts that growing children need.

Figure 25–9 Amino Acid Catabolism and Synthesis.

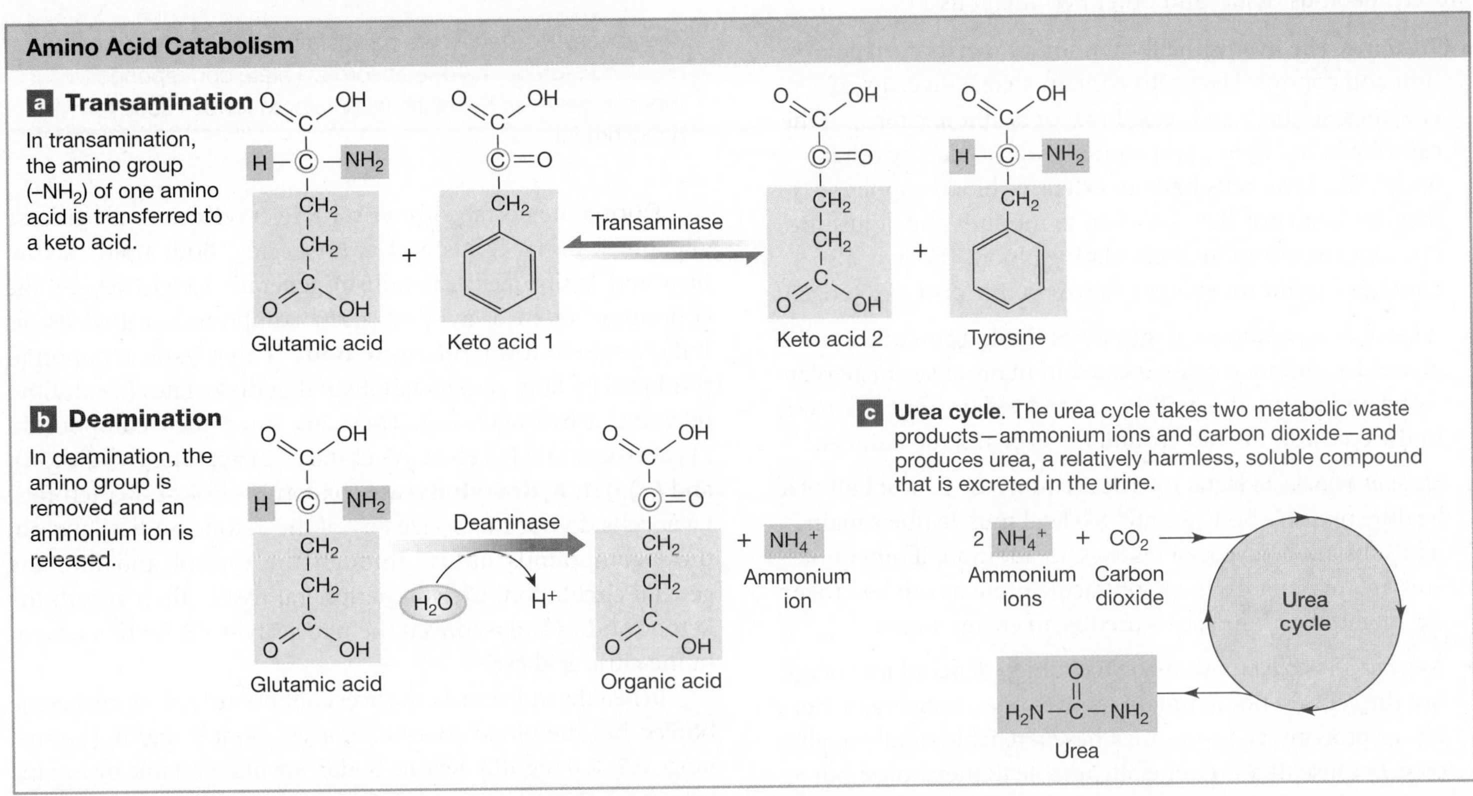
Amino Acid Catabolism
a Transamination
In transamination, the amino group (–NH2) of one amino acid is transferred to a keto acid.
Transaminase
Glutamic acid
Keto acid 1
Keto acid 2
Tyrosine
b Deamination
In deamination, the amino group is removed and an ammonium ion is released.
Deaminase
H2O
H+
Glutamic acid
Organic acid
NH4+
Ammonium ion
c Urea cycle. The urea cycle takes two metabolic waste products—ammonium ions and carbon dioxide—and produces urea, a relatively harmless, soluble compound that is excreted in the urine.
2 NH4+ + CO2
Ammonium ions
Carbon dioxide
Urea cycle
Urea
Amino Acid Synthesis
d Amination
In amination, an ammonium ion (NH4+) is used to form an amino group that is attached to a molecule, yielding an amino acid.
NH4+
H+
H2O
α–Ketoglutarate
Glutamic acid

Other amino acids are called **nonessential amino acids** because the body can make them on demand. Your body cells can readily synthesize their carbon frameworks. Then an amino group can be added by **amination**—using an ammonium ion as a reactant (Figure 25–9d).

Checkpoint

9. **Define *transamination* and *deamination*.**
10. **How would a diet that is deficient in pyridoxine (vitamin B_6) affect protein metabolism?**

See the blue Answers tab at the back of the book.

25-5 The body experiences two patterns of metabolic activity: energy storage in the absorptive state and energy release in the postabsorptive state

Learning Outcome **Differentiate between the absorptive and postabsorptive metabolic states, and summarize the characteristics of each.**

The nutrient requirements of each tissue vary with the types and quantities of enzymes present in the cytosol of cells. From a metabolic standpoint, we consider the body to have distinctive components: the liver, adipose tissue, skeletal muscle, nervous tissue, and other peripheral tissues:

- *The Liver.* The liver is the focal point of metabolic regulation and control. Liver cells contain a great diversity of enzymes, so they can break down or synthesize most of the carbohydrates, lipids, and amino acids needed by other body cells. Liver cells have an extensive blood supply, so they are in an excellent position to monitor and adjust the nutrient composition of circulating blood. The liver also contains significant energy reserves in the form of glycogen.
- *Adipose Tissue.* Adipose tissue stores lipids, primarily as triglycerides. Adipocytes are located in many areas: in areolar tissue, in mesenteries, within red and yellow bone marrows, in the epicardium, and around the eyes and the kidneys.
- *Skeletal Muscle.* Skeletal muscle accounts for almost half of a healthy person's body weight. Skeletal muscle fibers maintain substantial glycogen reserves. In addition, if other nutrients are unavailable, their contractile proteins can be broken down and the amino acids used as an energy source.
- *Nervous Tissue.* Nervous tissue has a high demand for energy, but the cells do not maintain reserves of carbohydrates, lipids, or proteins. Neurons must have a reliable supply of glucose, because they are generally unable to metabolize other molecules. If the blood glucose level becomes too low, nervous tissue in the central nervous system cannot continue to function, and the person falls unconscious.
- *Other Peripheral Tissues.* Other peripheral tissues do not maintain large metabolic reserves, but they are able to metabolize glucose, fatty acids, or other substrates. Their preferred source of energy varies according to instructions from the endocrine system.

To understand the interrelationships among these five components, let's consider events over a typical 24-hour period. During this time, the body experiences two broad patterns of metabolic activity: the *absorptive state* and the *postabsorptive state*. In the absorptive state that follows a meal, cells absorb nutrients to be used for growth, maintenance, and energy reserves. Hours later, in the postabsorptive state, metabolic reactions are focused on maintaining the blood glucose level that meets the needs of nervous tissue.

Spotlight Figure 25–10 summarizes the metabolic pathways for lipids, carbohydrates, and proteins. The diagrams show the reactions in a "typical" cell. Note, however, that no one cell can carry out all the anabolic and catabolic operations and interconversions required by the body as a whole. As cells differentiate, each type develops its own set of enzymes. These enzymes determine the cell's metabolic capabilities. In the face of such cellular diversity, homeostasis can be preserved only when the metabolic activities of tissues, organs, and organ systems are coordinated.

Tips & Tools

The absorptive state is like harvest time: Food is being gathered and stored. The postabsorptive state corresponds to the time between harvests, when stored food is used for nourishment.

During the postabsorptive state, liver cells conserve glucose and break down lipids and amino acids. Both lipid catabolism and amino acid catabolism generate acetyl-CoA. As the concentration of acetyl-CoA rises, compounds called *ketone bodies* begin to form. A **ketone body** is an organic compound produced by fatty acid metabolism that dissociates in solution, releasing a hydrogen ion. There are three such compounds: (1) **acetoacetate** (as-eh-tō-AS-eh-tāt), (2) **acetone** (AS-eh-tōn), and (3) **betahydroxybutyrate** (bā-tah-hī-droks-ē-BŪ-teh-rāt). Liver cells do not catabolize any of the ketone bodies. Instead, these compounds diffuse through the cytosol and into the general circulation. Cells in peripheral tissues then absorb the ketone bodies and reconvert them to acetyl-CoA for breakdown in the citric acid cycle.

In healthy individuals, the liver continuously produces ketone bodies, but the blood concentration is about 1 mg/dL (normal range 0.3–2.0 mg/dL). Ketone bodies are undetectable by routine urinalysis. During even a brief period of fasting, the increased production of ketone bodies results in *ketosis* (kē-TŌ-sis), a high concentration of ketone bodies in body fluids. The appearance of ketone bodies in the bloodstream—*ketonemia*—lowers blood pH, which must be controlled by buffers. During prolonged starvation, the level of ketone bodies continues to rise. Eventually, buffering capacities are exceeded and a dangerous drop in pH takes place. This acidification of the blood by ketone bodies is called ***ketoacidosis*** (kē-tō-as-ih-DŌ-sis). In severe ketoacidosis, the circulating concentration of ketone bodies can reach 200 mg/dL, and the pH may drop below 7.05. A pH that low can disrupt tissue activities and cause coma, cardiac arrhythmias, and death.

In summary, during the postabsorptive state, the liver acts to stabilize the blood glucose concentration. It does so first by the breakdown of glycogen reserves and later by gluconeogenesis. Over the remainder of the postabsorptive state, the combination of lipid and amino acid catabolism provides the necessary ATP. These processes generate large quantities of ketone bodies that diffuse into the bloodstream. Amino acid catabolism results in the release of toxic ammonium ions, which become tied up in the formation of urea.

Changes in the activity of the liver, adipose tissue, skeletal muscle, and other peripheral tissues ensure a steady supply of glucose to the nervous system. The supply remains steady despite daily or even weekly changes in nutrient availability. Only after a prolonged period of starvation does nervous tissue begin to metabolize ketone bodies and lactate molecules, as well as glucose.

✓ Checkpoint

11. **What process in the liver increases after you have eaten a high-carbohydrate meal?**
12. **Why does the blood level of urea increase during the postabsorptive state?**
13. **If a cell accumulates more acetyl-CoA than it can metabolize in the citric acid cycle, what products are likely to form?**

See the blue Answers tab at the back of the book.

25-6 Adequate nutrition allows normal physiological functioning

Learning Outcome Explain what makes up a balanced diet and why such a diet is important.

The postabsorptive state can be maintained for a considerable period. For homeostasis to be maintained indefinitely, however, the digestive tract must regularly absorb enough fluids, organic nutrients, minerals, and vitamins to keep pace with cellular demands. The absorption of nutrients from food is called **nutrition**.

The body's requirement for each nutrient varies from day to day and from person to person. *Nutritionists* attempt to analyze a diet in terms of its ability to meet the needs of a specific individual. Food supplies the necessary nutrients to ensure normal functions. Absorbed nutrients enter the bloodstream and lymph and are transported to body tissues. When a person is in good health and maintains a healthy diet, energy needs are met by absorbed nutrients. An important healthy diet consideration is *nitrogen balance*. Since proteins contain a great deal of nitrogen, nitrogen balance equals protein balance, or simply, protein anabolism equals protein catabolism. This section discusses the food groups, positive and negative nitrogen balance, and minerals and vitamins.

Food Groups and a Balanced Diet

A **balanced diet** contains all the ingredients needed to maintain homeostasis. Such a diet must include adequate essential amino acids and fatty acids, minerals and vitamins, and substrates for energy generation. In addition, the diet must include enough water to replace losses in urine, feces, and evaporation. A balanced diet prevents **malnutrition**, an unhealthy state resulting from inadequate or excessive absorption of one or more nutrients.

One way of avoiding malnutrition is to consume a balanced diet (Figure 25–11 and **Table 25–1**). The U.S. Department of Agriculture offers guidelines for healthy eating. In 2015, new dietary guidelines and the new MyPlate, MyWins Healthy Eating Solutions for Everyday Life were established. The color-coded food group slices indicate the proportions of food we should consume from **five food groups**: grains (orange), vegetables (green), fruits (red), dairy products (blue), and protein (purple) (see Figure 25–11). For more information, visit www.choosemyplate.gov/MyWins.

What is most important is that you take in nutrients in sufficient *quantity* (adequate to meet your energy needs) and *quality* (including all nutrients of a balance diet). The key is making intelligent choices about what you eat. Poor choices can lead to malnutrition even if all food groups are represented.

For example, consider the essential amino acids. ⮌ p. 956 The liver cannot synthesize any of these amino acids, so you must obtain them from your diet. Some foods in the dairy and protein groups—specifically, beef, fish, poultry, eggs, and milk—provide all the essential amino acids in sufficient quantities. They are said to contain **complete proteins**. Many plants also supply adequate *amounts* of protein but contain **incomplete proteins**, which are deficient in one or more of the essential amino acids. People who follow a *vegetarian diet*, which is largely restricted to the grains, vegetables, and fruits groups

Table 25–1 Basic Food Groups and Their General Effects on Health

Nutrient Group	Provides	Health Effects
Grains (recommended: at least half of the total eaten as whole grains)	Carbohydrates; vitamins E, thiamine, niacin, folate; calcium; phosphorus; iron; sodium; dietary fiber	Whole grains prevent rapid rise in blood glucose levels, and consequent rapid rise in insulin levels
Vegetables (recommended: especially dark-green and orange vegetables)	Carbohydrates; vitamins A, C, E, folate; dietary fiber; potassium	Reduce risk of cardiovascular disease; protect against colon cancer (folate) and prostate cancer (lycopene in tomatoes)
Fruits (recommended: a variety of fruit each day)	Carbohydrates; vitamins A, C, E, folate; dietary fiber; potassium	Reduce risk of cardiovascular disease; protect against colon cancer (folate)
Dairy (recommended: low-fat or fat-free milk, yogurt, and cheese)	Complete proteins; fats; carbohydrates; calcium; potassium; magnesium; sodium; phosphorus; vitamins A, B_{12}, pantothenic acid, thiamine, riboflavin	Whole milk: high in calories, may cause weight gain; saturated fats correlated with heart disease
Protein (recommended: lean meats, fish, poultry, eggs, dry beans, nuts, legumes)	Complete proteins; fats; calcium; potassium; phosphorus; iron; zinc; vitamins E, thiamine, B_6	Fish and poultry lower risk of heart disease and colon cancer (compared to red meat). Consumption of up to one egg per day does not appear to increase incidence of heart disease; nuts and legumes improve blood cholesterol ratios, lower risk of heart disease and diabetes

SPOTLIGHT

Figure 25–10 Absorptive and Postabsorptive States

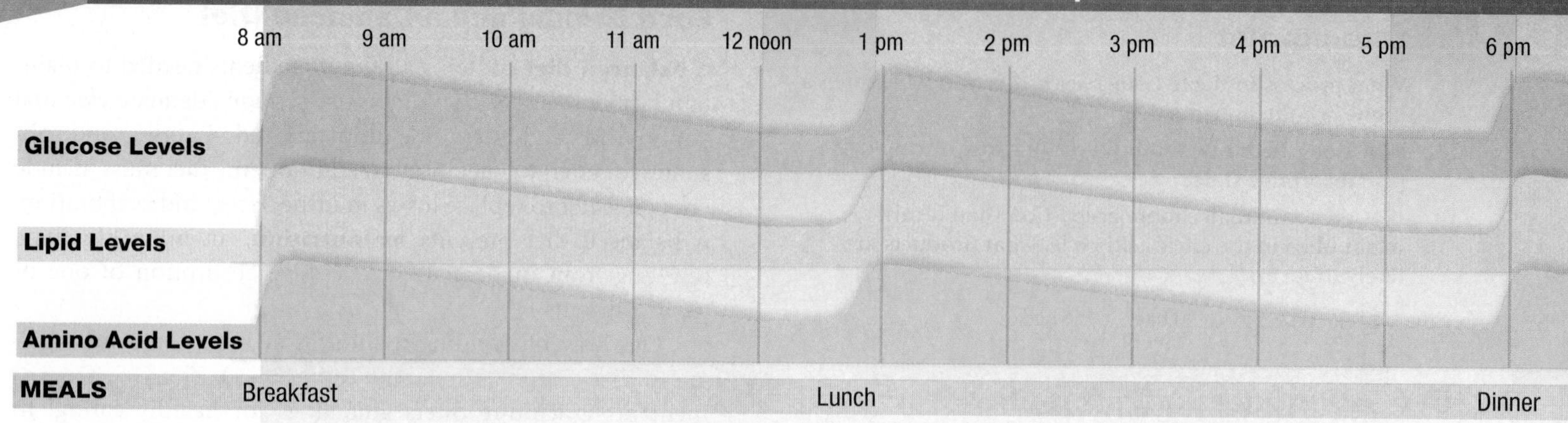
8 am
9 am
10 am
11 am
12 noon
1 pm
2 pm
3 pm
4 pm
5 pm
6 pm
Glucose Levels
Lipid Levels
Amino Acid Levels
MEALS
Breakfast
Lunch
Dinner

ABSORPTIVE STATE

The **absorptive state** is the time following a meal, when nutrient absorption is under way. Nutrient absorption was outlined in *Spotlight Figure 24–27.* After a typical meal, the absorptive state lasts for about 4 hours. If you are fortunate enough to eat three meals a day, you spend 12 out of every 24 hours in the absorptive state.

Insulin is the primary hormone of the absorptive state. Various other hormones stimulate amino acid uptake (growth hormone) and protein synthesis (growth hormone, androgens, and estrogens).

Blood glucose levels elevated
Insulin (binds to membrane receptor)
LIPIDS
CARBOHYDRATES
PROTEINS
Triglycerides
Glycogen
Proteins
Insulin
Glucose
Glycolysis
ATP
Insulin
Insulin
Androgens
Estrogens
Growth hormone
Blood lipid levels elevated
Fatty acids
Glycerol
Amino acids
Blood amino acids elevated
Insulin, Growth hormone
Pyruvate
CO_2
Insulin
KEY
Catabolic pathway
Anabolic pathway
Stimulation
Acetyl-CoA
ATP
Citric acid cycle
Coenzymes
Electron transport chain
O_2
O_2
MITOCHONDRIA
H_2O
CO_2

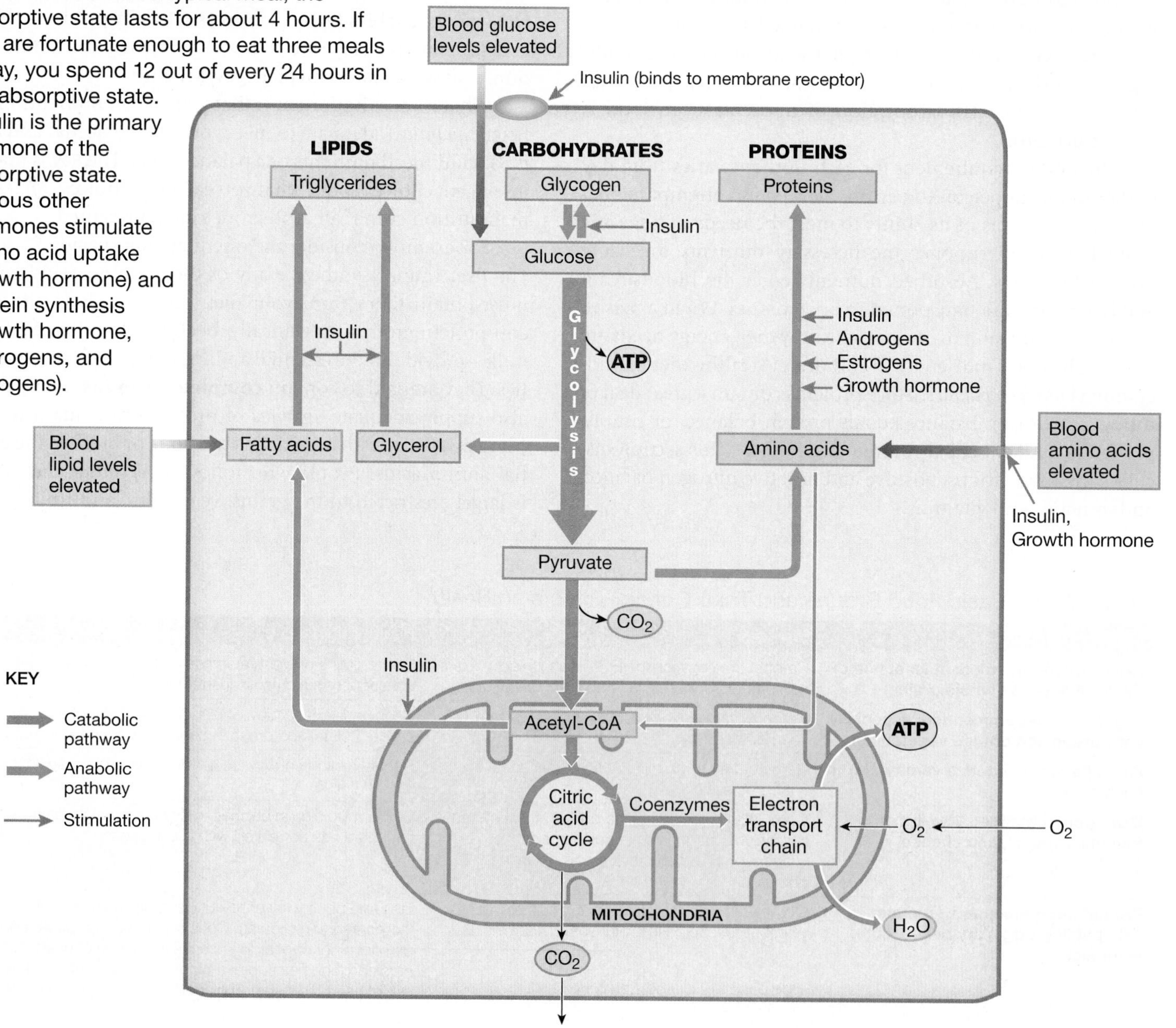

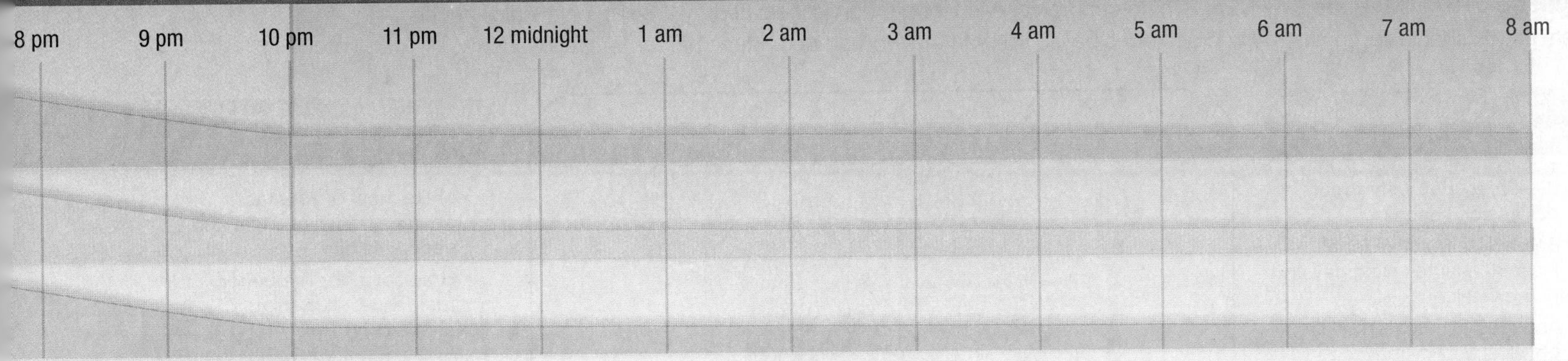
8 pm
9 pm
10 pm
11 pm
12 midnight
1 am
2 am
3 am
4 am
5 am
6 am
7 am
8 am

POSTABSORPTIVE STATE

The **postabsorptive state** is the time when nutrients are not being absorbed and your body must rely on internal energy reserves to meet its energy demands. You spend about 12 hours each day in the postabsorptive state. If you skip meals, however, you can extend that time considerably. Metabolic activity in the postabsorptive state is focused on mobilizing energy reserves and maintaining normal blood glucose levels. Several hormones coordinate these activities. These hormones include glucagon, epinephrine, glucocorticoids, and growth hormone.

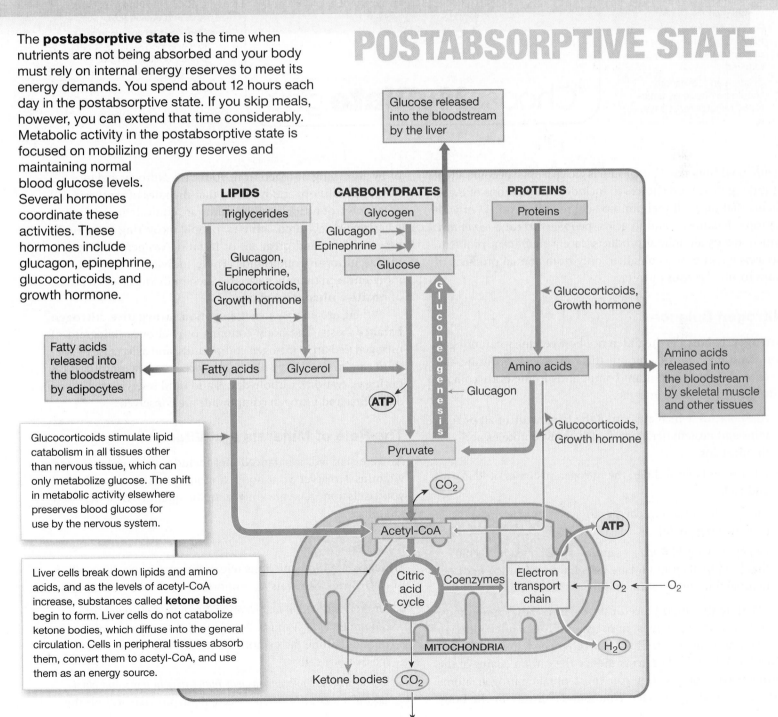

Glucose released into the bloodstream by the liver

LIPIDS

CARBOHYDRATES

PROTEINS

Triglycerides

Glycogen

Proteins

Glucagon

Epinephrine

Glucose

Glucagon, Epinephrine, Glucocorticoids, Growth hormone

Glucocorticoids, Growth hormone

Fatty acids released into the bloodstream by adipocytes

Fatty acids

Glycerol

Amino acids

Amino acids released into the bloodstream by skeletal muscle and other tissues

ATP

Gluconeogenesis

Glucagon

Glucocorticoids, Growth hormone

Pyruvate

CO_2

Acetyl-CoA

ATP

Citric acid cycle

Coenzymes

Electron transport chain

O_2

O_2

H_2O

MITOCHONDRIA

Ketone bodies

CO_2

Glucocorticoids stimulate lipid catabolism in all tissues other than nervous tissue, which can only metabolize glucose. The shift in metabolic activity elsewhere preserves blood glucose for use by the nervous system.

Liver cells break down lipids and amino acids, and as the levels of acetyl-CoA increase, substances called **ketone bodies** begin to form. Liver cells do not catabolize ketone bodies, which diffuse into the general circulation. Cells in peripheral tissues absorb them, convert them to acetyl-CoA, and use them as an energy source.

Figure 25–11 MyPlate, MyWins. The proportion of each food group is indicated by the size of its color-coded area. Foods should be consumed in proportions based on both the food group and the person's level of activity.

Find your balance between food and physical activity

- Be sure to stay within your daily calorie needs.
- Be physically active for at least 30 minutes most days of the week.
- About 60 minutes a day of physical activity may be needed to prevent weight gain.
- For sustaining weight loss, at least 60 to 90 minutes a day of physical activity may be required.
- Children and teenagers should be physically active for 60 minutes most days.

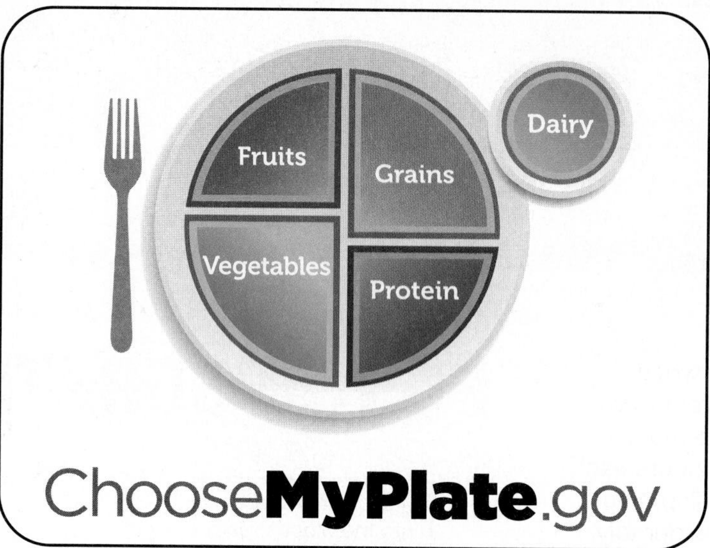
Fruits
Grains
Dairy
Vegetables
Protein
ChooseMyPlate.gov

Know the limits on fats, sugars, and salt (sodium)

- Make most of your fat sources from fish, nuts, and vegetable oils.
- Limit solid fats like butter, margarine, shortening, and lard.
- Check the nutrition facts label to keep saturated fats, trans fats, and sodium low.
- Choose food and beverages low in added sugars. Added sugars contribute calories with few, if any, nutrients.

(with or without the dairy products group), must become adept at varying their food choices to include combinations of ingredients that meet all their amino acid requirements. Even with a proper balance of amino acids, people who eat a *vegan diet*, which avoids all animal products, face a significant problem, because vitamin B_{12} is obtained only from animal products or from fortified cereals or tofu.

Nitrogen Balance

Nitrogen balance is the difference between the total nitrogen you consume and the amount that you excrete in urine and feces. A variety of important compounds in the body contain nitrogen atoms. These include:

- Amino acids, which are part of the framework of all proteins and protein derivatives, such as glycoproteins and lipoproteins
- Purines and pyrimidines, the nitrogenous bases of RNA and DNA
- *Creatine*, important in energy storage in muscle tissue (as creatine phosphate)
- *Porphyrins* (por-FĒR-inz), complex ring-shaped molecules that bind metal ions and are essential to the function of hemoglobin, myoglobin, and the cytochromes.

25

Despite the importance of nitrogen to these compounds, your body neither stores nitrogen nor maintains nitrogen reserves, as it does carbohydrates (glycogen) and lipids (triglycerides). Your body can synthesize the carbon chains of the nitrogen compounds, but you must obtain nitrogen atoms either by recycling nitrogen compounds already in the body or by absorbing nitrogen from your diet. Nitrogen balance is a normal condition, and it means that the rates of synthesis and breakdown of nitrogen compounds are equivalent.

Growing children, athletes, people recovering from an illness or injury, and pregnant or lactating women actively synthesize nitrogen compounds. These individuals must absorb more nitrogen than they excrete. Such individuals are in a state of **positive nitrogen balance**.

When excretion exceeds ingestion, a **negative nitrogen balance** exists. The body contains only about a kilogram of nitrogen tied up in nitrogen compounds, and a decrease of one-third can be fatal. Even when energy reserves are mobilized (as during starvation), carbohydrates and lipid reserves are broken down first and nitrogen compounds are conserved.

The Role of Minerals and Vitamins

As we explained, a balanced diet includes both minerals and vitamins. However, your body cannot synthesize minerals, and your cells can generate only a small quantity of a very few vitamins.

Minerals

Minerals are inorganic ions released through the dissociation of electrolytes. Minerals are important for three reasons:

- *Ions such as sodium and chloride determine the osmotic concentrations of body fluids.* Potassium ions are important in maintaining the osmotic concentration of the cytosol inside body cells.
- *Ions in various combinations play major roles in important physiological processes.* As we have seen, these processes include the

maintenance of membrane potentials; the construction and maintenance of the skeleton; muscle contraction; the generation of action potentials; the release of neurotransmitters; hormone production; blood clotting; the transport of respiratory gases; buffer systems; fluid absorption; and waste removal.

- *Ions are essential cofactors in a variety of enzymatic reactions.* For example, calcium-dependent ATPase in skeletal muscle also requires the presence of magnesium ions. Another ATPase required for the conversion of glucose to pyruvate needs both potassium and magnesium ions. Carbonic anhydrase is important in CO_2 transport, buffering systems, and gastric acid secretion. This enzyme requires the presence of zinc ions. Finally, components of the electron transport chain require an iron atom. The final cytochrome (a_3) of the electron transport chain must bind a copper ion as well.

The major minerals and a summary of their functional roles are listed in **Table 25–2**. Your body contains reserves of several important minerals, and these reserves help reduce the effects of variations in the dietary supply. Chronic dietary reductions can lead to a variety of clinical problems. Alternatively, a dietary excess of mineral ions can be equally dangerous, because storage capabilities are limited.

Problems involving iron are particularly common. The body of a healthy man contains about 3.5 g of iron in the ionic form Fe^{2+}. Of that amount, 2.5 g is bound to the hemoglobin in circulating red blood cells. The rest is stored in the liver and red bone marrow. In women, the total body iron content averages 2.4 g, with about 1.9 g incorporated into red blood cells. Thus, a woman's iron reserves consist of only 0.5 g, half that of a typical man. For this reason, if the diet contains inadequate amounts of iron, premenopausal women are more likely to develop signs of iron deficiency than are men.

Vitamins

A **vitamin** is an essential organic nutrient that functions as a coenzyme in vital enzymatic reactions. Vitamins are assigned to either of two groups based on their chemical structure and characteristics: *fat-soluble vitamins* or *water-soluble vitamins*.

Fat-Soluble Vitamins. Vitamins A, D, E, and K are the **fat-soluble vitamins**, which dissolve in lipids. You absorb these vitamins primarily from the digestive tract, along with the lipid contents of micelles. However, when exposed to sunlight, your skin can synthesize small amounts of vitamin D, and intestinal bacteria produce some vitamin K.

The mode of action of these vitamins is an area of active research. Vitamin A has long been recognized as a structural component of the visual pigment retinal. Its more general

Table 25–2 Minerals and Mineral Reserves

Mineral	Significance	Total Body Content	Primary Route of Excretion	Recommended Daily Allowance (RDA) in mg
BULK MINERALS				
Sodium	Major cation in body fluids; essential for normal membrane function	110 g, primarily in body fluids	Urine, sweat, feces	1500
Potassium	Major cation in cytoplasm; essential for normal membrane function	140 g, primarily in cytoplasm	Urine	4700
Chloride	Major anion in body fluids; functions in forming HCl	89 g, primarily in body fluids	Urine, sweat	2300
Calcium	Essential for normal muscle and neuron function and normal bone structure	1.36 kg, primarily in skeleton	Urine, feces	1000–1200
Phosphorus	In high-energy compounds, nucleic acids, and bone matrix (as phosphate)	744 g, primarily in skeleton	Urine, feces	700
Magnesium	Cofactor of enzymes, required for normal membrane functions	29 g (skeleton, 17 g; cytoplasm and body fluids, 12 g)	Urine	310–400
TRACE MINERALS				
Iron	Component of hemoglobin, myoglobin, and cytochromes	3.9 g (1.6 g stored as ferritin or hemosiderin)	Urine (traces)	8–18
Zinc	Cofactor of enzyme systems, notably carbonic anhydrase	2 g	Urine, hair (traces)	8–11
Copper	Required as cofactor for hemoglobin synthesis	127 mg	Urine, feces (traces)	0.9
Manganese	Cofactor for some enzymes	11 mg	Feces, urine (traces)	1.8–2.3
Cobalt	Cofactor for transaminations; mineral in vitamin B_{12} (cobalamin)	1.1 g	Feces, urine	0.0001
Selenium	Antioxidant	Variable	Feces, urine	0.055
Chromium	Cofactor for glucose metabolism	0.0006 mg	Feces, urine	0.02–0.035

Table 25–3 The Fat-Soluble Vitamins

Vitamin	Significance	Sources	Recommended Daily Allowance (RDA) in mg	Effects of Deficiency	Effects of Excess
A	Maintains epithelia; required for synthesis of visual pigments; supports immune system; promotes growth and bone remodeling	Leafy green and yellow vegetables	0.7–0.9	Retarded growth, night blindness, deterioration of epithelial membranes	Liver damage, skin paling, CNS effects (nausea, anorexia)
D	Required for normal bone growth, intestinal calcium and phosphorus absorption, and retention of these ions at the kidneys	Synthesized in skin exposed to sunlight	0.005–0.015*	Rickets, skeletal deterioration	Calcium deposits in many tissues, disrupting functions
E	Prevents breakdown of vitamin A and fatty acids	Meat, milk, vegetables	15	Anemia, other problems suspected	Nausea, stomach cramps, blurred vision, fatigue
K	Essential for liver synthesis of prothrombin and other clotting factors	Vegetables; production by intestinal bacteria	0.09–0.12	Bleeding disorders	Liver dysfunction, jaundice

* Unless exposure to sunlight is inadequate for extended periods and alternative sources are unavailable.

metabolic effects are not well understood. Vitamin D is ultimately converted to calcitriol. This hormone binds to cytoplasmic receptors within the intestinal epithelium and helps to increase the rate of intestinal absorption of calcium and phosphorus ions. Vitamin E stabilizes intracellular membranes. Vitamin K is essential to the synthesis of several proteins, including at least three blood clotting factors. Information about the fat-soluble vitamins is summarized in Table 25–3.

Water-Soluble Vitamins. Most of the **water-soluble vitamins** (Table 25–4) are components of coenzymes. For example, NAD is derived from niacin (vitamin B_3), FAD from vitamin B_2 (riboflavin), and coenzyme A from vitamin B_5 (pantothenic acid). The liver synthesizes niacin from the essential amino acid tryptophan.

Water-soluble vitamins are rapidly exchanged between the fluid compartments of the digestive tract and the circulating

Table 25–4 The Water-Soluble Vitamins

Vitamin	Significance	Sources	Recommended Daily Allowance (RDA) in mg	Effects of Deficiency	Effects of Excess
B_1 (thiamine)	Coenzyme in many pathways	Milk, meat, bread	1.1–1.2	Muscle weakness, CNS and cardiovascular problems, including heart disease; called *beriberi*	Hypotension
B_2 (riboflavin)	Part of FAD, involved in multiple pathways, including glycolysis and citric acid cycle	Milk, meat, eggs, and cheese	1.1–1.3	Epithelial and mucosal deterioration	Itching, tingling
B_3 (niacin)	Part of NAD, involved in multiple pathways	Meat, bread, potatoes	14–16	CNS, GI, epithelial, and mucosal deterioration; called *pellagra*	Itching, burning; vasodilation; death after large dose
B_5 (pantothenic acid)	Coenzyme A, in multiple pathways	Milk, meat	10	Retarded growth, CNS disturbances	None reported
B_6 (pyridoxine)	Coenzyme in amino acid and lipid metabolism	Meat, whole grains, vegetables, orange juice, cheese, and milk	1.3–1.7	Retarded growth, anemia, convulsions, epithelial changes	CNS alterations, perhaps fatal
B_9 (folic acid)	Coenzyme in amino acid and nucleic acid metabolism	Leafy vegetables, some fruits, liver, cereal, and bread	0.2–0.4	Retarded growth, anemia, gastrointestinal disorders, developmental abnormalities	Few noted, except at massive doses
B_{12} (cobalamin)	Coenzyme in nucleic acid metabolism	Milk, meat	0.0024	Impaired RBC production, causing *pernicious anemia*	Polycythemia
B_7 (biotin)	Coenzyme in many pathways	Eggs, meat, vegetables	0.03	Fatigue, muscular pain, nausea, dermatitis	None reported
C (ascorbic acid)	Coenzyme in many pathways	Citrus fruits	75–90; smokers add 35 mg	Epithelial and mucosal deterioration; called *scurvy*	Kidney stones

Clinical Note Vitamins

"If a little is good, a lot must be better" is a common—but dangerously incorrect—attitude about vitamins.

Too much of a vitamin can also have harmful effects. **Hypervitaminosis** (hī-per-vī-tuh-min-Ō-sis) occurs when dietary intake exceeds the body's ability to use, store, or excrete a particular vitamin. This condition usually involves one of the fat-soluble vitamins, because the excess is retained and stored in body lipids. Fat-soluble vitamins normally diffuse into plasma membranes and other lipids in the body, including the lipid inclusions in the liver and adipose tissue. As a result, your body contains a significant reserve of these vitamins. Normal metabolic operations can continue for several months after dietary sources have been cut off. For this reason, a dietary insufficiency of fat-soluble vitamins rarely causes signs and symptoms of **hypovitaminosis** (hī-pō-vi-ta-min-Ō-sis), or vitamin insufficiency. However, other factors can cause hypovitaminosis involving either fat-soluble or water-soluble vitamins. Problems may be due to an inability to absorb a vitamin from the digestive tract, inadequate storage, or excessive demand.

Clinical Note Alcohol by the Numbers

Alcohol is big business throughout the world. We see beer commercials on television, billboards advertising liquor, and TV or movie characters enjoying a drink. This cultural fondness for alcohol has serious medical consequences.

Consider these statistics provided by the Centers for Disease Control and Prevention (CDC) and the National Council on Alcoholism and Drug Dependence for people living in the United States:

- One in four children—over 7 million—grows up in a home where someone abuses alcohol.
- 17.6 million people—1 in every 12 adults—abuse alcohol.
- 79,000 deaths are attributed to alcohol use every year.
- Alcoholism is the third leading cause of lifestyle-related deaths.
- Alcohol is responsible for 60–90 percent of all liver disease in the United States.
- Women who consume 1 ounce of alcohol per day during pregnancy have a higher rate of spontaneous abortion and bear children with lower birth weights than do women who abstain.

Clinical Note Alcohol and Disease

Alcohol affects many organ systems. Major clinical symptoms of alcoholism include (1) disorientation and confusion (*nervous system*); (2) ulcers, diarrhea, and cirrhosis (*digestive system*); (3) cardiac arrhythmias, cardiomyopathy, and anemia (*cardiovascular system*); (4) depressed sexual drive and testosterone level (*reproductive system*); and (5) itching and angiomas (benign blood vessel or lymphatic vessel tumors) (*integumentary system*).

Alcohol affects developing fetuses. The toll on newborn infants has risen steadily since the 1960s as the number of female drinkers has increased. Women who drink heavily may have children with *fetal alcohol syndrome (FAS)*. Facial abnormalities, a small head, slow growth, and mental retardation characterize this irreversible condition.

Problems with alcohol are usually divided into those stemming from alcohol abuse and those involving alcoholism. The boundary between these conditions is hazy. *Alcohol abuse* is the general term for overuse and its behavioral and physical effects. *Alcoholism* is chronic alcohol abuse accompanied by the physiological changes associated with addiction to other CNS-active drugs. Alcoholism has received the most attention in recent years, but alcohol abuse—especially when combined with driving a car—is also in the spotlight. Alcohol abuse is considerably more widespread than alcoholism. The medical effects are less well documented, but they are clearly significant.

Several factors interact to produce alcoholism. The primary risk factors are sex (males are more likely to become alcoholics than are females) and a family history of alcoholism. There does appear to be a genetic component: A gene on chromosome 11 has been implicated in some inherited forms of alcoholism. The relative importance of genes versus social environment is difficult to assess. Both alcohol abuse and alcoholism probably are caused by a variety of factors.

blood. Excessive amounts are readily excreted in urine. For this reason, hypervitaminosis involving water-soluble vitamins is relatively uncommon, except among people who take large doses of vitamin supplements. Only vitamins B_{12} and C are stored in significant quantities. Insufficient intake of other water-soluble vitamins can lead to initial signs and symptoms of vitamin deficiency within a period of days to weeks.

The bacteria that live in the intestines help prevent deficiency diseases. They produce small amounts of five of the nine water-soluble vitamins, in addition to fat-soluble vitamin K. The intestinal epithelium can easily absorb all the water-soluble vitamins except B_{12}. Because the B_{12} molecule is large, it must be bound to *intrinsic factor* from the gastric mucosa before absorption can take place, as we discussed in Chapter 24. p. 904

+ Clinical Note Anorexia

Anorexia means lack of appetite. This is a common symptom of chronic disease. A patient with cancer loses his or her appetite and often loses body weight, too. The presence of anorexia can alert the health care worker to look for an underlying disease.

Anorexia nervosa, a more familiar term, refers to a specific mental health disorder in which an individual is obsessed with a desire to lose weight by refusing to eat. People with anorexia have a distorted body image: We see a skeletal appearance when we look at them, but they see themselves as fat. Repetitive starvation damages both body and brain. More common in females, anorexia nervosa has the highest mortality rate of any psychiatric condition.

+ Clinical Note Superfoods

Superfoods have star status! The health claims for superfoods attack us from advertisements everywhere: "Eat (*a specific food*) and banish (*a disease*)!" Oat bran is hyped for reversing high serum cholesterol, while pomegranates are touted to promote long life. What are we to make of these claims?

Some foods do in fact give more nutritional bang for the buck than other foods. Such foods may be nutrient dense. Raw almonds, for example, contain protein, anti-inflammatory omega-3 fats, and fiber. Other foods have high concentrations of a select nutrient. Red peppers, for example, are unusually high in vitamin C. The bright orange color of a sweet potato broadcasts its content of beta-carotene, a precursor of vitamin A. Dark leafy greens are rich plant sources of iron. Different foods confer different, and complementary, benefits.

However, the notion that foods have superpowers is more marketing ploy than nutrition science. A diet that incorporates a wide variety of whole foods is a powerful tool for nutritional health.

25

✓ Checkpoint

14. **Identify the two classes of vitamins.**
15. **Would an athlete in intensive training try to maintain a positive or a negative nitrogen balance?**
16. **How would a decrease in the amount of bile salts in the bile affect the amount of vitamin A in the body?**

See the blue Answers tab at the back of the book.

25-7 Metabolic rate is the average caloric expenditure, and thermoregulation involves balancing heat-producing and heat-losing mechanisms

Learning Outcome **Define metabolic rate, discuss the factors involved in determining an individual's BMR, and discuss the homeostatic mechanisms that maintain a constant body temperature.**

A person's daily energy expenditures vary widely with activity. For example, a person who leads a sedentary life may have minimal energy demands, but someone who includes a single hour of swimming has a significantly increased daily energy use. Your average daily energy use give a clue to your *metabolic rate*. If your daily energy intake exceeds your total energy demands, you store the excess energy, primarily as triglycerides in adipose tissue. If your daily energy expenditures exceed your intake, the result is a net reduction in your body's energy reserves and a corresponding loss in weight. In this section, first we consider aspects of energy intake and expenditure. Then we turn to the topic of *thermoregulation*, how the body maintains a consistent internal temperature.

Energy Gains and Losses

When chemical bonds are broken, energy is released. Inside cells, a significant amount of energy may be used to synthesize ATP, but much of it is lost to the environment as heat.

Measuring Energy and the Energy Content of Food

The process of *calorimetry* (kal-ō-RIM-eh-trē) measures the total amount of energy released when the bonds of organic molecules are broken. The unit of measurement is the **calorie** (KAL-ō-rē) (cal), defined as the amount of energy required to raise the temperature of 1 g of water 1 degree Celsius (°C). One gram of water is not a very practical measure when you are interested in the metabolic processes that keep a 70-kg human alive, so we use the **kilocalorie** (KIL-ō-kal-ō-rē) (kcal), or **Calorie** (with a capital C, also known as the "large calorie,"). One kilocalorie is the amount of energy needed to raise the temperature of 1 kilogram (1 Liter) of water 1°C. In human nutrition, the term *"calorie"* is commonly used instead of kilocalorie or Calorie to refer to a unit of food energy.

In cells, organic molecules are oxidized to carbon dioxide and water. Oxidation also takes place when something burns, and this process can be experimentally controlled. A known amount of food is placed in a chamber called a **calorimeter** (kal-ō-RIM-e-ter), which is filled with oxygen and surrounded by a known volume of water. Once the food is inside, the chamber is sealed and the contents are electrically ignited. When the material has completely oxidized and only ash remains in the chamber, the number of Calories released can be determined by comparing the water temperatures before and after the test.

The energy potential of food is usually expressed in kilocalories per gram (kcal/g). The catabolism of lipids releases a considerable amount of energy, roughly 9.46 kcal/g. The catabolism of carbohydrates or proteins is not as productive, because many of the carbon and hydrogen atoms are already bound to oxygen. Their average yields are comparable: 4.18 kcal/g for carbohydrates and 4.32 kcal/g for proteins. Most foods are mixtures of fats, proteins, and carbohydrates, so the calculated values listed in a "calorie counter" vary.

Measuring Energy Expenditure: Metabolic Rate

Clinicians can examine your metabolic state and determine how many calories you are using. The result can be expressed as calories per hour, calories per day, or calories per unit of body weight per day. The **basal metabolic rate (BMR)** is a measurement of the rate at which the body expends energy while at rest to maintain vital functions, such as breathing and keeping warm. On average, the BMR is 1 kcal of energy per hour for each kilogram of body weight. The metabolic rate changes and is affected by exercise, age, sex, hormones, endocrine activity, and climate. For instance, measurements taken while a person is sprinting are quite different from those taken while a person is sleeping. Figure 25–12 shows the caloric expenditures for various common activities.

In an attempt to reduce the variations in studies of energetics and metabolism, physiologists and clinicians standardize the testing conditions. Ideally, the BMR is the minimum resting energy expenditure of an awake, alert person. A direct method of determining the BMR involves monitoring respiratory activity. In resting individuals, energy use is proportional to oxygen consumption. If we assume that average amounts of carbohydrates, lipids, and proteins are being catabolized, 4.825 kcal are expended per liter of oxygen consumed.

Figure 25–12 Caloric Expenditures for Various Activities.

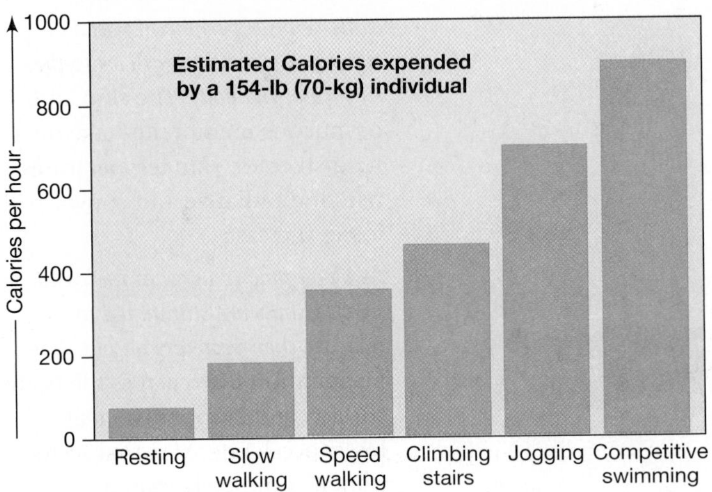
Estimated Calories expended by a 154-lb (70-kg) individual
Calories per hour
1000
800
600
400
200
0
Resting
Slow walking
Speed walking
Climbing stairs
Jogging
Competitive swimming

An average individual has a BMR of 70 kcal per hour, or about 1680 kcal per day. Although the test conditions are standardized, many uncontrollable factors can influence the BMR. These factors include physical condition, body weight, and genetic differences.

Regulation of Energy Intake

The control of appetite is poorly understood. Stretch receptors along the digestive tract, especially in the stomach, play a role, but other factors are probably more significant. Social factors, psychological pressures, and dietary habits are important. Evidence also indicates that complex hormonal stimuli interact to affect appetite. For example, the hormones *cholecystokinin (CCK)* and *adrenocorticotropic hormone (ACTH)* suppress appetite. The hormones leptin and ghrelin also interact to affect hunger. *Leptin* is released into the bloodstream by adipose tissues as they synthesize triglycerides. When leptin binds to CNS neurons that function in emotion and appetite control, the result is a sense of satiation and appetite suppression. Leptin levels are lower in thin people and higher in overweight people. Obese people have resistance to the appetite-suppressing effects of leptin. *Ghrelin* (GREL-in), a hormone secreted from the gastric mucosa, stimulates appetite. The ghrelin level declines when the stomach is full, and increases during the fasting state. Some research suggests that leptin plays a role in regulating ghrelin, showing that these "hunger hormones" are complex and warrant further study. p. 905

Obesity is defined as body weight more than 20 percent above the ideal weight for a given individual. Currently, obesity is considered an epidemic and is taking its toll on the health of adults and children. Diseases associated with obesity include heart disease, cancer, and diabetes. For this reason, calorie counting and exercise are important in a weight-control program.

Thermoregulation

The BMR estimates the rate of energy use by the body. The energy that cells do not capture and harness is released as heat. This heat serves an important homeostatic purpose. Humans are subject to vast changes in environmental temperatures, but our complex biochemical systems have a major limitation. Our enzymes operate over only a relatively narrow temperature range. Accordingly, our bodies have anatomical and physiological mechanisms that keep body temperatures within acceptable limits, regardless of environmental conditions. This homeostatic process is called **thermoregulation**. Failure to control body temperature can result in a series of physiological changes. For example, a body temperature below 36°C (97°F) or above 40°C (104°F) can cause disorientation. A body temperature above 42°C (108°F) can cause convulsions and permanent cell damage.

We continuously produce heat as a by-product of metabolism. When our energy use increases due to physical activity, or when our cells are more active metabolically (as they are during the absorptive state), additional heat is generated. The heat produced by biochemical reactions is retained by water, which makes up nearly two-thirds of body weight. Water is an excellent conductor of heat, so the heat produced in one region of the body is rapidly distributed by diffusion, as well as through the bloodstream. If body temperature is to remain constant, that heat must be lost to the environment at the same rate at which it is generated. When environmental conditions rise above or fall below "ideal," the body must control the gains or losses to maintain homeostasis.

Mechanisms of Heat Exchange

Heat exchange with the environment involves four basic processes: (1) *radiation*, (2) *convection*, (3) *evaporation*, and (4) *conduction* (Figure 25–13). Objects warmer than the environment lose heat as **radiation**. When you feel the heat from the sun, you are experiencing radiant heat. Your body loses heat the same way, but in proportionately smaller amounts. More than 50 percent of the heat you lose indoors is lost through radiation. The exact amount varies with both body temperature and skin temperature.

Convection is heat loss to the cooler air that moves across the surface of your body. As your body loses heat to the air next to your skin, that air warms and rises, moving away from the surface of the skin. Cooler air replaces it, and this air in turn becomes warmed. This pattern then repeats. Convection accounts for roughly 15 percent of the body's heat loss indoors but is insignificant as a mechanism of heat gain.

Figure 25–13 Mechanisms of Heat Exchange.

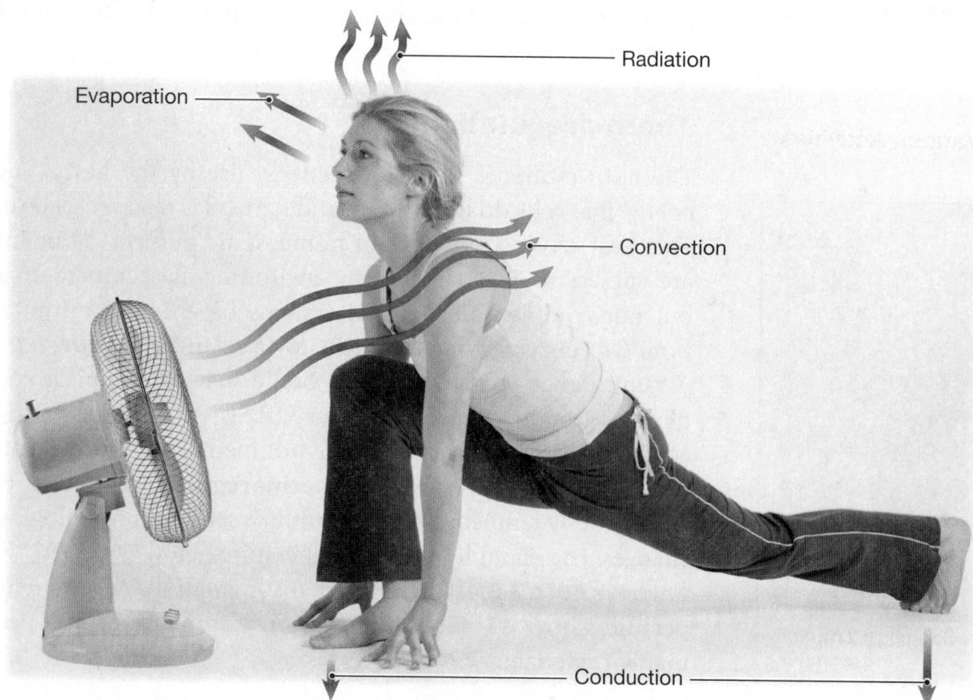
Radiation
Evaporation
Convection
Conduction

25

When water evaporates, it changes from a liquid to a vapor. **Evaporation** absorbs energy—about 0.58 kcal per gram of water evaporated—and cools the surface where evaporation occurs. The rate of evaporation at your skin is highly variable. Each hour, 20–25 mL of water crosses epithelia and evaporates from the alveolar surfaces and the surface of the skin. This *insensible water loss* remains relatively constant. At rest, it accounts for roughly 20 percent of your body's average indoor heat loss. The sweat glands responsible for *sensible perspiration* have a tremendous scope of activity, ranging from virtual inactivity to secretory rates of 2–4 liters (2.1–4.2 quarts) per hour. p. 157

Conduction is the direct transfer of energy through physical contact. When you come into an air-conditioned classroom and sit on a cold plastic chair, you are immediately aware of this process. Conduction is generally not an effective way of gaining or losing heat. Its impact depends on the temperature of the object and the amount of skin surface area involved. When you are lying on cool sand in the shade, conductive losses can be considerable. When you are standing on the same sand, conductive losses are negligible.

The Hypothalamus Regulates Heat Gain and Heat Loss

Heat loss and heat gain involve the activities of many systems. Those activities are coordinated by the **heat-loss center** and **heat-gain center**, respectively, in the pre-optic area of the anterior hypothalamus. p. 484 These centers modify the activities of other hypothalamic nuclei. The overall effect is to control temperature by influencing two processes: the rate of heat production and the rate of heat loss to the environment. Changes in behavior may also support these processes.

Mechanisms for Increasing Heat Loss. When the temperature at the pre-optic area rises above its set point, the heat-loss center is stimulated, producing major effects:

- *The inhibition of the vasomotor center causes peripheral vasodilation, and warm blood flows to the surface of the body.* The skin and/or mucous membranes take on a reddish color, skin temperatures rise, and radiative and convective losses increase.
- *As blood flow to the skin increases, sweat glands are stimulated to increase their secretory output.* The perspiration flows across the body surface, and evaporative heat losses accelerate. Maximal secretion, if completely evaporated, would remove 2320 kcal per hour.

- *The respiratory centers are stimulated, and the depth of respiration increases.* Often, the individual begins breathing through an open mouth rather than through the nasal passageways. This increases evaporative heat losses through the lungs.

Mechanisms for Promoting Heat Gain. The function of the heat-gain center of the brain is to prevent **hypothermia** (hī-pō-THER-mē-uh), or below-normal body temperature. When the temperature at the pre-optic area drops below the acceptable level, the heat-loss center is inhibited and the heat-gain center is activated.

To conserve heat, the sympathetic vasomotor center decreases blood flow to the dermis, thereby reducing losses by radiation, convection, and conduction and allowing the skin to cool. With blood flow restricted, the skin in lighter-skinned individuals may take on a bluish or pale color. The epithelial cells are not damaged, because they can tolerate extended periods at temperatures as low as 25°C (77°F) or as high as 49°C (120°F).

In addition, blood returning from the limbs is shunted into a network of deep veins. p. 766 Under warm conditions, blood flows in a superficial venous network (**Figure 25–14a**). In cold conditions, blood is diverted to a network of veins that lie deep to an insulating layer of subcutaneous fat (**Figure 25–14b**). This venous network wraps around the deep arteries, and heat is conducted from the warm blood flowing outward to the limbs to the cooler blood returning from the periphery (**Figure 25–14c**). This arrangement traps the heat close to the body core and restricts heat loss. Such exchange between fluids

Figure 25–14 Vascular Adaptations for Heat Loss and Conservation.

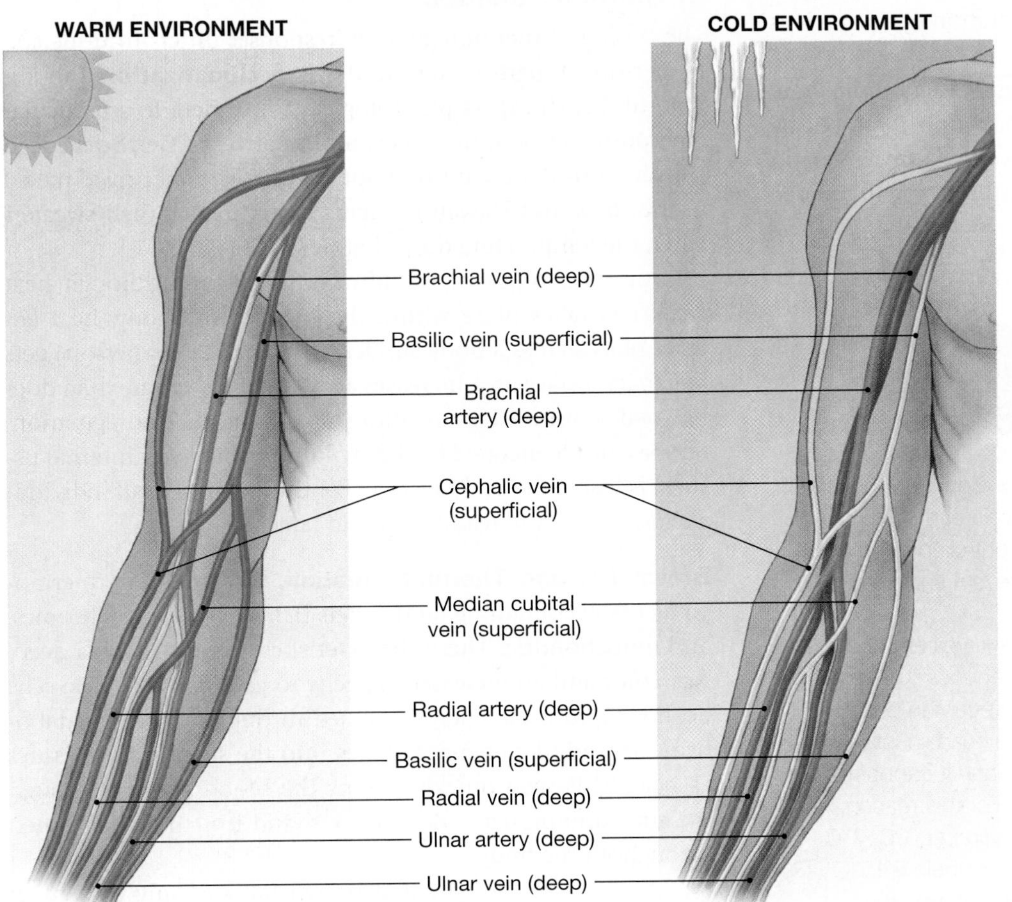
WARM ENVIRONMENT
COLD ENVIRONMENT
Brachial vein (deep)
Basilic vein (superficial)
Brachial artery (deep)
Cephalic vein (superficial)
Median cubital vein (superficial)
Radial artery (deep)
Basilic vein (superficial)
Radial vein (deep)
Ulnar artery (deep)
Ulnar vein (deep)

a Circulation through the blood vessels of the forearm in a warm environment. Blood enters the limb in a deep artery and returns to the trunk in a network of superficial veins that radiate heat to the environment through the overlying skin.

b Circulation through the blood vessels of the forearm in a cold environment. Blood now returns to the trunk by a network of deep veins that flow around the artery. The amount of heat loss is decreased, as shown in part c.

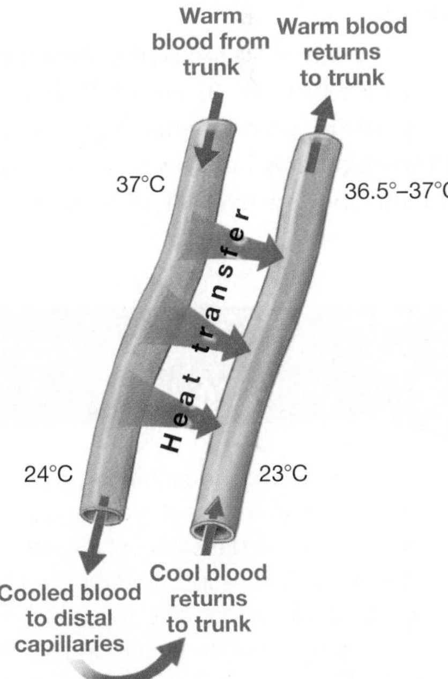
Warm blood from trunk
Warm blood returns to trunk
37°C
36.5°–37°C
Heat transfer
24°C
23°C
Cooled blood to distal capillaries
Cool blood returns to trunk

c Countercurrent heat exchange occurs as heat radiates from the warm arterial blood into the cooler venous blood flowing in the opposite direction. By the time the arterial blood reaches distal capillaries, where most heat loss to the environment occurs, it is already 13°C cooler than it was when it left the trunk. This mechanism decreases the rate of heat loss by conserving body heat within the trunk.

? In a warm environment, does more blood return to the trunk through deep or superficial veins? Why?

25

that are moving in opposite directions is called *countercurrent exchange*. (We return to this topic in Chapter 26.)

We can divide the mechanisms for generating heat into two broad categories: shivering thermogenesis and nonshivering thermogenesis. In **shivering thermogenesis** (ther-mō-JEN-eh-sis), a gradual increase in muscle tone increases the energy consumption of skeletal muscle tissue throughout your body. The more energy used, the more heat is produced. Both agonist and antagonist muscles are involved. The degree of stimulation varies with the demand.

If the heat-gain center is extremely active, muscle tone increases to the point at which stretch receptor stimulation will produce brief, oscillatory contractions of antagonistic muscles. In other words, you begin to **shiver**. Shivering increases the workload of the muscles and further elevates oxygen and energy consumption. The heat that is produced warms the deep vessels, to which blood has been shunted by the sympathetic vasomotor center. Shivering can elevate body temperature quite effectively, increasing the rate of heat generation by as much as 400 percent.

In contrast, *nonshivering thermogenesis* is a less immediate heat-gain process. In an adult, heat production increases by only 10–15 percent after a period of weeks. **Nonshivering thermogenesis** involves the release of hormones that increase the metabolic activity of all tissues:

- The heat-gain center stimulates the adrenal medullae through the sympathetic division of the autonomic nervous system. As a result, epinephrine is released. It increases the rates of glycogenolysis in liver and skeletal muscle, and the metabolic rate of most tissues. The effects are immediate.
- The pre-optic area regulates the production of thyrotropin-releasing hormone (TRH) by the hypothalamus. In children, when body temperatures are below normal, additional TRH is released. This hormone stimulates the release of thyroid -stimulating hormone (TSH) by the anterior lobe of the pituitary gland. In response to this release of TSH, the thyroid gland increases the rate of thyroxine release into the blood. Thyroxine increases not only the rate of carbohydrate catabolism, but also the rate of catabolism of all other nutrients. These effects develop gradually, over a period of days to weeks.

Clinical Note Hypothermia in the Operating Room

Hypothermia may be intentionally produced during surgery. The goal is to reduce the metabolic rate of a particular organ or of the patient's entire body. In controlled hypothermia, the patient is first anesthetized to prevent shivering, which would act to fight off hypothermia.

During open-heart surgery, the body is typically cooled to 25°C–32°C (79°F–89°F). This cooling reduces the metabolic demands of the body, which will be receiving blood from an external pump or oxygenator. The heart must be stopped completely during the operation, and it cannot be well supplied with blood over this period. For this reason, the heart is perfused with an *arresting solution* at 0°C–4°C (32°F–39°F) and maintained at a temperature below 15°C (60°F) during the operation. At these temperatures, the cardiac muscle can tolerate several hours of ischemia (inadequate blood supply) without damage.

When cardiac surgery is performed on infants, a deep hypothermia may be produced by cooling the entire body to temperatures as low as 11°C (52°F) for an hour or more. In effect, these conditions are similar to those experienced by victims of accidental drowning.

Sources of Individual Variation in Thermoregulation

The timing of thermoregulatory responses differs from person to person. A person may undergo **acclimatization** (ah-klī-mah-tih-ZĀ-shun)—a physiological adjustment to a particular environment over time. For example, natives of Tierra del Fuego (off the southernmost tip of South America) once lived naked in the snow, but Hawaii residents often unpack their sweaters when the temperature drops below 22°C (72°F).

Another source of variation is body size. Although heat *production* takes place within the mass of the body, heat *loss* takes place across a body surface. As an object (or person) gets larger, its surface area increases at a much slower rate than does its total volume. This relationship affects thermoregulation, because heat generated by the "volume" (that is, by internal tissues) is lost at the body surface. For this reason, small individuals lose heat more readily than do large individuals.

Brown Fat and Thermoregulation. Brown fat is thermogenic tissue containing fat droplets rich in heme, cytochromes, and mitochondria. These characteristics give the tissue a deep, rich color and an increased capacity to generate heat. The cells do not capture the energy released through fatty acid catabolism. Instead, the energy radiates into the surrounding tissues as heat. This heat quickly warms the blood passing through the surrounding network of vessels, and it is then distributed throughout the body.

It was once thought that brown fat was only present in infants, however, it is also found in adults. In newborns and infants, brown fat is important for thermoregulation, because their temperature-regulating mechanisms are not fully functional and they cannot shiver. Consequently, newborns must be dried promptly and kept bundled up. Babies born prematurely need a thermally regulated incubator. Infants also lose heat quickly as a result of their small size. So, the brown fat can very quickly speed up metabolic heat generation by 100 percent.

With increasing age and size, an infant's body temperature becomes more stable. As a result, brown fat becomes less important. With increased body size, skeletal muscle mass, and insulation, shivering thermogenesis becomes more effective in raising body temperature.

While we tend to lose brown fat as we age, everyone has brown fat nestled between areas of white fat, predominantly in the neck, upper chest, and shoulders. Although it is difficult to identify, PET scans are advancing the ability to recognize it. Interestingly, people who live in colder climates seem to have more brown fat than do people living in warmer climes, and thinner, younger people have more brown fat than do older, obese people. Hence, the role of brown fat—particularly its role in metabolism—is an area of active research.

Thermoregulatory Variations among Adults. Adults of a given body weight may differ in their thermal responses due to variations in body mass and tissue distribution. Adipose tissue is an excellent insulator. It conducts heat at only about one-third the rate of other tissues. As a result, people with a more substantial layer of subcutaneous fat may not begin to shiver until long after their thinner companions.

Two otherwise similar individuals may also differ in their response to temperature changes because their hypothalamic "thermostats" are at different settings. We all experience daily oscillations in body temperature. Temperatures peak during the day or early evening. They fall 1°C–2°C (1.8°F–3.6°F) at night. The ovulatory cycle also causes temperature fluctuations in women, as you will see in Chapter 28.

Individuals vary in the timing of their maximum temperature setting. Some have a series of peaks, with an afternoon low. The origin of these patterns is unclear. The patterns do not result from daily activity regimens. For example, the temperatures of people who work at night still peak during the same range of times as the rest of the population.

+ Clinical Note Excess Body Heat

Heat exhaustion and heat stroke are malfunctions of thermoregulatory mechanisms. With **heat exhaustion**, also known as *heat prostration*, excessive fluids are lost in perspiration, and the individual has difficulty maintaining blood volume. The heat-loss center in the hypothalamus stimulates sweat glands, whose secretions moisten the surface of the skin to provide evaporative cooling. As fluid losses mount, blood volume decreases. The resulting decline in blood pressure is not countered by peripheral vasoconstriction, because the heat-loss center is actively stimulating peripheral vasodilation. As blood flow to the brain declines, headache, nausea, and eventual collapse follow. Treatment is straightforward: provide fluids, salts, and a cooler place to be!

In **heat stroke**, body temperature rises uncontrollably because the thermoregulatory center stops functioning. The sweat glands are inactive, and the skin becomes hot and dry. Heat stroke is more serious and can follow an untreated case of heat exhaustion. Predisposing factors include any preexisting condition that affects peripheral circulation, such as heart disease or diabetes. Unless the problem is recognized in time, body temperature may climb to 41°C–45°C (106°F–113°F). Temperatures in this range quickly disrupt a variety of vital physiological systems and destroy brain, liver, skeletal muscle, and kidney cells. Effective treatment involves lowering the body temperature as rapidly as possible.

+ Clinical Note Deficient Body Heat

Hypothermia is below-normal body temperature. If body temperature drops significantly below the normal level, thermoregulatory mechanisms become less sensitive and less effective. Cardiac output and respiratory rate decrease. If the core temperature falls below 28°C (82°F), cardiac arrest is likely. The individual then has no heartbeat, no respiratory rate, and no response to external stimuli—even painful ones. The body temperature continues to decline, and the skin and/or mucous membranes turn blue or pale and cold.

At this point, people commonly assume that the person has died. But because metabolic activities have decreased system-wide, the victim may still be saved, even after several hours. Treatment consists of cardiopulmonary support and gradual rewarming, both external and internal. The skin can be warmed up to 45°C (113°F) without damage. Warm baths or blankets can be used. One effective method of raising internal temperatures involves the introduction of warm saline solution into the peritoneal cavity.

Hypothermia is a significant risk for people engaged in water sports. It may complicate the treatment of a drowning victim. Water absorbs heat about 27 times as fast as air does, and the body's heat-gain mechanisms are unable to keep pace over long periods or when faced with a large temperature gradient. But hypothermia in cold water does have a positive side. On several occasions, small children who have "drowned " in cold water have been successfully revived after periods of up to 4 hours. Children lose body heat quickly, and their systems soon stop functioning as their body temperature declines. This rapid drop in temperature prevents the oxygen starvation and tissue damage that would otherwise take place when breathing stops.

Resuscitation is not attempted if the person has actually frozen. Water expands roughly 7 percent during ordinary freezing. The process destroys plasma membranes throughout the body. Very small organisms can be frozen and subsequently thawed without ill effects, because their surface-to-volume ratio is enormous, and the freezing process occurs so rapidly that ice crystals never form.

Fevers

A **fever** is a body temperature greater than 37.2°C (99°F). We discussed fevers when we examined innate (nonspecific) immunity. p. 805 Fevers occur for a variety of reasons, not all of them pathological. In young children, transient fevers with no ill effects can result from exercise in warm weather. Similar exercise-related elevations were rarely seen in adults until running marathons became popular. Temperatures from 39°C to 41°C (103°F to 109°F) may result. For this reason, competitions are usually held when the air temperature is below 28°C (82°F).

Fevers can also result from the following factors:

- Abnormalities that affect the entire thermoregulatory mechanism, such as heat exhaustion or heat stroke
- Clinical problems that restrict blood flow, such as congestive heart failure
- Conditions that impair sweat gland activity, such as drug reactions and some skin conditions
- The resetting of the hypothalamic "thermostat" by circulating *pyrogens*—most notably, interleukin-1.

✓ Checkpoint

17. **How would the BMR of a pregnant woman compare with her BMR before she became pregnant?**
18. **What effect does peripheral vasoconstriction on a hot day have on an individual's body temperature?**
19. **Why do infants have greater problems with thermoregulation than adults?**

See the blue Answers tab at the back of the book.

25 Chapter Review

Study Outline

MasteringA&P™ Access more chapter study tools online in the MasteringA&P Study Area:

- Chapter Quizzes, Chapter Practice Test, MP3 Tutor Sessions, and Clinical Case Studies
- Practice Anatomy Lab PAL 3.0
- A&P Flix A&PFlix
- Interactive Physiology iP2
- PhysioEx PhysioEx 9.1

An Introduction to Metabolism, Nutrition, and Energetics p. 940

1. The nutrients in food support the metabolic activities of the body. Your body builds energy reserves when nutrients are abundant and mobilizes them when nutrients are in short supply.

25-1 Metabolism is the sum of all the catabolic and anabolic reactions in the body, and energetics is the flow and transformation of energy p. 940

2. All the available nutrients distributed in the blood form a **nutrient pool**. **Catabolism** converts larger molecules into smaller ones, and **anabolism** converts small molecules into larger ones. Various metabolic pathways are involved with catabolism and anabolism. *(Figure 25–1)*
3. Anabolic reactions (1) perform structural maintenance or repairs, (2) support growth, (3) produce secretions, and (4) build and store nutrient reserves. *(Figure 25–1)*
4. **Energetics** is the study of the flow of energy and its transformation, or change, from one form to another.
5. Energy changes in cells involve chemical reactions known as oxidation and reduction. **Oxidation** is the gain of oxygen, or loss of hydrogen or electrons, and **reduction** is the loss of oxygen, or gain of hydrogen or electrons. Oxidation–reduction reactions occur together and are referred to as *redox reactions*.

25-2 Carbohydrate metabolism generates ATP by glucose catabolism and forms glucose by gluconeogenesis p. 943

6. Most cells generate ATP and other high-energy compounds by the catabolism of carbohydrates, especially glucose.
7. **Glycolysis** and **aerobic metabolism**, or **cellular respiration**, provide most of the ATP used by typical cells. In the anaerobic process of glycolysis, each molecule of glucose yields 2 molecules of **pyruvic acid** (as **pyruvate** ions), a net gain of 2 molecules of ATP, and 2 **NADH** molecules. *(Figure 25–2)*
8. In the presence of oxygen, pyruvate enters the mitochondria, where it is broken down completely in the **citric acid cycle**. Carbon and oxygen atoms are lost as carbon dioxide (**decarboxylation**); hydrogen atoms are passed to coenzymes, which initiate the oxygen-consuming and ATP-generating reaction **oxidative phosphorylation**. *(Figure 25–3)*
9. The **electron transport chain (ETC)** is embedded in the inner mitochondrial membrane. It is made up of four respiratory protein complexes, coenzyme Q, and cytochromes that eventually pass along electrons to oxygen to form water. Energy released by the passage of the electrons is used to pump H^+ from the matrix to the intermembrane space. *(Spotlight Figure 25–4)*
10. ATP is formed through **chemiosmosis**. During this process, H^+ pass through H^+ channels in the inner membrane from the intermembrane space to the mitochondrial matrix. Energy from the passage of the H^+ is used by *ATP synthase* to form ATP from ADP and P_i. *(Spotlight Figure 25–4)*
11. For each glucose molecule processed through glycolysis, the citric acid cycle, and the ETC, most cells gain 30–32 molecules of ATP. *(Figure 25–5)*

12. **Gluconeogenesis**, the synthesis of glucose from noncarbohydrate molecules, enables a liver cell to synthesize glucose molecules when carbohydrate reserves are depleted. Glucose molecules are stored as **glycogen** in liver and skeletal muscle cells. **Glycogenesis** is the process of glycogen formation. Glycogen is an important energy reserve. **Glycogenolysis** is the breakdown of glycogen to glucose molecules. *(Figures 25–1, 25–6)*

25-3 Lipid metabolism provides long-term storage and release of energy p. 951

13. During **lipolysis** (lipid catabolism), lipids are broken down into pieces that can be converted into pyruvate or channeled into the citric acid cycle.
14. **Triglycerides**, the most abundant lipids in the body, are split into glycerol and fatty acids. The glycerol is converted to pyruvate, and the fatty acids enter the mitochondria.
15. **Beta-oxidation** is the breakdown of a fatty acid molecule into 2-carbon fragments that enter the citric acid cycle. The steps of beta-oxidation cannot be reversed, and the body cannot manufacture all the fatty acids needed for normal metabolic operations. *(Figure 25–7)*
16. Lipids cannot provide large amounts of ATP quickly. However, cells can shift to lipid-based energy production when glucose reserves are limited.
17. In **lipogenesis** (the synthesis of lipids), almost any organic substrate can be used to form glycerol. **Essential fatty acids** are those that cannot be synthesized and must be included in the diet.
18. Most lipids circulate as **lipoproteins** (lipid–protein complexes that contain large glycerides and cholesterol). The largest lipoproteins, chylomicrons, carry absorbed lipids from the intestinal tract to the bloodstream. All other lipoproteins are derived from the liver and carry lipids to and from various tissues of the body. *(Figure 25–8)*
19. Capillary walls of adipose tissue, skeletal muscle, cardiac muscle, and the liver contain **lipoprotein lipase**, an enzyme that breaks down lipids, releasing a mixture of fatty acids and monoglycerides into the interstitial fluid. *(Figure 25–8)*

25-4 Protein metabolism provides amino acids and synthesizes proteins p. 956

20. If other energy sources are inadequate, mitochondria can break down amino acids in the citric acid cycle to generate ATP. In the mitochondria, the amino group can be removed by either **transamination** or by **deamination**. *(Figure 25–9)*
21. Protein catabolism is impractical as a source of quick energy.
22. Roughly half the amino acids needed to build proteins can be synthesized. There are 10 **essential amino acids**, which must be acquired through the diet. **Amination**, the attachment of an amino group to a carbon framework, is an important step in the synthesis of **nonessential amino acids**. (Figure 25–9)

25-5 The body experiences two patterns of metabolic activity: energy storage in the absorptive state and energy release in the postabsorptive state p. 957

23. No one cell of a human can perform all the anabolic and catabolic operations necessary to support life. Homeostasis can be preserved only when metabolic activities of different tissues are coordinated.
24. The body has five metabolic components: the liver, adipose tissue, skeletal muscle, nervous tissue, and other peripheral tissues. The liver is the focal point for metabolic regulation and control. Adipose tissue stores lipids, primarily in the form of triglycerides. Skeletal muscle contains substantial glycogen reserves, and the contractile proteins can be degraded and the amino acids used as an energy source. Nervous tissue does not contain energy reserves; glucose must be supplied to it for energy. Other peripheral tissues are able to metabolize glucose, fatty acids, or other substrates under the direction of the endocrine system.
25. For about 4 hours after a meal, nutrients enter the blood as intestinal absorption proceeds. *(Spotlight Figure 25–10)*
26. The liver closely regulates the circulating levels of glucose and amino acids.
27. The **absorptive state** exists when nutrients are being absorbed by the digestive tract. Adipocytes remove fatty acids and glycerol from the bloodstream and synthesize new triglycerides to be stored for later use. *(Spotlight Figure 25–10)*
28. During the absorptive state, glucose molecules are catabolized and amino acids are used to build proteins. Skeletal muscles may also catabolize circulating fatty acids, and the energy obtained is used to increase glycogen reserves.
29. The **postabsorptive state** extends from the end of the absorptive state to the next meal. *(Spotlight Figure 25–10)*
30. When blood glucose levels fall, the liver begins breaking down glycogen reserves and releasing glucose into the bloodstream. As the time between meals increases, liver cells synthesize glucose molecules from smaller carbon fragments and from glycerol molecules. Fatty acids undergo beta-oxidation; the fragments enter the citric acid cycle or combine to form **ketone bodies**. *(Spotlight Figure 25–10)*
31. Some amino acids can be converted to pyruvate and used for gluconeogenesis; others, including most of the essential amino acids, are converted to acetyl-CoA and are either catabolized or converted to ketone bodies.
32. Nervous tissue continues to be supplied with glucose as an energy source until blood glucose levels become very low.

25-6 Adequate nutrition allows normal physiological functioning p. 959

33. **Nutrition** is the absorption of nutrients from food. A **balanced diet** contains all the ingredients needed to maintain homeostasis; a balanced diet prevents **malnutrition**.
34. The **basic food groups** include grains, vegetables, fruits, dairy, and protein. These are arranged on a *plate* to reflect the recommended daily food consumption balanced with daily physical activity. *(Figure 25–11; Table 25–1)*
35. Amino acids, purines, pyrimidines, creatine, and porphyrins are **nitrogen compounds**, which contain nitrogen atoms. An adequate dietary supply of nitrogen is essential, because the body does not maintain large nitrogen reserves. **Nitrogen balance** is a state in which the amount of nitrogen absorbed equals that lost in urine and feces.
36. **Minerals** act as cofactors in a variety of enzymatic reactions. They also contribute to the osmotic concentration of body fluids and play a role in membrane potentials, action potentials, the release of neurotransmitters, muscle contraction, skeletal construction and maintenance, gas transport, buffer systems, fluid absorption, and waste removal. *(Table 25–2)*
37. Vitamins are needed in very small amounts. Vitamins A, D, E, and K are **fat-soluble vitamins**; taken in excess, they can lead to **hypervitaminosis**. **Water-soluble vitamins** are not stored in the body; a lack of adequate dietary supplies may lead to **hypovitaminosis**, or vitamin insufficiency). *(Tables 25–3, 25–4)*
38. A balanced diet can improve one's general health.

25-7 Metabolic rate is the average caloric expenditure, and thermoregulation involves balancing heat-producing and heat-losing mechanisms p. 966

39. The energy content of food is usually expressed in **kilocalories (kcal)** per gram (kcal/g) or as **Calories** per gram (Cal/g).
40. The catabolism of lipids releases 9.46 kcal/g, about twice the amount released by equivalent weights of carbohydrates or proteins.
41. The **basal metabolic rate (BMR)** is a measurement of the rate at which the body expends energy while at rest to maintain vital functions, such as breathing and keeping warm. *(Figure 25–12)*
42. The homeostatic regulation of body temperature is **thermoregulation**. Heat exchange with the environment involves four processes: **radiation**, **convection**, **evaporation**, and **conduction**. *(Figure 25–13)*
43. The *pre-optic area* of the hypothalamus acts as the body's thermostat, affecting the **heat-loss center** and the **heat-gain center**.
44. Mechanisms for increasing heat loss include both physiological mechanisms (peripheral vasodilation, increased perspiration, and increased respiration) and behavioral modifications.
45. Responses that conserve heat include a decreased blood flow to the dermis and *countercurrent exchange*. *(Figure 25–14)*
46. Heat is generated by **shivering thermogenesis** and **nonshivering thermogenesis**.
47. Thermoregulatory responses differ among individuals. One important source of variation is **acclimatization** (a physiological adjustment to an environment over time).
48. **Fever**, a body temperature above 37.2°C (99°F), can result from problems with the thermoregulatory mechanism, circulation, or sweat gland activity, or from the resetting of the hypothalamic "thermostat" by circulating pyrogens.

Review Questions

See the blue Answers tab at the back of the book.

LEVEL 1 Reviewing Facts and Terms

1. Catabolism refers to **(a)** the creation of a nutrient pool, **(b)** the sum total of all chemical reactions in the body, **(c)** the production of organic compounds, **(d)** the breakdown of organic substrates.
2. The breakdown of glucose to pyruvate is an ________ process. **(a)** anaerobic, **(b)** aerobic, **(c)** anabolic, **(d)** oxidative, **(e)** both a and d.
3. The process that produces more than 90 percent of the ATP used by our cells is **(a)** glycolysis, **(b)** the citric acid cycle, **(c)** substrate-level phosphorylation, **(d)** oxidative phosphorylation.
4. The sequence of reactions responsible for the breakdown of fatty acid molecules is **(a)** beta-oxidation, **(b)** the citric acid cycle, **(c)** lipogenesis, **(d)** all of these.
5. The citric acid cycle must turn ________ time(s) to completely metabolize the pyruvate produced from one glucose molecule. **(a)** 1, **(b)** 2, **(c)** 3, **(d)** 4, **(e)** 5.
6. The largest metabolic reserves for the average adult are stored as **(a)** carbohydrates, **(b)** proteins, **(c)** amino acids, **(d)** triglycerides, **(e)** fatty acids.
7. Fill in the missing terms for the different mechanisms of heat transfer.

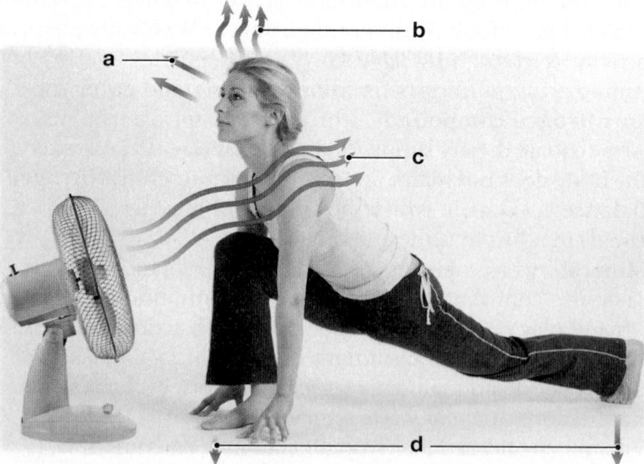

a
b
c
d

8. The vitamins generally associated with vitamin toxicity are **(a)** fat-soluble vitamins, **(b)** water-soluble vitamins, **(c)** the B complex vitamins, **(d)** vitamins C and B_{12}.
9. What is a lipoprotein? What are the major groups of lipoproteins, and how do they differ?

LEVEL 2 Reviewing Concepts

10. What is oxidative phosphorylation? Explain how the electron transport chain and chemiosmosis are involved in this process.
11. How are lipids catabolized in the body? How is beta-oxidation involved in lipid catabolism?
12. How do the absorptive and postabsorptive states maintain normal blood glucose levels?
13. Why is the liver the focal point for metabolic regulation and control?
14. Why are vitamins and minerals essential components of the diet?
15. Articles in the popular media often refer to "good cholesterol" and "bad cholesterol." To which types and functions of cholesterol do these terms refer? Explain your answer.

LEVEL 3 Critical Thinking and Clinical Applications

16. When blood levels of glucose, amino acids, and insulin are high, and glycogenesis is occurring in the liver, the body is in the ________ state. **(a)** fasting, **(b)** postabsorptive, **(c)** absorptive, **(d)** stress, **(e)** bulimic.
17. Charlie has a blood test that shows a normal level of LDLs but an elevated level of HDLs in his blood. Given that his family has a history of cardiovascular disease, he wonders if he should modify his lifestyle. What would you tell him?
18. Jill suffers from anorexia nervosa. One afternoon she is rushed to the emergency room because of cardiac arrhythmias. Her breath smells fruity, and her blood and urine samples contain high levels of ketone bodies. Why do you think she is having the arrhythmias?

+ CLINICAL CASE Wrap-Up The Miracle Supplement

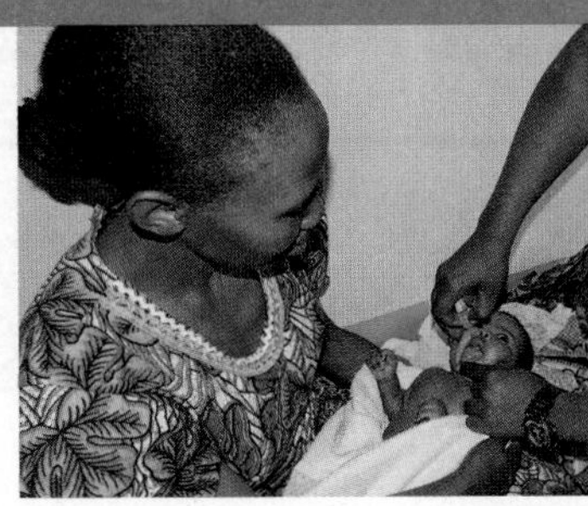

Vitamin A is the miracle vitamin supplement that can reduce overall childhood mortality by 23 percent in high-risk areas like rural Kenya. Preschool-age children and pregnant women in their last trimester are the most vulnerable to vitamin A deficiency.

Like other fat-soluble vitamins, vitamin A is an essential organic nutrient that functions as a coenzyme in vital enzymatic reactions. In the retina of the eye, *retinal* (made from vitamin A) combines with *opsin* to form *rhodopsin*, which is necessary for low-light and color vision. Vitamin A deficiency is the leading cause of preventable childhood blindness, affecting as many as half a million children worldwide. As a coenzyme, vitamin A is critical in many steps in the immune system. Vitamin A deficiency leads to a decreased immune response and increased vulnerability to malaria, respiratory infections, and diarrhea. Vitamin A is also critical for normal bone growth and development, and its deficiency can lead to stunting and skeletal deformities.

Vitamin A is found naturally in leafy green, yellow, and orange vegetables and fruits. Vitamin A is also available in milk, liver, eggs, and some fatty fish. None of these food items are available to Justine, who is among the very poor living in sub-Saharan Africa. But thanks to inexpensive vitamin A supplements, Justine's daughter will be protected from the health problems caused by vitamin A deficiency.

1. Other than diet and vitamin supplementation, is there any other available source of vitamin A?
2. If Justine's diet—which includes very little milk, liver, eggs, fish, leafy green, yellow, and orange vegetables, and fruits—is deficient in vitamin A, what other vitamin deficiencies might be expected?

See the blue Answers tab at the back of the book.

Related Clinical Terms

anorexia: Persistent loss of appetite.

anorexia nervosa: Eating disorder occurring primarily among girls and women characterized by a desire to lose, or not gain, weight through starvation, due to a distorted view of the person's own body. There are typically two types: strict diet and exercise, and binging and purging.

binge–purge syndrome: Eating disorder characterized by excessive eating followed by periods of fasting or self-induced vomiting.

bulimia: An eating disorder usually characterized by episodic binge eating that is followed by feelings of guilt, depression, and self-condemnation. It is often associated with steps taken to lose weight, such as self-induced vomiting, the use of laxatives, dieting, or fasting.

eating disorders: Psychological problems that result in inadequate or excessive food consumption. Examples include anorexia nervosa and bulimia.

familial hypercholesterolemia: The most common inherited type of hyperlipidemia characterized by high levels of lipids in one's blood. It affects 1 in every 500 children born, who then present with LDL levels in excess of 190.

heat cramps: A condition that usually follows heavy sweating due to strenuous activity that causes a loss of salt in the body and results in painful muscle spasms in the abdomen, arms, or legs.

hyperuricemia: A condition in which the plasma uric acid level exceeds 7.4 mg/dL. It can result in the condition called *gout*.

ketonuria: The presence of ketone bodies in the urine.

kwashiorkor: A form of malnutrition due to a protein deficiency in the diet that typically affects young children in tropical regions.

marasmus: Severe malnourishment that causes a child's weight to fall significantly below the standards set for children of similar age.

orexigenic: Having a stimulating effect on the appetite.

pica: Tendency or craving for substances other than normal food. Possible organic causes are iron deficiency, lead encephalopathy, pregnancy, and zinc deficiency.

protein-calorie malnutrition (PCM): A severe deficiency of protein in the diet in addition to inadequate caloric intake. It results in the condition termed *kwashiorkor*.

skin-fold test: Test to estimate the amount of body fat on a person.

26 The Urinary System

Learning Outcomes

These Learning Outcomes correspond by number to this chapter's sections and indicate what you should be able to do after completing the chapter.

- 26-1 ■ Identify the organs of the urinary system, and describe the functions of the system. p. 977
- 26-2 ■ Describe the location and structures of the kidneys, identify major blood vessels associated with each kidney, trace the path of blood flow through a kidney, and describe the structure of a nephron. p. 978
- 26-3 ■ Describe the basic processes that form urine. p. 987
- 26-4 ■ Describe the factors that influence glomerular filtration pressure and the rate of filtrate formation. p. 989
- 26-5 ■ Identify the types and functions of transport mechanisms found along each segment of the nephron and collecting system. p. 993
- 26-6 ■ Explain the role of countercurrent multiplication, describe hormonal influence on the volume and concentration of urine, and describe the characteristics of a normal urine sample. p. 1001
- 26-7 ■ Describe the structures and functions of the ureters, urinary bladder, and urethra, discuss the voluntary and involuntary regulation of urination, and describe the urinary reflexes. p. 1009
- 26-8 ■ Describe the effects of aging on the urinary system. p. 1013
- 26-9 ■ Give examples of interactions between the urinary system and other organ systems studied so far. p. 1014

CLINICAL CASE A Case of "Hidden" Bleeding

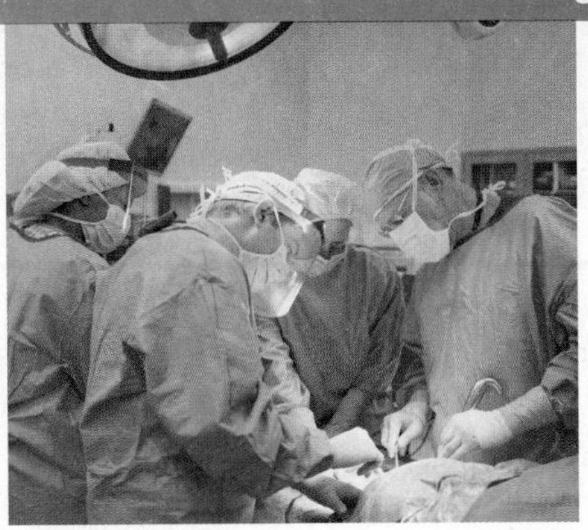

The motorcycle accident scene is horrific. The driver is dead, and his passenger—a 14-year-old high school freshman named Mike—is unconscious, flung more than 50 feet from the wreckage. When emergency responders deliver Mike to the trauma center, he has no measurable blood pressure, in spite of the 4 liters of fluids they have infused. The trauma team begins administering whole blood transfusions, but Mike is bleeding faster than replacement blood can be supplied.

Noting Mike's bloated abdomen and the abrasions on his left posterior chest and flank, the trauma team suspects that the massive blood loss is due to intra-abdominal injuries. An x-ray of Mike's chest and abdomen shows fractured left 8th through 12th ribs and fractures of the left transverse processes of L_1, L_2, and L_3 vertebrae. A nurse inserts a urinary catheter into Mike's bladder, which drains clear urine.

When the surgeon opens Mike's abdomen through a midline incision, he encounters a copious amount of free blood. Clearing this blood quickly, and moving the abdominal structures out of the way, he finds a ruptured spleen that is bleeding profusely. The surgeon performs an emergency splenectomy. Rinsing the abdominal structures reveals no further visible bleeding, but the surgical team cannot stabilize Mike's blood pressure, no matter how quickly they transfuse blood. **Something is still bleeding, but what is it?** **To find out, turn to the Clinical Case Wrap-Up on p. 1019.**

An Introduction to the Urinary System

The **urinary system** removes most of the metabolic wastes produced by the body's cells. In this chapter, we describe the organs that make up the urinary system, and explain how the *kidneys* remove metabolic wastes from the circulation to produce *urine*. We also discuss the major regulatory mechanisms controlling urine production and concentration, and consider how urine is eliminated from the body.

26-1 The organs of the urinary system function in excreting wastes and regulating body fluids

Learning Outcome **Identify the organs of the urinary system, and describe the functions of the system.**

This section introduces the organs of the urinary system, as well as their major functions.

Organs of the Urinary System

The organs of the urinary system are shown in Figure 26–1. The paired **kidneys** have excretory functions in the urinary system. These organs produce **urine**, which is a fluid that contains water, ions, and small soluble compounds. Urine leaving the kidneys flows along the *urinary tract*. The **urinary tract** consists of paired tubes called **ureters** (yū-RĒ-terz); the **urinary bladder**, a muscular sac for temporary urine storage; and a tube called the **urethra** (yū-RĒ-thrah).

When the bladder is full, urine is eliminated from the body in the process of **urination**, or **micturition** (mik-chū-RISH-un). In this process, the urinary bladder contracts and forces urine through the urethra, which conducts the urine to the exterior. ATLAS: Embryology Summary 20: The Development of the Urinary System

Figure 26–1 **An Overview of the Urinary System.** An anterior view of the urinary system, showing the positions of its organs.

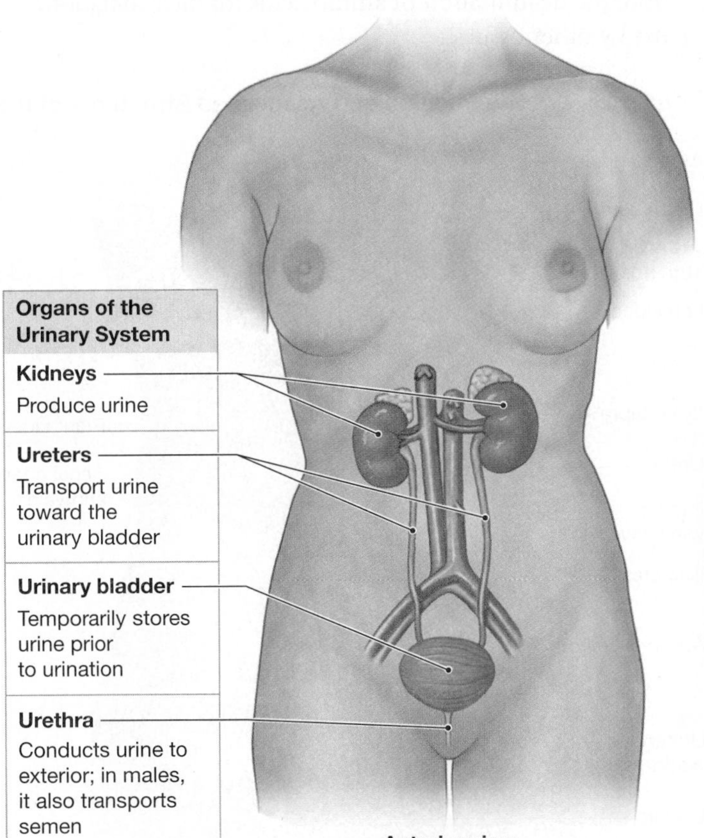
Organs of the Urinary System
Kidneys
Produce urine
Ureters
Transport urine toward the urinary bladder
Urinary bladder
Temporarily stores urine prior to urination
Urethra
Conducts urine to exterior; in males, it also transports semen
Anterior view

26

Urinary System Functions

We can see that the urinary system has the major functions of *excretion*, the removal of metabolic waste products from body fluids, and *elimination*, the discharge of these wastes out of the body. However, the urinary system also has a third major function: *homeostatic regulation* of the volume and solute concentration of blood. Essential to such homeostatic regulation are the following:

- *Regulating blood volume and blood pressure*, by adjusting the volume of water lost in urine, secreting erythropoietin, and releasing renin.
- *Regulating plasma concentrations of sodium, potassium, chloride, and other ions*, by influencing the quantities in urine. The kidneys also control the calcium ion level through the synthesis of calcitriol.
- *Helping to stabilize blood pH*, by controlling the concentrations of hydrogen and bicarbonate ions in urine.
- *Conserving valuable nutrients*, by preventing their loss in urine while removing metabolic wastes—especially the nitrogenous wastes *urea* and *uric acid*.
- *Assisting the liver* in detoxifying poisons and, during starvation, the deamination of amino acids for their metabolic use by other tissues.

These activities are carefully regulated to keep the composition of blood within acceptable limits. A disruption of any one of them has immediate consequences and can be fatal.

✓ Checkpoint

1. **Name the major functions of the urinary system.**
2. **Identify the organs of the urinary system.**

See the blue Answers tab at the back of the book.

26-2 Kidneys are highly vascular organs containing functional units called nephrons

Learning Outcome Describe the location and structures of the kidneys, identify major blood vessels associated with each kidney, trace the path of blood flow through a kidney, and describe the structure of a nephron.

The kidneys are major excretory organs located retroperitoneally in the superior lumbar region. Kidneys contain many functional units called nephrons.

Position and Associated Structures of the Kidneys

The kidneys are located on either side of the vertebral column, between vertebrae T_{12} and L_3 (Figure 26–2a). The left kidney

Figure 26–2 **The Position and Associated Structures of the Kidneys.** ATLAS: Plates 57a,b

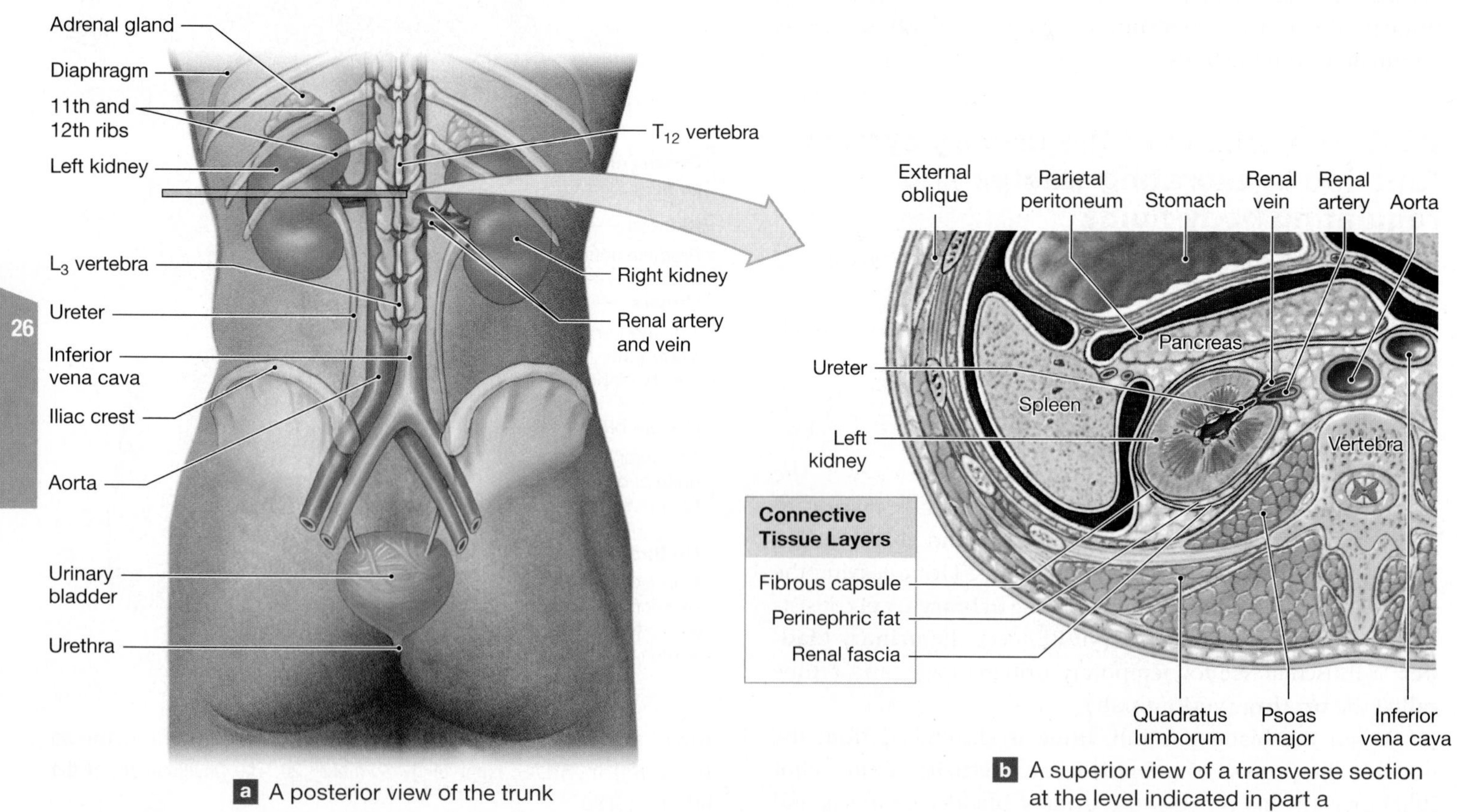
Adrenal gland
Diaphragm
11th and 12th ribs
Left kidney
L_3 vertebra
Ureter
Inferior vena cava
Iliac crest
Aorta
Urinary bladder
Urethra
T_{12} vertebra
Right kidney
Renal artery and vein
External oblique
Parietal peritoneum
Stomach
Renal vein
Renal artery
Aorta
Pancreas
Spleen
Ureter
Left kidney
Vertebra
Connective Tissue Layers
Fibrous capsule
Perinephric fat
Renal fascia
Quadratus lumborum
Psoas major
Inferior vena cava

a A posterior view of the trunk

b A superior view of a transverse section at the level indicated in part a

26

Go to MasteringA&P™ > Study Area > Menu > Lab Tools > PAL 3.0 > Human Cadaver and Anatomical Models > Urinary System lies slightly superior to the right kidney. The right kidney is slightly inferior due to the position of the right lobe of the liver.

The superior surface of each kidney is capped by an adrenal gland. The kidneys and adrenal glands lie between the muscles of the posterior body wall and the parietal peritoneum, in a retroperitoneal position (Figure 26–2b).

Tips & Tools

To visualize the kidneys' retroperitoneal positions, think of each kidney as a picture on the body wall that got covered over by wallpaper (the parietal peritoneum). To remember all the organs located entirely or partially retroperitoneally, use the mnemonic **SAD PUCKER** for **S**uprarenal (adrenal) glands, **A**orta and inferior vena cava, **D**uodenum, **P**ancreas, **U**reters, **C**olon, **K**idneys, **E**sophagus, **R**ectum.

The kidneys are held in position within the abdominal cavity by (1) the overlying peritoneum, (2) contact with adjacent visceral organs, and (3) supporting connective tissues. Three concentric layers of connective tissue protect and stabilize each kidney, which are, from deep to superficial (see Figure 26–2b):

1. The **fibrous capsule**, a layer of collagen fibers that covers the outer surface of the entire organ
2. The **perinephric fat**, a cushioning, thick layer of adipose tissue that surrounds the fibrous capsule
3. The **renal fascia**, a dense, fibrous outer layer that anchors the kidney to surrounding structures. Collagen fibers extend outward from the fibrous capsule through the perinephric fat to this layer. Posteriorly, the renal fascia fuses with the deep fascia surrounding the muscles of the body wall. Anteriorly, the renal fascia forms a thick layer that fuses with the peritoneum.

In effect, each kidney hangs suspended by collagen fibers from the renal fascia and is packed in a soft cushion of adipose tissue. This arrangement prevents the jolts and shocks of day-to-day living from disturbing normal kidney function. If the suspensory fibers break or become detached, a slight bump or hit can displace the kidney and stress the attached vessels and ureter. This condition is called a *floating kidney*. It may cause pain or other problems from the distortion of the ureter or blood vessels.

Gross Anatomy of the Kidneys

As shown in Figure 26–3, the kidneys are bean-shaped organs served by renal blood vessels, lymphatics, and nerves adjacent to the proximal region of the ureters.

Figure 26–3 **The Gross Anatomy of the Urinary System.** The abdominopelvic cavity (with the digestive organs removed), showing the kidneys, ureters, urinary bladder, and blood supply to the urinary structures. ATLAS: Plates 61a; 62a,b

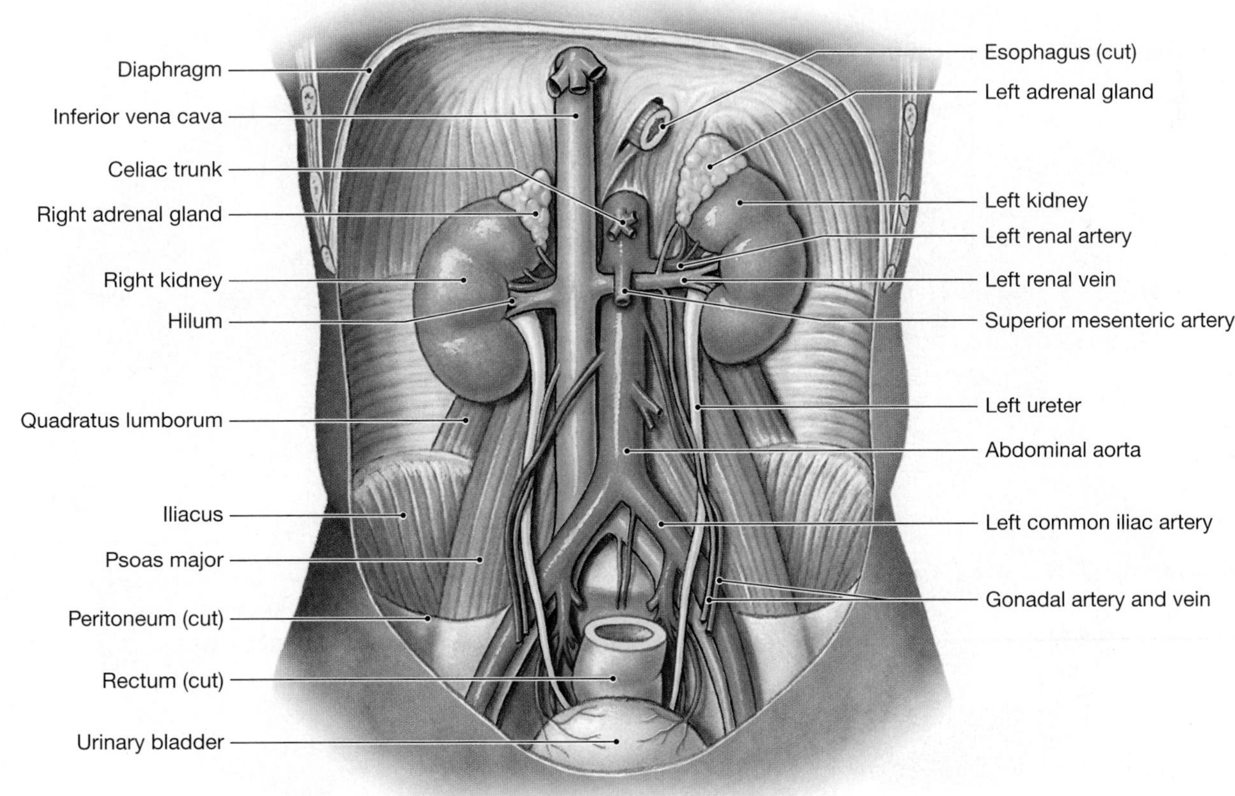
Diaphragm
Inferior vena cava
Celiac trunk
Right adrenal gland
Right kidney
Hilum
Quadratus lumborum
Iliacus
Psoas major
Peritoneum (cut)
Rectum (cut)
Urinary bladder
Esophagus (cut)
Left adrenal gland
Left kidney
Left renal artery
Left renal vein
Superior mesenteric artery
Left ureter
Abdominal aorta
Left common iliac artery
Gonadal artery and vein
Anterior view

26

External Anatomy of the Kidneys

A typical adult kidney is reddish-brown and about 10 cm (4 in.) long, 5.5 cm (2.2 in.) wide, and 3 cm (1.2 in.) thick (see Figure 26–3). Each kidney weighs about 150 g (5.25 oz). The **hilum**, a prominent medial indentation, is the point of entry for the *renal artery* and *renal nerves*. The hilum is also the point of exit for the *renal vein* and the ureter.

Internal Anatomy of the Kidneys

The fibrous capsule covering the outer surface of the kidney also lines the **renal sinus**, an internal cavity within the kidney (Figure 26–4). The fibrous capsule is bound to the outer surfaces of the structures within the renal sinus. In this way, it stabilizes the positions of the ureter, renal blood vessels, and nerves.

The kidney has an outer *cortex* and an inner *medulla*. The **renal cortex** is the superficial region of the kidney, in contact with the fibrous capsule. The cortex is reddish-brown and granular.

The **renal medulla** consists of 6 to 18 distinct triangular structures called **renal pyramids**. The base of each pyramid touches the cortex. The tip of each pyramid—a region known as the **renal papilla**—projects into the renal sinus. Each pyramid has a series of fine grooves that converge at the papilla. Bands of cortical tissue called **renal columns** extend into the medulla and separate adjacent renal pyramids. The columns have a distinctly granular texture, similar to that of the cortex. A **kidney lobe** consists of a renal pyramid, the overlying area of renal cortex, and adjacent tissues of the renal columns.

Urine is produced in the kidney lobes. Ducts within each renal papilla discharge urine into a cup-shaped drain called a **minor calyx** (KĀ-liks). Four or five minor calyces (KAL-ih-sēz) merge to form a **major calyx**, and two or three major calyces combine to form the **renal pelvis**, a large, funnel-shaped chamber. The renal pelvis fills most of the renal sinus and is connected to the ureter, which drains the kidney.

Urine production begins in microscopic, tubular structures called *nephrons* (NEF-ronz). Two types of nephrons found in the kidney are *cortical nephrons* (in the renal cortex) and *juxtamedullary nephrons* (in the renal medulla). Each kidney has about 1.25 million nephrons, with a combined length of about 145 km (85 miles).

Blood Supply and Innervation of the Kidneys

The kidneys receive 20–25 percent of the total cardiac output. In healthy individuals, about 1200 mL of blood flow through the kidneys each minute—a phenomenal amount of blood for organs with a combined weight of less than 300 g (10.5 oz)!

Each kidney receives blood through a **renal artery**. This vessel originates along the lateral surface of the abdominal aorta near the level of the superior mesenteric artery (look back at Figure 21–23a, p. 764). As it enters the renal sinus, the renal artery provides blood to the five **segmental arteries** (Figure 26–5a). Segmental arteries further divide into a series of **interlobar arteries**. These arteries radiate outward through the renal columns between the renal pyramids. The interlobar arteries supply blood to the **arcuate** (AR-kyū-āt)

Figure 26–4 The Internal Anatomy of the Kidney. ATLAS: Plates 57a,b; 61b

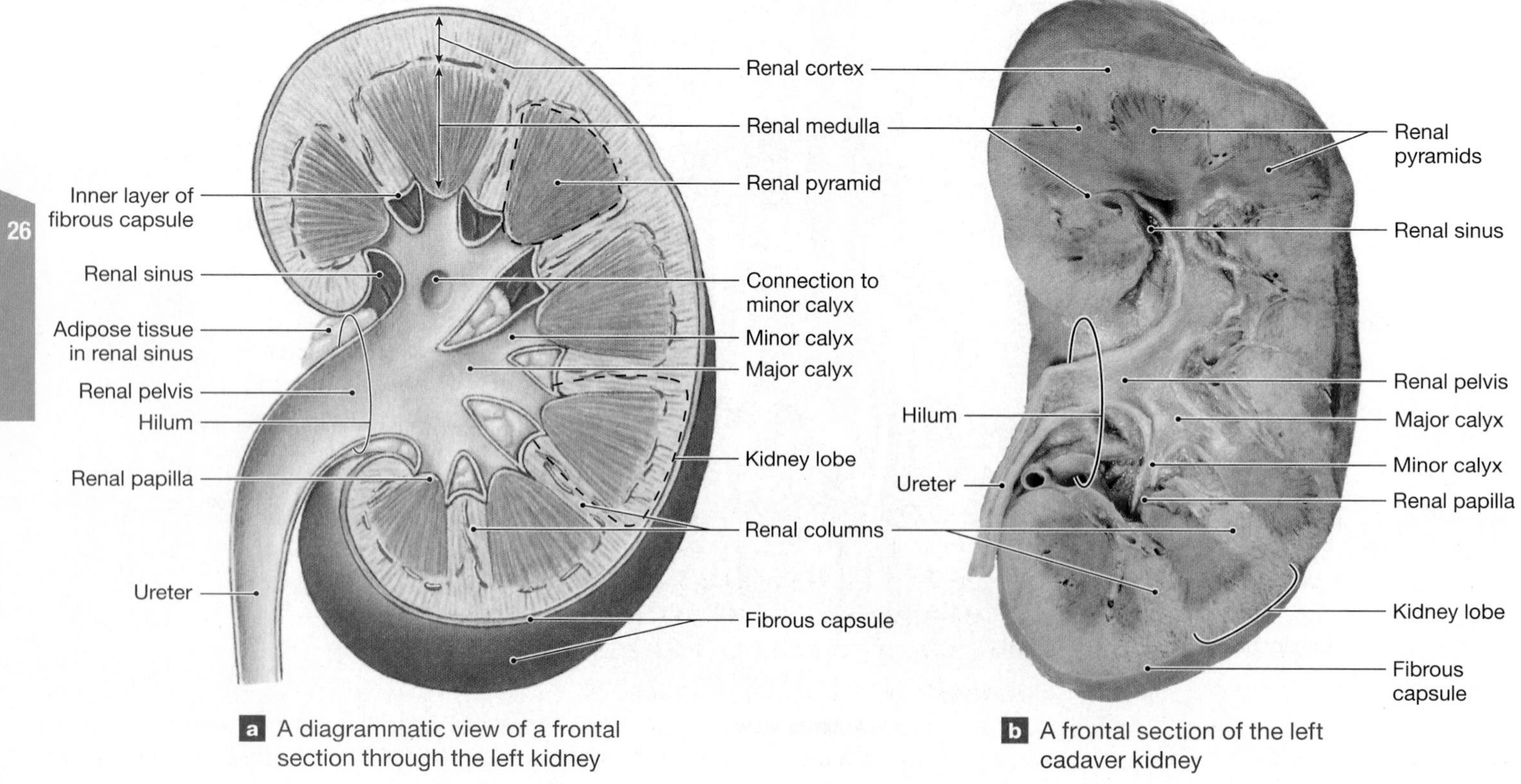
Renal cortex
Renal medulla
Renal pyramid
Inner layer of fibrous capsule
Renal sinus
Connection to minor calyx
Adipose tissue in renal sinus
Minor calyx
Major calyx
Renal pelvis
Hilum
Kidney lobe
Renal papilla
Renal columns
Ureter
Fibrous capsule
Renal pyramids
Renal sinus
Renal pelvis
Hilum
Major calyx
Minor calyx
Ureter
Renal papilla
Renal columns
Kidney lobe
Fibrous capsule

a A diagrammatic view of a frontal section through the left kidney

b A frontal section of the left cadaver kidney

Figure 26–5 **The Blood Supply to the Kidneys.** ATLAS: Plates 53c,d; 61a–c

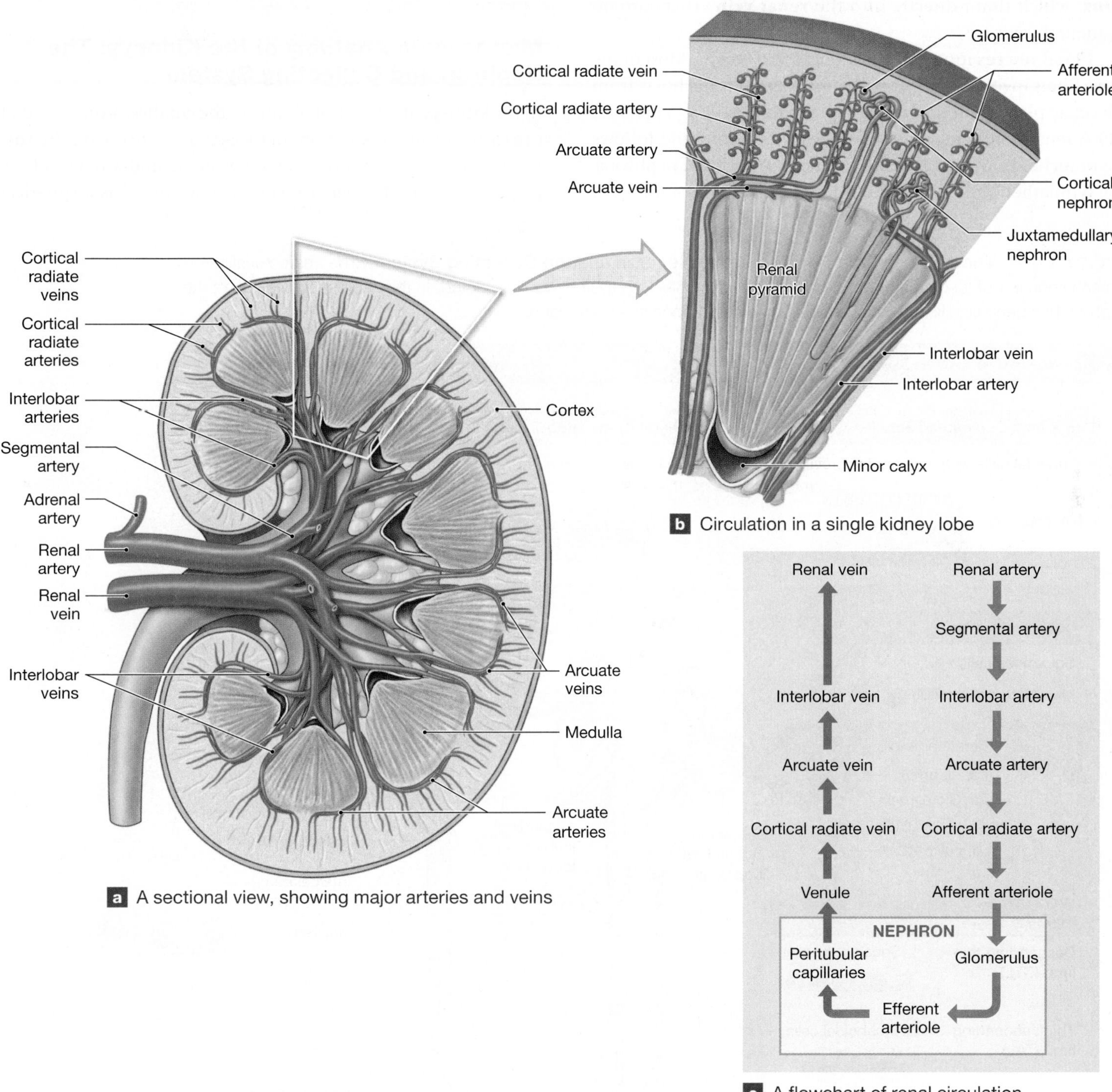
Cortical radiate veins
Cortical radiate arteries
Interlobar arteries
Segmental artery
Adrenal artery
Renal artery
Renal vein
Interlobar veins
Cortex
Arcuate veins
Medulla
Arcuate arteries
Cortical radiate vein
Cortical radiate artery
Arcuate artery
Arcuate vein
Glomerulus
Afferent arterioles
Cortical nephron
Juxtamedullary nephron
Renal pyramid
Interlobar vein
Interlobar artery
Minor calyx
Renal vein
Renal artery
Segmental artery
Interlobar vein
Interlobar artery
Arcuate vein
Arcuate artery
Cortical radiate vein
Cortical radiate artery
Venule
Afferent arteriole
NEPHRON
Peritubular capillaries
Glomerulus
Efferent arteriole

a A sectional view, showing major arteries and veins

b Circulation in a single kidney lobe

c A flowchart of renal circulation

? In which region of the kidney is a glomerulus located?

arteries, which arch along the boundary between the cortex and medulla of the kidney. Each arcuate artery gives rise to a number of **cortical radiate arteries**, also called *interlobular arteries*. They supply the cortical regions of the adjacent kidney lobes. Branching from each cortical radiate artery are numerous **afferent arterioles**. These vessels deliver blood to the capillaries supplying individual nephrons (Figure 26–5b,c).

After passing through the capillaries of the nephrons, blood enters a network of venules and small veins that converge on the **cortical radiate veins**, also called *interlobular veins* (see Figure 26–5a,c). The cortical radiate veins deliver blood to **arcuate veins**. These veins in turn empty into **interlobar veins**, which drain directly into the **renal vein**. There are no segmental veins.

Renal nerves innervate the kidneys and ureters. Most of the nerve fibers involved are sympathetic postganglionic fibers from the celiac plexus and the inferior splanchnic nerves. pp. 542, 549 A renal nerve enters each kidney at the hilum and follows the branches of the renal arteries to reach individual nephrons. The sympathetic innervation (1) adjusts rates of urine formation by changing blood flow at the nephron and (2) influences urine composition by stimulating release of the hormone *renin*.

Microscopic Anatomy of the Kidneys: The Nephron and Collecting System

In the kidneys, the functional units—the smallest structures that can carry out all the functions of the system—are the **nephrons**. Each nephron consists of a renal corpuscle and a renal tubule (Figure 26–6a). The *renal corpuscle* (KOR-pus-ul) is a spherical

Figure 26–6 **The Anatomy of a Representative Nephron and the Collecting System.** The characteristic cells of each segment of (a) the nephron (purple) and (b) the collecting system (tan) are noted. In this representative figure, the nephron has been significantly shortened and some components rearranged.

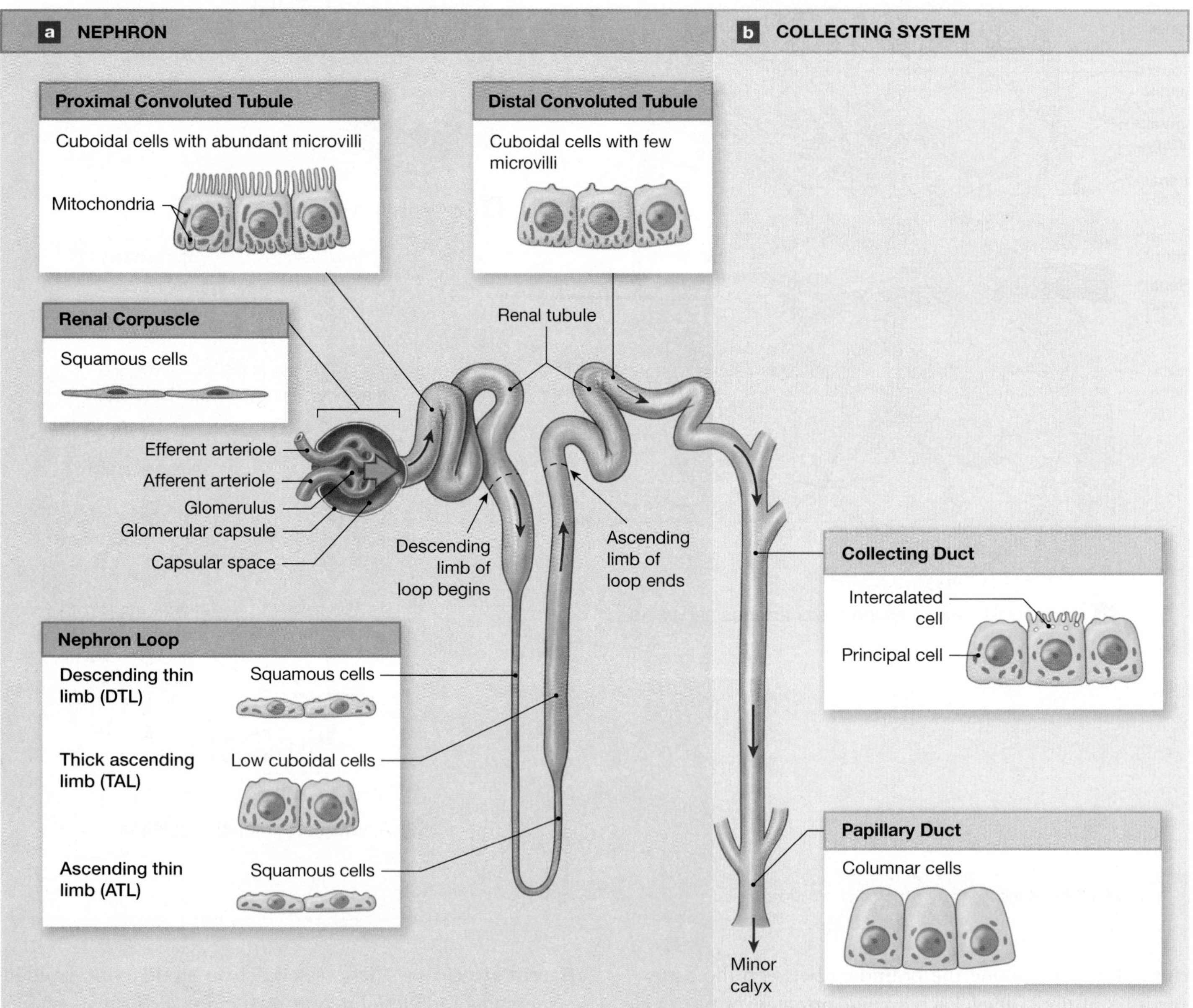
a NEPHRON
b COLLECTING SYSTEM
Proximal Convoluted Tubule
Cuboidal cells with abundant microvilli
Mitochondria
Distal Convoluted Tubule
Cuboidal cells with few microvilli
Renal Corpuscle
Squamous cells
Renal tubule
Efferent arteriole
Afferent arteriole
Glomerulus
Glomerular capsule
Capsular space
Descending limb of loop begins
Ascending limb of loop ends
Collecting Duct
Intercalated cell
Principal cell
Nephron Loop
Descending thin limb (DTL)
Squamous cells
Thick ascending limb (TAL)
Low cuboidal cells
Ascending thin limb (ATL)
Squamous cells
Papillary Duct
Columnar cells
Minor calyx

? What structure connects the proximal convoluted tubule to the distal convoluted tubule?

26

structure containing a capillary network that filters blood. The *renal tubule*, a long tubular passageway, begins at the renal corpuscle. Each renal tubule empties into the *collecting system*, a series of tubes that carry tubular fluid away from the nephron (Figure 26–6b). Let's examine the structure and basic functions of each segment of a representative nephron.

The Renal Corpuscle

Each **renal corpuscle** is 150–250 μm in diameter (Figure 26–7). It is made up of the *glomerular* (Bowman's) *capsule*, a cup-shaped chamber that envelopes the capillary network called the *glomerulus*.

The Glomerular Capsule and the Glomerulus. The outer wall of the renal corpuscle is formed by the **glomerular capsule**. This structure encapsulates the glomerular capillaries and is continuous with the initial segment of the renal tubule. The **glomerulus** (glo-MER-yū-lus; plural, *glomeruli*) consists of about 50 intertwined capillaries. Blood is delivered to the glomerulus by an afferent arteriole, and leaves the glomerulus in an **efferent arteriole** (see Figures 26–6, 26–7a).

The outer wall of the capsule is made up of a simple squamous epithelium and is called the **capsular outer layer** (see Figure 26–7a). This layer ends at the **visceral layer**, which covers the glomerular capillaries. The **capsular space** (*urinary space*) separates the capsular outer layer and visceral layer. The two layers are continuous where the glomerular capillaries are connected to the afferent arteriole and efferent arteriole. Therefore this space is continuous with the lumen of the renal tubule (look back at Figure 26–6a).

The visceral layer of the glomerulus consists of large cells with complex processes, or "feet," that wrap around the *dense layer*, a specialized basement membrane of the glomerular capillaries. These visceral cells of the capsule are called **podocytes** (PŌ-dō-sīts; *podos*, foot + *-cyte*, cell). Their feet are known as **foot processes**, or *pedicels* (see Figure 26–7b). There are narrow gaps called **filtration slits** between adjacent foot processes. These slits are only 6–9 nm wide.

Figure 26–7 The Renal Corpuscle.

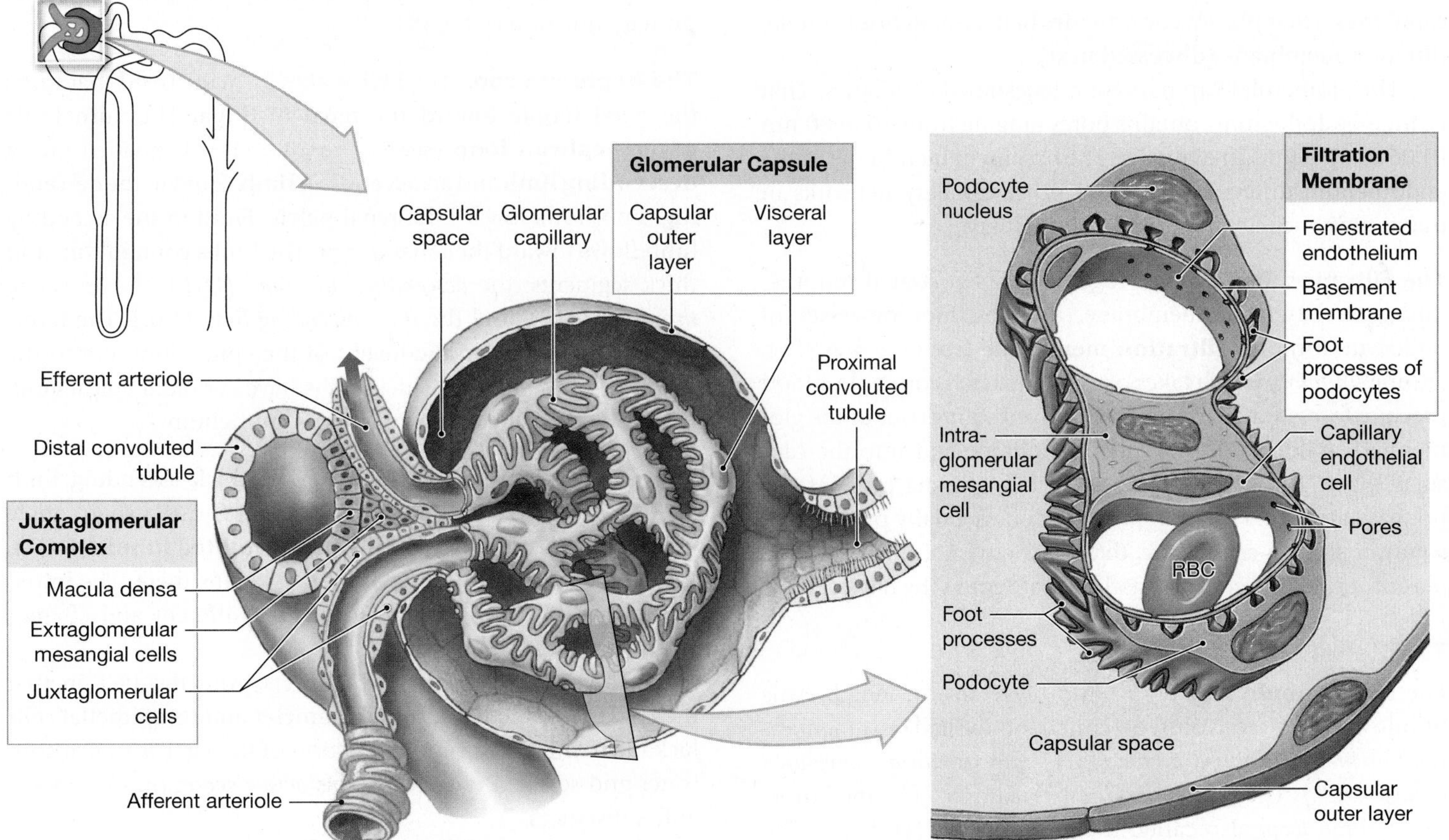

Glomerular Capsule
Capsular space
Glomerular capillary
Capsular outer layer
Visceral layer
Efferent arteriole
Distal convoluted tubule
Proximal convoluted tubule
Juxtaglomerular Complex
Macula densa
Extraglomerular mesangial cells
Juxtaglomerular cells
Afferent arteriole
Podocyte nucleus
Filtration Membrane
Fenestrated endothelium
Basement membrane
Foot processes of podocytes
Intra-glomerular mesangial cell
Capillary endothelial cell
Pores
RBC
Foot processes
Podocyte
Capsular space
Capsular outer layer

a Important structural features of a renal corpuscle.

b This cross section through a portion of the glomerulus shows the components of the filtration membrane of the nephron.

26

Clinical Note Glomerulonephritis

Glomerulonephritis (glo-mer-yū-lō-nef-RĪ-tis) is an inflammation of the glomeruli that impairs filtration in the kidneys. The condition is often an *immune complex disorder*. It may develop after an infection involving *Streptococcus* bacteria. The kidneys are not the sites of infection, but as the immune system responds, the number of circulating antigen–antibody complexes rapidly increases. These complexes are small enough to pass through the basement membrane, but too large to fit between podocyte foot processes of the filtration membrane. The complexes clog up the filtration mechanism. As a result, filtrate production decreases. Any condition that leads to a massive immune response, including viral infections and autoimmune disorders, can cause glomerulonephritis.

Intraglomerular mesangial cells are located among the glomerular capillaries (see Figure 26–7b). They are specialized cells derived from smooth muscle. Functions of these contractile cells include structural support, filtration, and phagocytosis. Their contraction decreases the luminal diameter of the capillaries. Their phagocytic activities help clear debris from the filtration membrane (discussed next).

The glomerular capillaries are fenestrated capillaries. That is, their endothelium contains pores ranging from 60 to 80 nm (0.06 to 0.08 μm) in diameter. The basement membrane of the endothelium differs from that in other capillary networks in that it may encircle more than one capillary.

The Filtration Membrane. Together, the fenestrated endothelium, the basement membrane, and the foot processes of podocytes form the **filtration membrane** (see Figure 26–7b). During *filtration*, which takes place in the renal corpuscle, blood pressure forces water and small dissolved solutes out of the glomerular capillaries through this membrane and into the capsular space. The larger solutes, especially plasma proteins, do not pass through. Filtration produces an essentially protein-free solution, known as a *filtrate*, that is otherwise similar to blood plasma. From the renal corpuscle, filtrate enters the renal tubule.

26

The Renal Tubule

The **renal tubule** may be 50 mm (1.97 in.) in length. This tubule has two convoluted (coiled or twisted) segments—the *proximal convoluted tubule (PCT)* and the *distal convoluted tubule (DCT)*. They are separated by a simple U-shaped tube, the *nephron loop*, also called the *loop of Henle* (HEN-lē). The convoluted segments lie in the cortex of the kidney, and the nephron loop dips at least partially into the medulla. For clarity, the nephron shown in Figure 26–6a has been shortened and straightened.

Tips & Tools

In the nephron, the terms *proximal tubule* and *distal tubule* refer to how far along the renal tubule these structures are located from the renal corpuscle. Think of *proximal* (nearer the renal corpuscle) as being first, and *distal* (farther from the renal corpuscle) as being last.

As the filtrate travels along the renal tubule, it is now called *tubular fluid*, and it gradually changes in composition. This is because the main functions of the renal tubule are reabsorption and secretion. When a substance is *reabsorbed*, it is "reclaimed," eventually reentering the blood. When a substance is *secreted*, it enters the tubular fluid from the blood. The changes that take place in the tubular fluid, and the characteristics of the urine that result, are due to the activities under way in each segment of the nephron.

The Proximal Convoluted Tubule. The **proximal convoluted tubule (PCT)** is the first segment (see Figure 26–6a). Its entrance lies almost directly opposite the point where the afferent and efferent arterioles connect to the glomerulus. The lining of the PCT is a simple cuboidal epithelium whose apical surfaces have microvilli. Reabsorption of critical ions is the primary function of the PCT.

The Nephron Loop. The PCT makes an acute bend that turns the renal tubule toward the renal medulla. This turn leads to the **nephron loop** (see Figure 26–6a). It is divided into a **descending limb** and an **ascending limb**. Fluid in the descending limb flows toward the renal pelvis. Fluid in the ascending limb flows toward the renal cortex. The limbs contain thin and thick segments: the *descending thin limb (DTL)*, the *ascending thin limb (ATL)*, and the *thick ascending limb (TAL)*. The terms *thick* and *thin* refer to the height of the epithelium, not to the diameter of the lumen. Thick limbs have cuboidal epithelium. Thin limbs are lined with squamous epithelium.

The Distal Convoluted Tubule. The thick ascending limb (TAL) of the nephron loop ends where it forms a sharp angle near the renal corpuscle. The **distal convoluted tubule (DCT)**, the third segment of the renal tubule, begins there. The initial portion of the DCT passes between the afferent and efferent arterioles (see Figure 26–7a).

In sectional view, the DCT differs from the PCT in that the DCT has a smaller luminal diameter and its epithelial cells lack microvilli. The primary function of the DCT is to reabsorb water and selected ions, as well as active secretion of undesirable substances.

The Juxtaglomerular Complex

The **juxtaglomerular complex (JGC)** is a structure that helps regulate blood pressure and filtrate formation. It consists of the *macula densa, juxtaglomerular cells* (granular cells), and *extraglomerular mesangial cells*. The **macula densa** (MAK-yū-lah DEN-sah) is a group of tall, closely packed epithelial cells in the distal tubular epithelium. They function as either chemoreceptors or baroreceptors. The **juxtaglomerular cells** are modified smooth muscle cells in the wall of the afferent arteriole that secrete renin. These cells function as baroreceptors that monitor blood pressure in the afferent arteriole. **Extraglomerular mesangial cells** are located in the triangular space between the afferent and efferent glomerular arterioles. These cells provide feedback control between the macula densa and the juxtaglomerular cells (see Figure 26–7a).

The Collecting System

The distal convoluted tubule, the last segment of the nephron, opens into the **collecting system** (see Figure 26–6b). Many individual nephrons drain their tubular fluid into a nearby **collecting duct**. Each collecting duct begins in the cortex and descends into the medulla. Several collecting ducts then converge into a larger **papillary duct**, which in turn empties into a *minor calyx*. The epithelium lining the papillary duct is typically columnar.

Two main types of cells are found in the collecting duct: *intercalated cells* and *principal cells*. Intercalated cells are cuboidal cells with microvilli. Type A intercalated cells and Type B intercalated cells make up the population of intercalated cells. Together, these cells regulate the acid–base balance in the blood. Principal cells are cuboidal cells that lack microvilli and reabsorb water and secrete potassium ions.

The collecting system transports tubular fluid from the nephrons to the renal pelvis. Along the way, it also adjusts the fluid's composition and determines the final osmotic concentration and volume of urine, the final product. We consider these activities of the collecting system later in the chapter.

Table 26–1 summarizes the regional specializations and general functions of the parts of the nephron and the collecting system.

Types of Nephrons

Nephrons from different locations within a kidney differ slightly in structure. Approximately 85 percent of all nephrons are **cortical nephrons**, located almost entirely within the superficial cortex of the kidney (Figure 26–8a,b). In a cortical nephron, the nephron loop is relatively short, and the efferent arteriole delivers blood to a network of **peritubular capillaries**, which surround the entire renal tubule. These capillaries drain

Figure 26–8 **The Locations and Structures of Cortical and Juxtamedullary Nephrons.**

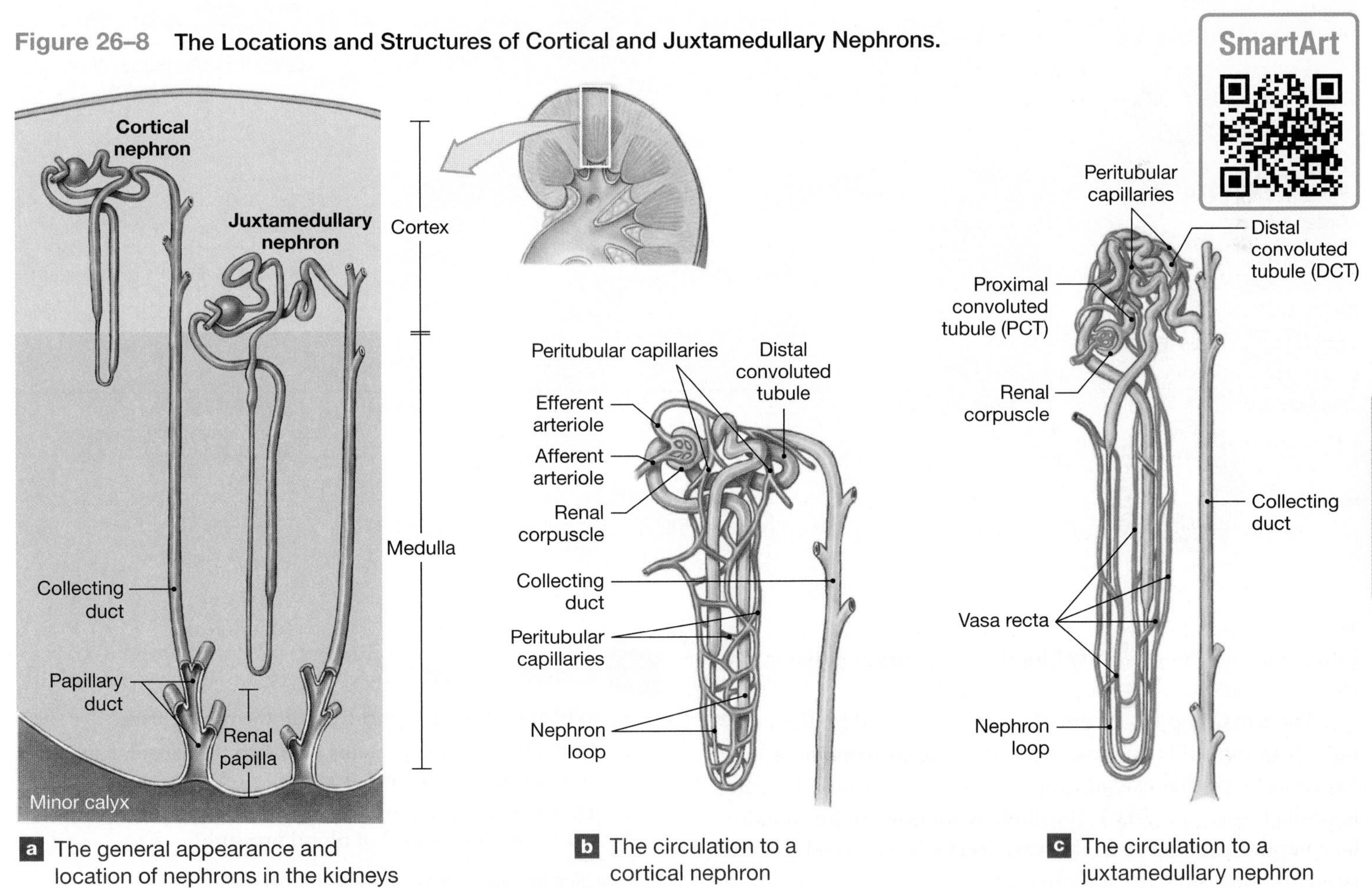

SmartArt
Cortical nephron
Juxtamedullary nephron
Cortex
Medulla
Collecting duct
Papillary duct
Renal papilla
Minor calyx
Peritubular capillaries
Distal convoluted tubule
Efferent arteriole
Afferent arteriole
Renal corpuscle
Collecting duct
Peritubular capillaries
Nephron loop
Peritubular capillaries
Distal convoluted tubule (DCT)
Proximal convoluted tubule (PCT)
Renal corpuscle
Collecting duct
Vasa recta
Nephron loop

a The general appearance and location of nephrons in the kidneys

b The circulation to a cortical nephron

c The circulation to a juxtamedullary nephron

26

Table 26–1 The Organization of the Nephron and Collecting System

Region	Histological Characteristics	Length	Diameter	Primary Function
NEPHRON				
Renal corpuscle	Glomerulus (capillary knot), intraglomerular mesangial cells, and basement membrane, enclosed by the glomerular capsule; visceral layer (podocytes) and capsular outer layer (simple squamous epithelium) separated by capsular space	150–250 µm (spherical)	150–250 µm	Filtration of blood plasma
Renal tubule				
Proximal convoluted tubule (PCT)	Cuboidal cells with microvilli	14 mm	60 µm	Reabsorption of ions, organic molecules, vitamins, water; secretion of drugs, toxins, acids
Nephron loop	Squamous or cuboidal cells	30 mm		
			15 µm	Descending limb: reabsorption of water from tubular fluid
			30 µm	Ascending limb: reabsorption of ions; assists in creation of a concentration gradient in the renal medulla
Distal convoluted tubule (DCT)	Cuboidal cells with few if any microvilli	5 mm	30–50 µm	Reabsorption of sodium ions and calcium ions; secretion of acids, ammonia, drugs, toxins
COLLECTING SYSTEM				
Collecting duct	Cuboidal cells	15 mm	50–100 µm	Reabsorption of water, sodium ions; secretion or reabsorption of bicarbonate ions or hydrogen ions
Papillary duct	Columnar cells	5 mm	100–200 µm	Conduction of tubular fluid to minor calyx; contributes to concentration gradient of the medulla

Capsular space
Capsular outer layer
Capillaries of glomerulus
Visceral layer
Microvilli
Intercalated cell
Principal cell

26

into small venules that carry blood to the cortical radiate veins (look back at **Figure 26–5c**).

The remaining 15 percent of nephrons, termed **juxtamedullary** (juks-tuh-MED-yū-lār-ē; *juxta*, near) **nephrons**, have long nephron loops that extend deep into the renal pyramids of the medulla (**Figure 26–8a,c**). The efferent arterioles of juxtamedullary nephrons connect to the **vasa recta** (*vasa*, vessel + *recta*, straight), long, straight capillaries that parallel the nephron loop.

✓ Checkpoint

3. **Which parts of a nephron are in the renal cortex?**
4. **Why don't plasma proteins pass into the capsular space under normal circumstances?**
5. **Damage to which part of a nephron would interfere with the hormonal control of blood pressure?**

See the blue Answers tab at the back of the book.

> Go to MasteringA&P™ > Study Area > IP2™ > Urinary System: Glomerular Filtration

26-3 Different segments of the nephron form urine by filtration, reabsorption, and secretion

Learning Outcome **Describe the basic processes that form urine.**

Although you may not have thought of it this way, the primary function of urine production is to maintain homeostasis by regulating the volume and composition of blood. This process involves the excretion of solutes—specifically, **metabolic wastes**.

Metabolic Wastes

Our bodies form three important metabolic wastes:

- *Urea.* **Urea** is the most abundant organic waste. You generate approximately 21 g of urea each day, most of it through the breakdown of amino acids.
- *Creatinine.* Skeletal muscle tissue generates **creatinine** through the breakdown of *creatine phosphate*, a high-energy compound that plays an important role in muscle contraction. ⮌ p. 320 Your body generates about 1.8 g of creatinine each day. Virtually all of it is excreted in urine.
- *Uric Acid.* **Uric acid** is a waste formed during the recycling of the nitrogenous bases from RNA molecules. You produce approximately 480 mg of uric acid each day.

These wastes are dissolved in the bloodstream. They can be eliminated only when dissolved in urine. For this reason, their removal involves an unavoidable water loss. The kidneys are usually capable of producing urine with an osmotic concentration more than four times that of plasma. If the kidneys were unable to concentrate the filtrate produced by filtration, fluid losses would lead to fatal dehydration in a matter of hours.

The kidneys also ensure that the fluid that *is* lost does not contain potentially useful substances that are present in blood plasma, such as sugars or amino acids. These valuable materials must be reabsorbed and retained for use by other tissues.

Basic Processes of Urine Formation

Let's look at the three distinct processes each nephron uses to produce urine:

1. *Filtration.* In **filtration**, blood pressure forces water and solutes across the walls of the glomerular capillaries and into the capsular space. Solute molecules small enough to pass through the filtration membrane are carried by the surrounding water molecules.
2. *Reabsorption.* **Reabsorption** is the removal of water and solutes from the filtrate, and their movement across the tubular epithelium and into the **peritubular fluid**, the interstitial fluid surrounding the renal tubule. Reabsorption takes place after the filtrate has left the renal corpuscle. Most of the reabsorbed substances are nutrients the body can use. Filtration is based solely on particle size, but reabsorption is a selective process. Reabsorption involves either simple diffusion or carrier proteins in the tubular epithelium. The reabsorbed substances in the peritubular fluid eventually reenter the blood. Water reabsorption takes place passively, through osmosis.
3. *Secretion.* **Secretion** is the transport of solutes from the peritubular fluid, across the tubular epithelium, and into the tubular fluid. Secretion is necessary because filtration does not force all the dissolved substances out of the plasma. Tubular secretion, which removes substances from the blood, can further lower the plasma concentration of undesirable materials. It provides a backup process for filtration. Secretion is often the primary method of preparing substances, including many drugs, for excretion.

Figure 26–9 summarizes where these processes occur in the various segments of the nephron and collecting system. Most regions carry out a combination of reabsorption and secretion. Note that the balance between the two processes shifts from one region to another.

Together, these processes produce urine, a fluid that is very different from other body fluids. Table 26–2 shows the functional efficiency of the kidneys by comparing the concentrations of solutes in urine and plasma.

Normal kidney physiology can continue only as long as filtration, reabsorption, and secretion function within relatively narrow limits. A disruption in kidney function has immediate effects on the composition of the circulating blood. If both kidneys are affected, death follows within a few days unless medical assistance is provided.

Table 26–2 Normal Laboratory Values for Solutes in Plasma and Urine

Solute	Plasma	Urine
IONS (mEq/L)		
Sodium (Na^+)	135–145	40–220
Potassium (K^+)	3.5–5.0	25–100
Chloride (Cl^-)	100–108	110–250
Bicarbonate (HCO_3^-)	20–28	1–9
METABOLITES AND NUTRIENTS (mg/dL)		
Glucose	70–110	0.009
Lipids	450–1000	0.002
Amino acids	40	0.188
Proteins	6–8 g/dL	0.000
NITROGENOUS WASTES (mg/dL)		
Urea	8–25	1800
Creatinine	0.6–1.5	150
Uric acid	2–6	40
Ammonia	<0.1	60

Figure 26–9 An Overview of Urine Formation.

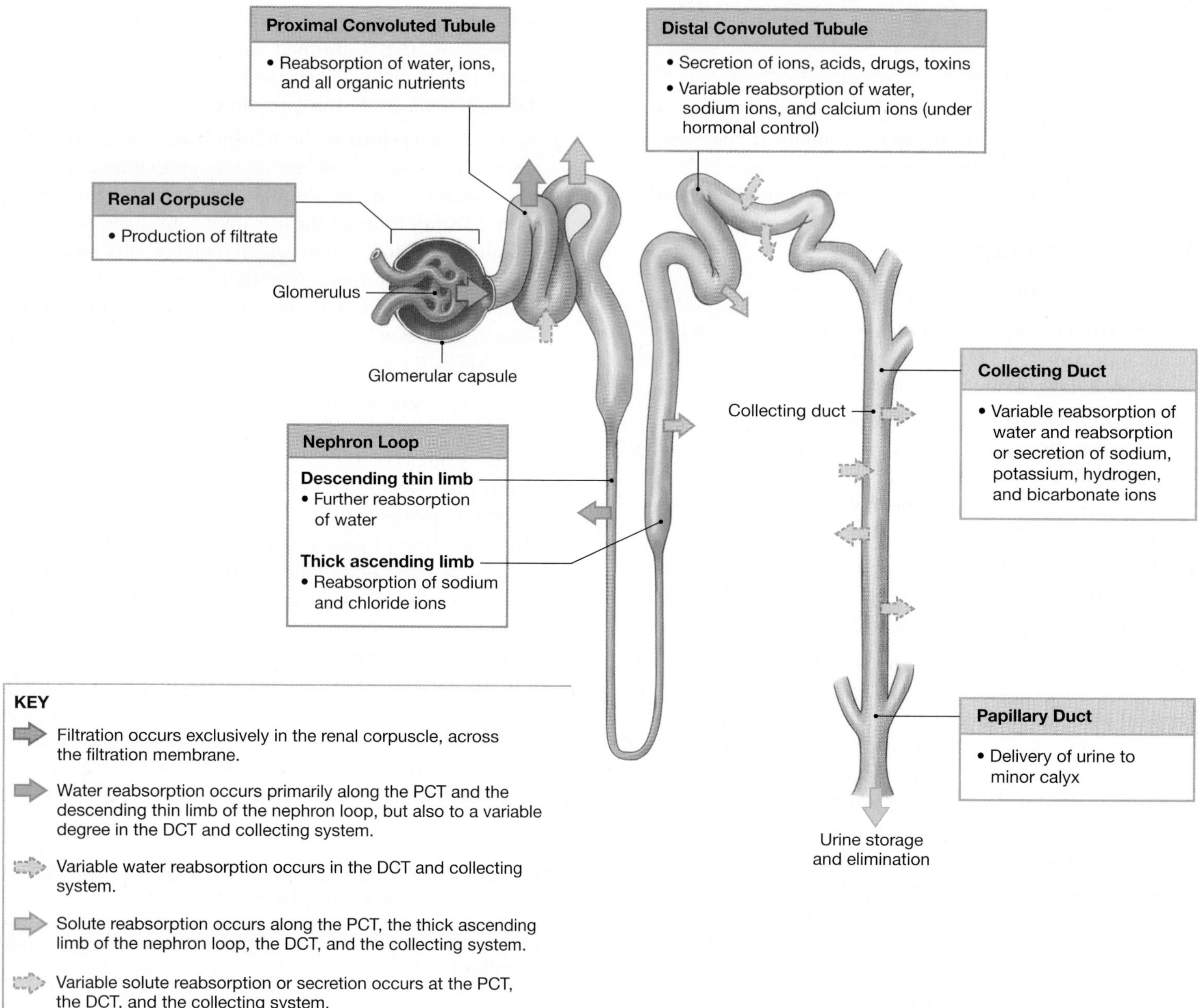
Proximal Convoluted Tubule
• Reabsorption of water, ions, and all organic nutrients
Distal Convoluted Tubule
• Secretion of ions, acids, drugs, toxins
• Variable reabsorption of water, sodium ions, and calcium ions (under hormonal control)
Renal Corpuscle
• Production of filtrate
Glomerulus
Glomerular capsule
Collecting duct
Collecting Duct
• Variable reabsorption of water and reabsorption or secretion of sodium, potassium, hydrogen, and bicarbonate ions
Nephron Loop
Descending thin limb
• Further reabsorption of water
Thick ascending limb
• Reabsorption of sodium and chloride ions
KEY
Filtration occurs exclusively in the renal corpuscle, across the filtration membrane.
Water reabsorption occurs primarily along the PCT and the descending thin limb of the nephron loop, but also to a variable degree in the DCT and collecting system.
Variable water reabsorption occurs in the DCT and collecting system.
Solute reabsorption occurs along the PCT, the thick ascending limb of the nephron loop, the DCT, and the collecting system.
Variable solute reabsorption or secretion occurs at the PCT, the DCT, and the collecting system.
Papillary Duct
• Delivery of urine to minor calyx
Urine storage and elimination

26

In the next sections, we proceed along the nephron to consider the formation of filtrate and its subsequent changes in composition and concentration as it passes along the individual segments of the renal tubule. Most of what follows applies equally to cortical and juxtamedullary nephrons. The functions of the renal corpuscle and of the proximal and distal convoluted tubules are the same in all nephrons. Cortical nephrons do most of the reabsorption and secretion in the kidneys because they are more numerous than juxtamedullary nephrons.

However, as you will see later in the chapter, it is the juxtamedullary nephrons that enable the kidneys to produce concentrated urine. Their long nephron loops play a key role in water conservation and the formation of concentrated urine. This process is vitally important, affecting the tubular fluid produced by every nephron in the kidney. For this reason, we use a juxtamedullary nephron as our example.

Checkpoint

6. **Identify the three distinct processes that form urine in the kidney.**

See the blue Answers tab at the back of the book.

26-4 The glomerulus filters blood through the filtration membrane to produce filtrate; several pressures determine the glomerular filtration rate

Learning Outcome **Describe the factors that influence glomerular filtration pressure and the rate of filtrate formation.**

In the body, the heart pushes blood around the cardiovascular system and generates *hydrostatic pressure*. As you have seen, this pressure causes filtration to occur across the walls of capillaries as water and dissolved materials are pushed into the interstitial fluids of the body (look back at Figure 21–10, p. 742). In kidney nephrons, this hydrostatic pressure drives *glomerular filtration*. Glomerular filtration is the vital first step for all other kidney functions. If filtration does not take place, wastes are not excreted, pH control is jeopardized, and an important mechanism for regulating blood volume is lost.

In this section we look at the pressures that affect the rate of glomerular filtration, and how that filtration is regulated. But first we revisit the filtration membrane, and take a closer look at how this process occurs.

Function of the Filtration Membrane

In **glomerular filtration**, blood plasma is forced through the pores of the specialized filtration membrane, and solute molecules small enough to pass through those pores are carried along. The filtration membrane restricts the passage of larger solutes and suspended materials from the plasma, producing filtrate.

We can see filtration in action in a drip coffee machine. Gravity forces hot water through the filter, and the water carries a variety of dissolved substances into the pot. The large coffee grounds never reach the pot, because they cannot fit through the pores of the filter. In other words, they are "filtered out" of the solution; the coffee we drink is the filtrate.

Recall that the filtration membrane has three parts: (1) the fenestrated endothelium, (2) the basement membrane, and (3) the foot processes of podocytes (Figures 26–7b and 26–10a). In glomerular filtration, water from the blood plasma is forced through all three parts of this membrane, with each filtering solutes of smaller size out of the final filtrate. The glomerular capillaries are made up of fenestrated endothelium with pores small enough to prevent the passage of blood cells, but too large to restrict the diffusion of solutes, even those the size of plasma proteins. The basement membrane is more selective: Only small plasma proteins, nutrients, and ions can cross it. The foot processes are the finest filters of all. The narrow gaps between the foot processes prevent the passage of most small plasma proteins. As a result, under normal circumstances only a few plasma proteins—such as albumin molecules, with an average diameter of 7 nm—can cross the filtration membrane and enter the capsular space. However, plasma proteins are all that stay behind, so the filtrate contains dissolved ions and small organic molecules in roughly the same concentrations as in plasma.

Filtration Pressures

We discussed the major forces that act across capillary walls in Chapters 21 and 22. (You may find it helpful to review Figures 21–10 and 21–11, pp. 742, 743, before you proceed.) The primary factor involved in glomerular filtration is basically the same one that regulates fluid and solute movement across capillaries throughout the body. This factor is the balance between **hydrostatic pressure**, or fluid pressure, and **colloid osmotic pressure**, or pressure due to materials in solution, on each side of the capillary walls.

Hydrostatic Pressure

Blood pressure is low in typical systemic capillaries. The reason is that capillary blood flows into the venous system, where resistance is fairly low. However, at the glomerulus, blood leaving the glomerular capillaries flows into an efferent glomerular arteriole, whose luminal diameter is *smaller* than that of the afferent glomerular arteriole. For this reason, the efferent arteriole offers considerable resistance. Relatively high pressures are needed to force blood into it. As a result, glomerular pressures are similar to those of small arteries. *Glomerular hydrostatic pressure* (*GHP*) averages about 50 mm Hg, instead of the 35 mm Hg typical of peripheral capillaries. GHP is shown with other pressures affecting glomerular filtration in Figure 26–10b.

Capsular hydrostatic pressure (*CsHP*) opposes glomerular hydrostatic pressure. The CsHP results from the resistance to flow along the nephron and the conducting system. (Before additional filtrate can enter the capsule, some of the filtrate already present must be forced into the PCT.) The CsHP averages about 15 mm Hg.

The *net hydrostatic pressure (NHP)* is the difference between the glomerular hydrostatic pressure, which tends to push water and solutes out of the bloodstream, and the capsular hydrostatic pressure, which tends to push water and solutes into the bloodstream. We can calculate net hydrostatic pressure as follows:

$$\text{NHP} = (\text{GHP} - \text{CsHP}) = (50\text{ mm Hg} - 15\text{ mm Hg}) = 35\text{ mm Hg}$$

Colloid Osmotic Pressure

The *blood colloid osmotic pressure (BCOP)* is the osmotic pressure resulting from suspended proteins in the blood (see Figure 26–10b). Under normal conditions, very few plasma proteins enter the capsular space, so no opposing colloid osmotic pressure exists within the capsule. However, if the glomeruli are damaged by disease or injury, and plasma proteins begin passing into the capsular space, a *capsular colloid osmotic pressure* is created that promotes filtration and increases fluid losses in urine.

26

Figure 26–10 Glomerular Filtration.

Efferent arteriole
Glomerulus
Afferent arteriole
Podocyte
Basement membrane
Capillary lumen
Filtration slit
Foot processes of podocytes
Pore
Capsular space
Filtration membrane

a The glomerular filtration membrane

Factors Controlling Glomerular Filtration

The **glomerular hydrostatic pressure (GHP)** is the blood pressure in the glomerular capillaries. This pressure tends to push water and solute molecules out of the plasma and into the filtrate. The GHP, which averages 50 mm Hg, is significantly higher than capillary pressures elsewhere in the systemic circuit, because the efferent arteriole is smaller in diameter than the afferent arteriole.

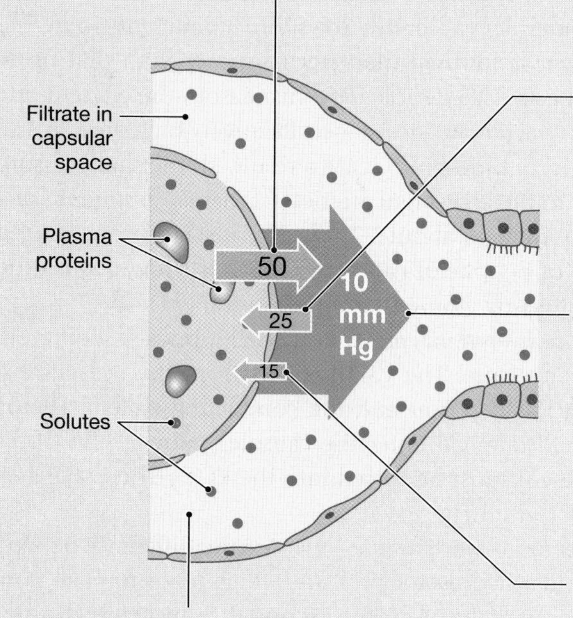
Filtrate in capsular space
Plasma proteins
50
25
15
10 mm Hg
Solutes

The **blood colloid osmotic pressure (BCOP)** tends to draw water out of the filtrate and into the plasma; it thus opposes filtration. Over the entire length of the glomerular capillary bed, the BCOP averages about 25 mm Hg.

The **net filtration pressure (NFP)** is the net pressure acting across the glomerular capillaries. It represents the sum of the hydrostatic pressures and the colloid osmotic pressures. Under normal circumstances, the net filtration pressure is approximately 10 mm Hg. This is the average pressure forcing water and dissolved substances out of the glomerular capillaries and into the capsular space.

Capsular hydrostatic pressure (CsHP) opposes GHP. CsHP, which tends to push water and solutes out of the filtrate and into the plasma, results from the resistance of filtrate already present in the nephron that must be pushed toward the renal pelvis. The difference between GHP and CsHP is the *net hydrostatic pressure (NHP)*.

The **capsular colloid osmotic pressure** is usually zero because few, if any, plasma proteins enter the capsular space.

b Net filtration pressure

Net Filtration Pressure

The *net filtration pressure* (*NFP*) in the glomerulus is the difference between the net hydrostatic pressure and the blood colloid osmotic pressure acting across the glomerular capillaries. Under normal circumstances, we can summarize this relationship as

$$\text{NFP} = \text{NHP} - \text{BCOP}$$

or, using the values in Figure 26–10b,

$$\text{NFP} = 35 \text{ mm Hg} - 25 \text{ mm Hg} = 10 \text{ mm Hg}$$

This is the average pressure forcing water and dissolved substances out of the glomerular capillaries and into the capsular spaces (see Figure 26–10b). Problems that affect the net filtration pressure can seriously disrupt kidney function and cause a variety of clinical signs and symptoms.

The Glomerular Filtration Rate (GFR)

The **glomerular filtration rate (GFR)** is the amount of filtrate the kidneys produce each minute. Each kidney contains about

6 m^2—some 64 square feet—of filtration surface, and the GFR averages an astounding *125 mL per minute*. This means that nearly 10 percent of the fluid delivered to the kidneys by the renal arteries leaves the bloodstream and enters the capsular spaces. In the course of a single day, the glomeruli generate about 180 liters (48 gal) of filtrate, approximately 70 times the total plasma volume.

The glomerular filtration rate depends on the net filtration pressure across the glomerular capillaries. Any factor that alters the net filtration pressure also alters the GFR and affects kidney function. One of the most significant factors would be a decrease in renal blood pressure. If blood pressure at the glomeruli decreases by 20 percent (from 50 mm Hg to 40 mm Hg), kidney filtration stops, because the net filtration pressure falls to 0 mm Hg. For this reason, the kidneys are sensitive to changes in blood pressure that have little or no effect on other organs. Hemorrhaging, shock, and dehydration are relatively common clinical conditions that can cause a dangerous decrease in the GFR and lead to *acute renal failure* (loss of kidney function) (p. 1014).

Regulation of the GFR

Filtration depends on adequate blood flow to the glomerulus and on the maintenance of normal filtration pressures. Three interacting levels of control ensure that the GFR remains within normal limits: (1) *autoregulation* occurring at the local level, (2) *hormonal regulation* initiated by the kidneys, and (3) *autonomic regulation* maintained primarily by the sympathetic division of the autonomic nervous system.

Autoregulation of the GFR

Autoregulation (local blood flow regulation) maintains an adequate GFR despite changes in local blood pressure and blood flow. *Myogenic mechanisms*—how arteries and arterioles react to an increase or decrease in blood pressure—play a role in the autoregulation of blood flow. Changes to the luminal diameters of afferent arterioles, efferent arterioles, and glomerular capillaries maintain GFR.

The most important regulatory mechanisms stabilize the GFR when systemic blood pressure decreases. One of these mechanisms relies on the supporting intraglomerular mesangial cells. Actin-like filaments in these cells enable them to contract. In this way, these cells control capillary luminal diameter and the rate of glomerular capillary blood flow. ⮌ p. 984

The GFR also remains relatively constant when systemic blood pressure increases. An increase in renal blood pressure stretches the walls of afferent glomerular arterioles, and the smooth muscle cells respond by contracting. The reduction in the luminal diameter of afferent glomerular arterioles decreases glomerular blood flow and keeps the GFR within normal limits.

If autoregulation is ineffective, the kidneys call for a wider response to increase the GFR (**Figure 26–11**).

Hormonal Regulation of the GFR

The GFR is regulated by the hormones of the **renin-angiotensin-aldosterone system (RAAS)** and the *natriuretic peptides*. We introduced these hormones and their actions in Chapters 18 and 21. ⮌ pp. 641, 643–644, 749–752 The release of **renin** ultimately restricts water and salt loss in the urine by stimulating reabsorption by the nephron. There are three triggers for the release of renin by the juxtaglomerular complex (JGC): (1) a decrease in blood pressure at the glomerulus as the result of a decrease in blood volume, a decrease in systemic pressures, or a blockage in the renal artery or its branches; (2) stimulation of juxtaglomerular cells by sympathetic innervation; or (3) a decrease in the osmotic concentration of the tubular fluid at the macula densa.

These triggers are often interrelated. For example, a decrease in systemic blood pressure reduces the glomerular filtration rate, while baroreceptor reflexes cause sympathetic activation. Meanwhile, a decrease in the GFR slows the movement of tubular fluid along the nephron. As a result, the tubular fluid is in the ascending limb of the nephron loop longer, and the concentration of sodium and chloride ions in the tubular fluid reaching the macula densa and DCT becomes abnormally low.

Figure 26–11 shows how the renin-angiotensin-aldosterone system responds to a decrease in GFR. A decrease in GFR leads to the release of renin by the juxtaglomerular complex. Renin converts the inactive plasma protein angiotensinogen to angiotensin I. Angiotensin I is also inactive but is then converted to angiotensin II by **angiotensin-converting enzyme (ACE)**. This conversion takes place primarily in the capillaries of the lungs. Angiotensin II acts at the nephron, adrenal glands, and in the CNS. Angiotensin II stimulates contraction of vascular smooth muscle in peripheral capillary beds, promotes aldosterone production, and stimulates the sympathetic nervous system to increase arterial pressures throughout the body. The combined effect is an increase in systemic blood pressure and blood volume and the restoration of normal GFR.

If blood volume increases, the GFR increases automatically. This increase promotes fluid losses that help return blood volume to the normal level. If the increase in blood volume is severe, hormonal factors further increase the GFR and speed up fluid losses in the urine. As noted in Chapter 18, the heart releases natriuretic peptides when increased blood volume or blood pressure stretches the walls of the heart. For example, the atria release *atrial natriuretic peptide (ANP)*. ⮌ p. 644 Among its other effects, this hormone triggers the dilation of afferent glomerular arterioles and the constriction of efferent glomerular arterioles. This mechanism increases glomerular pressures and increases the GFR. ANP also decreases sodium

Figure 26–11 The Response to a Decrease in the GFR.

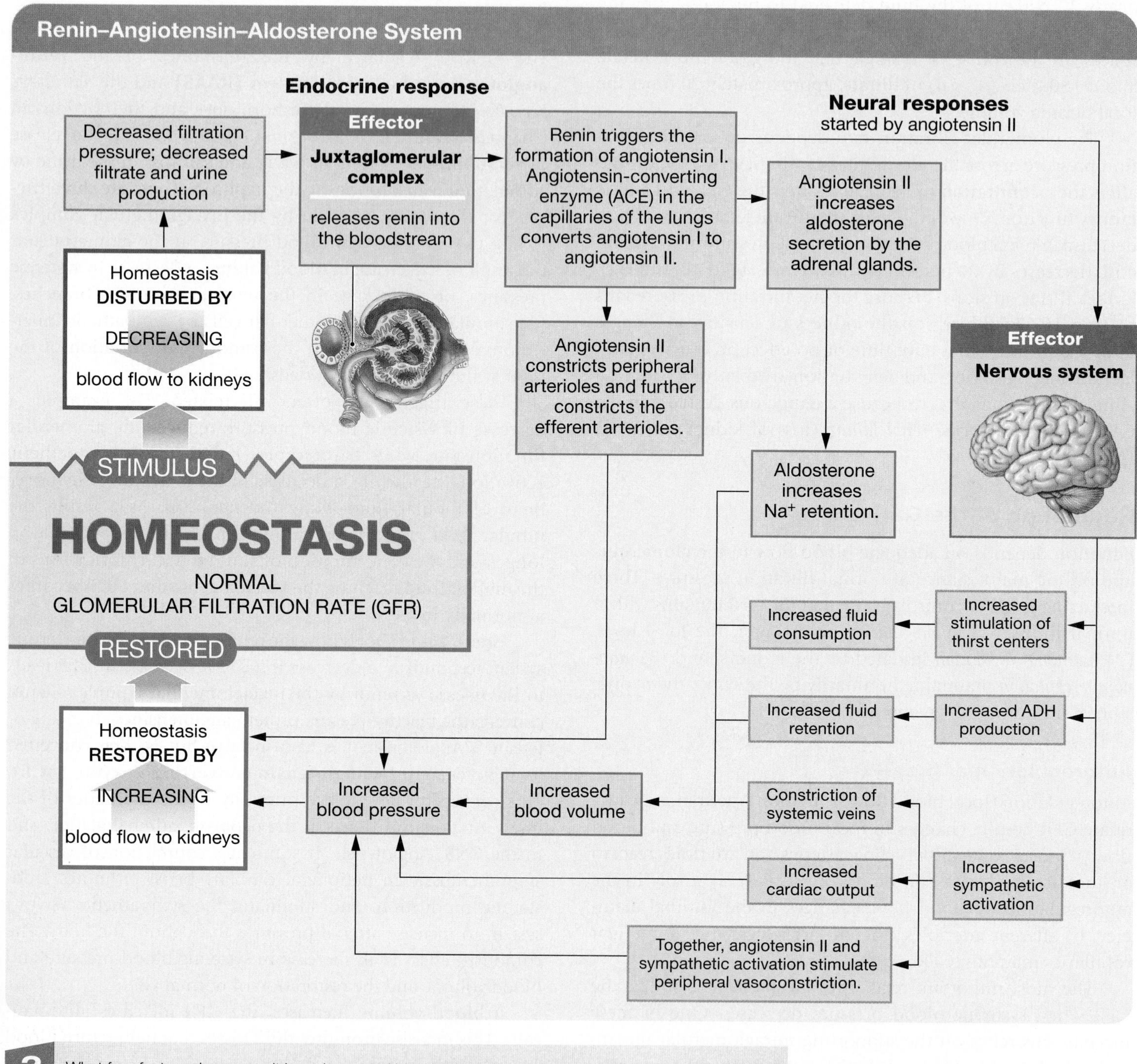
Renin–Angiotensin–Aldosterone System
Endocrine response
Decreased filtration pressure; decreased filtrate and urine production
Effector
Juxtaglomerular complex
releases renin into the bloodstream
Renin triggers the formation of angiotensin I. Angiotensin-converting enzyme (ACE) in the capillaries of the lungs converts angiotensin I to angiotensin II.
Neural responses
started by angiotensin II
Angiotensin II increases aldosterone secretion by the adrenal glands.
Homeostasis
DISTURBED BY
DECREASING
blood flow to kidneys
Angiotensin II constricts peripheral arterioles and further constricts the efferent arterioles.
Effector
Nervous system
STIMULUS
Aldosterone increases Na+ retention.
HOMEOSTASIS
NORMAL
GLOMERULAR FILTRATION RATE (GFR)
RESTORED
Increased fluid consumption
Increased stimulation of thirst centers
Increased fluid retention
Increased ADH production
Homeostasis
RESTORED BY
INCREASING
blood flow to kidneys
Increased blood pressure
Increased blood volume
Constriction of systemic veins
Increased cardiac output
Increased sympathetic activation
Together, angiotensin II and sympathetic activation stimulate peripheral vasoconstriction.

? What four factors shown result in an increased blood volume?

reabsorption at the renal tubules. The net result is increased urine production and decreased blood volume and pressure.

Autonomic Regulation of the GFR

Most of the autonomic innervation of the kidneys consists of sympathetic postganglionic fibers. Sympathetic activation has a direct effect on the GFR. It produces powerful vasoconstriction of afferent glomerular arterioles, which decreases the GFR and slows the production of filtrate. In this way, the sympathetic activation triggered by an acute decrease in blood pressure or a heart attack overrides the local regulatory mechanisms that act to stabilize the GFR. As the crisis passes and sympathetic tone decreases, the filtration rate gradually returns to normal.

26

When the sympathetic division alters regional patterns of blood circulation, blood flow to the kidneys is often affected. For example, the dilation of superficial dermal blood vessels in warm weather shunts blood away from the kidneys. As a result, glomerular filtration decreases temporarily. The effect becomes especially pronounced during strenuous exercise. As the blood flow to your skin and skeletal muscles increases, kidney perfusion gradually decreases.

At a maximal level of exertion, renal blood flow may be less than 25 percent of the normal resting level. This reduction can create problems for endurance athletes. Metabolic wastes build up over the course of a long event. *Proteinuria* (protein in the urine) commonly occurs after such events because the glomerular cells have been injured by prolonged *hypoxia* (a low oxygen level). If the damage is substantial, *hematuria* (blood in the urine) occurs. Hematuria develops in about 18 percent of marathon runners. The cause is trauma to the bladder epithelium from the shocks of running. Proteinuria and hematuria generally disappear within 48 hours as the glomerular tissues are repaired. However, a small number of marathon and ultramarathon runners experience acute renal failure.

✓ Checkpoint

7. **What parts of the glomerulus are involved in filtration?**
8. **List the factors that influence net filtration pressure.**
9. **List the factors that influence the rate of filtrate formation.**
10. **How would a decrease in blood pressure affect the GFR?**

See the blue Answers tab at the back of the book.

26-5 The renal tubule reabsorbs nutrients, ions, and water and secretes ions and wastes; the collecting system reabsorbs ions and water

Learning Outcome Identify the types and functions of transport mechanisms found along each segment of the nephron and collecting system.

Filtration by the renal corpuscle is both effective and energy efficient, but it has one major drawback: Useful substances such as glucose, free fatty acids, amino acids, vitamins, and other solutes enter the capsular space, in addition to metabolic wastes and excess ions.

The renal tubule has three specific functions: (1) reabsorbing all the useful organic nutrients in the filtrate; (2) reabsorbing more than 90 percent of the water in the filtrate; and (3) secreting into the tubule lumen any wastes that did not pass into the filtrate at the glomerulus. As filtrate passes through the renal tubules, about 99 percent of it is reabsorbed. Both reabsorption and secretion take place in every segment of the renal tubule and in the collecting system. However, their relative importance changes from segment to segment. Let's take a more detailed look at these critical processes, starting with the principles behind them.

Principles of Reabsorption and Secretion

The processes of reabsorption and secretion alter the composition of the filtrate produced by glomerular filtration. Tubular reabsorption returns nutrients, such as glucose, amino acids, water, and salt, from the tubular fluid to the blood. Tubular secretion is the reverse of reabsorption: it adds substances from the blood to the tubular fluid. We begin our discussion with reabsorption mechanisms.

Review of Carrier-Mediated Transport Mechanisms

The processes of reabsorption and secretion by the kidneys involve a combination of diffusion, osmosis, leak channels (channel-mediated diffusion), and carrier-mediated transport. We considered diffusion and osmosis in several other chapters. Here we briefly review *carrier-mediated transport* mechanisms.

Elsewhere in this text, we looked at four major types of carrier-mediated transport:

- In *facilitated diffusion*, a carrier protein transports a molecule across the plasma membrane without expending energy (look back at Figure 3–18, p. 96). Such transport always follows the concentration gradient for the ion or molecule transported.
- *Active transport* is driven by the hydrolysis of ATP to ADP on the inner plasma membrane surface (look back at Figure 3–19, p. 97). Exchange pumps and other carrier proteins are active along the renal tubules. Active transport can operate despite an opposing concentration gradient.
- In *cotransport*, carrier protein activity is not directly linked to the hydrolysis of ATP (look back at Figure 3–20, p. 98). Instead, two substrates (ions, molecules, or both) cross the membrane while bound to the carrier protein. The movement of the substrates always follows the concentration gradient of at least one of the transported substances. Cotransport is used for the reabsorption of organic and inorganic compounds from the tubular fluid.
- *Countertransport* resembles cotransport, except that the two transported ions move in *opposite* directions (look back at Figures 23–23, p. 869, and 24–14, p. 904). Countertransport operates in the PCT, DCT, and collecting system.

All carrier-mediated processes share the following characteristics, which are important to understanding kidney physiology:

- *A specific substrate binds to a carrier protein that facilitates movement across the membrane.* The movement may be active or passive.
- *A given carrier protein typically works in one direction only.* In facilitated diffusion, the concentration gradient of the

substance being transported determines that direction. In active transport, cotransport, and countertransport, the location and orientation of the carrier proteins determine whether a particular substance is reabsorbed or secreted. The carrier protein that transports amino acids from the tubular fluid to the cytosol of the tubule cells, for example, will not carry amino acids back into the tubular fluid.

- *The distribution of carrier proteins can vary in different regions of the cell surface.* Transport between tubular fluid and peritubular fluid involves two steps. First, the material enters the cell at its apical surface. Then the material leaves the cell at its basolateral surface and enters the peritubular fluid. Each step involves a different carrier protein. For example, the apical surfaces of cells along the proximal convoluted tubule contain carrier proteins that bring amino acids, glucose, and many other nutrients into these cells by sodium ion–linked cotransport. In contrast, the basolateral surfaces contain carrier proteins that move those nutrients out of the cell by facilitated diffusion.
- *The membrane of a single tubular cell contains many types of carrier proteins.* Each cell can have multiple functions. A cell that reabsorbs one compound can secrete another.

The Transport Maximum (T_m) and the Renal Threshold

Recall that when an enzyme is *saturated*, further increases in substrate concentration have no effect on the rate of reaction. p. 56 Likewise, when a carrier protein is saturated, further increases in substrate concentration have no effect on the rate of transport across the plasma membrane. For any substance, the concentration at saturation is called the **transport maximum (T_m)**, or *tubular maximum*. The T_m reflects the number of available carrier proteins in the renal tubules. In healthy people, carrier proteins involved in tubular secretion seldom become saturated. However, carriers involved in tubular reabsorption are often at risk of saturation, especially during the absorptive state following a meal. When the carrier proteins are saturated, excesses of that substrate are excreted.

26

Normally, any plasma proteins and nutrients, such as amino acids and glucose, are removed from the tubular fluid by cotransport or facilitated diffusion. If the concentrations of these nutrients rise in the tubular fluid, the rate of reabsorption increases until the carrier proteins are saturated. A concentration higher than the transport maximum exceeds the reabsorptive abilities of the nephron. In this case, some of the material will remain in the tubular fluid and appear in the urine. The transport maximum thus determines the **renal threshold**—the plasma concentration at which a specific substance or ion begins to appear in the urine.

The renal threshold varies with the substance involved. The renal threshold for glucose is approximately 180 mg/dL. When the plasma glucose concentration exceeds 180 mg/dL, glucose appears in urine; this condition is called *glucosuria* (glū-kō-SYŪ-rē-uh) or *glycosuria*. After you have eaten a meal rich in carbohydrates, your plasma glucose level may exceed the glucose T_m for a brief period. However, the liver quickly lowers the circulating glucose level, and very little glucose is lost in your urine. Chronically increased plasma and urinary glucose concentrations are abnormal. (Glycosuria is one of the key signs of diabetes mellitus. p. 642)

The renal threshold for amino acids is lower than that for glucose. Amino acids appear in urine when their plasma concentrations exceed 65 mg/dL. Plasma amino acid levels commonly exceed the renal threshold after you have eaten a protein-rich meal, causing some amino acids to appear in your urine. This condition is termed *aminoaciduria* (am-ih-nō-as-ih-DYŪ-rē-uh).

The T_m values for water-soluble vitamins are relatively low. As a result, you excrete excess quantities of these vitamins in urine. (This is typically the fate of water-soluble vitamins in daily supplements.)

Measuring Osmotic Concentration

The osmotic concentration, or *osmolarity*, of a solution is the total number of solute particles in each liter. p. 94 We usually express osmolarity in **osmoles** per liter (Osm/L) or **milliosmoles** per liter (mOsm/L). If a liter of a fluid contains 1 mole of dissolved particles, the solute concentration is 1 Osm/L, or 1000 mOsm/L. Body fluids have an osmotic concentration of about 300 mOsm/L. For comparison, seawater is about 1000 mOsm/L and fresh water is about 5 mOsm/L. In a clinical setting, the osmolarity of urine is often reported as *osmolality*, expressed as milliosmoles per kilogram of water (mOsm/kg H_2O). When discussing osmotic concentrations in body fluids, these terms are often used interchangeably (1 liter of water has a mass of 1 kilogram).

Ion concentrations are often reported in *milliequivalents* per liter (mEq/L). Milliequivalents indicate the number of positive or negative charges in solution, rather than the number of solute particles. To convert mmol/L to mEq/L, multiply mmol/L by the number of charges on each ion. For example, each sodium ion has one charge only, so for Na^+, 1 mmol/L = 1 mEq/L; for Ca^{2+}, each ion has two charges, so 1 mmol/L = 2 mEq/L. The concentrations of large organic molecules are usually reported in grams, milligrams, or micrograms per unit volume of solution (typically, per deciliter, or dL).

An Overview of Reabsorbed and Secreted Substances

Table 26–3 lists some substances that are actively reabsorbed by the renal tubules. It also includes some of the substances secreted into tubular fluid by the proximal and distal convoluted tubules.

Table 26–3 Tubular Reabsorption and Secretion

Reabsorbed	Secreted	No Transport Mechanism
Ions Na^+, Cl^-, K^+, Ca_2^+, Mg^{2+}, SO_4^{2-}, HCO_3^-	**Ions** K^+, H^+, Ca^{2+}, PO_4^{3-}	Urea Water Urobilinogen Bilirubin
	Wastes Creatinine Ammonia Metabolic acids and bases	
Metabolites Glucose Amino acids Proteins Vitamins		
	Miscellaneous Neurotransmitters (ACh, NE, E, dopamine) Histamine Drugs (penicillin, atropine, morphine, many others)	

Note that cells of the renal tubule ignore a number of substances in the tubular fluid, because there are no transport mechanisms for them. These substances tend to be metabolic wastes. As water and other substances are removed, the concentrations of these wastes in the tubular fluid gradually rise.

Reabsorption and Secretion along the PCT

The cells of the proximal convoluted tubule normally reabsorb 60–70 percent of the volume of the filtrate produced in the renal corpuscle. The tubular cells reabsorb organic nutrients, ions, water, and plasma proteins (if present) from the tubular fluid. The reabsorbed materials enter the peritubular fluid, diffuse into peritubular capillaries, and are quickly returned to the bloodstream. The epithelial cells can also secrete substances into the lumen of the renal tubule.

The PCT has the following major functions:

- *Reabsorption of Organic Nutrients.* Under normal circumstances, before the tubular fluid enters the nephron loop, the PCT reabsorbs more than 99 percent of the glucose, amino acids, and other organic nutrients in the fluid. This reabsorption involves a combination of facilitated diffusion and cotransport.
- *Active Reabsorption of Ions.* The PCT actively transports several ions, including sodium, potassium, and bicarbonate ions, plus magnesium, phosphate, and sulfate ions (Figure 26–12). Although the ion pumps involved are individually regulated, they may be influenced by circulating ion or hormone levels. For example, angiotensin II stimulates sodium ion (Na^+) reabsorption along the PCT. By absorbing carbon dioxide (CO_2), the PCT indirectly recaptures about 90 percent of the bicarbonate ions (HCO_3^-) from tubular fluid. Bicarbonate ions are important in stabilizing blood pH. We examine this process further in Chapter 27.
- *Reabsorption of Water.* The reabsorptive processes have a direct effect on the solute concentrations inside and outside the tubules. The filtrate entering the PCT has the same osmotic concentration as that of the surrounding peritubular fluid. As reabsorption proceeds, the solute concentration of tubular fluid decreases, and that of peritubular fluid and adjacent capillaries increases. Water then moves out of the tubular fluid and into the peritubular fluid by osmosis. Along the PCT, this mechanism results in the reabsorption of about 108 liters of water each day.
- *Passive Reabsorption of Ions.* As active reabsorption of ions takes place and water leaves the tubular fluid by osmosis, the concentration of other solutes in the tubular fluid increases above that in the peritubular fluid. If the tubular cells are permeable to them, those solutes move across the tubular cells and into the peritubular fluid by passive diffusion. Urea, chloride ions, and lipid-soluble materials may diffuse out of the PCT in this way. Such diffusion further decreases the solute concentration of the tubular fluid and promotes additional water reabsorption by osmosis.
- *Secretion.* Active secretion of hydrogen ions first takes place along the PCT (see **Table 26–4** p. 996). Because the PCT and DCT secrete similar substances, and the DCT carries out comparatively little reabsorption, we will consider secretory mechanisms when we discuss the DCT.

Sodium ion reabsorption plays an important role in all of these processes. Sodium ions may enter tubular cells by diffusion through sodium ion leak channels; by the sodium ion–linked cotransport of glucose, amino acids, or other organic solutes; or by countertransport for hydrogen ions (see Figure 26–12). Once inside the tubular cells, sodium ions diffuse toward the basement membrane. The plasma membrane in this area contains sodium–potassium exchange pumps that eject sodium ions in exchange for extracellular potassium ions (K^+). Reabsorbed sodium ions then diffuse through the peritubular fluid and into adjacent peritubular capillaries.

The reabsorption of ions and compounds along the PCT involves many different carrier proteins. Some people have an inherited inability to manufacture one or more of these carrier proteins. For this reason, these individuals are unable to recover specific solutes from tubular fluid. In *renal glucosuria,* for example, a defective glucose carrier protein makes it impossible for the PCT to reabsorb glucose from tubular fluid.

Reabsorption and Secretion along the Nephron Loop

Approximately 60–70 percent of the volume of filtrate produced by the glomerulus has been reabsorbed before the tubular fluid reaches the nephron loop. In the process, useful organic substrates and many mineral ions have been reclaimed.

Table 26–4 Renal Structures and Their Functions

Segment	General Functions	Specific Functions	Mechanisms
Renal corpuscle	*Filtration* of plasma; generates approximately 180 L/day of filtrate similar in composition to blood plasma without plasma proteins	*Filtration* of water, inorganic and organic solutes from plasma; retention of plasma proteins and blood cells	Glomerular hydrostatic (blood) pressure working across capillary endothelium, dense layer, and filtration slits
Proximal convoluted tubule (PCT)	*Reabsorption* of 60%–70% of the water (108–116 L/day), 99%–100% of the organic substrates, and 60%–70% of the sodium and chloride ions in the original filtrate	*Active reabsorption*: Glucose, other simple sugars, amino acids, vitamins, ions (including sodium, potassium, calcium, magnesium, phosphate, and bicarbonate)	Carrier-mediated transport, including facilitated diffusion (glucose, amino acids), cotransport (glucose, ions), and countertransport (with secretion of H^+)
		Passive reabsorption: Urea, chloride ions, lipid-soluble materials, water	Diffusion (solutes) or osmosis (water)
		Secretion: Hydrogen ions, ammonium ions, creatinine, drugs, and toxins (as at DCT)	Countertransport with sodium ions
Nephron loop	*Reabsorption* of 25% of the water (45 L/day) and 20%–25% of the sodium and chloride ions present in the original filtrate; creation of the concentration gradient in the medulla	*Reabsorption*: Sodium and chloride ions	Active transport by $Na^+–K^+/2\ Cl^-$ transporter
		Water	Osmosis
Distal convoluted tubule (DCT)	*Reabsorption* of a variable amount of water (usually 5%, or 9 L/day), under ADH stimulation, and a variable amount of sodium ions, under aldosterone stimulation	*Reabsorption*: Sodium and chloride ions	Cotransport
		Sodium ions (variable)	Countertransport with potassium ions; aldosterone-regulated
		Calcium ions (variable)	Carrier-mediated transport stimulated by parathyroid hormone and calcitriol
		Water (variable)	Osmosis; ADH-regulated
		Secretion: Hydrogen ions, ammonium ions	Countertransport with sodium ions
		Creatinine, drugs, toxins	Carrier-mediated transport
Collecting system	*Reabsorption* of a variable amount of water (usually 9.3%, or 16.8 L/day) under ADH stimulation, and a variable amount of sodium ions, under aldosterone stimulation	*Reabsorption*: Sodium ions (variable)	Countertransport with potassium or hydrogen ions; aldosterone-regulated
		Bicarbonate ions (variable)	Diffusion, generated within tubular cells
		Water (variable)	Osmosis; ADH-regulated
		Urea (distal segments only)	Diffusion
		Secretion: Potassium and hydrogen ions (variable)	Carrier-mediated transport
Peritubular capillaries	*Redistribution* of water and solutes reabsorbed in the cortex	Return of water and solutes to the general circulation	Osmosis and diffusion
Vasa recta	*Redistribution* of water and solutes reabsorbed in the medulla and stabilization of the concentration gradient of the medulla	Return of water and solutes to the general circulation	Osmosis and diffusion

The nephron loop reabsorbs about half of the remaining water and two-thirds of the remaining sodium and chloride ions.

Recall that the nephron loop is made up of ascending and descending limbs, with thin and thick segments. The ascending limb is impermeable to water, but passively and actively removes sodium and chloride ions from the tubular fluid. Its ascending thin limb is permeable to sodium ions, which diffuse into the surrounding peritubular fluid. The thick ascending limb (TAL) actively transports sodium and chloride ions out of the tubular fluid. The effect of this movement is most noticeable in the renal medulla, where the long ascending limbs of juxtamedullary nephrons create unusually high solute concentrations in the peritubular fluid.

The entire descending limb is freely permeable to water, but not to solutes, such as sodium and chloride ions. The descending thin limb has functions similar to those of the PCT: It reabsorbs sodium and chloride ions out of the tubular fluid. Because of the ion concentration gradients created by the thick ascending limb, water moves out of the descending limb, helping to increase the solute concentration of the tubular fluid. This process, which allows the production of concentrated urine, will be discussed in detail in Section 26-6.

Tips & Tools

To remember the names and order of the limbs of the nephron loop, think "descending thin limb turns and sprouts up as the ascending thin limb and grows TAL (thick ascending limb)."

Figure 26–12 Transport Activities at the PCT. Sodium ions may enter a tubular cell from the filtrate by diffusion, cotransport, or countertransport. The sodium ions are then pumped into the peritubular fluid by the sodium–potassium exchange pump. Other reabsorbed solutes may be ejected into the peritubular fluid by separate active transport mechanisms. The absorption of bicarbonate is indirectly associated with the reabsorption of sodium ions and the secretion of hydrogen ions.

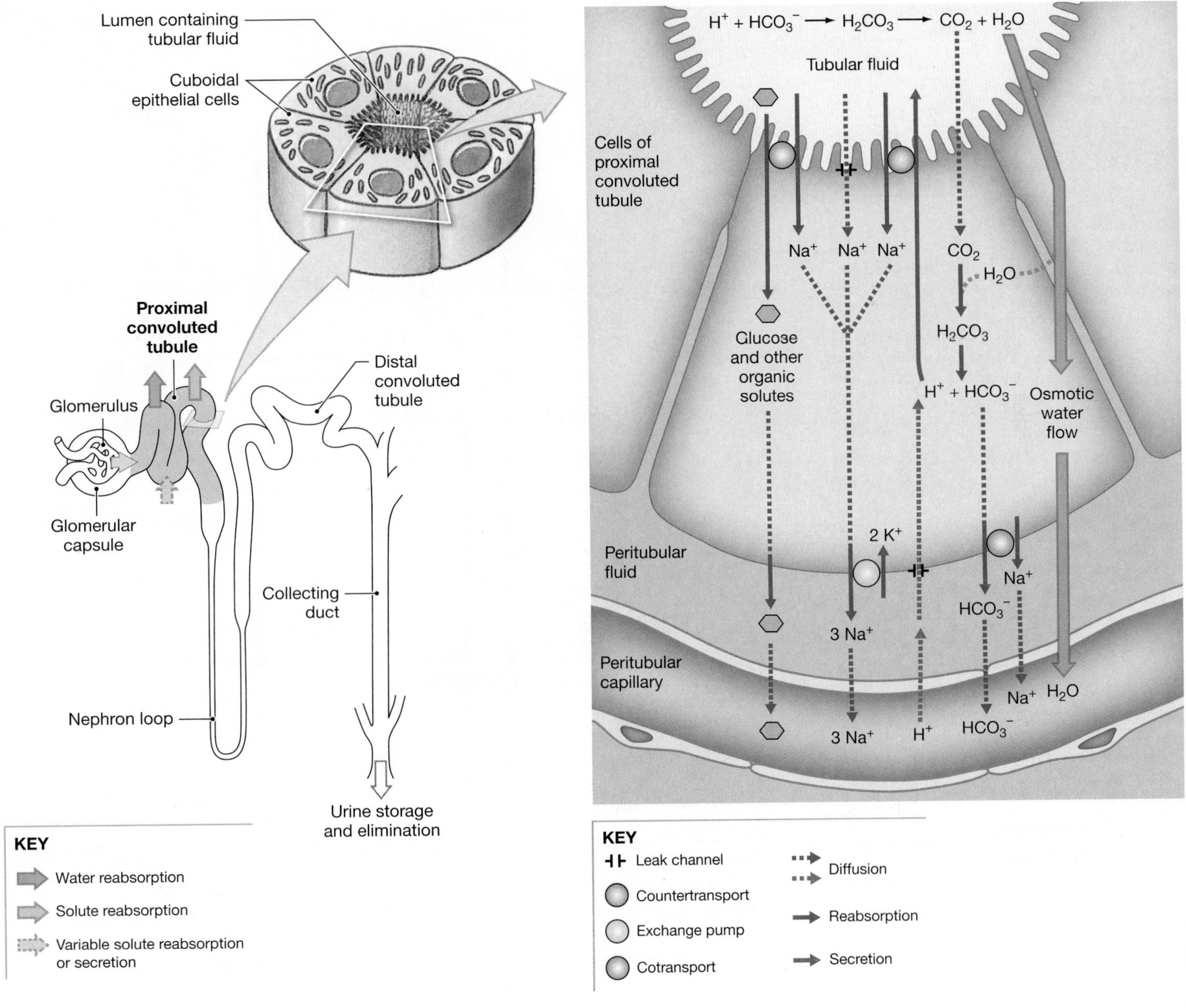
Lumen containing tubular fluid
Cuboidal epithelial cells
Proximal convoluted tubule
Glomerulus
Glomerular capsule
Distal convoluted tubule
Collecting duct
Nephron loop
Urine storage and elimination
KEY
Water reabsorption
Solute reabsorption
Variable solute reabsorption or secretion
H+ + HCO3− → H2CO3 → CO2 + H2O
Tubular fluid
Cells of proximal convoluted tubule
Na+ Na+ Na+
CO2
H2O
H2CO3
H+ + HCO3−
Glucose and other organic solutes
Osmotic water flow
2 K+
Peritubular fluid
Na+
HCO3−
3 Na+
Peritubular capillary
Na+ H2O
3 Na+ H+ HCO3−
KEY
Leak channel
Countertransport
Exchange pump
Cotransport
Diffusion
Reabsorption
Secretion

Reabsorption and Secretion along the DCT

As we have just seen, the composition and volume of tubular fluid change dramatically as it flows from the capsular space to the distal convoluted tubule. Only 15–20 percent of the initial filtrate volume reaches the DCT. The concentrations of electrolytes and metabolic wastes in the arriving tubular fluid no longer resemble the concentrations in blood plasma. Selective reabsorption or secretion, primarily along the DCT, makes the final adjustments in the solute composition and volume of the tubular fluid. The DCT is an important site for three vital processes: (1) the active secretion of ions, acids, drugs, and toxins into the tubule; (2) the selective reabsorption of sodium ions and calcium ions from tubular fluid; and (3) the selective reabsorption of water, which assists in concentrating the tubular fluid.

Reabsorption at the DCT

Throughout most of the DCT, the tubular cells actively transport sodium ions (Na^+) and chloride ions (Cl^-) out of the tubular

Figure 26–13 Tubular Secretion and Solute Reabsorption by the DCT.

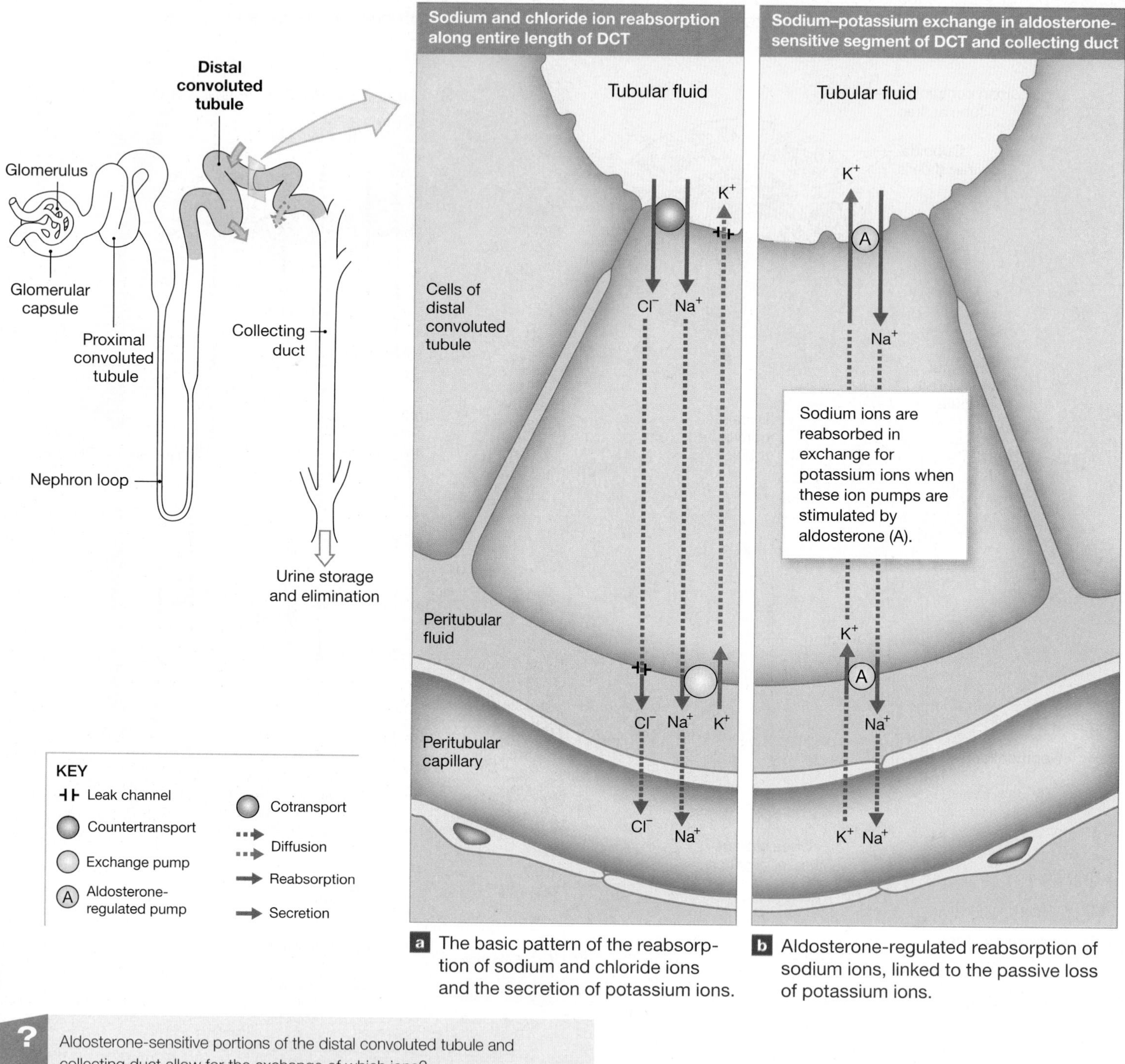
Distal convoluted tubule
Glomerulus
Glomerular capsule
Proximal convoluted tubule
Collecting duct
Nephron loop
Urine storage and elimination
KEY
Leak channel
Countertransport
Exchange pump
Aldosterone-regulated pump
Cotransport
Diffusion
Reabsorption
Secretion
Sodium and chloride ion reabsorption along entire length of DCT
Tubular fluid
Cells of distal convoluted tubule
Peritubular fluid
Peritubular capillary
Cl⁻ Na⁺ K⁺
Sodium–potassium exchange in aldosterone-sensitive segment of DCT and collecting duct
Sodium ions are reabsorbed in exchange for potassium ions when these ion pumps are stimulated by aldosterone (A).

a The basic pattern of the reabsorption of sodium and chloride ions and the secretion of potassium ions.

b Aldosterone-regulated reabsorption of sodium ions, linked to the passive loss of potassium ions.

? Aldosterone-sensitive portions of the distal convoluted tubule and collecting duct allow for the exchange of which ions?

fluid (Figure 26–13a). Tubular cells along the distal regions of the DCT also contain ion pumps that reabsorb tubular sodium ions in exchange for another cation (usually potassium ions [K^+]) (Figure 26–13b). The hormone *aldosterone*, produced by the adrenal cortex, controls the sodium ion channels and the ion pump. This hormone stimulates the synthesis and incorporation of sodium ion channels and sodium ion pumps in plasma membranes along the DCT and collecting duct. The net result is a reduction in the number of sodium ions lost in urine.

However, Na^+ conservation is associated with K^+ loss. Prolonged aldosterone stimulation can therefore produce

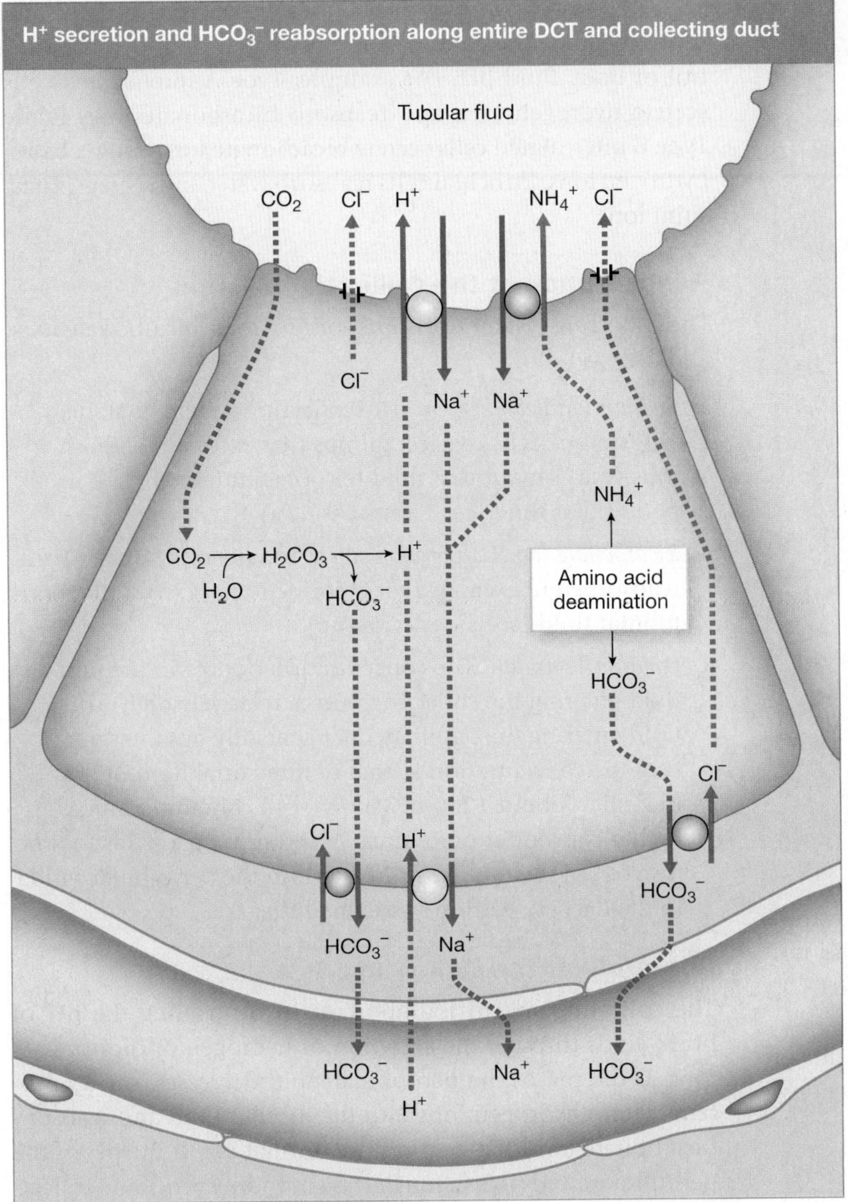
H⁺ secretion and HCO₃⁻ reabsorption along entire DCT and collecting duct
Tubular fluid
CO₂
Cl⁻
H⁺
NH₄⁺
Cl⁻
Cl⁻
Na⁺
Na⁺
NH₄⁺
CO₂
H₂CO₃
H⁺
H₂O
HCO₃⁻
Amino acid deamination
HCO₃⁻
Cl⁻
Cl⁻
H⁺
HCO₃⁻
HCO₃⁻
Na⁺
HCO₃⁻
Na⁺
HCO₃⁻
H⁺

c Hydrogen ion secretion and the acidification of urine occur by two routes. The central theme is the exchange of hydrogen ions in the cytosol for sodium ions in the tubular fluid, and the reabsorption of the bicarbonate ions generated in the process.

hypokalemia, a dangerous reduction in the plasma potassium ion concentration. The secretion of aldosterone and its actions on the DCT and collecting system are opposed by the natriuretic peptide ANP.

The DCT is also the primary site of calcium ion (Ca^{2+}) reabsorption. The circulating level of parathyroid hormone regulates this process. p. 633

Secretion at the DCT

The blood entering the peritubular capillaries still contains a number of potentially undesirable substances that did not cross the filtration membrane at the glomerulus. In most cases, the concentrations of these substances are too low to cause physiological problems. However, any ions or compounds in peritubular capillaries will diffuse into the peritubular fluid. If those concentrations become too high, the tubular cells may absorb these substances from the peritubular fluid and secrete them into the tubular fluid.

The rate of potassium and hydrogen ion secretion increases or decreases in direct response to changes in their concentrations in peritubular fluid. The higher their concentrations in the peritubular fluid, the higher the rate of secretion. Potassium and hydrogen ions merit special attention, because their concentrations in body fluids must be maintained within narrow limits.

Potassium Ion Secretion. Figure 26–13a,b diagrams the mechanism of potassium ion (K^+) secretion. In effect, tubular cells trade sodium ions (Na^+) in the tubular fluid for excess potassium ions in body fluids. Potassium ions are removed from the peritubular fluid in exchange for sodium ions from the tubular fluid. These potassium ions diffuse into the lumen of the DCT through potassium ion leak channels in the apical surfaces of the tubular cells.

Hydrogen Ion Secretion. Hydrogen ion (H^+) secretion is also associated with the reabsorption of sodium ions (Na^+). Figure 26–13c depicts two routes of secretion. Both involve the generation of carbonic acid (H_2CO_3) catalyzed by the enzyme *carbonic anhydrase.* pp. 865, 904 Hydrogen ions generated by the dissociation of the carbonic acid are secreted by sodium ion–linked countertransport in exchange for sodium ions in the tubular fluid. The bicarbonate ions (HCO_3^-) diffuse into the peritubular fluid and then into the bloodstream. There they help prevent changes in plasma pH.

Hydrogen ion secretion acidifies the tubular fluid while increasing the pH of the blood. Hydrogen ion secretion speeds up when the pH of the blood decreases. This can happen in metabolic acidosis (referred to as *lactic acidosis*), which can develop after exhaustive muscle activity, or *ketoacidosis*, which can develop in starvation or diabetes mellitus. p. 958 The combination of hydrogen ion removal and bicarbonate ion production by the kidneys plays an important role in the control of blood pH. Because one of the secretory pathways is aldosterone sensitive, aldosterone stimulates hydrogen ion secretion. Prolonged aldosterone stimulation can cause *alkalosis*, or abnormally high blood pH.

In Chapter 25, we noted that the production of lactate and ketone bodies during the postabsorptive state is associated with or could cause acidosis, respectively. Under these conditions,

+ Clinical Note Diuretics

Diuresis (dī-yū-RĒ-sis; *dia*, through + *ouresis*, urination) is the elimination of urine. *Urination* is an equivalent term in a general sense, but *diuresis* typically indicates the production of a large volume of urine. **Diuretics** (dī-yū-RET-iks) are drugs that promote water loss in urine. The usual goal in diuretic therapy is to decrease blood volume, blood pressure, extracellular fluid volume, or all three. The ability to control renal water losses with relatively safe and effective diuretics has saved the lives of many people, especially those with high blood pressure or congestive heart failure.

Diuretics have many mechanisms of action. However, all such drugs affect transport activities or water reabsorption along the nephron and collecting system. For example, consider the class of diuretics called *thiazides* (THĪ-uh-zīdz). These drugs increase water loss by decreasing sodium and chloride ion transport in the proximal and distal convoluted tubules.

Diuretic use for nonclinical reasons is on the rise. For example, some bodybuilders take large doses of diuretics to improve muscle definition temporarily. Some fashion models or horse jockeys do the same. Their goal is to reduce body weight for brief periods. This practice of "cosmetic dehydration" is extremely dangerous. It has caused several deaths due to electrolyte imbalance and consequent cardiac arrest.

the PCT and DCT deaminate amino acids in reactions that strip off the amino groups ($-NH_2$). The reaction sequence ties up hydrogen ions (H^+) and yields both **ammonium ions** (NH_4^+) and bicarbonate ions (HCO_3^-). As indicated in Figure 26–13c, the ammonium ions are then pumped into the tubular fluid by sodium ion–linked countertransport, and the bicarbonate ions enter the bloodstream by way of the peritubular fluid.

Tubular deamination thus has two major benefits. It provides carbon chains suitable for catabolism. It also generates bicarbonate ions that add to the buffering capacity of plasma.

26

Reabsorption and Secretion along the Collecting System

The collecting ducts receive tubular fluid from many nephrons and carry it toward the renal sinus. The normal amount of water and solute loss in the collecting system is hormonally regulated:

- By aldosterone, which controls sodium ion pumps along most of the DCT and the proximal portion of the collecting system. As we have noted, these actions are opposed by atrial natriuretic peptide (ANP).
- By ADH, which controls the permeability of the DCT and collecting system to water. The secretion of ADH is suppressed by ANP, and this—combined with the effects of atrial natriuretic peptide on aldosterone secretion and action—can dramatically increase urinary water losses.

The collecting system also has other reabsorptive and secretory functions. Many of them are important to the control of body fluid pH. For example, Type A intercalated cells secrete hydrogen ions and reabsorb bicarbonate ions, while Type B intercalated cells secrete bicarbonate ions and reabsorb hydrogen ions. Principal cells reabsorb water and secrete potassium ions.

Reabsorption at the Collecting System

The collecting system reabsorbs sodium ions, bicarbonate ions, and urea as follows:

- *Sodium Ion Reabsorption.* The collecting system contains aldosterone-sensitive ion pumps that exchange sodium ions (Na^+) in tubular fluid for potassium ions (K^+) in peritubular fluid (see Figure 26–13b).
- *Bicarbonate Ion Reabsorption.* Bicarbonate ions (HCO_3^-) are reabsorbed in exchange for chloride ions (Cl^-) in the peritubular fluid (see Figure 26–13c).
- *Urea Reabsorption.* The concentration of urea in the tubular fluid entering the collecting duct is relatively high. The fluid entering the papillary duct generally has the same osmotic concentration as that of interstitial fluid of the medulla—about 1200 mOsm/L—but contains a much higher concentration of urea. As a result, urea tends to diffuse out of the tubular fluid and into the peritubular fluid in the deepest portion of the medulla.

Secretion at the Collecting System

The collecting system is important in controlling the pH of body fluids through the secretion of hydrogen or bicarbonate ions. If the pH of the peritubular fluid decreases, carrier proteins pump hydrogen ions into the tubular fluid and reabsorb bicarbonate ions that help restore normal pH. If the pH of the peritubular fluid rises (a much less common event), the collecting system secretes bicarbonate ions and pumps hydrogen ions into the peritubular fluid. The net result is that the body eliminates a buffer and gains hydrogen ions that lower the pH. (We examine these responses in more detail in Chapter 27, when we consider acid–base balance.)

✓ Checkpoint

11. **What occurs when the plasma concentration of a substance exceeds its tubular maximum?**
12. **What effect would an increased amount of aldosterone have on the potassium ion concentration in urine?**
13. **What effect would a decrease in the sodium ion concentration of filtrate have on the pH of tubular fluid?**
14. **Why does a decrease in the amount of sodium ions in the distal convoluted tubule lead to an increase in blood pressure?**

See the blue Answers tab at the back of the book.

26-6 Countercurrent multiplication allows the kidneys to regulate the volume and concentration of urine

Learning Outcome Explain the role of countercurrent multiplication, describe hormonal influence on the volume and concentration of urine, and describe the characteristics of a normal urine sample.

The kidneys couple two countercurrent mechanisms to establish the conditions necessary to regulate the volume and concentration of urine. *Countercurrent* refers to the fact that the exchange takes place between fluids moving in opposite directions in two adjacent segments of the same tube. These two mechanisms are a *countercurrent multiplier* (filtrate flow in the nephron loop) and a *countercurrent exchanger* (blood flow in the vasa recta). These mechanisms establish and maintain an increasing osmotic gradient from the renal cortex through the medulla.

The Nephron Loop and Countercurrent Multiplication

The descending thin limb and the thick ascending limb of the nephron loop lie very close together. They are separated only by the peritubular fluid, which surrounds all the nephrons. The exchange of substances between these segments of a nephron is called **countercurrent multiplication**. Tubular fluid in the descending limb flows toward the renal pelvis, while tubular fluid in the ascending limb flows toward the cortex. *Multiplication* refers to the fact that the effect of the exchange increases as movement of the fluid continues.

Countercurrent multiplication performs two functions:

- It efficiently reabsorbs solutes and water before the tubular fluid reaches the DCT and collecting system.
- It establishes a concentration gradient in the peritubular fluid (called the *medullary osmotic gradient*) that permits the passive reabsorption of water from the tubular fluid in the collecting system. This allows the production of concentrated urine. The circulating level of antidiuretic hormone (ADH) regulates water reabsorption.

How does this occur? As we have indicated, the two parallel limbs of the nephron loop have very different permeability characteristics. The descending thin limb is permeable to water but relatively impermeable to solutes. The thick ascending limb is relatively impermeable to both water and solutes, but it contains active transport mechanisms that pump sodium and chloride ions from the tubular fluid into the peritubular fluid of the medulla.

Figure 26–14 shows an overview of countercurrent multiplication:

- Sodium ions (Na^+) and chloride ions (Cl^-) are pumped out of the thick ascending limb and into the peritubular fluid. This dilutes the tubular fluid.
- The pumping action increases the osmotic concentration in the peritubular fluid around the descending thin limb.
- This creates a small concentration difference between the tubular fluid and peritubular fluid in the renal medulla.
- The concentration difference results in an osmotic flow of water out of the descending thin limb and into the peritubular fluid. As a result, the solute concentration of the tubular fluid in the descending thin limb increases.
- The arrival of the highly concentrated tubular fluid in the thick ascending limb speeds up the transport of sodium and chloride ions into the peritubular fluid.

Solute pumping out of the tubular fluid in the thick ascending limb leads to higher solute concentrations in the tubular fluid in the descending thin limb, which then brings about increased solute pumping in the thick ascending limb. Notice that this process is a simple positive feedback loop that *multiplies* the concentration difference between the hypotonic tubular fluid in the thick ascending limb and the hypertonic peritubular fluid in the renal medulla.

Creation of the Medullary Osmotic Gradient

The concentration gradient created in the peritubular fluid of the medulla is called the **medullary osmotic gradient**. We can now take a closer look at how this is produced. Figure 26–14a diagrams ion transport across the epithelium of the thick ascending limb (TAL). Active transport at the apical surface moves sodium, potassium, and chloride ions out of the tubular fluid. The carrier is called a **$Na^+ – K^+/2\ Cl^-$ transporter**, because each cycle of the pump carries a sodium ion, a potassium ion, and two chloride ions into the tubular cell. Then cotransport carriers pump potassium and chloride ions into the peritubular fluid. However, potassium ions are removed from the peritubular fluid as the sodium–potassium exchange pump moves sodium ions out of the tubular cell. The potassium ions then diffuse back into the lumen of the tubule through potassium ion leak channels. The net result is that sodium and chloride ions enter the peritubular fluid of the renal medulla.

The removal of sodium and chloride ions from the tubular fluid in the ascending limb raises the osmotic concentration of the peritubular fluid around the descending thin limb (see Figure 26–14b). Recall that the descending thin limb is permeable to water but not to solutes. As tubular fluid travels deeper into the medulla within the descending thin limb, osmosis moves water into the peritubular fluid. Solutes remain behind. As a result, the tubular fluid at the turn of the nephron loop has a higher osmotic concentration than it did at the start.

The pumping mechanism of the TAL is highly effective. Almost two-thirds of the sodium and chloride ions that enter it are pumped out of the tubular fluid before that fluid reaches the DCT. In other tissues, differences in solute concentration are quickly resolved by osmosis. However, osmosis cannot take place

Figure 26–14 **Countercurrent Multiplication and Urine Concentration.**

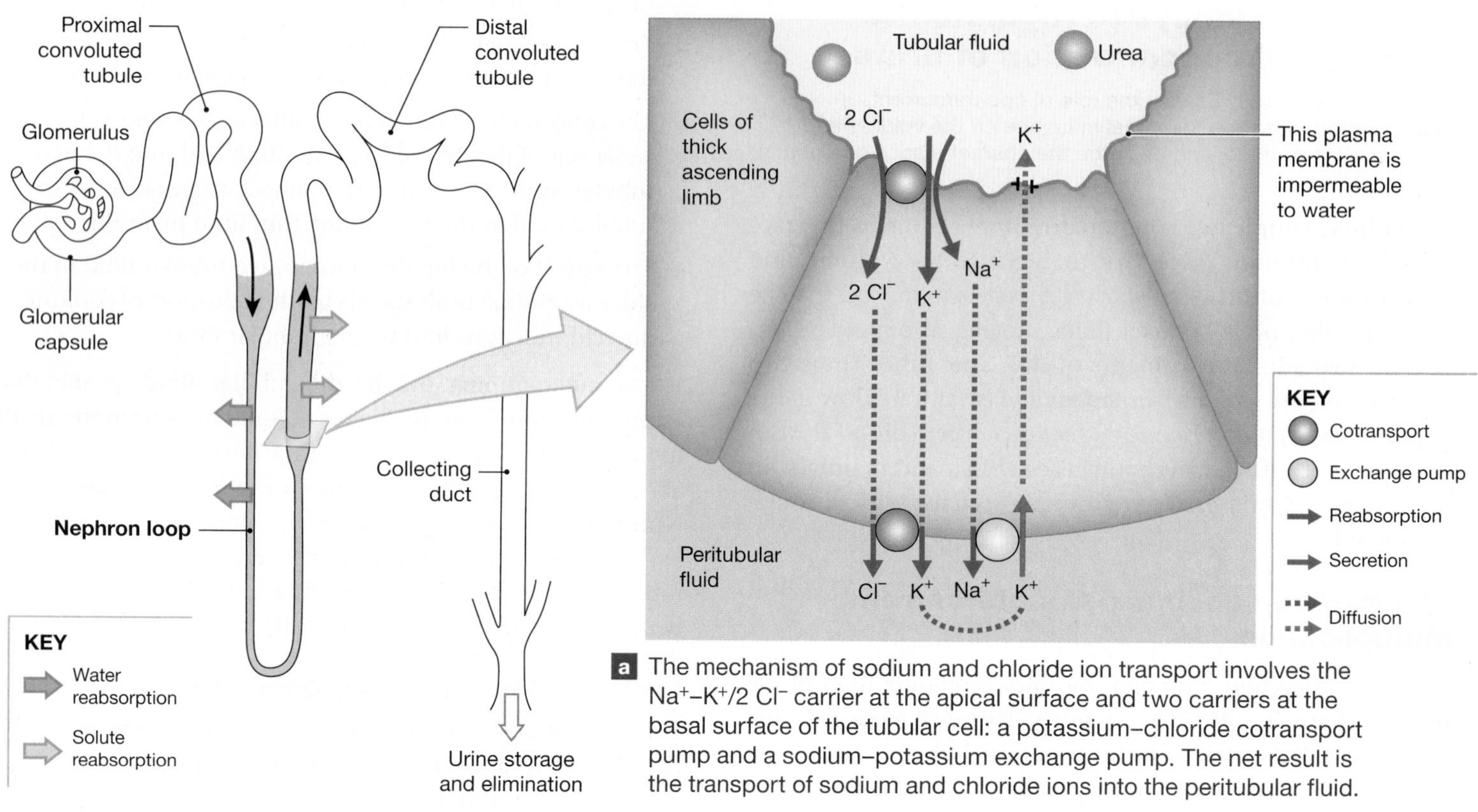
Proximal convoluted tubule
Distal convoluted tubule
Glomerulus
Glomerular capsule
Collecting duct
Nephron loop
KEY
Water reabsorption
Solute reabsorption
Urine storage and elimination
Tubular fluid
Urea
Cells of thick ascending limb
2 Cl⁻
K⁺
Na⁺
This plasma membrane is impermeable to water
Peritubular fluid
Cl⁻ K⁺ Na⁺ K⁺
KEY
Cotransport
Exchange pump
Reabsorption
Secretion
Diffusion

a The mechanism of sodium and chloride ion transport involves the Na^+–K^+/2 Cl^- carrier at the apical surface and two carriers at the basal surface of the tubular cell: a potassium–chloride cotransport pump and a sodium–potassium exchange pump. The net result is the transport of sodium and chloride ions into the peritubular fluid.

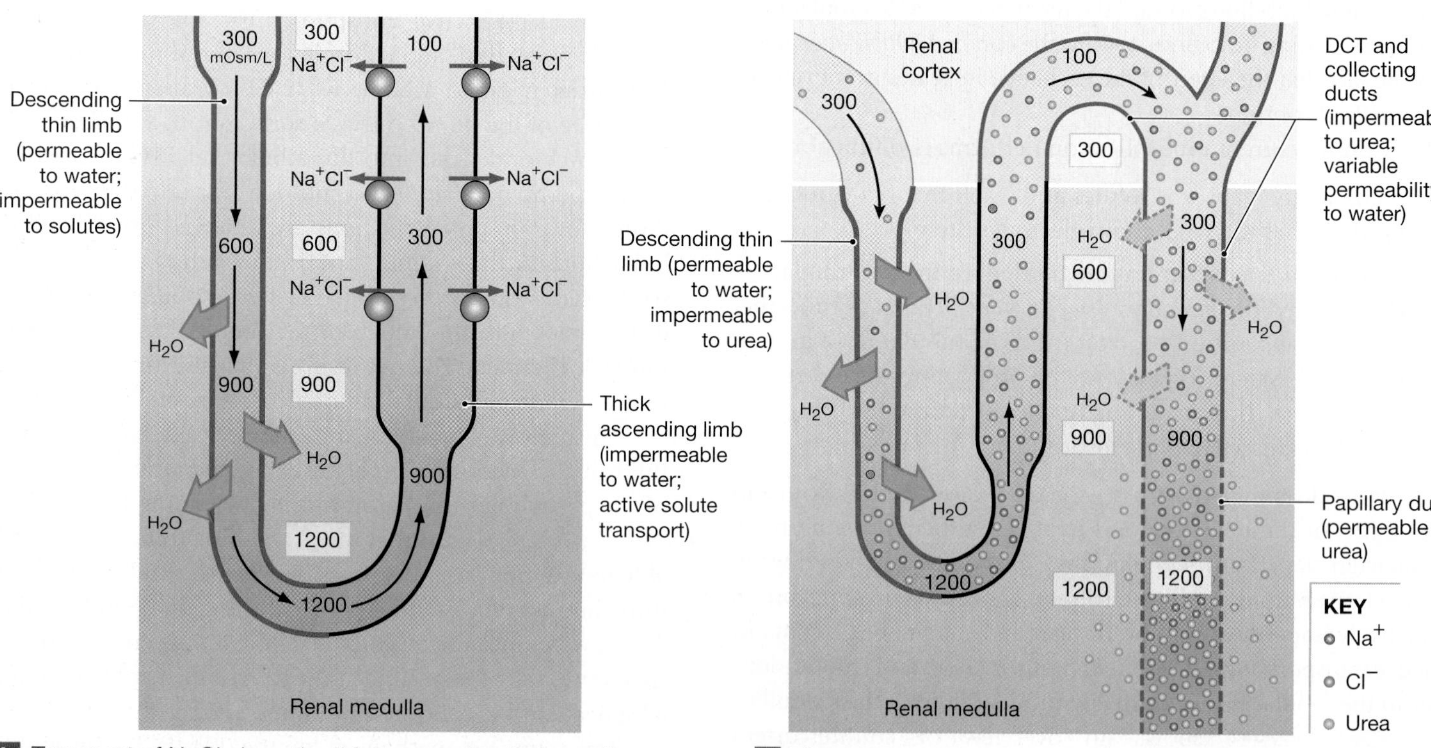
300 mOsm/L
300
100
Na⁺Cl⁻
Descending thin limb (permeable to water; impermeable to solutes)
600
300
H₂O
900
900
1200
1200
Thick ascending limb (impermeable to water; active solute transport)
Renal medulla
Renal cortex
DCT and collecting ducts (impermeable to urea; variable permeability to water)
Descending thin limb (permeable to water; impermeable to urea)
Papillary duct (permeable to urea)
KEY
Na⁺
Cl⁻
Urea

b Transport of NaCl along the thick ascending limb results in the movement of water from the descending thin limb.

KEY
- Impermeable to water
- Impermeable to solutes
- Impermeable to urea; variable permeability to water
- Permeable to urea

c The permeability characteristics of both the loop and the collecting duct tend to concentrate urea in the tubular fluid and in the medulla. The nephron loop, DCT, and collecting duct are impermeable to urea. As water reabsorption occurs, the urea concentration increases. Papillary duct permeability to urea results in urea making up nearly one-third of the solutes in the deepest portions of the medulla.

26

across the wall of the thick ascending limb, because the epithelium there is impermeable to water. So, as sodium and chloride ions are removed, the solute concentration in the tubular fluid decreases. Tubular fluid arrives at the DCT with an osmotic concentration of only about 100 mOsm/L. This value is one-third the concentration of the peritubular fluid of the renal cortex.

The rate of ion transport across the thick ascending limb is proportional to an ion's concentration in tubular fluid. As a result, more sodium and chloride ions are pumped into the medulla at the start of the thick ascending limb, where their ion concentrations are highest, than near the cortex. This regional difference in the rate of ion transport is the basis of the medullary osmotic gradient.

The Role of Urea

Normally, the maximum solute concentration of the peritubular fluid near the turn of the nephron loop is about 1200 mOsm/L. Sodium and chloride ions pumped out of the loop's thick ascending limb make up about two-thirds of that gradient (750 mOsm/L). The rest of the medullary osmotic gradient results from the presence of urea.

To understand how urea arrives in the medulla, let's look ahead to events in the last segments of the collecting system (Figure 26–14c). The thick ascending limb of the nephron loop, the DCT, and the collecting ducts are all impermeable to urea. As water is reabsorbed, the concentration of urea gradually increases in the tubular fluid. When the tubular fluid reaches the papillary duct, it typically contains urea at a concentration of about 450 mOsm/L. The papillary ducts are permeable to urea, so the urea concentration in the deepest parts of the medulla also averages 450 mOsm/L.

Summary of Countercurrent Multiplication

In summary, countercurrent multiplication is a way to either concentrate or dilute urine. The tubular fluid entering the descending limb of the nephron loop has an osmotic concentration of roughly 300 mOsm/L, due primarily to the presence of ions such as sodium and chloride. The concentration of organic wastes, such as urea, is low. About half of the tubular fluid entering the nephron loop is then reabsorbed along the descending thin limb. Two-thirds of the sodium and chloride ions are reabsorbed along the thick ascending limb. As a result, the DCT receives a reduced volume of tubular fluid with an osmotic concentration of about 100 mOsm/L. Urea and other organic wastes, which were not pumped out of the thick ascending limb, now represent a significant proportion of the dissolved solutes.

Regulation of Urine Volume and Osmotic Concentration: Production of Dilute and Concentrated Urine

Urine volume and osmotic concentration are regulated through the control of water reabsorption. Water is reabsorbed by osmosis along the proximal convoluted tubule and the descending limb of the nephron loop. The water permeabilities of these regions cannot be adjusted. As a result, water reabsorption takes place wherever the osmotic concentration of the peritubular fluid is greater than that of the tubular fluid. The ascending limb of the nephron loop is impermeable to water. In the distal convoluted tubule and collecting system, 1–2 percent of the volume of water in the original filtrate is recovered during sodium ion reabsorption. All these water movements represent *obligatory water reabsorption* because they cannot be prevented. This reabsorption usually recovers 85 percent of the volume of filtrate.

The volume of water lost in urine depends on how much of the remaining water in the tubular fluid is reabsorbed along the DCT and collecting system. (This remaining water represents 15 percent of the filtrate volume, or approximately 27 liters per day.) The amount reabsorbed can be precisely controlled by a process called *facultative water reabsorption*. Precise control is possible because these segments are relatively impermeable to water except when ADH is present. This hormone causes special *water channels*, or *aquaporins*, to be inserted into the apical plasma membranes. These water channels dramatically enhance the rate of osmotic water movement. The higher the circulating level of ADH, the greater the number of water channels, and the greater the water permeability of these segments.

As noted earlier in this chapter, the tubular fluid arriving at the DCT has an osmotic concentration of only about 100 mOsm/L. In the presence of ADH, osmosis takes place. Water moves out of the DCT until the osmotic concentration of the tubular fluid equals that of the surrounding cortex (roughly 300 mOsm/L).

The tubular fluid then flows along the collecting duct, which passes through the concentration gradient of the medulla. Additional water is then reabsorbed. The urine reaching the minor calyx has an osmotic concentration closer to 1200 mOsm/L. How closely the osmotic concentration approaches 1200 mOsm/L depends on how much ADH is present.

Figure 26–15 diagrams the effects of ADH on the DCT and collecting system. Without ADH (Figure 26–15a), water is not reabsorbed in these segments, so all the fluid reaching the DCT is lost in the urine. The person then produces large amounts of very dilute urine. That is just what happens in cases of *diabetes insipidus*. p. 625 With this condition, urinary water losses may reach 24 liters (6.3 gal) per day and the urine osmotic concentration range is 30–400 mOsm/L.

As the ADH level rises (Figure 26–15b), the DCT and collecting system become more permeable to water. As a result, the amount of water reabsorbed increases. At the same time, the osmotic concentration of the urine climbs. Under maximum ADH stimulation, the DCT and collecting system become so permeable to water that the osmotic concentration of the urine equals that of the deepest portion of the medulla.

Figure 26–15 **The Effects of ADH on the DCT and Collecting Duct.**

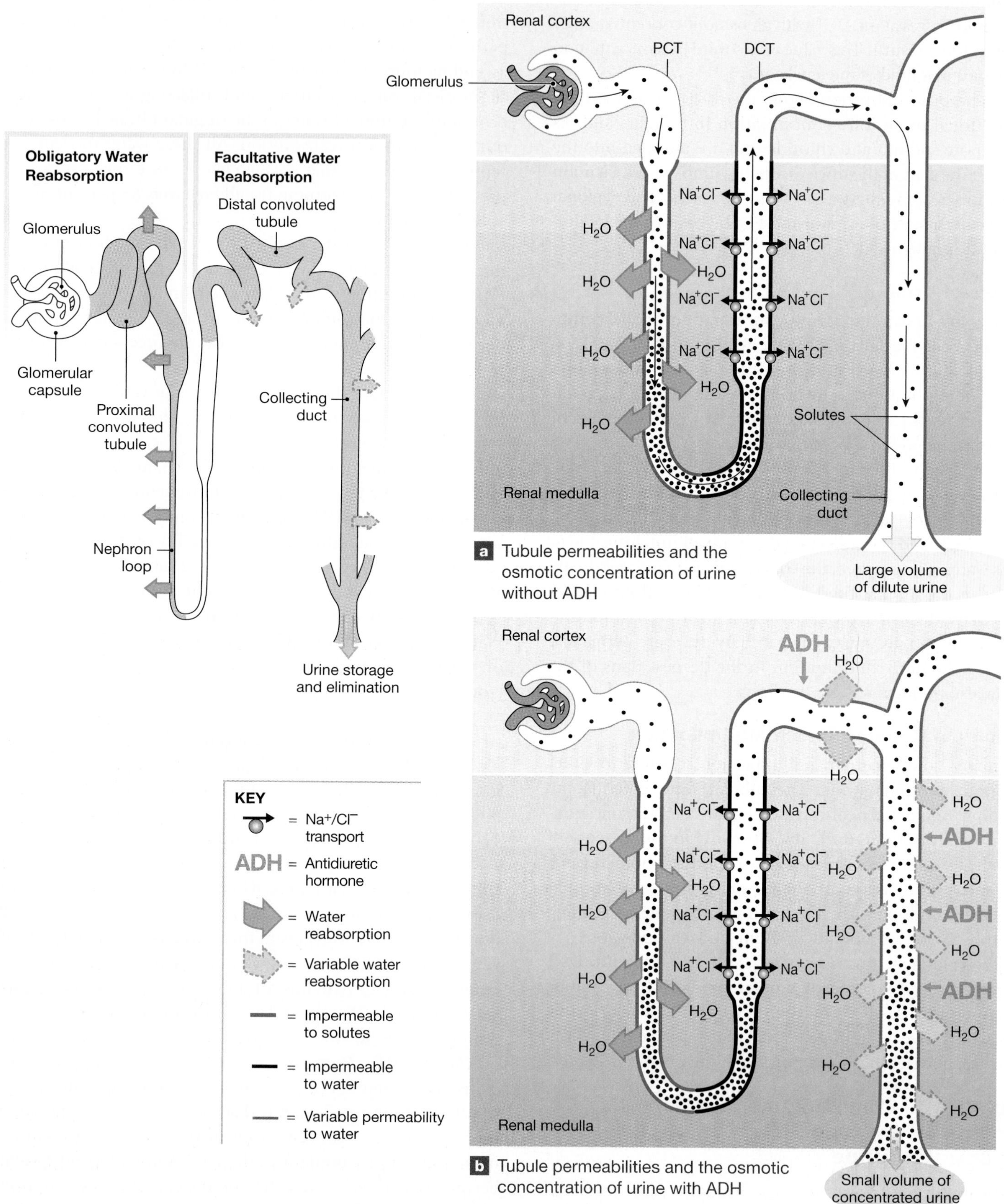
Obligatory Water Reabsorption
Glomerulus
Glomerular capsule
Proximal convoluted tubule
Nephron loop
Facultative Water Reabsorption
Distal convoluted tubule
Collecting duct
Urine storage and elimination
KEY
= Na+/Cl− transport
ADH = Antidiuretic hormone
= Water reabsorption
= Variable water reabsorption
= Impermeable to solutes
= Impermeable to water
= Variable permeability to water
Renal cortex
PCT
DCT
Glomerulus
Na+Cl−
H2O
Solutes
Renal medulla
Collecting duct
Large volume of dilute urine
ADH
Small volume of concentrated urine

a Tubule permeabilities and the osmotic concentration of urine without ADH

b Tubule permeabilities and the osmotic concentration of urine with ADH

? ADH creates a (*small* or *large*) volume of (*dilute* or *concentrated*) urine.

Note that the concentration of urine can never *exceed* that of the medulla, because the concentrating mechanism relies on osmosis.

The hypothalamus continuously secretes ADH at a low level. For this reason, the DCT and collecting system always have a significant degree of water permeability. At this low ADH level, the DCT reabsorbs approximately 9 liters of water per day, or about 5 percent of the original volume of filtrate produced by the glomeruli. At a normal ADH level, the collecting system reabsorbs about 16.8 liters per day, or about 9.3 percent of the original volume of filtrate. A healthy adult typically produces 1200 mL of urine per day (about 0.6 percent of the filtrate volume). Its osmotic concentration range is 855–1335 mOsm/L.

The effects of ADH are opposed by those of atrial natriuretic peptide (ANP). This hormone stimulates the production of a large volume of relatively dilute urine. This water loss reduces plasma volume to normal.

The Function of the Vasa Recta: Countercurrent Exchange

The solutes and water reabsorbed in the renal medulla must be returned to the bloodstream without disrupting the medullary osmotic gradient. This return is the function of the vasa recta, in a process called **countercurrent exchange**.

Recall that the vasa recta are long, straight capillaries that parallel the long nephron loops of juxtamedullary nephrons. Blood entering the vasa recta from the peritubular capillaries has an osmotic concentration of approximately 300 mOsm/L. As the blood descends into the medulla, it gradually increases in osmotic concentration as the solute concentration in the peritubular fluid increases. This increase in blood osmotic concentration involves both solute absorption and water loss. Solute absorption predominates, however, because the plasma proteins limit the osmotic flow of water out of the blood. ⮌ p. 742

Blood ascending toward the cortex gradually decreases in osmotic concentration as the solute concentration of the peritubular fluid decreases. Again, this decrease involves both solute diffusion and osmosis. In this case osmosis predominates, because the presence of plasma proteins does not oppose the osmotic flow of water into the blood.

The net results are that (1) some of the solutes absorbed in the descending segment of the vasa recta do not diffuse out in the ascending segment and (2) more water moves into the ascending portion of the vasa recta than moves out in the descending segment. Thus, the vasa recta carries both water and solutes out of the medulla. Under normal conditions, the removal of solutes and water by the vasa recta precisely balances the rates of solute reabsorption and osmosis in the medulla.

Table 26–5 General Characteristics of Normal Urine

Characteristic	Normal Range
pH	4.5–8 (average: 6.0)
Specific gravity	1.003–1.030
Osmotic concentration (osmolarity)	855–1335 mOsm/L
Water content	93%–97%
Volume	700–2000 mL/day
Color	Pale yellow
Clarity	Clear
Odor	Varies with composition
Bacterial content	None (sterile)

Spotlight Figure 26–16 provides a summary of kidney function showing the major steps in the reabsorption of water and the production of concentrated urine. Please revisit **Table 26–4** on p. 996 for a detailed summary of the functions of the various renal structures.

Urine Composition and Analysis

Urine is the fluid and dissolved substances excreted by the kidney. The analysis of that urine by physical, chemical, and microscopic means is called *urinalysis*. This section considers urine analysis.

The Composition of Normal Urine

The general characteristics of normal urine are listed in **Table 26–5**. However, the composition of the urine produced each day varies with the metabolic and hormonal events under way.

The composition and concentration of urine are two related but distinct properties. The *composition* of urine reflects the filtration, reabsorption, and secretion activities of the nephrons. Some substances (such as urea) are neither actively excreted nor reabsorbed along the nephron. In contrast, organic nutrients are completely reabsorbed. Other substances, such as creatinine, are missed by filtration but are actively secreted into the tubular fluid.

Filtration, reabsorption, and secretion determine the kinds and amounts of substances excreted in urine. The *concentration* of these substances in a given urine sample depends on the osmotic movement of water across the walls of the tubules and collecting ducts. Because the composition and concentration of urine vary independently, you can produce a small volume of concentrated urine or a large volume of dilute urine and still excrete the same amount of dissolved substances. For this reason, health care professionals who are interested in a detailed assessment of renal function commonly analyze the urine produced over a 24-hour period rather than a single urine sample.

SPOTLIGHT

Figure 26–16
Summary of Renal Function

1 The filtrate produced by the renal corpuscle has the same osmotic concentration as plasma—about 300 mOsm/L. It has the same composition as blood plasma but does not contain plasma proteins.

2 In the proximal convoluted tubule (PCT), the active removal of ions and organic nutrients results in a continuous osmotic flow of water out of the tubular fluid. This decreases the volume of filtrate but keeps the solutions inside and outside the tubule isotonic. Between 60 and 70 percent of the filtrate volume is absorbed here.

3 In the PCT and descending limb of the nephron loop, water moves into the surrounding peritubular fluid. This compulsory reabsorption of water results in a small volume of highly concentrated tubular fluid.

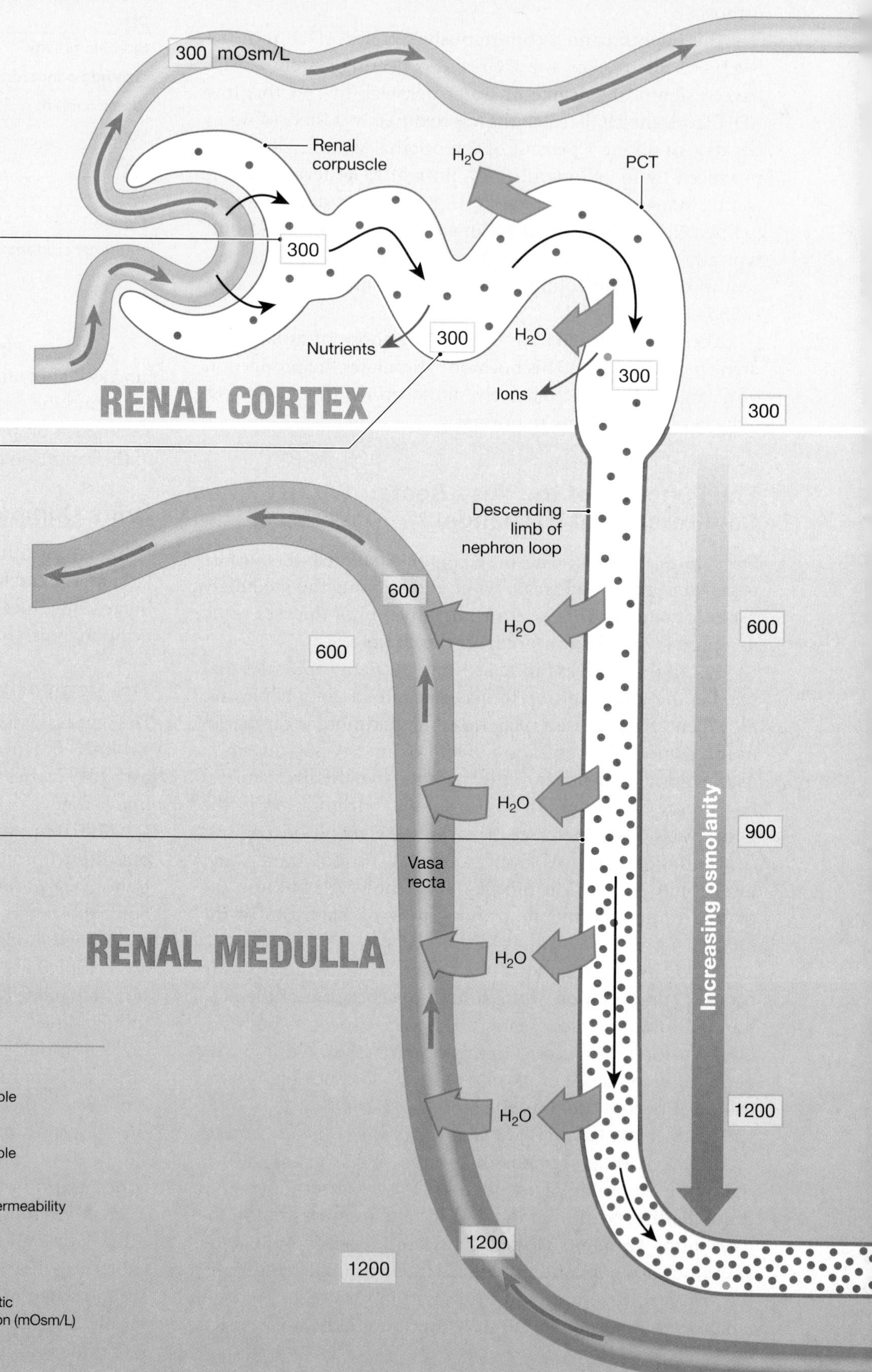

300 mOsm/L
Renal corpuscle
H_2O
PCT
300
Nutrients
300
H_2O
300
Ions
RENAL CORTEX
300
Descending limb of nephron loop
600
H_2O
600
600
H_2O
900
Vasa recta
RENAL MEDULLA
H_2O
Increasing osmolarity
H_2O
1200
1200
1200

KEY
= Water reabsorption
= Variable water reabsorption
= Na^+/Cl^- transport
(A) = Aldosterone-regulated pump
(U) = Urea transporter
= Impermeable to solutes
= Impermeable to water
= Variable permeability to water
= Solutes
300 = Fluid osmotic concentration (mOsm/L)

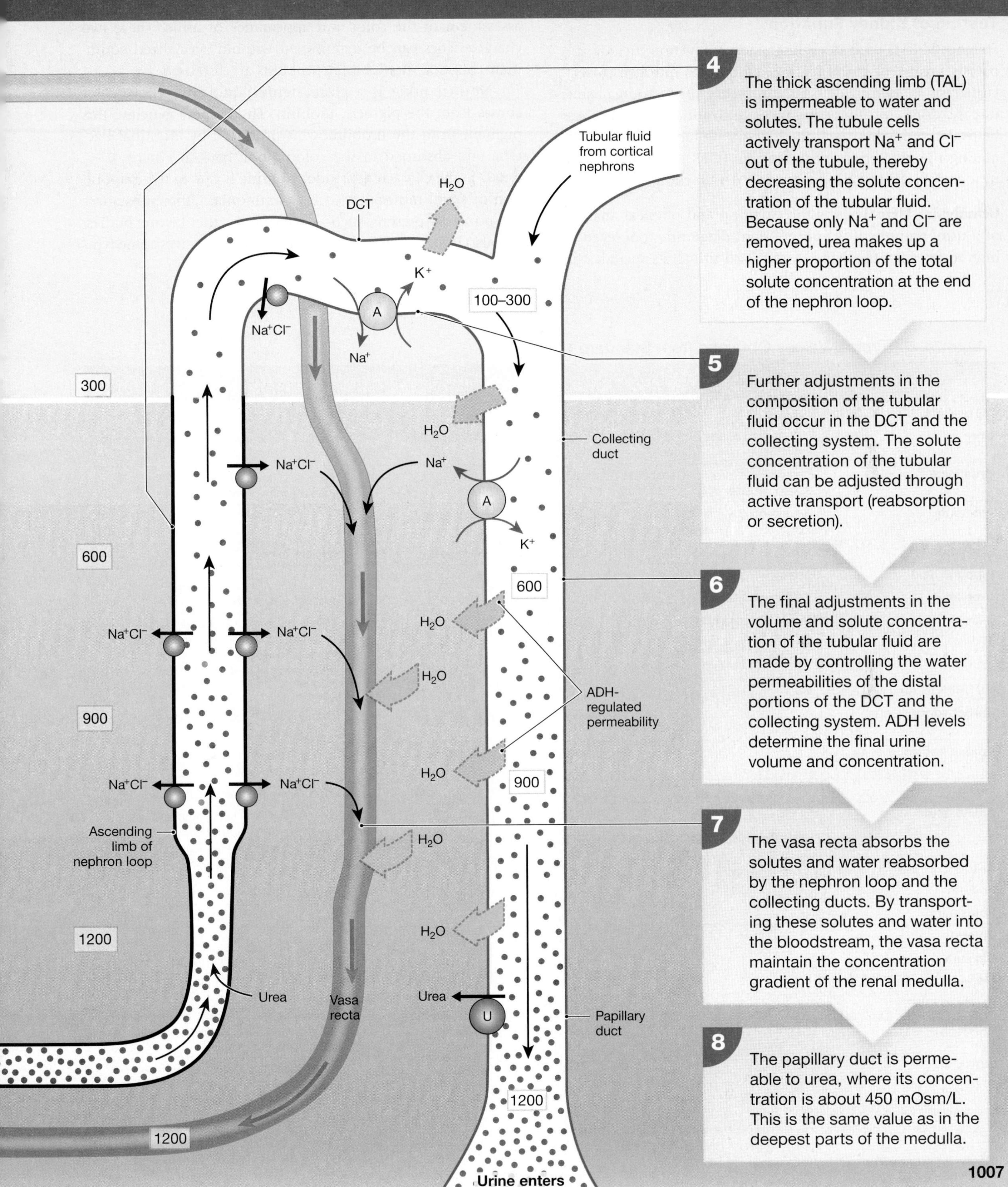

Tubular fluid from cortical nephrons
H_2O
DCT
K^+
A
Na^+Cl^-
Na^+
100–300
300
H_2O
Collecting duct
Na^+Cl^-
Na^+
A
K^+
600
600
Na^+Cl^-
Na^+Cl^-
H_2O
H_2O
ADH-regulated permeability
900
H_2O
900
Na^+Cl^-
Na^+Cl^-
Ascending limb of nephron loop
H_2O
1200
H_2O
Urea
Vasa recta
Urea
U
Papillary duct
1200
1200
Urine enters renal pelvis

4 The thick ascending limb (TAL) is impermeable to water and solutes. The tubule cells actively transport Na^+ and Cl^- out of the tubule, thereby decreasing the solute concentration of the tubular fluid. Because only Na^+ and Cl^- are removed, urea makes up a higher proportion of the total solute concentration at the end of the nephron loop.

5 Further adjustments in the composition of the tubular fluid occur in the DCT and the collecting system. The solute concentration of the tubular fluid can be adjusted through active transport (reabsorption or secretion).

6 The final adjustments in the volume and solute concentration of the tubular fluid are made by controlling the water permeabilities of the distal portions of the DCT and the collecting system. ADH levels determine the final urine volume and concentration.

7 The vasa recta absorbs the solutes and water reabsorbed by the nephron loop and the collecting ducts. By transporting these solutes and water into the bloodstream, the vasa recta maintain the concentration gradient of the renal medulla.

8 The papillary duct is permeable to urea, where its concentration is about 450 mOsm/L. This is the same value as in the deepest parts of the medulla.

Testing of Kidney Function

Diagnostic tests used to evaluate kidney function include urinalysis, creatinine clearance, and blood urea nitrogen (BUN). Urinalysis is used to detect the presence of microorganisms, disease, drugs, or other substances. A creatinine clearance test helps determine how well the kidneys are functioning by estimating the glomerular filtration rate (GFR). The blood urea nitrogen (BUN) test also measures renal function.

Urinalysis. Urinalysis is the chemical and physical analysis of a urine sample. It is an important diagnostic tool, even in high-technology medicine. A standard urinalysis includes an assessment of the color and appearance of urine. These two characteristics can be determined without specialized equipment, but quantitative analytical tests are also used.

Normal urine is a clear, sterile liquid. Its yellow color comes from the pigment urobilin. The kidneys generate this pigment from the urobilinogens produced by intestinal bacteria and absorbed in the colon (look back at **Figure 19–5**, p. 667). The characteristic odor of urine is due to the evaporation of small molecules, such as ammonia. Other substances not normally present, such as acetone or other ketone bodies, can also impart a distinctive smell. **Table 26–6** gives some typical values obtained from urinalysis.

Table 26–6 Typical Values Obtained from Standard Urinalysis

Compound	Primary Source	Daily Elimination*	Concentration	Remarks
NITROGENOUS WASTES				
Urea	Deamination of amino acids by liver and kidneys	21 g	1800 mg/dL	Increases if negative nitrogen balance exists
Creatinine	Breakdown of creatine phosphate in skeletal muscle	1.8 g	150 mg/dL	Proportional to muscle mass; decreases during atrophy or muscle disease
Ammonia	Deamination by liver and kidney, absorption from intestinal tract	0.68 g	60 mg/dL	
Uric acid	Breakdown of purines	0.53 g	40 mg/dL	Increases in gout, liver diseases
Hippuric acid	Breakdown of dietary toxins	4.2 mg	350 µg/dL	
Urobilin	Urobilinogens absorbed at colon	1.5 mg	125 µg/dL	Gives urine its yellow color
Bilirubin	Hemoglobin breakdown product	0.3 mg	20 µg/dL	Increase may indicate problem with liver elimination or excess production; causes yellowing of skin and mucous membranes in jaundice
NUTRIENTS AND METABOLITES				
Carbohydrates		0.11 g	9 µg/dL	Primarily glucose; *glycosuria* develops if T_m is exceeded
Ketone bodies		0.21 g	17 µg/dL	Ketonuria may occur during postabsorptive state
Lipids		0.02 g	0.002 mg/dL	May increase in some kidney diseases
Amino acids		2.25 g	188 µg/dL	Note relatively high loss compared with other metabolites due to low T_m; excess (*aminoaciduria*) indicates T_m problem
IONS				
Sodium		4.0 g	40–220 mEq/L	Varies with diet, urine pH, hormones, etc.
Potassium		2.0 g	25–100 mEq/L	Varies with diet, urine pH, hormones, etc.
Chloride		6.4 g	110–250 mEq/L	
Calcium		0.2 g	17 mg/dL	Hormonally regulated (PTH/CT)
Magnesium		0.15 g	13 mg/dL	
BLOOD CELLS‡				
RBCs		130,000/day	100/mL	Excess (*hematuria*) indicates vascular damage in urinary system
WBCs		650,000/day	500/mL	Excess (*pyuria*) indicates renal infection or inflammation

*Representative values for a 70-kg (154-lb) male.

‡Usually estimated by counting the cells in a sample of sediment after urine centrifugation.

26

Creatinine Clearance. Creatinine clearance compares the creatinine level in urine with the creatinine level in blood by estimating the GFR. It is used as an index of overall kidney function. For example, a GFR around 100 means the kidneys are working at 100%. Creatinine results from the breakdown of creatine phosphate in muscle tissue and is normally eliminated in urine. Creatinine enters the filtrate at the glomerulus and is not reabsorbed in significant amounts.

Consider, for example, a person who eliminates 84 mg of creatinine each hour and has a plasma creatinine concentration of 1.4 mg/dL. The GFR is equal to the amount secreted divided by the plasma concentration, so this person's GFR is

$$\frac{84 \text{ mg/h}}{1.4 \text{ mg/dL}} = 60 \text{ dL/h} = 100 \text{ mL/min}$$

The GFR is usually reported in milliliters per minute.

The value 100 mL/min is only an approximation of the GFR. The reason is that up to 15 percent of creatinine in the urine enters by means of active tubular secretion. When necessary, a more accurate GFR can be obtained by using the complex carbohydrate *inulin*. This compound is not metabolized in the body and is neither reabsorbed nor secreted by the kidney tubules.

BUN. A **BUN** (**blood urea nitrogen**) test measures the amount of urea in blood. Urea is created in the liver from the breakdown of proteins. As we have discussed, a critical role of the kidneys is to excrete nitrogenous wastes—urea, uric acid, and creatinine. We expect these wastes to appear in urine, but high levels of nitrogenous wastes should not remain in the blood.

Checkpoint

15. **How would a lack of juxtamedullary nephrons affect the volume and osmotic concentration of urine?**
16. **What effect would a high-protein diet have on the composition of urine?**

See the blue Answers tab at the back of the book.

26-7 Urine is transported by the ureters, stored in the bladder, and eliminated through the urethra by urinary reflexes

Learning Outcome Describe the structures and functions of the ureters, urinary bladder, and urethra, discuss the voluntary and involuntary regulation of urination, and describe the urinary reflexes.

Filtrate modification and urine production end when the fluid enters the renal pelvis. The urinary tract (the ureters, urinary bladder, and urethra) transports, stores, and eliminates urine.

Figure 26–17 **A Pyelogram.** This anterior-posterior view of urinary system structures is color enhanced. ATLAS: Plate 62b

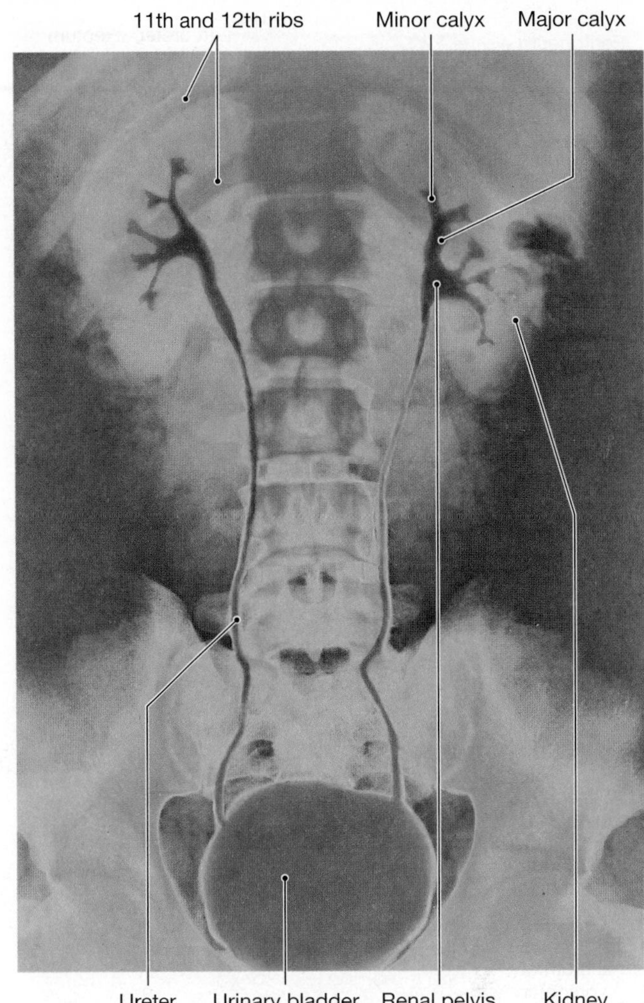
11th and 12th ribs
Minor calyx
Major calyx
Ureter
Urinary bladder
Renal pelvis
Kidney

A **pyelogram** (PĪ-el-ō-gram) is an image of the urinary system (Figure 26–17). It is obtained by taking an x-ray of the kidneys after a radiopaque dye has been administered intravenously. Such an image provides an orientation to the relative sizes and positions of the main structures. Note that the sizes of the minor and major calyces, the renal pelvis, the ureters, the urinary bladder, and the proximal portion of the urethra are somewhat variable. These regions are lined by a *transitional epithelium* that can tolerate cycles of distension and relaxation without damage. p. 122

26

The Ureters

The ureters are a pair of muscular tubes that extend from the kidneys to the urinary bladder—a distance of about 30 cm (12 in.). Each ureter begins at the funnel-shaped renal pelvis (look back at Figure 26–4). The ureters extend inferiorly and medially, passing over the anterior surfaces of the *psoas major* (look back

Figure 26–18 Organs for Conducting and Storing Urine. ATLAS: Plates 62b; 64; 65

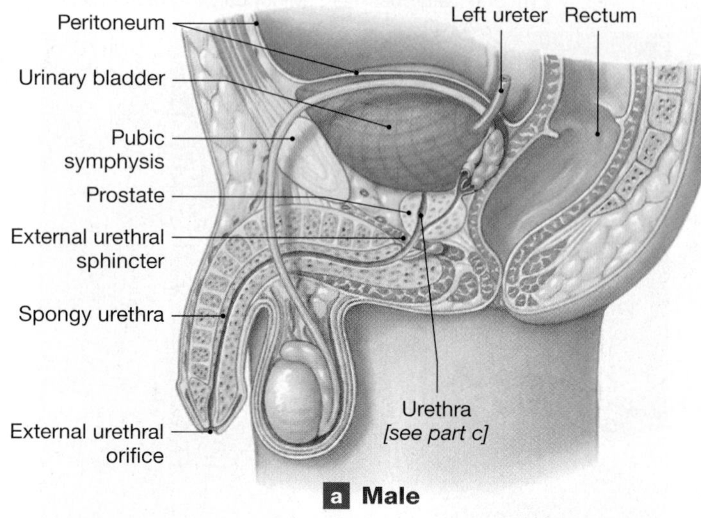
Peritoneum
Left ureter
Rectum
Urinary bladder
Pubic symphysis
Prostate
External urethral sphincter
Spongy urethra
External urethral orifice
Urethra [see part c]

a Male

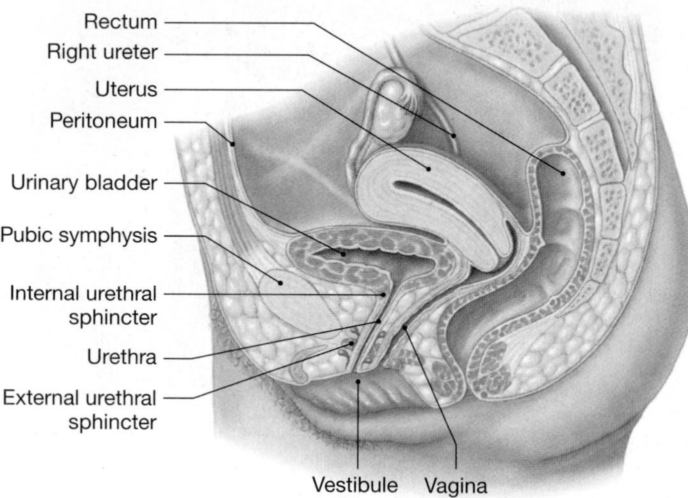
Rectum
Right ureter
Uterus
Peritoneum
Urinary bladder
Pubic symphysis
Internal urethral sphincter
Urethra
External urethral sphincter
Vestibule
Vagina

b Female

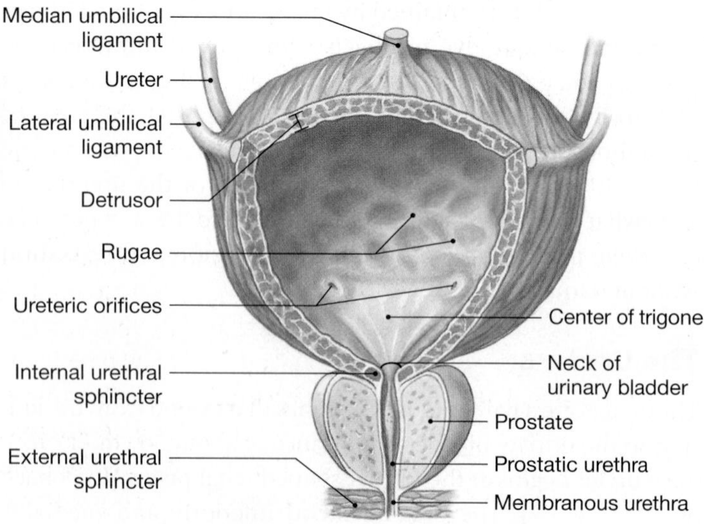
Median umbilical ligament
Ureter
Lateral umbilical ligament
Detrusor
Rugae
Ureteric orifices
Center of trigone
Internal urethral sphincter
Neck of urinary bladder
Prostate
External urethral sphincter
Prostatic urethra
Membranous urethra

c Urinary bladder in male

? The (*ureter* or *urethra*) transports urine *to* the bladder.

26

at Figure 26–3). The ureters are retroperitoneal and are firmly attached to the posterior abdominal wall. The paths taken by the ureters in men and women are different, due to variations in the nature, size, and position of the reproductive organs. In males, the base of the urinary bladder lies between the rectum and the pubic symphysis (Figure 26–18a). In females, the base of the urinary bladder sits inferior to the uterus and anterior to the vagina (Figure 26–18b).

The ureters penetrate the posterior wall of the urinary bladder without entering the peritoneal cavity. They pass through the bladder wall at an oblique angle. The **ureteric orifices** are slit-like rather than rounded (Figure 26–18c). This shape helps prevent the backflow of urine toward the ureter and kidneys when the urinary bladder contracts.

The wall of each ureter consists of three layers (Figure 26–19a): (1) an inner mucosa, made up of a transitional epithelium and the surrounding lamina propria; (2) a middle muscular layer made up of longitudinal and circular bands of smooth muscle; and (3) an outer connective tissue layer that is continuous with the fibrous capsule and peritoneum. About every 30 seconds, a peristaltic contraction begins at the renal pelvis. As it sweeps along the ureter, it forces urine toward the urinary bladder.

The Urinary Bladder

The urinary bladder is a hollow, muscular organ that serves as temporary storage for urine (see Figure 26–18c). The dimensions of the urinary bladder vary with its state of distension. A full urinary bladder can contain as much as a liter of urine.

A layer of peritoneum covers the superior surfaces of the urinary bladder. Several peritoneal folds assist in stabilizing its position. The **median umbilical ligament** extends from the anterior, superior border toward the umbilicus (navel). The **lateral umbilical ligaments** pass along the sides of the bladder to the umbilicus. These fibrous cords are the vestiges of the two *umbilical arteries*, which supplied blood to the placenta during embryonic and fetal development. ⮌ p. 775 The urinary bladder's posterior, inferior, and anterior surfaces lie outside the peritoneal cavity. In these areas, tough ligamentous bands anchor the urinary bladder to the pelvic and pubic bones.

In sectional view, the mucosa lining the urinary bladder usually has folds called **rugae** that disappear as the bladder fills. The triangular smooth area bounded by the openings of the ureters and the entrance to the urethra makes up a region called the **trigone** (TRĪ-gōn) of the urinary bladder. There, the mucosa is smooth and very thick. The trigone acts as a funnel that channels urine into the urethra when the urinary bladder contracts.

The urethral entrance lies at the apex of the trigone, at the most inferior point in the urinary bladder. The region

Figure 26–19 **The Histology of the Ureter, Urinary Bladder, and Urethra.**

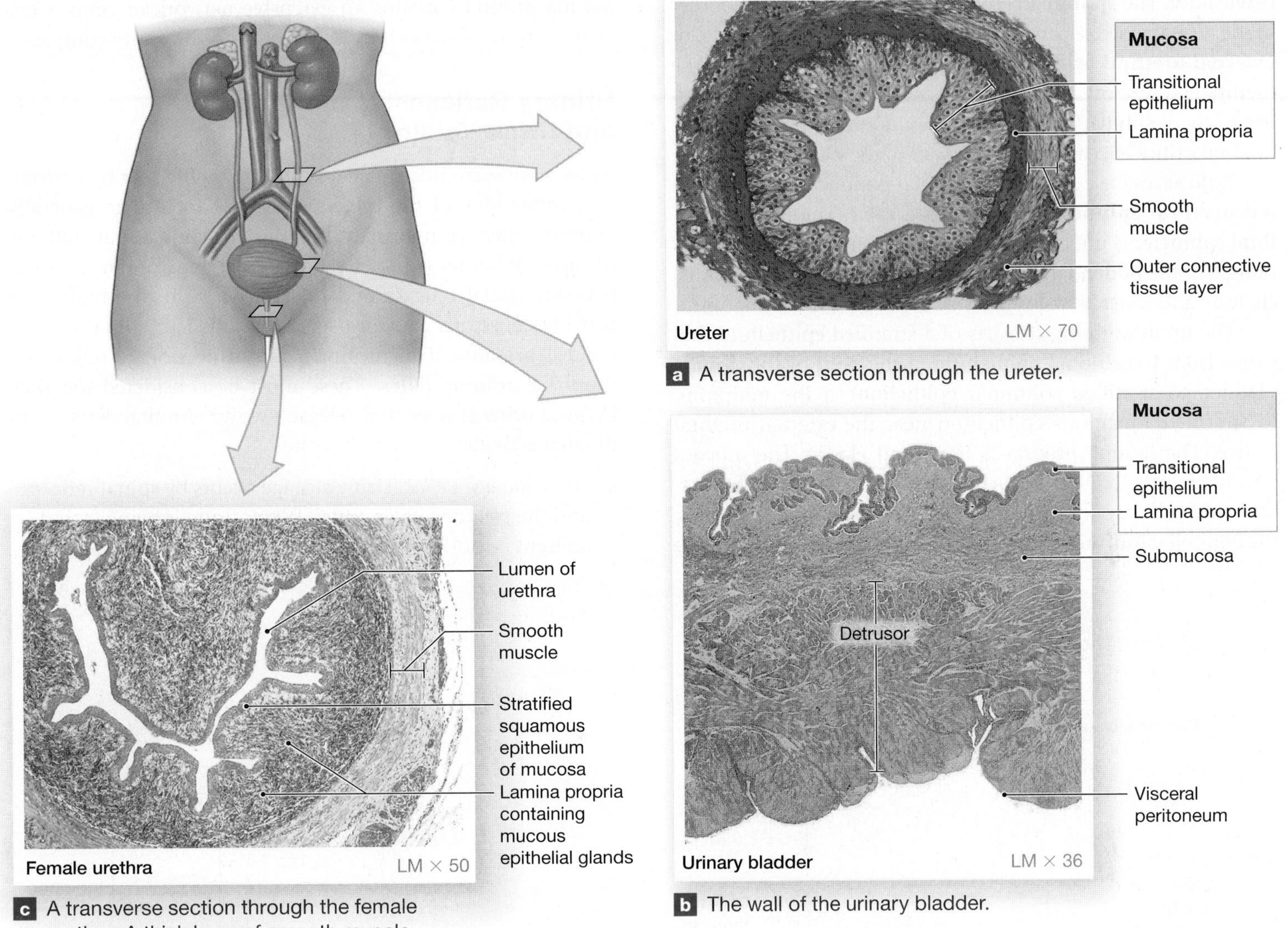
Mucosa
Transitional epithelium
Lamina propria
Smooth muscle
Outer connective tissue layer
Ureter
LM × 70
Mucosa
Transitional epithelium
Lamina propria
Submucosa
Detrusor
Visceral peritoneum
Urinary bladder
LM × 36
Lumen of urethra
Smooth muscle
Stratified squamous epithelium of mucosa
Lamina propria containing mucous epithelial glands
Female urethra
LM × 50

a A transverse section through the ureter.

b The wall of the urinary bladder.

c A transverse section through the female urethra. A thick layer of smooth muscle surrounds the lumen.

surrounding the urethral opening is known as the **neck** of the urinary bladder. It contains a muscular **internal urethral sphincter**. The smooth muscle fibers of this sphincter provide involuntary control over the discharge of urine from the bladder. The urinary bladder is innervated by postganglionic fibers from ganglia in the hypogastric plexus and by parasympathetic fibers from intramural ganglia that are controlled by branches of the pelvic nerves.

The wall of the urinary bladder contains mucosa, submucosa, and muscular layers (Figure 26–19b). The muscular layer consists of internal and external layers of longitudinal smooth muscle, with a circular layer between the two. Together, these layers form the powerful **detrusor** (de-TRŪ-sor), a muscle of the urinary bladder. When the detrusor contracts, it compresses the urinary bladder and expels urine into the urethra.

The Urethra

The urethra extends from the neck of the urinary bladder and transports urine to the exterior of the body. The male urethra and the female urethra differ in length and in function. In males, the urethra extends from the neck of the urinary bladder to the tip of the penis (look back at Figure 26–18a,c). This distance may be 18–20 cm (7–8 in.). The male urethra is subdivided into three segments: the prostatic urethra, the membranous urethra, and the spongy urethra. The **prostatic urethra** (*intermediate part of urethra*) passes through the center of the prostate. The **membranous urethra** includes the short segment that penetrates the deep transverse perineal muscle of the pelvic floor. The **spongy urethra** extends from the distal border of the deep transverse perineal muscle to the external opening, or **external urethral orifice**, at the tip of the penis. In females, the urethra is very short. It extends 3–5 cm (1–2 in.) from the bladder to the vestibule (look back at Figure 26–18b). The external urethral orifice is near the anterior wall of the vagina.

Both sexes have a skeletal muscular band, called the **external urethral sphincter**, that acts as a valve. The external urethral sphincter is under voluntary control through the perineal branch of the pudendal nerve. This sphincter has a resting muscle tone and must be voluntarily relaxed to permit urination.

The urethral lining consists of a stratified epithelium that varies from transitional epithelium at the neck of the urinary bladder, to stratified columnar epithelium at the midpoint, to stratified squamous epithelium near the external urethral orifice. The lamina propria is thick and elastic. The mucosa is folded into longitudinal creases (Figure 26–19c). Mucus-secreting cells are located in the epithelial pockets. In males, the epithelial mucous glands may form tubules that extend into the lamina propria. Connective tissues of the lamina propria anchor the urethra to surrounding structures. In females, the lamina propria contains an extensive network of veins. Concentric layers of smooth muscle surround the entire complex.

Urinary Reflexes: Urine Storage and Urine Voiding

As we have seen, urine reaches the urinary bladder by peristaltic contractions of the ureters. The urge to urinate generally appears when your urinary bladder contains about 200 mL of urine. Whether or not we urinate depends on an interplay between spinal reflexes and higher centers in the brain that provide conscious control over urination. Two spinal reflexes control urination (micturition): the urine storage reflex and the urine voiding reflex. These must be considered together, because urine storage and release involve simultaneous, coordinated activities (Figure 26–20).

- *Urine Storage Reflex.* Urine storage occurs by spinal reflexes and the *pontine storage center*. When urine is being stored, afferent impulses from stretch receptors in the urinary

Figure 26–20 The Control of Urination.

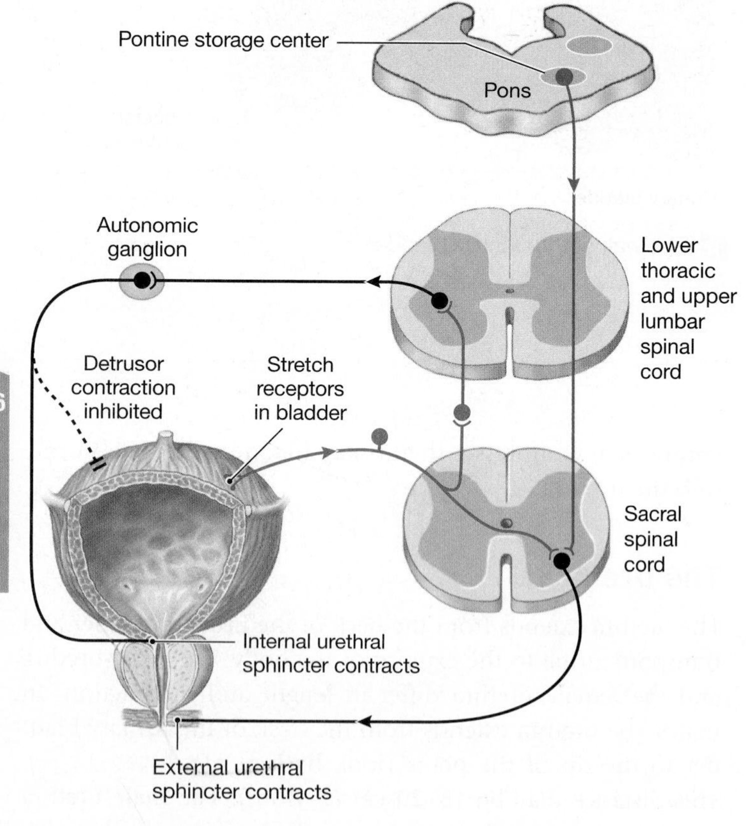
Pontine storage center
Pons
Autonomic ganglion
Lower thoracic and upper lumbar spinal cord
Detrusor contraction inhibited
Stretch receptors in bladder
Sacral spinal cord
Internal urethral sphincter contracts
External urethral sphincter contracts

a Urine storage reflex

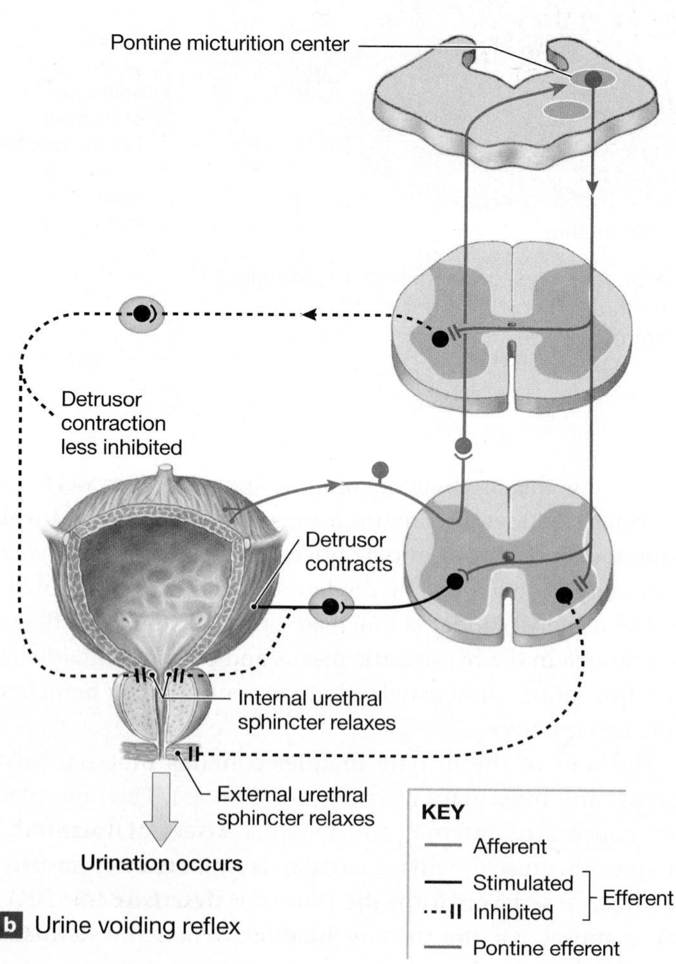
Pontine micturition center
Detrusor contraction less inhibited
Detrusor contracts
Internal urethral sphincter relaxes
External urethral sphincter relaxes
Urination occurs
KEY
Afferent
Stimulated
Inhibited
Efferent
Pontine efferent

b Urine voiding reflex

bladder stimulate sympathetic outflow to the detrusor and internal urethral sphincter, inhibiting contraction of the detrusor and stimulating contraction of the internal urethral sphincter (see **Figure 26–20a**). The pontine storage center inhibits urination by decreasing parasympathetic activity and increasing somatic motor nerve activity at the external urethral sphincter. Urine storage (continence) occurs as urine is unable to pass through the urethra.

- *Urine Voiding Reflex.* Urine voiding occurs by spinal reflexes and the *pontine micturition center*. The voiding reflex begins when afferent impulses from stretch receptors in the bladder stimulate interneurons that relay sensations to the pontine micturition center. This center initiates sacral spinal reflexes that (1) stimulate increased parasympathetic activity (detrusor contracts and internal urethral sphincter relaxes), (2) decrease sympathetic activity (detrusor contracts and internal urethral sphincter relaxes), and (3) decrease efferent somatic motor nerve activity (external urethral sphincter relaxes). Voiding (urination) occurs as urine passes through the urethra.

At the end of urination, less than 10 mL of urine remains in the urinary bladder.

Infants lack voluntary control over urination, because the necessary corticospinal connections have yet to be established. Accordingly, "toilet training" before age 2 often involves training the parent to anticipate the timing of the reflex rather than training the child to exert conscious control.

Incontinence (in-KON-tih-nens) is the inability to control urination voluntarily. Trauma to the internal or external urethral sphincter can contribute to incontinence in otherwise healthy adults. For example, some mothers develop *stress urinary incontinence* (SUI) if childbirth overstretches and damages the sphincter muscles. In this condition, increased intra-abdominal pressures—caused, for example, by a cough or sneeze—can overwhelm the sphincter muscles, causing urine to leak out. Incontinence can also develop in older people due to a general loss of muscle tone.

Damage to the central nervous system, the spinal cord, or the nerve supply to the urinary bladder or external urethral sphincter can also produce incontinence. For example, incontinence commonly accompanies Alzheimer's disease or spinal cord damage. In most cases, the affected individual develops an *automatic bladder*. The urinary reflexes remain intact, but voluntary control of the external urethral sphincter is lost, so the person cannot prevent the reflexive emptying of the urinary bladder. Damage to the pelvic nerves of the sacral spinal cord can destroy the urinary reflexes entirely, because those nerves carry both afferent and efferent fibers of both reflex arcs. The urinary bladder then becomes greatly distended with urine. It remains filled to capacity while the excess urine flows into the urethra in an uncontrolled stream. The insertion of a catheter is often needed to facilitate the discharge of urine.

+ Clinical Note Urinary Obstruction

Local blockages of the collecting ducts or ureters can result from *casts*—small blood clots, epithelial cells, lipids, or other materials—that form in the collecting ducts. Casts are commonly eliminated in urine. They are visible in microscopic analyses of urine samples. **Renal calculi** (KAL-kū-lī), or *kidney stones*, form within the urinary tract from calcium deposits, magnesium salts, or crystals of uric acid. The condition is called *nephrolithiasis* (nef-rō-lih-THĪ-uh-sis; *nephros*, kidney; *lithos*, stone). The blockage of the ureter by a stone or by other means (such as external compression) creates **urinary obstruction**. This problem is an emergency. In addition to causing extreme pain, it decreases or prevents filtration in the affected kidney by increasing the capsular hydrostatic pressure.

Calculi are generally visible on an x-ray. If urinary peristalsis and fluid pressures cannot dislodge them, they must be either surgically removed or destroyed. One nonsurgical procedure involves disintegrating the stones with a *lithotripter*, a device originally developed from machines used to de-ice airplane wings. Lithotripters focus sound waves on the stones, breaking them into smaller fragments that can be passed in the urine. Another nonsurgical approach is the insertion of a catheter armed with a laser that can shatter calculi with intense light beams.

Checkpoint

17. **Obstruction of a ureter by a kidney stone would interfere with the flow of urine between which two points?**
18. **The ability to control the urinary reflexes depends on your ability to control which muscle?**

See the blue Answers tab at the back of the book.

26-8 Age-related changes affect kidney function and urination

Learning Outcome **Describe the effects of aging on the urinary system.**

In general, aging is associated with an increased incidence of kidney problems. One example—*nephrolithiasis*, the formation of calculi, or kidney stones—is described in the Urinary Obstruction Clinical Note. Other age-related changes in the urinary system include the following:

- *A Decrease in the Number of Functional Nephrons.* The total number of kidney nephrons decreases by 30–40 percent between ages 25 and 85.
- *A Reduction in the GFR.* This reduction results from fewer glomeruli, cumulative damage to the filtration apparatus in the remaining glomeruli, and reduced renal blood flow.

Clinical Note Renal Failure and Kidney Transplant

Renal failure occurs when the kidneys cannot filter wastes from the blood and can no longer maintain homeostasis. When kidney filtration slows for any reason, urine production decreases. As the patient makes less and less urine, signs and symptoms of renal failure appear because water, ions, and metabolic wastes are retained. Virtually all systems in the body are affected. Blood pressure rises; anemia develops due to a decrease in erythropoietin production; and CNS problems can lead to sleeplessness, seizures, delirium, and even coma.

Acute renal failure occurs when filtration slows suddenly or stops due to renal ischemia, urinary obstruction, trauma, or nephrotoxic drugs. The reduction in kidney function takes place over a few days and may persist for weeks. Sensitized individuals can also develop acute renal failure after an allergic response to antibiotics or anesthetics.

In *chronic renal failure*, kidney function deteriorates gradually and problems accumulate over years. This condition generally cannot be reversed but only slowed. Eventually, symptoms of *end-stage renal failure* develop. *Hemodialysis* is a treatment that "cleanses" the blood by substituting for normal kidney functioning. It can relieve the symptoms of renal failure but is not a cure.

In terms of overall quality of life, the most satisfactory solution to end-stage renal failure is **kidney transplant**. This procedure involves implanting a new kidney from a living or deceased donor. According to the National Kidney Foundation, in January 2016, 83% of all people awaiting organ transplants were waiting for a kidney. Their median wait time is 3.6 years. The recipient's nonfunctioning kidney(s) may be removed, especially if an infection is present. The transplanted kidney and ureter are usually placed extraperitoneally in the pelvic cavity (within the iliac fossa). The ureter is connected to the recipient's urinary bladder. Patient survival is more than 90 percent at two years after the transplant.

- *A Reduced Sensitivity to ADH.* With age, the distal segments of the nephron and collecting system become less responsive to ADH. Water and sodium ions are reabsorbed at a reduced rate, and more sodium ions are lost in urine.
- *Problems with the Urinary Reflexes.* Three factors are involved in such problems: (1) The external sphincter loses muscle tone and become less effective at voluntarily retaining urine. This leads to incontinence, often involving a slow leakage of urine. (2) The ability to control urination can be lost due to a stroke, Alzheimer's disease, or other CNS problems affecting the cerebral cortex or hypothalamus. (3) In males, *urinary retention* may develop if the prostate enlarges and compresses the urethra, restricting the flow of urine.

Checkpoint

19. **List four age-related changes in the urinary system.**
20. **Define *nephrolithiasis*.**
21. **Describe how incontinence may develop in an elderly person.**

See the blue Answers tab at the back of the book.

26-9 The urinary system is one of several body systems involved in waste excretion

Learning Outcome **Give examples of interactions between the urinary system and other organ systems studied so far.**

The urinary system excretes wastes produced by other body systems, but it is not the only organ system involved in excretion. Indeed, the urinary, integumentary, respiratory, and digestive systems are together regarded as an anatomically diverse **excretory system** whose components perform all the excretory functions that affect the composition of body fluids:

- *Integumentary System.* Water losses and electrolyte losses in sensible perspiration can affect the volume and composition of the plasma. The effects are most apparent when losses are extreme, such as during peak sweat production. Small amounts of metabolic wastes, including urea, also are eliminated in perspiration.
- *Respiratory System.* The lungs remove the carbon dioxide generated by cells. Small amounts of other compounds, such as acetone and water, evaporate into the alveoli and are eliminated when you exhale.
- *Digestive System.* The liver excretes small amounts of metabolic waste products in bile. You lose a variable amount of water in feces.

These excretory activities have an impact on the composition of body fluids. For example, the removal of carbon dioxide by the respiratory system also affects the pH of blood plasma. Note that the excretory functions of these systems are not regulated as closely as are those of the kidneys. Under normal circumstances, the effects of integumentary and digestive excretory activities are minor compared with those of the urinary system.

Build Your Knowledge Figure 26–21 summarizes the functional relationships between the urinary system and other systems we have studied so far. We explore many of these relationships further in Chapter 27 when we consider major aspects of fluid, pH, and electrolyte balance.

Checkpoint

22. **Identify the role the urinary system plays for all other body systems.**
23. **Name the body systems that make up the excretory system.**

See the blue Answers tab at the back of the book.

Build Your Knowledge

Figure 26–21 **Integration of the URINARY system with the other body systems presented so far.**

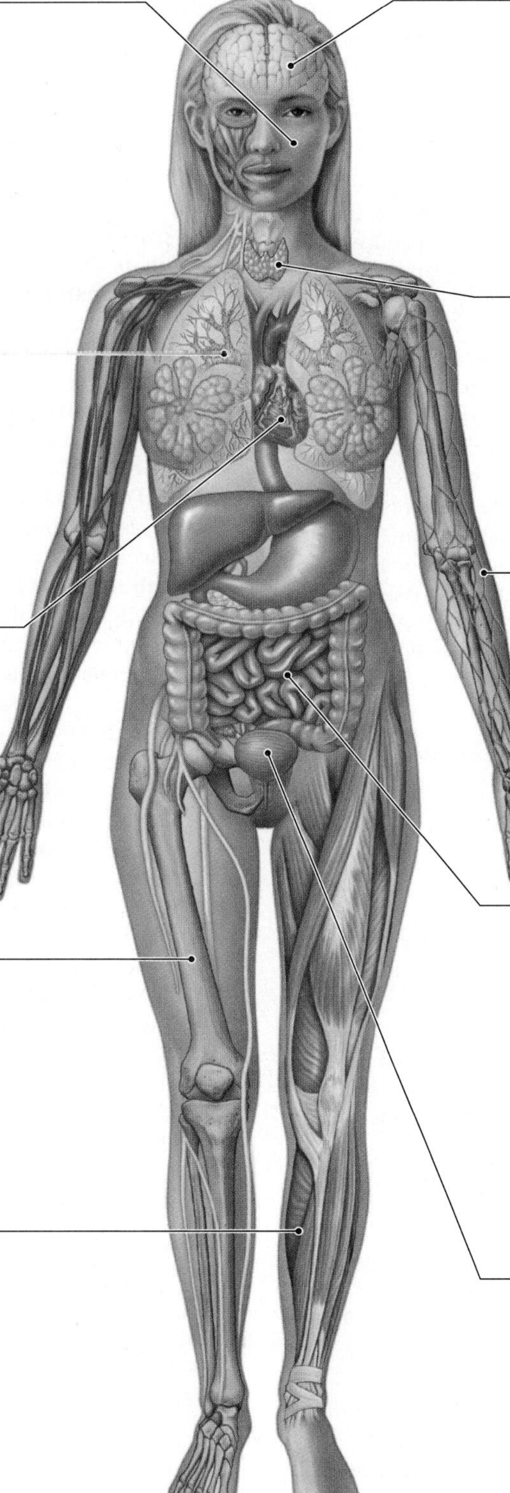

Integumentary System

- The Integumentary System prevents excessive fluid loss through skin surface; produces vitamin D_3, important for the renal production of calcitriol; sweat glands assist in elimination of water and solutes
- The urinary system excretes nitrogenous wastes; maintains fluid, electrolyte, and acid–base balance of blood that nourishes the skin

Respiratory System

- The Respiratory System assists in the regulation of pH by eliminating carbon dioxide
- The urinary system assists in the excretion of carbon dioxide; provides bicarbonate buffers that assist in pH regulation

Cardiovascular System

- The Cardiovascular System delivers blood to glomerular capillaries, where filtration occurs; accepts fluids and solutes reabsorbed during urine production
- The urinary system releases renin to elevate blood pressure and erythropoietin to accelerate red blood cell production

Skeletal System

- The Skeletal System provides some protection for kidneys and ureters with its axial divison; pelvis protects urinary bladder and proximal portion of urethra
- The urinary system conserves calcium and phosphate needed for bone growth

Muscular System

- The Muscular System controls urination by closing urethral sphincters. Muscle layers of trunk provide some protection for urinary organs
- The urinary system excretes waste products of muscle and protein metabolism; assists in regulation of calcium and phosphate concentrations

Nervous System

- The Nervous System adjusts renal blood pressure; monitors distension of urinary bladder and controls urination
- The urinary system eliminates nitrogenous wastes; maintains fluid, electrolyte, and acid–base balance of blood, which is critical for neural function

Endocrine System

- The Endocrine System produces aldosterone and antidiuretic hormone (ADH), which adjust rates of fluid and electrolyte reabsorption by kidneys
- The urinary system releases renin when local blood pressure drops and erythropoietin (EPO) when renal oxygen levels fall

Lymphatic System

- The Lymphatic System provides adaptive (specific) defense against urinary tract infections
- The urinary system excretes toxins and wastes generated by cellular activities; acid pH of urine provides innate (nonspecific) defense against urinary tract infections

Digestive System

- The Digestive System absorbs water needed to excrete wastes at kidneys; absorbs ions needed to maintain normal body fluid concentrations; liver removes bilirubin
- The urinary system excretes toxins absorbed by the digestive epithelium; excretes bilirubin and nitrogenous wastes from the liver; calcitriol production by kidneys aids calcium and phosphate absorption

Urinary System

The urinary system excretes metabolic wastes and maintains normal body fluid pH and ion composition.

It:

- regulates blood volume and blood pressure
- regulates plasma concentrations of sodium, potassium, chloride, and other ions
- helps to stabilize blood pH
- conserves valuable nutrients

26 Chapter Review

Study Outline

An Introduction to the Urinary System p. 977

1. The urinary system removes most physiological wastes and produces urine.

26-1 The organs of the urinary system function in excreting wastes and regulating body fluids p. 977

2. The urinary system includes two **kidneys**, two **ureters**, the **urinary bladder**, and the **urethra**. The kidneys produce **urine**, which is a fluid that contains water, ions, and soluble compounds. During **urination (micturition)**, urine is forced out of the body. *(Figure 26–1)*
3. The major functions of the **urinary system** are *excretion*, the removal of metabolic waste products from body fluids; *elimination*, the discharge of these waste products into the environment; and *homeostatic regulation* of the volume and solute concentration of blood plasma. Other homeostatic functions include regulating blood volume and pressure by adjusting the volume of water lost and releasing hormones; regulating plasma concentrations of ions; helping to stabilize blood pH; conserving nutrients; assisting the liver in detoxifying poisons; and, during starvation, deaminating amino acids so that they can be catabolized by other tissues.

26-2 Kidneys are highly vascular organs containing functional units called nephrons p. 978

4. The left kidney extends superiorly slightly more than the right kidney. Both kidneys and the adrenal gland superior to each kidney are retroperitoneal. *(Figures 26–1, 26–2, 26–3)*
5. The **hilum**, a medial indentation, provides entry for the *renal artery* and *renal nerves* and exit for the *renal vein* and the ureter. *(Figures 26–3, 26–4)*
6. The superficial region of the kidney, the cortex, surrounds the **medulla**. The ureter communicates with the **renal pelvis**, a chamber that branches into two **major calyces**. Each major calyx is connected to four or five **minor calyces**, which enclose the **renal papillae**. *(Figure 26–4)*
7. The blood supply to the kidneys includes the **renal**, **segmental**, **interlobar**, **arcuate**, and **cortical radiate arteries**. *(Figure 26–5)*
8. The **renal nerves**, which innervate the kidneys and ureters, are dominated by sympathetic postganglionic fibers.
9. The **nephron** is the basic functional unit in the kidney. It consists of the **renal corpuscle** and **renal tubule**. The renal tubule is long and narrow and divided into the *proximal convoluted tubule*, the *nephron loop*, and the *distal convoluted tubule*. **Filtrate** is produced at the renal corpuscle. The nephron empties **tubular fluid** into the **collecting system**, consisting of **collecting ducts** and **papillary ducts**. *(Figures 26–6, 26–7)*
10. The **proximal convoluted tubule (PCT)** actively reabsorbs nutrients, plasma proteins, and ions from the filtrate. These substances are released into the **peritubular fluid**, which surrounds the nephron. *(Figure 26–6)*
11. The **nephron loop**, also called the **loop of Henle**, includes a **descending limb** and an **ascending limb**. The *descending thin limb (DTL)* and the *ascending thin limb (ATL)* contain simple squamous epithelium and the *thick ascending limb (TAL)* contains cuboidal epithelium. The ascending limb delivers fluid to the **distal convoluted tubule (DCT)**, which actively secretes ions, toxins, and drugs, and reabsorbs sodium ions from the tubular fluid. *(Figures 26–6, 26–8)*
12. The renal tubule begins at the renal corpuscle, which includes a knot of intertwined capillaries called the **glomerulus**, surrounded by the **glomerular capsule**. Blood arrives at the glomerulus by the **afferent arteriole** and departs in the **efferent arteriole**. *(Figure 26–7)*
13. At the glomerulus, a **visceral layer** of **podocytes** cover the **basement membrane** of the capillaries that project into the **capsular space**. The **foot processes**, or *pedicels*, of the podocytes are separated by narrow **filtration slits**. Together, the fenestrated endothelium, basement membrane, and foot processes form the **filtration membrane**. **Intraglomerular mesangial cells** lie between adjacent capillaries and control capillary luminal diameter. *(Figure 26–7)*
14. Roughly 85 percent of the nephrons are **cortical nephrons**, located in the renal cortex. **Juxtamedullary nephrons** are closer to the renal medulla, with their *nephron loops* extending deep into the **renal pyramids**. *(Figure 26–8)*
15. Blood travels from the efferent arteriole to the **peritubular capillaries** and the **vasa recta**. *(Figure 26–8)*
16. Nephrons produce filtrate; reabsorb organic nutrients, water, and ions; and secrete various waste products into the tubular fluid. *(Table 26–1)*

26-3 Different segments of the nephron form urine by filtration, reabsorption, and secretion p. 987

17. Urine production maintains homeostasis by regulating blood volume and composition. In the process, metabolic

> MasteringA&P™ Access more chapter study tools online in the MasteringA&P Study Area:
> - Chapter Quizzes, Chapter Practice Test, MP3 Tutor Sessions, and Clinical Case Studies
> - Practice Anatomy Lab PAL 3.0
> - A&P Flix A&PFlix
> - Interactive Physiology iP2
> - PhysioEx PhysioEx 9.1

wastes —notably urea, creatinine, and uric acid—are excreted.

18. Urine formation involves **filtration**, **reabsorption**, and **secretion**.
19. Most regions of the nephron perform a combination of reabsorption and secretion. *(Figure 26–9; Table 26–2)*

26-4 The glomerulus filters blood through the filtration membrane to produce filtrate; several pressures determine the glomerular filtration rate p. 989

20. **Glomerular filtration** occurs as fluids and solutes move across the filtration membrane of the glomerulus into the capsular space in response to the **glomerular hydrostatic pressure (GHP)**—the hydrostatic (blood) pressure in the glomerular capillaries. This movement is opposed by the **capsular hydrostatic pressure (CsHP)** and by the **blood colloid osmotic pressure (BCOP)**. The **net filtration pressure (NFP)** at the glomerulus is the difference between the blood pressure and the opposing capsular and osmotic pressures. *(Figure 26–10)*
21. The **glomerular filtration rate (GFR)** is the amount of filtrate produced in the kidneys each minute. Any factor that alters the filtration pressure acting across the glomerular capillaries will change the GFR and affect kidney function.
22. A drop in filtration pressures stimulates the **juxtaglomerular complex (JGC)** to release renin and erythropoietin. *(Figure 26–11)*
23. Sympathetic activation (1) produces a powerful vasoconstriction of the afferent arterioles, decreasing the GFR and slowing the production of filtrate; (2) alters the GFR by changing the regional pattern of blood circulation; and (3) stimulates the release of renin by the juxtaglomerular complex. *(Figure 26–11)*
24. Glomerular filtration produces a filtrate with a composition similar to blood plasma, but with few, if any, plasma proteins.

26-5 The renal tubule reabsorbs nutrients, ions, and water, and secretes ions and wastes; the collecting system reabsorbs ions and water p. 993

25. Four types of *carrier-mediated transport (facilitated diffusion, active transport, cotransport,* and *countertransport)* are involved in modifying filtrate. The saturation limit of a carrier protein is its **transport maximum (T_m)**, which determines the **renal threshold**—the plasma concentration at which various substances will appear in urine. *(Table 26–2)*
26. The transport maximum determines the renal threshold for the reabsorption of substances in tubular fluid. *(Table 26–3)*
27. The cells of the PCT normally reabsorb sodium and other ions, water, and almost all the organic nutrients that enter the filtrate. These cells also secrete various substances into the tubular fluid. *(Figure 26–12)*
28. Water and ions are reclaimed from tubular fluid by the nephron loop. The entire descending limb of the nephron loop is freely permeable to water, but not to solutes, such as sodium and chloride ions. The ascending limb of the nephron loop is impermeable to water, but passively and actively removes sodium and chloride ions from the tubular fluid. Its ascending thin limb is permeable to sodium ions. The thick ascending limb (TAL) actively transports sodium and chloride ions out of the tubular fluid.
29. The DCT performs final adjustments by actively secreting or absorbing materials. Sodium ions are actively absorbed, in exchange for potassium or hydrogen ions discharged into tubular fluid. Aldosterone secretion increases the rate of sodium reabsorption and potassium loss. *(Figure 26–13)*
30. The amount of water and solutes in the tubular fluid of the collecting ducts is further regulated by aldosterone and ADH secretions. *(Figure 26–13)*

26-6 Countercurrent multiplication allows the kidneys to regulate the volume and concentration of urine p. 1001

31. The descending thin limb and the thick ascending limb of the nephron loop are close together and separated by peritubular fluid. The descending thin limb is permeable to water but relatively impermeable to solutes. The thick ascending limb is relatively impermeable to both water and solutes, but it contains active transport mechanisms that pump sodium and chloride ions from the tubular fluid into the peritubular fluid of the medulla.
32. The exchange of substances between the descending thin limb and the thick ascending limb of a nephron, called **countercurrent multiplication,** helps create the osmotic gradient in the medulla. The concentration gradient in the renal medulla encourages the osmotic flow of water out of the tubular fluid. As water is lost by osmosis and the volume of tubular fluid decreases, the urea concentration increases. *(Figure 26–14)*
33. Urine volume and osmotic concentration are regulated by controlling water reabsorption. Precise control over this occurs by *facultative water reabsorption*. *(Figure 26–15)*
34. Normally, the removal of solutes and water by the vasa recta precisely balances the rates of reabsorption and osmosis in the renal medulla.
35. More than 99 percent of the filtrate produced each day is reabsorbed before reaching the renal pelvis. *(Table 26–5)*
36. Each segment of the nephron and collecting system contributes to the production of concentrated urine. *(Spotlight Figure 26–16; Table 26–4)*
37. **Urinalysis** is the chemical and physical analysis of a urine sample. *(Table 26–6)*

26-7 Urine is transported by the ureters, stored in the bladder, and eliminated through the urethra by urinary reflexes p. 1009

38. Urine production ends when tubular fluid enters the renal pelvis. The rest of the urinary system transports, stores, and eliminates urine. *(Figure 26–17)*

39. The ureters extend from the renal pelvis to the urinary bladder. Peristaltic contractions by smooth muscles move the urine along the tract. *(Figures 26–18, 26–19)*
40. The urinary bladder is stabilized by the **middle umbilical ligament** and the **lateral umbilical ligaments**. Internal features include the **trigone**, the **neck**, and the **internal urethral sphincter**. The mucosal lining contains prominent **rugae** (folds). Contraction of the **detrusor** compresses the urinary bladder and expels urine into the urethra. *(Figures 26–18, 26–19)*
41. In both sexes, a circular band of skeletal muscles forms the **external urethral sphincter**, which is under voluntary control. *(Figure 26–18)*
42. **Urination** is coordinated by the **urine storage reflex** and the **urine voiding reflex**, which are initiated by stretch receptors in the wall of the urinary bladder. Voluntary urination involves coupling these reflexes with the voluntary relaxation of the external urethral sphincter, which allows the relaxation of the **internal urethral sphincter**. *(Figure 26–20)*

26-8 Age-related changes affect kidney function and urination p. 1013

43. Aging is generally associated with increased kidney problems. Age-related changes in the urinary system include (1) declining numbers of functional nephrons, (2) reduced GFR, (3) reduced sensitivity to ADH, and (4) problems with the urinary reflexes. (**Urinary retention** may develop in men whose prostate is enlarged.)

26-9 The urinary system is one of several body systems involved in waste excretion p. 1014

44. The urinary system is the major component of an anatomically diverse **excretory system** that includes the integumentary system, the respiratory system, and the digestive system. *(Figure 26–21)*

Review Questions

See the blue Answers tab at the back of the book.

LEVEL 1 Reviewing Facts and Terms

1. Identify the structures of the kidney in the following diagram.

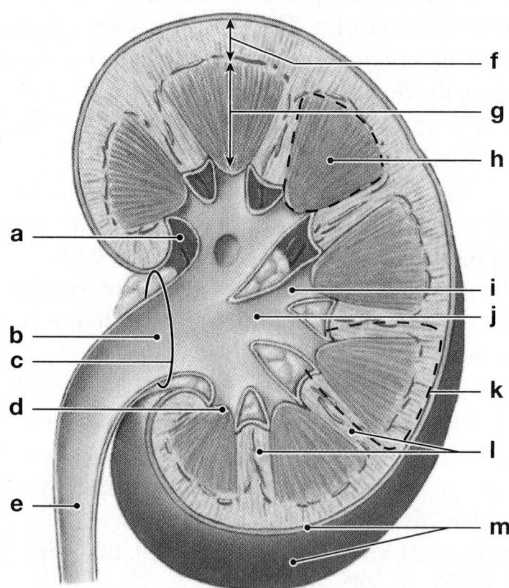
f
g
h
a
i
j
b
c
k
d
l
e
m

(a) ________ (b) ________
(c) ________ (d) ________
(e) ________ (f) ________
(g) ________ (h) ________
(i) ________ (j) ________
(k) ________ (l) ________
(m) ________

2. The basic functional unit of the kidney is the **(a)** nephron, **(b)** renal corpuscle, **(c)** glomerulus, **(d)** nephron loop, **(e)** filtration unit.
3. The process of urine formation involves all of the following, *except* **(a)** filtration of plasma, **(b)** reabsorption of water, **(c)** reabsorption of certain solutes, **(d)** secretion of wastes, **(e)** secretion of excess lipoprotein and glucose molecules.
4. The glomerular filtration rate is regulated by all of the following, *except* **(a)** autoregulation, **(b)** sympathetic neural control, **(c)** cardiac output, **(d)** angiotensin II, **(e)** the hormone ADH.
5. The distal convoluted tubule is an important site for **(a)** active secretion of ions, **(b)** active secretion of acids and other materials, **(c)** selective reabsorption of sodium ions from the tubular fluid, **(d)** all of these.
6. Changing the luminal diameters of the afferent and efferent arterioles to alter the GFR can be an example of **(a)** hormonal regulation, **(b)** autonomic regulation, **(c)** autoregulation, **(d)** all of these.
7. What are the primary functions of the urinary system?
8. What organs make up the urinary system?
9. Trace the pathway of the protein-free filtrate from where it is produced in the renal corpuscle until it drains into the renal pelvis in the form of urine. (Use arrows to indicate the direction of flow.)
10. Name the segments of the nephron distal to the renal corpuscle, and state the functions of each.
11. What is the function of the juxtaglomerular complex?
12. Trace the path of a drop of blood from its entry into the renal artery until its exit at a renal vein.
13. Name and define the three distinct processes involved in the production of urine.
14. What are the primary effects of angiotensin II on kidney function and regulation?
15. Which structures of the urinary system are responsible for the transport, storage, and elimination of urine?

LEVEL 2 Reviewing Concepts

16. When the renal threshold for a substance exceeds its tubular maximum **(a)** more of the substance will be filtered, **(b)** more of the substance will be reabsorbed, **(c)** more of the substance will be secreted, **(d)** the amount of the substance that exceeds the tubular maximum will be found in the urine, **(e)** both a and d occur.
17. Sympathetic activation of nerve fibers in the nephron causes **(a)** the regulation of glomerular blood flow and pressure, **(b)** the stimulation of renin release from the juxtaglomerular complex, **(c)** the direct stimulation of water and Na^+ reabsorption, **(d)** all of these.
18. Sodium reabsorption in the DCT and in the cortical portion of the collecting system is accelerated by the secretion of **(a)** ADH, **(b)** renin, **(c)** aldosterone, **(d)** erythropoietin.
19. When ADH levels rise, **(a)** the amount of water reabsorbed increases, **(b)** the DCT becomes impermeable to water, **(c)** the amount of water reabsorbed decreases, **(d)** sodium ions are exchanged for potassium ions.
20. The control of blood pH by the kidneys during acidosis involves **(a)** the secretion of hydrogen ions and reabsorption of bicarbonate ions from the tubular fluid, **(b)** a decrease in the amount of water reabsorbed, **(c)** hydrogen ion reabsorption and bicarbonate ion loss, **(d)** potassium ion secretion.
21. How are proteins excluded from filtrate? Why is this important?
22. What interacting controls stabilize the glomerular filtration rate (GFR)?
23. What primary changes occur in the composition and concentration of filtrate as a result of activity in the proximal convoluted tubule?
24. Describe two functions of countercurrent multiplication in the kidney.
25. Describe the urine voiding reflex.

LEVEL 3 Critical Thinking and Clinical Applications

26. In a normal kidney, which of the following conditions would cause an increase in the glomerular filtration rate (GFR)? **(a)** constriction of the afferent arteriole, **(b)** a decrease in the pressure of the glomerulus, **(c)** an increase in the capsular hydrostatic pressure, **(d)** a decrease in the concentration of plasma proteins in the blood, **(e)** a decrease in the net glomerular filtration process.
27. In response to *excess* water in the body, **(a)** antidiuretic hormone is secreted by the anterior lobe of the pituitary gland, **(b)** the active transport mechanisms in the ascending thick limb of the nephron loop cease functioning, **(c)** the permeability of the distal convoluted tubules and collecting ducts to water is decreased, **(d)** the permeability of the ascending limb of the nephron loop is increased, **(e)** the glomerular filtration rate is reduced.
28. Sylvia is suffering from severe edema in her arms and legs. Her physician prescribes a diuretic (a substance that increases the volume of urine produced). Why might this help alleviate Sylvia's problem?
29. David's grandfather suffers from hypertension. His doctor tells him that part of his problem stems from renal arteriosclerosis. Why would this cause hypertension?
30. *Mannitol* is a sugar that is filtered, but not reabsorbed, by the kidneys. What effect would drinking a solution of mannitol have on the volume of urine produced?
31. The drug *Diamox* is sometimes used to treat mountain sickness. Diamox inhibits the action of carbonic anhydrase in the proximal convoluted tubule. Polyuria (excessive urine production resulting in frequent urination) is a side effect associated with the medication. Why does polyuria occur?

+ CLINICAL CASE Wrap-Up A Case of "Hidden" Bleeding

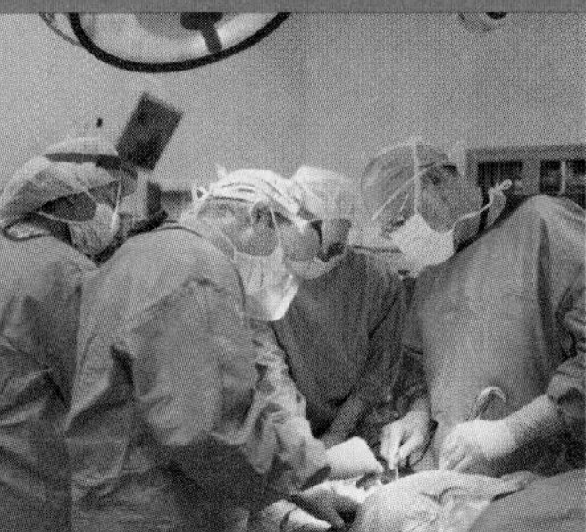

Normally, when a kidney is injured, the urine appears very bloody (*hematuria*). Mike has made very little urine since his injury and massive blood loss, but the urine in his catheter is clear. This reduces the initial suspicion of a renal injury.

However, when the surgeon opens the left retroperitoneal space, he finds another large collection of free blood. When the surgeon clears this blood away, the source of Mike's continued bleeding becomes apparent. His left renal artery, renal vein, and ureter have been completely torn away from the kidney at the hilum. The renal artery is briskly pumping what circulatory blood volume is left into the retroperitoneal space, where it is walled off from the abdominal space by the parietal peritoneum. Because the left ureter is completely torn from the damaged kidney, there is no visible blood in the urine.

This kidney is too damaged to heal. The surgeon quickly performs a left *nephrectomy* (kidney removal). If Mike survives this trauma, he will have to live with a solitary kidney. With the bleeding stopped, including the "hidden" bleeding, the trauma team is finally able to sustain an adequate circulating blood volume.

1. How could Mike's abdominal x-rays show that his kidneys might be injured?
2. What will likely happen to the glomerular filtration rate (GFR) of Mike's remaining right kidney if he survives this trauma?

See the blue Answers tab at the back of the book.

Related Clinical Terms

azotemia: The condition characterized by excessive urea or other nitrogen-containing compounds in the blood.

cystocele: Condition that occurs when the supportive tissue between a woman's bladder and vaginal wall weakens, stretches, and allows the bladder to bulge into the vagina.

cystoscopy: Diagnostic procedure using an optical instrument called a cystoscope that is inserted through the urethra to visually examine the bladder and lower urinary tract, to collect urine samples, or to view the prostate.

enuresis: Involuntary urination, especially that of a child while asleep.

nephroptosis: Condition in which the kidney is displaced downward from its usual and normal position; also called a *floating kidney*.

nephrotic syndrome: A kidney disorder that causes one to excrete excessive protein in the urine.

nephrotoxin: A toxin that has a specific harmful effect on the kidney.

polycystic kidney disease: An inherited abnormality that affects the development and structure of kidney tubules.

shock-wave lithotripsy: A noninvasive technique used to pulverize kidney stones by passing high-pressure shock waves through a water-filled tub in which the patient sits.

urologist: Physician who specializes in functions and disorders of the urinary system.

27 Fluid, Electrolyte, and Acid-Base Balance

Learning Outcomes

These Learning Outcomes correspond by number to this chapter's sections and indicate what you should be able to do after completing the chapter.

27-1 ■ **Explain what is meant by the terms fluid balance, electrolyte balance, and acid-base balance, and discuss their importance for homeostasis.** p. 1022

27-2 ■ **Compare the composition of ECF and ICF, explain the basic concepts involved in the regulation of fluids and electrolytes, and identify the hormones that play important roles in fluid and electrolyte regulation.** p. 1023

27-3 ■ **Describe the movement of fluid within the ECF, between the ECF and the ICF, and between the ECF and the environment.** p. 1027

27-4 ■ **Discuss the mechanisms by which sodium, potassium, calcium, magnesium, phosphate, and chloride ion concentrations are regulated to maintain electrolyte balance.** p. 1030

27-5 ■ **Explain the buffer systems that balance the pH of the ICF and ECF, and describe the compensatory mechanisms involved in maintaining acid-base balance.** p. 1036

27-6 ■ **Identify the most frequent disturbances of acid-base balance, and explain how the body responds when the pH of body fluids varies outside normal limits.** p. 1042

27-7 ■ **Describe the effects of aging on fluid, electrolyte, and acid-base balance** p. 1049

CLINICAL CASE When Treatment Makes You Worse

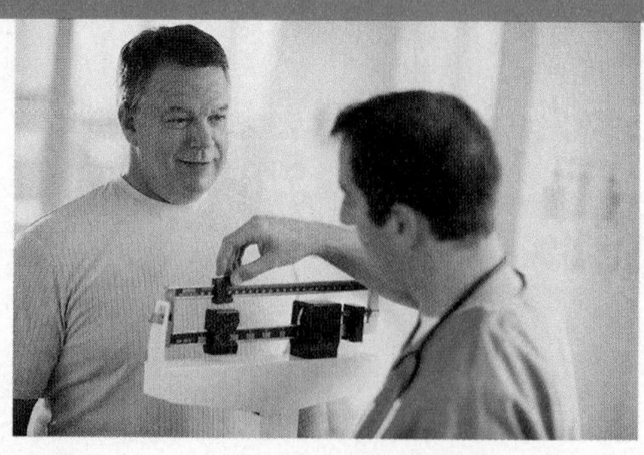

Marquis is a 43-year-old office worker who is slightly overweight. At his annual checkup, his blood pressure measures 150/96 mm Hg, which is considered moderately high. Marquis feels fine, but his doctor tells him it is time to treat his hypertension. Marquis's doctor advises him to improve his diet and get more exercise. He also prescribes a new medication, hydrochlorothiazide, or HCTZ, which he instructs him to take every morning. HCTZ is a diuretic that is commonly used to lower blood pressure.

Marquis begins taking HCTZ and continues to feel well, although he notices the medication makes him need to urinate more frequently. A few weeks later he and a group of coworkers eat some bad potato salad at the office picnic and come down with food poisoning. After 3 days of diarrhea, Marquis feels absolutely horrible. He is so weak he can barely get out of bed. His muscles ache and his legs and feet cramp a lot. Worst of all, he feels as if there is a frog jumping around in his chest. He is often dizzy when this happens.

After a week in bed, Marquis feels even weaker. He goes to see his doctor, who sends him to the lab for some tests. "I know why you feel so terrible," says his doctor when he sees the lab results. **What is happening to Marquis?** **To find out, turn to Clinical Case Wrap-Up on p. 1054.**

An Introduction to Fluid, Electrolyte, and Acid-Base Balance

The next time you see a small pond, think about the fish in it. They live their entire lives totally dependent on the quality of that isolated environment. Severe water pollution will kill them, but even subtle changes can have equally grave effects. Changes in the volume of the pond, for example, can be quite important. If evaporation removes too much water, the fish become overcrowded. The oxygen and food supplies will run out, and the fish will suffocate or starve. The ionic concentration of the water is also crucial. Most of the fish in a freshwater pond will die if the water becomes too salty. Fish in a saltwater pond will die if the water becomes too dilute. The pH of the pond water, too, is a vital factor. This is another reason that acid rain is such a problem.

Your cells live in a pond whose shores are the exposed surfaces of your skin. Most of your body weight is water. Water makes up about 99 percent of the volume of the fluid outside cells, the *extracellular fluid* (*ECF*), and it is also an essential ingredient of the cytosol inside cells, the *intracellular fluid* (*ICF*). All of a cell's operations rely on water as a diffusion medium for the distribution of gases, nutrients, and wastes. If the water volume of the body changes, cellular activities are jeopardized. For example, if the water volume reaches a very low level, proteins denature, enzymes stop functioning, and cells die. In this chapter we discuss the homeostatic mechanisms that regulate volume, ion concentrations, and pH in these fluids that are so critical to cells.

To survive, we must maintain a normal volume and composition of both the ECF and the ICF. The ionic concentrations and *pH* (hydrogen ion concentration) of these fluids are as important as their absolute quantities. If the concentration of calcium ions or potassium ions in the ECF becomes too high, cardiac arrhythmias develop and death can result. A pH outside the normal range can also lead to a variety of serious problems. Low pH—*acidity*—is especially dangerous, because hydrogen ions break chemical bonds, change the shapes of complex molecules, disrupt plasma membranes, and impair tissue functions.

Tips & Tools

The "p" in pH refers to power. Hence, pH refers to the **p**ower of **H**ydrogen.

Three interrelated processes stabilize the volumes, solute concentrations, and pH of the ECF and the ICF:

- *Fluid Balance.* You are in *fluid balance* when the amount of water you gain each day is equal to the amount you lose to the environment. Maintaining normal fluid balance involves regulating the content and distribution of body water in the ECF and the ICF. The digestive system is the main source of water gains. Metabolic activity generates a small amount of additional water. The urinary system is the main route for water loss under normal conditions, but as we saw in Chapter 25, sweating can become important when body temperature is elevated. ⮌ p. 968
- *Electrolyte Balance.* **Electrolytes** are ions released when inorganic compounds dissociate. They are so named because they can conduct an electrical current in a solution. ⮌ p. 41 Each day, your body fluids gain electrolytes from the food and drink you consume. Your body fluids also lose electrolytes in urine, sweat, and feces. For each ion,

27-1 Fluid balance, electrolyte balance, and acid-base balance are interrelated and essential to homeostasis

Learning Outcome **Explain what is meant by the terms fluid balance, electrolyte balance, and acid-base balance, and discuss their importance for homeostasis.**

daily gains must balance daily losses. For example, if you lose 500 mg of Na^+ in urine and sweat, you need to gain 500 mg of Na^+ from food and drink to remain in sodium ion balance. If the gains and losses for every electrolyte are in balance, you are said to be in *electrolyte balance*. Electrolyte balance primarily involves balancing the rates of absorption across the digestive tract with rates of loss by the kidneys, although losses at sweat glands and other sites can play a secondary role.

- *Acid-Base Balance.* You are in *acid-base balance* when the production of hydrogen ions (H^+) in your body is precisely offset by their loss. When acid-base balance exists, the pH of body fluids remains within normal limits. p. 43 Preventing a decrease in pH is the primary problem, because your body generates a variety of acids during normal metabolic operations. The kidneys play a major role by secreting hydrogen ions into the urine and generating *buffers* (compounds that stabilize the pH of a solution) that enter the bloodstream. Such secretion takes place primarily in the distal segments of the distal convoluted tubule (DCT) and along the collecting system. p. 999 The lungs also play a key role by eliminating carbon dioxide.

In this chapter, we explore the dynamics of exchange among the various body fluids, and between the body and the external environment. Much of the information in this chapter should be familiar to you from discussions in earlier chapters about aspects of fluid, electrolyte, or acid-base balance that affect specific systems. Here we provide an overview that integrates those discussions to highlight important functional patterns.

The subjects discussed in this chapter have wide-ranging clinical importance: Steps to restore normal fluid, electrolyte, and acid-base balances must be part of the treatment of any serious illness affecting the nervous, cardiovascular, respiratory, urinary, or digestive system. Because this chapter builds on information in earlier chapters, we have included many references to relevant discussions and figures that can provide a quick review.

Checkpoint

1. **Identify the three interrelated processes essential to stabilizing body fluid volumes.**

See the blue Answers tab at the back of the book.

27-2 Extracellular fluid (ECF) and intracellular fluid (ICF) are fluid compartments with differing solute concentrations that are closely regulated

Learning Outcome **Compare the composition of ECF and ICF, explain the basic concepts involved in the regulation of fluids and electrolytes, and identify the hormones that play important roles in fluid and electrolyte regulation.**

This section examines the composition of the ECF and ICF, how exchanges occur between them, and how these are regulated by hormones. But first let's take a look at the big picture of how much total water is found in the body.

Body Water Content

Figure 27–1a presents an overview of the body makeup of a 70-kg (154-lb) person with a minimum of body fat. The distribution is based on overall average values for males and females ages 18–40 years. In both sexes, the ICF contains a greater proportion of total body water than does ECF. However, there are some differences between the sexes in body water content (Figure 27–1b). Water makes up about 60 percent of the total body weight of an adult male, and 50 percent of that of an adult female. This difference between the sexes, which is mainly in the ICF, reflects the proportionately larger mass of adipose tissue in adult females, and the greater average muscle mass in adult males. (Adipose tissue is only 10 percent water, whereas skeletal muscle is 75 percent water.) Less striking differences occur in the ECF values, due to variations in the interstitial fluid volume of various tissues and the larger blood volume in males versus females.

The Fluid Compartments of the ECF and ICF

The **extracellular fluid**, or **ECF**, is made up mostly of the interstitial fluid of peripheral tissues and the plasma of circulating blood; the **intracellular fluid**, or **ICF**, consists of the cytosol inside cells. Minor components of the ECF include lymph, cerebrospinal fluid (CSF), synovial fluid, serous fluids (pleural, pericardial, and peritoneal fluids), aqueous humor, perilymph, and endolymph.

The ECF and ICF are called **fluid compartments** because they commonly behave as separate sections, maintaining different ionic compositions. The principal ions in the ECF are sodium, chloride, and bicarbonate (HCO_3^-). The ICF contains an abundance of potassium, magnesium, and phosphate ions (HPO_4^{2-}), plus large amounts of negatively charged proteins. Figure 27–2 compares the concentrations of the cations and anions of the ICF with those of the two major subdivisions of the ECF (plasma and interstitial fluid) in terms of milliequivalents per liter (mEq/L). This figure reflects the total number of positive and negative electrical charges in solution. Specific ions and substances, such as proteins, contribute more since they have more than a single electric charge (see Section 27-4).

The osmotic concentration, or osmolarity, of a solution depends on solute concentration, or the total number of solute particles per liter. p. 994 Despite the differences in the concentration of specific substances in Figure 27–2, the osmotic concentrations of the ICF and ECF are identical. This is because ions and proteins each count as one particle in solution, despite their number of electrical charges.

Figure 27–1 Composition of the Human Body.

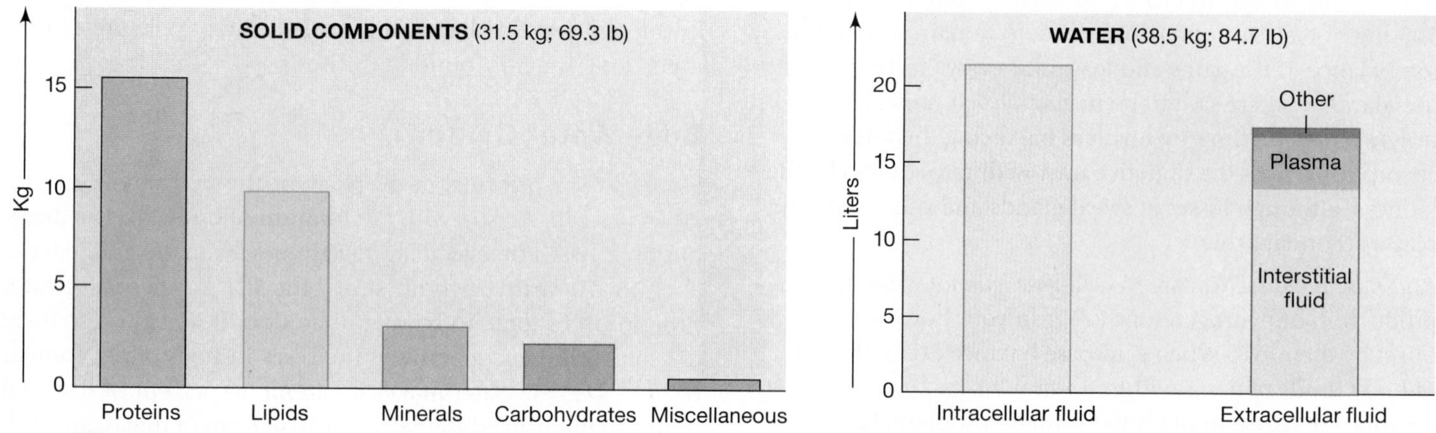
SOLID COMPONENTS (31.5 kg; 69.3 lb)
Kg
15
10
5
0
Proteins
Lipids
Minerals
Carbohydrates
Miscellaneous
WATER (38.5 kg; 84.7 lb)
Liters
20
15
10
5
0
Other
Plasma
Interstitial fluid
Intracellular fluid
Extracellular fluid

a The body composition (by weight, averaged for both sexes) and major body fluid compartments of a 70-kg (154-lb) person. For technical reasons, it is extremely difficult to determine the precise size of any of these compartments; estimates of their relative sizes vary widely.

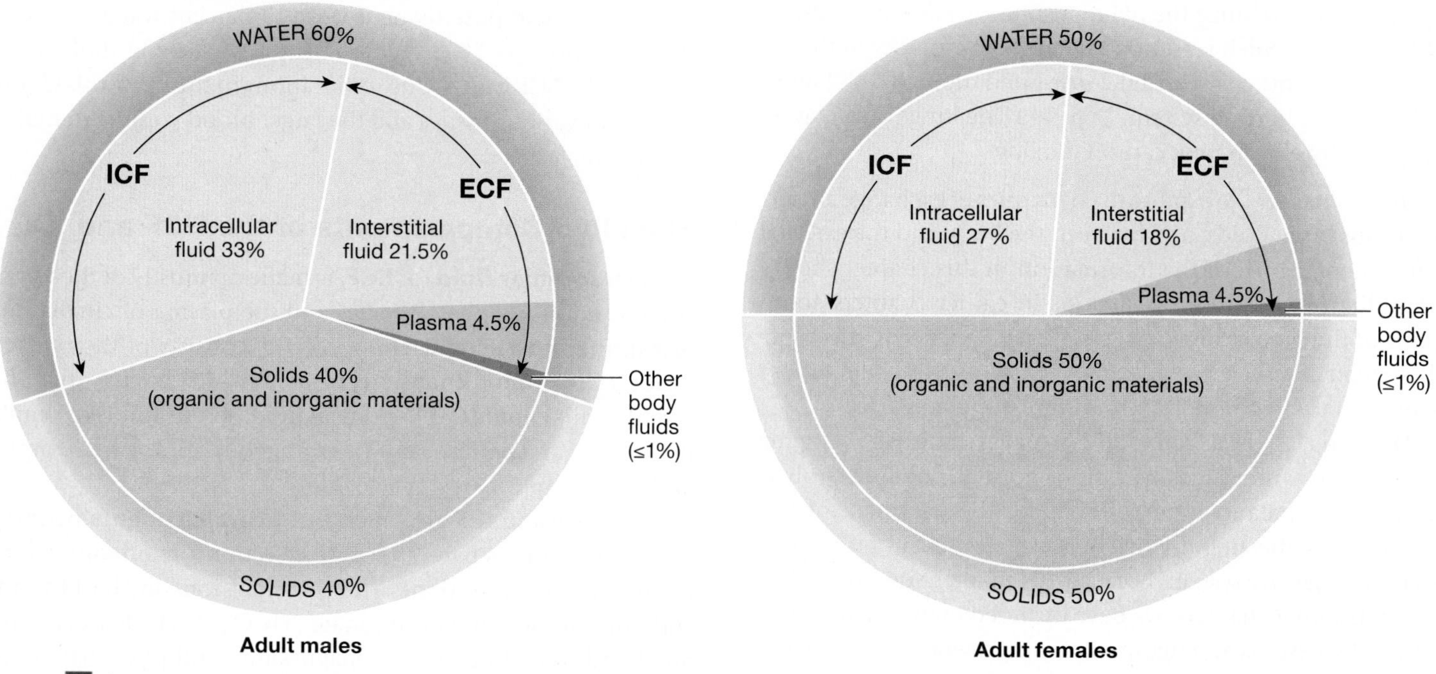
WATER 60%
ICF
ECF
Intracellular fluid 33%
Interstitial fluid 21.5%
Plasma 4.5%
Solids 40% (organic and inorganic materials)
Other body fluids (≤1%)
SOLIDS 40%
Adult males
WATER 50%
ICF
ECF
Intracellular fluid 27%
Interstitial fluid 18%
Plasma 4.5%
Other body fluids (≤1%)
Solids 50% (organic and inorganic materials)
SOLIDS 50%
Adult females

b A comparison of the body compositions of adult males and females, ages 18–40 years.

27

Osmosis eliminates minor differences in concentration almost at once, because most plasma membranes are freely permeable to water. (The only noteworthy exceptions are the apical surfaces of epithelial cells along the ascending limb of the nephron loop, the distal convoluted tubule, and the collecting system.)

Solute Exchanges between the ECF and the ICF

If the plasma membranes of cells were freely permeable, diffusion would continue until the ions in the ECF and ICF were evenly distributed across all of the membranes. But it does not, because plasma membranes are selectively permeable: Ions can enter or leave the cells only by specific membrane channels. Also, carrier mechanisms move specific ions into or out of the cells. Therefore, the exchange of ions between the ICF and the ECF takes place across plasma membranes by osmosis, diffusion, and carrier-mediated transport. (To review the mechanisms involved, look back at **Spotlight Figure 3–22**, p. 100.) The presence of a plasma membrane and active transport at the membrane surface enable cells to maintain internal environments with a composition that differs from their surroundings.

In addition to solute exchanges between the ECF and ICF, exchanges also occur among the subdivisions of the ECF. These exchanges occur primarily across the endothelial lining of

Figure 27–2 Cations and Anions in Body Fluids. Notice that each fluid compartment is electrically neutral and the ionic composition is markedly different between the ICF and ECF (plasma and interstitial fluid).

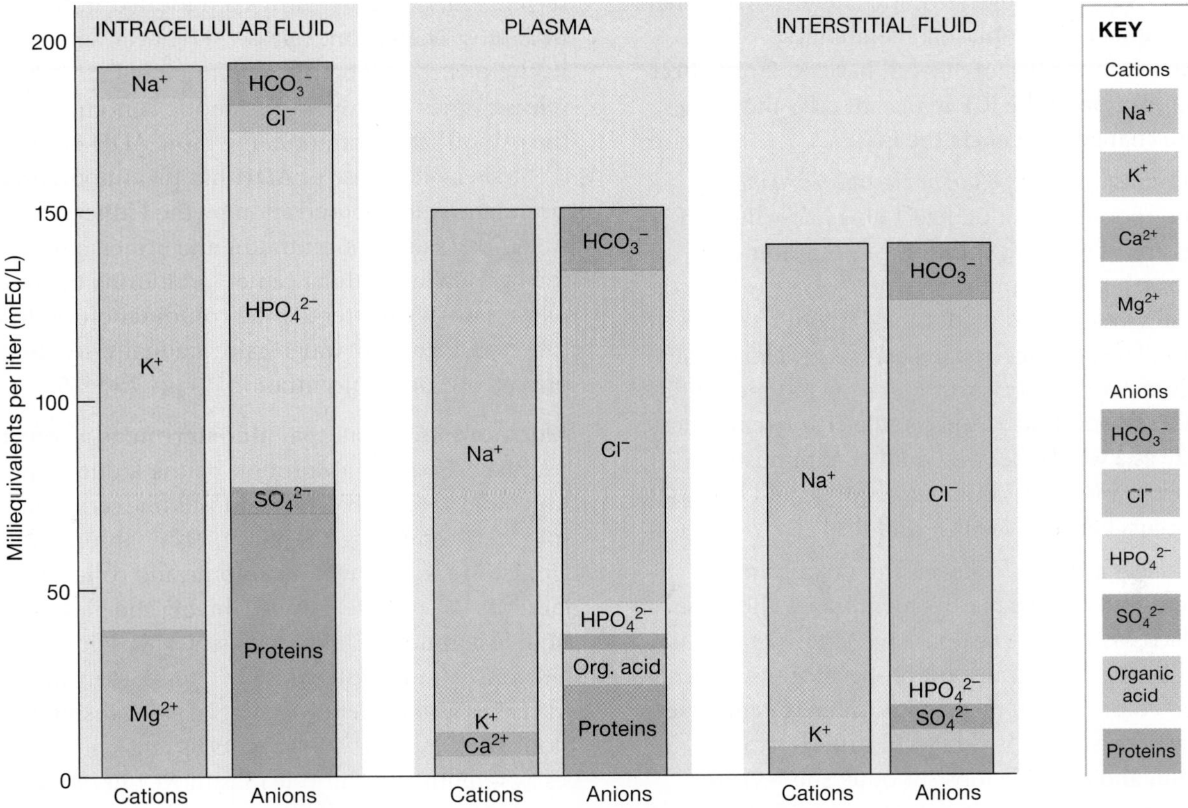
INTRACELLULAR FLUID
PLASMA
INTERSTITIAL FLUID
KEY
Cations
Na+
K+
Ca2+
Mg2+
Anions
HCO3−
Cl−
HPO4 2−
SO4 2−
Organic acid
Proteins
Milliequivalents per liter (mEq/L)
200
150
100
50
0
Na+
K+
Mg2+
HCO3−
Cl−
HPO4 2−
SO4 2−
Proteins
Na+
K+
Ca2+
HCO3−
Cl−
HPO4 2−
Org. acid
Proteins
Na+
K+
HCO3−
Cl−
HPO4 2−
SO4 2−
Cations
Anions
Cations
Anions
Cations
Anions

? Mg^{2+} and K^+ are found in higher concentrations in the (*intracellular*, *extracellular*) fluid, whereas Na^+ and Cl^- are found in higher concentrations in the (*intracellular*, *extracellular*) fluid.

capillaries. Fluid may also travel from the interstitial spaces to plasma through lymphatic vessels that drain into the venous system. p. 787 Such exchanges mean that the kinds and quantities of dissolved electrolytes, proteins, nutrients, and wastes in the ECF vary regionally. Still, the variations among the compartments of the ECF seem minor compared with the major differences between the ECF and the ICF (Figure 27–2).

In clinical situations, we usually estimate that two-thirds of the total body water is in the ICF and one-third in the ECF. This ratio underestimates the real volume of the ECF, because portions of the ECF—including the water in bone, in many dense connective tissues, and in many of the minor ECF components—are relatively isolated. Exchange between these fluid volumes and the rest of the ECF takes place more slowly than does exchange between plasma and other interstitial fluids. For this reason, they can be safely ignored in many cases. Clinical attention usually focuses on the rapid fluid and solute movements that accompany the administration of blood, plasma, or saline solutions to counteract blood loss or dehydration. Physiologists and clinicians pay particular attention to ionic distributions across membranes and to the electrolyte composition of body fluids. The Appendix at the end of the book reports normal physiological values in the units most often used in clinical reports.

An Overview of the Regulation of Fluid and Electrolyte Balance

The regulation of fluid balance and of electrolyte balance are tightly intertwined because changes in solute concentrations lead to immediate changes in water distribution. The following basic principles are key to understanding fluid and electrolyte balance:

- *All the homeostatic mechanisms that monitor and adjust the composition of body fluids respond to changes in the ECF, not in the ICF.* Receptors monitoring the composition of two key components of the ECF—plasma and cerebrospinal fluid (CSF)—detect significant changes in their composition or volume and trigger appropriate nervous and endocrine responses. This arrangement makes functional sense, because a change in one ECF component will spread rapidly throughout the entire ECF and affect all the body's cells. In contrast, the ICF is contained within trillions of individual cells that are physically and chemically isolated from one another by their plasma membranes. For this reason, changes in the ICF in one cell have no direct effect on the composition of the ICF in distant cells and tissues, unless those changes also affect the ECF.

- *No receptors directly monitor fluid or electrolyte balance.* In other words, receptors cannot detect how many liters of water, or grams of sodium, chloride, or potassium ions, the body contains. Nor can they count how many liters or grams we gain or lose throughout the day. But receptors *can* monitor *plasma volume* and *osmotic concentration*. The plasma volume and osmotic concentration are good indicators of the state of fluid balance and electrolyte balance for the body as a whole, because fluid continuously circulates between interstitial fluid and plasma, and because exchange occurs between the ECF and the ICF.
- *Cells cannot move water molecules by active transport.* All movement of water across plasma membranes and epithelia takes place passively, in response to osmotic gradients established by the active transport of specific ions, such as sodium and chloride. You may find it useful to remember that "water follows salt." As we saw in earlier chapters, when sodium and chloride ions (or other solutes) are actively transported across a membrane or epithelium, water follows by osmosis. p. 1001 This basic principle accounts for water absorption across the digestive epithelium, and for water conservation by the kidneys.
- *The body's content of water or electrolytes will increase if dietary gains exceed losses to the environment, and will decrease if losses exceed gains.* This basic rule is important when you consider the mechanics of fluid balance and electrolyte balance. Homeostatic adjustments affect the balance between urinary excretion and dietary absorption. As we saw in Chapter 26, circulating hormones regulate renal function. These hormones can also produce complementary changes in behavior, such as promoting thirst.

27

The Primary Hormones That Regulate Fluid and Electrolyte Balance

Three hormones mediate physiological adjustments to fluid balance and electrolyte balance: (1) *antidiuretic hormone (ADH)*, (2) *aldosterone*, and (3) the *natriuretic peptides (ANP* and *BNP)*.

Antidiuretic Hormone. Recall that **antidiuretic hormone (ADH)** is a pituitary hormone that promotes water retention by the kidneys. The hypothalamus contains specialized cells known as *osmoreceptors* that monitor the osmotic concentration of the ECF. These cells are very sensitive: A 2 percent change in osmotic concentration (approximately 6 mOsm/L) is enough to alter osmoreceptor activity.

The population of osmoreceptors includes neurons that secrete ADH. These neurons are located in the anterior hypothalamus. Their axons release ADH near fenestrated capillaries in the posterior lobe of the pituitary gland. The rate of ADH release varies directly with osmotic concentration: The higher the osmotic concentration, the more ADH is released.

Increased release of **ADH** has two important effects: (1) It stimulates water conservation by the kidneys, reducing urinary water losses and concentrating the urine; and (2) it stimulates the hypothalamic thirst center, promoting the intake of fluids. As we saw in Chapter 21, the combination of decreased water loss and increased water gain gradually restores the normal plasma osmotic concentration. pp. 750–752

Aldosterone. Recall that **aldosterone** is a corticosteroid that stimulates sodium absorption by the kidneys and so regulates water and sodium balance. Aldosterone secretion plays a major role in determining the rate of Na^+ absorption and K^+ loss along the distal convoluted tubule and collecting system of the kidneys. pp. 999–1000 The higher the plasma concentration of aldosterone, the more efficiently the kidneys conserve Na^+. Because "water follows salt," the conservation of Na^+ also increases water retention. As Na^+ is reabsorbed, Cl^- follows (look back at **Figure 26–13**, p. 998), and water follows by osmosis as sodium and chloride ions move out of the tubular fluid.

Aldosterone is secreted in response to an increasing K^+ or decreasing Na^+ level in the blood reaching the adrenal cortex, or in response to *renin-angiotensin-aldosterone system* activation. As we saw in earlier chapters, renin release occurs in response to any of three changes: (1) a decrease in plasma volume or blood pressure at the juxtaglomerular complex of the nephron; (2) a decrease in the osmotic concentration of tubular fluid at the DCT; or, as we will soon see, (3) a decreasing Na^+ or increasing K^+ concentration in the renal circulation.

Natriuretic Peptides. The **natriuretic** (*natrium*, sodium; *-uretic*, urine) **peptides**, *atrial natriuretic peptide (ANP)* and *B-type natriuretic peptide (BNP)*, are released by cardiac muscle cells in response to abnormal stretching of the heart walls. Increased blood pressure or increased blood volume causes this stretching. Among their other effects, these peptides reduce thirst and block the release of ADH and aldosterone that might otherwise lead to water and salt conservation. The resulting *diuresis* (fluid loss by the kidneys) decreases both blood pressure and plasma volume, eliminating the stretching.

The Interplay between Fluid Balance and Electrolyte Balance

At first glance, it can be difficult to distinguish between water balance and electrolyte balance. For example, when you lose body water, plasma volume decreases and electrolyte concentrations increase. Conversely, when you gain or lose excess electrolytes, you also gain or lose water due to osmosis.

However, it is often useful to consider fluid balance and electrolyte balance separately because the regulatory mechanisms involved are quite different. This distinction is absolutely vital in a clinical setting, where problems with fluid balance and electrolyte balance must be identified and corrected promptly.

Checkpoint

2. **List the components of extracellular fluid (ECF) and intracellular fluid (ICF).**
3. **Name three hormones that play a major role in adjusting fluid and electrolyte balance in the body.**
4. **What effect might drinking a half-gallon of distilled water (water without minerals) have on the ADH level?**

See the blue Answers tab at the back of the book.

27-3 Fluid balance involves the regulation and distribution of water gains and losses

Learning Outcome Describe the movement of fluid within the ECF, between the ECF and the ICF, and between the ECF and the environment.

As mentioned, **fluid balance** occurs when the amount of water gained each day is equal to the amount lost to the environment. We cannot easily determine the body's water content. However, the concentration of Na^+, the most abundant ion in the ECF, provides useful clues to the state of fluid balance (see Figure 27–2). When the body's water content increases, the Na^+ concentration of the ECF becomes abnormally low. And when the body's water content decreases, the Na^+ concentration becomes abnormally high. The body responds to such osmotic changes to ensure that fluid balance is maintained. Let's look at fluid gains and losses, and water movements.

Table 27–1 Water Gains and Losses

Gains	Daily Input (mL)
Water content of food	1000
Water consumed as liquid	1200
Metabolic water produced during catabolism	300
Total	**2500**
Losses	**Daily Output (mL)**
Urination	1200
Evaporation at skin	750
Evaporation at lungs	400
Loss in feces	150
Total	**2500**

Fluid Gains and Losses

Table 27–1 lists the major water gains and losses involved in fluid balance. Figure 27–3 highlights the routes of fluid exchange between the fluid compartments and the environment.

Water Gains

A water gain of about 2500 mL/day is required to balance your average water losses. This value amounts to about 40 mL/kg (about 1.35 fl oz/lb) of body weight per day. You obtain water through eating (1000 mL), drinking (1200 mL), and *metabolic generation* (300 mL).

Metabolic generation of water is the production of water within cells, primarily as a result of oxidative phosphorylation in mitochondria during aerobic metabolism, (In Chapter 25

Figure 27–3 Exchanges between Fluid Compartments and the Environment. Fluid movements that maintain fluid balance in a normal person. The volumes are drawn to scale: The ICF is about twice as large as the ECF.

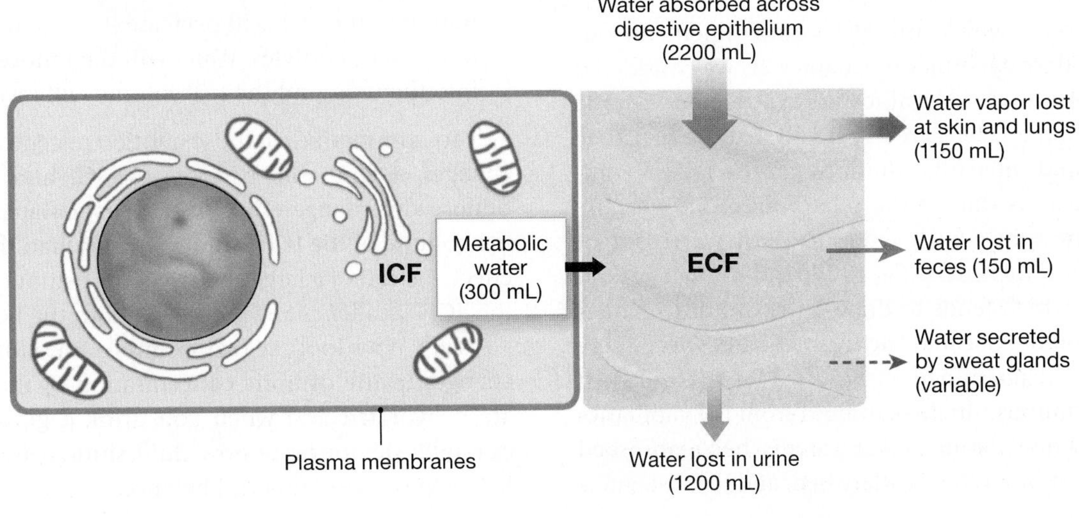
Water absorbed across digestive epithelium (2200 mL)
Water vapor lost at skin and lungs (1150 mL)
ICF
Metabolic water (300 mL)
ECF
Water lost in feces (150 mL)
Water secreted by sweat glands (variable)
Plasma membranes
Water lost in urine (1200 mL)

27

we described the synthesis of water at the end of the electron transport chain p. 947) When cells break down 1 g of lipid, 1.7 mL of water is generated. Breaking down proteins or carbohydrates yields much smaller amounts (0.41 mL/g and 0.55 mL/g, respectively).

A typical diet in the United States contains 46 percent carbohydrates, 40 percent lipids, and 14 percent protein. Such a diet produces approximately 300 mL of water per day, about 12 percent of your average daily requirement.

Water Losses

You lose about 2500 mL of water each day through urine, feces, and *insensible perspiration*—the evaporation of water across the epithelia of the skin and lungs. The losses due to *sensible perspiration*—evaporation due to the secretory activities of the sweat glands—vary with physical activity. Sensible perspiration can cause significant water deficits, with maximum perspiration rates reaching 4 liters per hour. p. 968 Fever can also increase water losses. For each degree that body temperature rises above normal, daily insensible water losses increase by 200 mL. The advice "Drink plenty of fluids" for anyone who is sick has a definite physiological basis.

Water Movement between Fluid Compartments

Recall that water movements occur between the ECF and ICF, and between the subdivisions of the ECF, but generally not within the ICF.

Water Movement between the ECF and ICF

Water can move between the ECF and the ICF. For example, a small amount of water moves from the ICF to the ECF each day, as the result of metabolic water generation (see Figure 27–3). However, under normal circumstances the ECF and the ICF are in osmotic equilibrium. For this reason, no large-scale circulation occurs between the two compartments.

27

Water Movement within the ECF

Water circulates freely within the ECF compartment. In our discussion of capillary dynamics in Chapter 21, we considered the basic principles that determine fluid movement among the subdivisions of the ECF. p. 742 Recall that the exchange between plasma and interstitial fluid, by far the largest components of the ECF, is due to the relationship between the net hydrostatic pressure, which tends to push water out of the plasma and into the interstitial fluid, and the net colloid osmotic pressure, which tends to draw water out of the interstitial fluid and into the plasma. The interaction between these opposing forces is diagrammed in Figure 21–11 (p. 743). It results in the continuous filtration of fluid from the capillaries into the interstitial fluid. Some of that water is then reabsorbed along the distal portion of the capillary bed, and the rest enters lymphatic vessels for transport to the venous circulation. At any moment, interstitial fluid and minor fluid compartments contain approximately 80 percent of the ECF volume, and plasma contains the other 20 percent.

Any factor that affects the net hydrostatic pressure or the net colloid osmotic pressure will alter the distribution of fluid within the ECF. The movement of abnormal amounts of water from plasma into interstitial fluid results in *edema* (swelling). Pulmonary edema, for example, can result from an increase in the blood pressure in pulmonary capillaries. Generalized edema can result from a decrease in blood colloid osmotic pressure, as seen in advanced starvation, when plasma protein concentrations decrease. Localized edema can result from damage to capillary walls (as occurs in bruising), the constriction of regional venous circulation, or a blockage of the lymphatic drainage (as happens with *lymphedema*, introduced in Chapter 22). p. 789

In addition to these water movements between major components of the ECF, water also moves continuously among the smaller ECF subdivisions. The volumes involved in these water movements are very small, and the volume and composition of the fluids are closely regulated.

Fluid Shifts between the ECF and ICF

A **fluid shift** is a rapid water movement between the ECF and the ICF in response to an osmotic gradient. Such a gradient is produced by changes in the osmotic concentration of the ECF (Figure 27–4). Fluid shifts occur rapidly, reaching equilibrium within minutes to hours. This process happens in the following two situations:

- *If the osmotic concentration of the ECF increases, that fluid will become hypertonic compared to the ICF.* The osmotic concentration of the ECF will increase if you lose water but retain electrolytes. Water will then move from the cells into the ECF until osmotic equilibrium is restored.
- *If the osmotic concentration of the ECF decreases, that fluid will become hypotonic compared to the ICF.* The osmotic concentration of the ECF will decrease if you gain water but do not gain electrolytes. Water will then move from the ECF into the cells, and the ICF volume will increase.

To summarize, if the osmotic concentration of the ECF changes, a fluid shift between the ICF and ECF will tend to oppose the change. The ICF acts as a water reserve, because the volume of the ICF is much greater than that of the ECF. In effect, instead of a large change in the osmotic concentration of the ECF, smaller changes occur in both the ECF and ICF.

Let's now look at two examples of situations that cause changes in the osmotic concentration of the ECF: when you are dehydrated and when you drink a glass of water. These examples demonstrate how fluid shifts between the ECF and ICF help to maintain fluid balance.

Figure 27–4 **Fluid Shifts between the ICF and ECF.**

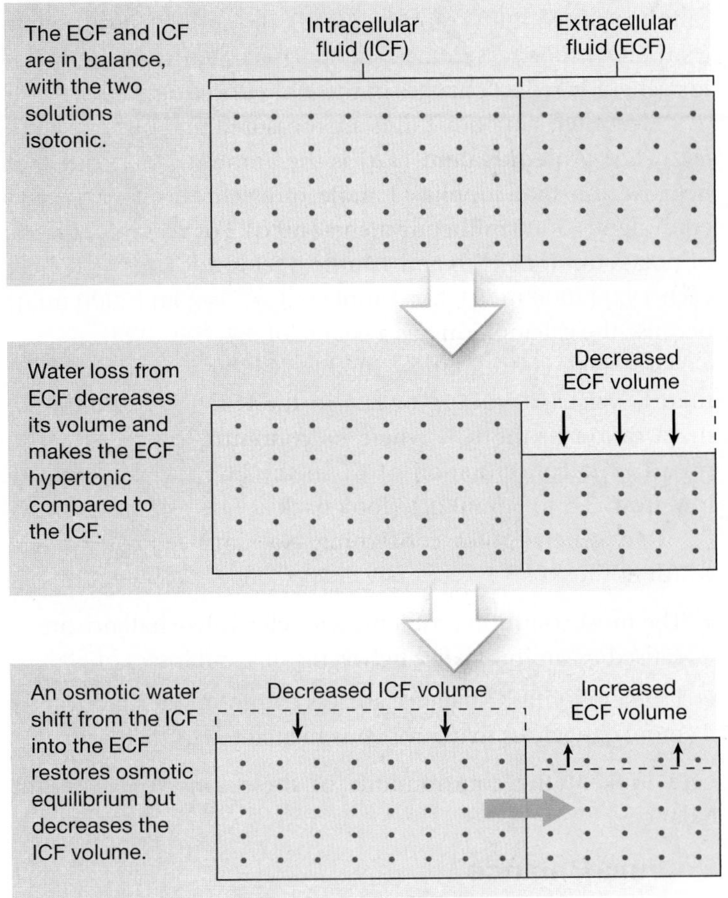
The ECF and ICF are in balance, with the two solutions isotonic.
Intracellular fluid (ICF)
Extracellular fluid (ECF)
Water loss from ECF decreases its volume and makes the ECF hypertonic compared to the ICF.
Decreased ECF volume
An osmotic water shift from the ICF into the ECF restores osmotic equilibrium but decreases the ICF volume.
Decreased ICF volume
Increased ECF volume

Clinical Note Water and Weight Loss

The safest way to lose weight is to reduce your intake of food while ensuring that all nutrient requirements are met. Water must be included on the list of nutrients, along with carbohydrates, fats, proteins, vitamins, and minerals. Nearly half of our normal water intake comes from food. For this reason, when you eat less, you become more dependent on drinking fluids and on whatever water is generated metabolically.

At the start of a diet, your body conserves water and catabolizes lipids. That is why the first week of dieting may seem rather unproductive. Over that week, the level of circulating ketone bodies gradually increases. In the weeks that follow, the rate of water loss by the kidneys increases in order to excrete wastes, such as the hydrogen ions released by ketone bodies and the urea and ammonia generated during protein catabolism. Your rate of fluid intake must also increase, or else you risk dehydration. While you are dieting, dehydration is especially serious, because as water is lost, the concentration of solutes in the ECF increases, further increasing the concentration of wastes and acids generated during the catabolism of energy reserves. This soon becomes a positive feedback loop: These wastes enter the filtrate at the kidneys, and their excretion draws additional water into the urine.

Allocation of Water Losses and Dehydration

Dehydration, or *water depletion*, develops when water losses outpace water gains. When you lose water but retain electrolytes, the osmotic concentration of the ECF increases. Water then moves out of the ICF and into the ECF by osmosis until the two solutions are again isotonic. At that point, both the ECF and ICF are somewhat more concentrated than normal, and both volumes are lower than they were before the fluid loss. Because the ICF has about twice the functional volume of the ECF, the net change in the ICF is small.

However, if the fluid imbalance continues unchecked, the loss of body water will produce severe thirst, dryness, and wrinkling of the skin. In the case of severe water losses such as excessive perspiration (as when you exercise in hot weather), inadequate water consumption, repeated vomiting, and diarrhea, water losses occur far in excess of electrolyte losses. Body fluids become increasingly concentrated, and the sodium ion concentration becomes abnormally high (a condition called *hypernatremia*). In addition, a significant decrease in plasma volume and blood pressure eventually occurs, and shock may develop.

Homeostatic responses to dehydration include physiological mechanisms (ADH and renin secretion) and behavioral changes (increasing fluid intake, preferably as soon as possible). Clinical therapies for acute dehydration include administering hypotonic fluids by mouth or intravenous infusion. These procedures rapidly increase ECF volume and promote the shift of water back into the ICF.

Distribution of Water Gains and Hyperhydration

When you drink a glass of water, your body's water content increases without a corresponding increase in the concentrations of electrolytes. As a result, the ECF increases in volume but becomes hypotonic compared to the ICF. A fluid shift then occurs, and the volume of the ICF increases at the expense of the ECF. Once again, the larger volume of the ICF limits the relative amount of osmotic change. After the fluid shift, the ECF and ICF have slightly larger volumes and slightly lower osmotic concentrations than they did originally.

Normally, this situation will be promptly corrected. The decreased plasma osmotic concentration decreases ADH secretion, discouraging fluid intake and increasing water losses in urine. If the situation is *not* corrected, a variety of clinical problems will develop as water shifts into the ICF, distorting cells, changing the solute concentrations around enzymes, and disrupting normal cell functions. This condition is called **hyperhydration**, or *water excess*. It can be caused by

(1) drinking a large volume of water or the infusion (injection into the bloodstream) of a hypotonic solution; (2) an inability to eliminate excess water in urine, due to chronic renal failure, heart failure, cirrhosis, or some other disorder; and (3) endocrine disorders, such as excessive ADH production.

The most obvious sign of hyperhydration is an abnormally low Na^+ concentration (*hyponatremia*). This reduction in Na^+ concentration in the ECF leads to a fluid shift into the ICF. The first signs are the effects on central nervous system (CNS) function. The individual initially acts drunk. This condition, called *water intoxication*, may sound odd, but is extremely dangerous. Untreated cases can rapidly progress from confusion to hallucinations, convulsions, coma, and then death. Treatment of severe hyperhydration generally involves administering diuretics and infusing a concentrated salt solution that promotes a fluid shift from the ICF to the ECF and returns the Na^+ concentration to a near-normal level.

Checkpoint

5. **Define** ***edema*****.**
6. **Describe a fluid shift.**
7. **Define** ***dehydration*****.**
8. **What effect would being in the desert without water for a day have on the osmotic concentration of your blood plasma?**

See the blue Answers tab at the back of the book.

27-4 In electrolyte balance, the concentrations of sodium, potassium, calcium, magnesium, phosphate, and chloride ions in body fluids are tightly regulated

Learning Outcome Discuss the mechanisms by which sodium, potassium, calcium, magnesium, phosphate, and chloride ion concentrations are regulated to maintain electrolyte balance.

27

As we discussed, **electrolyte balance** occurs when gains and losses are equal for each electrolyte in the body. Electrolyte balance is important for these reasons:

- *Total electrolyte concentrations directly affect water balance*, as previously described.
- *The concentrations of individual electrolytes can affect cell functions.* We saw many examples in earlier chapters, including the effect of an abnormal Na^+ concentration on neuron activity and the effects of high or low Ca^{2+} and K^+ concentrations on cardiac muscle tissue.

This section pays particular attention to sodium and potassium ion balance because (1) they are major contributors to the osmotic concentrations of the ECF and the ICF, respectively, and (2) they directly affect the normal functioning of all cells.

Sodium is the main cation in the ECF. More than 90 percent of the osmotic concentration of the ECF results from sodium salts, mainly sodium chloride (NaCl) and sodium bicarbonate ($NaHCO_3$). Changes in the osmotic concentration of body fluids generally reflect changes in the Na^+ concentration.

Electrolytes in body fluids are measured in terms of *equivalents* (Eq). An **equivalent (Eq)** is the amount of a positive or negative ion that supplies 1 mole of electrical charge, and 1 equivalent = 1000 milliequivalents (mEq). For example, 1 mole of potassium ions (K^+) and 1 mole of chloride ions (Cl^-) are each 1 Eq (1000 mEq), but 1 mole of Ca^{2+} is 2 Eq (2000 mEq) because the calcium ion has a charge of +2. p. 994

The Na^+ concentration in the ECF averages about 140 mEq/L, versus 10 mEq/L or less in the ICF. Potassium is the main cation in the ICF, where its concentration reaches 160 mEq/L. The concentration of K^+ in the ECF is generally very low, from 3.5 to 5.5 mEq/L (look back at Figure 27–2).

Two general rules concerning Na^+ and K^+ balance are worth noting:

- The most common problems with electrolyte balance are caused by an imbalance between gains and losses of Na^+.
- Problems with K^+ balance are less common but significantly more dangerous than problems related to Na^+ balance.

Let's look at the contributions of these ions to electrolyte balance.

Sodium Balance

The total amount of sodium ions in the ECF represents a balance between two factors:

- *Sodium Ion Uptake across the Digestive Epithelium.* Sodium ions enter the ECF by crossing the digestive epithelium through diffusion and carrier-mediated transport. The rate of absorption varies directly with the amount of sodium in the diet.
- *Sodium Ion Excretion by the Kidneys and Other Sites.* Sodium ion losses take place primarily by excretion in urine and through perspiration. The kidneys are the most important sites of Na^+ regulation. We discussed the mechanisms for Na^+ reabsorption by the kidneys in Chapter 26. pp. 995, 999

A person in sodium balance typically gains and loses 48–144 mEq (1.1–3.3 g) of Na^+ each day. When Na^+ gains exceed Na^+ losses, the total Na^+ content of the ECF increases. When losses exceed gains, the Na^+ content decreases. However, a change in the Na^+ content of the ECF does not produce a change in the Na^+ concentration. When Na^+ intake or output changes, a corresponding gain or loss of water tends to keep the Na^+ concentration constant. For example, if you eat a very salty meal, the osmotic concentration of the ECF will not increase. When Na^+ are pumped across the digestive epithelium, the solute concentration in that portion of the ECF increases, and that of the intestinal contents decreases. Osmosis then occurs: Additional water enters the ECF from the digestive tract, increasing the blood volume and blood pressure. For this reason, people with high blood pressure or a renal salt sensitivity are advised to restrict the amount of salt in their diets.

Sodium Balance and ECF Volume

The sodium ion regulatory mechanism is diagrammed in **Figure 27–5.** It changes the ECF volume but keeps the Na^+ concentration fairly stable. If you consume large amounts of salt *without* adequate fluid, as when you eat salty potato chips without taking a drink, the plasma Na^+ concentration increases temporarily. A change in ECF volume soon follows, however. Fluid will exit the ICF in a fluid shift, increasing ECF volume and decreasing the Na^+ concentration somewhat. The secretion of ADH restricts water loss and stimulates thirst, promoting additional water consumption. Due to the inhibition of water receptors in the pharynx, ADH secretion begins even before Na^+ absorption occurs. The ADH secretion rate increases more after Na^+ absorption, due to osmoreceptor stimulation. pp. 572, 625

When sodium ion losses exceed gains, the volume of the ECF decreases. This reduction occurs without a significant change in the osmotic concentration of the ECF. Thus, if you perspire heavily but consume only water, you will lose Na^+, and the osmotic concentration of the ECF will decrease briefly. However, as soon as the osmotic concentration drops by 2 percent or more, ADH secretion decreases, so water losses by your kidneys increase. As water leaves the ECF in a fluid shift, the osmotic concentration returns to normal.

Minor changes in ECF volume do not matter, because they do not cause adverse physiological effects. If, however, regulation of the Na^+ concentration results in a large change in ECF volume, the situation will be corrected by the same homeostatic mechanisms responsible for regulating blood volume and blood pressure. This is the case because when ECF volume changes, so does plasma volume and, in turn, blood volume. If ECF volume increases, blood volume increases,

Figure 27–5 Homeostatic Regulation of Sodium Ion Concentration in Body Fluids.

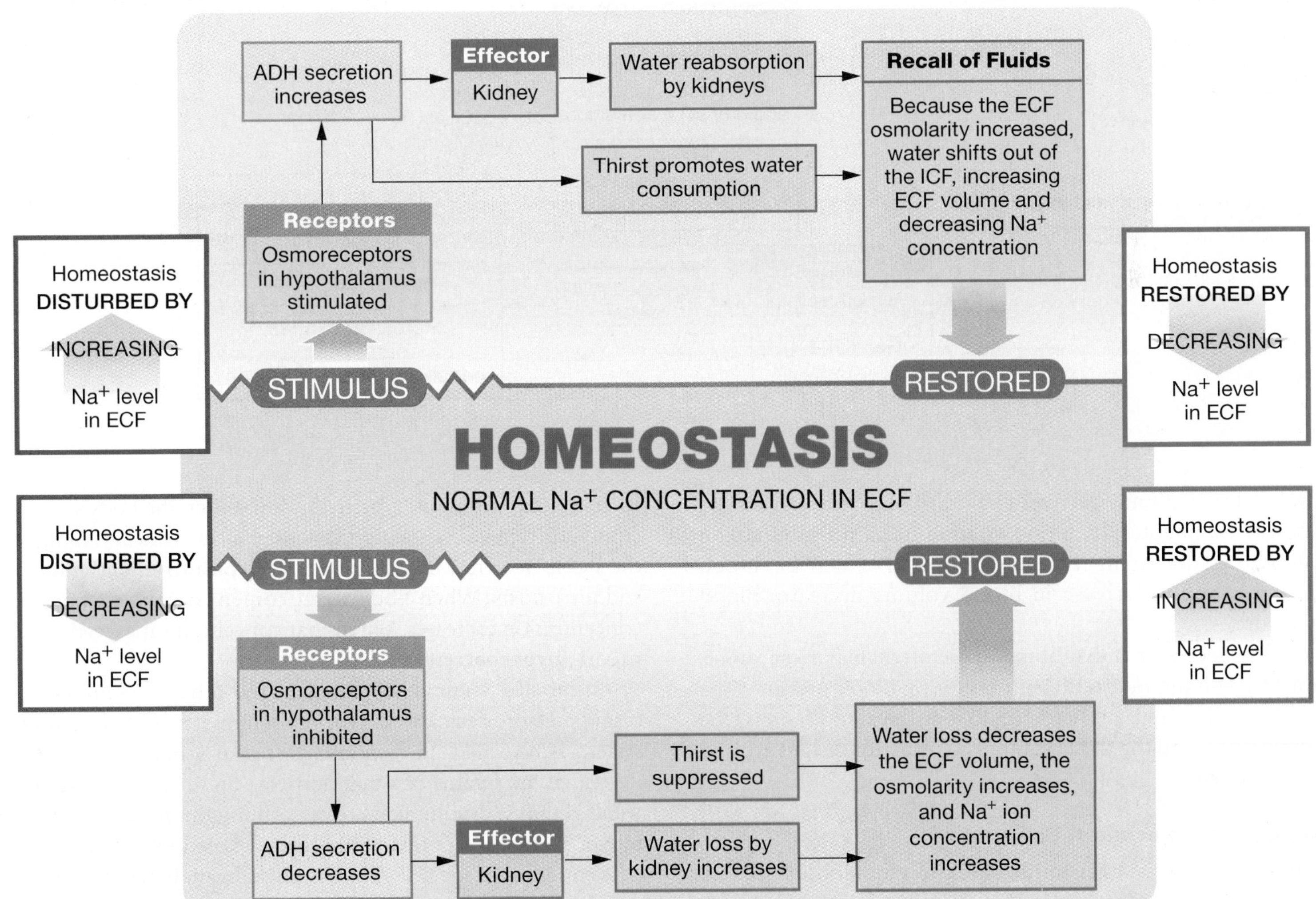
ADH secretion increases
Effector
Kidney
Water reabsorption by kidneys
Recall of Fluids
Because the ECF osmolarity increased, water shifts out of the ICF, increasing ECF volume and decreasing Na+ concentration
Thirst promotes water consumption
Receptors
Osmoreceptors in hypothalamus stimulated
Homeostasis DISTURBED BY INCREASING Na+ level in ECF
STIMULUS
RESTORED
Homeostasis RESTORED BY DECREASING Na+ level in ECF
HOMEOSTASIS
NORMAL Na+ CONCENTRATION IN ECF
Homeostasis DISTURBED BY DECREASING Na+ level in ECF
STIMULUS
RESTORED
Homeostasis RESTORED BY INCREASING Na+ level in ECF
Receptors
Osmoreceptors in hypothalamus inhibited
Thirst is suppressed
Water loss decreases the ECF volume, the osmolarity increases, and Na+ ion concentration increases
ADH secretion decreases
Effector
Kidney
Water loss by kidney increases

27

Figure 27–6 Integration of Fluid Volume Regulation and Sodium Ion Concentration in Body Fluids.

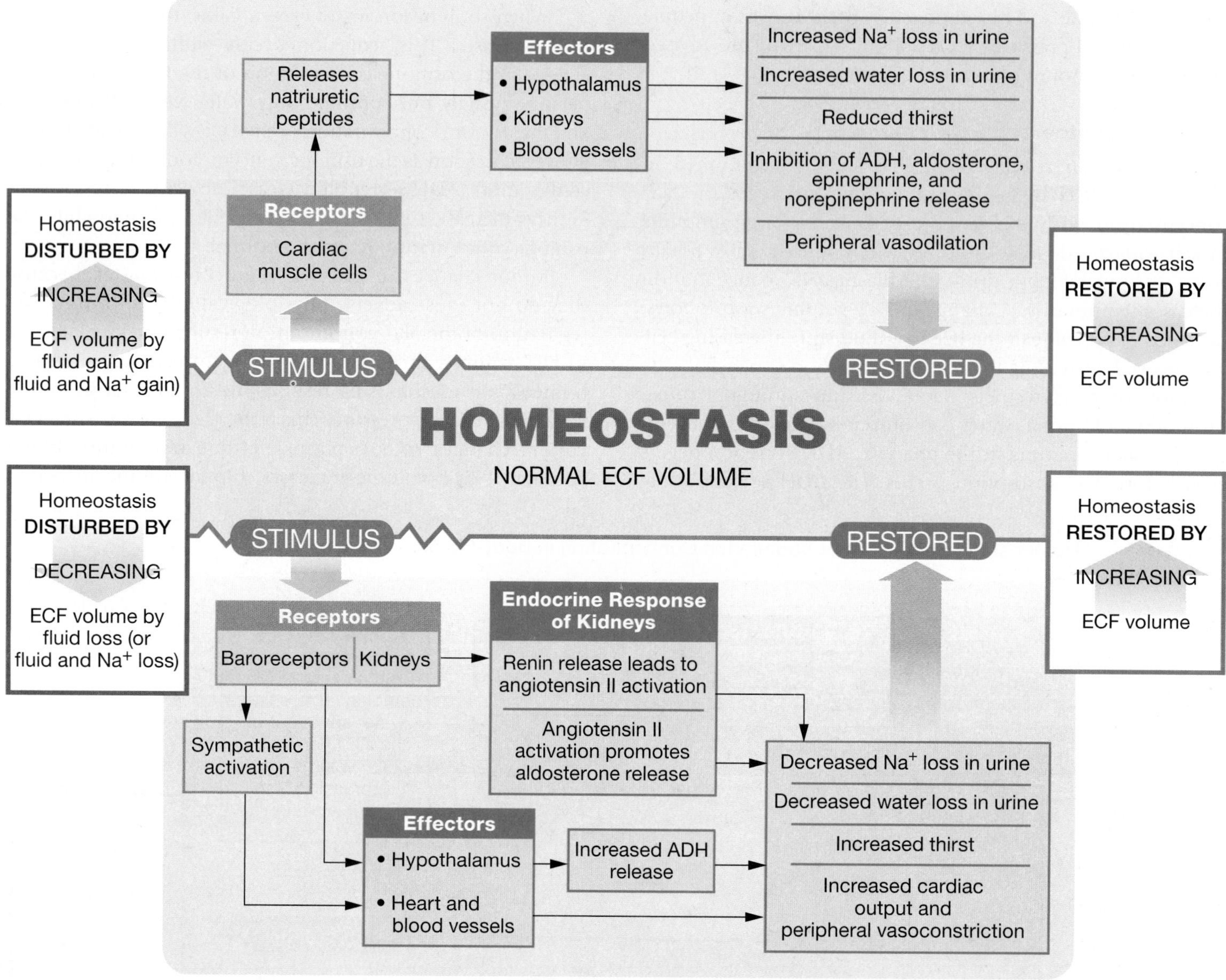
Homeostasis DISTURBED BY INCREASING ECF volume by fluid gain (or fluid and Na+ gain)
Receptors
Cardiac muscle cells
Releases natriuretic peptides
Effectors
• Hypothalamus
• Kidneys
• Blood vessels
Increased Na+ loss in urine
Increased water loss in urine
Reduced thirst
Inhibition of ADH, aldosterone, epinephrine, and norepinephrine release
Peripheral vasodilation
Homeostasis RESTORED BY DECREASING ECF volume
STIMULUS
RESTORED
HOMEOSTASIS
NORMAL ECF VOLUME
Homeostasis DISTURBED BY DECREASING ECF volume by fluid loss (or fluid and Na+ loss)
STIMULUS
RESTORED
Homeostasis RESTORED BY INCREASING ECF volume
Receptors
Baroreceptors
Kidneys
Endocrine Response of Kidneys
Renin release leads to angiotensin II activation
Angiotensin II activation promotes aldosterone release
Sympathetic activation
Effectors
• Hypothalamus
• Heart and blood vessels
Increased ADH release
Decreased Na+ loss in urine
Decreased water loss in urine
Increased thirst
Increased cardiac output and peripheral vasoconstriction

27

and if ECF volume decreases, blood volume decreases. As we saw in Chapter 21, blood volume has a direct effect on blood pressure. An increase in blood volume increases blood pressure, and a decrease in blood volume decreases blood pressure.

The net result is that homeostatic mechanisms can monitor ECF volume indirectly by monitoring blood pressure. The receptors involved are baroreceptors at the carotid sinus, the aortic sinus, and the right atrium. The regulatory steps involved are reviewed in Figure 27–6.

Disorders of Sodium Balance

Sustained abnormalities in the Na^+ concentration in the ECF occur only when there are severe problems with fluid balance, such as dehydration or hyperhydration. When the body's water content increases enough to decrease the Na^+ concentration of the ECF below 135 mEq/L, a state of **hyponatremia** (*natrium*, sodium) exists. When body water content decreases, the Na^+ concentration increases. When that concentration exceeds 145 mEq/L, **hypernatremia** exists.

If the ECF volume is inadequate, both blood volume and blood pressure decrease, and the renin-angiotensin-aldosterone system is activated. In response, losses of water and Na^+ are decreased, and gains of water and Na^+ are increased. The net result is that ECF volume increases. Although the total amount of Na^+ in the ECF is increasing (gains exceed losses), the Na^+ concentration in the ECF remains unchanged, because osmotic water movement accompanies Na^+ absorption.

If the plasma volume becomes abnormally large, venous return increases, stretching the atrial and ventricular walls and stimulating the release of natriuretic peptides (ANP and BNP). This in turn decreases thirst and blocks the secretion of ADH and aldosterone, which together promote water or salt conservation. As a result, salt and water loss by the kidneys increases and the volume of the ECF decreases.

Potassium Balance

About 98 percent of the body's potassium is in the ICF. Cells expend energy to recover K^+ as they diffuse across their plasma membranes and into the ECF. The K^+ concentration outside the cell is relatively low, and the concentration in the ECF at any moment represents a balance between (1) the rate of gain across the digestive epithelium and (2) the rate of loss into urine.

The K^+ concentration in the ECF is regulated by adjustments to K^+ losses in urine. This occurs by controlling the rate of active secretion by ion pumps along the distal convoluted tubule of the nephron and the collecting system. Whenever a Na^+ is reabsorbed from the tubular fluid, it generally is exchanged for a cation (typically K^+) from the peritubular fluid. p. 998 Urinary K^+ losses are usually limited to the amount gained by absorption across the digestive epithelium, typically 50–150 mEq (1.9–5.8 g) per day. (Potassium ion losses in feces and perspiration are negligible.)

The rate of tubular secretion of K^+ into urine varies in response to three factors:

- *Changes in the* K^+ *Concentration of the ECF.* In general, the higher the extracellular concentration of K^+, the higher the rate of secretion.
- *Changes in pH.* When the pH of the ECF decreases, so does the pH of peritubular fluid. The rate of K^+ secretion then decreases, because H^+, rather than K^+, are secreted in exchange for Na^+ in tubular fluid. The mechanisms for H^+ secretion were summarized in Figure 26–13c (p. 999).
- *The Level of Aldosterone.* The hormone aldosterone strongly affects the rate at which K^+ are lost in urine. The ion pumps that are sensitive to this hormone reabsorb Na^+ from tubular fluid in exchange for K^+ from peritubular fluid. Aldosterone secretion is stimulated by angiotensin II as part of the regulation of blood volume. A high plasma K^+ concentration also stimulates aldosterone secretion directly. Either way, under the influence of aldosterone, the amount of Na^+ conserved and the amount of K^+ excreted in urine are directly related.

Two disorders of K^+ imbalance are hypokalemia and hyperkalemia. **Hypokalemia** (*hypo-*, below + *kalium*, potassium) is a deficiency of K^+ in the bloodstream. **Hyperkalemia** (*hyper-*, above) is an elevated level of K^+ in the bloodstream (Figure 27–7).

+ Clinical Note Athletes and Salt Loss

Unfounded notions and rumors about water and salt requirements during exercise are common. Sweat is a hypotonic solution that contains Na^+ in lower concentration than the ECF. As a result, a person who is sweating profusely loses more water than salt. This loss leads to an increase in the Na^+ concentration of the ECF. The water content of the ECF decreases as the water loss occurs, so blood volume decreases. Clinically, this condition is often called *volume depletion*. Because volume depletion takes place at the same time that blood is being shunted away from the kidneys, kidney function is impaired. As a result, wastes accumulate in the blood.

To prevent volume depletion, exercising athletes should drink liquids at regular intervals. The primary problem in volume depletion is water loss. Research has revealed no basis for the rumor that cramps will result if you drink while exercising. The best choice for rehydration is plain water. Body reserves of electrolytes are sufficient to tolerate extended periods of strenuous activity. Problems with Na^+ balance are extremely unlikely, except during marathons or other activities that involve maximal exertion for more than 6 hours. However, both volume depletion (causing acute renal failure) and water intoxication (causing fatal hyponatremia) have occurred in marathon runners.

+ Clinical Note Sports Drinks

Some sports drinks contain sugars and vitamins as well as electrolytes. During endurance events (marathons, ultramarathons, and distance cycling), solutions containing less than 10 g/dL of glucose may improve performance if consumed late in the event, when metabolic reserves are exhausted. However, high sugar concentrations (above 10 g/dL) can cause cramps, diarrhea, and other problems. The benefit of "glucose polymers" (often cornstarch) in sports drinks has yet to be proved. Drinking beverages "fortified" with vitamins is actively discouraged: Vitamins are not lost during exercise, and the consumption of these beverages in large volumes could, over time, cause hypervitaminosis.

Tips & Tools

The chemical symbols for sodium (Na) and potassium (K) are derived from their Latin names, **Na**trium and **K**alium.

Figure 27–7 Major Factors Involved in Disturbances of Potassium Ion Balance.

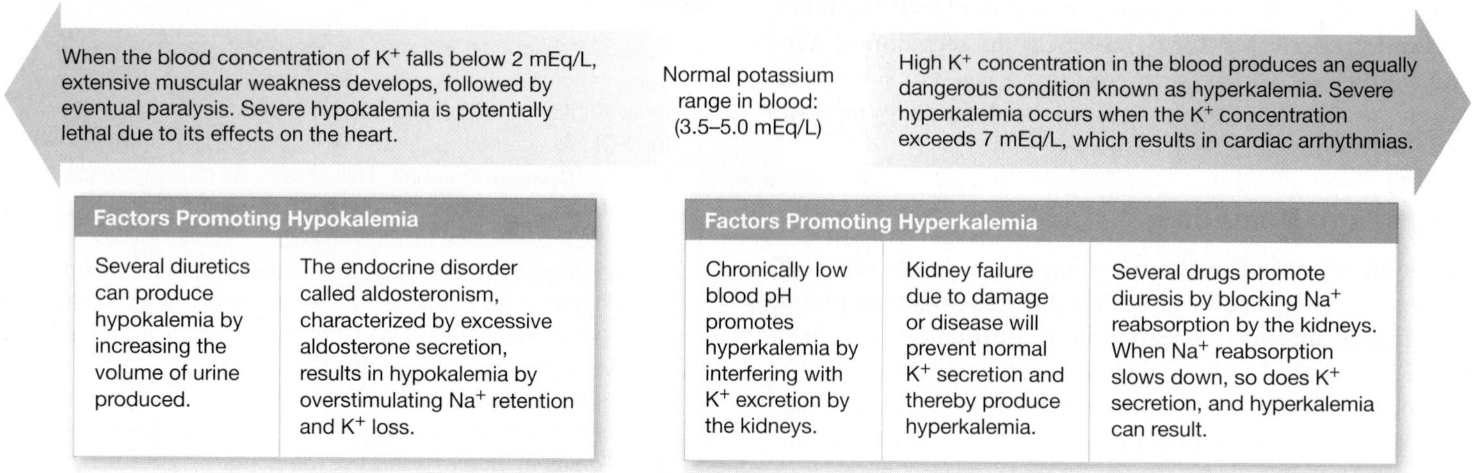

When the blood concentration of K^+ falls below 2 mEq/L, extensive muscular weakness develops, followed by eventual paralysis. Severe hypokalemia is potentially lethal due to its effects on the heart.

Normal potassium range in blood: (3.5–5.0 mEq/L)

High K^+ concentration in the blood produces an equally dangerous condition known as hyperkalemia. Severe hyperkalemia occurs when the K^+ concentration exceeds 7 mEq/L, which results in cardiac arrhythmias.

Factors Promoting Hypokalemia	
Several diuretics can produce hypokalemia by increasing the volume of urine produced.	The endocrine disorder called aldosteronism, characterized by excessive aldosterone secretion, results in hypokalemia by overstimulating Na^+ retention and K^+ loss.

Factors Promoting Hyperkalemia		
Chronically low blood pH promotes hyperkalemia by interfering with K^+ excretion by the kidneys.	Kidney failure due to damage or disease will prevent normal K^+ secretion and thereby produce hyperkalemia.	Several drugs promote diuresis by blocking Na^+ reabsorption by the kidneys. When Na^+ reabsorption slows down, so does K^+ secretion, and hyperkalemia can result.

What is a serious condition that could occur if the potassium ion concentration is >7 mEq/L?

Balance of Other Electrolytes

The ECF concentrations of other electrolytes are regulated as well. Here we will consider the most important ions involved: calcium, magnesium, phosphate, and chloride.

Calcium Balance

Calcium is the most abundant mineral in the body. A typical person has 1–2 kg (2.2–4.4 lb) of this element; ninety-nine percent of it is in the skeleton. In addition to forming the crystalline component of bone, calcium ions play key roles in controlling muscular and neural activities, in blood clotting, as cofactors for enzymatic reactions, and as second messengers.

As noted in Chapters 6 and 18, calcium homeostasis primarily reflects an interplay between the reserves in bone, the rate of absorption across the digestive tract, and the rate of loss by the kidneys. The hormones parathyroid hormone (PTH), calcitriol, and (to a lesser degree) calcitonin maintain calcium homeostasis in the ECF. Parathyroid hormone and calcitriol increase the Ca^{2+} concentration. Their actions are opposed by calcitonin. pp. 631–633

A small amount of Ca^{2+} is lost in the bile. Under normal circumstances, very few Ca^{2+} escape in urine or feces. To keep pace with these losses, an adult must absorb only 0.8–1.2 g/day of Ca^{2+}. That amount represents only about 0.03 percent of the calcium reserve in the skeleton. PTH from the parathyroid glands and calcitriol from the kidneys stimulate Ca^{2+} absorption by the digestive tract and Ca^{2+} reabsorption along the distal convoluted tubule of the kidneys.

Hypercalcemia exists when the Ca^{2+} concentration of the ECF exceeds 5.3 mEq/L. The primary cause of hypercalcemia in adults is *hyperparathyroidism*, a condition resulting from PTH oversecretion. Less common causes include malignant cancers of the breast, lung, kidney, and bone marrow, and excessive use of calcium or vitamin D supplements. Severe hypercalcemia (12–13 mEq/L) causes such signs and symptoms as fatigue, confusion, cardiac arrhythmias, and calcification of the kidneys and soft tissues throughout the body.

Hypocalcemia exists when the Ca^{2+} concentration is under 4.3 mEq/L. *Hypoparathyroidism* (undersecretion of PTH), vitamin D deficiency, or chronic renal failure is typically responsible for hypocalcemia. Signs and symptoms include muscle spasms, sometimes with generalized convulsions, weak heartbeats, cardiac arrhythmias, and osteoporosis.

Magnesium Balance

The adult body contains about 29 g of magnesium. Almost 60 percent of it is in the skeleton. The magnesium ions (Mg^{2+}) in body fluids are found primarily in the ICF. The cytosolic concentration of Mg^{2+} averages about 26 mEq/L. Magnesium ions are required as a cofactor for several important enzymatic reactions, including the phosphorylation of glucose within cells and the use of ATP by contracting muscle fibers.

The typical range of Mg^{2+} concentration in the ECF is 1.4–2.0 mEq/L, considerably lower than the level in the ICF. The proximal convoluted tubule reabsorbs magnesium ions very effectively. Keeping pace with the daily urinary loss requires a minimum dietary intake of only 24–32 mEq (0.3–0.4 g) per day.

Phosphate Balance

Phosphate ions (PO_4^{3-}) are required for bone mineralization. About 740 g are bound up in the mineral salts of the skeleton. In body fluids, the most important functions of PO_4^{3-} involve

Table 27–2 Electrolyte Balance for Average Adult

Ion and Normal ECF Range (mEq/L)	Disorder (mEq/L)	Signs and Symptoms	Causes	Treatments
Sodium (135–145)	Hypernatremia (>145)	Thirst, dryness and wrinkling of skin, reduced blood volume and pressure, eventual circulatory collapse	Dehydration; loss of hypotonic fluid	Ingestion of water or intravenous infusion of hypotonic solution
	Hyponatremia (<135)	Disturbed CNS function (water intoxication): confusion, hallucinations, convulsions, coma; death in severe cases	Infusion or ingestion of large volumes of hypotonic solution	Diuretic use and infusion of hypertonic salt solution
Potassium (3.5–5.0)	Hyperkalemia (>5.0)	Severe cardiac arrhythmias; muscle spasms	Renal failure; use of diuretics; chronic acidosis	Infusion of hypotonic solution; selection of different diuretics; infusion of buffers; dietary restrictions
	Hypokalemia (<3.5)	Muscular weakness and paralysis	Low-potassium diet; diuretics; hypersecretion of aldosterone; chronic alkalosis	Increase in dietary K^+ content; ingestion of K^+ tablets or solutions; infusion of potassium solution
Calcium (4.3–5.3)	Hypercalcemia (>5.3)	Confusion, muscle pain, cardiac arrhythmias, kidney stones, calcification of soft tissues	Hyperparathyroidism; cancer; vitamin D toxicity; calcium supplement overdose	Infusion of hypotonic fluid to lower $Ca^2{}_1$ levels; surgery to remove parathyroid gland; administration of calcitonin
	Hypocalcemia (<4.3)	Muscle spasms, convulsions, intestinal cramps, weak heartbeats, cardiac arrhythmias, osteoporosis	Poor diet; lack of vitamin D; renal failure; hypoparathyroidism; hypomagnesemia	Calcium supplements, administration of vitamin D
Magnesium (1.4–2.0)	Hypermagnesemia (>2.0)	Confusion, lethargy, respiratory depression, hypotension	Overdose of magnesium supplements or antacids (rare)	Infusion of hypotonic solution to lower plasma concentration
	Hypomagnesemia (<1.4)	Hypocalcemia, muscle weakness, cramps, cardiac arrhythmias, hypertension	Poor diet; alcoholism; severe diarrhea; kidney disease; malabsorption syndrome; ketoacidosis	Intravenous infusion of solution high in Mg^{2+}
Phosphate (1.8–3.0)	Hyperphosphatemia (>3.0)	No immediate symptoms; chronic elevation leads to calcification of soft tissues	High dietary phosphate intake; hypoparathyroidism	Dietary reduction; PTH supplementation
	Hypophosphatemia (<1.8)	Anorexia, dizziness, muscle weakness, cardiomyopathy, osteoporosis	Poor diet; kidney disease; malabsorption syndrome; hyperparathyroidism; vitamin D deficiency	Dietary improvement; vitamin D and/or calcitriol supplementation
Chloride (100–108)	Hyperchloremia (>108)	Acidosis, hyperkalemia	Dietary excess; increased chloride retention	Infusion of hypotonic solution to lower plasma concentration
	Hypochloremia (<100)	Alkalosis, anorexia, muscle cramps, apathy	Vomiting; hypokalemia	Diuretic use and infusion of hypertonic salt solution

the ICF. Inside cells these ions are required for the formation of high-energy compounds, the activation of enzymes, and the synthesis of nucleic acids.

The PO_4^{3-} concentration of the plasma is usually 1.8–3.0 mEq/L. Phosphate ions are reabsorbed from tubular fluid along the proximal convoluted tubule. Calcitriol stimulates PO_4^{3-} reabsorption along the PCT. Urinary and fecal losses of PO_4^{3-} amount to 30–45 mEq (0.8–1.2 g) per day.

Chloride Balance

Chloride ions are the most abundant anions in the ECF. The normal plasma concentration is 100–108 mEq/L. In the ICF, the Cl^- concentration is usually low (3 mEq/L). Chloride ions are absorbed across the digestive tract together with Na^+. Several carrier proteins along the renal tubules reabsorb Cl^- with Na^+. ↺ pp. 1001–1003 The amount of Cl^- loss is small. A gain of 48–146 mEq (1.7–5.1 g) per day will keep pace with losses in urine and perspiration.

Additional information about sodium, potassium, and these other ions is listed in **Table 27–2**. This table also includes information about disorders that occur if we have too much or too little of them in the body.

27

✓ Checkpoint

9. **Identify four physiologically important cations and two important anions in the ECF.**
10. **Why does prolonged sweating increase the plasma Na^+ level?**
11. **Which is more dangerous, disturbances of Na^+ balance or disturbances of K^+ balance? Explain your answer.**

See the blue Answers tab at the back of the book.

27-5 In acid-base balance, buffer systems as well as respiratory and renal compensation regulate pH changes in body fluids

Learning Outcome Explain the buffer systems that balance the pH of the ICF and ECF, and describe the compensatory mechanisms involved in maintaining acid-base balance.

We introduced the topic of **pH**—a measure of the acidity or alkalinity of a solution—and the chemical nature of acids, bases, and buffers in Chapter 2. Table 27–3 reviews key terms important to the discussion that follows. For a more detailed review, refer to the appropriate sections of Chapter 2 before you proceed. pp. 43–44

The pH of body fluids reflects interactions among all the acids, bases, and salts in solution in the body. Therefore, the pH of body fluids can be altered by adding either acids or bases. The pH of the ECF normally remains within narrow limits, usually 7.35–7.45. Any deviation from the normal range is extremely dangerous. Changes in the H^+ concentration disrupt the stability of plasma membranes, alter the structure of proteins, and change the activities of important enzymes. You could not survive for long with an ECF pH below 6.8 or above 7.7.

When the pH of blood falls below 7.35, *acidemia* exists. The physiological state that results is called **acidosis**. When the pH of blood rises above 7.45, *alkalemia* exists. The physiological state that results is called **alkalosis**.

Acidosis and alkalosis affect virtually all body systems, but the nervous and cardiovascular systems are particularly sensitive to pH fluctuations. For example, severe acidosis (pH below 7.0) can be deadly, because (1) central nervous system function deteriorates, and the person may become comatose; (2) cardiac contractions grow weak and irregular, and signs and symptoms of heart failure may develop; and (3) peripheral vasodilation produces a dramatic decrease in blood pressure, potentially producing circulatory collapse.

The control of pH, or **acid-base balance**, is a homeostatic process of great physiological and clinical significance. Several mechanisms contribute to this process, including *buffer systems* and compensation by the respiratory and urinary systems. The *combination* of buffer systems and respiratory and renal mechanisms maintains body pH within narrow limits. Both acidosis and alkalosis are dangerous, but in practice, problems with acidosis are much more common. This is so because normal cellular activities generate several types of acids.

Table 27–3 A Review of Important Terms Relating to Acid-Base Balance

Term	Definition
pH	The negative exponent (negative logarithm) of the hydrogen ion concentration $[H^+]$
Neutral	A solution with a pH of 7; the solution contains equal numbers of hydrogen ions and hydroxide ions
Acidic	A solution with a pH below 7; in this solution, hydrogen ions (H^+) predominate
Basic, or alkaline	A solution with a pH above 7; in this solution, hydroxide ions (OH^-) predominate
Acid	A substance that dissociates to release hydrogen ions, decreasing pH
Base	A substance that dissociates to release hydroxide ions or accept hydrogen ions, increasing pH
Salt	An ionic compound consisting of a cation other than hydrogen and an anion other than a hydroxide ion
Buffer	A substance that tends to oppose changes in the pH of a solution by removing or replacing hydrogen ions; in body fluids, buffers maintain blood pH within normal limits (7.35–7.45)

27

Types of Acids in the Body

There are three general categories of acids in the body: (1) **fixed acids**, which do not leave solution; (2) **metabolic acids**, which take part in or are produced by cellular metabolism; and (3) **volatile acids**, which can leave the body and enter the atmosphere through the lungs. Figure 27–8 gives examples of each. In general, fixed acids and metabolic acids do not greatly affect pH, so let's look at a volatile acid that might.

Carbonic acid (H_2CO_3) forms spontaneously from carbon dioxide (CO_2) and water in body fluids. This reaction proceeds much more rapidly in the presence of the enzyme *carbonic anhydrase*. This enzyme is found in the cytoplasm of red blood cells, liver and kidney cells, parietal cells of the stomach, and many other types of cells.

The partial pressure of carbon dioxide (P_{CO_2}) is the most important factor affecting the pH in blood. Most of the CO_2 in solution is converted to carbonic acid, and most of the carbonic acid dissociates into H^+ and **bicarbonate ions** (HCO_3^-). For this reason, the partial pressure of CO_2 and the pH are inversely related, as the seesaw illustrates in Figure 27–9. In other words, when the CO_2 level increases, additional H^+ and bicarbonate ions are released, so the pH decreases. (Recall that the pH is a *negative exponent*, so when the concentration of H^+ goes up, the pH value goes down.)

In contrast, as CO_2 leaves the bloodstream and diffuses into the alveoli, the number of H^+ and bicarbonate ions in the alveolar capillaries decreases, and blood pH increases.

Mechanisms of pH Control: Buffer Systems

To maintain acid-base balance over long periods of time, your body must balance gains and losses of H^+. Hydrogen ions are gained at the digestive tract and through metabolic activities within cells. Your kidneys eliminate H^+ by secreting them into urine. The lungs eliminate H^+ by forming water and CO_2 from H^+ and HCO_3^-. These sites of elimination are far removed from the sites of acid production. As the hydrogen ions travel through the body, they must be neutralized to avoid tissue damage.

Buffers in body fluids temporarily neutralize the acids produced by normal metabolic operations. To understand how

Figure 27–8 Three Classes of Acids Found in the Body.

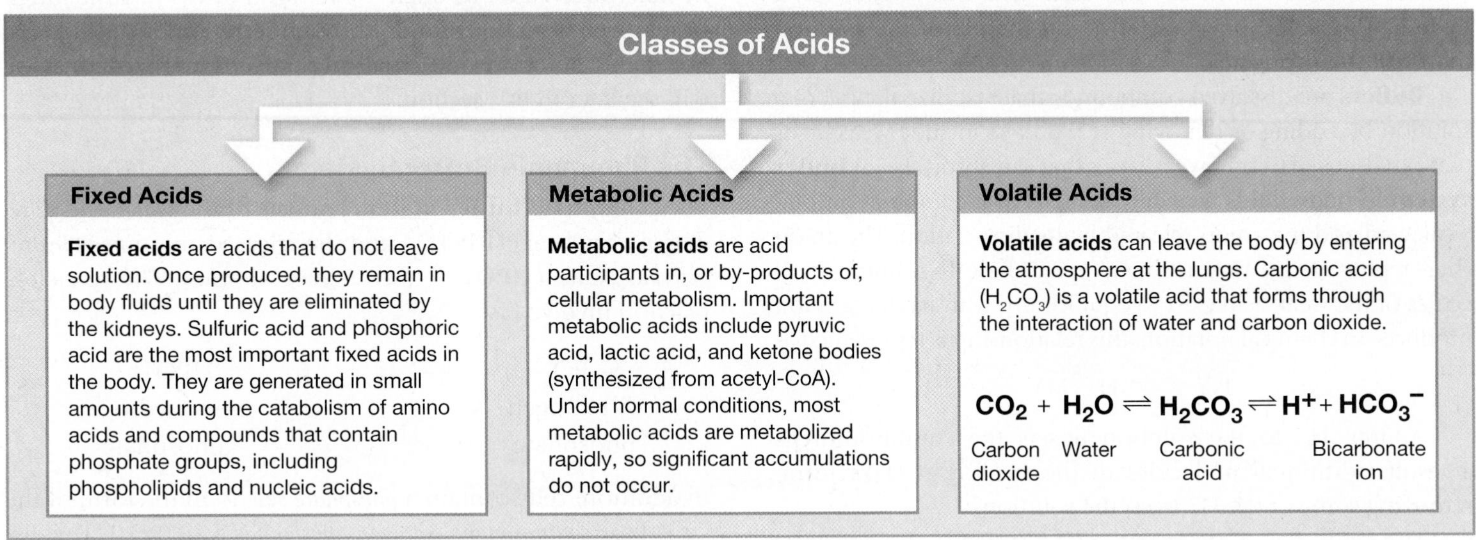

Classes of Acids
Fixed Acids
Fixed acids are acids that do not leave solution. Once produced, they remain in body fluids until they are eliminated by the kidneys. Sulfuric acid and phosphoric acid are the most important fixed acids in the body. They are generated in small amounts during the catabolism of amino acids and compounds that contain phosphate groups, including phospholipids and nucleic acids.
Metabolic Acids
Metabolic acids are acid participants in, or by-products of, cellular metabolism. Important metabolic acids include pyruvic acid, lactic acid, and ketone bodies (synthesized from acetyl-CoA). Under normal conditions, most metabolic acids are metabolized rapidly, so significant accumulations do not occur.
Volatile Acids
Volatile acids can leave the body by entering the atmosphere at the lungs. Carbonic acid (H_2CO_3) is a volatile acid that forms through the interaction of water and carbon dioxide.
$CO_2 + H_2O \rightleftharpoons H_2CO_3 \rightleftharpoons H^+ + HCO_3^-$
Carbon dioxide
Water
Carbonic acid
Bicarbonate ion

buffers work, we need a bit of review of acids and bases. Recall from Chapter 2 that we can categorize **acids** and **bases** as either *strong* or *weak*:

- **Strong acids** and **strong bases** dissociate completely in solution. For example, hydrochloric acid (HCl), a strong acid, dissociates in solution by the reaction

$$HCl \longrightarrow H^+ + Cl^-$$

- When **weak acids** or **weak bases** enter a solution, a significant number of molecules remain intact, so dissociation is not complete. For example, carbonic acid is a weak acid. At the normal pH of the ECF, an equilibrium exists, and the reaction can be diagrammed as follows:

$$\underset{\text{Carbonic acid}}{H_2CO_3} \rightleftharpoons \underset{\text{hydrogen ion}}{H^+} + \underset{\text{bicarbonate ion}}{HCO_3^-}$$

Figure 27–9 The Basic Relationship between P_{CO_2} and Blood pH. The P_{CO_2} is inversely related to the pH.

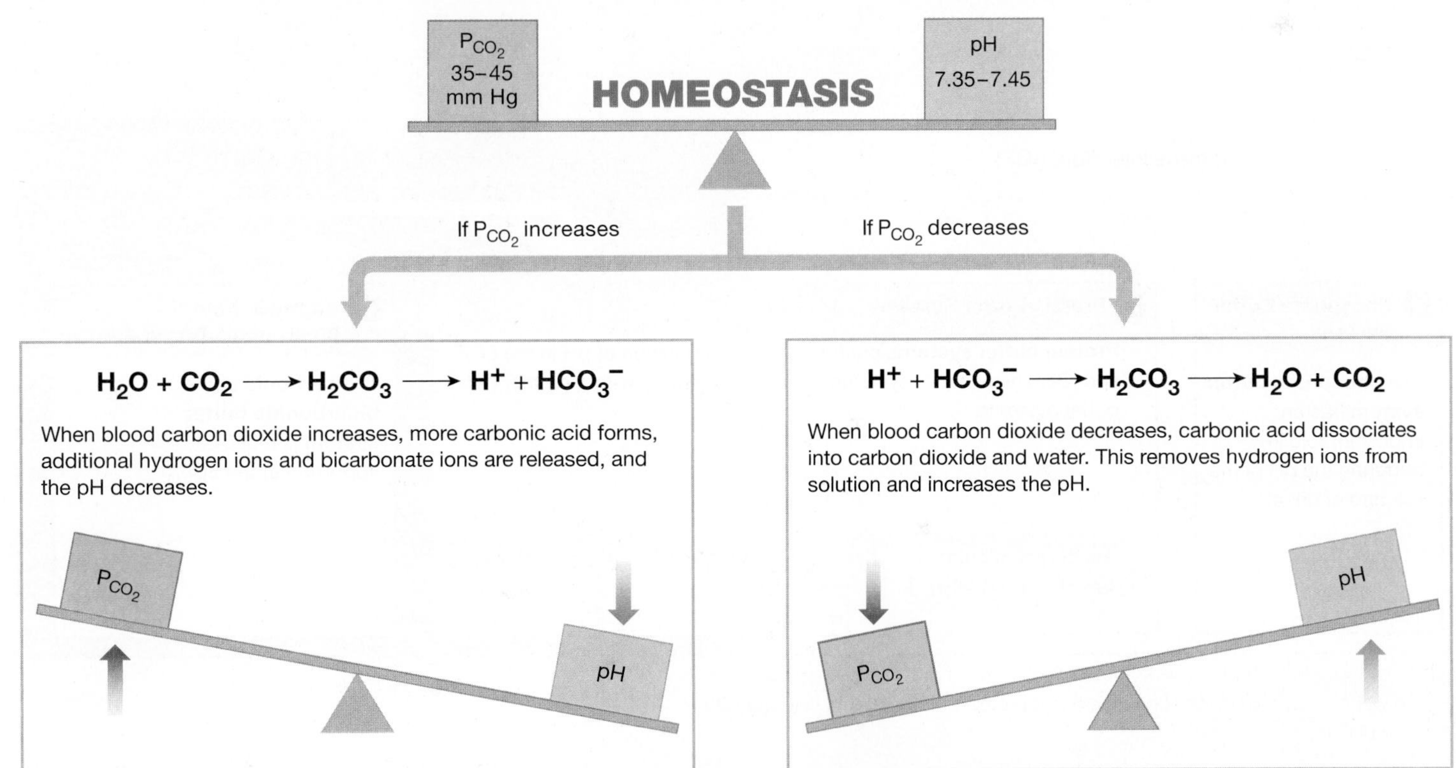

P_{CO_2} 35–45 mm Hg
HOMEOSTASIS
pH 7.35–7.45
If P_{CO_2} increases
If P_{CO_2} decreases
$H_2O + CO_2 \longrightarrow H_2CO_3 \longrightarrow H^+ + HCO_3^-$
When blood carbon dioxide increases, more carbonic acid forms, additional hydrogen ions and bicarbonate ions are released, and the pH decreases.
P_{CO_2}
pH
$H^+ + HCO_3^- \longrightarrow H_2CO_3 \longrightarrow H_2O + CO_2$
When blood carbon dioxide decreases, carbonic acid dissociates into carbon dioxide and water. This removes hydrogen ions from solution and increases the pH.
P_{CO_2}
pH

If you place the same concentration of a weak acid and a strong acid in separate solutions, the weak acid will release fewer hydrogen ions and have less effect on the pH of the solution than will the strong acid.

Buffers are dissolved compounds that stabilize the pH of a solution by adding or removing H^+. Buffers include weak acids that can donate H^+, and weak bases that can absorb H^+. A **buffer system** in body fluids generally consists of a combination of a weak acid and the anion released by its dissociation. The anion (here represented as Y) functions as a weak base. In solution, molecules of the weak acid exist in equilibrium with its dissociation products. In chemical notation, this relationship is represented as

$$HY \rightleftharpoons H^+ + Y^-$$

Adding H^+ to the solution upsets the equilibrium. As a result, additional molecules of the weak acid (HY) form, removing some of the H^+ from the solution.

The body has three major buffer systems: the *phosphate buffer system*, the *protein buffer systems*, and the *carbonic acid–bicarbonate buffer system*. Each has slightly different characteristics and distributions (Figure 27–10).

As we look more closely at these buffer systems, remember that they provide only a temporary solution to an acid-base imbalance. To preserve homeostasis, the H^+ that these systems capture must ultimately be either permanently tied up in water molecules, through the removal of CO_2 by the lungs, or removed from body fluids, through secretion by the kidneys. For that, respiratory and renal mechanisms are needed, as we will see later in this section.

The Phosphate Buffer System

The **phosphate buffer system** consists of the weak acid *dihydrogen phosphate* ($H_2PO_4^-$) and the released anion *monohydrogen phosphate* (HPO_4^{2-}) (see Figure 27–10a). The reversible reaction involved is

$$\underset{\text{dihydrogen phosphate}}{H_2PO_4^-} \rightleftharpoons H^+ + \underset{\text{monohydrogen phosphate}}{HPO_4^{2-}}$$

In addition, cells contain a *phosphate reserve* in the form of the weak base *sodium monohydrogen phosphate* (Na_2HPO_4). The dissociation of this weak base provides additional monohydrogen phosphate for use by this buffer system:

$$2Na^+ + \underset{\text{monohydrogen phosphate}}{HPO_4^{2-}} \rightleftharpoons \underset{\text{sodium monohydrogen phosphate}}{Na_2HPO_4}$$

Figure 27–10 **Buffer Systems in Body Fluids.**

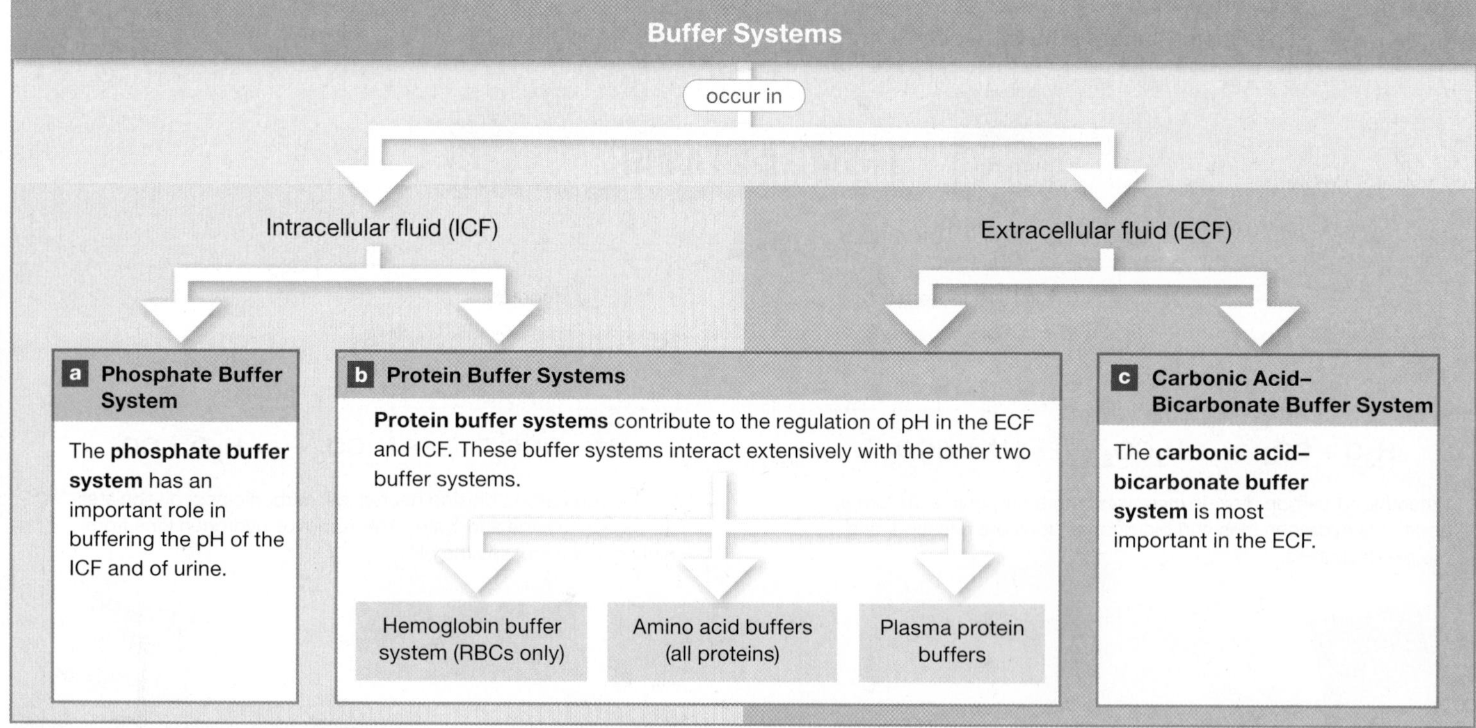
Buffer Systems
occur in
Intracellular fluid (ICF)
Extracellular fluid (ECF)
a Phosphate Buffer System
The phosphate buffer system has an important role in buffering the pH of the ICF and of urine.
b Protein Buffer Systems
Protein buffer systems contribute to the regulation of pH in the ECF and ICF. These buffer systems interact extensively with the other two buffer systems.
Hemoglobin buffer system (RBCs only)
Amino acid buffers (all proteins)
Plasma protein buffers
c Carbonic Acid–Bicarbonate Buffer System
The carbonic acid–bicarbonate buffer system is most important in the ECF.

? In which body fluid do the phosphate and protein buffer systems help regulate the pH?

27

The phosphate buffer system is quite important in buffering the pH of the ICF. This system is also important in stabilizing the pH of urine. In the ECF, the phosphate buffer system plays only a supporting role in the regulation of pH, primarily because the concentration of bicarbonate ions far exceeds that of dihydrogen phosphate.

Protein Buffer Systems

Protein buffer systems depend on the ability of amino acids to respond to pH changes by accepting or releasing H^+, that is, by acting as a base or an acid (see Figure 27–10b). Such molecules are said to be *amphoteric* (*ampho*, both).

The effects of protein buffer systems are very important in both the ECF and the ICF. In the ICF of active cells, structural and other proteins provide an extensive buffering capability. Their buffering prevents destructive changes in pH when cellular metabolism produces metabolic acids, such as lactic acid or pyruvic acid. Because exchange occurs between the ICF and the ECF, the protein buffer systems can also help stabilize the pH of the ECF. For example, when the pH of the ECF decreases, cells pump H^+ out of the ECF and into the ICF, where intracellular proteins can buffer them. When the pH of the ECF increases, pumps in plasma membranes move H^+ out of the ICF and into the ECF. These mechanisms can assist in buffering the pH of the ECF, but the process is slow. Hydrogen ions must be individually transported across the plasma membrane. For this reason, the protein buffer systems in most cells cannot make rapid, large-scale adjustments in the pH of the ECF.

All proteins function as *amino acid buffers*. In addition, there are two specialized protein buffer systems that help to maintain the pH of blood: *plasma protein buffers* (which are covered in Figure 27–10b) and the *hemoglobin buffer system*.

Amino Acid Buffers. The underlying buffering mechanism of amino acids is shown in Figure 27–11:

- If pH increases (becomes more basic), the carboxyl group ($-COOH$) of the amino acid can dissociate, acting as a weak acid and releasing an H^+. The carboxyl group then becomes a carboxylate ion ($-COO^-$). At the normal pH of body fluids (7.35–7.45), the carboxyl groups of most amino acids have already given up their H^+. (Proteins carry negative charges mainly for that reason.) However, some amino acids, notably histidine and cysteine, have R groups (side chains) that will donate H^+ if the pH increases outside the normal range.
- In a neutral solution, an amino acid carries both positive and negative charges. This state is called a *zwitterion* (the term is derived from German meaning *hybrid ion*).
- If pH decreases (becomes more acidic), the carboxylate ion and the amino group ($-NH_2$) can act as weak bases and accept additional H^+, forming a carboxyl group ($-COOH$) and an amino ion ($-NH_3{}^+$), respectively. In free amino acids, both the main structural chain and the side chain can act as buffers. In a protein, most of the carboxyl and amino groups in the main chain are tied up in peptide bonds. Only the exposed amino group and carboxyl group at either end of a protein are available as buffers. So most of the buffering capacity of proteins is provided by the R groups of amino acids.

The Hemoglobin Buffer System. The situation is somewhat different for red blood cells. These cells, which contain approximately 5.5 percent of the ICF, are normally suspended in the plasma. They are densely packed with hemoglobin, and their cytoplasm contains large amounts of carbonic anhydrase. Red blood cells have a significant effect on the pH of the ECF, because they absorb CO_2 from the plasma and convert it to carbonic acid. Carbon dioxide can diffuse across the RBC membrane very quickly, so no transport mechanism is needed. As the carbonic acid dissociates within an RBC, the bicarbonate ions diffuse into the plasma in exchange for chloride ions, a swap known as the *chloride shift*. p. 868 Hemoglobin molecules buffer the hydrogen ions (look back at Figure 23–22, p. 867). In the lungs, the entire reaction sequence diagrammed in Figure 23–23 on p. 869 proceeds in reverse. This mechanism is known as the **hemoglobin buffer system**.

Figure 27–11 The Role of Amino Acid Buffers in Protein Buffer Systems. An amino acid in solution has both a negative and positive charge. Depending on the pH of their surroundings, amino acids either donate an H^+ (as at left) or accept an H^+ (as at right). An amino acid in solution is a zwitterion and has both a negative and positive charge. Additionally, several different R groups may release or absorb H^+.

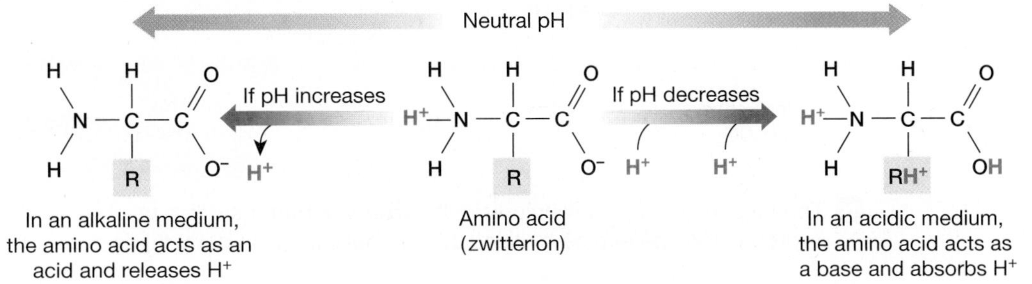

Neutral pH

If pH increases — If pH decreases

In an alkaline medium, the amino acid acts as an acid and releases H^+

Amino acid (zwitterion)

In an acidic medium, the amino acid acts as a base and absorbs H^+

The hemoglobin buffer system is the only intracellular buffer system that can have an immediate effect on the pH of the ECF. This buffer system helps prevent drastic changes in pH when the plasma P_{CO_2} is increasing or decreasing.

The Carbonic Acid–Bicarbonate Buffer System

Body cells generate CO_2 virtually 24 hours a day. (Exceptions are red blood cells, some cancer cells, and tissues temporarily deprived of oxygen.) As we have seen, most of the CO_2 is converted to carbonic acid, which then dissociates into a hydrogen ion and a bicarbonate ion (HCO_3^-). The carbonic acid and its dissociation products form the **carbonic acid–bicarbonate buffer system**. The primary role of this buffer system is to prevent changes in pH caused by metabolic acids and fixed acids in the ECF (see Figure 27–10c).

The carbonic acid–bicarbonate buffer system consists of the reaction introduced in our discussion of volatile acids (Figure 27–12a):

$$\underset{\text{carbon dioxide}}{CO_2} + \underset{\text{water}}{H_2O} \rightleftharpoons \underset{\text{carbonic acid}}{H_2CO_3} \rightleftharpoons \underset{\text{hydrogen ion}}{H^+} + \underset{\text{bicarbonate ion}}{HCO_3^-}$$

Because the reaction is freely reversible, a change in the concentration of any one component affects the concentrations of all others. For example, if H^+ are added, most of them will be removed by interactions with HCO_3^-, forming H_2CO_3. In the process, the HCO_3^- act as a weak base that buffers the excess H^+. The H_2CO_3 formed in this way in turn dissociates into CO_2 and H_2O (Figure 27–12b). The lungs can then excrete the extra CO_2. In effect, this reaction takes the H^+ released by a strong metabolic or fixed acid and generates a volatile acid that can easily be eliminated.

The carbonic acid–bicarbonate buffer system can also protect against increases in pH, although such changes are rare. If H^+ are removed from the plasma, the reaction is driven to the right: The P_{CO_2} decreases, and H_2CO_3 dissociates to replace the missing H^+.

The carbonic acid–bicarbonate buffer system has three important limitations:

- *It cannot protect the ECF from changes in pH that result from an increased or decreased level of CO_2.* A buffer system cannot protect against changes in the concentration of its own weak acid. As Figure 27–12a indicates, an equilibrium exists among the components of this buffer system. Thus, in this system, the addition of excess H^+ from an outside source would drive the reaction to the left. But the addition of excess CO_2 would form H_2CO_3 and drive the reaction to the right. The dissociation of H_2CO_3 would release H^+ and HCO_3^-, decreasing the pH of the blood.
- *It can function only when the respiratory system and the respiratory control centers are working normally.* Normally, the increase in P_{CO_2} that occurs when fixed acids or metabolic acids are buffered stimulates an increase in the respiratory rate. This increase accelerates the removal of CO_2 by the lungs. The efficiency of the buffer system will be reduced if the respiratory passageways are blocked, or blood flow

Figure 27–12 The Carbonic Acid–Bicarbonate Buffer System.

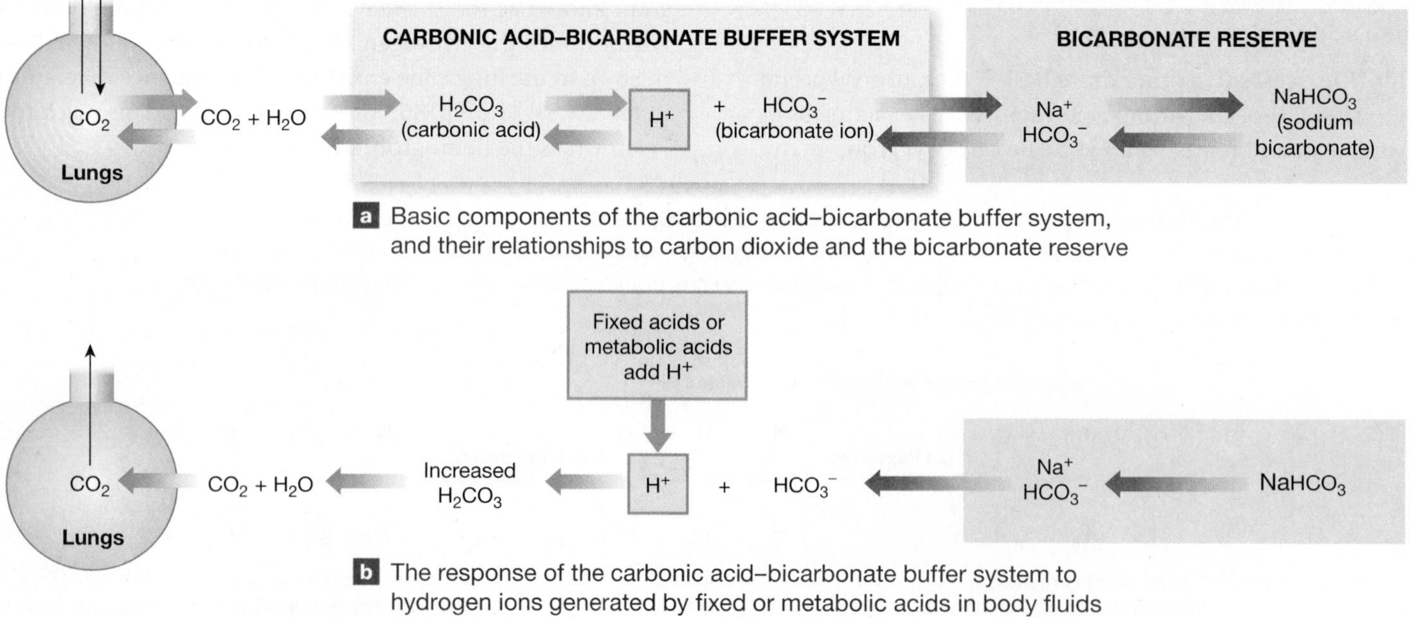

CARBONIC ACID–BICARBONATE BUFFER SYSTEM
BICARBONATE RESERVE
CO2
Lungs
CO2 + H2O
H2CO3 (carbonic acid)
H+
+ HCO3− (bicarbonate ion)
Na+ HCO3−
NaHCO3 (sodium bicarbonate)
Fixed acids or metabolic acids add H+
CO2
Lungs
CO2 + H2O
Increased H2CO3
H+
+ HCO3−
Na+ HCO3−
NaHCO3

a Basic components of the carbonic acid–bicarbonate buffer system, and their relationships to carbon dioxide and the bicarbonate reserve

b The response of the carbonic acid–bicarbonate buffer system to hydrogen ions generated by fixed or metabolic acids in body fluids

27

to the lungs is impaired, or the respiratory centers do not respond normally. This buffer system cannot eliminate H^+ and remove the threat to homeostasis unless the respiratory system is functioning normally.

- *The ability to buffer acids is limited by the availability of bicarbonate ions.* Every time an H^+ is removed from the plasma, an HCO_3^- goes with it. When all the HCO_3^- have been tied up, buffering capabilities are lost.

Problems due to a lack of bicarbonate ions are rare, for several reasons. First, body fluids contain a large reserve of HCO_3^-, mostly in the form of dissolved molecules of the weak base *sodium bicarbonate* ($NaHCO_3$). This readily available supply of HCO_3^- is known as the **bicarbonate reserve**. The reaction involved (see Figure 27–12a) is

$$\underset{}{Na^+} + \underset{\text{bicarbonate ion}}{HCO_3^-} \rightleftharpoons \underset{\text{sodium bicarbonate}}{NaHCO_3}$$

When H^+ enter the ECF, the HCO_3^- tied up in H_2CO_3 molecules are replaced by HCO_3^- from the bicarbonate reserve (see Figure 27–12b).

Second, additional HCO_3^- can be generated by the kidneys, through mechanisms described in Chapter 26 (look back at Figure 26–13c, p. 999). In the distal convoluted tubule of the nephron and collecting system, carbonic anhydrase converts CO_2 within tubular cells into H_2CO_3, which then dissociates. The H^+ is pumped into tubular fluid in exchange for a Na^+, and the HCO_3^- is transported into peritubular fluid in exchange for a Cl^-. In effect, tubular cells remove HCl from peritubular fluid in exchange for $NaHCO_3$.

Regulation of Acid-Base Balance

Buffer systems can tie up excess H^+, but they provide only a temporary solution to an acid-base imbalance. The H^+ are merely rendered harmless; they are not eliminated. To preserve homeostasis, the captured H^+ must ultimately be either permanently tied up in water molecules, through the removal of CO_2 by the lungs, or removed from body fluids, through secretion by the kidneys.

Suppose that a buffer molecule prevents a change in pH by binding an H^+ that enters the ECF. That buffer molecule is then tied up, reducing the capacity of the ECF to cope with any additional H^+. Eventually, all the buffer molecules are bound to H^+, and further pH control becomes impossible.

The situation can be resolved only by removing H^+ from the ECF (thereby freeing the buffer molecules) or by replacing the buffer molecules. Similarly, if a buffer provides an H^+ to maintain normal pH, homeostatic conditions will return only when either another H^+ has been obtained or the buffer has been replaced.

Regulating acid-base balance thus includes balancing H^+ gains and losses. This "balancing act" involves coordinating the actions of buffer systems with respiratory and renal compensation mechanisms. These mechanisms support the buffer systems by (1) secreting or absorbing H^+, (2) controlling the excretion of acids and bases, and, when necessary, (3) generating additional buffers. As we have noted, it is the combination of buffer systems and these respiratory and renal mechanisms that maintains body pH within narrow limits.

Respiratory Compensation

Respiratory compensation is a change in the respiratory rate that helps stabilize the pH of the ECF. Respiratory compensation takes place whenever body pH strays outside normal limits. Such compensation works because respiratory activity has a direct effect on the carbonic acid–bicarbonate buffer system. Increasing or decreasing the rate of respiration alters pH by decreasing or increasing the P_{CO_2}. When the P_{CO_2} increases, the pH decreases, because the addition of CO_2 drives the carbonic acid–bicarbonate buffer system to the right. When the P_{CO_2} falls, the pH increases because the removal of CO_2 drives that buffer system to the left.

We described the mechanisms responsible for the control of respiratory rate in Chapter 23. Here we present only a brief summary. (For a review, look back at **Spotlight Figure 23–25**, p. 872, and Figure 23–26, p. 875.)

Chemoreceptors of the carotid and aortic bodies are sensitive to the P_{CO_2} of circulating blood. Other receptors, located on the ventrolateral surfaces of the medulla oblongata, monitor the P_{CO_2} of the CSF. A rise in P_{CO_2} stimulates the chemoreceptors, leading to an increase in the respiratory rate. As the rate of respiration increases, more CO_2 is lost at the lungs, so the P_{CO_2} returns to a normal level. Conversely, when the P_{CO_2} of the blood or CSF decreases, the chemoreceptors are inhibited. Respiratory activity becomes depressed and the breathing rate decreases, causing an elevation of the P_{CO_2} in the ECF.

Renal Compensation

Renal compensation is a change in the rates of H^+ and HCO_3^- secretion or reabsorption by the kidneys in response to changes in plasma pH. Under normal conditions, the body generates enough metabolic and fixed acids to add about 100 mEq of H^+ to the ECF each day. An equivalent number of H^+ must be excreted in urine to maintain acid-base balance. In addition, the kidneys assist the lungs by eliminating any CO_2 that either enters the renal tubules during filtration or diffuses into the tubular fluid as it travels toward the renal pelvis.

Hydrogen ions are secreted into the tubular fluid along the proximal convoluted tubule (PCT), the distal convoluted tubule (DCT), and the collecting system.The basic mechanisms involved are shown in Figures 26–12 and 26–13c (pp. 997, 999). The ability to eliminate a large number of H^+ in a normal volume of urine depends on the presence of buffers in the urine. The secretion of H^+ can continue only until the pH of the tubular fluid reaches 4.0–4.5. (At that point, the H^+ concentration gradient is so great that H^+ leak out of the tubule as fast as they are pumped in.)

If the tubular fluid lacked buffers to absorb the H^+, the kidneys could secrete less than 1 percent of the acid produced each day before the pH reached this limit. To maintain acid balance under these conditions, the kidneys would have to produce about 1000 liters of urine each day just to keep pace with the generation of H^+ in the body! Buffers in tubular fluid are extremely important, because they keep the pH high enough for H^+ secretion to continue. Metabolic acids are being generated continuously. Without these buffering mechanisms, the kidneys could not maintain homeostasis.

The Process of pH Regulation by Kidney Tubule Cells. Figure 27–13 diagrams the primary routes of H^+ secretion and the buffering mechanisms that stabilize the pH of tubular fluid. The three major buffers involved are the previously discussed carbonic acid–bicarbonate buffer system and phosphate buffer system, and the *ammonia buffer system* (Figure 27–13a). Glomerular filtration puts components of the carbonic acid–bicarbonate buffer system and the phosphate buffer system into the filtrate. Tubule cells (mainly those of the PCT) generate ammonia.

Figure 27–13a shows the secretion of H^+, which relies on carbonic anhydrase activity within tubular cells. The H^+ generated may be pumped into the lumen in exchange for Na^+, individually or together with Cl^-. The net result is the secretion of H^+, accompanied by the removal of CO_2 (from the tubular fluid, the tubule cells, and the ECF), and the release of sodium bicarbonate into the ECF.

Figure 27–13b shows the generation of ammonia within the tubules. As tubule cells use the enzyme *glutaminase* to break down the amino acid *glutamine*, amino groups are released as either ammonium ions (NH_4^+) or ammonia (NH_3). The NH_4^+ are transported into the lumen in exchange for Na^+ in the tubular fluid. The NH_3, which is highly volatile and also toxic to cells, diffuses rapidly into the tubular fluid. There it reacts with an H^+, forming NH_4^+.

This reaction buffers the tubular fluid and removes a potentially dangerous compound from body fluids. The carbon chains of the glutamine molecules are ultimately converted to HCO_3^-, which is cotransported with Na^+ into the ECF. The generation of ammonia by tubule cells ties up H^+ in the tubular fluid and releases sodium bicarbonate into the ECF, where it contributes to the bicarbonate reserve. These mechanisms of H^+ secretion and buffering are always functioning, but their levels of activity vary widely with the pH of the ECF.

Renal Responses to Acidosis and Alkalosis. Acidosis (low blood pH) develops when the normal plasma buffer mechanisms are stressed by excessive hydrogen ions. The kidney tubules do not distinguish among the various acids that may cause acidosis. Whether the decrease in pH results from the production of volatile, fixed, or metabolic acids, the renal contribution remains limited to (1) the secretion of H^+, (2) the activity of buffers in the tubular fluid, (3) the removal of CO_2, and (4) the reabsorption of sodium bicarbonate.

Tubule cells thus bolster the capabilities of the carbonic acid–bicarbonate ion buffer system. They do so by increasing the concentration of HCO_3^- in the ECF, replacing those already used to remove H^+ from solution. In a starving person, tubule cells break down amino acids, yielding ammonium ions that are pumped into the tubular fluid, bicarbonate ions to help buffer ketone bodies in the blood, and carbon chains for catabolism (see Figure 27–13b).

When alkalosis (high blood pH) develops, (1) the rate of H^+ secretion by the kidneys declines, (2) tubule cells do not reclaim the HCO_3^- in tubular fluid, and (3) the collecting system transports HCO_3^- into tubular fluid while releasing H^+ and Cl^- into the peritubular fluid (Figure 27–13c). The concentration of HCO_3^- in blood decreases, promoting the dissociation of H_2CO_3 and the release of H^+. The additional H^+ generated by the kidneys help return the pH to a normal level.

✓ Checkpoint

12. **Identify the body's three major buffer systems.**
13. **What effect would a decrease in the pH of body fluids have on the respiratory rate?**
14. **Why must tubular fluid in nephrons be buffered?**

See the blue Answers tab at the back of the book.

27-6 Disorders of acid-base balance can be classified as respiratory or metabolic

Learning Outcome Identify the most frequent disturbances of acid-base balance, and explain how the body responds when the pH of body fluids varies outside normal limits.

For anyone considering a career in a health-related field, an understanding of acid-base dynamics is essential both in clinical diagnosis and patient management. Temporary shifts in the pH of body fluids occur frequently. For example, the *alkaline tide* following a meal causes a temporary increase in blood pH. p. 904 Rapid and complete return to normal pH involves a combination of buffer system activity and the respiratory and renal compensatory mechanisms (Figure 27–14). Together these mechanisms can generally control pH very precisely, so the pH of the ECF seldom varies more than 0.1 pH unit, from 7.35 to 7.45.

Figure 27–13 **pH Regulation of Tubular Fluid by Kidney Tubule Cells.**

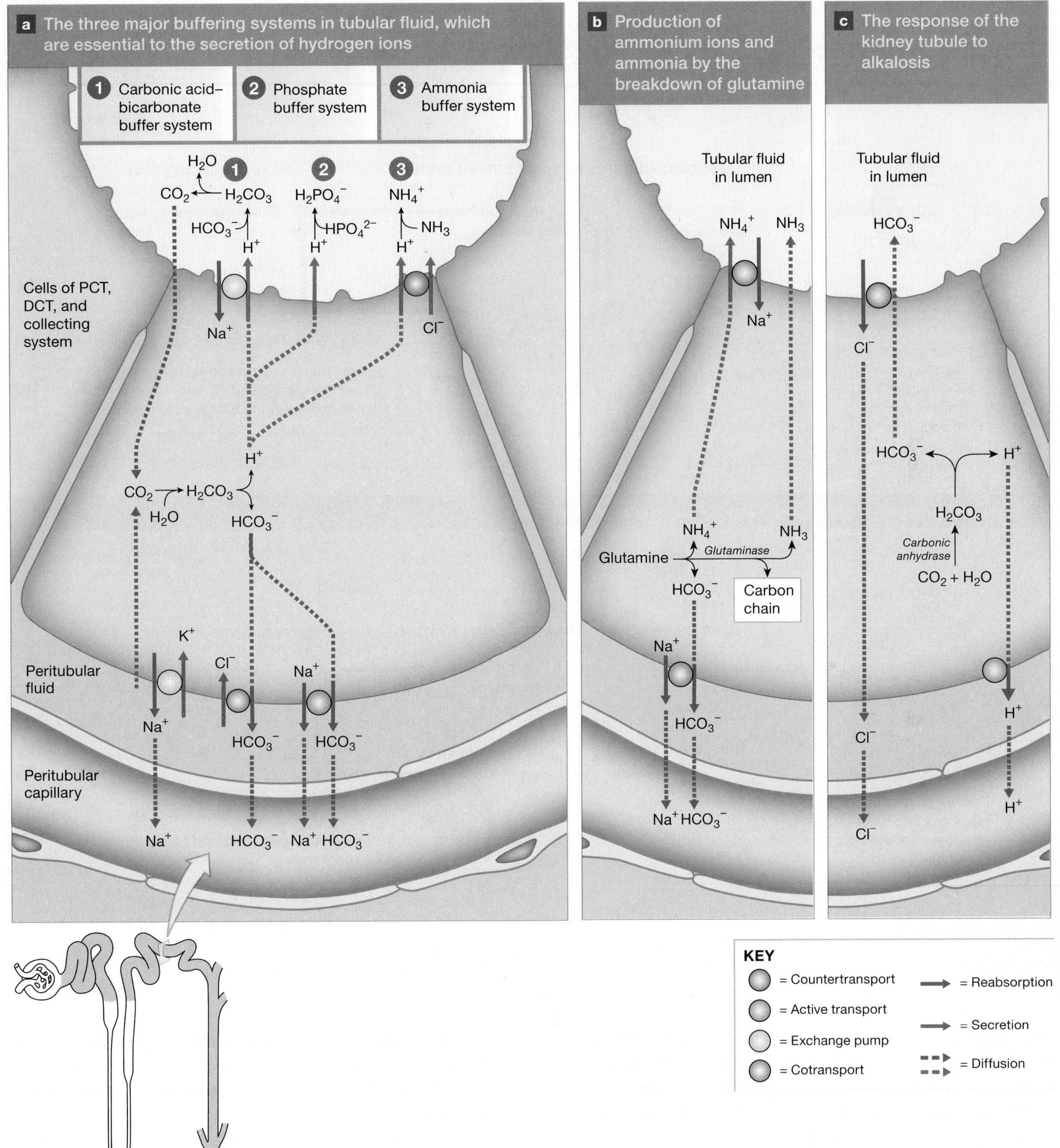
a The three major buffering systems in tubular fluid, which are essential to the secretion of hydrogen ions
1 Carbonic acid–bicarbonate buffer system
2 Phosphate buffer system
3 Ammonia buffer system
H_2O
CO_2
H_2CO_3
HCO_3^-
H^+
$H_2PO_4^-$
HPO_4^{2-}
NH_4^+
NH_3
Cells of PCT, DCT, and collecting system
Na^+
Cl^-
K^+
Peritubular fluid
Peritubular capillary
b Production of ammonium ions and ammonia by the breakdown of glutamine
Tubular fluid in lumen
Glutamine
Glutaminase
Carbon chain
$Na^+HCO_3^-$
c The response of the kidney tubule to alkalosis
Carbonic anhydrase
$CO_2 + H_2O$
KEY
= Countertransport
= Active transport
= Exchange pump
= Cotransport
= Reabsorption
= Secretion
= Diffusion
27

Figure 27–14 Interactions among the Carbonic Acid–Bicarbonate Buffer System and Respiratory and Renal Compensation Mechanisms in the Regulation of Plasma pH.

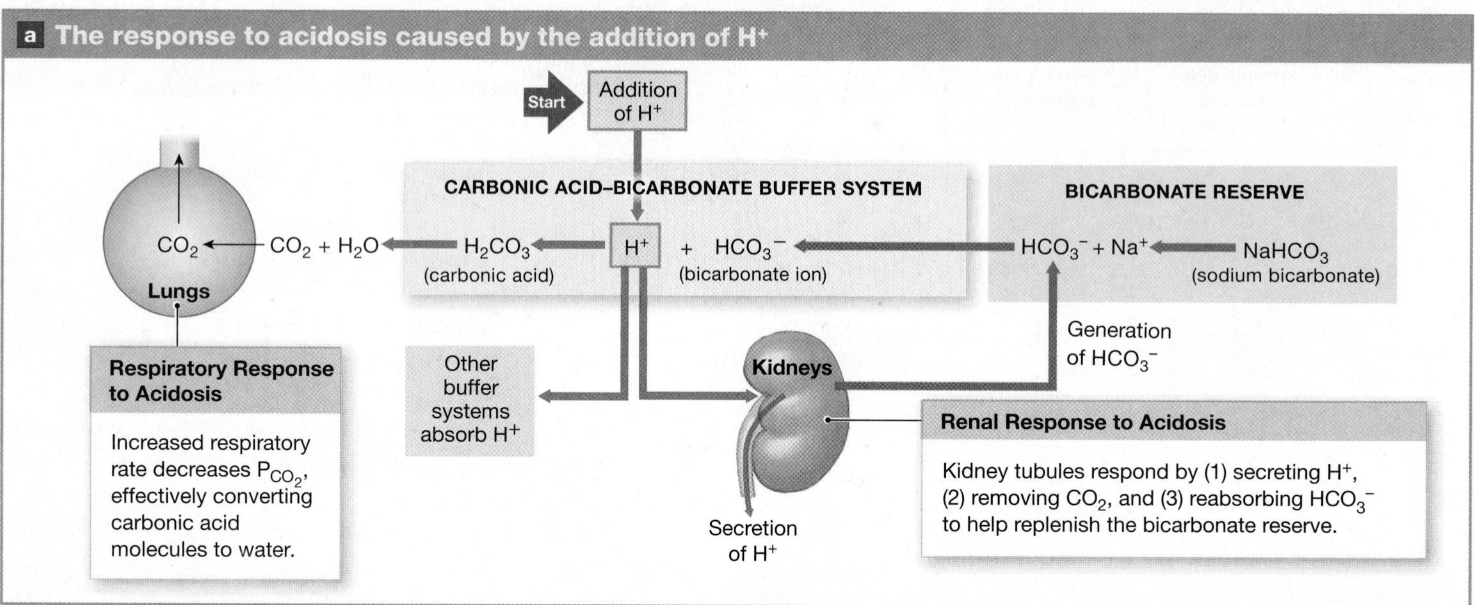

a The response to acidosis caused by the addition of H^+
Start
Addition of H^+
CARBONIC ACID–BICARBONATE BUFFER SYSTEM
BICARBONATE RESERVE
CO_2
$CO_2 + H_2O$
H_2CO_3 (carbonic acid)
H^+
+ HCO_3^- (bicarbonate ion)
$HCO_3^- + Na^+$
$NaHCO_3$ (sodium bicarbonate)
Lungs
Generation of HCO_3^-
Respiratory Response to Acidosis
Increased respiratory rate decreases P_{CO_2}, effectively converting carbonic acid molecules to water.
Other buffer systems absorb H^+
Kidneys
Renal Response to Acidosis
Kidney tubules respond by (1) secreting H^+, (2) removing CO_2, and (3) reabsorbing HCO_3^- to help replenish the bicarbonate reserve.
Secretion of H^+

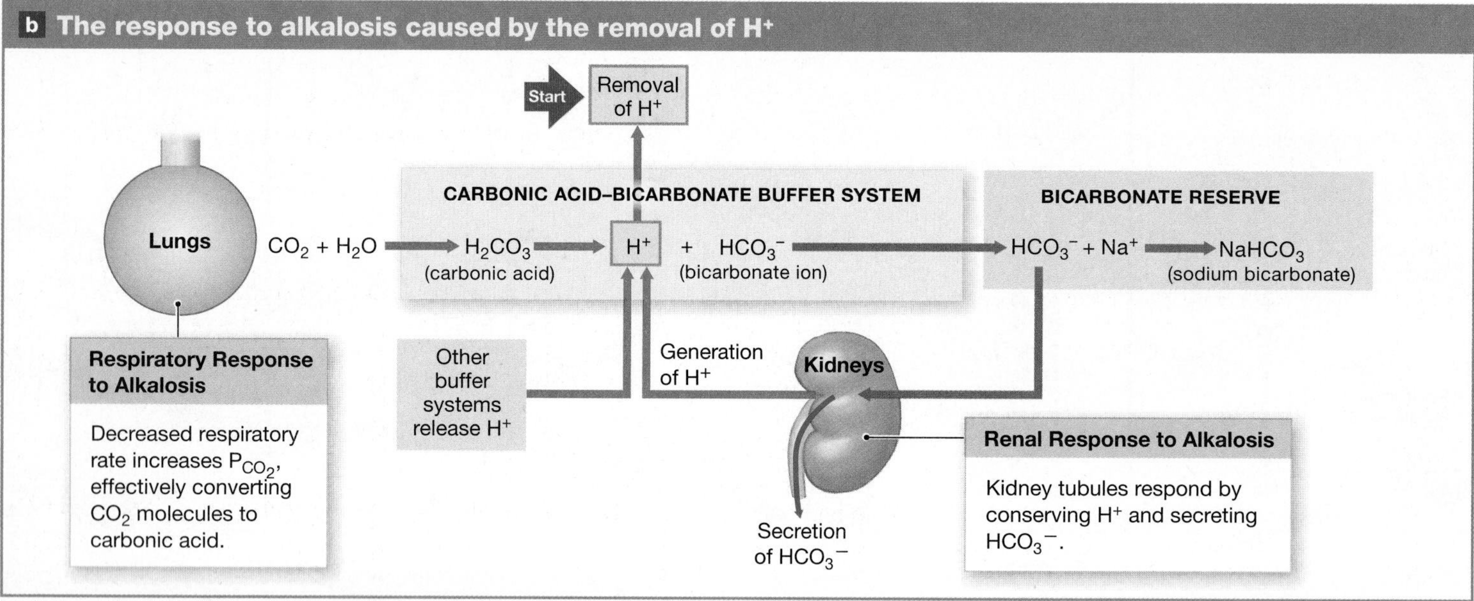

b The response to alkalosis caused by the removal of H^+
Start
Removal of H^+
CARBONIC ACID–BICARBONATE BUFFER SYSTEM
BICARBONATE RESERVE
Lungs
$CO_2 + H_2O$
H_2CO_3 (carbonic acid)
H^+
+ HCO_3^- (bicarbonate ion)
$HCO_3^- + Na^+$
$NaHCO_3$ (sodium bicarbonate)
Respiratory Response to Alkalosis
Decreased respiratory rate increases P_{CO_2}, effectively converting CO_2 molecules to carbonic acid.
Other buffer systems release H^+
Generation of H^+
Kidneys
Renal Response to Alkalosis
Kidney tubules respond by conserving H^+ and secreting HCO_3^-.
Secretion of HCO_3^-

27

? What is the kidney's response to alkalosis caused by the removal of H^+?

When buffering mechanisms are severely stressed, however, the pH drifts outside these limits, producing signs and symptoms of alkalosis or acidosis. More serious and prolonged disturbances of acid-base balance can result with the following conditions:

- *Disorders Affecting Circulating Buffers, Respiratory Performance, or Renal Function.* Several conditions, including *emphysema* and *renal failure,* are associated with dangerous changes in pH. pp. 877, 1014
- *Cardiovascular Conditions.* Conditions such as *heart failure* or *hypotension* can affect the pH of internal fluids by causing fluid shifts and by changing the glomerular filtration rate and respiratory efficiency. pp. 715, 740
- *Conditions Affecting the Central Nervous System.* Neural damage or disease that affects the CNS can affect the respiratory and cardiovascular reflexes that are essential to normal pH regulation.

Serious abnormalities in acid-base balance generally have an initial *acute phase,* in which the pH moves rapidly out of the normal range. If the condition persists, physiological adjustments take place. The person then enters the *compensated phase.* However, unless the underlying problem is corrected, compensation typically cannot return the blood pH to the normal range. Even if the pH is returned to normal, the P_{CO_2} or HCO_3^- concentration can still be abnormal.

The name given to the resulting condition usually indicates the primary source of the problem:

- *Respiratory acid-base disorders* result from a mismatch between CO_2 generation in peripheral tissues and CO_2 excretion by the lungs. When a respiratory acid-base disorder is present, the CO_2 level of the ECF is abnormal.
- *Metabolic acid-base disorders* are caused by the generation of metabolic acids or fixed acids or by conditions affecting the concentration of HCO_3^- in the ECF.

Respiratory compensation alone may restore normal acid-base balance in persons with respiratory acid-base disorders. In contrast, compensation mechanisms for metabolic acid-base disorders may be able to stabilize pH, but other aspects of acid-base balance (buffer system function, bicarbonate ion and P_{CO_2} levels) remain abnormal until the underlying metabolic cause is corrected. Let's look further into both respiratory and metabolic acid-base disorders.

Respiratory Acid-Base Disorders

We can subdivide the respiratory acid-base disturbances into *respiratory acidosis* and *respiratory alkalosis.*

Respiratory Acidosis

Respiratory acidosis develops when the respiratory system cannot eliminate all the CO_2 generated by peripheral tissues. The primary sign is low blood pH due to **hypercapnia**, an increased blood P_{CO_2}. The usual cause is *hypoventilation,* an abnormally low respiratory rate. When the P_{CO_2} in the ECF rises, H^+ and HCO_3^- concentrations also begin rising as H_2CO_3 forms and dissociates. Other buffer systems can tie up some of the H^+, but once the combined buffering capacity has been exceeded, the pH begins to fall rapidly. The effects are diagrammed in Figure 27–15a.

Respiratory acidosis is the most common challenge to acid-base equilibrium. Body tissues generate CO_2 rapidly. Even a few minutes of hypoventilation can cause acidosis, reducing the pH of the ECF to as low as 7.0. Under normal circumstances, the chemoreceptors that monitor the P_{CO_2} of blood and of CSF eliminate the problem by stimulating an increase in breathing rate.

If the chemoreceptors do not respond, if the breathing rate cannot be increased, or if the circulatory supply to the lungs is inadequate, the pH will continue to decrease. If the decline is severe, **acute respiratory acidosis** develops. Acute respiratory acidosis is an immediate, life-threatening condition. It is especially dangerous in people whose tissues are generating large amounts of CO_2, or in those who are incapable of normal respiratory activity. For this reason, the reversal of acute respiratory acidosis is probably the major goal in the resuscitation of cardiac arrest or drowning victims. Thus, first-aid, CPR, and lifesaving courses always stress the "ABCs" of emergency care: *A*irway, *B*reathing, and *C*irculation.

Chronic respiratory acidosis develops when normal respiratory function has been compromised, but the compensatory mechanisms have not failed completely. For example, normal respiratory compensation may not occur in response to chemoreceptor stimulation in persons with CNS injuries and those whose respiratory centers have been desensitized by drugs such as alcohol or barbiturates. As a result, these people are prone to developing acidosis due to chronic hypoventilation.

Even when respiratory centers are intact and functional, damage to some parts of the respiratory system can prevent increased pulmonary exchange. Examples of conditions fostering chronic respiratory acidosis include emphysema, congestive heart failure, and pneumonia (which typically cause alveolar damage or blockage). Pneumothorax and respiratory muscle paralysis have a similar effect, because they, too, limit adequate breathing rate.

When the pulmonary response is abnormal, the kidneys respond by increasing the rate of H^+ secretion into tubular fluid. This response slows the rate of pH change. However, renal mechanisms alone cannot return the pH to normal until the underlying respiratory or circulatory problems are corrected.

The primary problem in respiratory acidosis is that the rate of pulmonary exchange is inadequate to keep the arterial P_{CO_2} within normal limits. Breathing efficiency can typically be improved temporarily by inducing bronchodilation or by using mechanical aids that provide air under positive pressure. If breathing has stopped, artificial respiration or a mechanical ventilator is required. These measures may restore normal pH if the respiratory acidosis was not severe or prolonged. Treatment of acute respiratory acidosis is complicated by the fact that, as we will soon see, it causes a complementary metabolic acidosis as oxygen-starved tissues generate lactate and H^+.

Respiratory Alkalosis

Problems resulting from **respiratory alkalosis** are fairly uncommon (Figure 27–15b). Respiratory alkalosis develops when respiratory activity lowers the blood P_{CO_2} to a below-normal level, a condition called **hypocapnia** (*-capnia,* presence of CO_2). A temporary hypocapnia can be produced by *hyperventilation* when increased respiratory activity leads to a reduction in the arterial P_{CO_2}. Continued hyperventilation can increase the pH to a level as high as 8.0. This condition generally corrects itself, because the reduction in P_{CO_2} halts the stimulation of the chemoreceptors,

Figure 27–15 Homeostatic Regulation of Acid-Base Balance.

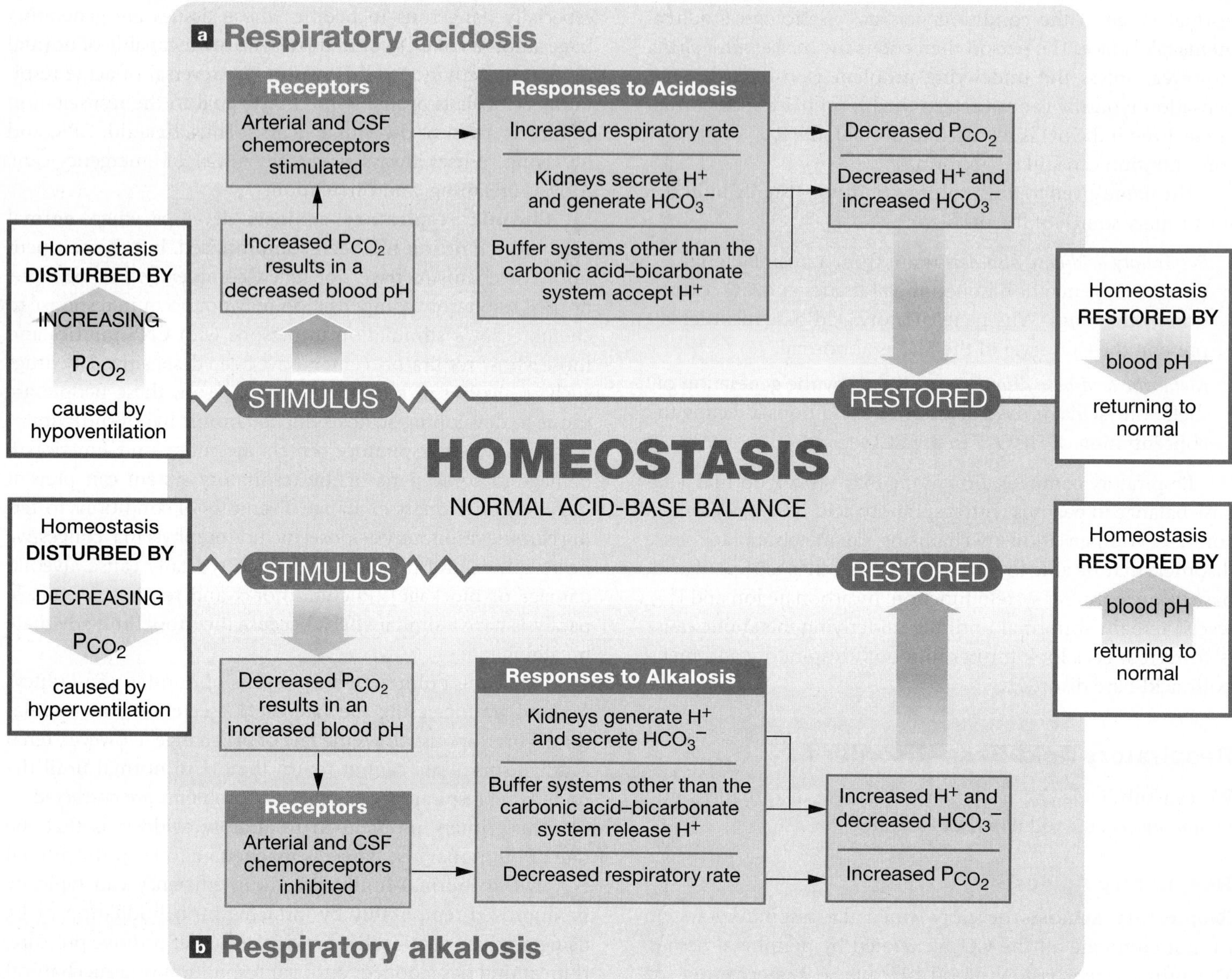
a Respiratory acidosis
Receptors
Arterial and CSF chemoreceptors stimulated
Responses to Acidosis
Increased respiratory rate
Kidneys secrete H^+ and generate HCO_3^-
Buffer systems other than the carbonic acid–bicarbonate system accept H^+
Decreased P_{CO_2}
Decreased H^+ and increased HCO_3^-
Increased P_{CO_2} results in a decreased blood pH
Homeostasis DISTURBED BY INCREASING P_{CO_2} caused by hypoventilation
STIMULUS
RESTORED
Homeostasis RESTORED BY blood pH returning to normal
HOMEOSTASIS
NORMAL ACID-BASE BALANCE
Homeostasis DISTURBED BY DECREASING P_{CO_2} caused by hyperventilation
STIMULUS
RESTORED
Homeostasis RESTORED BY blood pH returning to normal
Decreased P_{CO_2} results in an increased blood pH
Responses to Alkalosis
Kidneys generate H^+ and secrete HCO_3^-
Buffer systems other than the carbonic acid–bicarbonate system release H^+
Decreased respiratory rate
Receptors
Arterial and CSF chemoreceptors inhibited
Increased H^+ and decreased HCO_3^-
Increased P_{CO_2}
b Respiratory alkalosis

27

so the urge to breathe fades until the CO_2 level has returned to normal. Respiratory alkalosis caused by hyperventilation seldom lasts long enough to cause a clinical emergency.

Common causes of hyperventilation include physical stresses such as pain, or psychological stresses such as extreme anxiety. Hyperventilation gradually increases the pH of the CSF, and CNS function is affected. The initial symptoms involve tingling sensations in the hands, feet, and lips. The person may also feel light-headed. If hyperventilation continues, the person may lose consciousness. Unconsciousness removes any contributing psychological stimuli, and the breathing rate decreases. The P_{CO_2} then increases until the pH returns to normal.

A simple treatment for respiratory alkalosis caused by hyperventilation is to have the person rebreathe air exhaled into a small paper bag. As the P_{CO_2} in the bag increases, so do the person's alveolar and arterial CO_2 concentrations. This change eliminates the problem and restores the pH to a normal level. Other problems with respiratory alkalosis are rare and involve primarily (1) persons adapting to high altitudes, where the low P_{O_2} promotes hyperventilation; (2) patients on mechanical respirators; or (3) persons whose brainstem injuries render them incapable of responding to shifts in the plasma CO_2 concentration.

Metabolic Acid-Base Disorders

We can divide the metabolic acid-base disturbances into *metabolic acidosis* and *metabolic alkalosis*.

Metabolic Acidosis

Metabolic acidosis is the second most common type of acid-base imbalance (after respiratory acidosis). It has the following three major causes:

- The most widespread cause of metabolic acidosis is the production of a large number of fixed acids or metabolic acids. The H^+ released by these acids overload the carbonic acid–bicarbonate ion buffer system, so the pH begins to decrease (Figure 27–16a). We considered two examples of metabolic acidosis earlier:
  - **Lactic acidosis** can develop after strenuous exercise or prolonged tissue *hypoxia* (oxygen starvation) as cells rely on anaerobic metabolism (look back at Figure 10–20, p. 321).
  - **Ketoacidosis** results from the generation of large quantities of ketone bodies during the postabsorptive state of metabolism. Ketoacidosis is a problem in starvation, and a potentially lethal complication of poorly controlled diabetes mellitus. In either case, peripheral tissues cannot obtain adequate glucose from the bloodstream and they begin metabolizing lipids and ketone bodies.

Figure 27–16 Responses to Metabolic Acidosis.

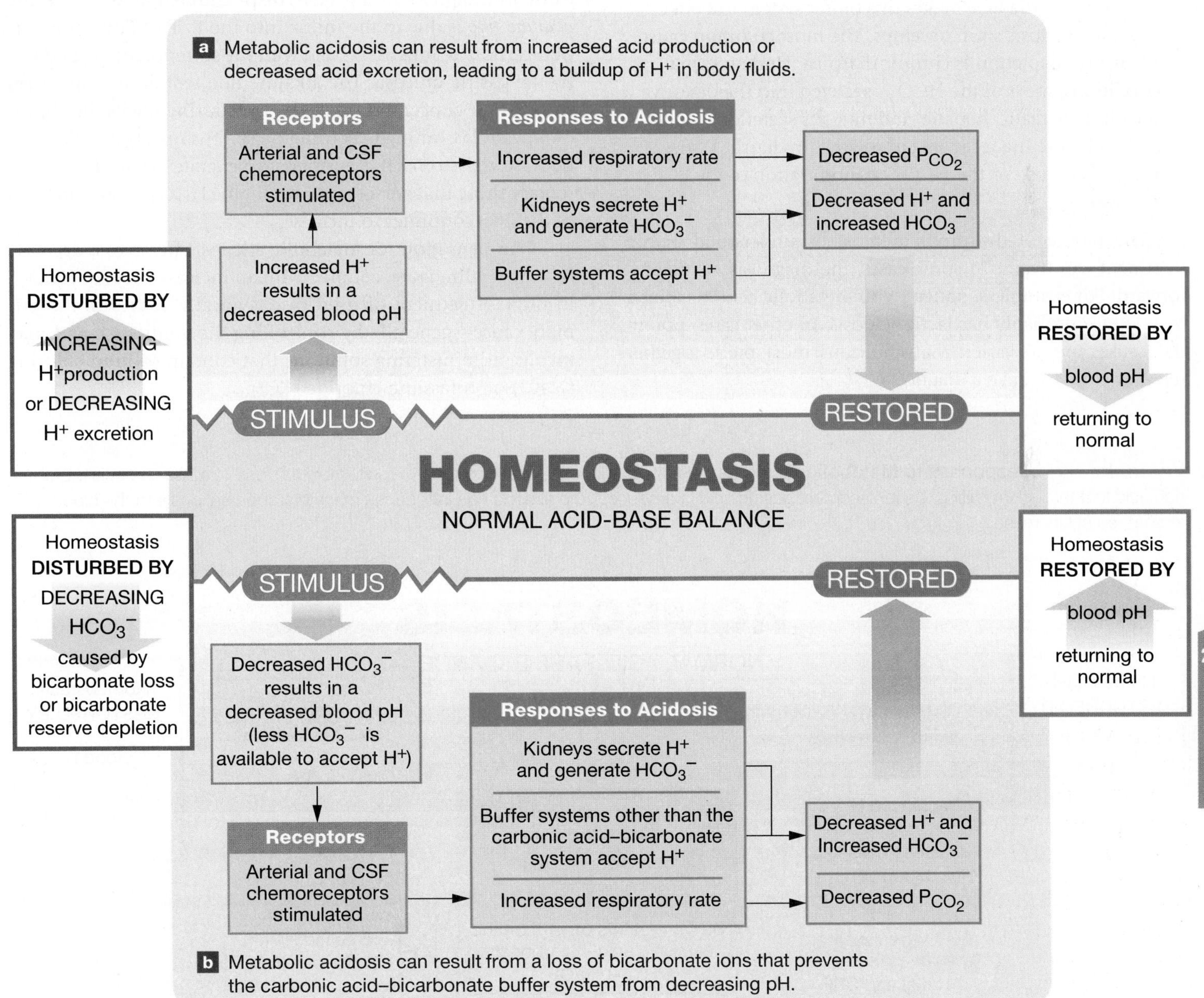
a Metabolic acidosis can result from increased acid production or decreased acid excretion, leading to a buildup of H+ in body fluids.
Receptors
Arterial and CSF chemoreceptors stimulated
Responses to Acidosis
Increased respiratory rate
Kidneys secrete H+ and generate HCO3−
Buffer systems accept H+
Decreased PCO2
Decreased H+ and increased HCO3−
Increased H+ results in a decreased blood pH
Homeostasis DISTURBED BY INCREASING H+production or DECREASING H+ excretion
STIMULUS
RESTORED
Homeostasis RESTORED BY blood pH returning to normal
HOMEOSTASIS
NORMAL ACID-BASE BALANCE
Homeostasis DISTURBED BY DECREASING HCO3− caused by bicarbonate loss or bicarbonate reserve depletion
STIMULUS
RESTORED
Homeostasis RESTORED BY blood pH returning to normal
Decreased HCO3− results in a decreased blood pH (less HCO3− is available to accept H+)
Receptors
Arterial and CSF chemoreceptors stimulated
Responses to Acidosis
Kidneys secrete H+ and generate HCO3−
Buffer systems other than the carbonic acid–bicarbonate system accept H+
Increased respiratory rate
Decreased H+ and Increased HCO3−
Decreased PCO2
b Metabolic acidosis can result from a loss of bicarbonate ions that prevents the carbonic acid–bicarbonate buffer system from decreasing pH.

? How do the lungs respond to metabolic acidosis?

27

- A less common cause of metabolic acidosis is an impaired ability to excrete H^+ by the kidneys (see Figure 27–16a). For example, conditions marked by severe kidney damage, such as *glomerulonephritis*, typically result in severe metabolic acidosis. p. 984 Diuretics that "turn off" the sodium–hydrogen transport system in the kidney tubules also cause metabolic acidosis. The secretion of H^+ is linked to the reabsorption of Na^+. When Na^+ reabsorption stops, so does H^+ secretion.
- Metabolic acidosis occurs after severe bicarbonate ion loss (Figure 27–16b). The carbonic acid–bicarbonate ion buffer system relies on bicarbonate ions to balance hydrogen ions that threaten pH balance. A decrease in the HCO_3^- concentration in the ECF makes this buffer system less effective, and acidosis soon develops. The most common cause of HCO_3^- depletion is chronic diarrhea. Under normal conditions, most of the HCO_3^- secreted into the digestive tract in pancreatic, hepatic, and mucous secretions are reabsorbed before the feces are eliminated. In diarrhea, these HCO_3^- are lost, so the HCO_3^- concentration of the ECF decreases.

The nature of the problem must be understood before treatment can begin. In some cases, the diagnosis is straightforward. For example, a patient with metabolic acidosis after a bicycle race probably has lactic acidosis. In other cases, potential causes are so varied that clinicians must piece together relevant clues to make a diagnosis.

Compensation for metabolic acidosis generally involves a combination of respiratory and renal mechanisms. Hydrogen ions interacting with bicarbonate ions form CO_2 molecules that are eliminated by the lungs. The kidneys excrete additional H^+ into the urine and generate HCO_3^- that are released into the ECF.

Metabolic Alkalosis

Metabolic alkalosis occurs when the HCO_3^- concentration is elevated (Figure 27–17). The HCO_3^- then interact with H^+ in solution, forming H_2CO_3. The resulting decrease in H^+ concentration causes signs of alkalosis.

Metabolic alkalosis is rare, but we noted one interesting cause in Chapter 24. p. 904 The phenomenon known as the *alkaline tide* is due to the influx into the ECF of large numbers of bicarbonate ions associated with the secretion of H^+ and Cl^- by the gastric mucosa. The alkaline tide temporarily increases the HCO_3^- concentration in the ECF during meals. But bouts of repeated vomiting may bring on serious metabolic alkalosis, because the stomach continues to generate stomach acids to replace those that are lost. As a result, the HCO_3^- concentration of the ECF continues to increase.

Compensation for metabolic alkalosis involves a decrease in the breathing rate, coupled with an increased loss of HCO_3^- in urine. Treatment of mild cases typically addresses the primary cause—generally by controlling the vomiting—and may involve administering solutions that contain sodium chloride (NaCl) or potassium chloride (KCl).

Figure 27–17 Responses to Metabolic Alkalosis. Metabolic alkalosis most commonly results from the loss of acids, especially stomach acid lost through vomiting. As replacement gastric acid is produced, the alkaline tide introduces many bicarbonate ions into the bloodstream, so pH increases.

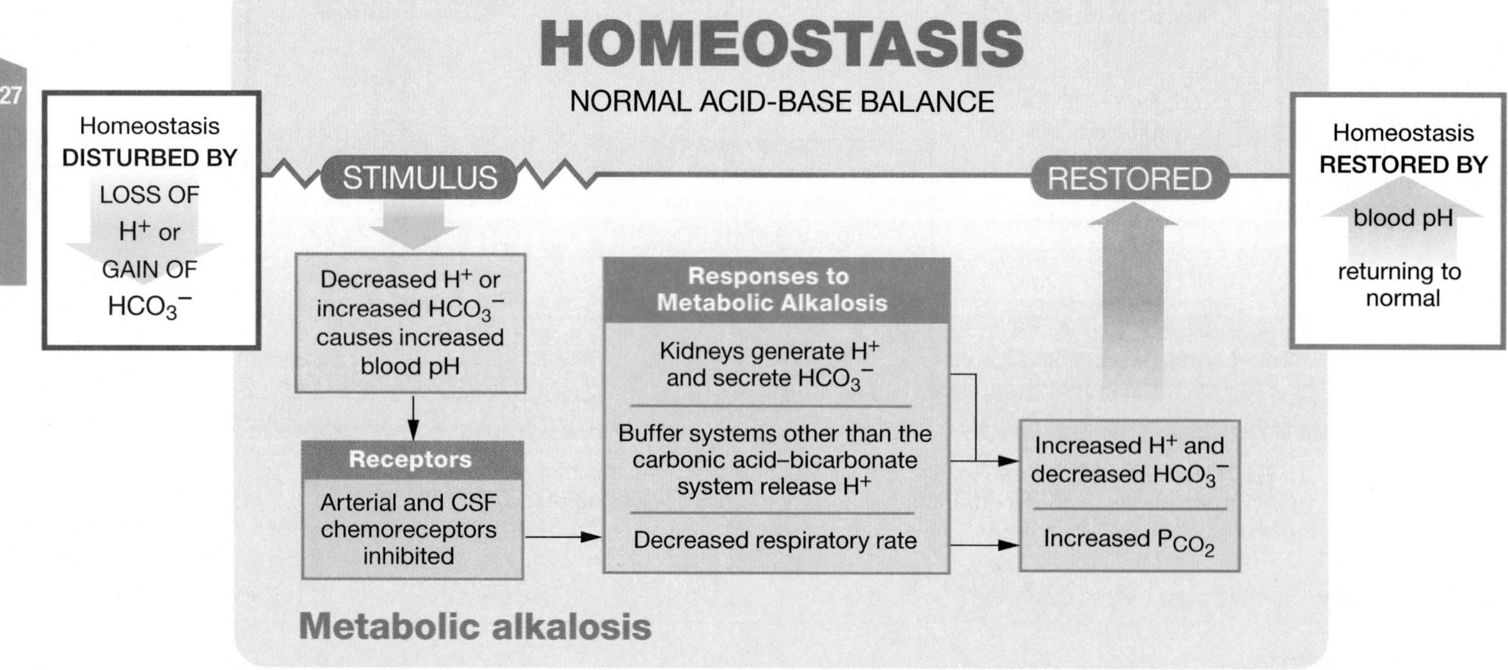
HOMEOSTASIS
NORMAL ACID-BASE BALANCE
Homeostasis DISTURBED BY
LOSS OF H+ or GAIN OF HCO3−
STIMULUS
RESTORED
Homeostasis RESTORED BY
blood pH returning to normal
Decreased H+ or increased HCO3− causes increased blood pH
Receptors
Arterial and CSF chemoreceptors inhibited
Responses to Metabolic Alkalosis
Kidneys generate H+ and secrete HCO3−
Buffer systems other than the carbonic acid–bicarbonate system release H+
Decreased respiratory rate
Increased H+ and decreased HCO3−
Increased PCO2
Metabolic alkalosis

27

Table 27–4 Changes in Blood Chemistry Associated with the Major Classes of Acid-Base Disorders

Disorder	pH (normal 7.35–7.45)	HCO_3^- (mEq/L) (normal = 21–28)	P_{CO_2} (mm Hg) (normal = 35–45)	Remarks	Treatments
Respiratory acidosis	Decreased (below 7.35)	Acute: normal Compensated: increased (above 28)	Increased (above 45)	Generally caused by hypoventilation and CO_2 buildup in tissues and blood	Improve ventilation; in some cases, with bronchodilation and mechanical assistance
Metabolic acidosis	Decreased (below 7.35)	Decreased (below 24)	Acute: normal Compensated: decreased (below 35)	Caused by buildup of metabolic or fixed acid, impaired H^+ elimination at kidneys, or HCO_3^- loss in urine or feces	Administration of bicarbonate (gradual), with other steps as needed to correct primary cause
Respiratory alkalosis	Increased (above 7.45)	Acute: normal Compensated: decreased (below 24)	Decreased (below 35)	Generally caused by hyperventilation and reduction in plasma CO_2 levels	Reduce respiratory rate, allow rise in P_{CO_2}
Metabolic alkalosis	Increased (above 7.45)	Increased (above 28)	Increased (above 45)	Generally caused by prolonged vomiting and associated acid loss	pH below 7.55: no treatment; pH above 7.55: may require administration of NH_4Cl

Treatment of acute cases of metabolic alkalosis may involve administering ammonium chloride (NH_4Cl). Metabolism of the ammonium ion in the liver liberates an H^+, so in effect the introduction of NH_4Cl leads to the internal generation of H^+ and Cl^-. As the H^+ diffuses into the bloodstream, the pH falls toward a normal level.

For a summary of the patterns that characterize the four major categories of acid-base disorders, please see Table 27–4.

Combined Respiratory and Metabolic Acidosis

Respiratory acidosis and metabolic acidosis are typically linked, because oxygen-starved tissues generate large quantities of lactic acid (lactate and H^+) and because sustained hypoventilation leads to decreased arterial P_{O_2}. The problem can be especially serious in cases of near drowning, in which body fluids have high P_{CO_2}, low P_{O_2}, and a large amount of lactic acid generated by the muscles of the struggling person. Prompt emergency treatment is essential. The usual procedure involves some form of artificial or mechanical respiratory assistance, coupled with intravenous infusion of an isotonic solution that contains sodium lactate, sodium gluconate, or sodium bicarbonate.

The Detection of Acidosis and Alkalosis

Spotlight Figure 27–18 shows the process of diagnosing acid-base disorders. Additional steps can help in identifying possible causes of the problem and in distinguishing compensated from uncompensated conditions. These steps may include determining the *anion gap* or using a diagnostic chart.

Checkpoint

15. **How would a prolonged fast affect the body's pH?**
16. **Why can prolonged vomiting produce metabolic alkalosis?**

See the blue Answers tab at the back of the book.

27-7 Aging affects fluid, electrolyte, and acid-base balance

Learning Outcome **Describe the effects of aging on fluid, electrolyte, and acid-base balance.**

Fetuses and infants have very different requirements for maintaining fluid, electrolyte, and acid-base balance than do adults. A fetus obtains the water, organic nutrients, and electrolytes it needs from the maternal bloodstream. Buffers in the fetal bloodstream provide short-term pH control, and the maternal kidneys eliminate the H^+ generated. A newborn's body water content is high: At birth, water makes up about 75 percent of body weight, compared with 50–60 percent in adults. Basic aspects of electrolyte balance are the same in newborns as in adults, but the effects of fluctuations in the diet are much more immediate in newborns because their reserves of minerals and energy sources are much smaller.

The descriptions of fluid, electrolyte, and acid-base balance in this chapter were based on the responses of healthy adults under age 40. Aging affects many aspects of fluid, electrolyte, and acid-base balance, including the following:

- Total body water content gradually decreases with age. Between ages 40 and 60, average total body water content declines slightly, to 55 percent for males and 47 percent for females. After age 60, the values decline further, to about 50 percent for males and 45 percent for females. Among other effects, each decrease reduces the dilution of wastes, toxins, and any drugs that have been administered.
- A decrease in the glomerular filtration rate and in the number of functional nephrons reduces the body's ability to regulate pH through renal compensation.
- The body's ability to concentrate urine declines, so more water is lost in urine. In addition, the rate of insensible perspiration increases as the skin becomes thinner and more delicate. Maintaining fluid balance therefore requires a

SPOTLIGHT

Figure 27–18

The Diagnosis of Acid-Base Disorders

ACID-BASE DISORDERS

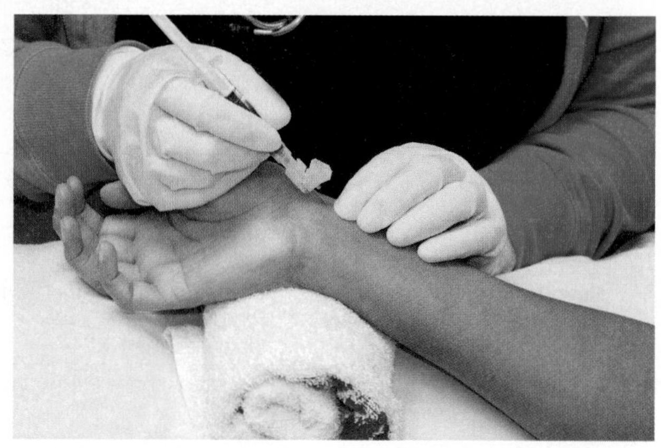

Virtually anyone who has a problem that affects the cardiovascular, respiratory, urinary, digestive, or nervous system may develop potentially dangerous acid-base imbalances. For this reason, an arterial blood gas test is done to evaluate the body's acid-base balance. Standard tests measure arterial blood pH, partial pressure of carbon dioxide (P_{CO_2}), partial pressure of oxygen (P_{O_2}), and bicarbonate (HCO_3^-) levels. These measurements make diagnosing acid-base disorders relatively straightforward. The flowchart below shows the diagnostic steps used to determine whether a person has an acid-base disorder, and to further determine what type of acidosis or alkalosis exists.

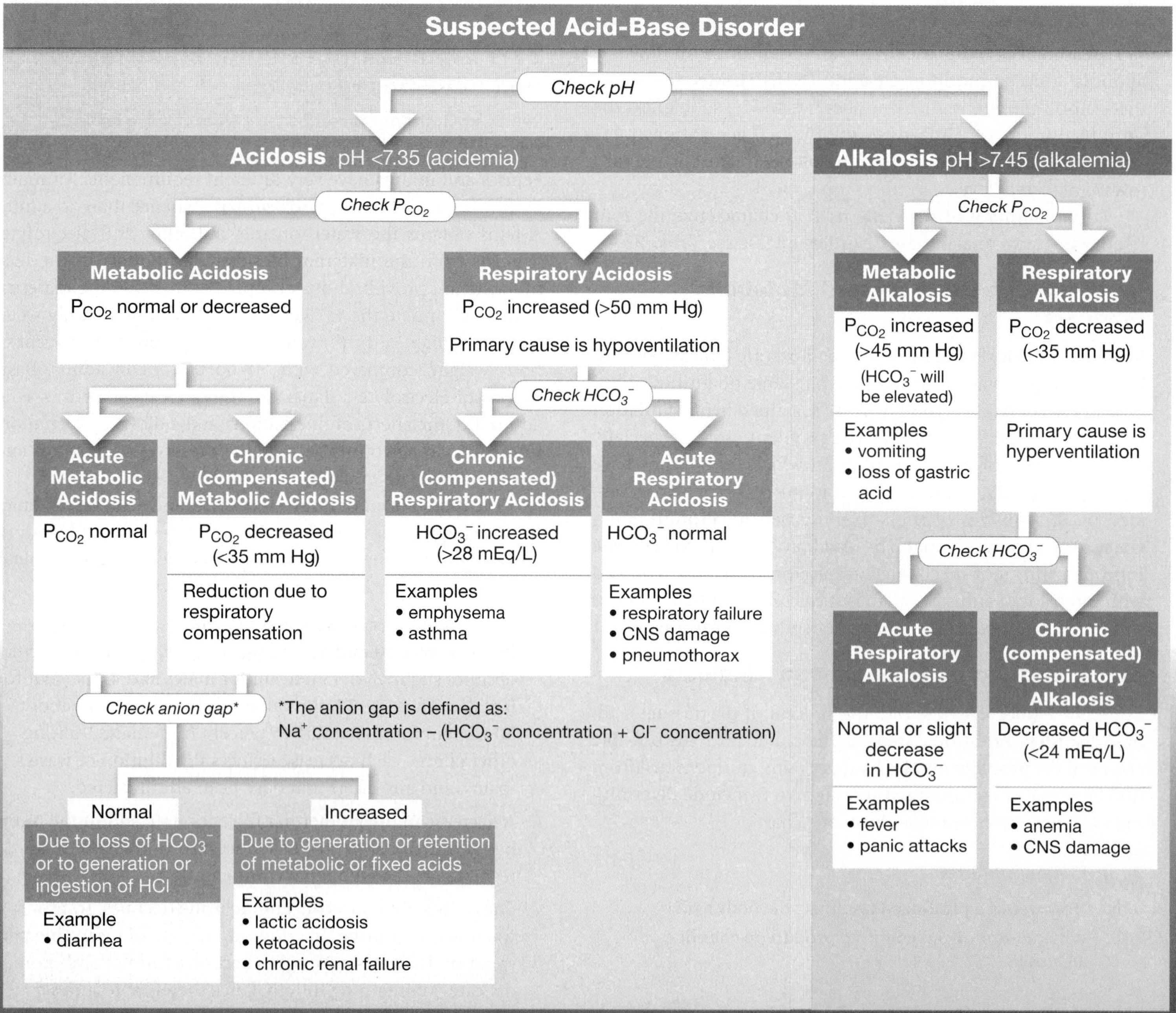

Suspected Acid-Base Disorder

Check pH

Acidosis pH <7.35 (acidemia)

Check P_{CO_2}

Metabolic Acidosis

P_{CO_2} normal or decreased

Acute Metabolic Acidosis

P_{CO_2} normal

Chronic (compensated) Metabolic Acidosis

P_{CO_2} decreased (<35 mm Hg)

Reduction due to respiratory compensation

*Check anion gap**

*The anion gap is defined as:
Na^+ concentration – (HCO_3^- concentration + Cl^- concentration)

Normal

Due to loss of HCO_3^- or to generation or ingestion of HCl

Example
- diarrhea

Increased

Due to generation or retention of metabolic or fixed acids

Examples
- lactic acidosis
- ketoacidosis
- chronic renal failure

Respiratory Acidosis

P_{CO_2} increased (>50 mm Hg)

Primary cause is hypoventilation

Check HCO_3^-

Chronic (compensated) Respiratory Acidosis

HCO_3^- increased (>28 mEq/L)

Examples
- emphysema
- asthma

Acute Respiratory Acidosis

HCO_3^- normal

Examples
- respiratory failure
- CNS damage
- pneumothorax

Alkalosis pH >7.45 (alkalemia)

Check P_{CO_2}

Metabolic Alkalosis

P_{CO_2} increased (>45 mm Hg)

(HCO_3^- will be elevated)

Examples
- vomiting
- loss of gastric acid

Respiratory Alkalosis

P_{CO_2} decreased (<35 mm Hg)

Primary cause is hyperventilation

Check HCO_3^-

Acute Respiratory Alkalosis

Normal or slight decrease in HCO_3^-

Examples
- fever
- panic attacks

Chronic (compensated) Respiratory Alkalosis

Decreased HCO_3^- (<24 mEq/L)

Examples
- anemia
- CNS damage

higher daily water intake. A reduction in ADH and aldosterone sensitivity makes older people less able than younger people to conserve body water when losses exceed gains.

- Many people over age 60 experience a net loss in body mineral content as muscle mass and skeletal mass decrease. This loss can be prevented, at least in part, by a combination of exercise and an increased dietary mineral supply.
- The reduction in vital capacity that accompanies aging reduces the body's ability to perform respiratory compensation, increasing the risk of respiratory acidosis. This problem can be compounded by arthritis, which can reduce vital capacity by limiting rib movements, and by emphysema, another condition that, to some degree, develops with aging.
- Disorders affecting major systems become more common with increasing age. Most, if not all, of these disorders have some effect on fluid, electrolyte, and/or acid-base balance.

✓ Checkpoint

17. **As a person ages, the glomerular filtration rate and the number of functional nephrons decreases. What effect would these changes have on pH regulation?**
18. **After the age of 40, does the total body water content increase or decrease?**

See the blue Answers tab at the back of the book.

27 Chapter Review

Study Outline

MasteringA&P™ Access more chapter study tools online in the MasteringA&P Study Area:

- Chapter Quizzes, Chapter Practice Test, MP3 Tutor Sessions, and Clinical Case Studies
- Practice Anatomy Lab PAL 3.0
- A&P Flix A&PFlix
- Interactive Physiology iP2
- PhysioEx PhysioEx 9.1

An Introduction to Fluid, Electrolyte, and Acid-base Balance p. 1022

1. The distribution of gases, nutrients, and wastes in the body are dependent on the homeostatic mechanisms that regulate volume, ion concentrations, and pH of body fluids.

27-1 Fluid balance, electrolyte balance, and acid-base balance are interrelated and essential to homeostasis p. 1022

2. Maintaining normal volume and composition of extracellular and intracellular fluids is vital to life. Three types of homeostasis are involved: **fluid balance**, **electrolyte balance**, and **acid-base balance**.

27-2 Extracellular fluid (ECF) and intracellular fluid (ICF) are fluid compartments with differing solute concentrations that are closely regulated p. 1023

3. The **intracellular fluid (ICF)** contains nearly two-thirds of the total body water. The **extracellular fluid (ECF)** contains the rest. Exchange occurs between the ICF and the ECF, but the two **fluid compartments** retain their distinctive characteristics. *(Figures 27–1, 27–2)*
4. Homeostatic mechanisms that monitor and adjust the composition of body fluids respond to changes in the ECF.
5. No receptors directly monitor fluid or electrolyte balance. Receptors involved in fluid balance and in electrolyte balance respond to changes in plasma volume and osmotic concentration.
6. Body cells cannot move water molecules by active transport. All movements of water across plasma membranes and epithelia occur passively, in response to osmotic gradients.
7. The body's content of water or electrolytes will increase if intake exceeds outflow and will decrease if losses exceed gains.
8. ADH promotes water reabsorption by the kidneys and stimulates thirst. Aldosterone increases the rate of sodium reabsorption by the kidneys. Natriuretic peptides (ANP and BNP) oppose those actions and promote fluid and electrolyte losses in urine.
9. The regulatory mechanisms of fluid balance and electrolyte balance are quite different, and the distinction is clinically important.

27-3 Fluid balance involves the regulation and distribution of water gains and losses p. 1027

10. Water circulates freely within the ECF compartment.
11. Water losses are normally balanced by gains through eating, drinking, and **metabolic generation**. *(Figure 27–3; Table 27–1)*
12. Water movement between the ECF and ICF is called a **fluid shift**. If the ECF becomes hypertonic relative to the ICF, water will move from the ICF into the ECF until osmotic equilibrium has been restored. If the ECF becomes hypotonic relative to the ICF, water will move from the ECF into the cells, and the volume of the ICF will increase. *(Figure 27–4)*

27-4 In electrolyte balance, the concentrations of sodium, potassium, calcium, magnesium, phosphate, and chloride ions in body fluids are tightly regulated p. 1030

13. An **equivalent** (Eq) is the amount of a positive ion or negative ion that supplies 1 mole of electrical charge, and 1 equivalent = 1000 milliequivalents (mEq).

14. Problems with electrolyte balance generally result from a mismatch between gains and losses of sodium. Problems with potassium balance are less common, but more dangerous.
15. The rate of sodium uptake across the digestive epithelium is directly proportional to the amount of sodium in the diet. Sodium losses occur mainly in urine and through perspiration. *(Figure 27–5)*
16. Shifts in sodium balance result in expansion or contraction of the ECF. Large variations in ECF volume are corrected by homeostatic mechanisms triggered by changes in blood volume. If the volume becomes too low, ADH and aldosterone are secreted; if the volume becomes too high, natriuretic peptides (ANP and BNP) are secreted. *(Figure 27–6)*
17. Potassium ion concentrations in the ECF are very low and not as closely regulated as are sodium ion concentrations. Potassium excretion increases as ECF concentrations rise, under aldosterone stimulation, and when the pH increases. Potassium retention occurs when the pH decreases. *(Figure 27–7)*
18. ECF concentrations of other electrolytes, such as calcium, magnesium, phosphate, and chloride, are also regulated. *(Table 27–2)*

27-5 In acid-base balance, buffer systems as well as respiratory and renal compensation regulate pH changes in body fluids p. 1036

19. Acids and bases are either **strong** or **weak**. *(Table 27–3)*
20. The pH of normal body fluids ranges from 7.35 to 7.45. Variations outside this relatively narrow range produce symptoms of **acidosis** or **alkalosis**.
21. **Volatile acids** can leave solution and enter the atmosphere. **Fixed acids** remain in body fluids until excreted by the kidneys. **Metabolic acids** are participants in, or are by-products of, cellular metabolism. *(Figure 27–8)*
22. Carbonic acid, a volatile acid, is the most important factor affecting the pH of the ECF. In solution, CO_2 reacts with water to form carbonic acid. An inverse relationship exists between pH and the partial pressure of CO_2. *(Figure 27–9)*
23. Sulfuric acid and phosphoric acid, the most important fixed acids, are generated during the catabolism of amino acids and compounds containing phosphate groups.
24. Metabolic acids include metabolic products such as lactic acid and ketone bodies.
25. A **buffer system** typically consists of a weak acid and the anion released by its dissociation. The anion functions as a weak base that can absorb H^+. The three major buffer systems are (1) the **phosphate buffer system** in the ICF and urine; (2) **protein buffer systems** in the ECF and ICF; and (3) the **carbonic acid–bicarbonate buffer system**, most important in the ECF. *(Figure 27–10)*
26. The phosphate buffer system plays a supporting role in regulating the pH of the ECF, but it is important in buffering the pH of the ICF and of urine.
27. In protein buffer systems, the initial, final, and R groups of component amino acids respond to changes in pH by accepting or releasing hydrogen ions. The **hemoglobin buffer system** is a protein buffer system that helps prevent drastic changes in pH when the P_{CO_2} is increasing or decreasing. *(Figure 27–11)*
28. The carbonic acid–bicarbonate buffer system prevents pH changes caused by metabolic acids and fixed acids in the ECF. The readily available supply of bicarbonate ions is the **bicarbonate reserve**. *(Figure 27–12)*
29. The lungs help regulate pH by affecting the carbonic acid–bicarbonate buffer system. A change in respiratory rate can raise or lower the P_{CO_2} of body fluids, affecting the body's buffering capacity. This process is called **respiratory compensation**.
30. In **renal compensation**, the kidneys vary their rates of hydrogen ion secretion and bicarbonate ion reabsorption, depending on the pH of the ECF. *(Figure 27–13)*

27-6 Disorders of acid-base balance can be classified as respiratory or metabolic p. 1042

31. Interactions among buffer systems, respiration, and renal function normally maintain tight control of the pH of the ECF, generally within a range of 7.35–7.45. *(Figure 27–14)*
32. *Respiratory acid-base disorders* result when abnormal respiratory function causes an extreme increase or decrease in CO_2 levels in the ECF. *Metabolic acid-base disorders* are caused by the generation of metabolic or fixed acids or by conditions affecting the concentration of bicarbonate ions in the ECF.
33. **Respiratory acidosis** results from excessive levels of CO_2 in body fluids. *(Figure 27–15a)*
34. **Respiratory alkalosis** is a relatively rare condition associated with hyperventilation. *(Figure 27–15b)*
35. **Metabolic acidosis** results from the depletion of the bicarbonate reserve, caused by an inability to excrete hydrogen ions by the kidneys, the production of large numbers of fixed and metabolic acids, or bicarbonate loss that accompanies chronic diarrhea. *(Figure 27–16)*
36. **Metabolic alkalosis** results when bicarbonate ion concentrations become elevated, as occurs during extended periods of vomiting. *(Figure 27–17)*
37. An arterial blood gas test evaluates blood pH, P_{CO_2}, P_{O_2}, and HCO_3^- levels to recognize and classify acidosis and alkalosis as either respiratory or metabolic in nature. *(Spotlight Figure 27–18; Table 27–4)*

27-7 Aging affects fluid, electrolyte, and acid-base balance p. 1049

38. Age-related changes that affect fluid, electrolyte, and acid-base balance include (1) reduced total body water content, (2) impaired ability to perform renal compensation, (3) increased water demands due to decreased ability to concentrate urine and reduced sensitivity to ADH and aldosterone, (4) a net loss of minerals, (5) reductions in respiratory efficiency that affect respiratory compensation, and (6) increased incidence of conditions that secondarily affect fluid, electrolyte, or acid-base balance.

Review Questions

See the blue Answers tab at the back of the book.

LEVEL 1 Reviewing Facts and Terms

1. The primary components of the ECF are **(a)** lymph and cerebrospinal fluid, **(b)** plasma and serous fluids, **(c)** interstitial fluid and plasma, **(d)** all of these.
2. The principal anions in the ICF are **(a)** phosphate and proteins (Pr^-), **(b)** phosphate and bicarbonate, **(c)** sodium and chloride, **(d)** sodium and potassium.
3. Osmoreceptors in the hypothalamus monitor the osmotic concentration of the ECF and secrete __________ in response to higher osmotic concentrations. **(a)** BNP, **(b)** ANP, **(c)** aldosterone, **(d)** ADH.
4. Write the missing names and molecular formulas for the following reactions between the carbonic acid–bicarbonate buffer system and the bicarbonate reserve.

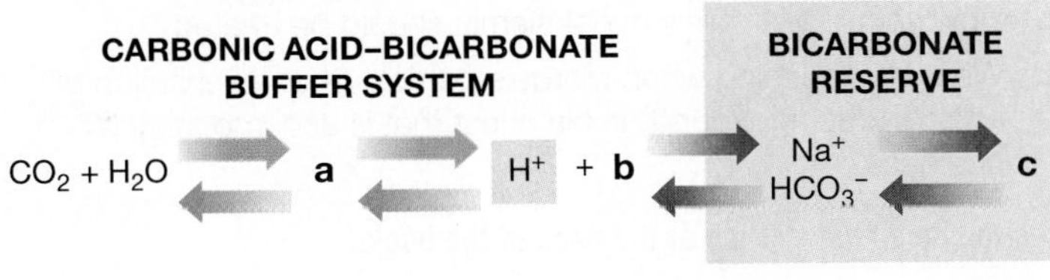
CARBONIC ACID–BICARBONATE BUFFER SYSTEM
BICARBONATE RESERVE
$CO_2 + H_2O$
a
H^+ + b
Na^+
HCO_3^-
c

(a) __________
(b) __________
(c) __________

5. Calcium homeostasis primarily reflects **(a)** a balance between absorption in the gut and excretion by the kidneys, **(b)** careful regulation of the blood calcium level by the kidneys, **(c)** an interplay between parathyroid hormone and aldosterone, **(d)** an interplay among reserves in the bones, the rate of absorption, and the rate of excretion, **(e)** hormonal control of calcium reserves in the bones.
6. The *most* important factor affecting the pH of body tissues is the concentration of **(a)** lactic acid, **(b)** ketone bodies, **(c)** metabolic acids, **(d)** carbon dioxide, **(e)** hydrochloric acid.
7. Changes in the pH of body fluids are compensated for by all of the following *except* **(a)** an increase in urine output, **(b)** the carbonic acid–bicarbonate buffer system, **(c)** the phosphate buffer system, **(d)** changes in the rate and depth of breathing, **(e)** protein buffers.
8. Respiratory acidosis develops when the blood pH is **(a)** increased due to a decreased blood P_{CO_2} level, **(b)** decreased due to an increased blood P_{CO_2} level, **(c)** increased due to an increased blood P_{CO_2} level, **(d)** decreased due to a decreased blood P_{CO_2} level.
9. Metabolic alkalosis occurs when **(a)** bicarbonate ion concentrations become elevated, **(b)** a severe bicarbonate loss occurs, **(c)** the kidneys fail to excrete hydrogen ions, **(d)** ketone bodies are generated in abnormally large quantities.
10. Identify four hormones that mediate major physiological adjustments affecting fluid and electrolyte balance. What are the primary effects of each hormone?

LEVEL 2 Reviewing Concepts

11. Drinking a solution hypotonic to the ECF causes the ECF to **(a)** increase in volume and become hypertonic to the ICF, **(b)** decrease in volume and become hypertonic to the ICF, **(c)** decrease in volume and become hypotonic to the ICF, **(d)** increase in volume and become hypotonic to the ICF.
12. The osmotic concentration of the ECF decreases if an individual gains water without a corresponding **(a)** gain of electrolytes, **(b)** loss of water, **(c)** fluid shift from the ECF to the ICF, **(d)** all of these.
13. When the pH of body fluids begins to *decrease*, free amino acids and proteins will **(a)** release a hydrogen from the carboxyl group, **(b)** release a hydrogen from the amino group, **(c)** release a hydrogen at the carboxyl group, **(d)** bind a hydrogen at the amino group.
14. In a protein buffer system, if the pH increases, **(a)** the protein acquires a hydrogen ion from carbonic acid, **(b)** hydrogen ions are buffered by hemoglobin molecules, **(c)** a hydrogen ion is released and a carboxylate ion is formed, **(d)** a chloride shift occurs.
15. Differentiate among fluid balance, electrolyte balance, and acid-base balance, and explain why each is important to homeostasis.
16. What are fluid shifts? What is their function, and what factors can cause them?
17. Why should a person with a fever drink plenty of fluids?
18. Define and give an example of (a) a volatile acid, (b) a fixed acid, and (c) a metabolic acid. Which represents the greatest threat to acid-base balance? Why?
19. What are the three major buffer systems in body fluids? How does each system work?
20. How do respiratory and renal mechanisms support the buffer systems?
21. Differentiate between respiratory compensation and renal compensation.
22. Distinguish between respiratory and metabolic disorders that disturb acid-base balance.
23. What is the difference between metabolic acidosis and respiratory acidosis? What can cause these conditions?
24. The most recent advice from medical and nutritional experts is to monitor one's intake of salt so that it does not exceed the amount needed to maintain a constant ECF volume. What effect does excessive salt ingestion have on blood pressure?
25. Exercise physiologists recommend that adequate amounts of fluid be ingested before, during, and after exercise. Why is fluid replacement during extensive sweating important?

LEVEL 3 Critical Thinking and Clinical Applications

26. After falling into an abandoned stone quarry filled with water and nearly drowning, a young boy is rescued. In assessing his condition, rescuers find that his body fluids have high P_{CO_2} and lactate levels, and low P_{O_2} levels. Identify the underlying problem and recommend the necessary treatment to restore homeostatic conditions.
27. Dan has been lost in the desert for 2 days with very little water. As a result of this exposure, you would expect to observe which of the following? **(a)** increased ADH levels, **(b)** decreased blood osmolarity, **(c)** normal urine production, **(d)** increased blood volume, **(e)** cells enlarged with fluid.
28. Mary, a nursing student, has been caring for burn patients. She notices that they consistently show elevated levels of potassium in their urine and wonders why. What would you tell her?
29. While visiting a foreign country, Milly inadvertently drinks some water, even though she had been advised not to. She contracts an intestinal disease that causes severe diarrhea. How would you expect her condition to affect her blood pH, urine pH, and pattern of ventilation?
30. Yuka is dehydrated, so her physician prescribes intravenous fluids. The attending nurse becomes distracted and erroneously gives Yuka a hypertonic glucose solution instead of normal saline. What effect will this mistake have on Yuka's plasma ADH levels and urine volume?
31. Refer to the diagnostic flowchart in **Spotlight Figure 27–18**. Use information from the blood test results in the accompanying table to categorize the suspected acid-base disorders of the patients represented in the table.

Results	Patient 1	Patient 2	Patient 3	Patient 4
pH	7.5	7.2	7.0	7.7
P_{CO_2} (mm Hg)	32	45	60	50
Na^- (mEq/L)	138	140	140	136
HCO_3^- (mEq/L)	22	20	28	34
Cl^- (mEq/L)	106	102	101	91
Anion gap* (mEq/L)	10	18	12	11

*Anion gap = Na^+ concentration − (HCO_3^- concentration + Cl^- concentration).

CLINICAL CASE Wrap-Up When Treatment Makes You Worse

"You're having trouble with your potassium," the doctor tells Marquis.

Marquis had been in a state of electrolyte balance, in which the rates of gain and loss for each electrolyte are equal every day, but he had high blood pressure before beginning treatment. The hydrochlorothiazide diuretic he is taking works by inhibiting reabsorption of Na^+ and K^+ in the distal convoluted tubules. Recall that "water follows salt." The increased excretion of Na^+ and K^+ causes increased excretion of water, which reduces blood volume and therefore decreases pressure on the arterial walls. Ninety-eight percent of the body's K^+ is in the ICF, leaving only a small amount in the ECF that is critically controlled. The fecal loss of K^+ is usually negligible, but diarrhea causes significant K^+ loss. Three days of diarrhea, combined with the new "potassium-wasting" diuretic, have pushed Marquis into a state of hypokalemia. His blood K^+ measures only 2.5 mEq/L (normal 3.5–5.0 mEq/L).

Potassium ions are critical for a variety of cell functions, particularly in electrically active cells in the nervous system, skeletal muscle, and cardiac muscle. When the blood K^+ level is low, cells cannot repolarize and are unable to fire repeatedly. This causes muscle weakness, aching, and cramping. In cardiac muscle, hypokalemia increases the resting membrane potential and increases both the duration of the action potential and the duration of the refractory period, leading to cardiac arrhythmias.

1. How should Marquis's hypokalemia should be treated?
2. What is happening when Marquis complains of a sensation of a "frog jumping around" in his chest that is accompanied by dizziness?

See the blue Answers tab at the back of the book.

Related Clinical Terms

antacid: A substance used to counteract or neutralize stomach acid.

bicarbonate loading: In sports, the act of ingesting bicarbonates prior to an athletic event to neutralize acidic by-products produced during strenuous physical activity involving anaerobic metabolism.

enema: A procedure in which a liquid or gas is injected into the rectum, usually with the intention of having the rectum dispel its contents, but sometimes used as a way to introduce drugs or to permit x-ray imaging.

fluid replacement therapy: Procedure conducted with the intent to replace body fluids lost due to disease or restricted intake; or to maintain a higher-than-normal rate of fluid excretion to ensure the removal of toxins; or possibly to administer therapeutic or anesthetic agents slowly over time.

potassium adaptation: The tolerance to increasing amounts of potassium.

syndrome of inappropriate secretion of ADH (SIADH): A condition associated with excessive ADH secretion that results in the excretion of concentrated urine. It disturbs fluid and electrolyte balance, and causes nausea, vomiting, muscle cramps, confusion, and convulsions. This syndrome can occur with some forms of cancer, such as small cell lung cancer, pancreatic cancer, prostate cancer, and Hodgkin disease, and possibly due to a number of other disorders.

total body water: The sum of fluids within all compartments.

28 The Reproductive System

Learning Outcomes

These Learning Outcomes correspond by number to this chapter's sections and indicate what you should be able to do after completing the chapter.

28-1 ■ List the basic structures of the human reproductive system, and summarize the functions of each. p. 1056

28-2 ■ Describe the structures of the male reproductive system, and specify the composition of semen. p. 1057

28-3 ■ Explain the differences between meiosis and mitosis, describe the process of spermatogenesis, and summarize the hormonal mechanisms that regulate male reproductive functions. p. 1065

28-4 ■ Describe the anatomy and functions of the ovaries, the uterine tubes, the uterus, the vagina, and the structures of the female external genitalia. p. 1072

28-5 ■ Describe the ovarian roles in oogenesis, explain the complete ovarian and uterine cycles, and summarize all aspects of the female reproductive cycle. p. 1082

28-6 ■ Discuss the physiology of sexual intercourse in males and females. p. 1091

28-7 ■ Describe the reproductive system changes that occur with development and aging. p. 1093

28-8 ■ Give examples of interactions between the reproductive system and each of the other organ systems. p. 1096

CLINICAL CASE And Baby Makes Three?

The Millers long to have a baby. They are both 30 years old, have been married for 5 years, and have been trying desperately to get pregnant.

Susan waits and watches. Each time her period comes, her heart drops. Her period has never been like clockwork. Now she worries about the STD (sexually transmitted disease) she had in college. At that time, she was admitted to the infirmary with a high fever and pelvic pain. A course of antibiotics finally turned things around.

Her husband, Tim, rates his health as a "9" on a 10-point scale. He's a distance runner and clocks an average of 40 miles per week. He says exercise helps him manage the stress of his sales job. Neither of them drinks nor smokes. **When should this couple seek help for infertility? To find out, turn to the Clinical Case Wrap-Up on p. 1101.**

An Introduction to the Reproductive System

Approximately 7.4 billion people currently live on the Earth. This is an astonishing number given that our *reproductive system* is the only system that is not essential to the life of an individual. Its activities do, however, affect other systems and ensure continuation of our species. In this chapter we discuss how the male and female reproductive organs produce and store specialized reproductive cells called *gametes.* We also look at how various reproductive organs secrete hormones that play major roles in maintaining normal sexual function, allowing those gametes to come together to form new individuals.

28-1 Male and female reproductive system structures produce gametes that combine to form a new individual

Learning Outcome List the basic structures of the human reproductive system, and summarize the functions of each.

The human **reproductive system** ensures the continued existence of the human species—by producing, storing, nourishing, and transporting functional male and female reproductive cells. These cells are called **gametes** (GAM-ēts).

The reproductive system includes the following basic structures:

- **Gonads** (GŌ-nadz; *gone,* seed), reproductive organs that produce gametes and hormones
- Ducts that receive and transport the gametes
- Accessory glands and organs that secrete fluids into the reproductive system ducts or into other excretory ducts
- Perineal structures that are collectively known as the **external genitalia** (jen-ih-TĀ-lē-uh).

In both males and females, the ducts are connected to chambers and passageways that open to the outside. The structures involved make up the *reproductive tract.*

The male and female reproductive systems are functionally quite different, however. In adult males, the **testes** (TES-tēz; singular, *testis*), or *testicles,* are male gonads that secrete sex hormones called *androgens.* The main androgen is *testosterone,* which was introduced in Chapter 18. p. 644 The testes also produce the male gametes, called *sperm.* Males produce about one-half billion sperm each day. During *emission,* mature sperm travel along a lengthy duct system, where they are mixed with the secretions of accessory glands. The mixture created is known as *semen* (SĒ-men). During *ejaculation,* semen is expelled from the body.

In adult females, the **ovaries**, or female gonads, each month release only one immature gamete, called an **oocyte** (Ō-ō-sī-t). The ovaries also secrete female sex hormones, including *estrogens.* The released oocyte travels along one of two short *uterine tubes,* which end in the muscular organ called the *uterus* (YŪ-ter-us). A short passageway, the *vagina* (va-JĪ-nuh), connects the uterus with the exterior.

During *sexual intercourse,* ejaculation introduces semen into the vagina, and the sperm then ascend the female reproductive tract. If a sperm reaches the oocyte and starts the process of *fertilization,* the oocyte matures into an *ovum* (plural, *ova*). The uterus will then enclose and support the developing *embryo* as it grows into a *fetus* and prepares for birth (a process we cover in Chapter 29).

In this chapter we examine the anatomy of the male and female reproductive systems. We consider the physiological and hormonal mechanisms that regulate reproductive function. Earlier chapters introduced the anatomical reference points used in the discussions that follow. You may find it helpful to review the figures on the pelvic girdle (**Figures 8–7, 8–8**, pp. 253, 254), perineal musculature (**Figure 11–13**, p. 360), pelvic innervation (Figure 13–12, p. 450), and regional blood supply (Figures 21–24, 21–28, pp. 766, 771).

Checkpoint

1. **Define *gamete*.**
2. **List the basic structures of the reproductive system.**
3. **Define *gonads*.**

See the blue Answers tab at the back of the book.

28-2 The structures of the male reproductive system consist of the testes, duct system, accessory glands, and penis

Learning Outcome Describe the structures of the male reproductive system, and specify the composition of semen.

Figure 28–1 shows the main structures of the male reproductive system. Starting from a *testis*, the sperm travel within the male *reproductive duct system*, which consists of the *epididymis* (ep-ih-DID-ih-mus), the *ductus deferens* (DUK-tus DEF-eh-renz), and the *urethra* before leaving the body. Accessory glands—the *seminal* (SEM-i-nal) *glands*, the *prostate* (PROS-tāt), and the *bulbo-urethral* (bul-bō-yū-RĒ-thral) *gland*—secrete various fluids into the duct system. The *male external genitalia* consist of the *scrotum* (SKRŌ-tum), that encloses the testes, the urethra, and the *penis* (PĒ-nis), an erectile organ. The distal portion of the urethra passes through the penis. This section introduces the anatomy and basic functions of these structures.

The Testes and Associated Structures

Each testis is about 5 cm (2 in.) long, 3 cm (1.2 in.) wide, and 2.5 cm (1 in.) thick. Each weighs 10–15 g (0.35–0.53 oz). The testes hang within the **scrotum**, a fleshy pouch suspended inferior to the perineum. The scrotum is anterior to the anus, and posterior to the base of the penis (see Figure 28–1).

The Spermatic Cords

The **spermatic cords** are paired structures extending between the abdominopelvic cavity and the testes (Figure 28–2). Each spermatic cord begins at the entrance to the *inguinal canal* (a passageway through the abdominal musculature). After passing through the inguinal canal, the spermatic cord descends into the scrotum.

Figure 28–1 Sagittal Section of the Male Reproductive System. ATLAS: Plate 64

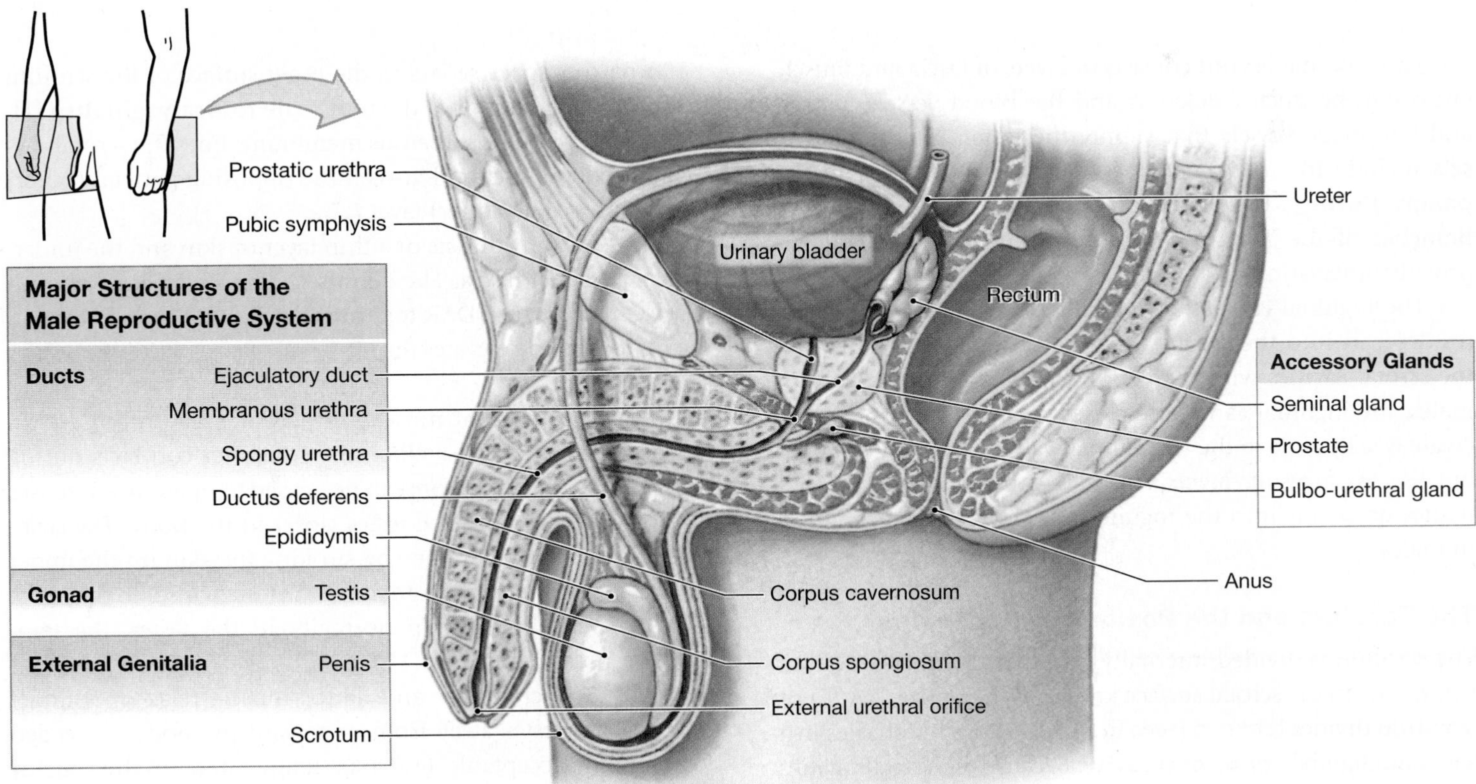

Prostatic urethra
Pubic symphysis
Urinary bladder
Rectum
Ureter
Major Structures of the Male Reproductive System
Ducts
Ejaculatory duct
Membranous urethra
Spongy urethra
Ductus deferens
Epididymis
Gonad
Testis
External Genitalia
Penis
Scrotum
Accessory Glands
Seminal gland
Prostate
Bulbo-urethral gland
Anus
Corpus cavernosum
Corpus spongiosum
External urethral orifice

28

Figure 28–2 Anterior View of the Male Reproductive System. In the anatomical position, the penis is erect, and the dorsal surface is closest to the navel. When the penis is flaccid, its dorsal surface faces anteriorly.

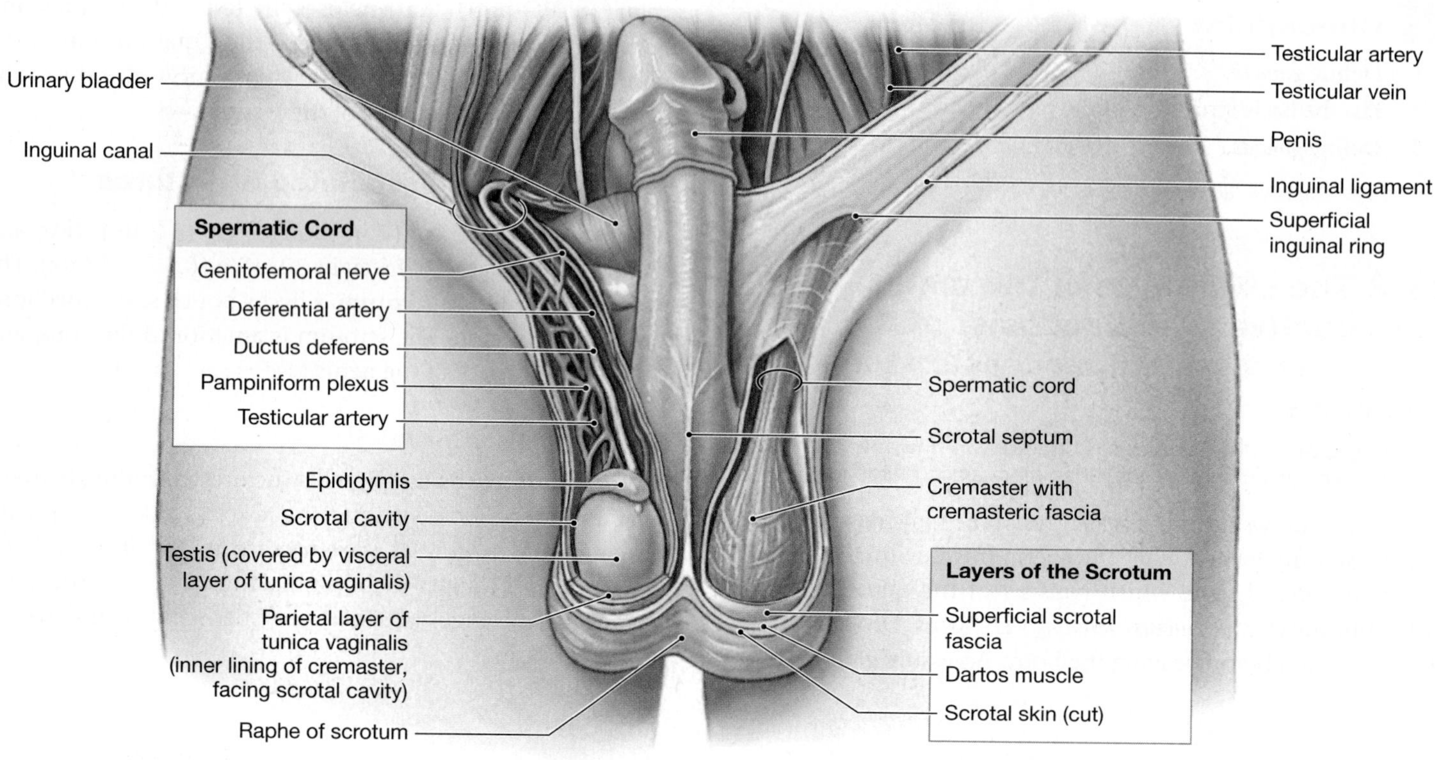
Urinary bladder
Inguinal canal
Spermatic Cord
Genitofemoral nerve
Deferential artery
Ductus deferens
Pampiniform plexus
Testicular artery
Epididymis
Scrotal cavity
Testis (covered by visceral layer of tunica vaginalis)
Parietal layer of tunica vaginalis (inner lining of cremaster, facing scrotal cavity)
Raphe of scrotum
Testicular artery
Testicular vein
Penis
Inguinal ligament
Superficial inguinal ring
Spermatic cord
Scrotal septum
Cremaster with cremasteric fascia
Layers of the Scrotum
Superficial scrotal fascia
Dartos muscle
Scrotal skin (cut)

Each spermatic cord consists of layers of fascia and muscle enclosing the ductus deferens and the blood vessels, nerves, and lymphatic vessels that supply the testes. The blood vessels include the *deferential artery*, a *testicular artery*, and the **pampiniform** (pam-PIN-ih-form) **plexus** of a *testicular vein*. Branches of the *genitofemoral nerve* from the lumbar plexus provide innervation.

The inguinal canals form during development as the testes descend into the scrotum. At that time, these canals link the scrotal cavities with the peritoneal cavity. In normal adult males, the inguinal canals are closed, but the spermatic cords create weak points in the abdominal wall that remain throughout life. As a result, *inguinal hernias*—protrusions of visceral tissues or organs into the inguinal canal—are fairly common in males.

The Scrotum and the Position of the Testes

The scrotum is divided internally into two chambers. A raised thickening in the scrotal surface known as the **raphe** (RĀ-fē) **of scrotum** divides it in two (see Figure 28–2). Each testis lies in a separate chamber, or **scrotal cavity**. Because the scrotal cavities are separated by a partition, infection or inflammation of one testis does not normally spread to the other.

A narrow space separates the inner surface of the scrotum from the outer surface of the testis. The **tunica vaginalis** (TU-nih-ka vaj-ih-NAL-is), a serous membrane, lines the scrotal cavity and reduces friction between the opposing parietal (outer) layer and visceral (inner) layer (Figure 28–3a).

The scrotum consists of a thin layer of skin and the underlying superficial fascia. The dermis contains a layer of smooth muscle, the **dartos** (DAR-tōs) **muscle**. Resting muscle tone in the dartos muscle elevates the testes and causes wrinkling of the scrotal surface.

A layer of skeletal muscle, the **cremaster** (krē-MAS-ter), lies deep to the dermis. When the cremaster contracts during sexual arousal or in response to decreased temperature, it tenses the scrotum and pulls the testes closer to the body. The *cremasteric reflex* can be initiated by stroking the skin on the upper thigh, causing the scrotum to move the testes closer to the body.

For sperm to develop normally in the testes, the temperature must be about 1.1°C (2°F) lower than elsewhere in the body. The cremaster and dartos muscle relax or contract to move the testes away from or toward the body as needed to maintain acceptable testicular temperatures. When air or body temperature increases, these muscles relax and the testes move away from the body. When the scrotum cools suddenly,

Figure 28–3 Anatomy of the Scrotum and the Testes.

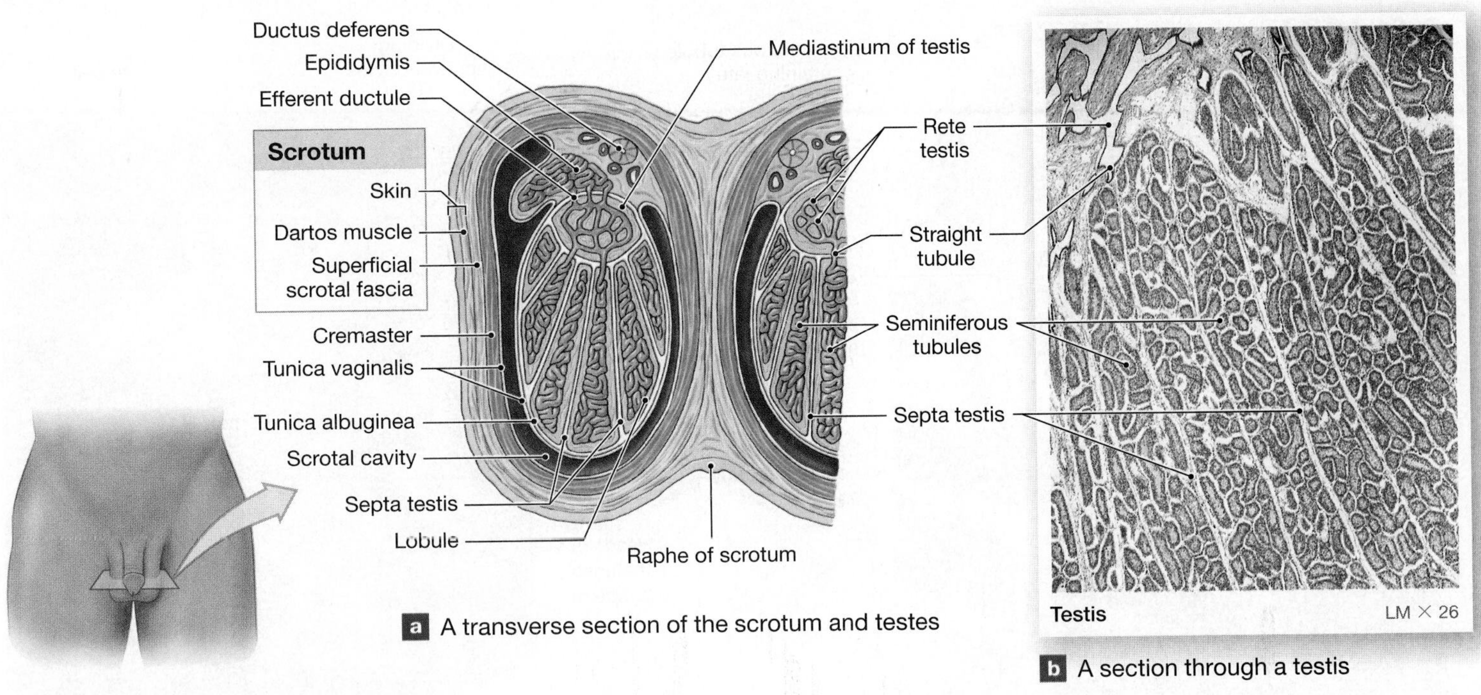
Ductus deferens
Epididymis
Efferent ductule
Scrotum
Skin
Dartos muscle
Superficial scrotal fascia
Cremaster
Tunica vaginalis
Tunica albuginea
Scrotal cavity
Septa testis
Lobule
Raphe of scrotum
Mediastinum of testis
Rete testis
Straight tubule
Seminiferous tubules
Septa testis
Testis
LM × 26

a A transverse section of the scrotum and testes

b A section through a testis

as happens when jumping into a cold swimming pool, these muscles contract to pull the testes closer to the body and keep testicular temperatures from decreasing.

Gross Anatomy of the Testes

Deep to the tunica vaginalis covering the testis is the **tunica albuginea** (al-bū-JIN-ē-uh), a dense layer of connective tissue rich in collagen fibers (see Figure 28–3a). These fibers are continuous with those surrounding the adjacent epididymis and extend into the testis. There they form fibrous partitions, or *septa testis*, that converge toward the region nearest the entrance to the epididymis. The connective tissues in this region support the blood vessels and lymphatic vessels that supply and drain the testis, and the *efferent ductules*, which transport sperm to the epididymis.

Histology of the Testes

The septa testis subdivides the testis into a series of **lobules** (see Figure 28–3a). Distributed among the lobules are approximately 800 slender, tightly coiled **seminiferous** (sem-ih-NIF-er-us) **tubules** (Figure 28–3a,b). Each tubule averages about 80 cm (32 in.) in length. A typical testis contains nearly one-half mile of seminiferous tubules. Sperm production takes place within these tubules.

Each lobule contains several seminiferous tubules that merge into **straight tubules** that enter the mediastinum of the testis. Straight tubules are extensively interconnected, forming a maze of passageways called the **rete** (RĒ-tē; *rete*, a net) **testis** (see Figure 28–3a,b). Fifteen to 20 **efferent ductules** connect the rete testis to the epididymis.

Because the seminiferous tubules are tightly coiled, most tissue slides show them in transverse section (Figure 28–4a). A delicate connective tissue capsule surrounds each tubule, and areolar tissue fills the spaces between the tubules (Figure 28–4b). Within those spaces are numerous blood vessels and large **interstitial endocrine cells** (*Leydig cells*). Interstitial endocrine cells produce androgens (androsterone and testosterone), the dominant sex hormones in males. Testosterone is the most important androgen.

Functional Anatomy of the Male Reproductive Duct System

The testes produce sperm that are immobile and not yet capable of successfully fertilizing an oocyte. The male reproductive duct system is responsible for the functional maturation, nourishment, storage, and transport of sperm.

The Epididymis

These immobile sperm are transported into the epididymis by fluid currents created by the cilia that line the efferent ductules. The **epididymis** (ep-ih-DID-ih-mis; *epi*, on + *didymos*, twin) is the start of the male reproductive tract (Figure 28–5). It is a coiled tube bound to the posterior border of each testis.

Figure 28–4 Anatomy of the Seminiferous Tubules.

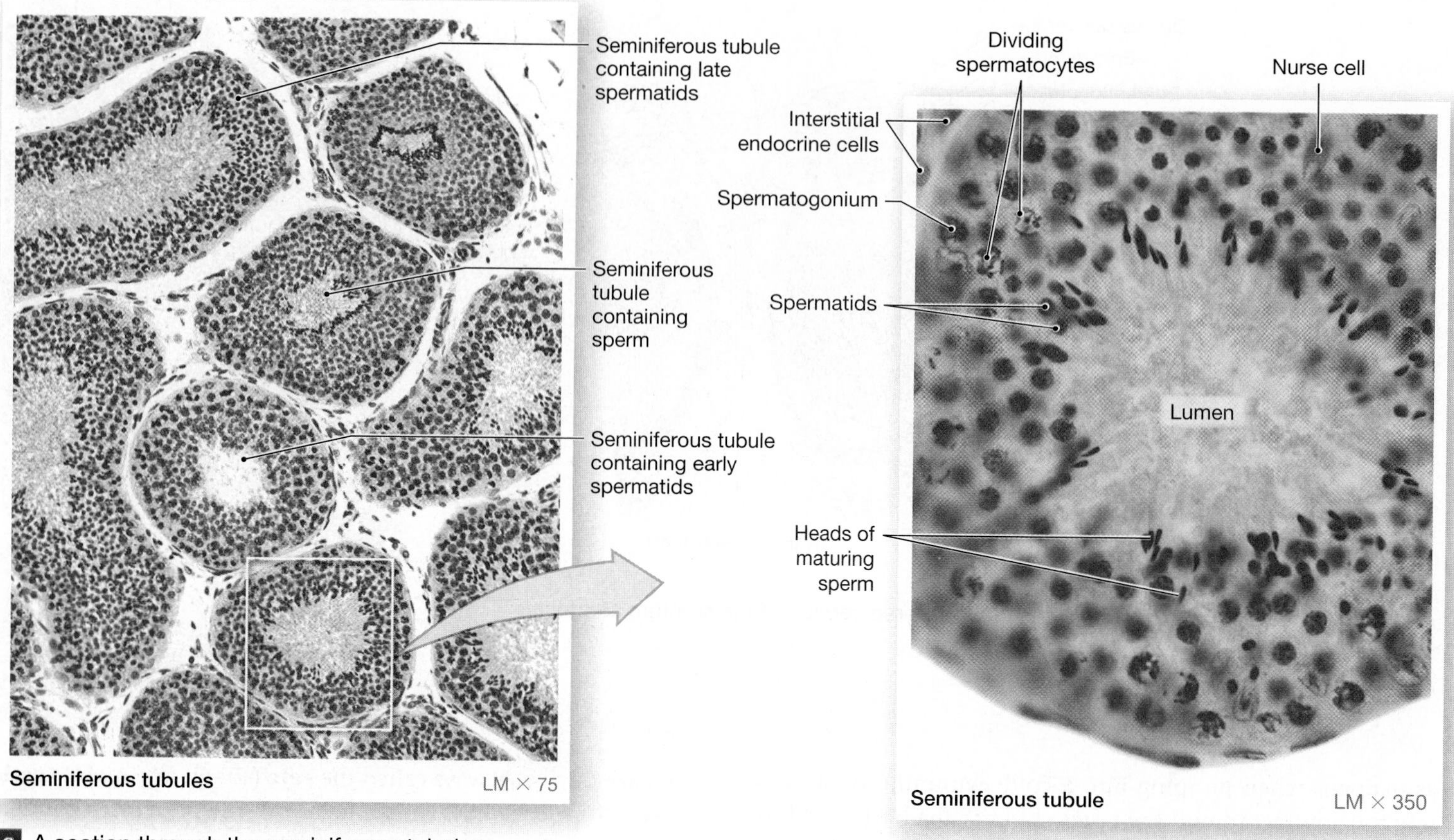
Seminiferous tubule containing late spermatids
Seminiferous tubule containing sperm
Seminiferous tubule containing early spermatids
Seminiferous tubules
LM × 75
Dividing spermatocytes
Nurse cell
Interstitial endocrine cells
Spermatogonium
Spermatids
Lumen
Heads of maturing sperm
Seminiferous tubule
LM × 350

a A section through the seminiferous tubules.

b A cross section through a single tubule.

The epididymides (ep-ih-dih-DIM-ih-dēz) can be felt through the scrotal skin. Each epididymis is almost 7 m (23 ft) long, but it is coiled and twisted so it takes up very little space. It has a head, a body, and a tail (Figure 28–5a,c). The superior **head** is the largest part of the epididymis. The head receives sperm from the efferent ductules.

The **body** is the middle part that extends inferiorly from the head to the tail of the epididymis on the posterior surface of the testis. Near the inferior border of the testis, the number of coils decreases, marking the start of the **tail**. The tail re-curves, ascends, and connects to the ductus deferens. Sperm are stored primarily within the tail of the epididymis.

The epididymis has the following functions:

- *It monitors and adjusts the composition of the fluid produced by the seminiferous tubules.* The pseudostratified columnar epithelial lining of the epididymis has distinctive stereocilia (Figure 28–5b). These stereocilia increase the surface area available for absorption from, and secretion into, the fluid in the tubule.
- *It acts as a recycling center for damaged sperm.* Cellular debris and damaged sperm are absorbed in the epididymis. The products of enzymatic breakdown are released into the surrounding interstitial fluids for pickup by the epididymal blood vessels.
- *It stores and protects sperm and facilitates their functional maturation.* It takes up to 3 weeks for a sperm to pass through the epididymis and complete its functional maturation. During this time, sperm exist in a sheltered environment that is regulated by the surrounding epithelial cells.

Transport along the epididymis involves a combination of fluid movement and peristaltic contractions of smooth muscle in the epididymis. After passing along the tail of the epididymis, the sperm enter the ductus deferens.

28

The Ductus Deferens

Each **ductus deferens** (plural, *ductus deferentia*), also called the *vas deferens*, is 40–45 cm (16–18 in.) long. It begins at the tail of the epididymis (see Figure 28–5a,c). As part of the spermatic cord, it ascends through the inguinal canal (look back at Figure 28–2). Inside the abdominal cavity, the ductus deferens passes posteriorly, curving inferiorly along the lateral surface of the urinary bladder toward the superior and posterior margin

Figure 28–5 Anatomy of the Epididymis. ATLAS: Plate 60a

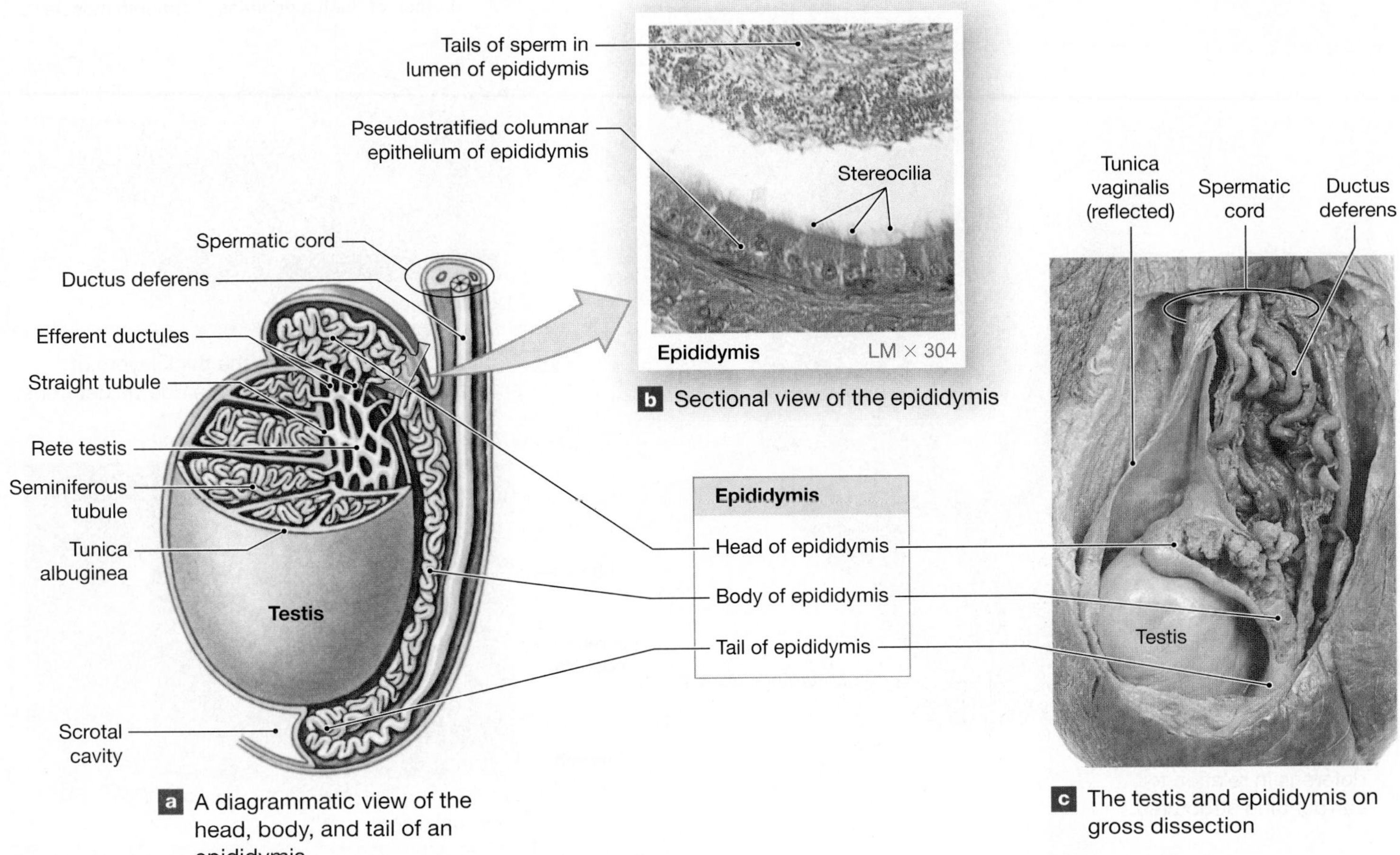

Tails of sperm in lumen of epididymis
Pseudostratified columnar epithelium of epididymis
Stereocilia
Epididymis
LM × 304
Spermatic cord
Ductus deferens
Efferent ductules
Straight tubule
Rete testis
Seminiferous tubule
Tunica albuginea
Testis
Scrotal cavity
Epididymis
Head of epididymis
Body of epididymis
Tail of epididymis
Tunica vaginalis (reflected)
Spermatic cord
Ductus deferens
Testis

a A diagrammatic view of the head, body, and tail of an epididymis

b Sectional view of the epididymis

c The testis and epididymis on gross dissection

of the prostate (see Figure 28–6a). Just before the ductus deferens reaches the prostate and seminal glands, its lumen enlarges. This expanded portion is known as the **ampulla** (am-PUL-luh) **of ductus deferens**.

The wall of the ductus deferens contains a thick layer of smooth muscle (see Figure 28–6b). Peristaltic contractions in this layer propel sperm and fluid along the duct, which is lined by a pseudostratified ciliated columnar epithelium. In addition to transporting sperm, the ductus deferens can store sperm in the ampulla for several months. During this time, the sperm remain in a temporary state of inactivity with low metabolic rates.

The Male Urethra

In males, the **urethra** is a passageway that extends 18–20 cm (7–8 in.) from the urinary bladder to the tip of the penis. It is divided into *prostatic*, *membranous*, and *spongy* regions (Figures 28–1, 28–7b). The male urethra is used by both the urinary and reproductive systems. It conducts urine to the exterior and introduces semen into the female's vagina during sexual intercourse.

The Accessory Glands

The fluids secreted by the seminiferous tubules and the epididymis account for only about 5 percent of the volume of semen. The fluid component of semen is a mixture of secretions—each with distinctive biochemical characteristics—from many glands. Important glands include the *seminal glands*, the *prostate*, and the *bulbo-urethral glands*. All of these occur only in males.

Among the major functions of these glands are (1) activating sperm; (2) providing the nutrients sperm need for motility; (3) propelling sperm and fluids along the reproductive tract, mainly by peristaltic contractions; and (4) producing buffers that counteract the acidity of the urethral and vaginal environments.

The Seminal Glands (Seminal Vesicles)

The ductus deferens on each side ends at the junction between the ampulla of ductus deferens and the duct that drains the seminal gland (see Figure 28–6a). The **seminal glands**, also called the *seminal vesicles*, are tubular glands embedded in connective tissue on each side of the midline, sandwiched between the posterior

Figure 28–6 Anatomy of the Ductus Deferens and Accessory Glands.

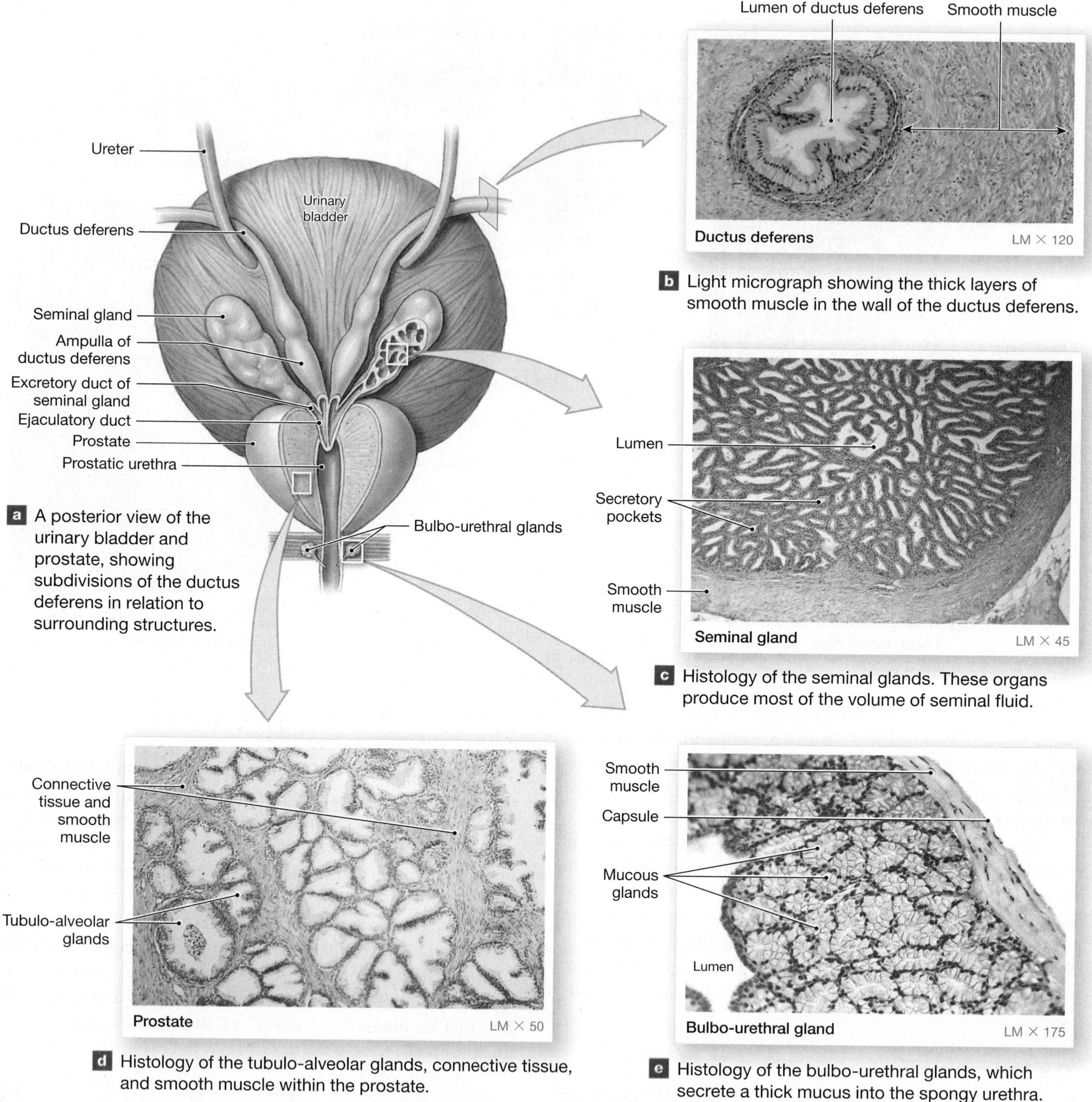
Ureter
Ductus deferens
Urinary bladder
Seminal gland
Ampulla of ductus deferens
Excretory duct of seminal gland
Ejaculatory duct
Prostate
Prostatic urethra
Bulbo-urethral glands
Lumen of ductus deferens
Smooth muscle
Ductus deferens
LM × 120
Lumen
Secretory pockets
Smooth muscle
Seminal gland
LM × 45
Connective tissue and smooth muscle
Tubulo-alveolar glands
Prostate
LM × 50
Smooth muscle
Capsule
Mucous glands
Lumen
Bulbo-urethral gland
LM × 175

a A posterior view of the urinary bladder and prostate, showing subdivisions of the ductus deferens in relation to surrounding structures.

b Light micrograph showing the thick layers of smooth muscle in the wall of the ductus deferens.

c Histology of the seminal glands. These organs produce most of the volume of seminal fluid.

d Histology of the tubulo-alveolar glands, connective tissue, and smooth muscle within the prostate.

e Histology of the bulbo-urethral glands, which secrete a thick mucus into the spongy urethra.

wall of the urinary bladder and the rectum. Each tubular seminal gland is about 15 cm (6 in.) in length. The body of the gland has many short side branches. The entire assemblage is coiled and folded into a compact, tapered mass roughly 5 cm × 2.5 cm (2 in. × 1 in.).

Seminal glands have an epithelial lining that contains extensive folds (Figure 28–6c), reflecting that they are extremely active secretory glands. The seminal glands produce the majority of the volume of semen. The secretions of the seminal glands are discharged into the **ejaculatory duct**. The junction of the duct of the seminal gland with the ampulla of ductus deferens marks the start of the ejaculatory duct. This short passageway is 2 cm (less than 1 in.). It penetrates the muscular wall of the prostate and empties into the prostatic urethra (see Figure 28–6a).

The Prostate

The **prostate** is a small, muscular, rounded organ about 4 cm (1.6 in.) in diameter. The prostate encircles the proximal portion of the urethra as it leaves the urinary bladder (see Figures 28–1, 28–6a). The glandular tissue of the prostate consists of a cluster of 30–50 compound tubulo-alveolar glands (Figure 28–6d). ⮌ p. 125 These glands are surrounded by a thick blanket of smooth muscle fibers.

The prostate produces **prostatic fluid**, a solution that makes up about a quarter of the volume of semen. This slightly acidic fluid is rich in enzymes, such as acid phosphatase and amylase, that prevent sperm coagulation in the vagina. Prostatic fluid is ejected into the prostatic urethra by peristaltic contractions of the muscular prostate wall.

Prostatic inflammation, or **prostatitis** (pros-ta-TĪ-tis), can occur in males at any age, but it most commonly afflicts older men. Prostatitis can result from bacterial infections but also occurs without pathogens. Signs and symptoms can resemble those of prostate cancer. Men with prostatitis may have pain in the lower back, perineum, or rectum. In some cases, the symptoms are accompanied by painful urination and mucus discharge from the external urethral orifice. Antibiotics are effective in treating most cases caused by bacterial infection.

The Bulbo-urethral Glands

The paired **bulbo-urethral glands**, or *Cowper's glands*, are compound tubular mucous glands located at the base of the penis (see Figure 28–7a). These glands are round, with diameters nearly 10 mm (less than 0.5 in.). The duct of each gland travels alongside the spongy urethra for 3–4 cm (1.2–1.6 in.) before emptying into the urethral lumen.

The bulbo-urethral glands secrete thick, alkaline mucus (Figure 28–6e). The secretion helps neutralize any urinary acids that may remain in the urethra, and it lubricates the tip of the penis.

Semen

A typical ejaculation releases 2–5 mL of **semen** (*ejaculate*). When sampled for analysis, semen is collected after a 36-hour period of sexual abstinence. Semen typically contains the following:

- *Sperm.* The normal sperm count ranges from 20 million to 100 million sperm per milliliter of semen. In addition, at least 60 percent of the sperm in the sample appear normal and will be actively swimming.
- *Seminal Fluid.* **Seminal fluid**, the liquid component of semen, is a mixture of glandular secretions with a distinct ionic and nutrient composition. A typical sample of seminal fluid contains the combined secretions of the seminal glands (60 percent), the prostate (30 percent), the nurse cells and epididymis (5 percent), and the bulbo-urethral glands (less than 5 percent). An abnormally low volume may indicate problems with the prostate or seminal glands. In particular, the secretion of the seminal glands contains (1) a higher concentration of fructose, which is easily metabolized by sperm; (2) prostaglandins, which can stimulate smooth muscle contractions along the male and female reproductive tracts; and (3) fibrinogen, which forms a temporary semen clot within the vagina. The glandular fluid the seminal glands produce generally has the same osmotic concentration as that of blood plasma, but the compositions of the two fluids are quite different. The secretions of the seminal glands are slightly alkaline, helping to neutralize acids in the secretions of the prostate and within the vagina.
- *Enzymes.* Several important enzymes are in seminal fluid, including (1) a protease that may help dissolve mucus in the vagina; (2) **seminalplasmin** (sem-i-nal-PLAZ-min), a protein with antibiotic properties that may help prevent urinary tract infections in males; (3) a prostatic enzyme that converts fibrinogen to fibrin, coagulating the semen within a few minutes after ejaculation; and (4) *fibrinolysin*, which liquefies the clotted semen after 15–30 minutes.

A complete chemical analysis of semen appears in the Appendix.

The Penis

The **penis** is a tubular organ for sexual intercourse and conducting urine to the exterior. Main structures include the root, body, and glans penis (Figure 28–7a). The **root of penis** is the fixed portion that attaches the penis to the body wall. It includes left crus of penis, right crus of penis, and the bulb of penis. This connection occurs immediately inferior to the pubic symphysis. The **body of penis** (*shaft*) is the tubular, movable portion of the organ. The **glans penis** (*head*) is the expanded distal end that surrounds the external urethral orifice. The *neck of glans* is the narrow portion of the penis between the body and the glans penis.

The skin overlying the penis resembles that of the scrotum. The dermis contains a layer of smooth muscle that is a continuation of the dartos muscle of the scrotum. The underlying areolar tissue allows the thin skin to move without distorting underlying structures. The subcutaneous layer also contains superficial arteries, veins, and lymphatic vessels.

A fold of skin called the **foreskin** or **prepuce** (PRĒ-pūs) surrounds the tip of the penis. (Figure 28–7b). The foreskin attaches to the relatively narrow neck of the penis and continues over the glans penis. **Preputial** (prē-PŪ-shē-al) **glands** in the skin of the neck and the inner surface of the foreskin secrete a waxy material known as *smegma* (SMEG-ma). Unfortunately,

Figure 28–7 **Anatomy of the Penis.** ATLAS: Plate 60b

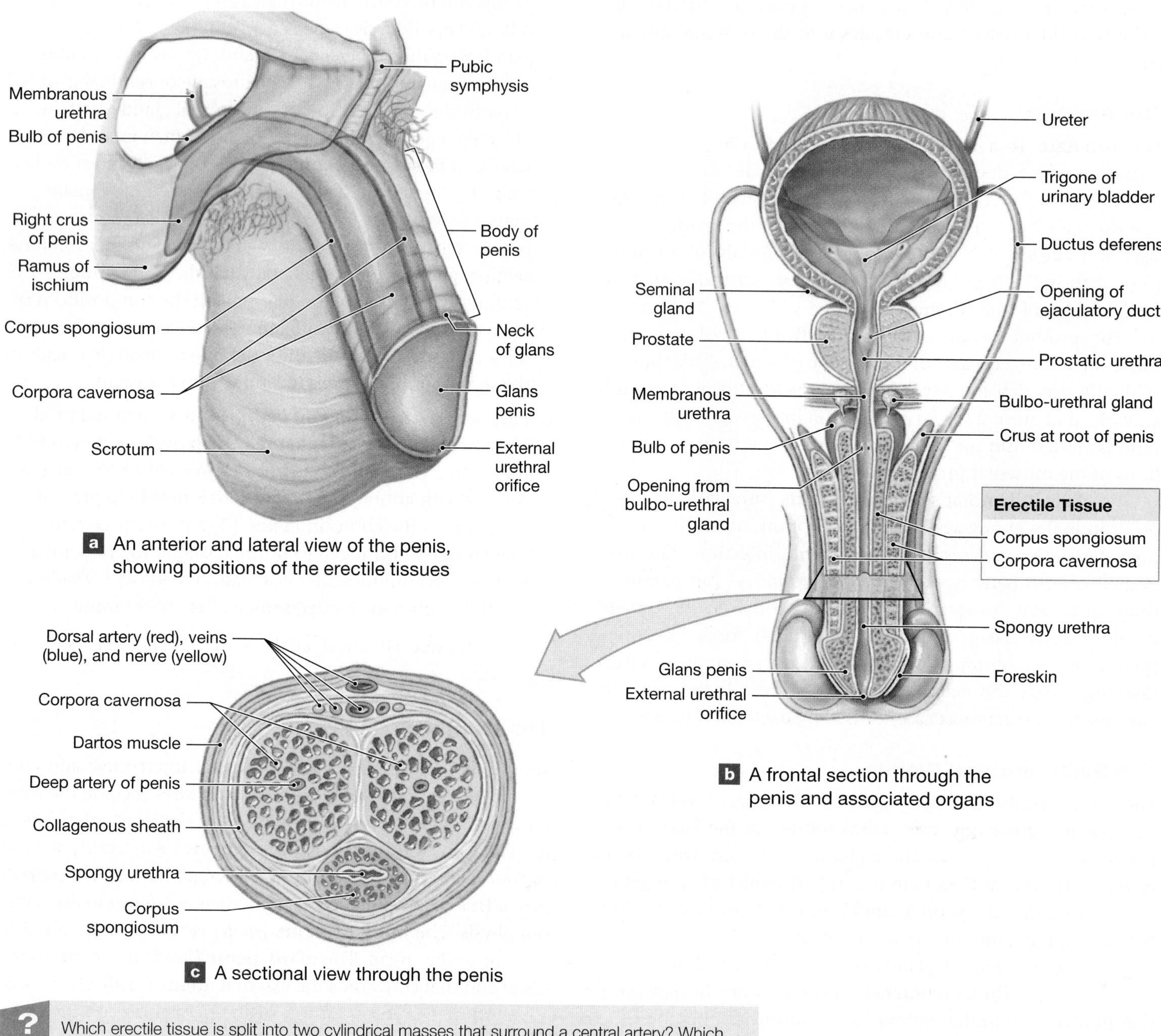

Membranous urethra
Bulb of penis
Pubic symphysis
Right crus of penis
Ramus of ischium
Body of penis
Corpus spongiosum
Neck of glans
Corpora cavernosa
Glans penis
Scrotum
External urethral orifice
Ureter
Trigone of urinary bladder
Ductus deferens
Seminal gland
Opening of ejaculatory duct
Prostate
Prostatic urethra
Membranous urethra
Bulbo-urethral gland
Bulb of penis
Crus at root of penis
Opening from bulbo-urethral gland
Erectile Tissue
Corpus spongiosum
Corpora cavernosa
Spongy urethra
Glans penis
Foreskin
External urethral orifice
Dorsal artery (red), veins (blue), and nerve (yellow)
Corpora cavernosa
Dartos muscle
Deep artery of penis
Collagenous sheath
Spongy urethra
Corpus spongiosum

a An anterior and lateral view of the penis, showing positions of the erectile tissues

b A frontal section through the penis and associated organs

c A sectional view through the penis

? Which erectile tissue is split into two cylindrical masses that surround a central artery? Which erectile tissue surrounds the urethra?

28 smegma can be an excellent nutrient source for bacteria. Mild inflammation and infections in this area are common, especially if the area is not washed thoroughly and frequently. One way to avoid such problems is **circumcision** (ser-kum-SIZH-un), the surgical removal of the foreskin. In Western societies (especially the United States), this procedure is often performed shortly after birth. Circumcision lowers the risks of developing urinary tract infections, HIV infection, and penile cancer. Because it is a surgical procedure with risk of bleeding, infection, and other complications, the practice remains controversial.

Deep to the areolar tissue, a dense network of elastic fibers encircles the internal structures of the penis. Most of the body of the penis consists of three cylindrical columns of **erectile tissue** (Figure 28–7b,c). Erectile tissue consists of a three-dimensional

+ Clinical Note Circumcision

Circumcision, or the removal of the foreskin of the penis, is an elective procedure for a baby boy. Certain religions conduct a ceremonial circumcision, as in the *bris* practiced in the Jewish faith when the infant is 8 days old. In the brief medical process of circumcision, the penis is anesthetized locally. A clamp or ring is applied to remove the prepuce. Circumcision in an adult may require a general anesthetic. A circumcised penis is easier to keep clean. The risk of developing a sexually transmitted disease, which can flourish under the foreskin, is decreased. However, these risks can also be managed by teaching the boy to practice careful personal hygiene.

maze of vascular channels incompletely separated by partitions of elastic connective tissue and smooth muscle fibers. In the resting state, the arterial branches are constricted and the muscular partitions are tense. This combination restricts blood flow into the erectile tissue. During **erection** the penis stiffens and elevates to an upright position, but the flaccid (nonerect) penis hangs inferior to the pubic symphysis and anterior to the scrotum.

The anterior surface of the flaccid penis covers two cylindrical masses of erectile tissue: the **corpora cavernosa** (KOR-por-uh ka-ver-NŌ-suh; singular, *corpus cavernosum*). The two are separated by a thin septum and encircled by a dense collagenous sheath (see Figure 28–7c). The corpora cavernosa diverge at their bases, forming the **crus of penis** (see Figure 28–7a). Each crus is bound to the ramus of the ischium and pubis by tough connective tissue ligaments. The corpora cavernosa extend along the length of the penis as far as its neck. The erectile tissue within each corpus cavernosum surrounds a central artery, or deep artery of the penis (see Figure 28–7c).

The penis contains the distal portion of the urethra, which the relatively slender **corpus spongiosum** (spon-jē-Ō-sum) surrounds (see Figure 28–7a,b). This erectile body extends to the tip of the penis, where it expands to form the glans penis. The sheath surrounding the corpus spongiosum contains more elastic fibers than does that of the corpora cavernosa, and the erectile tissue contains a pair of small arteries.

✓ Checkpoint

4. **Name the male reproductive structures.**
5. **On a warm day, would the cremaster be contracted or relaxed? Why?**
6. **Trace the pathway that a sperm travels from the site of its production to outside the body.**

See the blue Answers tab at the back of the book.

28-3 Spermatogenesis occurs in the testes, and hormones from the hypothalamus, pituitary gland, and testes control male reproductive functions

Learning Outcome **Explain the differences between meiosis and mitosis, describe the process of spermatogenesis, and summarize the hormonal mechanisms that regulate male reproductive functions.**

Overview of Mitosis and Meiosis

Mitosis, the production of genetically identical cells, and *meiosis*, the production of gametes, differ significantly in terms of the events that take place in the nucleus. Recall from Chapter 3 that human somatic cells contain 23 pairs of chromosomes, or 46 chromosomes altogether. Each pair consists of one chromosome provided by the father, and another provided by the mother, at the time of fertilization. The two members of each of the 22 pairs have similar sizes and genes and are known as **homologous** (huh-MOL-ō-gus) **chromosomes**.

Mitosis is part of the process of somatic cell division, producing two daughter cells each containing identical numbers and pairs of chromosomes. The pattern is illustrated in Figure 28–8a. Because daughter cells contain both members of each chromosome pair, they are called **diploid** (DIP-loyd; *diplo*, double) (2n = 46) cells. (You can review the description of mitosis and cell division in Chapter 3. ⮌ pp. 104–105)

Meiosis is a specialized form of cell division that produces only gametes. In contrast to mitosis, meiosis follows the pattern in Figure 28–8b. It involves two cycles of cell division (*meiosis I* and *meiosis II*) and produces four cells, each containing 23 *individual* chromosomes. These cells are called **haploid** (HAP-loyd; *haplo*, single) (n = 23) cells and contain only one member of each homologous pair of chromosomes. Because gametes contain half the number of chromosomes found in somatic cells, the fusion of the nuclei of a male gamete and a female gamete produces a cell that has the normal number of chromosomes, rather than twice that number. The events in the nucleus shown in Figure 28–8b are the same for the formation of sperm in males or oocytes in females.

As a cell prepares to begin meiosis, DNA replication occurs within the nucleus just as it does in a cell preparing for mitosis. The duplicated chromosomes condense and become visible with a light microscope. As in mitosis, each chromosome now consists of two duplicate *chromatids* (KRŌ-mah-tidz).

At this point, the close similarities between meiosis and mitosis end. In meiosis, the corresponding maternal (inherited from the mother) and paternal (inherited from the father) chromosomes now come together in an event known as **synapsis** (sih-NAP-sis). Synapsis involves 23 pairs of chromosomes, and each member of each pair consists of two chromatids.

28

Figure 28–8 A Comparison of Chromosomes in Mitosis and Meiosis.

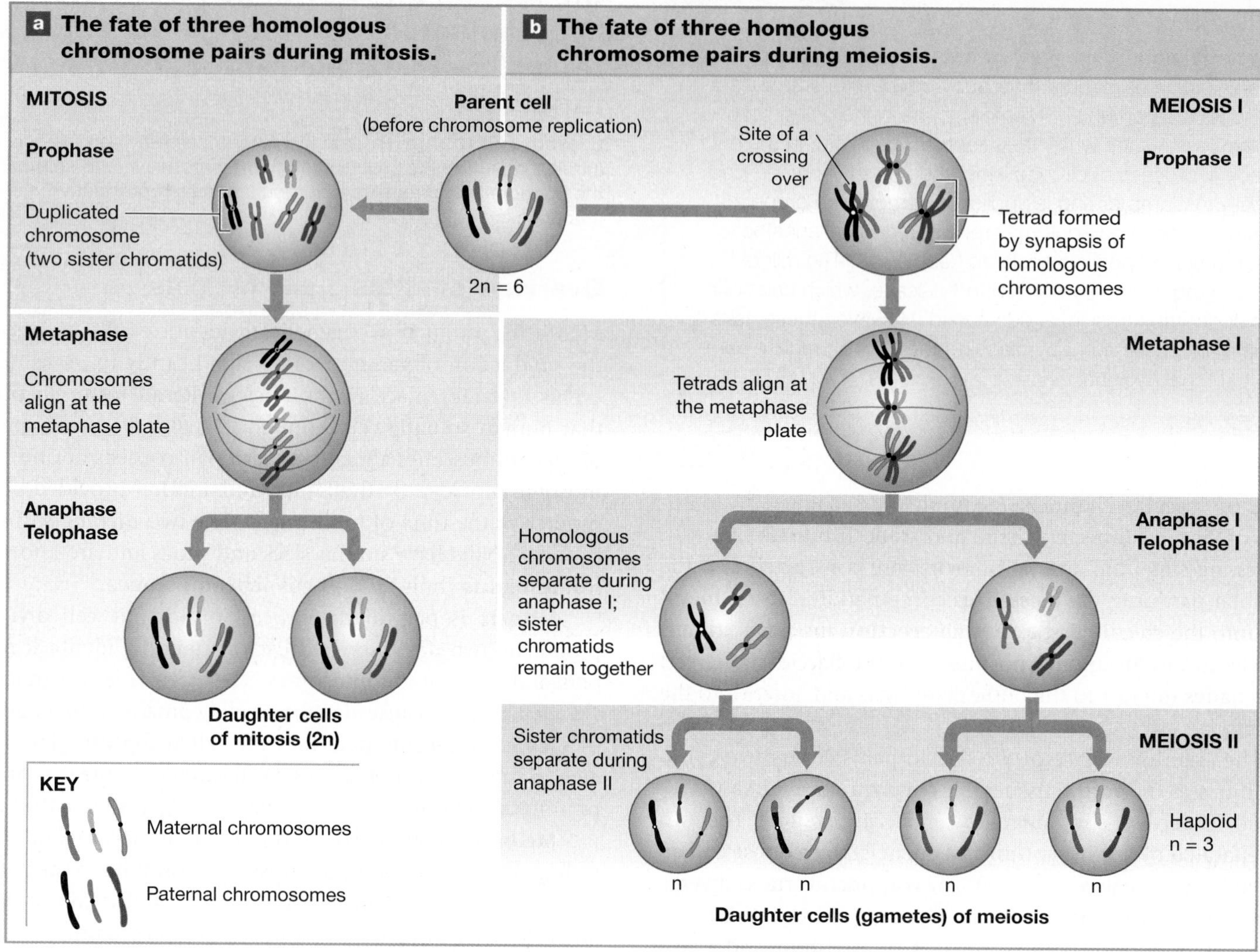
a The fate of three homologous chromosome pairs during mitosis.
b The fate of three homologus chromosome pairs during meiosis.
MITOSIS
Prophase
Duplicated chromosome (two sister chromatids)
Parent cell
(before chromosome replication)
2n = 6
Site of a crossing over
MEIOSIS I
Prophase I
Tetrad formed by synapsis of homologous chromosomes
Metaphase
Chromosomes align at the metaphase plate
Tetrads align at the metaphase plate
Metaphase I
Anaphase
Telophase
Homologous chromosomes separate during anaphase I; sister chromatids remain together
Anaphase I
Telophase I
Daughter cells of mitosis (2n)
Sister chromatids separate during anaphase II
MEIOSIS II
KEY
Maternal chromosomes
Paternal chromosomes
Haploid
n = 3
n
n
n
n
Daughter cells (gametes) of meiosis

A matched set of four chromatids is called a **tetrad** (TET-rad; *tetras*, four) (Figure 28–8b). Some genetic material can be exchanged between the maternal and paternal chromatids of a chromosome pair at this stage of meiosis. Such an exchange, called *crossing over*, increases genetic variation among offspring. We discuss genetics in Chapter 29.

Meiosis includes two division cycles, referred to as **meiosis I** and **meiosis II**. We identify the stages within each cycle with I or II, for example as prophase I or metaphase II.

Meiosis I

28 Meiosis I begins with prophase I; the nuclear envelope disappears at the end of prophase I. As metaphase I begins, the tetrads line up along the metaphase plate. As anaphase I begins, the tetrads break up—the maternal and paternal chromosomes separate. This is a major difference between mitosis and meiosis: In mitosis, each daughter cell receives a copy of every chromosome, maternal and paternal, but in meiosis I, each daughter cell receives two copies of *either* the maternal chromosome *or* the paternal chromosome. (Compare the two parts of Figure 28–8.)

As anaphase I proceeds, the maternal and paternal components are randomly and independently distributed. That is, as each tetrad splits, we cannot predict which daughter cell will receive copies of the maternal chromosome, and which will receive copies of the paternal chromosome. As a result, telophase I ends with the formation of two daughter cells containing unique combinations of maternal and paternal chromosomes. Both cells contain 23 chromosomes.

The first meiotic division is called a **reductional division** because it reduces the number of chromosomes from the diploid number ($2n = 46$) to the haploid number ($n = 23$). Each of these chromosomes still consists of two chromatids. The two chromatids will separate during meiosis II.

Meiosis II

The interphase between meiosis I and meiosis II is very brief. No DNA is replicated during that period. Each cell then proceeds through prophase II, metaphase II, and anaphase II. During anaphase II, the chromatids separate. Telophase II yields four haploid cells, each containing 23 chromosomes. Because the number of chromosomes is unchanged, meiosis II is an **equational division**.

The chromosomes are evenly distributed among the four cells, but the cytoplasm may not be. In males, in the process of *spermatogenesis*, meiosis produces four immature gametes that are identical in size. Each will develop into a functional sperm. In females, meiosis produces one relatively huge oocyte and three tiny, nonfunctional polar bodies. (If fertilization takes place, the oocyte completes meiosis II, yielding an ovum.) We examine the details of spermatogenesis next. We will consider the production of female oocytes, *oogenesis*, in Section 28-5.

Spermatogenesis

As we have indicated, **spermatogenesis** (sper-ma-tō-JEN-eh-sis) is the process of sperm formation (Figure 28–9). Spermatogenesis begins at *puberty* (sexual maturation) and continues until relatively late in life (past age 70).

The complete process of spermatogenesis takes about 64 days. It involves three steps:

1. *Mitosis.* **Spermatogonia** (sper-ma-tō-GŌ-nē-uh) are stem cells in the seminiferous tubules that undergo cell divisions throughout adult life. Some of these cells differentiate into **primary spermatocytes** (sper-MA-tō-sīts), which prepare to begin meiosis.

Figure 28–9 The Process of Spermatogenesis. The fates of three representative chromosome pairs are shown; for clarity, maternal and paternal chromatids are not identified.

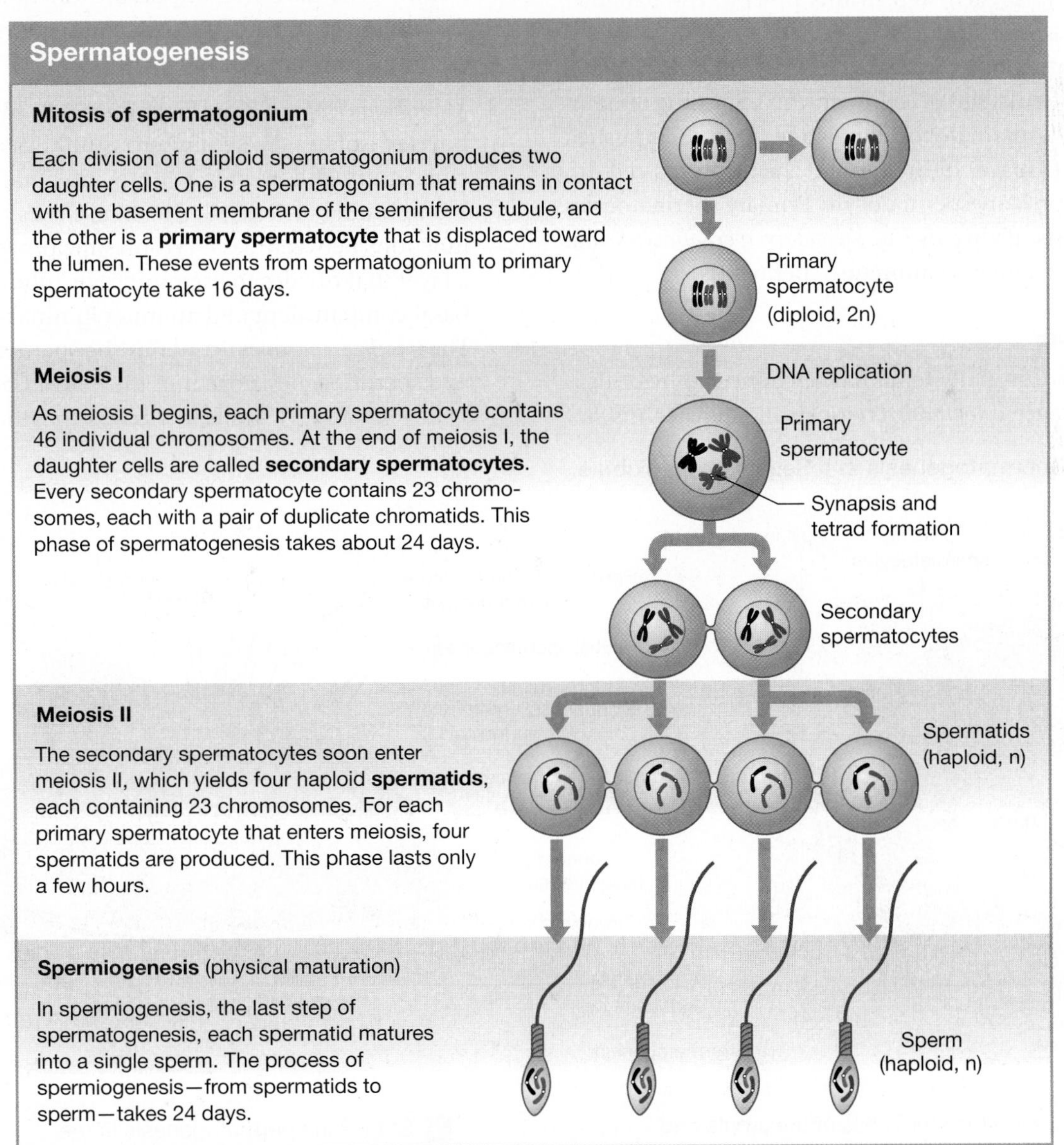
Spermatogenesis
Mitosis of spermatogonium
Each division of a diploid spermatogonium produces two daughter cells. One is a spermatogonium that remains in contact with the basement membrane of the seminiferous tubule, and the other is a primary spermatocyte that is displaced toward the lumen. These events from spermatogonium to primary spermatocyte take 16 days.
Primary spermatocyte (diploid, 2n)
Meiosis I
As meiosis I begins, each primary spermatocyte contains 46 individual chromosomes. At the end of meiosis I, the daughter cells are called secondary spermatocytes. Every secondary spermatocyte contains 23 chromosomes, each with a pair of duplicate chromatids. This phase of spermatogenesis takes about 24 days.
DNA replication
Primary spermatocyte
Synapsis and tetrad formation
Secondary spermatocytes
Meiosis II
The secondary spermatocytes soon enter meiosis II, which yields four haploid spermatids, each containing 23 chromosomes. For each primary spermatocyte that enters meiosis, four spermatids are produced. This phase lasts only a few hours.
Spermatids (haploid, n)
Spermiogenesis (physical maturation)
In spermiogenesis, the last step of spermatogenesis, each spermatid matures into a single sperm. The process of spermiogenesis—from spermatids to sperm—takes 24 days.
Sperm (haploid, n)

? What is the name of the cell at each stage of sperm development, from mitosis to spermiogenesis?

28

2. *Meiosis.* Primary spermatocytes then undergo meiotic divisions that produce **spermatids** (SPER-ma-tidz), undifferentiated male gametes.
3. *Spermiogenesis.* In *spermiogenesis,* spermatids differentiate into physically mature **sperm**.

Spermatogenesis is a continuous process, so all stages of meiosis are seen within the seminiferous tubules. Each seminiferous tubule contains spermatogonia, spermatocytes at various stages of meiosis, spermatids, and sperm. They also contain large *nurse cells,* which provide a microenvironment that supports spermatogenesis. The sperm eventually lose contact with the nurse cells of the seminiferous tubule and enter the fluid in the lumen.

Mitosis and Meiosis during Spermatogenesis

Spermatogenesis begins at the outermost layer of cells in the seminiferous tubules and proceeds toward the lumen (Figure 28–10a,b). At each step in this process, the daughter cells move closer to the lumen.

First, spermatogonia divide by mitosis to produce two daughter cells. One daughter cell from each division remains at that location as a spermatogonium, while the other is pushed toward the lumen (space) of the tubule. The displaced cell differentiates into a primary spermatocyte. Primary spermatocytes then begin meiosis, giving rise to *secondary spermatocytes* that divide and differentiate into immature spermatids.

Spermiogenesis

The four spermatids initially remain interconnected by bridges of cytoplasm because *cytokinesis* (cytoplasmic division) is not completed in meiosis I or meiosis II. These connections assist in transferring nutrients and hormones between the cells, helping ensure that the cells develop together. The bridges are not broken until the last stages of physical maturation.

In **spermiogenesis**, the last step of spermatogenesis, the spermatids gradually develop into mature sperm (Figures 28–9, 28–11). At *spermiation,* a sperm loses its attachment to the nurse cell and enters the lumen of the seminiferous tubule.

The Role of Nurse Cells

Developing spermatocytes undergoing meiosis, and spermatids undergoing spermiogenesis, are not free in the seminiferous tubules. Instead, they are surrounded by the cytoplasm of adjacent *nurse cells*. **Nurse cells**, which are also known as *Sertoli cells,* play a critical supportive role in spermatogenesis.

These cells have five important functions that directly or indirectly affect mitosis, meiosis, and spermiogenesis within the seminiferous tubules:

- *Maintenance of the Blood Testis Barrier.* A **blood testis barrier** isolates the seminiferous tubules from the general circulation. This barrier is comparable in function to the blood brain barrier. p. 473 Nurse cells are joined by tight junctions between basal cytoplasmic processes. They form a layer that divides the seminiferous tubule into an outer basal compartment and an inner luminal compartment. The *basal compartment* contains the spermatogonia. Meiosis and spermiogenesis occur in the *luminal compartment* (see Figure 28–10a,b). Transport across the nurse cells is tightly

Figure 28–10 Spermatogenesis in a Seminiferous Tubule.

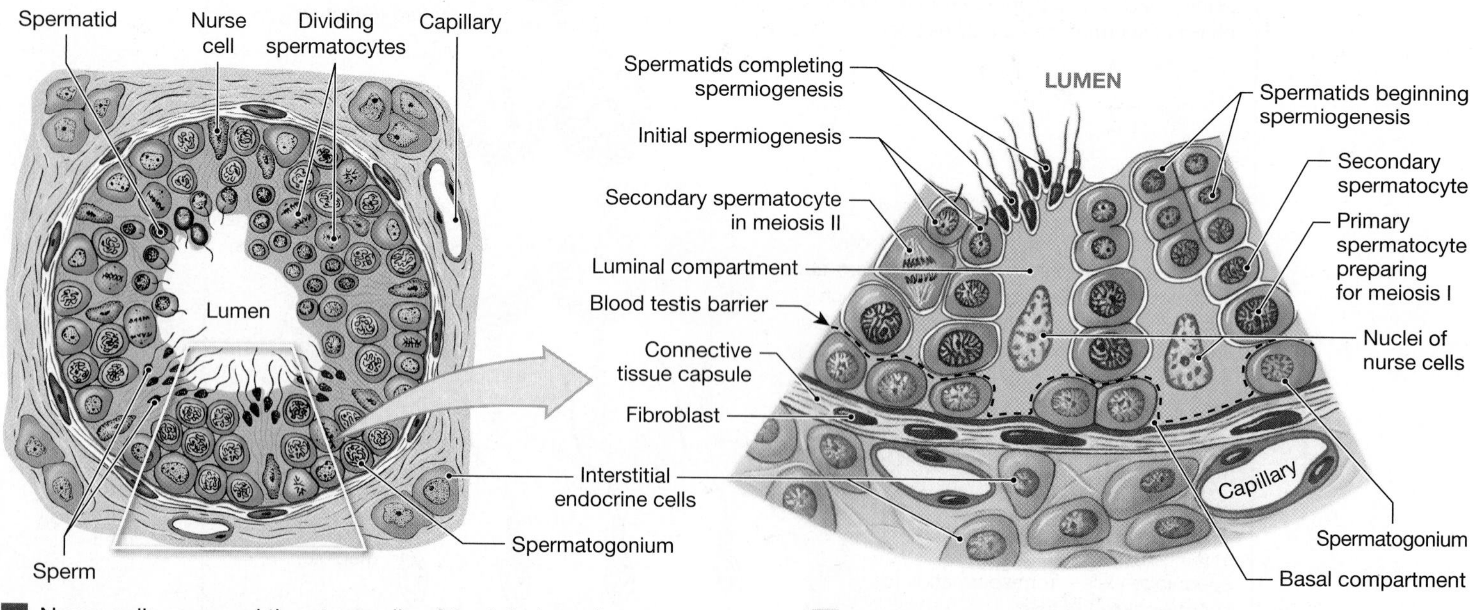
Spermatid
Nurse cell
Dividing spermatocytes
Capillary
Lumen
Sperm
Interstitial endocrine cells
Spermatogonium
Spermatids completing spermiogenesis
Initial spermiogenesis
Secondary spermatocyte in meiosis II
Luminal compartment
Blood testis barrier
Connective tissue capsule
Fibroblast
LUMEN
Spermatids beginning spermiogenesis
Secondary spermatocyte
Primary spermatocyte preparing for meiosis I
Nuclei of nurse cells
Capillary
Spermatogonium
Basal compartment

a Nurse cells surround the stem cells of the tubule and support the developing spermatocytes and spermatids.

b Stages in spermatogenesis in the wall of a seminiferous tubule.

Figure 28–11 The Process of Spermiogenesis and Anatomy of a Sperm. ATLAS: Plate 60a

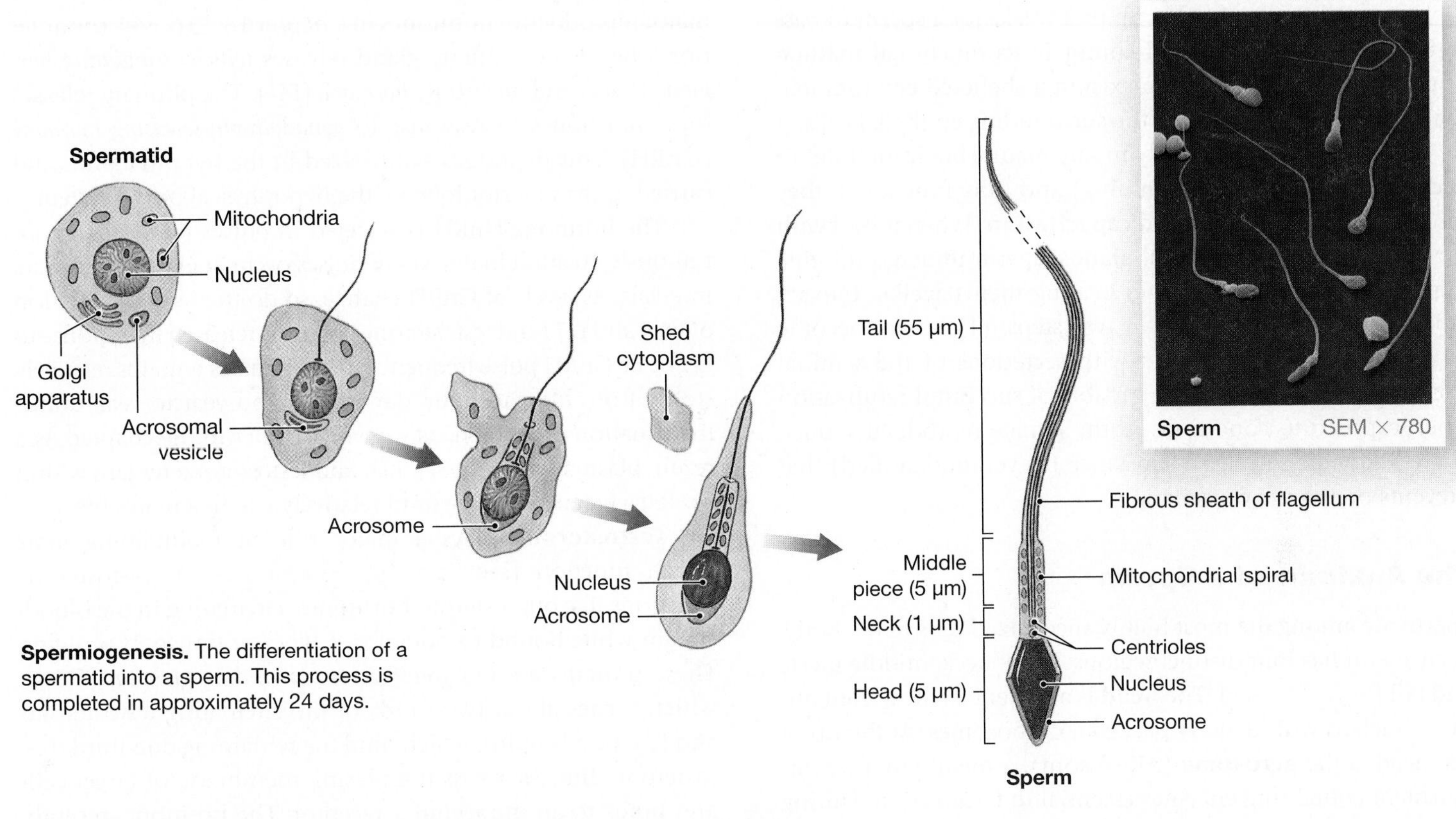

Spermatid
Mitochondria
Nucleus
Golgi apparatus
Acrosomal vesicle
Acrosome
Shed cytoplasm
Nucleus
Acrosome
Tail (55 μm)
Middle piece (5 μm)
Neck (1 μm)
Head (5 μm)
Fibrous sheath of flagellum
Mitochondrial spiral
Centrioles
Nucleus
Acrosome
Sperm
Sperm
SEM × 780

Spermiogenesis. The differentiation of a spermatid into a sperm. This process is completed in approximately 24 days.

? In which of the four parts of a sperm are the centrioles found?

regulated, so conditions in the luminal compartment remain very stable.

- The nurse cells produce the fluid in the lumen of a seminiferous tubule. They also regulate the fluid's composition. This luminal fluid is very different from the surrounding interstitial fluid. It is high in androgens, estrogens, potassium ions, and amino acids. The blood testis barrier is essential to preserving the differences between the luminal fluid and the interstitial fluid.
- In addition, this barrier prevents immune system cells from detecting and attacking the developing sperm. The plasma membranes of sperm contain sperm-specific antigens not found in somatic cell membranes, so they might be identified as "foreign."

- *Support of Mitosis and Meiosis.* Circulating follicle-stimulating hormone (FSH) and testosterone stimulate nurse cells. These stimulated nurse cells then promote the division of spermatogonia and the meiotic divisions of spermatocytes.
- *Support of Spermiogenesis.* Spermiogenesis requires nurse cells. These cells surround and enfold the spermatids, providing nutrients and chemical stimuli that promote their development. Nurse cells also phagocytize cytoplasm that is shed by spermatids as they develop into sperm.
- *Secretion of Inhibin.* Nurse cells secrete the peptide hormone *inhibin* (in-HIB-in) in response to factors released by developing sperm. Inhibin depresses the pituitary production of FSH, and perhaps the hypothalamic secretion of gonadotropin-releasing hormone (GnRH). The faster the rate of sperm production, the more inhibin is secreted. By regulating FSH and GnRH secretion, nurse cells provide negative feedback control of spermatogenesis.
- *Secretion of Androgen-Binding Protein. Androgen-binding protein* (ABP) binds androgens (primarily testosterone) in the luminal fluid of the seminiferous tubules. This protein is likely important for increasing androgen concentration and stimulating spermiogenesis. FSH stimulates the production of ABP.

Maturation of Sperm

The sperm that lie within the lumen of the seminiferous tubule have most of the physical characteristics of mature sperm, but are functionally immature and incapable of coordinated movement or fertilization. It takes up to 2 weeks for a sperm to pass through the epididymis and complete its functional maturation. During this time, sperm exist in a sheltered environment that is precisely regulated by the surrounding epithelial cells.

Sperm leaving the epididymis are mature but immobile. To become *motile* (actively swimming) and fully functional, they must undergo a process called **capacitation**. When mixed with the secretions of the seminal glands, sperm undergo the first step in capacitation and begin beating their flagella,. Capacitation normally takes place in two steps: (1) Sperm become motile when they are mixed with secretions of the seminal glands, and (2) they become capable of successful fertilization when exposed to conditions in the female reproductive tract. The epididymis secretes a substance (as yet unidentified) that prevents premature capacitation.

The Anatomy of a Sperm

Sperm are among the most highly specialized cells in the body. Each sperm has four distinct regions: head, neck, middle piece, and tail (Figure 28–11). The **head** is a flattened ellipse containing a nucleus with densely packed chromosomes. At the tip of the head is the **acrosome** (AK-rō-sōm), a membranous compartment containing enzymes essential to fertilization. During spermiogenesis, saccules of the spermatid's Golgi apparatus fuse and flatten into an *acrosomal vesicle*, which ultimately forms the acrosome.

A short **neck** attaches the head to the **middle piece.** The neck contains both centrioles of the original spermatid. The microtubules of the distal centriole are continuous with those of the middle piece and tail. Mitochondria in the middle piece are arranged in a spiral around the microtubules. Mitochondria provide the ATP required to move the tail.

The **tail** is the only flagellum in the human body. A *flagellum* is a whiplike organelle that moves a cell from one place to another. Cilia beat in a predictable, wavelike fashion, but the tail of a sperm has a whip-like, corkscrew motion.

Unlike other, less specialized cells, a mature sperm lacks an endoplasmic reticulum, a Golgi apparatus, lysosomes, peroxisomes, inclusions, and many other organelles. The loss of these organelles reduces the cell's size and mass. A sperm is essentially a mobile carrier for chromosomes, and extra weight would slow it down. Because a sperm lacks glycogen or other energy reserves, it must absorb nutrients (primarily fructose) from the surrounding fluid.

Hormonal Regulation of Male Reproductive Function

The hormonal interactions that regulate male reproductive function are diagrammed in **Spotlight Figure 28–12**. We introduced major reproductive hormones in Chapter 18. p. 644 The anterior lobe of the pituitary gland releases *follicle-stimulating hormone* (FSH) and *luteinizing hormone* (LH). The pituitary releases these hormones in response to *gonadotropin-releasing hormone* (GnRH). This peptide is synthesized in the hypothalamus and carried to the anterior lobe by the hypophyseal portal system.

The hormone GnRH is secreted in pulses rather than continuously. In adult males, small pulses occur at 60- to 90-minute intervals. As levels of GnRH change, so do the rates of secretion of FSH and LH (and testosterone, which is released in response to LH). The GnRH pulse frequency in adult males remains relatively steady from hour to hour, day to day, and year to year, unlike the situation in women, as we will see later in the chapter. As a result, plasma levels of FSH, LH, and testosterone remain within a relatively narrow range until relatively late in a man's life.

Testosterone plays a major role in maintaining male sexual function (see **Spotlight Figure 28–12**). Testosterone functions like other steroid hormones, circulating in the bloodstream while bound to one of two types of transport proteins. These proteins are (1) *gonadal steroid-binding globulin* (GBG), which carries about two-thirds of the circulating testosterone, and (2) the albumins, which bind the remaining one-third. Testosterone diffuses across the plasma membrane of target cells and binds to an intracellular receptor. The hormone–receptor complex then binds to the DNA in the nucleus.

+ Clinical Note Dehydroepiandrosterone (DHEA)

Dehydroepiandrosterone, or **DHEA**, is the primary androgen secreted by the zona reticularis of the adrenal cortex. p. 636 As noted in Chapter 18, these androgens, which are secreted in small amounts, are converted to testosterone (or estrogens) by other tissues. The significance of this small adrenal androgen secretion in both sexes remains unclear, but DHEA is being promoted as a wonder drug for increasing vitality, strength, and muscle mass. The effects of long-term high doses of DHEA remain largely unknown. However, high levels of DHEA in women have been linked to an increased risk of ovarian cancer as well as to masculinization, due to the conversion of DHEA to testosterone. The International Olympic Committee, NCAA, MLB, and NFL have banned the use of DHEA.

SPOTLIGHT

Figure 28–12
Hormonal Regulation of Male Reproduction

Male reproductive function is regulated by the complex interaction of hormones from the hypothalamus, anterior lobe of the pituitary gland, and the testes. Negative feedback systems keep testosterone levels within a relatively narrow range until late in life.

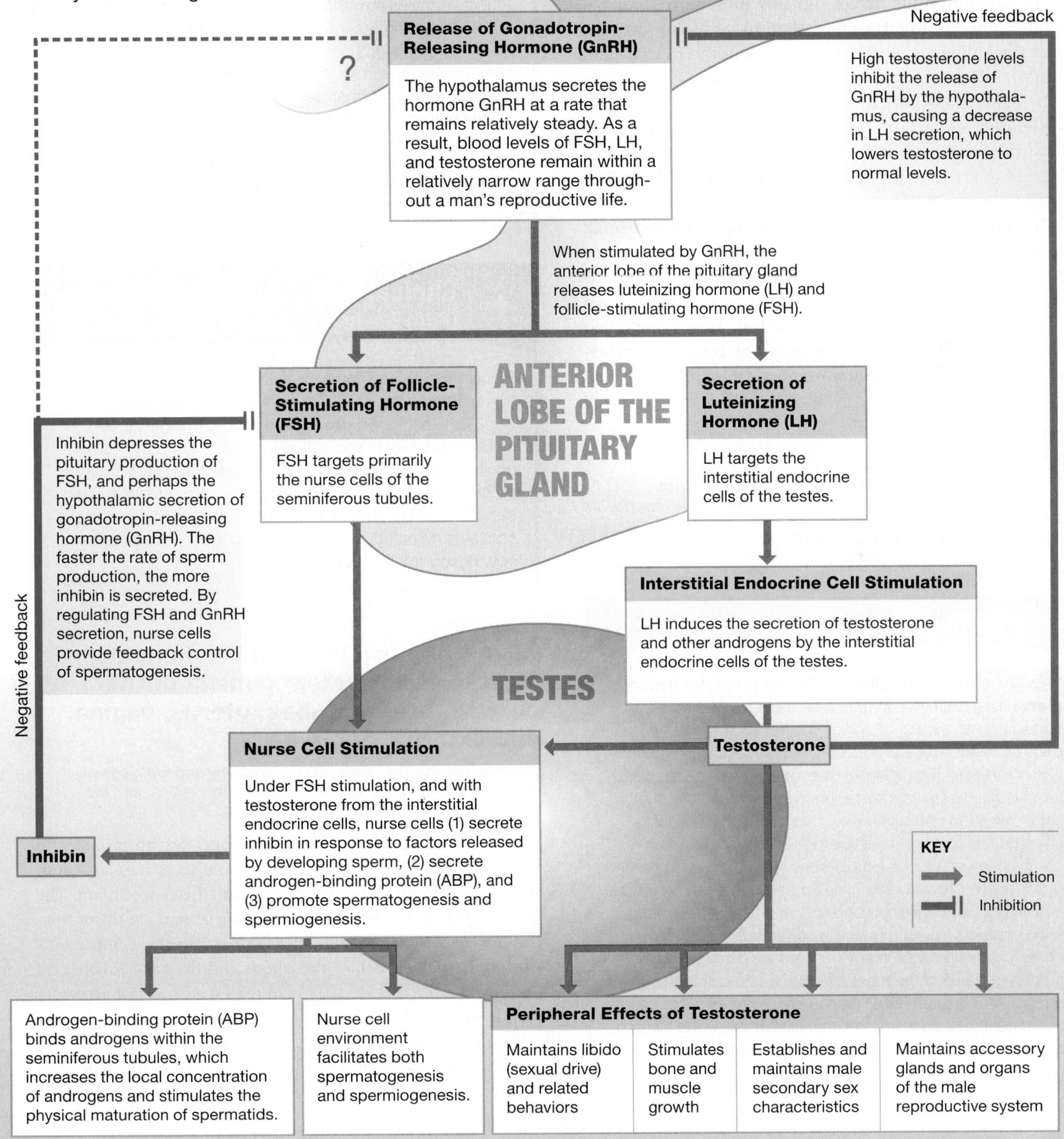

HYPOTHALAMUS

Release of Gonadotropin-Releasing Hormone (GnRH)

The hypothalamus secretes the hormone GnRH at a rate that remains relatively steady. As a result, blood levels of FSH, LH, and testosterone remain within a relatively narrow range throughout a man's reproductive life.

?

Negative feedback

High testosterone levels inhibit the release of GnRH by the hypothalamus, causing a decrease in LH secretion, which lowers testosterone to normal levels.

When stimulated by GnRH, the anterior lobe of the pituitary gland releases luteinizing hormone (LH) and follicle-stimulating hormone (FSH).

ANTERIOR LOBE OF THE PITUITARY GLAND

Secretion of Follicle-Stimulating Hormone (FSH)

FSH targets primarily the nurse cells of the seminiferous tubules.

Secretion of Luteinizing Hormone (LH)

LH targets the interstitial endocrine cells of the testes.

Negative feedback

Inhibin depresses the pituitary production of FSH, and perhaps the hypothalamic secretion of gonadotropin-releasing hormone (GnRH). The faster the rate of sperm production, the more inhibin is secreted. By regulating FSH and GnRH secretion, nurse cells provide feedback control of spermatogenesis.

Interstitial Endocrine Cell Stimulation

LH induces the secretion of testosterone and other androgens by the interstitial endocrine cells of the testes.

TESTES

Testosterone

Nurse Cell Stimulation

Under FSH stimulation, and with testosterone from the interstitial endocrine cells, nurse cells (1) secrete inhibin in response to factors released by developing sperm, (2) secrete androgen-binding protein (ABP), and (3) promote spermatogenesis and spermiogenesis.

Inhibin

KEY
Stimulation
Inhibition

Androgen-binding protein (ABP) binds androgens within the seminiferous tubules, which increases the local concentration of androgens and stimulates the physical maturation of spermatids.

Nurse cell environment facilitates both spermatogenesis and spermiogenesis.

Peripheral Effects of Testosterone

Maintains libido (sexual drive) and related behaviors	Stimulates bone and muscle growth	Establishes and maintains male secondary sex characteristics	Maintains accessory glands and organs of the male reproductive system

In many target tissues, some of the arriving testosterone is converted to **dihydrotestosterone (DHT)**. A small amount of DHT diffuses back out of the cell and into the bloodstream, and DHT levels are usually about 10 percent of circulating testosterone levels. Dihydrotestosterone can also enter peripheral cells and bind to the same hormone receptors targeted by testosterone. In addition, some tissues (notably those of the external genitalia) respond to DHT rather than to testosterone. Other tissues (including the prostate) are more sensitive to DHT than to testosterone.

In adult males, negative feedback controls the level of testosterone production. Above-normal testosterone levels inhibit the release of GnRH by the hypothalamus, causing a reduction in LH secretion and lowering testosterone levels (see **Spotlight Figure 28–12**).

The plasma of adult males also contains relatively small amounts of the estrogenic hormone *estradiol* (2 ng/dL versus 525 ng/dL of testosterone). Seventy percent of the estradiol is formed from circulating testosterone. Interstitial endocrine cells and nurse cells of the testes secrete the rest. An enzyme called aromatase converts testosterone to estradiol. For unknown reasons, estradiol production increases in older men.

Checkpoint

7. **What effect would a low FSH level have on sperm production?**

See the blue Answers tab at the back of the book.

Clinical Note Enlarged Prostate

Enlargement of the prostate, or **benign prostatic hyperplasia**, is so common in men over the age of 50 that it is considered an aspect of aging. The increase in size happens as testosterone production by the interstitial endocrine cells decreases. For unknown reasons, small masses called *prostatic concretions* may form within the glands. At the same time, the interstitial endocrine cells begin releasing small quantities of estrogens into the bloodstream. The combination of a lower testosterone level and the presence of estrogens probably stimulates prostatic growth. In severe cases, prostatic swelling constricts and blocks the urethra and constricts the rectum. If not corrected, the urinary obstruction can cause permanent kidney damage. Partial surgical removal is the most effective treatment. In the procedure known as a **TURP** (*transurethral prostatectomy*), an instrument pushed along the urethra cuts away the swollen prostatic tissue. Most of the prostate remains in place, and normal function returns.

Clinical Note Prostate Cancer

Prostate cancer is the second most common cancer and the second most lethal cancer in males. For reasons that are poorly understood, prostate cancer rates for Asian American males are relatively low compared with those of either Caucasian Americans or Black Americans. For all ages and ethnic groups, the rate of prostate cancer is rising sharply. The reason for the increase is not known. Aggressive diagnosis and treatment of localized prostate cancer in elderly patients is controversial because many of these men have nonmetastatic tumors (tumors that have not spread). In addition, even if their cancer is left untreated, these men are more likely to die of some other disease.

Clinical Note Prostate-Specific Antigen (PSA) Testing

Blood tests are used to screen for prostate cancer and to monitor the success of treatment. The most sensitive is a blood test called **prostate-specific antigen (PSA)**. An elevated level of this antigen, which is normally present in a low concentration, may indicate the presence of prostate cancer. Screening with periodic PSA tests is now being recommended for some men over age 50. If prostate cancer is detected, treatment decisions vary depending on how rapidly the PSA level is rising.

28-4 The structures of the female reproductive system consist of the ovaries, uterine tubes, uterus, vagina, and external genitalia

Learning Outcome **Describe the anatomy and functions of the ovaries, the uterine tubes, the uterus, the vagina, and the structures of the female external genitalia.**

A woman's reproductive system produces sex hormones and functional gametes. It must also be able to protect and support a developing embryo and nourish a newborn infant. The main organs of the female reproductive system are the *ovaries*, the *uterine tubes*, the *uterus*, the *vagina*, and the components of the external genitalia—the *clitoris* and the *labia majora* and *labia minora* (**Figure 28–13**). As in males, a variety of accessory glands—the *urethral glands* and the *greater vestibular glands*—release secretions into the female reproductive tract.

The ovaries, uterine tubes, and uterus are enclosed within an extensive mesentery known as the **broad ligament**

28

Figure 28–13 Sagittal Section of the Female Reproductive System. ATLAS: Plate 65

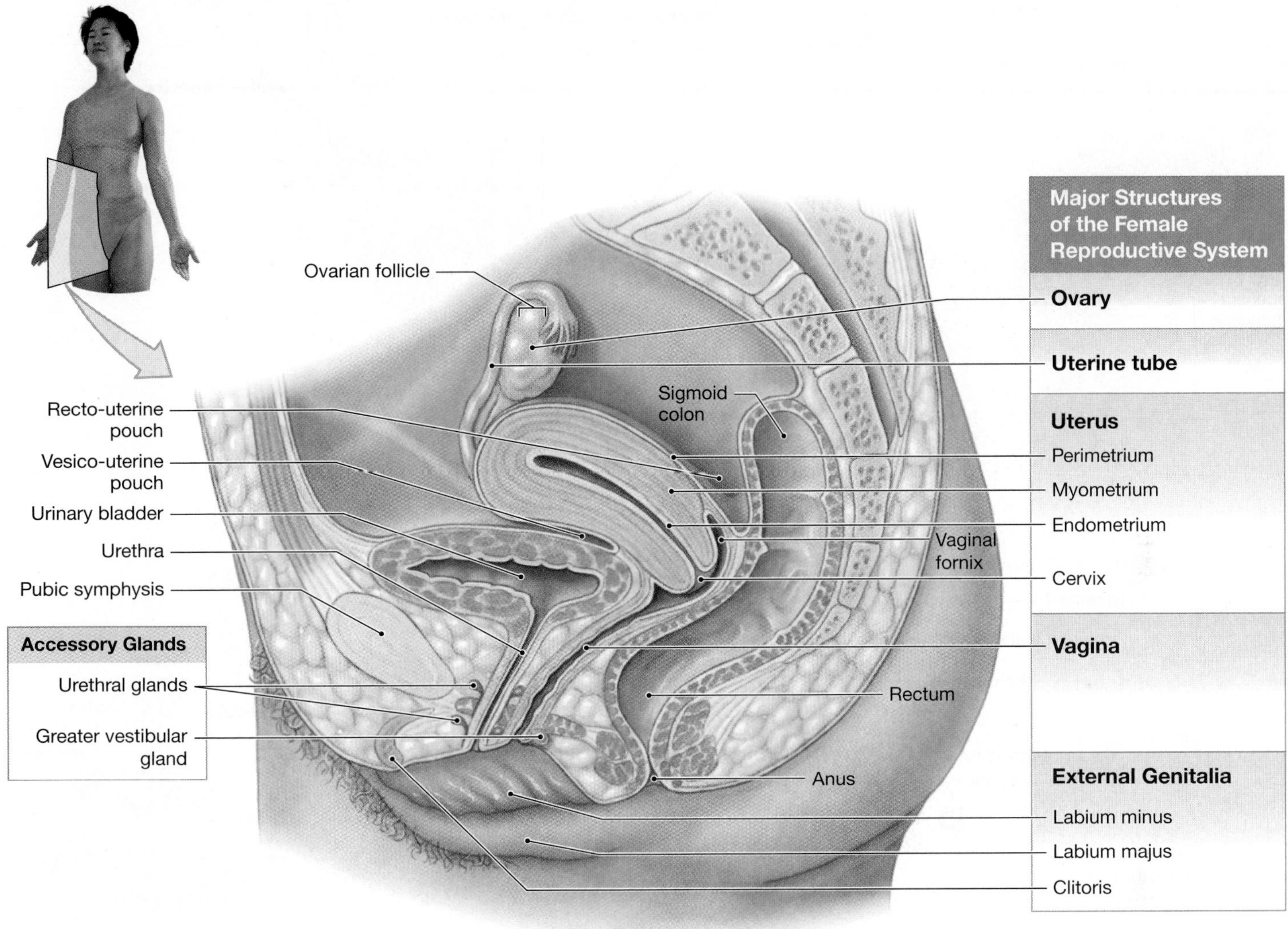
Ovarian follicle
Recto-uterine pouch
Vesico-uterine pouch
Urinary bladder
Urethra
Pubic symphysis
Accessory Glands
Urethral glands
Greater vestibular gland
Sigmoid colon
Vaginal fornix
Rectum
Anus
Major Structures of the Female Reproductive System
Ovary
Uterine tube
Uterus
Perimetrium
Myometrium
Endometrium
Cervix
Vagina
External Genitalia
Labium minus
Labium majus
Clitoris

(Figure 28–14a). The uterine tubes run along the superior border of the broad ligament and open into the pelvic cavity lateral to the ovaries. The broad ligament attaches to the sides and floor of the pelvic cavity, where it becomes continuous with the parietal peritoneum. The broad ligament subdivides this part of the peritoneal cavity.

The pocket formed between the posterior wall of the uterus and the anterior surface of the colon is the **recto-uterine** (rek-tō-YŪ-ter-in) **pouch**. The pocket formed between the uterus and the posterior wall of the bladder is the **vesico-uterine** (ves-ih-kō-YŪ-ter-in) **pouch**. These subdivisions are easily seen in sagittal section (see Figure 28–13).

Several other ligaments assist the broad ligament in supporting and stabilizing the uterus and associated reproductive organs. These ligaments lie within the mesentery sheet of the broad ligament and are connected to the ovaries or uterus. The broad ligament limits side-to-side movement and rotation, and the other ligaments (described in our discussion of the ovaries and uterus) prevent superior–inferior movement.

Let's look further at the structures and basic functions of the female reproductive system.

The Ovaries

The paired ovaries are small, lumpy, almond-shaped organs near the lateral walls of the pelvic cavity (Figure 28–14b). The ovaries perform three main functions. They (1) produce immature female gametes, or oocytes; (2) secrete female sex hormones, including *estrogens* and *progesterone*; and (3) secrete inhibin, involved in the feedback control of pituitary FSH production.

The **mesovarium** (mez-ō-VĀ-rē-um), a thickened fold of mesentery, supports and stabilizes the position of each ovary

Figure 28–14 Relationships between the Ovaries, Uterine Tubes, and Uterus. ATLAS: Plate 67

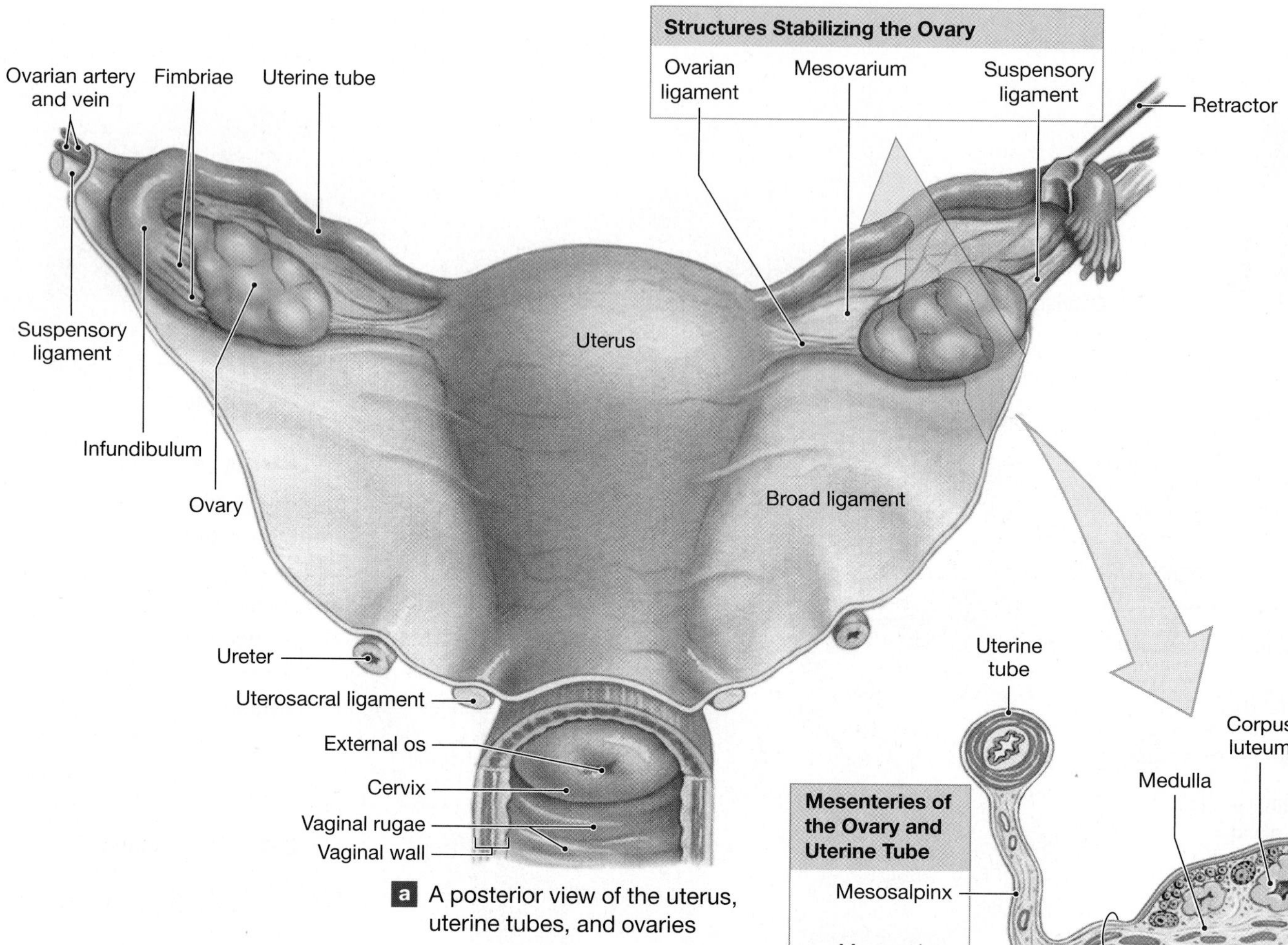

a A posterior view of the uterus, uterine tubes, and ovaries

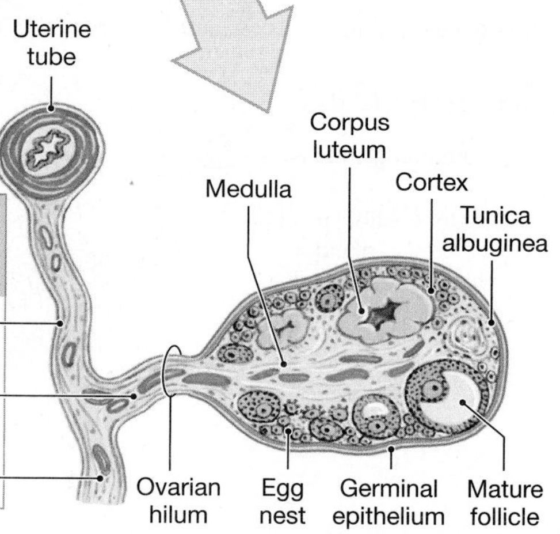

b A sectional view of the ovary, uterine tube, and associated mesenteries

(see Figure 28–14a). Each ovary is also stabilized by a pair of supporting ligaments: the ovarian ligament and the suspensory (*infundibulopelvic*) ligament. The **ovarian ligament** extends from the uterus, near the attachment of the uterine tube, to the medial surface of the ovary. The **suspensory ligament** extends from the lateral surface of the ovary past the open end of the uterine tube to the pelvic wall. The suspensory ligament contains the major blood vessels of the ovary: the **ovarian artery** and **ovarian vein**. These vessels are connected to the ovary at the **ovarian hilum**, where the ovary attaches to the mesovarium (see Figure 28–14b).

A typical ovary is about 5 cm long, 2.5 cm wide, and 8 mm thick (2 in. × 1 in. × 0.33 in.). It weighs 6–8 g (about 0.25 oz). An ovary is pink or yellowish and has a nodular consistency. The visceral peritoneum, or *germinal epithelium*, covers the surface of each ovary and consists of a layer of columnar epithelial cells that overlies a dense connective tissue layer called the **tunica albuginea** (al-bū-JIN-ē-ah) (see Figure 28–14b). We divide the interior tissues, or **stroma**, of the ovary into a superficial *cortex* and a deeper *medulla*. Gametes are produced in the cortex.

The Uterine Tubes

Each **uterine tube** (*fallopian tube* or *oviduct*) is a hollow, muscular cylinder measuring roughly 13 cm (5.2 in.) in length (see Figure 28–14a).

Regions of the Uterine Tubes

We divide each uterine tube into three regions, from lateral to medial (Figure 28–15a):

1. *The Infundibulum.* The end closest to the ovary forms an expanded funnel, or **infundibulum**, with numerous fingerlike projections that extend into the pelvic cavity. The projections are called **fimbriae** (FIM-brē-ē) (Figure 28–16a). Fimbriae drape over the surface of the ovary, but there is no physical connection between the two structures. The inner surfaces of the infundibulum are lined with cilia that beat toward the middle region of the uterine tube.
2. *The Ampulla.* The middle region is the called the **ampulla**. The thickness of its smooth muscle layers gradually increases as the tube approaches the uterus.
3. *The Isthmus.* The ampulla leads to the **isthmus** (IS-mus), a short region connected to the uterine wall.

+ Clinical Note Ovarian Cancer

Ovarian cancer is not the most common female reproductive cancer—it is the third most common—but it is the most dangerous, because it is so difficult to diagnose in its early stages. A woman in the United States has a 1-in-70 chance of developing ovarian cancer in her lifetime. The prognosis is relatively good for cancers that originate in the general ovarian tissues or from abnormal oocytes. These cancers generally respond well to some combination of chemotherapy, radiation, and surgery. However, 85 percent of ovarian cancers develop from epithelial cells, and sustained remission occurs in only about one-third of these cases.

Functional Anatomy of the Uterine Tube

The epithelium lining the uterine tube is composed of ciliated columnar epithelial cells with scattered mucin-secreting cells. Concentric layers of smooth muscle surround the mucosa (Figure 28–15b). A combination of ciliary movement and peristaltic contractions in the walls of the uterine tube transports oocytes toward the uterus (Figure 28–15c). In addition, the uterine tube provides a nutrient-rich environment that contains lipids and glycogen for both oocytes and sperm. If fertilization occurs, it typically takes place near the boundary between the ampulla and isthmus of the uterine tube.

In addition to ciliated cells, the epithelium lining the uterine tubes contains *peg cells*. The peg cells project into the lumen of the uterine tube. They secrete a fluid that both completes the capacitation of sperm and supplies nutrients to sperm and the pre-embryo.

The Uterus

The **uterus** protects, nourishes, and removes wastes for the developing *embryo* (weeks 1–8) and *fetus* (week 9 through delivery). In addition, contractions of the muscular uterus are important in expelling the fetus at birth.

The uterus is a small, pear-shaped organ (Figure 28–16). It is about 7.5 cm (3 in.) long with a maximum diameter of 5 cm (2 in.). It weighs 50–100 g (1.75–3.5 oz). In its normal position, the uterus bends anteriorly near its base, a condition known as *anteflexion* (an-tē-FLEK-shun) (look back at Figure 28–13). In this position, the uterus covers the superior and posterior surfaces of the urinary bladder. If the uterus bends

Figure 28–15 **Anatomy of the Uterine Tubes.**

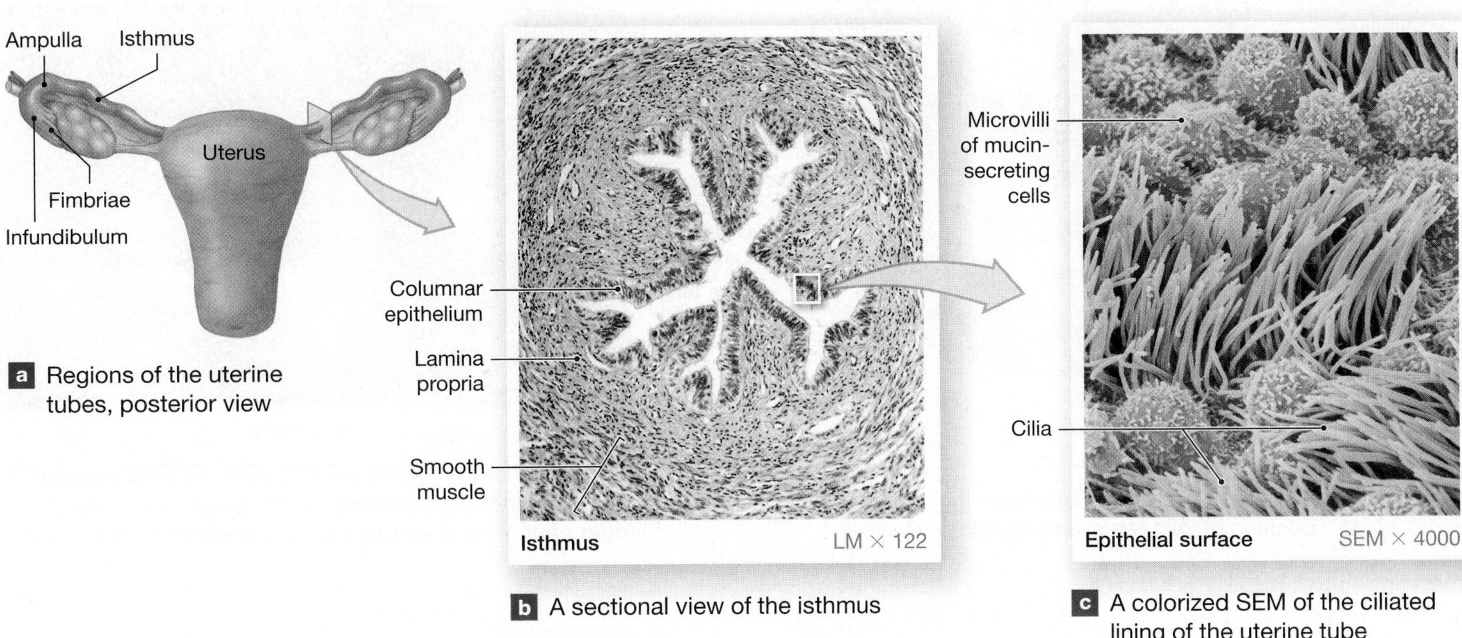
Ampulla
Isthmus
Uterus
Fimbriae
Infundibulum
Columnar epithelium
Lamina propria
Smooth muscle
Isthmus
LM × 122
Microvilli of mucin-secreting cells
Cilia
Epithelial surface
SEM × 4000

a Regions of the uterine tubes, posterior view

b A sectional view of the isthmus

c A colorized SEM of the ciliated lining of the uterine tube

Figure 28–16 **Gross Anatomy of the Uterus.** ATLAS: Plates 66; 67

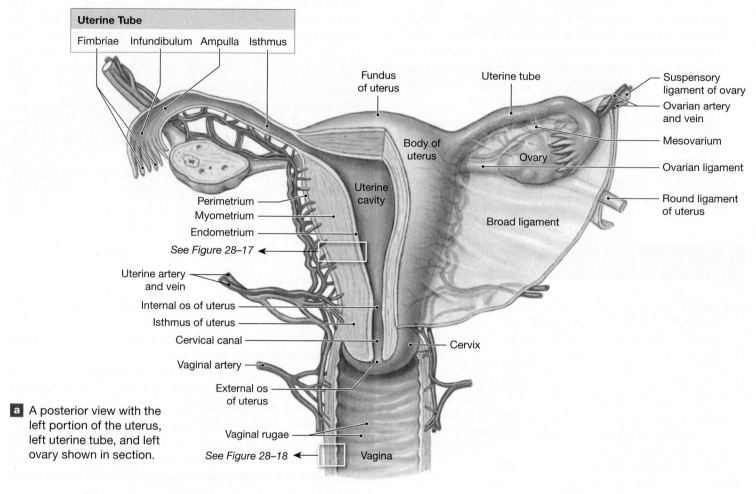

Uterine Tube
Fimbriae Infundibulum Ampulla Isthmus
Fundus of uterus
Uterine tube
Suspensory ligament of ovary
Ovarian artery and vein
Mesovarium
Body of uterus
Ovary
Ovarian ligament
Uterine cavity
Round ligament of uterus
Perimetrium
Myometrium
Broad ligament
Endometrium
See Figure 28–17
Uterine artery and vein
Internal os of uterus
Isthmus of uterus
Cervical canal
Cervix
Vaginal artery
External os of uterus
Vaginal rugae
See Figure 28–18
Vagina

a A posterior view with the left portion of the uterus, left uterine tube, and left ovary shown in section.

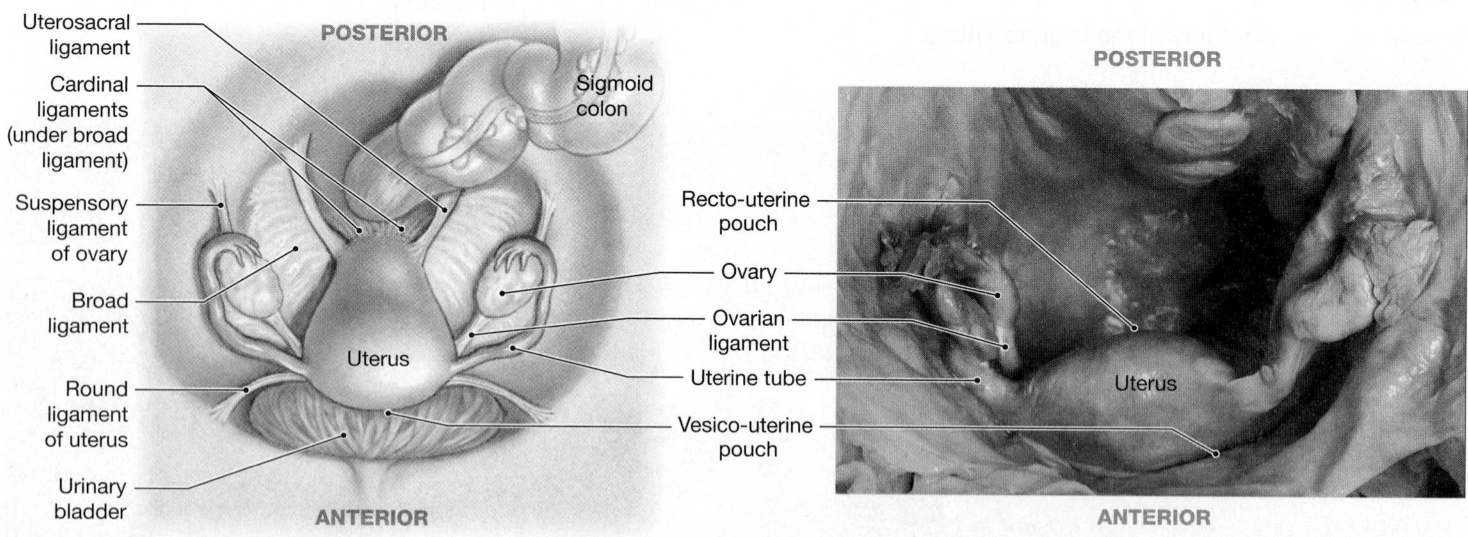
POSTERIOR
Uterosacral ligament
Cardinal ligaments (under broad ligament)
Sigmoid colon
Suspensory ligament of ovary
Broad ligament
Uterus
Round ligament of uterus
Urinary bladder
ANTERIOR
POSTERIOR
Recto-uterine pouch
Ovary
Ovarian ligament
Uterine tube
Vesico-uterine pouch
Uterus
ANTERIOR

b A superior view of the ligaments that stabilize the position of the uterus in the pelvic cavity.

c Superior view of the female pelvic cavity showing supporting ligaments of uterus and ovaries. In the photo, the urinary bladder cannot be seen because it is covered by peritoneum.

backward toward the sacrum, the condition is termed *retroflexion* (re-trō-FLEK-shun). Retroflexion occurs in about 20 percent of adult women and has no clinical significance. (A retroflexed uterus generally becomes anteflexed spontaneously during the third month of pregnancy.)

Suspensory Ligaments of the Uterus

In addition to the broad ligament, three pairs of suspensory ligaments stabilize the uterus and limit its range of movement (Figure 28–16b). The **uterosacral** (yū-te-rō-SĀ-krul) **ligaments** extend from the lateral surfaces of the uterus to the anterior face of the sacrum, keeping the body of the uterus from moving inferiorly and anteriorly. The **round ligaments** arise on the lateral margins of the uterus just posterior and inferior to the attachments of the uterine tubes. These ligaments extend through the inguinal canal and end in the connective tissues of the external genitalia. (Unlike the inguinal canals in males, these canals in females are very small, containing only the round ligaments and the *ilioinguinal nerves*; this makes inguinal hernias in women very rare.) The round ligaments restrict posterior movement of the uterus. The **cardinal ligaments** extend from the base of the uterus and vagina to the lateral walls of the pelvis. These ligaments tend to prevent inferior movement of the uterus. The muscles and fascia of the pelvic floor provide additional mechanical support.

Gross Anatomy of the Uterus

We divide the uterus into four main anatomical regions (Figure 28–16a). The uterine **body** is the largest portion of the uterus. The **fundus** is the rounded portion of the body superior to the attachment of the uterine tubes. The body ends at a constriction known as the **isthmus** of the uterus. The **cervix** (SER-viks) is the inferior portion of the uterus that extends from the isthmus to the vagina.

> **+ Clinical Note** Pap Smear
>
> For many young women, the **Pap smear** is the reason for a regular visit to the doctor. This important screening test evaluates the health of the *cervix*, the inferior portion of the uterus that projects into the vaginal canal. Early detection of cervical cancer, the most common reproductive cancer in women ages 15–34, is the main goal of the test. Developed by Dr. Georgios Papanikolaou (hence the name of the test), the simple, quick, and inexpensive Pap smear is a form of *exfoliative cytology*. The practitioner opens the patient's vagina with a speculum, and gently scrapes the cervix with a swab to collect epithelial cells. These cells are then studied in the pathology lab for any evidence of cancerous changes.

The tubular cervix projects about 1.25 cm (0.5 in.) into the vagina. Within the vagina, the distal end of the cervix forms a curving surface that surrounds the **external os** (*os*, an opening or mouth) of the uterus. The external os leads into the **cervical canal**, a constricted passageway that opens into the **uterine cavity** of the body at the **internal os**.

The uterus receives blood from branches of the **uterine arteries**. These vessels arise from branches of the *internal iliac arteries*, and from the *ovarian arteries*, which arise from the abdominal aorta inferior to the renal arteries. The arteries to the uterus are extensively interconnected, ensuring a reliable flow of blood to the organ despite changes in its position and shape during pregnancy. Numerous veins and lymphatic vessels also drain each portion of the uterus.

The organ is innervated by autonomic fibers from the hypogastric plexus (sympathetic) and from sacral segments S_3 and S_4 (parasympathetic). Sensory information reaches the central nervous system (CNS) within the posterior roots of spinal nerves T_{11} and T_{12}. The most delicate anesthetic procedures used during labor and delivery, known as *segmental blocks*, target only spinal nerves T_{10}–L_1.

The Uterine Wall

The dimensions of the uterus are highly variable. In women of reproductive age who have not given birth, the uterine wall is about 1.5 cm (0.6 in.) thick.

Layers of the Uterine Wall. The uterine wall is made up of three main layers. The fundus and the posterior surface of the uterine body and isthmus are covered by an outer serous membrane that is continuous with the peritoneal lining. This incomplete serosa is called the **perimetrium**. The wall has a thick, middle, muscular **myometrium** (mī-ō-MĒ-trē-um; *myo-*, muscle + *metra*, uterus) and a thin, inner, glandular **endometrium** (en-dō-MĒ-trē-um) (Figure 28–17a).

The myometrium is the thickest portion of the uterine wall. It makes up almost 90 percent of the mass of the uterus. Smooth muscle in the myometrium is arranged into longitudinal, circular, and oblique layers. The smooth muscle tissue of the myometrium provides much of the force needed to move a fetus out of the uterus and into the vagina.

The endometrium makes up about 10 percent of the mass of the uterus. The glandular and vascular tissues of the endometrium support the physiological demands of the growing fetus. Vast numbers of *uterine glands* open onto the endometrial surface and extend deep into the lamina propria, almost to the myometrium. Under the influence of estrogens, the uterine glands, blood vessels, and epithelium change with the phases of the monthly *uterine cycle*.

Histology of the Uterine Wall. The endometrium is divided into a **functional layer**—the layer closest to the uterine

Figure 28–17 Histology of the Uterine Wall.

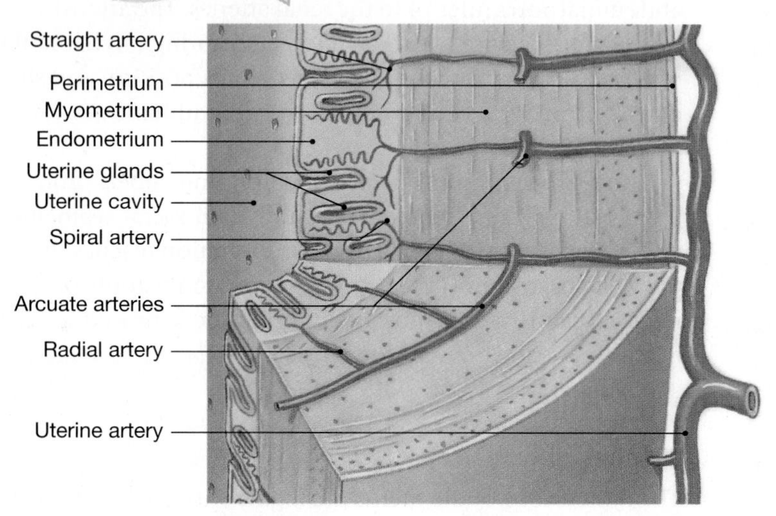
Straight artery
Perimetrium
Myometrium
Endometrium
Uterine glands
Uterine cavity
Spiral artery
Arcuate arteries
Radial artery
Uterine artery

a A diagrammatic sectional view of the uterine wall, showing the endometrial regions and the blood supply to the endometrium

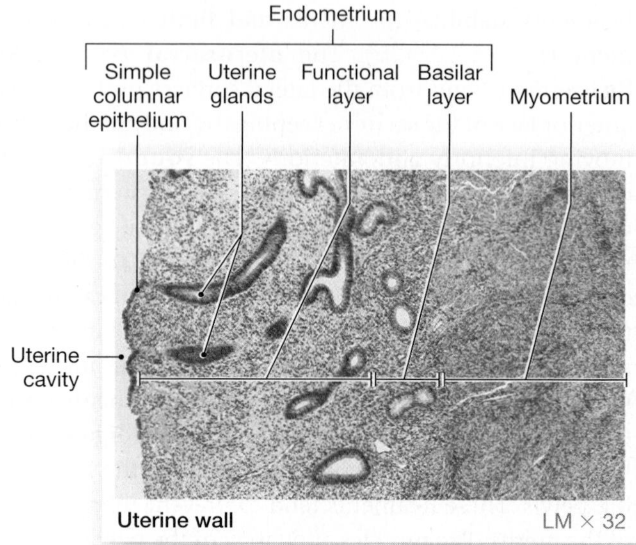
Endometrium
Simple columnar epithelium
Uterine glands
Functional layer
Basilar layer
Myometrium
Uterine cavity
Uterine wall
LM × 32

b The basic histological structure of the uterine wall

cavity—and a **basal layer**, adjacent to the myometrium (Figure 28–17b). The functional layer contains most of the **uterine glands** and contributes most of the endometrial thickness. This layer undergoes dramatic changes in thickness and structure during the uterine cycle. The basal layer attaches the endometrium to the myometrium and contains the terminal branches of the tubular uterine glands. The structure of the basal layer remains fairly constant over time, but the functional layer undergoes cyclical changes in response to sex hormone levels.

Within the myometrium, branches of the uterine arteries form **arcuate arteries**, which encircle the endometrium (see Figure 28–17a). **Radial arteries** supply **straight arteries**, which deliver blood to the basal layer of the endometrium, and **spiral arteries**, which supply the functional layer.

28

The Vagina

The **vagina** is an elastic, muscular tube that has three major functions. It (1) is a passageway for the elimination of menstrual fluids; (2) receives the penis during sexual intercourse, and holds sperm prior to their passage into the uterus; and (3) forms the inferior portion of the *birth canal*, through which the fetus passes during delivery.

Gross Anatomy of the Vagina

The vagina extends between the cervix and the body exterior (Figure 28–18a). At the proximal end of the vagina, the cervix projects into the **vaginal canal**. The shallow recess surrounding the cervical protrusion is known as the **vaginal fornix** (FOR-niks). The vagina is typically 7.5–9 cm (3–3.6 in.) long. Its diameter varies because it is highly distensible.

The two *bulbospongiosus* muscles extend along either side of the vaginal entrance. Contractions of the bulbospongiosus muscles constrict the vagina. ⮌ p. 361 These muscles cover the **bulb of vestibule**, a mass of erectile tissue on each side of the vagina (Figure 28–19). The bulb of vestibule has the same embryonic origins as the corpus spongiosum of the penis in males.

The vagina opens into the *vestibule*, a space bordered by the female external genitalia (see Figure 28–18a). The **hymen** (HĪ-men) is an elastic epithelial fold of variable size that partially blocks the entrance to the vagina. An intact hymen is typically stretched or torn during first sexual intercourse, tampon use, pelvic examination, or physical activity.

The vagina lies parallel to the rectum, and the two are in close contact posteriorly. Anteriorly, the urethra extends along

Figure 28–18 Anatomy of the Vagina.

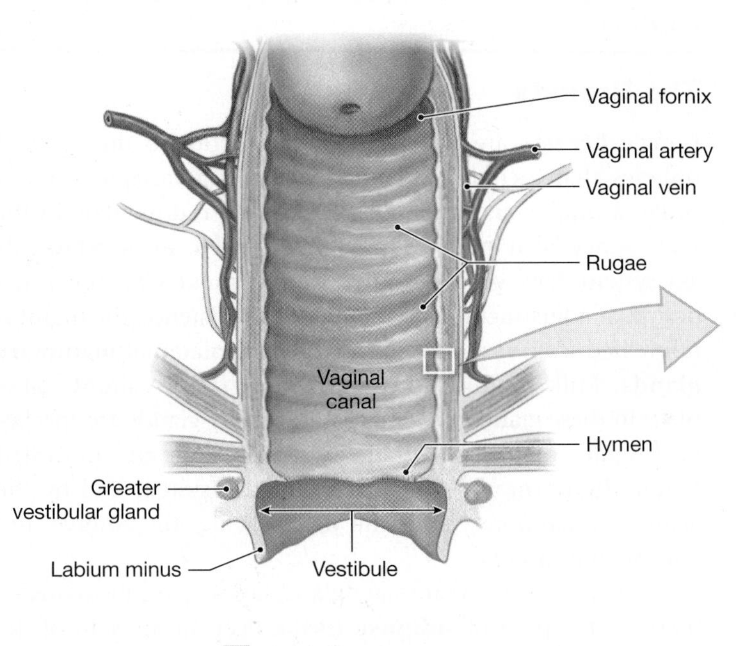
Vaginal fornix
Vaginal artery
Vaginal vein
Rugae
Vaginal canal
Hymen
Greater vestibular gland
Labium minus
Vestibule

a Vaginal structures

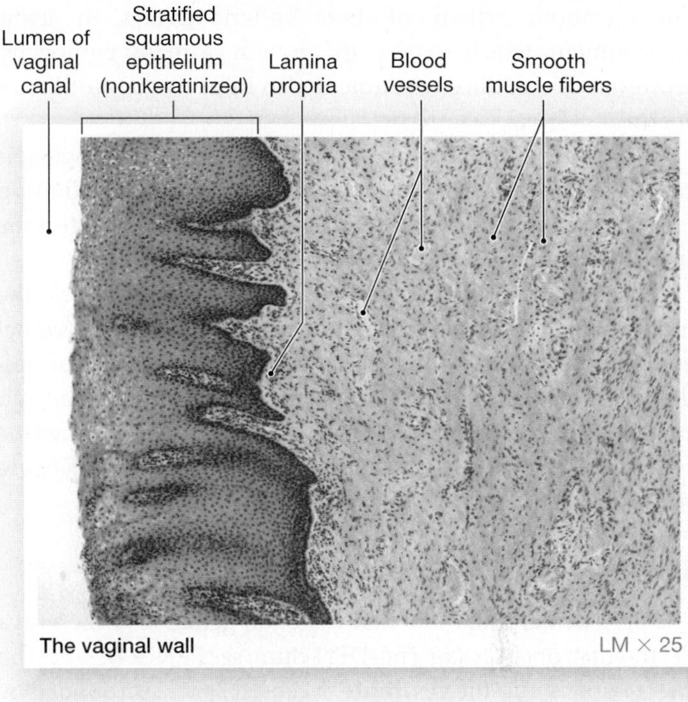
Lumen of vaginal canal
Stratified squamous epithelium (nonkeratinized)
Lamina propria
Blood vessels
Smooth muscle fibers
The vaginal wall
LM × 25

b Histology of vagina

Figure 28–19 Anatomy of the Female External Genitalia.

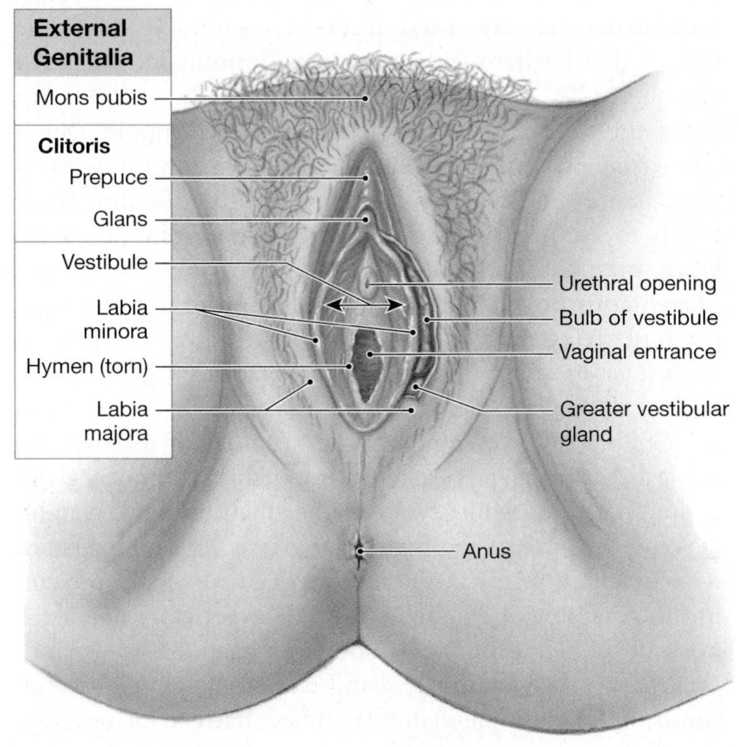
External Genitalia
Mons pubis
Clitoris
Prepuce
Glans
Vestibule
Labia minora
Hymen (torn)
Labia majora
Urethral opening
Bulb of vestibule
Vaginal entrance
Greater vestibular gland
Anus

the superior wall of the vagina from the urinary bladder to the external urethral orifice, which opens into the vestibule.

The primary blood supply of the vagina is by the **vaginal branches** of the internal iliac (or uterine) arteries and veins. Innervation is from the hypogastric plexus, sacral nerves S_2–S_4, and branches of the pudendal nerve. pp. 451, 764, 771

Histology of the Vagina

In sectional view, the lumen of the vagina appears constricted, forming a rough H-shape. The vaginal walls contain a network of blood vessels and layers of smooth muscle (Figure 28–18b). The lining is moistened by secretions of the cervical glands and by the movement of water across the permeable epithelium.

The vaginal lumen is lined by a nonkeratinized stratified squamous epithelium (see Figure 28–18b). In the relaxed state, this epithelium forms folds called *rugae* (RŪ-gē). The underlying lamina propria is thick and elastic, and it contains small blood vessels, nerves, and lymph nodes. The vaginal mucosa is surrounded by an elastic *muscular* layer consisting of layers of smooth muscle fibers arranged in circular and longitudinal bundles continuous with the uterine myometrium. The portion of the vagina adjacent to the uterus has a serosal covering that is continuous with the pelvic peritoneum. Along the rest of the vagina, the muscular layer is surrounded by an *adventitial layer* of fibrous connective tissue.

The vagina contains a population of resident bacteria, usually harmless, supported by nutrients in the cervical mucus. The metabolic activity of these bacteria creates an acidic environment, which restricts the growth of many pathogens. However, this acidic environment also inhibits the motility of sperm.

Vaginitis (vaj-ih-NĪ-tis), an inflammation of the vagina, is caused by fungal, bacterial, or parasitic infections. In addition to any discomfort that may result, the condition may affect the survival of sperm and thereby reduce fertility.

The hormonal changes associated with the *ovarian cycle*, the monthly cycle that produces an oocyte, also affect the vaginal epithelium. By examining a *vaginal smear*—a sample of epithelial cells shed at the surface of the vagina—a clinician can estimate the corresponding stages in the ovarian and uterine cycles. This diagnostic procedure is an example of *exfoliative cytology*. ⮌ p. 126

The Female External Genitalia

The area containing the female external genitalia is the **vulva** (VUL-vuh), or *pudendum* (pū-DEN-dum; see Figure 28–19). The vagina opens into the **vestibule**, a central space surrounded by small folds known as the **labia minora** (LĀ-bē-uh mi-NOR-uh; singular, *labium minus*). The labia minora are covered with smooth, hairless skin. The urethra opens into the vestibule just anterior to the vaginal entrance. The **urethral glands** discharge into the urethra near the external urethral opening.

Anterior to this opening, the **clitoris** (KLIT-ō-ris) projects into the vestibule. A small, rounded tissue projection, the clitoris is derived from the same embryonic structures as the penis in males, and has a major role in female sexual response. Internally, it contains erectile tissue called **corpus cavernosum of clitoris** (comparable to those structures of the penis) that form its **body**. A small erectile **glans** sits atop the body. The bulb of vestibule along the sides of the vestibule are comparable to the male corpus spongiosum. These erectile tissues engorge with blood during sexual arousal. Extensions of the labia minora encircle the body of the clitoris, forming its **prepuce**.

A variable number of small **lesser vestibular glands** discharge their secretions onto the exposed surface of the vestibule between the orifices of the vagina and urethra. During sexual arousal, the **greater vestibular glands**, located on each side of the distal portion of the vagina, secrete into the vestibule. These mucous glands keep the area moist and lubricated. The vestibular glands have the same embryonic origins as the bulbourethral glands in males.

The outer margins of the vulva are formed by the mons pubis and the labia majora. The **mons pubis** is a pad of adipose tissue covering the pubic symphysis. Adipose tissue also accumulates within the **labia majora** (singular, *labium majus*), prominent folds of skin that encircle and partially conceal the labia minora and adjacent structures. The outer margins of the labia majora and the mons pubis are covered with coarse hair, but the inner surfaces of the labia majora are hairless. Sebaceous glands and scattered apocrine sweat glands secrete onto the inner surface of the labia majora, moistening and lubricating them.

The Breasts

Female **breasts** are two projections anterior to the pectoral muscles that include a mammary gland, a variable amount of fat, a nipple, and an areola. Breasts are rudimentary in the male. A newborn infant cannot fend for itself, and several of its key systems have yet to complete development. Over the initial period of adjustment to an independent existence, the infant is nourished from the milk secreted by the maternal **mammary glands**. Milk production, or **lactation** (lak-TĀ-shun), takes place in these glands. In females, mammary glands are specialized organs of the integumentary system that are controlled mainly by hormones of the reproductive system and by the *placenta*, a temporary structure that provides the embryo and fetus with nutrients.

On each side, a mammary gland lies in the subcutaneous tissue of the pectoral **adipose tissue** deep to the skin of the chest (Figure 28–20a). Each breast has a **nipple**, a small conical projection where the ducts of the underlying mammary gland open onto the body surface. The reddish-brown skin around each nipple is the **areola** (a-RĒ-ō-luh). Large sebaceous glands deep to the areolar surface give it a grainy texture.

The glandular tissue of a mammary gland consists of separate lobes, each containing several secretory lobules. Ducts leaving the lobules converge, giving rise to a single **lactiferous** (lak-TIF-er-us) **duct** in each lobe. Near the nipple, that lactiferous duct enlarges, forming an expanded chamber called a **lactiferous sinus**. Typically, 15–20 lactiferous sinuses open onto the surface of each nipple. Dense connective tissue surrounds the duct system and forms partitions that extend between the lobes and the lobules. These bands of connective tissue, the *suspensory ligaments of the breast*, originate in the dermis of the overlying skin. A layer of areolar tissue separates the mammary gland complex from the underlying pectoralis muscles. Branches of the *internal thoracic artery* supply blood to each mammary gland (look back at Figure 21–20, p. 760).

Figure 28–20b,c compares the histology of inactive and active mammary glands. An inactive, or *resting*, mammary gland is dominated by a duct system rather than by active glandular cells. The size of the mammary glands in a nonpregnant woman reflects primarily the amount of adipose tissue present, not the amount of glandular tissue. The secretory apparatus normally does not complete its development unless pregnancy occurs. An active mammary gland is a tubulo-alveolar gland, consisting of multiple glandular tubes that end in secretory alveoli. We will discuss the hormonal mechanisms involved in lactation in Chapter 29.

Figure 28–20 **Anatomy of the Breast.** ATLAS: Plate 28

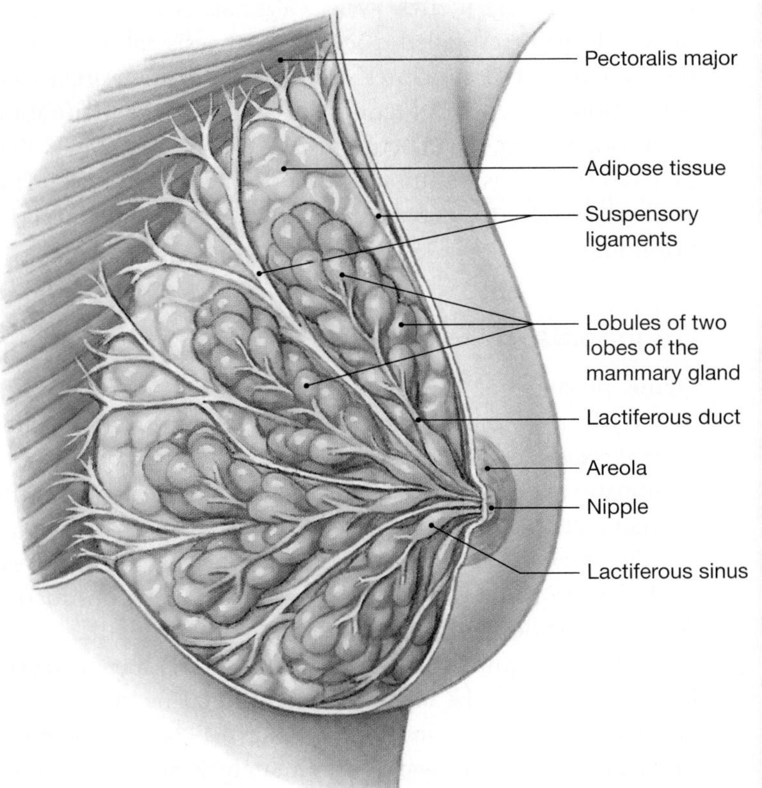

Pectoralis major
Adipose tissue
Suspensory ligaments
Lobules of two lobes of the mammary gland
Lactiferous duct
Areola
Nipple
Lactiferous sinus

a The mammary gland of the left breast

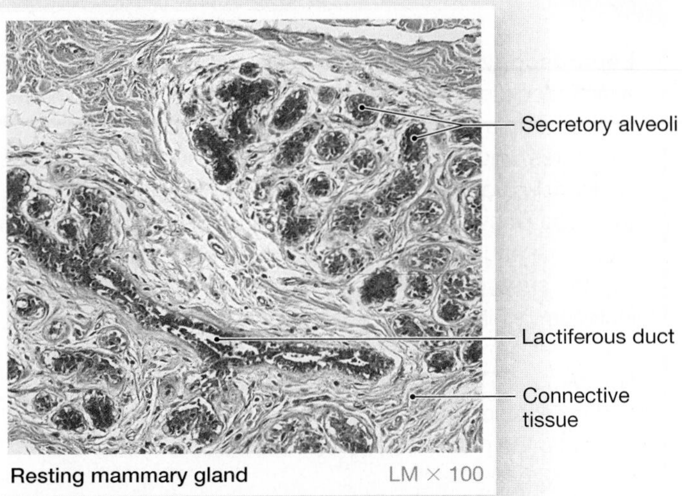

Secretory alveoli
Lactiferous duct
Connective tissue

Resting mammary gland LM × 100

b An inactive mammary gland of a nonpregnant woman

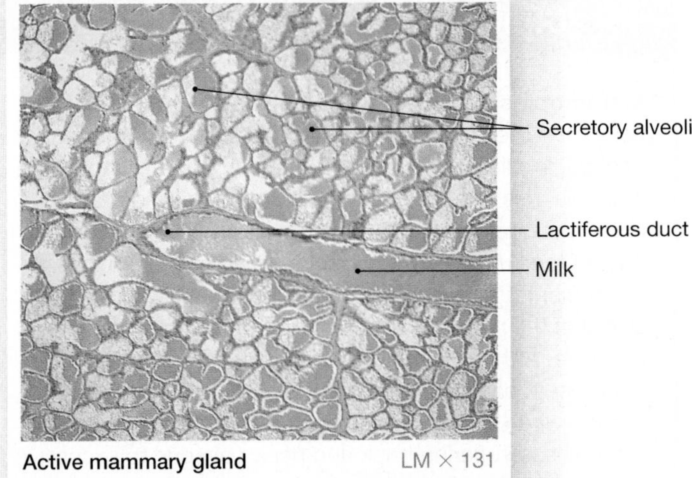

Secretory alveoli
Lactiferous duct
Milk

Active mammary gland LM × 131

c An active mammary gland of a nursing woman

Clinical Note Fibrocystic Disease and Breast Cancer

The uterine cycle also influences cyclical changes in the mammary glands. Occasional discomfort or even inflammation of mammary gland tissues can occur late in the cycle. If inflamed lobules become walled off by scar tissue, **cysts** are created. Clusters of cysts can be felt in the breast as discrete masses, a condition known as **fibrocystic disease**. Because its symptoms are similar to those of breast cancer, biopsies may be needed to distinguish between this benign condition and breast cancer.

Breast cancer is a malignant, metastasizing tumor of the mammary gland. It is the leading cause of death in women between ages 35 and 45, but it is most common in women over age 50. An estimated 12.4 percent of U.S. women will develop breast cancer at some point in their lifetime. The incidence is highest among Caucasian Americans, somewhat lower in Black Americans, and lowest in Asian Americans and Native Americans. Notable risk factors include (1) a family history of breast cancer, (2) a first pregnancy after age 30, and (3) early menarche (first menstruation) or late menopause (end of menstruation).

Despite repeated studies, no links have been proven between breast cancer and birth control pills, fat consumption, or alcohol use. It appears likely that multiple factors are involved. In some families an inherited genetic variation has been linked to higher-than-normal risk of developing the disease. However, most women never develop breast cancer—even women in families with a history of the disease. Mothers who breast-fed (nursed) their babies have a 20 percent lower incidence of breast cancer after menopause than do mothers who did not breast-feed. The reason is not known. Adding to the mystery, nursing does not appear to affect the incidence of premenopausal breast cancer.

+ Clinical Note Laparoscopy

Laparoscopy is a minimally invasive procedure that allows a direct look into the abdominopelvic cavity. The technique is often employed to assess the health of female pelvic structures when another diagnostic test, such as an x-ray, is inconclusive. The surgeon makes a tiny incision through the abdominal wall when the patient is under general anesthesia. Through this opening, carbon dioxide gas is injected into the abdomen in order to spread out the organs and increase visibility. Then the laparoscope is inserted. The scope is a flexible tube bearing a tiny light and an equally tiny camera. The operator can look around, take photos, and collect biopsies. Laparoscopy can determine, for example, the presence of a blocked uterine tube or a cyst on an ovary.

+ Clinical Note Mammoplasty

Mammoplasty includes three types of surgical alteration of the breast. *Breast augmentation* is most frequently desired. In this procedure, the size or contour of the mammary gland is modified by the placement of fluid-filled implants under the skin, and either superficial or deep to the pectoral muscles. The surgeon may be able to insert the implant through an incision along the margin of the areola of the nipple. A second approach is to enter from the axilla. A third option is to make a small incision directly underneath each breast. *Breast reduction* can be performed when a woman desires that cosmetic result, or when she is suffering back and neck pain due to excessive breast size. Finally, women who have undergone surgery for breast cancer may seek *breast reconstruction*. As a case in point, actor and activist Angelina Jolie went public in 2013 with her prophylactic bilateral mastectomy, or removal of the breasts in response to a high risk of breast cancer.

✓ Checkpoint

8. **Name the structures of the female reproductive system.**
9. **What effect would blockage of both uterine tubes by scar tissue (resulting from an infection such as gonorrhea) have on a woman's ability to conceive?**
10. **What benefit does the acidic pH of the vagina provide?**
11. **Would the blockage of a single lactiferous sinus interfere with the delivery of milk to the nipple? Explain.**

See the blue Answers tab at the back of the book.

28-5 Oogenesis occurs in the ovaries, and hormones from the hypothalamus, pituitary gland, and ovaries control female reproductive functions

Learning Outcome Describe the ovarian roles in oogenesis, explain the complete ovarian and uterine cycles, and summarize all aspects of the female reproductive cycle.

Three interrelated processes help to ensure the continuation of our species from the female perspective: oogenesis, the ovarian cycle, and the uterine cycle. *Oogenesis* is the production of female gametes called oocytes. Oogenesis takes place in the ovaries within structures called *follicles*. The *ovarian cycle* is the monthly series of events associated with the maturation of an oocyte. The *uterine cycle* involves a series of events that prepares the uterus for implantation of a fertilized oocyte. This section describes these events and how they are influenced by hormones.

Oogenesis

Ovum production, or **oogenesis** (ō-ō-JEN-eh-sis; *oon*, egg), begins before a woman's birth, accelerates at puberty, and ends at *menopause* (end of menstruation). Between puberty and menopause, oogenesis occurs on a monthly basis as part of the ovarian cycle.

Oogenesis is summarized in Figure 28–21. This process begins during fetal development, when female reproductive stem cells called **oogonia** (ō-ō-GŌ-nē-uh; singular, *oogonium*) undergo mitosis, producing diploid **primary oocytes** (2n). After 5 months of development, fetal ovaries contain an estimated 7 million primary oocytes. Not all survive. At birth the ovaries have approximately 2 million primary oocytes. However, most of these primary oocytes do not survive until puberty, degenerating in a process called *atresia* (ah-TRĒ-zē-uh). By puberty, their number has dropped to about 400,000.

The remaining primary oocytes begin meiosis I, proceed as far as prophase I, and then remain in that state until the female reaches puberty. At that point, some are stimulated to finish meiosis I, producing haploid **secondary oocytes** (n). Note that the cytoplasm of the primary oocyte is unevenly distributed during the two meiotic divisions (see Figure 28–21). Oogenesis produces one secondary oocyte, which contains most of the original cytoplasm, and two or three **polar bodies**, nonfunctional cells that later disintegrate.

The secondary oocyte then begins meiosis II, but is suspended at metaphase II. In the process of *ovulation*, the ovary releases such a secondary oocyte. This secondary oocyte will not complete meiosis II and become a mature ovum unless fertilization occurs.

Figure 28–21 The Process of Oogenesis. For clarity, maternal and paternal chromatids are not identified.

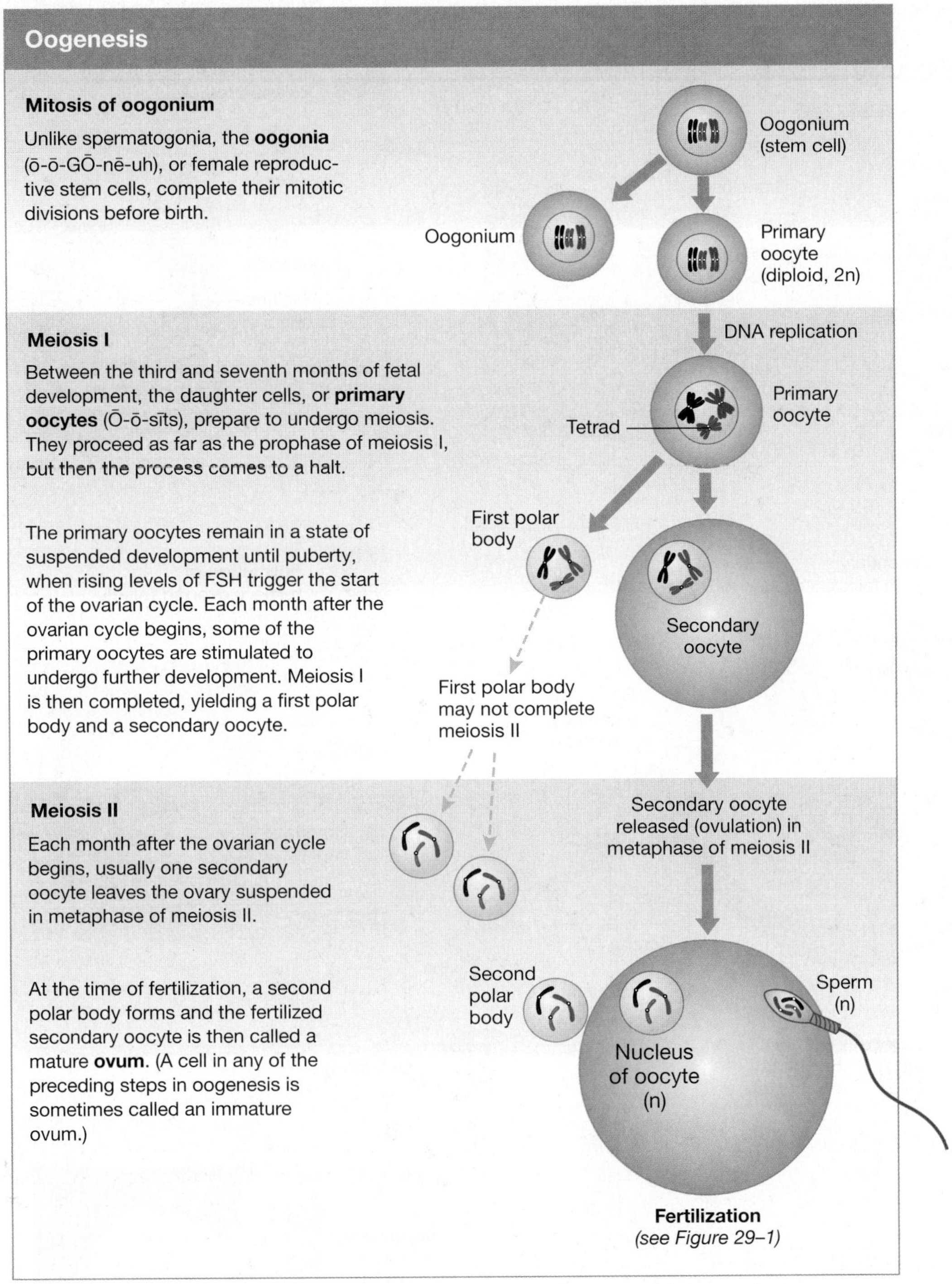

Oogenesis

Mitosis of oogonium

Unlike spermatogonia, the **oogonia** (ō-ō-GŌ-nē-uh), or female reproductive stem cells, complete their mitotic divisions before birth.

Oogonium (stem cell)

Oogonium

Primary oocyte (diploid, 2n)

DNA replication

Meiosis I

Between the third and seventh months of fetal development, the daughter cells, or **primary oocytes** (Ō-ō-sīts), prepare to undergo meiosis. They proceed as far as the prophase of meiosis I, but then the process comes to a halt.

Tetrad

Primary oocyte

The primary oocytes remain in a state of suspended development until puberty, when rising levels of FSH trigger the start of the ovarian cycle. Each month after the ovarian cycle begins, some of the primary oocytes are stimulated to undergo further development. Meiosis I is then completed, yielding a first polar body and a secondary oocyte.

First polar body

Secondary oocyte

First polar body may not complete meiosis II

Meiosis II

Each month after the ovarian cycle begins, usually one secondary oocyte leaves the ovary suspended in metaphase of meiosis II.

Secondary oocyte released (ovulation) in metaphase of meiosis II

At the time of fertilization, a second polar body forms and the fertilized secondary oocyte is then called a mature **ovum**. (A cell in any of the preceding steps in oogenesis is sometimes called an immature ovum.)

Second polar body

Nucleus of oocyte (n)

Sperm (n)

Fertilization
(see Figure 29–1)

1 Primordial Ovarian Follicles in Egg Nest. Primary oocytes are located in the outer portion of the ovarian cortex, near the tunica albuginea, in clusters called *egg nests.* A single squamous layer of *follicle cells* surrounds each primary oocyte within an egg nest. The primary oocyte and its follicle cells form a **primordial ovarian follicle.** At puberty, each ovary contains about 200,000 primordial follicles. Forty years later, few if any primary ovarian follicles remain, although only about 500 secondary oocytes will have been ovulated.

Beginning at puberty, primordial ovarian follicles are continuously activated to join other follicles already in development. The activating mechanism is unknown, but local hormones or growth factors within the ovary may be involved. The activated primordial ovarian follicle will either eventually mature and release a secondary oocyte or degenerate through follicular atresia.

2 Formation of Primary Ovarian Follicle. The preliminary steps in follicle development vary in length but may take almost a year to complete. Follicle development begins with the activation of primordial ovarian follicles into **primary ovarian follicles**. The follicular cells enlarge, divide, and form several layers of cells around the growing primary oocyte. The cells begin to produce sex hormones called estrogens. Microvilli from the surrounding follicle cells intermingle with microvilli originating at the surface of the oocyte. This region is called the **zona pellucida** (ZŌ-na pe-LŪ-sid-uh; *pellucidus,* translucent). The microvilli increase the surface area available for the transfer of materials from the follicular cells to the growing oocyte.

3 Formation of Secondary Ovarian Follicle. Many primordial ovarian follicles develop into primary ovarian follicles, but only a few mature further in the next four months or so. This

Follicle Development

Specialized structures in the cortex of the ovaries called **ovarian follicles** (ō-VAR-ē-an FOL-i-klz) are the sites of both oocyte growth and meiosis I of oogenesis. Follicle development and the ovarian cycle are shown in Figure 28–22.

Figure 28–22 Follicle Development and the Ovarian Cycle. The photomicrographs show the changes in an ovarian follicle as it develops. It enters the 28-day ovarian cycle as a tertiary ovarian follicle. The drawing of the ovary illustrates the sequence and relative sizes of the various stages in the development, ovulation, and degeneration of an ovarian follicle. Follicles do not physically move around the periphery of the ovary.

1 Primordial ovarian follicles in egg nest

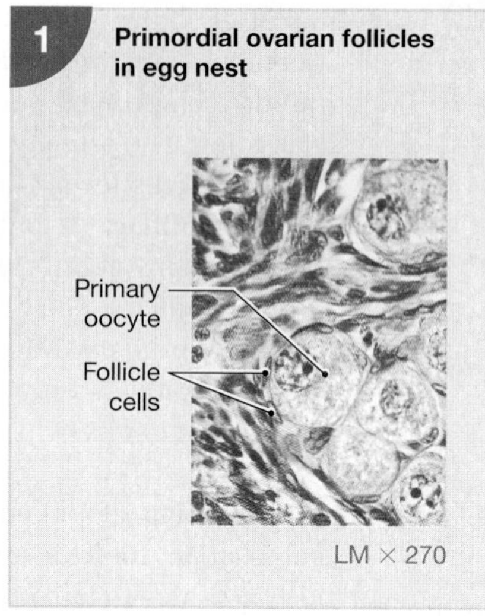
Primary oocyte
Follicle cells
LM × 270

2 Formation of primary ovarian follicle

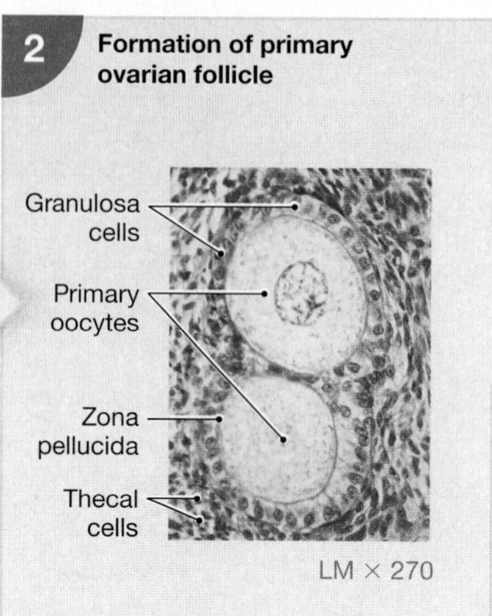
Granulosa cells
Primary oocytes
Zona pellucida
Thecal cells
LM × 270

3 Formation of secondary ovarian follicle

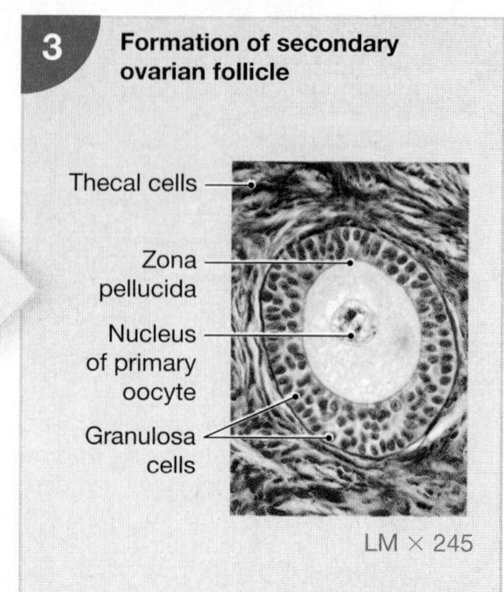
Thecal cells
Zona pellucida
Nucleus of primary oocyte
Granulosa cells
LM × 245

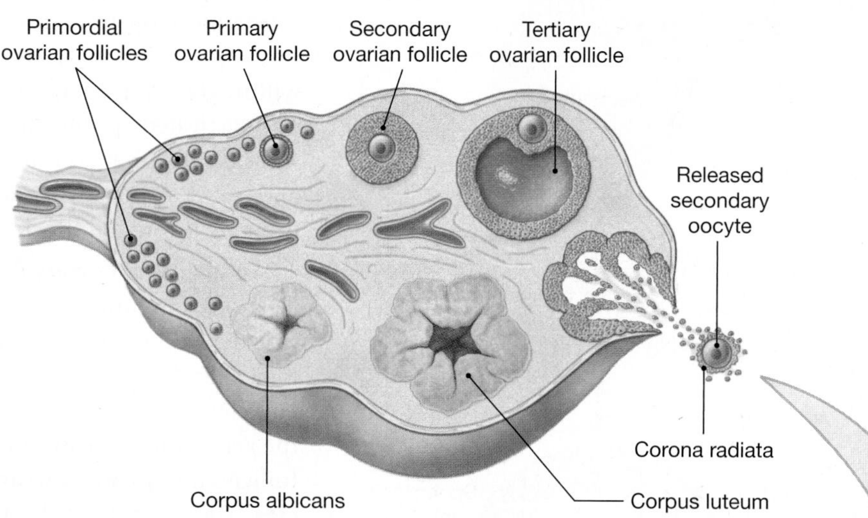
Primordial ovarian follicles
Primary ovarian follicle
Secondary ovarian follicle
Tertiary ovarian follicle
Released secondary oocyte
Corona radiata
Corpus albicans
Corpus luteum

4 Formation of tertiary ovarian follicle

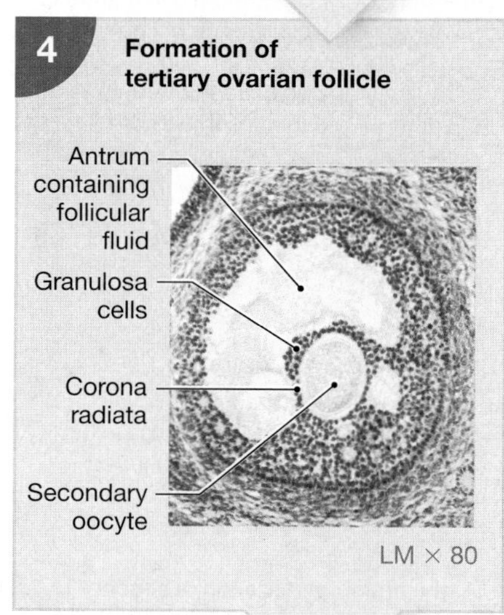
Antrum containing follicular fluid
Granulosa cells
Corona radiata
Secondary oocyte
LM × 80

5 Ovulation

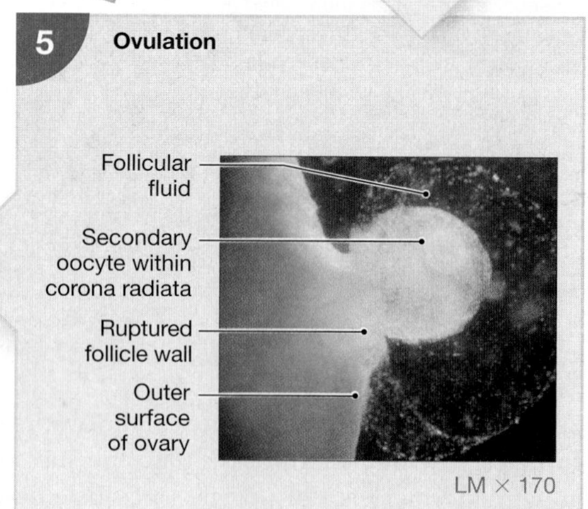
Follicular fluid
Secondary oocyte within corona radiata
Ruptured follicle wall
Outer surface of ovary
LM × 170

6 Formation of corpus luteum

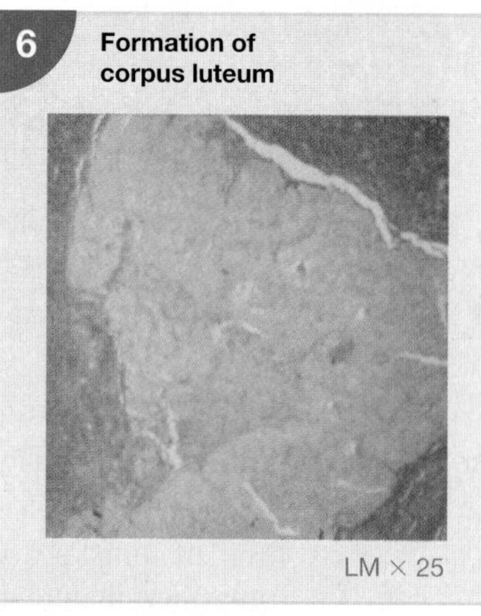
LM × 25

7 Formation of corpus albicans

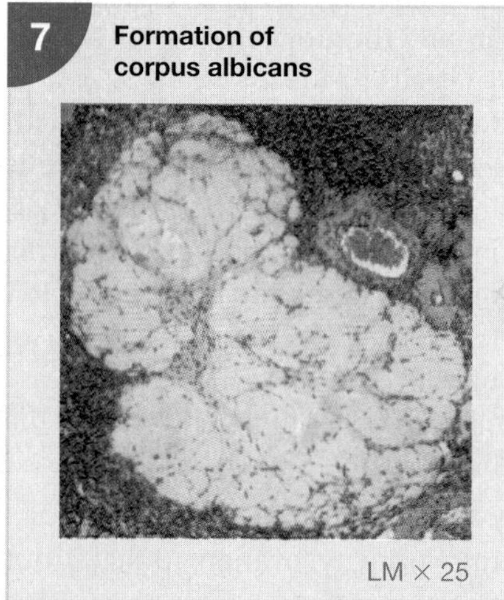
LM × 25

28

? At what stage of development is the oocyte during ovulation?

process is apparently under the control of a growth factor produced by the oocyte. The transformation begins as the wall of the follicle thickens. At this stage, the complex is known as a **secondary ovarian follicle**. The primary oocyte continues to grow slowly.

4 **Formation of Tertiary Ovarian Follicle.** During the next 2 to 3 months, the follicle wall continues to grow and the deeper follicular cells begin secreting *follicular fluid*. This fluid accumulates in small pockets that gradually expand between the inner and outer cellular layers of the follicle. The secondary ovarian follicle as a whole has doubled in size, and is now called a *tertiary ovarian follicle*. Tertiary ovarian follicles, also called *vesicular follicles*, continue to grow and accumulate follicular fluid.

The Ovarian Cycle

The tertiary follicles are then ready to complete their maturation as part of the 28-day ovarian cycle. The monthly process of the maturation, ovulation, and degeneration of a tertiary ovarian follicle is known as the **ovarian cycle** (see Figure 28–22). We can divide the ovarian cycle into a *follicular phase* (folliculogenesis) and a *luteal phase* (luteogenesis), each lasting about 14 days. The luteal phase involves the follicle after ovulation, when it becomes an endocrine structure called a *corpus luteum* (LŪ-tē-um) that secretes a sex hormone.

The Follicular Phase

In the **follicular phase**, or *preovulatory phase*, a tertiary ovarian follicle is ready to complete its maturation. Ovulation marks the end of this phase.

At the start of each ovarian cycle, an ovary contains only a few tertiary follicles destined for further development. Stimulated by FSH, one of these follicles becomes dominant by day 5 of the cycle and it rapidly increases in size. This remaining **tertiary ovarian follicle**, also called a *mature graafian* (GRAF-ē-an) *follicle*, is roughly 15–20 mm (0.6–0.8 in) in diameter. It is formed by days 10–14 of the ovarian cycle. Its large size creates a prominent bulge in the surface of the ovary. The oocyte and its covering of follicular cells projects into the expanded central chamber of the follicle, the **antrum** (AN-trum). The antrum is surrounded by a mass of **granulosa cells**. As the granulosa cells enlarge and multiply, adjacent cells in the ovarian stroma form a layer of **thecal endocrine cells** (*theca*, a box) around the follicle. Thecal endocrine cells and granulosa cells work together to produce the sex hormones called **estrogens**.

Until this time, the primary oocyte has been suspended in prophase of meiosis I. As the development of the remaining tertiary ovarian follicle ends, rising LH levels prompt the primary oocyte to complete meiosis I. The completion of the first meiotic division produces a secondary oocyte and a small, nonfunctional polar body (see Figure 28–21). The secondary oocyte begins meiosis II but stops short of dividing. Meiosis II will not be completed unless fertilization occurs.

Generally, on day 14 of a 28-day ovarian cycle, the secondary oocyte and its surrounding follicular cells lose their connections with the follicular wall and drift free within the antrum. The granulosa cells still associated with the secondary oocyte form a protective layer known as the **corona radiata** (kō-RŌ-nuh rā-dē-AH-tuh).

5 **Ovulation.** At **ovulation** (ōv-ū-LĀ-shun), the tertiary follicle releases the secondary oocyte. The distended follicular wall then ruptures, releasing the follicular contents, including the secondary oocyte, into the pelvic cavity. The sticky follicular fluid keeps the corona radiata attached to the surface of the ovary near the ruptured wall of the follicle. The oocyte is then moved into the uterine tube by contact with the fimbriae that extend from the tube's funnel-like opening (look back at Figure 28–14a), or by fluid currents produced by the cilia that line the tube.

Usually only a single oocyte is released into the pelvic cavity at ovulation, although many primordial ovarian follicles may have developed into primary ovarian follicles, and several primary ovarian follicles may have been converted to secondary ovarian follicles. These follicles undergo follicular atresia.

The Luteal Phase

After ovulation occurs, the **luteal phase**, or *postovulatory phase*, begins. This phase ends with the degeneration of the corpus luteum.

6 **Formation of Corpus Luteum.** The empty tertiary ovarian follicle initially collapses, and ruptured vessels bleed into the antrum. Under the stimulation of LH, the remaining granulosa cells then invade the area and proliferate to create the **corpus luteum**.

The corpus luteum is so named (*lutea*, yellow) because this is the color of the cholesterol it contains. This cholesterol is used to manufacture the steroid hormone **progesterone** (prō-JES-ter-ōn). Progesterone is the main hormone during the luteal phase. Progesterone's primary function is to prepare the uterus for pregnancy by stimulating the maturation of the uterine lining and the secretions of uterine glands. The corpus luteum also secretes moderate amounts of estrogens, but levels are not as high as they were at ovulation.

7 **Formation of Corpus Albicans.** The corpus luteum begins to degenerate about 12 days after ovulation (unless fertilization takes place). Progesterone and estrogen levels then decline markedly. Fibroblasts invade the nonfunctional corpus luteum, producing a knot of pale scar tissue called a **corpus albicans** (AL-bi-kanz).

The disintegration of the corpus luteum marks the end of the ovarian cycle. A new ovarian cycle then begins with another group of tertiary ovarian follicles.

The Uterine (Menstrual) Cycle

The **uterine cycle**, or *menstrual* (MEN-strū-ul) *cycle*, is a repeating series of changes in the structure of the endometrium (Figure 28–23). The uterine cycle averages 28 days in length, but it can range from 21 to 35 days in healthy women of reproductive age.

The uterine cycle begins at puberty. The first cycle, known as *menarche* (me-NAR-kē; *men*, month + *arche*, beginning), typically occurs at age 11–12. The cycles continue until *menopause* (MEN-ō-pawz), the termination of the uterine cycle, at age 45–55. Over the interim, the regular appearance of uterine cycles is interrupted only by circumstances such as illness, stress, starvation, or pregnancy.

The uterine cycle is divided into three phases: (1) the *menstrual phase*, (2) the *proliferative phase*, and (3) the *secretory phase.* The phases take place in response to hormones associated with the regulation of the ovarian cycle. The menstrual and proliferative phases occur during the follicular phase of the ovarian cycle. The secretory phase corresponds to the luteal phase of the ovarian cycle.

Figure 28–23 A Comparison of the Structure of the Endometrium during the Phases of the Uterine Cycle.

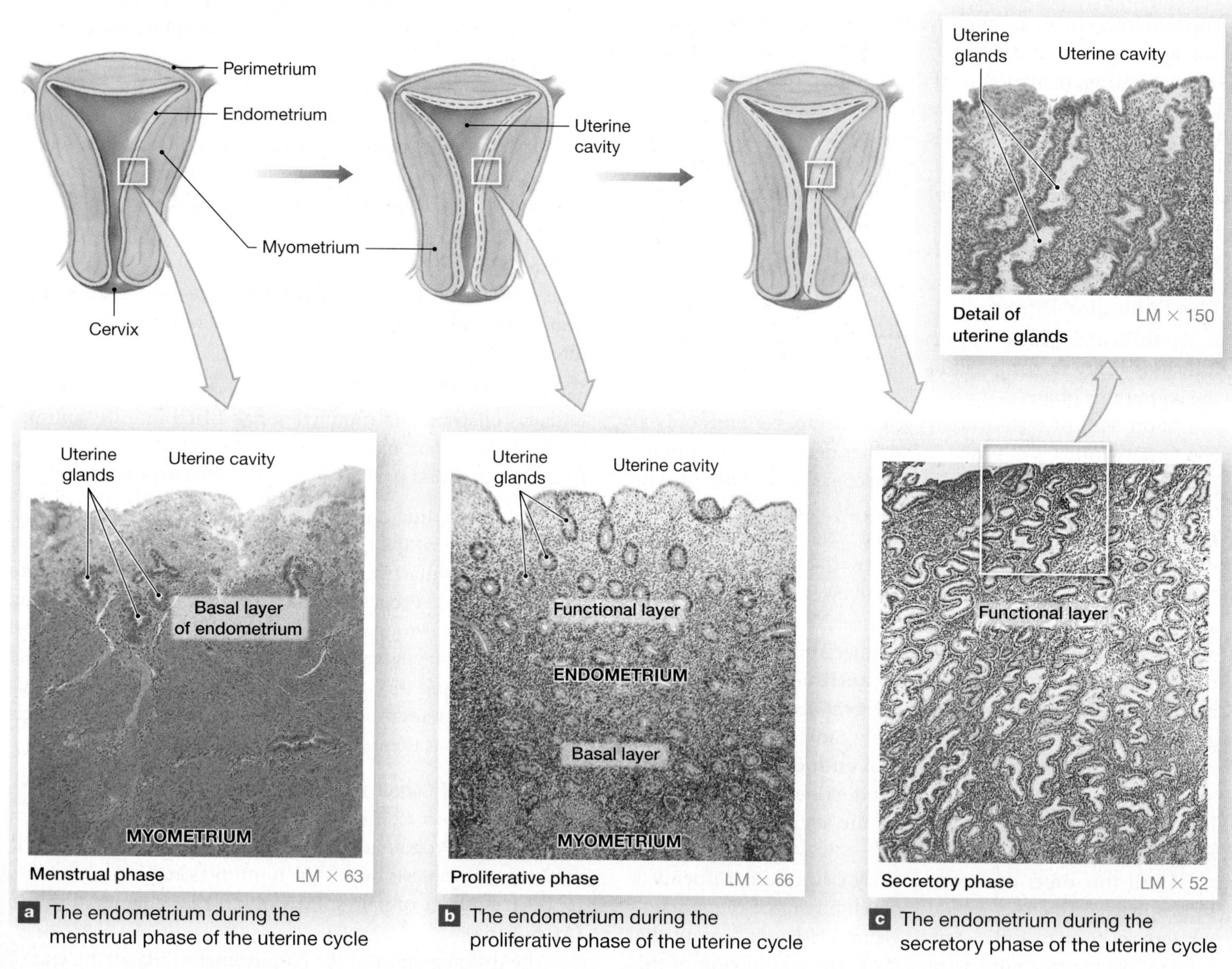
Perimetrium
Endometrium
Myometrium
Cervix
Uterine cavity
Uterine glands
Uterine cavity
Detail of uterine glands
LM × 150
Uterine glands
Uterine cavity
Basal layer of endometrium
MYOMETRIUM
Menstrual phase
LM × 63
Uterine glands
Uterine cavity
Functional layer
ENDOMETRIUM
Basal layer
MYOMETRIUM
Proliferative phase
LM × 66
Functional layer
Secretory phase
LM × 52

a The endometrium during the menstrual phase of the uterine cycle

b The endometrium during the proliferative phase of the uterine cycle

c The endometrium during the secretory phase of the uterine cycle

? Which layer of the endometrium is affected most during the uterine cycle?

Tips & Tools

Progesterone literally means a steroid (***-one***) that favors (***pro-***) gestation (***-gest***).

Menstrual Phase

The uterine cycle begins with the **menstrual phase**. This phase is marked by the degeneration and sloughing (shedding) of the endometrial functional layer, leading to **menstruation** (men-strū-Ā-shun), also called **menses** (MEN-sēz) (see Figure 28–23a). Endometrial degeneration occurs in patches. It is caused by constriction of the spiral arteries, which reduces blood flow to areas of the endometrium. Deprived of oxygen and nutrients, the secretory glands and other tissues in the functional layer begin to deteriorate. Eventually, the weakened arterial walls rupture, and blood pours into the connective tissues of the functional layer. Blood cells and degenerating tissues then break away and enter the uterine cavity, to be shed as they pass through the external os and into the vagina. Only the functional layer is affected, because the deeper, basal layer receives blood from the straight arteries, which remain unconstricted.

The sloughing of tissue is gradual, and repairs begin almost immediately at each site. Nevertheless, before menstruation has ended, the entire functional layer has been lost. The menstrual phase generally lasts from 1 to 7 days. During this time about 35–50 mL (1.2–1.7 oz) of blood are lost.

The process can be relatively painless. However, painful menstruation, or **dysmenorrhea**, can result from myometrial contractions ("cramps"), uterine inflammation, or from conditions involving adjacent pelvic structures.

The Proliferative Phase

The basal layer, including the basal parts of the uterine glands, survives menstruation intact. In the days after menstruation, the epithelial cells of the uterine glands multiply and spread across the endometrial surface, restoring the uterine epithelium (see Figure 28–23b). Further growth and vascularization completely restore the functional layer.

During this reorganization, the endometrium is in the **proliferative phase**. Restoration takes place at the same time as the tertiary ovarian follicles enlarge in the ovary. Estrogens secreted by the developing ovarian follicles stimulate and sustain the proliferative phase.

By the time ovulation occurs, the functional layer is several millimeters thick. Prominent mucous glands extend to its border with the basal layer. At this time, the uterine glands are manufacturing mucus that is rich in glycogen. This specialized mucus appears to be essential for the survival of the fertilized ovum through its earliest developmental stages. (We consider these stages in Chapter 29.) The entire functional layer is highly vascularized, with small arteries spiraling toward the endometrial surface from larger arteries in the myometrium.

The Secretory Phase

During the **secretory phase** of the uterine cycle, the uterine glands enlarge, accelerating their rates of secretion. The arteries that supply the uterine wall elongate and spiral through the tissues of the functional layer (see Figure 28–23c). This activity occurs under the combined stimulatory effects of progesterone and estrogens from the corpus luteum. The secretory phase begins at the time of ovulation and persists as long as the corpus luteum remains intact.

Secretory activities peak about 12 days after ovulation. Over the next day or two, the glands become less active. The uterine cycle ends as the corpus luteum stops producing stimulatory hormones. A new uterine cycle then begins with the onset of menstruation and the disintegration of the functional layer.

The secretory phase generally lasts 14 days. For this reason, you can identify the date of ovulation by counting backward 14 days from the first day of menstruation.

Hormonal Coordination of the Ovarian and Uterine Cycles

The female reproductive tract is under hormonal control that involves interplay between secretions of both the pituitary gland and the gonads. Circulating hormones control the **female reproductive cycle**, coordinating the ovarian and uterine cycles for proper reproductive function. If the two cycles are not properly coordinated, *infertility*, the inability to conceive or carry a pregnancy to term, results.

As in males, GnRH from the hypothalamus regulates reproductive function in females. However, in females, the GnRH pulse frequency and amplitude (amount secreted per pulse) change throughout the ovarian cycle. If the hypothalamus were a radio station, the pulse frequency would correspond to the radio frequency it is transmitting on, and the amplitude would be the volume. We will consider changes in pulse frequency, because their effects are both dramatic and reasonably well understood. Circulating levels of estrogens and progesterone primarily control changes in GnRH pulse frequency. Estrogens increase the GnRH pulse frequency, and progesterone decreases it.

The endocrine cells of the anterior lobe of the pituitary gland respond as if each group of endocrine cells is monitoring different frequencies. As a result, each group of cells is sensitive to some GnRH pulse frequencies and insensitive to others. For example, consider the *gonadotropes*, the cells that produce FSH and LH. At one pulse frequency, the gonadotropes respond preferentially and secrete FSH, but at another frequency, these cells release primarily LH. FSH and LH production also occurs in pulses that follow the rhythm of GnRH pulses. If GnRH is absent or is supplied at a constant rate (without pulses), FSH and LH secretion stops in a matter of hours.

Spotlight Figure 28–24 shows the changes in circulating hormone levels that accompany the ovarian and uterine cycles.

SPOTLIGHT

Figure 28–24

Hormonal Regulation of Female Reproduction

The ovarian and uterine cycles must operate in synchrony to ensure proper reproductive function. If the two cycles are not properly coordinated, infertility results. A female who doesn't ovulate cannot conceive, even if her uterus is perfectly normal. A female who ovulates normally, but whose uterus is not ready to support an embryo, will also be infertile.

As in males, GnRH from the hypothalamus regulates reproductive function in females. However, in females, GnRH levels change throughout the course of the ovarian cycle.

HYPOTHALAMUS
1 Release of Gonadotropin-Releasing Hormone (GnRH)
The cycle begins with the release of GnRH, which stimulates the production and secretion of FSH and the production—but not the secretion—of LH.
Release of GnRH
ANTERIOR LOBE OF PITUITARY GLAND
Production and secretion of FSH
Production of LH
Secretion of LH
2 Follicular Phase of the Ovarian Cycle
The follicular phase begins when FSH stimulates growth and development of a group of tertiary ovarian follicles. Usually only one follicle becomes dominant.
As tertiary ovarian follicles develop, FSH levels decline due to the negative feedback effects of inhibin.
Developing ovarian follicles also secrete estrogens, especially estradiol, the dominant hormone prior to ovulation.
In low concentrations, estrogens inhibit LH secretion. This inhibition gradually decreases as estrogen levels increase.
Negative feedback
OVARY
Before day 10
After day 10
• Ovarian follicle development
• Secretion of inhibin
• Secretion of estrogens
• Meisois I completion
• Ovulation
• Corpus luteum formation
Secretion of progesterone
3 Luteal Phase of the Ovarian Cycle
The combination of increased GnRH pulse frequency and elevated estrogen levels stimulates LH secretion.
On or around day 14, a massive surge in LH level triggers (1) the completion of meiosis I by the primary oocyte, (2) the forceful rupture of the follicular wall, (3) ovulation, roughly 9 hours after the LH peak, and (4) formation of the corpus luteum.
The corpus luteum secretes progesterone, which stimulates and sustains endometrial development.
After ovulation, progesterone levels rise and estrogen levels fall. This suppresses GnRH secretion. If pregnancy does not occur, the corpus luteum will degenerate after 12 days, and as progesterone levels decrease, GnRH secretion increases, and a new cycle begins.
Effects on CNS
Stimulation of bone and muscle growth
Establishment and maintenance of female secondary sex characteristics
Maintenance of accessory glands and organs
Stimulation of endometrial growth and secretion
KEY
Stimulation
Inhibition

This illustration combines the key events in the ovarian and uterine cycles. The monthly hormonal fluctuations cause physiological changes that affect core body temperature. During the follicular phase—when estrogens are the dominant hormones—the **basal body temperature**, or the resting body temperature measured upon awakening in the morning, is about 0.3°C (0.5°F) lower than it is during the luteal phase, when progesterone dominates.

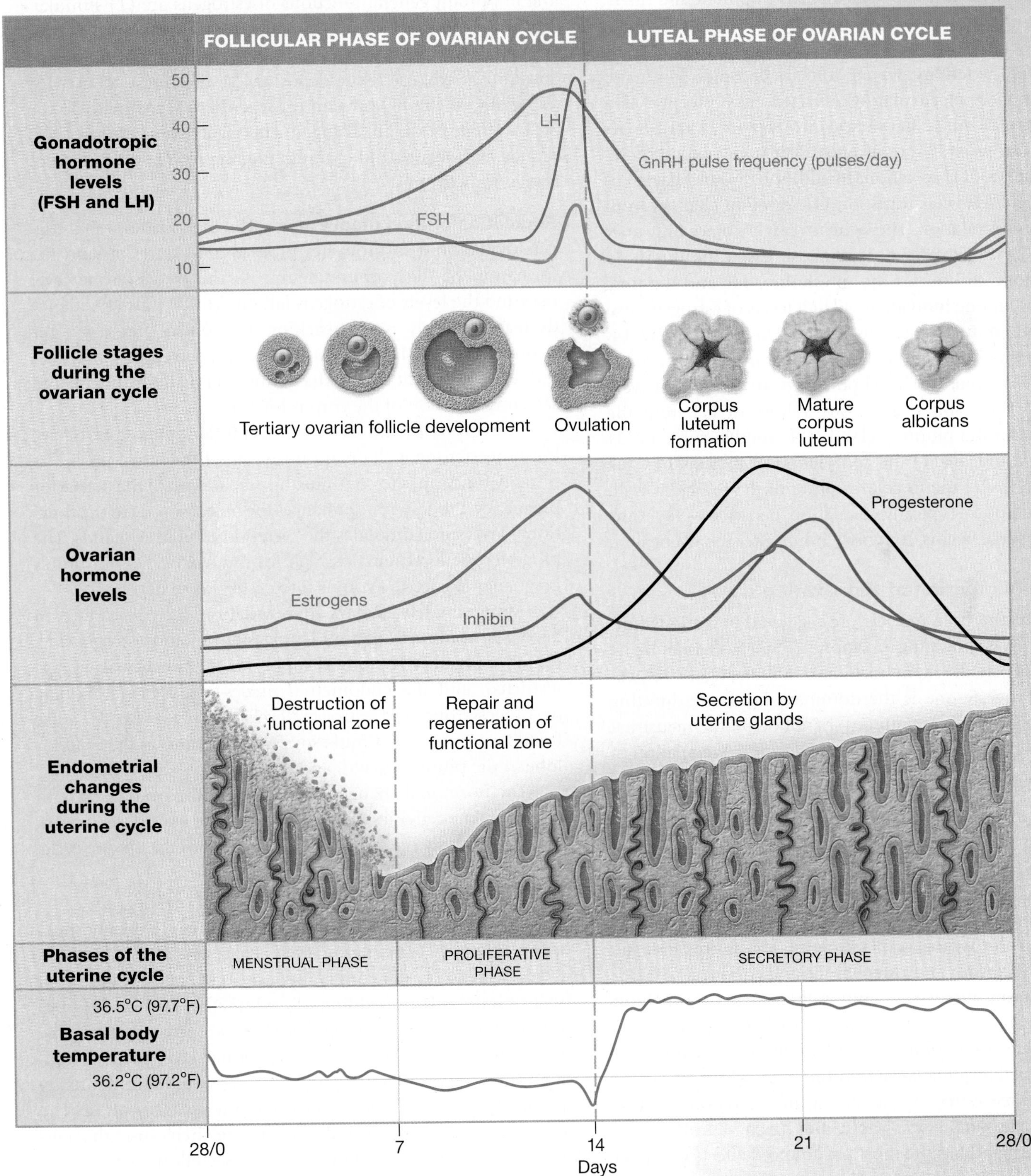

FOLLICULAR PHASE OF OVARIAN CYCLE
LUTEAL PHASE OF OVARIAN CYCLE
Gonadotropic hormone levels (FSH and LH)
50
40
30
20
10
LH
FSH
GnRH pulse frequency (pulses/day)
Follicle stages during the ovarian cycle
Tertiary ovarian follicle development
Ovulation
Corpus luteum formation
Mature corpus luteum
Corpus albicans
Ovarian hormone levels
Estrogens
Inhibin
Progesterone
Endometrial changes during the uterine cycle
Destruction of functional zone
Repair and regeneration of functional zone
Secretion by uterine glands
Phases of the uterine cycle
MENSTRUAL PHASE
PROLIFERATIVE PHASE
SECRETORY PHASE
Basal body temperature
36.5°C (97.7°F)
36.2°C (97.2°F)
28/0
7
14
21
28/0
Days

Early in the follicular phase of the ovarian cycle and prior to day 10, the levels of estrogens are low and the GnRH pulse frequency is 16–24 per day (one pulse every 60–90 minutes). At this frequency, FSH is the dominant hormone released by the anterior lobe of the pituitary gland. The estrogens released by developing follicles inhibit LH secretion. As tertiary ovarian follicles develop, the FSH level decreases due to the negative feedback effects of inhibin. Follicular development and maturation continue, however, supported by the combination of estrogens, FSH, and LH.

As one of the tertiary ovarian follicles becomes dominant, the concentration of circulating estrogens rises steeply. As a result, the GnRH pulse frequency increases to about 36 per day (one pulse every 30–60 minutes). The increased pulse frequency stimulates LH secretion. In addition, around day 10 of the cycle, the effect of estrogen on LH secretion changes from inhibition to stimulation. The switchover takes place only after the levels of estrogens have risen above a specific threshold value for about 36 hours. (The threshold value and the time required vary among individuals.) High levels of estrogens also increase gonadotrope sensitivity to GnRH. At about day 14, the estrogen levels have peaked, the gonadotropes are at maximum sensitivity, and the GnRH pulses are arriving about every 30 minutes. The result is a massive release of LH from the anterior lobe of the pituitary gland. This sudden surge in LH concentration triggers (1) the completion of meiosis I by the primary oocyte, (2) the forceful rupture of the follicular wall, and (3) ovulation. Typically, ovulation occurs 34–38 hours after the LH surge begins, roughly 9 hours after the LH peak.

Hormonal Regulation of the Ovarian Cycle

The phases of the ovarian cycle are regulated by various hormones. Follicle-stimulating hormone (FSH) and luteinizing hormone (LH) are the main hormones influencing the follicular phases. Progesterone is the dominant hormone directing the luteal phase. If implantation does not occur, the menstrual cycle begins. This section describes the hormonal regulation of the ovarian cycle.

Regulation of the Follicular Phase. Follicular development begins under FSH stimulation. Each month some of the tertiary ovarian follicles begin to grow but only one will become dominant. As the follicles enlarge, thecal endocrine cells start producing *androstenedione,* a steroid hormone that is a key intermediate in the synthesis of estrogens and androgens. The granulosa cells absorb androstenedione and convert it to estrogens. In addition, thecal endocrine cells scattered throughout the ovarian stroma secrete small quantities of estrogens. Circulating estrogens are bound primarily to albumins, with lesser amounts carried by gonadal steroid-binding globulin (GBG).

Of the three estrogens circulating in the bloodstream—*estradiol, estrone,* and *estriol*—**estradiol** (es-tra-DĪ-ol) is the most abundant and has the most pronounced effects on target tissues. It is the dominant hormone prior to ovulation. In estradiol synthesis, androstenedione is first converted to testosterone, which the enzyme aromatase converts to estradiol (**Figure 28–25**). The synthesis of both estrone and estriol proceeds directly from androstenedione.

Estrogens have multiple functions that affect the activities of many tissues and organs throughout the body. Among the important general functions of estrogens are (1) stimulating bone and muscle growth; (2) maintaining female secondary sex characteristics, such as body hair distribution and the location of adipose tissue deposits; (3) affecting CNS activity, especially in the hypothalamus, where estrogens increase the sexual drive; (4) maintaining functional accessory reproductive glands and organs; and (5) initiating the repair and growth of the endometrium.

Regulation of the Luteal Phase. The high LH level that triggers ovulation also promotes progesterone secretion and the formation of the corpus luteum. As the progesterone level rises and the levels of estrogens fall, the GnRH pulse frequency decreases sharply, soon reaching 1–4 pulses per day. This frequency of GnRH pulses stimulates LH secretion more than it does FSH secretion, and the LH maintains the structure and secretory function of the corpus luteum.

The corpus luteum secretes moderate amounts of estrogens, but progesterone is the main hormone of the luteal phase. Its primary function is to continue the preparation of the uterus for pregnancy. Progesterone enhances the blood supply to the functional layer and stimulates the secretion of uterine glands. The progesterone level remains high for the next week, but unless pregnancy occurs, the corpus luteum begins to degenerate.

Approximately 12 days after ovulation, the corpus luteum becomes nonfunctional, and progesterone and estrogen levels fall markedly. The blood supply to the functional layer is restricted, and the endometrial tissues begin to deteriorate. As progesterone and estrogen levels drop, the GnRH pulse frequency increases, stimulating FSH secretion by the anterior lobe of the pituitary gland, and the ovarian cycle begins again.

The hormonal changes involved with the ovarian cycle in turn affect the activities of other reproductive tissues and organs. At the uterus, the hormonal changes maintain the uterine cycle.

Hormonal Regulation of the Uterine Cycle

Spotlight Figure 28–24 also shows the changes in the endometrium during a single uterine cycle. Progesterone and estrogen levels decrease as the corpus luteum degenerates, resulting in menstruation. The shedding of endometrial tissue continues for several days, until rising levels of estrogens stimulate the repair and regeneration of the functional layer of the endometrium. The proliferative phase then continues until a rising progesterone level marks the arrival of the secretory phase. The combination of estrogen and progesterone then causes the uterine glands to enlarge and increase their secretions.

Figure 28–25 Pathways of Steroid Hormone Synthesis in Males and Females. All gonadal steroids are derived from cholesterol. In men, the pathway ends with the synthesis of testosterone, which may subsequently be converted to dihydrotestosterone. In women, an additional step after testosterone synthesis leads to estradiol synthesis. The synthesis of progesterone and estrogens other than estradiol involves alternative pathways.

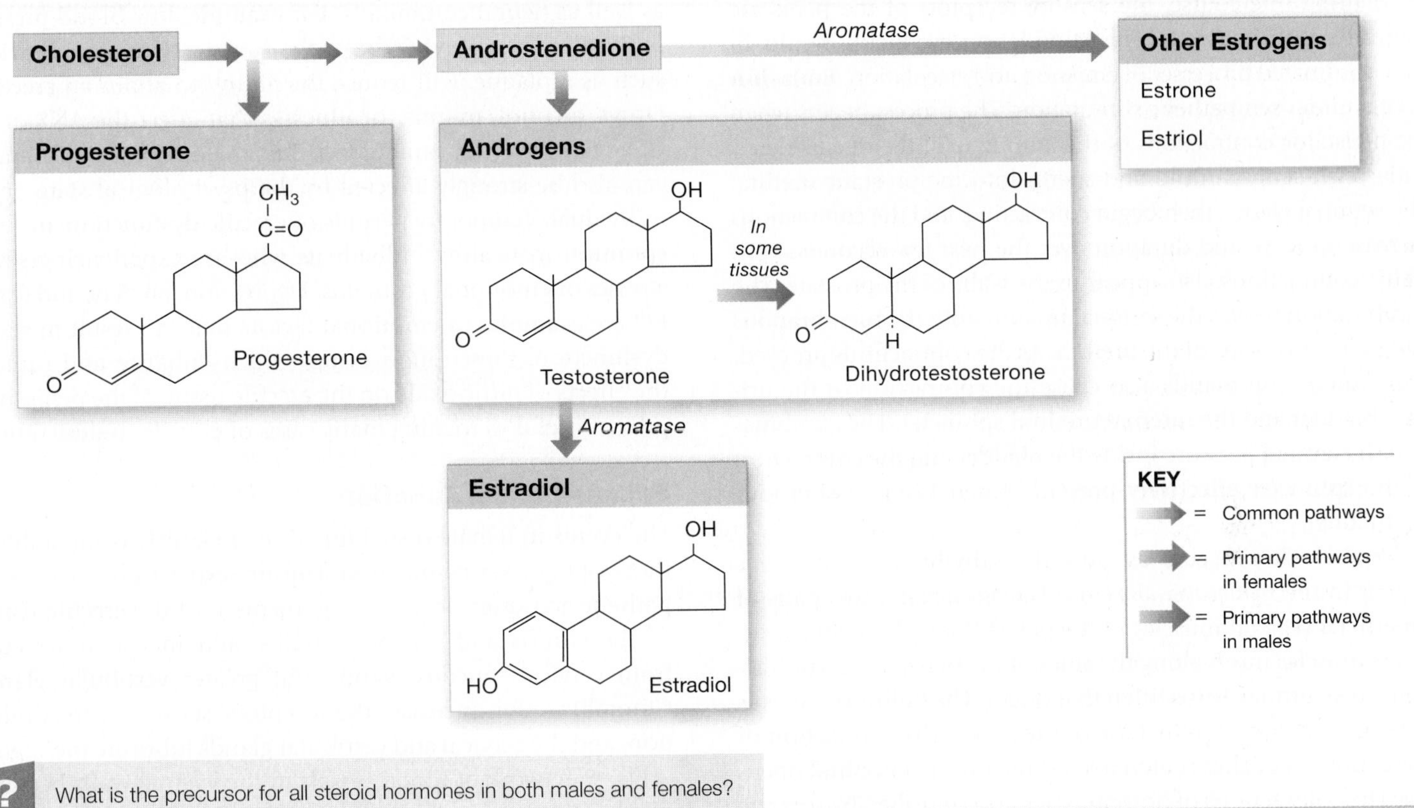
Cholesterol
Androstenedione
Aromatase
Other Estrogens
Estrone
Estriol
Progesterone
CH3
C=O
O
Progesterone
Androgens
OH
In some tissues
Testosterone
H
Dihydrotestosterone
Aromatase
Estradiol
HO
Estradiol
KEY
= Common pathways
= Primary pathways in females
= Primary pathways in males

? What is the precursor for all steroid hormones in both males and females?

Checkpoint

12. **Which layer of the uterus is sloughed off, or shed, during menstruation?**
13. **What changes would you expect to observe in the ovarian cycle if the LH surge did not occur?**
14. **What effect would a blockage of progesterone receptors in the uterus have on the endometrium?**
15. **What event in the uterine cycle occurs when the levels of estrogens and progesterone decrease?**

See the blue Answers tab at the back of the book.

28-6 The autonomic nervous system influences male and female sexual function

Learning Outcome **Discuss the physiology of sexual intercourse in males and females.**

Sexual intercourse introduces semen into the female reproductive tract. The process, in which the autonomic nervous system plays a critical part, affects both the male and female reproductive systems.

Sexual intercourse can lead to several consequences, including pregnancy (both wanted and unintended) and sexually transmitted infections. We also discuss these subjects in this section.

Human Sexual Function

Human sexual function refers to how the body reacts during the sexual response.

Male Sexual Function

Complex neural reflexes coordinate sexual function in males. The reflex pathways use the sympathetic and parasympathetic divisions of the autonomic nervous system (ANS). During sexual **arousal**, erotic thoughts, the stimulation of sensory nerves in the genital region, or both lead to an increase in parasympathetic outflow over the pelvic nerves. This outflow in turn leads to **erection** of the penis. The parasympathetic innervation of the penile arteries involves neurons that release nitric oxide at their axon terminals. The smooth muscles in the arterial walls relax when nitric oxide is released. At that time, the vessels dilate, blood flow increases, and the vascular channels become engorged with blood. The skin covering the glans penis contains numerous sensory receptors, and erection tenses the skin and increases sensitivity. Subsequent stimulation can initiate the secretion of the bulbo-urethral glands, providing **lubrication** for the spongy urethra and the surface of the glans penis.

During intercourse, the sensory receptors of the penis are rhythmically stimulated. This stimulation eventually results in the coordinated processes of emission and ejaculation. **Emission** occurs under sympathetic stimulation. The process begins when the peristaltic contractions of the ampullae of the ductus deferentia push seminal fluid and sperm into the prostatic urethra. The seminal glands then begin contracting, and the contractions increase in force and duration over the next few seconds. Peristaltic contractions also appear in the walls of the prostate. The combination moves the seminal mixture into the membranous and penile portions of the urethra. As the contractions proceed, sympathetic commands also cause the contraction of the urinary bladder and the internal urethral sphincter. The combination of elevated pressure inside the bladder and the contraction of the sphincter effectively prevent semen from passing into the bladder.

Ejaculation occurs as powerful, rhythmic contractions appear in the *ischiocavernosus* and *bulbospongiosus*, two pairs of superficial skeletal muscles of the pelvic floor. The ischiocavernosus muscles insert along the sides of the penis. Their contractions serve primarily to stiffen that organ. The bulbospongiosus muscles wrap around the base of the penis. The contraction of these muscles pushes semen toward the external urethral opening. The contractions of both muscles are controlled by somatic motor neurons in the inferior lumbar and superior sacral segments of the spinal cord. (The positions of these muscles are shown in Figure 11–13b, p. 360.) Contraction of the smooth muscle within the prostate acts to pinch off the urethra, preventing urine from passing through the erect penis.

Ejaculation is associated with intensely pleasurable sensations associated with perineal muscle contraction, an experience known as male **orgasm** (OR-gazm). Several other noteworthy physiological changes take place at this time, including pronounced but temporary increases in heart rate and blood pressure. After orgasm, a **resolution** phase occurs as heart rate and blood pressure decrease and blood begins to leave the erectile tissue and the erection begins to subside. This process of losing an erection, called **detumescence** (dē-tū-MES-ens), is mediated by the sympathetic nervous system. A *refractory period* occurs during the initial period of resolution in which it is generally impossible for another orgasm to occur.

In sum, arousal, erection, emission, ejaculation, orgasm, and resolution are controlled by a complex interplay between the sympathetic and parasympathetic divisions of the ANS. Higher centers, including the cerebral cortex, can facilitate or inhibit many of the important reflexes, modifying sexual function.

Any physical or psychological factor that affects a single component of the system can result in male sexual dysfunction. **Erectile dysfunction (ED)**, or *impotence*, is an inability to achieve or maintain an erection. Various physical causes may be responsible for ED, because erection involves vascular changes as well as neural commands. For example, low blood pressure in the arteries supplying the penis, due to a circulatory blockage such as a plaque, will reduce the ability to attain an erection. Drugs, alcohol, trauma, or illnesses that affect the ANS or the CNS can have the same effect. But male sexual performance can also be strongly affected by the psychological state of the individual. Temporary periods of erectile dysfunction are fairly common in healthy individuals who are experiencing severe stresses or emotional problems. Depression, anxiety, and fear of ED are examples of emotional factors that can result in sexual dysfunction. Prescription drugs, which enhance and prolong the effects of nitric oxide on the erectile tissue of the penis, have proven useful in treating many cases of erectile dysfunction.

Female Sexual Function

The events in female sexual function are largely comparable to those of male sexual function. During sexual arousal, parasympathetic activation leads to engorgement of the erectile tissues of the clitoris and vestibular bulbs, and increased secretion from cervical mucous glands and greater vestibular glands. Clitoral erection increases the receptors' sensitivity to stimulation, and the cervical and vestibular glands lubricate the vaginal walls. A network of blood vessels in the vaginal walls becomes filled with blood at this time, and the vaginal surfaces are also moistened by fluid that moves across the epithelium from underlying connective tissues. (This process accelerates during intercourse as the result of mechanical stimulation.) Parasympathetic stimulation also causes contraction of subcutaneous smooth muscle of the nipples, making them more sensitive to touch and pressure.

During sexual intercourse, rhythmic contact of the penis with the clitoris and vaginal walls—reinforced by touch sensations from the breasts and other stimuli (visual, olfactory, and auditory)—provides stimulation that can lead to orgasm. Female orgasm is accompanied by peristaltic contractions of the uterine and vaginal walls and, through impulses traveling over the pudendal nerves, rhythmic contractions of the bulbospongiosus and ischiocavernosus muscles. The latter contractions give rise to the intensely pleasurable sensations of orgasm. Unlike resolution and the refractory period in men, women may have little delay in achieving further orgasms.

Contraception and Infertility

Many different methods can be used to promote or avoid pregnancy. For example, at the time of ovulation, the basal body temperature (BBT) decreases noticeably, making the rise in temperature over the next day even more noticeable (see Spotlight Figure 28–24). Urine tests that detect LH are available, and testing daily for several days before expected ovulation can detect the LH surge more reliably than the BBT changes. Because fertilization typically occurs within a day of ovulation, this information can be used to time intercourse according to the wishes of the couple. Using this to avoid conception is called the *rhythm method*.

Other forms of **contraception** (methods to prevent pregnancy) include abstinence, barrier methods, intrauterine devices, chemical methods such as "the pill," and surgical methods.

Some couples find they have the opposite problem, and they are unable to conceive. **Infertility** is defined as the inability to conceive or carry a pregnancy to term after 1 year of unprotected intercourse. Infertility, which affects about 10 percent of the U.S. population, can happen in both males and females.

For men, the most common issues are low sperm count and motility. Most males with lower sperm counts are infertile, because too few sperm survive the ascent of the female reproductive tract to perform fertilization. A low sperm count may reflect inflammation of the epididymis, ductus deferens, or prostate. Common abnormalities in sperm are malformed heads and "twin" sperm that did not separate at the time of spermiation.

For women, all of their reproductive organs and hormonal regulation must work properly to conceive and carry a pregnancy to term. If menarche does not appear by age 16, or if the normal uterine cycle of an adult woman is interrupted for 6 months or more, the condition of **amenorrhea** (ā-men-ō-RĒ-uh) exists. *Primary amenorrhea* is the failure to initiate menstruation. This condition may indicate developmental abnormalities, such as nonfunctional ovaries, the absence of a uterus, or an endocrine or genetic disorder. It can also result from malnutrition: Puberty is delayed if the leptin level is too low. p. 646

Severe physical or emotional stresses can cause transient *secondary amenorrhea*. In effect, the reproductive system gets "switched off." Factors associated with amenorrhea include drastic weight loss, anorexia nervosa, and severe depression or grief. Amenorrhea has also been observed in marathon runners and other women engaged in training programs that require sustained high levels of exertion, which severely reduce body lipid reserves.

Sexually Transmitted Diseases (STDs)

Sexual activity carries with it the risk of infection with a variety of microorganisms. The consequences of such an infection may range from merely inconvenient to potentially lethal. **Sexually transmitted diseases (STDs)**, also called **sexually transmitted infections (STIs)**, are transferred from individual to individual, primarily or exclusively by sexual intercourse. At least two dozen bacterial, viral, and fungal infections are currently recognized as STIs. The bacterium *Chlamydia* can cause **pelvic inflammatory disease (PID)** and infertility. The *Zika virus* (*ZIKV*) is a cause of fetal brain defects, including microcephaly. Recent evidence indicates it can be spread through sexual contact. AIDS, caused by a virus, is deadly.

The incidence of STDs is a significant health challenge in the United States. According to the Centers for Disease Control and Prevention (CDC), an estimated 20 million new cases occur each year, almost 50 percent in people aged 15–24.

✓ Checkpoint

16. **What happens when the arteries within the penis dilate?**
17. **List the physiological events of sexual intercourse in both sexes, and indicate those that occur in males but not in females.**
18. **An inability to contract the ischiocavernosus and bulbospongiosus muscles would interfere with which part of the male sex act?**
19. **What changes occur in females during sexual arousal as the result of increased parasympathetic stimulation?**

See the blue Answers tab at the back of the book.

28-7 Changes in levels of reproductive hormones cause functional changes throughout the life span

Learning Outcome Describe the reproductive system changes that occur with development and aging.

Sex hormones have widespread effects on the body and begin early in development. They affect brain development and behavioral drives, muscle mass, bone mass and density, body proportions, and the patterns of hair and body fat distribution. As noted earlier in the chapter, these systems become fully functional at puberty. At puberty, secretion of sex hormones in both sexes accelerates markedly, initiating sexual maturation and the appearance of secondary sex characteristics. Later on, the aging process affects all body systems, including the reproductive systems of both men and women.

Development of the Genitalia

Male and female reproductive systems develop from an embryonic aggregation of cells, termed *primordial cells*. Primordial cells form the first trace of an organ. Both reproductive systems follow a similar pattern of development from the same primordial tissues. They become distinct from each other because of the influence of sex hormones.

Around weeks 5–6 of embryonic development, an embryo is said to be *sexually indifferent*, because the gonads and other internal and external reproductive structures are similar in both sexes. Whether male or female gonads develop depends on the embryo's genetic makeup. (We discuss the basis of sex determination in Chapter 29.) The male gonads begin to form during week 7. Female gonads begin forming about a week later. Once the testes have formed, they begin to secrete testosterone continuously throughout fetal development and until a few days after birth. Testosterone stimulates the differentiation and development of male reproductive structures. Without testosterone, the sexually indifferent embryo will develop female reproductive structures.

The development of male and female genitalia from a sexually indifferent stage is shown in Figure 28–26. At week 4 of development, a temporary cloacal opening marks the site of the future genitalia (Figure 28–26a). A *cloaca* is a cavity or chamber at the end of the digestive tract open to both excretory and reproductive products. By week 6, the cloacal opening becomes subdivided, thus isolating the digestive tract from future urogenital structures. An anterior urogenital membrane will eventually form the urethra and urinary bladder. The genital tubercle present in the sexually indifferent stage will give rise to the penis in the male and clitoris in the female (see Figure 28–26b,c).

During fetal development, the testes form inside the body cavity adjacent to the kidneys. A bundle of connective tissue fibers—called the *gubernaculum testis* (gū-bur-NAK-ū-lum TES-tis; plural, *gubernacula*)—extends from each testis to the posterior wall of a small anterior and inferior pocket of the peritoneum. As the fetus grows, the gubernacula do not get any longer, so they lock the testes in position. As a result, the position of each testis changes as the body enlarges. The testis gradually moves inferiorly and anteriorly toward the anterior abdominal wall. During the seventh developmental month, fetal growth continues rapidly, and circulating hormones stimulate a contraction of the gubernaculum testis. During this time, each testis moves through the abdominal musculature, along with small pockets of the peritoneal cavity. This process is called the *descent of the testes* into the scrotum (see Figure 28–26b). Descent of the female gonads is much less than that in males. The ovaries come to lie below the rim of the true pelvis. They are held in place by the suspensory ligament, ovarian ligament, and round ligament of the uterus (see Figure 28–16).

In *cryptorchidism* (krip-TOR-ki-dizm; *crypto*, hidden + *orchis*, testis), one or both of the testes have not descended into the scrotum by the time of birth. Typically, the cryptorchid (abdominal) testes are lodged in the abdominal cavity or within the inguinal canal. Cryptorchidism occurs in about 3 percent of full-term deliveries and in about 30 percent of premature births. In most instances, normal descent takes place a few weeks later.

The condition can be surgically corrected if it persists. Corrective measures should be taken before *puberty* (sexual maturation), because a cryptorchid testis will not produce sperm. If both testes are cryptorchid, the male will be *sterile (infertile)* and unable to father children. If the testes cannot be moved into the scrotum, they will usually be removed. This surgical procedure to remove the testes is called an *orchiectomy* (or-kē-EK-tō-mē). About 10 percent of males with uncorrected cryptorchid testes eventually develop testicular cancer.

As each testis moves through the body wall, the ductus deferens and the testicular blood vessels, nerves, and lymphatic vessels accompany it. Together, these structures form the body of the spermatic cord (see Figure 28–2, p. 1058).

Effects of Aging

As aging occurs, reductions in sex hormone levels affect appearance, strength, and a variety of physiological functions. The most striking age-related changes in the female reproductive system occur at menopause. Comparable age-related changes in the male reproductive system occur more gradually and over a longer period of time.

The Male Climacteric

Changes in the male reproductive system take place more gradually than do those in the female reproductive system. The period of declining reproductive function, which corresponds to perimenopause in women, is known as the **male climacteric** or *andropause*. The level of circulating testosterone begins to decrease between the ages of 50 and 60, and levels of circulating FSH and LH increase. Although sperm production continues (men well into their 80s can father children), older men experience a gradual reduction in sexual activity. This decrease may be linked to the declining testosterone level. Some clinicians suggest the use of testosterone replacement therapy to enhance the libido (sexual drive) of elderly men, but this may increase the risk of prostate disease.

Menopause

Menopause is usually defined as the time when ovulation and menstruation cease. Menopause typically occurs at age 45–55, but in the years immediately preceding it, an interval called *perimenopause*, the ovarian and uterine cycles become irregular. A shortage of primordial ovarian follicles is the underlying cause of the irregular cycles. With the arrival of perimenopause, the number of primordial ovarian follicles responding each month begins to decrease markedly. As their numbers decrease, the levels of estrogens decrease and may not increase enough to trigger ovulation. By age 50, there are often no primordial ovarian follicles left to respond to FSH. In **premature menopause**, this depletion occurs before age 40.

Menopause is accompanied by a decrease in circulating concentrations of estrogens and progesterone, and a sharp and sustained increase in the production of GnRH, FSH, and LH. The decrease in the levels of estrogens leads to reductions in

Figure 28–26 The Development of Male and Female Genitalia.

a

By 6 weeks, the cloacal opening has been subdivided into a posterior anal membrane and an anterior urogenital membrane. Urethral folds develop and prominent genital swellings form lateral to each urethral fold. The genital tubercle will develop into either the penis in males, or clitoris in females, depending on the presence or absence of testosterone.

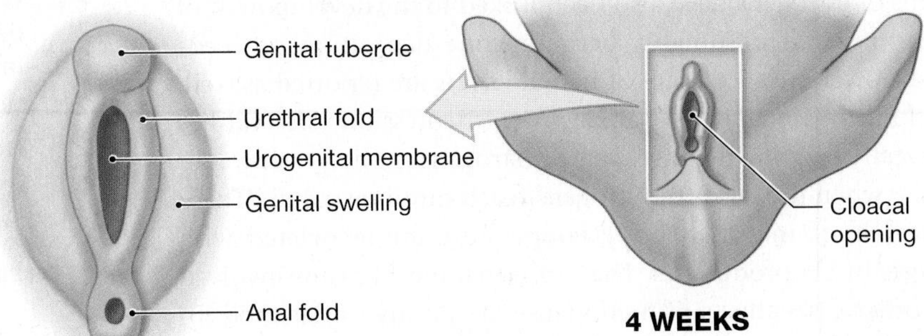
Genital tubercle
Urethral fold
Urogenital membrane
Genital swelling
Anal fold
6 WEEKS
Cloacal opening
4 WEEKS

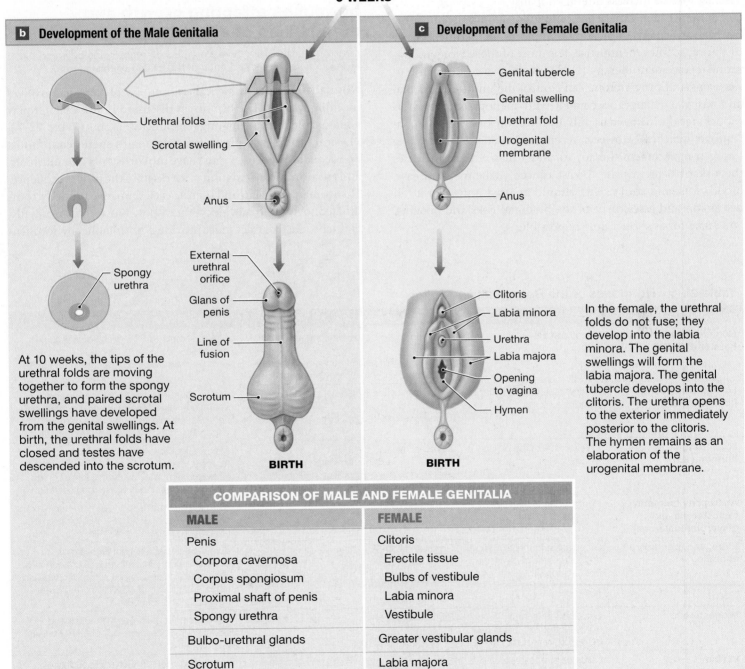

b Development of the Male Genitalia

Urethral folds
Scrotal swelling
Anus
Spongy urethra
External urethral orifice
Glans of penis
Line of fusion
Scrotum
BIRTH

At 10 weeks, the tips of the urethral folds are moving together to form the spongy urethra, and paired scrotal swellings have developed from the genital swellings. At birth, the urethral folds have closed and testes have descended into the scrotum.

c Development of the Female Genitalia

Genital tubercle
Genital swelling
Urethral fold
Urogenital membrane
Anus
Clitoris
Labia minora
Urethra
Labia majora
Opening to vagina
Hymen
BIRTH

In the female, the urethral folds do not fuse; they develop into the labia minora. The genital swellings will form the labia majora. The genital tubercle develops into the clitoris. The urethra opens to the exterior immediately posterior to the clitoris. The hymen remains as an elaboration of the urogenital membrane.

COMPARISON OF MALE AND FEMALE GENITALIA	
MALE	**FEMALE**
Penis Corpora cavernosa Corpus spongiosum Proximal shaft of penis Spongy urethra	Clitoris Erectile tissue Bulbs of vestibule Labia minora Vestibule
Bulbo-urethral glands	Greater vestibular glands
Scrotum	Labia majora

28

the size of the uterus and breasts, accompanied by a thinning of the urethral and vaginal epithelia. The decreased concentrations of estrogens have also been linked to the development of osteoporosis, presumably because bone deposition proceeds at a slower rate. A variety of neural effects are reported as well, including "hot flashes," anxiety, and depression. Hot flashes typically begin while the levels of estrogens are decreasing, and cease when the levels of estrogens reach minimal values. These intervals of increased body temperature are associated with surges in LH production. The hormonal mechanisms involved in other CNS effects of menopause are poorly understood. In addition, the risks of atherosclerosis and other forms of cardiovascular disease increase after menopause.

Some women experience only mild symptoms, but some experience acutely unpleasant symptoms in perimenopause or during or after menopause. For most of those women, hormone replacement therapy (HRT) involving a combination of estrogens and progesterone can control the unpleasant neural and vascular changes associated with menopause. The hormones are administered as pills, by injection, or by transdermal "estrogen patches." However, recent studies suggest that taking estrogen replacement therapy for more than 5 years increases the risk of heart disease, breast cancer, Alzheimer's disease, blood clots, and stroke. HRT should be used only after a full discussion and assessment of the potential risks and benefits, and taken for as short a time as possible.

✓ Checkpoint

20. **What structure in a sexually indifferent embryo may develop into a penis or clitoris and what hormone is involved in that development?**
21. **What is the male climacteric?**
22. **Define *menopause*.**
23. **Why does the level of FSH increase and remain high during menopause?**

See the blue Answers tab at the back of the book.

28-8 The reproductive system secretes hormones affecting growth and metabolism of all body systems

Learning Outcome Give examples of interactions between the reproductive system and each of the other organ systems.

Normal human reproduction is a complex process that involves multiple systems. The hormones discussed in this chapter play a major role in coordinating reproductive events (Table 28–1). Physical factors also play a role. The man's sperm count must be adequate, the semen must have the correct pH and nutrients, and erection and ejaculation must occur in the proper sequence. The woman's ovarian and uterine cycles must be properly coordinated; ovulation and oocyte transport must occur normally. Her reproductive tract must provide a hospitable environment

Table 28–1 Hormones of the Reproductive System

Hormone	Source	Regulation of Secretion	Primary Effects
Gonadotropin-releasing hormone (GnRH)	Hypothalamus	*Males:* inhibited by testosterone and possibly by inhibin *Females:* GnRH pulse frequency increased by estrogens, decreased by progesterone	*Males:* stimulates FSH secretion and LH synthesis *Females:* stimulates FSH secretion and LH synthesis
Follicle-stimulating hormone (FSH)	Anterior lobe of the pituitary gland	*Males:* stimulated by GnRH, inhibited by inhibin *Females:* stimulated by GnRH, inhibited by inhibin	*Males:* stimulates spermatogenesis and spermiogenesis through effects on nurse cells *Females:* stimulates follicle development, estrogen production, and oocyte maturation
Luteinizing hormone (LH)	Anterior lobe of the pituitary gland	*Males:* stimulated by GnRH *Females:* production stimulated by GnRH, secretion by the combination of high GnRH pulse frequencies and high estrogen levels	*Males:* stimulates interstitial cells to secrete testosterone *Females:* stimulates ovulation, formation of corpus luteum, and progesterone secretion
Androgens (primarily testosterone and dihydrotestosterone)	Interstitial endocrine cells of testes	Stimulated by LH	Establish and maintain male secondary sex characteristics and sexual behavior; promote maturation of sperm; inhibit GnRH secretion
Estrogens (primarily estradiol)	Granulosa and thecal endocrine cells of developing follicles; corpus luteum	Stimulated by FSH	Stimulate LH secretion (at high levels); establish and maintain female secondary sex characteristics and sexual behavior; stimulate repair and growth of endometrium; increase frequency of GnRH pulses
Progesterone	Granulosa cells from midcycle through functional life of corpus luteum	Stimulated by LH	Stimulate endometrial growth and glandular secretion; reduce frequency of GnRH pulses
Inhibin	Nurse cells of testes and granulosa cells of ovaries	Stimulated by factors released by developing sperm (male) and developing follicles (female)	Inhibits secretion of FSH (and possibly of GnRH)

Build Your Knowledge

Figure 28–27 Integration of the REPRODUCTIVE system with the other body systems presented so far.

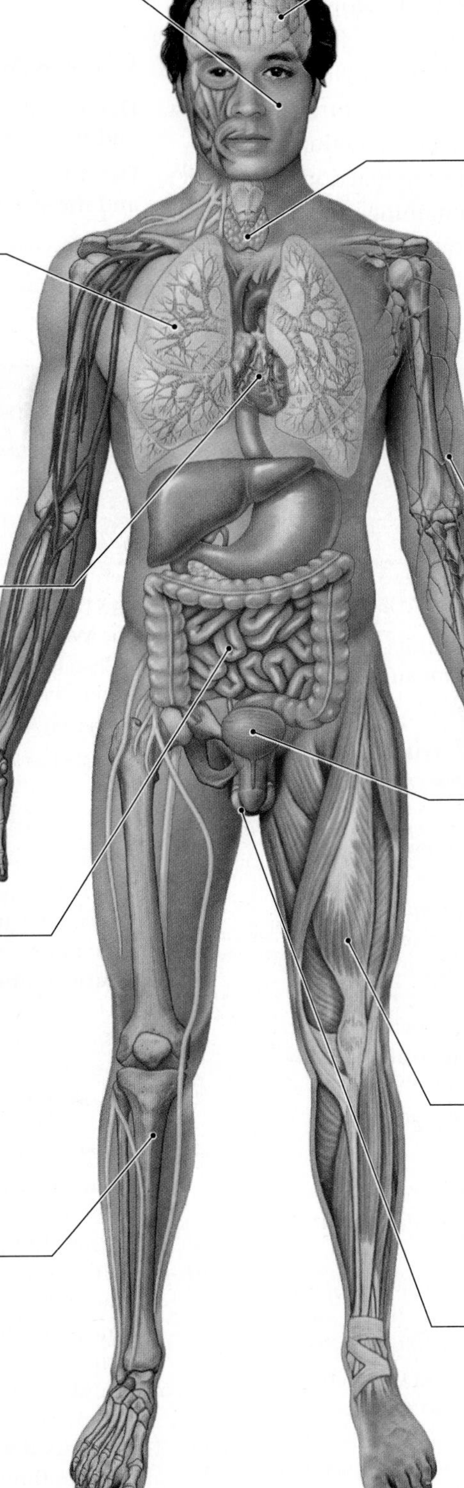

Integumentary System

- The Integumentary System covers external genitalia; provides sensations that stimulate sexual behaviors; mammary gland secretions nourish the newborn
- The reproductive system hormones affect the distribution of body hair and subcutaneous fat

Respiratory System

- The Respiratory System provides oxygen and removes carbon dioxide generated by tissues of reproductive system and (in pregnant women) by embryonic and fetal tissues
- The reproductive system changes respiratory rate and depth during sexual arousal, under control of the nervous system

Cardiovascular System

- The Cardiovascular System distributes reproductive hormones; provides nutrients, oxygen, and waste removal for fetus; local blood pressure changes responsible for physical changes during sexual arousal
- The reproductive system produces estrogens that may help maintain healthy vessels and slow development of atherosclerosis

Digestive System

- The Digestive System provides additional nutrients required to support gamete production and (in pregnant women) embryonic and fetal development
- The reproductive system in pregnant women with a developing fetus crowds digestive organs, causes constipation, and increases appetite

Skeletal System

- The Skeletal System (pelvic girdle) protects reproductive organs of females, portion of ductus deferens and accessory glands in males
- The reproductive system hormones stimulate bone growth and maintenance, and at puberty accelerate growth and closure of epiphyseal cartilages

Nervous System

- The Nervous System controls sexual behaviors and sexual function
- The reproductive system hormones affect CNS development and sexual behaviors

Endocrine System

- The Endocrine System produces hypothalamic regulatory hormones and pituitary hormones that regulate sexual development and function; oxytocin stimulates smooth muscle contractions in uterus and mammary glands
- The reproductive system produces steroid sex hormones and inhibin that inhibit secretory activities of hypothalamus and pituitary gland

Lymphatic System

- The Lymphatic System provides IgA for secretions by epithelial glands; assists in repairs and defense against infection
- The reproductive system secretes lysozymes and bactericidal chemicals that provide innate (nonspecific) defense against reproductive tract infections

Urinary System

- The Urinary System in males carries semen to exterior in urethra; kidneys excrete wastes generated by reproductive tissues and (in pregnant women) by a growing embryo and fetus
- The reproductive system secretions may have antibacterial activity that helps prevent urethral infections in males

Muscular System

- Muscular System contractions eject semen from male reproductive tract
- The reproductive system hormone testosterone accelerates skeletal muscle growth

Reproductive System

The reproductive system secretes hormones with effects on growth and metabolism. It:

- produces, stores, nourishes, and transports male and female gametes
- supports the developing embryo and fetus in the uterus

for the survival and movement of sperm, and for the subsequent fertilization of the oocyte. For these steps to occur, the reproductive, digestive, endocrine, nervous, cardiovascular, and urinary systems must all be functioning normally.

Even when all else is normal and fertilization occurs at the proper time and place, a healthy infant will not be produced unless the **zygote**—the union of an ovum with a sperm forming a single cell the size of a pinhead—manages to develop into a full-term fetus that typically weighs about 3 kg (6.6 lb). In Chapter 29 we will consider the process of development, focusing on the mechanisms that determine both the structure of the body and the distinctive characteristics of each individual.

Even though the reproductive system's primary function—producing offspring—doesn't play a role in maintaining homeostasis, reproduction depends on a variety of physical, physiological, and psychological factors. Many of these factors require intersystem cooperation. In addition, the hormones that control and coordinate sexual function have direct effects on the organs and tissues of other systems. For example, testosterone and estradiol affect both muscular development and bone density. Build Your Knowledge Figure 28–27 summarizes the functional relationships between the reproductive system and other body systems.

Checkpoint

24. **Describe the interaction between the reproductive system and the cardiovascular system.**
25. **Describe the interaction between the reproductive system and the skeletal system.**

See the blue Answers tab at the back of the book.

28 Chapter Review

Study Outline

An Introduction to the Reproductive System p. 1056

1. The *reproductive system* is the only system that is not essential to the life of an individual, but it ensures continuation of our species.

28-1 Male and female reproductive system structures produce gametes that combine to form a new individual p. 1056

2. The human **reproductive system** produces, stores, nourishes, and transports functional **gametes** (reproductive cells). **Fertilization** is the fusion of male and female gametes.
3. The reproductive system includes **gonads** (**testes** or **ovaries**), ducts, accessory glands and organs, and the **external genitalia**.
4. In males, the testes produce **sperm**, which are expelled from the body in **semen** during *ejaculation*. The ovaries of a sexually mature female produce **oocytes** (immature **ova**) that travel along *uterine tubes* toward the *uterus*. The *vagina* connects the uterus with the exterior of the body.

28-2 The structures of the male reproductive system consist of the testes, duct system, accessory glands, and penis p. 1057

5. Sperm travel along the *epididymis*, the *ductus deferens*, the *ejaculatory duct*, and the *urethra* before leaving the body. Accessory organs (notably the *seminal glands*, *prostate*, and *bulbo-urethral gland*) secrete fluids into the ejaculatory ducts and the urethra. The *scrotum* encloses the testes, and the *penis* is an erectile organ. *(Figure 28–1)*
6. The **raphe** marks the boundary between the two chambers in the **scrotum**. *(Figure 28–3)*

MasteringA&P™ Access more chapter study tools online in the MasteringA&P Study Area:

- Chapter Quizzes, Chapter Practice Test, MP3 Tutor Sessions, and Clinical Case Studies
- Practice Anatomy Lab PAL 3.0
- Interactive Physiology iP2
- A&P Flix
- PhysioEx PhysioEx 9.1

7. The **dartos** muscle tightens the scrotum, giving it a wrinkled appearance as it elevates the testes. The **cremaster** has skeletal muscle fibers that pull the testes close to the body.
8. The **tunica albuginea** surrounds each testis. Septa extend from the tunica albuginea to the region of the testis closest to the entrance to the epididymis, creating a series of **lobules**. *(Figure 28–3)*
9. **Seminiferous tubules** within each lobule are the sites of sperm production. From there, sperm pass through the **rete testis**. Seminiferous tubules connect to a **straight tubule**. **Efferent ductules** connect the rete testis to the epididymis. Between the seminiferous tubules are **interstitial cells**, which secrete sex hormones. *(Figures 28–3, 28–4)*
10. From the testis, the sperm enter the **epididymis**, an elongated tubule with **head**, **body**, and **tail** regions. The epididymis monitors and adjusts the composition of the fluid in the seminiferous tubules, serves as a recycling center for damaged sperm, stores and protects sperm, and facilitates their functional maturation. *(Figure 28–5)*
11. The **ductus deferens**, or *vas deferens*, begins at the epididymis and passes through the inguinal canal as part of the spermatic cord. Near the prostate, the ductus deferens enlarges to form

the **ampulla of ductus deferens**. The junction of the base of the seminal gland and the ampulla creates the **ejaculatory duct**, which empties into the urethra. *(Figures 28–5, 28–6)*

12. The **urethra** extends from the urinary bladder to the tip of the penis. The urethra can be divided into *prostatic, membranous,* and *spongy regions.*
13. Each **seminal gland (seminal vesicle)** is an active secretory gland that contributes about 60 percent of the volume of semen. Its secretions contain fructose (which is easily metabolized by sperm), bicarbonate ions, prostaglandins, and fibrinogen. The **prostate** secretes slightly acidic **prostatic fluid.** Alkaline mucus secreted by the **bulbo-urethral glands** has lubricating properties. *(Figures 28–6, 28–7)*
14. A typical ejaculation releases 2–5 mL of semen (**ejaculate**), which contains 20–100 million sperm per milliliter. The fluid component of semen is **seminal fluid**.
15. The skin overlying the **penis** resembles that of the scrotum. Most of the **body** of the penis consists of three masses of **erectile tissue**. Beneath the superficial fascia are two **corpora cavernosa** and a single **corpus spongiosum**, which surrounds the urethra. Dilation of the blood vessels within the erectile tissue produces an **erection**. *(Figure 28–7)*

28-3 Spermatogenesis occurs in the testes, and hormones from the hypothalamus, pituitary gland, and testes control male reproductive functions p. 1065

16. Seminiferous tubules contain **spermatogonia**, stem cells involved in **spermatogenesis** (the production of sperm).
17. **Nurse** (Sertoli) **cells** sustain and promote the development of sperm. *(Figures 28–8, 28–9, 28–10)*
18. Each **sperm** has a **head** capped by an **acrosome**, a **middle piece**, and a **tail**. *(Figure 28–11)*
19. Important regulatory hormones include **FSH** *(follicle-stimulating hormone)*, **LH** *(luteinizing hormone)*, and **GnRH** *(gonadotropin-releasing hormone)*. *Testosterone* is the most important androgen. *(Spotlight Figure 28–12)*

28-4 The structures of the female reproductive system consist of the ovaries, uterine tubes, uterus, vagina, and external genitalia p. 1072

20. Principal organs of the female reproductive system include the *ovaries, uterine tubes, uterus, vagina,* and external genitalia. *(Figure 28–13)*
21. The ovaries, uterine tubes, and uterus are enclosed within the **broad ligament**. The **mesovarium** supports and stabilizes each ovary. *(Figure 28–14)*
22. The ovaries are held in position by the **ovarian ligament** and the **suspensory ligament**. Major blood vessels enter the ovary at the **ovarian hilum**. Each ovary is covered by a **tunica albuginea**. *(Figure 28–14)*
23. Each **uterine tube** has an **infundibulum** with **fimbriae** (fingerlike projections), an **ampulla**, and an **isthmus**. Each uterine tube opens into the *uterine cavity.* For fertilization to occur, a secondary oocyte must encounter sperm during the first 12–24 hours of its passage from the infundibulum to the uterus. *(Figure 28–15)*
24. *Peg cells* lining the uterine tube secrete a fluid that completes the capacitation of sperm.
25. The **uterus** provides mechanical protection, nutritional support, and waste removal for the developing embryo and fetus. Normally, the uterus bends anteriorly near its base (*anteflexion*). The **broad ligament**, **uterosacral ligaments**, **round ligaments**, and **lateral ligaments** stabilize the uterus. *(Figure 28–16)*
26. Major anatomical landmarks of the uterus include the **body**, **isthmus**, **cervix**, **external os** *(external orifice)*, **uterine cavity**, **cervical canal**, and **internal os** *(internal orifice)*. The uterine wall consists of an inner **endometrium**, a muscular **myometrium**, and a superficial **perimetrium** (an incomplete serous layer). *(Figures 28–16, 28–17)*
27. The **vagina** is a muscular tube extending between the uterus and the external genitalia; it is lined by a nonkeratinized stratified squamous epithelium. A thin epithelial fold, the **hymen**, partially blocks the entrance to the vagina until physical distortion ruptures the membrane. *(Figures 28–18, 28–19)*
28. The components of the **vulva** are the **vestibule**, **labia minora**, **urethral glands**, **clitoris**, **mons pubis, labia majora**, and **lesser** and **greater vestibular glands**. *(Figure 28–19)*
29. A newborn infant is nourished from milk secreted by the maternal **mammary gland** in each breast. *(Figure 28–20)*

28-5 Oogenesis occurs in the ovaries, and hormones from the hypothalamus, pituitary gland, and ovaries control female reproductive functions p. 1082

30. Ovaries are the site of **oogenesis** (ovum production), which occurs in **ovarian follicles**. Follicle development proceeds from **primordial ovarian follicles** through **primary**, **secondary**, and **tertiary ovarian follicles**. A dominant tertiary ovarian follicle forms during the **ovarian cycle**. The **ovarian cycle** is divided into a **follicular** *(preovulatory)* **phase** and a **luteal** *(postovulatory)* **phase**. *(Figures 28–21, 28–22)*
31. At **ovulation**, a **secondary oocyte** and the attached follicular cells of the **corona radiata** are released through the ruptured ovarian wall. The follicular cells remaining within the ovary form the **corpus luteum**, which later degenerates into scar tissue called a **corpus albicans**. *(Figure 28–22)*
32. A typical 28-day **uterine**, or *menstrual*, **cycle** begins with the onset of menstruation, also called **menses**, and the destruction of the **functional layer** of the endometrium. This process of **menstruation** continues from 1 to 7 days. *(Figure 28–23)*
33. After the **menstrual phase**, the **proliferative phase** begins, and the functional layer thickens and undergoes repair. The proliferative phase is followed by the **secretory phase**, during which uterine glands enlarge. Menstruation begins at **menarche** and continues until **menopause**. *(Figure 28–23)*
34. Hormonal regulation of the **female reproductive cycle** involves the coordination of the ovarian and uterine cycles. *(Spotlight Figure 28–24)*
35. **Estradiol**, the most important *estrogen*, is the dominant hormone of the follicular phase. Ovulation occurs in response to a midcycle surge in LH. *(Spotlight Figure 28–24; Figure 28–25)*
36. The hypothalamic secretion of GnRH occurs in pulses that trigger the pituitary secretion of FSH and LH. FSH initiates follicular development, and activated follicles and ovarian interstitial cells produce estrogens. High estrogen levels stimulate LH secretion, increase pituitary sensitivity to GnRH, and increase the GnRH pulse frequency. **Progesterone** is the main hormone of the luteal phase. Changes in estrogen and progesterone levels are responsible for maintaining the uterine cycle. *(Spotlight Figure 28–24)*

28-6 The autonomic nervous system influences male and female sexual function p. 1091

37. During sexual **arousal** in males, erotic thoughts, sensory stimulation, or both lead to parasympathetic activity that produces erection. Stimuli accompanying **sexual intercourse** lead to **emission** and **ejaculation**. Contractions of the bulbospongiosus muscles are associated with **orgasm**. During **resolution**, the heart rate and blood pressure decrease and blood begins to leave the erectile tissue and the erection begins to subside.
38. The events of female sexual function resemble those of male sexual function, with parasympathetic arousal and skeletal muscle contractions associated with orgasm.
39. **Infertility** is the inability to conceive or carry a pregnancy to term after 1 year of unprotected intercourse.

28-7 Changes in levels of reproductive hormones cause functional changes throughout the life span p. 1093

40. At weeks 5–6 of embryonic development, an embryo is said to be **sexually indifferent**, because the gonads and other internal and external reproductive structures are similar in both sexes. The presence of testosterone stimulates the differentiation and development of male reproductive structures. Without testosterone, the sexually indifferent embryo will develop female reproductive structures. *(Figure 28–26)*
41. The **descent of the testes** through the *inguinal canals* begins during the seventh month of fetal development.
42. During the **male climacteric**, at ages 50–60, circulating testosterone levels decrease, and FSH and LH levels increase.
43. Menopause (the time that ovulation and menstruation stop) typically occurs at ages 45–55. The production of GnRH, FSH, and LH rises, whereas circulating concentrations of estrogen and progesterone decrease.

28-8 The reproductive system secretes hormones affecting growth and metabolism of all body systems p. 1096

44. Normal human reproduction depends on a variety of physical, physiological, and psychological factors, many of which require intersystem cooperation. *(Figure 28–27; Table 28–1)*

Review Questions

See the blue Answers tab at the back of the book.

LEVEL 1 Reviewing Facts and Terms

1. Developing sperm are nourished by **(a)** interstitial cells, **(b)** the seminal glands, **(c)** nurse cells, **(d)** Leydig cells, **(e)** the epididymis.
2. The ovaries are responsible for **(a)** the production of female gametes, **(b)** the secretion of female sex hormones, **(c)** the secretion of inhibin, **(d)** all of these.
3. In females, meiosis II is not completed until **(a)** birth, **(b)** puberty, **(c)** fertilization occurs, **(d)** uterine implantation occurs.
4. A sudden surge in LH secretion causes the **(a)** onset of menstruation, **(b)** rupture of the follicular wall and ovulation, **(c)** beginning of the proliferative phase, **(d)** end of the uterine cycle.
5. The main hormone of the postovulatory phase is **(a)** progesterone, **(b)** estradiol, **(c)** estrogen, **(d)** luteinizing hormone.
6. Which accessory structures contribute to the composition of semen? What are the functions of each structure?
7. What types of cells in the testes are responsible for functions related to reproductive activity? What are the functions of each cell type?
8. Identify the three regions of the male urethra.
9. List the functions of testosterone in males.
10. List and summarize the important steps in the ovarian cycle.
11. Describe the histology of the uterine wall.
12. What is the role of the clitoris in the female reproductive system?
13. Trace the path of milk flow from its site of production to outside the female.
14. Identify the main structures of the male reproductive system in the diagram on the right.

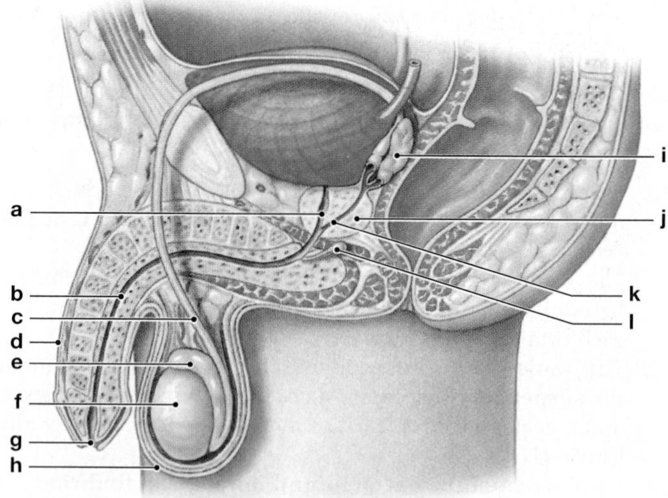
a
b
c
d
e
f
g
h
i
j
k
l

(a)________ **(b)**________ **(c)**________
(d)________ **(e)**________ **(f)**________
(g)________ **(h)**________ **(i)**________
(j)________ **(k)**________ **(l)**________

28

LEVEL 2 Reviewing Concepts

15. In the follicular phase of the ovarian cycle, the ovary **(a)** undergoes atresia, **(b)** forms a corpus luteum, **(c)** releases a mature ovum, **(d)** secretes progesterone, **(e)** matures a dominant tertiary ovarian follicle.
16. Summarize the roles of the hormones in the ovarian and uterine cycles.
17. What are the main differences in gamete production between males and females?
18. Describe the erectile tissues of the penis. How does erection occur?
19. Describe each of the three phases of a typical 28-day uterine cycle.
20. Describe the hormonal events associated with the ovarian cycle.
21. Describe the hormonal events associated with the uterine (menstrual) cycle.
22. Summarize the events that occur in sexual arousal and orgasm. Do these processes differ in males and females?
23. How does the aging process affect the reproductive systems of men and women?

LEVEL 3 Critical Thinking and Clinical Applications

24. Diane has peritonitis (an inflammation of the peritoneum), which her physician says resulted from a urinary tract infection. Why might this condition occur more readily in females than in males?
25. In a condition known as endometriosis, endometrial cells are believed to migrate from the body of the uterus into the uterine tubes or by way of the uterine tubes into the peritoneal cavity, where they become established. A major symptom of endometriosis is periodic pain. Why does such pain occur?
26. Birth control pills contain estradiol and progesterone, or progesterone alone, administered at programmed doses during the ovarian cycle to prevent tertiary ovarian follicle maturation and ovulation. Explain how such pills are effective.
27. Female bodybuilders and women with eating disorders such as anorexia nervosa commonly experience amenorrhea. What does this fact suggest about the relationship between body fat and menstruation? What effect would amenorrhea have on achieving a successful pregnancy?

+ CLINICAL CASE Wrap-Up And Baby Makes Three?

A basic rule of thumb is that a couple that has been trying to get pregnant for 1 year is a candidate for a fertility work-up. Fertility problems can originate in the female, in the male, in both, or for unknown reasons.

A woman's ability to get pregnant decreases with increasing age. Her fertility depends on how many oocytes she has left in her ovaries, how healthy the oocytes are, and—most importantly—if they are being released from the ovary. A healthy reproductive tract is needed for the oocyte to unite with a sperm, and then for the fertilized ovum to be conveyed to a healthy uterus to implant and complete gestation.

Is Susan ovulating? That is the first thing to check, as it is the most common reason for female infertility. At home she can take her morning temperature, which should spike with ovulation at day 14 and stay high through day 28 of her cycle under the influence of progesterone. Is her reproductive tract blocked by scarring from her earlier STI? If so, surgery can open a blocked uterine tube.

A man's infertility is most often due to problems with his sperm. The count may be too low or the motility of the sperm may be poor. Checking the quantity and quality of Tim's sperm is simple. He can provide a sample for analysis in the lab. Tim can change his running gear from briefs to boxers in order to provide a cooler environment in the testes for sperm production.

The Millers have youth and general good health on their side. However, they do have risk factors for infertility that need to be addressed. Baby Miller may yet join the family!

1. Trace the path of Susan's oocyte to meet Tim's sperm. Where will a fertilized ovum travel next?
2. When should Susan expect ovulation to occur during her cycle? How can we verify that she has ovulated?

See the blue Answers tab at the back of the book.

Related Clinical Terms

cervical dysplasia: Abnormal growth of epithelial cells in the uterine cervix; may progress to cancer.

endometriosis: The growth of endometrial tissue outside the uterus.

genital herpes: A sexually transmitted disease caused by a herpes virus and characterized by painful blisters in the genital area.

gonorrhea: A sexually transmitted bacterial disease caused by *Neisseria gonorrhoeae.* Commonly called "the clap."

gynecology: The branch of medicine that deals with the functions and diseases specific to women and girls affecting the reproductive system.

hydrocele: The accumulation of serous fluid in any body sac, but especially in the tunica vaginalis of the testis or along the spermatic cord.

hysterectomy: The surgical removal of the uterus.

menorrhagia: The condition of experiencing extremely heavy bleeding at menstruation.

oophorectomy: The surgical removal of one or both ovaries.

orchitis: Inflammation of one or both testicles.

ovarian cyst: A condition (usually harmless) in which fluid-filled sacs develop in or on the ovary.

polycystic ovary syndrome (PCOS): A condition in women that is characterized by irregular or no menstrual periods, acne, obesity, and excessive hair growth.

premature ejaculation: A common complaint of ejaculating semen sooner than the man desires while achieving orgasm during intercourse. An estimated 30 percent of men regularly experience the problem.

premenstrual dysphoric disorder (PMDD): A collection of physical and emotional symptoms that occurs 5 to 11 days before a woman's period begins, and goes away once menstruation starts. Over 150 signs and symptoms have been associated with the condition.

premenstrual syndrome: A condition occurring in the last half of the menstrual cycle after ovulation that is a combination of physical and mood disturbances that normally end with the onset of the menstrual flow.

salpingitis: Inflammation of a uterine tube.

uterine fibroids (leiomyomas): Benign tumors of the uterus composed of smooth muscle tissue that grows in the wall of the uterus of some women. Although not usually dangerous they can cause problems such as very heavy menstrual periods and pain.

uterine prolapse: Condition that occurs when a woman's pelvic floor muscles and ligaments stretch and weaken and provide inadequate support for the uterus, which then descends into the vaginal canal.

vasectomy: The surgical removal of a segment of each ductus (vas) deferens, making it impossible for sperm to reach the distal portions of the male reproductive tract.

vulvovaginal candidiasis: A common female vaginal infection caused by the yeast *Candida*, usually *Candida albicans.*

29 Development and Inheritance

Learning Outcomes

These Learning Outcomes correspond by number to this chapter's sections and indicate what you should be able to do after completing the chapter.

- 29-1 ■ **Explain the relationship between differentiation and development, and specify the various stages of development.** p. 1104
- 29-2 ■ **Describe the process of fertilization, and explain how developmental processes are regulated.** p. 1105
- 29-3 ■ **List the three stages of prenatal development, and describe the major events of each.** p. 1107
- 29-4 ■ **Explain how the three germ layers are involved in forming the extra-embryonic membranes, and discuss the importance of the placenta as an endocrine organ.** p. 1107
- 29-5 ■ **List several fetal changes that occur during the second and third trimesters.** p. 1115
- 29-6 ■ **Describe the interplay between the maternal organ systems and the developing fetus, and discuss the structural and functional changes in the uterus during gestation.** p. 1117
- 29-7 ■ **List and discuss the events that occur during labor and delivery.** p. 1123
- 29-8 ■ **Identify the features and physiological changes of the postnatal stages of life.** p. 1126
- 29-9 ■ **Relate basic principles of genetics to the inheritance of human traits.** p. 1131

CLINICAL CASE The Twins That Looked Nothing Alike

10
2

Annie and her twin, Ruth, are in Twinsburg, Ohio, for the Twins Day Festival, the largest annual celebration of twins in the world. There are thousands of identical twins in attendance, and a healthy representation of fraternal twins as well. Annie and Ruth are amazed to see so many twins who not only look alike but also move alike, sound alike, and smile alike.

As they walk through the festival grounds, Annie and Ruth notice tents set up to recruit twins for everything from FBI face-recognition software development to fingerprint and handwriting analysis. The festival is a popular event for "nature vs. nurture" researchers who are looking for twins to enlist in their studies.

Another centerpiece of the festival is the awarding of medals to twins in various categories, including the oldest twins, the youngest twins, the "most-alike" twins, and the "least-alike" twins. This year, the "least-alike" twins medals are awarded to two boys, one who is short and Caucasian-with blonde hair, the other who is tall and of African descent, with black hair. **Annie elbows Ruth and asks, "What do you think is the story with these twins?" To find out, turn to the Clinical Case Wrap-Up on p. 1143.**

An Introduction to Development and Inheritance

The physiological processes we have studied so far are relatively brief. Many last only a fraction of a second, and others may take hours. But some important processes are measured in months, years, or even decades. In the process of **development**, a human being forms in the womb for 9 months, and grows to maturity in 15 to 20 years. He or she may live the better part of a century, in that time always changing. Birth, growth, maturation, aging, and death are all parts of a single, continuous process. That process does not end with the individual, because humans can pass at least some of their characteristics on to their offspring. Each generation gives rise to a new generation that will repeat the cycle.

In this chapter, we explore how genetics, environmental factors, and various physiological processes affect the events following the union of male and female *gametes*—sperm and oocytes. The explanation begins with development before birth and continues through childhood and adolescence and into maturity and *senescence*, or aging.

29-1 Directed by inherited genes, a fertilized ovum differentiates during prenatal development to form an individual; postnatal development brings that individual to maturity

Learning Outcome Explain the relationship between differentiation and development, and specify the various stages of development.

The gradual modification of anatomical structures and physiological characteristics that occurs before birth is referred to as **prenatal** (*natus*, birth) **development**. **Postnatal development** begins at birth and continues to **maturity**, the state of full development or completed growth.

Prenatal development, the primary focus of this chapter, begins at *fertilization*, or *conception*, when the male gamete (a sperm) and the female gamete (oocyte) fuse. We divide development into stages characterized by specific anatomical changes. **Pre-embryonic development** involves the processes that occur in the first 2 weeks after fertilization, producing an *embryo*. **Embryonic development** includes the events during the third through eighth weeks. The study of these events is called **embryology** (em-brē-OL-ō -jē). The development of the **fetus**, or **fetal development**, begins at the start of the ninth week and continues until birth.

The changes that occur during development are truly remarkable. In a mere nine months, all the tissues, organs, and organ systems we have studied so far take shape and begin to function. What begins as a single egg cell slightly larger than the period at the end of this sentence becomes an individual like yourself whose body contains trillions of cells organized into a complex array of highly specialized structures. The formation of the different types of cells required in this process is called **differentiation**.

Differentiation occurs through selective changes in genetic activity. As development proceeds, some genes are turned off and others are turned on. Exactly which genes are turned off and on varies from one type of cell to another, and the patterns change over time.

All humans go through the same developmental stages, but this process produces individuals with distinctive characteristics. The term **inheritance**, or **heredity**, refers to the transfer of genetically determined characteristics from generation to generation. The study of the mechanisms responsible for inheritance is called **genetics**. In this chapter, we consider basic genetics as it applies to inherited characteristics, such as sex, hair color, and various diseases.

A basic understanding of human development gives important insights into anatomical structures. In addition, many of the processes of development and growth are similar to those that repair injuries. In this chapter, we focus on major aspects of development. We consider highlights of the developmental process rather than examine the events in great detail. We also consider the regulatory mechanisms involved, and how developmental patterns can be modified.

Checkpoint

1. **Define *differentiation*.**
2. **What event marks the onset of development?**
3. **Define *inheritance*.**

See the blue Answers tab at the back of the book.

29-2 Fertilization—the fusion of a secondary oocyte and a sperm—forms a zygote

Learning Outcome **Describe the process of fertilization, and explain how developmental processes are regulated.**

Fertilization involves the fusion of two haploid (n) gametes, each containing 23 chromosomes. This process produces a **zygote** that contains 46 chromosomes, the normal diploid (2n) number in a human somatic cell. Fertilization typically takes place near the junction between the ampulla and isthmus of the uterine tube. It generally occurs within a day after ovulation. Let's look at how this process takes place, and the different roles of the male and female gametes.

The Secondary Oocyte and Sperm before Fertilization

The functional roles and contributions of the male and female gametes are very different. The **sperm** delivers the paternal (father's) chromosomes to the site of fertilization. It is small, highly streamlined, and driven by a flagellum. In contrast, the female **oocyte** provides all the cellular organelles and inclusions, nourishment, and genetic programming necessary to support development of the embryo for nearly a week after conception. For this reason, the volume of this gamete is much greater than that of the sperm. At fertilization, the diameter of the secondary oocyte is more than twice the entire length of the sperm (Figure 29–1a). The ratio of their volumes is even more striking—approximately 2000:1.

Recall from Chapter 28 that ovulation releases a **secondary oocyte**. The secondary oocyte is surrounded by a thick, glycoprotein-rich envelope called the *zona pellucida*. Outside of that envelope is a protective layer of follicle cells called the **corona radiata** (Figure 29–1b). The cells of the corona radiata protect the secondary oocyte as it passes through the ruptured follicular wall, across the surface of the ovary, and into the infundibulum of the uterine tube. The oocyte then generally travels a few centimeters into the ampulla toward the isthmus.

The secondary oocyte begins meiosis II in the tertiary ovarian follicle, but the process is suspended in metaphase of meiosis II. The secondary oocyte is then ovulated in this state. The cell's metabolic operations await the stimulus of fertilization for further development. If fertilization does not take place, the oocyte disintegrates without completing meiosis.

To have a chance of fertilizing a secondary oocyte, sperm must travel the relatively long distance between the vagina and the ampulla of the uterine tube. The sperm deposited in the vagina are already motile, as a result of contact with secretions of the seminal glands—the first step of *capacitation*. ⮌ p. 1070 Capacitation is the functional maturation of the sperm in which the receptors are uncovered when the glycoprotein layer of the sperm membrane is removed. This also alters the acrosome, enabling the acrosome reaction, and allowing the sperm to penetrate and fertilize an oocyte. (An unidentified substance secreted by the epididymis appears to prevent premature capacitation.) The sperm, however, cannot accomplish fertilization until they have been exposed to conditions in the female reproductive tract. Uterine tube peg cell secretions help with capacitation, but the exact mechanism responsible for capacitation remains unknown. Furthermore, how sperm "find" the oocyte remains a rich field of investigation. A sperm can propel itself at speeds of only about 34 μm per second, or about 12.5 cm (5 in.) per hour. In theory it should take sperm several hours to reach the upper portions of the uterine tubes. The actual time, however, ranges from 2 hours to as little as 30 minutes. Contractions of the uterine musculature and ciliary currents in the uterine tubes may help the sperm travel from the vagina to the site of fertilization.

Of the nearly 200 million sperm introduced into the vagina in a typical ejaculation, only about 10,000 enter the uterine tube, and fewer than 100 reach the isthmus. In general, a male with a sperm count below 20 million per milliliter is functionally sterile because too few sperm survive to reach and fertilize an oocyte. While it is true that only one sperm fertilizes an oocyte, dozens of sperm are required for successful fertilization, as we will learn next.

The Process of Fertilization

Fertilization and the events that follow are diagrammed in Figure 29–1b. Before the physical process of fertilization can occur, a sperm must penetrate the corona radiata. This is aided by the *acrosome reaction*, which frees the enzymes contained in the acrosome of each sperm, including **hyaluronidase** (hī -uh-lū-RON-i-dās). Hyaluronidase breaks down the bonds between adjacent follicle cells. Dozens of sperm must release

Figure 29–1 Fertilization.

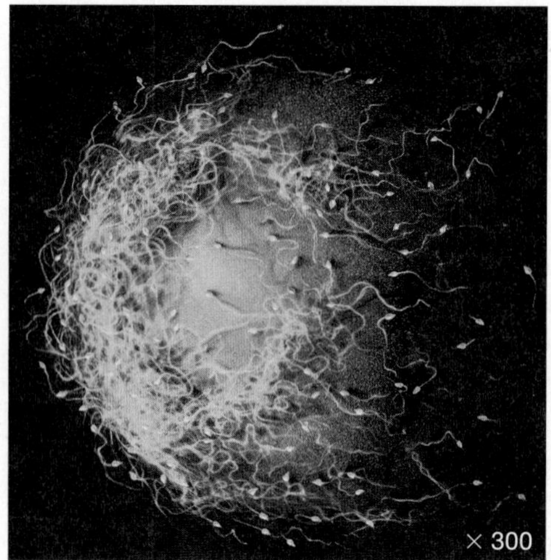
× 300

a A secondary oocyte and numerous sperm at the time of fertilization. Notice the difference in size between the gametes.

hyaluronidase before the connections between the follicle cells of the corona radiata break down enough to allow an intact sperm to reach the oocyte.

No matter how many sperm slip through the gap in the corona radiata, only a single sperm fertilizes the oocyte (1, Figure 29–1b). That sperm must have an intact acrosome. The first step is the binding of the sperm to *sperm receptors* in the zona pellucida. This binding triggers the rupture of the acrosome. The hyaluronidase and **acrosin**, another protein-digesting enzyme, then digest a path through the zona pellucida toward the surface of the oocyte. When the sperm contacts the oocyte membrane, the sperm and oocyte membranes begin to fuse.

Events after Fertilization

The fusion of these membranes triggers **oocyte activation**, which involves a series of changes in the metabolic activity of the oocyte. This process leads to increased permeability of the oocyte membrane to sodium ions; the entry of sodium ions causes the membrane to depolarize. This depolarization in turn causes the release of calcium ions from the smooth endoplasmic reticulum into the oocyte cytoplasm.

Oocyte at Ovulation

Ovulation releases a secondary oocyte and the first polar body; both are surrounded by the corona radiata. The oocyte is suspended in metaphase of meiosis II.

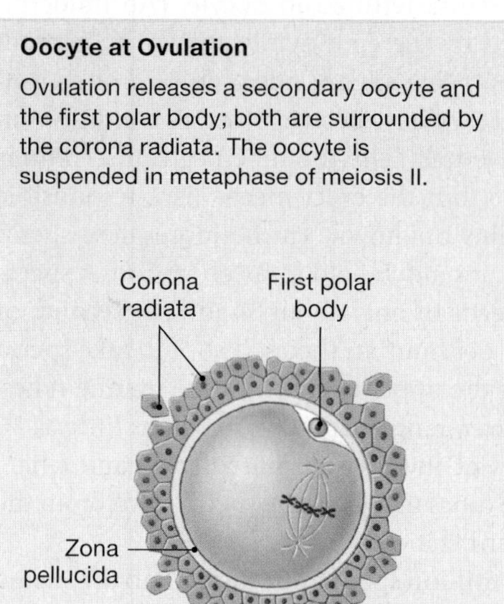
Corona radiata
First polar body
Zona pellucida

1 Fertilization and Oocyte Activation

Acrosomal enzymes from multiple sperm create gaps in the corona radiata. A single sperm then makes contact with the oocyte membrane, and membrane fusion occurs, triggering oocyte activation and the completion of meiosis.

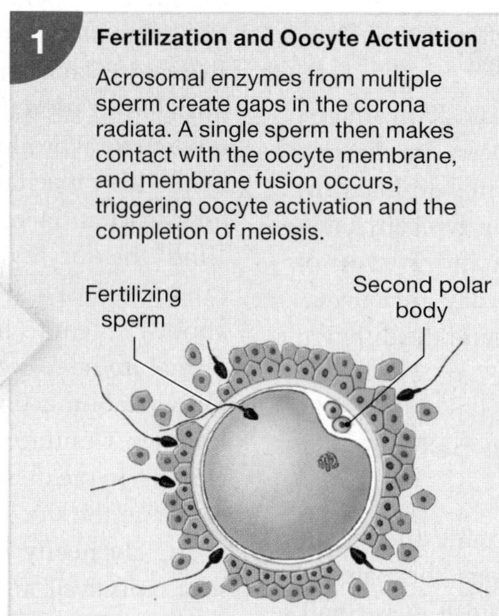
Fertilizing sperm
Second polar body

2 Pronuclei Develop and DNA Synthesis Occurs

The sperm is absorbed into the cytoplasm, and the female pronucleus develops. As the male pronucleus develops, most remaining sperm break down.

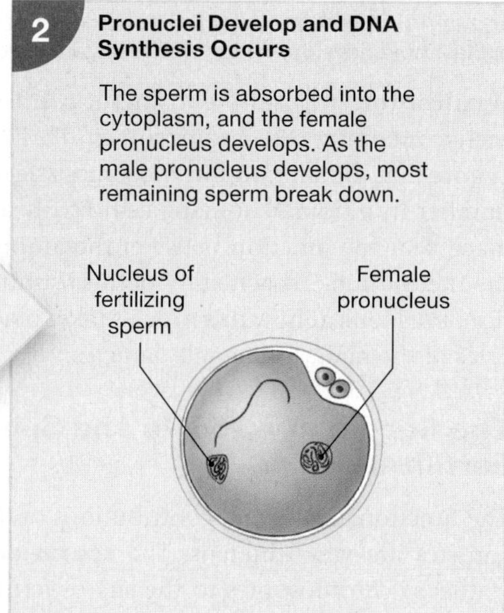
Nucleus of fertilizing sperm
Female pronucleus

3 Spindle Formation Begins

Pronuclei chromatin condenses into chromosomes, and spindle fibers appear in preparation for the first cell division.

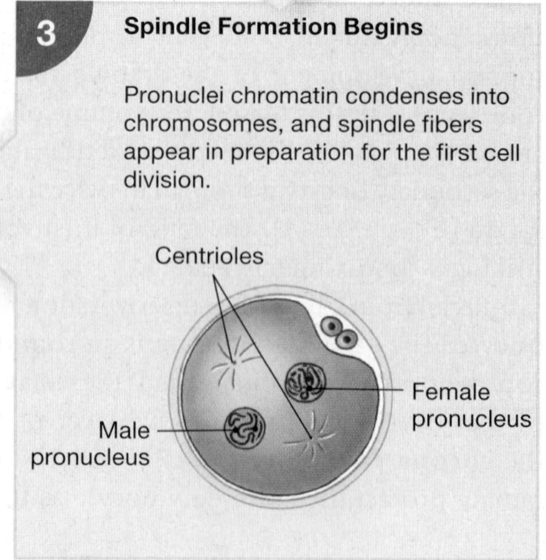
Centrioles
Female pronucleus
Male pronucleus

4 Amphimixis Occurs and Cleavage Begins

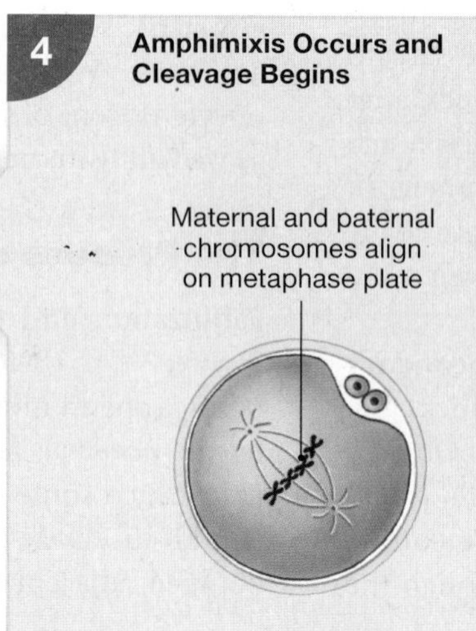
Maternal and paternal chromosomes align on metaphase plate

5 First Cleavage Forms Two Blastomeres

The first cleavage division nears completion about 30 hours after fertilization.

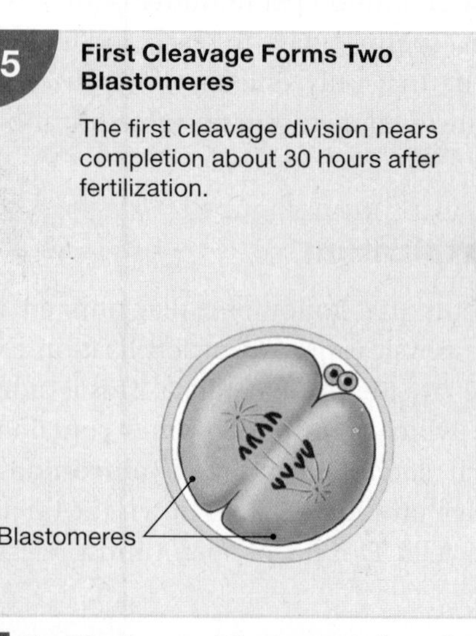
Blastomeres

b Fertilization and the preparations for cleavage.

29

The sudden increase in the calcium ion level has important effects, including the following:

- *Exocytosis of Vesicles Adjacent to the Oocyte Membrane.* This process, called the *cortical reaction*, releases enzymes called *zonal inhibiting proteins*, or *ZIPs*, that both inactivate the sperm receptors and harden the zona pellucida. This combination that prevents fertilization by more than one sperm is known as the **block to polyspermy**. Polyspermy would create a zygote incapable of normal development. (Prior to completion of the cortical reaction, depolarization of the oocyte membrane probably prevents fertilization by any other sperm that penetrate the zona pellucida.)
- *Completion of Meiosis II and Formation of the Second Polar Body.* Once the sperm enters the secondary oocyte, the oocyte can complete meiosis II. The fertilized oocyte is now called an **ovum**.
- *Activation of Enzymes That Cause a Rapid Increase in the Ovum's Metabolic Rate.* The cytoplasm of the ovum contains a large number of mRNA strands that have been inactivated by special proteins. The mRNA strands are now activated, so protein synthesis accelerates quickly. Most of the proteins synthesized are required for development to proceed.

After oocyte activation and the completion of meiosis, the nuclear material remaining within the ovum reorganizes into the **female pronucleus**. While these changes are under way, the nucleus of the sperm swells. As it forms the **male pronucleus**, the rest of the sperm breaks down (except for its centrosome). DNA synthesis then takes place in both the male and female pronuclei (2, Figure 29–1b). The male and female pronuclei move toward each other, their chromatin begins to condense into chromosomes (each consisting of two duplicate chromatids), and the male-donated centrioles duplicate and begin to form spindle fibers (3). The pronuclei membranes dissolve and their maternal and paternal chromosomes all line up on the metaphase plate. The combining of the male and female chromosomes is called *amphimixis* (am-fi-MIK-sis; "both mixed together") (4). The cell is now a zygote that contains the normal diploid complement of 46 chromosomes, and fertilization is complete. This is the "moment of conception." The first cleavage yields two daughter cells called *blastomeres* (5). The cell divisions of cleavage subdivide the cytoplasm of the zygote.

Checkpoint

4. **Name two sperm enzymes important to secondary oocyte penetration.**
5. **How many chromosomes are contained within a human zygote?**

See the blue Answers tab at the back of the book.

29-3 Gestation consists of three stages of prenatal development: the first, second, and third trimesters

Learning Outcome List the three stages of prenatal development, and describe the major events of each.

The time spent in prenatal development is known as **gestation** (jes-TĀ-shun). For convenience, we usually think of the gestation period as three integrated **trimesters**, each lasting 3 months:

1. The **first trimester** is the period of pre-embryonic, embryonic, and early fetal development. During this time, the beginnings of all the major organ systems appear.
2. The **second trimester** is dominated by the development of fetal organs and organ systems. This process nears completion by the end of the sixth month. During this time, body shape and proportions change. By the end of this trimester, the fetus looks distinctively human.
3. The **third trimester** is characterized by rapid fetal growth and deposition of adipose tissue. Early in the third trimester, most of the fetus's major organ systems become fully functional. An infant born 1 month or even 2 months prematurely has a reasonable chance of survival.

Checkpoint

6. **Define *gestation*.**
7. **Characterize the key features of each trimester.**

See the blue Answers tab at the back of the book.

29-4 The first trimester includes pre-embryonic and embryonic development, involving the processes of cleavage, implantation, placentation, and embryogenesis

Learning Outcome Explain how the three germ layers are involved in forming the extra-embryonic membranes, and discuss the importance of the placenta as an endocrine organ.

At the moment of conception, the fertilized ovum is a single cell about 0.135 mm (0.005 in.) in diameter. It weighs approximately 150 μg. By convention, pregnancies are clinically dated from the last menstrual period (LMP), which is usually 2 weeks before ovulation and conception. At the end of the first trimester (12 weeks from LMP, but only 10 developmental weeks), the fetus is almost 75 mm (3 in.) long and weighs perhaps 14 g (0.5 oz).

Many important and complex developmental events occur during the first trimester. Here we focus on four general processes: Two occur in the pre-embryonic period (cleavage and implantation) and two occur in the embryonic period (placentation and embryogenesis).

1. **Cleavage** (KLĒV-ij) is a sequence of cell divisions that begins immediately after fertilization (Figure 29–1b). During cleavage, the zygote becomes a **pre-embryo**, which develops into a multicellular complex known as a *blastocyst.* Cleavage ends when the blastocyst first contacts the uterine wall.
2. **Implantation** begins when the blastocyst attaches to the endometrium of the uterus. The process continues as the blastocyst invades maternal tissues. Important events during implantation set the stage for the formation of vital embryonic structures.
3. **Placentation** (plas-en-TĀ-shun) occurs as blood vessels form around the periphery of the blastocyst, and as the placenta develops. The **placenta** is a complex organ that permits exchange between maternal and embryonic blood. It supports the early fetus from its formation in the first trimester until just after birth, when the placenta stops functioning and is expelled from the uterus. From that point on, the newborn is physically independent of the mother.
4. **Embryogenesis** (em-brē-ō-JEN-e-sis) is the formation of a viable **embryo**. This process establishes the foundations for all major organ systems.

These processes are both complex and vital to the survival of the embryo. Perhaps because the events in the first trimester are so complex, it is the most dangerous prenatal stage. Only about 40 percent of conceptions produce embryos that survive the first trimester. For that reason, pregnant women are warned to avoid drugs, alcohol, tobacco, and other disruptive stresses during the first trimester, in the hope of preventing an error in the intricate processes that are under way. Let's look at each of these stages in turn.

The Pre-Embryonic Period

The pre-embryonic period is the first 14 days after fertilization, before implantation in the uterus. At this stage, tissue differentiation has not occurred.

Cleavage and Blastocyst Formation

Cleavage is a series of cell divisions that subdivides the cytoplasm of the zygote, producing an ever-increasing number of smaller and smaller daughter cells (Figure 29–2). Soon after the secondary oocyte completes meiosis, the chromosomes line up along a metaphase plate, and the cell prepares to divide. The first cleavage division is completed about 30 hours after fertilization. The first cleavage produces a pre-embryo consisting of two identical cells called **blastomeres** (see day 0 in Figure 29–2). Subsequent divisions take place at intervals of 10–12 hours. During the initial divisions, all the blastomeres divide simultaneously. As the number of blastomeres increases, the timing becomes less predictable.

After 3 days of cleavage, the pre-embryo is a solid ball of cells resembling a mulberry. This stage is called the **morula** (MOR-yū-luh; *morus*, mulberry) (see days 1–3). The morula typically reaches the uterus on day 4 or 5. Over the next 2 days, the blastomeres in the morula form a **blastocyst**, a hollow ball with an inner cavity known as the **blastocoele** (BLAS-tō-sēl) (see day 6). The blastomeres are now no longer identical in size and shape. The outer layer of cells is called the **trophoblast** (TRŌ-fō-blast). As the word *trophoblast* implies (*trophos*, food + *blast*, precursor), cells in this layer provide nutrients to the developing embryo. The trophoblast also releases enzymes that erode an opening in the zona pellucida, which is then shed in a process known as *hatching* (see Figure 29–2). In the blastocyst, a second group of cells, the **inner cell mass**, lies clustered at one end. These cells are exposed only to the blastocoele and insulated from contact with the outside uterine environment by the trophoblast. In time, the inner cell mass will form the embryo.

Implantation

Once the blastocyst is hatched, it is freely exposed to glycogen-rich fluid within the uterine cavity. The uterine glands secrete this fluid. Throughout the previous few days, the pre-embryo and early blastocyst had been absorbing fluid and nutrients from their surroundings. The process now accelerates, and the blastocyst enlarges. When fully formed, the blastocyst contacts the endometrium, and implantation takes place (Figure 29–3).

Implantation begins as the surface of the blastocyst closest to the inner cell mass adheres to the uterine lining (see day 7 in Figure 29–3). At the point of contact, the trophoblast cells divide rapidly, making the trophoblast several layers thick. The cells closest to the interior of the blastocyst remain intact, forming an inner cellular layer of the trophoblast called the **cytotrophoblast**. Near the endometrial wall, however, the plasma membranes separating the trophoblast cells disappear, creating a layer of cytoplasm with multiple nuclei (day 8). This outer layer is called the **syncytiotrophoblast**.

The syncytiotrophoblast erodes a path through the uterine epithelium by secreting hyaluronidase. This enzyme dissolves the proteoglycans between adjacent epithelial cells, just as hyaluronidase released by sperm dissolved the connections between cells of the corona radiata. At first, the erosion creates a gap in the uterine lining, but maternal epithelial cells migrate and divide and soon repair the surface. Once the blastocyst has lost contact with the uterine cavity, its further development occurs entirely within the functional layer of the endometrium.

As implantation proceeds, the syncytiotrophoblast continues to enlarge and spread into the surrounding endometrium (see day 9, Figure 29–3). The erosion of uterine glands releases nutrients that are absorbed by the syncytiotrophoblast. They

Figure 29-2 **Cleavage and Blastocyst Formation.**

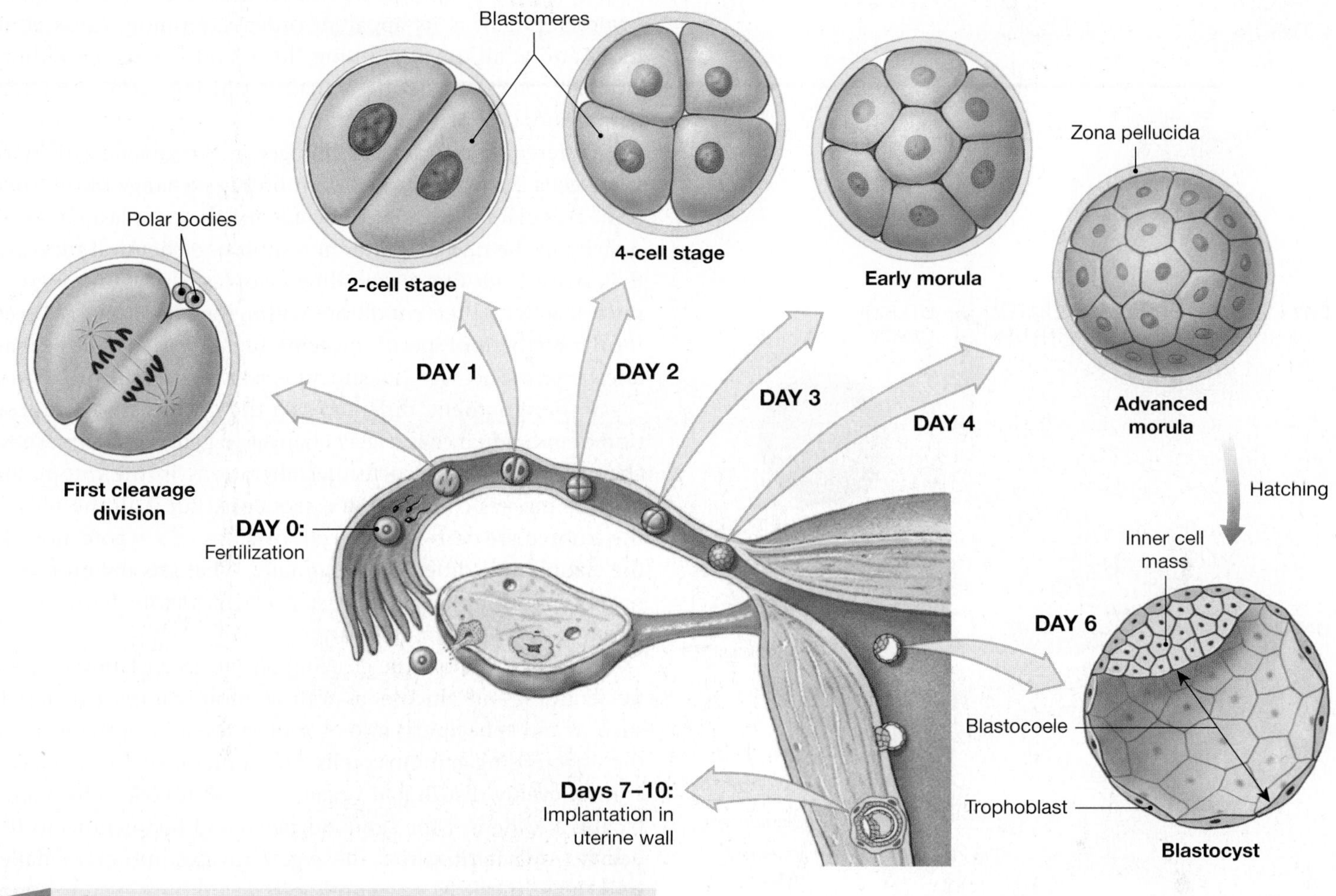

Blastomeres
Polar bodies
2-cell stage
4-cell stage
Early morula
Zona pellucida
Advanced morula
First cleavage division
DAY 0:
Fertilization
DAY 1
DAY 2
DAY 3
DAY 4
Hatching
Inner cell mass
DAY 6
Blastocoele
Trophoblast
Blastocyst
Days 7–10:
Implantation in uterine wall

? What is the name and function of the outer cell layer of a blastocyst?

then diffuse through the underlying cytotrophoblast to the inner cell mass. These nutrients provide the energy needed for the early stages of embryo formation. Trophoblastic extensions grow around endometrial capillaries. As the capillary walls are destroyed, maternal blood begins to percolate through trophoblastic channels known as **lacunae** (*lacuna*, singular). Fingerlike **villi** extend away from the trophoblast into the surrounding endometrium, gradually increasing in size and complexity until about day 21. As the syncytiotrophoblast spreads, it begins breaking down larger endometrial veins and arteries, and blood flow through the lacunae accelerates.

The inner cell mass has little apparent organization early in the blastocyst stage. Yet by the time of implantation, the inner cell mass has separated from the trophoblast. The separation gradually increases, creating a fluid-filled chamber called the **amniotic** (am-nē-OT-ik) **cavity** (see day 9 in Figure 29–3). When the amniotic cavity first appears, the cells of the inner cell mass are organized into an oval sheet that is two layers thick. A superficial layer faces the amniotic cavity, and a deeper layer is exposed to the fluid contents of the blastocoele.

In most cases, implantation occurs in the fundus or in the body of the uterus. In an **ectopic pregnancy**, implantation occurs somewhere other than within the uterus, such as in one of the uterine tubes. Approximately 0.6 percent of pregnancies are ectopic pregnancies. They do not produce a viable embryo and can be life threatening to the mother.

The Embryonic Period

The embryonic period comprises the events from implantation to week 9. During this time, hormonal changes that help maintain the pregnancy are occurring: Trophoblast cells and later the chorion secrete human chorionic gonadotropin (hCG), the corpus luteum secretes progesterone and estrogen, and by the end of the second month, the placenta begins secreting progesterone and estrogen.

Figure 29-3 Stages in Implantation.

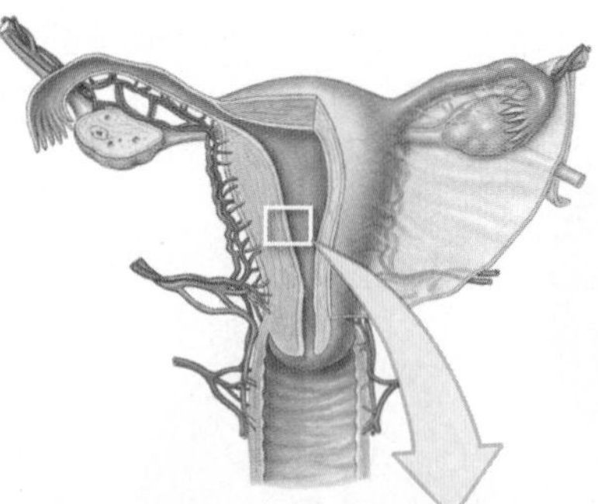

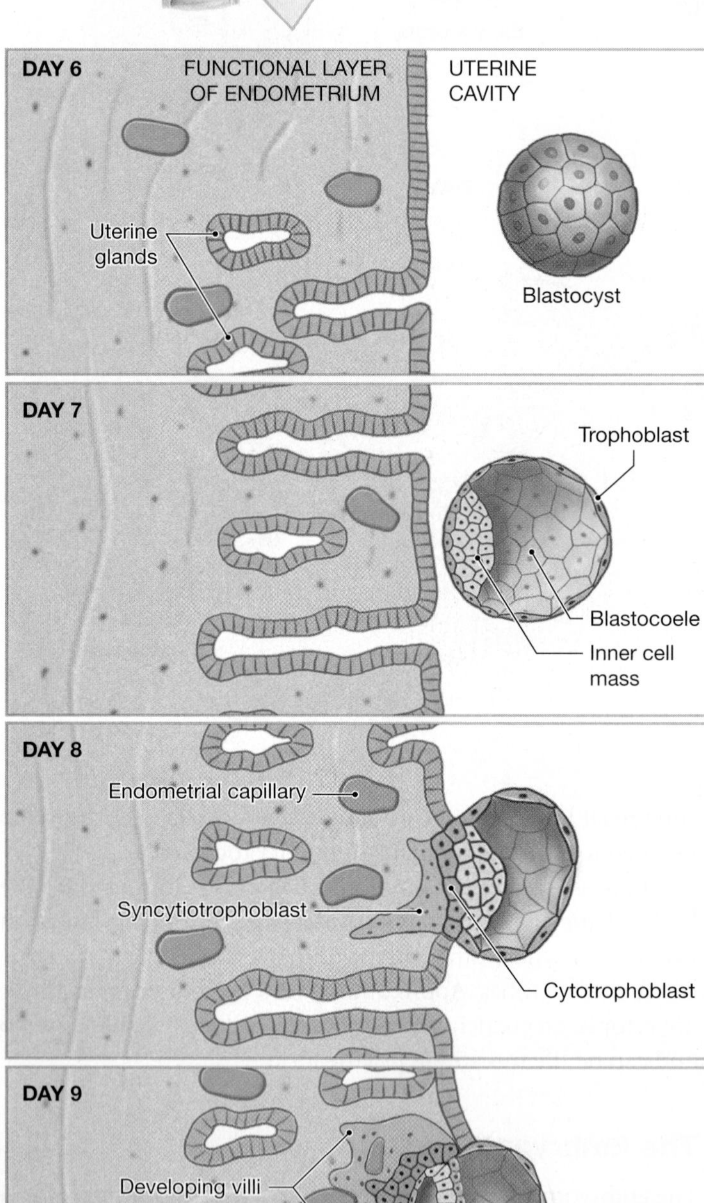
DAY 6
FUNCTIONAL LAYER OF ENDOMETRIUM
UTERINE CAVITY
Uterine glands
Blastocyst
DAY 7
Trophoblast
Blastocoele
Inner cell mass
DAY 8
Endometrial capillary
Syncytiotrophoblast
Cytotrophoblast

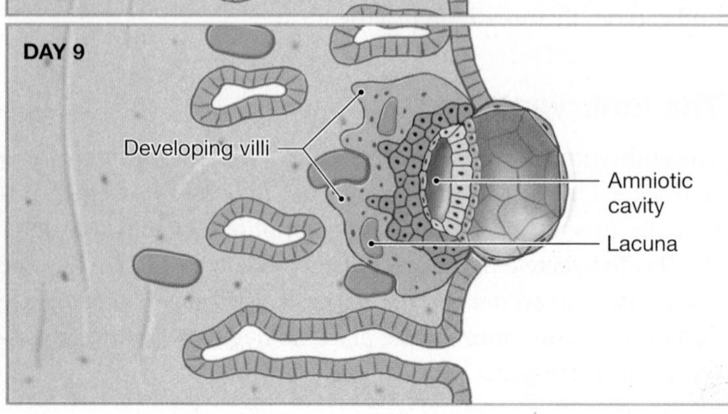
DAY 9
Developing villi
Amniotic cavity
Lacuna

Regulation of Differentiation

During prenatal development, a single cell ultimately forms a 3–4-kg (6.6–8.8-lb) infant. One of the most fascinating aspects of development is its apparent order. Continuity exists at all levels and at all times. Nothing "leaps" into existence without apparent precursors. Differentiation and increasing structural complexity take place hand in hand.

Differentiation involves changes in the genetic activity of some cells but not others. A continuous exchange of information takes place between the nucleus and the cytoplasm in a cell. Activity in the nucleus varies in response to chemical messages that arrive from the surrounding cytoplasm. In turn, ongoing nuclear activity alters conditions within the cytoplasm by directing the synthesis of specific proteins. In this way, the nucleus can affect enzyme activity, cell structure, and membrane properties.

In development, differences in the cytoplasmic composition of individual cells trigger changes in genetic activity. These changes in turn lead to further alterations in the cytoplasm, and the process continues in a sequence. But if all the cells of the embryo are derived from cell divisions of a zygote, how do the cytoplasmic differences originate? What sets the process in motion? The important first step occurs before fertilization, while the oocyte is in the ovary.

Before ovulation, the growing oocyte accepts amino acids, nucleotides, and glucose, as well as more complex materials such as phospholipids, mRNA molecules, and proteins, from the surrounding granulosa cells. The contents of the cytoplasm are not evenly distributed because not all follicle cells manufacture and deliver the same nutrients and instructions to the oocyte. After fertilization, the zygote divides into ever-smaller cells that differ from one another in cytoplasmic composition. These differences alter the genetic activity of each cell, creating cell lines with increasingly diverse fates.

As development proceeds, some of the cells release chemical substances, including RNA molecules, polypeptides, and small proteins that affect the differentiation of other embryonic cells. This type of chemical interplay among developing cells is called *induction* (in-DUK-shun). It works over very short distances, such as when two types of cells are in direct contact. It may also operate over longer distances, with the inducing chemicals functioning as hormones.

This type of regulation, which involves an integrated series of interacting steps, can control highly complex processes. The mechanism is not always error free, however. An abnormal or inappropriate inducer can change the course of development.

Early Embryogenesis: Formation of the Germ Layers and Extra-Embryonic Membranes

Embryogenesis is a phase of prenatal development in which embryonic structures are formed. It is usually regarded as the time from the end of the second week, when the embryonic disc is formed from the inner cell mass, to the end of the eighth week, when the embryo becomes a fetus. In week 3, the two-layered embryonic disc becomes a three-layered embryo as the primary germ layers (ectoderm, mesoderm, and endoderm) and the extra-embryonic membranes develop.

Gastrulation and Germ Layer Formation. Around day 15, a third layer of cells begins to form between the superficial and deep layers of the inner cell mass through a cell migration process called **gastrulation** (gas-trū-LĀ-shun) (Figure 29–4). Together, the three layers of cells are called **germ layers**. The layers are known as the **ectoderm**, **mesoderm**, and **endoderm**. The germ layers separate the trophoblast from the amniotic cavity. Table 29–1 contains a listing of the contributions each germ layer makes to form the body systems described in earlier chapters.

The **embryonic disc** becomes an oval, three-layered sheet during gastrulation. This disc will form the body of the embryo, whereas all other cells of the blastocyst will be part of the *extra-embryonic membranes*.

Shortly after gastrulation begins, the body of the embryo begins to separate itself from the rest of the embryonic disc. The body of the embryo and its internal organs now start to form. This process, embryogenesis, begins as folding and differential growth of the embryonic disc produce a bulge that projects into the amniotic cavity (Spotlight Figure 29–5, step 2).

Figure 29-4 The Inner Cell Mass and Gastrulation.

Day 10: Yolk Sac Formation

While cells from the superficial layer of the inner cell mass migrate around the amniotic cavity, forming the amnion, cells from the deeper layer migrate around the outer edges of the blastocoele. This is the first step in the formation of the yolk sac, a second extra-embryonic membrane. For about the next 2 weeks, the yolk sac is the primary nutrient source for the inner cell mass; it absorbs and distributes nutrients released into the blastocoele by the cytotrophoblast.

Syncytiotrophoblast
Cytotrophoblast
Blastocoele
Amniotic cavity
Yolk sac
Lacunae
Superficial layer
Deep layer

Day 15: Gastrulation

By day 15, superficial cells of the two-layered embryonic disc have begun to migrate toward a central line known as the **primitive streak**. Here, the migrating cells leave the surface and move between the two existing layers. This movement creates three distinct embryonic layers: (1) the **ectoderm**, consisting of superficial cells that did not migrate into the interior of the embryonic disc; (2) the **endoderm**, consisting of the cells that face the yolk sac; and (3) the **mesoderm**, consisting of the poorly organized layer of migrating cells between the ectoderm and the endoderm. Collectively, these three embryonic layers are called **germ layers**, and the migration process is called **gastrulation**. Through gastrulation, the **embryonic disc** becomes an oval, three-layered sheet. This disc will form the body of the embryo, whereas all other cells of the blastocyst will be part of the extra-embryonic membranes.

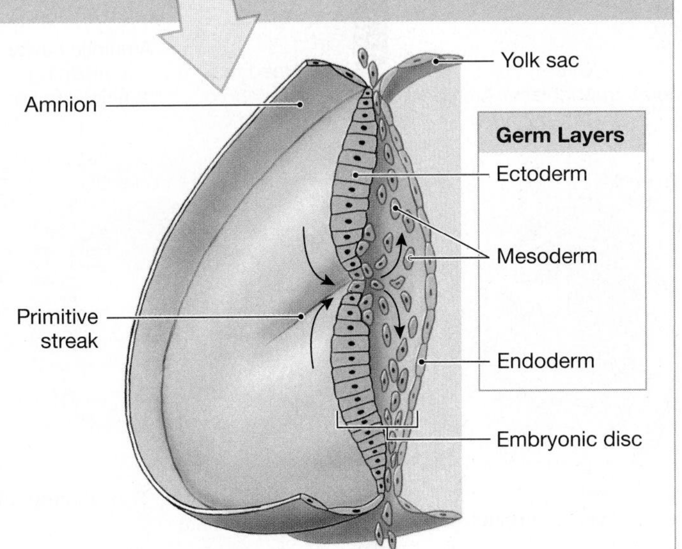
Amnion
Yolk sac
Germ Layers
Ectoderm
Mesoderm
Primitive streak
Endoderm
Embryonic disc

? What is the name of the the third germ layer that forms during gastrulation?

SPOTLIGHT

Figure 29-5
Extra-Embryonic Membranes and Placenta Formation

The germ layers introduced in Figure 29–4 also form four **extra-embryonic membranes: (1)** The **yolk sac** (endoderm and mesoderm), **(2)** the **amnion** (ectoderm and mesoderm), **(3)** the **allantois** (endoderm and mesoderm), and (4) the **chorion** (mesoderm and trophoblast). These membranes support embryonic and fetal development, but few traces of their existence remain in adults.

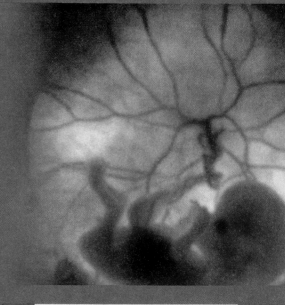

Yolk sac

The yolk sac begins as a layer of cells spreads out around the outer edges of the blastocoele to form a complete pouch. It is the primary nutrient source for early embryonic development, and becomes an important site for blood cell formation.

Amnion

Ectodermal cells spread over the inner surface of the amniotic cavity, soon followed by mesodermal cells. Amniotic fluid is produced, which cushions the developing embryo.

1 Week 2

Migration of mesoderm around the inner surface of the cytotrophoblast forms the chorion. Mesodermal migration around the outside of the amniotic cavity, between the ectodermal cells and the trophoblast, forms the amnion. Mesodermal migration around the endodermal pouch forms the yolk sac.

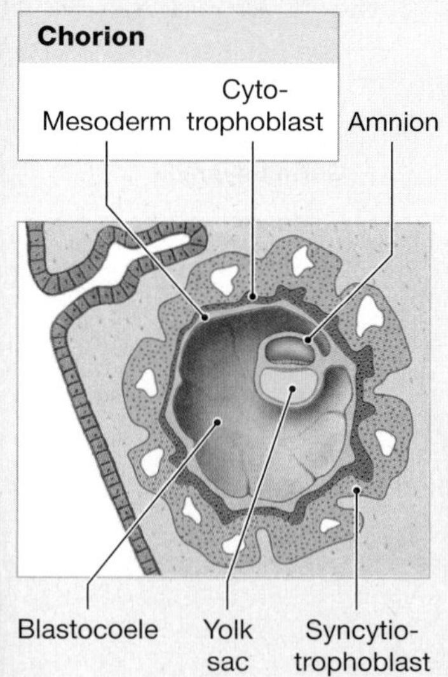

Chorion
Mesoderm
Cyto-trophoblast
Amnion
Blastocoele
Yolk sac
Syncytio-trophoblast

2 Week 3

The embryonic disc bulges into the amniotic cavity at the head fold. The allantois, an endodermal extension surrounded by mesoderm, extends toward the trophoblast.

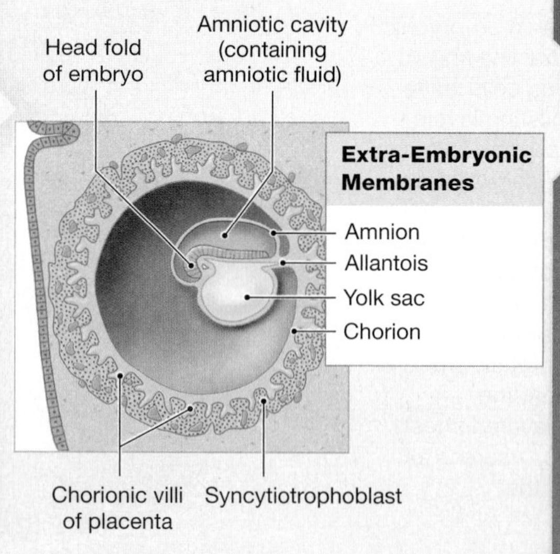

Head fold of embryo
Amniotic cavity (containing amniotic fluid)
Extra-Embryonic Membranes
Amnion
Allantois
Yolk sac
Chorion
Chorionic villi of placenta
Syncytiotrophoblast

3 Week 4

The embryo now has a head fold and a tail fold. Constriction of the connections between the embryo and the surrounding trophoblast narrows the yolk stalk and body stalk.

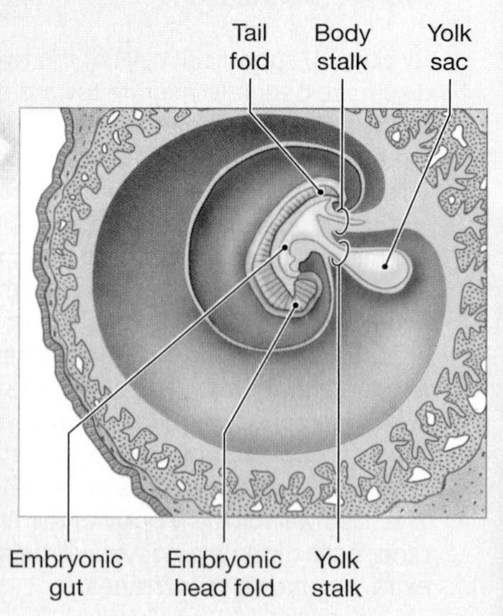

Tail fold
Body stalk
Yolk sac
Embryonic gut
Embryonic head fold
Yolk stalk

Allantois

The allantois begins as an outpocket of the endoderm near the base of the yolk sac. The free endodermal tip then grows toward the wall of the blastocyst, surrounded by a mass of mesodermal cells. The base of the allantois eventually gives rise to the urinary bladder.

Chorion

The mesoderm associated with the allantois spreads around the entire blastocyst, separating the cytotrophoblast from the blastocoele.
The appearance of blood vessels in the chorion is the first step in the creation of a functional placenta. By the third week of development, the mesoderm extends along the core of each trophoblastic villus, forming chorionic villi in contact with maternal tissues and blood vessels. These villi continue to enlarge and branch forming the placenta, the exchange platform between mother and fetus for nutrients, oxygen, and wastes.

4 Week 5

The developing embryo and extra-embryonic membranes bulge into the uterine cavity. The cytotrophoblast pushing out into the uterine cavity remains covered by endometrium but no longer participates in nutrient absorption and embryo support. The embryo moves away from the placenta, and the body stalk and yolk stalk fuse to form an umbilical stalk.

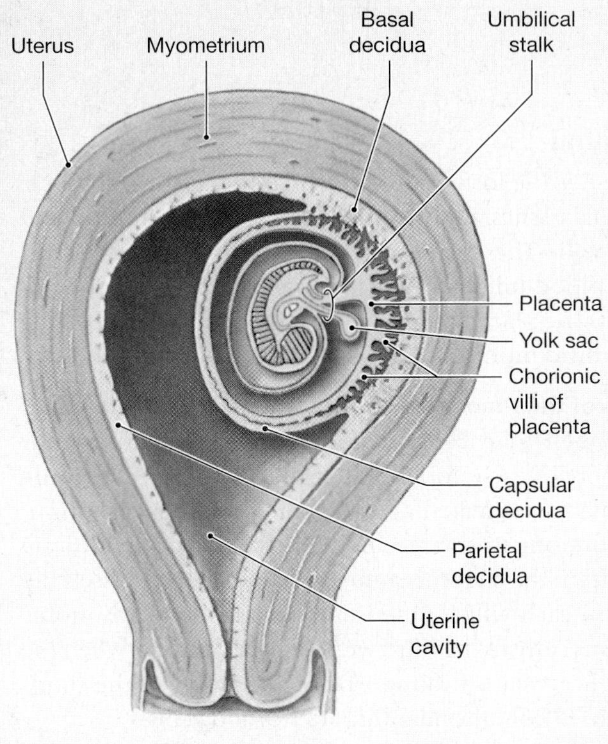
Uterus
Myometrium
Basal decidua
Umbilical stalk
Placenta
Yolk sac
Chorionic villi of placenta
Capsular decidua
Parietal decidua
Uterine cavity

5 Week 10

The amnion has expanded greatly, filling the uterine cavity. The fetus is connected to the placenta by an elongated umbilical cord that contains a portion of the allantois, blood vessels, and the remnants of the yolk stalk. A mucus plug forms, preventing bacteria from entering the uterus.

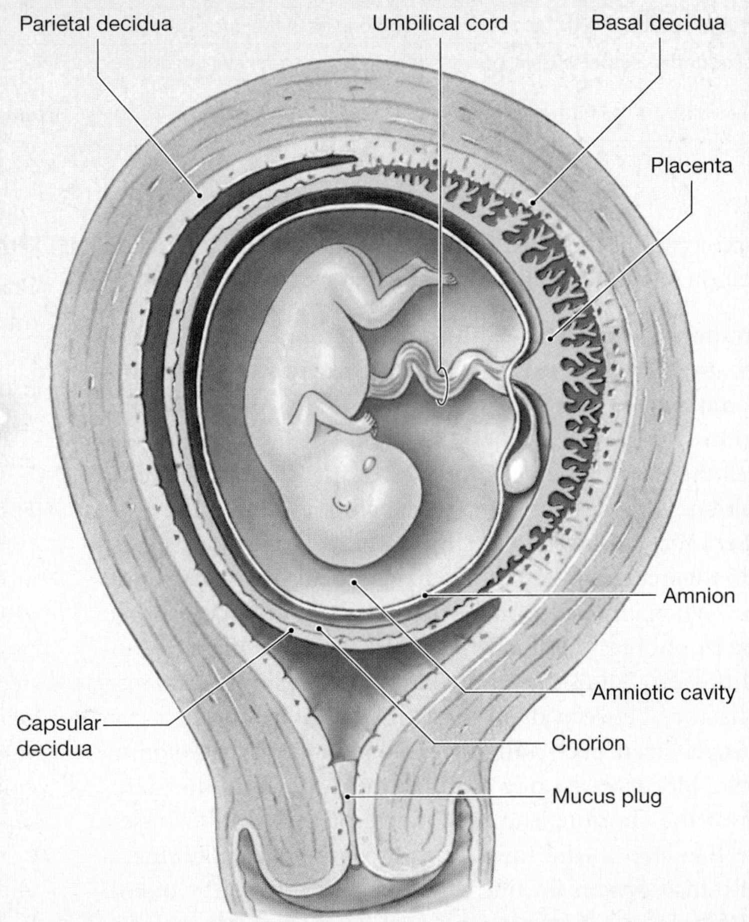
Parietal decidua
Umbilical cord
Basal decidua
Placenta
Amnion
Amniotic cavity
Chorion
Mucus plug
Capsular decidua

Table 29-1 The Fates of the Germ Layers

Ectodermal Contributions
Integumentary system: epidermis, hair follicles and hairs, nails, and glands communicating with the skin (sweat glands, mammary glands, and sebaceous glands)
Skeletal system: pharyngeal cartilages and their derivatives in adults (portion of sphenoid, the auditory ossicles, the styloid processes of the temporal bones, the cornu and superior rim of the hyoid bone)*
Nervous system: all neural tissue, including brain and spinal cord
Endocrine system: pituitary gland and adrenal medullae
Respiratory system: mucous epithelium of nasal passageways
Digestive system: mucous epithelium of mouth and anus, salivary glands
Mesodermal Contributions
Integumentary system: dermis and subcutaneous layer (hypodermis)
Skeletal system: all structures except some pharyngeal derivatives
Muscular system: all structures
Endocrine system: adrenal cortex, endocrine tissues of heart, kidneys, and gonads
Cardiovascular system: all structures
Lymphatic system: all structures
Urinary system: the kidneys, including the nephrons and the initial portions of the collecting system
Reproductive system: the gonads and the adjacent portions of the duct systems
Miscellaneous: the lining of the pleural, pericardial, and peritoneal cavities and the connective tissues that support all organ systems
Endodermal Contributions
Endocrine system: thymus, thyroid gland, and pancreas
Respiratory system: respiratory epithelium (except nasal passageways) and associated mucous glands
Digestive system: mucous epithelium (except mouth and anus), exocrine glands (except salivary glands), liver, and pancreas
Urinary system: urinary bladder and distal portions of the duct system
Reproductive system: distal portions of the duct system, stem cells that produce gametes

*The neural crest is derived from ectoderm and contributes to the formation of the skull and the skeletal derivatives of the embryonic pharyngeal arches.

This projection is the **head fold**. Similar movements lead to the formation of a **tail fold** (see **Spotlight Figure 29–5**, step 3).

Formation of the Extra-Embryonic Membranes. The germ layers also form four **extra-embryonic membranes**: the **yolk sac**, **amnion** (AM-nē-on), **allantois** (ah-LAN-tō-is), and **chorion** (KŌ-rē-on). **Spotlight Figure 29–5** illustrates and describes the developmental processes of these extra-embryonic membranes. Please study this figure before reading on.

Let's focus more on the chorion here because of its early role in the maternal–embryo exchange of nutrients, gases, and wastes. When implantation first occurs, the nutrients absorbed by the trophoblast can easily reach the inner cell mass by simple diffusion. But as the embryo and the trophoblast enlarge, the distance between them increases, so diffusion alone can no longer keep pace with the demands of the developing embryo. Blood vessels now begin to develop within the mesoderm of the chorion. The appearance of these blood vessels is the first step in the formation of a functional placenta, a rapid-transit system for nutrients that links the embryo with the trophoblast.

Placentation

Placentation is the formation of the placenta from embryonic and maternal tissues. Embryonic tissues form the chorion and chorionic villi. The maternal endometrial tissues that surround the blood-filled lacunae become restricted to a specialized region. The placenta is fully formed and functional by the end of the third month.

Formation of the Placenta. By the third week of development (see **Spotlight Figure 29–5,** step 2), the mesoderm extends along the core of each trophoblastic villus, forming **chorionic villi** in contact with maternal tissues and blood vessels. Chorionic villi continue to enlarge and branch, creating an intricate network within the endometrium. Embryonic blood vessels develop within each villus. Blood flow through those chorionic vessels begins early in the third week of development, when the embryonic heart starts beating. The blood supply to the chorionic villi arises from the allantoic arteries and veins.

At first, the entire blastocyst is surrounded by chorionic villi. The chorion continues to enlarge, expanding like a balloon within the endometrium. By week 4, the embryo, amnion, and yolk sac are suspended within an expansive, fluid-filled chamber (see **Spotlight Figure 29–5**, step 3). The **body stalk**, the connection between embryo and chorion, contains the distal portions of the allantois and blood vessels that carry blood to and from the placenta. The narrow connection between the endoderm of the embryo and the yolk sac is called the **yolk stalk**.

As the chorionic villi enlarge, more maternal blood vessels are eroded. The placenta does not continue to enlarge indefinitely. Regional differences in placental organization begin to develop as expansion of the placenta creates a prominent bulge in the endometrial surface. This relatively thin portion of the endometrium, called the **capsular decidua** (dē-SID-yū-uh; *deciduus*, a falling off), or *decidua capsularis*, no longer exchanges nutrients, and the chorionic villi in the region disappear (see **Spotlight Figure 29–5**, step 4), and **Figure 29–6a**). Placental functions are now concentrated in a disc-shaped area in the deepest portion of the endometrium, a region called the **basal decidua** (*decidua basalis*). The rest of the uterine endometrium, which has no contact with the chorion, is called the **parietal decidua** (*decidua parietalis*).

As the end of the first trimester approaches, the fetus moves farther from the placenta (see **Spotlight Figure 29–5**, steps 4 and 5). The fetus and placenta remain connected by the **umbilical cord**, which contains the allantois, the placental blood vessels, and the yolk stalk.

Tips & Tools

Like a deep-sea diver's air hose or a space-walking astronaut's tether, the umbilical cord supplies the fetus with life-sustaining substances and removes wastes.

Placental Circulation. **Figure 29–6a** illustrates placental circulation at the end of the first trimester. Blood flows from the fetus to the placenta through the paired **umbilical arteries** and returns in a single **umbilical vein**. p. 775

The chorionic villi provide the surface area for active and passive exchanges of gases, nutrients, and wastes between the fetal and maternal bloodstreams. The blood in the umbilical arteries is deoxygenated and contains wastes generated by fetal tissues. Maternal blood moves slowly through complex lacunae lined by the syncytiotrophoblast. Chorionic blood vessels pass close by, and gases and nutrients diffuse between the embryonic and maternal circulations across the layers of the trophoblast. Recall that fetal hemoglobin has a higher affinity for oxygen than does maternal hemoglobin, enabling fetal hemoglobin to strip oxygen from maternal hemoglobin. p. 866 At the placenta, oxygen supplies are replenished, nutrients added, and carbon dioxide and other wastes removed. Maternal blood then reenters the venous system of the mother through the broken walls of small uterine veins. No mixing of maternal and fetal blood takes place, because layers of trophoblast always separate the two.

The placenta places a considerable demand on the maternal cardiovascular system, and blood flow to the uterus and placenta is extensive. If the placenta is torn or otherwise damaged, the consequences may prove fatal to both fetus and mother.

Late Embryogenesis and Organogenesis

At about week 4 of development, the embryo is physically as well as developmentally distinct from the embryonic disc and the extra-embryonic membranes. The definitive orientation of the embryo can now be seen, complete with posterior and anterior surfaces and left and right sides. The changes in proportions and appearance that occur between the second developmental week and the end of the first trimester are summarized in **Figure 29–7**.

The first trimester is a critical period for development, because events in the first 12 weeks establish the basis for **organogenesis**, the process of organ formation. The rudiments of all the major organ systems have formed by the end of the first trimester. **Table 29–2** provides an overview of the development of the major organs and body systems in the first trimester (as well as subsequent development).

✓ Checkpoint

8. **What is the developmental fate of the inner cell mass of the blastocyst?**
9. **Improper development of which of the extra-embryonic membranes would affect the cardiovascular system?**
10. **What vessels carry blood between the fetus and the placenta?**

See the blue Answers tab at the back of the book.

29-5 During the second and third trimesters, fetal development involves growth and organ function

Learning Outcome **List several fetal changes that occur during the second and third trimesters.**

Over the second trimester, the fetus grows to a weight of about 0.6 kg (1.32 lb). Encircled by the amnion, the fetus grows faster than the surrounding placenta during this period. When the mesoderm on the outer surface of the amnion contacts the mesoderm on the inner surface of the chorion, these layers fuse, creating a compound *amniochorionic membrane*. **Figure 29–8a** shows a 4-month-old fetus, while **Figure 29–8b** shows a 6-month-old fetus.

During the third trimester, most organ systems become ready for normal functioning without maternal assistance. The rate of growth starts to slow, but weight gain is the greatest

Figure 29–6 Anatomy of the Placenta after the First Trimester.

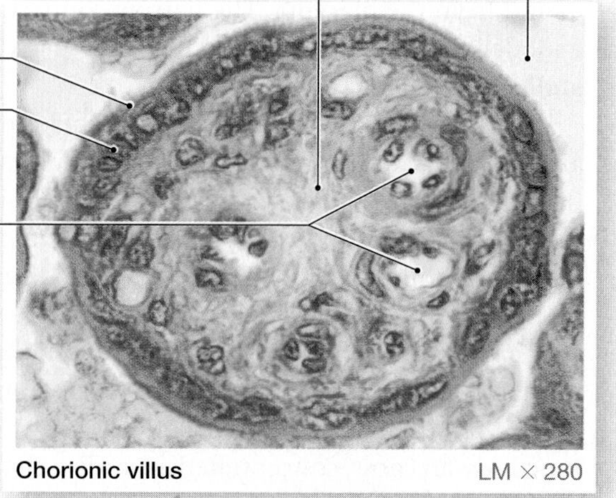

Embryonic connective tissue
Area filled with maternal blood
Syncytiotrophoblast
Cytotrophoblast
Fetal blood vessels
Chorionic villus
LM × 280

b A cross section through a chorionic villus, showing the syncytiotrophoblast exposed to the maternal blood space.

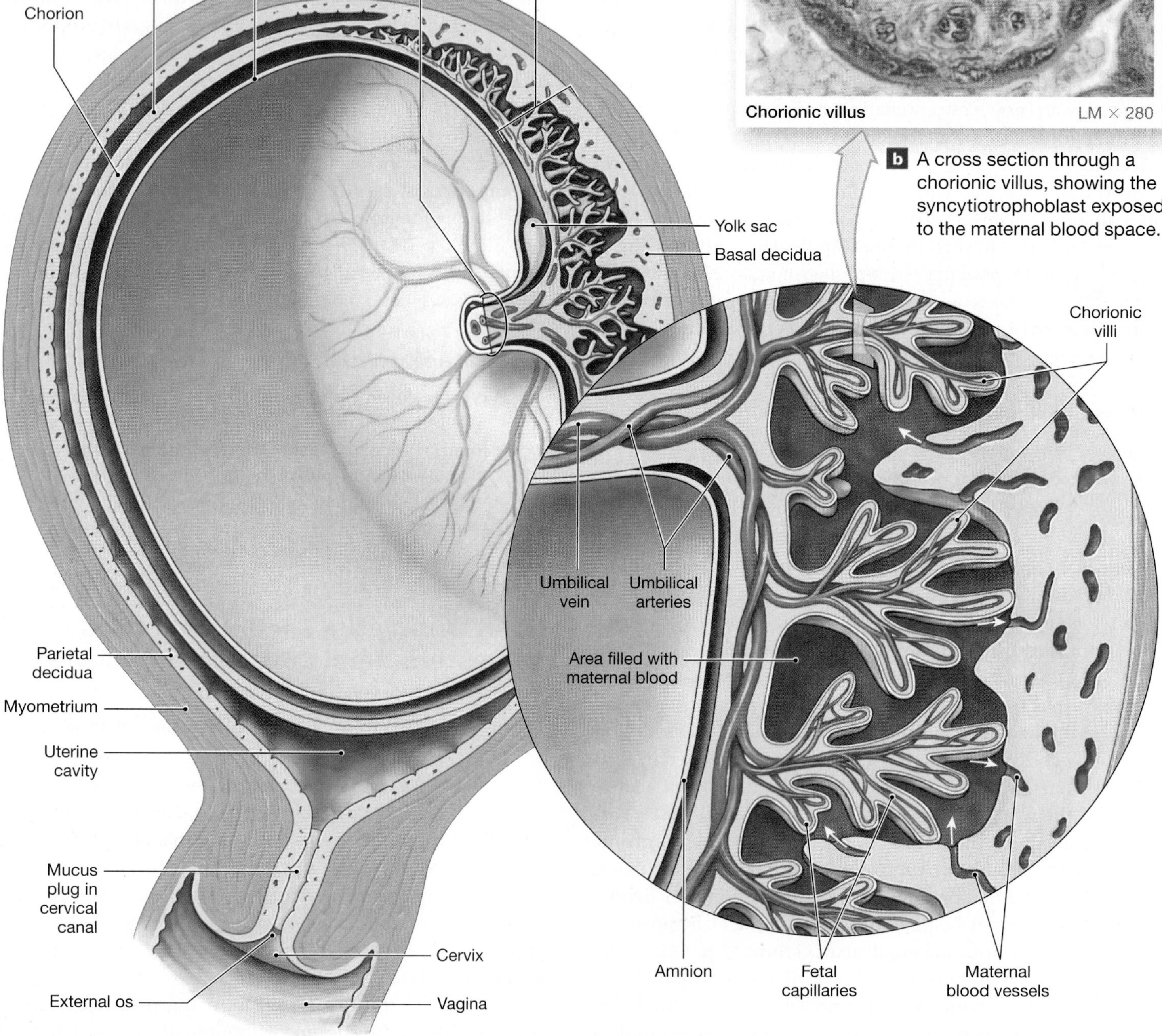

Capsular decidua
Amnion
Umbilical cord (cut)
Placenta
Chorion
Yolk sac
Basal decidua
Chorionic villi
Umbilical vein
Umbilical arteries
Area filled with maternal blood
Parietal decidua
Myometrium
Uterine cavity
Mucus plug in cervical canal
Cervix
External os
Vagina
Amnion
Fetal capillaries
Maternal blood vessels

a Placental circulation at the end of the first trimester. Arrows in the enlarged view indicate the direction of blood flow. Blood flows into the placenta through ruptured maternal arteries and then flows around chorionic villi, which contain fetal blood vessels.

Figure 29–7 The First 12 Weeks of Development. ATLAS: Plate 90a

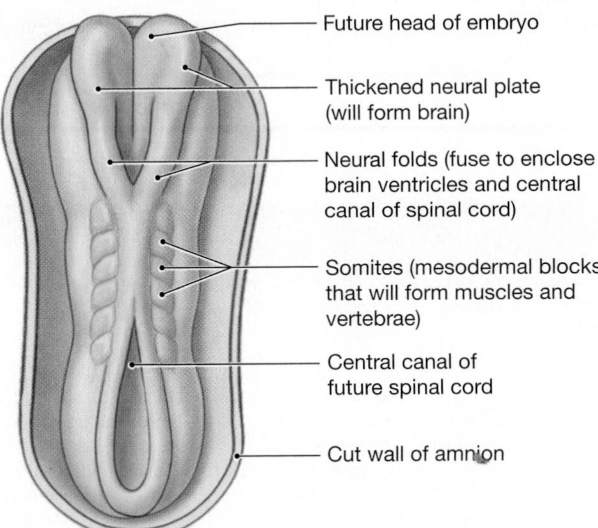
Future head of embryo
Thickened neural plate (will form brain)
Neural folds (fuse to enclose brain ventricles and central canal of spinal cord)
Somites (mesodermal blocks that will form muscles and vertebrae)
Central canal of future spinal cord
Cut wall of amnion

a **Week 3**. Diagram showing the formation of the CNS from neural plate ectoderm (about 2.25 mm in size).

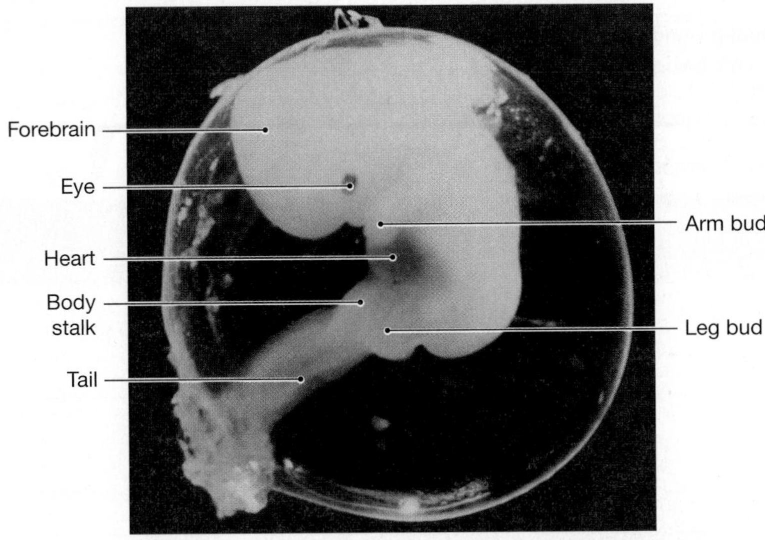
Forebrain
Eye
Heart
Body stalk
Tail
Arm bud
Leg bud

b **Week 4**. Fiberoptic view of human development at week 4 (about 5 mm in size).

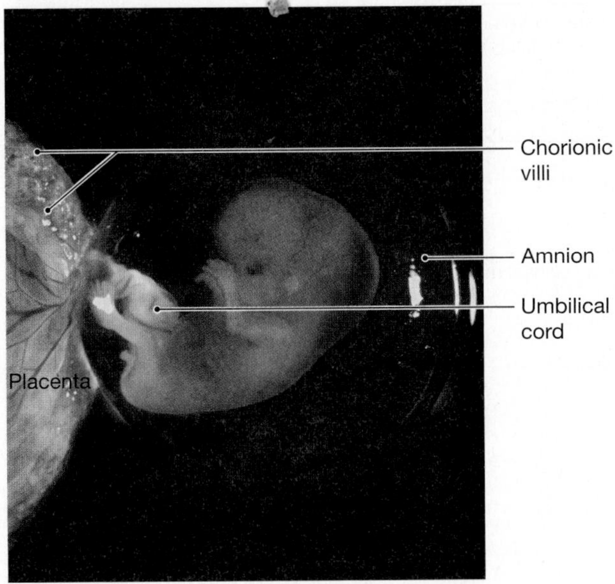
Chorionic villi
Amnion
Umbilical cord
Placenta

c **Week 8**. Fiberoptic view of human development at week 8 (about 1.6 cm in size).

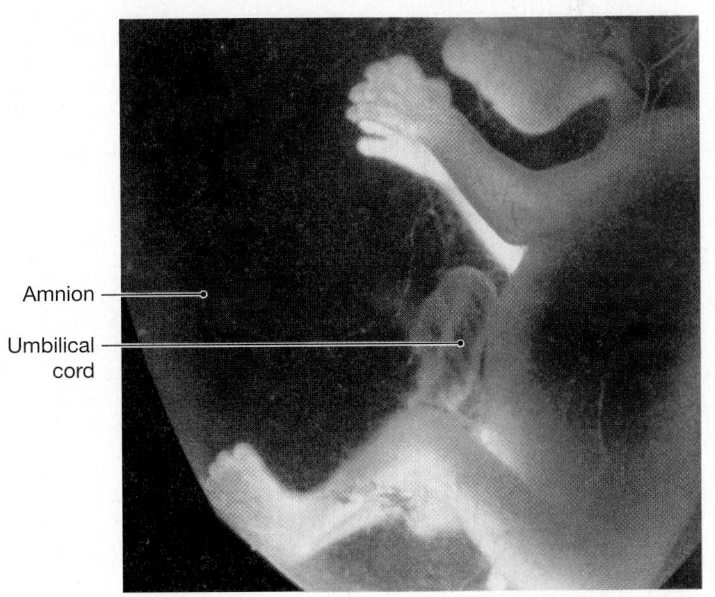
Amnion
Umbilical cord

d **Week 12**. Fiberoptic view of human development at week 12 (about 5.4 cm in size).

during this trimester. In the last 3 months of gestation, the fetus gains about 2.6 kg (5.7 lb), reaching a full-term weight of approximately 3.2 kg (7.05 lb).

Look at **Table 29–2** again for a complete summary of the development of different organ systems in the second and third trimesters.

 **Checkpoint**

11. **Explain how fetal size is measured between weeks 8 and 20 and between week 20 and birth.**

See the blue Answers tab at the back of the book.

29-6 During gestation, maternal organ systems support the developing fetus; the reproductive system undergoes structural and functional changes

Learning Outcome Describe the interplay between the maternal organ systems and the developing fetus, and discuss the structural and functional changes in the uterus during gestation.

Hormonal and anatomical changes are effects of pregnancy. Placental hormones play a role in maintaining the pregnancy, fetal growth, and breast maturation. Throughout the pregnancy, the uterus expands into most of the abdominal cavity.

Table 29-2 An Overview of Prenatal Development. From 8 weeks to about 20 weeks, the fetus is measured from crown to rump. From 20 weeks to birth, the baby is measured from crown to heel.

Background Material ATLAS: Embryology Summaries 1–4: The Formation of Tissues The Development of Epithelia The Origins of Connective Tissues The Development of Organ Systems		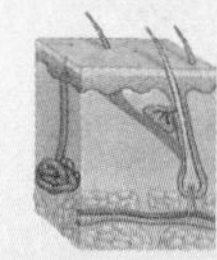	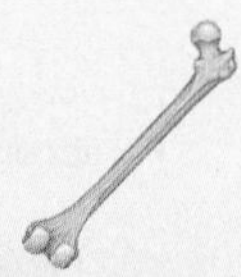		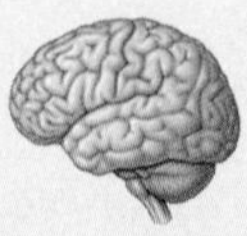	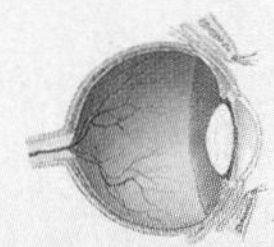
Gestational Age (embryonic age/fetal age plus 2 weeks) (months)	**Size and Weight**	**Integumentary System**	**Skeletal System**	**Muscular System**	**Nervous System**	**Special Sense Organs**
1	5 mm (0.2 in.), 0.02 g (0.0007 oz)		(b) Formation of somites	(b) Formation of somites	(b) Formation of neural tube	(b) Formation of eyes and ears
2	1.6 cm (0.63 in.), 1.0 g (0.04 oz)	(b) Formation of nail beds, hair follicles, sweat glands	(b) Formation of axial and appendicular cartilages	(c) Rudiments of axial musculature	(b) CNS, PNS organization, growth of cerebrum	(b) Formation of taste buds, olfactory epithelium
3	5.4 cm (2.13 in.), 14 g (0.5 oz)	(b) Epidermal layers appear	(b) Spreading of ossification centers	(c) Rudiments of appendicular musculature	(c) Basic spinal cord and brain structure	
4	11.6 cm (4.6 in.), 100 g (3.53 oz)	(b) Formation of hair, sebaceous glands (c) Sweat glands	(b) Articulations (c) Facial and palatal organization	Fetal movements can be felt by the mother	(b) Rapid expansion of cerebrum	(c) Basic eye and ear structure (b) Formation of peripheral receptors
5	16.4 cm (6.46 in.), 300 g (10.58 oz)	(b) Keratin production, nail production			(b) Myelination of spinal cord	
6	30 cm (11.81 in.), 600 g (1.32 lb)			(c) Perineal muscles	(b) Formation of CNS tract (c) Layering of cortex	
7	37.6 cm (14.8 in.), 1.005 kg (2.22 lb)	(b) Keratinization, formation of nails, hair				(c) Eyelids open, retinas sensitive to light
8	42.4 cm (16.69 in.), 1.702 kg (3.75 lb)		(b) Formation of epiphyseal cartilages			(c) Taste receptors functional
9	47.4 cm (18.66 in.), 3.2 kg (7.05 lb)					
Early postnatal development		Hair changes in consistency and distribution	Formation and growth of epiphyseal cartilages continue	Muscle mass and control increase	Myelination, layering, CNS tract formation continue	
Location of relevant text and illustrations		ATLAS: Embryology Summary 5	Ch. 6: pp. 190–193 Ch. 7: pp. 226–228 ATLAS: Embryology Summaries 6, 7, 8	ATLAS: Embryology Summary 9	Ch. 14: p. 468 ATLAS: Embryology Summaries 10, 11, 12	ATLAS: Embryology Summary 13

Note: (b) = beginning; (c) = completion.

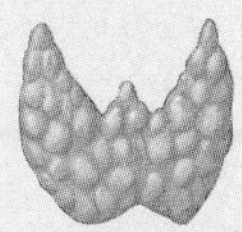 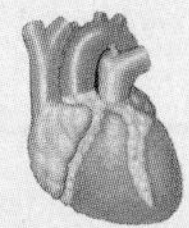 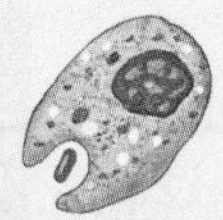 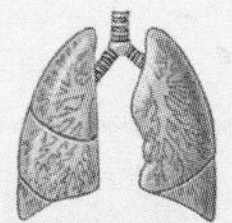 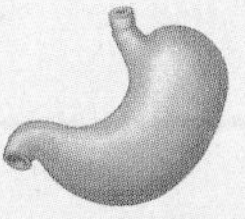 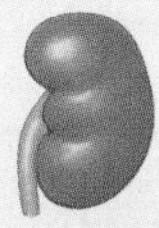 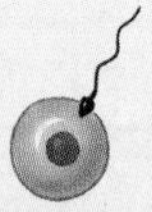

Gestational Age (embryonic age/fetal age plus 2 weeks) (months)	Endocrine System	Cardiovascular and Lymphatic Systems	Respiratory System	Digestive System	Urinary System	Reproductive System
1		(b) Heartbeat	(b) Formation of trachea and lungs	(b) Formation of intestinal tract, liver, pancreas (c) Yolk sac	(c) Allantois	
2	(b) Formation of thymus, thyroid, pituitary, adrenal glands	(c) Basic heart structure, major blood vessels, lymph nodes and ducts (b) Blood formation in liver	(b) Extensive bronchial branching into mediastinum (c) Diaphragm	(b) Formation of intestinal subdivisions, villi, salivary glands	(b) Formation of kidneys (metanephros)	(b) Formation of mammary glands
3	(c) Thymus, thyroid gland	(b) Tonsils, blood formation in bone marrow		(c) Gallbladder, pancreas		(b) Formation of gonads, ducts, genitalia; oogonia in female
4		(b) Migration of lymphocytes to lymphoid organs; blood formation in spleen			(b) Degeneration of embryonic kidneys (mesonephros)	
5		(c) Tonsils	(c) Nostrils open	(c) Intestinal subdivisions		
6	(c) Adrenal glands	(c) Spleen, liver, bone marrow	(b) Formation of alveoli	(c) Epithelial organization, glands		
7	(c) Pituitary gland			(c) Intestinal circular folds		(b) Descent of testes in male; primary oocytes in prophase I of meiosis in female
8			Complete pulmonary branching and alveolar structure		(c) Nephron formation	Descent of testes complete at or near time of birth
9						
Early postnatal development		Cardiovascular changes at birth; immune response gradually becomes fully operational				
Location of relevant text and illustrations	ATLAS: Embryology Summary 14	Ch. 19: pp. 664, 675 Ch. 21: pp. 775–777 ATLAS: Embryology Summaries 15, 16, 17	Ch. 23: p. 876 ATLAS: Embryology Summary 18	Ch. 24: pp. 887–889 ATLAS: Embryology Summary 19	Ch. 26: p. 1010 ATLAS: Embryology Summary 20	Ch. 28: pp. 1093–1095 ATLAS: Embryology Summary 21

Note: (b) = beginning; (c) = completion.

Figure 29–8 **Fetal Development in the Second and Third Trimesters.** ATLAS: Plate 90b

a A 4-month-old fetus, seen through a fiberoptic endoscope (about 13.3 cm in size)

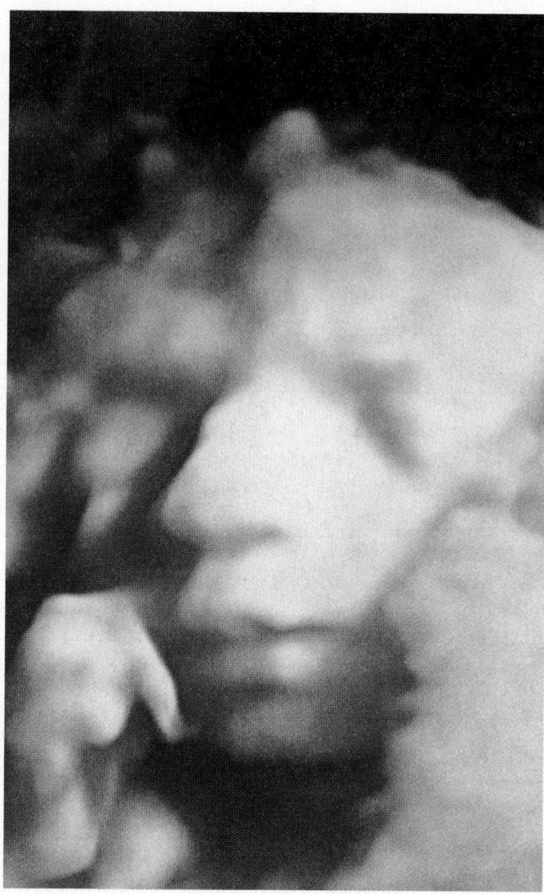

b Head of a 6-month-old fetus, revealed through ultrasound (about 30 cm in size)

Hormonal Regulation during Gestation

In addition to its role in the nutrition of the fetus, the placenta acts as an endocrine organ. Several hormones—including *human chorionic gonadotropin, human placental lactogen, placental prolactin, relaxin, progesterone,* and *estrogens*—are synthesized by the syncytiotrophoblast and released into the maternal bloodstream.

Human Chorionic Gonadotropin

29

The hormone **human chorionic gonadotropin (hCG)** appears in the maternal bloodstream soon after implantation. The presence of hCG in blood or urine samples is a reliable indication of pregnancy. Home pregnancy tests detect the presence of hCG in the urine.

In function, hCG resembles luteinizing hormone (LH), because it maintains the integrity of the corpus luteum and promotes the continued secretion of progesterone. As a result, the endometrial lining remains perfectly functional, and menstruation does not normally occur. Without hCG, the pregnancy ends, because another uterine cycle begins and the functional layer of the endometrium disintegrates.

With hCG, the corpus luteum persists for 3 to 4 months before gradually decreasing in size and secretory function. These changes do not trigger the return of uterine cycles, because by the end of the first trimester, the placenta actively secretes both estrogens and progesterone.

Human Placental Lactogen and Placental Prolactin

Human placental lactogen (hPL) helps prepare the mammary glands for milk production. It also has a stimulatory effect on other maternal tissues comparable to that of growth hormone (GH), ensuring that glucose and protein are available for the fetus. At the mammary glands, the conversion from inactive to active status requires placental hormones (hPL, **placental prolactin**, estrogens, and progesterone) and several maternal hormones (GH, prolactin, and thyroid hormones).

Relaxin

Relaxin is a peptide hormone secreted by the placenta and the corpus luteum during pregnancy. Relaxin (1) increases the flexibility of the pubic symphysis, permitting the pelvis to expand during delivery; (2) causes the cervix to dilate, making it easier for the fetus to enter the vaginal canal; and (3) suppresses the release of oxytocin by the hypothalamus and delays the onset of labor contractions.

Progesterone and Estrogens

After the first trimester, the placenta produces sufficient amounts of progesterone to maintain the endometrial lining and continue the pregnancy. As the end of the third trimester approaches, estrogen production by the placenta accelerates. As we will see in Section 29-7, the increasing levels of estrogens play a role in stimulating labor and delivery.

Changes in Maternal Organ Systems

Pregnancy affects the maternal respiratory, cardiovascular, digestive, and urinary systems. The tidal volume increases, blood volume and blood pressure increase, morning sickness and constipation are common, and urine production increases due to increased metabolism and fetal wastes.

Changes in the Respiratory, Cardiovascular, Digestive, and Urinary Systems

The developing fetus is totally dependent on maternal organ systems for nourishment, respiration, and waste removal. Maternal systems perform these functions in addition to their normal operations. For example, the mother must absorb enough oxygen, nutrients, and vitamins for herself *and* for her fetus, and she must eliminate all the wastes that are generated. This is not a burden over the initial weeks of gestation, but the demands placed on the mother become significant as the fetus grows. For the mother to survive under these conditions, maternal systems must compensate for changes introduced by the fetus. In practical terms, the mother must breathe, eat, and excrete for two. The major changes in maternal systems include the following:

- *Maternal respiratory rate increases and tidal volume increases.* As a result, the mother's lungs deliver the extra oxygen required, and remove the excess carbon dioxide generated by the fetus.
- *Maternal blood volume increases.* This increase occurs because blood flowing into the placenta decreases the volume in the rest of the systemic circuit, and because fetal metabolic activity both decreases blood P_{O_2} and increases P_{CO_2}. The latter combination stimulates the production of renin and erythropoietin, leading to an increase in maternal blood volume through mechanisms detailed in Chapter 21 (see Figure 21–15, p. 751). By the end of gestation, maternal blood volume has increased by almost 50 percent.
- *Maternal requirements for nutrients increase 10–30 percent.* Pregnant women tend to have increased appetites because they must nourish both themselves and their fetus.
- *Maternal glomerular filtration rate increases by approximately 50 percent.* This increase corresponds to the increase in blood volume and accelerates the excretion of metabolic wastes generated by the fetus. Pregnant women need to urinate frequently because the volume of urine produced increases and the weight of the uterus presses down on the urinary bladder.

Changes in the Reproductive System

At the end of gestation, a typical uterus has undergone a tremendous increase in size, from 7.5 cm (3 in.) in length and 30–40 g (1–1.4 oz) in weight to 30 cm (12 in.) in length and 1100 g (2.4 lb) in weight. This remarkable expansion occurs through the enlargement (hypertrophy) of existing cells, especially smooth muscle fibers, rather than by an increase in the total number of cells. Figure 29–9a,b,d shows the positions of the uterus, fetus, and placenta from 16 weeks to *full term* (9 months). The uterus may then contain 2 liters of fluid, plus fetus and placenta, for a total weight of about 6–7 kg (13–15.4 lb). When the pregnancy is at full term, the uterus and fetus push many of the maternal abdominal organs out of their normal positions (compare Figure 29–9c and Figure 29–9d).

In addition, the mammary glands increase in size, and secretory activity begins. Mammary gland development requires a combination of hormones, including human placental lactogen and placental prolactin from the placenta, and prolactin (PRL), estrogens, progesterone, GH, and thyroxine from maternal endocrine organs. By the end of the sixth month of pregnancy, the mammary glands are fully developed and begin to produce clear secretions that are stored in the duct system and may be expressed from the nipple.

✓ Checkpoint

12. **Sue's pregnancy test indicates an elevated level of the hormone hCG (human chorionic gonadotropin). Explain whether she is pregnant or not.**
13. **What respiratory changes occur in pregnant women?**
14. **Why does a mother's blood volume increase during pregnancy?**

See the blue Answers tab at the back of the book.

Figure 29–9 Growth of the Uterus and Fetus during Gestation.

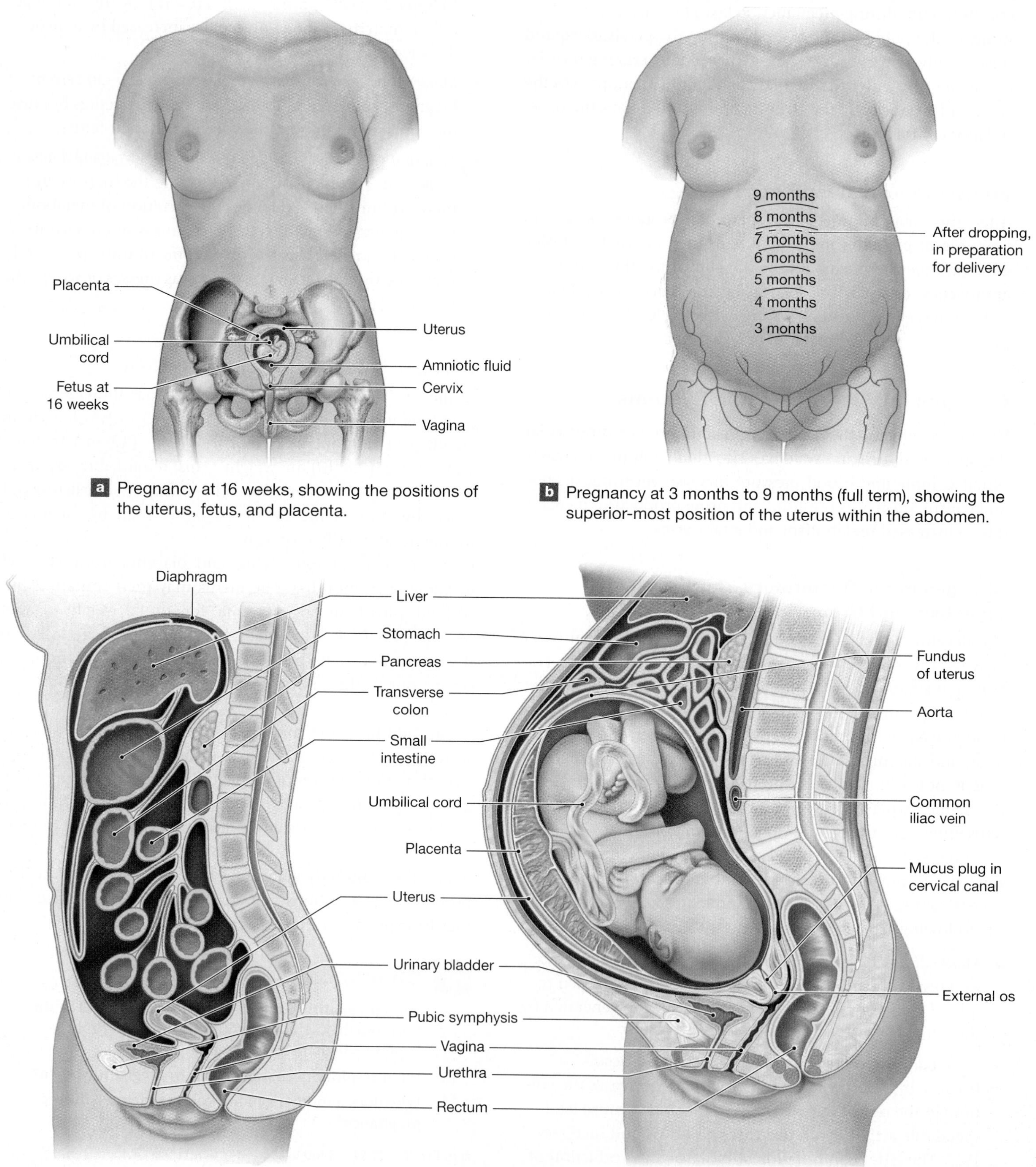

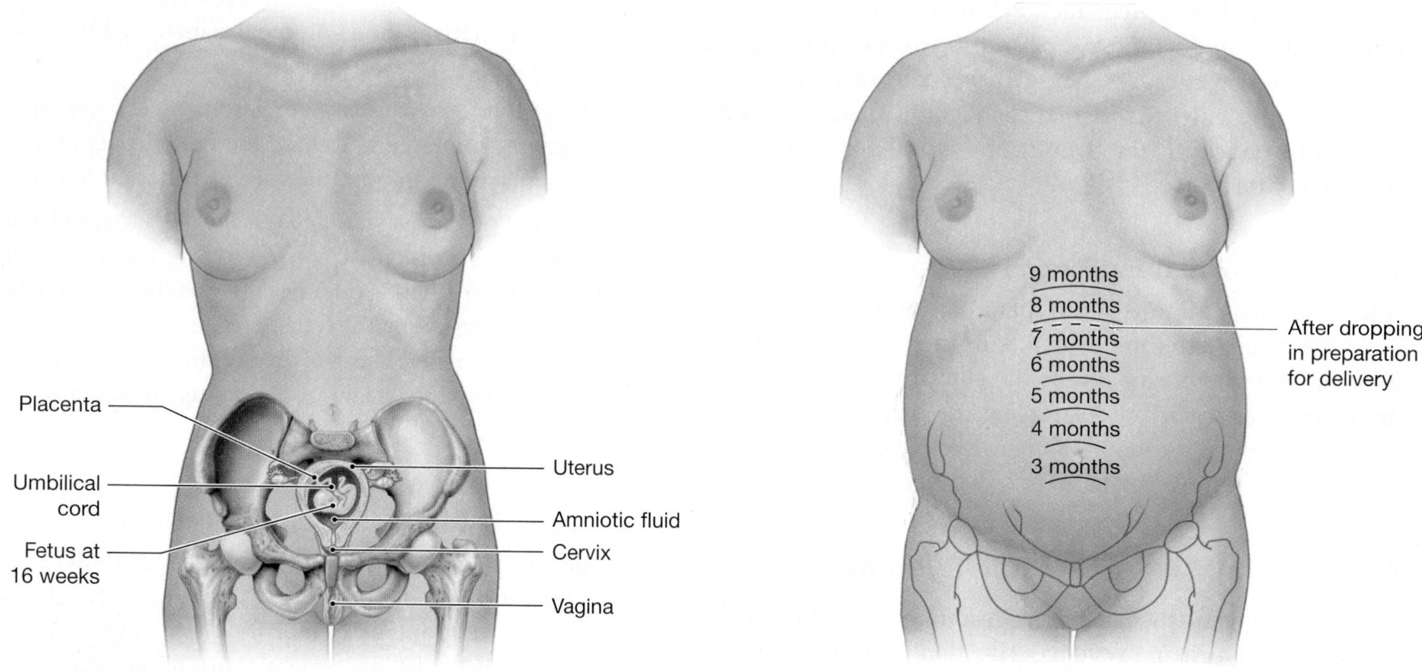
Placenta
Umbilical cord
Fetus at 16 weeks
Uterus
Amniotic fluid
Cervix
Vagina
9 months
8 months
7 months
6 months
5 months
4 months
3 months
After dropping, in preparation for delivery

a Pregnancy at 16 weeks, showing the positions of the uterus, fetus, and placenta.

b Pregnancy at 3 months to 9 months (full term), showing the superior-most position of the uterus within the abdomen.

Diaphragm
Liver
Stomach
Pancreas
Transverse colon
Small intestine
Umbilical cord
Placenta
Uterus
Urinary bladder
Pubic symphysis
Vagina
Urethra
Rectum
Fundus of uterus
Aorta
Common iliac vein
Mucus plug in cervical canal
External os

c A sectional view through the abdominopelvic cavity of a woman who is not pregnant.

d Pregnancy at full term. Note the positions of the uterus and full-term fetus within the abdomen, and the displacement of abdominal organs.

29

+ Clinical Note Abortion

Abortion is the termination of a pregnancy. Most references distinguish among spontaneous, therapeutic, and induced abortions. Most **spontaneous abortions**, or *miscarriages*, result from developmental problems (such as chromosomal defects in the embryo) or from hormonal problems, including abnormally low LH production by the maternal pituitary gland, reduced LH sensitivity at the corpus luteum, insufficient progesterone sensitivity in the endometrium, or insufficient placental production of hCG. Spontaneous abortions occur in at least 15 percent of recognized pregnancies. **Therapeutic abortions** are performed when continued pregnancy poses a threat to the life of the mother. **Induced abortions**, or *elective abortions*, are performed at the woman's request. Induced abortions remain the focus of considerable controversy.

29-7 Childbirth occurs through the process of labor, which consists of the dilation, expulsion, and placental stages

Learning Outcome **List and discuss the events that occur during labor and delivery.**

Parturition (par-tū-RISH-un), or *childbirth*, is the forcible expulsion of the fetus from the uterus through the cervix and vagina. This occurs by a series of strong, rhythmic uterine contractions called **labor**. This section looks at how labor starts, the three stages of the process, and difficulties that may occur during this process.

Initiation of Labor

The tremendous stretching of the uterus that occurs during pregnancy is associated with a gradual increase in the rate of spontaneous smooth muscle contractions in the myometrium. In the early stages of pregnancy, the contractions are weak, painless, and brief. Evidence indicates that progesterone released by the placenta has an inhibitory effect on uterine smooth muscle, preventing more extensive and more powerful contractions. Late in pregnancy, some women experience occasional spasms in the uterine musculature, but these contractions are not regular or persistent. Such contractions are called **Braxton Hicks contractions** (*false labor*).

Placental and fetal factors are involved in the initiation of **true labor** and delivery (Figure 29–10). When labor begins, the fetal pituitary gland secretes oxytocin, which is released into the maternal bloodstream at the placenta. This may be the actual trigger for the onset of true labor, as it increases myometrial contractions and prostaglandin production, on top of the priming effects of estrogens and maternal oxytocin.

The Stages of Labor

Once labor contractions have begun in the myometrium, positive feedback ensures that they will continue until delivery has been completed. During labor, each contraction begins near the superior portion of the uterus and sweeps in a wave toward the cervix. The contractions are strong and occur at regular intervals. As parturition approaches, the contractions increase in force and frequency, changing the position of the fetus and moving it toward the cervical canal. Labor has traditionally been divided into three stages: the *dilation stage*, the *expulsion stage*, and the *placental stage* (Figure 29–11).

The Dilation Stage

The **dilation stage** begins with the onset of true labor, as the cervix dilates and the fetus begins to shift toward the cervical canal, moved by gravity and uterine contractions (stage 1 in Figure 29–11). This stage is highly variable in length but typically lasts 8 or more hours. At the start of the dilation stage, labor contractions last up to half a minute and occur once every 10–30 minutes. Their frequency increases steadily. Late in this stage, the amniochorionic membrane ruptures, an event sometimes referred to as the woman having her "water break." If this event occurs before other events of the dilation stage, the life of the fetus may be at risk from infection. If the risk is sufficiently great, labor can be induced.

The Expulsion Stage

The **expulsion stage** begins as the cervix, pushed open by the approaching fetus, completes its dilation to about 10 cm (4 in.) (stage 2 in Figure 29–11). In this stage, contractions reach maximum intensity. They occur at perhaps 2- or 3-minute intervals and last a full minute. Expulsion continues until the fetus has emerged from the vagina. In most cases, the expulsion stage lasts less than 2 hours. The arrival of the newborn infant into the outside world is **delivery**, or birth.

If the vaginal canal is too small to permit the passage of the fetus, posing acute danger of perineal tearing, a clinician may temporarily enlarge the passageway by performing an **episiotomy** (eh-piz-ē-OT-ō-mē), an incision through the perineal musculature. After delivery, this surgical cut is repaired with sutures, a much simpler procedure than suturing the jagged edges associated with an extensive perineal tear.

Immediately after birth, clinicians assess the newborn's health in five areas: heart rate, breathing, skin color, muscle tone, and reflex response. This assessment is called an **Apgar score**. Each criterion is scored from 0 to 2, with 8–10 indicating a healthy baby.

The Placental Stage

During the **placental stage** of labor, muscle tension builds in the walls of the partially empty uterus, which gradually

Figure 29–10 Factors Involved in the Initiation of Labor and Delivery.

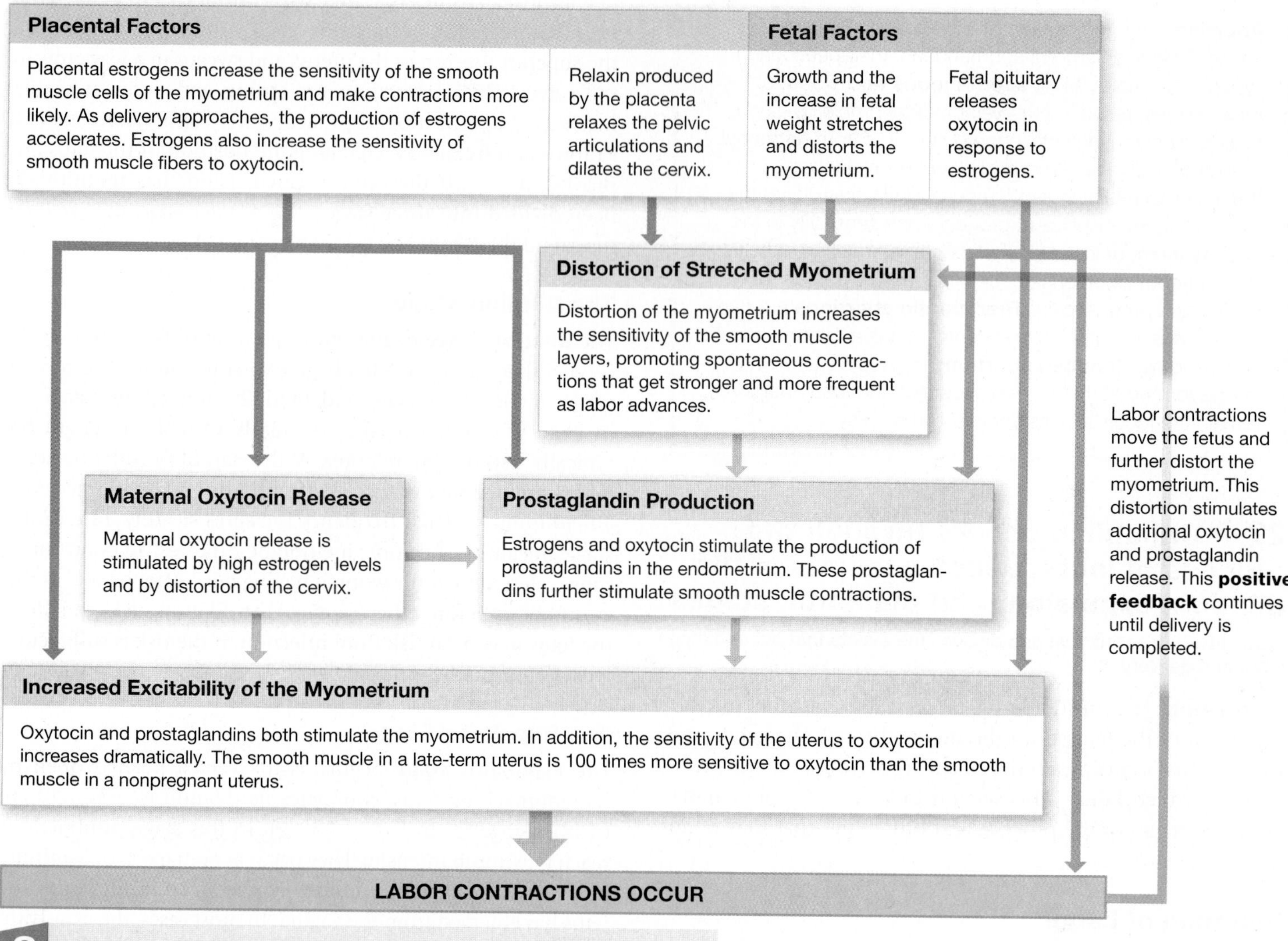
Placental Factors
Placental estrogens increase the sensitivity of the smooth muscle cells of the myometrium and make contractions more likely. As delivery approaches, the production of estrogens accelerates. Estrogens also increase the sensitivity of smooth muscle fibers to oxytocin.
Relaxin produced by the placenta relaxes the pelvic articulations and dilates the cervix.
Fetal Factors
Growth and the increase in fetal weight stretches and distorts the myometrium.
Fetal pituitary releases oxytocin in response to estrogens.
Distortion of Stretched Myometrium
Distortion of the myometrium increases the sensitivity of the smooth muscle layers, promoting spontaneous contractions that get stronger and more frequent as labor advances.
Labor contractions move the fetus and further distort the myometrium. This distortion stimulates additional oxytocin and prostaglandin release. This positive feedback continues until delivery is completed.
Maternal Oxytocin Release
Maternal oxytocin release is stimulated by high estrogen levels and by distortion of the cervix.
Prostaglandin Production
Estrogens and oxytocin stimulate the production of prostaglandins in the endometrium. These prostaglandins further stimulate smooth muscle contractions.
Increased Excitability of the Myometrium
Oxytocin and prostaglandins both stimulate the myometrium. In addition, the sensitivity of the uterus to oxytocin increases dramatically. The smooth muscle in a late-term uterus is 100 times more sensitive to oxytocin than the smooth muscle in a nonpregnant uterus.
LABOR CONTRACTIONS OCCUR

? What hormones stimulate and increase the excitability of the myometrium?

decreases in size (stage 3 in Figure 29–11). This uterine contraction tears the connections between the endometrium and the placenta. In general, within an hour of delivery, the placental stage ends with the ejection of the placenta, or *afterbirth*. The disruption of the placenta is accompanied by a loss of blood, but associated uterine contraction compresses the uterine vessels and usually restricts this flow. Because maternal blood volume has increased greatly during pregnancy, the blood loss can normally be tolerated without difficulty.

29

Difficulties of Labor and Delivery and Multiple Births

Common complications of labor include issues with the umbilical cord looping around the baby's neck, an abnormal fetal heart rate (normal is 110–160 beats per minute), and stalled labor. Others include premature ("preterm") labor, a condition in which uterine contractions cause the cervix to open too soon, or difficult deliveries, which can stress the mother and/or the baby and sometimes cause severe complications.

Premature Labor

Premature labor occurs when true labor begins before the fetus has completed normal development. The newborn's chances of surviving are directly related to its body weight at delivery. Even with massive supportive efforts, newborns weighing less than 400 g (14 oz) at birth will not survive, primarily because their respiratory, cardiovascular, and urinary systems are unable to support life without aid from maternal systems. As a result, the dividing line between spontaneous abortion and

Figure 29–11 The Stages of Labor.

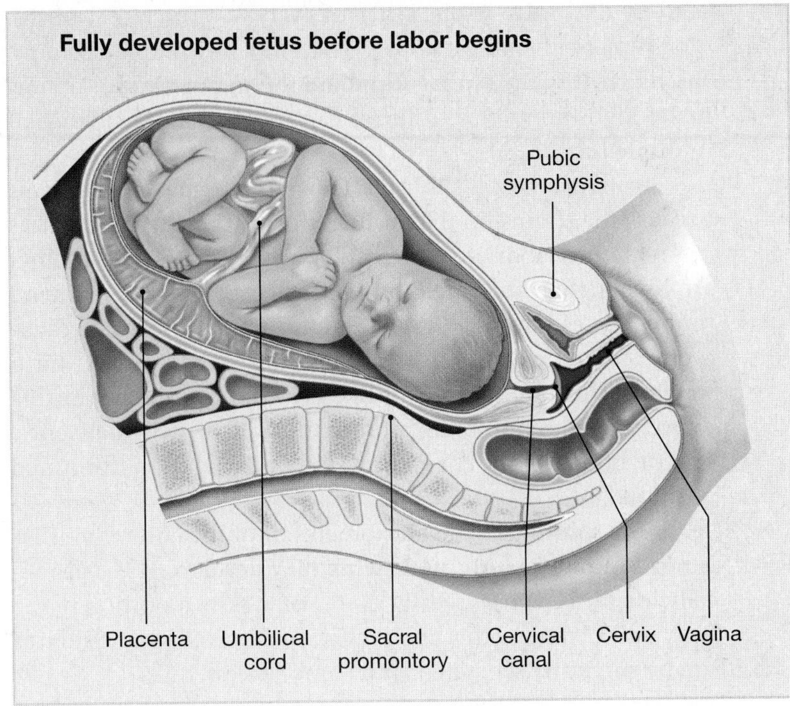
Fully developed fetus before labor begins
Pubic symphysis
Placenta
Umbilical cord
Sacral promontory
Cervical canal
Cervix
Vagina

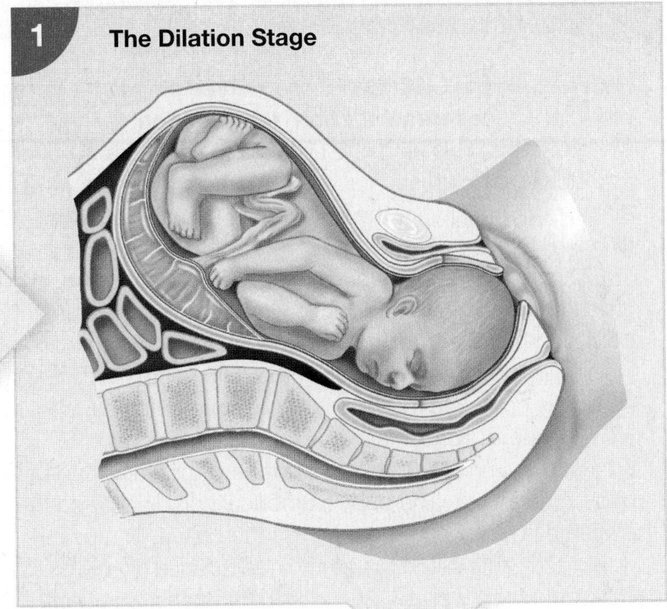
1
The Dilation Stage

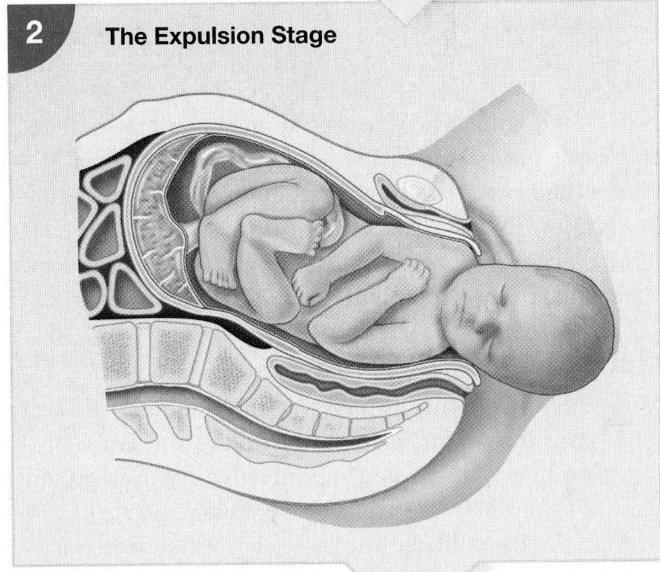
2
The Expulsion Stage

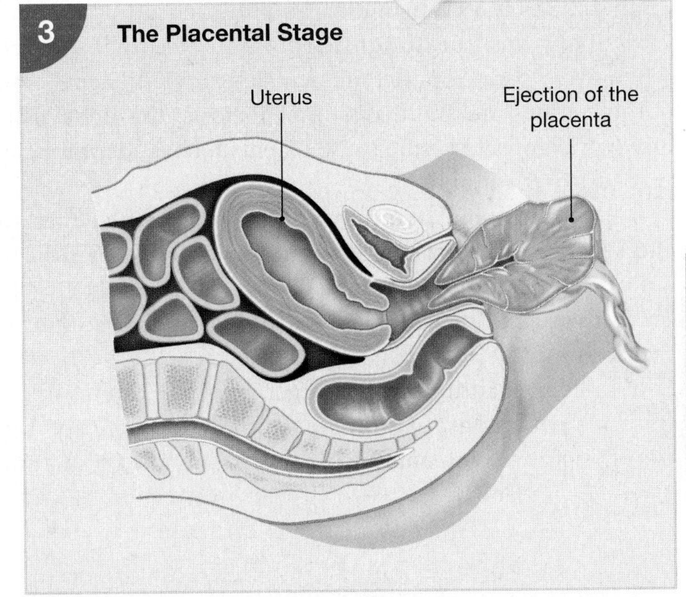
3
The Placental Stage
Uterus
Ejection of the placenta

immature delivery is usually set at 500 g (17.6 oz), the normal weight near the end of the second trimester.

Most fetuses born at 25–27 weeks of gestation (a birth weight under 600 g or 21.1 oz) die despite intensive neonatal care. Survivors have a high risk of developmental abnormalities. **Premature delivery** usually refers to birth at 28–36 weeks (a birth weight over 1 kg or 2.2 lb). With care, these newborns have a good chance of surviving and developing normally.

Difficult Deliveries

By the end of gestation in most pregnancies, the fetus has rotated within the uterus to transit the birth canal headfirst, facing the mother's sacrum. In about 6 percent of deliveries, the fetus faces the mother's pubis instead. These babies can be delivered normally, given enough time, but risks to infant and mother are reduced by a *forceps delivery*. Forceps resemble large, curved salad tongs that can be separated for insertion into the vaginal canal, one side at a time. Once in place, they are reunited and used to grasp the head of the fetus. An intermittent pull is applied, so that the forces on the head resemble those of normal delivery.

In 3–4 percent of deliveries, the legs or buttocks of the fetus enter the vaginal canal first. Such deliveries are **breech births**. Risks to the fetus are higher in breech births than in normal deliveries, because the umbilical cord can become constricted, cutting off placental blood flow.

The head is normally the widest part of the fetus. The mother's cervix may dilate enough to pass the baby's legs and

Clinical Note C-Section

According to the CDC, about one in three pregnant women give birth by **cesarean section**, or **C-section**, in the United States. This high rate appears to accompany new risk factors for birth such as older mothers, mothers with diabetes or obesity, or the occurrence of multiple births (twins or triplets).

During this procedure, which typically lasts 1 to 2 hours, the mother first receives regional anesthesia to the pelvis (an "epidural"). The surgeon makes a low horizontal incision in the abdominal wall at the "bikini line." (When healing is complete the scar is easily covered by her underwear.) Another horizontal cut is then made to open the uterus, the amniochorionic membrane is ruptured, and the baby is taken out. The baby's umbilical cord is cut and the placenta is removed. The uterus is then carefully repaired with dissolvable sutures, and the abdomen is often closed with removable staples. Afterward, the mother and baby will stay in the hospital for a few days to recover.

body, but not the head. This entrapment compresses the umbilical cord, prolongs delivery, and subjects the fetus to severe distress and potential injury. If attempts to reposition the fetus or promote further dilation are unsuccessful over the short term, delivery by *cesarean section*, or *C-section* (see Clinical Note: C-Section) may be required.

Multiple Births

Multiple births (twins, triplets, quadruplets, and so forth) can take place for several reasons, including the growing use of fertility drugs and assisted reproductive technology. Another reason is the rise in the average age of women give birth (older women are more likely to release more than one oocyte in a cycle). According to the CDC, the rate of twin births in the United States is 33.7 twin births out of every 1,000 deliveries. "Fraternal twins," or **dizygotic** (dī-zī-GOT-ik) **twins**, develop when two separate oocytes are ovulated and fertilized. Because chromosomes are shuffled during meiosis, the odds against any two zygotes from the same parents having identical genes exceed 1 in 8.4 million.

"Identical twins" or **monozygotic twins**, result either from the separation of blastomeres early in cleavage or from the splitting of the inner cell mass before gastrulation. In either event, the genetic makeup of the twins is identical because both formed from the same pair of gametes.

Triplets, quadruplets, and larger multiples can result from multiple ovulations, blastomere splitting, or some combination of the two. For unknown reasons, the rates of naturally occurring multiple births fall into a pattern: Twins occur in 1 of every 89 births, triplets in 1 of every 89^2 (or 7921) births, quadruplets in 1 of every 89^3 (704,969) births, and so forth. Taking fertility drugs that stimulate the maturation of abnormally high numbers of follicles can increase the incidence of multiple births.

Pregnancies with multiple fetuses pose special problems because the strains on the mother are multiplied. The chances of premature labor are increased, and the risks to the mother are higher than for single births. Increased risks also extend to the fetuses during gestation, and to the newborns, because even at full term such newborns have lower-than-average birth weights. They are also more likely to have problems during delivery. For example, in more than half of twin deliveries, one or both fetuses enter the vaginal canal in an abnormal position.

If the splitting of the blastomeres or of the embryonic disc is not complete, **conjoined twins** may develop. These genetically identical twins typically share some skin, a portion of the liver, and perhaps other internal organs as well. Some infants can be surgically separated with some success.

✓ Checkpoint

15. **Identify three major factors opposing the calming action of progesterone on the uterus.**
16. **Name the three stages of labor.**
17. **Differentiate between immature delivery and premature delivery.**
18. **Give the biological terms for fraternal twins and identical twins.**

See the blue Answers tab at the back of the book.

29-8 Postnatal stages are the neonatal period, infancy, childhood, adolescence, and maturity, followed by senescence and death

Learning Outcome Identify the features and physiological changes of the postnatal stages of life.

Developmental processes do not stop at delivery. Newborns have few of the anatomical, functional, or physiological characteristics of mature adults. The course of postnatal development typically includes five **life stages**: (1) the *neonatal period*, (2) *infancy*, (3) *childhood*, (4) *adolescence*, and (5) *maturity*. Each stage is typified by a distinctive combination of characteristics and abilities. These stages are familiar parts of human experience. Each stage has distinctive features, but the transitions between them are gradual, and the boundaries indistinct. At maturity, development ends and the process of aging, or *senescence*, begins, leading ultimately to death.

The Neonatal Period, Infancy, and Childhood

The **neonatal period** extends from birth to 1 month of age. **Infancy**, or babyhood, then continues through the first year of life. **Childhood** lasts from infancy until **adolescence** (puberty), the period of sexual and physical maturation. Two major events are under way during these developmental stages:

1. The organ systems (except those associated with reproduction) become fully operational and gradually acquire the functional characteristics of adult structures.
2. The individual grows rapidly, and body proportions change significantly.

Pediatrics is the medical specialty that focuses on postnatal development from infancy through adolescence. Infants and young children cannot clearly describe the problems they are experiencing, so pediatricians and parents must be skilled observers. Standardized tests are used to assess developmental progress relative to average values.

The Neonatal Period

The neonatal period is the 4-week period immediately after birth. During this time, the newborn adjusts to life outside the womb as extensive and ongoing system transitions occur.

Adjustments of the Neonate. Physiological and anatomical changes take place as the fetus completes the transition to the status of newborn, or **neonate**. The neonatal period is the time from birth through 28 days. Before delivery, dissolved gases, nutrients, wastes, hormones, and antibodies were transferred across the placenta. At birth, the neonate must become relatively self-sufficient, performing respiration, digestion, and excretion using its own specialized organs and organ systems. The transition from fetus to neonate can be summarized as follows:

- At birth, the lungs are collapsed and filled with fluid. Filling them with air requires a massive and powerful inhalation.
- When the lungs expand, the pattern of cardiovascular circulation changes due to alterations in blood pressure and flow rates. The ductus arteriosus closes, isolating the pulmonary and systemic trunks. Closure of the foramen ovale separates the atria of the heart, completing the separation of the pulmonary and systemic circuits.
- Typical neonatal heart rates (120–140 beats per minute) are lower than fetal heart rates (averaging 150 beats per minute). Both the neonatal heart rate and respiratory rate (30 breaths per minute) are considerably higher than in adults.
- Before birth, the digestive system is relatively inactive, although it does accumulate a mixture of bile secretions, mucus, and epithelial cells. This collection of debris, called *meconium*, is excreted during the first few days of life. During that period, the newborn begins to nurse.
- As wastes build up in the arterial blood, the kidneys excrete them. Glomerular filtration is normal, but the neonate's kidneys cannot concentrate urine to any significant degree. As a result, urinary water losses are high, and neonatal fluid requirements are proportionally much greater than those of adults.
- The neonate has little ability to control its body temperature, particularly in the first few days after delivery. As the infant grows larger and its insulating subcutaneous adipose "blanket" gets thicker, its metabolic rate also increases. Daily and even hourly shifts in body temperature continue throughout childhood. p. 970
- During embryonic and fetal development, the mother's body is at normal body temperature. At birth, the neonate has little ability to control its body temperature. Neonates also lose heat very quickly, because they have a high surface-area-to-volume ratio. To maintain a constant body temperature, their cellular metabolic rates must be high. In fact, the metabolic rate per unit of body weight in neonates is approximately twice that of adults.

Over the entire neonatal period, the newborn is dependent on nutrients contained in milk, typically breast milk secreted by the mother's mammary glands.

Maternal Adjustments, Including Lactation. By the end of the sixth month of pregnancy, the mammary glands are fully developed, and the gland cells begin to secrete **colostrum** (kō-LOS-trum). Ingested by the infant during the first 2 or 3 days of life, colostrum contains more proteins and far less fat than breast milk. Many of the proteins are antibodies that may help the infant ward off infections until its own immune system becomes functional. In addition, the mucins present in both colostrum and milk can inhibit the replication of a family of viruses *(rotaviruses)* that can cause dangerous forms of gastroenteritis and diarrhea in infants.

As colostrum production drops, the mammary glands convert to milk production. Breast milk consists of water, proteins, amino acids, lipids, sugars, and salts. It also contains large quantities of *lysozyme*, an enzyme with antibiotic properties. In terms of energy, human milk provides about 750 kilocalories per liter. The secretory rate varies with the demand, but a 5–6-kg (11–13-lb) infant usually requires about 850 mL (3.6 cups) of milk per day. (Maternal milk production throughout this period is maintained through the combined actions of several hormones, as detailed in Chapter 18. p. 624) 29

Milk becomes available to infants through the **milk ejection (milk let-down) reflex** (Figure 29–12). This reflex continues to function until *weaning*, the withdrawal of mother's milk, typically 1 to 2 years after birth. Milk production stops soon after, and the mammary glands gradually return to a resting state. Earlier weaning is a common practice in the United

States, where women take advantage of commercially prepared milk- or soy-based infant formulas that closely approximate the composition of natural breast milk. The major difference between such substitutes and natural milk is that the substitutes lack antibodies and lysozyme, which play important roles in maintaining the health of the infant. Consequently, early weaning is associated with an increased risk of infections and allergies in the infant.

Infancy and Childhood

The most rapid growth occurs during prenatal development, and the growth rate decreases after delivery. Growth during infancy and childhood occurs under the direction of circulating hormones, notably growth hormone, adrenal steroids, and thyroid hormones. These hormones affect each tissue and organ in specific ways, depending on the sensitivities of the individual cells. As a result, growth does not occur uniformly, so body proportions gradually change. The head, for example, is relatively large at birth but decreases in proportion to the rest of the body as the child grows to adulthood (Figure 29–13).

Figure 29–12 **The Milk Ejection Reflex.**

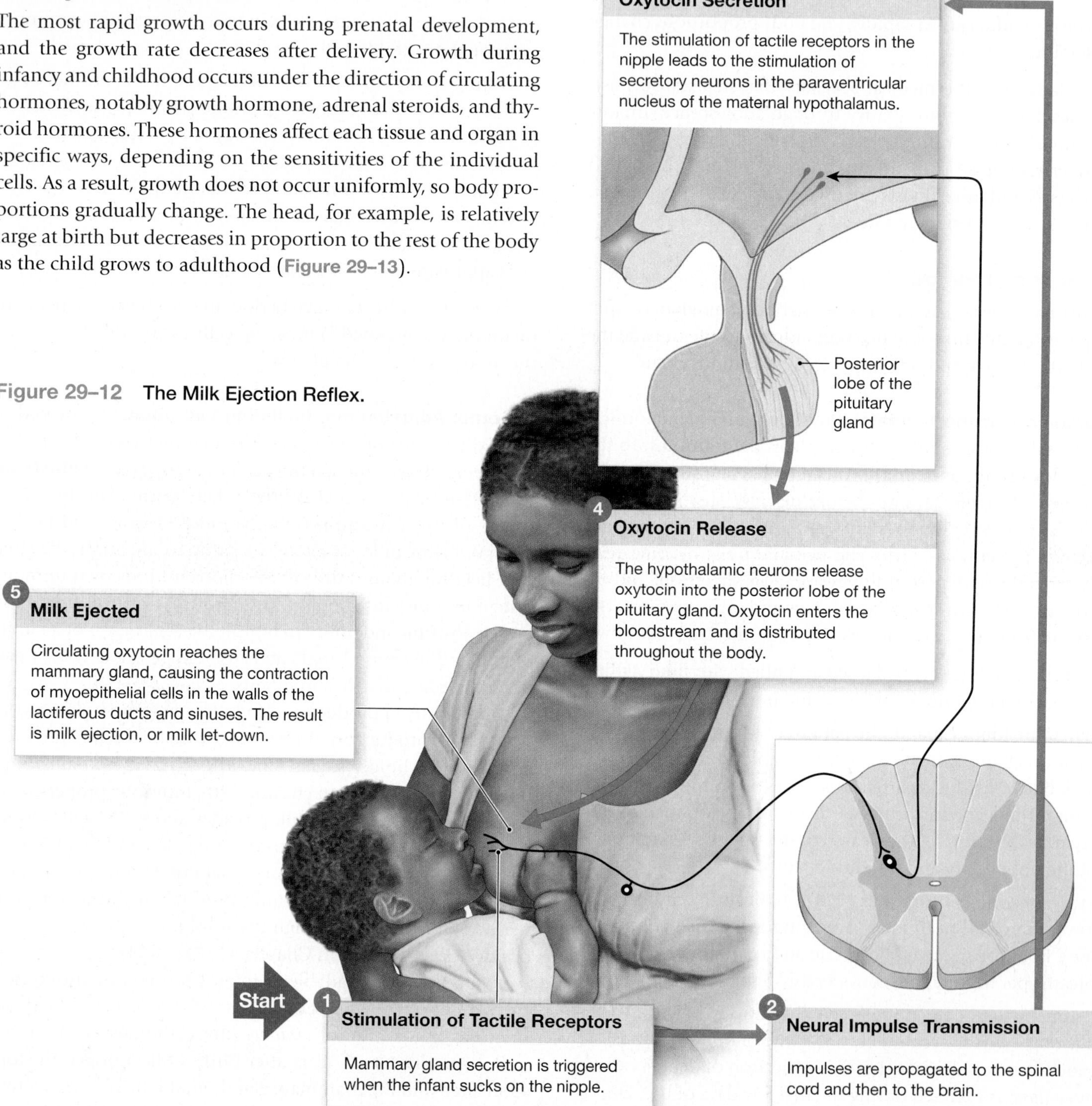
3
Oxytocin Secretion
The stimulation of tactile receptors in the nipple leads to the stimulation of secretory neurons in the paraventricular nucleus of the maternal hypothalamus.
Posterior lobe of the pituitary gland
4
Oxytocin Release
The hypothalamic neurons release oxytocin into the posterior lobe of the pituitary gland. Oxytocin enters the bloodstream and is distributed throughout the body.
5
Milk Ejected
Circulating oxytocin reaches the mammary gland, causing the contraction of myoepithelial cells in the walls of the lactiferous ducts and sinuses. The result is milk ejection, or milk let-down.
Start
1
Stimulation of Tactile Receptors
Mammary gland secretion is triggered when the infant sucks on the nipple.
2
Neural Impulse Transmission
Impulses are propagated to the spinal cord and then to the brain.

29

Figure 29–13 Prenatal and Postnatal Growth and Changes in Body Form and Proportion. The views at 4, 8, and 16 weeks of gestation are presented at actual size. Notice the changes in body form and proportions as development proceeds. For example, the head, which contains the brain and sense organs, is proportionately large at birth.

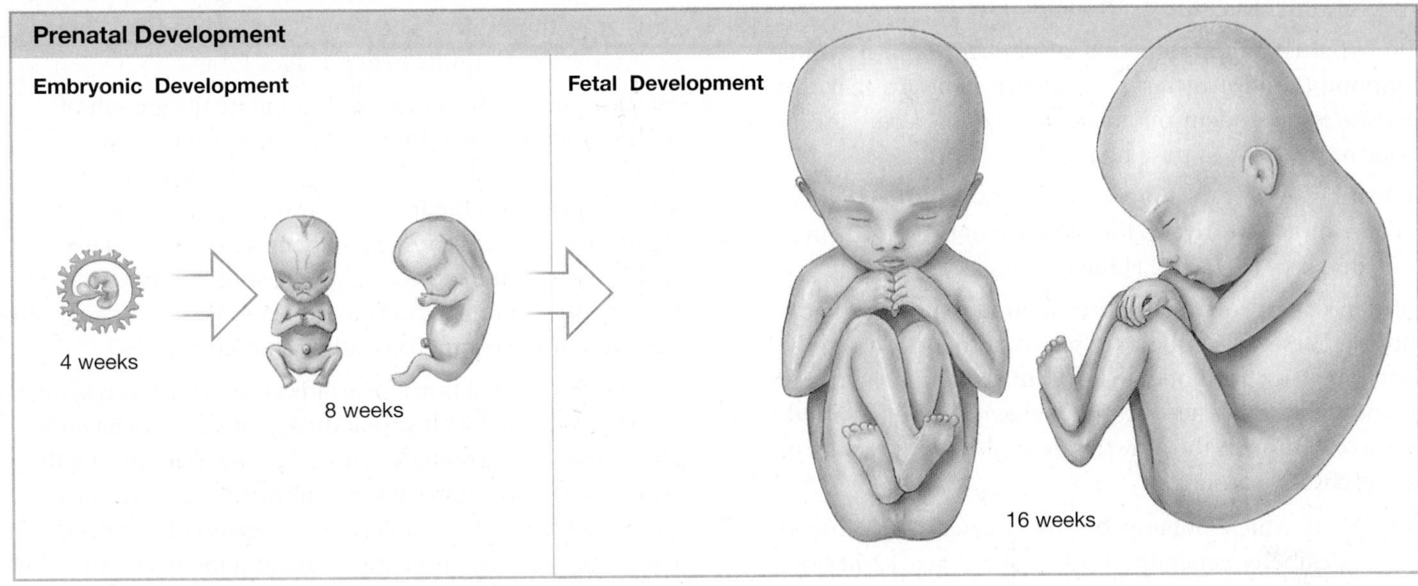
Prenatal Development
Embryonic Development
Fetal Development
4 weeks
8 weeks
16 weeks

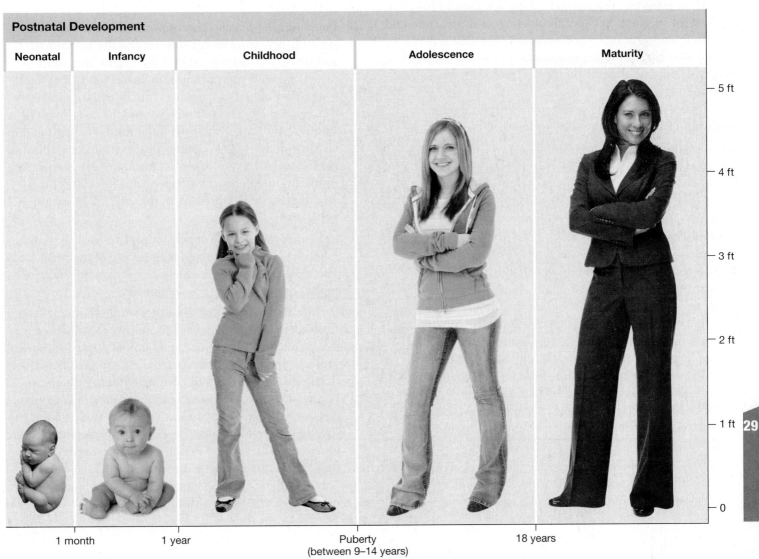

Postnatal Development
Neonatal
Infancy
Childhood
Adolescence
Maturity
5 ft
4 ft
3 ft
2 ft
1 ft
0
1 month
1 year
Puberty
(between 9–14 years)
18 years

Adolescence and Maturity

Adolescence begins at **puberty**, the period of sexual maturation, and ends when growth is completed. Three major hormonal events interact at the onset of puberty:

1. The hypothalamic production of gonadotropin-releasing hormone (GnRH) increases. Evidence indicates that this increase is dependent on adequate levels of *leptin*, a hormone released by adipose tissues. p. 646
2. Endocrine cells in the anterior lobe of the pituitary gland become more sensitive to the presence of GnRH, so circulating levels of FSH and LH increase rapidly.
3. Ovarian or testicular cells become more sensitive to FSH and LH. These cells initiate (1) gamete production; (2) the secretion of sex hormones that stimulate the appearance of secondary sex characteristics and behaviors; and (3) a sudden acceleration in the growth rate, ending with closure of the epiphyseal cartilages.

The age at which puberty begins varies. In the United States today, puberty generally starts at about age 12 in boys and 11 in girls, but the normal ranges are broad (10–15 in boys, 9–14 in girls). Sex-specific differences in structure and function develop as many body systems alter their activities in response to circulating sex hormones and to the presence of growth hormone, thyroid hormones, prolactin, and adrenocortical hormones.

At puberty, endocrine system changes induce characteristic changes in various body systems:

- *Integumentary System.* Testosterone stimulates the development of terminal hairs on the face and chest of males. Under estrogen stimulation those follicles continue to produce fine hairs in females. The hairline recedes under testosterone stimulation. Both testosterone and estrogen stimulate terminal hair growth in the axillae and in the genital area. Androgens, which are present in both sexes, also stimulate sebaceous gland secretion and may cause acne. Adipose tissues respond differently to testosterone than to estrogens, and this difference produces changes in the distribution of subcutaneous body fat. In women, the combination of estrogens, prolactin, growth hormone, and thyroid hormones promotes the initial development of the mammary glands. The duct system becomes more elaborate, but true secretory alveoli do not yet develop. Much of the growth of the breasts during this period reflects increased deposition of fat rather than glandular tissue.
- *Skeletal System.* Both testosterone and estrogen accelerate bone deposition and skeletal growth. In the process, they promote closure of the epiphyseal cartilages and thus place a limit on growth in height. Girls generally do not grow as tall as boys because estrogens cause more rapid epiphyseal cartilage closure than does testosterone, and the period of skeletal growth ends at an earlier age in girls than in boys. Girls grow most rapidly between ages 10 and 13, whereas boys grow most rapidly between ages 12 and 15.
- *Muscular System.* Sex hormones stimulate the growth of skeletal muscle fibers, increasing strength and endurance. The effects of testosterone greatly exceed those of the estrogens, and the increased muscle mass accounts for significant sex differences in body mass, even for males and females of the same height. The stimulatory effects of testosterone on muscle mass have led to the use of anabolic steroids among competitive athletes of both sexes.
- *Nervous System.* Sex hormones affect central nervous system centers concerned with sexual drive and sexual behaviors. These centers differentiated in sex-specific ways during the second and third trimesters, when the fetal gonads secrete either testosterone (in males) or estrogens (in females). The surge in sex hormone secretion at puberty activates the CNS centers.
- *Cardiovascular System.* Testosterone stimulates erythropoiesis, increasing blood volume and the hematocrit. In females whose uterine cycles have begun, the iron loss associated with menstruation increases the risk of developing iron-deficiency anemia. Late in each uterine cycle, estrogens and progesterone promote the movement of water from plasma into interstitial fluid, leading to an increase in tissue water content. Estrogens decrease plasma cholesterol levels and slow the formation of plaque. As a result, premenopausal women have a lower risk of atherosclerosis ("hardening of the arteries") than do adult men.
- *Respiratory System.* Testosterone stimulates growth of the larynx and a thickening and lengthening of the vocal cords. These changes cause a gradual deepening of the voice of males compared with those of females.
- *Reproductive System.* In males, testosterone stimulates the functional development of the accessory reproductive glands, such as the prostate and seminal glands, and helps promote spermatogenesis. In females, estrogens target the uterus, promoting a thickening of the myometrium, increasing blood flow to the endometrium, and stimulating cervical mucus production. Estrogens also promote the functional development of accessory reproductive organs in females. The first few uterine cycles may or may not be accompanied by ovulation. After the initial stage, the woman will be fertile, even though her growth and physical maturation will continue for several years.

After puberty, the continued background secretion of estrogens or androgens maintains sex-specific differences. In both sexes, growth continues at a slower pace until age 18–21. By that time most of the epiphyseal cartilages have closed. The boundary between adolescence and maturity is indistinct, because it has physical, emotional, and behavioral components. Adolescence is often said to be over when growth stops, in the late teens or early twenties. The individual is then considered physically mature.

Senescence and Death

Physical growth may stop at maturity, but physiological changes continue. The sex-specific differences produced at puberty remain, but further changes take place when sex hormone levels decline at menopause or the male climacteric (andropause). p. 1094 All these changes are part of the process of **senescence** (*senesco*, to grow old), or aging, which reduces the functional capabilities of the person. Even without disease or injury, senescence-related changes at the molecular level ultimately lead to **death**.

Table 29–3 summarizes the age-related changes in organ systems discussed in earlier chapters. Taken together, these changes both decrease the functional abilities of the individual and affect homeostatic mechanisms. As a result, elderly people are less able to make homeostatic adjustments in response to internal or environmental stresses. The risks of contracting a variety of infectious diseases are proportionately increased as immune function deteriorates. Age-induced changes in vital systems cause problems such as infections, cancers, heart disease, strokes, arthritis, senile dementia, and anemia.

Physicians can attempt to adjust homeostatic mechanisms, and treat cancers and infections. **Geriatrics** (*geras*, old age) is the medical specialty that deals with the problems associated with aging. Physicians trained in geriatrics are known as **geriatricians**.

Eventually, however, this deterioration leads to drastic physiological changes that affect all internal systems. Death ultimately occurs when the body's existing homeostatic mechanisms cannot counter some combination of stresses.

Checkpoint

19. Name the postnatal stages of development.
20. Describe the time frame for each of the following stages: neonatal period, infancy, and adolescence.
21. What is the difference between colostrum and breast milk?
22. Increases in the blood levels of GnRH, FSH, LH, and sex hormones mark the onset of which stage of development?

See the blue Answers tab at the back of the book.

Table 29–3 Effects of Aging on Organ Systems

The characteristic physical and functional changes that are part of the aging process affect all organ systems. Examples discussed in previous chapters include the following:

- A loss of elasticity in the skin that produces sagging and wrinkling. p. 175
- A decline in the rate of bone deposition, leading to weak bones, and degenerative changes in joints that make them less mobile. pp. 204, 285–286
- Reductions in muscular strength and ability. p. 325
- Impairment of coordination, memory, and intellectual function. pp. 558–559
- Reductions in the production of, and sensitivity to, circulating hormones. p. 647
- Appearance of cardiovascular problems and a reduction in peripheral blood flow that can affect a variety of vital organs. p. 778
- Reduced sensitivity and responsiveness of the immune system, leading to infection, cancer, or both. p. 827
- Reduced elasticity in the lungs, leading to decreased respiratory function. pp. 876–877
- Decreased peristalsis and muscle tone along the digestive tract. p. 931
- Decreased peristalsis and muscle tone in the urinary system, coupled with a reduction in the glomerular filtration rate. p. 1013
- Functional impairment of the reproductive system, which eventually becomes inactive when menopause or the male climacteric occurs. p. 1094

29-9 Genes and chromosomes determine patterns of inheritance

Learning Outcome Relate basic principles of genetics to the inheritance of human traits.

Recall that **chromosomes** contain **DNA** and proteins, and **genes** are functional segments of DNA. Each gene carries the information needed to direct the synthesis of a specific polypeptide. We introduced chromosome structure and the functions of genes in Chapter 3. pp. 83–90 We begin our discussion of genetics with the basic patterns of inheritance and their implications. We then examine the mechanisms responsible for regulating the activities of the genes during prenatal development.

Genotype and Phenotype

Every nucleated somatic cell in your body carries copies of the original 46 chromosomes present when you were a zygote. Those chromosomes and their genes make up your **genotype** (JĒN-ō-tīp; *geno-*, gene).

Through development and differentiation, the instructions contained in the genotype are **expressed**, or carried out, in many ways. No single cell or tissue uses all the information and instructions contained in the genotype. For example, in muscle fibers, the genes involved in the formation of excitable membranes and contractile proteins are active, but in cells of the pancreatic islets, a different set of genes is at work. Collectively, however, the instructions in your genotype determine the anatomical and physiological characteristics that make you a unique person. Those anatomical and physiological characteristics are your **phenotype** (FĒ-nō-tīp; *phaino*, to display). The phenotype results from the interaction between the person's genotype and the environment. In architectural terms, the genotype is a set of plans, and the phenotype is the finished building. Observable features, such as hair or eye color, are called phenotypic *characters*. Specific character variants, such as brown hair or eyes, are called phenotypic *traits*.

Your genotype is derived from the genotypes of your parents. Yet you are not an exact copy of either parent. Nor are you an easily identifiable mixture of their characteristics.

Homologous Chromosomes and Alleles

The 46 chromosomes carried by each human somatic cell occur in pairs. Every somatic cell contains 23 pairs of chromosomes. At amphimixis, the sperm supplies the paternal member of each pair, and the ovum supplies the maternal member. Recall that the two members of each of the 22 pairs are known as **homologous chromosomes.** ⮌ p. 1065 Those 22 chromosome pairs are called **autosomes.** Most of the genes of autosomes affect somatic, or body, characteristics, such as hair color and skin pigmentation. The chromosomes of the 23rd pair are not truly homologous and are known as the **sex chromosomes**. One of their functions is to determine whether the individual is genetically male or female. Figure 29–14 shows the **karyotype** (*karyon*, nucleus + *typos*, mark), or entire set of chromosomes, of a normal male. In the discussion that follows, we deal with the inheritance of traits carried on the autosomal chromosomes. We will examine the patterns of inheritance by the sex chromosomes later in this section.

The two chromosomes in an autosomal homologous pair have the same structure and carry genes that affect the same traits. For example, suppose that one member of the pair contains three genes in a row, with the first gene determining hair color, the second eye color, and the third skin pigmentation. The other chromosome, or *homolog*, carries genes that affect the same traits, and the genes are in the same sequence. The genes are also located at equivalent positions on their respective chromosomes. A gene's position on a chromosome is called a **locus** (LŌ-kus; plural, *loci* [LŌ-sī]).

Because the two chromosomes of a homologous pair have different origins, one paternal and the other maternal, they may not carry the same *form*, or version, of each gene, however. The various forms of a given gene are called **alleles** (uh-LĒLZ). These *alternative forms* determine the precise effect of the gene on your phenotype. If the two chromosomes of a homologous pair carry the same allele of a particular gene, you are **homozygous** (hō-mō-ZĪ-gus; *homos*, the same) for the trait affected by that gene. If you have two different alleles for the same gene, you are **heterozygous** (het-er-ō-ZĪ-gus; *heteros*, other) for the trait determined by that gene.

Figure 29–14 **A Human Male Karyotype.**

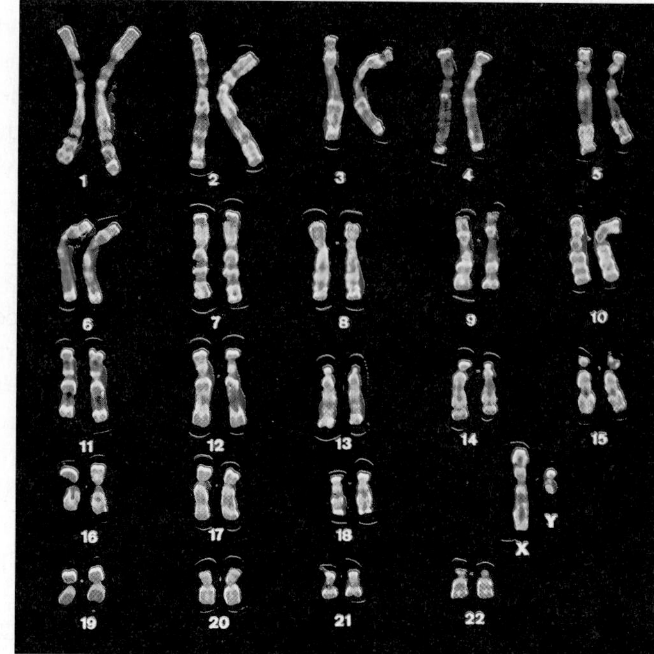
1
2
3
4
5
6
7
8
9
10
11
12
13
14
15
16
17
18
X
Y
19
20
21
22

? Explain why this is a karyotype of a male and not a female.

Autosomal Patterns of Inheritance

In **simple inheritance**, the phenotype is determined by interactions between a single pair of alleles. Some phenotypic traits are determined by interactions among several genes. Such interactions constitute **polygenic inheritance**. The potential interactions, which include examples of normal and abnormal phenotypic traits, are diagrammed in Figure 29–15.

Simple Inheritance

If you are homozygous for a particular trait, that allele will then indeed be expressed in your phenotype. For example, if you receive a gene for freckles from your father and a gene for freckles from your mother, you will be homozygous for freckles—and you will have freckles. About 80 percent of an individual's genome consists of homozygous alleles.

The phenotype that results from a heterozygous genotype depends on the nature of the interaction between the corresponding alleles. For example, if you received a gene for curly hair from your father, but a gene for straight hair from your

Figure 29–15 Major Patterns of Inheritance.

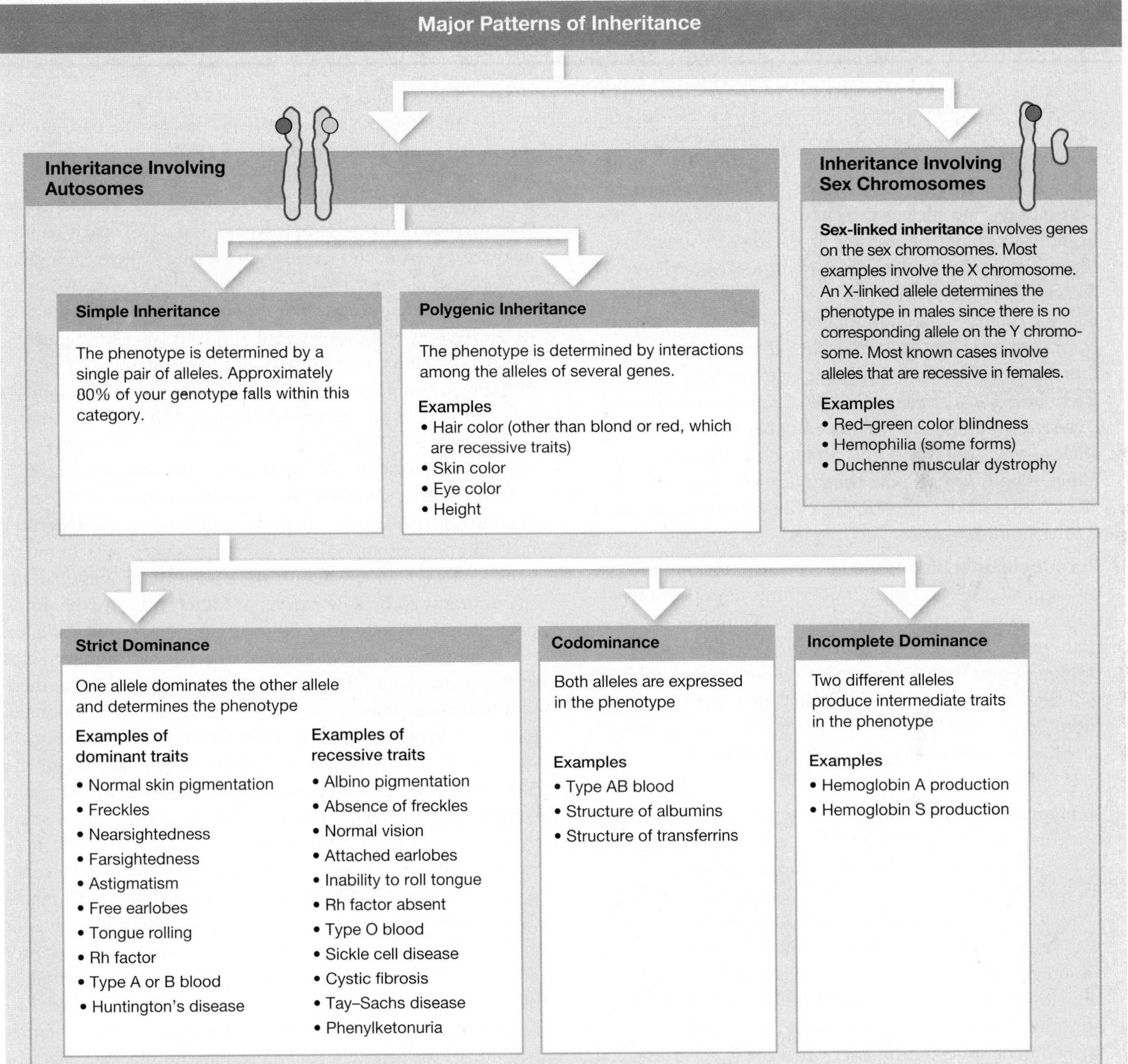
Major Patterns of Inheritance
Inheritance Involving Autosomes
Inheritance Involving Sex Chromosomes
Sex-linked inheritance involves genes on the sex chromosomes. Most examples involve the X chromosome. An X-linked allele determines the phenotype in males since there is no corresponding allele on the Y chromosome. Most known cases involve alleles that are recessive in females.
Examples
• Red–green color blindness
• Hemophilia (some forms)
• Duchenne muscular dystrophy
Simple Inheritance
The phenotype is determined by a single pair of alleles. Approximately 80% of your genotype falls within this category.
Polygenic Inheritance
The phenotype is determined by interactions among the alleles of several genes.
Examples
• Hair color (other than blond or red, which are recessive traits)
• Skin color
• Eye color
• Height
Strict Dominance
One allele dominates the other allele and determines the phenotype
Examples of dominant traits
• Normal skin pigmentation
• Freckles
• Nearsightedness
• Farsightedness
• Astigmatism
• Free earlobes
• Tongue rolling
• Rh factor
• Type A or B blood
• Huntington's disease
Examples of recessive traits
• Albino pigmentation
• Absence of freckles
• Normal vision
• Attached earlobes
• Inability to roll tongue
• Rh factor absent
• Type O blood
• Sickle cell disease
• Cystic fibrosis
• Tay–Sachs disease
• Phenylketonuria
Codominance
Both alleles are expressed in the phenotype
Examples
• Type AB blood
• Structure of albumins
• Structure of transferrins
Incomplete Dominance
Two different alleles produce intermediate traits in the phenotype
Examples
• Hemoglobin A production
• Hemoglobin S production

? What types of inheritance are involved in the following phenotypes-nearsightedness, normal vision, and red-green color blindness?

mother, whether *you* will have curly hair, straight hair, or even wavy hair depends on the relationship between the alleles for those traits. The relationship can be one of *strict dominance, codominance,* or *incomplete dominance,* as follows:

- In **strict dominance**, an allele that is **dominant** will be expressed in the phenotype, *regardless of any conflicting instructions carried by the other allele.* For instance, an individual with only one allele for freckles will have freckles, because that allele is dominant over the "nonfreckle" allele. An allele that is **recessive** will be expressed in the phenotype only if that same allele is present on *both chromosomes* of a homologous pair. For example, in Chapter 5 you learned that people with albinism cannot synthesize the yellow-brown pigment *melanin.* ⮌ p. 161 The presence of one allele that directs melanin production will result in normal color. Two recessive alleles must be present to produce an individual with albinism. A single gene can have many different alleles *in a population,* some dominant and others recessive. An individual can have a maximum of only two alleles—one from the mother and the other one from the father. If both parents have the same alleles, then an individual has only one kind of allele, but two copies of it, and is thus, homozygous.
- In **codominance**, an individual who is heterozygous (has different alleles) for a given trait exhibits both phenotypes for that trait. Blood type in humans is determined by codominance. The alleles for type A and type B blood are dominant over the allele for type O blood, but a person with one type A allele and one type B allele has type AB blood, not A or B. The red blood cells that make up type AB blood express *both* type A antigens and type B antigens. ⮌ p. 668
- In **incomplete dominance**, heterozygous alleles produce a phenotype that is intermediate (not completely dominant) to the phenotypes of individuals who are homozygous for one allele or the other. A good example is the sickling gene that affects the shape of red blood cells. Individuals with homozygous alleles that carry instructions for normal adult hemoglobin A have red blood cells of normal shape. Individuals with homozygous alleles for hemoglobin S, an abnormal form, have red blood cells that become sickle shaped in peripheral capillaries when the P_{O_2} decreases. These individuals develop *sickle cell disease.* ⮌ p. 686 Individuals who are heterozygous for this trait do not develop anemia but their red blood cells may sickle when tissue oxygen levels are extremely low. The distinction between incomplete dominance and codominance is not always clear-cut. For example, a person who has alleles for hemoglobin A and hemoglobin S shows incomplete dominance for RBC shape, but codominance for hemoglobin. Each red blood cell contains a mixture of hemoglobin A and hemoglobin S. Incomplete dominance is relatively rare in humans.

29

Predicting Simple Inheritance: The Punnett Square When an allele can be neatly characterized as dominant or recessive, you can predict the characteristics of individuals on the basis of their parents' alleles.

In such calculations, dominant alleles are traditionally indicated by capitalized abbreviations, and recessive alleles by lowercase abbreviations. Thus, for a given trait, the possible genotypes are indicated by *AA* (homozygous dominant), *Aa* (heterozygous), or *aa* (homozygous recessive). Each gamete involved in fertilization contributes a single allele for a given trait. That allele must be one of the two alleles contained by all cells in the parent's body. For example, let's consider the possible offspring of a mother with albinism (no skin pigmentation) and a father with normal skin pigmentation. The alleles of the maternal genotype are abbreviated *aa* because albinism is a homozygous recessive trait. No matter which of her oocytes is fertilized, it will carry the recessive *a* allele. The father has normal pigmentation, a dominant trait. His genotype therefore is either homozygous dominant *or* heterozygous for this trait, because both *AA* and *Aa* will produce the same phenotype: normal skin pigmentation. Every sperm produced by a homozygous dominant father will carry the *A* allele. In contrast, half the sperm produced by a heterozygous father will carry the dominant allele *A*, and the other half will carry the recessive allele *a*.

A simple box diagram known as a **Punnett square** helps us to predict the probabilities that children will have particular phenotypic traits by showing the various combinations of parental alleles they can inherit. In the Punnett squares shown in **Figure 29–16**, the maternal alleles for skin pigmentation are listed along the horizontal axis, and the paternal ones along the vertical axis. The combinations of alleles appear in the small boxes. **Figure 29–16a** shows the possible offspring of an *aa* mother and an *AA* father. All the children will have the genotype *Aa*, so all will have normal skin pigmentation. Compare these results with those of **Figure 29–16b**, for a heterozygous father (*Aa*) and an *aa* mother. The heterozygous father produces two types of gametes, *A* and *a*. The mother's secondary oocyte, carrying an *a* allele, may be fertilized by either one. As a result, the probability is 50 percent that a child of such a father will inherit the genotype *Aa* and so have normal skin pigmentation. The probability of inheriting the genotype *aa*, and thus having the albinism phenotype, is also 50 percent.

A Punnett square can also be used to draw conclusions about the identity and genotype of a parent. For example, in our scenario, a man with the genotype *AA* cannot be the father of an albino child (genotype *aa*).

We can also predict the frequency of appearance of any inherited disorder that results from simple inheritance by using a Punnett square. Although they are rare in terms of overall

Figure 29–16 **Predicting Phenotypic Traits by Using Punnett Squares.**

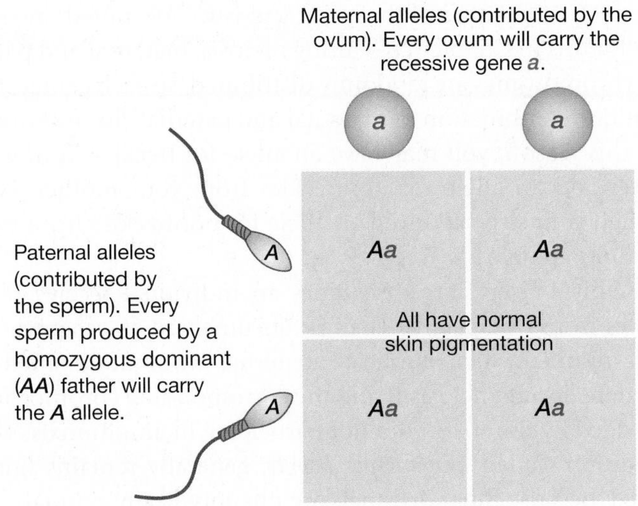
Maternal alleles (contributed by the ovum). Every ovum will carry the recessive gene a.
a
a
Paternal alleles (contributed by the sperm). Every sperm produced by a homozygous dominant (AA) father will carry the A allele.
A
A
Aa
Aa
All have normal skin pigmentation
Aa
Aa

a If the father is homozygous for normal pigmentation, all of the children will have the genotype Aa, and all will have normal skin pigmentation.

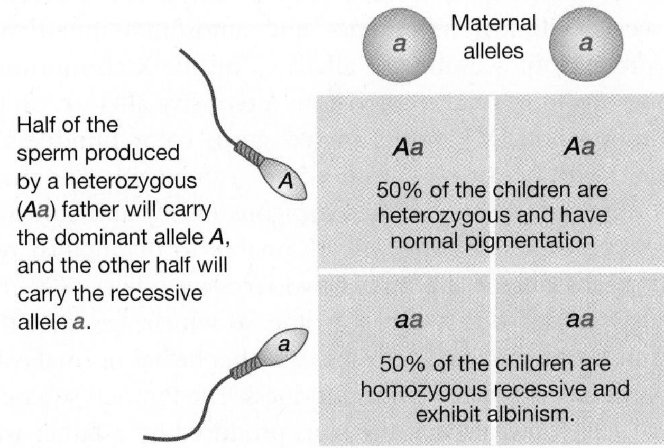
Maternal alleles
a
a
Half of the sperm produced by a heterozygous (Aa) father will carry the dominant allele A, and the other half will carry the recessive allele a.
A
a
Aa
Aa
50% of the children are heterozygous and have normal pigmentation
aa
aa
50% of the children are homozygous recessive and exhibit albinism.

b If the father is heterozygous for normal skin pigmentation, the probability that a child will have normal pigmentation is reduced to 50%.

numbers, more than 1200 inherited disorders have been identified that reflect the presence of one or two abnormal alleles for a single gene.

Polygenic Inheritance

We cannot predict the presence or absence of phenotypic traits inherited in a polygenic manner using a simple Punnett square, because the resulting phenotype depends not only on the nature of the alleles but also on how those alleles interact. In *suppression*, one gene suppresses the other. As a result, the second gene has no effect on the phenotype. In *complementary gene action*, dominant alleles on two genes interact to produce a phenotype different from that seen when one gene contains recessive alleles. The risks of developing several important adult disorders, including hypertension and coronary artery disease, are linked to polygenic inheritance.

Many of the developmental disorders responsible for fetal deaths and congenital malformations result from polygenic inheritance. In these cases, an individual's genetic composition does not by itself determine the onset of the disease. Instead, the conditions regulated by these genes establish a susceptibility to particular environmental influences. As a result, not every individual with the genetic tendency for a certain condition will develop that condition. For this reason, it is difficult to track polygenic conditions through successive generations. However, steps can be taken to prevent a crisis because many inherited polygenic conditions are *likely* (but not *guaranteed*) to occur. For example, you can reduce hypertension by controlling your diet and fluid volume, and you can prevent coronary artery disease by lowering your blood serum cholesterol levels.

Sex-Linked Patterns of Inheritance

Sex-linked inheritance involves genes on the sex chromosomes (see Figure 29–15). Unlike the other 22 chromosomal pairs, the sex chromosomes are not identical in appearance and gene content. There are two types of sex chromosomes: an **X chromosome** and a **Y chromosome**. X chromosomes are considerably larger and have more genes than do Y chromosomes. The smaller Y chromosome carries the *SRY gene* not found on the X chromosome. It specifies that an individual with that chromosome will be male. The normal pair of sex chromosomes in males is XY. Females do not have a Y chromosome. Their sex chromosome pair is XX.

All oocytes carry an X chromosome, because the only sex chromosomes females have are X chromosomes. But each sperm carries either an X or a Y chromosome, because males have one of each and can pass along either one during meiosis. As a Punnett square shows, the ratio of males to females in offspring should be 1:1. The worldwide birth statistics differ slightly from that prediction, with 107 males born for every 100 females. It has been suggested that more males are born because a sperm that carries the Y chromosome can reach the oocyte first, because that sperm does not have to carry the extra weight of the larger X chromosome.

The X chromosome also carries genes that affect somatic structures. Genes that are found on the X chromosome but not on the Y chromosome are called **X linked**. The inheritance of characteristics, or traits, regulated by these X-linked genes does not follow the pattern of alleles on autosomal chromosomes. (The SRY gene that determines "maleness" is an example of a Y-linked gene.)

The inheritance of color blindness shows the differences between sex-linked inheritance and autosomal inheritance. The presence of a dominant allele, *C*, on the X chromosome (X^C) results in normal color vision. A recessive allele, *c*, on the X chromosome (X^c) results in red–green color blindness. A woman, with her two X chromosomes, can be either homozygous dominant (X^CX^C) or heterozygous (X^CX^c) and still have normal color vision. She will be unable to distinguish reds from greens only if she carries two recessive alleles, X^cX^c. But a male has only one X chromosome, so whichever allele that X chromosome carries determines whether he has normal color vision or has red–green color blindness. The Punnett square in Figure 29–17 reveals that the sons produced by a father with normal vision and a heterozygous (carrier) mother have a 50 percent chance of being red–green color blind. Any daughters have normal color vision. Recessive alleles on X chromosomes produce genetic disorders in males at a higher frequency than in females.

A number of other clinical disorders noted earlier in the text are X-linked traits, including certain forms of hemophilia, diabetes insipidus, and muscular dystrophy. In several instances, advances in molecular genetics techniques have enabled geneticists to locate the specific genes on the X chromosome. These techniques provide a reasonably direct method of screening for the presence of a particular condition before any signs or symptoms appear, and even before birth.

Figure 29–17 Inheritance of an X-Linked Trait.

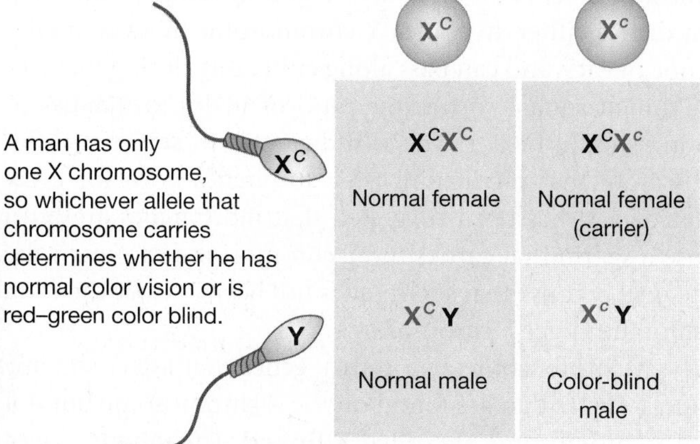
A woman—who has two X chromosomes—can be either homozygous dominant (X^CX^C) or heterozygous (X^CX^c) and still have normal color vision. She will be unable to distinguish reds from greens only if she carries two recessive alleles, X^cX^c.
X^C
X^c
A man has only one X chromosome, so whichever allele that chromosome carries determines whether he has normal color vision or is red–green color blind.
X^C
Y
X^CX^C
Normal female
X^CX^c
Normal female (carrier)
X^CY
Normal male
X^cY
Color-blind male

Sources of Individual Variation

Just as you are not a copy of either of your parents, you also are not a 50–50 mixture of their characteristics. We noted one reason for this in Chapter 28: During meiosis, maternal and paternal chromosomes are randomly distributed, so each gamete has a unique combination of maternal and paternal chromosomes. For this reason, you may have an allele for freckles from your father and an allele for nonfreckles from your mother, even though your sister received an allele for nonfreckles from each of your parents.

Only in very rare cases does an individual receive both alleles from one parent. The few documented cases appear to have resulted when duplicate maternal chromatids failed to separate during meiosis II and the corresponding chromosome provided by the sperm did not participate in amphimixis. This condition, called *uniparental disomy*, generally remains undetected, because the individuals are phenotypically normal.

Genetic Recombination

During meiosis, various changes can occur in chromosome structure, producing gametes with chromosomes that differ from those of each parent. **Genetic recombination** greatly increases the genetic variation among gametes, and thus among the genotypes formed when those gametes combine in fertilization.

In one normal form of recombination, parts of homologous chromosomes become rearranged during synapsis of meiosis (Figure 29–18). When tetrads form, adjacent chromatids may overlap. The chromatids may then break, and the overlapping segments trade places, an event called **crossing over**. In contrast, genetic recombination between nonhomologous chromosomes it is called **translocation**. During recombination, portions of chromosomes may also break away and be lost, or *deleted*. This

Figure 29–18 Crossing Over and Recombination.

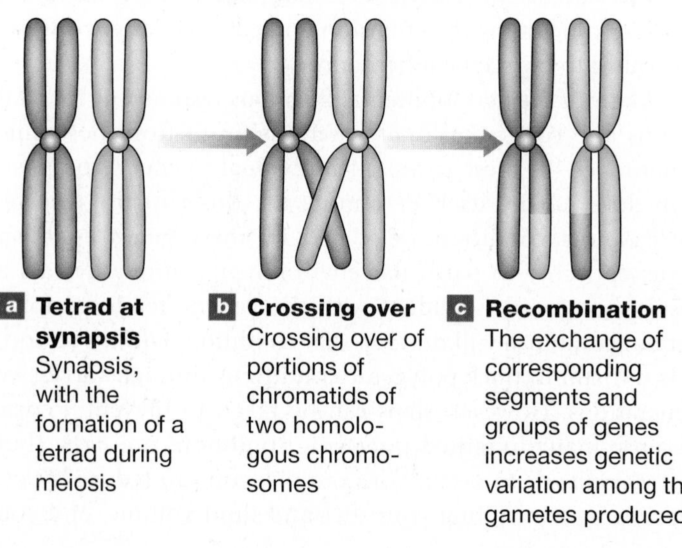
a Tetrad at synapsis
Synapsis, with the formation of a tetrad during meiosis
b Crossing over
Crossing over of portions of chromatids of two homologous chromosomes
c Recombination
The exchange of corresponding segments and groups of genes increases genetic variation among the gametes produced.

29

event is abnormal, and it can have severe or even lethal effects on the zygote, depending on the nature of the lost genes.

Recombination that results in abnormal chromosome shapes or numbers is lethal for the zygote in almost all cases. Roughly 10 percent of zygotes have **chromosomal abnormalities**—that is, damaged, broken, missing, or extra copies of chromosomes—but only about 0.5 percent of newborns have such abnormalities. Few individuals with chromosomal abnormalities survive to full term. *Down syndrome (trisomy 21)*, which involves one of the smallest chromosomes in humans, is an exception.

In addition to contributing to prenatal mortality, chromosomal abnormalities produce a variety of serious clinical conditions. The high mortality rate and the severity of the problems reflect the fact that large numbers of genes have been added or deleted. Women who become pregnant later in life run a higher risk of birth defects and miscarriage due to chromosomal abnormalities in the oocyte. It seems that the longer the oocyte remains suspended in meiosis I, the more likely are recombination errors when meiosis is completed.

However, a lack of specific gene expression in a zygote is not necessarily due to chromosomal aberrations or deletions. Genes can be present, yet be prevented from being fully expressed. A genetic process called **genomic imprinting** is particularly important in the early embryo. Imprinting does not change the DNA itself. Instead, it results in specific (and usually reversible) chemical modifications, or markings, of DNA and its associated proteins. These changes then dictate whether the gene is expressed or not (silenced). As gametes mature,

+ Clinical Note Amniocentesis

Many genetic conditions can be detected before birth through the analysis of fetal cells. In **amniocentesis**, a sample of **amniotic fluid** is removed and the fetal cells it contains are analyzed. This procedure permits the identification of many congenital conditions, including Down syndrome, sickle cell disease, and neural tube defects. The needle inserted to obtain a fluid sample is guided into position during an ultrasound procedure. ⮌ p. 17

Because the sampling procedure represents a potential threat to the health of both the fetus and the mother, amniocentesis is generally offered only to women who have a significant risk for genetic diseases.

Sampling cannot safely be performed until the volume of amniotic fluid is large enough that the fetus will not be injured during the process. The usual time for amniocentesis is between the 15th and 18th week of pregnancy. Results are usually available within 2–3 weeks.

+ Clinical Note Chromosomal Abnormalities

Embryos that have abnormal autosomal chromosomes (chromosomes 1–22) rarely survive. However, *translocation defects* and *trisomy* are two types of autosomal abnormalities that do not invariably result in prenatal death.

In a **translocation defect**, an exchange occurs between different (nonhomologous) chromosome pairs. For example, a piece of chromosome 8 may become attached to chromosome 14. The genes moved to their new position may function abnormally, becoming inactive or overactive. In a balanced translocation, where there is no net loss or gain of chromosomal material, embryos may survive.

In **trisomy**, a mistake occurs in meiosis. One of the gametes involved in fertilization carries an extra copy of one chromosome, so the zygote then has three copies of this chromosome rather than two. (The nature of the trisomy is indicated by the number of the chromosome involved. For example, individuals with trisomy 13 have three copies of chromosome 13.) Zygotes with extra copies of chromosomes seldom survive. The notable exception is trisomy 21.

Trisomy 21, or **Down syndrome**, is the most common viable chromosomal abnormality. According to the CDC, about 6000 babies, or about 1 in every 700 babies, are born with Down syndrome. The degree of developmental disability

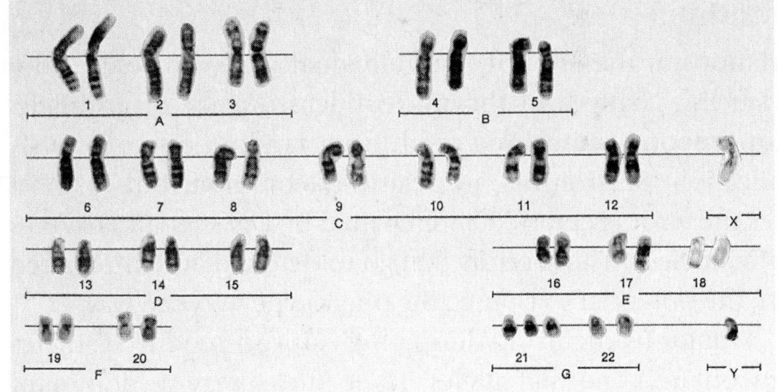
1
2
3
A
4
5
B
6
7
8
9
C
10
11
12
X
13
14
15
D
16
17
18
E
19
20
F
21
22
G
Y

ranges from moderate to severe. Anatomical problems affecting the cardiovascular system often prove fatal during childhood or early adulthood. Although some individuals survive to moderate old age, many develop Alzheimer's disease while still relatively young (before age 40).

For unknown reasons, there is a direct correlation between advancing maternal age and the risk of having a child with trisomy 21. This is becoming increasingly significant because many women are delaying childbearing until their mid-thirties or later.

a proportion of their genes develop characteristic, maternal and paternal specific, imprinting markings. These markings persist after fertilization and can regulate whether or not the maternally or the paternally derived gene is transcribed during embryonic development. Many of the genes known to be affected by genomic imprinting regulate many aspects of early development, including rates of prenatal and postnatal growth, behavior, and language development. Imprinting is a normal process that acts as a "volume control" for parental genes since it results in one active copy of an imprinted gene.

Incorrectly imprinted genes can have the same effect as deletion of the same gene. This is exemplified by two human genetic disorders linked to genes in a specific portion of chromosome 15 and their differential imprinting during sperm or oocyte formation. *Angelman syndrome*, which results in hyperactivity, severe developmental disability, and seizures, occurs when the maternal genes are inactive. *Prader–Willi syndrome*, which results in short stature, reduced muscle tone and skin pigmentation, underdeveloped gonads, and some degree of developmental disability, occurs when paternal genes are inactive.

Epigenetics is the study of inherited traits that are *not* due to changes in a person's genotype or DNA sequences. Similar to genomic imprinting, it acts at a level "above" the DNA by activating or inactivating specific genes. Such gene regulation occurs in a number of ways. One way is by chemically marking DNA and its histone proteins with methyl or acetyl groups, which permit or restrict transcription, respectively. Understanding the role of epigenetics in human clinical disorders, disease, and early development is an area of active study.

Mutations

Variations at the level of the individual gene can result from *mutations*—changes in the nucleotide sequence of an allele. **Spontaneous mutations** result from random errors in DNA replication. Such errors are relatively common, but in most cases the error is detected and repaired by DNA repair enzymes in the nucleus. Those errors that go undetected and unrepaired have the potential to change the phenotype in some way.

Mutations occurring during meiosis can produce gametes that contain abnormal alleles. These alleles may be dominant or recessive, and they may occur on autosomal chromosomes or on sex chromosomes. The vast majority of mutations make the zygote incapable of completing normal development. Mutation, rather than chromosomal abnormalities, is probably the primary cause of the high mortality rate among pre-embryos
29 and embryos. (Approximately 50 percent of all zygotes fail to complete cleavage, and another 10 percent fail to reach the fifth month of gestation.)

If the abnormal allele is dominant but does not affect gestational survival, the individual's phenotype will show the effects of the mutation. If the abnormal allele is recessive and is on an autosomal chromosome, it will not affect the individual's phenotype as long as the zygote contains a normal allele contributed by the other parent at fertilization. Over generations, a recessive autosomal allele can spread through the population, remaining undetected until a fertilization occurs in which the two gametes contribute identical recessive alleles. This individual, who will be homozygous for the abnormal allele, will be the first to show the phenotypic effects of the original mutation. Individuals who are heterozygous for an abnormal allele but do not show the effects of the mutation are called **carriers**. Available genetic tests can determine whether an individual is a carrier for any of several autosomal recessive disorders, including Tay–Sachs disease. The information obtained from these tests can be useful in counseling prospective parents. For example, if both parents are carriers of the same disorder, they have a 25 percent probability of producing a child with the disease. This information may affect their decision to conceive.

Effect of Environmental Factors: Penetrance and Expressivity

Differences in genotype lead to distinct variations in phenotype, but the relationships are not always predictable. The presence of a particular pair of alleles does not affect the phenotype in the same way in every individual. **Penetrance** is the percentage of individuals with a particular genotype that show the "expected" phenotype. In other individuals, the activity of other genes or environmental factors may override the effects of that genotype. For example, *emphysema*, a respiratory disorder discussed in Chapter 23, has been linked to a specific abnormal genotype. ⮌ p. 877 However, about 20 percent of the individuals with this genotype do not develop emphysema, making the penetrance of this genotype approximately 80 percent. The effects of environmental factors are apparent: Most people who develop emphysema are cigarette smokers.

If a given genotype *does* affect the phenotype, it can do so to various degrees, again depending on the activity of other genes or environmental stimuli. For example, even though identical twins have the same genotype, they do not have exactly the same fingerprints. The extent to which a particular allele is expressed when it is present is termed its **expressivity**.

Environmental effects on genetic expression are particularly evident during embryonic and fetal development. Drugs, including certain antibiotics, alcohol, and nicotine in cigarette smoke, can disrupt fetal development. Factors that result in abnormal development are called **teratogens** (TER-ah-tō-jenz).

The Human Genome

The **human genome** is the full set of genetic material (DNA) in our chromosomes. We now know the nucleotide sequences of its total 3.2 billion base pairs spread among our 22 autosomes and X and Y sex chromosomes. This knowledge has revealed that our genome contains 20,000–25,000 protein-coding genes. It

Figure 29–19 A Map of Human Chromosomes. The banding patterns of typical chromosomes in a male, and the general locations of the mutated genes associated with specific inherited disorders. The chromosomes are not drawn to scale.

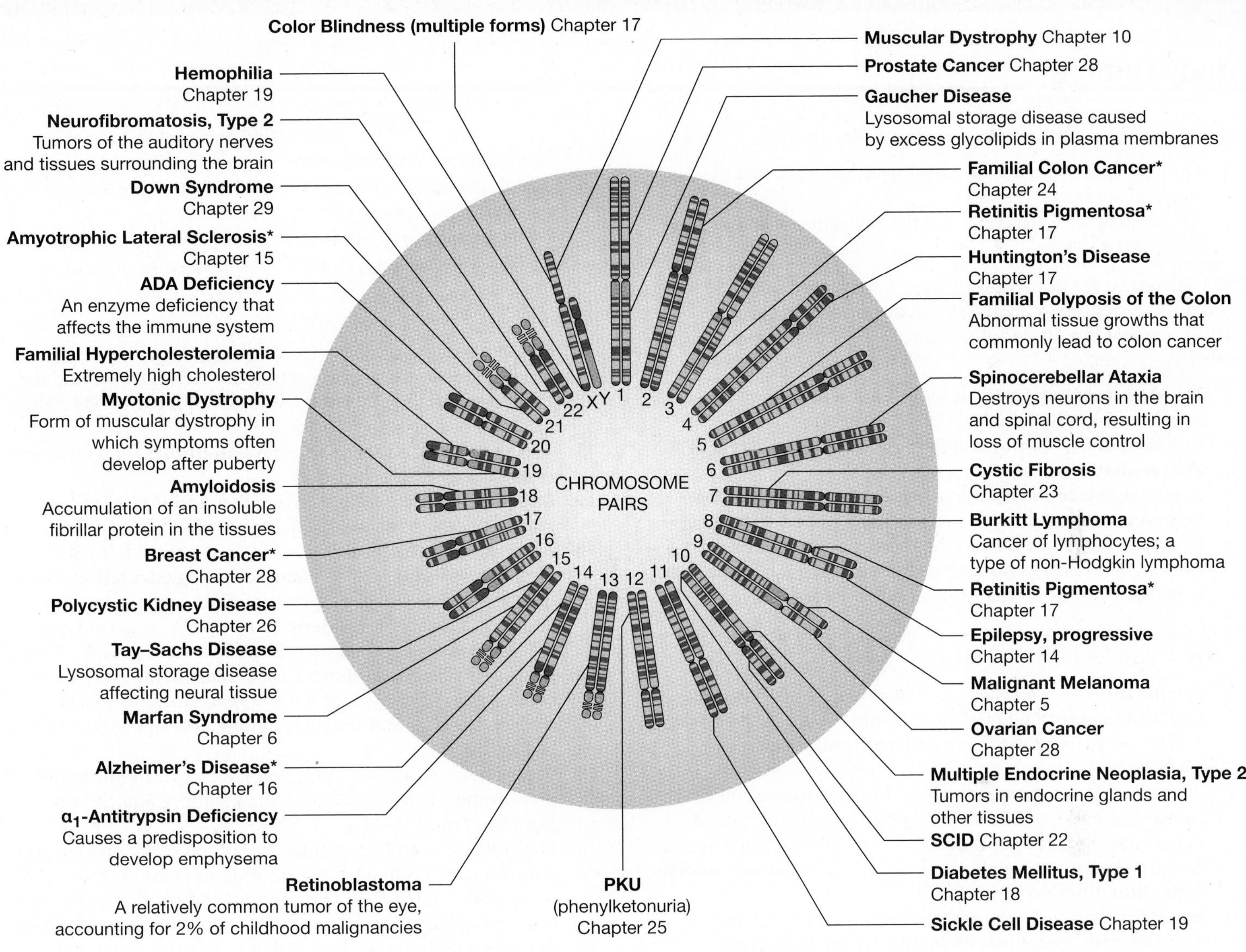
Color Blindness (multiple forms) Chapter 17
Hemophilia Chapter 19
Neurofibromatosis, Type 2 Tumors of the auditory nerves and tissues surrounding the brain
Down Syndrome Chapter 29
Amyotrophic Lateral Sclerosis* Chapter 15
ADA Deficiency An enzyme deficiency that affects the immune system
Familial Hypercholesterolemia Extremely high cholesterol
Myotonic Dystrophy Form of muscular dystrophy in which symptoms often develop after puberty
Amyloidosis Accumulation of an insoluble fibrillar protein in the tissues
Breast Cancer* Chapter 28
Polycystic Kidney Disease Chapter 26
Tay–Sachs Disease Lysosomal storage disease affecting neural tissue
Marfan Syndrome Chapter 6
Alzheimer's Disease* Chapter 16
α1-Antitrypsin Deficiency Causes a predisposition to develop emphysema
Retinoblastoma A relatively common tumor of the eye, accounting for 2% of childhood malignancies
PKU (phenylketonuria) Chapter 25
Muscular Dystrophy Chapter 10
Prostate Cancer Chapter 28
Gaucher Disease Lysosomal storage disease caused by excess glycolipids in plasma membranes
Familial Colon Cancer* Chapter 24
Retinitis Pigmentosa* Chapter 17
Huntington's Disease Chapter 17
Familial Polyposis of the Colon Abnormal tissue growths that commonly lead to colon cancer
Spinocerebellar Ataxia Destroys neurons in the brain and spinal cord, resulting in loss of muscle control
Cystic Fibrosis Chapter 23
Burkitt Lymphoma Cancer of lymphocytes; a type of non-Hodgkin lymphoma
Retinitis Pigmentosa* Chapter 17
Epilepsy, progressive Chapter 14
Malignant Melanoma Chapter 5
Ovarian Cancer Chapter 28
Multiple Endocrine Neoplasia, Type 2 Tumors in endocrine glands and other tissues
SCID Chapter 22
Diabetes Mellitus, Type 1 Chapter 18
Sickle Cell Disease Chapter 19
CHROMOSOME PAIRS
XY 1 2 3 4 5 6 7 8 9 10 11 12 13 14 15 16 17 18 19 20 21 22
* One form of the disease

has also led to a description of roughly 10,000 different single-gene disorders. Most are very rare, but collectively they may affect 1 in every 200 births. Several examples are included in Figure 29–19. Genetic screening and diagnostic tests for abnormal genes are now performed for many of these disorders.

We have identified the normal genetic composition of a "typical" human and have begun to reveal how genes are regulated. *Gene therapy* is a rapidly growing field of medicine. **Gene therapy** involves inserting corrective genes made in a laboratory into a patient's cells to treat disorders or diseases caused by genetic defects.

We all are variations on a basic theme. How do we decide what set of genes to accept as "normal"? Moreover, as we improve our abilities to modify the expression of a gene or correct an abnormal gene, we will face many additional ethical and legal dilemmas. In the years to come, we will have to find answers to these dilemmas that are acceptable to us all.

✓ Checkpoint

23. **Describe the relationship between genotype and phenotype.**
24. **Define *heterozygous*.**
25. **Tongue rolling is an autosomal dominant trait. What would be the phenotype of a person who is heterozygous for this trait?**
26. **Why are children not identical copies of their parents?**

See the blue Answers tab at the back of the book.

29

29 Chapter Review

Study Outline

MasteringA&P™ Access more chapter study tools online in the MasteringA&P Study Area:
- Chapter Quizzes, Chapter Practice Test, MP3 Tutor Sessions, and Clinical Case Studies
- Practice Anatomy Lab PAL 3.0
- A&P Flix A&PFlix
- Interactive Physiology iP2
- PhysioEx PhysioEx 9.1

An Introduction to Development and Inheritance p. 1104

1. Humans develop in the womb, grow to maturity, and live for most of a century.
2. An interplay between genes and the environment shapes physiological processes.

29-1 Directed by inherited genes, a fertilized ovum differentiates during prenatal development to form an individual; postnatal development brings that individual to maturity p. 1104

3. **Development** is the gradual modification of anatomical structures and physiological characteristics from conception to maturity. The formation of different types of cells is **differentiation**.
4. **Prenatal development** occurs before birth; **postnatal development** begins at birth and continues to **maturity**, when aging begins. **Inheritance** is the transfer of genetically determined characteristics from generation to generation. **Genetics** is the study of the mechanisms of inheritance.

29-2 Fertilization—the fusion of a secondary oocyte and a sperm—forms a zygote p. 1105

5. **Fertilization**, or *conception*, normally occurs in the uterine tube within a day after ovulation. Sperm cannot fertilize a secondary oocyte until they have undergone *capacitation*. *(Figure 29–1)*
6. The acrosomes of the sperm release **hyaluronidase** and **acrosin**, enzymes required for sperm to penetrate the corona radiata and zona pellucida of the oocyte. When a single sperm contacts the oocyte membrane, fertilization begins and **oocyte activation** follows. *(Figure 29–1)*
7. During activation, the oocyte completes meiosis II and thus becomes a functionally mature ovum. Membrane depolarization and the *cortical reaction* act as a **block to polyspermy**.
8. After activation, the **female pronucleus** and the **male pronucleus** replicate their DNA before fusing in a process called *amphimixis*. *(Figure 29–1)*

29-3 Gestation consists of three stages of prenatal development: the first, second, and third trimesters p. 1107

9. The 9-month **gestation** period can be divided into three **trimesters**.

29-4 The first trimester includes pre-embryonic and embryonic development, involving the processes of cleavage, implantation, placentation, and embryogenesis p. 1107

10. In the **first trimester**, **cleavage** subdivides the cytoplasm of the zygote in a series of mitotic divisions; the zygote becomes a **pre-embryo** and then a **blastocyst**. During **implantation**, the blastocyst becomes enclosed within the uterine endometrium. **Placentation** occurs as blood vessels form around the blastocyst and the **placenta** develops. **Embryogenesis** is the formation of a viable embryo.
11. The blastocyst consists of an outer **trophoblast** and an **inner cell mass**. *(Figure 29–2)*
12. Implantation occurs about 7 days after fertilization as the blastocyst adheres to the uterine lining. *(Figure 29–3)*
13. As the **syncytiotrophoblast** enlarges and spreads, maternal blood flows through open **lacunae**. After **gastrulation**, there is an **embryonic disc** composed of **endoderm**, **ectoderm**, and an intervening **mesoderm**. It is from these **germ layers** that the body systems differentiate. *(Figure 29–4; Table 29–1)*
14. During prenatal development, differences in the cytoplasmic composition of individual cells trigger changes in genetic activity. The chemical interplay among developing cells is **induction**.
15. Germ layers are involved in forming four **extra-embryonic membranes**: the yolk sac, amnion, allantois, and chorion. *(Spotlight Figure 29–5)*
16. The **yolk sac** is an important site of blood cell formation. The **amnion** encloses fluid that surrounds and cushions the developing embryo. The base of the **allantois** later gives rise to the urinary bladder. Circulation within the vessels of the **chorion** provides a rapid-transit system that links the embryo with the trophoblast. *(Spotlight Figure 29–5; Figure 29–6)*
17. **Chorionic villi** extend outward into the maternal tissues, forming an intricate, branching network through which maternal blood flows. As development proceeds, the **umbilical cord** connects the fetus to the placenta. *(Figure 29–6)*
18. The first trimester is critical, because events in the first 12 weeks establish the basis for **organogenesis** (organ formation). *(Figure 29–7; Table 29–2)*

29-5 During the second and third trimesters, fetal development involves growth and organ function p. 1115

19. In the **second trimester**, the organ systems increase in complexity. During the **third trimester**, many of the organ systems become fully functional. *(Figure 29–8; Table 29–2)*
20. The fetus undergoes its largest weight gain in the third trimester. At the end of gestation, the fetus and the enlarged uterus displace many of the mother's abdominal organs. *(Figure 29–9)*

29-6 During gestation, maternal organ systems support the developing fetus; the reproductive system undergoes structural and functional changes p. 1117

21. The syncytiotrophoblast synthesizes **human chorionic gonadotropin (hCG)**, **human placental lactogen (hPL)**, **placental prolactin**, **relaxin**, **progesterone**, and **estrogens**.
22. The developing fetus is totally dependent on maternal organs for nourishment, respiration, and waste removal. Maternal adaptations include increases in respiratory rate, tidal volume, blood volume, nutrient and vitamin intake, and glomerular filtration rate, as well as changes in the size of the uterus and mammary glands.

29-7 Childbirth occurs through the process of labor, which consists of the dilation, expulsion, and placental stages p. 1123

23. Progesterone produced by the placenta has an inhibitory effect on uterine muscles. Estrogens, oxytocin, and prostaglandins oppose its calming action. At some point, multiple factors interact to produce **labor contractions** in the uterine wall. *(Figure 29–10)*
24. The goal of **true labor** is **parturition**, the forcible expulsion of the fetus.
25. Labor can be divided into three stages: the **dilation stage**, the **expulsion stage**, and the **placental stage**. The Apgar score is used to assess the overall health of a newborn. *(Figure 29–11)*
26. **Premature labor** may result in **premature delivery**.
27. Difficult deliveries can include *forceps deliveries* and **breech births**—deliveries in which the legs or buttocks of the fetus, rather than the head, enter the vaginal canal first.
28. Twins are either **dizygotic** (fraternal) **twins** or **monozygotic** (identical) **twins**.

29-8 Postnatal stages are the neonatal period, infancy, childhood, adolescence, and maturity, followed by senescence and death p. 1126

29. Postnatal development involves a series of five **life stages**: the neonatal period, infancy, childhood, adolescence, and maturity. *Senescence*, or aging, begins at maturity and ends in the death of the individual.
30. The **neonatal period** extends from birth to 1 month. In the transition from fetus to **neonate**, the respiratory, circulatory, digestive, and urinary systems of the infant begin functioning independently. The newborn must also begin thermoregulation.
31. Mammary gland cells produce protein-rich **colostrum** during the neonate's first few days of life and then convert to milk production. These secretions are released as a result of the **milk ejection (milk let-down) reflex**. *(Figure 29–12)*
32. Body proportions gradually change during **infancy** (from age 1 month to 2 years) and during **childhood** (from age 2 years to puberty). *(Figure 29–13)*
33. **Adolescence** begins at **puberty**, when (1) the hypothalamus increases its production of GnRH, (2) circulating levels of FSH and LH rise rapidly, and (3) ovarian or testicular cells become more sensitive to FSH and LH. These changes initiate gamete formation, the production of sex hormones, and a sudden increase in the growth rate. The hormonal changes at puberty, especially changes in sex hormone levels, produce sex-specific differences in the structure and function of many systems; these differences will be retained. Adolescence continues until growth is completed. Further changes occur when sex hormone levels decline at menopause or the male climacteric.
34. **Senescence** then begins, producing gradual reductions in the functional capabilities of all systems. *(Table 29–3)*

29-9 Genes and chromosomes determine patterns of inheritance p. 1131

35. Every somatic cell carries copies of the original 46 chromosomes in the zygote; these chromosomes and their genes make up the individual's **genotype**. The physical expression of the genotype is the individual's **phenotype**.
36. Every somatic human cell contains 23 pairs of chromosomes; 22 pairs consist of **homologous chromosomes**. These 22 pairs are called **autosomes**. The chromosomes of the 23rd pair are the **sex chromosomes**; they differ between the sexes. *(Figure 29–14)*
37. Chromosomes contain DNA, and genes are functional segments of DNA. The various forms of a given gene are called **alleles**. If both homologous chromosomes carry the same allele of a particular gene, the individual is **homozygous**; if they carry different alleles, the individual is **heterozygous**.
38. In **simple inheritance**, phenotypic traits are determined by interactions between a single pair of alleles. **Polygenic inheritance** involves interactions among alleles on several genes. *(Figure 29–15)*
39. Alleles are either **dominant** or **recessive**, depending on how their traits are expressed.
40. Combining maternal and paternal alleles in a **Punnett square** helps us predict the characteristics of offspring. *(Figure 29–16)*
41. The two types of sex chromosomes are an **X chromosome** and a **Y chromosome**. The normal sex chromosome complement of males is XY; that of females is XX. **Sex-linked inheritance** involves genes on the sex chromosomes. The X chromosome carries **X-linked genes**, which affect somatic structures but have no corresponding alleles on the Y chromosome. *(Figure 29–17)*
42. **Genetic recombination**, the gene reshuffling in **crossing over** and **translocation** during meiosis, increases the genetic variation of male and female gametes. *(Figure 29–18)*
43. **Spontaneous mutations** are the result of random errors in DNA replication. Such mutations can cause the production of abnormal alleles.
44. The **human genome** contains some 20,000–25,000 protein-coding genes. Thousands of single-gene disorders have been identified, including some of those responsible for inherited disorders. *(Figure 29–19)*
45. **Gene therapy** involves inserting corrective genes made in a laboratory into a patient's cells to treat disorders or diseases caused by genetic defects.

Review Questions

See the blue Answers tab at the back of the book.

LEVEL 1 Reviewing Facts and Terms

1. The chorionic villi **(a)** form the umbilical cord, **(b)** form the umbilical vein, **(c)** form the umbilical arteries, **(d)** increase the surface area available for exchange between the placenta and maternal blood, **(e)** form the portion of the placenta called the capsular decidua.
2. Identify the two extra-embryonic membranes and the three different regions of the endometrium at week 10 of development in the following diagram.

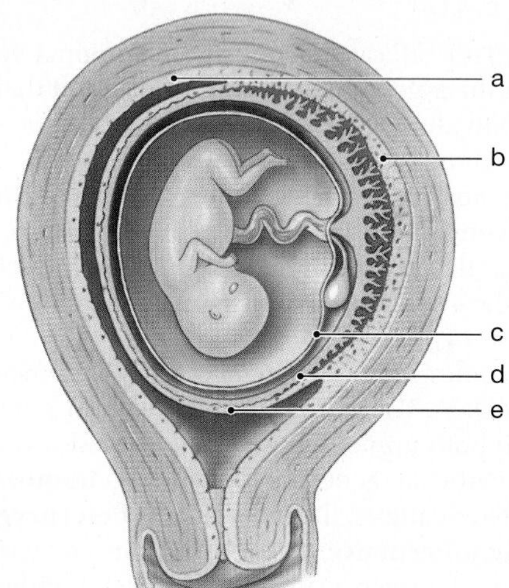
a
b
c
d
e

(a)__________ **(b)**__________
(c)__________ **(d)**__________
(e)__________

3. The hormone that is the basis for a pregnancy test is **(a)** LH, **(b)** progesterone, **(c)** human chorionic gonadotropin (hCG), **(d)** human placental lactogen (hPL), **(e)** either c or d, depending on the type of test.
4. Recessive X-linked traits **(a)** are passed from fathers to their sons, **(b)** are more likely to be expressed in males, **(c)** always affect some aspect of the reproductive system, **(d)** are never expressed in females, **(e)** cannot be passed from mothers to daughters.
5. The stage of development that follows cleavage is the **(a)** blastocyst, **(b)** morula, **(c)** trophoblast, **(d)** blastocoele.
6. The pre-embryo develops into a multicellular complex known as a **(a)** blastocyst, **(b)** trophoblast, **(c)** lacuna, **(d)** blastomere.
7. The structure(s) that allow(s) active and passive exchange between the fetal and maternal bloodstreams is/are the **(a)** yolk stalk, **(b)** chorionic villi, **(c)** umbilical veins, **(d)** umbilical arteries.
8. If an allele must be present on both the maternal and paternal chromosomes to affect the phenotype, the allele is said to be **(a)** dominant, **(b)** recessive, **(c)** complementary, **(d)** heterozygous.
9. Describe the changes that occur in the oocyte immediately after fertilization.
10. **(a)** What are the four extra-embryonic membranes? **(b)** From which germ layers do these membranes form, and what are their functions?
11. Identify the three stages of labor, and describe the events that characterize each stage.
12. List the factors involved in initiating labor contractions.
13. Identify the three life stages that occur between birth and approximately age 10. Describe the timing and characteristics of each stage.
14. What hormonal events are responsible for puberty? Which life stage does puberty initiate?

LEVEL 2 Reviewing Concepts

15. A normally pigmented woman whose father was an albino marries a normally pigmented man whose mother was an albino. What is the probability that they would have an albino child? **(a)** 50 percent, **(b)** 25 percent, **(c)** 12.5 percent, **(d)** 6.25 percent, **(e)** 100 percent.
16. If a sperm lacked hyaluronidase, it would *not* be able to **(a)** move its flagellum, **(b)** penetrate the corona radiata, **(c)** become capacitated, **(d)** survive the environment of the female reproductive tract, **(e)** metabolize fructose.
17. Problems involving the formation of the chorion would affect **(a)** the embryo's ability to produce blood cells, **(b)** the formation of limbs, **(c)** the embryo's ability to derive nutrition from the mother, **(d)** lung formation, **(e)** the urinary system.
18. After implantation, how does the developing embryo obtain nutrients? What structures and processes are involved?
19. Name the primary placental hormones and their functions.
20. Discuss the changes that occur in maternal systems during pregnancy. Why are these changes functionally significant?
21. During true labor, what physiological mechanisms ensure that uterine contractions continue until delivery has been completed?
22. What physiological adjustments must an infant make during the neonatal period in order to survive?
23. Distinguish between the following paired terms: **(a)** genotype and phenotype, **(b)** heterozygous and homozygous, **(c)** simple inheritance and polygenic inheritance.
24. Indicate the type of inheritance involved in each of the following situations. **(a)** Children who exhibit the trait have at least one parent who also exhibits it. **(b)** Children exhibit the trait even though neither parent does. **(c)** The trait is expressed more commonly in sons than in daughters. **(d)** The trait is expressed equally in daughters and sons.

LEVEL 3 Critical Thinking and Clinical Applications

25. Hemophilia A, a condition in which blood does not clot properly, is a recessive trait located on the X chromosome (X^h). Suppose that a woman who is heterozygous for this trait (XX^h) has children with a normal male (XY). What is the probability that the couple will have daughters with hemophilia? What is the probability that the couple will have sons with hemophilia?
26. Joe and Jane desperately want to have children, and although they have tried for 2 years, they have not been successful. Finally, each of them consults a physician, and it turns out that Joe suffers from oligospermia (a low sperm count). He confides to you that he doesn't understand why this would interfere with his ability to have children since he remembers from biology class that it only takes one sperm to fertilize an egg. What would you tell him?
27. Alina has just given birth to a little girl. When the nurses take the infant back to the nursery, she becomes cyanotic, a condition characterized by bluish discoloration of the skin. The episode passes, but when the infant is bathed, she becomes cyanotic again. Blood gas levels indicate that arterial blood is only 60 percent saturated. Physical examination reveals no structural deformities involving the respiratory or digestive system. What might be causing the problem?
28. Brianna gives birth to a baby with a congenital deformity of the stomach. Sally believes that her baby's affliction is the result of a viral infection she suffered during her third trimester. Is this a possibility? Explain.

CLINICAL CASE Wrap-Up The Twins That Looked Nothing Alike

10
2

The winners of the "least-alike" twins prize are named Hans and Willem. These charming boys traveled all the way from the Netherlands to win this award in Twinsburg.

Hans and Willem's parents used the services of a prestigious fertility clinic in the Netherlands in the 1990s. Their mother's ovaries were hormonally stimulated, and the secondary oocytes that were released were harvested laparoscopically (by abdominal surgery through small holes). These oocytes were then fertilized with sperm in the laboratory. On the sixth post-fertilization day, coinciding with the time a normally fertilized blastocyst would be finishing its journey through the uterine tube into the uterus, two blastocysts were transferred to the mother's uterus. Both had a successful implantation, placentation, embryogenesis, fetal development, and birth.

Why, then, do Hans and Willem look so different? As it turned out, the technician performing the *in vitro* fertilization mistakenly used a previously used pipette to transfer the sperm. The couple that visited the clinic just before Hans and Willem's parents were of African descent, and the African man's sperm was still present in the pipette. As a result, one oocyte was fertilized by sperm from a Caucasian man, and the other was fertilized by sperm from an African man.

1. Why were the fertilized blastocysts implanted in Hans and Willem's mother on her sixth day after ovulation?
2. Are Hans and Willem monozygotic twins or dizygotic twins?

See the blue Answers tab at the back of the book.

Related Clinical Terms

eclampsia: A condition in which one or more convulsions occur in a pregnant woman suffering from high blood pressure, often followed by coma and posing a threat to the health of mother and baby.

gamete intrafallopian transfer (GIFT): An assisted reproductive procedure in which a woman's eggs are removed, mixed with sperm, and replaced into the woman's uterine tube where the fertilization takes place, rather than in the laboratory.

in vitro fertilization: Fertilization outside the body, generally in a Petri dish.

infertility: The inability to achieve pregnancy after engaging in 1 year of appropriately timed intercourse.

neural tube defects (NTDs): Major birth defects caused by an abnormal development of the neural tube—the structure present during the embryonic stage that later becomes the central nervous system. These are very common birth defects that cause infant mortality and disability and include anencephaly and spina bifida.

placenta abruptio: Condition in which there is separation of the placenta from the uterine site of implantation before delivery of the baby.

placenta previa: Condition during pregnancy in which the placenta is abnormally placed so as to totally or partially cover the cervix.

pre-eclampsia: A condition in pregnancy characterized by sudden hypertension, albuminuria, and edema of the hands, feet, and face. It is the most common complication of pregnancy, affecting about 5 percent of pregnancies.

therapeutic cloning: A procedure that usually takes skin cells from a patient, and inserts a skin cell nucleus into a fertilized egg whose nucleus has been removed to create a new cell. That new cell divides repeatedly to form a blastocyst from which stem cells can be extracted to grow new tissue that is genetically matched to the patient.

Answers to Checkpoints, Review Questions, Clinical Case Wrap-Ups, and Figure-Based Questions

ANSWERS

Chapter 1

Answers to Checkpoints

PAGE 3 **1.** A learning outcome is an educational objective that uses key verbs to target specific skills, goals, aims, and achievements. In essence, learning outcomes describe what the reader should be able to do after completing a particular section. **2.** To enhance learning, the text and the art should be read together. That is, after reading the text, the reader should look at and study the accompanying art. Research has shown that reading the text and viewing the art together enhances learning. **PAGE 5** **3.** Anatomy is the study of internal and external body structures. **4.** Physiology is the study of how living organisms perform functions. **5.** Anatomy and physiology are closely related because all specific functions are performed by specific structures. **6.** Gross anatomy (often referred to as macroscopic anatomy) involves studying body structures that can be seen with the unaided eye. Microscopic anatomy is the study of body structures using a microscope to magnify the objects. **7.** Several specialties of physiology are cell physiology, organ physiology, systemic physiology, and pathological physiology. **8.** It is difficult to separate anatomy from physiology because the structures of body parts are so closely related to their functions; put another way, function follows form. **PAGE 6** **9.** The major levels of organization of the human body from the simplest to the most complex are as follows: chemical level → cellular level → tissue level → organ level → organ system level → organism level. **10.** Major organ systems (and some structures of each) are the integumentary system (skin, hair, sweat glands, and nails), skeletal system (bones, cartilages, associated ligaments, and bone marrow), muscular system (skeletal muscles and associated tendons), nervous system (brain, spinal cord, peripheral nerves, and sense organs), endocrine system (pituitary gland, thyroid gland, pancreas, adrenal glands, gonads, and other endocrine tissues), cardiovascular system (heart, blood, and blood vessels), lymphatic system (spleen, thymus, lymphatic vessels, lymph nodes, and tonsils), respiratory system (nasal cavities, sinuses, larynx, trachea, bronchi, lungs, and alveoli), digestive system (teeth, tongue, pharynx, esophagus, stomach, small intestine, large intestine, liver, gallbladder, and pancreas), urinary system (kidneys, ureters, urinary bladder, and urethra), male reproductive system (testes, epididymides, ductus deferentia, seminal glands, prostate gland, penis, and scrotum), and female reproductive system (ovaries, uterine tubes, uterus, vagina, labia, clitoris, and mammary glands). **11.** A histologist investigates structures and properties at the tissue level of organization. **PAGE 7** **12.** Medical terminology is the use of word roots, prefixes, suffixes, and combining forms to construct anatomical, physiological, or medical terms. **13.** An eponym is a commemorative name for a structure or clinical condition that was originally named for a real or mythical person. **14.** The book used as the international standard for anatomical vocabulary is *Terminologia Anatomica*. **PAGE 13** **15.** The purpose of anatomical terms is to provide a standardized frame of reference for describing the human body. **16.** In the anatomical position, an anterior view is seen from the front and a posterior view is from the back. **PAGE 18** **17.** Body cavities protect internal organs and cushion them from thumps and bumps that occur while walking, running, or jumping. Body cavities also permit the organs that they surround to change in size and shape without disrupting the activities of nearby organs. **18.** The thoracic cavity includes the pleural and pericardial cavities, which enclose the lungs and heart, respectively. The diaphragm forms the boundary between the superior thoracic cavity and the inferior abdominopelvic cavity. The abdominopelvic cavity contains the peritoneal cavity, which contains the superior abdominal cavity and the inferior pelvic cavity. **PAGE 19** **19.** Homeostasis refers to the existence of a stable internal environment. **20.** Extrinsic regulation is a type of homeostatic regulation resulting from activities of the nervous system or endocrine system. **21.** Physiological systems can function normally only under carefully controlled conditions. Homeostatic regulation prevents potentially disruptive changes in the body's internal environment. **PAGE 24** **22.** Negative feedback systems provide long-term control over the body's internal conditions—that is, they maintain homeostasis—by counteracting the effects of a stimulus or environmental change. **23.** When homeostasis fails, organ systems function less efficiently or even malfunction. The result is the state that we call disease. If the situation is not corrected, death can result. **24.** A positive feedback system amplifies or enhances the effects of a stimulus. **25.** Positive feedback is useful in processes that must be completed quickly, such as blood clotting. In contrast, it is harmful in situations in which a stable condition must be maintained, because it tends to increase any departure from the desired condition. Positive feedback in the regulation of body temperature, for example, would cause a slight fever to spiral out of control, with fatal results. For this reason, physiological systems are typically regulated by negative feedback, which tends to oppose any departure from the norm. **26.** Equilibrium is a dynamic (constantly changing) state in which two opposing forces or processes are in balance. **27.** When the body continuously adapts by using homeostatic mechanisms, it is said to be in a state of dynamic equilibrium.

Answers to Review Questions

PAGE 25

Level 1 Reviewing Facts and Terms

1. (a) superior **(b)** inferior **(c)** posterior or dorsal **(d)** anterior or ventral **(e)** cranial **(f)** caudal **(g)** lateral **(h)** medial **(i)** proximal **(j)** distal **2.** g **3.** d **4.** a **5.** j **6.** b **7.** l **8.** n **9.** f **10.** h **11.** e **12.** c **13.** o **14.** k **15.** m **16.** i **17.** b **18.** c **19.** d **20.** c **21.** b **22. (a)** pericardial cavity **(b)** peritoneal cavity **(c)** pleural cavity **(d)** abdominal (or abdominopelvic) cavity **23.** b **24.** a

Level 2 Reviewing Concepts

25. (a) Anatomy is the study of internal and external body structures and their physical relationships among other body parts. **(b)** Physiology is the study of how living organisms perform their vital functions. **26.** d **27.** Autoregulation occurs when the activities of a cell, tissue, organ, or organ system change automatically (that is, without neural or endocrine input) when faced with some environmental change. Extrinsic regulation results from the activities of the nervous or endocrine systems. It causes more extensive and potentially more effective adjustments in activities. **28.** The body is erect, and the hands are at the sides with the palms facing forward. **29.** b **30.** c

Level 3 Critical Thinking and Clinical Applications

31. Insulin is released when blood glucose levels are increased. This hormone should bring about a decrease in blood glucose levels, thus decreasing the stimulus for its own release. **32.** A lack of blood flow to the hypothalamus of the brain would cause brain cells there to die. Since this is where the body's thermoregulatory center lies, then control of body temperature would be affected.

Answers to Clinical Case Wrap-Up

PAGE 26 **1.** In addition to the liver and most of the large intestine, the abdominal cavity contains the stomach, spleen, and small intestine. **2.** A deep knife wound superior to the diaphragm would have entered the thoracic cavity, which contains the lungs and heart; associated organs of the respiratory, cardiovascular, and lymphatic systems; the inferior portions of the esophagus; and the thymus.

Answers to Figure-Based Questions

FIGURE 1–3 posterior

FIGURE 1–4 left upper quadrant and the epigastric region

FIGURE 1–5 distal; lateral

FIGURE 1–6 transverse/horizontal; frontal/coronal

FIGURE 1–9 increased blood flow to skin (blood vessels dilate) and sweating increases

Chapter 2

Answers to Checkpoints

PAGE 32 **1.** An atom is the smallest stable unit of matter. **2.** Atoms of the same element that have the same atomic number but different numbers of neutrons are called isotopes. **3.** Hydrogen has three isotopes: hydrogen-1, with a mass number of 1; deuterium, with a mass number of 2; and tritium, with a mass number of 3. The heavier sample must contain a higher proportion of one or both of the heavier isotopes. **PAGE 37** **4.** A chemical bond is an attractive force acting between two atoms that may be strong enough to hold them together in a molecule or compound. The strongest attractive forces result from the gain, loss, or sharing of electrons. Examples of such chemical bonds are ionic bonds and covalent bonds. In contrast, the weaker hydrogen bonds occur between molecules or compounds. **5.** The atoms in a water molecule are held together by polar covalent bonds. Water molecules are attracted to one another by hydrogen bonds. **6.** Atoms combine with each other to gain a complete set of electrons in their outer energy level or valence shell. Oxygen atoms do not have a full outer energy level, so they readily react with many other elements to attain this stable arrangement. Neon already has a full outer energy level so it cannot combine with other elements. **PAGE 39** **7.** Using chemical notation, an ion is represented by a superscript plus or minus sign following the symbol of an element. If more than one electron has been lost or gained, the electric charge on the ion is represented by a number preceding the plus or minus sign. **8.** The molecular formula for glucose, a compound composed of 6 carbon atoms, 12 hydrogen atoms, and 6 oxygen atoms, is $C_6H_{12}O_6$. **9.** Three types of chemical reactions important to the study of human physiology include decomposition reactions, synthesis reactions, and exchange reactions. In a decomposition reaction, a chemical reaction breaks a molecule into smaller fragments. A synthesis reaction assembles smaller molecules into larger ones. In an exchange reaction, parts of the reacting molecules are shuffled around to produce new products. **10.** This reaction involves a large molecule being broken down into two smaller ones, so it is a decomposition reaction. **PAGE 40** **11.** An enzyme is a protein that lowers the activation energy of a chemical reaction, which is the amount of energy required to start the reaction. **12.** Without enzymes, chemical reactions would proceed only under conditions that the cells could not tolerate (e.g., high temperatures). By lowering the activation energy, enzymes make it possible for chemical reactions to rapidly proceed under conditions compatible with life. **13.** Organic compounds always contain carbon and hydrogen. Inorganic compounds do not contain carbon and hydrogen atoms as structural components. **PAGE 43** **14.** Specific chemical properties of water that make life possible include its strong polarity (enables it to be an excellent solvent), its reactivity (it participates in many chemical reactions), its high heat capacity (it absorbs and releases heat slowly), and its ability to serve as a lubricant. **PAGE 44** **15.** The pH is a measure of the concentration of hydrogen ions in fluids. Acid and base concentrations are measured in pH, which is the negative logarithm of hydrogen ion concentration, expressed in moles per liter. On the pH scale, 7 represents a neutral solution; values below 7 indicate acidic solutions, and values above 7 indicate basic (alkaline) solutions. **16.** If the body is to maintain homeostasis and thus health, the pH of different body fluids must remain within a fairly narrow range. **17.** An acid is a compound whose dissociation in solution releases a hydrogen ion and an anion; a base is a compound whose dissociation releases a hydroxide ion (OH^-) or removes a hydrogen ion (H^+) from the solution; a salt is an inorganic compound consisting of a cation other than H^+ and an anion other than OH^-. **18.** Stomach discomfort is commonly the result of excessive stomach acidity ("acid indigestion"). Antacids contain a weak base that neutralizes the excess acid. **PAGE 45** **19.** The macromolecules important to life are carbohydrates, lipids, proteins, and nucleic acids. **20.** The functional group that acts as an acid is the carboxyl group —COOH. **PAGE 47** **21.** The monomer common to both plant starch and glycogen is glucose. **PAGE 51** **22.** Lipids are a diverse group of compounds that include fatty acids, eicosanoids, glycerides, steroids, phospholipids, and glycolipids. They are organic compounds that contain carbon, hydrogen, and oxygen in a ratio that does not approximate 1:2:1. **23.** Human plasma membranes contain mainly phospholipids, plus small amounts of cholesterol and glycolipids. **PAGE 57** **24.** Proteins are organic compounds formed from amino acids that contain a central carbon atom, a hydrogen atom, an amino group ($-NH_2$), a carboxyl group (–COOH), and an R group or variable side chain. Proteins function in support, movement, transport, buffering, metabolic regulation, coordination and control, and defense. **25.** The heat of boiling breaks bonds that maintain the protein's tertiary structure, quaternary structure, or both. The resulting structural change, known as denaturation, affects the ability of the protein molecule to perform its normal biological functions. **PAGE 59** **26.** A nucleic acid is a large organic molecule made of carbon, hydrogen, oxygen, nitrogen, and phosphorus. Nucleic acids regulate protein synthesis and make up the genetic material in cells. **27.** Both DNA (deoxyribonucleic acid) and RNA (ribonucleic acid) contain nitrogenous bases and phosphate groups, but the nucleic acid described is RNA because it contains the sugar ribose. **PAGE 60** **28.** Adenosine triphosphate

(ATP) is a high-energy compound consisting of adenosine to which three phosphate groups are attached. The second and third phosphate groups are each attached by a high-energy bond. **29.** Phosphorylation of an ADP molecule yields a molecule of ATP.

Answers to Review Questions

PAGE 63

Level 1 Reviewing Facts and Terms

1. (a)

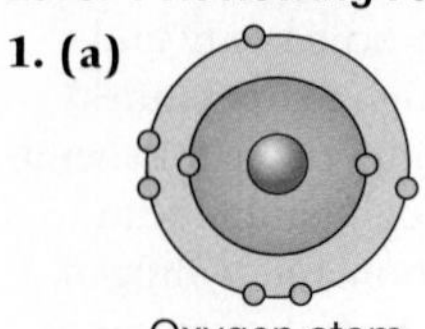
Oxygen atom

(b) Two more electrons can fit into the outermost energy level of an oxygen atom. **2.** hydrolysis **3.** a **4.** b **5.** d **6.** d **7.** b **8.** c **9.** d **10.** d **11.** b **12.** d **13.** a **14.** d **15.** b **16.** protons, neutrons, and electrons **17.** carbohydrates, lipids, proteins, and nucleic acids **18.** Triglycerides (1) provide a significant energy reserve, (2) serve as insulation and thus act in heat conservation, and (3) protect organs by cushioning them. **19.** (1) support (structural proteins); (2) movement (contractile proteins); (3) transport (transport proteins); (4) buffering; (5) metabolic regulation; (6) coordination and control (hormones and neurotransmitters); and (7) defense (antibodies) **20. (a)** DNA: deoxyribose, phosphate, and nitrogenous bases (A, T, C, G); **(b)** RNA: ribose, phosphate, and nitrogenous bases (A, U, C, G) **21.** (1) adenosine, (2) phosphate groups, and (3) appropriate enzymes

Level 2 Reviewing Concepts

22. c **23.** d **24.** Enzymes are specialized protein catalysts that lower the activation energy for chemical reactions. Enzymes speed up chemical reactions but are not used up or changed in the process. **25.** A salt is an ionic compound consisting of any cations other than hydrogen ions and any anions other than hydroxide ions. Acids dissociate and release hydrogen ions, while bases remove hydrogen ions from solution (usually by releasing hydroxide ions). **26.** Nonpolar covalent bonds have an equal sharing of electrons. Polar covalent bonds have an unequal sharing of electrons. Ionic bonds result from the loss or gain of electrons. **27.** e **28.** c **29.** The molecule is a nucleic acid. Carbohydrates and lipids do not contain nitrogen. Although both proteins and nucleic acids contain nitrogen, only nucleic acids normally contain phosphorus.

Level 3 Critical Thinking and Clinical Applications

30. (a) number of electrons = 20; **(b)** atomic number = 20; **(c)** atomic weight = 40; **(d)** 2 electrons in first energy level, 8 in second energy level, 8 in third energy level, and 2 in outer energy level (valence shell). **31.** Decreasing the amount of enzyme at the second step would slow down the remaining steps of the pathway because less substrate would be available for the next two steps. The net result would be a decrease in the amount of product. **32.** If a person exhales large amounts of CO_2, the level of H^+ in the blood will decrease. A decrease in the level of H^+ will cause the pH to increase.

Answers to Clinical Case Wrap-Up

PAGE 64 **1.** Baby Sean's stools contained undigested lipids and carbohydrates, which gave them a fatty or greasy appearance. **2.** Digestive enzymes aid in the decomposition reactions that break down lipids and carbohydrates in food, so they can be absorbed from the digestive tract. If these large molecules cannot be broken down, they cannot be used as energy sources or chemical building blocks.

Answers to Figure-Based Questions

FIGURE 2–3 The second energy level can hold 8 electrons when completely filled.

FIGURE 2–8 Hydrogen bonds act between adjacent molecules and between atoms of the same molecule. Hydrogen bonds form between a slight positive charge on the hydrogen atom of a polar covalent bond and a slight negative charge on an oxygen atom of another polar covalent bond.

FIGURE 2–9 2; 4

FIGURE 2–12 Six oxygen atoms are shown in each glucose structure.

FIGURE 2–15 An unsaturated fatty acid contains at least one double covalent bond between its carbon atoms.

FIGURE 2–17 The double bond between the two carbon atoms makes fatty acid 3 unsaturated.

FIGURE 2–24 Adenine and guanine are purines, which means they are double-ringed molecules. Cytosine, thymine, and uracil are pyrimidines, which means they are single-ringed molecules.

FIGURE 2–26 AMP undergoes two phosphorylations, which adds two phosphate groups (for a total of three phosphates) to form ATP.

Chapter 3

Answers to Checkpoints

PAGE 71 **1.** The general functions of the plasma membrane include physical isolation, regulation of exchange with the environment, sensitivity to the environment, and structural support. **2.** The components of the plasma membrane that allow it to perform its characteristic functions are membrane lipids, membrane proteins, and membrane carbohydrates. **3.** The phospholipid bilayer of the plasma membrane forms a physical barrier between the cell's internal and external environments. **4.** Channel proteins are integral proteins that allow water, ions, and small water-soluble solutes to pass through the plasma membrane. **PAGE 82** **5.** Cytoplasm is the material between the plasma membrane and the nuclear membrane. Cytosol is the fluid portion of the cytoplasm; cytosol is also called intracellular fluid. It is a mixture of water and various dissolved and insoluble materials, in which the organelles and inclusions are suspended. **6.** Cytosol has a higher concentration of potassium ions and suspended proteins, and a lower concentration of sodium ions, than does extracellular fluid. Cytosol also includes small quantities of carbohydrates and large reserves of amino acids and lipids. **7.** The nonmembranous organelles and their functions include (1) centriole = essential for movement of chromosomes during cell division; organization of microtubules in cytoskeleton; (2) cilia = movement of materials over cell surface (motile cilia), environmental sensor (primary cilium); flagella = propel sperm (3) cytoskeleton = strength and support; movement of cellular structures and materials; cell movement; (4) microvilli = increase surface area to facilitate absorption of extracellular materials; (5) proteasomes = breakdown and recycling of intracellular proteins; (6) ribosomes = protein synthesis. **PAGE 83** **8.** The membranous organelles and their functions include (1) endoplasmic reticulum = synthesis of secretory products; intracellular storage and transport; (2) smooth ER = lipid and carbohydrate synthesis; (3) rough ER = modification and packaging of newly synthesized proteins; (4) Golgi apparatus = storage, modification, and packaging of secretory products and lysosomal enzymes; (5) lysosomes = intracellular removal of damaged organelles or pathogens; (6) peroxisomes = neutralization of toxic compounds; (7) mitochondria = production of 95 percent of the ATP required by the cell. **9.** The SER synthesizes lipids, such as steroids. Ovaries and testes produce large amounts of steroid hormones, which are lipids, and thus need large amounts of SER. **10.** Mitochondria produce energy, in the form of ATP molecules, for the cell. A large number of mitochondria in a cell indicates a high demand for energy. **PAGE 85** **11.** The nucleus is a cellular organelle that contains DNA, RNA, enzymes, and proteins. The nuclear envelope is a double membrane that surrounds the nucleus; the perinuclear space is the region between this double membrane. Nuclear pores enable chemical communication between the nucleus and the cytosol. **12.** A gene is a portion of a DNA strand that functions as a hereditary unit. Each gene is located at a particular site on a specific chromosome and codes for a specific protein. **PAGE 90** **13.** Gene activation is the process of uncoiling the segment of DNA containing the nucleotide sequence of that gene, and temporarily removing histones, so that the gene can affect the cell. **14.** Transcription is the synthesis of RNA from a DNA template. Translation uses the information from an mRNA strand to form a linear chain of amino acids, or protein. **15.** A cell that lacked the enzyme RNA polymerase would not be able to transcribe RNA from DNA. **PAGE 94** **16.** A selectively permeable membrane allows the passage of some substances and restricts the passage of others. It lies between two extremes: impermeable, which allows no substances to pass, and freely permeable, which permits the passage of any substance. **17.** Diffusion is the passive movement of a substance from an area of higher concentration to an area of lower concentration. Diffusion proceeds until the concentration gradient is eliminated (equilibrium is reached). **18.** Five factors that influence the diffusion of substances in the body are (1) distance, (2) ion or molecule size, (3) temperature, (4) concentration gradient, and (5) electrical forces. **19.** Diffusion is driven by a concentration gradient. The steeper the concentration gradient, the faster the rate of diffusion, and the less steep the concentration gradient, the slower the rate of diffusion. If the concentration of oxygen in the lungs were to decrease, the concentration gradient between oxygen in the lungs and oxygen in the blood would decrease (as long as the oxygen level of the blood remained constant). Thus, oxygen would diffuse more slowly into the blood. **20.** Osmosis is the diffusion of water across a selectively permeable membrane from one solution to another solution that contains a higher solute concentration. **21.** The 10 percent salt solution is hypertonic with respect to the cells lining the nasal cavity, because it contains a higher salt (solute) concentration than do the cells. The hypertonic solution would draw water out of the cells, causing the cells to shrink and adding water to the mucus, thereby diluting it and relieving congestion. **PAGE 99** **22.** In carrier-mediated transport, integral proteins bind specific ions or organic substrates and carry them across the plasma membrane. All forms of carrier-mediated transport are specific, have saturation limits, and are regulated (usually by hormones). **23.** An active transport process must be involved because it takes an expenditure of energy to move the hydrogen ions against their concentration gradient—that is, from a region where they are less concentrated (the cells lining the stomach) to a region where they are more concentrated (the interior of the stomach). **24.** Endocytosis is the movement of relatively large volumes of extracellular material into the cytoplasm by the formation of a membranous vesicle at the cell surface. Types of endocytosis are receptor-mediated endocytosis, the clathrin-coated vesicle-mediated movement into the cytoplasm of specific target molecules (ligands); pinocytosis, the vesicle-mediated movement into the cytoplasm of extracellular fluid and its contents; and phagocytosis, the vesicle-mediated movement into the cytoplasm of extracellular solids, especially bacteria and debris. **25.** Exocytosis is the ejection of cytoplasmic materials by the fusion of a membranous vesicle with the plasma membrane. **26.** The process by which certain white blood cells engulf bacteria is called phagocytosis. **PAGE 102** **27.** The membrane potential of a cell is the difference in electrical potential that results from the uneven distribution of positive and negative charges across the plasma membrane. It is expressed in millivolts (mV). **28.** If the plasma membrane of a cell were freely permeable to sodium ions, more of these positively charged ions would move into the cell, and the membrane potential would become less negative (move closer to zero). **PAGE 103** **29.** The biological term for cellular reproduction is cell division, and the term for cell death is apoptosis. **30.** Important enzymes for DNA replication include helicases, DNA polymerase, and ligases. **31.** Interphase is a stage in a cell's life when it performs all its normal functions and, if necessary, prepares for cell division. The stages of interphase include G_1, S, G_2, and G_0. A cell in G_0 is not preparing for cell division. **32.** This cell is likely in the G_1 phase of its life cycle. **33.** Mitosis is the essential step in cell division in which a single cell nucleus divides to produce two identical daughter cell nuclei. The four stages of mitosis are prophase, metaphase, anaphase, and telophase. **34.** If spindle fibers failed to form during mitosis, the cell would not be able to separate the chromosomes into two sets. If cytokinesis occurred, the result would be one cell with two sets of chromosomes and one cell with none. **PAGE 107** **35.** A growth factor is a natural substance, such as a peptide or hormone, that can stimulate the division of specific cell types. Representative growth factors include M-phase promoting factor (maturation-promoting factor), growth hormone, prolactin, fibroblast growth factor (FGF), nerve growth factor (NGF), epidermal growth factor (EGF), erythropoietin, thymosins and related compounds, and chalones. **PAGE 109** **36.** An illness characterized by mutations that disrupt normal controls and produce potentially malignant cells is termed cancer. **37.** Metastasis is the spread of cancer cells to distant tissues and organs, leading to the establishment of secondary tumors. **38.** Cellular differentiation is the development of specific cellular characteristics and functions that are different from those of the original cell. It results from genes in the cell being switched off, restricting the cell's potential functions.

Answers to Review Questions

PAGE 112

Level 1 Reviewing Facts and Terms

1. a. isotonic b. hypotonic c. hypertonic **2.** b **3.** c **4.** d **5.** a **6.** c **7.** a **8.** c **9.** a **10.** b **11.** (1) Cells are the building blocks of all plants and animals. (2) Cells are produced by the division of preexisting cells. (3) Cells are the smallest units that perform all vital physiological functions. (4) Each cell maintains homeostasis at the cellular level. **12.** Four general functions of the plasma membrane are (1) physical isolation, (2) regulation of exchange with the environment, (3) sensitivity, and (4) structural support. **13.** Membrane proteins function as receptors, channels, carriers, enzymes, anchors, and identifiers. **14.** The major transport mechanisms are (1) diffusion, (2) carrier-mediated transport, and (3) vesicular transport. **15.** Factors that affect diffusion rate are (1) distance, (2) steepness of the concentration gradient, (3) molecule size, (4) temperature, and

ANSWERS

(5) electrical forces. **16.** Major functions of the ER are (1) synthesis of proteins, carbohydrates, and lipids; (2) storage of absorbed or synthesized molecules; (3) transport of materials; and (4) detoxification of drugs or toxins.

Level 2 Reviewing Concepts

17. b **18.** d **19.** b **20.** c **21.** c **22.** d **23.** b **24.** G_0: normal cell functions, nondividing state; G_1: cell growth, duplication of organelles, and protein synthesis; S: DNA replication and synthesis of histones; G_2: protein synthesis **25.** Prophase: chromatin condenses and chromosomes become visible; centrioles migrate to opposite poles of the cell and spindle fibers develop; and the nuclear membrane disintegrates. Metaphase: chromatids attach to spindle fibers and line up along the metaphase plate. Anaphase: chromatids separate and migrate toward opposite poles of the cell. Telophase: the nuclear membrane forms; chromosomes disappear as they uncoil back to chromatin; and nucleoli appear. **26. (a)** Cytokinesis is the cytoplasmic movement that separates two daughter cells. **(b)** Cytokinesis completes the process of cell division.

Level 3 Critical Thinking and Clinical Applications

27. This process is facilitated diffusion, which requires a carrier molecule but not cellular energy. The energy for this process is provided by the concentration gradient of the substance being transported. When all the carriers are actively involved in transport, the rate of transport reaches a saturation point. **28.** Solution A must have initially had more solutes than solution B. As a result, water moved by osmosis across the selectively permeable membrane from side B to side A, increasing the fluid level on side A. **29.** c **30.** The isolation of the internal contents of membrane-bound organelles allows them to manufacture or store secretions, enzymes, or toxins that could adversely affect the cytoplasm in general. Another benefit is the increased efficiency of having specialized enzyme systems concentrated in one place. For example, the concentration of enzymes necessary for energy production in the mitochondrion increases the efficiency of cellular respiration.

Answers to Clinical Case Wrap-Up

PAGE 113 **1.** The other cell that might also be affected is the sperm cell, the only human cell with a tail-like flagellum, which has the same microtubule structure as motile cilia. **2.** Because Jackson's sperm cells will be affected, he is likely to experience difficulties in fathering children. Also, normal ciliary action is necessary to move sperm from the testes into the male reproductive tract.

Answers to Figure-Based Questions

FIGURE 3–3 The three components that make up the cytoskeleton in all body cells are microfilaments, intermediate filaments, and microtubules.

FIGURE 3–9 The two reactants needed are oxygen and short carbon chains (such as the pyruvate molecule shown). The products from this reaction are carbon dioxide (CO_2), water (H_2O), and ATP.

FIGURE 3–11 In a dividing cell, DNA is tightly coiled and organized as chromosomes. In a nondividing cell, DNA is loosely coiled and organized as chromatin.

FIGURE 3–15 Small water-soluble molecules and ions diffuse through membrane channels. Lipid-soluble molecules diffuse directly through the plasma membrane.

FIGURE 3–17 In part b, the intracellular fluid is hypertonic to the extracellular fluid. In part c, the intracellular fluid is hypotonic to the extracellular fluid.

FIGURE 3–18 The concentration gradient determines the direction in which glucose molecules will be transported. Facilitated diffusion is a passive process, so glucose molecules will travel from high to low concentration.

FIGURE 3–19 Breaking the high-energy bond of ATP provides the energy for the sodium–potassium exchange pump.

Chapter 4

Answers to Checkpoints

PAGE 115 **1.** Histology is the study of tissues. **2.** The four major types of tissues that form all body structures are epithelial, connective, muscle, and nervous tissue. **PAGE 120** **3.** Epithelial tissue provides physical protection, controls permeability, provides sensation, and produces specialized secretions. **4.** Epithelial tissue is characterized by polarity, cellularity, attachment, avascularity, and regeneration. **5.** An epithelium whose cells have microvilli is probably involved in absorption or secretion. The microvilli greatly increase the cellular surface area available for these processes. **6.** Epithelial cell junctions include tight junctions, gap junctions, spot desmosomes, and hemidesmosomes. **7.** Gap junctions allow small molecules and ions to pass from cell to cell. When connecting epithelial cells, they help coordinate such functions as the beating of cilia. In cardiac and smooth muscle tissues, they are essential for coordinating muscle cell contractions. **PAGE 126** **8.** The three cell shapes that are characteristic of epithelial cells are squamous (thin and flat), cuboidal (box shaped), and columnar (tall and slender). **9.** When classifying epithelial tissue, one with a single layer of cells is called simple, whereas one with multiple layers of cells is called stratified. **10.** No, this is not a sample from the skin surface. A simple squamous epithelium does not provide enough protection against infection, abrasion, or dehydration. The skin surface has a stratified squamous epithelium. **11.** All these regions are subject to mechanical trauma and abrasion—by food (pharynx and esophagus), by feces (anus), and by intercourse or childbirth (vagina). **12.** The two primary types of glandular epithelia are endocrine glands and exocrine glands. **13.** Sebaceous glands exhibit holocrine secretion. **14.** An endocrine gland releases its secretions directly into the interstitial fluid. **PAGE 128** **15.** Functions of connective tissues include (1) establishing a structural framework for the body; (2) transporting fluids and dissolved materials; (3) protecting delicate organs; (4) supporting, surrounding, and interconnecting other types of tissue; (5) storing energy reserves; and (6) defending the body from invading microorganisms. **16.** The three categories of connective tissues are connective tissue proper, fluid connective tissues, and supporting connective tissues. **PAGE 135** **17.** Cells found in connective tissue proper are adipocytes, fibroblasts, fibrocytes, lymphocytes, macrophages, mast cells, melanocytes, mesenchymal cells, and microphages. **18.** The reduced collagen production resulting from a lack of vitamin C in the diet would cause connective tissue to be weak and prone to damage. **19.** Mast cells and basophils produce the histamine, which antihistamines block. **20.** The type of connective tissue that contains triglycerides is adipose (fat) tissue. **21.** The three layers of fascia are superficial fascia (areolar and adipose tissue), deep fascia (dense regular connective tissue), and subserous fascia (areolar tissue). **PAGE 136** **22.** Blood and lymph each have a fluid matrix and are fluid connective tissues. **23.** The recirculation of fluid in the body is a continuous process from the cardiovascular system, through the interstitial fluid, to the lymph, and then back to the cardiovascular system. **PAGE 140** **24.** The two types of supporting connective tissue are cartilage and bone. **25.** Bone heals faster than cartilage because, unlike cartilage, bone has a direct blood supply, which is necessary for proper and rapid healing to occur. **26.** The type of cartilage damaged in a herniated intervertebral disc is fibrocartilage. **PAGE 142** **27.** The four types of tissue membranes found in the body are the cutaneous membrane, mucous membranes, serous membranes, and synovial membranes. **28.** The pleural, peritoneal, and pericardial cavities are all lined by serous membranes. **29.** Mucous membranes typically line body passageways that open to the exterior. **PAGE 144** **30.** The three types of muscle tissue in the body are skeletal muscle, cardiac muscle, and smooth muscle. **31.** Muscle tissue that lacks striations is smooth muscle. Both cardiac and skeletal muscles are striated. **32.** Skeletal muscle is repaired by the division of myosatellite cells, which are stem cells that persist in adult skeletal muscle tissue. **PAGE 145** **33.** The cells are most likely neurons. **PAGE 146** **34.** The two phases in the response to tissue injury are inflammation and regeneration. **35.** The four common signs of inflammation are redness, heat (warmth), swelling, and pain. **36.** With advancing age, the speed and effectiveness of tissue repair decrease, the rate of energy consumption in general decreases, hormonal activity is altered, and other factors contribute to changes in structure and chemical composition. **37.** Longer life is associated with a longer period of potential exposure to chemicals and other environmental factors.

Answers to Review Questions

PAGE 150

Level 1 Reviewing Facts and Terms

1. (a) simple squamous epithelium, **(b)** simple cuboidal epithelium, **(c)** simple columnar epithelium, **(d)** stratified squamous epithelium, **(e)** stratified cuboidal epithelium, **(f)** stratified columnar epithelium. **2.** b **3.** b **4.** d **5.** c **6.** c **7.** b **8.** d **9.** b **10.** d **11.** c **12.** a **13.** b **14.** e **15.** a **16.** b **17.** Epithelial tissue (1) provides physical protection, (2) controls permeability, (3) provides sensations, and (4) produces specialized secretions. **18.** Endocrine glands secrete hormones onto the surface of the gland or directly into the surrounding fluid; exocrine glands secrete through ducts. **19.** Exocrine glandular epithelial cells release their secretions by (1) merocrine secretion, (2) apocrine secretion, or (3) holocrine secretion. **20.** Connective tissues contain (1) specialized cells, (2) extracellular protein fibers, and (3) a fluid ground substance. **21.** The four membranes in the body are (1) mucous membranes, (2) serous membranes, (3) the cutaneous membrane (skin), and (4) synovial membranes. **22.** Nervous tissue contains (1) neurons, which propagate electrical impulses in the form of reversible changes in the membrane potential, and (2) neuroglia, which include several kinds of supporting cells and play a role in providing nutrients to neurons.

Level 2 Reviewing Concepts

23. Exocrine secretions are secreted onto a surface or outward through a duct. Endocrine secretions are secreted by ductless glands into surrounding tissues. Endocrine secretions are called hormones, which usually diffuse into the bloodstream for distribution to other parts of the body. **24.** Tight junctions block the passage of water or solutes between cells. In the digestive system, these junctions keep enzymes, acids, and wastes from damaging delicate underlying tissues. **25.** Fluid connective tissues have a watery matrix. They differ from supporting connective tissues in that they have many soluble proteins in the matrix, and they do not include insoluble fibers. **26.** The extensive connections between cells formed by cell junctions, proteoglycans, and physical interlocking hold skin cells together and can deny access to chemicals or pathogens that cover their free surfaces. If the skin is damaged and the connections are broken, infection can occur. **27.** b **28.** b **29.** Similarities: actin and myosin interactions produce contractions. Differences: skeletal muscle tissue cells are relatively large, multinucleate, striated, and contract only under neural stimulation; cardiac muscle tissue cells have one to five nuclei, are interconnected in a branching network, and contract in response to pacemaker cell activity; smooth muscle tissue cells are small, spindle shaped, nonstriated, with only one nucleus, and are not under voluntary control.

Level 3 Critical Thinking and Clinical Applications

30. Because apocrine secretions are released by pinching off a portion of the secreting cell, you could test for the presence of plasma membranes, specifically for the phospholipids in plasma membranes. Merocrine secretions do not contain a portion of the secreting cell, so they would lack plasma membrane components. **31.** Skeletal muscle tissue would be made up of densely packed fibers running in the same direction, but since muscle fibers are composed of cells, they would have many nuclei and mitochondria. Skeletal muscle tissue also has an obvious banding pattern, or striations, due to the arrangement of the actin and myosin filaments within the cell. The student is probably looking at a slide of tendon (dense connective tissue). The small nuclei would be those of fibroblasts. **32.** You would expect the skin in the area of the injury to become red and warm. It would also swell, and Jim would experience a painful sensation. These changes occur as a result of inflammation, the body's first response to injury. Injury to the epithelium and underlying connective tissue would trigger the release of chemicals such as histamine and heparin from mast cells in the area. These chemicals in turn initiate the changes that we observe.

Answers to Clinical Case Wrap-Up

PAGE 151 **1.** The collagen fibers within the matrix of connective tissue are defective in Ehlers-Danlos syndrome. **2.** Connective tissues throughout Anne Marie's body are affected by Ehlers-Danlos syndrome. This includes structures and organs composed of dense regular connective tissue (tendons, ligaments, aponeuroses), dense irregular connective tissue (skin, organ capsules of liver, kidneys, and spleen), and reticular tissue (spleen, liver, lymph nodes) as well.

Answers to Figure-Based Questions

FIGURE 4–3 Only gap junctions allow the diffusion of ions and molecules between cells. Tight junctions, hemidesmosomes, and spot desmosomes do not.

FIGURE 4–6 Columnar describes cell shape. Simple describes the number of cell layers; here, simple means one layer.

FIGURE 4–8 Holocrine secretion destroys superficial gland cells. Merocrine secretion uses exocytosis to release product from vesicles, and apocrine secretion sheds the apical cytoplasm to produce secretions.

FIGURE 4–11 Adipose tissue provides padding, cushions shock, reduces heat loss, and stores energy. It is most commonly found deep to the skin of the sides, buttocks, and breasts.

FIGURE 4–14 White blood cells contain a nucleus. Red bloods cells and platelets do not.

FIGURE 4–16 Chondrocytes (cartilage cells) are the only type of cell found in cartilage matrix. Chondrocytes are found within small chambers in the matrix called lacunae.

FIGURE 4–18 The epithelium that lines the peritoneal cavity is called the mesothelium.

FIGURE 4–19 Skeletal muscle tissue consists of multinucleate cells.

Chapter 5

Answers to Checkpoints

PAGE 158 **1.** The layers of the epidermis are the stratum basale, stratum spinosum, stratum granulosum, stratum lucidum, and stratum corneum. **2.** Dandruff consists of cells from the stratum corneum. **3.** This splinter is lodged in the stratum granulosum. **4.** Wrinkles in the tips of the fingers (and toes) are due to the constriction of blood vessels in the underlying skin, which causes the skin to shrink. This effect is controlled by the autonomic nervous system, a division of the nervous system that acts outside our awareness. **5.** Sanding the tips of the fingers will not permanently remove fingerprints. The ridges of the fingerprints are formed in layers of the skin that are constantly regenerated, so these ridges will eventually reappear. The pattern of the ridges is determined by the arrangement of tissue in the dermis, which is not affected by sanding. **6.** Salivary glands and duodenal glands (glands of the duodenum) produce epidermal growth factor (EGF). **7.** Epidermal growth factor (EGF) promotes the divisions of basal cells in the stratum basale and stratum spinosum, accelerates the production of keratin in differentiating keratinocytes, stimulates epidermal development and epidermal repair after injury, and stimulates synthetic activity and secretion by epithelial glands. **PAGE 160** **8.** The dermis (a connective tissue layer) is located between the epidermis and the subcutaneous layer (hypodermis). **9.** The capillaries and sensory nerve fibers that supply the epidermis are located in the papillary layer of the dermis. **10.** The presence of elastic fibers and the flexibility and resilience of skin turgor allow the dermis to undergo repeated cycles of stretching and recoiling (returning to its original shape). **PAGE 161** **11.** The tissue that connects the dermis to underlying tissues is the subcutaneous layer or hypodermis. **12.** The subcutaneous layer is a layer of loose connective tissue (areolar tissue) and adipose tissue below (deep to) the dermis. The subcutaneous layer is not considered a part of the integument, but it is important in stabilizing the position of the skin in relation to underlying tissues. **13.** Subcutaneous fat provides insulation to help reduce heat loss, serves as an energy reserve, and acts as a shock absorber for the body. **PAGE 163** **14.** The two major pigments in the epidermis are melanin, which ranges in color from red-yellow (pheomelanin) to brown-black (eumelanin), and carotene, an orange-yellow pigment. **15.** When exposed to the ultraviolet (UV) radiation in sunlight, melanocytes in the epidermis and dermis synthesize the pigment melanin, darkening the skin. **16.** When skin gets warm, arriving oxygenated blood is diverted to the superficial dermis for the purpose of eliminating heat. The oxygenated blood gives a reddish coloration to the skin. **PAGE 165** **17.** In the presence of ultraviolet radiation in sunlight, epidermal cells in the stratum spinosum and stratum basale convert a cholesterol-related steroid compound into cholecalciferol, or vitamin D_3. **18.** Cholecalciferol (vitamin D_3) is needed to form strong bones and teeth. When the body surface is covered, UV light cannot penetrate to the stratum basale in the skin to begin vitamin D_3 production, resulting in fragile bones. **PAGE 168** **19.** A typical hair is a keratinous strand produced by epithelial cells of the hair follicle. **20.** The contraction of the arrector pili muscle pulls the hair follicle erect. The result is known as "goose bumps." **21.** Even though hair is a derivative of the epidermis, the hair follicles are in the dermis. Where the epidermis and deep dermis are destroyed, new hair will not grow. **PAGE 171** **22.** Two types of exocrine glands found in the skin are sebaceous (oil) glands and sweat glands. **23.** The functions of sebaceous secretions (called sebum) are to lubricate and protect the keratin of the hair shaft, lubricate and condition the surrounding skin, and inhibit the growth of bacteria. **24.** Deodorants are used to mask the odor of apocrine sweat gland secretions, which contain several kinds of organic compounds. Some of these compounds have an odor, and others produce an odor when metabolized by skin bacteria. **25.** Apocrine sweat glands enlarge and increase secretory activity in response to the increase in sex hormones that occurs at puberty. **26.** Keratin is the substance that makes fingernails hard. **27.** The area of thickened stratum corneum under the free edge of a nail is called the hyponychium. **28.** Nail growth occurs at the nail root, an epidermal fold that is not visible from the surface. **PAGE 173** **29.** The combination of fibrin clots, fibroblasts, and the extensive network of capillaries in tissue that is healing is called granulation tissue. **30.** Skin can regenerate effectively even after considerable damage because stem cells persist in both the epithelial and connective tissue components of skin. When injury occurs, cells of the stratum basale replace epithelial cells, and mesenchymal cells replace cells lost from the dermis. **PAGE 175** **31.** As a person ages, the blood supply to the dermis decreases and eccrine sweat glands become less active. These changes make it more difficult for older individuals to cool themselves in hot weather. **32.** With advancing age, melanocyte activity decreases, leading to gray or white hair.

Answers to Review Questions

PAGE 178

Level 1 Reviewing Facts and Terms

1. **(a)** epidermis **(b)** dermis **(c)** papillary layer **(d)** reticular layer **(e)** subcutaneous layer (hypodermis) **2.** a **3.** c **4.** d **5.** d **6.** c **7.** a **8.** d **9.** b **10.** b **11.** d **12.** d **13.** b **14.** a **15.** Epidermal cell division occurs in the stratum basale. **16.** These smooth muscles cause hairs to stand erect when stimulated. **17.** Epidermal growth factor promotes the divisions of basal cells in the stratum basale and stratum spinosum, accelerates the production of keratin in differentiating epidermal cells, stimulates both epidermal development and epidermal repair after injury, and stimulates secretory product synthesis and secretion by epithelial glands. **18.** The dermis is made up of (1) the papillary layer, which consists of areolar connective tissue and contains capillaries and sensory nerve fibers and receptors, and (2) the reticular layer, which consists of dense irregular connective tissue and bundles of collagen fibers. Both layers contain networks of blood vessels, lymphatic vessels, and nerve fibers. **19.** Regeneration of injured skin involves (1) bleeding and inflammation (inflammation phase), (2) scab and granulation tissue formation (migration phase), (3) loss of granulation and undermining of the scab (proliferation phase), and (4) scarring (scarring phase).

Level 2 Reviewing Concepts

20. Insensible perspiration is water loss by evaporation through the stratum corneum. Sensible perspiration is produced by active sweat glands. **21.** The integument is the first barrier to substances entering the body. This method of delivering drugs requires the substance to move across the epidermal plasma membranes. The lipid bilayer of the plasma membranes creates a barrier that prevents water from rapidly entering or leaving the cells. Fat-soluble (lipophilic) substances, such as oil containing a dissolved drug, can easily pass through this barrier to underlying connective tissue and enter the circulation. Water-soluble drugs are lipophobic and thus do not readily penetrate the integument. **22.** A tan develops when the skin is exposed to the sun. The tan is a result of the synthesis of melanin in the skin. Melanin, in turn, helps prevent excessive skin damage by absorbing UV radiation before it reaches the deep layers of the epidermis and dermis. Within the epidermal cells, melanin concentrates around the nucleus, so it absorbs the UV light before it can damage nuclear DNA. Some sun exposure, however, is beneficial because sunlight causes epidermal cells to convert a cholesterol-related steroid compound into cholecalciferol (vitamin D_3), which is important to bone maintenance and growth. **23.** Incisions along tension lines—which represent the orientation of dermal collagen and elastin fibers—are more likely to remain closed, and thus heal more quickly, than incisions that cut across tension lines. **24.** c **25.** c **26.** d

Level 3 Critical Thinking and Clinical Applications

27. a **28.** The puncture wound by the nail has a greater chance of becoming infected. Whereas the knife cut will bleed freely, washing many of the bacteria from the wound site, the nail can carry bacteria beneath the surface of the skin, where oxygen is limited (anaerobic), and past the skin's protective barriers. **29. (a)** Ultraviolet radiation in sunlight converts a cholesterol-related steroid compound into vitamin D_3, or cholecalciferol. This compound is then converted to the hormone calcitriol, which is essential for absorbing the calcium and phosphate ions by the small intestine that is necessary for normal bone maintenance and growth. **(b)** The child can drink more milk. Milk is routinely fortified with cholecalciferol, normally identified as "vitamin D," which is easily absorbed by the intestines. **30.** Sweating from eccrine sweat glands is precisely regulated, and one influencing factor is emotional state. Presumably, a person who is lying is nervous and sweats profusely; the lie detector machine detects a change in electrical skin conductance caused by this sweating. **31.** The chemicals in hair dyes break the protective covering of the cortex, allowing the dyes to stain the medulla of the shaft. Dying is not permanent because the cortex remains damaged, allowing shampoo and UV rays from the sun to enter the medulla and affect the color. Also, the living portion of the hair remains unaffected, so that when the shaft is replaced the color will be lost.

Answers to Clinical Case Wrap-Up

PAGE 179 **1.** Ichthyosis affects keratinization, also called cornification. This disorder causes abnormal function of keratinocytes, the most numerous cells of the epidermis. Dysfunction begins in the stratum granulosum and continues through the stratum corneum. Keratinocytes in the granulosum layer make defective keratohyalin, which does a poor job of linking keratin fibers into the waterproof web desired for protection of the skin. As dying keratinocytes are pushed up, they bring this fragmented web to the surface, where it dries, peels, and scales. **2.** Sweat glands are connected to the epidermal surface through ducts. In ichthyosis, the thick scales block the sweat pores. Grandpa can't easily secrete sweat onto his body surface to cool his body temperature through evaporation. The thermoregulatory function of his integument is impaired, so he must guard against overheating.

Answers to Figure-Based Questions

FIGURE 5–1 Sebaceous glands and sweat glands are found in the skin.

FIGURE 5–6 The subpapillary plexus supplies blood to the capillary loops that follow the epidermis–dermis boundary.

FIGURE 5–8 Melanocytes are found in the stratum basale.

FIGURE 5–10 Sunlight and food are the body's sources of vitamin D_3.

FIGURE 5–13 A sebaceous gland discharges its secretion into a hair follicle. A sebaceous follicle discharges its secretion directly onto the epidermis.

Chapter 6

Answers to Checkpoints

PAGE 182 **1.** Major functions of the skeletal system are support, storage of minerals and lipids, blood cell production, protection, and leverage. **PAGE 183** **2.** The different broad categories for classifying bones according to shape are sutural bones, irregular bones, short bones, flat bones, long bones, and sesamoid bones. **3.** Bone markings (surface features) are characteristics of a bone's surface that have specific functions, such as joint formation, muscle attachment, or the passage of nerves and blood vessels. **PAGE 187** **4.** Mature bone cells are known as osteocytes, bone-building cells are called osteoblasts, and osteoclasts are bone-resorbing cells. **5.** If the ratio of collagen to hydroxyapatite in a bone increased, the bone would become less strong (as well as more flexible). **6.** Because osteoclasts break down or demineralize bone, the bone would have a decreased mineral content (less mass); as a result, it would also be weaker. **PAGE 190** **7.** Compact bone consists of osteons (Haversian systems) with little space between them. Compact bone lies over spongy bone and makes up most of the diaphysis. It functions to protect, support, and resist stress. Spongy bone consists of trabeculae with numerous red marrow–filled spaces. Spongy bone makes up most of the structure of short, flat, and irregular bones and is also found at the epiphyses of long bones. Spongy bone functions in storing marrow and providing some support. **8.** The presence of lamellae that are not arranged in osteons is indicative of spongy bone, which is located in an epiphysis. **PAGE 194** **9.** In endochondral ossification, cells of the inner layer of the perichondrium differentiate into osteoblasts, and a cartilage model is gradually replaced by bone. **10.** During intramembranous ossification, fibrous connective tissue is replaced by bone. **11.** Long bones of the body, such as the femur, have an epiphyseal cartilage, a plate of cartilage that separates the epiphysis from the diaphysis so long as the bone is still growing lengthwise. An x-ray would indicate whether the epiphyseal cartilage is still present. If it is, growth is still occurring. If it is not, the bone has reached its adult length and, thus, the person has reached full height. **PAGE 196** **12.** Bone remodeling refers to the process whereby old bone is continuously being destroyed by osteoclasts while new bone is being constructed by osteoblasts. **13.** The biochemistry of some heavy-metal ions, such as strontium, cobalt, uranium, and plutonium, is very similar to that of calcium. Osteoblasts cannot differentiate these heavy-metal ions from normal calcium ions, so the heavy-metal ions become incorporated into the bone matrix. Over time, these dangerous ions can be released into the bloodstream during normal bone remodeling. **PAGE 197** **14.** The larger arm muscles of the weight lifter would apply more mechanical stress to the bones of the upper limbs. In response to that stress, the bones would grow thicker. **15.** Growth continues throughout childhood. At puberty, a growth spurt takes place and is followed by the closure of the epiphyseal cartilages. The later puberty begins, the taller the child will be when the growth spurt begins, so the taller the person will be when growth is completed. **16.** Increased levels of growth hormone prior to puberty will result in excessive bone growth, making the child taller. **PAGE 200** **17.** Parathyroid hormone (PTH) indirectly stimulates osteoclasts to release calcium ions from bone and enhances calcitriol's effect on the intestinal absorption of calcium. Increased PTH secretion would result in an increase in the level of calcium ions in the blood. **18.** The bones of children who have rickets are poorly mineralized and as a result are quite flexible. Under the weight of the body, the leg bones bend. The instability makes walking difficult and can lead to other problems of the legs and feet. **PAGE 201** **19.** Immediately following a fracture, extensive bleeding occurs at the injury site. After several hours, a large blood clot called a fracture hematoma develops. Next, an internal callus forms as a network of spongy bone unites the inner edges, and an external callus of cartilage and bone stabilizes the outer edges. The cartilaginous external callus is eventually replaced by bone, and the struts of new spongy bone now unite the broken ends. With time, the swelling that initially marks the location of the fracture is remodeled, new compact bone is formed, and there is little evidence that a break occurred. **20.** An external callus forms early in the healing process, when cells from the endosteum and periosteum migrate to the area of the fracture. These cells form an enlarged collar (external callus) that encircles the bone in the area of the fracture. **PAGE 204** **21.** Osteopenia is inadequate ossification and is common during the aging process. It results from decreasing osteoblast activity accompanied by normal osteoclast activity. **22.** In women, the sex hormone known as estrogen plays an important role in moving calcium into bones. After menopause, the level of estrogen decreases dramatically. As a result, older women have difficulty replacing the calcium in bones that is being lost due to normal aging. In men, the level of testosterone does not decrease until much later in life.

Answers to Review Questions

PAGE 206

Level 1 Reviewing Facts and Terms

1. b **2.** a **3.** a **4.** c **5.** b **6.** b **7.** **(a)** long bone **(b)** flat bone **(c)** sutural bone **(d)** irregular bone **(e)** short bones **(f)** sesamoid bone. **8.** a **9.** a **10.** c **11.** (1) support, (2) storage of minerals and lipids, (3) blood cell production, (4) protection, and (5) leverage **12.** (1) osteogenic cells, (2) osteoblasts, (3) osteocytes, and (4) osteoclasts **13.** (1) diaphysis (shaft), (2) epiphysis, (3) epiphyseal cartilage/line, (4) articular cartilage, (5) medullary canal, (6) periosteum, (7) endosteum **14.** In endochondral ossification, bone replaces a cartilage model. In intramembranous ossification, bone replaces mesenchyme or fibrous connective tissue. **15.** organic = collagen; inorganic = hydroxyapatite crystals **16.** **(a)** calcium salts, phosphate salts, and vitamins A, C, and D_3; **(b)** calcitriol, growth hormone, thyroxine, estrogen (in females) or testosterone (in males), parathyroid hormone, and calcitonin. **17.** The three organs or tissues are the bones, the intestines of the digestive tract, and the kidneys. **18.** Parathyroid hormone stimulates osteoclast activity, increases the rate of intestinal absorption of calcium ions, and decreases the rate of excretion of calcium ions by the kidneys. These effects increase blood calcium levels.

Level 2 Reviewing Concepts

19. Nutrients reach the osteocytes in spongy bone by diffusing along canaliculi that open onto the surface of the trabeculae. **20.** The osteons are aligned parallel to the long axis of the shaft, which does not bend when forces are applied to either end. Stresses or impacts to the side of the shaft can lead to a fracture. **21.** The lack of physical stress during inactivity leads to the removal of calcium salts from bones. Up to one-third of the bone mass can be lost in this manner, causing the bones to become thin and brittle. **22.** The digestive and urinary (kidneys) systems play important roles in providing the calcium and phosphate minerals needed for bone growth. In return, the skeleton provides protection and acts as a reserve of calcium, phosphate, and other minerals that can compensate for changes in the dietary supplies of these ions. **23.** There are many long bones in the hand, each of which has an epiphyseal cartilage (plate). Measuring the width of these plates will provide clues to the hormonal control of growth in the child. **24.** Once a bone fracture has been repaired, the bone at the fracture site tends to be stronger and thicker than normal. **25.** b **26.** Bone markings give clues as to the size, age, sex, and general appearance of an individual.

Level 3 Critical Thinking and Clinical Applications

27. The fracture might have damaged the epiphyseal cartilage in Sally's right leg. Even though the bone healed properly, the damaged leg did not produce as much cartilage as did the undamaged leg. The result would be a shorter bone on the side of the injury. **28.** d **29.** a **30.** The matrix of bone will absorb traces of minerals from the diet. These minerals can be identified hundreds of years later. A diet rich in calcium and vitamin B will produce denser bones than will a diet deficient in them. Cultural practices such as the binding of appendages or the wrapping of infant heads will manifest in misshapen bones. Heavy muscular activity will result in larger bone markings, indicating an athletic or physically demanding lifestyle.

Answers to Clinical Case Wrap-Up

PAGE 207 **1.** The compact bone in Alex's tibia is composed of osteons that have plenty of calcified matrix (for hardness) but few "supporting beams" of collagen. The spongy bone of his epiphyses is likewise missing collagen reinforcement in the trabeculae. Therefore, the bony architecture of Alex's tibia is intrinsically weak. The bone won't be able to bend with forces placed on it during daily activities, especially in sports, and will be prone to snap. **2.** Alex should not continue playing soccer, and he should avoid contact sports. As this incident shows, Alex is at much higher risk for fractures than any other soccer player on the field. He should be encouraged to pursue swimming, music, writing, art, or other stimulating, yet less physical, activities.

Answers to Figure-Based Questions

FIGURE 6–3 In long bones, spongy bone is found in the epiphyses. In flat bones, spongy bone is found between two layers of compact bone.

FIGURE 6–5 An osteogenic cell divides to produce osteoblasts. Osteoblasts become osteocytes when they are completely surrounded by bone matrix.

FIGURE 6–7 Lamellae are organized concentrically around an osteon, interstitially between osteons, and circumferentially around the entire bone.

FIGURE 6–10 The periosteum covers the superficial layer of compact bone and is made up of the fibrous and cellular layers.

FIGURE 6–16a The parathyroid gland secretes parathyroid hormone when the blood calcium level is too low.

Chapter 7

Answers to Checkpoints

PAGE 209 **1.** The axial skeleton includes the skull (8 cranial bones and 14 facial bones), the bones associated with the skull (6 auditory ossicles and the hyoid bone), the thoracic cage (the sternum and 24 ribs), and the vertebral column (24 vertebrae, the sacrum, and the coccyx). **2.** The axial skeleton serves as a framework that supports and protects organs, provides an extensive surface area for muscle attachment, and stabilizes and positions parts of the appendicular skeleton that support the limbs. **PAGE 225** **3.** The foramen magnum is located in the base of the occipital bone. **4.** Tomás has fractured his right parietal bone. **5.** The sphenoid contains the sella turcica, which in turn contains the pituitary gland. **6.** The 14 facial bones include 2 inferior nasal conchae, 2 lacrimal bones, 1 mandible, 2 maxillae (maxillary bones), 2 nasal bones, 2 palatine bones, 1 vomer, and 2 zygomatic bones. **7.** The mental foramen is found in the mandible. Structures passing through this opening include the mental nerve, which carries sensory information to the brain, and mental vessels, which supply the chin and lips. **8.** The optic canal is found in the sphenoid. The optic nerve and ophthalmic artery pass through this structure. **9.** The foramina in the ethmoid are the olfactory foramina. **PAGE 226** **10.** The bones forming the orbital complex are the frontal, sphenoid, zygomatic, palatine, maxilla, lacrimal, and ethmoid. **11.** The bones forming the nasal complex are the frontal, ethmoid, nasal, maxilla, palatine, and sphenoid. **12.** The sphenoid, ethmoid, frontal, and paired palatine and maxillary bones contain the paranasal sinuses. **PAGE 227** **13.** A fontanelle is a relatively soft, flexible, fibrous region between two flat bones in the developing skull. The major fontanelles are the anterior fontanelle, posterior fontanelle, sphenoidal fontanelles, and mastoid fontanelles. **14.** Because they are not ossified at birth, fontanelles permit the skull to change shape during childbirth, and they allow for enlargement of the brain during infancy and early childhood. **PAGE 229** **15.** The secondary curves of the spine allow us to balance our body weight on our lower limbs with minimal muscular effort. Without the secondary curves, we would not be able to stand upright for extended periods. **16.** The adult vertebral column consists of 26 bones: the vertebrae (24), the sacrum (1), and the coccyx. **PAGE 237** **17.** The adult vertebral column has fewer vertebrae than a newborn because, in an adult, the five sacral vertebrae fuse to form a single sacrum, and the four coccygeal vertebrae fuse to form a single coccyx. **18.** The axis is fractured. The dens is a tooth-like process projecting upward from the body of the axis, or second cervical vertebra, which is located in the neck (cervical region). **19.** The presence of transverse foramina indicates that this vertebra is from the cervical region of the vertebral column. These foramina are passageways for important blood vessels that supply the brain. **20.** The bodies of the lumbar vertebrae are larger because they must support a great deal more weight than do vertebrae that are more superior in the spinal column. The larger vertebral bodies allow the weight to be distributed over a larger area. **PAGE 239** **21.** False ribs (ribs 8–12) are the five lower ribs on either side that do not articulate with the sternum directly. **22.** Improper compression of the chest during CPR can—and commonly does—result in a fracture of the xiphoid process of the sternum or the ribs. **23.** True ribs (ribs 1–7) attach directly to the sternum by separate costal cartilages; floating ribs (ribs 11–12) are the two lower ribs on either side that are not attached anteriorly to the sternum.

Answers to Review Questions

PAGE 241

Level 1 Reviewing Facts and Terms

1. (a) occipital bone **(b)** parietal bone **(c)** frontal bone **(d)** temporal bone **(e)** sphenoid **(f)** ethmoid **(g)** vomer **(h)** mandible **(i)** lacrimal bone **(j)** nasal bone **(k)** zygomatic bone **(l)** maxilla **2.** a **3.** b **4.** d **5.** d **6.** a **7. (a)** cervical **(b)** thoracic **(c)** lumbar **8.** b **9.** c **10.** d **11.** (1) occipital bone; (2) frontal bone; (3) sphenoid; (4) ethmoid; (5) paired parietal bones; and (6) paired temporal bones **12.** (1) sphenoid; (2) frontal bone; (3) ethmoid; (4) lacrimal bone; (5) maxilla; (6) palatine bone; (7) zygomatic bone **13.** The vomer forms the anterior, inferior portion of the bony nasal septum that separates the right and left nasal cavities. **14.** The fibrocartilage discs between adjacent vertebrae make the vertebral column more flexible.

Level 2 Reviewing Concepts

15. The petrous part of the temporal bone encloses the structures of the internal ear. The middle ear is located in the tympanic cavity within the petrous part. The external acoustic meatus (auditory canal) ends at the tympanic membrane (eardrum), which leads into the middle ear. Mastoid air cells within the mastoid process are connected to the tympanic cavity. **16.** The ethmoid forms the superior surface of the nasal cavity. The olfactory foramina within the cribriform plate of the ethmoid allow neurons associated with the sense of smell to extend into the nasal cavity. **17.** The ribs raise and lower to increase and decrease the volume of the thoracic cage. When they rise, the thoracic cage expands and we breathe in. When the ribs are lowered to their original position, the volume of the thoracic cage decreases and we breathe out. **18.** Keeping your back straight keeps the weight aligned along the axis of your vertebral column, where it can be transferred to your lower limbs. Bending your back would strain the muscles and ligaments of the back, increasing the risk of injury. **19.** d **20.** Fontanelles are fibrous connections between cranial bones. They help ease birth by permitting the skull to change shape without damage during delivery. **21.** a **22.** c **23.** e

Level 3 Critical Thinking and Clinical Applications

24. d **25.** The large bones of a child's cranium are not yet fused; they are connected by fontanelles, areas of fibrous tissue. By examining the bones, the archaeologist could readily see if sutures had formed. By knowing approximately how long it takes for the various fontanelles to close and by determining their sizes, she could estimate the age of the person at death. **26.** Women in later stages of pregnancy develop lower back pain because of changes in the lumbar curve of the spine. The increased mass of the pregnant uterus shifts the woman's center of gravity. To compensate, the lumbar curve is exaggerated, and the lumbar region supports more of her body weight than normal. This results in sore muscles that produce lower back pain.

Answers to Clinical Case Wrap-Up

PAGE 242 **1.** The seven bones that make up the orbital complex are the frontal bone (forming the roof), the zygomatic bone (forming the lateral wall), the maxilla (forming most of the floor), the lacrimal bone, the ethmoidal labyrinth, the sphenoid, and the palatine bone (forming the thin medial wall and floor). **2.** The four paranasal sinuses surrounding the orbital complex are the frontal, maxillary, and sphenoidal sinuses, and ethmoidal cells.

Answers to Figure-Based Questions

FIGURE 7–3 The skull contains four major sutures: the sagittal suture, the lambdoid suture, the squamous suture, and the coronal suture.

FIGURE 7–16 The infant skull has six fontanelles: sphenoidal (paired), mastoid (paired), anterior, and posterior. As the skull develops, bones ossify and fuse. In an adult skull, all of the bones are complete and fused, so there aren't any fontanelles on the adult skull.

FIGURE 7–17 The thoracic and the sacral regions of the spine form primary curves.

FIGURE 7–23 The last two pairs of ribs (11 and 12) are the floating ribs and do not attach to the sternum. The other false ribs are indirectly attached to the sternum by costal cartilage.

ANSWERS

Chapter 8

Answers to Checkpoints

PAGE 247 **1.** Each of the two pectoral girdles consists of a clavicle (collarbone) and a scapula (shoulder blade). Each arm articulates with the trunk at the pectoral girdle. **2.** The clavicle attaches the scapula to the sternum, and restricts the scapula's range of motion (mobility). When the clavicle is broken, the scapula has a greater than normal range of motion and is less stable. **3.** The head of the humerus articulates with the scapula at the glenoid cavity.
PAGE 252 **4.** The bones of an upper limb are the humerus, ulna, radius, carpal (wrist) bones (scaphoid, lunate, triquetrum, pisiform, trapezium, trapezoid, capitate, and hamate), 5 metacarpals (palm bones), and 14 phalanges (fingers). **5.** The two rounded projections on either side of the elbow are the lateral and medial epicondyles of the humerus. **6.** The radius is lateral to the ulna while in the anatomical position; or the ulna is medial to the radius while in the anatomical position. **7.** The first distal phalanx is located at the tip of the pollex (thumb), so Bill's thumb is broken.
PAGE 254 **8.** The pelvic girdle is composed of the paired hip bones, also known as the coxal bones or pelvic bones. **9.** The three bones that make up a hip bone are the ilium, ischium, and pubis. **10.** When you are seated, your body weight is supported by the ischial tuberosities. **PAGE 260** **11.** The bones of the lower limb are the femur (thigh), patella (kneecap), tibia and fibula (leg), tarsal (ankle) bones (talus, calcaneus, cuboid, navicular, medial cuneiform, intermediate cuneiform, and lateral cuneiform), 5 metatarsals, and 14 phalanges. **12.** The fibula is not part of the knee joint and does not bear weight, but it is an important point of attachment for many leg muscles. When the fibula is fractured, these muscles cannot function properly to move the leg, so walking is difficult and painful. The fibula also helps stabilize the ankle joint. **13.** Joey has most likely fractured the calcaneus (heel bone). **14.** The talus transfers the weight of the body from the tibia toward the toes. **15.** In general, the bones of males tend to be heavier than those of females, and bone markings are more prominent in males than in females.
16. Bones can reveal information about a person's sex, age, muscular development, nutritional state, handedness, and occupation, plus other information relative to the medical history. **17.** The female pelvis is adapted for supporting the weight of the developing fetus and enabling its passage through the pelvic outlet during delivery. Compared to males, such adaptations of the female pelvis include an enlarged pelvic outlet, a sacrum and coccyx with less curvature, a pelvic inlet that is wider and more circular, a relatively broad shape, ilia that project farther laterally, and an inferior angle between the pubic bones that is greater than 100 degrees (as opposed to 90 degrees or less for males).

Answers to Review Questions

PAGE 263

Level 1 Reviewing Facts and Terms

1. **(a)** anterior view; **(b)** lateral view; **(c)** posterior view; **(d)** acromion; **(e)** coracoid process; **(f)** spine; **(g)** glenoid cavity **2.** d
3. **(a)** sacrum; **(b)** coccyx; **(c)** ilium; **(d)** pubis; **(e)** ischium; **(f)** hip bone **4.** a **5.** b **6.** d **7.** b **8.** interosseus membrane **9.** ilium, ischium, and pubis **10.** (1) talus; (2) calcaneus; (3) cuboid; (4) navicular; and (5–7) three cuneiform bones

Level 2 Reviewing Concepts

11. d **12.** a **13.** e **14.** d **15.** d **16.** d **17.** c **18.** The pelvic girdle consists of the two hip bones (coxal bones or pelvic bones). The pelvis is a composite structure; it consists of the two hip bones of the appendicular skeleton and the sacrum and coccyx of the axial skeleton. **19.** The clavicles are small and fragile, so they are easy to break. Once this part of a pectoral girdle is broken, the attacker would no longer have efficient use of his arms. **20.** d **21.** e
22. The slender fibula parallels the tibia of the leg and provides an important site for muscle attachment. It does not help in transferring weight to the ankle and foot because it is excluded from the knee joint. **23.** e

Level 3 Critical Thinking and Clinical Applications

24. In osteoporosis, a decrease in the calcium content of the bones leads to bones that are weak and brittle. Hip fractures involve the femur, not the hip bones. Because the hip joint must help support the body's weight, any weakening of the femoral bones may result in their breaking under the weight of the body. The shoulder, by contrast, is not a load-bearing joint and is not subject to the same great stresses or strong muscle contractions as the hip joint. As a result, it is less likely to become broken. **25.** Fred probably dislocated his shoulder, which is quite a common injury due to the weak nature of the shoulder (glenohumeral) joint. **26.** Several characteristics are important in determining a person's sex from the pelvis, including its general appearance, shape of the pelvic inlet, depth of the iliac fossa, characteristics of the ilium, angle inferior to the pubic symphysis, position of the acetabulum, shape of the obturator foramen, and characteristics of the ischium. The person's age can be estimated by the bone's size, degree of mineralization (mineral content), and various markings. The person's general appearance can be reconstructed from the markings where muscles attach to the bones, which can indicate the size and shape of the muscles and thus the person's general body contours.

Answers to Clinical Case Wrap-Up

PAGE 264 **1.** The clavicle makes up the anterior part of the pectoral, or shoulder, girdle. It anchors the upper extremity to the axial skeleton by articulating with the acromion of the scapula laterally and the manubrium of the sternum medially. **2.** When his clavicle was fractured, Ken's shoulder was destabilized, resulting in his upper extremity dangling uselessly on that side.

Answers to Figure-Based Questions

FIGURE 8–1 The appendicular skeleton is composed of the pectoral girdles (clavicles and scapulae), upper limbs (humeri, radii, ulnae, carpal bones, metacarpals, and phalanges), pelvic girdle (hip bones), and lower limbs (femurs, patellae, tibias, fibulas, tarsal bones, metatarsals, and phalanges).

FIGURE 8–6 Metacarpal IV articulates with the hamate and capitate.

FIGURE 8–8 The opening formed by the pubis and the ischium is the obturator foramen.

FIGURE 8–12 The fibula does not articulate with the femur. The fibula does articulate with the tibia.

Chapter 9

Answers to Checkpoints

PAGE 268 **1.** The three types of joints as classified by range of motion are (1) an immovable joint or synarthrosis, (2) a slightly movable joint or amphiarthrosis, and (3) a freely movable joint or diarthrosis. A synarthrosis can be fibrous or cartilaginous, depending on the nature of the connection, or it can be a bony fusion, which develops over time. An amphiarthrosis is either fibrous or cartilaginous, depending on the nature of the connection. A diarthrosis is a synovial joint, which permits the greatest range of motion. **2.** Both synarthrotic joints (except synostoses) and amphiarthrotic joints consist of bony regions separated by fibrocartilage or cartilaginous connective tissue. **3.** These are fibrous joints known as sutures. **PAGE 270** **4.** Components of a synovial joint include a joint capsule (articular capsule), articular cartilages, synovial fluid, and various accessory structures (menisci, fat pads, ligaments, tendons, and bursae). Synovial joints are freely movable joints that have a joint (synovial) cavity, which is the space between the articulating surfaces of two bones in a joint. Articular cartilages resembling hyaline cartilage cover the articulating bone surfaces. The fibrous joint capsule surrounds the joint, and a synovial membrane, which secretes synovial fluid, lines the walls of the joint cavity. Within the cavity, synovial fluid lubricates, distributes nutrients, and absorbs shocks. Menisci are pads of fibrocartilage that allow for variation in the shapes of the articulating surfaces. Fat pads protect the cartilages. Ligaments are cords of fibrous tissue that support, strengthen, and reinforce the joint. Tendons passing across or around the joint limit the range of motion and provide mechanical support. Bursae are synovial fluid-filled pockets that reduce friction and absorb shocks. **5.** Because articular cartilages are avascular (lack a blood supply), they rely on synovial fluid to supply nutrients and eliminate wastes. Impairing the circulation of synovial fluid would have the same effect as impairing a tissue's blood supply: Nutrients would not be delivered to meet the tissue's needs, and wastes would accumulate. Tissue damage or death could ultimately result. **6.** A dislocation is a complete loss of contact between the articulating surfaces within a joint. **PAGE 276** **7.** Based on the shapes of the articulating surfaces, synovial joints are classified as plane (gliding), hinge, pivot, condylar (ellipsoid), saddle, and ball-and-socket joints. **8.** When doing jumping jacks, you move your lower limbs away from the body's midline; this movement is abduction. When you bring the lower limbs back together, the movement is adduction. **9.** Flexion and extension are the movements associated with hinge joints. **PAGE 277** **10.** The type of joint that only permits slight movements is an amphiarthrosis **11.** Intervertebral discs are not found between the first and second cervical vertebrae, between sacral vertebrae in the sacrum, or between coccygeal vertebrae in the coccyx. An intervertebral disc between the first and second cervical vertebrae would prohibit rotation. The vertebrae in the sacrum and coccyx are fused to provide a firm attachment for muscles and ligaments. **12.** The vertebral movement involved in (a) bending forward is flexion. The movement involved in (b) bending to the side is lateral flexion. The movement involved in (c) moving the head to signify "no" is rotation. **PAGE 281** **13.** Terry has most likely damaged his annular ligament. **14.** Damage to the menisci of the knee joint decreases the joint's stability, so the person would have a difficult time locking the knee in place while standing and would have to use muscle contractions to stabilize the joint. If the person had to stand for a long time, the muscles would fatigue and the knee would "give out." It is also likely that the person would feel pain. **15.** Like members of the clergy, carpet layers and roofers kneel a lot (and they also slide along on their knees), causing inflammation of the bursae in the knee joint. **16.** The bones involved with the knee joint are the femur, tibia, and patella. The fibula is not part of the knee joint. **PAGE 285** **17.** Ligaments and muscles provide most of the stability for the shoulder joint. **18.** The subscapular bursa is located in the shoulder joint, so the tennis player would be more likely to develop inflammation of this structure (bursitis). The condition is associated with repetitive motion that occurs at the shoulder, such as swinging a tennis racket. The jogger would be more at risk for injuries to the knee joint. **19.** A shoulder separation is an injury involving partial or complete dislocation of the acromioclavicular joint. **20.** The bones making up the shoulder joint are the humerus and scapula. **21.** The iliofemoral, pubofemoral, and ischiofemoral ligaments are at the hip joint. **PAGE 286** **22.** Rheumatism is a general term describing any painful condition of the bones, muscles, and joints (musculoskeletal system). One of several forms of rheumatism is arthritis. **23.** Arthritis applies to all the rheumatic diseases that affect synovial joints. Of the many different types of arthritis, osteoarthritis is the most common. **24.** Osteoblast activity must equal osteoclast activity or else the integrity of bone structure is compromised. For example, if osteoclast activity outpaces osteoblast activity, the result is thin, brittle bones. **25.** The skeletal system provides structural support for all body systems and stores energy, calcium, and phosphate. The integumentary system synthesizes vitamin D_3, which is essential for calcium and phosphorus absorption. Calcium and phosphorus are needed for bone growth and maintenance.

Answers to Review Questions

PAGE 289

Level 1 Reviewing Facts and Terms

1. (a) joint capsule; **(b)** synovial membrane; **(c)** articular cartilage; **(d)** joint cavity (contains synovial fluid) **2.** d **3.** b **4.** c **5.** c **6.** d **7.** c **8.** b **9.** c **10.** b **11.** b **12.** d **13.** c **14.** c **15.** a **16.** b **17.** a **18.** d **19.** b

Level 2 Reviewing Concepts

20. b **21.** d **22.** A meniscus subdivides a synovial cavity, channels the flow of synovial fluid, and allows variations in the shape of the articular surfaces. Menisci also act as cushions and shock absorbers. **23.** Partial or complete dislocation of the acromioclavicular joint is called a shoulder separation. **24.** Articular cartilages lack a perichondrium, and their matrix contains more water than does the matrix of other cartilages. **25.** In a bulging disc, the nucleus pulposus does not extrude. In a herniated disc, the nucleus pulposus breaks through the annulus fibrosus. **26.** Height decreases during adulthood in part as a result of osteoporosis in the vertebrae, and in part as a result of the decrease in water content of the nucleus pulposus region of intervertebral discs. **27.** (1) plane (gliding): clavicle and sternum; (2) hinge: elbow; (3) condylar: radius and carpal bones; (4) saddle: thumb; (5) pivot: atlas and axis; (6) ball-and-socket: shoulder

Level 3 Critical Thinking and Clinical Applications

28. The problem is probably a sprained ankle. The ligaments have been damaged but not ruptured, and the joint remains unaffected. **29.** Cartilage does not contain blood vessels, so the chondrocytes rely on diffusion to gain nutrients and eliminate wastes. The synovial fluid is very important in supplying nutrients to the articular cartilages that it bathes and in removing wastes. If the circulation of the synovial fluid is impaired or stopped, the cells will not get enough nutrients or be able to get rid of their waste products. This combination of factors can lead to the death of the chondrocytes and the breakdown of the cartilage. **30.** Shoulder dislocations would occur more frequently than hip dislocations because the shoulder is a more mobile joint. Because the shoulder joint is not bound tightly by ligaments or other structures, it is easier to dislocate when excessive forces are applied. By contrast, the hip joint, although mobile, is less easily dislocated because it is stabilized by four heavy ligaments, and the bones fit together snugly in the joint. Additionally, the synovial capsule of the hip joint is larger than that of the shoulder, and its range of motion is decreased.

Answers to Clinical Case Wrap-Up

PAGE 290 **1.** Only diarthroses, or synovial joints, are affected by juvenile rheumatoid arthritis. This is a disease of the synovial lining, so it does not affect joints that have no synovial membrane.
2. Normal synovial membranes are thin, delicate, and smooth. The inflamed synovial membrane in a joint affected by rheumatoid arthritis would look thickened, red, excessively fluid filled, swollen, and overgrown.

Answers to Figure-Based Questions

FIGURE 9–1 Articular cartilage covers the bony surfaces of a synovial joint. The articular cartilage on each bony surface is separated by a thin layer of synovial fluid.

FIGURE 9–3 Angular movements include flexion/extension/hyperextension, abduction/adduction, and circumduction.

FIGURE 9–6 The annulus fibrosus provides the tough outer layer of fibrocartilage and the nucleus pulposus provides the soft inner core for resiliency and shock absorption.

FIGURE 9–9 The acromioclavicular ligament connects the scapula (specifically, the acromion of the scapula) and the clavicle.

Chapter 10

Answers to Checkpoints

PAGE 293 **1.** The three types of muscle tissue are skeletal muscle, cardiac muscle, and smooth muscle. Skeletal muscle tissue moves the body by pulling on our bones, cardiac muscle tissue pumps blood through the cardiovascular system, and smooth muscle tissue pushes fluids and solids along the digestive tract and other internal organs, and regulates the diameters of small arteries. **2.** Muscle tissues share the common properties of excitability (the ability to receive and respond to a stimulus), contractility (the ability of a muscle cell to shorten when it is stimulated), extensibility (stretching movement of a muscle), and elasticity (the ability of a muscle to recoil to its resting length). **3.** Skeletal muscles produce skeletal movement, maintain posture and body position, support soft tissues, guard body entrances and exits, maintain body temperature, and store nutrients. **4.** The epimysium is a dense layer of collagen fibers that surrounds the entire muscle; the perimysium divides the skeletal muscle into a series of compartments, each containing a bundle of muscle fibers called a fascicle, and the endomysium surrounds individual skeletal muscle cells (muscle fibers). The collagen fibers of the epimysium, perimysium, and endomysium come together to form either bundles known as tendons, or broad sheets called aponeuroses. Tendons and aponeuroses generally attach skeletal muscles to bones. **5.** Because tendons attach muscles to bones, severing the tendon would disconnect the muscle from the bone, and so the muscle could not move a body part. **PAGE 302** **6.** Sarcomeres, the smallest contractile units of a striated muscle cell, are segments of myofibrils. Each sarcomere has a dark A band and light I bands. The A band contains the M line, the H band, and the zone of overlap. Each I band contains thin filaments, but not thick filaments. Z lines bisect the I bands and mark the boundaries between adjacent sarcomeres. **7.** Skeletal muscle fibers appear striated when viewed through a light microscope because the Z lines and thick filaments of the myofibrils within the muscle fibers are aligned. **8.** You would expect the greatest concentration of calcium ions in a resting skeletal muscle fiber to be in the terminal cisternae of the sarcoplasmic reticulum. **PAGE 311** **9.** The neuromuscular junction, is the synapse between a motor neuron and a skeletal muscle cell (fiber). This connection enables communication between the nervous system and a skeletal muscle fiber. **10.** Acetylcholine (ACh) release from the axon terminal is necessary for skeletal muscle contraction, because it serves as the first step in the process that leads to the formation of cross-bridges in the sarcomeres. A muscle's ability to contract depends on the formation of cross-bridges between the myosin heads and actin myofilaments. A drug that blocks ACh release would interfere with this cross-bridge formation and prevent muscle contraction. **11.** If the sarcolemma of a resting skeletal muscle suddenly became permeable to Ca^{2+}, the cytosolic concentration of Ca^{2+} would increase, and the muscle would contract. In addition, because the amount of calcium ions in the cytosol must decrease for relaxation to occur, the increased permeability of the sarcolemma to Ca^{2+} might prevent the muscle from relaxing completely. **12.** Without acetylcholinesterase (AChE), the motor end plate would be stimulated continuously by acetylcholine, locking the muscle in a state of contraction. **PAGE 314** **13.** A muscle's ability to contract depends on the formation of cross-bridges between the myosin and actin myofilaments in the muscle. In a muscle that is overstretched, the myofilaments would overlap very little, so very few cross-bridges between myosin and actin could form and, thus, the contraction would be weak. If the myofilaments did not overlap at all, then no cross-bridges would form and the muscle could not contract.
PAGE 319 **14.** Yes, a skeletal muscle can contract without shortening. The muscle can shorten (isotonic concentric contraction), elongate (isotonic eccentric contraction), or remain the same length (isometric contraction), depending on the relationship between the load (resistance) and the tension produced by actin–myosin interactions. **PAGE 323** **15.** Muscle fibers synthesize ATP continuously by utilizing creatine phosphate (CP) and metabolizing glycogen and fatty acids. Most cells generate ATP only through aerobic metabolism in the mitochondria and through glycolysis in the cytosol. **16.** Oxygen debt is the amount of oxygen required to restore normal, pre-exertion conditions in muscle tissue. **PAGE 327** **17.** The three types of skeletal muscle fibers are (1) fast fibers (also called white muscle fibers, fast-twitch glycolytic fibers, Type II-B fibers, and fast fatigue fibers); (2) slow fibers (also called red muscle fibers, slow-twitch oxidative fibers, Type I fibers, and slow oxidative fibers); and (3) intermediate fibers (also called fast-twitch oxidative fibers,

Type II-A fibers, and fast resistant fibers). **18.** Muscle fatigue is a muscle's reduced ability to contract due to low pH (hydrogen ion buildup), low ATP levels, or other problems. **19.** General age-related effects on skeletal muscles include decreased skeletal muscle fiber diameters, diminished muscle elasticity, decreased tolerance for exercise, and a decreased ability to recover from muscular injuries **20.** A sprinter requires large amounts of energy for a short burst of activity. To supply this energy, the sprinter's muscles rely on anaerobic metabolism. Anaerobic metabolism is less efficient in producing energy than aerobic metabolism, and the process also produces acidic wastes; this combination contributes to muscle fatigue. Conversely, marathon runners derive most of their energy from aerobic metabolism, which is more efficient and produces fewer wastes than anaerobic metabolism does. **21.** Activities that require short periods of strenuous activity produce a greater oxygen debt, because such activities rely heavily on energy production by anaerobic metabolism. Because lifting weights is more strenuous over the short term than swimming laps, which is an aerobic activity, weight lifting would likely produce a greater oxygen debt than would swimming laps. **22.** People who excel at endurance activities have a higher than normal percentage of slow fibers. Slow fibers are physiologically better adapted to this type of activity than are fast fibers, which are less vascular and fatigue faster. **PAGE 328** **23.** Compared to skeletal muscle tissue, cardiac muscle tissue (1) has relatively small cells; (2) has cells with a centrally located nucleus (some may contain two or more nuclei); (3) has T tubules that are short and broad and do not form triads; (4) has an SR that lacks terminal cisternae and has tubules that contact the plasma membrane as well as the T tubules; (5) has cells that are nearly totally dependent on aerobic metabolism as an energy source; and (6) contains intercalated discs that assist in stabilizing tissue structure and spreading action potentials. **24.** Cardiac muscle cells are joined by gap junctions, which allow ions and small molecules to flow directly between cells. As a result, action potentials generated in one cell spread rapidly to adjacent cells. Thus, all the cells contract simultaneously, as if they were a single unit (a syncytium). **PAGE 331** **25.** Smooth muscle cells lack sarcomeres, and thus smooth muscle tissue is nonstriated. Additionally, the thin filaments are anchored to dense bodies. **26.** Skeletal muscle contractions are least affected by changes in extracellular Ca^{2+} concentrations. In skeletal muscle, most of the calcium ions come from the sarcoplasmic reticulum. Most of the calcium ions that trigger a contraction in cardiac and smooth muscles come from the extracellular fluid. **27.** The looser organization of actin and myosin filaments in smooth muscle allows smooth muscle to contract over a wider range of resting lengths than skeletal muscle.

Answers to Review Questions

PAGE 334

Level 1 Reviewing Facts and Terms

1. (a) sarcolemma **(b)** sarcoplasm **(c)** mitochondria **(d)** myofibril **(e)** thin filament **(f)** thick filament **(g)** sarcoplasmic reticulum **(h)** T tubules **2.** d **3.** c **4.** c **5.** a **6.** d **7.** d **8.** d **9.** (1) skeletal muscle; (2) cardiac muscle; and (3) smooth muscle **10.** (1) epimysium: surrounds entire muscle; (2) perimysium: surrounds bundles of muscle fibers (fascicles); and (3) endomysium: surrounds individual skeletal muscle fibers (cells) **11.** a **12.** The transverse (T) tubules propagate action potentials into the interior of the cell. **13.** (1) exposure of active sites; (2) attachment of cross-bridges; (3) pivoting of myosin heads (power stroke); (4) detachment of cross-bridges; and (5) reactivation of myosin heads (recocking myosin head) **14.** Both the frequency of motor unit stimulation and the number of motor units involved affect the amount of tension produced when a skeletal muscle contracts. **15.** Resting skeletal muscle fibers contain ATP, creatine phosphate, and glycogen. **16.** Glycolysis and aerobic metabolism generate ATP from glucose in muscle cells. **17.** The calcium-binding protein in smooth muscle tissue is calmodulin.

Level 2 Reviewing Concepts

18. d **19.** b **20.** a **21.** e **22.** In an initial latent period (after the stimulus arrives and before tension begins to increase), an action potential generated in the muscle fiber triggers the release of calcium ions from the SR. In the contraction phase, calcium ions bind to troponin (cross-bridges form) and tension begins to increase. In the relaxation phase, tension decreases because cross-bridges have detached and because calcium ion levels have decreased; the active sites are once again covered by the troponin–tropomyosin complex. **23.** (1) O_2 for aerobic metabolism is consumed by liver cells, which must make a great deal of ATP to convert lactate to pyruvate, and pyruvate to glucose; (2) O_2 for aerobic metabolism is consumed by skeletal muscle fibers as they restore ATP, creatine phosphate, and glycogen concentrations to their former levels; and (3) the normal O_2 concentration in blood and peripheral tissues is replenished. **24.** The timing of cardiac muscle contractions is determined by specialized cardiac muscle fibers called pacemaker cells; this property of cardiac muscle tissue is termed *automaticity*. **25.** If atracurium blocked the binding of ACh to ACh receptors at the motor end plates of neuromuscular junctions, the muscle's ability to contract would be inhibited. **26.** In rigor mortis, the membranes of the dead cells are no longer selectively permeable; the SR is no longer able to retain calcium ions. As calcium ions enter the cytosol, a sustained contraction develops, making the body extremely stiff. Contraction persists because the dead muscle cells can no longer make the ATP required for cross-bridge detachment from the active sites. Rigor mortis begins a few hours after death and ends after one to six days, or when decomposition begins. Decomposition begins when the lysosomal enzymes released by autolysis break down the myofilaments. **27.** c

Level 3 Critical Thinking and Clinical Applications

28. Because organophosphates block the action of acetylcholinesterase (AChE), ACh released into the synaptic cleft would not be removed. It would continue to stimulate the motor end plate, causing a state of persistent contraction (spastic paralysis). If the muscles of respiration were affected (which is likely), Ivan would die of suffocation. Prior to death, the most obvious sign would be uncontrolled tetanic contractions of skeletal muscles. **29.** CK is the enzyme creatine kinase. It functions in the anaerobic reaction that transfers phosphate from creatine phosphate to ADP in muscle cells. The presence of cardiac troponin, a form found only in cardiac muscle cells, provides direct evidence that cardiac muscle cells have been severely damaged. **30.** A skeletal muscle not regularly stimulated by a motor neuron will lose muscle tone and mass and become weak (it will atrophy). While his leg was immobilized, it did not receive sufficient stimulation to maintain proper muscle tone. It will take a while for Bill's muscles to regain enough strength to support his weight.

Answers to Clinical Case Wrap-Up

PAGE 335 **1.** A drug that inhibits acetylcholinesterase (AChE) is inhibiting the enzyme that breaks down the neurotransmitter acetylcholine (ACh). As a result, ACh is present longer in the neuromuscular junction (NMJ). Because a patient with myasthenia gravis does not have enough functioning ACh receptors, a greater amount of ACh is better for strength of muscular response. In short, by inhibiting the breakdown of ACh, the drug strengthens the patient's muscular response. **2.** Myasthenia gravis affects voluntary, skeletal muscle at the neuromuscular junction (NMJ). Cardiac muscle tissue contracts without neural stimulation, a property known as automaticity. There is no neural stimulation, no NMJ, no ACh neurotransmitter or receptor required for cardiac muscle contraction. For these reasons, myasthenia gravis would not affect cardiac muscle.

Answers to Figure-Based Questions

FIGURE 10–3 The terminal cisternae are part of the sarcoplasmic reticulum.

FIGURE 10–6 A skeletal muscle is surrounded by epimysium. A muscle fascicle is surrounded by perimysium. A muscle fiber is surrounded by endomysium. A myofibril is surrounded by sarcoplasmic reticulum.

FIGURE 10–14 If the zone of overlap is decreased, the sarcomere length is increased.

FIGURE 10–21 Slow fibers contain more mitochondria than fast fibers.

Chapter 11

Answers to Checkpoints

PAGE 339 **1.** Based on the patterns of fascicle arrangement, the four types of skeletal muscles are parallel muscles, convergent muscles, pennate muscles, and circular muscles. **2.** Contraction of a pennate muscle generates more tension than would contraction of a parallel muscle of the same size because a pennate muscle contains more muscle fibers, and thus more myofibrils and sarcomeres, than does a parallel muscle of the same size. **3.** The anal opening between the large intestine and the exterior would be guarded by a circular muscle, or sphincter. The concentric circles of muscle fibers found in sphincters are ideally suited for opening and closing passageways and for acting as valves in the body. **4.** A lever is a rigid structure—such as a board, a pry bar, or a bone—that moves on a fixed joint called a fulcrum. There are three classes of levers: In a first-class lever, the fulcrum lies between the applied force and load; in a second-class lever, the load lies between the applied force and fulcrum; and in a third-class lever, the applied force or pull exerted is between the fulcrum and the load. Third-class levers are the most common type in the body. **5.** The joint between the occipital bone and the first cervical vertebra is part of a first-class lever system. The joint between the two bones (the fulcrum) lies between the skull (which provides the load) and the neck muscles (which provide the applied force). **PAGE 341** **6.** The origin of a muscle is the end that is less movable. Because the gracilis moves the tibia, the origin of this muscle must be on the pelvis (pubis and ischium). **7.** Muscles A and B are antagonists to each other, because they perform opposite actions. **8.** A synergist is a muscle that helps a larger prime mover (or agonist—a muscle that is responsible for a specific movement) perform its actions more efficiently. **PAGE 346** **9.** Names of skeletal muscles are based on several factors, including region of the body; position, direction, or fascicle arrangement; origin and insertion; shape and size; and action. **10.** The name *flexor carpi radialis longus* tells you that this muscle is a long muscle (longus) that lies next to the radius (radialis) and flexes (flexor) the wrist (carpi). **PAGE 347** **11.** Axial muscles arise on the axial skeleton. They position the head and vertebral column, move the rib cage, and form the pelvic floor. **12.** Appendicular muscles stabilize or move structures of the appendicular skeleton. These muscles move and support the pectoral (shoulder) girdles and the upper and lower limbs. **PAGE 362** **13.** Contracting the masseter elevates the mandible, and relaxing this muscle depresses the mandible. You would probably be chewing something. **14.** The buccinator, which positions the mouth for blowing, is well developed in a trumpet player. **15.** Swallowing involves contractions of the palatal muscles, which elevate the soft palate as well as portions of the superior pharyngeal wall. Elevation of the superior portion of the pharynx enlarges the opening to the auditory tube, permitting airflow to the middle ear and the inside of the eardrum. Making this opening larger by swallowing facilitates airflow into or out of the middle ear cavity. **16.** Damage to the intercostal muscles would interfere with breathing. **17.** A hit in the rectus abdominis would cause that muscle to contract forcefully, resulting in flexion of the vertebral column. In other words, you would "double over." **18.** The sore muscles in Joe's back are most likely the erector spinae, especially the longissimus and the iliocostalis of the lumbar region. These muscles would have to contract with more tension than normal to counterbalance the increased anterior weight when carrying heavy boxes. **PAGE 383** **19.** When you shrug your shoulders, you are contracting your levator scapulae. **20.** The muscles involved in a rotator cuff injury are the supraspinatus, infraspinatus, teres minor, and subscapularis (SITS). **21.** Injury to the flexor carpi ulnaris would impair your ability to flex and adduct the wrist. **22.** Injury to the obturator would impair your ability to laterally rotate the hip. **23.** A "pulled hamstring" refers to a strain affecting one or more of the three muscles that collectively flex the knee: the biceps femoris, semitendinosus, and semimembranosus. **24.** A torn calcaneal tendon would make plantar flexion difficult because this tendon attaches the soleus and gastrocnemius to the calcaneus (heel bone). **25.** The muscular system generates heat that maintains normal body temperature. **26.** Exercise affects the cardiovascular system by increasing heart rate and dilating blood vessels. With exercise, the rate and depth of breathing increases, and sweat gland secretion increases. The physiological effects that result from exercise are directed and coordinated by the nervous and endocrine systems.

Answers to Review Questions

PAGE 386

Level 1 Reviewing Facts and Terms

1. (a) deltoid (multipennate); **(b)** extensor digitorum (unipennate); **(c)** rectus femoris (bipennate) **2. (a)** supraspinatus; **(b)** infraspinatus; **(c)** teres minor **3.** b **4.** a **5.** a **6.** a **7.** b **8.** c **9.** a **10.** a **11.** b **12.** d **13.** a **14.** The four patterns of fascicle arrangement are (1) parallel, (2) convergent, (3) pennate, and (4) circular. **15.** An aponeurosis is a collagenous sheet connecting two muscles. The epicranial aponeurosis and the linea alba are examples. **16.** The axial musculature includes (1) muscles of the head and neck, (2) muscles of the vertebral column, (3) oblique and rectus muscles, and (4) muscles of the pelvic floor. **17.** The muscles of the pelvic floor (1) support the organs of the pelvic cavity, (2) flex joints of the sacrum and coccyx, and (3) control movement of materials through the urethra and anus. **18.** The supraspinatus, infraspinatus, and teres minor originate on the posterior body of the scapula, and the subscapularis originates on the anterior body of the scapula. All four muscles insert on the humerus. **19.** The functional muscle groups in the lower limbs are (1) muscles that move the thigh, (2) muscles that move the leg, and (3) muscles that move the foot and toes.

Level 2 Reviewing Concepts

20. b **21.** a **22.** hernia **23.** synovial tendon sheaths **24.** The vertebral column does not have a massive series of anterior flexors because many of the large trunk muscles flex the vertebral column when they contract. In addition, most of the body weight lies anterior to the vertebral column, and gravity tends to flex the intervertebral joints. **25.** In a convergent muscle, the direction of pull can be changed by stimulating only one group of muscle cells at any one time. When all the fibers contract at once, they do not pull as hard on the tendon as would a parallel muscle of the same size, because the muscle fibers on opposite sides of the tendon are pulling in different directions rather than working together. **26.** A pennate muscle contains more muscle fibers, and thus more myofibrils and sarcomeres, than does a parallel muscle of the same size, resulting in a contraction that generates more tension. **27.** Lifting heavy objects becomes easier as the elbow approaches a 90° angle. As you increase the angle at or near full extension, tension production decreases, so movement becomes more difficult. **28.** When the hamstrings are injured, flexion at the knee and extension at the hip are affected.

Level 3 Critical Thinking and Clinical Applications

29. Mary is not happy to see Jill. Contraction of the frontalis would wrinkle Mary's brow, contraction of the procerus would flare her nostrils, and contraction of the levator labii on the right side would raise the right side of her lip, as in sneering. **30.** b **31.** Although the pectoralis is located across the chest, it inserts on the greater tubercle of the humerus, the bone of the arm. When this muscle contracts, it contributes to flexion, adduction, and medial rotation of the humerus at the shoulder joint. All of these arm movements would be impaired if the muscle were damaged.

Answers to Clinical Case Wrap-Up

PAGE 387 **1.** Downward-Facing Dog engages muscles throughout the body: extensors of the neck and vertebral column (erector spinae group), hip flexors (iliopsoas group and tensor fasciae latae), extensors of the knee (quadriceps femoris group), ankle flexors (tibialis anterior), extensors of the shoulder (latissimus dorsi and teres major), extensors of the elbow (triceps brachii) and wrist (extensor carpi group), and abductors of the fingers (abductor pollicis brevis and interossei). **2.** The muscles of the "core" support good posture. These include the abdominal muscles (rectus abdominis, external and internal obliques, transversus abdominis); muscles of the pelvic floor; the erector spinae group, especially the longissimus thoracis; and the diaphragm, a muscle that is often overlooked when considering skeletal muscles.

Answers to Figure-Based Questions

FIGURE 11–5 The term *levator anguli oris* means "to raise the mouth." This is one of the muscles that contributes to a smile.

FIGURE 11–6 From the lateral or the medial surface you can see five of the six eye muscles. From the lateral surface, you cannot see the medial rectus and from the medial surface you cannot see the lateral rectus.

FIGURE 11–10 The digastric, geniohyoid, omohyoid, sternohyoid, stylohyoid, and thyrohyoid all insert on the hyoid bone. The mylohyoid inserts on the hyoid bone by a band of connective tissue.

FIGURE 11–17 The triceps brachii has three heads, and each head has a different origin. The lateral head's origin is on the superior lateral surface of the humerus. The long head's origin is on the scapula. The medial head's origin is on the posterior surface of the humerus.

FIGURE 11–19 The adductor pollicis and abductor pollicis brevis are an agonist-antagonist pair.

FIGURE 11–21 The vastus intermedius is not visible from the superficial anterior view of the thigh. This muscle and the vastus lateralis, vastus medialis, and rectus femoris make up the quadriceps femoris.

Chapter 12

Answers to Checkpoints

PAGE 392 **1.** The three anatomical divisions of the nervous system are the central nervous system (CNS), consisting of the brain and spinal cord, the peripheral nervous system (PNS), consisting of all nervous tissue outside the CNS and ENS, and the enteric nervous system (ENS), consisting of the network of neurons and nerve networks in the walls of the digestive tract. **2.** The two functional divisions of the peripheral nervous system are the afferent division, which brings sensory information to the CNS from receptors in peripheral tissues and organs, and the efferent division, which carries motor commands from the CNS to muscles, glands, and adipose tissue. **3.** The two components of the efferent division of the PNS and their effectors are the somatic nervous system (SNS) and the autonomic nervous system (ANS). The SNS effectors are skeletal muscles and the ANS effectors include smooth and cardiac muscle, glands, and adipose tissue.
4. Damage to the afferent division of the PNS, which is composed of nerves that carry sensory information to the brain and spinal cord, would interfere with a person's ability to experience a variety of sensory stimuli. **PAGE 395** **5.** The structural components of a typical neuron include a cell body or soma (which contains the nucleus and perikaryon), dendrites (some with dendritic spines), an axon, telodendria, Nissl bodies, neurofilaments, intermediate neurotubules, neurofibrils, axoplasm, axolemma, initial segment, axon hillock, and collaterals. **6.** According to structure, neurons are classified as anaxonic, bipolar, unipolar, or multipolar. **7.** According to function, neurons are classified as sensory neurons, motor neurons, or interneurons. **8.** Because most sensory neurons of the PNS are unipolar, these neurons most likely function as sensory neurons. **PAGE 401** **9.** Central nervous system neuroglia include astrocytes, ependymal cells, oligodendrocytes, and microglia. **10.** Peripheral nervous system neuroglia include satellite cells and Schwann cells (neurolemmocytes). **11.** The small phagocytic cells called microglia occur in increased numbers in infected (and damaged) areas of the CNS **PAGE 409** **12.** The resting membrane potential is the membrane potential of a normal, unstimulated cell. **13.** If the voltage-gated sodium ion channels in the plasma membrane of a neuron could not open, sodium ions could not flood into the neuron and it would not be able to depolarize. **14.** If the extracellular concentration of potassium ions decreased, more potassium would leave the cell, and the membrane potential would become more negative. This condition is called hyperpolarization. **PAGE 416** **15.** An action potential is a propagated change in the membrane potential of excitable cells, initiated by a change in the membrane permeability to sodium ions. **16.** The four steps involved in the generation of action potentials are (1) depolarization to threshold; (2) activation of voltage-gated sodium ion channels and rapid depolarization; (3) inactivation of voltage-gated sodium ion channels and activation

ANSWERS

of voltage-gated potassium ion channels; and (4) closing of voltage-gated potassium ion channels and return to normal permeability. **17.** The presence of myelin greatly increases the propagation speed of action potentials. **18.** Action potentials travel along myelinated axons (by saltatory propagation) at much higher speeds than along unmyelinated axons. The axon with a propagation speed of 50 meters per second must be the myelinated axon. Typical propagation speeds for unmyelinated axons are about 1 m/sec. **PAGE 420** **19.** The general structural components of a synapse, the site where a neuron communicates with another cell, are a presynaptic cell and a postsynaptic cell, whose plasma membranes are separated by a narrow synaptic cleft. **20.** If a synapse involves direct physical contact between cells, it is termed *electrical*; if the synapse involves a neurotransmitter, it is termed *chemical*. **21.** If the voltage-gated calcium ion channels at a cholinergic synapse were blocked, Ca^{2+} could not enter the axon terminal and trigger the release of ACh into the synapse, so no communication would take place across the synapse. **22.** Because of synaptic delay, the pathway with fewer neurons (in this case, 3) will transmit impulses more rapidly. **PAGE 424** **23.** Both neurotransmitters and neuromodulators are substances that are released by one neuron and affect another neuron. A neurotransmitter alters the membrane potential of the other neuron, whereas a neuromodulator alters the other neuron's response to specific neurotransmitters. **24.** Neurotransmitters and neuromodulators are either (1) compounds that have a direct effect on membrane potential, (2) compounds that have an indirect effect on membrane potential through G proteins, or (3) lipid-soluble gases that have an indirect effect on membrane potential through intracellular enzymes. **PAGE 428** **25.** No action potential will be generated. **26.** Yes, an action potential will be generated. **27.** Temporal summation would occur if the two EPSPs happened at the same time.

Answers to Review Questions

PAGE 430

Level 1 Reviewing Facts and Terms

1. (a) dendrite; **(b)** nucleolus; **(c)** nucleus; **(d)** axon hillock; **(e)** initial segment; **(f)** axolemma; **(g)** axon; **(h)** telodendria; **(i)** axon terminals **2.** c **3.** c **4.** d **5.** b **6.** a **7.** b **8.** a **9. (a)** the CNS: brain and spinal cord **(b)** the PNS: all the nervous tissue outside the CNS and ENS divided between the efferent division (which consists of the somatic nervous system and the autonomic nervous system) and the afferent division (which consists of receptors and sensory neurons) **(c)** the ENS: network of neurons and nerve networks in the walls of the digestive tract. **10.** Neuroglia in the PNS are (1) satellite cells and (2) Schwann cells. **11.** (1) sensory neurons: transmit impulses from the PNS to the CNS; (2) motor neurons: transmit impulses from the CNS to peripheral effectors; and (3) interneurons: analyze sensory inputs and coordinate motor outputs.

Level 2 Reviewing Concepts

12. b **13.** Neurons lack centrioles and therefore cannot divide and replace themselves. **14.** Anterograde flow is the movement of materials from the cell body to the axon terminals. Retrograde flow is the movement of materials toward the cell body. **15.** Chemically gated (ligand-gated) ion channels open or close when they bind specific extracellular chemicals (ligands). Voltage-gated ion channels open or close in response to changes in the membrane potential. Mechanically gated ion channels open or close in response to physical distortion of the membrane surface. **16.** The all-or-none principle of action potentials states that any depolarization event sufficient to reach threshold will generate an action potential of the same strength, regardless of the amount of stimulation above threshold. **17.** The membrane depolarizes to threshold. Next, voltage-gated sodium ion channels are activated, and the membrane rapidly depolarizes. These sodium channels are then inactivated, and voltage-gated potassium ion channels are activated. Finally, normal permeability returns. The voltage-gated sodium channels become activated once the repolarization is complete; the voltage-gated potassium channels begin closing as the membrane potential reaches the normal resting membrane potential. **18.** In saltatory propagation, which occurs in myelinated axons, only the nodes along the axon can respond to a depolarizing stimulus. In continuous propagation, which occurs in unmyelinated axons, an action potential appears to move across the membrane surface in a series of tiny steps. **19.** Type A fibers are myelinated and carry action potentials very quickly (120 m/sec). Type B fibers are also myelinated, but carry action potentials more slowly due to their smaller diameter. Type C fibers are extremely slow due to their small diameter and lack of myelination. **20.** (1) The action potential arrives at the axon terminal, depolarizing it; (2) extracellular calcium enters the axon terminal, triggering the exocytosis of ACh; (3) ACh binds to the postsynaptic membrane and depolarizes the next neuron in the chain; (4) ACh is removed by AChE. **21.** Temporal summation is the addition of stimuli that arrive at a single synapse in rapid succession. Spatial summation occurs when simultaneous stimuli at multiple synapses have a cumulative effect on the membrane potential.

Level 3 Critical Thinking and Clinical Applications

22. Harry's kidney condition is causing the retention of potassium ions. As a result, the K^+ concentration of the extracellular fluid is higher than normal. Under these conditions, less potassium diffuses from heart muscle cells than normal, resulting in a resting membrane potential that is less negative (more positive). This change in the resting membrane potential moves the membrane potential closer to threshold, so it is easier to stimulate the muscle. The ease of stimulation accounts for the increased number of contractions evident in the rapid heart rate. **23.** To reach threshold, the postsynaptic membrane must receive enough neurotransmitter to produce an EPSP of +20 mV (+10 mV to reach threshold and +10 mV to cancel the IPSPs produced by the 5 inhibitory neurons). Each neuron releases enough neurotransmitter to produce a change of +2 mV, so at least 10 of the 15 excitatory neurons must be stimulated to produce this effect by spatial summation. **24.** Action potentials travel faster along fibers that are myelinated than fibers that are nonmyelinated. Destruction of the myelin sheath increases the time it takes for motor neurons to communicate with their effector muscles. This delay in response results in varying degrees of uncoordinated muscle activity. The situation is very similar to that of a newborn, who cannot control its arms and legs very well because the myelin sheaths are still being laid down. Since not all motor neurons to the same muscle may be demyelinated to the same degree, there would be some fibers that are slow to respond while others are responding normally, producing contractions that are erratic and poorly controlled. **25.** The absolute refractory period limits the number of action potentials that can travel along an axon in a given unit of time. During the absolute refractory period, the membrane cannot conduct an action potential, so a new depolarization event cannot occur until after the absolute refractory period has passed. If the absolute refractory period for a particular axon is 0.001 sec, then the maximum frequency of action potentials propagated by this axon would be 1000/sec.

Answers to Clinical Case Wrap-Up

PAGE 431 **1.** The polio virus attacks the motor neurons located in the spinal cord. **2.** Guillain-Barré syndrome is an autoimmune disorder that attacks the myelin covering of peripheral nerves in adults. This causes both sensory and motor signs and symptoms because peripheral nerves are mixed nerves.

Answers to Figure-Based Questions

FIGURE 12–2 The axon hillock is a region between the cell body and the axon of a neuron.

FIGURE 12–4 Oligodendrocytes myelinate CNS axons and Schwann cells myelinate PNS axons.

FIGURE 12–16 Choline undergoes reuptake by the axon terminal. The acetate part of acetylcholine does not undergo reuptake.

Chapter 13

Answers to Checkpoints

PAGE 435 **1.** The central nervous system (CNS) is made up of the brain and spinal cord, while cranial nerves and spinal nerves make up the peripheral nervous system (PNS). **2.** A spinal reflex is a rapid, automatic response triggered by specific stimuli that is controlled in the spinal cord. **PAGE 440** **3.** The three spinal meninges are the dura mater, arachnoid mater, and pia mater. **4.** Damage to the anterior root of a spinal nerve, which contains both visceral and somatic motor fibers, would interfere with motor function. **5.** The cerebrospinal fluid that surrounds the spinal cord is located in the subarachnoid space, which lies between the epithelium of the arachnoid mater and the pia mater. **PAGE 442** **6.** Sensory nuclei receive and relay sensory information from peripheral receptors. Motor nuclei issue motor commands to peripheral effectors. **7.** The polio virus–infected neurons would be in the anterior horns of the spinal cord, where the cell bodies of somatic motor neurons are located. **8.** A disease that damages myelin sheaths would affect the columns in the white matter of the spinal cord, because the columns are composed of bundles of myelinated axons. **PAGE 452** **9.** The major nerve plexuses are the cervical, brachial, lumbar, and sacral. **10.** An anesthetic that blocks the function of the posterior rami of the cervical spinal nerves would affect the skin and muscles of the back of the neck and of the shoulders. **11.** Damage to the cervical plexus—or more specifically to the left and right phrenic nerves, which originate in this plexus and innervate the diaphragm—would greatly interfere with the ability to breathe and might even be fatal. **12.** Compression of the sciatic nerve produces the sensation that your leg has "fallen asleep." **PAGE 453** **13.** A neuronal pool is a functional group of organized interneurons within the CNS. **14.** The five neuronal pool circuit patterns are divergence, convergence, serial processing, parallel processing, and reverberation. **PAGE 457** **15.** A reflex is a rapid, automatic response to a specific stimulus. It is an important mechanism for maintaining homeostasis. **16.** The minimum number of neurons required for a reflex arc is two. One must be a sensory neuron that brings impulses to the CNS, and the other a motor neuron that transmits a response to the effector. **17.** The suck reflex is an innate reflex. **PAGE 460** **18.** All polysynaptic reflexes involve pools of interneurons; are intersegmental in distribution; and involve reciprocal inhibition, reverberating circuits, and the cooperation of several reflexes. **19.** When intrafusal fibers are stimulated by gamma motor neurons, the muscle spindles become more sensitive. As a result, little if any stretching stimulus would be needed to stimulate the contraction of the quadriceps muscles in the patellar reflex. For this reason, the reflex response would appear more quickly. **20.** This response is the tendon reflex. **21.** During a withdrawal reflex, the limb on the opposite side is extended. This response is called a crossed extensor reflex. **PAGE 461** **22.** Reinforcement is an enhancement of spinal reflexes. It occurs when the postsynaptic neuron enters a state of generalized facilitation caused by chronically active excitatory synapses. **23.** A Babinski reflex is abnormal in adults. It indicates possible damage to descending tracts in the spinal cord.

Answers to Review Questions

PAGE 463

Level 1 Reviewing Facts and Terms

1. (a) white matter; **(b)** anterior root; **(c)** posterior root; **(d)** pia mater; **(e)** arachnoid mater; **(f)** gray matter; **(g)** spinal nerve; **(h)** posterior root ganglion; **(i)** dura mater **2.** d **3.** a **4.** d **5.** c **6.** c **7.** c **8.** c **9.** a **10.** d **11.** a **12.** b **13.** c **14.** c **15.** (a) 1 (b) 7 (c) 3 (d) 5 (e) 4 (f) 6 (g) 2 (h) 8

Level 2 Reviewing Concepts

16. The vertebral column continues to grow, extending beyond the spinal cord. The end of the spinal cord is visible as the conus medullaris near L_1, and the cauda equina extends the remainder of the vertebral column. **17.** (1) arrival of stimulus and activation of receptor; (2) activation of sensory neuron; (3) information processing; (4) activation of a motor neuron; and (5) response by an effector (muscle, gland, or adipose tissue) **18.** d **19.** The first cervical nerve exits superior to vertebra C_1 (between the skull and vertebra); the last cervical nerve exits inferior to vertebra C_7 (between the last cervical vertebra and the first thoracic vertebra). So, there are 8 cervical nerves but only 7 cervical vertebrae. **20.** The cell bodies of spinal motor neurons are located in the anterior horns, so damage to these horns would result in a loss of motor control. **21.** Cerebrospinal fluid fills the subarachnoid space between the arachnoid mater and the pia mater. CSF acts as a shock absorber and a diffusion medium for dissolved gases, nutrients, chemicals, and wastes. **22.** (1) involvement of pools of interneurons; (2) intersegmental distribution; (3) involvement of reciprocal inhibition; (4) motor response prolonged by reverberating circuits; and (5) cooperation of reflexes to produce a coordinated, controlled response **23.** Transection of the spinal cord at segment C_7 would most likely result in paralysis from the neck down. Transection at segment T_{10} would produce paralysis and eliminate sensory input in the lower half of the body only. **24.** a **25.** b **26.** a **27.** Stimulation of the sensory neuron will increase muscle tone.

Level 3 Critical Thinking and Clinical Applications

28. The median nerve is involved. **29.** The radial nerve is involved. **30.** The individual would still exhibit a defecation (bowel) and urination (urinary bladder) reflex because these spinal reflexes are processed at the level of the spinal cord. Efferent impulses from the organs would stimulate specific interneurons in the sacral region that synapse with the motor neurons controlling the sphincters, thus

ANSWERS

bringing about emptying when the organs began to fill. (This is the same situation that exists in a newborn infant who has not yet fully developed the descending tracts required for conscious control.) However, an individual with the spinal cord transection at L_1 would lose voluntary control of the bowel and bladder because these functions rely on impulses carried by motor neurons in the brain that must travel down the cord and synapse with the interneurons and motor neurons involved in the reflex. **31.** The anterior horn in the lumbar region of the spinal cord contains somatic motor neurons that direct the activity of skeletal muscles of the hip, lower limb, and foot. As a result of the injury, Karen would be expected to have poor control of most muscles of the lower limbs, causing difficulty walking (if she could walk at all) and problems maintaining balance (if she could stand).

Answers to Clinical Case Wrap-Up

PAGE 464 **1.** No. Recovery of motor activity or sensation is not expected inferior to the level of Joe's complete injury at T_{11}. Central nervous tissue (brain and spinal cord) has very limited healing ability. **2.** Joe would exhibit Babinski reflexes with toes going up (dorsiflexed), which is abnormal in adults (but not in infants). The descending motor pathways to the feet have been completely disrupted at T_{11}.

Answers to Figure-Based Questions

FIGURE 13–3 The area between the arachnoid mater and the pia mater is the subarachnoid space.

FIGURE 13–5 Somatic motor nuclei are located in the anterior horn.

FIGURE 13–9 The axillary, radial, and ulnar nerves branch from the brachial plexus.

FIGURE 13–15 Reflexes can be classified by development (innate or acquired), response (somatic or visceral), complexity (monosynaptic or polysynaptic), and processing site (spinal or cranial).

Chapter 14

Answers to Checkpoints

PAGE 469 **1.** The four major regions of the brain are the cerebrum, cerebellum, diencephalon, and brainstem. **2.** The brainstem is made up of the midbrain, pons, and medulla oblongata. **3.** The rhombencephalon or hindbrain develops into the cerebellum, pons, and medulla oblongata. **PAGE 474** **4.** The layers of the cranial meninges are the outer dura mater, the middle arachnoid mater, and the inner pia mater. **5.** If the normal circulation or resorption of cerebrospinal fluid (CSF) became blocked, CSF would continue to be produced at the choroid plexus in each ventricle, but the fluid would remain there, causing the ventricles to swell—a condition known as hydrocephalus. **6.** If diffusion across the arachnoid granulations decreased, less CSF would reenter the bloodstream, and CSF would accumulate in the ventricles. The increased pressure within the ventricles due to accumulated CSF could damage the brain. **7.** Many water-soluble molecules are rare or absent in the extracellular (ECF), or cerebrospinal fluid (CSF), of the brain because the blood brain barrier regulates the movement of such molecules from the blood to the ECF of the brain. **PAGE 477** **8.** The gracile nucleus and cuneate nucleus are responsible for relaying somatic sensory information to the thalamus. **9.** Damage to the medulla oblongata can be lethal because it contains many vital autonomic reflex centers, including those that control breathing and regulate heart rate and force of contraction, which affect blood pressure. **10.** The pons contains (1) sensory and motor nuclei of cranial nerves (V, VI, VII, and VIII), (2) nuclei involved with the control of respiration, (3) nuclei and tracts that process and relay information sent to or from the cerebellum, and (4) ascending, descending, and transverse pontine fibers. **11.** Damage to the respiratory centers of the pons could result in loss of ability to modify the respiratory rhythmicity centers of the medulla oblongata, which set the basic pace for respiratory movements. **PAGE 478** **12.** Two pairs of sensory nuclei make up the corpora quadrigemina: the superior colliculi and inferior colliculi. **13.** The superior colliculi of the midbrain control reflexive movements of the eyes, head, and neck to visual stimuli, such as a bright light. **PAGE 481** **14.** Components of the cerebellar gray matter include the cerebellar cortex and cerebellar nuclei. **15.** The arbor vitae, which is the internal, white matter part of the cerebellum, connects the cerebellar cortex and nuclei with the cerebellar peduncles. **PAGE 484** **16.** The main components of the diencephalon are the epithalamus, thalamus, and hypothalamus. **17.** Damage to the lateral geniculate body would interfere with the sense of sight. **18.** Changes in body temperature stimulate the pre-optic area of the hypothalamus, a component of the diencephalon. **PAGE 486** **19.** The limbic system is responsible for processing memories and creating emotional states, drives, and associated behaviors. **20.** Damage to the amygdaloid body would interfere with the sympathetic ("fight or flight") division of the autonomic nervous system (ANS). **PAGE 506** **21.** Projection fibers link the cerebral cortex to the spinal cord, passing through the diencephalon, brainstem, and cerebellum. **22.** Damage to the basal nuclei would result in decreased muscle tone and the loss of coordination of learned movement patterns. **23.** The primary motor cortex is located in the precentral gyrus of the frontal lobe of the cerebrum. **24.** Damage to the temporal lobes of the cerebrum would interfere with the processing of olfactory (smell) and auditory (sound) sensations. **25.** The stroke has damaged Broca's area (motor speech area), located in the left cerebral hemisphere. **26.** The temporal lobe of the cerebrum is probably involved, specifically the hippocampus and the amygdaloid body. His problems may also involve other parts of the limbic system that act as a gate for loading and retrieving long-term memories. **27.** Cranial reflexes are automatic responses to stimuli that involve the sensory and motor fibers of cranial nerves. Both monosynaptic and polysynaptic reflex arcs occur. Cranial reflex testing is often used to assess damage to cranial nerves or to the associated processing centers in the brain. **28.** Shining a light in a person's eye tests the pupillary reflex and the consensual light reflex, which are visceral reflexes.

Answers to Review Questions

PAGE 509

Level 1 Reviewing Facts and Terms

1. **(a)** cerebrum; **(b)** diencephalon; **(c)** midbrain; **(d)** pons; **(e)** medulla oblongata; **(f)** cerebellum **2.** **(a)** periosteal cranial dura; **(b)** dural sinus; **(c)** meningeal cranial dura; **(d)** subdural space; **(e)** arachnoid mater; **(f)** subarachnoid space; **(g)** pia mater **3.** d **4.** b **5.** c **6.** d **7.** b **8.** c **9.** a **10.** b **11.** a **12.** a **13.** a **14.** (1) cushioning delicate neural structures; (2) supporting the brain; and (3) transporting nutrients, chemical messengers, and waste products **15.** (1) portions of the hypothalamus where the capillary endothelium is extremely permeable; (2) capillaries in the pineal gland; and (3) capillaries at the choroid plexus **16.** CN I: olfactory nerve; CN II: optic nerve; CN III: oculomotor nerve; CN IV:

trochlear nerve; CN V: trigeminal nerve; CN VI: abducens nerve; CN VII: facial nerve; CN VIII: vestibulocochlear nerve; CN IX: glossopharyngeal nerve; CN X: vagus nerve; CN XI: accessory nerve; and CN XII: hypoglossal nerve

Level 2 Reviewing Concepts

17. The brain can respond with greater versatility because it includes many more interneurons, pathways, and connections than the tracts of the spinal cord. **18.** The cerebellum adjusts voluntary and involuntary motor activities based on sensory information and stored memories of previous experiences. **19.** d **20.** In Parkinson's disease, the substantia nigra is inhibited from secreting the neurotransmitter dopamine at the basal nuclei. This is generally due to the loss of dopamine-producing neurons in the substantia nigra.
21. Roles of the hypothalamus include (1) secretion of two hormones, (2) regulation of body temperature, (3) control of autonomic function, (4) coordination between voluntary and autonomic functions, (5) coordination of activities of the nervous and endocrine systems, (6) regulation of circadian rhythms, (7) subconscious control of skeletal muscle contractions, and (8) production of emotions and behavioral drives. **22.** Stimulation of the feeding and thirst centers of the hypothalamus would produce sensations of hunger and thirst. **23.** This nucleus is the hippocampus, which is part of the limbic system. **24.** The left cerebral hemisphere contains the specialized language areas and speech centers and is responsible for analytical tasks, logical decision making, and language-based skills (reading, writing, and speaking). The right hemisphere analyzes sensory information and relates the body to the sensory environment. Interpretive centers in this hemisphere permit the identification of familiar objects by touch, smell, sight, and taste. The right cerebral hemisphere is also important in understanding three-dimensional relationships and in analyzing the emotional context of a conversation. **25.** d **26.** c **27.** Lesions in Wernicke's area cause defective visual and auditory comprehension of language, repetition of spoken sentences, and defective naming of objects. Lesions in Broca's area result in hesitant and distorted speech.

Level 3 Critical Thinking and Clinical Applications

28. Sensory innervation of the nasal mucosa occurs by the maxillary nerve (V_2), a division of the trigeminal nerve (V). Irritation of the nasal lining by ammonia increases the frequency of action potentials along the maxillary nerve through the semilunar ganglion to reach centers in the midbrain, which in turn excite the neurons of the reticular activating system (RAS). Increased activity by the RAS can bring the cerebrum back to consciousness. **29.** The officer is testing the function of Bill's cerebellum. Many drugs, including alcohol, have pronounced effects on cerebellar function. A person who is under the influence of alcohol cannot properly anticipate the range and speed of limb movement, because processing and correction by the cerebellum are slow. Because of disturbance in muscular coordination (ataxia), Bill might have a difficult time walking a straight line or touching his finger to his nose. **30.** Increasing pressure in the cranium could compress important blood vessels, leading to further brain damage in areas not directly affected by the hematoma. Pressure on the brainstem could disrupt vital respiratory, cardiovascular, and vasomotor functions and possibly cause death. Pressure on the motor nuclei of the cranial nerves would lead to drooping eyelids and dilated pupils. Pressure on descending motor tracts would impair muscle function and decrease muscle tone in the affected areas of the body. **31.** In any inflamed tissue, edema occurs in the area of inflammation. The accumulation of fluid in the subarachnoid space can cause damage by pressing against neurons. If the intracranial pressure is excessive, brain damage can occur, and if the pressure involves vital autonomic reflex areas, death could occur. **32.** Most of the functional problems observed in shaken baby syndrome are the result of trauma to the cerebral hemispheres due to contact between the brain and the inside of the skull. Damage to and distortion of the brainstem and medulla oblongata can cause death.

Answers to Clinical Case Wrap-Up

PAGE 510 **1.** The speech center, also called *Broca's area,* or the *motor speech area,* is usually located in the left hemisphere. Dr. Taylor was unable to speak or understand speech after her stroke. The primary motor cortex that directs voluntary movements of the right side of the body is located in the left hemisphere, in the precentral gyrus of the cerebral cortex. Dr. Taylor was unable to use her right arm after her stroke. For these reasons, you would be able to determine that the left side of her brain was affected by the stroke. **2.** Neuroplasticity is the ability of the nerve cells in the brain to make new connections to other nerve cells. Following trauma to the brain, as in Dr. Taylor's stroke, brain cells that are still alive but dormant can make new connections with other nerve cells. We can think of these changes as healing after a stroke or other trauma to the brain.

Answers to Figure-Based Questions

FIGURE 14–1 The brainstem is made up of the midbrain, the pons, and the medulla oblongata.

FIGURE 14–3 The maters surrounding the brain (from deepest to most superficial) are the pia mater, the arachnoid mater, and the dura mater (meningeal cranial dura and periosteal cranial dura).

FIGURE 14–9 The cerebellar cortex and cerebellar nuclei make up the gray matter in the cerebellum.

FIGURE 14–13 The central sulcus divides the anterior frontal lobe from the more posterior parietal lobe.

FIGURE 14–15 The globus pallidus of the lentiform nucleus is more medial and the putamen of the lentiform nucleus is more lateral.

FIGURE 14–20 The lateral geniculate body of the thalamus receives visual information before it reaches the visual cortex.

FIGURE 14–22 The divisions of the trigeminal nerve are the ophthalmic nerve (V_1), the maxillary nerve (V_2), and the mandibular nerve (V_3).

Chapter 15

Answers to Checkpoints

PAGE 514 **1.** The specialized cells that monitor specific conditions in the body or the external environment are sensory receptors.
2. Yes, it is possible for somatic motor commands to occur at the subconscious (involuntary) level. They also occur at the conscious (voluntary) level. **PAGE 517** **3.** Receptor A provides more precise sensory information because it has a smaller receptive field on the skin surface. **4.** Adaptation is a decrease in receptor sensitivity or a decrease in perception during constant stimulation. **PAGE 522**
5. The major types of general sensory receptors (and the stimuli that excite them) are nociceptors (pain), thermoreceptors (temperature), mechanoreceptors (physical distortion), and chemoreceptors (chemical concentration). **6.** The three classes of mechanoreceptors

are tactile receptors, baroreceptors, and proprioceptors. **7.** If proprioceptors in your legs could not relay information about limb position and movement to the CNS (especially the cerebellum), your movements would be uncoordinated and you likely could not walk. **PAGE 527** **8.** The tract being compressed is the gracile fasciculus in the posterior column of the spinal cord, which carries information about touch and pressure from the lower limbs to the brain. **9.** The tracts that carry action potentials generated by nociceptors (pain receptors) are the lateral spinothalamic tracts. **10.** The left cerebral hemisphere (specifically, the primary somatosensory cortex) receives impulses carried by the right gracile fasciculus. **PAGE 532** **11.** The anatomical basis for opposite-side motor control is crossing over (decussation) of axons, so the motor fibers of the corticospinal pathway innervate lower motor neurons on the opposite side of the body. **12.** An injury involving the superior portion of the motor cortex would affect the ability to control the muscles in the upper limb and the proximal portion of the lower limb. **13.** Increased stimulation of the motor neurons of the red nucleus would increase stimulation of the skeletal muscles in the upper limbs, thereby increasing their muscle tone.

Answers to Review Questions

PAGE 533

Level 1 Reviewing Facts and Terms

1. c **2.** Phasic **3.** c **4.** d

5.

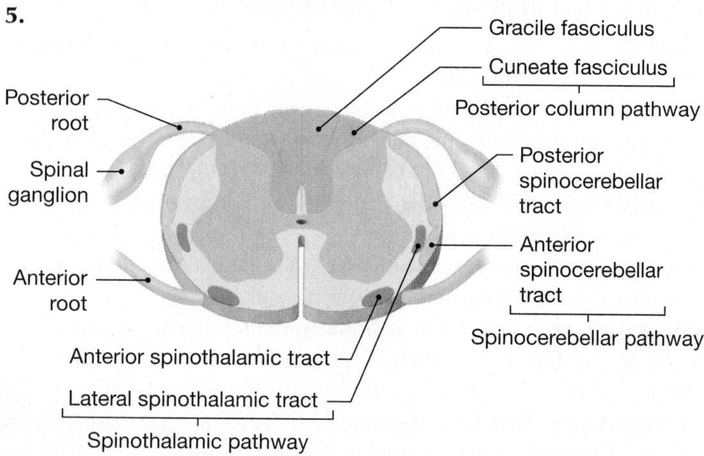
Gracile fasciculus
Cuneate fasciculus
Posterior column pathway
Posterior root
Spinal ganglion
Posterior spinocerebellar tract
Anterior spinocerebellar tract
Spinocerebellar pathway
Anterior root
Anterior spinothalamic tract
Lateral spinothalamic tract
Spinothalamic pathway

6. (1) free nerve endings: sensitive to touch and pressure; (2) root hair plexus: monitors distortions and movements across the body surface; (3) tactile discs: detect fine touch and pressure; (4) bulbous (Ruffini) corpuscles: sensitive to pressure and distortion of the skin; (5) lamellar (pacinian) corpuscles: sensitive to pulsing or vibrating stimuli (deep pressure); and (6) tactile (Meissner) corpuscles: detect fine touch and pressure **7.** (1) tactile receptors; (2) baroreceptors; and (3) proprioceptors **8.** (1) spinothalamic pathway: provides conscious sensations of poorly localized ("crude") touch, pressure, pain, and temperature; (2) posterior column pathway: provides conscious sensations of highly localized ("fine") touch, pressure, vibration, and proprioception; and (3) spinocerebellar pathway: carries proprioceptive information about the position of skeletal muscles, tendons, and joints to the cerebellum **9.** (1) corticobulbar tracts; (2) lateral corticospinal tracts; and (3) anterior corticospinal tracts **10.** (1) vestibulospinal tract; (2) tectospinal tract; and (3) reticulospinal tract **11.** The cerebellum (1) integrates proprioceptive sensations with visual information from the eyes and equilibrium-related sensations from the internal ear, and (2) adjusts the activities of the voluntary and involuntary motor centers on the basis of sensory information and the stored memories of previous experiences. **12.** a **13.** (1) An arriving stimulus alters the membrane potential of the receptor cell. (2) The receptor potential directly or indirectly affects a sensory neuron. (3) Action potentials travel to the CNS along an afferent fiber.

Level 2 Reviewing Concepts

14. A tonic receptor is always active; a phasic receptor is normally inactive and becomes active only when a change occurs in the condition (modality) being monitored. **15.** A motor homunculus, a mapped-out area of the primary motor cortex, provides an indication of the degree of fine motor control available. A sensory homunculus indicates the relative density of peripheral sensory receptors. **16.** A sensory neuron that delivers sensations to the CNS is a first-order neuron. Within the CNS, the axon of the first-order neuron synapses on a second-order neuron, which is an interneuron located in the spinal cord or brainstem. The second-order neuron synapses on a third-order neuron in the thalamus. The axons of third-order neurons synapse on neurons of the primary somatosensory cortex of the cerebral hemispheres. **17.** Damage to the posterior spinocerebellar tract on the left side of the spinal cord at the L_1 level would interfere with the coordinated movement of the left leg. **18.** Injury to the primary motor cortex affects the ability to exert fine control over motor units. Gross movements are still possible, however, because they are controlled by the basal nuclei that use the reticulospinal or rubrospinal tracts. Thus, walking and other voluntary and involuntary movements can be performed with difficulty, and the movements are imprecise and awkward. **19.** Muscle tone is controlled by the basal nuclei, cerebellum, and red nuclei through commands distributed by the reticulospinal and rubrospinal tracts. **20.** Strong pain sensations arriving at a particular segment of the spinal cord can cause stimulation of the interneurons of the spinothalamic pathway. This stimulation is interpreted by the somatosensory cortex as originating in the region of the body surface associated with the origin of that same labeled line or pathway.

Level 3 Critical Thinking and Clinical Applications

21. Kayla's tumor is most likely adjacent to the corticobulbar tracts. The axons of those tracts carry action potentials to motor nuclei of the cranial nerves, which control eye muscles and muscles of facial expression. **22.** Injuries to the primary motor cortex eliminate the ability to produce fine control of motor units. However, as long as the cerebral nuclei are functional, gross movements would still be possible. Harry should still be able to walk, maintain his balance, and perform voluntary and involuntary movements using the rubrospinal and reticulospinal tracts in place of the corticospinal tracts. Although these movements may be awkward or difficult, they will still be possible. **23.** Denzel is experiencing phantom pain. Since pain perception occurs in the somatosensory cortex of the brain, he can still feel pain in his fingers if the brain projects feeling to that area. When he bumps the arm at the elbow, sensory receptors are stimulated to send impulses to the somatosensory cortex. The brain perceives a sensation from a general area, and projects that feeling to a body part. Since more sensory information reaches the brain from the hands and fingers, it is not unusual for the brain to project to this area.

Answers to Clinical Case Wrap-Up

PAGE 534 **1.** All the principal descending motor pathways carry information that would have been affected by the brain damage causing Michael's cerebral palsy. They include the corticospinal pathway, the medial pathway, and the lateral pathway. **2.** Poor motor control of the muscles used in swallowing can contribute to an increased risk of choking and poor infant nutrition, which can affect development and even be life threatening.

Answers to Figure-Based Questions

FIGURE 15–1 The involuntary motor pathway directs a faster response than the voluntary motor pathway.

FIGURE 15–2 The area monitored by a single receptor cell is its receptive field. The branching free nerve endings extend across the area and are receptive to stimuli.

FIGURE 15–4 Root hair plexuses, bulbous (Ruffini) corpuscles, and lamellar (pacinian) corpuscles are located in the dermis.

FIGURE 15–7 The lateral spinothalamic tracts sense pain and temperature.

FIGURE 15–10 The lateral pathway has one tract (the rubrospinal tract), the corticospinal pathway has two tracts (the lateral and anterior corticospinal tracts), and the medial pathway has three tracts (the reticulospinal, the tectospinal, and the vestibulospinal tracts).

Chapter 16

Answers to Checkpoints

PAGE 538 **1.** The two major divisions of the ANS are the sympathetic division and the parasympathetic division. **2.** Two neurons are needed to carry an action potential from the spinal cord to smooth muscles in the intestine. The first neuron carries the action potential from the spinal cord to the autonomic ganglion, and a second neuron carries the action potential from the autonomic ganglion to the smooth muscles in the intestinal wall. **3.** The sympathetic division of the ANS is responsible for the physiological changes that occur in response to stress (confronting an angry dog) and increased activity (running). **PAGE 543** **4.** Sympathetic chain ganglia lie on each side of the vertebral column; they are also called paravertebral ganglia. Collateral ganglia are anterior to the vertebral column; they are also called prevertebral ganglia. **5.** The nerves that synapse in collateral ganglia originate in the inferior thoracic and superior lumbar regions (between T_1 and L_2) of the spinal cord. The preganglionic fibers they contain pass through the sympathetic chain ganglia to the collateral ganglia. **PAGE 545** **6.** Because preganglionic fibers of the sympathetic nervous system release acetylcholine (ACh), a drug that stimulates cholinergic receptors would stimulate the postganglionic fibers of sympathetic nerves, resulting in increased sympathetic activity. **7.** Blocking the beta receptors on cells would decrease or prevent sympathetic stimulation of tissues containing those cells. Heart rate, force of contraction of cardiac muscle, and contraction of smooth muscle in the walls of blood vessels would decrease, lowering blood pressure. **PAGE 546** **8.** The vagus nerve (X) carries preganglionic parasympathetic fibers that innervate the lungs, heart, stomach, liver, pancreas, and parts of the small and large intestines (as well as several other visceral organs). **9.** The parasympathetic division is sometimes referred to as the anabolic system because parasympathetic stimulation leads to a general increase in the nutrient content of the blood. Cells throughout the body respond to the increase by absorbing the nutrients and using them to support growth and other anabolic activities. **PAGE 547** **10.** The neurotransmitter acetylcholine (ACh) is released by all parasympathetic neurons. **11.** The two types of cholinergic receptors on the postsynaptic membranes of parasympathetic neurons are nicotinic receptors and muscarinic receptors. **12.** Stimulation of muscarinic receptors, a type of cholinergic receptor located in postganglionic synapses of the parasympathetic nervous system, would cause K^+ channels to open, resulting in hyperpolarization of cardiac plasma membranes and a decreased heart rate. **PAGE 548** **13.** In the sympathetic division, axons emerge from the thoracic and lumbar segments of the spinal cord and innervate ganglia relatively close to the spinal cord, whereas in the parasympathetic division, axons emerge from the brainstem and sacral segments of the spinal cord and innervate ganglia very close to (or within) target organs. **14.** Divergence refers to the number of synapses a single preganglionic fiber may have with ganglionic neurons. The sympathetic division has a much higher degree of divergence than the parasympathetic division. A greater amount of divergence results in more complex and coordinated responses. **PAGE 551** **15.** General responses to increased sympathetic activity include heightened mental alertness, increased metabolic rate, reduced digestive and urinary functions, activation of energy reserves, increased respiratory rate and dilation of respiratory passageways, elevated heart rate and blood pressure, and activation of sweat glands. General responses to increased parasympathetic activity include decreased metabolic rate, decreased heart rate and blood pressure, increased secretion by salivary and digestive glands, increased motility and blood flow in the digestive tract, and stimulation of urination and defecation. **16.** Most blood vessels receive sympathetic stimulation, so a loss of sympathetic tone would relax the smooth muscles lining the vessels; the resulting vasodilation would increase blood flow to the tissue. **17.** In anxious individuals, an increase in sympathetic stimulation would probably cause some or all of the following changes: a dry mouth; increased heart rate, blood pressure, and rate of breathing; cold sweats; an urge to urinate or defecate; a change in the motility of the digestive tract (that is, "butterflies in the stomach"); and dilated pupils. **PAGE 553** **18.** A visceral reflex is an autonomic reflex initiated in the viscera. It is an automatic motor response that can be modified, facilitated, or inhibited by higher centers, especially those of the hypothalamus. **19.** A brain tumor that interferes with hypothalamic function would also interfere with autonomic function, because centers in the hypothalamus are involved with autonomic (visceral) function. **PAGE 558** **20.** Higher-order functions require action by the cerebral cortex, involve both conscious and unconscious information processing, and are subject to modification and adjustment over time. **21.** Test-taking involves short-term memory, although your instructor would like you to transfer this information to long-term memory. **22.** If your RAS were suddenly stimulated while you were asleep, it would rouse the cerebrum to a state of consciousness and you would wake up. **23.** A drug that increases the amount of serotonin released in the brain would produce a heightened perception of certain sensory stimuli (e.g., auditory or visual stimuli) and hallucinations. **24.** Amphetamines stimulate the secretion of dopamine. **PAGE 559** **25.** Some possible reasons for slower recall and for loss of memory in older people include a loss of neurons (possibly those involved in specific memories), changes in synaptic organization of the brain, changes in the neurons themselves, and decreased blood flow, which would affect the metabolic rate of neurons and perhaps slow the retrieval of information from memory. **26.** Common age-related anatomical changes in the nervous system include a reduction in brain size and weight, a reduction in the number of neurons, a decrease in blood flow to the brain, changes in the synaptic organization of the brain, and intracellular and extracellular changes in CNS neurons. **27.** Alzheimer's disease is the most common form of senile dementia. **28.** The nervous system controls the actions of the arrector pili muscles and sweat glands of the integumentary system. It also controls skeletal muscle contractions of the muscular system, which, in turn, affects the thickening of bones of the skeletal system. It also coordinates muscular activities associated with the respiratory and cardiovascular systems.

Answers to Review Questions

PAGE 563

Level 1 Reviewing Facts and Terms

1.

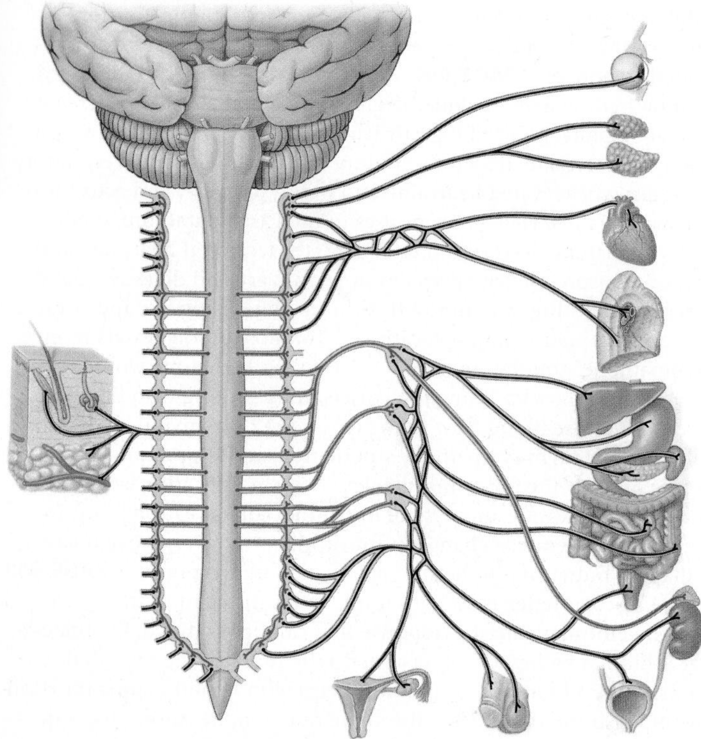

2. d **3.** a **4.** b **5.** d **6.** preganglionic neuron T_5–L_2 → collateral ganglia → postganglionic fibers → visceral effector in abdominopelvic cavity **7.** (1) ciliary ganglion; (2) pterygopalatine ganglion; (3) submandibular ganglion; and (4) otic ganglion **8.** Visceral reflex arcs include a receptor, a sensory neuron, an interneuron (may or may not be present), and two visceral motor neurons. **9.** Increased neurotransmitter release, facilitation of synapses, and the formation of additional synaptic connections are thought to be involved in memory formation and storage. **10.** During non-REM sleep, the entire body relaxes, and activity at the cerebral cortex is at a minimum; heart rate, blood pressure, respiratory rate, and energy utilization decrease. During REM sleep, active dreaming occurs, accompanied by alterations in blood pressure and respiratory rates; muscle tone decreases markedly, and response to outside stimuli decreases.
11. Aging causes a reduction in brain volume and weight, a reduction in the number of neurons, a decrease in blood flow to the brain, changes in synaptic organization, and intracellular and extracellular changes in CNS neurons. **12.** c **13.** d **14.** Sympathetic preganglionic fibers emerge from the thoracolumbar area (T_1 through L_2) of the spinal cord. Parasympathetic fibers emerge from the brainstem and the sacral region of the spinal cord (craniosacral). **15.** (1) celiac ganglion; (2) superior mesenteric ganglion; and (3) inferior mesenteric ganglion **16.** Stimulation of sympathetic ganglionic neurons causes (1) release of norepinephrine at specific locations and (2) secretion of epinephrine (and modest amounts of norepinephrine) into the bloodstream. **17.** The four pairs of cranial nerves are CN III, CN VII, CN IX, and CN X.
18. (1) cardiac plexus: heart rate increases (sympathetic)/decreases (parasympathetic); force of heart contraction increases (sympathetic)/decreases (parasympathetic); blood pressure increases (sympathetic)/decreases (parasympathetic); (2) pulmonary plexus: respiratory passageways dilate (sympathetic)/constrict (parasympathetic); (3) esophageal plexus: respiratory rate increases (sympathetic)/decreases (parasympathetic); (4) celiac plexus: digestion inhibited (sympathetic)/stimulated (parasympathetic); (5) inferior mesenteric plexus: digestion inhibited (sympathetic)/stimulated (parasympathetic); and (6) hypogastric plexus: defecation inhibited (sympathetic)/stimulated (parasympathetic); urination inhibited (sympathetic)/stimulated (parasympathetic); sexual organs: stimulation of secretion (sympathetic)/erection (parasympathetic)
19. Higher-order functions (1) are performed by neurons of the cerebral cortex and involve complex interactions between areas of the cortex and between the cerebral cortex and other parts of the brain; (2) involve both conscious and unconscious information processing; and (3) are subject to modification and adjustment over time.

Level 2 Reviewing Concepts

20. a **21.** c **22.** The preganglionic fibers innervating the cervical ganglia originate in the anterior roots of the thoracic segments, which are undamaged. **23.** c **24.** b **25.** d **26.** Due to the stimulation of the sympathetic division, you would experience increased respiratory rate, increased peripheral vasoconstriction and elevation of blood pressure, increased heart rate and force of contraction, and an increased rate of glucose release into the bloodstream. **27.** If autonomic motor neurons maintain a background level of activity at all times, they can either increase or decrease their activity, providing a greater range of control options. **28.** Cholinergic receptors are found in all the ganglia of the ANS, so nicotine would stimulate both sympathetic and parasympathetic responses in cardiovascular tissues. Although increased sympathetic stimulation increases heart rate and force of contraction, increased parasympathetic stimulation simultaneously decreases blood flow to the heart muscle. In addition to elevating heart rate and force of contraction, sympathetic stimulation also constricts peripheral blood vessels, all of which contribute to increased blood pressure. **29.** The upsetting situation would be processed by the higher centers of the CNS and relayed to the hypothalamus. The hypothalamus could suppress the vasomotor center of the medulla oblongata, resulting in fewer sympathetic impulses to peripheral blood vessels. This would cause a decrease in sympathetic tone in the smooth muscle of the blood vessels, resulting in vasodilation. The vasodilation would cause blood to pool in the limbs, decreasing the amount of blood returning to the heart and producing shock.

Level 3 Critical Thinking and Clinical Applications

30. Epinephrine would be more effective, because it would reduce inflammation and relax the smooth muscle of the airways, making it easier for Phil to breathe. **31.** The molecule is probably mimicking NE and binding to alpha-1 receptors.

Answers to Clinical Case Wrap-Up

PAGE 564 **1.** The first part of Helen's brain to be affected is the hippocampus in the limbic system where memory is encoded. Later progressive losses in cognition involve wide areas of her cerebral cortex. **2.** First, Helen loses orientation to time, then to place, and finally to person. The memory of person is the last to go because it is a deep-seated, long-term, associative memory. **3.** A cholinesterase inhibitor will inactivate acetylcholinesterase, the enzyme whose role is to break down the neurotransmitter acetylcholine at the cholinergic synapse. This will help maintain ACh levels in the early phases of Alzheimer's disease.

Answers to Figure-Based Questions

FIGURE 16–1 The autonomic nervous system affects smooth muscles, glands, cardiac muscle, and adipocytes.

FIGURE 16–3 Collateral ganglionic neurons innervate visceral organs in the abdominopelvic cavity.

FIGURE 16–4 Norepinephrine (NE) is released from most varicosities in the sympathetic division.

FIGURE 16–7 The short reflex bypasses the central nervous system.

FIGURE 16–11 Visual input arrives via CN II and auditory input arrives via CN VIII.

Chapter 17

Answers to Checkpoints

PAGE 570 **1.** Olfaction is the sense of smell; it involves olfactory sensory neurons in paired olfactory organs responding to airborne chemical stimuli. **2.** Axons from the olfactory epithelium collect into bundles that reach the olfactory bulb. In the olfactory pathway, axons leaving the olfactory bulb then travel along the olfactory tract to the olfactory cortex, hypothalamus, and portions of the limbic system. **3.** By the end of the lab period, central adaptation has occurred. Inhibition of synapses along the olfactory pathway reduces the amount of information reaching the olfactory cortex, even though the olfactory neurons remain active. **PAGE 572** **4.** Gustation is the sense of taste, provided by gustatory epithelial cells (taste receptors) responding to dissolved chemical stimuli. **5.** Gustatory epithelial cells (taste receptors in taste buds) are sensitive only to molecules and ions that are in solution. If you dry the surface of your tongue, the salt ions or sugar molecules have no moisture in which to dissolve, so they will not stimulate the cells. **6.** Your grandfather is experiencing the effects of several age-related gustatory and olfactory changes. The number of taste buds decreases dramatically after age 50, and those that remain are not as sensitive as they once were. In addition, the loss of olfactory sensory neurons contributes to the perception of fewer flavors in foods. **PAGE 580** **7.** The conjunctiva would be the first layer of the eye affected by inadequate tear production. Drying of the conjunctiva would produce an irritated, scratchy feeling. **8.** Sue will likely be unable to see at all. The fovea centralis (fovea) contains only cones, which need high-intensity light to be stimulated. The dimly lit room contains light that is too weak to stimulate the cones. **9.** If the scleral venous sinus (canal of Schlemm) were blocked, the aqueous humor could not drain, producing an eye condition called glaucoma. Accumulation of this fluid increases the pressure within the eye, distorting soft tissues and interfering with vision. If untreated, the condition would ultimately cause blindness. **PAGE 583** **10.** The focal point is the point at which the light rays from an object intersect on the retina. **11.** When the lens becomes more rounded, you are looking at an object that is close to you. **PAGE 591** **12.** If you were born without cones, you would still be able to see—so long as you had functioning rods—but you would lack color vision and see in black and white only. **13.** A vitamin A deficiency would reduce the quantity of retinal the body could produce, thereby interfering with night vision (which operates at the body's threshold ability to respond to light). **14.** Vision would be impaired. Decreased phosphodiesterase activity would increase intracellular cGMP levels, which, by keeping gated sodium ion channels open, would prevent both hyperpolarization of the photoreceptor and a decrease in its neurotransmitter release. There would be no signal to the bipolar cell that a photon had been absorbed. **PAGE 605** **15.** If the round window could not move, the perilymph would not be moved by the vibration of the stapes at the oval window, reducing or eliminating the perception of sound. **16.** The loss of stereocilia (as a result of constant exposure to loud noises, for instance) would reduce hearing sensitivity and could lead to deafness. **17.** If the auditory tube were blocked, it would not be possible to equalize the pressure on both sides of the tympanic membrane. If external pressure then decreases, the pressure in the middle ear would be greater than that on the outside, forcing the tympanic membrane outward and producing pain.

Answers to Review Questions

PAGE 608

Level 1 Reviewing Facts and Terms

1. **(a)** vascular layer; **(b)** iris; **(c)** ciliary body; **(d)** choroid; **(e)** inner layer (retina); **(f)** neural layer; **(g)** pigmented layer; **(h)** fibrous layer; **(i)** cornea; **(j)** sclera **2.** d **3.** c **4.** e **5.** c **6.** c **7.** b **8.** d **9.** b **10.** **(a)** auricle; **(b)** external acoustic meatus; **(c)** tympanic membrane; **(d)** auditory ossicles; **(e)** semicircular canals; **(f)** vestibule; **(g)** auditory tube; **(h)** cochlea; **(i)** vestibulocochlear nerve (VIII) **11.** d **12.** d **13.** c **14.** (1) filiform papillae; (2) fungiform papillae; (3) foliate papillae; and (3) vallate papillae **15.** The fibrous layer **(a)** is composed of the sclera and the cornea and **(b)** provides mechanical support and some physical protection, serves as an attachment site for the extrinsic eye muscles, and contains structures that assist in the focusing process. **16.** The vascular layer consists of the iris, ciliary body, and choroid. **17.** The malleus, incus, and stapes transmit and amplify a mechanical vibration from the tympanic membrane to the oval window.

Level 2 Reviewing Concepts

18. Axons leaving the olfactory epithelium collect into 20 or more bundles that penetrate the cribriform plate of the ethmoid to reach the olfactory bulbs of the cerebrum. Axons leaving the olfactory bulb travel along the olfactory tract to reach the olfactory cortex, hypothalamus, and portions of the limbic system. **19.** Olfactory sensations are long lasting and important to memories because the sensory information reaches the cerebral cortex by the hypothalamus and the limbic system without first being filtered through the thalamus. **20.** An infected sebaceous gland of an eyelash or tarsal gland usually becomes a sty, a painful swelling. **21.** a **22.** c **23.** a

Level 3 Critical Thinking and Clinical Applications

24. Your medial rectus muscles would contract, directing your gaze more medially. In addition, your pupils would constrict and the lenses would become more spherical. **25.** Myopia is corrected by **(a)** concave lenses. **26.** In removing the polyps, some of the olfactory epithelium was probably damaged or destroyed, decreasing the area available for the solution of odorants and reducing the intensity of the stimulus. As a result, after the surgery it would take a larger stimulus to provide the same level of smell. **27.** The rapid descent in the elevator causes the otoliths of the saccular maculae to slide upward, producing the sensation of downward vertical motion. After the elevator abruptly stops, it takes a few seconds for the otoliths of the maculae to come to rest in the normal position. So long as the otoliths are displaced, you will perceive movement. **28.** When Juan closes his eyes, visual cues are gone, and his brain must rely solely on proprioceptive information from the equilibrium centers of the internal ear to maintain normal posture. Because either the internal ear receptors or the sensory nerves are not functioning normally, he is unstable. The most likely reason for his drift to the left is that he is getting inappropriate sensations from equilibrium receptors, either at the maculae (affecting his ability to determine which way is "down") or one of the horizontal semicircular ducts (making him attempt to compensate for a perceived roll to the right).

ANSWERS

Answers to Clinical Case Wrap-Up

PAGE 609 **1.** A visual acuity of 20/200 means Makena must be 20 feet away from an object to see it as clearly as someone with normal vision would see it at 200 feet. In the United States she would be considered legally blind. **2.** If Makena's vision is corrected to a state of emmetropia, her vision would be corrected to normal.

Answers to Figure-Based Questions

FIGURE 17–4 The lacrimal gland is located superior and lateral to the eye.

FIGURE 17–7 Amacrine cells are found where bipolar cells synapse with ganglion cells. Horizontal cells are found where photoreceptors (rods and cones) synapse with bipolar cells.

FIGURE 17–21 The tympanic cavity and auditory ossicles are found in the middle ear.

FIGURE 17–28 The scala vestibuli and scala tympani contain perilymph (a fluid similar in composition to cerebrospinal fluid). The cochlear duct contains endolymph (a fluid with electrolyte concentrations that differ from other bodily fluids).

Chapter 18

Answers to Checkpoints

PAGE 613 **1.** A hormone is a chemical messenger that is secreted by one cell and travels through the bloodstream to affect the activities of cells in other parts of the body. **2.** Paracrine communication is the use of chemical messengers (paracrines) to transfer information from cell to cell within a single tissue. **3.** The five mechanisms of intercellular communication are direct, paracrine, autocrine, endocrine, and synaptic. **PAGE 618** **4.** A cell's hormonal sensitivities are determined by the presence or absence of extracellular and intracellular receptors that bind a given hormone. **5.** A substance that inhibits adenylate cyclase, the enzyme that converts ATP to cAMP, would block the action of any hormone that requires cAMP as a second messenger. **6.** Humoral stimuli are those involving changes in the composition of the extracellular fluid. Such a stimulus triggers the secretion of a hormone, and the direct or indirect effects of the hormone reduce the intensity of the stimulus. **PAGE 627** **7.** The two lobes of the pituitary gland are the anterior lobe and the posterior lobe. **8.** In dehydration, blood osmotic concentration is increased, which would stimulate the posterior lobe of the pituitary gland to release more ADH. **9.** Somatomedins mediate the action of growth hormone. Elevated levels of growth hormone typically accompany elevated levels of somatomedins. **10.** An elevated circulating level of cortisol would inhibit the endocrine cells that control the release of ACTH from the pituitary gland, so the ACTH level would decrease. This is an example of a negative feedback mechanism. **PAGE 632** **11.** Thyroxine (T_4), triiodothyronine (T_3), and calcitonin (CT) are hormones associated with the thyroid gland. **12.** A person whose diet lacks iodine would be unable to form the thyroid hormones thyroxine (T_4) and triiodothyronine (T_3). As a result, you would expect to see signs and symptoms associated with their deficiency: decreased metabolic rate, decreased body temperature, a poor response to physiological stress, and an increase in the size of the thyroid gland (goiter). **13.** Most of the body's reserves of the thyroid hormones T_4 and T_3 are bound to blood-borne proteins called thyroid-binding globulins. Because these compounds represent such a large reservoir of T_4 and T_3, it takes several days after removal of the thyroid gland for the blood level of T_4 and T_3 to decrease. **PAGE 634** **14.** The parathyroid glands are embedded in the posterior surfaces of the lateral lobes of the thyroid gland. **15.** The hormone secreted by the parathyroid glands is parathyroid hormone (PTH). **16.** The removal of the parathyroid glands would result in a decrease in the blood concentration of calcium ions. **PAGE 637** **17.** The two regions of the adrenal gland are the cortex and medulla. The cortex secretes mineralocorticoids (mainly aldosterone), glucocorticoids (mainly cortisol, corticosterone, and cortisone), and androgens; the medulla secretes epinephrine and norepinephrine. **18.** The three zones of the adrenal cortex from superficial to deep are the zona glomerulosa, zona fasciculata, and zona reticularis. **19.** One function of cortisol is to decrease the cellular use of glucose while increasing both the available glucose (by promoting the breakdown of glycogen) and the conversion of amino acids to carbohydrates. Therefore, the net result of elevated cortisol level would be an elevation of blood glucose. **20.** Pinealocytes are the hormone-secreting cells of the pineal gland. **21.** Increased amounts of light would inhibit the production (and release) of melatonin from the pineal gland, which receives neural input from the optic tracts. Melatonin secretion is influenced by day–night patterns. **22.** Melatonin influences circadian rhythms, inhibits reproductive functions, and protects against free radical damage. **PAGE 641** **23.** The cells of the pancreatic islets (and their hormones) are alpha cells (glucagon), beta cells (insulin), delta cells (GH–IH), and pancreatic polypeptide cells (pancreatic polypeptide). **24.** A person with Type 1 or Type 2 diabetes has such a high blood glucose level that the kidneys cannot reabsorb all the glucose; some glucose is lost in urine. Because the urine contains a high concentration of glucose, less water can be reclaimed by osmosis, so the volume of urine production increases. The water losses reduce blood volume and elevate blood osmotic pressure, promoting thirst and triggering the secretion of ADH. **25.** An increased blood level of glucagon stimulates the conversion of glycogen to glucose in the liver, which would in turn decrease the amount of glycogen in the liver. **PAGE 646** **26.** Two hormones secreted by the kidneys are erythropoietin (EPO) and calcitriol. **27.** Leptin is a hormone released by adipose tissue. **28.** Once released into the bloodstream, renin functions as an enzyme, catalyzing the conversion of angiotensinogen to angiotensin I.

PAGE 649 **29.** The type of hormonal interaction in which two hormones have opposite effects on their target tissues is called antagonism. **30.** A lack of GH, thyroid hormone, PTH, and the gonadal hormones would inhibit the formation and development of the skeletal system. **31.** During the resistance phase of the general adaptation syndrome, there is a high demand for glucose, especially by the nervous system. The hormones GH–RH and CRH increase the levels of GH and ACTH, respectively. Growth hormone (GH) mobilizes fat reserves and promotes the catabolism of protein; ACTH increases cortisol, which stimulates both the conversion of glycogen to glucose and the catabolism of fat and protein. Together, these metabolic processes conserve glucose for use by nervous tissue. **32.** The endocrine system adjusts metabolic rates and substrate use, and regulates growth and development, in all other body systems. **33.** Hormones of the endocrine system adjust muscle metabolism, energy production, and growth; hormones also regulate calcium and phosphate levels, which are critical to normal muscle functioning. Skeletal muscles protect some endocrine organs.

Answers to Review Questions

PAGE 653

Level 1 Reviewing Facts and Terms

1. b **2.** c **3.** d **4.** a **5.** d **6.** d **7.** d **8.** **(a)** hypothalamus; **(b)** pituitary gland; **(c)** thyroid gland; **(d)** thymus; **(e)** adrenal glands; **(f)** pineal gland; **(g)** parathyroid glands; **(h)** heart; **(i)** kidneys; **(j)** adipose tissue; **(k)** digestive tract; **(l)** pancreas (pancreatic islets); **(m)** gonads **9.** (1) The hypothalamus releases ADH and oxytocin into the bloodstream at the posterior lobe of the pituitary gland. (2) The hypothalamus produces regulatory hormones that control secretion by endocrine cells in the anterior lobe of the pituitary gland. (3) The hypothalamus contains autonomic centers that exert direct neural control over the endocrine cells of the adrenal medullae. These hypothalamic activities are adjusted through negative feedback loops involving hormones released by peripheral endocrine tissues and organs. **10.** The anterior lobe of the pituitary gland releases (1) thyroid-stimulating hormone (TSH); (2) adrenocorticotropic hormone (ACTH); (3) follicle-stimulating hormone (FSH); (4) luteinizing hormone (LH); (5) prolactin (PRL); (6) growth hormone (GH); and (7) melanocyte-stimulating hormone (MSH). **11.** Growth is affected by (1) growth hormone, (2) thyroid hormones, (3) insulin, (4) parathyroid hormone, (5) calcitriol, and (6) the reproductive hormones. **12.** Effects of thyroid hormones are (1) increased rate of energy consumption and utilization in cells; (2) accelerated production of sodium–potassium ATPase; (3) activation of genes coding for the synthesis of enzymes involved in glycolysis and energy production; (4) accelerated ATP production by mitochondria; and (5) in growing children, normal development of the skeletal, muscular, and nervous systems. **13.** Parathyroid hormone causes an increase in the concentration of calcium ions in blood. **14.** (1) zona glomerulosa: mineralocorticoids; (2) zona fasciculata: glucocorticoids; and (3) zona reticularis: androgens **15.** The kidneys release (1) erythropoietin, which stimulates the production of RBCs by the red bone marrow, and (2) calcitriol, which stimulates calcium and phosphate absorption along the digestive tract. **16.** Natriuretic peptides (1) promote the loss of sodium ions and water at the kidneys; (2) inhibit the secretion of water-conserving hormones, such as ADH and aldosterone; (3) suppress thirst; and (4) inhibits the effects of angiotensin II and norepinephrine and decreases blood pressure. Angiotensin II opposes these actions by stimulating aldosterone secretion by the adrenal cortex and ADH by the posterior lobe of the pituitary gland, and further by retaining salt and water by the kidneys. Angiotensin II also stimulates thirst and elevates blood pressure. **17.** (1) alpha (α) cells: glucagon; (2) beta (β) cells: insulin; (3) delta (δ) cells: growth hormone–inhibiting hormone (GH–IH, or somatostatin); and (4) pancreatic polypeptide cells: pancreatic polypeptide

Level 2 Reviewing Concepts

18. The primary difference involves speed and duration. In the nervous system, the source and destination of communication are quite specific, and the effects are extremely quick and short-lived. In endocrine communication, the effects are slow to appear and commonly persist for days. A single hormone can alter the metabolic activities of multiple tissues and organs simultaneously. **19.** Hormones can (1) direct the synthesis of an enzyme (or other protein) not already present in the cytoplasm, (2) turn an existing enzyme "on" or "off," and (3) increase the rate of synthesis of a particular enzyme or other protein. **20.** Inactivation of phosphodiesterase (PDE), which converts cAMP to AMP, would prolong the effect of the hormone. **21.** The adrenal medulla is controlled by the sympathetic nervous system, whereas the adrenal cortex is stimulated by the release of ACTH from the anterior lobe of the pituitary gland. **22.** b **23.** b **24.** b

Level 3 Critical Thinking and Clinical Applications

25. Extreme thirst and frequent urination are characteristic of both diabetes insipidus and diabetes mellitus. To distinguish between the two, glucose levels in the blood and urine could be measured. A high glucose concentration would indicate diabetes mellitus. **26.** Julie's poor diet would not supply enough Ca^{2+} for her developing fetus, which would remove large amounts of Ca^{2+} from the maternal blood. A lowering of the mother's blood Ca^{2+} would lead to an increase in parathyroid hormone levels and increased release of stored Ca^{2+} from maternal bones. **27.** Sherry's signs and symptoms suggest hyperthyroidism. Blood tests could be performed to assay the levels of TSH, T_3, and T_4. From these tests, the physician could make a positive diagnosis (hormone levels would be elevated in hyperthyroidism) and also determine whether the condition is primary (a problem with the thyroid gland) or secondary (a problem with hypothalamo-pituitary control of the thyroid gland). **28.** One benefit of a portal system is that it ensures that the regulatory hormones will be delivered directly to the target cells. Secondly, because the regulatory hormones go directly to their target cells without first passing through the general circulation, they are not diluted. The hypothalamus can control the cells of the anterior lobe of the pituitary gland with much smaller amounts of releasing and inhibiting hormones than would be necessary if the hormones had to first go through the circulatory pathway before reaching the pituitary. **29.** The natural effects of testosterone are to increase muscle mass, increase endurance, and enhance the "competitive spirit." Side effects in women include hirsutism (abnormal hair growth on the face or body), enlargement of the laryngeal cartilages, premature closure of the epiphyseal cartilages, liver dysfunction, and irregular menstrual periods.

Answers to Clinical Case Wrap-Up

PAGE 654 **1.** In addition to mobilizing Ca^{2+} from bone, PTH enhances the reabsorption of Ca^{2+} by the kidneys, decreasing urinary losses and keeping blood concentrations high. PTH also stimulates the

ANSWERS

secretion of calcitriol by the kidneys, which enhances absorption of Ca^{2+} in the digestive tract, which also increases blood concentrations of Ca^{2+}. **2.** Calcitonin, produced by the thyroid, decreases blood Ca^{2+} levels indirectly. Calcitonin does this by causing an increased excretion of Ca^{2+} by the kidneys.

Answers to Figure-Based Questions

FIGURE 18–6 The endocrine cells whose secretion is controlled by regulatory hormones from the hypothalamus are located in the anterior lobe of the pituitary gland.

FIGURE 18–8 In a typical regulation pattern of endocrine secretion, hormone 2 (which is released from the target organ) is responsible for negative feedback.

FIGURE 18–9 Antidiuretic hormone (ADH) and oxytocin (OXT) are released from the posterior lobe of the pituitary gland. Posterior pituitary hormones are released directly from the hypothalamus, and regulatory hormones from the hypothalamus control the release of anterior pituitary hormones.

FIGURE 18–14 The zona glomerulosa produces mineralocorticoids. The primary mineralocorticoid is aldosterone.

FIGURE 18–17 Alpha cells secrete glucagon. Glucagon increases blood glucose level by (1) increasing the breakdown of glycogen to glucose in liver and skeletal muscle, (2) increasing the breakdown of fat to fatty acids in adipose tissue, and (3) increasing the synthesis and release of glucose by the liver.

ANSWERS

Chapter 19

Answers to Checkpoints

PAGE 659 **1.** Five major functions of blood are transporting dissolved gases, nutrients, hormones, and metabolic wastes; regulating the pH and ion composition of interstitial fluids; restricting fluid losses at injury sites; defending against toxins and pathogens; and stabilizing body temperature. **2.** The three types of formed elements in blood are red blood cells, white blood cells, and platelets. **3.** The three major types of plasma proteins are albumins, globulins, and fibrinogen. **4.** A decrease in the amount of plasma proteins in the blood would lower plasma osmotic pressure, reduce the ability to fight infection, and decrease the transport and binding of some ions, hormones, and other molecules. **PAGE 666** **5.** Hemoglobin is a protein composed of four globular subunits, each bound to a heme molecule, which gives red blood cells the ability to transport oxygen in the blood. **6.** After a significant loss of blood, the hematocrit—the amount of formed elements (mostly red blood cells) as a percentage of the total blood—would be reduced. **7.** The hematocrit would increase, because reduced blood flow to the kidneys triggers the release of erythropoietin, which stimulates an increase in erythropoiesis (red blood cell formation). **8.** Bilirubin would accumulate in the blood, producing jaundice, because diseases that damage the liver, such as hepatitis or cirrhosis, impair the liver's ability to excrete bilirubin in the bile. **PAGE 670** **9.** Surface antigens on RBCs are glycoproteins in the plasma membrane; they determine blood type. **10.** Only type O blood can be safely transfused into a person whose blood type is O. **11.** If a person with type A blood receives a transfusion of type B blood, which contains anti-A antibodies, the red blood cells will agglutinate (clump), potentially blocking blood flow to various organs and tissues. **PAGE 678** **12.** The five types of white blood cells are neutrophils, eosinophils, basophils, monocytes, and lymphocytes. **13.** An infected cut would contain a large number of neutrophils, because these phagocytic white blood cells are the first to arrive at the site of an injury. **14.** The blood of a person fighting a viral infection would contain elevated numbers of lymphocytes, because B cells (B lymphocytes) produce circulating antibodies. **15.** Basophils respond to an injury by releasing a variety of chemicals, including histamine and heparin. Histamine dilates blood vessels and heparin prevents blood clotting. Basophils also release other chemicals that attract eosinophils and other basophils to the injured area. **16.** Thrombocytopoiesis is the term for platelet production. **17.** Platelets release chemicals important to the clotting process, they form a temporary patch in the walls of damaged blood vessels, and they reduce the size of a break in a vessel wall. **PAGE 683** **18.** A decreased number of megakaryocytes would interfere with the blood's ability to clot properly, because fewer megakaryocytes would produce fewer platelets. **19.** Vitamin K is necessary for blood clotting, and fats are required for vitamin K absorption. So, if a person did not eat foods containing fat, this would lead to a vitamin K deficiency, which would, in turn, result in a decreased production of several clotting factors—most notably, prothrombin. As a result, clotting time would increase. **20.** The activation of Factor XII initiates the intrinsic pathway.

Answers to Review Questions

PAGE 685

Level 1 Reviewing Facts and Terms

1. **(a)** neutrophil; **(b)** eosinophil; **(c)** basophil; **(d)** monocyte; **(e)** lymphocyte **2.** c **3.** c **4.** a **5.** d **6.** d **7.** b **8.** d **9.** a **10.** Blood (1) transports dissolved gases, nutrients, hormones, and metabolic wastes; (2) regulates pH and electrolyte composition of interstitial fluids throughout the body; (3) restricts fluid losses through damaged vessels or at other injury sites; (4) defends against toxins and pathogens; and (5) stabilizes body temperature. **11.** Major types of plasma proteins are (1) albumins, which maintain the osmotic pressure of plasma and are important in the transport of fatty acids; (2) globulins, which (a) bind small ions, hormones, or compounds that might otherwise be filtered out of the blood at the kidneys or have very low solubility in water (transport globulins), or (b) attack foreign proteins and pathogens (immunoglobulins); and (3) fibrinogen, which functions in blood clotting. **12.** **(a)** anti-B antibodies; **(b)** anti-A antibodies; **(c)** neither anti-A nor anti-B antibodies; **(d)** both anti-A and anti-B antibodies **13.** WBCs exhibit (1) emigration (diapedesis), squeezing between adjacent endothelial cells in the capillary wall; (2) amoeboid movement, a gliding movement that transports the cell; (3) positive chemotaxis, the attraction to specific chemical stimuli, and (4) phagocytosis (engulfing particles for neutrophils, eosinophils, and monocytes). **14.** Neutrophils, eosinophils, basophils, and monocytes function in nonspecific defenses. **15.** The primary lymphocytes are (1) T cells, which are responsible for cell-mediated immunity; (2) B cells, which are responsible for humoral immunity; and (3) NK cells, which are responsible for immune surveillance. **16.** Prothrombin is an inactive precursor that is converted to thrombin during coagulation. Thrombin is an enzyme that causes the clotting of blood by converting fibrinogen to fibrin. **17.** Erythropoietin is released (1) during anemia, (2) when blood flow to the kidneys declines, (3) when the oxygen content of air in the lungs declines, and (4) when the respiratory surfaces of the lungs are damaged. **18.** Initiation of the common pathway requires the activation of Factor X by the extrinsic and/or intrinsic pathways.

Level 2 Reviewing Concepts

19. a **20.** d **21.** c **22.** c **23.** Red blood cells are biconcave discs that lack mitochondria, ribosomes, and nuclei, and they contain a large amount of hemoglobin. RBCs transport oxygen, while WBCs are involved in immunity. The five types of white blood cells vary in size from slightly larger to twice the diameter of an RBC, contain a prominent nucleus, and may contain granules with distinct staining properties. **24.** White blood cells defend against toxins and pathogens. Neutrophils, eosinophils, and monocytes engulf and digest bacteria, protozoa, fungi, viruses, and cellular debris. Lymphocytes are specialized to attack and destroy specific foreign cells, proteins, and cancerous cells, directly or through the production of antibodies. **25.** Blood stabilizes and maintains body temperature by absorbing and redistributing the heat produced by active skeletal muscles. **26.** Each molecule of hemoglobin consists of four protein subunits, each of which contains a single molecule of heme, a nonprotein ring surrounding an iron ion. These central iron ions are what actually bind and release oxygen molecules. **27.** Aspirin helps prevent vascular problems by inhibiting clotting. It inhibits platelet enzymes involved in the production of thromboxane A_2 and prostaglandins, thereby inhibiting clotting. It also prolongs bleeding time.

Level 3 Critical Thinking and Clinical Applications

28. A prolonged prothrombin time and a normal partial thromboplastin time indicate a deficiency in the extrinsic system but not in the intrinsic system or common pathway. Factor VII would be deficient. **29.** As the spleen enlarges, so does its capacity to store additional red blood cells, leading to fewer red blood cells in circulation, producing anemia. The decreased number of RBCs in circulation decreases the body's ability to deliver oxygen to the tissues, slowing their metabolism and producing the tired feeling and lack of energy. Because there are fewer RBCs than normal, the blood circulating through the skin is not as red, producing a pale complexion. **30.** Taking a broad-spectrum antibiotic kills a wide range of bacteria, both pathogenic and nonpathogenic, including many of the normal flora, or microbiota, of the intestine. Reducing the intestinal flora substantially decreases the amount of vitamin K they make available to the liver to produce prothrombin, a vital component of the common pathway. With decreased amounts of prothrombin in the blood, normal minor breaks in the vessels of the nasal passageways do not seal off as quickly, producing nosebleeds. **31.** Removal of most of Randy's stomach eliminated the production of intrinsic factor, which is essential for the absorption of vitamin B_{12} by intestinal cells. Thus Randy was prescribed vitamin B_{12} to prevent pernicious anemia, and he needed injections of vitamin B_{12} because if taken orally, his intestines could no longer absorb it.

Answers to Clinical Case Wrap-Up

PAGE 686 **1.** In Clare's sickle cell crisis, the sickled RBCs block blood vessels in her legs, producing severe pain. Additionally, sickled RBCs are piling up and blocking the circulation to her feet. The surrounding tissues are being starved of blood with its oxygen supply and its warmth (blood is 1 °C warmer than body temperature). So, her feet are freezing and the skin of her legs shows redness and swelling from local inflammation. **2.** Hydration is critical in sickle cell disease because water helps to maintain the proper volume of blood. When Clare's red blood cells become concentrated, the impact of the sickling process is increased.

Answers to Figure-Based Questions

FIGURE 19–3 Each molecule of heme contains an ion of iron (Fe^{2+}).

FIGURE 19–5 Red blood cells are produced in the red bone marrow.

FIGURE 19–6 Type A blood has RBCs with surface antigen A only and contains anti-B antibodies.

FIGURE 19–10 Granulocytes consist of basophils, eosinophils, and neutrophils. Granulocytes differentiate from myeloblasts.

Chapter 20

Answers to Checkpoints

PAGE 702 **1.** Damage to the semilunar valve of the right ventricle, or pulmonary valve, would affect blood flow to the pulmonary trunk. **2.** Contraction of the papillary muscles (just before the rest of the ventricular myocardium contracts) pulls on the chordae tendineae, which prevent the AV valves from opening back into the atria. **3.** The left ventricle is more muscular than the right ventricle because the left ventricle must generate enough force to propel blood throughout the body, except the lungs, whereas the right ventricle must generate only enough force to propel blood a few centimeters to the lungs. **PAGE 711** **4.** Autorhythmicity is the ability of cardiac muscle tissue to contract without neural or hormonal stimulation. **5.** The sinoatrial (SA) node is known as the cardiac pacemaker or the natural pacemaker. **6.** If the cells of the SA node did not function, the AV node would act as the pacemaker and set the heart rhythm. The heart would still continue to beat, but at a slower rate. **7.** If the impulses from the atria were not delayed at the AV node, they would be conducted through the ventricles so quickly by the bundle branches and Purkinje cells that the ventricles would begin contracting immediately, before the atria had finished their contraction. As a result, the ventricles would not be as full of blood as they could be, and the pumping of the heart would not be as efficient, especially during physical activity. **PAGE 716** **8.** The technical term for heart contraction is systole, and the term for heart relaxation is diastole. **9.** The phases of the cardiac cycle are atrial systole, atrial diastole, ventricular systole, and ventricular diastole. **10.** No. When pressure in the left ventricle first rises, the heart is contracting but no blood is leaving the heart. During this initial phase of contraction, both the AV valves and the semilunar valves are closed. The increase in pressure is the result of increased tension as the cardiac muscle contracts. When the blood pressure in the ventricle exceeds the blood pressure in the aorta, the aortic semilunar valves are forced open, and blood is rapidly ejected from the ventricle. **11.** An increase in the size of the QRS complex indicates a larger-than-normal amount of electrical activity during ventricular depolarization. One possible cause is an enlarged heart. Because more cardiac muscle is depolarizing, the magnitude of the electrical event would be greater. **PAGE 723** **12.** Cardiac output is the amount of blood pumped by the left ventricle in 1 minute. **13.** Caffeine acts directly on the conducting system and contractile cells of the heart, increasing the rate at which they depolarize. Drinking large amounts of caffeinated drinks would increase the heart rate. **14.** The heart pumps in proportion to the amount of blood that enters. A heart that beats too rapidly does not have sufficient time to fill completely between beats. Thus, when the heart beats too fast, very little blood leaves the ventricles and enters the

circulation, so tissues suffer damage from inadequate blood supply. **15.** Stimulating the acetylcholine receptors of the heart would slow the heart rate (parasympathetic neurons release acetylcholine). Since cardiac output is the product of stroke volume and heart rate, a reduction in heart rate will lower the cardiac output (assuming that the stroke volume remains the same or doesn't increase). **16.** The venous return fills the heart with blood, stretching the heart muscle. According to the Frank–Starling principle, the more the heart muscle is stretched, the more forcefully it will contract (to a point). The more forceful the contraction, the more blood the heart will eject with each beat (stroke volume). Therefore, increased venous return would increase the stroke volume (if all other factors are constant).
17. SV = EDV − ESV, so SV = 125 mL − 40 mL = 85 mL

Answers to Review Questions

PAGE 724

Level 1 Reviewing Facts and Terms

1. c **2.** b **3.** d **4.** d **5.** d **6.** a and b **7.** b **8.** **(a)** superior vena cava; **(b)** auricle of right atrium; **(c)** right ventricle; **(d)** left ventricle; **(e)** aortic arch; **(f)** left pulmonary artery; **(g)** pulmonary trunk; **(h)** auricle of left atrium **9.** **(a)** ascending aorta; **(b)** opening of coronary sinus; **(c)** right atrium; **(d)** cusp of tricuspid valve; **(e)** chordae tendineae; **(f)** right ventricle; **(g)** pulmonary valve; **(h)** left pulmonary veins; **(i)** left atrium; **(j)** aortic valve; **(k)** cusp of mitral valve; **(l)** left ventricle; **(m)** interventricular septum **10.** a **11.** a **12.** During ventricular contraction, tension in the papillary muscles pulls against the chordae tendineae, which keep the cusps of the AV valve from swinging into the atrium. This action prevents the backflow, or regurgitation, of blood into the atrium as the ventricle contracts. **13.** (1) The epicardium is the visceral layer of the serous pericardium, which covers the outer surface of the heart. (2) The myocardium is the muscular wall of the heart, which forms both atria and ventricles. It contains cardiac muscle tissue and associated connective tissues, blood vessels, and nerves. (3) The endocardium is made up of a simple squamous epithelium and an areolar layer that covers the inner surfaces of the heart, including the valves. **14.** The tricuspid valve (right atrioventricular valve) and the mitral valve (left atrioventricular valve) prevent the backflow of blood from the ventricles into the atria. The pulmonary and aortic semilunar valves prevent the backflow of blood from the pulmonary trunk and aorta into the right and left ventricles. **15.** SA node → AV node → AV bundle → right and left bundle branches → Purkinje fibers (into the mass of ventricular muscle tissue) **16.** The cardiac cycle comprises the events in a complete heartbeat, including a contraction–relaxation period for both atria and ventricles. The cycle begins with atrial systole as the atria contract and push blood into the relaxed ventricles. As the atria relax (atrial diastole), the ventricles contract (ventricular systole), forcing blood through the semilunar valves into the pulmonary trunk and aorta. The ventricles then relax (ventricular diastole). For the rest of the cardiac cycle, both the atria and ventricles are in diastole; passive filling occurs. **17.** The factors that regulate stroke volume are (1) preload, the stretch on the heart before it contracts; (2) contractility, the force of contraction of individual ventricular contractile cells; and (3) afterload, the pressure that must be exceeded before blood can be ejected from the ventricles.

Level 2 Reviewing Concepts

18. c **19.** a **20.** d **21.** a **22.** The SA node, which is composed of cells that exhibit rapid pacemaker potential, is the pacemaker of the heart. The AV node slows the impulse that signals contraction, because its cells are smaller than those of the conduction pathway. **23.** The first sound ("lubb"), which marks the start of ventricular contraction, is produced as the AV valves close and the semilunar valves open. The second sound ("dupp") occurs when the semilunar valves close and the AV valves open, marking the start of ventricular diastole. The third heart sound is associated with blood flow into the ventricles, and the fourth sound is associated with atrial contraction. Listening to the heart sounds (auscultation) is a simple and effective diagnostic tool. **24.** Stroke volume (SV) is the volume of blood ejected by a ventricle in a single contraction. Cardiac output (CO) is the amount of blood pumped by the left ventricle in 1 minute: CO (in mL/min) = HR (in beats/min) × SV (in mL/beat). **25.** Stroke volume and heart rate influence cardiac output. **26.** Sympathetic activation increases the heart rate and the force of contractions; parasympathetic stimulation decreases the heart rate and the force of contractions. **27.** All these hormones have positive inotropic effects, which means that they increase the strength of cardiac contraction (contractility).

Level 3 Critical Thinking and Clinical Applications

28. During tachycardia (an abnormally fast heart rate), there is less time between contractions for the heart to fill with blood again. Thus, over time the heart fills with less and less blood, and pumps less blood out. As the stroke volume decreases, the cardiac output decreases. When cardiac output decreases to the point where not enough blood reaches the brain, loss of consciousness occurs. **29.** Harvey probably has a regurgitating mitral valve, or left atrioventricular (AV) valve. When an AV valve fails to close properly, blood flowing back into the atrium produces a murmur. A murmur at the beginning of systole implicates the AV valve because this is the period when the valve has just closed and the blood in the ventricle is under increasing pressure; thus the likelihood of backflow is the greatest. A sound heard at the end of systole or the beginning of diastole would implicate a regurgitating semilunar valve—in this case, the aortic valve. **30.** Using CO = HR × SV, person 1 has a cardiac output of 4500 mL, and person 2 has a cardiac output of 8550 mL. According to the Frank–Starling principle, in a normal heart the cardiac output is directly proportional to the venous return. Thus, person 2 has the greater venous return. Ventricular filling decreases with increased heart rate; person 1 has the lower heart rate and therefore the longer ventricular filling time. **31.** By blocking calcium channels, verapamil will decrease the force of cardiac contraction, which directly lowers Karen's stroke volume.

Answers to Clinical Case Wrap-Up

PAGE 726 **1.** The "lubb" heard when the AV valves close, and the "dupp" heard when the semilunar valves close, would be muffled and difficult to hear because the sounds were transmitted through a pericardial cavity filled with blood. This dampens the sounds when heard through a stethoscope. **2.** The right atrium and right ventricle sit anteriorly, just deep to the sternum, and would likely be injured in this collision. Delicate structures that could also be injured include the anterior interventricular artery and great cardiac vein, running in the anterior interventricular sulcus. With a hard enough impact, the papillary muscles and chordae tendineae could rupture, leading to failure of the AV valves.

Answers to Figure-Based Questions

FIGURE 20–1 The **left** side of the heart supplies blood to the systemic circuit, while the **right** side of the heart supplies blood to the pulmonary circuit.

FIGURE 20–5 Blood will pass through the tricuspid valve (right AV valve), the pulmonary valve, the mitral valve (left AV valve), and finally the aortic valve.

FIGURE 20–11 When the electrical impulse reaches the AV node it is delayed about 100 msec.

FIGURE 20–15 The entry of sodium ions (Na^+) causes rapid depolarization. The entry of calcium ions (Ca^{2+}) causes the plateau. The exit of potassium ions (K^+) causes repolarization.

FIGURE 20–21 The sympathetic nervous system would increase heart rate. The cardioacceleratory center controls sympathetic neurons that increase heart rate.

FIGURE 20–24 Normal cardiac output at rest is about 5–7.5 L/min.

Chapter 21

Answers to Checkpoints

PAGE 736 **1.** The five general classes of blood vessels are arteries, arterioles, capillaries, venules, and veins. **2.** Blood vessels with thin walls and very little smooth muscle tissue in the tunica media are veins. Arteries and arterioles have a large amount of smooth muscle tissue in a thick, well-developed tunica media. **3.** In the arterial system, pressures are high enough to keep the blood moving forward. In the venous system, blood pressure is too low to keep the blood moving on toward the heart. Valves in veins prevent blood from flowing back toward the capillaries whenever the venous pressure drops. **4.** Fenestrated capillaries are located where fluids and small solutes move freely into and out of the blood, including endocrine glands, the choroid plexus of the brain, absorptive areas of the intestine, and filtration areas of the kidneys. **PAGE 745** **5.** The factors that contribute to total peripheral resistance are vascular resistance, vessel length, vessel luminal diameter, blood viscosity, and turbulence. **6.** In a healthy person, blood pressure is greater at the aorta than at the inferior vena cava. Blood, like other fluids, moves along a pressure gradient from areas of high pressure to areas of low pressure. If the pressure were higher in the inferior vena cava than in the aorta, the blood would flow backward. **7.** While Lailah was standing for a period of time, blood pooled in her lower limbs, which decreased venous return to her heart. In turn, cardiac output decreased, so less blood reached her brain, causing light-headedness and fainting. A hot day adds to this effect, because the loss of body water through sweating reduces blood volume. **8.** Mike's mean arterial pressure (MAP) is approximately 88.3 mm Hg; $70 + (125 - 70)/3 = 70 + 18.3 = 88.3$. **PAGE 752** **9.** Vasodilators promote the dilation of precapillary sphincters. Local vasodilators, such as decreased O_2 level or increased CO_2 level, act on tissues to accelerate blood flow. **10.** Pressure on the common carotid artery would decrease blood pressure at the baroreceptors in the carotid sinus. This decrease would lower the frequency of action potentials along the glossopharyngeal nerve (IX) to the medulla oblongata, and more sympathetic impulses would be sent to the heart. The net result would be an increase in the heart rate. **11.** Vasoconstriction of the renal artery would decrease both blood flow and blood pressure at the kidney. In response, the kidney would increase the amount of renin it releases, which in turn would lead to an increase in the level of angiotensin II. The angiotensin II would bring about increased blood pressure and increased blood volume. **PAGE 756** **12.** Blood pressure increases during exercise because (1) cardiac output increases and (2) resistance in visceral tissues increases. **13.** The immediate problem during hemorrhaging is maintaining adequate blood pressure and peripheral blood flow; the long-term problem is restoring normal blood volume. **14.** Both aldosterone and ADH promote fluid retention and reabsorption at the kidneys, preventing further reductions in blood volume. **PAGE 757** **15.** The two circuits of the cardiovascular system are the pulmonary circuit and the systemic circuit. **16.** The major patterns are as follows: (1) The peripheral distributions of arteries and veins on the body's left and right sides are generally identical, except near the heart, where the largest vessels connect to the atria or ventricles; (2) a single vessel may have several names as it crosses specific anatomical boundaries, making accurate anatomical descriptions possible; and (3) tissues and organs are usually serviced by several arteries and veins. **PAGE 758** **17.** The pulmonary arteries enter the lungs carrying deoxygenated blood, and the pulmonary veins leave the lungs carrying oxygenated blood. **18.** The path of blood through the lungs is right ventricle → pulmonary trunk → left and right pulmonary arteries → pulmonary arterioles → alveolar capillary network → pulmonary venules → pulmonary veins → left atrium. **PAGE 775** **19.** A blockage of the left subclavian artery would interfere with blood flow to the left arm. **20.** Compression of the common carotid arteries would decrease blood pressure at the carotid sinus and cause a rapid reduction in blood flow to the brain, resulting in a loss of consciousness. An immediate reflexive increase in heart rate and blood pressure would follow. **21.** Rupture of the celiac trunk would most directly affect the stomach, spleen, liver, and pancreas. **22.** The vein that is bulging in Noah's neck is the external jugular vein. **23.** A blockage of the popliteal vein would interfere with blood flow in the tibial and fibular veins (which form the popliteal vein) and the small saphenous vein (which joins the popliteal vein). **PAGE 778** **24.** Two umbilical arteries supply blood to the placenta, and one umbilical vein returns blood from the placenta. The umbilical vein then drains into the ductus venosus within the fetal liver. (Remember, arteries carry blood away from the heart, and veins carry blood to the heart.) **25.** This blood sample was taken from the umbilical vein, which carries oxygenated, nutrient-rich blood from the placenta to the fetus. **26.** Structures specific to the fetal circulation include two umbilical arteries, an umbilical vein, the ductus venosus, the foramen ovale, and the ductus arteriosus. In the newborn, the foramen ovale closes and persists as the fossa ovalis, a shallow depression; the ductus arteriosus persists as the ligamentum arteriosum, a fibrous cord; and the umbilical vessels and ductus venosus persist throughout life as fibrous cords. **27.** Components of the cardiovascular system affected by age include the blood, heart, and blood vessels. **28.** A thrombus is a stationary blood clot within the lumen of a blood vessel. **29.** An aneurysm is the ballooning out of a weakened arterial wall resulting from sudden pressure increases. **30.** The cardiovascular system provides other body systems with oxygen, hormones, nutrients, and white blood cells in blood, while removing carbon dioxide and metabolic wastes; it also transfers heat to body tissues. **31.** The skeletal system provides calcium needed for normal cardiac muscle contraction, and it protects developing blood cells in the red bone marrow. The cardiovascular system provides calcium and phosphate for bone deposition, delivers erythropoietin to red bone marrow, and transports parathyroid hormone and calcitonin to osteoblasts and osteoclasts.

Answers to Review Questions

PAGE 782

Level 1 Reviewing Facts and Terms

1. **(a)** brachiocephalic trunk; **(b)** brachial; **(c)** radial; **(d)** external iliac; **(e)** anterior tibial; **(f)** right common carotid; **(g)** left subclavian; **(h)** common iliac; **(i)** femoral **2.** b **3.** e **4.** c **5.** b **6.** b **7.** b **8.** d **9.** **(a)** external jugular; **(b)** brachial; **(c)** median cubital; **(d)** radial; **(e)** great saphenous; **(f)** internal jugular; **(g)** superior vena cava; **(h)** left and right common iliac; **(i)** femoral **10.** b **11.** b **12.** d **13.** c **14.** c **15.** c **16.** c **17.** **(a)** Capillary hydrostatic pressure forces fluid out of a capillary at the arterial end. **(b)** Blood colloid osmotic pressure causes the movement of fluid back into a capillary at its venous end. **18.** When an infant takes its first breath, the lungs expand and pulmonary vessels dilate. The smooth muscles in the ductus arteriosus contract, due to increased venous return from the lungs, isolating the pulmonary and aortic trunks, and blood begins flowing through the pulmonary circuit. As pressure increases in the left atrium, the valvular flap closes the foramen ovale, completing the vascular remodeling.

Level 2 Reviewing Concepts

19. b **20.** b **21.** a **22.** Artery walls are generally thicker and contain more smooth muscle and elastic fibers, enabling them to resist and adjust to the pressure generated by the heart. Venous walls are thinner; the pressure in veins is less than that in arteries. Arteries constrict more than veins do when not expanded by blood pressure, due to a greater degree of elastic tissue. Finally, the endothelial lining of an artery has a pleated appearance because it cannot contract and so forms folds. The lining of a vein looks like a typical endothelial layer. **23.** Capillary walls are thin, so distances for diffusion are short. Continuous capillaries have small gaps between adjacent endothelial cells that permit the diffusion of water and small solutes into the surrounding interstitial fluid but prevent the loss of blood cells and plasma proteins. Fenestrated capillaries contain pores that permit very rapid exchange of fluids and solutes between interstitial fluid and plasma. The walls of arteries and veins are several cell layers thick and are not specialized for diffusion. **24.** Contraction of the surrounding skeletal muscles squeezes venous blood toward the heart. This mechanism, the muscular pump, is assisted by the presence of valves in the veins, which prevent backflow of the blood. The respiratory pump, which results from the increase in internal pressure of the thoracic cavity during exhalation, pushes venous blood into the right atrium.
25. Cardiac output and peripheral blood flow are directly proportional to blood pressure. Blood pressure is closely regulated by a combination of neural and hormonal mechanisms. The resistance of the vascular system opposes the movement of blood, so blood flow is inversely proportional to the resistance. Sources of peripheral resistance include vascular resistance, blood viscosity, and turbulence. **26.** The brain receives arterial blood from four arteries that form anastomoses within the cranium. An interruption of any one vessel will not compromise the blood flow to the brain.
27. The cardioacceleratory and vasomotor centers are stimulated when general sympathetic activation occurs. The result is an increase in cardiac output and blood pressure. When the parasympathetic division is activated, the cardioinhibitory center is stimulated, reducing cardiac output.

Level 3 Critical Thinking and Clinical Applications

28. Fluid loss lowers blood volume, leading to sympathetic stimulation, which elevates blood pressure. So, Bob's blood pressure is high instead of low. **29.** Antihistamines and decongestants are sympathomimetic drugs; they have the same effects on the body as does stimulation of the sympathetic nervous system. In addition to the desired effects of counteracting the symptoms of the allergy, these medications can produce an increased heart rate, increased stroke volume, and increased peripheral resistance, all of which will contribute to elevating blood pressure. In a person with hypertension (high blood pressure), these drugs would aggravate this condition, with potentially hazardous consequences. **30.** When Jolene stood up rapidly, gravity caused her blood volume to move to the lower parts of her body away from the heart, decreasing venous return. The decreased venous return resulted in a decreased end-diastolic volume (EDV), leading to a decreased stroke volume and cardiac output. In turn, blood flow to the brain decreased, so the diminished oxygen supply caused her to be light-headed and feel faint. This reaction doesn't happen all the time because as soon as the pressure drops due to inferior movement of blood, baroreceptors in the aortic arch and carotid sinuses trigger the baroreceptor reflex. Action potentials are carried to the medulla oblongata, where appropriate responses are integrated. In this case, we would expect an increase in peripheral resistance to compensate for the decreased blood pressure. If this doesn't compensate enough for the drop, then an increase in heart rate and force of contraction would occur. Normally, these responses occur so quickly that changes in pressure following changes in body position go unnoticed.

Answers to Clinical Case Wrap-Up

PAGE 783 **1.** Calcification follows the degeneration of smooth muscle in the tunica media of arteries. Calcification in arteries is related to atherosclerotic injury of the vessels. Calcification does not occur in normal, young, healthy blood vessel walls.
2. Atherosclerosis is never seen in the venous system because veins are never subjected to the pressures of the arterial system. In the average person the systemic arterial pressure is over 100 mm Hg. This causes wear and tear in the arterial walls, leading to gradual degeneration of the smooth muscle in the tunica media, followed by deposition of calcium salts, and then by deposition of atherosclerotic plaques. The pressure within the venous system is always quite low (usually around 16 mm Hg). Damage to the endothelial surfaces does not get started.

Answers to Figure-Based Questions

FIGURE 21–2 (a) Veins have larger lumens than arteries. (b) Arteries, specifically muscular arteries, have the thickest tunica media.

FIGURE 21–7 Factors that would increase vascular resistance would be increased vessel length and turbulence. Increased vessel luminal diameter would decrease vascular resistance.

FIGURE 21–12 For blood flow to be autoregulated, physical stress (such as trauma or high temperature), chemical changes, or increased tissue activity are stimuli that signal inadequate local blood pressure and blood flow.

FIGURE 21–15 For short-term regulation of a decrease in blood pressure and blood volume, sympathetic activation and the release of the adrenal hormones epinephrine (E) and norepinephrine (NE) increase cardiac output and peripheral vasoconstriction, which will increase blood pressure.

FIGURE 21–21 The superficial temporal, maxillary, occipital, facial, and lingual arteries are all branches of the external carotid artery.

FIGURE 21–29 The right and left brachiocephalic veins, the mediastinal veins, and the azygos vein directly drain into the superior vena cava.

Chapter 22

Answers to Checkpoints

PAGE 796 **1.** A pathogen is any disease-causing organism, such as a virus, bacterium, fungus, or parasite that can survive and even thrive inside the body. **2.** The components of the lymphatic system are lymph, lymphatic vessels, lymphoid cells (lymphocytes and other cells), primary lymphoid tissues and organs (red bone marrow and the thymus), and secondary lymphoid tissues and organs (lymph nodes, tonsils, MALT, the appendix, and the spleen). **3.** A blockage of the thoracic duct would impair the drainage of lymph from inferior to the diaphragm and from the left side of the head and thorax, slowing the return of lymph to the venous blood and promoting the accumulation of fluid in the limbs (lymphedema). **4.** A lack of thymic hormones would drastically reduce the population of T lymphocytes. **5.** Lymph nodes enlarge during some infections because lymphocytes and phagocytes in the nodes multiply to defend against the infectious agent. **PAGE 797** **6.** Immunity is the body's ability to resist infectious organisms or other substances that could damage tissues and organs. Resistance is the ability of the body to maintain its immunity. **7.** Innate (nonspecific) immunity is a type of body defense that you are born with and does not distinguish among different threats. Adaptive (specific) immunity is acquired after birth and after exposure to a particular antigen. **PAGE 805** **8.** A decrease in the number of monocyte-forming cells in red bone marrow would result in fewer macrophages of all types, including stellate macrophages (Kupffer cells) of the liver, microglia of the central nervous system, and alveolar macrophages. **9.** A rise in the level of interferon suggests a viral infection. Interferon does not help an infected cell, but "interferes" with replication of the virus and thus the virus's ability to infect other cells. **10.** Pyrogens increase body temperature (produce a fever) by stimulating the temperature control area of the pre-optic nucleus of the hypothalamus. **PAGE 809** **11.** In cell-mediated (cellular) immunity, T cells defend against abnormal cells and pathogens inside cells. In antibody-mediated (humoral) immunity, B cells secrete antibodies that defend against antigens and pathogens in body fluids. **12.** The four major types of T cells are cytotoxic T cells, helper T cells, regulatory T cells, and memory T cells. **13.** The two forms of active immunity are naturally acquired active immunity and artificially acquired active immunity; the two forms of passive immunity are naturally acquired passive immunity and artificially acquired passive immunity. **14.** The four general properties of adaptive immunity are specificity, versatility, memory, and tolerance. **PAGE 816** **15.** Abnormal peptides in the cytoplasm of a cell can become attached to MHC (major histocompatibility complex) proteins and then be displayed on the surface of the cell's plasma membrane. The recognition of such displayed peptides by T cells can initiate an immune response. **16.** A decrease in the number of cytotoxic T cells would affect cell-mediated immunity, reducing the effectiveness of cytotoxic T cells in killing foreign cells and virus-infected cells. **PAGE 821** **17.** Without helper T cells—which promote B cell division, the maturation of plasma cells, and antibody production by plasma cells—the antibody-mediated immune response would probably not occur. **18.** Sensitization is the process by which a B cell becomes able to react with a specific antigen. **19.** Plasma cells produce and secrete antibodies, so observing an elevated number of plasma cells would lead us to expect increasing levels of antibodies in the blood. **20.** An antibody molecule consists of two parallel pairs of polypeptide chains: one pair of heavy chains and one pair of light chains. Each chain contains both constant segments and variable segments. **21.** Because production of a secondary response depends on the presence of memory B cells and memory T cells formed during a primary response, the secondary response would be more negatively affected by a lack of memory B cells for a particular antigen. **PAGE 826** **22.** A developing fetus is protected primarily by naturally acquired passive immunity, through the maternal production of IgG antibodies that have crossed the placenta from the mother's bloodstream. **23.** Stress can interfere with the immune response by depressing inflammation, reducing the number and activity of phagocytes, and inhibiting interleukin secretion. **24.** An autoimmune disorder is a condition that results when the immune system's sensitivity to normal cells and tissues causes the production of autoantibodies. **PAGE 829** **25.** Elderly people are more susceptible to viral and bacterial infections because the number of helper T cells declines with age and B cells are less responsive, so antibody levels rise more slowly after antigen exposure. **26.** As one ages, immune surveillance declines, so cancerous cells are not eliminated as effectively. **27.** Glucocorticoids released by the endocrine system have anti-inflammatory effects; thymic hormones stimulate the development and maturation of lymphocytes; and many hormones affect immune function. The thymus secretes thymic hormones, and cytokines affect cells throughout the body. **28.** The lymphatic system provides defenses against infection; performs immune surveillance to eliminate cancer cells; and returns tissue fluid to the circulation.

Answers to Review Questions

PAGE 832

Level 1 Reviewing Facts and Terms

1. **(a)** tonsil; **(b)** cervical lymph nodes; **(c)** right lymphatic duct; **(d)** thymus; **(e)** cisterna chyli; **(f)** lumbar lymph nodes; **(g)** appendix; **(h)** lymphatics of lower limb; **(i)** lymphatics of upper limb; **(j)** axillary lymph nodes; **(k)** thoracic duct; **(l)** lymphatics of mammary gland; **(m)** spleen; **(n)** mucosa-associated lymphoid tissue (MALT); **(o)** pelvic lymph nodes; **(p)** inguinal lymph nodes; **(q)** red bone marrow **2.** b **3.** d **4.** c **5.** c **6.** c **7.** a **8.** c **9.** d **10.** d **11.** c **12.** Primary lymphoid tissues and organs: (1) red bone marrow: maintain normal lymphocyte populations and other defense cells, such as monocytes and macrophages; (2) thymus: production of mature T cells and hormones that promote immune function. Secondary lymphoid tissues and organs: (1) spleen: filtration of blood, recycling of red blood cells, detection of blood-borne pathogens or toxins; (2) lymph nodes: filtration of lymph, detection of pathogens, initiation of immune response; (3) lymphoid nodules and MALT: defense of entrance and passageways of digestive tract and protection of epithelia lining the respiratory, urinary, and reproductive tracts against pathogens and foreign proteins or toxins. **13.** **(a)** lymphocytes responsible for cell-mediated immunity; **(b)** stimulate the activation and function of T cells and B cells; **(c)** inhibit the activation and function of both T cells and B cells; **(d)** produce and secrete antibodies; **(e)** recognize and destroy abnormal cells; **(f)** produce interleukin-7, which promotes the differentiation of B cells; **(g)** regulate T cell development and function; **(h)** interfere with viral replication inside the cell and stimulate the activities of macrophages and NK cells; **(i)** reset the body's thermostat, causing a rise in body temperature (fever); **(j)** provide cell-mediated immunity, which defends against abnormal cells and pathogens inside cells; **(k)** provide humoral immunity, which defends against antigens and pathogens in the body (but not inside cells); **(l)** enhance innate (nonspecific) defenses and increase T cell sensitivity and stimulate B cell activity; **(m)** slow tumor growth and kill sensitive tumor cells; **(n)** stimulate the production of blood cells in the red bone marrow and lymphocytes in lymphoid tissues and organs **14.** (1) T cells, derived from the thymus; (2) B cells, derived from red bone marrow; and (3) NK cells, derived from red bone marrow **15.** Innate (nonspecific) immunity: (1) physical barriers; (2) phagocytic cells; (3) immune surveillance; (4) interferons; (5) complement system; (6) inflammation; and (7) fever

Level 2 Reviewing Concepts

16. c **17.** c **18.** b **19.** b **20.** Complement activation can rupture the target cell by forming a membrane attack complex (MAC). It also enhances phagocytosis (opsonization) and inflammation. Interferon interferes with viral replication inside virus-infected cells by triggering the production of antiviral proteins. **21.** Cytotoxic T cells kill by rupturing the target cell's plasma membrane, by stimulating lymphotoxin secretion, or by activating genes in the nucleus that program cell death (apoptosis). **22.** Formation of an antigen–antibody complex eliminates antigens by neutralization; by agglutination and precipitation; by activating complement; by attracting phagocytes; by opsonization; by stimulating inflammation; or by preventing bacterial and viral adhesion. **23.** Innate immunity is genetically programmed; an example is immunity to fish diseases. Naturally acquired active immunity develops after birth from contact with pathogens; an example is exposure to chickenpox in grade school. Artificially acquired active immunity develops after intentional exposure to a pathogen; an example is administration of measles vaccine. Artificially acquired passive immunity is temporary immunity provided by injection with antibodies produced in another organism, such as antibodies against rabies. Naturally acquired passive immunity is gained by acquiring antibodies from mother's milk or placental exchange. **24.** The injections are timed to trigger the primary and secondary responses of the immune system. Upon first exposure to hepatitis antigens, B cells produce daughter cells that differentiate into plasma cells and memory B cells. The plasma cells begin producing antibodies, which represent the primary response to exposure. However, the primary response does not maintain elevated antibody levels for long periods, so the second and third injections are necessary to trigger secondary responses, when memory B cells differentiate into plasma cells and produce antibody concentrations that remain high much longer.

Level 3 Critical Thinking and Clinical Applications

25. Yes, the crime lab could determine whether the sample is blood plasma, which contains IgM, IgG, IgD, and IgE, or semen, which contains only IgA. **26.** Ted cannot yet know whether he'll come down with the measles. Ted's elevated blood IgM level and antibody titer indicates that he is in the early stages of a primary response to the measles virus. If his immune response proves unable to control and then eliminate the virus, Ted will develop the measles. **27.** In a radical mastectomy, lymph nodes in the nearby axilla and surrounding region are removed along with the cancerous breast to try to prevent the spread of cancer cells by the lymphatic system. Lymphatic vessels from the limb on the affected side are tied off, and because there is no place for the lymph to drain, over time lymphedema causes swelling of the limb. **28.** A key characteristic of cancer cells is their ability to metastasize—to break free from a tumor and form new tumors in other tissues. The primary route of metastasis is the lymphatic system, including the lymph nodes. Examination of regional lymph nodes for the presence of cancer cells can help the physician determine if the cancer was caught in an early stage or whether it has started to spread to other tissues. **29.** Allergies occur when allergens bind to specific IgE antibodies that are bound to the surface of mast cells and basophils. A person becomes allergic when he or she develops IgE antibodies for a specific allergen. Theoretically, at least, a molecule that would bind to the specific IgE for ragweed allergen and prevent the allergen from binding should help to relieve the signs and symptoms of the allergy.

Answers to Clinical Case Wrap-Up

PAGE 833 **1.** Innate immunity is genetically determined and present at birth. For example, skin and mucous membranes provide an innate physical and chemical barrier to infecting organisms. All humans are susceptible to chickenpox, an infectious disease. The immunity developed after a chickenpox vaccination is artificially acquired adaptive immunity, developed to a specific antigen, the varicella-zoster virus (VZV). **2.** Baby Ruthie has developed lifelong immunity to chickenpox as a result of having the disease. Her immune memory will provoke a secondary immune response if she is exposed to the varicella-zoster virus again. The recommended schedule (according to the Centers for Disease Control and Prevention, CDC) for varicella-zoster vaccination is 12–15 months of age. However, there is no need for Ruthie to get a chickenpox vaccination after having had the disease.

Answers to Figure-Based Questions

FIGURE 22–3 Lymphatic vessels have valves, similar to veins, which allow lymph to flow in only one direction.

FIGURE 22–8 Lymphocytes are found in the white pulp of the spleen, whereas red blood cells are found in the red pulp of the spleen.

FIGURE 22–12 Interferon alpha attracts and stimulates NK cells as a way to enhance viral resistance to viral infection.

FIGURE 22–17 Artificially acquired active immunity develops after receiving a vaccine.

FIGURE 22–25 During the primary response, IgM peaks sooner than IgG. During the secondary response, the level of IgG is higher than the level of IgM.

FIGURE 22–26 After the appearance of bacteria, NK cells are present for the shortest amount of time (typically less than 1 week).

Chapter 23

Answers to Checkpoints

PAGE 838 **1.** Functions of the respiratory system include providing an extensive surface area for gas exchange between air and circulating blood; moving air to and from exchange surfaces along the respiratory passageways; protecting respiratory surfaces from dehydration, temperature changes, and pathogens; producing sounds; and detecting odors. **2.** The two anatomical divisions of the respiratory system are the upper respiratory system and the lower respiratory system. **3.** The respiratory mucosa lines the conducting portion of the respiratory tract. **PAGE 841** **4.** The upper respiratory system consists of the nose, nasal cavity, paranasal sinuses, and pharynx. **5.** The rich vascularization of the nose delivers body heat to the nasal cavity, so inhaled air is warmed before it leaves the nasal cavity and moves toward the lungs. The heat also evaporates moisture from the epithelium to humidify the incoming air. **6.** The lining of the nasopharynx, which receives air from only the nasal cavity, is the same as that of the nasal cavity: a pseudostratified ciliated columnar epithelium. Because the oropharynx and laryngopharynx receive air from the nasal cavity and potentially abrasive food from the oral cavity, they have a more protective lining: a stratified squamous epithelium. **PAGE 846** **7.** The unpaired laryngeal cartilages include the thyroid cartilage, cricoid cartilage, and epiglottis. The paired cartilages are the arytenoid cartilages, corniculate cartilages, and cuneiform cartilages. **8.** The trachea conveys air between the larynx and main bronchi; cilia and the mucus produced by epithelial cells also protect the respiratory tree by trapping inhaled debris and sweeping it toward the pharynx, where it is removed through coughing or swallowing. **9.** The tracheal cartilages are C-shaped to allow space for expansion of the esophagus when food or liquid is swallowed. **10.** Objects are more likely to be lodged in the right main bronchus because it is slightly larger and more vertical than the left main bronchus. **PAGE 848** **11.** Without surfactant, the alveoli would collapse as a result of surface tension in the thin layer of water that moistens the alveolar surfaces. **12.** Air passing through the glottis flows into the larynx and through the trachea. From there, the air flows into a main bronchus, which supplies the lungs. In the lungs, the air passes to lobar bronchi, segmental bronchi, bronchioles, a terminal bronchiole, a respiratory bronchiole, an alveolar duct, an alveolar sac, an alveolus, and ultimately to the blood air barrier. **PAGE 851** **13.** The pulmonary arteries supply blood to the gas-exchange surfaces; the bronchial capillaries, supplied by the bronchial arteries, which branch from the thoracic aorta, supply the conducting portions of the respiratory system. **14.** The pleura is a serous membrane that secretes pleural fluid, which lubricates the opposing parietal and visceral surfaces to prevent friction during breathing. **PAGE 852** **15.** External respiration includes all the processes involved in the exchange of oxygen and carbon dioxide between the body's interstitial fluids and the external environment. Internal respiration is the absorption of oxygen and the release of carbon dioxide by the body's cells. **16.** The integrated steps involved in external respiration are pulmonary ventilation (breathing), gas diffusion across the blood air barrier, and the transport of oxygen and carbon dioxide. **PAGE 859** **17.** Compliance is the ease with which the lungs expand. Factors affecting compliance include (a) the connective tissue structure of the lungs, (b) the level of surfactant production, and (c) the mobility of the thoracic cage. **18.** The measurable pulmonary volumes are tidal volume (V_T), expiratory reserve volume (ERV), residual volume, and inspiratory reserve volume (IRV). **19.** When the rib penetrates Mark's chest wall, it also penetrates the left pleural cavity, allowing atmospheric air to enter and increase the intrapleural pressure within the pleural cavity (a condition called pneumothorax). As a result, the natural elasticity of the left lung may cause the lung to collapse, a condition called atelectasis. **20.** Because the fluid produced in pneumonia takes up space that would normally be occupied by air, vital capacity would decrease. **PAGE 864** **21.** Dalton's law states that each gas in a mixture exerts a pressure equal to its relative abundance. **22.** Henry's law states that at a constant temperature, the quantity of a particular gas that will dissolve in a liquid is directly proportional to the partial pressure of that gas. **PAGE 868** **23.** Carbon dioxide is transported in the bloodstream as bicarbonate ions, bound to hemoglobin, or dissolved in the plasma. **24.** During exercise, active skeletal muscles generate heat, lactate, and hydrogen ions. A combination of increased body temperature and decreased pH causes hemoglobin to release more oxygen when you are exercising than when you are resting. **25.** Blockage of the trachea would interfere with the body's ability to gain oxygen and eliminate carbon dioxide. Because most carbon dioxide is transported in blood as bicarbonate ions formed from the dissociation of carbonic acid, an inability to eliminate carbon dioxide would result in an excess of hydrogen ions, which decreases blood pH. **PAGE 876** **26.** Exciting the pneumotaxic centers would inhibit the inspiratory and apneustic centers, which would result in shorter and more rapid breaths. **27.** Peripheral chemoreceptors are more sensitive to the carbon dioxide level than to the oxygen level. When carbon dioxide dissolves, it forms carbonic acid, which dissociates and releases hydrogen ions, thereby decreasing pH and altering cell or tissue activity. **28.** Johnny isn't likely to turn blue and die. When he holds his breath, the blood carbon dioxide level is increasing, causing increased stimulation of the inspiratory center and forcing him to breathe again. **PAGE 877** **29.** Aging results in deterioration of elastic tissue (reducing compliance), decreased vital capacity, arthritic changes in rib articulations, decreased flexibility at costal cartilages, and some degree of emphysema. **PAGE 879** **30.** The respiratory system provides oxygen and eliminates carbon dioxide for all body systems. **31.** The respiratory system provides alveolar phagocytes and respiratory defenses to trap pathogens and protect deeper tissues. Tonsils within the lymphatic system protect against respiratory infection, and lymph drainage from the lungs mobilizes defenses to ward off infection.

Answers to Review Questions

PAGE 882

Level 1 Reviewing Facts and Terms

1. d **2.** c **3.** a **4.** c **5.** c **6.** **(a)** nasal cavity; **(b)** pharynx; **(c)** right lung; **(d)** nose; **(e)** larynx; **(f)** trachea; **(g)** bronchus; **(h)** bronchioles; **(i)** left lung **7.** Since the air Brad is breathing is dry, large amounts of moisture are leaving the mucus in his respiratory tract to humidify inhaled air. Drying makes the mucus viscous and makes it difficult for the cilia to move, so mucus builds up, producing nasal congestion by morning. When Brad showers and drinks fluids, body water is replaced, so the mucus loosens up and can be moved along as usual. **8.** The upper respiratory system consists of the nose, nasal cavity, paranasal sinuses, and pharynx. The lower respiratory system consists of the larynx, trachea, bronchi, bronchioles, and alveoli of the lungs. **9.** The regions of the pharynx are the superior nasopharynx, where the nasal cavity opens into the pharynx; the middle oropharynx, posterior to the oral cavity; and the inferior laryngopharynx, which is posterior to the hyoid bone and glottis. **10.** The thyroid cartilage forms the anterior walls of the larynx; the cricoid cartilage protects the glottis and the entrance to the trachea; the epiglottis forms a lid over the glottis; the arytenoid cartilages and the corniculate cartilages are involved in the formation of sound; and the cuneiform cartilages are found in the folds of the larynx. **11.** The steps of external respiration are **(a)** pulmonary ventilation (breathing); **(b)** gas diffusion across the blood air barrier and between blood and interstitial fluids; and **(c)** the transport of oxygen and carbon dioxide between alveolar and peripheral capillaries. **12.** Fetal hemoglobin has a higher affinity for oxygen than does adult hemoglobin. Thus, it binds more of the oxygen that is present, enabling it to "steal" oxygen from the maternal hemoglobin. **13.** Carbon dioxide is transported in the blood as bicarbonate ions, bound to hemoglobin, and dissolved in the plasma.

Level 2 Reviewing Concepts

14. c **15.** d **16.** c **17.** b **18.** The nasal cavity cleanses, moistens, and warms inhaled air, whereas the mouth does not. Drier air entering through the mouth can irritate the trachea and cause throat soreness. **19.** Smooth muscle tissue in the walls of bronchioles allows changes in airway diameter (bronchodilation or bronchoconstriction), which provides control of the flow and distribution of air within the lungs, just as vasodilation and vasoconstriction of the arterioles control blood flow and blood distribution. **20.** Pneumocytes type II produce surfactant, which reduces surface tension in the water coating the alveolar surface. Without surfactant, the surface tension would be so high that the alveoli would collapse. **21.** Pulmonary ventilation, the physical movement of air into and out of the respiratory tract, maintains adequate alveolar ventilation. Alveolar ventilation, air movement into and out of the alveoli, prevents the buildup of carbon dioxide in the alveoli and ensures a continuous supply of oxygen that keeps pace with absorption by the bloodstream. **22. (a)** Boyle's law describes the inverse relationship between gas pressure and volume: If volume decreases, pressure rises; if volume increases, pressure falls. It is the basis for the direction of air movement in pulmonary ventilation. **(b)** Dalton's law states that each of the gases that make up a mixture of gases contributes to the total pressure in proportion to its relative abundance; that is, all the partial pressures added together equal the total pressure exerted by the gas mixture. This relationship is the basis for the calculation of the partial pressures of oxygen and carbon dioxide, and their exchange between blood and alveolar air. **(c)** Henry's law states that, at a given temperature, the amount of a particular gas that dissolves in a liquid is directly proportional to the partial pressure of that gas. Henry's law underlies the diffusion of gases between capillaries, alveoli, and interstitial fluid. **23.** Both sneezing and coughing involve a temporary cessation of respiration, known as apnea. **24.** Pulmonary volumes are determined experimentally and include tidal volume (averaging 500 mL), expiratory reserve volume (approximately 1200 mL), residual volume (averaging 1200 mL), minimal volume (30–120 mL), and inspiratory reserve volume (approximately 3300 mL). Respiratory capacities include inspiratory capacity, functional residual capacity, vital capacity, and total lung capacity. The difference between the two measures is that respiratory capacities are the sums of various pulmonary volumes. **25.** The DRG is the inspiratory center that contains neurons that control lower motor neurons innervating the external intercostal muscles and the diaphragm. The DRG functions in every respiratory cycle, whether quiet or forced. The VRG functions only during forced respiration—active exhalation and maximal inhalation. The neurons involved with active exhalation are sometimes said to form an expiratory center.

Level 3 Critical Thinking and Clinical Applications

26. AVR = respiratory rate × (tidal volume – dead space). In this case, the dead air space is 200 mL (the anatomical dead air space [150 mL] plus the volume of the snorkel [50 mL]); therefore, AVR = respiratory rate × (500 – 200). To maintain an AVR of 6.0 L/min, or 6000 mL/min, the respiratory rate must be 6000/(500 – 200), or 20 breaths per minute. **27.** A person with chronic emphysema has constantly elevated blood levels of P_{CO_2} due to an inability to eliminate CO_2 efficiently because of physical damage to the lungs. Over time, the brain ignores the stimulatory signals produced by the increased CO_2 and begins to rely on information from peripheral chemoreceptors to set the pace of breathing. (In other words, accommodation has occurred.) The peripheral chemoreceptors also accommodate to the elevated CO_2 and respond primarily to the level of O_2 in the blood, increasing breathing when O_2 levels are low and decreasing breathing when O_2 levels are high. When pure O_2 was administered, chemoreceptors responded with fewer action potentials to the medulla oblongata, so Mr. B. stopped breathing. **28.** Cary's hyperventilation resulted in abnormally low P_{CO_2}. This reduced his urge to breathe, so he stayed underwater longer, unaware that his P_{CO_2} was dropping to the point of loss of consciousness. **29.** In anemia, the blood's ability to carry oxygen is decreased due to the lack of functional hemoglobin, red blood cells, or both. Anemia does not interfere with the exchange of carbon dioxide within the alveoli, nor with the amount of oxygen that will dissolve in the plasma. Because chemoreceptors respond to dissolved gases and pH, as long as the pH and the concentrations of dissolved carbon dioxide and oxygen are normal, ventilation patterns should not change significantly. **30.** The obstruction in Doris's right lung would prevent gas exchange. Thus, blood moving through the right lung would not be oxygenated and would retain carbon dioxide, which would lead to a lower blood pH than that of blood leaving the left lung. The lower pH for blood in the right lung would shift the oxygen–hemoglobin saturation curve to the left (the Bohr effect) as compared with the curve for the left lung.

Answers to Clinical Case Wrap-Up

PAGE 883 **1.** Breathing can be obstructed in the nose (a bad cold, growths such as nasal polyps, a broken nasal septum), in the sinuses (a seasonal allergy or sinus infection), in the oral cavity (an enlarged or swollen tongue), or in the pharynx (a swollen, sore throat). **2.** Children who have frequent URT infections often have huge, infected tonsils whose locations (*lingual, palatine,* and *pharyngeal*) crowd the back of the throat. These children can become mouth-breathers. A tonsillectomy (surgical removal of the tonsils) may be indicated to help the child breathe better and sleep through the night.

Answers to Figure-Based Questions

FIGURE 23–2 The conducting portion of the respiratory tract is lined with pseudostratified ciliated columnar epithelium.

FIGURE 23–8 Smooth muscle wraps around a respiratory bronchiole and can change the diameter of the airway.

FIGURE 23–9 The right lung has three lobes, the left lung has two lobes, and the left lung has a cardiac notch.

FIGURE 23–14 The primary muscles of respiration are the diaphragm and the external intercostal muscles.

FIGURE 23–16 The inspiratory capacity is a total of the tidal volume and inspiratory reserve volume.

FIGURE 23–20 More oxygen is released when the pH is *lower* than normal and when temperature is *higher* than normal.

FIGURE 23–26 Inhibition of medulla oblongata chemoreceptors and respiratory muscles will decrease respiratory rate, the elimination of CO_2 at alveoli, and increase **arterial** P_{CO_2}.

Chapter 24

Answers to Checkpoints

PAGE 892 **1.** Organs of the digestive system include the oral cavity, pharynx, esophagus, stomach, small intestine, large intestine, and accessory organs (teeth, tongue, salivary glands, liver, pancreas, and gallbladder). **2.** The main function of the digestive system is to break down food, absorb the nutrients that can be used by body cells, and eliminate the remaining waste. **3.** The six primary processes of the digestive system include the following: (1) ingestion = entry of food; (2) mechanical digestion and propulsion = crushing and shearing of food to make it more susceptible to enzymes and the propelling of food along the digestive tract; (3) chemical digestion = the chemical breakdown of food into smaller products for absorption; (4) secretion = the release of water, acids, and other substances by the epithelium of the digestive tract and by glandular organs; (5) absorption = movement of digested particles across the digestive epithelium and into the interstitial fluid of the digestive tract; and (6) defecation= the elimination of wastes as feces from the body. **4.** The mesenteries—sheets consisting of two layers of serous membrane connected by their loose connective tissue—support and stabilize the organs in the abdominopelvic cavity and provide a route for the associated blood vessels, nerves, and lymphatic vessels. **5.** The layers of the gastrointestinal tract, beginning at the surface of the lumen, are the mucosa (adjacent to the lumen), submucosa, muscular layer, and serosa. **6.** The waves of contractions responsible for peristalsis are more efficient in propelling intestinal contents than segmentation, which is basically a churning action that mixes intestinal contents with digestive fluids. **7.** A drug that blocks parasympathetic stimulation, which increases muscle tone and activity in the digestive tract, would slow peristalsis. **PAGE 898** **8.** Structures associated with the oral cavity include the tongue, teeth, and salivary glands. **9.** Stratified squamous epithelium lines the oral cavity. It protects against friction or abrasion by food. **10.** Damage to the parotid glands, which secrete the carbohydrate-digesting enzyme salivary amylase, would interfere with the digestion of complex carbohydrates. **11.** The incisors are the teeth most useful for chopping (or cutting or shearing) pieces of relatively rigid food, such as raw vegetables. **12.** The fauces are the dividing line between the oral cavity and the pharynx. **PAGE 900** **13.** The pharynx is an anatomical space that receives a food bolus or liquid and passes it to the esophagus as part of the swallowing process. Air also passes through the pharynx. **14.** The structure connecting the pharynx to the stomach is the esophagus. **15.** The muscular layer of the esophagus is an unusual segment of the digestive tract because it (1) contains skeletal muscle cells along most of the length of the esophagus and (2) is surrounded by an adventitia. **16.** Muscles associated with the pharynx and deglutition are the palatal muscles, the palatopharyngeus, stylopharyngeus, and pharyngeal constrictors. **17.** When the soft palate and larynx elevate and the glottis closes, swallowing (deglutition) is occurring. **PAGE 905** **18.** The four major regions of the stomach are the cardia, fundus, body, and pyloric part. **19.** The low pH of the stomach creates an acidic environment that kills most microorganisms ingested with food, denatures proteins and inactivates most enzymes in food, helps break down plant cell walls and meat connective tissue, and activates pepsin. **20.** Large meals, especially meals with a high protein content, stimulate increased stomach acid secretion. When gastric glands are secreting, bicarbonate ions enter the bloodstream and increase the blood pH. This phenomenon is referred to as the alkaline tide. **21.** The vagus nerves (X) contain parasympathetic motor fibers that can stimulate gastric secretions, even if food is not present in the stomach (the cephalic phase of gastric activity). Cutting the branches of the vagus nerves that supply the stomach would prevent this type of secretion from occurring and thereby reduce the likelihood of ulcer formation. **PAGE 915** **22.** Damage to the exocrine pancreas would most impair the digestion of fats (lipids), because it is the primary source of lipases. Even though such damage would also reduce carbohydrate and protein digestion, enzymes for digesting these nutrients are produced by other digestive system structures, including the salivary glands (carbohydrates), the small intestine (carbohydrates and proteins), and the stomach (proteins). **23.** The portal triad is a region of a hepatic lobule made up of the interlobular artery, interlobular vein, and interlobular bile duct. **24.** The gallbladder concentrates and stores bile before secreting it into the small intestine. **PAGE 921** **25.** The three regions of the small intestine from proximal to distal are the duodenum, jejunum, and ileum. **26.** The small intestine has several adaptations that increase its surface area and thus its absorptive capacity for nutrients. The walls of the small intestine have folds called circular folds. The mucosa that covers the circular folds has fingerlike projections, the villi. The epithelial cells that cover the villi have an exposed surface covered by small fingerlike projections, the microvilli. In addition, the small intestine has a very rich supply of blood vessels and lymphatic vessels, which transport the nutrients that are absorbed. **27.** A high-fat meal would raise the cholecystokinin level in the blood. **28.** The hormone secretin, among other things, stimulates the pancreas to release fluid high in buffers to neutralize the chyme that enters the duodenum from the stomach. If the small intestine did not produce secretin, the pH of the intestinal contents would be lower than normal. **PAGE 926** **29.** The four regions of the colon are the ascending colon, transverse colon, descending colon, and sigmoid colon. **30.** The large intestine is larger in luminal diameter than the small intestine. The thin walls of the large intestine lack villi and have an abundance of mucous cells and intestinal glands. **31.** In haustral churning, segmentation movements mix the contents of nearby haustra. In mass movements, which occur a few times per day, strong peristaltic contractions move material from the transverse colon through the rest of the large intestine. **PAGE 931** **32.** The nutrients the body requires are carbohydrates, lipids, proteins, vitamins, electrolytes, minerals, and water. **33.** Because chylomicrons are formed from the fats digested in a meal, fats increase the number of chylomicrons in the lacteals. **34.** Removal of the stomach would interfere with the absorption of vitamin B_{12}. Parietal cells of the stomach are the source of intrinsic factor, which is required for the vitamin's absorption. **35.** When someone with diarrhea loses fluid and electrolytes faster than they can be replaced, the resulting dehydration can be fatal. Although constipation can be quite uncomfortable, it does not interfere with any life-supporting processes; the few toxic wastes normally eliminated by the digestive system can move into the blood and be eliminated by the kidneys. **PAGE 932** **36.** General age-related digestive system changes include decreased mitotic activity of epithelial cells, decreased secretory mechanisms, decreased gastric and intestinal motility, and loss of muscle tone associated with tract sphincters; cumulative damage becomes more apparent, cancer rates increase, and dehydration occurs as a result of decreased osmoreceptor sensitivity. **37.** The digestive system absorbs the organic substrates, vitamins, ions, and water required by cells of all other body systems. **38.** The skeletal, muscular, nervous, endocrine, and cardiovascular systems may all be affected by inadequate absorption of calcium ions.

Answers to Review Questions

PAGE 936

Level 1 Reviewing Facts and Terms

1. d **2.** a **3.** d **4.** c **5.** **(a)** oral cavity (mouth); **(b)** liver; **(c)** gallbladder; **(d)** pancreas; **(e)** large intestine; **(f)** salivary glands; **(g)** pharynx; **(h)** esophagus; **(i)** stomach; **(j)** small intestine; **(k)** anus **6.** **(a)** mucosa; **(b)** submucosa; **(c)** muscular layer; **(d)** serosa **7.** e **8.** a **9.** a **10.** **(a)** duodenum; **(b)** jejunum; **(c)** ileum **11.** d **12.** c **13.** a **14.** The six main processes of the digestive system are: (1) ingestion; (2) mechanical digestion and propulsion; (3) chemical digestion; (4) secretion; (5) absorption; and (6) defecation. **15.** Layers of the digestive tract from the innermost layer nearest the lumen to the outermost layer are (1) the mucosa: the epithelial layer that performs chemical digestion and absorption of nutrients; (2) the submucosa: the connective tissue layer containing lymphatic and blood vessels and the submucosal nerve plexus; (3) the muscular layer: the smooth muscle layer containing the myenteric nerve plexus; and (4) the serosa: the outermost layer, epithelium and connective tissue that forms the visceral peritoneum (or connective tissue that forms the adventitia). **16.** Activities of the digestive tract are regulated by local, neural, and hormonal mechanisms. **17.** The three phases of swallowing—the buccal, pharyngeal, and esophageal phases—are controlled by the swallowing center of the medulla oblongata by the trigeminal and glossopharyngeal cranial nerves. The motor commands originating at the swallowing center are distributed by cranial nerves V, IX, X, and XII. Along the esophagus, primary peristaltic contractions are coordinated by afferent and efferent fibers within the glossopharyngeal and vagus cranial nerves, but secondary peristaltic contractions occur in the absence of CNS instructions. **18.** The pancreas provides digestive enzymes, plus bicarbonate ions that elevate the pH of the chyme. The liver produces bile and is also the primary organ involved in regulating the composition of circulating blood. The gallbladder stores and releases bile, which contains additional buffers and bile salts that aid in the digestion and absorption of lipids. **19.** The hormones include the following: enterocrinin, which stimulates the submucosal glands of the duodenum; secretin, which stimulates the pancreas and liver to increase the secretion of water and bicarbonate ions; cholecystokinin (CCK), which causes an increase in the release of pancreatic secretions and bile into the duodenum, inhibits gastric activity, and appears to have CNS effects that reduce the sensation of hunger; gastric inhibitory peptide (GIP), which stimulates insulin release at pancreatic islets and the activity of the duodenal submucosal glands; vasoactive intestinal peptide (VIP), which stimulates the secretion of intestinal glands, dilates regional capillaries, and inhibits acid production in the stomach; gastrin, which is secreted by G cells in the duodenum when they are exposed to large quantities of incompletely digested proteins; and, in small quantities, motilin, which stimulates intestinal contractions; villikinin, which promotes the movement of villi and associated lymph flow; and somatostatin, which inhibits gastric secretion. **20.** The large intestine reabsorbs water and compacts the intestinal contents into feces, absorbs important vitamins liberated by bacterial action, and stores fecal material prior to defecation. **21.** Positive feedback loops in the defecation reflex involve (1) stretch receptors in the rectal walls, which promote a series of peristaltic contractions in the colon and rectum, moving feces toward the anus in the intrinsic myenteric defecation reflex; and (2) the sacral parasympathetic defecation reflex, also activated by the stretch receptors, which stimulates peristalsis by motor commands.

Level 2 Reviewing Concepts

22. e **23.** e **24.** a **25.** A root canal involves drilling through the enamel and the dentin. **26.** The stomach is protected from digestion by mucous secretions of its epithelial lining and by neural and hormonal control over the times and rates of acid secretion. **27.** (1) The cephalic phase of gastric secretion begins with the sight or thought of food. Directed by the CNS, this phase prepares the stomach to receive food. (2) The gastric phase begins with the arrival of food in the stomach; this phase is initiated by distension of the stomach, an increase in the pH of the gastric contents, and the presence of undigested materials in the stomach. (3) The intestinal phase begins when chyme starts to enter the small intestine. This phase controls the rate of gastric emptying and ensures that the secretory, digestive, and absorptive functions of the small intestine can proceed reasonably efficiently. **28.** After a heavy meal, bicarbonate ions pass from the parietal cells of the stomach into the interstitial fluid. The diffusion of bicarbonate ions from the interstitial fluid into the bloodstream increases blood pH. This sudden influx of bicarbonate ions into the bloodstream has been called the alkaline tide.

Level 3 Critical Thinking and Clinical Applications

29. If a gallstone is small enough, it can pass through the bile duct and block the pancreatic duct. Enzymes from the pancreas then cannot reach the small intestine. As the enzymes accumulate, they irritate the duct and ultimately the exocrine pancreas, producing pancreatitis. **30.** The darker color of his urine is probably due to increased amounts of the pigment urobilin, which gives urine its normal yellow color. Urobilin is derived from urobilinogen, which is formed in the large intestine by the action of intestinal bacteria on bile pigments. In an intestinal obstruction, the bile pigments cannot be eliminated by their normal route, so a larger-than-normal amount diffuses into the bloodstream, where the kidneys eliminate it. **31.** If an individual cannot digest lactose, this sugar passes into the large intestine in an undigested form. The presence of extra sugar in the chyme increases its osmolarity, so less water is reabsorbed by the intestinal mucosa. The bacteria that inhabit the large intestine can metabolize the lactose, and in the process they produce large amounts of carbon dioxide. This gas overstretches the intestine, which stimulates local reflexes that increase peristalsis. The combination of more-fluid contents and increased peristalsis causes diarrhea. The overexpansion of the intestine by gas, which is directly related to increased gas production by the bacteria, causes the severe pain and abdominal cramping. **32.** The primary effect of such surgeries would be a reduction in the volume of food (and thus in the amount of calories) consumed because the person feels full after eating a small amount. This can result in significant weight loss.

Answers to Clinical Case Wrap-Up

PAGE 938 **1.** If Mirasol's *Clostridium difficile* colitis is not successfully treated with a fecal transplant, she is likely to continue to experience bloody diarrhea. This may result in progressive dehydration that can eventually lead to her death. She may also continue to experience abdominal pain and blood loss. Eventually she may suffer from a lack of vitamin K (required for clotting factor production), biotin, and vitamin B_5. **2.** If you could look at the lining of Mirasol's large intestine with a colonoscope, the surface would look inflamed, swollen, reddened. You would likely see areas of raw, bleeding mucosa. You might even see areas of white cells, accumulated in pockets of pus, as Mirasol's immune system tries to fight this *C. difficile* infection.

Answers to Figure-Based Questions

FIGURE 24–4 Circular muscles contract behind the bolus to force the bolus forward. Longitudinal muscles contract ahead of the bolus to shorten that segment of the digestive tract.

FIGURE 24–9 The submandibular glands produce glycoproteins called mucins.

FIGURE 24–12 The cardia is the region of the stomach that connects to the esophagus.

FIGURE 24–23 Chyme in the duodenum stimulates the secretion of GIP, secretin, CCK, and VIP.

FIGURE 24–26 Feces moving into the rectum cause distension, which stimulates stretch receptors.

Chapter 25

Answers to Checkpoints

PAGE 942 **1.** Metabolism is the sum of all biochemical processes under way within the human body; it includes anabolism and catabolism. Catabolism is the breakdown of organic molecules into simpler components, accompanied by the release of energy. Anabolism is the synthesis of new organic molecules from simpler components, and it requires energy. **2.** Oxidation is a gain of oxygen, or loss of hydrogen or electrons, from an atom or molecule. Reduction is the loss of oxygen, or gain of hydrogen or electrons, to an atom or molecule. **PAGE 951** **3.** The primary role of the citric acid cycle in ATP production is to transfer electrons from substrates to coenzymes. These electrons provide energy for the production of ATP by the electron transport chain. **4.** The process of oxidative phosphorylation releases energy from high-energy electrons that powers proton (hydrogen ion) pumps. These pumps move hydrogen ions from the mitochondrial matrix into the intermembrane space, creating an electrochemical proton gradient. In chemiosmosis, protons pass through hydrogen ion channels providing the energy necessary to bind ADP and P_i together into ATP. **5.** A decrease in cytosolic NAD would reduce ATP production in mitochondria. Decreased NAD would reduce the amount of pyruvate produced by glycolysis; less pyruvate means that the citric acid cycle could produce less ATP in mitochondria. **PAGE 955** **6.** Beta-oxidation is fatty acid catabolism that produces molecules of acetyl-CoA. **7.** The four major groups of lipoproteins are chylomicrons, very low density lipoproteins (VLDLs), low-density lipoproteins (LDLs), and high-density lipoproteins (HDLs). **8.** High-density lipoproteins (HDLs) are considered beneficial to health because they reduce the amount of cholesterol in the bloodstream by transporting it to the liver for storage or excretion in the bile. **PAGE 957** **9.** Transamination is the removal and transfer of an amino group from one molecule to another, especially from an amino acid to a keto acid. Deamination is the removal of an amino group from an amino acid. **10.** A diet deficient in pyridoxine (vitamin B_6), an important coenzyme in deaminating and transaminating amino acids in cells, would interfere with the body's ability to metabolize proteins. **PAGE 959** **11.** Glycogenesis (the formation of glycogen) in the liver increases after a high-carbohydrate meal. **12.** Blood levels of urea increase during the postabsorptive state because the deamination of many amino acids at that time yields ammonium ions, which are then converted to urea by the liver. **13.** Accumulated acetyl-CoA is likely to be converted into ketone bodies. **PAGE 966** **14.** The two classes of vitamins are fat soluble and water soluble. **15.** An athlete who is adding muscle mass through extensive training would try to maintain a positive nitrogen balance. **16.** A decrease in the amount of bile salts, which are necessary for digesting and absorbing fats and fat-soluble vitamins (including vitamin A), would result in less vitamin A in the body, and perhaps a vitamin A deficiency. **PAGE 972** **17.** The BMR of a pregnant woman would be higher than her own BMR when she is not pregnant, due to both the increased metabolism associated with supporting the fetus and the contribution of fetal metabolism. **18.** The vasoconstriction of peripheral vessels would decrease blood flow to the skin and thus the amount of heat the body can lose. As a result, body temperature would increase. **19.** Infants have higher surface-to-volume ratios than do adults, and the body's temperature-regulating mechanisms are not fully functional at birth. As a result, infants must expend more energy to maintain body temperature, and they get cold more easily than do healthy adults.

Answers to Review Questions

PAGE 974

Level 1 Reviewing Facts and Terms

1. d **2.** e **3.** d **4.** a **5.** b **6.** d **7.** **(a)** evaporation; **(b)** radiation; **(c)** convection; **(d)** conduction **8.** a **9.** Lipoproteins are lipid–protein complexes that contain large insoluble glycerides and cholesterol, with a superficial coating of phospholipids and proteins. The major groups are chylomicrons, which consist of 95 percent triglyceride, are the largest lipoproteins, and carry absorbed lipids from the intestinal tract to the bloodstream; very low density lipoproteins (VLDLs), which consist of triglyceride, phospholipid, and cholesterol, and transport triglycerides to peripheral tissues; low-density lipoproteins (LDLs, or "bad cholesterol"), which are mostly cholesterol and deliver cholesterol to peripheral tissues; and high-density lipoproteins (HDLs, or "good cholesterol"), which are equal parts protein and lipid (cholesterol and phospholipids) and transport excess cholesterol to the liver for storage or excretion in bile.

Level 2 Reviewing Concepts

10. Oxidative phosphorylation is the generation of ATP as the result of the transfer of electrons from the coenzymes NADH and $FADH_2$ to oxygen by a sequence of electron carriers within mitochondria. The electrons are carried by the electron transport chain in the inner mitochondrial membrane. Energy from the stepwise passage of electrons (from H atoms) along the electron transport chain is used to pump hydrogen ions from the matrix into the intermembrane space. This forms an electrochemical proton (hydrogen ion) gradient across the inner mitochondrial membrane. Through the process of chemiosmosis, the hydrogen ions diffuse back through ATP synthase and generate ATP. The electrons, hydrogen ions, and oxygen combine to produce water as a by-product. **11.** In lipid catabolism, a triglyceride is hydrolyzed, yielding glycerol and fatty acids. Glycerol is converted to pyruvate and enters the citric acid cycle. Fatty acids are broken into 2-carbon fragments by beta-oxidation inside mitochondria. The 2-carbon compounds then enter the citric acid cycle. **12.** During the absorptive state, insulin prevents a large surge in blood glucose after a meal by causing the liver to remove glucose from the hepatic portal circulation. During the postabsorptive state, blood glucose begins to decline, triggering the liver to release glucose produced through glycogenolysis and gluconeogenesis. **13.** Liver cells can break down or synthesize most carbohydrates, lipids, and amino acids. Because the liver has an extensive blood supply, it can easily monitor blood composition of these nutrients and regulate them accordingly. The liver also stores energy in the form of glycogen. **14.** Vitamins and minerals are essential components of the diet because the body cannot synthesize most of the vitamins and minerals it requires. **15.** These terms refer to lipoproteins in the blood that transport cholesterol. "Good cholesterol" (high-density lipoproteins, or HDLs) transports excess cholesterol to the liver for storage or breakdown, whereas "bad cholesterol" (low-density lipoproteins, or LDLs) transports cholesterol to peripheral tissues, which unfortunately may include the arteries. The buildup of cholesterol in the arteries is linked to cardiovascular disease.

Level 3 Critical Thinking and Clinical Applications

16. c **17.** Based just on the information given, Charlie would appear to be in good health, at least relative to his diet and probable exercise. Problems are associated with elevated levels of LDLs, which carry cholesterol to peripheral tissues and make it available for the formation of atherosclerotic plaques in blood vessels. High levels of HDLs indicate that a considerable amount of cholesterol is being removed from the peripheral tissues and carried to the liver for disposal. You would encourage Charlie not to change, and keep up the good work. **18.** It appears that Jill is suffering from ketoacidosis as a consequence of her anorexia. Because she is literally starving herself, her body is metabolizing large amounts of fatty acids and amino acids to provide energy, and in the process is producing large quantities of ketone bodies (normal metabolites from these catabolic processes). One of the ketones formed is acetone, which can be eliminated through the lungs. Acetone has a fruity aroma, so her breath would also smell fruity. The ketones are also converted into keto acids such as acetic acid. In large amounts, this lowers the body's pH. This is probably the cause of her arrhythmias.

Answers to Clinical Case Wrap-Up

PAGE 975 **1.** Diet and vitamin supplementation are the only known sources of vitamin A. Unlike some other vitamins that can be made by our skin (vitamin D_3) or by intestinal bacteria [vitamin K, biotin (B_7), pantothenic acid(B_5)], vitamin A is a true *essential* vitamin that must be consumed orally. **2.** Justine's vitamin-deficient diet is likely to lead to deficiencies of all the fat-soluble and water-soluble vitamins, including D, E, K, all the B vitamins, and vitamin C.

Answers to Figure-Based Questions

FIGURE 25–2 Glycolysis produces 2 pyruvate molecules with a net gain of 2 ATP molecules.

FIGURE 25–5 The energy produced from aerobic metabolism comes from the electron transport chain and the citric acid cycle.

FIGURE 25–7 Beta-oxidation, the breakdown of fatty acids, takes place within the mitochondria.

FIGURE 25–8 The liver manufactures very low density lipoproteins (VLDLs) and low-density lipoproteins (LDLs).

FIGURE 25–14 In a warm environment, more blood returns to the trunk through superficial veins. This allows heat to radiate to the environment through the skin instead of conserving heat.

Chapter 26

Answers to Checkpoints

PAGE 978 **1.** The three major functions of the urinary system include (1) the excretion of metabolic wastes from body fluids, (2) the elimination of body wastes to the exterior, and (3) the regulation of homeostasis by maintaining the volume and solute concentration of blood. **2.** The urinary system consists of two kidneys, two ureters, a urinary bladder, and a urethra. **PAGE 986** **3.** The renal corpuscle, proximal convoluted tubule, distal convoluted tubule, and the proximal part of the nephron loop and collecting duct are all in the renal cortex. (In a cortical nephron, most of the nephron loop is in the renal cortex; in a juxtamedullary nephron, most of the nephron loop is in the renal medulla.) **4.** Plasma proteins are too large to pass through the pores of the glomerular capillaries; only the smallest plasma proteins can pass between the foot processes of the podocytes. **5.** Damage to the juxtaglomerular complex of a nephron would interfere with blood pressure regulation. **PAGE 988** **6.** The three distinct processes that form urine in the kidney are filtration, reabsorption, and secretion. **PAGE 993** **7.** The parts of the glomerulus involved in filtration are the glomerular capillaries, the basement membrane, and the foot processes of the podocytes. **8.** The factors that influence net filtration pressure are net hydrostatic pressure and blood colloid osmotic pressure. **9.** The factors that influence the rate of filtrate formation are the filtration pressure across glomerular capillaries, plus interactions among the responses by autoregulation, hormonal regulation (endocrine), and autonomic regulation (neural). **10.** A decrease in blood pressure would decrease the GFR by decreasing the blood hydrostatic pressure within the glomerulus. **PAGE 1000** **11.** When the plasma concentration of a substance exceeds its tubular maximum, the excess is not reabsorbed, so it is excreted in the urine. **12.** Increased amounts of aldosterone, which promotes Na^+ retention and K^+ secretion by the kidneys, would increase the K^+ concentration of urine. **13.** If the concentration of Na^+ in the filtrate decreased, fewer hydrogen ions could be secreted by the countertransport mechanism involving these two ions. As a result, the pH of the tubular fluid would increase. **14.** When the amount of Na^+ in the tubular fluid passing through the distal convoluted tubule decreases, the cells of the macula densa are stimulated to release renin. The resulting activation of angiotensin II increases blood pressure. **PAGE 1009** **15.** Without juxtamedullary nephrons, a steep osmotic concentration gradient could not exist in the medulla, and the kidneys would be unable to form concentrated urine. **16.** Digestion of a high-protein diet would lead to increased production of urea, a nitrogenous waste formed from the metabolism of amino acids during the breakdown of proteins. As a result, the urine would contain more urea, and urine volume might also increase as a result of the need to flush the excess urea. **PAGE 1013** **17.** An obstruction of a ureter would interfere with the passage of urine between the renal pelvis and the urinary bladder. **18.** The ability to control the urinary reflexes depend on the ability to control the external urethral sphincter, a ring of skeletal muscle, which acts as a valve. **PAGE 1014** **19.** Four general age-related changes in the urinary system are a decrease in the number of functional nephrons, a reduction in the GFR, a reduced sensitivity to ADH, and problems with the urinary reflex. **20.** Nephrolithiasis is the formation of renal calculi, or kidney stones. **21.** Incontinence may develop in elderly individuals as a result of either decreased muscle tone of the sphincter muscles controlling micturition or disorders of the CNS affecting the ability to control urination. **PAGE 1014** **22.** For all systems, the urinary system excretes wastes collected from blood and maintains normal body fluid pH and ion composition. **23.** Body systems that make up the body's excretory system include the urinary, integumentary, respiratory, and digestive systems.

Answers to Review Questions

PAGE 1018

Level 1 Reviewing Facts and Terms

1. **(a)** renal sinus; **(b)** renal pelvis; **(c)** hilum; **(d)** renal papilla; **(e)** ureter; **(f)** renal cortex; **(g)** renal medulla; **(h)** renal pyramid; **(i)** minor calyx; **(j)** major calyx; **(k)** kidney lobe; **(l)** renal columns; **(m)** fibrous capsule **2.** a **3.** e **4.** c **5.** d **6.** d **7.** The urinary system excretes the metabolic wastes generated by cells throughout the body and eliminates them from the body in urine. It also regulates the volume and solute concentration of body fluids. **8.** The kidneys, ureters, urinary bladder, and urethra are the organs of the urinary system. **9.** renal corpuscle (glomerulus/glomerular capsule) → proximal convoluted tubule → nephron loop → distal convoluted tubule → collecting duct → papillary duct → renal pelvis **10.** proximal convoluted tubule: reabsorbs all the useful organic substrates from the filtrate; nephron loop: reabsorbs over 90 percent of the water in the filtrate; and distal convoluted tubule: secretes wastes into the tubular fluid that were not filtered or reabsorbed **11.** The juxtaglomerular complex secretes the enzyme renin and the hormone erythropoietin. **12.** renal artery → segmental arteries → interlobar arteries → arcuate arteries → cortical radiate arteries → afferent arterioles → glomerulus → venules → cortical radiate veins → arcuate veins → interlobar veins → renal vein **13.** Processes in urine production are (1) filtration: the selective removal of large solutes and suspended materials from a solution on the basis of size; requires a filtration membrane and hydrostatic pressure, as provided by gravity or by blood pressure; (2) reabsorption: the removal of water and solute molecules from the filtrate after it enters the renal tubules; and (3) secretion: the transport of solutes from the peritubular fluid, across the tubular epithelium, and into the tubular fluid **14.** In peripheral capillary beds, angiotensin II causes powerful vasoconstriction of precapillary sphincters, increasing pressures in the renal arteries and their branches. At the nephron, angiotensin II causes the efferent arteriole to constrict, increasing glomerular pressures and filtration rates. At the PCT, it stimulates the reabsorption of sodium ions and water. In the CNS, angiotensin II triggers the release of ADH, stimulating the reabsorption of water in the distal portion of the DCT and the collecting system, and it causes the sensation of thirst. At the adrenal gland, angiotensin II stimulates the secretion of aldosterone by the cortex. Aldosterone accelerates sodium reabsorption in the DCT and the cortical portion of the collection system. **15.** Structures responsible for the transport, storage, and elimination of urine are the ureters, urinary bladder, and urethra.

Level 2 Reviewing Concepts

16. d **17.** d **18.** c **19.** a **20.** a **21.** Proteins are excluded from filtrate because they are too large to fit through the filtration slits between adjacent foot processes (pedicels) of podocytes. Keeping proteins in the plasma ensures that blood colloid osmotic pressure will oppose filtration and return water to the plasma. **22.** Controls on GFR are autoregulation at the local level, hormonal regulation initiated by the kidneys, and autonomic regulation (by the sympathetic division of the ANS). **23.** As a result of facilitated diffusion and cotransport mechanisms in the PCT, 99 percent of the glucose, amino acids, and other nutrients are reabsorbed before the filtrate leaves the PCT. A reduction of the solute concentration of the tubular fluid occurs due to active ion reabsorption of sodium, potassium, calcium, magnesium, bicarbonate, phosphate, and sulfate ions. The passive diffusion of urea, chloride ions, and lipid-soluble materials further reduces the solute concentration of the tubular fluid and promotes additional water reabsorption. **24.** Countercurrent multiplication (1) is an efficient way to reabsorb solutes and water before the tubular fluid reaches the DCT and collecting system, and (2) establishes an osmotic concentration gradient in the medulla that permits the passive reabsorption of water from urine in the collecting system. **25.** The urge to urinate usually appears when the urinary bladder contains about 200 mL of urine. The urine voiding reflex begins to function when the stretch receptors have provided adequate stimulation to the pontine micturition center, which stimulates parasympathetic motor neurons. The activity in the motor neurons generates action potentials that reach the detrusor in the wall of the urinary bladder. These efferent impulses travel over the pelvic nerves of the sacral spinal cord, producing a sustained contraction of the urinary bladder and relaxation of the internal urethral sphincter. Urination occurs when the external urethral sphincter relaxes.

Level 3 Critical Thinking and Clinical Applications

26. d **27.** c **28.** Increasing the volume of urine produced decreases the total blood volume of the body, which in turn leads to a decreased blood hydrostatic pressure. Edema frequently results when the hydrostatic pressure of the blood exceeds the opposing forces at the capillaries in the affected area. Depending on the actual cause of the edema, decreasing the blood hydrostatic pressure would decrease edema formation and possibly cause some of the fluid to move from the interstitial spaces back to the blood. **29.** Renal arteriosclerosis restricts blood flow to the kidneys and produces renal ischemia. Decreased blood flow and ischemia trigger the juxtaglomerular complex to produce more renin, which leads to elevated levels of angiotensin II and aldosterone. Angiotensin II causes vasoconstriction, increased peripheral resistance, and thus increased blood pressure. The aldosterone promotes sodium retention, which leads to more water retained by the body and an increase in blood volume. This too contributes to a higher blood pressure. Additionally, in response to tissue hypoxia, erythropoietin release is increased, stimulating the formation of red blood cells, which leads to increased blood viscosity and again contributes to hypertension. **30.** Because mannitol is filtered but not reabsorbed, drinking a mannitol solution would lead to an increase in the osmolarity of the filtrate. Less water would be reabsorbed, and an increased volume of urine would be produced. **31.** Carbonic anhydrase catalyzes the reaction that forms carbonic acid, a source of hydrogen ions that are excreted by the kidneys. Hydrogen ion excretion occurs by countertransport in which sodium ions are exchanged for hydrogen ions. Fewer hydrogen ions would be available, so less sodium would be reabsorbed, causing an increased osmolarity of the filtrate. In turn, an increased volume of urine and more frequent urination would result.

Answers to Clinical Case Wrap-Up

PAGE 1019 **1.** Mike's x-rays show fractures of the left 11th and 12th ribs, overlying the left kidney. They also show fractures of the transverse processes of the first three lumbar vertebrae on the left. The left renal hilum lies between the fractured ribs and the fractured lumbar vertebral transverse processes. These visible fractures are a strong indication that the soft tissues and organs in this area are also injured. **2.** If Mike survives this trauma, the GFR of his solitary right kidney is likely to increase to compensate for the loss of his left kidney. This is to be expected if there has been no trauma to his right kidney.

Answers to Figure-Based Questions

FIGURE 26–5 A glomerulus is located in the renal cortex.

FIGURE 26–6 The proximal convoluted tubule is connected to the distal convoluted tubule by the nephron loop, which consists of the descending thin limb (DTL), ascending thin limb (ATL), and the thick ascending limb (TAL).

FIGURE 26–11 An increased blood volume results from increased Na^+ retention, increased fluid consumption, increased fluid retention, and constriction of systemic veins.

FIGURE 26–13 Aldosterone-sensitive portions of the distal convoluted tubule and collecting duct allow for sodium ions to be reabsorbed in exchange for potassium ions to be passively lost.

FIGURE 26–15 ADH creates a small volume of concentrated urine.

FIGURE 26–18 The ureter transports urine *to* the bladder.

ANSWERS

Chapter 27

Answers to Checkpoints

PAGE 1023 **1.** The three interrelated processes essential to stabilizing body fluid volumes are fluid balance, electrolyte balance, and acid-base balance. **PAGE 1027** **2.** The components of extracellular fluid (ECF) are interstitial fluid, plasma, and other body fluids; the intracellular fluid (ICF) is the cytosol of a cell. **3.** Three hormones affecting fluid and electrolyte balance are antidiuretic hormone (ADH), aldosterone, and natriuretic peptides (ANP and BNP). **4.** Drinking a half-gallon of distilled water might temporarily decrease your blood osmolarity (osmotic concentration). This decrease in osmolarity would lead to a decrease in the level of ADH in your blood, because ADH release is triggered by increases in osmolarity. **PAGE 1030** **5.** Edema (swelling) results from the movement of abnormally large amounts of water from plasma into interstitial fluid. **6.** A fluid shift is a rapid movement of water between the ECF and ICF in response to an osmotic gradient. **7.** Dehydration is a decrease in the water content of the body that develops when water losses outpace water gains, and this threatens homeostasis. **8.** Being in the desert without water, you would lose fluid through perspiration, urination, and breathing. As a result, the osmotic concentration of your blood plasma (and other body fluids) would increase. **PAGE 1035** **9.** In the ECF, four important cations are sodium, potassium, calcium, and magnesium; two important anions are phosphate and chloride. **10.** Sweat is a hypotonic solution with lower sodium ion concentration than the ECF. Sweating causes a greater loss of water than of sodium, increasing the blood plasma sodium ion level. **11.** Potassium ion imbalances are more dangerous than sodium ion imbalances because they can lead to extensive muscle weakness or even paralysis when blood concentrations are too low, and to cardiac arrhythmias when the levels are too high. Disturbances in sodium balance, by contrast, produce dehydration or tissue edema. **PAGE 1042** **12.** The body's three major buffer systems are the phosphate buffer system, the protein buffer systems, and the carbonic acid–bicarbonate buffer system. **13.** A decrease in the pH of body fluids accompanies an increase in the partial pressure of carbon dioxide. Chemoreceptors sensitive to P_{CO_2} would stimulate the respiratory centers of the medulla oblongata, resulting in an increase in the respiratory rate. **14.** Tubular fluid in nephrons must be buffered to avoid decreasing the pH below approximately 4.5, at which point H^+ secretion cannot continue because the H^+ concentration gradient becomes too steep. The hydrogen ions would leak out of the tubule as fast as they are pumped in. **PAGE 1049** **15.** In a prolonged fast, fatty acids are catabolized, producing large numbers of ketone bodies, which are metabolic acids that decrease the blood pH. (The decreased pH would eventually lead to ketoacidosis.) **16.** When a person has prolonged vomiting, large amounts of stomach hydrochloric acid (HCl) are lost from the body. More H^+ and Cl^- are then formed by the parietal cells of the stomach. The H^+ that are released were produced by the dissociation of carbonic acid and exchange of bicarbonate ions with chloride ions from the blood. The alkaline tide of released bicarbonate ions increases the blood pH, leading to metabolic alkalosis. **PAGE 1051** **17.** A decrease in the glomerular filtration rate and the number of functional nephrons reduces the body's ability to regulate pH through renal compensation. **18.** Total body water content gradually decreases with age.

Answers to Review Questions

PAGE 1053

Level 1 Reviewing Facts and Terms

1. c **2.** a **3.** d **4.** **(a)** carbonic acid (H_2CO_3); **(b)** bicarbonate ion (HCO_3^-); **(c)** sodium bicarbonate ($NaHCO_3$) **5.** d **6.** d **7.** a **8.** b **9.** a **10.** Four major hormones involved in fluid and electrolyte balance are (1) antidiuretic hormone (ADH): stimulates water conservation by the kidneys and stimulates the thirst center; (2) aldosterone: determines the rate of sodium reabsorption and potassium secretion along the DCT and collecting system of the kidney; and the natriuretic peptides, (3) ANP and (4) BNP: reduce thirst, promote the loss of Na^+ and water by the kidneys, and block the release of ADH and aldosterone.

Level 2 Reviewing Concepts

11. d **12.** a **13.** d **14.** c **15.** Fluid balance is a state in which the amount of water gained each day is equal to the amount lost to the environment. It is vital that the water content of the body remain stable, because water is an essential ingredient of cytoplasm and accounts for about 99 percent of ECF volume. Electrolyte balance exists when there is neither a net gain nor a net loss of any ion in body fluids. It is important that the ionic concentrations in body water remain within normal limits; if levels of calcium or potassium become too high, for instance, cardiac arrhythmias can develop. Acid-base balance exists when the production of hydrogen ions precisely offsets their loss. The pH of body fluids must remain within a relatively narrow range; variations outside this range can be life threatening. **16.** Fluid shifts are rapid water movements between the ECF and the ICF that occur in response to increases or decreases in the osmotic concentration of the ECF. Such water movements dampen extreme shifts in electrolyte balance. **17.** A person with a fever should increase fluid intake because for each degree (Celsius) the body temperature increases above normal, daily water loss increases by 200 mL. **18.** **(a)** Acids that can leave solution and enter the atmosphere, such as carbonic acid, are volatile acids. **(b)** Acids that do not leave solution, such as sulfuric acid, are fixed acids. **(c)** Acids produced during metabolism, such as pyruvic acid, lactic acid, and ketones, are metabolic acids. Volatile acids are the greatest threat because of the large amounts generated by normal cellular processes. **19.** (1) Phosphate buffer system: This buffer system consists of dihydrogen phosphate ($H_2PO_4^-$) a weak acid that, in solution, reversibly dissociates into a hydrogen ion and monohydrogen phosphate (HPO_4^{2-}). The phosphate buffer system plays a relatively small role in regulating the pH of the ECF, because the ECF contains far higher concentrations of bicarbonate ions than phosphate ions; however, it is important in buffering the pH of the ICF. (2) Protein buffer systems: These depend on the ability of amino acids to respond to changes in pH by accepting or releasing hydrogen ions. If the pH rises, the carboxyl group of the amino acid dissociates to release a hydrogen ion; if the pH drops, the amino group accepts an additional hydrogen ion to form an amino ion (NH_3^+) and the carboxylate ion can accept a hydrogen ion to form a carboxyl group. Plasma proteins contribute to the buffering capabilities of the blood; inside cells, protein buffer systems stabilize the pH of the ECF by absorbing extracellular hydrogen ions or exchanging intracellular hydrogen ions for extracellular potassium. (3) Carbonic acid–bicarbonate system: Most carbon dioxide generated in tissues is converted to carbonic acid, which dissociates into a hydrogen ion and a bicarbonate ion. Hydrogen ions released by dissociation of organic or fixed acids combine with bicarbonate ions, elevating the P_{CO_2}; additional CO_2 is lost at the lungs. **20.** Respiratory and renal mechanisms support buffer systems by secreting or absorbing hydrogen ions, by controlling the excretion of acids and bases, and by generating additional buffers. **21.** Respiratory compensation is a change in the respiratory rate that helps stabilize the pH of the ECF. Increasing or decreasing the rate of respiration alters pH by decreasing or increasing the P_{CO_2}. When the P_{CO_2} decreases, the pH increases; when the P_{CO_2} increases, the pH decreases. Renal compensation is a change in the rates of hydrogen and bicarbonate ion secretion or reabsorption in response to changes in blood pH. The secretion of hydrogen ions into tubular fluid results in the diffusion of bicarbonate ions into the ECF. **22.** Respiratory disorders result from abnormal CO_2 levels in the ECF. An imbalance exists between the rate of CO_2 removal by the lungs and its generation in other tissues. Metabolic disorders are caused by the generation of metabolic or fixed acids or by conditions affecting the concentration of bicarbonate ions in the ECF. **23.** Metabolic acidosis, which occurs when bicarbonate ion levels decrease, can result from overproduction of fixed or metabolic acids, impaired ability to secrete H^+ ions by the kidney, or severe bicarbonate loss. Respiratory acidosis, which results from an abnormally high level of carbon dioxide (hypercapnia), is usually caused by hypoventilation. **24.** Excessive salt intake causes an increase in total blood volume and blood pressure due to an obligatory increase in water absorption across the intestinal lining and recall of fluid from the ICF. **25.** Since sweat is usually hypotonic, the loss of a large volume of sweat causes hypertonicity in body fluids. The loss of fluid volume is primarily from the interstitial space, which leads to a decrease in plasma volume and an increase in the hematocrit. Severe dehydration can cause the blood viscosity to increase substantially, resulting in an increased workload on the heart, ultimately increasing the probability of heart failure.

Level 3 Critical Thinking and Clinical Applications

26. The young boy has metabolic and respiratory acidosis. The metabolic acidosis resulted primarily from the large amounts of lactate and H^+ generated by the boy's muscles as he struggled in the water. Sustained hypoventilation during drowning contributed to both tissue hypoxia and respiratory acidosis. Respiratory acidosis developed as the P_{CO_2} increased in the ECF, increasing the production of carbonic acid and its dissociation into H^+ and HCO_3^-. Prompt emergency treatment is essential. The usual procedure involves some form of artificial or mechanical respiratory assistance (to increase the respiratory rate and decrease in the ECF) coupled with the intravenous infusion of a buffered isotonic solution containing sodium lactate, sodium gluconate, or sodium bicarbonate that would absorb the H^+ in the ECF and increase body fluid pH. **27.** a **28.** When tissues are burned, cells are destroyed and their cytoplasmic contents leak into the interstitial fluid and then move into the plasma. Since potassium ions are normally found within the cell, damage to a large number of cells releases relatively large amounts of potassium ions into the blood. The elevated potassium level stimulates cells of the adrenal cortex to produce aldosterone and cells of the juxtaglomerular complex to produce renin. Renin activates the renin-angiotensin-aldosterone system. Ultimately, angiotensin II stimulates increased aldosterone secretion, which promotes sodium retention and potassium secretion by the kidneys, thereby accounting for the elevated potassium levels in the patient's urine. **29.** Digestive secretions contain high levels of bicarbonate, so individuals with diarrhea can lose significant amounts of this important ion, leading to acidosis. We would expect Milly's blood pH to be lower than 7.35, and that of her urine to be low (due to increased renal excretion of hydrogen ions). We would also expect an increase in the rate and depth of breathing as the respiratory system tries to compensate by eliminating carbon dioxide. **30.** The hypertonic solution will cause fluid to move from the ICF to the ECF, further aggravating Yuka's dehydration. The slight increase in pressure and osmolarity of the blood should lead to an increase in ADH, even though ADH levels are probably quite high already. Despite the high ADH levels, urine volume would probably increase, because the kidneys could not reabsorb much of the glucose. The remaining glucose would increase the osmolarity of the tubular filtrate, decreasing water reabsorption and increasing urine volume. **31.** Patient 1 has compensated respiratory alkalosis. Patient 2 has acute metabolic acidosis due to the generation or retention of metabolic or fixed acids. Patient 3 has acute respiratory acidosis. Patient 4 has metabolic alkalosis.

Answers to Clinical Case Wrap-Up

PAGE 1054 **1.** Marquis should be treated with potassium replacement. He can take K^+ orally, or it can be given intravenously (IV) for faster replacement. His doctor should consider changing his blood pressure medication to one that does not cause a loss of K^+. (These are often called potassium-sparing diuretics.) **2.** When Marquis complains of a "frog jumping around" in his chest with dizziness, he is likely describing a cardiac arrhythmia. This can be the most serious complication of hypokalemia and can lead to sudden cardiac death.

Answers to Figure-Based Questions

FIGURE 27–2 Mg^{2+} and K^+ are found in higher concentrations in the intracellular fluid, while Na^+ and Cl^- are found in higher concentrations in the extracellular fluid.

FIGURE 27–7 If the K^+ concentration rises above 7 mEq/L (hyperkalemia), a serious condition that could occur is a cardiac arrhythmia.

FIGURE 27–10 The phosphate and protein buffer systems help regulate the pH of the intracellular fluid (ICF).

FIGURE 27–14 The kidney's response to alkalosis is to conserve H^+ and secrete HCO_3^-.

FIGURE 27–16 In metabolic acidosis, arterial and CSF chemoreceptors are stimulated, which results in increased respiratory rate and a decrease in P_{CO_2}.

Chapter 28

Answers to Checkpoints

PAGE 1057 **1.** A gamete is a functional male or female reproductive cell. **2.** Basic structures of the reproductive system are gonads (reproductive organs), ducts (which receive and transport gametes), accessory glands (which secrete fluids), and external genitalia (perineal structures). **3.** Gonads are reproductive organs that produce gametes and hormones. **PAGE 1065** **4.** Male reproductive structures are the scrotum, testes, epididymides, ductus deferentia, ejaculatory duct, urethra, seminal glands, prostate, bulbo-urethral glands, and penis. **5.** On a warm day, the cremaster (as well as the dartos muscle) would be relaxed so that the scrotum could descend away from the warmth of the body, thereby cooling the testes. **6.** The pathway of sperm is as follows: seminiferous tubule → straight tubule → rete testis → efferent ductules → epididymis → ductus deferens → ejaculatory duct → urethra. **PAGE 1072** **7.** Low FSH levels would lead to low levels of testosterone in the seminiferous tubules, decreasing both the sperm production rate and sperm count. **PAGE 1082** **8.** Structures of the female reproductive system include the ovaries, uterine tubes, uterus, vagina, and external genitalia (mons pubis, the clitoris and the labia majora and labia minora). **9.** The blockage of both uterine tubes would cause sterility. **10.** The acidic pH of the vagina helps prevent bacterial, fungal, and parasitic infections in this region. **11.** Blockage of a single lactiferous sinus would not interfere with the delivery of milk to the nipple because the mammary gland of each breast generally has 15–20 lactiferous sinuses. **PAGE 1091** **12.** The functional layer of the endometrium is sloughed off, or shed, during menstruation. **13.** If the LH surge did not occur during an ovarian cycle, ovulation and corpus luteum formation would not occur. **14.** Blockage of progesterone receptors in the uterus would inhibit the development of the endometrium, making the uterus unprepared for pregnancy. **15.** A decrease in the levels of estrogens and progesterone signals the beginning of menstruation, the end of the uterine cycle. **PAGE 1093** **16.** When the arteries within the penis dilate, the increased blood flow causes the erectile tissues of the corpora cavernosa and corpus spongiosum to engorge with blood, producing an erection. **17.** The physiological events of sexual intercourse in both sexes are arousal, erection, lubrication, orgasm, and resolution. Emission and ejaculation are additional phases that occur only in males. **18.** An inability to contract the ischiocavernosus and bulbospongiosus muscles would interfere with a male's ability to ejaculate and to experience orgasm. **19.** Parasympathetic stimulation in females during sexual arousal causes **(a)** engorgement of the erectile tissue of the clitoris and vestibular bulbs, **(b)** increased secretion of cervical and greater vestibular glands, **(c)** increased blood flow to the wall of the vagina, and **(d)** engorgement of the blood vessels in the nipples. **PAGE 1096** **20.** The genital tubercle of a sexually indifferent embryo will develop into a penis in the presence of testosterone or a clitoris without testosterone. **21.** The male climacteric, or andropause, is a period of declining reproductive function in men, typically between the ages of 50 and 60. **22.** Menopause is the time when ovulation and menstruation cease, typically around ages 45–55. **23.** At menopause, circulating estrogen levels begin to decrease. Estrogen has an inhibitory effect on FSH (and on GnRH). As the level of estrogen decreases, the levels of FSH increase and remain high. **PAGE 1098** **24.** The cardiovascular system distributes reproductive hormones; provides nutrients, oxygen, and waste removal for the fetus; and produces local blood pressure changes responsible for the physical changes that occur during sexual intercourse. The reproductive system supplies estrogens that may help maintain healthy blood vessels and slow the development of atherosclerosis. **25.** Pelvic bones protect reproductive organs in females and portions of the ductus deferens and accessory glands in males; sex hormones stimulate growth and maintenance of bones, and accelerate growth and closure of epiphyseal cartilages at puberty.

Answers to Review Questions

PAGE 1100

Level 1 Reviewing Facts and Terms

1. c **2.** d **3.** c **4.** b **5.** a **6.** The accessory structures include the seminal glands (seminal vesicles), which provide the nutrients that sperm need for motility; prostaglandins that stimulate smooth muscle contractions along the male and female reproductive tracts thereby propelling sperm and fluids; fibrinogen that temporarily clots the ejaculate within the vagina; buffers that counteract the acidity of the prostatic secretions and urethral and vaginal contents; the prostate gland, which aids the activation of the sperm (with seminal gland secretions); and the bulbo-urethral glands, which buffer acids in the penile urethra and lubricate the glans penis. The fluids secreted by the accessory glands make up about 95 percent of the volume of semen. **7.** Interstitial endocrine cells (Leydig cells) produce male sex hormones (androgens), the most important of which is testosterone; nurse cells maintain the blood testis barrier, support spermatogenesis and spermiogenesis, and secrete inhibin, and androgen-binding protein. **8.** The three regions of the male urethra are the prostatic urethra, membranous urethra, and spongy urethra. **9.** In males, testosterone stimulates spermatogenesis and promotes the functional maturation of sperm; maintains the male accessory reproductive organs; determines male secondary sex characteristics; stimulates metabolic pathways, especially those concerned with protein synthesis and muscle growth; and affects CNS function, by stimulating sexual drive and sexual behaviors. **10.** Steps of the ovarian cycle are (1) the development of a dominant tertiary follicle, or mature graafian follicle, (2) ovulation, and (3) the formation of the corpus luteum and its degeneration into the corpus albicans. **11.** The myometrium is the middle, muscular layer of the uterine wall; the endometrium is the inner, glandular layer; and the perimetrium is an incomplete outer serosal layer. **12.** The clitoris—a part of the external genitalia of females that is derived from the same embryonic structures as the penis—contains erectile tissue that becomes engorged with blood during sexual arousal and provides pleasurable sensations. **13.** Path of milk flow: secretory lobules of glandular tissue (lobes) → ducts → lactiferous duct → lactiferous sinus, which opens onto the surface of the nipple. **14.** **(a)** prostatic urethra; **(b)** spongy urethra; **(c)** ductus deferens; **(d)** penis; **(e)** epididymis; **(f)** testis; **(g)** external urethral orifice; **(h)** scrotum; **(i)** seminal gland; **(j)** prostate; **(k)** ejaculatory duct; **(l)** bulbo-urethral gland.

Level 2 Reviewing Concepts

15. e **16.** The hypothalamic secretion of GnRH triggers the pituitary secretion of FSH and LH. FSH initiates follicular development, and activated follicles and ovarian thecal endocrine cells produce estrogens. High estrogen levels stimulate LH secretion and increase anterior pituitary lobe sensitivity to GnRH, causing the release of LH. Progesterone is the main hormone of the luteal phase. Changes in estrogen and progesterone levels are responsible for maintaining the uterine cycle. **17.** Males produce gametes from puberty until death; females produce gametes only from menarche to menopause. Males produce many gametes at a time; females typically produce one or two per 28-day cycle. Males release mature gametes that have completed meiosis; females release secondary oocytes held in metaphase of meiosis II. **18.** The corpora cavernosa extend along the length of the penis as far as the neck of the penis, and the erectile tissue within each corpus cavernosum surrounds a central artery. The slender corpus spongiosum surrounds the urethra and extends from the superficial fascia of the urogenital diaphragm to the tip of the penis, where it expands to form the glans penis. The sheath surrounding the corpus spongiosum contains more elastic fibers than the corpora cavernosa, and the erectile tissue contains a pair of arteries. When parasympathetic neurons innervating the penile arteries release nitric oxide, smooth muscles in the arterial walls relax, dilating the vessels and increasing blood flow; the resulting engorgement of the vascular channels with blood causes erection of the penis. **19.** (1) The menstrual phase is the interval marked by the degeneration and loss of the functional layer of the endometrium, lasts 1–7 days, and 35–50 mL (1.2–1.7 oz) of blood is lost. (2) The proliferative phase features growth and vascularization, resulting in the complete restoration of the functional layer; it lasts from the end of menstruation until the beginning of ovulation, around day 14. (3) During the secretory phase, the uterine glands enlarge, accelerating their rates of secretion, and the arteries elongate and spiral through the tissues of the functional layer; this phase begins at ovulation, occurs under the combined stimulatory effects of progesterone and estrogens from the corpus luteum, and persists as long as the corpus luteum remains intact. **20.** As follicular development proceeds, the concentration of circulating estrogen rises and the level of circulating inhibin rises. The rising estrogen and inhibin levels inhibit hypothalamic secretion of GnRH and pituitary production and release of FSH. (GnRH pulse frequency is increased by estrogens and decreased by progesterone.) As tertiary ovarian follicles develop and estrogen levels rise, the pituitary output of LH gradually increases. Estrogens, FSH, and LH continue to support follicular development and maturation despite a gradual decline in FSH levels. In the second week of the ovarian cycle, estrogen levels sharply increase, and the dominant tertiary ovarian follicle enlarges in preparation for ovulation. By day 14, estrogen levels peak, triggering a massive outpouring of LH from the anterior lobe of the pituitary gland. The rupture of the follicular wall results in ovulation. Next, LH stimulates the formation of the corpus luteum, which secretes moderate amounts of estrogens but large amounts of progesterone, the principal hormone of the postovulatory period. About 12 days after ovulation, declining progesterone and estrogen levels stimulate hypothalamic receptors and GnRH production increases, leading to increased FSH and LH production in the anterior lobe of the pituitary gland; the cycle begins again. **21.** The corpus luteum degenerates and progesterone and estrogen levels decrease, causing degeneration of the endometrium and the onset of menstruation. Next, rising levels of FSH, LH, and estrogen stimulate the repair and regeneration of the functional layer of the endometrium. During the postovulatory phase, the combination of estrogen and progesterone causes the enlargement of the uterine glands and an increase in their secretory activity. **22.** During sexual arousal, erotic thoughts or physical stimulation of sensory nerves in the genital region increases the parasympathetic outflow over the pelvic nerve, leading to erection of the clitoris or penis. Orgasm is the intensely pleasurable sensation associated with perineal muscle contraction and ejaculation in males, and with uterine and vaginal contractions and perineal muscle contraction in females. These processes are comparable in males and females, but only males undergo emission and ejaculation. **23.** Men aged 50–60 experience the male climacteric, a time when circulating testosterone levels begin to decline and circulating levels of FSH and LH rise. Although sperm production continues, a gradual reduction in sexual activity occurs in older men. Women ages 45–55 experience menopause—the time that ovulation and menstruation cease, accompanied by a sharp and sustained rise in the production of GnRH, FSH, and LH and a drop in the concentrations of circulating estrogen and progesterone. The decline in estrogen levels leads to reductions in the size of the uterus and breasts, accompanied by a thinning of the urethral and vaginal walls. In addition to neural and cardiovascular effects, reduced estrogen concentrations have been linked to the development of osteoporosis, presumably because bone deposition proceeds at a slower rate.

Level 3 Critical Thinking and Clinical Applications

24. Women more frequently experience peritonitis stemming from a urinary tract infection because infectious organisms exiting the urethral orifice can readily enter the nearby vagina. From there, they can then proceed to the uterus, into the uterine tubes, and finally into the peritoneal cavity. No such direct path of entry into the abdominopelvic cavity exists in men. **25.** Regardless of their location, endometrial cells have receptors that respond to estrogen and progesterone. Under the influence of estrogen at the beginning of the menstrual cycle, any endometrial cells in the peritoneal cavity proliferate and begin to develop glands and blood vessels, which then further develop under the control of progesterone. The dramatic increase in size of this tissue presses on neighboring abdominal tissues and organs, causing periodic painful sensations.
26. Slightly elevated levels of estrogen and progesterone inhibit both GnRH release at the hypothalamus and the release of FSH and LH from the anterior lobe of the pituitary gland. Without FSH, tertiary ovarian follicles do not begin to develop, and levels of estrogen remain low. An LH surge, triggered by the peaking of estradiol, is necessary for ovulation to occur. If the level of estradiol is not allowed to rise above the critical level, the LH surge will not occur, and thus ovulation will not occur, even if a tertiary ovarian follicle managed to develop to a stage at which it could ovulate. Any mature follicles would ultimately degenerate, and no new follicles would mature to take their place. Although the ovarian cycle is interrupted, the level of hormones is still adequate to regulate a normal menstrual cycle. **27.** These observations suggest that a certain amount of body fat is necessary for menstrual cycles to occur. The nervous system appears to respond to circulating levels of the adipose tissue hormone leptin; when leptin levels fall below a certain set point, menstruation ceases. Because a woman lacking adequate fat reserves might not be able to have a successful pregnancy, the body prevents pregnancy by shutting down the ovarian cycle, and thus the menstrual cycle. Once sufficient energy reserves become available, the cycles begin again.

Answers to Clinical Case Wrap-Up

PAGE 1101 **1.** Susan's secondary oocyte leaves the tertiary ovarian follicle and is picked up by the fimbriae of the uterine tube. The oocyte is moved by cilia through the infundibulum, ampulla, and isthmus of the uterine tube. Tim's spermatozoa will meet and fertilize the oocyte at the junction of ampulla and isthmus. The fertilized oocyte, now an ovum, will continue into the uterine cavity where it will attach to the endometrium. **2.** Ovulation occurs at the midpoint, or day 14, of a 28-day cycle. Susan might experience "mittelschmerz" (German, "middle pain"), a discrete painful sensation as the secondary oocyte bursts out of the ovary. Blood tests will show a surge in LH from the anterior lobe of the pituitary gland. Her basal body temperature will rise and stay high until the end of her cycle.

Answers to Figure-Based Questions

FIGURE 28–7 The corpora cavernosa is split into two cylindrical masses and surrounds a central artery. The corpus spongiosum surrounds the urethra.

FIGURE 28–9 During mitosis, the spermatogonium divides, forming another spermatogonium and a primary spermatocyte. During meiosis I, the primary spermatocyte divides into two secondary spermatocytes. During meiosis II, each secondary spermatocytes divides into two spermatids. During spermiogenesis (the physical maturation), the four spermatids develop into four sperm.

FIGURE 28–11 The centrioles are found in the neck of the sperm.

FIGURE 28–22 During ovulation, the oocyte is a secondary oocyte.

FIGURE 28–23 The functional layer of the endometrium is affected most during the uterine cycle. During menstruation, the entire functional layer is sloughed off and lost.

FIGURE 28–25 Cholesterol is the precursor for all steroid hormones in both males and females.

Chapter 29

Answers to Checkpoints

PAGE 1105 **1.** Differentiation is the formation of different types of cells during development. **2.** Development begins at fertilization (conception), the union of a male gamete (a sperm) and a female gamete (oocyte). **3.** Inheritance refers to the transfer of genetically determined characteristics from one generation to the next. **PAGE 1107** **4.** Two sperm enzymes important to secondary oocyte penetration are hyaluronidase and acrosin. **5.** A normal human zygote contains 46 chromosomes. **6.** Gestation is the period of prenatal development. It consists of three trimesters. **7.** The first trimester is the time of pre-embryonic, embryonic, and early fetal development. The second trimester is a time of organ and organ system development. By the end of this stage, the fetus appears distinctly human. The third trimester is characterized by rapid fetal growth and the deposition of adipose tissue. **PAGE 1115** **8.** The inner cell mass of the blastocyst eventually develops into the embryo. **9.** Improper development of the yolk sac—the mesoderm-derived structure that gives rise to blood vessels and is an important site of blood cell formation—would affect the development and function of the cardiovascular system. **10.** Blood flows from the fetus to the placenta through paired umbilical arteries and returns in a single umbilical vein. **PAGE 1117** **11.** From 8 weeks to about 20 weeks, the fetus is measured from crown to rump. From 20 weeks to birth, the baby is measured from crown to heel. **PAGE 1121** **12.** Yes, Sue is pregnant. After fertilization, the developing trophoblast (and later, the placenta) produce and release the hormone hCG. **13.** Pregnant women experience an increase in respiratory rate and tidal volume. **14.** A mother's blood volume increases during pregnancy to compensate for the reduction in maternal blood volume resulting from blood flow through the placenta. **PAGE 1126** **15.** Three factors opposing the calming action of progesterone on the uterus are increasing estrogen levels, increasing oxytocin levels, and prostaglandin production. **16.** The three stages of labor are the dilation stage, expulsion stage, and placental stage. **17.** Immature delivery is the birth of a fetus weighing at least 500 g (17.6 oz), which is the normal weight near the end of the second trimester. Premature delivery usually refers to birth at 28–36 weeks at a weight of more than 1 kg (2.2 lb). **18.** Fraternal twins are dizygotic twins, and identical twins are monozygotic twins. **PAGE 1131** **19.** The postnatal stages of development are the neonatal period, infancy, childhood, adolescence, and maturity. These stages are followed by senescence, or aging, and death. **20.** The neonatal period is the time from birth to 1 month. Infancy continues from the neonatal period to age 1. Adolescence is the period after childhood in which sexual and physical maturity occur. **21.** Colostrum is produced by the mammary glands from the end of the sixth month of pregnancy until a few days after birth. After that, the glands begin producing breast milk, which contains fewer proteins (including antibodies) and far more fat than colostrum. **22.** Increases in the blood levels of GnRH, FSH, LH, and sex hormones mark the onset of puberty. **PAGE 1139** **23.** Genotype is a person's genetic makeup. Phenotype is a person's physical and physiological characteristics. It results from the interaction between the person's genotype and the environment. **24.** Heterozygous refers to a genotype possessing two different alleles at corresponding sites of a chromosome pair. For a heterozygous trait, one or both of the alleles determines the individual's phenotype. **25.** The phenotype of a person who is heterozygous for tongue rolling—who has one dominant allele for tongue-rolling and one recessive allele for non-tongue-rolling—would be "tongue roller." **26.** One reason children are not identical copies of their parents is that during meiosis, parental chromosomes are randomly distributed such that each gamete has a unique set of chromosomes. Additionally, crossing over and mutations during meiosis introduce new variations in gametes. Besides gene interactions, environmental factors can also affect the expression of genes and the resulting phenotype through epigenetic mechanisms.

Answers to Review Questions

PAGE 1142

Level 1 Reviewing Facts and Terms

1. d **2.** **(a)** parietal decidua; **(b)** basal decidua; **(c)** amnion; **(d)** chorion; **(e)** capsular decidua **3.** c **4.** b **5.** b **6.** a **7.** b **8.** b **9.** When a sperm contacts the secondary oocyte, their plasma membranes fuse. The oocyte is then activated: Its metabolic rate rises; it completes meiosis II; and the cortical reaction prevents additional sperm from entering. (Vesicles beneath the oocyte surface fuse with the plasma membrane and discharge their contents.) The male and female pronuclei replicate their DNA, fuse (amphimixis), and the zygote begins preparing for the first cleavage division. **10.** **(a)** The four extra-embryonic membranes are the yolk sac, amnion, allantois, and chorion. **(b)** The yolk sac forms from endoderm and mesoderm; it is an important site of blood cell formation. The amnion forms from ectoderm and mesoderm; it encloses the fluid that surrounds and cushions the developing embryo and fetus. The allantois forms from endoderm and mesoderm; its base gives rise to the urinary bladder. The chorion forms from mesoderm and trophoblast; circulation through chorionic vessels provides a "rapid-transit system" for blood and nutrients. **11.** The dilation stage begins with the onset of true labor, as the cervix dilates and the fetus begins to move toward the cervical canal; late in this stage, the amniochorionic membrane ruptures. The expulsion stage begins as the cervix dilates completely and continues until the fetus has completely emerged from the vagina (delivery). In the placental stage, the uterus gradually contracts, tearing the connections between the endometrium and the placenta and ejecting the placenta. **12.** The fetal pituitary secretes oxytocin, which is released into the maternal bloodstream at the placenta. This release may be the trigger for the onset of true labor. **13.** (1) Neonatal period (birth to 1 month): The newborn becomes relatively self-sufficient and begins performing respiration (breathing), digestion, and excretion on its own. Heart rates and fluid requirements are higher than those of adults. Neonates have little ability to thermoregulate. (2) Infancy (1 month to 1 year): Major organ systems (other than those related to reproduction) become fully operational. (3) Childhood (1 year to puberty): Growth continues; body proportions change significantly. **14.** Three events interact to promote increased hormone production and sexual maturation at puberty: (1) The hypothalamus increases its production of GnRH; (2) the anterior lobe of the pituitary gland becomes more sensitive to the presence of GnRH, and circulating levels of FSH and LH rise rapidly; and (3) ovarian or testicular cells become more sensitive to FSH and LH. Puberty initiates adolescence, which includes gametogenesis (gamete production) and the production of sex hormones that stimulate the appearance of secondary sexual characteristics and behaviors.

Level 2 Reviewing Concepts

15. b **16.** b **17.** c **18.** The post-implantation embryo obtains nutrients through the chorionic villi and later the placenta. The placenta develops during placentation. **19.** The placenta produces (1) human chorionic gonadotropin, which maintains the integrity of the corpus luteum and promotes the continued secretion of progesterone (keeping the endometrial lining functional); and (2) human placental lactogen and placental prolactin, which help prepare the mammary glands for milk production. Human placental lactogen also ensures adequate levels of glucose and protein for the developing fetus. (3) The placenta also produces relaxin, which increases the flexibility of the pubic symphysis, causes dilation of the cervix, and suppresses the release of oxytocin by the hypothalamus, delaying the onset of labor contractions. (4) The placenta becomes the main source of progesterone to maintain the endometrial lining after the first trimester. (5) Placental estrogen production rapidly increases in the third trimester and plays a role in stimulating labor and delivery. **20.** Respiratory rate and tidal volume increase, allowing the lungs to obtain the extra oxygen the fetus needs, and to remove the excess carbon dioxide the fetus generates. Maternal blood volume increases, compensating for blood flowing into the placenta. Nutrient and vitamin requirements increase 10–30 percent, reflecting the fact that some of the mother's nutrients go to nourish the fetus. Glomerular filtration rate increases by about 50 percent, which corresponds to the increased blood volume, and this accelerates the excretion of metabolic wastes generated by the fetus. The uterus and mammary glands increase in size, and secretory activity begins in the mammary glands. **21.** Positive feedback mechanisms between increasing levels of oxytocin and increased uterine distortion ensure that labor contractions continue until delivery has been completed. **22.** A neonate must fill its lungs (which are collapsed and filled with fluid at birth) with air, which alters the pattern of cardiovascular circulation due to changes in blood pressure and flow rates. It must excrete the mixture of debris (meconium) that has collected in the fetal digestive system. The neonate must obtain nourishment from a new source—the mother's mammary glands. Neonatal fluid requirements are high, because the newborn cannot concentrate its urine significantly. The infant also has little ability to thermoregulate, although as it grows its insulating adipose tissue increases, and its metabolic rate also rises. **23.** **(a)** Genotype refers to an individual's genetic makeup; phenotype refers to the individual's physical and physiological characteristics. **(b)** A heterozygous genotype carries different alleles for a given gene; a homozygous genotype carries identical alleles for that gene. **(c)** In simple inheritance, phenotypes are determined by interactions between a single pair of alleles; in polygenic inheritance, interactions occur among multiple genes. **24.** **(a)** dominant; **(b)** recessive; **(c)** X-linked; **(d)** autosomal

Level 3 Critical Thinking and Clinical Applications

25. The probability that this couple's daughters will have hemophilia is 0, because each daughter will receive a normal allele from her father. There is a 50 percent chance that a son will have hemophilia, because each son has a 50 percent chance of receiving the mother's normal allele and a 50 percent chance of receiving the mother's recessive allele. **26.** Although technically it only takes one sperm to fertilize an egg, the probability of this occurring is very low if too few sperm are present. Most sperm that enter the female reproductive tract are killed or disabled before they reach the uterus. Many of the sperm reaching the uterus are incapable of reaching the secondary oocyte, which is in a uterine tube. Once at the oocyte, the sperm must penetrate the corona radiata, which requires the combined acrosomal enzymes of dozens of sperm. If too few sperm are present in the vagina, the number reaching the uterine tube is too low for any one sperm to penetrate the corona radiata. **27.** The most obvious possibility is that the infant has a problem with the cardiovascular supply to the lungs. A good guess would be a patent ductus arteriosus (the ductus arteriosus has failed to close off completely). When the baby is not being stressed (bathing creates heat loss and thermal stress) or eating (less air is entering the lungs), the infant appears normal. Because some of the blood flow to the lungs is being shunted over to the aorta during stress and eating, too little blood is being oxygenated, so the infant becomes cyanotic. **28.** It is very unlikely that the baby's condition results from a viral infection contracted during the third trimester, because organ systems develop during the first trimester and are fully formed by the end of the second trimester.

Answers to Clinical Case Wrap-Up

PAGE 1143 **1.** Implanting the blastocysts on the sixth day after ovulation matches the fertility cycle in nature. This timing ensures the blastocysts are available for implantation in the uterine wall by days 7 through 10 after ovulation. **2.** Hans and Willem are not monozygotic twins because they did not result from the fertilization of a single oocyte with a single sperm, forming a single zygote that subsequently split. Neither are they technically dizygotic twins.

They are the developmental result of two separate oocytes from the same mother, but unlike dizygotic twins, they have different *biological* fathers. In terms of biological inheritance, they are half brothers. Socially they are fraternal twins.

Answers to Figure-Based Questions

FIGURE 29–2 The outer cell layer of a blastocyst is called the trophoblast. The function of the trophoblast is to provide nutrients to the developing embryo.

FIGURE 29–4 The third germ layer that forms during gastrulation is called mesoderm.

FIGURE 29–10 Oxytocin and prostaglandins stimulate and increase the excitability of the myometrium.

FIGURE 29–14 The 23rd pair of chromosomes in this karyotype are the sex chromosomes. They are characteristic of a male because both X and Y sex chromosomes are present. A female karyotype would have two X sex chromosomes.

FIGURE 29–15 Simple inheritance is involved in determining nearsightedness (dominant trait) and normal vision (recessive trait), and sex-linked inheritance is involved in determining red-green color blindness.

Appendices

Appendix A Normal Physiological Values

Tables 1 and 2 present normal averages or ranges for the chemical composition of body fluids. These values are approximations rather than absolute values, because test results vary from laboratory to laboratory as a result of differences in procedures, equipment, normal solutions, or other factors. Blanks in the tabular data appear where data are not available. The following locations in the text contain additional information about body fluid analysis:

Table 19–3 (p. 676) presents data on the formed elements of the blood.

Table 26–2 (p. 987) compares the normal values for solutes in plasma and urine.

Table 26–5 (p. 1005) gives the general characteristics of normal urine, and Table 26–6 (p. 1008) shows the typical values of substances in a standard urinalysis.

Table 1 The Composition of Body Fluids

	Normal Averages or Ranges					
Test	Perilymph	Endolymph	Synovial Fluid	Sweat	Saliva	Semen
pH			7.4	4–6.8	6.4*	7.19
Specific gravity			1.008–1.015	1.001–1.008	1.007	1.028
Electrolytes (mEq/L)						
Potassium	5.5–6.3	140–160	4.0	4.3–14.2	21	31.3
Sodium	143–150	12–16	136.1	0–104	14*	117
Calcium	1.3–1.6	0.05	2.3–4.7	0.2–6	3	12.4
Magnesium	1.7	0.02		0.03–4	0.6	11.5
Bicarbonate	17.8–18.6	20.4–21.4	19.3–30.6		6*	24
Chloride	121.5	107.1	107.1	34.3	17	42.8
Proteins (total) (mg/dL)	200	150	1.72 g/dL	7.7	386†	4.5 g/dL
Metabolites (mg/dL)						
Amino acids				47.6	40	1.26 g/dL
Glucose	104		70–110	3.0	11	224 (fructose)
Urea				26–122	20	72
Lipids (total)	12		20.9	‡	25–500§	188

*Increases under salivary stimulation.
†Primarily alpha-amylase, with some lysozymes.
‡Not present in eccrine (merocrine) secretions.
§Cholesterol.

Table 2 The Chemistry of Blood, Cerebrospinal Fluid (CSF), and Urine

Test	Normal Averages or Ranges		
	Blood*	CSF	Urine
pH	S: 7.35–7.45	7.31–7.34	4.5–8.0
Osmolarity (mOsm/L)	S: 280–295	292–297	855–1335
Electrolytes	|——(mEq/L unless noted)——|		(urinary loss, mEq per 24-hour period[†])
Bicarbonate	P: 20–28	20–24	0
Calcium	S: 8.5–10.3	2.1–3.0	6.5–16.5
Chloride	P: 97–107	100–108	110–250
Iron	S: 50–150 μg/L	23–52 μg/L	40–150 μg
Magnesium	S: 1.4–2.1	2–2.5	6.0–10.0
Phosphorus	S: 1.8–2.9	1.2–2.0	0.4–1.3 g
Potassium	P: 3.5–5.0	2.7–3.9	25–125
Sodium	P: 135–145	137–145	40–220
Sulfate	S: 0.2–1.3		1.07–1.3 g
Metabolites	|——(mg/dL unless noted)——|		(urinary loss, mg per 24-hour period[‡])
Amino acids	P/S: 2.3–5.0	10.0–14.7	41–133
Ammonia	P: 20–150 μg/dL	25–80 μg/dL	340–1200
Bilirubin	S: 0.5–1.0	<0.2	0
Creatinine	P/S: 0.6–1.5	0.5–1.9	770–1800
Glucose	P/S: 70–110	40–70	0
Ketone bodies	S: 0.3–2.0	1.3–1.6	10–100
Lactic acid	WB: 0.7–2.5 mEq/L[§]	10–20	100–600
Lipids (total)	P: 450–1000	0.8–1.7	0.002
Cholesterol (total)	S: 150–300	0.2–0.8	1.2–3.8
Triglycerides	S: 40–150	0–0.9	0
Urea	P: 8–25	12.0	1800
Uric acid	P: 2.0–6.0	0.2–1.5	250–750
Proteins	(g/dL)	(mg/dL)	(urinary loss, mg per 24-hour period[‡])
Total	P: 6.0–8.0	2.0–4.5	0–8
Albumin	S: 3.2–4.5	10.6–32.4	0–3.5
Globulins (total)	S: 2.3–3.5	2.8–15.5	7.3
Immunoglobulins	S: 1.0–2.2	1.1–1.7	3.1
Fibrinogen	P: 0.2–0.4	0.65	0

*S = serum, P = plasma, WB = whole blood.

[†]Because urinary output averages just over 1 liter per day, these electrolyte values are comparable to mEq/L.

[‡]Because urinary metabolite and protein data approximate mg/L or g/L, these data must be divided by 10 for comparison with CSF or blood concentrations.

[§]Venous blood sample.

Appendix B Gas Pressure Measurements and Cell Turnover Times

To understand the principles of blood flow, capillary exchange, respiratory gas exchange, and the mechanics of breathing, we must know the pressures of blood and gases in peripheral tissues, in the bloodstream, and inside and outside the respiratory tract. Blood and gas pressures are measured and reported in several ways. Table 3 shows four common reporting methods.

Table 3 Four Common Methods of Reporting Gas Pressure

Measurement Units	Description
millimeters of mercury (mm Hg)	This is the most common method of reporting blood pressure and gas pressures. Normal atmospheric pressure is approximately 760 mm Hg.
torr	This unit of measurement is preferred by many respiratory therapists; it is also commonly used in Europe and in some technical journals. Note that 1 torr is equivalent to 1 mm Hg; in other words, normal atmospheric pressure is equal to 760 torr.
centimeters of water (cm H_2O)	In a hospital setting, anesthetic gas pressures and oxygen pressures are commonly measured in centimeters of water. Note that 1 cm H_2O is equivalent to 0.735 mm Hg; normal atmospheric pressure is 1033.6 cm H_2O.
pounds per square inch (psi)	Pressures in compressed gas cylinders and other industrial applications are generally reported in psi. Normal atmospheric pressure at sea level is approximately 15 psi.

Appendix B Turnover Times for Selected Cells

Most of the organic molecules in a cell are replaced at intervals ranging from hours to months. The average time between synthesis and breakdown is known as the *turnover time.* This table lists the turnover times of the organic components of representative cells.

Table 4 Turnover Times of the Organic Components of Four Cell Types

Cell Type	Component	Average Recycling Time*
Liver	Total protein	5–6 days
	Enzymes	1 hour to several days, depending on the enzyme
	Glycogen	1–2 days
	Cholesterol	5–7 days
Muscle	Total protein	30 days
	Glycogen	12–24 hours
Nerve	Phospholipids	200 days
	Cholesterol	100+ days
Fat	Triglycerides	15–20 days

*Most values were obtained from studies on mammals other than humans.

Appendix C Codon Chart

A codon is a sequence of three consecutive nucleotides in mRNA that codes for a particular amino acid or signals to stop protein synthesis. Because each mRNA codon consists of three nucleotides, the four nucleotides in mRNA (A, U, G, and C) can produce 64 different combinations. Of these, 61 codons correspond to amino acids and 3 act as stop signals of protein synthesis. Only 20 different amino acids are used to synthesize proteins, so most amino acids are represented by more than one codon. One codon, AUG, has a dual role. It codes for the amino acid methionine and also acts as the start signal for protein synthesis. It is always the first codon in a strand of mRNA.

Table 5 The Genetic Code

Genetic Code (mRNA codons)

First nucleotide	Second nucleotide: U	Second nucleotide: C	Second nucleotide: A	Second nucleotide: G	Third nucleotide
U	UUU Phenylalanine	UCU Serine	UAU Tyrosine	UGU Cysteine	U
U	UUC Phenylalanine	UCC Serine	UAC Tyrosine	UGC Cysteine	C
U	UUA Leucine	UCA Serine	UAA **Stop**	UGA **Stop**	A
U	UUG Leucine	UCG Serine	UAG **Stop**	UGG Tryptophan	G
C	CUU Leucine	CCU Proline	CAU Histidine	CGU Arginine	U
C	CUC Leucine	CCC Proline	CAC Histidine	CGC Arginine	C
C	CUA Leucine	CCA Proline	CAA Glutamine	CGA Arginine	A
C	CUG Leucine	CCG Proline	CAG Glutamine	CGG Arginine	G
A	AUU Isoleucine	ACU Threonine	AAU Asparagine	AGU Serine	U
A	AUC Isoleucine	ACC Threonine	AAC Asparagine	AGC Serine	C
A	AUA Isoleucine	ACA Threonine	AAA Lysine	AGA Arginine	A
A	AUG **Start** *or* Met*	ACG Threonine	AAG Lysine	AGG Arginine	G
G	GUU Valine	GCU Alanine	GAU Aspartic acid	GGU Glycine	U
G	GUC Valine	GCC Alanine	GAC Aspartic acid	GGC Glycine	C
G	GUA Valine	GCA Alanine	GAA Glutamic acid	GGA Glycine	A
G	GUG Valine	GCG Alanine	GAG Glutamic acid	GGG Glycine	G

*Abbreviation for methionine

Appendix D The Periodic Table

The periodic table presents the known elements in order of their atomic weights. Each horizontal row represents a single electron shell. The number of elements in that row is determined by the maximum number of electrons that can be stored at that energy level. The element at the left end of each row contains a single electron in its outermost electron shell; the element at the right end of the row has a filled outer electron shell. This organization highlights similarities that reflect the composition of the outer electron shell, or valence shell. These similarities are evident when you examine the vertical columns. All the elements of the right-most column—helium, neon, argon, krypton, xenon, and radon—have full electron shells; each is a gas at normal atmospheric temperature and pressure, and none reacts readily with other elements. These elements, highlighted in blue, are known as the noble, or inert, gases. In contrast, the elements of the left-most column below hydrogen—lithium, sodium, potassium, rubidium, caesium, and francium—are silvery, soft metals that are so highly reactive that pure forms cannot be found in nature. The fourth and fifth electron levels can each hold up to 18 electrons. Table inserts are used for the so-called lanthanoids (atomic numbers 57–71) and actinoids (atomic numbers 89–103) to save space, as higher levels can store up to 32 electrons. Elements of particular importance to our discussion of human anatomy and physiology are highlighted in pink.

Key: Atomic number — 1; Chemical symbol — H; Element name — Hydrogen; Atomic weight — 1.01

1 H Hydrogen 1.01																	2 He Helium 4.00
3 Li Lithium 6.94	4 Be Beryllium 9.01											5 B Boron 10.81	6 C Carbon 12.01	7 N Nitrogen 14.01	8 O Oxygen 16.00	9 F Fluorine 19.00	10 Ne Neon 20.18
11 Na Sodium 22.99	12 Mg Magnesium 24.31											13 Al Aluminum 26.98	14 Si Silicon 28.09	15 P Phosphorus 30.97	16 S Sulfur 32.07	17 Cl Chlorine 35.45	18 Ar Argon 39.95
19 K Potassium 39.10	20 Ca Calcium 40.08	21 Sc Scandium 44.96	22 Ti Titanium 47.87	23 V Vanadium 50.94	24 Cr Chromium 52.00	25 Mn Manganese 54.94	26 Fe Iron 55.85	27 Co Cobalt 58.93	28 Ni Nickel 58.69	29 Cu Copper 63.55	30 Zn Zinc 65.38	31 Ga Gallium 69.72	32 Ge Germanium 72.63	33 As Arsenic 74.92	34 Se Selenium 78.96	35 Br Bromine 79.90	36 Kr Krypton 83.80
37 Rb Rubidium 85.47	38 Sr Strontium 87.62	39 Y Yttrium 88.91	40 Zr Zirconium 91.22	41 Nb Niobium 92.91	42 Mo Molybdenum 95.96	43 Tc Technetium	44 Ru Ruthenium 101.1	45 Rh Rhodium 102.9	46 Pd Palladium 106.4	47 Ag Silver 107.9	48 Cd Cadmium 112.4	49 In Indium 114.8	50 Sn Tin 118.7	51 Sb Antimony 121.8	52 Te Tellurium 127.6	53 I Iodine 126.9	54 Xe Xenon 131.3
55 Cs Caesium 132.9	56 Ba Barium 137.3	57–71 Lanthanoids	72 Hf Hafnium 178.5	73 Ta Tantalum 180.9	74 W Tungsten 183.8	75 Re Rhenium 186.2	76 Os Osmium 190.2	77 Ir Iridium 192.2	78 Pt Platinum 195.1	79 Au Gold 197.0	80 Hg Mercury 200.6	81 Tl Thallium 204.3	82 Pb Lead 207.2	83 Bi Bismuth 209.0	84 Po Polonium	85 At Astatine	86 Rn Radon
87 Fr Francium	88 Ra Radium	89–103 Actinoids	104 Rf Rutherfordium	105 Db Dubnium	106 Sg Seaborgium	107 Bh Bohrium	108 Hs Hassium	109 Mt Meitnerium	110 Ds Darmstadtium	111 Rg Roentgenium	112 Cn Copernicium	113 Nh Nihonium*	114 Fl Flerovium	115 Mc Moscovium*	116 Lv Livermorium	117 Ts Tennessine*	118 Og Oganesson*

57 La Lanthanum 138.9	58 Ce Cerium 140.1	59 Pr Praseodymium 140.9	60 Nd Neodymium 144.2	61 Pm Promethium	62 Sm Samarium 150.4	63 Eu Europium 152.0	64 Gd Gadolinium 157.3	65 Tb Terbium 158.9	66 Dy Dysprosium 162.5	67 Ho Holmium 164.9	68 Er Erbium 167.3	69 Tm Thulium 168.9	70 Yb Ytterbium 173.1	71 Lu Lutetium 175.0
89 Ac Actinium	90 Th Thorium 232.0	91 Pa Protactinium 231.0	92 U Uranium 238.0	93 Np Neptunium	94 Pu Plutonium	95 Am Americium	96 Cm Curium	97 Bk Berkelium	98 Cf Californium	99 Es Einsteinium	100 Fm Fermium	101 Md Mendelevium	102 No Nobelium	103 Lr Lawrencium

*Pending formal approval

Glossary of Key Terms

This glossary is an alphabetical list of key terms related to topics covered in the textbook. Brief explanations, along with the chapter number(s) where the topic was covered, are given.

A

abdomen: The region of the trunk between the inferior margin of the rib cage and the superior margin of the pelvis. (1)

abdominopelvic cavity: The term used to refer to the general region bounded by the abdominal wall and the pelvis; it contains the peritoneal cavity and visceral organs. (1)

abducens nerve: Cranial nerve VI, which innervates the lateral rectus muscle of the eye. (14)

abduction: Movement away from the midline of the body, as viewed in the anatomical position. (9)

abortion: The premature loss or expulsion of an embryo or fetus. (29)

abscess: A localized collection of pus within a damaged tissue. (4, 22)

absorption: The active or passive uptake of gases, fluids, or solutes. (24, 25)

accommodation: An alteration in the curvature of the lens of the eye to focus an image on the retina. (17)

acetabulum: The fossa on the lateral aspect of the pelvis that accommodates the head of the femur. (8)

acetylcholine (ACh): A chemical neurotransmitter in the brain and peripheral nervous system; the dominant neurotransmitter in the peripheral nervous system, released at neuromuscular junctions and synapses of the parasympathetic division. (10, 12, 16)

acetylcholinesterase (AChE): An enzyme found in the synaptic cleft, bound to the postsynaptic membrane, and in tissue fluids; breaks down and inactivates acetylcholine molecules; also called *cholinesterase*. (10, 12)

acetyl-CoA: An acetyl group bound to coenzyme A, a participant in the anabolic and catabolic pathways for carbohydrates, lipids, and many amino acids. (25)

acetyl group: —CH_3CO; an acetic acid molecule without the hydroxyl group. (25)

acid: A compound whose dissociation in solution releases a hydrogen ion and an anion; an acidic solution has a pH below 7.0 and has an excess of hydrogen ions. (2, 27)

acidosis: An abnormal physiological state characterized by blood pH below 7.35. (2, 25, 26, 27)

acinus/acini: A histological term referring to a blind pocket, pouch, or sac.

acoustic: Pertaining to sound or the sense of hearing. (17)

acquired immunodeficiency syndrome (AIDS): A disease caused by the human immunodeficiency virus (HIV); characterized by the destruction of helper T cells and a resulting severe impairment of the immune response. (22)

acromegaly: A condition caused by the overproduction of growth hormone in adults, characterized by a thickening of bones and an enlargement of cartilages and other soft tissues. (6, 18)

acromion: A continuation of the scapular spine that projects superior to the capsule of the glenohumeral joint. (8)

acrosome: A membranous sac at the tip of a sperm that contains hyaluronidase; also called *acrosomal cap*. (28)

actin: The protein component of microfilaments that forms thin filaments in skeletal muscles and produces contractions of all muscles through interaction with thick (myosin) filaments; *see also* **sliding-filament theory**. (3, 10)

action potential: A propagated change in the membrane potential of excitable cells, initiated by a change in the membrane permeability to sodium ions; *see also* **nerve impulse**. (10, 12)

active transport: The ATP-dependent absorption or secretion of solutes across a plasma membrane. (3, 26)

acute: Sudden in onset, typically severe in intensity, and brief in duration.

adaptation: A decrease in receptor sensitivity or perception after chronic stimulation (15); a change in pupillary size in response to changes in light intensity (17); physiological responses that produce acclimatization. (25)

Addison disease: A condition resulting from the hyposecretion of glucocorticoids; characterized by lethargy, weakness, hypotension, and increased skin pigmentation. (5, 18)

adduction: Movement toward the axis or midline of the body, as viewed in the anatomical position. (9)

adenine (A): A purine; one of the nitrogenous bases in the nucleic acids RNA and DNA. (2)

adenohypophysis: The anterior lobe of the pituitary gland. (18)

adenoid: The pharyngeal tonsil. (22, 23)

adenosine: A compound consisting of adenine and ribose. (2)

adenosine diphosphate (ADP): A compound consisting of adenosine with two phosphate groups attached. (2, 25)

adenosine monophosphate (AMP): A nucleotide consisting of adenosine plus a phosphate group (PO_4^{3-}); also called *adenosine phosphate*. (2)

adenosine triphosphate (ATP): A high-energy compound consisting of adenosine with three phosphate groups attached; the third is attached by a high-energy bond. (2, 10, 25)

adenylate cyclase: An enzyme bound to the inner surfaces of plasma membranes that can convert ATP to cyclic AMP; formerly called *adenylyl cyclase*. (12, 18)

adhesion: The fusion of two mesenterial layers after damage or irritation of their opposing surfaces; this process restricts relative movement of the organs involved (4); the binding of a phagocyte to its target (22).

adipocyte: A fat cell. (4)

adipose tissue: Loose connective tissue dominated by adipocytes. (4, 18, 28)

adrenal cortex: The superficial region of the adrenal gland that produces steroid hormones; also called the *suprarenal cortex*. (18)

adrenal gland: A small endocrine gland that secretes steroids and catecholamines and is located superior to each kidney; also called *suprarenal gland*. (18)

adrenal medulla: A modified sympathetic ganglion that secretes catecholamines into the blood during sympathetic activation; also called *suprarenal medulla* (16); the core of the adrenal gland (18).

adrenergic: An axon terminal that, when stimulated, releases norepinephrine. (12)

adrenocortical hormone: Any steroid produced by the adrenal cortex. (18)

adrenocorticotropic hormone (ACTH): The hormone that stimulates the production and secretion of glucocorticoids by the zona fasciculata of the adrenal cortex; released by the anterior lobe of the pituitary gland (adenohypophysis) in response to corticotropin-releasing hormone. (18)

adventitia: The superficial layer of connective tissue surrounding an internal organ; fibers are continuous with those of surrounding tissues, providing support and stabilization. (24)

aerobic: Requiring oxygen.

aerobic metabolism: The complete breakdown of organic substrates into carbon dioxide and water; a process that yields large amounts of ATP but requires mitochondria and oxygen. (3, 10, 25)

afferent: Conducting toward a center.

afferent arteriole: An arteriole that carries blood to a glomerulus of the kidney. (26)

afferent fiber: An axon that carries sensory information to the central nervous system. (12)

agglutination: The aggregation of red blood cells due to interactions between surface antigens and plasma antibodies. (19, 22)

agglutinins: Immunoglobulins (antibodies) in plasma that react with antigens on the surfaces of foreign red blood cells when donor and recipient differ in blood type. (19)

agglutinogens: Genetically determined surface antigens on red blood cells that stimulate the formation of specific agglutinins. (19)

aggregated lymphoid nodules: Lymphoid nodules beneath the epithelium of the small intestine; also called *Peyer's patches*. (22)

agonist: A muscle responsible for a specific movement; also called a *prime mover*. (11)

agranular: Without granules; *agranular leukocytes* are monocytes and lymphocytes. (19)

AIDS: *See* **acquired immunodeficiency syndrome**. (22)

aldosterone: A mineralocorticoid produced by the zona glomerulosa of the adrenal cortex; stimulates the kidneys to conserve sodium and water; secreted in response to the presence of angiotensin II. (18, 26, 27)

alkalosis: The condition characterized by a blood pH greater than 7.45; associated with a relative deficiency of hydrogen ions or an excess of bicarbonate ions. (2, 27)

alpha (α) receptors: Membrane receptors sensitive to norepinephrine or epinephrine; stimulation normally results in target cell excitation. (16)

alveolar sac: An air-filled chamber at the terminal end of the alveolar duct that gives rise to the alveoli in the lung. (23)

alveolus/alveoli: Terminal pockets at the end of the respiratory tree, lined by a simple squamous epithelium and surrounded by a capillary network; sites of gas exchange with the blood (23); a bony socket that holds the root of a tooth. (24)

Alzheimer's disease: A disorder resulting from degenerative changes in populations of neurons in the cerebrum, causing dementia characterized by problems with attention, short-term memory, and emotions. (16)

GLOSSARY

amination: The attachment of an amino group to a carbon chain; performed by a variety of cells and important in the synthesis of amino acids. (25)
amino acids: Organic compounds made up of carbon, hydrogen, oxygen, and nitrogen; building block of protein; chemical structure can be summarized as R—$CHNH_2$—COOH. (2, 25)
amino group: —NH_2. (2)
amnion: One of the four extra-embryonic membranes; surrounds the developing embryo or fetus. (29)
amniotic fluid: Fluid that fills the amniotic cavity; cushions and supports the embryo or fetus. (4, 29)
amphiarthrosis: A joint (articulation) that permits a small degree of independent movement; *see* **interosseous membrane** (8) and **pubic symphysis**. (9)
ampulla/ampullae: A localized dilation in the lumen of a canal or passageway. (17, 24, 28)
amygdaloid body: A basal nucleus that is a component of the limbic system and acts as an interface between the limbic system, the cerebrum, and sensory systems; also called *amygdala*. (14)
amylase: An enzyme that breaks down polysaccharides; produced by the salivary glands and pancreas. (24)
anabolism: The synthesis of complex organic compounds from simpler precursors. (2, 25)
anaerobic: Without oxygen.
analgesic: A substance that relieves pain. (15)
anal triangle: The posterior subdivision of the perineum. (11)
anaphase: The mitotic stage in which the paired chromatids separate and move toward opposite ends of the spindle apparatus. (3)
anaphylaxis: A hypersensitivity reaction due to the binding of antigens to immunoglobulins (IgE) on the surfaces of mast cells; the release of histamine, serotonin, and prostaglandins by mast cells then causes widespread inflammation; a sudden decrease in blood pressure may occur, producing anaphylactic shock. (22)
anastomosis: The joining of two tubes, usually referring to a connection between two peripheral vessels without an intervening capillary bed. (21)
anatomical position: An anatomical reference position; the body viewed from the anterior surface with the palms facing forward. (1)
anatomy: The study of internal and external body structures and their physical relationships among other body parts. (1)
androgen: A steroid sex hormone mainly produced by the interstitial cells of the testis and manufactured in small quantities by the adrenal cortex in both sexes. (18, 28)
anemia: The condition marked by a decrease in the hematocrit, the hemoglobin content of the blood, or both. (19)
aneurysm: A weakening in the arterial wall causing an outpouching or enlargement of the artery. (21)
angiotensin-converting enzyme (ACE): A chemical that converts angiotensin I (a hormone) to angiotensin II (a vasoconstrictor), which causes an increase in blood pressure. (26)
angiotensin I: The hormone produced by the activation of angiotensinogen by renin (18); angiotensin-converting enzyme (ACE) converts angiotensin I into angiotensin II in lung capillaries. (26)
angiotensin II: A hormone that causes an increase in systemic blood pressure, stimulates the secretion of aldosterone, promotes thirst, and causes the release of antidiuretic hormone (18); angiotensin-converting enzyme (ACE) in lung capillaries converts angiotensin I into angiotensin II. (21, 26)
angiotensinogen: The blood protein produced by the liver that is converted to angiotensin I by the enzyme renin. (18)
anion: An ion having a negative charge. (2, 27)
anoxia: Inadequate oxygen reaching body tissues. (23)
antagonist: A muscle that opposes the movement of an agonist. (11)
antebrachium: The forearm. (8)
anterior: On or near the front, or ventral surface, of the body.
antibiotic: A chemical agent that selectively kills pathogens, primarily bacteria. (20)
antibody: A globular protein produced by plasma cells that will bind to specific antigens and promote their destruction or removal from the body. (19, 22)
antibody-mediated immunity: The form of immunity resulting from the presence of circulating antibodies produced by plasma cells; also called *humoral immunity*. (22)
anticodon: Three nitrogenous bases on a tRNA molecule that interact with a complementary codon on a strand of mRNA. (3)
antidiuretic hormone (ADH): A hormone synthesized in the hypothalamus and secreted at the posterior lobe of the pituitary gland (neurohypophysis); causes the kidneys to retain water and an increase in blood pressure. (18, 21, 26, 27)
antigen: A substance capable of inducing an immune response that may include the production of antibodies. (22)
antigen–antibody complex: The combination of an antigen and a specific antibody. (22)
antigenic determinant site: A region of an antigen that can interact with an antibody molecule. (22)
antigen-presenting cell (APC): A cell that processes antigens and displays them, bound to class II MHC proteins; essential to the initiation of a normal immune response. (22)
antihistamine: A chemical that blocks the action of histamine on peripheral tissues. (22)
antrum: A chamber or pocket. (28)
anulus: A cartilage or bone shaped like a ring; also spelled *annulus*. (9)
anus: The external opening of the anal canal. (24)
aorta: The large, elastic artery that carries blood away from the left ventricle and into the systemic circuit. (20)
aortic bodies: Receptors in the aortic arch that are sensitive to changes in oxygen, carbon dioxide, and pH levels in the blood. (21)
aortic sinus: Space between the superior portion of each of the three aortic valve cusps and the dilated portion of the wall of the ascending aorta.(21)
apnea: Cessation of breathing. (23)
apocrine secretion: A mode of secretion in which the glandular cell sheds portions of its cytoplasm. (4, 5)
aponeurosis/aponeuroses: Broad tendinous sheet(s) that may serve as the origin or insertion of a skeletal muscle. (4, 6, 10)
apoptosis: Genetically programmed cell death. (3)
appendicular: Pertaining to the upper or lower limbs. (7, 8)
appendix: A blind sac connected to the cecum of the large intestine; also called *vermiform appendix*. (24)
appositional growth: The enlargement of a cartilage or bone by the addition of cartilage or bony matrix at its surface. (4, 6)
aqueous humor: A fluid similar to perilymph or cerebrospinal fluid that fills the anterior chamber of the eye. (17)
arachidonic acid: One of the essential fatty acids. (2, 18)
arachnoid granulations: Processes of the arachnoid mater that project into the superior sagittal sinus; sites where cerebrospinal fluid enters the venous circulation. (14)
arachnoid mater: The middle meninx (layer) that encloses cerebrospinal fluid and protects the central nervous system. (13, 14)
arbor vitae: The central, branching mass of white matter inside the cerebellum. (14)
arcuate: Curving.
areolar: Containing minute spaces, as in areolar tissue.
areolar tissue: Loose connective tissue with an open framework. (4)
arrector pili: Smooth muscles whose contractions force hairs to stand erect. (5)
arrhythmias: Abnormal patterns of cardiac contractions. (20)
arteriole: A small arterial branch that delivers blood to a capillary network. (21)
artery: A blood vessel that carries blood away from the heart and toward a peripheral capillary. (4, 20, 21)
articular: Pertaining to a joint.
articular capsule: The dense collagen fiber sleeve that surrounds a joint and provides protection and stabilization; also called *joint capsule*. (6, 9)
articular cartilage: The cartilage pad that covers the surface of a bone inside a joint cavity. (6, 9)
articulation: A joint. (9)
arytenoid cartilages: A pair of small cartilages in the larynx. (23)
ascending tract: A tract carrying information from the spinal cord to the brain. (13, 14)
association areas: Cortical areas of the cerebrum that are responsible for the integration of sensory inputs and/or motor commands. (14)
association neuron: *See* **interneuron**. (12)
astrocyte: One of the four types of neuroglia in the central nervous system; responsible for maintaining the blood brain barrier by the stimulation of endothelial cells. (12)
atherosclerosis: The formation of fatty plaques in the walls of arteries, restricting blood flow. (21)
atom: The smallest stable unit of matter. (1, 2)
atomic number: The number of protons in the nucleus of an atom. (2)
atomic weight: The average of the different atomic masses and isotope proportions of a particular element. (2)
atria: Thin-walled chambers of the heart that receive venous blood from the pulmonary or systemic circuit.
atrial natriuretic peptide (ANP): *See* **natriuretic peptides**. (20)
atrial reflex: The reflexive increase in heart rate after an increase in venous return; due to mechanical and neural factors; also called *Bainbridge reflex*. (20, 21)
atrioventricular (AV) bundle: Specialized conducting cells in the interventricular septum that carry the contracting stimulus from the AV node to bundle branches and then to Purkinjie fibers; also called *bundle of His*. (20)
atrioventricular (AV) node: Specialized cardiac muscle cells that relay the contractile stimulus to the atrioventricular bundle (bundle of His), the bundle branches, the Purkinje fibers, and the ventricular myocardium; located at the boundary between the atria and ventricles. (20)
atrioventricular (AV) valve: One of the valves that prevents backflow into the atria during ventricular systole (contraction). (20)
atrophy: The wasting away of tissues from a lack of use, ischemia, or nutritional abnormalities. (10)
auditory: Pertaining to the sense of hearing. (17)
auditory ossicles: The bones of the middle ear: malleus, incus, and stapes. (7, 17)
auditory tube: A passageway that connects the nasopharynx with the middle ear cavity; also called *Eustachian tube* or *pharyngotympanic tube*. (17)

auricle: A broad, flattened process that resembles the external ear; in the ear, the expanded, projecting portion that surrounds the external auditory meatus; also called *pinna* (17); in the heart, the externally visible flap formed by the collapse of the outer wall of a relaxed atrium (20).
autoantibodies: Antibodies that react with antigens on the surfaces of a person's own cells and tissues. (22)
autoimmunity: The immune system's sensitivity to one's own tissues, resulting in the production of autoantibodies. (22)
autolysis: The destruction of a cell due to the rupture of lysosomal membranes in its cytoplasm. (3)
automaticity: The spontaneous depolarization to threshold, characteristic of cardiac pacemaker cells. (10, 20)
autonomic ganglion: A collection of visceral motor neurons outside the central nervous system. (16)
autonomic nerve: A peripheral nerve consisting of preganglionic or postganglionic autonomic fibers. (16)
autonomic nervous system (ANS): Centers, nuclei, tracts, ganglia, and nerves involved in the unconscious regulation of visceral functions; includes components of the central nervous system and the peripheral nervous system. (12, 16)
autoregulation: Changes in activity that maintain homeostasis in direct response to changes in the local environment; does not require neural or endocrine control. (1, 21, 26)
autosome: A chromosome other than the X or Y sex chromosome. (29)
avascular: Without blood vessels. (4)
axilla: The armpit. (1, 8)
axolemma: The plasma membrane of an axon, continuous with the plasma membrane of the cell body and dendrites and distinct from any neuroglial coverings. (12)
axon: The elongated extension of a neuron that conducts an action potential. (4, 12)
axon hillock: In a multipolar neuron, the portion of the cell body adjacent to the initial segment. (12)
axoplasm: The cytoplasm within an axon. (12)

GLOSSARY

B

bacteria: Single-celled microorganisms, some pathogenic, that are common in the environment and in and on the body. (22)
Bainbridge reflex: The reflexive increase in heart rate after an increase in venous return; due to mechanical and neural factors; also called *atrial reflex*. (20, 21)
baroreception: The ability to detect changes in pressure. (15, 23)
baroreceptor reflex: A reflexive change in cardiac activity in response to changes in blood pressure. (21)
baroreceptors: The nerve endings that sense pressure changes. (15, 21)
basal nuclei: Nuclei of the cerebrum that are important in the subconscious control of skeletal muscle activity. (14, 15)
base: A compound whose dissociation releases a hydroxide ion (OH^-) or removes a hydrogen ion (H^1) from the solution. (2, 27)
basement membrane: A layer of filaments and fibers that attach an epithelium to the underlying connective tissue. (4)
basophils: Circulating granulocytes (white blood cells) similar in size and function to tissue mast cells. (19)
B cells: Lymphocytes capable of differentiating into plasma cells, which produce antibodies. (19, 22)
benign: Not malignant. (3)
beta (β) cells: Cells of the pancreatic islets that secrete insulin in response to increased blood sugar concentrations. (18)
beta-oxidation: Fatty acid catabolism that produces molecules of acetyl-CoA. (25)
beta (β) receptors: Membrane receptors sensitive to epinephrine; stimulation may result in the excitation or inhibition of the target cell. (16)
bicarbonate ion: HCO_3^-; anion component of the carbonic acid-bicarbonate buffer system. (26, 27)
bicuspid: Having two cusps or points; refers to a premolar tooth, which has two roots, or to the left AV valve, which has two cusps. (24)
bicuspid valve: *See* **mitral valve** (20).
bifurcate: To branch into two parts.
bile: The exocrine secretion of the liver; stored in the gallbladder and ejected into the duodenum. (24)
bile duct: The duct formed by the union of the cystic duct from the gallbladder and the common hepatic duct from the liver; terminates at the duodenal ampulla, where it meets the pancreatic duct. (24)
bile salts: Steroid derivatives in bile; responsible for the emulsification of ingested lipids. (2, 24)
bilirubin: A pigment that is the by-product of hemoglobin catabolism. (19)
biopsy: The removal of a small sample of tissue for pathological analysis. (4, 13)
bladder: A muscular sac that distends as fluid is stored and whose contraction ejects the fluid at an appropriate time; used alone, the term usually refers to the urinary bladder. (26)
blastocyst: An early stage in the developing embryo, consisting of an outer trophoblast and an inner cell mass. (29)
blockers/blocking agents: Drugs that block membrane pores or prevent binding to membrane receptors. (16)
blood brain barrier (BBB): The isolation of the central nervous system from the general circulation; primarily the result of astrocyte regulation of capillary permeabilities. (12, 14)
blood clot: *See* **clot**. (19)
blood CSF barrier: The isolation of the cerebrospinal fluid from the capillaries of the choroid plexus; primarily the result of specialized ependymal cells. (14)
blood pressure (BP): A force exerted against vessel walls by the blood in the vessels, due to the push exerted by cardiac contraction and the elasticity of the vessel walls; usually measured along one of the muscular arteries, with systolic pressure measured during ventricular systole (contraction) and diastolic pressure during ventricular diastole (relaxation). (21)
blood testis barrier: The isolation of the interior of the seminiferous tubules from the general circulation, due to the activities of the nurse (sustentacular) cells. (28)
Bohr effect: The increased oxygen release by hemoglobin in the presence of increased carbon dioxide levels. (23)
bolus: A compact mass; usually refers to compacted ingested material (food) on its way to the stomach. (23, 24)
bone tissue: A strong connective tissue containing specialized cells and a mineralized matrix of crystalline calcium phosphate and calcium carbonate; also called *osseous tissue*.
bony labyrinth: A series of cavities within the petrous part of the temporal bone that is filled with perilymph and in which the membranous labyrinth is suspended.(17)
bowel: The intestinal tract. (24)
Bowman's capsule: *See* **glomerular capsule**. (26)
brachial: Pertaining to the arm.
brachial plexus: A network formed by branches of spinal nerves C_5–T_1 en route to innervating the upper limb. (13)
brachium: The arm. (11)
bradycardia: An abnormally slow heart rate, usually below 50 bpm. (20)
brainstem: The brain minus the cerebrum, diencephalon, and cerebellum; midbrain, pons, and medulla oblongata. (14)
brevis: Short. (11)
Broca's area: The speech center of the brain, normally located on the neural cortex of the left cerebral hemisphere. (14)
bronchial tree: The trachea, bronchi, and bronchioles. (23)
bronchodilation: The dilation of the bronchial passages; can be caused by sympathetic stimulation. (23)
bronchus/bronchi: Branch/branches of the bronchial tree between the trachea and bronchioles. (23)
B-type natriuretic peptide (BNP): *See* **natriuretic peptides**. (27)
buccal: Pertaining to the cheeks. (24)
buffer: A compound that stabilizes the pH of a solution by removing or releasing hydrogen ions. (2, 27)
buffer system: Interacting compounds that prevent increases or decreases in the pH of body fluids; includes the carbonic acid–bicarbonate buffer system, the phosphate buffer system, and the protein buffer systems. (27)
bulbar: Pertaining to the brainstem. (14)
bulbo-urethral glands: Mucous glands at the base of the penis that secrete into the penile urethra; the equivalent of the greater vestibular glands of females; also called *Cowper's glands*. (28)
bundle branches: Specialized conducting cells in the ventricles that carry the contractile stimulus from the atrioventricular bundle (bundle of His) to the Purkinje fibers. (20)
bundle of His: *See* **atrioventricular (AV) bundle**. (20)
bursa: A small sac filled with synovial fluid that cushions adjacent structures and reduces friction. (9)

C

calcaneal tendon: The large tendon that inserts on the calcaneus; tension on this tendon produces extension (plantar flexion) of the foot; also called *Achilles tendon*. (8, 11)
calcaneus: The heel bone, the largest of the tarsal bones. (8)
calcification: The deposition of calcium salts within a tissue. (4, 6)
calcitonin (CT): The hormone secreted by C cells of the thyroid when calcium ion concentrations are abnormally high; restores homeostasis by increasing the rate of bone deposition (mostly in childhood) and the rate of calcium loss by the kidneys. (6, 18)
calculus/calculi: Solid mass/masses of insoluble materials that form within body fluids, especially the gallbladder, kidneys, or urinary bladder. (26)

callus: A localized thickening of the epidermis due to chronic mechanical stresses (5); a thickened area that forms at the site of a bone break as part of the repair process. (6)

calorigenic effect: The stimulation of energy production and heat loss by thyroid hormones. (18)

canaliculi: Microscopic passageways between cells; bile canaliculi carry bile to bile ducts in the liver (24); in bone, canaliculi permit the diffusion of nutrients and wastes to and from osteocytes. (4, 6)

cancer: An illness caused by mutations leading to the uncontrolled growth and replication of the affected cells. (3)

cannula: A tube that can be inserted into the body; commonly placed in blood vessels prior to transfusion or dialysis. (19)

capacitation: The activation process that must occur before a sperm can successfully fertilize an oocyte; occurs in the vagina after ejaculation. (28, 29)

capillary: A small blood vessel, located between an arteriole and a venule, whose thin wall permits the diffusion of gases, nutrients, and wastes between plasma and interstitial fluids. (4, 19, 20, 21)

capitulum: A general term for a small, elevated articular process; refers to the rounded distal surface of the humerus that articulates with the head of the radius. (8)

caput: The head. (7)

carbaminohemoglobin: Hemoglobin bound to carbon dioxide molecules. (19, 23)

carbohydrase: An enzyme that breaks down carbohydrate molecules. (24)

carbohydrate: An organic compound containing carbon, hydrogen, and oxygen in a ratio that approximates 1:2:1. (2, 25)

carbon dioxide: CO_2; a compound produced by the decarboxylation reactions of aerobic metabolism. (2, 23)

carbonic anhydrase: An enzyme that catalyzes the reaction $H_2O + CO_2 \rightarrow H_2CO_3$; important in carbon dioxide transport (23), gastric acid secretion (24), and renal pH regulation (26).

carcinogenic: Stimulating cancer formation in affected tissues. (3)

cardia: The region of the stomach surrounding its connection with the esophagus. (24)

cardiac: Pertaining to the heart. (10, 20)

cardiac cycle: One complete heartbeat, including atrial and ventricular systole and diastole. (20)

cardiac output (CO): The amount of blood ejected by the left ventricle each minute; normally about 5 liters. (20)

cardiac reserve: The potential percentage increase in cardiac output above resting levels. (20)

cardiac tamponade: A compression of the heart due to fluid accumulation in the pericardial cavity. (20)

cardiovascular: Pertaining to the heart, blood, and blood vessels. (19, 20, 21)

cardiovascular (CV) centers: Poorly localized area in the reticular formation of the medulla oblongata of the brain; includes cardioacceleratory, cardioinhibitory, and vasomotor centers. (14, 21)

carotene: A yellow-orange pigment, found in carrots, sweet potatoes, and other vegetables, that the body can convert to vitamin A. (5)

carotid artery: The main artery of the neck, servicing cervical and cranial structures; one branch, the internal carotid, provides a major blood supply to the brain. (21)

carotid body: A group of receptors, adjacent to the carotid sinus, that are sensitive to changes in the carbon dioxide levels, pH, and oxygen concentrations of arterial blood. (15, 21)

carotid sinus: A dilated segment at the base of the internal carotid artery whose walls contain baroreceptors sensitive to changes in blood pressure. (21)

carotid sinus reflex: Reflexive changes in blood pressure that maintain homeostatic pressures at the carotid sinus, stabilizing blood flow to the brain. (21)

carpus/carpal: The wrist. (8, 11)

cartilage: A connective tissue with a gelatinous matrix that contains an abundance of fibers. (4)

catabolism: The breakdown of complex organic molecules into simpler components, accompanied by the release of energy. (2, 25)

catalyst: A substance that accelerates a specific chemical reaction but that is not altered by the reaction. (2)

catecholamines: Epinephrine, norepinephrine, dopamine, and related compounds. (18)

catheter: A tube surgically inserted into a body cavity or along a blood vessel or excretory passageway for the collection of body fluids, monitoring of blood pressure, or introduction of medications or radiographic dyes. (20)

cation: An ion having a positive charge. (2, 27)

cauda equina: Bundle of spinal nerve roots arising from the lumbosacral enlargement and medullary cone of the adult spinal cord; they extend caudally inside the vertebral canal en route to lumbar and sacral segments. (13)

caudal/caudally: Closest to or toward the tail (coccyx).

caudate nucleus: One of the basal nuclei involved with the subconscious control of skeletal muscular activity. (14)

cell: The smallest living structural unit in the human body. (1, 3)

cell body: The body of a neuron; also called *soma*. (4, 12)

cell-mediated immunity: Resistance to disease by sensitized T cells that destroy antigen-bearing cells by direct contact or through the release of lymphotoxins; also called *cellular immunity*. (22)

center of ossification: The site in a connective tissue where bone formation begins. (6)

central canal: Longitudinal canal in the center of an osteon that contains blood vessels and nerves; also called *Haversian canal* (6); a passageway along the longitudinal axis of the spinal cord that contains cerebrospinal fluid. (13, 14)

central nervous system (CNS): The brain and spinal cord. (12, 13)

centriole: A cylindrical intracellular organelle composed of nine groups of microtubules, three in each group (9 + 0 array); functions in mitosis or meiosis by organizing the microtubules of the spindle apparatus. (3)

centromere: The localized region where two chromatids remain connected after the chromosomes have replicated; site of spindle fiber attachment. (3)

centrosome: A region of cytoplasm that contains a pair of centrioles oriented at right angles to one another. (3)

cerebellum: The posterior portion of the metencephalon, containing the cerebellar hemispheres; includes the arbor vitae, cerebellar nuclei, and cerebellar cortex. (14, 15)

cerebral cortex: An extensive area of neural cortex covering the surfaces of the cerebral hemispheres. (14)

cerebral hemispheres: A pair of expanded portions of the cerebrum covered in neural cortex. (14)

cerebrospinal fluid (CSF): Fluid bathing the internal and external surfaces of the central nervous system; secreted by the choroid plexus. (12, 13, 14)

cerebrovascular accident (CVA): The occlusion of a blood vessel that supplies a portion of the brain, resulting in damage to the dependent neurons; also called *stroke*. (14)

cerebrum: The largest portion of the brain, composed of the cerebral hemispheres; includes the cerebral cortex, the basal nuclei, and the internal capsule. (14)

cerumen: The waxy secretion of the ceruminous glands along the external acoustic meatus. (5, 17)

ceruminous glands: Integumentary glands that secrete cerumen. (5, 17)

cervix: The inferior portion of the uterus. (28)

chemoreception: The detection of changes in the concentrations of dissolved compounds or gases (mainly CO_2 and O_2). (15, 17, 21, 23, 25)

chemotaxis: The attraction of phagocytic cells to the source of abnormal chemicals in tissue fluids. (22)

chemotherapy: The treatment of illness through the administration of specific chemicals.

chief cells: Gastric cells near the base of gastric glands that secrete pepsinogen. (24)

chloride shift: The movement of plasma chloride ions into red blood cells in exchange for bicarbonate ions generated by the intracellular dissociation of carbonic acid. (23, 27)

cholecystokinin (CCK): A duodenal hormone that stimulates the contraction of the gallbladder and the secretion of enzymes by the exocrine pancreas. (24)

cholesterol: A steroid component of plasma membranes and a substrate for the synthesis of steroid hormones and bile salts. (2, 25)

choline: A breakdown product or precursor of acetylcholine. (12)

cholinergic synapse: A synapse where the presynaptic membrane releases acetylcholine on stimulation. (12, 16)

chondrocyte: A cartilage cell. (4)

chondroitin sulfate: The predominant proteoglycan in cartilage, responsible for the gelatinous consistency of the matrix. (4)

chordae tendineae: Fibrous cords that stabilize the position of the AV valves in the heart, preventing backflow during ventricular systole (contraction). (20)

chorion/chorionic: An extra-embryonic membrane, consisting of the trophoblast and underlying mesoderm, that forms the placenta. (29)

choroid: The middle, vascular layer in the wall of the eye. (17)

choroid plexus: The vascular complex in the roof of the third and fourth ventricles of the brain, responsible for the production of cerebrospinal fluid. (14)

chromatid: One complete copy of a DNA strand and its associated nucleoproteins. (3, 28, 29)

chromatin: A histological term referring to the grainy material visible in cell nuclei during interphase; the appearance of the DNA content of the nucleus when the chromosomes are uncoiled. (3)

chromosomes: Dense structures, composed of tightly coiled DNA strands and associated histones, that become visible in the nucleus when a cell prepares to undergo mitosis or meiosis; normal human somatic cells each contain 46 chromosomes. (3, 28, 29)

chronic: Habitual or long term.

chylomicrons: Relatively large droplets that may contain triglycerides, phospholipids, and cholesterol in association with proteins; synthesized and released by intestinal cells and transported to the venous blood by the lymphatic system. (24, 25)

ciliary body: A thickened region of the choroid that encircles the lens of the eye; includes the ciliary muscle and the ciliary processes that support the *ciliary zonule* (suspensory ligament) of the lens. (17)

cilium/cilia: Slender organelle/organelles that extends above the free surface of an epithelial cell. Multiple *motile cilia* generally undergo cycles of movement and are composed of a basal body and microtubules in a 9 + 2 array; a nonmotile, solitary *primary cilium* acts as an environmental sensor, and has microtubules arranged in a 9 + 0 array. (3)

circulatory system: The network of blood vessels and lymphatic vessels that facilitate the distribution and circulation of extracellular fluid. (21, 22)

circumduction: A movement at a synovial joint in which the distal end of the bone moves in a circular direction, but the shaft does not rotate. (9)

cisterna: An expanded or flattened chamber derived from and associated with the endoplasmic reticulum. (3, 10)

citric acid cycle: The reaction sequence that occurs in the matrix of mitochondria; in the process, organic molecules are broken down, carbon dioxide molecules are released, and hydrogen molecules are transferred by coenzymes to the electron transport chain. (3, 10, 25)

clot: A network of fibrin fibers and trapped blood cells; also called a *thrombus* if it occurs within blood vessels. (19)

clotting factors: Plasma proteins, synthesized by the liver, that are essential to the clotting response. (19)

clotting response: The series of events that result in the formation of a clot. (19)

coccygeal ligament: The fibrous extension of the dura mater and filum terminale; provides longitudinal stabilization to the spinal cord. (13)

coccyx: The terminal portion of the spinal column, consisting of relatively tiny, fused vertebrae. (7)

cochlea: The spiral portion of the bony labyrinth of the internal ear that surrounds the organ of hearing. (17)

cochlear duct: *See* **scala media**. (17)

codon: A sequence of three nitrogenous bases along an mRNA strand that will specify the location of a single amino acid in a peptide chain. (3)

coelom: The body cavity lined by a serous membrane and subdivided during fetal development into the pleural, pericardial, and peritoneal cavities.

coenzymes: Complex organic cofactors; most are structurally related to vitamins. (2, 25)

cofactor: Ions or molecules that must be attached to the active site before an enzyme can function; examples include mineral ions and several vitamins. (2)

collagen fibers: Strong, insoluble protein fibers common in connective tissues. (4)

collateral ganglion: A sympathetic ganglion located anterior to the spinal column and separate from the sympathetic chain. (12, 16)

colliculus/colliculi: Little mound/mounds; in the brain, refers to one of the thickenings in the roof of the mesencephalon; the superior colliculi are associated with the visual system, and the inferior colliculi with the auditory system. (14, 15, 17)

colloid/colloidal suspension: A solution containing large organic molecules in suspension. (2, 26)

colon: The large intestine from the cecum to the rectum. (24)

coma: An unconscious state from which an individual cannot be aroused, even by strong stimuli. (16)

comminuted: Broken or crushed into small pieces.

commissure: A crossing over from one side to another.

common pathway: The final clotting pathway that begins with the activation of clotting factor X by enzyme complexes from either the extrinsic or intrinsic pathways. The common pathway ends with the formation of a blood clot when the enzyme thrombin converts soluble fibrinogen to insoluble fibrin. (19)

compact bone: Dense bone that contains parallel osteons. (6)

complement system: A group of distinct plasma proteins that interact in a chain reaction after exposure to activated antibodies or the surfaces of certain pathogens; complement proteins promote cell lysis, phagocytosis, and other defense mechanisms. (22)

compliance: Expandability; the ability of certain organs to tolerate changes in volume; indicates the presence of elastic fibers and smooth muscles. (23)

compound: A pure chemical substance made up of atoms of two or more different elements in a fixed proportion, regardless of the type of chemical bond joining them. (2)

concentration: The amount (in grams) or number of atoms, ions, or molecules (in moles) per unit volume. (2, 3, 25, 26)

concentration gradient: Regional differences in the concentration of a particular substance. (3, 25, 26)

conception: Fertilization. (29)

concha/conchae: Three pairs of thin, scroll-like bones that project into the nasal cavities; the superior and middle conchae are part of the ethmoid, and the inferior nasal conchae articulate with the ethmoid, lacrimal, maxilla, and palatine bones; also called *turbinate(s)*. (7)

condyle: A rounded articular projection on the surface of a bone. (8)

cone: A photoreceptor responsible for sharp vision and color vision. (17)

congenital: Present at birth.

congestive heart failure (CHF): The failure to maintain adequate cardiac output due to cardiovascular problems or myocardial damage. (23)

conjunctiva: A layer of stratified squamous epithelium that covers the inner surfaces of the eyelids and the anterior surface of the eye to the edges of the cornea. (17)

connective tissue: One of the four primary tissue types; provides a structural framework that stabilizes the relative positions of the other tissue types; includes connective tissue proper, cartilage, bone, and blood; contains cell products, cells, and ground substance. (4)

continuous propagation: The propagation of an action potential along an unmyelinated axon or a muscle plasma membrane, wherein the action potential affects every portion of the membrane surface. (12)

contractility: The ability to contract; possessed by skeletal, smooth, and cardiac muscle cells. (4, 20)

contralateral reflex: A reflex that affects the opposite side of the body from the stimulus. (13)

conus medullaris: The conical tip of the spinal cord that gives rise to the filum terminale. (13)

convergence: In the nervous system, the innervation of a single neuron by axons from several neurons; most common along motor pathways. (13)

coracoid process: A hook-shaped process of the scapula that projects above the anterior surface of the capsule of the shoulder joint. (8)

Cori cycle: The metabolic exchange of lactate from skeletal muscle for glucose from the liver; occurs during the recovery period after muscular exertion. (10)

cornea: The transparent portion of the fibrous layer of the anterior surface of the eye. (17)

corniculate cartilages: A pair of small laryngeal cartilages. (23)

cornu: Horn-shaped.

coronoid: Hooked or curved. (8)

corpora quadrigemina: The superior and inferior colliculi of the mesencephalic tectum (roof) in the brain. (14)

corpus/corpora: Body/bodies.

corpus callosum: A large bundle of axons that links centers in the left and right cerebral hemispheres. (14)

corpus luteum: The progesterone-secreting mass of follicle cells that develops in the ovary after ovulation. (18, 28)

cortex: The outer layer or region of an organ (5) or bone. (6)

corticobulbar tracts: Descending tracts that carry information or commands from the cerebral cortex to nuclei and centers in the brainstem. (15)

corticospinal tracts: Descending tracts that carry motor commands from the cerebral cortex to the anterior gray horns of the spinal cord. (15)

corticosteroid: A steroid hormone produced by the adrenal (suprarenal) cortex. (2, 18)

corticosterone: A corticosteroid secreted by the zona fasciculata of the adrenal (suprarenal) cortex; a glucocorticoid. (18)

corticotropin-releasing hormone (CRH): The releasing hormone, secreted by the hypothalamus, that stimulates secretion of adrenocorticotropic hormone (ACTH) by the anterior lobe of the pituitary gland (adenohypophysis). (18)

cortisol: A corticosteroid secreted by the zona fasciculata of the adrenal (suprarenal) cortex; a glucocorticoid. (18)

cortisone: A glucocorticoid produced by the adrenal cortex that exerts its effects after it is converted to cortisol. (18)

costa/costae: Rib/ribs. (7, 23)

cotransport: The membrane transport of a nutrient, such as glucose, in company with the movement of an ion, normally sodium; transport requires a carrier protein but does not involve direct ATP expenditure and can occur regardless of the concentration gradient for the nutrient. (3, 24, 26)

countercurrent exchange: The transfer of heat, water, or solutes between two fluids that travel in opposite directions. (25, 26)

countercurrent multiplication: Active transport between two limbs of a loop that contains a fluid moving in one direction in which the effect of the transport increases; responsible for the concentration of urine in the kidney tubules. (26)

covalent bond: A chemical bond between atoms that involves the sharing of electrons. (2)

coxal bone: Hip bone. (7, 8)

cranial: Pertaining to the head. (7)

cranial nerves: Peripheral nerves originating at the brain. (12)

craniosacral division: The parasympathetic division of the autonomic nervous system. (16)

cranium: The braincase; the skull bones that surround and protect the brain. (7)

creatine: A nitrogenous compound, synthesized in the body, that can form a high-energy bond by connecting to a phosphate group and that serves as an energy reserve. (10)

creatine phosphate (CP): A high-energy compound in muscle cells; during muscle activity, the phosphate group is donated to ADP, regenerating ATP; also called *phosphorylcreatine*. (10)
creatinine: A breakdown product of creatine metabolism. (26)
crenation: Cellular shrinkage due to an osmotic movement of water out of the cytoplasm. (3)
cribriform plate: A portion of the ethmoid that contains the foramina used by the axons of olfactory receptors en route to the olfactory bulbs of the cerebrum. (7)
cricoid cartilage: A ring-shaped cartilage that forms the inferior margin of the larynx. (23)
crista/cristae: A ridge-shaped collection of hair cells in the ampulla of a semicircular duct; the crista and cupula form a receptor complex sensitive to movement along the plane of the semicircular canal. (3, 17)
cross-bridge: The binding of a myosin head that projects from the surface of a thick filament at the active site of a thin filament in the presence of calcium ions. (10)
cuneiform cartilages: A pair of small cartilages in the larynx. (23)
cupula: A gelatinous mass that is located in the ampulla of a semicircular duct in the internal ear and whose movement stimulates the hair cells of the crista. (17)
Cushing disease: A condition caused by the oversecretion of adrenal steroids. (18)
cutaneous membrane: The epidermis and papillary layer of the dermis. (4, 5)
cuticle: The layer of dead, keratinized cells that surrounds the shaft of a hair; for nails, *see* **eponychium**. (5)
cyanosis: An abnormal bluish coloration of the skin due to the presence of deoxygenated blood in vessels near the body surface. (5)
cystic duct: A duct that carries bile between the gallbladder and the common bile duct. (24)
cytochrome: A protein-pigment component of the electron transport chain; a structural relative of heme. (25)
cytokinesis: The cytoplasmic movement that separates two daughter cells at the completion of mitosis. (3)
cytology: The study of the internal structures, physiology, and chemistry of cells. (1, 3)
cytoplasm: The material between the plasma membrane and the nuclear membrane; cell contents. (3)
cytosine (C): A pyrimidine; one of the nitrogenous bases in the nucleic acids RNA and DNA. (2)
cytoskeleton: A network of microtubules and microfilaments in the cytoplasm. (3)
cytosol: The fluid portion of the cytoplasm; also called *intracellular fluid*. (3, 27)
cytotoxic: Poisonous to cells. (22)
cytotoxic T cells: Lymphocytes involved in cell-mediated immunity that kill target cells by direct contact or by the secretion of lymphotoxins; also called *killer T cells* and *T_C cells*. (22)

D

daughter cells: Genetically identical cells produced by somatic cell division. (3, 28)
deamination: The removal of an amino group from an amino acid. (25, 26)
decomposition reaction: A chemical reaction that breaks a molecule into smaller fragments. (2)
decussate: To cross over to the opposite side, usually referring to the crossover of the descending tracts of the corticospinal pathway on the ventral surface of the medulla oblongata. (15)
defecation: The elimination of fecal wastes. (24)
degradation: Breakdown, catabolism. (2, 25)
dehydration: A reduction in the water content of the body that threatens homeostasis. (27)
dehydration synthesis reaction: The joining of two molecules associated with the removal of a water molecule. (2)
demyelination: The loss of the myelin sheath of an axon, normally due to chemical or physical damage to Schwann cells or oligodendrocytes. (12)
denaturation: A temporary or permanent change in the three-dimensional structure of a protein. (2)
dendrite: A sensory process of a neuron. (4, 12)
deoxyribonucleic acid (DNA): A nucleic acid consisting of a double chain of nucleotides that contains the sugar deoxyribose and the nitrogenous bases adenine, guanine, cytosine, and thymine. (2)
deoxyribose: A five-carbon sugar resembling ribose but lacking an oxygen atom. (2)
depolarization: A change in the membrane potential from a negative value toward 0 mV. (12, 20)
depression: Inferior (downward) movement of a body part. (9)
dermatitis: An inflammation of the skin. (5)
dermatome: A sensory region monitored by the dorsal rami of a single spinal segment. (13)
dermis: The connective tissue layer beneath the epidermis of the skin. (5)
detrusor: Collectively, the three layers of smooth muscle in the wall of the urinary bladder. (26)
detumescence: The loss of a penile erection. (28)
development: Growth and the acquisition of increasing structural and functional complexity; includes the period from conception to maturity. (29)
diabetes insipidus: Disorder of the pituitary gland characterized by polyuria (excessive urination) and polydipsia (excessive thirst) that results from inadequate production of antidiuretic hormone (ADH). (18)
diabetes mellitus: Disorder characterized by polyuria (excessive urination) and glycosuria (glucose in the urine), most commonly due to the inadequate production or diminished sensitivity to insulin with a resulting increase of blood glucose levels. (18)
diapedesis: The movement of white blood cells through the walls of blood vessels by migration between adjacent endothelial cells. (19, 22)
diaphragm: Any muscular partition; the respiratory muscle that separates the thoracic cavity from the abdominopelvic cavity. (1, 11, 23)
diaphysis: The shaft of a long bone. (6)
diarthrosis: A synovial joint. (9)
diastole: A period of relaxation in a chamber of the heart, as part of the cardiac cycle. (20)
diastolic pressure: Pressure measured in the walls of a muscular artery when the left ventricle is in diastole (relaxation). (21)
diencephalon: A division of the brain that includes the epithalamus, thalamus, and hypothalamus. (14)
differential count: An estimate of the number of each type of white blood cell on the basis of a random sampling of 100 white blood cells. (19)
differentiation: The gradual appearance of characteristic cellular specializations during development as the result of gene activation or repression. (3, 29)
diffusion: Passive molecular movement from an area of higher concentration to an area of lower concentration. (3, 21, 23, 26)
digestion: The chemical breakdown of ingested food into simple molecules that can be absorbed by the cells of the digestive tract. (24)
digestive system: The digestive tract and associated glands. (24)
digestive tract: An internal passageway that begins at the mouth, ends at the anus, and is lined by a mucous membrane; also called *gastrointestinal(GI) tract* or *alimentary canal*. (24)
dilate: To increase in diameter; to enlarge or expand.
disaccharide: A compound formed by the joining of two simple sugars by dehydration synthesis. (2)
dissociation: *See* **ionization**. (2)
distal: A direction away from the point of attachment or origin; for a limb, away from its attachment to the trunk. (1, 8)
distal convoluted tubule (DCT): The segment of the nephron closest to the connecting tubules and collecting duct; an important site of active secretion. (26)
diuresis: Fluid loss by the kidneys; the production of unusually large volumes of urine. (26)
divergence: In nervous tissue, the spread of information from one neuron to many neurons; an organizational pattern common along sensory pathways of the central nervous system. (13)
diverticulum: A sac or pouch in the wall of the colon or other organ. (24)
DNA molecule: Two DNA strands wound in a double helix and held together by hydrogen bonds between complementary nitrogenous base pairs. (3)
dopamine: An important neurotransmitter in the central nervous system. (12)
dorsal: Toward the back, posterior.
dorsal root ganglion: *See* **posterior root ganglion**. (13)
dorsiflexion: Upward movement of the foot through flexion at the ankle. (9)
Down syndrome: A genetic abnormality resulting from the presence of three copies of chromosome 21; people with this condition have characteristic physical and intellectual deficits; also called *trisomy 21*. (16, 29)
duct: A passageway that delivers exocrine secretions to an epithelial surface. (4)
ductus arteriosus: A vascular connection between the pulmonary trunk and the aorta that functions throughout fetal life; normally closes at birth or shortly thereafter and persists as the ligamentum arteriosum. (21, 23)
ductus deferens: A passageway that carries sperm from the epididymis to the ejaculatory duct; also called the *vas deferens*. (28)
duodenal ampulla: A chamber that receives bile from the common bile duct and pancreatic secretions from the pancreatic duct. (24)
duodenal papilla: A conical projection from the inner surface of the duodenum that contains the opening of the duodenal ampulla. (24)
duodenum: The proximal 25 cm (9.8 in.) of the small intestine that contains short villi and submucosal glands. (24)
dura mater: The outermost membrane of the cranial and spinal meninges. (13, 14)

E

eccrine sweat glands: Sweat glands of the skin that produce a watery secretion. (5)
ectoderm: One of the three primary germ layers; covers the surface of the embryo and gives rise to the nervous system, the epidermis and associated glands, and a variety of other structures. (29)
ectopic: Outside the normal location.
effector: A peripheral gland or muscle cell innervated by a motor neuron. (1, 12)

efferent: Conducting a fluid or nerve impulse away from an organ or structure.
efferent arteriole: An arteriole carrying blood away from a glomerulus of the kidney. (26)
efferent fiber: An axon that carries impulses away from the central nervous system. (12)
ejaculation: The ejection of semen from the penis as the result of muscular contractions of the bulbospongiosus and ischiocavernosus muscles. (28)
ejaculatory ducts: Short ducts that pass within the walls of the prostate gland and connect the ductus deferens with the prostatic urethra. (28)
elastase: A pancreatic enzyme that breaks down elastin fibers. (24)
elastin: Connective tissue fibers that stretch and recoil, providing elasticity to connective tissues. (4, 5)
electrical coupling: A connection between adjacent cells that permits the movement of ions and the transfer of graded or propagated changes in the membrane potential from cell to cell. (12)
electrocardiogram (ECG, EKG): A graphic record of the electrical activities of the heart, as monitored at specific locations on the body surface. (20)
electroencephalogram (EEG): A graphic record of the electrical activities of the brain. (14)
electrolytes: Soluble inorganic compounds whose ions will conduct an electrical current in solution. (2, 27)
electron: One of the three fundamental subatomic particles; has a negative charge and normally orbits the protons of the nucleus. (2, 25)
electron transport chain (ETC): The system made up of four respiratory protein complexes, coenzyme Q, and electron carriers made up of a series of cytochrome molecules responsible for aerobic energy production in cells; bound to the inner mitochondrial membrane. (25)
element: All the atoms with the same atomic number. (2)
elevation: Movement in a superior, or upward, direction. (9)
elimination: The ejection of wastes from the body through urination or defecation. (24, 26)
embolism: The obstruction of a vessel, typically by a blood clot. (19)
embolus: An air bubble, fat globule, or blood clot drifting in the bloodstream. (19)
embryo: The developmental stage beginning at fertilization and ending at the start of the third developmental month. (28, 29)
embryology: The study of embryonic development, focusing on the first 2 months after fertilization. (1, 28, 29)
endocardium: The simple squamous epithelium that lines the heart and is continuous with the endothelium of the great vessels. (20)
endochondral ossification: The replacement of a cartilaginous model with bone; the characteristic mode of formation for skeletal elements other than the bones of the cranium, the clavicles, and sesamoid bones. (6)
endocrine gland: A gland that secretes hormones into the blood. (4, 18)
endocrine system: The endocrine (ductless) glands and organs of the body. (18)
endocytosis: The movement of relatively large volumes of extracellular material into the cytoplasm by the formation of a membranous vesicle at the cell surface; includes pinocytosis and phagocytosis. (3)
endoderm: One of the three primary germ layers; the layer on the undersurface of the embryonic disc; gives rise to the epithelia and glands of the digestive system, the respiratory system, and portions of the urinary system. (29)
endogenous: Produced within the body.
endolymph: The fluid contents of the membranous labyrinth (the saccule, utricle, semicircular ducts, and cochlear duct) of the internal ear. (17)
endometrium: The mucous membrane lining the uterus. (28)
endomysium: A delicate network of connective tissue fibers that surrounds individual muscle cells. (10)
endoneurium: A delicate network of connective tissue fibers that surrounds individual nerve fibers. (13)
endoplasmic reticulum: A network of membranous channels in the cytoplasm of a cell that function in intracellular transport, synthesis, storage, packaging, and secretion. (3)
endorphins: Neuromodulators, produced in the central nervous system, that inhibit activity along pain pathways. (12)
endosteum: An incomplete cellular lining on the inner (medullary) surfaces of bones. (6)
endothelium: The simple squamous epithelial cells that line blood and lymphatic vessels. (4, 19, 21)
enkephalins: Neuromodulators, produced in the central nervous system, that inhibit activity along pain pathways. (12)
enteric nervous system (ENS): One of the three main divisions of the nervous system; it is an extensive nerve network in the walls of the digestive tract that initiates and coordinates digestive motility and secretions. (16)
enterocrinin: A hormone secreted by the lining of the duodenum after exposure to chyme; stimulates the secretion of the submucosal glands. (24)
enteroendocrine cells: Endocrine cells scattered among the epithelial cells that line the digestive tract. (24)
enterogastric reflex: The reflexive inhibition of gastric secretion; initiated by the arrival of chyme in the small intestine. (24)
enterohepatic circulation: The excretion of bile salts by the liver, followed by the absorption of bile salts by intestinal cells for return to the liver by the hepatic portal vein. (24)
enteropeptidase: An enzyme in the lumen of the small intestine that activates the proenzymes secreted by the pancreas; formerly called *enterokinase*. (24)
enzyme: A protein that catalyzes a specific biochemical reaction. (2)
eosinophil: A type of white blood cell with a lobed nucleus and red-staining granules; participates in the immune response and is especially important during allergic reactions. (19)
ependymal cells: The layer of cells lining the ventricles and central canal of the central nervous system. (12)
epicardium: A serous membrane covering the outer surface of the heart; also called *visceral pericardium*. (20)
epidermis: The epithelium covering the surface of the skin. (5)
epididymis: A coiled duct that connects the rete testis to the ductus deferens; site of functional maturation of sperm. (28)
epidural space: The space between the spinal dura mater and the walls of the vertebral foramen; contains blood vessels and adipose tissue; a common site of injection for regional anesthesia. (13)
epiglottis: A blade-shaped flap of tissue, reinforced by cartilage, that is attached to the posterior and superior surface of the thyroid cartilage; folds over the entrance to the larynx during swallowing. (23)
epimysium: A dense layer of collagen fibers that surrounds a skeletal muscle and is continuous with the tendons/aponeuroses of the muscle and with the perimysium. (10)
epineurium: A dense layer of collagen fibers that surrounds a peripheral nerve. (13)
epiphyseal cartilage: The cartilaginous region between the epiphysis and diaphysis of a growing bone; also called *epiphyseal plate*. (6)
epiphysis: The head of a long bone. (6)
epithelium: One of the four primary tissue types; a layer of cells that forms a superficial covering or an internal lining of a body cavity or vessel. (4, 24)
eponychium: A narrow zone of stratum corneum that extends across the surface of a nail at its exposed base; also called the *cuticle*. (5)
eponym: A commemorative name for a structure or clinical condition that was originally named for a real or mythical person. (1)
equilibrium: A dynamic state in which two opposing forces or processes are in balance. (1, 15, 17)
erection: The stiffening of the penis due to the erectile tissues of the corpora cavernosa and corpus spongiosum filling with blood. (28)
erythema: Redness and inflammation at the surface of the skin. (5, 22)
erythrocyte: A red blood cell (RBC); has no nucleus and contains large quantities of hemoglobin. (4, 19)
erythropoiesis: The formation of red blood cells. (19)
erythropoietin (EPO): A hormone released by most tissues, and especially by the kidneys, when oxygen levels decrease; stimulates erythropoiesis (red blood cell formation) in red bone marrow. (18, 19, 21)
Escherichia coli: A normal bacterial resident of the large intestine. (24)
esophagus: A muscular tube that connects the pharynx to the stomach. (24)
essential amino acids: Amino acids that cannot be synthesized in the body in adequate amounts and must be obtained from the diet. (25)
essential fatty acids: Fatty acids that cannot be synthesized in the body and must be obtained from the diet. (25)
estrogens: A class of steroid sex hormones that includes estradiol. (2, 18, 28)
evaporation: A movement of molecules from the liquid state to the gaseous state. (25)
eversion: A turning outward. (9)
excitable membranes: Membranes that propagate action potentials, a characteristic of muscle cells and nerve cells. (10, 12)
excitatory postsynaptic potential (EPSP): The depolarization of a postsynaptic membrane by a chemical neurotransmitter released by the presynaptic cell. (12)
excretion: The removal of wastes from the blood, tissues, or organs. (24)
exocrine gland: A gland that secretes onto the body surface or into a passageway connected to the exterior. (4)
exocytosis: The ejection of cytoplasmic materials by the fusion of a membranous vesicle with the plasma membrane. (3)
expiration: Exhalation; breathing out.
extension: An increase in the angle between two articulating bones; the opposite of flexion. (8, 9)
external acoustic meatus: A passageway in the temporal bone that leads to the tympanic membrane of the middle ear. (7, 17)
external ear: The auricle, external acoustic meatus, and tympanic membrane. (17)

external nares: The nostrils; the external openings into the nasal cavity. (23)
external respiration: The diffusion of gases between the alveolar air and the alveolar capillaries and between the systemic capillaries and peripheral tissues. (23)
exteroceptors: General sensory receptors in the skin, mucous membranes, and special sense organs that provide information about the external environment and about our position within it. (12)
extracellular fluid (ECF): All body fluids other than that contained within cells; includes plasma and interstitial fluid. (3, 27)
extra-embryonic membranes: The yolk sac, amnion, chorion, and allantois. (29)
extrafusal muscle fibers: Contractile muscle fibers (as opposed to the sensory intrafusal fibers, or muscle spindles). (13)
extrinsic pathway: A clotting pathway that begins with damage to blood vessels or surrounding tissues and ends with the formation of tissue factor complex. (19)
eyelids: Movable folds that cover the front of the eyeballs. *See* **palpebrae.** (17)

F

fabella: A sesamoid bone commonly located in the gastrocnemius. (11)
facilitated: Brought closer to threshold, as in the depolarization of a nerve plasma membrane toward threshold; making the cell more sensitive to depolarizing stimuli. (12)
facilitated diffusion: The passive movement of a substance across a plasma membrane by means of a protein carrier. (3, 24, 26)
falciform ligament: A sheet of mesentery that contains the ligamentum teres, the fibrous remains of the umbilical vein of the fetus. (24, 29)
falx: Sickle shaped.
falx cerebri: The curving sheet of dura mater that extends between the two cerebral hemispheres; encloses the superior sagittal sinus. (7, 14)
fasciae: Connective tissue fibers, primarily collagen, that form sheets or bands beneath the skin to attach, stabilize, enclose, and separate muscles and other internal organs. (4)
fasciculus: A small bundle; usually refers to a collection of nerve axons or muscle fibers. (10, 15)
fatty acids: Hydrocarbon chains that end in a carboxyl group. (2)
fauces: The passage from the mouth to the pharynx, bounded by the palatal arches, the soft palate, and the uvula. (24)
febrile: Characterized by or pertaining to a fever. (22, 25)
feces: Waste products eliminated by the digestive tract at the anus; contains indigestible residue, bacteria, mucus, and epithelial cells. (24)
fenestra/fenestrae: Opening(s).
fertilization: The fusion of a secondary oocyte and a sperm to form a zygote. (28, 29)
fetus: The developmental stage lasting from the start of the third developmental month to delivery. (28, 29)
fibrin: Insoluble protein fiber that forms the basic framework of a blood clot. (19)
fibrinogen: A plasma protein that is the soluble precursor of the insoluble protein fibrin. (19)
fibroblasts: Cells of connective tissue proper that produce extracellular fibers and secrete the organic substances of the extracellular matrix. (4)
fibrocartilage: Cartilage containing an abundance of collagen fibers; located around the edges of joints, in the intervertebral discs, and the menisci of the knee; also referred to as *fibrous cartilage*. (4)
fibrocytes: Mature fibroblasts; maintain connective tissue fibers of connective tissue proper. (4)
fibrous layer: The outermost layer of the eye, composed of the sclera and cornea; also called *fibrous tunic*. (17)
fibula: The lateral, slender bone of the leg. (8)
filariasis: A condition resulting from infection by mosquito-borne parasites; can cause elephantiasis. (21, 22)
filiform papillae: Slender conical projections from the dorsal surface of the anterior two-thirds of the tongue. (17)
filtrate: The fluid produced by filtration at a glomerulus in the kidney. (26)
filtration: The movement of a fluid across a membrane whose pores restrict the passage of solutes on the basis of size. (21, 26)
filtration pressure: The hydrostatic pressure responsible for filtration. (21, 26)
filum terminale: A fibrous extension of the spinal cord, from the conus medullaris to the coccygeal ligament. (13)
fimbriae: Fringes; the fingerlike processes that surround the entrance to the uterine tube. (28)
fissure: An elongated groove or opening. (7, 14)
fistula: An abnormal passageway between two organs or from an internal organ or space to the body surface.
flaccid: Limp, soft, flabby; a muscle without muscle tone.
flagellum/flagella: An organelle that is structurally similar to a motile cilium but is used to propel a cell through a fluid; found on sperm. (3, 28)
flatus: Intestinal gas. (24)
flexion: A movement that decreases the angle between two articulating bones; the opposite of extension. (8, 9)
flexor: A muscle that produces flexion. (11)
flexor reflex: A reflex contraction of the flexor muscles of a limb in response to a painful stimulus. (13)
flexure: A bending.
folia: Leaflike folds; the slender folds in the surface of the cerebellar cortex. (14)
follicle: A small secretory sac or gland.
follicle-stimulating hormone (FSH): A hormone secreted by the adenohypophysis (anterior lobe of the pituitary gland); stimulates oogenesis (female) and spermatogenesis (male). (18, 28)
fontanelle: A relatively soft, flexible, fibrous region between two flat bones in the developing skull; also spelled *fontanel*. (7)
foramen/foramina: Opening/openings or passage/passages through a bone. (7, 20)
forearm: The distal portion of the upper limb between the elbow and wrist. (8)
forebrain: The cerebrum. (14)
fornix: An arch or the space bounded by an arch; in the brain, an arching tract that connects the hippocampus with the mammillary bodies (14); in the eye, a slender pocket located where the epithelium of the ocular conjunctiva folds back on itself as the palpebral conjunctiva (17); in the vagina, the shallow recess surrounding the protrusion of the cervix. (28)
fossa: A shallow depression or furrow in the surface of a bone. (8, 20)
fourth ventricle: An elongated ventricle of the metencephalon (pons and cerebellum) and the myelencephalon (medulla oblongata) of the brain; the roof contains a region of choroid plexus. (14)
fovea centralis: The portion of the retina within the macula that provides the sharpest vision because it has the highest concentration of cones; also called *fovea*. (17)
fracture: A break or crack in a bone. (6)
frenulum: A bridle; usually referring to a band of tissue that restricts movement, e.g., *lingual frenulum*. (24)
frontal (coronal) plane: A vertical plane that divides the body into anterior and posterior portions. (1)
fructose: A hexose (six-carbon simple sugar) in foods and in semen. (2, 28)
fundus: The base of an organ such as the stomach, uterus, or gallbladder.

G

gallbladder: The pear-shaped reservoir for bile after it is secreted by the liver. (24)
gametes: Reproductive cells (sperm or oocytes) that contain half the normal chromosome complement. (28, 29)
gametogenesis: The formation of gametes. (28)
gamma-aminobutyric acid (GABA): A neurotransmitter of the central nervous system whose effects are generally inhibitory. (12)
gamma motor neurons: Motor neurons that adjust the sensitivities of muscle spindles (intrafusal fibers). (13)
ganglion/ganglia: A collection of neuron cell bodies in the peripheral nervous system. (12, 16)
gangliosides: Glycolipids that are important components of plasma membranes in the central nervous system. (12)
gap junctions: Connections between cells that permit electrical coupling. (4)
gaster: The stomach (24); the body, or belly, of a skeletal muscle. (11)
gastric: Pertaining to the stomach. (24)
gastric glands: The tubular glands of the stomach whose cells produce acid, enzymes, intrinsic factor, and hormones. (24)
gastrointestinal (GI) tract: *See* **digestive tract**. (24)
gene: A portion of a DNA strand that functions as a hereditary unit, is located at a particular site on a specific chromosome, and codes for a specific protein or polypeptide. (3, 29)
genetic engineering: Research and experiments involving the manipulation of the genetic makeup of an organism. (29)
genetics: The study of heredity and genetic variations. (29)
geniculate body: The medial geniculate body and the lateral geniculate body are in the thalamus of the brain. (14)
genitalia: The reproductive organs. (28)
genotype: Genetic makeup of an organism. (29)
germinal centers: Pale regions in the interior of lymphoid tissues or lymphoid nodules, where cell divisions occur that produce additional lymphocytes. (22)
gestation: The period of intrauterine development; pregnancy. (29)
gland: Cells that produce exocrine or endocrine secretions. (4)
glenoid cavity: A rounded depression that forms the articular surface of the scapula at the shoulder joint. (8)
glial cells: *See* **neuroglia**. (4, 12)
globular proteins: Proteins whose tertiary structure makes them rounded and compact. (2)
glomerular capsule: The expanded initial portion of the nephron that surrounds the glomerulus; also called *Bowman's capsule*. (26)

glomerular filtration rate (GFR): The rate of filtrate formation at the glomerulus. (26)
glomerulus: A knot of capillaries that projects into the enlarged, proximal end of a nephron; the site of filtration, the first step in the production of urine. (26)
glossopharyngeal nerve: Cranial nerve IX that provides sensations to the pharynx and posterior third of the tongue; it also carries motor fibers to the stylopharyngeus muscle. (14)
glottis: Structure within the larynx that consists of the vocal folds and the rima glottidis. (23)
glucagon: A hormone secreted by the alpha cells of the pancreatic islets; increases blood glucose concentrations. (18)
glucocorticoids: Hormones secreted by the zona fasciculata of the adrenal (suprarenal) cortex to modify glucose metabolism; cortisol and corticosterone are important examples. (18)
gluconeogenesis: The synthesis of glucose from noncarbohydrate precursors (e.g., lactate, glycerol, or amino acids). (25)
glucose: A six-carbon sugar, $C_6H_{12}O_6$; the preferred energy source for most cells and normally the only energy source for neurons. (2, 10, 18, 25)
glycerides: Lipids composed of glycerol bound to fatty acids. (2)
glycogen: A polysaccharide that is an important energy reserve; a polymer consisting of a long chain of glucose molecules. (2, 10, 25)
glycogenesis: The synthesis of glycogen from glucose molecules. (25)
glycogenolysis: Glycogen breakdown and the liberation of glucose molecules. (25)
glycolipids: Compounds created by the combination of carbohydrate and lipid components. (2)
glycolysis: The anaerobic cytosolic breakdown of glucose into two 3-carbon molecules of pyruvate, with a net gain of 2 ATP molecules. (3, 10, 25)
glycoprotein: A compound containing a relatively small carbohydrate group attached to a large protein. (2, 18)
glycosuria: The presence of glucose in urine; also called *glucosuria*. (18, 26)
goblet (mucous) cell: A goblet-shaped, mucus-producing, unicellular gland in certain epithelia of the digestive and respiratory tracts; they are called goblet cells in the mucosa of the small intestine, large intestine, terminal bronchioles, and conjunctiva; they are called mucous cells when they are found in the stomach mucosa, respiratory mucosa, and salivary glands; (4, 23, 24)
Golgi apparatus: A cellular organelle consisting of a series of membranous plates that give rise to lysosomes and secretory vesicles. (3)
gomphosis: A fibrous synarthrosis that binds a tooth to the bone of the jaw; *see* **periodontal ligament**. (24)
gonadotropin-releasing hormone (GnRH): A hypothalamic releasing hormone that causes the secretion of both follicle-stimulating hormone and luteinizing hormone by the anterior lobe of the pituitary gland (adenohypophysis). (18, 28)
gonadotropins: Follicle-stimulating hormone and luteinizing hormone, hormones that stimulate gamete development and sex hormone secretion. (18, 28)
gonads: Reproductive organs that produce gametes and sex hormones. (28)
granulocytes: White blood cells containing granules that are visible with the light microscope; includes eosinophils, basophils, and neutrophils; also called *granular leukocytes*. (19)
gray matter: Areas in the central nervous system that are dominated by neuron cell bodies, neuroglia, and unmyelinated axons. (12, 13, 14)
gray ramus communicans: A bundle of postganglionic sympathetic nerve fibers distributed to effectors in the body wall, skin, and limbs by way of a spinal nerve. (13)
greater omentum: A large fold of the dorsal mesentery of the stomach; hangs anterior to the intestines. (24)
groin: The inguinal region. (11)
gross anatomy: The study of the structural features of the body without the aid of a microscope; also called *macroscopic anatomy*. (1)
growth hormone (GH): An adenohypophysis (anterior lobe of the pituitary gland) hormone that stimulates tissue growth and anabolism when nutrients are abundant and restricts tissue glucose dependence when nutrients are in short supply. (18)
growth hormone–inhibiting hormone (GH–IH): A hypothalamic regulatory hormone that inhibits growth hormone secretion by the anterior lobe of the pituitary gland (adenohypophysis); also called *somatostatin*. (18)
growth hormone–releasing hormone (GH–RH): A peptide released by the hypothalamus that causes the release of growth hormone from the anterior pituitary gland. (18)
guanine (G): A purine; one of the nitrogenous bases in the nucleic acids RNA and DNA. (2)
gustation: Sense of taste. (15, 17)
gyrus: A prominent fold or ridge of neural cortex on the surfaces of the cerebral hemispheres. (14)

H

hair: A keratinous strand produced by epithelial cells of the hair follicle. (5)
hair cells: Sensory cells of the internal ear. (17)
hair follicle: An accessory structure of the integument; a tube lined by a stratified squamous epithelium that begins at the surface of the skin and ends at the hair papilla. (5)
hallux: The big toe; also called the *great toe*. (8)
haploid: Possessing half the normal number of chromosomes; a characteristic of gametes. (28, 29)
hard palate: The bony roof of the oral cavity, formed by the maxillae and palatine bones. (23, 24)
helper T cells: Lymphocytes whose secretions and other activities coordinate cell-mediated and antibody-mediated immunities; also called *T_H cells*. (22)
hematocrit: The percentage of formed elements in a sample of blood; also called *volume of packed red cells (VPRC)* or *packed cell volume (PCV)*. (19)
hematoma: A tumor or swelling filled with blood.
hematuria: The abnormal presence of red blood cells in urine. (19, 26)
heme: A porphyrin ring containing a central iron atom that can reversibly bind oxygen molecules; a component of the hemoglobin molecule. (19)
hemocytoblasts: Stem cells whose divisions produce each of the various populations of blood cells; also called *hematopoietic stem cells*. (19)
hemoglobin (Hb or Hgb): A protein composed of four globular subunits, each bound to a heme molecule; gives red blood cells the ability to transport oxygen in the blood. (5, 19, 23, 27)
homozygous: Having identical alleles at corresponding sites on a chromosome pair. (29)
hemolysis: The breakdown of red blood cells. (3)
hemopoiesis: Blood cell formation and differentiation. (19)
hemorrhage: Blood loss; to bleed. (21)
hemostasis: The stoppage of bleeding. (19)
heparin: An anticoagulant released by activated basophils and mast cells. (4, 19)
hepatic duct: The duct that carries bile away from the liver lobes and toward the union with the cystic duct. (24)
hepatic portal vein: The vessel that carries blood from the intestinal capillaries to the sinusoids of the liver. (21)
hepatocyte: A liver cell. (24)
heterozygous: Possessing two different alleles at corresponding sites on a chromosome pair; a person's phenotype is determined by one or both of the alleles. (29)
hexose: A six-carbon simple sugar. (2)
hiatus: A gap, cleft, or opening.
high-density lipoprotein (HDL): A lipoprotein with a relatively small lipid content; responsible for the movement of cholesterol from peripheral tissues to the liver. (25)
hilum: A localized region where blood vessels, lymphatic vessels, nerves, and/or other anatomical structures are attached to an organ. (22, 23, 26)
hippocampus: A region, deep to the floor of a lateral ventricle, involved with emotional states and the conversion of short-term to long-term memories. (12, 14)
histamine: The chemical released by stimulated mast cells or basophils to initiate or enhance inflammation. (4, 12)
histology: The study of tissues. (1, 4)
histones: Proteins associated with the DNA of the nucleus; the DNA strands are wound around them. (3)
holocrine secretion: A form of exocrine secretion in which the secretory cell becomes swollen with vesicles and then ruptures. (4)
homeostasis: The maintenance of a relatively constant internal environment. (1)
hormone: A chemical that is secreted by one cell and travels through the bloodstream to affect the activities of cells in another part of the body. (2, 4, 6, 18, 21, 24, 28, 29)
human chorionic gonadotropin (hCG): The placental hormone that maintains the corpus luteum for the first 3 months of pregnancy. (29)
human immunodeficiency virus (HIV): The infectious agent that causes acquired immunodeficiency syndrome (AIDS). (22)
human leukocyte antigen (HLA): *See* **MHC protein**. (22)
human placental lactogen (hPL): The placental hormone that stimulates the functional development of the mammary glands. (29)
humoral immunity: *See* **antibody-mediated immunity**. (22)
hyaluronan: A carbohydrate component of proteoglycans in the matrix of many connective tissues; also called *hyaluronic acid*. (4)
hyaluronidase: An enzyme that breaks down the bonds between adjacent follicle cells; produced by some bacteria and found in the acrosome of a sperm. (29)
hydrogen bond: A weak interaction between the hydrogen atom on one molecule and a negatively charged portion of another molecule. (2)
hydrogen ion: A hydrogen atom that has lost an electron; a proton. (2)
hydrolysis reaction: The breakage of a chemical bond through the addition of a water molecule; the reverse of dehydration synthesis. (2)
hydrophilic: Freely associating with water; readily entering into solution; water loving. (2)
hydrophobic: Incapable of freely associating with water molecules; insoluble; water fearing. (2)
hydrostatic pressure: Fluid pressure. (3, 21, 26)
hydroxide ion: OH^-. (2)

hypercapnia: High blood carbon dioxide concentrations, commonly as a result of hypoventilation or inadequate tissue perfusion. (23, 27)
hyperextension: Extension of a body part past the anatomical position. (9)
hyperplasia: An abnormal increase in the number of cells in a tissue or organ. (3)
hyperpolarization: The movement of the membrane potential away from the normal resting potential and farther from 0 mV. (12)
hypersecretion: The overactivity of glands that produce exocrine or endocrine secretions. (18)
hypertension: Abnormally high blood pressure. (21)
hypertonic: In comparing two solutions, the solution with the higher osmolarity. (3)
hypertrophy: An increase in tissue size without cell division. (10)
hyperventilation: A rate of respiration sufficient to decrease blood P_{CO_2} to levels below normal. (23, 27)
hypocapnia: An abnormally low blood P_{CO_2}; commonly results from hyperventilation. (23, 27)
hypodermic needle: A needle inserted through the skin to introduce drugs into the subcutaneous layer. (5)
hypodermis: The layer of loose connective tissue deep to the dermis; also called *subcutaneous layer* (4, 5)
hypophyseal portal system: The network of vessels that carries blood from capillaries in the hypothalamus to capillaries in the anterior lobe of the pituitary gland (adenohypophysis). (18)
hypophysis: *See* **pituitary gland**. (18)
hyposecretion: Abnormally low exocrine or endocrine secretion. (18)
hypothalamus: The floor of the diencephalon; the region of the brain containing centers involved with the subconscious regulation of visceral functions, emotions, drives, and the coordination of neural and endocrine functions. (14, 18)
hypothermia: An abnormally low body temperature. (25)
hypothesis: A prediction that can be subjected to scientific analysis and review.
hypotonic: In comparing two solutions, the solution with the lower osmolarity. (3)
hypoventilation: A respiratory rate that is insufficient to keep blood P_{CO_2} within normal levels. (23, 27)
hypoxia: A low tissue oxygen concentration. (19, 23)

I

ileum: The distal segment of the small intestine. (24)
ilium: The largest of the three bones whose fusion creates a coxal bone. (8)
immunity: Resistance to infection and disease caused by foreign substances, toxins, or pathogens. (22)
immunization: The production of immunity by the deliberate exposure to antigens under conditions that prevent the development of illness but stimulate the production of memory B cells. (22)
immunoglobulin (Ig): A circulating antibody. (19, 22)
implantation: The attachment of a blastocyst into the endometrium of the uterine wall. (29)
inclusions: Aggregations of insoluble pigments, nutrients, or other materials in cytoplasm. (3)
incus: The central auditory ossicle, located between the malleus and the stapes in the middle ear cavity. (17)
inducer: A stimulus that promotes the activity of a specific gene. (29)
inexcitable: Incapable of propagating an action potential. (12)
infarct: An area of dead cells that results from an interruption of blood flow. (19, 20)
infection: The invasion and colonization of body tissues by pathogens. (4)
inferior: Below, in reference to a particular structure, with the body in the anatomical position.
inferior vena cava (IVC): The vein that carries blood from the parts of the body inferior to the heart to the right atrium. (20, 21)
infertility: The inability to conceive; also called *sterility*. (28, 29)
inflammation: A nonspecific defense mechanism that operates at the tissue level; characterized by swelling, redness, heat (warmth), pain, and sometimes loss of function. (4, 22)
infundibulum: A tapering, funnel-shaped structure; in the brain, the connection between the pituitary gland and the hypothalamus (14, 18); in the uterine tube, the entrance bounded by fimbriae that receives the oocyte at ovulation. (28)
ingestion: The introduction of food into the digestive tract by way of the mouth; eating. (24)
inguinal canal: A passage through the abdominal wall that marks the path of testicular descent and that contains the testicular arteries, veins, and ductus deferens. (11, 28)
inguinal region: The area of the abdominal wall near the junction of the trunk and the thighs that contains the external genitalia; the groin. (28)
inhibin: A hormone, produced by nurse cells of the testes and follicular cells of the ovaries, that inhibits the secretion of follicle-stimulating hormone by the adenohypophysis (anterior lobe of the pituitary gland). (18, 28)
inhibitory postsynaptic potential (IPSP): A hyperpolarization of the postsynaptic membrane after the arrival of a neurotransmitter. (12)
initial segment: The proximal portion of the axon where an action potential first appears. (12)
injection: The forcing of fluid into a body part or organ.
inner cell mass: Cells of the blastocyst that will form the body of the embryo. (29)
innervation: The distribution of sensory and motor nerves to a specific region or organ. (11, 16)
insensible perspiration: Evaporative water loss by diffusion across the epithelium of the skin or evaporation across the alveolar surfaces of the lungs. (5, 27)
insertion: A point of attachment of a muscle; the end that is easily movable. (11)
insoluble: Incapable of dissolving in solution. (2)
inspiration: Inhalation; the movement of air into the respiratory system. (23)
insulin: A hormone secreted by beta cells of the pancreatic islets; causes a decrease in blood glucose concentrations. (18)
integument: The skin. (5)
intercalated discs: Regions where adjacent cardiac muscle cells interlock and where gap junctions permit electrical coupling between the cells. (4, 10, 20)
intercellular fluid: *See* **interstitial fluid**. (3)
interferons (IFNs): Peptides released by virus-infected cells, especially lymphocytes, that slow viral replication and make other cells more resistant to viral infection. (22)
interleukins: Peptides, released by activated monocytes and lymphocytes, that assist in the coordination of cell-mediated and antibody-mediated immunities. (22)
internal capsule: The collection of afferent and efferent fibers of the white matter of the cerebral hemispheres, visible on gross dissection of the brain. (14)
internal ear: The membranous labyrinth that contains the organs of hearing and equilibrium. (17)
internal respiration: The diffusion of gases between interstitial fluid and cytoplasm. (23)
interneuron: An association neuron; central nervous system neurons that are between sensory and motor neurons. (12)
interoceptors: Sensory receptors monitoring the functions and status of internal organs and systems. (12)
interosseous membrane: The fibrous connective tissue membrane between the shafts of the tibia and fibula and between the radius and ulna; an example of a fibrous amphiarthrosis. (8)
interphase: The stage in the life cycle of a cell during which the chromosomes are uncoiled and all normal cellular functions except mitosis are under way. (3, 29)
intersegmental reflex: A reflex that involves several segments of the spinal cord. (13)
interstitial fluid: The fluid in the tissues that fills the spaces between cells. (3)
interstitial growth: A form of cartilage growth in which the cartilage expands from within through the growth, mitosis, and secretion of chondrocytes in the matrix. (4)
interventricular foramen: The opening that permits fluid movement between the lateral and third ventricles of the brain. (14)
intervertebral disc: A fibrocartilage pad between the bodies of successive vertebrae that absorbs shocks. (7, 9)
intestinal gland: A tubular epithelial pocket that is lined by secretory cells and opens into the lumen of the digestive tract; also called *intestinal crypt*. (24)
intestine: The tubular organ of the digestive tract. (18, 24)
intracellular fluid (ICF): The cytosol. (27)
intrafusal muscle fibers: Muscle spindle fibers. (13)
intramembranous ossification: The formation of bone within a connective tissue without the prior development of a cartilaginous model. (6)
intrinsic factor: A glycoprotein secreted by the parietal cells of the stomach that facilitates the intestinal absorption of vitamin B_{12}. (19, 24, 25)
intrinsic pathway: A pathway of the clotting system that begins with the activation of platelets and ends with the formation of factor X activator complex. (19)
inversion: A turning inward. (9)
in vitro: Outside the body, in an artificial environment.
in vivo: In the living body.
involuntary: Not under conscious control.
ion: An atom or molecule having a positive or negative charge due to the loss or gain, respectively, of one or more electrons. (2, 26, 27)
ionic bond: A molecular bond created by the attraction between ions with opposite charges. (2)
ionization: the dissociation of an atom, molecule, or substance into ions; also called *dissociation*. (2)
ipsilateral: A reflex response that affects the same side as the stimulus. (13)
iris: A contractile structure made up of smooth muscle that forms the colored portion of the eye. (17)
ischemia: An inadequate blood supply to a region of the body. (11)
ischium: One of the three bones whose fusion creates a coxal bone. (8)
islets of Langerhans: *See* **pancreatic islets**. (18, 24)
isotonic: A solution with an osmolarity that does not result in water movement across plasma membranes. (3, 10)

isotopes: Forms of an element whose atoms contain the same number of protons but different numbers of neutrons (and thus differ in atomic mass). (2)
isthmus: A narrow band of tissue connecting two larger masses.

J

jejunum: The middle part of the small intestine. (24)
joint: An area where adjacent bones interact; also called *articulation*. (9)
juxtaglomerular cells: Modified smooth muscle cells between the walls of the distal convoluted tubule and the afferent and efferent arterioles adjacent to the glomerulus and the macula densa. (26)
juxtaglomerular complex (JGC): The macula densa, extraglomerular mesangial cells, and the juxtaglomerular cells; a complex responsible for the release of renin and erythropoietin. (26)

K

keratin: The tough, fibrous protein component of nails, hair, calluses, and the general integumentary surface. (5)
keto acid: The carbon chain that remains after the deamination or transamination of an amino acid. (25)
ketoacidosis: A condition characterized by a decrease in blood pH due to the presence of large numbers of ketone bodies. (25, 26, 27)
ketone bodies: Keto acids produced during the catabolism of lipids and ketogenic amino acids; specifically, acetone, acetoacetate, and beta-hydroxybutyrate. (25)
kidney: A component of the urinary system; an organ functioning in the regulation of blood composition, including the excretion of wastes and the maintenance of normal fluid and electrolyte balances. (18, 26)
killer T cells: *See* **cytotoxic T cells**. (22)
Krebs cycle: *See* **citric acid cycle**. (3, 10, 25)
Kupffer cell: *See* **stellate macrophage.** (22, 24)

L

labium/labia: Lip/lips; the labia majora and labia minora are structures of the female external genitalia. (28)
labrum: A lip or rim.
labyrinth: A maze of passageways; the structures of the internal ear. (17)
lacrimal gland: A tear gland on the dorsolateral surface of the eye. (17)
lactase: An enzyme that breaks down the milk sugar, lactose. (24)
lactate: An anion released by the dissociation of lactic acid, or formed from the reduction of pyruvate. (10)
lactation: The production of milk by the mammary glands. (28)
lacteal: A terminal lymphatic within an intestinal villus. (24)
lacuna/lacunae: A small pit or cavity. (4, 6, 29)
lambdoid suture: The synarthrosis between the parietal and occipital bones of the cranium. (7)
lamellae: Concentric layers; the concentric layers of bone within an osteon. (6)
lamellated corpuscle: A receptor sensitive to vibration. (15)
lamina: A thin sheet or layer.
lamina propria: The areolar tissue that underlies a mucous epithelium and forms part of a mucous membrane. (4, 23, 24)
Langerhans cells: Cells in the epithelium of the skin (15) and digestive tract (24) that participate in the immune response by presenting antigens to T cells; also called *dendritic cells*.
large intestine: The terminal portions of the intestinal tract, consisting of the colon, the rectum, and the anal canal. (24)
laryngopharynx: The division of the pharynx that is inferior to the epiglottis and superior to the esophagus. (23)
larynx: A complex cartilaginous structure that surrounds and protects the glottis and vocal cords; the superior margin is bound to the hyoid bone, and the inferior margin is bound to the trachea. (23)
latent period: The time between the stimulation of a muscle and the start of the contraction phase. (10)
lateral: Pertaining to the side.
lateral apertures: Openings in the roof of the fourth ventricle that permit the circulation of cerebrospinal fluid into the subarachnoid space. (14)
lateral ventricle: A fluid-filled chamber within a cerebral hemisphere. (14)
lens: The transparent refractive structure of the eye that is between the iris and the vitreous humor. (17)
lesion: A localized abnormality in tissue organization. (4)
lesser omentum: A double-layered peritoneal fold formed by the mesentery that connects the lesser curvature of the stomach to the liver. (24)
leukocyte: A white blood cell. (4, 19)
ligament: A dense band of connective tissue fibers that attaches one bone to another. (4, 9)
ligamentum arteriosum: The fibrous strand in adults that is the remnant of the ductus arteriosus of the fetal stage. (21)
ligamentum nuchae: An elastic ligament between the vertebra prominens and the occipital bone. (7)
ligamentum teres: The fibrous strand in the falciform ligament of adults that is the remnant of the umbilical vein; also called the *round ligament of the liver*. (24)
ligate: To tie off.
limbic system: The nuclei and centers in the cerebrum and diencephalon that are involved with emotional states, memories, and behavioral drives. (14)
lingual: Pertaining to the tongue. (17, 24)
lipid: An organic compound containing carbon, hydrogen, and oxygen in a ratio that does not approximate 1:2:1; includes fats, oils, and waxes. (2, 24, 25)
lipogenesis: The synthesis of lipids from nonlipid precursors. (25)
lipolysis: The catabolism of lipids as a source of energy. (25)
lipoprotein: A compound containing a relatively small lipid bound to a protein. (25)
liver: An organ of the digestive system that has varied and vital functions, including the production of plasma proteins, the excretion of bile, the storage of energy reserves, the detoxification of poisons, and the interconversion of nutrients. (24)
lobule: Histologically, the basic organizational unit of the liver. (24)
local hormone: *See* **prostaglandin**. (2, 18)
loop of Henle: *See* **nephron loop**. (26)
loose connective tissue: A loosely organized, easily distorted connective tissue that contains several fiber types, a varied population of cells, and a viscous ground substance. (4)
lumbar: Pertaining to the lower back. (7, 13)
lumen: The central space within a duct or other internal passageway. (4)
lungs: The paired organs of breathing enclosed by the pleural cavities. (23)
luteinizing hormone (LH): A hormone produced by the adenohypophysis (anterior lobe of the pituitary gland). In females, it assists FSH in follicle stimulation, triggers ovulation, and promotes the maintenance and secretion of endometrial glands. In males, it was formerly called *interstitial cell-stimulating hormone* because it stimulates testosterone secretion by the interstitial endocrine cells of the testes. (18, 28)
lymph: The fluid contents of lymphatic vessels, similar in composition to interstitial fluid. (4, 22)
lymphatic vessels: The vessels of the lymphatic system; also called *lymphatics*. (4, 22)
lymphedema: Sewlling as a result of lymphatic vessel obstruction. (22)
lymph nodes: Lymphoid organs that monitor the composition of lymph. (22)
lymphocyte: A cell of the lymphatic system that plays a role in the immune response. (4, 19, 22)
lymphokines: Chemicals secreted by activated lymphocytes. (22)
lymphopoiesis: The production of lymphocytes from lymphoid stem cells. (19, 22)
lymphotoxin: A secretion of lymphocytes that kills the target cells. (22)
lysis: The destruction of a cell through the rupture of its plasma membrane. (3)
lysosome: An intracellular vesicle containing digestive enzymes. (3)
lysozyme: An enzyme, present in some exocrine secretions, that has antibiotic properties. (17)

M

macrophage: A phagocytic cell of the monocyte–macrophage system. (4, 22)
macula: A receptor complex, located in the saccule or utricle of the internal ear, that responds to linear acceleration or gravity. (17)
macula (of retina)**:** The oval area in the retina whose center (the fovea centralis) is the region of sharpest vision; also called *macula lutea*. (17)
macula densa: A group of specialized secretory cells that is located in a portion of the distal convoluted tubule, adjacent to the glomerulus, extraglomerular mesangial cells, and the juxtaglomerular cells; a component of the juxtaglomerular complex. (26)
major histocompatibility complex: *See* **MHC protein**. (22)
malignant tumor: A form of cancer characterized by rapid cell growth and the spread of cancer cells throughout the body. (3)
malleus: An auditory ossicle, bound to the tympanic membrane and the incus. (17)
malnutrition: An unhealthy state produced by inadequate dietary intake or absorption of nutrients. (25)
mammillary bodies: Nuclei in the hypothalamus that affect eating reflexes and behaviors; a component of the limbic system. (14)
mammary glands: Milk-producing glands of the female breasts. (5, 28)
manus: The hand; manual. (8, 11)
marrow: A tissue that fills the internal cavities in bone; dominated by hemopoietic cells (red bone marrow) or by adipose tissue (yellow bone marrow). (6, 19)
mass: A physical property that determines the weight of an object in Earth's gravitational field; the amount of material in mater. (2)
mast cell: A connective tissue cell that, when stimulated, releases histamine, serotonin, and heparin, initiating the inflammatory response. (4)
mastication: Chewing. (11, 24)

GLOSSARY

mastoid sinus: Air-filled spaces in the mastoid process of the temporal bone. (7)
matrix: The extracellular fibers and ground substance of a connective tissue. (3, 4)
matter: Anything that takes up space and has mass. (2)
maxillary sinus: One of the paranasal sinuses; an air-filled chamber lined by a respiratory epithelium that is located in a maxilla and opens into the nasal cavity. (7)
meatus: An opening or entrance into a passageway. (23, 26)
mechanoreception: The detection of mechanical stimuli, such as touch, pressure, or vibration. (15)
medial: Toward the midline of the body.
mediastinum: The central tissue mass that divides the thoracic cavity into two pleural cavities. (1, 20)
medulla: The inner layer or core of an organ. (5)
medulla oblongata: The most caudal of the brain regions; also called the *myelencephalon*. (14)
medullary cavity: The space within a bone that contains the marrow. (6)
megakaryocytes: Bone marrow cells responsible for the formation of platelets. (19)
meiosis: Cell division that produces gametes with half the normal somatic chromosome complement. (3, 28)
melanin: The red-yellow or brown-black pigments produced by the melanocytes of the skin. (4, 5)
melanocyte: A specialized cell in the deeper layers of the stratified squamous epithelium of the skin; responsible for the production of melanin. (4, 5, 18)
melanocyte-stimulating hormone (MSH): A hormone, produced by the pars intermedia of the anterior lobe of the pituitary gland (adenohypophysis), that stimulates melanin production. (18)
melatonin: A hormone secreted by the pineal gland; inhibits secretion of MSH and GnRH. (14, 18)
membrane: Any sheet or partition; a layer consisting of an epithelium and the underlying connective tissue. (2)
membrane flow: The continuous movement and exchange of membrane segments. (3)
membrane potential: The potential difference, measured across a plasma membrane and expressed in millivolts, that results from the uneven distribution of positive and negative ions across the plasma membrane; also called *transmembrane potential*. (3, 10, 12)
membranous labyrinth: Endolymph-filled tubes that enclose the receptors of the internal ear. (17)
memory: The ability to recall information or sensations; can be divided into short-term and long-term memories. (16, 22)
memory T cells: T lymphocytes that provide immunologic memory, enabling an enhanced immune response when reexposed to a specific antigen. (22)
meninges: Three membranes that surround the surfaces of the central nervous system; the dura mater, the pia mater, and the arachnoid mater. (13)
meniscus: A fibrocartilage pad between opposing surfaces in a joint. (9)
menstruation: The phase of the uterine cycle in which the endometrial functional layer is shed; also called *menses*. (28)
merocrine secretion: A method of secretion in which the cell ejects materials from secretory vesicles through exocytosis. (4, 5)
mesencephalon: The midbrain; the region between the diencephalon and pons. (14)
mesenchyme: Embryonic or fetal connective tissue. (4)
mesentery: A double layer of serous membrane that supports and stabilizes the position of an organ in the abdominopelvic cavity and provides a route for the associated blood vessels, nerves, and lymphatic vessels. (24)
mesoderm: The middle germ layer, between the ectoderm and endoderm of the embryo. (29)
mesothelium: A simple squamous epithelium that lines the body cavities enclosing the lungs, heart, and abdominal organs. (4)
messenger RNA (mRNA): RNA formed at transcription to direct protein synthesis in the cytoplasm. (2, 3)
metabolic turnover: The continuous breakdown and replacement of organic materials within cells. (2, 25)
metabolism: The sum of all biochemical processes under way within the human body at any moment; includes anabolism and catabolism. (2, 25)
metabolites: Compounds produced in the body as a result of metabolic reactions. (2)
metacarpals: The five bones of the palm of the hand. (8)
metalloproteins: Proteins containing a metal ion cofactor; examples include plasma transport proteins, storage proteins, and enzymes. (19, 25)
metaphase: The stage of mitosis in which the chromosomes line up along the equatorial plane of the cell. (3)
metaphysis: The region of a long bone between the epiphysis and diaphysis, corresponding to the location of the epiphyseal cartilage of the developing bone. (6)
metastasis: The spread of cancer cells from one organ to another, leading to the establishment of secondary tumors. (3)
metatarsal: One of the five bones of the foot that articulate with the tarsal bones (proximally) and the phalanges (distally). (8)
metencephalon: The pons and cerebellum of the brain. (14)
MHC protein: A surface antigen that is important to the recognition of foreign antigens and that plays a role in the coordination and activation of the immune response; also called *major histocompatibility complex (MHC) protein* or *human leukocyte antigen (HLA)*. (22)
micelle: A droplet with hydrophilic portions on the outside; a spherical aggregation of bile salts, monoglycerides, and fatty acids in the lumen of the intestinal tract. (2, 24)
microfilaments: Fine protein filaments visible with the electron microscope; components of the cytoskeleton. (3)
microglia: Phagocytic neuroglia in the central nervous system. (12, 22)
microphages: Neutrophils and eosinophils. (4, 19, 22)
microtubules: Microscopic tubules that are part of the cytoskeleton and are a component in cilia, flagella, the centrioles, and spindle fibers. (3)
microvilli: Small, fingerlike extensions of the exposed plasma membrane of an epithelial cell. (3)
micturition: Urination. (26)
midbrain: The mesencephalon. (14)
middle ear: The space between the external and internal ears that contains auditory ossicles. (17)
midsagittal plane: A plane passing through the midline of the body that divides it into left and right halves. (1)
mineralocorticoids: Corticosteroids produced by the zona glomerulosa of the adrenal cortex; steroids such as aldosterone that affect mineral metabolism. (18)
mitochondrion: An intracellular organelle responsible for generating most of the ATP required for cellular operations. (3, 25)
mitosis: The division of a single cell nucleus that produces two identical daughter cell nuclei; an essential step in cell division. (3, 28)
mitral valve: The left atrioventricular (AV) valve, also called *bicuspid valve*. (20)
mixed gland: A gland that contains exocrine and endocrine cells, or an exocrine gland that produces serous and mucous secretions. (4)
mixed nerve: A peripheral nerve that contains sensory and motor fibers. (13)
mole: A quantity of an element or compound having a mass in grams equal to the element's atomic weight or to the compound's molecular weight. (2)
molecular weight: The sum of the atomic weights of all the atoms in a molecule or compound. (2, 3)
molecule: A chemical structure containing two or more atoms that are bonded together by shared electrons. (1, 2, 3)
monoclonal antibodies: Antibodies produced by genetically identical cells under laboratory conditions. (22)
monocytes: Phagocytic agranulocytes (white blood cells) in the circulating blood. (19)
monoglyceride: A lipid consisting of a single fatty acid bound to a molecule of glycerol. (2)
monokines: Secretions released by activated cells of the monocyte–macrophage system to coordinate various aspects of the immune response. (22)
monomer: A molecule that can be bonded to other identical molecules to form a polymer. (2)
monosaccharide: A simple sugar, such as glucose or ribose. (2, 24)
monosynaptic reflex: A reflex in which the sensory afferent neuron synapses directly on the motor efferent neuron. (13)
motor unit: All the muscle cells controlled by a single motor neuron. (10)
mucins: Proteoglycans responsible for the lubricating properties of mucus. (2, 24)
mucosa: A mucous membrane; the epithelium plus the lamina propria; also called *mucous membrane*. (4, 24)
mucosa-associated lymphoid tissue (MALT): The extensive collection of lymphoid tissues linked with the epithelia of the digestive, respiratory, urinary, and reproductive tracts. (22)
mucous (adjective)**:** Indicating the presence or production of mucus.
mucous (goblet) cell: A goblet-shaped, mucus-producing, unicellular gland in certain epithelia of the digestive and respiratory tracts; they are called mucous cells when they are found in the stomach mucosa, respiratory mucosa, and salivary glands; they are called goblet cells in the mucosa of the small intestine, large intestine, terminal bronchioles, and conjunctiva. (4, 23, 24)
mucous membrane: *See* **mucosa**. (4, 24)
mucus (noun)**:** A lubricating fluid that is composed of water and mucins and is produced by unicellular and multicellular glands along the digestive, respiratory, urinary, and reproductive tracts. (2, 4)
multipolar neuron: A neuron with many dendrites and a single axon; the typical form of a motor neuron. (12)
multiunit smooth muscle: A smooth muscle tissue whose muscle cells are innervated in motor units. (10)
muscarinic receptors: Membrane receptors sensitive to acetylcholine and to muscarine, a toxin produced by certain mushrooms; located at all parasympathetic neuromuscular and neuroglandular junctions and at a few sympathetic neuromuscular and neuroglandular junctions. (16)

muscle: A contractile organ composed of muscle tissue, blood vessels, nerves, connective tissues, and lymphatic vessels. (10, 11)
muscle tissue: A tissue characterized by the presence of cells capable of contraction; includes skeletal, cardiac, and smooth muscle tissues. (4, 10)
muscular layer: Concentric layers of smooth muscle responsible for peristalsis. (24)
muscularis mucosae: The layer of circular and longitudinal smooth muscle deep to the lamina propria; responsible for moving the mucosal surface. (24)
mutagens: Chemical agents that induce mutations and may be carcinogenic. (3)
mutation: A change in the nucleotide sequence of the DNA in a cell. (3)
myelencephalon: *See* **medulla oblongata**. (14)
myelin: An insulating sheath around an axon; consists of multiple layers of neuroglial membrane; significantly increases the nerve impulse propagation rate along the axon. (12)
myelination: The formation of myelin. (12)
myenteric plexus: Parasympathetic motor neurons and sympathetic postganglionic fibers located between the circular and longitudinal layers of the muscularis externa. (24)
myocardial infarction: A heart attack; damage to the heart muscle due to an interruption of regional coronary circulation. (20)
myocardium: The cardiac muscle tissue of the heart. (20)
myofibril: Organized collections of myofilaments in skeletal and cardiac muscle cells. (10)
myofilaments: Fine protein filaments composed primarily of the proteins actin (thin filaments) and myosin (thick filaments). (10)
myoglobin: An oxygen-binding pigment that is especially common in slow skeletal muscle fibers and cardiac muscle cells. (2, 10)
myogram: A recording of the tension produced by muscle fibers on stimulation. (10)
myometrium: The thick layer of smooth muscle in the wall of the uterus. (28)
myosatellite cells: Cells that are the precursors to skeletal muscle cells (fibers) (10)
myosin: The protein component of thick filaments. (3, 10)

N

nail: A keratinous structure produced by epithelial cells of the nail root. (5)
nasal cavity: A chamber in the skull that is bounded by the internal and external nares. (7, 17, 23)
nasolacrimal duct: The passageway that transports tears from the nasolacrimal sac to the nasal cavity. (7, 17)
nasolacrimal sac: A chamber that receives tears from the lacrimal ducts. (17)
nasopharynx: A region that is posterior to the internal nares and superior to the soft palate and ends at the oropharynx. (23)
natriuretic peptides: Hormones released by specialized cardiac muscle cells when they are stretched by an abnormally large venous return; promotes fluid loss and reductions in blood pressure and in venous return. Includes atrial natriuretic peptide (ANP) and B-type natriuretic peptide (BNP). (18, 21, 26, 27)
natural killer (NK) cells: Type of lymphocyte that can kill target cells without previous sensitivity. (19)
N compound: An organic compound containing nitrogen atoms. (25)
necrosis: The death of cells or tissues from disease or injury. (4, 22)
negative feedback: A corrective mechanism that opposes or negates a variation from normal limits. (1, 18)
neonate: A newborn infant, or baby. (29)
neoplasm: A tumor, or mass of abnormal tissue. (3)
nephron: The basic functional unit of the kidney. (26)
nephron loop: The segment of the nephron that creates the concentration gradient in the renal medulla; also called *loop of Henle*. (26)
nerve impulse: An action potential in a neuron plasma membrane. (12)
net filtration pressure (NFP): The difference between the net hydrostatic pressure and the net osmotic pressure. (21, 26)
neural cortex: An area of gray matter at the surface of the central nervous system. (13)
neurofibrils: Microfibrils in the cytoplasm of a neuron. (12)
neurofilaments: Microfilaments in the cytoplasm of a neuron. (12)
neuroglandular junction: A cell junction at which a neuron controls or regulates the activity of a secretory (gland) cell. (12)
neuroglia: Cells of the central nervous system and peripheral nervous system that support and protect neurons; also called *glial cells*. (4, 12)
neurohypophysis: The posterior lobe of the pituitary gland, or pars nervosa; contains the axons of hypothalamic neurons, which release OXT and ADH. (18)
neurolemma: The outer surface of a Schwann cell that encircles an axon in the peripheral nervous system. (12)
neuromodulator: A compound, released by a neuron, that adjusts the sensitivities of another neuron to specific neurotransmitters. (12)
neuromuscular junction (NMJ): A synapse between a neuron and a muscle cell. (10, 12)
neuron: A cell in nervous tissue that is specialized for intercellular communication through (1) changes in membrane potential and (2) synaptic connections. (4, 12, 15, 16)
neuroplasticity: The ability of nerve cells to make new connections. (14)
neurotransmitter: A chemical compound released by one neuron to affect the membrane potential of another. (10, 12, 16)
neurotubules: Microtubules in the cytoplasm of a neuron. (12)
neutron: A fundamental particle that does not have a positive or a negative charge. (2)
neutrophil: A white blood cell that is very numerous and normally the first of the mobile phagocytic cells to arrive at an area of injury or infection. (19)
nicotinic receptors: Acetylcholine receptors on the surfaces of sympathetic and parasympathetic ganglion cells; respond to the compound nicotine. (16)
nipple: An elevated epithelial projection on the surface of the breast; contains the openings of the lactiferous sinuses. (28)
Nissl bodies: The ribosomes, Golgi apparatus, rough endoplasmic reticulum, and mitochondria of the perikaryon of a typical neuron. (12)
nitrogenous wastes: Organic waste products of metabolism that contain nitrogen, such as urea, uric acid, and creatinine. (25)
nociception: Pain perception. (15)
nodes: The areas between adjacent neuroglia where the myelin covering of an axon is incomplete; also called *nodes of Ranvier*. (12)
nodose ganglion: A sensory ganglion of cranial nerve X; also called *inferior ganglion*. (14)
noradrenaline: *See* **norepinephrine (NE)**. (12, 18)
norepinephrine (NE): A catecholamine neurotransmitter in the peripheral nervous system and central nervous system, released at most sympathetic neuromuscular and neuroglandular junctions, and a hormone secreted by the adrenal (suprarenal) medulla; also called *noradrenaline*. (12, 18)
nucleic acid: A polymer of nucleotides that contains a pentose sugar, a phosphate group, and one of four nitrogenous bases that regulate the synthesis of proteins and make up the genetic material in cells. (2)
nucleolus: The dense region in the nucleus that is the site of RNA synthesis. (3)
nucleoplasm: The fluid content of the nucleus. (3)
nucleoproteins: Proteins of the nucleus that are generally associated with DNA. (3)
nucleotide: A compound consisting of a nitrogenous base, a simple sugar, and a phosphate group. (2)
nucleus: The central region of an atom. (1) A cellular organelle that contains DNA, RNA, and proteins; in the central nervous system, a mass of gray matter. (3)
nucleus pulposus: The gelatinous central region of an intervertebral disc. (9)
nurse cells: Supporting cells of the seminiferous tubules of the testis; responsible for the differentiation of spermatids, the maintenance of the blood testis barrier, and the secretion of inhibin, androgen-binding protein, and Müllerian-inhibiting factor; also called *Sertoli cells*. (18, 28)
nutrient: An inorganic or organic substance that can be used by the body as a cofactor or to produce energy. (2, 25)
nystagmus: An unconscious, continuous movement of the eyes as if to adjust to constant motion. (17)

O

obesity: Body weight more than 20 percent above the ideal weight for a given individual. (25)
occlusal surface: The opposing surfaces of the teeth that come into contact when chewing food. (24)
ocular: Pertaining to the eye. (17)
oculomotor nerve: Cranial nerve III, which controls the extraocular muscles other than the superior oblique and the lateral rectus muscles. (14)
olecranon: The proximal end of the ulna that forms the prominent point of the elbow. (8)
olfaction: The sense of smell. (15, 17, 23)
olfactory bulb: The expanded ends of the olfactory tracts (17); the sites where the axons of the first cranial nerves (I) synapse on central nervous system interneurons that lie inferior to the frontal lobes of the cerebrum. (14)
oligodendrocytes: Central nervous system neuroglia that maintain cellular organization within gray matter and provide a myelin sheath in areas of white matter. (12)
oligopeptide: A short chain of amino acids. (2)
oocyte: A cell whose meiotic divisions will produce a single ovum and three polar bodies. (3, 28, 29)
oogenesis: Formation and development of an oocyte. (28)
opsonization: An effect of coating an object with antibodies; the attraction and enhancement of phagocytosis. (22)
optic chiasm: The crossing point of the optic nerves. (14, 17)

optic nerve: The second cranial nerve (II), which carries signals from the retina of the eye to the optic chiasm. (14)
optic tract: The tract over which nerve impulses from the retina are transmitted between the optic chiasm and the thalamus. (14)
orbit: The bony recess of the skull that contains the eyeball. (7)
organ: Combinations of tissues that perform complex functions. (1)
organelle: An intracellular structure with a specific function or group of functions. (3)
organic compound: A compound containing carbon, hydrogen, and in most cases oxygen. (2)
organogenesis: The formation of organs during embryonic and fetal development. (29)
organ systems: Groups of organs that function together in a coordinated manner. (1)
origin: In a skeletal muscle, the point of attachment that does not change position when the muscle contracts; usually defined in terms of movements from the anatomical position. (11)
oropharynx: The middle portion of the pharynx, bounded superiorly by the nasopharynx, anteriorly by the oral cavity, and inferiorly by the laryngopharynx. (23)
osmolarity: The total concentration of dissolved materials in a solution, regardless of their specific identities, expressed in moles; also called *osmotic concentration*. (3, 26, 27)
osmoreceptor: A receptor sensitive to changes in the osmolarity of plasma. (27)
osmosis: The movement of water across a selectively permeable membrane from one solution to another solution that contains a higher solute concentration. (3, 21, 26, 27)
osmotic pressure (OP): The force of osmotic water movement; the pressure that must be applied to prevent osmosis across a membrane. (3, 21, 26, 27)
osseous tissue: A strong connective tissue containing specialized cells and a mineralized matrix of crystalline calcium phosphate and calcium carbonate; also called *bone tissue*. (4, 6)
ossicles: Small bones; middle ear bones. (17)
ossification: The formation of bone; osteogenesis. (6)
osteoblast: A cell that produces the fibers and matrix of bone. (6)
osteoclast: A cell that dissolves the fibers and matrix of bone. (6)
osteocyte: A bone cell responsible for the maintenance and turnover of the mineral content of the surrounding bone. (4, 6)
osteogenic layer: The inner, cellular layer of the periosteum that aids in bone growth and repair. (6)
osteolysis: The breakdown of the mineral matrix of bone. (6)
osteon: The basic histological unit of compact bone, consisting of osteocytes organized around a central canal and separated by concentric lamellae. (6)
osteoporosis: A reduction in bone mass that causes brittle, fragile bones and compromises normal function. (6)
otic: Pertaining to the ear. (17)
otoliths: Calcium carbonate crystals embedded in a gelatinous matrix; located on each macula of the vestibule. (17)
oval window: An opening in the bony labyrinth where the stapes attaches to the membranous wall of the scala vestibuli (vestibular duct). (17)
ovarian cycle: The monthly process of the maturation, ovulation, and degeneration of a tertiary ovarian follicle. (28)
ovary: The female reproductive organ that produces oocytes (gametes). (18, 28)
ovulation: The release of a secondary oocyte, surrounded by cells of the corona radiata, after the rupture of the wall of a tertiary ovarian follicle. (28, 29)
ovum/ova: The functional product of meiosis II, produced after the fertilization of a secondary oocyte. (28, 29)
oxidation: The gain of oxygen, or the loss of hydrogen or electrons. (25)
oxytocin (OXT): A hormone produced by hypothalamic cells and secreted into capillaries at the posterior lobe of the pituitary gland (neurohypophysis); stimulates smooth muscle contractions of the uterus or mammary glands in females and the prostate gland in males. (18)

P

pacemaker cells: Cells of the sinoatrial node that set the pace of cardiac contraction. (4, 10, 20)
palate: The horizontal partition separating the oral cavity from the nasal cavity and nasopharynx; divided into an anterior bony (hard) palate and a posterior fleshy (soft) palate. (7, 23, 24)
palatine: Pertaining to the palate. (24)
palpate: To examine by touch.
palpebrae: *See* **eyelids**. (17)
pancreas: A digestive organ containing exocrine and endocrine tissues; the exocrine portion secretes pancreatic juice, and the endocrine portion secretes hormones, including insulin and glucagon. (18, 24)
pancreatic duct: A tubular duct that carries pancreatic juice from the pancreas to the duodenum. (18, 24)
pancreatic islets: Aggregations of endocrine cells in the pancreas; also called *islets of Langerhans*. (18, 24)
pancreatic juice: A mixture of buffers and digestive enzymes that is discharged into the duodenum under the stimulation of the enzymes secretin and cholecystokinin. (18, 24)
Papanicolaou test (Pap smear): A test for the detection of malignancies based on the cytological appearance of epithelial cells, especially those of the cervix and uterus. (28)
papilla: A small, conical projection.
paralysis: The loss of voluntary motor control over a portion of the body. (13)
paranasal sinuses: Bony chambers, lined by respiratory epithelium, that open into the nasal cavity; the frontal, ethmoidal, sphenoidal, and maxillary sinuses. (7)
parasagittal plane: A plane that parallels the midsagittal plane but that does not pass along the midline. (1)
parasympathetic division: One of the two divisions of the autonomic nervous system; generally responsible for activities that conserve energy and lower the metabolic rate; the "rest and digest" division; also called *craniosacral division*. (16)
parathyroid glands: Four small glands embedded in the posterior surface of the thyroid gland that secrete parathyroid hormone. (6, 18)
parathyroid hormone (PTH): A hormone secreted by the parathyroid glands when blood calcium levels decrease below the normal range; causes increased osteoclast activity, increased intestinal calcium uptake, and decreased calcium ion loss by the kidneys. (6, 18)
parenchyma: The cells of a tissue or organ that are responsible for fulfilling its functional role; distinguished from the stroma of that tissue or organ. (4)
paresthesia: A sensory abnormality that produces a tingling sensation.
parietal: Relating to the parietal bone (7); referring to the wall of a cavity. (23)
parietal cells: Cells of the gastric glands that secrete hydrochloric acid and intrinsic factor. (24)
parotid glands: Large salivary glands that secrete saliva with high concentrations of salivary (alpha) amylase. (24)
pars distalis: The large, anterior portion of the anterior lobe of the pituitary gland (adenohypophysis). (18)
pars intermedia: The portion of the anterior lobe of the pituitary gland (adenohypophysis) that is immediately adjacent to the posterior lobe of the pituitary gland (neurohypophysis) and the infundibulum. (18)
pars nervosa: The posterior lobe of the pituitary gland (neurohypophysis). (18)
pars tuberalis: The portion of the anterior lobe of the pituitary gland (adenohypophysis) that wraps around the infundibulum superior to the posterior lobe (neurohypophysis). (18)
patella: The sesamoid bone of the knee; also called the *kneecap*. (8)
pathogen: A disease-causing organism. (1, 22)
pathologist: A physician specializing in the identification of diseases on the basis of characteristic structural and functional changes in tissues and organs.
pelvic cavity: The inferior subdivision of the abdominopelvic cavity; encloses the urinary bladder, the sigmoid colon and rectum, and male or female reproductive organs. (1, 8)
pelvis: A bony complex created by the articulations among the coxal bones, the sacrum, and the coccyx. (8, 11)
penis: A component of the male external genitalia; a copulatory organ that surrounds the urethra and introduces semen into the female vagina; the developmental equivalent of the female clitoris. (28)
peptide: A chain of amino acids linked by peptide bonds. (2, 18)
peptide bond: A covalent bond between the amino group of one amino acid and the carboxyl group of another. (2)
perforating canals: Passageways within compact bone that extend perpendicular to the surface. (6)
pericardial cavity: The space between the parietal pericardium and the epicardium (visceral pericardium) that encloses the outer surface of the heart. (1, 20)
pericardium: The fibrous sac that surrounds the heart; its inner, serous lining is continuous with the epicardium. (4, 20)
perichondrium: The layer that surrounds a cartilage, consisting of an outer fibrous region and an inner cellular region. (4)
perikaryon: The cytoplasm that surrounds the nucleus in the cell body of a neuron. (12)
perilymph: A fluid similar in composition to cerebrospinal fluid; located in the spaces between the bony labyrinth and the membranous labyrinth of the internal ear. (17)
perimetrium: The incomplete serosa of the uterus. (28)
perimysium: A connective tissue partition that separates adjacent fasciculi in a skeletal muscle. (10)
perineum: The pelvic floor and its associated structures. (11)
perineurium: A connective tissue partition that separates adjacent bundles of nerve fibers in a peripheral nerve. (13)
periodontal ligament: Collagen fibers that anchor a tooth within its alveolus. (24)

periosteum: The layer that surrounds a bone, consisting of an outer fibrous region and inner cellular region. (4, 6)

peripheral nervous system (PNS): All nervous tissue outside the central nervous system and the enteric nervous system. (12, 13)

peripheral resistance: The resistance to blood flow; primarily caused by friction with the vessel walls. (21)

peristalsis: A wave of smooth muscle contractions that propels materials along the lumen of a tube such as the digestive tract (24), the ureters (26), or the ductus deferens. (28)

peritoneal cavity: The potential space within the abdominopelvic cavity lined by the peritoneum. (1)

peritoneum: The serous membrane that lines the peritoneal cavity. (4, 24, 28)

peritubular capillaries: A network of capillaries that surrounds the proximal and distal convoluted tubules of the kidneys. (26)

permeability: The ease with which dissolved materials can cross a membrane; if the membrane is freely permeable, any molecule can cross it; if impermeable, nothing can cross; most biological membranes are selectively permeable (allow some and restrict others). (3)

peroxisome: A membranous vesicle containing enzymes that break down hydrogen peroxide (H_2O_2). (3)

pes: The foot. (8, 11)

petrosal ganglion: A sensory ganglion of the glossopharyngeal nerve (N IX). (14, 15)

petrous: Stony; usually refers to the thickened portion of the temporal bone that encloses the internal ear. (17)

pH: The negative exponent (negative logarithm) of the hydrogen ion concentration, expressed in moles per liter. (2, 27)

phagocyte: A cell that performs phagocytosis. (22)

phagocytosis: The engulfing of extracellular materials or pathogens; the movement of extracellular materials into the cytoplasm by enclosure in a membranous vesicle. (3, 19, 22)

phalanx/phalanges: Bone(s) of the finger(s) or toe(s). (8)

pharmacology: The study of drugs, their physiological effects, and their clinical uses.

pharynx: The throat; a muscular passageway shared by the digestive and respiratory tracts. (11, 23, 24)

phasic response: A pattern of response to stimulation by sensory neurons that are normally inactive; stimulation causes a burst of neural activity that ends when the stimulus either stops or stops changing in intensity. (15)

phenotype: Physical characteristics that are genetically determined. (29)

phosphate group: PO_4^{3-}; a functional group that can be attached to an organic molecule; required for the formation of high-energy bonds. (2, 25, 27)

phospholipid: An important membrane lipid whose structure includes both hydrophilic and hydrophobic regions. (2, 3)

phosphorylation: The addition of a high-energy phosphate group to a molecule. (2, 25)

photoreception: Sensitivity to light. (17)

physiology: The study of how living organisms perform their functions. (1)

pia mater: The innermost layer of the meninges bound to the underlying nervous tissue. (13, 14)

pineal gland: Nervous tissue in the posterior portion of the roof of the diencephalon; secretes melatonin. (14, 18)

pinna: *See* **auricle**. (17)

pinocytosis: The introduction of fluids into the cytoplasm by enclosing them in membranous vesicles at the cell surface. (3)

pituitary gland: An endocrine organ that is located in the sella turcica of the sphenoid bone and is connected to the hypothalamus by the infundibulum; includes the posterior lobe (neurohypophysis) and the anterior lobe (adenohypophysis); also called the *hypophysis*. (14, 18)

placenta: A temporary structure in the uterine wall that permits diffusion between the fetal and maternal circulatory systems. (29)

plantar: Referring to the sole of the foot (1); muscles (11); plantar reflex. (13)

plantar flexion: Ankle extension; toe pointing. (8, 9, 11)

plasma: The fluid ground substance of whole blood; what remains after the cells have been removed from a sample of whole blood. (4, 19)

plasma cell: An activated B cell that secretes antibodies. (4, 19, 22)

plasma membrane: A cell membrane; also called a *plasmalemma*. (3)

platelets: Small packets of cytoplasm that contain enzymes important in the clotting response; manufactured in bone marrow by megakaryocytes. (4, 19)

pleura: The serous membrane that lines the pleural (lung) cavities. (4, 23)

pleural cavities: Body cavities of the thoracic region that surround the lungs. (1, 23)

plexus: A network or braid.

polar body: A nonfunctional packet of cytoplasm that contains chromosomes eliminated from an oocyte during meiosis. (28, 29)

polarized: Referring to cells that have regional differences in organelle distribution or cytoplasmic composition along a specific axis, such as between the basement membrane and free surface of an epithelial cell. (4)

pollex: The thumb. (8)

polymer: A large molecule consisting of a long chain of monomer subunits. (2)

polypeptide: A chain of amino acids strung together by peptide bonds; those containing more than 100 peptides are called *proteins*. (2)

polyribosome: Several ribosomes linked by their translation of a single mRNA strand. (3)

polysaccharide: A complex sugar, such as glycogen or a starch. (2)

polysynaptic reflex: A reflex in which interneurons are interposed between the sensory fiber and the motor neuron(s). (13)

polyunsaturated fats: Fatty acids containing carbon atoms that are linked by double bonds. (1, 2)

pons: The portion of the metencephalon that is anterior to the cerebellum. (14)

popliteal: Pertaining to the back of the knee. (9, 11, 21)

porphyrins: Ring-shaped molecules that form the basis of important respiratory and metabolic pigments, including heme and the cytochromes. (23)

positive feedback: A mechanism that increases a deviation from normal limits after an initial stimulus. (1)

postcentral gyrus: The primary sensory cortex, where touch, vibration, pain, temperature, and taste sensations arrive and are consciously perceived. (14)

posterior: Toward the back; dorsal.

posterior root ganglion: A peripheral nervous system ganglion containing the cell bodies of sensory neurons; *see* **dorsal root ganglion**. (13)

postganglionic neuron: An autonomic neuron in a peripheral ganglion, whose activities control peripheral effectors. (16)

postsynaptic membrane: The portion of the plasma membrane of a postsynaptic cell that is part of a synapse. (12)

potential difference: The separation of opposite charges; requires a barrier that prevents ion migration. (3, 12)

precentral gyrus: The primary motor cortex of a cerebral hemisphere, located anterior to the central sulcus. (14)

prefrontal cortex: The anterior portion of each cerebral hemisphere; thought to be involved with higher intellectual functions, predictions, and calculations. (14)

preganglionic neuron: A visceral motor neuron in the central nervous system whose output controls one or more ganglionic motor neurons in the peripheral nervous system. (16)

premotor cortex: The motor association area between the precentral gyrus and the prefrontal area. (14)

pre-optic body: The hypothalamic nucleus that coordinates thermoregulatory activities. (14)

presynaptic membrane: The synaptic surface where neurotransmitter release occurs. (12)

prevertebral ganglion: *See* **collateral ganglion**. (12, 16)

prime mover: A muscle that performs a specific action; also called an *agonist*. (11)

proenzyme: An inactive enzyme secreted by an epithelial cell. (19)

progesterone: The most important hormone secreted by the corpus luteum after ovulation. (18, 28)

prognosis: A prediction about the possible course or outcome from a specific disease.

projection fibers: Axons carrying information from the thalamus to the cerebral cortex. (14)

prolactin (PRL): The hormone that stimulates functional development of the mammary glands in females; a secretion of the anterior lobe of the pituitary gland (adenohypophysis). (18)

pronation: The rotation of the forearm that makes the palm face posteriorly. (9)

prone: Lying face down with the palms facing the floor. (1)

pronucleus: An enlarged ovum or sperm nucleus that forms after fertilization but before amphimixis. (29)

prophase: The initial phase of mitosis; characterized by the appearance of chromosomes, the breakdown of the nuclear membrane, and the formation of the spindle apparatus. (3)

proprioceptor: Sensory organ that monitors the position and movement of skeletal muscles and joints. (12)

proprioception: The awareness of the positions of bones, joints, and muscles. (15)

prostaglandin: A fatty acid secreted by one cell that alters the metabolic activities or sensitivities of adjacent cells; also called *local hormone*. (2, 18)

prostate gland: An accessory gland of the male reproductive tract, contributing about one-third of the volume of semen. (28)

prosthesis: An artificial substitute for a body part.

protease: An enzyme that breaks down proteins into peptides and amino acids. (2, 3, 24)

protein: A large polypeptide with a complex structure. (2, 25)

proteoglycan: A substance containing a large polysaccharide complex linked by polypeptide chains; examples include hyaluronan (hyaluronic acid) and chondroitin sulfate. (2, 4)

proton: A fundamental subatomic particle having a positive charge. (2)

protraction: Movement anteriorly in the horizontal plane. (9)
proximal: A direction toward the point of attachment or origin; for a limb, toward its attachment to the trunk. (1, 8)
proximal convoluted tubule (PCT): The segment of the nephron between the glomerular capsule (Bowman's capsule) and the nephron loop; the major site of active reabsorption from filtrate. (26)
pseudopodia: Temporary cytoplasmic extensions typical of mobile or phagocytic cells. (3)
pseudostratified epithelium: An epithelium that contains several layers of nuclei but whose cells are all in contact with the underlying basement membrane. (4)
puberty: A period of rapid growth, sexual maturation, and the appearance of secondary sexual characteristics; normally occurs at ages 10–15 years. (18, 28, 29)
pubic symphysis: The fibrocartilaginous amphiarthrosis between the pubic bones of the hip. (8, 9)
pubis: The anterior, inferior component of the hip bone. (8)
pudendum: The external genitalia. (28)
pulmonary circuit: Blood vessels between the pulmonary semilunar valve of the right ventricle and the entrance to the left atrium; the blood flow through the lungs. (20)
pulmonary ventilation: The movement of air into and out of the lungs. (23)
pulvinar nucleus: The thalamic nucleus involved in the integration of sensory information prior to projection to the cerebral hemispheres. (14)
pupil: The opening in the center of the iris through which light enters the eye. (17)
purine: A nitrogen compound with a double ring-shaped structure; examples include adenine and guanine, two nitrogenous bases that are common in nucleic acids. (2, 12)
Purkinje cell layer: A large, branching neuron of the cerebellar cortex. (14)
Purkinje fibers: Specialized conducting cardiac muscle cells in the ventricles of the heart. (20)
pus: An accumulation of debris, fluid, dead and dying cells, and necrotic tissue. (4, 20, 22)
pyloric part: The gastric region between the body of the stomach and the duodenum; includes the muscular pylorus and the pyloric sphincter. (24)
pyloric sphincter: A ring of smooth muscle that regulates the passage of chyme from the stomach to the duodenum. (24)
pyrimidine: A nitrogen compound with a single ring-shaped structure; examples include cytosine, thymine, and uracil, nitrogenous bases that are common in nucleic acids. (2)
pyruvate: The anion formed by the dissociation of pyruvic acid, a three-carbon compound produced by glycolysis. (25)

Q

quaternary structure: The three-dimensional protein structure produced by interactions between protein subunits. (2)

R

radiodensity: The relative resistance to the passage of x-rays. (1)
radiographic techniques: Methods of visualizing internal structures by using various forms of radiational energy. (1)
radiopaque: Having a high radiodensity. (1)
rami communicantes: Axon bundles that link the spinal nerves with the ganglia of the sympathetic chain. (13)
ramus/rami: Branch/branches.
raphe: A seam. (11, 28)
receptive field: The area monitored by a single sensory receptor. (15)
rectum: The inferior 15 cm (6 in.) of the digestive tract. (24)
rectus: Straight. (11)
red blood cell (RBC): *See* **erythrocyte**. (4, 19)
reduction: The loss of oxygen, or the gain of hydrogen or electrons. (25)
reductional division: The first meiotic division, which reduces the chromosome number from 46 to 23. (28)
reflex: A rapid, automatic response to a stimulus. (12, 13, 16, 18, 21)
reflex arc: The receptor, sensory neuron, motor neuron, and effector involved in a particular reflex; interneurons may be present, depending on the reflex considered. (13)
refractory period: The time between the initiation of an action potential and the restoration of the normal resting membrane potential; during this period, the membrane will not respond normally to stimulation. (12, 20)
regulatory T cells: Population of T lymphocytes that suppress the immune response. (22)
relaxation phase: The period after a contraction when the tension in the muscle fiber returns to resting levels. (10)
relaxin: A hormone that loosens the pubic symphysis; secreted by the placenta. (29)
renal: Pertaining to the kidneys. (26)
renal corpuscle: The initial segment of the nephron, consisting of an expanded chamber that encloses the glomerulus. (26)
renin: The enzyme released by cells of the juxtaglomerular complex when renal blood flow decreases; converts angiotensinogen to angiotensin I. (18, 26)
rennin: A gastric enzyme that breaks down milk proteins. (24)
replication: Duplication. (29)
repolarization: The movement of the membrane potential away from a positive value and toward the resting potential. (12, 20)
respiration: The exchange of gases between cells and the environment; includes pulmonary ventilation, external respiration, and internal respiration. (23, 27)
respiratory minute volume ($\dot{V}_E$): The amount of air moved into and out of the respiratory system each minute. (23)
respiratory pump: A mechanism by which changes in the intrapleural pressures during the respiratory cycle assist the venous return to the heart; also called *thoracoabdominal pump*. (21, 23)
respiratory rhythmicity center: The center in the medulla oblongata that sets the background pace of respiration; includes inspiratory and expiratory centers. (14, 23)
resting membrane potential: The membrane potential of a normal cell under homeostatic conditions. (3, 12)
rete: An interwoven network of blood vessels or passageways. (28)
reticular activating system (RAS): The midbrain portion of the reticular formation; responsible for arousal and the maintenance of consciousness. (16)
reticular formation: A diffuse network of gray matter that extends the entire length of the brainstem (medulla oblongata to the midbrain). (14)
reticulospinal tracts: Descending tracts of the medial pathway that carry involuntary motor commands issued by neurons of the reticular formation. (15)
retina: The innermost layer of the eye, lining the vitreous chamber; also called *inner layer*. (17)
retinal: A visual pigment derived from vitamin A. (17)
retraction: Movement posteriorly in the horizontal plane. (9)
retroperitoneal: Behind or outside the peritoneal cavity. (1, 24)
reverberation: A positive feedback along a chain of neurons such that they remain active once stimulated. (13)
rheumatism: A general term used to describe pain in muscles, tendons, bones, or joints. (9)
Rh factor: A surface antigen that may be present (Rh positive) or absent (Rh negative) from the surfaces of red blood cells. (19)
rhodopsin: The visual pigment in the membrane discs of the distal segments of rods. (17)
ribonucleic acid: A nucleic acid consisting of a chain of nucleotides that contain the sugar ribose and the nitrogenous bases adenine, guanine, cytosine, and uracil. (2, 3)
ribose: A five-carbon sugar that is a structural component of RNA. (2, 3)
ribosome: An organelle that contains rRNA and proteins and is essential to mRNA translation and protein synthesis. (2, 3)
rod: A photoreceptor responsible for vision in dim lighting. (17)
rough endoplasmic reticulum (RER): A membranous organelle that is a site of protein synthesis and storage. (3)
round window: An opening in the bony labyrinth of the internal ear that exposes the membranous wall of the tympanic duct to the air of the middle ear cavity. (17)
rubrospinal tracts: Descending tracts of the lateral pathway that carry involuntary motor commands issued by the red nucleus of the midbrain. (15)
rugae: Mucosal folds in the lining of the empty stomach that disappear as gastric distension occurs (24); folds in the urinary bladder (26).

S

saccule: A portion of the vestibular apparatus of the internal ear; contains a macula important for providing sensations of gravity and linear acceleration in a vertical dimension. (17)
sagittal plane: A vertical plane that divides the body into left and right portions. (1)
salt: An inorganic compound consisting of a cation other than H^+ and an anion other than OH^-. (2)
saltatory propagation: The relatively rapid propagation of an action potential between successive nodes of a myelinated axon. (12)
sarcolemma: The plasma membrane of a muscle cell. (10)
sarcomere: The smallest contractile unit of a striated muscle cell. (10)
sarcoplasm: The cytoplasm of a muscle cell. (10)
scala media: The central membranous tube within the cochlea that is filled with endolymph and contains the spiral organ (*organ of Corti*); also called *cochlear duct*. (17)
scala tympani: The perilymph-filled chamber of the internal ear, adjacent to the basilar membrane; pressure changes there distort the round window; also called *tympanic duct*. (17)

scala vestibuli: The perilymph-filled chamber of the internal ear, adjacent to the vestibular membrane; pressure waves are induced by movement of the stapes at the oval window; also called *vestibular duct*. (17)
scar tissue: The thick, collagenous tissue that forms at an injury site. (5)
Schwann cells: Neuroglia responsible for the neurolemma that surrounds axons in the peripheral nervous system. (12)
sciatic nerve: A nerve innervating the posteromedial portions of the thigh and leg. (13)
scientific method: A system of advancing knowledge that begins by proposing a hypothesis to answer a question, and then testing that hypothesis with data collected through observation and experimentation. (1)
sclera: The fibrous, outer layer of the eye that forms the white area of the anterior surface; a portion of the fibrous layer of the eye. (17)
sclerosis: A hardening and thickening that commonly occurs secondary to tissue inflammation. (4)
scrotum: The loose-fitting, fleshy pouch that encloses the testes of the male. (28)
sebaceous glands: Glands that secrete sebum; normally associated with hair follicles. (5)
sebum: A waxy secretion that coats the surfaces of hairs. (5)
secondary sex characteristics: Physical characteristics that appear at puberty in response to sex hormones but are not involved in the production of gametes. (28)
secretin: A hormone, secreted by the duodenum, that stimulates the production of buffers by the pancreas and inhibits gastric activity. (24)
sectional plane: A view or slice along a two-dimensional flat surface. (1)
semen: The fluid ejaculate that contains sperm and the secretions of accessory glands of the male reproductive tract. (28)
semicircular ducts: The tubular components of the membranous labyrinth of the internal ear; respond to rotational movements of the head. (17)
semilunar valve: A three-cusped valve guarding the exit from one of the cardiac ventricles; the pulmonary and aortic valves. (20)
seminal glands: Glands of the male reproductive tract that produce roughly 60 percent of the volume of semen; also called *seminal vesicles*. (28)
seminiferous tubules: Coiled tubules where sperm production occurs in the testes. (28)
senescence: Aging. (29)
sensible perspiration: Water loss due to secretion by sweat glands. (5, 27)
septa: Partitions that subdivide an organ. (20, 22)
serosa: *See* **serous membrane**. (4, 24)
serotonin: A neurotransmitter in the central nervous system; a substance that enhances inflammation and is released by activated mast cells and basophils. (12, 19)
serous cell: A cell that produces a serous secretion. (4)
serous membrane: A squamous epithelium and the underlying loose connective tissue; the lining of the pericardial, pleural, and peritoneal cavities; also called a *serosa*. (4, 24)
serous secretion: A watery secretion that contains high concentrations of enzymes; produced by serous cells. (4)
Sertoli cells: *See* **nurse cells**. (18, 28)
serum: The ground substance of blood plasma from which clotting agents have been removed. (19)
sesamoid bone: A bone that forms within a tendon. (6)
sigmoid colon: The S-shaped region of the colon between the descending colon and the rectum. (24)
sign: The visible, objective evidence of the presence of a disease. (1)
simple epithelium: An epithelium containing a single layer of cells superficial to the basement membrane. (4)
sinoatrial (SA) node: The natural pacemaker of the heart; located in the wall of the right atrium. (20)
sinus: A chamber or hollow in a tissue; a large, dilated vein. (6, 7, 20)
sinusoid: An exchange vessel that is similar in general structure to a fenestrated capillary. The two differ in size (sinusoids are larger and more irregular in cross section), continuity (sinusoids have gaps between endothelial cells), and support (sinusoids have thin basement membranes, if present at all). (20, 21)
skeletal muscle: A contractile organ of the muscular system. (10)
skeletal muscle tissue: A contractile tissue dominated by skeletal muscle fibers; characterized as striated, voluntary muscle. (4, 10)
sliding-filament theory: The concept that a sarcomere shortens as the thick and thin filaments slide past one another. (10)
small intestine: The duodenum, jejunum, and ileum; the digestive tract between the stomach and the large intestine. (24)
smooth endoplasmic reticulum (SER): A membranous organelle in which lipid and carbohydrate synthesis and storage occur. (3)
smooth muscle tissue: Muscle tissue in the walls of many visceral organs; characterized as nonstriated, involuntary muscle. (4, 10, 24, 26)
soft palate: The fleshy posterior extension of the hard palate, separating the nasopharynx from the oral cavity. (24)
solute: Any materials dissolved in a solution. (2, 21, 26)
solution: A fluid containing dissolved materials. (2, 21)
solvent: The fluid component of a solution. (2, 21)
somatic: Pertaining to the body.
somatic nervous system (SNS): The efferent division of the nervous system that innervates skeletal muscles. (12, 15, 16)
somatomedins: Substances stimulating tissue growth; released by the liver after the secretion of growth hormone; also called *insulin-like growth factors*. (18)
somatotropin: Growth hormone (GH); produced by the anterior lobe of the pituitary gland (adenohypophysis) in response to growth hormone–releasing hormone (GH–RH). (18)
sperm: Male gamete(s). (3, 28, 29)
spermatic cord: Collectively, the spermatic vessels, nerves, lymphatic vessels, and the ductus deferens, extending between the testes and the proximal end of the inguinal canal. (28)
spermatocyte: A cell of the seminiferous tubules that is engaged in meiosis. (28)
spermatogenesis: Sperm production. (28)
sphincter: A muscular ring that contracts to close the entrance or exit of an internal passageway. (10, 11, 26)
spinal nerve: One of 31 pairs of nerves that originate on the spinal cord from anterior and posterior roots. (12, 13)
spindle apparatus: Microtubule-based structure that distributes duplicated chromosomes to opposite ends of a dividing cell during mitosis. (3)
spinocerebellar tracts: Ascending tracts that carry sensory information to the cerebellum. (15)
spinothalamic tracts: Ascending tracts that carry poorly localized touch, pressure, pain, vibration, and temperature sensations to the thalamus. (15)
spinous process: The prominent posterior projection of a vertebra; formed by the fusion of two laminae. (7)
spiral organ: A receptor complex in the scala media of the cochlea that includes the inner and outer hair cells, supporting cells and structures, and the tectorial membrane; provides the sensation of hearing; also called the *organ of Corti*. (17)
spleen: A lymphoid organ important for the phagocytosis of red blood cells, the immune response, and lymphocyte production. (22)
spongy bone: Bone that consists of an open network of struts and plates that resembles a three-dimensional garden lattice. (6)
squama: A broad, flat surface.
squamous: Flattened.
squamous epithelium: An epithelium whose superficial cells are flattened and platelike. (4)
stapes: The auditory ossicle attached to the tympanic membrane. (17)
stellate macrophage: Phagocytic cell of the liver sinusoids; also called *Kupffer cell*. (22, 24)
stenosis: A constriction or narrowing of a passageway.
stereocilia: Elongated microvilli characteristic of the epithelium of the epididymis, sections of the ductus deferens (28), and the internal ear. (17)
steroid: A ring-shaped lipid structurally related to cholesterol. (2, 18)
stimulus: An environmental change that produces a change in cellular activities; often used to refer to events that alter the membrane potentials of excitable cells. (15)
stratified epithelium: An epithelium containing several layers. (4)
stratum/strata: Layer/layers. (4)
stretch receptors: Sensory receptors that respond to stretching of the surrounding tissues. (13)
stretch reflex: Tonic muscle contraction in response to stimulation of muscle proprioceptors. (13)
stroke volume (SV): The amount of blood pumped out of one heart ventricle in a single heart beat. (20)
stroma: The connective tissue framework of an organ; distinguished from the functional cells (parenchyma) of that organ.
subarachnoid space: A meningeal space containing cerebrospinal fluid; the area between the arachnoid membrane and the pia mater. (13, 14)
subclavian: Pertaining to the region immediately posterior and inferior to the clavicle.
submucosa: The region between the muscularis mucosae and the muscularis externa. (23, 24)
subserous fascia: The loose connective tissue layer deep to the serous membrane that lines body cavities. (4)
substrate: A participant (product or reactant) in an enzyme-catalyzed reaction. (2)
sulcus: A groove or furrow. (14)
summation: The temporal or spatial addition of contractile force or neural stimuli. (10, 12)
superficial fascia: The layer of loose connective tissue below the dermis. (4, 5)
superior: Above, in reference to a portion of the body in the anatomical position.
superior vena cava (SVC): The vein that carries blood to the right atrium from parts of the body that are superior to the heart. (20, 21)
supination: The rotation of the forearm such that the palm faces anteriorly. (9)
supine: Lying face up, with palms facing anteriorly. (1)

GLOSSARY

suprarenal cortex: *See* **adrenal cortex**. (18)
suprarenal gland: *See* **adrenal gland**. (18)
suprarenal medulla: *See* **adrenal medulla**. (16)
surfactant: A lipid secretion that coats the alveolar surfaces of the lungs and prevents their collapse. (23)
sutural bones: Irregular bones that form in fibrous tissue between the flat bones of the developing cranium; also called *Wormian bones*. (6)
suture: A fibrous joint between flat bones of the skull. (7, 9)
sympathetic division: The division of the autonomic nervous system that is responsible for "fight or flight" reactions; primarily concerned with the elevation of metabolic rate and increased alertness; also called the *thoracolumbar division*. (12, 16)
symphysis: A fibrous amphiarthrosis, such as that between adjacent vertebrae or between the pubic bones of the coxal bones. (9)
symptom: An abnormality of function as a result of disease; subjective experience of patient. (1)
synapse: The site of communication between a nerve cell and some other cell; if the other cell is not a neuron, the term *neuromuscular junction* or *neuroglandular junction* is often used. (12, 16, 18)
synaptic delay: The period between the arrival of an impulse at the presynaptic membrane and the initiation of an action potential in the postsynaptic membrane. (12)
syncytium: A multinucleate mass of cytoplasm, produced by the fusion of cells or repeated mitoses without cytokinesis. (29)
syndrome: A discrete set of signs and symptoms that occur together.
synergist: A muscle that assists a prime mover in performing its primary action. (11)
synovial cavity: A fluid-filled chamber in a synovial joint. (4, 9)
synovial fluid: The substance secreted by synovial membranes that lubricates joints. (4, 9)
synovial joint: A freely movable joint where the opposing bone surfaces are separated by synovial fluid; a diarthrosis. (4, 9)
synovial membrane: An incomplete layer of fibroblasts confronting the synovial cavity, plus the underlying loose connective tissue. (4)
synthesis: Manufacture; anabolism. (23)
system: An interacting group of organs that performs one or more specific functions.
systemic circuit: The vessels between the aortic valve and the entrance to the right atrium; the system other than the vessels of the pulmonary circuit. (20)
systole: A period of contraction in a chamber of the heart, as part of the cardiac cycle. (20)
systolic pressure: The peak arterial pressure measured during ventricular systole. (20, 21)

T

tachycardia: Rapid heart beat, usually over 90 beats per minute. (20)
tactile: Pertaining to the sense of touch. (15)
tarsal bones: The bones of the ankle (the talus, calcaneus, navicular, and cuneiform bones). (8)
tarsus: The ankle. (8)
TCA (tricarboxylic acid) cycle: *See* **citric acid cycle**. (3, 10, 25)
T cells: Lymphocytes responsible for cell-mediated immunity and for the coordination and regulation of the immune response; includes cytotoxic T cells, helper T cells, regulatory T cells, and memory T cells. (19, 22)
tectospinal tracts: Descending tracts of the medial pathway that carry involuntary motor commands issued by the colliculi. (15)
telodendria: Terminal axonal branches that end in axon terminals. (12)
telomeres: The distal ends of chromosomes. (3)
telophase: The final stage of mitosis, characterized by the disappearance of the spindle apparatus, the reappearance of the nuclear membrane, the disappearance of the chromosomes, and the completion of cytokinesis. (3)
temporal: Pertaining to time (temporal summation) or to the temples (temporal bone). (7)
tendon: A collagenous band that connects a skeletal muscle to an element of the skeleton. (4, 10)
teres: Round. (11)
terminal: Toward the end.
tertiary structure: The protein structure that results from interactions among distant portions of the same molecule; complex coiling and folding. (2)
testes: The male gonads, sites of gamete production and hormone secretion; also called *testicles*. (18, 28)
testosterone: The main androgen produced by the interstitial endocrine cells of the testes. (2, 18, 28)
tetraiodothyronine: T_4, or thyroxine, a thyroid hormone. (18)
thalamus: The walls of the diencephalon. (14)
therapy: The treatment of disease.
thermoreception: Sensitivity to temperature changes. (15)
thermoregulation: Homeostatic maintenance of body temperature. (1, 25)
thick filament: A cytoskeletal filament in a skeletal or cardiac muscle cell; composed of myosin, with a core of titin. (3, 10)
thin filament: A cytoskeletal filament in a skeletal or cardiac muscle cell; consists of actin, troponin, nebulin, and tropomyosin. (3, 10)
thoracolumbar division: The sympathetic division of the autonomic nervous system. (16)
thorax: The chest. (7)
threshold: The membrane potential at which an action potential begins. (12)
thrombin: The enzyme that converts soluble fibrinogen to insoluble fibrin. (19)
thymine (T): A pyrimidine; one of the nitrogenous bases in the nucleic acid DNA. (2)
thymosins: Thymic hormones essential to the development and differentiation of T cells. (18, 22)
thymus: A lymphoid organ, the site of T cell development and maturation. (18, 22)
thyroglobulin: A circulating transport globulin that binds thyroid hormones. (18)
thyroid gland: An endocrine gland whose lobes are lateral to the thyroid cartilage of the larynx. (18)
thyroid hormones: Thyroxine (T_4) and triiodothyronine (T_3), hormones of the thyroid gland; stimulate tissue metabolism, energy utilization, and growth. (18)
thyroid-stimulating hormone (TSH): The hormone, produced by the adenohypophysis (anterior lobe of the pituitary gland), that triggers the secretion of thyroid hormones by the thyroid gland. (18)
thyroxine: A thyroid hormone; also called T_4 or *tetraiodothyronine*. (18)
tidal volume (V_T): The volume of air moved into and out of the lungs during a normal quiet respiratory cycle. (23)
tight junction: Connection between cells formed by the fusion of membrane proteins. (4)
tissue: A collection of specialized cells and cell products that performs a specific function. (1, 4)
tonic response: An increase or decrease in the frequency of action potentials by sensory receptors that are chronically active. (15)
tonsil: A lymphoid nodule in the wall of the pharynx; the palatine, pharyngeal, and lingual tonsils. (22)
topical: Applied to the body surface.
toxic: Poisonous.
trabecula: A connective tissue partition that subdivides an organ. (22)
trachea: The windpipe; an airway extending from the larynx to the primary bronchi. (23)
tract: A bundle of axons in the central nervous system. (13, 14)
transamination: The reversible chemical reaction between an amino acid and a keto acid in which the amino group from the amino acid is transferred to the keto acid. (25)
transcription: The encoding of genetic instructions on a strand of mRNA. (3)
transection: The severing or cutting of an object in the transverse plane.
translation: The process of peptide formation from the instructions carried by an mRNA strand. (3)
transmembrane potential: *See* **membrane potential**. (3, 10, 12)
transudate: A fluid that diffuses across a serous membrane and lubricates opposing surfaces. (4)
transverse plane: A horizontal plane that divides the body into superior and inferior portions. (1)
transverse tubules: The transverse, tubular extensions of the sarcolemma that extend deep into the sarcoplasm, contacting cisternae of the sarcoplasmic reticulum; also called *T tubules*. (10)
tricuspid valve: The right atrioventricular valve, which prevents the backflow of blood into the right atrium during ventricular systole. (20)
trigeminal nerve: Cranial nerve V, which provides sensory information from the lower portions of the face (including the upper and lower jaws) and delivers motor commands to the muscles of mastication. (14)
triglyceride: A lipid that is composed of a molecule of glycerol attached to three fatty acids. (2, 25)
triiodothyronine: T_3, a thyroid hormone. (18)
trisomy: The abnormal possession of three copies of a chromosome; trisomy 21 is responsible for Down syndrome. (29)
trochanter: Large process near the head of the femur. (8)
trochlea: A pulley; the spool-shaped medial portion of the condyle of the humerus. (8)
trochlear nerve: Cranial nerve IV, controlling the superior oblique muscle of the eye. (14)
tropomyosin: Fibrous muscle protein that covers active sites on G actin and prevents actin-myosin interaction. (10)
troponin: Globular muscle protein that binds to tropomyosin. (10)
trunk: The thoracic and abdominopelvic regions (1); a large bundle of spinal nerve axons (13); a major arterial branch (21).
T tubules: *See* **transverse tubules**.
tuberculum: A small, localized elevation on a bony surface. (7)
tuberosity: A large, roughened elevation on a bony surface. (6, 8)
tumor: A tissue mass formed by the abnormal growth and replication of cells. (3)
tunica: A layer or covering.
twitch: A single stimulus–contraction–relaxation cycle in a skeletal muscle fiber. (10)
tympanic duct: *See* **scala tympani**.

tympanic membrane: The membrane that separates the external acoustic meatus from the middle ear; the membrane whose vibrations are transferred to the auditory ossicles and ultimately to the oval window; also called *eardrum* or *tympanum*. (17)
Type A fibers: Large myelinated axons. (12)
Type B fibers: Small myelinated axons. (12)
Type C fibers: Small unmyelinated axons. (12)

U

umbilical cord: The connecting stalk between the fetus and the placenta; contains the allantois, the umbilical arteries, and the umbilical vein. (21, 29)
umbilicus: The navel. (29)
unicellular gland: Mucous cells. (4)
unipolar neuron: A sensory neuron whose cell body is in a dorsal root ganglion or a sensory ganglion of a cranial nerve. (12)
unmyelinated axon: An axon whose neurolemma does not contain myelin and across which continuous propagation occurs. (12)
uracil (U): A pyrimidine; one of the nitrogenous bases in the nucleic acid RNA. (2)
ureters: Muscular tubes, lined by transitional epithelium, that carry urine from the renal pelvis to the urinary bladder. (26)
urethra: A muscular tube that carries urine from the urinary bladder to the exterior. (26, 28)
urinary bladder: The muscular, distensible sac that stores urine prior to micturition. (26)
urination: The elimination of urine; also called *micturition*. (26)
uterus: The muscular organ of the female reproductive tract in which implantation, placenta formation, and fetal development occur. (28)
utricle: The largest chamber of the vestibular apparatus of the internal ear; contains a macula important for providing sensations of gravity and linear acceleration in a horizontal dimension. (17)

V

vagina: A muscular tube extending between the uterus and the vestibule. (28)
vallate papilla: One of the large, dome-shaped papillae on the superior surface of the tongue that forms a V, separating the body of the tongue from the root. (17)
vascular: Pertaining to blood vessels. (19)
vas deferens: *See* **ductus deferens.** (28)
vasoconstriction: A reduction in the diameter of arterioles due to the contraction of smooth muscles in the tunica media; increases peripheral resistance; may occur in response to local factors, through the action of hormones, or from the stimulation of the vasomotor center. (21)
vasodilation: An increase in the diameter of arterioles due to the relaxation of smooth muscles in the tunica media; decreases peripheral resistance; may occur in response to local factors, through the action of hormones, or after decreased stimulation of the vasomotor center. (21)
vasomotion: Rhythmic changes in the pattern of blood flow through a capillary bed due to the alternate contraction and relaxation of precapillary sphincters. (21)
vasomotor center: The center in the medulla oblongata whose stimulation produces vasoconstriction and an increase of peripheral resistance. (14)
vein: A blood vessel carrying blood from a capillary bed toward the heart. (20, 21)
vena cava: One of the major veins delivering systemic blood to the right atrium; superior and inferior venae cavae. (20, 21)
ventilation: Air movement into and out of the lungs. (23)
ventral: Pertaining to the anterior surface.
ventricle: A fluid-filled chamber; in the heart, one of the large chambers discharging blood into the pulmonary or systemic circuits (20); in the brain, one of four fluid-filled interior chambers. (14)
venule: Thin-walled veins that receive blood from capillaries. (21)
vermiform appendix: *See* **appendix**. (24)
vertebral canal: The passageway that encloses the spinal cord; a tunnel bounded by the neural arches of adjacent vertebrae. (7)
vertebral column: The cervical, thoracic, and lumbar vertebrae, the sacrum, and the coccyx. (7, 11)
vesicle: A membranous sac in the cytoplasm of a cell. (3)
vestibular duct: *See* **scala vestibuli**.
vestibular nucleus: The processing center for sensations that arrive from the vestibular apparatus of the internal ear, located near the border between the pons and the medulla oblongata. (14, 17)
vestibulospinal tracts: Descending tracts of the medial pathway that carry involuntary motor commands issued by the vestibular nucleus to stabilize the position of the head. (15)
villus/villi: Slender, finger-shaped projection(s) of a mucous membrane. (24, 29)
virus: A noncellular pathogen. (22)
viscera: Internal organs of the thoracic and abdominopelvic cavities. (1)
visceral: Pertaining to viscera (internal organs) or their outer coverings. (1)
visceral smooth muscle: A smooth muscle tissue that forms sheets or layers in the walls of visceral organs; the cells may not be innervated, and the layers often show automaticity (rhythmic contractions). (10, 24)
viscosity: The resistance to flow that a fluid exhibits as a result of molecular interactions within the fluid. (21)
viscous: Thick, syrupy.
vitamin: An essential organic nutrient that functions as a coenzyme in vital enzymatic reactions. (24, 25)
vitreous humor: The fluid component of the vitreous body, the gelatinous mass in the posterior cavity (vitreous chamber) of the eye. (17)
voluntary: Controlled by conscious thought processes.

W

white blood cells (WBCs): The granulocytes and agranulocytes of whole blood; also called *leukocytes*. (4, 19)
white matter: Regions in the central nervous system that are dominated by myelinated axons. (12, 13, 14)
white ramus communicans: A nerve bundle containing the myelinated preganglionic axons of sympathetic motor neurons en route to the sympathetic chain or to a collateral ganglion. (13)
Wormian bones: *See* **sutural bones**. (6)

X

X chromosome: One of two sex chromosomes; females have two X chromosomes. (29)
xiphoid process: The slender, inferior extension of the sternum. (7)

Y

Y chromosome: The sex chromosome whose presence indicates that the individual is a genetic male. (29)

Z

zona fasciculata: The region of the adrenal cortex that secretes glucocorticoids. (18)
zona glomerulosa: The region of the adrenal cortex that secretes mineralocorticoids. (18)
zona reticularis: The region of the adrenal cortex that secretes androgens. (18)
zygote: The fertilized ovum, prior to the start of cleavage. (28, 29)

Credits

Chapter 1 CO-1 Rubberball/Nicole Hill/Getty Images; 1-CC-1 Rafael Macia/Science Source 1-CN-1 Sebastian Pfuetze/Fancy/AGE Fotostock 1-CN-2 top Digital Vision/Getty Images 1-CN-3a Science Source 1-CN-3b Dr. Kathleen Welch 1-CN-3c Science Source 1-CN-3e Dr. Kathleen Welch 1-CN-3f ICVI-CCN/Voisin/Science Source 1-CN-3g Alexander Tsiaras/Photo Researchers, Inc./Science Source 1-CN-3h Chad Ehlers/Alamy Stock Photo 1-CN-3i Science Source 1-CN-3d Sciencefoto.De-Dr. Andre Kempe/Stockbyte/Getty Images 1.4a–c Sciencefoto.De-Dr. Andre Kempe/Stockbyte/Getty Images
Chapter 2 CO-2 PNC/Getty Images 2-CN-1 Nicholas Kostin/Sygma/Getty Images 2.5c Gyssels/BSIP/Alamy Stock Photo 2-CC-1,2 Monkey Business/Fotolia 2-CN-2 U.S. Food and Drug Administration FDA
Chapter 3 CO-3 View Stock/Getty Images 3-CC-1 John Giustina/Getty Images 3.03a Heuse Hirokara; Fawcett/Science Source 3.03b Torsten Wittmann/Science Source 3.04a Kalnins; de Harven; Fawcett/Science Source 3.05b Don W. Fawcett/Science Source 3.06b Biophoto Associates/Science Source 3.07 Dr. Birgit H. Satir//Science Source 3.09a CNRI/Science Source 3.10a Don W. Fawcett M.D. 3.10c Biophoto Associates/Science Source 3.17a Steve Geschmeissner/Science Source 3.17b David M. Phillips/Science Source 3.17c Steve Geschmeissner/Science Source 3.24b–f Ed Reschke 3-CC-2 John Giustina/Getty Images 3.24a Pete Draper/Getty Images 3.24h Center for Disease Control
Chapter 4 CO-4 Biophoto Associates/Science Source 4.1a Frederic H. Martini 4.1b Pearson Education, Inc.. 4.1c Robert B. Tallitsch 4.1d Pearson Education, Inc. 4.4a Robert B. Tallitsch 4.4b Frederic H. Martini 4.5a Robert B. Tallitsch 4.5b Gregory N. Fuller, MD, PhD/Frederic H. Martini 4.5c,d Robert B. Tallitsch 4.6a William C. Ober 4.6b Frederic H. Martini 4.6c Robert B. Tallitsch 4.6d William C. Ober 4.6e Courtesy of Dr. Gregory N. Fuller, Chief Neuropathologist, UT MD Anderson Cancer Center 4.8a H. Jastrow from Dr. Jastrow's electron microscopic atlas on the www, http://www.drjastrown.de 4.10a,b Robert B. Tallitsch 4.11a Pearson Education, Inc. 4.11b Robert B. Tallitsch 4.11c Ward's Science 4.13 Anita Impagliazzo 4.15b William C. Ober 4.15c Photo Researchers, Inc./Science Source 4.15d, 4.17, 4.19a–c Robert B. Tallitsch 4.20 Pearson Education, Inc. 4.21a Pixelmaker/Fotolia 4.21b Leah-Anne Thompson/Shutterstock 4-CC-1 Biophoto Associates/Science Source
Chapter 5 CO-5 Ariel Skelley/Getty Images 5.2a–c, 5.3a,b Robert B. Tallitsch 5.3c Cloud Hill Imaging 5.4, 5.7 Frederic H. Martini 5.8a Robert B. Tallitsch 5.9 John F. Wilson, MD/Science Source 5.11 Biophoto Associates/Science Source 5-CN-1 Luanateutzi/Shutterstock 5-CN-2a,b Frederic H. Martini 5-CN-3 Mediscan/Alamy 5.12d Michael Abbey/Science Source 5.13, 5.14a,b Frederic H. Martini 5-CC-1 Mediscan/Alamy 5.16 William C. Ober 5.17 Dr. M.A. Ansary/Photo Researchers, Inc. 5-CC-2 Mediscan/Alamy Stock Photo
Chapter 6 CO-6 Peopleimages/E+/Getty Images 6.3 Ralph T. Hutchings 6.4 Seeleve 6.6 Andrew Syred/Science Source 6.8, 6.11 Ralph T. Hutchings 6.12a Matthew Borkoski/Getty Images 6.12b SMC Images/Getty Images 6-CN-1a Mustafa Ozer/Newcom 6-CN-1b Joeff Davis 6.17a Alliance/Shutterstock 6.17b Dr. Kathleen Welch 6.17c Du Cane Medical Imaging Ltd./Science Source 6.17d Zephyr/Science Source 6.17e Dr. P. Marazzi/Science Source 6.17f Dan Zika 6.17g Mark Aiken 6.17h Frederic H. Martini 6.17i Scott Camazine/Science Source 6.17j Living Art Enterprises, LLC/Science Source 6.17k Kammeel/Fotolia; 6.18a William C. Ober 6.18b Prof. P. Motta/Science Source Library/Science Source 6.18c Lisa F. Young/Shutterstock 6.18d Prof. P. Motta/Science Source Library/Science Source 6-CC-1, 6-CC-2 Kris Timken/Blend Images/Alamy
Chapter 7 CO-7 Clicknique/E+/Getty Images 7-CC-1 Ruth Anne O'Keefe 7.1b(1,2) Ralph T. Hutchings 7.2a Ralph T. Hutchings 7.2b VideoSurgery/Science Source 7.5a,b, 7.6a,b, 7.7a,b Ralph T. Hutchings 7.7c Michael J. Timmons 7.8a,b, 7.9a,b, 7.10a, 7.11, 7.12a,b, Ralph T. Hutchings 7.13a William C. Ober 7.13b, 7.14 Ralph T. Hutchings 7.15a William C. Ober 7-CN-1a,b William C. Ober 7.17 William C. Ober 7-CN-2a–c William C. Ober 7-CC-2 Ruth Anne O'Keefe 7.19a–c, 7.T1a–c, 7.20a–c, 7.21a–c, 7.22a–c, 7.23a,b, 7.24a,b, 7-EOC.01a–c Ralph T. Hutchings
Chapter 8 CO-8 Shapecharge/E+/Getty Images 8.2a,b, 8.3a–c, 8.4a–d, 8.5a–c, 8.7a,b, 8.8a,b 8.10a,b, 8.11a–c, 8.12a,b, 8.13a–c Ralph T. Hutchings 8-CN-2 Jim Stevenson/Science Source 8-EOC-01a–c Ralph T. Hutchings 8-CC-1, 8-CC-2 Steve Mann/Shutterstock
Chapter 9 CO-9 Steve Prezant/Image Source/Getty Images; 9.3a Anita Impagliazzo 9.3b Ralph T. Hutchings 9.3c Anita Impagliazzo 9.3d–g Ralph T. Hutchings 9.4a Anita Impagliazzo 9.4b Ralph T. Hutchings; 9.4c William C. Ober 9.5a–g Ralph T. Hutchings 9.8b,d Ralph T. Hutchings 9-CC-1, 9-CC-2 Jamie Grill/JGI/Getty Images 9-CN-2 Soupstock/Fotolia.com
Chapter 10 CO-10 Echo/Cultura/Getty Images 10.1 William C. Ober 10.2 Michael Abbey/Science Source 10.4 Don W. Fawcett/Science Source 10.11 Pkline/E+/Getty Images 10.21a,b Frederic H. Martini 10.21c Lippincott Williams & Wilkins; 10.22 Robert B. Tallitsch 10.23 Frederic H. Martini 10-CC-1, 10-CC-2 Oliver Gutfleisch/ImageBROKER/Alamy Stock Photo
Chapter 11 CO-11 Henrik Sorensen/The Image Bank/Getty Images 11-CC-1 Visage/Stockbyte/Getty Images 11.9, 11.10 William C. Ober 11.14a,b, 11.15b, 11.16a,b Anita Impagliazzo 11.17a,b, 11.18a–e, 11.22 William C. Ober
Chapter 12 CO-12 Marc Romanelli/Blend Images/Getty Images 12.1 William C. Ober 12.6a Biophoto Associates/Science Source 12.6b Don W. Fawcett/Science Source 12-CC-1, 12-CC-2 Margaret "Daisy" Suckley/Franklin Delano Roosevelt Library 12-CN-1 Steve Gschmeissner/Science Source
Chapter 13 CO-13 Caroline Schiff/DigitalVision/Getty Images 13-CC-1 Victor Biro/Alamy Stock Photo 13.4 Ralph T. Hutchings 13.5 Michael J. Timmons 13.6b Steve Gschmeissner/Science Source 13-CN-1, 13.9, 13.10 William C. Ober 13-CC-2 Victor Biro/Alamy Stock Photo
Chapter 14 CO-14 Wavebreakmedia/Shutterstock 14-CC-1 Dr. Jill Bolte Taylor 14.5d, 14.8b,c Karen Krabbenhoft/Pearson Education 14.9a Ralph T. Hutchings 14.9b Volker Steger/Science Source 14.11, 14.13 Ralph T. Hutchings 14.25, 14.26 William C. Ober 14.27 Anita Impagliazzo 14-CC-2 Dr. Jill Bolte Taylor 14.17 Science Photo Library/Getty Images 14.18 Ralph T. Hutchings
Chapter 15 CO-15 Jun Takahashi/The Image Bank/Getty Images 15-CC-1, 15-CC-2 Ruth Anne O'Keefe; Anne Marie O'Keefe 15.4d,e Frederic H. Martini 15-CN-1 M. Spuhler/Pearson Education, Inc.
Chapter 16 CO-16 Swissmediavision/E+/Getty Images 16.6 Pearson Education, Inc. 16.10 Michael Pettigrew/Shutterstock 16-CC-1, 16-CC-2 Nick White/ImageSource/AGE Fotostock
Chapter 17 CO-17 Geri Lavrov/Photographer's Choice/Getty Images 17-CC-1 Center for Vision in the Developing World 17.2a Robnroll/Shutterstock 17.2b Andresr/Shutterstock 17.3a Robert B. Tallitsch 17.4a,b Ralph T Hutchings 17.6a,b Diane Hirsch/Fundamental Photographs 17.7a Ed Reschky 17.7c Thalerngsak Mongkolsin/Shutterstock 17.13a,b William C. Ober 17.13c ZUMA Press/Alamy Stock Photo 17.22b Lennart Nilsson/Tidningarnas Telelgrambyra AB 17.22c Ralph T. Hutchings 17.25c(1,2) Pearson Education, Inc. 17.28b Michael J. Timmons 17.29b Biophoto Associates/Science Source 17-CC-2 Center for Vision in the Developing World
Chapter 18 CO-18 Hero Images/Hero Images/Getty Images 18-CC-1 Phovoir/Shutterstock 18.2a Bloody/Shutterstock 18.2b Bochkarev Photography/Shutterstock 18.2c James Steidl/Shutterstock 18.5 Robert B. Tallitsch 18.10a William C. Ober 18.10b Robert B. Tallitsch 18.10c Frederic H. Martini 18.12b Frederic H. Martini 18.12c Pearson Education, Inc. 18.14a Pearson Education, Inc. 18.14c Biophoto Associates/Science Source 18.16b Robert B. Tallitsch 18.18 Mik122/Getty Images 18-CC-2 Phovoir/Shutterstock 18.20 Jake Lyell/Alamy Stock Photo
Chapter 19 CO-19 Jan Wolak-Dyszynski/Getty Images 19.1 Dizzy/Getty Images 19.2a Frederic H. Martini 19.2b Cheryl Power/Science Source 19.2d Ed Reschke 19.7 Karen E. Petersen 19.9a–d Ralph T. Hutchings 19.9e Pearson Education, Inc. 19.11 Mediscan/Alamy Stock Photo 19-CC-1 Rook76/Fotolia 19-CC-2 Eye of Science/Science Source 19-EOC.01a–e Ralph T. Hutchings
Chapter 20 CO-20 Image Source/Photodisc/Getty Images 20.2 Pearson Education, Inc. 20.3c Karen Krabbenhoft/Pearson Education, Inc. 20.5b Philippe Plailly/Look at Sciences/Science Source 20.5c Karen Krabbenhoft/Pearson Education, Inc. 20.7a Biophoto Associates/Science Source 20.7b SPL/Science Source 20.8c Ralph T. Hutchings 20.9a Phanie/Voisin/Alamy Stock Photo 20.9b,c Howard Sochurek/Corbis Images/Getty Images 20.9d Maciej Olesky/Shutterstock 20.9e Serg64/Shutterstock 20.9f BSIP/Science Source 20.9g kalewa/Shutterstock 20.12a Science Source 20.14 Ralph T. Hutchings 20.18 Pearson Education, Inc. 20-CC-1, 20-CC-2 Dr. Guillermo Perez Chagerben/Clinica Santa Gema
Chapter 21 CO-21 PeopleImages.com/DigitalVision/Getty Images 21-CC-1, 21-CC-2 Jim Bourg/Reuters 21.1, 21.4 Biophoto Associates/Science Source 21.18 Pearson Education, Inc. 21.21 William C. Ober 21.23a Pearson Education, Inc. 21.27 William C. Ober 21.28 Pearson Education, Inc. 21-CN-1a Martin M. Rotker/Science Source 21-CN-1b Richard Kirby/David Spears Ltd./Science Source 21-CN-1c Martin M. Rotker/Science Source
Chapter 22 CO-22 Mika/Corbis/Getty Images 22-CC-1, 22-CC-2 Mariana Nakashima/Robert Hutchinson 22.3 Frederic H. Martini 22.5a Biophoto Associates/Science Source 22.5b Robert B. Tallitsch 22.7a Pearson Education, Inc. 22.7c,d Robert B. Tallitsch 22.8a Pearson Education, Inc. 22.8c Robert B. Tallitsch 22.18b Amnis Corporation 22-CN-1 Dr. P. Marazzi/Science Source 22-CN-2 C. Goldsmith/Centers for Disease Control and Prevention (CDC)
Chapter 23 CO-23 Saro17/E+/Getty Images 23.1, 23.2 left Pearson Education, Inc. 23.2 right Photo Insolite Realite/Science Source; 23.3a, 23.3b left William C. Ober 23.3b right Ralph T. Hutchings 23.3c William C. Ober 23.4 Pearson Education, Inc. 23.5 CNRI/Science Source 23.7d Image Source/Alamy Stock Photo 23.8 Robert B. Tallitsch 23.9a left Pearson Education, Inc. 23.9a right Ralph T. Hutchings 23.10a left Pearson Education, Inc. 23.10a right Ralph T. Hutchings 23.11 Pearson Education, Inc. 23-CN-1 William C. Ober 23-CC-1, 23-CC-2 ChameleonsEye/Shutterstock 23-EOC-01 Pearson Education, Inc.
Chapter 24 CO-24 Sam Diephuis/Blend Images/Getty Images 24-CC-1 Ampyang/Shutterstock 24.1, 24.2 Pearson Education, Inc. 24.8 Ralph T. Hutchings 24.9a Pearson Education, Inc. 24.9b Robert B. Tallitsch 24.10a left Pearson Education, Inc. 24.10a right Alfred Pieska/Getty Images 24.10b Robert B. Tallitsch 24.12a Karen Krabbenhoft/Pearson Education, Inc. 24.16c Frederic H. Martini 24.17 Pearson Education, Inc. 24.18c Robert B. Tallitsch 24.19c Frederic H. Martini 24.20b Karen Krabbenhoft/Pearson Education, Inc. 24.21d M. I. Walker/Science Source 24.24a Pearson Education, Inc. 24.24b Ralph T. Hutchings; 24.25 Robert B. Tallitsch 24.27 Monkey Business Images/iStock/Getty Images 24-CC-2 Ampyang/Shutterstock 24-EOC-01 Pearson Education, Inc.
Chapter 25 CO-25 Zave Smith/Image Source/Getty Images 25-CC-1 Ruth Anne O'Keefe 25.1 Adapted from L.L. Langley, Homeostasis (New York: Reinhold, 1965). 25.13 Fotandy/Thinkstock/Getty Images 25-CC-2 Ruth Anne O'Keefe 25.11 Logo from ChooseMyPlate.gov. Copyright © 2016, Used by permission of U.S. Department of Agriculture.
Chapter 26 CO-26 Dave & Les Jacobs/Blend Images/Getty Images 26-CC-1 Blend Images/ERproductions Ltd 26.4b Ralph T. Hutchings 26.17 Photo Researchers, Inc./Science Source 26.19a Ed Reschke/Oxford Scientific/

CREDITS

Getty Images **26.19b** Frederic Martini **26.19c** Pearson Education, Inc. **26-CC-2** Blend Images/ERproductions Ltd/ Getty Images

Chapter 27 CO-27 David Buffington/Getty Images 27-CC-1, 27-CC-2 Tetra Images/AGE Fotostock **27.18** Alexander Raths/Shutterstock

Chapter 28 CO-28 Peathegee Inc/Getty Images **28.3** Frederic H. Martini **28.4a,b** Robert B. Tallitsch **28.5b** Frederic H. Martini **28.5c** Ralph T. Hutchings **28.6b** Ward's Natural Science Establishment **28.6c–e** Frederic H. Martini **28.11** Eye of Science/Science Source **28.13** Pearson Education, Inc. **28.15b** Frederic H. Martini **28.15c** Steve Gschmeissner/Science Source **28.16c** Karen Krabbenhoft/Pearson Education, Inc. **28.17b, 28.18, 28.20b** Robert B. Tallitsch **28.20c** Pearson Education, Inc. **28.22(1–4)** Frederic H. Martini **28.22(5)** C. Edelmann/La Villete/Photo Researchers/Science Source **28.22(6)** BSIP/AGE Fotostock **28.22(7)** Biophoto Associates/Science Source **28.23a,b** Frederic H. Martini **28.23c(1)** Michael J. Timmons **28.23c(2)** Huhman, Heather **28-CC-1, 28-CC-2** Kupicoo/Getty Images

Chapter 29 CO-29 Getty Images **29-CC-1** Barcroft Media/Getty Images **29.5** Nestle/Petit Format/Science Source **29.6b** Frederic H. Martini **29.7b** Pr. M. Forest, CNRI/Science Source **29.7c** Claude Edelmann/Science Source **29.8a** Nestle/Petit Format/Science Source **29.8b** Kate Bowers **29.13a** Victoria Penafiel/Moment/Getty Images **29.13b,c** Jaroslaw Wojcik/Getty Images **29.13d** Justin Horrocks/E+/Getty Images **29.13e** Jacob Wackerhausen/iStock/Getty Images **29.14** CNRI/Science Source

FM **FM-01** PK Martini **FM-02** Cesere Brothers Photography **FM-03** Amy Heflinger **FM-04** William C. Ober **FM-05** Claire E. Ober **FM-06** PK Martini **FM-07** Ralph T. Hutchings **FM-08** Christine Boudrie, M.D. **FM-09** Anne Marie O'Keefe **FM-10** Kevin Petti, PhD **VWT:** [laptop] Artjazz/Shutterstock; [phone] Nik Merkulov/ Shutterstock; [tablet] Designsstock/Shutterstock.

Index

INDEX

INDEX

INDEX

INDEX

INDEX

D

INDEX

INDEX

INDEX

INDEX

INDEX